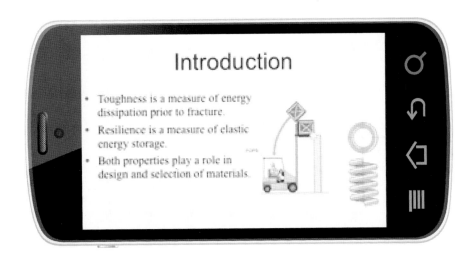

Redeem access to
Video Solutions, Pearson eText, and more:

www.pearsonhighered.com/kalpakjian

Scratch here to reveal your access code.
This access code may only be used by the original purchaser.

Use a coin to scratch off the coating and reveal your student access code.
Do not use a knife or other sharp object as it may damage the code.

More than 120 QR codes appear throughout the textbook, enabling you to use your smartphone or tablet to watch video solutions and manufacturing processes. Here's how to get started:

- Download a QR code reader. You can download free apps from your app store or use a built-in code reader if your device has one.

- Scan the code using the QR code reader.
- View content online. You will automatically be redirected to exclusive online extras. (Note: Data usage charges may apply.)

Manufacturing Engineering and Technology

SEVENTH
EDITION

Serope Kalpakjian
Illinois Institute of Technology

Steven R. Schmid
The University of Notre Dame

PEARSON

Upper Saddle River Boston Columbus San Francisco New York
Indianapolis London Toronto Sydney Singapore Tokyo Montreal
Dubai Madrid Hong Kong Mexico City Munich Paris Amsterdam Cape Town

Vice President and Editorial Director, ECS: *Marcia J. Horton*
Executive Editor: *Holly Stark*
Editorial Assistant: *Carlin Heinle*
Executive Marketing Manager: *Tim Galligan*
Marketing Assistant: *Jon Bryant*
Permissions Project Manager, Photos: *Karen Sanatar*
Permissions Project Manager, Text: *Alison Bruckner*
Senior Managing Editor: *Scott Disanno*
Production Project Manager: *Clare Romeo*
Operations Specialist: *Lisa McDowell*
Cover Designer: *Black Horse Designs*
Cover Photo: *Jason Richards, Oak Ridge National Laboratory/Science Photo Library/Photo Researchers*
Manager, Rights and Permissions: *Mike Lackey*
Composition: *Jouve India*
Project Manager: *Pavithra Jayapaul, Jouve India*
Printer/Binder: *LSC Communications*
Cover Printer: *LSC Communications*
Typeface: *10/12 Sabon*

About the Cover Photo: Next generation hydraulically controlled robotic hand fabricated from titanium using electron beam melting. The hand was designed and manufactured at the Department of Energy's Manufacturing Demonstration Facility at Oak Ridge National Laboratory.

Pearson Education Ltd., *London*
Pearson Education Singapore, Pte. Ltd
Pearson Education Canada, Inc.
Pearson Education—Japan
Pearson Education Australia PTY, Limited
Pearson Education North Asia, Ltd., *Hong Kong*
Pearson Educación de Mexico, S.A. de C.V.
Pearson Education Malaysia, Pte. Ltd.
Pearson Education, Inc., Upper Saddle River, New Jersey.

Library of Congress Cataloging-in-Publication Data

Kalpakjian, Serope, 1928-
   Manufacturing engineering and technology / Serope Kalpakjian, Illinois Institute of Technology, Steven R. Schmid, The University of Notre Dame.—Seventh edition.
      pages cm
   ISBN-13: 978-0-13-312874-1
   ISBN-10: 0-13-312874-1
   1. Production engineering. 2. Manufacturing processes. I. Schmid, Steven R. II. Title.
   TS176.K34 2012
   670.42—dc23   2012035148

11 2022

ISBN-13: 978-0-13-312874-1
ISBN-10:   0-13-312874-1

To the memory of

Margaret Jean Kalpakjian

*"And ever has it been known that love knows not its own depth
until the hour of separation."*

Khalil Gibran

Contents

3 Physical Properties of Materials   88

4 Metal Alloys: Their Structure and Strengthening by Heat Treatment   101

5   Ferrous Metals and Alloys: Production, General Properties, and Applications   128

6   Nonferrous Metals and Alloys: Production, General Properties, and Applications   150

7   Polymers: Structure, General Properties, and Applications   169

Part II: Metal-casting Processes and Equipment   235

10 Fundamentals of Metal Casting   237

11 Metal-casting Processes and Equipment   256

12 Metal Casting: Design, Materials, and Economics   294

19 Plastics and Composite Materials: Forming and Shaping 494

20 Rapid-prototyping Processes and Operations 535

Part IV: Machining Processes and Machine Tools   563

21 Fundamentals of Machining   566

22 Cutting-tool Materials and Cutting Fluids   600

23 Machining Processes: Turning and Hole Making   625

24 Machining Processes: Milling, Broaching, Sawing, Filing, and Gear Manufacturing   668

25 Machining Centers, Machine-tool Structures, and Machining Economics   703

26 Abrasive Machining and Finishing Operations   729

27   Advanced Machining Processes and Equipment   769

Part V: Micromanufacturing and Fabrication of Microelectronic Devices   797

28   Fabrication of Microelectronic Devices   800

32 Brazing, Soldering, Adhesive-bonding, and Mechanical Fastening Processes 934

Part VII: Surface Technology 961

33 Surface Roughness and Measurement; Friction, Wear, and Lubrication 963

34 Surface Treatments, Coatings, and Cleaning 985

Part VIII: Engineering Metrology, Instrumentation, and Quality Assurance   1007

35   Engineering Metrology and Instrumentation   1008

36   Quality Assurance, Testing, and Inspection   1030

Case Studies

Preface

•••

In preparing the seventh edition of this book, our goal continues to be to provide a comprehensive and state-of-the-art manufacturing engineering and technology textbook, with the additional aims of motivating and challenging students in studying this important engineering discipline. As in the previous editions, the book attempts to present a broad overview, emphasizing a largely qualitative coverage of the science, technology, and practice of manufacturing. We have included detailed descriptions of the fundamentals of manufacturing processes, operations, and the manufacturing enterprise.

The book continues to address the various challenges and issues in modern manufacturing processes and operations, ranging from traditional topics such as casting, forming, machining, and joining processes, to advanced topics such as the fabrication of microelectronic devices and microelectromechanical systems and nanomanufacturing. The book provides numerous examples and case studies, as well as comprehensive and up-to-date coverage of all topics relevant to modern manufacturing, as a solid background for students as well as for professionals.

New to This Edition

In response to the suggestions by several of our colleagues and their students, numerous changes have been made to this edition of the book:

- Recognizing the proliferation of intelligent phones and the intention of Internet browsing ability in these phones and tablet devices, *QR Codes* have been introduced with this edition. Each QR Code is a link to a video solution to problems or a manufacturing process video. See sample QR Code to the right.
- The 56 video solutions are complete, step-by-step solution walkthroughs of representative problems from the text. The problems featured in video solutions provide additional assistance for students with homework or in preparing for an exam or quiz.
- The 65 manufacturing videos provide students with real-world context and allow students to watch an interactive demonstration of relevant issues or problem-solving strategies.

  Please note: Users must download a QR code reader to their smartphone or tablet. Data and roaming charges may also apply.

- Wherever appropriate, *illustrations* and *examples* have been replaced with newer ones to indicate recent advances in manufacturing.
- To provide a better perspective of the topics covered, the text now contains more *cross-references* to other relevant chapters, sections, tables, and illustrations in the book.
- The *questions* and *problems* and *projects* for class discussions, at the end of each chapter have been significantly expanded.
- The *bibliographies* at the end of each chapter have been thoroughly updated.

Video Solution 1.1 Calculating the Number of Grains in a Ball Bearing

- Every attempt has been made to ensure that *figures* and *tables* are placed on the same page as they are referenced in the text; this has been made possible by rearranging the page layout, including the use of margins on the pages.
- An *eText* version of this edition, and *videolinks* are provided on the web at www.pearsonhighered.com/kalpakjian.
- A *Solutions Manual* is, as always, available for use by instructors.

The text has been thoroughly edited and updated, as can be noted by page-by-page comparison with the previous edition. Specific revisions in the book include:

1. Expansion of the discussions regarding carbon in its useful forms, such as nanotubes and graphene (Section 8.6)
2. Expansion of product design considerations for manufacturing processes (various chapters)
3. Servo presses for forging and sheet-metal forming (Section 14.8)
4. Hot stamping of sheet metals (Section 16.11)
5. Expanded discussion on sintering, material properties, explosive compaction, roll densification, and combustion synthesis in powder metallurgy (Chapter 17)
6. Discussion of laser-engineered net shaping and self-replicating machines in rapid prototyping (Chapter 20)
7. Through-the-tool cooling systems in machining (Section 22.12)
8. Electrolytic laser MicroJet and Blue Arc machining (Chapter 27)
9. Immersion and pitch splitting lithography (Section 28.7)
10. MolTun (Section 29.3.4) and an expanded discussion of MEMS devices (Chapter 29)
11. Visual sensing (Section 37.7)
12. Production flow analysis (Section 38.8)
13. Development of MTConnect (Section 39.7)
14. Energy consumption in manufacturing (Section 40.5)

Audience

As in the previous editions, this seventh edition has been written for students in mechanical, manufacturing, industrial, aerospace, metallurgical and materials, and biomedical engineering programs. The book is also appropriate for use in associate degree programs at junior and community colleges. It is hoped that by studying this book, students will begin to appreciate the critical role of manufacturing engineering and technology in our daily lives and professional activities, and also view it as an academic subject that is as exciting, challenging, and as important as any other engineering discipline.

We would be very grateful for comments and suggestions from instructors and students regarding the nature and contents of the large number of topics presented in this book, as well as informing us of any errors in the text and in the illustrations that may have escaped our attention during the preparation of this text. Please send your comments or suggestions to Steven R. Schmid, schmid.2@nd.edu or to Serope Kalpakjian, kalpakjian@iit.edu.

Translations

Various editions of this book and our other textbook, *Manufacturing Processes for Engineering Materials*, are available in Chinese, Korean, Spanish, Italian, and German.

Acknowledgments

It gives us great pleasure to acknowledge the assistance of the following colleagues in the preparation and publication of this edition of the book: K. Scott Smith, University of North Carolina at Charlotte; Steven Hayashi, General Electric Corp.; James Adams, Metal Powder Industries Federation; Timotius Pasang, Auckland University of Technology, New Zealand; Miguel Sellés Cantó, Escola Politècnica Superior d'Alcoi, Universitat Politecnica de València, Spain; and Megan McGann and Holly Weiss, University of Notre Dame. We also acknowledge Kent M. Kalpakjian, Micron Technology, as the original author of Fabrication of Microelectronic Devices (Chapter 28), and would like to thank Flora Derminjan, Micron Technology, and Vahagn Sargsyan, Carl Zeiss Nano Technology Systems, for reviewing this chapter for this edition. Thank you also to the reviewers for this edition: Keng Hsu, University of Illinois at Urbana-Champaign; Wayne Hung, Texas A&M University; Stephen Jenkins, Wor-Wic Community College; Edward Red, Brigham Young University; David Veazie, Southern Polytechnic State University; Yan Wang, Worcester Polytechnic Institute; and Y. Lawrence Yao, Columbia University.

We would also like to thank our editor, Holly Stark, at Prentice Hall for her enthusiastic support and guidance; and Clare Romeo, Sr. Production Editor, for her meticulous editorial and production supervision. We would like to acknowledge the help of the following individuals in the production of this edition of the book; Renata Butera, Media Editor; Carlin Heinle, Editorial Assistant; and Black Horse Designs, Cover Design.

We are happy to present below a cumulative list of all those individuals who generously have made numerous contributions to various editions of our two books.

B.J. Aaronson	A. Cinar	P. Grigg	J. Lewandowski
R. Abella	R.O. Colantonio	M. Grujicic	X.Z. Li
D. Adams	P. Cotnoir	P.J. Guichelaar	Z. Liang
S. Arellano	P.J. Courtney	B. Harriger	B.W. Lilly
D.D. Arola	P. Demers	D. Harry	D.A. Lucca
R.A. Arlt	D. Descoteaux	M. Hawkins	M. Madou
V. Aronov	M.F. DeVries	R.J. Hocken	S. Mantell
A. Bagchi	R.C. Dix	E.M. Honig, Jr.	L. Mapa
E.D. Baker	M. Dollar	Y. Huang	R.J. Mattice
J. Barak	D.A. Dornfeld	S. Imam	C. Maziar
J. Ben-Ari	H.I. Douglas	R. Ivester	T. McClelland
G.F. Benedict	M. Dugger	R. Jaeger	L. McGuire
S. Bhattacharyya	D.R. Durham	C. Johnson	K.E. McKee
JT. Black	D. Duvall	D. Kalisz	K.P. Meade
W. Blanchard	S.A. Dynan	J. Kamman	M.H. Miller
C. Blathras	K.F. Ehmann	S.G. Kapoor	T.S. Milo
G. Boothroyd	J. El Gomayel	R.L. Kegg	D.J. Morrison
D. Bourell	M.G. Elliott	W.J. Kennedy	S. Mostovoy
B. Bozak	N.X. Fang	R. Kerr	C. Nair
N.N. Breyer	E.C. Feldy	T. Kesavadas	P.G. Nash
C.A. Brown	J. Field	J.E. Kopf	J. Nazemetz
R.G. Bruce	G.W. Fischer	R.J. Koronkowski	E.M. Odom
J. Cesarone	D.A. Fowley	J. Kotowski	U. Pal
T.-C. Chang	R.L. French	S. Krishnamachari	S.J. Parelukar
R.L. Cheaney	B.R. Fruchter	K.M. Kulkarni	T. Pasang
A. Cheda	D. Furrer	T. Lach	Z.J. Pei
S. Chelikani	R. Giese	L. Langseth	J. Penaluna
S. Chen	E. Goode	M. Levine	M. Philpott
S.-W. Choi	K.L. Graham	B.S. Levy	M. Pradheeradhi

J.M. Prince
D.W. Radford
W.J. Riffe
R.J. Rogalla
Y. Rong
A.A. Runyan
G.S. Saletta
M. Salimian
M. Savic
W.J. Schoech
S.A. Schwartz
M. Selles

S. Shepel
R. Shivpuri
M.T. Siniawski
J.E. Smallwood
J.P. Sobczak
L. Soisson
P. Stewart
J. Stocker
L. Strom
A.B. Strong
K. Subramanian
T. Sweeney

T. Taglialavore
M. Tarabishy
K.S. Taraman
R. Taylor
B.S. Thakkar
A. Trager
A. Tseng
C. Tszang
M. Tuttle
J. Vigneau
G.A. Volk
G. Wallace

J.F. Wang
K.J. Weinmann
R. Wertheim
K. West
J. Widmoyer
K.R. Williams
G. Williamson
B. Wiltjer
P.K. Wright
N. Zabaras

We are also grateful to numerous organizations, companies, and individuals who supplied us with many of the illustrations and case studies. These contributions have been specifically acknowledged throughout the text.

SEROPE KALPAKJIAN
STEVEN R. SCHMID

About the Authors

● ●

Serope Kalpakjian is a professor emeritus of mechanical and materials engineering at the Illinois Institute of Technology, Chicago. He is the author of *Mechanical Processing of Materials* (Van Nostrand, 1967) and co-author of *Lubricants and Lubrication in Metalworking Operations* (with E.S. Nachtman, Dekker, 1985). Both of the first editions of his books *Manufacturing Processes for Engineering Materials* (Addison-Wesley, 1984) and *Manufacturing Engineering and Technology* (Addison-Wesley, 1989) have received the M. Eugene Merchant Manufacturing Textbook Award of SME. He has conducted research in several areas in manufacturing processes; is the author of numerous technical papers and articles in professional journals, handbooks, and encyclopedias; and has edited several conference proceedings. He also has been editor and co-editor of various technical journals and has served on the editorial board of *Encyclopedia Americana*.

Professor Kalpakjian has received the *Forging Industry Educational and Research Foundation Best Paper Award* (1966), the *Excellence in Teaching Award* from the Illinois Institute of Technology (1970), the *ASME Centennial Medallion* (1980), the *SME International Education Award* (1989), a *Person of the Millennium Award* from IIT (1999), and the *Albert Easton White Outstanding Teacher Award* from ASM International (2000); the *SME Outstanding Young Manufacturing Engineer Award* for 2002 was named after him. Professor Kalpakjian is a life fellow of ASME, fellow of SME, fellow and life member of ASM International, fellow emeritus of The International Academy for Production Engineering (CIRP), and a past president and founding member of the North American Manufacturing Research Institution/SME. He is a high honors graduate of Robert College (Istanbul), Harvard University, and the Massachusetts Institute of Technology.

Steven R. Schmid is an associate professor in the Department of Aerospace and Mechanical Engineering at the University of Notre Dame, where he teaches and conducts research in the general areas of manufacturing, machine design, and tribology. He received his bachelor's degree in mechanical engineering from the Illinois Institute of Technology (with Honors) and master's and Ph.D. degrees, both in mechanical engineering, from Northwestern University. He has received numerous awards, including the *John T. Parsons Award* from SME (2000), the *Newkirk Award* from ASME (2000), the *Kaneb Center Teaching Award* (2000 and 2003), and the *Ruth and Joel Spira Award for Excellence in Teaching* (2005). He is also the recipient of a *National Science Foundation CAREERS Award* (1996) and an *ALCOA Foundation Award* (1994).

Professor Schmid is the author of over 100 technical papers, has co-authored *Fundamentals of Machine Elements* (McGraw-Hill), *Fundamentals of Fluid Film Lubrication* (Dekker), *Manufacturing Processes for Engineering Materials* (Prentice Hall), and has contributed two chapters to the *CRC Handbook of Modern Tribology*. He is a registered professional engineer, a certified manufacturing engineer of SME, a member of the North American Research Institution, and a fellow of the ASME. In 2012, he was named an ASME Foundation Swanson Fellow and served as Assistant Director for Research Partnerships at the Advanced Manufacturing National Program Office, National Institute for Science and Technology.

General Introduction

I.1 What is Manufacturing?

As you begin to read this chapter, take a few moments to inspect various objects around you: mechanical pencil, light fixture, chair, cell phone, and computer. You soon will note that all these objects, and their numerous individual components, are made from a variety of materials and have been produced and assembled into the items that you now see. You also will note that some objects, such as a paper clip, nail, spoon, and door key, are made of a single component. As shown in Table I.1, however, the vast majority of objects around us consist of numerous individual pieces that are built and assembled by a combination of processes called **manufacturing**. This is well illustrated by the tractor shown in Fig. I.1, with several components and their materials highlighted.

The word *manufacture* first appeared in English in 1567, and is derived from the Latin *manu factus*, meaning "made by hand." The word *manufacturing* first appeared in 1683, and the word *production*, which is often used interchangeably with the word *manufacturing*, first appeared sometime during the 15th century.

Manufacturing is concerned with making products. A manufactured product may, in turn, itself be used to make other products, such as (a) a large press, to shape flat sheet metal into automobile bodies; (b) a drill, for producing holes; (c) industrial sewing machines, for making clothing at high rates; and (d) machinery, to produce an endless variety of individual items, ranging from thin wire for guitars and electric motors to crankshafts and connecting rods for automotive engines.

Items such as nails, bolts, screws, nuts, washers, and paper clips are *discrete products*, meaning individual items. By contrast, a spool of wire, metal or plastic tubing, and a roll of aluminum foil are *continuous products*, which are then cut into individual pieces of various lengths for specific purposes.

Because a manufactured item typically begins with raw materials, which are then subjected to a sequence of processes to make individual products, it has a certain value. For example, clay has some value as mined, but when made into a product such as cookware, pottery, an electrical insulator, or a cutting tool, value is *added* to the clay. Similarly, a nail has a value over and above the cost of the short piece of wire or rod from which it is made. Products such as computer chips, electric motors, medical implants, machine tools, and aircraft are known as *high-value-added products*.

A Brief History of Manufacturing. Manufacturing dates back to the period 5000–4000 B.C. (Table I.2), thus it is older than recorded history (which dates to the Sumerians around 3500 B.C.). Primitive cave drawings, as well as markings on clay tablets and stone, needed (a) some form of a brush and some sort of pigment, as in the prehistoric cave paintings in Lascaux, France, estimated to be 16,000 years old; (b) some means of first scratching the clay tablets and then baking them, as in cuneiform scripts and pictograms of 3000 B.C.; and (c) simple

TABLE I.1

Approximate Number of Parts in Products	
Common pencil	4
Rotary lawn mower	300
Grand piano	12,000
Automobile	15,000
Boeing 747–400	6,000,000

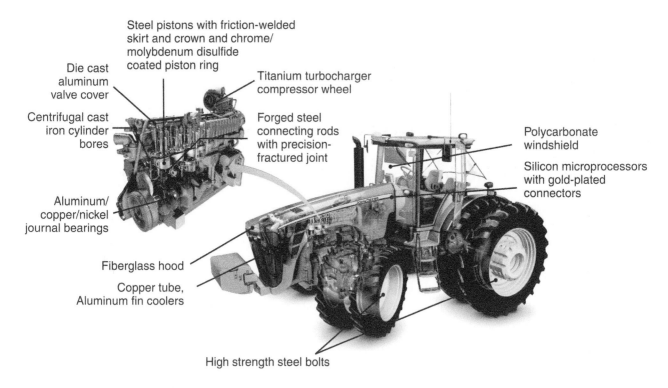

Steel pistons with friction-welded skirt and crown and chrome/molybdenum disulfide coated piston ring

Die cast aluminum valve cover

Titanium turbocharger compressor wheel

Centrifugal cast iron cylinder bores

Forged steel connecting rods with precision-fractured joint

Polycarbonate windshield

Silicon microprocessors with gold-plated connectors

Aluminum/copper/nickel journal bearings

Fiberglass hood

Copper tube, Aluminum fin coolers

High strength steel bolts

FIGURE I.I   Model 8430 John Deere tractor, with detailed illustration of its diesel engine, showing the variety of materials and processes incorporated. *Source:* Courtesy of John Deere Company.

tools for making incisions and carvings on the surfaces of stone, as in the hieroglyphs in ancient Egypt.

The manufacture of items for specific uses began with the production of various household artifacts, typically made of wood, stone, or metal. The materials first used in making utensils and ornamental objects included gold, copper, and iron, followed by silver, lead, tin, bronze, and brass. The processing methods first employed involved mostly *casting* and *hammering*, because they were relatively easy to perform. Over the centuries, these simple processes gradually began to be developed into more and more complex operations, at increasing rates of production and at higher levels of product quality. Note, for example, from Table I.2 that lathes for cutting screw threads already were available during the period from 1600 to 1700, but it was not until some three centuries later that automatic screw machines were developed.

Although ironmaking began in the Middle East in about 1100 B.C., a major milestone was the production of steel in Asia during the period 600–800 A.D. A wide variety of materials continually began to be developed. Today, countless metallic and nonmetallic materials with unique properties are available, including **engineered materials** and various other advanced materials. Among the materials now available are high-tech or industrial ceramics, reinforced plastics, composite materials, and nanomaterials that are now used in an extensive variety of products, ranging from prosthetic devices and computers to supersonic aircraft.

Until the **Industrial Revolution**, which began in England in the 1750s (also called the *First Industrial Revolution*), goods had been produced in batches, which required high reliance on manual labor in all phases of production. The **Second Industrial Revolution** is regarded by some as having begun in the mid-1900s, with the

TABLE I.2

Historical Development of Materials, Tools, and Manufacturing Processes

Period	Dates	Metals and casting	Various materials and composites	Forming and shaping	Joining	Tools, machining, and manufacturing systems
Egypt: ~3100 B.C.–300 B.C.	~4000 B.C.	Gold, copper, meteoric iron	Earthenware, glazing, natural fibers	Hammering		Tools of stone, flint, wood, bone, ivory, composite tools
	3000	Copper casting, stone and metal molds, lost-wax process, silver, lead, tin, bronze		Stamping, jewelry	Soldering (CuAu, CuPb, PbSn)	Corundum (alumina, emery)
Greece: ~1100 B.C.–146 B.C.	2000	Bronze casting and drawing, gold leaf	Glass beads, potter's wheel, glass vessels	Wire by slitting sheet metal	Riveting, brazing	Hoe making, hammered axes, tools for ironmaking and carpentry
	1000	Wrought iron, brass				
Roman Empire: ~500 B.C.–476 A.D.	1 B.C.	Cast iron, cast steel	Glass pressing and blowing	Stamping of coins	Forge welding of iron and steel, gluing	Improved chisels, saws, files, woodworking lathes
	1 A.D.	Zinc, steel	Venetian glass	Armor, coining, forging, steel swords		Etching of armor
Middle ages: ~476–1492	1000	Blast furnace, type metals, casting of bells, pewter	Crystal glass	Wire drawing, gold-and silversmith work		Sandpaper, windmill driven saw
Renaissance: 1400–1600	1500	Cast-iron cannon, tinplate	Cast plate glass, flint glass	Waterpower for metalworking, rolling mill for coinage strips		Hand lathe for wood
	1600	Permanent-mold casting, brass from copper and metallic zinc	Porcelain	Rolling (lead, gold, silver), shape rolling (lead)		Boring, turning, screw-cutting lathe, drill press

(continued)

TABLE I.2 (*continued*)

Historical Development of Materials, Tools, and Manufacturing Processes

Period	Dates	Metals and casting	Various materials and composites	Forming and shaping	Joining	Tools, machining, and manufacturing systems
First Industrial Revolution: ~1780–1850	1700	Malleable cast iron, crucible steel (iron bars and rods)		Extrusion (lead pipe), deep drawing, rolling		
	1800	Centrifugal casting, Bessemer process, electrolytic aluminum, nickel steels, babbitt, galvanized steel, powder metallurgy, open-hearth steel	Window glass from slit cylinder, light bulb, vulcanization, rubber processing, polyester, styrene, celluloid, rubber extrusion, molding	Steam hammer, steel rolling, seamless tube, steel-rail rolling, continuous rolling, electroplating		Shaping, milling, copying lathe for gunstocks, turret lathe, universal milling machine, vitrified grinding wheel
WWI	1900		Automatic bottle making, bakelite, borosilicate glass	Tube rolling, hot extrusion	Oxyacetylene; arc, electrical-resistance, and thermit welding	Geared lathe, automatic screw machine, hobbing, high-speed-steel tools, aluminum oxide and silicon carbide (synthetic)
WWII	1920	Die casting	Development of plastics, casting, molding, polyvinyl chloride, cellulose acetate, polyethylene, glass fibers	Tungsten wire from metal powder	Coated electrodes	Tungsten carbide, mass production, transfer machines
Second Industrial Revolution: 1947–	1940	Lost-wax process for engineering parts	Acrylics, synthetic rubber, epoxies, photosensitive glass	Extrusion (steel), swaging, powder metals for engineering parts	Submerged arc welding	Phosphate conversion coatings, total quality control
	1950	Ceramic mold, nodular iron, semiconductors, continuous casting	Acrylonitrile-butadiene-styrene, silicones, fluorocarbons, polyurethane, float glass, tempered glass, glass ceramics	Cold extrusion (steel), explosive forming, thermomechanical processing	Gas metal arc, gas tungsten arc, and electroslag welding; explosion welding	Electrical and chemical machining, automatic control

TABLE I.2

Historical Development of Materials, Tools, and Manufacturing Processes

Period	Dates	Metals and casting	Various materials and composites	Forming and shaping	Joining	Tools, machining, and manufacturing systems
Space age	1960–1970	Squeeze casting, single-crystal turbine blades	Acetals, polycarbonate, cold forming of plastics, reinforced plastics, filament winding	Hydroforming, hydrostatic extrusion, electroforming	Plasma-arc and electron-beam welding, adhesive bonding	Titanium carbide, synthetic diamond, numerical control, integrated circuit chip
	1970–1990	Compacted graphite, vacuum casting, organically bonded sand, automation of molding and pouring, rapid solidification, metal-matrix composites, semisolid metalworking, amorphous metals, shape-memory alloys (smart materials), computer simulation	Adhesives, composite materials, semiconductors, optical fibers, structural ceramics, ceramic-matrix composites, biodegradable plastics, electrically conducting polymers	Precision forging, isothermal forging, superplastic forming, dies made by computer-aided design and manufacturing, net-shape forging and forming, computer simulation	Laser beam, diffusion bonding (also combined with superplastic forming), surface-mount soldering	Cubic boron nitride, coated tools, diamond turning, ultraprecision machining, computer-integrated manufacturing, industrial robots, machining and turning centers, flexible-manufacturing systems, sensor technology, automated inspection, expert systems, artificial intelligence, computer simulation and optimization
Information age	1990–2010	Rheocasting, computer-aided design of molds and dies, rapid tooling, TRIP and TWIP steels	Nanophase materials, metal foams, advanced coatings, high-temperature superconductors, machinable ceramics, diamondlike carbon, carbon nanotubes, graphene	Rapid prototyping, rapid tooling, environmentally friendly metalworking fluids, digital manufacturing	Friction stir welding, lead-free solders, laser butt-welded (tailored) sheet-metal blanks, electrically conducting adhesives, linear friction welding	Micro- and nano fabrication, LIGA (a German acronym for a process involving lithography, electroplating, and molding), dry etching, linear motor drives, artificial neural networks, six sigma, three-dimensional computer chips, blue-arc machining, soft lithography

5

development of solid-state electronic devices and computers (Table I.2). **Mechanization** began in England and other countries of Europe, basically with the development of textile machinery and machine tools for cutting metal. This technology soon moved to the United States, where it continued to be further developed.

A major advance in manufacturing began in the early 1800s, with the design, production, and use of **interchangeable parts**, conceived by the American manufacturer and inventor E. Whitney (1765–1825). Prior to the introduction of interchangeable parts, much hand fitting was necessary, because no two parts could be made exactly alike. By contrast, it is now taken for granted that a broken bolt can easily be replaced with an identical one produced decades after the original. Further developments soon followed, resulting in countless consumer and industrial products that we now cannot imagine being without.

Beginning in the early 1940s, several milestones were reached in all aspects of manufacturing, as can be observed by a detailed review of Table I.2. Note particularly the progress that has been made during the 20th century, compared with that achieved during the 40-century period from 4000 b.c. to 1 b.c.

For example, in the Roman Empire (around 500 b.c. to 476 a.d.), factories were available for the mass production of glassware; however, the methods used were generally very slow, and much manpower was required in handling the parts and operating the machinery. Today, production methods have advanced to such an extent that (a) aluminum beverage cans are made at rates higher than 500 per minute, with each can costing about four cents to make, (b) holes in sheet metal are punched at rates of 800 holes per minute, and (c) light bulbs are made at rates of more than 2000 bulbs per minute.

The period approximately from the 1940s to the 1990s was characterized by **mass production** and expanding **global** markets. Initially, the United States had a dominant position, as it was the only developed nation with an intact infrastructure following World War II; however, this advantage dissipated by the 1960s. The **quality** revolution began to change manufacturing in the 1960s and 1970s, and programmable computers started becoming widely applied in the 1980s.

The era of **digital manufacturing** can be considered to have begun around 1990. As a fundamental change in manufacturing operations, powerful computers and software have now been fully integrated across the design and manufacturing enterprise. Communications advances, some Internet-based, have led to further improvements in organization and capabilities. The effects are most striking when considering the origin and proliferation of rapid prototyping. Prior to 1990, a prototype of a part could be produced only through intensive effort and costly manufacturing approaches, requiring significant operator skill. Today, a part can first be drafted in a CAD program, and then produced in a matter of minutes or hours (depending on size and part complexity) without the need for tools or skilled labor. Over time, prototyping systems have become more economical and faster, and use improved raw materials. The term *digital manufacturing* has been applied to reflect the notion that the manufacture of components can take place completely through such computer-driven CAD and production machinery.

I.2 Product Design and Concurrent Engineering

Product design involves the creative and systematic prescription of the shape and characteristics of an artifact to achieve specified objectives, while simultaneously satisfying several constraints. Design is a critical activity, because it has been estimated that as much as 80% of the cost of product development and manufacture is determined by the decisions made in the *initial* stages of design. The product design process has been studied extensively; it is briefly introduced here because of the strong interactions among manufacturing and design activities.

Innovative approaches are essential in successful product design, as are clearly specified functions and a clear statement of the performance expected of the product. The market for the product, which may be new or a modified version of an existing product, and its anticipated use or uses also must be defined at this stage. This aspect also involves the assistance of market analysts and sales personnel who will bring valuable and timely input to the manufacturer, especially regarding market needs and trends.

The Design Process.   Traditionally, design and manufacturing activities have taken place *sequentially*, as shown in Fig. I.2a. This methodology may, at first, appear to be straightforward and logical; in practice, however, it is wasteful of resources. Consider the case of a manufacturing engineer who, for example, determines that, for a variety of reasons, it would be more desirable to (a) use a different material, such as a polymer or a ceramic instead of a metal, (b) use the same material but in a different condition, such as a softer instead of a harder material or a material with a smoother surface finish, or (c) modify the design of a component in order to make it easier, faster, and less costly to manufacture. Note that these decisions must take place at the sixth box from the top in Fig. I.2a.

Each of the modifications just described will necessitate a repeat of the design analysis stage (the third box from the top in Fig. I.2a) and the subsequent stages. This approach is to ensure that the product will still meet all specified requirements and will function satisfactorily. A later change from, say, a forged, cast, or machined component will, likewise, necessitate a repeat analysis. Such iterations obviously waste both time and the resources of a company.

Concurrent Engineering.   Driven primarily by the consumer electronics industry, a continuing trend has been to bring products to the marketplace as rapidly as possible, so as to gain a higher percentage share of the market and thus higher profits. An important methodology aimed at achieving this end is *concurrent engineering*, which involves the product-development approach shown in Fig. I.2b.

Although this concept, also called **simultaneous engineering**, still has the same general product-flow sequence as in the traditional approach shown in Fig. I.2a, it now contains several deliberate modifications. From the earliest stages of product design and engineering, all relevant disciplines are now *simultaneously* considered. As a result, any iterations that may have to be made will require a smaller effort, thus resulting in much less wasted time than occurs in the traditional approach to design. It should be apparent that a critical feature of this approach is the recognition of the importance of *communication* among and within all relevant disciplines.

Concurrent engineering can be implemented in companies large or small; this is particularly significant because 98% of all U.S. manufacturing companies, for example, have fewer than 500 employees. Such companies are generally referred to as *small businesses*. As an example of the benefits of concurrent engineering, one automotive company reduced the number of components in one of its engines by 30%, decreased the engine weight by 25%, and reduced its manufacturing time by 50%.

Life Cycle.   In concurrent engineering, the design and manufacture of products are integrated with a view toward optimizing all elements involved in the *life cycle* of the

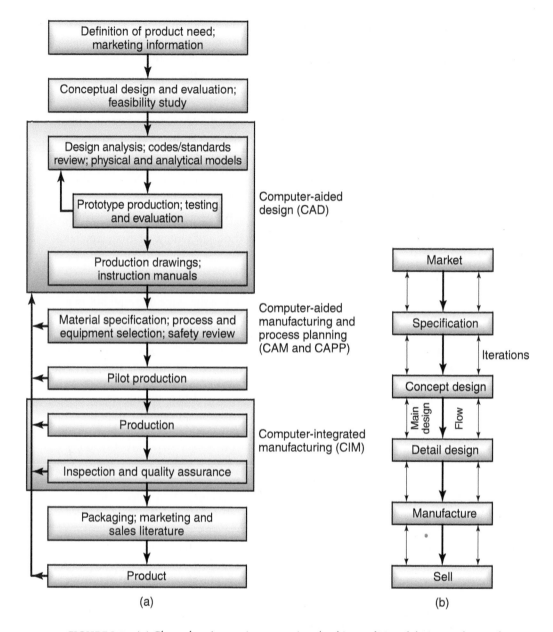

FIGURE I.2   (a) Chart showing various steps involved in *traditional* design and manufacture of a product. Depending on the complexity of the product and the type of materials used, the time span between the original concept and the marketing of the product may range from a few months to several years. (b) Chart showing general product flow in *concurrent engineering*, from market analysis to marketing the product. *Source: After S. Pugh.*

product (see Section I.4). The life cycle of a new product generally consists of four stages:

1. Product start-up
2. Rapid growth of the product in the marketplace
3. Product maturity
4. Decline

Consequently, **life-cycle engineering** requires that the *entire life* of a product be considered, beginning with the design stage and on through production, distribution, product use, and, finally, recycling or disposal of the product.

Role of Computers in Product Design.   Typically, product design first requires the preparation of *analytical* and *physical models* of the product, for the purposes of visualization and engineering analysis. Although the need for such models depends on product complexity, constructing and studying these models are now highly simplified through the use of **computer-aided design** (CAD) and **computer-aided engineering** (CAE) techniques.

CAD systems are capable of rapid and thorough analyses of designs, whether it is a simple shelf bracket or a gear in large and complex structures. The Boeing 777 passenger airplane, for example, was designed completely by computers, in a process known as **paperless design,** with 2000 workstations linked to eight design servers. Unlike previous mock-ups of aircraft, no prototypes or mock-ups were built and the 777 was constructed and assembled *directly* from the CAD/CAM software that had been developed.

Through computer-aided engineering, the performance of structures subjected, for example, to static or fluctuating loads or to temperature gradients also can be simulated, analyzed, and tested, rapidly and accurately. The information developed is stored and can be retrieved, displayed, printed, and transferred anytime and anywhere within a company's organization. Design modifications can be made and optimized (as is often the practice in engineering, especially in the production of large structures such as aircraft) directly, easily, and at any time.

Computer-aided manufacturing involves all phases of manufacturing, by utilizing and processing the large amount of information on materials and processes gathered and stored in the organization's database. Computers greatly assist in organizing the information developed and performing such tasks as (a) programming for numerical-control machines and for robots for material-handling and assembly operations, (b) designing tools, dies, molds, fixtures, and work-holding devices, and (c) maintaining quality control throughout the total operation.

On the basis of the models developed and analyzed in detail, product designers then finalize the geometric features of each of the product's components, including specifying their dimensional tolerances and surface-finish characteristics. Because all components, regardless of their size, eventually have to be *assembled* into the final product, dimensional tolerances are a major consideration in manufacturing. Indeed, dimensional tolerances are equally important for small products as well as for car bodies or aircraft. The models developed also allow the specification of the mechanical and physical properties required, which in turn affect the selection of materials.

Prototypes.   A *prototype* is a physical model of an individual component or product. The prototypes are carefully reviewed for possible modifications to the original design, materials, or production methods. An important and continuously evolving technology is **rapid prototyping**. Using CAD/CAM and various specialized technologies, designers can now make prototypes rapidly and at low cost, from metallic or nonmetallic materials such as plastics and ceramics.

Rapid prototyping can significantly reduce costs and the associated product-development times. Rapid-prototyping techniques are now advanced to such a level that they also can be used for low-volume (in batches typically of fewer than 100 parts) economical production of a variety of actual and functional parts, later to be assembled into products.

Virtual Prototyping. *Virtual prototyping* is a software-based method that uses advanced graphics and virtual-reality environments to allow designers to view and examine a part in detail. This technology, also known as **simulation-based design,** uses CAD packages to render a part such that, in a 3-D interactive virtual environment, designers can observe and evaluate the part as it is being developed. Virtual prototyping has been gaining importance, especially because of the availability of low-cost computers and simulation and analysis tools.

I.3 Design for Manufacture, Assembly, Disassembly, and Service

Design for manufacture (DFM) is a *comprehensive* approach to integrating the design process with production methods, materials, process planning, assembly, testing, and quality assurance. DFM requires a fundamental understanding of (a) the characteristics and capabilities of materials, manufacturing processes, machinery, equipment, and tooling and (b) variability in machine performance, dimensional accuracy and surface finish of the workpiece, processing time, and the effect of processing methods employed on product quality. Establishing *quantitative* relationships is essential in order to be able to analyze and optimize a design for ease of manufacturing and assembly at the lowest cost.

The concepts of **design for assembly** (DFA), **design for manufacture and assembly** (DFMA), and **design for disassembly** (DFD) are all important considerations in manufacturing. Methodologies and computer software are available for design for assembly, utilizing 3-D conceptual designs and solid models. Subassembly, assembly, and disassembly times and costs can now be minimized, while product integrity and performance are maintained. Experience has indicated that a product which is easy to assemble is, usually, also easy to disassemble.

Assembly is an important phase of manufacturing, requiring considerations of the ease, speed, and cost of putting together the numerous individual components of a product (Fig. I.3). Depending on the type of product, assembly costs in manufacturing can be substantial, typically ranging from 10 to 60% of the total product cost. *Disassembly* of a product is an equally important consideration, for such activities as maintenance, servicing, and recycling of individual components.

There are several methods of assembly of components, including the use of a wide variety of fasteners, adhesives, or joining techniques such as welding, brazing, or soldering. As is the case in all types of manufacturing, each of these assembly operations has its own specific characteristics, times, advantages and limitations, associated costs, and special design considerations. Individual parts may be assembled by hand or by a variety of automatic equipment and industrial robots. The choice depends on several factors, such as product complexity, the number of components to be assembled, the care and protection required to prevent damage to the individual parts, and the relative cost of labor compared with the cost of machinery required for automated assembly.

Design for Service. In addition to design for assembly and for disassembly, *design for service* is an important aspect of product design. Products often have to be disassembled to varying degrees in order to service them and, if necessary, repair them. The design should take into account the concept that, for ease of access, components that are most likely to be in need of servicing be placed, as much as possible, at the outer layers of the product. This methodology can be appreciated by anyone who has had the experience of servicing machinery.

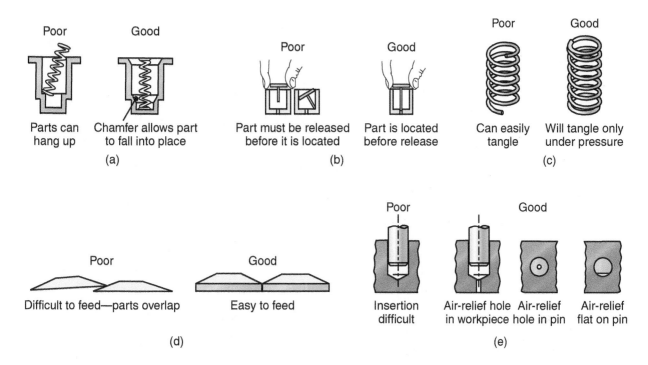

FIGURE I.3 Redesign of parts to facilitate assembly. *Source:* After G. Boothroyd and P. Dewhurst.

I.4 Green Design and Manufacturing

In the United States alone, 9 million passenger cars, 300 million tires, 670 million compact fluorescent lamps, and more than 5 billion kg of plastic products are discarded each year. Every 3 months, industries and consumers discard enough aluminum to rebuild the U.S. commercial air fleet. Note that, as indicated below, the term *discarding* suggests that the product has reached the end of its useful life; it does not necessarily mean that it has to be dumped into landfills.

The particular manufacturing process and the operation of machinery can each have a significant environmental impact. Manufacturing operations generally produce some waste, such as:

1. Chips from machining and trimmed materials from sheet forming, casting, and molding operations
2. Slag from foundries and welding operations
3. Additives in sand used in sand-casting operations
4. Hazardous waste and toxic materials used in various products
5. Lubricants and coolants in metalworking and machining operations
6. Liquids from processes such as heat treating and plating
7. Solvents from cleaning operations
8. Smoke and pollutants from furnaces and gases from burning fossil fuels

The adverse effects of these activities, their damage to our environment and to the Earth's ecosystem, and, ultimately, their effect on the quality of human life are now widely recognized and appreciated. Major concerns involve global warming, greenhouse gases (carbon dioxide, methane, and nitrous oxide), acid rain, ozone depletion, hazardous wastes, water and air pollution, and contaminant seepage into water sources. One measure of the adverse impact of human activities is called the **carbon footprint**, which quantifies the amount of greenhouse gases produced in our daily activities. In 2011, the amount of carbon dioxide released worldwide into the air was estimated to be 2.4 million pounds per second; China being the most polluter.

The term **green design and manufacturing** has become common in all industrial activities, with a major emphasis on **design for the environment** (DFE). Also called **environmentally conscious design and manufacturing**, this approach considers all possible adverse environmental impacts of materials, processes, operations, and products, so that they can all be taken into account at the earliest stages of design and production.

These goals, which now have become global, also have led to the concept of **design for recycling** (DFR). Recycling may involve one of two basic activities:

- **Biological cycle:** Organic materials degrade naturally, and in the simplest version of a biological cycle, they lead to new soil that can sustain life; thus, product design involves the use of (usually) organic materials. The products function well for their intended life and can then be safely discarded.
- **Industrial cycle:** The materials in the product are recycled and reused continuously; for example, aluminum beverage cans are recycled and the metal is reused. To demonstrate the economic benefits of this approach, it has been estimated that producing aluminum from scrap, instead of from bauxite ore, reduces production costs by as much as 66% and reduces energy consumption and pollution by more than 90%.

One of the basic principles of design for recycling is the use of materials and product-design features that facilitate biological or industrial recycling. In the U.S. automotive industry, for example, about 75% of automotive parts (mostly metal) are now recycled, and there are continuing plans to recycle the rest as well, including plastics, glass, rubber, and foam. About 80% of the discarded automobile tires are reused in various ways.

Cradle-to-cradle Production.   Also called C2C, *cradle-to-cradle* production considers the impact of each stage of a product's life cycle, from the time natural resources are mined and processed into raw materials, through each stage of manufacturing products, their use and, finally, recycling. *Cradle-to-grave* production, also called *womb-to-tomb* production, has a similar approach, but it does not necessarily consider or take on the responsibility of recycling.

Cradle-to-cradle production especially emphasizes:

1. Sustainable and efficient manufacturing activities, using clean technologies
2. Waste-free production
3. Using recyclable and nonhazardous materials
4. Reducing energy consumption
5. Using renewable energy, such as wind and solar
6. Maintaining ecosystems by minimizing the environmental impact of all manufacturing activities
7. Using materials and energy sources that are available locally, so as to reduce energy use associated with their transport which, by and large, has an inherently high carbon footprint
8. Continuously exploring the reuse and recycling of materials, thus perpetually trying to recirculate materials; also included is investigating the composting of materials, whenever appropriate or necessary, instead of dumping them into landfills

Guidelines for Green Design and Manufacturing.   In reviewing the various activities described thus far, it can be noted that there are overarching relationships among the basic concepts of DFMA, DFD, DFE, and DFR. These relationships can be summarized as guidelines, now rapidly accepted worldwide:

1. Reduce waste of materials, by refining product design, reducing the amount of materials used in products, and selecting manufacturing processes that minimize scrap (such as forming instead of machining).
2. Reduce the use of hazardous materials in products and processes.
3. Investigate manufacturing technologies that produce environmentally friendly and safe products and by-products.
4. Make improvements in methods of recycling, waste treatment, and reuse of materials.
5. Minimize energy use and, whenever possible, encourage the use of renewable sources of energy; selection of materials can have a major impact on the latent energy in products.
6. Encourage recycling by using materials that are a part of either industrial or biological cycling, but not both in the same product assembly. Ensure proper handling and disposal of all waste of materials that are used in products, but are not appropriate for industrial or biological cycling.

I.5   Selection of Materials

An increasingly wide variety of materials are now available, each type having its own specific properties and manufacturing characteristics, advantages and limitations, and costs. The selection of materials for products (consumer or industrial) and their components is typically made in consultation with materials engineers, although design engineers may also be sufficiently experienced and qualified to do so.

The general types of materials used, either individually or in combination with other materials, are:

- **Ferrous metals:** Carbon, alloy, stainless, and tool and die steels (Chapter 5)
- **Nonferrous metals:** Aluminum, magnesium, copper, nickel, titanium, superalloys, refractory metals, beryllium, zirconium, low-melting-point alloys, and precious metals (Chapter 6)
- **Plastics (polymers):** Thermoplastics, thermosets, and elastomers (Chapter 7)
- **Ceramics, glasses, glass ceramics, graphite, diamond, and diamondlike materials** (Chapter 8)
- **Composite materials:** Reinforced plastics and metal-matrix and ceramic-matrix composites (Chapter 9)
- **Nanomaterials** (Section 8.8)
- **Shape-memory alloys** (*smart materials*), **amorphous alloys, semiconductors,** and **superconductors** (Chapters 6, 18 and 28)

As new developments continue, the selection of an appropriate material for a particular application from a very large variety of options has become even more challenging. Furthermore, there are continuously shifting trends in the substitution of materials, driven not only by technological considerations but also by economics.

Properties of Materials.   *Mechanical properties* of interest in manufacturing generally include strength, ductility, hardness, toughness, elasticity, fatigue, and creep resistance (Chapter 2). *Physical properties* are density, specific heat, thermal expansion and conductivity, melting point, and electrical and magnetic properties (Chapter 3). Optimum designs often require a consideration of a combination of mechanical and physical properties. A typical example is the *strength-to-weight* and *stiffness-to-weight* ratios of materials for minimizing the weight of structural members; weight minimization is particularly important for aerospace and automotive applications, in order to improve performance and fuel economy.

Chemical properties include oxidation, corrosion, degradation, toxicity, and flammability; these properties play a significant role under both hostile (such as corrosive) and normal environments. *Manufacturing properties* indicate whether a particular material can be cast, formed, machined, joined, and heat treated with relative ease. As Table I.3 illustrates, no one material has the same manufacturing characteristics. Another consideration is *appearance*, which includes such characteristics as surface texture, color, and feel, all of which can play a significant role in a product's acceptance by the public.

Availability.   As emphasized throughout this book, the economic aspect of material selection is as important as technological considerations (Chapter 40); thus, the availability of materials is a major concern in manufacturing. Furthermore, if materials are not available in the desired shapes, dimensions, surface texture, and quantities, then materials substitution or additional processing of a particular material may well be required, all of which can contribute significantly to product cost.

Reliability of supply is important in order to meet production schedules; in automotive industries, for example, materials must arrive at a plant at appropriate time intervals. (See also *just-in-time*, Section I.7.) Reliability of supply is also important, because most countries import numerous raw materials. The United States, for example, imports most of the cobalt, titanium, chromium, aluminum, nickel, natural rubber, and diamond that it needs. Consequently, a country's self-reliance on resources, especially energy, is an often-expressed political goal, but challenging to achieve. *Geopolitics* (defined briefly as the study of the influence of a nation's physical geography on its foreign policy) must thus be a consideration, particularly during periods of global instability or hostility.

Service Life.   Everyone has directly experienced a shortened service life of a product, which often can be traced to one or more of the following: (a) improper selection of materials, (b) improper selection of production methods, (c) insufficient control of processing variables, (d) defective raw materials or parts, or manufacturing-induced

TABLE I.3

General Manufacturing Characteristics of Various Materials			
Alloy	Castability	Weldability	Machinability
Aluminum	Excellent	Fair	Excellent–good
Copper	Good–fair	Fair	Good–fair
Gray cast iron	Excellent	Difficult	Good
White cast iron	Good	Very poor	Very poor
Nickel	Fair	Fair	Fair
Steels	Fair	Excellent	Fair
Zinc	Excellent	Difficult	Excellent

The ratings shown depend greatly on the particular material, its alloys, and its processing history.

defects, (e) poor maintenance of machinery or equipment, and (f) improper use of the product.

Generally, a product is considered to have failed when it

- Stops functioning, due to the failure of one or more of its components, such as a broken shaft, gear, bolt, or turbine blade, or a burned-out electric motor
- Does not function properly or perform within required specifications, due, for example, to worn gears or bearings
- Becomes unreliable or unsafe for further use, as in the erratic behavior of a switch, poor connections in a printed-circuit board, or delamination of a composite material

Material Substitution in Products.   For a variety of reasons, numerous substitutions are made in materials, as evidenced by a routine inspection and comparison of common products, such as home appliances, sports equipment, or automobiles. As a measure of the challenges faced in material substitution, consider the following examples: (a) metal versus wooden handle for a hammer, (b) aluminum versus cast-iron lawn chair, (c) copper versus aluminum electrical wire, (d) plastic versus steel car bumper, and (e) alloy steel versus titanium submarine hull.

The following two case studies describe some details of the major factors involved in material substitution in common products.

CASE STUDY I.1   Baseball Bats

Baseball bats for the major and minor leagues are generally made of wood from the northern white ash tree, a material that has high dimensional stability, high elastic modulus and strength-to-weight ratio, and high shock resistance. Wooden bats can break, however, and may cause serious injury. (This is especially true of the relatively new trend of using maple wood in baseball bats.) Wooden bats are made on semiautomatic lathes (Section 23.3), followed by finishing operations for appearance and labeling. The straight uniform grain required for such bats has become increasingly difficult to find, particularly when the best wood comes from ash trees that are at least 45 years old.

For the amateur market and for school and college players, aluminum bats (top portion of Fig. I.4) have been made since the 1970s as a cost-saving alternative to wood. The bats are made by various metalworking operations, described throughout Part III. Although, at first, their performance was not as good as that of wooden bats, the technology has advanced to a great extent. Metal bats are now mostly made from high-strength aluminum tubing, but can incorporate titanium. The bats are designed to have the same center of percussion (known as the

FIGURE I.4   Cross-sections of baseball bats made of aluminum (top two) and composite material (bottom two).

(*continued*)

sweet spot, as in tennis racquets) as wooden bats, and are usually filled with polyurethane or cork for improved sound damping and for controlling the balance of the bat.

Metal bats possess such desirable performance characteristics as lower weight than wooden bats, optimum weight distribution along the bat's length, and superior impact dynamics. Also, as documented by scientific studies, there is a general consensus that metal bats outperform wooden bats. Further

developments in bat materials include composite materials (Chapter 9), consisting of high-strength graphite or glass fibers embedded in an epoxy resin matrix. The inner woven sleeve (lower portion of Fig. I.4) is made of Kevlar fibers (an aramid), which add strength to the bat and dampen its vibrations; these bats perform and sound much like wooden bats.

Source: Mizuno USA, Inc.

CASE STUDY I.2   U.S. Pennies

Billions of pennies are produced and put into circulation each year by the U.S. Mint. The materials used have undergone significant changes throughout its history, largely because of periodic material shortages and the resulting fluctuating cost of appropriate materials. The following table shows the chronological development of material substitutions in pennies:

1793–1837	100% copper
1837–1857	95% copper, 5% tin and zinc
1857–1863	88% copper, 12% nickel
1864–1962	95% copper, 5% tin and zinc
1943 (WWII years)	Steel, plated with zinc
1962–1982	95% copper, 5% zinc
1982–present	97.5% zinc, plated with copper

I.6   Selection of Manufacturing Processes

There is often more than one method that can be employed to produce a component for a product from a given material. The following broad categories of manufacturing methods are all applicable to metallic as well as nonmetallic materials:

1. Casting (Fig. I.5a): Expendable mold and permanent mold (Part II)
2. Forming and shaping (Figs. I.5b–d): Rolling, forging, extrusion, drawing, sheet forming, powder metallurgy, and molding (Part III)
3. Machining (Fig. I.5e): Turning, boring, drilling, milling, planing, shaping, broaching; grinding; ultrasonic machining; chemical, electrical, and electrochemical machining; and high-energy-beam machining (Part IV); this broad category also includes micromachining for producing ultraprecision parts (Part V)
4. Joining (Fig. I.5f): Welding, brazing, soldering, diffusion bonding, adhesive bonding, and mechanical joining (Part VI)
5. Finishing: Honing, lapping, polishing, burnishing, deburring, surface treating, coating, and plating (Chapters 26 and 34)
6. Microfabrication and nanofabrication: Technologies that are capable of producing parts with dimensions at the micro (one-millionth of a meter) and nano (one-billionth of a meter) levels; fabrication of microelectromechanical systems (MEMS) and nanoelectromechanical systems (NEMS), typically involving processes such as lithography, surface and bulk micromachining, etching, LIGA, and various other specialized processes (Chapters 28 and 29)

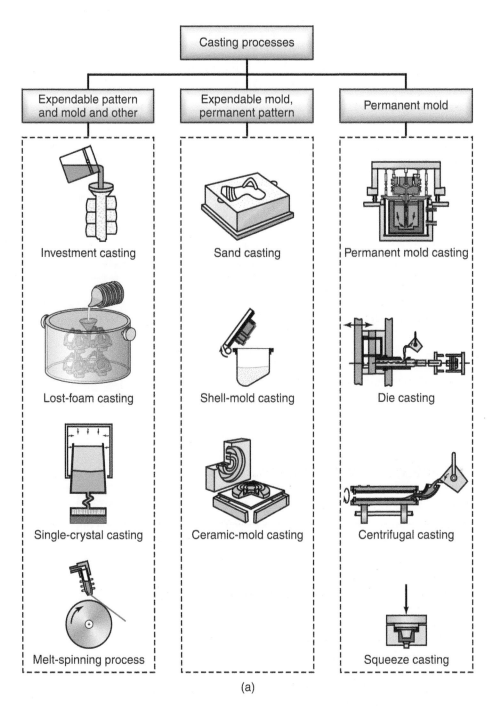

(a)

FIGURE I.5a   Schematic illustrations of various casting processes.

Process Selection.   The selection of a particular manufacturing process or, more often, sequence of processes, depends on the geometric features of the parts to be produced, including the dimensional tolerances and surface texture required, and on numerous factors pertaining to the particular workpiece material and

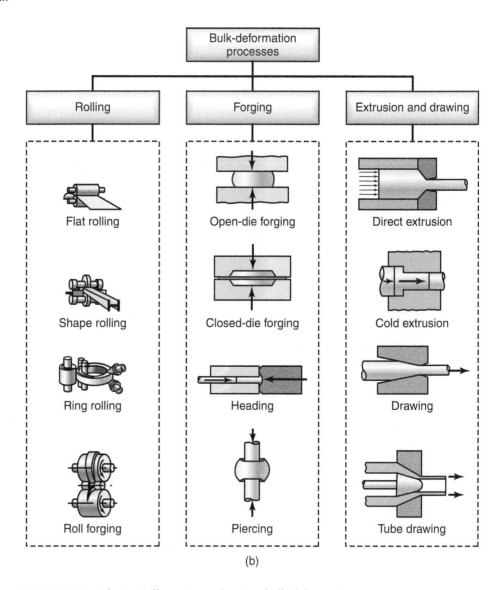

(b)

FIGURE I.5b   Schematic illustrations of various bulk-deformation processes.

its manufacturing properties. To emphasize the challenges involved, consider the following two cases:

1. Brittle and hard materials cannot be shaped or formed without the risk of fracture, unless they are performed at elevated temperatures, whereas these materials can be cast, machined, or ground with relative ease.
2. Metals that have been preshaped at room temperature become less formable during subsequent processing, which, in practice, is often necessary to finish the part; this is because the metals have become stronger, harder, and less ductile than they were prior to processing them further.

There is a constant demand for new approaches to production challenges and, especially, for manufacturing cost reduction; for example, sheet-metal parts traditionally have been cut and fabricated using common mechanical tools, such as punches and dies. Although still widely used, some of these operations have been replaced by

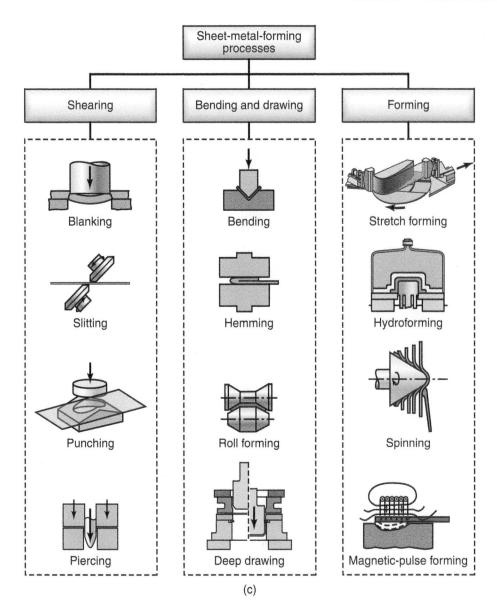

FIGURE I.5c   Schematic illustrations of various sheet-metal forming processes.

laser cutting, as shown in Fig. I.6. This method eliminates the need for hard tools, which typically have only fixed shapes, and can be expensive and time consuming to make.

The laser path in this cutting operation is computer controlled, thereby increasing the operation's flexibility and its capability of producing an infinite variety of shapes accurately, repeatably, and economically. Because of the high heat involved in using lasers, however, the surfaces produced after cutting have very different characteristics (such as surface texture and discoloration) than those produced by traditional methods. This difference can have significant adverse effects, not only on appearance but especially on its subsequent processing and in the service life of the product. Moreover, the inherent flexibility of the laser cutting process is countered by the fact that it is slower than traditional punching operations.

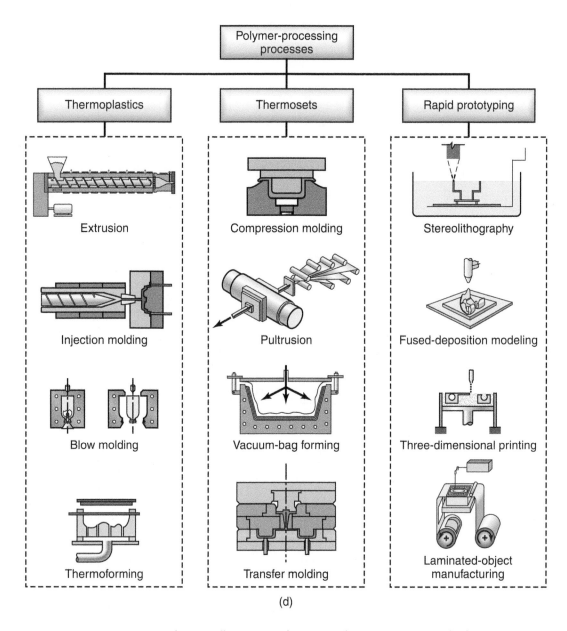

(d)

FIGURE I.5d Schematic illustrations of various polymer-processing methods.

Several factors can have a major role in process selection, including part size, shape complexity, and dimensional accuracy and surface finish required. For example:

- Flat parts and thin cross-sections can be difficult to cast.
- Complex parts generally cannot be shaped easily and economically by such metalworking techniques as forging, whereas, depending on part size and level of complexity, the parts may be precision cast, fabricated and assembled from individual pieces, or produced by powder metallurgy techniques.
- Dimensional tolerances and surface finish in hot-working operations are not as fine as those obtained in operations performed at room temperature (cold

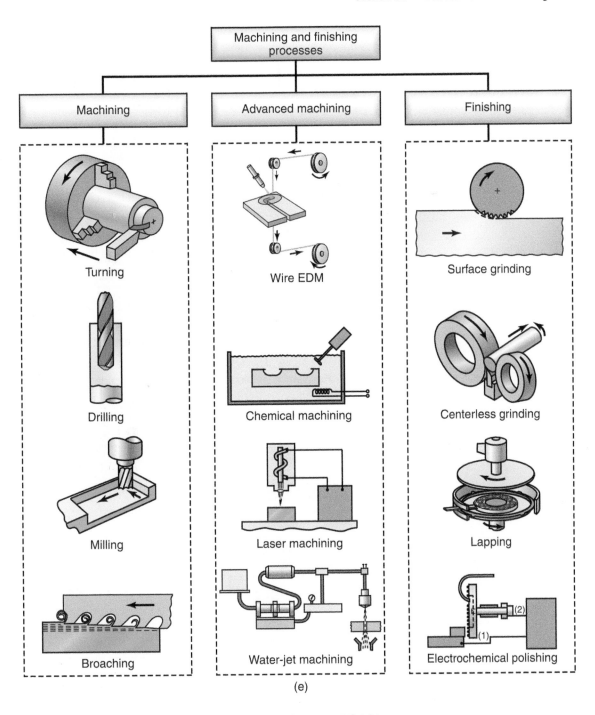

Machining and finishing processes

Machining | Advanced machining | Finishing

Turning

Drilling

Milling

Broaching

Wire EDM

Chemical machining

Laser machining

Water-jet machining

Surface grinding

Centerless grinding

Lapping

Electrochemical polishing

(e)

FIGURE I.5e Schematic illustrations of various machining and finishing processes.

working), because of the dimensional changes, distortion, warping, and surface oxidation due to the elevated temperatures employed.

The size of manufactured products, and the machinery and equipment involved in processing them, vary widely, ranging from microscopic gears and mechanisms of micrometer size, as illustrated in Fig. I.7, to (a) the main landing gear for the Boeing

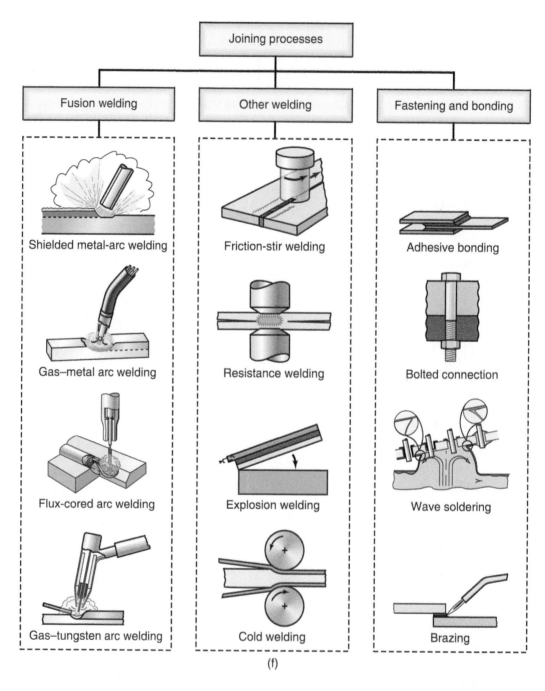

FIGURE I.5f Schematic illustrations of various joining processes.

777 aircraft, which is 4.3 m (14 ft) tall and includes three axles and six wheels; (b) the runner for the turbine for a hydroelectric power plant, which is 4.6 m (180 in.) in diameter and weighs 50,000 kg (110,000 lb); and (c) a large steam turbine rotor, weighing 300,000 kg (700,000 lb).

Process Substitution.   It is common practice in industry that, for a variety of reasons and after a review of all appropriate and applicable processes, a particular production method (that may have been employed in the past) may well have to be substituted with another. Consider, for example, the following products that can be produced by any of the sets of the following processes: (a) cast versus forged crankshaft, (b) stamped sheet-metal versus forged or cast automobile wheels, (c) cast versus stamped sheet-metal frying pan, (d) injection molded versus extruded or cast polymer bracket, and (e) welded versus riveted sheet-metal safety hood for a machine.

Criteria for their selection include factors such as cost, the maintenance required, whether the product is for industrial or consumer use, the parameters to which the product will be subjected (such as external forces, impact, temperatures, and chemicals), environmental concerns that have to be addressed, and the product's appeal to the customer.

Net-shape and Near-net-shape Manufacturing.   *Net-shape* and *near-net-shape manufacturing* together constitute an important methodology, by which a part is made in only one operation and at or close to the final desired dimensions, tolerances, and surface finish. The difference between net shape and near net shape is a matter of degree of how close the product is to its final dimensional and surface finish characteristics.

The necessity for, and benefits of, net-shape manufacturing can be appreciated from the fact that, in the majority of cases, more than one additional manufacturing operation or step is often necessary to produce the part. For example, a cast or forged crankshaft generally will not have the necessary dimensional surface finish characteristics, and will typically require additional processing, such as machining or grinding. These additional operations can contribute significantly to the cost of a product.

Typical examples of net-shape manufacturing include precision casting (Chapter 11), forging (Chapter 14), forming sheet

FIGURE I.6   Cutting sheet metal with a laser beam. *Source:* Courtesy of Rofin-Sinar, Inc., and Society of Manufacturing Engineers.

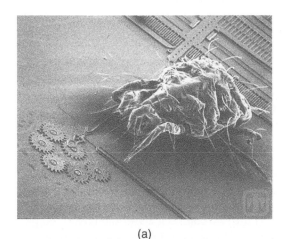

(a)

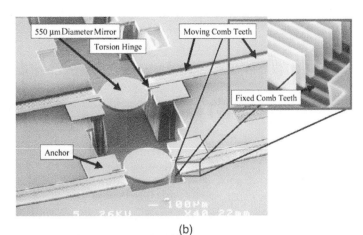

(b)

FIGURE I.7   (a) Microscopic gears with dust mite. *Source:* Courtesy of Sandia National Laboratory. Printed with permission; (b) a movable micromirror component of a light sensor; note the scale at the bottom of the figure. *Source:* Courtesy of R. Mueller, University of California at Berkeley.

metal (Chapter 16), powder metallurgy and injection molding of metal powders (Chapter 17), and injection molding of plastics (Chapter 19).

Ultraprecision Manufacturing.   Dimensional accuracies for some modern equipment and instrumentation have now reached the magnitude of the atomic lattice (below 1 nm). Several techniques, including the use of highly sophisticated technologies (see *micromechanical and microelectromechanical device fabrication*, Chapter 29), are rapidly being developed to attain such extreme accuracy. Also, mirrorlike surfaces on metals can now be produced by machining, using a very sharp diamond, with a nose radius of 250 μm, as the cutting tool. The equipment is highly specialized, with very high stiffness (to minimize deflections, as well as vibration and chatter, during machining). It is operated in a room where the ambient temperature is controlled to within 1°C, in order to avoid thermal distortions of the machine.

Types of Production.   The number of parts to be produced (such as the annual quantity) and the production rate (the number of pieces made per unit time) are important economic considerations in determining the appropriate processes and the types of machinery required. Note, for example, that light bulbs, beverage cans, fuel-injection nozzles, and hubcaps are produced in numbers and at rates that are much higher than those for jet engines.

A brief outline of the general types of production, in increasing order of annual quantities produced, are:

1. **Job shops:** Small lot sizes, typically less than 100, using general-purpose machines, such as lathes, milling machines, drill presses, and grinders, many now typically equipped with computer controls.
2. **Small-batch production:** Quantities from about 10 to 100, using machines similar to those in job shops.
3. **Batch production:** Lot sizes typically between 100 and 5000, using more advanced machinery with computer control.
4. **Mass production:** Lot sizes generally over 100,000, using special-purpose machinery, known as dedicated machines, and various automated equipment in a plant for transferring materials and parts in progress.

CASE STUDY I.3   Saltshaker and Pepper Mill

The saltshaker and pepper mill set shown in Fig. I.8 consists of metallic as well as nonmetallic components. The main parts (the body) of the set are made by injection molding of a thermoplastic (Chapter 19), such as an acrylic, which has both transparency and other desirable characteristics for this application and is easy to mold. The round metal top of the saltshaker is made of sheet metal, has punched holes (Chapter 16), and is electroplated for improved appearance (Section 34.9).

The knob on the top of the pepper mill is made by machining (Chapter 23) and is threaded on the inside to allow it to be screwed and unscrewed. The square rod connecting the top portion of the pepper mill to the two pieces shown at the bottom of the figure is made by a rolling operation (Chapter 13). The two grinder components shown at the bottom of the figure are made of stainless steel. A design for manufacturing analysis indicated that casting or machining the two components would be too costly; consequently, it was determined that an appropriate and economical method would be the powder metallurgy technique (Chapter 17).

FIGURE I.8   A saltshaker and pepper mill set. The two metal pieces (at the bottom) for the pepper mill are made by powder metallurgy techniques. *Source:* Metal Powder Industries Federation.

I.7   Computer-integrated Manufacturing

Computer-integrated manufacturing (CIM), as the name suggests, integrates the software and hardware needed for computer graphics, computer-aided modeling, and computer-aided design and manufacturing activities, from initial product concept through its production and distribution in the marketplace. This comprehensive and integrated approach began in the 1970s and has been particularly effective because of its capability of making possible the following tasks:

- Responsiveness to rapid changes in product design modifications and to varying market demands
- Better use of materials, machinery, and personnel
- Reduction in inventory
- Better control of production and management of the total manufacturing operation

The following is a brief outline of the various elements in CIM, all described in detail in Chapters 38 and 39:

1. **Computer numerical control** (CNC). First implemented in the early 1950s, this is a method of controlling the movements of machine components by the direct insertion of coded instructions in the form of numerical data.
2. **Adaptive control** (AC). The processing parameters in an operation are automatically adjusted to optimize the production rate and product quality and to minimize manufacturing costs. For example, in machining, the forces, temperature, surface finish, and the dimensions of the part can be constantly monitored. If they move outside a specified range, the system automatically adjusts the appropriate variables until the parameters are within the specified range.
3. **Industrial robots.** Introduced in the early 1960s, industrial robots (Fig. I.9) have rapidly been replacing humans, especially in operations that are repetitive, dangerous, and boring. As a result, variability in product quality is decreased and productivity improved. Robots are particularly effective in material movement

FIGURE I.9   Automated spot welding of automobile bodies in a mass-production line.

and assembly operations; *intelligent* robots have been developed with sensory-perception capabilities and movements that simulate those of humans.

4. **Automated materials handling.** Computers have made possible highly efficient handling of materials and components in various stages of completion (work in progress), as in moving a part from one machine to another, and then to points of inspection, to inventory, and finally, to shipment.

5. **Automated assembly systems.** These systems have been developed to replace assembly by humans, although they still have to perform some operations. Depending on the type of product, assembly costs can be high; consequently, products must be designed such that they can be assembled more easily and faster by automated machinery.

6. **Computer-aided process planning** (CAPP). By optimizing process planning, this system is capable of improving productivity, product quality, and consistency, and thus reducing costs. Functions such as cost estimating and monitoring work standards (time required to perform a certain operation) are also incorporated into the system.

7. **Group technology** (GT). The concept behind group technology is that numerous parts can be grouped and produced by classifying them into families according to similarities in (a) design and (b) the manufacturing processes employed to produce them. In this way, part designs and process plans can be standardized and new parts, based on similar parts made previously, can be produced efficiently and economically.

8. **Just-in-time production** (JIT). The principle behind JIT is that (a) supplies of raw materials and parts are delivered to the manufacturer just in time to be used, (b) parts and components are produced just in time to be made into subassemblies, and (c) products are assembled and finished just in time to be delivered to the customer. As a result, inventory carrying costs are low, defects in components are detected right away, productivity is increased, and high-quality products are made and at low cost.

9. **Cellular manufacturing** (CM). This system utilizes workstations that consist of a number of *manufacturing cells*, each containing various production machinery,

all controlled by a central robot, with each machine performing a specific operation on the part, including inspection.

10. **Flexible manufacturing systems** (FMS). These systems integrate manufacturing cells into a large production facility, with all of the cells interfaced with a central computer. Although very costly, flexible manufacturing systems are capable of producing parts efficiently (although in relatively small quantities, because hard automation is still most efficient for mass production) and of quickly changing manufacturing sequences required for making different parts. Flexibility enables these systems to meet rapid changes in market demand for all types of products.

11. **Expert systems** (ES). Consisting basically of complex computer programs, these systems have the capability of performing various tasks and solving difficult real-life problems (much as human experts would), including expediting the traditional iterative process.

12. **Artificial intelligence** (AI). Computer-controlled systems are capable of learning from experience and of making decisions that optimize operations and minimize costs, ultimately replacing human intelligence.

13. **Artificial neural networks** (ANN). These networks are designed to simulate the thought processes of the human brain, with such capabilities as modeling and simulating production facilities, monitoring and controlling manufacturing operations, diagnosing problems in machine performance, and conducting financial planning and managing a company's manufacturing strategy.

CASE STUDY I.4   Mold for Making Sunglass Frames

The metal mold used for injection molding of plastic sunglass frames is made on a computer numerical-control milling machine, using a cutter (called a ball-nosed end mill) as illustrated in Fig. I.10. First, a model of the sunglass is made using a computer-aided design software package, from which a model of the mold is automatically generated. The geometric information is sent to the milling machine, and the machining steps are planned.

Next, an offset is added to each surface to account for the nose radius of the end mill during machining, thus determining the cutter path (the path followed by the center of rotation of the machine spindle). The numerical-control programming software executes this machining program on the milling machine, producing the die cavity with appropriate dimensions and tolerances. Electrical-discharge machining (Section 27.5) can also be used to make this mold; however, it was determined that the operation was about twice as expensive as machining the mold by computer numerical control, and it also produced molds that had lower dimensional accuracy.

Source: Based on Mastercam/CNC Software, Inc.

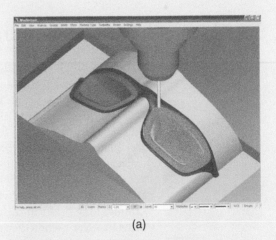

(a)

FIGURE I.10   Machining a mold cavity for making sunglasses. (a) Computer model of the sunglasses as designed and viewed on the monitor. *Source:* Courtesy of Mastercam/CNC Software, Inc.

(*continued*)

(b)

(c)

FIGURE I.10 (*continued*) (b) Machining of the die cavity, using a computer numerical-control milling machine. (c) Final product produced from the mold. *Source:* Courtesy of Mastercam/CNC Software, Inc.

I.8   Quality Assurance and Total Quality Management

Product quality is one of the most critical considerations in manufacturing, because it directly influences customer satisfaction, thus playing a crucial role in determining a product's success in the marketplace (Chapter 36). The traditional approach of inspecting products after they are made has largely been replaced by the recognition that *quality must be built into the product,* from its initial design through all subsequent stages of manufacture and assembly.

Because products are typically made through several manufacturing steps and operations, each step can involve its own significant variations in performance; this variation can occur even within a relatively short time. A production machine, for example, may perform differently when it is first turned on than after it warms up during its use, or when the ambient temperature in the plant fluctuates. Consequently, *continuous control of processes* (known as *online monitoring*) is a critical factor in maintaining product quality, the objective being to *control processes, not products.*

Quality assurance and *total quality management* (TQM) are widely recognized as being the responsibility of everyone involved in the design and manufacture of products and their components. *Product integrity* is a term generally used to define the degree to which a product

- Functions reliably during its life expectancy (Table I.4)
- Is suitable for its intended purposes
- Can be maintained with relative ease

Producing and marketing defective products can be very costly to the manufacturer, with costs varying by orders of magnitude, as shown in Table I.5.

Pioneers in quality control, particularly W.E. Deming (1900–1993), J.M. Juran (1904–2008), and G. Taguchi (1924–2012), all emphasized the importance of management's commitment to (a) product quality, (b) pride of workmanship at all levels of production, and (c) the necessity of using **statistical process control** (SPC) and **control charts** (Chapter 36). They also pointed out the importance of online monitoring and rapidly identifying the *sources of quality problems* in production, before even another defective part is produced. The major goal of control is to *prevent* defective parts from ever being made, rather than to inspect, detect, and reject defective parts after they have been made.

As an example of strict quality control, computer chips are now produced with such high quality that only a few out of a million chips may be defective. The level of defects is identified in terms of **standard deviation**, denoted by the symbol σ (the Greek letter *sigma*). Three sigma would result in 2700 defective parts per million, which is much too high in modern manufacturing. In fact, it has been estimated that at this level, no modern computer would function reliably. At **six sigma**, defective parts are reduced to only 3.4 per million parts made; this level has been reached through major improvements in manufacturing *process capabilities* in order to *reduce variability* in product quality.

Important developments in quality assurance include the implementation of **experimental design**, a technique in which the factors involved in a manufacturing operation and their interactions are studied simultaneously. For example, the variables affecting dimensional accuracy or surface finish in a machining operation can be identified readily, thus making it possible for appropriate preventive on-time adjustments to be taken.

Quality Standards.   Global manufacturing and competitiveness have led to an obvious need for international conformity and consensus in establishing quality control methods. This need resulted in the establishment of the ISO 9000 standards series on quality management and quality assurance standards, as well as of the QS 9000 standards (Section 36.6), introduced in 1994. A company's registration for these standards, which is a *quality process certification* and not a product certification, means that the company conforms to consistent practices as specified by its own quality system. ISO 9000 and QS 9000 have permanently influenced the manner in which companies conduct business in world trade, and they are now the world standard for quality.

Human-factors Engineering.   This topic deals with human–machine interactions, and thus is an important aspect of manufacturing operations in a plant, as well as of products in their expected use. The human-factors approach is essential in the design and manufacture of safe products; it emphasizes **ergonomics**, defined as the study of how a workplace and the machinery and equipment in it can best be designed and arranged for comfort, safety, efficiency, and productivity.

Some examples of the need for proper ergonomic considerations are: (a) a mechanism that is difficult to operate manually, causing injury to the worker, (b) a poorly

TABLE I.4

Average Life Expectancy of Various Products

Type of product	Life expectancy (years)
U.S. dollar bill	1.5
Personal computer	2
Car battery	4
Hair dryer	5
Automobile	8
Dishwasher	10
Kitchen disposal unit	10
Vacuum cleaner	10
Water heater (gas)	12
Clothes dryer (gas)	13
Clothes washer	13
Air-conditioning unit (central)	15
Manufacturing cell	15
Refrigerator	17
Furnace (gas)	18
Machinery	30
Nuclear reactor	40

Note: Significant variations can be expected, depending on the quality of the product and how well it has been maintained.

TABLE I.5

Relative Cost of Repair at Various Stages of Product Development and Sale

Stage	Relative cost of repair
When the part is being made	1
Subassembly of the product	10
Assembly of the product	100
Product at the dealership	1000
Product at the customer	10,000

designed keyboard that causes pain to the user's hands and arms during its normal use (known as *repetitive stress syndrome*), and (c) a control panel on a machine that is difficult to reach or use safely and comfortably.

Product Liability.   Designing and manufacturing safe products is an essential aspect of a manufacturer's responsibilities. All those involved with product design, manufacture, and marketing must fully recognize the consequences of a product's failure, including failure due to foreseeable misuse of the product.

A product's malfunction or failure can cause bodily injury or even death, as well as financial loss to an individual, a bystander, or an organization. This important topic is referred to as *product liability*. The laws governing it generally vary from state to state and from country to country. Among the numerous examples of products that could involve liability are:

- A grinding wheel shatters and causes injury to a worker
- A cable supporting a platform snaps, allowing the platform to drop and cause bodily harm or death
- Automotive brakes suddenly become inoperative because of the failure of a particular component of the brake system
- Production machinery lacks appropriate safety guards
- Electric and pneumatic tools lack appropriate warnings and instructions for their safe use

I.9   Lean Production and Agile Manufacturing

Lean production (Section 39.6) is a methodology that involves a thorough assessment of each activity of a company, with the basic purpose of minimizing waste at all levels and calling for the elimination of unnecessary operations that do not provide any added value to the product being made. This approach, also called *lean manufacturing*, identifies all of a manufacturer's activities and optimizes the processes used in order to *maximize added value*. Lean production focuses on (a) the efficiency and effectiveness of each and every manufacturing step, (b) the efficiency of the machinery and equipment used, and (c) the activities of the personnel involved in each operation. This methodology also includes a comprehensive analysis of the costs incurred in each activity and those for productive and for nonproductive labor.

The lean production strategy requires a fundamental change in corporate culture, as well as having an understanding of the importance of *cooperation and teamwork* among the company's workforce and management. Lean production does not necessarily require cutting back on a company's physical or human resources; rather, it aims at *continually* improving efficiency and profitability by removing all waste in the company's operations and dealing with any problems as soon as they arise.

Agile Manufacturing.   The principle behind *agile manufacturing* is ensuring *agility*, hence *flexibility*, in the manufacturing enterprise, so that it can respond rapidly and effectively to changes in product demand and the needs of the customer. Flexibility can be achieved through people, equipment, computer hardware and software, and advanced communications systems. As an example of this approach, it has been demonstrated that the automotive industry can configure and build a car in three days and that, eventually, the traditional assembly line will be replaced by a system in which a nearly custom-made car will be produced by combining several individual modules.

The methodologies of both lean and agile production require that a manufacturer **benchmark** its operations. Benchmarking involves assessing the competitive position of other manufacturers with respect to one's own position (including product quality, production time, and manufacturing costs) and setting realistic goals for the future. Benchmarking thus becomes a *reference point* from which various measurements can be made and to which they can be compared.

I.10   Manufacturing Costs and Global Competition

Always critically important, the economics of manufacturing has become even more so with (a) ever-increasing global competition and (b) the demand for high-quality products, generally referred to as *world-class manufacturing*, and at low prices. Typically, the *manufacturing cost* of a product represents about 40% of its *selling price*, which often is the overriding consideration in a product's marketability and general customer satisfaction. An approximate, but typical, breakdown of costs in modern manufacturing is given in Table I.6; the percentages indicated can, however, vary significantly depending on product type.

The *total cost* of manufacturing a product generally consists of the following components:

TABLE I.6

Typical Cost Breakdown in Manufacturing

Design	5%
Materials	50%
Manufacturing	
Direct labor	15%
Indirect labor	30%

1. **Materials.** Raw-material costs depend on the material itself and on supply and demand for the material. Low cost may not be the deciding factor if the cost of processing a particular material is higher than that for a more expensive material. For example, a low-cost piece of metal may require more time to machine, or form, than one of higher cost, thus increasing production costs.

2. **Tooling.** Tooling costs include those for cutting tools, dies, molds, work-holding devices, and fixtures. Some cutting tools cost as little as $2 and as much as about $100 for materials such as cubic boron nitride and diamond. Depending on their size and the materials involved in making them, molds and dies can cost from only a few hundred dollars to over $2 million for a set of dies for stamping sheet metal to make automobile fenders.

3. **Fixed.** Fixed costs include costs for energy, rent for facilities, insurance, and real-estate taxes.

4. **Capital.** Production machinery, equipment, buildings, and land are typical capital costs. Machinery costs can range from a few hundred to millions of dollars. Although the cost of computer-controlled machinery can be very high, such an expenditure may well be warranted if it reduces labor costs.

5. **Labor.** Labor costs consist of direct and indirect costs. Direct labor, also called productive labor, concerns the labor that is directly involved in manufacturing products. Indirect labor, also called nonproductive labor or overhead, pertains to servicing of the total manufacturing operation.

Direct-labor costs may be only 10 to 15% of the total cost (Table I.6), but it can be as much as 60% for labor-intensive products. Reductions in the direct-labor share of manufacturing costs can be achieved by such means as extensive automation, computer control of all aspects of manufacturing, implementation of modern technologies, and increased efficiency of operations.

As expected and shown in Table I.7, there continues to be a worldwide disparity in labor costs, by an order of magnitude. It is not surprising that today numerous consumer products are manufactured or assembled in Pacific Rim countries, especially China. Likewise, software and information technologies are often much less costly to develop in countries such as India and China than in the United States or

TABLE I.7

Approximate Relative Hourly Compensation for Workers in Manufacturing in 2010 (United States = 100)			
Norway	166	Italy	96
Switzerland	153	Japan	92
Belgium	146	Spain	76
Denmark	131	New Zealand	59
Germany	126	Israel	58
Sweden	126	Singapore	55
Finland	122	Korea (South)	48
Austria	118	Argentina, Slovakia	36
Netherlands, Australia	118	Portugal	34
France	117	Czech Republic	33
Ireland	104	Poland	23
United States	100	Mexico	18
Canada	97	China, India, Philippines	6

Note: Compensation can vary significantly with benefits. Data for China and India are estimates, they use different statistical measures of compensation, and are provided here for comparison purposes only. *Source:* U.S. Department of Labor.

Europe. As living standards continue to rise, however, labor costs, too, are beginning to rise significantly in these countries.

Outsourcing.   *Outsourcing* is defined as the purchase by a company of parts and/or labor from an outside source, either from another company or another country, in order to reduce design and manufacturing costs. In theory, this approach allows companies to concentrate on their core competencies, and be able to optimize their critical technologies. Outsourcing, however, has several drawbacks, including its social impact and political implications of any ensuing lowered employment, especially in the European Union countries and the United States. In recent years, the costs of shipping and transport have increased and have become more uncertain, and also manufacturers often prefer to be located near their customers and/or suppliers. As a result, a **reshoring** trend has begun, which involves relocating manufacturing activities to a few critical locations, usually near the customers.

I.II   Trends in Manufacturing

Several trends regarding various aspects of modern manufacturing are:

1. Product variety and complexity continue to increase.
2. Product life cycles are becoming shorter.
3. Markets continue to become multinational and global competition is increasing rapidly.
4. Customers are consistently demanding high-quality, reliable, and low-cost products.
5. Developments continue in the quality of materials and their selection for improved recyclability.
6. Weight savings continue with the use of materials with higher strength-to-weight and stiffness-to-weight ratios, particularly in the automotive, aerospace, and sporting industries.

7. Improvements are being made in predictive models of the effects of material-processing parameters on product integrity, applied during a product's design stage.

8. Developments in ultraprecision manufacturing, micromanufacturing, and nanomanufacturing, approaching the level of atomic dimensions.

9. Computer simulation, modeling, and control strategies are being applied to all areas of manufacturing.

10. Rapid-prototyping technologies are increasingly being applied to the production of tooling and direct digital manufacturing.

11. Advances in optimization of manufacturing processes and production systems are making them more agile.

12. Lean production and information technology are being implemented as powerful tools to help meet global challenges.

13. Manufacturing activities are viewed not as individual, separate tasks, but as making up a large system, with all its parts interrelated.

14. It has become common to build quality into the product at each stage of its production.

15. The most economical and environmentally friendly (green) manufacturing methods are being increasingly pursued; energy management is increasingly important.

16. Continued efforts are aimed at achieving higher levels of productivity and eliminating or minimizing waste with optimum use of an organization's resources.

Fundamentals of Materials: Behavior and Manufacturing Properties

Part I of this text begins by describing the behavior and properties of materials, their manufacturing characteristics, and their applications, as well as their advantages and limitations that influence their selection in the design and manufacture of products.

In order to emphasize the importance of the topics to be described, consider a typical automobile as an example of a common product that utilizes a wide variety of materials (Fig. I.1). These materials were selected not only because they possess the desired properties and characteristics for the intended functions of specific parts, but also they were the ones that could be manufactured at the lowest cost.

For example, steel was chosen for parts of the body because it is strong, easy to shape, and inexpensive. Plastics were used in many components because of characteristics such as light weight, resistance to corrosion, availability in a wide variety of colors, and ease of manufacturing into complex shapes and at low cost. Glass was chosen for all the windows because it is transparent, hard (hence scratch resistant), easy to shape, and easy to clean. Numerous similar observations can be made about each component of an automobile, ranging from very small screws to wheels. In recent years, fuel efficiency and the need for improved performance have driven the

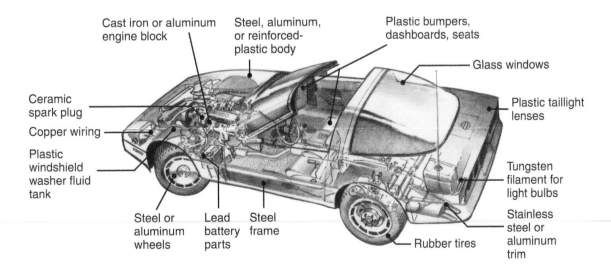

FIGURE I.1 An outline of the topics described in Part I.

QR Code 1.1 Demonstration of a tension test. (*Source:* Courtesy of Instron®)

substitution of such materials as aluminum, magnesium, and plastics for steel, and the use of composite materials for structural (load-bearing) components.

As stated in the General Introduction, the selection of materials for individual components in a product requires a thorough understanding and assessment of their properties, functions, and manufacturing costs. A typical automobile is an assemblage of some 15,000 individual parts; consequently, by saving just one cent on the cost per part, such as by selecting a different material or manufacturing process, the cost of an automobile would be reduced by $150. This task thus becomes very challenging, especially with the ever-increasing variety of materials and manufacturing processes that are now available, as outlined in Fig. I.2.

A general outline of the topics described in Part I of this text is given in Fig. I.3. The fundamental knowledge presented on the behavior, properties, and characteristics of materials will help understand their significance and relevance to all the manufacturing processes described in Parts II through V.

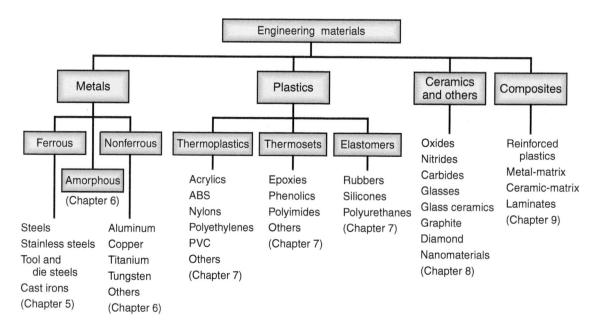

FIGURE I.2   An outline of the engineering materials described in Part I.

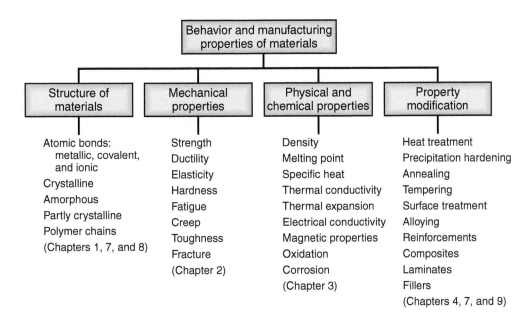

FIGURE I.3 An outline of the behavior and the manufacturing properties of materials described in Part I.

1

The Structure of Metals

- This chapter describes the crystalline structure of metals and explains how they determine their properties and behavior.
- It begins with a review of the types of atomic bonds and their characteristics: ionic, covalent, and metallic.
- Metal structures and the arrangement of atoms within the structure are then examined, and the types of imperfections in the crystal structure and their effects are presented.
- The effects of grains and grain boundaries are examined, followed by a discussion of strain hardening and anisotropy of metals.

1.1 Introduction

Why are some metals hard and others soft? Why are some brittle, while others are ductile and can be shaped easily without fracture? Why is it that some metals can withstand high temperatures, while others cannot? Why a piece of sheet metal may behave differently when stretched in one direction versus another?

These questions can be answered by studying the **atomic structure** of metals—that is, the arrangement of the atoms within metals. This knowledge then serves as a guide to controlling and predicting the behavior and performance of metals in various manufacturing processes. Understanding the structure of metals also allows us to predict and evaluate their **properties** (such as strength and stiffness), thus help us to make appropriate selections for specific applications. For example, single-crystal turbine blades (Fig. 1.1) for use in jet engines have properties that are better than those for conventional blades. In addition to atomic structure, several other factors also influence the properties and behavior of metals. They include the composition of the particular metal, impurities and vacancies in their atomic structure, grain size, grain boundaries, environment, size and surface condition of the metal, and the methods by which they are made into products.

The topics described in this chapter and their sequence are outlined in Fig. 1.2. The structure and general properties of materials other than metals are described in Chapter 7 (polymers), Chapter 8 (ceramics and glasses), and Chapter 9 (composite materials). The structure of metal alloys, the control of their structure, and heat-treatment processes are described in Chapter 4.

(a)   (b)   (c)

FIGURE 1.1   Turbine blades for jet engines, manufactured by three different methods: (a) conventionally cast, (b) directionally solidified, with columnar grains as can be seen from the vertical streaks, and (c) single crystal. Although more expensive, single-crystal blades have properties at high temperatures that are superior to those of other blades. *Source:* Courtesy of United Technologies, Pratt & Whitney Division.

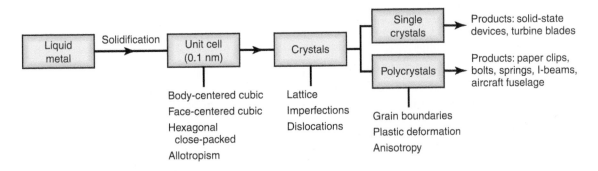

FIGURE 1.2   An outline of the topics described in this chapter.

1.2 Types of Atomic Bonds

All matter is made up of **atoms** consisting of a nucleus of *protons* and *neutrons* and surrounding clouds or orbits of electrons. The number of protons in the nucleus determines whether a particular atom will be metallic, nonmetallic, or semimetallic. An atom with a balanced charge has the same number of electrons as protons; when there are too many or too few electrons, the atom is called an **ion**. An excess of electrons results in a negatively charged atom, referred to as an **anion**, while too few electrons results in a positively charged atom, called a **cation**. The number of electrons in the outermost orbit of an atom determines the chemical affinity of that atom for other atoms.

Atoms can transfer or share electrons; in doing so, multiple atoms combine to form **molecules**. Molecules are held together by attractive forces called **bonds**, which act through electron interaction. The basic types of atomic attraction associated with electron transfer, called **primary bonds** or **strong bonds**, are:

- **Ionic bonds.** When one or more electrons from an outer orbit are transferred from one material to another, a strong attractive force develops between the two ions. An example is that of sodium (Na) and chlorine (Cl) in common table salt; the salt consists of Na^+ and Cl^- ions (hence the term *ionic bond*), which are strongly attracted to each other. Also, the attraction is between all adjacent ions, allowing crystalline structures to be formed, as discussed in Section 1.3. Molecules with ionic bonds generally have low ductility and low thermal and electrical conductivity. Ionic bonding is the predominant bond in ceramic materials.
- **Covalent bonds.** In a covalent bond, the electrons in outer orbits are shared by atoms to form molecules. The number of electrons shared is reflected by terms such as "single bond," "double bond," etc. Polymers consist of large molecules that are covalently bonded together. Solids formed by covalent bonding typically have low electrical conductivity and can have high hardness; diamond, a form of covalently bonded carbon, is an example.
- **Metallic bonds.** Metals have relatively few electrons in their outer orbits, and thus they cannot complete the outer shell when self-mated. Instead, metals and alloys form *metallic bonds*, whereby the available electrons are shared by all atoms in contact. The resulting electron cloud provides attractive forces to hold the atoms together and results in generally high thermal and electrical conductivity.

In addition to the strong attractive forces associated with electrons, weak or **secondary bonds** or attractions occur between molecules. Also referred to as **van der Waals** forces, these forces arise from the attraction of opposite charges without electron transfer. Water molecules, for example, consist of one oxygen atom and two smaller hydrogen atoms, located around 104° from each other. Although each molecule has a balanced, or neutral, charge, there are more hydrogen atoms on one side of the molecule (i.e., it is a *dipole*), so that the molecule develops a weak attraction to nearby oxygen atoms on that side.

1.3   The Crystal Structure of Metals

When metals solidify from a molten state (Chapter 10), the atoms arrange themselves into various orderly configurations, called **crystals**; this atomic arrangement is called **crystal structure** or **crystalline structure**. The smallest group of atoms showing the characteristic **lattice structure** of a particular metal is known as a **unit cell**.

The following are the three basic atomic arrangements in metals:

1. **Body-centered cubic (bcc)**; alpha iron, chromium, molybdenum, tantalum, tungsten, and vanadium.
2. **Face-centered cubic (fcc)**; gamma iron, aluminum, copper, nickel, lead, silver, gold, and platinum.
3. **Hexagonal close-packed (hcp)**; beryllium, cadmium, cobalt, magnesium, alpha titanium, zinc, and zirconium.

These structures are represented by the illustrations given in Figs. 1.3–1.5, in which each sphere represents an atom. The distance between the atoms in these

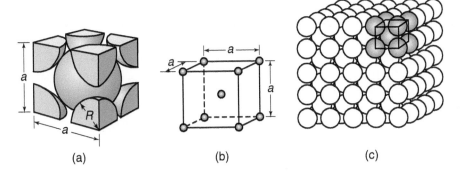

FIGURE 1.3   The body-centered cubic (bcc) crystal structure: (a) hard-ball model, (b) unit cell, and (c) single crystal with many unit cells.

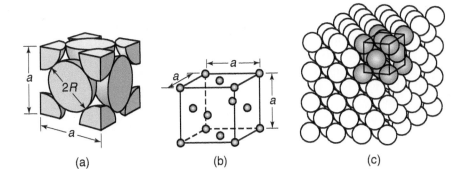

FIGURE 1.4   The face-centered cubic (fcc) crystal structure: (a) hard-ball model, (b) unit cell, and (c) single crystal with many unit cells.

crystal structures is on the order of 0.1 nm (10^{-8} in.). The models shown are known as **hard-ball** or **hard-sphere** models, and can be likened to tennis balls arranged in various configurations in a box.

In the three structures illustrated, the hcp crystals have the most densely packed configurations, followed by fcc and then bcc. In the hcp structure, the top and bottom planes are called **basal planes**. All three arrangements can be modified by adding atoms of some other metal or metals, known as **alloying**, often to improve various properties of the metal.

The appearance of more than one type of crystal structure in metals is known as **allotropism** or **polymorphism** (meaning "many shapes"). Because the properties and behavior of a particular metal depend greatly on its crystal structure, allotropism is an important factor in the heat treatment of metals, as well as in metalworking and welding operations, described in Parts III and VI, respectively. Single crystals of metals are now produced as ingots in sizes on the order of 1 m (40 in.) long and up to 300 mm (12 in.) in diameter, with applications such as turbine blades and semiconductors (see Sections 11.5 and 28.3). Most metals used in manufacturing are, however, polycrystalline, as described in Section 1.5.

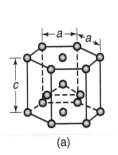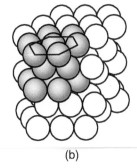

FIGURE 1.5   The hexagonal close-packed (hcp) crystal structure: (a) unit cell and (b) single crystal with many unit cells.

1.4   Deformation and Strength of Single Crystals

When a single crystal is subjected to an external force, it first undergoes **elastic deformation** (Chapter 2); that is, it returns to its original shape when the force is removed. A simple analogy to this type of behavior is a helical spring that stretches when loaded and returns to its original shape when the load is removed. If the force is increased sufficiently, the crystal undergoes **plastic deformation** or **permanent deformation**; that is, it does not return to its original shape when the force is removed.

There are two basic mechanisms by which plastic deformation takes place in crystal structures. One mechanism involves a plane of atoms slipping over an adjacent plane (called the **slip plane**) under a **shear stress** (Fig. 1.6a); note that this behavior is much like sliding of a set of playing cards against each other. *Shear stress* is defined as the ratio of the applied shearing force to the cross-sectional area being sheared.

Just as it takes a certain force to slide playing cards against each other, a single crystal requires a certain magnitude of shear stress (called **critical shear stress**) to undergo permanent deformation. Thus, there must be a shear stress of sufficient magnitude to cause plastic deformation; otherwise the deformation remains elastic.

The shear stress required to cause slip in single crystals is directly proportional to the ratio b/a in Fig. 1.6a, where a is the spacing of the atomic planes and b is inversely proportional to the atomic density in the atomic plane. As the ratio b/a decreases, the shear stress required to cause slip decreases. Thus, slip in a single crystal takes

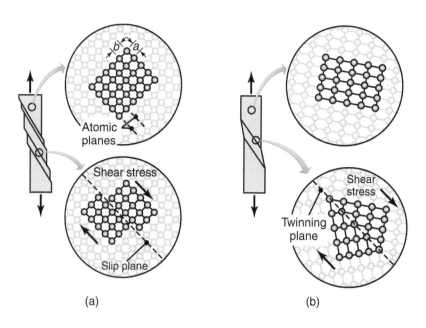

(a)   (b)

FIGURE 1.6 Permanent deformation of a single crystal under a tensile load; the highlighted grid of atoms emphasizes the movement that occurs within the lattice. (a) Deformation by slip. The b/a ratio influences the magnitude of the shear stress required to cause slip. (b) Deformation by twinning, involving the generation of a "twin" around a line of symmetry subjected to shear. Note that the tensile load results in a shear stress in the plane illustrated.

place along planes of *maximum atomic density*; in other words, slip takes place in closely packed planes and in closely packed directions.

Because the b/a ratio varies for different directions within the crystal, a single crystal exhibits different properties when tested in different directions, a property called **anisotropy**. An example is the behavior of plywood, which is much stronger in the planar direction than along its thickness direction.

The second, and less common, mechanism of plastic deformation in crystals is **twinning**, in which a portion of the crystal forms a mirror image of itself across the *plane of twinning* (Fig. 1.6b). Twins form abruptly and are the cause of the creaking sound (called "tin cry") that occurs when a rod of tin or zinc is bent at room temperature. Twinning usually occurs in hcp metals.

Slip Systems. The combination of a slip plane and slip direction is known as a *slip system*. In general, metals with five or more slip systems are ductile.

1. In **body-centered cubic** crystals, there are 48 possible slip systems; therefore, the probability is high that an externally applied shear stress will operate on one of these systems and cause slip. Because of the relatively high b/a ratio in this type of crystal, however, the required shear stress is high. Metals with bcc structures (such as titanium, molybdenum, and tungsten) generally have good strength and moderate ductility, but can have high ductility at elevated temperatures.

2. In **face-centered cubic** crystals, there are 12 slip systems. The probability of slip is moderate, and the shear stress required is low because of the relatively low b/a ratio. These metals (such as aluminum, gold, copper, and silver) generally have moderate strength and good ductility.

3. The **hexagonal close-packed** crystal has three slip systems, and therefore has a low probability of slip; however, more slip systems become active at elevated temperatures. Metals with hcp structures (such as beryllium, magnesium, and zinc) are generally brittle at room temperature.

Note in Fig. 1.6a that the portions of the single crystal that have slipped have rotated from their original angular position toward the direction of the tensile force; note also that slip has taken place only along certain planes. It can be observed using electron microscopy that what appears to be a single slip plane is actually a **slip band**, consisting of a number of slip planes (Fig. 1.7).

1.4.1 Imperfections in the Crystal Structure of Metals

The actual strength of metals is approximately one to two orders of magnitude lower than the strength levels obtained from theoretical calculations. This discrepancy is explained in terms of **defects** and **imperfections** in the crystal structure. Unlike in idealized models described earlier, actual metal

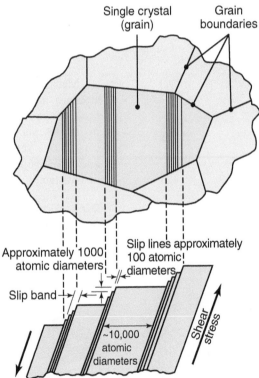

FIGURE 1.7 Schematic illustration of slip lines and slip bands in a single crystal (grain) subjected to a shear stress. A slip band consists of a number of slip planes. The crystal at the center of the upper illustration is an individual grain surrounded by several other grains.

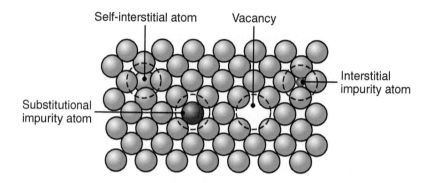

FIGURE 1.8 Schematic illustration of types of defects in a single-crystal lattice: self-interstitial, vacancy, interstitial, and substitutional.

crystals contain a large number of defects and imperfections, which generally are categorized as:

1. *Point defects*, such as a **vacancy** (missing atom), an **interstitial atom** (extra atom in the lattice), or an **impurity** (foreign atom that has replaced the atom of the pure metal) (Fig. 1.8).
2. *Linear, or one-dimensional, defects*, called **dislocations** (Fig. 1.9).
3. *Planar, or two-dimensional, imperfections*, such as **grain boundaries** and **phase boundaries** (Section 1.5).
4. *Volume, or bulk, imperfections*, such as **voids, inclusions** (nonmetallic elements such as oxides, sulfides, and silicates), other **phases**, or **cracks**.

Mechanical and electrical properties of metals, such as yield stress, fracture strength, and electrical conductivity, are adversely affected by the presence of defects; these properties are known as **structure sensitive**. By contrast, physical and chemical properties, such as melting point, specific heat, coefficient of thermal expansion, and elastic constants, such as modulus of elasticity and modulus of rigidity (Sections 2.2.1 and 2.4), are not sensitive to these defects; these properties are known as **structure insensitive**.

Dislocations. First observed in the 1930s, *dislocations* are defects in the orderly arrangement of a metal's atomic structure. Because a slip plane containing a dislocation (Fig. 1.10) requires much lower shear stress to allow slip than does a plane in a perfect lattice, dislocations are the most significant defects that explain the discrepancy between the actual and theoretical strengths of metals.

There are two types of dislocations: **edge** and **screw** (Fig. 1.9). An analogy to the movement of an edge dislocation is the progress of an earthworm, which moves forward by means of a hump that starts at the tail and moves toward the head. Another analogy is moving a large carpet on a floor by first forming a hump at one end and moving the hump gradually to the other end. (Recall that the force required to move a carpet in this way is much lower than that required to slide the whole carpet along the floor.) Screw dislocations are so named because the atomic planes form a spiral ramp, like the threads on a screw or bolt.

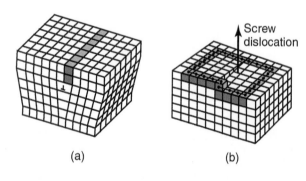

(a)   (b)

FIGURE 1.9 Types of dislocations in a single crystal: (a) edge dislocation and (b) screw dislocation.

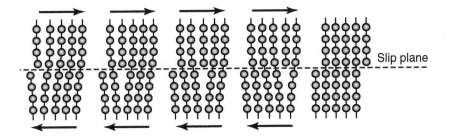

FIGURE 1.10   Movement of an edge dislocation across the crystal lattice under a shear stress. Dislocations help explain why the actual strength of metals is much lower than that predicted by theory.

1.4.2   Work Hardening (Strain Hardening)

Although the presence of a dislocation lowers the shear stress required to cause slip, dislocations can be

1. Entangled and interfere with each other
2. Impeded by barriers, such as grain boundaries, impurities, and inclusions in the material

The higher shear stress required to overcome entanglements and impediments results in an increase in the overall strength and hardness of the metal, and is known as **work hardening** or **strain hardening**. The greater the deformation, the greater is the number of entanglements and hence the higher the increase in the metal's strength. Work hardening is a mechanism used extensively for strengthening of metals in metalworking processes at low to moderate temperatures. Typical examples are producing sheet metal for automobile bodies and aircraft fuselages by cold rolling (Chapter 13), producing the head of a bolt by forging (Chapter 14), and strengthening wire by reducing its cross-section by drawing it through a die (Chapter 15).

1.5   Grains and Grain Boundaries

When a mass of molten metal begins to solidify, crystals form independently of each other at various locations within the liquid mass, and thus have random and unrelated orientations (Fig. 1.11). Each of these crystals eventually grows into a crystalline structure, or *grain*; each grain consists of either a single crystal (for pure metals) or a polycrystalline aggregate (for alloys).

The number and size of the grains developed in a unit volume of the metal depends on the *rate* at which **nucleation** (the initial stage of crystal formation) takes place. The *median size* of the grains developed depends on (a) the number of different sites at which individual crystals begin to form (note that there are seven in Fig. 1.11a) and (b) the rate at which these crystals grow. If the nucleation rate is high, the number of grains in a unit volume of metal will be large, and thus grain size will be small. Conversely, if the crystal growth rate is high (as compared with their nucleation rate), there will be fewer grains per unit volume, and thus grain size will be larger. Generally, rapid cooling produces smaller grains, whereas slow cooling produces larger grains.

Note in Fig. 1.11d that the growing grains eventually interfere with and impinge upon one another; the interfaces that separate the individual grains are called **grain boundaries**. Note also that the crystallographic orientation changes abruptly from one grain to the next across the grain boundaries. Recall, from Section 1.4, that the

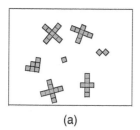
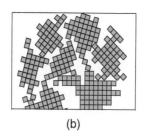
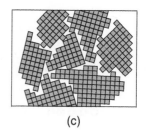
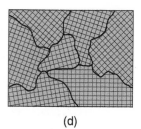

(a)   (b)   (c)   (d)

FIGURE 1.11   Schematic illustration of the stages during the solidification of molten metal; each small square represents a unit cell. (a) Nucleation of crystals at random sites in the molten metal; note that the crystallographic orientation of each site is different. (b) and (c) Growth of crystals as solidification continues. (d) Solidified metal, showing individual grains and grain boundaries; note the different angles at which neighboring grains meet each other.

behavior of a single crystal or a single grain is anisotropic. Thus, because its many grains have random crystallographic orientations, the behavior of a polycrystalline metal is essentially **isotropic**; that is, its properties do not vary with direction.

1.5.1   Grain Size

Grain size has a major influence on the mechanical properties of metals. At room temperature, for example, a large grain size is generally associated with low strength, low hardness, and low ductility. Grains can be so large as to be visible with the naked eye; zinc grains on the surface of galvanized sheet steels are an example. Large grains also cause a rough surface appearance after the material has been plastically deformed, particularly in the stretching of sheet metals (see **orange peel**, Section 1.7).

The yield strength, Y, of the metal is the most sensitive property and is related to grain size by the empirical formula (known as the *Hall–Petch equation*)

$$Y = Y_i + kd^{-1/2}, \tag{1.1}$$

where Y_i is the yield stress for a large grained material, k is a constant, and d is the mean grain diameter. Equation (1.1) is valid below the recrystallization temperature of the material.

Grain size is usually measured by counting the number of grains in a given area, or by counting the number of grains that intersect a prescribed length of a line randomly drawn on an enlarged photograph of the grains (taken under a microscope on a polished and etched specimen). Grain size may also be determined by comparing such a photograph against a standard chart.

The ASTM (American Society for Testing and Materials) grain size number, n, is related to the number of grains, N, per square inch at a magnification of $100\times$ (equal to 0.0645 mm^2 of actual area) by the formula

$$N = 2^{n-1}. \tag{1.2}$$

Because grains are typically extremely small, many grains can occupy a very small volume of metal (Table 1.1). Grain sizes between 5 and 8 are generally considered fine grained. A grain size of 7 is typically acceptable for sheet metals for making car bodies, appliances, and kitchen utensils (Chapter 16).

TABLE 1.1

Grain sizes		
ASTM No.	Grains/mm^2	Grains/mm^3
−3	1	0.7
−2	2	2
−1	4	5.6
0	8	16
1	16	45
2	32	128
3	64	360
4	128	1020
5	256	2900
6	512	8200
7	1024	23,000
8	2048	65,000
9	4096	185,000
10	8200	520,000
11	16,400	1,500,000

EXAMPLE 1.1   Number of Grains in a Paper Clip

Given: A paper clip is made of wire that is 120 mm long and 0.75 mm in diameter, with an ASTM grain size of 9.

Find: Calculate the number of grains in the paper clip.

Solution: A metal with an ASTM grain size of 9 has 185,000 grains per mm^3 (see Table 1.1). The volume of the paper clip is

$$V = \frac{\pi}{4}d^2 l = \frac{\pi}{4}(0.75)^2 (120) = 53.0 \text{ mm}^3.$$

The total number of grains is calculated by multiplying the volume by the grains per mm^3, or

$$\text{No. grains} = (53.0 \text{ mm}^3)(185,000 \text{ grains/mm}^3)$$
$$= 9.81 \text{ million.}$$

1.5.2 Influence of Grain Boundaries

Grain boundaries have an important influence on the strength and ductility of metals; they interfere with dislocation movement and thus also influence strain hardening. The magnitude of these effects depends on temperature, deformation rate, and the type and amount of impurities present along grain boundaries.

Because the atoms along the grain boundaries are more disordered and hence packed less efficiently, grain boundaries are more reactive than the grains themselves. As a result, the boundaries have lower energy than the atoms in the orderly lattice within the grains; thus, they can be more easily removed or chemically bonded to another atom. As a result, for example, the surface of a metal piece becomes rougher when etched or is subjected to corrosive environments (see also *end grains in forging*, in Section 4.11).

At elevated temperatures, and in metals whose properties depend on the rate at which they are deformed, plastic deformation also takes place by means of grain-boundary sliding. The **creep** mechanism (deformation under stress over time, usually at elevated temperatures) involves *grain-boundary sliding* (Section 2.8).

Grain-boundary embrittlement.   When exposed to certain low-melting-point metals, a normally ductile and strong metal can crack when subjected to very low external stresses. Examples of such behavior are (a) aluminum wetted with a mercury–zinc amalgam or with liquid gallium and (b) copper at elevated temperature wetted with lead or bismuth; these elements weaken the grain boundaries of the metal by **embrittlement**. The term **liquid-metal embrittlement** is used to describe such phenomena, because the embrittling element is in a liquid state. However, embrittlement can also occur at temperatures well below the melting point of the embrittling element, a phenomenon known as **solid-metal embrittlement.**

Another embrittlement phenomenon, called **hot shortness,** is caused by local melting of a constituent or of an impurity along the grain boundary at a temperature below the melting point of the metal itself. When subjected to plastic deformation at elevated temperatures (*hot working*), a piece of metal crumbles along its grain boundaries; examples are (a) antimony in copper, (b) leaded steels (Section 21.7.1), and (c) leaded brass. To avoid hot shortness, the metal is usually worked at a lower temperature to prevent softening and melting along the grain boundaries. **Temper embrittlement** in alloy steels is another form of embrittlement, caused by segregation (movement) of impurities to the grain boundaries (Section 4.11).

1.6 Plastic Deformation of Polycrystalline Metals

When a polycrystalline metal with uniform *equiaxed grains* (grains having equal dimensions in all directions) is subjected to plastic deformation at room temperature (called *cold working*), the grains become deformed and elongated, as shown schematically in Fig. 1.12. Deformation may be carried out by, for example, compressing the metal piece, as is done in a forging operation to make a turbine disk (Chapter 14) or by subjecting it to tension, as is done in stretch forming of sheet metal (Section 16.6). The deformation within each grain takes place by the mechanisms described in Section 1.4 for a single crystal.

During plastic deformation, the grain boundaries remain intact and mass continuity is maintained. The deformed metal exhibits higher strength, because of the entanglement of dislocations with grain boundaries and with each other. The increase in strength depends on the degree of deformation (*strain*) to which the metal is subjected; the higher the deformation, the stronger the metal becomes. The strength is higher for metals with smaller grains, because they have a larger grain-boundary surface area per unit volume of metal and hence more entanglement of dislocations.

Anisotropy (Texture).   Note in Fig. 1.12b that, as a result of plastic deformation, the grains have elongated in one direction and contracted in the other. Consequently, this piece of metal has become *anisotropic*, and thus its properties in the vertical direction are different from those in the horizontal direction. The degree of anisotropy depends on the temperature at which deformation takes place and on how uniformly the metal is deformed. Note from the crack direction in Fig. 1.13, for example, that the ductility of the cold-rolled sheet in the transverse direction is lower than in its rolling direction (see also Section 16.5).

Anisotropy influences both mechanical and physical properties of metals, described in Chapter 3. For example, sheet steel for electrical transformers is rolled in such a manner that the resulting deformation imparts anisotropic magnetic properties to the sheet. This operation

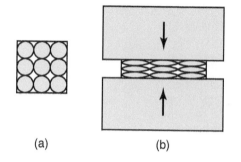

(a)   (b)

FIGURE 1.12 Plastic deformation of idealized (equiaxed) grains in a specimen subjected to compression (such as occurs in the forging or rolling of metals): (a) before deformation and (b) after deformation. Note the alignment of grain boundaries along a horizontal direction, an effect known as *preferred orientation*.

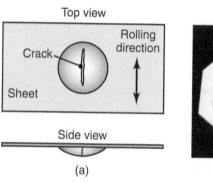

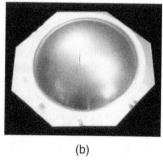

(a)   (b)

FIGURE 1.13 (a) Schematic illustration of a crack in sheet metal that has been subjected to bulging (caused, for example, by pushing a steel ball against the sheet). Note the orientation of the crack with respect to the rolling direction of the sheet; this sheet is anisotropic. (b) Aluminum sheet with a crack (vertical dark line at the center) developed in a bulge test; the rolling direction of the sheet was vertical. *Source:* Courtesy of J.S. Kallend, Illinois Institute of Technology.

reduces magnetic-hysteresis losses and thus improves the efficiency of transformers. (See also *amorphous alloys*, Section 6.14.) There are two general types of anisotropy in metals: preferred orientation and mechanical fibering.

Preferred Orientation.   Also called **crystallographic anisotropy**, *preferred orientation* can be best described by referring to Fig. 1.6a. When a single-crystal metal piece is subjected to tension, the sliding blocks rotate toward the direction of the tensile force; as a result, slip planes and slip bands tend to align themselves with the general direction of deformation. Similarly, for a polycrystalline metal, with grains in random orientations, all slip directions tend to align themselves with the direction of the tensile force. By contrast, slip planes under compression tend to align themselves in a direction perpendicular to the direction of the compressive force.

Mechanical Fibering.   This is a type of anisotropy that results from the alignment of inclusions (*stringers*), impurities, and voids in the metal during deformation. Note that if the spherical grains in Fig. 1.12a were coated with impurities, these impurities would align themselves in a generally horizontal direction after deformation. Because impurities weaken the grain boundaries, this piece of metal will now be weaker and less ductile when tested in the vertical direction. As an analogy, consider plywood, which is strong in tension along its planar direction, but splits easily when pulled in tension in its thickness direction.

1.7   Recovery, Recrystallization, and Grain Growth

Recall that plastic deformation at room temperature causes (a) distortion of the grains and grain boundaries, leading to anisotropic behavior; (b) a general increase in strength; and (c) a decrease in ductility. These effects can be reversed, and the properties of the metal can be brought back to their original levels, by heating the metal to a specific temperature range for a given period of time—a process called **annealing** (described in detail in Section 4.11). Three events take place consecutively during the annealing process:

1. **Recovery.** During *recovery*, which occurs at a certain temperature range below the **recrystallization temperature** of the metal (described next), the stresses in the highly deformed regions of the metal are relieved. Subgrain boundaries begin to form (called **polygonization**), with no significant change in mechanical properties such as hardness and strength (Fig. 1.14).

2. **Recrystallization.** This is the process in which, within a certain temperature range, new equiaxed and strain-free grains are formed, replacing the older grains. The temperature required for recrystallization ranges approximately between 0.3 and $0.5T_m$, where T_m is the melting point of the metal on the absolute scale.

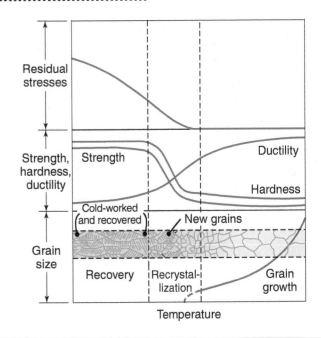

FIGURE 1.14   Schematic illustration of the effects of recovery, recrystallization, and grain growth on mechanical properties and on the shape and size of grains; note the formation of small new grains during recrystallization. *Source:* After G. Sachs.

Generally, the recrystallization temperature is defined as the temperature at which complete recrystallization occurs within approximately one hour. Recrystallization decreases the density of dislocations, lowers the strength, and raises the ductility of the metal (Fig. 1.14). Lead, tin, cadmium, and zinc recrystallize at about room temperature; consequently, they do not normally work harden.

The recrystallization temperature depends on the degree of prior cold work (work hardening): the more the cold work, the lower the temperature required for recrystallization. The reason is that, as the amount of cold work increases, the number of dislocations and hence the amount of energy stored in dislocations (**stored energy**) also increase. This energy supplies some of the work required for recrystallization. Recrystallization is also a function of time, because it involves **diffusion**—the movement and exchange of atoms across grain boundaries.

The effects of temperature, time, and plastic deformation by cold working on recrystallization are:

a. For a constant amount of deformation by cold working, the time required for recrystallization decreases with increasing temperature.
b. The more the prior cold work, the lower the temperature required for recrystallization.
c. The higher the amount of deformation, the smaller the grain size becomes during recrystallization; this effect is a commonly used method of converting a coarse-grained structure to one having a finer grain, and thus one with improved properties.
d. Some anisotropy due to preferred orientation usually persists after recrystallization; to restore isotropy, a temperature higher than that required for recrystallization may be necessary.

3. **Grain growth.** If the temperature is raised further, the grains begin to grow (see lower part of Fig. 1.14) and their size may eventually exceed the original grain size; called *grain growth*, this phenomenon adversely affects mechanical properties (Fig. 1.14). Large grains also produce a rough surface appearance on sheet metals, called **orange peel**, when they are stretched to form a part, or on the surfaces of a piece of metal when subjected to bulk deformation, such as compression in forging.

1.8 Cold, Warm, and Hot Working

Cold working refers to plastic deformation that is usually carried out at room temperature; when deformation occurs above the recrystallization temperature, it is called **hot working.** "Cold" and "hot" are relative terms, as can be seen from the fact that deforming lead at room temperature is a hot-working process, because the recrystallization temperature of lead is about room temperature. As the name implies, **warm working** is carried out at intermediate temperatures; thus, warm working is a compromise between cold and hot working. The important technological differences in products that are processed by cold, warm, and hot working are described in Part III.

The temperature ranges for these three categories of plastic deformation are given in Table 1.2 in terms of a ratio, T/T_m, where T is the working temperature and T_m is the melting point of the metal, both on the absolute scale. Although it is a dimensionless quantity, this ratio is known as the **homologous temperature**.

TABLE 1.2

Homologous Temperature Ranges for Various Processes

Process	T/T_m
Cold working	<0.3
Warm working	0.3–0.5
Hot working	>0.6

SUMMARY

- There are three basic crystal structures in metals: body-centered cubic (bcc), face-centered cubic (fcc), and hexagonal close-packed (hcp). Grains made of these crystals typically contain various defects and imperfections, such as dislocations, vacancies, impurities, inclusions, and grain boundaries. Polycrystalline metals consist of many crystals, or grains, in random orientations.

- Plastic deformation in metals takes place by a slip mechanism. Although the theoretical shear stress required to cause slip is very high, actual stresses are much lower because of the presence of dislocations (edge or screw type). Dislocations become entangled with one another or are impeded by barriers such as grain boundaries, impurities, and inclusions. As a result, the shear stress required to cause further slip is increased; consequently, the overall strength and hardness of the metal is also increased (through work hardening or strain hardening).

- Grain size has a significant effect on the strength of metals: the smaller the size, the stronger is the metal, and the larger the size, the more ductile is the metal; however, excessively large grains are generally associated with brittle behavior.

- Grain boundaries have a major influence on the behavior of metals. Boundaries can undergo embrittlement, severely reducing ductility at elevated temperatures (hot shortness); they are also responsible for the creep phenomenon, which is due to grain boundary sliding.

- Metals may be plastically deformed at room, warm, or high temperatures; their behavior and workability depend largely on whether deformation takes place below or above the recrystallization temperature of the metal. Deformation at room temperature (cold working) results in higher strength, but reduced ductility; generally, it also causes anisotropy (preferred orientation or mechanical fibering), whereby the properties are different in different directions.

- The effects of cold working can be reversed by annealing the metal; that is, heating it within a certain temperature range for a given period of time, thereby allowing the successive processes of recovery, recrystallization, and grain growth to take place.

KEY TERMS

Allotropism	Grain boundaries	Nucleation	Slip plane
Anisotropy	Grain growth	Orange peel	Slip system
Basal plane	Grain size	Plastic deformation	Strain hardening
Body-centered cubic	Hall–Petch equation	Polycrystalline	Structure insensitive
Cold working	Hexagonal close-packed	Polygonization	Structure sensitive
Covalent bond	Homologous temperature	Polymorphism	Texture
Creep	Hot shortness	Preferred orientation	Twinning
Crystals	Hot working	Primary bond	Unit cell
Dislocations	Imperfections	Recovery	Vacancy
Elastic deformation	Ionic bond	Recrystallization	van der Waals force
Embrittlement	Lattice structure	Secondary bond	Warm working
Face-centered cubic	Mechanical fibering	Shear stress	Work hardening
Grains	Metallic bond	Slip band	

BIBLIOGRAPHY

Ashby, M.F., **Materials Selection in Mechanical Design**, 4th ed., Butterworth-Heinemann, 2010.

Ashby, M.F., and Jones, D.R.H., **Engineering Materials**, Vol. 1, *An Introduction to Properties, Applications, and Microstructure*, 4th ed., 2012; Vol. 2: *An Introduction to Microstructures and Processing*. Butterworth-Heinemann, 2012.

Ashby, M., Shercliff, H., and Cebon, D., **Materials: Engineering, Science, Processing and Design**,   2nd ed., Butterworth-Heinemann, 2009.

Askeland, D.R., Fulay, P.P., and Wright, W.J., **The Science and Engineering of Materials**, 6th ed., CL Engineering, 2010.

Callister, W.D., Jr., and Rethwisch, D.G., **Materials Science and Engineering: An Introduction**, 8th ed., Wiley, 2010.

Shackelford, J.F., **Introduction to Materials Science for Engineers**, 7th ed., Prentice Hall, 2008.

REVIEW QUESTIONS

1.1 What is the difference between an atom and a molecule? A molecule and a crystal?

1.2 Describe ionic, covalent, and metallic bonds.

1.3 Explain the difference between a unit cell and a single crystal.

1.4 In tables on crystal structures, iron is listed as having both a bcc and an fcc structure. Why?

1.5 Define anisotropy. What is its significance?

1.6 What effects does recrystallization have on the properties of metals?

1.7 What is strain hardening, and what effects does it have on the properties of metals?

1.8 Explain what is meant by structure-sensitive and structure-insensitive properties of metals.

1.9 Make a list of each of the major types of imperfection in the crystal structure of metals, and describe them.

1.10 What influence does grain size have on the mechanical properties of metals?

1.11 What is the relationship between the nucleation rate and the number of grains per unit volume of a metal?

1.12 What is a slip system, and what is its significance?

1.13 Explain the difference between recovery and recrystallization.

1.14 What is hot shortness, and what is its significance?

1.15 Explain the advantages and limitations of cold, warm, and hot working, respectively.

1.16 Describe what the orange peel effect is. Explain why we may have to be concerned with the orange-peel effect on metal surfaces.

1.17 Some metals, such as lead, do not become stronger when worked at room temperature. Explain the reason.

1.18 Describe the difference between preferred orientation and mechanical fibering.

1.19 Differentiate between stress relaxation and stress relieving.

1.20 What is twinning? How does it differ from slip?

QUALITATIVE PROBLEMS

1.21 Explain your understanding of why the study of the crystal structure of metals is important.

1.22 What is the significance of the fact that some metals undergo allotropism?

1.23 Is it possible for two pieces of the same metal to have different recrystallization temperatures? Is it possible for recrystallization to take place in some regions of a part before it does in other regions of the same part? Explain.

1.24 Describe your understanding of why different crystal structures exhibit different strengths and ductilities.

1.25 A cold-worked piece of metal has been recrystallized. When tested, it is found to be anisotropic. Explain the probable reason.

1.26 What materials and structures can you think of (other than metals) that exhibit anisotropic behavior?

1.27 Two parts have been made of the same material, but one was formed by cold working and the other by hot working. Explain the differences you might observe between the two.

1.28 Do you think it might be important to know whether a raw material to be used in a manufacturing process has anisotropic properties? What about anisotropy in the finished product? Explain.

1.29 Explain why the strength of a polycrystalline metal at room temperature decreases as its grain size increases.

1.30 Describe the technique you would use to reduce the orange-peel effect on the surface of workpieces.

1.31 What is the significance of the fact that such metals as lead and tin have a recrystallization temperature that is about room temperature?

1.32 It was stated in this chapter that twinning usually occurs in hcp materials, but Fig. 1.6b shows twinning in a rectangular array of atoms. Can you explain the discrepancy?

1.33 It has been noted that the more a metal has been cold worked, the less it strain hardens. Explain why.

1.34 Is it possible to cold work a metal at temperatures above the boiling point of water? Explain.

1.35 Comment on your observations regarding Fig. 1.14.

1.36 Is it possible for a metal to be completely isotropic? Explain.

QUANTITATIVE PROBLEMS

1.37 How many atoms are there in a single repeating cell of an fcc crystal structure? How many in a repeating cell of an hcp structure?

1.38 The atomic weight of copper is 63.55, meaning that 6.023×10^{23} atoms weigh 63.55 g. The density of copper is 8970 kg/m^3, and pure copper forms fcc crystals. Estimate the diameter of a copper atom.

1.39 Plot the data given in Table 1.1 in terms of grains/mm^2 versus grains/mm^3, and discuss your observations.

1.40 A strip of metal is reduced from 30 mm in thickness to 20 mm by cold working; a similar strip is reduced from 40 to 30 mm. Which of these cold-worked strips will recrystallize at a lower temperature? Why?

1.41 The ball of a ballpoint pen is 1 mm in diameter and has an ASTM grain size of 10. How many grains are there in the ball?

Video Solution 1.1 Calculating the Number of Grains in a Ball Bearing

1.42 How many grains are there on the surface of the head of a pin? Assume that the head of a pin is spherical with a 1-mm diameter and has an ASTM grain size of 12.

1.43 The unit cells shown in Figs. 1.3–1.5 can be represented by tennis balls arranged in various configurations in a box. In such an arrangement, the *atomic packing factor* (APF) is defined as the ratio of the sum of the volumes of the atoms to the volume of the unit cell. Show that the APF is 0.68 for the bcc structure and 0.74 for the fcc structure.

Video Solution 1.2 Atomic Packing Factor for Hexagonal Close-packed Materials

1.44 Show that the lattice constant a in Fig. 1.4a is related to the atomic radius by the formula $a = 2\sqrt{2}R$, where R is the radius of the atom as depicted by the tennis-ball model.

1.45 Show that, for the fcc unit cell, the radius r of the largest hole is given by $r = 0.414R$. Determine the size of the largest hole for the iron atoms in the fcc structure.

1.46 A technician determines that the grain size of a certain etched specimen is 8. Upon further checking, it is found that the magnification used was 125×, instead of the 100× that is required by the ASTM standards. Determine the correct grain size.

1.47 If the diameter of the aluminum atom is 0.28 nm, how many atoms are there in a grain of ASTM grain size 8?

1.48 The following data are obtained in tension tests of brass:

Grain size (μm)	Yield stress (MPa)
15	150
20	140
50	105
75	90
100	75

Does the material follow the Hall–Petch equation? If so, what is the value of k?

1.49 Assume that you are asked to submit a quantitative problem for a quiz. Prepare such a question, supplying the answer.

1.50 The atomic radius of iron is 0.125 nm, while that of a carbon atom is 0.070 nm. Can a carbon atom fit inside a steel bcc structure without distorting the neighboring atoms?

1.51 Estimate the atomic radius for the following materials and data: (a) Aluminum (atomic weight = 26.98 g/mol, density = 2700 kg/m^3); (b) tungsten (atomic weight = 183.85 g/mol, density = 19,300 kg/m^3); and (c) magnesium (atomic weight = 24.31 g/mol, density = 1740 kg/m^3).

1.52 A simple cubic structure consists of atoms located at the cube corners that are in contact with each other along the cube edges. Make a sketch of a simple cubic structure, and calculate its atomic packing factor.

1.53 Same as Prob. 1.39, but ASTM no. versus grains/mm^3.

SYNTHESIS, DESIGN, AND PROJECTS

1.54 By stretching a thin strip of polished metal, as in a tension-testing machine, demonstrate and comment on what happens to its reflectivity as the strip is being stretched.

1.55 Draw some analogies to mechanical fibering—for example, layers of thin dough sprinkled with flour or melted butter between each layer.

1.56 Draw some analogies to the phenomenon of hot shortness.

1.57 Obtain a number of small balls made of plastic, wood, marble, or metal, and arrange them with your hands or glue them together to represent the crystal structures shown in Figs. 1.3–1.5. Comment on your observations.

1.58 Take a deck of playing cards, place a rubber band around it, and then slip the cards against each other to represent Figs. 1.6a and 1.7. If you repeat the same experiment with more and more rubber bands around the same deck, what are you accomplishing as far as the behavior of the deck is concerned?

1.59 Give examples in which anisotropy is scale dependent. For example, a wire rope can contain annealed wires that are isotropic on a microscopic scale, but the rope as a whole is anisotropic.

1.60 The movement of an edge dislocation was described in Section 1.4.1, by means of an analogy involving a hump in a carpet on the floor and how the whole carpet can eventually be moved by moving the hump forward. Recall that the entanglement of dislocations was described in terms of two humps at different angles. Use a piece of cloth placed on a flat table to demonstrate these phenomena.

Mechanical Behavior, Testing, and Manufacturing Properties of Materials

- This chapter examines the effects of external forces on the behavior of materials, and the test methods employed in determining their mechanical properties.
- The tension test, described first, is commonly used for quantifying several material parameters, including elastic modulus, yield stress, ultimate strength, ductility, and toughness.
- Compression tests, described next, are useful because they more closely simulate some metalworking processes; nonetheless, they have the unavoidable drawback of contributing friction to the test results.
- Bending tests are particularly useful for brittle materials; three- and four-point tests are in common use.
- Hardness and the variety of hardness tests and their range of applicability are then described.
- Fatigue involves the failure of material due to cyclic or repeating loads, whereas creep is deformation due to the application of a constant load over an extended period; these concepts are also discussed.
- The chapter ends with descriptions of the types of and factors involved in failure and fracture of materials.

2.1 Introduction

In manufacturing operations, parts and components of products are formed into a wide variety of shapes by applying external forces to the workpiece, typically by means of various tools and dies. Common examples of such operations are forging of turbine disks, extruding various components for aluminum ladders, drawing wire for making nails, and rolling metal to make sheets for car bodies, appliances, and office equipment. Forming operations may be carried out at room temperature or elevated temperatures and at a low or a high rate of deformation. Many of these operations are also used in forming and shaping nonmetallic materials such as plastics, ceramics, and composite materials, as described throughout this text.

As indicated in Table I.2, a wide variety of metallic and nonmetallic materials is now available, with an equally wide range of properties and characteristics, as shown qualitatively in Table 2.1. This chapter covers those aspects of mechanical properties and behavior of metals that are relevant to the design and manufacturing of products, and includes commonly used test methods employed in assessing various properties.

TABLE 2.1

Relative Mechanical Properties of Various Materials at Room Temperature (in Decreasing Order). Metals Are in Their Alloy Form				
Strength	Hardness	Toughness	Stiffness	Strength/Density
Glass fibers	Diamond	Ductile metals	Diamond	Reinforced plastics
Carbon fibers	Cubic boron nitride	Reinforced plastics	Carbides	Titanium
Kevlar fibers	Carbides	Thermoplastics	Tungsten	Steel
Carbides	Hardened steels	Wood	Steel	Aluminum
Molybdenum	Titanium	Thermosets	Copper	Magnesium
Steels	Cast irons	Ceramics	Titanium	Beryllium
Tantalum	Copper	Glass	Aluminum	Copper
Titanium	Thermosets		Ceramics	Tantalum
Copper	Magnesium		Reinforced plastics	
Reinforced thermosets	Thermoplastics		Wood	
Reinforced thermoplastics	Tin		Thermosets	
Thermoplastics	Lead		Thermoplastics	
Lead			Rubbers	

2.2 Tension

The **tension test** is the most commonly used method for determining the *mechanical properties* of materials, such as strength, ductility, toughness, elastic modulus, and strain-hardening capability. The test first requires the preparation of a **test specimen**, as shown in Fig. 2.1a; although most tension-test specimens are solid and round, they can also be flat or tubular. The specimen is prepared generally according to

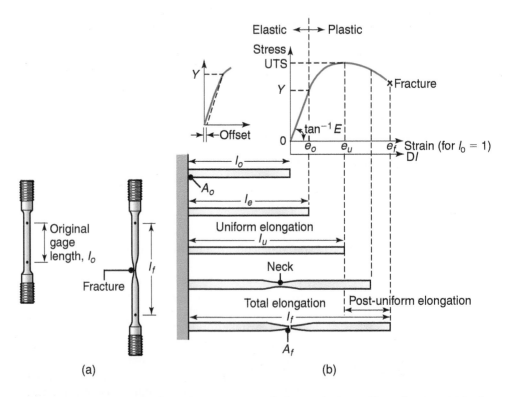

(a)   (b)

FIGURE 2.1   (a) A standard tensile-test specimen before and after pulling, showing original and final gage lengths. (b) Stages in specimen behavior in a tension test.

ASTM specifications, although various other specifications also are available from corresponding organizations around the world.

The specimen has an **original gage length,** l_o, generally 50 mm (2 in.), and a cross-sectional area, A_o, usually with a diameter of 12.5 mm (0.5 in.). It is mounted in the jaws of a tension-testing machine, equipped with various accessories and controls so that the specimen can be tested at different temperatures and rates of deformation.

2.2.1   Stress–strain Curves

A typical sequence in a tension test is shown in Fig. 2.1b. When the load is first applied, the specimen elongates in proportion to the load, called **linear elastic behavior** (Fig. 2.2). If the load is removed, the specimen returns to its original length and shape, in a manner similar to stretching a rubber band and releasing it.

The **engineering stress,** also called **nominal stress,** is defined as the ratio of the applied load, P, to the original cross-sectional area, A_o, of the specimen:

$$\sigma = \frac{P}{A_o}. \tag{2.1}$$

The **engineering strain** is defined as

$$e = \frac{l - l_o}{l_o}, \tag{2.2}$$

where l is the instantaneous length of the specimen.

As the load is increased, the specimen begins to undergo *nonlinear* elastic deformation, at a stress called the *proportional limit*. At that point, the stress and strain are no longer proportional, as they were in the linear elastic region; but when unloaded, the specimen still returns to its original shape. **Permanent (plastic) deformation** occurs when the **yield stress,** Y, of the material is reached. The yield stress and other properties of various metallic and nonmetallic materials are given in Table 2.2.

For soft and ductile materials, it may not be easy to determine the exact location on the stress–strain curve at which yielding occurs, because the slope of the curve

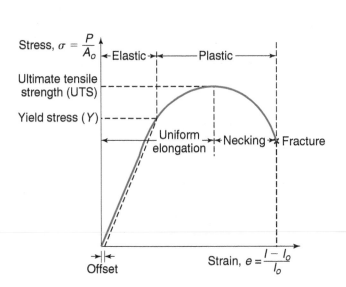

FIGURE 2.2 A typical stress–strain curve obtained from a tension test, showing various features.

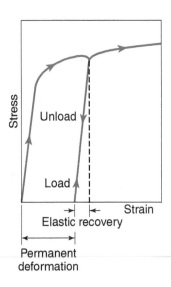

FIGURE 2.3 Schematic illustration of the loading and the unloading of a tensile-test specimen. Note that, during unloading, the curve follows a path parallel to the original elastic slope.

TABLE 2.2

Mechanical Properties of Various Materials at Room Temperature

Material	Elastic modulus (GPa)	Yield strength (MPa)	Ultimate tensile strength (MPa)	Elongation in 50 mm (%)	Poisson's ratio, v
Metals (wrought)					
Aluminum and its alloys	69–79	35–550	90–600	45–4	0.31–0.34
Copper and its alloys	105–150	76–110	140–1310	65–3	0.33–0.35
Lead and its alloys	14	14	20–55	50–9	0.43
Magnesium and its alloys	41–45	130–305	240–380	21–5	0.29–0.35
Molybdenum and its alloys	330–360	80–2070	90–2340	40–30	0.32
Nickel and its alloys	180–214	105–1200	345–1450	60–5	0.31
Steels	190–210	205–1725	415–1750	65–2	0.28–0.33
Titanium and its alloys	80–130	344–1380	415–1450	25–7	0.31–0.34
Tungsten and its alloys	350–400	550–690	620–760	0	0.27
Zinc and its alloys	50	25–180	240–550	65–5	0.27
Nonmetallic materials					
Ceramics	70–1000	—	140–2600	0	0.2
Diamond	820–1050	—	60,000	—	0.2
Glass and porcelain	70–80	—	140	0	0.24
Silicon carbide (SiC)	200–500	—	310–400	—	0.19
Silicon nitride (Si_2N_4)	280–310	—	160–580	—	0.26
Rubbers	0.01–0.1	—	—	—	0.5
Thermoplastics	1.4–3.4	—	7–80	1000–5	0.32–0.40
Thermoplastics, reinforced	2–50	—	20–120	10–1	0–0.5
Thermosets	3.5–17	—	35–170	0	0.34–0.5
Boron fibers	380	—	3500	0	0.27
Carbon fibers	275–415	—	2000–3000	0	0.21–0.28
Glass fibers	73–85	—	3500–4600	0	0.22–0.26
Kevlar fibers	62–117	—	2800	0	0.36
Spectra Fibers	73–100	—	2400–2800	3	0.46

Note: In the upper part of the table the lowest values for E, Y, and UTS and the highest values for elongation are for pure metals. Multiply gigapascals (GPa) by 145,000 to obtain pounds per square in. (psi), and megapascals (MPa) by 145 to obtain psi.

begins to decrease slowly above the proportional limit. For such materials, Y is usually defined by drawing a line with the same slope as the linear elastic curve but that is **offset** by a strain of 0.002, or 0.2% elongation. The yield stress is then defined as the stress where this offset line intersects the stress–strain curve; this simple procedure is shown on the left side in Fig. 2.2.

As the specimen begins to elongate under a continuously increasing load, its cross-sectional area decreases **permanently** and **uniformly** within its gage length. If the specimen is unloaded (from a stress level higher than the yield stress), the curve follows a straight line downward and parallel to the original slope of the curve, as shown in Fig. 2.3. As the load is increased further, the engineering stress eventually reaches a maximum and then begins to decrease (Fig. 2.2). The maximum engineering stress is called the **tensile strength** or **ultimate tensile strength** (**UTS**) of the material. Values for UTS for various materials are given in Table 2.2.

If the specimen is loaded beyond its ultimate tensile strength, it begins to **neck**, or *neck down*; the cross-sectional area of the specimen is no longer uniform along the gage length and is smaller in the necked region. As the test progresses, the engineering stress drops further and the specimen finally fractures at the necked region (Fig. 2.1a); the engineering stress at fracture is known as the **breaking** or **fracture stress**.

The ratio of stress to strain in the elastic region is the **modulus of elasticity, E,** or **Young's modulus** (after the British scientist T. Young, 1773–1829):

$$E = \frac{\sigma}{e}. \tag{2.3}$$

This linear relationship is known as **Hooke's law** (after the British physicist R. Hooke, 1635–1703).

Note in Eq. (2.3) that, because engineering strain is dimensionless, E has the same units as stress. The modulus of elasticity is the slope of the elastic portion of the curve and hence indicates the **stiffness** of the material. The higher the elastic modulus, the higher is the load required to stretch the specimen to the same extent, and thus the stiffer is the material. Compare, for example, the stiffness of metal wire with that of a rubber band or plastic sheet when they are stretched.

The elongation of the specimen under tension is accompanied by lateral contraction; this effect can easily be observed by stretching a rubber band. The absolute value of the ratio of the lateral strain to the longitudinal strain is known as **Poisson's ratio** (after the French mathematician S.D. Poisson, 1781–1840) and is denoted by the symbol ν.

2.2.2 Ductility

An important behavior observed during a tension test is **ductility**—the extent of plastic deformation that the material undergoes prior to fracture. There are two common measures of ductility. The first is the **total elongation** of the specimen, given by

$$\text{Elongation} = \frac{l_f - l_o}{l_o} \times 100, \tag{2.4}$$

where l_f and l_o are measured as shown in Fig. 2.1a. Note that the elongation is based on the original gage length of the specimen and that it is calculated as a percentage.

The second measure of ductility is the **reduction of area**, given by

$$\text{Reduction of area} = \frac{A_o - A_f}{A_o} \times 100, \tag{2.5}$$

where A_o and A_f are, respectively, the original and final (fracture) cross-sectional areas of the test specimen. Thus, the ductility of a piece of chalk is zero, because it does not stretch at all or reduce in cross-section; by contrast, a ductile specimen, such as a pure metal or thermoplastic, stretches and necks considerably before it fails.

2.2.3 True Stress and True Strain

Recall that engineering stress is based on the original cross-sectional area, A_o, of the specimen. However, the instantaneous cross-sectional area of the specimen becomes smaller as it elongates, just as the area of a rubber band does; thus, engineering stress does not represent the *actual* (or true) stress to which the specimen is subjected.

True stress is defined as the ratio of the load, P, to the actual (instantaneous, hence *true*) cross-sectional area, A, of the specimen:

$$\sigma = \frac{P}{A}. \tag{2.6}$$

For determining **true strain**, first consider the elongation of the specimen as consisting of increments of instantaneous change in length. Then, using calculus, it can be shown that the true strain (*natural* or *logarithmic strain*) is calculated as

$$\epsilon = \ln\left(\frac{l}{l_o}\right). \tag{2.7}$$

Note from Eqs. (2.2) and (2.7) that, for small values of strain, the engineering and true strains are approximately equal; however, they diverge rapidly as the strain increases. For example, when $e = 0.1$, $\epsilon = 0.095$, and when $e = 1$, $\epsilon = 0.69$.

Unlike engineering strains, true strains are consistent with actual physical phenomena in the deformation of materials. For example, consider a hypothetical situation where a specimen 50 mm (2 in.) in height is compressed, between flat platens, to a final height of zero; in other words, the specimen is deformed infinitely. According to their definitions, the engineering strain that the specimen undergoes is $(0 - 50)/50 = -1$, but the true strain is $-\infty$; note that the answer will be the same regardless of the original height of the specimen. Clearly, then, true strain describes the extent of deformation correctly, since the deformation is indeed infinite.

2.2.4 Construction of Stress–strain Curves

The procedure for constructing an engineering stress–strain curve is to take the load–elongation curve (Fig. 2.4a, and also Fig. 2.2) and then to divide the load (vertical

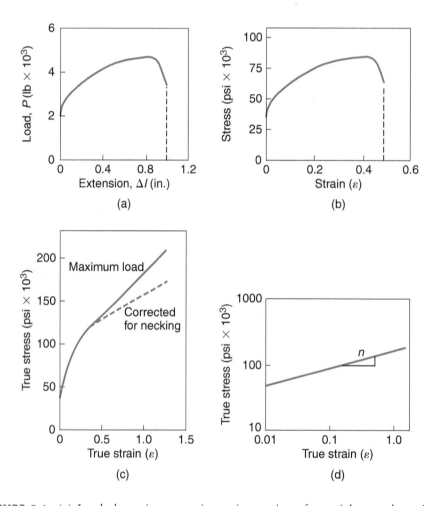

FIGURE 2.4 (a) Load–elongation curve in tension testing of a stainless steel specimen. (b) Engineering stress–engineering strain curve, drawn from the data in Fig. 2.4a. (c) True stress–true strain curve, drawn from the data in Fig. 2.4b. Note that this curve has a positive slope, indicating that the material is becoming stronger as it is strained. (d) True stress–true strain curve plotted on log–log paper and based on the corrected curve in Fig. 2.4c. The correction is due to the triaxial state of stress that exists in the necked region of the specimen.

axis) by the original cross-sectional area, A_o, and the elongation (horizontal axis) by the original gage length, l_o. Because A_o and l_o are constants, the engineering stress–strain curve obtained (shown in Fig. 2.4b) has the same shape as the load–elongation curve shown in Fig. 2.4a. (In this example, $A_o = 0.056$ in^2 and $A_f = 0.016$ in^2.)

True stress–true strain curves are obtained similarly, by dividing the load by the instantaneous cross-sectional area, with the true strain calculated from Eq. (2.7); the result is shown in Fig. 2.4c. Note the *correction* to the curve, reflecting the fact that the specimen's necked region is subjected to three-dimensional tensile stresses, as described in more advanced texts. This stress state gives higher stress values than the actual true stress; hence, to compensate, the curve must be corrected downward.

The true stress–true strain curve in Fig. 2.4c can be represented by the equation

$$\sigma = K\epsilon^n, \qquad (2.8)$$

where K is the **strength coefficient** and n is the **strain-hardening** or **work-hardening exponent**. Typical values for K and n for several metals are given in Table 2.3.

When the curve shown in Fig. 2.4c is a log–log plot, it will be found that the curve is approximately a straight line (Fig. 2.4d); the slope of the curve is the exponent n. Thus, the higher the slope, the greater is the strain-hardening capacity of the material—that is, the stronger and harder it becomes as it is strained. True stress–true strain curves for a variety of metals are given in Fig. 2.5. When reviewed in detail, some differences between Table 2.3 and Fig. 2.5 will be noted. These discrepancies result from the fact that different sources of data and different specimens have been involved in obtaining them.

Note that the elastic regions in the curves have been deleted, because the slope in this region is very high. Consequently, the point of intersection of each curve with the vertical axis in this figure can be considered to be the yield stress, Y, of the material.

The area under the true stress–true strain curve at a particular strain is the energy per unit volume (**specific energy**) of the material deformed, and indicates the work required to plastically deform a unit volume of the material to that strain. The area under the true stress–true strain curve up to fracture is known as the material's **toughness**, that is, the amount of energy per unit volume that the material dissipates prior to fracture. Note that toughness involves both the height and width of the stress–strain curve of the material, whereas strength is related only to the *height* of the curve and ductility is related only to the *width* of the curve.

2.2.5   Strain at Necking in a Tension Test

As stated earlier, the onset of necking in a tension-test specimen corresponds to the ultimate tensile strength of the material. Note that the slope of the load–elongation curve at this point is zero, and it is there that the specimen begins to neck. The specimen cannot support the load being applied because the cross-sectional area of the neck is becoming smaller at a rate that is higher than the rate at which the material becomes stronger (strain hardens).

The true strain at the onset of necking is numerically equal to the strain-hardening exponent, n, of the material. Thus, the higher the value of n, the higher the strain

TABLE 2.3

Typical Values for K and n for Selected Metals		
Material	K (MPa)	n
Aluminum		
1100-O	180	0.20
2024-T4	690	0.16
5052-O	202	0.13
6061-O	205	0.20
6061-T6	410	0.05
7075-O	400	0.17
Brass		
70-30, annealed	900	0.49
85-15, cold rolled	580	0.34
Cobalt-based alloy, heat treated	2070	0.50
Copper, annealed	315	0.54
Steel		
Low-C, annealed	530	0.26
1020, annealed	745	0.20
4135, annealed	1015	0.17
4135, cold rolled	1100	0.14
4340, annealed	640	0.15
304 stainless, annealed	1275	0.45
410 stainless, annealed	960	0.10
Titanium		
Ti-6Al-4V, annealed, 20°C	1400	0.015
Ti-6Al-4V, annealed, 200°C	1040	0.026
Ti-6Al-4V, annealed, 600°C	650	0.064
Ti-6Al-4V, annealed, 800°C	350	0.146

Video Solution 2.1 Toughness and Resilience of Power Law Materials

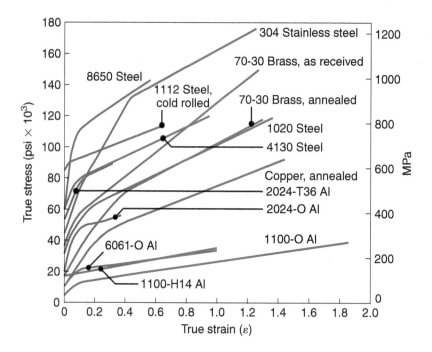

FIGURE 2.5 True stress–true strain curves in tension at room temperature for various metals; the curves start at a finite level of stress. The elastic regions have too steep a slope to be shown in this figure; thus, each curve starts at the yield stress, Y, of the material.

that a piece of material can experience before it begins to neck. This observation is important, particularly in regard to sheet-metal forming operations that involve stretching of the workpiece material (Chapter 16). It can be seen in Table 2.3 that annealed copper, brass, and stainless steel have high n values, meaning that they can be stretched uniformly to a greater extent than can the other metals listed in the table.

EXAMPLE 2.1   Calculation of Ultimate Tensile Strength

Given: This example shows that the UTS of a material can be calculated from its strength coefficient, K, and strain hardening exponent, n. Assume that a material has a true stress–true strain curve given by

$$\sigma = 100{,}000\epsilon^{0.5} \text{ psi.}$$

Find: Calculate the true ultimate tensile strength and the engineering UTS of this material.

Solution: Recall that the necking strain corresponds to the strain hardening exponent; the necking strain for this material is

$$\epsilon = n = 0.5,$$

therefore the *true* ultimate tensile strength is

$$\sigma = Kn^n = 100{,}000(0.5)^{0.5} = 70{,}710 \text{ psi.}$$

The true area at the onset of necking is obtained from

$$\ln\left(\frac{A_o}{A_{\text{neck}}}\right) = n = 0.5.$$

Thus,

$$A_{\text{neck}} = A_o\epsilon^{-0.5},$$

and the maximum load, P, is

$$P = \sigma A_{\text{neck}} = \sigma A_o e^{-0.5},$$

where σ is the true UTS. Hence,

$$P = (70{,}710)(0.606)(A_o) = 42{,}850 A_o \text{ lb.}$$

Since $\text{UTS} = P/A_o$,

$$\text{UTS} = 42{,}850 \text{ lb.}$$

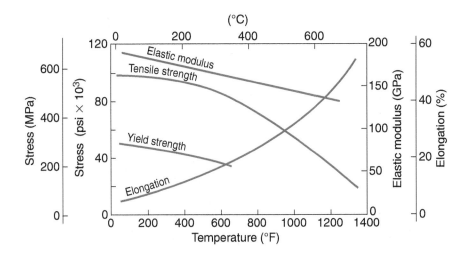

FIGURE 2.6   Effect of temperature on mechanical properties of a carbon steel; most materials display similar temperature sensitivity for elastic modulus, yield strength, ultimate strength, and ductility.

2.2.6   Temperature Effects

Increasing the temperature generally has the following effects on stress–strain curves (Fig. 2.6):

a. The ductility and toughness increase
b. The yield strength and modulus of elasticity decrease

Temperature also affects the strain-hardening exponent, n, of most metals, in that it increases with increasing temperature. The influence of temperature is, however, best described in conjunction with the rate of deformation because increasing strain rate tends to decrease n, as described below.

2.2.7   Effects of Rate of Deformation and Strain Rate

Just as a balloon can be inflated or a rubber band stretched at different rates, a piece of material in a manufacturing process can be shaped at different speeds. Some machines, such as hydraulic presses, form materials at low speeds, while others, such as mechanical presses, shape materials at high speeds.

The **deformation rate** in a tension test is the speed at which a specimen is being stretched, in units such as m/s or ft/min. The **strain rate**, on the other hand, is a function of the specimen's length. For example, consider two rubber bands, one 20 mm and the other 100 mm long, respectively, that are stretched by 10 mm within a period of 1 s. The engineering strain in the shorter specimen is $\frac{10}{20} = 0.5$; the strain in the longer is $\frac{10}{100} = 0.1$. Thus, the strain rates are 0.5 s^{-1} and 0.1 s^{-1}, respectively; thus, the short band is being subjected to a strain rate five times higher than that for the long band, even though they are both being stretched at the same deformation rate.

Deformation rates typically employed in various testing as well as metalworking processes, and the true strains involved, are given in Table 2.4. Because of the wide range encountered in practice, strain rates are usually stated in terms of orders of magnitude, such as 10^2 s^{-1}, 10^4 s^{-1}, etc.

The typical effects that temperature and strain rate jointly have on the strength of metals are shown in Fig. 2.7. Note that increasing the strain rate increases the

TABLE 2.4

Typical Ranges of Strain and Deformation Rate in Manufacturing Processes		
Process	True Strain	Deformation rate (m/s)
Cold working		
Forging, rolling	0.1–0.5	0.1–100
Wire and tube drawing	0.05–0.5	0.1–100
Explosive forming	0.05–0.2	10–100
Hot working and warm working		
Forging, rolling	0.1–0.5	0.1–30
Extrusion	2–5	0.1–1
Machining	1–10	0.1–100
Sheet-metal forming	0.1–0.5	0.05–2
Superplastic forming	0.2–3	10^{-4}–10^{-2}

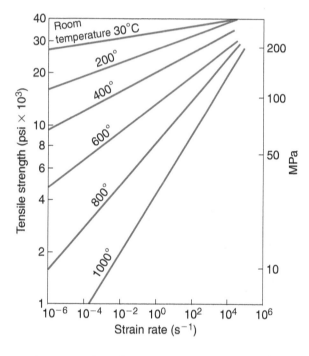

FIGURE 2.7 The effect of strain rate on the ultimate tensile strength for aluminum. Note that, as the temperature increases, the slopes of the curves increase; thus, strength becomes more and more sensitive to strain rate as temperature increases. *Source:* J.H. Hollomon.

strength of the material, called **strain-rate hardening**. The slope of these curves is called the **strain-rate sensitivity exponent, *m***. The value of *m* is obtained from log–log plots, provided that the vertical and horizontal scales are the same (unlike those shown in Fig. 2.7). A slope of 45° would therefore indicate a value of $m = 1$. The relationship is given by the equation

$$\sigma = C\dot{\epsilon}^{m}, \tag{2.9}$$

where *C* is the **strength coefficient** and $\dot{\epsilon}$ is the *true strain rate*, defined as the true strain that the material undergoes per unit time. Note that *C* has the units of stress and is similar to, but not to be confused with, the strength coefficient *K* in Eq. (2.8).

From Fig. 2.7, it can be seen that the sensitivity of strength to strain rate increases with temperature; in other words, *m* increases with increasing temperature. Also note that the slope is relatively flat at room temperature; that is, *m* is very low. This condition is true for most metals, but not for those that recrystallize at room temperature, such as lead and tin. Typical ranges of *m* for metals are up to 0.05 for cold-working, 0.05 to 0.4 for hot-working, and 0.3 to 0.85 for superplastic materials (see below).

The magnitude of the strain-rate sensitivity exponent significantly influences necking in a tension test. With increasing *m*, the material stretches farther before it fails; thus, increasing *m* delays necking. Ductility enhancement caused by the high strain-rate sensitivity of some materials has been exploited in **superplastic forming** of sheet metal, as described in Section 16.10.

Superplasticity.   The term *superplasticity* refers to the capability of some materials to undergo large uniform elongation prior to necking and fracture in tension. The elongation ranges from a few hundred percent to as much as 2000%. Common nonmetallic materials exhibiting superplastic behavior are bubble gum, glass (at elevated temperatures), and thermoplastics; as a result, glass and thermoplastics can successfully be formed into a wide variety of complex shapes. Among metals exhibiting superplastic behavior are very fine grained (10 to 15 μm) titanium alloys and zinc–aluminum alloys; when heated, they can elongate to several times their original length.

2.2.8 Hydrostatic Pressure Effects

Various tests have been performed to determine the effect of hydrostatic pressure on mechanical properties of materials. Test results at pressures up to 3.5 GPa (500 ksi) indicate that increasing the hydrostatic pressure substantially increases the strain at fracture, both for ductile and for brittle materials. This beneficial effect of hydrostatic pressure has been exploited in metalworking processes, particularly in hydrostatic extrusion (Section 15.4.2) and in compaction of metal powders (Section 17.3).

2.2.9 Radiation Effects

In view of the use of various metals and alloys in nuclear applications, extensive studies have been conducted on radiation's effects on mechanical properties. Typical changes in the properties of steels and other metals exposed to doses of high radiation are increased yield stress, tensile strength, and hardness, and decreased ductility and toughness.

2.3 Compression

Numerous operations in manufacturing, such as forging, rolling, and extrusion (Part III), are performed with the workpiece subjected to compressive forces. The **compression test,** in which the specimen is subjected to a compressive load, gives information that is helpful in estimating forces and power requirements in these processes. This test is usually carried out by compressing a solid cylindrical specimen between two well-lubricated flat dies (platens). Because of friction between the specimen and the platens, the specimen's cylindrical surface bulges, called **barreling** (Fig. 2.8). The height-to-diameter ratio of the specimen is typically less than 3:1 in order to avoid buckling during the test. (See also Section 14.4 on *heading*.)

Because of barreling, the specimen's cross-sectional area varies along its height, and thus obtaining the stress–strain curves in compression can be difficult. Furthermore, since friction dissipates energy, the compressive force is higher than it otherwise would be in order to overcome friction. With effective lubrication, however, friction can be minimized and thus a reasonably constant cross-sectional area can be maintained during the test.

When the results of compression and tension tests on *ductile* metals are compared, it will be seen that the true stress–true strain curves coincide. This behavior, however, does not hold true for *brittle* materials, which are generally stronger and more ductile in compression than in tension. (See Table 8.2.)

If a specimen is first subjected to tension and deformed plastically and then the load is released and a compressive load is applied, the yield stress in compression is found to be lower than that in tension. This behavior is known as the **Bauschinger effect** (after the German engineer J. Bauschinger, reported in 1881); it is exhibited in varying

FIGURE 2.8 Barreling in compressing a round solid cylindrical specimen (7075-O aluminum) between flat dies. Barreling is caused by friction at the die–specimen interfaces, which retards the free flow of the material. See also Fig. 14.3.

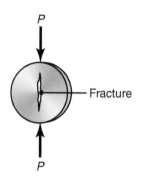

P

P

FIGURE 2.9 Disk test on a brittle material, showing the direction of loading and the fracture path.

degrees by all metals and alloys. The phenomenon is also called **strain softening** or **work softening**, because of the lowered yield stress in the direction opposite that of the original load application.

Disk Test. For brittle materials such as ceramics and glasses (Chapter 8), the **disk test** can be used, in which a disk is subjected to diametral compression forces between two hardened flat platens (Fig. 2.9). When the specimen is loaded as shown, tensile stresses develop perpendicular to the vertical centerline along the disk; fracture initiates and the disk splits vertically in half.

The *tensile stress*, σ, in the disk is uniform along the centerline and can be calculated from the formula

$$\sigma = \frac{2P}{\pi dt}, \tag{2.10}$$

where P is the load at fracture, d is the diameter of the disk, and t is its thickness. In order to avoid premature failure at the contact points, thin strips of soft metal are placed between the disk and the two platens; these strips also protect the platens from being damaged during the test. The phenomenon of fracture at the center of the specimen has been utilized in the manufacture of *seamless tubing* (Section 13.5).

2.4 Torsion

In addition to undergoing tensile and compressive forces, a workpiece may also be subjected to shear strains (Fig. 2.10), such as in the punching of holes in sheet metals (Section 16.2), swaging (Section 14.4), and machining (Section 21.2). The method generally used to determine properties of materials in shear is the **torsion test**. This test is usually performed on a thin tubular specimen in order to obtain an approximately uniform stress and strain distribution along its cross-section.

The test specimen usually has a reduced cross-section in order to confine the deformation to a narrow zone. The **shear stress**, τ, can be calculated from the formula

$$\tau = \frac{T}{2\pi r^2 t}, \tag{2.11}$$

where T is the torque applied, r is the average radius of the tube, and t is the thickness of the tube at its narrow section.

The **shear strain**, γ, can be calculated from the formula

$$\gamma = \frac{r\phi}{l}, \tag{2.12}$$

where l is the length of the tube section and ϕ the **angle of twist** in radians.

The ratio of the shear stress to the shear strain in the elastic range is known as the **shear modulus**, or **modulus of rigidity**, G. The shear modulus is a quantity related to the modulus of elasticity, E, by the formula

$$G = \frac{E}{2(1 + \nu)}. \tag{2.13}$$

The *angle of twist*, ϕ, to fracture in the torsion of solid round bars at elevated temperatures has been found to be useful in estimating the forgeability of metals (Section 14.5). The greater the number of twists prior to failure, the better is the forgeability.

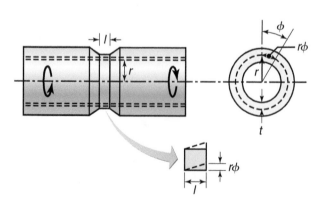

FIGURE 2.10 A typical torsion-test specimen; it is mounted between the two heads of a testing machine and twisted; note the shear deformation of an element in the reduced section of the specimen.

2.5   Bending (Flexure)

Preparing specimens from brittle materials can be challenging because of the difficulties involved in shaping or machining them to proper dimensions. Furthermore, the specimens are sensitive to surface defects (such as scratches and notches), and clamping brittle specimens for testing is difficult. Also, improper alignment of the test specimen can result in a nonuniform stress distribution along its cross-section.

A commonly used test method for brittle materials is the **bend** or **flexure test**, which typically involves a specimen that has a rectangular cross-section and is supported in a manner shown in Fig. 2.11. The load is applied vertically, at either one point or two points; consequently, these tests are referred to as **three-point** and **four-point bending,** respectively. The longitudinal stresses in the specimens are tensile at their lower surfaces and compressive at their upper surfaces; these stresses can be calculated using simple beam equations described in texts on the mechanics of solids.

The stress at fracture in bending is known as the **modulus of rupture,** or **transverse rupture strength** (see Table 8.2). Note that, because of the larger volume of material subjected to the same bending moment in Fig. 2.11b, there is a higher probability that defects exist within this volume than in that shown in Fig. 2.11a. Consequently, the four-point test gives a lower modulus of rupture than the three-point test.

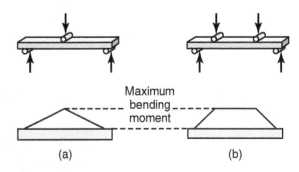

FIGURE 2.11   Two bend-test methods for brittle materials: (a) three-point bending and (b) four-point bending. The areas over the beams represent the bending-moment diagrams, described in texts on the mechanics of solids. Note the region of constant maximum bending moment in (b); by contrast, the maximum bending moment occurs only at the center of the specimen in (a).

2.6   Hardness

Hardness is generally defined as *resistance to permanent indentation*; thus, steel is harder than aluminum and aluminum is harder than lead. Hardness, however, is not a fundamental property, because the resistance to indentation depends on the shape of the indenter and on the load applied. Hardness is a commonly used property; it gives a general indication of the strength of the material and of its resistance to scratching and wear.

QR Code 2.1 Instron 5544 3-point bend test. (*Source:* Courtesy of Instron®)

2.6.1   Hardness Tests

Several test methods, using different indenter materials and shapes (Fig. 2.12), have been developed to measure the hardness of materials. The most commonly used hardness tests are described below.

Brinell Test.   Introduced by J.A. Brinell in 1900, this test involves pressing a steel or tungsten-carbide ball 10 mm (0.4 in.) in diameter against a surface, with a load of 500, 1500, or 3000 kg (Fig. 2.13). The *Brinell hardness number* (HB) is defined as the ratio of the applied load, P, to the curved surface area of the indentation. The harder the material tested, the smaller the impression; a 1500-kg or 3000-kg load is usually recommended in order to obtain impressions sufficiently large for accurate measurement of hardness.

Depending on the condition of the material tested, one of two types of impression develops on the surface after the performance of a hardness test (Fig. 2.14). The impressions in annealed metals generally have a rounded profile along the periphery (Fig. 2.14a); in cold-worked metals, they usually have a sharp profile

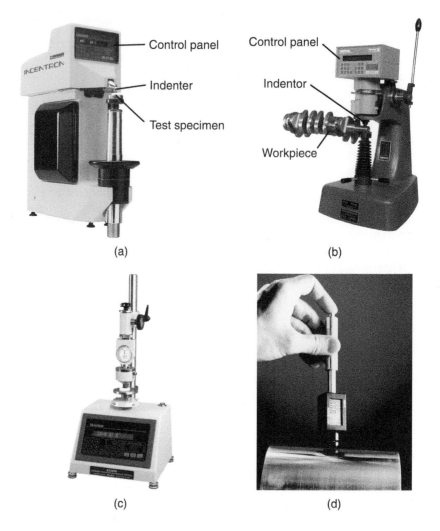

FIGURE 2.12   A selection of hardness testers. (a) A Micro Vickers hardness tester, (b) Rockwell hardness tester (the support for the part has been removed for clarity), (c) Durometer, and (d) Leeb tester. *Source:* (a) through (c) Courtesy of Newage Testing Instruments, Inc. (d) Courtesy of Wilson® Instruments.

(Fig. 2.14b). The correct method of measuring the indentation diameter, d, is shown in the figure.

The indenter has a finite elastic modulus, and thus undergoes *elastic* deformation under the applied load; as a result, hardness measurements may not be as accurate as expected, depending on the indenter material. One method for minimizing this effect is using tungsten-carbide balls (Section 22.4); because of their higher modulus of elasticity, they distort less than steel balls do. These indenters are usually recommended for materials with a Brinell hardness number higher than 500.

Rockwell Test.   Developed by S.P. Rockwell in 1922, this test measures the *depth* of penetration instead of the diameter of the indentation. The indenter is pressed onto the surface, first with a *minor load* and then with a *major load*; the difference in the depths of penetration is a measure of the hardness of the material. Figure 2.13 shows some of the more common Rockwell hardness scales and the indenters used

QR Code 2.2 Video of a Rockwell hardness test. (*Source:* Courtesy of Instron®)

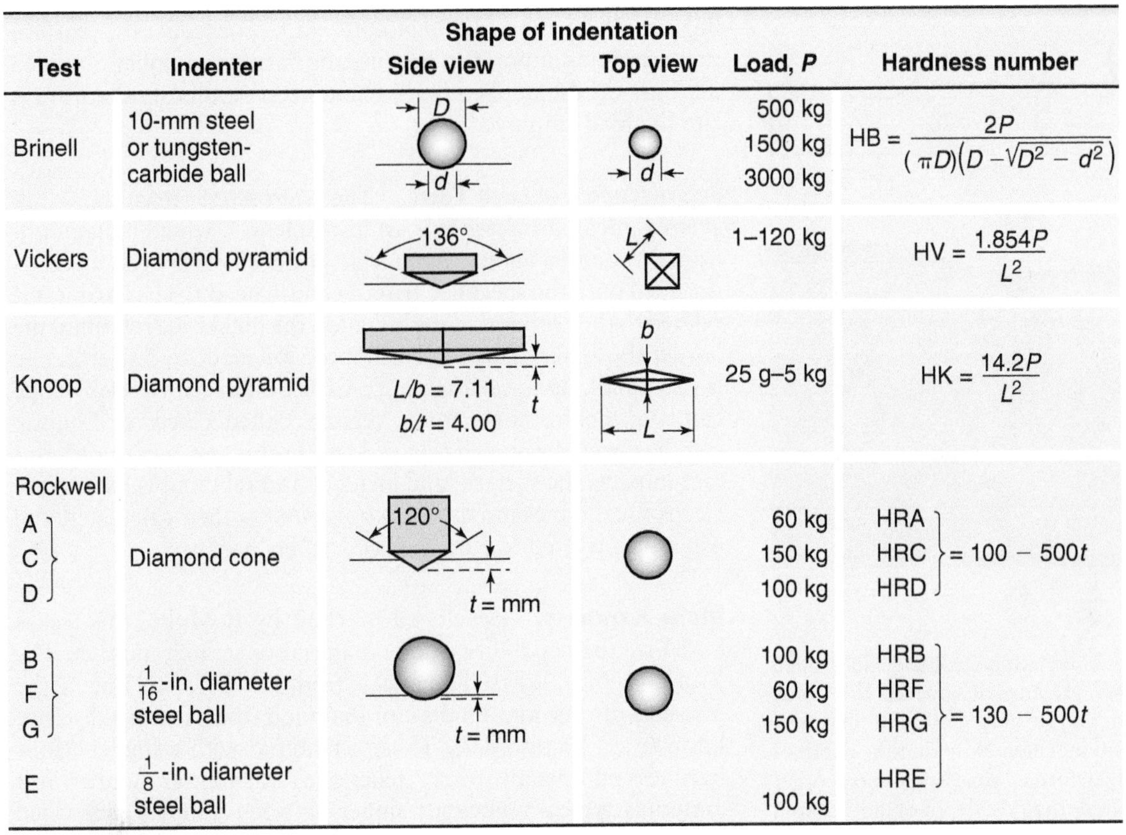

Test	Indenter	Shape of indentation Side view	Shape of indentation Top view	Load, P	Hardness number
Brinell	10-mm steel or tungsten-carbide ball	D ⟷, d ⟷	d ⟷	500 kg 1500 kg 3000 kg	$HB = \dfrac{2P}{(\pi D)\left(D - \sqrt{D^2 - d^2}\right)}$
Vickers	Diamond pyramid	136°	L	1–120 kg	$HV = \dfrac{1.854P}{L^2}$
Knoop	Diamond pyramid	$L/b = 7.11$ $b/t = 4.00$ t	b $\leftarrow L \rightarrow$	25 g–5 kg	$HK = \dfrac{14.2P}{L^2}$
Rockwell A C D	Diamond cone	120° t = mm		60 kg 150 kg 100 kg	HRA HRC HRD $\Big\} = 100 - 500t$
B F G	$\frac{1}{16}$-in. diameter steel ball	t = mm		100 kg 60 kg 150 kg	HRB HRF HRG $\Big\} = 130 - 500t$
E	$\frac{1}{8}$-in. diameter steel ball			100 kg	HRE

FIGURE 2.13   General characteristics of hardness-testing methods and formulas for calculating hardness.

in practice. Rockwell **superficial hardness** tests, using the same type of indenters but at lighter loads, also have been developed.

Vickers Test.   This test, developed by the Vickers Ltd. company in 1922 and formerly known as the *diamond pyramid hardness* test, uses a pyramid-shaped diamond indenter (Fig. 2.13) and a load that ranges from 1 to 120 kg. The *Vickers hardness number* is indicated by HV. The impressions obtained are typically less than 0.5 mm (0.020 in.) on the diagonal, and penetration depths can be as low as 20 nm. This test gives essentially the same hardness number regardless of the load, and is suitable for testing materials with a wide range of hardness, including heat-treated steels. Test procedures also have been developed to perform tests using atomic force microscopes and nanoindenters.

Knoop Test.   This test, developed by F. Knoop in 1939, uses a diamond indenter in the shape of an elongated pyramid (Fig. 2.13); the applied loads range generally from 25 g to 5 kg. The *Knoop hardness number* is indicated by HK. Because of the light loads applied, this test is a **microhardness** test, and is therefore suitable for very small or very thin specimens and for brittle materials such as carbides, ceramics, and glass.

This test is also used for measuring the hardness of the individual grains and components in a metal alloy. The size of the indentation is generally in the range from

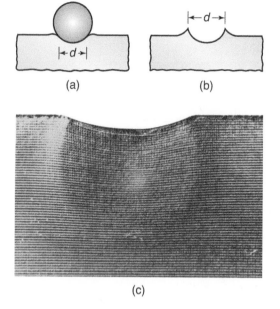

FIGURE 2.14 Indentation geometry in Brinell hardness testing: (a) annealed metal, (b) work-hardened metal, and (c) deformation of mild steel under a spherical indenter. Note that the depth of the permanently deformed zone is about one order of magnitude larger than the depth of indentation; for a hardness test to be valid, this zone should be fully developed in the material. *Source:* After M.C. Shaw and C.T. Yang.

0.01 to 0.10 mm (0.0004–0.004 in.); consequently, surface preparation is important. Because the hardness number obtained depends on the applied load, Knoop test results should always cite the load employed.

Scleroscope and Leeb Tests.   The *scleroscope* (from the Greek *skleros*, meaning "hard") is an instrument in which a diamond-tipped indenter (called *hammer*), enclosed in a glass tube, is dropped onto the specimen from a certain height. The hardness is related to the *rebound* of the indenter: the higher the rebound, the harder the material tested. The impression made by a scleroscope is very small. Because obtaining reliable results with a scleroscope can be difficult, an electronic version, called a *Leeb* or Equotip test, has been developed (Fig. 2.12d). In this test, a carbide hammer impacts the surface, and incident and rebound velocities are electronically measured. A *Leeb number* is then calculated and usually converted to Rockwell or Vickers hardness.

Mohs Hardness.   Developed in 1822 by F. Mohs, this test is based on the capability of one material to scratch another. The Mohs hardness is based on a scale from 1 to 10, with 1 being the measure for talc and 10 that for diamond (the hardest substance known; see Section 8.7). Thus, a material with a higher Mohs hardness number always scratches one with a lower number. Soft materials typically have a number between 2 and 3, hardened steels about 6, and aluminum oxide (used for cutting tools and as an abrasive in grinding wheels) of 9. Although the Mohs scale is qualitative and is used mainly by mineralogists, it correlates well with Knoop hardness.

Shore Test and Durometer.   The hardness of materials such as rubbers, plastics, and similar soft and elastic nonmetallic materials is generally measured by a Shore test, with an instrument called a *durometer* (from the Latin *durus*, meaning "hard"). An indenter is first pressed against the surface and then a constant load is rapidly applied. The *depth* of penetration is measured after one second; the hardness is inversely related to the penetration. There are two different scales for this test. Type A has a blunt indenter and an applied load of 1 kg; the test is typically used for softer materials. Type D has a sharper indenter and a load of 5 kg, and is used for harder materials. The hardness numbers in these tests range from 0 to 100.

Hot Hardness.   The hardness of materials at elevated temperatures (see Fig. 22.1) is important in applications such as cutting tools in machining and dies in hot-working and casting operations. Hardness tests can be performed at elevated temperatures with conventional testers, with some modifications such as enclosing the specimen and indenter in a small electric furnace.

2.6.2   Hardness and Strength

Video Solution 2.2   Hardness Test Indentation Dimensions

Because hardness is the resistance to *permanent* indentation, it can be likened to performing a compression test on a small volume on the surface of a material (Fig. 2.14c). Studies have shown that the hardness of a cold-worked metal is about three times its yield stress Y (in the same units); for annealed metals, the hardness is about five times Y.

A relationship has been established between the UTS and the Brinell hardness (HB) for steels, measured for a load of 3000 kg. In SI units, the relationship is given by

$$\text{UTS} = 3.5(\text{HB}), \qquad\qquad (2.14)$$

where UTS is in MPa. In traditional units,

$$\text{UTS} = 500(\text{HB}), \qquad\qquad (2.15)$$

where UTS is in psi.

2.6.3   Hardness-testing Procedures

For a hardness test to be meaningful and reliable, the **zone of deformation** under the indenter (see Fig. 2.14c) must be allowed to develop freely. Consequently, the *location* of the indenter (with respect to the location of the *edges* of the specimen to be tested) and the *thickness* of the specimen are important considerations. Generally, the location should be at least two diameters of the indentation from the edge of the specimen, and the thickness of the specimen should be at least 10 times the depth of penetration of the indenter. Also, successive indentations on the same surface of the workpiece should be far enough apart so as not to interfere with each other.

Moreover, the indentation should be sufficiently large to give a representative hardness value for the bulk material. If hardness variations need to be detected in a small area, or if the hardness of individual constituents in a matrix or an alloy is to be determined, the indentations should be very small, such as those obtained in Knoop or Vickers tests, using light loads. While *surface preparation* is not critical for the Brinell test, it is important for the Rockwell test and even more important for the other hardness tests, because of the small sizes of the indentations. Surfaces may have to be polished to allow correct measurement of the impression's dimensions.

The values obtained from different hardness tests, on different scales, can be interrelated and can be converted using Fig. 2.15. Care should be exercised in using these charts because of the variables in material characteristics and in the shape of the indentation.

EXAMPLE 2.2   Calculation of Modulus of Resilience from Hardness

Given: A piece of steel is highly deformed at room temperature; its hardness is found to be 300 HB.

Find: Estimate the area of the elastic portion of the stress–strain curve up to the yield point (that is, the *resilience*) for this material if the yield strength is one-third the Brinell hardness.

Solution: Since the steel has been subjected to large strains at room temperature, it may be assumed that its stress–strain curve has flattened considerably, thus approaching the shape of a perfectly-plastic curve. Since the yield strength is one-third the Brinell hardness,

$$Y = \frac{300}{3} = 100 \text{ kg/mm}^2 = 142{,}250 \text{ psi.}$$

The area under the stress–strain curve is

$$\text{Modulus of resilience} = \frac{Y^2}{2E}.$$

From Table 2.2, $E = 210$ and GPa = 30×10^6 psi for steel. Hence,

$$\text{Modulus of resilience} = \frac{(142{,}250)^2}{2\,(30 \times 10^6)}$$

$$= 337 \text{ in.-lb/in}^3.$$

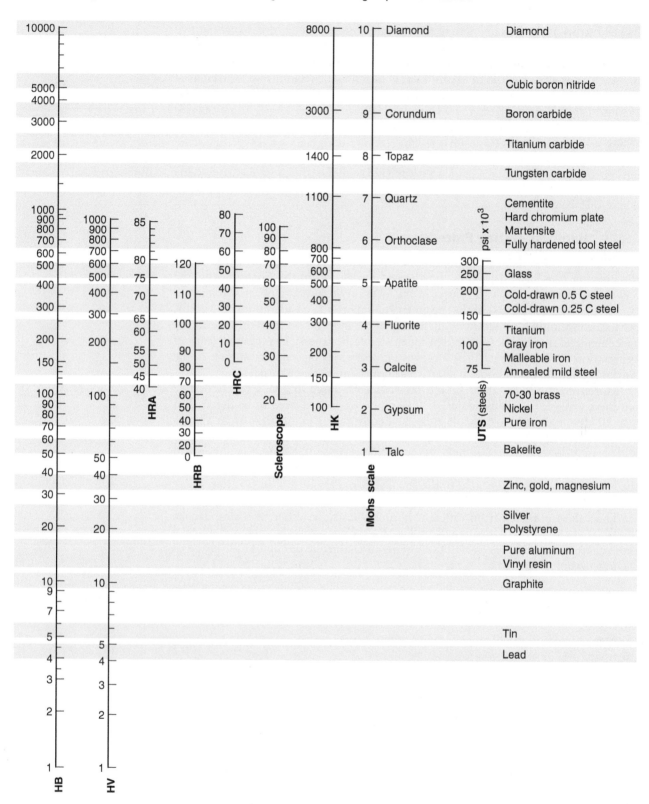

FIGURE 2.15 Chart for converting various hardness scales; note the limited range of most of the scales. Because of the many factors involved, these conversions are approximate.

2.7 Fatigue

Numerous components, such as tools, dies, gears, cams, shafts, and springs, are subjected to rapidly fluctuating (cyclic or periodic) loads, in addition to static loads. **Cyclic stresses** may be caused by fluctuating mechanical loads, such as (a) on gear teeth or reciprocating sliders, (b) by rotating machine elements under constant bending stresses, as is commonly encountered in shafts, or (c) by thermal stresses, as when a room-temperature die comes into repeated contact with hot workpieces, and then begins to cool between successive contacts. Under these conditions, the component fails at a stress level *below* that at which failure would occur under static loading. Upon inspection, failure is found to be associated with cracks that grow with every stress cycle and that propagate through the material until a critical crack length is reached when the material fractures. Known as **fatigue failure**, this phenomenon is responsible for the majority of failures in mechanical components.

Fatigue test methods involve testing specimens under various states of stress, usually in a combination of tension and bending. The test is carried out at various *stress amplitudes* (S), and the number of cycles (N) it takes to cause total failure of the specimen or part is then recorded. Stress amplitude is defined as the maximum stress, in tension and compression, to which the specimen is subjected. Typical plots, called **S–N curves**, are shown in Fig. 2.16. These curves are based on complete reversal of the stress—that is, maximum tension, then maximum compression, then maximum tension, and so on—such as that imposed by bending a rectangular eraser or a piece of wire alternately in one direction and then the other.

Tests can also be performed on a rotating shaft in four-point bending (see Fig. 2.11b). With some materials, the S–N curve becomes horizontal at low stress levels, indicating that the material will not fail at stresses below this limit. The maximum stress to which the material can be subjected without fatigue failure, regardless of the number of cycles, is known as the **endurance limit** or **fatigue limit**.

Although many materials, especially steels, have a definite endurance limit, others, such as aluminum alloys, do not have one and the S–N curve continues its downward trend. For metals exhibiting such behavior, the fatigue strength is specified at a certain number of cycles, such as 10^7; in this way, the useful service life of the component can be specified. The endurance limit for metals can be approximately related to their

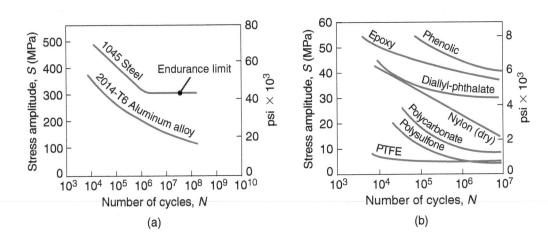

FIGURE 2.16 (a) Typical S–N curves for two metals; note that, unlike steel, aluminum does not have an endurance limit. (b) S–N curves for some polymers.

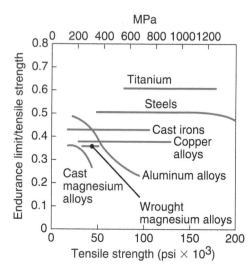

FIGURE 2.17 Ratio of endurance limit to tensile strength for various metals, as a function of tensile strength. Because aluminum does not have an endurance limit, the correlations for aluminum are based on a specific number of cycles, as is seen in Fig. 2.16.

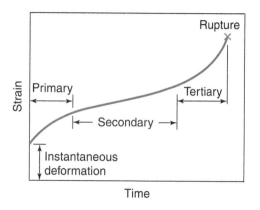

FIGURE 2.18 Schematic illustration of a typical creep curve; the linear segment of the curve (secondary) is used in designing components for a specific creep life.

UTS (Fig. 2.17). For carbon steels, the endurance limit is usually 0.4–0.5 times the tensile strength.

2.8 Creep

Creep is the permanent deformation of a component under a static load maintained for a period of time. This phenomenon occurs in metals and some nonmetallic materials, such as thermoplastics and rubbers, and it can occur at any temperature; lead, for example, creeps under a constant tensile load at room temperature. However, for metals and their alloys, creep of any significance occurs at elevated temperatures, beginning at about 200°C (400°F) for aluminum alloys and at about 1500°C (2800°F) for refractory alloys. The mechanism of creep at elevated temperature in metals is generally attributed to **grain-boundary sliding** (Section 1.5).

Creep is especially important in high-temperature applications, such as gas-turbine blades and components in jet engines and rocket motors. High-pressure steam lines, nuclear-fuel elements, and furnace components are likewise subject to creep. Creep can also occur in tools and dies that are subjected to high stresses at elevated temperatures during hot-working operations, such as forging and extrusion.

The *creep test* typically consists of subjecting a specimen to a constant tensile load (hence, constant engineering stress) at elevated temperature and measuring the changes in length at various time increments. A creep curve usually consists of *primary, secondary,* and *tertiary stages* (Fig. 2.18). During the test, the specimen eventually fails by necking and fracture, called *rupture* or *creep rupture*. As expected, the creep rate increases with specimen temperature and applied load.

Design against creep usually requires knowledge of the secondary (linear) range and its slope, because the creep rate can be determined reliably only when the curve has a constant slope. Generally, resistance to creep increases with the melting temperature of a material. Stainless steels, superalloys, and refractory metals and alloys are thus commonly used in applications where resistance to creep is required.

Stress Relaxation.   *Stress relaxation* is closely related to creep. In stress relaxation, the stresses resulting from external loading of a structural component decrease in magnitude over a period of time, even though the dimensions of the component remain constant. An example is the decrease in tensile stress of a wire in tension between two fixed ends (as in the wires in a piano or violin). Other examples include stress relaxation in rivets, bolts, guy wires, and various similar parts under either tension, compression, or flexure. Stress relaxation is particularly common and important in thermoplastics (Section 7.3).

2.9 Impact

In numerous machinery components and manufacturing operations, materials are subjected to **impact**, or **dynamic, loading**; for example, in high-speed metalworking processes such as heading to shape nails and bolt heads (Section 14.4).

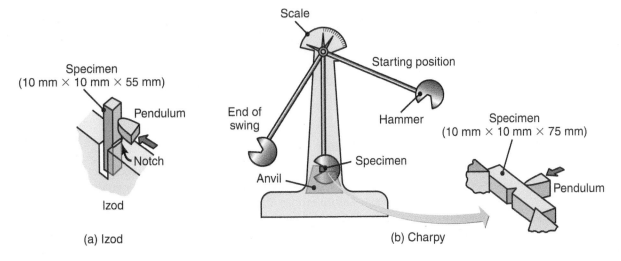

FIGURE 2.19   Impact test specimens.

A typical *impact test* for determining impact properties of materials consists of placing a *notched* specimen in an impact tester and breaking the specimen with a swinging pendulum (Fig. 2.19).

In the **Charpy test** (after the French scientist G. Charpy), the specimen is supported at both ends, while in the **Izod test** (after the English engineer E.G. Izod), it is supported at one end like a cantilever beam. From the swing of the pendulum, the energy dissipated in breaking the specimen can be obtained; this energy is the **impact toughness** of the material. Unlike hardness-test conversions (Fig. 2.15), no quantitative relationships have yet been established between Charpy and the Izod tests. Materials that have high impact resistance generally have high strength, high ductility, and, hence, high toughness. Sensitivity to surface defects (**notch sensitivity**) is important, as it significantly lowers impact toughness, particularly in heat-treated metals, ceramics, and glasses. Impact tests are particularly useful in determining the ductile–brittle transition temperature of materials (Section 2.10.1).

2.10   Failure and Fracture of Materials

Failure is one of the most important aspects of material behavior, because it directly influences the selection of a material for a particular application and the method(s) of manufacturing, and determining the service life of the component. Because of the several factors involved, failure and fracture of materials is a complex area of study. This section focuses only on those aspects of failure that are of particular significance to selecting and processing materials.

There are two general types of failure:

1. **Fracture,** through either internal or external cracking; fracture is further subclassified into two general categories: *ductile* and *brittle* (Figs. 2.21 and 2.22).
2. **Buckling,** as shown in Fig. 2.20b.

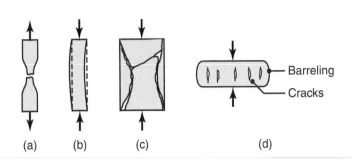

FIGURE 2.20   Schematic illustration of types of failures in materials: (a) necking and fracture of ductile materials, (b) buckling of ductile materials under a compressive load, (c) fracture of brittle materials in compression, and (d) cracking on the barreled surface of ductile materials in compression.

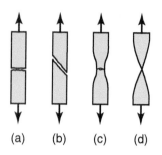

(a)   (b)   (c)   (d)

FIGURE 2.21 Schematic illustration of the types of
fracture in tension: (a) brittle fracture in polycrystalline
metals, (b) shear fracture
in ductile single crystals—
see also Fig. 1.6a, (c) ductile cup-and-cone fracture in
polycrystalline metals, and
(d) complete ductile fracture in polycrystalline metals,
with 100% reduction of area.

Although failure of materials is generally regarded as undesirable, it should be
noted that some products are designed in such a way that failure is essential for their
functioning. Typical examples are (a) food and beverage containers with tabs or entire
tops which are removed by tearing; (b) shear pins on shafts to prevent machinery
damage in the case of overloads; (c) perforated sheet to ease tearing along a path;
and (d) metal or plastic screw caps for beverage bottles.

2.10.1 Ductile Fracture

Ductile fracture is characterized by *plastic deformation*, which precedes failure
(Fig. 2.20a). In a tension test, highly ductile materials such as gold and lead may
neck down to a point before failing (Fig. 2.21d); most metals and alloys, however,
neck down to a finite area and then fail. Ductile fracture generally takes place along
planes on which the shear stress is a maximum. Thus, in torsion, for example, a
ductile metal fractures along a plane perpendicular to the axis of twist; that is, the
plane on which the shear stress is a maximum. Fracture in shear, by contrast, is a
result of extensive slip along slip planes within the grains (see Fig. 1.7).

Close examination of the surface of ductile fracture (Fig. 2.22) shows a *fibrous*
pattern with *dimples*, as if a number of very small tension tests have been carried out
over the fracture surface. Failure is initiated with the formation of tiny *voids*, usually
around small inclusions or pre-existing voids, which, in turn, *grow* and *coalesce*,
developing into microcracks which then grow in size and eventually lead to fracture.

In a tension-test specimen, fracture begins at the center of the necked region
resulting in the growth and coalescence of cavities (Fig. 2.23). The central region
becomes one large crack, as can be seen in the midsection of the tension-test specimen
in Fig. 2.23d; this crack then propagates to the periphery of the necked region and
results in total failure. Because of its appearance, the fracture surface of a tension-test
ductile specimen is called a **cup-and-cone fracture**.

Effects of Inclusions.   Because they are nucleation sites for voids, *inclusions* have
an important influence on ductile fracture and, consequently, on the workability of
metals. Inclusions typically consist of impurities of various kinds
and of second-phase particles, such as oxides, carbides, and sulfides. The extent of their influence depends on such factors as their
shape, hardness, distribution, and their fraction of the total volume;
the higher the volume fraction of inclusions, the lower will be the
ductility of the material.

Voids and porosity can also develop during processing of metals, such as porosity in casting (Section 10.6.1), and metalworking
processes such as drawing and extrusion, described in Chapter 15.
Two factors affect void formation:

1. The strength of the bond at the interface between an inclusion
   and the matrix. If the bond is strong, there is less tendency for
   void formation during plastic deformation.
2. The hardness of the inclusion. If the inclusion is soft, such as
   manganese sulfide, it will conform to the overall shape change
   of the workpiece during plastic deformation. If the inclusion
   is hard, as, for example, in carbides and oxides (see also
   Section 8.2), it could lead to void formation (Fig. 2.24). Hard
   inclusions may also break up into smaller particles during
   plastic deformation, because of their brittle nature.

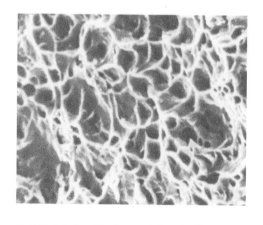

FIGURE 2.22 Surface of ductile fracture in
low-carbon steel, showing dimples. Fracture
is usually initiated at impurities, inclusions,
or pre-existing voids (microporosity) in the
metal. *Source:* Courtesy of K.-H. Habig and
D. Klaffke.

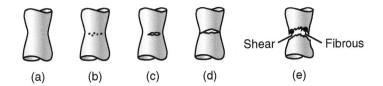

(a)   (b)   (c)   (d)   (e)

Shear   Fibrous

FIGURE 2.23 Sequence of events in the necking and fracture of a tensile-test specimen: (a) early stage of necking; (b) small voids begin to form within the necked region; (c) voids coalesce, producing an internal crack; (d) the rest of the cross-section begins to fail at the periphery, by shearing; and (e) the final fracture, known as a cup- (top fracture surface) and cone- (bottom surface) fracture, surfaces.

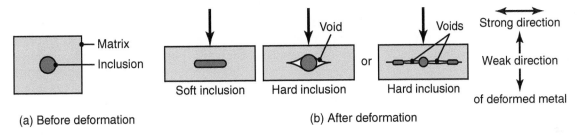

Matrix
Inclusion
Soft inclusion   Void   Hard inclusion   Voids   Hard inclusion   or

Strong direction
Weak direction
of deformed metal

(a) Before deformation   (b) After deformation

FIGURE 2.24 Schematic illustration of the deformation of soft and hard inclusions and of their effect on void formation in plastic deformation. Note that, because they do not conform to the overall deformation of the ductile matrix, hard inclusions can cause internal voids.

The alignment of inclusions during plastic deformation leads to **mechanical fibering** (Section 1.6). Subsequent processing of such a material must therefore involve considerations of the proper direction of working the material, in order to develop maximum ductility and strength.

Transition Temperature. Metals may undergo a sharp change in ductility and toughness across a narrow temperature range, called the *transition temperature* (Fig. 2.25). This phenomenon occurs mostly in body-centered cubic and in some hexagonal close-packed metals; it is rarely exhibited by face-centered cubic metals. The transition temperature depends on such factors as (a) the composition, microstructure, and grain size of the material, (b) surface finish and shape of the specimen, and (c) the deformation rate. High rates, abrupt changes in workpiece shape, and the presence of surface notches raise the transition temperature.

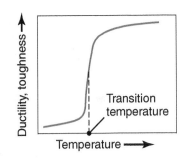

FIGURE 2.25 Schematic illustration of transition temperature in metals.

Strain Aging. *Strain aging* is a phenomenon in which carbon atoms in steels segregate to dislocations, thereby pinning the dislocations and, in this way, increasing the resistance to their movement; the result is increased strength and reduced ductility. Instead of taking place over several days at room temperature, strain aging can occur in just a few hours at a higher temperature; it is then called *accelerated strain aging*. An example of accelerated strain aging in steels is **blue brittleness**, so named because it occurs in the blue-heat range where the steel develops a bluish oxide film. Blue brittleness causes a significant decrease in ductility and toughness, and an increase in the strength of plain-carbon and some alloy steels.

2.10.2   Brittle Fracture

Brittle fracture occurs with little or no gross plastic deformation. In tension, fracture takes place along the crystallographic plane (**cleavage plane**) on which the normal

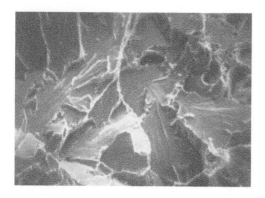

FIGURE 2.26 Fracture surface of steel that has failed in a brittle manner; the fracture path is transgranular (through the grains). Magnification: 200×. *Source:* Courtesy of B.J. Schulze and S.L. Meiley and Packer Engineering Associates, Inc.

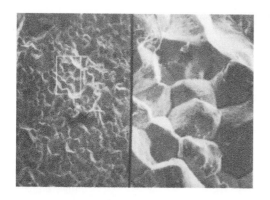

FIGURE 2.27 Intergranular fracture, at two different magnifications; grains and grain boundaries are clearly visible in this micrograph. The fracture path is along the grain boundaries. Magnification: left, 100×; right, 500×. *Source:* Courtesy of B.J. Schulze and S.L. Meiley and Packer Engineering Associates, Inc.

tensile stress is a maximum. Face-centered cubic metals usually do not fail by brittle fracture, whereas body-centered cubic and some hexagonal close-packed metals fail by cleavage. In general, low temperature and a high rate of deformation promote brittle fracture.

In a polycrystalline metal under tension, the fracture surface has a bright granular appearance (unlike the fibrous appearance in ductile fracture) because of the changes in the direction of the cleavage planes as the crack propagates from one grain to another (Fig. 2.26). Brittle fracture in compression is more complex; fracture may even follow a path that is theoretically at an angle of 45° to the direction of the applied force.

Examples of fracture along a cleavage plane are the splitting of rock salt and the peeling of layers of mica. Tensile stresses normal to the cleavage plane, caused by pulling, initiate and control the propagation of fracture. Another example is the behavior of brittle materials, such as chalk, gray cast iron, and concrete; in tension, they fail in the manner shown in Fig. 2.21a; in torsion, they fail along a plane at an angle of 45° to the axis of twist (Fig. 2.10)—that is, along a plane on which the tensile stress is a maximum.

Defects. An important factor in fracture is the presence of *defects*, such as scratches, flaws, and pre-existing external or internal cracks. Under tension, the sharp tip of a crack is subjected to high tensile stresses, which then lead the crack to propagate rapidly.

The presence of defects explains why brittle materials exhibit weakness in tension as compared with their strength in compression. (See Table 8.2.) For example, the ratio of compressive to tensile strength is on the order of 10 for rocks and similar materials, about 5 for glass, and about 3 for gray cast iron. Under tensile stresses, cracks propagate rapidly, causing what is known as *catastrophic failure*.

In polycrystalline metals, the fracture paths most commonly observed are **transgranular** (*transcrystalline* or *intragranular*); that is, the crack propagates through the grain. In **intergranular** fracture, the crack propagates along the grain boundaries (Fig. 2.27); it generally occurs when the grain boundaries are soft, contain a brittle phase, or they have been weakened by liquid- or solid-metal embrittlement (Section 1.5.2).

Fatigue Fracture. *Fatigue fracture* typically occurs in a brittle manner. Minute external or internal cracks develop at pre-existing flaws or defects in the material; these cracks then propagate over time and eventually lead to sudden and total failure of the part. The surface in fatigue fracture is generally characterized by the term **beach marks**, because of its appearance (Fig. 2.28). Under high magnification (typically more than 1000×), however, a series of **striations** can be seen on fracture surfaces, each beach mark consisting of several striations.

Improving Fatigue Strength. Fatigue life is greatly influenced by the method of preparation of the surfaces of the part (Fig. 2.29). The fatigue strength of manufactured products can be improved overall by the following methods:

1. Inducing compressive residual stresses on surfaces—for example, by shot peening or by roller burnishing (Section 34.2).
2. Case hardening (surface hardening) by various means (Section 4.10).
3. Providing a fine surface finish, thereby reducing the detrimental effects of notches and other surface imperfections.
4. Selecting appropriate materials and ensuring that they are free from significant amounts of inclusions, voids, and impurities.

Conversely, the following factors and processes can *reduce* fatigue strength:

1. Tensile residual stresses on the surface
2. Decarburization
3. Surface pits (such as due to corrosion), that act as stress raiser
4. Hydrogen embrittlement
5. Galvanizing
6. Electroplating

Stress–corrosion Cracking. An otherwise ductile metal can fail in a brittle manner by *stress–corrosion cracking* (also called **stress cracking** or **season cracking**). Parts that are free from defects may develop cracks, either over time or even soon after being manufactured. Crack propagation may be either intergranular or transgranular. The susceptibility of metals to stress–corrosion cracking depends mainly on the material, the presence and magnitude of *tensile residual stresses*, and the environment (corrosive media such as salt water or chemicals).

Brass and austenitic stainless steels are among metals that are highly susceptible to stress cracking. The usual procedure to avoid stress–corrosion cracking is to *stress relieve* the part just after it is formed. Full annealing (Section 4.11) may also be done, but this treatment reduces the strength of cold-worked parts.

Hydrogen Embrittlement. The presence of hydrogen can reduce ductility and cause severe embrittlement and premature failure in metals, alloys, and nonmetallic materials. Called *hydrogen embrittlement*, this phenomenon is especially severe in high-strength steels. Possible sources of hydrogen arise during melting of the metal for casting, pickling (removing of surface oxides by chemical or electrochemical reaction), and electrolysis in electroplating. Other sources of hydrogen are water vapor in the atmosphere and moisture on electrodes and in fluxes used during welding. Oxygen can also cause embrittlement, particularly in copper alloys.

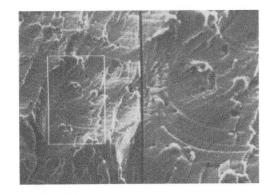

FIGURE 2.28 Typical fatigue-fracture surface on metals, showing beach marks. Magnification: left, 500×; right, 1000×. *Source:* Courtesy of B.J. Schulze and S.L. Meiley and Packer Engineering Associates, Inc.

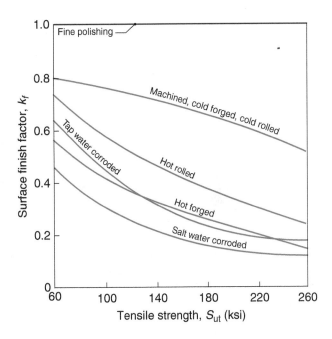

FIGURE 2.29 Reductions in the fatigue strength of cast steels subjected to various surface-finishing operations. Note that the reduction becomes greater as the surface roughness and the strength of the steel increase. *Source:* Reprinted by permission of CRC Press, Inc. *Fundamentals of Machine Elements,* 3rd edition by Schmid, Hamrock and Jacobson; © 2013 Taylor & Francis LLC.

2.11 **Residual Stresses**

Residual stresses may develop when workpieces are subjected to plastic deformation that is not uniform throughout the part; these are stresses that remain within a part after it has been formed and all the external forces (applied through tools and dies) are removed. A typical example is the bending of a metal bar (Fig. 2.30). Note that the external bending moment first produces a linear elastic stress distribution (Fig. 2.30a); as the moment is increased, the outer fibers in the bar reach a stress level high enough to cause yielding. For a typical strain-hardening material, the stress distribution shown in Fig. 2.30b is eventually reached, and the bar has now undergone permanent bending.

Let's now remove the external bending moment on the bar. This operation is equivalent to applying an equal but opposite moment to the bar; thus, the moments of the areas *oab* and *oac* in Fig. 2.30c must be equal. Line *oc*, which represents the opposite bending moment, is linear, because all unloading and recovery are *elastic* (see Fig. 2.3). The difference between the two stress distributions gives the residual stress pattern within the bar, as shown in Fig. 2.30d.

Note the presence of compressive residual stresses in layers *ad* and *oe*, and tensile residual stresses in layers *do* and *ef*. Because there are now no external forces applied to the bar, the internal forces resulting from these residual stresses must be in static equilibrium. It should be noted that although this example involves residual stresses in the longitudinal direction of the bar only, in most cases residual stresses are three dimensional and hence more difficult to analyze.

The removal of a layer of material from the surfaces of the bar, such as by machining or grinding, will disturb the equilibrium of the residual stresses shown in Fig. 2.30d. The bar will then acquire a new radius of curvature in order to balance the internal forces. Such disturbances of residual stresses cause **warping** of parts (Fig. 2.31). The equilibrium of residual stresses may also be disturbed by *relaxation* of these stresses over a period of time; see below.

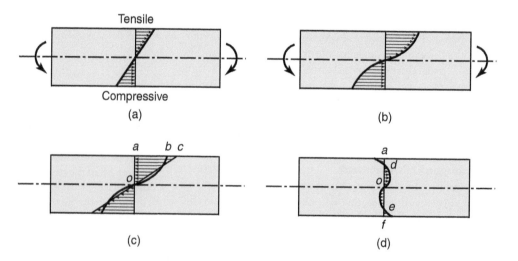

FIGURE 2.30 Residual stresses developed in bending a beam having a rectangular cross-section; note that the horizontal forces and moments caused by residual stresses in the beam must be balanced internally. Because of nonuniform deformation, especially during cold-metalworking operations, most parts develop residual stresses.

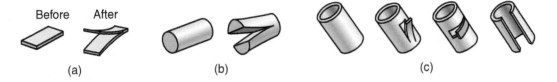

FIGURE 2.31 Distortion of parts with residual stresses after cutting or slitting: (a) flat sheet or plate, (b) solid round rod, and (c) thin-walled tubing or pipe.

Residual stresses can also be developed by *temperature gradients* within the part, such as occur during cooling of a casting or a hot forging. The local expansions and contractions caused by temperature gradients within the part will produce a nonuniform deformation, such as described in the permanent bending of a beam.

Tensile residual stresses on the surface of a part are generally undesirable, as they lower the fatigue life and fracture strength of the part. This is because a surface with tensile residual stresses cannot sustain additional tensile stresses (from external forces) as high as those that a surface free from residual stresses can. This reduction in strength is particularly characteristic of brittle or less ductile materials, in which fracture takes place with little or no plastic deformation preceding fracture.

Tensile residual stresses can also lead, over a period of time, to *stress cracking* or *stress–corrosion cracking* of parts (Section 2.10.2). Compressive residual stresses on a surface, on the other hand, are generally desirable. In fact, in order to increase the fatigue life of components, compressive residual stresses can be imparted to surfaces by such techniques as shot peening and surface rolling (Section 34.2).

Reduction and Elimination of Residual Stresses. Residual stresses can be reduced or eliminated either by *stress-relief annealing* (Section 4.11) or by a further *plastic deformation* of the part, such as stretching it. Given sufficient time, residual stresses may also diminish at room temperature, by *relaxation* of residual stresses. The time required for relaxation can be greatly reduced by raising the temperature of the workpiece.

2.12 Work, Heat, and Temperature

Almost all the mechanical work in plastic deformation is converted into **heat**. However, this conversion is not complete, because a portion of this work is stored within the deformed material as elastic energy, known as **stored energy** (Section 1.7). This is generally 5–10% of the total energy input; in some alloys, however, it may be as high as 30%.

In a simple frictionless deformation process, and assuming that the work is completely converted into heat, the theoretical (adiabatic) *temperature rise*, ΔT, in the workpiece is given by

$$\Delta T = \frac{u}{\rho c}, \tag{2.16}$$

where u is the **specific energy** (work of deformation per unit volume), ρ is the density, and c is the specific heat of the material. It can be seen that higher temperatures are associated with large areas under the stress–strain curve and with smaller values of specific heat. However, such physical properties as specific heat and thermal conductivity (Chapter 3) may themselves also depend on temperature; thus, they must be taken into account in the calculations.

The temperature rise for a true strain of 1 (such as occurs in a 27-mm-high specimen when it is compressed down to 10 mm) can be calculated to be: for aluminum, 75°C (165°F); copper, 140°C (285°F); low-carbon steel, 280°C (535°F); and titanium 570°C (1060°F). In actual metalworking operations, heat is lost to the environment, to tools and dies, and to lubricants or coolants used, if any, in the process. If deformation is performed rapidly, the heat losses will be relatively small over that brief period; if, on the other hand, the process is carried out slowly, the actual temperature rise will be only a fraction of the calculated value.

SUMMARY

- Numerous manufacturing processes involve shaping materials by plastic deformation; consequently, such mechanical properties as strength (yield strength, Y, and ultimate tensile strength, UTS); modulus of elasticity, E; ductility (total elongation and reduction of area); hardness; and the energy required for plastic deformation are important factors. These properties depend, to various extents, on the particular material and on its condition, temperature, deformation rate, surface condition, and the environment.

- Because of its relative simplicity, the tensile test is the most commonly used to determine mechanical properties. From these tests, true stress–true strain curves are constructed that are needed to determine the strength coefficient (K), the strain-hardening exponent (n), the strain-rate sensitivity exponent (m), and the toughness of materials.

- Compression tests are subject to inaccuracy due to the presence of friction and to resulting barreling of the specimen. Torsion tests are conducted on tubular specimens and subjected to twisting. Bending or flexure tests are commonly used for brittle materials to determine their modulus of rupture or the transverse rupture strength.

- Several hardness tests may be used to determine the resistance of a material to permanent indentation or to scratching. Hardness is related to strength and wear resistance of a material, but it is not a fundamental property.

- Fatigue tests indicate the endurance limit or fatigue limit of materials—that is, the maximum stress to which a material can be subjected without fatigue failure, regardless of the number of cycles. Some materials have no endurance limit; their allowable stress must be reported with respect to the number of loading cycles.

- Creep is the permanent deformation of a component under a static load maintained for a period of time. In tension, the specimen eventually fails by rupture (necking and fracturing).

- Impact tests determine the energy required to completely break a specimen, called the impact toughness of the material. Impact tests are also useful for determining the transition temperature of materials.

- Failure and fracture constitute an important aspect of a material's behavior when it is subjected to deformation during metalworking operations. Ductile fracture is characterized by plastic deformation preceding fracture, and it requires a considerable amount of energy. Brittle fracture can be catastrophic, because it is not preceded by plastic deformation; however, it requires much less energy than does ductile fracture. Impurities, inclusions, and voids play a major role in the fracture of materials.

- Residual stresses are those that remain in a workpiece after it has been plastically deformed and then has had all external forces removed. Surface tensile residual stresses are generally undesirable; they may be reduced or eliminated by stress-relief annealing, further plastic deformation, or by relaxation over a period of time.

KEY TERMS

Bauschinger effect	Engineering strain	Modulus of rupture	Strength coefficient
Blue brittleness	Engineering stress	Poisson's ratio	Stress–corrosion cracking
Brittle fracture	Fatigue	Reduction of area	Stress relaxation
Buckling	Fatigue failure	Residual stresses	Superplasticity
Charpy test	Flexural strength	Rupture	Tension
Compression	Fracture	Shear	Torsion test
Creep	Hardness	Shear modulus	Toughness
Defects	Impact loading	Shore test	Transition temperature
Deformation rate	Inclusions	Strain aging	True strain
Disk test	Izod test	Strain-hardening exponent	True stress
Ductile fracture	Leeb test	Strain rate	Ultimate tensile strength
Ductility	Microhardness	Strain-rate sensitivity exponent	Yield stress
Durometer	Modulus of elasticity		
Elongation	Modulus of rigidity	Strain softening	

BIBLIOGRAPHY

Ashby, M.F., Materials Selection in Mechanical Design, 4th ed., Butterworth-Heinemann, 2010.

Ashby, M.F., and Jones, D.R.H., Engineering Materials, Vol. 1: An Introduction to Properties, Applications, and Microstructure, 4th ed., 2012; Vol. 2: An Introduction to Microstructures and Processing. Butterworth-Heinemann, 2012.

Ashby, M., Shercliff, H., and Cebon, D., Materials: Engineering, Science, Processing and Design, 2nd ed., Butterworth-Heinemann, 2009.

Askeland, D.R., Fulay, P.P., and Wright, W.J., The Science and Engineering of Materials, 6th ed., CL Engineering, 2010.

ASM Handbook, Vol. 8: Mechanical Testing and Evaluation. ASM International, 2000.

Budinski, K.G., and Budinski, M.K., Engineering Materials: Properties and Selection, 9th ed., Prentice Hall, 2009.

Callister, W.D., Jr., and Rethwisch, D.G., Materials Science and Engineering: An Introduction, 8th ed., Wiley, 2010.

Chandler, H., Hardness Testing, 2nd ed., ASM International, 1999.

Courtney, T.H., Mechanical Behavior of Materials, 2nd ed., Waveland Press, 2005.

Davis, J.R. (ed.), Tensile Testing, 2nd ed., ASM International, 2004.

Dowling, N.E., Mechanical Behavior of Materials: Engineering Methods for Deformation, Fracture, and Fatigue, 3rd ed., Prentice Hall, 2006.

Herrmann, K. (ed.), Hardness Testing: Principles and Applications, ASM International, 2011.

Hosford, W.F., Mechanical Behavior of Materials, Cambridge, 2005.

Shackelford, J.F., Introduction to Materials Science for Engineers, 7th ed., Prentice Hall, 2008.

Tamarin, Y., Atlas of Stress–Strain Curves, 2nd ed., ASM International, 2002

Wulpi, D.J., Understanding How Components Fail, 2nd ed., ASM International, 1999.

REVIEW QUESTIONS

2.1 Distinguish between engineering stress and true stress.

2.2 In a stress–strain curve, what is the proportional limit? Is it different than the yield point?

2.3 Describe the events that take place when a specimen undergoes a tension test. Sketch a plausible stress–strain curve, and identify all significant regions and points between them. Assume that loading continues up to fracture.

2.4 What is ductility, and how is it measured?

2.5 In the equation $\sigma = K\epsilon^n$, which represents the true stress–strain curve for a material, what is the significance of the exponent n?

2.6 What is strain-rate sensitivity, and how is it measured?

2.7 What test can measure the properties of a material undergoing shear strain?

2.8 What testing procedures can be used to measure the properties of brittle materials, such as ceramics and carbides?

2.9 Describe the differences between brittle and ductile fracture.

2.10 What is hardness? Explain.

2.11 Describe the features of a Rockwell hardness test.

2.12 What is a Leeb test? How is it different from a Rockwell A test?

2.13 Differentiate between stress relaxation and creep.

2.14 Describe the difference between elastic and plastic behavior.

2.15 Explain what uniform elongation means in tension testing.

2.16 Describe the difference between deformation rate and strain rate. What unit does each one have?

2.17 Describe the difficulties involved in conducting a compression test.

2.18 What is Hooke's law? Young's modulus? Poisson's ratio?

2.19 Describe the difference between transgranular and intergranular fracture.

2.20 What is the reason that yield strength is generally defined as a 0.2% offset strength?

2.21 Why does the fatigue strength of a specimen or part depend on its surface finish?

2.22 If striations are observed under microscopic examination of a fracture surface, what do they suggest regarding the mode of fracture?

2.23 What is an Izod test? Why are Izod tests useful?

2.24 Why does temperature increase during plastic deformation?

2.25 What is residual stress? How can residual stresses be removed?

QUALITATIVE PROBLEMS

2.26 On the same scale for stress, the tensile true stress–true strain curve is higher than the engineering stress–engineering strain curve. Explain whether this condition also holds for a compression test.

2.27 What are the similarities and differences between deformation and strain?

2.28 Can a material have a negative Poisson's ratio? Give a rationale for your answer.

2.29 It has been stated that the higher the value of m, the more diffuse the neck is, and likewise, the lower the value of m, the more localized the neck is. Explain the reason for this behavior.

2.30 Explain why materials with high m values, such as hot glass and Silly Putty®, when stretched slowly, undergo large elongations before failure. Consider events taking place in the necked region of the specimen.

2.31 With a simple sketch, explain whether it is necessary to use the offset method to determine the yield stress, Y, of a material that has been highly cold worked.

2.32 Explain why the difference between engineering strain and true strain becomes larger as strain increases. Does this difference occur for both tensile and compressive strains? Explain.

2.33 Consider an elastomer, such as a rubber band. This material can undergo a large elastic deformation before failure, but after fracture it recovers completely to its original shape. Is this material brittle or ductile? Explain.

2.34 If a material (such as aluminum) does not have an endurance limit, how then would you estimate its fatigue life?

2.35 What role, if any, does friction play in a hardness test? Explain.

2.36 Which hardness tests and scales would you use for very thin strips of metal, such as aluminum foil? Explain.

2.37 Consider the circumstance where a Vickers hardness test is conducted on a material. Sketch the resulting indentation shape if there is a residual stress on the surface.

2.38 Which of the two tests, tension or compression, would require a higher capacity of testing machine, and why?

2.39 List and explain briefly the conditions that induce brittle fracture in an otherwise ductile metal.

2.40 List the factors that you would consider in selecting a hardness test. Explain why.

2.41 On the basis of Fig. 2.5, can you calculate the percent elongation of the materials listed? Explain.

2.42 If a metal tension-test specimen is rapidly pulled and broken, where would the temperature be highest, and why?

2.43 Comment on your observations regarding the contents of Table 2.2.

2.44 Will the disk test be applicable to a ductile material? Why or why not?

2.45 What hardness test is suitable for determining the hardness of a thin ceramic coating on a piece of metal?

2.46 Wire rope consists of many wires that bend and unbend as the rope is run over a sheave. A wire-rope failure is investigated, and it is found that some of the wires, when examined under a scanning electron microscope, display cup-and-cone failure surfaces, while others display transgranular fracture surfaces. Comment on these observations.

2.47 A statistical sampling of Rockwell C hardness tests are conducted on a material, and it is determined that the material is defective because of insufficient hardness. The supplier claims that the tests are flawed because the diamond-cone indenter was probably dull. Is this a valid claim? Explain.

2.48 In a Brinell hardness test, the resulting impression is found to be elliptical. Give possible explanations for this result.

2.49 Some coatings are extremely thin—some as thin as a few nanometers. Explain why even the Knoop test is not able to give reliable results for such coatings. Recent investigations have attempted to use highly polished diamonds (with a tip radius around 5 nm) to indent such coatings in atomic force microscopes. What concerns would you have regarding the appropriateness of the test results?

2.50 Select an appropriate hardness test for each of the following materials, and justify your answer:

a. Cubic boron nitride
b. Lead
c. Cold-drawn 0.5%C steel
d. Diamond
e. Caramel candy
f. Granite

QUANTITATIVE PROBLEMS

2.51 A paper clip is made of wire 0.5 mm in diameter. If the original material from which the wire is made is a rod 25 mm in diameter, calculate the longitudinal engineering and true strains that the wire has undergone during processing.

2.52 A 250-mm-long strip of metal is stretched in two steps, first to 300 mm and then to 400 mm. Show that the total true strain is the sum of the true strains in each step; in other words, the true strains are additive. Show that, in the case of engineering strains, the strains cannot be added to obtain the total strain.

2.53 Identify the two materials in Fig. 2.5 that have the lowest and the highest uniform elongations. Calculate these quantities as percentages of the original gage lengths.

2.54 Plot the ultimate strength vs. stiffness for the materials listed in Table 2.2, and prepare a three-dimensional plot for these materials where the third axis is their maximum elongation in 50 mm.

2.55 If you remove the layer of material *ad* from the part shown in Fig. 2.30d (for instance, by machining or grinding), which way will the specimen curve? (*Hint:* Assume that the part shown in sketch *d* in the figure is composed of four horizontal springs held at the ends. Thus, from the top down, you have compression, tension, compression, and tension springs.)

2.56 Prove that the true strain at necking equals the strain hardening exponent.

2.57 Percent elongation is always defined in terms of the original gage length, such as 50 mm or 2 in. Explain how percent elongation would vary as the gage length of the tensile-test specimen increases. (*Hint:* Recall that necking is a *local* phenomenon.)

2.58 Make a sketch showing the nature and distribution of residual stresses in Fig. 2.31a and b, prior to the material's being cut. (*Hint:* Assume that the split parts are free from any stresses; then force these parts back to the shape they originally had.)

2.59 You are given the K and n values of two different metals. Is this information sufficient to determine which metal is tougher? If not, what additional information do you need?

2.60 A cable is made of two strands of different materials, A and B, and cross-sections, as follows:

For material A, $K = 60,000$ psi, $n = 0.5$, $A_o = 0.6$ in^2;

for material B, $K = 30,000$ psi, $n = 0.5$, $A_o = 0.3$ in^2.

Calculate the maximum tensile force that this cable can withstand prior to necking.

2.61 On the basis of the information given in Fig. 2.5, calculate the ultimate tensile strength (engineering) of 304 stainless steel.

2.62 In a disk test performed on a specimen 1.00 in. in diameter and 1/4 in. thick, the specimen fractures at a stress of 40,000 psi. What was the load on it?

2.63 A piece of steel has a hardness of 300 HB. Calculate its tensile strength, in MPa and in psi.

2.64 A metal has the following properties: UTS = 70,000 psi and $n = 0.20$. Calculate its strength coefficient, K.

2.65 Using only Fig. 2.5, calculate the maximum load in tension testing of an annealed copper specimen with an original diameter of 5 mm.

2.66 Estimate the modulus of resilience for a highly cold worked piece of steel having a hardness of 250 HB, and for a piece of highly cold-worked copper with a hardness of 100 HRB.

2.67 A metal has a strength coefficient $K = 100,000$ psi and $n = 0.25$. Assuming that a tensile-test specimen made from this metal begins to neck at a true strain of 0.25, show that the ultimate tensile strength is 59,340 psi.

2.68 Plot the true stress–true strain curves for the materials listed in Table 2.3.

2.69 The design specification for a metal requires a minimum hardness of 80 HRA. If a Rockwell test is performed and the depth of penetration is 60 μm, is the material acceptable?

2.70 Calculate the major and minor pyramid angles for a Knoop indenter, and compare your results with those obtained from Vickers and Rockwell A indenters.

2.71 If a material has a target hardness of 300 HB, what is the expected indentation diameter?

2.72 A Rockwell A test was conducted on a material and a penetration depth of 0.15 mm was recorded. What is the hardness of the material? What material would typically have such a hardness value? If a Brinell hardness test were to be conducted on this material, give an estimate of the indentation diameter if the load used was 1500 kg.

2.73 For a cold-drawn 0.5% carbon steel, will a Rockwell C test or a Brinell test at 500 kg result in a deeper penetration?

2.74 A material is tested in tension. Over a 1-in. gage length, the engineering strain measurements are 0.01, 0.02, 0.03,

0.04, 0.05, 0.1, 0.15, 0.2, 0.5, and 1.0. Plot the true strain versus engineering strain for these readings.

2.75 A horizontal rigid bar c–c is subjecting specimen a to tension and specimen b to frictionless compression such that the bar remains horizontal. (See Fig. P2.75.) The force F is located at a distance ratio of 2:1. Both specimens have an original cross-sectional area of 1 in^2 and the original lengths are $a = 8$ in. and $b = 4.5$ in. The material for specimen a has a true stress–true strain curve of $\sigma = 100,000\epsilon^{0.5}$. Plot the true stress–true strain curve that the material for specimen b should have for the bar to remain horizontal.

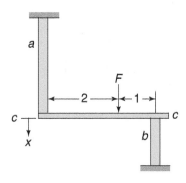

FIGURE P2.75

SYNTHESIS, DESIGN, AND PROJECTS

2.76 List and explain the desirable mechanical properties of (a) an elevator cable, (b) a paper clip, (c) a leaf spring for a truck, (d) a bracket for a bookshelf, (e) piano wire, (f) a wire coat hanger, (g) the clip for a pen, and (h) a staple.

2.77 When making a hamburger, you may have observed the type of cracks shown in Fig. 2.20d. What would you do to avoid such cracks? [*Note:* Test hamburger patties by compressing them at different temperatures, and observe the crack path (i.e., the path through the fat particles, the meat particles, or their interface).]

2.78 An inexpensive claylike material called Silly Putty is generally available in stores that sell toys and games. Obtain a sample and perform the following experiments: (a) Shape it into a ball, and drop it onto a flat surface. (b) Reround the ball and place a heavy book on it for one minute. (c) Shape the putty into a long rod, and pull on it—first slowly, then very quickly. Describe your observations, referring to the specific sections in this chapter where each particular observation is relevant.

2.79 Make individual sketches of the mechanisms of testing machines that, in your opinion, would be appropriate for tension, for torsion, and for compression testing of specimens at different rates of deformation. What modifications

would you make on these machines to include the effects of temperature on material properties?

2.80 In tension testing of specimens, mechanical and electronic instruments are typically used to measure elongation. Make sketches of instruments that would be suitable for this purpose, commenting on their accuracy. What modifications would you make to these instruments to include the use of specimens at elevated temperatures?

2.81 Obtain small pieces of different metallic and nonmetallic materials, including stones. Rub them against each other, observe the scratches made, and order them in a manner similar to the Mohs hardness numbering system.

2.82 Demonstrate the stress-relaxation phenomenon by tightly stretching thin plastic strings between two nails placed at the ends of a long piece of wood. Pluck the strings frequently, to test the tension as a function of time. Repeat the test at a higher temperature by placing the fixture in an oven set on low.

2.83 Demonstrate the impact toughness of a piece of round chalk by first using a triangular file to produce a V-notch on the cylindrical surface (as shown in Fig. 2.19a) and then bending the chalk to break it.

2.84 Using a large rubber band and a set of weights, obtain the force–displacement curve for the rubber band. Is the result different from the stress–strain curves shown in Fig. 2.4? Comment.

2.85 Design a test protocol to obtain the work of plastic deformation by measuring the temperature rise in a workpiece, assuming that there is no heat loss and that the temperature distribution is uniform throughout. If the specific heat of the material decreases with increasing temperature, will the work of deformation calculated using the specific heat at room temperature be higher or lower than the actual work done? Explain.

2.86 Find or prepare some solid circular pieces of brittle materials, such as chalk, ceramics, etc. and subject them to the type of test shown in Fig. 2.9 by using the jaws of a simple vise. Describe your observations as to how the materials fracture. Repeat the tests, using ductile materials, such as clay, soft metals, etc., and describe your observations.

2.87 Take several rubber bands and pull them at different temperatures, including from a frozen state. Comment on their behavior, such as ductile or brittle.

2.88 Devise a simple fixture for conducting the bend tests shown in Fig. 2.11. Test sticks of various brittle materials by loading them with dead weights until they break. Verify the statement in the text that the specimens on the right in the figure will fracture sooner than the ones on the left.

2.89 By pressing a small ball bearing against the top surfaces of various materials, such as clay and dough, observe the shape of the indentation with a magnifier, referring to those shapes shown in Fig. 2.14a and b.

2.90 Describe your observations regarding Fig. 2.14c.

2.91 Embed a small steel ball in a soft block of material such as clay, and compress the clay as shown in Fig. 2.24a. Then slice the clay carefully along the center plane and observe the deformation of the material. Repeat the experiment by embedding a small round jelly bean in the clay and deforming the material. Comment on your observations.

2.92 Devise a simple experiment, and perform tests on materials commonly found around the house by bending them at different temperatures, for a qualitative assessment of their transition temperature, as shown in Fig. 2.25.

2.93 Obtain some solid and some tubular metal pieces, and slit them as shown in Fig. 2.31. Comment on whether there were any residual stresses in the parts prior to slitting them.

2.94 Explain how you would obtain an estimate of the hardness for a carbon nanotube. (See Section 8.6.2.)

2.95 Without using the words "stress" or "strain," define elastic modulus.

2.96 We know that it is relatively easy to subject a specimen to hydrostatic compression, such as by using a chamber filled with a liquid. Devise a means whereby the specimen (say, in the shape of a cube or a round disk) can be subjected to hydrostatic tension, or one approaching this state of stress. (Note that a thin-walled, internally pressurized spherical shell is not a correct answer, because it is subjected only to a state of plane stress.)

3 Physical Properties of Materials

- Physical properties can have several significant roles in the selection, processing, and use of materials. They can be key factors in determining a material's suitability for specific applications, especially when considered simultaneously with mechanical properties.
- Strength-to-weight and stiffness-to-weight ratios, as examples, are discussed in the context of lightweight design, an important consideration in aerospace and automotive industries.
- Thermal, electrical, magnetic, and optical properties are then presented.
- The importance of corrosion and corrosion-resistant materials are described.
- Design and manufacturing implications of all physical properties are considered, with various specific examples given.

3.1 Introduction

Why is electrical wiring generally made of copper? Why are aluminum, stainless steel, and copper commonly used in cookware? Why are the handles of cookware usually made of wood or plastic, while other types of handles are made of metal? What type of material should be chosen for the heating elements in toasters? Why does aluminum feel colder to the touch than plastic, when both are at room temperature? Why are the metallic components in some machines being replaced with ceramics? Why are commercial airplane bodies generally made of aluminum, and why are some now being replaced gradually with various composite materials, including reinforced plastics?

It is apparent from these questions that an important criterion in material selection is consideration of **physical properties**, such as density, melting point, specific heat, thermal conductivity, thermal expansion, electrical and magnetic properties, and resistance to oxidation and corrosion. Combinations of mechanical and physical properties, such as the strength-to-weight and stiffness-to-weight ratios of materials, are equally important, particularly for aircraft and aerospace structures. Also, high-speed equipment, such as textile and printing machinery, and forming and cutting machines for high-speed operations, require lightweight components to reduce inertial forces, and thus prevent the machines from being subjected to excessive vibration. Several other examples of the importance of physical properties are described in this chapter.

3.2 Density
••

The **density** of a material is its mass per unit volume. Another term is **specific gravity**, which expresses a material's density relative to that of water, therefore specific gravity has no units. The range of densities for a variety of materials at room temperature, along with other properties, is given in Tables 3.1 and 3.2.

Weight saving is particularly important for aircraft and aerospace structures, automotive bodies and components, and for various other products where energy

TABLE 3.1

Physical Properties of Selected Materials at Room Temperature

Material	Density (kg/m^3)	Melting point (°C)	Specific heat (J/kg-K)	Thermal conductivity (W/m-K)	Coefficient of thermal expansion (μm/m-°C)	Electrical resistivity (Ω-m)
Metallic						
Aluminum	2700	660	900	222	23.6	2.8×10^{-8}
Aluminum alloys	2630–2820	476–654	880–920	121–239	23.0–23.6	$2.8–4.0 \times 10^{-8}$
Beryllium	1854	1278	1884	146	8.5	4.0×10^{-8}
Niobium(columbium)	8580	2468	272	52	7.1	15×10^{-8}
Copper	8970	1082	385	393	16.5	1.7×10^{-8}
Copper alloys	7470–8940	885–1260	377–435	219–234	16.5–20	$1.7–5.9 \times 10^{-8}$
Gold	19,300	1063	129	317	19.3	2.4×10^{-8}
Iron	7860	1537	460	74	11.5	9.5×10^{-8}
Lead	11,350	327	130	35	29.4	20.6×10^{-8}
Lead alloys	8850–11,350	182–326	126–188	24–46	27.1–31.1	$20.6–24 \times 10^{-8}$
Magnesium	1745	650	1025	154	26.0	4.5×10^{-8}
Magnesium alloys	1770	610–621	1046	75–138	26.0	$4.5–15.9 \times 10^{-8}$
Molybdenum alloys	10,210	2610	276	142	5.1	5.3×10^{-8}
Nickel	8910	1453	440	92	13.3	6.2×10^{-8}
Nickel alloys	7750–8850	1110–1454	381–544	12–63	12.7–18.4	$6.2–110 \times 10^{-8}$
Platinum	2145	1768	133	71.6	8.8	10.5×10^{-8}
Silicon	2330	1423	712	148	7.63	1.0×10^{-3}
Silver	10,500	961	235	429	19.3	1.6×10^{-8}
Steels	6920–9130	1371–1532	448–502	15–52	11.7–17.3	17.0×10^{-8}
Tantalum alloys	16,600	2996	142	54	6.5	13.5×10^{-8}
Tin	7310	232	217	67	22	11.5×10^{-8}
Titanium	4510	1668	519	17	8.35	42×10^{-8}
Titanium alloys	4430–4700	1549–1649	502–544	8–12	8.1–9.5	$40–171 \times 10^{-8}$
Tungsten	19,290	3410	138	166	4.5	5.0×10^{-8}
Zinc	7140	419	385	113	32.5	5.45×10^{-8}
Zinc alloys	6640–7200	386–525	402	105–113	32.5–35	$6.06–6.89 \times 10^{-8}$
Nonmetallic						
Ceramics	2300–5500	—	750–950	10–17	5.5–13.5	—
Glasses	2400–2700	580–1540	500–850	0.6–1.7	4.6–70	—
Graphite	1900–2200	—	840	5–10	7.86	—
Plastics	900–2000	110–330	1000–2000	0.1–0.4	72–200	—
Wood	400–700	—	2400–2800	0.1–0.4	2–60	—

TABLE 3.2

Physical Properties of Materials (in Descending Order)					
Density	Melting point	Specific heat	Thermal conductivity	Thermal expansion	Electrical conductivity
Platinum	Tungsten	Wood	Silver	Plastics	Silver
Gold	Tantalum	Beryllium	Copper	Lead	Copper
Tungsten	Molybdenum	Porcelain	Gold	Tin	Gold
Tantalum	Niobium	Aluminum	Aluminum	Magnesium	Aluminum
Lead	Titanium	Graphite	Magnesium	Aluminum	Magnesium
Silver	Iron	Glass	Graphite	Copper	Tungsten
Molybdenum	Beryllium	Titanium	Tungsten	Steel	Beryllium
Copper	Copper	Iron	Beryllium	Gold	Steel
Steel	Gold	Copper	Zinc	Ceramics	Tin
Titanium	Silver	Molybdenum	Steel	Glass	Graphite
Aluminum	Aluminum	Tungsten	Tantalum	Tungsten	Ceramics
Beryllium	Magnesium	Lead	Ceramics		Glass
Glass	Lead		Titanium		Plastics
Magnesium	Tin		Glass		Quartz
Plastics	Plastics		Plastics		

consumption and power limitations are major concerns. Substitution of materials for weight savings and economy is a major factor in the design of advanced equipment and machinery and consumer products, such as automobiles, aircraft, sporting goods, laptop computers, and bicycles.

A significant role that density plays is in the **strength-to-weight ratio** (**specific strength**) and **stiffness-to-weight ratio** (**specific stiffness**) of materials. Figure 3.1 shows the ratio of maximum yield stress to density for a variety of metal alloys.

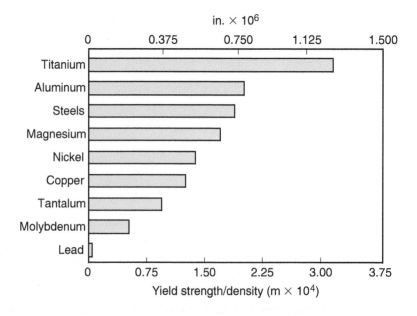

FIGURE 3.1   Ratio of maximum yield stress to density for selected metal alloys.

Note that titanium and aluminum are at the top of the list; consequently, and as described in Chapter 6, they are among the most commonly used metals for various applications.

The specific tensile strength and specific stiffness at room temperature for a variety of metallic and nonmetallic materials is given in Fig. 3.2. Note the positions of composite materials, as compared with those of metals, with respect to these properties; these advantages have led composites to become among the most important materials, as described in Chapter 9. At elevated temperatures, specific strength and specific stiffness are likewise important considerations, especially for components operating at these temperatures, such as automotive and jet engines, gas turbines, and furnaces. Typical ranges for a variety of materials as a function of temperature are given in Fig. 3.3.

Density is also an important factor in the selection of materials for high-speed equipment, such as magnesium in printing and textile machinery, many components of which typically operate at very high speeds. To obtain exposure times of 1/4000 s in cameras without sacrificing accuracy, the focal plane shutters of some high-quality digital cameras are made of titanium. Aluminum is used with some digital cameras for better performance in cold weather. Because of their low density, ceramics (Chapter 8) are being used for components in high-speed automated machinery and in machine tools.

On the other hand, there are applications where weight is desirable. Examples are counterweights for various mechanisms (using lead or steel), flywheels, ballasts on yachts and aircraft, and weights on golf clubs (using high-density materials such as tungsten).

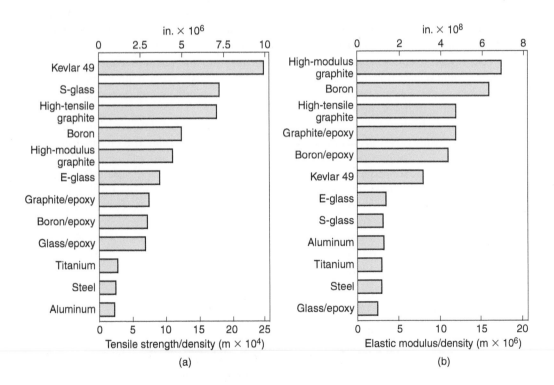

FIGURE 3.2 Specific strength (tensile strength/density) and specific stiffness (elastic modulus/density) for various materials at room temperature. (See also Chapter 9.)

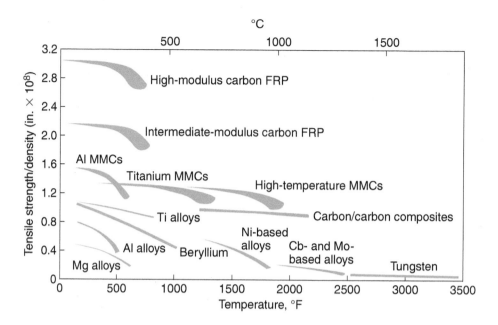

FIGURE 3.3 Specific strength (tensile strength/density) for a variety of materials as a function of temperature; note the useful temperature range for these materials and the high values for composite materials. MMC = metal-matrix composite; FRP = fiber-reinforced plastic.

3.3 Melting Point

The temperature range within which a component or structure is designed to function is an important consideration in the selection of materials. Plastics, for example, have the lowest useful temperature range, while graphite and refractory-metal alloys have the highest useful range. Note also that pure metals have a definite melting point, whereas the melting temperature of a metal alloy can have a wide range (Table 3.1) depending on its composition.

The melting point has a number of indirect effects on manufacturing operations. Because the recrystallization temperature of a metal is related to its melting point (Section 1.7), operations such as annealing and heat treating (Chapter 4) and hot working (Part III) require a knowledge of the melting points of the materials involved. These considerations are also important in the selection of tool and die materials. Melting point also plays a major role in the selection of the equipment and the melting practice employed in metal-casting operations (Part II) and in the electrical-discharge machining process (Section 27.5), where the melting points of metals are related to the rate of material removal and of electrode wear.

3.4 Specific Heat

A material's **specific heat** is the energy required to raise the temperature of a unit mass by one degree. Alloying elements have a relatively minor effect on the specific heat of metals. The temperature rise in a workpiece, such as those resulting from forming or machining operations (Parts III and IV, respectively), is a function of the work done and of the specific heat of the workpiece material (Section 2.12). An excessive temperature rise in a workpiece can

a. Decrease product quality by adversely affecting its surface finish and dimensional accuracy
b. Cause excessive tool and die wear
c. Result in undesirable metallurgical changes in the material

3.5 Thermal Conductivity

Thermal conductivity indicates the rate at which heat flows within and through a material. Metallically bonded materials (metals) generally have high thermal conductivity, while ionically or covalently bonded materials (ceramics and plastics) have poor conductivity (Table 3.2). Alloying elements can have a significant effect on the thermal conductivity of alloys, as can be seen by comparing pure metals with their alloys in Table 3.1. In general, materials with high electrical conductivity also have high thermal conductivity.

Thermal conductivity is an important consideration in many applications. For example, high thermal conductivity is desirable in cooling fins, cutting tools, and die-casting molds, to extract heat faster. In contrast, materials with low thermal conductivity are used, for instance, in furnace linings, insulation, coffee cups, and handles for pots and pans. One function of a lubricant in hot metalworking is to serve as an insulator to keep the workpiece hot and thus formable.

Video Solution 3.1 Selection of Materials for Thermal Insulation

3.6 Thermal Expansion

The **thermal expansion** of materials can have several significant effects, particularly the relative expansion or contraction of different materials in assemblies, such as electronic and computer components, glass-to-metal seals, struts on jet engines, coatings on cutting tools (Section 22.5), and moving parts in machinery that require certain clearances for proper functioning. The use of ceramic components in cast-iron engines, for example, also requires consideration of their relative expansion during their operation. *Shrink fits* utilize thermal expansion and contraction; a shrink fit is a part, often a sleeve or a hub, that is to be installed over a shaft. The part is first heated and then slipped over the shaft or spindle; when allowed to cool, the hub shrinks and the assembly becomes an integral component.

Typical coefficients of thermal expansion are given in Table 3.1 (see also *Invar* below). Generally, the coefficient of thermal expansion is inversely proportional to the melting point of the material. Alloying elements have a relatively minor effect on the thermal expansion of metals.

Thermal expansion, in conjunction with thermal conductivity, plays the most significant role in the development of **thermal stresses** (due to *temperature gradients*), both in manufactured components and in tools and dies, and molds for casting operations. This consideration is particularly important in, for example, a forging operation during which hot workpieces are repeatedly placed over a relatively cool die, thus subjecting the die surfaces to thermal cycling. To reduce thermal stresses, a combination of high thermal conductivity and low thermal expansion is desirable. Thermal stresses can also be caused by **anisotropy of thermal expansion**; that is, the material expands differently in different directions. This property is generally observed in hexagonal close-packed metals, ceramics, and composite materials.

Thermal expansion and contraction can lead to cracking, warping, or loosening of components during their service life, as well as cracking of ceramic parts and in tools and dies made of relatively brittle materials. **Thermal fatigue** results from thermal cycling and causes a number of surface cracks, especially in tools and dies for casting and metalworking operations (*heat checking*). **Thermal shock** is the term generally used to describe development of a crack or cracks after being subjected to a single thermal cycle.

To alleviate some of the problems caused by thermal expansion, a family of iron–nickel alloys with very low thermal expansion coefficients are available, called **low-expansion alloys.** The low thermal-expansion characteristic of these alloys is often referred to as the **Invar effect,** after the metal *Invar*. Their thermal coefficient of expansion is generally in the range of 2×10^{-6} to 9×10^{-6} per °C (compare with those given in Table 3.1). Typical compositions are 64% Fe–36% Ni for Invar and 54% Fe–28% Ni–18% Co for Kovar.

Low-expansion alloys also have good thermal-fatigue resistance, and because of their good ductility, they can easily be formed into various shapes. Applications include (a) bimetallic strips, consisting of a low-expansion alloy metallurgically bonded to a high-expansion alloy (the strip bends when subjected to temperature variations) and (b) glass-to-metal seals, in which the thermal expansions of the two materials are matched.

3.7 Electrical, Magnetic, and Optical Properties

Electrical conductivity and the *dielectric* properties of materials are important not only in electrical equipment and machinery but also in such manufacturing processes as magnetic-pulse forming (Section 16.12), resistance welding (Section 31.5), and the electrical-discharge machining and electrochemical grinding of hard and brittle materials (Chapter 27). The units of electrical conductivity are mho/m or mho/ft, where mho is the reverse of ohm, the unit of electrical resistance. Alloying elements have a major effect on the electrical conductivity of metals; the higher the conductivity of the alloying element, the higher is the electrical conductivity of the alloy.

Dielectric Strength.   An electrically insulating material's *dielectric strength* is the largest electric field to which it can be subjected without degrading or losing its insulating properties. This property is defined as the voltage required per unit distance for electrical breakdown and has the units of V/m or V/ft.

Conductors.   Materials with high electrical conductivity, such as metals, are generally referred to as *conductors*. **Electrical resistivity** is the inverse of electrical conductivity; materials with high electrical resistivity are referred to as **dielectrics** or **insulators.**

Superconductors.   *Superconductivity* is the phenomenon of near-zero electrical resistivity that occurs in some metals and alloys below a critical temperature, often near absolute zero (0 K, –273°C, or –460°F). The highest temperature at which superconductivity has been exhibited to date is for a mercury–barium–calcium–copper compound ($HgBa_2Ca_2Cu_3O_x$), which may be as high as –109°C (–164°F), under pressure; advances in high temperature superconductivity continue to be made.

The main application of superconductors is for high-power magnets, and are the enabling technology for magnetic resonance imaging used for medical imaging. Other applications proposed for superconductors include magnetic levitation (maglev) trains, efficient power transmission lines, and components for extremely fast computers.

Semiconductors.   The electrical properties of *semiconductors*, such as single-crystal silicon, germanium, and gallium arsenide, are extremely sensitive to temperature and the presence and type of minute impurities. Thus, by controlling the concentration and type of impurities (called **dopants**), such as phosphorus and boron in silicon, electrical conductivity can be controlled. This property is utilized in semiconductor (solid-state) devices, used extensively in miniaturized electronic circuitry (Chapter 28).

Ferromagnetism and Ferrimagnetism.   *Ferromagnetism* is a phenomenon characterized by high permeability and permanent magnetization that are due to the alignment of iron, nickel, and cobalt atoms. It is important in such applications as electric motors, electric generators, electric transformers, and microwave devices. *Ferrimagnetism* is a permanent and large magnetization exhibited by some ceramic materials, such as cubic ferrites.

Piezoelectric Effect.   The *piezoelectric effect* (*piezo* from Greek, meaning "to press") is exhibited by what are called **smart materials**. Two basic behaviors are involved: (a) When subjected to an electric current, these materials undergo a reversible change in shape, by as much as 4% and (b) when deformed by an external force, the materials emit a small electric current. Piezoelectric materials include quartz crystals and some ceramics and polymers. The piezoelectric effect is utilized in making transducers, which are devices that convert the strain from an external force into electrical energy. Typical applications are sensors, force or pressure transducers, inkjet printers, strain gages, sonar detectors, and microphones. As an example, an air bag in an automobile has a sensor that, when subjected to an impact force, sends an electric charge which deploys the bag.

Magnetostriction.   The phenomenon of expansion or contraction of a material when it is subjected to a magnetic field is called *magnetostriction*. Materials such as pure nickel and some iron–nickel alloys exhibit this behavior. Magnetostriction is the principle behind ultrasonic machining equipment (Section 26.6).

Magnetorheostatic and Electrorheostatic Effects.   When subjected to magnetic or electric fields, some fluids undergo a major and reversible change in their viscosity within a fraction of a second, turning from a liquid to an almost solid state. For example, magnetorheostatic behavior is attained by mixing very fine iron filings with oil. Called **smart fluids**, these materials are being developed for such applications as vibration dampeners, engine mounts, prosthetic devices, clutches, and valves.

Optical Properties.   Among various other properties, color and opacity are particularly relevant to polymers and glasses. These two properties are described in Sections 7.2.2 and 8.4.3, respectively.

3.8   Corrosion Resistance

Metals, ceramics, and plastics are all subject to forms of **corrosion**. The word *corrosion* itself usually refers to the deterioration of metals and ceramics, while similar phenomena in plastics (Chapter 7) are generally called **degradation**. Corrosion not only leads to surface deterioration of components and structures, such as bridges and ships, but also reduces their strength and structural integrity. The direct cost of corrosion to the U.S. economy alone has been estimated to be over $400 billion per year, approximately 3% of the gross domestic product. Indirect costs of corrosion are estimated at twice this amount.

Corrosion resistance is an important aspect of material selection for applications in the chemical, food, and petroleum industries, as well as in manufacturing operations. In addition to various possible chemical reactions from the elements and compounds present, the environmental oxidation and corrosion of components and structures is a major concern, particularly at elevated temperatures and in automobiles and other transportation vehicles.

Resistance to corrosion depends on the composition of the material and on the particular environment. Corrosive media may be chemicals (acids, alkalis, and salts) and the environment (oxygen, moisture, pollution, and acid rain), including water (fresh or salt water). Nonferrous metals, stainless steels, and nonmetallic materials generally have high corrosion resistance, whereas steels and cast irons generally have poor resistance and must be protected by various coatings and surface treatments (Chapter 34).

Corrosion can occur over an entire surface or it can be *localized*, called **pitting**, a term that is also used for fatigue wear or failure of gears and in forging; see Section 33.2. Corrosion can also occur along grain boundaries of metals as intergranular corrosion, and at the interface of bolted or riveted joints as **crevice corrosion**.

Two dissimilar metals may form a **galvanic cell** (after the Italian physician L. Galvani, 1737–1798); that is, two electrodes in an electrolyte in a corrosive environment that includes moisture and cause **galvanic corrosion**. Two-phase alloys (Chapter 4) are more susceptible to galvanic corrosion (because of the physical separation of the two different metals involved) than are single-phase alloys or pure metals; as a result, heat treatment can have a significant beneficial influence on corrosion resistance.

Stress-corrosion cracking (Section 2.10.2) is an example of the effect of a corrosive environment on the integrity of a product that, as manufactured, had residual stresses. Likewise, cold-worked metals are likely to have residual stresses, hence they are more susceptible to corrosion than are hot worked or annealed metals.

Tool and die materials also can be susceptible to chemical attack by lubricants and by coolants; the chemical reaction alters their surface finish and adversely influences the metalworking operation. One example is that of carbide tools and dies having cobalt as a binder (Section 22.4); the cobalt is attacked by elements in the metalworking fluid (**selective leaching**). Thus, compatibility of the tool, die, and workpiece materials with the metalworking fluid, under actual operating conditions is an important consideration in the selection of materials.

However, chemical reactions should not be regarded as having only adverse effects. Advanced machining processes such as chemical and electrochemical machining (Chapter 27) are indeed based on controlled chemical reactions. These processes remove material by chemical action, in a manner similar to the etching of metallurgical specimens.

The usefulness of some level of **oxidation** is demonstrated by the corrosion resistance of aluminum, titanium, and stainless steel (Section 33.2). Aluminum develops a thin (a few atomic layers), strong, and adherent hard-oxide film (Al_2O_3) that better protects the surface from further environmental corrosion. Titanium develops a film of titanium oxide (TiO_2). A similar phenomenon occurs in stainless steels which, because of the chromium present in the alloy, develop a protective chromium oxide film on their surfaces. These processes are known as **passivation**. When the protective film is scratched and exposes the metal underneath, a new oxide film begins to form.

CASE STUDY 3.1   Selection of Materials for Coins

There are six general criteria in the selection of materials for coins (Fig. 3.4).

1. *Subjective factors*, such as the *appearance* of the coin, its color, weight, and its ring (the sound made when striking). Also included in this criterion is the *feel* of the coin, a term that is similar in effect to the feel of a fine piece of wood, polished stone, or tableware. It is difficult to quantify because it combines several human factors.

2. The intended *life* of the coin is also a consideration; this duration will reflect resistance to corrosion and wear (Chapter 33) while the coin is in circulation. These two factors basically determine the span over which the surface imprint of the coin will remain identifiable, as well as the ability of the coin to retain its original luster.

3. *Manufacturing* of the coin includes factors such as the formability of the candidate coin materials, the life of the dies used in the coining operation (Section 14.4), and the capability of the materials and processes to resist counterfeiting.

4. Another consideration is the *suitability for use* in coin-operated devices, such as vending machines and turnstiles. These machines are generally equipped with detection devices that test the coins—first, for proper diameter, thickness, and surface condition, and second, for

FIGURE 3.4 A selection of U.S. coins, manufactured from different metal alloys of copper, nickel, tin, zinc, and aluminum. Valuable metals such as gold and silver are used for coins, but are not used for general currency.

electrical conductivity and density. The coin is rejected if it fails any of these tests.

5. *Health* issues must be considered. For example, given the large portion of the population with nickel allergies, Euro coins are minted from nickel-free alloys.

6. A final consideration is the *cost* of raw materials and processing, and whether there is a sufficient *supply* of the coin materials. For example, Canada recently decided it would eliminate the penny because of the high cost of production and its limited currency value. The United States has similar concerns, since a penny (one cent) costs around 1.6 cents to manufacture.

SUMMARY

- Physical properties can have several important influences on materials selection, manufacturing, and on the service life of components. These properties and other relevant material characteristics must be considered because of their possible effects on product design, service requirements, and compatibility with other materials, including tools, dies, and workpieces.

- The combined properties of strength-to-weight and stiffness-to-weight ratios are important factors in selecting materials for lightweight and high-performance structures.

- Thermal conductivity and thermal expansion are major factors in the development of thermal stresses and thermal fatigue and shock, effects that are important in tool and die life in manufacturing operations.

- Chemical reactions, including oxidation and corrosion, are important factors in material selection, design, and manufacturing, as well as in the service life of components. Passivation and stress-corrosion cracking are two other phenomena to be considered.

• Certain physical properties are utilized in manufacturing processes and their control, such as the magnetostriction effect (for ultrasonic machining of materials) and the piezoelectric effect (for force transducers and various sensors).

KEY TERMS

Conductors	Ferromagnetism	Passivation	Specific strength
Corrosion	Galvanic corrosion	Piezoelectric effect	Stress-corrosion
Degradation	Heat checking	Selective leaching	cracking
Density	Invar effect	Semiconductors	Superconductivity
Dielectric	Magnetorheostatic	Smart fluids	Thermal conductivity
Electrical conductivity	Magnetostriction	Smart materials	Thermal expansion
Electrical resistivity	Melting point	Specific heat	Thermal fatigue
Electrorheostatic	Oxidation	Specific stiffness	Thermal stresses

BIBLIOGRAPHY

ASM Handbook, Vol. 13A: **Corrosion: Fundamentals, Testing, and Protection,** 2003; Vol. 13B: **Corrosion: Materials,** 2005; Vol. 13C: **Corrosion: Environments and Industries,** 2006, ASM International.

Hummel, R.E., **Electronic Properties of Materials,** 4th ed., Springer, 2011.

Pollock, D.D., **Physical Properties of Materials for Engineers,** 2nd ed., CRC Press, 1993.

Schweitzer, P.A., **Encyclopedia of Corrosion Technology,** 3rd ed., Marcel Dekker, 2004.

Solymar, L., and Walsh, D., **Electrical Properties of Materials,** Oxford, 2004.

White, M.A., **Physical Properties of Materials,** 2nd ed., CRC Press, 2011.

REVIEW QUESTIONS

3.1 List several reasons that density is an important material property.

3.2 Explain why the melting point of a material can be an important factor in material selection.

3.3 What adverse effects can be caused by thermal expansion of materials? Give some examples.

3.4 Is thermal cracking the same as thermal shock? Why or why not?

3.5 What is the piezoelectric effect?

3.6 Describe the factors that can lead to the corrosion of a metal.

3.7 What is a superconductor? Describe two applications of superconducting materials.

3.8 What is the difference between thermal conductivity and thermal expansion?

3.9 What is corrosion? How can it be prevented or accelerated?

3.10 Explain stress-corrosion cracking. Why is it also called season cracking?

3.11 What is the difference between a superconductor and a semiconductor?

3.12 What are smart materials?

QUALITATIVE PROBLEMS

3.13 What is the fundamental difference between mechanical properties of materials discussed in Chapter 2, and physical properties of materials, described in this chapter?

3.14 Describe the significance of structures and machine components made of two materials with different coefficients of thermal expansion.

3.15 Which of the properties described in this chapter are important for (a) pots and pans, (b) cookie sheets for baking, (c) rulers, (d) paper clips, (e) music wire, and (f) beverage cans? Explain your answers.

3.16 Note in Table 3.1 that the properties of the alloys of metals have a wide range compared with the properties of the pure metals. Explain why.

3.17 Rank the following in order of increasing thermal conductivity: aluminum, copper, silicon, titanium, ceramics, and plastics. Comment on how this ranking influences applications of these materials.

3.18 Does corrosion have any beneficial effects? Explain.

3.19 Explain how thermal conductivity can play a role in the development of residual stresses in metals.

3.20 What material properties are desirable for heat shields such as those placed on the space shuttle?

3.21 List examples of products where materials that are transparent are desired. List applications for opaque materials.

3.22 Refer to Fig. 3.2 and explain why the trends seen are to be expected.

3.23 Two physical properties that have a major influence on the cracking of workpieces, tools, or dies during thermal cycling are thermal conductivity and thermal expansion. Explain why.

3.24 Which of the materials described in this chapter has the highest (a) density, (b) electrical conductivity, (c) thermal conductivity, (d) specific heat, (e) melting point, and (f) cost.

3.25 Which properties described in this chapter can be affected by applying a coating?

QUANTITATIVE PROBLEMS

3.26 If we assume that all the work done in plastic deformation is converted into heat, the temperature rise in a workpiece is (1) directly proportional to the work done per unit volume and (2) inversely proportional to the product of the specific heat and the density of the workpiece. Using Fig. 2.5, and letting the areas under the curves be the unit work done, calculate the temperature rise for (a) 8650 steel, (b) 304 stainless steel, and (c) 1100-H14 aluminum.

3.27 The natural frequency, f, of a cantilever beam is given by

$$f = 0.56\sqrt{\frac{EIg}{wL^4}},$$

where E is the modulus of elasticity, I is the moment of inertia, g is the gravitational constant, w is the weight of the beam per unit length, and L is the length of the beam. How

does the natural frequency of the beam change, if at all, as its temperature is increased? Assume that the material is steel.

3.28 Plot the following for the materials described in this chapter: elastic modulus versus density, yield stress versus density, thermal conductivity versus density. Comment on the implications of these plots.

3.29 It can be shown that thermal distortion in precision devices is low for high values of thermal conductivity divided by the thermal expansion coefficient. Rank the materials in Table 3.1 according to their ability to resist thermal distortion.

3.30 Add a column to Table 3.1 that lists the volumetric heat capacity of the materials listed, expressed in units of J/cm^3-K. Compare the results to the value for liquid water (4.184 J/cm^3-K). Note that the volumetric heat capacity of a material is the product of its density and specific heat.

SYNTHESIS, DESIGN, AND PROJECTS

3.31 Conduct a literature search and add the following materials to Table 3.1: cork, cement, ice, sugar, lithium, graphene, and chromium.

3.32 From your own experience, make a list of parts, components, or products that have corroded and have had to be replaced or discarded.

3.33 List applications where the following properties would be desirable: (a) high density, (b) low density, (c) high melting point, (d) low melting point, (e) high thermal conductivity, and (f) low thermal conductivity.

3.34 Describe several applications in which both specific strength and specific stiffness are important.

3.35 Design several mechanisms or instruments based on utilizing the differences in thermal expansion of materials, such as bimetallic strips that develop a curvature when heated.

3.36 For the materials listed in Table 3.1, determine the specific strength and specific stiffness. Describe your observations.

3.37 The maximum compressive force that a lightweight column can withstand before buckling depends on the ratio of the square root of the stiffness to the density for the material. For the materials listed in Table 2.2, determine (a) the ratio of tensile strength to density and (b) the ratio of elastic modulus to density. Comment on the suitability of each for being made into lightweight columns.

3.38 Describe possible applications and designs using alloys exhibiting the Invar effect of low thermal expansion.

3.39 Collect some pieces of different metallic and nonmetallic materials listed in Table 3.2. Using simple tests and/or instruments, determine the validity of the descending order of the physical properties shown in the table.

3.40 Design an actuator to turn on a switch when the temperature drops below a certain level. Use two materials with different coefficients of thermal expansion in your design.

3.41 Conduct an Internet and technical literature review and write a one-page paper highlighting applications of piezoelectric materials.

3.42 It has been widely reported that mechanical properties such as strength and ductility can be very different for micro-scale devices than are measured at normal length scales. Explain whether or not you would expect the physical properties described in this chapter to be scale dependent.

3.43 If you were given a metal (not an alloy) and asked to identify it, list (in order) the experiments or measurements you would perform. Explain what influence the shape of the metal would have on your prioritization.

Metal Alloys: Their Structure and Strengthening by Heat Treatment

CHAPTER 4

- This chapter examines the structures of metal alloys, including solid solutions, intermetallic compounds, and two-phase systems.
- Phase diagrams show graphically the various phases that develop as a function of alloy composition and temperature.
- The system of iron and carbon and the phases involved are described in detail.
- Heat treatment of metals is a common method to improve mechanical properties; it involves establishing a desired phase at elevated temperatures, followed by controlled cooling before the microstructure can transform into a different phase.
- Some metals, such as aluminum and stainless steels, can be heat treated only by precipitation hardening or aging.
- Improving the ductility of a material is at the expense of properties such as strength or hardness.
- The chapter ends with a discussion of the characteristics of heat-treating equipment.

4.1 Introduction

The properties and behavior of metals and alloys during manufacturing into a product and their performance during their service life depend on their composition, structure, and their processing history and the heat treatment to which they have been subjected. Important properties such as strength, hardness, ductility, toughness, and resistance to wear are greatly influenced by alloying elements and the heat-treatment processes employed. The properties of non-heat-treatable alloys are improved by mechanical working, such as rolling, forging, and extrusion (Part III).

The most common example of a process that improves properties is *heat treatment* (Sections 4.7–4.10), which modifies microstructures. Several mechanical properties that are important to manufacturing then develop, such as improved formability, machinability, or increased strength and hardness for better performance of tools and dies. These properties also enhance the service performance of gears (Fig. 4.1), cams, shafts, tools, dies, and molds.

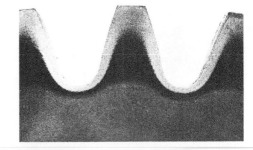

FIGURE 4.1 Cross-section of gear teeth showing induction-hardened surfaces. *Source:* TOCCO Div., Park-Ohio Industries, Inc.

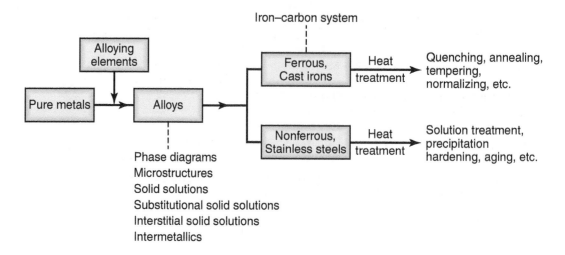

FIGURE 4.2   Outline of topics described in this chapter.

This chapter follows the outline shown in Fig. 4.2, beginning with the role of various alloying elements, the solubility of one element in another, phases, equilibrium phase diagrams, and the influence of composition, temperature, and time. The chapter also discusses methods and techniques of heating, quenching, tempering, and annealing of metals and alloys, and describes the characteristics of the equipment involved.

4.2   Structure of Alloys

When describing the basic crystal structure of metals in Chapter 1, it was noted that the atoms are all of the *same* type, except for the presence of rare impurity atoms; these metals are known as **pure metals**, even though they may not be completely pure. *Commercially pure* metals are used for various purposes, such as aluminum for foil, copper for electrical conductors, nickel or chromium for plating, and gold for electrical contacts. Pure metals have somewhat limited properties but they can be enhanced and modified by **alloying**. An **alloy** consists of two or more chemical elements, at least one of which is a metal; the majority of metals used in engineering applications are some form of alloy. Alloying consists of two basic forms: *solid solutions* and *intermetallic compounds*.

4.2.1   Solid Solutions

Two terms are essential in describing alloys: **solute** and **solvent**. The solute is the *minor* element (such as salt or sugar) that is added to the solvent, which is the *major* element (such as water). In terms of the elements in a crystal structure, the solute (composed of *solute atoms*) is the element that is added to the solvent (composed of *host atoms*). When the particular crystal structure of the solvent is maintained during alloying, the alloy is called a *solid solution*.

Substitutional Solid Solutions.   If the size of the solute atom is similar to that of the solvent atom, the solute atoms can replace solvent atoms and form a *substitutional solid solution* (see Fig. 1.8). An example is brass, which is an alloy of zinc and copper in which zinc (the solute atom) is introduced into the lattice of copper (solvent atoms). The properties of brass can thus be altered by controlling the amount of zinc in copper.

Interstitial Solid Solutions. If the size of the solute atom is much smaller than that of the solvent atom, each solute atom can occupy an *interstitial* position; such a process forms an *interstitial solid solution.*

An important family of interstitial solid solutions is **steel** (Chapter 5), an alloy of iron and carbon in which the carbon atoms are present in interstitial positions between iron atoms. The atomic radius of carbon is 0.071 nm, which is very small compared to the 0.124-nm radius of the iron atom. The properties of carbon steels can be varied over a wide range by adjusting the ratio of carbon to iron. The ability to control this ratio is a major reason why steel is such a versatile and useful material, with a very wide variety of properties and applications.

4.2.2   Intermetallic Compounds

Intermetallic compounds are complex structures consisting of two metals in which solute atoms are present among solvent atoms in certain proportions. Typical examples are the aluminides of titanium (Ti_3Al), nickel (Ni_3Al), and iron (Fe_3Al). Some intermetallic compounds have solid solubility, and the type of atomic bond may range from metallic to ionic. Intermetallic compounds are strong, hard, and brittle. Because of their high melting points, strength at elevated temperatures, good oxidation resistance, and relatively low density, they are candidate materials for such applications as advanced gas-turbine engines.

4.2.3   Two-phase Systems

Recall that a solid solution is one in which two or more elements form a single homogeneous solid phase, in which the elements are uniformly distributed throughout the solid mass. Such a system has a maximum concentration of solute atoms in the solvent-atom lattice, just as there is a solubility limit for sugar in water. Most alloys consist of two or more solid phases and may be regarded as mechanical mixtures; such a system with two solid phases is known as a *two-phase system.*

A **phase** is defined as a physically distinct and homogeneous portion in a material. Each phase is a homogeneous part of the total mass, and has its own characteristics and properties. Consider a mixture of sand and water as an example of a two-phase system. These two different components have their own distinct structures, characteristics, and properties; there is a clear boundary in this mixture between the water (one phase) and the sand particles (the second phase). Another example is ice in water: the two phases are the same chemical compound of exactly the same chemical elements (hydrogen and oxygen), even though their properties are very different. Note that it is not necessary that one phase be a liquid; for example, sand suspended in ice is also a two-phase system.

A typical example of a two-phase system in metals occurs when lead is added to copper in the molten state. After the mixture solidifies, the structure consists of two phases: one having a small amount of lead in solid solution in copper, the other having lead particles (roughly spherical in shape) *dispersed* throughout the structure (Fig. 4.3a). The lead particles are analogous to the sand particles in water, described above. This copper–lead alloy has properties that are different from those of either copper or lead alone.

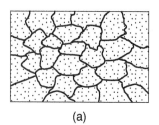

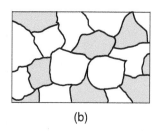

(a)   (b)

FIGURE 4.3   (a) Schematic illustration of grains, grain boundaries, and particles dispersed throughout the structure of a two-phase system, such as a lead–copper alloy. The grains represent lead in solid solution in copper, and the particles are lead as a second phase. (b) Schematic illustration of a two-phase system consisting of two sets of grains: dark and light. The colored and white grains have different compositions and properties.

Alloying with finely dispersed particles (**second-phase particles**) is an important method of strengthening metal alloys and controlling their properties. In two-phase alloys, the second-phase particles become obstacles to dislocation movement and thus increase the strength of the alloy. Another example of a two-phase alloy is the aggregate structure shown in Fig. 4.3b, where there are two sets of grains, each with its own composition and properties. The darker grains in the figure may have a different structure from the lighter grains; they may, for example, be brittle, while the lighter grains are ductile.

Defects may develop during metalworking operations such as forging or extrusion (as described in Chapters 14 and 15, respectively). Such flaws may be due to the lack of ductility of one of the phases in the alloy. In general, two-phase alloys are stronger and less ductile than solid solutions.

4.3   Phase Diagrams

Pure metals have clearly defined melting or freezing points, and solidification takes place at a *constant temperature*. When the temperature of a molten metal is reduced to the freezing point, the energy or the *latent heat of solidification* is given off while the temperature remains constant. Eventually, solidification is complete and the solid metal continues cooling to ambient (room) temperature.

Unlike pure metals, alloys solidify over a range of temperatures (Fig. 4.4). Solidification begins when the temperature of the molten metal drops below the **liquidus**; it is completed when the temperature reaches the **solidus**. Within this temperature range, the alloy is in a *mushy* or *pasty* state; its composition and state are then described by the particular alloy's phase diagram.

A phase diagram, also called an **equilibrium** or constitutional diagram, shows the relationships among temperature, composition, and the phases present in a particular alloy system at equilibrium. *Equilibrium* means that the state of a system does not vary with time; the word *constitutional* indicates the relationships among the structure, the composition, and the physical makeup of the alloy. As described in detail below, types of phase diagrams include those for (a) complete solid solutions; (b) eutectics, such as cast irons; and (c) eutectoids, such as steels.

One example of a phase diagram is shown in Fig. 4.4 for the copper–nickel alloy; it is called a **binary phase diagram** because there are two elements (copper and nickel) present in the system. The left boundary of this phase diagram (100% Ni) indicates the melting point of pure nickel; the right boundary (100% Cu) indicates the melting point of pure copper. (All percentages are by weight, not by number of atoms.)

Lever Rule.   The composition of various phases in a phase diagram can be determined by a procedure called the *lever rule*. As shown in the lower portion of Fig. 4.4, we first construct a lever between the solidus and liquidus lines (called *tie line*), which is balanced (on the triangular support) at the nominal weight composition C_o of the alloy. The left end of the lever represents the composition C_S of the solid phase and the right end of the composition C_L of the liquid phase. Note from the graduated scale in the figure that the liquid fraction is also indicated along this tie line, ranging from 0 at the left (fully solid) to 1 at the right (fully liquid).

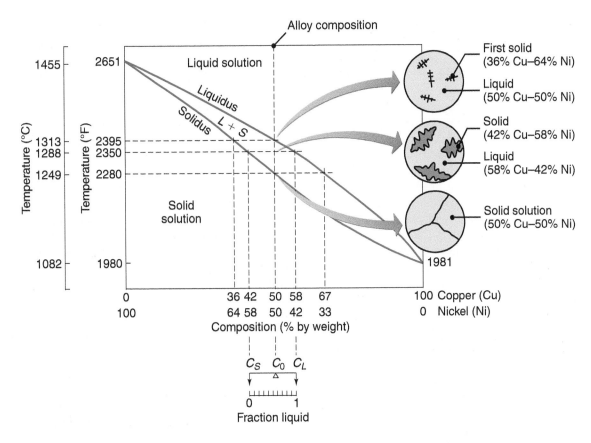

FIGURE 4.4   Phase diagram for copper–nickel alloy system obtained at a slow rate of solidification. Note that pure nickel and pure copper each have one freezing or melting temperature. The top circle on the right depicts the nucleation of crystals. The second circle shows the formation of dendrites (see Section 10.2.2). The bottom circle shows the solidified alloy, with grain boundaries.

The lever rule states that the **weight fraction of solid** is proportional to the distance between C_o and C_L:

$$\frac{S}{S+L} = \frac{C_o - C_L}{C_S - C_L}.\tag{4.1}$$

Likewise, the **weight fraction of liquid** is proportional to the distance between C_S and C_o, hence

$$\frac{L}{S+L} = \frac{C_S - C_o}{C_S - C_L}.\tag{4.2}$$

Note that these quantities are fractions and must be multiplied by 100 to obtain percentages.

From inspection of the tie line in Fig. 4.4 (and for a nominal alloy composition of $C_o = 50\%$ Cu–50% Ni) it can be noted that, because C_o is closer to C_L than it is to C_S, the solid phase contains less copper than does the liquid phase. By measuring on the phase diagram and using the lever-rule equations, it can be seen that the composition of the solid phase is 42% Cu and of the liquid phase is 58% Cu, as stated in the middle circle at the right in Fig. 4.4. These calculations refer to copper. If we now reverse the phase diagram in the figure, so that the left boundary is 0% nickel (whereby nickel now becomes the alloying element in copper), these calculations will give us the compositions of the solid and liquid phases in terms of nickel. The lever rule is also known as the *inverse lever rule* because, as indicated by Eqs. (4.1) and (4.2), the amount of each phase is proportional to the length of the opposite end of the lever.

The *completely solidified* alloy in the phase diagram shown in Fig. 4.4 is a *solid solution*, because the alloying element, Cu (the solute atom), is completely dissolved in the host metal, Ni (the solvent atom), and each grain has the same composition. The atomic radius of copper is 0.128 nm and that of nickel is 0.125 nm, and both elements are of face-centered cubic structure; thus, they readily form solid solutions.

The mechanical properties of solid solutions of Cu–Ni alloy depend on their composition (Fig. 4.5). The properties of pure copper are, up to a point, improved upon by increasing the nickel content; thus, there is an optimal percentage of nickel that gives the highest strength and hardness to the Cu–Ni alloy.

Figure 4.5 also shows how zinc, as an alloying element in copper, affects the mechanical properties of the alloy. Note the maximum of 40% solid solubility for zinc (solute) in copper (solvent), whereas copper and nickel are completely soluble in each other. The improvements in properties are due to *pinning* (blocking) of dislocations (Section 1.4.1) at substitutional nickel or zinc atoms, which may also be regarded as impurity atoms. As a result, dislocations cannot move as freely and thus the strength of the alloy increases.

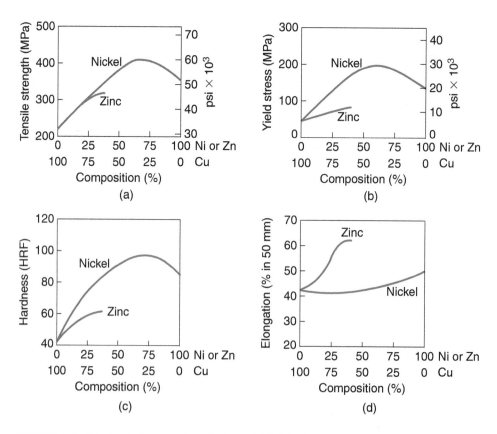

FIGURE 4.5 Mechanical properties of copper–nickel and copper–zinc alloys as a function of their composition. The curves for zinc are short, because zinc has a maximum solid solubility of 40% in copper.

4.4 The Iron–carbon System

Steels and cast irons are represented by the iron–carbon binary system. Commercially pure iron contains up to 0.008% C, steels up to 2.11% C, and cast irons up to 6.67% C, although most cast irons contain less than 4.5% C. In this section, the iron–carbon system is described, including the techniques employed to evaluate and modify the properties of these important materials for specific applications.

The **iron–iron-carbide phase diagram** is shown in Fig. 4.6. Although this diagram can be extended to the right—to 100% C (pure graphite; see Fig. 4.10)—the range that is significant to engineering applications is up to 6.67% C, because Fe_3C is a stable phase. Pure iron melts at a temperature of 1538°C (2798°F), as shown at the left boundary in Fig. 4.6. As iron cools, it first forms delta ferrite, then austenite, and finally alpha ferrite.

Ferrite. **Alpha ferrite,** also denoted α-ferrite or simply **ferrite,** is a solid solution of body-centered cubic (bcc) iron; it has a maximum solid solubility of 0.022% C at a temperature of 727°C (1341°F). Just as there is a solubility limit for salt in water (with any extra amount precipitating as solid salt at the bottom of a container), there is a solid solubility limit for carbon in iron as well.

Ferrite is relatively soft and ductile; it is magnetic from room temperature to 768°C (1414°F), the so-called *Curie temperature* (after the Polish physicist and chemist M. Curie, 1867–1934). Although very little carbon can dissolve interstitially in bcc iron, the amount of carbon can significantly affect the mechanical properties of ferrite. Furthermore, significant amounts of chromium, manganese, nickel, molybdenum, tungsten, and silicon can be contained in iron in solid solution, imparting special properties.

Austenite. As shown in Fig. 4.6, within a certain temperature range iron undergoes a **polymorphic transformation** from a bcc to an fcc structure, becoming *gamma iron*

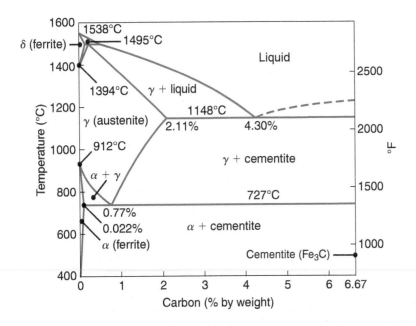

FIGURE 4.6   The iron–iron-carbide phase diagram.

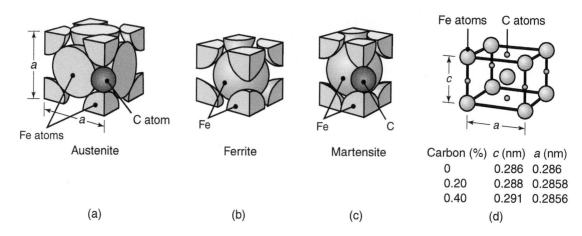

FIGURE 4.7   The unit cells for (a) austenite, (b) ferrite, and (c) martensite. The effect of percentage of carbon (by weight) on the lattice dimensions for martensite is shown in (d); note the interstitial position of the carbon atoms. (See Fig. 1.8.) Note also the increase in dimension c with increasing carbon content; this effect causes the unit cell of martensite to be in the shape of a rectangular prism.

(γ-iron), or, more commonly, **austenite** (after the British metallurgist W.C. Roberts–Austen, 1843–1902). This structure has a solid solubility of up to 2.11% C at 1148°C (2098°F). Because the fcc structure has more interstitial positions, the solid solubility of austenite is about two orders of magnitude higher than that of ferrite, with the carbon occupying interstitial positions, as shown in Fig. 4.7a.

Austenite is an important phase in the heat treatment of steels (Section 4.7). It is denser than ferrite, and its single-phase fcc structure is ductile at elevated temperatures; consequently, it possesses good formability. Large amounts of nickel and manganese can also be dissolved in fcc iron to impart various properties. Steel is non-magnetic in the austenitic form, either at high temperatures or, for austenitic stainless steels, at room temperature.

Cementite.   The right boundary of Fig. 4.6 represents **cementite**, which is 100% iron carbide (Fe_3C), having a carbon content of 6.67%. Cementite (from the Latin *caementum*, meaning "stone chips") is also called **carbide**, but should not be confused with other carbides used as dies, cutting tools, and abrasives, such as tungsten carbide, titanium carbide, and silicon carbide (Chapters 8 and 22). Cementite is a very hard and brittle intermetallic compound, with a significant influence on the properties of steels. It can include other alloying elements, such as chromium, molybdenum, and manganese.

4.5   The Iron–iron-carbide Phase Diagram and the Development of Microstructures in Steels

The region of the iron–iron-carbide phase diagram that is significant for steels is shown in Fig. 4.8 (an enlargement of the lower left-hand portion of Fig. 4.6). Various microstructures can be developed, depending on the (a) carbon content, (b) amount of plastic deformation (working), and (c) method of heat treatment. For example, consider the **eutectic point** of iron with a 0.77% C content, while it is being cooled very slowly from a temperature of, say, 1100°C (2000°F) in the austenite phase. The reason for very slow cooling is to maintain equilibrium.

At 727°C (1341°F), a reaction takes place in which austenite is transformed into alpha ferrite (bcc) and cementite. Because the solid solubility of carbon in ferrite is only 0.022%, the extra carbon forms cementite. This reaction is called a **eutectoid** (meaning *eutecticlike*) reaction, which means that, at a certain temperature, a single solid phase (austenite) is transformed into two other solid phases (ferrite and cementite).

The structure of eutectoid steel is called pearlite because, at low magnifications, it resembles mother of pearl (Fig. 4.9). The microstructure of pearlite consists of alternating layers (lamellae) of ferrite and cementite; consequently, the mechanical properties of pearlite are intermediate between those of ferrite (soft and ductile) and cementite (hard and brittle).

4.5.1   Effects of Alloying Elements in Iron

Although carbon is the basic element that transforms iron into steel, other elements are also added to impart a variety of desirable properties. The main effect of these alloying elements on the iron–iron-carbide phase diagram is to shift the eutectoid temperature and eutectoid composition (percentage of carbon in steel at the eutectoid point); these elements shift other phase boundaries as well.

The eutectoid temperature may be raised or lowered, from 727°C (1341°F), depending on the particular alloying element. On the other hand, alloying elements always lower the eutectoid composition; that is, its carbon content is lower than 0.77%. Lowering the eutectoid temperature means increasing the austenite range; as a result, an alloying element such as nickel is known as an **austenite former**. Because nickel has an fcc structure, it favors the fcc structure of austenite. Conversely, chromium and molybdenum have a bcc structure, thus favoring the bcc structure of ferrite; these elements are known as **ferrite stabilizers**.

4.6   Cast Irons

The term **cast iron** refers to a family of ferrous alloys composed of iron, carbon (ranging from 2.11% to about 4.5%), and silicon (up to about 3.5%). Cast irons are usually classified according to their solidification morphology from the eutectic temperature (see also Section 12.3.2):

1. Gray cast iron, or gray iron
2. Ductile cast iron, also called nodular cast iron or spheroidal graphite cast iron
3. White cast iron
4. Malleable iron
5. Compacted graphite iron

Cast irons are also classified by their structure: ferritic, pearlitic, quenched and tempered, or austempered.

The equilibrium phase diagram relevant to cast irons is shown in Fig. 4.10, in which the right boundary is 100% C—that is, pure graphite. Because the eutectic

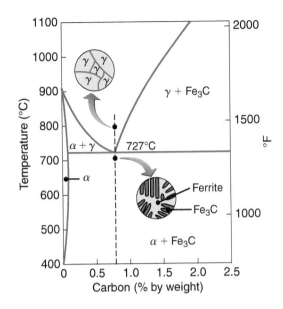

FIGURE 4.8 Schematic illustration of the microstructures for an iron–carbon alloy of eutectoid composition (0.77% carbon), above and below the eutectoid temperature of 727°C (1341°F).

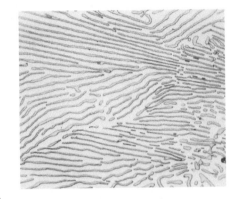

FIGURE 4.9 Microstructure of pearlite in 1080 steel, formed from austenite of eutectoid composition. In this lamellar structure, the lighter regions are ferrite and the darker regions are carbide. Magnification: 2500×.

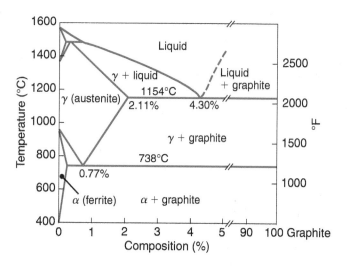

FIGURE 4.10 Phase diagram for the iron–carbon system with graphite (instead of cementite) as the stable phase; note that this figure is an extended version of Fig. 4.6.

temperature is 1154°C (2109°F), cast irons are completely liquid at temperatures lower than those required for liquid steels. Consequently, iron with high carbon content can be cast (see Part II) at lower temperatures than can steels.

Cementite is not completely stable; it is **meta-stable**, with an extremely low rate of decomposition. It can, however, be made to decompose into alpha ferrite and graphite. The formation of graphite (**graphitization**) can be controlled, promoted, and accelerated by modifying the composition, the rate of cooling, and by the addition of silicon.

Gray Cast Iron. In this structure, graphite exists largely in the form of *flakes* (Fig. 4.11a). It is called **gray cast iron**, or **gray iron** because, when it is broken, the fracture path is along the graphite flakes and has a gray, sooty appearance. These flakes act as stress raisers; as a result, gray iron has negligible ductility and is weak in tension, although strong in compression. On the other hand, the presence of graphite flakes gives this material the capacity to dampen vibrations (by internal friction). This capacity makes gray cast iron a suitable and commonly used material for constructing machine-tool bases and machinery structures (Section 25.3).

Three types of gray cast iron are **ferritic, pearlitic,** and **martensitic;** because of the different structures, each has different properties and applications. In ferritic gray iron (also known as *fully gray iron*), the structure consists of graphite flakes in an alpha-ferrite matrix. Pearlitic gray iron has a structure of graphite in a matrix of pearlite, and although still brittle, it is stronger than fully gray iron. Martensitic gray iron is obtained by austenitizing a pearlitic gray iron and then quenching it rapidly to produce a structure of graphite in a martensite matrix; as a result, this cast iron is very hard.

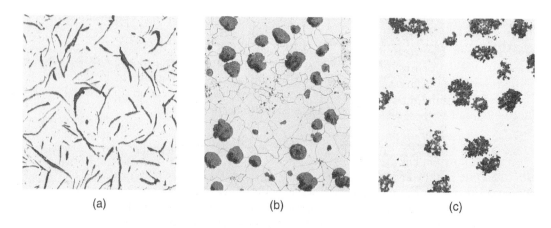

(a)   (b)   (c)

FIGURE 4.11 Microstructure for cast irons. Magnification: 100×. (a) Ferritic gray iron with graphite flakes. (b) Ferritic ductile iron (nodular iron), with graphite in nodular form. (c) Ferritic malleable iron; this cast iron solidified as white cast iron, with the carbon present as cementite, and was heat treated to graphitize the carbon.

Ductile (Nodular) Iron.   In the ductile-iron structure, graphite is in a **nodular** or **spheroid** form (Fig. 4.11b), which permits the material to be somewhat ductile and shock resistant. The shape of graphite flakes can be changed into nodules (spheres), by small additions of magnesium and/or cerium to the molten metal prior to pouring. Ductile iron can be made ferritic or pearlitic by heat treatment; it can also be heat treated to obtain a structure of tempered martensite (Section 4.7).

White Cast Iron.   *White cast iron* is obtained either by cooling gray iron rapidly or by adjusting the composition by keeping the carbon and silicon content low; it is also called *white iron* because of the white crystalline appearance of the fracture surface. The white cast iron structure is very hard, wear resistant, and brittle, because of the presence of large amounts of iron carbide (instead of graphite).

Malleable Iron.   *Malleable iron* is obtained by annealing white cast iron in an atmosphere of carbon monoxide and carbon dioxide, at between 800°C and 900°C (1470°F and 1650°F), for up to several hours, depending on the size of the part. During this process, the cementite decomposes (*dissociates*) into iron and graphite. The graphite exists as *clusters* or *rosettes* (Fig. 4.11c) in a ferrite or pearlite matrix; consequently, malleable iron has a structure similar to that of nodular iron. This structure promotes good ductility, strength, and shock resistance—hence, the term *malleable* (from the Latin *malleus* meaning "it can be hammered").

Compacted-graphite Iron.   The graphite in this structure is in the form of short, thick, interconnected flakes having undulating surfaces and rounded extremities. The mechanical and physical properties of this cast iron are intermediate between those of flake-graphite and nodular-graphite cast irons.

4.7 Heat Treatment of Ferrous Alloys

The various microstructures described thus far can be modified by **heat-treatment** techniques—that is, by controlled heating and cooling of the alloys at various rates. These treatments induce **phase transformations**, which greatly influence such mechanical properties as strength, hardness, ductility, toughness, and wear resistance. The effects of thermal treatment depend on the particular alloy, its composition and microstructure, the degree of prior cold work, and the rates of heating and cooling during heat treatment.

This section focuses on the microstructural changes in the iron–carbon system. Because of their technological significance, the structures considered are pearlite, spheroidite, bainite, martensite, and tempered martensite. The heat-treatment processes described are annealing, quenching, and tempering.

Pearlite.   If the ferrite and cementite lamellae in the pearlite structure of the eutectoid steel, shown in Fig. 4.9, are thin and closely packed, the microstructure is called **fine pearlite**; if they are thick and widely spaced, it is called **coarse pearlite**. The difference between the two depends on the rate of cooling through the eutectoid temperature, which is the site of a reaction in which austenite is transformed into pearlite. If the rate of cooling is relatively high (as in air), fine pearlite is produced; if cooling is slow (as in a furnace), coarse pearlite is produced.

Spheroidite.   When pearlite is heated to just below the eutectoid temperature and then held at that temperature for a period of time, such as at 700°C (1300°F)

QR Code 4.1 Hardening chains with induction heating. (*Source:* Courtesy of GH Induction Atmospheres)

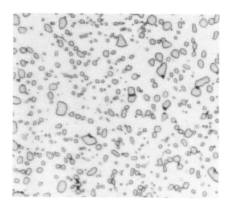

FIGURE 4.12 Microstructure of eutectoid steel. Spheroidite is formed by tempering the steel at 700°C (1292°F). Magnification: 1000×.

for a day, the cementite lamellae transform to roughly spherical shapes (Fig. 4.12). Unlike the lamellar shapes of cementite, which act as stress raisers, **spheroidites** (spherical particles) have smaller stress concentrations because of their rounded shapes. Consequently, this structure has higher toughness and lower hardness than the pearlite structure. It can be cold worked, because the ductile ferrite has high toughness and the spheroidal carbide particles prevent the initiation of cracks within the material.

Bainite.   Visible only through electron microscopy, *bainite* is a very fine microstructure, consisting of ferrite and cementite, similar to pearlite, but having a different morphology. Bainite can be produced in steels with alloying elements and at cooling rates that are higher than those required for pearlite. This structure, called **bainitic steel** (after the American metallurgist E.C. Bain, 1891–1971), is generally stronger and more ductile than pearlitic steels at the same hardness level.

Martensite.   When austenite is cooled at a high rate, such as by quenching in water, its fcc structure is transformed into a **body-centered tetragonal** (bct) structure, which can be described as a body-centered rectangular prism that is slightly elongated along one of its principal axes (see Fig. 4.7d). This microstructure is called *martensite* (after the German metallurgist A. Martens, 1850–1914). Because martensite does not have as many slip systems as a bcc structure, and the carbon is in interstitial positions, it is extremely hard and brittle (Fig. 4.13). Martensite transformation takes place almost instantaneously, because it involves not the diffusion process but a slip mechanism and thus involves plastic deformation. This is a time-dependent phenomenon that is the mechanism in other transformations as well.

Retained Austenite.   If the temperature to which the alloy is quenched is not sufficiently low, only a portion of the structure is transformed to martensite. The rest is *retained austenite*, which is visible as white areas in the structure, along with the dark, needlelike martensite. Retained austenite can cause dimensional instability and cracking, and lower the hardness and strength of the alloy.

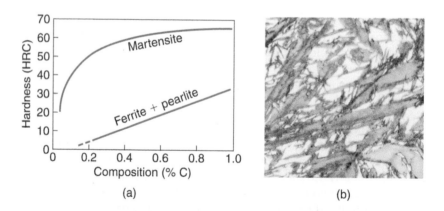

(a)   (b)

FIGURE 4.13   (a) Hardness of martensite as a function of carbon content. (b) Micrograph of martensite containing 0.8% carbon. The gray platelike regions are martensite; they have the same composition as the original austenite (white regions). Magnification: 1000×.

Tempered Martensite. Martensite is tempered in order to improve its mechanical properties. *Tempering* is a heating process by which hardness is reduced and toughness is increased. The body-centered tetragonal martensite is heated to an intermediate temperature, typically 150°–650°C (300°–1200°F), where it decomposes to a two-phase microstructure, consisting of bcc alpha ferrite and small particles of cementite.

With increasing tempering time and temperature, the hardness of tempered martensite decreases (Fig. 4.14). The reason is that the cementite particles coalesce and grow, and the distance between the particles in the soft ferrite matrix increases as the less stable and smaller carbide particles dissolve.

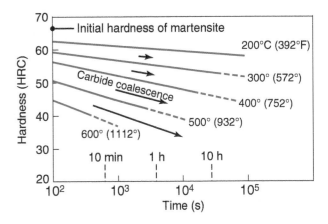

FIGURE 4.14 Hardness of tempered martensite as a function of tempering time for 1080 steel quenched to 65 HRC. Hardness decreases because the carbide particles coalesce and grow in size, thereby increasing the interparticle distance of the softer ferrite.

4.7.1   Time–temperature–transformation Diagrams

The percentage of austenite transformed into pearlite is a function of temperature and time (Fig. 4.15a). This transformation is best illustrated by Figs. 4.15b and c in diagrams called **isothermal transformation (IT) diagrams**, or *time–temperature–transformation* (TTT) *diagrams*, constructed from the data given in Fig. 4.15a. The higher the temperature or the longer the time, the more austenite is transformed into pearlite. Note that, for each temperature, there is a minimum time for the transformation to begin. This time period defines the critical cooling rate; with longer times, austenite begins to transform into pearlite, as can be traced in Figs. 4.15b and c.

The TTT diagrams shown allow the design of heat-treatment schedules to obtain desirable microstructures. For example, consider the TTT curves shown in Fig. 4.15c; the steel can be raised to a very high temperature (above the eutectic temperature) to start with a state of austenite. If the material is cooled rapidly, it can follow the 140°C/s cooling rate trajectory shown, resulting in complete martensite. On the other hand, it can be more slowly cooled (in a molten salt bath) to develop pearlite- or bainite-containing steels. If tempered martensite is desired, the heat treat and quench stages is followed by a tempering process.

The differences in hardness and toughness of the various structures obtained are shown in Fig. 4.16. Fine pearlite is harder and less ductile than coarse pearlite. The effects of various percentages of carbon, cementite, and pearlite on other mechanical properties of steels are shown in Fig. 4.17.

Video Solution 4.1 Application of TTT Diagrams

4.8   Hardenability of Ferrous Alloys

The capability of an alloy to be hardened by heat treatment is called its **hardenability**, and is a measure of the *depth* of hardness that can be obtained by heating and subsequent quenching. The term "hardenability" should not be confused with "hardness," which is the resistance of a material to indentation or scratching (Section 2.6). From the discussion thus far, it can be seen that hardenability of ferrous alloys depends on the (a) carbon content, (b) grain size of the austenite, (c) alloying elements present in the material, and (d) cooling rate.

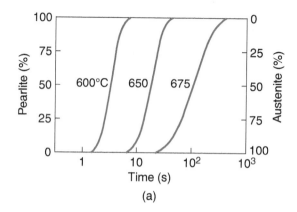

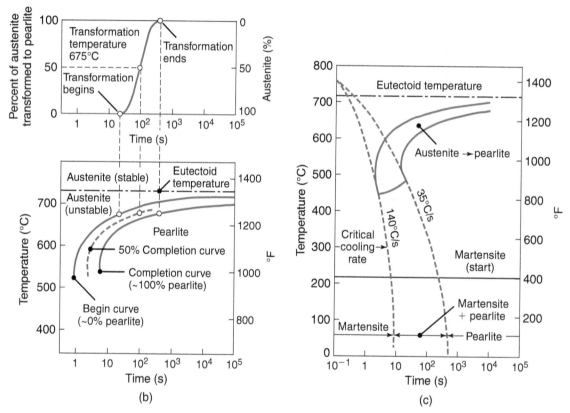

FIGURE 4.15 (a) Austenite-to-pearlite transformation of iron–carbon alloy as a function of time and temperature. (b) Isothermal transformation diagram obtained from (a) for a transformation temperature of 675°C (1247°F). (c) Microstructures obtained for a eutectoid iron–carbon alloy as a function of cooling rate.

4.8.1   The End-quench Hardenability Test

In this commonly used **Jominy test** (after the American metallurgist W.E. Jominy, 1893–1976), a round test bar 100 mm (4 in.) long, made from a particular alloy, is **austenitized**—that is, heated to the proper temperature to form 100% austenite. It is then quenched directly at one end (Fig. 4.18a) with a stream of water at 24°C (75°F). The cooling rate thus varies throughout the length of the bar, the rate being

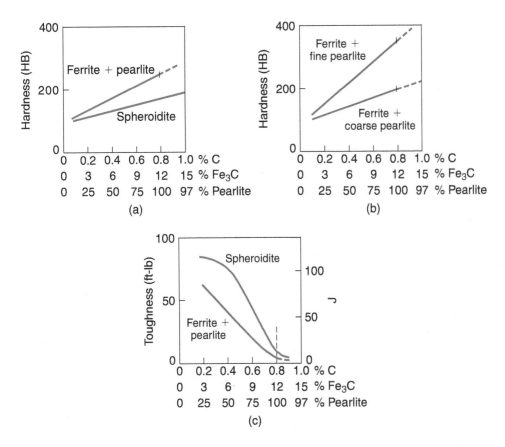

FIGURE 4.16   (a) and (b) Hardness and (c) toughness for annealed plain-carbon steels as a function of carbide shape. Carbides in the pearlite are lamellar. Fine pearlite is obtained by increasing the cooling rate. The spheroidite structure has sphere-like carbide particles.

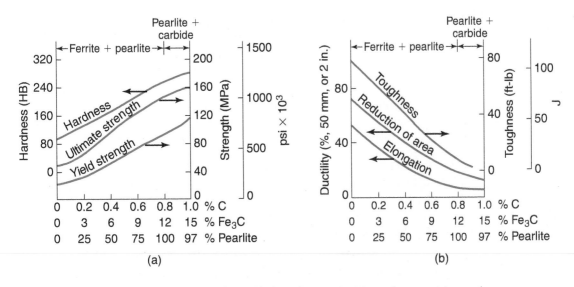

FIGURE 4.17   Mechanical properties of annealed steels as a function of composition and microstructure. Note in (a) the increase in hardness and strength, and in (b), the decrease in ductility and toughness, with increasing amounts of pearlite and iron carbide.

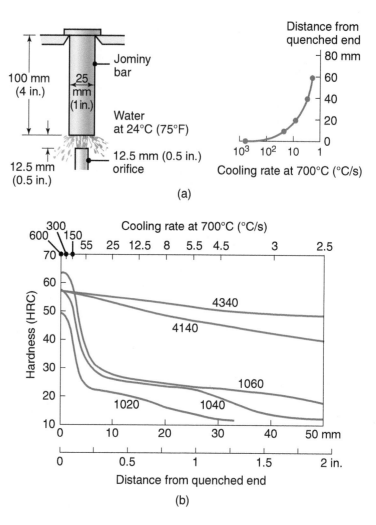

FIGURE 4.18 (a) End-quench test and cooling rate. (b) Hardenability curves for five different steels, as obtained from the end-quench test. Small variations in composition can change the shape of these curves. Each curve is actually a band, and its exact determination is important in the heat treatment of metals, for better control of properties.

highest at the lower end, being in direct contact with the water. The hardness along the length of the bar is then measured at various distances from the quenched end.

As expected from the discussion of the effects of cooling rates in Section 4.7, hardness decreases away from the quenched end of the bar (Fig. 4.18b). The greater the depth to which hardness increases, the greater the hardenability of the alloy. Each composition of an alloy has its particular **hardenability band**. Note that the hardness at the quenched end increases with increasing carbon content, and that 1040, 4140, and 4340 steels have the same carbon content (0.40%) and thus they have the same hardness (57 HRC) at the quenched end.

4.8.2   Quenching Media

The fluid used for quenching the heated specimen also has an effect on hardenability. Quenching may be carried out in water, brine (salt water), oil, molten salt, or air; caustic solutions, and polymer solutions; gases may also be used. Because of the

differences in thermal conductivity, specific heat, and heat of vaporization of these media, the rate of cooling of the specimen (**severity of quench**) is also different. In relative terms and in decreasing order, the cooling capacities of several quenching media are: agitated brine, 5; still water, 1; still oil, 0.3; cold gas, 0.1; and still air, 0.02.

Agitation also is a significant factor in the rate of cooling; the more vigorous the agitation, the higher is the rate of cooling. In tool steels, the quenching medium is specified by a letter (see Table 5.7), such as W for water hardening, O for oil hardening, and A for air hardening. The cooling rate also depends on the surface-area-to-thickness or surface-area-to-volume ratio of the part to be quenched; the higher this ratio, the higher is the cooling rate. For example, a thick plate cools more slowly than a thin plate with the same surface area. These considerations are also significant in the cooling of metals and of plastics in casting and in molding processes (Sections 10.5.1 and 19.3).

Water is a common medium for rapid cooling. However, the heated specimen may form a **vapor blanket** along its surfaces due to the water-vapor bubbles that form when water boils at the metal–water interface. This blanket creates a barrier to heat conduction, because of the lower thermal conductivity of the vapor. Agitating the fluid or the part helps to reduce or eliminate the blanket; also, water may be sprayed onto the part under high pressure. Brine is an effective quenching medium, because salt helps to nucleate bubbles at the interfaces, which improves agitation; note, however, that brine can corrode the part.

Polymer quenchants can be used for ferrous as well as for nonferrous alloys; they have cooling characteristics that generally are between those of water and petroleum oils. Typical polymer quenchants are polyvinyl alcohol, polyalkaline oxide, polyvinyl pyrrolidone, and polyethyl oxazoline. These quenchants have such advantages as better control of hardness, elimination of fumes and fire (as may occur when oils are used as a quenchant), and reduction of corrosion (as may occur when water is used).

4.9   Heat Treatment of Nonferrous Alloys and Stainless Steels

Nonferrous alloys and some stainless steels cannot be heat treated by the techniques described for ferrous alloys. The reason is that nonferrous alloys do not undergo phase transformations like those in steels; the hardening and strengthening mechanisms for these alloys are fundamentally different. Heat-treatable aluminum alloys, copper alloys, and martensitic and some other stainless steels are hardened and strengthened by a process called **precipitation hardening**. In this process, small particles of a different phase, called **precipitates**, are uniformly dispersed in the matrix of the original phase (Fig. 4.3a). Precipitates form because the solid solubility of one element (one component of the alloy) in the other is exceeded.

Three stages are involved in precipitation hardening, which can best be described by reference to the phase diagram for the aluminum–copper system (Fig. 4.19a). For a composition of 95.5% Al–4.5% Cu, a single-phase (kappa phase) substitutional solid solution of copper (solute) in aluminum (solvent) exists between 500°C and 570°C (930°F and 1060°F). This *kappa phase* is aluminum rich, has an fcc structure, and is ductile. Below the lower temperature (that is, below the lower solubility curve) there are two phases: *kappa* (κ) and *theta* (θ) which is a hard intermetallic compound of $CuAl_2$. This alloy can be heat treated and its properties modified by two different methods: *solution treatment* and *precipitation hardening*.

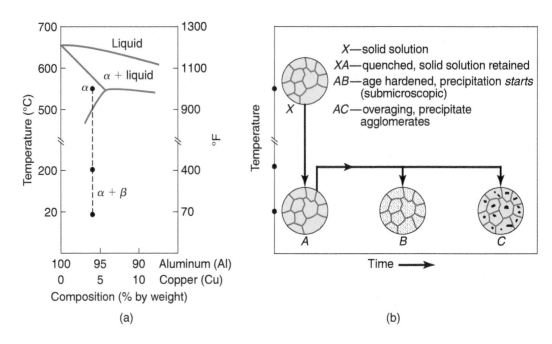

FIGURE 4.19   (a) Phase diagram for the aluminum–copper alloy system. (b) Various micro-structures obtained during the age-hardening process.

4.9.1   Solution Treatment

In *solution treatment*, the alloy is heated to within the solid-solution kappa phase (say, 540°C; 1000°F) and then cooled rapidly, such as by quenching it in water. The structure obtained soon after quenching (*A* in Fig. 4.19b) consists only of the single phase kappa. This alloy has moderate strength and considerable ductility.

4.9.2   Precipitation Hardening

The structure obtained in *A* in Fig. 4.19b can be made stronger by *precipitation hardening*. In this process, the alloy is first reheated to an intermediate temperature and then held there for a period of time, during which precipitation takes place. The copper atoms diffuse to nucleation sites and combine with aluminum atoms. This process produces the theta phase, which forms as submicroscopic precipitates (shown in *B* by the small dots within the grains of the kappa phase). The resulting structure is stronger than that in *A*, although it is less ductile. The increase in strength is due to increased resistance to dislocation movement in the region of the precipitates.

Aging.   Because the precipitation process is one of time and temperature, it is also called aging, and the property improvement is known as **age hardening**. If carried out above room temperature, the process is called **artificial aging**. Several aluminum alloys harden and become stronger over a period of time at room temperature; this process is then called **natural aging**. Such alloys are first quenched and then, if desired, they are shaped by plastic deformation at room temperature. Finally, they are allowed to develop strength and hardness by aging naturally. The rate of natural aging can be slowed by refrigerating the quenched alloy (**cryogenic treatment**).

In the precipitation process, if the reheated alloy is held at the elevated temperature for an extended period of time, the precipitates begin to coalesce and grow. They become larger but fewer in number, as is shown by the larger dots in C in Fig. 4.19b. This process is called **over-aging,** and the resulting alloy is softer and weaker.

There is an optimal time–temperature relationship in the aging process that must be followed in order to obtain the desired properties (Fig. 4.20). It is apparent that an aged alloy can be used only up to a certain maximum temperature in service, otherwise it will over-age and lose its strength and hardness. Although weaker, an over-aged part has better dimensional stability.

Maraging.   This is a precipitation-hardening treatment for a special group of high-strength iron-base alloys. The word *maraging* is derived from *martensite age hardening*, a process in which one or more intermetallic compounds are precipitated in a matrix of low-carbon martensite. A typical maraging steel may contain 18% Ni, in addition to other elements, and aging takes place at 480°C (900°F). Because hardening by maraging does not depend on the cooling rate, uniform and full hardness can be obtained throughout large parts and with minimal distortion. Typical uses of maraging steels are in dies and tooling for casting, molding, forging, and extrusion (Parts II and III).

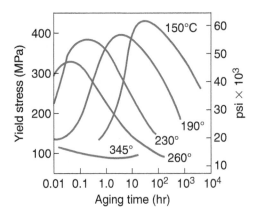

FIGURE 4.20   The effect of aging time and temperature on the yield stress of 2014-T4 aluminum alloy; note that, for each temperature, there is an optimal aging time for maximum strength.

4.10   Case Hardening

The heat-treatment processes described thus far involve microstructural alterations and property changes in the bulk of the component by means of *through hardening.* It is not always desirable to through harden parts, because a hard part lacks the required toughness for some applications. For example, a small surface crack could propagate rapidly through a part and cause sudden and total failure. In many cases, modification of only the *surface properties* of a part (hence, the term *surface* or *case hardening*) is desirable. This widely used method is particularly useful for improving resistance to surface indentation, fatigue, and wear. Typical applications for case hardening are gear teeth, cams, shafts, bearings, fasteners, pins, automotive clutch plates, tools, and dies.

Several case-hardening processes are available (Table 4.1):

1. **Carburizing** (gas, liquid, and pack carburizing)
2. **Carbonitriding**
3. **Cyaniding**
4. **Nitriding**
5. **Boronizing**
6. **Flame hardening**
7. **Induction hardening**
8. **Laser-beam hardening**

Basically, these are operations where the component is heated in an atmosphere containing elements (such as carbon, nitrogen, or boron) that alter the composition, microstructure, and properties of surface layers. For steels with sufficiently high carbon content, surface hardening takes place without using any of these additional elements; only the heat-treatment processes described in Section 4.7 are needed to alter the microstructures, usually by either flame hardening or induction hardening, as outlined in Table 4.1.

TABLE 4.1

Outline of Heat-treatment Processes for Surface Hardening

Process	Metals hardened	Element added to surface	Procedure	General characteristics	Typical applications
Carburizing	Low-carbon steel (0.2% C), alloy steels (0.08–0.2% C)	C	Heat steel at 870°C–950°C (1600°F–1750°F) in an atmosphere of carbonaceous gases (gas carburizing) or carbon-containing solids (pack carburizing); then quench.	A hard, high-carbon surface is produced. Hardness = 55–65 HRC. Case depth = <0.5–1.5 mm (<0.020–0.060 in.); some distortion of part during heat treatment.	Gears, cams, shafts, bearings, piston pins, sprockets, clutch plates
Carbonitriding	Low-carbon steel	C and N	Heat steel at 700°C–800°C (1300°F–1600°F) in an atmosphere of carbonaceous gas and ammonia; then quench in oil.	Surface hardness = 55–62 HRC. Case depth = 0.07–0.5 mm (0.003–0.020 in.); less distortion than in carburizing.	Bolts, nuts, gears
Cyaniding	Low-carbon steel (0.2% C), alloy steels (0.08–0.2% C)	C and N	Heat steel at 760°C–845°C (1400°F–1550°F) in a molten bath of solutions of cyanide (e.g., 30% sodium cyanide) and other salts.	Surface hardness up to 65 HRC. Case depth = 0.025–0.25 mm (0.001–0.010 in.); some distortion.	Bolts, nuts, screws, small gears
Nitriding	Steels (1% Al, 1.5% Cr, 0.3% Mo), alloy steels (Cr, Mo), stainless steels, high-speed tool steels	N	Heat steel at 500°C–600°C (925°F–1100°F) in an atmosphere of ammonia gas or mixtures of molten cyanide salts; no further treatment.	Surface hardness up to 1100 HV. Case depth = 0.1–0.6 mm (0.005–0.030 in.) and 0.02–0.07 mm (0.001–0.003 in.) for high-speed steel.	Gears, shafts, sprockets, valves, cutters, boring bars, fuel-injection pump parts
Boronizing	Steels	B	Part is heated using boron-containing gas or solid in contact with part.	Extremely hard and wear-resistant surface. Case depth = 0.025–0.075 mm (0.001–0.003 in.).	Tool and die steels
Flame hardening	Medium-carbon steels, cast irons	None	Surface is heated with an oxyacetylene torch, then quenched with water spray or other quenching methods.	Surface hardness = 50–60 HRC. Case depth = 0.7–6 mm (0.030–0.25 in.); little distortion.	Gear and sprocket teeth, axles, crankshafts, piston rods, lathe beds and centers
Induction hardening	Same as above	None	Metal part is placed in copper induction coils and is heated by high frequency current; then quenched.	Same as above	Same as above

Laser beams and **electron beams** (Sections 27.6 and 27.7) are used effectively to harden small as well as large surfaces, such as gears, valves, punches, and engine cylinders. The depth of the case-hardened layer is usually less than 2.5 mm (0.1 in.). These methods are also used for through hardening of relatively small parts. The main advantages of laser surface hardening are close control of power input, low part distortion, and the ability to reach areas that may be inaccessible by other means.

Because case hardening involves a localized layer, case-hardened parts have a *hardness gradient*. Typically, the hardness is maximum at the surface and decreases inward, the rate of decrease depending on the composition and physical properties of the metal and processing variables. Surface-hardening techniques can also be used for *tempering* (Section 4.11) and for modifying the properties of surfaces that have been subjected to heat treatment. Several other processes and techniques for surface hardening, such as shot peening and surface rolling, to improve wear resistance and other characteristics, are described in Section 34.2.

Decarburization is the phenomenon in which alloys lose carbon from their surfaces as a result of heat treatment or of hot working in a medium, usually oxygen, that reacts with the carbon. Decarburization is undesirable because it affects the hardenability of the surfaces of a part, by lowering its carbon content; it also adversely affects the hardness, strength, and fatigue life of steels, significantly lowering their endurance limit.

4.11 Annealing

Annealing is a general term used to describe the restoration of a cold-worked or heat-treated alloy to its original properties—for instance, to increase ductility (and hence formability) and to reduce hardness and strength, or to modify the microstructure of the alloy. The annealing process is also used to relieve residual stresses in a part, as well as to improve machinability and dimensional stability. The term "annealing" also applies to the thermal treatment of glass and similar products (Section 18.4), castings, and weldments.

The annealing process consists of the following steps:

1. Heating the workpiece to a specific temperature in a furnace
2. Holding it at that temperature for a period of time (soaking)
3. Cooling the workpiece, in air or in a furnace

The annealing process may be carried out in an inert or a controlled atmosphere, or it may be performed at lower temperatures to minimize or prevent surface oxidation.

The *annealing temperature* may be higher than the material's recrystallization temperature, depending on the degree of cold work. For example, the recrystallization temperature for copper ranges between 200° and 300°C (400° and 600°F), whereas the annealing temperature required to fully recover the original properties ranges from 260° to 650°C (500° to 1200°F), depending on the degree of prior cold work (see also Section 1.7).

Full annealing is a term applied to ferrous alloys. The steel is heated to above A_1 or A_3 (Fig. 4.21) and the cooling takes place slowly [typically at 10°C (20°F) per hour], in a furnace, after which it is turned off. The structure obtained through full annealing is coarse pearlite, which is soft and ductile and has small, uniform grains.

To avoid excessive softness from annealing of steels, the cooling cycle may be done completely in still air. This process is called **normalizing**, to indicate that

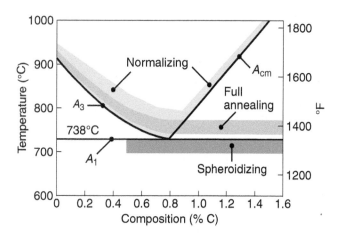

FIGURE 4.21 Heat-treating temperature ranges for plain-carbon steels, as indicated on the iron–iron carbide phase diagram.

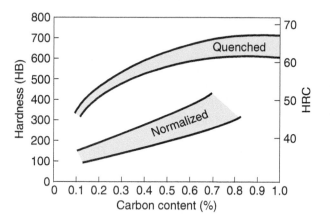

FIGURE 4.22 Hardness of steels in the quenched and normalized conditions as a function of carbon content.

the part is heated to a temperature above A_3 or A_{cm} in order to transform the structure to austenite. Normalizing results in somewhat higher strength and hardness, and lower ductility, than does full annealing (Fig. 4.22). The structure obtained is fine pearlite, with small, uniform grains. Normalizing is generally carried out to refine the grain structure, obtain uniform structure (homogenization), decrease residual stresses, and improve machinability.

The structure of spheroidites and the procedure for obtaining it are described in Section 4.7 and shown in Figs. 4.12 and 4.21. *Spheroidizing annealing* improves the cold workability (Section 14.5) and the machinability of steels (Section 21.7).

Stress-relief Annealing. To reduce or eliminate residual stresses, a workpiece is generally subjected to *stress-relief annealing*, or **stress relieving**. The temperature and time required for this process depend on the material and on the magnitude of the residual stresses present. The residual stresses may have been induced during forming, machining, or caused by volume changes during phase transformations.

Tempering. If steels are hardened by heat treatment, *tempering* (also called **drawing**) is used in order to reduce brittleness, increase ductility and toughness, and reduce residual stresses. The term "tempering" is also used for glasses (Section 8). In tempering, the steel is heated to a specific temperature, depending on its composition, and then cooled at a prescribed rate. The results of tempering for an oil-quenched AISI 4340 steel are shown in Fig. 4.23.

Alloy steels may undergo **temper embrittlement**, which is caused by the segregation of impurities along the grain boundaries, at temperatures between 480°C and 590°C (900°F and 1100°F).

Austempering. In *austempering*, the heated steel is quenched from the austenitizing temperature rapidly, to avoid formation of ferrite or pearlite. It is held at a certain temperature until isothermal transformation from austenite to bainite is complete. It is then cooled to room temperature, usually in still air and at a moderate rate in order to avoid thermal gradients within the part. The quenching medium most commonly used is molten salt, at temperatures ranging from 160°C to 750°C (320°F to 1380°F).

Austempering is often substituted for conventional quenching and tempering, either to reduce the tendency for cracking and distortion during quenching or to improve ductility and toughness while maintaining hardness. Because of the shorter cycle time involved, this process is also economical.

Martempering (Marquenching). In *martempering*, steel or cast iron is first quenched from the austenitizing temperature in a hot fluid medium, such as hot oil or molten salt. Next, it is held at that temperature until the temperature is uniform throughout the part. It is cooled at a moderate rate, such as in air, in order to avoid excessive temperature gradients within the part. Usually, the part is subsequently tempered, because the structure obtained is otherwise primarily untempered martensite, and thus not suitable for most applications.

Martempered steels have lower tendency to crack, distort, or develop residual stresses during heat treatment. In **modified martempering**, the quenching temperature is lower, and thus the cooling rate is higher. The process is suitable for steels with lower hardenability.

Ausforming. In *ausforming*, also called **thermomechanical processing**, the steel is formed into desired shapes, within controlled ranges of temperature and time to avoid formation of nonmartensitic transformation products. The part is then cooled at various rates to obtain the desired microstructures.

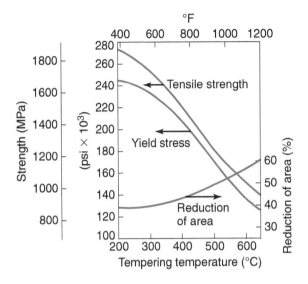

FIGURE 4.23 Mechanical properties of oil-quenched 4340 steel as a function of tempering temperature.

4.12 Heat-treating Furnaces and Equipment

Two basic types of furnaces are used for heat treating: batch furnaces and continuous furnaces. Because they consume much energy, their insulation and efficiency are important design considerations, as are their initial cost, the personnel needed for their operation and maintenance, and their safe use.

Uniform temperature and accurate control of temperature–time cycles are important. Modern furnaces are equipped with various electronic controls, including computer-controlled systems, programmed to run through a complete heat-treating cycle repeatably and with reproducible accuracy. The fuels used are usually natural gas, oil, or electricity (for resistance or induction heating); the type of fuel affects the furnace's atmosphere. Unlike electric heating, gas or oil introduces combustion products into the furnace. Electrical heating, however, has a slower start-up time and is more difficult to adjust and control.

Batch Furnaces. In a *batch furnace*, the parts to be heat treated are loaded into and unloaded from the furnace in individual batches. The furnace basically consists of an insulated chamber, a heating system, and an access door or doors. Batch furnaces are of the following basic types:

1. A **box furnace** is a horizontal rectangular chamber, with one or two access doors through which parts are loaded.
2. A **pit furnace** is a vertical pit below ground level into which the parts are lowered.
3. A **bell furnace** is a round or rectangular box furnace without a bottom, and is lowered over stacked parts that are to be heat treated; this type of furnace is particularly suitable for coils of wire, rods, and sheet metal.
4. In an **elevator furnace**, the parts to be heat treated are loaded onto a car platform, rolled into position, and then raised into the furnace.

Continuous Furnaces. In this type of furnace, the parts to be heat treated move continuously through the furnace on conveyors of various designs.

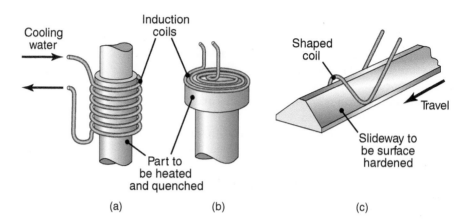

FIGURE 4.24   Types of coils used in induction heating of various surfaces of parts.

Salt-bath Furnaces.   Because of their high heating rates and better control of uniformity of temperature, *salt baths* are commonly used in various heat-treating operations, particularly for nonferrous strip and wire. Heating rates are high because of the higher thermal conductivity of liquid salts as compared with that of air or gases.

Fluidized Beds.   Dry, fine, and loose solid particles, usually aluminum oxide, are heated and suspended in a chamber by an upward flow of hot gas at various speeds. The parts to be heat treated are then placed within the floating particles, hence the term *fluidized bed*.

Induction Heating.   In this method, the part is heated rapidly by the electromagnetic field generated by an *induction coil* carrying alternating current, which induces eddy currents in the part. The coil, which can be shaped to fit the contour of the part to be heat treated (Fig. 4.24), is made of copper or of a copper-base alloy. The coil, which is usually water cooled, may be designed to quench the part as well.

Furnace Atmospheres.   The atmospheres in furnaces can be controlled so as to avoid oxidation, tarnishing, and decarburization of ferrous alloys heated to elevated temperatures. Oxygen causes corrosion, rusting, and scaling. Carbon dioxide, which has various effects, may be neutral or decarburizing, depending on its concentration in the furnace atmosphere. Nitrogen is a common neutral atmosphere, and a vacuum provides a completely neutral atmosphere. Water vapor in the furnace causes oxidation of steels, resulting in a blue color. The term **bluing** is used to describe formation of a thin, blue film of oxide on finished parts to improve their appearance and their resistance to oxidation.

QR Code 4.3 Demonstration of induction heating. (*Source:* Courtesy of GH Induction Atmospheres)

4.13   Design Considerations for Heat Treating

In addition to metallurgical factors, successful heat treating involves design considerations for avoiding such problems as cracking, distortion, and nonuniformity of properties throughout and among heat-treated parts. The rate of cooling during quenching may not be uniform, particularly in complex shapes having varying cross-sections and thicknesses, producing severe temperature gradients in the part.

Nonuniformity can lead to variations in contraction, resulting in thermal stresses that may cause warping or cracking of the part. Nonuniform cooling also causes residual stresses in the part, which then can lead to stress-corrosion cracking. The quenching method selected, the care taken during the operation, and the selection of a proper quenching medium and temperature also are important considerations.

As a general guideline for part design for heat treating.

- Sharp internal or external corners should be avoided, as otherwise stress concentrations at these corners may raise the level of stresses high enough to cause cracking.
- The part should have its thicknesses as nearly uniform as possible.
- The transition between regions of different thicknesses should be made smooth.
- Parts with holes, grooves, keyways, splines, and asymmetrical shapes may be difficult to heat treat, because they may crack during quenching.
- Large surfaces with thin cross-sections are likely to warp.
- Hot forgings and hot steel-mill products may have a *decarburized skin* which may not respond successfully to heat treatment.

SUMMARY

- Commercially pure metals generally do not have sufficient strength for most engineering applications; consequently, they must be alloyed with various elements which alter their structures and properties. Important concepts in alloying are the solubility of alloying elements in a host metal and the phases present at various ranges of temperature and composition.
- Alloys basically have two forms: solid solutions and intermetallic compounds. Solid solutions may be substitutional or interstitial. There are certain conditions pertaining to the crystal structure and atomic radii that have to be met in order to develop these structures.
- Phase diagrams show the relationships among the temperature, composition, and phases present in a particular alloy system. As temperature is decreased at various rates, correspondingly various transformations take place, resulting in microstructures that have widely different characteristics and properties.
- Among binary systems, the most important is the iron–carbon system, which includes a wide range of steels and cast irons. Important components in this system are ferrite, austenite, and cementite. The basic types of cast irons are gray iron, ductile (nodular) iron, white iron, malleable iron, and compacted-graphite iron.
- The mechanisms for hardening and strengthening metal alloys basically involve heating the alloy and subsequently quenching it at varying cooling rates. Important phase transformations then take place, producing structures such as pearlite (fine or coarse), spheroidite, bainite, and martensite. Heat treating of nonferrous alloys and stainless steels involves solution treatment and precipitation hardening.
- The furnace atmosphere, the quenchants used, the control and characteristics of the equipment, and the shape of the parts to be heat treated are important heat-treatment considerations
- Hardenability is the capability of an alloy to be hardened by heat treatment. The Jominy end-quench hardenability test is a method commonly used to determine hardenability bands for alloys.

- Case hardening is an important process for improving the wear and fatigue resistance of parts. Several methods are available, such as carburizing, nitriding, induction hardening, and laser hardening.
- Annealing includes normalizing, process annealing, stress relieving, tempering, austempering, and martempering, each with the purpose of enhancing the ductility and toughness of heat-treated parts.

KEY TERMS

Age hardening	Distortion	Jominy test	Solid solution
Aging	End-quench test	Maraging	Solute
Alloy	Equilibrium diagram	Martempering	Solution treatment
Annealing	Eutectic point	Martensite	Solvent
Austempering	Eutectoid reaction	Normalizing	Spheroidite
Austenite	Ferrite	Over-aging	Stress relieving
Bainite	Furnaces	Pearlite	Tempered martensite
Case hardening	Hardenability	Phase diagram	Tempering
Cast iron	Heat treatment	Phase transformations	Time–temperature–
Cementite	Intermetallic	Precipitation hardening	transformation
Curie temperature	compounds	Pure metals	diagrams
Decarburization	Iron–carbon system	Retained austenite	Two-phase systems

BIBLIOGRAPHY

ASM Handbook, Vol. 3: **Alloy Phase Diagrams**, ASM International, 1992.

ASM Handbook, Vol. 4: **Heat Treating**, ASM International, 1991.

ASM Handbook, Vol. 9: **Metallography and Microstructures**, ASM International, 2004.

Brooks, C.R., **Principles of the Heat Treatment of Plain Carbon and Low Alloy Steel**, ASM International, 1996.

Bryson, W.E., **Heat Treatment, Selection, and Application of Tool Steels**, 2nd ed., Hanser Gardner, 2005.

Campbell, F.C., (ed.), **Phase Diagrams: Understanding the Basics**, ASM International, 2012.

Heat Treater's Guide: Practices and Procedures for Irons and Steels, 2nd ed., ASM International, 1995.

Heat Treater's Guide: Practices and Procedures for Nonferrous Alloys, ASM International, 1996.

Krauss, G., **Steels: Processing, Structure, and Performance**, ASM International, 2005.

Totten, G.E., **Steel Heat Treatment Handbook**, 2 Vols., 2nd ed., CRC Press/Taylor & Francis, 2007.

REVIEW QUESTIONS

4.1 Describe the difference between a solute and a solvent.

4.2 What is a solid solution?

4.3 What are the conditions for obtaining (a) substitutional and (b) interstitial solid solutions?

4.4 Describe the difference between a single-phase and a two-phase system.

4.5 What is an induction heater? What kind of part shapes can be heated by induction heating?

4.6 Describe the major features of a phase diagram.

4.7 What do the terms "equilibrium" and "constitutional," as applied to phase diagrams, indicate?

4.8 What is the difference between "eutectic" and "eutectoid"?

4.9 What is tempering? Why is it performed?

4.10 Explain what is meant by "severity of quenching."

4.11 What are precipitates? Why are they significant in precipitation hardening?

4.12 What is the difference between natural and artificial aging?

4.13 Describe the characteristics of ferrite, austenite, and cementite.

4.14 What is the purpose of annealing?

4.15 What is a time–temperature–transformation diagram? How is it used?

QUALITATIVE PROBLEMS

4.16 You may have seen some technical literature on products stating that certain parts in those products are "heat treated." Describe briefly your understanding of this term and why the manufacturer includes it.

4.17 Describe the engineering significance of the existence of a eutectic point in phase diagrams.

4.18 What is the difference between hardness and hardenability?

4.19 Referring to Table 4.1, explain why the items listed under typical applications are suitable for surface hardening.

4.20 It generally is not desirable to use steels in their as-quenched condition. Explain why.

4.21 Describe the differences between case hardening and through hardening, insofar as engineering applications of metals are concerned.

4.22 Describe the characteristics of (a) an alloy, (b) pearlite, (c) austenite, (d) martensite, and (e) cementite.

4.23 Explain why carbon, among all elements, is so effective in imparting strength to iron in the form of steel.

4.24 How does the shape of graphite in cast iron affect its properties?

4.25 In Section 4.8.2, several fluids are listed in terms of their cooling capacity in quenching. Which physical properties of these fluids influence their cooling capacity?

4.26 Why is it important to know the characteristics of heat-treating furnaces? Explain.

4.27 Explain why, in the abscissa of Fig. 4.16c, the percentage of pearlite begins to decrease after 0.8% carbon content is reached.

4.28 What is the significance of decarburization? Give some examples.

4.29 Explain your understanding of size distortion and shape distortion in heat-treated parts, and describe their causes.

4.30 Comment on your observations regarding Fig. 4.18b.

QUANTITATIVE PROBLEMS

4.31 Design a heat-treating cycle for carbon steel, including temperature and exposure times, to produce (a) pearlite–martensite steels and (b) bainite–martensite steels.

4.32 Using Fig. 4.4, estimate the following quantities for a 75% Cu–25% Ni alloy: (a) the liquidus temperature, (b) the solidus temperature, (c) the percentage of nickel in the liquid at 1150°C (2102°F), (d) the major phase at 1150°C, and (e) the ratio of solid to liquid at 1150°C.

4.33 Extrapolating the curves in Fig. 4.14, estimate the time that it would take for 1080 steel to soften to 40 HRC at (a) 300°C and (b) 400°C.

4.34 A typical steel for tubing is AISI 1040, and one for music wire is 1085. Considering their applications, explain the reason for the difference in carbon content.

SYNTHESIS, DESIGN, AND PROJECTS

4.35 It was stated in this chapter that, in parts design, sharp corners should be avoided in order to reduce the tendency toward cracking during heat treatment. If it is essential for a part to have sharp corners for functional purposes, and it still requires heat treatment, what method would you recommend for manufacturing this part?

4.36 The heat-treatment processes for surface hardening are summarized in Table 4.1. Each of these processes involves different equipment, procedures, and cycle times; as a result, each incurs different costs. Review the available literature, contact various companies, and then make a similar table outlining the costs involved in each process.

4.37 It can be seen that, as a result of heat treatment, parts can undergo size distortion and shape distortion to various degrees. By referring to the Bibliography at the end of this chapter, make a survey of the technical literature and report quantitative data regarding the distortions of parts having different shapes.

4.38 Figure 4.18b shows hardness distributions in end-quench tests, as measured along the length of the round bar. Make a simple qualitative sketch showing the hardness distribution across the diameter of the bar. Would the shape of the curve depend on the bar's carbon content? Explain.

4.39 Throughout this chapter, you have seen specific examples of the importance and the benefits of heat-treating parts or certain regions of parts. Refer to the Bibliography at the end of this chapter, make a survey of the heat-treating literature, and then compile several examples and illustrations of parts that have been heat treated.

4.40 Refer to Fig. 4.24, and think of a variety of other part shapes to be heat treated, and design coils that are appropriate for these shapes. Describe how different your designs would be if the parts have varying shapes along their length (such as from a square at one end to a round shape at the other end).

4.41 Inspect various parts in your car or home, and identify those that are likely to have been case hardened. Explain your reasons.

CHAPTER 5

Ferrous Metals and Alloys: Production, General Properties, and Applications

- Ferrous metals and alloys are the most widely used structural materials, generally because of their properties and performance, ease of manufacturing, and low cost.
- The chapter opens with a brief outline of iron and steel production from ore, including descriptions of furnaces and foundry operations.
- The casting of ingots is described, followed by continuous casting operations, which are now into wide use for producing high-quality metals in large volumes.
- The chapter then discusses in detail the properties and applications of ferrous metals, including cast irons, carbon and alloy steels, stainless steels, and tool and die steels.

5.1 Introduction

By virtue of their relatively low cost and wide range of mechanical, physical, and chemical properties, **ferrous metals and alloys** are among the most useful of all metals. They contain iron as their base metal and are generally classified as *carbon and alloy steels*, *stainless steels*, *tool and die steels*, *cast irons*, and *cast steels*. **Steel** refers to a ferrous alloy, which can be as simple as an alloy of iron and carbon, but also often containing a number of alloying elements to impart various properties. Ferrous alloys are produced as

- Sheet steel, for automobiles, appliances, and containers
- Plates, for boilers, ships, and bridges
- Structural members, such as I-beams, bar products, axles, crankshafts, and railroad rails
- Tools, dies, and molds
- Rods and wire, for fasteners such as bolts, rivets, nuts, and staples

Carbon steels are the least expensive of all structural metals. As an example of their widespread use, ferrous metals make up 70–85% by weight of structural members and mechanical components. The average U.S. passenger vehicle (including trucks and sport utility vehicles) contains about 1200 kg (2700 lb) of steel, accounting for about 60% of its total weight.

The use of iron and steel as structural materials has been one of the most important technological developments. Primitive ferrous tools, which first appeared about 4000 to 3000 B.C., were made from meteoritic iron, obtained from meteorites that had struck the earth. True ironworking began in Asia Minor in about 1100 B.C. and signaled the advent of the *Iron Age*. Invention of the blast furnace in about 1340 A.D. made possible the production of large quantities of high-quality iron and steel. (See Table I.2.)

5.2 Production of Iron and Steel

5.2.1 Raw Materials

The three basic materials used in iron- and steelmaking are **iron ore, limestone,** and **coke.** Although it does not occur in a free state in nature, *iron* is one of the most abundant elements (in the form of various ores) in the world, making up about 5% of the earth's crust. The principal iron ores are (a) *taconite* (a black flintlike rock), (b) *hematite* (an iron-oxide mineral), and (c) *limonite* (an iron oxide containing water). After it is mined, the ore is crushed into fine particles, the impurities are removed (by various means such as magnetic separation), and the ore is formed into pellets, balls, or briquettes, using water and various binders. Typically, pellets are about 65% pure iron and about 25 mm (1 in.) in diameter. The concentrated iron ore is referred to as *beneficiated* (as are other concentrated ores). Some iron-rich ores are used directly, without pelletizing.

Coke is obtained from special grades of bituminous coal (a soft coal rich in volatile hydrocarbons and tars) that are heated in vertical ovens to temperatures of up to 1150°C (2100°F), and then cooled with water in quenching towers. Coke has several functions in steelmaking, including (a) generating the high level of heat required for the chemical reactions in ironmaking to take place and (b) producing carbon monoxide (a reducing gas, meaning that it removes oxygen), to reduce iron oxide to iron. The chemical by-products of coke are used in the synthesis of plastics and of chemical compounds; the gases evolved during the conversion of coal to coke are used as fuel for plant operations.

The function of *limestone* (calcium carbonate) is to remove impurities from the molten iron. The limestone reacts chemically with impurities, acting like a **flux** (meaning to flow as a fluid) that causes the impurities to melt at a low temperature. The limestone combines with the impurities and forms a **slag** (which, being light, floats over the molten metal and subsequently is removed). *Dolomite* (an ore of calcium magnesium carbonate) also is used as a flux. The slag is later used in making cement, fertilizers, glass, building materials, rock-wool insulation, and road ballast.

5.2.2 Ironmaking

The three raw materials described above are dumped into the top of a **blast furnace** (Fig. 5.1), an operation called *charging the furnace*. A blast furnace is basically a large steel cylinder lined with refractory (heat-resistant) brick; it has the height of about a 10-story building. The charge mixture is melted in a reaction at 1650°C (3000°F), with air preheated to about 1100°C (2000°F) and *blasted* into the furnace (hence the term "blast furnace"), through nozzles called *tuyeres*. Although a number of other reactions may take place, the basic reaction is that of oxygen combining with carbon to produce carbon monoxide, which, in turn, reacts with the iron oxide, reducing it to **iron.** Preheating the incoming air is necessary because the burning coke alone does not produce sufficiently high temperatures for these reactions to occur.

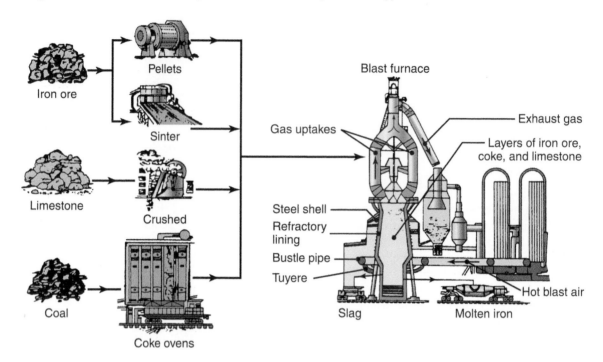

FIGURE 5.1   Schematic illustration of a blast furnace.

The molten metal accumulates at the bottom of the blast furnace, while the impurities float to the top. At intervals of four to five hours, the molten metal is drawn off (*tapped*) into ladle cars, each holding as much as 160 tons of molten iron. The molten metal at this stage is called **pig iron**, or simply **hot metal**, and has a typical composition of 4% C, 1.5% Si, 1% Mn, 0.04% S, 0.4% P, with the rest being iron. The word **pig** comes from the early practice of pouring the molten iron into small sand molds arranged around a main channel, reminding early ironworkers of a litter of small pigs crowding against their mother sow. The solidified metal is then used in making iron and steels.

5.2.3   Steelmaking

Steel was first produced in China and Japan about 600 to 800 A.D. The steelmaking process is essentially one of **refining** the pig iron by (a) reducing the percentages of manganese, silicon, carbon, and other elements and (b) controlling the composition of the output through the addition of various elements. The molten metal from the blast furnace is transported into one of four types of furnaces: **open-hearth, electric, vacuum,** or **basic-oxygen**. The name "open-hearth" derives from the shallow hearth shape that is open directly to the flames that melt the metal. Developed in the 1860s, the open-hearth furnace essentially has been replaced by electric furnaces and by the basic-oxygen process, because the latter two are more efficient and produce steels of better quality.

Electric Furnace.   The source of heat in this type of furnace is a continuous electric arc that is formed between the electrodes and the charged metal (Figs. 5.2a and b); temperatures as high as 1925°C (3500°F) are generated. There are usually three

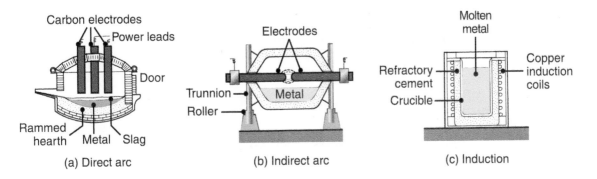

Carbon electrodes
Power leads
Door
Rammed hearth   Metal   Slag
(a) Direct arc

Electrodes
Trunnion
Metal
Roller
(b) Indirect arc

Molten metal
Refractory cement
Crucible
Copper induction coils
(c) Induction

FIGURE 5.2 Schematic illustration of types of electric furnaces.

graphite electrodes, and they can be as large as 750 mm (30 in.) in diameter and 1.5–2.5 m (5–8 ft) in length. Their height in the furnace can be adjusted in response to the amount of metal present and the amount of wear of the electrodes.

Steel scrap and a small amount of carbon and limestone are first dropped into the electric furnace through the open roof. The roof is then closed and the electrodes are lowered. The power is turned on, and within a period of about two hours, temperatures increase sufficiently to melt the metal. The current is then shut off, the electrodes are raised, the furnace is tilted, and the molten metal is poured into a *ladle* (a receptacle used for transferring and pouring molten metal). Electric-furnace capacities range from 60 to 90 tons of steel per day, and the quality of steel produced is better than that from either the open-hearth or the basic-oxygen process (see below).

For smaller quantities, electric furnaces can be of the **induction** type (Fig. 5.2c). The metal is placed in a **crucible,** a large pot made of refractory material and surrounded with a copper coil through which alternating current is passed. The induced current in the charge generates heat and melts the metal.

Basic-oxygen Furnace.   The basic-oxygen furnace (BOF) is the fastest and by far the most common steelmaking furnace. Typically, 200 tons of molten pig iron and 90 tons of scrap are charged into a vessel (Fig. 5.3a); some units can hold as much as 350 tons. Pure oxygen is then blown into the furnace, for about 20 min, through a water-cooled *lance* (a long tube) and under a pressure of about 1250 kPa (180 psi), as shown in Fig. 5.3b. Fluxing agents (such as calcium or magnesium oxide) are added through a chute. The process is known as *basic* because of the pH of these fluxing agents.

The vigorous agitation of the oxygen refines the molten metal by an oxidation process, in which iron oxide is produced. The oxide reacts with the carbon in the molten metal, producing carbon monoxide and carbon dioxide. The lance is then retracted, and the furnace is tapped by tilting it (note the opening in Fig. 5.3c for the molten metal); then the slag is removed by tilting the furnace in the opposite direction. The BOF process is capable of refining 250 tons of steel in 35–50 min. Most BOF steels, which have low impurity levels and are of better quality than open-hearth furnace steels, are processed into plates, sheets, and various structural shapes, such as I-beams and channels (see Fig. 13.1).

Vacuum Furnace.   Steel may also be melted in induction furnaces from which the air has been removed, similar to the one shown in Fig. 5.2c. Cooling is accomplished by injecting an inert gas (typically argon) at high pressure into the furnace. Because the operation removes gaseous impurities from the molten metal and prevents oxidation,

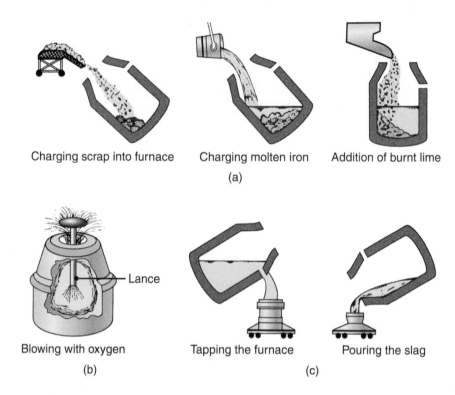

Charging scrap into furnace   Charging molten iron   Addition of burnt lime

(a)

Blowing with oxygen

(b)

— Lance

Tapping the furnace   Pouring the slag

(c)

FIGURE 5.3   Schematic illustrations showing charging, melting, and pouring of molten iron in a basic-oxygen process.

vacuum furnaces produce high-quality steels. They are also commonly used for heat treating (Section 4.7) and brazing (Section 32.2).

5.3   Casting of Ingots

QR Code 5.1 Animation of ingot casting. (*Source:* Courtesy of Sandvik Coromant)

Traditionally, the steelmaking process involves the shaping of the molten steel into a solid form (**ingot**), for such further processing as rolling it into shapes, casting into semifinished forms, or forging. The molten metal is poured (*teemed*) from the ladle into ingot molds, in which the metal solidifies. Molds usually are made of cupola iron or blast-furnace iron with 3.5% carbon, and are tapered in order to facilitate the removal of the solidified metal. The bottoms of the molds may be closed or open; if they are open, the molds are placed on a flat surface. The cooled ingots are then stripped from the molds and lowered into **soaking pits**, where they are reheated to a uniform temperature of about 1200°C (2200°F) prior to subsequent processing.

Certain reactions take place during the solidification of an ingot that have an important influence on the quality of the steel produced. For example, significant amounts of oxygen and other gases can dissolve in the molten metal during steelmaking. Most of these gases are rejected during the solidification of the metal, because the solubility limit of the gases in the metal decreases sharply as its temperature decreases (see Fig. 10.15). The rejected oxygen combines with carbon to form carbon monoxide, which causes porosity in the solidified ingot.

Depending on the amount of gas evolved during solidification, three types of steel ingots can be produced: killed, semi-killed, and rimmed.

1. **Killed Steel.** The term *killed* comes from the fact that the steel lies quietly after being poured into the mold. **Killed steel** is a fully deoxidized steel; that is, oxygen is removed and the associated porosity is thus eliminated. In the deoxidation process, the oxygen dissolved in the molten metal is made to react with elements such as aluminum, silicon, manganese, and vanadium that have been added to the melt. These elements have an affinity for oxygen and form metallic oxides. If aluminum is used, the product is called *aluminum-killed steel* (see Table 16.4).

   If they are sufficiently large, the oxide inclusions in the molten bath float out and adhere to, or are dissolved in, the slag. A fully killed steel thus is free of any porosity caused by gases; it also is free of any **blowholes** (large spherical holes near the surfaces of the ingot). Consequently, the chemical and mechanical properties of a killed-steel ingot are relatively uniform throughout. Because of shrinkage during solidification, however, an ingot of this type develops a **pipe** at the top (also called a **shrinkage cavity**); it has the appearance of a funnel-like shape. This pipe can take up a substantial volume of the ingot, and it has to be cut off and scrapped.

2. **Semi-killed Steel.** **Semi-killed steel** is a *partially deoxidized steel.* It contains some porosity (generally in the upper central section of the ingot), but it has little or no pipe. Although the piping in semi-killed steels is less, this advantage is offset by the presence of porosity in that region. Semi-killed steels are economical to produce.

3. **Rimmed Steel.** In a **rimmed steel**, which generally has a carbon content of less than 0.15%, the evolved gases are only partially killed (or controlled) by the addition of other elements, such as aluminum. The gases produce blowholes along the outer rim of the ingot—hence the term *rimmed.* These steels have little or no piping, and they have a ductile skin with good surface finish; however, if not controlled properly, blowholes may break through the skin. Furthermore, impurities and inclusions tend to segregate toward the center of the ingot. Products made from this steel may thus be defective, hence thorough inspection is essential.

Refining. The properties and manufacturing characteristics of ferrous alloys are affected adversely by the amount of impurities, inclusions, and other elements present (see Section 2.10). The removal of impurities is known as *refining*. Most refining is done in melting furnaces or in ladles, by the addition of various elements.

Refining is particularly important in producing high-grade steels and alloys for high-performance and critical applications, such as aircraft components, automobile structural elements, medical devices, and cutlery. Moreover, warranty periods on shafts, camshafts, crankshafts, and similar parts can be increased significantly by using higher quality steels. Such steels are subjected to **secondary refining** in ladles (**ladle metallurgy**) and ladle refining (**injection refining**), which generally consists of melting and processing the steel in a vacuum. Several processes using controlled atmospheres have been developed, such as electron-beam melting, vacuum-arc remelting, argon–oxygen decarburization, and vacuum-arc double-electrode remelting.

5.4 Continuous Casting

Conceived in the 1860s, **continuous** or **strand casting** was first developed for casting nonferrous metal strips. The process is now used widely for steel production, with major productivity improvements and cost reductions. One system for continuous casting is shown schematically in Fig. 5.4a. The molten metal in the ladle is cleaned,

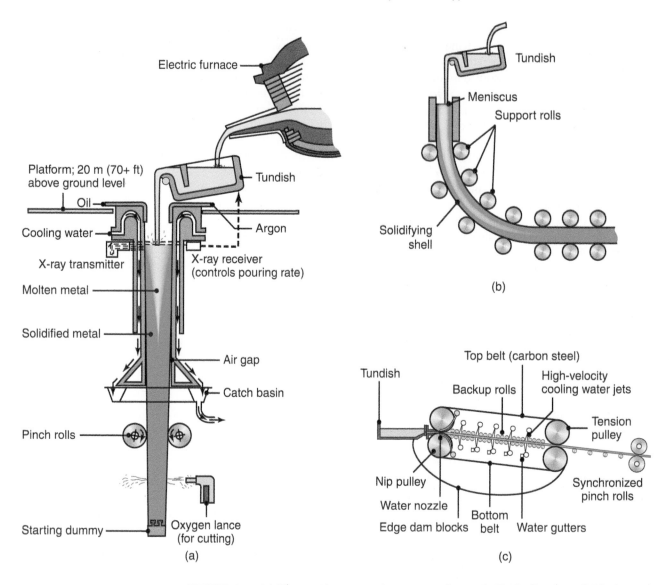

FIGURE 5.4 (a) The continuous-casting process for steel. Typically, the solidified metal descends at a speed of 25 mm/s (1 in./s); note that the platform is about 20 m (65 ft) above ground level. *Source:* Figure adapted from *Metalcaster's Reference and Guide* (c. 1989, p. 41), American Foundrymen's Society. (b) Continuous casting using support or guide rollers to allow transition from a vertical pour zone to horizontal conveyors. (c) Continuous strip casting of nonferrous metal strip. *Source:* Courtesy of Hazelett Corporation.

QR Code 5.2 Animation of continuous casting. (*Source:* Courtesy of Sandvik Coromant)

then it is equalized in temperature by blowing nitrogen gas through it for 5–10 min. The metal is then poured into a refractory-lined intermediate pouring vessel (**tundish**), where impurities are skimmed off. The tundish holds as much as three tons of metal. The molten metal is then tapped from the tundish, travels downward through water-cooled copper molds, and begins to solidify as it is drawn through the molds at a constant velocity by rollers (called *pinch rolls*).

Before starting the casting operation, a solid *starter bar* (*dummy bar*) is inserted into the bottom of the mold. When the molten metal is first poured, it solidifies onto the dummy bar. The bar is withdrawn at the same rate at which the metal is poured. The cooling rate is such that the metal develops a solidified skin (*shell*), so as to support itself during its travel downward, typically at speeds of about 25 mm/s (1 in./s). The shell thickness at the exit end of the mold is about 12–18 mm (0.5–0.75 in.). Additional cooling is provided by water sprays along the travel path of the solidifying metal. The molds generally are coated with graphite or similar solid lubricants, in order to reduce both friction and adhesion at the mold–metal interfaces; also, the molds are vibrated in order to reduce friction and sticking.

The continuously cast metal may be cut into desired lengths by shearing or computer-controlled torch cutting, or it may be fed directly into a rolling mill for further reduction in thickness and for the shaping of products, such as channels and I-beams. In addition to lower cost, continuously cast metals have more uniform compositions and properties than those obtained by ingot casting.

Modern facilities include computer-controlled operations on continuously cast strands, with final sheet thicknesses on the order of 2–6 mm (0.08–0.25 in.) for carbon and stainless steels. They have capabilities for a rapid switchover from one type of steel to another. Steel plates or other shapes undergo one or more further processing, such as (a) cleaning and pickling by chemicals, to remove surface oxides, (b) cold rolling, to improve strength and surface finish, (c) annealing, and (d) galvanizing or aluminizing, to improve resistance to corrosion.

In **strip casting**, thin slabs, or strips, are produced from molten metal. The metal solidifies in similar manner to strand casting, but the hot solid is then rolled to produce the final shape (Fig. 5.4b). The compressive stresses in rolling (see Section 13.2) serve to reduce porosity and to provide better material properties. In effect, strip casting eliminates a hot-rolling operation in the production of metal strips or slabs.

5.5   Carbon and Alloy Steels

Steel is an alloy that consists primarily of iron, and has a carbon content between 0.2 and 2.1%, by weight. Alloys with higher than 2.1% carbon are known as cast irons (discussed in Section 12.3.2), and have a lower melting point than other steels and good castability. Carbon and alloy steels are among the most commonly used metals and have a wide variety of compositions, processing options, and applications (Table 5.1). These steels are available in a variety of basic product shapes: plate, sheet, strip, bar, wire, tube, castings, and forgings.

5.5.1   Effects of Various Elements in Steels

Various elements are added to steels in order to impart specific properties, such as hardenability, strength, hardness, toughness, wear resistance, workability, weldability, and machinability. These elements are listed in Table 5.2, with summaries of their beneficial and detrimental effects. Generally, the higher the percentages of these elements in steels, the greater are

TABLE 5.1

Applications for Selected Carbon and Alloy Steels	
Product	Steel
Aircraft forgings, tubing, fittings	4140, 8740
Automobile bodies	1010
Axles	1040, 4140
Ball bearings and races	52100
Bolts	1035, 4042, 4815
Camshafts	1020, 1040
Chains (transmission)	3135, 3140
Coil springs	4063
Connecting rods	1040, 3141, 4340
Crankshafts (forged)	1045, 1145, 3135, 3140
Differential gears	4023
Gears (car and truck)	4027, 4032
Landing gear	4140, 4340, 8740
Lock washers	1060
Nuts	3130
Railroad rails and wheels	1080
Springs (coil)	1095, 4063, 6150
Springs (leaf)	1085, 4063, 9260, 6150
Tubing	1040
Wire	1045, 1055
Wire (music)	1085

TABLE 5.2

Effect of Various Elements in Steels	
Element	Effect
Aluminum	Deoxidizes nitriding steels, limits austenite grain growth, increases hardness of nitriding steels
Bismuth	Improves machinability
Boron	Improves hardness without loss of (and perhaps some improvement in) machinability and formability
Calcium	Deoxidizes steel; improves toughness; may improve formability and machinability
Carbon	Improves hardenability, strength, hardness, and wear resistance; reduces ductility, weldability, and toughness
Cerium, magnesium, zirconium	Deoxidizes steel, improves toughness in HSLA steels; controls shape of inclusions
Chromium	Improves toughness, hardenability, wear and corrosion resistance, and high-temperature strength; promotes carburization and depth of hardening in heat treatment
Cobalt	Improves strength and hardness at elevated temperatures
Copper	Improves resistance to atmospheric corrosion; can increase strength without loss in ductility; adversely affects hot workability and surface quality
Lead	Improves machinability; can cause liquid metal embrittlement
Manganese	Deoxidizes steel; improves hardenability, strength, abrasion resistance, and machinability; reduces hot shortness; decreases weldability
Molybdenum	Improves hardenability, wear resistance, toughness, elevated-temperature strength, creep resistance, and hardness; minimizes temper embrittlement
Nickel	Improves strength, toughness, corrosion resistance, and hardenability
Niobium, tantalum	Improves strength and impact toughness; lowers transition temperature; may decrease hardenability
Phosphorus	Improves strength, hardenability, corrosion resistance, and machinability; severely reduces ductility and toughness
Selenium	Improves machinability
Silicon	Improves strength, hardness, corrosion resistance, and electrical conductivity; decreases machinability and cold formability
Sulfur	Improves machinability when combined with manganese; decreases impact strength, ductility, and weldability
Tellurium	Improves machinability, formability, and toughness
Titanium	Deoxidizes steel; improves hardenability
Tungsten	Improves hardness, especially at elevated temperature
Vanadium	Improves strength, toughness, abrasion resistance, and hardness at elevated temperatures; inhibits grain growth during heat treatment

the particular properties that they impart. For example, the higher the carbon content, the greater the hardenability of the steel and the higher its strength, hardness, and wear resistance. On the other hand, ductility, weldability, and toughness are reduced with increasing carbon content.

Some *residual elements*, called **trace elements**, may remain after production, refining, and processing of steels. Although the elements in Table 5.2 may also be considered as residuals, the following generally are considered unwanted residual elements:

Antimony and **arsenic** cause temper embrittlement.

Hydrogen severely embrittles steels; however, heating during processing drives out most of the hydrogen.

Nitrogen improves strength, hardness, and machinability; in aluminum-deoxidized steels, it controls the size of inclusions. Nitrogen can increase or decrease strength, ductility, and toughness, depending on the presence of other elements.

Oxygen slightly increases the strength of rimmed steels; it severely reduces toughness.

Tin causes hot shortness and temper embrittlement.

5.5.2   Designations for Steels

Traditionally, the American Iron and Steel Institute (AISI) and the Society of Automotive Engineers (SAE) have designated carbon and alloy steels by four digits. The first two digits indicate the alloying elements and their percentages, and the last two digits indicate the carbon content by weight.

The American Society for Testing and Materials (ASTM) has a designation system, which incorporates the AISI and SAE designations and includes standard specifications for steel products. For ferrous metals, the designation consists of the letter "A" followed by a few numbers (generally three). The current standard numbering system is known as the *Unified Numbering System* (UNS), and has been widely adopted by both ferrous and nonferrous industries. It consists of a letter, indicating the general class of the alloy, followed by five digits, designating its chemical composition. Typical letter designations are:

G — AISI and SAE carbon and alloy steels
J — Cast steels
K — Miscellaneous steels and ferrous alloys
S — Stainless steels and superalloys
T — Tool steels

Two examples are: G41300 for AISI 4130 alloy steel, and T30108 for AISI A-8 tool steel.

5.5.3   Carbon Steels

Carbon steels generally are classified by their proportion, by weight, of carbon content. The general mechanical properties of carbon and alloy steels are given in Table 5.3, and the effect of carbon on the properties of steel is shown in Fig. 5.5, and summarized as:

- **Low-carbon steel,** also called **mild steel,** has less than 0.30% C. It often is used for common industrial products (such as bolts, nuts, sheets, plates, and tubes) and for machine components that do not require high strength.
- **Medium-carbon steel** has 0.30–0.60% C. It generally is used in applications requiring higher strength than is available in low-carbon steels, such as in machinery, automotive and agricultural parts (gears, axles, connecting rods, and crankshafts), railroad equipment, and parts for metalworking machinery.
- **High-carbon steel** has more than 0.60% C. Generally, high-carbon steel is used for applications requiring strength, hardness, and wear resistance, such as cutting tools, cables, music wire, springs, and cutlery. After being manufactured into shapes, the parts usually are heat treated and tempered. The higher

TABLE 5.3

		Ultimate tensile strength (MPa)	Yield strength (MPa)	Elongation in 50 mm (%)	Reduction of area (%)	Typical hardness (HB)
Typical Mechanical Properties of Selected Carbon and Alloy Steels						
AISI	Condition					
1020	As-rolled	448	346	36	59	143
	Normalized	441	330	35	67	131
	Annealed	393	294	36	66	111
1080	As-rolled	1010	586	12	17	293
	Normalized	965	524	11	20	293
	Annealed	615	375	24	45	174
3140	Normalized	891	599	19	57	262
	Annealed	689	422	24	50	197
4340	Normalized	1279	861	12	36	363
	Annealed	744	472	22	49	217
8620	Normalized	632	385	26	59	183
	Annealed	536	357	31	62	149

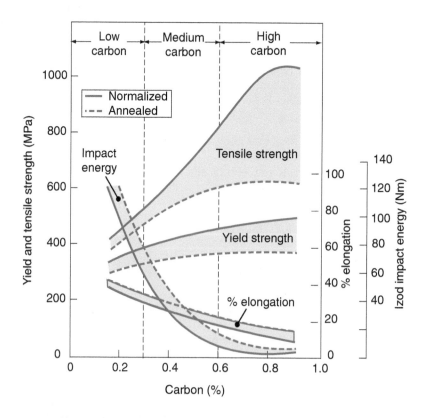

FIGURE 5.5   Effect of carbon content on the mechanical properties of carbon steel.

the carbon content of the steel, the higher is its hardness, strength, and wear resistance after heat treatment.

- Carbon steels containing sulfur and phosphorus are known as **resulfurized** carbon steels (11xx series) and **rephosphorized and resulfurized** carbon steels (12xx series). For example, 1112 steel is a resulfurized steel with a carbon content of 0.12%. These steels have improved machinability, as described in Section 21.7.

5.5.4   Alloy Steels

Steels containing significant amounts of alloying elements are called **alloy steels**. **Structural-grade alloy steels** are used mainly in the construction and transportation industries, because of their high strength. Other types of alloy steels are used in applications where strength, hardness, creep and fatigue resistance, and toughness are required; they can be heat treated to obtain specific desired properties.

5.5.5   High-strength Low-alloy Steels

In order to improve the strength-to-weight ratio of steels, a number of **high-strength, low-alloy steels** (HSLA) have been developed. These steels have a low carbon content (usually less than 0.30%) and are characterized by a microstructure consisting of fine-grain ferrite as one phase and a hard second phase of martensite and austenite. Mechanical properties for selected HSLA steels are given in Table 5.4. The steels have high strength and energy-absorption capabilities as compared to conventional steels. The ductility, formability, and weldability of HSLA steels are, however, generally inferior to those of conventional low-alloy steels (see Fig. 5.6). To improve these properties, several ultra-high-strength steels have been developed, as described in Section 5.5.6.

Sheet products of HSLA steels typically are used for parts of automobile bodies and other transportation equipment (in order to reduce weight and, hence, fuel consumption) and in mining, agricultural, and various other industrial applications. Plates are used in ships, bridges, building construction, and for such shapes as I-beams, channels, and angles used in buildings and in various structures.

Designations.   Three categories compose the system of AISI designations for high-strength sheet steel (Table 5.5): (a) *Structural quality* (S) includes the elements C, Mn, P, and N; (b) *low alloys* (X) contain Nb, Cr, Cu, Mo, Ni, Si, Ti, V, and Zr, either singly or in combination; (c) *weathering steels* (W) have environmental-corrosion resistance that is approximately four times higher than that of conventional low-carbon steels

TABLE 5.4

Mechanical Properties of Selected Advanced High-strength Steels			
Steel	Minimum ultimate strength (MPa)	Minimum yield strength (MPa)	Elongation in 50 mm (%)	Strain-hardening exponent, n
BH 210/340	340	210	36	0.18
BH 260/370	370	260	32	0.13
HSLA 350/450	450	450	25	0.14
DP 350/600	600	600	27	0.14
DP 500/800	800	800	17	0.14
DP 700/1000	1,000	1,000	15	0.13
TRIP 450/800	800	450	29	0.24
TRIP 400/600	600	400	30	0.23
CP 700/800	800	700	12	0.13
MART 950/1200	1200	950	6	0.07
MART 1250/1520	1520	1250	5	0.065
27MnCrB5, as rolled	967	478	12	0.06
27MnCrB5, hot stamped	1350	1097	5	0.06
37MnB4, as rolled	810	580	12	0.06
37MnB4, hot stamped	2040	1378	4	0.06

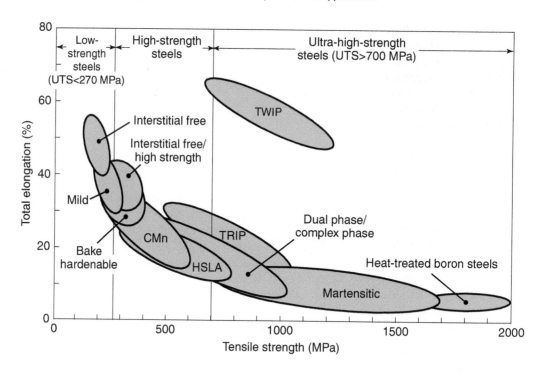

FIGURE 5.6   Comparison of advanced high-strength steels.

TABLE 5.5

AISI Designations for High-strength Sheet Steel			
Yield strength			
ksi	MPa	Chemical composition	Deoxidation practice
35	240	S = structural quality	F = killed plus sulfide inclusion control
40	275	X = low alloy	K = killed
45	310	W = weathering	O = nonkilled
50	350	D = dual-phase	
60	415		
70	485		
80	550		
100	690		
120	830		
140	970		
Example: 50XF			
50×10^3 psi min yield strength		Low alloy	Killed plus sulfide inclusion control

and contain Si, P, Cu, Ni, and Cr in various combinations. In addition, the formability of these sheet steels is graded by the letters F (excellent), K (good), and O (fair).

Another designation scheme in wide use is that defined by the Ultralight Steel Auto Body Consortium (ULSAB). The ULSAB practice is to define both the type of a steel and its yield and tensile strengths, in a compact designation, in the form XX aaa/bbb, where XX is the type of steel, aaa is the yield strength in MPa, and bbb is the ultimate tensile strength, in MPa. The types of steel are:

BH   — Bake-hardenable
HSLA — High-strength low-alloy
DP   — Dual-phase
TRIP   — Transformation-induced plasticity
TWIP   — Twinning-induced plasticity
MART — Martensitic
CP   — Complex phase

Thus, for example; HSLA 350/450 would be a high-strength low-alloy steel with a minimum yield strength of 350 MPa and a minimum ultimate tensile strength of 450 MPa.

Microalloyed Steels.   These steels provide superior properties and can eliminate the need for heat treatment. They have a ferrite–pearlite microstructure, with fine dispersed particles of carbonitride. When subjected to carefully controlled cooling (usually in air), these steels develop improved and uniform strength. Compared to medium-carbon steels, **microalloyed steels** also can provide cost savings of as much as 10%, since the steps of quenching, tempering, and stress relieving are not required.

Nanoalloyed Steels.   These steels have extremely small grain sizes (10–100 nm), and are produced using metallic glasses (Section 6.14) as a precursor. The metallic glass is subjected to a carefully controlled vitrification (crystallization) process at a high nucleation rate, resulting in fine nanoscale phases. (See also Section 8.8.)

5.5.6   Ultra-high-strength Steels

Ultra-high-strength steels are defined by AISI as those with an ultimate tensile strength higher than 700 MPa (100 ksi). There are five important types of ultra-high-strength steel: dual-phase, TRIP, TWIP, complex phase, and martensitic. The main application of these steels is for crashworthy design of automobiles. The use of stronger steels allows for smaller cross-sections in structural components, thus resulting in weight savings and increases in fuel economy without compromising safety. The significant drawbacks of all these steels are higher springback, tool and die wear, and forming loads.

Dual-phase steels are processed specially to have a mixed ferrite and martensite structure. They have a high work-hardening exponent, n, in Eq. (2.8), which improves their ductility and formability.

TRIP steels consist of a ferrite–bainite matrix and 5–20% retained austenite. During forming, the austenite progressively transforms into martensite. TRIP steels have both excellent ductility (because of the austenite) and high strength after forming. As a result, these steels can be used to produce more complex parts than other high-strength steels.

TWIP steels (from *TWinning-Induced Plasticity*) are austenitic and have high manganese content (17–20%). These steels derive their properties from the generation of twins during deformation (see Section 1.4) without a phase change, resulting in very high strain hardening and avoiding thinning during processing. As shown in Fig. 5.5, TWIP steels combine high strength with high formability.

Complex-phase grades (CP grades) are very-fine-grained microstructures of ferrite and a high volume fraction of hard phases (martensite and bainite). These steels have ultimate tensile strengths as high as 800 MPa (115 ksi), and are therefore of interest for automotive crash applications, such as bumpers and roof supports. *Martensitic grades* also are available, consisting of high fractions of martensite to attain tensile strengths as high as 1500 MPa (217 ksi).

CASE STUDY 5.1   Advanced High-strength Steels in Automobiles

Increasing fuel economy in automobiles has received considerable attention in recent years, for both environmental and economic reasons. Regulatory requirements call for automobile manufacturers to achieve *corporate average fuel economy* (CAFE) standards. To achieve higher fuel economy without compromising performance or safety, manufacturers have increasingly applied advanced high-strength steels in structural elements of automobiles; for example, the application of steel in the Ford 500 automobile shown in Fig. 5.7. Note that although 60% of the steel in this automobile is mild steel, and is associated with body panels and transmission and engine components, structural components are exploiting the higher strength-to-weight ratios of advanced high-strength steels.

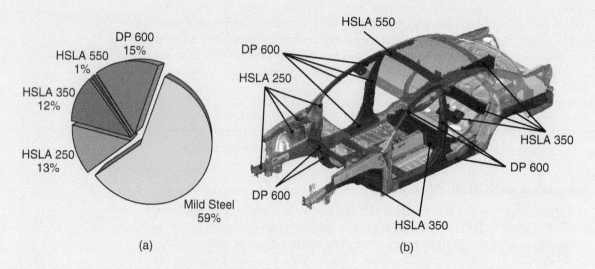

FIGURE 5.7   Advanced high-strength steel applications in the Ford 500 automobile. (a) Use of advanced high-strength steels by weight percent; (b) structural components and alloys used. *Source:* Courtesy of American Iron and Steel Institute.

5.6   Stainless Steels

Stainless steels are characterized primarily by their corrosion resistance, high strength, and ductility. They are called *stainless* because, in the presence of oxygen (air), they develop a thin, hard, and adherent film of chromium oxide that protects the metal from corrosion (*passivation*; see Section 3.8); this protective film builds up again in the event that the surface is scratched. For passivation to occur, the minimum chromium content should be 10–12% by weight. In addition to chromium and carbon, other alloying elements in stainless steels typically are nickel, molybdenum, copper, titanium, silicon, manganese, columbium, aluminum, nitrogen, and sulfur.

The higher the carbon content, the lower is the corrosion resistance of stainless steels. The reason is that the carbon combines with the chromium in the steel and forms chromium carbide; the reduced presence of chromium lowers the passivity of the steel. In addition, the chromium carbide introduces a second phase and thereby promotes galvanic corrosion.

Developed in the early 1900s, stainless steels are produced in electric furnaces or by the basic-oxygen process, and using techniques similar to those used in other types of steelmaking. The level of purity is controlled through various refining techniques. Stainless steels are available in a wide variety of shapes. Typical applications include cutlery, kitchen equipment, health care and surgical equipment, and applications in the chemical, food-processing, and petroleum industries.

Stainless steels generally are divided into five types (see also Table 5.6).

Austenitic (200 and 300 series).   These steels generally are composed of chromium, nickel, and manganese in iron. They are nonmagnetic and have excellent corrosion resistance, but are susceptible to stress-corrosion cracking (Section 3.8). Austenitic stainless steels, which are hardened by cold working, are the most ductile of all stainless steels and can be formed easily. They are used in a wide variety of applications, such as kitchenware, fittings, welded construction, lightweight transportation equipment, furnace and heat-exchanger parts, and components for severe chemical environments.

Ferritic (400 series).   These steels have a high chromium content. They are magnetic and have good corrosion resistance, but have lower ductility than austenitic stainless steels. Ferritic stainless steels are hardened by cold working. They generally are used for nonstructural applications, such as kitchen equipment and automotive trim.

Martensitic (400 and 500 series).   Most martensitic stainless steels do not contain nickel and are hardenable by heat treatment. These steels are magnetic and have high strength, hardness, and fatigue resistance, good ductility, and moderate corrosion resistance. Martensitic stainless steels typically are used for cutlery, surgical tools, instruments, valves, and springs.

Precipitation-hardening (PH).   These stainless steels contain chromium and nickel, along with copper, aluminum, titanium, or molybdenum. They have good corrosion

TABLE 5.6

	Mechanical Properties and Typical Applications of Selected Annealed Stainless Steels at Room Temperature			
AISI (UNS) designation	Ultimate tensile strength (MPa)	Yield strength (MPa)	Elongation in 50 mm (%)	Characteristics and typical applications
303 (S30300)	550–620	240–260	53–50	Screw machine products (shafts, valves, bolts, bushings, and nuts) and aircraft fittings (bolts, nuts, rivets, screws, and studs)
304 (S30400)	5–620	240–290	60–55	Chemical and food-processing equipment, brewing equipment, cryogenic vessels, gutters, downspouts, and flashings
316 (S31600)	50–590	210–290	60–55	High corrosion resistance and high creep strength, chemical- and pulp-handling equipment, photographic equipment, brandy vats, fertilizer parts, ketchup-cooking kettles, and yeast tubs
410 (S41000)	480–520	240–310	35–25	Machine parts, pump shafts, bolts, bushings, coal chutes, cutlery, tackle, hardware, jet engine parts, mining machinery, rifle barrels, screws, and valves
416 (S41600)	480–520	275	30–20	Aircraft fittings, bolts, nuts, fire extinguisher inserts, rivets, and screws

resistance and ductility, and have high strength at elevated temperatures. Their main application is in aircraft and aerospace structural components.

Duplex Structure. These stainless steels have a mixture of austenite and ferrite. They have good strength and higher resistance to both corrosion (in most environments) and stress-corrosion cracking than do the 300 series of austenitic steels. Typical applications are in water-treatment plants and in heat-exchanger components.

CASE STUDY 5.2   Stainless Steels in Automobiles

The types of stainless steel usually selected by materials engineers for use in automobile parts are 301, 409, 430, and 434. Because of its good corrosion resistance and mechanical properties, type 301 is used for wheel covers; cold working during the forming process increases its yield strength and gives the wheel cover a springlike action.

Type 409 is used extensively for catalytic converters. Type 430 had been used for automotive trim, but it is not as resistant as type 434 is to the road deicing salts used in colder climates in winter; as a result, its use is now limited. In addition to being more corrosion resistant, type 434 closely resembles the color of chromium plating, thus becoming an attractive alternative to 430.

Stainless steels also are well suited for use in other automobile components. Examples include: exhaust manifolds (replacing cast-iron manifolds, to reduce weight, increase durability, provide higher thermal conductivity, and reduce emissions), mufflers and tailpipes (for better corrosion protection in harsh environments), and brake tubing.

5.7   Tool and Die Steels

Tool and die steels are specially alloyed steels (Tables 5.7 and 5.8), designed for tool and die requirements, such as high strength, impact toughness, and wear resistance at room and elevated temperatures. They commonly are used in forming and machining of metals (Parts III and IV).

5.7.1   High-speed Steels

High-speed steels (HSS) are the most highly alloyed tool and die steels. First developed in the early 1900s, they maintain their hardness and strength at elevated operating temperatures. There are two basic types of high-speed steels: the **molybdenum type** (M-series) and the **tungsten type** (T-series).

The **M-series** steels contain up to about 10% molybdenum, with chromium, vanadium, tungsten, and cobalt as other alloying elements. The **T-series** steels contain 12–18% tungsten, with chromium, vanadium, and cobalt as other alloying elements. The M-series steels generally have higher abrasion resistance than T-series steels, undergo less distortion in heat treatment, and are less expensive. The M-series constitutes about 95% of all the high-speed steels produced in the United States. High-speed steel tools can be coated with titanium nitride and titanium carbide for improved wear resistance.

TABLE 5.7

Basic Types of Tool and Die Steels

Type	AISI
High speed	M (molybdenum base)
	T (tungsten base)
Hot work	H1 to H19 (chromium base)
	H20 to H39 (tungsten base)
	H40 to H59 (molybdenum base)
Cold work	D (high carbon, high chromium)
	A (medium alloy, air hardening)
	O (oil hardening)
Shock resisting	S
Mold steels	P1 to P19 (low carbon)
	P20 to P39 (others)
Special purpose	L (low alloy)
	F (carbon–tungsten)
Water hardening	W

TABLE 5.8

Processing and Service Characteristics of Common Tool and Die Steels							
AISI designation	Resistance to decarburization	Resistance to cracking	Approximate hardness (HRC)	Machinability	Toughness	Resistance to softening	Resistance to wear
M2	Medium	Medium	60–65	Medium	Low	Very high	Very high
H11, 12, 13	Medium	Highest	38–55	Medium to high	Very high	High	Medium
A2	Medium	Highest	57–62	Medium	Medium	High	High
A9	Medium	Highest	35–56	Medium	High	High	Medium to high
D2	Medium	Highest	54–61	Low	Low	High	High to very high
D3	Medium	High	54–61	Low	Low	High	Very high
H21	Medium	High	36–54	Medium	High	High	Medium to high
P20	High	High	28–37	Medium to high	High	Low	Low to medium
P21	High	Highest	30–40	Medium	Medium	Medium	Medium
W1, W2	Highest	Medium	50–64	Highest	High	Low	Low to medium

5.7.2 Die Steels

Hot-work steels (H-series) are designed for use at elevated temperatures. They have high toughness, as well as high resistance to wear and cracking. The alloying elements generally are tungsten, molybdenum, chromium, and vanadium. **Cold-work steels** (A-, D-, and O-series) are used for cold-working operations. They generally have high resistance to wear and cracking, and are available as oil-hardening or air-hardening types. **Shock-resisting steels** (S-series) are designed for impact toughness and are used in applications such as header dies, punches, and chisels. Various other tool and die materials for a variety of manufacturing applications are given in Table 5.9.

SUMMARY

- The major categories of ferrous metals and alloys are carbon steels, alloy steels, stainless steels, and tool and die steels. Their wide range of properties, availability, and generally low cost have made them among the most useful of all metallic materials.

- Steelmaking processes increasingly involve continuous-casting and secondary-refining techniques, resulting in higher quality steels and higher efficiency and productivity.

- Carbon steels generally are classified as low-carbon (mild steel), medium-carbon, and high-carbon steels. Alloy steels contain several alloying elements, particularly chromium, nickel, and molybdenum.

- High-strength low-alloy (HSLA) steels have a low carbon content and consist of fine-grained ferrite as one phase and a second phase of martensite and austenite. Micro- and nanoalloyed steels are fine-grained, high-strength low-alloy steels that provide superior properties without the need for heat treatment.

- Stainless steels have chromium as the major alloying element; they are called stainless because they form a passivating chromium–oxide layer on their surface. These steels are generally classified as austenitic, ferritic, martensitic, and precipitation-hardening steels.

TABLE 5.9

Typical Tool and Die Materials for Metalworking Processes	
Process	Material
Die casting	H13, P20
Powder metallurgy	
Punches	A2, S7, D2, D3, M2
Dies	WC, D2, M2
Molds for plastics and rubber	S1, O1, A2, D2, 6F5, 6F6, P6, P20, P21, H13
Hot forging	6F2, 6G, H11, H12
Hot extrusion	H11, H12, H13, H21
Cold heading	W1, W2, M1, M2, D2, WC
Cold extrusion	
Punches	A2, D2, M2, M4
Dies	O1, W1, A2, D2
Coining	52100, W1, O1, A2, D2, D3, D4, H11, H12, H13
Drawing	
Wire	WC, diamond
Shapes	WC, D2, M2
Bar and tubing	WC, W1, D2
Rolls	
Rolling	Cast iron, cast steel, forged steel, WC
Thread rolling	A2, D2, M2
Shear spinning	A2, D2, D3
Sheet metals	
Cold shearing	D2, A2, A9, S2, S5, S7
Hot shearing	H11, H12, H13
Pressworking	Zinc alloys, 4140 steel, cast iron, epoxy composites, A2, D2, O1
Deep drawing	W1, O1, cast iron, A2, D2
Machining	Carbides, high-speed steels, ceramics, diamond, cubic boron nitride

Notes: Tool and die materials usually are hardened 55–65 HRC for cold working and 30–55 HRC for hot working. Tool and die steels contain one or more of the following major alloying elements: chromium, molybdenum, tungsten, and vanadium. (For further details, see the bibliography at the end of this chapter.)

- Tool and die steels are among the most important metallic materials, and are used widely in casting, forming, and machining operations. They generally consist of high-speed steels, hot- and cold-work steels, and shock-resisting steels.

KEY TERMS

Alloy steels	Electric furnace	Nanoalloyed steels	Stainless steels
Basic-oxygen furnace	High-strength low-alloy steels	Open-hearth furnace	Steel
Blast furnace		TWIP steels	Strand casting
Carbon steels	Ingot	Pig iron	Tool and die steels
Complex-phase steels	Killed steel	Refining	Trace elements
Continuous casting	Martensitic steel	Rimmed steel	TRIP steels
Dual-phase steels	Microalloyed steels	Semi-killed steel	TWIP steels

BIBLIOGRAPHY

ASM Handbook, Vol. 1: **Properties and Selection: Iron, Steels, and High-Performance Alloys**, ASM International, 1990.

ASM Specialty Handbook, **Carbon and Alloy Steels**, ASM International, 1995.

ASM Specialty Handbook, **Stainless Steels**, ASM International, 1994.

ASM Specialty Handbook, **Tool Materials**, ASM International, 1995.

Beddoes, J., and Parr, J.G., **Introduction to Stainless Steels**, 3rd ed., ASM International, 1999.

Bhadeshia, H., and Honeycombe, R., **Steels: Microstructure and Properties**, 3rd ed., Butterworth-Heinemann, 2006.

Bryson, W.E., **Heat Treatment, Selection and Application of Tool Steels**, 2nd ed., Hanser Gardner, 2005.

Krauss, G., **Steels: Processing, Structure, and Performance**, ASM International, 2005.

Llewellyn, D.T., and Hudd, R.C., **Steels: Metallurgy and Applications**, 3rd ed., Butterworth-Heinemann, 1999.

McGuire, M.F., **Stainless Steels for Design Engineers**, ASM International, 2008.

Reed, C., **The Superalloys: Fundamentals and Applications**, Cambridge University Press, 2008.

Roberts, G.A., Krauss, G., and Kennedy, R., **Tool Steels**, 5th ed., ASM International, 1998.

Schneider, W., **Continuous Casting**, Wiley, 2006.

REVIEW QUESTIONS

5.1 What are the major categories of ferrous alloys?

5.2 Why is steel so commonly used?

5.3 List the basic raw materials used in making iron and steel, and explain their functions.

5.4 List the types of furnaces commonly used in steelmaking, and describe their characteristics.

5.5 List and explain the characteristics of the types of steel ingots.

5.6 What does refining mean? How is it done?

5.7 What is continuous casting? What advantages does continuous casting have over casting into ingots?

5.8 What is the role of a tundish in continuous casting?

5.9 Name the four alloying elements that have the greatest effect on the properties of steels.

5.10 What are trace elements?

5.11 What are the percentage carbon contents of low-carbon, medium-carbon, and high-carbon steels?

5.12 How do stainless steels become stainless?

5.13 What are the major alloying elements in tool and die steels and in high-speed steels?

5.14 How does chromium affect the surface characteristics of stainless steels?

5.15 What kinds of furnaces are used to refine steels?

5.16 What is high-speed steel?

5.17 What are TRIP and TWIP?

5.18 What are the applications of advanced high-strength steels?

5.19 What characteristics are common among die steels?

5.20 What effect does carbon content have on mechanical properties of steel? What effects does it have on physical properties?

QUALITATIVE PROBLEMS

5.21 Identify several different products that are made of stainless steel, and explain why they are made of that material.

5.22 Professional cooks generally prefer carbon-steel to stainless-steel knives, even though the latter are more popular with consumers. Explain the reasons for those preferences.

5.23 Why is the control of the structure of an ingot important?

5.24 Explain why continuous casting has been such an important technological advancement.

5.25 Describe applications in which you would not want to use carbon steels.

5.26 Explain what would happen if the speed of the continuous-casting process shown in Fig. 5.4a is (a) higher or (b) lower than that indicated, typically 25 mm/s.

5.27 The cost of mill products of metals increases with decreasing thickness and section size. Explain why.

5.28 Describe your observations regarding the information given in Table 5.9.

5.29 How do trace elements affect the ductility of steels?

5.30 Comment on your observations regarding Table 5.1.

5.31 In Table 5.9, D2 steel is listed as a more common tool and die material for most applications. Why is this so?

5.32 List the common impurities in steel. Which of these are the ones most likely to be minimized if the steel is melted in a vacuum furnace?

5.33 Explain the purpose of the oil shown at the top left of Fig. 5.4a given that the molten-steel temperatures are far above the ignition temperatures of the oil.

5.34 Recent research has identified mold-surface textures that will either (a) inhibit a solidified steel from separating from the mold or (b) force it to stay in contact in continuous casting. What is the advantage of a mold that maintains intimate contact with the steel?

5.35 Identify products that cannot be made of steel, and explain why this is so. (For example, electrical contacts commonly are made of gold or copper, because their softness results in low contact resistance, whereas for steel, the contact resistance would be very high.)

5.36 List and explain the advantages and disadvantages of using advanced high-strength steels.

QUANTITATIVE PROBLEMS

5.37 Conduct an Internet search and determine the chemical composition of (a) TRIP 450/800; (b) 304 stainless steel; and (c) 4140 steel. If a foundry ladle will pour 50,000 kg, calculate the weight of each element in the ladle.

5.38 Refer to the available literature, and estimate the cost of the raw materials for (a) an aluminum beverage can, (b) a stainless-steel two-quart cooking pot, and (c) the steel hood of a car.

5.39 In Table 5.1, more than one type of steel is listed for some applications. Refer to data available in the technical literature listed in the bibliography, and determine the range of properties for these steels in various conditions, such as cold worked, hot worked, and annealed.

5.40 Some soft drinks are now available in steel cans (with aluminum tops) that look similar to aluminum cans. Obtain one of each type, weigh them when empty, and determine their respective wall thicknesses.

5.41 Using strength and density data, determine the minimum weight of a 1-m-long tension member that must support a load of 4 kN, manufactured from (a) annealed 303 stainless steel, (b) normalized 8620 steel, (c) as-rolled 1080 steel, (d) any two aluminum alloys, (e) any brass alloy, and (f) pure copper.

5.42 The endurance limit (fatigue life) of steel is approximately one-half the ultimate tensile strength (see Fig. 2.16), but never higher than 100 ksi (700 MPa). For iron, the endurance limit is 40% of the ultimate strength, but never higher than 24 ksi (170 MPa). Plot the endurance limit versus the ultimate strength for the steels described in this chapter and for the cast irons shown in Table 12.3. On the same plot, show the effect of surface finish by plotting the endurance limit, assuming that the material is in the as-cast state. (See Fig. 2.29.)

5.43 Using the data given in Table 5.4, obtain the power-law curves for the advanced high-strength steels shown and plot the curves. Compare these materials with those given in Table 2.3.

SYNTHESIS, DESIGN, AND PROJECTS

5.44 Based on the information given in Section 5.5.1, make a table with columns for each improved property, such as hardenability, strength, toughness, and machinability. In each column, list the elements that improve that particular property and identify the element that has the most influence.

5.45 Assume that you are in charge of public relations for a large steel-producing company. Outline all of the attractive characteristics of steels that you would like your customers to be informed about.

5.46 Assume that you are in competition with the steel industry and are asked to list all of the characteristics of steels that are not attractive. Make a list of those characteristics and explain their relevance to engineering applications.

5.47 Section 5.5.1 noted the effects of various individual elements, such as lead alone or sulfur alone, on the properties and characteristics of steels. What was not discussed, however, was the role of combinations of these elements (such as lead and sulfur together). Review the technical literature, and prepare a table indicating the combined effects of several elements on steels.

5.48 In the past, waterfowl hunters used lead shot in their shotguns, but this practice resulted in lead poisoning of unshot birds that ingested lead pellets (along with gravel) to help them digest food. Steel and tungsten are being used as replacement materials. If all pellets have the same velocity upon exiting the shotgun barrel, what concerns would you have regarding this substitution of materials? Consider both performance and environmental effects.

5.49 Aluminum is being used as a substitute material for steel in automobiles. Describe your concerns, if any, in purchasing an aluminum automobile.

5.50 In the 1940s (the Second World War), the *Yamato* and its sister ship, the *Musashi*, were the largest battleships ever built. Find out the weight of these ships, and estimate the number of automobiles that could have been built from the steel used in just one such ship. Estimate the time it would take to cast that much steel by continuous casting.

5.51 Search the technical literature, and add more parts and materials to those shown in Table 5.1.

5.52 Referring to Fig. 5.4a, note that the mold has cooling channels incorporated to remove heat. Can continuous casting be done without such cooling channels? Can it be done with a heated mold? Explain your answer.

Nonferrous Metals and Alloys: Production, General Properties, and Applications

- Nonferrous metals include a wide variety of materials, ranging from aluminum to zinc, with special properties that are indispensable in most products.

- This chapter introduces each class of nonferrous metal and its alloys, and briefly describes their methods of production.

- Their physical and mechanical properties are then summarized, along with general guidelines for their selection and applications.

- Shape-memory alloys, amorphous alloys, and metal foams are also described, with examples of their unique applications.

6.1   Introduction

Nonferrous metals and alloys cover a very wide range, from the more common metals (such as aluminum, copper, and magnesium) to high-strength, high-temperature alloys (such as those of tungsten, tantalum, and molybdenum). Although generally more expensive than ferrous metals (Table 6.1), nonferrous metals have numerous important applications because of such properties as good corrosion resistance, high thermal and electrical conductivity, low density, and ease of fabrication (Table 6.2).

Typical examples of nonferrous metal and alloy applications are aluminum for cooking utensils and aircraft bodies, copper wire for electrical power cords, zinc for galvanized sheet metal for car bodies, titanium for jet-engine turbine blades and for orthopedic implants, and tantalum for rocket engine components. As an example, the turbofan jet engine (Fig. 6.1) for the Boeing 757 aircraft typically contains the following nonferrous metals and alloys: 38% Ti, 37% Ni, 12% Cr, 6% Co, 5% Al, 1% Nb, and 0.02% Ta.

This chapter introduces the general properties, production methods, and important engineering applications for nonferrous metals and alloys. The manufacturing properties of these materials (such as formability, machinability, and weldability) are described in various chapters throughout this text.

6.2   Aluminum and Aluminum Alloys

The important characteristics of **aluminum** (Al) and its alloys are their high strength-to-weight ratios, resistance to corrosion by many chemicals, high thermal and electrical conductivities, nontoxicity, reflectivity, appearance, and ease of formability and machinability; they are also nonmagnetic.

The principal uses of aluminum and its alloys, in decreasing order of consumption, are in containers and packaging (aluminum beverage cans and foil), architectural and structural applications, transportation (aircraft and aerospace applications, buses, automobiles, railroad cars, and marine craft), electrical applications (as economical and nonmagnetic electrical conductors), consumer durables (appliances, cooking utensils, and furniture), and portable tools (Tables 6.3 and 6.4). Nearly all high-voltage transmission wiring is made of aluminum.

In its structural (load-bearing) components, 82% of a Boeing 747 aircraft and 70% of a 777 aircraft is aluminum. The Boeing 787 Dreamliner (first placed into service in late 2012) is well recognized for its carbon-fiber reinforced composite fuselage, although it still uses 20% aluminum by weight, as compared to 15% titanium. The frame and the body panels of the Rolls Royce Phantom coupe are made of aluminum, improving the car's strength-to-weight and torsional rigidity-to-weight ratios.

Aluminum alloys are available as mill products—that is, as wrought products made into various shapes by rolling, extrusion, drawing, forging, and sheet forming (Chapters 13 through 16). Aluminum ingots are available for casting, as is aluminum in powder form for powder metallurgy applications (Chapter 17). Most aluminum alloys can be machined, formed, and welded with relative ease. There

TABLE 6.1

Approximate Cost-per-unit-volume for Wrought Metals and Plastics Relative to the Cost of Carbon Steel

Material	Relative cost
Gold	30,000
Silver	600
Molybdenum alloys	75–100
Nickel	20
Titanium alloys	20–40
Copper alloys	8–10
Zinc alloys	1.5–3.5
Stainless steels	2–9
Magnesium alloys	4–6
Aluminum alloys	2–3
High-strength low-alloy steels	1.4
Gray cast iron	1.2
Carbon steel	1
Nylons, acetals, and silicon rubber*	1.1–2
Other plastics and elastomers*	0.2–1

*As molding compounds.
Note: Costs vary significantly with quantity of purchase, supply and demand, size and shape, and various other factors.

TABLE 6.2

General Characteristics of Nonferrous Metals and Alloys

Material	Characteristics
Nonferrous alloys	More expensive than steels and plastics; wide range of mechanical, physical, and electrical properties; good corrosion resistance; high-temperature applications
Aluminum	Alloys have high strength-to-weight ratio; high thermal and electrical conductivity; good corrosion resistance; good manufacturing properties
Magnesium	Lightest metal; good strength-to-weight ratio
Copper	High electrical and thermal conductivity; good corrosion resistance; good manufacturing properties
Superalloys	Good strength and resistance to corrosion at elevated temperatures; can be iron-, cobalt-, and nickel-based alloys
Tin	Good corrosion resistance and bright appearance; used also in solders and as bearing materials
Titanium	Highest strength-to-weight ratio of all metals; good strength and corrosion resistance at high temperatures
Refractory metals	Molybdenum, niobium, tungsten, and tantalum; high strength at elevated temperatures
Precious metals	Gold, silver, and platinum; generally good corrosion resistance and aesthetic characteristics
Zinc	Very good corrosion resistance; commonly used in castings and galvanizing steel sheet for corrosion protection

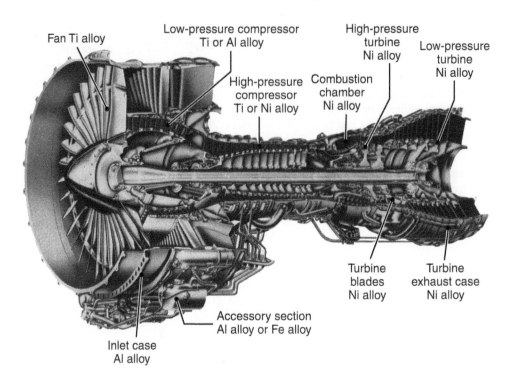

Fan Ti alloy

Low-pressure compressor
Ti or Al alloy

High-pressure
turbine
Ni alloy

Low-pressure
turbine
Ni alloy

High-pressure
compressor
Ti or Ni alloy

Combustion
chamber
Ni alloy

Turbine
blades
Ni alloy

Turbine
exhaust case
Ni alloy

Accessory section
Al alloy or Fe alloy

Inlet case
Al alloy

FIGURE 6.1   Cross-section of a jet engine (PW2037), showing various components and the alloys used in manufacturing them. *Source:* Courtesy of United Technologies, Pratt & Whitney Division.

TABLE 6.3

Properties of Selected Aluminum Alloys at Room Temperature

Alloy (UNS)	Temper	Ultimate tensile strength (MPa)	Yield strength (MPa)	Elongation in 50 mm (%)
1100 (A91100)	O	90	35	35–45
	H14	125	120	9–20
2024 (A92024)	O	190	75	20–22
	T4	470	325	19–20
3003 (A93003)	O	110	40	30–40
	H14	150	145	8–16
5052 (A95052)	O	190	90	25–30
	H34	260	215	10–14
6061 (A96061)	O	125	55	25–30
	T6	310	275	12–17
7075 (A97075)	O	230	105	16–17
	T6	570	500	11

are two types of wrought alloys of aluminum: (a) alloys that can be hardened by mechanical processing and are not heat treatable and (b) alloys that can be hardened by heat treatment.

Unified Numbering System.   As is the case with steels, aluminum and other non-ferrous metals and alloys are identified internationally by the Unified Numbering System (UNS), consisting of a letter, indicating the general class of the alloy, followed

TABLE 6.4

Manufacturing Characteristics and Typical Applications of Selected Wrought Aluminum Alloys				

| | Characteristics* | | | |
Alloy	Corrosion resistance	Machinability	Weldability	Typical Applications
1100	A	C–D	A	Sheet-metal work, spun hollowware, tin stock
2024	C	B–C	B–C	Truck wheels, screw machine products, aircraft structures
3003	A	C–D	A	Cooking utensils, chemical equipment, pressure vessels, sheet-metal work, builders' hardware, storage tanks
5052	A	C–D	A	Sheet-metal work, hydraulic tubes, and appliances; bus, truck, and marine uses
6061	B	C–D	A	Heavy-duty structures where corrosion resistance is needed; truck and marine structures, railroad cars, furniture, pipelines, bridge railings, hydraulic tubing
7075	C	B–D	D	Aircraft and other structures, keys, hydraulic fittings

*A, excellent; D, poor.

by five digits, indicating its chemical composition. For example, A is for aluminum, C for copper, N for nickel alloys, P for precious metals, and Z for zinc; in the UNS designation, 2024 wrought aluminum alloy is A92024.

Production. Aluminum was first produced in 1825. It is the most abundant metallic element, making up about 8% of the Earth's crust, and is produced in a quantity second only to that of iron. The principal ore for aluminum is *bauxite*, which is a hydrous (water-containing) aluminum oxide and includes various other oxides. After the clay and dirt are washed off, the ore is crushed into powder and treated with hot caustic soda (sodium hydroxide) to remove impurities. Next, alumina (aluminum oxide) is extracted from this solution and then dissolved in a molten sodium-fluoride and aluminum-fluoride bath at 940° to 980°C (1725° to 1800°F). This mixture is then subjected to direct-current electrolysis. Aluminum metal forms at the cathode (negative pole), while oxygen is released at the anode (positive pole). *Commercially pure aluminum* is up to 99.99% Al. The production process consumes a great deal of electricity, which contributes significantly to the cost of aluminum.

Porous Aluminum. Blocks of *porous aluminum* are produced that are 37% lighter than solid aluminum and have uniform permeability (*microporosity*). This characteristic allows their use in applications where a vacuum or differential pressure has to be maintained. Examples are the (a) vacuum holding of fixtures for assembly and automation (Section 37.8) and (b) vacuum forming or thermoforming of plastics (Section 19.6). These blocks are 70–90% aluminum powder; the rest is epoxy resin. They can be machined with relative ease and can be joined using adhesives.

CASE STUDY 6.1   An All-aluminum Automobile

Aluminum use in automobiles and in light trucks has been increasing steadily. As recently as 1990, there were no aluminum-structured passenger cars in production, but by 1997 there were seven, including the Plymouth Prowler and the Audi A8 (Fig. 6.2). With weight savings of up to 47% over steel vehicles, such cars use less fuel, create less pollution, and are recyclable.

(a)

(b)

FIGURE 6.2   (a) The aluminum body structure, showing various components made by extrusion, sheet forming, and casting processes. (b) The Audi A8 automobile, which has an all-aluminum body structure. *Source:* (a) Courtesy of National Institute of Standards and Technology.

New alloys and new design and manufacturing methodologies had to be developed to enable the use of aluminum in automobiles. For example, welding and adhesive bonding procedures had to be refined, the structural frame design had to be optimized, and new tooling designs had to be developed. Because of these new technologies, the desired environmental savings were realized without an accompanying drop in performance or safety. In fact, the Audi A8 was the first luxury-class car to earn a dual five-star (highest safety) rating for both driver and front-seat passenger in the National Highway Transportation Safety Administration (NHSTA) New Car Assessment Program.

6.3 Magnesium and Magnesium Alloys

Magnesium (Mg) is the lightest engineering metal, and has good vibration-damping characteristics. Its alloys are used in structural and nonstructural applications wherever weight is of primary importance. Magnesium is also an alloying element in various nonferrous metals.

Typical uses of magnesium alloys are in aircraft and missile components, material-handling equipment, portable power tools, ladders, luggage, bicycles, sporting goods, and general lightweight components. Like aluminum, magnesium is finding increased use in the automotive sector, mainly in order to achieve weight savings. Magnesium alloys are available either as castings (such as die-cast camera frames) or as wrought products (such as extruded bars and shapes, forgings, and rolled plates and sheets). Magnesium alloys are also used in printing and textile machinery to minimize inertial forces in high-speed components.

Because it is not sufficiently strong in its pure form, magnesium is alloyed with various elements (Table 6.5) in order to impart certain specific properties, particularly a high strength-to-weight ratio. A variety of magnesium alloys have good casting, forming, and machining characteristics. Because they oxidize rapidly (i.e., they are *pyrophoric*), however, a fire hazard exists, and hence precautions must be taken when machining, grinding, or sand-casting magnesium alloys. Products made of magnesium and its alloys are not a fire hazard during normal use.

Magnesium is easy to cast but difficult to form. Efforts have been made to promote the increased use of magnesium in automobiles through improved welding and sheet formability. Alloys ZEK100, AZ31, and ZE10 are of current high interest.

TABLE 6.5

Properties and Typical Forms of Selected Wrought Magnesium Alloys

Alloy	Nominal composition	Condition	Ultimate tensile strength (MPa)	Yield strength (MPa)	Elongation in 50 mm (%)	Typical forms
AZ31B	3.0 Al, 1.0 Zn, 0.2 Mn	F	260	200	15	Extrusions
		H24	290	220	15	Sheet and plate
AZ80A	8.5 Al, 0.5 Zn, 0.2 Mn	T5	380	275	7	Extrusions and forgings
HK31A	0.7 Zr, 3 Th	H24	255	200	8	Sheet and plates
ZE10	1.0 Zn, 1.0 Ce	F	263	163	16	Sheet and plates
ZEK199	1.0 Zn, 0.3 Zr, 1.0 Ce	F	311	308	19	Extrusions and sheet
ZK60A	5.7 Zn, 0.55 Zr	T5	365	300	11	Extrusions and forgings

Production. Magnesium is the third most abundant metallic element (2%) in the Earth's crust, after iron and aluminum. Most magnesium comes from seawater, which contains 0.13% magnesium, in the form of magnesium chloride. First produced in 1808, magnesium metal can be obtained either electrolytically or by thermal reduction. In the *electrolytic method*, seawater is mixed with lime (calcium hydroxide) in settling tanks. Magnesium hydroxide precipitates to the bottom, is filtered and mixed with hydrochloric acid. The resulting solution is subjected to electrolysis (as is done with aluminum), producing magnesium metal, which is then cast into ingots for further processing into various shapes.

In the *thermal-reduction method*, magnesium ores (dolomite, magnesite, and other rocks) are broken down with reducing agents (such as powdered ferrosilicon, an alloy of iron and silicon) by heating the mixture in a vacuum chamber. As a result of this reaction, vapors of magnesium form, and they condense into magnesium crystals, which are then melted, refined, and poured into ingots to be further processed into various shapes.

6.4   Copper and Copper Alloys

First produced in about 4000 B.C., **copper** (Cu, from the Latin *cuprum*) and its alloys have properties somewhat similar to those of aluminum and its alloys. In addition, they have good corrosion resistance and are among the best conductors of electricity and heat (Tables 3.1 and 3.2). Copper and its alloys can be processed easily by various forming, machining, casting, and joining techniques.

Copper alloys often are attractive for applications in which a combination of electrical, mechanical, nonmagnetic, corrosion-resistant, thermally conductive, and wear-resistant qualities are required. Applications include electrical and electronic components, springs, coins, plumbing components, heat exchangers, marine hardware, and consumer goods (such as cooking utensils, jewelry, and other decorative objects). Although aluminum is the most common material for dies in polymer injection molding (Section 19.3), copper often is used because of its better thermal properties. Also, pure copper can be used as a solid lubricant in hot metal-forming operations (Section 33.7.6).

Copper alloys can develop a wide variety of properties by the addition of alloying elements and by heat treatment, to improve their manufacturing characteristics. The most common copper alloys are brasses and bronzes. **Brass** (an alloy of copper and zinc) is one of the earliest alloys developed and has numerous applications, including decorative objects (Table 6.6). **Bronze** is an alloy of copper and tin (Table 6.7); there are also other bronzes, such as aluminum bronze (an alloy of copper and aluminum) and tin bronzes. Beryllium copper (or beryllium bronze) and phosphor bronze have good strength and hardness, with applications such as springs and bearings; other major copper alloys are copper nickels and nickel silvers.

Production. Copper is found in several types of ores, the most common being sulfide ores. The ores are generally of low grade and usually are obtained from open-pit mines. The ore is ground into fine particles in ball mills (rotating cylinders with metal balls inside to crush the ore, as illustrated in Fig. 17.6b); the resulting particles are then suspended in water to form a slurry. Chemicals and oil are added, and the mixture is agitated. The mineral particles form a *froth*, which is scraped and dried. The dry copper concentrate (as much as one-third of which is copper)

TABLE 6.6

Properties and Typical Applications of Selected Wrought Copper and Brasses

Type and UNS number	Nominal composition (%)	Ultimate tensile strength (MPa)	Yield strength (MPa)	Elongation in 50 mm (%)	Typical applications
Electrolytic tough-pitch copper (C11000)	99.90 Cu, 0.04 O	220–450	70–365	55–4	Downspouts, gutters, roofing, gaskets, auto radiators, bus bars, nails, printing rolls, rivets
Red brass, 85% (C23000)	85.0 Cu, 15.0 Zn	270–725	70–435	55–3	Weather stripping, conduits, sockets, fasteners, fire extinguishers, condenser and heat-exchanger tubing
Cartridge brass, 70% (C26000)	70.0 Cu, 30.0 Zn	300–900	75–450	66–3	Radiator cores and tanks, flashlight shells, lamp fixtures, fasteners, locks, hinges, ammunition components, plumbing accessories
Free-cutting brass (C36000)	61.5 Cu, 3.0 Pb, 35.5 Zn	340–470	125–310	53–18	Gears, pinions, automatic high-speed screw machine parts
Naval brass (C46400 to C46700)	60.0 Cu, 39.25 Zn, 0.75 Sn	380–610	170–455	50–17	Aircraft: turnbuckle barrels, balls, bolts; marine hardware: propeller shafts, rivets, valve stems, condenser plates

TABLE 6.7

Properties and Typical Applications of Selected Wrought Bronzes

Type and UNS number	Nominal composition (%)	Ultimate tensile strength (MPa)	Yield strength (MPa)	Elongation in 50 mm (%)	Typical applications
Architectural bronze (C38500)	57.0 Cu, 3.0 Pb, 40.0 Zn	415 (as extruded)	140	30	Architectural extrusions, storefronts, thresholds, trim, butts, hinges
Phosphor bronze, 5% A (C51000)	95.0 Cu, 5.0 Sn, trace P	325–960	130–550	64–2	Bellows, clutch disks, cotter pins, diaphragms, fasteners, wire brushes, chemical hardware, textile machinery
Free-cutting phosphor bronze (C54400)	88.0 Cu, 4.0 Pb, 4.0 Zn, 4.0 Sn	300–520	130–435	50–15	Bearings, bushings, gears, pinions, shafts, thrust washers, valve parts
Low-silicon bronze, B (C65100)	98.5 Cu, 1.5 Si	275–655	100–475	55–11	Hydraulic pressure lines, bolts, marine hardware, electrical conduits, heat-exchanger tubing
Nickel silver, 65–10 (C74500)	65.0 Cu, 25.0 Zn, 10.0 Ni	340–900	125–525	50–1	Rivets, screws, slide fasteners, hollowware, nameplates

is traditionally **smelted** (melted and fused) and refined; this process is known as **pyrometallurgy,** because heat is used to refine the metal. For applications such as electrical conductors, the copper is further refined electrolytically to a purity of at least 99.95% (OFEC, *oxygen-free electrolytic copper*). Copper is also processed by **hydrometallurgy,** involving both chemical and electrolytic reactions.

6.5 Nickel and Nickel Alloys

Nickel (Ni) is a silver-white metal and a major alloying element in metals that imparts strength, toughness, and corrosion resistance, and is used extensively in stainless steels and nickel-based alloys (also called **superalloys**). Nickel alloys are used in high-temperature applications (such as jet engine components, rockets, and nuclear power plants), food-handling and chemical-processing equipment, coins, and marine applications. Because nickel is magnetic, its alloys also are used in electromagnetic applications, such as solenoids.

The principal use of nickel as a metal is in the electroplating of parts for their appearance and the improvement of their corrosion and wear resistance. Nickel alloys have high strength and corrosion resistance at elevated temperatures. Alloying elements in nickel are chromium, cobalt, and molybdenum. The behavior of nickel alloys in machining, forming, casting, and welding can be modified by various other alloying elements.

A variety of nickel alloys, with a wide range of strengths at different temperatures, have been developed (Table 6.8). Although trade names are still in wide use, nickel alloys are identified in the UNS system with the letter N. Thus, Hastelloy G is N06007; other common trade names are:

- **Monel** is a nickel–copper alloy
- **Hastelloy** (also a nickel–chromium alloy) has good corrosion resistance and high strength at elevated temperatures
- **Nichrome** (an alloy of nickel, chromium, and iron) has high electrical resistance and high resistance to oxidation, and is used for electrical heating elements
- **Invar** and **Kovar** (alloys of iron and nickel) have relatively low sensitivity to temperature changes (Section 3.6)

Production.   The main sources of nickel are sulfide and oxide ores, all of which have low concentrations of nickel. The metal is produced by sedimentary and thermal processes, followed by electrolysis; this sequence yields 99.95% pure nickel. Nickel

TABLE 6.8

Properties and Typical Applications of Selected Nickel Alloys (All Are Trade Names)

Type and UNS number	Nominal composition (%)	Ultimate tensile strength (MPa)	Yield strength (MPa)	Elongation in 50 mm (%)	Typical applications
Nickel 200 (annealed)	–	380–550	100–275	60–40	Chemical and food processing industry, aerospace equipment, electronic parts
Duranickel 301 (age hardened)	4.4 Al, 0.6 Ti	1300	900	28	Springs, plastics extrusion equipment, molds for glass, diaphragms
Monel R-405 (hot rolled)	30 Cu	525	230	35	Screw-machine products, water meter parts
Monel K-500 (age hardened)	29 Cu, 3 Al	1050	750	30	Pump shafts, valve stems, springs
Inconel 600 (annealed)	15 Cr, 8 Fe	640	210	48	Gas turbine parts, heat-treating equipment, electronic parts, nuclear reactors
Hastelloy C-4 (solution treated and quenched)	16 Cr, 15 Mo	785	400	54	Parts requiring high-temperature stability and resistance to stress-corrosion cracking

also is present in the ocean bed in significant amounts, but undersea mining does not currently account for significant nickel production.

6.6 Superalloys

Superalloys are important in high-temperature applications, hence they are also known as **heat-resistant** or **high-temperature alloys**. Superalloys generally have good resistance to corrosion, mechanical and thermal fatigue, mechanical and thermal shock, and creep and erosion at elevated temperatures. Major applications of superalloys are in jet engines and gas turbines; other applications are in reciprocating engines, rocket engines, tools and dies for hot working of metals, and in the nuclear, chemical, and petrochemical industries.

Generally, superalloys are identified by trade names or by special numbering systems, and are available in a variety of shapes. Most superalloys have a maximum service temperature of about 1000°C (1800°F) in structural applications. For non-load-bearing components, temperatures can be as high as 1200°C (2200°F).

Superalloys are referred to as *iron-based, cobalt-based,* or *nickel-based.*

- **Iron-based superalloys** generally contain from 32-67% Fe, 15-22% Cr, and 9-38% Ni. Common alloys in this group are the *Incoloy* series.
- **Cobalt-based superalloys** generally contain from 35-65% Co, 19-30% Cr, and up to 35% Ni. These superalloys are not as strong as nickel-based superalloys, but they retain their strength at higher temperatures.
- **Nickel-based superalloys** are the most common of the superalloys and are available in a wide variety of compositions (Table 6.9). The proportion of nickel is from 38-76% and also contain up to 27% Cr and 20% Co. Common alloys in this group are the *Hastelloy, Inconel, Nimonic, René, Udimet, Astroloy,* and *Waspaloy* series.

TABLE 6.9

Properties and Typical Applications of Selected Nickel-based Superalloys at 870°C (1600°F) (All Are Trade Names)

Alloy	Condition	Ultimate tensile strength (MPa)	Yield strength (MPa)	Elongation in 50 mm (%)	Typical applications
Astroloy	Wrought	770	690	25	Forgings for high-temperature use
Hastelloy X	Wrought	255	180	50	Jet engine sheet parts
IN-100	Cast	885	695	6	Jet engine blades and wheels
IN-102	Wrought	215	200	110	Superheater and jet engine parts
Inconel 625	Wrought	285	275	125	Aircraft engines and structures, chemical processing equipment
Inconel 718	Wrought	340	330	88	Jet engine and rocket parts
MAR-M 200	Cast	840	760	4	Jet engine blades
MAR-M 432	Cast	730	605	8	Integrally cast turbine wheels
René 41	Wrought	620	550	19	Jet engine parts
Udimet 700	Wrought	690	635	27	Jet engine parts
Waspaloy	Wrought	525	515	35	Jet engine parts

6.7   Titanium and Titanium Alloys

Titanium (Ti, named after the Greek god Titan) is a silvery-white metal, discovered in 1791 but not produced commercially until the 1950s. Although titanium is expensive, its high strength-to-weight ratio and corrosion resistance at room and elevated temperatures make it attractive for numerous applications, including aircraft; jet engines (see Fig. 6.1); racing cars; golf clubs; chemical, petrochemical, and marine components; submarine hulls; armor plate; and medical applications, such as orthopedic implants (Table 6.10). Titanium alloys can be used for service at 550°C (1000°F) for long periods of time and at up to 750°C (1400°F) for shorter periods.

Unalloyed titanium, known as *commercially pure titanium*, has excellent corrosion resistance for applications where strength considerations are secondary. Aluminum, vanadium, molybdenum, manganese, and other alloying elements impart special properties, such as improved workability, strength, and hardenability.

The properties and manufacturing characteristics of titanium alloys are extremely sensitive to small variations in both alloying and residual elements. Control of composition and processing are therefore important, especially the prevention of surface contamination by hydrogen, oxygen, or nitrogen during processing. These elements cause embrittlement of titanium and, consequently, reduce toughness and ductility.

The body-centered cubic structure of titanium (*beta-titanium*) is above 880°C (1600°F) and is ductile, whereas its hexagonal close-packed structure (*alpha-titanium*) is somewhat brittle and is very sensitive to stress corrosion. A variety of other structures (alpha, near-alpha, alpha–beta, and beta) can be obtained by alloying and heat treating, so that the properties can be optimized for specific applications. **Titanium aluminide intermetallics** (TiAl and Ti_3Al; see Section 4.2.2) have higher stiffness and lower density than conventional titanium alloys, and can withstand higher temperatures.

Production.   Ores containing titanium are first reduced to titanium tetrachloride in an arc furnace, then converted to titanium chloride in a chlorine atmosphere. The compound is reduced further to titanium metal by distillation and leaching (dissolving). This sequence forms *sponge titanium*, which is then pressed into billets, melted, and poured into ingots to be later processed into various shapes. The complexity of these multistep thermochemical operations (the *Kroll process*, after

TABLE 6.10

Properties and Typical Applications of Selected Wrought Titanium Alloys at Various Temperatures

UNS number	Nominal composition (%)	Condition	Temperature (°C)	Ultimate tensile strength (MPa)	Yield strength (MPa)	Elongation (%)	Reduction in area (%)
R50250	99.5 Ti	Annealed	25	330	240	30	55
			300	150	95		
R54520	5 Al, 2.5 Sn	Annealed	25	860	810	16	40
			300	565	450		
R56400	6 Al, 4 V	Annealed	25	1000	925	14	30
			300	725	650		
		Solution + age	25	1175	1100	10	20
			300	980	900		
R58010	13 V, 11 Cr, 3 Al	Solution + age	25	1275	1210	8	—
			425	1100	830		

the Luxembourg metallurgist W.J. Kroll) adds considerably to the cost of titanium. Recent developments in electrochemical extraction processes have reduced the number of steps involved and the energy consumption, thereby reducing the cost of producing titanium.

6.8   Refractory Metals and Alloys

There are four **refractory metals**: molybdenum, niobium, tungsten, and tantalum; they are called *refractory* because of their high melting points, and are important alloying elements in steels and superalloys. More than most other metals and alloys, refractory metals maintain their strength at elevated temperatures, and thus are of great importance in rocket engines, gas turbines, and various other aerospace applications; in the electronic, nuclear-power, and chemical industries; and as tool and die materials. The temperature range for some of these applications is on the order of 1100° to 2200°C (2000° to 4000°F), where strength and oxidation are of major concern.

6.8.1   Molybdenum

Molybdenum (Mo, from the Greek meaning lead) is a silvery-white metal and has a high melting point, high modulus of elasticity, good resistance to thermal shock, and good electrical and thermal conductivity. Molybdenum is used in greater amounts than any other refractory metal, in applications such as solid-propellant rockets, jet engines, honeycomb structures, electronic components, heating elements, and dies for die casting. The principal alloying elements are titanium and zirconium. Molybdenum is itself also an important alloying element in cast and wrought alloy steels and in heat-resistant alloys, imparting strength, toughness, and corrosion resistance. A major limitation of molybdenum alloys is their low resistance to oxidation at temperatures above 500°C (950°F), which necessitates the need for protective coatings.

Production.   The main source of molybdenum is the mineral *molybdenite* (molybdenum disulfide). The ore is first processed and the molybdenum is concentrated; it is then chemically reduced, first with oxygen and then with hydrogen. Powder metallurgy techniques also are used to produce ingots for further processing into various shapes.

6.8.2   Niobium (Columbium)

Niobium (Nb, for niobium, after Niobe, the daughter of the mythical Greek king Tantalus), also called **columbium** (after its source mineral, *columbite*), possesses good ductility and formability, and has higher oxidation resistance than other refractory metals. With various alloying elements, niobium alloys can have moderate strength and good fabrication characteristics. These alloys generally are used in rockets and missiles and in nuclear, chemical, and superconductor applications. Niobium is also an alloying element in various alloys and superalloys. The metal is processed from ores by reduction and refinement, and from powder by melting and shaping into ingots.

6.8.3   Tungsten

Tungsten (W, for *wolfram*, its European name, and from its source mineral *wolframite*; in Swedish, *tung* means "heavy" and *sten* means "stone") is the most abundant of all the refractory metals. Tungsten has the highest melting point of any metal (3410°C, 6170°F) and is notable for its high strength at elevated temperatures. However, it has high density, hence it is used for balancing weights and counterbalances in mechanical systems, including self-winding watches. It is brittle at low temperatures and has poor resistance to oxidation. As an alloying element, tungsten imparts elevated-temperature strength and hardness to steels.

Tungsten alloys are used for applications involving temperatures above 1650°C (3000°F), such as nozzle throat liners in missiles and rocket engines, circuit breakers, welding electrodes, tooling for electrical-discharge machining, and spark-plug electrodes. Tungsten carbide, with cobalt as a binder for the carbide particles, is one of the most important tool and die materials. Tungsten is processed from ore concentrates by chemical decomposition, then reduced, and further processed by powder metallurgy techniques in a hydrogen atmosphere.

6.8.4 Tantalum

Tantalum (Ta, after the mythical Greek king Tantalus) is characterized by its high melting point (3000°C, 5425°F), high density, good ductility, and resistance to corrosion; however, it has poor chemical resistance at temperatures above 150°C (300°F). Tantalum is used extensively in electrolytic capacitors and in various components in the electrical, electronic, and chemical industries. It also is used for thermal applications, such as in furnaces and in acid-resistant heat exchangers. A variety of tantalum-based alloys are available in various forms for use in missiles and aircraft. Tantalum also is used as an alloying element. It is processed by techniques similar to those used for processing niobium.

6.9 Beryllium

Steel-gray in color, **beryllium** (Be, from the ore *beryl*) has a high strength-to-weight ratio. Unalloyed beryllium is used in rocket nozzles, space and missile structures, aircraft disk brakes, and precision instruments and mirrors; it is also used in nuclear and X-ray applications because of its low neutron absorption. Beryllium is also an alloying element, and its alloys of copper and nickel are used in various applications, including springs (*beryllium copper*), electrical contacts, and nonsparking tools for use in such explosive environments as mines and metal-powder production. Beryllium and its oxide are toxic, and should be handled accordingly.

6.10 Zirconium

Zirconium (Zr, from the mineral *zircon*) is silvery in appearance; it has good strength and ductility at elevated temperatures and has good corrosion resistance because of an adherent oxide film. Zirconium is used in electronic components and in nuclear-power reactor applications because of its low neutron absorption.

6.11 Low-melting Alloys

Low-melting alloys are so named because of their relatively low melting points. The major metals in this category are lead, zinc, tin, and their alloys.

6.11.1   Lead

Lead (Pb, after *plumbum*, the root of the word "plumber") has the properties of high density, resistance to corrosion (by virtue of the stable lead-oxide layer that forms to protect the surface), softness, low strength, ductility, and good workability. Alloying lead with various elements (such as antimony and tin) enhances its desirable properties, making it suitable for piping, collapsible tubing, bearing alloys (Babbitt), cable sheathing, foil (as thin as 0.01 mm), roofing, and lead–acid storage batteries. Lead also is used for damping vibrations, radiation shielding against X-rays, ammunition, and as a solid lubricant for hot-metal-forming operations. The oldest known lead artifacts were made in about 3000 b.c. Lead pipes made by the Romans and installed in the Roman baths in Bath, England, two millennia ago, are still in use.

Because of its toxicity, major efforts are being made to replace lead with other elements, such as *lead-free solders* (Section 32.3.1). The most important mineral source of lead is *galena* (PbS); it is mined, smelted, and refined by chemical treatments.

6.11.2   Zinc

Zinc (Zn, from the Latin *zincum*, although its origins are not clear) is bluish-white in color and is the metal that is fourth most utilized industrially, after iron, aluminum, and copper. It has three major uses: for galvanizing iron, steel sheet, and wire; as an alloy in other metals; and as a metal for castings. In **galvanizing**, zinc serves as an anode and protects steel (cathode) from corrosive attack should the coating be scratched or punctured. Zinc is also used as an alloying element; brass, for example, is an alloy of copper and zinc. In zinc-based alloys the major alloying elements are aluminum, copper, and magnesium, imparting strength and providing dimensional control during casting of the metal.

Zinc-based alloys are used extensively in die casting, for making such products as fuel pumps and grills for automobiles, components for household appliances such as vacuum cleaners and washing machines, kitchen equipment, machinery parts, and photoengraving equipment. Another use for zinc is in superplastic alloys (Section 2.2.7). A very fine grained 78% Zn–22% Al sheet is a common example of a superplastic zinc alloy that can be formed by methods used for forming plastics or metals.

Production. The principal mineral source for zinc is zinc sulfide, also called *zincblende*. The ore is first roasted in air and converted to zinc oxide. It is then reduced to zinc, either electrolytically (using sulfuric acid) or by heating it in a furnace with coal, which causes the molten zinc to separate.

6.11.3   Tin

Although used in small amounts compared with iron, aluminum, or copper, **tin** (Sn, from the Latin *stannum*) is an important metal. The most extensive use of tin (a silver-white, lustrous metal) is as a protective coating on steel sheets (*tin plates*) used in making containers (*tin cans*), for food and various other products. The low shear strength of the tin coatings on steel sheet improves its deep drawability (Section 16.7.1). Unlike galvanized steels, if this coating is punctured or destroyed, the steel corrodes because the tin is cathodic.

Unalloyed tin is used in such applications as a lining for water distillation plants and as a molten layer of metal in the production of float glass plate (Section 18.3.1). Tin-based alloys (also called **white metals**) generally contain copper, antimony, and lead; these alloying elements impart hardness, strength, and corrosion resistance.

Tin itself is an alloying element for dental alloys and for bronze (copper–tin alloy), titanium, and zirconium alloys. Tin–lead alloys are common soldering materials (Section 32.3), with a wide range of compositions and melting points.

Because of their low friction coefficients (which result from low shear strength and low adhesion), some tin alloys are used as journal-bearing materials. Known as **babbitts** (after the American goldsmith I. Babbitt, 1799–1862), these alloys contain tin, copper, and antimony. **Pewter,** an alloy of tin, copper, and antimony, is used for tableware, hollowware, and decorative artifacts. Tin alloys are also used in making organ pipes. The most important tin mineral is *cassiterite* (tin oxide), which is of low grade. The ore is first mined, then concentrated by various techniques, smelted, refined, and cast into ingots for further processing.

6.12   Precious Metals

The most important precious (costly) metals, also called **noble metals,** are the following:

- **Gold** (Au, from the Latin *aurum*) is soft and ductile, and has good corrosion resistance at any temperature. Typical applications include jewelry, coinage, reflectors, gold leaf for decorative purposes, dental work, electroplating, and electrical contacts and terminals.
- **Silver** (Ag, from the Latin *argentum*) is ductile and has the highest electrical and thermal conductivity of any metal (see Table 3.2); however, it develops an oxide film that adversely affects its surface characteristics and appearance. Typical applications for silver include tableware, jewelry, coinage, electroplating, solders, bearing linings, and food and chemical equipment. *Sterling silver* is an alloy of silver and 7.5% copper.
- **Platinum** (Pt) is a soft, ductile, grayish-white metal that has good corrosion resistance, even at elevated temperatures. Platinum alloys are used as electrical contacts; for spark-plug electrodes; as catalysts for automobile pollution-control devices; in filaments and nozzles; in dies for extruding glass fibers (Section 18.3.4), in thermocouples; and in jewelry and dental work.

6.13   Shape-memory Alloys (Smart Materials)

Shape-memory alloys are unique in that, after being plastically deformed into various shapes at room temperature, they return to their original shape upon heating. For example, a piece of straight wire, made of such a material, can be wound into the shape of a helical spring; when heated, the spring uncoils and returns to its original straight shape. Shape-memory alloys can be used to generate motion and/or force in temperature-sensitive actuators. The behavior of these alloys, also called **smart materials,** can be reversible; that is, the shape can switch back and forth repeatedly upon application and removal of heat.

A typical shape-memory alloy is 55% Ni–45% Ti (*Nitinol*); other alloys are copper–aluminum–nickel, copper–zinc–aluminum, iron–manganese–silicon, and titanium–nickel–hafnium. Shape-memory alloys generally also have such properties as good ductility, corrosion resistance, and high electrical conductivity.

Applications of shape-memory alloys include sensors, stents for blocked arteries, relays, pumps, switches, connectors, clamps, fasteners, and seals. As an example, a nickel–titanium valve has been made to protect people from being scalded in sinks, tubs, and showers. It is installed directly into the piping system and brings the water flow down to a trickle within 3 s after the water temperature reaches 47°C (116°F). More recent developments include thin-film shape-memory alloys deposited on polished silicon substrates for use in microelectromechanical (MEMS) devices (Chapter 29).

6.14 Amorphous Alloys (Metallic Glasses)

A class of metal alloys that, unlike metals, do not have a long-range crystalline structure, is called **amorphous alloys**; they have no grain boundaries, and their atoms are packed randomly and tightly. The amorphous structure was first obtained in the late 1960s by **rapid solidification** of a molten alloy (Section 11.6). Because their structure resembles that of glasses, these alloys are also called **metallic glasses**.

Amorphous alloys typically contain iron, nickel, and chromium, which are alloyed with carbon, phosphorus, boron, aluminum, and silicon, and are available as wire, ribbon, strip, and powder. One application is for faceplate inserts on golf-club heads; the alloy has a composition of zirconium, beryllium, copper, titanium, and nickel and is made by die casting. Another application is in hollow aluminum baseball bats, coated with a composite of amorphous metal by thermal spraying, and is said to improve the performance of the bat.

Amorphous alloys exhibit excellent corrosion resistance, good ductility, high strength, and very low magnetic hysteresis (utilized in magnetic steel cores for transformers, generators, motors, lamp ballasts, magnetic amplifiers, and linear accelerators). The low magnetic hysteresis loss provides greatly improved efficiency; however, fabrication costs are significant. Amorphous steels have been demonstrated to have strengths twice those of high-strength steels, and with potential applications in large structures; however, they are presently cost prohibitive. A major application for the superalloys of rapidly solidified powders is the consolidation into near-net shapes for parts used in aerospace engines.

6.15 Metal Foams

Metal foams are structures where the metal consists of only 5–20% of the structure's volume, as shown in Fig. 6.3. Usually made of aluminum alloys (but also of titanium, tantalum, and others), metal foams can be produced by blowing air into molten metal and tapping the froth that forms at the surface; this froth then solidifies into a foam. Other approaches to producing **metal foam** include (a) chemical vapor deposition (Section 34.6.2) onto a carbon foam lattice, (b) depositing metal powders from a slurry onto a polymer foam lattice, followed by sintering (Section 17.4) to fuse the metals and burn off the polymer, (c) doping molten or powder metals with titanium hydride (TiH_2), which then releases hydrogen gas at the elevated casting or sintering temperatures, and (d) pouring molten metal into a porous salt and, upon cooling, leaching out the salt with acid.

Metal foams have unique combinations of strength-to-density and stiffness-to-density ratios, although these ratios are not as high as the base metals themselves. However, metal foams are very lightweight and thus are attractive materials, especially for aerospace applications. Because of their porosity, other applications of metal foams include filters and orthopedic implants. More recent developments include nickel–manganese–gallium metal foams with shape-memory characteristics.

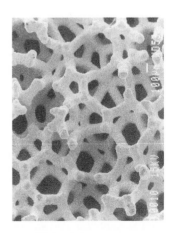

FIGURE 6.3 Structure of a metal foam used in orthopedic implants to encourage bone ingrowth. *Source:* Courtesy of Zimmer, Inc.

SUMMARY

- Nonferrous metals and alloys include a very broad range of materials. The most common are aluminum, magnesium, and copper and their alloys, with a wide range of applications. For high temperature service, nonferrous metals include nickel, titanium, refractory alloys (molybdenum, niobium, tungsten, tantalum), and superalloys. Other nonferrous metal categories include low-melting alloys (lead, zinc, tin) and precious metals (gold, silver, platinum).

- Nonferrous alloys have a wide variety of desirable properties, such as strength, toughness, hardness, and ductility; resistance to high temperature, creep, and oxidation; a wide range of physical, thermal, and chemical properties; and high strength-to-weight and stiffness-to-weight ratios (particularly for aluminum and titanium). Nonferrous alloys can be heat treated to impart certain specific properties.

- Shape-memory alloys (smart materials) have unique properties, with numerous applications in a variety of products as well as in manufacturing operations.

- Amorphous alloys (metallic glasses) have properties that are superior to other materials; available in various forms, they have numerous applications.

- Metal foams are very lightweight and thus are attractive for aerospace and various other applications.

- As with all materials, the selection of a nonferrous material for a particular application requires a careful consideration of several factors, including design and service requirements, long-term effects, chemical affinity to other materials, environmental attack, and cost.

KEY TERMS

Amorphous alloys	Low-melting alloys	Pewter	Shape-memory alloys
Babbitts	Metal foam	Precious metals	Smart materials
Brass	Metallic glasses	Pyrometallurgy	Smelted
Bronze	Nonferrous	Refractory metals	Superalloys
Galvanizing			

BIBLIOGRAPHY

ASM Handbook, Vol. 2: **Properties and Selection: Nonferrous Alloys and Special-Purpose Materials**, ASM International, 1990.

ASM Specialty Handbook, **Aluminum and Aluminum Alloys**, ASM International, 1993.

ASM Specialty Handbook, **Copper and Copper Alloys**, ASM International, 2001.

ASM Specialty Handbook, **Heat-Resistant Materials**, ASM International, 1997.

ASM Specialty Handbook, **Magnesium and Magnesium Alloys**, ASM International, 1999.

ASM Specialty Handbook, **Nickel, Cobalt, and Their Alloys**, ASM International, 2000.

Donachie, M.J. (ed.), **Titanium: A Technical Guide**, 2nd ed., ASM International, 2000.

Donachie, M.J., and Donachie, S.J., **Superalloys: A Technical Guide**, 2nd ed., ASM International, 2002.

Fremond, M., and Miyazaki, S., **Shape-Memory Alloys**, Springer Verlag, 1996.

Geddes, B., H. Leon, and Huang, X., **Superalloys: Alloying and Performance**, ASM International, 2010.

Kaufman, J.G., **Introduction to Aluminum Alloys and Tempers**, ASM International, 2000.

Lagoudas, D.C. (ed.), **Shape Memory Alloys: Modeling and Engineering Applications**, Springer, 2008.

Leo, D.J., **Engineering Analysis of Smart Material Systems**, Wiley, 2007.

Lutjering, G., and Williams, J.C., **Titanium**, 2nd ed., Springer, 2007.

Russel, A., and Lee, K.L., **Structure-Property Relations in Nonferrous Metals**, Wiley-Interscience, 2005.

Schwartz, M., **Smart Materials**. CRC Press, 2008.

REVIEW QUESTIONS

6.1 Given the abundance of aluminum in the Earth's crust, explain why it is more expensive than steel.

6.2 Why is magnesium often used as a structural material in power hand tools? Why are its alloys used instead of pure magnesium?

6.3 What are the major uses of copper? What are the alloying elements in brass and bronze, respectively?

6.4 What are superalloys? Why are they so named?

6.5 What properties of titanium make it attractive for use in race-car and jet-engine components? Why is titanium not used widely for engine components in passenger cars?

6.6 Which properties of each of the major refractory metals define their most useful applications?

6.7 What are metallic glasses? Why is the word "glass" used for these materials?

6.8 What is the composition of (a) babbitts, (b) pewter, and (c) sterling silver?

6.9 Name the materials described in this chapter that have the highest (a) density, (b) electrical conductivity, (c) thermal conductivity, (d) strength, and (e) cost.

6.10 What are the major uses of gold and silver, other than in jewelry?

6.11 Describe the advantages to using zinc as a coating for steel.

6.12 What are nanomaterials? Why are they being developed?

6.13 Why are aircraft fuselages made of aluminum alloys, even though magnesium is the lightest metal?

6.14 How is metal foam produced?

6.15 What metals have the lowest melting points? What applications for these metals take advantage of their low melting points?

QUALITATIVE PROBLEMS

6.16 Explain why cooking utensils generally are made of stainless steels, aluminum, or copper.

6.17 Would it be advantageous to plot the data in Table 6.1 in terms of cost per unit weight rather than cost per unit volume? Explain and give some examples.

6.18 Compare the contents of Table 6.3 with those in various other tables and data on materials in this book, then comment on which of the two hardening processes (heat treating and work hardening) is more effective in improving the strength of aluminum alloys.

6.19 What factors other than mechanical strength should be considered in selecting metals and alloys for high-temperature applications? Explain.

6.20 Assume that, for geopolitical reasons, the price of copper increases rapidly. Name two metals with similar mechanical and physical properties that can be substituted for

copper. Comment on your selection and any observations you make.

6.21 If aircraft, such as a Boeing 757, are made of 79% aluminum, why are automobiles made predominantly of steel?

6.22 Portable (notebook) computers and digital cameras can have their housing made of magnesium. Why?

6.23 Most household wiring is made of copper wire. By contrast, grounding wire leading to satellite dishes and the like is made of aluminum. Explain the reason.

6.24 The example in this chapter showed the benefits of making cars from aluminum alloys. However, the average amount of steel in cars has increased in the past decade. List reasons to explain these two observations.

6.25 If tungsten is the highest melting-point metal, why are no high temperature parts in Fig. 6.1 made from tungsten?

QUANTITATIVE PROBLEMS

6.26 A simply supported rectangular beam is 25 mm wide and 1 m long, and it is subjected to a vertical load of 10 kg at its center. Assume that this beam could be made of any of the materials listed in Table 6.1. Select three different materials,

and for each, calculate the beam height that would cause each beam to have the same maximum deflection. Calculate the ratio of the cost for each of the three beams.

6.27 Obtain a few aluminum beverage cans, cut them, and measure their wall thicknesses. Using data in this chapter and simple formulas for thin-walled, closed-end pressure vessels, calculate the maximum internal pressure these cans can withstand before yielding. (Assume that the can is a thin-walled, closed-end, internally pressurized vessel.)

6.28 Beverage cans usually are stacked on top of each other in stores. Use the information from Problem 6.24, and, referring to textbooks on the mechanics of solids, estimate the crushing load each of these cans can withstand.

6.29 Using strength and density data, determine the minimum weight of a 1 m-long tension member that must support 3000 N if it is manufactured from (a) 3003-O aluminum, (b) 5052-H34 aluminum, (c) AZ31B-F magnesium, (d) any brass alloy, and (e) any bronze alloy.

6.30 Plot the following for the materials described in this chapter: (a) yield strength versus density, (b) modulus of elasticity versus strength, (c) modulus of elasticity versus relative cost, and (d) electrical conductivity versus density.

SYNTHESIS, DESIGN, AND PROJECTS

6.31 Because of the number of processes involved in making metals, the cost of raw materials depends on the condition (hot or cold rolled), shape (plate, sheet, bar, tubing), and size of the metals. Make a survey of the technical literature, obtain price lists or get in touch with suppliers, and prepare a list indicating the cost per 100 kg of the nonferrous materials described in this chapter, available in different conditions, shapes, and sizes.

6.32 The materials described in this chapter have numerous applications. Make a survey of the available literature in the bibliography, and prepare a list of several specific parts or components and applications, indicating the types of materials used.

6.33 Name products that would not have been developed to their advanced stages (as we find them today) if alloys having high strength, high corrosion resistance, and high creep resistance (all at elevated temperatures) had not been developed.

6.34 Assume that you are the technical sales manager of a company that produces nonferrous metals. Choose any one of the metals and alloys described in this chapter, and

prepare a brochure, including some illustrations, for use as sales literature by your staff in their contact with potential customers.

6.35 Give some applications for (a) amorphous metals, (b) precious metals, (c) low-melting alloys, and (d) nanomaterials.

6.36 Describe the advantages of making products with multilayer materials. (For example, aluminum bonded to the bottom of stainless-steel pots.)

6.37 In the text, magnesium was described as the lightest engineering metal. Is it also the lightest metal? Explain.

6.38 Review the technical literature and the Internet and summarize the rare earth metals, their sources, and their main applications.

6.39 Review the technical literature, and write a detailed description of how magnesium is produced from sea water.

6.40 If you were to design an implant for use in the human body, what materials would you exclude? Which metals are possible for such applications? Of these, list three that you feel are best.

Polymers: Structure, General Properties, and Applications

- Polymers display a wide range of properties and have several advantages over metallic materials, including low cost and ease of manufacturing; for these reasons, polymers continue to be among the most commonly used materials.

- This chapter first describes the structure of polymers, the polymerization process, crystallinity, and the glass-transition temperature.

- Mechanical properties and their dependence on temperature and deformation rate are then discussed.

- There are two basic types of polymers: thermoplastics and thermosets. Thermoplastics follow a basic manufacturing strategy of heating them until they soften or melt, and then shaping them into the desired product. Thermosets involve precursors that are formed to a desired shape and set through polymerization or cross-linking between polymer chains.

- The chapter also describes the properties and uses of elastomers, or rubbers.

- The general properties, typical applications, advantages, and limitations of polymers are discussed throughout the chapter, with several specific examples.

7.1 Introduction

The word **plastics** was first used as a noun in 1909 and is commonly employed as a synonym for **polymers**, a term first used in 1866. Plastics are unique in that they have extremely large molecules (*macromolecules* or *giant molecules*). Consumer and industrial products made of plastics include food and beverage containers, packaging, signs, housewares, housings for computers and monitors, textiles (clothing), medical devices, foams, paints, safety shields, toys, appliances, lenses, gears, electronic and electrical products, and automobile and aircraft bodies and components.

Because of their many unique and diverse properties, polymers increasingly have replaced metallic components in such applications as automobiles, civilian and military aircraft, sporting goods, toys, appliances, and office equipment. These substitutions reflect the advantages of polymers in terms of the following characteristics:

- Relatively low cost (see Table 6.1)
- Corrosion resistance and resistance to chemicals
- Low electrical and thermal conductivity
- Low density
- High strength-to-weight ratio, particularly when reinforced

- Noise reduction
- Wide choice of colors and transparencies
- Complexity of design possibilities and ease of manufacturing
- Other characteristics that may or may not be desirable (depending on the application), such as low strength and stiffness (Table 7.1), high coefficient of thermal expansion, low useful-temperature range—up to about 350°C (660°F)—and lower dimensional stability in service over a period of time.

The word *plastic* is from the Greek *plastikos*, meaning "capable of being molded and shaped." Plastics can be formed, cast, machined, and joined into various shapes with relative ease. Minimal additional surface-finishing operations, if any at all, are required; this characteristic provides an important advantage over metals. Plastics are available commercially as film, sheet, plate, rods, and tubing of various cross-sections.

Video Solution 7.1 Selection of Materials for a Lightweight Beam.

TABLE 7.1

Range of Mechanical Properties for Various Engineering Plastics at Room Temperature

Material	Ultimate tensile strength (MPa)	Elastic modulus (GPa)	Elongation (%)	Poisson's ratio, v
Thermoplastics:				
Acrylonitrile-butadiene-styrene (ABS)	28–55	1.4–2.8	75–5	—
ABS, reinforced	100	7.5	—	0.35
Acetal	55–70	1.4–3.5	75–25	—
Acetal, reinforced	135	10	—	0.35–0.40
Acrylic	40–75	1.4–3.5	50–5	—
Cellulosic	10–48	0.4–1.4	100–5	—
Fluorocarbon	7–48	0.7–2	300–100	0.46–0.48
Nylon	55–83	1.4–2.8	200–60	0.32–0.40
Nylon, reinforced	70–210	2–10	10–1	—
Polycarbonate	55–70	2.5–3	125–10	0.38
Polycarbonate, reinforced	110	6	6–4	—
Polyester	55	2	300–5	0.38
Polyester, reinforced	110–160	8.3–12	3–1	—
Polyethylene	7–40	0.1–1.4	1000–15	0.46
Polypropylene	20–35	0.7–1.2	500–10	—
Polypropylene, reinforced	40–100	3.5–6	4–2	—
Polystyrene	14–83	1.4–4	60–1	0.35
Polyvinyl chloride	7–55	0.014–4	450–40	—
Thermosets:				
Epoxy	35–140	3.5–17	10–1	—
Epoxy, reinforced	70–1400	21–52	4–2	—
Phenolic	28–70	2.8–21	2–0	—
Polyester, unsaturated	30	5–9	1–0	—
Elastomers:				
Chloroprene (neoprene)	15–25	1–2	100–500	0.5
Natural rubber	17–25	1.3	75–650	0.5
Silicone	5–8	1–5	100–1100	0.5
Styrene-butadiene	10–25	2–10	250–700	0.5
Urethane	20–30	2–10	300–450	0.5

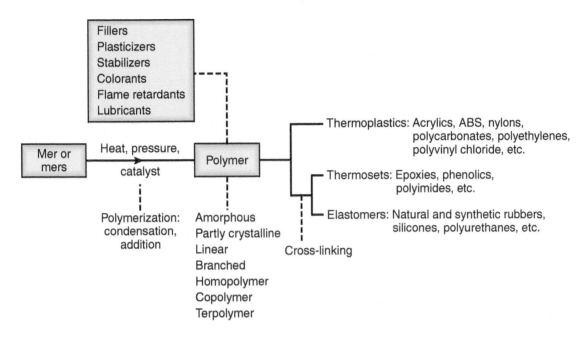

FIGURE 7.1 Outline of the topics described in this chapter.

An outline of the basic process for making synthetic polymers is given in Fig. 7.1. In polyethylene, only carbon and hydrogen atoms are involved, but other polymer compounds can be obtained by including chlorine, fluorine, sulfur, silicon, nitrogen, and oxygen. As a result, an extremely wide range of polymers—having among them an equally wide range of properties—has been developed.

7.2 The Structure of Polymers

The properties of polymers depend largely on the structures of individual polymer molecules, molecule shape and size, and the arrangement of molecules to form a polymer structure. Polymer molecules are characterized by their very large size, a feature that distinguishes them from most other organic chemical compositions. Polymers are **long-chain molecules** that are formed by *polymerization*, that is, by the linking and cross-linking of different monomers. A **monomer** is the basic building block of a polymer. The word mer (from the Greek *meros*, meaning "part") indicates the smallest repetitive unit, thus the term is similar to that of *unit cell* in crystal structures of metals (Section 1.3).

The word **polymer** means "many mers," generally repeated hundreds or thousands of times in a chainlike structure. Most monomers are *organic materials*, in which carbon atoms are joined in *covalent* (electron-sharing) bonds with other atoms (such as hydrogen, oxygen, nitrogen, fluorine, chlorine, silicon, and sulfur). An ethylene molecule (Fig. 7.2) is an example of a simple monomer, consisting of carbon and hydrogen atoms.

7.2.1 Polymerization

Monomers can be linked in repeating units to make longer and larger molecules by a chemical process called a **polymerization reaction**. Although there are several

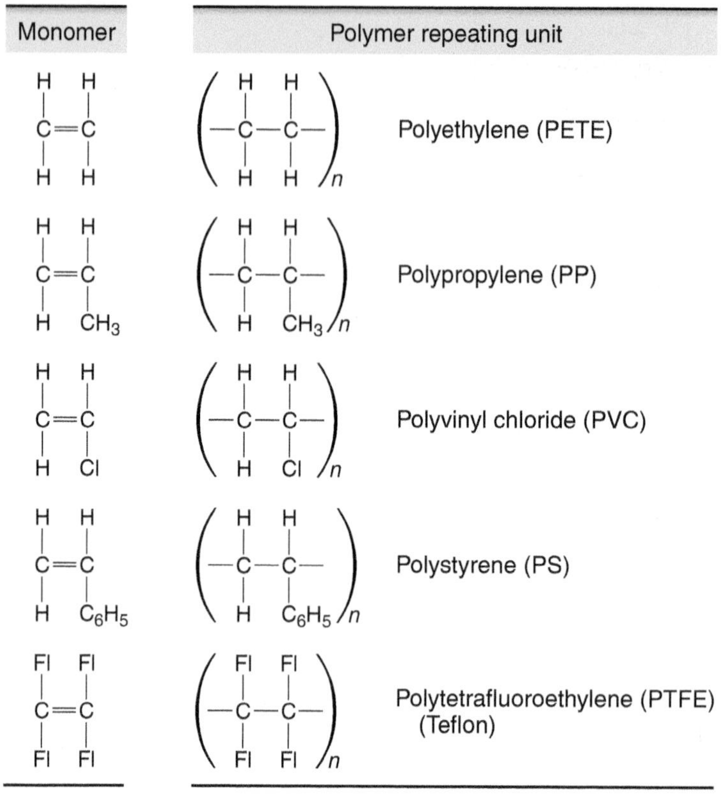

Monomer	Polymer repeating unit	
H H | | C＝C | | H H	⎛ H H ⎞ ⎜ —C—C— ⎟ ⎝ H H ⎠$_n$	Polyethylene (PETE)
H H | | C＝C | | H CH$_3$	⎛ H H ⎞ ⎜ —C—C— ⎟ ⎝ H CH$_3$ ⎠$_n$	Polypropylene (PP)
H H | | C＝C | | H Cl	⎛ H H ⎞ ⎜ —C—C— ⎟ ⎝ H Cl ⎠$_n$	Polyvinyl chloride (PVC)
H H | | C＝C | | H C$_6$H$_5$	⎛ H H ⎞ ⎜ —C—C— ⎟ ⎝ H C$_6$H$_5$ ⎠$_n$	Polystyrene (PS)
Fl Fl | | C＝C | | Fl Fl	⎛ Fl Fl ⎞ ⎜ —C—C— ⎟ ⎝ Fl Fl ⎠$_n$	Polytetrafluoroethylene (PTFE) (Teflon)

FIGURE 7.2 Molecular structure of various polymers; these are examples of the basic building blocks for plastics.

variations, two polymerization processes are important: condensation and addition polymerization.

In **condensation polymerization** (Fig. 7.3a), polymers are produced by the formation of bonds between two types of reacting mers. A characteristic of this reaction is that reaction by-products (such as water) are condensed out (hence the word *condensation*). This process is also known as **step-growth** or **step-reaction polymerization**, because the polymer molecule grows step-by-step until all of one reactant is consumed.

In **addition polymerization**, also called **chain-growth** or **chain-reaction polymerization**, bonding takes place without reaction by-products, as shown in Fig. 7.3b. It is called *chain reaction* because of the high rate at which long molecules form simultaneously, usually within a few seconds. This rate is much higher than that in condensation polymerization. In addition polymerization, an *initiator* is added to open the double bond between two carbon atoms, which then begins the linking process by adding several more monomers to a growing chain. For example, ethylene monomers (Fig. 7.3b) link to produce *polyethylene*; other examples of addition-formed polymers are given in Fig. 7.2.

Molecular Weight.   The sum of the molecular weights of the mers in a representative chain is known as the *molecular weight* of the polymer; the higher the molecular weight of a given polymer, the greater the average chain length. Most commercial polymers have a molecular weight between 10,000 and 10,000,000. Because polymerization is a random event, the polymer chains produced are not all of equal

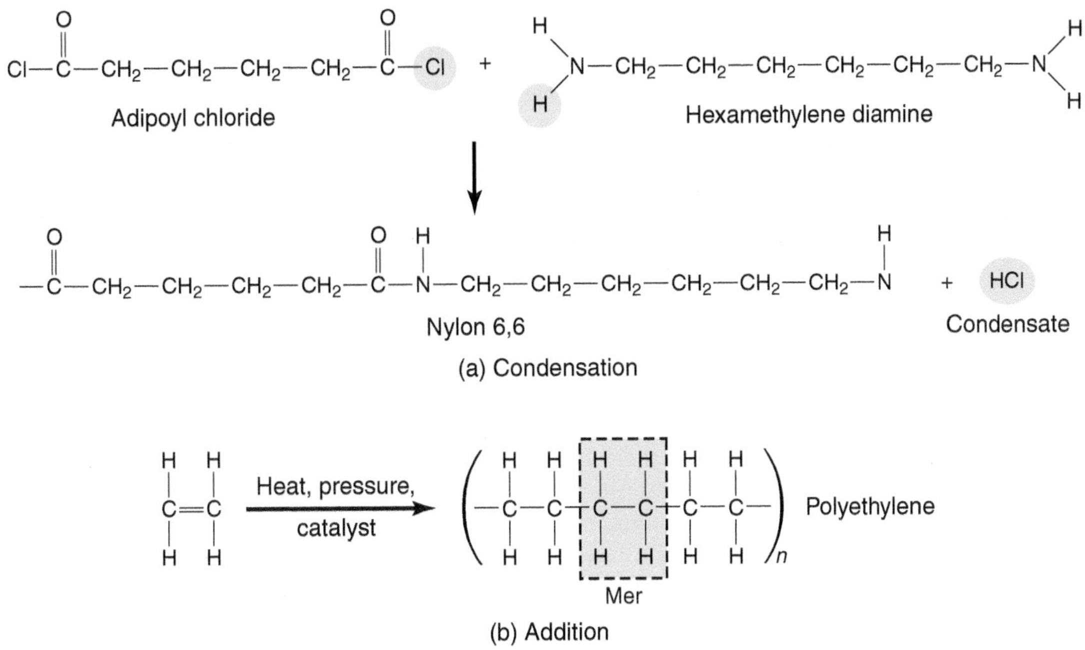

$$Cl-\overset{\overset{\displaystyle O}{\|}}{C}-CH_2-CH_2-CH_2-CH_2-\overset{\overset{\displaystyle O}{\|}}{C}-Cl \quad + \quad \overset{H}{\underset{H}{\diagup}}N-CH_2-CH_2-CH_2-CH_2-CH_2-CH_2-N\overset{H}{\underset{H}{\diagdown}}$$

Adipoyl chloride   Hexamethylene diamine

↓

$$-\overset{\overset{\displaystyle O}{\|}}{C}-CH_2-CH_2-CH_2-CH_2-\overset{\overset{\displaystyle O}{\|}}{C}-\overset{\overset{\displaystyle H}{|}}{N}-CH_2-CH_2-CH_2-CH_2-CH_2-CH_2-\overset{\overset{\displaystyle H}{|}}{N} \quad + \quad HCl$$

Nylon 6,6   Condensate

(a) Condensation

$$\overset{\overset{\displaystyle H}{|}\ \overset{\displaystyle H}{|}}{\underset{\underset{\displaystyle H}{|}\ \underset{\displaystyle H}{|}}{C=C}} \quad \xrightarrow[\text{catalyst}]{\text{Heat, pressure,}} \quad \left(-\overset{\overset{\displaystyle H}{|}}{\underset{\underset{\displaystyle H}{|}}{C}}-\overset{\overset{\displaystyle H}{|}}{\underset{\underset{\displaystyle H}{|}}{C}}\!\!\left|\overset{\overset{\displaystyle H}{|}}{\underset{\underset{\displaystyle H}{|}}{C}}-\overset{\overset{\displaystyle H}{|}}{\underset{\underset{\displaystyle H}{|}}{C}}\right|\overset{\overset{\displaystyle H}{|}}{\underset{\underset{\displaystyle H}{|}}{C}}-\overset{\overset{\displaystyle H}{|}}{\underset{\underset{\displaystyle H}{|}}{C}}-\right)_{\!\!n} \quad \text{Polyethylene}$$

Mer

(b) Addition

FIGURE 7.3 Examples of polymerization. (a) Condensation polymerization of nylon 6,6 and (b) addition polymerization of polyethylene molecules from ethylene mers.

length, although the chain lengths produced fall into a traditional distribution curve (as described in Section 36.7). The molecular weight of a polymer is determined on a statistical basis by averaging.

The spread of the molecular weights in a chain is called the **molecular weight distribution**. A polymer's molecular weight and its distribution have a major influence on its properties. For example, the tensile and the impact strength, the resistance to cracking, and the viscosity (in the molten state) of the polymer all increase with increasing molecular weight (Fig. 7.4).

Degree of Polymerization.   It is convenient to express the size of a polymer chain in terms of the *degree of polymerization* (DP), defined as the ratio of the molecular weight of the polymer to the molecular weight of the repeating unit. For example, polyvinyl chloride (PVC) has a mer weight of 62.5; thus, the DP of PVC with a molecular weight of 50,000 is $50{,}000/62.5 = 800$. In terms of polymer processing (described in Chapter 19), the higher the DP, the higher is the polymer's viscosity or its resistance to flow (Fig. 7.4). On the one hand, high viscosity adversely affects the ease of shaping and, thus, raises the overall cost of processing; moreover, high DP can result in stronger polymers.

Bonding.   During polymerization, the monomers are linked together by **covalent bonds** (Section 1.2), forming a polymer chain. Because of their strength, covalent bonds also are called **primary bonds.** The polymer chains are, in turn, held together by **secondary bonds,** such as van der Waals bonds, hydrogen bonds, and ionic bonds. Secondary bonds are weaker than primary bonds by one to two orders of magnitude.

In a given polymer, the increase in strength and viscosity with molecular weight is due, in part, to the fact that the longer the polymer chain, the

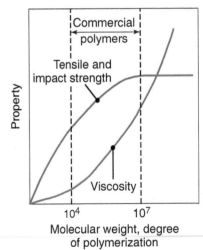

FIGURE 7.4 Effect of molecular weight and degree of polymerization on the strength and viscosity of polymers.

greater is the energy required to overcome the combined strength of the secondary bonds. For example, ethylene polymers having DPs of 1, 6, 35, 140, and 1350 at room temperature are, respectively, in the form of gas, liquid, grease, wax, and hard plastic.

Linear Polymers.   The chainlike polymers shown in Fig. 7.2 are called *linear polymers* because of their sequential structure (Fig. 7.5a); however, a linear molecule is not necessarily straight in shape. In addition to those shown in the figure, other linear polymers are polyamides (nylon 6,6) and polyvinyl fluoride. Generally, a polymer consists of more than one type of structure; thus, a linear polymer may contain some branched and some cross-linked chains. As a result of branching and cross-linking, the polymer's properties are changed significantly.

Branched Polymers.   The properties of a polymer depend not only on the type of monomers, but also on their arrangement in the molecular structure. In *branched polymers* (Fig. 7.5b), side-branch chains are attached to the main chain during synthesis of the polymer. Branching interferes with the relative movement of the molecular chains, and as a result, their resistance to deformation and stress cracking is increased. The density of branched polymers is lower than that of linear-chain polymers, because the branches interfere with the packing efficiency of polymer chains.

The behavior of branched polymers can be compared to that of linear-chain polymers, by making an analogy with a pile of tree branches (*branched polymers*) and a bundle of straight logs (*linear polymers*). Note that it is more difficult to move a branch within the pile of branches than to move a log within its bundle. The three-dimensional entanglements of branches make their movements more difficult, a phenomenon akin to increased strength of the polymer.

Cross-linked Polymers.   Generally three-dimensional in structure, *cross-linked polymers* have adjacent chains linked by covalent bonds (Fig. 7.5c). Polymers with a

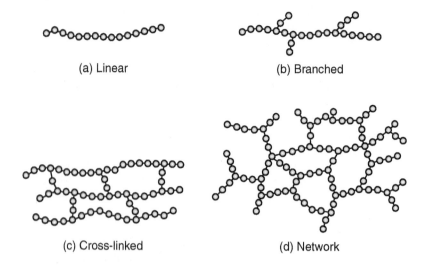

(a) Linear   (b) Branched

(c) Cross-linked   (d) Network

FIGURE 7.5   Schematic illustration of polymer chains. (a) Linear structure—thermoplastics such as acrylics, nylons, polyethylene, and polyvinyl chloride have linear structures. (b) Branched structure, such as in polyethylene. (c) Cross-linked structure—many rubbers, or elastomers, have this structure, and the vulcanization of rubber produces this structure. (d) Network structure, which is basically highly cross-linked—examples are thermosetting plastics, such as epoxies and phenolics.

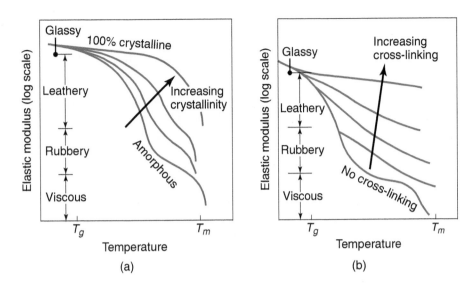

FIGURE 7.6 Behavior of polymers as a function of temperature and (a) degree of crystallinity and (b) cross-linking. The combined elastic and viscous behavior of polymers is known as viscoelasticity.

cross-linked structure are called **thermosets** or **thermosetting plastics,** such as epoxies, phenolics, and silicones. Cross-linking has a major influence on the properties of polymers, generally imparting hardness, strength, stiffness, brittleness, and better dimensional stability (see Fig. 7.6), and the on **vulcanization** of rubber (Section 7.9).

Network Polymers.   These polymers consist of spatial (three-dimensional) networks consisting of three or more active covalent bonds (Fig. 7.5d). A highly cross-linked polymer also is considered a *network polymer.* Thermoplastic polymers that already have been shaped can be cross-linked, to obtain higher strength by subjecting them to high-energy radiation, such as ultraviolet light, X-rays, or electron beams. However, excessive radiation can cause degradation of the polymer.

Copolymers and Terpolymers.   If the repeating units in a polymer chain are all of the same type, the molecule is called a *homopolymer.* However, as with solid-solution metal alloys (Section 4.2), two or three different types of monomers can be combined to develop certain properties and characteristics, such as improved strength, toughness, and formability of the polymer. (a) *Copolymers* contain two types of polymers; for example, styrene-butadiene, widely used for automobile tires. (b) *Terpolymers* contain three types; for example, acrylonitrile-butadiene-styrene (ABS), used for helmets, telephones, and refrigerator liners.

CASE STUDY 7.1   **Dental and Medical Bone Cement**

Polymethylmethacrylate (PMMA) is an acrylic polymer commonly used in dental and medical applications as an adhesive, often referred to as bone cement. There are several forms of PMMA, but the adhesive is one common form involving an addition-polymerization reaction. PMMA is delivered in two parts: a powder and a liquid, which are hand-mixed. The liquid wets and partially dissolves

(continued)

the powder, resulting in a liquid with a viscosity similar to that of vegetable oil. The viscosity increases markedly until a doughy state is reached, in about five minutes. The dough fully hardens in an additional five minutes.

The powder consists of high-molecular-weight poly[(methylmethacrylate)-costyrene] particles, about 50 μm in diameter, and contain a small volume fraction of benzoyl peroxide. The liquid consists of methyl methacrylate (MMA) monomer, with a small amount of dissolved n,n dimethylptoluidine (DMPT). When the liquid and powder are mixed, the DMPT cleaves the benzoyl peroxide molecule into two parts, to form a catalyst with a free electron (sometimes referred to as a free radical). This catalyst causes rapid growth of PMMA from the MMA mers, so that the final material is a composite of high-molecular-weight PMMA particles interconnected by PMMA chains. An illustration of a fully set bone cement is given in Fig. 7.7.

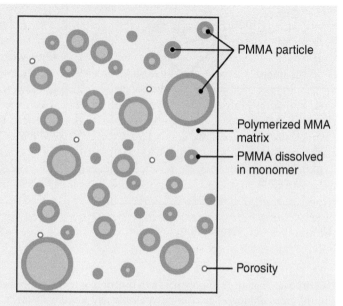

FIGURE 7.7 Schematic illustration of the microstructure of polymethylmethacrylate cement used in dental and medical applications.

7.2.2 Crystallinity

Polymers such as PMMA, polycarbonate, and polystyrene are generally **amorphous**; that is, the polymer chains exist without long-range order. (See also *amorphous alloys*, Section 6.14.) The amorphous arrangement of polymer chains is often described as being like a bowl of spaghetti, or like worms in a bucket, all intertwined with each other. In some polymers, however, it is possible to impart some crystallinity and thereby modify their characteristics. This arrangement may be fostered either during the synthesis of the polymer or by deformation during its subsequent processing.

The crystalline regions in polymers are called **crystallites** (Fig. 7.8). They are formed when the long molecules arrange themselves in an orderly manner, similar to the folding of a fire hose in a cabinet or of facial tissues in a box. A partially crystalline (**semicrystalline**) polymer can be regarded as a two-phase material, one phase being crystalline and the other amorphous.

By controlling the chain structure as well as the rate of solidification during cooling, it is possible to impart different **degrees of crystallinity** to polymers, although never 100%. Crystallinity ranges from an almost complete crystal (up to about 95% by volume in the case of polyethylene) to slightly crystallized (and mostly amorphous) polymers. The degree of crystallinity is also affected by branching. A linear polymer can become highly crystalline; a highly branched polymer cannot, although it may develop some low level of crystallinity. It will never achieve a high crystallite content, because the branches interfere with the alignment of the chains into a regular crystal array.

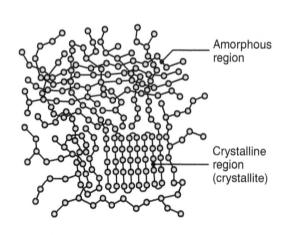

FIGURE 7.8 Amorphous and crystalline regions in a polymer; the crystalline region (crystallite) has an orderly arrangement of molecules. The higher the crystallinity, the harder, stiffer, and less ductile the polymer.

Effects of Crystallinity. The mechanical and physical properties of polymers are greatly influenced by the degree of crystallinity. As crystallinity increases, polymers become stiffer, harder, less ductile, more dense, less rubbery, and more resistant to solvents and heat (Fig. 7.6). The increase in density with increasing crystallinity is called *crystallization shrinkage*, and is caused by more efficient packing of the molecules in the crystal lattice. For example, the highly crystalline form of polyethylene, known as high-density polyethylene (HDPE), has a specific gravity in the range of 0.941–0.970 (80–95% crystalline). It is stronger, stiffer, tougher, and less ductile than low-density polyethylene (LDPE), which is about 60–70% crystalline and has a specific gravity in the range of 0.910–0.925.

Optical properties of polymers also are affected by the degree of crystallinity. The reflection of light from the boundaries between the crystalline and the amorphous regions in the polymer (Fig. 7.8) causes opaqueness. Furthermore, because the index of refraction is proportional to density, the greater the density difference between the amorphous and crystalline phases, the greater is the opaqueness of the polymer. Polymers that are completely amorphous can be transparent, such as polycarbonate and acrylics.

7.2.3   Glass-transition Temperature

Although amorphous polymers do not have a specific melting point, they undergo a distinct change in their mechanical behavior across a narrow range of temperatures. At low temperatures, they are hard, rigid, brittle, and glassy; at high temperatures, they are rubbery or leathery. The temperature at which a transition occurs is called the **glass-transition temperature** (T_g), also called the *glass point* or *glass temperature*. The term "glass" is used in this description because glasses, which are amorphous solids, behave in the same manner. (See *metallic glasses*, Section 6.14 and *glass*, Section 8.4.) Although most amorphous polymers exhibit this behavior, an exception is polycarbonate, which is neither rigid nor brittle below its glass-transition temperature. Polycarbonate is tough at ambient temperatures and is thus used for safety helmets and shields.

To determine T_g, a plot of the specific volume of the polymer as a function of temperature is produced; T_g occurs where there is a sharp change in the slope of the curve (Fig. 7.9). In the case of highly cross-linked polymers, the slope of the curve changes gradually near T_g, thus it can be difficult to determine T_g for these polymers. The glass-transition temperature varies with the type of polymer (Table 7.2), and it can be above or below room temperature. Unlike amorphous polymers, partly crystalline polymers have a distinct melting point, T_m (Fig. 7.9; see also Table 7.2). Because of the structural changes (called first-order changes) that occur, the specific volume of the polymer drops rapidly as its temperature is reduced.

7.2.4   Polymer Blends

The brittle behavior of amorphous polymers below their glass-transition temperature can be reduced by *blending* them, usually with small quantities of an **elastomer** (Section 7.9). The tiny particles that make up the elastomer are dispersed

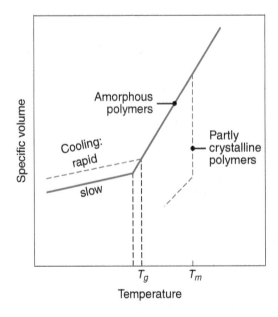

FIGURE 7.9 Specific volume of polymers as a function of temperature. Amorphous polymers, such as acrylic and polycarbonate, have a glass-transition temperature, but do not have a specific melting point. Partly crystalline polymers, such as polyethylene and nylons, contract sharply while passing through their melting temperatures during cooling.

TABLE 7.2

Glass-transition and Melting Temperatures of Some Polymers		
Material	T_g (°C)	T_m (°C)
Nylon 6,6	57	265
Polycarbonate	150	265
Polyester	73	265
Polyethylene		
High density	−90	137
Low density	−110	115
Polymethylmethacrylate	105	—
Polypropylene	−14	176
Polystyrene	100	239
Polytetrafluoroethylene	−90	327
Polyvinyl chloride	87	212
Rubber	−73	—

throughout the amorphous polymer, enhancing its toughness and impact strength by improving its resistance to crack propagation. These polymer blends are known as **rubber-modified polymers**.

Blending involves combining several components, creating **polyblends** that utilize the favorable properties of different polymers. **Miscible blends** (mixing without separation of two phases) are produced by a process, similar to the alloying of metals, that enables polymer blends to become more ductile. Polymer blends account for about 20% of all polymer production.

7.3   Thermoplastics

It was noted above that within each molecule, the bonds between adjacent long-chain molecules (secondary bonds) are much weaker than the covalent bonds between mers (primary bonds). It is the strength of the secondary bonds that determines the overall strength of the polymer. Linear and branched polymers have weak secondary bonds.

As the temperature is raised above the glass-transition temperature, T_g, or melting point, T_m, certain polymers become easier to form or mold into desired shapes. When the polymer is cooled, it returns to its original hardness and strength; in other words, the process is reversible. Polymers that exhibit this behavior are known as **thermoplastics**, common examples of which are acrylics, cellulosics, nylons, polyethylenes, and polyvinyl chloride.

The behavior of thermoplastics also depends on other variables, including their structure and composition; among the most important are temperature and rate of deformation. Below the glass-transition temperature, most polymers are *glassy* (brittle) and behave like an elastic solid. The relationship between stress and strain is linear, as shown in Fig. 2.2. The behavior depends on the particular polymer. For example, PMMA is glassy below its T_g, whereas polycarbonate is not glassy below its T_g. When the applied stress is increased further, polycarbonate eventually fractures, just as a piece of glass does at ambient temperature.

Typical stress–strain curves for some thermoplastics and thermosets at room temperature are shown in Fig. 7.10. Their various behavior may be described as rigid, soft, brittle, flexible, and so on. As can be noted from the mechanical properties of the polymers listed in Table 7.1, thermoplastics are about two orders of magnitude less stiff than metals, and their ultimate tensile strength is about one order of magnitude lower than that of metals (see Table 2.2).

Effects of Temperature.   If the temperature of a thermoplastic polymer is raised above its T_g, it first becomes *leathery* and then, with increasing temperature, *rubbery* (Fig. 7.6). Finally, at higher temperatures (above T_m for crystalline thermoplastics), the polymer becomes a *viscous fluid*, and its viscosity decreases with increasing temperature. As a viscous fluid, it can be softened, molded into shapes, resolidified, remelted, and remolded several times. In practice, however, repeated heating and cooling causes **degradation** or **thermal aging** of thermoplastics.

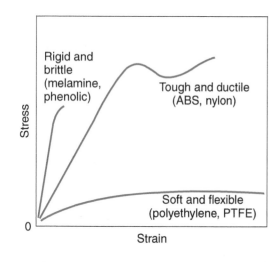

FIGURE 7.10 General terminology describing the behavior of three types of plastics. PTFE (polytetrafluoroethylene) has *Teflon* as its trade name. *Source:* After R.L.E. Brown.

As with metals, the strength and the modulus of elasticity of thermoplastics decrease with increasing temperature while the ductility increases (see Figs. 2.6 and 7.11). The effect of temperature on impact strength is shown in Fig. 7.12; note the large difference in the impact behavior among various polymers.

Effect of Rate of Deformation.   When deformed rapidly, the behavior of thermoplastics is similar to metals, as was shown by the strain-rate sensitivity exponent, m, in Eq. (2.9). Thermoplastics, in general, have high m values, indicating that they can undergo large *uniform deformation* in tension before fracture. Note in Fig. 7.13 how, unlike in ordinary metals, the necked region of the specimen elongates considerably.

This phenomenon can be easily demonstrated by stretching a piece of the plastic holder for a 6-pack of beverage cans, and observing the sequence of the necking and stretching behavior shown in Fig. 7.13a. This characteristic, which is the same in the superplastic metals (Section 2.2.7), enables the thermoforming of thermoplastics (Section 19.6) into such complex shapes as candy trays, lighted signs, and packaging.

Orientation.   When thermoplastics are deformed (say, by stretching), the long-chain molecules tend to align in the general direction of the elongation: a behavior called *orientation.* As in metals, the polymer becomes *anisotropic* (see also Section 1.6): the specimen becomes stronger and stiffer in the elongated (stretched) direction than in its transverse direction. Stretching is an important technique for enhancing the strength and toughness of polymers, and is especially exploited in producing high-strength fibers for use in composite materials, as described in Chapter 9.

Creep and Stress Relaxation.   Because of their viscoelastic behavior, thermoplastics are particularly susceptible to creep and stress relaxation, and to a larger extent than metals (Section 2.8). The extent of these phenomena depends on the particular polymer, stress level, temperature, and time. Thermoplastics exhibit creep and stress relaxation at room temperature, whereas most metals do so only at elevated temperatures.

Crazing.   When subjected to tensile or bending stresses, some thermoplastics (such as polystyrene and PMMA) develop localized, wedge-shaped narrow regions of highly deformed material, a behavior called *crazing.* Although they may appear to be like cracks, crazes are spongy material, typically containing about 50% voids. With increasing tensile load on the specimen, these voids coalesce to form a crack, which eventually can lead to fracture of the polymer. Crazing has been observed both in transparent, glassy polymers and in other types. The environment (particularly the presence of solvents, lubricants, or water vapor) can enhance the formation of crazes, called **environmental-stress cracking** and **solvent crazing.** Residual stresses in the material also contribute to crazing and cracking of the polymer.

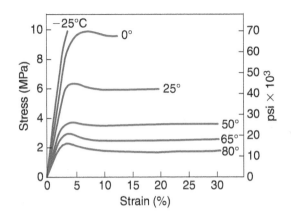

FIGURE 7.11   Effect of temperature on the stress–strain curve for cellulose acetate, a thermoplastic. Note the large drop in strength and the large increase in ductility with a relatively small increase in temperature. *Source:* After T.S. Carswell and H.K. Nason.

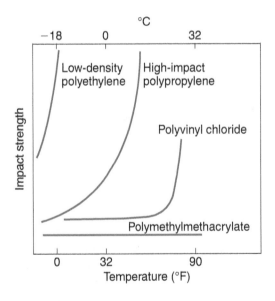

FIGURE 7.12   Effect of temperature on the impact strength of various plastics. Small changes in temperature can have a significant effect on impact strength. *Source:* After P.C. Powell.

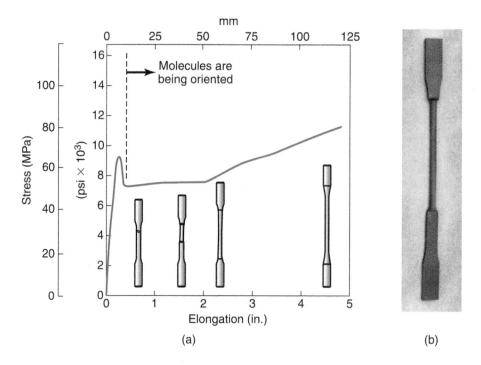

FIGURE 7.13 (a) Stress-elongation curve for polycarbonate, a thermoplastic. *Source:* Courtesy of R.P. Kambour and R.E. Robertson. (b) High-density polyethylene tensile-test specimen, showing uniform elongation (the long, narrow region in the specimen).

A behavior related to crazing is **stress whitening.** When subjected to tensile stresses (such as those caused by bending or folding), the polymer becomes lighter in color, a phenomenon usually attributed to the formation of microvoids in the material. As a result, the polymer becomes less translucent (transmits less light), or more opaque. This behavior can easily be demonstrated by bending plastic components commonly found in colored binder strips for report covers, household products, and toys.

Water Absorption.   An important characteristic of some polymers, such as nylons, is their ability to absorb water, known as *hygroscopic* behavior; water acts as a plasticizing agent, making the polymer more plastic. (See Section 7.5.) In a sense, it lubricates the chains in the amorphous regions. With increasing moisture absorption, the glass-transition temperature, the yield stress, and the elastic modulus of the polymer typically becomes rapidly lower. Dimensional changes also occur, especially in a humid environment.

Thermal and Electrical Properties.   Compared to metals, plastics generally are characterized by low thermal and electrical conductivity, low specific gravity (ranging from 0.90 to 2.2), and a high coefficient of thermal expansion (about an order of magnitude higher, as shown in Tables 3.1 and 3.2). Because most polymers have low electrical conductivity, they can be used for insulators and as packaging material for electronic components.

The electrical conductivity of some polymers can be increased by **doping** (introducing impurities, such as metal powders, salts, and iodides, into the polymer). Discovered in the late 1970s, **electrically conducting polymers** include polyethylene oxide, polyacetylene, polyaniline, polypyrrole, and polythiophene. The electrical conductivity of polymers increases with moisture absorption; their electronic properties also can be changed by irradiation. Applications for conducting polymers include adhesives, microelectronic devices, rechargeable batteries, capacitors, catalysts, fuel cells, fuel-level sensors, deicer panels, radar dishes, antistatic coatings, and thermoactuating motors (used in linear-motion applications, such as for power antennae, sun roofs, and power windows).

Thermally conducting polymers are being developed for applications requiring dimensional stability and heat transfer (such as heat sinks), as well as for reducing cycle times in molding and processing of thermoplastics. These polymers are typically thermoplastics (such as polypropylene, polycarbonate, and nylon), embedded with thermally conducting particles. Their conductivity can be as much as 100 times that of conventional plastics. (See also *sprayed-metal tooling*, Section 20.6.1.)

Shape-memory Polymers.   Polymers also can behave in a manner similar to the shape-memory alloys, described in Section 6.13. The polymers can be stretched or compressed to very large strains, and then, when subjected to heat, light, or a chemical environment, they recover their original shape. The potential applications for these polymers are similar to those for shape-memory alloys, such as in opening blocked arteries, probing neurons in the brain, and improving the toughness of spines.

CASE STUDY 7.2   Use of Electrically Conducting Polymers in Rechargeable Batteries

One of the earliest applications of conducting polymers was in rechargeable batteries. Modern rechargeable batteries use lithium, or an oxide of lithium, as the cathode and lithium carbide ($Li_y C_6$) as the anode, separated by a conducting polymer layer. Lithium is the lightest of all metals and has a high electrochemical potential, so that its energy per volume is highest.

The polymer, usually polyethylene oxide (PEO), with a dissolved lithium salt, is placed between the cathode and anode. During discharge of the battery, $Li_y C_6$ is oxidized, emitting free electrons and lithium ions. The electrons are available to power devices, and the Li^+ ions are stored in the polymer. When the cathode is depleted, the battery must be recharged to restore the cathode. During charging, Li^+ is transferred through the polymer electrolytes to the cathode. Lithium-ion batteries have good capacity, can generate up to 4.2 V each, and can be placed in series to obtain higher voltages, such as 18 V for hand tools. Battery cells are now being developed in which both electrodes are made of conducting polymers; one has been constructed with a capacity of 3.5 V.

7.4   Thermosetting Plastics

When the long-chain molecules in a polymer are cross-linked in a three-dimensional arrangement, the structure in effect becomes one *giant molecule*, with strong covalent bonds. These polymers are called **thermosetting polymers** or **thermosets** because, during polymerization, the network is completed and the shape of the part is permanently set; this **curing** (**cross-linking**) reaction, unlike that of thermoplastics, is irreversible. The response of a thermosetting plastic to a sufficiently elevated temperature can be likened to what happens when baking a cake or boiling an egg: Once the cake is baked and cooled, or the egg boiled and cooled, reheating it will not change its shape.

Some thermosets (such as epoxy, polyester, and urethane) cure at room temperature, because the heat produced by the exothermic reaction is sufficient to cure the plastic. A typical thermoset is **phenolic**, which is a product of the reaction between phenol and formaldehyde. Common products made from this polymer are the handles and knobs on cooking pots and pans and components of light switches and outlets.

The polymerization process for thermosets generally takes place in two stages. The first occurs at the chemical plant, where the molecules are partially polymerized into linear chains. The second stage occurs during the final step of part production, where cross-linking is completed under *heat* and *pressure* during molding and shaping of the part (Chapter 19).

Thermosetting polymers do not have a sharply defined glass-transition temperature. Because of the nature of the bonds, the strength and hardness of a thermoset are not affected by temperature or by rate of deformation, unlike those for thermoplastics. If the temperature is increased sufficiently, a thermosetting polymer instead begins to burn, degrade, and char. Thermosets generally possess better mechanical, thermal, and chemical properties; electrical resistance; and dimensional stability than do thermoplastics.

7.5 Additives in Plastics

Polymers usually are compounded with *additives* which modify and improve certain characteristics, such as stiffness, strength, color, weatherability, flammability, arc resistance (for electrical applications), and ease of subsequent processing. Additives may consist of:

- **Plasticizers.** These are added to polymers to impart *flexibility* and *softness* by lowering their glass-transition temperature. *Plasticizers* are low-molecular-weight solvents with high boiling points (nonvolatility); they reduce the strength of the secondary bonds between the long-chain molecules, and thus make the polymer flexible and soft. The most common use of plasticizers is in PVC, which remains flexible during its numerous uses; other applications are in thin sheets, films, tubing, shower curtains, and clothing materials.

- **Colorants.** The wide variety of colors available in plastics is obtained by adding *colorants*, which are either *dyes* (organic) or *pigments* (inorganic).

- **Fillers.** Because of their low cost, *fillers* are important in reducing the overall cost of polymers. Depending on their type, fillers may also improve the strength, hardness, toughness, abrasion resistance, dimensional stability, or stiffness of plastics. These properties are greatest at specific percentages of different types of polymer-filler combinations. Fillers are generally wood flour (fine sawdust), silica flour (fine silica powder), clay, powdered mica, talc, calcium carbonate, and short fibers of cellulose, glass, or asbestos.

- **Flame Retardants.** If the temperature is sufficiently high, most polymers will ignite and burn. The **flammability** (ability to support combustion) of polymers varies considerably, depending on their composition (especially on their chlorine and fluorine content). The flammability of polymers can be reduced either by making them from less flammable raw materials or by adding flame retardants, such as compounds of chlorine, bromine, and phosphorus. Cross-linking also reduces polymer flammability.

- **Other Additives.** Most polymers are affected adversely by *ultraviolet radiation* (such as from sunlight) and by *oxygen*, which weaken and break the primary bonds and cause the scission (splitting) of the long-chain molecules. The polymer then degrades and becomes stiff and brittle. A typical example of protection against ultraviolet radiation is the compounding of certain plastics and rubber with **carbon black** (soot). Protection against degradation caused by oxidation, particularly at elevated temperatures, is achieved by adding **antioxidants** to the polymer.

7.6   General Properties and Applications of Thermoplastics

The general characteristics and typical applications of major classes of thermoplastics, particularly as they relate to the manufacturing and service life of plastic products and components, are outlined in this section. General recommendations for various plastics applications are given in Table 7.3, and Table 7.4 lists some of the more common trade names for thermoplastics.

Acetals (from *acetic* and *alcohol*) have good strength and stiffness, and good resistance to creep, abrasion, moisture, heat, and chemicals. Typical applications include mechanical parts and components requiring high performance over a long period (e.g., bearings, cams, gears, bushings, and rolls), impellers, wear surfaces, pipes, valves, shower heads, and housings.

TABLE 7.3

General Recommendations for Plastic Products		
Design requirement	Typical applications	Plastics
Mechanical strength	Gears, cams, rolls, valves, fan blades, impellers, pistons	Acetals, nylon, phenolics, polycarbonates, polyesters, polypropylenes, epoxies, polyimides
Wear resistance	Gears, wear strips and liners, bearings, bushings, roller blades	Acetals, nylon, phenolics, polyimides, polyurethane, ultrahigh-molecular-weight polyethylene
Frictional properties		
High	Tires, nonskid surfaces, footware, flooring	Elastomers, rubbers
Low	Sliding surfaces, artificial joints	Fluorocarbons, polyesters, polyimides
Electrical resistance	All types of electrical components and equipment, appliances, electrical fixtures	Polymethylmethacrylate, ABS, fluorocarbons, nylon, polycarbonate, polyester, polypropylenes, ureas, phenolics, silicones, rubbers
Chemical resistance	Containers for chemicals, laboratory equipment, components for chemical industry, food and beverage containers	Acetals, ABS, epoxies, polymethylmethacrylate, fluorocarbons, nylon, polycarbonate, polyester, polypropylene, ureas, silicones
Heat resistance	Appliances, cookware, electrical components	Fluorocarbons, polyimides, silicones, acetals, polysulfones, phenolics, epoxies
Functional and decorative	Handles, knobs, camera and battery cases, trim moldings, pipe fittings	ABS, acrylics, cellulosics, phenolics, polyethylenes, polypropylenes, polystyrenes, polyvinyl chloride
Functional and transparent	Lenses, goggles, safety glazing, signs, food-processing equipment, laboratory hardware	Acrylics, polycarbonates, polystyrenes, polysulfones
Housings and hollow shapes	Power tools, housings, sport helmets, telephone cases	ABS, cellulosics, phenolics, polycarbonates, polyethylenes, polypropylene, polystyrenes

TABLE 7.4

Trade Names for Thermoplastic Polymers

Acetal: Delrin, Duracon, Lupital, Ultraform
Acrylic: Lucite, Acrylite, Acrysteel, Cyrolite, Diakon, Implex, Kamax, Korad, Plexiglass, XT, Zylar
Acrylic-polyvinyl chloride: Kydex
Acrylonitrile-butadiene-styrene: Cycolac, Delta, Denka, Magnum, Novodur, Royalite, Terluran
Aramid: Kevlar
Fluorocarbon: Teflon (polytetrafluoroethylene)
Polyamide: Capron, Celanese, Durethan, Grilamid, Maranyl, Nylon, Rilsan, Ultramid, Vespel, Vydyne, Zytel
Polycarbonate: APEC, Calibre, Hyzod, Lexan, Makrolon, Merlon
Polyester: Dacron, Eastpac, Ektar, Kodel, Mylar, Rynite
Polyetherimide: Ultem
Polyethylene: Alathon, Dowlex, Forar, Fortiflex, Hostalen, Marlex, Petrothene
Polyimide: Aurum, Avimid, Estamid, Envex, Kapton, Lenzing, VTEC
Polyphenylene: Forton, Fortron, Noryl
Polypropylene: Fortilene, Oleplate, Olevac, Pro-Fax
Polystyrene: Dylene, Fosta Tuf-Flex, Fostalite, Fostarene, Lustrex, Polystrol, Styron, Syrofoam
Polysulfone: Mindel, Udel
Polyurethane: Estane, Isoplast, Pellethane
Polyvinyl chloride: Fiberloc, Geon, Saran, Sintra, Tygon
Polyvinylidene fluoride: Foraflon, Kynar
Styrene-methylmethacrylate: Zerlon

Acrylics (such as PMMA) possess moderate strength, good optical properties, and weather resistance. They are transparent (but can be made opaque), are generally resistant to chemicals, and have good electrical resistance. Typical applications include optical lenses, lighted signs, displays, window glazing, skylights, automotive headlight lenses, windshields, lighting fixtures, and furniture.

Acrylonitrile-butadiene-styrene is rigid and dimensionally stable. It has good impact, abrasion, and chemical resistance; good strength and toughness; good low-temperature properties; and high electrical resistance. Typical applications include pipes, fittings, chrome-plated plumbing supplies, helmets, tool handles, automotive components, boat hulls, telephones, luggage, housing, appliances, refrigerator liners, and decorative panels.

Cellulosics have a wide range of mechanical properties, depending on their composition. They can be made rigid, strong, and tough; however, they weather poorly and are affected by heat and chemicals. Typical applications include tool handles, pens, knobs, frames for eyeglasses, safety goggles, machine guards, helmets, tubing and pipes, lighting fixtures, rigid containers, steering wheels, packaging film, signs, billiard balls, toys, and decorative parts.

Fluorocarbons possess good resistance to high temperature [e.g., a melting point of 327°C (621°F) for *Teflon*], chemicals, weather, and electricity. They also have unique nonadhesive properties and low friction. Typical applications include linings for chemical-processing equipment, nonstick coatings for cookware, electrical insulation for high-temperature wire and cable, gaskets, low-friction surfaces, bearings, and seals.

Polyamides (from the words *poly*, *amine*, and *carboxyl acid*) are available in two main types: *nylons* and *aramids*:

- **Nylons** have good mechanical properties and abrasion resistance; they also are self-lubricating and resistant to most chemicals. All nylons are *hygroscopic* (absorb water); the moisture absorption reduces desirable mechanical properties and increases part dimensions. Typical applications include gears, bearings, bushings, rolls, fasteners, zippers, electrical parts, combs, tubing, wear-resistant surfaces, guides, and surgical equipment.
- **Aramids** (aromatic polyamides) have very high tensile strength and stiffness. Typical applications include fibers for reinforced plastics, bulletproof vests, cables, and radial tires.

Polycarbonates are versatile; they have good mechanical and electrical properties, high impact resistance, and they can be made resistant to chemicals. Typical applications include safety helmets, optical lenses, bullet-resistant window glazing, signs, bottles, food-processing equipment, windshields, load-bearing electrical components, electrical insulators, medical apparatus, business machine components, guards for machinery, and parts requiring dimensional stability.

Polyesters (thermoplastic polyesters; see also Section 7.7) have good mechanical, electrical, and chemical properties; good abrasion resistance; and low friction. Typical applications include gears, cams, rolls, load-bearing members, pumps, and electromechanical components.

Polyethylenes possess good electrical and chemical properties; their mechanical properties depend on composition and structure. Three major polyethylene classes are: (1) *low density* (LDPE), (2) *high density* (HDPE), and (3) *ultrahigh molecular weight* (UHMWPE). Typical applications for LDPE and HDPE are housewares, bottles, garbage cans, ducts, bumpers, luggage, toys, tubing, bottles, and packaging materials. UHMWPE is used in parts requiring high-impact toughness and resistance to abrasive wear; examples include artificial knee and hip joints.

Polyimides have the structure of a thermoplastic, but the nonmelting characteristic of a thermoset. (See also Section 7.7.)

Polypropylenes have good mechanical, electrical, and chemical properties and good resistance to tearing. Typical applications include automotive trim and components, medical devices, appliance parts, wire insulation, TV cabinets, pipes, fittings, drinking cups, dairy-product and juice containers, luggage, ropes, and weather stripping.

Polystyrenes generally have average properties and are somewhat brittle, but inexpensive. Typical applications include disposable containers; packaging; trays for meats, cookies, and candy; foam insulation; appliances; automotive and radio/TV components; housewares; and toys and furniture parts (as a substitute for wood).

Polysulfones have excellent resistance to heat, water, and steam; they have dielectric properties that remain virtually unaffected by humidity, are highly resistant to some chemicals, but are attacked by organic solvents. Typical applications include steam irons, coffeemakers, hot-water containers, medical equipment that requires sterilization, power-tool and appliance housings, aircraft cabin interiors, and electrical insulators.

Polyvinyl chloride has a wide range of properties, is inexpensive and water resistant, and can be made rigid or flexible; it is not suitable for applications requiring strength and heat resistance. *Rigid* PVC is tough and hard; it is used for signs and in the construction industry (e.g., in pipes and conduits). *Flexible* PVC is used in wire and cable coatings, in low-pressure flexible tubing and hoses, and in footwear, imitation leather, upholstery, records, gaskets, seals, trim, film, sheet, and coatings.

7.7 General Properties and Applications of Thermosetting Plastics

This section outlines the general characteristics and typical applications of the major thermosetting plastics.

Alkyds (from *alkyl*, meaning alcohol, and *acid*) possess good electrical insulating properties, impact resistance, dimensional stability, and low water absorption. Typical applications are in electrical and electronic components.

Aminos have properties that depend on composition; generally, they are hard, rigid, and resistant to abrasion, creep, and electrical arcing. Typical applications include small-appliance housings, countertops, toilet seats, handles, and distributor caps. **Urea** typically is used for electrical and electronic components; and **melamine** for dinnerware.

Epoxies have excellent mechanical and electrical properties, good dimensional stability, strong adhesive properties, and good resistance to heat and chemicals. Typical applications include electrical components requiring mechanical strength and high insulation, tools and dies, and adhesives. **Fiber-reinforced epoxies** have excellent mechanical properties and are used in pressure vessels, rocket-motor casings, tanks, and similar structural components.

Phenolics are rigid (though brittle) and dimensionally stable; they have high resistance to heat, water, electricity, and chemicals. Typical applications include knobs, handles, laminated panels, and telephones; bonding material to hold abrasive grains together in grinding wheels; and electrical components (such as wiring devices, connectors, and insulators).

Polyesters (thermosetting polyesters; see also Section 7.6) have good mechanical, chemical, and electrical properties; they generally are reinforced with glass (or other) fibers and also are available as casting resins. Typical applications include boats, luggage, chairs, automotive bodies, swimming pools, and materials for impregnating cloth and paper.

Polyimides possess good mechanical, physical, and electrical properties at elevated temperatures; they also have good creep resistance, low friction, and low wear characteristics. Polyimides have the non-melting characteristic of a thermoset, but the structure of a thermoplastic. Typical applications include pump components (bearings, seals, valve seats, retainer rings, and piston rings), electrical connectors for high-temperature use, aerospace parts, high-strength impact-resistant structures, sports equipment, and safety vests.

Silicones have properties that depend on their composition; generally, they weather well, possess excellent electrical properties over a wide range of humidity and temperature, and resist chemicals and heat (Section 7.9). Typical applications include electrical components requiring strength at elevated temperatures, oven gaskets, heat seals, and waterproof materials.

Health Hazards.   There is increasing concern that some of the chemicals used in polymers may present health hazards, especially in such products as polycarbonate water containers and baby bottles, and also medical devices, sports safety equipment, and eating utensils. The chemical that is of particular concern is bisphenol A (BPA), a widely used chemical; several worldwide investigations are being conducted to determine whether there is any link to human diseases, specifically heart disease and diabetes.

CASE STUDY 7.3   Materials for a Refrigerator Door Liner

In the selection of candidate materials for a refrigerator door liner (where eggs, butter, salad dressings, and small bottles are typically stored) the following factors should be considered:

1. *Mechanical requirements*: strength, toughness (to withstand impacts, door slamming, and racking), stiffness, resilience, and resistance to scratching and wear at operating temperatures.
2. *Physical requirements*: dimensional stability and electrical insulation.
3. *Chemical requirements*: resistance to staining, odor, chemical reactions with food and beverages, and cleaning fluids.
4. *Appearance*: color, stability of color over time, surface finish, texture, and feel.
5. *Manufacturing properties*: methods of manufacturing and assembly, effects of processing on

material properties and behavior over a period of time, compatibility with other components in the door, and cost of materials and manufacturing.

An extensive study, considering all of the factors involved, identified two candidate materials for door liners: ABS and HIPS (high-impact polystyrene). One aspect of the study involved the effect of vegetable oils, such as from salad dressing stored in the door shelf, on the strength of those plastics. Experiments showed that the presence of vegetable oils significantly reduced the load-bearing capacity of HIPS. It was found that HIPS becomes brittle in the presence of oils (solvent stress cracking), whereas ABS is not affected to any significant extent.

7.8 Biodegradable Plastics

Plastic wastes contribute about 16% of municipal solid waste by weight, and make up 50–80% of waste littering beaches, oceans, and sea beds; on a volume basis, they contribute between two and three times their weight. Only about one-third of plastic production goes into disposable products, such as bottles, packaging, and garbage bags. With the growing use of plastics and great concern over environmental issues regarding the disposal of plastic products and the shortage of landfills, major efforts continue to develop completely biodegradable plastics.

Traditionally, most plastic products have been made from synthetic polymers that are derived from nonrenewable natural resources, are not biodegradable, and are difficult to recycle. **Biodegradability** means that microbial species in the environment (e.g., microorganisms in soil and water) will degrade all or part of the polymeric material under the proper environmental conditions, without producing toxic by-products. The end products of the degradation of the biodegradable portion of the material are carbon dioxide and water. Because of the variety of constituents in biodegradable plastics, these plastics can be regarded as composite materials; consequently, only a portion of them may be truly biodegradable.

Three different *biodegradable plastics* have thus far been developed. They have different degradability characteristics, and they degrade over different periods of time (anywhere from a few months to a few years).

1. The **starch-based system** is the farthest along in terms of production capacity. Starch may be extracted from potatoes, wheat, rice, or corn. The starch granules are processed into a powder, which is heated and becomes a sticky liquid; the liquid is then cooled, shaped into pellets, and processed in conventional plastic-processing equipment. Various additives and binders are blended with the starch to impart specific characteristics to the bioplastic materials; for example, a

composite of polyethylene and starch is produced commercially as degradable garbage bags.

2. In the **lactic-based system**, fermenting feedstocks produce lactic acid, which is then polymerized to form a polyester resin. Typical uses include medical and pharmaceutical applications.

3. In the **fermentation of sugar**, organic acids are added to a sugar feedstock. The resulting reaction produces a highly crystalline and very stiff polymer, which, after further processing, behaves in a manner similar to polymers developed from petroleum.

Studies continue to be conducted on producing fully biodegradable plastics, by using various agricultural waste (*agrowastes*), plant carbohydrates, plant proteins, and vegetable oils. Typical applications of this approach include

- Disposable tableware made from a cereal substitute, such as rice grains or wheat flour
- Plastics made almost entirely from starch extracted from potatoes, wheat, rice, or corn
- Plastic articles made from coffee beans and rice hulls, dehydrated and molded under high pressure and temperature
- Water-soluble and compostable polymers for medical and surgical use
- Food and beverage containers made from potato starch, limestone, cellulose, and water, that can dissolve in storm sewers and oceans without affecting wildlife or marine life

Recycling of Plastics.   Much effort continues to be expended globally on the collecting and recycling of used plastic products. Thermoplastics are recycled by melting, blending, and reforming them into other products. The products carry *recycling symbols* (for efficient separation of different types), in the shape of a triangle outlined by three clockwise arrows and with a number in the middle. These numbers developed in 1988 and used internationally, correspond to the following plastics:

1. PETE (polyethylene terephthalate)
2. HDPE (high-density polyethylene)
3. V (vinyl) or PVC (polyvinyl chloride)
4. LDPE (low-density polyethylene)
5. PP (polypropylene)
6. PS (polystyrene)
7. Other, or O (acrylic, nylon, polycarbonate, etc.)

Recycled plastics are increasingly being used for a variety of products. For example, a recycled polyester (filled with glass fibers and minerals) is used for the engine cover for an F-series Ford pickup truck; it has the appropriate stiffness, chemical resistance, and shape retention up to 180°C (350°F).

7.9   Elastomers (Rubbers)

Elastomers (derived from the words *elastic* and *mer*) consist of a large family of amorphous polymers (Section 7.2.1) with a low glass-transition temperature. They have the characteristic ability of undergoing large elastic deformations without rupture; also, they are soft and have a low elastic modulus.

The structure of elastomer molecules is highly kinked (tightly twisted or curled) when stretched, but then return to their original shape after the load is removed

(Fig. 7.14). They can also be cross-linked, the best example of which is the elevated-temperature **vulcanization** of rubber with sulfur, discovered by the American inventor C. Goodyear in 1839 and named for Vulcan, the Roman god of fire. Once the elastomer is cross-linked, it cannot be reshaped; for example, an automobile tire, which is one giant molecule, cannot be softened and reshaped.

The terms *elastomer* and *rubber* often are used interchangeably. Generally, however, an **elastomer** is defined as being capable of recovering substantially in shape and size after the load has been removed; a **rubber** is defined as being capable of recovering from large deformations quickly.

The hardness of elastomers, which is measured with a durometer (Section 2.6.1), increases with the cross-linking of the molecular chains. As with plastics, a variety of additives can be blended into elastomers to impart specific properties. Elastomers have a wide range of applications, such as high-friction and nonskid surfaces, protection against corrosion and abrasion, electrical insulation, and shock and vibration insulation. Examples include tires, hoses, weather stripping, footwear, linings, gaskets, seals, printing rolls, and flooring.

One unique property of elastomers is their *hysteresis loss* in stretching or compression (Fig. 7.14). The clockwise loop indicates energy loss, whereby mechanical energy is converted into heat; this property is important for absorbing vibrational energy (damping) and sound insulation.

Natural Rubber. The base for *natural rubber* is **latex**, a milk-like sap obtained from the inner bark of a tropical tree. Natural rubber has good resistance to abrasion and fatigue, and high friction, but low resistance to oil, heat, ozone, and sunlight. Typical applications are tires, seals, couplings, and engine mounts.

Synthetic Rubbers. Examples of *synthetic rubbers* are butyl, styrene butadiene, polybutadiene, and ethylene propylene. Compared to natural rubber, they have better resistance to heat, gasoline, and chemicals, and have a higher range of useful temperatures. Synthetic rubbers that are resistant to oil are neoprene, nitrile, urethane, and silicone. Typical applications of synthetic rubbers are tires, shock absorbers, seals, and belts.

Silicones. *Silicones* (Section 7.7) have the highest useful temperature range of elastomers (up to 315°C; or 600°F), but other properties, such as strength and resistance to wear and oils, generally are inferior to those of other elastomers. Typical applications of silicones are seals, gaskets, thermal insulation, high-temperature electrical switches, and electronic apparatus.

Polyurethane. This elastomer has very good overall properties of high strength, stiffness, and hardness, and it has exceptional resistance to abrasion, cutting, and tearing. Typical applications of *polyurethane* are seals, gaskets, cushioning, diaphragms for the rubber forming of sheet metals (Section 16.8), and auto body parts.

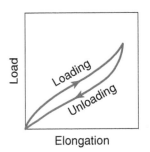

FIGURE 7.14 Typical load-elongation curve for rubbers; the clockwise loop, indicating the loading and the unloading paths, displays the hysteresis loss. Hysteresis gives rubbers the capacity to dissipate energy, damp vibration, and absorb shock loading, as is necessary in automobile tires and in vibration dampers placed under machinery.

SUMMARY

- Polymers are a major class of materials, and possess a very wide range of mechanical, physical, chemical, and optical properties. Compared to metals, polymers are generally characterized by (a) lower density, strength, elastic modulus, thermal and electrical conductivity, cost; (b) higher strength-to-weight ratio, higher resistance to corrosion, higher thermal expansion, (c) wide choice of colors and transparencies; and (d) greater ease of manufacture into complex shapes.

- Plastics are composed of polymer molecules and various additives. The smallest repetitive unit in a polymer chain is called a mer. Monomers are linked by polymerization processes (condensation or addition) to form larger molecules. The glass-transition temperature separates the region of brittle behavior in polymers from that of ductile behavior.

- The properties of polymers depend on their molecular weight, structure (linear, branched, cross-linked, or network), degrees of polymerization and crystallinity, and on the additives present in their formulation. Additives have such functions as improving strength, flame retardation, lubrication, imparting flexibility and color, and providing stability against ultraviolet radiation and oxygen. Polymer structures can be modified by various means to impart a wide range of desirable properties.

- Two major classes of polymers are thermoplastics and thermosets. Thermoplastics become soft and easy to form at elevated temperatures; their behavior includes such phenomena as creep and stress relaxation, crazing, and water absorption. Thermosets, which are obtained by cross-linking polymer chains, do not become soft to any significant extent with increasing temperature; they are much more rigid and harder than thermoplastics.

- Elastomers have a characteristic ability to undergo large elastic deformations and then return to their original shapes when unloaded. Consequently, they have important applications in tires, seals, footwear, hoses, belts, and shock absorbers.

- Among important considerations in polymers are their recyclability and biodegradability. Several formulations of biodegradable plastics are available, and others are under continued development.

KEY TERMS

Additives	Degree of crystallinity	Molecular weight	Rubber
Biodegradable	Degree of polymerization	Monomer	Secondary bonds
Blends	Doping	Network polymers	Shape-memory polymers
Bonding	Elastomer	Orientation	Silicones
Branched polymers	Fillers	Plasticizers	Stress whitening
Colorants	Flame retardants	Plastics	Thermal aging
Crazing	Glass-transition temperature	Polyblends	Thermoplastics
Cross-linked polymers	Latex	Polymer	Thermosets
Crystallinity	Linear polymers	Polymerization	Vulcanization
Curing	Lubricants	Primary bonds	
Degradation	Mer	Recycling	

BIBLIOGRAPHY

Bhowmick, A.K., and Stephens, H.L., Handbook of Elastomers, 2nd ed., CRC Press, 2000.

Buckley, C.P., Bucknall, C.B., and McCrum, N.G., Principles of Polymer Engineering, 2nd ed., Oxford University Press, 1997.

Campo, E.A., Selection of Polymeric Materials, William Andrew, Inc., 2008.

Chanda, M., and Roy, S.K., Plastics Technology Handbook, 4th ed., Marcel Dekker, 2006.

Drobny, J.G., Handbook of Thermoplastic Elastomers, William Andrew, Inc., 2007.

Fink, J.K., High Performance Polymers, William Andrew, Inc., 2008.

Goodship, V., **Introduction to Plastics Recycling**, 2nd ed., Smithers Rapra Press, 2008.

Harper, C., **Handbook of Plastics, Elastomers, and Composites**, 4th ed., McGraw-Hill, 2003.

Harper, C.A., **Modern Plastics Handbook**, McGraw-Hill, 2000.

Khemani, K., and Scholz, C., **Degradable Polymers and Materials: Principles and Practice**, American Chemical Society, 2006.

Kutz, M., **Applied Plastics Engineering Handbook: Processing and Materials**, Wilhelm Andrew, Inc., 2011.

MacDermott, C.P., and Shenoy, A.V., **Selecting Thermoplastics for Engineering Applications**, 2nd ed., Marcel Dekker, 1997.

Margolis, J., **Engineering Plastics Handbook**, McGraw-Hill, 2006.

Mark, J.E. (ed.), **Physical Properties of Polymers Handbook**, 2nd ed., Springer, 2006.

Michler, G.H., and Balta-Calleja, F.J., **Mechanical Properties of Polymers Based on Nanostructure and Morphology**, CRC Press, 2006.

Mills, N., **Plastics: Microstructure and Engineering Applications**, 3rd ed., Butterworth-Heinemann, 2006.

Mittal, V. (ed.), **High Performance Polymers and Engineering Plastics**, Wiley Scrinever, 2011.

Owald, T.A., and Menges, G., **Materials Science of Polymers for Engineers**, 2nd ed., Hamser, 2003,

Properties and Behavior of Polymers, 2 vols., Wiley, 2011.

Salamone, J.C. (ed.), **Concise Polymeric Materials Encyclopedia**, CRC Press, 1999.

Strong, A.B., **Plastics: Materials and Processing**, 3rd ed., Prentice Hall, 2005.

Ward, I.M., and Sweeny, J., **An Introduction to the Mechanical Properties of Solid Polymers**, 2nd ed., Wiley, 2004.

Xanthos, M., **Functional Fillers for Plastics**, 2nd ed., Wiley, 2010.

Young, R.J., and Lovell, P., **Introduction to Polymers**, 3rd ed., CRC Press, 2008.

Zweifel, H., Maier, H., and Schiller, M., **Plastics Additives Handbook**, 6th ed., Hanser, 2009.

REVIEW QUESTIONS

7.1 Summarize the important mechanical and physical properties of plastics.

7.2 What are the major differences between the (a) mechanical and (b) physical properties of plastics and metals?

7.3 List properties that are influenced by the degree of polymerization.

7.4 What is the difference between condensation polymerization and addition polymerization?

7.5 Explain the differences between linear, branched, and cross-linked polymers.

7.6 What is the glass-transition temperature?

7.7 List and explain the additives commonly used in plastics.

7.8 What is crazing?

7.9 What are polyblends?

7.10 List the major differences between thermoplastics and thermosets.

7.11 What is an elastomer?

7.12 What effects does a plasticizing agent have on a polymer?

7.13 Define the following abbreviations: PMMA, PVC, ABS, HDPE, UHDPE, and LDPE.

7.14 Explain why it would be advantageous to produce a polymer with a high degree of crystallinity.

7.15 What are the differences and similarities of addition and condensation polymerization?

7.16 Are molecular weight and degree of polymerization related? Explain.

7.17 Why do polymers need to be dried before processing?

QUALITATIVE PROBLEMS

7.18 What characteristics of polymers make them attractive for clothing?

7.19 Do polymers strain harden more than metals or vice versa? Explain.

7.20 Inspect various plastic components in an automobile, and state whether they are made of thermoplastic materials or of thermosetting plastics.

7.21 Give applications for which flammability of plastics would be of major importance.

7.22 What characteristics make polymers advantageous for applications such as gears? What characteristics are drawbacks in such applications?

7.23 What properties do elastomers have that thermoplastics in general do not have?

7.24 Do you think that the substitution of plastics for metals in products traditionally made of metal may be viewed negatively by the public at large? If so, why?

7.25 Is it possible for a material to have a hysteresis behavior that is the opposite of that shown in Fig. 7.14, so that the two arrows run counterclockwise? Explain.

7.26 Observe the behavior of the specimen shown in Fig. 7.13, and state whether the material has a high or a low strain-rate sensitivity exponent, m (see Section 2.2.7).

7.27 Add more to the applications column in Table 7.3.

7.28 Discuss the significance of the glass-transition temperature, T_g, in engineering applications.

7.29 Describe how a rechargeable lithium battery works.

7.30 Explain how cross-linking improves the strength of polymers.

7.31 Describe the methods by which the optical properties of polymers can be altered.

7.32 How can polymers be made to conduct electricity? Explain.

7.33 Explain the reasons for which elastomers were developed.

7.34 Give several examples of plastic products or components in which creep and stress relaxation would be important considerations.

7.35 Describe your opinions regarding the recycling of plastics versus the development of plastics that are biodegradable.

7.36 Explain how you would go about determining the hardness of plastics.

7.37 Compare the values of the elastic modulus, given in Table 7.1, to the values for metals given in Chapters 2, 5, and 6. Comment on your observations.

7.38 Why is there so much variation in the stiffness of products made of polymers? Explain.

7.39 Explain why thermoplastics are easier to recycle than thermosets.

7.40 Give an example where crazing is desirable.

7.41 Describe the principle behind shrink wrapping.

7.42 List and explain some environmental pros and cons of using plastic shopping bags versus paper bags.

7.43 List the characteristics required of a polymer for (a) a bucket, (b) a golf ball, (c) an automobile dashboard, (d) clothing, (e) flooring, and (f) fishing nets.

7.44 How can you tell whether a part is made of a thermoplastic or a thermoset?

7.45 As you know, there are plastic paper clips available in various colors. Why are there no plastic staples?

7.46 By incorporating small amounts of a blowing agent, it is possible to manufacture hollow polymer fibers with gas cores. List possible applications for such fibers.

7.47 In injection-molding operations (Section 19.3), it is common practice to remove the part from its runner, to place the runner into a shredder, and to recycle the resultant pellets. List the concerns you would have in using such recycled pellets as opposed to so-called virgin pellets.

7.48 From an environmental standpoint, do you feel it is best to incorporate polymers or metals into designs? Explain.

QUANTITATIVE PROBLEMS

7.49 Calculate the areas under the stress–strain curve (toughness) for the materials shown in Fig. 7.11, plot them as a function of temperature, and describe your observations.

7.50 Note in Fig. 7.11 that, as expected, the elastic modulus of the polymer decreases as temperature increases. Using the stress–strain curves in the figure, make a plot of the modulus of elasticity versus the temperature. Comment on the shape of the curve.

7.51 A rectangular cantilever beam 75 mm high, 20 mm wide, and 1 m long is subjected to a concentrated load of 50 kg at its end. From Table 7.1, select three unreinforced and three reinforced materials and calculate the maximum deflection of the beam in each case. Then select aluminum and steel for the same beam dimensions, calculate the maximum deflection, and compare the results.

7.52 Estimate the number of molecules in a typical automobile tire, then estimate the number of atoms in the tire.

7.53 Using strength and density data, determine the minimum weight of a 1-m-long tension member that must support a load of 5000 N, if it is manufactured from (a) high-molecular-weight polyethylene, (b) polyester, (c) rigid PVC, (d) ABS, (e) polystyrene, and (e) reinforced nylon.

7.54 Plot the following for any five polymers described in this chapter: (a) UTS versus density and (b) elastic modulus versus UTS. Where appropriate, plot a range of values.

SYNTHESIS, DESIGN, AND PROJECTS

7.55 Conduct an Internet search, and describe differential scanning calorimetry. What does this technique measure?

7.56 Describe the design considerations involved in replacing a metal beverage container with one made of plastic.

7.57 Assume that you are manufacturing a product in which all of the gears are made of metal. A salesperson visits you and asks you to consider replacing some of these metal gears with plastic ones. Make a list of the questions that you would raise before making a decision.

7.58 Assume you work for a company that produces polymer gears. You have arranged to meet with a potential new customer, who currently uses gears made of metal. Make a list of the benefits that plastic gears present, and prepare a presentation for the meeting.

7.59 Sections 7.6 and 7.7 list several plastics and their applications. Rearrange this information, making a table of products (gears, helmets, luggage, electrical parts, etc.) that shows the types of plastic that can be used to make these products.

7.60 Make a list of products or parts that currently are not made of plastics, and offer possible reasons why they are not.

7.61 Review the three curves shown in Fig. 7.10, and give some applications for each type of behavior. Explain your choices.

7.62 Repeat Problem 7.61 for the curves shown in Fig. 7.12.

7.63 In order to use a steel or aluminum container for an acidic liquid, such as tomato sauce, a polymeric barrier is usually placed between the container and its contents. Describe possible methods of producing such a barrier.

7.64 Conduct a study of plastics used for some products. Measure the hardness and stiffness of these plastics. (For example, dog chew toys use plastics with a range of properties.) Describe your observations.

7.65 Add a column to Table 7.1 that describes the appearance of these plastics, including available colors and opaqueness.

7.66 With Table 7.3 as a guide, inspect various products, both in a typical kitchen and in an automobile, and describe the types of plastics that were used or could be used in making their individual components.

8

Ceramics, Glass, Graphite, Diamond, and Nanomaterials: Structure, General Properties, and Applications

- Ceramics, glass, and various forms of carbon possess unique combinations of mechanical and physical properties that cannot be obtained with other metallic or nonmetallic materials.
- Ceramic materials are first described, in terms of their chemistry, microstructure, mechanical and physical properties, and applications.
- The basic types of ceramics include oxide ceramics, such as aluminum or zirconium oxide, and carbides and nitrides.
- Glasses have numerous formulations, all containing at least 50% silica. Their general properties and typical uses are described.
- Various forms of carbon are commercially important; graphite is the most common, with numerous uses, including as reinforcement in composite materials, electrodes for electrical discharge machining, and solid lubricant.
- Diamond is the hardest material known and, as such, is used for precision and abrasive machining and for polishing operations.
- Nanomaterials, such as carbon nanotubes, are becoming increasingly important, with numerous applications for nanoscale electrical and microelectronic devices.

8.1 Introduction

The various types of materials described in the preceding chapters are not all suitable for certain applications, including

1. An electrical insulator for use at high temperatures
2. Floor tiles to resist scuffing and abrasion
3. A transparent baking dish
4. Small ball bearings that are light, rigid, hard, and resist high temperatures
5. Automobile windshields that are hard, abrasion resistant, and transparent

It is apparent from these examples that the properties required include high-temperature strength; hardness; inertness to chemicals, foods, and the environment; resistance to wear and corrosion; and low electrical and thermal conductivity.

The general characteristics and applications of those ceramics, glasses, and glass ceramics that are of importance in engineering applications and in manufacturing are first described. Because of their unique properties and uses, the various forms of carbon (graphite, diamond, and carbon nanotubes) are described next. The manufacturing of ceramic and of glass components and various shaping and finishing operations are detailed in Chapter 18; composites, which contain the materials described, are described in Chapter 9.

8.2   The Structure of Ceramics

Ceramics are compounds of metallic and nonmetallic elements. The term *ceramics* (from the Greek *keramos*, meaning "potter's clay," and *keramikos*, meaning "clay products") refers both to the material and to the ceramic product itself. Because of the large number of possible combinations of elements, a wide variety of ceramics is now available for a broad range of consumer and industrial applications. The earliest use of ceramics was in pottery and bricks, dating back to before 4000 B.C. They have become increasingly important in tool and die materials, heat engines, and automotive components, such as exhaust-port liners, automotive spark plugs, coated pistons, and cylinder liners.

Ceramics can be divided into two general categories:

1. **Traditional ceramics**, such as whiteware, tiles, brick, sewer pipe, pottery, and abrasive wheels
2. **Industrial ceramics**, also called **engineering, high-tech**, or **fine ceramics**, such as automotive, turbine, structural, and aerospace components (Fig. 8.1), heat exchangers, semiconductors, seals, prosthetics, and cutting tools

The structure of ceramic crystals, containing various atoms of different sizes, is among the most complex of all material structures. The bonding between these atoms

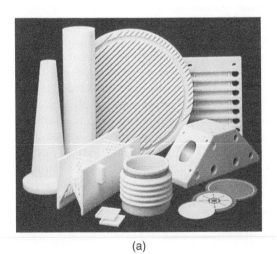
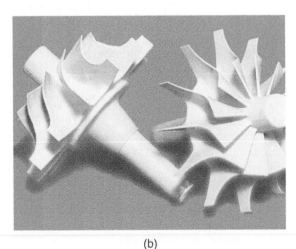

(a)   (b)

FIGURE 8.1 A variety of ceramic components. (a) High-strength alumina for high-temperature applications. (b) Gas-turbine rotors made of silicon nitride. *Source:* Courtesy of Wesgo Div., GTE.

is generally **covalent** or **ionic** (see Section 1.2), and as such are much stronger than metallic bonds. Consequently, properties such as hardness and thermal and electrical resistance are significantly higher in ceramics than in metals (Tables 3.1 and 3.2). Ceramics are available in *single-crystal* or in *polycrystalline* form. Grain size has a major influence on the strength and properties of ceramics; the finer the grain size (hence the term **fine ceramics**), the higher the strength and toughness.

8.2.1   Raw Materials

Among the oldest of the raw materials used for making ceramics is **clay**, which has a fine-grained sheetlike structure. The most common example is *kaolinite* (from Kaoling, a hill in China); it is a white clay consisting of silicate of aluminum, with alternating weakly bonded layers of silicon and aluminum ions (Fig. 8.2). When added to kaolinite, water attaches itself to these layers (*adsorption*); this makes the layers slippery and gives wet clay both its well-known softness and the plastic properties (*hydroplasticity*) that make it easily formable.

Other major raw materials for ceramics that are found in nature are **flint** (a rock composed of very fine-grained silica, SiO_2) and **feldspar** (a group of crystalline minerals consisting of aluminum silicates and potassium, calcium, or sodium). **Porcelain** is a **white ceramic**, composed of kaolin, quartz, and feldspar; its largest use is in appliances and kitchen and bath ware. In their natural state, these raw materials generally contain impurities of various kinds, which have to be removed prior to their further processing into useful products.

8.2.2   Oxide Ceramics

There are two major types of oxide ceramics: alumina and zirconia (Table 8.1).

Alumina.   Also called **corundum** or **emery**, *alumina* (aluminum oxide, Al_2O_3) is the most widely used *oxide ceramic*, either in pure form or as a raw material to be blended with other oxides. It has high hardness and moderate strength. Although alumina exists in nature, it contains varying levels of impurities and possesses nonuniform properties; as a result, its performance also varies. Aluminum oxide, silicon carbide, and most other ceramics are now manufactured almost totally synthetically, so that their quality can be controlled at a consistently high level. First made in 1893, synthetic aluminum oxide is obtained from the fusion of molten bauxite (an aluminum-oxide ore, which is the principal source of aluminum), iron filings, and coke, in electric furnaces. The cooled product is then crushed and graded by size, by passing it through standard screens. Aluminum oxide can be blended with small amounts of other ceramics, such as titanium oxide and titanium carbide.

Structures containing alumina and various other oxides are known as **mullite** and **spinel**, used as refractory materials for high-temperature applications. The mechanical and physical properties of alumina are suitable particularly in such applications as electrical and thermal insulation and in cutting tools and abrasives.

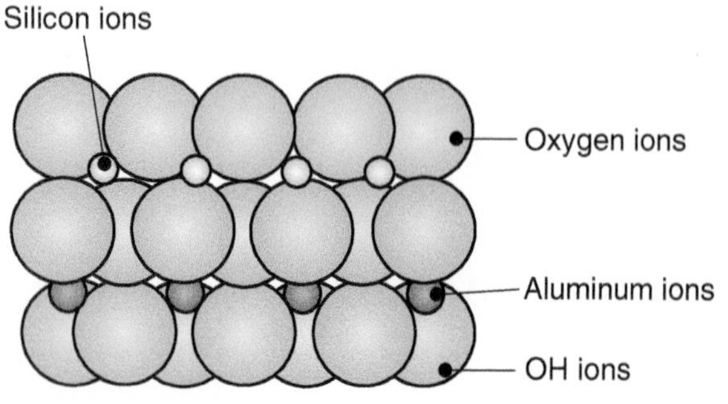

Zirconia.   Zirconia (zirconium oxide, ZrO_2, white in color) has good toughness, good resistance to thermal shock, wear, and corrosion, low thermal conductivity, and a low friction coefficient. **Partially stabilized zirconia**

Silicon ions

Oxygen ions

Aluminum ions

OH ions

FIGURE 8.2   The crystal structure of kaolinite, commonly known as clay; compare with Figs. 1.3–1.5 for metals.

TABLE 8.1

Types, General Characteristics, and Principal Uses of Ceramics	
Type	General characteristics and uses
Oxide ceramics	
Alumina	High hardness and moderate strength; most widely used ceramic; cutting tools; abrasives; electrical and thermal insulation
Zirconia	High strength and toughness; thermal expansion close to cast iron; suitable for high-temperature applications such as metallurgical furnace linings, jet-engine components, and nuclear fuel cladding
Carbides	
Tungsten carbide	Hardness, strength, and wear resistance depend on cobalt binder content; commonly used for dies and cutting tools
Titanium carbide	Not as tough as tungsten carbide; has nickel and molybdenum as the binder; used as cutting tools
Silicon carbide	High-temperature strength and wear resistance; used for heat engines and as abrasives in grinding wheels
Nitrides	
Cubic boron nitride	Second-hardest substance known, after diamond; used as abrasives and cutting tools
Titanium nitride	Gold in color; used as coatings because of low frictional characteristics
Silicon nitride	High resistance to creep and thermal shock; used in high-temperature applications such as turbocharger components, rolling element bearings and cutting tools
Sialon	Consists of silicon nitrides and other oxides and carbides; used as cutting tools and feed tubes and linings for non-ferrous metal casting
Cermets	Consist of oxides, carbides, and nitrides; used in high-temperature applications such as cutting tools and composite armor for military applications
Silica	High-temperature resistance; quartz exhibits piezoelectric effect; silicates containing various oxides are used in nonstructural applications such as fiber glass, plate glass, and optical glass
Glasses	Contain at least 50% silica; amorphous structures; several types available with a wide range of mechanical and physical properties
Glass ceramics	Have a high crystalline component to their structure; good thermal-shock resistance and strong. Typical applications include glass-ceramic cooking tops for stoves and cookware
Graphite	Crystalline form of carbon; high electrical and thermal conductivity; good thermal-shock resistance; used for structural reinforcement in composite materials, electrical discharge machining electrodes, piston rings
Diamond	Hardest substance known; available as single crystal or in polycrystalline form; used as cutting tools and abrasives and as dies for drawing fine wire
Carbon nanotubes	Unique crystalline form of graphite, with high electrical and thermal conductivity; under investigation for MEMS and microelectronics applications and in composite materials
Nanophase ceramics	Stronger and easier to fabricate and machine than conventional ceramics; used in automotive and jet-engine applications

(PSZ) has higher strength and toughness and better reliability in performance than does zirconia. It is obtained by doping zirconia with oxides of calcium, yttrium, or magnesium; this process forms a material with fine particles of tetragonal zirconia in a cubic lattice. Typical applications include dies for the hot extrusion of metals, and zirconia beads used as grinding and polishing media for aerospace coatings, for automotive primers, paint, and fine glossy print on flexible food packaging.

Two important characteristics of PSZ are its high coefficient of thermal expansion (which is only about 20% lower than that of cast iron), and its low thermal conductivity (which is about one-third that of other ceramics). Consequently, PSZ is very suitable for heat-engine components, such as cylinder liners and valve bushings, to help keep the cast-iron engine assembly intact. **Transformation-toughened zirconia** (TTZ) has higher toughness, because of dispersed tough phases in the ceramic matrix.

***

CASE STUDY 8.1   Ceramic Knives

Generally made of zirconium oxide, ceramic knives are produced by a process described in Section 18.2. It starts with a ceramic powder mixed with various binders, and compacted (molded) into blanks under high pressure. The blanks are then fired (sintered) at temperatures above 1000°C (1830°F) for several days. An optional hot isostatic pressing operation (Section 17.3.2) can be used to densify and toughen the ceramic. Next, the blanks are ground and polished on a diamond wheel to form a sharp edge, and the handle is attached. The Mohs hardness (Section 2.6) of the zirconium oxide ceramic is 8.2, as compared to 6 for hardened steel and a maximum of 10 for diamond.

Among the advantages of ceramic knives over steel knives are: (a) Because of their very high hardness and wear resistance, ceramic knives can last months and even years before resharpening. (b) They are chemically inert; consequently, they do not stain

and food does not stick to them, hence they are easy to clean, and leave no metallic taste or smell. (c) Because they are lightweight, they are easier to use.

The knives should, however, be stored in wooden knife blocks and handled carefully. Sharp impact against other objects (such as dishes or dropping it on its edge on a hard surface) should be avoided, as the sharp edges of the knife can chip. Also, they should be used only for cutting (not for prying); in cutting meat, contact with bones is not advisable. Furthermore, the knives have to be sharpened at the factory to a precise edge, using diamond grinding wheels. Ceramic knives are comparable in cost to high-quality steel knives, typically ranging from $20–80 for a 3-in. paring knife to $50–200 for a 6-in. serrated knife.

Source: Courtesy of Kyocera Corporation.

8.2.3   Other Ceramics

Carbides.   *Carbides* are typically used as cutting tools and die materials, and as an abrasive, especially in grinding wheels. Common examples of carbides are:

- **Tungsten carbide** (WC) consists of tungsten-carbide particles with cobalt as a binder. The amount of binder has a major influence on the material's properties; toughness increases with cobalt content, whereas hardness, strength, and wear resistance decrease.
- **Titanium carbide** (TiC) has nickel and molybdenum as the binder, and is not as tough as tungsten carbide.
- **Silicon carbide** (SiC) has good resistance to wear (thus suitable for use as an abrasive), thermal shock, and corrosion. It has a low friction coefficient and retains strength at elevated temperatures, and thus it is suitable for high-temperature components in heat engines. First produced in 1891, synthetic silicon carbide is made from silica sand, coke, and small amounts of sodium chloride and sawdust; the process is similar to that for making synthetic aluminum oxide (Section 8.2.2).

Nitrides.   Examples of *nitrides* are:

- **Cubic boron nitride** (cBN) is the second-hardest known substance (after diamond), and has special applications, such as in cutting tools and as abrasives in grinding wheels. It does not exist in nature, and was first made synthetically in the 1970s, using techniques similar to those used in making synthetic diamond (Section 8.7).

- **Titanium nitride** (TiN) is used widely as a coating on cutting tools; it improves tool life by virtue of its low friction characteristics.
- **Silicon nitride** (Si_3N_4) has high resistance to creep at elevated temperatures, low thermal expansion, and high thermal conductivity, thus it resists thermal shock (Section 3.6). It is suitable for high-temperature structural applications, such as components in automotive-engine and gas-turbine, cam-follower rollers, bearings, sandblast nozzles, and components for the paper industry.

Sialon.   Derived from the words *si*licon, *al*uminum, *o*xygen, and *n*itrogen, *sialon* consists of silicon nitride, with various additions of aluminum oxide, yttrium oxide, and titanium carbide. It has higher strength and thermal-shock resistance than silicon nitride, and is used primarily as a cutting-tool material.

Cermets.   *Cermets* are combinations of a *ceramic* phase bonded with a *metallic* phase. Introduced in the 1960s and also called **black ceramics** or **hot-pressed ceramics,** they combine the high-temperature oxidation resistance of ceramics with the toughness, thermal-shock resistance, and ductility of metals. A common application of cermets is in cutting tools, with a typical composition being 70% Al_2O_3 and 30% TiC; other cermets contain various oxides, carbides, and nitrides.

Cermets have been developed for high-temperature applications, such as nozzles for jet engines and brakes for aircraft, as well as electrical resistors and capacitors that may experience high temperatures. Cermets can be regarded as *composite* materials and can be used in various combinations of ceramics and metals bonded by powder metallurgy techniques (Chapter 17).

8.2.4   Silica

Abundant in nature, *silica* is a polymorphic material—that is, it can have different crystal structures. The cubic structure is found in refractory bricks, used for high-temperature furnace applications. Most glasses contain more than 50% silica. The most common form of silica is **quartz,** a hard, abrasive hexagonal crystal, used extensively in communications applications as an oscillating crystal of fixed frequency, because it exhibits the piezoelectric effect (Section 3.7).

Silicates are products of the reaction of silica with oxides of aluminum, magnesium, calcium, potassium, sodium, and iron; examples are clay, asbestos, mica, and silicate glasses. **Lithium aluminum silicate** has very low thermal expansion and thermal conductivity, and good thermal-shock resistance. However, it also has very low strength and fatigue life, thus it is suitable only for nonstructural applications, such as catalytic converters, regenerators, and heat-exchanger components.

8.2.5   Nanoceramics and Composites

In order to improve the ductility and manufacturing properties of ceramics, the particle size in ceramics has been reduced by means of various techniques, most commonly gas condensation, use of sol-gels, or combustion synthesis. Called *nanoceramics* or **nanophase ceramics,** the structure of these materials consists of atomic clusters, each containing a few thousand atoms. Control of particle size, distribution, and contamination are important in nanoceramics, which exhibit ductility at significantly lower temperatures than do conventional ceramics, and are stronger and easier to fabricate and machine with fewer flaws. Applications are in automotive components such as valves, rocker arms, turbocharger rotors, and cylinder liners, and in jet-engine components.

Nanocrystalline second-phase particles (on the order of 100 nm or less) and fibers also are used as reinforcements in *composites* (Chapter 9). These composites have such enhanced properties as better tensile strength and creep resistance.

8.3 General Properties and Applications of Ceramics

Compared with metals, ceramics typically have the following relative characteristics: brittleness; high strength, elastic modulus, and hardness at elevated temperatures; low toughness, density, and thermal expansion; and low thermal and electrical conductivity. Because of the wide variety of material compositions and grain size, the mechanical and physical properties of ceramics vary considerably. Properties of ceramics can also vary widely because of their sensitivity to flaws, defects, and surface or internal cracks. The presence of different types and levels of impurities; and different methods of manufacturing also affect their properties.

8.3.1 Mechanical Properties

The mechanical properties of selected engineering ceramics are given in Fig. 8.3 and Table 8.2. Note that their strength in tension (transverse rupture strength, Section 2.5) is approximately one order of magnitude lower than their compressive strength, because of their sensitivity to cracks, impurities, and **porosity**. Such defects lead to the initiation and propagation of cracks under tensile stresses, and thus significantly reduce the tensile strength of the ceramic; thus, reproducibility and reliability are important aspects in the service life of ceramic components.

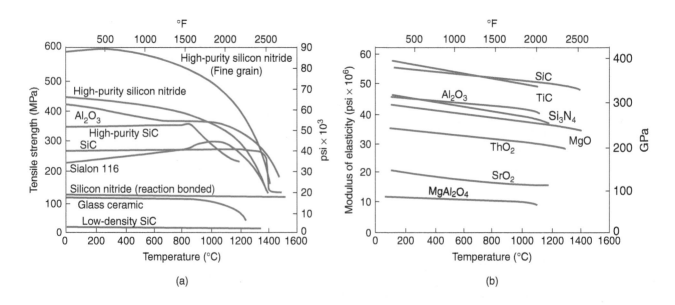

(a)   (b)

FIGURE 8.3 (a) Effect of temperature on the strength of various engineering ceramics; note that much of the strength is maintained at high temperatures; compare with Fig. 5.5. (b) Effect of temperature on the modulus of elasticity for various ceramics; compare with Fig. 2.6.

TABLE 8.2

Properties of Various Ceramics at Room Temperature

Material	Symbol	Transverse rupture strength (MPa)	Compressive strength (MPa)	Elastic modulus (GPa)	Hardness (HK)	Poisson's ratio, v	Density (kg/m^3)
Aluminum oxide	Al_2O_3	140–240	1000–2900	310–410	2000–3000	0.26	4000–4500
Cubic boron nitride	cBN	725	7000	850	4000–5000	—	3480
Diamond	—	1400	7000	830–1000	7000–8000	—	3500
Silica, fused	SiO_2	—	1300	70	550	0.25	—
Silicon carbide	SiC	100–750	700–3500	240–480	2100–3000	0.14	3100
Silicon nitride	Si_3N_4	480–600	—	300–310	2000–2500	0.24	3300
Titanium carbide	TiC	1400–1900	3100–3850	310–410	1800–3200	—	5500–5800
Tungsten carbide	WC	1030–2600	4100–5900	520–700	1800–2400	—	10,000–15,000
Partially stabilized zirconia	PSZ	620	—	200	1100	0.30	5800

Note: These properties vary widely depending on the condition of the material.

The tensile strength of polycrystalline ceramic increases with decreasing grain size and porosity. This relationship is represented approximately by the expression

$$UTS = UTS_o e^{-nP}, \qquad (8.1)$$

where P is the volume fraction of pores in the solid, thus if the porosity is 15%, $P = 0.15$, UTS_o is the tensile strength at zero porosity; and the exponent n ranges between 4 and 7. The modulus of elasticity of ceramics is related to porosity by the expression

$$E \simeq E_o \left(1 - 1.9P + 0.9P^2\right), \qquad (8.2)$$

where E_o is the elastic modulus at zero porosity.

Unlike most metals and thermoplastics, ceramics generally lack impact toughness and thermal-shock resistance, because of their inherent lack of ductility; once initiated, a crack propagates rapidly. In addition to undergoing fatigue failure under cyclic loading, ceramics exhibit a phenomenon called **static fatigue**, also exhibited by glasses (Section 8.4). When subjected to a static tensile load over time, these materials may suddenly fail; this phenomenon occurs in environments where water vapor is present. Static fatigue, which does not occur in a vacuum or in dry air, has been attributed to a mechanism similar to the stress–corrosion cracking of metals (Section 2.10.2).

Ceramic components that are to be subjected to tensile stresses may be *prestressed*, in much the same way that concrete is prestressed. Prestressing the shaped ceramic components subjects them to compressive stresses; the methods used include

- Heat treatment and chemical tempering (Section 18.4)
- Laser treatment of surfaces (Section 34.8)
- Coating with ceramics having different thermal-expansion coefficients (Section 3.6)
- Surface-finishing operations, such as grinding, in which compressive residual stresses are induced on the surfaces (Section 26.3)

Major advances have been made in improving the toughness and other properties of ceramics, including the development of **machinable** and **grindable ceramics**

(Section 18.2.5). Among these advances are the proper selection and processing of raw materials, the control of purity and structure, and the use of reinforcements—with particular emphasis on advanced methods of stress analysis during the design of ceramic components.

8.3.2   Physical Properties

Most ceramics have a relatively low specific gravity, ranging from about 3 to 5.8 for oxide ceramics as compared to 7.86 for iron (Table 3.1). They have very high melting or decomposition temperatures.

Thermal conductivity in ceramics varies by as much as three orders of magnitude, depending on their composition, whereas in metals it varies by only one order. Like that of other materials, the thermal conductivity of ceramics decreases with increasing temperature and porosity, because air is a poor thermal conductor. The thermal conductivity, k, is related to porosity by the expression

$$k = k_o (1 - P), \tag{8.3}$$

where k_o is the thermal conductivity at zero porosity and P is the porosity, as a fraction of the total volume.

Thermal expansion and thermal conductivity induce internal stresses that can lead to thermal shock or to thermal fatigue in ceramics. The tendency toward **thermal cracking** (called **spalling**, when a small piece or a layer from the surface breaks off) is lower with the combination of low thermal expansion and high thermal conductivity. For example, fused silica has high thermal-shock resistance, because of its virtually zero thermal expansion. (See also Sections 3.5 and 3.6.)

The *optical properties* of ceramics can be controlled by using various formulations and by controlling the structure. These methods make possible the imparting of different degrees of transparency and translucency, and of different colors. For example, single-crystal sapphire is completely transparent, zirconia is white, and fine-grained polycrystalline aluminum oxide is a translucent gray. Porosity influences the optical properties of ceramics in much the same way as air trapped in ice cubes, making them less transparent and giving a white appearance. Although ceramics are basically resistors, they can be made *electrically conducting* by alloying them with certain elements in order to make the ceramic behave like a semiconductor or even like a superconductor.

8.3.3   Applications

Ceramics have numerous consumer and industrial applications. Various types of ceramics are used in the electrical and electronics industries, because they have high electrical resistivity, high dielectric strength (voltage required for electrical breakdown per unit thickness), and magnetic properties suitable for such applications as magnets for speakers.

The capability of ceramics to maintain their strength and stiffness at elevated temperatures makes them very attractive for high-temperature applications. The higher operating temperatures made possible by the use of ceramic components mean more efficient combustion of fuel and reduction of emissions in automobiles. Currently, internal combustion engines are only about 30% efficient, but with the use of ceramic components, the operating performance can be improved by at least 30%.

Ceramics that are being used successfully, especially in automotive gas-turbine engine components (such as rotors), are: silicon nitride, silicon carbide, and partially stabilized zirconia. Other attractive properties of ceramics are their low density and high elastic modulus; they enable product weight to be reduced and allow the inertial forces generated by moving parts to be lower. Ceramic turbochargers, for example, are about 40% lighter than conventional ones. High-speed components for machine tools also are candidates for ceramics (Section 25.3). Furthermore, the high elastic modulus of ceramics makes them attractive for improving the stiffness of machines, while reducing the weight. Their high resistance to wear also makes them suitable for such applications as cylinder liners, bushings, seals, bearings, and liners for gun barrels. Coating metal with ceramics is another application, often done to reduce wear, prevent corrosion, or provide a thermal barrier.

CASE STUDY 8.2   Ceramic Gun Barrels

The wear resistance and low density of ceramics have led to research into their use as liners for gun barrels. Their limited success has led to more recent developments in making composite ceramic gun barrels, which have improved performance over traditional steel barrels. For example, a 50-caliber zirconia ceramic barrel is formed in several separate segments, each 150–200 mm (6–8 in.) long and with a wall thickness of 3.75 mm (0.150 in.), using the shaping and sintering processes described in Chapter 17.

The segments are subsequently machined to the required dimensions and surface finish. Zirconia is used because of its high toughness, flexural strength, specific heat, operating temperature, and very low thermal conductivity. The thermal properties are important for gun performance.

The separate ceramic segments are then assembled, and the barrel is wrapped with a carbon-fiber/polymer-matrix composite that subjects the ceramic barrel to a compressive stress of 100,000 psi, thus greatly improving its capacity to withstand tensile stresses developed during firing. Finally, the inside of the barrel is rifled (cut to produce internal spiral grooves that give rotation to the exiting bullet for gyroscopic stability) and fitted to a breech.

Source: Courtesy of K.H. Kohnken, Surface Conversion Technologies, Inc., Cumming, Georgia.

CASE STUDY 8.3   Ceramic Ball and Roller Bearings

Silicon-nitride ceramic ball and roller bearings (Fig. 8.4) are used when high temperature, high speed, or marginally lubricated conditions exist. The bearings can be made entirely from ceramics, or just the ball and rollers are ceramic and the races are metal, in which case they are referred to as *hybrid bearings*. Examples of applications for ceramic and hybrid bearings include high-performance machine tool spindles, metal-can seaming heads, high-speed flow meters, and bearings for motorcycles, go karts, and snowmobiles.

The ceramic spheres have a diametral tolerance of 0.13 μm (5 μin.) and a surface roughness of 0.02 μm (0.8 μin.) They have high wear resistance, high fracture toughness, low density, and perform well with little or no lubrication. The balls have a coefficient of thermal expansion one-fourth that of steel, and they can withstand temperatures of up to 1400°C (2550°F).

Produced from titanium and carbon nitride by powder metallurgy techniques, the full-density titanium carbonitride (TiCN) or silicon nitride (Si_3N_4) bearing-grade material can be twice as hard as chromium steel and 40% lighter. Components up to 300 mm (12 in.) in diameter have been produced.

(continued)

(a)   (b)

FIGURE 8.4   A selection of ceramic bearings and races. *Source:* Courtesy of The Timken Company.

Bioceramics.   Because of their strength and inertness, ceramics are also used as bio-materials (*bioceramics*), to replace joints in the human body, as prosthetic devices, and in dental work. Commonly used bioceramics are aluminum oxide, hydrox-yapatite (a naturally occurring mineral), tricalcium phosphate, silicon nitride, and various compounds of silica. Ceramic implants can be made porous, so that bone can grow into the porous structure (as is the case with porous titanium implants) and develop a strong bond with structural integrity.

8.4   Glasses

Glass is an amorphous solid with the structure of a liquid (Fig. 8.5), a condition that is obtained by supercooling (cooling at a rate too high to allow crystals to form). Technically, glass is defined as an inorganic product of fusion that has cooled to a

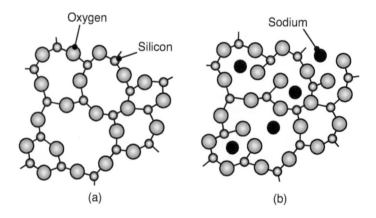

(a)   (b)

FIGURE 8.5   Schematic illustration of the structure of silica glass. (a) Pure silica glass, in the form of $(SiO_2)_n$ random structure and (b) partially depolymerized glass; note that a fourth bond for each silicon is outside the plane shown.

rigid state without crystallizing. Glass has no distinct melting or freezing point; thus, its behavior is similar to that of amorphous alloys (see metallic glasses, Section 6.14) and amorphous polymers (Section 7.2.2).

Glass beads were first produced in about 2000 B.C., and the art of glassblowing began in about 200 B.C. Silica solely was used for all glass products until the late 1600s. Rapid developments in glasses began in the early 1900s. There are about 750 different types of commercially available glasses, with applications ranging from window glass to glass for containers, cookware, lighting, and mobile phones, and to glasses with special mechanical, electrical, high-temperature, antichemical, corrosion, and optical characteristics. Special glasses are used in fiber optics (for communication by light with little loss in signal power) and in **glass fibers**, with very high strength (for use in reinforced plastics, Section 9.2.1).

All glasses contain at least 50% silica, which is known as a **glass former**. The composition and properties of glasses can be modified by the addition of oxides of aluminum, sodium, calcium, barium, boron, magnesium, titanium, lithium, lead, and potassium. Depending on their function, these oxides are known as **intermediates** or **modifiers**.

8.4.1 Types of Glasses

Almost all *commercial glasses* are categorized by the following types (Table 8.3):

- **Soda-lime glass** (the most common type)
- **Lead-alkali glass**
- **Borosilicate glass**
- **Aluminosilicate glass**
- **96%-silica glass**
- **Fused silica glass**

Glasses also are classified as colored, opaque (white and translucent), multiform (a variety of shapes), optical, photochromatic (darkens when exposed to light, as in some sunglasses), photosensitive (changing from clear to opaque), fibrous (drawn into long fibers, as in fiberglass), and foam or cellular (containing bubbles, thus a good thermal insulator). Glasses also can be referred to as **hard** or **soft**, usually in the sense of a thermal rather than mechanical property (see also hardness of glasses, Section 8.4.2); thus, a soft glass softens at a lower temperature than

TABLE 8.3

Properties of Various Glasses					
Property	Soda-lime glass	Lead-alkali glass	Borosilicate glass	96% silica	Fused silica
Density	High	Highest	Medium	Low	Lowest
Strength	Low	Low	Moderate	High	Highest
Resistance to thermal shock	Low	Low	Good	Better	Best
Electrical resistivity	Moderate	Best	Good	Good	Good
Hot workability	Good	Best	Fair	Poor	Poorest
Heat treatability	Good	Good	Poor	None	None
Chemical resistance	Poor	Fair	Good	Better	Best
Impact-abrasion resistance	Fair	Poor	Good	Good	Best
Ultraviolet-light transmission	Poor	Poor	Fair	Good	Good
Relative cost	Lowest	Low	Medium	High	Highest

does a hard glass. Soda-lime and lead-alkali glasses are considered soft, and the rest as hard.

8.4.2 Mechanical Properties

The behavior of glass, like that of most ceramics, is generally regarded as perfectly elastic and brittle. The modulus of elasticity for commercial glasses ranges from 55 to 90 GPa (8–13 million psi), and their Poisson's ratio from 0.16 to 0.28. The hardness of glasses, as a measure of resistance to scratching, ranges from 5 to 7 on the Mohs scale, which is equivalent to a range from around 350–500 HK. (See Fig. 2.15.)

Glass in **bulk** form generally has a strength lower than 140 MPa (20 ksi). The relatively low strength of bulk glass is attributed to the presence of small flaws and microcracks on its surface, some or all of which may be introduced during normal handling of the glass by inadvertently abrading it. These defects reduce the strength of glass by two to three orders of magnitude, compared to its ideal (defect free) strength. Glasses can be strengthened by thermal or chemical treatments to obtain high strength and toughness (Section 18.4).

The strength of glass theoretically can reach 35 GPa (5 million psi). When molten glass is drawn into fibers (**fiberglass**), its tensile strength ranges from 0.2 to 7 GPa (30–1000 ksi), with an average of about 2 GPa (300 ksi). These fibers are stronger than steel, and are used to reinforce plastics in such applications as boats, automobile bodies, furniture, and sporting equipment (see Tables 2.2 and 9.2).

8.4.3 Physical Properties

Glasses are characterized by low thermal conductivity and high electrical resistivity and dielectric strength. Their coefficient of thermal expansion is lower than those for metals and plastics, and may even approach zero. For example, titanium silicate glass (a clear, synthetic high-silica glass) has a near-zero coefficient of thermal expansion. Fused silica (a clear, synthetic amorphous silicon dioxide of very high purity) also has a near-zero coefficient of expansion. The optical properties of glasses, such as reflection, absorption, transmission, and refraction, can be modified by varying their composition and treatment. Glasses generally are resistant to chemical attack, and are ranked by their resistance to corrosion by acids, alkalis, or water.

8.5 Glass Ceramics

Although glasses are amorphous, *glass ceramics* have a high crystalline component to their microstructure. Glass ceramics, such as *Pyroceram* (a trade name), contain large proportions of several oxides; thus, their properties are a combination of those for glass and those for ceramics. Most glass ceramics are stronger than glass. The products are first shaped and then heat treated, whereby **devitrification** (recrystallization) of the glass occurs. Glass ceramics are generally white or gray in color.

The hardness of glass ceramics ranges approximately from 520 to 650 HK. Because they have a near-zero coefficient of thermal expansion, they also have high thermal-shock resistance, and because of the absence of the porosity usually found in conventional ceramics, they are strong. The properties of glass ceramics can be improved by modifying their composition and by heat-treatment techniques. First developed in 1957, glass ceramics are typically used for cookware, heat exchangers in gas-turbine engines, radomes (housings for radar antennas), and electrical and electronics components.

8.6 Graphite

Graphite is a crystalline form of carbon, and has a *layered structure*, with basal planes or sheets of close-packed carbon atoms (see Fig. 1.5); consequently, graphite is weak when sheared along the layers. This characteristic, in turn, gives graphite its low frictional properties, and explains its use as a lubricant, especially at elevated temperatures. However, its frictional properties are low only in an environment of air or moisture; in a vacuum, it is abrasive and thus a poor lubricant. Unlike with other materials, strength and stiffness of graphite increase with temperature. Amorphous graphite is known as **lampblack** (black soot) and is used as a pigment. Ordinary pencil "lead" is a mixture of graphite and clay; graphite deposits found in the early 16th century were first thought of as being a form of lead.

Although brittle, graphite has high electrical and thermal conductivity and good resistance to thermal shock and to high temperature, although it begins to oxidize at 500°C (930°F). It is an important material for applications such as electrodes, heating elements, brushes for motors, high-temperature fixtures and furnace parts, mold materials (such as crucibles for the melting and casting of metals), and seals (Fig. 8.6). A characteristic of graphite is its resistance to chemicals, thus it is used in filters for corrosive fluids; its low absorption cross-section and high scattering cross-section for thermal neutrons make graphite also suitable for nuclear applications.

Graphite Fibers.   An important use of graphite is as *fibers* in reinforced plastics and composite materials, as described in Section 9.2.

Carbon and Graphite Foams.   These foams have high service temperatures, chemical inertness, low thermal expansion, and thermal and electrical properties that can be tailored to specific applications. *Carbon foams* are available in either graphitic or nongraphitic structures. *Graphitic foams* (typically produced from petroleum, coal tar, and synthetic pitches) have low density, high thermal and electrical conductivity, but lower mechanical strength, and are much more expensive than nongraphitic foams (produced from coal or organic resins), which are highly amorphous.

These foams have a cellular microstructure, with interconnected pores, thus their mechanical properties depend on density (see also Section 8.3). Blocks of foam

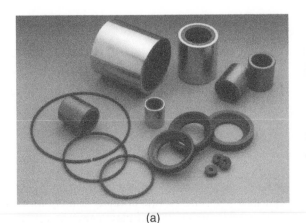

(a)   (b)

FIGURE 8.6   (a) Various engineering components made of graphite. *Source:* Courtesy of Entegris, Inc. (b) Examples of graphite electrodes for electrical discharge machining (Section 27.5). *Source:* Courtesy of Unicor, Inc.

can easily be machined into various complex shapes. Applications of carbon foams include their use as core materials for aircraft and ship interior panels, structural insulation, sound-absorption panels, substrates for spaceborne mirrors, lithium-ion batteries, and for fire and thermal protection.

8.6.1 Fullerenes

Carbon molecules (typically C60) are now produced in the shape of soccer balls, called *fullerenes* or **buckyballs**, after B. Fuller (1895–1983), the inventor of the geodesic dome. These chemically–inert spherical molecules are produced from soot, and act much like solid lubricant particles. When mixed with metals, fullerenes can become superconductors at low temperatures (around 40 K). Despite their promise, no commercial applications of buckyballs currently exist.

8.6.2 Nanotubes

Carbon nanotubes can be thought of as tubular forms of graphite, and are of interest for the development of nanoscale devices. (See also *nanomaterials*, Section 8.8.) *Nanotubes* are produced by laser ablation of graphite, carbon-arc discharge, and, most often, by chemical vapor deposition (CVD, Section 34.6.2). They can be single-walled (SWNT) or multi-walled (MWNT) and can be doped with various species.

Carbon nanotubes have exceptional strength, thus making them attractive as reinforcing fibers for composite materials; however, because they have very low adhesion with most materials, delamination with a matrix can limit their reinforcing effectiveness. It is difficult to disperse nanotubes properly because they have a tendency to clump, and this limits their effectiveness as a reinforcement. A few products have used carbon nanotubes, such as bicycle frames, specialty baseball bats, golf clubs, and tennis racquets. Nanotubes provide only a fraction of the reinforcing material (by volume), graphite fibers playing the major role.

An additional characteristic of carbon nanotubes is their very high electrical-current carrying capability. They can be made as semiconductors or conductors, depending on the orientation of the graphite in the nanotube (Fig. 8.7). Armchair nanotubes are theoretically capable of carrying a current density higher than 1,000

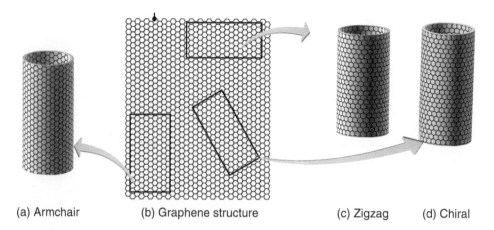

(a) Armchair   (b) Graphene structure   (c) Zigzag   (d) Chiral

FIGURE 8.7 Forms of carbon nanotubes produced from a section of graphene: armchair, zigzag, and chiral. Armchair nanotubes are noteworthy for their high electrical conductivity, whereas zigzag and chiral nanotubes are semiconductors.

times that for silver or copper, thus making them attractive for electrical connections in nanodevices (Section 29.5). Carbon nanotubes have been incorporated into polymers to improve their static-electricity discharge capability, especially in fuel lines for automotive and aerospace applications.

Among the numerous proposed uses for carbon nanotubes are storage of hydrogen for use in hydrogen-powered vehicles, flat-panel displays, human tissue engineering, electrical cables for nano-scale circuitry, catalysts, and X-ray and microwave generators. Highly sensitive sensors using aligned carbon nanotubes are now being developed for detecting deadly gases, such as sarin.

8.6.3   Graphene

Graphene can be considered to be a single sheet of graphite, or an unwrapped nanotube, as shown in Fig. 8.7b. It is one of the most commonly encountered materials, but its direct observation in transmission electron microscopes dates only to the early 1960s. A number of methods have been developed for producing graphene, including epitaxy (Section 28.5) on silicon carbide or metal substrates and by chemical reduction of graphite. Research interest in graphene has grown considerably in the past few years, and applications of graphene as a transistor in integrated circuits (Chapter 28) and in solar cells have been suggested.

8.7   Diamond

Diamond is a form of carbon, with a covalently bonded structure. It is the hardest substance known (7000–8000 HK); however, it is brittle and begins to decompose in air at about 700°C (1300°F), but resists higher temperatures in a nonoxidizing environment.

Synthetic or **industrial diamond** was first made in 1955. A common method of manufacturing it is to subject graphite to a hydrostatic pressure of 14 GPa (2 million psi) and a temperature of 3000°C (5400°F), referred to as *high-pressure, high-temperature* (HPHT) synthesis. An alternative is to produce diamonds through a chemical vapor deposition process (CVD; Section 34.6.2), depositing carbon onto a starting *seed* of diamond powder. The CVD process is used most often for synthetic gemstones. Synthetic diamond has identical, and sometimes slightly superior, mechanical properties as natural diamond, because of the presence of fewer and smaller impurities. The gemstones have a characteristic orange or yellow tint due to impurities, resulting from the CVD process, whereas laser treatment of the diamond can change the tint to pink or blue. However, since most of a gemstone's cost is attributed to grinding and finishing (Chapter 26) to achieve a desired shape, synthetic diamonds are only slightly less expensive than natural ones.

Synthetic diamond is available in a variety of sizes and shapes; for use in abrasive machining, the most common grit size is 0.01 mm (0.0004 in.) in diameter. Diamond particles can be coated with nickel, copper, or titanium for improved performance in grinding operations. **Diamond-like carbon** also has been developed and is used as a diamond film coating, described in Section 34.13.

In addition to its use in jewelry, gem-quality synthetic diamond has applications as heat sinks for computers, in telecommunications and integrated-circuit industries, and in high-power lasers. Its electrical conductivity is 50 times higher than that of natural diamond, and it is 10 times more resistant to laser damage.

Because of its favorable characteristics, diamond has numerous important applications, such as:

- Cutting-tool materials, as a single crystal or in polycrystalline form
- Abrasives in grinding wheels, for hard materials
- Dressing of grinding wheels (e.g., sharpening of the abrasive grains)
- Die inserts, for drawing wire less than 0.06 mm (0.0025 in.) in diameter
- Coatings for cutting tools and dies

8.8 Nanomaterials

Important developments continue to take place in the production of materials as particles, fibers, wire, tube, films, and composites, with features typically on the order of 1 nm to up to 100 nm. First investigated in the early 1980s and generally called *nanomaterials* or *nanostructured, nanocrystalline,* or *nanophase* materials, they have certain properties that are often superior to traditional materials. These characteristics include high strength, hardness, ductility, toughness, resistance to wear and corrosion, and suitable for structural (load bearing) and nonstructural applications in combination with unique electrical, magnetic, thermal, and optical properties.

The composition of a nanomaterial can be any combination of chemical elements; among the more important compositions are carbides, oxides, nitrides, metals and their alloys, organic polymers, semiconductors, and various composites. *Nanometal-polymer hybrid nanomaterials* have been developed for very lightweight components. More recent investigations include the development of *nanopaper*, with very high strength and toughness, produced from wood pulp, with fibers rearranged into an entangled porous mesh.

Production methods for nanomaterials include inert-gas condensation, sputtering, plasma synthesis, electrode position, sol–gel synthesis, and mechanical alloying or ball milling. The synthesized powders are consolidated into bulk materials by various techniques, including compaction and sintering. Nanoparticles have a very high surface area-to-volume ratio, thus affecting their behavior in such processes as diffusion and agglomeration. Because the synthesis of nanomaterials is at atomic levels, their purity is on the order of 99.9999%, and their homogeneity and uniformity of their microstructure are highly controlled. As a result, their mechanical, electrical, magnetic, optical, and chemical properties also can be controlled precisely. Nanomaterials are very expensive to produce and process them into products, thus their cost-effectiveness is under continued study.

Applications of Nanomaterials.   The unique properties of nanomaterials enable manufacturing of products that are strong and light. The following are some current and potential applications for nanomaterials:

1. Cutting tools and inserts, made of nanocrystalline carbides and other ceramics
2. Nanophase ceramics, that are ductile and machinable
3. Specialty bicycle frames, baseball bats, and tennis racquets, using carbon nanotubes (see also Section 8.6.2)
4. Next-generation computer chips, using nanocrystalline starting materials with very high purity, better thermal conductivity, and more durable interconnections

5. Flat-panel displays for laptop computers and televisions, made by synthesizing nanocrystalline phosphorus to improve screen resolution
6. Spark-plug electrodes, igniters and fuels for rockets, medical implants, high-sensitivity sensors, catalysts for elimination of pollutants, high-power magnets, and high-energy-density batteries
7. Switches, valves, motor, and pumps
8. Coatings made of nanomaterials are being investigated for improved wear, abrasion, corrosion resistance, and thermal insulation; nanocrystalline materials; and nanophase materials because of their lower thermal conductivity

Health Hazards. Nanoparticles can present various health hazards by virtue of their very small size and absorption through the skin, lungs, or the digestive track; they can also penetrate human cells. There is also increasing evidence that nanoparticles can pollute air, water, and the ground. Consequently, there is growing research on the risks of nanoparticles to humans and the environment.

SUMMARY

- Ceramics, glasses, and various forms of carbon are of major importance in engineering applications and in manufacturing processes. Ceramics, which are compounds of metallic and nonmetallic elements, generally are characterized by high hardness, high compressive strength, high elastic modulus, low thermal expansion, high temperature resistance, good chemical inertness, low density, and low thermal and electrical conductivity. They are brittle and have low toughness.

- Ceramics are generally classified as either traditional ceramics or industrial (or high-tech) ceramics; the latter are particularly attractive for applications such as engine components, cutting tools, and components requiring resistance against wear and corrosion. Ceramics of importance in design and manufacturing are the oxide ceramics (alumina and zirconia), tungsten and silicon carbides, nitrides, and cermets.

- Glasses are supercooled liquids and are available in a wide variety of compositions and mechanical, physical, and optical properties. Glass ceramics are predominantly crystalline in structure, and have properties that are more desirable than those of glasses.

- Glass in bulk form has relatively low strength, but it can be strengthened by thermal and chemical treatments. Glass fibers are used widely as a reinforcement in composite materials.

- Graphite, fullerenes, carbon nanotubes, and diamond are forms of carbon that display unique combinations of properties. Graphite has high-temperature use and electrical applications; graphite fibers are used to reinforce plastics and other composite materials.

- Diamond is used as cutting tools for precision machining operations, as dies for drawing of thin wire and as abrasives for grinding wheels. Diamond-like carbon has applications as a coating material for improved wear resistance.

- Nanomaterials have physical, mechanical, optical, chemical, and thermal properties, with several unique applications. Carbon nanotubes are of continued research interest, particularly because of their applications in nanoscale electrical and electromechanical systems.

KEY TERMS

Alumina	Diamond	Industrial ceramics	Sialon
Bioceramics	Diamond-like carbon	Industrial diamond	Silica
Buckyballs	Feldspar	Nanoceramics	Static fatigue
Carbides	Flint	Nanophase ceramics	Transformation-
Carbon	Fullerenes	Nanotubes	toughened zirconia
Carbon foam	Glass	Nitrides	White ceramics
Carbon nanotubes	Glass ceramics	Oxide ceramics	Zirconia
Ceramics	Glass fibers	Partially stabilized	
Cermets	Glass former	zirconia	
Clay	Graphene	Porcelain	
Devitrification	Graphite	Porosity	

BIBLIOGRAPHY

Bansal, N.P. (ed.), **Handbook of Ceramic Composites,** Springer, 2005.

Barsoum, M.W., **Fundamentals of Ceramics,** Institute of Physics Publishing, 2003.

Bhushan, B. (ed.), **Springer Handbook of Nanotechnology,** 3rd ed., Springer, 2010.

Buchanon, R.C. (ed.), **Ceramic Materials for Electronics: Processing, Properties and Applications,** 3rd ed., Dekker, 2004.

Carter, C.B., and Norton, M.G., **Ceramic Materials: Science and Engineering,** Springer, 2008.

Edinsinghe, M.J., **An Introduction to Structural Engineering Ceramics,** Ashgate Pub. Co., 1997.

Gogotsi, Y., **Nanomaterials Handbook,** CRC press, 2006.

Green, D.J., **An Introduction to the Mechanical Properties of Ceramics,** Cambridge University Press, 1998.

Harper, C.A. (ed.), **Handbook of Ceramics, Glasses, and Diamonds,** McGraw-Hill, 2001.

Holand, W., and Beall, G.H., **Design and Properties of Glass Ceramics,** 2nd ed., Wiley-American Chemical Society, 2012.

Mitura, S., **Nanomaterials,** Elsevier, 2000.

O'Connell, M.J. (ed.), **Carbon Nanotubes: Properties and Applications,** CRC Press, 2006.

Parinov, I., **Microstructure and Properties of High-Temperature Superconductors,** Springer, 2008.

Park, J., **Bioceramics: Properties, Characterization and Applications,** Springer, 2008.

Prelas, M.A., Popovichi, G., and Bigelow, L.K. (eds.), **Handbook of Industrial Diamonds and Diamond Films,** Marcel Dekker, 1998.

Rice, R.W., **Porosity of Ceramics,** CRC Press, 1998.

Richerson, D.W., **Modern Ceramic Engineering: Properties, Processing, and Use in Design,** 3rd ed., Marcel Dekker, 2005.

Shelby, J.E., **Introduction to Glass Science and Technology,** Royal Society of Chemistry, 2005.

Somiya, S., Aldiner, F., Spriggs, R., Uchino, K., Kuomoto, K., and Kaneno, M. (eds.) **Handbook of Advanced Ceramics: Materials, Applications, Processing and Properties,** Academic Press, 2003.

Vollath, D., **Nanomaterials: An Introduction to Synthesis, Properties and Applications,** Wiley, 2008.

Wachtman, J.B., Cannon, W.R., Matthewson, M.J., **Mechanical Properties of Ceramics,** Wiley, 2009.

REVIEW QUESTIONS

8.1 What is a ceramic?

8.2 List the major differences between the properties of ceramics and those of metals and plastics.

8.3 List the major types of ceramics that are useful in engineering applications.

8.4 What do the following materials typically consist of: (a) carbides, (b) cermets, and (c) sialon?

8.5 What is porcelain?

8.6 What is glass? Why is it called a supercooled material?

8.7 How is glass different from a glass ceramic?

8.8 What is devitrification?

8.9 List the major types of glasses and their applications.

8.10 What is static fatigue? What is its significance?

8.11 Describe the major uses of graphite.

8.12 How are alumina ceramics produced?

8.13 What features of PSZ differentiate it from other ceramics?

8.14 What are buckyballs?

8.15 List the major uses of diamond.

8.16 What is a carbon nanotube? Explain why they are not as prevalent as other forms of carbon.

8.17 What is graphene? How is it related to graphite?

8.18 What do the terms "armchair," "zigzag," and "chiral" have in common?

QUALITATIVE PROBLEMS

8.19 Explain why ceramics are weaker in tension than in compression.

8.20 What are the advantages of cermets? Suggest applications in addition to those given in this chapter.

8.21 Explain why the electrical and thermal conductivity of ceramics decreases with increasing porosity.

8.22 Explain why the mechanical property data given in Table 8.2 have such a broad range. What is the significance of this in engineering practice?

8.23 Describe the reasons that have encouraged the development of synthetic diamond.

8.24 Explain why the mechanical properties of ceramics generally differ from those of metals.

8.25 Explain how ceramics can be made tougher.

8.26 List and describe situations in which static fatigue can be important.

8.27 What properties are important in making heat-resistant ceramics for use on oven tops? Why?

8.28 A large variety of glasses is now available. Why is this so?

8.29 What is the difference between the structure of graphite and that of diamond? Is it important? Explain.

8.30 List and explain materials that are suitable for use as a coffee cup.

8.31 Aluminum oxide and PSZ are described as white in appearance. Can they be colored? If so, how would you accomplish this?

8.32 Why does the strength of a ceramic part depend on its size?

8.33 In old castles and churches in Europe, the glass windows display pronounced ripples and are thicker at the bottom than at the top. Explain.

8.34 Is a carbide an example of a composite material? Explain.

8.35 Ceramics are hard and strong in both compression and shear. Why, then, are they not used as nails or other fasteners? Explain.

8.36 Perform an Internet search and determine the chemistry of glass used for (a) fiber-optic communication lines, (b) crystal glassware, and (c) high-strength glass fibers.

8.37 Investigate and list the ceramics used for high-temperature superconductor applications.

8.38 Explain why synthetic diamond gemstones are not appreciably less expensive than natural diamond gemstones.

QUANTITATIVE PROBLEMS

8.39 In a fully dense ceramic, $UTS_o = 200$ MPa and $E_o = 330$ GPa. What are these properties at 15% porosity for values of $n = 4, 5, 6$, and 7, respectively?

8.40 Plot the UTS, E, and k values for ceramics as a function of porosity P. Describe and explain the trends that you observe in their behavior.

8.41 What would be the tensile strength and the modulus of elasticity of the ceramic in Problem 8.39 for porosities of 25% and 50%, for the four n values given?

8.42 Calculate the thermal conductivities for ceramics at porosities of 10%, 20%, and 40% for $k_o = 0.7$ W/mK.

8.43 A ceramic has $k_o = 0.80$ W/mK. If this ceramic is shaped into a cylinder with a porosity distribution given by $P = 0.1(x/L)(1 - x/L)$, where x is the distance from one end of the cylinder and L is the total cylinder length, plot the porosity as a function of distance, evaluate the average porosity, and calculate the average thermal conductivity.

8.44 It can be shown that the minimum weight of a column which will support a given load depends on the ratio of the material's stiffness to the square root of its density. Plot this property for a ceramic as a function of porosity.

SYNTHESIS, DESIGN, AND PROJECTS

8.45 Make a list of the ceramic parts that you can find around your house or in your car. Give reasons why those parts are made of ceramics.

8.46 Assume that you are working in technical sales and are fully familiar with all the advantages and limitations of ceramics. Which of the markets traditionally using nonceramic materials do you think ceramics can penetrate? What would you like to talk about to your potential customers during your sales visits? What questions do you think they may ask you about ceramics?

8.47 Describe applications in which a ceramic material with a near-zero coefficient of thermal expansion would be desirable.

8.48 The modulus of elasticity of ceramics is typically maintained at elevated temperatures. What engineering applications could benefit from this characteristic?

8.49 List and discuss the factors that you would take into account when replacing a metal component with a ceramic component in a specific product.

8.50 Obtain some data from the technical literature in the Bibliography of this chapter, and show quantitatively the effects of temperature on the strength and the modulus of elasticity of several ceramics. Comment on how the shape of these curves differs from those for metals.

8.51 Conduct a literature search and write a brief paper summarizing the properties and potential applications of graphene.

8.52 It was noted in Section 8.4.1 that there are several basic types of glasses available. Make a survey of the technical literature and prepare a table for these glasses, indicating various mechanical, physical, and optical properties.

8.53 Ceramic pistons are being considered for high-speed combustion engines. List the benefits and concerns that you would have regarding this application.

8.54 It has been noted that the strength of brittle materials (such as ceramics and glasses) is very sensitive to surface defects, such as scratches (known as *notch sensitivity*). Obtain several pieces of these materials, scratch them, and test them by carefully clamping them in a vise and bending them. Comment on your observations.

8.55 Electric space heaters for home use commonly utilize a ceramic filament as the heating element. List the required properties for this filament, explain why a ceramic is a suitable material, and perform an Internet search to determine the specific ceramic material actually utilized in this application.

Composite Materials: Structure, General Properties, and Applications

CHAPTER 9

- With their attractive properties, especially high strength-to-weight and stiffness-to-weight ratios, composites are among the most important engineered materials.
- Composites are widely used as structural components, especially in the aerospace industry, where weight savings are a major consideration.
- This chapter describes the major types of composite materials, the characteristics of the commonly used reinforcing fibers, and their effect in improving mechanical properties.
- The role of the matrix is then described, and the three principal classes of matrix materials (plastic, metal, and ceramic) are examined.
- The chapter ends with a discussion of the selection and applications of a variety of reinforced plastics and composites.

9.1 Introduction

A **composite material** is a combination of two or more chemically distinct and insoluble phases with a recognizable interface, in such a manner that its properties and structural performance are superior to those of the constituents acting independently. These combinations are known as **polymer-matrix, metal-matrix**, and **ceramic-matrix composites**. As shown in Table 7.1, fiber reinforcements significantly improve the strength, stiffness, and creep resistance of plastics, particularly their strength-to-weight and stiffness-to-weight ratios. Composite materials have found increasingly wider applications in aircraft (Fig. 9.1), space vehicles, satellites, offshore structures, piping, electronics, automobiles, boats, and sporting goods.

The oldest example of composites, dating back to 4000 B.C., is the addition of straw to clay to make bricks stronger. In this combination, the straws are the reinforcing fibers and the clay is the matrix. Another example of a composite material is reinforced concrete, which was developed in the 1800s. By itself, concrete is brittle and has little or no useful tensile strength; reinforcing steel rods (*rebar*) impart the necessary tensile strength to the concrete.

Composites, in a general sense, can include a wide variety of materials: cermets (Section 8.2.3), two-phase alloys (Section 4.2), natural materials such as wood and bone, and reinforced or combined materials such as steel-wire reinforced automobile tires. This chapter describes the structure, properties, and applications of

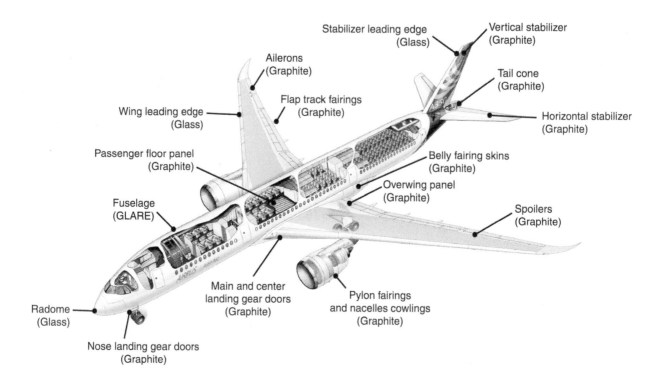

FIGURE 9.1   Application of advanced composite materials in the Airbus 350. The reinforcement type is shown, with the fuselage made of GLARE (a glass-reinforced polymer/aluminum laminate). *Source:* Reuse courtesy of Flight Global, http://www.flightglobal.com/news/articles/paris-air-show-a350-xwb-ready-to-rock-327500/.

composite materials; the processing and shaping of composite materials are described in Chapter 19.

9.2   The Structure of Reinforced Plastics

Reinforced plastics, also known as **polymer-matrix composites** (PMC) and **fiber-reinforced plastics** (FRP), consist of **fibers** (the discontinuous, or dispersed, phase) in a polymer **matrix** (the continuous phase), as shown in Fig. 9.2. These fibers are strong and stiff (Table 9.1), and they have high specific strength (strength-to-weight ratio) and specific stiffness (stiffness-to-weight ratio); see Fig. 9.3. In addition, reinforced-plastic structures have improved fatigue resistance, and higher toughness and creep resistance than those made of unreinforced plastics.

The fibers in reinforced plastics, by themselves, have little structural value; they are stiff in their longitudinal direction but have no transverse stiffness or strength. Although the polymer matrix is less strong and less stiff than the fibers, it is tougher and often more chemically inert; thus, reinforced plastics combine the advantages of each of the two constituents. The percentage of fibers (by volume) in reinforced plastics usually ranges between 10 and 60%.

9.2.1   Reinforcing Fibers

Glass, carbon, ceramics, aramids, and boron are the most common reinforcing fibers for PMCs (Table 9.2).

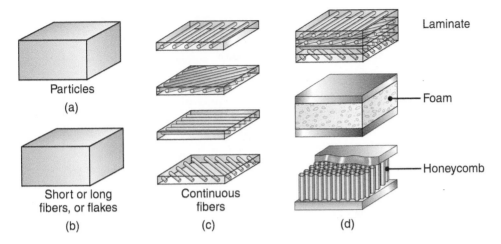

FIGURE 9.2 Schematic illustration of methods of reinforcing plastics (matrix) with (a) particles, (b) short or long fibers or flakes, and (c) continuous fibers. The laminate structures shown in (d) can be produced from layers of continuous fibers or sandwich structures using a foam or honeycomb core (see also Fig. 16.58).

TABLE 9.1

Types and General Characteristics of Composite Materials

Material	Characteristics
Fibers	
Glass	High strength, low stiffness, high density; lowest cost; E (calcium aluminoborosilicate) and S (magnesia aluminosilicate) types commonly used
Carbon	Available as high modulus or high strength; low cost; less dense than glass; sometimes used in combination with carbon nanotubes (see Section 8.6.2)
Boron	High strength and stiffness; highest density; highest cost; has tungsten filament at its center
Aramids (Kevlar)	Highest strength-to-weight ratio of all fibers; high cost
Other fibers	Nylon, silicon carbide, silicon nitride, aluminum oxide, boron carbide, boron nitride, tantalum carbide, steel, tungsten, molybdenum
Matrix materials	
Thermosets	Epoxy and polyester, with the former most commonly used; others are phenolics, fluorocarbons, polyethersulfone, silicon, and polyimides
Thermoplastics	Polyetheretherketone; tougher than thermosets, but lower resistance to temperature
Metals	Aluminum, aluminum–lithium, magnesium, and titanium; fibers are carbon, aluminum oxide, silicon carbide, and boron
Ceramics	Silicon carbide, silicon nitride, aluminum oxide, and mullite; fibers are various ceramics

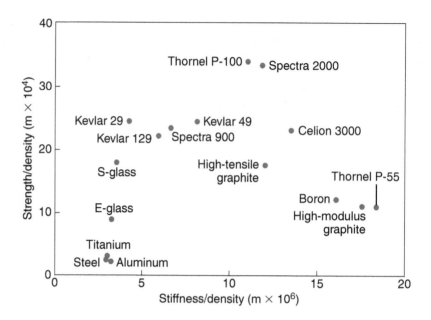

FIGURE 9.3 Specific tensile strength (tensile-strength-to-density ratio) and specific tensile modulus (modulus-of-elasticity-to-density ratio) for various fibers used in reinforced plastics; note the wide range of specific strength and stiffness available.

TABLE 9.2

Typical Properties of Reinforcing Fibers

Type	Tensile strength (MPa)	Elastic modulus (GPa)	Density (kg/m^3)	Relative cost
Boron	3500	380	2380	Highest
Carbon				
High strength	3000	275	1900	Low
High modulus	2000	415	1900	Low
Glass				
E-type	3500	73	2480	Lowest
S-type	4600	85	2540	Lowest
Kevlar				
29	2920	70.5	1440	High
49	3000	112.4	1440	High
129	3200	85	1440	High
Nextel				
312	1700	150	2700	High
610	2770	328	3960	High
Spectra				
900	2270	64	970	High
1000	2670	90	970	High
2000	3240	115	970	High
Alumina (Al$_2$O$_3$)	1900	380	3900	High
Silicon carbide	3500	400	3200	High

Note: These properties vary significantly depending on the material and method of preparation.

Glass Fibers.   Glass fibers are the most widely used and the least expensive of all fibers. The composite material is called **glass-fiber reinforced plastic** (GFRP), and may contain between 30% and 60% glass fibers. The fibers are made by drawing molten glass through small openings in a platinum die (see Section 18.3.4); they are then elongated, cooled, and wound on a roll. The glass fibers are later treated with silane (a silicon hydride), as described in Section 9.3. The principal types of glass fibers are:

- **E-type:** a calcium aluminoborosilicate glass, the type most commonly used
- **S-type:** a magnesia aluminosilicate glass, offering higher strength and stiffness, but at a higher cost
- **E-CR-type:** a high-performance glass fiber, with higher resistance to elevated temperatures and acid corrosion than does the E-glass

Carbon Fibers.   Carbon fibers (Fig. 9.4a), although more expensive than glass fibers, have a combination of low density, high strength, and high stiffness. The composite is called **carbon-fiber reinforced plastic** (CFRP). Although the words are often used interchangeably, the difference between *carbon* and *graphite* depends on the purity of the material and the temperature at which it was processed. Carbon fibers are at least 90% carbon; graphite fibers are usually more than 99% carbon. A typical carbon fiber contains amorphous (noncrystalline) carbon and graphite (crystalline carbon). These fibers are classified by their elastic modulus, which ranges from 35 to 800 GPa, as *low, intermediate, high,* and *very high modulus*. Some trade names for carbon fibers are Celion and Thornel (see Fig. 9.3). Carbon nanotubes also have been used as reinforcement in composite materials, as described in Section 8.6.2.

All carbon fibers are made by **pyrolysis** of organic **precursors**, commonly *polyacrylonitrile* (PAN) because of its low cost. *Rayon* and *pitch* (the residue from catalytic crackers in petroleum refining) also can be used as precursors. Pyrolysis is the process of inducing chemical changes by heat–for instance, by burning a length of yarn and causing the material to carbonize and become black in color. With PAN, the fibers are partially cross-linked at a moderate temperature (in order to prevent melting during subsequent processing steps), and are simultaneously elongated. At this stage, the fibers are *carburized*; they are exposed to elevated temperatures to expel the hydrogen (dehydrogenation) and the nitrogen (denitrogenation) from the PAN. The temperatures for carbonizing range up to about 1500°C (2730°F); for graphitizing, up to 3000°C (5400°F).

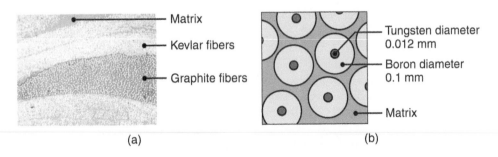

(a)   (b)

FIGURE 9.4   (a) Cross-section of a tennis racket, showing graphite and aramid (Kevlar) reinforcing fibers. *Source:* Courtesy of F. Garrett, Wilson Sporting Goods Co. (b) Cross-section of boron-fiber reinforced composite material.

Conductive Graphite Fibers. These fibers are produced to make it possible to enhance the electrical and thermal conductivity of reinforced plastic components. The fibers are coated with a metal (usually nickel), using a continuous electroplating process. The coating is typically 0.5-μm thick on a 7-μm-diameter graphite fiber core. Available in chopped or continuous form, the conductive fibers are incorporated directly into injection-molded plastic parts (Section 19.3). Applications include electromagnetic and radio-frequency shielding and lightning-strike protection.

Ceramic Fibers. Ceramic fibers are advantageous for high-temperature applications and in metal-matrix composites (Section 9.5). These fibers have low elongation, low thermal conductivity, and good chemical resistance, in addition to being suitable for high-temperature applications. One family of ceramic fibers is *Nextel*, a trade name; the fibers are oval in cross-section and consist of alumina, silica, and boric oxide. Typical mechanical properties of ceramic are given in Table 9.2.

Polymer Fibers. Polymer fibers may be made of aramids, nylon, rayon, or acrylics; the most common are **aramid fibers**. Aramids (Section 7.6), such as **Kevlar**, are among the toughest fibers, with very high specific strength (Fig. 9.3). Aramids can undergo some plastic deformation prior to fracture and, hence, have higher toughness than brittle fibers. However, aramids absorb moisture (*hygroscopic*), thus degrading their properties.

A high-performance polyethylene fiber is *Spectra* (a trade name); it has ultra-high molecular weight and high molecular-chain orientation. Spectra, bright white in color, has better abrasion resistance and flexural-fatigue strength than aramid fibers, at a similar cost. In addition, because of its lower density (970 kg/m^3), it has a higher specific strength and specific stiffness than aramid fibers (see Table 9.2). However, a low melting point and poor adhesion characteristics, as compared to other polymers, are its major limitations to applications. (The manufacture of polymer fibers is described in Section 19.2.2.)

Boron Fibers. These fibers consist of tungsten fibers with a layer of boron, deposited by chemical vapor-deposition techniques (Fig. 9.4b); boron also can be deposited onto carbon fibers. Boron fibers have such desirable properties as high strength and stiffness, both in tension and in compression, and resistance to high temperatures. However, because of the high density of tungsten these fibers are heavy, and expensive.

Other Fibers. Among other fibers used in composites are silicon carbide, silicon nitride, aluminum oxide, sapphire, steel, tungsten, molybdenum, boron carbide, boron nitride, and tantalum carbide. **Whiskers** also are used as reinforcing fibers (see Section 22.10). Whiskers are tiny needlelike single crystals that grow to 1–10 μm (40–400 μin.) in diameter, with high aspect ratios (the ratio of fiber length to its diameter), ranging from 100 to $15,000$. Because of their small size, whiskers are either free of imperfections or the imperfections they contain do not significantly affect their strength, which approaches the theoretical strength of the material. The elastic moduli of whiskers range between 400 and 700 GPa, and their tensile strength is on the order of 15 to 20 GPa, depending on the material.

9.2.2   Fiber Size and Length

Fibers are very strong and stiff in tension, because (a) the molecules in the fibers are oriented in the longitudinal direction, and (b) their cross-sections are so small, usually less than 0.01 mm (0.0004 in.) in diameter, that the probability is low for any significant defects to exist in the fiber. Glass fibers can have tensile strengths as high as 4600 MPa (650 ksi), whereas the strength of glass in bulk form (Section 8.4.2) is much lower (see Table 2.2).

Fibers generally are classified as **short (discontinuous)** or **long (continuous)**. The designations "short" and "long" are, in general, based on the following distinction: In a given type of fiber, if the mechanical properties improve as a result of increasing average fiber length, it is called a *short fiber*; if there is no such improvement in fiber properties, it is called a *long fiber*. Short fibers typically have aspect ratios between 20 and 60, long fibers between 200 and 500. Reinforcing elements in composites may also be in the form of *chopped fibers, particles, flakes*, or in the form of continuous *roving* (slightly twisted strands) fibers, *woven fabric* (similar to cloth), *yarn* (twisted strands), and *mats* of various combinations.

9.2.3 Matrix Materials

The matrix in reinforced plastics has three principal functions:

1. Support the fibers in place and transfer the stresses to them, so that the fibers can carry most of the load (see Example 9.1)
2. Protect the fibers against physical damage and the environment
3. Slow the propagation of cracks in the composite, by virtue of the higher ductility and toughness of the polymer matrix

Matrix materials usually are *thermoplastics* or *thermosets*, which commonly consist of epoxy, polyester, phenolic, fluorocarbon, polyethersulfone, or silicon. The most commonly used materials are epoxies (in 80% of all reinforced plastics) and polyesters (less expensive than the epoxies). Polyimides, which resist exposure to temperatures in excess of 300°C (575°F), are available for use as a matrix with carbon fibers. Some thermoplastics, such as polyetheretherketone (PEEK), also are used as matrix materials. They generally have higher toughness than thermosets, but their resistance to temperature is lower, being limited to 100° to 200°C (200° to 400°F).

9.3 Properties of Reinforced Plastics

The mechanical and physical properties of reinforced plastics depend on the type, shape, and orientation of the reinforcing material, the length of the fibers, and the volume fraction (percentage) of the reinforcing material. Short fibers are less effective than long fibers (Fig. 9.5), and their properties are strongly influenced by temperature and time under load. Long fibers transmit the load through the matrix better, and are less likely to pull out of the matrix (caused by shear failure of the fiber–matrix interface); thus, they are commonly used in critical applications, particularly at elevated temperatures. The physical properties of reinforced plastics and their resistance to fatigue, creep, and wear depend greatly on the type and amount of reinforcement. Composites can be tailored to impart specific properties (such as permeability and dimensional stability), make processing easier, and reduce production costs.

A critical factor in reinforced plastics is the strength of the bond between the fiber and the polymer matrix, because the load is transmitted through the fiber–matrix interface. Weak interfacial bonding can cause **fiber pullout** and **delamination** of the composite, particularly under adverse environmental conditions. Adhesion at

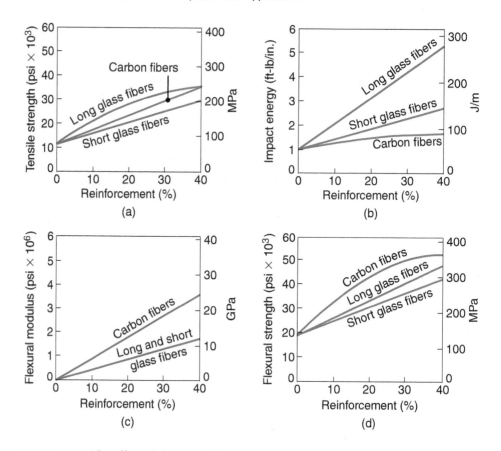

FIGURE 9.5 The effect of the type of fiber on various properties of fiber-reinforced nylon (6,6). *Source:* Courtesy of NASA.

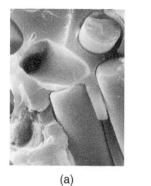

(a)   (b)

FIGURE 9.6 (a) Fracture surface of a glass-fiber reinforced epoxy composite; the fibers are 10 μm (400 μin.) in diameter and have random orientation. (b) Fracture surface of a graphite-fiber reinforced epoxy composite; the fibers, 9–11 μm in diameter, are in bundles and are all aligned in the same direction. *Source:* Courtesy of L.J. Broutman.

the interface can be improved by special surface treatments, such as coatings and coupling agents. Glass fibers, for example, are treated with **silane** (a silicon hydride) for improved wetting and bonding between the fiber and the matrix. The importance of proper bonding can be appreciated by inspecting the fracture surfaces of reinforced plastics; note in Figs. 9.6a and b, for example, the separation between the fibers and the matrix.

Generally, the highest stiffness and strength in reinforced plastics are obtained when the fibers are aligned in the direction of the tension force. The composite is then highly anisotropic (Fig. 9.7); that is, it has properties, such as strength and stiffness, that depend on direction. As a result, other properties, such as creep resistance, thermal and electrical conductivity, and thermal expansion, also are anisotropic. The transverse properties of such a unidirectionally reinforced structure are much lower than the longitudinal properties. For example, note how strong fiber-reinforced packaging tape is when pulled in tension, yet how easily it can split and tear when pulling it in the width direction.

Because it is an **engineered material**, a part made of reinforced plastic can be given an optimal configuration for a specific service condition. For example, if the part is to be subjected to stresses in different directions, such as in thin-walled, pressurized vessels, (a) the fibers can be criss-crossed in the matrix or (b) layers of fibers oriented in different directions can be built up into a laminate having improved properties in more than one direction. (See *filament winding*, Section 19.13.3.) For example, a composite flywheel rotor has been produced using a special weaving technique, in which the reinforcing fibers (E-glass) are aligned in the radial direction as well as in the hoop direction. Designed for mechanical-energy storage systems in low-emission electric and hybrid vehicles, the flywheel can operate at speeds up to 50,000 rpm.

9.3.1   Strength and Elastic Modulus of Reinforced Plastics

The strength and elastic modulus of a reinforced plastic, with unidirectional fibers, can be determined in terms of the (a) strengths and moduli of the fibers and matrix and (b) volume fraction of fibers in the composite. In the following equations, c refers to the composite, f to the fiber, and m to the matrix. The total tensile load, P_c, on the composite is shared by the fiber (P_f) and the matrix (P_m). Thus,

$$P_c = P_f + P_m, \tag{9.1}$$

which can be written as

$$\sigma_c A_c = \sigma_f A_f + \sigma_m A_m, \tag{9.2}$$

where A_c, A_f, and A_m are the cross-sectional areas of the composite, the fiber, and the matrix, respectively; thus, $A_c = A_f + A_m$. Let's now denote x as the area fraction of the fibers in the composite. (Note that x also represents the volume fraction, because the fibers are uniformly longitudinal in the matrix.) Then Eq. (9.2) can be written as

$$\sigma_c = x\sigma_f + (1 - x)\sigma_m. \tag{9.3}$$

The fraction of the total load carried by the fibers can now be calculated. First, note that in the composite under a tensile load, the strains sustained by the fibers and the matrix are the same; that is, $e_c = e_f = e_m$. Next, recall from Section 2.2 that

$$e = \frac{\sigma}{E} = \frac{P}{AE}.$$

Consequently,

$$\frac{P_f}{P_m} = \frac{A_f E_f}{A_m E_m}. \tag{9.4}$$

Since the relevant quantities for a specific situation are known, by using Eq. (9.1), the fraction P_f/P_c can be found. Then, using the foregoing relationships, the elastic modulus, E_c, of the composite can be calculated, by replacing σ in Eq. (9.3) with E. Thus,

$$E_c = xE_f + (1 - x)E_m. \tag{9.5}$$

FIGURE 9.7   The tensile strength of glass-reinforced polyester as a function of fiber content and fiber direction in the matrix.

Video Solution 9.1 Design of a Composite Material

EXAMPLE 9.1   Calculation of Stiffness of a Composite and Load Supported by Fibers

Given: Assume that a graphite–epoxy reinforced plastic with longitudinal fibers contains 20% graphite fibers. The elastic modulus of the fibers is 300 GPa, and that of the epoxy matrix is 100 GPa.

Find: Calculate the elastic modulus of the composite and the fraction of the load supported by the fibers.

Solution: The data given are $x = 0.2$, $E_f = 300$ GPa, and $E_m = 100$ GPa. Using Eq. (9.5),

$$E_c = 0.2(300) + (1 - 0.2)100$$

$$= 60 + 80 = 140 \text{ GPa}.$$

From Eq. (9.4), the load fraction P_f/P_m is found to be

$$\frac{P_f}{P_m} = \frac{0.2(300)}{0.8(100)} = 0.75.$$

Because

$$P_c = P_f + P_m \quad \text{and} \quad P_m = \frac{P_f}{0.75},$$

we obtain,

$$P_c = P_f + \frac{P_f}{0.75} = 2.33 P_f, \quad \text{or} \quad P_f = 0.43 P_c.$$

Thus, the fibers support 43% of the load, even though they occupy only 20% of the cross-sectional area (and hence volume) of the composite.

Video Solution 9.2 Mechanical Properties of Composites

9.4   Applications of Reinforced Plastics

The first engineering application of reinforced plastics was in 1907, for an acid-resistant tank made of a phenolic resin with asbestos fibers. In the 1920s, *Formica* (a trade name) was developed, and used commonly for countertops. Epoxies first were used as a matrix material in the 1930s. Beginning in the 1940s, boats were made with fiberglass, and reinforced plastics were used for aircraft, electrical equipment, and sporting goods. Major developments in composites began in the 1970s, resulting in materials that are now called **advanced composites**. Glass or carbon-fiber reinforced hybrid plastics are available for high-temperature applications, with continuous use ranging up to about 300°C (550°F).

Reinforced plastics are typically used in commercial and military aircraft, rocket components, helicopter blades, automobile bodies, leaf springs, drive shafts, pipes, ladders, pressure vessels, sporting goods, helmets, boat hulls, and various other structures and components. About 50% (by weight) of the Boeing 787 Dreamliner is made of composites. By virtue of the resulting weight savings, reinforced plastics have reduced fuel consumption in aircraft by about 2%. The Airbus jumbo jet A380, with a capacity of up to 700 passengers, has horizontal stabilizers, ailerons, wing boxes and leading edges, secondary mounting brackets of the fuselage, and a deck structure made of composites with carbon fibers, thermosetting resins, and thermoplastics. The upper fuselage is made of alternating layers of aluminum and glass-fiber reinforced epoxy prepregs (see Section 19.13).

The contoured frame of the Stealth bomber is made of composites, consisting of carbon and glass fibers, epoxy-resin matrices, high-temperature polyimides, and other advanced materials. Boron-fiber reinforced composites are used in military aircraft, golf-club shafts, tennis rackets, fishing rods, and sailboards (Fig. 9.8). Another example is the development of a small, all-composite ship (twin-hull catamaran design) for the U.S. Navy, capable of speeds of 50 knots (58 mph). More recent developments include (a) reinforcing bars for concrete, replacing steel bars, thus

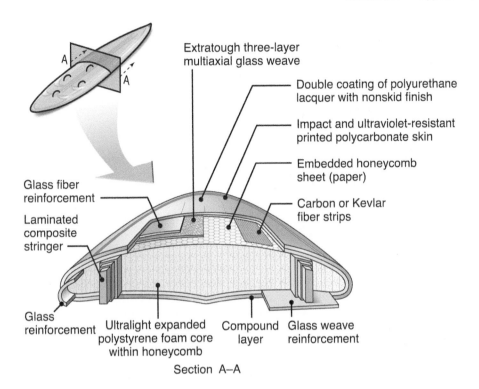

Extratough three-layer multiaxial glass weave

Double coating of polyurethane lacquer with nonskid finish

Impact and ultraviolet-resistant printed polycarbonate skin

Embedded honeycomb sheet (paper)

Carbon or Kevlar fiber strips

Glass fiber reinforcement

Laminated composite stringer

Glass reinforcement

Ultralight expanded polystyrene foam core within honeycomb

Compound layer

Glass weave reinforcement

Section A–A

FIGURE 9.8 Cross-section of a composite sailboard, an example of advanced materials construction. *Source:* K. Easterling, *Tomorrow's Materials*, 2nd ed., Institute of Metals, 1990. Courtesy of Maney Publishing. www.maneypublishing.com.

lowering the costs involved due to their corrosion and (b) rollers for papermaking and similar industries, with lower deflections as compared to traditional steel rollers.

CASE STUDY 9.1   Composite Military Helmets and Body Armor

Personal protective equipment, in the form of body armor and composite helmets, have become widespread for military and police use. Body armor relies on high-strength woven fibers to prevent the penetration of projectiles. To stop a bullet, a composite material must first deform or flatten it, a process that occurs when the bullet's tip comes into contact with as many individual fibers of the composite as possible, without the fibers being pushed aside. The momentum associated with projectiles is felt by the user of the armor, but successful designs will contain bullets and shrapnel and thus prevent serious and fatal injuries.

There are two basic types of body armor: (a) *soft armor*, which relies upon several layers of high-strength, woven fibers, and is designed mainly to contain handgun bullets, and (b) *hard armor*, which uses a metal, ceramic, or polymer plate, in addition to the woven fiber, and is intended to provide protection against rifle rounds and shrapnel. A schematic of a body armor is shown in Fig. 9.9.

Several types of fiber meshes have been used in body armor applications. Different suppliers employ different combinations of fiber meshes, and may include additional layers to provide protection against blunt trauma. The first fiber used for flexible body armor was Kevlar 29 (an aramid), which has been improved through a number of versions. Other forms include Kevlar 49, Kevlar 129, and Kevlar Protera, where tensile strength

(continued)

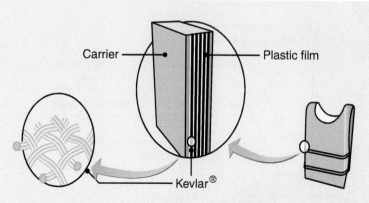

FIGURE 9.9 Schematic illustration of body armor, showing the layers of woven fibers.

and energy-absorbing capabilities have been improved through the development of advanced spinning processes to produce the fibers. Aramid fibers are used very commonly in flexible body armor. Honeywell also produces an aramid-fiber based body armor, but other designs, such as Twaron aramid fiber, use over a thousand finely spun filaments that interact with each other to dissipate the impact energy.

Spectra fiber is used to make a composite for use in body armor. A layer of Spectra Shield composite consists of two unidirectional layers of Spectra fiber, arranged to cross each other at 0° and 90° angles and held in place by a flexible resin. Both the fiber and resin layers are sealed between two thin sheets of polyethylene film, which is similar in appearance to plastic food wrap. Hard armor uses several designs, but typically it consists of steel, ceramic (usually aluminum oxide and silica), or polyethylene plates, strategically located to prevent penetration of ballistic particles to critical areas.

Designs currently being evaluated utilize fluids, with suspended nanoparticles of silica; at low strain rates, these fluids are inviscid and flow readily. At the high strain rates typical of ballistic particles, these fluids are very resistant to deformation and can provide additional protection (see also Section 2.2.7). The fluid is contained by the woven fiber mesh (acting like a sponge holding the fluid in place) and is contained by the outer fabric.

A composite military helmet also has been developed that, although weighing about the same as conventional manganese–steel helmets, covers more of the head and offers twice the ballistic and fragmentation protection. This helmet has a nonwoven fiber construction made with Spectra fibers in a thermosetting polymer matrix, which effectively stops the bullet by flattening it as it strikes the first layer of material.

Source: Courtesy of Pinnacle Armor, Allied Signal Corp., and CGS Gallet SA.

9.5 Metal-matrix Composites

Matrix materials in *metal-matrix composites* (MMC) are usually aluminum, aluminum–lithium alloy (lighter than aluminum), magnesium, copper, titanium, or superalloys (Fig. 9.10). Fiber materials are graphite, aluminum oxide, silicon carbide, boron, molybdenum, or tungsten. The elastic modulus of nonmetallic fibers ranges between 200 and 400 GPa, with tensile strengths in the range from 2000 to 3000 MPa. The advantages of a metal matrix over a polymer matrix are higher elastic modulus, toughness, ductility, and higher resistance to elevated temperatures; the limitations are higher density and a greater difficulty in processing the composite parts. Typical compositions and applications for MMC are given in Table 9.3.

FIGURE 9.10   Examples of metal-matrix composite parts. *Source:* Courtesy of 3M Speciality Materials Division.

TABLE 9.3

Metal-matrix Composite Materials and Applications

Fiber	Matrix	Applications
Graphite	Aluminum	Satellite, missile, and helicopter structures
	Magnesium	Space and satellite structures
	Lead	Storage-battery plates
	Copper	Electrical contacts and bearings
Boron	Aluminum	Compressor blades and structural supports
	Magnesium	Antenna structures
	Titanium	Jet-engine fan blades
Alumina	Aluminum	Superconductor restraints in fission power reactors
	Lead	Storage-battery plates
	Magnesium	Helicopter transmission structures
Silicon carbide	Aluminum, titanium	High-temperature structures
	Superalloy (cobalt base)	High-temperature engine components
Molybdenum, tungsten	Superalloy	High-temperature engine components

CASE STUDY 9.2   Aluminum-matrix Composite Brake Calipers

One of the trends in automobile design and manufacture is the increased effort toward lighter-weight designs in order to realize improved performance and/or fuel economy. This trend also can be seen in the development of MMC brake calipers. Traditional brake calipers are made from cast iron, and can weigh around 3 kg (6.6 lb) each in a small car and up to 14 kg (30 lb) in a truck. The cast-iron caliper could be redesigned completely, using aluminum to achieve weight savings, but that would require a larger volume since the nominal strength of aluminum is lower than the cast iron, and the space available between the wheel and rotor is very constrained.

A new brake caliper was designed, using an aluminum alloy locally reinforced with precast composite inserts using continuous ceramic fiber. The fiber is a nanocrystalline alumina, with a diameter of 10–12 μm and a fiber volume fraction of 65%. The fiber and composite properties are summarized

(*continued*)

in Table 9.4. Finite element analysis confirmed the placement and amount of reinforcement, leading to a design that exceeded minimum design requirements, and matched deflections of cast-iron calipers in a packaging-constrained environment. The new brake caliper is shown in Fig. 9.11. It has a weight savings of 50%, with the additional benefits of corrosion resistance and ease of recyclability.

TABLE 9.4

Summary of Fiber and Composite Properties for an Automotive Brake Caliper

Property	Alumina fiber	Alumina-reinforced composite material
Tensile strength	3.1 GPa (450 ksi)	1.5 GPa (220 ksi)
Elastic modulus	380 GPa (55 Mpsi)	270 GPa (39 Mpsi)
Density	3.9 g/cm^3	3.48 g/cm^3

FIGURE 9.11 Aluminum-matrix composite brake caliper using nanocrystalline alumina fiber reinforcement. *Source:* Courtesy of 3M Speciality Materials Division.

9.6 Ceramic-matrix Composites

Ceramic-matrix composites (CMC) are characterized by their resistance to high temperatures and corrosive environments. As described in Section 8.3.1, ceramics are strong and stiff; they resist high temperatures, but generally lack toughness. Matrix materials that retain their strength up to 1700°C (3000°F) are silicon carbide, silicon nitride, aluminum oxide, and mullite (a compound of aluminum, silicon, and oxygen). Carbon–carbon-matrix composites retain much of their strength up to 2500°C (4500°F), although they lack oxidation resistance at high temperatures. Fiber materials are usually carbon and aluminum oxide. Applications of CMC include jet and automotive engine components, deep-sea mining equipment, pressure vessels, structural components, cutting tools, and dies for the extrusion and drawing of metals.

9.7 Other Composites

Composites also may consist of *coatings* of various types, applied on base metals or substrates (Chapter 34). Examples are:

- Plating of aluminum or other metals over plastics, generally for decorative purposes
- Enamels, for wear resistance, hardness, and decorative purposes
- Vitreous (glasslike) coatings on metal surfaces, for various functional or ornamental purposes

Composites are made into cutting tools and dies, such as cemented carbides and cermets. Other composites are grinding wheels, made of aluminum oxide, silicon carbide, diamond, or cubic-boron-nitride abrasive particles, all held together with various organic, inorganic, or metallic binders. A composite, used in machine-tool beds for some precision grinders (see Section 25.3.1), consists of granite particles in an epoxy matrix; it has high strength, good vibration-damping capacity (better than gray cast iron), and good frictional characteristics.

CASE STUDY 9.3 Composites in the Aircraft Industry

High fuel prices have significantly affected the operations of the aircraft industry; any design advantages that lead to increased efficiency or fuel economy continue to be aggressively pursued by aircraft manufacturers. One area where this effect is most dramatic is the increased composite content in commercial aircraft, as shown in Fig. 9.12.

In addition to the amount of composite materials used, there are a number of design innovations in the types and applications of composite materials, including:

- GLARE is a GLAss-REinforced aluminum, consisting of several layers of glass-fiber reinforced

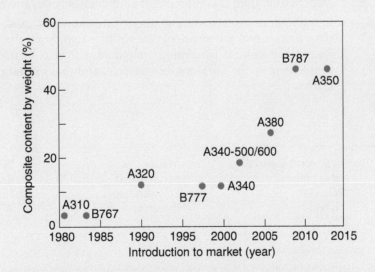

FIGURE 9.12 Composite content in selected commercial aircraft as a function of time (measured by date first introduced into the market). A = Airbus, B = Boeing.

(*continued*)

polymer and sandwiched between thin sheets of aluminum. It is used on the upper fuselage of the Airbus A380 and the leading edges of the tail of the plane, and has been credited with over 500 kg of weight savings, as compared to previously used materials. GLARE also provides improved fatigue strength and corrosion resistance.

• The Boeing 787 Dreamliner has an all-composite fuselage, constructed mainly from CFRP. In addition to weight savings, the fuselage is constructed in one piece and joined end to end, eliminating the need for an estimated 50,000 fasteners. Composites make up around 50% of the weight of the Dreamliner, as compared to 12% on the 777 aircraft, first introduced in 1994.

SUMMARY

- Composites are an important class of engineered materials, with numerous attractive properties. Three major categories are fiber-reinforced plastics, metal-matrix composites, and ceramic-matrix composites. They have a wide range of applications in the aircraft, aerospace, and transportation industries, sporting goods, and for structural components.

- In fiber-reinforced plastics, the fibers are usually glass, graphite, aramids, or boron. Polyester and epoxies commonly are used as the matrix material. These composites have particularly high toughness and high strength-to-weight and stiffness-to-weight ratios.

- In metal-matrix composites, the fibers are typically graphite, boron, aluminum oxide, silicon carbide, molybdenum, or tungsten. Matrix materials generally consist of aluminum, aluminum–lithium alloy, magnesium, copper, titanium, or superalloys.

- In ceramic-matrix composites, the fibers usually are carbon and aluminum oxide, and the matrix materials are silicon carbide, silicon nitride, aluminum oxide, carbon, or mullite (a compound of aluminum, silicon, and oxygen).

- In addition to the type and quality of the materials used, important factors in the structure and properties of composite materials are the size and length of the fibers, their volume percentage compared with that of the matrix, the strength of the bond at the fiber–matrix interface, and the orientation of the fibers in the matrix.

KEY TERMS

Advanced composites	Engineered materials	Matrix	Pyrolysis
Ceramic matrix	Fiber pullout	Metal matrix	Reinforced plastics
Composite materials	Fibers	Polymer matrix	Silane
Delamination	Hybrid	Precursor	Whiskers

BIBLIOGRAPHY

Agarwal, B.D., Broutman, L.J., and Chandrashekhara, K., **Analysis and Performance of Fiber Composites**, 3rd ed., Wiley, 2006.

ASM Handbook, Vol. 21: **Composites**, ASM International, 2001.

Bansal, N.P. (ed.), **Handbook of Ceramic Composites**, Springer, 2004.

Campbell, F.C., **Structural Composite Materials**, ASM International, 2010.

Cantor, B., Dunne, F.P.E., and Stone, I.C. (eds.), **Metal and Ceramic Matrix Composites**, Taylor & Francis, 2003.

Chawla, K.K., **Composite Materials: Science and Engineering**, 3rd ed., Springer, 2008.

Chung, D.D.L., **Composite Materials: Science and Applications**, 2nd ed., Springer, 2010.

Daniel, I.M., and Ishai, O., **Engineering Mechanics of Composite Materials**, 2nd ed., Oxford, 2005.

Fitzer, E., and Manocha, L.M., **Carbon Reinforcements and Carbon/Carbon Composites**, Springer, 1998.

Gay, D., and Hoa, S.V., **Composite Materials: Design and Applications**, 2nd ed., CRC Press, 2007.

Krenker, W. (ed.), **Ceramic Matrix Composites: Fiber Reinforced Materials and Their Applications**, Wiley, 2008.

Strong, A.B., **Fundamentals of Composites Manufacturing: Materials, Methods and Applications**, 2nd ed., Society of Manufacturing Engineers, 2007.

REVIEW QUESTIONS

9.1 Distinguish between composites and metal alloys.

9.2 Describe the functions of the matrix and the reinforcing fibers. What fundamental differences are there in the characteristics of the two materials?

9.3 Name the reinforcing fibers generally used to make composites. Which type of fiber is the strongest? Which type is the weakest?

9.4 What is the range in length and diameter of typical reinforcing fibers?

9.5 List the important factors that determine the properties of reinforced plastics.

9.6 Comment on the advantages and limitations of metal-matrix composites, reinforced plastics, and ceramic-matrix composites, respectively.

9.7 What are the most commonly used matrix materials? Why?

9.8 Describe the advantages of hybrid composites over other composites.

9.9 What material properties are improved by the addition of reinforcing fibers?

9.10 Describe the purpose of the matrix material.

9.11 What are the most common types of glass fibers?

9.12 Explain the difference between a carbon fiber and a graphite fiber.

9.13 How can a graphite fiber be made electrically and thermally conductive?

9.14 What is a whisker? What is the difference between a whisker and a fiber?

9.15 Describe the composition of boron fibers. Why are they heavy?

9.16 Give a succinct definition of fiber, yarn, and fabric, respectively.

QUALITATIVE PROBLEMS

9.17 How do you think the use of straw mixed with clay originally came about in making brick for dwellings?

9.18 What products have you personally seen that are made of reinforced plastics? How can you tell?

9.19 Describe applications that are not well suited for composite materials. Explain.

9.20 Is there a difference between a composite material and a coated material? Explain.

9.21 Identify metals and alloys that have strengths comparable to those of reinforced plastics.

9.22 What limitations or disadvantages do composite materials have? What suggestions would you make to overcome the limitations?

9.23 Give examples of composite materials other than those described in this chapter.

9.24 Explain why the behavior of the materials depicted in Fig. 9.5 is as shown.

9.25 Explain why fibers are so capable of supporting a major portion of the tensile load in composite materials.

9.26 Do metal-matrix composites have any advantages over reinforced plastics? Explain.

9.27 Give reasons for the development of ceramic-matrix composites. Name some applications, and explain why they should be effective.

9.28 Explain how you would go about determining the hardness of reinforced plastics and of composite materials. Are hardness measurements on these types of materials meaningful? Does the size of the indentation make any difference? Explain.

9.29 How would you go about trying to determine the strength of a fiber?

9.30 Glass fibers are said to be much stronger than bulk glass. Why is this so?

9.31 Describe situations in which a glass could be used as a matrix material.

9.32 When the American Plains states were settled, no trees existed for the construction of housing. Pioneers cut bricks from sod—basically, prairie soil as a matrix and grass and its root system as reinforcement. Explain why this approach was successful.

9.33 By incorporating small amounts of a blowing agent, it is possible to manufacture hollow polymer fibers with gas cores. List possible applications for such fibers.

9.34 Referring to Fig. 9.2c, would there be an advantage in using layers of cloth (woven fibers) instead of continuous fiber stacks without weaving? Explain.

9.35 Is it possible to design a composite material that has a Poisson's ratio of zero in a desired direction? Explain. Can a composite material be designed that has a thermal conductivity of zero in a desired direction? Explain.

QUANTITATIVE PROBLEMS

9.36 Calculate the average increase in the properties of the plastics given in Table 7.1 as a result of their reinforcement, and describe your observations.

9.37 In Example 9.1, what would be the percentage of the load supported by the fibers if their strength were 1000 MPa and the matrix strength were 200 MPa? What would be the answer if the fiber stiffness were doubled and the matrix stiffness were halved?

9.38 Calculate the percent increase in the mechanical properties of reinforced nylon from the data shown in Fig. 9.5.

9.39 Plot E/ρ and $E/\rho^{0.5}$ for the composite materials listed in Table 9.1, and compare your results with the properties of the materials described in Chapters 4 through 8. (See also Table 9.2.)

9.40 Calculate the stress in the fibers and in the matrix in Example 9.1. Assume that the cross-sectional area is 0.25 in^2 and $P_c = 500$ lb.

9.41 Repeat the calculations in Example 9.1 if (a) Nextel 610 fiber is used and (b) Spectra 2000 is used.

9.42 Refer to the properties listed in Table 7.1. If acetal is reinforced with E-type glass fibers, what is the range of fiber content in glass-reinforced acetal?

9.43 Plot the elastic modulus and strength of an aluminum metal-matrix composite with high-modulus carbon fibers, as a function of fiber content.

9.44 For the data in Example 9.1, what should be the fiber content so that the fibers and the matrix fail simultaneously? Use an allowable fiber stress of 200 MPa and a matrix strength of 30 MPa.

9.45 It is desired to obtain a composite material with a target stiffness of 10 GPa. If a high strength carbon fiber is to be used, determine the required fiber volume if the matrix is (a) nylon, (b) polyester, (c) acetal, and (d) polyethylene.

9.46 A rectangular cantilever beam, 100 mm high, 20 mm wide, and 1 m long, is subjected to a concentrated load of 50 kg at its end. (a) Consider a polymer reinforced with high modulus carbon fibers, with a fiber volume ratio of $x = 10\%$. What is the maximum deflection of the beam if the matrix material is polyester? (b) Obtain the deflection of the beam if aluminum or steel was used, for the same beam dimensions. (c) What fiber volume ratio is needed to produce the same deflection as the aluminum or steel beams? (d) Determine the weight of the beams considered in parts (b) and (c), and compare them.

SYNTHESIS, DESIGN, AND PROJECTS

9.47 What applications for composite materials can you think of in addition to those given in Section 9.4? Why do you think your applications would be suitable for these materials?

9.48 Using the information given in this chapter, develop special designs and shapes for possible new applications of composite materials.

9.49 Would a composite material with a strong and stiff matrix and a soft and flexible reinforcement have any practical uses? Explain.

9.50 Make a list of products for which the use of composite materials could be advantageous because of their anisotropic properties.

9.51 Inspect Fig. 9.1 and explain what other components of an aircraft, including the cabin, could be made of composites.

9.52 Name applications in which both specific strength and specific stiffness are important.

9.53 What applications for composite materials can you think of in which high thermal conductivity would be desirable? Explain.

9.54 As with other materials, the mechanical properties of composites are obtained by preparing appropriate specimens and then testing them. Explain what problems you might encounter in preparing such specimens for testing in tension. Suggest methods for making appropriate specimens, including their shape and how they would be clamped into the jaws of testing machines.

9.55 Developments are taking place in techniques for three-dimensional reinforcement of composites. Describe (a) applications in which strength in the thickness direction of the composite is important and (b) your ideas on how to achieve this strength. Include simple sketches of the structure utilizing such reinforced plastics.

9.56 Design and describe a test method to determine the mechanical properties of reinforced plastics in their thickness direction. (Note, for example, that plywood is weak in its thickness direction.)

9.57 As described in this chapter, reinforced plastics can be adversely affected by the environment—in particular, moisture, chemicals, and temperature variations. Design and describe test methods to determine the mechanical properties of composite materials subjected to these environmental conditions.

9.58 Comment on your observations on the design of the sailboard illustrated in Fig. 9.8.

9.59 Make a survey of various sports equipment and identify the components made of composite materials. Explain the reasons for and the advantages of using composites in these specific applications.

9.60 Several material combinations and structures were described in this chapter. In relative terms, identify those that would be suitable for applications involving each of the following: (a) very low temperatures, (b) very high temperatures, (c) vibrations, and (d) high humidity.

9.61 Obtain a textbook on composite materials, and investigate the effective stiffness of a continuous fiber-reinforced polymer. Plot the stiffness of such a composite as a function of orientation with respect to the fiber direction.

9.62 It is possible to make fibers or whiskers with a varying cross-section, or a "wavy" fiber. What advantages would such fibers have?

9.63 Describe how you can produce some simple composite materials using raw materials that are available around a home. Explain.

9.64 *Gel spinning* is a specialized process used in making fibers with high strength or special properties. Search the technical literature, and write a brief paper on this subject.

9.65 Figure P9.65 shows a section of a three-dimensional weave that uses a binder yarn to tie layers of fibers together. Conduct a literature search, and determine the advantages and limitations of using three-dimensional weaves as reinforcements in composite materials.

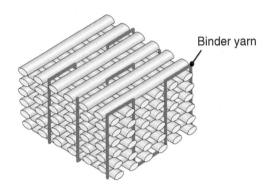

Binder yarn

FIGURE P9.65

Metal-casting Processes and Equipment

As described throughout the rest of this book, several methods are available to shape metals into products. One of the oldest processes is **casting**, which basically involves pouring molten metal into a mold cavity. Upon solidification, the metal takes the shape of the cavity; two examples of cast parts are shown in Fig. II.1. Casting was first used around 4000 B.C. to make ornaments, arrowheads, and various other objects. A wide variety of products can be cast, and the process is capable of producing intricate shapes in one piece, including those with internal cavities, such as engine blocks. Figure II.2 shows cast components in a typical automobile, a product that was used in the introduction to Part I to illustrate the selection and use of a variety of materials. The common casting processes, developed over the years, are shown in Fig. II.3.

As in all manufacturing operations, each casting process has its own characteristics, applications, advantages, limitations, and costs involved. Casting is most often selected over other manufacturing methods because:

(a)

(b)

FIGURE II.1 Examples of cast parts. (a) A die-cast aluminum transmission housing. (b) A tree of rings produced through investment casting. *Source:* (b) Courtesy of Romanoff, Inc.

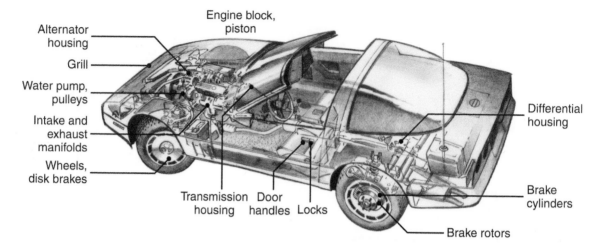

FIGURE II.2   Cast parts in a typical automobile.

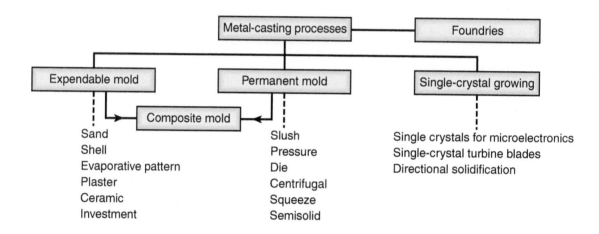

FIGURE II.3   Outline of metal-casting processes described in Part II.

- Casting can produce complex shapes and can incorporate internal cavities or hollow sections
- Very large parts can be produced in one piece
- Casting can utilize materials that are difficult or uneconomical to process by other means, such as hard metals that are difficult to machine or plastically deform
- The casting process can be competitive with other manufacturing processes

   Almost all metals can be cast in, or nearly in, the final shape desired, often requiring only minor finishing operations. This capability places casting among the most important *net-shape manufacturing* technologies, along with net-shape forging (Chapter 14), stamping of sheet metal (Chapter 16), and powder metallurgy and metal-injection molding (Chapter 17). With modern processing techniques and control of chemical composition, mechanical properties of castings can equal those made by other processes.

Fundamentals of Metal Casting

- First used about 6000 years ago, casting continues to be an important manufacturing process for producing very small, as well as very large and complex, parts.
- The first topic described is solidification of molten metals, including the differences between solidification of pure metals and alloys.
- Fluid flow in casting is then described, with Bernoulli's and the continuity equations being applied to establish a framework for analyzing molten metal flow through the cavities of a mold.
- The importance of turbulence versus laminar flow is introduced.
- Heat transfer and shrinkage of castings are also discussed, including Chvorinov's rule for solidification time.
- The chapter ends with a description of the causes of porosity in castings and common methods of reducing them to improve cast-metal properties.

10.1 Introduction

The **casting** process basically involves (a) pouring molten metal into a mold, patterned after the part to be cast, (b) allowing it to solidify, and (c) removing the part from the mold. As with all other manufacturing processes, an understanding of the underlying science is essential for the production of good quality and economical parts, and for establishing proper techniques for mold design and casting practice.

Important considerations in casting operations are:

- Flow of the molten metal into the mold cavity, and design of gating systems or pathways for molten metal to fill the cavity
- Solidification and cooling of the metal in the mold
- Influence of the mold material

This chapter describes relationships among the many factors involved in casting. The flow of molten metal into the mold cavity is first described, in terms of mold design and fluid-flow characteristics. Solidification and cooling of metals in the mold are affected by several factors, including the metallurgical and thermal properties of the metal. The type of mold also has an important influence, because it affects the rate of cooling. The chapter concludes with a description of the factors influencing defect formation in castings.

Metal-casting processes, design considerations, and casting materials are described in Chapters 11 and 12. The casting of ceramics and plastics, which involves methods and procedures somewhat similar to those for metal, are described in Chapters 18 and 19, respectively.

10.2 Solidification of Metals

After molten metal is poured into a **mold**, a sequence of events takes place during solidification and cooling of the metal to ambient temperature. These events greatly influence the size, shape, uniformity, and chemical composition of the grains formed throughout the casting, which, in turn, influence the overall properties of the casting. The significant factors affecting these events are the type of metal cast, the thermal properties of both the metal and the mold, the geometric relationship between volume and surface area of the casting, and the shape of the mold.

10.2.1 Pure Metals

Because a pure metal has a clearly defined melting, or freezing, point, it solidifies at a constant temperature, as shown in Fig. 10.1. Pure aluminum, for example, solidifies at 660°C (1220°F), iron at 1537°C (2798°F), and tungsten at 3410°C (6170°F). (See also Table 3.1.) After the temperature of the molten metal drops to its freezing point, its temperature remains constant while the *latent heat of fusion* is given off. The *solidification front* (the solid–liquid interface) moves through the molten metal from the mold walls in toward the center. The solidified metal, called the *casting*, is then removed from the mold and allowed to cool to ambient temperature.

As shown in Fig. 10.1b and described in greater detail in Section 10.5.2, metals shrink while cooling and, generally, also shrink when they solidify. This is an

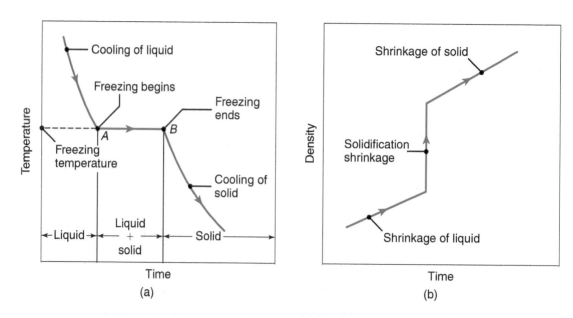

FIGURE 10.1   (a) Temperature as a function of time for the solidification of pure metals; note that freezing takes place at a constant temperature. (b) Density as a function of time.

important consideration, because shrinkage can lead to microcracking and the associated porosity, which can adversely affect the mechanical properties of the casting.

As an example of the grain structure that develops in a casting, Fig. 10.2a shows a cross-section of a box-shaped mold. At the mold walls, which are at ambient temperature at first, or typically are much cooler than the molten metal, the metal cools rapidly, producing a solidified **skin**, or *shell*, of fine equiaxed grains. The grains generally grow in a direction opposite to that of the heat transfer out through the mold.

Those grains that have favorable orientation grow preferentially, and are called **columnar grains** (Fig. 10.3). Those grains that have substantially different orientations are blocked from further growth. As the driving force of the heat transfer decreases away from the mold walls, the grains become equiaxed and coarse. This sequence of grain development is known as **homogenous nucleation,** meaning that the grains (crystals) grow upon themselves, starting at the mold wall.

10.2.2 Alloys

Solidification in alloys begins when the temperature drops below the *liquidus*, T_L, and is complete when it reaches the *solidus*, T_S (Fig. 10.4). Within this temperature range, the alloy is in a *mushy* or *pasty* state, consisting of **columnar dendrites** (from the Greek *dendron*, meaning "akin to," and *drys*, meaning "tree"). Note the presence of liquid metal between the dendrite *arms*.

Dendrites have three-dimensional arms and branches (*secondary arms*), which eventually interlock, as can be seen in Fig. 10.5. The study of dendritic structures, although complex, is important, because such structures can contribute to detrimental factors, such as compositional variations, segregation, and microporosity within a cast part.

The width of the **mushy zone,** where both liquid and solid phases are present, is an important factor during solidification. This zone is described in terms of a temperature difference, known as the **freezing range,** as

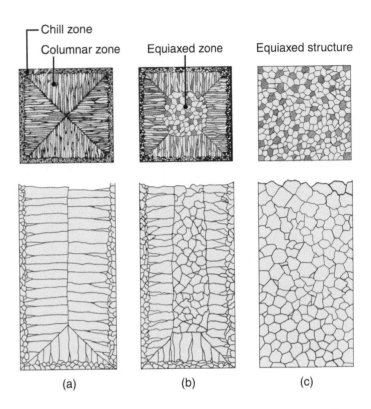

FIGURE 10.2 Schematic illustration of three cast structures of metals solidified in a square mold: (a) pure metals; (b) solid–solution alloys; and (c) structure obtained by using nucleating agents. *Source:* After G.W. Form, J.F. Wallace, J.L. Walker, and A. Cibula.

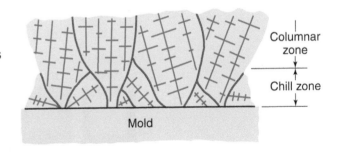

FIGURE 10.3 Development of a preferred texture at a cool mold wall; note that only favorably oriented grains grow away from the surface of the mold surface.

$$\text{Freezing range} = T_L - T_S. \qquad (10.1)$$

It can be seen in Fig. 10.4 that pure metals have a freezing range that approaches zero, and that the solidification front moves as a plane without developing a mushy zone. Eutectics (Section 4.3) solidify in a similar manner, with an essentially plane

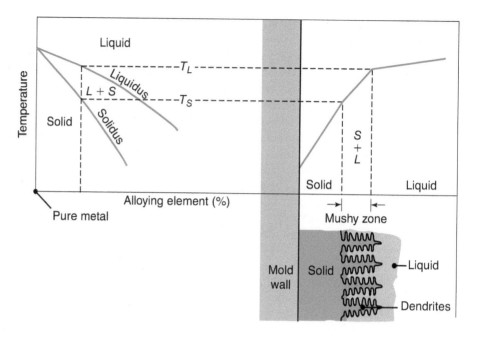

FIGURE 10.4 Schematic illustration of alloy solidification and temperature distribution in the solidifying metal; note the formation of dendrites in the mushy zone.

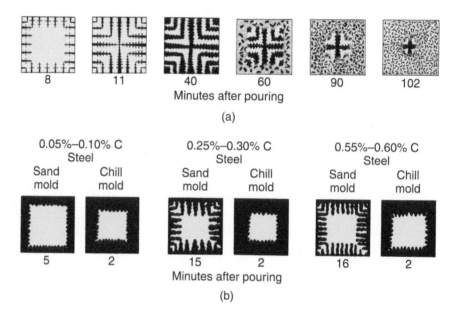

FIGURE 10.5 (a) Solidification patterns for gray cast iron in a 180-mm (7-in.) square casting. Note that after 11 min of cooling, dendrites reach each other, but the casting is still mushy throughout; it takes about 2 h for this casting to solidify completely. (b) Solidification of carbon steels in sand and chill (metal) molds; note the difference in solidification patterns as the carbon content increases. *Source:* After H.F. Bishop and W.S. Pellini.

front. The structure developed upon solidification depends on the composition of the eutectic. In alloys with a nearly symmetrical phase diagram (see Fig. 4.4), the structure is generally lamellar, with two or more solid phases present, depending on the alloy system. When the volume fraction of the minor phase of the alloy is less than about 25%, the structure generally becomes fibrous. These conditions are particularly significant for cast irons.

For alloys, a *short freezing range* generally involves a temperature difference of less than 50°C (90°F), and for a *long freezing range*, more than 110°C (200°F). Ferrous castings generally have narrow mushy zones, whereas aluminum and magnesium alloys have wide mushy zones. Consequently, these alloys are in a mushy state throughout most of their solidification process.

Effects of Cooling Rates.   Slow cooling rates, on the order of 10^2 K/s, or long local solidification times, result in *coarse* dendritic structures, with large spacing between dendrite arms. For higher cooling rates, on the order of 10^4 K/s, or for short local solidification times, the structure becomes *finer*, with smaller dendrite arm spacing. For still higher cooling rates, on the order of from 10^6 to 10^8, the structures developed are *amorphous*, as described in Section 6.14.

The structures developed and the resulting grain size have an influence on the properties of the casting. As grain size decreases, the strength and ductility of the cast alloy increase, microporosity (*interdendritic shrinkage voids*) in the casting decreases, and the tendency for the casting to crack (*hot tearing*, see Fig. 10.12) during solidification decreases. Lack of uniformity in grain size and grain distribution result in castings that have *anisotropic properties*.

A criterion describing the kinetics of the liquid–solid interface is the ratio G/R, where G is the *thermal gradient* and R is the *rate* at which the interface moves. Typical values for G range from 10^2 to 10^3 K/m, and for R the range is from 10^{-3} to 10^{-4} m/s. Dendritic-type structures (Figs. 10.6a and b) typically have a G/R ratio in the range from 10^5 to 10^7, whereas ratios of 10^{10} to 10^{12} produce a plane front, nondendritic liquid–solid interface (Fig. 10.7).

10.2.3   Structure–property Relationships

Because all castings are expected to possess certain properties to meet design and service requirements, the relationships between properties and the structures developed

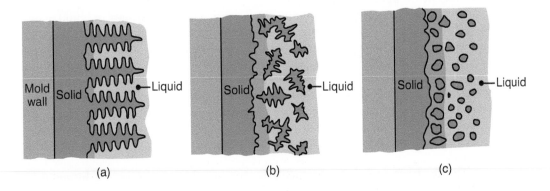

FIGURE 10.6   Schematic illustration of three basic types of cast structures: (a) columnar dendritic; (b) equiaxed dendritic; and (c) equiaxed nondendritic. *Source:* Courtesy of D. Apelian.

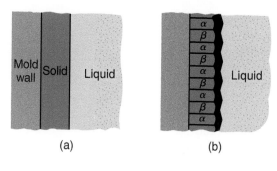

FIGURE 10.7 Schematic illustration of cast structures in (a) plane front, single phase, and (b) plane front, two phase. *Source:* Courtesy of D. Apelian.

during solidification are important aspects of casting. This section describes these relationships in terms of dendrite morphology and the concentration of alloying elements in various regions within the casting.

The compositions of dendrites and the liquid metal are given by the *phase diagram* of the particular alloy. When the alloy is cooled very slowly, each dendrite develops a uniform composition. However, under the normally higher cooling rates encountered in practice, **cored dendrites** are formed. These dendrites have a surface composition different from that at their centers, a difference referred to as *concentration gradient*. The surface of the dendrite has a higher concentration of alloying elements than at its core, due to solute rejection from the core toward the surface during solidification of the dendrite (**microsegregation**). The darker shading in the interdendritic liquid near the dendrite roots shown in Fig. 10.6 indicates that these regions have a higher solute concentration. Microsegregation in these regions is much more pronounced than in others.

There are several types of **segregation**. In contrast to microsegregation, **macrosegregation** involves differences in composition throughout the casting itself. In situations where the solidification front moves away from the surface of a casting as a plane (Fig. 10.7), lower melting-point constituents in the solidifying alloy are driven toward the center (**normal segregation**). Consequently, such a casting has a higher concentration of alloying elements at its center than at its surfaces. In dendritic structures, such as those found in solid–solution alloys (Fig. 10.2b), the opposite occurs; that is, the center of the casting has a lower concentration of alloying elements (**inverse segregation**) than does at its surface. The reason is that the liquid metal (having a higher concentration of alloying elements) enters the cavities developed from solidification shrinkage in the dendrite arms, which have solidified sooner.

Another form of segregation is due to gravity. **Gravity segregation** involves a process whereby higher-density inclusions or compounds sink while lighter elements (such as antimony in an antimony–lead alloy) float to the surface.

A typical cast structure of a solid–solution alloy, with an inner zone of equiaxed grains, is shown in Fig. 10.2b. This inner zone can be extended throughout the casting, as shown in Fig. 10.2c, by adding an **inoculant** (*nucleating agent*) to the alloy. The inoculant induces nucleation of the grains throughout the liquid metal (**heterogeneous nucleation**).

Because of the presence of *thermal gradients* in a solidifying mass of liquid metal, and due to gravity and the resultant density differences, *convection* has a strong influence on the structures developed. Convection involves heat transfer by the movement of matter, and in a casting it usually is associated with the flow of the liquid metal. Convection promotes the formation of an outer chill zone, refines grain size, and accelerates the transition from columnar to equiaxed grains. The structure shown in Fig. 10.6b also can be obtained by increasing convection within the liquid metal, whereby dendrite arms separate (**dendrite multiplication**). Conversely, reducing or eliminating convection results in coarser and longer columnar dendritic grains.

The dendrite arms are not particularly strong and can be broken up by agitation or mechanical vibration in the early stages of solidification (as in **semisolid metal forming** and **rheocasting**, described in Section 11.4.7). This process results in finer grain size, with equiaxed nondendritic grains distributed more uniformly throughout the casting (Fig. 10.6c). A side benefit is the *thixotropic* behavior of alloys (that is, the viscosity decreases when the liquid metal is agitated), leading to improved castability of the metal. Another form of semisolid metal forming is **thixotropic casting**, where a solid billet is first heated to a semisolid state and then injected into a die-casting mold (Section 11.4.5).

10.3 Fluid Flow

To emphasize the importance of fluid flow in casting, consider a basic gravity casting system, as shown in Fig. 10.8. The molten metal is poured through a **pouring basin** or **cup**; it then flows through the **gating system** (consisting of sprue, runners, and gates) into the mold cavity. As also illustrated in Fig. 11.3, the **sprue** is a tapered vertical channel through which the molten metal flows downward in the mold. **Runners** are the channels that carry the molten metal from the sprue into the mold cavity or connect the sprue to the **gate** (that portion of the runner through which the molten metal enters the mold cavity). **Risers**, also called **feeders**, serve as reservoirs of molten metal to supply any molten metal necessary to prevent porosity due to shrinkage during solidification.

Although such a gating system appears to be relatively simple, successful casting requires proper design and control of the solidification process to ensure adequate fluid flow in the system. For example, an important function of the gating system in sand casting is to trap contaminants (such as oxides and other inclusions) and remove them from the molten metal, by having the contaminants adhere to the walls of the gating system, thereby preventing them from reaching the mold cavity. Furthermore, a properly designed gating system helps avoid or minimize such problems as premature cooling, turbulence, and gas entrapment. Even before it reaches the mold cavity, the molten metal must be handled carefully to avoid the formation of oxides on molten-metal surfaces from exposure to the environment or the introduction of impurities into the molten metal.

Two basic principles of fluid flow are relevant to gating design: Bernoulli's theorem and the law of mass continuity.

Bernoulli's Theorem.   This theorem is based on the principle of the conservation of energy, and relates pressure, velocity, the elevation of the fluid at any location in the system, and the frictional losses in a system that is full of liquid. The Bernoulli equation is

$$h + \frac{p}{\rho g} + \frac{v^2}{2g} = \text{constant}, \qquad (10.2)$$

where h is the elevation above a certain reference level, p is the pressure at that elevation, v is the velocity of the liquid at that elevation, ρ is the density of the fluid (assuming that it is incompressible), and g is the gravitational constant.

Conservation of energy requires that, at a particular location in the system, the following relationship be satisfied:

$$h_1 + \frac{p_1}{\rho g} + \frac{v_1^2}{2g} = h_2 + \frac{p_2}{\rho g} + \frac{v_2^2}{2g} + f, \qquad (10.3)$$

where the subscripts 1 and 2 represent two different locations in the system and f represents the frictional loss in the liquid as it travels through the system. The frictional loss includes

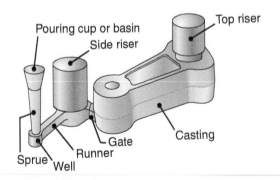

FIGURE 10.8   Schematic illustration of a typical riser-gated casting. Risers serve as reservoirs, supplying molten metal to the casting as it shrinks during solidification.

such factors as energy loss at the liquid–mold wall interfaces and turbulence in the liquid.

Mass Continuity.   The law of mass continuity states that, for incompressible liquids and in a system with impermeable walls, the rate of flow is constant. Thus,

$$Q = A_1 v_1 = A_2 v_2, \tag{10.4}$$

where Q is the volume rate of flow (such as m^3/s), A is the cross-sectional area of the liquid stream, and v is the average velocity of the liquid in that cross-section. The subscripts 1 and 2 refer to two different locations in the system. According to this law, the flow rate must be maintained everywhere in the system. The wall permeability is important, because otherwise some liquid will escape through the walls (as occurs in sand molds); thus, the flow rate will decrease as the liquid moves through the system. Coatings often are used to inhibit such behavior in sand molds. A small amount of permeability is useful to allow escape of gases and can aid in heat transfer.

Sprue Design.   An application of the Bernoulli and mass continuity equations is the traditional tapered design of sprues shown in Fig. 10.8. Note that in a free-falling liquid (such as water from a faucet), the cross-sectional area of the stream decreases as the liquid gains velocity downward. Thus, if a sprue has a constant cross-sectional area and molten metal is poured into it, regions can develop where the liquid loses contact with the sprue walls. As a result, **aspiration** (a process whereby air is drawn in or entrapped in the liquid) may take place.

One of two basic alternatives is used to prevent aspiration: (a) A tapered sprue is used to prevent molten metal separation from the sprue wall or (b) straight-sided sprues are supplied with a **choking** mechanism at the bottom, consisting of either a choke core or a runner choke, as shown in Fig. 11.3. The choke slows the flow sufficiently to prevent aspiration in the sprue.

The specific shape of a tapered sprue that prevents aspiration can be determined from Eqs. (10.3) and (10.4). Assuming that the pressure at the top of the sprue is equal to the pressure at the bottom, and that there are no frictional losses, the relationship between height and cross-sectional area at any point in the sprue is given by the parabolic relationship

Video Solution 10.1 De-sign of a Sprue

$$\frac{A_1}{A_2} = \sqrt{\frac{h_2}{h_1}}, \tag{10.5}$$

where, for example, the subscript 1 denotes the top of the sprue and 2 denotes the bottom. The distances h_1 and h_2 are measured from the liquid level in the pouring cup or basin (Fig. 10.8), so that h_2 is larger than h_1. Moving downward from the top, the cross-sectional area of the sprue must therefore decrease. The area at the bottom of the sprue, A_2, is selected to allow for desired flow rates, as discussed below, and the profile produced is according to Eq. (10.5).

Depending on the assumptions made, expressions other than Eq. (10.5) also can be obtained. For example, assume a certain molten-metal velocity, V_1, at the top of the sprue; then, using Eqs. (10.3) and (10.4), an expression can be obtained for the ratio A_1/A_2 as a function of h_1, h_2, and V_1.

Modeling.   Another application of the foregoing equations is in the *modeling of mold filling*. For example, consider the situation shown in Fig. 10.8 where molten metal is poured into a pouring cup or basin; it flows through a sprue to a runner and a gate, and fills the mold cavity. If the pouring basin has a much larger cross-sectional area than the sprue bottom, then the velocity of the molten metal at the top of the pouring basin is very low and can be taken to be zero. If frictional losses are due to a viscous dissipation of energy, then f in Eq. (10.3) can be taken to be a function of the vertical distance, and is often approximated as a linear function. The velocity of the molten metal leaving the gate is obtained from Eq. (10.3) as

$$v = c\sqrt{2gh},$$

where h is the distance from the sprue base to the liquid metal height and c is a friction factor. For frictionless flow, c equals unity, and for flows with friction, c is always between 0 and 1. The magnitude of c varies with mold material, runner layout, and channel size, and can include energy losses due to turbulence, as well as viscous effects.

If the liquid level has reached a height of x at the gate, then the gate velocity is

$$v = c\sqrt{2g}\sqrt{h - x}.$$

The flow rate through the gate will then be the product of this velocity and the gate area, according to Eq. (10.4). The shape of the casting will determine the height as a function of time. Integrating Eq. (10.4) gives the mean fill time and flow rate, and dividing the casting volume by this mean flow rate gives the mold fill time.

Simulation of mold filling assists designers in the specification of the runner diameter, as well as the size and number of sprues and pouring basins. To ensure that the runners stay open, the fill time must be a small fraction of the solidification time, but the velocity should not be so high as to erode the mold material (referred to as *mold wash*) or to result in too high of a **Reynolds number** (see below); otherwise, turbulence and associated air entrainment results. Several computational tools are now available to evaluate gating designs and assist in the sizing of components, such as Magmasoft, ProCast, Quikcast, SolidCast, Star-cast, SutCast, and PASSAGE/Powercast.

Flow Characteristics.   An important consideration of fluid flow in gating systems is the presence of **turbulence**, as opposed to the *laminar flow* of fluids. Turbulence is flow that is highly chaotic; in casting systems such flow can lead to aspiration. The **Reynolds number**, Re, is used to quantify this aspect of fluid flow; it represents the ratio of the *inertia* to the *viscous* forces in fluid flow, and is defined as

$$\text{Re} = \frac{vD\rho}{\eta}, \tag{10.6}$$

where v is the velocity of the liquid, D is the diameter of the channel, and ρ and η are the density and viscosity of the liquid, respectively.

The higher the Reynolds number, the greater the tendency for turbulent flow to occur. In gating systems, Re typically ranges from 2000 to 20,000, where a value of up to 2000 represents laminar flow. Between 2000 and 20,000, it represents a mixture of laminar and turbulent flow; such a mixture generally is regarded as harmless in gating systems. However, Re values in excess of 20,000 represent severe turbulence, resulting in significant air entrainment and the formation of *dross* (the scum that forms on the surface of molten metal), from the reaction of the liquid metal with air and other gases. Techniques for minimizing turbulence generally involve avoidance of sudden changes in local flow direction and in the geometry of channel cross-sections in gating system design.

Dross or slag can be eliminated only by *vacuum casting* (Section 11.4.2). Conventional atmospheric casting mitigates dross or slag by (a) skimming, (b) using properly

Video Solution 10.2 Reynolds Number in Gating Systems

designed pouring basins and runner systems, or (c) using filters, which also can eliminate turbulent flow in the runner system. Filters usually are made of ceramics, mica,
or fiberglass; their proper location and placement are important for effective filtering
of dross and slag.

10.4   Fluidity of Molten Metal

The capability of molten metal to fill mold cavities is called *fluidity*, which consists of
two basic factors: (1) characteristics of the molten metal and (2) casting parameters.
Although complex, the term **castability** generally is used to describe the ease with
which a metal can be cast to produce a part with good quality. Castability includes
not only fluidity, but the nature of casting practices as well.

The following characteristics of molten metal influence fluidity:

1. Viscosity.   As viscosity and its sensitivity to temperature increase, fluidity
decreases.

2. Surface Tension.   A high surface tension of the liquid metal reduces fluidity.
Because of this, oxide films on the surface of the molten metal have a significant
adverse effect on fluidity; for example, an oxide film on the surface of pure molten
aluminum triples the surface tension.

3. Inclusions.   Because they are insoluble, inclusions can have a significant adverse
effect on fluidity. This effect can be verified by observing the viscosity of a liquid
(such as oil) with and without sand particles in it; the liquid with sand in it has a
higher viscosity and, hence, lower fluidity.

4. Solidification Pattern of the Alloy.   The manner in which solidification takes
place (Section 10.2) can influence fluidity. Moreover, fluidity is inversely proportional
to the freezing range: The shorter the range (as in pure metals and eutectics), the
higher the fluidity. Conversely, alloys with long freezing ranges (such as solid–solution
alloys) have lower fluidity.

The following casting parameters influence fluidity and also influence the fluid
flow and thermal characteristics of the system:

1. Mold Design.   The design and dimensions of the sprue, runners, and risers all
influence fluidity.

2. Mold Material and Its Surface Characteristics.   The higher the thermal conductivity of the mold and the rougher its surfaces, the lower the fluidity of the molten
metal. Although heating the mold improves fluidity, it slows down solidification of
the metal; thus, the casting develops coarse grains and hence has lower strength.

3. Degree of Superheat.   *Superheat* (defined as the increment of temperature of
an alloy above its melting point) improves fluidity by delaying solidification. The
pouring temperature often is specified instead of the degree of superheat, because it
can be specified more easily.

4. Rate of Pouring.   The slower the rate of pouring molten metal into the mold, the
lower the fluidity, because of the higher rate of cooling when poured slowly.

5. Heat Transfer.   This factor directly affects the viscosity of the liquid metal (see
below).

10.4.1   Tests for Fluidity

Several tests have been developed to quantify fluidity, although none is accepted universally. In one such common test, the molten metal is made to flow along a channel that is at room temperature (Fig. 10.9); the distance the metal flows before it solidifies and stops flowing is a measure of its fluidity. Obviously, this length is a function of the thermal properties of the metal and the mold, as well as of the design of the channel. Still, such fluidity tests are useful and simulate casting situations to a reasonable degree.

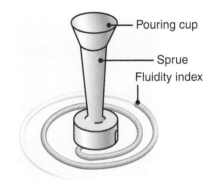

FIGURE 10.9 A test method for fluidity using a spiral mold. The fluidity index is the length of the solidified metal in the spiral passage; the greater the length of the solidified metal, the greater is the metal's fluidity.

10.5   Heat Transfer

The heat transfer during the complete cycle (from pouring, to solidification, and to cooling to room temperature) is another important consideration in metal casting. Heat flow at different locations in the system is a complex phenomenon and depends on several factors relating to the material cast, the mold, and process parameters. For instance, in casting thin sections, the metal flow rates must be high enough to avoid premature chilling and solidification. On the other hand, the flow rate must not be so high as to cause excessive turbulence—with its detrimental effects on the casting process.

A typical temperature distribution at the mold liquid–metal interface is shown in Fig. 10.10. Heat from the liquid metal is given off through the mold wall and to the surrounding air. The temperature drop at the air–mold and mold–metal interfaces is caused by the presence of boundary layers and imperfect contact at these interfaces. The shape of the curve depends on the thermal properties of the molten metal and the mold.

10.5.1   Solidification Time

During the early stages of solidification, a thin skin begins to form at the relatively cool mold walls, and as time passes, the thickness of the skin increases (Fig. 10.11). With flat mold walls, the thickness is proportional to the square root of time; thus, doubling the time will make the skin $\sqrt{2} = 1.41$ times or 41% thicker.

The **solidification time** is a function of the volume of a casting and its surface area (*Chvorinov's rule*):

$$\text{Solidification time} = C \left(\frac{\text{Volume}}{\text{Surface area}} \right)^n, \quad (10.7)$$

where C is a constant that reflects (a) the mold material, (b) the metal properties (including latent heat), and (c) the temperature. The parameter n has a value between 1.5 and 2, but usually is taken as 2. Thus, a large solid sphere will solidify and cool to ambient temperature at a much slower rate than will a smaller solid sphere. The reason for this is that the volume of a sphere is proportional to the cube of its diameter, whereas the surface area is proportional to the square of its diameter.

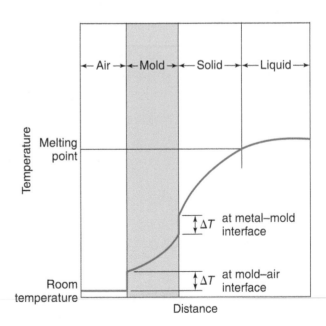

FIGURE 10.10 Temperature distribution at the interface of the mold wall and the liquid metal during the solidification of metals in casting.

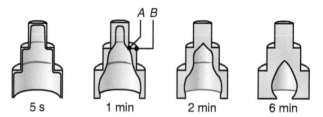

5 s   1 min   2 min   6 min

FIGURE 10.11 Solidified skin on a steel casting. The remaining molten metal is poured out at the times indicated in the figure. Hollow ornamental and decorative objects are made by a process called *slush casting*, which is based on this principle. *Source:* After H.F. Taylor, J. Wulff, and M.C. Flemings.

Video Solution 10.3   Application of Chvorinov's Rule

Similarly, it can be shown that molten metal in a cube-shaped mold will solidify faster than in a spherical mold of the same volume (see Example 10.1).

The effects of mold geometry and elapsed time on skin thickness and shape are shown in Fig. 10.11. As illustrated, the unsolidified molten metal has been poured from the mold at time intervals ranging from 5 s to 6 min. Note that (as expected) the skin thickness increases with elapsed time, and that the skin is thinner at internal angles (location A in the figure) than at external angles (location B). The latter condition is caused by slower cooling at internal angles than at external angles.

EXAMPLE 10.1   Solidification Times for Various Shapes

Given: Three metal pieces being cast have the same volume, but different shapes: One is a sphere, one a cube, and the other a cylinder with its height equal to its diameter. Assume that $n = 2$.

Find: Which piece will solidify the fastest, and which one the slowest?

Solution: The volume of the piece is taken as unity. Thus from Eq. (10.7),

$$\text{Solidification time} \propto \frac{1}{(\text{Surface area})^2}.$$

The respective surface areas are:

Sphere:

$$V = \left(\frac{4}{3}\right)\pi r^3, \quad r = \left(\frac{3}{4\pi}\right)^{1/3},$$

$$A = 4\pi r^2 = 4\pi \left(\frac{3}{4\pi}\right)^{2/3} = 4.84.$$

Cube:

$$V = a^3, \quad a = 1, \quad \text{and} \quad A = 6a^2 = 6.$$

Cylinder:

$$V = \pi r^2 h = 2\pi r^3, \quad r = \left(\frac{1}{2\pi}\right)^{1/3},$$

$$A = 2\pi r^2 + 2\pi rh$$

$$= 6\pi r^2 = 6\pi \left(\frac{1}{2\pi}\right)^{2/3} = 5.54.$$

The respective solidification times are

$$t_{\text{sphere}} = 0.043C, \quad t_{\text{cube}} = 0.028C,$$

$$t_{\text{cylinder}} = 0.033C.$$

Hence, the cube-shaped piece will solidify the fastest, and the spherical piece will solidify the slowest.

10.5.2 Shrinkage

Because of their thermal expansion characteristics, metals usually shrink (contract) during solidification and while cooling to room temperature. *Shrinkage*, which causes dimensional changes and sometimes warping and cracking, is the result of the following three sequential events:

1. Contraction of the molten metal as it cools prior to its solidification
2. Contraction of the metal during phase change from liquid to solid
3. Contraction of the solidified metal (the casting) as its temperature drops to ambient temperature

The largest shrinkage occurs during the phase change of the material from liquid to solid, but this can be reduced or eliminated through the use of risers or pressure-feeding of molten metal. The amount of contraction during the solidification of various metals is shown in Table 10.1; note that some metals (such as gray cast iron) expand. The reason is that graphite has a relatively high specific volume, and when it precipitates as graphite flakes during solidification of the gray cast iron, it causes a net expansion of the metal. Shrinkage, especially that due to thermal contraction, is further discussed in Section 12.2.1 in connection with design considerations in casting.

TABLE 10.1

Volumetric Solidification Contraction or Expansion for Various Cast Metals			
Contraction (%)		Expansion (%)	
Aluminum	7.1	Bismuth	3.3
Zinc	6.5	Silicon	2.9
Al–4.5% Cu	6.3	Gray iron	2.5
Gold	5.5		
White iron	4–5.5		
Copper	4.9		
Brass (70–30)	4.5		
Magnesium	4.2		
90% Cu–10% Al	4		
Carbon steels	2.5–4		
Al–12% Si	3.8		
Lead	3.2		

10.6 Defects

Various defects can develop during manufacturing, depending on such factors as the quality of raw materials, casting design, and control of processing parameters. While some defects affect only the appearance, others can have major adverse effects on the structural integrity of the parts made.

Several defects can develop in castings, as illustrated in Figs. 10.12 and 10.13. Because different terms have been used in the past to describe the same defect, the International Committee of Foundry Technical Associations has developed a standardized nomenclature, consisting of seven basic categories of casting defects, identified with boldface capital letters:

A—**Metallic projections,** consisting of fins, flash, or projections, such as swells and rough surfaces.

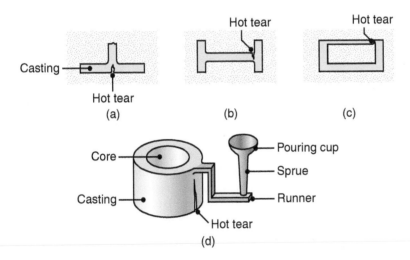

FIGURE 10.12 Examples of hot tears in castings. These defects occur because the casting cannot shrink freely during cooling, owing to constraints in various portions of the molds and cores. Exothermic (heat-producing) compounds may be used (as exothermic padding) to control cooling at critical sections to avoid hot tearing.

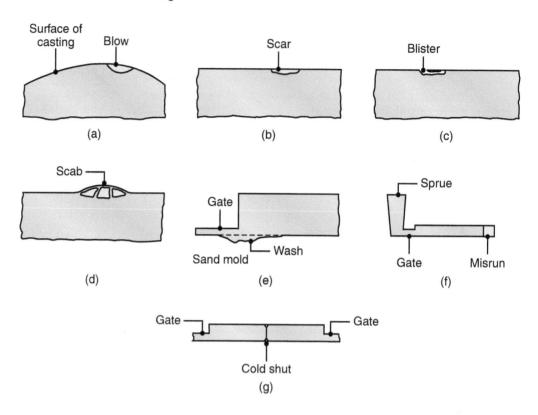

FIGURE 10.13 Examples of common defects in castings; these defects can be minimized or eliminated by proper design and preparation of molds and control of pouring procedures. *Source:* After J. Datsko.

B—Cavities, consisting of rounded or rough internal or exposed cavities, including blowholes, pinholes, and shrinkage cavities (see *porosity*, Section 10.6.1).

C—Discontinuities, such as cracks, cold or hot tearing, and cold shuts. If the solidifying metal is constrained from shrinking freely, cracking and tearing may occur. Although several factors are involved in tearing, coarse grain size and the presence of low-melting-point segregates along the grain boundaries of the metal increase the tendency for hot tearing. *Cold shut* is an interface in a casting that lacks complete fusion, because of the meeting of two streams of liquid metal from different gates.

D—Defective surface, such as surface folds, laps, scars, adhering sand layers, and oxide scale.

E—Incomplete casting, such as *misruns* (due to premature solidification), insufficient volume of the metal poured, and *runout* (due to loss of metal from the mold after pouring). Incomplete castings also can result from the molten metal being at too low a temperature or from pouring the metal too slowly.

F—Incorrect dimensions or shape, due to such factors as improper shrinkage allowance, pattern mounting error, irregular contraction, deformed pattern, or warped casting.

G—Inclusions, which form during melting, solidification, and molding. Generally nonmetallic, they are regarded as harmful, because they act as stress raisers, and thus reduce the strength of the casting. Inclusions may form during melting when the molten metal reacts with the environment (usually oxygen), with the crucible or the mold material. Chemical reactions among components in the molten metal itself may produce inclusions. Slags and other foreign material entrapped in the molten metal also become inclusions, although filtering the molten metal can remove particles as small as 30 μm. Finally, spalling of the mold and core surfaces can produce inclusions, thus indicating the importance of the quality of molds and their maintenance.

10.6.1   Porosity

Porosity in a casting may be caused by *shrinkage*, entrained or dissolved *gases*, or both. Porous regions can develop in castings because of **shrinkage** of the solidified metal. Thin sections in a casting solidify sooner than thicker regions; as a result, molten metal flows into the thicker regions that have not yet solidified. Porous regions may develop at their centers because of contraction as the surfaces of the thicker region begin to solidify first. *Microporosity* also can develop when the liquid metal solidifies and shrinks between dendrites and between dendrite branches.

Porosity is detrimental to the strength and ductility of a casting and its surface finish, potentially making the casting permeable, thus affecting the pressure tightness of a cast pressure vessel.

Porosity caused by shrinkage can be reduced or eliminated by various means, including the following:

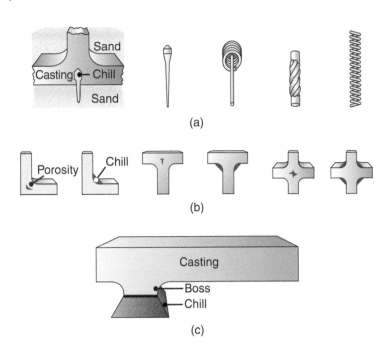

FIGURE 10.14   Various types of (a) internal and (b) external chills (dark areas at corners) used in castings to eliminate porosity caused by shrinkage. Chills are placed in regions where there is a larger volume of metal, as shown in (c).

- Adequate liquid metal should be provided to prevent cavities caused by shrinkage.
- Internal or external chills, as those used in sand casting (Fig. 10.14), also are an effective means of reducing shrinkage porosity. The function of chills is to increase the rate of solidification in critical regions. Internal chills are usually made from the same material as the casting itself, and are left in the casting. However, problems may arise that involve proper fusion of the internal chills with the casting; thus, foundries generally avoid the use of internal chills. External chills may be made from the same material as the casting or may be made of iron, copper, or graphite.
- With alloys, porosity can be reduced or eliminated by high temperature gradients, that is, by increasing the cooling rate; for example, mold materials with higher thermal conductivity may be used.
- Subjecting the casting to hot isostatic pressing is another method of reducing porosity (see Section 17.3.2).

Because *gases* are more soluble in liquid metals than in solid metals (Fig. 10.15), when a metal begins to solidify, the dissolved gases are expelled from the solution; gases also may be due to reactions of the molten metal with the mold materials. Gases either accumulate in regions of existing porosity (such as in interdendritic regions) or cause microporosity in the casting,

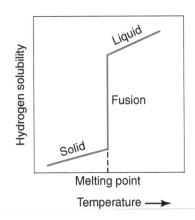

FIGURE 10.15   Solubility of hydrogen in aluminum; note the sharp decrease in solubility as the molten metal begins to solidify.

particularly in cast iron, aluminum, and copper. Dissolved gases may be removed from the molten metal by *flushing* or *purging* with an inert gas or by melting and pouring the metal in a vacuum. If the dissolved gas is oxygen, the molten metal can be *deoxidized*. Steel usually is deoxidized with aluminum, silicon, copper-based alloys with phosphorus, copper, titanium, or zirconium-bearing materials.

Whether microporosity is a result of shrinkage or is caused by gases may be difficult to determine. If the porosity is spherical and has smooth walls (similar to the shiny holes in Swiss cheese), it is generally from gases. If the walls are rough and angular, porosity is likely from shrinkage between dendrites. Gross porosity is from shrinkage and usually is called a **shrinkage cavity**.

SUMMARY

- Casting is a solidification process in which molten metal is poured into a mold and allowed to cool. The metal may flow through a variety of passages (pouring basins, sprues, runners, risers, and gating systems) before reaching the mold cavity. Bernoulli's theorem, the continuity law, and the Reynolds number are the analytical tools used in designing castings, with the goals of achieving an appropriate flow rate and eliminating defects associated with fluid flow.

- Solidification of pure metals takes place at a constant temperature, whereas solidification of alloys occurs over a range of temperatures. Phase diagrams are important tools for identifying the solidification point or points for technologically important metals.

- The composition and cooling rates of the molten metal both affect the size and shape of the grains and the dendrites in the solidifying alloy; in turn, the size and structure of grains and dendrites influence properties of the solidified casting. Solidification time is a function of the volume and surface area of a casting (Chvorinov's rule).

- The grain structure of castings can be controlled by various means to obtain desired properties. Because most metals contract during solidification and cooling, cavities can form in the casting. Porosity caused by gases evolved during solidification can be a significant problem, particularly because of its adverse effect on the properties of castings. Various defects also can develop in castings from lack of control of material and process variables.

- Dimensional changes and cracking (hot tearing) are difficulties that can arise during solidification and cooling. Several basic categories of casting defects have been identified.

KEY TERMS

Aspiration	Freezing range	Mold	Segregation
Bernoulli's theorem	Gate	Mushy zone	Shrinkage
Casting	Gating system	Normal segregation	Skin
Chills	Heterogeneous	Porosity	Solidification
Columnar dendrite	nucleation	Pouring basin	Sprue
Columnar grain	Homogenous nucleation	Reynolds number	Turbulence
Cored dendrite	Inoculant	Rheocasting	
Dendrite	Macrosegregation	Riser	
Fluidity	Microsegregation	Runner	

BIBLIOGRAPHY

Analysis of Casting Defects, American Foundrymen's Society, 2002.

ASM Handbook, Vol. 15: **Casting**, ASM International, 2008.

Blair, M., Stevens, T.L., and Linskey, B. (eds.), **Steel Castings Handbook**, 6th ed., ASM International, 1995.

Campbell, J., **Castings**, Butterworth-Heinemann, 2nd ed., 2003.

Campbell, J., **Complete Casting Handbook: Metal Casting Processes, Techniques and Design**, Butterworth-Heinemann, 2011.

Cantor, B., and O'Reilly, K., **Solidification and Casting**, Taylor & Francis, 2002.

Glicksmann, M.E., **Principles of Solidification: An Introduction to Modern Casting and Crystal Growth Concepts**, Springer, 2010.

Poirer, D.R., and Poirer, E.J., **Heat Transfer Fundamentals for Metal Casting**, Wiley, 1998.

Reikher, A., and Barkhudarov, M., **Casting: An Analytical Approach**, Springer, 2008.

Stefanescu, D.M., **Science and Engineering of Casting Solidification**, 2nd ed., Springer, 2008.

REVIEW QUESTIONS

10.1 Explain why casting is an important manufacturing process.

10.2 Why do most metals shrink when they are cast?

10.3 What are the differences between the solidification of pure metals and metal alloys?

10.4 What are dendrites? Why are they called so?

10.5 Describe the difference between short and long freezing ranges.

10.6 What is superheat? Is it important? What are the consequences of excessive superheat?

10.7 Define shrinkage and porosity. How can you tell whether cavities in a casting are due to porosity or to shrinkage?

10.8 What is the function of chills? What are they made of?

10.9 Why is the Reynolds number important in casting?

10.10 What is a sprue? What shape should a sprue have if a mold has no other choking means?

10.11 How is fluidity defined? Why is it important?

10.12 Explain the reasons for hot tearing in castings.

10.13 Why is it important to remove dross or slag during the pouring of molten metal into the mold? What methods are used to remove them?

10.14 Why is Bernoulli's equation important in casting?

10.15 Describe thixocasting and rheocasting.

10.16 What is Chvorinov's Rule?

10.17 How is a blister related to a scab?

QUALITATIVE PROBLEMS

10.18 Is there porosity in a chocolate bar? In an ice cube? Explain.

10.19 Describe the stages involved in the contraction of metals during casting.

10.20 Explain the effects of mold materials on fluid flow and heat transfer in casting operations.

10.21 It is known that pouring metal at a high rate into a mold can have certain disadvantages. Are there any disadvantages to pouring it very slowly?

10.22 Describe the events depicted in Fig. 10.5.

10.23 Would you be concerned about the fact that portions of internal chills are left within the casting? Explain.

10.24 Review Fig. 10.8 and make a summary, explaining the purpose of each feature shown and the consequences of omitting the feature from the mold design.

10.25 Make a sketch of volume vs. temperature for a metal that shrinks when it cools from the liquid state to room temperature. On the graph, mark the area where shrinkage is compensated by risers.

10.26 What practical demonstrations can you suggest to indicate the relationship of the solidification time to the volume and surface area of a casting?

10.27 Explain why a casting may have to be subjected to various heat treatments.

10.28 List and explain the reasons why porosity can develop in a casting.

10.29 Why does porosity have detrimental effects on the mechanical properties of castings? Would physical properties, such as thermal and electrical conductivity, also be adversely affected by porosity? Explain.

10.30 A spoked handwheel is to be cast in gray iron. In order to prevent hot tearing of the spokes, would you insulate the spokes or chill them? Explain.

10.31 Which of the following considerations are important for a riser to function properly? Must it: (a) have a surface area larger than the part being cast, (b) be kept open to atmospheric pressure, and/or (c) solidify first? Explain.

10.32 Explain why the constant C in Eq. (10.7) depends on mold material, metal properties, and temperature.

10.33 Are external chills as effective as internal chills? Explain.

10.34 Explain why, as shown in Table 10.1, gray cast iron undergoes expansion rather than contraction during solidification.

10.35 Referring to Fig. 10.11, explain why internal corners, such as A, develop a thinner skin than external corners, such as B, during solidification.

10.36 Note the shape of the two risers shown in Fig. 10.8, and discuss your observations with respect to Eq. (10.7).

10.37 Is there any difference in the tendency for shrinkage void formation in metals with short and long freezing ranges, respectively? Explain.

10.38 What is the influence of the cross-sectional area of the spiral channel, shown in Fig. 10.9, on fluidity test results? What is the effect of sprue height? If this test is run with the entire test setup heated to elevated temperatures, would the results be more useful? Explain.

10.39 It has long been observed that (a) low pouring temperatures (i.e., low superheat) promote the formation of equiaxed grains over columnar grains and (b) equiaxed grains become finer as the pouring temperature decreases. Explain these two phenomena.

10.40 In casting metal alloys, what would you expect to occur if the mold were agitated (vibrated) aggressively after the molten metal had been in the mold for a sufficient period of time to form a skin?

10.41 If you inspect a typical cube of ice, you are likely to see air pockets and cracks in the cube. Some ice cubes, however, are tubular in shape and do not have noticeable air pockets or cracks in their structure. Explain this phenomenon.

10.42 How can you tell whether cavities in a casting are due to shrinkage or entrained air bubbles?

10.43 Describe the drawbacks to having a riser that is (a) too large and (b) too small.

10.44 Reproduce Fig. 10.2 for a casting that is spherical in shape.

10.45 List the process variables that affect the fluidity index shown in Fig. 10.9.

10.46 Assume that you have a method of measuring porosity in a casting. Could you use this information to accurately predict the strength of the casting? Explain.

QUANTITATIVE PROBLEMS

10.47 A round casting is 0.2 m (7.9 in.) in diameter and 0.75 m (29.5 in.) in length. Another casting of the same metal is elliptical in cross-section, with a major-to-minor axis ratio of 2, and has the same length and cross-sectional area as the round casting. Both pieces are cast under the same conditions. What is the difference in the solidification times of the two castings?

10.48 A cylinder with a diameter of 2.0 in. and a height of 3 in. solidifies in three minutes in a sand casting operation. What is the solidification time if the cylinder height is doubled? What is the time if the diameter is doubled?

10.49 The constant C in Chvorinov's rule is given as 2.5 s/mm^2, and is used to produce a cylindrical casting with a diameter of 50 mm and height of 125 mm. Estimate the time for the casting to fully solidify. The mold can be broken safely when the solidified shell is at least 20 mm. Assuming that the cylinder cools evenly, how much time must transpire after pouring the molten metal before the mold can be broken?

10.50 Pure copper is poured into a sand mold. The metal level in the pouring basin is 10 in. above the metal level in the mold, and the runner is circular with a 0.4-in. diameter. What are the velocity and rate of the flow of the metal into the mold? Is the flow turbulent or laminar?

10.51 For the sprue described in Problem 10.58, what runner diameter is needed to ensure a Reynolds number of 2000? How long will a 15 in^3 casting take to fill with such a runner?

10.52 Assume that you are an instructor covering the topics described in this chapter, and you are giving a quiz on the numerical aspects to test the understanding of the students. Prepare two quantitative problems and supply the answers to them.

10.53 When designing patterns for casting, pattern makers use special rulers that automatically incorporate solid shrinkage allowances into their designs; therefore, a 100-mm patternmaker's ruler is longer than 100 mm. How long should a patternmaker's ruler be for (1) aluminum castings, (2) malleable cast iron, and (3) high manganese steel?

SYNTHESIS, DESIGN, AND PROJECTS

10.54 Can you devise fluidity tests other than that shown in Fig. 10.9? Explain the features of your test methods.

10.55 Figure P10.55 indicates various defects and discontinuities in cast products. Review each defect and offer solutions to prevent it.

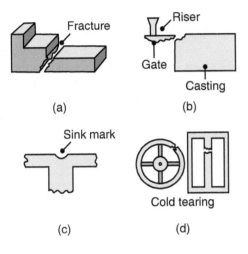

FIGURE P10.55

10.56 The fluidity test shown in Fig. 10.9 illustrates only the principle of this test. Design a setup for such a test, showing the type of materials and the equipment to be used. Explain the method by which you would determine the length of the solidified metal in the spiral passage.

10.57 Utilizing the equipment and materials available in a typical kitchen, design an experiment to reproduce results similar to those shown in Fig. 10.11. Comment on your observations.

10.58 One method of relieving stress concentrations in a part is to apply a small, uniform plastic deformation to it. Make a list of your concerns and recommendations if such an approach is suggested for a casting.

10.59 Describe the effects on mold design, including the required change in the size of the risers, runners, chokes, and sprues, for a casting of a given shape that is to be doubled in volume.

10.60 Small amounts of slag often persist after skimming and are introduced into the molten-metal flow in casting. Recognizing that the slag is much less dense than the metal, design mold features that will remove small amounts of slag before the metal reaches the mold cavity.

10.61 Figure II.2 shows a variety of components in a typical automobile that are produced by casting. Think of other products, such as power tools and small appliances, and prepare an illustration similar to the figure.

10.62 Design an experiment to measure the constants C and n in Chvorinov's rule, Eq. (10.7). Describe the features of your design, and comment on any difficulties that might be encountered in running such an experiment.

Metal-casting Processes and Equipment

- Building upon the fundamentals of solidification, fluid flow, and heat transfer described in the preceding chapter, this chapter describes the industrial casting processes.
- Casting processes are generally categorized as permanent-mold and expendable-mold processes; expendable-mold processes are further categorized as permanent-mold and expendable-pattern processes.
- The characteristics of each process are described, together with typical applications, advantages, and limitations.
- Special casting processes that produce single-crystal components as well as amorphous alloys are then described.
- The chapter ends with a description of inspection techniques for castings.

Typical products made by casting: Engine blocks, crankshafts, power tool housings, turbine blades, plumbing parts, zipper teeth, dies and molds, gears, railroad wheels, propellers, office equipment, and statues. Casting is extremely versatile and suitable for a wide variety of products.

Alternative processes: Forging, powder metallurgy, machining, rapid prototyping, and fabrication.

11.1 Introduction

The first metal castings were made during the period from 4000 to 3000 B.C., using stone or metal molds for casting copper. Various casting processes have been developed over time, each with its own characteristics and applications (see also Fig. I.5a), to meet specific design requirements (Table 11.1). A very wide variety of parts and components are made by casting (Fig. 11.1), such as engine blocks, crankshafts, automotive components and powertrains, agricultural and railroad equipment, pipes and plumbing fixtures, power-tool housings, gun barrels, frying pans, jewelry, orthopedic implants, and very large components for hydraulic turbines.

Two trends have had a major impact on the casting industry. The first is the mechanization and automation of casting operations, which has led to significant changes in the use of equipment and labor. Advanced machinery and automated process-control systems have replaced traditional methods of casting. The second major trend has been the increasing demand for high-quality castings with close dimensional tolerances.

This chapter is organized around the major classifications of casting practices (given in Fig. II.3 in the Introduction to Part II). These classifications are related to

TABLE 11.1

Summary of Casting Processes		
Process	Advantages	Limitations
Sand	Almost any metal can be cast; no limit to part size, shape, or weight; low tooling cost	Some finishing required; relatively coarse surface finish; wide tolerances
Shell mold	Good dimensional accuracy and surface finish; high production rate	Part size limited; expensive patterns and equipment
Evaporative pattern	Most metals can be cast, with no limit to size; complex part shapes	Patterns have low strength and can be costly for low quantities
Plaster mold	Intricate part shapes; good dimensional accuracy and surface finish; low porosity	Limited to nonferrous metals; limited part size and volume of production; mold-making time relatively long
Ceramic mold	Intricate part shapes; close-tolerance parts; good surface finish; low cooling rate	Limited part size
Investment	Intricate part shapes; excellent surface finish and accuracy; almost any metal can be cast	Part size limited; expensive patterns, molds, and labor
Permanent mold	Good surface finish and dimensional accuracy; low porosity; high production rate	High mold cost; limited part shape and complexity; not suitable for high-melting-point metals
Die	Excellent dimensional accuracy and surface finish; high production rate	High die cost; limited part size; generally limited to nonferrous metals; long lead time
Centrifugal	Large cylindrical or tubular parts with good quality; high production rate	Expensive equipment; limited part shape

mold materials, pattern production, molding processes, and methods of feeding the mold with molten metal. The major categories are:

1. **Expendable molds,** typically made of sand, plaster, ceramics, and similar materials, and generally mixed with various binders (*bonding agents*) for improved properties. A typical sand mold consists of 90% sand, 7% clay, and 3% water. As described in Chapter 8, these materials are *refractories* (i.e., they are capable of withstanding the high temperatures of molten metals). After the casting has solidified, the mold is broken up to remove the casting, hence the word *expendable*.

   The mold is produced from a pattern; in some processes, and although the mold is expendable, the pattern is reused to produce several molds. Such processes are referred to as *expendable-mold, permanent-pattern casting processes*. On the other hand, investment casting requires a pattern for each mold produced, and is an example of an *expendable-mold, expendable-pattern process*.

2. **Permanent molds,** made of metals that maintain their strength at high temperatures. As the name implies, the molds are used repeatedly, and are designed in such a way that the casting can be removed easily and the mold used for the next casting. Metal molds are better heat conductors than expendable nonmetallic molds (see Table 3.1), hence the solidifying casting is subjected to a higher rate of cooling, which in turn affects the microstructure and grain size within the casting.

3. **Composite molds,** made of two or more different materials (such as sand, graphite, and metal), combining the advantages of each material. These molds have a permanent and an expendable portion, and are used in various casting processes to improve mold strength, control the cooling rate, and optimize the overall economics of the casting operation.

(a)

(b)

(c)

(d)

FIGURE 11.1   (a) Typical gray-iron castings used in automobiles, including the transmission valve body (left) and the hub rotor with disk-brake cylinder (front). *Source:* Courtesy of Central Foundry Division of General Motors Corporation. (b) Die-cast magnesium housing for the Olympus E-3 camera; *Source:* Courtesy of Olympus Inc. (c) A cast transmission housing. (d) Cast aluminum impellers for automotive turbochargers. *Source:* (c) and (d) Courtesy of American Foundry Society.

The general characteristics of sand casting and other casting processes are summarized in Table 11.2. As it can be seen, almost all commercial metals can be cast. The surface finish obtained is largely a function of the mold material, and can be very good, although, as expected, sand castings generally have rough, grainy surfaces. Dimensional tolerances generally are not as good as those in machining and other net-shape processes; however, intricate shapes, such as engine blocks and very large propellers for ocean liners, can be made by casting.

Because of their unique characteristics and applications, particularly in manufacturing microelectronic devices (described in Part V), basic crystal-growing techniques also are included in this chapter, which concludes with a brief overview of modern foundries.

TABLE 11.2

General Characteristics of Casting Processes

	Sand	Shell	Evaporative pattern	Plaster	Investment	Permanent mold	Die	Centrifugal
Typical materials cast	All	All	All	Nonferrous (Al, Mg, Zn, Cu)	All	All	Nonferrous (Al, Mg, Zn, Cu)	All
Weight (kg):								
Minimum	0.01	0.01	0.01	0.01	0.001	0.1	<0.01	0.01
Maximum	No limit	100+	100+	50+	100+	300	50	5000+
Typical surface finish (R_a in μm)	5–25	1–3	5–25	1–2	0.3–2	2–6	1–2	2–10
Porosity[1]	3–5	4–5	3–5	4–5	5	2–3	1–3	1–2
Shape complexity[1]	1–2	2–3	1–2	1–2	1	2–3	3–4	3–4
Dimensional accuracy[1]	3	2	3	2	1	1	1	3
Section thickness (mm):								
Minimum	3	2	2	1	1	2	0.5	2
Maximum	No limit	—	—	—	75	50	12	100
Typical dimensional tolerance (mm)	1.6–4 mm (0.25 mm for small parts)	±0.003		±0.005–0.010	±0.005	±0.015	± 0.001–0.005	0.015
Equipment	3–5	3	2–3	3–5	3–5	2	1	1
Pattern/die	3–5	2–3	2–3	3–5	2–3	2	1	1
Labor	1–3	3	3	1–2	1–2	3	5	5
Typical lead time[2]	Days	Weeks	Weeks	Days	Weeks	Weeks	Weeks to months	Months
Typical production rate[2] (parts/mold-hour)	1–20	5–50	1–20	1–10	1–1000	5–50	2–200	1–1000
Minimum quantity[2]	1	100	500	10	10	1000	10,000	10–10,000

Notes: 1. Relative rating, from 1 (best) to 5 (worst). For example, die casting has relatively low porosity, mid to low shape complexity, high dimensional accuracy, high equipment and die costs, and low labor costs. These ratings are only general; significant variations can occur, depending on the manufacturing methods used.
2. Approximate values without the use of rapid-prototyping technologies; minimum quantity is 1 when applying rapid prototyping.
Source: Data taken from J.A. Schey, *Introduction to Manufacturing Processes*, 3rd ed., McGraw-Hill, 2000.

11.2 Expendable-mold, Permanent-pattern Casting Processes

The major categories of expendable-mold, permanent-pattern casting processes are sand, shell mold, plaster mold, ceramic mold, and vacuum casting.

11.2.1 Sand Casting

The traditional method of casting metals is in sand molds and has been used for millennia. Sand casting is still the most prevalent form of casting; in the United States alone, about 15 million metric tons of metal are cast by this method each year. Typical applications of sand casting include machine bases, large turbine impellers, propellers, plumbing fixtures, and a wide variety of other products and components. The capabilities of sand casting are given in Table 11.2.

Sand casting basically consists of (a) placing a pattern, having the shape of the part to be cast, in sand to make an imprint, (b) incorporating a gating system, (c) removing the pattern and filling the mold cavity with molten metal, (d) allowing the metal to cool until it solidifies, (e) breaking away the sand mold, and (f) removing the casting (Fig. 11.2).

Sands.   Most sand-casting operations use silica sand (SiO_2) as the mold material. Sand is inexpensive and is suitable as a mold material because of its high-temperature characteristics and high melting point. There are two general types of sand: **naturally bonded** (*bank sand*) and **synthetic** (*lake sand*). Because its composition can be controlled more accurately, synthetic sand is preferred by most foundries.

Several factors are important in the selection of sand for molds, and certain trade-offs with respect to properties have to be considered. Sand having fine, round grains can be packed closely and thus forms a smooth mold surface. Although fine-grained sand enhances mold strength, the fine grains also lower mold *permeability* (where fluids and gases can pass through pores). Good permeability of molds and cores allows gases and steam evolved during the casting process to escape easily. The mold also

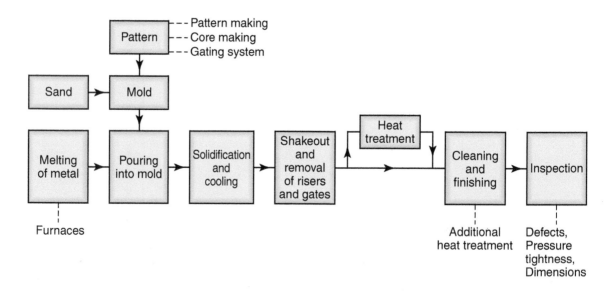

FIGURE 11.2   Outline of production steps in a typical sand-casting operation.

should have good *collapsibility*, in order to allow the casting to shrink while it is cooling, and thus prevent defects in the casting, such as hot tearing and cracking shown in Fig. 10.12.

Types of Sand Molds. Sand molds (Fig. 11.3) are characterized by the types of sand that comprise them and by the methods used to produce them. There are three basic types of sand molds: (a) green-sand, (b) cold-box, and (c) no-bake molds. The most common mold material is **green molding sand**, a mixture of sand, clay, and water. The term "green" refers to the fact that the sand in the mold is moist or damp while the metal is poured into it.

Green-sand molding is the least expensive method of making molds, and the sand is recycled easily for subsequent reuse. In the *skin-dried* method, the mold surfaces are dried, either by storing the mold in air or by drying it with torches. Because of their higher strength, these molds are generally used for large castings.

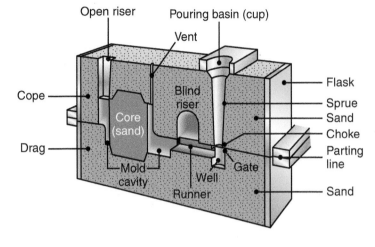

FIGURE 11.3 Schematic illustration of a sand mold, showing various features.

In the **cold-box mold** process, various organic and inorganic *binders* are blended into the sand to bond the grains chemically for greater strength. These molds are more dimensionally accurate than green-sand molds, but are more expensive to make. In the no-bake mold process, a synthetic liquid resin is mixed with the sand, and the mixture hardens at room temperature. Because the bonding of the mold in this and in the cold-box process takes place without heat, they are called **cold-setting processes.**

Sand molds can be oven dried (*baked*) prior to pouring the molten metal; they then become stronger than green-sand molds and impart better dimensional accuracy and surface finish to the casting. However, this method has drawbacks in that (a) distortion of the mold is greater; (b) the castings are more susceptible to hot tearing, because of the lower collapsibility of the mold; and (c) the production rate is lower, because of the considerable drying time required.

The major features of molds in sand casting are:

1. The **flask**, which supports the mold itself. Two-piece molds consist of a **cope** on top and a **drag** on the bottom; the seam between them is the *parting line*. When more than two pieces are used in a sand mold, the additional parts are called *cheeks*.
2. A **pouring basin** or **pouring cup**, into which the molten metal is poured.
3. A **sprue**, through which the molten metal flows downward.
4. The **runner system**, which has channels that carry the molten metal from the sprue to the mold cavity. **Gates** are the inlets into the mold cavity.
5. **Risers**, which supply additional molten metal to the casting as it shrinks during solidification. Two types of risers—a *blind riser* and an *open riser*—are shown in Fig. 11.3.
6. **Cores**, which are inserts made from sand; they are placed in the mold to form hollow regions or otherwise define the interior surface of the casting. Cores also are used on the outside of the casting to form features, such as lettering or numbering.

QR Code 11.1 Sand core production. (*Source:* Courtesy of Alcast Technologies, Ltd.)

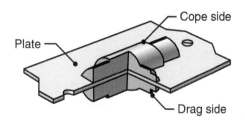

FIGURE 11.4 A typical metal match-plate pattern used in sand casting.

QR Code 11.2 Sand casting using manual molding methods. (*Source:* Courtesy of Alcast Technologies, Ltd.)

7. **Vents,** which are placed in molds to carry off gases produced when the molten metal comes into contact with the sand in the mold and the core. Vents also exhaust air from the mold cavity as the molten metal flows into the mold.

Patterns.   *Patterns* are used to mold the sand mixture into the shape of the casting, and may be made of wood, plastic, or metal. The selection of a pattern material depends on the size and shape of the casting, the dimensional accuracy and the quantity of castings required, and the molding process. Because patterns are used repeatedly to make molds, the strength and durability of the material selected for a pattern must reflect the number of castings that the mold will produce. Patterns may also be made of a combination of materials to reduce wear in critical regions; they usually are coated with a **parting agent** to facilitate the removal of the pattern from the molds.

Patterns can be designed with a variety of features for specific applications and economic requirements. **One-piece patterns,** also called *loose* or *solid patterns,* are generally used for simpler shapes and low-quantity production; they generally are made of wood and are inexpensive. **Split patterns** are two-piece patterns, made such that each part forms a portion of the cavity for the casting; in this way, castings with complicated shapes can be produced. **Match-plate patterns** are a common type of mounted pattern in which two-piece patterns are constructed by securing each half of one or more split patterns to the opposite sides of a single plate (Fig. 11.4). In such constructions, the gating system can be mounted on the drag side of the pattern. This type of pattern is used most often in conjunction with molding machines and for large production runs to produce smaller castings.

An important development in mold and pattern making is the application of **rapid prototyping** (Chapter 20). In sand casting, for example, a pattern can be fabricated in a rapid-prototyping machine, and fastened to a backing plate at a fraction of the time and the cost of machining a pattern. There are several rapid-prototyping techniques with which these tools can be produced quickly. These technologies are best suited for small production runs.

Pattern design is a critical aspect of the total casting operation. The design should provide for **metal shrinkage,** permit proper metal flow in the mold cavity, and allow the pattern to be easily removed from the sand mold, by means of a taper or draft (Fig. 11.5) or some other geometric feature. (These topics are described in greater detail in Chapter 12.)

Cores.   For castings with internal cavities or passages, such as those found in automotive engine blocks or valve bodies, *cores* are utilized. Cores are placed in the mold cavity to form the interior surfaces of the casting, and are removed from the finished part during shakeout and further processing. Like molds, cores must possess strength, permeability, collapsibility, and the ability to withstand heat; hence cores are made of sand aggregates.

The core is anchored by **core prints,** which are geometric features added to the pattern to locate and support the core and to provide vents for the escape of gases (Fig. 11.6a). A common problem with cores is that, for some casting requirements, as in the case where a recess is required, they may

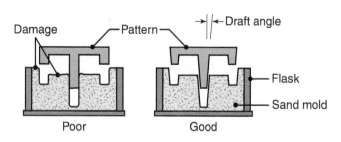

FIGURE 11.5 Taper on patterns for ease of removal from the sand mold.

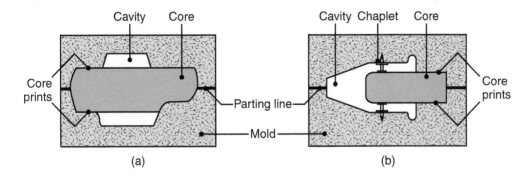

FIGURE 11.6   Examples of sand cores, showing core prints and chaplets to support the cores.

lack sufficient structural support in the cavity. To keep the core from shifting, metal supports (**chaplets**) may be used to anchor the core in place (Fig. 11.6b).

Cores are generally made in a manner similar to that used in sand moldmaking; the majority are made with shell (see Section 11.2.2), no-bake, or cold-box processes. Cores are shaped in *core boxes*, which are used in much the same way that patterns are used to form sand molds.

Sand-molding Machines.   The oldest known method of molding, which is still used for simple castings and for small production runs, is to compact the sand by hand hammering (tamping) or ramming it around the pattern. For most operations, however, the sand mixture is compacted around the pattern by *molding machines*. These machines eliminate arduous labor, manipulate the mold in a controlled manner, offer high-quality casting by improving the application and distribution of forces, and increase production rate.

In **vertical flaskless molding**, the pattern halves form a vertical chamber wall against which sand is blown and compacted (Fig. 11.7). Then, the mold halves are packed horizontally with the parting line oriented vertically, and moved along a pouring conveyor. This is a simple operation and eliminates the need to handle

QR Code 11.3 Sand casting using automated molding methods. (*Source:* Courtesy of Alcast Technologies, Ltd.)

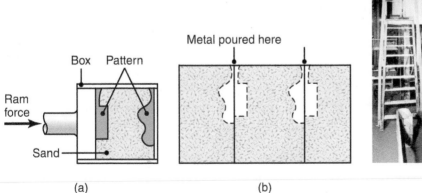

FIGURE 11.7   Vertical flaskless molding. (a) Sand is squeezed between two halves of the pattern. (b) Assembled molds pass along an assembly line for pouring. (c) A photograph of a vertical flaskless molding line. *Source:* Courtesy of American Foundry Society.

flasks, allowing for very high production rates, particularly when other aspects of the operation (such as coring and pouring) are automated.

Sandslingers fill the flask uniformly with sand under a high-pressure stream; they are used to fill large flasks and are often automated. An impeller in the machine throws sand from its blades or cups at such high speeds that the machine not only places the sand, but also rams it appropriately.

In **impact molding**, the sand is compacted by a controlled explosion or instantaneous release of compressed gases. This method produces molds with uniform strength and good permeability.

In **vacuum molding** (also known as the *V process*), shown in Fig. 11.8, the pattern is covered tightly with a thin sheet of plastic. A flask is placed over the covered pattern and is filled with dry, binderless sand. A second sheet of plastic then is placed on top of the sand, and a vacuum action compacts the sand; the pattern can then be removed. Both halves of the mold are made in this manner and subsequently assembled. During pouring of the molten metal, the mold remains under vacuum, but the casting cavity does not. When the metal has solidified, the vacuum is turned off and the sand falls away, releasing the casting.

As shown in Fig. 11.8, vacuum molding does not require a draft in the part, and can be very economical because of the low tooling costs, long pattern life, and absence of binders in the sand (which also simplifies sand recovery and reuse). Vacuum molding produces castings with high-quality surface detail and dimensional accuracy; it is suited especially well for large, relatively flat (plane) castings.

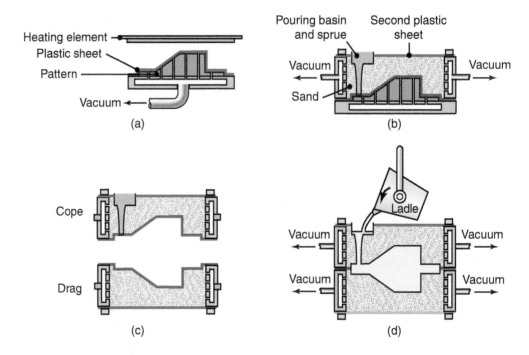

FIGURE 11.8 The vacuum molding process. (a) A plastic sheet is thermoformed (see Section 19.6) over a pattern; (b) a vacuum flask is placed over the pattern, a pouring basin/sprue insert is located, and the flask is filled with sand. A second sheet is located on the top of the sand mold, and vacuum is applied to tightly compact the sand against the pattern. (c) A drag is also produced, along with cheeks, cores, etc., as in conventional sand casting; the cope and drag can be carefully transported without vacuum applied. (d) After the mold halves are joined, vacuum is applied to ensure mold strength, and molten metal is poured into the mold.

The Sand-casting Operation.   After the mold has been shaped and the cores have been placed in position, the two mold halves (cope and drag) are closed, clamped, and weighted down, to prevent the separation of the mold sections under the pressure exerted when the molten metal is poured into the mold cavity. A complete sequence of operations in sand casting is shown in Fig. 11.9.

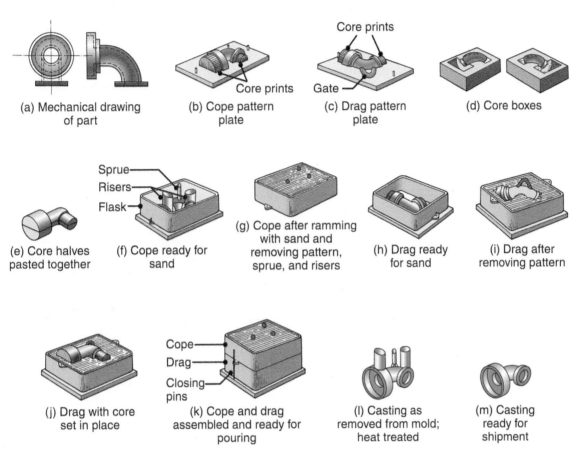

FIGURE 11.9   Schematic illustration of the sequence of operations for sand casting. (a) A mechanical drawing of the part is used to generate a design for the pattern. Considerations such as part shrinkage and draft must be built into the drawing. (b) and (c) Patterns have been mounted on plates equipped with pins for alignment; note the presence of core prints designed to hold the core in place. (d) and (e) Core boxes produce core halves, which are pasted together; the cores will be used to produce the hollow area of the part shown in (a). (f) The cope half of the mold is assembled by securing the cope pattern plate to the flask with aligning pins and attaching inserts to form the sprue and risers. (g) The flask is rammed with sand, and the plate and inserts are removed. (h) The drag half is produced in a similar manner with the pattern inserted; a bottom board is placed below the drag and aligned with pins. (i) The pattern, flask, and bottom board are inverted, and the pattern is withdrawn, leaving the appropriate imprint. (j) The core is set in place within the drag cavity. (k) The mold is closed by placing the cope on top of the drag and securing the assembly with pins; the flasks are then subjected to pressure to counteract buoyant forces in the liquid, which might lift the cope. (l) After the metal solidifies, the casting is removed from the mold. (m) The sprue and risers are cut off and recycled, and the casting is cleaned, inspected, and heat treated (when necessary). *Source:* Courtesy of Steel Founders' Society of America.

After it solidifies, the casting is shaken out of its mold, and the sand and oxide layers adhering to the casting are removed (by vibration, using a shaker, or by sand blasting). Castings are also cleaned by blasting with steel shot or grit (*shot blasting*; Section 26.8). The risers and gates are cut off by oxyfuel-gas cutting, sawing, shearing, or abrasive wheels; or they are trimmed in dies. Gates and risers on steel castings also may be removed with air carbon-arc cutting (Section 30.8) or torches. Castings may be cleaned further by electrochemical means or by pickling with chemicals to remove surface oxides.

The casting subsequently may be *heat treated* to improve certain properties required for its intended use; heat treatment is particularly important for steel castings. *Finishing operations* may involve machining, straightening, or forging with dies (sizing) to obtain final dimensions. *Inspection* is an important final step, and is carried out to ensure that the casting meets all design and quality-control requirements.

Rammed-graphite Molding.   In this process, rammed graphite (Section 8.6) is used to make molds for casting reactive metals, such as titanium and zirconium. Sand cannot be used because these metals react vigorously with silica. The molds are packed like sand molds, air dried, baked at 175°C (350°F), fired at 870°C (1600°F), and then stored under controlled humidity and temperature. The casting procedures are similar to those for sand molds.

Mold Ablation.   Ablation has been used to improve the mechanical properties and production rates in sand casting. In this process, a sand mold is filled with molten metal, and the mold is then immediately sprayed with a liquid and/or gas solvent, to progressively erode the sand. As the mold is removed, the liquid stream causes rapid and directional solidification of the metal. With proper risers, mold ablation results in significantly lower porosity than conventional sand casting, leading to higher strength and ductility, and has therefore been applied to normally difficult-to-cast materials or metal-matrix composites. Since ablation speeds up solidification and also removes cores, significant productivity improvements can also be achieved.

11.2.2   Shell Molding

Shell molding, first developed in the 1940s, has grown significantly because it can produce many types of castings, with close dimensional tolerances and good surface finish, and at low cost. Shell-molding applications include small mechanical parts requiring high precision, such as gear housings, cylinder heads, and connecting rods. The process is also used widely in producing high-precision molding cores. The capabilities of shell-mold casting are given in Table 11.2.

In this process, a mounted pattern, made of a ferrous metal or aluminum, is (a) heated to a range of 175°–370°C (350°–700°F), (b) coated with a parting agent (such as silicone), and (c) clamped to a box or chamber. The box contains fine sand, mixed with 2.5–4% of a thermosetting resin binder (such as phenol-formaldehyde), which coats the sand particles. Either the box is rotated upside down (Fig. 11.10) or the sand mixture is blown over the pattern, allowing it to form a coating.

The assembly is then placed in an oven for a short period of time to complete the curing of the resin. In most shell-molding machines, the oven consists of a metal box, with gas-fired burners that swing over the shell mold to cure it. The shell hardens around the pattern and is removed from the pattern using built-in ejector pins. Two half-shells are made in this manner and are bonded or clamped together to form a mold.

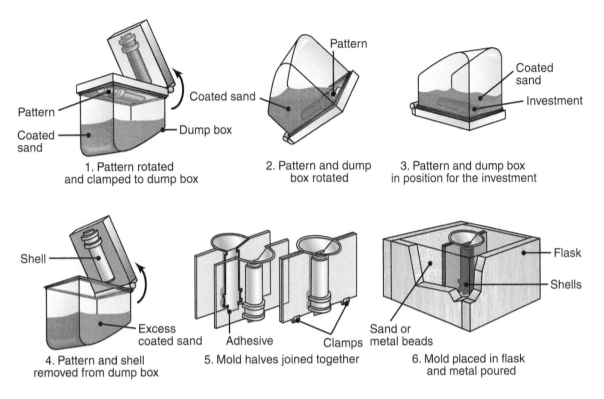

FIGURE 11.10   The shell-molding process, also called the dump-box technique.

The thickness of the shell can be determined accurately by controlling the time that the pattern is in contact with the mold. In this way, the shell can be formed with the required strength and rigidity, in order to hold the weight of the molten liquid. The shells are light and thin—usually 5 to 10 mm (0.2–0.4 in.)—and, consequently, their thermal characteristics are different from those for thicker molds.

Shell sand has a much smaller grain size, and therefore a lower permeability than the sand used for green-sand molding. The decomposition of the shell-sand binder also produces a high volume of gas. Thus, unless the molds are vented properly, trapped air and gas can result in defects in the shell molding of ferrous castings. The high quality of the finished casting can reduce cleaning, machining, and other finishing costs significantly. Complex shapes can be produced with less labor, and the process can be automated fairly easily.

11.2.3 Plaster-mold Casting

This process, and the ceramic-mold and investment casting processes described in Sections 11.2.4 and 11.3.2, are known as **precision casting**, because of the high dimensional accuracy and good surface finish obtained. Typical parts made are lock components, gears, valves, fittings, tooling, and ornaments. The castings are typically in the range of 125–250 g (0.25–0.50 lb), although parts as light as 1 g (0.035 oz) have been made. The capabilities of plaster-mold casting are given in Table 11.2.

In the *plaster-molding process*, the mold is made of plaster of paris (gypsum or calcium sulfate), with the addition of talc and silica powder to improve strength and to control the time required for the plaster to set. These components are mixed with water, and the resulting slurry is poured over the pattern. After the plaster sets (usually within 15 min), it is removed and the mold is dried at a temperature range of 120°–260°C (250°–500°F); higher drying temperatures may be used, depending on the type of plaster. The mold halves are assembled to form the mold cavity and are preheated to about 120°C (250°F). The molten metal is then poured into the mold.

Because plaster molds have very low permeability, gases evolved during solidification of the metal cannot escape; consequently, the molten metal is poured either in a vacuum or under pressure. Mold permeability can be increased substantially by the *Antioch process*, in which the molds are dehydrated in an autoclave (pressurized oven) for 6–12 h, and then rehydrated in air for 14 h. Another method of increasing the permeability of the mold is to use foamed plaster, containing trapped air bubbles.

Patterns for plaster molding generally are made of aluminum alloys, thermosetting plastics, brass, or zinc alloys. Wood patterns are not suitable for making a large number of molds, because they are repeatedly in contact with the water-based plaster slurry and thus warp or degrade quickly. Since there is a limit to the maximum temperature that the plaster mold can withstand (generally about 1200°C; 2200°F), plaster-mold casting is used only for aluminum, magnesium, zinc, and some copper-based alloys. The castings have good surface finish with fine details. Because plaster molds have lower thermal conductivity than other mold materials, the castings cool slowly, and thus a more uniform grain structure is obtained, with less warpage. The wall thickness of the cast parts can be as thin as 1 to 2.5 mm (0.04–0.1 in.).

11.2.4   Ceramic-mold Casting

The *ceramic-mold casting* process (also called *cope-and-drag investment casting*) is similar to the plaster-mold process, except that it uses refractory mold materials suitable for high-temperature exposure. Typical parts made are impellers, cutters for machining operations, dies for metalworking, and molds for making plastic and rubber components. Parts weighing as much as 700 kg (1500 lb) have been cast by this process. The slurry is a mixture of fine-grained zircon ($ZrSiO_4$), aluminum oxide, and fused silica, which are mixed with bonding agents and poured over the pattern (Fig. 11.11) which has been placed in a flask.

The pattern may be made of wood or metal. After setting, the molds (ceramic facings) are removed, dried, ignited to burn off volatile matter, and baked. The molds are clamped firmly and used as all-ceramic molds. In the *Shaw process*, the ceramic facings are backed by fireclay (which resists high temperatures) to give strength to the mold. The facings are then assembled into a complete mold, ready to be poured.

The high-temperature resistance of the refractory molding materials allows these molds to be used for casting ferrous and other high-temperature alloys, stainless

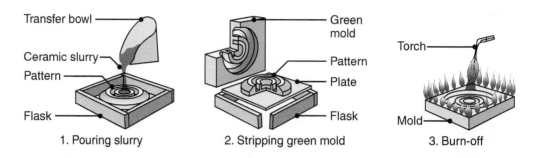

FIGURE 11.11   Sequence of operations in making a ceramic mold. *Source: Metals Handbook,* Vol. 5, 8th ed., ASM International, 1970.

steels, and tool steels. Although the process is somewhat expensive, the castings have good dimensional accuracy and surface finish over a wide range of sizes and intricate shapes.

11.3   Expendable-mold, Expendable-pattern Casting Processes

Evaporative-pattern and investment casting also are referred to as *expendable-pattern* casting processes or *expendable mold–expendable pattern* processes. They are unique in that a mold and a pattern must be produced for each casting, whereas the patterns in the processes described in the preceding section are reusable. Typical applications are cylinder heads, engine blocks, crankshafts, brake components, manifolds, and machine bases.

11.3.1   Evaporative-pattern Casting (Lost-foam Process)

The *evaporative-pattern casting* (EPC) process uses a polystyrene pattern, which evaporates upon contact with molten metal to form a cavity for the casting; this process is also known as *lost-foam casting*, or *full-mold casting* (FMC) process. It has become one of the more important casting processes for ferrous and nonferrous metals, particularly for the automotive industry.

In this process, polystyrene beads, containing 5–8% pentane (a volatile hydro-carbon), are placed in a preheated die, usually made of aluminum. Complex patterns may be made by bonding various individual pattern sections using hot-melt adhesive (Section 32.4.1). Polymethylmethacrylate (PMMA) and polyalkylene carbonate also may be used as pattern materials for ferrous castings.

The polystyrene expands and takes the shape of the die cavity; additional heat is applied to fuse and bond the beads together. The die is then cooled and opened, and the polystyrene pattern is removed. The pattern is then coated with a water-based refractory slurry, dried, and placed in a flask. The flask is filled with loose, fine sand, which surrounds and supports the pattern (Fig. 11.12), and may be dried or mixed with bonding agents to give it additional strength. The sand is compacted periodically, without removing the polystyrene pattern; then the molten metal is poured into the mold. The molten metal vaporizes the pattern and fills the mold cavity, completely replacing the space previously occupied by the polystyrene. Any degradation products from the polystyrene are vented into the surrounding sand.

Because the polymer requires considerable energy to degrade, large thermal gradients are present at the metal–polymer interface; in other words, the molten metal cools faster than it would if it were poured directly into an empty cavity. Consequently, fluidity is less than in sand casting. This has important effects on the microstructure throughout the casting, and also leads to directional solidification of the metal.

The evaporative-pattern process has several advantages over other casting methods:

- The process is relatively simple, because there are no parting lines, cores, or riser systems.
- Inexpensive flasks are used for the process.
- Polystyrene is inexpensive, and can be processed easily into patterns having complex shapes, various sizes, and fine surface detail.
- The casting requires minimal finishing and cleaning operations.

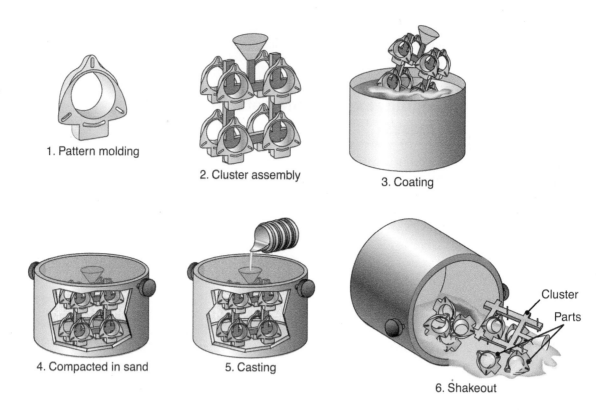

1. Pattern molding

2. Cluster assembly

3. Coating

4. Compacted in sand

5. Casting

6. Shakeout

Cluster

Parts

FIGURE 11.12 Schematic illustration of the expendable-pattern casting process, also known as lost-foam or evaporative-pattern casting.

- The process can be automated and is economical for long production runs; however, the cost to produce the die and the need for two sets of tooling are significant factors.

In a modification of the evaporative-pattern process, called the *Replicast® C-S process*, a polystyrene pattern is surrounded by a ceramic shell; then the pattern is burned out prior to pouring the molten metal into the mold. Its principal advantage over investment casting (which uses wax patterns, Section 11.3.2) is that carbon pickup into the metal is avoided entirely. Further developments in EPC include the production of metal-matrix composites (Sections 9.5 and 19.14). During molding of the polymer pattern, fibers or particles are embedded throughout, which then become an integral part of the casting. Other techniques include the modification and grain refinement of the casting by using grain refiners and modifier master alloys.

CASE STUDY 11.1   Lost-foam Casting of Engine Blocks

One of the most important components in an internal combustion engine is the engine block. Industry trends have focused upon high-quality, low-cost, and lightweight designs. Economic benefits can be attained through casting more complex geometries and by incorporating multiple components into one part. Recognizing that evaporative-pattern casting can simultaneously satisfy all of these requirements, Mercury Castings built a lost-foam casting line to produce aluminum engine blocks and cylinder heads.

One example of a part produced through lost-foam casting is a 45-kW (40-hp), three-cylinder engine block used for marine applications, such as an outboard motor on a small boat illustrated in Fig. 11.13. Previously manufactured as eight separate die castings, the block was converted to a single 10-kg (22-lb) casting, with a weight and cost savings of 1 kg (2 lb) and $25, respectively, on each block. Lost-foam casting also allowed consolidation of the engine's cylinder head and the exhaust and cooling systems into the block, thus eliminating the associated machining operations and fasteners required in sand-cast or die-cast designs. In addition, since the pattern contained holes, which could be cast without the use of cores, numerous drilling operations were eliminated.

Mercury Marine also was in the midst of developing a new V6 engine, utilizing a new corrosion-resistant aluminum alloy with increased wear resistance. This engine design also required a cylinder block and head integration, featuring hollow sections for water jacket cooling that could not be cored out in die casting or semipermanent mold processes (which were used for its other V6 blocks). Based on the success the foundry had with the three-cylinder lost-foam block, engineers applied this process for casting the V6 die block (Fig. 11.13b). The new engine block involves only one casting, and is lighter and less expensive than the previous designs. Produced with an integrated cylinder head and exhaust and cooling system, this component is cast hollow to develop more efficient water jacket cooling of the engine during its operation.

The company also has developed a pressurized lost-foam process. First, a foam pattern is made, placed in a flask, and surrounded by sand. Then the flask is inserted into a pressure vessel, where a robot pours molten aluminum onto the polystyrene pattern. A lid on the pressure vessel is closed, and a pressure of 1 MPa (150 psi) is applied to the casting until it solidifies, in about 15 min. The result is a casting with better dimensional accuracy, lower porosity, and improved strength, as compared to conventional lost-foam casting.

Source: Courtesy of Mercury Marine.

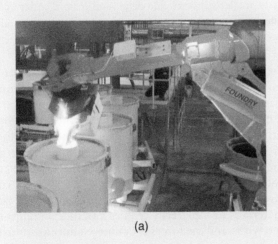

(a)   (b)   (c)

FIGURE 11.13   (a) Metal is poured into a mold for lost-foam casting of a 40-hp, three-cylinder marine engine; (b) finished engine block; and (c) completed outboard motor. *Source:* Mercury Marine.

11.3.2 Investment Casting

The *investment-casting* process, also called the **lost-wax process**, was first used during the period from 4000 to 3000 B.C. Typical parts made are components for office equipment and mechanical components, such as gears, cams, valves, and ratchets. Parts up to 1.5 m (60 in.) in diameter and weighing as much as 1140 kg (2500 lb) have been cast successfully by this process. The capabilities of investment casting are given in Table 11.2.

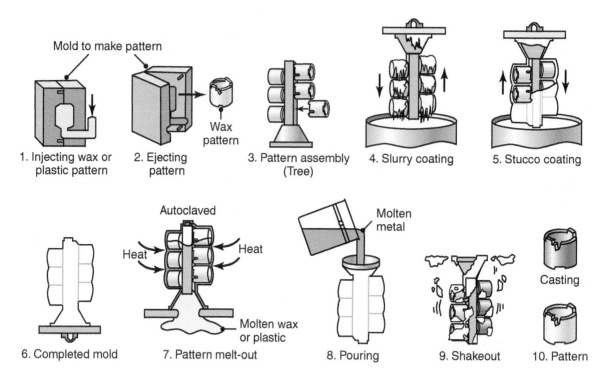

FIGURE 11.14 Schematic illustration of the investment-casting (lost-wax) process. Castings produced by this method can be made with very fine detail and from a variety of metals. *Source:* Courtesy of Steel Founders' Society of America.

Video Solution 11.1 Cast-in-place Gems

QR Code 11.4 Investment casting of sculptures. (*Source:* Courtesy of the National Sculpture Society)

The sequence involved in investment casting is shown in Fig. 11.14. The pattern is made of wax, or of a plastic such as polystyrene, by molding or rapid-prototyping techniques (Chapter 20). It is then dipped into a slurry of refractory material such as very fine silica and binders, including water, ethyl silicate, and acids. After this initial coating has dried, the pattern is coated repeatedly to increase its thickness, for higher strength. Note that smaller particles can be used for the initial coating to develop a better surface finish in the casting; subsequent layers use larger particles and are intended to increase the coating thickness quickly.

The term *investment* derives from the fact that the pattern is invested (surrounded) with the refractory material. Wax patterns require careful handling, because they are not strong enough to withstand the forces encountered during mold making; unlike plastic patterns, however, wax can be recovered and reused.

The one-piece mold is dried in air and heated to a temperature of 90°–175°C (200°–375°F); it is held in an inverted position for a few hours to melt out the wax. The mold is then fired to 650°–1050°C (1200°–1900°F) for about four hours (depending on the metal to be cast), to drive off the water of crystallization (chemically combined water) and to burn off any residual wax. After the metal has been poured and has solidified, the mold is broken up and the casting is removed.

A number of patterns can be joined to make one mold, called a **tree** (Fig. 11.14), significantly increasing the production rate. For small parts, the tree can be inserted into a permeable flask and filled with a liquid slurry investment. The investment is then placed into a chamber and evacuated (to remove the air bubbles in it) until the mold solidifies. The flask is usually placed in a vacuum-casting machine, so that the molten metal is drawn into the permeable mold producing fine detail.

Although the mold materials and labor involved make the lost-wax process costly, it is suitable for casting high-melting-point alloys, with good surface finish and close dimensional tolerances. Few or no finishing operations are required, which otherwise would add significantly to cost of the casting. The process is capable of producing intricate shapes from a wide variety of ferrous and nonferrous metals and alloys, with parts weighing from 1 g to 35 kg (0.035 oz to 75 lb). Advances include the casting of titanium aircraft-engine and structural airframe components, with wall thicknesses on the order of 1.5 mm (0.060 in.), thus competing with previously used sheet-metal structures.

FIGURE 11.15   Investment casting of an integrally cast rotor for a gas turbine. (a) Wax pattern assembly. (b) Ceramic shell around wax pattern. (c) Wax is melted out and the mold is filled, under a vacuum, with molten superalloy. (d) The cast rotor, produced to net or near-net shape. *Source:* Courtesy of Howmet Corporation.

Ceramic-shell Investment Casting.   A variation of the investment-casting process is *ceramic-shell casting*. It uses the same type of wax or plastic pattern, which is dipped first in ethyl silicate gel and subsequently into a fluidized bed (see Section 4.12) of fine-grained fused silica or zircon powder. The pattern is then dipped into coarser grained silica, to build up additional coatings and develop a proper thickness so that the pattern can withstand the thermal shock during pouring. The rest of the procedure is similar to investment casting. The process is economical and is used extensively for the precision casting of steels and high-temperature alloys.

The sequence of operations involved in making a turbine disk by this method is shown in Fig. 11.15. If ceramic cores are used in the casting, they are removed by leaching using caustic solutions under high pressure and temperature. The molten metal may be poured in a vacuum to extract evolved gases and reduce oxidation, thus improving the casting quality. To further reduce microporosity, the castings made by this, as well as other processes, are subjected to hot isostatic pressing.

CASE STUDY 11.2   Investment-cast Superalloy Components for Gas Turbines

Investment-cast superalloys have been replacing wrought counterparts in high-performance gas turbines since the 1960s. The microstructure of an integrally investment-cast gas-turbine rotor is shown in the upper half of Fig. 11.16; note the fine, uniform equiaxed grains throughout the rotor cross-section. Casting procedures include the use of a nucleant addition to the molten metal, as well as close control of its superheat, pouring techniques, and the cooling rate of the casting (see Section 10.2).

In contrast, note the coarse-grained structure in the lower half of the figure, showing the same type of rotor cast conventionally; this rotor has inferior properties as compared with the fine-grained rotor. Due to further developments in these processes, the proportion of cast parts to other parts in aircraft engines has increased from 20% to about 45% by weight.

(continued)

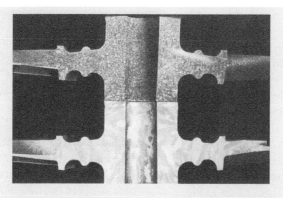

FIGURE 11.16 Cross-section and microstructure of two rotors: (top) investment cast; (bottom) conventionally cast. *Source:* Courtesy of ASM International.

CASE STUDY 11.3   Investment Casting of Total Knee Replacements

With major advances in medical care in the past few decades, life expectancies have increased significantly, but so have expectations for the quality of life in later years to remain high. One of the reasons for improvements in the past 40 years has been the great success of orthopedic implants. Hip, knee, shoulder, spine, and other implants have resulted in increased activity and reduced pain for millions worldwide.

An example of an orthopedic implant that has greatly improved quality of life is the *total knee replacement* (TKR), as shown in Fig. 11.17a. TKRs are very popular and reliable for the relief of osteoarthritis, a chronic and painful degenerative condition of the knee joint, that typically sets in after middle age. TKRs consist of multiple parts, including femoral, tibial, and patellar components. Typical materials used include cobalt alloys, titanium alloys, and ultrahigh-molecular-weight polyethylene (UHMWPE; Section 7.6). Each material is chosen for specific properties that are important in the application of the implant.

This case study describes the investment casting of femoral components of TKRs, which are produced from cobalt–chrome alloy (Section 6.6). The manufacturing process begins with injection molding of the patterns, which are then hand assembled onto trees, as shown in Fig. 11.17b. The patterns are spaced properly on a central wax sprue, and then are welded in place by dipping them into molten wax and pressing them against the sprue, until the patterns are held in place. The final assembled tree, shown in Fig. 11.18a, contains 12 knee implants arranged into four rows.

The completed trees are then placed in a rack, where they form a queue and are then taken in order by an industrial robot (Section 37.6). The robot follows a set sequence in building up the mold. It first dips the pattern into a dilute slurry, and then rotates it under a sifting of fine particles. Next, the robot moves the tree beneath a blower to quickly dry the ceramic coating, and then it repeats the cycle. After a few cycles of such exposure to dilute slurry and fine particles, the details of the patterns are well produced, and good surface finish is ensured. The robot then dips the pattern into a thicker slurry, that quickly builds up the mold thickness (Fig. 11.18c). The trees are then dried and placed into a furnace to melt out and burn the wax. They are placed into another furnace to preheat them in preparation for the casting process.

A mold, ready for investment casting, is placed into a casting machine. The mold is placed upside down on the machine, directly over a measured volume of molten cobalt chrome. The machine then rotates so that the metal flows into the mold, as shown in Fig. 11.17d. The tree is then allowed to cool and the mold is removed. The cast parts are machined from the tree and are further machined and polished to the required dimensional tolerance and surface finish. Figure 11.18 shows the progression of investment casting, from tree, to investment, to casting. The parts are then removed from the tree and subjected to finishing operations.

Source: Courtesy of M. Hawkins, Zimmer, Inc.

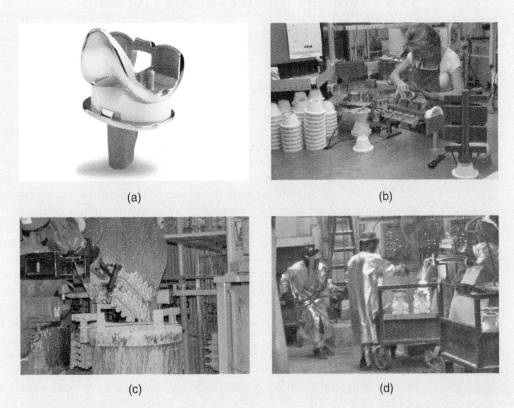

(a)

(b)

(c)

(d)

FIGURE 11.17 Manufacture of total knee replacements. (a) The Zimmer NexGen mobile-bearing knee (MBK); the femoral portion of the total knee replacement is the subject of this case study. (b) Assembly of patterns onto a central tree. (c) Dipping of the tree into slurry to develop a mold from investment. (d) Pouring of metal into a mold. *Source:* Courtesy of M. Hawkins, Zimmer, Inc.

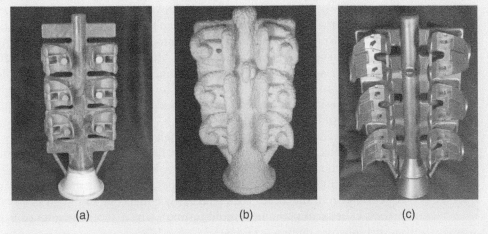

(a)

(b)

(c)

FIGURE 11.18 Progression of the tree. (a) After assembly of blanks onto the tree; (b) after coating with investment; (c) after removal from the mold. *Source:* Courtesy of M. Hawkins, Zimmer, Inc.

11.4   Permanent-mold Casting Processes

Permanent-mold casting processes have certain advantages over other casting processes, as described in this section.

11.4.1   Permanent-mold Casting

In *permanent-mold casting* (also called *hard-mold casting*), two halves of a mold are made from materials with high resistance to erosion and thermal fatigue, such as cast iron, steel, bronze, graphite, or refractory metal alloys. Typical parts made are automobile pistons, cylinder heads, connecting rods, gear blanks for appliances, and kitchenware. Parts that can be made economically generally weigh less than 25 kg (55 lb), although special castings, weighing a few hundred kilograms, have been made using this process. The capabilities of permanent-mold casting are given in Table 11.2.

The mold cavity and gating system are machined into the mold and thus become an integral part of it. To produce castings with internal cavities, cores made of metal or sand aggregate are placed in the mold prior to casting. Typical core materials are oil-bonded or resin-bonded sand, plaster, graphite, gray iron, low-carbon steel, and hot-work die steel. Gray iron is used most commonly, particularly for large molds used for aluminum and magnesium casting. Inserts also are used in various locations of the mold.

In order to increase the life of permanent molds, the surfaces of the mold cavity are usually coated with a refractory slurry, such as sodium silicate and clay, or are sprayed with graphite every few castings. These coatings also serve as parting agents and as thermal barriers, thus controlling the rate of cooling of the casting. Mechanical ejectors (such as pins located in various parts of the mold) may be required for the removal of complex castings. Ejectors usually leave small round impressions, which generally are not significant.

The molds are clamped together by mechanical means, and heated to about 150°–200°C (300°–400°F) to facilitate metal flow and reduce thermal damage to the dies. Molten metal is then poured through the gating system. After solidification, the molds are opened and the casting is removed. The mold often incorporates special cooling features, such as a means for pumping cooling water through the channels located in the mold and the use of cooling fins. Although the permanent-mold casting operation can be performed manually, it is often automated for large production runs.

The process is used mostly for aluminum, magnesium, and copper alloys, as well as for gray iron because of their generally lower melting points, although steels also can be cast using graphite or heat-resistant metal molds. Permanent-mold casting produces castings with good surface finish, close dimensional tolerances, uniform and good mechanical properties, and at high production rates.

Although equipment costs can be high, because of high die costs, labor costs are kept low through automation. The process is not economical for small production runs and is not suitable for intricate shapes, because of the difficulty in removing the casting from the mold. However, easily collapsible sand cores can be used (in a process called **semipermanent mold casting**), which are then removed from castings, leaving intricate internal cavities.

11.4.2   Vacuum Casting

A schematic illustration of the *vacuum-casting* process, also called *countergravity low-pressure* (CL) *process* (not to be confused with the vacuum-molding process

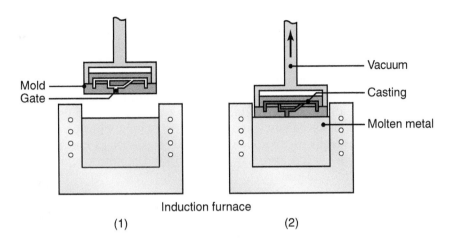

FIGURE 11.19 Schematic illustration of the vacuum-casting process; note that the mold has a bottom gate. (a) Before and (b) after immersion of the mold into the molten metal. *Source: After R. Blackburn.*

described in Section 11.2.1), is shown in Fig. 11.19. Vacuum casting is an alternative to investment, shell-mold, and green-sand casting, and is suitable particularly for thin-walled complex shapes, with uniform properties. Typical parts made are superalloy gas-turbine components with walls as thin as 0.5 mm (0.02 in.).

In this process, a mixture of fine sand and urethane is molded over metal dies, and cured with amine vapor. The mold is then held with a robot arm and immersed partially into molten metal held in an induction furnace. The metal may be melted in air (*CLA process*) or in a vacuum (*CLV process*). The vacuum reduces the air pressure inside the mold to about two-thirds of atmospheric pressure, thus drawing the molten metal into the mold cavities through a gate in the bottom of the mold. The metal in the furnace is usually at a temperature of 55°C (100°F) above the liquidus temperature of the alloy; consequently, it begins to solidify within a very short time.

The process can be automated, with production costs that are similar to those for green-sand casting. Carbon, low- and high-alloy steel, and stainless steel parts, weighing as much as 70 kg (155 lb), have been vacuum cast by this method. CLA castings are made easily at high volume and relatively low cost, and CLV parts usually involve reactive metals, such as aluminum, titanium, zirconium, and hafnium.

11.4.3 Slush Casting

It was noted in Fig. 10.11 that a solidified skin develops in a casting, which becomes thicker with time. Thin-walled hollow castings can be made by permanent-mold casting using this principle, in a process called *slush casting*. The molten metal is poured into the metal mold. After the desired thickness of solidified skin is obtained, the mold is inverted (or slung) and the remaining liquid metal is poured out. The mold halves are then opened and the casting is removed. Note that this operation is similar to making hollow chocolate shapes, eggs, and other confectionaries. Slush casting is suitable for small production runs, and is generally used for making ornamental and decorative objects (such as lamp bases and stems) and toys from low-melting-point metals such as zinc, tin, and lead alloys.

11.4.4 Pressure Casting

In the two permanent-mold processes described previously, the molten metal flows into the mold cavity by gravity. In *pressure casting* (also called *pressure pouring* or *low-pressure casting*), the molten metal is forced upward by gas pressure into a graphite or metal mold. The pressure is maintained until the metal has solidified completely in the mold. The molten metal also may be forced upward by a vacuum, which also removes dissolved gases and produces a casting with lower porosity. Pressure casting is generally used for high-quality castings, such as steel railroad-car wheels, although these wheels also may be cast in sand molds or semipermanent molds made of graphite and sand.

11.4.5 Die Casting

The *die-casting* process, developed in the early 1900s, is a further example of permanent-mold casting. The European term for this process is *pressure die casting*, and should not be confused with pressure casting, described above. Typical parts made by die casting are housings for transmissions, business-machine and appliance components, hand-tool components, and toys. The weight of most castings generally ranges from less than 90 g (3 oz) to about 25 kg (55 lb). Equipment costs, particularly the cost of dies, are somewhat high, but labor costs are generally low, because the process is semi- or fully automated. Die casting is economical for large production runs. The capabilities of die casting are given in Table 11.2.

In the die-casting process, molten metal is forced into the die cavity at pressures ranging from 0.7 to 700 MPa (0.1–100 ksi). There are two basic types of die-casting machines: *hot-* and *cold-chamber*.

The **hot-chamber process** (Fig. 11.20) involves the use of a piston, which forces a specific volume of molten metal into the die cavity through a gooseneck and nozzle. Pressures range up to 35 MPa (5000 psi), with an average of about 15 MPa (2000 psi). The metal is held under pressure until it solidifies in the die. To improve die life and to aid in rapid metal cooling (thereby reducing cycle time), dies are usually cooled by circulating water or oil, through various passageways in the die block. Low-melting-point alloys (such as zinc, magnesium, tin, and lead) commonly are cast using this process. Cycle times usually range from 200 to 300 shots (individual injections) per hour, for zinc, although very small components, such as zipper teeth, can be cast at rates of 18,000 shots per hour.

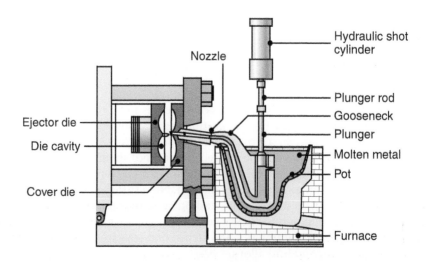

FIGURE 11.20 Schematic illustration of the hot-chamber die-casting process.

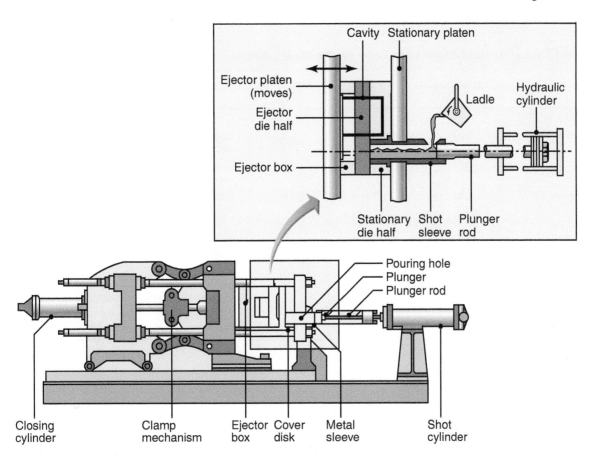

FIGURE 11.21 Schematic illustration of the cold-chamber die-casting process. These machines are large compared to the size of the casting, because high forces are required to keep the two halves of the dies closed under pressure during casting.

In the **cold-chamber process** (Fig. 11.21), molten metal is poured into the injection cylinder (*shot chamber*). The chamber is not heated, hence the term *cold chamber*. The metal is forced into the die cavity at pressures usually ranging from 20 to 70 MPa (3–10 ksi), although they may be as high as 150 MPa (20 ksi).

The machines may be horizontal (as shown in the figure), or vertical, in which case the shot chamber is vertical. High-melting-point alloys of aluminum, magnesium, and copper normally are cast using this method, although ferrous and other metals also can be cast. Molten-metal temperatures begin at about 600°C (1150°F) for aluminum and some magnesium alloys, and increase considerably for copper-based and iron-based alloys.

Process Capabilities and Machine Selection.   Die casting has the capability for rapid production of high-quality parts with complex shapes, especially with aluminum, brass, magnesium, and zinc (Table 11.3). It also produces good dimensional accuracy and surface details, so that parts require little or no subsequent machining or finishing operations (net-shape forming). Because of the high pressures involved, walls as thin as 0.38 mm (0.015 in.) are produced, which are thinner than those obtained by other casting methods. However, ejector marks remain on part surfaces, as may small amounts of flash (thin material squeezed out between the dies) at the die parting line.

A typical part made by die casting is the aluminum impeller shown in Fig. 11.1d; note the intricate shape and fine surface detail. For certain parts, die casting can compete favorably with other manufacturing methods (such as sheet-metal stamping

TABLE 11.3

Properties and Typical Applications of Some Common Die-casting Alloys				
Alloy	Ultimate tensile strength (MPa)	Yield strength (MPa)	Elongation in 50 mm (%)	Applications
Aluminum				
380 (3.5 Cu–8.5 Si)	320	160	2.5	Appliances, automotive components, electrical motor frames and housings
13 (12 Si)	300	150	2.5	Complex shapes with thin walls, parts requiring strength at elevated temperatures
Brass 858 (60 Cu)	380	200	15	Plumbing fixtures, lock hardware, bushings, ornamental castings
Magnesium				
AZ91 B (9 Al–0.7 Zn)	230	160	3	Power tools, automotive parts, sporting goods
Zinc				
No. 3 (4 Al)	280	—	10	Automotive parts, office equipment, household utensils, building hardware, toys
No. 5 (4 Al–1 Cu)	320	—	7	Appliances, automotive parts, building hardware, business equipment

Source: Images provided by: North American Die Casting Association, Wheeling, Illinois.

and forging) or other casting processes. In addition, because the molten metal chills rapidly at the die walls, the casting has a fine-grained, hard skin with high strength. Consequently, the strength-to-weight ratio of die-cast parts increases with decreasing wall thickness. With good surface finish and dimensional accuracy, die casting can produce smooth surfaces for bearings that otherwise would normally have to be machined.

Components such as pins, shafts, and threaded fasteners can be die cast integrally; called **insert molding**, this process is similar to placing wooden sticks in popsicles prior to freezing (see also Section 19.3). For good interfacial strength, insert surfaces may be knurled (see Fig. 23.1l), grooved, or splined. Steel, brass, and bronze inserts are commonly used in die-casting alloys. In selecting insert materials, the possibility of galvanic corrosion should be taken into account; to avoid this potential problem, the insert can be insulated, plated, or surface treated.

Because of the high pressures involved, dies for die casting have a tendency to separate unless clamped together tightly (see Fig. 11.21). Die-casting machines are thus rated according to the clamping force that can be exerted, to keep the dies closed during casting. The capacities of commercially available machines range from about 25 to 3000 tons. Other factors involved in the selection of die-casting machines are die size, piston stroke, shot pressure, and cost.

Die-casting dies (Fig. 11.22) may be *single cavity, multiple cavity* (several identical cavities), *combination cavity* (several different cavities), or *unit dies* (simple, small dies that can be combined in two or more units in a master holding die). Typically, the ratio of die weight to part weight is 1000 to 1. Thus, for example, the die for a casting weighing 2 kg would weigh about 2000 kg. The dies are usually made of hot-work die steels or mold steels (see Section 5.7). Die wear increases with the temperature of the molten metal. **Heat checking** of dies (surface cracking from repeated heating and cooling of the die, described in Section 3.6) can be a problem. When the materials are selected and maintained properly, however, dies can last more than a half million shots before any significant die wear takes place.

 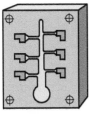 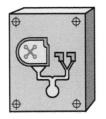 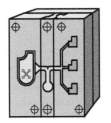

(a) Single-cavity die   (b) Multiple-cavity die   (c) Combination die   (d) Unit die

FIGURE 11.22 Various types of cavities in a die-casting die. *Source:* Images provided by: North American Die Casting Association, Wheeling, Illinois.

CASE STUDY 11.4 Die Casting of a Headlight Mount

Figure 11.23 shows a die-cast aluminum component of a daytime running lamp and turn signal for an automobile. Aluminum was preferable to plastic, because of its higher heat sink characteristics and rigidity, and also because tight tolerances were required for mounting and providing wiring access to LED bulbs. The fin size, thickness, and spacing were determined from a heat transfer analysis. Given this constraint, the fins were tapered to allow for easy removal from a die, and the corner radii were designed to prevent distortion during ejection. The part was then oriented so that mounting holes and pockets were coplanar to the die parting line to simplify die fabrication.

Heating channels were incorporated into the die near the thin sections to slow cooling, while cooling channels were incorporated near the thick sections. The resulting thermal balance led to lower distortion in the final product. The final product was cast from

380 aluminum, and measures 100 mm × 75 mm × 100 mm for the turn signal and 250 mm × 100 mm × 50 mm for the daytime running light sub-assembly.

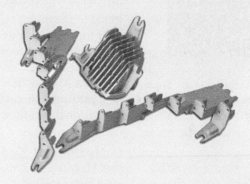

FIGURE 11.23 Die-cast running light and turn signal mounts produced from die-cast aluminum. *Source:* Courtesy of American Foundry Society.

11.4.6 Centrifugal Casting

As its name implies, the *centrifugal-casting* process utilizes *inertia* (caused by rotation) to force the molten metal into the mold cavities, a method that was first suggested in the early 1800s. The capabilities of centrifugal casting are given in Table 11.2. There are three types of centrifugal casting: true centrifugal casting, semicentrifugal casting, and centrifuging.

True Centrifugal Casting. In *true centrifugal casting*, hollow cylindrical parts (such as pipes, gun barrels, bushings, engine-cylinder liners, bearing rings with or without flanges, and street lampposts) are produced by the technique shown in Fig. 11.24. In this process, molten metal is poured into a rotating mold; the axis of rotation

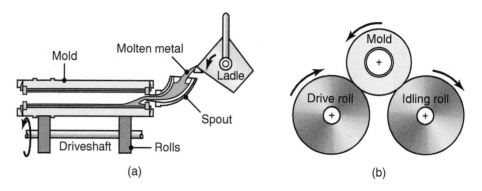

FIGURE 11.24 (a) Schematic illustration of the centrifugal-casting process; pipes, cylinder liners, and similarly shaped parts can be cast with this process. (b) Side view of the machine.

Video Solution 11.2 Force and Energy in Centrifugal Casting

is usually horizontal, but can be vertical for short workpieces. Molds are made of steel, iron, or graphite, and may be coated with a refractory lining to increase mold life. The mold surfaces can be shaped so that pipes with various external designs can be cast. The inner surface of the casting remains cylindrical, because the molten metal is distributed uniformly by the centrifugal forces. However, because of density differences, lighter elements (such as dross, impurities, and pieces of the refractory lining in the mold) tend to collect on the inner surface of the casting; consequently, the properties of the casting can vary throughout its thickness.

Cylindrical parts ranging from 13 mm (0.5 in.) to 3 m (10 ft) in diameter and 16 m (50 ft) long can be cast centrifugally, with wall thicknesses ranging from 6 to 125 mm (0.25–5 in.). The pressure generated by the centrifugal force is high; such high pressure is necessary for casting thick-walled parts. Castings with good quality, dimensional accuracy, and external surface detail are produced by this process.

Semicentrifugal Casting. An example of semicentrifugal casting is shown in Fig. 11.25(a). This method is used to cast parts with rotational symmetry, such as a wheel with spokes.

Centrifuging. In *centrifuging*, also called *centrifuge casting*, mold cavities are placed at a certain distance from the axis of rotation. The molten metal is poured from the center, and is forced into the mold by centrifugal forces (Fig. 11.25b). The properties of the castings can vary by distance from the axis of rotation, as in true centrifugal casting.

11.4.7 Squeeze Casting and Semisolid-metal Forming

Two casting processes that incorporate the features of both casting and forging (Chapter 14) are squeeze casting and semisolid-metal forming.

Squeeze Casting. The *squeeze-casting* or *liquid-metal forging* process was invented in the 1930s, but developed for industrial applications in the 1960s, and involves the solidification of molten metal under high pressure (Fig. 11.26). Typical products made are automotive components and mortar bodies (a short-barreled cannon). The machinery includes a die, punch, and ejector pin. The pressure applied by the punch keeps the entrapped gases in solution, and the contact under high pressure at the

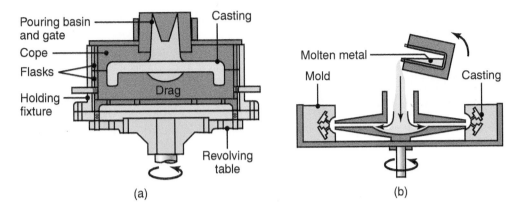

FIGURE 11.25   (a) Schematic illustration of the semicentrifugal casting process; wheels with spokes can be cast by this process. (b) Schematic illustration of casting by centrifuging; the molds are placed at the periphery of the machine and the molten metal is forced into the molds by centrifugal force.

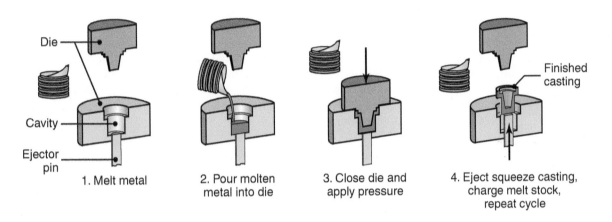

FIGURE 11.26   Sequence of operations in the squeeze-casting process; this process combines the advantages of casting and forging.

die–metal interface promotes rapid heat transfer, thus resulting in a casting with a fine microstructure and good mechanical properties.

The application of pressure also overcomes feeding difficulties that may arise when casting metals with a long freezing range (Section 10.2.2). Complex parts can be made to near-net shape, with fine surface detail using both nonferrous and ferrous alloys.

Semisolid-metal Forming.   *Semisolid-metal forming* (also called *mushy-state processing*; see Fig. 10.4) was developed in the 1970s. When it enters the die, the metal (consisting of liquid and solid components) is stirred so that all of the dendrites are crushed into fine solids; when cooled in the die, it develops into a fine-grained structure. The alloy exhibits thixotropic behavior, described in Section 10.2.3; hence the process is also called **thixoforming** or **thixomolding**, meaning its viscosity decreases when agitated. Thus, at rest and above its solidus temperature, the molten alloy has the consistency of butter, but when agitated vigorously, its consistency becomes more like motor oil.

Processing metals in their mushy state also has led to developments in *mushy-state extrusion*, similar to injection molding (described in Section 19.3), *forging*, and *rolling* (hence the term *semisolid metalworking*). These processes are also used in making parts with specially designed casting alloys, wrought alloys, and metal-matrix composites (Section 9.5). They also have the capability for blending granules of different alloys, called *thixoblending*, for specific applications.

Thixotropic behavior also has been utilized in developing technologies that combine casting and forging of parts, using cast billets that are forged when the metal is 30–40% liquid. Parts made include automotive control arms, brackets, and steering components. Processing steels by thixoforming has not yet reached the same stage as with aluminum and magnesium, largely because of the high temperatures involved which adversely affect die life and the difficulty in making complex shapes.

The advantages of semisolid metal forming over die casting are: (a) the structures developed are homogeneous, with uniform properties, lower porosity, and high strength; (b) both thin and thick parts can be made; (c) casting alloys as well as wrought alloys can be used; (d) parts can subsequently be heat treated; and (e) the lower superheat results in shorter cycle times. However, material and overall costs are higher than those for die casting.

Rheocasting.   This technique, first investigated in the 1960s, is used for forming metals in the semisolid state. The metal is heated to just above its solidus temperature, and poured into a vessel to cool it down to the semisolid state; the slurry is then mixed and delivered to the mold or die. This process is being used successfully with aluminum and magnesium alloys.

11.4.8   Composite-mold Casting Operations

Composite molds are made of two or more different materials and are used in shell molding and various other casting processes; they are generally employed in casting complex shapes, such as impellers for turbines. Composite molds increase the strength of the mold, improve the dimensional accuracy and surface finish of the castings, and can help reduce overall costs and processing time. Molding materials commonly used are shells (made as described in Section 11.2.2), plaster, sand with binder, metal, and graphite. These molds also may include cores and chills to control the rate of solidification in critical areas of castings.

11.5   Casting Techniques for Single-crystal Components

This section describes the techniques used to cast single-crystal components, such as gas turbine blades which generally are made of nickel-based superalloys, and used in the hot stages of the engine.

Conventional Casting of Turbine Blades.   In the *conventional-casting process*, the molten metal is poured into a ceramic mold, and begins to solidify at the mold walls. The grain structure developed is polycrystalline, similar to that shown in

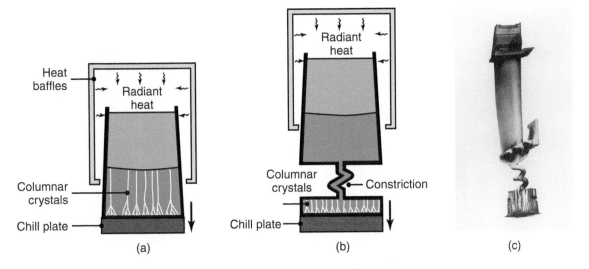

FIGURE 11.27 Methods of casting turbine blades: (a) directional solidification; (b) method to produce a single-crystal blade; and (c) a single-crystal blade with the constriction portion still attached. (See also Fig. 1.1) *Source:* (a) and (b) After B.H. Kear, (c) Courtesy of ASM International.

Fig. 10.2c. However, the presence of grain boundaries makes this structure suscep-tible to creep and cracking along the boundaries, under the centrifugal forces and elevated temperatures commonly encountered in an operating gas turbine.

Directionally Solidified Blades.   The *directional-solidification process* (Fig. 11.27a) was first developed in 1960. The ceramic mold, supported by a water-cooled chill plate, is preheated by radiant heating; after the metal is poured into the mold, the chill-plate assembly is lowered slowly. Crystals begin to grow at the chill-plate surface and on upward, like the *columnar grains* shown in Fig. 10.3. The blade is solidified directionally, with longitudinal but no transverse grain boundaries. The blade is thus stronger in the direction of centrifugal forces developed in the gas turbine.

Single-crystal Blades.   In *crystal growing*, developed in 1967, the mold has a con-striction in the shape of a corkscrew or helix (Figs. 11.27b and c), with a cross-section so small that it allows only one crystal to fit through. The mechanism of crys-tal growth is such that only the most favorably oriented crystals are able to grow through the helix (a situation similar to that shown in Fig. 10.3), because all others are intercepted by the walls of the helical passage.

As the assembly is slowly lowered, a single crystal grows upward through the constriction and begins to grow in the mold; strict control of the rate of movement is essential. Although single-crystal blades are more expensive than other types, the absence of grain boundaries makes them resistant to creep and thermal shock, hence they have a longer and more reliable service life.

Single-crystal Growing.   Single-crystal growing is a major activity in the semicon-ductor industry, in the manufacture of the silicon wafers for microelectronic devices (Chapter 28). There are two basic methods of crystal growing:

- In the **crystal-pulling method,** also known as the **Czochralski (CZ) process** (Fig. 11.28), a seed crystal is dipped into the molten metal, and then pulled

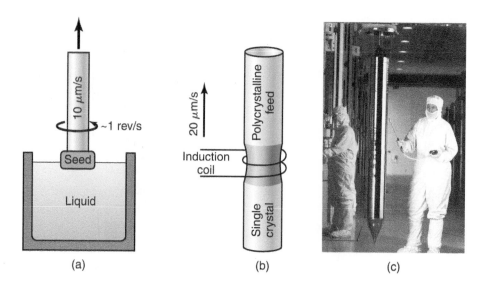

FIGURE 11.28 Two methods of crystal growing: (a) crystal pulling (Czochralski process) and (b) the floating-zone method. Crystal growing is especially important in the semiconductor industry (Chapter 28). (c) A single-crystal ingot produced by the Czochralski process. *Source:* Courtesy of Intel Corporation.

up slowly (at a rate of about 10 μm/s) while being rotated. The liquid metal begins to solidify on the seed, and the crystal structure of the seed continues throughout. *Dopants* (alloying elements, Section 28.3) may be added to the liquid metal to impart specific electrical properties. Single crystals of silicon, germanium, and various other elements are grown using this process. Single-crystal ingots up to 400 mm (16 in.) in diameter and over 2 m (80 in.) in length have been produced by this technique, although 300 mm (12 in.) ingots are more common in the production of silicon wafers for integrated circuit manufacture (Part V).

- The **floating-zone method** (Fig. 11.28b) starts with a rod of polycrystalline silicon resting on a single crystal; an induction coil then heats these two pieces while the coil moves slowly upward. The single crystal grows upward, while maintaining its orientation. Thin wafers are then cut from the rod, cleaned, and polished for use in microelectronic device fabrication. This process is suitable for producing diameters under 150 mm (6 in.), with very low levels of impurities.

11.6 Rapid Solidification

The properties of *amorphous alloys*, also known as *metallic glasses*, were described in Section 6.14. The technique for making these alloys (called *rapid solidification*) involves cooling the molten metal, at rates as high as 10^6 K/s, so that it does not have sufficient time to crystallize (see also Fig. 1.11). Rapid solidification results in a significant extension of solid solubility (Section 4.2), grain refinement, and reduced microsegregation (see Section 10.2.3), among other effects.

In a common method, called **melt spinning** (Fig. 11.29), the alloy is melted by induction in a ceramic crucible. It is then propelled, under high gas pressure, against a rotating copper disk (chill block), which chills the alloy rapidly (**splat cooling**).

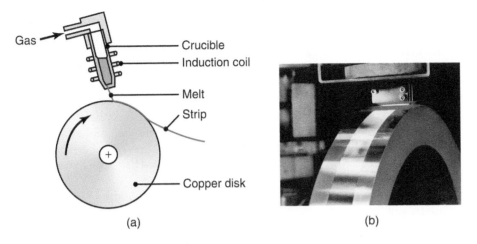

FIGURE 11.29 (a) Schematic illustration of melt spinning to produce thin strips of amorphous metal. (b) Photograph of nickel-alloy production through melt spinning. *Source: Courtesy of Siemens AG.*

11.7   Inspection of Castings

Several methods can be used to inspect castings to determine their quality and the presence and types of any defects. Castings can be inspected *visually*, or *optically*, for surface defects. Subsurface and internal defects are investigated using various nondestructive techniques, described in Section 36.10. In *destructive testing* (Section 36.11), specimens are removed from various sections of a casting, and tested for strength, ductility, and other mechanical properties, and to determine the presence, location, and distribution of porosity and any other defects.

Pressure tightness of cast components (such as valves, pumps, and pipes) is usually determined by sealing the openings in the casting, then pressurizing it with water, oil, or air. For leak tightness requirements in critical applications, pressurized helium or specially scented gases, with detectors (*sniffers*), are used. The casting is then inspected for leaks while the pressure is maintained; unacceptable or defective castings are remelted for reprocessing.

11.8   Melting Practice and Furnaces

Melting practice is an important aspect of casting operations, because it has a direct bearing on the quality of castings. Furnaces are charged with *melting stock*, consisting of metal, alloying elements, and various other materials such as **flux** and slag-forming constituents. Fluxes are inorganic compounds that refine the molten metal by removing dissolved gases and various impurities; they may be added manually or can be injected automatically into the molten metal.

Melting Furnaces.   The melting furnaces commonly used in foundries are electric-arc furnaces, induction furnaces, crucible furnaces, and cupolas.

1. **Electric arc** furnaces, described in Section 5.2.3 and illustrated in Fig. 5.2, are used extensively in foundries, because of their high rate of melting (thus high-production rate), much less pollution than other types, and their ability to

hold the molten metal (*i.e.*, keeping it at a constant temperature for a period of time) for alloying purposes.

2. **Induction** furnaces (Fig. 5.2c) are especially useful in smaller foundries, and produce smaller composition-controlled melts. There are two basic types. The *coreless induction furnace* consists of a crucible, surrounded with a water-cooled copper coil through which high-frequency current passes. Because there is a strong electromagnetic stirring action during induction heating, this type of furnace has excellent mixing characteristics for alloying and adding a new charge of metal into the furnace. The other type, called a *core* or *channel furnace*, uses low-frequency current (as low as 60 Hz), and has a coil that surrounds only a small portion of the unit.

   These furnaces are commonly used in nonferrous foundries, and are particularly suitable for (a) superheating (heating above normal casting temperature to improve fluidity), (b) holding, which makes it suitable for die-casting applications, and (c) duplexing (using two furnaces—for instance, melting the metal in one furnace and transferring it to another).

3. **Crucible** furnaces (Fig. 11.30a), which have been used extensively throughout history, are heated with various fuels, such as commercial gases, fuel oil, and fossil fuel, and with electricity. Crucible furnaces may be stationary, tilting, or movable.

4. **Cupolas** are basically vertical, refractory-lined steel vessels, charged with alternating layers of metal, coke, and flux (Fig. 11.30b). Although they require major investments and are increasingly replaced by induction furnaces, cupolas operate continuously, have high melting rates, and produce large amounts of molten metal.

5. **Levitation melting** involves *magnetic suspension* of the molten metal. An induction coil simultaneously heats a solid billet and stirs and confines the melt, thus eliminating the need for a crucible (which could contaminate the molten metal with oxide inclusions). The molten metal flows downward into an investment-casting mold, placed directly below the coil. Investment castings made by this method are free of refractory inclusions and of gas porosity, and have a uniform fine-grained structure.

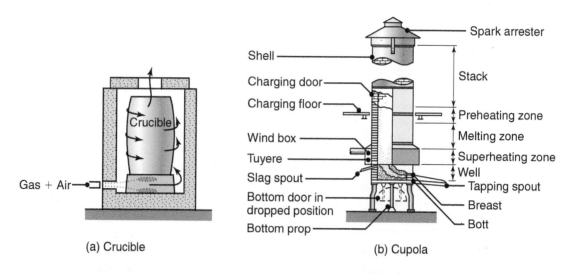

(a) Crucible   (b) Cupola

FIGURE 11.30 Two types of melting furnaces used in foundries.

11.9   Foundries and Foundry Automation

Casting operations are usually carried out in **foundries** (from the Latin *fundere*, meaning "melting and pouring"). Although these operations traditionally have involved much manual labor, modern foundries have incorporated automation and computer integration of all aspects of their operations.

As outlined in Fig. 11.2, foundry operations initially involve two separate groups of activities. The first group is pattern and moldmaking. Computer-aided design and manufacturing (Chapter 38) and rapid-prototyping techniques (Chapter 20) are now widely used to minimize trial and error, and thus improve efficiency. A variety of automated machinery is used to minimize labor costs, which can be significant in the production of castings.

The second group of activities involves melting the metals, and controlling their composition and impurities. Operations such as pouring into molds (some carried along conveyors), shakeout, cleaning, heat treatment, and inspection, also are automated. Automation minimizes labor, reduces the possibility of human error, increases the production rate, and attains higher quality levels. Industrial robots (Section 37.6) are now used extensively in foundry operations, such as for cleaning, riser cutting, mold venting, mold spraying, pouring, sorting, and inspection. Other operations involve automatic storage and retrieval systems for cores and patterns, using automated guided vehicles (Section 37.5).

SUMMARY

- Expendable-mold, permanent-pattern processes include sand, shell-mold, plaster-mold, and ceramic-mold casting. These processes require the destruction of the mold for each casting produced, but mold production is facilitated by reusable patterns.

- Expendable-mold, expendable-pattern processes include lost-foam and investment casting. In these processes, a pattern is consumed for each mold produced, and the mold is destroyed after each casting.

- Permanent-mold processes have molds or dies that can be used to produce castings at high production rates. Common permanent-mold processes include slush casting, pressure casting, die casting, and centrifugal casting.

- The molds used in permanent-mold casting are made of metal or graphite, and are used repeatedly to produce a large number of parts. Because metals are good heat conductors but do not allow gases to escape, permanent molds have fundamentally different effects on castings than sand or other aggregate mold materials.

- In permanent-mold casting, die and equipment costs are relatively high, but the processes are economical for large production runs. Scrap loss is low, dimensional accuracy is relatively high, and good surface detail can be achieved.

- Other casting processes include squeeze casting (a combination of casting and forging), semisolid-metal forming, rapid solidification (for the production of amorphous alloys), and casting of single-crystal components (such as turbine blades and silicon ingots for making wafers in integrated-circuit manufacture).

- Melting processes and their control also are important factors in casting operations. They include proper melting of the metals, preparation for alloying and removal of slag and dross, and pouring the molten metal into the molds. Inspection of castings for possible internal or external defects also is essential.

- Castings are generally subjected to subsequent processing, such as heat treatment and machining operations, to produce the final desired shapes, surface characteristics, and the required surface finish and dimensional accuracy.

KEY TERMS

Binders	Expendable mold	Parting agent	Rheocasting
Centrifugal casting	Expendable-pattern	Patterns	Sand casting
Ceramic-mold casting	casting	Permanent mold	Semisolid-metal forming
Chaplets	Fluxes	Permanent-mold casting	Shell-mold casting
Composite mold	Foundry	Plaster-mold casting	Slush casting
Core print	Green molding sand	Precision casting	Squeeze casting
Cores	Insert molding	Pressure casting	Thixotropy
Crystal growing	Investment casting	Rammed-graphite	Vacuum casting
Die casting	Levitation melting	molding	
Evaporative-pattern	Lost-foam process	Rapid prototyping	
casting	Lost-wax process	Rapid solidification	

BIBLIOGRAPHY

Analysis of Casting Defects, American Foundrymen's Society, 2002.

ASM Handbook, Vol. 15: **Casting**, ASM International, 2008.

Beeley, P., **Foundry Technology**, Butterworth-Heinemann, 2002.

Blair, M., Stevens, T.L., and Linskey, B. (eds.), **Steel Castings Handbook**, 6th ed., ASM International, 1995.

Campbell, J., **Complete Casting Handbook: Metal Casting Processes, Techniques and Design**, Butterworth-Heinemann, 2011.

Investment Casting Handbook, Investment Casting Institute, 1997.

Kaufman, J.G., and Rooy, E.L., **Aluminum Alloy Castings Properties, Processes and Applications**, ASM International, 2004.

Kirkwood, D.H., Suery, M., Kapranos, P., and Atkinson, H.V., **Semi-solid Processing of Alloys**, Springer, 2009.

Martin, A., **The Essential Guide to Mold Making & Slip Casting**, Lark Books, 2007.

Sias, F.R., **Lost-Wax Casting**, Woodsmere Press, 2006.

Vinarcik, E.J., **High Integrity Die Casting**, Wiley, 2002.

Young, K.P., **Semi-solid Processing**, Chapman & Hall, 1997.

REVIEW QUESTIONS

11.1 Describe the differences between expendable and permanent molds.

11.2 Name the important factors in selecting sand for molds.

11.3 What are the major types of sand molds? What are their characteristics?

11.4 List important considerations when selecting pattern materials.

11.5 What is the function of a core?

11.6 What is the difference between sand-mold and shell-mold casting?

11.7 What are composite molds? Why are they used?

11.8 Describe the features of plaster-mold casting.

11.9 Name the type of materials typically used for permanent-mold casting processes.

11.10 What are the advantages of pressure casting over other processes?

11.11 List the advantages and limitations of die casting.

11.12 What is the purpose of a riser? What is a blind riser?

11.13 Explain the purpose of a vent and a runner in a casting mold.

11.14 How are shell molds produced?

11.15 What keeps the mold together in vacuum casting?

11.16 What is squeeze casting? What are its advantages?

11.17 What are the advantages of the lost-foam casting process?

11.18 How are single-crystal turbine blades produced?

QUALITATIVE PROBLEMS

11.19 What are the reasons for the large variety of casting processes that have been developed over the years? Explain with specific examples.

11.20 Why are risers not as useful in die casting as they are in sand casting?

11.21 Describe the drawbacks to having a riser that is (a) too large and (b) too small.

11.22 Why can blind risers be smaller than open-top risers?

11.23 Why does die casting produce the smallest cast parts?

11.24 Why is the investment-casting process capable of producing fine surface detail on castings?

11.25 What differences, if any, would you expect in the properties of castings made by permanent-mold versus sand-casting processes?

11.26 Recently, cores for sand casting have been produced from salt. What advantages and disadvantages would you expect from using salt cores?

11.27 Would you recommend preheating the molds used in permanent-mold casting? Would you remove the casting soon after it has solidified? Explain your reasons.

11.28 Give reasons for, and examples of, using die inserts.

11.29 Referring to Fig. 11.3, do you think it is necessary to weigh down or clamp the two halves of the mold? Explain your reasons. Do you think that the kind of metal cast, such as gray cast iron versus aluminum, should make a difference in the clamping force? Explain.

11.30 Explain why squeeze casting produces parts with better mechanical properties, dimensional accuracy, and surface finish than do expendable-mold processes.

11.31 How are the individual wax patterns attached on a "tree" in investment casting?

11.32 Describe the measures that you would take to reduce core shifting in sand casting.

11.33 You have seen that, even though die casting produces thin parts, there is a limit to how thin they can be. Why can't even thinner parts be made by this process?

11.34 How are hollow parts with various cavities made by die casting? Are cores used? If so, how? Explain.

11.35 It was stated that the strength-to-weight ratio of die-cast parts increases with decreasing wall thickness. Explain why.

11.36 How are risers and sprues placed in sand molds? Explain, with appropriate sketches.

11.37 In shell-mold casting, the curing process is critical to the quality of the finished mold. In this stage of the process, the shell-mold assembly and cores are placed in an oven for a short period of time to complete the curing of the resin binder. List probable causes of unevenly cured cores or of uneven core thicknesses.

11.38 Why does the die-casting machine shown in Fig. 11.21 have such a large mechanism to close the dies? Explain.

11.39 Chocolate forms are available in hollow shapes. What process should be used to make these chocolates?

11.40 What are the benefits to heating the mold in investment casting before pouring in the molten metal? Are there any drawbacks? Explain.

11.41 The "slushy" state of alloys refers to that state between the solidus and liquidus temperatures, as described in Section 10.2.2. Pure metals do not have such a slushy state. Does this mean that pure metals cannot be slush cast? Explain.

11.42 Can a chaplet also act as a chill? Explain.

11.43 Rank the casting processes described in this chapter in terms of their solidification rate. That is, which processes extract heat the fastest from a given volume of metal?

QUANTITATIVE PROBLEMS

11.44 Estimate the clamping force for a die-casting machine in which the casting is rectangular with projected dimensions of 100 mm × 150 mm (4 in. × 6 in.). Would your answer depend on whether it is a hot-chamber or cold-chamber process? Explain.

11.45 The blank for the spool shown in Fig. P11.45 is to be sand cast out of A-319, an aluminum casting alloy. Make a sketch of the wooden pattern for this part, and include all necessary allowances for shrinkage and machining.

11.46 Repeat Problem 11.45, but assume that the aluminum spool is to be cast by expendable-pattern casting. Explain the important differences between the two patterns.

11.47 In sand casting, it is important that the cope-mold half be weighted down with sufficient force to keep it from floating when the molten metal is poured in. For the casting shown in Fig. P11.47, calculate the minimum amount of weight necessary to keep the cope from floating up as the molten metal is poured in. (*Hint:* The buoyancy force exerted by the molten metal on the cope is dependent on the effective height of the metal head above the cope.)

11.48 If an acceleration of 100 g is necessary to produce a part in true centrifugal casting and the part has an inner diameter of 8 in., a mean outer diameter of 14 in., and a length of 20 ft, what rotational speed is needed?

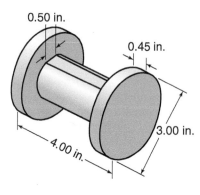

FIGURE P11.45

0.50 in.

0.45 in.

3.00 in.

4.00 in.

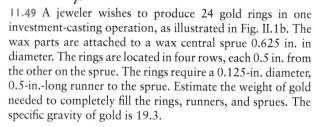

11.49 A jeweler wishes to produce 24 gold rings in one investment-casting operation, as illustrated in Fig. II.1b. The wax parts are attached to a wax central sprue 0.625 in. in diameter. The rings are located in four rows, each 0.5 in. from the other on the sprue. The rings require a 0.125-in. diameter, 0.5-in.-long runner to the sprue. Estimate the weight of gold needed to completely fill the rings, runners, and sprues. The specific gravity of gold is 19.3.

11.50 Assume that you are an instructor covering the topics described in this chapter, and you are giving a quiz on the numerical aspects of casting processes to test the understanding of the students. Prepare two quantitative problems and supply the answers.

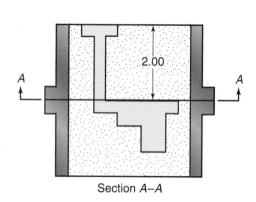

Section *A–A*

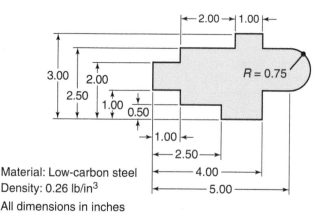

Material: Low-carbon steel
Density: 0.26 lb/in³
All dimensions in inches

FIGURE P11.47

SYNTHESIS, DESIGN, AND PROJECTS

11.51 Describe the procedures that would be involved in making a large outdoor bronze statue. Which casting process(es) would be suitable? Why?

11.52 The optimum shape of a riser is spherical to ensure that it cools more slowly than the casting it feeds; however, spherically shaped risers are difficult to cast. (a) Sketch the shape of a blind riser that is easy to mold, but also has the smallest possible surface-area-to-volume ratio. (b) Compare the solidification time of the riser in part (a) with that of a riser shaped like a right circular cylinder. Assume that the volume of each riser is the same and the height of each is equal to the diameter. (See Example 10.1.)

11.53 Sketch and describe a casting line consisting of machinery, conveyors, robots, sensors, etc., that automatically could perform the expendable-pattern casting process.

11.54 Outline the casting processes that would be most suitable for making small toys. Explain your choices.

11.55 Make a list of the mold and die materials used in the casting processes described in this chapter. Under each type of material, list the casting processes that are employed, and explain why these processes are suitable for that particular mold or die material.

11.56 Write a brief paper on the permeability of molds and the techniques that are used to determine permeability.

11.57 Light metals commonly are cast in vulcanized rubber molds. Conduct a literature search and describe the mechanics of this process.

11.58 It sometimes is desirable to cool metals more slowly than they would be if the molds were maintained at room temperature. List and explain the methods you would use to slow down the cooling process.

11.59 The part shown in Fig. P11.59 is a hemispherical shell used as an acetabular (mushroom-shaped) cup in a total hip replacement. Select a casting process for making this part, and provide a sketch of all the patterns or tooling needed if it is to be produced from a cobalt–chrome alloy.

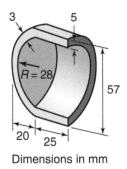

Dimensions in mm

FIGURE P11.59

11.60 Porosity that has developed in the boss of a casting is illustrated in Fig. P11.60. Show that the porosity can be eliminated simply by repositioning the parting line of this casting.

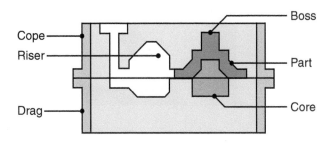

FIGURE P11.60

11.61 In Fig. II.1b the gemstones have been cast in place. Design a ring with a means of securing a gemstone in the wax pattern, such that it will remain in the mold as the wax is being melted. Could such an approach be used in lost-foam casting?

12

Metal Casting: Design, Materials, and Economics

- This final chapter on metal casting serves as a general guide to the interrelationships among product design, material, process selection, and economical considerations in casting.
- The chapter describes in detail the design considerations for casting operations, and discusses the general guidelines for successful casting.
- The characteristics and applications of the most common ferrous and nonferrous alloys are then described.
- The chapter ends with a discussion of casting economics.

12.1 Introduction

In the preceding two chapters, it was noted that successful casting practice requires the proper control of a large number of variables. These variables pertain to the particular characteristics of the metals and alloys cast, method of casting, mold and die materials, mold design, and processing parameters. Factors such as the flow of the molten metal in the mold cavities, the gating systems, the rate of cooling, and the gases evolved all influence the quality of a casting.

This chapter describes general design considerations for metal casting and presents guidelines for avoiding defects. It then describes the characteristics of the metals and alloys that are commonly cast, together with their typical applications. Because the economics of casting operations are just as important as their technical aspects, this chapter also briefly outlines the basic economic factors that are relevant to casting operations.

12.2 Design Considerations in Casting

As in all manufacturing operations, certain *design principles* pertaining also to casting have been developed over many years. Although these principles have been established primarily through experience, analytical methods, process simulation and modeling, and computer-aided design and manufacturing techniques all have come into wide use as well, improving the quality of castings and productivity, and resulting in significant cost savings.

All casting processes share some basic characteristics. Consequently, a number of design considerations apply equally to, for example, sand casting and die casting; however, each process will have its own particular design considerations. Sand casting will require consideration of mold erosion and associated sand inclusions in the casting, whereas die casting will not have this concern, although it has others, such as heat checking of dies, which reduces die life.

Troubleshooting the causes of defects in cast products is often complicated, and the considerations presented in this chapter are to serve only as guidelines. Furthermore, defects frequently are random and can be difficult to reproduce, thus complicating the implementation of corrective measures. In most cases, a given mold design will produce mostly good parts, as well as some defective parts. For these reasons, strict quality control procedures are implemented, especially for critical applications. (See Chapter 36.)

12.2.1 General Design Considerations for Castings

There are two types of design issues in casting: (a) geometric features, tolerances, etc. that should be incorporated into the part and (b) mold features that are needed to produce the desired casting. Robust design of castings usually involves the following steps:

1. Design the part so that the shape is cast as easily as possible. Several design considerations are given in this chapter to assist in such efforts.
2. Select a casting process and a material suitable for the part, size, required production quantity, mechanical properties, and so on. Often, steps 1 and 2 in this list have to be specified simultaneously, which can be a demanding design challenge.
3. Locate the parting line of the mold in the part.
4. Design and locate the gates to allow uniform feeding of the mold cavity with molten metal.
5. Select an appropriate runner geometry for the system.
6. Locate mold features, such as sprue, screens, and risers, as appropriate.
7. Make sure proper controls and good practices are in place.

Design of Parts to be Cast.   The following considerations are important in designing castings, as outlined in Fig. 12.1:

1. **Corners, angles, and section thickness.** Sharp corners, angles, and fillets should be avoided as much as possible, because they act as stress raisers and may cause cracking and tearing of the metal (as well as of the dies) during solidification. Fillet radii should be selected to minimize stress concentrations and to ensure proper molten-metal flow during pouring. Fillet radii usually range from 3 to 25 mm (1/8–1 in.), although smaller radii may be permissible in small castings and for specific applications. On the other hand, if the fillet radii are too large, the volume of the material in those regions also is large, and consequently the cooling rate is lower.

   Section changes in castings should be blended smoothly into each other. The location of the largest circle that can be inscribed in a particular region (Figs. 12.2a and b) is critical so far as shrinkage cavities are concerned. Because the cooling rate in regions with larger circles is lower, these regions are called **hot spots,** and can cause **shrinkage cavities** and **porosity** (Figs. 12.2c and d).

   Cavities at hot spots can be eliminated by using small cores, and although they produce cored holes in the casting (Fig. 12.2e), these holes do not affect strength significantly. It is also important to try to maintain uniform cross-sections and wall thicknesses throughout the casting, in order to avoid or minimize shrinkage cavities. Although they increase the production cost, *metal paddings* or *chills* in the mold can eliminate or minimize hot spots (see Fig. 10.14).

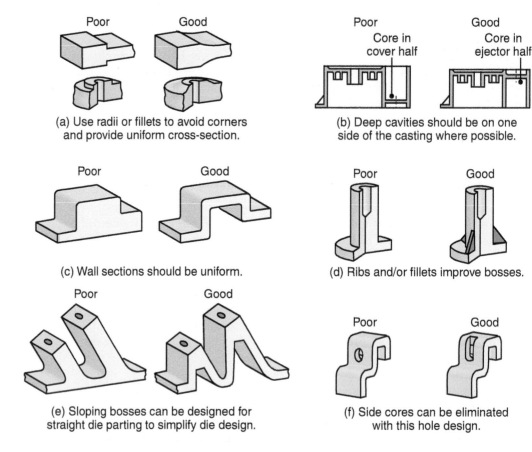

FIGURE 12.1 Suggested design modifications to avoid defects in castings. *Source:* Images provided by: North American Die Casting Association, Wheeling, Illinois.

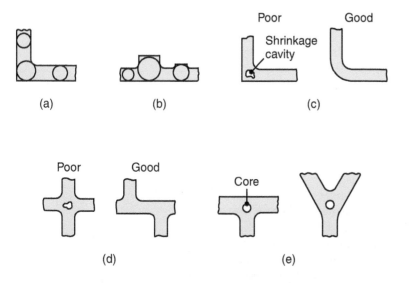

FIGURE 12.2 Examples of designs showing the importance of maintaining uniform cross-sections in castings to avoid hot spots and shrinkage cavities.

2. **Flat areas.** Large flat areas (plane surfaces) should be avoided, since (a) they may warp during cooling because of temperature gradients or (b) cause poor surface finish because of uneven flow of the metal during pouring. One of the common techniques for avoiding these problems is to break up flat surfaces with staggered ribs and serrations, as described below.

3. **Ribs.** One method of producing uniform thickness parts is to eliminate large, bulky volumes in the casting, as shown in Fig. 12.1. However, this can result in a loss in stiffness and, especially with flat regions, can lead to warping. One solution to these problems is to use ribs or support structure on the casting, as shown in Fig. 12.3. These are usually placed on the surface

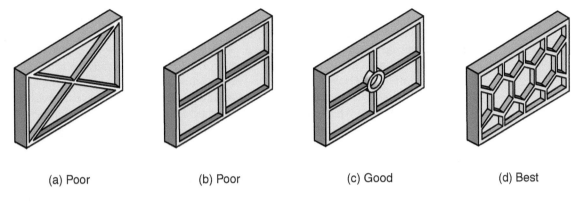

FIGURE 12.3 Rib designs for use on thin sections or flat surfaces to reduce or eliminate warping.

that is less visible. Ribs should, in general, have a thickness around 80% of the adjoining member thickness, and should be deeper than their strut thickness. It usually is beneficial to have the ribs solidify before the members they adjoin. Ribbing should not be used on both sides of a casting, and ribs should not meet at acute angles, because of complications to molding.

4. **Shrinkage.** To avoid cracking of the casting during cooling, there should be allowances for shrinkage during solidification and/or cooling to room temperature. In castings with intersecting ribs, the tensile stresses developed can be reduced by staggering the ribs or by changing the intersection geometry. Pattern dimensions also should allow for shrinkage of the metal during solidification and cooling. Allowances for shrinkage, known as **patternmaker's shrinkage allowances,** usually range from about 10 to 20 mm/m (1/8–1/4 in./ft). Table 12.1 gives the normal shrinkage allowance for metals commonly sand cast.

5. **Draft.** A small draft (taper) is typically provided in sand-mold patterns to enable removal of the pattern without damaging the mold (see Fig. 11.5). Drafts generally range from 5 to 15 mm/m (1/16–3/16 in./ft). Depending on the quality of the pattern, draft angles usually range from 0.5° to 2°. The angles on inside surfaces typically are twice this range; they have to be higher than those for outer surfaces because the casting shrinks inward toward the core as it cools.

6. **Dimensional tolerances.** Dimensional tolerances depend on the particular casting process employed, size of the casting, and type of pattern used. Tolerances should be as wide as possible, within the limits of good part performance, as otherwise the cost of the casting increases. In commercial practice, tolerances are typically in the range of ±0.8 mm (1/32 in.) for small castings, and increase with the size of the castings. Tolerances for large castings, for instance, may be as much as ±6 mm (0.25 in.).

7. **Lettering and markings.** It is common practice to include some form of part identification, such as lettering, numbers, or corporate logos, in castings. These features can be sunk into the casting or can protrude from the surface; which one is more desirable depends on the method of producing the molds. For example, in sand casting, a pattern plate is produced by machining on a computer numerically controlled milling machine (Section 24.2), because it is simpler to

TABLE 12.1

Normal Shrinkage Allowance for Some Metals Cast in Sand Molds

Metal	Shrinkage allowance (%)
Cast irons	
Gray cast iron	0.83–1.3
White cast iron	2.1
Malleable cast iron	0.78–1.0
Aluminum alloys	1.3
Magnesium alloys	1.3
Copper alloys	
Yellow brass	1.3–1.6
Phosphor bronze	1.0–1.6
Aluminum bronze	2.1
High-manganese steel	2.6

machine letters into the pattern plate. On the other hand, in die casting, it is simpler to machine letters into the mold.

8. **Finishing operations.** In designing a casting, it is important to consider the subsequent machining and finishing operations, if any, that may be required. For example, if a hole is to be drilled in a casting, it is better to locate the hole on a flat surface rather than on a curved surface, in order to prevent the drill from wandering. An even better design would incorporate a small dimple on the curved surface as a starting point for the drilling operation. Castings should also include features that allow them to be clamped easily in machine tools, if secondary machining operations are necessary.

Selecting the Casting Process.   Casting processes cannot be selected separately from economic considerations, as described in Section 12.4. Table 11.1 lists some of the advantages and limitations of casting processes that have an impact on casting design.

Locating the Parting Line.   A part should be oriented in a mold so that the large portion of the casting is relatively low and the height of the casting is minimized. Part orientation also determines the distribution of porosity. For example, in casting aluminum, hydrogen is soluble in liquid metal but is not soluble as the aluminum solidifies (see Fig. 10.15). Thus, hydrogen bubbles can form during the casting of aluminum, which float upward due to buoyancy and cause a higher porosity in the top regions of castings; critical surfaces should be oriented so that they face downward.

A properly oriented casting then can have the parting line specified; the parting line is the line or plane separating the upper (cope) and lower (drag) halves of molds (see Fig. 11.4). In general, the parting line should be along a flat plane rather than be contoured. Whenever possible, the parting line should be at the corners or edges of castings, rather than on flat surfaces in the middle of the casting, so that the **flash** at the **parting line** (material squeezing out between the two halves of the mold) will not be as visible. The location of the parting line is important because it influences mold design, ease of molding, number and shape of cores required, method of support, and the gating system.

The parting line should be placed as low as possible (relative to the casting) for less dense metals (such as aluminum alloys) and located at around midheight for denser metals (such as steels). However, the molten metal should not be allowed to flow vertically, especially when unconstrained by a sprue. The placement of the parting line has a large effect on the remainder of the mold design; for example, in sand casting, it is typical that the runners, gates, and sprue well are all placed in the drag on the parting line. Also, the placement of the parting line and orientation of the part determine the number of cores needed, especially when it is preferable to avoid the use of cores, whenever practical.

Locating and Designing Gates.   Gates are the connections between the runners and the part to be cast. Some considerations in designing gating systems are:

- Multiple gates often are preferable, and are necessary for large parts. Multiple gates have the benefits of allowing lower pouring temperature and reducing the temperature gradients in the casting.
- Gates should feed into thick sections of castings.

- A fillet should be used where a gate meets a casting; this feature produces less turbulence than abrupt junctions.
- The gate closest to the sprue should be placed sufficiently away from the sprue, so that the gate can be easily removed. This distance may be as small as a few mm for small castings and up to 500 mm for large ones.
- The minimum gate length should be three to five times the gate diameter, depending on the metal being cast. The gate cross-section should be large enough to allow the filling of the mold cavity, and should be smaller than the runner cross-section.
- Curved gates should be avoided; when necessary, a straight section in the gate should be located immediately adjacent to the casting.

Runner Design.   The runner is a horizontal distribution channel that receives molten metal from the sprue and delivers it to the gates. Runners are used to trap dross (a mixture of oxide and metal that forms on the surface of metals) and keep it from entering the gates and mold cavity. Commonly, dross traps are placed at the ends of runners, and the runner projects above the gates to ensure that the metal in the gates is tapped from below the surface. A single runner is used for simple parts, but two-runner systems may be necessary for more complicated castings.

Designing Various Mold Features.   The main goal in designing a *sprue* (described in Section 10.3) is to achieve the required molten-metal flow rates, while preventing aspiration (entrainment of air) or excessive dross formation. Flow rates are determined such that turbulence is avoided, but so that the mold is filled quickly as compared to the solidification time required. A *pouring basin* can be used to ensure that the metal flow into the sprue is uninterrupted; also, if molten metal is maintained in the pouring basin during pouring, the dross will float and will not enter the mold cavity. *Filters* are used to trap large contaminants; they also serve to reduce the metal velocity and make the flow more laminar. *Chills* can be used to speed solidification of the metal in a particular region of a casting.

Establishing Good Practices.   It has been widely observed that a given mold design can produce acceptable parts as well as defective ones, and rarely will produce only good or only defective castings. To check for defective ones, quality control procedures are necessary. Some common concerns are:

- Starting with a high-quality molten metal is essential for producing superior castings. Pouring temperature, metal chemistry, gas entrainment, and handling procedures all can affect the quality of metal being poured into a mold.
- The pouring of metal should not be interrupted, because it can lead to dross entrainment and turbulence. The meniscus of the molten metal in the mold cavity should experience a continuous, uninterrupted, and upward advance.
- The different cooling rates within the body of a casting cause residual stresses; thus, stress relieving (Section 4.11) may be necessary to avoid distortions of castings in critical applications.

12.2.2 Design for Expendable-mold Casting

Expendable-mold processes have certain specific design requirements, involving mainly the mold material, size of parts, and the manufacturing method. Note that a casting in an expendable-mold process (such an investment casting) will cool much more slowly than it would in, say, die casting. This has important implications in the layout of molds. Important design considerations for expendable-mold casting are:

Mold Layout.   The features in the mold must be placed logically and compactly, using gates as necessary. One of the most important goals in mold layout is to have solidification initiate at one end of the mold and progress in a uniform front across the casting, with the risers solidifying last. Traditionally, mold layout has been based on experience and on considerations of fluid flow and heat transfer. Commercial computer programs have now become widely available, that assist in the analysis of fluid flow and heat transfer in casting. These programs simulate mold filling and allow the rapid evaluation and design of mold layouts.

Riser Design.   A major concern in the design of castings is the size of risers and their placement. Risers are very useful in affecting the solidification-front progression across a casting, and are an essential feature in the mold layout, described previously. Blind risers are good design features and maintain heat longer than open risers do.

Risers are designed according to the following basic rules:

1. The riser must not solidify before the casting does. This rule usually is satisfied by avoiding the use of small risers and by using cylindrical risers with small aspect ratios (i.e., small ratios of height to cross-section). Spherical risers are the most efficient shape, but are difficult to work with.
2. The riser volume must be large enough to provide a sufficient amount of molten metal to compensate for shrinkage in the casting.
3. Junctions between the casting and the riser should not develop hot spots, where shrinkage porosity can occur.
4. Risers must be placed such that the molten metal can be delivered to locations where it is most needed.
5. There must be sufficient pressure to drive the molten metal into locations in the mold where it is needed. Thus, risers are not as useful for metals with low density (such as aluminum alloys) as they are for those with higher density (such as steel and cast irons).
6. The pressure head from the riser should suppress cavity formation and encourage complete filling of the mold cavity.

Machining Allowance.   Most expendable-mold castings require some additional finishing operations, such as machining and grinding; thus, allowances have to be made in casting design for these operations. Machining allowances, which are included in pattern dimensions, depend on the type of casting operation, and they increase with the size and section thickness of the casting. Allowances usually range from about 2 to 5 mm (0.1–0.2 in.) for small castings to more than 25 mm (1 in.) for large castings.

12.2.3   Design for Permanent-mold Casting

Typical design guidelines for permanent-mold casting are discussed in Example 12.1. Special considerations are generally included in designing tooling for die casting. Although designs may be modified to eliminate the draft for better dimensional accuracy, a draft angle of 0.5° or even 0.25° is usually required; otherwise, galling (localized seizure or sticking of two surfaces, Section 33.5) may take place between the part and the dies, and cause distortion of the casting. Die-cast parts are nearly net shaped, typically requiring only the removal of gates and minor trimming to remove flashing and other minor defects. The surface finish and dimensional accuracy of die-cast parts are very good (see Table 11.2), and in general, they do not require a machining allowance.

CASE STUDY 12.1   Illustrations of Poor and Good Casting Designs

Several examples of poor and good designs in permanent-mold and die casting are illustrated in Fig. 12.4. The significant differences in design are outlined here for each example:

(a) The lower portion of the design on the left has a thin wall, with no apparent function. This location of the part may fracture if subjected to high forces or impact. The good design eliminates this problem, and also may simplify die and mold manufacturing.

(b) Large flat surfaces always present difficulties in casting metals (as well as nonmetallic materials,

described in Part III), as they tend to warp and develop uneven surfaces. A common practice to avoid this situation is to break up the surface with ribs (see Fig. 12.3) and serrations on the reverse side of the casting. This approach greatly reduces distortion, while not adversely affecting the appearance and function of the flat surface. In addition to ribs, it is beneficial to use a textured surface, as shown in Fig. 12.4b, since very smooth surfaces are difficult to cast without objectionable aesthetic features.

(c) This example of poor and good design is relevant not only to castings, but also to parts that

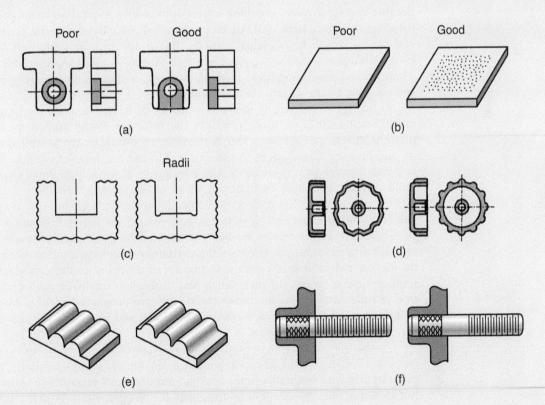

FIGURE 12.4 Examples of undesirable (poor) and desirable (good) casting designs. *Source:* Images provided by: North American Die Casting Association, Wheeling, Illinois.

(*continued*)

are machined or ground. It is difficult to produce sharp internal radii or corners that may be required for functional purposes, such as inserts designed to reach the bottom of the part cavity. Also, in the case of lubricated cavities, the lubricant can accumulate at the bottom and, because it is incompressible, prevent full insertion of an insert. The placement of a small radius at the corners or periphery at the bottom of the part eliminates this problem.

(d) A part could function, for instance, as a knob to be gripped and rotated, hence the outer features along its periphery. Note in the design on the left that the inner periphery of the knob also has features which are not functional but help save material. The die for the good design is easier to manufacture.

(e) Note that the poor design has sharp fillets at the base of the longitudinal grooves, meaning that the die has sharp (knife-edge) protrusions. It is thus possible that, with overextended use of the die, these edges may chip off.

(f) The poor design on the left has threads reaching the right face of the casting. It then is possible that, during casting, some molten metal will penetrate this region, forming a flash and interfering with the function of the threaded insert, such as when a nut is used. The good design incorporates an offset on the threaded rod, eliminating this problem. This design consideration is also applicable for injection molding of plastics, an example of which is shown in Fig. 19.9.

12.2.4 Computer Modeling of Casting Processes

Because casting involves complex interactions among several material and process variables, a quantitative study of these interactions is essential to the proper design and production of high-quality castings. Rapid advances in computers and modeling techniques have led to important innovations in modeling casting processes. These include fluid flow, heat transfer, and the microstructures developed during solidification under various casting conditions.

Modeling of *fluid flow* in molds is based on Bernoulli's and the continuity equations (Section 10.3). A model predicts the behavior of the molten metal during pouring into the gating system and its travel into the mold cavity, as well as the velocity and pressure distributions in the molds. Modern software can couple fluid flow and *heat transfer* and the effects of such parameters as surface conditions (roughness, mold permeability, etc.), thermal properties of the materials involved, and natural and forced convection on cooling rate. Recall that the surface conditions vary during solidification, as a layer of air develops between the casting and the mold wall due to shrinkage. Similar studies are being conducted on modeling the development of *microstructures* in casting. These studies encompass heat flow, temperature gradients, nucleation and growth of crystals, formation of dendritic and equiaxed structures, impingement of grains on each other, and movement of the liquid–solid interface during solidification. Several commercial software programs, such as Magmasoft, SOLIDCast, CAP, NovaFlow, Flow 3-D, WinCast, and Star–Cast, are now available for modeling casting processes.

The models are capable of predicting, for example, the width of the mushy zone (see Fig. 10.4) during solidification and the grain size in castings. Similarly, the capability to calculate isotherms (lines of equal temperature) gives insight into possible hot spots and the subsequent development of shrinkage cavities. With the availability of user-friendly software and advances in computer-aided design and manufacturing (Chapter 38), modeling techniques are becoming easier to implement. The benefits of this approach are improved quality, easier planning and cost estimating, increased productivity, and faster response to design modifications.

12.3   Casting Alloys

The general properties and applications of ferrous and nonferrous metals and alloys were presented in Chapters 5 and 6, respectively. This section describes the properties and applications of cast metals and alloys; their properties and casting and manufacturing characteristics are summarized in Fig. 12.5 and Tables 12.2–12.5. In addition to their casting characteristics, other important considerations in casting alloys include their machinability and weldability, as alloys typically are assembled with other components to produce the entire part.

The most commonly used casting alloy (in tonnage) is gray iron, followed by ductile iron, aluminum, zinc, lead, copper, malleable iron, and magnesium. Shipments of castings, in the United States alone, are around 13 million tons per year.

12.3.1   Nonferrous Casting Alloys

Aluminum-based Alloys.   Aluminum alloys have a wide range of mechanical properties, mainly because of various hardening mechanisms and heat treatments that can be used (Section 4.9). Parts made of aluminum and magnesium alloys are known as **light-metal** castings. These alloys have high electrical conductivity and generally good atmospheric corrosion resistance; however, their resistance to some acids and all alkalis is poor, and care must be taken to prevent galvanic corrosion. Aluminum alloys are lightweight, nontoxic, and have good machinability. Except for alloys containing silicon, they generally have low resistance to wear and abrasion. Aluminum-based alloys have numerous applications, including architectural and decorative uses. An increasing trend is their use in automobiles, for components such as engine blocks, cylinder heads, intake manifolds, transmission cases, suspension components, wheels, and brakes.

Magnesium-based Alloys.   The lowest density of all commercial casting alloys are those in the magnesium-based group; they have good corrosion resistance and moderate strength, depending on the particular heat treatment used. Typical applications include automotive wheels, housings, and air-cooled engine blocks. Because of their light weight, magnesium castings are being increasingly used in automobiles to increase fuel economy.

Copper-based Alloys.   Copper-based alloys have the advantages of good electrical and thermal conductivity, corrosion resistance, and nontoxicity, as well as wear properties thus making them suitable as bearing materials. A wide variety of copper-based alloys is available, including brasses, aluminum bronzes, phosphor bronzes, and tin bronzes.

Zinc-based Alloys.   A low-melting-point alloy group, zinc-based alloys have good corrosion resistance, good fluidity, and sufficient strength for structural applications. These alloys are commonly used in die casting, particularly for parts with intricate shapes and thin walls.

Tin-based Alloys.   Although low in strength, these alloys have good corrosion resistance and are typically used for linings or bearing surfaces.

Lead-based Alloys.   These alloys have applications similar to tin-based alloys, but the toxicity of lead is a major drawback to their wider application.

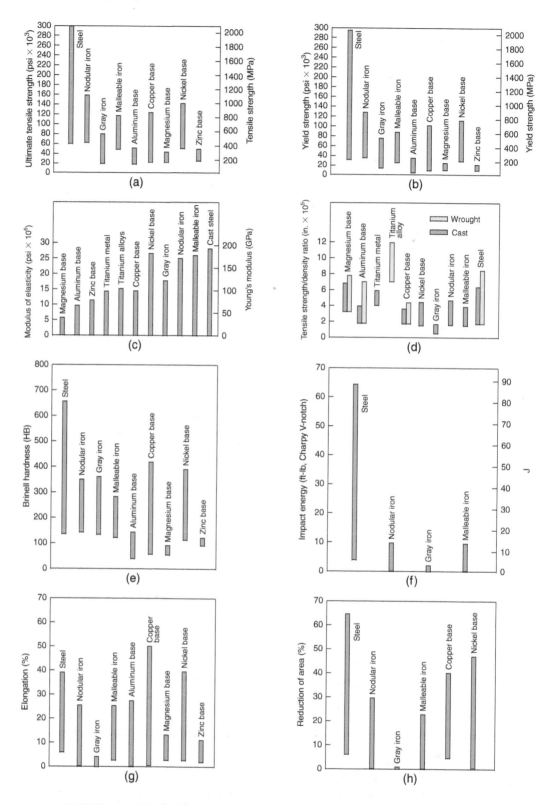

FIGURE 12.5 Mechanical properties for various groups of cast alloys; note that even within the same group, the properties vary over a wide range, particularly for cast steels. *Source:* Courtesy of Steel Founders' Society of America.

TABLE 12.2

Typical Applications for Castings and Casting Characteristics

Type of alloy	Castability*	Weldability*	Machinability*	Typical applications
Aluminum	E	F	G–E	Pistons, clutch housings, intake manifolds
Copper	F–G	F	F–G	Pumps, valves, gear blanks, marine propellers
Iron				
Ductile	G	D	G	Crankshafts, heavy-duty gears
Gray	E	D	G	Engine blocks, gears, brake disks and drums, machine bases
Malleable iron	G	D	G	Farm and construction machinery, heavy-duty bearings, railroad rolling stock
White iron	G	VP	VP	Mill liners, shot-blasting nozzles, railroad brake shoes, crushers, and pulverizers
Magnesium	G–E	G	E	Crankcase, transmission housings
Nickel	F	F	F	Gas turbine blades, pump and valve components for chemical plants
Steel				
Carbon and low alloy	F	E	F	Die blocks, heavy-duty gear blanks, aircraft undercarriage members, railroad wheels
High alloy	F	E	F	Gas-turbine housings, pump and valve components, rock-crusher jaws
Zinc	E	D	E	Door handles, radiator grills

*E = Excellent; G = Good; F = Fair; VP = Very poor; D = Difficult.

TABLE 12.3

Properties and Typical Applications of Cast Irons

Cast iron	Type	Ultimate tensile strength (MPa)	Yield strength (MPa)	Elongation in 50 mm (%)	Typical applications
Gray	Ferritic	170	140	0.4	Pipe, sanitary ware
	Pearlitic	275	240	0.4	Engine blocks, machine tools
	Martensitic	550	550	0	Wear surfaces
Ductile (Nodular)	Ferritic	415	275	18	Pipe, general service
	Pearlitic	550	380	6	Crankshafts, highly stressed parts
	Tempered martensite	825	620	2	High-strength machine parts, wear-resistant parts
Malleable	Ferritic	365	240	18	Hardware, pipe fittings, general engineering service
	Pearlitic	450	310	10	Railroad equipment, couplings
	Tempered martensite	700	550	2	Railroad equipment, gears, connecting rods
White	Pearlitic	275	275	0	Wear-resistant parts, mill rolls

High-temperature Alloys.   High-temperature alloys have a wide range of properties, and typically require temperatures of up to 1650°C (3000°F) for casting titanium and superalloys, and even higher for refractory alloys (Mo, Nb, W, and Ta). Special techniques are used to cast these alloys for nozzles and various jet- and rocket-engine components. Some high-temperature alloys are more suitable and economical

TABLE 12.4

Mechanical Properties of Gray Cast Irons				
ASTM class	Ultimate tensile strength (MPa)	Compressive strength (MPa)	Elastic modulus (GPa)	Hardness (HB)
20	152	572	66–97	156
25	179	669	79–102	174
30	214	752	90–113	210
35	252	855	100–119	212
40	293	965	110–138	235
50	362	1130	130–157	262
60	431	1293	141–162	302

TABLE 12.5

Properties and Typical Applications of Nonferrous Cast Alloys					
Alloys (UNS)	Condition	Ultimate tensile strength (MPa)	Yield strength (MPa)	Elongation in 50 mm (%)	Typical applications
Aluminum alloys					
195 (AO1950)	Heat treated	220–280	110–220	8.5–2	Sand castings
319 (AO3190)	Heat treated	185–250	125–180	2–1.5	Sand castings
356 (AO3560)	Heat treated	260	185	5	Permanent mold castings
Copper alloys					
Red brass (C83600)	Annealed	235	115	25	Pipe fittings, gears
Yellow brass (C86400)	Annealed	275	95	25	Hardware, ornamental
Manganese bronze (C86100)	Annealed	480	195	30	Propeller hubs, blades
Leaded tin bronze (C92500)	Annealed	260	105	35	Gears, bearings, valves
Gun metal (C90500)	Annealed	275	105	30	Pump parts, fittings
Nickel silver (C97600)	Annealed	275	175	15	Marine parts, valves
Magnesium alloys					
AZ91A	F	230	150	3	Die castings
AZ63A	T4	275	95	12	Sand and permanent mold castings
AZ91C	T6	275	130	5	High-strength parts
EZ33A	T5	160	110	3	Elevated-temperature parts
HK31A	T6	210	105	8	Elevated-temperature parts
QE22A	T6	275	205	4	Highest-strength parts

for casting than for shaping by other manufacturing methods, such as forging and powder metallurgy techniques.

12.3.2   Ferrous Casting Alloys

Cast Irons.   Cast irons represent the largest quantity of all metals cast; they can be cast easily into intricate shapes, and generally possess several desirable properties, such as wear resistance, high hardness, and good machinability. The term *cast iron* refers to a family of alloys, and as described in Section 4.6, they are classified as gray cast iron (gray iron), ductile (nodular or spheroidal) iron, white cast iron, malleable iron, and compacted-graphite iron. Their general properties and typical applications are given in Tables 12.3 and 12.4.

1. **Gray cast iron.** Gray iron castings have relatively few shrinkage cavities and low porosity. Various forms of gray cast iron are *ferritic, pearlitic,* and *martensitic,* and because of differences in their structures, each type has different properties (Table 12.4). Gray cast irons are specified by a two-digit ASTM designation; thus, for example, class 20 specifies that the material must have a minimum tensile strength of 20 ksi (140 MPa). Typical uses of gray cast iron are in engine blocks, electric-motor housings, pipes, and wear surfaces for machines; also, because of its high damping capacity, gray iron is used widely for machine-tool bases (Section 25.3).

2. **Ductile (nodular) iron.** Typically used for machine parts, housings, gears, pipe, rolls for rolling mills, and automotive crankshafts, ductile irons are specified by a set of two-digit numbers. For example, class or grade 80-55-06 indicates that the material has a minimum tensile strength of 80 ksi (550 MPa), a minimum yield strength of 55 ksi (380 MPa), and 6% elongation in 2 in. (50 mm).

3. **White cast iron.** Because of its very high hardness and wear resistance, white cast iron is used typically for rolls for rolling mills, railroad-car brake shoes, and liners in machinery for processing abrasive materials.

4. **Malleable iron.** The principal use of malleable iron is for railroad equipment and various types of hardware, fittings, and components for electrical applications. Malleable irons are specified by a five-digit designation; for example, 35018 indicates that the yield strength of the material is 35 ksi (240 MPa) and its elongation is 18% in 2 in.

5. **Compacted-graphite iron.** First produced commercially in 1976, compacted-graphite iron (CGI) has properties that are between those of gray irons and ductile irons. Gray iron has good damping and thermal conductivity, but low ductility, whereas ductile iron has poor damping and thermal conductivity, but high tensile strength and fatigue resistance. CGI has damping and thermal properties similar to gray iron, and strength and stiffness comparable to those of ductile iron. Because of its strength, castings made of CGI can be smaller and thus lighter; it is easy to cast and has consistent properties throughout the casting. Also, its machinability is better than that of ductile iron, an important consideration since CGI is used for automotive engine blocks and cylinder heads, which require extensive machining on their surfaces.

Cast Steels.   Because of the high temperatures required to melt steels (up to about 1650°C, or 3000°F), casting them requires special considerations. The high temperatures involved present difficulties in the selection of mold materials, particularly in view of the high reactivity of steels with oxygen during the melting and pouring of the metal. Steel castings possess properties that are more uniform (isotropic) than those made by mechanical working processes (Part III). Although cast steels can be welded, welding alters the cast microstructure in the heat-affected zone (see Fig. 30.20), thus influencing the strength, ductility, and toughness of the base metal. Subsequent heat treatment would be required to restore the mechanical properties of the casting. Cast weldments have gained importance for assembling large machines and structures. Cast steels have important applications in mining, chemical plants, oil fields, heavy construction, and equipment for railroads.

Cast Stainless Steels.   Casting of stainless steels involves considerations similar to those for steels. Stainless steels generally have long freezing ranges and high melting temperatures. They can develop several structures, depending on their composition and processing parameters. Cast stainless steels are available in several compositions, and they can be heat treated and welded; the parts have high heat and corrosion resistance, especially in the chemical and food industries. Nickel-based casting alloys are used for very corrosive environments and for very high temperature service.

12.4   Economics of Casting

As is the case with all manufacturing processes, the cost of each cast part (**unit cost**) depends on several factors, including materials, equipment, and labor. Of the various casting processes described in Chapter 11, some require more labor than others, some require expensive dies and machinery, and some require long production times to produce the castings (Table 12.6). Each of these individual factors thus affects the overall cost of a casting operation and to varying degrees. As can be seen in Table 12.6, relatively little cost is involved in making molds for sand casting, whereas molds for various casting processes and dies for die casting require expensive materials and manufacturing operations. There are also major costs involved in making patterns for casting, although (as stated in Section 11.2.1) much progress is being made in utilizing rapid-prototyping techniques (Chapter 20) to reduce costs and time.

Costs are also incurred in melting and pouring the molten metal into molds, and in heat treating, cleaning, and inspecting the castings. Heat treating is an important part of the production of many alloy groups (especially ferrous castings), and may be necessary for improved mechanical properties. However, heat treating may also introduce another set of production problems (such as scale formation on casting surfaces and warpage of the part) that can be a significant aspect of production costs.

The labor and skills required for these operations can vary considerably, depending on the particular casting operation and level of automation in the foundry. Investment casting, for example, requires much labor because of the several steps involved in the operation. Some automation is possible, such as using robots (Fig. 11.13a), whereas operations such as a highly automated die-casting process can maintain high production rates with little labor required.

TABLE 12.6

General Cost Characteristics of Casting Processes

Casting process	Cost*			Production rate (pieces/h)
	Die	Equipment	Labor	
Sand	L	L	L–M	< 20
Shell mold	L–M	M–H	L–M	< 10
Plaster	L–M	M	M–H	< 10
Investment	M–H	L–M	H	< 1000
Permanent mold	M	M	L–M	< 60
Die	H	H	L–M	< 200
Centrifugal	M	H	L–M	< 50

*L = Low; M = Medium; H = High.

Note that the equipment cost per casting will decrease as the number of parts cast increases; sustained high production rates can justify the high cost of dies and machinery. However, if demand is relatively small, the cost per casting increases rapidly, and it then becomes more economical to manufacture the parts by other casting processes described in this chapter, or by other manufacturing processes described in detail in Parts III and IV.

SUMMARY

- General guidelines have been established to aid in the production of castings that are free from defects, and meet dimensional tolerances, surface finish, service requirements, and various specifications and standards. The guidelines concern the shape of the casting and various techniques to minimize hot spots that could lead to shrinkage cavities. Because of the large number of variables involved, close control of all parameters is essential, particularly those related to the nature of liquid-metal flow into the molds and dies, and the rate of cooling in different regions of the mold.

- Numerous nonferrous and ferrous casting alloys are available, with a wide range of properties, casting characteristics, and applications. Because many castings are designed and produced to be assembled with other mechanical components and structures (subassemblies), several other considerations, such as weldability, machinability, and surface characteristics, also are important.

- Within the limits of good performance, the economics of casting is just as important as the technical considerations. Factors affecting the overall cost are the cost of materials, molds, dies, equipment, and labor, each of which varies with the particular casting operation.

KEY TERMS

Cast iron	Draft	Parting line	Shrinkage cavities
Compacted-graphite iron	Flash	Patternmaker's shrinkage allowance	Unit cost
Design principles	Hot spots	Porosity	
	Machining allowance		

BIBLIOGRAPHY

Analysis of Casting Defects, American Foundrymen's Society, 2002.

ASM Handbook, Vol. 15: Casting, ASM International, 2008.

ASM Specialty Handbook: Cast Irons, ASM International, 1996.

Campbell, J., Complete Casting Handbook: Metal Casting Processes, Techniques and Design, Butterworth-Heinemann, 2011.

Casting Design and Performance, ASM International, 2008.

Casting Design Handbook, ASM International, 2012.

Laird, G., Gundlach, R., and Rohrig, K., Abrasion-Resistant Cast Iron Handbook, American Foundry Society, 2000.

Powell, G.W., Cheng, S.-H., and Mobley, C.E., Jr., A Fractography Atlas of Casting Alloys, Battelle Press, 1992.

Product Design for Die Casting, Diecasting Development Council, 1988.

Steel Castings Handbook, 6th ed., ASM International, 1995.

REVIEW QUESTIONS

12.1 Why are steels more difficult to cast than cast irons?

12.2 What is the significance of hot spots in metal casting?

12.3 What is shrinkage allowance? Machining allowance?

12.4 Explain the reason for drafts in molds.

12.5 Why are ribs useful for flat surfaces?

12.6 What are light castings and where are they used most commonly?

12.7 Name the types of cast irons generally available, and list their major characteristics and applications.

12.8 Comment on your observations regarding Fig. 12.5.

12.9 Describe the difference between a runner and a gate.

12.10 What is the difference between machining allowance and dimensional tolerance?

12.11 What is dross? Can it be eliminated?

QUALITATIVE PROBLEMS

12.12 Describe your observation concerning the design modifications shown in Fig. 12.1.

12.13 If you need only a few castings of the same design, which three processes would be the most expensive per piece cast?

12.14 Do you generally agree with the cost ratings in Table 12.6? If so, why?

12.15 Describe the nature of the design differences shown in Fig. 12.4. What general principles do you observe in this figure?

12.16 Note in Fig. 12.5 that the ductility of some cast alloys is very low. Do you think this should be a significant concern in engineering applications of castings? Explain.

12.17 Do you think that there will be fewer defects in a casting made by gravity pouring versus one made by pouring under pressure? Explain.

12.18 Explain the difference in the importance of drafts in green-sand casting versus permanent-mold casting.

12.19 What type of cast iron would be suitable for heavy-machine bases, such as presses and machine tools? Why?

12.20 Explain the advantages and limitations of sharp and rounded fillets, respectively, in casting design.

12.21 Explain why the elastic modulus, E, of gray cast iron varies so widely, as shown in Table 12.4.

12.22 If you were to incorporate lettering or numbers on a sand-cast part, would you make them protrude from the surface or recess them into the surface? What if the part were to be made by investment casting? Explain.

12.23 The general design recommendations for a well in sand casting (see Fig. 11.3) are that (a) its diameter should be at least twice the exit diameter of the sprue and (b) its depth should be approximately twice the depth of the runner. Explain the consequences of deviating from these guidelines.

12.24 The heavy regions of parts typically are placed in the drag in sand casting and not in the cope. Explain why.

12.25 What are the benefits and drawbacks to having a pouring temperature that is much higher than the metal's melting temperature? What are the advantages and disadvantages in having the pouring temperature remain close to the melting temperature?

QUANTITATIVE PROBLEMS

12.26 When designing patterns for casting, patternmakers use special rulers that automatically incorporate solid shrinkage allowances into their designs. For example, a 12-in. patternmaker's ruler is longer than 1 ft. How long should a patternmaker's ruler be for making patterns for (a) aluminum castings and (b) high-manganese steel?

12.27 Using the information given in Table 12.2, develop approximate plots of (a) castability versus weldability and (b) castability versus machinability, for at least five of the materials listed in the table.

12.28 The part in Figure P12.28 is to be cast of 10% Sn bronze, at the rate of 100 parts per month. To find an appropriate casting process, consider all casting processes, then reject those that are (a) technically inadmissible, (b) technically feasible but too expensive for the purpose, and (c) identify the most economical one. Write a rationale using common-sense assumptions about cost.

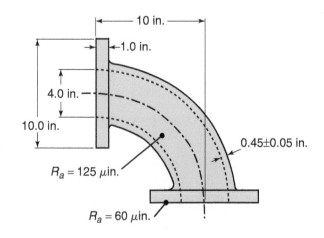

FIGURE P12.28

SYNTHESIS, DESIGN, AND PROJECTS

12.29 Describe the general design considerations pertaining to metal casting.

12.30 Add more examples of applications to those shown in Fig. 12.2.

12.31 Explain how ribs and serrations are helpful in casting flat surfaces that otherwise may warp. Give a specific illustration.

12.32 List casting processes that are suitable for making hollow parts with (a) complex external features, (b) complex internal features, and (c) both complex external and complex internal features. Explain your choices.

12.33 Small amounts of slag and dross often persist after skimming, and are introduced into the molten metal flow in casting. Recognizing that slag and dross are less dense than the molten metal, design mold features that will remove small amounts of slag before the metal reaches the mold cavity.

12.34 If you need only a few units of a particular casting, which process(es) would you use? Why?

12.35 For the cast metal wheel illustrated in Fig. P12.35, show how (a) riser placement, (b) core placement, (c) padding, and (d) chills may be used to help feed molten metal and eliminate porosity in the isolated hub boss.

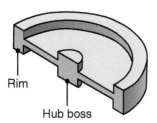

FIGURE P12.35

12.36 Assume that the introduction to this chapter is missing. Write a brief introduction to highlight the importance of the topics covered in it.

12.37 In Fig. P12.37, the original casting design shown in (a) was resized and modified to incorporate ribs in the design shown in (b). The casting is round and has a vertical axis of symmetry. What advantages do you think the new design has as a functional part over the old one?

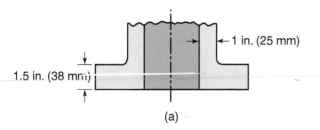

FIGURE P12.37

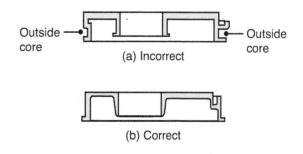

FIGURE P12.37 (*continued*)

12.38 An incorrect and a correct design for casting are shown in Fig. P12.38. Review the changes made and comment on their advantages.

(a) Incorrect

(b) Correct

FIGURE P12.38

12.39 Using the method of inscribed circles, shown in Fig. 12.2, justify the trend shown in Fig. 12.3.

12.40 A growing trend is the production of patterns and molds through rapid-prototyping approaches, described in Chapter 20. Consider the case of an injection molding operation, where the patterns are produced by rapid prototyping, and then hand assembled onto trees and processed in traditional fashion. What design rules discussed in this chapter would still be valid, and which would not be as important in this case?

12.41 Repeat Problem 12.40 for the case where (a) a pattern for sand casting is produced by rapid prototyping and (b) a sand mold for sand casting is produced.

Forming and Shaping Processes and Equipment

●●●

When inspecting the numerous products all around us, we soon realize that a wide variety of materials and processes have been used in making them, as can be seen from the example of an automobile shown in Fig. III.1. It will also be noted that some products consist of a few parts, such as eyeglasses, pencils, and light fixtures, while others consist of hundreds or thousands of parts, such as clocks, automobiles, and computers, or even millions of parts, as in airplanes and ships.

Some products have simple shapes, with smooth curvatures, yet others have complex configurations and detailed surface features. Some are used in critical applications, such as elevator cables and turbine blades, whereas others are used for routine applications, such as paper clips, forks, and door keys. Some are very thin, such as aluminum foil and plastic film, whereas others are very thick, as in ship hulls, boiler plates, and large machine bases.

Note that the words **forming** and **shaping** are both used in the title of this part of the book. Although there are not always clear distinctions between the two terms, *forming* generally indicates changing the shape of an existing solid body. Thus, in **forming processes**, the starting material (usually called the workpiece, stock, or blank) may be in the shape of a plate, sheet, bar, rod, wire, or tubing. For example, an ordinary wire coat hanger is made by forming a straight piece of wire, by bending

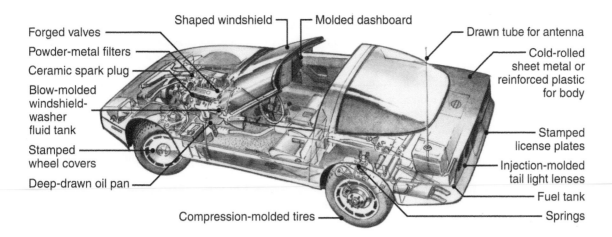

FIGURE III.1   Formed and shaped parts in a typical automobile.

and twisting it into the shape of a hanger. Likewise, the sheet metal body for an automobile is generally made of flat, cold-rolled steel sheet, which is then formed into various shapes, such as hood, roof, trunk, and door panels, using a pair of large dies.

Shaping processes typically involve molding and casting, producing a part that generally is at or near the final desired shape. A plastic coat hanger, for example, is made by forcing molten plastic into a two-piece mold, with a cavity in the shape of the hanger. Telephone and computer housings, refrigerator-door liners, some auto-body parts, and countless other plastic products are likewise shaped by forcing molten polymer into a mold, and removing it after it solidifies.

Some forming and shaping operations produce long *continuous* products, such as plates, sheets, tubing, wire, and rod and bars with various cross-sections, which then are shaped into specific products. Rolling, extrusion, and drawing processes (Chapters 13 and 15) are capable of making such long products, which then are cut into desired lengths. On the other hand, processes such as forging (Chapter 14), sheet metal forming and stamping (Chapter 16), powder metallurgy (Chapter 17), ceramic casting and glass pressing (Chapter 18), and processes involving plastics and reinforced plastics (Chapter 19), typically produce *discrete* products.

The initial raw material used in forming and shaping metals is usually molten metal, which is *cast* into individual *ingots* or *continuously cast* into slabs, rods, or pipes. Cast structures are converted to *wrought structures* by plastic-deformation processes. The raw material used also may consist of *metal powders*, which then are pressed and sintered (heated without melting) into individual parts. For plastics, the

TABLE III.1

General Characteristics of Forming and Shaping Processes

Process	Characteristics
Rolling	
Flat	Production of flat plate, sheet, and foil at high speeds; good surface finish, especially in cold rolling; very high capital investment; low-to-moderate labor cost
Shape	Production of various structural shapes (such as I-beams and rails) at high speeds; includes thread rolling; requires shaped rolls and expensive equipment; low-to-moderate labor cost; requires moderate operator skill
Forging	Production of discrete parts with a set of dies; some finishing operations usually required; usually performed at elevated temperatures, but also cold for smaller parts; die and equipment costs are high; moderate-to-high labor cost; requires moderate-to-high operator skill
Extrusion	Production of long lengths of solid or hollow shapes with constant cross-section; product is then cut into desired lengths; usually performed at elevated temperatures; cold extrusion has similarities to forging and is used to make discrete products; moderate-to-high die and equipment cost; low-to-moderate labor cost; requires low-to-moderate operator skill
Drawing	Production of long rod and wire with various cross-sections; good surface finish; low-to-moderate die, equipment, and labor costs; requires low-to-moderate operator skill
Sheet-metal forming	Production of a wide variety of shapes with thin walls and simple or complex geometries; generally low-to-moderate die, equipment, and labor costs; requires low-to-moderate operator skill
Powder metallurgy	Production of simple or complex shapes by compacting and sintering metal powders; moderate die and equipment cost; low labor cost and skill
Processing of plastics and composite materials	Production of a wide variety of continuous or discrete products by extrusion, molding, casting, and fabricating processes; moderate die and equipment costs; requires high operator skill in processing of composite materials
Forming and shaping of ceramics	Production of discrete products by various shaping, drying, and firing processes; low-to-moderate die and equipment cost; requires moderate-to-high operator skill

starting material is usually pellets, flakes, or powder; for ceramics, it is clays and oxides, obtained from ores or produced synthetically.

In this part of the text, the important factors involved in each forming and shaping process are described, along with how material properties and processes affect the quality and integrity of the product made (Table III.1). Detailed mathematical models of processes are available, as can be found in the Bibliographies at the end of the chapters; this book will provide only simple models for the various forming and shaping processes considered.

We also explain why some materials can be processed only by certain specific manufacturing methods, and why parts with particular shapes can only be processed by certain techniques and not by others. Also included are the characteristics of the machinery and equipment used, which can significantly influence product quality, production rate, and the economics of a particular manufacturing operation.

13

Metal-rolling Processes and Equipment

- This chapter describes the rolling of metals, which is the most important metal-forming operation based on volume of metals rolled.

- The chapter begins with a description of the flat-rolling process and analyzes the force, torque, and power required, in terms of relevant material and process parameters; it also includes a review of defects and their causes in rolled products.

- Shape-rolling processes are then described, where workpieces pass through a series of shaped rolls.

- Special rolling processes, such as cross rolling, ring rolling, thread rolling, tube rolling, and tube piercing, are also described.

- The chapter ends with a description of the characteristics of rolling mills and roll arrangements for making specific products.

Typical products made by various rolling processes: Plates for ships, bridges, structures, machines; sheet metal for car bodies, aircraft fuselages, appliances, containers; foil for packaging; I-beams, railroad rails, architectural shapes, large rings, seamless pipe and tubing; bolts, screws, and threaded components.

Alternative processes: Continuous casting, extrusion, drawing, machining of threaded components.

13.1 Introduction

Rolling is the process of reducing the thickness or changing the cross-section of a long workpiece by compressive forces applied through a set of **rolls** (Fig. 13.1), a process that is similar to rolling dough with a rolling pin. Rolling, which accounts for about 90% of all metals produced by metalworking processes, was first developed in the late 1500s. Modern steelmaking practices and the production of various ferrous and nonferrous metals and alloys now generally integrate *continuous casting* with rolling processes; this method greatly improves productivity and lowers production costs, as described in Section 5.4. Nonmetallic materials also are rolled to reduce their thickness and enhance their properties.

Rolling is first carried out at elevated temperatures (*hot rolling*). During this phase, the coarse-grained, brittle, and porous structure of the ingot or the continuously cast metal is broken down into a *wrought structure*, having a finer grain size and enhanced properties, such as increased strength and hardness. Subsequently,

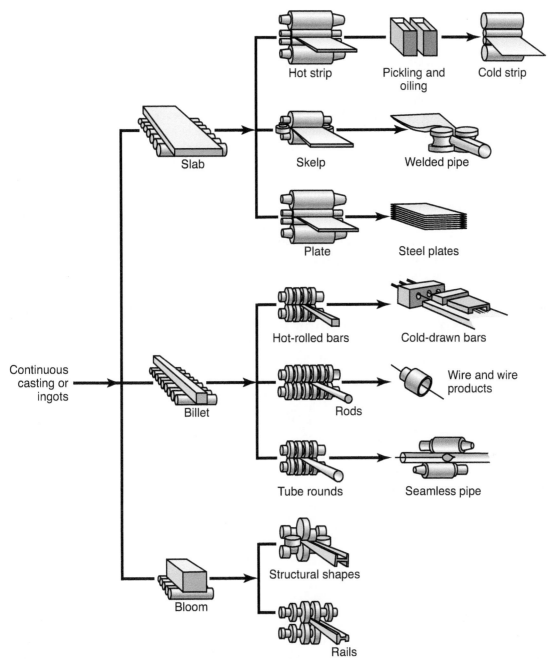

FIGURE 13.1   Schematic outline of various flat-rolling and shape-rolling processes. *Source:* Courtesy of American Iron and Steel Institute.

rolling can be performed at room temperature (*cold rolling*), whereby the rolled sheet has higher strength and hardness, and better surface finish. Cold rolling will, however, result in a product with anisotropic properties, due to preferred orientation or mechanical fibering, described in Section 1.6.

   Plates generally have a thickness of more than 6 mm (0.25 in.), and are used for structural applications, such as ship hulls, boilers, bridges, and heavy machinery.

QR Code 13.1 Animation of hot rolling of billets. (*Source:* Courtesy of Sandvik Coromant)

Plates can be as thick as 300 mm (12 in.) for large structural supports, 150 mm (6 in.) for reactor vessels, and 100 to 125 mm (4–5 in.) for machinery frames and warships.

Sheets are generally less than 6 mm thick, and are typically provided to manufacturing facilities as coils, weighing as much as 30,000 kg (33 tons), or as flat sheets for further processing into a wide variety of sheet-metal products. Sheets are used for trucks and aircraft bodies, appliances, food and beverage containers, and kitchen and office equipment. Commercial aircraft fuselages and trailer bodies usually are made of aluminum–alloy sheets, with a minimum thickness of 1-mm (0.04 in.). The skin thickness of a Boeing 747 fuselage, for example, is 1.8 mm (0.07 in.); for a Lockheed L1011 it is 1.9 mm (0.075 in.).

Steel sheets used for automobile and appliance bodies typically are about 0.7 mm (0.03 in.) thick. Aluminum beverage cans are made from sheets 0.28 mm (0.01 in.) thick, which, after processing into a can (Section 16.7), become a cylindrical body with a wall thickness of 0.1 mm (0.004 in.). Aluminum **foil** typically has a thickness of 0.008 mm (0.0003 in.), although thinner foils, down to 0.003 mm (0.0001 in.), also can be produced.

13.2 The Flat-rolling Process

A schematic illustration of the *flat-rolling* process is shown in Fig. 13.2. A metal strip, of thickness h_o, enters the **roll gap** and is reduced to thickness h_f by a pair of rotating rolls, each powered individually by electric motors. The surface speed of the rolls is V_r. The velocity of the strip increases from its entry value of V_o as it moves through the roll gap, and is highest at the exit from the roll, where it is denoted as V_f. The metal accelerates in the roll gap, in the same manner as an incompressible fluid flows through a converging channel.

Because the surface speed of the rigid roll is constant, there is *relative sliding* between the roll and the strip along the contact length, L. At one point, called the **neutral point** or **no-slip point**, the velocity of the strip is the same as that of the roll. To the left of this point, the roll moves faster than the strip; to the right of this point, the strip moves faster than the roll; consequently, the frictional forces act on the strip as shown in Fig. 13.2b.

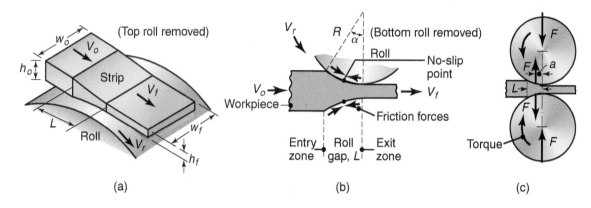

(a)   (b)   (c)

FIGURE 13.2 (a) Schematic illustration of the flat-rolling process. (b) Friction forces acting on strip surfaces. (c) Roll force, F, and torque, T, acting on the rolls. The width of the strip, w, usually increases during rolling, as shown in Fig. 13.5.

Forward slip in rolling is defined in terms of the exit velocity of the strip, V_f, and the surface speed of the roll, V_r, as

$$\text{Forward slip} = \frac{V_f - V_r}{V_r}, \qquad (13.1)$$

and is a measure of the relative velocities involved in the toll gap. Forward slip can easily be obtained by measuring the roll and workpiece velocities on a rolling mill, and gives a real-time indication of the neutral point location. Forward slip also correlates with the surface finish of the rolled strip, with low values being preferable to high values.

The rolls pull the material into the roll gap through a *frictional force* on the material; thus, a net frictional force must exist and be to the right in Fig. 13.2b; this also means that the frictional force to the left of the neutral point must be higher than the friction force to the right. Although friction is essential to enable rolling (just as it is in driving a car on a road), energy is dissipated in overcoming friction; note that increasing friction also increases rolling forces and power requirements. Furthermore, high friction could damage the surface of the rolled product or cause sticking. A compromise is therefore made in practice through lubricant selection, leading to low and controlled levels of friction.

The maximum possible **draft** is defined as the difference between the initial and final strip thicknesses, or $(h_o - h_f)$; a large draft could cause the rolls to slip. It can be shown that maximum draft is a function of the roll radius, R, and the coefficient of friction, μ, as:

$$h_o - h_f = \mu^2 R. \qquad (13.2)$$

Thus, as expected, the higher the friction and the larger the roll radius, the greater the maximum possible draft. This is a situation similar to the use of large tires (high R) and rough treads (high μ) on farm tractors and off-road earthmoving equipment, which allows the vehicles to travel over rough terrain without skidding.

13.2.1   Roll Force, Torque, and Power Requirements

The rolls apply pressure on the flat strip, resolved into a *roll force*, F, as shown in Fig. 13.2c. Note that in the figure this force appears to be perpendicular to the plane of the strip, rather than at an angle; this is because, in practice, the arc of contact is very small compared to the roll radius, thus it can be assumed that the roll force is perpendicular to the strip.

The roll force in flat rolling can be estimated from the expression

$$F = Lw Y_{avg}, \qquad (13.3)$$

where L is the roll-strip contact length, w is the width of the strip, and Y_{avg} is the average true stress (see Section 2.2.3) of the strip in the roll gap. Equation (13.3) is for a *frictionless* condition; however, an estimate of the *actual roll force*, including friction, may be made by increasing this calculated force by about 20%.

The *torque* on the roll is the product of F and a. The power required per roll can then be estimated by assuming that F acts in the middle of the arc of contact; thus in Fig. 13.2c, $a = L/2$. Therefore, the *total power* (for two rolls), in S.I. units, is

$$\text{Power (kW)} = \frac{2\pi F L N}{60{,}000}, \qquad (13.4)$$

Video Solution 13.1 Draft and Force in Metal Rolling

where F is in newtons, L is in meters, and N is the revolutions per minute of the roll. In traditional English units, the total power can be expressed as

$$\text{Power (hp)} = \frac{2\pi FLN}{33,000}, \tag{13.5}$$

where F is in pounds and L is in feet.

EXAMPLE 13.1   Calculation of Roll Force and Torque in Flat-rolling

Given: An annealed copper strip 9 in. (228 mm) wide and 1.00 in. (25 mm) thick is being rolled to a thickness of 0.80 in. (20 mm), in one pass. The roll radius is 12 in. (300 mm), and the rolls rotate at 100 rpm.

Find: Calculate the roll force and the power required in this operation.

Solution: The roll force is determined from Eq. (13.3), in which L is the roll-strip contact length. It can be shown, from simple geometry, that this length is given approximately by

$$L = \sqrt{R(h_o - h_f)} = \sqrt{12(1.00 - 0.80)} = 1.55 \text{ in.}$$

The average true stress, Y_{avg}, for annealed copper is determined as follows: First note that the absolute value of the true strain that the strip undergoes in this operation is

$$\epsilon = \ln\left(\frac{1.00}{0.80}\right) = 0.223.$$

Referring to Fig. 2.5, annealed copper has a true stress of about 12,000 psi in the unstrained condition, and at a true strain of 0.223, the true stress is

40,000 psi. Hence, the average true stress in the roll gap is $(12,000 + 40,000)/2 = 26,000$ psi. Thus, the roll force is

$$F = LwY_{avg} = (1.55)(9)(26,000)$$
$$= 363,000 \text{ lb} = 1.6 \text{ MN}$$

The total power is calculated from Eq. (13.5), with $N = 100$ rpm. Thus,

$$\text{Power} = \frac{2\pi FLN}{33,000} = \frac{2\pi(363,000)(1.55/12)(100)}{33,000}$$
$$= 898 \text{ hp} = 670 \text{ kW}$$

Exact calculation of the force and the power requirements in rolling can be difficult, because of the uncertainties involved in (a) determining the exact contact geometry between the roll and the strip and (b) accurately estimating both the coefficient of friction and the strength of the material in the roll gap. For hot rolling, the sensitivity of the strength of the material to strain rate, as described in Section 2.2.7, must be taken into account.

Video Solution 13.2 Force and Power in Rolling of Aluminum Foil

Reducing Roll Force. Roll force can cause significant deflection and flattening of the rolls, as it does in a rubber tire. Such changes will, in turn, affect the rolling process, its efficiency, and its ability to produce a uniform thickness in the rolled sheet (known as *gage control*). Also, the columns of the **roll stand** (including the housing, chocks, and bearings, as shown in Fig. 13.3c) would deflect under high roll forces to such an extent that the roll gap may open up significantly. Consequently, the rolls have to be set closer than originally calculated in order to compensate for this deflection and to obtain the desired final thickness.

Roll forces can be reduced by the following means (see Fig. 13.3 and Sections 13.3 and 13.4):

- Using smaller diameter rolls, to reduce the contact area
- Taking smaller reductions per pass, to reduce the contact area

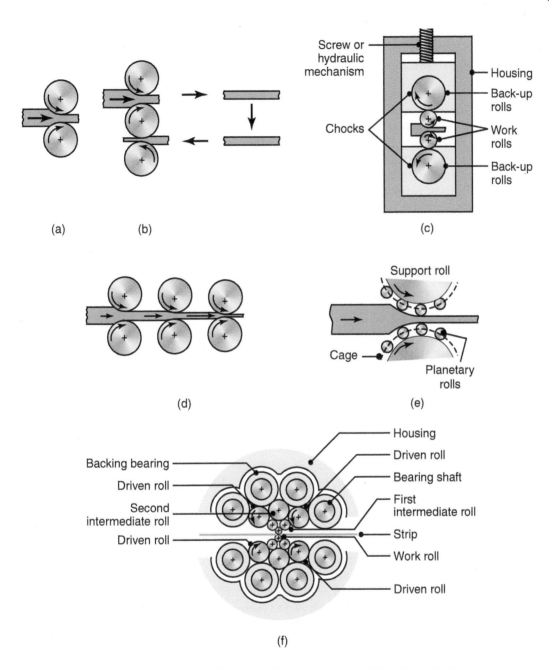

FIGURE 13.3 Schematic illustration of various roll arrangements: (a) Two-high mill; if a two-high mill is used for thick but short workpieces, it will commonly roll a billet back and forth in multiple passes, known as a *reversing mill*. (b) Three-high mill with elevator for multiple passes. (c) Four-high rolling mill showing various features; the stiffness of the housing, the rolls, and the roll bearings are all important in controlling and maintaining the thickness of the rolled strip. (d) Tandem rolling, with three stands. (e) Planetary mill, and (f) Cluster mill, also known as a *Sendzimir* or Z-mill.

- Rolling at elevated temperatures, to lower the strength of the material
- Applying back and/or front tensions to the strip, to reduce the roll pressure
- Reducing friction at the roll–workpiece interface

An effective method of reducing roll forces is to apply longitudinal **tension** to the strip during rolling, as a result of which the compressive stresses required to plastically deform the material become lower (known as the apparent decrease in the yield stress of the material; see Bibliography of Chapter 2). Because they require high roll forces, tensions are important, particularly in rolling high-strength metals. Tensions can be applied to the strip at either the entry zone (**back tension**), the exit zone (**front tension**), or both. Back tension is applied to the sheet by a braking action to the reel that supplies the sheet into the roll gap (*pay-off reel*). Front tension is applied by increasing the rotational speed of the *take-up reel*. Although it has limited and specialized applications, rolling also can be carried out by front tension only, with no power supplied to the rolls, known as **Steckel rolling**.

13.2.2 Geometric Considerations

Because of the forces acting on the rolls, they undergo shape changes during rolling. Just as a straight beam deflects under a transverse load, roll forces tend to *elastically* bend the rolls during rolling, as shown in Fig. 13.4a; as a result, the rolled strip will be thicker at its center than at its edges, known as **crown**. A common method of avoiding this problem is to grind the rolls in such a way that their diameter at the center is slightly larger than that at their edges (called **camber**). Thus, when the rolls bend, the strip being rolled will have a constant thickness along its width (Fig. 13.4b).

For rolling sheet metals, the radius of the maximum camber is generally 0.25 mm (0.01 in.) greater than the radius at the ends of the roll; however, a particular camber is correct only for a certain load and strip width. To reduce deflection, the rolls can also be subjected to external bending, by applying moments at their bearings (as can be demonstrated by simply bending a wooden stick at its ends to simulate camber). Note also that the higher the elastic modulus of the roll material, the smaller the roll deflection (see also Section 13.4).

Because of the heat generated, due to the work of plastic deformation during rolling, rolls can become slightly barrel shaped, known as **thermal camber**. Unless compensated for by some means, this condition can produce strips that are thinner at the center than their edges. Thermal camber can be controlled by adjusting the location of coolants and their flow rate along the length of the rolls.

Roll forces also tend to *flatten* the rolls elastically, producing an effect much like the flattening of automobile tires. Flattening is undesirable because it results, in effect, in a larger roll radius which, in turn, means a larger contact area for the same draft, and thus the roll force increases because of the now larger contact area.

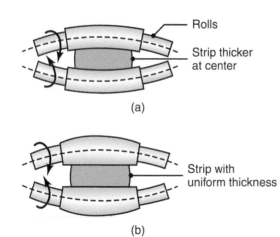

Rolls

Strip thicker
at center

(a)

Strip with
uniform thickness

(b)

FIGURE 13.4 (a) Bending of straight cylindrical rolls caused by roll forces. (b) Bending of rolls ground with camber, producing a strip with uniform thickness through the strip width. (Roll deflections have been exaggerated for clarity.)

Spreading.   In rolling plates and sheets with high width-to-thickness ratios, the width of the strip remains effectively constant during rolling. However, with smaller ratios (such as a bar with a square cross-section), its width increases significantly as it passes through the rolls, an effect that can easily be observed in the rolling of dough with a rolling pin. The increase in width is called *spreading* (Fig. 13.5).

It can be shown that spreading increases with (a) decreasing width-to-thickness ratio of the entering strip, because of reduction in the width constraint, (b) increasing friction, and (c) decreasing ratio of roll radius to strip thickness. Spreading can also be prevented by using vertical rolls, in contact with the strip edges; known as *edger mills*, the vertical rolls provide a physical constraint to spreading.

13.2.3   Vibration and Chatter

Vibration and *chatter* can have significant adverse effects on product quality and productivity in manufacturing operations. Generally defined as *self-excited vibration*, chatter in rolling leads to periodic variations in the rolled sheet thickness and surface finish, and may lead to excessive scrap (see Table 40.3). Chatter in rolling is found predominantly in tandem mills (Fig. 13.3d). Chatter is very detrimental to productivity; it has been estimated, for example, that modern rolling mills could operate at up to 50% higher speeds were it not for chatter.

Chatter is a complex phenomenon (see also Section 25.4), and results from interactions between the structural dynamics of the mill stand and the dynamics of the rolling operation. Rolling speed and lubrication are found to be the two most significant parameters. Although not always practical to implement, it also has been suggested that chatter can be reduced by (a) increasing the distance between the stands of the rolling mill, (b) increasing the strip width, (c) decreasing the reduction per pass (draft), (d) increasing the roll radius, (e) increasing the strip-roll friction, and (f) incorporating external dampers in the roll supports.

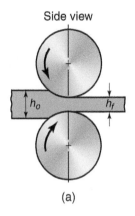

Side view

(a)

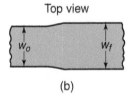

Top view

(b)

FIGURE 13.5 Spreading in flat rolling; note that similar spreading can be observed when dough is rolled with a rolling pin.

13.3   Flat-rolling Practice

The initial rolling steps (*breaking down*) of the material is usually done by **hot rolling**, above the recrystallization temperature of the metal (Section 1.7). As described in Section 10.2 and illustrated in Fig. 10.2, a **cast structure** typically is dendritic, consisting of coarse and nonuniform grains; this structure is usually brittle, and may also be porous. Hot rolling converts the cast structure to a **wrought structure** (Fig. 13.6), with finer grains and enhanced ductility, both of which result from the breaking up of brittle grain boundaries and the closing up of internal defects, including porosity. Typical temperature ranges for hot rolling are about 450°C (850°F) for aluminum alloys, up to 1250°C (2300°F) for alloy steels, and up to 1650°C (3000°F) for refractory alloys (see also Table 14.3).

The rolled product of the first hot-rolling operation is called **bloom, slab**, or **billet** (see Fig. 13.1). A bloom usually has a square cross-section, at least 150 mm (6 in.) on the side, whereas a slab is usually rectangular in cross-section. Blooms are further processed by *shape rolling* into structural shapes, such as I-beams and railroad rails (Section 13.5). Slabs are rolled into plates and sheets. Billets usually are square (with a cross-sectional area smaller than that for blooms), and are later rolled into various shapes, such as round rods and bars, using shaped rolls. Hot-rolled round rods, called **wire rods**, are used as the starting material for rod- and wire-drawing operations (Chapter 15).

In hot rolling of blooms, billets, and slabs, the surface of the material is usually **conditioned** (prepared for a subsequent operation) prior to rolling them. Conditioning is often done by means of a torch (*scarfing*), which removes heavy

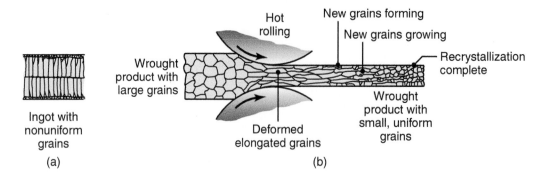

FIGURE 13.6   Changes in the grain structure of cast or of large-grain wrought metals during hot rolling. Hot rolling is an effective way to reduce grain size in metals for improved strength and ductility; cast structures of ingots or continuous casting (Section 5.4) are converted to a wrought structure by hot working.

scale, or by rough grinding, which smoothens surfaces. Prior to cold rolling, the scale developed during hot rolling may be removed by *pickling* with acids (acid etching), and by such mechanical means as blasting with water or by grinding.

Cold rolling is carried out at room temperature and, compared with hot rolling, produces sheets and strips with a much better surface finish, better dimensional tolerances, and enhanced mechanical properties (because of strain hardening).

Pack rolling is a flat-rolling operation in which two or more layers of sheet are rolled together, thus increasing productivity. Aluminum foil, for example, is pack rolled in two layers, thus only the top and bottom outer surfaces are in contact with the rolls. Note that one side of aluminum foil is matte, while the other side is shiny. The foil-to-foil side has a matte and satiny finish, whereas the foil-to-roll side is shiny and bright (because the polished surface is impressed onto the foil during rolling).

Rolled mild steel, when subsequently stretched during sheet-forming operations, undergoes *yield-point elongation* (Section 16.3), a phenomenon that causes surface irregularities, called *stretcher strains* or *Lüder's bands*. To prevent this situation, the sheet metal is subjected to a final light pass of 0.5–1.5% reduction (known as **temper rolling** or **skin pass**) shortly before stretching it in a subsequent forming operation.

A rolled sheet may not be sufficiently flat as it exits the roll gap, due to factors such as variations in the incoming material or in the processing parameters during rolling. To improve *flatness*, the rolled strip typically passes through a series of **leveling rolls**. Several roll arrangements can be used, as shown in Fig. 13.7, in which the sheet is flexed in opposite directions as it passes through the sets of rolls.

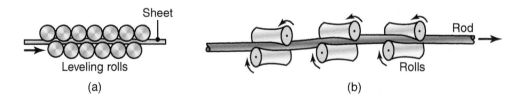

FIGURE 13.7   (a) A method of roller leveling to flatten rolled sheets. (b) Roller leveling to straighten drawn rods.

13.3.1   Defects in Rolled Plates and Sheets

Defects may be present on the surfaces of rolled plates and sheets, or there may be internal structural defects. Defects are undesirable not only because they adversely affect surface appearance, but also because they may affect strength, formability, and other manufacturing characteristics of the rolled sheets. Several surface defects (scale, rust, scratches, gouges, pits, and cracks) have been identified in sheet metals. These defects may be caused by inclusions and impurities in the original cast material or by various other conditions related to material preparation and to the rolling operation.

Wavy edges on sheets (Fig. 13.8a) are due to roll bending. The strip becomes thinner along its edges than at its center (see Fig. 13.4a), thus, because of volume constancy in plastic deformation, the edges have to elongate more than the material at the center. Consequently, the edges buckle because they are constrained by the central region from expanding freely in the longitudinal (rolling) direction.

The **cracks** shown in Figs. 13.8b and c are usually the result of low material ductility at the rolling temperature. Because the quality of the edges of the sheet is important in subsequent forming operations, edge defects in rolled sheets may have to be removed by shearing and slitting operations (Section 16.2). **Alligatoring** (Fig. 13.8d) is typically caused by nonuniform bulk deformation of the billet during rolling or by the presence of defects in the original cast material.

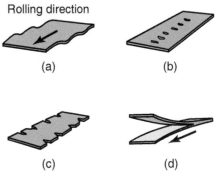

Rolling direction

(a)   (b)

(c)   (d)

FIGURE 13.8 Schematic illustration of typical defects in flat rolling: (a) wavy edges; (b) zipper cracks in the center of the strip; (c) edge cracks; and (d) alligatoring.

13.3.2   Other Characteristics of Rolled Metals

Residual Stresses.   Because of nonuniform deformation of the material in the roll gap, residual stresses can develop in rolled plates and sheets, especially in cold rolling. Small-diameter rolls or small thickness reductions per pass tend to plastically deform the metal to a higher degree at its surfaces than in its bulk (Fig. 13.9a). This situation results in compressive residual stresses on the surfaces and tensile stresses in the bulk. Conversely, large-diameter rolls or high reductions per pass tend to deform the bulk more than its surfaces (Fig. 13.9b). This is due to the higher frictional constraint

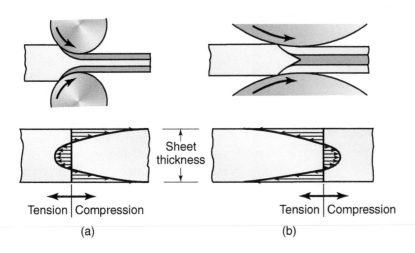

Tension | Compression   Tension | Compression

(a)   (b)

FIGURE 13.9   (a) Residual stresses developed in rolling with small-diameter rolls or at small reductions in thickness per pass. (b) Residual stresses developed in rolling with large-diameter rolls or at high reductions per pass; note the reversal of the residual stress patterns.

at the surfaces along the arc of contact, a situation that produces residual stress distributions that are the opposite of those with small-diameter rolls.

Dimensional Tolerances. Thickness tolerances for cold-rolled sheets usually range from ±0.01 to 0.05 mm (±0.0004 to 0.002 in.); tolerances are much greater for hot-rolled plates. *Flatness tolerances* are usually within ±15 mm/m (±3/16 in./ft) for cold rolling and ±55 mm/m (±5/8 in./ft) for hot rolling.

Surface Roughness. The ranges of surface roughness in cold and hot rolling are given in Fig. 23.14, which, for comparison, also includes other manufacturing processes. Note that cold rolling can produce a very fine surface finish; thus, products made of cold-rolled sheets may not require additional finishing operations. Note also in the figure that hot rolling and sand casting produce the same range of surface roughness.

Gage Numbers. The thickness of a sheet is usually identified by a *gage number*: the smaller the number, the thicker the sheet. Several numbering systems are used in industry, depending on the type of sheet metal. Rolled sheets of copper and brass are generally identified by thickness changes during rolling, such as 14 hard, 12 hard, and so on.

13.4 Rolling Mills

Several types of *rolling mills* and equipment are available, with a range of sizes and diverse roll arrangements. Although the designs of equipment for hot and cold rolling are essentially the same, there are important differences in the roll materials, process parameters, lubricants, and cooling systems. The design, construction, and operation of rolling mills (Fig. 13.10) require major investments. Highly automated mills now produce close-tolerance, high-quality plates and sheets, at high production rates and at low cost per unit weight, particularly when integrated with continuous casting (Section 5.4). The width of rolled products may range up to 5 m (200 in.), and rolling speeds are up to 40 m/s (130 ft/s).

Two-high rolling mills (Fig. 13.3a) are used for hot rolling in initial breakdown passes (*primary roughing* or **cogging mills**) on cast ingots or in continuous casting, with roll diameters ranging from 0.6 to 1.4 m (24–55 in.). In the **three-high mill** (*reversing mill*, Fig. 13.3b) the direction of material movement through the rolls is reversed after each pass, using an elevator mechanism and various manipulators.

Four-high mills (Fig. 13.3c) and **cluster mills** (Sendzimir or Z mill, Fig. 13.3f) use small-diameter work rolls to lower roll forces (because of smaller roll-strip contact area), and thus lower power requirements and reduced spreading. Moreover, when worn or

Coil storage   Take-up reel   Mill stands

Operator controls

FIGURE 13.10   View of a rolling mill. *Source:* Courtesy of Ispat Inland.

broken, small rolls can be replaced at lower cost than can large ones. On the other hand, small rolls deflect more under roll forces, and thus have to be supported by other larger—diameter rolls. Although the cost of a Sendzimir mill facility is very high, it is particularly suitable for cold rolling thin sheets of high-strength metals and alloys. Common rolled widths in this mill are 0.66 m (26 in.), with a maximum of 1.5 m (60 in.).

In **tandem rolling**, the strip is rolled continuously, through a number of **stands**, to thinner gages with each pass (Fig. 13.11). Each stand consists of a set of rolls, with its own housing and controls; a group of stands is called a *train*. The control of the strip thickness and the speed at which the strip travels through each roll gap is critical. Extensive electronic and computer controls are used in these operations, particularly in precision rolling at high speeds.

Roll Materials.   The basic requirements for roll materials are strength and resistance to wear (see also Table 5.8). Common roll materials are cast iron, cast steel, and forged steel; tungsten carbide is also used for small-diameter rolls, such as the work roll in the cluster mill (Fig. 13.3f). Forged-steel rolls, although more costly than cast rolls, have higher strength, stiffness, and toughness than cast-iron rolls. Rolls for cold rolling are ground to a fine finish, and for special applications, they are also polished. Rolls made for cold rolling should not be used for hot rolling, because they may crack due to thermal cycling (*heat checking*) or *spall* (cracking or flaking of surface layers).

Lubricants.   Hot rolling of ferrous alloys is usually carried out without lubricants, although graphite may be used to reduce friction. Water-based solutions may be used to cool the rolls and to break up the scale on the rolled material. Nonferrous

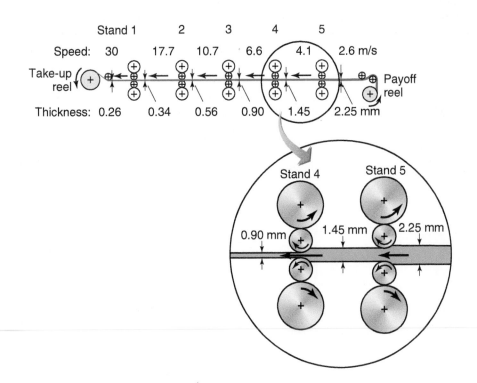

FIGURE 13.11   An example of a tandem-rolling operation.

alloys are hot rolled using a variety of compounded oils, emulsions, and fatty acids. Cold rolling is carried out with water-soluble oils or low-viscosity lubricants, such as mineral oils, emulsions, paraffin, and fatty oils. (See also Section 33.7.)

13.4.1   Integrated Mills and Minimills

Integrated Mills.   These mills are large facilities that involve complete integration of the activities, from the production of hot metal in a blast furnace to the casting and rolling of finished products, ready to be shipped to the customer.

Minimills.   Competition in the steel industry has led to the development of *minimills*, in which scrap metal is (a) melted in electric-arc furnaces, (b) cast continuously, and (c) rolled directly into specific lines of products. Each minimill produces essentially one type of rolled product (rod, bar, or structural sections such as angle iron), from basically one type of metal or alloy. The scrap metal, obtained locally to reduce transportation costs, is typically old machinery, cars, and farm equipment. Minimills have the economic advantage of lower capital equipment costs for each type of metal and product line, with low labor and energy costs. The products typically are aimed at markets in the mill's particular geographic location.

13.5   Various Rolling Processes and Mills

Several rolling processes and mills have been developed to produce a variety of product shapes.

QR Code 13.2 Animation of shape rolling. (*Source: Courtesy of Sandvik Coromant*)

Shape Rolling.   Straight and long structural shapes (such as channels, I-beams, railroad rails, and solid bars) are formed by *shape rolling (profile rolling)*, in which the heated stock passes through pairs of specially designed rolls (Fig. 13.12; see also Fig. 13.1). *Cold shape rolling* also can be done, with the starting materials in the shape of rod, with various cross-sections. Because the entering material's cross-section is reduced nonuniformly, the proper design of a roll sequence, called **roll-pass design**, is critical in order to prevent the formation of external and internal defects, hold dimensional tolerances, and to reduce roll wear.

Roll Forging.   In this operation, also called *cross rolling*, the cross-section of a round bar is shaped by passing it through a pair of rolls with specially profiled grooves (Fig. 13.13). This process is typically used to produce tapered shafts and leaf springs, table knives, and hand tools. Roll forging also may be used as a preliminary forming operation, to be followed by other forging processes (Chapter 14).

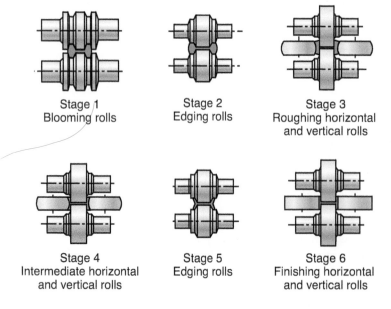

| Stage 1 Blooming rolls | Stage 2 Edging rolls | Stage 3 Roughing horizontal and vertical rolls |

Stage 4 Intermediate horizontal and vertical rolls   Stage 5 Edging rolls   Stage 6 Finishing horizontal and vertical rolls

FIGURE 13.12   Steps in the shape rolling of an I-beam. Various other structural sections, such as channels and rails, also are rolled by this process.

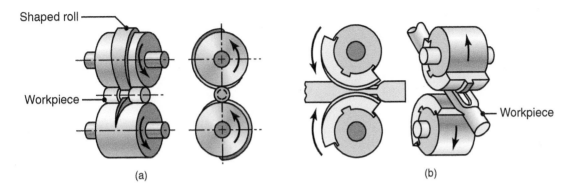

FIGURE 13.13 Two examples of the roll-forging operation, also known as cross rolling; tapered leaf springs and knives can be made by this process. *Source: After J. Holub.*

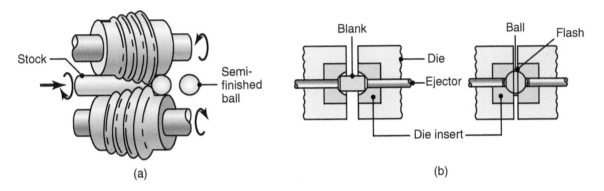

FIGURE 13.14 (a) Production of steel balls by the skew-rolling process. (b) Production of steel balls by upsetting a cylindrical blank: note the formation of flash. The balls made by these processes are subsequently ground and polished for use in ball bearings.

Skew Rolling. A process similar to roll forging is *skew rolling*, typically used for making ball bearings (Fig. 13.14a). Round wire or rod is fed into the roll gap, and spherical blanks are formed continuously by the action of the rotating rolls. (Another method of forming ball bearings is illustrated in Fig. 13.14b, which is basically a forging operation, described in Fig. 14.12.) The balls, which require further finishing, are subsequently ground and polished in special machinery (see Fig. 26.15).

Ring Rolling. In *ring rolling*, a ring-shaped blank is placed between two rolls, one of which is driven while the other idles (Fig. 13.15a). The thickness of the ring is reduced by bringing the rolls closer together as they rotate. Since the volume of the ring remains constant during deformation (volume constancy), the reduction in ring thickness results in its increase in diameter. The process can be carried out at room temperature or at an elevated temperature, depending on the size (which can be up to 10 ft, or 3 m, in diameter), and strength and ductility of the material.

Depending on its size, the blank may be produced by such means as cutting from a plate, piercing, or sawing or shearing a thick-walled pipe. Various shapes can be ring rolled using shaped rolls (Figs. 13.15b–d). The thickness of rings also can be reduced by an open-die forging process, as illustrated in Fig. 14.4c; however, dimensional control and surface finish will not be as good as in ring rolling.

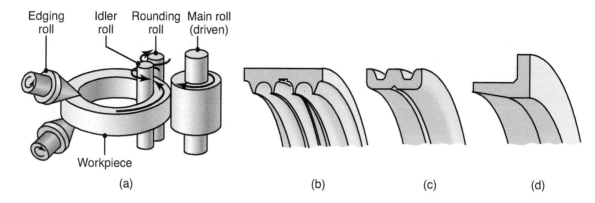

(a)   (b)   (c)   (d)

FIGURE 13.15   (a) Schematic illustration of a ring-rolling operation: thickness reduction results in an increase in the part diameter. (b) through (d) Examples of cross-sections that can be formed by ring rolling.

QR Code 13.3 Ring-rolling operations. (*Source:* Courtesy of the Forging Industry Association, www .forging.org)

Typical applications of ring rolling are body casings for rockets and jet-engine, ball- and roller-bearing races, flanges, and reinforcing rings for pipes. Compared with other manufacturing processes that are capable of producing the same part, the advantages of ring rolling are short production times, low scrap, close dimensional tolerances, and favorable grain flow in the product, thus enhancing its strength in the desired direction.

Thread Rolling.   *Thread rolling* is a cold-forming process by which straight or tapered threads are formed on round rods or wire. The threads are formed on the rod or wire with each stroke of a pair of flat reciprocating dies (Fig. 13.16a). In another method, threads are formed using two rolls (Fig. 13.16b) or *rotary* or *planetary dies* (Fig. 13.16c), at production rates as high as 80 pieces per second. Typical parts made are screws, bolts, and similar threaded parts. Depending on die design, the major diameter of a rolled thread may or may not be larger than a machined thread (Fig. 13.17a), that is, the same as the blank diameter.

The thread-rolling process has the advantages of generating threads at high production rates and without any scrap. The surface finish produced is very smooth, and the process induces compressive residual stresses on the surfaces, thus improving fatigue life. Thread rolling is superior to other methods of thread manufacture (notably thread cutting, as illustrated in Fig. 23.1k), because machining the threads cuts through the grain-flow lines of the material, whereas rolling the threads results in a grain-flow pattern that improves thread strength (Figs. 13.17b and c). **Internal thread rolling** can be carried out with a fluteless **forming tap** (Section 23.7). This is an operation that is similar to external thread rolling, and it produces accurate internal threads and with good strength.

Spur and helical gears can be produced by a cold-rolling process similar to thread rolling (see also Section 24.7); the operation may be carried out on solid cylindrical blanks or on precut gears. Cold rolling of gears has extensive applications in automatic transmissions and in power tools.

Lubrication is important in thread-rolling operations in order to obtain good surface finish and surface integrity and to minimize defects. Lubrication affects the manner in which the material deforms during deformation, an important consideration because of the possibility of internal defects being developed. Typically made of hardened steel, rolling dies are expensive because of their complex shape. The

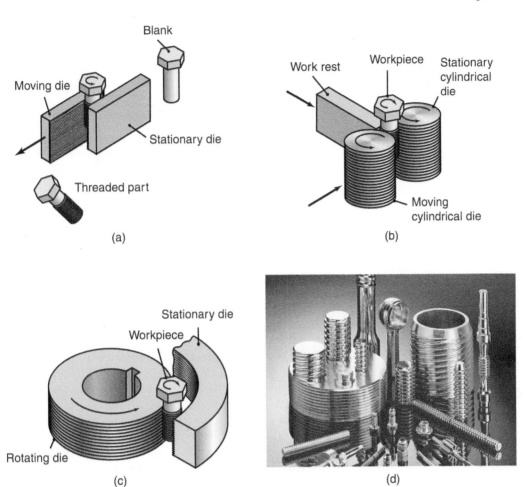

FIGURE 13.16   Thread-rolling processes: (a) reciprocating flat dies used to produce a threaded fastener; (b) two-roll dies; (c) rotary or planetary die set; and (d) a collection of thread-rolled parts made economically at high production rates. *Source:* (d) courtesy of Tesker Manufacturing Corp., Saukville, Wisconsin.

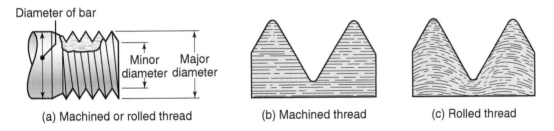

FIGURE 13.17   (a) Features of a machined or rolled thread. Grain flow in (b) machined and (c) rolled threads. Unlike machining, which cuts through the grains of the metal, the rolling of threads imparts improved strength because of cold working and favorable grain flow.

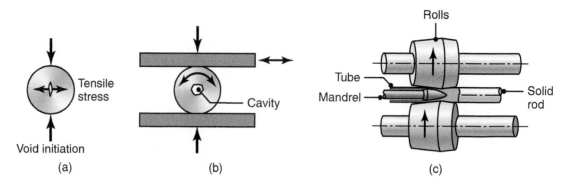

FIGURE 13.18 Cavity formation in a solid, round bar and its utilization in the rotary tube piercing process for making seamless pipe and tubing. (See also Fig. 2.9.)

dies usually cannot be reground after they are worn. With proper die materials and preparation, however, die life may range up to millions of pieces.

Rotary Tube Piercing. Also known as the **Mannesmann process**, this is a hot-working operation for making long, thick-walled *seamless pipe and tubing* (Fig. 13.18). Developed in the 1880s, this process is based on the principle that when a round bar is subjected to radial compressive forces, tensile stresses develop at its center (see Fig. 2.9). When continuously subjected to these cyclic compressive stresses (Fig. 13.18b), the bar begins to develop a small cavity at its center, which then begins to grow. This phenomenon can be demonstrated with a short piece of round eraser, by rolling it back and forth on a hard flat surface, as shown in Fig. 13.18b.

Rotary tube piercing is carried out using an arrangement of rotating rolls (Fig. 13.18c). The axes of the rolls are *skewed* in order to pull the round bar through

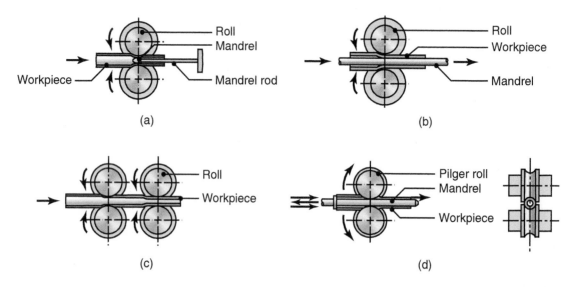

FIGURE 13.19 Schematic illustration of various tube-rolling processes: (a) with a fixed mandrel; (b) with a floating mandrel; (c) without a mandrel; and (d) pilger rolling over a mandrel and a pair of shaped rolls. Tube diameters and thicknesses also can be changed by other processes, such as drawing, extrusion, and spinning.

the rolls, by the axial component of the rotary motion. An internal mandrel assists the operation by expanding the hole and sizing the inside diameter of the tube. The mandrel may be held in place by a long rod or it may be a floating mandrel, without a support (see Fig. 15.21c for a similar floating mandrel, used in drawing). Because of the severe deformation that the bar undergoes, the round blank must be of high quality and free of inclusions.

Tube Rolling.   The diameter and thickness of pipes and tubing can be reduced by *tube rolling*, which utilizes shaped rolls, arranged in various configurations (Fig. 13.19). These operations can be carried out either with or without an internal mandrel. In the *pilger mill*, the tube and an internal mandrel undergo a reciprocating motion; the rolls are specially shaped and are rotated continuously. During the gap cycle on the roll, the tube is advanced and rotated, starting another cycle of tube reduction, whereby the tube undergoes a reduction in both diameter and wall thickness. Steel tubing 265 mm (10.5 in.) in diameter has been produced by this process. Other operations for tube manufacturing are described in Chapter 15.

SUMMARY

- Rolling is the process of reducing the thickness or changing the cross-section of a long strip by compressive forces applied through a set of rolls. Shape rolling is used to make products with various cross-sections. Other rolling operations include ring rolling and thread rolling.

- Rolling may be carried out at room temperature (cold rolling) or at elevated temperatures (hot rolling). The process involves several material and process variables, including roll diameter (relative to material thickness), reduction per pass, speed, lubrication, and temperature. Spreading, bending, and flattening are important considerations for controlling the dimensional accuracy of the rolled stock.

- Rolling mills have a variety of roll configurations, such as two-high, three-high, four-high, cluster (Sendzimir), and tandem. Front and/or back tension may be applied to the material to reduce roll forces.

- Continuous casting and rolling of ferrous and nonferrous metals into semifinished products are now in common practice because of the economic benefits.

- Integrated mills are large facilities involving the complete sequence of activities, from the production of hot metal in a blast furnace to the casting and the rolling of finished products ready to be shipped to the customer. On a much smaller scale, minimills utilize scrap metal that is melted in electric-arc furnaces, cast, and continuously rolled into specific lines of products.

KEY TERMS

Alligatoring	Chatter	Foil	Neutral point
Back tension	Cogging mill	Front tension	Pack rolling
Billet	Cold rolling	Gage number	Pilger mill
Bloom	Crown	Hot rolling	Plate
Camber	Draft	Mannesmann process	Ring rolling
Cast structure	Flat rolling	Minimill	Roll

Roll forging	Sendzimir mill	Spreading	Thread rolling
Roll stand	Shape rolling	Stand	Tube rolling
Rolling	Sheet	Steckel rolling	Wrought structure
Rolling mill	Skew rolling	Tandem rolling	
Rotary tube piercing	Slab	Temper rolling	

BIBLIOGRAPHY

Ginzburg, V.B., and Ballas, R., **Flat Rolling Fundamentals,** CRC Press, 2001.

Hosford, W.F., and Caddell, R.M., **Metal Forming: Mechanics and Metallurgy,** 4th ed., Prentice Hall, 2011.

Lee, Y., **Rod and Bar Rolling: Theory and Applications.** CRC Press, 2004.

Lenard, J.G., **Primer on Flat Rolling,** Elsevier, 2007.

Pittner, J., and Simaan, M.A., **Tandem Cold Metal Rolling Mill Control: Using Practical Advanced Methods,** Springer, 2010.

Tschaetch, H., **Metal Forming Practise: Processes, Machines, Tools,** Springer, 2007.

REVIEW QUESTIONS

13.1 What is the difference between a plate and a sheet?

13.2 Define roll gap, neutral point, and draft.

13.3 What factors contribute to spreading in flat rolling?

13.4 What is forward slip? Why is it important?

13.5 Describe the types of deflections that rolls undergo.

13.6 Describe the difference between a bloom, a slab, and a billet.

13.7 Why may roller leveling be a necessary operation?

13.8 List the defects commonly observed in flat rolling.

13.9 What are the advantages of tandem rolling? Pack rolling?

13.10 How are seamless tubes produced?

13.11 Why is the surface finish of a rolled product better in cold rolling than in hot rolling?

13.12 What is a Sendzimir mill? What are its important features?

13.13 What is the Mannesmann process? How is it different from tube rolling?

13.14 Describe ring rolling. Is there a neutral plane in ring rolling?

13.15 How is back tension generated?

QUALITATIVE PROBLEMS

13.16 Explain why the rolling process was invented and developed.

13.17 Flat rolling reduces the thickness of plates and sheets. It is possible, instead, to reduce their thickness simply by stretching the material? Would this be a feasible process? Explain.

13.18 Explain how the residual stress patterns shown in Fig. 13.9 become reversed when the roll radius or reduction-per-pass is changed.

13.19 Explain whether it would be practical to apply the roller-leveling technique shown in Fig. 13.7a to thick plates.

13.20 Describe the factors that influence the magnitude of the roll force, F, in Fig. 13.2c.

13.21 Explain how you would go about applying front and back tensions to sheet metals during rolling. How would you go about controlling these tensions?

13.22 What typically is done to make sure that the product in flat rolling is not crowned?

13.23 Make a list of parts that can be made by (a) shape rolling and (b) thread rolling.

13.24 Describe the methods by which roll flattening can be reduced. Which property or properties of the roll material can be increased to reduce roll flattening?

13.25 It was stated that spreading in flat rolling increases with (a) a decreasing width-to-thickness ratio of the entering material, (b) decreasing friction, and (c) a decreasing ratio of the roll radius to the strip thickness. Explain why.

13.26 Flat rolling can be carried out by front tension only, using idling rolls (Steckel rolling). Since the torque on the rolls is now zero, where, then, is the energy coming from to supply the work of deformation in rolling?

13.27 Explain the consequence of applying too high a back tension in rolling.

13.28 Note in Fig. 13.3f that the driven rolls (powered rolls) are the third set from the work roll. Why isn't power supplied through the work roll itself? Is it even possible? Explain.

13.29 Describe the importance of controlling roll speeds, roll gaps, temperature, and other process variables in a tandem-rolling operation, as shown in Fig. 13.11. Explain how you would go about determining the optimum distance between the stands.

13.30 In Fig. 13.9a, if you remove the top compressive layer by, say, grinding, will the strip remain flat? If not, which way will it curve and why?

13.31 Name several products that can be made by each of the operations shown in Fig. 13.1.

13.32 List the possible consequences of rolling at (a) too high of a speed and (b) too low of a speed.

13.33 It is known that in thread rolling, as illustrated in Fig. 13.16, a workpiece must make roughly six revolutions to form the thread. Under what conditions (process parameters, thread geometry or workpiece properties) can deviation from this rule take place?

13.34 If a rolling mill encounters chatter, what process parameters would you change, and in what order? Explain.

13.35 Can the forward slip ever become negative? Why or why not?

QUANTITATIVE PROBLEMS

13.36 In Example 13.1, calculate the roll force and the power for the case in which the workpiece material is 1100-O aluminum and the roll radius, R, is 8 in.

13.37 Calculate the individual drafts in each of the stands in the tandem-rolling operation shown in Fig. 13.11.

13.38 Estimate the roll force, F, and the torque for an AISI 1020 carbon-steel strip that is 200 mm wide, 10 mm thick, and rolled to a thickness of 7 mm. The roll radius is 200 mm, and it rotates at 200 rpm.

13.39 A rolling operation takes place under the conditions shown in Fig. P13.39. What is the position, x_n, of the neutral point? Note that there are front and back tensions that have not been specified. Additional data are as follows: Material is 5052-O aluminum; hardened steel rolls; surface roughness of the rolls = 0.02 μm; rolling temperature = 210°C.

13.40 Estimate the roll force and power for annealed low-carbon steel strip 200 mm wide and 10 mm thick, rolled to a thickness of 6 mm. The roll radius is 200 mm, and the roll rotates at 200 rpm; use $\mu = 0.1$.

13.41 A flat-rolling operation is being carried out where $h_o = 0.20$ in., $h_f = 0.15$ in., $w_o = 10$ in., $R = 8$ in., $\mu = 0.25$, and the average flow stress of the material is 40,000 psi. Estimate the roll force and the torque; include the effects of roll flattening.

13.42 It can be shown that it is possible to determine μ in flat rolling without measuring torque or forces. By inspecting the equations for rolling, describe an experimental procedure to do so. Note that you are allowed to measure any quantity other than torque or forces.

13.43 Assume that you are an instructor covering the topics described in this chapter, and you are giving a quiz on the numerical aspects to test the understanding of the students. Prepare two quantitative problems and supply the answers.

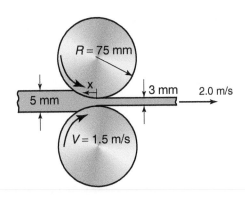

FIGURE P13.39

SYNTHESIS, DESIGN, AND PROJECTS

13.44 A simple sketch of a four-high mill stand is shown in Fig. 13.3c. Make a survey of the technical literature and present a more detailed sketch for such a stand, showing the major components.

13.45 Obtain a piece of soft, round rubber eraser, such as that at the end of a pencil, and duplicate the process shown in Fig. 13.18b. Note how the central portion of the eraser will begin to disintegrate, producing a rough hole.

13.46 If you repeat the experiment in Problem 13.45 with a harder eraser, such as that used for erasing ink, you will note that the whole eraser will begin to crack and crumble. Explain why.

13.47 Design a set of rolls to produce cross-sections other than those shown in Fig. 13.12.

13.48 Design an experimental procedure for determining the neutral point in a flat rolling operation.

13.49 Using a rolling pin and any available dough (bread, cookie, etc), measuring 100 mm by 100 mm by 8 mm, quantify the spreading in flat rolling for different reductions in thickness.

13.50 Derive an expression for the thickest workpiece that can be drawn between two rolls as a function of roll gap, roll radius, and coefficient of friction.

CHAPTER 14

Metal-forging Processes and Equipment

- This chapter describes the fundamentals of forging and related processes, including design and economic considerations.

- Open-die forging operations for producing simple shapes are discussed first, followed by impression-die and closed-die forging operations for producing more intricate shapes.

- Various forging operations, such as heading, piercing, coining, swaging, and cold extrusion, are then introduced.

- Factors involved in forging defects and die failures are explained.

- The economics of forging, as it relates to process selection, is also discussed.

- The chapter ends with a review of the design of parts to be forged, guidelines for die design and manufacturing, and selection of die materials and lubricants in forging operations.

Typical parts made by forging and related processes: Shafts, gears, bolts, turbine blades, hand tools, dies, and components for machinery, transportation, and farm equipment.

Alternative processes: Casting, powder metallurgy, machining, and fabrication.

14.1 Introduction

Forging is a basic process in which the workpiece is shaped by compressive forces applied through various dies and tooling. One of the oldest and most important metalworking operations, dating back at least to 4000 B.C., forging was first used to make jewelry, coins, and various implements, by hammering metal with tools made of stone. Forged parts now include large rotors for turbines, gears, cutlery (Fig. 14.1a), hand tools, components for machinery, aircraft (Fig. 14.1b), railroads, and transportation equipment.

Unlike rolling operations, described in Chapter 13, that generally produce continuous plates, sheets, strips, and various structural cross-sections, forging operations produce discrete parts. Because the metal flow in a die and the material's grain structure can be controlled, forged parts have good strength and toughness, and are very reliable for highly stressed and critical applications (Fig. 14.2). Simple forging operations can be performed with a heavy hammer and an anvil, as has been done traditionally by blacksmiths. Most forgings require a set of dies and such equipment as press or powered hammers.

QR Code 14.1 What is Forging. (*Source:* Courtesy of the Forging Industry Association, www.forging.org)

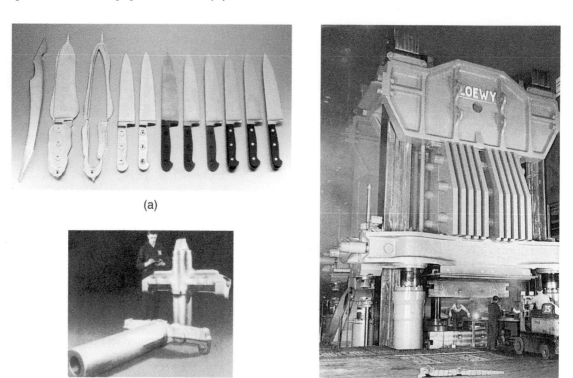

FIGURE 14.1 (a) Illustration of the steps involved in forging a knife. (b) Landing-gear components for the C5A and C5B transport aircraft, made by forging. (c) General view of a 445-MN (50,000-ton) hydraulic press. *Source:* (a) Courtesy of Mundial, LLC. (b) and (c) Courtesy of Wyman-Gordon Company.

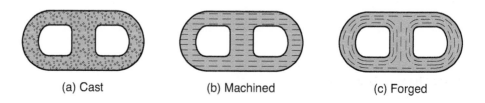

FIGURE 14.2 Schematic illustration of a part (dragline chain link, approximately 2 m long) made by three different processes and showing grain flow. Each process has its own advantages and limitations regarding external and internal characteristics, material properties, dimensional accuracy, surface finish, and the economics of production. *Source:* Courtesy of the Forging Industry Association.

Forging may be carried out at room temperature (*cold forging*) or at elevated temperatures (*warm* or *hot forging*), depending on the homologous temperature, described in Section 1.8. Cold forging requires higher forces, because of the higher strength of the workpiece material, which also must possess sufficient ductility at room temperature to be able to undergo the required deformation without cracking. Cold-forged parts have good surface finish and dimensional accuracy. Hot forging requires lower forces, but the dimensional accuracy and surface finish of the parts are not as good as those in cold forging.

Forgings generally are subjected to additional finishing operations, such as heat treating to modify properties and machining for accuracy in final dimensions and good surface finish. The finishing operations can be minimized by *precision forging*, which is an important example of *net-shape* or *near-net-shape* forming processes. As described throughout this book, parts that can be forged successfully also may be manufactured economically by other methods, such as casting (Chapter 11), powder metallurgy (Chapter 17), or machining (Part IV). Each of these methods will produce a part having different characteristics, particularly with regard to strength, toughness, dimensional accuracy, surface finish, and the possibility of internal or external defects.

14.2 Open-die Forging

Open-die forging is the simplest forging operation (Table 14.1). Although most open-die forgings generally weigh 15 to 500 kg (30–1000 lb), forgings as heavy as 300 tons have been made. Part sizes may range from very small (such as pins, nails, and bolts) to very large [up to 23 m (75 ft) long shafts for ship propellers]. Open-die forging can be simply described by a metal workpiece (blank), placed between two flat dies, and reduced in height by compressing it (Fig. 14.3), a process called **upsetting** or **flat-die forging**. The die surfaces may have shallow cavities or features to produce relatively simple shapes.

The deformation of a workpiece under *frictionless* conditions is shown in Fig. 14.3b. Because constancy of volume is maintained, any reduction in height increases the diameter of the forged part. Note in the figure that the workpiece is deformed uniformly; in actual operations, however, there is friction at the die–workpiece interfaces, and the part develops a barrel shape (Fig. 14.3c), a deformation mode also called *pancaking*.

Barreling is caused primarily by frictional forces that oppose the outward flow of the workpiece at the die interfaces; thus, it can be minimized by using an effective

QR Code 14.2 Open-die forging operations. (*Source:* Courtesy of the Forging Industry Association, www .forging.org)

QR Code 14.3 Animation of open-die forging. (*Source:* Courtesy of Sandvik Coromant)

TABLE 14.1

General Characteristics of Forging Processes		
Process	Advantages	Limitations
Open die	Simple and inexpensive dies; wide range of part sizes; good strength characteristics; generally for small quantities	Limited to simple shapes; difficult to hold close tolerances; machining to final shape necessary; low production rate; relatively poor utilization of material; high degree of skill required
Closed die	Relatively good utilization of material; generally better properties than open-die forgings; good dimensional accuracy; high production rates; good reproducibility	High die cost, not economical for small quantities; machining often necessary
Blocker	Low die costs; high production rates	Machining to final shape necessary; parts with thick webs and large fillets
Conventional	Requires much less machining than blocker type; high production rates; good utilization of material	Higher die cost than blocker type
Precision	Close dimensional tolerances; very thin webs and flanges possible; machining generally not necessary; very good material utilization	High forging forces, intricate dies, and provision for removing forging from dies

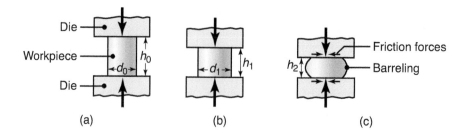

FIGURE 14.3 (a) Solid cylindrical billet upset between two flat dies. (b) Uniform deformation of the billet without friction. (c) Deformation with friction; note barreling of the billet caused by friction forces at the billet–die interfaces.

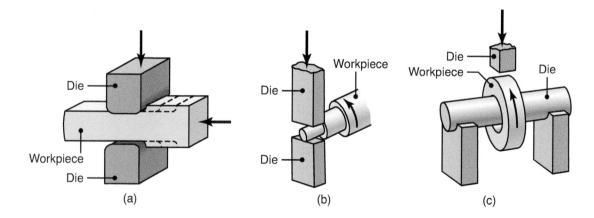

FIGURE 14.4 (a) Schematic illustration of a cogging operation on a rectangular bar. Blacksmiths use this process to reduce the thickness of bars by hammering the part on an anvil; reduction in thickness is accompanied by barreling. (b) Reducing the diameter of a bar by open-die forging; note the movements of the dies and the workpiece. (c) The thickness of a ring being reduced by open-die forging.

lubricant. Barreling also can develop in upsetting hot workpieces between cold dies. The material at the die surfaces cools rapidly, while the rest remains relatively hot; consequently, the material at the top and bottom of the workpiece has higher resistance to deformation than the material at the center. As a result, the central portion of the workpiece expands laterally to a greater extent than do the ends. Barreling from thermal effects can be reduced or eliminated by using heated dies. Thermal barriers, such as glass cloth at the die–workpiece interfaces also can be used for this purpose.

Cogging, also called *drawing out*, is an open-die forging operation in which the thickness of a bar is reduced by successive forging steps (*bites*) at specific intervals (Fig. 14.4a). The thickness of bars and rings also can be reduced by similar open-die forging techniques, as illustrated in Figs. 14.4b and c. Because the contact area between the die and the workpiece is small, a long section of a bar can thus be reduced in thickness without requiring large forces or heavy machinery. Note that blacksmiths have been performing such operations for centuries on hot workpieces, using a hammer and an anvil. Cogging of larger workpieces is usually done using mechanized equipment and computer controls, in which lateral and vertical movements of the dies are coordinated to produce the desired part shape.

Forging Force. The *forging force*, F, in an *open-die forging operation* on a solid cylindrical workpiece can be estimated from the formula

$$F = Y_f \pi r^2 \left(1 + \frac{2\mu r}{3h} \right),$$

(14.1)

where Y_f is the *flow stress* of the material (see Example 14.1), μ is the coefficient of friction between the workpiece and the die, and r and h are, respectively, the instantaneous radius and height of the workpiece.

Friction Hill. Consider the upsetting of a solid cylinder, as depicted in Fig. 14.3. If the workpiece-die interface is frictionless, then the die pressure is equal to the flow stress of the material. If friction is present, as is the case in actual forging operations, then calculating the die pressure is more involved. For upsetting of a cylinder with outer radius r_o, height, h, and coefficient of friction, μ, the die pressure at any radius can be expressed as

$$p = Ye^{2\mu(r_o - r)/h}.$$

(14.2)

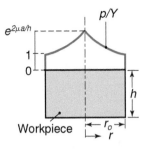

FIGURE   14.5 Distribution of die pressure in upsetting with sliding friction. Note that the pressure at the outer radius is equal to the flow stress, Y, of the material. Sliding friction means that the frictional stress is directly proportional to the normal stress.

The die-pressure distribution is plotted in Fig. 14.5. Note that the pressure is maximum at the center of the workpiece, and can be very high, especially if the diameter-to-height ratio of the workpiece is large. Because of its shape, the pressure-distribution curve in Fig. 14.5 is referred to as the *friction hill*.

EXAMPLE 14.1   Calculation of Forging Force in Upsetting

Given: A solid cylindrical workpiece made of 304 stainless steel is 150 mm (6 in.) in diameter and 100 mm (4 in.) high. It is reduced in height by 50%, at room temperature, by open-die forging with flat dies. Assume that the coefficient of friction is 0.2.

Find: Calculate the forging force at the end of the stroke.

Solution: The forging force at the end of the stroke is calculated using Eq. (14.1), in which the dimensions pertain to the final dimensions of the forging. The final height is $h = 100/2 = 50$ mm, and the final radius, r, is determined from volume constancy by equating the volumes before and after deformation. Hence,

$$(\pi)(75)^2(100) = (\pi)(r)^2(50).$$

Thus, $r = 106$ mm (4.17 in.).

The quantity Y_f in Eq. (14.1) is the flow stress of the material, which is the stress required to continue plastic deformation of the workpiece at a particular true strain. The absolute value of the true strain that the workpiece has undergone at the end of the stroke in this operation is

$$\epsilon = \ln\left(\frac{100}{50}\right) = 0.69.$$

The flow stress can be determined by referring to Eq. (2.8) and noting from Table 2.3 that, for 304 stainless steel, $K = 1275$ MPa and $n = 0.45$. Thus, for a true strain of 0.69, the flow stress is calculated to be 1100 MPa. Another method is to refer to Fig. 2.5 and note that the flow stress for 304 stainless steel at a true strain of 0.69 is about 1000 MPa (140 ksi). The small difference between the two values is due to the fact that the data in Table 2.3 and Fig. 2.5 are from different sources.

Taking the latter value for flow stress, the forging force can now be calculated, noting that in this problem the units in Eq. (14.1) must be in N and m. Thus,

$$F = \left(1000 \times 10^6 \right)(\pi)(0.106)^2(1) + \frac{(2)(0.2)(0.106)}{(3)(0.050)}$$

$$= 4.5 \times 10^7 \text{ N} = 45 \text{ MN} = 10^7 \text{ lb} = 5000 \text{ tons}.$$

14.3 Impression-die and Closed-die Forging

In *impression-die forging*, the workpiece takes the shape of the die cavity while being forged between two shaped dies (Figs. 14.6a–c). This process is usually carried out at elevated temperatures, in order to lower the forging forces and develop enhanced ductility of the workpiece. Note in Fig. 14.6c that, during deformation, some of the material flows outward and forms a **flash**.

The flash has an important role in impression-die forging: The high pressure, and the resulting high frictional resistance in the flash, presents a severe constraint on any radially outward flow of the material in the die. This is due to the friction hill effect,

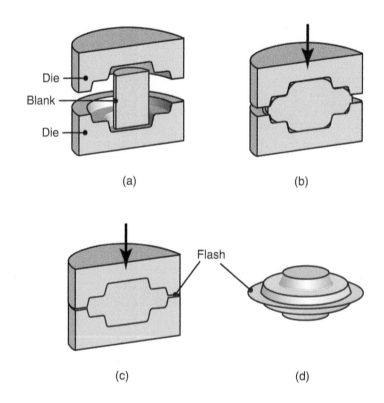

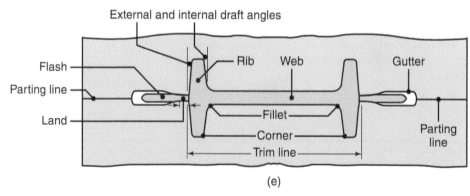

FIGURE 14.6 (a) through (d) Stages in impression-die forging of a solid round billet; note the formation of flash, which is excess metal that is subsequently trimmed off. (e) Standard terminology for various features of a forging die.

described in Section 14.2. Thus, based on the principle that, in plastic deformation, the material flows in the direction of least resistance (because it requires less energy), the material flows preferentially into the die cavity, eventually filling it completely.

The standard terminology for a typical forging die is shown in Fig. 14.6e. Instead of being made as one piece, dies may be made of several pieces (segmented), including die inserts (Fig. 14.7), and particularly for complex part shapes. The inserts can easily be replaced in the case of wear or failure in a particular region of the die, and are usually made of stronger and harder materials.

The blank to be forged is prepared by (a) *cropping* (shearing, Section 16.2) from an extruded or drawn bar stock; (b) *powder metallurgy* or *casting*; or (c) a preformed blank from a prior forging operation. The blank is placed on the lower die, and as the upper die begins to descend, its shape gradually changes (Fig. 14.8a).

Preforming operations (Figs. 14.8b and c) are typically used to enhance the distribution of the material into various regions of the blank, using simple dies of various contours. In **fullering**, material is distributed away from a region in the dies. In **edging**, it is gathered into a localized region. The part is then formed into a rough shape by a process called **blocking**, using *blocker dies*. The final operation is the finishing of the forging in *impression dies*, which give the forging its final shape; the flash is later removed by a trimming operation (Fig. 14.9).

Forging Force.   The *forging force*, F, required in an *impression-die forging* operation can be estimated from the formula

$$F = kY_f A, \qquad (14.3)$$

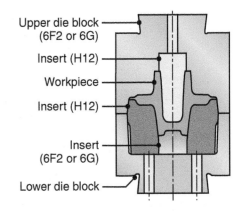

FIGURE 14.7 Die inserts used in forging an automotive axle housing. (See Section 5.7 for die materials.)

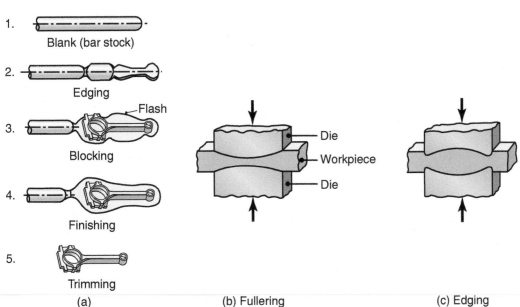

FIGURE 14.8 (a) Stages in forging a connecting rod for an internal combustion engine; note the amount of flash required to ensure proper filling of the die cavities. (b) Fullering and (c) edging operations to distribute the material properly when preshaping the blank for forging.

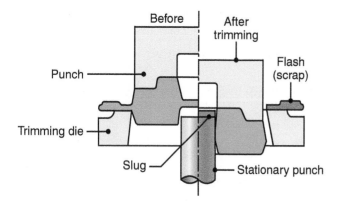

FIGURE 14.9 Trimming flash from a forged part; note that the thin material at the center (slug) is removed by punching.

TABLE 14.2

Range of *k* Values for Eq. (14.3)	
Shape	*k*
Simple shapes, without flash	3–5
Simple shapes, with flash	5–8
Complex shapes, with flash	8–12

where k is a multiplying factor, obtained from Table 14.2, Y_f is the flow stress of the material at the forging temperature, and A is the projected area of the forging, including the flash area.

In hot-forging operations, the actual forging pressure for most metals typically ranges from 550 to 1000 MPa (80–140 ksi). As an example, assume that the flow stress of a material at the forging temperature is 100,000 psi, and a part (such as that shown in Fig. 14.8a) has a projected area (with flash) of 60 in². Taking a value of $k = 10$ from Table 14.2, the forging force would be $F = (10)(100,000)(60) = 60 \times 10^6$ lb, or 26.8 tons.

Closed-die Forging. The process shown in Fig. 14.6 is also referred to as *closed-die forging*. In true closed-die forging, however, a flash does not form (hence the term *flashless forging*), and the workpiece completely fills the die cavity (see right side of Fig. 14.10b). The accurate control of the blank volume and proper die design are essential to producing a forging with the required dimensional tolerances. Undersized blanks prevent the complete filling of the die cavity; conversely, oversized blanks generate excessive pressures and may cause dies to fail prematurely or the machine to jam.

Precision Forging. In order to reduce the number of additional finishing operations (hence cost), the trend continues toward greater precision in forged products (net-shape forming). Typical precision-forged products are gears, connecting rods, and turbine blades. Precision forging requires (a) special and more complex dies, (b) precise control of the blank's volume and shape, and (c) accurate positioning of the blank in the die cavity. Also, because of the higher forces required to obtain fine details on the part, precision forging requires higher capacity equipment. Although steel and titanium also can be precision forged, aluminum and magnesium alloys are particularly suitable, because of the relatively low forging loads and temperatures that they require; however, steels and titanium also can be precision forged.

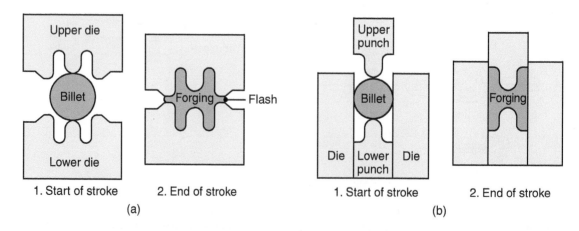

FIGURE 14.10 Comparison of (a) closed-die forging with flash and (b) precision or flashless forging of a round billet. *Source:* After H. Takemasu, V. Vazquez, B. Painter, and T. Altan.

Forging Practice and Product Quality.   A forging operation typically involves the following sequence of steps:

1. A slug, billet, or preform is prepared; if necessary, cleaned by such means as shot blasting.
2. For hot forging, the workpiece is heated in a suitable furnace and then, if necessary, descaled with a wire brush, water jet, steam, or by scraping. Some descaling also may occur during the initial stages of forging, when the thick, brittle scale falls off during forging.
3. For hot forging, the dies are preheated and lubricated; for cold forging, the blank is lubricated.
4. The billet is forged in appropriate dies and in the proper sequence. If necessary, any excess material is removed (especially flash) by trimming, machining, or grinding.
5. The forging is cleaned, checked for dimensions and, if necessary, machined or ground to final dimensions and specified tolerances and surface finish.
6. Additional operations are performed, such as straightening and heat treating (for improved mechanical properties). Any additional finishing operations that may be required are performed.
7. The forging is inspected for any external and internal defects.

The quality, dimensional tolerance, and surface finish of a forging depend on how well these operations have been performed. Generally, dimensional tolerances range between ±0.5% and ±1% of the dimensions of the forging. In good practice, tolerances for hot forging of steel are usually less than ±6 mm (0.25 in.); in precision forging, they can be as low as ±0.25 mm (0.01 in.). Other factors that contribute to dimensional inaccuracies are draft angles, radii, fillets, die wear, die closure (whether the dies have closed properly), and mismatching of the dies.

14.4 Various Forging Operations

Several other operations related to the basic forging process are described below.

Coining.   A closed-die forging process, *coining* was originally used in the minting of coins, medallions, and jewelry (Fig. 14.11), but currently is used to produce a wide range of parts with high accuracy, such as gears, industrial seals, and medical devices. The blank or slug is coined in a completely closed-die cavity, in order to produce fine details, such as in coins; the pressures required can be as high as five or six times the strength of the material. On some parts, several coining operations may be required. Lubricants should not be applied in coining, because they can become entrapped in the die cavities and, being incompressible, prevent the full reproduction of die-surface details and surface finish.

Marking parts with letters and numbers also can be done rapidly through coining. **Sizing** is a process used in forging, and other processes, to improve surface finish and to impart the desired dimensional accuracy with little or no change in part size.

Heading.   Also called **upset forging,** *heading* is essentially an upsetting operation, performed on the end of a rod or wire in order to increase the cross-section. Typical products are nails, bolt heads, screws, rivets, and various other fasteners (Fig. 14.12). Heading can be carried out cold, warm, or hot. An important consideration in heading is the tendency for the bar to *buckle* if its unsupported length-to-diameter ratio is too high. This ratio usually is limited to less than 3:1, but with appropriate dies,

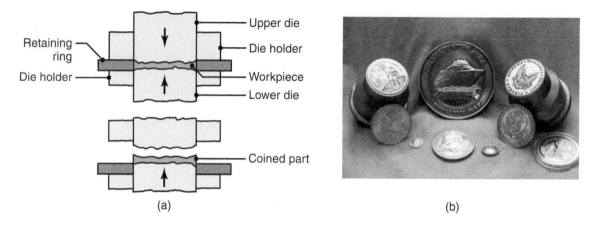

(a)   (b)

FIGURE 14.11 (a) Schematic illustration of the coining process; the earliest coins (see Table I.2) were made by open-die forging and lacked precision and sharp details. (b) An example of a modern coining operation, showing the coins and tooling; note the detail and superior surface finish that can be achieved in this process. *Source:* Courtesy of C & W Steel Stamp Company Inc.

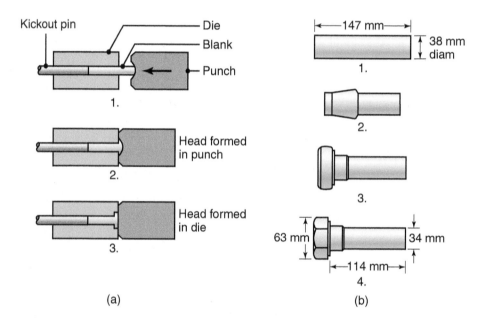

(a)   (b)

FIGURE 14.12 (a) Heading operation to form heads on fasteners, such as nails and rivets. (b) Sequence of operations in producing a typical bolt head by heading.

it can be higher; for example, higher ratios can be accommodated if the diameter of the die cavity is not more than 1.5 times the bar diameter.

Heading operations are performed on machines called **headers,** which usually are highly automated, with production rates of hundreds of pieces per minute for small parts. Hot heading on larger parts typically are performed on **horizontal upsetters.** Heading operations can be combined with cold-extrusion processes to make various parts, as described in Section 15.4.

Punch

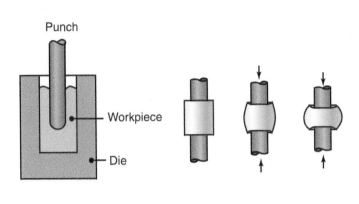

Workpiece

Die

(a)

(b)

FIGURE 14.13   (a) Examples of piercing operations. (b) A pierced round billet showing grain-flow pattern (see also Fig. 14.2c). *Source:* Courtesy of Ladish Co., Inc.

Piercing.   This is a process of indenting (but not breaking through) the surface of a workpiece with a punch, in order to produce a cavity or an impression (Fig. 14.13). The workpiece may be confined in a container, such as a die cavity or may be unconstrained. The extent of surface deformation of the workpiece depends on how much it is constrained from flowing freely as the punch descends. A common example of piercing is the indentation of the hexagonal cavity in bolt heads. Piercing may be followed by punching to produce a hole in the part. (For a similar depiction of this situation, see the slug above the stationary punch in the central portion of Fig. 14.9.) Piercing also is performed to produce hollow regions in forgings, using side-acting auxiliary equipment.

The *piercing force* depends on (a) the cross-sectional area and the tip geometry of the punch, (b) the strength of the workpiece, and (c) friction at the punch–workpiece interfaces. The pressure may range from three to five times the strength of the material, which is about the same level of stress required to make an indentation in hardness testing (Section 2.6).

CASE STUDY 14.1   Manufacture of a Stepped Pin by Heading and Piercing Operations

Figure 14.14a shows a stepped pin made from SAE 1008 steel, and used as a portion of a roller assembly to adjust the position of a car seat. The part is fairly complex and must be produced in a progressive manner, in order to produce the required details and fill the die completely.

The cold-forging steps used to produce this part are shown in Fig. 14.14b. First, a solid, cylindrical blank is extruded (Chapter 15) in two operations, followed by upsetting. The upsetting operation uses a conical cross-section in the die to produce the preform, and is oriented such that material is concentrated at the top of the part in order to ensure proper die filling. After impression-die forming, a piercing operation is performed which forms the bore.

(continued)

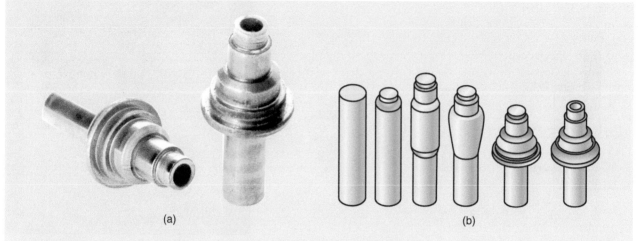

FIGURE 14.14   (a) The stepped pin used in Case Study 14.1. (b) Illustration of the manufacturing steps used to produce the stepped pin. *Source:* Courtesy of National Machinery, LLC.

Hubbing.   This process consists of pressing a hardened punch, with a particular tip geometry, into the surface of a block of metal. The cavity produced is subsequently used as a die for forming operations, such as those employed in making tableware. The die cavity usually is shallow, but for deeper cavities, some material may first be removed from the surface of the block by machining prior to hubbing (see Figs. 24.2c and d). The *hubbing force* can be estimated from the equation

$$\text{Hubbing force} = 3(\text{UTS})(A), \tag{14.4}$$

where UTS is obtained from Table 2.2, and A is the projected area of the impression. For example, for high-strength steel with UTS = 1500 MPa and a part with a projected area of 400 mm^2, the hubbing force is $(3)(1500 \text{ N/mm}^2)(400 \text{ mm}^2) = 1.8$ MN $= 179$ tons.

Orbital Forging.   In this process, the upper die moves along an orbital path and forms the part *incrementally*, an operation that is similar to the action of a mortar and pestle, used for crushing herbs and seeds. Typical components that may be forged by this process are disk-shaped and conical parts, such as bevel gears and gear blanks. The forging force is relatively small, because at any particular instant, the die contact is concentrated onto a small area of the workpiece (see also *incremental forging* below). The operation is relatively quiet, and parts can be formed within 10–20 cycles of the orbiting die.

Incremental Forging.   In this process, a tool forges a blank into a particular shape in several small steps. The operation is somewhat similar to cogging (Fig. 14.4a), in which the die deforms the blank to a different extent at different positions. Because of the smaller area of contact with the die, the process requires much lower forces as compared with conventional impression-die forging, and the tools are simpler and less costly.

Isothermal Forging.   Also known as **hot-die forging**, the dies in this process are heated to the same temperature as that of the hot workpiece (see also Table 14.3). Because the workpiece remains hot (essentially no heat is lost to the dies), its flow strength and high ductility are maintained during forging; thus, the forging load is low, and the material flow within the die cavity is improved. Complex parts can be isothermally forged, with good dimensional accuracy and to near-net shape by one stroke in a hydraulic press. The dies for hot forging usually are made of nickel or molybdenum alloys (because of their resistance to high temperature), but steel dies can be used for aluminum alloys.

The process is expensive and the production rate is low. It can, however, be economical for specialized, intricate forgings, made of such materials as aluminum, titanium, and superalloys, provided that the quantity of forgings to be made is sufficiently high to justify high die costs.

Rotary Swaging.   In this process, also known as *radial forging*, *rotary forging*, or simply *swaging*, a solid rod or tube is subjected to radial impact forces using a set of reciprocating dies (Figs. 14.15a and b). The dies are activated by means of a set of rollers within a cage, in an action similar to that of a roller bearing. The workpiece is stationary and the dies rotate (while moving radially in their slots), striking the workpiece at rates as high as 20 strokes per second. In **die-closing swaging machines,** die movements are through the reciprocating motion of wedges (Fig. 14.15c). The dies can be opened wider than those in rotary swagers, thereby accommodating large-diameter or variable-diameter parts. In another type of machine, the dies do not rotate but move radially in and out. Typical products made by this machine are screwdriver blades and soldering-iron tips.

Swaging also can be used to *assemble* fittings over cables and wire; in such cases, the tubular fitting is swaged directly onto the cable. The process also is used for operations such as pointing (tapering the tip of a cylindrical part) and *sizing* (finalizing the dimensions of a part).

Swaging generally is limited to a maximum workpiece diameter of about 150 mm (6 in.), and parts as small as 0.5 mm (0.02 in.) have been swaged; dimensional tolerances range from ±0.05 to ±0.5 mm (0.002–0.02 in.). The process is suitable for medium-to-high rates of production, with rates as high as 50 parts per minute possible, depending on part complexity. Swaging is a versatile process and is limited in length only by the length of the bar supporting the mandrel, if one is needed (see Fig. 14.15b).

Tube Swaging.   In this process, the internal diameter and/or the thickness of the tube is reduced, with or without the use of *internal mandrels* (Figs. 14.16a and b). For small-diameter tubing, high-strength wire can be used as a mandrel. Mandrels also can be made with longitudinal grooves, to allow swaging of internally shaped tubes (Fig. 14.16c). For example, the *rifling* in gun barrels (internal spiral grooves to give gyroscopic effect to bullets) can be produced by swaging a tube over a mandrel with spiral grooves. Special machinery can swage gun barrels, and other tubular parts, with starting diameters as large as 350 mm (14 in.).

TABLE 14.3

Forgeability of Metals, in Decreasing Order (See also Table 15.1)

Metal or alloy	Approximate range of hot-forging temperatures (°C)
Aluminum alloys	400–550
Magnesium alloys	250–350
Copper alloys	600–900
Carbon- and low-alloy steels	850–1150
Martensitic stainless steels	1100–1250
Austenitic stainless steels	1100–1250
Titanium alloys	700–950
Iron-based superalloys	1050–1180
Cobalt-based superalloys	1180–1250
Tantalum alloys	1050–1350
Molybdenum alloys	1150–1350
Nickel-based superalloys	1050–1200
Tungsten alloys	1200–1300

14.5   Forgeability of Metals; Forging Defects

Forgeability is generally defined as the capability of a material to undergo deformation in forging without cracking. Various tests have been developed to quantify forgeability; however, because of their complex nature, only two simple tests have had general acceptance: upsetting and hot twist.

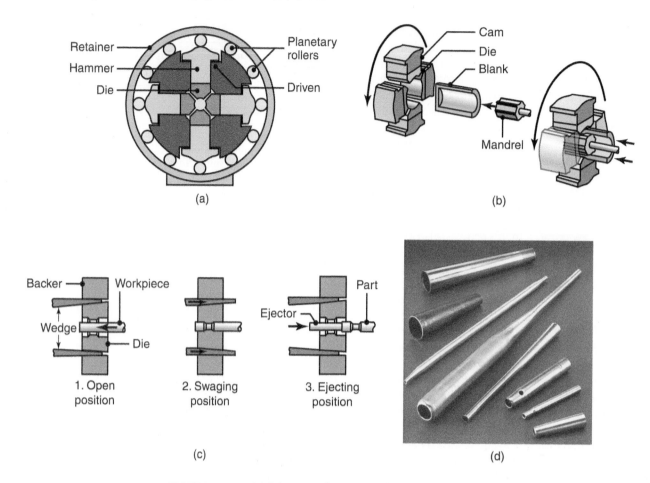

FIGURE 14.15 (a) Schematic illustration of the rotary-swaging process. (b) Forming internal profiles on a tubular workpiece by swaging. (c) A die-closing swaging machine, showing forming of a stepped shaft. (d) Typical parts made by swaging. *Source:* (d) Courtesy of J. Richard Industries.

In the **upsetting test**, a solid, cylindrical specimen is upset between flat dies, and the reduction in height at which cracks on the barreled surfaces begin to develop (see also Fig. 2.20d); the greater the deformation prior to cracking, the greater the forgeability of the metal. The second method is the **hot-twist test**, in which a round specimen is twisted continuously and in the same direction until it fails. This test is performed on a number of specimens and at different temperatures; the number of complete turns that each specimen undergoes before failure at each temperature is then plotted. The temperature at which the maximum number of turns occurs becomes the forging temperature for maximum forgeability. This test has been found to be useful particularly for steels.

The forgeability of various metals and alloys is given in Table 14.3, in decreasing order. More comprehensively, forgeability is rated on considerations such as (a) ductility and strength of the material, (b) forging temperature required, (c) frictional behavior between die and workpiece, and (d) the quality of the forgings produced. The ratings should be regarded only as general guidelines. Typical *hot-forging temperature* ranges for various metals and alloys are included in Table 14.3; for *warm* forging, temperatures range from 200° to 300°C (400° to 600°F) for aluminum alloys, and 550° to 750°C (1000° to 1400°F) for steels.

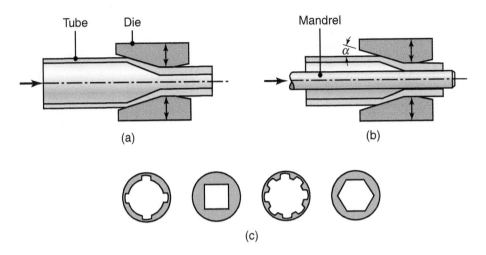

FIGURE 14.16   (a) Swaging of tubes without a mandrel; note the increase in wall thickness in the die gap. (b) Swaging with a mandrel; note that the final wall thickness of the tube depends on the mandrel diameter. (c) Examples of cross-sections of tubes produced by swaging on shaped mandrels; rifling (internal spiral grooves) in small gun barrels can be made by this process.

Forging Defects.   In addition to surface cracking, other defects can develop during forging as a result of the material flow pattern in the die, as described next in Section 14.6 regarding die design. For example, if there is an insufficient volume of material to fill the die cavity completely, the web may buckle during forging and develop laps (Fig. 14.17a). Conversely, if the web is too thick, the excess material flows past the already formed portions of the forging and develops internal cracks (Fig. 14.17b).

The various radii in the forging-die cavity can significantly influence the formation of such defects. Internal defects also may develop from (a) nonuniform deformation of the material in the die cavity, (b) temperature gradients developed throughout the workpiece during forging, and (c) microstructural changes caused by phase transformations. The *grain-flow pattern* of the material in forging also is important. The flow lines may reach a surface perpendicularly, as shown in Fig. 14.13b, known as **end grains**; the grain boundaries become directly exposed to the environment and can be attacked by it, developing a rough surface and acting as stress raisers.

Forging defects can cause fatigue failures, corrosion, and wear during the service life of the forging. The importance of inspecting forgings prior to their placement in service, particularly in critical applications, is obvious. Inspection techniques for manufactured parts are described in Chapter 36.

14.6   Die Design, Die Materials, and Lubrication

The design of forging dies requires considerations of (a) the shape and complexity of the workpiece, (b) forgeability, (c) strength and its sensitivity to deformation rate, (d) temperature, (e) frictional characteristics at the die–workpiece interfaces, and (f) die distortion under the forging loads. The most important rule in die design is

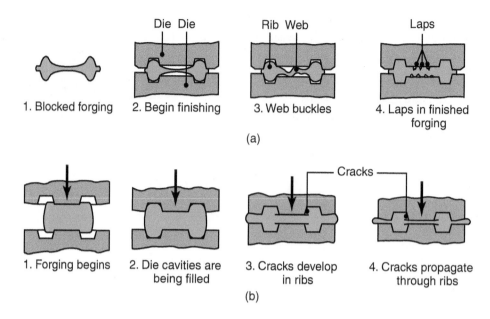

FIGURE 14.17 Examples of defects in forged parts. (a) Laps due to web buckling during forging; web thickness should be increased to avoid this problem. (b) Internal defects caused by an oversized billet; die cavities are filled prematurely, and the material at the center flows past the filled regions as the dies close.

the fact that the part will flow in the direction of least resistance. Workpiece *intermediate shapes* should be considered so that die cavities are filled properly, and without defects; an example of the intermediate shapes for a connecting rod is given in Fig. 14.8a.

With continuing advances in reliable simulation of all types of metalworking operations, software is available to help predict material flow in forging-die cavities (see Fig. 14.18). The simulation incorporates various conditions, such as workpiece temperature and heat transfer to dies, frictional conditions at die–workpiece contact surfaces, and forging speed. Such software has become essential in die design, especially in eliminating defective forgings (see also Section 38.7).

Preshaping. The requirements for preshaping a workpiece are: (a) the material should not flow easily into the flash area, as otherwise die filling will be incomplete, (b) the grain flow pattern should be favorable for the products' strength and reliability, and (c) sliding at the die–workpiece interface should be minimized in order to reduce die wear. The selection of preshapes involves calculations of cross-sectional areas at each location in the forging; computer modeling and simulation techniques are very useful in such calculations.

Die Features. The terminology for forging dies is shown in Fig. 14.6e. For most forgings, the **parting line** is located at the largest cross-section of the part. For simple symmetrical shapes, the parting line is usually a single plane and at the center of the forging; for more complex shapes, the line may not lie in a single plane. The dies are then designed in such a way that they make proper contact with the workpiece, while avoiding side thrust forces and maintaining die alignment during forging.

After sufficiently constraining lateral flow to ensure proper die filling, the flash material is allowed to flow into a **gutter**, so that the extra flash does not increase the forging load excessively. A general guideline for flash thickness is 3% of the maximum thickness (vertical dimension) of the forging. The length of the **land** is usually two to five times the flash thickness.

Draft angles are necessary in almost all forging dies in order to facilitate removal of the forging. Upon cooling, the forging shrinks both radially and longitudinally; internal draft angles (about 7°–10°) are therefore made larger than external ones (about 3°–5°).

Selection of the proper radii for corners and fillets is important in ensuring smooth flow of the metal into the die cavity and improving die life. Small radii generally are undesirable, because of their adverse effect on metal flow and their tendency to wear rapidly (as a result of stress concentration and thermal cycling). Small fillet radii also can cause fatigue cracking of the dies. As a general rule, these radii should be as large as can be permitted by the design of the forging. As with the patterns used in casting (Section 12.2.1), *allowances* are provided in forging-die design, when machining or grinding of the forging is necessary in order to impart final desired dimensions and surface finish. Machining allowance should be provided at flanges, holes, and mating surfaces.

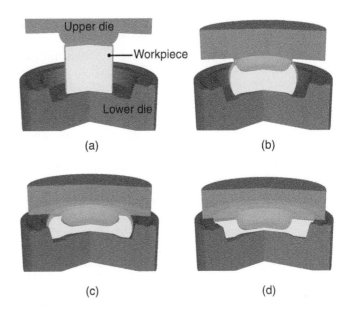

(a)   (b)

(c)   (d)

FIGURE 14.18 Deformation of a blank during forging as predicted by the software program DEFORM, based on the finite-element method of analysis. *Source:* Courtesy of Scientific Forming Technologies Corporation.

Die Materials.   General requirements for die materials are:

- Strength and toughness, especially at elevated temperatures
- Hardenability and ability to harden uniformly
- Resistance to mechanical and thermal shock
- Wear resistance, particularly resistance to abrasive wear, because of the presence of scale in hot forging

Common die materials are tool and die steels containing chromium, nickel, molybdenum, and vanadium (see Tables 5.7–5.9). Dies are made from die blocks, which themselves are forged from castings, and then machined and finished to the desired shape and surface finish.

Lubrication.   A wide variety of metalworking fluids are available for use in forging, as described in Section 33.7. Lubricants greatly influence friction and wear, in turn affecting the forces required [see Eq. (14.1)], die life, and the manner in which the material flows into the die cavities. Lubricants can also act as a *thermal barrier* between the hot workpiece and the relatively cool dies, thus slowing the rate of cooling of the workpiece and improving metal flow. Another important function of the lubricant is to act as a *parting agent*, preventing the forging from sticking to the dies and helping release it from the die.

14.7   Die-manufacturing Methods and Die Failure

Dies have an important impact on the overall economics in forging, because of the significant cost and the lead time required to produce them, as some dies can take months to make. Equally important are the proper maintenance of dies and their modifications as parts are first produced and evaluated.

Several manufacturing methods, either singly or in combination, can be used to make dies for forging. These methods include casting, forging, machining, grinding, electrical and electrochemical methods [particularly electrical-discharge machining (EDM) and wire EDM], and laser beams for making small dies. An important and continuing trend is the production of tools and dies by **rapid tooling**, using rapid-prototyping techniques, described in Section 20.6.

Producing a die cavity in a die block is called **die sinking**. Recall that the process of **hubbing** (Section 14.4), either cold or hot, also may be used to make small dies with shallow cavities. Usually, dies are subsequently heat treated, for higher hardness and wear resistance (Chapter 33). If necessary, their surface profile and finish are further improved by finish grinding and polishing, usually using computer numerically controlled machines.

The choice of a die manufacturing method depends on die size and shape, and the particular operation in which the die is to be used, such as casting, forging, extrusion, powder metallurgy, or plastics molding. As in all manufacturing operations, cost often dictates the process selected, because tool and die costs can be significant in manufacturing operations. Dies of various sizes and shapes can be cast from steels, cast irons, and nonferrous alloys; the processes used for preparing them may range from sand casting (for large dies, weighing several tons) to shell molding (for casting small dies). Cast steels generally are preferred for large dies, because of their strength and toughness as well as the ease with which the steel composition, grain size, and other properties can be controlled and modified.

Most commonly, dies are *machined* from forged die blocks, using processes such as high-speed milling, turning, grinding, and electrical discharge and electrochemical machining (see Part IV). Such an operation is shown in Fig. I.10b for making molds for eyeglass frames. For high-strength and wear-resistant die materials that are hard or are heat treated (and thus difficult to machine), processes such as hard machining and electrical and electrochemical machining are in common practice.

Typically, a die is machined by milling on computer-controlled machine tools, using various software packages that have the capability of optimizing the cutting-tool path; thus, for example, the best surface finish can be obtained in the least possible machining time. Equally important is the setup for machining, because dies should be machined as much as possible in one setup, without having to remove them from their fixtures and reorient them for subsequent machining operations.

After heat treating, to achieve the desired mechanical properties, dies usually are subjected to *finishing operations* (Section 26.7), such as grinding, polishing, and chemical and electrical processes, to obtain the desired surface finish and dimensional accuracy. This also may include *laser surface treatments* and *coatings* (Chapter 34) to improve die life. Laser beams also may be used for die repair and reconfiguration of the worn regions of dies (see also Fig. 33.11).

Die Costs.   From the preceding discussion, it is evident that the cost of a die depends greatly on its size, shape complexity, and surface finish required, as well as the die material and manufacturing, heat treating, and finishing methods employed. Some qualitative ranges of tool and die costs are given throughout this book, such as in Table 12.6. Even small and relatively simple dies can cost hundreds of dollars to make, and the cost of a set of dies for automotive body panels can be on the order of $2 million. On the other hand, because a large number of parts usually are made from one set of dies, *die cost per piece made* is generally a small portion of a part's manufacturing cost (see also Section 40.9). The *lead time* required to produce dies also can have a significant impact on the overall manufacturing cost of parts made.

Die Failures.   Failure generally results from one or more of the following causes:

- Improper die design
- Improper selection of die material
- Improper manufacturing, heat treatment, and finishing operations
- Overheating and heat checking (cracking caused by temperature cycling)
- Excessive die wear
- Overloading (excessive force on the die)
- Improper alignment of die components or segments
- Misuse
- Improper handling of the die

Other Considerations.   In order to withstand the forces involved, a die must have sufficiently large cross-sections and clearances (to prevent jamming). Abrupt changes in cross-section, sharp corners, radii, fillets, and a coarse surface finish (including grinding marks and their orientation on die surfaces) act as stress raisers, and thus reduce die life. For improved strength and to reduce the tendency for cracking, dies may be made in segments and assembled into a complete die, with rings that prestress the dies. Proper handling, installation, assembly, and alignment of dies are essential. Overloading of tools and dies can cause premature failure. A common cause of damage to dies is the failure of the operator, or of a programmable robot, to remove a formed part from the die before another blank is placed (loaded) into the die.

14.8 Forging Machines

Various types of forging machines are available, with a wide range of capacities (tonnage), speeds, and speed–stroke characteristics (Fig. 14.19 and Table 14.4).

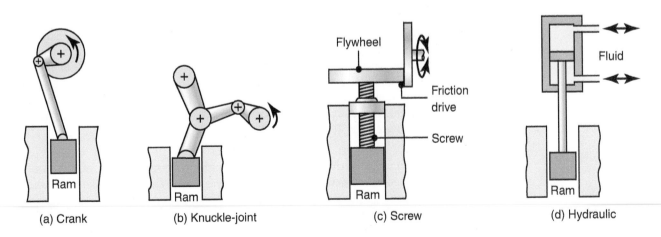

(a) Crank   (b) Knuckle-joint   (c) Screw   (d) Hydraulic

FIGURE 14.19 Schematic illustration of the principles of various forging machines. (a) Crank press with an eccentric drive; the eccentric shaft can be replaced by a crankshaft to give up-and-down motion to the ram. (b) Knuckle-joint press. (c) Screw press. (d) Hydraulic press.

TABLE 14.4

Typical Speed Ranges of Forging Equipment	
Equipment	m/s
Hydraulic press	0.06–0.30
Mechanical press	0.06–1.5
Screw press	0.6–1.2
Gravity drop hammer	3.6–4.8
Power drop hammer	3.0–9.0
Counterblow hammer	4.5–9.0

Mechanical Presses.   These presses are basically of either the *crank* or the *eccentric* type (Fig. 14.19a). The speed varies from a maximum at the center of the stroke to zero at the bottom of the stroke, thus mechanical presses are *stroke limited*. The energy in a mechanical press is generated by a large flywheel powered by an electric motor. A clutch engages the flywheel to an eccentric shaft; a connecting rod then translates the rotary motion into a reciprocating linear motion. A *knuckle-joint* mechanical press is shown in Fig. 14.19b; because of the linkage design, very high forces can be applied in this type of press (see also Fig. 11.21).

The force available in a mechanical press depends on the stroke position, and becomes extremely high at the end of the stroke; thus, proper setup is essential to avoid breaking the dies or equipment components. Mechanical presses have high production rates, are easier to automate, and require less operator skill than do other types of machines. Press capacities generally range from 2.7 to 107 MN (300–12,000 tons). Mechanical presses are preferred for forging parts requiring high precision.

Screw Presses.   These presses (Fig. 14.19c) derive their energy from a flywheel, hence they are *energy limited*. The forging load is transmitted through a large vertical screw, and the ram comes to a stop when the flywheel energy has been dissipated. If the dies do not close at the end of the cycle, the operation is repeated until the forging is completed. Screw presses are used for various open-die and closed-die forging operations. They are particularly suitable for small production quantities and for thin parts with high precision, such as turbine blades. Press capacities range from 1.4 to 280 MN (160–31,500 tons).

Hydraulic Presses.   These presses (Fig. 14. 19d) operate at constant speeds and are *load limited* (load restricted), whereby the press stops if the load required exceeds its capacity. Large amounts of energy can be transmitted from the press to the workpiece by a constant load throughout the stroke, the speed of which can be controlled. Because forging in a hydraulic press takes longer than in the other types of forging machines, described next, the workpiece may cool rapidly unless the dies are heated (see *isothermal forging*, Section 14.4). Compared with mechanical presses, hydraulic presses are slower and involve higher initial costs, but they require less maintenance.

A hydraulic press typically consists of a frame with two or four columns, pistons, cylinders, rams, and hydraulic pumps driven by electric motors. The ram speed can be varied during the stroke. Press capacities range up to 125 MN (14,000 tons) for open-die forging, and up to 730 MN (82,000 tons) for closed–die forging. The main landing-gear support beam for the Boeing 747 aircraft is forged in a 450-MN (50,000-ton) hydraulic press; this part is made of a titanium alloy and weighs approximately 1350 kg (3000 lb).

Hammers.   Hammers derive their energy from the potential energy of the ram, which is converted into kinetic energy; thus they are *energy limited*. Unlike hydraulic presses, hammers operate at high speeds, minimizing the cooling of a hot forging. Low cooling rates allow the forging of complex shapes, particularly those with thin and deep recesses. To complete the forging, several successive blows are usually made in the same die. Hammers are available in a variety of designs, and are the most versatile and the least expensive type of forging equipment.

Drop Hammers. In *power drop hammers*, the ram's downstroke is accelerated by steam, air, or hydraulic pressure. Ram weights range from 225 to 22,500 kg (500–50,000 lb), with energy capacities reaching 1150 kJ (850,000 ft-lb). In the operation of *gravity drop hammers*, a process called **drop forging**, the energy is derived from the free-falling ram. The available energy is the product of the ram's weight and the height of its drop. Ram weights range from 180 to 4500 kg (400–10,000 lb), with energy capacities ranging up to 120 kJ (90,000 ft-lb).

Counterblow Hammers. These hammers have two rams that simultaneously approach each other horizontally or vertically to forge the part. As in open-die forging operations, the workpiece may be rotated between blows in shaping the workpiece during forging. Counterblow hammers operate at high speeds and transmit less vibration to their bases; capacities range up to 1200 kJ (900,000 ft-lb).

High-energy-rate Forging Machines. In these machines, the ram is accelerated rapidly by inert gas at high pressure, and the part is forged in one blow at a very high speed. Although there are several types of these machines, various problems associated with their operation and maintenance, as well as die breakage and safety considerations, have greatly limited their use in industry.

Servo Presses. A recent development is the use of servo presses for forging and stamping applications (Fig. 14.20). These presses utilize servo drives along with

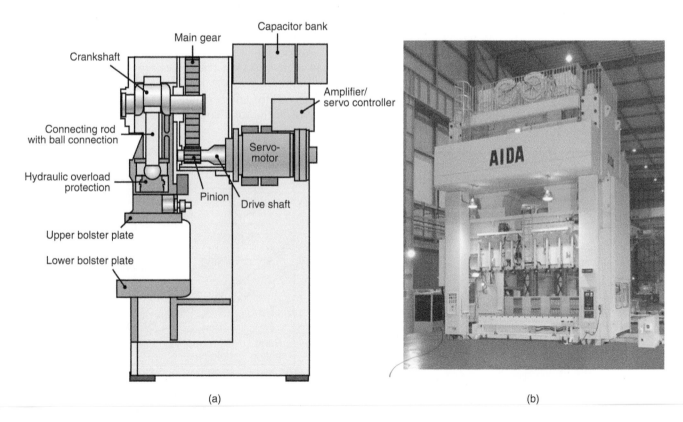

(a)   (b)

FIGURE 14.20 (a) Schematic illustration of a servo press, with the power source and transmission components highlighted. (b) An example of a servo press, with a 23,000 kN (2500 ton) capacity. *Source:* Courtesy of Aida Engineering, Inc.

linkage mechanisms, as in mechanical, knuckle joint, or screw presses. There are no clutches or brakes; instead, the desired velocity profile is achieved through a servo motor controller. The servo drive thus allows considerable flexibility regarding speeds and stroke heights, which simplifies setup and allows an optimized velocity profile for forging difficult materials or products; in addition, servo presses can produce parts with as little as 10% of the energy consumption of other presses, attributable mainly to their low energy costs when not producing parts (see Section 40.4). Servo presses can develop forces up to 25,000 kN (2800 tons); larger forces can be developed by hybrid machines that combine servo drives with energy storage in a flywheel.

QR Code 14.5 The Forging Advantage. (*Source:* Courtesy of the Forging Industry Association, www .forging.org)

14.9 Economics of Forging

Several factors are involved in the cost of forgings, depending on the complexity of the forging and tool and die costs, which range from moderate to high. As in other manufacturing operations, these costs are spread out over the number of parts forged with that particular die set. Thus, referring to Fig. 14.21, even though the cost of workpiece material per forging is constant, setup and tooling costs per piece decrease as the number of pieces forged increases.

The ratio of the cost of the die material to the total cost of forging the part increases with the weight of forgings: The more expensive the material, the higher the cost of the material relative to the total cost. Because dies must be made and forging operations must be performed regardless of the size of the forging, the cost of dies and of the forging operation relative to material cost is high for small parts; by contrast, die material costs are relatively low.

The size of forgings also has some effect on cost. Sizes range from small forgings (such as utensils and small automotive components) to large ones (such as gears, crankshafts, and connecting rods for large engines). As the size of the forging increases, the share of material cost in the total cost also increases, but at a lower

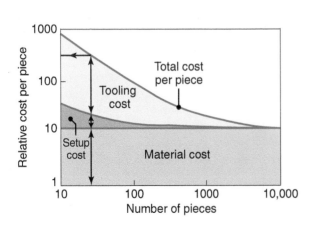

FIGURE 14.21 Typical cost per piece (*unit cost*) in forging; note how the setup and the tooling costs per piece decrease as the number of pieces forged increases (if all pieces use the same die).

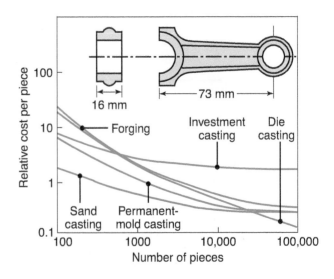

FIGURE 14.22 Relative costs per piece of a small connecting rod made by various forging and casting processes; note that, for large quantities, forging is more economical, and sand casting is the most economical process for fewer than about 20,000 pieces.

rate. This is because (a) the incremental increase in die cost for larger dies is relatively small, (b) the machinery and operations involved are essentially the same, regardless of forging size, and (c) the labor involved per piece made is not that much higher.

The total cost involved in a forging operation is not influenced to any major extent by the type of materials forged. Because they now have been reduced significantly by automated and computer-controlled operations, labor costs in forging generally are moderate. Moreover, die design and manufacturing are now mostly performed by computer-aided design and manufacturing techniques (Chapter 38), resulting in major savings in time and effort.

The cost of forging a part, compared to that of producing it by various processes such as casting, powder metallurgy, machining, or other methods, is an important consideration. For example, for shorter production runs and all other factors being the same, manufacturing a certain part by, say, expendable-mold casting may well be more economical than producing it by forging (Fig. 14.22). Recall that this casting method does not require expensive molds and tooling, whereas forging typically requires expensive dies. The competitive aspects of manufacturing and process selection are discussed in greater detail in Chapter 40.

CASE STUDY 14.2   Suspension Components for the Lotus Elise Automobile

The automotive industry increasingly has been subjected to a demanding set of performance, cost, safety, fuel efficiency, and environmental regulations. One of the main strategies in improving vehicle design with respect to all of these possibly conflicting constraints is to reduce vehicle weight while using advanced materials and manufacturing processes. Previous design optimization for this car has shown that weight savings of up to 34% can be realized on suspension system components, a significant savings since suspensions make up approximately 12% of a car's mass.

Weight savings could be achieved largely by developing optimum designs, utilizing advanced analytical tools, and using net-shape or near-net-shape steel forgings, instead of cast-iron components. In addition, it has been demonstrated that significant cost savings can be achieved for many parts when using steel forgings, as opposed to aluminum castings and extrusions.

The Lotus Elise is a high-performance sports car (Fig. 14.23a), designed for superior ride and handling. The Lotus group investigated the use of steel forgings, instead of extruded-aluminum suspension uprights, in order to reduce cost and improve reliability and performance. Their development efforts consisted of two phases, shown in Figs. 14.23b and c. The first phase involved the development of a forged-steel component that could be used on the existing Elise sports car; the second phase involved the production of a suspension upright for a new model. A new design was developed using an iterative process, with advanced software tools, to reduce the number of components and to determine the optimum geometry. The material selected for the upright was an air-cooled forged steel, which gives uniform grain size and microstructure, and uniform high strength, without the need for heat treatment. These materials also have approximately 20% higher fatigue strengths than traditional carbon steels, such as AISI 1548-HT used for similar applications.

The revised designs are summarized in Table 14.5. As can be seen, the optimized new forging design (Fig. 14.23d) resulted in significant cost savings. Although it also resulted in a small weight increase, when compared to the aluminum-extrusion design, the weight penalty is recognized as quite small. Furthermore, the use of forged steel for such components is especially advantageous in fatigue-loading conditions, constantly encountered by suspension components. The new design also had certain performance advantages, in that the component stiffness is now higher, which registered as improved customer satisfaction and better "feel" during driving. The new design also reduced the number of parts required, thus satisfying another fundamental principle in design.

(continued)

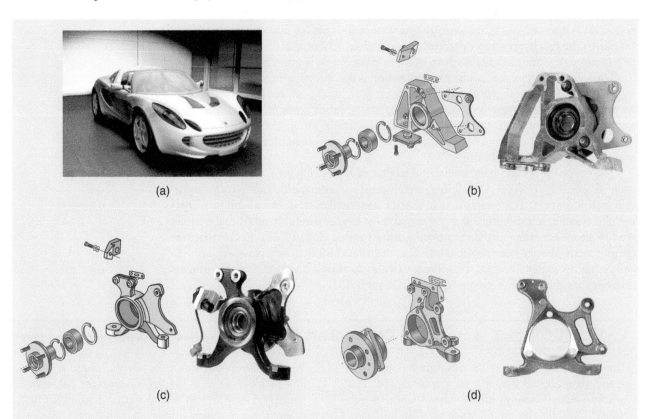

FIGURE 14.23 (a) The Lotus Elise Series 2 automobile; (b) illustration of the original design for the vertical suspension uprights, using an aluminum extrusion; (c) retrofit design, using a steel forging; and (d) optimized steel forging design for new car models. *Source:* (a) Courtesy of Fox Valley Motorcars; (b) through (d) Courtesy of American Iron and Steel Institute.

TABLE 14.5

Comparison of Suspension Upright Designs for the Lotus Elise Automobile

Fig. 14.23 sketch	Material	Application	Mass (kg)	Cost ($)
(b)	Aluminum extrusion, steel bracket, steel bushing, housing	Original design	2.105	85
(c)	Forged steel	Phase I	2.685 (+28%)	27.7 (−67%)
(d)	Forged steel	Phase II	2.493 (+18%)	30.8 (−64%)

Source: Courtesy of American Iron and Steel Institute

SUMMARY

- Forging denotes a family of metalworking processes in which deformation of the workpiece is carried out by compressive forces applied through a set of dies. Forging is capable of producing a wide variety of structural parts, with favorable characteristics such as higher strength, improved toughness, dimensional accuracy, and reliability in service.

- The forging process can be carried out at room, warm, or high temperatures. Workpiece material behavior during deformation, friction, heat transfer, and material-flow characteristics in the die cavity are important considerations, as are the proper selection of die materials, lubricants, workpiece and die temperatures, forging speeds, and equipment.

- Various defects can develop if the process is not designed or controlled properly. Computer-aided design and manufacturing techniques are now used extensively in die design and manufacturing, preform design, predicting material flow, and avoiding the possibility of internal and external defects during forging.

- A variety of forging machines is available, each with its own capabilities and characteristics. Forging operations are now highly automated, using industrial robots and computer controls.

- Swaging is a type of rotary forging in which a solid rod or a tube is reduced in diameter by the reciprocating radial movement of a set of two or four dies. The process is suitable for producing short or long lengths of bar or tubing, with various internal or external profiles.

- Because die failure has a major economic impact, die design, material selection, and production method are of great importance. A variety of die materials and manufacturing methods is available, including advanced material-removal and finishing processes.

KEY TERMS

Barreling	Forgeability	Impression-die forging	Precision forging
Closed-die forging	Forging	Incremental forging	Presses
Cogging	Fullering	Isothermal forging	Sizing
Coining	Hammers	Net-shape forging	Swaging
Edging	Heading	Open-die forging	Upsetting
End grain	Hot-twist test	Orbital forging	
Flash	Hubbing	Piercing	

BIBLIOGRAPHY

Altan, T., Ngaile, G., and Shen, G. (eds.), **Cold and Hot Forging: Fundamentals and Applications**, ASM International, 2005.

ASM Handbook, Vol. 14A: **Metalworking: Bulk Forming**, ASM International, 2005.

Boljanovic, V., **Metal Shaping Processes**, Industrial Press, 2009.

Byrer, T.G. (ed.), **Forging Handbook**, Forging Industry Association, 1985.

Dieter, G.E., Kuhn, H.A., and Semiatin, S.L. (eds.), **Handbook of Workability and Process Design**, ASM International, 2003.

Hosford, W.F., and Caddell, R.M., **Metal Forming: Mechanics and Metallurgy**, 4th ed., Cambridge, 2011.

Product Design Guide for Forging, Forging Industry Association, 1997.

Spitler, D., Lantrip, J., Nee, J., and Smith, D.A., **Fundamentals of Tool Design**, 5th ed., Society of Manufacturing Engineers, 2003.

Tschaetch, H., **Metal Forming Practice: Processes, Machines, Tools**, Springer, 2007.

REVIEW QUESTIONS

14.1 What is the difference between cold, warm, and hot forging?

14.2 Explain the difference between open-die and impression-die forging.

14.3 Explain the difference between fullering, edging, and blocking.

14.4 What is flash? What is its function?

14.5 Why is the intermediate shape of a part important in forging operations?

14.6 Describe the features of a typical forging die.

14.7 Explain what is meant by "load limited," "energy limited," and "stroke limited" as these terms pertain to forging machines.

14.8 What type of parts can be produced by rotary swaging?

14.9 Why is hubbing an attractive alternative to producing simple dies?

14.10 What is the difference between piercing and punching?

14.11 What is a hammer? What are the different kinds of hammers?

14.12 Why is there barreling in upsetting?

14.13 What are the advantages and disadvantages of isothermal forging?

14.14 Why are draft angles required in forging dies?

14.15 Is a mandrel needed in swaging?

QUALITATIVE PROBLEMS

14.16 Describe and explain the factors that influence spread in cogging operations on long square billets.

14.17 How can you tell whether a certain part is forged or cast? Explain the features that you would investigate.

14.18 Identify casting design rules, described in Section 12.2, that also can be applied to forging.

14.19 Describe the factors involved in precision forging.

14.20 Why is control of the volume of the blank important in closed-die forging?

14.21 Why are there so many types of forging machines available? Describe the capabilities and limitations of each.

14.22 What are the advantages and limitations of cogging operations? Should cogging be performed hot or cold? Explain.

14.23 What are the advantages and limitations of using die inserts? Give some examples.

14.24 Review Fig. 14.6e and explain why internal draft angles are larger than external draft angles. Is this also true for permanent-mold casting?

14.25 Comment on your observations regarding the grain-flow pattern in Fig. 14.13b.

14.26 Describe your observations concerning the control of the final tube thickness in Fig. 14.16a.

14.27 By inspecting some forged products, such as hand tools, you will note that the lettering on them is raised rather than sunk. Offer an explanation as to why they are made that way.

14.28 Describe the difficulties involved in defining the term "forgeability" precisely.

14.29 Describe the advantages of servo presses for forging and stamping.

14.30 List the general recommendations you would make for forging materials with limited ductility.

14.31 Which would you recommend, (a) hot forging and heat treating a workpiece or (b) cold forging it and relying upon strain hardening for strengthening? Explain.

QUANTITATIVE PROBLEMS

14.32 Take two solid, cylindrical specimens of equal diameter, but different heights, and compress them (frictionless) to the same percent reduction in height. Show that the final diameters will be the same.

14.33 Calculate the room-temperature forging force for a solid, cylindrical workpiece made of 5052-O aluminum that is 3.5 in. high and 5 in. in diameter and is to be reduced in height by 30%. Let the coefficient of friction be 0.15.

14.34 Using Eq. (14.2), estimate the forging force for the workpiece in Problem 14.33, assuming that it is a complex forging and that the projected area of the flash is 30% greater than the projected area of the forged workpiece.

14.35 To what thickness can a solid cylinder of 1020 steel that is 1 in. in diameter and 2 in. high be forged in a press that can generate 100,000 lb?

14.36 In Example 14.1, calculate the forging force, assuming that the material is 1100-O aluminum and that the coefficient of friction is 0.10.

14.37 Using Eq. (14.1), make a plot of the forging force, F, as a function of the radius, r, of the workpiece. Assume that the flow stress, Y_f, of the material is constant. Recall that the volume of the material remains constant during forging; thus, as h decreases, r increases.

14.38 How would you go about estimating the punch force required in a hubbing operation, assuming that the material is mild steel and the projected area of the impression is 0.5 in²? Explain. (*Hint:* See Section 2.6 on hardness.)

14.39 A mechanical press is powered by a 30-hp motor and operates at 40 strokes per minute. It uses a flywheel, so that the crankshaft speed does not vary appreciably during the stroke. If the stroke is 6 in., what is the maximum constant force that can be exerted over the entire stroke length?

14.40 A solid cylindrical specimen, made of a perfectly plastic material, is being upset between flat dies with no friction. The process is being carried out by a falling weight, as in a drop hammer. The downward velocity of the hammer is at a maximum when it first contacts the workpiece, and becomes zero when the hammer stops at a certain height of the specimen. Establish quantitative relationships between workpiece height and velocity, and make a qualitative sketch of the velocity profile of the hammer. (*Hint:* The loss in the kinetic energy of the hammer is the plastic work of deformation; thus, there is a direct relationship between workpiece height and velocity.)

14.41 Assume that you are an instructor covering the topics described in this chapter and you are giving a quiz on the numerical aspects to test the understanding of the students. Prepare two quantitative problems and supply the answers.

SYNTHESIS, DESIGN, AND PROJECTS

14.42 Devise an experimental method whereby you can measure only the force required for forging the flash in impression-die forging.

14.43 Assume that you represent the forging industry and that you are facing a representative of the casting industry. What would you tell that person about the merits of forging processes?

14.44 Figure P14.44 shows a round impression-die forging made from a cylindrical blank, as illustrated on the left. As described in this chapter, such parts are made in a sequence of forging operations. Suggest a sequence of intermediate forging steps to make the part on the right, and sketch the shape of the dies needed.

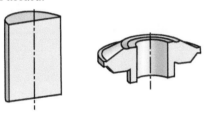

FIGURE P14.44

14.45 In comparing forged parts with cast parts, we have noted that the same part may be made by either process. Comment on the pros and cons of each process, considering such factors as part size, shape complexity, design flexibility, mechanical properties developed, and performance in service.

14.46 From the data given in Table 14.3, obtain the approximate value of the yield strength of the materials listed at their hot-forging temperatures.

14.47 Review the sequence of operations in the production of the stepped pin shown in Fig. 14.14. If the conical-upsetting step is not performed, how would the final part be affected?

14.48 Using a flat piece of wood, perform simple cogging operations on pieces of clay, and make observations regarding the spread of the pieces as a function of the original cross-sections (for example, square or rectangular with different thickness-to-width ratios).

14.49 Discuss the possible environmental concerns regarding the operations described in this chapter.

14.50 List the advantages and disadvantages in using a lubricant in forging operations.

15

Metal Extrusion and Drawing Processes and Equipment

- Extrusion and drawing involve, respectively, pushing or pulling a material through a die, for the purpose of modifying its cross-section.
- The chapter begins by describing the basic types of extrusion processes, and how the extrusion force can be estimated from material and processing parameters.
- Hot and cold extrusion are then discussed, including die design, and describing how cold extrusion is often done in combination with forging to produce specific shapes of parts.
- Extrusion practices and die designs that avoid common defects also are presented.
- The drawing of rod, wire, and tubing is then examined in a similar manner, along with die design considerations.
- The equipment characteristics for these processes also are described.

Typical parts made by extrusion and drawing: Long pieces having a wide variety of constant cross-sections, rods, shafts, bars for machinery and automotive power-train applications, aluminum ladders, collapsible tubes, and wires for numerous electrical and mechanical applications and musical instruments.

Alternative processes: Machining, powder metallurgy, shape rolling, roll forming, pultrusion, and continuous casting.

15.1 Introduction

Extrusion and drawing have numerous applications in manufacturing continuous as well as discrete products from a wide variety of metals and alloys. In extrusion, a usually cylindrical billet is forced through a die (Fig. 15.1), in a manner similar to squeezing toothpaste from a tube. A wide variety of solid or hollow cross-sections can be produced by extrusion, which essentially are semifinished products.

A characteristic of extrusion (from the Latin *extrudere*, meaning "to force out") is that large deformations can take place without fracture (see Section 2.2.8), because the material is under high triaxial compressive stresses. Since the die geometry remains unchanged throughout the process, extruded products typically have a constant cross-section.

Typical products made by extrusion are railings for sliding doors, window frames, tubing, aluminum ladder frames, and structural and architectural shapes. Extrusions can be cut into desired lengths, which then become discrete parts, such as brackets, gears, and coat hangers (Fig. 15.2). Commonly extruded metals are aluminum, copper, steel, magnesium, and lead; other metals and alloys also can be extruded, with various levels of difficulty.

Each billet is extruded individually, thus extrusion is a batch or semicontinuous operation. The process can be economical for large as well as short production runs. Tool costs generally are low, particularly for producing simple, solid cross-sections. Depending on the required ductility of the material, the process is carried out at room or at elevated temperatures. Extrusion at

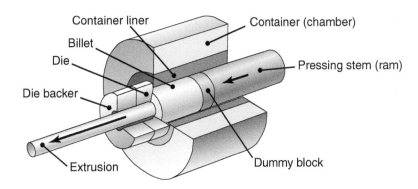

FIGURE 15.1   Schematic illustration of the direct-extrusion process.

room temperature is often combined with forging operations, in which case it is generally called **cold extrusion** (see also Section 14.4), with numerous applications, such as fasteners and components for automobiles, bicycles, motorcycles, heavy machinery, and transportation equipment.

In **drawing**, developed between 1000 and 1500 A.D., the cross-section of a solid rod, wire, or tubing is reduced or changed in shape by pulling it through a die. Drawn rods are used for shafts, spindles, and small pistons, and as the raw material for fasteners such as rivets, bolts, and screws. In addition to round rods, various profiles also can be drawn.

QR Code 15.1 Aluminum Extrusion. (*Source:* Courtesy of PBC Linear, a Pacific Bearing Company)

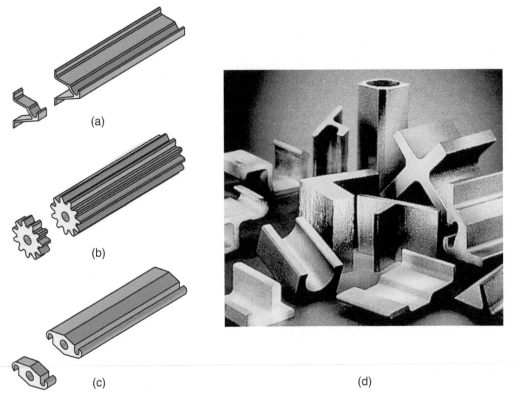

(a)

(b)

(c)

(d)

FIGURE 15.2   Extrusions and examples of products made by sectioning off extrusions. *Source:* (d) Courtesy of Plymouth Extruded Shapes. (For extruding plastics, see Section 19.2.)

The distinction between the terms **rod** and **wire** is somewhat arbitrary, with rod taken to be larger in cross-section than wire. In industry, wire is generally defined as a rod that has been drawn through a die at least once, or that its diameter is small enough so that it can be coiled. Wire drawing involves much smaller diameters than rod drawing, with sizes down to 0.01 mm (0.0004 in.) for magnet wire, and even smaller for use in very low current fuses.

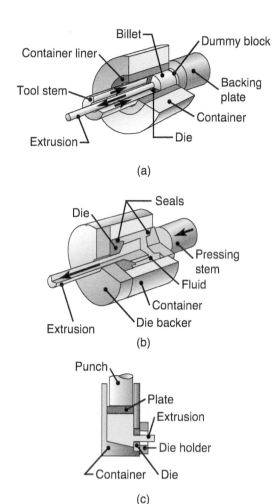

FIGURE 15.3 Types of extrusion: (a) indirect; (b) hydrostatic; and (c) lateral.

15.2 The Extrusion Process

There are three basic types of extrusion. In **direct** or **forward extrusion**, a billet is placed in a container (*chamber*) and forced through a die, as shown in Fig. 15.1. The die opening may be round, or it may have various shapes, depending on the desired profile. The function of the dummy block, shown in the figure, is to protect the tip of the pressing stem (punch), particularly in hot extrusion.

In **indirect** extrusion, also called *reverse, inverted,* or *backward extrusion,* the die moves toward the unextruded billet (Fig. 15.3a). Indirect extrusion has the advantage of having no billet–container friction, since there is no relative motion; thus, it is used on materials with very high friction, such as hot extrusion of high strength and stainless steels.

In **hydrostatic extrusion** (Fig. 15.3b), the billet is smaller in diameter than the container (which is filled with a fluid), and the pressure is transmitted to the fluid by a ram. The fluid pressure imparts triaxial compressive stresses acting on the workpiece and thus has improved formability (Section 2.10). Furthermore, there is much less workpiece–container friction than in direct extrusion. A less common type of extrusion is *lateral* or *side extrusion* (Fig. 15.3c).

As can be seen in Fig. 15.4, the geometric variables in extrusion are the die angle, α, and the ratio of the cross-sectional area of the billet to that of the extruded part, A_o/A_f, called the **extrusion ratio**, R. Other processing variables are the billet temperature, the speed at which the ram travels, and the type of lubricant used.

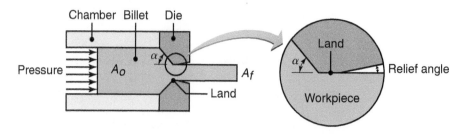

FIGURE 15.4 Process variables in direct extrusion; the die angle, reduction in cross-section, extrusion speed, billet temperature, and lubrication all affect the extrusion pressure.

Extrusion Force.   The force required for extrusion depends on (a) the strength of the billet material, (b) extrusion ratio, (c) friction between the billet, container, and die surfaces, and (d) process variables. For a small die angle, α, it has been shown that the extrusion pressure can be approximated as

$$p = Y\left(1 + \frac{\tan\alpha}{\mu}\right)\left(R^{\mu\cot\alpha} - 1\right), \qquad (15.1)$$

where μ is the coefficient of friction, Y is the yield stress of the billet material, and R is the extrusion ratio. The extrusion force can then be obtained by multiplying the pressure by the billet area, and can be simplified as

$$F = A_o k \ln\left(\frac{A_o}{A_f}\right), \qquad (15.2)$$

where k is the extrusion constant, determined experimentally; thus, k is a measure of the strength of the material being extruded and the frictional conditions. Figure 15.5 gives the values of k for several metals for a range of extrusion temperatures.

Video Solution 15.1   Forces in Extrusion

EXAMPLE 15.1   Calculation of Force in Hot Extrusion

Given: A round billet made of 70–30 brass is extruded at a temperature of 1250°F (675°C). The billet diameter is 5 in. (125 mm), and the diameter of the extrusion is 2 in. (50 mm).

Find: Calculate the extrusion force required.

Solution: The extrusion force is calculated using Eq. (15.2), in which the extrusion constant, k, is

obtained from Fig. 15.5. For 70–30 brass, $k = 35,000$ psi (250 MPa) at the given extrusion temperature. Thus,

$$F = \pi(2.5)^2(35,000)\ln\left[\frac{\pi(2.5)^2}{\pi(1.0)^2}\right] = 1.26 \times 10^6 \text{ lb}$$

$$= 630 \text{ tons} = 5.5 \text{ MN}.$$

Metal Flow in Extrusion.   The metal flow pattern in extrusion, as in other forming processes, is important because of its influence on the quality and the properties of the extruded product. The material flows longitudinally, much like an incompressible fluid flow in a channel; thus, extruded products have an elongated grain structure (*preferred orientation*, Section 1.6). Improper metal flow during extrusion can produce various defects in the extruded product, as described in Section 15.5.

A common technique for investigating the flow pattern is to cut the round billet lengthwise in half and mark one face with a square grid pattern. The two halves are placed in the chamber together and are extruded. Figure 15.6 shows typical flow patterns obtained by this technique, for the case of direct extrusion with square dies (90° die angle). The conditions under which these different flow patterns occur are described in the caption of Figs. 15.6. Note the **dead-metal zone** in Figs. 15.6b and c, where the metal at the corners is essentially stationary; a situation similar to the stagnation of fluid flow in channels that have sharp angles or turns.

Process Parameters.   In practice, extrusion ratios, R, usually range from about 10 to 100. They may be higher for special applications (400 for softer nonferrous metals)

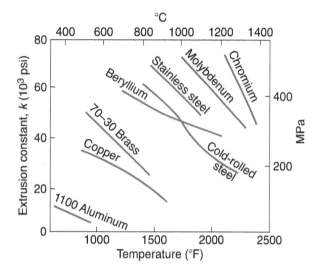

FIGURE 15.5 Extrusion constant k for various metals at different temperatures, as determined experimentally. *Source:* After P. Loewenstein.

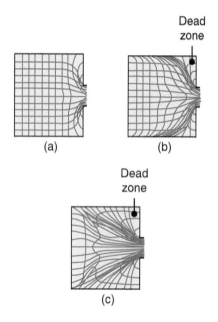

FIGURE 15.6 Types of metal flow in extruding with square dies. (a) Flow pattern obtained at low friction or in indirect extrusion. (b) Pattern obtained with high friction at the billet–chamber interfaces. (c) Pattern obtained at high friction or with cooling of the outer regions of the hot billet in the chamber; this type of pattern, observed in metals whose strength increases rapidly with decreasing temperature, leads to a defect known as pipe (or extrusion) defect.

or lower for less ductile materials, although the ratio usually has to be at least 4 to deform the material plastically through the bulk of the workpiece. Extruded products are usually less than 7.5 m (25 ft) long, because of the difficulty in handling greater lengths, but they can be as long as 30 m (100 ft). Ram speeds range up to 0.5 m/s (100 ft/min). Generally, lower speeds are preferred for aluminum, magnesium, and copper, higher speeds for steels, titanium, and refractory alloys. Dimensional tolerances in extrusion are usually in the range from ±0.25 to 2.5 mm (±0.01 − 0.1 in.), and they increase with increasing cross-section.

Because they have high ductility, aluminum, copper, and magnesium and their alloys, as well as steels and stainless steels, are extruded with relative ease into numerous shapes. Other metals, such as titanium and refractory metals, also can be extruded, but only with some difficulty and significant die wear. Most extruded products, particularly those with small cross-sections, require straightening and twisting. This is typically done by stretching and twisting the extruded product, usually in a hydraulic stretcher equipped with jaws.

The presence of a die angle causes a small portion of the end of the billet to remain in the chamber after the stroke of the ram. This portion, called *scrap* or the *butt end*, is later removed by cutting it off. Alternatively, another billet or a graphite block may be placed in the chamber to extrude the piece remaining from the previous extrusion.

In **coaxial extrusion**, or **cladding**, coaxial billets are extruded together, provided that the strength and ductility of the two metals are compatible. An example is copper clad with silver. *Stepped extrusions* also are produced, by extruding the billet partially in one die and then in successively larger dies (see also *cold extrusion*, Section 15.4). *Lateral extrusion* (Fig. 15.3c) is used for the sheathing of wire and the coating of electric wire with plastic.

15.3 Hot Extrusion

For metals and alloys that do not have sufficient ductility at room temperature, or in order to reduce the forces required, extrusion can be carried out at elevated temperatures (Table 15.1). As in all other elevated-temperature operations, hot extrusion has special requirements, because of the high operating temperatures involved. For example, die wear can be excessive, and cooling of the hot-billet surfaces (in the cooler container) and the die can result in highly nonuniform deformation of the billet, as shown in Fig. 15.6c. Thus, extrusion dies may be preheated, as is done in hot-forging operations (Chapter 14).

Because the billet is hot, it develops an oxide film, unless it is heated in an inert atmosphere. Oxide films can be abrasive (see

Section 33.2), and can affect the flow pattern of the material. Their presence also results in an extruded product that may be unacceptable when good surface finish is required. In order to avoid the formation of oxide films on the hot extruded product, the dummy block placed ahead of the ram (Fig. 15.1) is made a little smaller in diameter than the container. As a result, a thin shell (*skull*), consisting mainly of the outer oxidized layer of the billet, is left in the container. The skull is later removed from the chamber.

Die Design.   Die design requires considerable experience, as can be appreciated by reviewing Fig. 15.7. *Square dies*, also called *shear dies*, are used in extruding nonferrous metals, especially aluminum. These dies develop *dead-metal zones*, which in turn form a "die angle" (see Figs. 15.6b and c) along which the material flows. These zones produce extrusions with bright finishes, because of the burnishing (Section 16.2) that takes places as the material flows past the "die angle" surface and land in the die.

Tubing can be extruded from a solid or hollow billet (Fig. 15.8). Wall thickness is usually limited to 1 mm (0.040 in.) for aluminum, 3 mm (0.125 in.) for carbon steels, and 5 mm (0.20 in.) for stainless steels. When solid billets are used, the ram is fitted with a mandrel that pierces a hole into the billet. Billets with a previously

TABLE 15.1

Typical Extrusion Temperature Ranges for Various Metals and Alloys. (See also Table 14.3.)

Material	Extrusion temperature (°C)
Lead	200–250
Aluminum and its alloys	375–475
Copper and its alloys	650–975
Steels	875–1300
Refractory alloys	975–2200

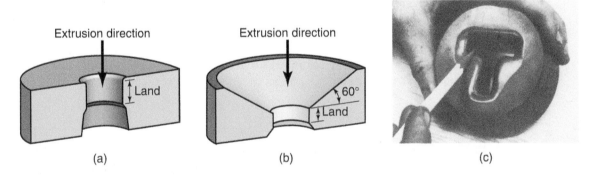

FIGURE 15.7   Typical extrusion die shapes: (a) die for nonferrous metals; (b) die for ferrous metals; and (c) die for a T-shaped extrusion made of hot-work die steel and used with molten glass as a lubricant. *Source:* (c) Courtesy of LTV Steel Company.

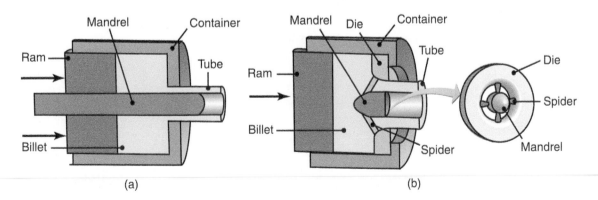

FIGURE 15.8   Extrusion of a seamless tube (a) using an internal mandrel that moves independently of the ram: an alternative arrangement has the mandrel integral with the ram and (b) using a spider die (see Fig. 15.9) to produce seamless tubing.

pierced hole also may be extruded in this manner. Because of friction and the severity of deformation, thin-walled extrusions are more difficult to produce than those with thick walls.

Hollow cross-sections (Fig. 15.9a) can be extruded by *welding-chamber* methods and using various dies known as a **porthole die, spider die,** and **bridge die** (Figs. 15.9b to d). During extrusion, the metal divides and flows around the supports for the internal mandrel into strands; this is a condition much like that of air flowing around a moving car and rejoining downstream, or water flowing around large rocks in a river and rejoining. The strands being extruded then become rewelded, under the high pressure in the welding chamber, before exiting the die. The rewelded surfaces have good strength, because they have not been exposed to the environment; as otherwise, they would develop oxides on their surfaces, thereby inhibiting good welding. The welding-chamber process is suitable only for aluminum and some of its alloys, because they can develop a strong weld under pressure, as described in Section 31.2. Lubricants cannot be used, because they prevent rewelding of the metal surfaces in the die.

Die Materials. Die materials for hot extrusion usually are hot-work die steels (Section 5.7). Coatings, such as partially stabilized zirconia (PSZ), may be applied to the dies to extend their life. Dies made of PSZ (Section 8.2.2) also are used for hot extrusion of tubes and rods. However, they are not suitable for dies for extruding complex shapes, because of the severe stress gradients that develop in the die, possibly leading to their premature failure.

Lubrication. Lubrication is important in hot extrusion, because of its effects on (a) material flow during extrusion, (b) surface finish and integrity, (c) product quality, and (d) extrusion forces. *Glass* (Section 8.4) is an excellent lubricant for steels,

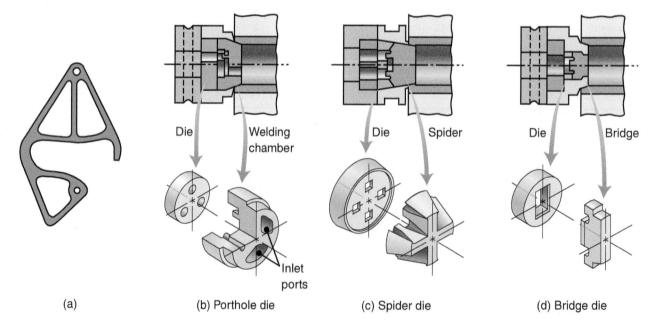

(a)   (b) Porthole die   (c) Spider die   (d) Bridge die

FIGURE 15.9 (a) An extruded 6063-T6 aluminum-ladder lock for aluminum extension ladders; this part is 8 mm (5/16 in.) thick and is sawed from the extrusion (see Fig. 15.2). (b) through (d) Components of various dies for extruding intricate hollow shapes.

stainless steels, and high-temperature metals and alloys. In a process developed in the 1940s and known as the **Séjournet process** (after J. Séjournet), a circular glass or fiberglass pad is placed in the chamber at the die entrance. The hot billet conducts heat to the glass pad, whereupon a thin layer of glass melts and acts as a lubricant. Before the hot billet is placed in the chamber, its surface is coated with a layer of powdered glass, to develop a thin glass lubricant layer at the billet–chamber interface.

For metals that have a tendency to stick or even weld to the container and the die, the billet can be enclosed in a *jacket*, a thin-walled container made of a softer and lower strength metal, such as copper or mild steel. This procedure is called **jacketing** or **canning**. In addition to acting as a low-friction interface, the jacket prevents contamination of the billet by the environment. For billet materials that are toxic or radioactive, the jacket also prevents it from contaminating the environment.

CASE STUDY 15.1   Manufacture of Aluminum Heat Sinks

Aluminum is used widely to transfer heat for both cooling and heating applications, because of its very high thermal conductivity. In fact, on a weight-to-cost basis, no other material conducts heat as efficiently as does aluminum.

Hot extrusion of aluminum is attractive for heat-sink applications, such as those in the electronics industry. Figure 15.10a shows an extruded heat sink, used for removing heat from a transformer on a printed circuit board. Heat sinks usually are designed with a large number of fins, that maximize the surface area and are evaluated

from a thermodynamics standpoint, using computer simulations. The fins are very difficult and expensive to machine, forge, or roll form, but they can be made economically by hot extrusion, using dies made by electrical-discharge machining (Section 27.5).

Figure 15.10b shows a die and a hot-extruded cross-section, suitable to serve as a heat sink. The shapes shown also could be produced through a casting operation, but extrusion is preferred, because there is no internal porosity in the part and its thermal conductivity is thus higher.

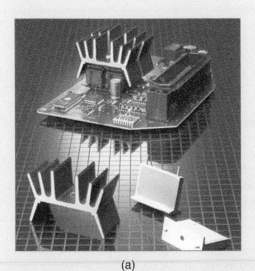

(a)

(b)

FIGURE 15.10   (a) Aluminum extrusion used as a heat sink for a printed circuit board. (b) Extrusion die and extruded heat sinks. *Source:* Courtesy of Aluminum Extruders Council.

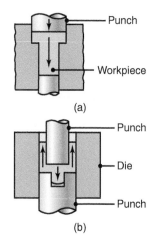

(a)

(b)

FIGURE 15.11 Two examples of cold extrusion; the arrows indicate the direction of metal flow during extrusion.

15.4 Cold Extrusion

Developed in the 1940s, *cold extrusion* is a general term often denoting a *combination* of operations, such as a combination of direct and indirect *extrusion and forging* (Fig. 15.11). Cold extrusion is used widely for components in automobiles, motorcycles, bicycles, appliances, and in transportation and farm equipment.

The cold-extrusion process uses slugs cut from cold-finished or hot-rolled bars, wire, or plates. Slugs that are less than about 40 mm (1.5 in.) in diameter are sheared (*cropped*), and, if necessary, their ends are squared off by processes such as upsetting, machining, or grinding. Larger diameter slugs are machined from bars into specific lengths. Cold-extruded parts weighing as much as 45 kg (100 lb) and having lengths of up to 2 m (80 in.) can be made, although most parts weigh much less. Powder metal slugs (*preforms*) also may be cold extruded (Section 17.3.3).

The *force*, F, in cold extrusion may be estimated from the formula

$$F = 1.7 A_o Y_{avg} \epsilon, \tag{15.3}$$

where A_o is the cross-sectional area of the blank, Y_{avg} is the average flow stress of the metal, and ϵ is the true strain that the piece undergoes, based on its original and final cross-sectional area. For example, assume that a round slug 10 mm in diameter and made of a metal with $Y_{avg} = 50,000$ psi is reduced to a final diameter of 7 mm by cold extrusion. The force would be

$$F = 1.7(\pi)\left(10^2/4\right)(50,000)\left[\ln(10/7)^2\right] = 4.8 \times 10^6 \text{ lb} = 2140 \text{ tons}.$$

Cold extrusion has the following advantages over hot extrusion:

- Improved mechanical properties, resulting from work hardening, provided that the heat generated by plastic deformation and friction does not recrystallize the extruded metal.
- Good control of dimensional tolerances, thus reducing the need for subsequent machining or finishing operations.
- Improved surface finish, due partly to the absence of an oxide film and provided that lubrication is effective.
- Production rates and costs are competitive with those of other methods of producing the same part. Some machines are capable of producing more than 2000 parts per hour.

The magnitude of the stresses on the tooling in cold extrusion, on the other hand, is very high (especially with steel and specialty-alloy workpieces), being on the order of the hardness of the workpiece material. The punch hardness usually ranges between 60 and 65 HRC, and the die hardness between 58 and 62 HRC. Punches are a critical component, as they must possess not only sufficient strength but also high toughness and resistance to wear and fatigue failure. *Lubrication* is critical, especially with steels, because of the possibility of sticking (*seizure*) between the workpiece and the tooling, in case of lubricant breakdown. The most effective means of lubrication is the application of a *phosphate-conversion coating* on the workpiece, followed by a coating of soap or wax, as described in Section 34.10.

Tooling design and the selection of appropriate tool and die materials are essential to the success of cold extrusion. Also important are the selection and control of the workpiece material with regard to its quality and the repeat accuracy of the slug dimensions and its surface condition.

CASE STUDY 15.2   Cold-extruded Part

A typical cold-extruded part, similar to the metal component of an automotive spark plug, is shown in Fig. 15.12. First, a slug is sheared off the end of a round rod (Fig. 15.12, left). It then is cold extruded (Fig. 15.12, middle) in an operation similar to those shown in Fig. 15.11, but with a blind hole. Then the material at the bottom of the blind hole is punched out, producing the small slug shown. Note the respective diameters of the slug and the hole at the bottom of the sectioned part.

Investigating material flow during the deformation of the slug helps avoid defects and leads to improvements in punch and die design. The part usually is sectioned in the midplane, and then polished and etched to display the grain flow, as shown in Fig. 15.13 (see also Fig. 14.13).

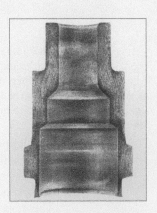

FIGURE 15.13 A cross-section of the metal part in Fig. 15.12, showing the grain-flow pattern (see also Fig 14.13b). *Source:* Courtesy of National Machinery Company.

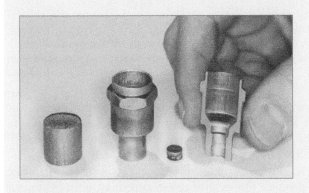

FIGURE 15.12 Production steps for the metal portion of a cold-extruded spark plug. *Source:* Courtesy of National Machinery Company.

15.4.1   Impact Extrusion

Impact extrusion is similar to indirect extrusion, and the process often is included in the cold-extrusion category. The punch descends rapidly on the blank (slug), which is extruded backward (Fig. 15.14). Because of volume constancy, the thickness of the tubular extruded region is a function of the clearance between the punch and the die cavity.

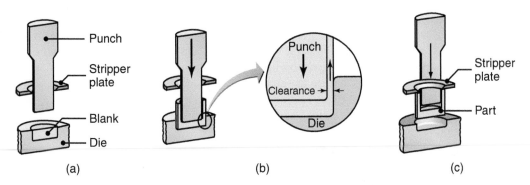

(a)   (b)   (c)

FIGURE 15.14 Schematic illustration of the impact-extrusion process; the extruded parts are stripped by using a stripper plate.

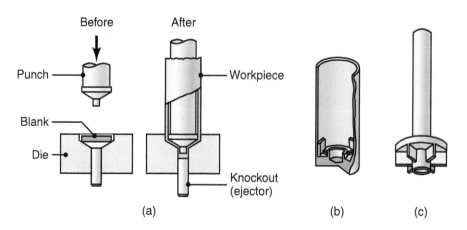

FIGURE 15.15 (a) Impact extrusion of a collapsible tube by the *Hooker process*. (b) and (c) Two examples of products made by impact extrusion. These parts also may be made by casting, forging, or machining; the choice depends on the materials part dimensions and wall thickness, and the properties desired. Economic considerations also are important in final process selection.

Typical products made by this process are shown in Fig. 15.15; other examples are collapsible tubes (similar to those used for toothpaste), light fixtures, automotive parts, and small pressure vessels. Most nonferrous metals can be impact extruded in vertical presses and at production rates as high as two parts per second.

The maximum diameter of the parts made is about 150 mm (6 in.). The impact-extrusion process can produce thin-walled tubular sections, with thickness-to-diameter ratios as small as 0.005. Consequently, the symmetry of the part and the concentricity of the punch and the blank are important.

15.4.2 Hydrostatic Extrusion

In *hydrostatic extrusion*, the pressure required in the chamber is supplied via a piston and through an incompressible fluid medium surrounding the billet (Fig. 15.3b). Pressures are typically on the order of 1400 MPa (200 ksi). The high pressure in the chamber transmits some of the fluid to the die surfaces, where it significantly reduces friction. Hydrostatic extrusion is usually carried out at room temperature, typically using vegetable oils as the fluid.

Brittle materials can be extruded successfully by this method, because the hydrostatic pressure, along with low friction and the use of small die angles and high extrusion ratios, increases the ductility of the material. Long wires also have been extruded from an aluminum billet, at room temperature and an extrusion ratio of 14,000; this means that a 1-m billet becomes a 14-km-long wire. In spite of the success obtained, hydrostatic extrusion has had limited industrial applications, mainly because of the complex nature of the tooling, the design of specialized equipment, and the long cycle times required—all of which make the process uneconomical for most materials and applications.

15.5 Extrusion Defects

Depending on workpiece material condition and process variables, extruded products can develop several types of defects that can affect significantly their strength and product quality. Some defects are visible to the naked eye, while others can be detected only by the nondestructive techniques described in Section 36.10. There are three principal *extrusion defects*: surface cracking, pipe, and internal cracking.

Surface Cracking.   If extrusion temperature, friction, or speed is too high, surface temperatures can become excessive, which may cause surface cracking and tearing (*fir-tree cracking* or *speed cracking*). These cracks are intergranular (along the grain boundaries; see Fig. 2.27), and usually are caused by **hot shortness** (Section 1.5.2). These defects occur especially in aluminum, magnesium, and zinc alloys, and can be avoided by lowering the billet temperature and the extrusion speed.

Surface cracking also may occur at lower temperatures, attributed to periodic sticking of the extruded part along the die land. Because of its similarity in appearance to the surface of a bamboo stem, it is known as a **bamboo defect**. The explanation is that, when the product being extruded temporarily sticks to the die land (see Fig. 15.7), the extrusion pressure increases rapidly; shortly thereafter, it moves forward again, and the pressure is released. The cycle is repeated continually, producing periodic circumferential cracks on the surface.

Pipe.   The type of metal-flow pattern in extrusion shown in Fig. 15.6c tends to draw surface oxides and impurities toward the center of the billet, much like a funnel. This defect is known as *pipe defect*, *tailpipe*, or *fishtailing*; as much as one-third of the length of the extruded product may contain this type of defect, and has to be cut off as scrap. Piping can be minimized by modifying the flow pattern to be more uniform, such as by controlling friction and minimizing temperature gradients within the part. Another method is to machine the billet's surface prior to extrusion, so that scale and surface impurities are removed, or by chemical etching of the surface oxides prior to extrusion.

Internal Cracking.   The center of the extruded product can develop cracks, called *center cracking*, *center-burst*, *arrowhead fracture*, or *chevron cracking* (Fig. 15.16a). They are attributed to a state of hydrostatic tensile stress that develops at the centerline in the deformation zone in the die (Fig. 15.16b), a condition similar to the

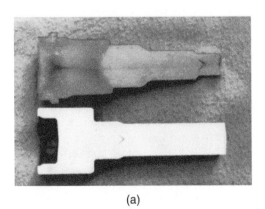

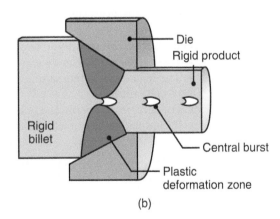

(a)   (b)

FIGURE 15.16   (a) Chevron cracking (central burst) in extruded round steel bars; unless the products are inspected, such internal defects may remain undetected and later cause failure of the part in service; this defect can also develop in the drawing of rod, wire, and tubes. (b) Schematic illustration of rigid and plastic zones in extrusion; the tendency toward chevron cracking increases if the two plastic zones do not meet; note that the plastic zone can be made larger either by decreasing the die angle, by increasing the reduction in cross-section, or both. *Source:* After B. Avitzur.

necked region in a tensile-test specimen (see Fig. 2.23). These cracks also have been observed in tube extrusion and in tube spinning (see Figs. 16.49b and c), appearing on the *inside* surfaces of tubes. The tendency for center cracking (a) increases with increasing die angle, (b) increases with increasing amount of impurities in the material, and (c) decreases with increasing extrusion ratio and friction.

15.6   Design Considerations

Extrusion of *constant cross-sections* is often a more economical method of producing a part than by forging, casting, or machining. While the designer has considerable freedom in designing cross-sections, there are several general rules that should be followed to simplify manufacturing and reduce defects. Before laying out the cross-section, the designer should consider the following:

- Some guidelines for proper die design in extrusion are illustrated in Fig. 15.10. Note the (a) importance of symmetry of cross-section, (b) avoiding sharp corners, (c) maintaining uniform wall thickness, and (d) avoiding severe changes in die dimensions within the cross-section.
- Solid shapes are the easiest to extrude. When possible, the cross-section should avoid hollow sections, but when necessary, such sections can be extruded using porthole, spider, or bridge dies, as illustrated in Fig. 15.9.
- If there is a critical dimension in a cross-section, it should not be located at the end of a gap. Figure 15.17 shows the use of a metal web to decrease the tolerance on a critical dimension. Note that this design approach requires the extrusion of a hollow cross-section; if the cross-section is complex, it can be extruded in two sections, and then assembled using the geometries shown in Fig. 15.18.
- Extrusions will usually develop some curvature, which may have to be straightened. Wide, thin sections can be difficult to straighten, hence the need to use ribs, as shown in the lower half of Fig. 15.17.

Impact extrusions should incorporate the following considerations:

- Impact extrusions should be symmetrical about the punch. External and internal bosses can be used as long as they are in the part axis.
- The maximum length-to-diameter ratio should not exceed eight or so, to avoid punch failure.
- For reverse extrusion, the outer radius can be small, but the inner radius should be as small as possible, and should preferably incorporate a chamfer.

15.7   Extrusion Equipment

The basic equipment for extrusion is a *horizontal hydraulic press* (Fig. 15.19; see also Fig. 14.19d). These presses are suitable for extrusion because the stroke and speed of the operation can be controlled, and are capable of applying a constant force over a long stroke. Consequently, long billets can be used, correspondingly larger extrusions are produced per setup, and the production rate is thus increased. Hydraulic presses with a ram-force capacity as high as 120 MN (14,000 tons) have been built, particularly for hot extrusion of large-diameter billets.

Vertical hydraulic presses typically are used for cold extrusion, and generally have lower capacity than those used for hot extrusion, but they take up less floor space.

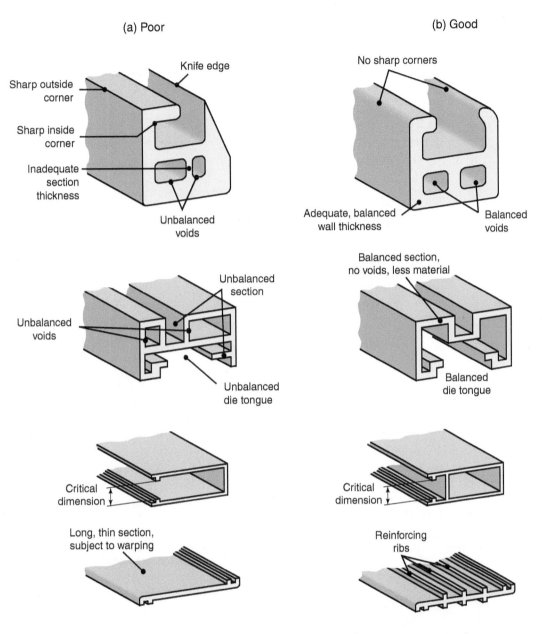

(a) Poor

(b) Good

Knife edge

No sharp corners

Sharp outside corner

Sharp inside corner

Inadequate section thickness

Unbalanced voids

Adequate, balanced wall thickness

Balanced voids

Unbalanced section

Unbalanced voids

Unbalanced die tongue

Balanced section, no voids, less material

Balanced die tongue

Critical dimension

Critical dimension

Long, thin section, subject to warping

Reinforcing ribs

FIGURE 15.17 Examples of poor and good design practices for extrusion; note the importance of eliminating sharp corners and keeping section thicknesses uniform.

Crank and *knuckle-joint* mechanical presses (Figs. 14.19a and b) are used for cold extrusion and for impact extrusion, to mass-produce small components. Multistage operations, where the cross-sectional area is reduced in a number of individual steps, are carried out on specially designed presses.

15.8 The Drawing Process

In *drawing*, the cross-section of a rod or wire is reduced or changed in shape by pulling (hence the term drawing) it through a die called a *draw die* (Fig. 15.20). The difference between drawing and extrusion is that in extrusion, the material is pushed through a

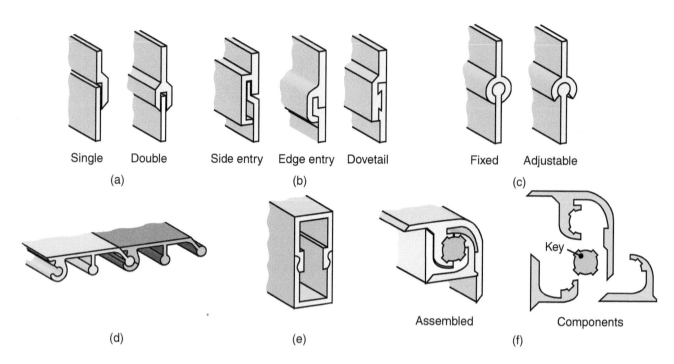

Single   Double   Side entry   Edge entry   Dovetail   Fixed   Adjustable

(a)   (b)   (c)

Key

Assembled   Components

(d)   (e)   (f)

FIGURE 15.18   Examples of part geometries that allow assembly of extruded sections. (a) Lap joints; (b) lap-lock joints; (c) cylindrical sliding fits; (d) cylindrical sliding lock joints; (e) snap fit and (f) keyed assembly.

die, whereas in drawing, it is pulled through it. Drawn rod and wire products cover a very wide range of applications, including shafts for power transmission, machine and structural components, blanks for bolts and rivets, electrical wiring, cables, tension-loaded structural members, welding electrodes, springs, paper clips, spokes for bicycle wheels, and stringed musical instruments.

The major processing variables in drawing are similar to those in extrusion, that is, reduction in cross-sectional area, die angle, friction along the die–workpiece interface, and drawing speed.

FIGURE 15.19   General view of a 9-MN (1000-ton) hydraulic-extrusion press. *Source:* Courtesy of Jones & Laughlin Steel Corporation.

Drawing Force.   The expression for the *drawing force*, F, under *ideal and frictionless* conditions is similar to that for extrusion, and is given by the equation

$$F = Y_{\text{avg}} A_f \ln \left(\frac{A_o}{A_f} \right), \qquad (15.4)$$

where Y_{avg} is the average true stress of the material in the die gap. Since more work has to be done to overcome friction, the force increases with increasing friction. Furthermore, because of nonuniform deformation within the die zone, additional energy, known as the *redundant work of deformation*, is required. Although several equations have been developed over the years to estimate the force (described in greater detail in advanced texts), a useful formula that includes friction and the redundant work is

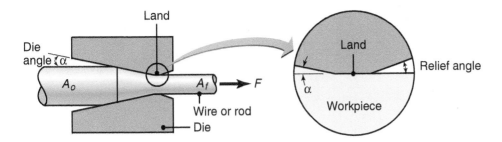

Land

Die angle α

A_o

A_f → F

Wire or rod

Die

Land

Relief angle

α

Workpiece

FIGURE 15.20 Process variables in wire drawing; the die angle, reduction in cross-sectional area per pass, speed of drawing, temperature, and lubrication condition all affect the drawing force, F.

$$ F = Y_{avg}A_f \left[\left(1 + \frac{\mu}{\alpha}\right) \ln\left(\frac{A_o}{A_f}\right) + \frac{2}{3}\alpha \right], \qquad (15.5) $$

where α is the die angle, in radians.

As can be seen from the two equations above, the drawing force increases as reduction increases. However, there has to be a limit to the magnitude of the force, because when the tensile stress reaches the yield stress of the metal drawn, the wire will simply yield and, eventually, break. It can be shown that, *ideally* and *without friction*, the maximum reduction in cross-sectional area per pass is 63%. Thus, for example, a 10-mm-diameter rod can be reduced to a diameter of 6.1 mm in one pass without failure.

It can also be shown that, for a certain reduction in diameter and a certain frictional condition, there is an *optimum die angle* at which the drawing force is a minimum. Often, however, the die force is not the major product quality concern, and the actual die angle may deviate from this optimum value.

Video Solution 15.2 Forces in Drawing

Drawing of Other Shapes.   Various solid cross-sections can be produced by drawing through dies with different profiles. Proper die design and the proper selection of reduction sequence per pass require considerable experience to ensure proper material flow in the die, reduce the development of internal or external defects, and improve surface quality.

The wall thickness, diameter, or shape of tubes that have been produced by extrusion or by other processes, described in this book, can be further reduced by *tube drawing* processes (Fig. 15.21). Tubes as large as 0.3 m (12 in.) in diameter can be drawn by these techniques.

Wedge-shaped dies are used for the drawing of *flat strips*. Although used only in specific applications, the principle behind this process is the fundamental deformation mechanism in **ironing**, used extensively in making aluminum beverage cans, as shown in Fig. 16.31.

15.9 Drawing Practice

Successful drawing requires proper selection of process parameters. In drawing, reductions in the cross-sectional area per pass range up to about 45%, and usually, the smaller the initial cross-section, the smaller the reduction per pass. Fine wires are drawn at 15–25% reduction per pass, and larger sizes at 20–45%. Reductions of higher than 45% may result in lubricant breakdown, leading to deterioration of surface finish. Although most drawing is done at room temperature, drawing

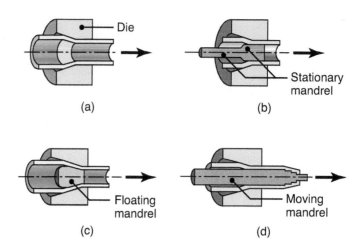

FIGURE 15.21 Examples of tube-drawing operations, with and without an internal mandrel; note that a variety of diameters and wall thicknesses can be produced from the same initial tube stock (which has been made by other processes).

large solid or hollow sections can be done at elevated temperatures, in order to reduce forces.

A light reduction, known as **sizing pass**, may be taken on rods to improve their surface finish and dimensional accuracy. However, because they basically deform only the surface layers, light reductions usually produce highly nonuniform deformation of the material and its microstructure. Consequently, the properties of the material will vary with radial distance within the cross-section.

Note in Fig. 15.20 that a rod or wire has to have its tip reduced in cross-section in order to be fed through the die opening, and be pulled. This typically is done by **swaging** the tip of the rod or wire in a manner similar to that shown in Figs. 14.15a and b, in an operation called *pointing*. Drawing speeds depend on the material and the reduction in cross-sectional area. They may range from 1 to 2.5 m/s (200–500 ft/min) for heavy sections to as much as 50 m/s (165 ft/s) for very fine wire, such as that used for electromagnets. Because the product does not have sufficient time to dissipate the heat generated in drawing, temperatures can rise significantly at high drawing speeds, and can have detrimental effects on product quality, such as surface finish and dimensional tolerances associated with thermal warping.

Drawn copper and brass wires are designated by their *temper* (such as 1/4 hard, 1/2 hard, etc.), because of work hardening. Intermediate annealing between passes may be necessary to maintain sufficient ductility of the material during cold drawing. High-carbon steel wires for springs and for musical instruments are made by **patenting**, a heat treating operation on the drawn wire, whereby the microstructure developed is fine pearlite (see Fig. 4.9). These wires have ultimate tensile strengths as high as 5 GPa (700 ksi), and a tensile reduction of area of about 20%.

Bundle Drawing.   Although very fine wire can be made by drawing, the cost can be high because the volume of metal produced per unit time is low. One method employed to increase productivity is to draw several wires simultaneously as a *bundle*. The interfaces between a hundred or more such wires are kept separate from one another by a suitable metallic material with similar properties, but with lower chemical resistance, so that it subsequently can be leached out from the drawn-wire surfaces.

Bundle drawing produces wires that are somewhat polygonal, rather than round, in cross-section. The wires produced can be as small as 4 μm (0.00016 in.) in diameter and can be made from such materials as stainless steels, titanium, and high-temperature alloys. In addition to producing continuous lengths, techniques have been developed to produce fine wire that is subsequently broken or chopped into various sizes and shapes. These wires are then used in applications such as electrically conductive plastics, heat-resistant and electrically conductive textiles, filter media, radar camouflage, and medical implants.

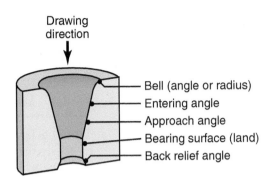

FIGURE 15.22 Terminology pertaining to a typical die used for drawing a round rod or wire.

Die Design.   The characteristic features of a typical drawing die are shown in Fig. 15.22. Note that there are two angles (entering and approach) in a typical die. Approach angles usually range from 6° to 15°, with the entering angle usually larger. The bell and the entering angles are used to control lubricant supply and film thickness. The purpose of the bearing surface (land) is to set the final diameter of the product (sizing), and to maintain this diameter even if the die–workpiece interface wears away.

A set of dies is required for **profile drawing**, which involves various stages of deformation to produce the final profile. The dies may be made in one piece or, depending on the complexity of the cross-sectional profile, with several segments, held together in a retaining ring. Computer-aided design techniques are implemented to design dies, for smooth material flow and to minimize the formation of defects.

A set of idling cylindrical or shaped rolls also may be used in drawing rods or bars of various shapes. Such an arrangement (called a *Turk's head*) is more versatile than that in common draw dies, because the rolls can be adjusted to different positions and angles for drawing specific profiles.

Die Materials.   Die materials for drawing (Table 5.8) typically are tool steels and carbides. For hot drawing, cast-steel dies are used, because of their high resistance to wear at elevated temperatures. Diamond dies are used for drawing fine wire, with diameters ranging from 2 μm to 1.5 mm (0.0001–0.06 in.). They may be made from a *single-crystal* diamond or in *polycrystalline* form, with diamond particles embedded in a metal matrix, called *compacts*. Because of their very low tensile strength and toughness, carbide and diamond dies are typically used as **inserts** or **nibs**, which are supported in a steel casing (Fig. 15.23).

Lubrication.   Lubrication is essential in drawing operations in order to improve die life and product surface finish, and to reduce drawing forces and temperature rise. Lubrication is critical, particularly in tube drawing, because of the difficulty of maintaining a sufficiently thick lubricant film at the mandrel–tube interface. In the drawing of rods, a common method of lubrication uses phosphate conversion coatings.

The following are the basic methods of lubrication used in wire drawing (see also Section 33.7):

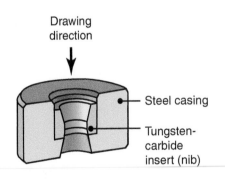

- **Wet drawing:** the dies and the rod are immersed completely in the lubricant.
- **Dry drawing:** the surface of the rod to be drawn is coated with a lubricant, by passing it through a box filled with the lubricant (*stuffing box*).
- **Metal coating:** the rod or wire is coated with a soft metal, such as copper or tin, that acts as a solid lubricant.

FIGURE 15.23 Tungsten-carbide die insert in a steel casing; diamond dies, used in drawing thin wire, are encased in a similar manner.

- **Ultrasonic vibration:** vibrations of the dies and mandrels improve surface finish and die life, and reduce drawing forces, thus allowing higher reductions per pass, without failure.

15.10 Drawing Defects and Residual Stresses

Typical defects in a drawn rod or wire are similar to those observed in extrusion, especially **center cracking** (see Fig. 15.16). Another major type of defect in drawing is **seams**, which are longitudinal scratches or folds in the drawn product (see also Section 33.2). Seams may open up during subsequent forming operations, such as upsetting, heading, thread rolling, or bending of the rod or wire, and may cause serious quality-control problems. Various other surface defects, such as scratches and die marks, may be due to improper selection of process parameters, poor lubrication, or poor die condition.

Because they undergo nonuniform deformation during drawing, cold-drawn products usually have *residual stresses*. For light reductions, such as only a few percent, the longitudinal-surface residual stresses are compressive, while the bulk is in tension, and fatigue life is thus improved. Conversely, heavier reductions induce tensile surface stresses, while the bulk is in compression. Residual stresses can be significant in causing stress-corrosion cracking of the part over time. Moreover, they cause the component to *warp*, if a layer of material is subsequently removed (see Fig. 2.30), such as by slitting, machining, or grinding.

Rods and tubes that are not sufficiently straight, or are supplied as coil, can be straightened by passing them through an arrangement of rolls placed at different axes, a process similar to roller leveling, shown in Fig. 13.7b.

15.11 Drawing Equipment

Although available in several designs, the equipment for drawing is basically of two types: the draw bench and the bull block.

A **draw bench** contains a single die, and its design is similar to that of a long, horizontal tension-testing machine (Fig. 15.24). The drawing force is supplied by a chain drive or hydraulic cylinder. Draw benches are used for a single-length drawing of straight rods and tubes, with diameters larger than 20 mm (0.75 in.) and lengths

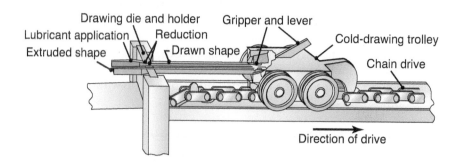

FIGURE 15.24 Cold drawing of an extruded channel on a draw bench to reduce its cross-section; individual lengths of straight rods or of cross-sections are drawn by this method.

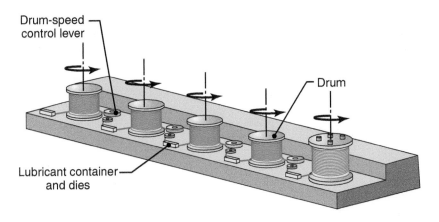

FIGURE 15.25 An illustration of multistage wire drawing typically used to produce copper wire for electrical wiring. Shown is a five bull block configuration; wire drawing machines can incorporate 15 or more of these drums, depending on the material and wire size. *Source:* After H. Auerswald.

up to 30 m (100 ft.). Machine capacities reach 1.3 MN (300 klb) of pulling force, with a speed range of 6 to 60 m/min (20–200 ft/min).

Very long rods and wire (several miles or kilometers) and wire of smaller cross-sections, usually less than 13 mm (0.5 in.), are drawn by a rotating *drum* (**bull block** or **capstan**, Fig. 15.25). The tension in this setup provides the force required for drawing the wire, usually through multiple dies (tandem drawing).

SUMMARY

- Extrusion is the process of pushing a billet through a die, to reduce its cross-section or to produce various solid or hollow cross-sections. This process is generally carried out at elevated temperatures, in order to reduce the extrusion force and improve the ductility of the material.

- Important factors in extrusion are die design, extrusion ratio, billet temperature, lubrication, and extrusion speed. Although the term "cold extrusion" applies to extrusion at room temperature, it is also the name for a combination of extrusion and forging operations. Cold extrusion is capable of economically producing discrete parts in various shapes and with good mechanical properties and dimensional tolerances.

- Rod, wire, and tube drawing operations basically involve pulling the material through a die or a set of dies in tandem. The cross-sections of most drawn products are round, but other shapes also can be drawn. Drawing tubular products, to reduce either their diameter or their thickness, usually requires the use of internal mandrels.

- Die design, reduction in cross-sectional area per pass, and selection of die materials and lubricants are all important parameters in making drawn products of high quality and with a good surface finish. External and internal defects can develop both in extrusion and in drawing. The significant factors are the die angle, reduction per pass, and quality of the workpiece material.

KEY TERMS

Bamboo defect

Bridge die

Bull block

Bundle drawing

Canning

Capstan

Center cracking

Chevron cracking

Cold extrusion

Conversion coating

Dead-metal zone

Draw bench

Drawing

Extrusion

Extrusion constant

Extrusion defects

Extrusion ratio

Fir-tree cracking

Hydrostatic extrusion

Impact extrusion

Ironing

Jacketing

Patenting

Pipe defect

Porthole die

Rod

Seam

Séjournet process

Shear die

Sizing pass

Speed cracking

Spider die

Turk's head

Wire

BIBLIOGRAPHY

Altan, T., Ngaile, G., and Shen, G. (eds.), *Cold and Hot Forging: Fundamentals and Applications*, ASM International, 2004.

ASM Handbook, Vol. 14A: **Metalworking: Bulk Forming**, ASM International, 2005.

Bauser, M., Sauer, G., and Siegert, K. (eds.), **Extrusion**, 2nd ed., ASM International, 2006.

Hosford, W.F., and Caddell, R.M., **Metal Forming: Mechanics and Metallurgy**, 4th ed., Cambridge, 2010.

Saha, P., **Aluminum Extrusion Technology**, ASM International, 2000.

Sheppard, T., **Extrusion of Aluminum Alloys**, Springer, 2010.

Tschaetch, H., **Metal Forming Practice: Processes, Machines, Tools**, Springer, 2007.

Wright, R.N., **Wire Technology: Process Engineering and Metallurgy**, Butterworth-Heinemann., 2010.

REVIEW QUESTIONS

15.1 How does extrusion differ from rolling and forging?

15.2 Explain the difference between extrusion and drawing.

15.3 What is a spider die? What is it used for?

15.4 Why are wires sometimes drawn in bundles?

15.5 What is a dead-metal zone?

15.6 Define the terms (a) cladding, (b) dummy block, (c) shear dies, (d) skull, and (e) canning.

15.7 Why is glass a good lubricant in hot extrusion?

15.8 What types of defects may occur in (a) extrusion and (b) drawing?

15.9 Describe the difference between direct and reverse extrusion.

15.10 What is land? What is its function in a die?

15.11 How are tubes extruded? Can they also be drawn? Explain.

15.12 It is possible to extrude straight gears; can helical gears also be extruded? Explain.

15.13 What is the difference between piping and bambooing?

15.14 What is impact extrusion?

15.15 What is the pipe defect in extrusion?

QUALITATIVE PROBLEMS

15.16 List the similarities and differences between direct extrusion and drawing.

15.17 Explain why extrusion is a batch, or semicontinuous, process. Do you think it can be made into a continuous process? Explain.

15.18 The extrusion ratio, die geometry, extrusion speed, and billet temperature all affect the extrusion pressure. Explain why.

15.19 Explain why cold extrusion is an important manufacturing process.

15.20 What is the function of a stripper plate in impact extrusion?

15.21 Explain the different ways by which changing the die angle affects the extrusion process.

15.22 Glass is a good lubricant in hot extrusion. Would you use glass for impression-die forging also? Explain.

15.23 How would you go about avoiding center-cracking defects in extrusion? Explain why your methods would be effective.

15.24 Table 15.1 gives temperature ranges for extruding various metals. Describe the possible consequences of extruding at a temperature (a) below and (b) above these ranges.

15.25 Will the force in direct extrusion vary as the billet becomes shorter? If so, why?

15.26 Comment on the significance of metal flow patterns shown in Fig. 15.6.

15.27 In which applications could you use the type of impact-extruded parts shown in Fig. 15.15?

15.28 What is the purpose of the land in a drawing die? Is there a limit to the size of the land that should be used? Explain.

15.29 Can spur gears be made by (a) drawing and (b) extrusion? Can helical gears? Explain.

15.30 How would you prepare the end of a wire in order to be able to feed it through a die so that a drawing operation can commence?

15.31 What is the purpose of a dummy block in extrusion? Explain.

15.32 Describe your observations concerning Fig. 15.9.

15.33 Occasionally, steel wire drawing will take place within a sheath of a soft metal, such as copper or lead. What is the purpose of this sheath?

15.34 Explain the advantages of bundle drawing.

15.35 Under what circumstances would backward extrusion be preferable to direct extrusion?

15.36 Why is lubrication detrimental in extrusion with a porthole die?

15.37 In hydrostatic extrusion, complex seals are used between the ram and the container, but not between the billet and the die. Explain why.

15.38 Describe the purpose of a container liner in direct extrusion, as shown in Fig. 15.1. What is the liner's function in reverse extrusion?

QUANTITATIVE PROBLEMS

15.39 Estimate the force required in extruding 70–30 brass at 700°C if the billet diameter is 200 mm and the extrusion ratio is 30.

15.40 Assuming an ideal drawing process, what is the smallest final diameter to which a 60-mm diameter rod can be drawn?

15.41 If you include friction in Problem 15.40, would the final diameter be different? Explain.

15.42 Calculate the extrusion force for a round billet 250 mm in diameter, made of 304 stainless steel, and extruded at 1000°C to a diameter of 70 mm.

15.43 A planned extrusion operation involves steel at 1000°C with an initial diameter of 100 mm and a final diameter of 25 mm. Two presses, one with capacity of 20 MN

and the other with a capacity of 10 MN, are available for the operation. Is the smaller press sufficient for this operation? If not, what recommendations would you make to allow the use of the smaller press?

15.44 A round wire made of a perfectly plastic material with a yield stress of 30,000 psi is being drawn from a diameter of 0.1–0.07 in. in a draw die of 15°. Let the coefficient of friction be 0.15. Using both Eqs. (15.4) and (15.5), estimate the drawing force required. Comment on the differences in your answer.

15.45 Assume that you are an instructor covering the topics described in this chapter and you are giving a quiz on the numerical aspects to test the understanding of the students. Prepare two quantitative problems and supply the answers.

SYNTHESIS, DESIGN, AND PROJECTS

15.46 Assume that the summary to this chapter is missing. Write a one-page summary of the highlights of the wire-drawing process.

15.47 Review the technical literature, and make a detailed list of the manufacturing steps involved in making common metallic hypodermic needles.

15.48 Figure 15.2 shows examples of discrete parts that can be made by cutting extrusions into individual pieces. Name several other products that can be made in a similar fashion.

15.49 The parts shown in Fig. 15.2 are economically produced by extrusion, but difficult to produce otherwise. List the processes that could be used to produce these parts, and explain why they are not as attractive as extrusion.

15.50 Survey the technical literature, and explain how external vibrations can be applied to a wire-drawing operation to

reduce friction. Comment also on the possible directions of vibration, such as longitudinal or torsional.

15.51 How would you go about making a stepped extrusion that has increasingly larger cross-sections along its length? Is it possible? Would your process be economical and suitable for high production runs? Explain.

15.52 List the processes that are suitable for producing an aluminum tube. For each process in your list, make a sketch of the grain structure you would expect to see in the finished product.

15.53 Assume that you are the technical director of trade associations of (a) extruders and (b) rod- and wire-drawing operations. Prepare a technical leaflet for potential customers, stating all of the advantages of these processes.

16

Sheet-metal Forming Processes and Equipment

- This chapter describes the important characteristics of sheet metals and the forming processes employed to produce a wide variety of products.
- The chapter opens with a description of the shearing operation, to cut sheet metal into blanks of desired shapes or to remove portions of the material, such as for holes or slots.
- A discussion of sheet-metal formability follows, with special emphasis on the specific metal properties that affect formability.
- The chapter then presents various bending operations for sheets, plates, and tubes, as well as such operations as stretch forming, rubber forming, spinning, peen forming, and superplastic forming.
- Deep drawing is then described, along with drawability, as it relates to the production of containers with thin walls.
- The chapter ends with a discussion of sheet-metal parts design, equipment characteristics, and the economic considerations for all these operations.

Typical parts made by sheet-metal forming: Car bodies, aircraft fuselages, trailers, office furniture, appliances, fuel tanks, and cookware.

Alternative process: Die casting, thermoforming, pultrusion, injection molding, blow molding.

16.1 Introduction

Products made of **sheet metals** are all around us. They include a very wide range of consumer and industrial products, such as beverage cans, cookware, file cabinets, metal desks, appliances, car bodies, trailers, and aircraft fuselages (Fig. 16.1). Sheet forming dates back to about 5000 B.C., when household utensils and jewelry were made by hammering and stamping gold, silver, and copper. Compared to those made by casting and by forging, sheet-metal parts offer the advantages of lightweight and versatile shapes.

As described throughout this chapter, there are numerous processes employed for making sheet-metal parts. The terms **pressworking** or **press forming** are commonly used in industry to describe these operations, because they typically are performed on *presses* (described in Sections 14.8 and 16.15), using a set of dies. A sheet-metal part produced in presses is called a **stamping** (after the word *stamp*, first used around 1200 A.D., and meaning "to force downward" or "to pound"). Low-carbon steel is the most commonly used sheet metal, because of its low cost and generally good

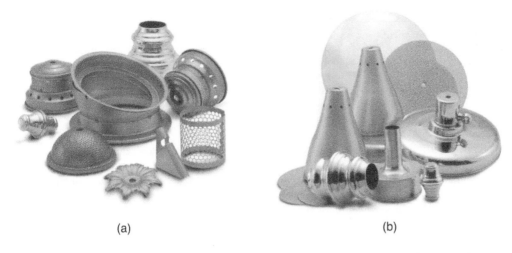

(a)   (b)

FIGURE 16.1 Examples of sheet-metal parts. (a) Stamped parts. (b) Parts produced by spinning. *Source:* Courtesy of Williamsburg Metal Spinning & Stamping Corp.

strength and formability characteristics. More recently developed alloys, such as TRIP and TWIP steels (see Section 5.5.6), have become more common for automotive applications because of their high strength. They are also well suited for providing good crash protection in a lightweight design. Aluminum is the most common material for such applications as beverage cans, packaging, kitchen utensils, and where corrosion resistance is an important factor. The common metallic materials for aircraft and aerospace applications are aluminum and titanium, although they are being replaced increasingly with composite materials, as described in Chapters 9 and 19.

Most manufacturing processes involving sheet metal are performed at room temperature. Hot stamping is occasionally performed in order to increase formability and decrease forming loads on machinery. Typical sheet metals in hot-stamping operations are titanium alloys and various high-strength steels.

This chapter first describes the methods by which blanks are cut from large rolled sheets, then processed further into desired shapes. The chapter also includes discussions on the characteristic features of sheet metals, the techniques employed to determine their formability, and the construction of forming-limit diagrams (FLDs). All of the major processes of sheet forming and the equipment also are described, as outlined in Table 16.1

16.2 Shearing

All sheet-metal forming operations begin with a **blank** of suitable dimensions and removed from a large sheet (usually from a *coil*) by **shearing**. Shearing subjects the sheet to shear stresses, generally using a punch and a die (Fig. 16.2a). The typical features of the sheared edges of the sheet metal and of the slug are shown in Figs. 16.2b and c, respectively. Note that, in this illustration, the edges are not smooth nor are they perpendicular to the plane of the sheet.

Shearing generally starts with the formation of cracks on both the top and bottom edges of the workpiece, at points *A* and *B*, and *C* and *D* in Fig. 16.2a. These cracks eventually meet each other, and complete separation occurs. The rough *fracture surfaces* are due to the cracks; the smooth and shiny *burnished surfaces* on the hole and the slug are from the contact and rubbing of the sheared edge against the walls of the punch and die, respectively.

TABLE 16.1

General Characteristics of Sheet-metal Forming Processes (in alphabetic order)	
Forming process	Characteristics
Drawing	Shallow or deep parts with relatively simple shapes, high production rates, high tooling and equipment costs
Explosive	Large sheets with relatively simple shapes, low tooling costs but high labor cost, low-quantity production, long cycle times
Incremental	Simple to moderately complex shapes with good surface finish; low production rates, but no dedicated tooling required; limited materials
Magnetic-pulse	Shallow forming, bulging, and embossing operations on relatively low strength sheets, requires special tooling
Peen	Shallow contours on large sheets, flexibility of operation, generally high equipment costs, process also used for straightening formed parts
Roll	Long parts with constant simple or complex cross-sections, good surface finish, high production rates, high tooling costs
Rubber	Drawing and embossing of simple or relatively complex shapes, sheet surface protected by rubber membranes, flexibility of operation, low tooling costs
Spinning	Small or large axisymmetric parts; good surface finish; low tooling costs, but labor costs can be high unless operations are automated
Stamping	Includes a wide variety of operations, such as punching, blanking, embossing, bending, flanging, and coining; simple or complex shapes formed at high production rates; tooling and equipment costs can be high, but labor cost is low
Stretch	Large parts with shallow contours, low-quantity production, high labor costs, tooling and equipment costs increase with part size
Superplastic	Complex shapes, fine detail and close dimensional tolerances, long forming times (hence production rates are low), parts not suitable for high-temperature use

The major processing parameters in shearing are:

- The shape of the punch and die
- The clearance, c, between the punch and the die
- The speed of punching
- Lubrication

The **clearance** is a major factor in determining the shape and the quality of the sheared edge. As clearance increases, the deformation zone (Fig. 16.3a) becomes larger and the sheared-edge surface becomes rougher. With excessive clearances, the sheet tends to be pulled into the die cavity, and the perimeter or edges of the sheared zone become rougher. Unless such edges are acceptable as produced, secondary operations may be necessary to make them smoother, which will increase the production cost. (See also *fine blanking* in Section 16.2.1.)

Edge quality can be improved with increasing punch speed, which may be as high as 10 to 12 m/s (30–40 ft/s). As shown in Fig. 16.3b, sheared edges can undergo severe cold working due to the high shear strains involved. Work hardening of the edges then will reduce the ductility of the edges, thus adversely affecting the formability of the sheet during subsequent forming operations, such as bending and stretching.

The ratio of the burnished area to the rough areas along the sheared edge increases with increasing ductility of the sheet metal, and decreases with increasing sheet thickness and clearance. The extent of the deformation zone, shown in Fig. 16.3, depends on the punch speed. With increasing speed, the heat generated by plastic

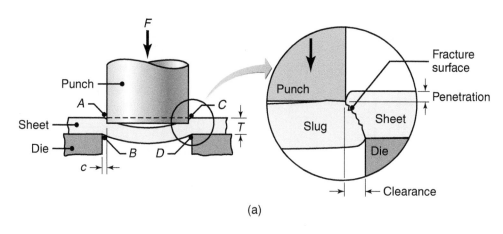

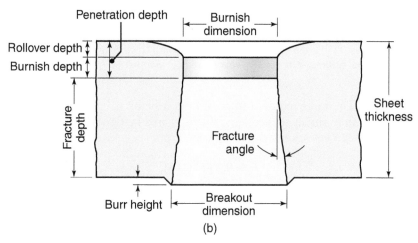

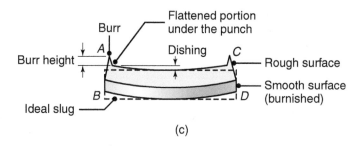

FIGURE 16.2 (a) Schematic illustration of shearing with a punch and die, indicating some of the process variables. Characteristic features of (b) a punched hole and (c) the slug; note that the scales of (b) and (c) are different.

deformation becomes confined to a smaller and smaller zone. Consequently, the sheared zone becomes narrower, and the sheared surface is smoother and exhibits less burr formation.

A **burr** is a thin edge or ridge, as shown in Figs. 16.2b and c. Burr height increases with increasing clearance and ductility of the sheet metal. Dull tool edges contribute greatly to large burr formation. The height, shape, and size of the burr can significantly affect subsequent forming operations. Several **deburring** processes are described in Section 26.8.

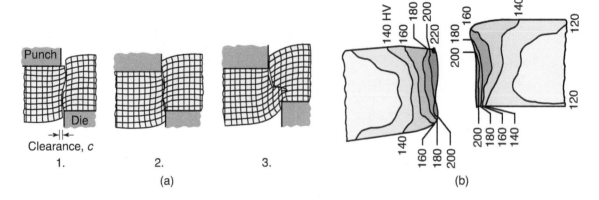

FIGURE 16.3 (a) Effect of the clearance, *c*, between punch and die on the deformation zone in shearing; as the clearance increases, the material tends to be pulled into the die rather than be sheared. In practice, clearances usually range between 2 and 10% of sheet thickness. (b) Microhardness (HV) contours for a 6.4-mm (0.25-in.) thick AISI 1020 hot-rolled steel in the sheared region. *Source:* After H.P. Weaver and K.J. Weinmann.

Punch Force.   The force required to punch out a blank is basically the product of the shear strength of the sheet metal and the total area being sheared. The *maximum punch force*, *F*, can be estimated from the equation

$$F = 0.7TL(\text{UTS}), \tag{16.1}$$

where *T* is the sheet thickness, *L* is the total length sheared (such as the perimeter of a hole), and UTS is the ultimate tensile strength of the material. It has been observed that as the clearance increases, the punch force decreases, and the wear on dies and punches also is reduced. (The effects of punch shape and die shape on punch forces are described in Section 16.2.3.)

Friction between the punch and the workpiece increases the punch force significantly. Furthermore, a force is required to strip the punch from the sheet during its return stroke. This force, which is in opposite direction of the punch force, is difficult to estimate because of the several factors involved in the punching operation.

Video Solution 16.1   Forces in Shearing

EXAMPLE 16.1   Calculation of Punch Force

Given: A 1-in. (25-mm) diameter hole is to be punched through a $\frac{1}{8}$-in. (3.2-mm) thick annealed titanium-alloy Ti-6Al-4V sheet at room temperature.

Find: Estimate the force required.

Solution: The force is estimated from Eq. (16.1), where the UTS for this alloy is found, from

Table 6.10, to be 1000 MPa, or 140,000 psi. Thus,

$$F = 0.7\left(\frac{1}{8}\right)(\pi)(1)(140,000) = 38,500 \text{ lb}$$

$$= 19.25 \text{ tons} = 0.17 \text{ MN}.$$

16.2.1   Shearing Operations

The most common shearing operations are **punching** [where the sheared slug is scrap (Fig. 16.4a) or may be used for some other purpose] and **blanking** (where the slug

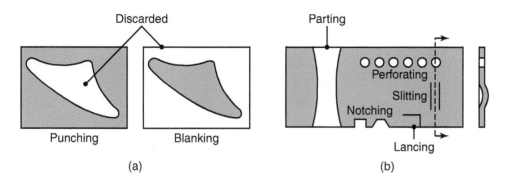

FIGURE 16.4 (a) Punching (piercing) and blanking. (b) Examples of various die-cutting operations on sheet metal; lancing involves slitting the sheet to form a tab.

is the part to be used and the rest is scrap). The shearing operations described next, as well as those described throughout the rest of this chapter, are often carried out on computer-numerical-controlled machines with quick-change toolholders (see Section 16.15).

Die Cutting.   This is a shearing operation that consists of the following basic processes, as shown in Fig. 16.4b:

- *Perforating:* punching a number of holes in a sheet
- *Parting:* shearing the sheet into two or more pieces
- *Notching:* removing pieces from edges
- *Lancing:* producing a tab without removing any material

Parts produced by these processes have various uses, particularly in assembly with other sheet-metal components. Perforated sheet metals, for example, with hole diameters ranging from 1 mm (0.040 in.) to 75 mm (3 in.) have uses as filters, as screens, in ventilation, as guards for machinery, in noise abatement, and in weight reduction of fabricated parts and structures. They are punched in crank presses (see Fig. 14.19a), at rates as high as 300,000 holes per minute, using special dies and equipment.

Fine Blanking.   Square edges with very smooth sheared surfaces can be produced by *fine blanking* (Fig. 16.5a). One basic die design is shown in Fig. 16.5b. A V-shaped stinger or impingement mechanically locks the sheet tightly in place, and thus prevents the type of distortion of the material shown in Figs. 16.2b and 16.3. The fine-blanking process involves clearances on the order of 1% of the sheet thickness, which may range from 0.5 to 13 mm (0.02–0.5 in.) in most applications. Dimensional tolerances typically are on the order of ±0.05 mm (0.002 in).

Slitting.   Shearing operations can be carried out by means of a pair of circular blades, similar to those in a can opener (Fig. 16.6). In *slitting*, the blades follow either a straight line, a circular path, or a curved path. A slit edge normally has a burr, which may be folded over the sheet surface by rolling it (flattening) between two cylindrical rolls. If not performed properly, slitting operations can cause various distortions of the sheared edges.

Steel Rules.   Soft metals, as well as paper, leather, and rubber, can be blanked with a *steel-rule die.* Such a die consists of a thin strip of hardened steel bent into the

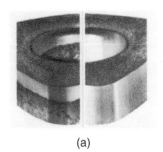

(a)

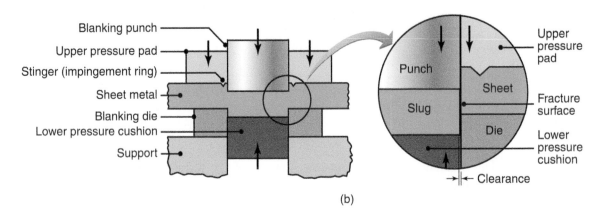

(b)

FIGURE 16.5   (a) Comparison of sheared edges produced by conventional (left) and by fine-blanking (right) techniques. (b) Schematic illustration of one setup for fine blanking. *Source:* Reprinted by permission of Feintool U.S. Operations.

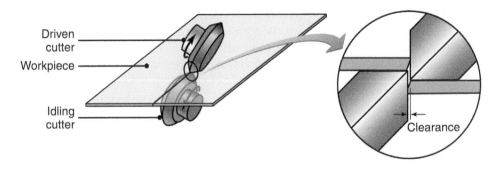

FIGURE 16.6   Slitting with rotary knives, a process similar to opening cans.

shape to be produced, and held on its edge on a flat wood or polymer base. The die is pressed against the sheet, which rests on the flat surface, and it shears the sheet along the shape defined by the steel rule.

Nibbling.   In *nibbling*, a machine called a *nibbler* moves a small straight punch up and down rapidly into a die. A sheet is fed through the gap and several overlapping holes are made. With manual or automatic control, sheets can be cut along any desired path. In addition to its flexibility, an advantage of nibbling is that intricate slots and notches, such as those shown in Fig. 16.4b, can be produced using standard punches. Because no special dies are required, the process is economical for small production runs.

Scrap in Shearing.   The amount of *scrap* (*trim loss*) produced in shearing operations can be significant, and can be as high as 30% on large stampings (see Table 40.3). Scrap, which can be a significant factor in manufacturing costs, can be reduced substantially by efficient arrangement of the shapes on the sheet to be cut (**nesting**, see Fig. 16.59). Computer-aided design techniques are now available to minimize scrap.

16.2.2   Tailor-welded Blanks

In the sheet-metal-forming processes to be described throughout this chapter, the blank is usually a one-piece sheet of constant thickness, and cut (blanked) from a large sheet. An important variation from this practice involves *laser-beam butt welding* (see Section 30.7) of two or more pieces of sheet metal with different shapes and thicknesses. The strips are welded to obtain a locally thicker sheet or add a different material. (See Case Study 16.1.)

Because of the small thicknesses involved, the proper alignment of the sheets prior to welding is important. The welded assembly is subsequently formed into a final shape. This technique has become increasingly important, particularly to the automotive industry. Because each piece now can have a different thickness, composition, coating, or other characteristics, the use of tailor-welded blanks has the following advantages:

- Reduction in scrap
- Elimination of the need for subsequent spot welding operations (as in making a car body; see Fig. I.9)
- Better control of dimensions
- Increased productivity

CASE STUDY 16.1   Tailor-welded Sheet Metal for Automotive Applications

An example of the use of tailor-welded sheet metals in automobile bodies is shown in Fig. 16.7. Note that five different pieces are first blanked, which includes cutting by laser beams. Four of these pieces are 1 mm thick, and one is 0.8 mm thick. The pieces are laser butt welded (Section 30.7) and then stamped into the final shape. In this manner, the blanks can be tailored to a particular application, not only as to shape and thickness, but also by using different-quality sheets, with or without coatings.

Laser-welding techniques are highly developed and the joints are very strong and reliable. The combination of welding and forming sheet-metal pieces makes possible significant flexibility in product design, formability, structural stiffness, and crash behavior of an automobile. It also makes possible the use of different materials in one product, weight savings, and cost reductions in materials, scrap, assembly, equipment, and labor.

The various components shown in Fig. 16.8 utilize the advantages outlined above. For example, note in Fig. 16.8b that the strength and stiffness required for the support of the shock absorber are achieved by welding a round piece onto the surface of the large sheet. The sheet thickness in such components varies (depending on its location and on its contribution to such characteristics as stiffness and strength), resulting in significant weight savings without loss of structural strength and stiffness.

More recent advances include the use of *friction stir welding* (Section 31.4) to produce the tailor-welded blank, and in the production of *tailor-welded coils*, where the material and/or thickness can be varied at a given location in the sheet. Tailor-welded blanks have also been used in hot stamping of automotive space frame pillars (Section 16.11). In this application, a steel grade is used to minimize deflections and protect occupants, but a more ductile steel that absorbs energy (see *toughness*, Sections 2.2.4 and 2.10) is used where the pillar is attached to the car frame.

(continued)

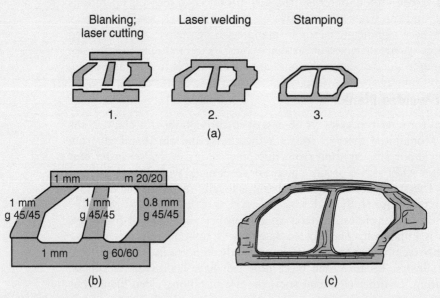

Legend:
g 60/60 (45/45)   Hot-galvanized alloy steel sheet. Zinc amount: 60/60 (45/45) g/m^2.
m 20/20   Double-layered iron–zinc alloy electroplated steel sheet. Zinc amount 20/20 g/m^2.

FIGURE 16.7   Production of an outer side panel of a car body by laser butt welding and stamping. *Source:* After M. Geiger and T. Nakagawa.

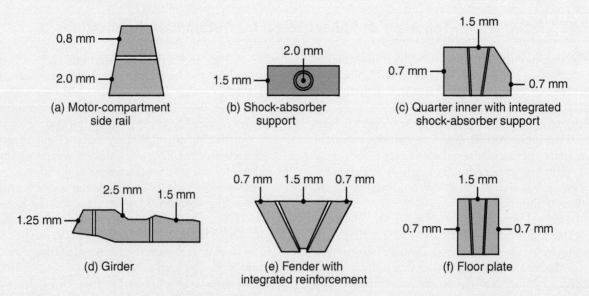

FIGURE 16.8   Examples of laser butt-welded and stamped automotive-body components. *Source:* After M. Geiger and T. Nakagawa.

16.2.3   Characteristics and Types of Shearing Dies

Clearance.   Because the formability of the sheared part can be influenced by the quality of its sheared edges, clearance control is important. The appropriate clearance depends on

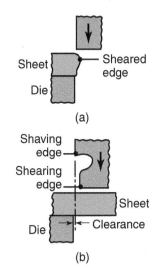

- Type of material and its temper
- Thickness and size of the blank
- Proximity to the edges of other sheared edges or the edges of the original blank

Clearances generally range between 2 and 8% of the sheet thickness, although they may be as small as 1% (as in *fine blanking*, Section 16.2.1) or as large as 30%. The smaller the clearance, the better is the quality of the edge. If the sheared edge is rough and not acceptable, it can be subjected to a process called **shaving** (Fig. 16.9a), whereby the extra material from the edge is trimmed by cutting, as also depicted in Fig. 21.3.

As a general guideline, (a) clearances for soft materials are less than those for harder grades; (b) the thicker the sheet, the larger the clearance must be; and (c) as the ratio of hole-diameter to sheet-thickness decreases, clearances must be larger. In using larger clearances, attention must be paid to the rigidity and the alignment of the presses, the dies, and their setups.

FIGURE 16.9 Schematic illustrations of the shaving process. (a) Shaving a sheared edge. (b) Shearing and shaving combined in one stroke.

Punch and Die Shape.   Note in Fig. 16.2a that the surfaces of the punch and of the die are both flat. Because the entire thickness is sheared at the same time, the punch force increases rapidly during shearing. The location of the regions being sheared at any particular instant can be controlled by *beveling* the punch and die surfaces (Fig. 16.10). This shape is similar to that of some paper punches, which can be observed by inspecting the tip of the punch. Beveling is suitable particularly for shearing thick sheets, because it reduces the force at the beginning of the stroke and the operation's noise level.

Note in Fig. 16.10c that the punch tip is symmetrical, and in Fig. 16.10d the die is symmetrical, thus there are no lateral forces acting on the punch to cause distortion. By contrast, the punch in Fig. 16.10b has a single taper, and thus it is subjected to a lateral force. Consequently, the punch and press setups must both have sufficient lateral stiffness, so that they neither produce a hole that is located improperly, nor allow the punch to hit the edge of the lower die and cause damage (as it might at point B or D in Fig. 16.2a).

Compound Dies.   Several operations may be performed on the same sheet in one stroke, and at one station, with a *compound die* (Fig. 16.11). Such combined operations usually are limited to relatively simple shapes, because (a) the process is somewhat slow and (b) the dies rapidly become much more expensive, especially for complex dies, to produce than those for individual shearing operations.

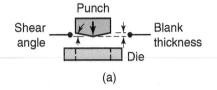
(a)

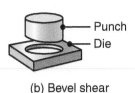
(b) Bevel shear

(c) Double-bevel shear

(d) Convex shear

FIGURE 16.10   Examples of shear angles on punches and dies.

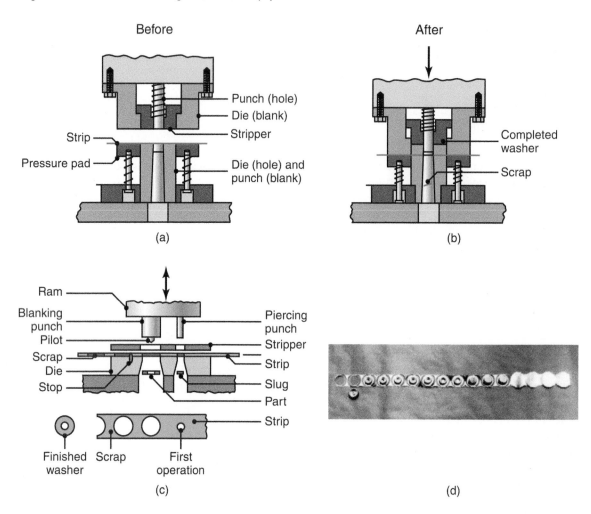

FIGURE 16.11 Schematic illustrations (a) before and (b) after blanking a common washer in a compound die; note the separate movements of the die (for blanking) and the punch (for punching the hole in the washer). (c) Schematic illustration of making a washer in a progressive die. (d) Forming of the top piece of an aerosol spray can in a progressive die; the part is attached to the strip until the last operation is completed.

Progressive Dies. Parts requiring multiple forming operations can be made, at high production rates, using *progressive dies*. The sheet metal is fed through as a coil strip, and a different operation (such as punching, blanking, and notching) is performed at the same station of the machine, with each stroke using a series of punches (Fig. 16.11c). An example of a part made in progressive dies is shown in Fig. 16.11d. The part is the small round metal piece that supports the plastic tip in spray cans.

Transfer Dies. In a *transfer die* setup, the sheet metal undergoes different operations at different stations of the machine, arranged along a straight line or a circular path. After each step in a station, the part is transferred to the next station for further operations.

Tool and Die Materials. Tool and die materials for shearing generally are tool steels and carbides (for high production rates). (See Tables 5.7–5.9.) Lubrication is important for reducing tool and die wear, thus maintaining edge quality.

16.2.4   Miscellaneous Methods of Cutting Sheet Metal

There are several other methods of cutting metal sheets and plates:

- **Laser-beam cutting** is an important process (Section 27.6), and typically used with computer-controlled equipment to cut a variety of shapes consistently, in various thicknesses, and without the use of dies. The process can also be combined with punching and shearing operations. Some parts with certain features may be produced best by one process, while others, with various features, may best be produced by the other process. Combination machines, incorporating both capabilities, have been designed and built for this reason. (See also Example 27.1.)
- **Water-jet cutting** is effective on metallic as well as nonmetallic materials (Section 27.8).
- Cutting with a **band saw**; this is a chip-removal process.
- **Friction sawing** involves a disk or blade that rubs against the sheet or plate at high surface speeds, thus raising the temperature and separating it into two pieces (Section 24.5).
- **Flame cutting** is another common method, particularly for thick plates; it is used widely in shipbuilding and on heavy structural components (Section 30.8).

16.3   Sheet-metal Characteristics and Formability

After a blank is cut from a larger sheet or coil, it is formed into various shapes by several processes, described in the rest of this chapter. This section presents a brief review of those characteristics of sheet metals that have significant effects on forming operations, as outlined in Table 16.2.

Elongation.   Sheet-metal forming processes rarely involve simple uniaxial stretching, as in a tension test. However, observations from tensile testing are useful and necessary for understanding the behavior of metals in these operations. Recall from Section 2.2 that a specimen subjected to tension first undergoes **uniform elongation**, and that when the load exceeds the ultimate tensile strength, the specimen begins to neck and elongation is no longer uniform.

Because in sheet forming the material usually is being stretched, high uniform elongation is essential for good formability. The true strain at which necking begins is numerically equal to the *strain-hardening exponent*, n, shown in Eq. (2.8). Thus, a high n value indicates large uniform elongation (see also Table 2.3). Necking may be *localized* or it may be *diffused*, depending on the *strain-rate sensitivity*, m, of the material, as given in Eq. (2.9). The higher the value of m, the more diffuse the neck becomes. A diffuse neck is desirable in sheet-forming operations. In addition to uniform elongation and necking, the **total elongation** of the specimen (in terms of that for a 50-mm gage length) also is a significant factor in the formability of sheet metals.

Yield-point Elongation.   Low-carbon steels and some aluminum–magnesium alloys exhibit a behavior called *yield-point elongation*, having both upper and lower yield points, shown in Fig. 16.12a. This phenomenon results in **Lüder's bands** (also called *stretcher-strain marks* or *worms*) on the sheet (Fig. 16.12b), which are elongated depressions on the surface of the sheet, such as can be found on the bottom of steel cans for common household products (Fig. 16.12c). The marks may be objectionable in the formed product, because coarseness on the surface degrades appearance and may cause difficulties in subsequent coating and painting operations.

TABLE 16.2

Important Metal Characteristics for Sheet-metal Forming Operations

Characteristic	Importance
Elongation	Determines the capability of the sheet metal to stretch without necking and failure; high strain-hardening exponent (n) and strain-rate sensitivity exponent (m) are desirable
Yield-point elongation	Typically observed with mild-steel sheets (also called Lüder's bands or stretcher strains); results in depressions on the sheet surface; can be eliminated by temper rolling, but sheet must be formed within a certain time after rolling
Anisotropy (planar)	Exhibits different behavior in different planar directions, present in cold-rolled sheets because of preferred orientation or mechanical fibering, causes earing in deep drawing, can be reduced or eliminated by annealing but at lowered strength
Anisotropy (normal)	Determines thinning behavior of sheet metals during stretching, important in deep drawing
Grain size	Determines surface roughness on stretched sheet metal; the coarser the grain, the rougher is the appearance (like an orange peel); also affects material strength and ductility
Residual stresses	Typically caused by nonuniform deformation during forming, results in part distortion when sectioned, can lead to stress-corrosion cracking, reduced or eliminated by stress relieving
Springback	Due to elastic recovery of the plastically deformed sheet after unloading, causes distortion of part and loss of dimensional accuracy, can be controlled by techniques such as overbending and bottoming of the punch
Wrinkling	Caused by compressive stresses in the plane of the sheet; can be objectionable; depending on its extent, can be useful in imparting stiffness to parts by increasing their section modulus; can be controlled by proper tool and die design
Quality of sheared edges	Depends on process used; edges can be rough, not square, and contain cracks, residual stresses, and a work-hardened layer, which are all detrimental to the formability of the sheet; edge quality can be improved by fine blanking, reducing the clearance, shaving, and improvements in tool and die design and lubrication
Surface condition of sheet	Depends on sheet-rolling practice; important in sheet forming, as it can cause tearing and poor surface quality

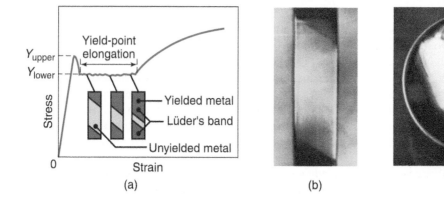

FIGURE 16.12   (a) Yield-point elongation in a sheet-metal specimen. (b) Lüder's bands in a low-carbon steel sheet. (c) Stretcher strains at the bottom of a steel can for household products. *Source:* (b) Courtesy of Caterpillar, Inc.

The usual method of avoiding Lüder's bands is to eliminate or reduce yield-point elongation by reducing the thickness of the sheet 0.5 to 1.5% by cold rolling, known as **temper** or **skin rolling**. Because of *strain aging*, however, the yield-point elongation reappears after a few days at room temperature, or after a few hours at higher temperatures; thus, the material should be formed within a certain time limit (which depends on the material).

Anisotropy.   An important factor that influences sheet-metal forming is *anisotropy* (*directionality*) of the sheet (see Fig. 16.17). Recall that anisotropy is acquired during the thermomechanical processing of the sheet, and that there are two types of anisotropy: *crystallographic anisotropy* (preferred orientation of the grains) and *mechanical fibering* (alignment of impurities, inclusions, and voids throughout the thickness of the sheet). The relevance of anisotropy is discussed further in Section 16.4.

Grain Size.   As described in Section 1.5, grain size affects mechanical properties and influences the surface appearance of the formed part (*orange peel*). The smaller the grain size, the stronger is the metal, and the coarser the grain, the rougher is the surface appearance. An ASTM grain size of 7 or finer (Table 1.1) is preferred for general sheet-forming operations.

Dent Resistance of Sheet Metals.   Dents are commonly found on cars, appliances, and office furniture. They usually are caused by dynamic forces from moving objects hitting the sheet metal. In typical automotive panels, for example, velocities at impact range up to 45 m/s (150 ft/s). Thus, it is the *dynamic yield stress* (yield stress under high rates of deformation), rather than the static yield stress, that is the significant strength parameter.

The factors significant in dent resistance can be shown to be the yield stress, Y, the sheet metal thickness, T, and the shape of the panel. *Dent resistance* is then expressed by a combination of material and geometrical parameters, as

$$\text{Dent resistance} = \frac{Y^2 T^4}{S}, \tag{16.1a}$$

where S is the panel stiffness, which, in turn, is defined as

$$S = ET^a(\text{shape}), \tag{16.1b}$$

where the value of a ranges from 1 to 2 for most panels. As for shape, the flatter the panel, the greater is its dent resistance, because of the sheet's flexibility. Thus, dent resistance (1) increases with increasing strength and thickness of the sheet, (2) decreases with increasing elastic modulus and stiffness, and (3) decreases with decreasing curvature of the sheet. Consequently, panels rigidly held at their edges have lower dent resistance (because of their higher stiffness) than those held with a set of springs.

Dynamic forces tend to cause *localized dents*, whereas static forces tend to *diffuse* the dented area. This phenomenon may be demonstrated by trying to dent a piece of flat sheet metal, by pushing a ball-peen hammer against it versus by striking it with the hammer; note how localized the dent will be in the latter case.

16.4 Formability Tests for Sheet Metals

Sheet-metal formability is generally defined as the ability of the sheet metal to undergo the required shape change without failure, such as by cracking, wrinkling, necking, or tearing. As will be noted throughout the rest of this chapter, depending on part shape, sheet metals may undergo two basic modes of deformation: (1) *stretching* and

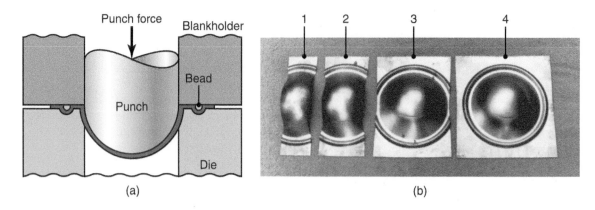

FIGURE 16.13 (a) A cupping test (the Erichsen test) to determine the formability of sheet metals. (b) Bulge-test results on steel sheets of various widths; the specimen farthest left is subjected to, basically, simple tension. The specimen that is farthest right is subjected to equal biaxial stretching. *Source:* Courtesy of (a) Arcelor Mittal and (b) Inland Steel Company.

(2) *drawing*. There are important distinctions between these two modes, and different parameters are involved in determining formability under different conditions. This section describes the methods that generally are used to predict formability.

Cupping Tests.   The earliest tests developed to predict sheet-metal formability were cupping tests (Fig. 16.13a). In the *Erichsen* test, the sheet specimen is clamped between two circular flat dies, and a steel ball (or a round punch) is forced into the sheet until a crack begins to appear on the stretched specimen. The *punch depth*, d, at which a crack appears is a measure of the formability of the sheet. Although this and other similar tests are easy to perform, they do not simulate the exact conditions of actual forming operations, and hence are not particularly reliable, especially for complex parts.

Forming-limit Diagrams.   An important approach for determining the formability of sheet metals is the development of *forming-limit diagrams* (FLD), as shown in Fig. 16.14. For a particular sheet metal, this diagram is constructed by first marking the flat sheet with a grid pattern of circles (Fig. 16.15), using chemical or photo-printing techniques. The blank is then stretched over a punch (Fig. 16.13a), and the deformation of the circles is observed and measured in the region where failure (*necking* or *tearing*) has occurred. Although the circles typically are 2.5 to 5 mm (0.1–0.2 in.) in diameter, for improved accuracy of measurement, they should be made as small as is practical.

In order to simulate the typically unequal stretching encountered in actual sheet-forming operations, the flat specimens are cut to varying widths (Fig. 16.13b), and then tested. Note that a square specimen (farthest right in the figure) produces *equal biaxial stretching* (such as that achieved in blowing up a spherical balloon), whereas a narrow specimen (farthest left in the figure) basically undergoes a state of *uniaxial stretching* (that is, simple tension). After a series of such tests is performed on a particular sheet metal, and at different widths, an FLD is constructed, showing the boundaries between failure and safe zones (Fig. 16.14b).

In order to develop an FLD, the major and minor engineering strains, as measured from the deformation of the original circles, are obtained. Note in Fig. 16.14a that an original circle has deformed into an ellipse, the *major axis* of which represents the major direction and magnitude of stretching. The major strain is the

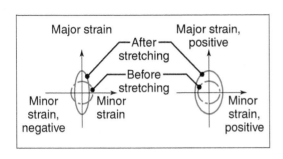

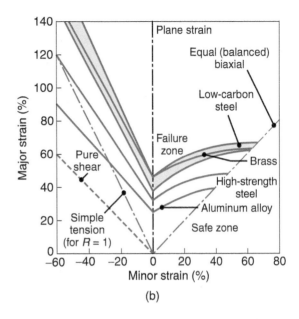

(a)	(b)

FIGURE 16.14 (a) Strains in deformed circular grid patterns. (b) Forming-limit diagrams (FLD) for various sheet metals. Although the major strain is always positive (stretching), the minor strain may be either positive or negative. R is the normal anisotropy of the sheet, as described in Section 16.7. *Source:* After S.S. Hecker and A.K. Ghosh.

engineering strain in this direction, and is always *positive* (because the sheet is being stretched). The *minor axis* of the ellipse represents the minor direction and magnitude of strain in the *transverse* direction, which may have undergone stretching or shrinking.

Note that the minor strain can be either *positive* or *negative*. For example, if a circle is placed in the center of a tensile-test specimen, and then stretched uniaxially (simple tension), the specimen becomes narrower as it is stretched (due to the Poisson effect, Section 2.2.1); thus the minor strain is negative. This behavior can be demonstrated easily by stretching a rubber band and observing the dimensional changes it undergoes. On the other hand, if we place a circle on a spherical rubber balloon and inflate it, the minor and major strains are both positive and equal in magnitude.

By comparing the surface areas of the original circle and the deformed circle on the formed sheet, we also can determine whether the thickness of the sheet has changed during deformation. Because in plastic deformation the volume remains constant, we know that if

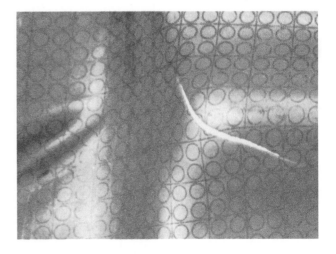

FIGURE 16.15 The deformation of the grid pattern and the tearing of sheet metal during forming; the major and minor axes of the circles are used to determine the coordinates on the forming-limit diagram in Fig. 16.14b. *Source:* After S.P. Keeler.

the area of the deformed circle is larger than the original circle, the sheet has become thinner. This phenomenon can be demonstrated easily by blowing up a balloon and noting that it becomes more translucent as it is stretched, because it is becoming thinner.

The data thus obtained from different locations in each of the samples shown in Fig. 16.13b are then plotted as shown in Fig. 16.14b. The curves represent the boundaries between failure zones and *safe zones* for each type of metal, and as can be noted, the higher the curve, the better is the formability of that particular sheet metal. As expected, different materials and conditions, such as cold worked or heat treated, have different FLDs. Taking the aluminum alloy in Fig. 16.14b as an example, if a circle in a particular location on the sheet has undergone major and minor strains of +20% and −10%, respectively, there would be no tear in that location of the specimen. On the other hand, if at another location on the sheet the major and minor strains were +80% and −40%, respectively, there would be a tear in that particular location of the specimen. An example of a formed sheet-metal part with a grid pattern is shown in Fig. 16.15; note the deformation of the circular patterns in the vicinity of the tear on the formed sheet.

It is important to note in FLDs that a compressive minor strain of, say, 20% is associated with a higher major strain than is a tensile (positive) minor strain of the same magnitude. In other words, it is desirable for the minor strain to be negative (i.e., shrinking in the minor direction). In forming complex parts, special tooling or clamps can be designed to take advantage of the beneficial effect of negative minor strains on formability. The effect of sheet thickness on FLDs is to raise the curves in Fig. 16.14b.

Friction and lubrication at the interface between the punch and the sheet metal also are important factors in the test results. With well-lubricated interfaces, the strains in the sheet are distributed more uniformly over the punch. Also, as expected, and depending on the material and surface defects such as notch sensitivity, surface scratches (see *notch sensitivity*, Section 2.9), deep gouges, and blemishes can significantly reduce formability and thereby lead to premature tearing and failure of the part.

A procedure that has been followed with some success to improve sheet-metal formability is to carefully control and vary process parameters during forming. For example, deep drawability (Section 16.7.1) can be improved by varying the blankholder force (see Fig. 16.32) during deep drawing. This force can be changed with position in the die if multiple actuators are used for the blankholder, or it can be changed with respect to time. Carefully optimized velocity profiles can be programmed into servo presses (described in Section 14.8) also improve formability.

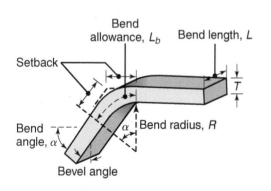

FIGURE 16.16 Bending terminology; note that the bend radius is measured to the inner surface of the bent part.

16.5 Bending Sheets, Plates, and Tubes

Bending is one of the most common forming operations, as evidenced by observing automobile bodies, exhaust pipes, appliances, paper clips, or file cabinets. Bending also imparts stiffness to the part, by increasing its moment of inertia. Note, for example, how corrugations, flanges, beads, and seams improve the stiffness of structures without adding weight. As a specific example, observe the diametral stiffness of a can with and without circumferential beads (see also Section 16.7).

The terminology used in bending sheet or plate is given in Fig. 16.16. Note that the outer fibers of the material are in tension, while the inner fibers are in compression. Because of the Poisson effect, the width of the part (*bend length*, L) has become smaller in the outer region, and larger in the inner region than the original width (as can be seen in Fig. 16.17c). This phenomenon may easily

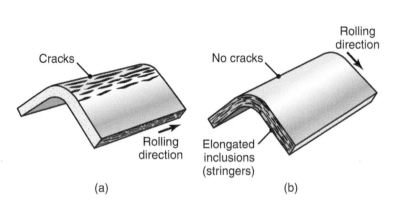

(a)   (b)   (c)

FIGURE 16.17 (a) and (b) The effect of elongated inclusions (stringers) on cracking as a function of the direction of bending with respect to the original rolling direction of the sheet. (c) Cracks on the outer surface of an aluminum strip bent to an angle of 90°. Note also the narrowing of the top surface in the bend area (due to the Poisson effect).

be observed by bending a rectangular rubber eraser, and observing the changes in its cross-section.

As shown in Fig. 16.16, the **bend allowance**, L_b, is the length of the *neutral axis* in the bend; it is used to determine the length of the blank for a part to be bent. The position of the neutral axis, however, depends on bend radius and bend angle, as described in texts on mechanics of materials. An approximate formula for the bend allowance is

$$L_b = \alpha \left(R + kT \right), \tag{16.3a}$$

where α is the bend angle (in radians), T is the sheet thickness, R is the bend radius, and k is a constant, which in practice typically ranges from 0.33 (for $R < 2T$) to 0.5 (for $R > 2T$). Note that for the ideal case, the neutral axis is at the center of the sheet thickness, $k = 0.5$, and hence,

$$L_b = \alpha \left[R + \left(\frac{T}{2} \right) \right]. \tag{16.3b}$$

Minimum Bend Radius. The radius at which a crack first appears at the outer fibers of a sheet being bent is referred to as the *minimum bend radius*. It can be shown that the engineering strain on the outer and inner fibers of a sheet during bending is given by the expression

$$e = \frac{1}{(2R/T) + 1}. \tag{16.4}$$

Thus, as R/T decreases (i.e., as the ratio of the bend radius to the thickness becomes smaller), the tensile strain at the outer fiber increases, and the material eventually develops cracks (Fig. 16.17). The bend radius usually is expressed (reciprocally) in terms of the thickness, such as $2T$, $3T$, $4T$, and so on (see Table 16.3). Thus, a $3T$ minimum bend radius indicates that the smallest radius to which the sheet can be bent, without cracking, is three times its thickness.

TABLE 16.3

Minimum Bend Radius for Various Metals at Room Temperature

	Condition	
Material	Soft	Hard
Aluminum alloys	0	6T
Beryllium copper	0	4T
Brass (low-leaded)	0	2T
Magnesium	5T	13T
Steels		
Austenitic stainless	0.5T	6T
Low-carbon, low-alloy, and HSLA	0.5T	4T
Titanium	0.7T	3T
Titanium alloys	2.6T	4T

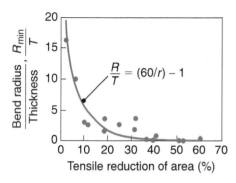

FIGURE 16.18 Relationship between R_{min}/T and tensile reduction of area for sheet metals. Note that sheet metal with a 50% tensile reduction of area can be bent over itself in a process like the folding of a piece of paper without cracking. *Source:* After J. Datsko and C.T. Yang.

It has been shown that there is an inverse relationship between *bendability* and the tensile reduction of the area, r, of the material (Fig. 16.18). The *minimum bend radius*, R_{min}, is, approximately,

$$R_{min} = T \left(\frac{50}{r} - 1 \right). \tag{16.5}$$

Thus, for $r = 50$, the minimum bend radius is zero; that is, the sheet can be folded over itself, called *hemming* (see Fig. 16.23), in much the same way as a piece of paper is folded. To increase the bendability of metals, their tensile reduction of area can be increased either by heating or by bending in a high-pressure environment, which improves the ductility of the material (see *hydrostatic stress*, Section 2.2.8).

Bendability also depends on the *edge condition* of the sheet. Since rough edges are points of stress concentration, bendability decreases as edge roughness increases. Another significant factor in edge cracking is the amount, shape, and hardness of *inclusions* present in the sheet metal and the amount of cold working that the edges undergo during shearing. Because of their pointed shape, inclusions in the form of stringers are more detrimental than globular-shaped inclusions (see also Fig. 2.24). The resistance to edge cracking during bending can be significantly improved by removing the cold-worked regions, by shaving or machining the edges of the part (see Fig. 16.9) or by annealing the sheet to improve its ductility.

Anisotropy of the sheet is another important factor in bendability. Cold rolling results in anisotropy of the sheet by *preferred orientation* or by *mechanical fibering*, due to the alignment of any impurities, inclusions, and voids that may be present, (see also Fig. 1.12). Prior to laying out or *nesting* the blanks (see Fig. 16.59) for subsequent bending or forming, caution should be exercised to cut them, as much as possible, in the proper direction from a rolled sheet.

Springback. Because all materials have a finite modulus of elasticity, plastic deformation is always followed by some elastic recovery when the load is removed (see Fig. 2.3). In bending, this recovery is called *springback*, which can easily be observed by bending and then releasing a piece of sheet metal or wire. As noted in Fig. 16.19, the final bend angle of a sheet metal after springback is smaller than the angle to which the sheet was bent, and the final bend radius is larger than before springback.

Springback can be calculated approximately in terms of the radii R_i and R_f (Fig. 16.19) as

$$\frac{R_i}{R_f} = 4 \left(\frac{R_i Y}{ET} \right)^3 - 3 \left(\frac{R_i Y}{ET} \right) + 1. \tag{16.6}$$

Note from this formula that springback increases as the R/T ratio and the yield stress, Y, of the material increase, and as the elastic modulus, E, decreases.

In V-die bending (Figs. 16.20 and 16.21), it is possible for the material to also exhibit *negative springback*. This is a condition caused by the nature of the deformation occurring within the sheet metal just when the punch completes the bending operation at the end of the stroke. Negative springback does not occur in *air bending*, shown in Fig. 16.22a (also called *free bending*), because of the absence of constraints that a V-die imposes on the bend area.

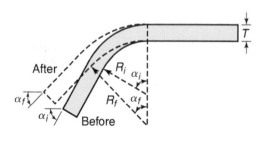

FIGURE 16.19 Springback in bending; the part tends to recover elastically after bending, and its bend radius becomes larger. Under certain conditions, it is possible for the final bend angle to be smaller than the original angle (negative springback).

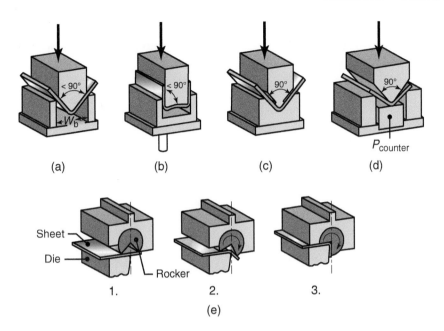

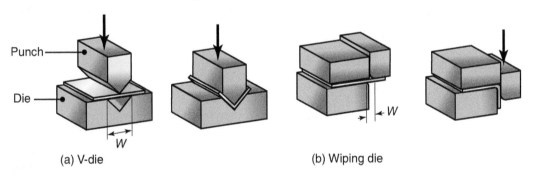

FIGURE 16.20   Methods of reducing or eliminating springback in bending operations.

FIGURE 16.21   Common die-bending operations showing the die-opening dimension, W, used in calculating bending forces.

Compensation for Springback.   Springback in forming operations usually is compensated by overbending the part (Figs. 16.20a and b), although several trials may be necessary to obtain the desired results. Another method is to *coin* the bend area by subjecting it to highly localized compressive stresses between the tip of the punch and the die surface (Figs. 16.20c and d), a technique called *bottoming the punch*. In another method, the part is subjected to *stretch bending*, in which it is under external tension while being bent (see also *stretch forming*, Section 16.6). In Fig. 16.20e, the upper die rotates clockwise as the dies close.

Video Solution 16.2 Springback

Bending Force.   The bending force for sheets and plates can be estimated by assuming that the process is one of simple bending of a rectangular beam (as described in texts on mechanics of solids). Thus, the bending force is a function of the strength of the material, the length, L, of the bend, the thickness, T, of the sheet, and the die opening, W (see Fig. 16.21). Excluding friction, the *maximum bending force*, P, is

$$P = \frac{kYLT^2}{W}, \tag{16.7}$$

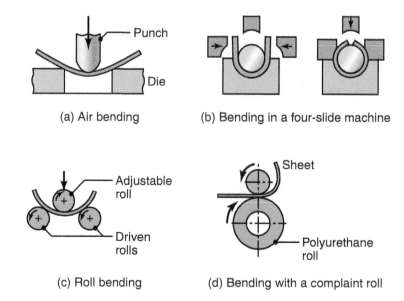

(a) Air bending

(b) Bending in a four-slide machine

(c) Roll bending

(d) Bending with a complaint roll

FIGURE 16.22   Examples of various bending operations.

where the factor k ranges from about 0.3 for a wiping die, to about 0.7 for a U-die, to about 1.3 for a V-die (Fig. 16.21), and Y is the yield stress of the material.

$$P = \frac{(\text{UTS})LT^2}{W},\qquad (16.8)$$

where UTS is the ultimate tensile strength of the sheet metal. This equation applies well to situations in which the punch-tip radius and the sheet thickness are relatively small, as compared to the die opening, W.

The force in die bending varies throughout the bending cycle. It increases from zero to a maximum, and it may even decrease as the bend is being completed, but then it increases sharply as the punch reaches the bottom of its stroke. In air bending (Fig. 16.22a), however, the force does not increase again after it begins to decrease, because the sheet has no resistance to its free movement downward.

16.6   Miscellaneous Bending and Related Forming Operations

Press-brake Forming.   Sheet metal or plate can easily be bent with simple fixtures using a press. Sheets or narrow strips that are 7 m (20 ft) or even longer usually are bent in a *press brake* (Fig. 16.23). The machine utilizes long dies, in a mechanical or hydraulic press, and is particularly suitable for small production runs. As can be seen in Fig. 16.23, the tooling is simple, the motions are only up and down, and is easily adaptable to a wide variety of part shapes. The operation can be automated easily for low-cost, high-production runs. Die materials for press brakes range from hardwood (for low-strength materials and small-production runs) to carbides for strong and abrasive sheet materials. For most applications, carbon-steel or gray-iron dies are generally used.

Bending in a Four-slide Machine.   Bending relatively short pieces can be done on a machine such as that shown in Fig. 16.22b. The lateral movements of the dies are

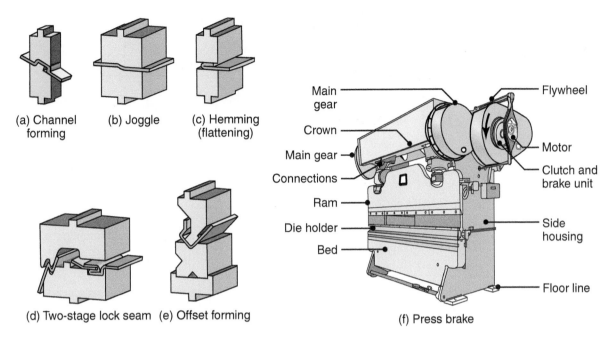

(a) Channel forming
(b) Joggle
(c) Hemming (flattening)
(d) Two-stage lock seam
(e) Offset forming
(f) Press brake

Main gear
Crown
Main gear
Connections
Ram
Die holder
Bed

Flywheel
Motor
Clutch and brake unit
Side housing
Floor line

FIGURE 16.23 (a) through (e) Schematic illustrations of various bending operations in a press brake. (f) Schematic illustration of a press brake. *Source:* Enprotech Industrial Technologies Inc.

controlled and synchronized with the vertical die movement to form the part. This process is typically used for making seamed tubing and conduits, bushings, fasteners, and various machinery components.

Roll Bending. In this process (Fig. 16.22c), plates are bent using a set of rolls. By adjusting the distance between the three rolls, various curvatures can be obtained. The process is flexible, and is used widely for bending plates for applications such as boilers, cylindrical pressure vessels, and various curved structural members. Figure 16.22d shows the bending of a strip, with a compliant roll made of polyurethane, which conforms to the shape of the strip as the hard upper roll presses upon it.

Beading. In *beading*, the periphery of the sheet metal is bent in the cavity of a die (Fig. 16.24). The bead imparts stiffness to the part by increasing the moment of inertia of that section. Also, beads improve the appearance of the part and eliminate exposed sharp edges that may be hazardous.

Flanging. This is a process of bending the edges of sheet metals, usually to 90° (see also Section 16.7). In **shrink flanging** (Fig. 16.25a), the flange is subjected to compressive hoop stresses which, if excessive, can cause the flange periphery to wrinkle. The wrinkling tendency increases with decreasing radius of curvature of the flange. In *stretch flanging*, the flange periphery is subjected to tensile stresses which, if excessive, can lead to cracking along the periphery (see Fig. 16.25c).

Roll Forming. Also called *contour-roll forming* or *cold-roll forming*, this process is used for forming continuous lengths of sheet metal and for large production runs. As it passes through a set of driven rolls, the metal strip is bent in consecutive stages (Fig. 16.26). The formed strip is then sheared into specific lengths and stacked.

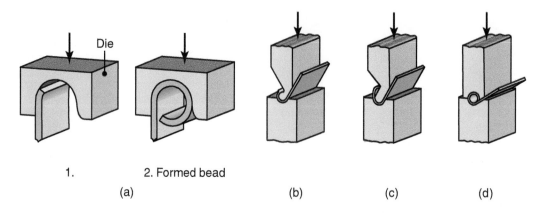

FIGURE 16.24 (a) Bead forming with a single die. (b) through (d) Bead forming with two dies in a press brake.

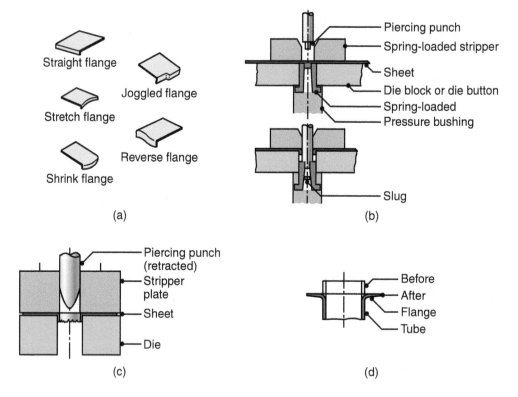

FIGURE 16.25 Various flanging operations. (a) Flanges on flat sheet. (b) Dimpling. (c) The piercing of sheet metal to form a flange. In this operation, a hole does not have to be prepunched before the punch descends; note the rough edges along the circumference of the flange. (d) The flanging of a tube; note the thinning of the edges of the flange.

Typical roll-formed products are door and picture frames, panels, channels, gutters, siding, and pipes and tubing with lock seams (see Section 32.5). The length of the part is limited only by the amount of sheet metal supplied to the rolls from a coiled stock. Sheet thickness usually ranges from about 0.125 to 20 mm (0.005–0.75 in.). Forming speeds are generally below 1.5 m/s (300 ft/min), although they can be much higher for special applications.

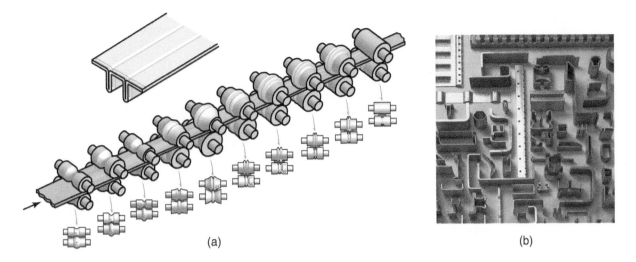

FIGURE 16.26 (a) Schematic illustration of the roll-forming process. (b) Examples of roll-formed cross-sections. *Source:* (b) Courtesy of Sharon Custom Metal Forming, Inc.

In designing the rolls and their sequence, dimensional tolerances and springback, as well as tearing and buckling of the strip, have to be considered. The rolls generally are made of carbon steel or gray iron, and they may be chromium plated for a better surface finish of the formed product and to reduce wear of the rolls. Lubricants may be used to reduce wear, improve surface finish, and cool the rolls and the sheet being formed.

Tube Bending and Forming.   Bending and forming tubes and other hollow sections requires special tooling because of the tendency for buckling and folding, as one can note while trying to bend a piece of copper tubing or even a plastic soda straw. The oldest method of bending a tube or pipe is to first fill it with loose particles (commonly sand), and then bend it in a suitable fixture. The function of the filler is to prevent the tube from buckling inward; after the tube has been bent, the sand is shaken out. Tubes also can be plugged with various flexible internal mandrels (Fig. 16.27), for the same purpose as sand. Note that, because of its lower tendency for buckling, a relatively thick tube can be bent safely without using fillers or plugs. (See also *tube hydroforming*, Section 16.8.)

The beneficial effect of forming metals under highly compressive stresses is demonstrated in Fig. 16.28 for bending a tube with relatively sharp corners. Note that, in this operation, the tube is subjected to longitudinal compressive stresses, which reduce the stresses in the outer fibers in the bend area, thus improving the bendability of the material.

Dimpling, Piercing, and Flaring.   In *dimpling* (Fig. 16.25b), a hole is first punched and then expanded into a flange. Flanges also may be produced by *piercing* with a shaped punch (Fig. 16.25c), and tube ends can be flanged by a similar process (Fig. 16.25d). When the bend angle is less than 90°, as in fittings with conical ends, the process is called *flaring*. The condition of the sheared edges (see Fig. 16.3) is important in these operations, because stretching the material causes high tensile stresses along the periphery (tensile hoop stresses), which can lead to cracking and tearing of the flange.

As the ratio of flange diameter to hole diameter increases, the strains increase proportionately. Depending on the roughness of the edge, there will therefore be a tendency for cracking along the outer periphery of the flange. To reduce this

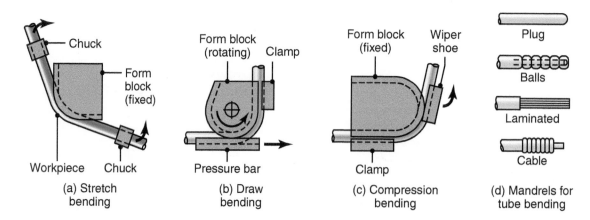

(a) Stretch bending

(b) Draw bending

(c) Compression bending

(d) Mandrels for tube bending

FIGURE 16.27 Methods of bending tubes. Internal mandrels or filling of tubes with particulate materials such as sand are often necessary to prevent collapse of the tubes during bending. Tubes also can be bent by a technique in which a stiff, helical tension spring is slipped over the tube. The clearance between the outer diameter of the tube and the inner diameter of the spring is small; thus, the tube cannot kink and the bend is uniform.

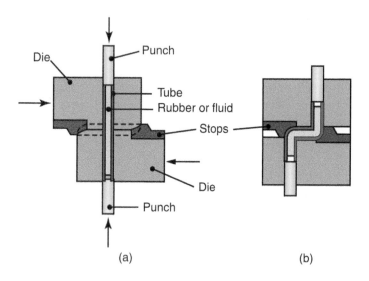

FIGURE 16.28 A method of forming a tube with sharp angles, applying an axial compressive force; compressive stresses are beneficial in forming operations because they delay fracture. Note that the tube is supported internally with rubber or fluid to avoid collapsing during forming. *Source:* After J.L. Remmerswaal and A. Verkaik.

possibility, sheared or punched edges could be shaved off with a sharp tool (see Fig. 16.9) to improve the surface finish of the edge.

Hemming and Seaming. In the *hemming* process, also called *flattening*, the edge of the sheet is folded over itself (Fig. 16.23c). Hemming increases the stiffness of the part, improves its appearance, and eliminates sharp edges. *Seaming* involves joining two edges of sheet metal by hemming (Fig. 16.23d). Double seams are made by a similar process, using specially shaped rolls for watertight and airtight joints, such as those in food and beverage containers.

Bulging. This process involves placing a tubular, conical, or curvilinear part into a split-female die, and then expanding the part, usually with a polyurethane plug (Fig. 16.29a). The punch is then retracted, the plug returns to its original shape (by total elastic recovery), and the formed part is removed by opening the split dies. Typical products made are coffee or water pitchers, beer barrels, and beads on oil drums. For parts with complex shapes, the plug, instead of being cylindrical, may be shaped, in order to be able to apply higher pressures at critical regions of the part. The major advantages of using polyurethane plugs are that they are highly resistant to abrasion and wear, and do not damage the surface finish of the part being formed.

Segmented Dies. These dies consist of individual segments that are placed inside the part to be formed and expanded mechanically in a generally radial direction. The segments are then retracted to remove the formed part. These dies are relatively inexpensive, and they can be used for large production runs.

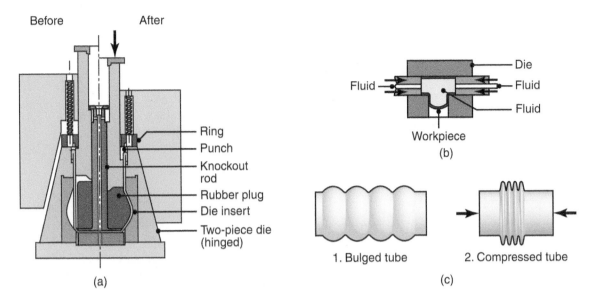

Before   After

— Ring
— Punch
— Knockout rod
— Rubber plug
— Die insert
— Two-piece die (hinged)

(a)

Die
Fluid —
Fluid
Fluid

Workpiece
(b)

1. Bulged tube   2. Compressed tube

(c)

FIGURE 16.29 (a) The bulging of a tubular part with a flexible plug; water pitchers can be made by this method. (b) Production of fittings for plumbing by expanding tubular blanks under internal pressure; the bottom of the piece is then punched out to produce a "T." (c) Steps in manufacturing bellows. *Source:* (b) After J.A. Schey, Introduction to Manufacturing Processes, 3rd ed., 2000, McGraw-Hill, p. 425. ISBN No. 0-07-031136-6.

Stretch Forming. In *stretch forming*, the sheet metal is clamped along its edges and then stretched over a male die, called *form block* or *form punch*. The die can move upward, downward, or sideways, depending on the particular design of the machine (Fig. 16.30). Stretch forming is used primarily to make aircraft wing-skin panels, fuselages, and boat hulls. Aluminum skins for the Boeing 767 and 757 aircraft, for example, are made by stretch forming, with a tensile force of 9 MN (2 million lb). The rectangular sheets are 12 m × 2.5 m × 6.4 mm (40 ft × 8.3 ft × 0.25 in.). Although this process is generally used for low-volume production, it is versatile and economical, particularly for applications in the aerospace industry.

In most operations, the blank is a rectangular sheet, clamped along its narrower edges and stretched lengthwise, thus allowing the material to shrink in its width. Controlling the amount of stretching is important in order to prevent tearing. Stretch forming cannot produce parts with sharp contours or with reentrant corners. Various accessory equipment can be used in conjunction with stretch forming, including further forming with both male and female dies while the part is under tension. Dies for stretch forming generally are made of zinc alloys, steel, plastics, or wood. Most applications require little or no lubrication.

16.7   Deep Drawing
··

It can be noted that numerous sheet metal parts are *cylindrical* or *box shaped*, such as pots and pans, all types of containers for food and beverages (Fig. 16.31), stainless-steel kitchen sinks, canisters, and automotive fuel tanks. Such parts usually are made by a process in which a punch forces a flat sheet-metal blank into a die cavity, as shown in Fig. 16.32a. Deep drawing is one of the most important metalworking processes because of its widespread use.

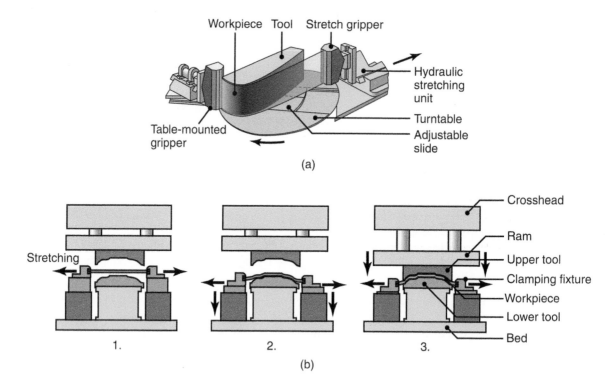

FIGURE 16.30 Schematic illustration of a stretch-forming process; aluminum skins for aircraft can be made by this method. *Source:* (a) Courtesy of Cyril Bath Co.

Consider the case of a round sheet-metal blank placed over a circular die opening, and held in place with a **blankholder**, or *hold-down ring* (Fig. 16.32b). The punch travels downward, forcing the blank into the die cavity, thus forming a cup. The major variables in this process are (a) properties of the sheet metal, (b) ratio of blank diameter, D_o, (c) punch diameter, D_p, (d) clearance, c, between punch and die; (e) punch radius, R_p, (f) die-corner radius, R_d, (g) blankholder force; and (h) friction and lubrication between all contacting interfaces.

During the drawing operation, the movement of the blank into the die cavity induces compressive circumferential (hoop) stresses in the flange, which tend to cause the flange to wrinkle during drawing. This phenomenon can be demonstrated simply by trying to force a circular piece of paper into a round cavity. Wrinkling can be reduced or eliminated if a blankholder is pressed downward with a certain force. In order to improve performance, the magnitude of this force can be computer controlled as a function of punch travel or location in the blankholder.

Because of the several variables involved, the *punch force*, F, is difficult to calculate directly. It has been shown, however, that the *maximum punch force*, F_{max}, can be estimated from the formula

$$F_{max} = \pi D_p T \, (\text{UTS}) \left[\left(\frac{D_o}{D_p} \right) - 0.7 \right], \qquad (16.9)$$

where the nomenclature is the same as that in Fig. 16.32b. It can be seen that the force increases with increasing blank diameter, sheet thickness, strength, and the ratio (D_o/D_p). The wall of the cup being drawn is subjected principally to a longitudinal (vertical) tensile stress, due to the punch force. Elongation under this stress causes the cup wall to become thinner and, if excessive, can cause *tearing* of the cup.

Process	Process illustration	Result

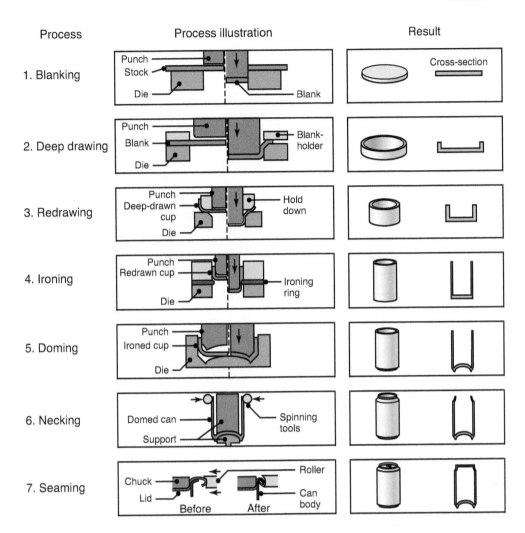

FIGURE 16.31 The metal-forming processes employed in manufacturing two-piece aluminum beverage cans.

16.7.1 Deep Drawability

In a deep-drawing operation, failure generally is a result of *thinning* of the cup wall under the high longitudinal tensile stresses due to the action of the punch. Following the movement of the material as it flows into the die cavity, it can be seen that the sheet metal (a) must be capable of undergoing a reduction in width due to a reduction in diameter and (b) must also resist thinning under the longitudinal tensile stresses in the cup wall.

Deep drawability is generally expressed by the **limiting drawing ratio** (LDR) as

$$\text{LDR} = \frac{\text{Maximum blank diameter}}{\text{Punch diameter}} = \frac{D_o}{D_p}. \qquad (16.10)$$

Whether a particular sheet metal can be deep drawn successfully into a round cup has been found to be a function of the **normal anisotropy**, R (also called *plastic anisotropy*), of the sheet metal. Normal anisotropy is defined in terms

Video Solution 16.3 Limiting Drawing Ratio

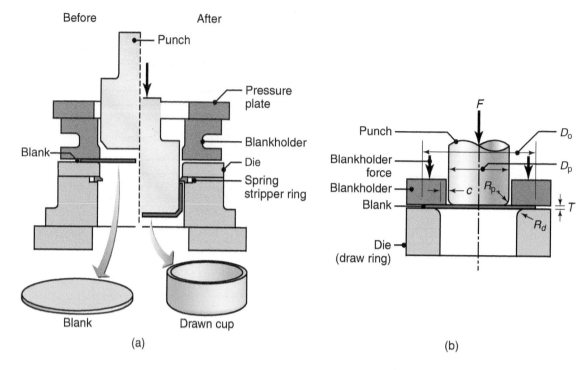

Before   After

Punch

Pressure plate

Blankholder

Blank

Die

Spring stripper ring

Blank

Drawn cup

(a)

F

Punch

Blankholder force

Blankholder

Blank

Die (draw ring)

D_o

D_p

R_p

c

R_d

T

(b)

FIGURE 16.32 (a) Schematic illustration of the deep-drawing process on a circular sheet-metal blank; the stripper ring facilitates the removal of the formed cup from the punch. (b) Process variables in deep drawing. Except for the punch force, F, all the parameters indicated in the figure are independent variables.

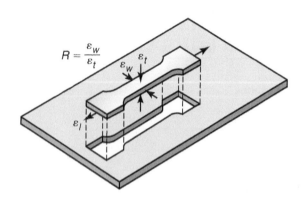

$R = \dfrac{\epsilon_w}{\epsilon_t}$

FIGURE 16.33 Strains on a tensile-test specimen removed from a piece of sheet metal; these strains are used in determining the normal and planar anisotropy of the sheet metal.

of the true strains that a tensile-test specimen undergoes (Fig. 16.33):

$$R = \frac{\text{Width strain}}{\text{Thickness strain}} = \frac{\epsilon_w}{\epsilon_t}. \quad (16.11)$$

In order to determine the magnitude of R, a specimen is first prepared and subjected to an elongation of 15–20%. The true strains that the sheet undergoes are then calculated, in the manner described in Section 2.2. Because cold-rolled sheets are generally anisotropic in their *planar* direction, the R value of a specimen cut from a rolled sheet will depend on its orientation with respect to the rolling direction of the sheet. An average value, R_{avg}, is calculated from the equation

$$R_{avg} = \frac{R_0 + 2R_{45} + R_{90}}{4}, \quad (16.12)$$

where the subscripts are the angles with respect to the rolling direction of the sheet. Some typical R_{avg} values are given in Table 16.4.

The experimentally determined relationship between R_{avg} and the LDR is shown in Fig. 16.34. It has been established that no other mechanical property of sheet metal shows a more consistent relationship to LDR as does R_{avg}. Thus, by using a simple tensile-test result and obtaining the normal anisotropy of the sheet metal, the LDR of a material can be determined.

Earing. In deep drawing, the edges of cups may become wavy, called *earing* (Fig. 16.35). Ears are objectionable on deep-drawn cups because they have to be trimmed off, as they serve no useful purpose, and interfere with further processing of the cup, resulting in scrap. Earing is caused by the **planar anisotropy** of the sheet metal, and the number of ears produced may be two, four, or eight, depending on the processing history and microstructure of the material. If the sheet is stronger in the rolling direction than transverse to the rolling direction, and the strength varies uniformly with respect to orientation, then two ears will form. If the sheet has high strength at different orientations, then more ears will form.

The planar anisotropy of the sheet, indicated by ΔR, is defined in terms of directional R values, from the equation

$$\Delta R = \frac{R_0 - 2R_{45} + R_{90}}{2}. \qquad (16.13)$$

When $\Delta R = 0$ no ears form, and the height of the ears increases as ΔR increases.

It can be seen that deep drawability is enhanced by a high R_{avg} value and a low ΔR. Generally, however, sheet metals with high R_{avg} also have high ΔR values. Sheet-metal textures continue to be developed to improve drawability, by controlling the type of alloying elements in the material and by adjusting various processing parameters during cold rolling of the sheet.

16.7.2 Deep-drawing Practice

Certain guidelines have been established over the years for successful deep-drawing practice. The blankholder pressure is chosen generally as 0.7–1.0% of the sum of the yield strength and the ultimate tensile strength of the sheet metal. Too high a blankholder force increases the punch force and causes the cup wall to tear. On the other hand, if the blankholder force is too low, wrinkling of the cup flange will occur.

Clearances are usually 7–14% greater than sheet thickness. If they are too small, the blank may be pierced or sheared by the punch. The corner radii of the punch and of the die are also important parameters. If they are too small, they can cause fracture at the corners; if they are too large, the cup wall may wrinkle, called *puckering*.

Draw beads (Fig. 16.36) are often necessary to control the flow of the blank into the die cavity. They restrict the free flow of the sheet metal by bending and unbending it during the drawing cycle, thereby increasing the force required to pull the sheet into the die cavity. Draw beads also help to reduce the necessary blankholder forces, because the beaded sheet has a higher stiffness (due to its higher moment of inertia) and, thus, less tendency to wrinkle. Draw-bead diameters may range from 13 to 20 mm (0.50–0.75 in.), the latter applicable to large stampings, such as automotive panels.

Draw beads also are useful in drawing *box-shaped* and *nonsymmetric* parts (Figs. 16.36b and c). Note in Fig. 16.36c, for example, that various regions of the part

TABLE 16.4

Typical Ranges of Average Normal Anisotropy, for Various Sheet Metals

Material	Range of R_{avg}
Zinc alloys	0.4–0.6
Hot-rolled steel	0.8–1.0
Cold-rolled, rimmed steel	1.0–1.4
Cold-rolled, aluminum-killed steel	1.4–1.8
Aluminum alloys	0.6–0.8
Copper and brass	0.6–0.9
Titanium alloys (alpha)	3.0–5.0
Stainless steels	0.9–1.2
High-strength, low-alloy steels	0.9–1.2

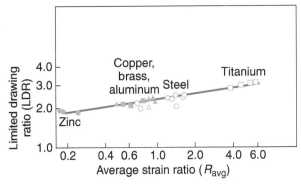

FIGURE 16.34 The relationship between average normal anisotropy and the limiting drawing ratio for various sheet metals. *Source:* After M. Atkinson.

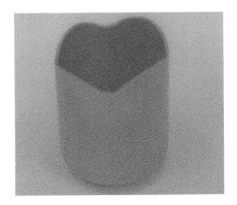

FIGURE 16.35 Earing in a drawn steel cup, caused by the planar anisotropy of the sheet metal.

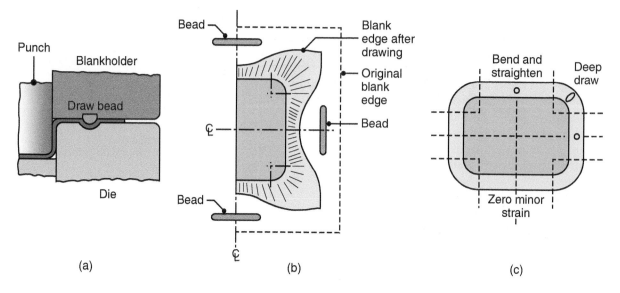

FIGURE 16.36 (a) Schematic illustration of a draw bead. (b) Metal flow during the drawing of a box-shaped part while using beads to control the movement of the material. (c) Deformation of circular grids in the flange in deep drawing.

being drawn undergo different types of deformation during drawing. (Recall also the fundamental principle that the material flows in the direction of least resistance.)

In order to avoid tearing of the sheet metal during forming, it often is necessary to incorporate the following:

- Proper design and location of draw beads
- Large die radii
- Effective lubrication
- Proper blank size and shape
- Cutting off of corners of square or rectangular blanks at 45° to reduce tensile stresses that develop during drawing
- Using blanks that are free of internal and external defects, including burrs

Ironing. If the clearance between the punch and the die is sufficiently large, the drawn cup will have thicker walls at its rim than at its base (see Fig. 16.32). The reason for this is that the cup rim consists of material from the outer diameter of the blank; hence, it has been reduced in diameter more, and thus becomes thicker, than the material constituting the rest of the cup wall. As a result, the cup will develop a nonuniform wall thickness.

The thickness of the wall can be controlled by *ironing*, a process in which a drawn cup is pushed through one or more ironing rings (see Fig. 16.31). The clearance between the punch and ironing rings is less than the cup wall thickness, thus the drawn cup essentially has a constant wall thickness. Aluminum beverage cans, for example, are pushed through a set of two or three ironing rings, in one stroke.

Redrawing. Containers that are too difficult to draw in one operation generally undergo *redrawing* (see Fig. 16.37). Because of the volume constancy of the metal, the cup becomes longer as it is redrawn to smaller diameters. In *reverse redrawing*, the cup is placed upside down in the die, and thus undergoes bending in the direction opposite to its original configuration.

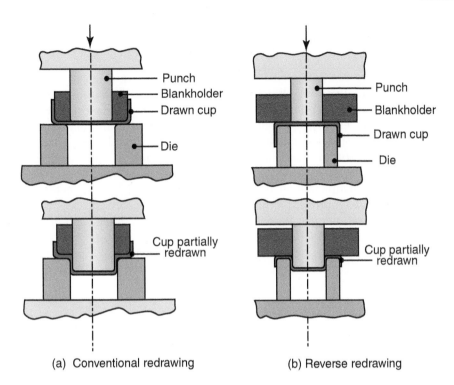

(a) Conventional redrawing   (b) Reverse redrawing

FIGURE 16.37 Reducing the diameter of drawn cups by redrawing operations: (a) conventional redrawing and (b) reverse redrawing. Small-diameter deep containers may undergo several redrawing operations.

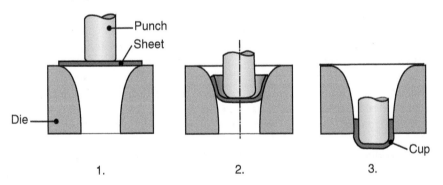

FIGURE 16.38 Stages in deep drawing without a blankholder, using a *tractrix* die profile.

Drawing without Blankholder. Deep drawing also may be carried out without a blankholder. The dies are specially contoured for this operation to prevent wrinkling; one example is shown in Fig. 16.38. The sheet metal must be sufficiently thick to prevent wrinkling. The following formula is a general guide:

$$D_o - D_p < 5T, \qquad (16.14)$$

where T is the sheet thickness. Thus, the thicker the sheet, the larger the blank diameter, and the deeper the cup that can be made, without wrinkling.

Embossing. This is an operation consisting of shallow or moderate drawing, made with male and female matching shallow dies (Fig. 16.39). Embossing is used

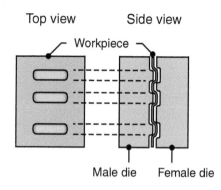

Top view   Side view

Workpiece

Male die   Female die

FIGURE 16.39 An embossing operation with two dies; letters, numbers, and designs on sheet-metal parts can be produced by this process.

principally for the stiffening of flat sheet-metal panels and for decorating, numbering, and lettering.

Tooling and Equipment for Drawing. The most common tool and die materials for deep drawing are tool steels and cast irons, and include dies made from ductile-iron castings produced by the lost-foam process (Section 11.3.1). Other materials, such as carbides, also may be used (see Table 5.7). Die-manufacturing methods are described in Section 14.7. Because of the generally axisymmetric shape of the punch and die components, such as for making cylindrical cans and containers, they can be manufactured on such equipment as high-speed machining on computer-controlled lathes (Section 25.5).

The equipment for deep drawing is usually a *double-action hydraulic press* or a *mechanical press*, the latter generally being favored because of its higher operating speed. In the double-action hydraulic press, the punch and the blankholder are controlled independently. Punch speeds generally range between 0.1 and 0.3 m/s (20–60 ft/min).

CASE STUDY 16.2   Manufacturing of Food and Beverage Cans

Can manufacturing is a major and competitive industry, with approximately 100 billion beverage cans and 30 billion food cans produced each year in the United States alone. These containers are strong and lightweight, typically weighing less than 0.5 oz, and they are under an internal pressure of 90 psi, reliably and without leakage of their contents. There are stringent requirements for the surface finish of the can, since brightly decorated and shiny cans are preferred over dull-looking containers. Considering all of these features, metal cans are very inexpensive. Can makers charge approximately $40 per 1000 cans, or about 4 cents per can.

Food and beverage cans may be produced in a number of ways, the most common ones being two-piece and three-piece cans. Two-piece cans consist of the can body and the lid (Fig. 16.40a). The body is made of one piece that has been drawn and ironed, hence the industry practice of referring to this style as D&I (drawn and ironed) cans. Three-piece cans are produced by attaching a lid and a bottom to a sheet-metal cylindrical body, typically made by forming a seam on a sheet metal blank.

Drawn and ironed can bodies are produced from a number of alloys, but the most common are 3004-H19 aluminum (Section 6.2) and electrolytic tin-plated ASTM A623 steel. Aluminum lids are used for both steel and aluminum cans, and are produced from 5182-H19 or 5182-H48. The lid has

a demanding set of design requirements, as can be appreciated by reviewing Fig. 16.40b. Not only must the can lid be *scored* easily (curved grooves around the tab), but an integral rivet is formed and headed (Section 14.4) in the lid to hold the tab in place. Aluminum alloy 5182 has the unique characteristics of having sufficient formability to enable forming of the integral rivet without cracking, and has the ability to be scored. The lids basically are stamped from 5182 aluminum sheet, the pop-top is scored, and a plastic seal is placed around the inside periphery of the lid. The polymer layer seals the can's contents after the lid is seamed to the can body, as described next.

The traditional method of manufacturing the can bodies is shown in Fig. 16.31. The process starts with 5.5-in.-diameter blanks produced from rolled sheet stock. The blanks are (a) *deep drawn* to a diameter of about 3.5 in., (b) *redrawn* to the final diameter of around 2.6 in., (c) *ironed* through two or three ironing rings in one pass, and (d) *domed* for shaping the can bottom. The deep-drawing and ironing operations are performed in a special type of press, typically producing cans at speeds over 400 strokes per minute. Following this series of operations, a number of additional processes take place.

Necking of the can body is performed either through *spinning* (Section 16.9) or by *die necking* (a forming operation similar to that shown in Fig. 15.21a, where a thin-walled tubular part is

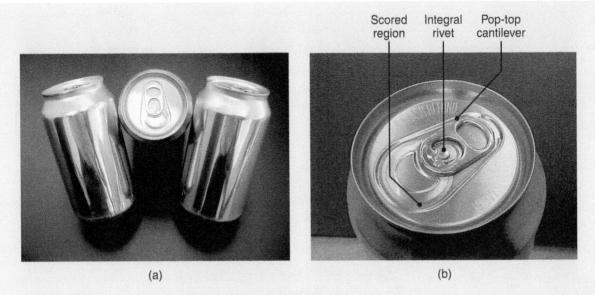

FIGURE 16.40   (a) Aluminum beverage cans; note the smooth surface. (b) Detail of the can lid, showing the integral rivet and scored edges for the pop-top.

pushed into the die), and then spin flanged. The reason for necking the can top is that the 5182 aluminum used for the lid is relatively expensive; thus, by tapering the top of the can, a smaller volume of material is needed, thereby reducing the material cost. It should also be noted that the cost of a can often is calculated to millionths of a dollar, hence any design feature that reduces cost will be exploited by this competitive industry.

Source: Printed with permission of J.E. Wang, Texas A&M University.

16.8   Rubber Forming and Hydroforming

In the processes described in the preceding sections, it has been noted that the dies generally are made of solid materials, such as steels and carbides. In *rubber forming* (also known as the *Guerin process*), one of the dies in a set is made of a flexible material, typically a polyurethane membrane. Polyurethanes are used widely because of their abrasion resistance, fatigue life, and resistance to cutting or tearing (Section 7.9).

In bending and embossing of sheet metal by this process, the female die is replaced with a rubber pad (Fig. 16.41). Note that the outer surface of the sheet is protected from damage or scratches, because it is not in direct contact with a hard metal surface during forming. Pressures in rubber forming are typically on the order of 10 MPa (1500 psi).

In the **hydroform** or *fluid-forming process* (Fig. 16.42), the pressure over the rubber membrane is controlled throughout the forming cycle, with a maximum pressure of up to 100 MPa (15,000 psi). This procedure allows close control of the part during forming, and prevents wrinkling or tearing. Deeper draws are obtained than in conventional deep drawing, because the pressure around the rubber membrane forces the cup against the punch. As a result, the friction at the punch–cup interface increases, which then reduces the longitudinal tensile stresses in the cup, and thus delays fracture.

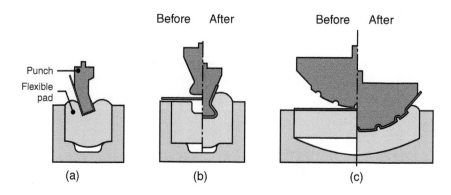

FIGURE 16.41 Examples of the bending and embossing of sheet metal with a metal punch and with a flexible pad serving as the female die. *Source:* Courtesy of Polyurethane Products Corporation.

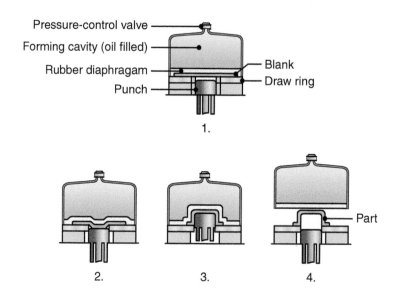

FIGURE 16.42 The hydroform (or fluid-forming) process; note that, in contrast to the ordinary deep-drawing process, the pressure in the dome forces the cup walls against the punch. The cup travels with the punch; in this way, deep drawability is improved.

The control of frictional conditions in rubber forming, as well as other sheet-forming operations, can be a critical factor in making parts successfully. The use of proper lubricants and their method of application is also important.

In **tube hydroforming** (Fig. 16.43a), metal tubing is formed in a die and pressurized internally by a fluid, usually water. This process, which now is being applied more widely, can form either simple tubes or various intricate hollow shapes (Fig. 16.43b). Parts made include automotive-exhaust and tubular structural components.

When selected properly, rubber-forming and hydroforming processes have the advantages of (a) the capability to form complex shapes, (b) forming parts with laminated sheets made of various materials and coatings, (c) flexibility and ease of operation, (d) avoiding damage to the surfaces of the sheet, (e) low die wear, and (f) low tooling cost.

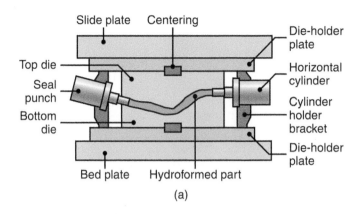

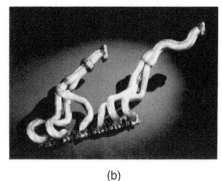

(a)   (b)

FIGURE 16.43   (a) Schematic illustration of the tube-hydroforming process. (b) Example of tube-hydroformed parts. Automotive-exhaust and structural components, bicycle frames, and hydraulic and pneumatic fittings are produced through tube hydroforming. *Source:* Courtesy of Schuler GmBH.

CASE STUDY 16.3   **Tube Hydroforming of an Automotive Radiator Closure**

The conventional assembly used to support an automotive radiator, or radiator closure, is constructed through stamping of the components, which are subsequently welded together. To simplify the design and to achieve weight savings, a hydroformed assembly was designed, as shown in Fig. 16.44. Note that this design uses varying cross-sections, an important feature to reduce weight and provide surfaces to facilitate assembly and mounting of the radiator.

A typical tube hydroforming processing sequence consists of the following steps:

1. Bending of tube to the desired configuration
2. Tube hydroforming to achieve the desired shape
3. Finishing operations, such as end shearing and inspection
4. Assembly, including welding of components

The operations performed on one of the tube components of the closure is shown in Fig. 16.45. The tube, constructed of steel with a 300 MPa (43.5 ksi) yield strength, is bent to shape (see Fig. 16.27). The bent tube is then placed in a hydroforming press and the end caps are attached.

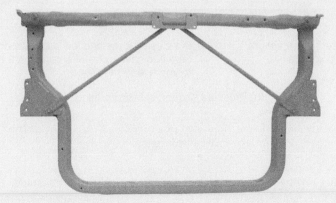

FIGURE 16.44   Hydroformed automotive radiator closure, which serves as a mounting frame for the radiator.

(*continued*)

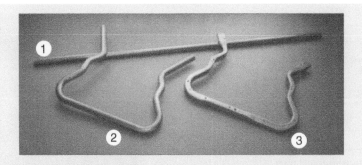

FIGURE 16.45 Sequence of operations in producing a tube-hydroformed component: (1) tube as cut to length; (2) after bending; and (3) after hydroforming.

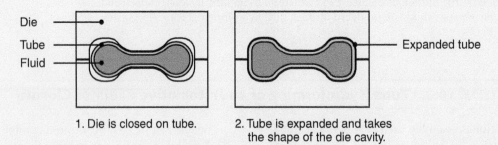

(a) Conventional hydroforming

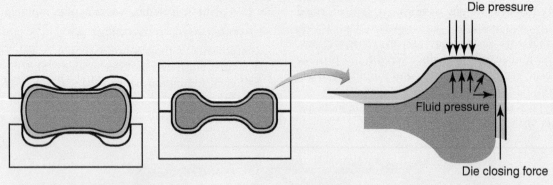

(b) Pressure sequence hydroforming

FIGURE 16.46 Schematic illustration of expansion of a tube to a desired cross-section through (a) conventional hydroforming and (b) pressure sequence hydroforming.

Conventional hydroforming involves closing the die onto the tube, followed by internal pressurization to force the tube to the desired shape. Figure 16.46a shows a typical cross-section. Note that as the tube is expanded, there is significant wall thinning, especially at the corners, because of friction at the tube–die interface. A sequence of pressures that optimize corner formation is therefore used, as shown in Fig. 16.46b.

In this approach, a first pressure stage (prepressure stage) is applied as the die is closing, causing the tube to partially fill the die cavity and form

FIGURE 16.47 View of the tube-hydroforming press, with bent tube in place in the forming die.

the cross-sectional corners. After the die is completely closed, the internal pressure is increased to lock-in the form and provide support needed for hole piercing. This sequence has the benefit of forming the sharp corners in the cross-section by bending, as opposed to pure stretching as in conventional hydroforming. The resulting wall thickness is much more uniform, producing a more structurally sound component. Figure 16.47 shows a part being hydroformed in a press.

The assembly shown in Fig. 16.44 has 76 holes that are pierced inside the hydroforming die; the ends are then sheared to length. The 10 components in the hydroformed closure are then assembled through robotic gas-metal arc welding (see Section 30.4.3), using threaded fasteners to aid in serviceability.

Compared to the original stamped design, the hydroformed design has four fewer components, uses only 20 welds as opposed to 174 for the stamped design, and weighs 10.5 kg (23 lb) versus 14.1 kg (31.1 lb). Furthermore, the stiffness of the enclosure and the water cooling areas are both significantly increased.

Source: Courtesy of B. Longhouse, Vari-Form, Inc.

16.9 Spinning

Spinning is a process that involves forming of axisymmetric parts over a mandrel, using various tools and rollers; a process similar to that of shaping clay on a potter's wheel.

Conventional Spinning.   In *conventional spinning*, a circular blank of flat or preformed sheet metal is placed and held against a mandrel, and rotated while a rigid tool shapes the material over the mandrel (Fig. 16.48a). The tool may be activated either manually or, for higher production rates, by computer-controlled mechanisms. The process typically involves a sequence of passes, and requires considerable skill. Conventional spinning is particularly suitable for making conical and curvilinear shapes (Fig. 16.48b), with part diameters ranging up to 6 m (20 ft), which otherwise would be difficult or uneconomical to produce. Although most spinning is performed at

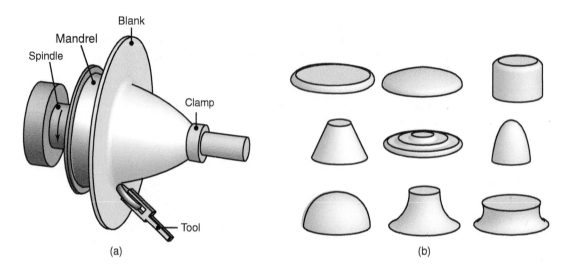

FIGURE 16.48 (a) Schematic illustration of the conventional spinning process. (b) Types of parts conventionally spun. All parts are axisymmetric.

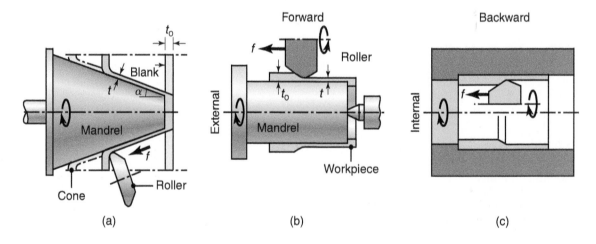

FIGURE 16.49 (a) Schematic illustration of the shear-spinning process for making conical parts; the mandrel can be shaped so that curvilinear parts can be spun. (b) and (c) Schematic illustrations of the tube-spinning process.

room temperature, thick parts and metals with high strength or low ductility require spinning at elevated temperatures.

Shear Spinning. Also known as *power spinning*, *flow turning*, *hydrospinning*, and *spin forging*, this operation produces an axisymmetric conical or curvilinear shape, reducing the sheet's thickness while maintaining its blank diameter (Fig. 16.49a). A single forming roll can be used, but two rolls are preferable in order to balance the radial forces acting on the mandrel. Typical parts made are rocket motor casings and missile nose cones. Parts up to 3 m (10 ft) in diameter can be formed by shear spinning. This operation produces little waste of material, and it can be completed in a relatively short time, in some cases in as little as a few seconds. Various shapes can be spun with fairly simple tooling, which generally is made of tool steel.

The *spinnability* of a metal in this process is generally defined as the maximum reduction in thickness to which a part can be subjected by spinning without fracture. Spinnability is found to be related to the tensile reduction of area of the material, just as is bendability (see Fig. 16.18). Thus, if a metal has a tensile reduction of area of 50% or higher, its thickness can be reduced, in one pass, by as much as 80%. For metals with low ductility, the operation is carried out at elevated temperatures, by heating the blank in a furnace and transferring it to the mandrel.

Tube Spinning.   In *tube spinning*, the thickness of hollow, cylindrical blanks is reduced or shaped by spinning them on a round mandrel, using rollers (Figs. 16.49b and c). This operation is capable of producing various external and internal profiles, from cylindrical blanks with constant wall thickness. The parts may be spun *forward* or *backward*. The maximum thickness reduction per pass in tube spinning is related to the tensile reduction of area of the material, as it is in shear spinning. Tube spinning can be used to make rocket, missile, and jet-engine parts, pressure vessels, and automotive components, such as car and truck wheels.

Incremental Forming.   *Incremental forming* is a term applied to a class of processes that are related to conventional metal spinning. The simplest version is *incremental stretch expanding* (shown in Fig. 16.50a), wherein a rotating blank is deformed by a steel rod with a smooth hemispherical tip, to produce axisymmetric parts. No special tooling or mandrel is used, and the motion of the rod determines the final part shape, made in one or more passes. Proper lubrication is essential.

CNC incremental forming uses a computer numerical control machine tool (see Section 37.3) programmed to follow contours at different depths across the sheet-metal surface. The blank is clamped and is stationary, and the forming tool rotates. Tool paths are calculated in a manner similar to machining (Part IV), using a CAD model of the desired shape as the starting point (see Fig. 20.2). Figure 16.50b depicts an example of a part that is produced by CNC incremental forming. Note that the part does not have to be axisymmetric. The main advantages of CNC incremental forming are low tooling costs and high flexibility in the shapes that can be produced. This process has been used for rapid prototyping of sheet-metal parts (see Chapter 20). The main drawbacks to incremental forming include low production rates and limitations on materials that can be formed.

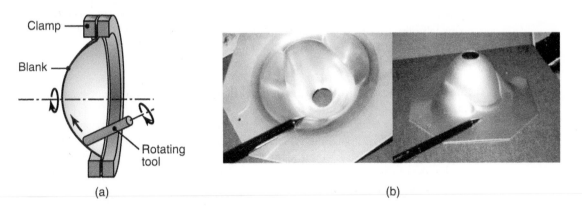

(a)   (b)

FIGURE 16.50   (a) Illustration of an incremental-forming operation; note that no mandrel is used and that the final part shape depends on the path of the rotating tool. (b) An automotive headlight reflector produced through CNC incremental forming. Note that the part does not have to be axisymmetric. *Source:* Courtesy of J. Jeswiet, Queen's University, Ontario.

16.10 Superplastic Forming

The superplastic behavior of certain metals and alloys, described in Section 2.2.7, involves tensile elongations of up to 2000%, exhibited within certain temperature ranges. Examples of such materials are zinc–aluminum and titanium alloys, with very fine grains, typically less than 10–15 μm (see Table 1.1). Superplastic alloys can be formed into complex shapes by *superplastic forming* (SPF), a process that employs common metalworking techniques, as well as by polymer-processing techniques (such as thermoforming, vacuum forming, and blow molding, described in Chapter 19). The behavior of the material in SPF is similar to that of bubble gum or hot glass, which, when blown, expands many times its original diameter before it bursts.

Superplastic alloys, particularly Zn-22Al and Ti-6Al-4V, can also be formed by bulk-deformation processes, including closed-die forging, coining, hubbing, and extrusion. Commonly used die materials in SPF are low-alloy steels, cast tool steels, ceramics, graphite, and plaster of paris. Their selection depends on the forming temperature and the strength of the superplastic alloy.

The very high ductility and relatively low strength of superplastic alloys offer the following advantages:

- Complex shapes can be formed from one piece, with fine detail, close tolerances, and elimination of secondary operations.
- Weight and material savings can be significant, because of the good formability of the materials.
- Little or no residual stresses are present in the formed parts.
- Because of the low strength of the material at forming temperatures, the tooling can be made of materials that have lower strength than those in other metalworking processes, thus tooling costs are lower.

On the other hand, SPF has the following limitations:

- The material must not be superplastic at service temperatures, as otherwise the part will undergo shape changes during use.
- Because of the high strain-rate sensitivity of the superplastic material (see Section 2.2.7), it must be formed at sufficiently low strain rates, typically 10^{-4} to 10^{-2}/s. Forming times range anywhere from a few seconds to several hours; cycle times are thus much longer than those of conventional forming processes; consequently, SPF is a batch-forming process.

Diffusion Bonding/Superplastic Forming.   Fabricating complex sheet-metal structures by combining *diffusion bonding* with *superplastic forming* (SPF/DB) is an important process, particularly in the aerospace industry. Typical structures made are shown in Fig. 16.51, in which flat sheets are *diffusion bonded* (Section 31.7) and formed. In this process, selected locations of the sheets are first diffusion bonded while the rest of the interfaces remains unbonded, using a layer of material (*stop-off*) to prevent bonding. The structure is then expanded in a mold (thus taking the shape of the mold), typically by using pressurized neutral (argon) gas. These structures have high stiffness-to-weight ratios, because they are thin and, by design, have high section moduli. This important feature makes this process particularly attractive in aircraft and aerospace applications.

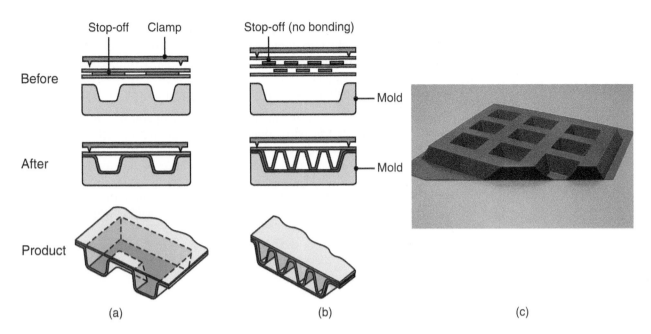

FIGURE 16.51 Types of structures made by superplastic forming and diffusion bonding of sheet metals; such structures have a high stiffness-to-weight ratio. *Source:* (a) and (b) Courtesy of the Boeing Company. Printed with permission. (c) courtesy of Triumph Group, Inc.

The SPF/DB process improves productivity by eliminating mechanical fasteners, and produces parts with good dimensional accuracy and low residual stresses. The technology is well advanced for titanium structures for aerospace applications. In addition to various aluminum alloys being developed using this technique, other metals for SPF include various nickel alloys.

16.11 Hot Stamping

Increasing fuel economy in automobiles has received considerable attention in recent years for both environmental and economic reasons. To achieve increased fuel economy, without compromising performance or safety, manufacturers have increasingly applied advanced materials in automobiles. Die-cast magnesium or extruded aluminum components are examples, but these materials are not sufficiently stiff or as well suited as steel for occupant safety. Thus, there has been a recent trend to consider hot stamping of advanced high-strength steels.

As discussed in Section 5.5.5, high-strength TRIP and TWIP steels have been developed, with yield strengths and ultimate strengths that can exceed 1300 MPa and 2000 MPa, respectively (see Table 5.4). Conventional sheet metal forming of these materials would be difficult or impossible, because of the high forces required and the excessive springback after forming. For these reasons, the sheet metal is preheated to above 900°C (usually 1000°−1200°C) and hot stamped. To extend die life and to quench the material within the die (as discussed below), the tooling is maintained at a much lower temperature, typically 400°−500°C.

Hot stamping allows exploitation of steel phases to facilitate forming and maximize part strength. Basically, the steel is maintained at elevated temperatures to form austenite (see Section 4.4), which has a ductile fcc structure at elevated temperatures. When formed and brought into contact with the much cooler tooling, the steel is rapidly quenched to form martensite, which is a very hard and strong but brittle form of steel (Section 4.7).

A typical hot-stamping sequence involves the following steps:

1. The material is heated up to the austenization temperature, and allowed to *dwell* or *soak* for a sufficiently long time to ensure that quenching will occur quickly when it contacts the die, but not before. Three basic means are used to heat blanks prior to stamping: roller hearth furnaces, induction heating coils, and resistive heating. The last two methods have the advantage of shorter soak times, but may not lead to uniform temperatures throughout the part. The soak time must be optimized in order to ensure proper quenching while minimizing the cycle time.

2. In order to avoid cooling of the part before forming it, the blank must be transferred to the forming dies as quickly as possible. Forming must be performed quickly, before the beginning of transformation of austenite into martensite.

3. Once the part is formed, the dies remain closed while the part is quenched, which takes from 2 to 10 s, depending on sheet thickness, temperature of sheet and die, and workpiece material. The cooling rate must be higher than 27°C/s to obtain martensite. Thus, forming is performed in steel tools that have cooling channels incorporated in them, in order to maintain proper tooling temperature. A complete transformation into martensite results in the high strengths given in Table 5.4. It should be noted that quenching from austenite to martensite results in an increase in volume, which influences the residual stress distribution and workpiece distortion in forming.

A more recent development is to use *pressurized hot gas* (air or nitrogen) as a working media to form the material, similar to hydroforming. This method improves formability, and with proper process control, allows for more uniform blank and tooling temperatures, and thus lower residual stresses and warping.

Because the workpiece is hot and quenching must be done very rapidly, hot stamping is usually performed without a lubricant, and often shot blasting (Section 26.8) is required after forming to remove scale from part surfaces. The steel may also be coated with an aluminum-silicon layer, to prevent oxidation and eliminate the grit blasting step. In such a case, the coating requires a slightly longer soak time, in order to properly bond to the steel substrate.

Hot stamping is not restricted to steels. Magnesium alloys ZEK100, AZ31, and ZE10 are also of great interest, because of their lightweight; however, these materials have limited formability at room temperature, and are therefore stamped at up to 300°C. Also, some advanced aluminum-alloy sheets are formed at elevated temperatures in order to attain improved ductility, and even develop superplastic behavior.

16.12 Specialized Forming Processes

Although not as commonly used as the other processes described thus far, several other sheet-forming processes are used for specialized applications.

Explosive Forming.   Explosives generally are used for demolition in construction, in road building, and for destructive purposes. However, controlling their quantity and shape makes it possible to use explosives as a source of energy for sheet-metal

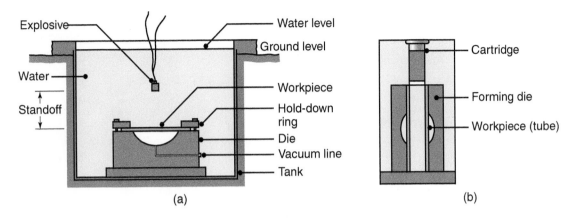

FIGURE 16.52   (a) Schematic illustration of the explosive-forming process. (b) Illustration of the confined method of the explosive bulging of tubes.

forming. In *explosive forming*, first utilized to form metals in the early 1900s, the sheet-metal blank is clamped over a die, and the entire assembly is lowered into a tank, filled with water (Fig. 16.52a). The air in the die cavity is then evacuated, an explosive charge is placed at a certain height, and the charge is detonated.

The explosive generates a shock wave, developing a pressure that is sufficient to form sheet metals. The *peak pressure*, p, generated in water is given by the expression

$$p = K \left(\frac{\sqrt[3]{W}}{R} \right)^a ,   (16.15)$$

where p is in psi, K is a constant, which depends on the type of explosive, such as 21,600 for TNT (trinitrotoluene), W is the weight of the explosive, in pounds, R is the distance of the explosive from the sheet-metal surface (called the *standoff*), in feet, and a is a constant, generally taken as 1.15.

A variety of shapes can be formed by explosive forming, provided that the material is sufficiently ductile at the high rates of deformation encountered in this process (see Table 2.4). The process is versatile, as there is virtually no limit to the size of the sheet or plate. It is suitable particularly for low-quantity production runs of large parts, such as those used in aerospace applications. Steel plates 25 mm (1 in.) thick and 3.6 m (12 ft) in diameter have been formed by this method, as have tubes with wall thicknesses as much as 25 mm (1 in.).

The explosive-forming method also can be used at a much smaller scale, as shown in Fig. 16.52b. In this case, a *cartridge* (canned explosive) is used as the source of energy. The process can be useful in bulging and expanding of thin-walled tubes, for specialized applications.

The mechanical properties of parts made by explosive forming are basically similar to those of others made by conventional forming methods. Depending on the number of parts to be produced, dies may be made of aluminum alloys, steel, ductile iron, zinc alloys, reinforced concrete, wood, plastics, or composite materials.

Electromagnetically Assisted Forming.   In *electromagnetically assisted forming*, also called *magnetic-pulse forming*, the energy stored in a capacitor bank is discharged rapidly through a magnetic coil. In a typical example, a ring-shaped coil is placed over a tubular workpiece. The tube is then collapsed by magnetic forces over a solid piece, thus making the assembly an integral part (Fig. 16.53).

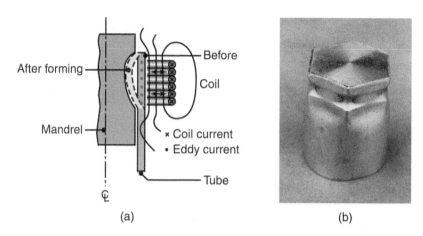

FIGURE 16.53   (a) Schematic illustration of the magnetic-pulse-forming process used to form a tube over a plug. (b) Aluminum tube collapsed over a hexagonal plug by the magnetic-pulse-forming process.

The mechanics of this process is based on the fact that a magnetic field, produced by the coil (Fig. 16.53a), crosses the metal tube (which is an electrical conductor) and generates *eddy currents* in the tube. In turn, these currents produce their own magnetic field. The forces produced by the two magnetic fields oppose each other. The repelling force generated between the coil and the tube then collapses the tube over the inner piece. The higher the electrical conductivity of the workpiece, the higher the magnetic forces. Note that it is not necessary for the workpiece material to have magnetic properties, but it must be electrically conducting.

It has been shown that the basic advantages of this process are that the formability of the material is increased, dimensional accuracy is improved, and springback and wrinkling are reduced. Magnetic coil design is an important factor in the success of the operation. Flat magnetic coils also can be made for use in such operations as embossing and shallow drawing of sheet metals.

First used in the 1960s, this process has been demonstrated to be particularly effective for aluminum alloys. Electromagnetically assisted forming has been applied to (a) collapsing thin-walled tubes over rods, cables, and plugs; (b) compression-crimp sealing of automotive oil filter canisters; (c) specialized sheet-forming operations; (d) bulging and flaring operations; and (e) swaging end fittings onto torque tubes for the Boeing 777 aircraft.

Peen Forming.   As shown in Fig. 16.54, peen forming is used to produce curvatures on thin sheet metals by *shot peening* (see Section 34.2) one surface of the sheet. As a result, the surface of the sheet is subjected to compressive stresses, which tend to expand the surface layer. Because the material below the peened surface remains rigid, the surface expansion causes the sheet to develop a curvature. The process also induces compressive surface residual stresses, which improve the fatigue strength of the sheet metal.

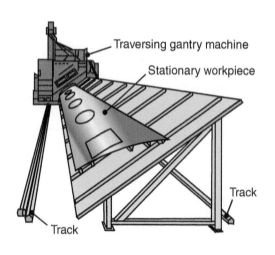

FIGURE 16.54   Schematic illustration of a peen-forming machine to shape a large sheet-metal part, as an aircraft-skin panel; note that the sheet is stationary and the peening head travels along its length. *Source:* Metal Improvement Company.

Peening is done with cast-iron or steel shot, discharged either from a rotating wheel or by an air blast from a nozzle. Peen forming is used by the aircraft industry to generate smooth and complex curvatures on aircraft wing skins. Cast-steel shot about 2.5 mm (0.1 in.) in diameter, traveling at speeds of 60 m/s (200 ft/s), have been used to form wing panels 25 m (80 ft) long. For heavy sections, shot diameters as large as 6 mm (1/4 in.) may be used. The peen-forming process also is used for *straightening* twisted or bent parts, including out-of-round rings to make them round.

Laser Beam Forming.   This process involves the application of laser beams as a localized heat source over specific regions of the sheet metal. The steep thermal gradients developed through the thickness of the sheet produce thermal stresses, which are sufficiently high to cause localized plastic deformation of the sheet. With this method, a sheet, for example, can be bent permanently without using dies. In **laser-assisted forming**, the laser acts as a localized heat source, thus reducing the strength of the sheet metal at specific locations, improving formability and increasing process flexibility. Applications include straightening, bending, embossing, and forming of complex tubular or flat components.

Microforming.   This is a more recent development and includes a family of processes that are used to produce very small metallic parts and components. Examples of *miniaturized products* include a wristwatch with an integrated digital camera and a one-gigabyte computer storage component. Typical components made by microforming include small shafts for micromotors, springs, screws, and a variety of cold-headed, extruded, bent, embossed, coined, punched, or deep-drawn parts. Dimensions are typically in the submillimeter range, and part weights are on the order of milligrams.

Electrohydraulic Forming.   Also called *underwater spark* or *electric-discharge forming*, the source of energy in this process is a spark between two electrodes, connected with a short, thin wire. The rapid discharge of the energy, from a capacitor bank, through the wire generates a shock wave in the water, similar to those created by explosives. The pressure developed in the water medium is sufficiently high to form the part. The energy levels are lower than those in explosive forming, being typically a few kJ. Electrohydraulic forming is a batch process and can be used in making various small parts.

CASE STUDY 16.4   Cymbal Manufacture

Cymbals (Fig. 16.55a) are an essential percussion instrument for all forms of music. Modern drum-set cymbals cover a wide variety of sounds, from deep, dark, and warm to bright, high-pitched, and cutting. Some cymbals sound "musical," while others are "trashy." A wide variety of sizes, shapes, weights, hammerings, and surface finishes (Fig. 16.55b) are available to achieve the desired performance.

Cymbals are produced from metals, such as B20 bronze (80% Cu–20% Sn, with a trace of silver), B8 bronze (92% Cu–8% Sn), nickel–silver alloy, and brass. The manufacturing sequence for producing a bronze cymbal is shown in Fig. 16.56. The B20 metal is first cast into mushroom-shaped ingots, and then cooled in ambient temperature. The ingot is then rolled successively, up to 14 times, with water cooling the metal with each pass through the rolling mill. Special care is taken to roll the bronze at a different angle with each pass, in order to minimize anisotropy and develop an even, round shape. The as-rolled blanks are then reheated and stretch formed (pressed) into the cup or bell shape, which determines the cymbal's overtones. The cymbals are then center drilled or punched, to create hang holes, and trimmed on a rotary shear to approximate final diameters. This operation is followed by another stretch-forming step, to achieve the characteristic "Turkish dish" form that controls the cymbal's pitch.

Automatic peen-forming is done on machinery (Fig. 16.57) and without templates, since the

(*continued*)

(a)   (b)

FIGURE 16.55 (a) Selected common cymbals. (b) Detailed view of different surface textures and finishes of cymbals. *Source:* Courtesy of W. Blanchard, Sabian Ltd.

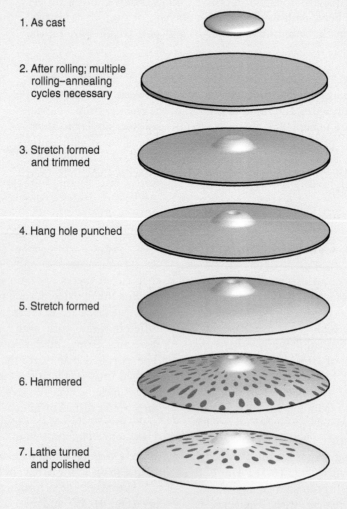

1. As cast

2. After rolling; multiple rolling–annealing cycles necessary

3. Stretch formed and trimmed

4. Hang hole punched

5. Stretch formed

6. Hammered

7. Lathe turned and polished

FIGURE 16.56 Manufacturing sequence for the production of cymbals. *Source:* Courtesy of W. Blanchard, Sabian Ltd.

FIGURE 16.57   Automated hammering of a cymbal on a peening machine. *Source:* Courtesy of W. Blanchard, Sabian Ltd.

cymbals have already been pressed into shape, but the peening pattern is controllable and uniform. The size and pattern of the peening operations depend on the desired response, such as tone, sound, response, and pitch of the cymbal. The cymbals are then hammered to impart a distinctive character to each instrument. Hammering can be done by hand, which involves placing the bronze blank on a steel anvil, where the cymbals then are struck manually by hand hammers.

Several finishing operations are performed on the cymbals. These can involve merely cleaning and printing of identifying information, as some musicians prefer the natural surface appearance and sound of formed, hot-rolled bronze. More commonly, the cymbals are turned on a lathe (without using any machining fluid) in order to remove the oxide surface and reduce the thickness of the cymbal to create the desired weight and sound. As a result, the surface finish becomes lustrous and, in some cases, also develops a favorable microstructure. Some cymbals are polished to a glossy "brilliant finish." In many cases, the surface indentations from peening persist after finishing; this is recognized as an essential performance feature of the cymbal, and it is also an aesthetic feature appreciated by musicians. Various surface finishes associated with modern cymbals are shown in Fig. 16.55b.

Source: Courtesy of W. Blanchard, Sabian Ltd.

16.13   Manufacturing of Metal Honeycomb Structures

A *honeycomb structure* basically consists of a core of honeycomb, or other corrugated shapes, bonded to two thin outer skins (Fig. 16.58). The most common example of such a structure is corrugated cardboard, which has a high stiffness-to-weight ratio and is used extensively in packaging for shipping consumer and industrial goods. Because of their lightweight and high resistance to bending, metal honeycomb structures are used for aircraft and aerospace components, in buildings, and in transportation equipment. The chassis of the Koenigsegg (Swedish) sports car, for example, is made partly of aluminum honeycomb with an integrated fuel tank. Honeycomb structures also may be made of nonmetallic materials, such as polymers and various composite materials.

Honeycomb structures are made most commonly of 3000-series aluminum, but may also be made of titanium, stainless steels, and nickel alloys, for specialized applications and corrosion resistance. Reinforced plastics, such as aramid-epoxy, also are used to make these structures.

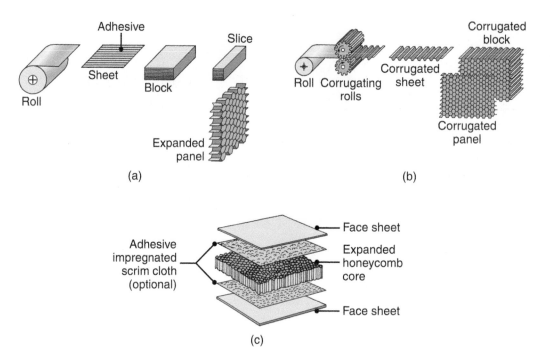

FIGURE 16.58 Methods of manufacturing honeycomb structures: (a) expansion process; (b) corrugation process; and (c) assembling a honeycomb structure into a laminate.

There are two basic methods of manufacturing honeycomb materials. In the **expansion process,** which is the more common method (Fig. 16.58a), sheets are first cut from a coil, and an *adhesive* (see Section 32.4) is applied at intervals (node lines) on their surfaces. The sheets are then stacked and cured in an oven, developing strong bonds at the adhesive surfaces. The block is then cut into slices of the desired dimensions, and stretched to produce a honeycomb structure.

In the **corrugation process** (Fig. 16.58b) the sheet metal first passes through a pair of specially designed rolls, becoming a corrugated sheet; it is then cut into desired lengths. Adhesive is applied to the node lines, the corrugated sheets are stacked into a block, and the block is cured. Because the sheets are already preformed, no expansion process is involved. The honeycomb is finally made into a sandwich structure (Fig. 16.58c), using face sheets that are joined by adhesives (or *brazed*; see Section 32.2) to the top and bottom surfaces.

16.14 Design Considerations in Sheet-metal Forming

As with most other processes described throughout this book, certain design guidelines and practices have evolved with time. Careful design using the best established design practices, computational tools, and manufacturing techniques is the best approach to achieving high-quality designs and realizing cost savings. The following guidelines apply to sheet-metal-forming operations, with the most significant design issues identified.

Blank Design.   Material scrap is the primary concern in blanking operations. (See also Table 40.6.) Poorly designed parts will not *nest* properly, and there can be considerable scrap produced (Fig. 16.59).

Bending. The main concerns in bending operations are material fracture, wrinkling, and the inability to properly form the bend. As shown in Fig. 16.60, a sheet-metal part with a flange will force the flange to undergo compression, which may cause buckling (see also *flanging*, Section 16.6). Buckling can be controlled with a relief notch, cut to limit the stresses developed during bending, or else a design modification as shown in the figure can be made. Right-angle bends have similar difficulties, and relief notches can be used to avoid tearing (Fig. 16.61).

Because the bend radius is a highly stressed area, all stress concentrations should be removed from the bend-radius location, such as holes near bends. It is advantageous to move the hole away from the bend area, but when this is not possible, a crescent slot or ear can be used (Fig. 16.62a). Similarly, in bending flanges, tabs and notches should be avoided, since their stress concentrations will greatly reduce formability. When tabs are necessary, large radii should be used to reduce stress concentration (Fig. 16.62b).

If notches are to be used, it is important to orient them properly with respect to the grain direction of the sheet metal. As shown in Fig. 16.17, bends ideally should be perpendicular to the rolling direction of the sheet (or oblique, if this is not possible) in order to avoid cracking. Bending to sharp radii can be accomplished by scoring or embossing (Fig. 16.63), but this operation can result in fracture. Burrs should not be present in a bend allowance (see Fig. 16.16), because they are less ductile (due to strain hardening) and can lead to crack initiation and propagation into the rest of the sheet.

Roll Forming. The process should, in general, be designed so as to control springback. Also, it is not difficult to include perforating rolls in the forming line, so that periodic holes, notches, or embossings can be located on the roll-formed shape

Stamping and Progressive-die Operations. In progressive dies, the cost of the tooling and the number of stations are determined by the number and spacing of the features on a part. Thus, it is advantageous to keep the number of features to a minimum, in order to minimize tooling costs. Closely-spaced features may provide insufficient clearance for punches, and may require two punches. Narrow cuts and protrusions also may present difficulties in forming with a single set of punch and die.

Deep Drawing. After a cup is deep drawn, it invariably will spring back, slightly toward its original shape. For this reason, designs that require a vertical wall may be

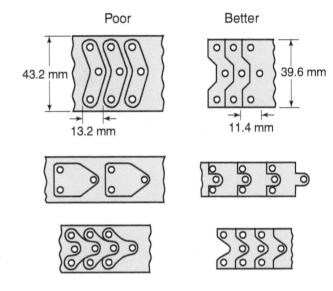

FIGURE 16.59 Efficient nesting of parts for optimum material utilization in blanking. *Source:* Reuse with permission from Society of Manufacturing Engineers in *Die Design Handbook*, 3rd edition, edited by David Smith.

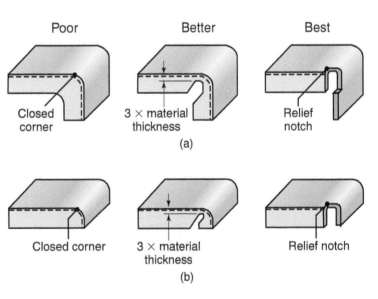

FIGURE 16.60 Control of tearing and buckling of two different flanges in a right-angle bend. *Source:* Reuse with permission from Society of Manufacturing Engineers in *Die Design Handbook*, 3rd edition, edited by David Smith.

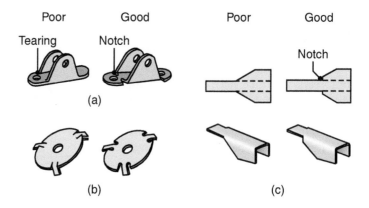

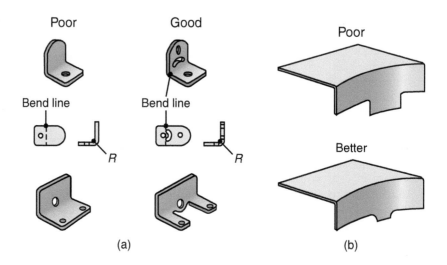

FIGURE 16.61 Application of notches to avoid tearing and wrinkling in right-angle bending operations. *Source:* Reuse with permission from Society of Manufacturing Engineers in *Die Design Handbook*, 3rd edition, edited by David Smith.

FIGURE 16.62 Stress concentrations near bends. (a) Use of a crescent or ear for a hole near a bend. (b) Reduction of severity of tab in flange. *Source:* Reuse with permission from Society of Manufacturing Engineers in *Die Design Handbook*, 3rd edition, edited by David Smith.

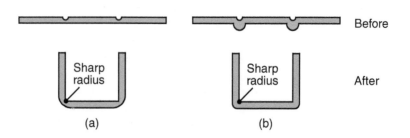

FIGURE 16.63 Application of (a) scoring or (b) embossing to obtain a sharp inner radius in bending. Unless properly designed, these features can lead to fracture. *Source:* Reuse with permission from Society of Manufacturing Engineers in *Die Design Handbook*, 3rd edition, edited by David Smith.

difficult to draw. Relief angles of at least 3° on each wall are easier to produce. Cups with sharp internal radii are difficult to produce, and deep cups will often require one or more ironing operations.

16.15 Equipment for Sheet-metal Forming

For most general pressworking operations, the basic equipment consists of mechanical, hydraulic, pneumatic, or pneumatic–hydraulic presses, with a wide variety of designs, features, capacities, and computer controls. Recently, servo presses (see Section 14.8) are being used for sheet-metal forming, because of their ability to vary speed and forces in a controlled manner during forming. Typical designs for press frames are shown in Fig. 16.64 (see also Figs. 14.19 and 16.23f). The proper design,

   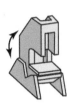 

(a) Basic C-frame design   (b) Wide design   (c) Adjustable bed   (d) Open-back inclinable   (e) Pillar   (f) Double column

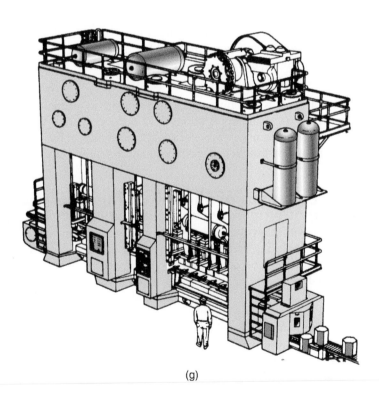

(g)

FIGURE 16.64 (a) through (f) Schematic illustrations of types of press frames for sheet-forming operations; each type has its own characteristics of stiffness, capacity, and accessibility. (g) A large stamping press. *Source:* (g) Printed with permission from Enprotech Industrial Technologies, Inc.

stiffness, and construction of such equipment is essential to the efficient operation of the system, and to achieving high production rate, good dimensional control, and high product quality.

The traditional **C-frame** structure (Fig. 16.64a) has been used widely for ease of tool and workpiece accessibility, but it is not as stiff as the **box-type pillar** (Fig. 16.64e) or the **double-column frame** structure (Fig. 16.64f). Accessibility to working areas in presses has become less important, due to advances in automation and in the use of industrial robots and computer controls.

Press selection for sheet-metal forming operations depends on several factors:

1. Type of forming operation, the size and shape of the dies, and the tooling required
2. Size and shape of the parts
3. Length of stroke of the slide, the number of strokes per minute, the operating speed, and the shut height (the distance from the top of the bed to the bottom of the slide with the stroke down)
4. Number of slides: single-action presses have one reciprocating slide; double-action presses have two slides, reciprocating in the same direction, and typically are used for deep drawing, one slide for the punch and the other for the blankholder; triple-action presses have three slides and generally are used for reverse redrawing and for other complicated forming operations
5. Maximum force required (press capacity and tonnage rating)
6. Type and level of mechanical, hydraulic, and computer controls
7. Features for changing the dies. Because the time required for changing dies in presses can be significant (as much as a few hours), and thus affect productivity, rapid die-changing systems have been developed; in a system called *single-minute exchange of die* (SMED), die setups can be changed in less than 10 min, by using computer-controlled hydraulic or pneumatic systems
8. Safety features

Because a press is a major capital investment, its present and future use for a broad variety of parts and applications must be investigated. Versatility and multiple use are important factors in press selection, particularly for product modifications and for making new products to respond to continually changing markets.

16.16 Economics of Sheet-forming Operations

Sheet-metal forming involves economic considerations that are similar to those for the other metalworking processes. Sheet-forming operations are very versatile, and a number of different processes can be considered to produce the same part. The costs involved (see also Chapter 40) depend on the particular operations, such as die and equipment costs and labor. For small and simple parts, die costs and lead times to make the dies are relatively low. On the other hand, for large-scale operations, such as stretch forming of aircraft panels and boat hulls, these costs are very high. Furthermore, because the number of such parts required is low, the cost per piece can be very high (see Fig. 14.21).

Deep drawing requires expensive dies and tooling, but a very high number of parts, such as containers, cans, and similar household products, can be produced with the same setup. These costs for other processes, such as punching, blanking, bending, and spinning, vary considerably.

Equipment costs vary widely, depending largely on the complexity of the forming operation, part loading and unloading features, part size and shape, and level of automation and computer controls required. Automation, in turn, directly affects the labor and the skill level required. Note that the higher the extent of automation, the lower the skill level required. Furthermore, many sheet-metal parts generally require some finishing operations, one of the most common being deburring of the edges of part made, which generally is labor intensive. However, significant advances have been made in automated deburring, which itself requires computer-controlled equipment, hence it can be costly.

As an example of the versatility of sheet-forming operations and the costs involved, recall that a cup-shaped part can be formed by deep drawing, spinning, rubber forming, or explosive forming. The part also may be formed by impact extrusion, casting, or fabrication by assembling or welding together different pieces. Each of these methods involves different costs. The part shown in Fig. 16.65, for example, can be made either by deep drawing or by conventional spinning, but the die costs for the two processes are significantly different.

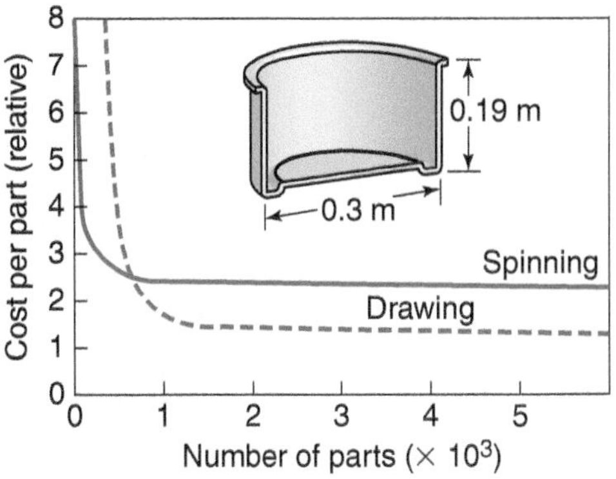

Deep-drawing dies have several components, and they cost much more than the relatively simple mandrels and tools employed in a process such as spinning. Consequently, the die cost per part in drawing will be high, especially if only a few parts are required. This part also can be formed by deep drawing and in a much shorter time than by spinning, even if the latter operation is automated and computer controlled. Also, spinning generally requires more skilled labor. Considering these factors, the break-even point for this part is around 700 parts, and for quantities greater than that, deep drawing is more economical.

FIGURE 16.65 Cost comparison for manufacturing a round sheet-metal container either by conventional spinning or by deep drawing; note that for small quantities, spinning is more economical.

SUMMARY

- Sheet-metal forming processes are among the most versatile of all metalworking operations. They generally are used on workpieces having high ratios of surface area to thickness. Unlike bulk deformation processes, such as forging and extrusion, sheet-metal forming operations often prevent the sheet thickness from being reduced.

- Important material parameters are the quality of the sheared edge of the blank, the capability of the sheet to stretch uniformly, the material's resistance to thinning, its normal and planar anisotropy, grain size, and for low-carbon steels, yield-point elongation.

- The forces and energy required in forming processes are transmitted to the sheet through solid tools and dies, by flexible rubber or polyurethane members, or by electrical, chemical, magnetic, and gaseous means.

- Because of the relatively thin materials used, springback, buckling, and wrinkling are significant factors in sheet forming. These difficulties can be eliminated or reduced by proper tool and die design, minimizing the unsupported length of the sheet during processing, and controlling the thickness and surface finish of the incoming sheet and its mechanical properties.

- Superplastic forming of diffusion-bonded sheets is an important process for making complex sheet-metal structures, particularly for aerospace applications in which high stiffness-to-weight ratios are important.

- Several test methods have been developed for predicting the formability of sheet metals.
- For general stamping operations, forming-limit diagrams are very useful, because they establish quantitative relationships among the major and minor principal strains that limit safe forming.

KEY TERMS

Beading	Drawing	Laser-assisted forming	Roll forming
Bendability	Earing	Limiting drawing ratio	Rubber forming
Bend allowance	Electrohydraulic	Lüder's bands	Shaving
Bending	forming	Magnetic-pulse forming	Shearing
Blankholder	Embossing	Microforming	Slitting
Blanking	Explosive forming	Minimum bend radius	Spinning
Bulging	Fine blanking	Nesting	Springback
Burnished surface	Flanging	Nibbling	Steel rule
Burr	Formability	Normal anisotropy	Stretch forming
Clearance	Forming-limit diagram	Peen forming	Superplastic forming
Compound dies	Hemming	Planar anisotropy	Tailor-welded blanks
Deburring	Honeycomb structures	Plastic anisotropy	Transfer dies
Deep drawing	Hot stamping	Press brake	Wrinkling
Dent resistance	Hydroform process	Progressive dies	
Dimpling	Incremental forming	Punching	
Draw bead	Ironing	Redrawing	

BIBLIOGRAPHY

Altan, T., and Tekkaya, T. (eds.), **Sheet Metal Forming: Fundamentals**, ASM International, 2012.

Altan, T., Tekkaya, T. (eds.), **Sheet Metal Forming: Processes and Applications**, ASM International, 2012.

ASM Handbook, Vol. 14B: **Metalworking: Sheet Forming**, ASM International, 2006.

Boljanovic, V., **Sheet Metal Forming Process and Die Design**, Industrial Press, 2004.

Davies, G., **Materials for Automobile Bodies**, Butterworth-Heinemann, 2003.

Hosford, W.F., and Caddell, R.M., **Metal Forming: Mechanics and Metallurgy**, 4th ed., Cambridge, 2011.

Hu, J., Marciniak, Z., and Duncan, J., **Mechanics of Sheet Metal Forming**, 2nd ed., Butterworth-Heinemann, 2002.

Pearce, R., **Sheet Metal Forming**, Springer, 2006.

Rapien, B.L., **Fundamentals of Press Brake Tooling**, Hanser Gardner, 2005.

Spitler, D., Lantrip, J., Nee, J., and Smith, D.A., **Fundamentals of Tool Design**, 5th ed., Society of Manufacturing Engineers, 2005.

Suchy, I., **Handbook of Die Design**, 2nd ed., McGraw-Hill, 2005.

Szumera, J.A., **The Metal Stamping Process**, Industrial Press, 2003.

Tschaetch, H., **Metal Forming Practise: Processes, Machines, Tools**, Springer, 2007.

REVIEW QUESTIONS

16.1 How does sheet-metal forming differ from rolling, forging, and extrusion?

16.2 What causes burrs? How can they be reduced or eliminated?

16.3 Explain the difference between punching and blanking.

16.4 Describe the difference between compound, progressive, and transfer dies.

16.5 Describe the characteristics of sheet metals that are important in sheet-forming operations. Explain why they are important.

16.6 Describe the features of forming-limit diagrams (FLDs).

16.7 List the properties of materials that influence springback. Explain why and how they do so.

16.8 Give one specific application for each of the common bending operations described in this chapter.

16.9 Why do tubes buckle when bent? What is the effect of the tube thickness-to-diameter ratio?

16.10 Define normal anisotropy, and explain why it is important in determining the deep drawability of a material.

16.11 Describe earing and why it occurs.

16.12 What are the advantages of rubber forming? Which processes does it compete with?

16.13 Explain the difference between deep drawing and redrawing.

16.14 How is roll forming fundamentally different from rolling?

16.15 What is nesting? What is its significance?

16.16 Describe the differences between compound, progressive, and transfer dies.

16.17 What is microforming?

16.18 Explain the advantages of superplastic forming.

16.19 What is hot stamping? For what materials is it used?

16.20 What is springback? What is negative springback?

QUALITATIVE PROBLEMS

16.21 Explain the differences that you have observed between products made of sheet metals and those made by casting and forging.

16.22 Take any three topics from Chapter 2, and, with specific examples for each, show their relevance to the topics covered in this chapter.

16.23 Do the same as for Problem 16.22, but for Chapter 3.

16.24 Identify the material and process variables that influence the punch force in shearing, and explain how each of them affects this force.

16.25 Explain why springback in bending depends on yield stress, elastic modulus, sheet thickness, and bend radius.

16.26 Explain why cupping tests may not predict well the formability of sheet metals in actual forming processes.

16.27 Identify the factors that influence the deep-drawing force, F, in Fig. 16.32b, and explain why they do so.

16.28 Why are the beads in Fig. 16.36b placed in those particular locations?

16.29 A general rule for dimensional relationships for successful drawing without a blankholder is given by Eq. (16.14). Explain what would happen if this limit were exceeded.

16.30 Section 16.2 stated that the punch stripping force is difficult to estimate because of the many factors involved. Make a list of these factors, with brief explanations about why they would affect the stripping force.

16.31 Is it possible to have ironing take place in an ordinary deep-drawing operation? What is the most important factor?

16.32 Note the roughness of the periphery of the flanged hole in Fig. 16.25c, and comment on its possible effects when the part is used in a product.

16.33 What recommendations would you make in order to eliminate the cracking of the bent piece shown in Fig. 16.17c? Explain your reasons.

16.34 It has been stated that the quality of the sheared edges can influence the formability of sheet metals. Explain why.

16.35 Give several specific examples from this chapter in which friction is desirable and several in which it is not desirable.

16.36 As you can see, some of the operations described in this chapter produce considerable scrap. Describe your thoughts regarding the reuse, recycling, or disposal of this scrap. Consider its size, shape, and contamination by metalworking fluids during processing.

16.37 Through changes in clamping or die design, it is possible for a sheet metal to undergo a negative minor strain. Explain how this effect can be advantageous.

16.38 How would you produce the part shown in Fig. 16.43b other than by tube hydroforming?

16.39 It has been stated that the thicker the sheet metal, the higher is the curve in the forming-limit diagram. Explain why.

16.40 If a cupping test (see Fig. 16.13) were to be performed using a pressurized lubricant instead of a spherical die, would you expect the forming limit diagram to change? Why or why not?

QUANTITATIVE PROBLEMS

16.41 Calculate R_{avg} for a metal where the R values for the 0°, 45°, and 90° directions are 0.9, 1.7, and 1.8, respectively. What is the limiting drawing ratio (LDR) for this material?

16.42 Calculate the value of ΔR in Problem 16.41. Will any ears form when this material is deep drawn? Explain.

16.43 Estimate the limiting drawing ratio for the materials listed in Table 16.4.

16.44 Using Eq. (16.15) and the K value for TNT, plot the pressure as a function of weight, W, and R, respectively. Describe your observations.

16.45 Section 16.5 states that the k values in bend allowance depend on the relative magnitudes of R and T. Explain why this relationship exists.

16.46 For explosive forming, calculate the peak pressure in water for 0.25 lb of TNT at a standoff distance of 4 ft. Comment on whether or not the magnitude of this pressure is sufficiently high to form sheet metals.

16.47 Measure the respective areas of the solid outlines in Fig. 16.14a, and compare them with the areas of the original circles. Calculate the final thicknesses of the sheets, assuming that the original sheet is 1 mm thick.

16.48 Plot Eq. (16.6) in terms of the elastic modulus, E, and the yield stress, Y, of the material, and describe your observations.

16.49 What is the minimum bend radius for a 1.0-mm-thick sheet metal with a tensile reduction of area of 30%? Does the bend angle affect your answer? Explain.

16.50 Survey the technical literature and explain the mechanism by which negative springback can occur in V-die bending. Show that negative springback does not occur in air bending.

16.51 Using the data in Table 16.3 and referring to Eq. (16.5), calculate the tensile reduction of area for the materials and the conditions listed in the table.

16.52 What is the force required to punch a square hole 50 mm on each side in a 0.1-mm-thick 5052-O aluminum sheet by using flat dies? What would be your answer if beveled dies are used?

16.53 In Case Study 16.2, it was stated that the reason for reducing the tops of cans (necking) is to save material for making the lid. How much material will be saved if the lid diameter is reduced by 5%? By 20%?

16.54 A cup is being drawn from a sheet metal that has a normal anisotropy of 3. Estimate the maximum ratio of cup height to cup diameter that can be drawn successfully in a single draw. Assume that the thickness of the sheet throughout the cup remains the same as the original blank thickness.

16.55 Estimate the percent scrap in producing round blanks if the clearance between blanks is one tenth of the radius

of the blank. Consider single and multiple-row blanking, as sketched in Fig. P16.55.

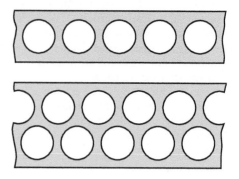

FIGURE P16.55

16.56 Plot the final bend radius as a function of initial bend radius in bending for (a) 5052-O aluminum; (b) 5052-H34 aluminum; (c) C24000 brass; and (d) AISI 304 stainless steel.

16.57 Figure P16.57 shows a parabolic profile that will define the mandrel shape in a spinning operation. Determine the equation of the parabolic surface. If a spun part will be produced from a 10-mm thick blank, determine the minimum required blank diameter.

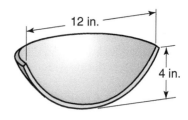

FIGURE P16.57

16.58 Assume that you are an instructor covering the topics described in this chapter and you are giving a quiz on the numerical aspects to test the understanding of the students. Prepare two quantitative problems and supply the answers.

SYNTHESIS, DESIGN, AND PROJECTS

16.59 Examine some of the products in your home or in an automobile that are made of sheet metal, and discuss the process or combination of processes by which you think they were made.

16.60 Consider several shapes to be blanked from a large sheet (such as oval, triangular, L-shaped, and so forth) by laser-beam cutting, and sketch a nesting layout to minimize scrap generation.

16.61 Give several specific product applications for (a) hemming and (b) seaming.

16.62 Many axisymmetric missile bodies are made by spinning. What other methods could you use if spinning processes were not available?

16.63 Give several structural designs and applications in which diffusion bonding and superplastic forming can be used jointly. Comment on whether this combination is capable of producing parts at high volume.

16.64 Metal cans are either two-piece (in which the bottom and sides are integral) or three-piece (in which the sides, the bottom, and the top are each separate pieces). For a three-piece can, should the vertical seam in the can body be (a) in the rolling direction, (b) normal to the rolling direction, or (c) oblique to the rolling direction? Prove your answer.

16.65 The design shown in Fig. P16.65 is proposed for a metal tray, the main body of which is made from cold-rolled sheet steel. Noting its features and that the sheet is bent in two different directions, comment on various manufacturing considerations. Include factors such as anisotropy of the rolled sheet, its surface texture, the bend directions, the nature of the sheared edges, and the way the handle is snapped in for assembly.

FIGURE P16.65

16.66 Suggest consumer-product designs that could utilize honeycomb structures. For example, an elevator can use a honeycomb laminate as a stiff and lightweight floor material.

16.67 How would you produce the part shown in Fig. 16.44 other than by tube hydroforming? Give two options.

16.68 Using a ball-peen hammer, strike the surface of aluminum sheets of various thicknesses until they develop a curvature. Describe your observations about the shapes produced.

16.69 Inspect a common paper punch and observe the shape of the punch tip. Compare it with those shown in Fig. 16.10 and comment on your observations.

16.70 Obtain an aluminum beverage can and slit it in half lengthwise with a pair of tin snips. Using a micrometer, measure the thickness of the can bottom and the wall. Estimate the thickness reductions in ironing and the diameter of the original blank.

16.71 In order to improve its ductility, a coil of sheet metal is placed in a furnace and annealed. However, it is observed that the sheet has a lower limiting drawing ratio than it had before being annealed. Explain the reasons for this behavior.

16.72 With automotive parts, it is often advantageous to have a part with tailored properties. For example, a pillar that provides structural support for the operator's compartment may be strong but less ductile at the center, but more ductile and less strong where the pillar attaches to the remainder of the car structure. List ways of producing such tailored properties in hot stampings.

16.73 Give three examples of sheet metal parts that (a) can and (b) cannot be produced by incremental forming.

16.74 Conduct a literature search and obtain the equation for a tractrix curve, as used in Fig. 16.38.

16.75 On the basis of experiments, it has been suggested that concrete, either plain or reinforced, can be a suitable material for dies in sheet-metal forming operations. Describe your thoughts regarding this suggestion, considering die geometry and any other factors that may be relevant.

16.76 Investigate methods for determining optimum shapes of blanks for deep-drawing operations. Sketch the optimally shaped blanks for drawing rectangular cups, and optimize their layout on a large sheet of metal.

16.77 Design a box that will contain a 4-in. × 6-in. × 3-in. volume. The box should be produced from two pieces of sheet metal and require no tools or fasteners for assembly.

16.78 Repeat Problem 16.77, but design the box from a single piece of sheet metal.

17

Powder Metal Processes and Equipment

- This chapter describes the powder metallurgy process for producing net-shape parts from metal powders.
- The chapter begins by examining methods of producing and blending metal powders, and investigates the shapes that powders will develop based on the process employed to make them.
- A number of secondary operations are then presented, such as compaction to consolidate the powder into a desired shape, and sintering to fuse the particles to achieve the required strength.
- Additional processes particular to powder metallurgy are then discussed, and design rules are presented.
- The chapter ends with a discussion of process capabilities and economics as compared with other competing manufacturing operations.

Typical products made: Connecting rods, piston rings, gears, cams, bushings, bearings, cutting tools, surgical implants, magnets, and metal filters.

Alternative processes: Casting, forging, and machining.

QR Code 17.1 Introduction to Powder-Metallurgy. (*Source:* Courtesy of Metal Powder Industries Federation)

17.1 Introduction

In the manufacturing processes described thus far, the raw materials used have been metals and alloys, either in a molten state (casting) or in solid form (metalworking). This chapter describes **powder metallurgy** (PM), in which metal powders are compacted into desired and often complex shapes and sintered (heated without melting) to form a solid piece. This process first was used in Egypt in about 3000 B.C. to make iron tools. One of its first modern uses was in the early 1900s to make the tungsten filaments for incandescent light bulbs. The availability of a wide range of metal powder compositions, the ability to produce parts to net dimensions (**net-shape forming**), and the overall favorable economics of the operation give this process its numerous attractive and expanding applications.

A wide range of parts and components are made by powder metallurgy techniques (Fig. 17.1): balls for ballpoint pens; automotive components, which now constitute about 70% of the PM market, such as piston rings, connecting rods, brake pads, gears, cams, and bushings; tool steels, tungsten carbides, and cermets as tool and die materials, graphite brushes impregnated with copper for electric motors; magnetic materials; metal filters and oil-impregnated bearings with controlled porosity; metal foams; and surgical implants. Advances in PM now permit *structural* parts of aircraft,

(a)

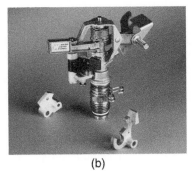
(b)

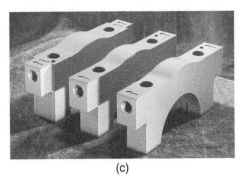
(c)

FIGURE 17.1   (a) Examples of typical parts made by powder metallurgy processes. (b) Upper trip lever for a commercial irrigation sprinkler made by PM; this part is made of an unleaded brass alloy; it replaces a die-cast part with a 60% cost savings. (c) Main-bearing metal powder caps for 3.8- and 3.1-liter General Motors automotive engines. *Source:* (a) and (b) Reproduced with permission from *Success Stories on PM Parts*, Metal Powder Industries Federation, Princeton, New Jersey, 1998. (c) Courtesy of GKN Sinter Metals, Auburn Hills, Michigan.

such as landing gear components, engine-mount supports, engine disks, impellers, and engine nacelle frames.

Powder metallurgy has become competitive with processes such as casting, forging, and machining, particularly for relatively complex parts made of high-strength and hard alloys. Although most parts weigh less than 2.5 kg (5 lb), they can weigh as much as 50 kg (100 lb). It has been shown that PM parts can be mass-produced economically in quantities as small as 5000 per year, and as much as 100 million per year for vibrator weights for cell phones.

The most commonly used metals in PM are iron, copper, aluminum, tin, nickel, titanium, and the refractory metals. For parts made of brass, bronze, steels, and stainless steels, *prealloyed powders* are used, where each powder particle itself is an alloy. The sources for metals are generally bulk metals and alloys, ores, salts, and other compounds.

17.2   Production of Metal Powders

The powder metallurgy process basically consists of the following operations, in sequence (Fig. 17.2):

1. *Powder production*
2. *Blending*
3. *Compaction*
4. *Sintering*
5. *Finishing operations*

17.2.1   Methods of Powder Production

There are several methods of producing metal powders, and most powders can be produced by more than one method; the choice depends on the requirements of the end product. The microstructure, bulk and surface properties, chemical purity, porosity, shape, and size distribution of the particles depend on the particular process used (Figs. 17.3 and 17.4). These characteristics are important because they significantly

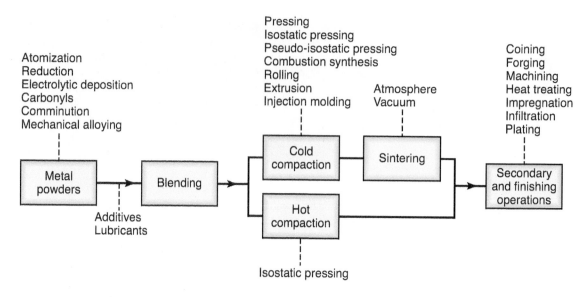

FIGURE 17.2   Outline of processes and operations involved in producing powder metallurgy parts.

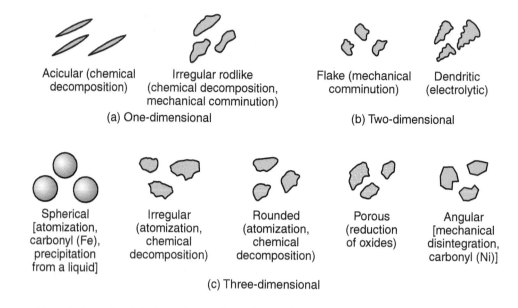

FIGURE 17.3   Particle shapes in metal powders, and the processes by which they are produced; iron powders are produced by many of these processes (see also Fig. 17.4).

affect the flow and permeability during compaction, and in subsequent sintering operations. Particle sizes produced range from 0.1 to 1000 μm (4 μin.–0.04 in.).

Atomization.   Atomization involves a liquid-metal stream produced by injecting molten metal through a small orifice. The stream is broken up by jets of inert gas or air (Fig. 17.5a) or water (Fig. 17.5b), known as *gas* or *water atomization*, respectively. The size and shape of the particles formed depend on the temperature of the molten metal, rate of flow, nozzle size, and jet characteristics. The use of water results in a slurry of metal powder and liquid at the bottom of the atomization chamber. Although the powders must be dried before they can be used, the water allows for more rapid cooling of the particles and thus higher production rates. Gas atomization usually results in more spherical particles (see Fig. 17.3c).

In *centrifugal atomization*, the molten-metal stream drops onto a rapidly rotating disk or cup; the centrifugal forces break up the stream and generate particles (Fig. 17.5c). In a variation of this method, a consumable electrode is rotated rapidly (about 15,000 rev/min) in a helium-filled chamber (Fig. 17.5d). The centrifugal force breaks up the molten tip of the electrode into metal particles.

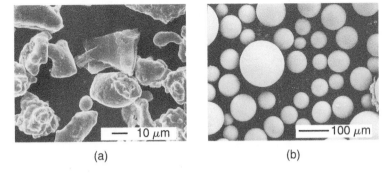

(a)   (b)

FIGURE 17.4 (a) Scanning-electron microscope image of iron-powder particles made by atomization. (b) Nickel-based superalloy (Udimet 700) powder particles made by the rotating electrode process; see Fig. 17.5d. *Source:* Courtesy of P.G. Nash, Illinois Institute of Technology, Chicago.

Reduction. The *reduction* (removal of oxygen) of metal oxides uses gases, such as hydrogen and carbon monoxide, as reducing agents. By this means, very fine metallic oxides are reduced to the metallic state. The powders produced are spongy and porous, and have uniformly sized spherical or angular shapes.

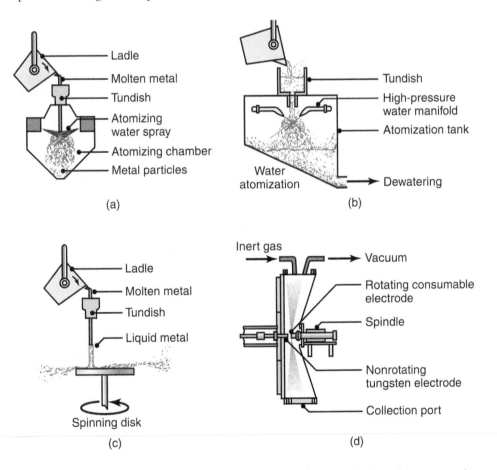

FIGURE 17.5 Methods of metal powder production by atomization: (a) gas atomization; (b) water atomization; (c) centrifugal atomization with a spinning disk or cup; and (d) atomization with a rotating consumable electrode.

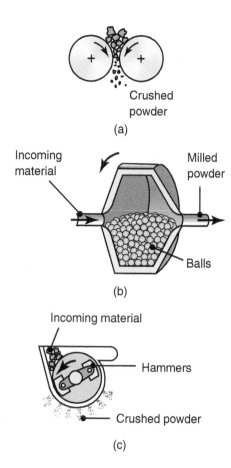

Crushed powder

(a)

Incoming material

Milled powder

Balls

(b)

Incoming material

Hammers

Crushed powder

(c)

FIGURE 17.6 Methods of mechanical comminution to obtain fine particles: (a) roll crushing; (b) ball mill; and (c) hammer milling.

Electrolytic Deposition. *Electrolytic deposition* utilizes either aqueous solutions or fused salts. The powders produced are among the purest available.

Carbonyls. *Metal carbonyls*, such as iron carbonyl [$Fe(CO)_5$] and nickel carbonyl [$Ni(CO)_4$], are formed by allowing iron or nickel to react with carbon monoxide. The reaction products are then decomposed to iron and nickel, and they turn into small, dense, uniformly spherical particles with high purity.

Comminution. *Mechanical comminution* (*pulverization*) involves crushing (Fig. 17.6), milling in a ball mill, or grinding brittle or less ductile metals into small particles. A *ball mill* (Fig. 17.6b) is a machine with a rotating hollow cylinder, partly filled with steel or white cast-iron balls. The powder or particles placed into a ball mill are impacted by the balls as the cylinder is rotated, or its contents may be agitated. This action has two effects: (a) the particles periodically fracture, resulting in smaller particles and (b) the shape of the particles is affected. With brittle materials, the particles produced have angular shapes; with ductile metals, they are flaky and not particularly suitable for powder metallurgy applications.

Mechanical Alloying. In *mechanical alloying*, powders of two or more pure metals are mixed in a ball mill, as illustrated in Fig. 17.7. Under the impact of the hard balls, the powders fracture and bond together by diffusion, entrapping the second phase and forming alloy powders. The dispersed phase can result in strengthening of the particles, or can impart special electrical or magnetic properties to the powder.

Miscellaneous Methods. Less commonly used methods for making powders are:

- **Precipitation** from a chemical solution
- Production of fine metal chips by **machining**
- **Vapor condensation**

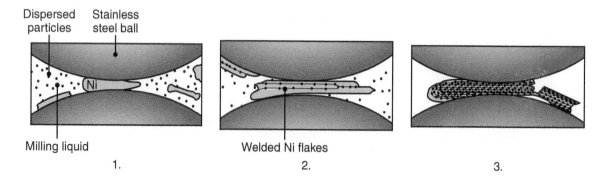

Dispersed particles   Stainless steel ball

Ni

Milling liquid

1.

Welded Ni flakes

2.

3.

FIGURE 17.7 Sequence of mechanical alloying of nickel particles with dispersed smaller particles. As nickel particles are flattened between two balls, the second, smaller phase is impressed into the nickel surface and eventually is dispersed throughout the particle due to successive flattening, fracture, and welding.

Developments include techniques based on *high-temperature extractive metal-lurgical processes*, which, in turn, are based on the reaction of volatile halides (a compound of halogen and an electropositive element) with liquid metals and controlled reduction and reduction/carburization of solid oxides.

Nanopowders. More recent developments include the production of *nanopowders* of copper, aluminum, iron, titanium, and various other metals (see also *nanomaterials*, Section 8.8). Because these powders are *pyrophoric* (ignite spontaneously), or are contaminated readily when exposed to air, they are shipped as thick slurries under hexane gas (which itself is highly volatile and combustible). When the material is subjected to large plastic deformation by compression and shear, at stress levels of 5500 MPa (800 ksi) encountered during processing of the powders, the particle size is reduced and the material becomes pore free, thus possessing enhanced properties.

Microencapsulated Powders. These metal powders are completely coated with a binder. For electrical applications, such as magnetic components of ignition coils and other pulsed AC and DC applications, the binder acts like an insulator, preventing electricity from flowing between the particles, and thus reducing eddy-current losses. The powders are compacted by warm pressing, and are used with the binder still in place. (See also *powder-injection molding*, Section 17.3.3.)

17.2.2 Particle Size, Shape, and Distribution

Particle size is generally controlled by *screening*, that is, by passing the metal powder through screens (*sieves*) of various mesh sizes. The screens are stacked vertically, with the mesh size becoming finer as the powder flows downward through the screens. The larger the mesh size, the smaller is the opening in the screen. A mesh size of 30, for example, has an opening of 600 μm, size 100 has 150 μm, and size 400 has 38 μm. (This method is similar to the numbering of abrasive grains. The larger the number, the smaller is the size of the abrasive particle: see Fig. 26.7.)

Several other methods also are available for particle-size analysis:

1. **Sedimentation,** which involves measuring the rate at which particles settle in a fluid
2. **Microscopic analysis,** which may include the use of transmission and scanning-electron microscopy
3. **Light scattering** from a laser that illuminates a sample, consisting of particles suspended in a liquid medium; the particles cause the light to be scattered, and a detector then digitizes the signals and computes the particle-size distribution
4. **Optical methods,** such as particles blocking a beam of light, in which the particle is sensed by a photocell
5. **Suspending particles** in a liquid and detecting particle size and distribution by electrical sensors

Particle Shape. A major influence on processing characteristics, particle shape usually is described in terms of aspect ratio or shape factor (SF). *Aspect ratio* is the ratio of the largest dimension to the smallest dimension of the particle, and ranges from unity, for a spherical particle, to about 10, for flakelike or needlelike particles.

Shape Factor. Also called the *shape index*, *shape factor* (SF) is a measure of the ratio of the surface area of the particle to its volume, normalized by reference to a spherical particle of equivalent volume. Thus, the SF for a flake is higher than that for a sphere.

Size Distribution.   The size distribution of particles is an important consideration, because it affects the processing characteristics of the powder. The distribution of particle size is given in terms of a *frequency-distribution plot* (see Section 36.7 for details). The maximum is called the *mode size*.

Other properties of metal powders that have an effect on their behavior in processing are (a) *flow properties* when the powders are being filled into dies, (b) *compressibility* when they are being compacted, (c) *density*, as defined in various terms such as theoretical density, apparent density, and the density, when the powder is shaken or tapped in the die cavity.

17.2.3   Blending Metal Powders

Blending (mixing) powders is the next step in powder metallurgy processing, and is carried out for the following purposes:

- Powders of different metals and other materials can be blended or mixed in order to impart special physical and mechanical properties and characteristics to the PM product. Mixtures of metals can be produced by alloying the metal before producing a powder, or else blends can be produced. Proper mixing is essential to ensure the uniformity of mechanical properties throughout the part.
- Even when a single metal is used, the powders may vary significantly in size and shape, hence they must be blended to ensure uniformity from part to part. An ideal mix is one in which all of the particles of each material, and of each size and morphology, are distributed uniformly.
- *Lubricants* can be mixed with the powders to improve their flow characteristics. They reduce friction between the metal particles, improve flow of the powder metals into the dies, and improve die life. Lubricants typically are stearic acid or zinc stearate, in a proportion of 0.25–5% by weight.
- Other additives, such as *binders* (as in sand molds, Section 11.2.1), are used to impart sufficient *green strength* (see Section 17.3); additives also can be used to facilitate sintering.

Powder mixing must be carried out under controlled conditions, in order to avoid contamination or deterioration. Deterioration is caused by excessive mixing, which may alter the shape of the particles and cause work hardening, thus making subsequent compaction more difficult. Powders can be mixed in air, in inert atmospheres (to avoid oxidation), or in liquids (which act as lubricants and make the mix more uniform). Several types of blending equipment are available (Fig. 17.8).

Hazards.   Because of their high surface area-to-volume ratio, metal powders can be explosive, particularly aluminum, magnesium, titanium, zirconium, and thorium. Great care must therefore be exercised, both during blending and in storage and handling. Precautions include (a) grounding equipment, (b) preventing sparks, by using nonsparking tools, (c) avoiding friction as a source of heat, and (d) avoiding dust clouds and exposed ignition sources, such as open flames.

17.3   Compaction of Metal Powders

Compaction is the step in which the blended powders are pressed into dies, as shown in sequence in Fig. 17.9. The purposes of compaction are to (a) obtain the required shape, density, and particle-to-particle contact and (b) make the part sufficiently

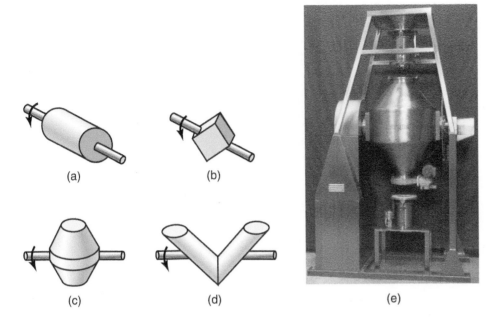

FIGURE 17.8 (a)–(d) Some common bowl geometries for mixing or blending powders. (e) A mixer suitable for blending metal powders. Since metal powders are abrasive, mixers rely on the rotation or tumbling of enclosed geometries, as opposed to using aggressive agitators. *Source:* Courtesy of Kemutec Group, Inc.

strong for further processing. The powder, called *feedstock*, is fed into the die by a *feed shoe*, and the upper punch descends into the die. The presses used are actuated either hydraulically or mechanically, and the process generally is carried out at room temperature, although it can also be done at elevated temperatures for high melting point metals.

The stages of powder compaction are shown in Fig. 17.10. Initially, the powder is loosely packed, and thus there is significant porosity. With low applied pressure, the powder rearranges, filling the voids and producing a denser powder; however, the stresses at points among the powders are still low. Continued compaction causes increased contact stress and plastic deformation of the powders, resulting in increased powder adhesion.

The pressed powder is known as **green compact**, because it has low strength, just as is seen in green parts in slip casting (Section 18.2.1). The green parts are very fragile, and can easily crumble or become damaged; a situation that is exacerbated by poor pressing practices, such as rough handling or insufficient compaction. For higher green strengths, the powder must be fed properly into the die cavity, and sufficient pressure must be developed throughout the part.

The *density* of the green compact depends on the compacting pressure (Fig. 17.11a). As the pressure is increased, the compact density approaches that of the metal in its bulk form. An important factor in density is the size *distribution* of the particles. If all of the particles are of the same size, there will always be some porosity when packed together. Theoretically, the porosity is at least 24% by volume. (Observe, for example, a box filled with rice, where there are always open spaces between the individual grains.) Introducing smaller particles into the powder mix will fill the spaces between the larger powder particles, and thus result in a higher density of the compact (see also *porous aluminum*, Section 6.2).

Video Solution 17.1 Compaction of a PM Part

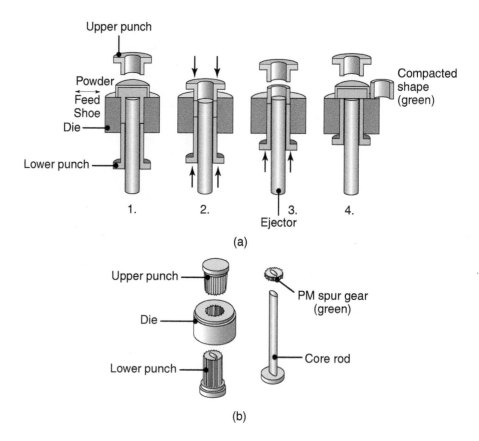

FIGURE 17.9 (a) Compaction of metal powder to form a bushing; the pressed powder part is called green compact. (b) A typical tool and die set for compacting a spur gear. *Source:* Reprinted with permission from Metal Powder Industries Federation, Princeton, NJ, USA.

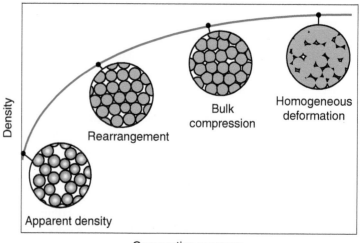

FIGURE 17.10 Compaction of metal powders; at low compaction pressures, the powder rearranges without deforming, leading to a high rate of density increase. Once the powders are more closely packed, plastic deformation occurs at their interfaces, leading to further density increases but at lower rates. At very high densities, the powder behaves like a bulk solid.

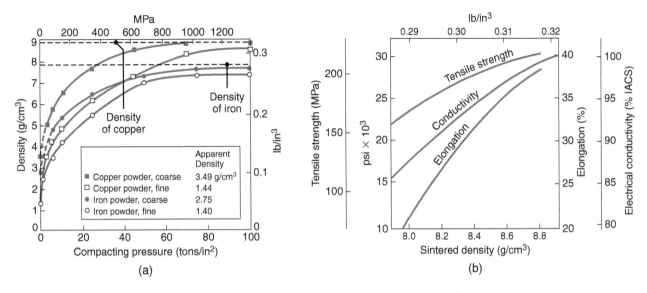

FIGURE 17.11   (a) Density of copper- and iron-powder compacts as a function of compacting pressure; density greatly influences the mechanical and physical properties of PM parts. (b) Effect of density on tensile strength, elongation, and electrical conductivity of copper powder. *Source:* (a) After F.V. Lenel. (b) After the International Annealed Copper Standard (IACS) for electrical conductivity.

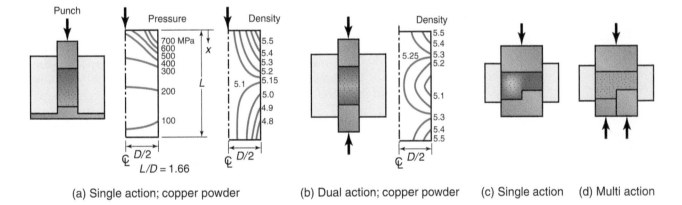

FIGURE 17.12   Density variation in compacting metal powders in various dies.

The higher the density of the compacted part, the higher are its strength and elastic modulus (Fig. 17.11b). The reason is that the higher the density, the higher the amount of solid metal in the same volume, and hence the higher its strength. Because of friction between (a) the metal particles in the powder and (b) the punch surfaces and die walls, the density within the part can vary considerably. This variation can be minimized by proper punch and die design and by control of friction. Thus, it may be necessary to use multiple punches, each with separate movements, in order to ensure that the density is more uniform throughout the part (Fig. 17.12). Recall a similar discussion regarding the compaction of sand in mold making (see Fig. 11.7). On the other hand, in some compacted parts, such as gears, and cams, density variations may be desirable. For example, densities can be increased in critical locations where high strength and wear resistance are important.

TABLE 17.1

Compacting Pressures for Various Powders

Material	Pressure (MPa)
Metals	
Aluminum	70–275
Brass	400–700
Bronze	200–275
Iron	350–800
Tantalum	70–140
Tungsten	70–140
Other materials	
Aluminum oxide	110–140
Carbon	140–165
Cemented carbides	140–400
Ferrites	110–165

FIGURE 17.13 A 7.3-MN (825-ton) mechanical press for compacting metal powder. *Source:* Courtesy of Cincinnati Incorporated.

Pressure Distribution during Compaction. As can be seen in Fig. 17.12, the pressure during compaction decays rapidly away from tooling surfaces. The pressure distribution along the length of the compact in a single-action press can be shown to be

$$p_x = p_o e^{-4\mu kx/D}, \qquad (17.1)$$

where μ is the coefficient of friction between particles and the container wall, k is a factor indicating the interparticle friction during compaction, D is the compact diameter, and p is the pressure in the compacting direction, x. Note that the pressure on the bottom of the punch is p_o. Equation (17.1) also includes a variable to account for friction between particles, given by

$$\sigma_r = kp_x,$$

where σ_r is the stress in the radial direction. If there is no friction between the particles, $k = 1$, the powder behaves like a fluid, and thus $\sigma_r = p_x$, signifying a state of hydrostatic pressure. If there is very high friction, $k = 0$, and the pressure will be low near the punch. It can be seen from Eq. (17.1) that the pressure within the compact decays as the coefficient of friction, the parameter k, and the length-to-diameter ratio increase. The pressure required for pressing metal powders typically ranges from 70 MPa (10 ksi) for aluminum to 800 MPa (120 ksi) for high-density iron parts (see Table 17.1).

17.3.1 Equipment

Press capacities for PM are generally around 1.8 to 2.7 MN (200–300 tons), although presses with much higher capacities are used for special applications. Most applications actually require less than 100 tons. For small tonnage, crank- or eccentric-type mechanical presses are used; for higher capacities, toggle- or knuckle-joint presses are employed (see Fig. 14.19b). Hydraulic presses (Fig. 17.13) with capacities as high as 45 MN (5000 tons) can be used for large parts. Press selection depends on part size and the configuration, density requirements, and production rate. However, the higher the pressing speed, the greater is the tendency for the press to trap air in the die cavity, and thus prevent proper compaction.

17.3.2 Isostatic Pressing

Green compacts may be subjected to *hydrostatic pressure* in order to achieve more uniform compaction and, hence, density. Typical applications include automotive cylinder liners and high-quality parts, such as turbine shafts, oil pipeline end swivels and pump manifolds, valves, and bearings. In **cold isostatic pressing** (CIP), the metal powder is placed in a flexible rubber mold (Fig. 17.14), typically made of neoprene rubber, urethane, polyvinyl chloride, or another elastomer (Section 7.9). The assembly then is pressurized hydrostatically in a chamber, usually using water. The most common pressure is 400 MPa (60 ksi), although pressures of up to 1000 MPa (150 ksi) may be used. The ranges for CIP and other compacting methods in terms of the size and complexity of a part are shown in Fig. 17.15.

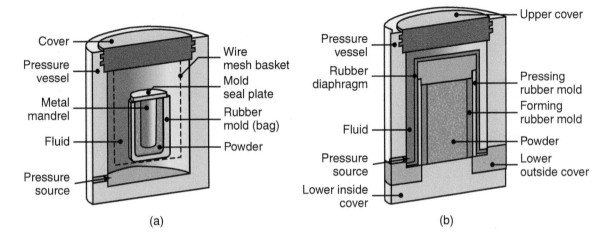

FIGURE 17.14   Schematic diagrams of cold isostatic pressing; pressure is applied isostatically inside a high-pressure chamber. (a) The wet bag process to form a cup-shaped part; the powder is enclosed in a flexible container around a solid-core rod. (b) The dry bag process used to form a PM cylinder.

In **hot isostatic pressing** (HIP), the container is generally made of a high-melting-point sheet metal, generally of mild or stainless steel, and the pressurizing medium is high-temperature inert gas or a vitreous (glasslike) fluid (Fig. 17.16). Common conditions for HIP are pressures as high as 100 MPa (15 ksi), although they can be three times as high and temperatures of 1200°C (2200°F). The main advantage of HIP is its ability to produce compacts having almost 100% density, good metallurgical bonding of the particles, and good mechanical properties.

The HIP process is used mainly to produce superalloy components for the aircraft and aerospace industries and in military, medical, and chemical applications. It also is used (a) to close internal porosity, (b) to improve properties in superalloy and titanium-alloy castings for the aerospace industry, and (c) as a final densification step for tungsten-carbide cutting tools and PM tool steels (Chapter 22).

The main advantages of HIP over conventional PM are:

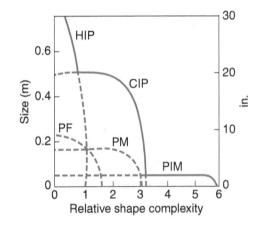

FIGURE 17.15   Capabilities, with respect to part size and shape complexity, available from various PM operations. PF = powder forging. *Source:* Reprinted with permission from Metal Powder Industries Federation, Princeton, NJ, USA.

- Because of the uniformity of pressure from all directions and the absence of die-wall friction, it produces fully dense compacts of practically uniform grain structure and density, irrespective of part shape (thus the properties are *isotropic*). Parts with high length-to-diameter ratios have been produced, with very uniform density, strength, toughness, and good surface detail.
- HIP is capable of handling much larger parts than those in other compacting processes; on the other hand, if has some limitations, such as:

  - Dimensional tolerances are higher than those in other compacting methods.
  - Equipment costs are higher and production time is longer than those in other processes.

- HIP is applicable only to relatively small production quantities, typically less than 10,000 parts per year.

QR Code 17.2 Hot Isostatic Pressing. (*Source:* Courtesy of Metal Powder Industries Federation)

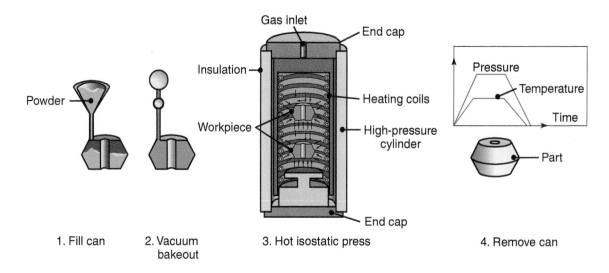

1. Fill can   2. Vacuum bakeout   3. Hot isostatic press   4. Remove can

FIGURE 17.16   Schematic illustration of hot isostatic pressing; the pressure and temperature variations versus time are shown in the diagram (not to scale).

CASE STUDY 17.1   Hot Isostatic Pressing of a Valve Lifter

An HIP-clad valve lifter, used in a full range of medium- to heavy-duty truck diesel engines, is shown in Fig. 17.17. The 0.2-kg (0.45-lb) valve lifter rides on the camshaft, and opens and closes the engine valves. Consequently, it is desirable to have a tungsten-carbide (WC) face for wear resistance, and a steel shaft for fatigue resistance. Before the HIP valve lifter was developed, these parts were produced through furnace brazing (Section 32.2), but resulted in occasional field failures and relatively high scrap rates. Because the required annual production of these parts is over 400,000, high scrap rates are particularly objectionable.

The new part product consists of a (a) 9% cobalt bonded tungsten-carbide face made from powder (pressed and sintered), (b) steel sheet-metal cap fitted over the WC disk, (c) copper-alloy foil interlayer, and (d) steel shaft. The steel cap is electron-beam welded to the steel shaft, then the assembly is hot isostatically pressed to provide a very strong bond. The HIP takes place at 1010°C (1850°F) and at a pressure of 100 MPa (15,000 psi). The tungsten-carbide surface

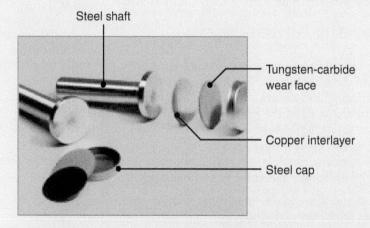

FIGURE 17.17   A valve lifter for heavy-duty diesel engines produced from a hot-isostatic-pressed carbide cap on a steel shaft. *Source:* Courtesy of the Metal Powder Industries Federation.

has a density of 14.52–14.72 g/cm^3, a hardness of 90.8 $\pm$ 5 HRA, and a minimum transverse rupture strength of 2450 MPa (355,000 psi).

Secondary operations are limited to grinding the face to remove any protruding sheet-metal cap, and to expose the wear-resistant tungsten-carbide face. The high reliability of the HIP bond greatly reduced scrap rates, to under 0.2%. No field failures have been experienced in over four years of full production. Also, production costs were substantially reduced, because of the HIP employed.

Source: Reprinted with permission from Metal Powder Industries Federation, Princeton, NJ, USA.

17.3.3 Miscellaneous Compacting and Shaping Processes

Powder-injection Molding. In *powder-injection molding* (PIM), also called **metal-injection molding** (MIM), very fine metal powders ($< 10 \mu m$) are blended with a 25%–45% polymer or a wax-based binder. The mixture then undergoes a process similar to die casting (Section 11.4.5; see also *injection molding of plastics* in Section 19.3), where it is injected into the mold, at a temperature of 135° to 200°C (275°–400°F). Parts generally have sprues and runners, as with injected molded parts (Fig. 17.18), and hence they are carefully separated before additional processing. The molded green parts are placed in a low-temperature oven, to burn off the plastic (*debinding*), or the binder is removed by solvent extraction. Often, a small amount of binder may be retained to provide sufficient green strength for transfer to a sintering furnace, where the parts are sintered (Section 17.4), at temperatures as high as 1375°C (2500°F). Subsequent operations, such as hole tapping, metal infiltration, and heat treating, also may be performed as required.

Generally, metals that are suitable for PIM are those that melt at temperatures above 1000°C (1830°F), such as carbon and stainless steels, tool steels, copper, bronze, and titanium. Typical parts made are components for watches, small-caliber gun barrels, scope rings for rifles, door hinges, impellers for sprinkler systems, and surgical knives.

The major advantages of PIM over conventional compaction are:

- Complex shapes, with wall thicknesses as small as 5 mm (0.2 in.), can be molded, then removed easily from the dies.
- Mechanical properties are nearly the same as those for wrought products.
- Dimensional tolerances are good.
- High production rates can be achieved by using multicavity dies (see Figs. 11.22 and 19.10).
- Parts produced by the PIM process compete well against small investment-cast parts and forgings, and complex machined parts. However, the PIM process does not compete well with zinc and aluminum die casting (Section 11.4.5), or with screw machining (Section 23.3.4).

The major limitations of PIM are the high cost for small production runs and the need for fine metal powders.

An example where the advantages of metal injection molding are apparent is in the production of light-duty gears, such as office equipment, where load and power is low. An inexpensive gear can be produced directly from metal injection molding, instead of producing a blank, such as from casting or forging, followed by machining and finishing operations (Section 24.7). Avoiding the high machining

FIGURE 17.18 A single shot of four metal injection molded components, with sprue, runners, and gates (see also Fig. 19.10). *Source:* Courtesy HARBEC, Inc.

costs thus results in significant cost savings; however, this approach may not be suitable for more demanding applications, such as power transmissions for automobiles or in gear pumps.

CASE STUDY 17.2   Mobile Phone Components Produced by Metal Injection Molding

Figure 17.19 shows the PM components used on mobile phones to allow them to flip open. These components were produced from 17-4 PH stainless steel (Section 5.6), with a final density of 7.6 g/cm^3, a hardness of 30 HRC, and a tensile strength of 172 ksi. The parts shown are complex, making them ideal candidates for MIM. The complexity arises because multiple components have to be assembled to make these parts, thus high production costs. The parts are produced within a maximum of +0.23% of the linear tolerance.

These components are integral to the unique opening mechanism deployed by the Motorola PEBL mobile phone. The dual hinge consists of the center barrel and a collar that extends to house intricate slots, and ends with the two straight legs. This feature enables the single-movement opening of the clamshell mobile phone. To accomplish this feature, the components must exhibit both the capability of design flexibility and material strength. The only other process option was machining, which would have required considerably higher material utilization and added cost, estimated to be five times the MIM component cost. The parts, as designed, have successfully completed accelerated life testing.

Source: Reprinted with permission from Metal Powder Industries Federation, Princeton, NJ, USA.

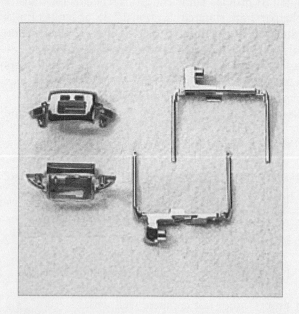

FIGURE 17.19 Powder metal components for mobile phones to achieve a flip-open feature.

Forging.   In *powder forging* (PF), the part produced from compaction and sintering serves as the *preform* in a hot-forging operation. The forged products have almost fully dense, and have good surface finish, good dimensional tolerances, and uniform and fine grain size. The superior properties obtained make PF particularly suitable for such applications as highly stressed automotive parts, such as connecting rods, and jet-engine components.

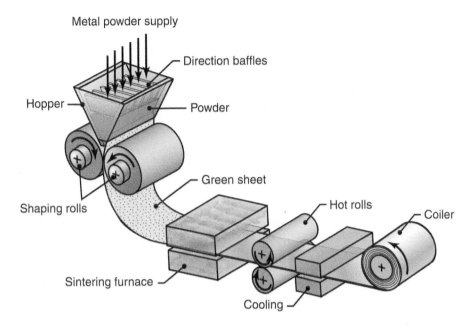

FIGURE 17.20   An illustration of metal powder rolling.

Rolling.   In *powder rolling*, also called *roll compaction*, the metal powder is fed into the roll gap in a two-high rolling mill (Fig. 17.20), and is compacted into a continuous strip, at speeds of up to 0.5 m/s (100 ft/min). The rolling operation can be carried out at room or elevated temperatures. Sheet metal for electrical and electronic components and for coins can be made by this process.

Extrusion.   Powders can be compacted by *extrusion*, where the powder is encased in a metal container and hot extruded. After sintering, the parts may be reheated and forged in a closed die (Section 14.3) to their final shape. Superalloy powders, for example, are hot extruded for enhanced properties.

Pressureless Compaction.   In this operation, the die is gravity filled with metal powder, and the powder is then sintered directly in the die. Because of the resulting low density, pressureless compaction is used principally for porous metal parts, such as filters.

Spray Deposition.   This is a shape-generation process (Fig. 17.21) involving (a) an atomizer, (b) a spray chamber, with an inert atmosphere, and (c) a mold for producing preforms. The mold may be made in various shapes, such as billets, tubes, disks, and cylinders. Although there are several variations, the best known is the *Osprey process*, shown in Fig. 17.21. After the metal is atomized, it is deposited onto a cooled preform mold, usually made of copper or ceramic, where it solidifies. The metal particles bond together, developing a density that usually is above 99% of the solid-metal density. Spray-deposited forms may be subjected to additional shaping and consolidation processes, such as forging, rolling, and extrusion. The grain size of the part is fine, and its mechanical properties are comparable to those of wrought products made of the same alloy.

Ceramic Molds.   *Ceramic* molds for shaping metal powders are made by the same technique used in investment casting (Section 11.3.2). After the mold is made, it is

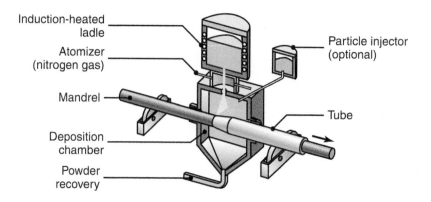

FIGURE 17.21 Spray deposition (*Osprey process*) in which molten metal is sprayed over a rotating mandrel to produce seamless tubing and pipe.

filled with metal powder and placed in a steel container. The space between the mold and the container is filled with particulate material. The container is then evacuated, sealed, and subjected to HIP. Titanium-alloy compressor rotors for missile engines have been made by this process.

Dynamic and Explosive Compaction. Some metal powders that are difficult to compact with sufficient green strength can be compacted rapidly to near full-density, using the setup shown in Fig. 17.22. The explosive drives a mass into green powder at high velocities, generating a shock wave that develops pressures up to 30 GPa. The shock wave traverses across the powder metal part at speeds up to 6 km/s. Preheating of the powder is often practiced to prevent fracture.

Combustion Synthesis. *Combustion synthesis* takes advantage of the highly combustible nature of metal powders, by placing a lightly compacted powder into a pressure vessel. An ignition source is then introduced, such as an arc from a tungsten

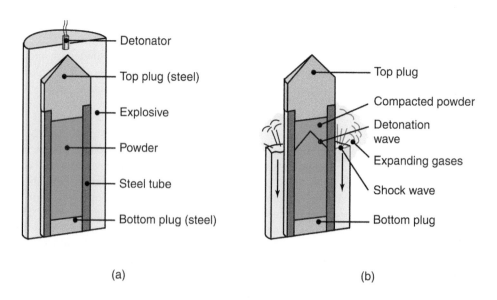

(a)   (b)

FIGURE 17.22 Schematic illustration of explosive compaction. (a) A tube filled with powder is surrounded by explosive media inside a container, typically cardboard or wood. (b) After detonation, a compression wave follows the detonation wave, resulting in a compacted metal powder part.

electrode, igniting the powder. The explosion produces a shock wave that travels across the compact, developing heat and pressure that is sufficient for compaction of the powder metal.

Pseudo-isostatic Pressing.   In *pseudo-isostatic pressing*, a preform is preheated, surrounded by hot ceramic or graphite granules, and placed in a container. A mechanical press compacts the granules and the preform. (Note that the granules are large and cannot penetrate the pores of the PM part.) The compaction is uniaxial, but because of the presence of the ceramic, the loading on the preform is multiaxial. This technique has cycle times shorter than HIP, but because the pressure is not strictly hydrostatic, dimensional changes during compaction are not uniform.

Selective Laser Sintering.   Some PM parts can be produced by selective laser sintering, a *rapid-prototyping operation* described in detail in Section 20.3.4.

17.3.4   Punch and Die Materials

The selection of punch and die materials for PM depends on the abrasiveness of the powder metal and the number of parts to be produced. Most common die materials are air- or oil-hardening tool steels, such as D2 or D3, with a hardness range from 60 to 64 HRC (Table 5.8). Because of their higher hardness and wear resistance, tungsten-carbide dies are used for more severe applications. Punches generally are made of similar materials.

Close control of die and punch dimensions is essential for proper compaction and die life. Too large a clearance between the punch and the die will allow the metal powder to enter the gap, where it will severely interfere with the operation and cause eccentric parts. Diametral clearances generally are less than 25 μm (0.001 in.). Die and punch surfaces must be lapped or polished, in the direction of tool movements in the die, for improved die life and overall performance.

17.4   Sintering

As described in Section 17.3, the green compact is brittle and its *green strength* is low. *Sintering* is the process whereby green compacts are heated, in a controlled-atmosphere furnace, to a temperature below the melting point of the metal, but sufficiently high to allow bonding (fusion) of the individual particles to impart strength to the part. The nature and strength of the bond between the particles, and thus that of the sintered compact, involve the complex mechanisms of diffusion, plastic flow, evaporation of volatile materials in the compact, recrystallization, grain growth, and extent of pore shrinkage.

The principal variables in sintering are temperature, time, and the furnace atmosphere. Sintering temperatures (Table 17.2) are generally within 70–90% of the melting point of the metal or alloy (see Table 3.1). Sintering times (Table 17.2) range from a minimum of about 10 min for iron and copper alloys to as much as eight hours for tungsten and tantalum.

Continuous-sintering furnaces, which are used for most production, have three chambers:

1. *Burn-off chamber*, for volatilizing the lubricants in the green compact, in order to improve bond strength and prevent cracking
2. *High-temperature chamber*, for sintering
3. *Cooling chamber*

TABLE 17.2

Sintering Temperature and Time for Various Metals		
Material	Temperature (°C)	Time (min)
Copper, brass, and bronze	760–900	10–45
Iron and iron graphite	1000–1150	8–45
Nickel	1000–1150	30–45
Stainless steels	1100–1290	30–60
Alnico alloys (for permanent magnets)	1200–1300	120–150
Ferrites	1200–1500	10–600
Tungsten carbide	1430–1500	20–30
Molybdenum	2050	120
Tungsten	2350	480
Tantalum	2400	480

To obtain optimum properties, proper control of the furnace atmosphere is essential for successful sintering. An oxygen-free atmosphere is necessary to control the carburization and decarburization of iron and iron-based compacts, and to prevent oxidation of the powders. A vacuum is generally used for sintering refractory-metal alloys and stainless steels. The gases most commonly used for sintering are hydrogen, dissociated or burned ammonia, partially combusted hydrocarbon gases, and nitrogen.

Sintering mechanisms depend on the composition of the metal particles, as well as on the processing parameters. The mechanisms are *diffusion, vapor-phase transport,* and *liquid-phase sintering.* As the temperature increases, two adjacent powder particles begin to form a bond by a **diffusion mechanism** (*solid-state bonding,* Fig. 17.23a). As a result, the strength, density, ductility, and thermal and electrical conductivities of the compact increase. At the same time, however, the compact shrinks, thus allowances must be made for shrinkage, as are done in casting.

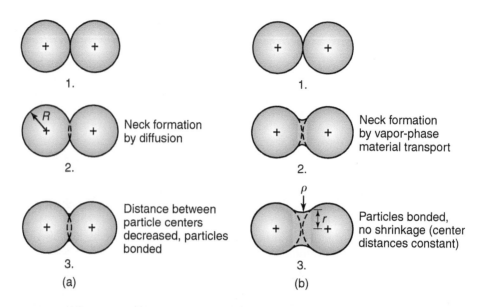

FIGURE 17.23 Schematic illustration of two mechanisms for sintering metal powders: (a) solid-state material transport and (b) vapor-phase material transport. R = particle radius, r = neck radius, and ρ = neck-profile radius.

A second sintering mechanism is **vapor-phase transport** (Fig. 17.23b). Because the material is heated to close to its melting temperature, metal atoms will be released to the vapor phase from the particles. At convergent geometries (the interface of two particles), the melting temperature is locally higher and the vapor phase resolidifies. Thus, the interface grows and strengthens while each particle shrinks as a whole.

If two adjacent particles are of different metals, alloying can take place at the interface of the two particles. If one of the particles has a lower melting point than the other, the particle will melt and, because of surface tension, surround the particle that has not melted (Fig. 17.24). An example of this mechanism, known as *liquid-phase sintering*, is cobalt in tungsten-carbide tools and dies (see Section 22.4), and stronger and denser parts can be obtained in this way. In *spark sintering*, loose metal powders are placed in a graphite mold, heated by electric current, subjected to a high-energy discharge, and compacted, all in one step. Another technique is *microwave sintering*, which reduces sintering time, thereby prevents grain growth, which can adversely affect strength.

Mechanical Properties.   Depending on temperature, time, and the processing history, different structures and porosities can be obtained in a sintered compact, and thus affect its properties. Porosity cannot be eliminated completely because (a) voids remain after compaction and (b) gases evolve during sintering. Porosity may consist either of a *network* of interconnected pores or of *closed holes*. Generally, if the density of the material is less than 80% of its bulk density, the pores are interconnected. Although porosity reduces the strength of the PM product, it is an important characteristic for making metal filters and bearings, and to allow for infiltration with liquid lubricants by surface tension.

Typical mechanical properties for several sintered PM alloys are given in Table 17.3. The differences in mechanical properties of wrought versus PM metals are given in Table 17.4. To further evaluate the differences between the properties of PM, wrought, and cast metals and alloys, compare these tables with the ones given in Parts I and II.

The effects of various manufacturing processes on the mechanical properties of a titanium alloy are shown in Table 17.5. Note that hot isostatic pressed titanium has properties that are similar to those for cast and forged titanium. It should be remembered, however, that unless they are precision forged, forgings generally require some additional machining or finishing operations that a PM component may not require.

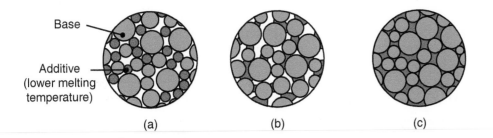

FIGURE 17.24 Schematic illustration of liquid phase sintering using a mixture of two powders. (a) Green compact of a higher melting point base metal and lower temperature additive; (b) liquid melting, wetting and reprecipitation on surfaces; and (c) fully sintered solid material.

TABLE 17.3

Mechanical Properties of Selected PM Materials

Material	Yield strength (MPa)	Ultimate tensile strength (MPa)	Elastic modulus (GPa)	Hardness	Elongation in 25 mm (%)	Density (g/cm^3)	Notes
Ferrous							
F-0008-20	170	200	85	35 HRB	<1	5.8	F-008 is often most
F-0008-35	260	390	140	70 HRB	1	7.0	cost effective
F-0008-55HT		450	115	22 HRC	<1	6.3	
F-0008-85HT		660	150	35 HRC	<1	7.1	
FC-0008-30	240	240	85	50 HRB	<1	5.8	Copper added for strength,
FC-0008-60	450	520	155	84 HRB	<1	7.2	hardness, and wear
FC-0008-95		720	150	43 HRC	<1	7.1	resistance
FN-0205-20	170	280	115	44 HRB	1	6.6	Good heat-treated strength,
FN-0205-35	280	480	170	78 HRB	5	7.4	impact energy
FN-0205-180HT		1280	170	78 HRB	<1	7.4	
FX-1005-40	340	530	160	82 HRB	4	7.3	Copper infiltrated steel
FX-1005-110HT		830	160	38 HRC	<1	7.3	Copper infiltrated steel
Stainless steel							
SS-303N1-38	310	470	115	70 HRB	5	6.9	Good machinability
SS-304N1-30	260	300	105	61 HRB	<1	6.4	High corrosion resistance
SS-316N1-25	230	280	105	59 HRB	<1	6.4	Good general-purpose alloy
SS-316N2-38	310	480	140	65 HRB	131	6.9	
Copper and copper alloys							
CZ-1000-9	70	120	80	65 HRH	9	7.6	General purpose structural parts
CZ-1000-11	80	160	100	80 HRH	12	8.1	General purpose structural parts
CZP-3002-14	110	220	90	88 HRH	16	8.0	High strength structural parts
CT-1000-13	110	150	60	82 HRH	4	7.2	Common self-lubricated bearing material
Aluminum alloys							
Ax 123-T1	200	270	—	47 HRB	3	2.7	General purpose
Ax 123-T6	390	400	—	72 HRB	<1	2.7	structural parts
Ax 231-T6	200	220	—	55 HRB	1	2.7	High wear resistance
Ax 231-T6	310	320	—	77 HRB	<1	2.7	
Ax 431-T6	270	300	—	55 HRB	5	2.8	High strength structural parts
Ax 431-T6	440	470	—	80 HRB	2	2.8	
Titanium alloys							
Ti-6Al-4V (HIP)	917	827	—	—	—	13	
Superalloys							
Stellite 19	—	1035	—	—	49 HRC	<1	

17.5 Secondary and Finishing Operations

In order to further improve the properties of sintered PM products or to impart special characteristics, several additional operations may be carried out following sintering:

1. **Coining** and **sizing** are compacting operations, performed under high pressure in presses. The purposes of these operations are to impart dimensional accuracy to the sintered part and to improve its strength and surface finish by further densification.

2. *Preformed and sintered* alloy-powder compacts subsequently may be cold or hot **forged** to the desired final shapes, and sometimes by *impact forging*. These

TABLE 17.4

Comparison of Mechanical Properties of Selected Wrought and Equivalent PM Metals (as Sintered)

Metal	Condition	Relative density[a] (%)	Ultimate tensile strength[a] (MPa)	Elongation in 50 mm (%)
Aluminum				
2014-T6	Wrought (W)	100	480	20
	PM	94	330	2
6061-T6	W	100	310	15
	PM	94	250	2
Copper, OFHC[b]	W, annealed	100	235	50
	PM	89	160	8
Brass, 260	W, annealed	100	300	65
	PM	89	255	26
Steel, 1025	W, hot rolled	100	590	25
	PM	84	235	2
Stainless steel, 303	W, annealed	100	620	50
	PM	82	360	2

Notes: [a] The density and strength of PM materials greatly increase with further processing, such as forging, isostatic pressing, and heat treatments. [b] OFHC = oxygen-free, high conductivity.

TABLE 17.5

Mechanical Property Comparisons for Ti-6AL-4V Titanium Alloy

Process*	Relative density (%)	Yield strength (MPa)	Ultimate tensile strength (MPa)	Elongation (%)	Reduction of area (%)
Cast	100	840	930	7	15
Cast and forged	100	875	965	14	40
Blended elemental (P + S)	98	786	875	8	14
Blended elemental (HIP)	>99	805	875	9	17
Prealloyed (HIP)	100	880	975	14	26
Electron-beam melting	100	910	970	16	—

*P + S = pressed and sintered, HIP = hot isostatically pressed. *Source:* Courtesy of R.M. German and Stratasys, Inc.

products have good surface finish and dimensional tolerances, and uniform and fine grain size. The superior properties obtained make this technology particularly suitable for such applications as highly stressed automotive and jet-engine components.

3. Powder metal parts may be subjected to other finishing operations, such as

- **Machining,** for producing various geometric features by milling, drilling, and tapping
- **Grinding,** for improving dimensional accuracy and surface finish
- **Plating,** for improving appearance and resistance to wear and corrosion
- **Heat treating,** for increasing hardness and strength

4. The inherent porosity of PM components can be utilized by **impregnating** them with a fluid. Bearings and bushings that are lubricated internally, with up to 30% oil by volume, are made by immersing the sintered bearing in heated oil.

Subsurface porosity

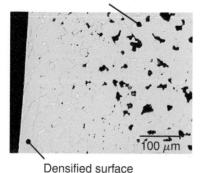

Densified surface

FIGURE 17.25 Micrograph of a PM material surface after roll densification; note the low porosity near the surface, increasing the materials ability to support contact stresses and resist fatigue. *Source:* Courtesy of Capstan Atlantic Corp.

The bearings have a continuous supply of lubricant during their service lives, due to capillary action (also referred to as *permanently lubricated*). Universal joints, for example, are made by means of grease-impregnated PM techniques, thus no longer requiring traditional grease fittings on the joints.

5. **Infiltration** is a process whereby a slug of a lower-melting-point metal is placed in contact with the sintered part. The assembly is then heated to a temperature sufficiently high to melt the slug. The molten metal infiltrates the pores by capillary action, producing a relatively pore-free part having good density and strength. The most common application is the infiltration of iron-based compacts by copper or bronze. The advantages of infiltration are that the hardness and tensile strength of the part are improved and the pores are filled, thus preventing moisture penetration, which could cause corrosion. Moreover, because some porosity is essential when an infiltrant is used, the part may be sintered only partially, thus resulting in lower thermal warpage.

6. **Electroplating** (Section 34.9) can be applied on PM parts, but special care is required to thoroughly remove the electrolytic fluid, since it presents health hazards. Under some conditions, electroplating can seal a part and eliminate its permeability.

7. **Densification**, or **roll densification**, is similar to roller burnishing (Section 34.2), where a small-diameter hard roll is pressed against a PM part, resulting in sufficiently high contact pressures to cause plastic deformation of its surface layers. Thus, instead of cold working the part, the effect is to cause an increase in density, or densification, of the surface layers (Fig. 17.25). Powder metallurgy gears and bearing races are generally treated by roll densification, since the surface layer is more fatigue resistant and better able to support higher contact stresses than untreated components.

17.6 Design Considerations

Because of the unique properties of metal powders, their flow characteristics in the die and the brittleness of green compacts, there are certain design principles that should be followed (Figs. 17.26–17.28):

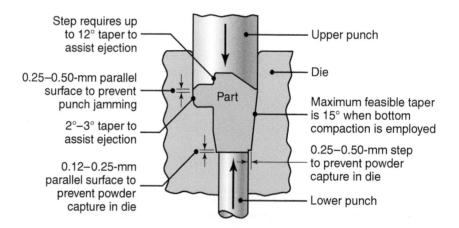

FIGURE 17.26 Die geometry and design features for powder metal compaction. *Source:* Reprinted with permission from Metal Powder Industries Federation, Princeton, NJ, USA.

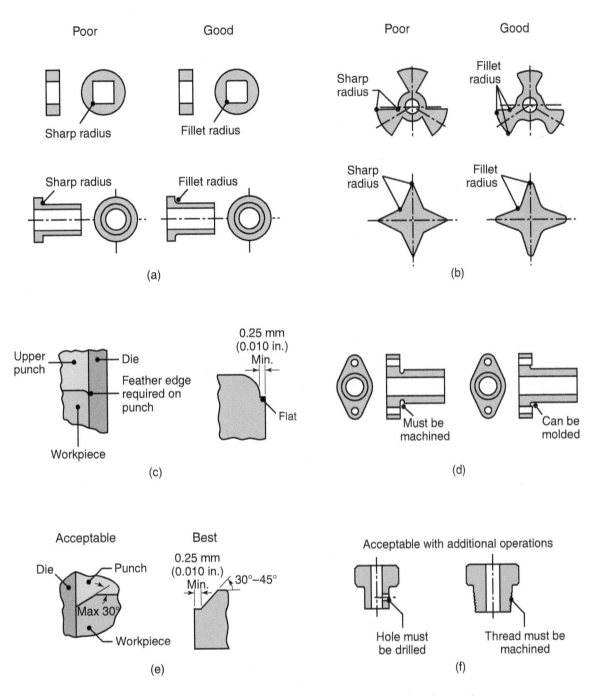

FIGURE 17.27 Examples of PM parts showing poor and good designs; note that sharp radii and reentry corners should be avoided and that threads and transverse holes have to be produced separately by additional machining operations. *Source:* Reprinted with permission from Metal Powder Industries Federation, Princeton, NJ, USA.

1. The shape of the compact must be kept as simple and uniform as possible. Sharp changes in contour, thin sections, variations in thickness, and high length-to-diameter ratios should be avoided.
2. Provision must be made for ejection of the green compact from the die without damaging the compact. Thus, holes or recesses should be parallel to the axis of

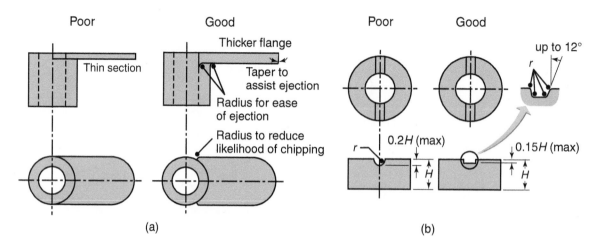

FIGURE 17.28   (a) Design features for use with unsupported flanges. (b) Design features for use with grooves. *Source:* Reprinted with permission from Metal Powder Industries Federation, Princeton, NJ, USA.

punch travel. Chamfers also should be provided to avoid damage to the edges during part ejection.

3. PM parts should be made with the widest acceptable dimensional tolerances, consistent with their intended applications, in order to increase tool and die life and reduce production costs.

4. Part walls generally should not be less than 1.5 mm (0.060 in.) thick; however, with special care, walls as thin as 0.34 mm (0.0135 in.) can be pressed successfully on components as little as 1 mm (0.04 in.) in length. Walls with length-to-thickness ratios greater than 8:1 are difficult to press, and density variations are virtually unavoidable.

5. Steps in parts can be produced if they are simple and their size doesn't exceed 15% of the overall part length. Larger steps can be pressed, but they require more complex, multiple-motion tooling.

6. Letters and numbers can be pressed if they are oriented perpendicular to the direction of pressing, and these can be raised or recessed. Raised letters are more susceptible to damage in the green stage, and also they prevent stacking during sintering.

7. Flanges or overhangs can be produced by a step in the die. However, long flanges can be broken during ejection, and may thus require more elaborate tooling. A long flange should incorporate a draft around the flange, a radius at the bottom edge, and a radius at the juncture of the flange and/or component body, in order to reduce stress concentrations and thus the likelihood of fracture.

8. A true radius cannot be pressed into the edge of a part because it would require the punch to be feathered (gently tapered) to a zero thickness, as shown in Fig. 17.27c. Chamfers or flats are preferred for pressing, and a 45° angle in a 0.25-mm (0.010-in.) flat is a common design practice.

9. Keys, keyways, and holes, used for transmitting torques on gears and pulleys, can be formed during powder compaction. Bosses (see Fig. 10.14c) also can be produced, provided that proper drafts are used, and their length is small compared to the overall component dimension.

10. Notches and grooves can be made if they are oriented perpendicular to the pressing direction. It is recommended that circular grooves not exceed a depth of 20% of the overall component, and rectangular grooves should not exceed 15%.

11. Parts produced by PIM have design constraints similar to those of parts produced by injection molding of polymers (Section 19.3). With PIM, wall thicknesses should be uniform, in order to minimize distortion during sintering. Also, molds should be designed with smooth transitions, to prevent powder accumulation and to allow uniform distribution of metal powder.

12. Dimensional tolerances of sintered PM parts are usually on the order of ± 0.05 to 0.1 mm (± 0.002–0.004 in.). Tolerances improve significantly with subsequent operations, such as sizing, machining, and grinding.

17.7 Economics of Powder Metallurgy

Because PM can produce parts at net or near-net shape, and thus eliminate many secondary manufacturing and assembly operations, it increasingly has become competitive with casting, forging, and machining. On the other hand, the high initial cost of punches, dies, and equipment for PM processing means that production volume must be sufficiently high to warrant this expenditure. Although there are exceptions, the process generally is economical for quantities over 10,000 pieces.

As in other metalworking operations, the cost of dies and tooling in PM depends on the part complexity and the method of processing the metal powders. Thus, tooling costs for processes such as HIP and PIM are higher than the more conventional powder processing. Because it is a near-net shape-manufacturing method, the cost of finishing operations in PM is low compared to other processes. However, if there are certain features to the part, such as threaded holes, undercuts, and transverse cavities and holes, then finishing costs will increase. Consequently, following design guidelines in PM to minimize or avoid such additional operations can be more important in this process than in others.

Equipment costs for conventional PM processing are somewhat similar to those for bulk deformation processing of metals, such as forging. However, the cost increases significantly when using methods such as HIP and PIM. Although the cost of materials has increased significantly (see Table 6.1), it has actually improved the economic viability of PM, since tooling and equipment costs are a smaller fraction of the total cost of production.

Labor costs for PM are not as high as those in other processes, primarily because the individual operations, such as powder blending, compaction, and sintering, are performed on highly automated equipment; thus, the skills required are not as high.

The near-net-shape capability of PM significantly reduces or eliminates scrap. Weight comparisons for aircraft components, produced by forging and by PM processes, are shown in Table 17.6. Note that the PM parts are subjected to further machining processes; thus, the final parts weigh less than those made by either of the two processes alone.

TABLE 17.6

Forged and PM Titanium Parts and Cost Savings				
	Weight (kg)			
Part	Forged billet	PM	Final part	Cost savings (%)
F-14 Fuselage brace	2.8	1.1	0.8	50
F-18 Engine mount support	7.7	2.5	0.5	20
F-18 Arrestor hook support fitting	79.4	25	12.9	25
F-14 Nacelle frame	143	82	24.2	50

CASE STUDY 17.3   Powder Metallurgy Parts in a Snowblower

Some of the parts in the freewheeling steering system of a commercial snowblower are shown in Fig. 17.29. Among the 16 PM components, the sprocket is the largest, at around 140 mm (5.5 in.) in diameter.

The final assembly incorporates a stamped steel frame, bronze and plastic bearings, and a wrought-steel axle, to produce a highly functional and low-cost machine. Unique features compatible with PM manufacturing were incorporated into the design of these parts to enhance their functionality.

The PM components in the assembly range from single-level parts with fixed features on punch faces and core rods, to intricate multilevel parts with complex die geometry, core rods, and transfer punches. These are unique features and they manage the powder for local density control. The clutch pawl, for example, is produced to a net-shape peripheral geometry that is not practical or economical with other manufacturing technologies. The material used is FLC4608-70 steel (a prealloyed powder of iron, with 1.9% NI, 0.56% Mo, and 0.8% C mixed in with 2% Cu), with a tensile strength of 70 ksi and a density of 6.8 g/cm^3 (0.23 lb/in^3).

Part numbers are pressed into the face of the components, as a simple means of identifying them. Two of the components are made with especially close tolerances: The pawl latch gear has a 0.15-mm (0.006-in.) tolerance on the pitch diameter (PD), with 0.11 mm (0.0045 in.) PD to ID run-out and 0.025 mm (0.001 in.) tolerance on the bore. The 32-tooth sprocket has a thin-walled 57.75 mm (2.274 in.) ID with a 0.05-mm (0.002-in.) tolerance. Both the pawl latch gear and the sprocket acquire a density of 6.7 g/cm^3 (0.21 lb/in^3) and a tensile strength of 690 MPa (100 ksi).

All components shown passed normal life-cycle testing and product-life testing, including shock loading by engaging the drive in reverse, while traveling at maximum forward speed down an incline. Clutch components, which were also subjected to salt-spray corrosion resistance, and proper operation in subzero temperatures, experienced no failures. No machining is required on these parts, as these are sufficiently net-shape components. The only additional operations, prior to final assembly, are vibratory deburring and honing of the 32-tooth sprocket, in order to produce a close-tolerance bore and surface finish. The clutch pawls, produced with sinter-hardened steel, are quenched in an atmosphere so that the porosity present can be filled with a lubricant, to provide lubricity at the interface of mating parts (see also Section 33.6).

Source: Reprinted with permission from Metal Powder Industries Federation, Princeton, NJ, USA.

FIGURE 17.29   Powder metallurgy parts in a commercial snowblower.

SUMMARY

- Powder metallurgy is a net-shape forming process consisting of metal powder production, blending, compaction in dies, and sintering in order to impart strength, hardness, and toughness. Although the size and the weight of PM products are limited, the process is capable of producing relatively complex parts economically, in net-shape form, to close dimensional tolerances, and from a wide variety of metal and alloy powders.

- Secondary and finishing operations may be performed on PM parts, to improve their dimensional accuracy, surface finish, mechanical and physical properties, and appearance. These operations include forging, heat treating, machining, grinding, plating, impregnation (as with oil), and infiltration (as with low-melting-point metals).

- Control of powder shape and quality, process variables, and sintering atmospheres are important considerations in product quality. Density and mechanical and physical properties can be controlled by tooling design and by adjusting the compacting pressure.

- An important PM process is powder-injection molding, which involves mixing the very fine metal powders with a polymer, to make them flow more easily into molds of complex shape.

- Design considerations for PM include the shape of the part, the ability to eject the green compact from the die, and the dimensional tolerances that are acceptable for the particular application.

- The PM process is suitable for medium- to high-volume production runs and for relatively small parts. It has some competitive advantages over other methods of production, such as casting, forging, and machining.

KEY TERMS

Atomization	Electrolytic deposition	Mechanical alloying	Screening
Blending	Green compact	Metal injection molding	Shape factor
Carbonyls	Green strength	Powder-injection molding	Sintering
Cold isostatic pressing	Hot isostatic pressing	Powder metallurgy	Spark sintering
Comminution	Impregnation	Pressing	
Compaction	Infiltration	Pressureless compaction	
Diffusion	Injection molding	Reduction	

BIBLIOGRAPHY

Allen, T., **Powder Sampling and Particle Size Determination,** Elsevier, 2003.

ASM Handbook, Vol. 7: **Powder Metal Technologies and Applications,** ASM International, 1998.

Atkinson, H.V., and Rickinson, B.A., **Hot Isostatic Pressing,** Adam Hilger, 1991.

Douvard, D. (ed.), **Powder Metallurgy,** ISTE Publishing, 2009.

Fayed, M., and Otten, L. (eds.), **Handbook of Powder Science and Technology,** 2nd ed., Chapman & Hall, 1997.

German, R.M., **A-Z of Powder Metallurgy,** Elsevier, 2007.

German, R.M., **Powder Metallurgy and Particulate Materials Processing,** Metal Powder Industry, 2006.

Kang, S.-J., **Sintering: Densification, Grain Growth and Microstructure,** Butterworth-Heinemann, 2005.

Klar, E., and Samal, P.K., **Powder Metallurgy Stainless Steels: Processing, Microstructures and Properties,** ASM International, 2008.

Leander, F., and West, W.G., **Fundamentals of Powder Metallurgy,** Metal Powder Industries Federation, 2002.

Mohan, T.R.R., and Ramakrishnan, P. (eds.), **Powder Metallurgy in Automotive Applications-II**, Science Publishers, 2002.

Neikov, O.D., Naboychenko, S., Moura chova, I.B., and Gopienko, (eds.), **Handbook of Non-Ferrous Metal Powders: Technologies and Applications**, Elsevier, 2009.

Powder Metallurgy Design Manual, 2nd ed., Metal Powder Industries Federation, 1995.

Selesca, S.M., Selesca, M., and Danninger, S., **Machinability of Powder Metallurgy Steels**, Cambridge International Science, 2005.

Upadhyaya, A., and Upadhyaya, G.S., **Powder Metallurgy: Science, Materials, and Technology**, Universities Press, 2011.

Upadhyaya, G.S., **Sintered Metallic and Ceramic Materials: Preparation, Properties and Applications**, Wiley, 2000.

REVIEW QUESTIONS

17.1 Describe briefly the production steps involved in making powder metallurgy parts.

17.2 Name the various methods of powder production and explain the types of powders produced.

17.3 Explain why metal powders may be blended.

17.4 Describe the methods used in metal powder compaction.

17.5 What is isostatic pressing? How is it different from pseudo-isostatic pressing?

17.6 What hazards are involved in PM processing? Explain their causes.

17.7 Describe what occurs to metal powders during sintering.

17.8 Describe the wet-bag and dry-bag techniques.

17.9 Why might secondary and finishing operations be performed on PM parts?

17.10 Explain the difference between impregnation and infiltration. Give some applications of each.

17.11 What is roll densification? Why is it done?

17.12 What is mechanical alloying? What are its advantages over the conventional alloying of metals?

17.13 What is the Osprey process?

17.14 What is screening of metal powders? Why is it done?

17.15 Why are protective atmospheres necessary in sintering? What would be the effects on the properties of PM parts if such atmospheres were not used?

QUALITATIVE PROBLEMS

17.16 Why is there density variation in the compacting of powders? How is it reduced?

17.17 What is the magnitude of the stresses and forces involved in powder compaction?

17.18 List the similarities and differences of forging and compacting metal powders.

17.19 Give some reasons that powder-injection molding is an important process.

17.20 How does the equipment used for powder compaction vary from those used in other metalworking operations in the preceding chapters?

17.21 Explain why the mechanical and physical properties of PM parts depend on their density.

17.22 What are the effects of the different shapes and sizes of metal particles in PM processing?

17.23 Describe the relative advantages and limitations of cold and hot isostatic pressing.

17.24 How different, if any, are the requirements for punch and die materials in powder metallurgy from those for forging and extrusion operations? Explain.

17.25 The powder metallurgy process can be competitive with processes such as casting and forging. Explain why this is so.

17.26 What are the reasons for the shapes of the curves shown in Fig. 17.11 and for their relative positions on the charts?

17.27 Should green compacts be brought up to the sintering temperature slowly or rapidly? Explain your reasoning.

17.28 Because they undergo special processing, metal powders are more expensive than the same metals in bulk form, especially powders used in powder-injection molding. How is the additional cost justified in processing powder metallurgy parts?

17.29 In Fig. 17.12, it can be seen that the pressure is not uniform across the diameter of the compact at a particular distance from the punch. What is the reason for this variation?

17.30 Why do the compacting pressure and the sintering temperature depend on the type of powder metal?

17.31 What will be stronger: a blend of stainless steel and copper powder that is compacted and sintered, or a stainless steel powder that is compacted, sintered, and infiltrated by copper? Explain.

17.32 Name the various methods of powder production and sketch the morphology of the powders produced.

QUANTITATIVE PROBLEMS

17.33 Estimate the maximum tonnage required to compact a brass slug 3.0 in. in diameter. Would the height of the slug make any difference in your answer? Explain your reasoning.

17.34 Refer to Fig. 17.11a. What should be the volume of loose, fine iron powder in order to make a solid cylindrical compact 25 mm in diameter and 20 mm high?

17.35 Determine the shape factors for (a) a cylinder with a dimensional ratio of 1:1:1 and (b) a flake with a ratio of 1:12:12.

17.36 Estimate the number of particles in a 500-g sample of iron powder if the particle size is 50 μm.

17.37 Assume that the surface of a copper particle is covered by an oxide layer 0.1 mm in thickness. What is the volume (and the percentage of volume) occupied by this layer if the copper particle itself is 75 μm in diameter?

17.38 A coarse copper powder is compacted in a mechanical press at a pressure of 20 tons/in^2. During sintering, the green part shrinks an additional 8%. What is the final density?

17.39 A gear is to be manufactured from iron powders. It is desired that it have a final density 90% that of cast iron, and it is known that the shrinkage in sintering will be approximately 5%. For a gear that is 2.5 in. in diameter and has a 0.75-in. hub, what is the required press force?

17.40 What volume of powder is needed to make the gear in Problem 11.39?

17.41 The axisymmetric part shown in Fig. P17.41 is to be produced from fine copper powder and is to have a tensile strength of 175 GPa. Determine the compacting pressure and the initial volume of powder needed.

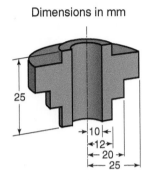

Dimensions in mm

FIGURE P17.41

17.42 Assume that you are an instructor covering the topics described in this chapter and you are giving a quiz on the numerical aspects to test the understanding of the students. Prepare two quantitative problems and supply the answers.

SYNTHESIS, DESIGN, AND PROJECTS

17.43 Prepare an illustration similar to Fig. 13.1, showing the variety of PM manufacturing options.

17.44 Make sketches of PM products in which density variations (see Fig. 17.12) would be desirable. Explain why in terms of the functions of these parts.

17.45 Compare the design considerations for PM products with those for (a) casting and (b) forging. Describe your observations.

17.46 Are there applications in which you, as a manufacturing engineer, would not recommend a PM product? Explain.

17.47 Describe other methods of manufacturing the parts shown in Fig. 17.1.

17.48 Using the Internet, locate suppliers of metal powders and compare the cost of the powder with the cost of ingots for five different materials.

17.49 Explain why powder metal parts are commonly used for machine elements requiring good frictional and wear characteristics and for mass-produced parts.

17.50 It was stated that powder-injection molding competes well with investment casting and small forgings for various materials, but not with zinc and aluminum die castings. Explain why.

17.51 Describe how the information given in Fig. 17.15 would be helpful in designing PM parts.

17.52 It was stated that, in the process shown in Fig. 17.21, shapes produced are limited to axisymmetric parts. Do you think it would be possible to produce other shapes as well? Describe how you would modify the design of the setup to produce other shapes, and explain the difficulties that may be encountered.

17.53 It has been noted that PM gears are very common for low-cost office equipment, such as the carriage mechanism of inkjet printers. Review the design requirements of these gears and list the advantages of PM manufacturing approaches for these gears.

17.54 The axisymmetric parts shown in Fig. P17.54 are to be produced through PM. Describe the design changes that you would recommend.

17.55 Assume you are working in technical sales. What applications currently using non-PM parts would you attempt to develop? What would you say to your potential customers during your sales visits? What kind of questions do you think they would ask?

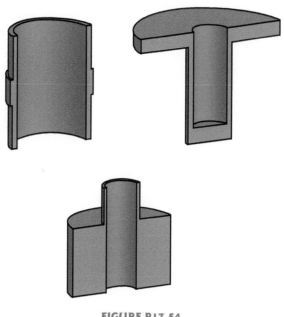

FIGURE P17.54

Ceramics, Glasses, and Superconductors: Processing and Equipment

- This chapter presents the manufacturing processes associated with ceramics, glass, and superconductors.
- It first describes the preparation of ceramic powders, followed by operations that produce discrete parts through the basic processes of casting, pressing, extrusion, and molding.
- Drying and firing, followed by finishing operations for ceramics, also are discussed.
- Glass manufacture involves production of continuous shapes, such as plate, tube, and bars, through drawing, rolling, or floating methods; the operations for discrete products, typically consist of molding, blowing, or pressing.
- The chapter ends with the processing of superconductors, which are produced mainly through the oxide-powder-in-tube process.

Typical products made: Ceramics: electrical insulators, rotors for gas turbines, lightweight components for high-speed machines, ball and roller bearings, seals, furnace components, ovenware, and tiles. Glass: glazing, laminated glass, bulletproof glass, bulbs, lenses, bottles, glass fibers, rods, and tubing. Superconductors: MRI magnets.

Alternative processes: Casting, forging, powder injection molding, blow molding, and rapid prototyping.

18.1 Introduction

The properties and applications of ceramics and glasses are described in Chapter 8. These materials have important characteristics, such as high-temperature strength and hardness, low electrical and thermal conductivity, chemical inertness, and resistance to wear and corrosion. The wide range of applications for these materials (Fig. 18.1) include parts such as electrical insulators, spark plugs, ball bearings, floor tiles, and dishes.

The processing methods employed for ceramics consist of (a) crushing the raw materials, (b) shaping them by various means, (c) drying and firing, and (d) finishing operations, as needed, to impart the required dimensional tolerances and surface finish. For glasses, the processes involve (a) mixing and melting the raw materials in a furnace and (b) shaping them in molds using various techniques, depending on the

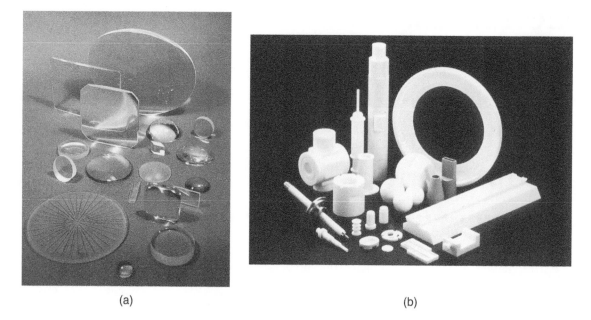

(a)   (b)

FIGURE 18.1 Examples of (a) glass parts and (b) ceramic parts. *Source:* (a) Courtesy of Commercial Optical Manufacturing, Inc. (b) Courtesy of Kyocera.

shape and size of the part. Both discrete products, such as bottles, and continuous products, such as flat glass, rods, tubing, and fibers, can be produced. Glasses can be strengthened by thermal or chemical means, as well as by laminating them with polymer sheets, as is done with windshields and bulletproof glass.

18.2 Shaping Ceramics

Several techniques are available for processing ceramics into useful products (Table 18.1), depending on the type of ceramics involved and their shapes. Production of some ceramic parts, such as pottery, ovenware, or floor tiles, generally does not

TABLE 18.1

General Characteristics of Ceramics Processing

Process	Advantages	Limitations
Slip casting	Large parts, complex shapes, low equipment cost	Low production rate, limited dimensional accuracy
Extrusion	Hollow shapes and small diameters, high production rate	Parts have constant cross-section, limited thickness
Dry pressing	Close tolerances, high production rates (with automation)	Density variation in parts with high length-to-diameter ratios, dies require abrasive-wear resistance, equipment can be costly
Wet pressing	Complex shapes, high production rate	Limited part size and dimensional accuracy, tooling costs can be high
Hot pressing	Strong, high-density parts	Protective atmospheres required, die life can be short
Isostatic pressing	Uniform density distribution	Equipment can be costly
Jiggering	High production rate with automation, low tooling cost	Limited to axisymmetric parts, limited dimensional accuracy
Injection molding	Complex shapes, high production rate	Tooling can be costly

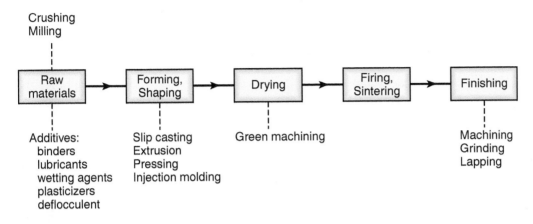

FIGURE 18.2 Processing steps involved in making ceramic parts.

involve the same level of control of materials and processes as do high-tech parts made of such structural ceramics as silicon nitride and silicon carbide and cutting tools made, for example, of aluminum oxide. Generally, the procedure is similar and involves the following steps (Fig. 18.2):

1. Crushing or grinding the raw materials into very fine particles
2. Mixing them with additives to impart certain desirable characteristics
3. Shaping, drying, and firing the material

The first step in processing ceramics is *crushing*, also called *comminution* or *milling*, of the raw materials. Crushing is generally done in a *ball mill* (see Fig. 17.6b), either dry or wet. Wet crushing is more effective, because it keeps the particles together, and also prevents the fine particles from contaminating the environment. The particles then may be *sized*, by passing them through a sieve, and then filtered and washed.

The ground particles are then mixed with *additives*, such as one or more of the following:

- **Binder,** for holding ceramic particles together
- **Lubricant,** to reduce internal friction between particles during molding and to help remove the part from the mold
- **Wetting agent,** to improve mixing
- **Plasticizer,** to make the mix more plastic and easy to shape
- **Agents,** to control foaming and sintering
- **Deflocculent,** to make the ceramic–water suspension more uniform, by changing the electrical charges on the particles of clay, so that the particles repel rather than attract each other; water is added to make the mixture more pourable and thus less viscous; typical deflocculents are Na_2CO_3 and Na_2SiO_3, in amounts of less than 1%

The three basic shaping processes for ceramics are casting, plastic forming, and pressing. The parts made also may be subjected to additional processing, such as machining and grinding, for better control of their dimensions and surface finish.

18.2.1 Casting

The most common casting process is **slip casting**, also called **drain casting**, as illustrated in Fig. 18.3. A **slip** consists of ceramic particles suspended in an immiscible liquid, generally water. The slip is poured into a porous mold, typically made

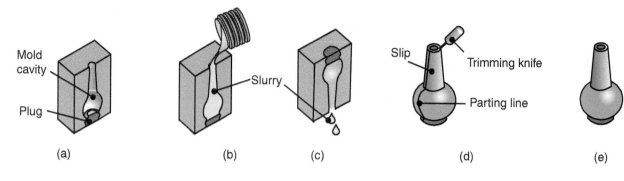

FIGURE 18.3 Sequence of operations in slip casting a ceramic part. (a) Mold is assembled and plug attached; some plugs incorporate draining features; (b) slurry, mixed from ceramic particles, binder, and water, is poured into the mold; (c) the mold is inverted and the slurry is poured from the mold, leaving a thin coating over the mold cavity; (d) after an initial drying period, the slip is removed from the mold, and features such as parting lines and sprue lips are removed; and (e) the slip is ready to be dried and fired in an oven, to develop strength and hardness.

of plaster of paris, and may consist of several components, as in other shaping processes.

The slip must have sufficient fluidity and low viscosity for it to flow easily into the mold, much like the importance of fluidity of molten metals in casting operations, described in Section 10.3. Pouring the slip must be done properly to avoid air entrapment, which can be a significant problem during casting.

After the mold has absorbed some of the water from the outer layers of the suspension, it is inverted and the remaining suspension is poured out. The product is a hollow object, as in the slush casting of metals, described in Section 11.4.3. The top of the part is then trimmed (note the trimming knife in Fig. 18.3d), the mold is opened, and the part is removed.

Large and complex parts, such as plumbing ware or art objects, can be made by slip casting. Although mold and equipment costs are low, dimensional control is poor and the production rate is low. In some applications, components of the product, such as handles for cups and pitchers, are made separately and then joined, using the slip as an adhesive. Molds also may consist of multiple components.

For solid-ceramic parts, the slip is supplied continuously into the mold to replenish the absorbed water, as otherwise the part will shrink. At this stage, the part is described as either a soft solid or semirigid. The higher the concentration of solids in the slip, the less water has to be removed. The part removed from the mold is referred to as a *green part*, as in powder metallurgy.

While the parts are still green, they may be machined to produce certain features or for better dimensional accuracy. Because of the delicate nature of the green compacts, however, machining usually is done manually with simple tools. For example, the flashing in a slip casting may be removed gently with a fine wire brush, or holes can be drilled in the mold.

Doctor-blade and Other Processes. Thin sheets of ceramics, less than 1.5 mm (0.06 in.) thick, can be made by the *doctor-blade process* (Fig. 18.4). The slip is cast over a moving plastic belt, while its thickness is being controlled by a blade. Ceramic sheets also may be produced by such methods as (a) *rolling* the slip between pairs of rolls and (b) *casting* the slip over a paper tape, which subsequently burns off during firing.

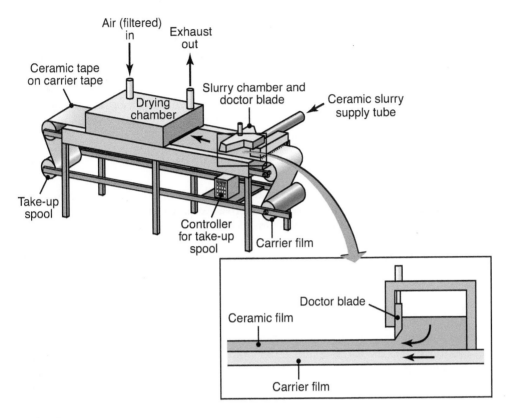

FIGURE 18.4   Production of ceramic sheets through the doctor-blade process.

18.2.2   Plastic Forming

Plastic forming, also called *soft*, *wet*, or *hydroplastic forming*, can be carried out by various methods, such as extrusion, injection molding, or molding, and by *jiggering* (Fig. 18.5). Plastic forming tends to orient the layered structure of clay along the direction of material flow, and thus tends to cause anisotropic behavior

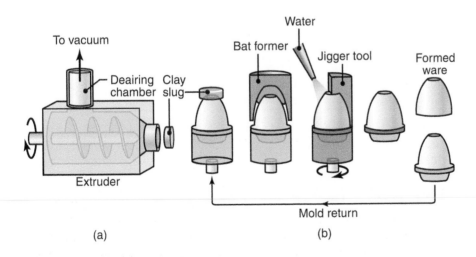

FIGURE 18.5   (a) Extruding and (b) jiggering operations. *Source:* After R.F. Stoops.

of the material, both in subsequent processing and in the final properties of the ceramic product.

In *extrusion*, the clay mixture (containing 20–30% water) is forced through a die opening, by a screw type equipment (see, for example, Fig. 19.2). In this process, the cross-section of the extruded product is constant, and there are limitations to wall thickness for hollow extrusions, because it may fracture during firing (Section 18.2.4). The products may be subjected to additional shaping operations. Tooling costs are low and production rates are high.

18.2.3   Pressing

Dry Pressing.   This is a technique similar to powder metal compaction, as described in Section 17.3. *Dry pressing* is used for relatively simple shapes, such as whiteware, refractories for furnaces, and abrasive products. The moisture content of the mixture generally is below 4%, but it may be as high as 12%. Organic and inorganic binders (such as stearic acid, wax, starch, and polyvinyl alcohol) are usually added to the mixture; these additives give strength and also act as lubricants to aide compaction. Dry pressing has the same high production rates and close control of dimensional accuracy as does powder metallurgy.

The pressing pressure ranges from 35 to 200 MPa (5–30 ksi). Density can vary significantly in dry-pressed ceramics, as in PM compaction (see Fig. 17.12), because of friction among the particles and at the mold walls. Density variations cause warping during firing, which is particularly severe for parts having high length-to-diameter ratios, the recommended maximum ratio being 2:1. Several methods may be used to minimize density variations, including (a) proper design of tooling, (b) vibratory pressing and impact forming, particularly for nuclear-reactor fuel elements, and (c) isostatic pressing. Modern presses for dry pressing are highly automated. The dies, usually made of carbides or hardened steel, must have high wear resistance in order to withstand the abrasive ceramic particles.

Wet Pressing.   In *wet pressing*, the part is formed in a mold while under high pressure in a hydraulic or mechanical press. Moisture content usually ranges from 10 to 15%. Production rates are high; however, (a) part size is limited, (b) dimensional control is difficult to achieve because of shrinkage during drying, and (c) tooling costs can be high. Wet pressing is generally used for making parts with intricate shapes, such as filters and electronic packaging.

Isostatic Pressing.   This process is used for ceramics in order to obtain a uniform density distribution throughout the part during compaction (see Section 17.3.2). The white insulators for automotive spark-plugs, for example, are made by this method at room temperature. Silicon-nitride vanes for high-temperature applications (see Fig. 8.1) are made by *hot isostatic pressing*.

Jiggering.   Ceramic dinner plates are made by a series of steps (Fig. 18.5). First, clay slugs are extruded and formed into a bat over a plaster mold; they are then jiggered on a rotating mold. *Jiggering* is a motion in which the clay bat is formed by means of templates or rollers. The part is then dried and fired. The jiggering process is confined to axisymmetric parts, and has limited dimensional accuracy.

Injection Molding.   *Injection molding* is used extensively for precision forming of ceramics, in high-technology applications such as for rocket-engine components. The raw material is mixed with a binder, such as a thermoplastic polymer (polypropylene, low-density polyethylene, or ethylene vinyl acetate) or wax, and injection molded.

The binder is usually removed by pyrolysis (inducing chemical changes by heat), and the part is fired.

The injection-molding process can produce thin sections, typically less than 10 to 15 mm (0.4–0.6 in.) thick, from most engineering ceramics, such as alumina, zirconia, silicon nitride, silicon carbide, and sialon (see Chapter 8). Thicker sections require careful control of the materials used and of the processing parameters, in order to avoid such defects as internal voids and cracks.

Hot Pressing.   In this process, also called *pressure sintering*, pressure and heat are applied simultaneously, thereby reducing porosity in the part, thus making it denser and stronger. Graphite is commonly used as a punch and die material, and protective atmospheres usually are employed during pressing.

Hot isostatic pressing also may be used, particularly to improve shape accuracy and the quality of high-technology ceramics, such as silicon carbide and silicon nitride. Glass-encapsulated HIP processing has been shown to be effective for this purpose.

18.2.4   Drying and Firing

The next step in ceramic processing is to dry and fire the part to give it the proper strength and hardness. *Drying* is a critical stage, because of the tendency for the part to warp or crack from variations in its moisture content and thickness. Control of atmospheric humidity and ambient temperature is important in order to reduce warping and cracking.

Loss of moisture during drying causes shrinkage of the part, by as much as 20% from the original, moist size (Fig. 18.6). In a humid environment, the evaporation rate is low, and thus the moisture gradient across the thickness of the part is lower than that in a dry environment. The low moisture gradient prevents a large, uneven gradient in shrinkage from the surface to the interior, reducing the tendency for excessive warping or cracking.

A ceramic part that has been shaped by any of the methods described thus far is in the green state. It can be machined in order to bring it closer to a near-net shape. Although the green part should be handled carefully, machining it is not particularly difficult, because of the relative softness of the materials.

Firing, also called **sintering**, involves heating the part to an elevated temperature in a controlled environment. Although some shrinkage occurs during firing, the ceramic part becomes stronger and harder. This improvement in mechanical properties is due to (a) the development of strong bonds among the complex oxide particles in the ceramic body and (b) reduced porosity. A more recent technology is **microwave sintering** of ceramics, conducted in furnaces with generators producing microwaves with frequencies in excess of 2 GHz. Microwave sintering can be significantly faster and less expensive than conventional sintering, and requires less energy per part.

Nanophase ceramics, described in Section 8.2.5, can be sintered at lower temperatures than those used for conventional ceramics. They are easier to fabricate, because they can be (a) compacted at room temperature to high densities, (b) hot pressed to theoretical density, and (c) formed into net-shaped parts without using binders or sintering aids.

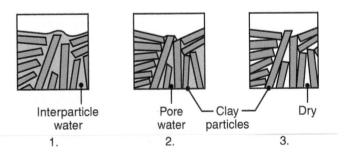

1.   Interparticle water
2.   Pore water — Clay particles
3.   Dry

FIGURE 18.6   Shrinkage of wet clay caused by the removal of water during drying; shrinkage may be as much as 20% by volume. *Source:* After F.H. Norton.

EXAMPLE 18.1   Dimensional Changes during the Shaping of Ceramic Components

Given: A solid, cylindrical ceramic part is to be made, with a final length, L, of 20 mm. For this material, it has been established that linear shrinkages during drying and firing are 7 and 6%, respectively, based on the dried dimension, L_d.

Find: Calculate (a) the initial length, L_o, of the part and (b) the dried porosity, P_d, if the porosity of the fired part, P_f, is 3%.

Solution:

a. On the basis of the information given and noting that firing is preceded by drying, we can write

$$\frac{(L_d - L)}{L_d} = 0.06,$$

or

$$L = (1 - 0.06)\, L_d.$$

Hence,

$$L_d = \frac{20}{0.94} = 21.28 \text{ mm}$$

and

$$L_o(1 + 0.07)L_d = (1.07)(21.28) = 22.77 \text{ mm}.$$

b. Since the final porosity is 3%, the actual volume, V_a, of the solid material in the part is

$$V_a = (1 - 0.03)\, V_f = 0.97 V_f,$$

where V_f is the volume of the part after firing. Because the linear shrinkage during firing is 6%, we can determine the dried volume, V_d, of the part as

$$V_d = \frac{V_f}{(1 - 0.06)^3} = 1.2 V_f.$$

Hence,

$$\frac{V_a}{V_d} = \frac{0.97}{1.2}, \text{ or } 81\%.$$

Therefore, the porosity, P_d, of the dried part is 19%.

18.2.5   Finishing Operations

Because firing causes dimensional changes, additional operations may be performed to (a) give the ceramic part its final shape, (b) remove any surface flaws, and (c) improve its surface finish and dimensional accuracy. Although they are hard and brittle, major advances have been made in producing **machinable ceramics** and **grindable ceramics**, thus enabling the production of ceramic components with high dimensional accuracy and good surface finish. An example is silicon carbide, which can be machined into final shapes from sintered blanks.

The finishing processes employed can be one or more of the following operations, described in detail in various sections in Part IV:

1. *Grinding*, using a diamond wheel
2. *Lapping* and *honing*
3. *Ultrasonic machining*
4. *Drilling*, using a diamond-coated drill
5. *Electrical-discharge machining*
6. *Laser-beam machining*
7. *Abrasive water-jet cutting*
8. *Tumbling*, to remove sharp edges and grinding marks

Process selection is an important consideration because of the brittle nature of most ceramics and the additional costs involved in using some of these processes. The effect of the finishing operation on the properties of the product also must be considered. For instance, because of notch sensitivity, the finer the surface finish of the part, the higher its strength and load-carrying capacity, particularly its fatigue strength. Ceramic parts also can undergo *static fatigue*, as described for glass in Section 18.5.

To improve their appearance and strength and to make them impermeable, ceramic products often are coated with a **glaze** or **enamel** (Section 34.12), which forms a glassy coating after firing.

18.3 Forming and Shaping of Glass

Glass is processed by melting and shaping it, either in molds or by blowing. The shapes produced include flat sheets and plates, rods, tubing, glass fibers, and discrete products, such as bottles, light bulbs, headlights, lenses, and cookware. Glass products may be as thick as those for large telescope mirrors, and as thin as those for holiday tree ornaments. The strength of glass can be improved by thermal and chemical treatments, which induce compressive surface residual stresses, or by laminating it with a thin sheet of tough plastic.

Glass products generally can be categorized as

1. **Flat sheets** or **plates**, ranging in thickness from about 0.8 to 10 mm (0.03–0.4 in.), and used as window glass, glass doors, and tabletops
2. **Rods** and **tubing**, used for neon lights, decorative artifacts, and for processing and handling chemicals
3. **Discrete products**, such as bottles, vases, and eyeglasses
4. **Glass fibers**, as reinforcements in composite materials (Section 9.2.1) and for use in fiber optics

All forming and shaping processes begin with molten glass, at a temperature typically in the range of 1000°–1200°C (1830°–2200°F), and has the appearance of a red-hot, viscous liquid.

18.3.1 Flat-sheet and Plate Glass

Flat-sheet glass can be made by any of the following three methods from the molten state, with glass supplied from a melting furnace or tank:

1. In the **float method** (Fig. 18.7), molten glass from the furnace is fed into a long bath, in which the glass, under a controlled atmosphere and at a temperature of 1150°C (2100°F), floats over a bath of molten tin. The glass then moves at a temperature of about 650°C (1200°F) over rollers into another chamber (*lehr*), where it solidifies. *Float glass* has smooth (*fire-polished*) surfaces, thus further finishing operations, such as grinding or polishing, are not necessary. The width can be as much as 4 m (13 ft). Both thin and plate glass are made by this process.

2. The **drawing** process for making flat sheets or plates involves passing the molten glass through a pair of rolls (Fig. 18.8a). The solidifying glass is squeezed between these two rolls, forming it into a sheet; it then moves forward over a set of smaller rolls.

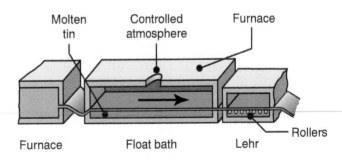

FIGURE 18.7 The float method of forming sheet glass.

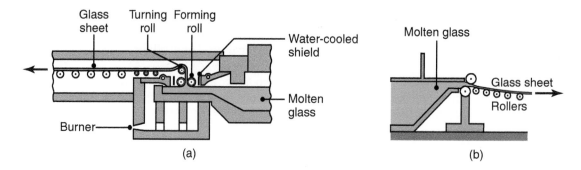

FIGURE 18.8 (a) Drawing process for drawing sheet glass from a molten bath. (b) Rolling process. *Source: After W.D. Kingery.*

3. In the **rolling** process (Fig. 18.8b), the molten glass is squeezed between powered rollers, thereby forming a sheet, with a surface that is somewhat rough. The surfaces of the glass may also be embossed with a pattern, using textured roller surfaces. In this way, the glass surface becomes a replica of the roll surface.

18.3.2 Tubing and Rods

Glass *tubing* is produced by the process shown in Fig. 18.9. Molten glass is wrapped around a rotating (cylindrical or cone-shaped) hollow mandrel, and is drawn out by a set of rolls. Air is blown through the mandrel to prevent the glass tube from collapsing. The machines may be horizontal, vertical, or slanted downward. This method is also used in making glass tubes for fluorescent bulbs.

An alternative method for making tubes is by *extruding* a strip of glass (with a thin rectangular cross-section), which is then wrapped obliquely (at an angle) around a rotating mandrel. The molten glass strips bond together along the edges, forming a continuous tube, which is then drawn off the mandrel in a continuous manner.

Glass *rods* are extruded or drawn directly from a molten bath, without the internal pressurization needed to make tubing.

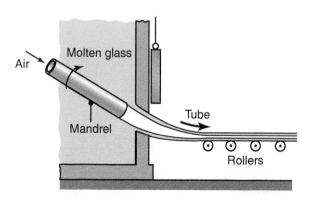

FIGURE 18.9 Manufacturing process for glass tubing; air is blown through the mandrel to keep the tube from collapsing; glass tubes for fluorescent bulbs are made by this method.

18.3.3 Discrete Glass Products

Blowing. Hollow and thin-walled glass items, such as bottles, vases, and flasks, are made by *blowing*, a process that is similar to blow molding of thermoplastics (Section 19.4). The steps involved in the production of an ordinary glass bottle by the blowing process are shown in Fig. 18.10. Blown air expands a hollow gob of heated glass against the inner walls of the mold. The mold surfaces are usually coated with a parting agent, such as oil or emulsion, to prevent the glass from sticking to the mold. Blowing may be followed by a second blowing operation to finalize the product shape, called the *blow and blow* process.

The surface finish of parts made by blowing is acceptable for most applications such as bottles and jars. It is difficult to precisely control the wall thickness of the product, because of the lack of an inner mold, but this process is economical for high-rate production.

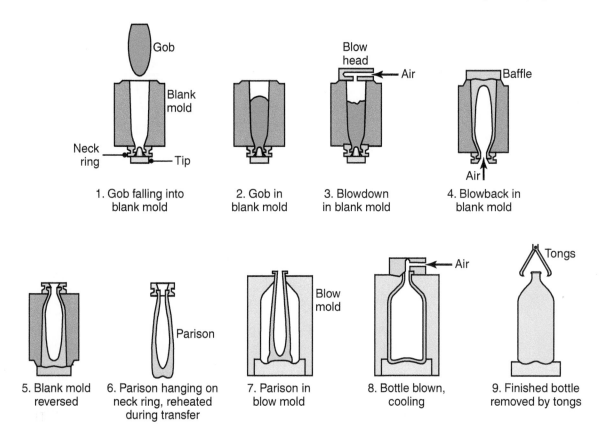

FIGURE 18.10   Steps in manufacturing an ordinary glass bottle. *Source:* After F.H. Norton.

Pressing.   In the *pressing* method, a gob of molten glass is placed into a mold, and is pressed into a confined cavity with a plunger, thus the process is similar to closed-die forging. The mold may be made in one piece, such as that shown in Fig. 18.11, or it may be a split mold (Fig. 18.12). After pressing, the solidifying glass acquires the shape of the mold-plunger cavity. Because of the confined environment, the product has better dimensional accuracy than can be obtained with blowing.

Pressing in one-piece molds cannot be used for (a) shapes of parts from which the plunger cannot be retracted or (b) thin-walled items, because of high force requirements and distortion upon part removal. For example, split molds are used for bottles, while, for thin-walled items, pressing can be combined with blowing. In the latter process, known as *press and blow*, the pressed part is subjected to air pressure (hence the term blow), which further expands the molten glass into the mold.

QR Code 18.1 Demonstration of glass bottle production. (*Source:* Courtesy of Owens-Illinois, Inc.)

| 1. Empty mold | 2. Loaded mold | 3. Glass pressed | 4. Finished part |

FIGURE 18.11   Manufacturing a glass item by pressing molten glass into a mold. *Source:* Based on data from Corning Glass Works.

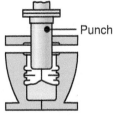

Punch

1. Empty mold   2. Loaded mold   3. Glass pressed   4. Finished part

FIGURE 18.12   Pressing molten glass into a split mold. *Source:* After E.B. Shand.

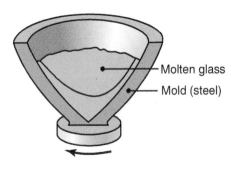

Molten glass
Mold (steel)

FIGURE 18.13 Centrifugal casting of glass; large telescope lenses and television-tube funnels are made by this process. *Source:* Based on data from Corning Glass Works.

Centrifugal Casting. Also known, in the glass industry, as **spinning** (Fig. 18.13), this process is similar to that used for metals (see Section 11.4.6). The centrifugal force pushes the molten glass against the mold wall, where it begins to solidify. Typical products made are large lenses for research telescopes and architectural shapes.

Sagging. Shallow dish-shaped or lightly embossed glass parts can be made by the *sagging* process. A sheet of hot glass is placed over a mold and heated, whereby the glass sags by its own weight and takes the shape of the mold. The process is similar to the thermoforming of thermoplastics (Section 19.6), but no pressure or vacuum is involved. Typical parts made are dishes, sunglass lenses, mirrors for telescopes, and lighting panels.

Glass Ceramics Manufacture. Glass ceramics (trade names: *Pyroceram, Corningware*) contain large proportions of several oxides, as noted in Section 8.5. Their manufacture involves a combination of the methods used for ceramics and glasses. Glass ceramics are shaped into discrete products, such as dishes and baking pans, and then heat treated, whereby glass becomes *devitrified* (recrystallized).

18.3.4   Glass Fibers

Continuous glass fibers are drawn through multiple orifices (200–400 holes) in heated platinum plates, at speeds as high as 500 m/s (1700 ft/s). Fibers as small as 2 μm (80 μin.) in diameter can be produced by this method. In order to protect their surfaces, fibers are subsequently coated with chemicals known as *sizing*, which are mainly silane compounds in water, but many sizing blends are possible. Short fibers (*chopped*) are produced by subjecting long fibers to compressed air or steam as they leave the orifice.

  Glass wool (short glass fibers), used as a thermal insulating material and for acoustic insulation, is made by a *centrifugal spraying process*, in which molten glass is ejected (*spun*) from a rotating head. The diameter of the fibers typically ranges from 20 to 30 μm (800–1200 μin.).

18.4   Techniques for Strengthening and Annealing Glass

Glass can be strengthened by a number of processes, and discrete glass products may be subjected to annealing and other finishing operations to impart desired properties and surface characteristics.

Thermal Tempering. In this process, also called *physical tempering* or *chill tempering*, the surfaces of the hot glass are cooled rapidly by a blast of air (Fig. 18.14). As a result, the surfaces shrink and, at first, tensile stresses develop on the surfaces. The bulk of the glass then begins to cool, and because it contracts, the already solidified surfaces of the glass also are forced to contract. Consequently, residual compressive surface stresses develop on the surfaces, while the interior develops tensile stresses (see also Section 2.11). Compressive surface stresses improve the strength of the glass in the same way that they do in metals and other materials.

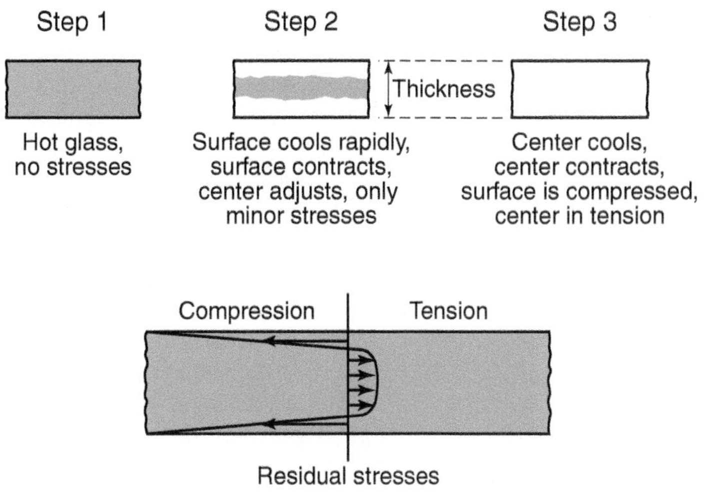

The higher the coefficient of thermal expansion of the glass and the lower its thermal conductivity, the higher will be the level of residual stresses developed, and hence, the stronger the glass becomes. Thermal tempering takes a relatively short time (minutes), and can be applied to most glasses. Because of the high amount of energy stored in residual stresses, **tempered glass** shatters into numerous pieces when broken. The broken pieces are not as sharp and hazardous as those from ordinary window glass, which has a sharp jagged fracture path.

FIGURE 18.14 (a) The stages involved in inducing compressive surface residual stresses for improved strength. (b) Residual stresses in a tempered glass plate. *Source:* Courtesy of Corning Glass Works.

Chemical Tempering. In this process, the glass is heated in a bath of molten KNO_3, K_2SO_4, or $NaNO_3$, depending on the type of glass. Ion exchanges then take place, with larger atoms replacing the smaller atoms on the surface of the glass. As a result, residual compressive stresses develop on the surface. This condition is similar to that created by forcing a wedge between two bricks in a brick wall.

Chemical tempering may be performed at various temperatures. At low temperatures, part distortion is minimal, thus complex part shapes can be tempered. At elevated temperatures, there may be some distortion of the part, but the product can then be used at higher temperatures, without loss of strength. The time required for chemical tempering is much longer than that for thermal tempering.

Laminated Glass. Laminated glass is a product of another strengthening method. Called *laminate strengthening*, it consists of two pieces of flat glass with a thin sheet of tough plastic in between. Thus, when laminated glass cracks, the pieces are held together by the plastic sheet and it becomes far less hazardous, a phenomenon commonly observed in a shattered windshield of an automobile.

Flat glass for glazing windows and doors can be strengthened with wire netting (such as *chicken wire*, with a hexagonal mesh), embedded in the glass during its production. When a hard object strikes the surface, the glass shatters, but the pieces are held together because of the embedded wire, which has both strength and ductility, hence toughness (see Section 2.2.4).

Bulletproof Glass. Laminated glass has considerable ballistic impact resistance, and can prevent the full penetration of solid objects, because of the presence of a tough polymer film in between the two layers of glass. *Bulletproof glass* (used in some automobiles, armored bank vehicles, and buildings) is a more challenging design, because of the very high speed and energy level of the bullet and the small size and the shape of the bullet tip, thus representing a small contact area and high localized stresses. Depending on the caliber of the weapon, bullet speeds range from about 350 to 950 m/s (1150–3100 ft/s).

Bulletproof glass, also called *bullet-resistant glass*, ranges in thickness from 7 to 75 mm (0.3–3 in.). The thinner glass plates are designed for resistance to handguns, and thicker ones to rifles. Although there are several designs, bulletproof glass basically consists of glass laminated with a polymer sheet (usually polycarbonate). The capacity of a bulletproof glass to stop a bullet depends on (a) the type and thickness of the glass; (b) the size, shape, weight, and speed of the bullet; and (c) the properties and thickness of the polymer sheet.

Polycarbonate sheets are commonly used for bulletproof glass. As a material widely used for safety helmets, windshields, and guards for machinery, polycarbonate is a tough and flexible polymer. Laminated with a thick glass, it can stop a bullet, although the glass itself develops a circular shattered region. Proper bonding of the glass–polymer interfaces also is an important consideration, as there usually is more than one round fired during encounters. Moreover, in order to maintain the transparency of the glass and minimize distortion, the *index of refraction* of the glass and the polymer must be nearly identical.

If a polymer sheet is bonded only on one side of the glass, it is known as a *one-way* bulletproof glass. In a vehicle, the polymer layer is on the inside surface of the glass. An external bullet will not penetrate the window, because the bullet will strike the glass first, shattering it. The glass absorbs some of the energy of the bullet, thus slowing the bullet down. The remaining energy is dissipated in the polymer sheet, which then stops the bullet. This arrangement allows someone inside a vehicle to fire back. A bullet from inside penetrates the polymer sheet and forces the glass to break outward, allowing the bullet to go through. Thus, a one-way glass stops a bullet fired from outside but allows a bullet to be fired from inside.

A more recent design for bulletproof glass consists of two adjacent layers of thermoplastic polymer sheet over the same surface of the glass, and is based on a somewhat different principle:

- The outermost layer (the side where the bullet enters) is an acrylic sheet (polymethylmethacrylate, PMMA). This sheet *dulls* the tip of the bullet, thus slowing the bullet and its ability to penetrate easily, because of its now-blunt tip. In addition, the acrylic film has high weather resistance, making it suitable as the outer layer, which is exposed to the elements.
- The next layer after the glass is a polycarbonate sheet. Because it has high toughness, the polycarbonate layer stops the bullet, which has already been dulled by first penetrating the acrylic sheet. The glass shatters in the same manner as in the other designs.

18.4.1 Finishing Operations

As in metal products, residual stresses can develop in glass products if they are not cooled at a sufficiently low rate. In order to ensure that the product is free from these stresses, it is **annealed,** by a process similar to the stress-relief annealing of metals (Section 4.11). The glass is first heated to a certain temperature, and then cooled slowly. Depending on the size, the thickness, and the type of the glass, annealing times may range from a few minutes to as long as 10 months, as in the case of a 600-mm (24-in.) mirror for a telescope in an observatory.

In addition to annealing, glass products may be subjected to further operations, such as cutting, drilling, grinding, and polishing. Sharp edges and corners can be made smooth by (a) grinding (as can be seen in glass tops for desks and shelves) or

(b) holding a torch against the edges (**fire polishing**), which rounds them by localized softening of the glass and subsequent effect of surface tension.

In all finishing operations on glass, and other brittle materials, care should be exercised to ensure that there is no surface damage, especially stress raisers such as rough surface finish and scratches. Because of its notch sensitivity, even a single scratch on glass can cause premature failure, especially if the scratch is in a direction where the tensile stresses are a maximum (see also Fig. 16.17).

18.5 Design Considerations for Ceramics and Glasses

Ceramic and glass products require careful selection of composition, processing methods, finishing operations, and methods of assembly with other components. With such properties as poor tensile strength, sensitivity to internal and external defects, low impact toughness, and static fatigue, the consequences of part failure always are a significant factor in designing ceramic and glass products. On the other hand, these limitations must be balanced against such desirable material characteristics as hardness, scratch resistance, compressive strength at room and elevated temperatures, and a wide range of diverse physical properties.

As noted in Section 8.3.1, ceramics and glasses undergo a phenomenon called *static fatigue*, whereby they can suddenly fracture under a static load after a period of time. Although this behavior does not occur in a vacuum or in dry air, provisions must be made to prevent such failure. A general rule is that, in order for a glass item to withstand a load of 1000 h or longer, the maximum stress that can be applied is about one-third of the maximum stress that it can withstand during the first second of loading. The control of processing parameters and of the quality and level of impurities in the raw materials is important.

Dimensional changes, warping, and the possibility of cracking during processing and service life are significant factors in selecting methods for shaping these materials. When a ceramic or glass is a component of a larger assembly, its compatibility with other components is another important consideration. Particularly significant are the type of external forces and thermal expansion, such as in seals and windows with metal frames. Recall from Table 3.1 the wide range of coefficients of thermal expansion for various metallic and nonmetallic materials. Thus, when a plate glass fits tightly in a metal window frame, temperature variations (such as the sun shining on only a portion of the window) can cause thermal stresses that may lead to cracking, a phenomenon often observed in some tall buildings. A common solution is placing rubber seals between the glass and the window frame, to allow for dimensional changes.

18.6 Processing of Superconductors

Although *superconductors* (Section 3.7) have major energy-saving potential in the generation, storage, and distribution of electrical power, their processing into useful shapes and sizes for practical applications has presented significant difficulties. The following are two basic types of superconductors:

1. Metals, called *low-temperature superconductors* (LTSC), include combinations of niobium, tin, and titanium. For example, niobium–tin alloys, cooled by liquid helium, constitute the superconducting magnet used in most *magnetic resonance imaging* (MRI) scanners for medical imaging.

2. Ceramics, called **high-temperature superconductors** (HTSCs), include various copper oxides. Here, "high" temperature means closer to ambient temperature, although the commercially important HTSCs maintain superconductivity above the boiling point of liquid nitrogen ($-196°C$, or $-321°F$).

Ceramic superconducting materials are available in powder form. The difficulty in manufacturing them is their (a) inherent brittleness and (b) anisotropy, making it difficult to align the grains in the proper direction to achieve high efficiency. The smaller the grain size, the more difficult it is to align the grains.

The basic manufacturing process for superconductors consists of the following steps:

1. Preparing the powders, blending them, and milling them in a ball mill down to a grain size of 0.5–10 μm
2. Forming the powder into the desired shape
3. Heat treating the product, to enhance properties

The most common forming process is the **oxide-powder-in-tube** (OPIT) method. The powder is first packed into silver tubes (because silver has the highest electrical conductivity of any metal; see Table 3.1), and sealed at both ends. The tubes are then deformed, by such processes as swaging, drawing, extrusion, isostatic pressing, and rolling. The final product shape may be wire, tape, coil, or bulk.

Other methods of processing superconductors are (a) coating silver wire with superconducting material, (b) depositing superconductor films by laser ablation, (c) forming by the doctor-blade process (Section 18.2.1), (d) explosive cladding, and (e) chemical spraying. The formed part subsequently may be heat treated to improve the grain alignment of the superconducting powder.

CASE STUDY 18.1   Production of High-temperature Superconducting Tapes

Two bismuth-based oxides are preferred as superconducting ceramic materials for various military and commercial applications, such as electrical propulsion for ships and submarines, shallow-water and ground minesweeping systems, transmission cable generators, and superconducting magnetic energy storage (SMES). Several different processing methods have been explored to produce wires and multifilament tapes. The *powder-in-tube* process (Fig. 18.15) has been used successfully to fabricate long lengths of bismuth-based wires and tapes, with desirable properties. The following example

illustrates this method for the production of high-temperature superconducting multifilament tapes:

1. First, a composite billet is produced, using a silver casing and ceramic powder. The casing is made of an annealed high-purity silver, filled with the bismuth-ceramic powder in an inert atmosphere. A steel ram is used to compact the casing in several increments, to a 30% relative density. In order to minimize density gradients, such as those shown in Fig. 17.12, about one gram of powder is added to the billet for each

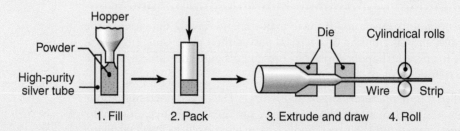

FIGURE 18.15 Schematic illustration of the powder-in-tube process.

stroke of the ram. Each billet is weighed and measured to verify the initial packing density. The billet ends are then sealed with a silver alloy, to avoid contamination during subsequent processing.

2. The billet is then extruded and drawn, to reduce its diameter and increase the powder density. Billets are drawn on a draw bench (see Fig. 15.24) down to a wire with a final diameter of 1.63 mm. It takes 12 passes, with a 20% reduction per pass, to perform this task. The dies have a semicone angle of 8°, and the drawing speed is approximately 1.4 m/min. A semisoluble oil or zinc-stearate spray is used as lubricant.

3. Following the drawing process, the wire is transformed progressively into tape in a single-stand rolling mill in a two-high or four-high configurations. For the four-high case, the diameter of the backup rolls (which are the work rolls for the two-high configuration) is 213 mm, and the diameter of the work rolls is 63.5 mm. The final tape dimensions are 100–200 μm in thickness and 2–3 mm in width, with a ceramic core ranging from 40 to 80 μm in thickness and 1.0 to 1.5 mm in width.

Source: Based on data from S. Vaze and M. Pradheeradhi, Concurrent Technologies Corporation.

SUMMARY

- Ceramic products are shaped by various casting, plastic forming, or pressing techniques. The parts are then dried and fired to impart strength and hardness. Finishing operations, such as machining and grinding, may be performed to give the part its final shape and dimensional accuracy, or to subject it to surface treatments. Because of their inherent brittleness, ceramics are processed with due consideration of distortion and cracking. The control of raw-material quality and processing parameters also is important.

- Glass products are made by several shaping processes, similar to those used for ceramics and plastics. They are available in a wide variety of forms, compositions, and mechanical, physical, and optical properties. Their strength can be improved by thermal and chemical treatments.

- Continuous methods of glass processing are drawing, rolling, and floating. Discrete glass products can be manufactured by blowing, pressing, centrifugal casting, or sagging. The parts subsequently may be annealed to relieve residual stresses.

- Design considerations for ceramics and glasses are guided by such factors as their low tensile strength and toughness, and their sensitivity to external and internal defects. Warping and cracking during production are important considerations.

- Manufacturing superconductors into useful products can be challenging, because of the anisotropy and inherent brittleness of the materials involved. Although new processes are being developed, the basic process consists of packing the powder into a silver tube and forming it into desired shapes.

KEY TERMS

Binder	Deflocculent	Gob	Laminated glass
Blow and blow	Doctor-blade process	High-temperature	Low-temperature
Blowing	Drawing	superconductors	superconductors
Bulletproof glass	Fire polishing	Hot pressing	Microwave sintering
Centrifugal casting	Firing	Injection molding	Oxide-powder-in-tube
Chemical tempering	Float glass	Jiggering	process

Plastic forming   Pressing   Slip casting   Thermal tempering
Plasticizer   Sagging   Static fatigue   Wetting agent
Press and blow   Slip   Tempered glass

BIBLIOGRAPHY

Advanced Ceramic Technologies & Products, Springer, 2012.
Bansal, N.P., and Boccaccini, A.R., (eds.), Ceramics and Composites: Processing Methods, Wiley-American Ceramic Society, 2012.
Barsoum, M.W., Fundamentals of Ceramics, Taylor & Francis, 2002.
Basu, B., and Balani, K., Advanced Structural Ceramics, Wiley-American Ceramic Society, 2011.
Bengisu, M., Engineering Ceramics, Springer, 2010.
Carter, C.B., and Norton, W.G., Ceramic Materials: Science and Engineering, Springer, 2007.
Engineered Materials Handbook, Vol. 4: Ceramics and Glasses, ASM International, 1991.
Handbook of Advanced Ceramics: Materials, Applications, Processing and Properties. Academic Press, 2004.
Harper, C.A. (ed.), Handbook of Ceramics, Glasses, and Diamonds, McGraw-Hill, 2001.

Holand, W., and Beall, G.H., Glass Ceramic Technology, 2nd ed., Wiley-American Ceramic Society, 2012.
King, A.G., Ceramics Processing and Technology, Noyes Publishing, 2001.
Rahaman, M.N., Ceramics Processing, CRC Press, 2007.
Richerson, D.W., Modern Ceramic Engineering: Properties, Processing, and Use in Design, 3rd ed., Marcel Dekker, 2005.
Riley, F.L., Structural Ceramics: Fundamentals and Case Studies, Cambridge, 2009.
Shackelford, G.F., and Doremus, R.H., Ceramic and Glass Materials: Structure, Properties and Processing, Springer, 2008.
Shelby, J.E., Introduction to Glass Science and Technology, 2nd ed., Royal Society of Chemistry, 2005.

REVIEW QUESTIONS

18.1 Outline the steps involved in processing (a) ceramics and (b) glasses.

18.2 List and describe the functions of additives in ceramics.

18.3 Describe the doctor-blade process.

18.4 Explain the advantages of isostatic pressing.

18.5 What is jiggering? What shapes does it produce?

18.6 Name the parameters that are important in drying ceramic products.

18.7 What types of finishing operations are used on ceramics? On glass? Why?

18.8 Describe the methods by which sheet glass is made.

18.9 What is float glass?

18.10 What is a gob?

18.11 How is glass tubing produced?

18.12 What is the difference between physical and chemical tempering of glass?

18.13 What is the structure of laminated glass? Bulletproof glass?

18.14 How are glass fibers made? What are their sizes?

18.15 Describe the processes of chemical and thermal tempering of glass.

18.16 What is a superconductor? How are superconductors made into a wire?

18.17 Is diamond a ceramic? Explain.

QUALITATIVE PROBLEMS

18.18 Inspect various products; noting their shape, color, and transparency, and identify those that are made of (a) ceramic, (b) glass, and (c) glass ceramics.

18.19 Describe the differences and similarities in processing metal powders vs. ceramics.

18.20 Which property of glasses allows them to be expanded to large dimensions by blowing? Can metals undergo such behavior? Explain.

18.21 Explain why ceramic parts may distort or warp during drying. What precautions should be taken to avoid this situation?

18.22 What properties should plastic sheets have to be used in laminated glass? Why?

18.23 It is stated that the higher the coefficient of thermal expansion of a glass and the lower its thermal conductivity, the higher the level of the residual stresses developed. Explain why.

18.24 Are any of the processes used for making discrete glass products similar to ones described in preceding chapters? Describe them.

18.25 Injection molding is a process that is used for powder metals, polymers, and ceramics. Explain why is this so.

18.26 Explain the phenomenon of static fatigue and how it affects the service life of a ceramic or glass component.

18.27 Describe and explain the differences in the manner in which each of the following would fracture when struck with a heavy piece of rock: (a) ordinary window glass, (b) tempered glass, and (c) laminated glass.

18.28 Is there any flash that develops in slip casting? How would you propose to remove such flash?

18.29 Explain the difficulties involved in making large ceramic components. What recommendations would you make to improve the process?

QUANTITATIVE PROBLEMS

18.30 Using Example 18.1, calculate (a) the porosity of the dried part if the porosity of the fired part is to be 9% and (b) the initial length, L_o, of the part if the linear shrinkages during drying and firing are 8 and 7%, respectively.

18.31 What would be the answers to Problem 18.30 if the quantities given were halved?

18.32 Assume that you are an instructor covering the topics described in this chapter and you are giving a quiz on the numerical aspects to test the understanding of the students. Prepare two quantitative problems and supply the answers.

SYNTHESIS, DESIGN, AND PROJECTS

18.33 List similarities and differences between the processes described in this chapter and those in (a) Part II on metal casting and (b) Part III on forming and shaping.

18.34 Consider some ceramic products with which you are familiar, and outline a sequence of processes that you think were used to manufacture them.

18.35 Make a survey of the technical literature, and describe the differences, if any, between the quality of glass fibers made for use in reinforced plastics and those made for use in fiber-optic communications. Comment on your observations.

18.36 How different, if any, are the design considerations for ceramics from those for other materials? Explain.

18.37 Visit a ceramics/pottery shop, and investigate the different techniques used for coloring and decorating a ceramic part. What are the methods of applying a metallic finish to the part?

18.38 Give examples of designs and applications in which static fatigue should be taken into account.

18.39 Construct a table that describes the approach for manufacturing dinner plate from (a) metals, (b) thermoplastics,

(c) ceramics, (d) powder metal, and (e) glass. Include descriptions of process capabilities and shortcomings in your descriptions.

18.40 Pyrex cookware displays a unique phenomenon: it functions well for a large number of cycles and then shatters into many pieces. Investigate this phenomenon, list the probable causes, and discuss the manufacturing considerations that may alleviate or contribute to such failures.

18.41 It has been noted that the strength of brittle materials such as ceramics and glasses are very sensitive to surface defects, such as scratches (notch sensitivity). Obtain some pieces of these materials, make scratches on them, and test them by carefully clamping in a vise and bending them. Comment on your observations.

18.42 Describe your thoughts on the processes that can be used to make (a) small ceramic statues, (b) whiteware for bathrooms, (c) common brick, and (d) floor tile.

18.43 Perform a literature search, and make a list of automotive parts or components that are made of ceramics. Explain why they are made of ceramics.

19

Plastics and Composite Materials: Forming and Shaping

- This chapter describes the manufacturing processes involved in producing polymers and composite materials.
- Extrusion is widely used to produce rods and tubing, and pellets as base stock for producing plastic parts, sheet, and film.
- The chapter describes several molding operations for producing discrete parts, including injection and reaction-injection molding, transfer molding, rotational molding, and compression molding.
- The manufacturing processes associated with shaping composite materials are then described, including compression and vacuum molding, contact molding, pultrusion, and filament winding.
- The chapter ends with a description of the characteristics of the machinery used, mold design principles, and economic considerations in polymer processing.

Typical parts made: Extensive variety of consumer and industrial products with a range of colors and characteristics.

Alternative processes: Casting, forming, powder metallurgy, and machining.

19.1 Introduction

The processing of plastics and elastomers involves operations similar to those used in the forming and shaping of metals, described in preceding chapters. The processing of rubbers and elastomers began in the 1800s, with the discovery of vulcanization by C. Goodyear in 1839. Plastics began to be developed in the 1920s, and rapid progress in the 1940s and onward led to important advances in materials, design, and manufacturing to make numerous consumer and industrial products in large quantities and at low cost. In the 1970s, reinforced plastics began to be introduced, leading the way for rapid progress in the use of composite materials, with unique properties and applications, and their associated challenges in producing them.

As noted in Chapter 7, thermoplastics melt and thermosets cure at relatively low temperatures. Hence, unlike metals, they are relatively easy to handle, and require much less force and energy to process them. Plastics, in general, can be molded, cast, shaped, and machined into complex shapes, in few operations, with relative ease, and at high production rates (Table 19.1). They also can be joined by various techniques (Section 32.6) and coated by various processes (Chapter 34). Plastics are shaped into discrete products or as continuous products such as sheets, plates, rods, and tubing; they may then be formed by secondary processes into a variety of discrete products.

TABLE 19.1

General Characteristics of Forming and Shaping Processes for Plastics and Composite Materials	
Process	Characteristics
Extrusion	Continuous, uniformly solid or hollow, and complex cross-sections; high production rates; relatively low tooling costs; wide tolerances
Injection molding	Complex shapes of various sizes; thin walls; very high production rates; costly tooling; good dimensional accuracy
Structural foam molding	Large parts with high stiffness-to-weight ratio; less expensive tooling than in injection molding; low production rates
Blow molding	Hollow, thin-walled parts and bottles of various sizes; high production rates; relatively low tooling costs
Rotational molding	Large, hollow items of relatively simple shape; relatively low tooling costs; relatively low production rates
Thermoforming	Shallow or relatively deep cavities; low tooling costs; medium production rates
Compression molding	Parts similar to impression-die forging; expensive tooling; medium production rates
Transfer molding	More complex parts than compression molding; higher production rates; high tooling costs; some scrap loss
Casting	Simple or intricate shapes made with rigid or flexible low-cost molds; low production rates
Processing of composite materials	Long cycle times; expensive operation; tooling costs depend on process

The types and properties of polymers, and the shape and complexity of the parts that can be produced, are greatly influenced by their manufacturing and processing characteristics.

Plastics are usually shipped to manufacturing plants as pellets, granules, or powders, and are melted (for thermoplastics) just before the shaping process. Liquid plastics that cure into solid form also are used, especially in making thermosets and reinforced-plastic parts. With increasing awareness of the environment, raw materials also may consist of reground, chopped, or melted plastics, obtained from recycling centers. As expected, however, product quality may not be as high as when made from the materials.

Following the outline shown in Fig. 19.1, this chapter describes the basic processes, operations, machinery, and the economics of forming and shaping plastics. The processing techniques for reinforced plastics and metal-matrix and ceramic-matrix composites also are described. The chapter begins with melt-processing techniques, starting with extrusion, and continuing on to molding processes.

19.2 Extrusion

In *extrusion*, which produces the largest volume of plastics, raw materials in the form of thermoplastic pellets, granules, or powder are placed into a *hopper* and fed into the barrel of a *screw extruder* (Fig. 19.2). The barrel is equipped with a helical screw, which builds up pressure in the barrel, blends the pellets, and conveys them down the barrel toward the die. The barrel heaters and the internal friction from the mechanical action of the screw heat the pellets and liquify them.

Screws have three distinct sections:

1. *Feed section*: Conveys the material from the hopper into the central region of the barrel

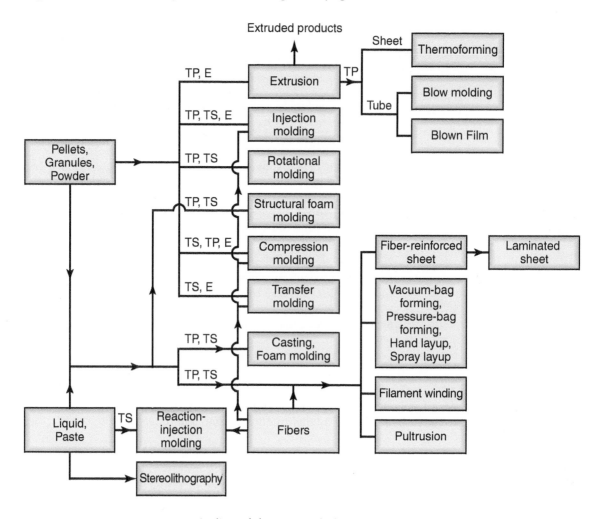

FIGURE 19.1 Outline of forming and shaping processes for plastics, elastomers, and composite materials. (TP = Thermoplastic; TS = Thermoset; E = Elastomer.)

2. *Melt section*, also called *compression* or *transition section*: Where the heat generated by the viscous shearing of the plastic pellets and by the external heaters around the barrel cause melting to begin

3. *Metering* or *pumping section*: Where additional shearing and melting occur, with pressure building up at the die entrance

The lengths of these individual sections can be changed to accommodate the melting characteristics of different types of plastics. A metal-wire filter screen (Fig. 19.2a) usually is placed just before the die, to filter out unmelted or congealed resin. This screen, which is replaced periodically, also causes back pressure in the barrel, which needs to be overcome by the extruder screw. Between the screen and the die is a *breaker plate*, which has several small holes in it and helps improve mixing of the polymer prior to its entering the die. The extruded product is cooled, generally by exposing it to blowing air or by passing it through a water-filled channel (trough).

Controlling the rate and uniformity of cooling is important in extruding, in order to minimize product shrinkage and distortion. In addition to single-screw extruders, other designs include *twin* (two parallel screws side by side) and *multiple screws*, for polymers that are difficult to extrude (see also *reciprocating screw*, Section 19.3).

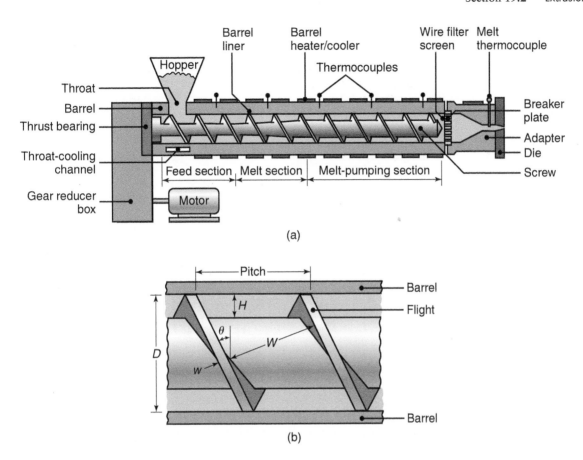

FIGURE 19.2   (a) Schematic illustration of a typical screw extruder. (b) Geometry of an extruder screw. Complex shapes can be extruded with relatively simple and inexpensive dies.

A typical helical screw is shown in Fig. 19.2b, indicating the important parameters that affect the mechanics of polymer extrusion. At any point in time, the molten plastic is in the shape of a helical ribbon, with thickness H and width W, and is conveyed toward the extruder outlet by the rotating screw *flights*. The shape, pitch, and flight angle of the helical screw are important parameters, as they affect the flow of the polymer through the extruder. The ratio of the barrel length, L, to its diameter, D, is also important. In typical commercial extruders, the L/D ratio ranges from 5 to 30, and barrel diameters generally are in the range from 25 to 200 mm (1–8 in.).

Video Solution 19.1   Extruder Characteristics

Process Characteristics.   Because there is a continuous supply of raw material from the hopper, long products such as solid rods, sections, channels, sheet, tubing, pipe, and architectural components can be extruded continuously by extrusion. Complex shapes with constant cross-sections also can be extruded with relatively inexpensive tooling. Some common die profiles are shown in Fig. 19.3b. Polymers usually undergo much greater and uneven shape recovery than is encountered in metal extrusion. Since the polymer will *swell* at the exit of the die, the openings shown in Figs. 19.3b and c are smaller than the extruded cross-sections. After it has cooled, the extruded product may subsequently be drawn (*sized*) by a puller and coiled or cut into desired lengths.

The control of processing parameters, such as extruder-screw rotational speed, barrel-wall temperatures, die design, and rate of cooling, and drawing speeds are all important, in order to ensure product integrity and uniform dimensional accuracy. *Defects* observed in extruding plastics are similar to those observed in metal

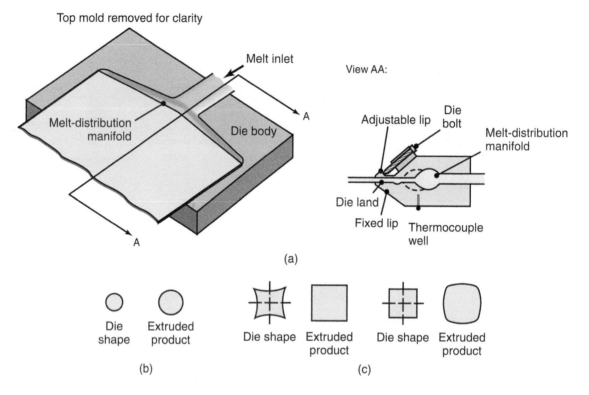

FIGURE 19.3 Common extrusion die geometries: (a) coat-hanger die for extruding sheet; (b) round die for producing rods; and (c) and (d) nonuniform recovery of the part after it exits the die.

extrusion, described in Section 15.5. Die shape is important, as it can induce high stresses in the product, causing it to develop surface fractures, as also occur with metals. Other surface defects are *bambooing* and *sharkskin defects*, due to a combination of friction at the die–polymer interfaces, elastic recovery, and nonuniform deformation of the outer layers of the product with respect to its bulk during extrusion.

Extruders generally are rated by the diameter, D, of the barrel and the length-to-diameter (L/D) ratio of the barrel. Machinery costs can be on the order of $300,000, including the cost for the equipment for downstream cooling and winding of the extruded product.

19.2.1   Miscellaneous Extrusion Processes

There are several variations of the basic extrusion process for producing a number of different polymer products.

Plastic Tubes and Pipes.   These are produced in an extruder with a *spider die*, as shown in Fig. 19.4a (see also Fig. 15.8 for details). Woven fiber or wire reinforcements also may be fed through specially designed dies in this operation, for the production of reinforced products to withstand higher pressures, such as garden hose. The extrusion of tubes is also a first step for related processes, such as extrusion blow molding and blown film.

Rigid Plastic Tubing.   Extruded by a process in which the die is *rotated*, rigid plastic tubing causes the polymer to be sheared and biaxially oriented during extrusion. As a result, the tube has a higher crushing strength and a higher strength-to-weight ratio than conventionally extruded tubing.

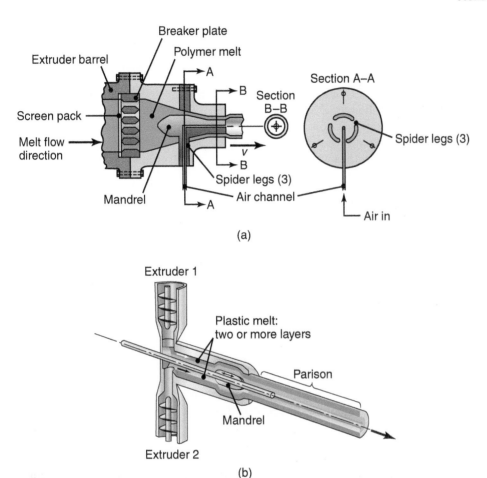

FIGURE 19.4 Extrusion of tubes. (a) Extrusion using a spider die (see also Fig. 15.8) and pressurized air. (b) Coextrusion for producing a parison for a plastic bottle.

Coextrusion. Shown in Fig. 19.4b, coextrusion involves simultaneous extrusion of two or more polymers through a single die; the product cross-section thus contains different polymers, each with its own characteristics and function. Coextrusion is commonly performed in shapes such as flat sheets, films, and tubes, and used especially in food packaging where different layers of polymers have different functions. These are: (a) providing inertness for food, (b) serving as barriers to fluids, such as water or oil, and (c) labeling of the product.

Plastic-coated Electrical Wire. Electrical wire, cable, and strips are simultaneously extruded and coated with plastic. The wire is fed into the die opening, at a controlled rate with the extruded plastic, in order to have a uniform coating on the wire. To ensure proper insulation, extruded electrical wires are checked continuously for their electrical resistance as they exit the die; the wire is also marked with ink using a roller to identify the specific type of wire. Plastic-coated wire *paper clips* also are made by this coating process.

Polymer Sheets and Films. Sheet and film can be produced by using a specially designed flat extrusion die, such as that shown in Fig. 19.3a. Also known as a *coat-hanger die*, it is designed to distribute the polymer melt evenly throughout the width of the die opening. The polymer is extruded by forcing it through the die, after which the extruded sheet is taken up by rolls, first on water-cooled rolls (to cool the sheets), and then by a pair of rubber-covered pull-off rolls. Generally, polymer *sheet* is considered to be thicker than 0.5 mm, and *film* is thinner than 0.5 mm.

Thin Polymer Films.   Common *plastic bags* and other thin polymer film products are made from **blown film,** which itself is made from a thin-walled tube produced by an extruder. In this process, a tube is extruded vertically (Fig. 19.5), continuously pulled upward and expanded into a balloon shape, by blowing air through the center of the extrusion die until the desired film thickness is reached. Because of the molecular orientation of thermoplastics (Section 7.3), a *frost line* develops on the balloon, which reduces its transparency.

The balloon usually is cooled by air from a cooling ring around it, which can also act as a physical barrier to further diametral expansion of the balloon, thus controlling its dimensions. The cooled bubble is then slit lengthwise, becoming *wrapping film,* or it is pinched/welded and cut off, becoming a plastic bag. The width of the film produced after slitting can be on the order of 6 m (20 ft) or more.

The ratio of the blown diameter to the extruded tube diameter is called the *blow ratio,* which is about 3:1 in Fig. 19.5. Note that, as described in Section 2.2.7, the polymer must have a high strain-rate sensitivity exponent, m, to successfully be blown by this process without tearing. (See also Example 9.1.)

Plastic Films.   Plastic films, especially polytetrafluoroethylene (PTFE; trade name: *Teflon*), can be produced by *shaving* the circumference of a solid round plastic billet, with specially designed knives and in a manner similar to producing veneer from a large piece of round wood, in a process called **skiving** (see also Section 24.4).

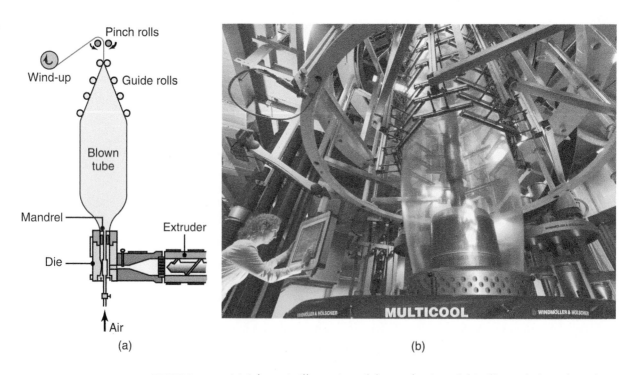

(a)   (b)

FIGURE 19.5   (a) Schematic illustration of the production of thin film and plastic bags from tube, first produced by an extruder and then blown by air. (b) A blown-film operation; this process is well developed, producing inexpensive and very large quantities of plastic film and shopping bags. *Source:* (b) Courtesy of Windmoeller & Hoelscher Corp.

Pellets. Used as raw material for other plastic-processing methods described in this chapter, pellets also are made by extrusion. A small-diameter, solid rod is extruded continuously, and then chopped into short lengths (pellets). With some modifications, extruders also can be used as simple melters for other shaping processes, such as injection molding and blow molding.

EXAMPLE 19.1   Blown Film

Given: A typical plastic shopping bag made by blown film has a lateral dimension (width) of 400 mm.

Find: a. Determine the extrusion-die diameter. b. These bags are relatively strong in use; how is this strength achieved?

Solution:

a. The perimeter of the flat bag is $(2)(400) = 800$ mm. Since the original cross-section of the film is round, the blown diameter should be $\pi D = 800$, thus $D = 255$ mm. Recall that in this process, a tube is expanded from 1.5 to 2.5 times the extrusion-die diameter. Taking the maximum value of 2.5, the die diameter is $255/2.5 = 100$ mm.

b. Note in Fig. 19.5a that, after extrusion, the balloon is being pulled upward by the pinch rolls. Thus, in addition to diametral stretching and the attendant molecular orientation, the film is *stretched* and *oriented* in the longitudinal direction. The resulting biaxial orientation of the polymer molecules significantly improves the strength and toughness of the plastic bag.

19.2.2   Production of Polymer Reinforcing Fibers

Polymer fibers have numerous important applications. In addition to their use as reinforcement in composite materials, these fibers are used in a wide variety of consumer and industrial products, including clothing, carpeting, fabrics, rope, and packaging tape.

Most synthetic fibers used in reinforced plastics are polymers, extruded through tiny holes of a device called a *spinneret* (resembling a shower head), to form continuous filaments of semisolid polymer. The extruder forces the polymer through the spinneret, which may have from one to several hundred holes. If the polymers are thermoplastics, they first are melted in the extruder, as described in Section 19.2. Thermosetting polymers also can be formed into fibers, by first dissolving or chemically treating them so that they can be extruded. These operations are performed at high production rates and with very high reliability.

As the filaments emerge from the holes in the spinneret, the liquid polymer is first converted to a rubbery state, and then it solidifies. This process of extrusion and solidification of continuous filaments is called **spinning**, a term also used for the production of natural textiles, such as cotton or wool. There are four methods of spinning fibers: melt, wet, dry, and gel spinning.

1. In **melt spinning** (Fig. 19.6), the polymer melt is extruded through the spinneret and then solidified directly by cooling. A typical spinneret for this operation is around

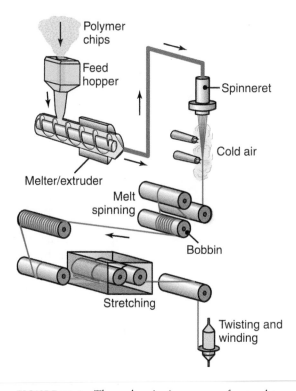

FIGURE 19.6 The melt-spinning process for producing polymer fibers, used in a variety of applications, including fabrics and as reinforcements for composite materials; in the stretching box, the right roll rotates faster than the left roll.

5 mm (0.2 in.) thick and has about 50 holes, about 0.25 mm (0.01 in.) in diameter. The fibers that emerge from the spinneret are cooled by forced-air convection, and are simultaneously pulled, so that their final diameter becomes much smaller than the spinneret opening. Polymers such as nylon, olefin, polyester, and PVC are produced in this manner.

Melt-spun fibers also can be extruded from the spinneret in various other cross-sections, such as trilobal (a triangle with curved sides), pentagonal, octagonal, and hollow shapes. Hollow fibers trap air, and thus provide additional thermal insulation.

2. **Wet spinning**, the oldest process for fiber production, is used for polymers that have been dissolved in a solvent, by submerging the spinnerets in a chemical bath. As the filaments emerge, they precipitate in the bath, producing a fiber that is then wound onto a *bobbin* (*spool*). The term "wet" refers to the use of a precipitating liquid bath, resulting in wet fibers, that require drying before they can be used. Acrylic, rayon, and aramid fibers are produced by this process.

3. **Dry spinning** is used for thermosets that are dissolved by a fluid. Instead of precipitating the polymer by dilution, as in wet spinning, solidification is achieved by evaporating the solvent fluid in a stream of air or inert gas. Thus, the filaments do not come in contact with a precipitating liquid, eliminating the need for drying. Dry spinning is used for the production of acetate, triacetate, polyether-based elastane, and acrylic fibers.

4. **Gel spinning** is a special process, used to obtain high strength or special fiber properties. Some polyethylene and aramid fibers are produced by gel spinning. The polymer is not melted completely, or dissolved in a liquid, but the molecules bond together at various points in liquid-crystal form. This operation produces strong interchain forces in the resulting filaments, that can significantly increase the tensile strength of the fibers. Moreover, the liquid crystals are aligned along the fiber axis, by the strain encountered during extrusion. Thus, the filaments emerge from the spinneret with an unusually high degree of orientation relative to each other, further enhancing their strength. This process is also called *dry wet spinning*, because the filaments first pass through air and are then cooled further in a liquid bath.

A necessary step in the production of most fibers is the application of significant *stretching*, to induce orientation of the polymer molecules in the fiber direction. This orientation is the main reason for the high strength of the fibers, as compared with the polymer in bulk form. The stretching can be done while the polymer is still pliable (just after extrusion from the spinneret) or it can be performed as a cold-drawing operation. The strain induced can be as high as 800%.

Graphite fibers are produced from polymer fibers by *pyrolysis*. In this operation, controlled heat, in the range from 1500° to 3000°C (2730°–5400°F), is applied to the polymer fiber (typically polyacrylonitrile, PAN) to drive off all elements except the carbon. The fiber is under tension in order to develop a high degree of orientation in the resulting fiber structure. (See also Section 9.2.1 on the properties of graphite fibers and other details.)

19.3   Injection Molding

Injection molding is similar to hot-chamber die casting (Fig. 19.7; see also Section 11.4.5). The pellets or granules are fed into the heated cylinder, and the melt is forced into the mold, either by a hydraulic *plunger* or by the *rotating screw*

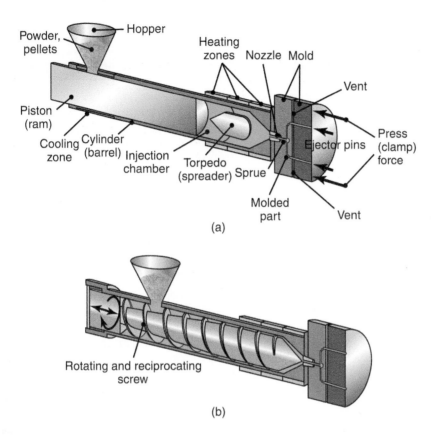

FIGURE 19.7 Schematic illustration of injection molding with (a) a plunger and (b) a reciprocating rotating screw.

system of an extruder. As in plastic extrusion, the barrel (cylinder) is heated externally to promote melting of the polymer. In injection-molding machines, however, a far greater portion of the heat transferred to the polymer is due to frictional heating.

Modern machines are of the *reciprocating* or *plasticating screw* type (Fig. 19.7b), with the sequence of operations shown in Fig. 19.8. As the pressure builds up at the mold entrance, the rotating screw begins to move backward under pressure to a predetermined distance. This movement controls the volume of material to be injected. The screw then stops rotating, and is pushed forward hydraulically, forcing the molten plastic into the mold cavity. The pressures developed usually range from 70 to 200 MPa (10,000–30,000 psi).

Some injection-molded products are shown in Fig. 19.9; other products include cups, containers, housings, tool handles, knobs, toys, plumbing fixtures, and components for electrical and communication-equipment. For thermoplastics, the molds are kept relatively cool, at about 90°C (190°F). Thermoset parts are molded in heated molds, at about 200°C (400°F), where *polymerization* and *cross-linking* take place.

After the part has cooled sufficiently (for thermoplastics) or cured (for thermosets), the molds are opened, and the part is ejected. The molds are then closed again, and the process is repeated automatically. Because the material is molten when injected into the mold, complex shapes with good dimensional accuracy can be obtained. However, because of uneven cooling of the part in the mold, residual stresses develop. Elastomers also are injection molded into discrete products by these processes.

Video Solution 19.2 Clamp Pressure in Injection Molding

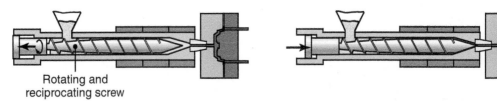

Rotating and
reciprocating screw

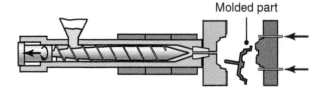

1. Build up polymer in front of sprue bushing; pressure pushes the screw backward. When sufficient polymer has built up, rotation stops.

2. When the mold is ready, the screw is pushed forward by a hydraulic cylinder, filling the sprue bushing, sprue, and mold cavity with polymer. The screw begins rotating again to build up more polymer.

Molded part

3. After polymer is solidified/cured, the mold opens, and ejector pins remove the molded part.

FIGURE 19.8 Sequence of operations in the injection molding of a part with a reciprocating screw; this process is used widely for numerous consumer and commercial products, such as toys, containers, knobs, and electrical equipment (see Fig. 19.9).

(a)

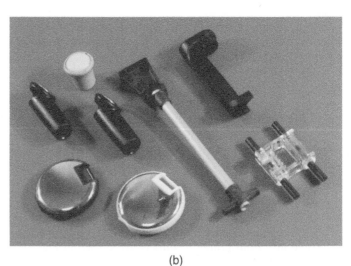

(b)

FIGURE 19.9 Typical products made by injection molding, including examples of insert molding. *Source:* (a) Courtesy of Plainfield Molding, Inc. (b) Courtesy of Rayco Mold and Mfg. LLC.

Molds with moving and unscrewing mandrels also are used in injection molding, as they allow the molding of parts having multiple cavities or internal and external threaded features. To accommodate part design, molds may have several components (Fig. 19.10), including runners (such as those used in metal-casting dies), cores, cavities, cooling channels, inserts, knockout pins, and ejectors. There are three basic types of molds for injection molding:

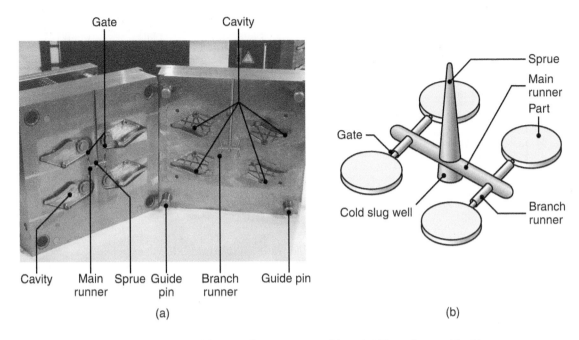

FIGURE 19.10 Illustration of mold features for injection molding. (a) Two-plate mold with important features identified. (b) Schematic illustration of the features in a mold. *Source: Courtesy of Tooling Molds West, Inc.*

1. **Cold-runner, two-plate mold:** This design is the simplest and most common (Fig. 19.11a).
2. **Cold-runner, three-plate mold:** The runner system is separated from the part when the mold is opened (Fig. 19.11b).
3. **Hot-runner mold,** also called **runnerless mold:** The molten plastic is kept hot in a heated runner plate (Fig. 19.11c).

In cold-runner molds, the solidified plastic, remaining in the channels connecting the mold cavity to the end of the barrel, must be removed, usually by trimming; later, this scrap can be chopped and recycled. In hot-runner molds, which are more expensive, there are no gates, runners, or sprues attached to the molded part. Cycle times are shorter, because only the molded part must be cooled and ejected.

Multicomponent injection molding. Also called *co-injection* or *sandwich molding*, this process allows the forming of parts that have a combination of various colors and shapes. An example is the molding of automobile rear-light covers, made of different materials and colors, such as red, amber, and white. For some parts, printed film also can be placed in the mold cavity, so they need not be decorated or labeled after molding.

Insert molding involves metallic components (such as screws, pins, and strips) that are placed in the mold cavity prior to injection and become an integral part of the molded product (Fig. 19.9). The most common examples of such combinations are hand tools, where the handle is insert molded onto a metal component; other examples include electrical and automotive components and faucet parts.

Overmolding. This is a process for making such products as hinge joints and ball-and-socket joints in one operation, and thus without requiring postmolding assembly. Two different plastics usually are used to ensure that no bonds will form between the molded halves of the joint, as otherwise motion would be impeded.

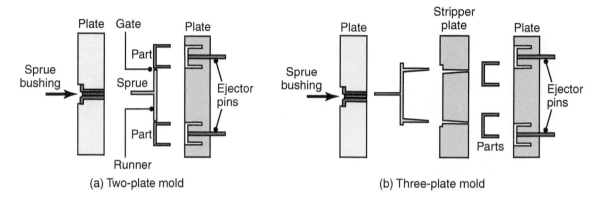

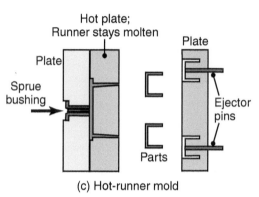

FIGURE 19.11   Types of molds used in injection molding.

In **ice-cold molding,** the same type of plastic is used for both components of the joint. The operation is carried out in a conventional injection-molding machine and in one cycle. A two-cavity mold is used, with cooling inserts positioned in the area of contact between the first and second molded components of the joint. In this way, no bonds develop between the two pieces, and thus the two components have free movements, as in a hinge or a sliding mechanism.

Process Capabilities.   Injection molding is a high-rate production process and permits good dimensional control. Although most parts generally weigh from about 100 to 600 g (3–20 oz), they also can be much heavier, such as automotive body panels and exterior components. Typical cycle times range from 5 to 60 seconds, although they can be several minutes for thermosetting materials.

Injection molding is a versatile process and capable of producing complex shapes, with good dimensional accuracy. As in other forming processes, mold design and the control of material flow in the die cavities are important factors in the quality of the product, and thus in avoiding defects. Because of the basic similarities to metal casting regarding material flow and heat transfer, *defects* observed in injection molding are somewhat similar to those in casting. For example:

- In Fig. 10.13g, the molten metal flows in from two opposite runners, and then meets in the middle of the mold cavity. Thus, a cold shut in casting is equivalent to *weld lines* in injection molding.
- If the runner cross-sections are too small, the polymer may solidify prematurely, thus preventing full filling of the mold cavity. Solidification of the outer layers

in thick sections can cause *porosity* or *voids* due to shrinkage, as in the metal parts shown in Fig. 12.2.

- If, for some reason, the dies do not close completely or because of die wear, *flash* will form, in a manner similar to flash formation in impression-die forging (see Figs. 14.6 and 19.17c).
- A defect known as *sink marks* (or *pull-in*), similar to that shown in Fig. 19.32c, also is observed in injection-molded parts.
- Methods of avoiding defects consist of proper control of temperatures, pressures, and mold design modifications, using simulation software.

Modeling techniques and *simulation software* continue to be developed for studying optimum gating systems, mold filling, mold cooling, and part distortion. *Software programs* are now available to expedite the design process for molding parts and with good dimensions and characteristics. The programs take into account such factors as injection pressure, temperature, heat transfer, and the condition of the polymer.

Machines.   Injection-molding machines are usually horizontal (Fig. 19.12); vertical machines are used for making small, close-tolerance parts and for insert molding. The clamping force on the dies is typically supplied by hydraulic means, although electrical means (weighing less and more quiet than hydraulic machines) also are used. Modern machines are equipped with microprocessors in a control panel, and monitor all aspects of the molding operation.

Injection-molding machines are rated according to the capacity of the mold and the clamping force. In most machines, this force ranges from 0.9 to 2.2 MN (100–250 tons). The largest machine in operation has a capacity of 45 MN (5000 tons), and it can produce parts weighing 25 kg (55 lb). The cost of a 100-ton machine ranges from about $60,000 to about $90,000 and of a 300-ton machine from about $85,000 to about $140,000. Die costs typically range from $20,000 to

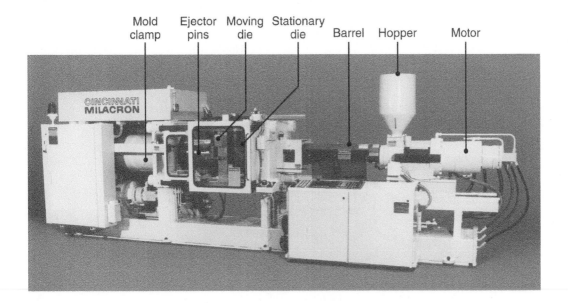

FIGURE 19.12 A 2.2-MN (250-ton) injection-molding machine; the tonnage is the force applied to keep the dies closed during the injection of molten plastic into the mold cavities, and hold it there until the parts are cool and stiff enough to be removed from the die. *Source:* Courtesy of Cincinnati Milacron, Plastics Machinery Division.

$200,000. Consequently, high-volume production is essential to justify such high expenditure.

The molds generally are made of tool steels, beryllium–copper, or aluminum. They may have multiple cavities, so that more than one part can be molded in one cycle (see also Fig. 11.22). Large molds can cost on the order of $100,000. Mold life may be on the order of 2 million cycles for steel molds, but it can be only about 10,000 cycles for aluminum molds.

EXAMPLE 19.2   Force Required in Injection Molding

Given: A 250-ton injection-molding machine is to be used to make spur gears, 4.5 in. in diameter and 0.5 in. thick. The gears have a fine-tooth profile.

Find: How many gears can be injection molded in one set of molds? Does the thickness of the gears affect the force?

Solution: Because of the detail involved (fine gear teeth), assume that the pressure required in the mold cavity will be on the order of 100 MPa (15 ksi). The cross-sectional (projected) area of the

gear is $\pi(4.5)^2/4 = 15.9 \ \text{in}^2$. Assuming that the parting plane of the two halves of the mold is in the mid-plane of the gear, the force required is $(15.9)(15,000) = 238,500$ lb.

Since the capacity of the machine is 250 tons, the clamping force available is $(250)(2000) = 500,000$ lb. Thus, the mold can accommodate two cavities and produce two gears per cycle. Because it does not influence the cross-sectional area of the gear, gear thickness does not directly affect the pressures involved and thus does not change the answer.

19.3.1   Reaction-injection Molding

In the *reaction-injection molding* (RIM) process, a monomer (Section 7.2) and two or more reactive fluids are forced, at high speed, into a mixing chamber at a pressure of 10 to 20 MPa (1400–2800 psi), and then into the mold cavity (Fig. 19.13). Chemical reactions take place rapidly in the mold, and the polymer solidifies. Typical polymers

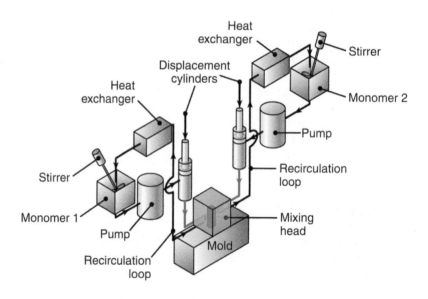

FIGURE 19.13   Schematic illustration of the reaction-injection molding process; typical parts made are automotive-body panels, water skis, and thermal insulation for refrigerators and freezers.

used are polyurethane, nylon, and epoxy. Cycle times may range up to about 10 min, depending on the materials, part size, and shape.

Major applications of this process include automotive parts (such as bumpers, fenders, steering wheels, and instrument panels), thermal insulation for refrigerators and freezers, water skis, and stiffeners for structural components. Parts made may range up to about 50 kg (110 lb). Reinforcing fibers (such as glass or graphite) also may be used to improve the product's strength and stiffness. Depending on the number of parts to be made and the part quality required, molds can be made of materials such as steel or aluminum.

19.4 Blow Molding

Blow molding is a modified extrusion- and injection-molding process. In **extrusion blow molding**, a tube or preform (usually oriented so that it is vertical) is first extruded. It is then clamped into a mold, with a cavity much larger than the tube diameter, and blown outward to fill the mold cavity (Fig. 19.14a). Depending on the material, the blow ratio may be as high as 7:1. Blowing usually is done with a hot-air blast, at a pressure ranging from 350 to 700 kPa (50–100 psi). Plastic drums, with a volume as large as 2000 liters (530 gallons), can be made by this process. Typical die materials are steel, aluminum, and beryllium copper.

In some operations, the extrusion is continuous and the molds move with the tubing. The molds close around the tubing, sealing off one end, breaking the long tube into individual sections, and moving away as air is injected into the tubular piece. The part is then cooled and ejected from the mold. Corrugated-plastic pipe and tubing are made by continuous blow molding, in which the pipe or tubing is extruded horizontally and then blown into moving molds.

In **injection blow molding**, a short tubular piece (**parison**) is injection molded (Fig. 19.14b) in cool dies. The dies are then opened, and the parison is transferred to a blow-molding die using an indexing mechanism (Fig. 19.14c). Hot air is injected into the parison, expanding it to contact the walls of the mold cavity. Typical products made are plastic beverage bottles (generally made of polyethylene or polyetheretherketone, PEEK) and small, hollow containers. A related process is **stretch blow molding**, in which the parison is expanded and elongated simultaneously, subjecting the polymer to biaxial stretching, and thus enhancing its properties.

Multilayer blow molding involves the use of coextruded tubes or parisons, and thus permits the production of a multilayer structure (see Fig. 19.4b). A typical example of such a product is plastic packaging for food and beverages, having such characteristics as odor and permeation barrier, taste and aroma protection, scuff resistance, the capability of being printed, and the ability to be filled with hot fluids. Other applications of this process are for containers in the cosmetics and the pharmaceutical industries.

19.5 Rotational Molding

Most thermoplastics and some thermosets can be formed into large, hollow parts by *rotational molding*. In this process, a thin-walled metal mold is made in two pieces (split-female mold), and is designed to be rotated about two perpendicular axes (Fig. 19.15). For each part cycle, a premeasured quantity of powdered plastic material is placed inside the warm mold. (The powder is obtained from a polymerization process that precipitates a powder from a liquid.) The mold is then heated, usually in a large oven, and is rotated continuously about its two principal axes.

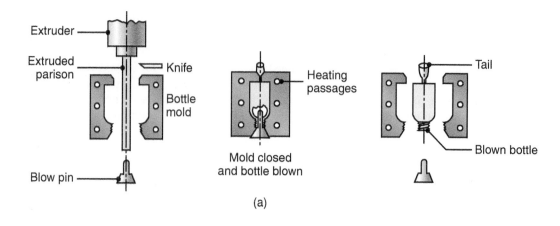

(a)

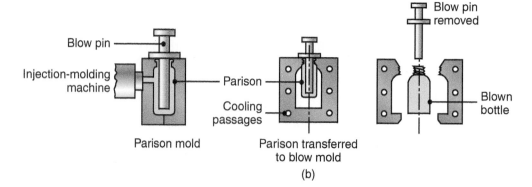

(b)

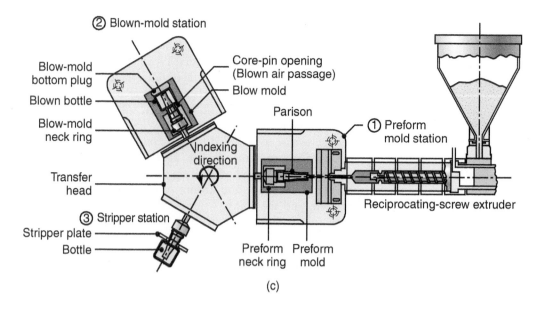

(c)

FIGURE 19.14 Schematic illustrations of (a) the extrusion blow-molding process for making plastic beverage bottles; (b) the injection blow-molding process; and (c) a three-station injection blow-molding machine for making plastic bottles.

This action tumbles the powder against the mold, where the heat fuses the powder, without melting it. For thermosetting parts, a chemical agent is added to the powder; cross-linking occurs after the part is formed in the mold. The machines are highly automated, with parts moved by an indexing mechanism, similar to that shown in Fig. 19.14c.

A large variety of parts are made by rotational molding, such as storage tanks, trash cans, boat hulls, buckets, housings, large hollow toys, carrying cases, and footballs. The outer surface finish of the part is a replica of the surface finish of the inside mold walls. Various metallic or plastic inserts or components also may be molded integrally into the parts.

In addition to powders, liquid polymers (**plastisols**) can be used in rotational molding, PVC plastisols being the most common material. In this operation, called **slush molding** or *slush casting*, the mold is heated and rotated simultaneously. Due to the tumbling action, the polymer is forced against the inside walls of the mold, where it melts and coats the mold walls. The part is cooled while it is still rotating, and removed by opening the mold. Parts made are typically thin-walled products, such as boots, buckets for aerial cranes, and toys.

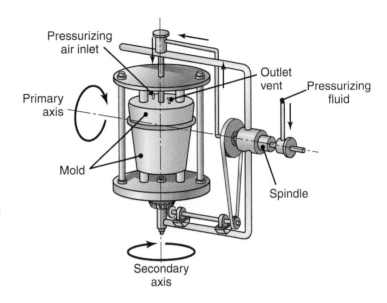

FIGURE 19.15   The rotational molding (rotomolding or rotocasting) process; trash cans, buckets, and plastic footballs can be made by this process.

Process Capabilities.   Rotational molding can produce parts with complex, hollow shapes with wall thicknesses as small as 0.4 mm (0.016 in.). Parts as large as 1.8 m × 1.8 m × 3.6 m (6 ft × 6 ft × 12 ft), with a volume as large as 80,000 liters (21,000 gallons) have been produced. Cycle times are longer than in other molding processes. Quality-control considerations usually involve accurate weight of the powder, proper rotational speed of the mold, and temperature–time relationships during the oven cycle.

19.6 Thermoforming

Thermoforming is a process for forming thermoplastic sheets or films over a mold through the application of heat and pressure (Fig. 19.16). In this process, a sheet is (a) clamped and heated to the *sag point* (above the *glass-transition temperature*, T_g, of the polymer; Table 7.2), usually by radiant heating and (b) forced against the mold surfaces by applying a vacuum or air pressure. The sheets are available as a coiled strip or as blanks, with lengths and widths of various sizes.

The mold is generally at room temperature, thus the shape produced becomes set upon contact with the mold. Because of the low strength of the materials formed, the pressure difference caused by a vacuum usually is sufficient for forming. Thicker and more complex parts require air pressure, which may range from about 100 to 2000 kPa (15–300 psi), depending on the type of material and its thickness. Mechanical means, such as the use of plugs, also may be employed to help form the parts. Variations of the basic thermoforming process are shown in Fig. 19.16.

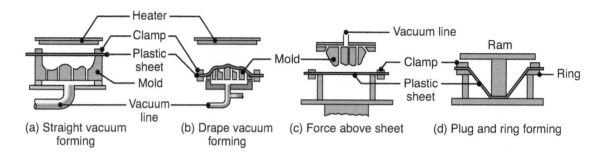

(a) Straight vacuum forming   (b) Drape vacuum forming   (c) Force above sheet   (d) Plug and ring forming

FIGURE 19.16 Various thermoforming processes for a thermoplastic sheet; these processes are commonly used in making advertising signs, cookie and candy trays, panels for shower stalls, and packaging.

Process Capabilities. Typical parts made by thermoforming are packaging, trays (such as for cookies and candy), advertising signs, refrigerator liners, appliance housings, and panels for shower stalls. Parts with openings or holes cannot be formed by this process, because the pressure difference cannot be maintained during forming. Because thermoforming is a combination of *drawing* and *stretching* operations, much like in some sheet-metal forming processes, the material must exhibit high uniform elongation, as otherwise it will neck and tear. Thermoplastics have high capacities for uniform elongation, by virtue of their high strain-rate sensitivity exponent, *m*, as described in Section 2.2.7.

Molds for thermoforming usually are made of aluminum, because high strength is not required; thus, tooling is relatively inexpensive. The molds have small through-holes in order to aid vacuum forming. These holes typically are less than 0.5 mm (0.02 in.) in diameter, as otherwise they may leave circular marks on the parts formed. *Defects* encountered in thermoforming include (a) tearing of the sheet during forming, (b) excessive nonuniform wall thickness, (c) improperly filled molds, (d) poor part definition, and (e) lack of surface details.

19.7 Compression Molding

In *compression molding*, a preshaped charge of material or a premeasured volume of powder, or a viscous mixture of liquid-resin and filler material, is placed directly into a heated mold cavity, which typically is around 200°C (400°F) but can be much higher. Forming is done under pressure from a plug or from the upper half of the die (Fig. 19.17); thus, the process is somewhat similar to closed-die forging of metals (Section 14.3). Polymers also can be molded by cold or hot isostatic pressing, described in Sec. 17.3.2.

Pressures range from about 10 to 150 MPa (1400–22,000 psi). As can be seen in Fig. 19.17, there is a flash formed, which subsequently is removed by trimming or some other means such as grinding. Typical parts made are dishes, handles, container caps, fittings, electrical and electronic components, washing-machine agitators, and housings. Fiber-reinforced parts with chopped fibers also are formed by this process.

Compression molding is mainly used with thermosetting plastics, with the original material being in a partially polymerized state; thermoplastics and elastomers are also processed by compression molding. Curing times are in the range of 0.5–5 min, depending on the material and part thickness and its shape. The thicker the material, the longer the time required to cure.

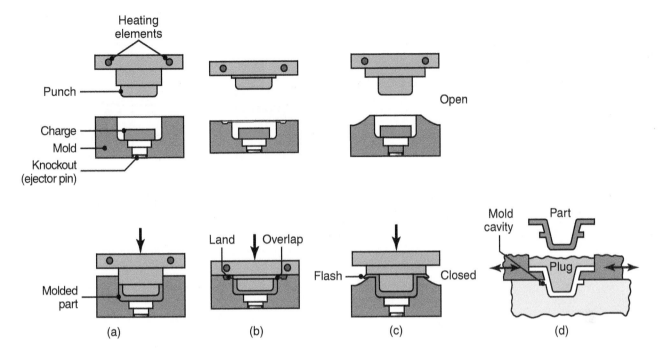

FIGURE 19.17 Types of compression molding, a process similar to forging: (a) positive; (b) semipositive; and (c) flash, in which the flash is later trimmed off. (d) Die design for making a compression-molded part with external undercuts.

Process Capabilities.   Three types of compression molds are made:

- *Flash type*, for shallow or flat parts
- *Positive type*, for high-density parts
- *Semipositive type*, for quality production

Undercuts in parts are not recommended; however, dies can be designed to open sideways (Fig. 19.17d) to allow removal of the molded part. In general, the complexity of parts produced is less than that from injection molding, but the dimensional control is better. Surface areas of compression-molded parts may range up to about 2.5 m^2 (8 ft^2). Because of their relative simplicity, dies for compression molding generally are less costly than those for injection molding. Die materials typically are tool steels and may be chrome plated or polished, for improved surface finish of the molded product.

19.8 Transfer Molding

Transfer molding is a further development of compression molding. The uncured thermosetting resin is placed in a heated transfer pot or chamber (Fig. 19.18), heated, and injected into heated closed molds. Depending on the type of machine used, a ram, plunger, or rotating-screw feeder forces the material to flow through the narrow channels into the mold cavity, at pressures up to 300 MPa (43,000 psi). The viscous flow generates considerable heat, which raises the temperature of the material and homogenizes it; curing takes place by cross-linking. Because the resin is in a molten state as it enters the molds, the complexity of the parts made and their dimensional control approach those of injection molding.

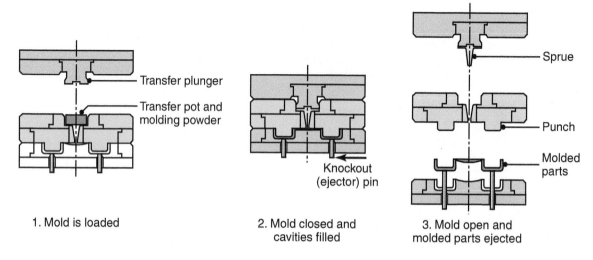

1. Mold is loaded   2. Mold closed and cavities filled   3. Mold open and molded parts ejected

FIGURE 19.18 Sequence of operations in transfer molding for thermosetting plastics; this process is suitable particularly for intricate parts with varying wall thickness.

Process Capabilities.   Typical parts made by transfer molding are electrical connectors and electronic components, rubber and silicone parts, and the encapsulation of microelectronic devices. The process is especially suitable for intricate shapes with varying wall thicknesses. The molds tend to be more expensive than those for compression molding, and some excess material is left in the channels of the mold during filling, which is later removed.

19.9 Casting

Some thermoplastics, such as nylons and acrylics, and thermosetting plastics, such as epoxies, phenolics, polyurethanes, and polyester, can be *cast* into a variety of shapes, using either rigid or flexible molds (Fig. 19.19a). Compared to other methods of processing plastics, casting is a slow, but simple and inexpensive, process. Also, the polymer must have sufficiently low viscosity in order to flow easily into the mold. Typical parts cast are gears (especially nylon), bearings, wheels, thick sheets, lenses, and components requiring resistance to abrasive wear.

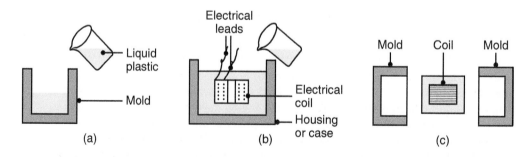

FIGURE 19.19 Schematic illustration of (a) casting, (b) potting, and (c) encapsulation processes for plastics and for electrical assemblies, where the surrounding plastic serves as a dielectric.

In basic conventional casting of thermoplastics, a mixture of monomer, catalyst, and various additives (*activators*) is heated to above its melting point, T_m, and poured into the mold. The part is shaped after polymerization takes place at ambient pressure. Intricate shapes can be produced using *flexible molds*, which are then peeled off (in a manner similar to using rubber gloves) and reused. As with metals (Section 5.4), thermoplastics may be continuously cast, with the polymer carried over continuous stainless-steel belts and polymerized by external heat.

Centrifugal Casting.   This process is similar to centrifugal metal casting, described in Section 11.4.6, and is used with thermoplastics, thermosets, and reinforced plastics with short fibers.

Potting and Encapsulation.   As a variation of casting, particularly important to the electrical and electronics industry, *potting* and *encapsulation* involve casting the plastic material (typically a liquid resin, such as expoxy) around an electrical component (such as a transformer) to embed it in the plastic. *Potting* (Fig. 19.19b) is carried out in a housing or case, which becomes an integral part of the component. In *encapsulation* (Fig. 19.19c), the component is coated with a layer of the plastic, surrounding it completely and then solidifying.

In both of these processes, the plastic material can serve as a *dielectric* (nonconductor); consequently, it must be free of moisture and porosity, which would require processing in a vacuum. Mold materials may be metal, glass, or various polymers. Small structural members, such as hooks and studs, may be encapsulated partially by dipping them in a hot thermoplastic. A wide variety of polymer colors and hardnesses are available.

19.10   Foam Molding

Products such as Styrofoam cups, food containers, thermally insulating blocks, and shaped packaging materials, such as for shipping appliances, computers, and electronics, are made by *foam molding*, using expandable **polystyrene beads** as the raw material. These products have a **cellular structure**, wherein it may have *open and interconnected porosity* (for polymers with low viscosity) or have *closed cells* (for polymers with high viscosity).

There are several techniques that can be used in foam molding. In the basic operation, polystyrene beads, obtained by polymerization of styrene monomer, are placed in a mold with a blowing agent, typically pentane (a volatile hydrocarbon) or inert gas (nitrogen), and exposed to heat, usually by steam. As a result, the beads expand to as much as 50 times their original size, taking the shape of the mold cavity. The amount of expansion can be controlled by varying the temperature and time. Various other particles, including hollow glass beads or plastic spheres, may be added to impart specific structural characteristics to the foam produced.

Polystyrene beads are available in three sizes: (a) small, for cups with a finished part density of about 50 kg/m^3 (3 lb/ft^3); (b) medium, for molded shapes; and (c) large, for molding insulating blocks, with a finished part density of about 15–30 kg/m^3 (which can then be cut to size). The bead size selected also depends on the minimum wall thickness of the product; the smaller the size, the thinner the part. The beads can be colored prior to expansion, thus making a part that is integrally colored. Both thermoplastics and thermosets can be used for foam molding, but thermosets are in a liquid-processing form, and are thus in a condition similar to that of polymers in RIM.

FIGURE 19.20 Cross-section of a structural foam molding, showing a dense skin and porous core. *Source:* Courtesy of M&T Industries.

A common method of foam molding is to use *pre-expanded polystyrene beads,* in which the beads are expanded partially by steam (hot air, hot water, or an oven also can be used) in an open-top chamber. The beads are then placed in a storage bin, and allowed to stabilize for a period of 3–12 h. They then can be molded into desired shapes, in the same manner described previously.

Structural Foam Molding. This process is used to make plastic products with a *solid outer skin* and a *cellular core structure* (Fig. 19.20). Typical products made are furniture components, computer and business-machine housings, and moldings (replacing more expensive wood moldings). In this process, thermoplastics are mixed with a blowing agent (usually an inert gas such as nitrogen) and injection molded into cold molds of desired shapes. The rapid cooling against the cold mold surfaces produces a skin that is rigid, which can be as much as 2 mm (0.08 in.) thick, and a core of the part that is cellular in structure. The overall part density can be as low as 40% of the density of the solid plastic. Thus, with a rigid skin and a less dense bulk, molded parts have a high stiffness-to-weight ratio (see also Fig. 3.2).

Polyurethane Foam Processing. Products such as furniture cushions and insulating blocks are made by this process. Basically, the operation starts with mixing two or more components; chemical reactions then take place after the mixture is (a) poured into molds of various shapes or (b) sprayed over surfaces, with a spray gun, to provide sound and thermal insulation. Various types of low-pressure and high-pressure machines are available, having computer controls to ensure proper mixing. The mixture solidifies into a cellular structure, the characteristics of which depend on the type and proportion of the components used.

19.11   Cold Forming and Solid-phase Forming

Processes that have been used in the cold working of metals (such as rolling, closed-die forging, coining, deep drawing, and rubber forming—all described in Part III) also can be used to form thermoplastics at room temperature (*cold forming*). Typical materials formed are polypropylene, polycarbonate, ABS, and rigid PVC. Important considerations regarding this process are that (a) the polymer must be sufficiently ductile at room temperature, thus polystyrenes, acrylics, and thermosets cannot be formed and (b) its deformation must be nonrecoverable, in order to minimize springback and creep of the formed part.

The advantages of cold forming over other methods of shaping plastics are:

- Strength, toughness, and uniform elongation are increased
- Plastics with high molecular weights (Section 7.2) can be used to make parts with superior properties
- Forming speeds are not affected by part thickness because, unlike other processing methods, there is no heating or cooling involved; also, cycle times generally are shorter than those in molding processes

Solid-phase Forming. Also called *solid-state forming,* this process is carried out at a temperature 10° to 20°C (20°–40°F) below the melting temperature of the plastic (for a crystalline polymer). Thus, the forming operation takes place while the polymer is still in a solid state. The main advantages of this process over cold forming are that forming forces and springback are lower. These processes are not used as widely as hot-processing methods and generally are restricted to special applications.

19.12 Processing Elastomers

Recall from Section 7.9 that, in terms of its processing characteristics, an *elastomer* is a polymer; in terms of its function and performance, it is a *rubber*. The raw material to be processed into various shapes is basically a compound of rubber and various additives and fillers. The additives include carbon black, an important element that enhances such properties as tensile and fatigue strength, abrasion and tear resistance, ultraviolet protection, and resistance to chemicals.

These materials are then mixed to break them down and to lower their viscosity; the mixture is subsequently *vulcanized*, using sulfur as the vulcanizing agent. This compound is then ready for further processing, such as calendering, extrusion, and molding, which may also include placing reinforcements, in such forms as fibers and fabric. During processing, the part becomes cross-linked, imparting the desirable properties that are associated with rubber products, ranging from rubber boots to pneumatic tires.

Elastomers can be shaped by a variety of processes that also are used for shaping thermoplastics. Thermoplastic elastomers are commonly shaped by extrusion or injection molding, the former being the more economical and faster process; they also can be formed by blow molding or thermoforming. Thermoplastic polyurethane, for example, can be shaped by any of the conventional methods. It also can be blended with thermoplastic rubbers, polyvinyl chloride compounds, ABS, and nylon to obtain specific properties.

The temperatures for elastomer extrusion are typically in the range from 170° to 230°C (340°–440°F), and for molding are up to 60°C (140°F). Dryness of the materials is important for product integrity. Reinforcements are used in conjunction with extrusion, to impart greater strength. Examples of extruded elastomer products are tubing, hoses, moldings, and inner tubes. Injection-molded elastomer products cover a broad range of applications, such as components for automobiles and appliances.

Rubber and some thermoplastic sheets are formed by the **calendering** process (Fig. 19.21), wherein a warm mass of the compound is fed into a series of rolls, and **masticated** (compressed and kneaded into a pulp). The thickness produced is typically 0.3 to 1 mm (0.01–0.040 in.), but can be made less by stretching the material. It then is stripped off, at speeds on the order of 2 m/s (6.5 ft/s) to form a sheet, which may be as wide as 3 m (10 ft). The calendered rubber then may be molded into various products, such as tires and belts for machinery. The rubber or thermoplastics also may be formed over both surfaces of a tape, paper, fabric, or plastics, thus making them permanently *laminated*. Roll surfaces may also be textured to produce a rubber sheet with various patterns and designs.

Discrete rubber products, such as gloves, balloons, swim caps, etc., are made by *dipping* or **dip molding** a solid metal form, such as in the shape of a hand for making gloves, repeatedly into a liquid compound that adheres to the form. A typical compound is *latex*, a milk-like sap obtained from the inner bark of a tropical tree. The compound is then vulcanized (cross-linked), usually in steam, and stripped from the form, thus becoming a discrete product.

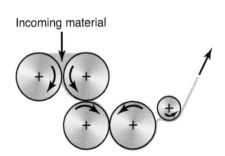

Incoming material

FIGURE 19.21 Schematic illustration of calendering; sheets produced by this process subsequently are used in thermoforming; this process also is used in the production of various elastomer and rubber products.

FIGURE 19.22   Reinforced-plastic components for a Honda motorcycle; the parts shown are front and rear forks, a rear swing arm, a wheel, and brake disks.

19.13   Processing Polymer-matrix Composites

As described in Chapter 9, *polymer-matrix composites* (PMC), also called **reinforced plastics**, are *engineered materials* with unique mechanical properties, especially high strength-to-weight ratio, stiffness-to-weight ratio, fatigue strength, creep resistance, and directional properties. Because of their complex structure, however, reinforced plastics require special methods to shape them into consumer and industrial products (Fig. 19.22).

Fabrication to ensure reliable properties in composite parts and structures, particularly over the long range of their service life, can be challenging, because of the presence of two or more types of materials. The matrix and the reinforcing fibers in the composite have, by design, very different properties and characteristics, and consequently have different responses to the methods of processing (Section 9.2).

The several steps required for manufacturing reinforced plastics, and the time and care involved, make processing costs very high, and generally not competitive with fabricating more common materials and shapes. This situation has necessitated the proper assessment and integration of design and manufacturing processes (*concurrent engineering*), in order to take advantage of the unique properties of these composites. This must be done by minimizing manufacturing costs while maintaining long-range product integrity, reliability, and production rate. An important safety and environmental concern in reinforced plastics is the dust generated during processing. For example, airborne carbon fibers are known to remain in the work area long after the fabrication of parts has been completed.

19.13.1   Fiber Impregnation

For good bonding between the reinforcing fibers and the polymer matrix, and to protect them during handling, fibers are surface treated by impregnation (*sizing*). When impregnation is carried out as a separate step, the resulting partially cured sheets are called by various terms, as described below.

Prepregs.   In a typical procedure for making fiber-reinforced plastic *prepregs* (meaning pre-impregnated with resin), the continuous fibers are aligned and subjected to a surface treatment to enhance the adhesion to the polymer matrix (Fig. 19.23a). They then are coated by dipping them in a resin bath and are made into a tape (Fig. 19.23b), typically in widths of 75 to 150 mm (3–6 in.). Individual segments of prepreg tape are then cut and assembled into *laminated structures* (Fig. 19.24a), such as for the horizontal stabilizer for the F-14 fighter aircraft.

Typical composites made from prepregs are flat or corrugated architectural paneling, panels for construction and electrical insulation, and structural components of aircraft, requiring good property retention over a period of time and under adverse conditions (including fatigue strength under hot or wet conditions), typically encountered by military aircraft.

Because the process of laying prepreg tapes is a time-consuming and labor-intensive operation, special and highly automated *computer-controlled tape-laying machines* have been built for this purpose (Fig. 19.24b). The prepreg tapes automatically are cut from a reel and placed on a mold in the desired patterns, with much

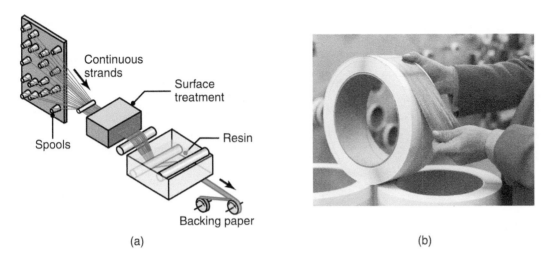

(a)

(b)

FIGURE 19.23 (a) Manufacturing process for polymer-matrix composite tape. (b) Boron-epoxy prepreg tape. These tapes are then used in making reinforced plastic parts and components with high strength-to-weight and stiffness-to-weight ratios, particularly important for aircraft and aerospace applications and sports equipment. *Source:* (a) After T.W. Chou, R.L. McCullough, and R.B. Pipes. (b) Courtesy of Avco Specialty Materials/Textron.

(a)

(b)

FIGURE 19.24 (a) Single-ply layup of boron-epoxy tape for the horizontal stabilizer for an F-14 fighter aircraft. (b) A 10-axis computer-numerical-controlled tape-laying system; this machine is capable of laying up 75- and 150-mm (3- and 6-in.) wide tapes on contours of up to and at speeds of up to 0.5 m/s (1.7 ft/s). *Source:* (a) Courtesy of Grumman Aircraft Corporation. (b) Courtesy of The Ingersoll Milling Machine Company.

better dimensional control than can be achieved by hand. The layout patterns can be modified easily and quickly for a variety of parts, by computer control and with high repeatability.

Sheet-molding Compound.   In making *sheet-molding compound* (SMC), continuous strands of reinforcing fiber are first chopped into short fibers (Fig. 19.25), and deposited in random orientations over a layer of resin paste. Generally, the paste is a polyester mixture (which may contain fillers, such as various mineral powders), and is carried on a polymer film, such as polyethylene. A second layer of resin paste is then deposited on top, and the sheet is pressed between rolls.

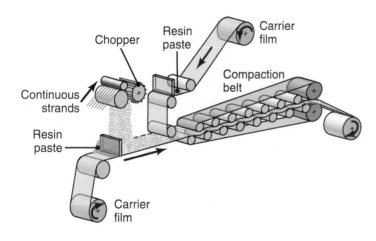

FIGURE 19.25 Schematic illustration of the manufacturing process for producing fiber-reinforced plastic sheets; the sheet still is viscous at this stage and later can be shaped into various products. *Source:* After T.-W. Chou, R.L. McCullough, and R.B. Pipes.

The product is then gathered into rolls, or placed into containers in several layers, and stored until it has undergone a maturation period and has reached the desired viscosity. The maturing process takes place under controlled conditions of temperature and humidity, that usually takes about one day.

The molding compounds should be stored at a temperature sufficiently low to delay curing. They have a limited shelf life, usually about 30 days, and hence must be processed within this period. Alternatively, the resin and the fibers can be mixed together only at the time they are to be placed into the mold.

Bulk-molding Compound. *Bulk-molding compounds* (BMCs) are in the shape of billets (hence the term "bulk"), and generally are up to 50 mm (2 in.) in diameter. They are made in the same manner as SMCs and are extruded to produce a bulk form. When processed into products, BMCs have flow characteristics that are similar to those of dough, thus they also are called *dough-molding compounds.*

Thick-molding Compound. *Thick-molding compounds* (TMCs) combine a characteristic of BMCs (lower cost) with one of SMCs (higher strength). They are generally injection molded, using chopped fibers of various lengths. One application is in electrical components because of the high dielectric strength of TMCs.

19.13.2   Molding of Reinforced Plastics

There are several molding processes used for reinforced plastics.

Compression Molding. The material is placed between two molds, and pressure is applied. The molds may be either at room temperature or heated to accelerate hardening of the part. The material may be a bulk-molding compound, which is a viscous, sticky mixture of polymers, fibers, and additives, powder, or it can be an uncured thermoset with a dough-like consistency. Generally, it is molded into the shape of a log, which subsequently is cut or sliced into the desired shape. Fiber lengths generally range from 3 to 50 mm (0.125–2 in.), although longer fibers of 75 mm (3 in.) also may be used.

Sheet-molding compounds also can be processed by compression molding; they are similar to BMC, except that the resin–fiber mixture is laid between plastic sheets, to make a sandwich that can be handled easily. The sheets have to be removed prior to placing the SMC in the mold.

Vacuum-bag Molding. In this process (Fig. 19.26a), prepregs are laid in a mold to form the desired shape. The pressure required, to shape the product and to ensure good bonding, is applied by covering the layup with a plastic bag and creating a vacuum. Curing takes place at room temperature or in an oven.

A variation of this process is **pressure-bag molding** (Fig. 19.26b). A flexible bag is placed over the resin and reinforcing fiber mixture, then pressure is applied over the mold and at a range typically from 200 to 400 kPa (30–50 psi). If higher heat and pressure are needed to produce parts with higher density and fewer

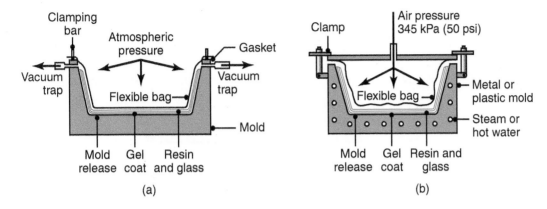

FIGURE 19.26 Schematic illustration of (a) vacuum-bag molding and (b) pressure-bag molding; these processes are used in making discrete reinforced plastic parts. *Source:* After T.H. Meister.

voids, the entire assembly is placed into an *autoclave* (a chamber under heat and pressure).

Care should be exercised to maintain fiber orientation if specific directional properties are desired. In chopped-fiber materials, no specific orientation is intended. In order to prevent the resin from sticking to the vacuum bag, and also to facilitate removal of excess resin, several sheets of various materials, called *release cloth* or *bleeder cloth*, are placed on top of the prepreg sheets.

The molds can be made of metal, usually aluminum, but more often they are made from the same resin (with reinforcement) as the material to be cured. This practice eliminates any difficulties caused by the difference in thermal expansion between the mold and the part.

Contact Molding.   Also referred to as *open-mold processing*, this is a series of processes that uses a single male or female mold, made of such materials as reinforced plastics, wood, metal, or plaster (Fig. 19.27). The operation is a wet method, in which the materials are applied in layers, and the reinforcement is impregnated with the resin at the time of molding. Contact molding is used in making *laminated products*, with high surface area-to-thickness ratios, hence the process is also called *contact lamination*. Typical examples of products are backyard swimming pools, boat hulls, automotive-body panels, tub and shower units, and housings.

The simplest method of contact molding is **hand layup.** The materials are placed in proper order (resins and reinforcements), brushed with liquid monomer, and shaped in the mold by hand with a roller (Fig. 19.27a). The squeezing action of the roller expels any trapped air bubbles and compacts the part. The reinforcements placed in the mold may consist of various shapes, including prepregs, and their orientation in the final product can be controlled.

In **spray layup,** molding is done by spraying the materials into the mold. As seen in Fig. 19.27b, both the resin and the chopped fibers are sprayed over the mold surfaces. Rolling the deposited materials (as in hand layup) to remove any porosity may be necessary. Because the chopped fibers have random orientations, directional properties cannot be imparted in products made by spray layup. Note also that only the mold-side surface of the formed part is smooth, from being in contact with the mold surfaces.

Both hand layup and spray layup are relatively slow operations, have high labor costs, and require significant time and labor in finishing operations. Also, the choice

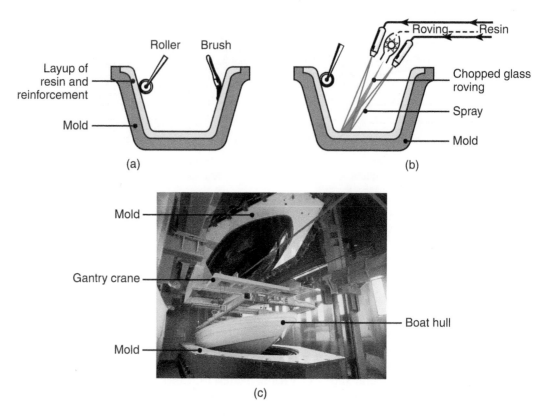

FIGURE 19.27 Manual methods of processing reinforced plastics: (a) hand layup and (b) spray layup. Note that, even though the process is slow, only one mold is required. The figures show a female mold, but male molds are used as well; these methods also are called *open-mold processing*. (c) A boat hull being made by these processes. *Source:* Courtesy of VEC Technology, LLC.

of materials that can be used is limited. However, both processes are simple to perform and the tooling is inexpensive.

Resin-transfer Molding. This process is based on transfer molding of plastics (Section 19.8). A resin is mixed with a catalyst, and is forced by a piston-type, positive-displacement pump into the mold cavity, which is filled with a fiber reinforcement. The process is a viable alternative to hand layup, spray layup, and compression molding for low- or intermediate-volume production.

Transfer/Injection Molding. This is an automated operation that combines compression-molding, injection-molding, and transfer-molding processes. This combination has the good surface finish, dimensional stability, and mechanical properties obtained in compression molding, as well as the high-automation capability and low cost of injection molding and transfer molding.

19.13.3 Filament Winding, Pultrusion, and Pulforming

Filament Winding. This is a process in which the resin and fibers are combined at the time of curing (Fig. 19.28a), in order to develop a composite structure. Axisymmetric parts, such as pipes and storage tanks, and even nonsymmetric parts, are produced on a rotating mandrel. The reinforcing filament, tape, or roving is wrapped

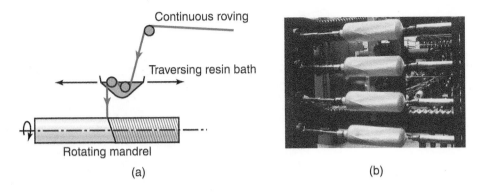

Continuous roving

Traversing resin bath

Rotating mandrel

(a)

(b)

FIGURE 19.28   (a) Schematic illustration of the filament-winding process; (b) fiberglass being wound over aluminum liners for slide-raft inflation vessels for the Boeing 767 aircraft. The products made by this process have a high strength-to-weight ratio and also serve as lightweight pressure vessels. *Source:* Courtesy of Brunswick Corporation.

continuously around the form. The reinforcements are impregnated by passing them through a polymer bath. Filament winding can be modified by using a prepreg material, such as a tape, to wrap the mandrel.

The products made by filament winding are very strong, because of their highly reinforced structure. Parts as large as 4.5 m (15 ft) in diameter and 20 m (65 ft) long have been made by this process. The process also has been used for strengthening cylindrical or spherical pressure vessels (Fig. 19.28b), made of materials such as aluminum and titanium, where the presence of a metal inner lining makes the part impermeable. Filament winding also can be used directly over solid-rocket propellant forms. Seven-axis computer-controlled machines have been developed that automatically dispense several unidirectional prepregs, to also make such nonsymmetric parts as aircraft engine ducts, fuselages, propellers, blades, and struts.

Pultrusion.   Long parts with various uniform cross-sections, such as rods, profiles, flat strips, and tubing, are made continuously by the pultrusion process. The sequence of operations is shown in Fig. 19.29. The continuous reinforcement, glass roving, or fabric (typically made of E type calcium aluminosilicate glass fiber; see Section 9.2.1) is supplied through several bobbins. The bundle is pulled first through a thermosetting polymer bath (usually polyesters), then through a preforming die, and finally through a heated steel die.

The product is cured during its travel through the heated die, which has a length of up to 1.5 m (5 ft) and a speed that allows sufficient time for the polymer to set. Note that this is an operation similar to continuously baking bread and cookies or making resin-bonded grinding wheels. The exiting material is then cut into desired lengths. Cross-sections as large as 1.5 m × 0.3 m (60 in. × 12 in.) have been made by this process. Typical products made by pultrusion, which may contain up to about 75% reinforcing fiber, are golf clubs, ski poles, fishing poles, driveshafts, and such structural members as ladders, walkways, and handrails.

Pulforming.   Continuously reinforced products, other than those with constant cross-sectional profiles, are made by *pulforming*. After being pulled through the polymer bath, the composite is clamped between the two halves of a die and cured into a finished product. Commonly made products are hammer handles reinforced by glass fibers and curved automotive leaf springs.

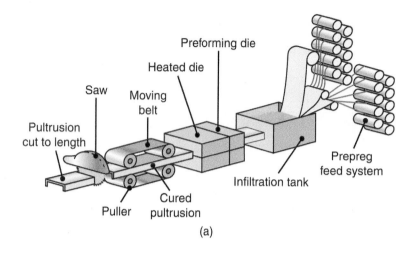

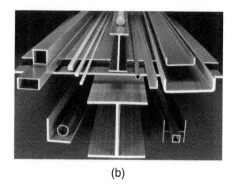

(a)

(b)

FIGURE 19.29   (a) Schematic illustration of the pultrusion process. (b) Examples of parts made by pultrusion. The major components of fiberglass ladders (used especially by electricians) are made by this process; they are available in different colors, but are heavier because of the presence of glass fibers. *Source:* Courtesy of Strongwell Corporation.

CASE STUDY 19.1   Polymer Automotive-body Panels Shaped by Various Processes

Polymeric materials are commonly used for automobile bodies; this example outlines typical applications of polymers. Three commonly used and competing processing methods are: (a) injection-molding of thermoplastics and elastomers, (b) RIM of polyurea/polyurethanes, and (c) compression-molding of SMC with resin-transfer-molded polyester and vinylester.

Typical examples of parts made are: (a) body panels and other large exterior components made by injection molding, (b) front fenders and rear quarter panels made of polyphenylene-ether/nylon or thermoplastic polyester, (c) outer door panels made of polycarbonate/ABS, and (d) fascias made of thermoplastic polyolefin. These materials are selected for design flexibility, impact strength and toughness, corrosion resistance, high durability, and low mass. Vertical panels and fascias are made in multicavity molds on large injection-molding machines, then assembled mechanically to a steel frame.

Large exterior-body parts also are made of reaction-injection molded polyurethane, although polyureas are important for body panels and bumpers. Thermoset fascias are made of reinforced RIM polyurethane and polyureas because of their higher thermal stability, low-temperature toughness, and lower cycle times. Large horizontal exterior-body panels, such as hoods, roofs, and rear decks, are made of reinforced polyester or vinylester in the form of compression-molded SMCs. Lower volume parts are made by resin-transfer molding.

Environmental and recycling considerations in material and process selection for automobiles, as well as other products, continue to be important. For example, polyphenylene oxide is being replaced with polycarbonate, which is made of 100% recycled or reclaimed materials.

19.13.4   Quality Considerations in Processing Reinforced Plastics

The major quality considerations in the processes described thus far concern internal voids and gaps between successive layers of material. Volatile gases, which develop during processing, must be allowed to escape from the layup through the vacuum bag, in order to avoid porosity due to trapped gases. Microcracks may develop during improper curing or during the transportation and handling of parts. These defects can be detected using ultrasonic scanning and other techniques described in Section 36.10.

CASE STUDY 19.2   Manufacture of Head Protector® Tennis Racquets

Competitive tennis is a demanding sport, and as a result, there is a strong demand to produce exceptionally lightweight and stiff racquets to improve performance. A tennis racquet consists of a number of regions, as shown in Fig. 19.30. Of particular interest is the sweet spot; when the tennis ball is struck at the sweet spot, the player has optimum control and power, and vibration is minimized. Several innovative racquet-head designs have been developed over the years to maximize the size of the sweet spot. A stiff composite material, typically with high-modulus graphite fibers in an epoxy matrix (see Chapter 9) is used in the manufacture of the racquet head. The orientation of the fibers varies in different locations of the racquet. For example, the main tube for the racquet consists of carbon-epoxy prepreg, oriented at $\pm 30°$ from layer to layer.

The advantages to such materials are obvious, in that stiff racquets allow higher forces to be applied to the ball. However, the use of these advanced materials has led to an increased frequency of tennis elbow, a painful condition associated with the tendons, that anchor muscles to the bones at the elbow. The condition is due not only to the higher forces involved, but also to the associated greater vibration of the racquet encountered with each stroke, especially when balls are struck away from the sweet spot.

An innovative design for a racquet, the Protector (made by Head Sport AG) uses lead zirconate titanate (PZT) fibers as an integral layer of the composite racquet frames. PZT is well known as a piezoelectric material (see Section 3.7); that is, it produces an electric response when deformed. Modules of the fibers, called Intellifibers®, are integrated into the throat on all sides of the racket; that is, left, right, front, and back. The module consists of about 50 PZT fibers, each approximately 0.3 mm in diameter, sandwiched between two polyamide layers, with printed electrodes for generating the potential difference when the fibers are bent.

During impact, the vibrations constantly excite the Intellifibers, generating a very high voltage potential but at low current. The energy is stored in coils on the printed circuit board (Chipsystem®) incorporated in the racquet handle in real time, and released back to the Intellifibers, in the optimal phase and waveform for the most efficient damping. The

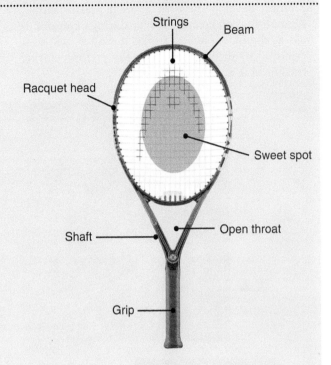

FIGURE 19.30 A Head Protector tennis racquet. *Source: Courtesy of Head Sport AG.*

stored energy is sent back to the Intellifibers in a phase that causes a mechanical force opposite to the vibration, thereby reducing it. The Chipsystem is tuned to the first natural frequency of the racket, and can damp vibrations only within a range of its design frequency.

The manufacture of a Protector tennis racquet involves a number of steps. First, a carbon-epoxy prepreg is produced, as described in Section 19.13.1. The prepreg is cut to the proper size and placed on a flat, heated bench to make the matrix material tackier, resulting in better adhesion to adjacent layers. A polyamide sleeve (or bladder) is then placed over a rod, and the prepreg is rolled over the sleeve. When the bar is removed, the result is a tube of carbon-epoxy prepreg with a polyamide sleeve, that can be placed in a mold and internally pressurized to develop the desired cross-section.

The throat piece is molded separately by wrapping the prepreg around sand-filled polyamide preforms or expandable foam. Since there is no

(continued)

easy way to provide air pressure to the throat, the preform develops its own internal pressurization, because of the expansion of air during exposure to elevated molding temperatures. If sand is used, it is removed by drilling holes into the preform during the finishing operation.

Prior to molding, all the components are assembled onto a template, and final prepreg pieces are added to strategic areas. The main tube is bent around the template, and the ends are pressed together and wrapped with a prepreg layer to form the handle. The PZT fibers are incorporated

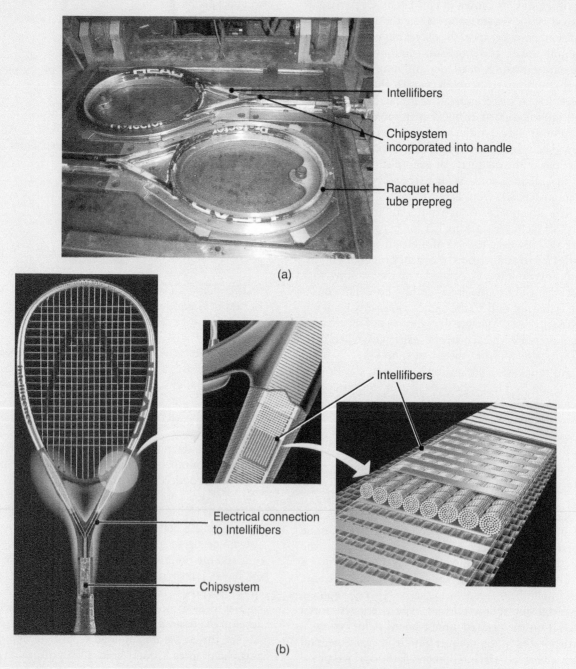

(a)

(b)

FIGURE 19.31   (a) The composite Head Protector racquet immediately after molding and (b) a completed Head Protector racquet, highlighting the incorporation of piezoelectric Intellifibers. *Source:* Courtesy of J. Kotze and R. Schwenger, Head Sport AG.

as the outer layer in the racquet in the throat area, and the printed electrodes are connected to the Chipsystem. The racquet is then placed into the mold, internally pressurized, and allowed to cure. Note that this operation is essentially an internally-pressurized, pressure-bag molding process (see Fig. 19.26b). A racquet as it appears directly after molding is shown in Fig. 19.31a.

The racquet then undergoes a number of finishing operations, including flash removal, drilling of

holes to accommodate strings, and finishing of the handle, including wrapping it with a special grip material. A completed Head Protector racquet is shown in Fig. 19.31b. This design has been found to reduce racquet vibrations by up to 50%, resulting in clinically proven reductions in tennis elbow, without any compromise in performance.

Source: Courtesy of J. Kotze and R. Schwenger, Head Sport AG.

19.14 Processing Metal-matrix and Ceramic-matrix Composites

Metal-matrix composites can be made into near-net shaped parts by the following processes:

- **Liquid-phase processing** basically consists of casting together the liquid-matrix material (such as aluminum or titanium) and the solid reinforcement (such as graphite, aluminum oxide, or silicon carbide) by conventional casting processes or by pressure-infiltration casting. In the latter process, pressurized gas forces the liquid-metal matrix into a preform, usually shaped out of wire or sheet and made of reinforcing fibers.
- **Solid-phase processing** utilizes powder metallurgy techniques (Chapter 17), including cold and hot isostatic pressing. Proper mixing is important for homogeneous distribution of the fibers throughout the part. An example is the production of tungsten-carbide tools and dies, with cobalt as the matrix material.
- **Two-phase (liquid–solid) processing** involves technologies that consist of rheocasting (Section 11.4.7) and the techniques of *spray atomization* and *deposition*. In the latter two processes, the reinforcing fibers are mixed with a matrix that contains both liquid and solid phases of the metal.

In making complex metal-matrix composite parts with whisker or fiber reinforcement, die geometry and control of process variables are very important, for ensuring the proper distribution and orientation of the fibers within the part. MMC parts made by powder metallurgy techniques generally are heat treated for optimum properties.

CASE STUDY 19.3 Metal-matrix Composite Brake Rotors and Cylinder Liners

Some brake rotors are made of composites consisting of an aluminum-based matrix, reinforced with 20% silicon-carbide particles. First, the particles are stirred into molten aluminum alloys, and the mixture is cast into ingots. The ingots are then remelted

and cast into shapes, by such casting processes as green-sand, bonded-sand, investment, permanent-mold, and squeeze casting. These rotors (a) are about one-half the weight of those made of gray cast iron, (b) conduct heat three times faster, (c) add stiffness

(continued)

and wear-resistance characteristics of ceramics, and (d) reduce noise and vibration, because of internal damping in the rotors.

To improve the wear- and heat-resistance of cast-iron cylinder liners in aluminum engine blocks, aluminum-matrix liners are also available. The metal-matrix layer consists of 12% aluminum-oxide fiber and 9% graphite fiber, and has a thickness that ranges from 1.5 to 2.5 mm (0.06–0.1 in.).

19.14.1 Processing Ceramic-matrix Composites

Several processes, including techniques such as melt infiltration, controlled oxidation, and hot-press sintering, are used to make ceramic-matrix composites; other processes are:

- **Slurry infiltration** is the most common process for making ceramic-matrix composites. It involves the preparation of a fiber preform, which is first hot pressed and then impregnated with a combination of slurry (containing the matrix powder), a carrier liquid, and an organic binder. High strength, toughness, and uniform structure are obtained by this process, but the product has limited high-temperature properties. A further improvement on the process is *reaction bonding* or *reaction sintering* of the slurry.
- **Chemical-synthesis** processes involve the sol-gel and the polymer-precursor techniques. In the *sol-gel process*, a *sol* (a colloidal fluid having the liquid as its continuous phase) that contains fibers is converted to a *gel*, which is then subjected to heat treatment to produce a ceramic-matrix composite. The *polymer-precursor method* is analogous to the process used in making ceramic fibers with aluminum oxide, silicon nitride, and silicon carbide.
- In **chemical-vapor infiltration**, a porous fiber preform is infiltrated with the matrix phase, using the chemical vapor deposition technique (Section 34.6). The product has very good high-temperature properties, but the process is time consuming and costly.

19.15 Design Considerations

Design considerations in forming and shaping plastics are similar to those for casting metals (Section 12.2). The selection of appropriate materials from an extensive list requires considerations of (a) service requirements, (b) possible long-range effects on properties and behavior, such as dimensional stability and wear, and (c) ultimate disposal of the product after its life cycle. (Some of these issues are described in Sections I.4 and I.6 in the General Introduction, and Section 7.8.)

Outlined below are the general design guidelines for the production of plastic and composite-material parts.

1. The processes for plastics have inherent flexibility, thus a wide variety of part shapes and sizes can be produced. Complex parts with internal and external features can be produced with relative ease, and at high production rates. Consequently, a process such as injection molding competes well with powder-injection molding and die casting. All are capable of producing complex shapes and having thin walls. In process substitutions, it is essential to consider the materials involved and their characteristics that may be very different, each having its own properties that are critical to a particular method of production.

2. Compared with metals, plastics have much lower stiffness and strength; thus, section size, shape, and thickness must be selected accordingly. Depending

on the application, a high section modulus can be achieved on the basis of design principles common to I-beams and tubes. Large, flat surfaces can be stiffened by such simple means as specifying curvatures on parts; for example, observe the stiffness of very thin, but gently curved slats in venetian blinds. Reinforcement with fibers or particles also are effective in achieving stiff and lightweight designs.

3. The overall part shape and thickness often determine the particular shaping or molding process to be selected. Even after a particular process is chosen, the design of parts and the dies should be appropriate for a particular shape generation (Fig. 19.32), dimensional control, and surface finish. Because of low stiffness and thermal effects, dimensional tolerances, especially for thermoplastics, are not as small as in metalworking processes. For example, dimensional tolerances are much smaller in injection molding than they are in thermoforming. As in casting metals and alloys, the control of material flow in the mold cavities is important. The effects of molecular orientation during the processing of the polymer also must be considered, especially in extrusion, thermoforming, and blow molding.

4. Large variations in cross-sectional areas and section thicknesses, as well as abrupt changes in geometry, should be avoided. Note, for example, that the *sink marks* (pull-in) shown in the top piece in Fig. 19.32c are due to the fact that thick sections in a part solidify last. Moreover, contractions in larger cross-sections during cooling tend to cause *porosity* in plastic parts, as they do in metal casting (see Fig. 12.2), thus affecting product integrity and quality. By contrast, a lack of stiffness may make it more difficult to remove thin parts from molds after shaping them.

5. The low elastic moduli of plastics further requires that shapes be selected properly for improved stiffness of the component (Fig. 19.32b), particularly when saving material is an important factor. Note that these considerations are similar to those applicable to the design of metal castings and forgings, as is the need for drafts (typically less than 1° for polymers) to enable removal of the part from molds and dies. Generally, the recommended part thickness ranges from about 1 mm (0.04 in.) for small parts to about 3 mm (0.12 in.) for large parts.

6. Physical properties, especially a high coefficient of thermal expansion, are important factors. Improper part design can lead to uneven shrinking (Fig. 19.32a)

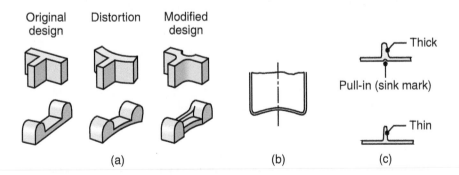

FIGURE 19.32 Examples of design modifications to eliminate or minimize distortion in plastic parts: (a) suggested design changes to minimize distortion; (b) stiffening the bottoms of thin plastic containers by doming, a technique similar to the process used to shape the bottoms of aluminum beverage cans (see Fig. 16.31); and (c) design change in a rib to minimize pull-in (*sink mark*) caused by shrinkage during the cooling of thick sections in molded parts.

and distortion (warping). Plastics can easily be molded around metallic parts and inserts; however, their interfacial strength and compatibility with metals when so assembled is an important consideration.

7. The properties of the final product depend on the original material and its processing history. For example, cold working of polymers improves their strength and toughness. On the other hand, because of the nonuniformity of deformation, even in simple rolling, residual stresses develop in polymers, just as they do in metals. These stresses also can be due to the thermal cycling of the part during processing. The magnitude and direction of residual stresses, however produced, are important factors, such as in stress cracking over time. Furthermore, these stresses can relax over a period of time and cause distortion of the part during its service life.

8. A major design advantage of reinforced plastics is the directional nature of the strength of the composite (see, for example, Fig. 9.7). External forces applied to the part are transferred by the matrix to the fibers, which are much stronger and stiffer than the matrix. When all of the fibers are oriented in one direction, the resulting composite material is exceptionally strong in the fiber direction, but weak in the transverse direction. To achieve strength in two principal directions, individual unidirectional layers are laid at controlled angles to each other, as is done in filament winding. If strength in the third (thickness) direction is required, a different type of reinforcement, such as woven fiber, is used to form a sandwich structure.

19.16 Economics of Processing Plastics and Composite Materials

General characteristics of processing of plastics and composite materials are given in Table 19.2. Note the wide range of equipment and tooling costs, and the economic production quantities. As described throughout this chapter, it will be noted that there is some relationship between equipment costs and tool and die costs.

The most expensive machines are for injection molding, followed by compression molding and transfer molding; tool and die costs also are high for these operations.

TABLE 19.2

Comparative Production Characteristics of Various Molding Methods

Molding method	Equipment and tooling cost*	Production rate*	Economical production quantity*
Extrusion	M–L	VH–H	VH
Injection molding	VH	VH	VH
Rotational molding	M	M–L	M
Blow molding	M	H–M	H
Compression molding	H–M	M	H–M
Transfer molding	H	M	VH
Thermoforming	M–L	M–L	H–M
Casting	M–L	M–L	L
Centrifugal casting	H–M	M–L	M–L
Pultrusion	H–M	H	H
Filament winding	H–M	L	L
Spray layup and hand layup	L–VL	L–VL	L

*VH = very high; H = high; M = medium; L = low; VL = very low.

Thus, in an operation like injection molding, the size of the die, and the optimum number of cavities in the die for producing more and more parts in one cycle are important considerations, as they are in an operation like die casting. Larger dies may be considered in order to accommodate several cavities (with runners to each cavity), but at the expense of increasing die cost even further. On the other hand, more parts will be produced per machine cycle, thus the production rate will increase.

A detailed analysis is thus required in order to determine the overall die size, the number of cavities in the die, and the machine capacity required to optimize the total operation, and to produce parts at minimum cost. Similar considerations also apply to all other processing methods described throughout this chapter.

SUMMARY

- Thermoplastics can be shaped by a variety of processes, including extrusion, molding, casting, and thermoforming, as well as by some of the processes used in metalworking. The raw material usually is in the form of pellets, granules, and powders.

- The high strain-rate sensitivity of thermoplastics allows extensive stretching in forming operations; thus, complex and deep parts can be produced easily. Thermosetting plastics generally are molded or cast, and they have better dimensional accuracy than forming thermoplastics.

- Fiber-reinforced plastics are processed into structural components using liquid plastics, prepregs, and bulk- and sheet-molding compounds. Fabricating techniques include various molding methods, filament winding, pultrusion, and pulforming. The type and orientation of the fibers and the strength of the bond between fibers and matrix and between layers of materials are important considerations.

- The design of plastic parts must take into account their low strength and stiffness, and such physical properties as high thermal expansion and generally low resistance to temperature. Inspection techniques are available to determine the integrity of these products.

- Processing of metal-matrix and ceramic-matrix composites continues to undergo significant developments, to ensure product integrity, reliability, and reduced costs. Metal-matrix composites are processed by liquid-phase, solid-phase, and two-phase processes. Ceramic-matrix composites can be processed by slurry infiltration, chemical synthesis, or chemical-vapor infiltration.

- The relevant factors in the economics of the operations described include the costs of the machinery, the level of controls, tooling, cycle times, and production rate and volume.

KEY TERMS

Blow molding	Coextrusion	Foam molding	Open-mold processing
Blow ratio	Cold forming	Hand layup	Overmolding
Bulk-molding compound	Compression molding	Ice-cold forming	Parison
Calendering	Contact molding	Injection molding	Pellets
Casting	Encapsulation	Insert molding	Plastisols
Chemical synthesis	Extrusion	Liquid-phase processing	Potting
Chemical-vapor infiltration	Extrusion blow molding	Masticated	Prepregs
Coat-hanger die	Filament winding	Melt spinning	Pulforming

Pultrusion
Reaction-injection molding
Resin-transfer molding
Rotational molding
Sheet-molding compound
Sink marks

Sizing
Slurry infiltration
Slush molding
Solid-phase forming
Solid-phase
    processing

Spinneret
Spinning
Spray layup
Structural foam molding
Swell
Thermoforming

Thick-molding compound
Transfer molding
Two-phase processing
Vacuum-bag molding

BIBLIOGRAPHY

Campbell, F., **Manufacturing Processes for Advanced Composites**, Elsevier, 2004.

Chanda, J.M., and Roy, S.K., **Plastics Technology Handbook**, 4th ed., CRC Press, 2006.

Dave, R.S., and Loos, A.C., **Processing of Composites**, Hanser Gardner, 2000.

Gerdeen, J.C., Lord, H.W., and Rorrer, R.A.L., **Engineering Design with Polymers and Composites**, CRC Press, 2006.

Gordon, Jr., M.J., **Industrial Design of Plastics Products**, Wiley, 2002.

Harper, C, **Handbook of Plastics, Elastomers & Composites**, 4th ed., McGraw-Hill, 2002.

Harper, C., **Handbook of Plastics Technologies: The Complete Guide to Processes and Performance**, 2nd ed., McGraw-Hill, 2006.

Johnson, P.S., **Rubber Processing: An Introduction**, Hanser Gardner, 2001.

Kutz, M., **Applied Plastics Engineering Handbook: Processing and Materials**, William Andrew, 2011.

Mallick, P.K. (ed.), **Fiber-reinforced Composites: Materials, Manufacturing and Design**, 3rd ed., Dekker, 2008.

Mazumdar, S.K., **Composites Manufacturing: Materials, Products and Process Engineering**, CRC Press, 2001.

Strong, A.B., **Fundamentals of Composites Manufacturing**, 2nd ed., Society of Manufacturing Engineers, 2007.

Strong, A.B., **Plastics: Materials and Processing**, 3rd ed., Prentice Hall, 2005.

Tadmore, Z., and Gogos, C.G., **Principles of Polymer Processing**, 2nd ed., Wiley, 2006.

REVIEW QUESTIONS

19.1 What are the forms of raw materials for processing plastics into products?

19.2 What is extrusion? What products are produced by polymer extrusion?

19.3 Describe the features of an extruder screw and their functions.

19.4 How are injection-molding machines rated?

19.5 What is (a) a parison, (b) a plastisol, and (c) a prepreg?

19.6 How is thin plastic film produced?

19.7 List several common products that can be made by thermoforming.

19.8 What similarities and differences are there between compression molding and closed-die forging?

19.9 Explain the difference between potting and encapsulation.

19.10 What is thermoforming?

19.11 Describe runner, gate, sprue, and well.

19.12 Describe the advantages of cold-forming plastics over other plastic-processing methods.

19.13 What are the characteristics of filament-wound products? Explain why they are desirable.

19.14 Describe the methods that can be used to make tubular plastic products.

19.15 What is pultrusion? Pulforming?

19.16 How are very thin plastic film produced?

19.17 What process is used to make foam drinking cups?

19.18 If a polymer is in the form of a thin sheet, is it a thermoplastic or thermoset? Why?

19.19 How are polymer fibers made? Why are they much stronger than bulk forms of the polymer?

19.20 What are the advantages of coextrusion?

19.21 Explain how latex rubber gloves are made.

QUALITATIVE PROBLEMS

19.22 Describe the features of a screw extruder and its functions.

19.23 Explain why injection molding is capable of producing parts with complex shapes and fine detail.

19.24 Describe the advantages of applying the traditional metal-forming techniques, described in Chapters 13 through 16, to making (a) thermoplastic and (b) thermoset products.

19.25 Explain the reasons that some plastic-forming processes are more suitable for certain polymers than for others. Give examples.

19.26 Describe the problems involved in recycling products made from reinforced plastics.

19.27 Can thermosetting plastics be used in injection molding? Explain.

19.28 Inspect some plastic containers, such as those containing talcum powder, and note that the integral lettering on them is raised rather than depressed. Explain.

19.29 An injection-molded nylon gear is found to contain small pores. It is recommended that the material be dried before molding it. Explain why drying will solve this problem.

19.30 Explain why operations such as blow molding and film-bag making are performed vertically.

19.31 Comment on the principle of operation of the tape-laying machine shown in Fig. 19.24b.

19.32 Typical production rates are given in Table 19.2. Comment on your observations and explain why there is such a wide range.

19.33 What determines the cycle time for (a) injection molding, (b) thermoforming, and (c) compression molding? Explain.

19.34 Does the pull-in defect (sink marks) shown in Fig. 19.32c also occur in metal-forming and casting processes? Explain.

19.35 What determines the intervals at which the indexing head in Fig. 19.14c rotates from station to station?

19.36 Identify processes that would be suitable for small production runs on plastic parts, of, say, 100.

19.37 Identify processes that are capable of producing parts with the following fiber orientations in each: (a) uniaxial, (b) cross-ply, (c) in-plane random, and (d) three-dimensional random.

19.38 Inspect several electrical components, such as light switches, outlets, and circuit breakers, and describe the process or processes used in making them.

19.39 Inspect several similar products that are made of metals and plastics, such as a metal bucket and a plastic bucket of similar shape and size. Comment on their respective thicknesses, and explain the reasons for their differences, if any.

19.40 What are the advantages of using whiskers as a reinforcing material?

19.41 Construct a table that lists the main manufacturing processes described in this chapter. Indicate those that can be used for (a) thermoplastics, (b) thermosets, and (c) composite materials.

QUANTITATIVE PROBLEMS

19.42 Estimate the die-clamping force required for injection molding five identical 8-in.-diameter disks in one die. Include the runners of appropriate length and diameter.

19.43 A 2-liter plastic beverage bottle is made by blow molding a parison 5 in. long and with a diameter that is the same as that of the threaded neck of the bottle. Assuming uniform deformation during molding, estimate the wall thickness of the tubular portion of the parison.

19.44 Consider a Styrofoam drinking cup. Measure the volume of the cup and its weight. From this information, estimate the percent increase in volume that the polystyrene beads have undergone.

19.45 In Fig. 19.2, what flight angle, θ, should be used so that one flight translates to a distance equal to the barrel diameter with each revolution?

SYNTHESIS, DESIGN, AND PROJECTS

19.46 Make a survey of a variety of sports equipment, such as bicycles, tennis racquets, golf clubs, and baseball bats, and identify the components made of composite materials. Explain the reasons for and advantages of using composites for these specific applications.

19.47 Explain the design considerations involved in replacing a metal beverage can with one made completely of plastic.

19.48 Give examples of several parts suitable for insert molding. How would you manufacture these parts if insert molding were not available?

19.49 Give other examples of design modifications in addition to those shown in Fig. 19.32.

19.50 With specific examples, discuss the design issues involved in making products out of plastics vs. reinforced plastics.

19.51 Die swell in extrusion is radially uniform for circular cross-sections, but is not uniform for other cross-sections. Recognizing this fact, make a qualitative sketch of a die profile that will produce (a) square, (b) triangular, (c) elliptical, and (d) gear-shaped cross-sections of extruded polymer.

19.52 Inspect various plastic components in a typical automobile, and identify the processes that could have been used in making them.

19.53 Write a brief paper on how plastic coatings are applied to (a) electrical wiring; (b) sheet-metal panels; (c) wire baskets, racks; and similar structures; and (d) handles for electricians tools, such as wire cutters and pliers requiring electrical insulation.

19.54 It is well known that plastic forks, spoons, and knives are not particularly rigid. What suggestions would you have to make them better? Describe processes that could be used for producing them.

19.55 Some plastic products have lids with integral hinges; that is, no other material or part is used at the junction of the two parts. Identify such products, and describe a method for making them.

19.56 Make a survey of the technical literature, and describe how different types of (a) pneumatic tires, (b) automotive hoses, and (c) garden hoses are manufactured.

19.57 Obtain a boxed kit for assembling a model car or airplane. Examine the injection-molded parts provided, and describe your thoughts on the layout of the molds to produce these parts.

19.58 In injection-molding operations, it is common practice to remove the part from its runner, place the runner in a shredder, and recycle the runner by producing pellets. List the concerns you may have in using such recycled pellets for products, as against virgin pellets.

19.59 An increasing environmental concern is the very long period required for the degradation of polymers in landfills. Noting the information given in Section 7.8 on biodegradable plastics, conduct a literature search on the trends and developments in the production of these plastics.

19.60 Examine some common and colorful plastic poker chips and give an opinion on how they were manufactured.

19.61 Obtain different styles of toothpaste tubes, carefully cut them across, and comment on your observations regarding (a) the type of materials used and (b) how the tubes were produced.

19.62 By incorporating small amounts of blowing agent, it is possible to manufacture polymer fibers with gas cores. List some applications for such fibers.

Rapid-prototyping Processes and Operations

- This chapter describes the technologies associated with rapid prototyping, sharing the characteristics of computer integration, production without the use of traditional tools and dies, and the ability to rapidly produce a single part on demand; all have the basic characteristics of producing individual parts layer by layer.
- The chapter describes the nonmetallic and metallic materials used in rapid prototyping, and covers the commercially important rapid-prototyping technologies, including fused-deposition modeling, stereolithography, polyjet modeling, three-dimensional printing, and selective laser sintering.
- The chapter ends with a description of the practice of applying rapid-prototyping techniques to the production of tooling that can be used in other manufacturing processes.

Typical parts made: A wide variety of metallic and nonmetallic parts for product design analysis, evaluation and finished products.

Alternative processes: Machining, casting, molding, and fabricating.

20.1 Introduction

In the development of a new product, there is invariably a need to produce a sample, or **prototype,** of a designed part or system before allocating large amounts of capital to new production facilities or assembly lines. The main reasons for this need are that the capital is very high and production tooling takes considerable time to prepare. Consequently, a working prototype is needed for design evaluation and troubleshooting before a complex product or system is ready to be produced and marketed.

A typical product development process is outlined in Fig. I.2 in the General Introduction. An iterative process occurs naturally when (a) errors are discovered or (b) more efficient or better design solutions are gleaned from the study of an earlier generation prototype. The main problem with this approach, however, is that the production of a prototype can be very time consuming. Tooling can take several months to prepare, and the production of a single complex part by conventional manufacturing operations can be very difficult. Furthermore, during the time that a prototype is being prepared, facilities and staff still generate costs.

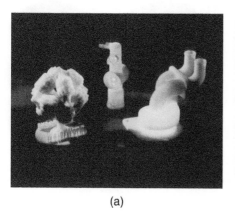

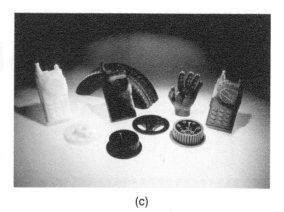

(a)   (b)   (c)

FIGURE 20.1   Examples of parts made by rapid-prototyping processes: (a) selection of parts from fused-deposition modeling; (b) stereolithography model of a communications device; and (c) selection of parts from three-dimensional printing. *Source:* (a) Courtesy of Stratasys, Inc. (b) and (c) Courtesy of 3D Systems, Inc.

QR Code 20.1 3D printing of a wrench with moving parts. (*Source:* Courtesy of Stratasys, Inc.)

An even more important concern is the speed with which a product flows from concept to a marketable item. In a competitive marketplace, it is well known that products that are introduced before those of their competitors generally are more profitable, and enjoy a larger share of the market. At the same time, there are important concerns regarding the production of high-quality products, and premature introduction of products can ruin a brand by causing frustration associated with poor performance or lost productivity due to repairs and upgrades. For these reasons, there is a concerted effort to bring high-quality products to market quickly.

A technology that speeds up the iterative product-development process considerably is the concept and practice of **rapid prototyping** (RP), also called **desktop manufacturing**, **digital manufacturing**, or **solid free-form fabrication**. Examples of rapid-prototyped parts are shown in Fig. 20.1.

CASE STUDY 20.1   Functional Rapid Prototyping

Toys are examples of mass-produced items that have universal appeal. Because some toys are actually quite complex, the function and benefits of a computer-aided design (CAD) cannot be ensured until prototypes are produced. Figure 20.2 shows a CAD model and a rapid-prototyped version of a water squirt gun (Super Soaker Power Pack Back Pack water gun), which was produced on a fused-deposition modeling machine. Each component was produced separately and assembled into the squirt gun; the prototype could actually hold and squirt water. The alternative would be to produce components on computer-numerical-control (CNC) milling machines or fabricate them in some fashion; however, this can be done only at much higher cost.

By producing a prototype, interference issues and assembly problems can be assessed and corrected, if necessary. Furthermore, from an aesthetic standpoint, the elaborate decorations on such a toy can be more effectively evaluated from a prototype than on a CAD file, and can be adjusted to improve the toy's appeal. Each component, having its design verified, then has its associated tooling produced, with better certainty that the tooling, as ordered, will produce the parts desired.

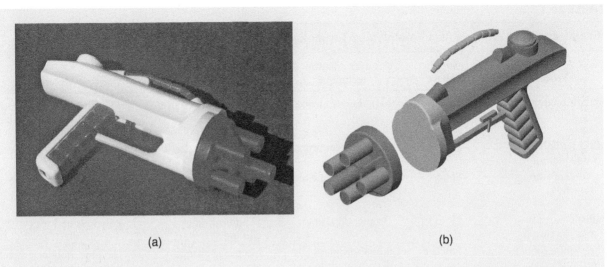

(a)   (b)

FIGURE 20.2   Rapid prototyping of a Super Soaker squirt gun. (a) Fully functional toy produced through fused-deposition modeling and (b) original CAD description. *Source:* Courtesy of Rapid Models and Prototypes, Inc., and Stratasys, Inc.

Developments in rapid prototyping began in the mid-1980s. The advantages of this technology include:

- Physical models of parts produced from CAD data files can be manufactured in a matter of hours, and thus allow the rapid evaluation of manufacturability and design effectiveness. In this way, rapid prototyping serves as an important tool for visualization and for product concept verification.
- With suitable materials, the prototype itself can be used in subsequent manufacturing operations, to produce the final parts. Also called *direct prototyping*, this approach can serve as an important manufacturing technology.
- Rapid-prototyping operations can be used in some applications to produce actual tooling for manufacturing operations (**rapid tooling**, see Section 20.6.1); thus, one can make tooling in a matter of a few days.

Rapid-prototyping processes can be classified into three major groups:

1. **Subtractive,** removing material from a workpiece that is larger than the final part
2. **Additive,** building up a part by adding material incrementally
3. **Virtual,** using advanced computer-based visualization technologies

Almost all materials can be used as a workpiece through one or more rapid-prototyping operations, as outlined in Table 20.1. However, because their properties are more suitable for these operations, polymers are the most commonly used material today, followed by metals and ceramics (see Table 20.2). Still, new processes are being introduced continually. The rest of this chapter serves as a general introduction to the most common rapid-prototyping operations, describe their advantages and limitations, and explore the present and future applications of these processes.

TABLE 20.1

Characteristics of Additive Rapid-prototyping Technologies

Process	Supply phase	Layer creation technique	Type of phase change	Materials
Stereolithography	Liquid	Liquid layer curing	Photopolymerization	Photopolymers (acrylates, epoxies, colorable resins, and filled resins)
Multi Jet/PolyJet modeling	Liquid	Liquid layer curing	Photopolymerization	Photopolymers
Fused-deposition modeling	Solid	Extrusion of melted polymer	Solidification by cooling	Polymers (such as ABS, polycarbonate, and polysulfone)
Ballistic-particle manufacturing	Liquid	Droplet deposition	Solidification by cooling	Polymers and wax
Three-dimensional printing	Powder	Binder-droplet deposition onto powder layer	No phase change	Ceramic, polymer, metal powder, and sand
Selective laser sintering	Powder	Layer of powder	Sintering or melting	Polymers, metals with binder, metals, ceramics and sand with binder
Electron-beam melting	Powder	Layer of powder	Melting	Titanium and titanium alloys, cobalt chrome
Laminated-object manufacturing	Solid	Deposition of sheet material	No phase change	Paper and polymers
Laser-engineered net shaping	Powder	Injection of powder stream	No phase change	Titanium, stainless steel, aluminum

20.2 Subtractive Processes

Making a prototype traditionally has involved a series of processes using a variety of tooling and machines, usually taking anywhere from weeks to months, depending on part complexity and size. This approach requires skilled operators and uses a series of *material removal by machining and finishing operations* (as described in detail in Part IV) until the prototype is completed. To speed the process, subtractive processes increasingly use computer-based technologies, such as:

- **Computer-based drafting packages,** which can produce three-dimensional representations of parts
- **Interpretation software,** which can translate the CAD file into a format usable by manufacturing software
- **Manufacturing software,** which is capable of planning the operations required to produce the desired part shape
- **Computer-numerical-control (CNC) machinery,** with capabilities necessary to produce the parts

When a prototype is required only for the purpose of shape verification, a soft material (usually a polymer or wax) is used as the workpiece, in order to reduce or avoid any machining difficulties. The material intended for use in the actual application also may be machined, but it may be more time consuming. Depending on part complexity and the level of machining capabilities in a plant, prototypes can be

TABLE 20.2

Mechanical Properties of Selected Materials for Rapid Prototyping

Process	Material	Tensile strength (MPa)	Elastic modulus (GPa)	Elongation in 50 mm (%)	Characteristics
Stereolithography	Accura 60	68	3.10	5	Transparent; good general-purpose material for rapid prototyping
	Somos 9920	9	1.35–1.81	15–26	Transparent amber; good chemical resistance; good fatigue properties; used for producing patterns in rubber molding
	WaterClear Ultra	56	2.9	6–9	Optically clear resin with ABS-like properties
	WaterShed 11122	47.1–53.6	2.65–2.88	11–20	Optically clear with a slight green tinge; mechanical properties similar to those of ABS; used for rapid tooling
	DMX-SL 100	32	2.2–2.6	12–28	Opaque beige; good general-purpose material for rapid prototyping
PolyJet	FC720	60.3	2.87	20	Transparent amber; good impact strength, good paint adsorption and machinability
	FC830	49.8	2.49	20	White, blue, or black; good humidity resistance; suitable for general-purpose applications
	FC 930	1.4	0.185	218	Semiopaque, gray, or black; highly flexible material used for prototyping of soft polymers or rubber
Fused-deposition modeling	Polycarbonate	52	2.0	3	White; high-strength polymer suitable for rapid prototyping and general use
	Ultem 9085	71.64	2.2	5.9	Opaque tan, high-strength FDM material, good flame, smoke and toxicity rating
	ABS-M30i	36	2.4	4	Available in multiple colors, most commonly white; a strong and durable material suitable for general use; biocompatible
	PC	68	2.28	4.8	White; good combination of mechanical properties and heat resistance
Selective laser sintering	WindForm XT	77.85	7.32	2.6	Opaque black polymide and carbon; produces durable heat- and chemical-resistant parts; high wear resistance
	Polyamide PA 3200GF	45	3.3	6	White; glass-filled polyamide has increased stiffness and is suitable for higher temperature applications
	SOMOS 201	—	0.015	110	Multiple colors available; mimics mechanical properties of rubber
	ST-100c	305	137	10	Bronze-infiltrated steel powder
Electron-beam melting	Ti-6Al-4V	970–1030	120	12–16	Can be heat treated by HIP to obtain up to 600-MPa fatigue strength

produced in a few days to a few weeks. Subtractive systems can take many forms; they are similar in approach to manufacturing cells, described in Section 39.2. Operators may or may not be involved in the operation, although the handling of parts is usually a human task.

20.3   Additive Processes

Additive rapid-prototyping operations all build parts in *layers*, and are summarized in Table 20.1. These processes consist of *stereolithography, Multi Jet/PolyJet modeling, fused-deposition modeling, ballistic-particle manufacturing, three-dimensional printing, selective laser sintering, electron-beam melting,* and *laminated-object manufacturing.* All of the processes described in this section *build parts layer by layer.* In order to visualize the methodology employed, it is beneficial to think of the construction of a loaf of bread, by stacking and bonding individual slices of bread on top of each other (hence the term *additive*). The main difference between the various additive processes lies in the method of producing the individual slices, which are typically 0.1–0.5 mm (0.004–0.020 in.) thick, although they can be thicker in some systems.

All additive operations require dedicated software. As an example, note the solid part shown in Fig. 20.3a. The first step is to obtain a CAD file description of the part; the computer then constructs slices of the three-dimensional part (Fig. 20.3b). Each slice is analyzed separately, and a set of instructions is compiled in order to provide the rapid-prototyping machine with detailed information regarding the manufacture of the part. Figure 20.3d shows the paths of the extruder in one slice, using the fused-deposition-modeling operation, described in Section 20.3.1.

This method requires operator input, both in the setup of the proper computer files and in the initiation of the production process. Following that stage, the machines generally operate unattended and provide a rough part after a few hours. The part can then be subjected to a series of manual finishing operations (such as sanding and painting) in order to complete the rapid-prototyping process.

The setup and finishing operations are very labor intensive and the production time is only a portion of the time required to obtain a prototype. In general, however, additive processes are much faster than subtractive processes, taking as little as a few minutes to a few hours to produce a part.

20.3.1   Fused-deposition Modeling

In the *fused-deposition-modeling* (FDM) process (Fig. 20.4), a gantry-robot controlled extruder head moves in two principal directions over a table, which can be raised and lowered as required. The extruder head is heated, and extrudes polymer filament at a constant rate through a small orifice. The head follows a predetermined path (see Fig. 20.3d); the extruded polymer bonds to the previously deposited layer. (The initial layer is placed on a foam foundation.) When the first layer is completed, the table is lowered so that subsequent layers can be superimposed over the first layer. When the part is finished, it can be easily removed from the foam base.

Complex parts, such as the one shown in Fig. 20.5a, may be difficult to build up directly, because once the part has been constructed up to height *a*, the next slice would require the filament to be placed at a location where no material exists underneath to support it. The solution is to extrude a support material separately from the modeling material, as shown in Fig. 20.5b. Note that the use of such support structures allows all of the layers to be supported by the material directly beneath them. The support material is produced with a less dense filament spacing on a layer,

QR Code 20.2 Fused deposition modeling. (*Source: Courtesy of Stratasys, Inc.*)

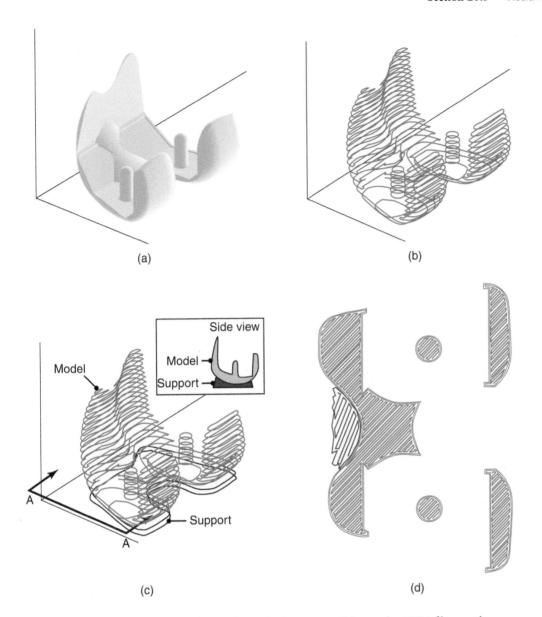

(a)

(b)

(c)

(d)

FIGURE 20.3 The computational steps in producing a stereolithography (STL) file; see also Fig. 38.5. (a) Three-dimensional description of part. (b) The part is divided into slices: only 1 in 10 is shown. (c) Support material is planned. (d) A set of tool directions is determined to manufacture each slice. Also shown is the extruder path at section A–A from (c) for a fused-deposition modeling operation.

so that it is weaker than the model material and can thus be broken off easily after the part is completed.

The layers in an FDM model are determined by the extrusion-die diameter, which typically ranges from 0.05 to 0.12 mm (0.002–0.005 in.). This thickness represents the best achievable tolerance in the vertical direction. In the x–y plane, dimensional accuracy can be as fine as 0.025 mm (0.001 in.) as long as a filament can be extruded into the feature. A variety of polymers are available for different applications. *Flat wire metal deposition* uses a metal wire, instead of a polymer filament, but also requires a laser to heat and bond the deposited wire to build parts.

Video Solution 20.1 Manufacture of a Klein bottle.

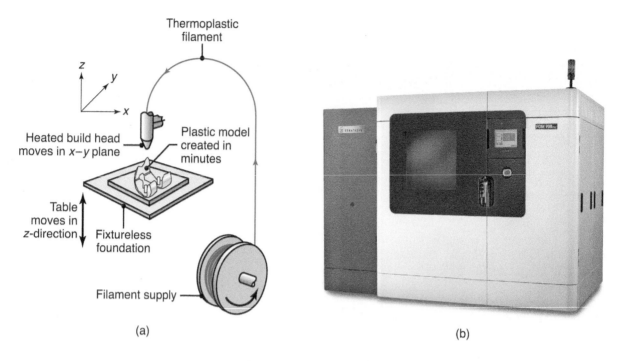

FIGURE 20.4 (a) Schematic illustration of the fused-deposition modeling process. (b) The FDM 900mc, a fused-deposition-modeling machine. *Source:* Courtesy of Stratasys, Inc.

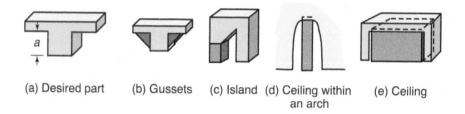

(a) Desired part   (b) Gussets   (c) Island   (d) Ceiling within an arch   (e) Ceiling

FIGURE 20.5 (a) A part with a protruding section that requires support material. (b) through (e) Common support structures used in rapid-prototyping machines. The gray areas are support material. *Source:* Reused with permission from Society of Manufacturing Engineers.

Close examination of an FDM-produced part will indicate that a stepped surface exists on oblique exterior planes. If this surface roughness is objectionable, a heated tool can be used to smoothen the surface, or it can be hand sanded, or a coating can be applied, often in the form of a polishing wax. However, the overall tolerances may then be compromised unless care is taken in these finishing operations.

Although some commercial FDM machines can be obtained for around $20,000, others can cost as much as $300,000. Some self-replicating machines (Section 20.5) are based on FDM and can cost around $1000. The main differences among them are the maximum size of the parts that can be produced and the numbers and types of materials that can be used.

20.3.2   Stereolithography

A common rapid-prototyping process, one that actually was developed prior to FDM, is *stereolithography* (STL). This process (Fig. 20.6) is based on the principle of *curing* (hardening) of a liquid photopolymer into a specific shape. A vat, containing a mechanism whereby a platform can be lowered and raised, is filled with a photocurable liquid-acrylate polymer. The liquid is a mixture of acrylic monomers, oligomers (polymer intermediates), and a photoinitiator (a compound that undergoes a reaction upon absorbing light).

At the highest position of the platform (*a* in Fig. 20.6), a shallow layer of liquid exists above the platform. A *laser*, generating an ultraviolet (UV) beam, is focused upon a selected surface area of the photopolymer, and then moved around in the *x–y* plane. The beam cures that portion of the photopolymer (say, a ring-shaped portion), and thereby produces a layer of the solid body. The platform is then lowered sufficiently to cover the cured polymer with another layer of liquid polymer, and the sequence is repeated. The process is repeated until level *b* in Fig. 20.6 is reached. Thus far, a cylindrical part with a constant wall thickness has been generated. Note that the platform is now lowered by a vertical distance *ab*.

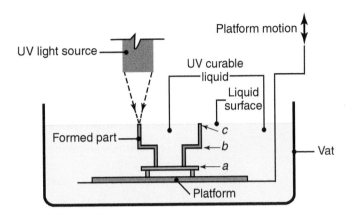

FIGURE 20.6 Schematic illustration of the stereolithography process.

At level *b*, the *x–y* movements of the beam define a wider geometry, thus a flange-shaped portion is being produced over the previously formed segment. After the proper thickness of the liquid has been cured, the process is repeated, producing another cylindrical section between levels *b* and *c*. Note that the surrounding liquid polymer is still fluid (because it has not been exposed to the UV beam), and that the part has been produced from the bottom up in individual "slices." The unused portion of the liquid polymer can be used again to make another part or another prototype.

Note that the term "stereolithography," as used to describe this process, comes from the facts that the movements are three dimensional (hence the word stereo) and the process is similar to lithography (see Section 28.7), in which the image to be printed on a flat surface is ink receptive and the blank areas are ink repellent. Note also that, as in FDM, stereolithography can require a weaker support material. In stereolithography, this support takes the form of perforated structures.

After its completion, the part is removed from the platform, blotted, and cleaned ultrasonically and with an alcohol bath. Then the support structure is removed, and the part is subjected to a final curing cycle in an autoclave. The smallest tolerance that can be achieved in stereolithography depends on the sharpness of the focus of the laser; typically, it is around 0.0125 mm (0.0005 in.). Oblique surfaces also can be produced with high quality.

Solid parts can be made by applying special laser-scanning patterns to speed up production. For example, by spacing scan lines in stereolithography, volumes or pockets of uncured polymer can be formed within cured solid shells. When the part is later placed in a postprocessing oven, the pockets cure and a solid part forms. Similarly, parts that are to be investment cast (Section 11.3.2) will have a drainable honeycomb structure, which permits a significant fraction of the part to remain uncured.

Total cycle times in stereolithography range from a few hours to a day, without post-processing steps such as sanding and painting. Depending on their capacity, the cost of the machines is in the range from $100,000 to $400,000. The cost of the liquid polymer is on the order from $300 per gallon. The maximum part size that can be produced is 0.5 m × 0.5 m × 0.6 m (19 in. × 19 in. × 24 in.).

Stereolithography has been used with highly focused lasers to produce parts with micrometer-sized features. The use of the optics required to produce such features necessitates thinner layers and lower volumetric cure rates. When stereolithography is used to fabricate micromechanical systems (Chapter 29), it is called **microstereolithography**.

20.3.3   MultiJet/PolyJet Modeling

The *MultiJet modeling* (MJM) or *PolyJet process* is similar to ink-jet printing, where print heads deposit the photopolymer on the build tray. Ultraviolet bulbs, alongside the jets, instantly cure and harden each layer, thus eliminating the need for any postmodeling curing (that is needed in stereolithography). The result is a smooth surface of layers as thin as 16 μm (0.0006 in.) that can be handled immediately after the process is completed.

Two different materials are used: one for the actual model, and a second gel-like resin for support, such as these shown in Fig. 20.5. Each material is simultaneously jetted and cured, layer by layer. When the model is completed, the support material is removed, with an aqueous solution. Build sizes have an envelope of up to 500 mm × 400 mm × 200 mm (20 in. × 16 in. × 8 in.). MJM has capabilities similar to those of stereolithography and uses similar resins (Table 20.2). The main advantages are the capabilities of avoiding part cleanup and lengthy post-process curing operations, and the much thinner layers produced, thus allowing for better resolution.

CASE STUDY 20.2   Coffeemaker Design

Alessi Corporation is well known for its high-end kitchen products. Although it makes products from a wide range of materials, it is best known for its highly polished stainless-steel designs. An example is the Cupola coffeemaker, a market favorite that was to be redesigned from the bottom up, while preserving the general characteristics of the established design.

Alessi engineers used MJM to produce prototypes of components of the coffeemaker, as shown in Fig. 20.7. The prototypes allowed engineers to evaluate the ease and security of mechanical assembly, but a significant effort was expended on the design of the coffeemakers lip in order to optimize the pouring of coffee. A large number of lip prototypes were constructed and evaluated, to obtain the most robust and aesthetically pleasing design. The ability to compare physical prototypes to the existing product was deemed essential to evaluating the new designs.

After a final design was selected from the numerous prototypes produced, it was found that a five to six week time savings was achieved in product development. The time savings resulted in cost savings, as well as assuring timely market launch of the redesigned product.

FIGURE 20.7 Coffeemaker prototypes produced through MultiJet modeling and final product (at right). *Source:* Courtesy Alessi Corporation, and 3D Systems, Inc.

20.3.4 Selective Laser Sintering

Selective laser sintering (SLS) is a process based on the sintering (Section 17.4) of non-metallic or, less commonly, metallic powders selectively into an individual object. The basic elements in this process are shown in Fig. 20.8. The bottom of the processing chamber is equipped with two cylinders:

1. A powder-feed cylinder, which is raised incrementally to supply powder to the part-build cylinder, through a roll mechanism
2. A part-build cylinder, which is lowered incrementally as the part is being formed

A thin layer of powder is first deposited in the part-build cylinder. Then a laser beam, guided by a process-control computer using instructions generated by the three-dimensional CAD program of the desired part, is focused on that layer, tracing and sintering a particular cross-section into a solid mass. The powder in other areas remains loose, but it supports the sintered portion. Another layer of powder is then deposited; this cycle is repeated again and again until the entire three-dimensional part has been produced. The loose particles are shaken off, and the part is recovered. The part does not require further curing, unless it is a ceramic, which has to be fired to develop strength.

A variety of materials can be used in this process, including polymers (such as ABS, PVC, nylon, polyester, polystyrene, and epoxy), wax, metals, and ceramics, with appropriate binders. It is most common to use polymers because of the smaller, less expensive, and less complicated lasers required for sintering. With ceramics and metals, it is a common practice to sinter only a polymer binder that has been blended with the ceramic or metal powders. If desired, the part can be sintered again in a furnace and infiltrated with another metal (see also Section 17.5).

QR Code 20.4 The selective laser sintering (SLS) process. (*Source:* Courtesy of 3D Systems)

QR Code 20.5 Selective laser sintering of metal. (*Source:* Production Systems Group, NIST)

20.3.5 Electron-beam Melting

A process similar to selective laser sintering and electron-beam welding (Section 30.6), *electron-beam,* or *E-beam, melting* uses the energy source associated with an electron

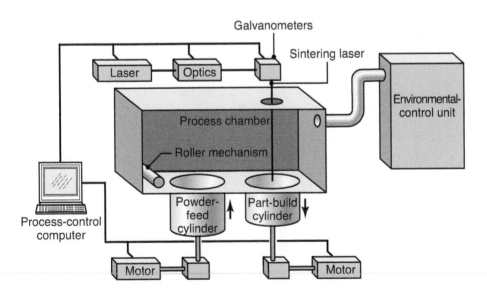

FIGURE 20.8 Schematic illustration of the selective-laser-sintering process. *Source:* After C. Deckard and P.F. McClure.

beam to melt titanium or cobalt-chrome powder to make metal prototypes. The workpiece is produced in a vacuum, thus part build size is limited to around 200 mm × 200 mm × 180 mm (8 in. × 8 in. × 6.3 in). Electron-beam melting (EBM) is up to 95% efficient from an energy standpoint, as compared with 10–20% efficiency for selective laser sintering, so that the (most commonly) titanium powder is actually melted, and fully dense parts can be produced. A volume build rate of up to 60 cm^3/h (3.7 in^3/h) can be obtained, with individual layer thicknesses of 0.050 to 0.200 mm (0.002–0.008 in.). Hot isostatic pressing (Section 17.3.2) also can be performed on parts, to improve their fatigue strength. Although applied mainly to titanium and cobalt-chrome to date, the process is being developed for stainless steels, aluminum, and copper alloys.

20.3.6   Three-dimensional Printing

QR Code 20.6 Color rapid prototyping with three-dimensional printing. (*Source:* Courtesy of 3D Systems)

In the *three-dimensional-printing* (3DP) process, a print head deposits an inorganic binder material onto a layer of polymer, ceramic, or metallic powder, as shown in Fig. 20.9. A piston, supporting the powder bed, is lowered incrementally, and with each step, a layer is deposited and then fused by the binder.

Multi Jet modeling and PolyJet processes (Section 20.3.3) are sometimes referred to as three-dimensional printing technologies, because they operate similarly to ink-jet printers, but incorporate a third (thickness) direction. In fact, 3DP has been used interchangeably with rapid prototyping or digital manufacturing to include all rapid prototyping operations; however, 3DP is most commonly associated with printing a binder onto powder.

Three-dimensional printing allows considerable flexibility in the materials and binders used. Commonly used powder materials are blends of polymers and fibers, foundry sand, and metals. Furthermore, since multiple binder print heads can be incorporated into one machine, it is possible to produce full-color prototypes by having different-color binders (see Case Study 20.3). The effect is a three-dimensional analog to printing photographs, using three ink colors (red, cyan, and blue) on an ink-jet printer.

A common part produced by 3DP from ceramic powder is a ceramic-casting shell (see Section 11.2.4), in which an aluminum-oxide or aluminum-silica powder is fused

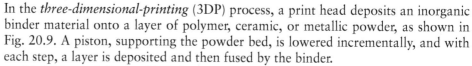

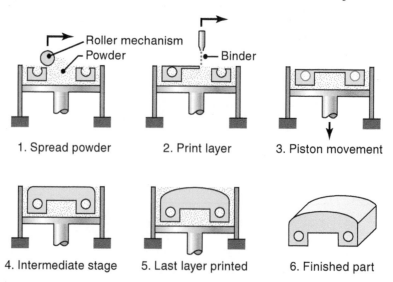

1. Spread powder   2. Print layer   3. Piston movement

4. Intermediate stage   5. Last layer printed   6. Finished part

FIGURE 20.9 Schematic illustration of the three-dimensional-printing process. *Source:* After E. Sachs and M. Cima.

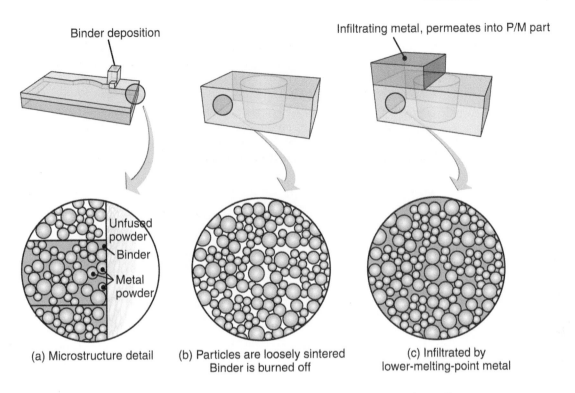

Binder deposition

Infiltrating metal, permeates into P/M part

(a) Microstructure detail

Unfused powder
Binder
Metal powder

(b) Particles are loosely sintered
Binder is burned off

(c) Infiltrated by
lower-melting-point metal

FIGURE 20.10 Three-dimensional printing using (a) part-build; (b) sinter; and (c) infiltration steps to produce metal parts. *Source:* Courtesy of The ExOne Company.

with a silica binder. The molds are post-processed in two steps: (a) curing at around 150°C (300°F) and (b) firing at 1000° to 1500°C (1840°–2740°F).

The parts produced through the 3DP process are somewhat porous, and thus may lack strength. Three-dimensional printing of metal powders also can be combined with sintering and metal infiltration (see Section 17.5) to produce fully dense parts, using the sequence shown in Fig. 20.10. Here, the part is produced as before by directing the binder onto powders. However, the build sequence is then followed by sintering to burn off the binder and partially fuse the metal powders, just as in powder injection molding, described in Section 17.3.3. Common metals used in 3DP are stainless steels, aluminum, and titanium. Infiltrating materials typically are copper and bronze, which provide good heat-transfer capabilities as well as wear resistance. This approach represents an efficient strategy for rapid tooling (Section 20.6.1).

In a related **ballistic-particle manufacturing process**, a stream of a material (such as plastic, ceramic, metal, or wax) is ejected through a small orifice and deposited on a surface (target), using an ink-jet type mechanism. A powder is not involved; the material deposited is used to build the prototype. The ink-jet head is guided by a three-axis robot to produce three-dimensional prototypes.

CASE STUDY 20.3   Production of Second Life Avatars

Second Life and World of Warcraft are examples of virtual worlds accessed through a website. To participate, users create an "avatar" that depicts

their alter ego in a fictional world. Many modern computer games (such as Rock Band 2) also allow users to produce very detailed avatars, with a unique

(continued)

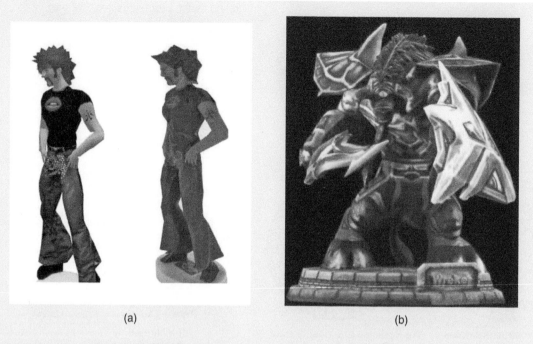

(a)   (b)

FIGURE 20.11 Rapid-prototyped versions of user-defined characters, or avatars, produced from geometric descriptions within popular websites or games. (a) Second Life avatar, as appears on a computer screen (left) and after printing (right) and (b) an avatar known as "Wreker" from World of Warcraft. *Source:* (a) Courtesy of Z Corporation. (b) Courtesy of Figure Prints and Fabjectory, Inc.

appearance and unique personalities. Avatars contain three-dimensional geometry data that describes their appearance, which can be translated to a file format suitable for rapid prototyping.

Avatars can be printed in full color to a 150-mm (6-in.) high figurine, with Z-Corp Spectrum Z510 or ZPrinter 450 three-dimensional printers (Fig 20.11). Users can order their avatar prototypes on the web, which are then printed and shipped to the user within days.

CASE STUDY 20.4   Fuselage Fitting for Helicopters

Sikorsky Aircraft Company needed to produce a limited number of the fuselage fittings shown in Fig. 20.12a through forging. Sikorsky wanted to produce the forging dies by means of three-dimensional-printing technologies. A die was designed, using the CAD part description. Forging allowances were incorporated and flashing accommodated by the die design.

The dies were produced using a ProMetal three-dimensional printer, as shown in Fig. 20.12b. The prototype was made by 0.178-mm (0.007-in.) layers, with stainless-steel powder as the workpiece media. The total time spent in the 3DP machine was just under 45 h. This was followed by curing of the binder (10 h, plus 5 h for cooldown), sintering (40 h, plus 17 h for cooldown), and infiltration (27 h, plus 15 h for cooldown). The dies then were finished and positioned in a die holder, and the part was forged in an 800-ton hydraulic press, with a die temperature of around 300°C. An as-forged part is shown in Fig. 20.12c, and requires trimming of the flash before it can be used. The dies were produced in just over six days, as compared with the several months required for conventional die production (Section 14.7).

Source: Courtesy of The ExOne Company.

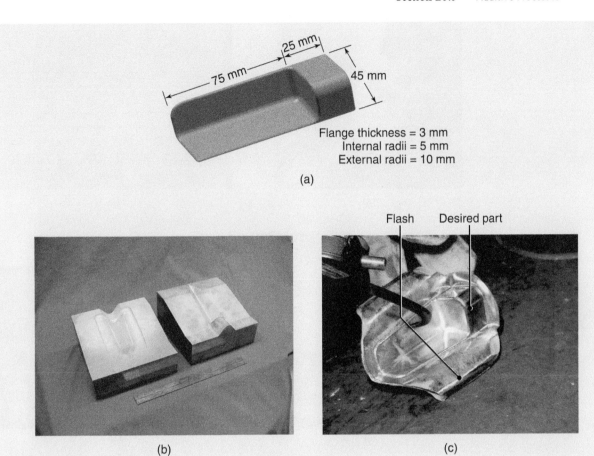

FIGURE 20.12   A fitting for a helicopter fuselage. (a) CAD representation with added dimensions. (b) Dies produced by three-dimensional printing. (c) Final forged workpiece. *Source:* (a) Courtesy of The ExOne Company; (b) and (c) Courtesy of Kennametal Extrude Hone Corporation.

20.3.7   Laminated-object Manufacturing

Lamination involves laying down layers that are bonded adhesively to one another. Several variations of *laminated-object manufacturing* (LOM) are available. The simplest and least expensive versions of LOM involve using control software and vinyl cutters to produce the prototype. Vinyl cutters are simple CNC machines that cut shapes from vinyl or paper sheets. Each sheet has a number of registration holes, which allow proper alignment and placement onto a build fixture. Figure 20.13 illustrates the manufacture of a prototype by LOM, with manual assembly. LOM systems are highly economical and are popular in schools and universities, because of the hands-on demonstration of additive manufacturing and production of parts by layers.

LOM systems can also be elaborate, where the more advanced systems use layers of paper or plastic, with a heat-activated glue on one side to produce parts. The shapes are burned into the sheet with a laser, and the parts are built layer by layer (Fig. 20.14). On some systems, the excess material must be removed manually once the part is completed. Removal is simplified by programming the laser to burn perforations in crisscrossed patterns. The resulting grid lines make the part appear as if it had been constructed from gridded paper, with squares printed on it, similar to graph paper.

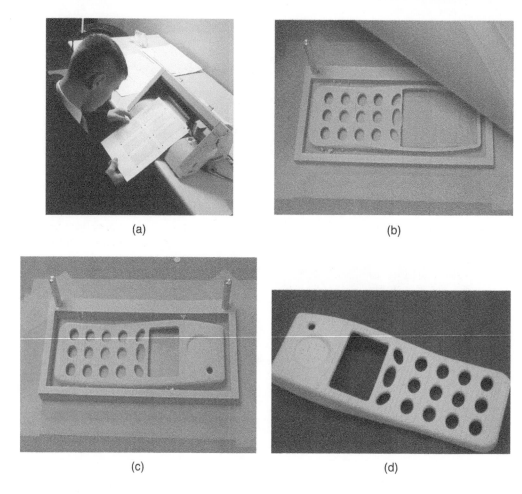

(a)

(b)

(c)

(d)

FIGURE 20.13 Production of a prototype through laminated-object manufacturing. (a) Layers are obtained from a vinyl cutter; (b) layers are manually stacked to form the part; (c) completed laminated assembly; and (d) final part prototype. *Source:* Courtesy of P. Barraclough, Boxford Ltd.

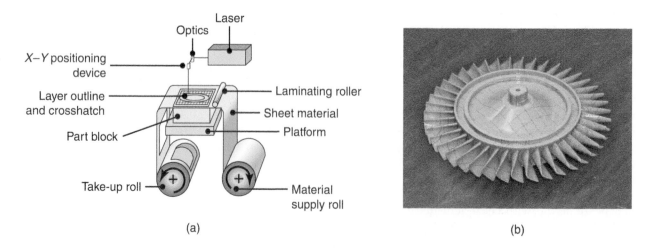

(a)

(b)

FIGURE 20.14 (a) Schematic illustration of the laminated-object-manufacturing process. (b) Turbine disk prototype made by LOM. *Source:* Courtesy of M. Feygin, Cubic Technologies, Inc.

20.3.8 Laser-engineered Net Shaping

More recent developments in additive manufacturing processes involve the principle of using a laser beam to melt and deposit metal powder or wire, again layer by layer, over a previously molten layer. The patterns of deposited layers are controlled by a CAD file. This near-net-shaping process is called *laser-engineered net shaping* (LENS, a trade name), and is based on the technologies of laser-beam welding and cladding (Sections 30.7 and 34.3). The heat input and cooling are controlled precisely to develop a favorable microstructure.

The deposition process is carried out inside a closed area and in an argon environment, to avoid the adverse effects of oxidation, particularly on aluminum. It is suitable for a wide variety of metals and specialty alloys, for the direct manufacturing of parts, including fully dense tools and molds. It can also be used for repairing thin and delicate components. There are other, similar processing methods using lasers, including *controlled-metal buildup* (CMB) and *precision-metal deposition* (PMD).

20.3.9 Solid-ground Curing

The *solid-ground curing* (SGC) process is unique in that entire slices of a part are manufactured at one time; as a result, a large throughput is achieved, compared with that from other rapid-prototyping processes. However, SGC is among the most expensive processes; hence, its adoption has been much less common than that of other types of rapid prototyping, and new machines are not available. Basically, the method consists of the following steps:

1. A slice is first created by the computer software, a mask of the slice then is printed on a glass sheet, by an electrostatic printing process similar to that used in laser printers. A mask is required because the area of the slice (where the solid material is desired) remains transparent.
2. While the mask is being prepared, a thin layer of photoreactive polymer is deposited on the work surface and is spread evenly.
3. The photomask is placed over the work surface, and a UV floodlight is projected through the mask. Wherever the mask is clear, the light shines through to cure the polymer, causing the desired slice to harden.
4. The unaffected resin, still liquid, is vacuumed off the surface.
5. Water-soluble liquid wax is then spread across the work area, filling the cavities previously occupied by the unexposed liquid polymer. Since the workpiece is on a chilling plate and it remains cool, the wax hardens quickly.
6. The layer is then milled (Section 24.2) to achieve the correct thickness and flatness.
7. This process is repeated, layer by layer, until the part is completed.

Solid-ground curing has the advantage of high production rate, because entire slices are produced at once, and the two glass screens are used concurrently; that is, while one mask is being used to expose the polymer, the next mask already is being prepared, and it is ready as soon as the milling operation is completed.

20.4 Virtual Prototyping

Virtual prototyping is a purely software form of prototyping; it uses advanced graphics and virtual-reality environments to allow designers to examine a part. This technology is used by common, conventional CAD packages, to render a part so that the designer can observe and evaluate it as drawn. However, the classification of virtual-prototyping is usually restricted to the most advanced rendering systems.

The simplest forms of such systems use complex software and three-dimensional graphics routines, to allow viewers to change the view of the parts on a computer screen. More complicated versions use virtual-reality headgear and gloves with appropriate sensors, to let the user observe a computer-generated prototype of the desired part in a completely virtual environment.

Virtual prototyping provides an instantaneous rendering of parts for evaluation, but the more advanced systems are costly. Because familiarity with software interfaces is a prerequisite to their application, these systems have steep learning curves. Moreover, many manufacturing and design practitioners prefer a physical prototype to evaluate, rather than a video-screen rendering. They often perceive virtual-reality prototypes to be inferior to mechanical prototypes, even though designers debug as many or more errors in the virtual environment.

There have been some important examples of complex products manufactured without any physical prototypes (**paperless design**). Perhaps the best known example is the Boeing 777 aircraft, for which mechanical fits and interferences were evaluated on a CAD system, and any difficulties were corrected before the first production model was manufactured (see Section 38.5).

QR Code 20.7 Features of a Cube. (*Source:* Courtesy of 3D Systems)

20.5 Self-replicating Machines

With the expiration of patent protection for many of the rapid prototyping processes, a movement, based on open-source sharing of software and machine designs, has now developed. Taking advantage of Internet-based data-sharing tools, inexpensive machines, based on FDM, have been developed and are available as kits, fully constructed, or with merely the plans for construction available for download.

Extending the *open-source* software traditions to hardware have encouraged the development of **self-replicating machines,** where the rapid prototyping machine creates parts that are used to produce another identical rapid prototyping machine. Not all parts are replicated (control hardware and metal structural parts, for example, are separate items), but up to 60% of the components can be produced.

This has led to a proliferation of low-cost rapid prototyping machines. For example, RepRaps can be constructed from components that cost a few hundred dollars unassembled, Thing-o-Matic kits cost around $1000, and a complete Cube is around $1300 (see Fig. 20.15).

Materials are limited for such machines, and the build spaces are smaller and resolutions coarser than those for more established commercial equipment. However, as with open-source software, the field is

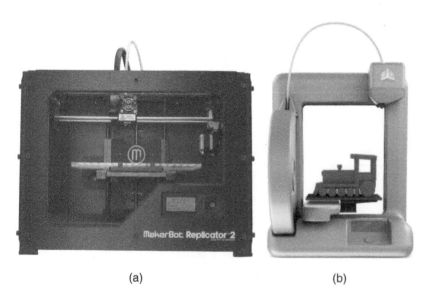

(a)   (b)

FIGURE 20.15 Examples of low-cost rapid-prototyping systems, based on fused-deposition modeling. (a) The MakerBot® Replicator® 2 Desktop 3D printer, based on fused-deposition modeling and open-source software, with a build volume of up to 110 mm × 110 mm × 120 mm, using either ABS or PLA (polylactic acid) polymers and (b) the Cube, with a build space of up to 140 mm × 140 mm × 140 mm. *Source:* (a) Courtesy of MakerBot, Inc. (b) Courtesy of 3D systems.

developing rapidly. In addition to continuing hardware developments, several web-based design libraries also have been established for part sharing. File sharing is performed using the Internet, and collaborative design is possible. This so-called **maker movement** allows rapid and inexpensive manufacture of prototypes.

20.6   Direct Manufacturing and Rapid Tooling

While extremely beneficial as a demonstration and visualization tool, rapid-prototyping processes also have been used as a manufacturing step in actual production. There are two basic methodologies used:

1. Direct production of engineering metal, ceramic, and polymer components or parts, by rapid prototyping
2. Production of tooling or patterns by rapid prototyping, for use in various manufacturing operations

Not only are the polymer parts, obtained from various rapid-prototyping operations, useful for design evaluation and troubleshooting, occasionally these processes also can be used to manufacture parts directly, referred to as *direct manufacturing*. Thus, the component is generated directly to a near-net shape, from a computer file containing part geometry. The main limitations to the widespread use of rapid prototyping for direct manufacturing, or *rapid manufacturing*, are:

QR Code 20.8 Direct manufacturing: 3D Printed Magic Arms. (*Source:* Courtesy of Stratasys, Inc.)

- Raw-material costs are high, and the time required to produce each part is too long to be viable for large production runs; however, there are many applications in which production runs are sufficiently small to justify direct manufacturing.
- The long-term and consistent performance of rapidly manufactured parts (as compared with the more traditional methods of manufacturing them) may be suspect, especially with respect to fatigue, wear, and life cycle of the parts made.

Much progress is being made to address these concerns, in order to make rapid manufacturing a more competitive and viable option in manufacturing. The future of these processes remains challenging though promising, especially in view of the fact that rapid manufacturing is now being regarded as a method of producing a product on demand. Thus, customers will be able to order a particular part, which will be produced within a relatively short waiting time.

CASE STUDY 20.5   Invisalign® Orthodontic Aligners

Orthodontic braces have been available to straighten teeth for more than 50 years. The braces involve metal, ceramic, or plastic brackets, that are bonded adhesively to teeth with fixtures for attachment to a wire, which then forces compliance on the teeth and straightens them to the desired shape within a few years. Conventional orthodontic braces are a well known and successful technique for ensuring long-term dental health. However, there are several drawbacks to conventional braces, including the facts that (a) they are aesthetically unappealing; (b) the sharp wires and brackets can be painful;

(c) they trap food, leading to premature tooth decay; (d) brushing and flossing teeth are more difficult and less effective with braces in place; and (e) certain foods must be avoided, because they will damage the braces.

One solution is the Invisalign system, made by Align Technology, Inc. It consists of a series of aligners, each of which the person wears for approximately two weeks. Each aligner (Fig. 20.16) consists of a precise geometry that incrementally moves the teeth to the desired positions. Because the aligners can be removed for eating, brushing, and flossing,

(continued)

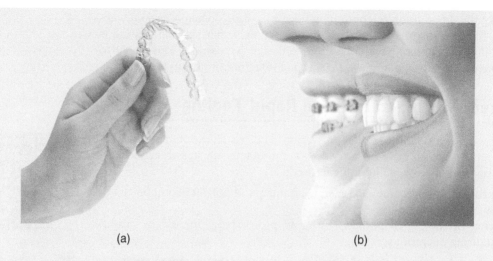

FIGURE 20.16 (a) An aligner for orthodontic use, manufactured by a combination of rapid tooling and thermoforming. (b) Comparison of conventional orthodontic braces using transparent aligners. *Source:* Courtesy of Align Technology, Inc.

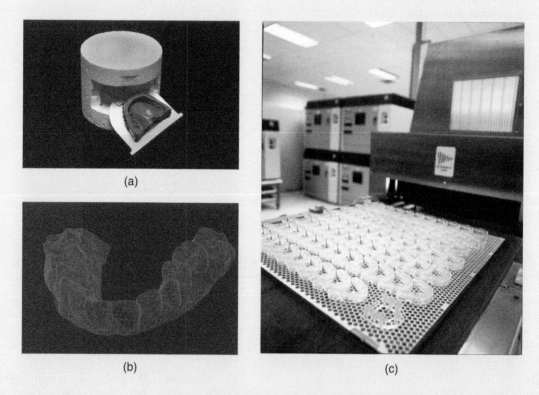

FIGURE 20.17 The manufacturing sequence for Invisalign orthodontic aligners. (a) Creation of a polymer impression of the patient's teeth. (b) Computer modeling to produce CAD representations of desired tooth profiles. (c) Production of incremental models of desired tooth movement; an aligner is produced by thermoforming a transparent plastic sheet against this model. *Source:* Courtesy of Align Technology, Inc.

most of the drawbacks of conventional braces are eliminated. Furthermore, since they are produced from a transparent plastic, the aligners do not seriously affect the appearance of the person's teeth.

The Invisalign product uses a combination of advanced technologies in the production process, shown in Fig. 20.17. The treatment begins with an orthodontist, or a general dentist, making a polymer impression of the patient's teeth (Fig. 20.17a). These impressions then are used to create a three-dimensional CAD representation of the patient's teeth (Fig. 20.17b). Proprietary CAD software then assists in the development of a treatment strategy for moving the teeth in an optimal manner.

Once the treating orthodontist has approved the treatment plan and it has been developed, the computer-based information is used to produce the aligners. This is done through a novel application of stereolithography. Although several materials are available for stereolithography, they have a characteristic yellow-brown shade to them, and therefore are unsuitable for direct application as an orthodontic product. Instead, the Align process uses a stereolithography machine that produces patterns of the desired incremental positions of the teeth (Fig. 20.17c). A sheet of clear polymer is then thermoformed (Section 19.6) over these patterns to produce the aligners, which are then sent to the treating orthodontist. With the doctors' supervision, patients are instructed to change the next set of aligners every two weeks.

The Invisalign product has proven to be popular for patients who wish to promote dental health, and to preserve their teeth long into their lives. The use of stereolithography, to produce accurate tools quickly and inexpensively, allows this orthodontic treatment to be an economically viable choice.

Source: Used with permission from Align Technology, Inc.

20.6.1 Rapid Tooling

Several methods have been devised for the rapid production of tooling (RT) by means of rapid-prototyping processes. The advantages to rapid tooling include:

1. The high cost of labor and short supply of skilled patternmakers can be overcome.
2. There is a major reduction in lead time.
3. Hollow designs can be adopted easily so that lightweight castings can be produced more easily.
4. The integral use of CAD technologies allows the use of modular dies, with base-mold tooling (match plates) and specially fabricated inserts; this technique can further reduce tooling costs.
5. Chill- and cooling-channel placement in molds can be optimized more easily, leading to reduced cycle times.
6. Shrinkage due to solidification or thermal contraction can be compensated for automatically, through software to produce tooling of the proper size and, in turn, to produce the desired parts.

The main shortcoming of rapid tooling is the potentially reduced tool or pattern life, as compared to those obtained from machined tool and die materials, such as tool steels or tungsten carbides (Chapter 21).

The simplest method of applying rapid-prototyping operations to other manufacturing processes is in the direct production of patterns or molds. As an example, Fig. 20.18 shows an approach for investment casting. Here, the individual patterns are made in a rapid-prototyping operation (in this case, stereolithography), and then used as patterns in assembling a tree for investment casting (Fig. 11.14). Note that this approach requires a polymer that will completely melt and burn from the ceramic mold; such polymers are available for all forms of polymer rapid-prototyping operations. Furthermore, as drawn in CAD programs, the parts are usually software

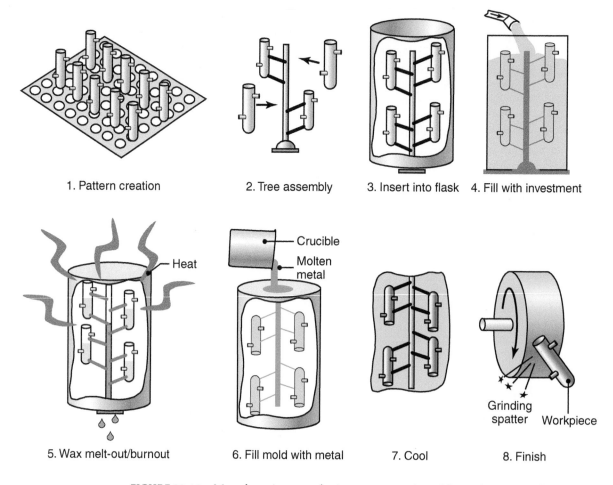

1. Pattern creation 2. Tree assembly 3. Insert into flask 4. Fill with investment

Crucible

Molten metal

Heat

Grinding spatter Workpiece

5. Wax melt-out/burnout 6. Fill mold with metal 7. Cool 8. Finish

FIGURE 20.18 Manufacturing steps for investment casting with rapid-prototyped wax parts as blanks; this method uses a flask for the investment, but a shell method also can be used. *Source:* Courtesy of 3D Systems, Inc.

modified to account for shrinkage, and it is the modified part that is produced in rapid-prototyping machinery.

As another example, 3DP can easily produce a ceramic-mold casting shell (Section 11.2.2) or a sand mold (Section 11.2.1), in which an aluminum-oxide or aluminum-silica powder is fused with a silica binder. The molds have to be post-processed in two steps: curing at around 150°C (300°F) and then firing at 1000° to 1500°C (1840°–2740°F).

Another common application of rapid tooling is injection molding of polymers (Section 19.3), in which the mold or, more typically, a *mold insert* is manufactured by rapid prototyping. Molds for slip casting of ceramics (Section 18.2.1) also can be produced in this manner. To produce individual molds, rapid-prototyping processes are used directly, and the molds will be shaped with the desired permeability. For example, in fused-deposition modeling, this requirement mandates that the filaments be placed onto the individual slices, with a small gap between adjacent filaments. The filaments are then positioned at right angles in adjacent layers.

The advantage of rapid tooling is the capability to produce a mold or a mold insert that can be used to manufacture components, without the time lag (typically several months) traditionally required for the procurement of tooling. Moreover, the design is simplified, because the designer need only analyze a CAD file of the desired part; software then produces the tool geometry and automatically compensates for shrinkage.

In addition to the straightforward application of rapid-prototyping technology to tool or pattern production, other rapid-tooling approaches, based on rapid-prototyping technologies, have been developed.

Room-temperature vulcanizing (RTV) molding/urethane casting can be performed by preparing a pattern of a part by any rapid-prototyping operation. The pattern is coated with a parting agent, and may or may not be modified to define mold parting lines. Liquid RTV rubber is poured over the pattern, and cures (usually within a few hours) to produce mold halves. The mold is then used with liquid urethanes in injection molding or reaction-injection molding operations (Section 19.3). One main limitation to this approach is a lower mold life, because the polyurethane present in the mold undergoes progressive damage and the mold may be suitable for as few as 25 parts.

Epoxy and *aluminum-filled epoxy* molds also can be produced, but mold design then requires special care. With RTV rubber, the mold flexibility allows it to be peeled off the cured part. With epoxy molds, the high stiffness precludes this method of part removal, and mold design is more complicated. Thus, for example, drafts are needed, and undercuts and other design features that can be produced by RTV molding must be avoided.

Acetal clear epoxy solid (ACES) injection molding, also known as *direct AIM*, refers to the use of rapid prototyping, usually stereolithography, to directly produce molds suitable for injection molding. The molds are shells, with an open end to allow filling with a material such as epoxy, aluminum-filled epoxy, or a low-melting-point metal. Depending on the polymer used in injection molding, mold life may be as few as 10 parts, although 100 parts per mold are possible.

Sprayed-metal tooling. In this process, shown in Fig. 20.19, a pattern is created through rapid prototyping. A metal spray operation (Section 34.5) then coats the pattern surface with a zinc–aluminum alloy. The metal coating is placed in a flask, and potted with an epoxy or an aluminum-filled epoxy material. In some applications, cooling lines can be incorporated into the mold before the epoxy is applied. The pattern is removed, and two such mold halves are used as in injection-molding operations. Mold life is highly dependent on the material used and temperatures involved, and can vary from a few to thousands of parts.

Keltool process. In the *Keltool process*, an RTV rubber mold is produced, based on a rapid-prototyped pattern, as described earlier. The mold is then filled with a mixture of powdered A6 tool steel (Section 5.7), tungsten carbide, and polymer binder, and is allowed to cure. The so-called *green* tool (green, as in ceramics and powder metallurgy) is fired to burn off the polymer and fuse the steel and the tungsten-carbide powders. The tool is then infiltrated with copper in a furnace to produce the final mold. The mold can subsequently be machined or polished to impart a superior surface finish and good dimensional tolerances. Keltool molds are limited in size to around 150 mm × 150 mm × 150 mm (6 in. × 6 in. × 6 in.); a mold insert, suitable for high-volume molding operations, can be made and installed. Depending on the material and processing conditions, mold life can range from 100,000 to 10 million parts.

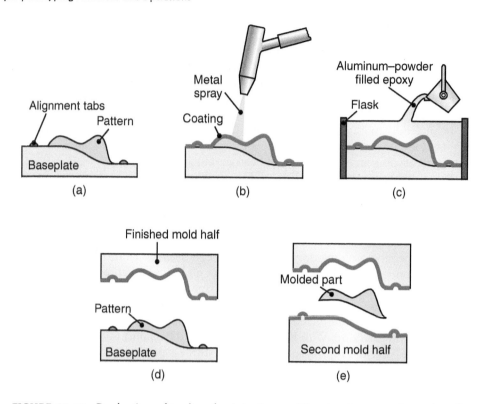

FIGURE 20.19 Production of tooling for injection molding by the sprayed-metal tooling process. (a) A pattern and baseplate are prepared through a rapid-prototyping operation; (b) a zinc–aluminum alloy is sprayed onto the pattern (see Section 34.5); (c) the coated baseplate and pattern assembly are placed together in a flask and backfilled with aluminum-impregnated epoxy; (d) after curing, the baseplate is removed from the finished mold; and (e) a second mold half suitable for injection molding is prepared.

CASE STUDY 20.6 Casting of Plumbing Fixtures

A global manufacturer of plumbing fixtures and accessories for baths and kitchens used rapid tooling to transform its product development process. One of the company's major product lines is decorative water faucets, produced from brass castings that are subsequently polished to achieve the desired surface finish. The ability to produce prototypes from brass is essential for quickly evaluating designs and identifying processing difficulties that may occur.

A new faucet design was prepared in a CAD program; the finished product is shown in Fig. 20.20. As part of the product development cycle, it was decided to produce prototypes of the faucet in order to confirm the aesthetics of the design. Since such

faucets are typically produced by sand casting, it was also essential to validate the design through a sand-casting process, followed by polishing. This approach allowed evaluation of the cast parts in terms of porosity and various other casting defects, and also would identify processing difficulties that might arise in the finishing stages.

A sand mold was first produced, as shown in Fig. 20.21. The mold material was a blend of foundry sand, plaster, and other additives that were combined to provide strong molds with good surface finish (see also Section 11.2.1). A binder was printed onto the sand mixture to produce the mold. The mold could be produced as one piece, with an integral core (see Figs. 11.3 and 11.6), but in

FIGURE 20.20 A new faucet design, produced by casting from rapid-prototyped sand molds.

FIGURE 20.21 Sand molds produced through three-dimensional printing.

practice, it is often desired to smoothen the core and assemble it later onto core prints. In addition, slender cores may become damaged, as support powder is removed from the mold, especially for complex casting designs. Therefore, the core is produced separately and then assembled into the two-part mold.

Using 3D printing, the operation produced brass prototypes of the faucets in five days, which included the time required for mold design, printing, metal casting, and finishing. The actual print time of the mold was just under three hours, and the material cost was approximately $280. The production of pattern plates for sand casting is, in general, too expensive for producing prototypes, and would cost over $10,000 and add several months to the lead time. The incorporation of 3D printing into the design process provided new capabilities that confirmed the design aesthetics and function, as well as manufacturing robustness and reliability.

Source: Courtesy of Z Corporation.

SUMMARY

- Rapid prototyping continues to grow into a unique manufacturing discipline. As a physical-model producing technology, it is a useful technique for identifying and correcting design errors. Several techniques have been developed for producing parts through rapid prototyping.

- Fused-deposition modeling consists of a computer-controlled extruder, through which a polymer filament is deposited to produce a part slice by slice.

- Stereolithography involves a computer-controlled laser-focusing system, that cures a liquid thermosetting polymer containing a photosensitive curing agent.

- MultiJet and PolyJet modeling use mechanisms similar to ink-jet printer heads to eject photopolymers to directly build prototypes.

- Laminated-object manufacturing uses a laser beam or vinyl cutter to first cut the slices on paper or plastic sheets (laminations); then it applies an adhesive layer, if necessary, and finally stacks the sheets to produce the part.

- Three-dimensional printing uses an ink-jet mechanism to deposit liquid droplets of the liquid binder onto polymer, metal, or ceramic powders. The related process of ballistic particle manufacturing directly deposits the build material. Using multiple printheads, three-dimensional printing can also produce full-color prototypes.

- Selective laser sintering uses a high-powered laser beam to sinter powders or coatings on the powders in a desired pattern. Selective laser sintering has been applied to polymers, sand, ceramics, and metals. Electron-beam melting uses the power of an electron beam to melt powders and form fully-dense functional parts.

- Rapid-prototyping techniques have made possible much faster product development times, and they are having a major effect on other manufacturing processes. When appropriate materials are used, rapid-prototyping machinery can produce blanks for investment casting or similar processes, so that metallic parts can now be obtained quickly and inexpensively, even for lot sizes as small as one part. Such technologies also can be applied to producing molds for operations (such as injection molding, sand and shell mold casting, and even forging), thereby significantly reducing the lead time between design and manufacture.

KEY TERMS

ACES	Electron-beam melting	Photopolymer	Solid-ground curing
Additive processes	Free-form fabrication	Polyjet	Sprayed–metal tooling
Ballistic-particle manufacturing	Fused-deposition modeling	Prototype	Stereolithography
Desktop manufacturing	Keltool	Rapid tooling	Subtractive processes
Digital manufacturing	Laminated-object manufacturing	RTV molding/urethane casting	Three-dimensional printing
Direct AIM	Multijet modeling	Selective laser sintering	Virtual prototyping
Direct manufacturing		Self-replicating machine	

BIBLIOGRAPHY

Bocking, C.E., Rennie, A., and Jacobson, D., **Rapid and Virtual Prototyping and Applications**, Wiley, 2003.

Chua, C.K., Leong, K.F., and Lim, C.S., **Rapid Prototyping: Principles and Applications**, 3rd ed., World Scientific Publishing, 2010.

Gebhardt, A., **Laser Rapid Prototyping**, Springer, 2005.

Gebhardt, A., **Understanding Additive Manufacturing: Rapid Prototyping, Rapid Tooling, Rapid Manufacturing**, Hanser, 2012.

Gibson, I., Rosen, D.W., and Stucker, B., **Additive Manufacturing Technologies: Rapid Prototyping to Direct Digital Manufacturing**, Springer, 2009.

Grimm, T., **Users Guide to Rapid Prototyping**, Society of Manufacturing Engineers, 2004.

Grimm, T., **Engineering Design and Rapid Prototyping**, Springer, 2010.

Hopkinson, N., Hague, R., and Dickens, R., (eds.), **Rapid Prototyping: An Industrial Revolution for the Digital Age**, Wiley, 2006.

Kamrani, K., and Nasr, E.A. (eds.), **Rapid Prototyping: Theory and Practice**, Springer, 2006.

Liou, F.W., **Rapid Prototyping and Engineering Applications: A Toolbox of Prototype Development**, Dekker, 2007.

Noorani, R.I., **Rapid Prototyping: Principles and Applications**, Wiley, 2006.

Pham, D.T., and Dimov, S.S., **Rapid Manufacturing: The Technologies and Applications of Rapid Prototyping and Rapid Tooling**, Springer, 2001.

Venuvinod, P.K., and Ma, W., **Rapid Prototyping: Laser-Based and Other Technologies**, Springer, 2010.

Wang, W., Stoll, H., and Conley, J.G., **Rapid Tooling Guidelines for Sand Casting**, Springer, 2010.

REVIEW QUESTIONS

20.1 What is the basic difference between additive manufacturing and rapid prototyping?

20.2 What is stereolithography?

20.3 What is virtual prototyping, and how does it differ from additive methods?

20.4 What is fused-deposition modeling?

20.5 Explain what is meant by rapid tooling.

20.6 Why are photopolymers essential for stereolithography?

20.7 Explain what each of the following means: (a) 3DP, (b) LOM, (c) STL, (d) SGC, (e) FDM, and (f) LENS.

20.8 What starting materials can be used in fused-deposition modeling? In three-dimensional printing?

20.9 What are the cleaning and finishing operations in rapid-prototyping processes? Why are they necessary?

20.10 Which rapid-prototyping technologies do not require a laser?

20.11 What are the advantages of electron beam melting?

20.12 What is the Keltool process?

20.13 What is a self-replicating machine?

20.14 Which rapid-prototyping operations can produce transparent workpieces?

20.15 Which rapid-prototyping operations can produce multicolored workpieces?

QUALITATIVE PROBLEMS

20.16 How can a mold for sand casting be produced using rapid prototyping techniques? Explain.

20.17 Examine a ceramic coffee cup and determine in which orientation you would choose to produce the part if you were using (a) fused-deposition manufacturing or (b) laminated-object manufacturing.

20.18 How would you rapidly manufacture tooling for injection molding? Explain any difficulties that may be encountered.

20.19 Explain the significance of rapid tooling in manufacturing.

20.20 List the processes described in this chapter that are best suited for the production of ceramic parts. Explain.

20.21 Few parts in commercial products today are directly manufactured through rapid-prototyping operations. Explain the possible reasons.

20.22 Can rapid-prototyped parts be made of paper? Explain.

20.23 Careful analysis of a rapid-prototyped part indicates that it is made up of layers with a distinct filament outline visible on each layer. Is the material a thermoset or a thermoplastic? Explain.

20.24 Why are the metal parts in three-dimensional printing often infiltrated by another metal?

20.25 Make a list of the advantages and limitations of each of the rapid-prototyping operations described in this chapter.

20.26 In making a prototype of a toy automobile, list the post-rapid-prototyping finishing operations that you think would be necessary. Explain.

QUANTITATIVE PROBLEMS

20.27 Using an approximate cost of $400 per gallon for the liquid polymer, estimate the material cost of a rapid-prototyped rendering of a typical computer mouse.

20.28 The extruder head in a fused-deposition modeling setup has a diameter of 1.25 mm (0.05 in.) and produces layers that are 0.25 mm (0.01 in.) thick. If the extruder head and polymer extrudate velocities are both 40 mm/s, estimate the production time for the generation of a 38-mm (1.5-in.) solid cube. Assume that there is a 10-second delay between layers as the extruder head is moved over a wire brush for cleaning.

20.29 Using the data for Problem 20.28 and assuming that the porosity for the support material is 50%, calculate the production rate for making a 100-mm (4-in.) high cup with an outside diameter of 90 mm (3.5 in.) and a wall thickness of 4 mm (0.16 in.). Consider the cases (a) with the closed end up and (b) with the closed end down.

20.30 Inspect Table 20.2 and compare the numerical values given with those for metals and other materials, as can be found in Part I of this text. Comment on your observations.

SYNTHESIS, DESIGN, AND PROJECTS

20.31 Rapid-prototyping machines represent a large capital investment; consequently, few companies can justify the purchase of their own system. Thus, service companies that produce parts based on their customers drawings have become common. Conduct an informal survey of such service companies, identify the classes of rapid-prototyping machines that they use, and determine the percentage use of each class.

20.32 One of the major advantages of stereolithography is that it can use transparent polymers, so that internal details of parts can readily be discerned. List and describe several parts in which this feature is valuable.

20.33 A manufacturing technique is being proposed that uses a variation of fused-deposition modeling, in which there are two polymer filaments that are melted and mixed prior to being extruded to make the part. What advantages does this method have?

20.34 Identify the rapid-prototyping processes described in this chapter that can be performed with materials available in your home or that you can purchase easily at low cost. Explain how you would go about it. Consider materials such as thin plywood, thick paper, glue, and butter, as well as the use of various tools and energy sources.

20.35 Design a machine that uses rapid-prototyping technologies to produce ice sculptures. Describe its basic features,

commenting on the effect of size and shape complexity on your design.

20.36 Because of relief of residual stresses during curing, long unsupported overhangs in parts made by stereolithography tend to curl. Suggest methods of controlling or eliminating this problem.

20.37 Describe methods that would allow the use of reinforced polymers to be used in rapid prototyping.

20.38 Conduct an Internet and literature study and write a two-page paper on developments of producing artificial organs through rapid prototyping related processes.

20.39 A current topic of research involves producing parts from rapid-prototyping operations and then using them in experimental stress analysis, in order to infer the strength of final parts produced by means of conventional manufacturing operations. List your concerns with this approach, and outline means of addressing these concerns.

20.40 Outline the approach you would use to produce prototypes of small gears from plastic. Assume the gears are 100 mm in diameter, 25 mm thick, and have 25 teeth. Explain how your preferred method of production would change if you needed to produce (a) one gear; (b) 100 per month; (c) 100 per day; and (d) 100 per hour.

Machining Processes and Machine Tools

Parts manufactured by the casting, forming, and shaping processes described in Parts II and III, including many parts made by near-net or net-shape methods, often require further operations before the product is ready for use. Consider, for example, the following features on parts and whether they could be produced by the processes described thus far:

- Smooth and shiny surfaces, such as the bearing surfaces of the crankshaft shown in Fig. IV.1
- Small-diameter deep holes in a part, such as the fuel–injector nozzle shown in Fig. IV.2
- Parts with sharp features, a threaded section, and specified close dimensional tolerances, such as the part shown in Fig. IV.3
- A threaded hole or holes on different surfaces of a part for assembly with other components
- Demanding, complex geometries, often in hard or high-performance materials that cannot be easily or economically produced in the quantities desired through the processes described earlier in the book (see Fig. 25.1)
- Special surface finish and texture for functional purposes or for appearance

It soon will become clear that none of the processes described in the preceding chapters is capable of producing parts with the specific characteristics outlined above. Thus, the parts will require further processing, generally referred to as *secondary* and *finishing operations*. **Machining** is a general term describing a group of processes that consist of the **removal** of material and **modification** of the workpiece surfaces after it has been produced by various methods. The very wide variety of shapes produced by machining can be seen in an automobile, as shown in Fig. IV.4.

In reviewing the contents of Parts II and III of this text, it will be recalled that some parts may indeed be produced to final shape (net shape) and at high quantities. However, machining may be more economical, provided that the number of parts required is relatively small or the material and shape allow the parts to be machined at

Before   After

FIGURE IV.1 A forged crankshaft before and after machining the bearing surfaces. The shiny bearing surfaces of the part on the right cannot be made to their final dimensions and surface finish by any of the processes described in previous chapters of this book. *Source:* Courtesy of Wyman-Gordon Company.

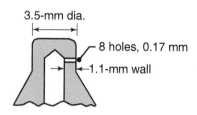

FIGURE IV.2 Cross-section of a fuel-injection nozzle, showing a small hole made by the electrical-discharge machining process, as described in Section 27.5. The material is heat-treated steel.

high rates, quantities, and with high dimensional accuracy. A good example is the production of brass screw-machine parts on multiple-spindle automatic screw machines.

In general, however, resorting to machining often suggests that a part could not have been produced to the final desired specifications by the primary processes used in making them, and that additional operations are necessary. We again emphasize the importance of net-shape manufacturing, as described in Section I.5, to avoid these additional steps and reduce production costs.

Furthermore, in spite of their advantages, material-removal processes have certain disadvantages; they

- *Waste material*, even though the amount may be relatively small
- Generally take *longer* than other manufacturing processes
- Generally require *more energy* than do forming and shaping operations
- Can have *adverse effects* on the surface quality and properties of the product

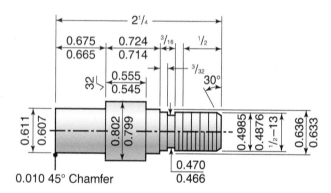

FIGURE IV.3 A machined and threaded part, showing various dimensions and tolerances; all dimensions are in inches; note that some tolerances are only a few thousandths of an inch.

As outlined in Fig. I.5e in the General Introduction, machining consists of several types of material-removal processes:

- **Cutting,** typically involving single-point or multi-point cutting tools, each with a clearly defined shape (Chapters 23 through 25)
- **Abrasive processes,** such as grinding and related processes (Chapter 26)
- **Advanced machining processes,** typically utilizing electrical, chemical, laser, thermal, and hydrodynamic methods (Chapter 27)

The machines on which these operations are carried out are called **machine tools.** As can be noted in Table I.2 in the General Introduction, the first primitive tools, dating back several millennia, were made for

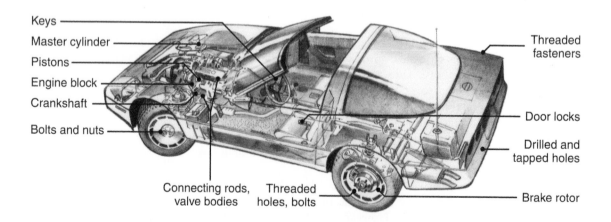

FIGURE IV.4 Typical parts in an automobile that require machining operations to impart desirable shapes, surface characteristics, dimensions, and tolerances

the main purpose of chipping away and cutting all types of natural materials, such as wood, stone, vegetation, and hunted livestock. Note also that it was not until the 1500s that developments began on making products by machining operations, particularly with the introduction of the lathe. Compared to the rather simple machinery and tools employed, a wide variety of computer-controlled machine tools and advanced techniques are available for making functional parts as small as tiny insects and with cross-sections much smaller than a human hair.

As in all other manufacturing operations, it is essential to view machining operations as a *system*, consisting of the (a) workpiece, (b) cutting tool, (c) machine tool, and (d) operator.

In the next seven chapters, the basic mechanics of chip formation in machining are described; these include tool forces, power requirements, temperature, tool wear, surface finish and integrity of the part machined, cutting tools, and cutting fluids. Specific machining processes are then described, including their capabilities, limitations, and typical applications, and important machine-tool characteristics for such basic operations as turning, milling, boring, drilling, and tapping.

The features of *machining centers*, which are versatile machine tools controlled by computers and capable of efficiently performing a variety of operations, are then described. The next group of processes described are those in which the removal of material is carried out by **abrasive processes** and related operations, and to very high dimensional accuracy and surface finish. For technical and economic reasons, some parts cannot be machined satisfactorily by cutting or abrasive processes. Since the 1940s, important developments have taken place in **advanced machining processes**, including chemical, electrochemical, electrical-discharge, laser-beam, electron-beam, abrasive-jet, and hydrodynamic machining, described throughout the rest of Part IV.

21

Fundamentals
of Machining

- This chapter is an introduction to the fundamentals of machining processes, and presents the basic concepts relevant to all machining operations.
- The chapter opens with a description of the mechanics of chip formation in machining, and includes the model typically used for the basic cutting operations allowing the calculation of force and power in machining.
- Temperature rise and its importance on the workpiece and cutting tool performance, and the mechanism of tool wear, are then discussed.
- The chapter concludes with a discussion of surface finish, the integrity of parts produced by machining, and the factors involved in the machinability of metallic and nonmetallic materials.

21.1 Introduction

Cutting processes remove material from the various surfaces of a workpiece by producing **chips**. Some of the more common cutting processes, illustrated in Fig. 21.1 (see also Fig. I.5e), are:

- **Turning,** in which the workpiece is rotated and a cutting tool removes a layer of material as the tool moves along its length, as shown in Fig. 21.1a
- **Cutting off,** in which the tool moves radially inward, and separates the piece on the right in Fig. 21.1b from the blank
- **Slab milling,** in which a rotating cutting tool removes a layer of material from the surface of the workpiece (Fig. 21.1c)
- **End milling,** in which a rotating cutter travels along a certain depth in the workpiece and produces a cavity (Fig. 21.1d)

In the turning process, illustrated in greater detail in Fig. 21.2, the cutting tool is set at a certain *depth of cut* (mm or in.), and travels to the left with a certain speed as the workpiece rotates. The *feed,* or *feed rate,* is the distance the tool travels per unit revolution of the workpiece (mm/rev or in./rev);

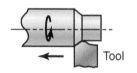

(a) Straight turning

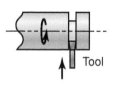

(b) Cutting off

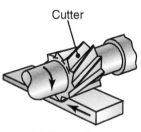

(c) Slab milling

(d) End milling

FIGURE 21.1 Some examples of common machining operations.

this movement of the cutting tool produces a chip, which moves up the face of the tool.

In order to analyze this basic process in greater detail, a two-dimensional model of it is presented in Fig. 21.3a. In this *idealized* model, a cutting tool moves to the left along the workpiece at a constant velocity, V, and a depth of cut, t_o. A chip is produced ahead of the tool by plastic deformation, and shears the material continuously along the *shear plane*. This phenomenon can easily be demonstrated by slowly scraping the surface of a stick of butter lengthwise with a sharp knife, and observing how a chip is produced. Chocolate shavings used as decorations on cakes and pastries also are produced in a similar manner.

In comparing Figs. 21.2 and 21.3, note that the feed in turning is equivalent to t_o, and the depth of cut in turning is equivalent to the width of cut (i.e., the dimension perpendicular to the page). These dimensional relationships can be visualized by rotating Fig. 21.3 clockwise by 90°. With this brief introduction as a background, the cutting process will now be described in greater detail.

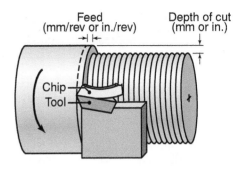

FIGURE 21.2 Schematic illustration of the turning operation, showing various features; surface finish is exaggerated to show feed marks.

21.2 **Mechanics of Cutting**

The factors that influence the cutting process are outlined in Table 21.1. In order to appreciate the contents of this table, note that the major *independent variables* in the basic cutting process are: (a) tool material and coatings, if any; (b) tool shape, surface finish, and sharpness; (c) workpiece material and its processing history; (d) cutting speed, feed, and depth of cut; (e) cutting fluids, if any; (f) characteristics of the machine tool; and (g) the type of work-holding device and fixturing.

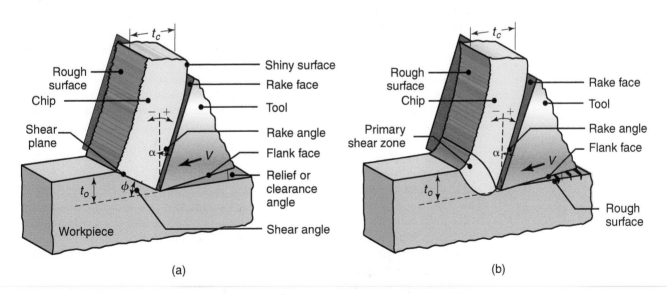

(a)   (b)

FIGURE 21.3 Schematic illustration of a two-dimensional cutting process, also called orthogonal cutting: (a) Orthogonal cutting with a well-defined shear plane, also known as the M.E. Merchant model. Note that the tool shape, the depth of cut, t_o, and the cutting speed, V, are all independent variables. (b) Orthogonal cutting without a well-defined shear plane.

TABLE 21.1

Factors Influencing Machining Operations	
Parameter	Influence and interrelationship
Cutting speed, depth of cut, feed, cutting fluids	Forces, power, temperature rise, tool life, type of chip, surface finish, and integrity
Tool angles	As above; influence on chip flow direction; resistance to tool wear and chipping
Continuous chip	Good surface finish; steady cutting forces; undesirable, especially in modern machine tools
Built-up edge chip	Poor surface finish and integrity; if thin and stable, edge can protect tool surfaces
Discontinuous chip	Desirable for ease of chip disposal; fluctuating cutting forces; can affect surface finish and cause vibration and chatter
Temperature rise	Influences tool life, particularly crater wear and dimensional accuracy of workpiece; may cause thermal damage to workpiece surface
Tool wear	Influences surface finish and integrity, dimensional accuracy, temperature rise, and forces and power
Machinability	Related to tool life, surface finish, forces and power, and type of chip produced

Dependent variables in cutting are those that are influenced by changes made in the independent variables listed above. These include: (a) type of chip produced; (b) force and energy dissipated during cutting; (c) temperature rise in the workpiece, the tool, and the chip; (d) tool wear and failure; and (e) surface finish and surface integrity of the workpiece.

The importance of establishing *quantitative* relationships among the independent and dependent variables in machining can best be appreciated by considering some typical questions to be posed: Which of the independent variables should be changed first, and to what extent, if (a) the surface finish of the workpiece being machined is unacceptable, (b) the cutting tool wears rapidly and becomes dull, (c) the workpiece becomes very hot, and (d) the tool begins to vibrate and chatter.

In order to understand these phenomena and respond to the question posed, it is essential to first study the mechanics of chip formation. The subject of chip formation mechanics has been studied extensively since the early 1940s. Several models, with varying degrees of complexity, have been proposed describing the cutting process. As is being done in many other manufacturing operations, advanced machining models are being continuously developed, including especially *computer simulation* of the basic machining process. Studying the complex interactions among the numerous variables involved, in turn, helps develop the capabilities to optimize machining operations and minimize costs.

The simple model shown in Fig. 21.3a, referred to as the M.E. Merchant model, and developed in the early 1940s, is sufficient for the purposes of this introduction. This model is known as **orthogonal cutting**, because it is two dimensional and the forces involved are perpendicular to each other. The cutting tool has a **rake angle**, α (positive as shown in the figure), and a **relief** or **clearance angle**.

Microscopic examination of chips produced in actual machining operations has revealed that they are produced by *shearing* (as modeled in Fig. 21.4a), a phenomenon similar to the movement in a deck of cards sliding against each other (see also Fig. 1.6). Shearing takes place within a **shear zone** (usually along a well-defined plane, referred to as the **shear plane**) at an angle ϕ (called the **shear angle**). Below the shear plane, the workpiece remains undeformed; above it, the chip (already formed) moves up the rake face of the tool. The dimension d is highly exaggerated in the figure to show the mechanism involved in chip formation. This dimension, in reality, has been found to be only on the order of 10^{-2} to 10^{-3} mm (10^{-3} to 10^{-4} in.).

Some materials, notably cast irons at low speeds, do not shear along a well-defined plane, but instead shear within a zone, as shown in Fig. 21.3b. The shape and size of this zone is important in the machining operation, as will be discussed in Section 21.2.1.

Cutting Ratio.   It can be seen from Fig. 21.3a that the chip thickness, t_c, can be determined from the depth of cut, t_o, the rake angle, α, and the shear angle, ϕ. The ratio of t_o/t_c is known as the **cutting ratio** (or *chip-thickness ratio*), r, and is related to the two angles ϕ and α by the following relationships:

$$\tan \phi = \frac{r \cos \alpha}{1 - r \sin \alpha} \tag{21.1}$$

and

$$r = \frac{t_o}{t_c} = \frac{\sin \phi}{\cos (\phi - \alpha)}. \tag{21.2}$$

Because the chip thickness is always greater than the depth of cut, the value of r is always less than unity. The reciprocal of r is known as the *chip-compression ratio* or *chip-compression factor*, and is a measure of how thick the chip has become as compared with the depth of cut; thus, the chip-compression ratio always is greater than unity. As may be visualized by reviewing Fig. 21.3a, the depth of cut is also referred to as the *undeformed chip thickness*.

The cutting ratio is an important and useful parameter for evaluating cutting conditions. Since the undeformed chip thickness, t_o, is easily specified as a machine setting, and is therefore known, the cutting ratio can be calculated by measuring the chip thickness with a micrometer. With the rake angle also known for a particular cutting operation (it is a function of the tool and workpiece geometry in use), Eq. (21.1) allows calculation of the shear angle.

Although t_o is referred to as the *depth of cut*, note that in a machining process such as turning (shown in Fig. 21.2), this quantity is the *feed* or *feed rate*, expressed in distance traveled per revolution of the workpiece. Assume, for instance, that the workpiece in Fig. 21.2 is a thin-walled tube, and the width of the cut is the same as the thickness of the tube. Then, by rotating Fig. 21.3 clockwise by 90°, the figure is now similar to the view in Fig. 21.2.

Shear Strain.   Referring to Fig. 21.4a, it can now be seen that the **shear strain**, γ, that the material undergoes can be expressed as

$$\gamma = \frac{AB}{OC} = \frac{AO}{OC} + \frac{OB}{OC},$$

or

$$\gamma = \cot \phi + \tan (\phi - \alpha). \tag{21.3}$$

Note that large shear strains are associated with (a) low shear angles or (b) with low or negative rake angles. Shear strains of five or higher have been observed in

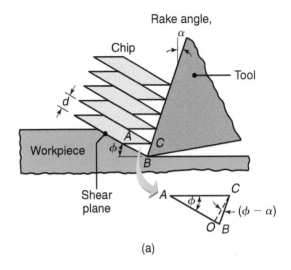
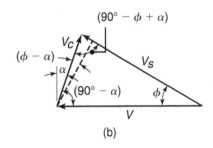

FIGURE 21.4   (a) Schematic illustration of the basic mechanism of chip formation by shearing. (b) Velocity diagram showing angular relationships among the three speeds in the cutting zone.

QR Code 21.1 Measuring strains during metal cutting. (*Source:* Courtesy of the National Institute of Standards and Technology)

actual cutting operations. Compared to forming and shaping processes, the work-piece material undergoes greater deformation during cutting, as is also seen in Table 2.4. Furthermore, deformation in cutting generally takes place within a very narrow zone; in other words, the dimension $d = OC$ in Fig. 21.4a is very small. Thus, the rate at which shearing takes place is high. (The nature and size of the deformation zone is further discussed in Section 21.3.)

The shear angle has a major significance in the mechanics of cutting operations, as it influences force and power requirements, chip thickness, and temperature rise in machining. One of the earliest analyses regarding the shear angle was based on the assumption that the shear angle adjusts itself to minimize the cutting force, or that the shear plane is a plane of maximum shear stress. This analysis yielded the expression

$$\phi = 45° + \frac{\alpha}{2} - \frac{\beta}{2}, \tag{21.4}$$

where β is the **friction angle**, and is related to the *coefficient of friction*, μ, at the tool–chip interface by the expression $\mu = \tan\beta$. Among several shear-angle relationships that have been developed, another approximate but useful formula is

$$\phi = 45° + \alpha - \beta. \tag{21.5}$$

The coefficient of friction in metal cutting has been found to generally range from about 0.5 to 2 (see also Section 33.4), indicating that the chip encounters considerable frictional resistance as it moves up the rake face of the tool. Experiments have shown that μ varies considerably along the tool–chip interface, because of large variations in contact pressure and temperature. Consequently, μ is also called the *apparent mean coefficient of friction*.

Equation (21.4) indicates that (a) as the rake angle decreases or the friction at the tool–chip interface increases, the shear angle decreases and the chip becomes thicker; (b) thicker chips mean more energy dissipation, because the shear strain is higher, as can be noted from Eq. (21.2); and (c) because the work done during cutting is converted into heat, the temperature rise is also higher.

Velocities in the Cutting Zone.   Note in Fig. 21.3 that since the chip thickness is greater than the depth of cut, the chip velocity, V_c, has to be lower than the cutting speed, V. Because mass continuity has to be maintained,

$$Vt_o = V_c t_c \quad \text{or} \quad V_c = Vr,$$

hence,

$$V_c = \frac{V \sin\phi}{\cos(\phi - \alpha)}. \tag{21.6}$$

A *velocity diagram* also can be constructed, as shown in Fig. 21.4b, in which, from trigonometric relationships, it can be shown that

$$\frac{V}{\cos(\phi - \alpha)} = \frac{V_s}{\cos\alpha} = \frac{V_c}{\sin\phi}, \tag{21.7}$$

where V_s is the velocity at which shearing takes place in the shear plane; note also that

$$r = \frac{t_o}{t_c} = \frac{V_c}{V}. \tag{21.8}$$

Video Solution 21.1   Mechanics of Cutting

These velocity relationships will be utilized further in Section 21.3 when describing power requirements in cutting operations.

21.2.1   Types of Chips Produced in Metal Cutting

The types of metal chips commonly observed in practice and their photomicrographs are shown in Fig. 21.5. The four main types are:

- Continuous
- Built-up edge
- Serrated or segmented
- Discontinuous

QR Code 21.2 Cutting steel showing BUE. (*Source:* Courtesy of the National Institute of Standards and Technology)

Note that a chip has two surfaces. One surface has been in contact with the rake face of the tool, and has a shiny and burnished appearance caused by sliding as the chip moves up the tool face. The other surface is from the exterior surface

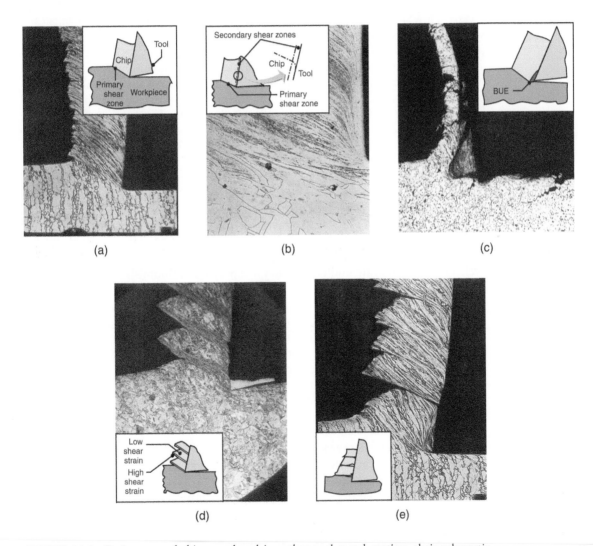

FIGURE 21.5   Basic types of chips produced in orthogonal metal cutting, their schematic representation, and photomicrographs of the cutting zone: (a) continuous chip, with narrow, straight, and primary shear zone; (b) continuous chip, with secondary shear zone at the chip–tool interface; (c) built-up edge; (d) segmented or nonhomogeneous chip; and (e) discontinuous chip. *Source:* After M.C. Shaw, P.K. Wright, and S. Kalpakjian.

of the workpiece. It has a jagged, rough appearance, as can be seen on the chips in Figs. 21.3 and 21.5, caused by the shearing mechanism shown in Fig. 21.4a.

Continuous Chips.   *Continuous chips* usually are formed with ductile materials, machined at high cutting speeds and/or at high rake angles (Fig. 21.5a). Deformation of the material takes place along a narrow shear zone, called the *primary shear zone*. Continuous chips may develop a *secondary shear zone* (Fig. 21.5b) because of high friction at the tool–chip interface. This zone becomes thicker as friction increases.

Deformation in continuous chips also may take place along a wide primary shear zone with *curved boundaries* (see Fig. 21.3b), unlike that shown in Fig. 21.5a. Note that the lower boundary of the deformation zone in Fig. 21.3b projects *below* the machined surface, subjecting it to distortion, as depicted by the distorted vertical lines in the machined subsurface. This situation occurs generally in machining soft metals at low speeds and low rake angles. It usually results in a poor surface finish and surface residual stresses, which may be detrimental to the properties of the machined part in their service life.

Although they generally produce a good surface finish, continuous chips are not necessarily desirable, as they tend to become tangled around the toolholder, the fixturing, and the workpiece. They also interfere with chip-disposal systems, described in Section 23.3.7. This problem can be alleviated with **chip breakers** (see Figure 21.7), by changing processing parameters, such as cutting speed, feed, and depth of cut, or by using cutting fluids.

Built-up Edge Chips.   A *built-up edge* (BUE) consists of layers of material from the workpiece that gradually are deposited on the tool tip, hence the term *built-up* (Fig. 21.5c). As it grows larger, the BUE becomes unstable, and eventually breaks apart. A portion of the BUE material is carried away by the tool side of the chip; the rest is deposited randomly on the workpiece surface. In effect, a BUE changes the geometry of the cutting edge and dulls it, as can be noted in Fig. 21.6a. The cycle of BUE formation and destruction is repeated continuously during the cutting operation.

Built-up edge is a major factor that adversely affects surface finish, as can be seen in Figs. 21.5c and 21.6b and c. On the other hand, a thin, stable BUE is usually regarded as desirable, because it reduces tool wear by protecting its rake face. Cold-worked metals generally have a lower tendency to form BUE than those in their annealed condition. Because of work hardening and deposition of successive layers of material, the BUE hardness increases significantly (Fig. 21.6a).

The tendency for BUE formation can be reduced by one or more of the following means:

- Increase the cutting speed
- Decrease the depth of cut
- Increase the rake angle
- Use a cutting tool that has lower chemical affinity for the workpiece material or use a sharp tool
- Use an effective cutting fluid

Serrated Chips.   *Serrated chips*, also called *segmented* or *nonhomogeneous chips* (Fig. 21.5d), are semicontinuous chips with large zones of low shear strain and small zones of high shear strain, hence the latter zone is called *shear localization*. The chips

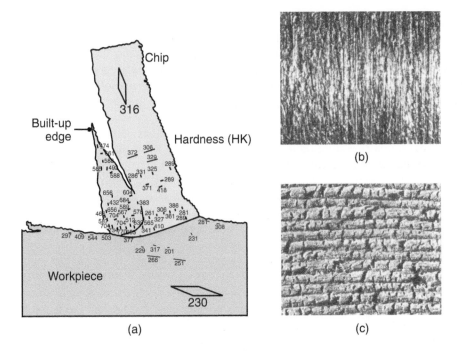

(a)

(b)

(c)

FIGURE 21.6   (a) Hardness distribution in a built-up edge in the cutting zone (material: 3115 steel); note that some regions in the built-up edge are as much as three times harder than the bulk metal of the workpiece. (b) Surface finish produced in turning 5130 steel with a built-up edge. (c) Surface finish on 1018 steel in face milling. Magnifications: 15×. *Source:* Courtesy of Metcut Research Associates, Inc.

have a sawtooth-like appearance. (This type of chip should not be confused with the illustration in Fig. 21.4a, in which the dimension d is highly exaggerated.) Metals that have low thermal conductivity and have strength that decreases sharply with temperature (called *thermal softening*) exhibit this behavior, and is most notably observed with titanium and its alloys.

Discontinuous Chips.   *Discontinuous chips* consist of segments, attached either firmly or loosely to each other (Fig. 21.5e). Discontinuous chips usually form under the following conditions:

- Brittle workpiece materials, because they do not have the capacity to undergo the high shear strains encountered in machining
- Workpiece materials that contain hard inclusions and impurities, or have structures such as the graphite flakes in gray cast iron (see Fig. 4.13a)
- Very low or very high cutting speed, V
- Large depth of cut, d
- Low rake angle, α
- Lack of an effective cutting fluid (Section 22.12)
- Low stiffness of the toolholder or the machine tool, thus allowing vibration and chatter to occur (Section 25.4)

Because of the discontinuous nature of chip formation, cutting forces continually vary during machining. Consequently, the stiffness or rigidity of the cutting-tool holder, the work-holding devices, and the machine tool (Chapters 23 through 25) are significant factors in machining with serrated or discontinuous chips. If it is not sufficiently rigid, the machine tool may begin to vibrate and chatter, as discussed in detail in Section 25.4. This, in turn, adversely affects the surface finish and dimensional accuracy of the machined part, and may cause premature wear or damage to the cutting tool. Even the components of the machine tool may be damaged if the vibration is excessive.

Chip Curl.   In all cutting operations on metals, as well as on nonmetallic materials, chips develop a curvature (*chip curl*) as they leave the workpiece surface (Fig. 21.5). Among the factors affecting the chip curl are:

- The distribution of stresses in the primary and secondary shear zones
- Thermal effects in the cutting zone
- Work-hardening characteristics of the workpiece material
- The geometry of the cutting tool
- Processing variables
- Cutting fluids

The first four items above are complex phenomena, and beyond the scope of this text. As for the effects of processing variables, as the depth of cut decreases, the radius of curvature of the chip generally decreases (the chip becomes curlier). Cutting fluids can make chips become more curly, thus reducing the tool–chip contact area (see Fig. 21.7a) and thus concentrating the heat closer to the tip of the tool (Section 21.4); as a result, tool wear increases.

Chip Breakers.   As stated above, continuous and long chips are undesirable in machining operations, as they tend to become entangled, severely interfere with machining operations, and can also become a potential safety hazard. If all of the processing variables are under control, the usual procedure employed to avoid such a situation is to break the chip intermittently, with shapes of cutting tools that have *chip-breaker* features, as shown in Fig. 21.7.

The basic principle of a chip breaker on a tool's rake face is to bend and break the chip periodically. Cutting tools and inserts (see Fig. 22.2) now have built-in chip-breaker features of various designs (Fig. 21.7). Chips also can be broken by changing the tool geometry to control chip flow, as in the turning operations shown in Fig. 21.8. Experience indicates that the ideal chip size to be broken is in the shape of either the letter C or the number 9, and fits within a 25-mm (1-in.) square space.

Controlled Contact on Tools.   Cutting tools can be designed so that the tool–chip contact length is reduced by recessing the rake face of the tool some distance away from its tip. This reduction in contact length affects the chip-formation mechanics. Primarily, it reduces the cutting forces, and thus the energy and temperature. Determining an optimum length is important, as too small a contact length would concentrate the heat at the tool tip, thus increasing wear.

Cutting Nonmetallic Materials.   A variety of chips are encountered in cutting thermoplastics (Section 7.3), depending on the type of polymer and process parameters, such as depth of cut, tool geometry, and cutting speed. The discussions concerning metals also are generally applicable to polymers. Because they are brittle, thermosetting plastics (Section 7.4) and ceramics (Chapter 8) generally produce discontinuous chips. The characteristics of other machined materials are described in Section 21.7.3.

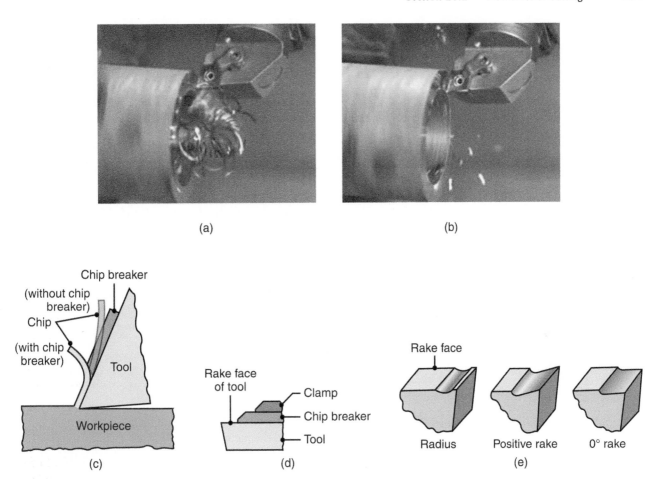

FIGURE 21.7 (a) Machining of aluminum using an insert without a chip breaker; note the long chips that can interfere with the tool and present a safety hazard. (b) Machining of aluminum with a chip breaker. (c) Schematic illustration of the action of a chip breaker; note that the chip breaker decreases the radius of curvature of the chip and eventually breaks it. (d) Chip breaker clamped on the rake face of a cutting tool. (e) Grooves in cutting tools acting as chip breakers. Most cutting tools now used are inserts with built-in chip-breaker features. *Source:* (a) and (b) Courtesy of Kennametal Inc.

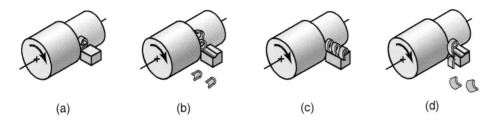

FIGURE 21.8 Chips produced in turning: (a) tightly curled chip; (b) chip hits workpiece and breaks off; (c) continuous chip moving radially away from workpiece; and (d) chip hits tool shank and breaks off.

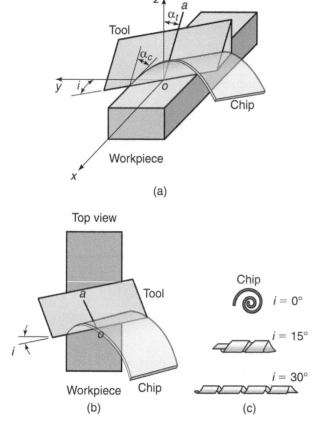

(a)

Top view

(b)

(c)

FIGURE 21.9   (a) Schematic illustration of cutting with an oblique tool; note the direction of chip movement. (b) Top view, showing the inclination angle, i. (c) Types of chips produced with tools at increasing inclination angles.

21.2.2   Oblique Cutting

The majority of machining operations involve tool shapes that are three dimensional, whereby the cutting is *oblique*. The basic difference between oblique and orthogonal cutting can be seen in Figs. 21.9a and c. In orthogonal cutting, the chip slides directly up the face of the tool and becomes like a spiral, whereas in *oblique cutting*, the chip becomes helical and leaves the workpiece surface at an angle i, called the **inclination angle** (Fig. 21.9b). Note that the lateral direction of chip movement in oblique cutting is similar to the action of a snowplow blade, whereby the snow is thrown sideways.

Note that the chip in Fig. 21.9a flows up the rake face of the tool at angle α_c (called the **chip flow angle**), and is measured in the plane of the tool face. Angle α_i is the **normal rake angle**, and is a basic geometric property of the tool; this is the angle between line oz normal to the workpiece surface and line oa on the tool face in the figure.

In oblique cutting, the workpiece material approaches the cutting tool at a velocity V and leaves the surface (as a chip) with a velocity V_c. The *effective rake angle*, α_e, is calculated in the plane of these two velocities. Assuming that the chip flow angle, α_c, is equal to the inclination angle, i (an assumption that has been verified experimentally), the effective rake angle, α_e, is

$$\alpha_e = \sin^{-1}\left(\sin^2 i + \cos^2 i \sin \alpha_n\right). \qquad (21.9)$$

Since both i and α_n can be measured directly, the effective rake angle can now be calculated. Note that, as i increases, the effective rake angle increases, the chip becomes thinner and longer, and, as a consequence, the cutting force decreases. The influence of the inclination angle on chip shape is shown in Fig. 21.9c.

A typical single-point turning tool used on a lathe is shown in Fig. 21.10a. Note the various angles involved, each of which has to be selected properly for efficient cutting. Although these angles have traditionally been produced by grinding (Chapter 26), the majority of cutting tools are now widely available as **inserts**, as shown in Fig. 21.10b and described in detail in Chapter 22. Various three-dimensional cutting tools, including those for drilling, tapping, milling, planing, shaping, broaching, sawing, and filing, are described in greater detail in Chapters 23 and 24.

Shaving and Skiving.   Thin layers of material can be removed from straight or curved surfaces by a process similar to the use of a plane to shave wood. *Shaving* is particularly useful in improving the surface finish and dimensional accuracy of sheared sheet metals and punched slugs, as shown in Fig. 16.9. Another common application of shaving is in finishing gears, using a cutter that has the shape of the gear tooth (see Section 24.7). Parts that are long or have complex shapes are shaved by *skiving*, using a specially shaped cutting tool that moves tangentially across the length of the workpiece.

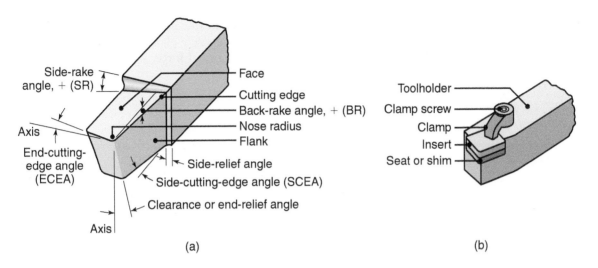

FIGURE 21.10 (a) Schematic illustration of a right-hand cutting tool. The various angles on these tools and their effects on machining are described in Section 23.2. Although these tools traditionally have been produced from solid tool–steel bars, they have been replaced largely with (b) inserts made of carbides and other materials of various shapes and sizes.

21.3   Cutting Forces and Power

Studying the *cutting forces* and *power* involved in machining operations is important for the following reasons:

- Data on cutting forces is essential so that:

  a. Machine tools can be properly designed to minimize distortion of the machine components, maintain the desired dimensional accuracy of the machined part, and help select appropriate toolholders and work-holding devices.

  b. The workpiece is capable of withstanding these forces without excessive distortion.

- Power requirements must be known in order to enable the selection of a machine tool with adequate electrical power.

The forces acting in orthogonal cutting are shown in Fig. 21.11a. The **cutting force**, F_c, acts in the direction of the cutting speed, V, and supplies the energy required for cutting. The ratio of the cutting force to the cross-sectional area being cut (i.e., the product of width of cut and depth of cut) is referred to as the *specific cutting force*.

The **thrust force**, F_t, acts in a direction normal to the cutting force. These two forces produce the **resultant force**, R, as can be seen from the force circle diagram shown in Fig. 21.11b. Note that the resultant force can be resolved into two components on the tool face: a **friction force**, F, along the tool–chip interface, and a **normal force**, N, perpendicular to it. It can also be shown that

$$F = R \sin \beta \tag{21.10}$$

and

$$N = R \cos \beta. \tag{21.11}$$

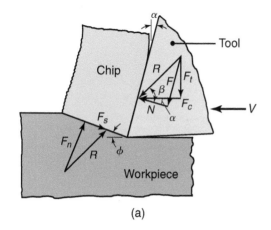

(a)

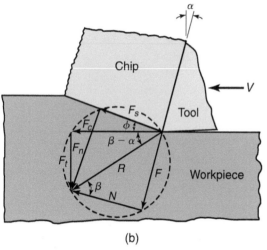

(b)

FIGURE 21.11 (a) Forces acting in the cutting zone during two-dimensional cutting; note that the resultant force, R, must be colinear to balance the forces. (b) Force circle to determine various forces acting in the cutting zone.

Note that the resultant force is balanced by an equal and opposite force along the shear plane, and is resolved into a **shear force, F_s,** and a **normal force, F_n.** These forces can be expressed, respectively, as

$$F_s = F_c \cos\phi - F_t \sin\phi \qquad (21.12)$$

and

$$F_n = F_c \sin\phi + F_t \cos\phi. \qquad (21.13)$$

Because the area of the shear plane can be calculated by knowing the shear angle and the depth of cut, the shear and normal stresses in the shear plane can be determined.

The ratio of F to N is the **coefficient of friction, μ,** at the tool–chip interface, and the angle β is the **friction angle** (as in Fig. 21.11). The magnitude of μ can be determined as

$$\mu = \frac{F}{N} = \frac{F_t + F_c \tan\alpha}{F_c - F_t \tan\alpha}. \qquad (21.14)$$

Although the magnitude of forces in actual cutting operations is generally on the order of a few hundred newtons, the *local stresses* in the cutting zone and the *pressure* on the cutting tool are very high, because the contact areas are very small. For example, the tool–chip contact length (see Fig. 21.3) is typically on the order of 1 mm (0.04 in.); consequently, the tool tip is subjected to very high stresses, which lead to wear, as well as chipping and fracture of the tool.

Thrust Force. The *thrust force* in cutting is important because the toolholder, the work-holding devices, and the machine tool itself must be sufficiently stiff to support that force, with minimal deflections. For example, if the thrust force is too high or if the machine tool is not sufficiently stiff, the tool will be pushed away from the workpiece surface being machined. This movement will, in turn, reduce the depth of cut, resulting in poor dimensional accuracy of the machined part.

The effect of rake angle and friction angle on the magnitude and direction of thrust force can be determined by noting from Fig. 21.11b that

$$F_t = R \sin(\beta - \alpha), \qquad (21.15)$$

or

$$F_t = F_c \tan(\beta - \alpha). \qquad (21.16)$$

The magnitude of the cutting force, F_c, is always positive, as shown in Fig. 21.11, because it is this force that supplies the work required in cutting. However, the sign of the thrust force, F_t, can be either positive or negative, depending on the relative magnitudes of β and α. Note that when $\beta > \alpha$, the sign of F_t is positive (*downward*), and when $\beta < \alpha$, the sign is negative (*upward*). Thus, it is possible to have an upward thrust force under the conditions of (a) high rake angles, (b) low friction at the tool–chip interface, or (c) both. A negative thrust force can have important implications in the design of machine tools and work holders and in the stability of the cutting process.

Power.   It can be seen from Fig. 21.11 that the power input in cutting is

$$\text{Power} = F_c V. \tag{21.17}$$

This power is dissipated mainly in the shear zone (due to the energy required to shear the material) and on the rake face of the tool (due to tool–chip interface friction).

From Figs. 21.4b and 21.11, the power dissipated in the shear plane is

$$\text{Power for shearing} = F_s V_s. \tag{21.18}$$

Denoting the width of cut as w, the **specific energy for shearing**, u_s, is given by

$$u_s = \frac{F_s V_s}{w t_o V}. \tag{21.19}$$

Similarly, the power dissipated in friction is

$$\text{Power for friction} = F V_c, \tag{21.20}$$

and the **specific energy for friction**, u_f, is

$$u_f = \frac{F V_c}{w t_o V} = \frac{F r}{w t_o}. \tag{21.21}$$

The **total specific energy**, u_t, is thus

$$u_t = u_s + u_f. \tag{21.22}$$

Because of the numerous factors involved, reliable prediction of cutting forces and power still is based largely on experimental data, such as those given in Table 21.2. The wide range of values seen in the table can be attributed to differences in strength within each material group, and to various other factors, such as friction, use of cutting fluids, and a wide range in processing variables. The sharpness of the tool tip also influences forces and power; because the tip rubs against the machined surface and makes the deformation zone ahead of the tool larger, duller tools require higher forces and power.

Video Solution 21.2 Power in Cutting

Measuring Cutting Forces and Power.   Cutting forces can be measured using a **force transducer** (typically with quartz piezoelectric sensors), a **dynamometer**, or a **load cell** (with resistance-wire strain gages placed on octagonal rings) mounted on the cutting-tool holder. Transducers have a much higher natural frequency and stiffness than dynamometers, which are prone to excessive deflection and vibration. It is also possible to calculate the cutting force from the **power consumption** during cutting from Eq. (21.4).

It should be recognized that Eq. (21.4) represents the power in the machining process, and the machine tool will require more power in order to overcome friction. Thus, to obtain the cutting force from the measured machine power consumption, the mechanical efficiency of the machine tool must be known. The *specific energy* in cutting, such as that shown in Table 21.2, also can be used to estimate cutting forces.

TABLE 21.2

Approximate Range of Energy Requirements in Cutting Operations at the Drive Motor of the Machine Tool (for Dull Tools, Multiply by 1.25)

Material	Specific energy	
	W-s/mm^3	hp-min/in^3
Aluminum alloys	0.4–1	0.15–0.4
Cast irons	1.1–5.4	0.4–2
Copper alloys	1.4–3.2	0.5–1.2
High-temperature alloys	3.2–8	1.2–3
Magnesium alloys	0.3–0.6	0.1–0.2
Nickel alloys	4.8–6.7	1.8–2.5
Refractory alloys	3–9	1.1–3.5
Stainless steels	2–5	0.8–1.9
Steels	2–9	0.7–3.4
Titanium alloys	2–5	0.7–2

EXAMPLE 21.1   Relative Energies in Cutting

Given: In an orthogonal cutting operation, $t_o = 0.005$ in., $V = 400$ ft/min, $\alpha = 10°$, and the width of cut is 0.25 in. It is observed that $t_c = 0.009$ in., $F_c = 125$ lb, and $F_t = 50$ lb.

Find: Calculate the percentage of the total energy that goes into overcoming friction at the tool–chip interface.

Solution: The percentage of the energy can be expressed as

$$\frac{\text{Friction energy}}{\text{Total energy}} = \frac{FV_c}{F_c V} = \frac{Fr}{F_c},$$

where

$$r = \frac{t_o}{t_c} = \frac{5}{9} = 0.555,$$

$$F = R \sin \beta,$$

$$F_c = R \cos(\beta - \alpha),$$

and

$$R = \sqrt{F_t^2 + F_c^2} = \sqrt{50^2 + 125^2} = 135 \text{ lb}.$$

Thus,

$$125 = 135 \cos(\beta - 10°),$$

and $\beta = 32°$ and

$$F = 135 \sin 32° = 71.5 \text{ lb}.$$

Hence,

$$\text{Percentage} = \frac{(71.5)(0.555)}{125} = 32,$$

or 32%.

21.4   Temperatures in Cutting

QR Code 21.3 Measuring temperatures during metal cutting. (*Source:* Courtesy of the National Institute of Standards and Technology)

As in all metalworking processes where plastic deformation is involved (Chapters 13 through 16), the energy dissipated in cutting is converted into *heat* which, in turn, raises the temperature in the cutting zone and the workpiece surface. *Temperature rise* is a very important factor in machining because of its major adverse effects:

- Excessive temperature lowers the strength, hardness, stiffness, and wear resistance of the cutting tool; tools also may soften and undergo plastic deformation, thus altering the tool shape.
- Increased heat causes uneven dimensional changes in the part being machined, depending on the physical properties of the material (Chapter 3), thus making it difficult to control its dimensional accuracy and tolerances.
- An excessive temperature rise can induce thermal damage and metallurgical changes (Chapter 4) in the machined surface, adversely affecting its properties.

The main sources of heat in machining are: (a) the work done in shearing in the primary shear zone, (b) energy dissipated as friction at the tool–chip interface, and (c) heat generated as the tool rubs against the machined surface, especially for dull or worn tools. Much effort has been expended in establishing relationships among temperature and various material and process variables in cutting. It can be shown that, in *orthogonal cutting*, the *mean temperature*, T_{mean}, in °F is

$$T_{\text{mean}} = \frac{1.2Y_f}{\rho c} \sqrt[3]{\frac{Vt_o}{K}}, \tag{21.23}$$

where Y_f is the flow stress in psi, ρc is the volumetric specific heat in in.-lb/in.^{3}-°F, and K is the thermal diffusivity (ratio of thermal conductivity to volumetric specific heat) in in.^{2}/s. Because the material parameters in this equation also depend on temperature, it is important to use appropriate values that are compatible with the predicted temperature range. It can be seen from Eq. (21.23) that the mean cutting temperature increases with workpiece strength, cutting speed, and depth of cut, and decreases with increasing specific heat and thermal conductivity of the workpiece material.

A simple expression for the *mean temperature in turning* on a lathe is given by

$$T_{\text{mean}} \propto V^a f^b, \qquad (21.24)$$

where V is the cutting speed and f is the feed of the tool, as shown in Fig. 21.2. Approximate values of the exponents a and b are $a = 0.2$ and $b = 0.125$ for *carbide* tools and $a = 0.5$ and $b = 0.375$ for *high-speed steel* tools.

Temperature Distribution. Because the sources of heat generation in machining are concentrated in the primary shear zone and at the tool–chip interface, it is to be expected that there will be severe temperature gradients in the cutting zone. A typical temperature distribution is shown in Fig. 21.12; note the presence of severe gradients and that the maximum temperature is about halfway up the tool–chip interface.

The temperatures developed in a *turning* operation on 52100 steel are shown in Fig. 21.13. The temperature distribution along the *flank surface* of the tool is shown in Fig. 21.13a, for $V = 60, 90,$ and 170 m/min, as a function of the distance from the tip of the tool. The temperature distributions at the *tool–chip interface* for the same three cutting speeds are shown in Fig. 21.13b, as a function of the fraction of the contact length. Thus, zero on the abscissa represents the tool tip, and 1.0 represents the end of the tool–chip contact length.

Note in Eq. 21.23 that the temperature increases with cutting speed, and that the highest temperature is almost 1100°C (2000°F). The presence of such high temperatures in machining can be verified simply by observing the dark-bluish color of the chips (caused by oxidation) typically produced at high cutting speeds. Chips can indeed become red hot, and hence create a safety hazard for the operator.

From Eq. (21.24) and the values for the exponent a, it can be seen that the cutting speed, V, greatly influences temperature. The explanation is that, as speed increases, the time for heat dissipation decreases, and hence the temperature rises, eventually becoming almost an *adiabatic* process. This effect of speed can be simulated easily by rubbing your hands together faster and faster.

As can be seen from Fig. 21.14, the chip carries away most of the heat generated. It has been estimated that in a typical machining operation, 90% of the energy is removed by the chip, with the rest going into the tool and the workpiece. Note in this figure that, as the cutting speed increases, a larger proportion of the total heat generated is carried away by the chip, and less heat goes elsewhere. This is one reason for the continued desire to increase machining speeds (see *high-speed machining*, Section 25.5). The other main benefit of higher cutting speeds is associated with the favorable economics in reducing machining time, as described in Section 25.8.

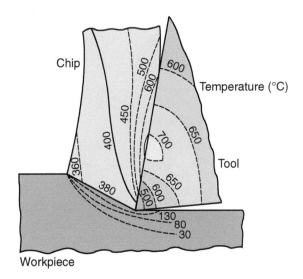

FIGURE 21.12 Typical temperature distribution in the cutting zone; note the severe temperature gradients within the tool and the chip, and that the workpiece remains relatively cool. *Source:* After G. Vieregge.

Video Solution 21.3 Temperatures in Turning

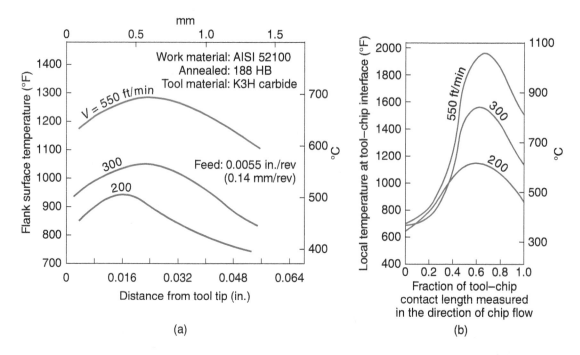

FIGURE 21.13 Temperatures developed in turning 52100 steel: (a) flank temperature distribution and (b) tool–chip interface temperature distribution. *Source:* After B.T. Chao and K.J. Trigger.

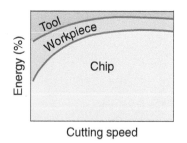

FIGURE 21.14 Proportion of the heat generated in cutting transferred to the tool, workpiece, and chip as a function of the cutting speed; note that the chip removes most of the heat.

Techniques for Measuring Temperature. Temperatures and their distribution in the cutting zone may be determined from **thermocouples** embedded in the tool or the workpiece. This technique has been used successfully, although it involves considerable effort. It is easier to determine the *mean* temperature with the **thermal emf** (electromotive force) at the tool–chip interface, which acts as a *hot junction* between two different materials (tool and chip). **Infrared radiation** from the cutting zone may also be monitored with a *radiation pyrometer*; however, this technique indicates only surface temperatures and its accuracy depends on the emissivity of the surfaces, which is difficult to determine accurately.

21.5   Tool Life: Wear and Failure

The previous sections have shown that cutting tools are subjected to (a) high localized stresses at the tip of the tool; (b) high temperatures, especially along the rake face; (c) sliding of the chip at high speeds along the rake face; and (d) sliding of the tool along the newly machined workpiece surface. These conditions induce **tool wear**, a major consideration in all machining operations, as are mold and die wear in casting and metalworking processes. Tool wear adversely affects tool life, the quality of the machined surface and its dimensional accuracy, and, consequently, the economics of cutting operations.

Wear is a gradual process (see Section 33.5), much like the wear of the tip of an ordinary pencil. The *rate* of tool wear (i.e., volume worn per unit time) depends on

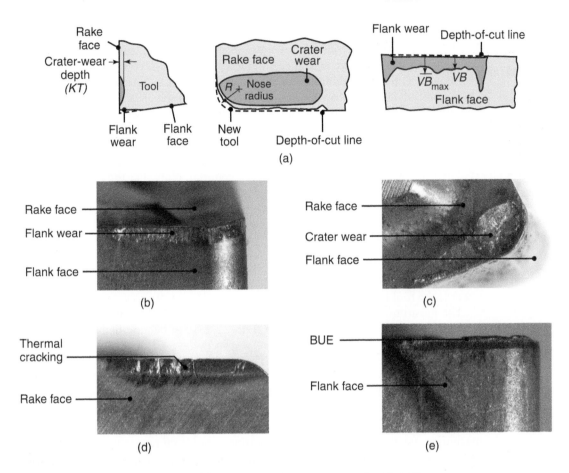

FIGURE 21.15 (a) Features of tool wear in a turning operation; the VB indicates average flank wear. (b) through (e) Examples of wear in cutting tools: (b) flank wear, (c) crater wear, (d) thermal cracking, and (e) flank wear and built-up edge. *Source:* (a) Terms and definitions reproduced with the permission of the International Organization for Standardization, ISO, copyright remains with ISO. (b) through (e) Courtesy of Kennametal Inc.

workpiece material, tool material and its coatings, tool geometry, process parameters, cutting fluids, and the characteristics of the machine tool. Tool wear and the resulting changes in tool geometry (Fig. 21.13) are generally classified as *flank wear, crater wear, nose wear, notching, plastic deformation, chipping,* and *gross fracture.*

21.5.1 Flank Wear

Flank wear occurs on the relief (flank) face of the tool, as shown in Figs. 21.15a, b, and e. It generally is attributed to (a) rubbing of the tool along the machined surface, thereby causing adhesive or abrasive wear and (b) high temperatures, which adversely affect tool-material properties.

In a classic study by F.W. Taylor (1856–1915) on the machining of steels conducted in the early 1890s, the following approximate relationship for tool life, known as the *Taylor tool-life equation*, was established:

$$VT^n = C, \tag{21.25}$$

TABLE 21.3

Ranges of n Values for the Taylor Equation (21.25) for Various Tool Materials	
High-speed steels	0.08–0.2
Cast alloys	0.1–0.15
Carbides	0.2–0.5
Coated carbides	0.4–0.6
Ceramics	0.5–0.7

where V is the cutting speed, T is the time (in minutes) that it takes to develop a certain flank **wear land** (shown as VB in Fig. 21.15a), n is an exponent that depends on tool and workpiece materials and cutting conditions, and C is a constant. Each combination of workpiece and tool materials and each cutting condition have their own n and C values, both of which are determined experimentally and often are based on surface finish requirements. Also, the Taylor equation is often applied even when flank wear is not the dominant wear mode (see Fig. 21.15), or if a different criterion (such as required machining power) is used to define C and n. Generally, n depends on the tool material, as shown in Table 21.3, and C on the workpiece material. Note that the magnitude of C is the cutting speed at $T = 1$ min.

To appreciate the importance of the exponent n, Eq. (21.25) can be rewritten as

$$T = \left(\frac{C}{V}\right)^{1/n}, \tag{21.26}$$

where it can be seen that for a constant value of C, the smaller the value of n, the lower is the tool life.

The most important variable associated with tool life is cutting speed, followed by depth of cut and feed, f. For turning, Eq. (21.25) can be modified as

$$VT^n d^x f^y = C, \tag{21.27}$$

where d is the depth of cut and f is the feed in mm/rev or in./rev, as shown in Fig. 21.2. The exponents x and y must be determined experimentally for each cutting condition.

Taking $n = 0.15$, $x = 0.15$, and $y = 0.6$ as typical values encountered in machining practice, it can be seen that cutting speed, feed rate, and depth of cut are of decreasing importance. Equation (21.27) can be rewritten as

$$T = C^{1/n} V^{-1/n} d^{-x/n} f^{-y/n}, \tag{21.28}$$

or, using typical values for the exponents, as

$$T \approx C^7 V^{-7} d^{-1} f^{-4}. \tag{21.29}$$

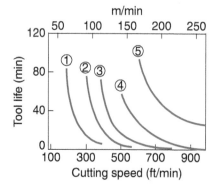

FIGURE 21.16 Effect of workpiece hardness and microstructure on tool life in turning ductile cast iron; note the rapid decrease in tool life (approaching zero) as the cutting speed increases. Tool materials have been developed that resist high temperatures, such as carbides, ceramics, and cubic boron nitride, as described in Chapter 22.

To obtain a constant tool life, the following observations can be made from Eq. (21.29):

- If the feed or the depth of cut is increased, the cutting speed must be decreased, and vice versa, and
- Depending on the exponents, a reduction in speed can result in an increase in the volume of the material removed, because of the increased feed or depth of cut.

Tool-life Curves. *Tool-life curves* are plots of experimental data obtained from cutting tests on various materials and under different cutting conditions, such as cutting speed, feed, depth of cut, tool material and geometry, and cutting fluids. Note in Fig. 21.16, for example, that (a) tool life decreases rapidly as the cutting speed increases, (b) the condition of the workpiece material has a strong influence on tool life, and (c) there is a large difference in tool life for different microstructures of the workpiece material (Chapter 4).

Heat treatment of the workpiece is important, due largely to increasing workpiece hardness. For example, ferrite has a hardness of about 100 HB, pearlite 200 HB, and martensite 300–500 HB. Impurities and hard constituents in the material or on the surface of the workpiece, such as rust, scale, and slag, also are important factors, because their abrasive action reduces tool life.

The exponent n can be determined from tool-life curves (Fig. 21.17). Note that the smaller the value of n, the faster the tool life decreases with increasing cutting speed. Although tool-life curves are somewhat linear over a limited range of cutting speeds, they rarely are linear over a wide range. Moreover, the exponent n can indeed become *negative* at low cutting speeds, meaning that tool-life curves actually can reach a maximum and then curve downward. Because of this possibility, caution should be exercised in using tool-life equations beyond the range of cutting speeds to which they are applicable.

Because temperature has a major influence on the physical and mechanical properties of materials (see Chapters 2 and 3), it is to be expected that it also strongly influences wear. Thus, as temperature increases, wear increases.

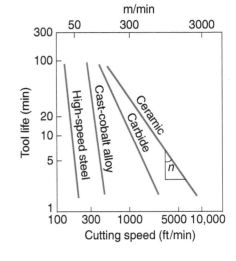

FIGURE 21.17 Tool-life curves for a variety of cutting-tool materials. The negative reciprocal of the slope of these curves is the exponent n in the Taylor tool-life equation (21.25), and C is the cutting speed at $T = 1$ min, ranging from about 200 to 10,000 ft/min in this figure.

EXAMPLE 21.2   Increasing Tool Life by Reducing the Cutting Speed

Given: Assume that for a given tool and workpiece combination, $n = 0.5$ and $C = 400$.

Find: Calculate the percentage increase in tool life when the cutting speed is reduced by 50%, using the Taylor equation for tool life.

Solution: Since $n = 0.5$, the Taylor equation can be rewritten as $VT^{0.5} = 400$. Denote V_1 as the initial speed and V_2 as the reduced speed; thus, $V_2 = 0.5V_1$. Because C is a constant at 400, we have the relationship

$$0.5V_1\sqrt{T_2} = V_1\sqrt{T_1}.$$

Simplifying this equation,

$$\frac{T_2}{T_1} = \frac{1}{0.25} = 4.$$

Thus the change in tool life is

$$\frac{T_2 - T_1}{T_1} = \left(\frac{T_2}{T_1}\right) - 1 = 4 - 1 = 3,$$

or that tool life is increased by 300%. Note that a *reduction* in cutting speed has resulted in a major *increase* in tool life. Note also that, for this problem, the magnitude of C is not relevant.

Video Solution 21.4 Taylor Tool-life Equation

Allowable Wear Land. A knife or a pair of scissors has to be sharpened when the quality of the cut deteriorates or the forces required become too high. Similarly, cutting tools need to be resharpened or replaced when (a) the surface finish of the machined workpiece begins to deteriorate, (b) cutting forces increase significantly, or (c) the temperature rises significantly. The *allowable wear land*, indicated as *VB* in Fig. 21.15a, for various machining conditions is given in Table 21.4. For improved dimensional accuracy and surface finish, the allowable wear land may be smaller than the values given in the table. The *recommended cutting speed* for a high-speed steel tool (see Section 22.2) is generally the one that yields a tool life of 60–120 min, and for a carbide tool (Section 22.4), it is 30–60 min.

Optimum Cutting Speed. Recall that as cutting speed increases, tool life is reduced rapidly. On the other hand, if the cutting speed is low, tool life is long, but the rate at which material is removed is also low. Thus, there is an *optimum cutting speed*, based on economic or production considerations, where the tool life is long and production speeds are reasonably high. Because it involves several other parameters, this topic is described further in Section 25.8.

EXAMPLE 21.3   Effect of Cutting Speed on Material Removal

The effect of cutting speed on the volume of metal removed between tool changes (or resharpenings) can be appreciated by analyzing Fig. 21.16. Assume that a material is being machined in the as-cast condition, with a hardness of 265 HB. Note that when the cutting speed is 60 m/min, tool life is about 40 min. Therefore, the tool travels a distance of 60 m/min × 40 min = 2400 m before it has to be replaced. However, when the cutting speed is increased to 120 m/min, the tool life is reduced to

about 5 min, and thus the tool travels 120 m/min × 5 min = 600 m before it has to be replaced.

Since the volume of material removed is directly proportional to the distance the tool has traveled, it can be seen that by *decreasing* the cutting speed, *more* material is removed between tool changes. It is important to note, however, that the lower the cutting speed, the longer is the time required to machine a part, which has a significant economic impact on the operation (see Section 25.8).

TABLE 21.4

Allowable Average Wear Land (See VB in Fig. 21.15a) for Cutting Tools in Various Machining Operations

Operation	Allowable wear land (mm)	
	High-speed steel tools	Carbide tools
Turning	1.5	0.4
Face milling	1.5	0.4
End milling	0.3	0.3
Drilling	0.4	0.4
Reaming	0.15	0.15

Note: Allowable wear for ceramic tools is about 50% higher; allowable notch wear (see Section 21.5.3), VB_{max}, is about twice that for *VB*.

21.5.2 Crater Wear

Crater wear occurs on the rake face of the tool, as shown in Figs. 21.15a and c, and Fig. 21.18, which illustrates various types of tool wear and failures. It readily can be seen that crater wear changes the tool–chip interface contact geometry. The most significant factors that influence crater wear are (a) the temperature at the tool–chip interface and (b) the chemical affinity of the tool and workpiece materials. Additionally, the same factors influencing flank wear also may affect crater wear.

Crater wear generally is attributed to a **diffusion mechanism**; that is, the movement of atoms across the tool–chip interface. Since diffusion rate increases with increasing temperature, crater wear increases as temperature increases. Note in Fig. 21.19, for example, how rapidly crater wear increases with temperature within a narrow range. Applying protective *coatings* to tools is an effective means of slowing the diffusion process, and thus reducing crater wear. Typical tool coatings are titanium nitride, titanium

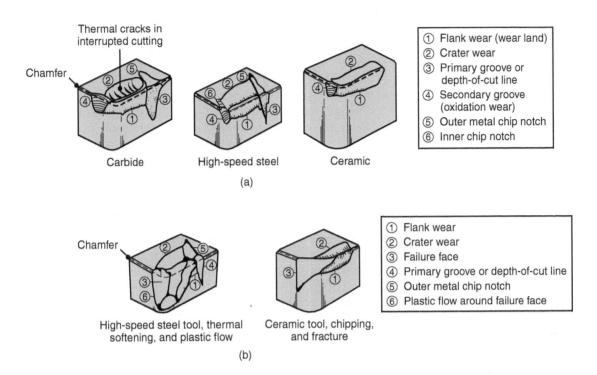

(a)

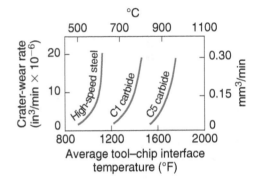

(b)

FIGURE 21.18 (a) Schematic illustrations of types of wear observed on various cutting tools. (b) Schematic illustrations of catastrophic tool failures. A wide range of parameters influence these wear and failure modes. *Source:* Printed by permission of V.C. Venkatesh.

carbide, titanium carbonitride, and aluminum oxide, and are described in greater detail in Section 22.6.

In comparing Figs. 21.12 and 21.15a, it can be seen that the location of the *maximum depth* of crater wear, *KT*, coincides with the location of the *maximum temperature* at the tool–chip interface. An actual cross-section of this interface, for steel machined at high speeds, is shown in Fig. 21.20. Note that the crater-wear pattern on the tool coincides with its discoloration pattern, an indication of the presence of high temperatures.

21.5.3   Other Types of Wear, Chipping, and Fracture

Nose wear (Fig. 21.15a) is the rounding of a sharp tool due to mechanical and thermal effects. It dulls the tool, affects chip formation, and causes rubbing of the tool over the workpiece, raising its temperature and inducing residual stresses on the machined surface. A related phenomenon is **edge rounding**, as shown in Fig. 21.15a.

An increase in temperature is particularly detrimental to high-speed steel tools, as can be appreciated from Fig. 22.1. Tools also may undergo **plastic deformation** because of temperature rises in the cutting zone, where temperatures can easily reach 1000°C (1800°F) in machining steels, and can be higher depending on the strength of the material machined.

Notches or **grooves** observed on cutting tools, as shown in Figs. 21.15a and 21.18, have been attributed to the fact that the region where they occur is the boundary where the chip is no longer in contact with the tool. Known as the **depth-of-cut line**

FIGURE 21.19 Relationship between crater-wear rate and average tool–chip interface temperature. Note how rapidly crater-wear rate increases with an incremental increase in temperature. *Source:* After B.T. Chao and K.J. Trigger.

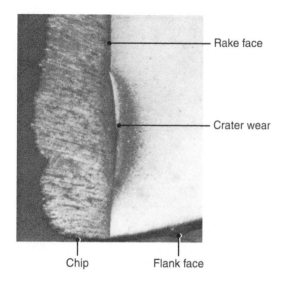

Chip   Flank face

FIGURE 21.20 Interface of a cutting tool (right) and chip (left) in machining plain-carbon steel; the discoloration of the tool indicates the presence of high temperatures. (Compare this figure with the temperature profiles shown in Fig. 21.12.) *Source:* Courtesy of P.K. Wright.

(DOC) (see Fig. 21.15a), this boundary oscillates, because of inherent variations in the cutting operation. In orthogonal cutting or with low feed rates, this region is at least partially in contact with the newly generated machined surface; the thin work-hardened layer that can develop in the workpiece will contribute to the formation of the wear groove. If sufficiently deep, the groove can lead to gross chipping of the tool tip, because of (a) its now reduced cross-section and (b) the notch sensitivity of the tool material.

Scale and *oxide layers* on a workpiece surface also contribute to notch wear, because these layers are hard and abrasive; thus, light cuts should not be taken on such workpieces. In Fig. 21.3, for example, the depth of cut, t_o, should be greater than the thickness of the scale on the workpiece.

In addition to being subjected to wear, cutting tools may undergo **chipping**, in which a small fragment from the cutting edge of the tool breaks away. This phenomenon, which typically occurs in brittle tool materials, such as ceramics, is similar to chipping the tip of a pencil if it is too sharp. The chipped fragments from the cutting tool may be very small (called **microchipping** or **macrochipping**, depending on its size), or they may be relatively large, in which case they are variously called **gross chipping, gross fracture**, and **catastrophic failure** (Fig. 21.18).

Chipping also may occur in a region of the tool where a small crack or defect already exists during its production. Unlike wear, which is a gradual process, chipping is a sudden loss of tool material, thus changing the tool's shape. As can be expected, chipping has a major detrimental effect on surface finish, surface integrity, and the dimensional accuracy of the workpiece.

Two main causes of chipping are:

- **Mechanical shock,** such as impact due to interrupted cutting, as in turning a splined shaft on a lathe
- **Thermal fatigue,** due to cyclic variations in the temperature of the tool in interrupted cutting

Thermal cracks usually are perpendicular to the cutting edge of the tool, as shown on the rake face of the carbide tool in Figs. 21.15d and 21.18a. Major variations in the composition or structure of the workpiece material also may cause chipping, due to differences in their thermal properties.

Chipping can be reduced by selecting tool materials with high impact and thermal-shock resistance, as described in Chapter 22. High positive rake angles can contribute to chipping because of the small included angle of the tool tip, as can be visualized from Fig. 21.3. Also, it is possible for the crater-wear region to progress toward the tool tip, thus weakening the tip because of reduced volume of material.

21.5.4 Tool-condition Monitoring

With rapid advances in computer-controlled machine tools and automated manufacturing, the reliable and repeatable performance of cutting tools has become a major consideration. As described in Chapters 23 through 25, modern machine tools operate with little direct supervision by a machine operator, and they generally are enclosed, making it impossible or difficult to monitor the machining operation and the condition of the cutting tool. It is therefore essential to continuously and indirectly monitor the condition of the cutting tool so as to note, for example, whether excessive wear, chipping, or gross tool failure is occurring. In modern machine tools, tool-condition monitoring systems are integrated into computer numerical control and programmable logic controllers.

Techniques for tool-condition monitoring typically fall into two general categories: direct and indirect. The **direct method** for observing the condition of a cutting tool involves *optical* measurements of wear, such as periodic observation of changes in the tool profile. This is a common and reliable technique, and is done using a microscope (*toolmakers' microscope*). However, this method requires that the cutting operation be stopped for tool observation. Another direct method involves programming the tool to contact a sensor after each machining cycle; this approach allows the measurement of wear and/or the detection of broken tools. Usually, the sensor involves a touch probe that must be depressed by the tool tip.

Indirect methods involve the correlation of the tool condition with parameters such as cutting forces, power, temperature rise, workpiece surface finish, vibration, and chatter. A common technique is **acoustic emission** (AE), which utilizes a *piezoelectric transducer* mounted on a toolholder. The transducer picks up acoustic emissions (typically above 100 kHz) which result from the stress waves generated during cutting. By analyzing the signals, tool wear and chipping can be monitored. This technique is particularly effective in precision-machining operations, where cutting forces are low (because of the small amounts of material removed). Another effective use of AE is in detecting the fracture of small carbide tools at high cutting speeds.

A similar indirect tool-condition monitoring system consists of **transducers** that are installed in original machine tools, or are retrofitted on existing machines. The system continually monitors torque and forces during cutting. The signals are preamplified, and a microprocessor analyzes and interprets their content. The system is capable of differentiating the signals that come from different sources, such as tool breakage, tool wear, a missing tool, overloading of the machine tool, or colliding with machine components. The system also can compensate automatically for tool wear, and thus improve the dimensional accuracy of the part being machined.

The design of transducers must be such that they are (a) nonintrusive to the machining operation, (b) accurate and repeatable in signal detection, (c) resistant to abuse and robust for the shop-floor environment (see Sections 36.5.1 and 40.7), and (d) cost effective. Continued progress is being made in the development of **sensors**, including the use of *infrared* and *fiber-optic techniques* for temperature measurement during machining.

In lower-cost computer numerical-control machine tools, monitoring is done by **tool-cycle time**. In a production environment, once the life expectancy of a cutting tool or insert has been determined, it can be entered into the machine control unit, so that the operator is prompted to make a tool change when that time is reached. This system is inexpensive and fairly reliable, although not totally so, because of the inherent statistical variation in tool life.

21.6 Surface Finish and Integrity

Surface finish influences not only the dimensional accuracy of machined parts but also their properties and their performance in service. The term *surface finish* describes the geometric features of a surface (see Chapter 33), and *surface integrity* pertains to material properties, such as fatigue life and corrosion resistance, that are strongly influenced by the nature of the surface produced.

(a)

(b)

FIGURE 21.21 Machined surfaces produced on steel (highly magnified), as observed with a scanning-electron microscope: (a) turned surface and (b) surface produced by shaping. *Source:* Courtesy of JT Black and S. Ramalingam.

With its significant effect on changing the tool-tip profile, the *built-up edge* (see Fig. 21.6) has the greatest influence on surface finish. The surfaces produced in two different cutting operations are shown in Fig. 21.21. Note the considerable damage to the surfaces from BUE; its damage can be noted by the scuffing marks, which deviate from the straight grooves that would result from normal machining, as seen in Fig. 21.2. Ceramic and diamond tools generally produce a better surface finish than other tools, largely because of their much lower tendency to form a BUE.

A *dull* tool has a large radius along its edges, just like the tip of a dull pencil or the cutting edge of a knife. Figure 21.22 illustrates the relationship between the radius of the cutting edge and the depth of cut in orthogonal cutting. Note that at small depths of cut, the rake angle effectively can become negative, and the tool simply may ride over the workpiece surface instead of cutting it and producing chips. This is a phenomenon similar to trying to scrape a thin layer from the surface of a stick of butter with a dull knife.

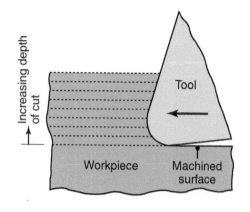

FIGURE 21.22 Schematic illustration of a dull tool with respect to the depth of cut in orthogonal machining (exaggerated); note that the tool has a positive rake angle, but as the depth of cut decreases, the rake angle effectively can become negative. The tool then simply rides over the workpiece (without cutting) and burnishes its surface; this action raises the workpiece temperature and causes surface residual stresses and metallurgical changes.

If the tip radius of the tool (not to be confused with the radius R in Fig. 21.15a) is large in relation to the depth of cut, the tool simply will rub over the machined surface. Rubbing will generate heat and induce residual surface stresses, which in turn may cause surface damage, such as tearing and cracking. Consequently, the depth of cut should be greater than the radius on the cutting edge.

In a turning operation, as in other cutting processes described in the rest of Part IV of this text, the tool leaves a spiral profile (called **feed marks**) on the machined surface as it moves across the workpiece, as shown in Figs. 21.2 and 21.23. It can be noted that the higher the feed, f, and the smaller the tool-nose radius, R, the more prominent the feed marks will be. It can be shown that the surface roughness, for such a case, is given by

$$R_t = \frac{f^2}{8R}, \tag{21.30}$$

where R_t is the *roughness height*, as described in Section 33.3. Although not significant in rough machining operations, feed marks are important in finish machining. (Further details on surface roughness are given for individual machining processes as they are described.)

Vibration and chatter are described in detail in Section 25.4. For now, it should be recognized that if the tool vibrates or chatters during cutting, it will adversely affect the workpiece surface finish. The reason is that a vibrating tool periodically changes the dimensions of the cut. Excessive chatter also can cause chipping and premature failure of the more brittle cutting tools, such as ceramics and diamond.

Factors influencing surface integrity are:

- Temperatures generated during processing and possible metallurgical transformations
- Surface residual stresses
- Severe plastic deformation and strain hardening of the machined surfaces, tearing, and cracking

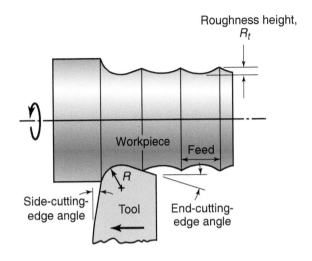

FIGURE 21.23   Schematic illustration of feed marks on a surface being turned (exaggerated).

Each of these factors can have major adverse effects on the machined part, although they can be taken care of by careful selection and maintenance of cutting tools and control of process variables.

The difference between **finish machining** and **rough machining** should be emphasized. In finish machining, it is important to consider the surface finish to be produced, whereas in rough machining the main purpose is to remove a large amount of material at a high rate. Surface finish is not a primary consideration, since it will be improved during finish machining. It is important that there be no subsurface-damage resulting from rough machining that cannot be removed during finish machining (see Fig. 21.21).

21.7   Machinability

The *machinability* of a material is usually defined in terms of four factors:

1. Surface finish and surface integrity of the machined part
2. Tool life
3. Force and power requirements
4. The level of difficulty in chip control after it is generated

Thus, good machinability indicates good surface finish and surface integrity, a long tool life, and low force and power requirements. As for chip control, and as stated earlier regarding continuous chips, chips that are long, thin, stringy, and curled can severely interfere with the machining operation by becoming entangled in the cutting zone (see Fig. 21.7).

Because of the complex nature of cutting operations, it is difficult to establish relationships that quantitatively define the machinability of a particular material. In machining practice, tool life and surface roughness generally are considered to be the most important factors in machinability. Approximate **machinability ratings** (indexes) have been available for many years for each type of material and its condition; however, they are not particularly useful or reliable because of their qualitative nature. In subsequent chapters, several tables are presented in which, for various groups of materials, *specific recommendations* are given regarding such parameters as cutting speed, feed, depth of cut, cutting tools and their shape, and type of cutting fluids.

21.7.1   Machinability of Ferrous Metals

This section describes the machinability of steels, alloy steels, stainless steels, and cast irons.

Steels.   Because steels are among the most important engineering materials, as also noted in Chapter 5, their machinability has been studied extensively. Carbon steels have a wide range of machinability, depending on their ductility and hardness. If a carbon steel is too ductile, a BUE can develop, leading to poor surface finish. If the steel is too hard, it can cause abrasive wear of the tool, because of the presence of carbides in the steel. Cold-worked carbon steels are thus desirable from a machinability standpoint.

An important group of steels is **free-machining steels**, containing sulfur and phosphorus. *Sulfur* forms manganese-sulfide inclusions (*second-phase particles*, Section 4.2.3), which act as stress raisers in the primary shear zone. As a result, the chips produced break up easily and are small, thus improving machinability. The size, shape, distribution, and concentration of these inclusions significantly influence machinability. Elements such as *tellurium* and *selenium*, both of which are chemically similar to sulfur, act as *inclusion modifiers* in resulfurized steels.

Phosphorus in steels has two major effects: (a) It strengthens the ferrite, causing increased hardness and resulting in better chip formation and surface finish and (b) it increases hardness and thus causes the formation of short chips instead of continuous stringy ones, thereby improving machinability. Soft steels can be difficult to machine because of their tendency for BUE formation and the resulting poor surface finish.

In **leaded steels**, a high percentage of lead solidifies at the tips of manganese-sulfide inclusions. In nonresulfurized grades of steel, lead takes the form of dispersed fine particles. Lead is insoluble in iron, copper, and aluminum and their alloys, and because of its low shear strength, it acts as a *solid* lubricant (see Section 33.7.6) and is smeared over the tool–chip interface during machining.

When the temperature developed is sufficiently high, such as at high cutting speeds and feeds, the lead melts directly in front of the tool, acting as a liquid lubricant. In addition to having this effect, lead lowers the shear stress in the primary shear zone, thus reducing cutting forces and power consumption. Lead can be used with every grade of steel and is identified by the letter "L" between the second and third numerals in steel identification (e.g., 10L45). In stainless steels, a similar use of the letter L means "low carbon," which improves their corrosion resistance.

Because lead is a well-known *toxin* and a pollutant, there are serious environmental concerns about its use in steels (estimated at 4500 tons of lead consumption every year in the production of steels). Consequently, there is a continuing trend toward eliminating the use of lead in steels (**lead-free steels**). *Bismuth* and *tin* are substitutes for lead in steels, but are not as effective in improving machinability.

Calcium-deoxidized steels contain oxide flakes of calcium silicates (CaSO) that reduce the strength of the secondary shear zone, and decrease tool–chip interface friction and wear. Because temperature increase is reduced correspondingly, these steels produce less crater wear, especially at high cutting speeds.

Alloy steels can have a wide variety of compositions and hardnesses, thus their machinability cannot be generalized. An important trend in machining these steels is *hard turning*, described in detail in Section 25.6. Alloy steels at hardness levels of 45–65 HRC can be machined with polycrystalline cubic-boron-nitride (cBN) cutting tools (see Section 22.7), producing good surface finish, integrity, and dimensional accuracy.

Effects of Various Elements in Steels.   The presence of *aluminum* and *silicon* in steels is always harmful, because these elements combine with oxygen and form aluminum oxide and silicates, which are hard and abrasive. As a result, tool wear increases and machinability is reduced.

Carbon and *manganese* have various effects on the machinability of steels, depending on their composition. Plain low-carbon steels (less than 0.15% C) can produce poor surface finish, by forming a BUE. Cast steels can be abrasive, although their machinability is similar to that of wrought steels. Tool and die steels are very difficult to machine, and usually require annealing prior to machining. The machinability of most steels is improved by cold working, which hardens the material and reduces the tendency for BUE formation.

Other alloying elements, such as *nickel, chromium, molybdenum,* and *vanadium,* that improve the properties of steels also generally reduce machinability. The effect of *boron* is negligible. Gaseous elements such as *hydrogen* and *nitrogen* can have particularly detrimental effects on the properties of steel. Oxygen has been shown to have a strong effect on the aspect ratio of the manganese-sulfide inclusions: The higher the oxygen content, the lower the aspect ratio, and the higher the machinability.

In improving the machinability of steels, it is important to also consider the possible detrimental effects of the alloying elements on the properties and strength of machined parts in service. At elevated temperatures, for example, lead causes *embrittlement* of steels (liquid-metal embrittlement and hot shortness; see Section 1.5.2), although at room temperature it has no effect on mechanical properties.

Sulfur can reduce the hot workability of steels severely, because of the formation of iron sulfide, unless sufficient manganese is present to prevent such formation. At room temperature, the mechanical properties of resulfurized steels depend on the orientation of the deformed manganese-sulfide inclusions. Rephosphorized steels are significantly less ductile, and are produced solely for the purpose of improving machinability.

Stainless Steels.   Austenitic (300 series) steels generally are difficult to machine. Chatter can be a problem, thus necessitating machine tools with high stiffness. Ferritic stainless steels (300 series) have good machinability. Martensitic (400 series) steels are abrasive, tend to form a BUE, and require tool materials with high hot hardness and crater-wear resistance. *Precipitation-hardening stainless steels* are strong and abrasive, thus requiring hard and abrasion-resistant tool materials.

Cast Irons.   *Gray irons* generally are machinable, although they can be abrasive, depending on composition, especially pearlite. Free carbides in castings reduce their machinability, and can cause tool chipping or fracture. *Nodular* and *malleable irons* are machinable, using hard tool materials.

21.7.2   Machinability of Nonferrous Metals

The following is a summary of the machinability of nonferrous metals and alloys, in alphabetic order:

- **Aluminum** is generally very easy to machine, although the softer grades tend to form a BUE, resulting in poor surface finish; thus, high cutting speeds, high rake angles, and high relief angles are recommended. Wrought aluminum alloys with high silicon content and cast aluminum alloys are generally abrasive, hence they require harder tool materials. Dimensional tolerance control may be a problem in machining aluminum, because it has a high thermal expansion coefficient and a relatively low elastic modulus.

- **Beryllium** generally is machinable, but because the fine particles produced during machining are toxic, it requires machining in a controlled environment.
- **Cobalt-based alloys** are abrasive and highly work hardening; they require sharp, abrasion-resistant tool materials and low feeds and speeds.
- **Copper,** in the wrought condition, can be difficult to machine, because of BUE formation; cast copper alloys are easy to machine. *Brasses* are easy to machine, especially with the addition of lead (*leaded free-machining brass*); note, however, the toxicity of lead and the associated environmental concerns. *Bronzes* are more difficult to machine than brass.
- **Magnesium** is very easy to machine, with good surface finish and prolonged tool life; however, care should be exercised because of its high rate of oxidation (*pyrophoric*) and the danger of fire.
- **Molybdenum** is ductile and work hardening; it can produce poor surface finish, thus sharp tools are essential.
- **Nickel-based alloys and superalloys** are work hardening, abrasive, and strong at high temperatures; their machinability depends on their condition and improves with annealing.
- **Tantalum** is very work hardening, ductile, and soft; it produces a poor surface finish, and tool wear is high.
- **Titanium** and its alloys have very poor thermal conductivity (the lowest of all metals, see Table 3.2), thus causing a significant temperature rise and BUE; they are highly reactive and can be difficult to machine.
- **Tungsten** is brittle, strong, and very abrasive; hence, its machinability is low, although it improves greatly at elevated temperatures.
- **Zirconium** has good machinability, but it requires a coolant-type cutting fluid because of the danger of explosion and fire.

21.7.3 Machinability of Miscellaneous Materials

Thermoplastics generally have low thermal conductivity and low elastic modulus, and they are thermally softening. Consequently, machining them requires sharp tools with positive rake angles (to reduce cutting forces), large relief angles, small depths of cut and feed, relatively high speeds, and proper support of the workpiece. External cooling of the cutting zone may be necessary, to keep the chips from becoming gummy and sticking to the cutting tools. Cooling usually can be achieved with a jet of air, a vapor mist, or using water-soluble oils.

Thermosetting plastics are brittle and sensitive to thermal gradients during cutting; machining conditions generally are similar to those of thermoplastics.

Polymer-matrix composites are very abrasive, because of the fibers that are present; hence, they are difficult to machine. Fiber tearing, pulling, and edge delamination are significant problems, and can lead to severe reduction in the load-carrying capacity of machined components. Machining of these materials requires careful handling and removal of debris, in order to avoid contact with and inhaling of the fibers.

Metal-matrix and ceramic-matrix composites can be difficult to machine, depending on the properties of the matrix material and the reinforcing fibers.

Graphite is abrasive; it requires sharp, hard, and abrasion-resistant tools.

Ceramics now have a steadily improved machinability, particularly with the development of *machinable ceramics* and nanoceramics (Section 8.2.5), and with the selection of appropriate processing parameters, such as ductile-regime cutting (described in Section 25.7).

Wood is an *orthotropic* material, with properties varying with its grain direction; consequently, the type of chips and the surfaces produced also vary significantly, depending on the type of wood and its condition. Woodworking, which dates back to 3000 B.C., remains largely an art. The basic requirements are generally sharp tools and high cutting speeds.

21.7.4 Thermally Assisted Machining

Metals and alloys that are difficult to machine at room temperature can be machined more easily at elevated temperatures. In thermally assisted machining, also called hot machining, a source of heat (such as a torch, induction coil, electric current, laser-beam, electron-beam, or plasma arc) is focused onto an area just ahead of the cutting tool. First investigated in the early 1940s, this operation typically is carried out above the *homologous temperature* of $T/T_m = 0.5$ (see Section 1.7, and Tables 1.2 and 3.1); thus, steels are hot machined above the temperature range of 650°–750°C (1200°–1400°F).

Although difficult and complicated to perform in production plants, the general advantages of hot machining are: (a) reduced cutting forces, (b) increased tool life, (c) higher material-removal rates, and (d) a reduced tendency for vibration and chatter.

SUMMARY

- Machining processes are often necessary to impart the desired dimensional accuracy, geometric features, and surface-finish characteristics to components, particularly those with complex shapes that cannot be produced economically using other shaping techniques. On the other hand, machining generally takes more time, wastes some material in the form of chips, doesn't affect the bulk properties of the workpiece, and may have adverse effects on surfaces produced.

- Commonly observed chip types in machining are continuous, built-up edge, discontinuous, and serrated. Important process variables in machining are tool shape and tool material; cutting conditions such as speed, feed, and depth of cut; the use of cutting fluids; and the characteristics of the workpiece material and the machine tool. Parameters influenced by these variables are forces and power consumption, tool wear, surface finish and surface integrity, temperature rise, and dimensional accuracy of the workpiece.

- Temperature rise in machining is an important consideration, since it can have adverse effects on tool life, as well as on the dimensional accuracy and surface integrity of the machined part.

- Two principal types of tool wear are flank wear and crater wear. Tool wear depends on workpiece and tool material characteristics; cutting speed, feed, depth of cut, and cutting fluids; and the characteristics of the machine tool. Tool failure also may occur by notching, chipping, and gross fracture.

- The surface finish of machined components can adversely affect product integrity. Important variables are the geometry and condition of the cutting tool, the type of chip produced, and process variables.

- Machinability generally is defined in terms of surface finish, tool life, force and power requirements, and chip control. The machinability of materials depends on their composition, properties, and microstructure. Proper selection and control of process variables are important.

KEY TERMS

Acoustic emission	Depth-of-cut line	Notch wear	Shear plane
Allowable wear land	Diffusion	Oblique cutting	Skiving
Built-up edge	Discontinuous chip	Orthogonal cutting	Specific energy
Chip	Feed marks	Primary shear zone	Surface finish
Chip breaker	Flank wear	Rake angle	Surface integrity
Chip curl	Friction angle	Relief angle	Taylor equation
Chipping of tool	Hot machining	Rephosphorized steel	Thrust force
Clearance angle	Inclination angle	Resulfurized steel	Tool-condition monitoring
Continuous chip	Machinability	Secondary shear zone	Tool life
Crater wear	Machinability ratings	Serrated chip	Turning
Cutting force	Machine tool	Shaving	Wear land
Cutting ratio	Machining	Shear angle	

BIBLIOGRAPHY

ASM Handbook, Vol. 16: **Machining**, ASM International, 1989.

Astakhov, V.P., **Metal Cutting Mechanics**, CRC Press, 1998.

Boothroyd, G., and Knight, W.A., **Fundamentals of Metal Machining and Machine Tools**, 3rd ed., Marcel Dekker, 2006.

Childs, T.H.C., Maekawa, K., Obikawa, T., and Yamane, Y., **Metal Machining: Theory and Applications**, Butterworth-Heinemann, 2000.

Cormier, D., **McGraw-Hill Machining and Metalworking Handbook**, McGraw-Hill 2005.

Davim, J.P. (ed.), **Machining: Fundamentals and Recent Advances**, Springer, 2010.

Davim, J.P. (ed.), **Surface Integrity in Machining**, Springer, 2010.

Shaw, M.C., **Metal Cutting Principles**, 2nd ed., Oxford, 2005.

Stephenson, D.A., and Agapiou, J.S., **Metal Cutting: Theory and Practice**, 2nd ed., CRC Press, 2005.

Trent, E.M., and Wright, P.K., **Metal Cutting**, 4th ed., Butterworth-Heinemann, 2000.

Tschatsch, H., **Applied Machining Technology**, Springer, 2009.

REVIEW QUESTIONS

21.1 Explain why continuous chips are not necessarily desirable.

21.2 Name the factors that contribute to the formation of discontinuous chips.

21.3 What is the cutting ratio? Is it always less than 1? Explain.

21.4 Explain the difference between positive and negative rake angles. What is the importance of the rake angle?

21.5 Explain how a dull tool can lead to negative rake angles.

21.6 Comment on the role and importance of the relief angle.

21.7 Explain the difference between discontinuous chips and segmented chips.

21.8 Why should we be interested in the magnitude of the thrust force in cutting?

21.9 What are the differences between orthogonal and oblique cutting?

21.10 What is a BUE? Why does it form?

21.11 Is there any advantage to having a built-up edge on a tool? Explain.

21.12 What is the function of chip breakers? How do they function? Do you need a chip breaker to eliminate continuous chips in oblique cutting? Explain.

21.13 Identify the forces involved in a cutting operation. Which of these forces contribute to the power required?

21.14 Explain the characteristics of different types of tool wear.

21.15 List the factors that contribute to poor surface finish in cutting.

21.16 Explain what is meant by the term machinability and what it involves. Why does titanium have poor machinability?

21.17 What is shaving in machining? When would it be used?

QUALITATIVE PROBLEMS

21.18 List reasons that machining operations may be required, and provide an example for each reason.

21.19 Are the locations of maximum temperature and crater wear related? If so, explain why.

21.20 Is material ductility important for machinability? Explain.

21.21 Explain why studying the types of chips produced is important in understanding cutting operations.

21.22 Why do you think the maximum temperature in orthogonal cutting is located at about the middle of the tool–chip interface? (*Hint:* Note that the two sources of heat are (a) shearing in the primary shear plane and (b) friction at the tool–chip interface.)

21.23 Tool life can be almost infinite at low cutting speeds. Would you then recommend that all machining be done at low speeds? Explain.

21.24 Explain the consequences of allowing temperatures to rise to high levels in cutting.

21.25 The cutting force increases with the depth of cut and decreasing rake angle. Explain why.

21.26 Why is it not always advisable to increase the cutting speed in order to increase the production rate?

21.27 What are the consequences if a cutting tool chips?

21.28 What are the effects of performing a cutting operation with a dull tool? A very sharp tool?

21.29 To what factors do you attribute the difference in the specific energies in machining the materials shown in Table 21.2? Why is there a range of energies for each group of materials?

21.30 Explain why it is possible to remove more material between tool resharpenings by lowering the cutting speed.

21.31 Noting that the dimension d in Fig. 21.4a is very small, explain why the shear strain rate in metal cutting is so high.

21.32 Explain the significance of Eq. (21.9).

21.33 Comment on your observations regarding Figs. 21.12 and 21.13.

21.34 Describe the consequences of exceeding the allowable wear land (Table 21.4) for various cutting-tool materials.

21.35 Comment on your observations regarding the hardness variations shown in Fig. 21.6a.

21.36 Why does the temperature in cutting depend on the cutting speed, feed, and depth of cut? Explain in terms of the relevant process variables.

21.37 You will note that the values of a and b in Eq. (21.24) are higher for high-speed steels than for carbides. Why is this so?

21.38 As shown in Fig. 21.14, the percentage of the total cutting energy carried away by the chip increases with increasing cutting speed. Why?

21.39 Describe the effects that a dull tool can have on cutting operations.

21.40 Explain whether it is desirable to have a high or low (a) n value and (b) C value in the Taylor tool-life equation.

21.41 The Taylor tool-life equation is directly applicable to flank wear. Explain whether or not it can be used to model tool life if other forms of wear are dominant.

21.42 The tool-life curve for ceramic tools in Fig. 21.17 is to the right of those for other tool materials. Why?

21.43 Why are tool temperatures low at low cutting speeds and high at high cutting speeds?

21.44 Can high-speed machining be performed without the use of a cutting fluid?

21.45 Given your understanding of the basic metal-cutting process, what are the important physical and chemical properties of a cutting tool?

21.46 Explain why the power requirements in cutting depend on the cutting force but not the thrust force.

21.47 State whether or not the following statements are true, explaining your reasons: (a) For the same shear angle, there are two rake angles that give the same cutting ratio. (b) For the same depth of cut and rake angle, the type of cutting fluid used has no influence on chip thickness. (c) If the cutting speed, shear angle, and rake angle are known, the chip velocity can be calculated. (d) The chip becomes thinner as the rake angle increases. (e) The function of a chip breaker is to decrease the curvature of the chip.

QUANTITATIVE PROBLEMS

21.48 Let $n = 0.5$ and $C = 400$ in the Taylor equation for tool wear. What is the percent increase in tool life if the cutting speed is reduced by (a) 50% and (b) 75%?

21.49 Assume that, in orthogonal cutting, the rake angle is 15° and the coefficient of friction is 0.2. Using Eq. (21.4), determine the percentage increase in chip thickness when the friction is doubled.

21.50 Derive Eq. (21.14).

21.51 Taking carbide as an example and using Eq. (21.24), determine how much the feed should be reduced in order to keep the mean temperature constant when the cutting speed is doubled.

21.52 Using trigonometric relationships, derive an expression for the ratio of shear energy to frictional energy in orthogonal cutting, in terms of angles α, β, and ϕ only.

21.53 An orthogonal cutting operation is being carried out under the following conditions: $t_o = 0.1$ mm, $t_c = 0.2$ mm, width of cut = 4 mm, $V = 3$ m/s, rake angle = 10°, $F_c = 500$ N, and $F_t = 200$ N. Calculate the percentage of the total energy that is dissipated in the shear plane.

21.54 Explain how you would go about estimating the C and n values for the four tool materials shown in Fig. 21.17.

21.55 Derive Eqs. (21.1) and (21.3).

21.56 Assume that, in orthogonal cutting, the rake angle, α, is 20° and the friction angle, β, is 35° at the chip–tool interface. Determine the percentage change in chip thickness when the friction angle is 45°. [*Note:* do not use Eq. (21.4) or Eq. (21.5).]

21.57 Show that, for the same shear angle, there are two rake angles that give the same cutting ratio.

21.58 With appropriate diagrams, show how the use of a cutting fluid can change the magnitude of the thrust force, F_t, in Fig. 21.11. Consider both heat transfer and lubrication effects.

21.59 In a cutting operation using a −5° rake angle, the measured forces were $F_c = 1330$ N and $F_t = 740$ N. When a cutting fluid was used, these forces were $F_c = 1200$ N and $F_t = 710$ N. What is the change in the friction angle resulting from the use of a cutting fluid?

21.60 For a turning operation using a ceramic cutting tool, if the speed is increased by 50%, by what factor must the feed rate be modified to obtain a constant tool life? Use $n = 0.5$ and $y = 0.6$.

21.61 In Example 21.3, if the cutting speed V is doubled, will the answer be different? Explain.

21.62 Using Eq. (21.30), select an appropriate feed for $R = 1$ mm and a desired roughness of 0.5 μm. How would you adjust this feed to allow for nose wear of the tool during extended cuts? Explain your reasoning.

21.63 With a carbide tool, the temperature in a cutting operation is measured as 1200°F when the speed is 300 ft/min and the feed is 0.002 in./rev. What is the approximate temperature if the speed is doubled? What speed is required to lower the maximum cutting temperature to 900°F?

21.64 The following flank wear data were collected in a series of machining tests, using C6 carbide tools on 1045 steel (HB = 192). The feed rate was 0.015 in./rev, and the width of cut was 0.030 in. (a) Plot flank wear as a function of cutting time. Using a 0.015 in. wear land as the criterion of tool failure, determine the lives for the two cutting speeds. (b) Plot your results on log–log plot and determine the values of n and C in the Taylor tool-life equation. (Assume a straight line relationship.) (c) Using these results, calculate the tool life for a cutting speed of 300 ft/min.

Cutting speed, ft/min	Cutting time, min	Flank wear, in.
400	0.5	0.0014
	2.0	0.0023
	4.0	0.0030
	8.0	0.0055
	16.0	0.0082
	24.0	0.0112
	54.0	0.0150
600	0.5	0.0018
	2.0	0.0035
	4.0	0.0060
	8.0	0.0100
	13.0	0.0145
	14.0	0.0160
800	0.5	0.0050
	2.0	0.0100
	4.0	0.0140
	5.0	0.0160
1000	0.5	0.0100
	1.0	0.0130
	1.8	0.0150
	2.0	0.0160

21.65 The following data are available from orthogonal cutting experiments. In both cases depth of cut (feed) $t_o = 0.13$ mm, width of cut $b = 2.5$ mm, rake angle $\alpha = -5°$, and cutting speed $V = 2$ m/s.

	Workpiece material	
	Aluminum	Steel
Chip thickness, t_c, mm	0.23	0.58
Cutting force, F_c, N	430	890
Thrust force, F_t, N	280	800

Determine the shear angle ϕ, friction coefficient μ, shear stress τ, shear strain γ on the shear plane, chip velocity V_c, and shear velocity V_s, as well as energies u_f, u_s, and u_t.

21.66 Estimate the cutting temperatures for the conditions of Problem 21.65 if the following properties apply:

	Workpiece material	
	Aluminum	Steel
Cutting energy, u, N-mm/mm^3	1320	2740
Thermal diffusivity, K, mm^2/s	97	14
Volumetric specific heat, ρc, N/mm^2°C	2.6	3.3

21.67 Assume that you are an instructor covering the topics described in this chapter, and you are giving a quiz on the numerical aspects to test the understanding of the students. Prepare two quantitative problems and supply the answers.

SYNTHESIS, DESIGN, AND PROJECTS

21.68 Tool life is increased greatly when an effective means of cooling and lubrication is implemented. Design methods of delivering this fluid to the cutting zone, and discuss the advantages and limitations of your design.

21.69 Design an experimental setup whereby orthogonal cutting can be simulated in a turning operation on a lathe.

21.70 Describe your thoughts on whether chips produced during machining can be used to make useful products. Give some examples of possible products, and comment on their characteristics and differences if the same products were made by other manufacturing processes. Which types of chips would be desirable for this purpose?

21.71 Recall that cutting tools can be designed so that the tool–chip contact length is reduced by recessing the rake face of the tool some distance away from its tip. Explain the possible advantages of such a tool.

21.72 Recall that the chip-formation mechanism also can be observed by scraping the surface of a stick of butter with a sharp knife. Using butter at different temperatures, including frozen butter, conduct such an experiment. Keep the depth of cut constant and hold the knife at different angles (to simulate the tool rake angle), including oblique scraping. Describe your observations regarding the type of chips produced. Also, comment on the force that your hand feels while scraping and whether you observe any chatter when the butter is very cold.

21.73 Experiments have shown that it is possible to produce thin, wide chips, such as 0.08-mm (0.003-in.) thick and 10-mm (4-in.) wide, which would be similar to the dimensions of a rolled sheet. Materials have been aluminum, magnesium, and stainless steel. A typical setup would be similar to orthogonal cutting, by machining the periphery of a solid round bar with a straight tool moving radially inward. Describe your thoughts regarding producing thin metal sheets by this method, taking into account the metal's surface characteristics and properties.

21.74 Describe your thoughts regarding the recycling of chips produced during machining in a plant. Consider chips produced by dry cutting versus those produced by machining with a cutting fluid.

21.75 List products that can be directly produced from metal chips or shavings.

21.76 Obtain a wood planer and some wood specimens. Show that the chips produced depend on the direction of cut with respect to the wood grain. Explain why.

21.77 It has been noted that the chips from certain carbon steels are noticeably magnetic, even if the original workpiece is not. Research the reasons for this effect and write a one-page paper explaining the important mechanisms.

21.78 As we have seen, chips carry away the majority of the heat generated during machining. If chips did not have this capacity, what suggestions would you make in order to be able to carry out machining processes without excessive heat? Explain.

22

Cutting-tool Materials and Cutting Fluids

- Continuing the coverage of the fundamentals of machining in the preceding chapter, this chapter describes two essential elements in machining operations: cutting-tool materials and cutting fluids.

- The chapter opens with a discussion of the types and characteristics of cutting-tool materials, including high-speed steels, carbides, ceramics, cubic boron nitride, diamond, and coated tools.

- The types of cutting fluids in common use are then described, including their functions and how they affect the machining operation.

- Trends in near-dry and dry machining, and in methods for cutting fluid application are also described, and their significance with respect to environmentally friendly machining operations are explained.

22.1 Introduction

The selection of cutting-tool materials for a particular application is among the most important factors in machining operations. This chapter describes the relevant properties and performance characteristics of all major types of cutting-tool materials, as a guide to tool selection. However, because of its complex nature, this subject does not readily render itself to the precise determination of appropriate tool materials for a particular application. Consequently, general guidelines and recommendations have been established in industry over many years. More detailed information on recommendations for specific workpiece materials and machining operations are presented beginning with Chapter 23.

As noted in the preceding chapter, the cutting tool is subjected to (a) high temperatures, (b) high contact stresses, and (c) rubbing along the tool–chip interface and along the machined surface. Consequently, the cutting-tool material must possess the following characteristics:

- **Hot hardness,** so that the hardness, strength, and wear resistance of the tool are maintained at the temperatures encountered in machining operations. This property ensures that the tool does not undergo any plastic deformation, and thus retains its shape and sharpness. Tool-material hardness is a function of temperature, as shown in Fig. 22.1; note the wide response of these materials, how rapidly carbon tool steels lose their hardness (meaning that they cannot be used for many operations), and how well ceramics maintain their hardness at high temperatures. Carbon tool steels (Section 5.7) were commonly used as tool materials until the development of high-speed steels in the early

1900s; high speed meaning that machining speeds can be high, hence higher productivity.

- **Toughness** and **impact strength** (mechanical shock resistance, Section 2.9), so that impact forces on the tool encountered repeatedly in interrupted cutting operations (such as milling and turning a splined shaft on a lathe) or forces due to vibration and chatter during machining do not chip or fracture the tool.
- **Thermal shock resistance,** to withstand the rapid temperature cycling (Section 3.6), as encountered in interrupted cutting.
- **Wear resistance** (Section 33.5), so that an acceptable tool life is obtained before replacement is necessary.
- **Chemical stability** and **inertness** with respect to the workpiece material, to avoid or minimize any adverse reactions, adhesion, and tool–chip diffusion that would contribute to tool wear.

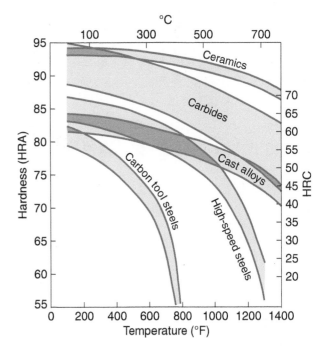

FIGURE 22.1   The hardness of various cutting-tool materials as a function of temperature (hot hardness); the wide range in each group of materials is due to the variety of tool compositions and treatments available for that group.

To respond to these demanding requirements, a variety of cutting-tool materials, with a wide range of mechanical, physical, and chemical properties, have been developed over the years, as shown in Table 22.1. The properties listed in the first column of this table are useful in determining desirable tool-material characteristics for a particular application. For example:

- Hardness and strength are important with respect to the mechanical properties of the workpiece material to be machined.
- Impact strength is important in making interrupted cuts in machining, such as in milling.
- Melting temperature of the tool material is important, especially compared to the temperatures developed in the cutting zone.
- Thermal conductivity and coefficient of thermal expansion are important in determining the resistance of the tool materials to thermal fatigue and shock.

Video Solution 22.1 Wear Resistance of Cutting–tool Materials

It will be recognized that a particular tool material may not have *all* of the desired properties for a particular machining operation. This situation can readily be seen from Table 22.2, by observing the opposite directions of the long horizontal arrows showing trends. Note, for example, that (a) high-speed steels are tough, but they have limited hot hardness and (b) ceramics have high resistance to temperature and wear, but they are brittle and can easily chip. Note also how the cost of tools increases from HSS to diamond.

The operating characteristics of tool materials are shown in Table 22.3, listed in the order in which they were developed and implemented in industry. Note that many of these materials also are used for dies and molds in casting, forming, and shaping metallic and nonmetallic materials.

1. High-speed steels
2. Cast-cobalt alloys
3. Carbides

TABLE 22.1

General Characteristics of Tool Materials

Property	High-speed steels	Cast-cobalt alloys	Carbides		Ceramics	Cubic boron nitride	Single-crystal diamond*
			WC	TiC			
Hardness	83–86 HRA	82–84 HRA 46–62 HRC	90–95 HRA 1800–2400 HK	91–93 HRA 1800–3200 HK	91–95 HRA 2000–3000 HK	4000–5000 HK	7000–8000 HK
Compressive strength,							
$\quad$MPa	4100–4500	1500–2300	4100–5850	3100–3850	2750–4500	6900	6900
$\quad$psi $\times 10^3$	600–650	220–335	600–850	450–560	400–650	1000	1000
Transverse rupture strength,							
$\quad$MPa	2400–4800	1380–2050	1050–2600	1380–1900	345–950	700	1350
$\quad$psi $\times 10^3$	350–700	200–300	150–375	200–275	50–135	105	200
Impact strength,							
$\quad$J	1.35–8	0.34–1.25	0.34–1.35	0.79–1.24	<0.1	<0.5	<0.2
$\quad$in.-lb	12–70	3–11	3–12	7–11	<1	<5	<2
Modulus of elasticity,							
$\quad$GPa	200	—	520–690	310–450	310–410	850	820–1050
$\quad$psi $\times 10^6$	30	—	75–100	45–65	45–60	125	120–150
Density,							
$\quad$kg/m^3	8600	8000–8700	10,000–15,000	5500–5800	4000–4500	3500	3500
$\quad$lb/in^3	0.31	0.29–0.31	0.36–0.54	0.2–0.22	0.14–0.16	0.13	0.13
Volume of hard phase, %	7–15	10–20	70–90	—	100	95	95
Melting or decomposition temperature,							
$\quad$°C	1300	—	1400	1400	2000	1300	700
$\quad$°F	2370	—	2550	2550	3600	2400	1300
Thermal conductivity, W/m-K	30–50	—	42–125	17	29	13	500–2000
Coefficient of thermal expansion, $\times 10^{-6}$/°C	12	—	4–6.5	7.5–9	6–8.5	4.8	1.5–4.8

*The values for polycrystalline diamond are generally lower, except for impact strength, which is higher.

TABLE 22.2

General Characteristics of Cutting-tool Materials (These Materials Have a Wide Range of Compositions and Properties; Overlapping Characteristics Exist in Many Categories of Tool Materials)

	High-speed steels	Cast-cobalt alloys	Uncoated carbides	Coated carbides	Ceramics	Polycrystalline cubic boron nitride	Diamond
Hot hardness	───►						
Toughness	◄───						
Impact strength	◄───						
Wear resistance	───►						
Chipping resistance	◄───						
Cutting speed	───►						
Thermal-shock resistance	◄───						
Tool material cost	───►						
Depth of cut	Light to heavy	Light to heavy	Light to heavy	Light to heavy	Light to heavy	Light to heavy	Very light for single-crystal diamond
Processing method	Wrought, cast, HIP* sintering	Cast and HIP sintering	Cold pressing and sintering	CVD or PVD**	Cold pressing and sintering or HIP sintering	High-pressure, high-temperature sintering	High-pressure, high-temperature sintering

Source: After R. Komanduri.
*Hot-isostatic pressing.
**Chemical-vapor deposition, physical-vapor deposition.

TABLE 22.3

General Operating Characteristics of Cutting-tool Materials

Tool materials	General characteristics	Modes of tool wear or failure	Limitations
High-speed steels	High toughness, resistance to fracture, wide range of roughing and finishing cuts, good for interrupted cuts	Flank wear, crater wear	Low hot hardness, limited hardenability, and limited wear resistance
Uncoated carbides	High hardness over a wide range of temperatures, toughness, wear resistance, versatile, wide range of applications	Flank wear, crater wear	Cannot use at low speeds because of cold welding of chips and microchipping
Coated carbides	Improved wear resistance over uncoated carbides, better frictional and thermal properties	Flank wear, crater wear	Cannot use at low speeds because of cold welding of chips and microchipping
Ceramics	High hardness at elevated temperatures, high abrasive wear resistance	Depth-of-cut line notching, microchipping, gross fracture	Low strength and low thermomechanical fatigue strength
Polycrystalline cubic boron nitride (cBN)	High hot hardness, toughness, cutting-edge strength	Depth-of-cut line notching, chipping, oxidation, graphitization	Low strength, and lower chemical stability than ceramics at higher temperature
Diamond	High hardness and toughness, abrasive wear resistance	Chipping, oxidation, graphitization	Low strength, and low chemical stability at higher temperatures

Source: After R. Komanduri and other sources.

4. Coated tools
5. Alumina-based ceramics
6. Cubic boron nitride
7. Silicon-nitride-based ceramics
8. Diamond
9. Whisker-reinforced materials and nanomaterials

Carbon steels are the oldest tool materials, and have been used widely for drills, taps, broaches, and reamers since the 1880s. Low-alloy and medium-alloy steels were developed later for similar applications, but with longer tool life. Although inexpensive and easily shaped and sharpened, these steels do not have sufficient hot hardness and wear resistance for machining at high speeds, where the temperature rises significantly. Their use is limited to very low speed cutting operations, particularly in woodworking, hence they are not of any particular significance in modern machining operations.

In this chapter, the following topics are described:

- The characteristics, applications, and limitations of cutting-tool materials, including the required characteristics and costs
- The applicable range of process variables for optimal performance
- The types and characteristics of cutting fluids and their specific applications in a wide variety of machining operations

22.2   High-speed Steels

High-speed steel (HSS) tools are so named because they were developed to machine at higher speeds than was previously possible. First produced in the early 1900s, high-speed steels are the most highly alloyed of the tool steels (Section 5.7). They can be hardened to various depths, have good wear resistance, and are relatively inexpensive. Because of their toughness, and associated high resistance to fracture, high-speed steels are suitable especially for (a) high positive rake-angle tools (those with small included angles), (b) interrupted cuts, (c) machine tools with low stiffness that are subject to vibration and chatter, and (d) complex tools, such as drills, reamers, taps, and gear cutters. Their most important limitation, due to their lower hot hardness, is that the cutting speeds are low compared with those of carbide tools, as can be seen in Fig. 22.1.

There are two basic types of high-speed steels: **molybdenum** (M-series) and **tungsten** (T-series). The M-series contains up to about 10% Mo, with Cr, V, W, and Co as alloying elements. The T-series contains 12–18% W, with Cr, V, and Co as alloying elements. Carbides formed in the steel constitute about 10–20% by volume. The M-series generally has higher abrasion resistance than the T-series, undergoes less distortion during heat treating (Section 4.7), and is less expensive. Consequently, 95% of all high-speed steel tools are made of the M-series steels. Table 5.6 lists three of these steels and their characteristics.

High-speed steel tools are available in wrought (rolled or forged), cast, and powder metallurgy (sintered) forms. They can be **coated** for improved performance, as described in Section 22.5. High-speed steel tools also may be subjected to *surface treatments* (such as case hardening for improved hardness and wear resistance; see Section 4.10) or steam treatment at elevated temperatures to develop a hard, black oxide layer (*bluing*) for improved performance, including a lower tendency for built-up edge formation.

The major alloying elements in HSS are chromium, vanadium, tungsten, cobalt, and molybdenum. To appreciate their role in cutting tools, refer to Table 5.2 on the effects of various elements in steels and note the following:

- *Chromium* improves toughness, wear resistance, and high-temperature strength
- *Vanadium* improves toughness, abrasion resistance, and hot hardness
- *Tungsten* and *cobalt* have similar effects, namely, improved strength and hot hardness
- *Molybdenum* improves wear resistance, toughness, and high-temperature strength and hardness

22.3 Cast-cobalt Alloys

Introduced in 1915, *cast-cobalt alloys* have the following composition ranges: 38%–53% Co, 30%–33% Cr, and 10%–20% W. Because of their high hardness, typically 58–64 HRC, they have good wear resistance and can maintain their hardness at elevated temperatures. They are not as tough as high-speed steels and are sensitive to impact forces; consequently, they are less suitable than high-speed steels for interrupted cutting operations. Commonly known as *Stellite* tools, these alloys are cast and ground into relatively simple shapes. They are now used only for special applications that involve deep, continuous *roughing cuts* at relatively high feeds and speeds, by as much as twice the rates possible with high-speed steels.

22.4 Carbides

The two groups of tool materials just described possess the required toughness, impact strength, and thermal shock resistance, but they also have important limitations, particularly with respect to strength and hot hardness. Consequently, they cannot be used as effectively where high cutting speeds, hence high temperatures, are involved; such speeds often are necessary to improve plant productivity.

To meet the challenge for increasingly higher cutting speeds, *carbides*, also known as *cemented* or *sintered carbides*, were introduced in the 1930s. Because of their high hardness over a wide range of temperatures (Fig. 22.1), high elastic modulus, high thermal conductivity, and low thermal expansion, carbides are among the most important, versatile, and cost-effective tool and die materials for a wide range of applications. The two major groups of carbides used for machining are *tungsten carbide* and *titanium carbide*. In order to differentiate them from the coated tools described in Section 22.5, plain-carbide tools usually are referred to as **uncoated carbides**.

22.4.1 Tungsten Carbide

Tungsten carbide (WC) typically consists of tungsten-carbide particles bonded together in a cobalt matrix. These tools are manufactured using powder metallurgy techniques, hence the term *sintered carbides* or *cemented carbides*, as described in Chapter 17. Tungsten-carbide particles are first combined typically with cobalt, resulting in a composite material with a cobalt matrix surrounding the carbide particles. These particles, which are 1–5 μm (40–200 μin.) in size, are then pressed and sintered into the desired *insert* (see Section 22.4.3 and Fig. 22.2) shapes. Tungsten

FIGURE 22.2 Typical cutting-tool inserts with various shapes and chip-breaker features; round inserts also are available, as can be seen in Figs. 22.3c and 22.4; the holes in the inserts are standardized for interchangeability in toolholders. *Source:* Courtesy of Kennametal Inc.

QR Code 22.1 Production of inserts. (*Source:* Courtesy of Sandvik Coromant)

carbides frequently are also compounded with *titanium carbide* and *niobium carbide*, to impart special properties to the material.

The amount of cobalt present, ranging typically from 6 to 16%, significantly affects the properties of tungsten-carbide tools. As the cobalt content increases, the strength, hardness, and wear resistance of WC decrease, while its toughness increases, because of the higher toughness of cobalt. Tungsten-carbide tools generally are used for cutting steels, cast irons, and abrasive nonferrous materials, and largely have replaced HSS tools because of their better performance.

Micrograin Carbides. Cutting tools also are made of submicron and ultra-fine-grained (*micrograin*) carbides, including tungsten carbide, titanium carbide, and tantalum carbide. The grain size is typically in the range from 0.2 to 0.8 μm (8–30 μin.). Compared with the traditional carbides, described previously, these tool materials are stronger, harder, and more wear resistant, thus improving productivity. In one application, microdrills, with diameters on the order of 100 μm (0.004 in.), are being made from micrograin carbides, and used in the fabrication of circuit boards (Section 28.13).

Functionally Graded Carbides. In these tools, the composition of the carbide in the insert has a *gradient* through its near-surface depth, instead of being uniform as it is in common carbide inserts. The gradient has a smooth distribution of compositions and phases, with functions similar to those described as desirable properties of coatings on cutting tools. Graded mechanical properties eliminate stress concentrations and promote tool life and performance; they are, however, more expensive and cannot be justified for all applications.

22.4.2 Titanium Carbide

Titanium carbide (TiC) consists of a nickel–molybdenum matrix. It has higher wear resistance than tungsten carbide but is not as tough. Titanium carbide is suitable for machining hard materials, mainly steels and cast irons, and for machining at speeds higher than those appropriate for tungsten carbide.

22.4.3 Inserts

Although a supply of sharp, or resharpened, tools is usually maintained in plants, tool-changing operations can be time consuming and thus inefficient. The need for a more effective method has led to the development of *inserts*, which are individual cutting tools with several cutting points (Fig. 22.2). Thus, a square insert has eight cutting points, and a triangular insert has six. Inserts usually are clamped on the *toolholder*, with various locking mechanisms (Fig. 22.3); when one point of the insert is worn, it is **indexed** (rotated in its holder) to make another cutting point

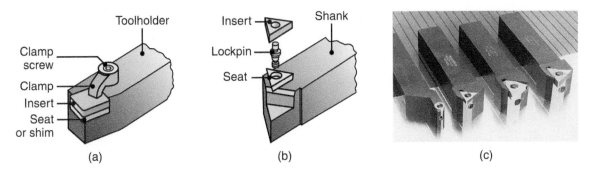

FIGURE 22.3   Methods of mounting inserts on toolholders: (a) clamping and (b) wing lockpins. (c) Examples of inserts mounted with threadless lockpins, which are secured with side screws. *Source:* Courtesy of Valenite.

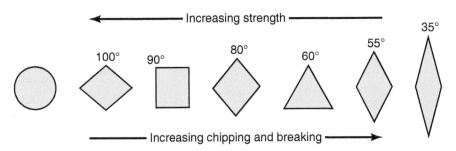

FIGURE 22.4   Relative edge strength and tendency for chipping of inserts with various shapes; strength refers to the cutting edge indicated by the included angles. *Source:* Courtesy of Kennametal Inc.

available. In addition to the examples in this figure, a wide variety of other toolholders is available for specific applications, including those with quick insertion and removal features.

Carbide inserts are available in a variety of shapes, such as square, triangle, diamond, and round. The strength of the cutting edge of an insert depends on its shape; the smaller the included angle (see top of Fig. 22.4), the lower is the strength of the edge. In order to further improve edge strength and prevent chipping, insert edges usually are honed, chamfered, or produced with a negative land (Fig. 22.5). Most inserts are honed to a radius of about 0.025 mm (0.001 in.).

Chip-breaker features (see Fig. 21.7 and Section 21.2.1) on inserts are for the purposes of (a) controlling chip flow during machining, (b) eliminating long chips, (c) reducing heat generated, and (d) reducing the tendency for vibration and chatter. Carbide inserts are commercially available with a wide variety of complex chip-breaker features, typical examples of which are shown in Fig. 22.2. The selection of a particular chip-breaker feature depends on the feed and depth of cut of the operation, the workpiece material, the type of chip produced during cutting, and whether it is a roughing or finishing machining. Optimum chip-breaker geometries continue to be developed by computer-aided design and finite-element analysis techniques.

Stiffness of the machine tool (Section 25.3) is of major importance in using carbide tools. Light feeds,

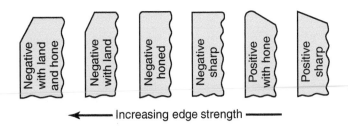

FIGURE 22.5   Edge preparation for inserts to improve edge strength. *Source:* Courtesy of Kennametal Inc.

low speeds, and chatter are detrimental, because they tend to damage the tool's cutting edge. Light feeds, for example, concentrate the forces and temperature closer to the edges of the tool, increasing the tendency for the edges to chip off.

22.4.4 Classification of Carbides

Carbide tool grades are classified using the letters P, M, K, N, S, and H (as shown in Tables 22.4 and 22.5) for a range of applications, including the traditional C grades used in the United States. Because of the wide variety of carbide compositions available and the broad range of machining applications and workpiece materials involved, efforts at ISO classification continue to be a difficult task.

TABLE 22.4

ISO Classification of Carbide Cutting Tools According to Use				
			Designation in order of decreasing wear and toughness in each category (in increments of 5)	
Symbol	Workpiece material	Color code	Uncoated	Coated
P	Ferrous metals with long chips	Blue	P01, P05–P20	P20–P50
M	Stainless steels with long or short chips	Yellow	M10–M20	M20–M40
K	Cast iron with short chips	Red	K05–K20	K05–K30
N	Nonferrous metals	Green	N10–20	N05–N30
S	High-temperature alloys	Orange	S10–20	S20–S30
H	Hardened materials	Gray	—	H10

TABLE 22.5

Classification of Tungsten Carbides According to Selected Machining Applications						
ISO standard	ANSI classification number (grade)	Materials to be machined	Machining operation	Type of carbide	Characteristics of	
					Cut	Carbide
K30–K40	C1	Cast iron, nonferrous metals, and nonmetallic materials requiring abrasion resistance	Roughing	Wear-resistant grades; generally straight WC–Co with varying grain sizes	Increasing cutting speed ↓ Increasing feed rate	Increasing hardness and wear resistance ↓ ↑ Increasing strength and binder content
K20	C2		General purpose			
K10	C3		Light finishing			
K01	C4		Precision finishing			
P30–P50	C5	Steels requiring crater and deformation resistance	Roughing	Crater-resistant grades; various WC–Co compositions with TiC and/or TaC alloys	Increasing cutting speed ↓ Increasing feed rate	Increasing hardness and wear resistance ↓ ↑ Increasing strength and binder content
P20	C6		General purpose			
P10	C7		Light finishing			
P01	C8		Precision finishing			

Note: The ISO and ANSI comparisons are approximate.

22.5 Coated Tools

As described in Part I, new metal alloys and engineered materials are being developed continuously, particularly since the 1960s. These materials have high strength and toughness, but generally are abrasive and chemically reactive with tool materials. The difficulty of machining these materials efficiently and the need for improving their performance have led to important developments in *coated tools*. Compared with the tool materials themselves, coatings have advantageous properties, such as:

- Lower friction
- Higher resistance to wear and cracking
- Higher hot hardness and impact resistance
- Acting as a diffusion barrier between the tool and the chip

Coated tools can last 10 times more than those of uncoated tools, thus allowing for high cutting speeds and reducing both the time required for machining operations and production costs. As can be seen from Fig. 22.6, machining time has been reduced steadily by a factor of more than 100 since 1900. This improvement has had a major impact on the economics of machining operations, in conjunction with continued improvements in the design and construction of modern machine tools and their computer controls (see Chapter 25 and Part IX). As a result, coated tools now are used in as much as 80% of all machining operations, particularly turning, milling, and drilling.

22.5.1 Coating Materials and Coating Methods

Commonly used coating materials are *titanium nitride* (TiN), *titanium carbide* (TiC), *titanium carbonitride* (TiCN), and *aluminum oxide* (Al_2O_3). These coatings, typically in the thickness range from 2 to 15 μm (80–600 μin.), are applied on

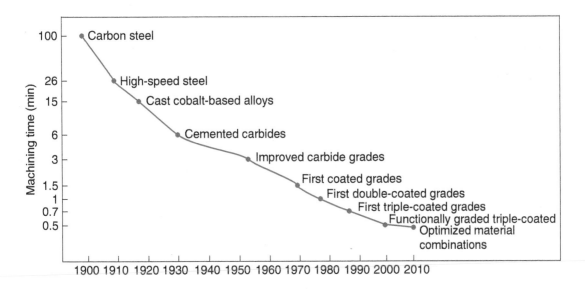

FIGURE 22.6 Relative time required to machine with various cutting-tool materials, indicating the year the tool materials were first introduced; note that machining time has been reduced by two orders of magnitude within a 100 years. *Source:* Courtesy of Sandvik.

cutting tools and inserts by two principal techniques, described in greater detail in Section 34.6:

1. **Chemical-vapor deposition** (CVD), including **plasma-assisted chemical-vapor deposition**
2. **Physical-vapor deposition** (PVD)

The CVD process is the most commonly used method for carbide tools with multiphase and ceramic coatings, both of which are described later in this section. However, the PVD-coated carbides with TiN coatings have higher cutting-edge strength, lower friction, and a lower tendency to form a built-up edge, and the coatings are smoother and more uniform in thickness, which generally is in the range from 2 to 4 μm (80–160 μin.). Another technology, used particularly for multiphase coatings, is **medium-temperature chemical-vapor deposition** (MTCVD), developed to machine ductile (nodular) iron and stainless steels, and to provide higher resistance to crack propagation than CVD coatings provide.

Coatings for cutting tools and dies should have the following general characteristics:

- **High hardness** at elevated temperatures, to resist wear
- **Chemical stability** and **inertness** to the workpiece material, in order to reduce wear
- **Low thermal conductivity**, to prevent temperature rise in the substrate
- **Compatibility** and **good bonding**, to prevent flaking or spalling from the substrate, which may be carbide or high-speed steel
- **Little or no porosity**, to maintain its integrity and strength

The effectiveness of coatings is enhanced by the hardness, toughness, and high thermal conductivity of the substrate. Honing (Section 26.7) of the cutting edges is an important procedure for the maintenance of coating strength; otherwise, the coating may peel or chip off at sharp edges and corners.

Titanium-nitride Coatings.   Titanium-nitride coatings have low friction coefficients, high hardness, resistance to high temperature, and good adhesion to the substrate. Consequently, they greatly improve the life of high-speed steel tools, as well as the lives of carbide tools, drill bits, and cutters. Titanium-nitride-coated tools (gold in color) perform well at higher cutting speeds and feeds. Flank wear is significantly lower than that of uncoated tools (Fig. 22.7), and flank surfaces can be reground after use, since regrinding the flank face does not remove the coating on the rake face of the tool. However, these coated tools do not perform as well at low cutting speeds, because the coating can be worn off by chip adhesion, thus the use of appropriate cutting fluids to minimize adhesion is important.

Titanium-carbide Coatings.   Titanium-carbide coatings on tungsten-carbide inserts have high flank-wear resistance in machining abrasive materials.

Ceramic Coatings.   Because of their chemical inertness, low thermal conductivity, resistance to high temperature, and resistance to flank and crater wear, ceramics are suitable coating materials for cutting tools. The most commonly used ceramic coating is *aluminum oxide* (Al_2O_3).

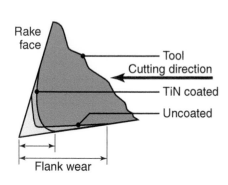

FIGURE 22.7 Schematic illustration of typical wear patterns on uncoated high-speed steel tools and titanium-nitride-coated tools; note that flank wear is significantly lower for the coated tool.

However, because they are very stable (not chemically reactive), oxide coatings generally bond weakly to the substrate.

Multiphase Coatings.   The desirable properties of the coatings just described can be combined and optimized using *multiphase coatings*. Carbide tools are available with two or more layers of such coatings, and are particularly effective in machining cast irons and steels. For example, TiC can first be deposited over the substrate, followed by Al_2O_3, and then TiN. The first layer should bond well with the substrate, the outer layer should resist wear and have low thermal conductivity, and the intermediate layer should bond well and be compatible with both layers.

Typical applications of multiple-coated tools are:

- High-speed, continuous cutting: TiC/Al_2O_3
- Heavy-duty, continuous cutting: $TiC/Al_2O_3/TiN$
- Light, interrupted cutting: TiC/TiC + TiN/TiN

Coatings also are available in **alternating multiphase layers**; the thickness of these layers is on the order of 2–20 μm, thinner than regular multiphase coatings (Fig. 22.8). The reason for using thinner coatings is that coating hardness increases with decreasing grain size, a phenomenon similar to the increase in the strength of metals with decreasing grain size (see Section 1.5.1); thus, thinner layers are harder than thicker layers.

A typical multiphase-coated carbide tool may consist of the following layers, starting from the top, along with their primary functions:

1. TiN: low friction
2. Al_2O_3: high thermal stability
3. TiCN: fiber reinforced, with a good balance of resistance to flank wear and crater wear, effective particularly for interrupted cutting
4. A thin-carbide substrate: high fracture toughness
5. A thick-carbide substrate: hard and resistant to plastic deformation at high temperatures

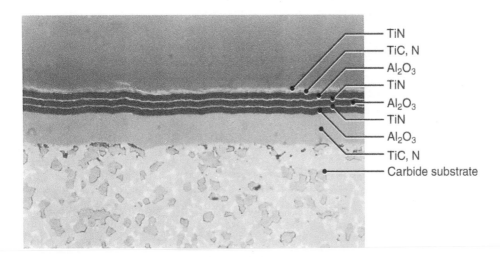

FIGURE 22.8   Multiphase coatings on a tungsten-carbide substrate; three alternating layers of aluminum oxide are separated by very thin layers of titanium nitride. Inserts with as many as 13 layers of coatings have been made. Coating thicknesses are typically in the range from 2 to 20 μm. *Source:* Courtesy of Kennametal Inc.

Diamond Coatings. The properties and applications of diamond, diamond coatings, and *diamondlike carbon* are described in Sections 8.7 and 34.13, and the use of these materials as cutting tools is given in Section 22.9. *Polycrystalline diamond* is used widely as a coating for cutting tools, particularly on tungsten-carbide and silicon-nitride inserts. Diamond-coated tools are particularly effective in machining (a) nonferrous metals, (b) abrasive materials, such as aluminum alloys containing silicon, (c) fiber-reinforced and metal-matrix composite materials, and (d) graphite. As much as 10-fold improvements in tool life have been obtained over the lives of other coated tools.

Diamond-coated inserts, available commercially, have thin films deposited on substrates through PVD or CVD techniques. Thick diamond films are obtained by growing a large sheet of pure diamond, which is then laser cut to shape and brazed to a carbide insert. *Multilayer nanocrystal diamond coatings* also are being developed, with interlocking layers of diamond that give strength to the coating. As with all coatings, it is essential to develop good adhesion of the diamond film to the substrate, and to minimize the difference in thermal expansion between the diamond and substrate materials (see Section 3.6).

22.5.2 Miscellaneous Coating Materials

Major advances are taking place in further improving the performance of coated tools. **Titanium carbonitride** (TiCN) and **titanium-aluminum nitride** (TiAlN) are effective in cutting stainless steels. TiCN (deposited through physical-vapor deposition) is harder and tougher than TiN, and can be used on carbides and high-speed steel tools. TiAlN is effective in machining aerospace alloys. Chromium-based coatings, such as **chromium carbide** (CrC), have been found to be effective in machining softer metals that have a tendency to adhere to the cutting tool, such as aluminum, copper, and titanium. Other coating materials include **zirconium nitride** (ZrN) and hafnium nitride (HfN).

More recent developments include (a) **nanolayer coatings,** such as carbide, boride, nitride, oxide, or some combination thereof (see also Section 8.8) and (b) **composite coatings,** using a variety of materials. The hardness of some of these coatings approaches that of cubic boron nitride (see Fig. 2.15).

22.5.3 Ion Implantation

In this process, ions are introduced into the surface of the cutting tool, improving its surface properties (Section 34.7); the process does not change the dimensions of tools. **Nitrogen-ion** implanted carbide tools have been used successfully on alloy steels and stainless steels. **Xenon-ion** implantation of tools is also under development.

22.6 Alumina-based Ceramics

Ceramic tool materials, introduced in the early 1950s, consist primarily of fine-grained, high-purity **aluminum oxide** (Section 8.2). They are cold pressed into insert shapes under high pressure, then sintered at high temperature; the end product is referred to as **white (cold-pressed) ceramics.** Additions of titanium carbide and zirconium oxide help improve properties, such as toughness and thermal-shock resistance.

Alumina-based ceramic tools have very high abrasion resistance and hot hardness (Fig. 22.9). Chemically, they are more stable than high-speed steels and carbides, so they have less tendency to adhere to metals during machining, and a correspondingly lower tendency to form a built-up edge. Consequently, in machining cast irons and steels, good surface finish is obtained using ceramic tools. On the other hand, ceramics generally lack toughness, and their use can result in premature tool failure, by chipping or in catastrophic failure.

Ceramic inserts are available in shapes similar to those of carbide inserts (Section 22.4.3). They are effective in high-speed, uninterrupted cutting operations, such as finishing or semifinishing. To reduce thermal shock, cutting should be performed either dry or with a copious amount of cutting fluid, applied in a steady stream (Section 22.12). Improper or intermittent applications of the fluid can cause thermal shock and fracture of the ceramic tool.

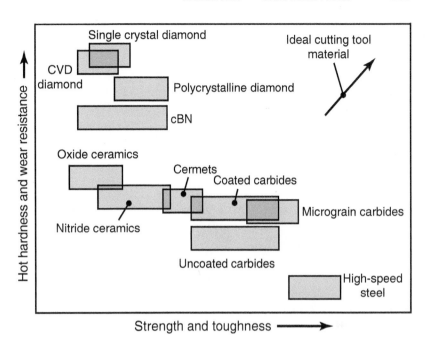

FIGURE 22.9 Ranges of mechanical properties for various groups of tool materials. HIP = hot isostatically pressed. (See also Tables 22.1–22.5.)

Ceramic tool shape and setup are important. Negative rake angles (large included angles) generally are preferred in order to avoid chipping, due to the poor tensile strength of ceramics. Tool failure can be reduced by increasing the stiffness and damping capacity of machine tools, mountings, and work-holding devices, thus reducing vibration and chatter.

Cermets.   *Cermets* (from the words *ceramic* and *metal*) consist of ceramic particles in a metallic matrix. They were introduced in the 1960s and are referred to as *black* or *hot-pressed ceramics* (carboxides). A typical cermet consists of 70% aluminum oxide and 30% titanium carbide; other cermets contain molybdenum carbide, niobium carbide, and tantalum carbide. Although they have chemical stability and resistance to built-up edge formation, the brittleness and high cost of cermets have been a limitation to their wider use.

Further developments and refinements of these tools have resulted in improved strength, toughness, and reliability. Their performance is somewhere between that of ceramics and carbides, and has been particularly suitable for light roughing cuts and high-speed finishing cuts. Chip-breaker features are important for cermet inserts. Although cermets can be coated, the benefits of coated cermets are somewhat controversial, as the improvement in wear resistance appears to be marginal.

22.7 Cubic Boron Nitride

Next to diamond, *cubic boron nitride* (cBN) is the hardest material available. Introduced in 1962 under the trade name *Borazon*, cubic boron nitride is made by bonding a 0.5–1 mm (0.02–0.04 in.) layer of **polycrystalline cubic boron nitride** to a carbide substrate, by sintering under high pressure and high temperature. While the carbide provides shock resistance, the cBN layer provides very high wear resistance and cutting-edge strength (Fig. 22.10).

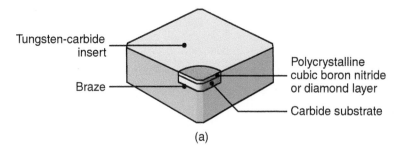

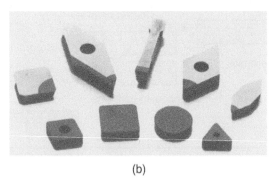

FIGURE 22.10 Cubic boron nitride inserts. (a) An insert of a polycrystalline cubic boron nitride or a diamond layer on tungsten carbide. (b) Inserts with polycrystalline cubic boron nitride tips (top row), and solid-polycrystalline cBN inserts (bottom row). *Source:* (b) Courtesy of Valenite.

The thermochemical stability of cBN is a significant advantage; it can be used safely up to 1200°C (2200°F). Also, at elevated temperatures, cBN maintains high chemical inertness to iron and nickel, thus there is no wear due to diffusion. Its resistance to oxidation is high, making it particularly suitable for machining hardened ferrous and high-temperature alloys (see *hard machining*, Section 25.6) and for high-speed machining operations (Section 25.5).

cBN also is used as an abrasive; however, because these tools are brittle, the stiffness of the machine tool and the fixturing is important in order to avoid vibration and chatter. Furthermore, in order to avoid chipping and cracking, due to thermal shock, machining generally should be performed dry, particularly in interrupted cutting operations, such as milling, which repeatedly subject the tool to thermal cycling.

22.8   Silicon-nitride-based Ceramics

Developed in the 1970s, *silicon-nitride* (SiN)–*based ceramic* tool materials consist of silicon nitride with various additions of aluminum oxide, yttrium oxide, and titanium carbide. These tools have high toughness, hot hardness, and good thermal-shock resistance. An example of an SiN-based material is **sialon**, named after the elements of which it is composed: *si*licon, *al*uminum, *o*xygen, and *n*itrogen. Sialon has higher thermal-shock resistance than silicon nitride, and is recommended for machining cast irons and nickel-based superalloys, at intermediate cutting speeds. Because of their chemical affinity to iron at elevated temperatures, however, SiN-based tools are not suitable for machining steels.

22.9   Diamond

Of all known materials, the hardest substance is diamond, described in Section 8.7. As a cutting tool, it has highly desirable properties such as low friction, high wear resistance, and the ability to maintain a sharp cutting edge. Diamond is used when good surface finish and dimensional accuracy are required, particularly when machining soft nonferrous alloys and abrasive nonmetallic and metallic materials, especially some aluminum–silicon alloys. *Synthetic* or *industrial diamond* is widely used because natural diamond has flaws, and thus its performance can be unpredictable.

Although **single-crystal diamond** of various carats (for precious stones, 1 carat = 200 mg) can be used for special applications, they have been replaced largely by **polycrystalline diamond** (PCD) tools, called **compacts** (also used as dies

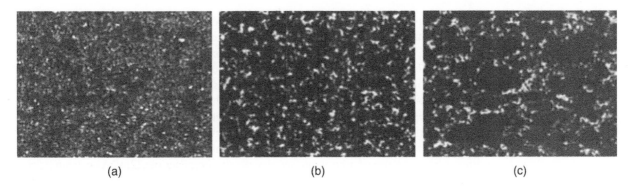

(a)   (b)   (c)

FIGURE 22.11 Microphotographs of diamond compacts. (a) Fine-grained diamond, with mean grain size around 2 μm; (b) Medium grain, with mean grain size around 10 μm; (c) Coarse grain, with grain size around 25 μm. Grain sizes ranging from 0.5 to 30 μm are commercially available. *Source:* Courtesy of Kennametal Inc.

for fine wire drawing, Section 15.7). These diamond tools consist of very small synthetic crystals (Fig. 22.11), fused by a high-pressure, high-temperature process, to a thickness of about 0.5–1 mm (0.02–0.04 in.), and bonded to a carbide substrate; this product is similar to cBN tools (Fig. 22.10). Fine grains are used when a high cutting edge quality and higher strength are required; coarse grains are preferred for increased abrasion resistance. The random orientation of the diamond crystals prevents the propagation of cracks through the structure, thus significantly improving its toughness.

Because diamond is brittle, tool shape and sharpness are important. Low rake angles generally are used to provide a strong cutting edge, because of the larger included angles. Special attention should be given to proper mounting and crystal orientation in order to obtain optimum tool life. Wear may occur through microchipping (caused by thermal stresses and oxidation) and through transformation to carbon (caused by the heat generated during machining). Diamond tools can be used satisfactorily at almost any speed, but are most suitable for light, uninterrupted finishing cuts. In order to minimize tool fracture, the single-crystal diamond must be resharpened as soon as it becomes dull. Because of its strong chemical affinity at elevated temperatures (resulting in diffusion), diamond is not recommended for machining plain-carbon steels or for titanium, nickel, and cobalt-based alloys.

22.10   Whisker-reinforced Materials and Nanomaterials

In order to further improve the performance and wear resistance of cutting tools, particularly in machining new materials and composites, continued progress is being made in developing new tool materials with enhanced properties, such as:

- High fracture toughness
- Resistance to thermal shock
- Cutting-edge strength
- Creep resistance
- Hot hardness

Advances include the use of **whiskers** as reinforcing fibers in composite cutting-tool materials. Examples of *whisker-reinforced cutting tools* include

(a) silicon-nitride-based tools reinforced with silicon-carbide whiskers and (b) aluminum-oxide-based tools reinforced with 25–40% silicon-carbide whiskers, sometimes with the addition of *zirconium oxide* (ZrO_2). Silicon-carbide whiskers are typically 5–100 μm long and 0.1–1 μm in diameter. However, the high reactivity of silicon carbide with ferrous metals makes SiC-reinforced tools unsuitable for machining irons and steels.

Nanomaterials also are becoming important in advanced cutting-tool materials (see Section 8.8); suitable nanomaterials are carbides and ceramics. Often, nanomaterials are applied as a thin coating, usually in an attempt to obtain a reasonable tool life without the use of a coolant (see *dry machining*, Section 22.12.1) or to machine at high speeds (see Section 25.5).

22.11   Tool Costs and Reconditioning of Tools

Tool costs vary widely, depending on the tool material, size, shape, chip-breaker features, and quality. The approximate cost for a typical 0.5-in. (12.5-mm) *insert* is approximately (a) $10–$15 for uncoated carbides, (b) $10–$25 for coated carbides, (c) $30–$50 for ceramics, (d) $50–$90 for diamond-coated carbides, (e) $130–$180 for cubic boron nitride, and (f) $150–$200 for a diamond-tipped insert.

After reviewing the costs involved in machining and considering all of the aspects involved in the total operation, it can be seen that the cost of an individual insert is relatively insignificant. Tooling costs in machining have been estimated to be on the order of 2–4% of the manufacturing costs. This small amount is due to the fact that a single cutting tool, for example, can perform a large amount of material removal before it is indexed to use all the cutting edges of an insert, and eventually recycled. Recall from Section 21.5 that the expected tool life can be in the range of 30–60 min; thus, considering that a square insert has eight cutting edges, a tool can last many hours before it is removed from the machine tool and replaced.

Cutting tools can be **reconditioned** by resharpening them, using tool and cutter grinders with special fixtures (Section 26.4). This operation may be carried out by hand or on computer-controlled tool and cutter grinders. Advanced methods of shaping cutting tools also are available, as described in Chapter 27. Reconditioning of coated tools also is done by recoating them, usually in special facilities available for these purposes; it is important that reconditioned tools have the same geometric features as the original tools. Often, a decision has to be made whether further reconditioning of tools is economical, especially when the costs of typical small inserts are not a major contribution to total costs. *Recycling* of tools is always a significant consideration, especially if they contain expensive and strategically important materials, such as tungsten and cobalt.

22.12   Cutting Fluids

Cutting fluids are used extensively in machining operations for the following purposes:

- Reduce friction and wear, thus improving the tool life and surface finish of the workpiece
- Cool the cutting zone, thus improving tool life and reducing the temperature and thermal distortion of the workpiece
- Reduce forces and energy consumption

- Flush away the chips from the cutting zone, thus preventing the chips from interfering with the cutting operation, particularly in drilling and tapping
- Protect the machined surface from environmental corrosion

Depending on the type of machining operation, the cutting fluid needed may be a **coolant**, a **lubricant**, or both. The effectiveness of cutting fluids depends on a number of factors, such as the type of machining operation, tool and workpiece materials, cutting speed, and the method of application. Water is an excellent coolant, and can effectively reduce the high temperatures developed in the cutting zone; however, water is not an effective lubricant and hence, it does not reduce friction and can cause corrosion of workpieces and machine-tool components.

The necessity for a cutting fluid depends on the *severity* of the particular machining operation, defined as (a) the temperatures and forces encountered and the ability of the tool materials to withstand them, (b) the tendency for built-up edge formation, (c) the ease with which chips produced can be removed from the cutting zone, and (d) how effectively the fluids can be supplied to the proper region at the tool–chip interface. The relative severities of specific machining processes, in increasing order of severity, are sawing, turning, milling, drilling, gear cutting, thread cutting, tapping, and internal broaching.

There are operations, however, in which the cooling action of cutting fluids can be detrimental. It has been shown, for example, that cutting fluids may cause the chip to become *more curly* (see Fig. 21.9c), and thus concentrate the heat closer to the tool tip, reducing tool life. Moreover, in interrupted cutting operations, such as milling with multiple-tooth cutters, cooling of the cutting zone leads to thermal cycling of the cutter teeth, which can cause *thermal cracks*, by the mechanisms of thermal fatigue or thermal shock.

Cutting-fluid Action.   The basic mechanisms of lubrication in metalworking operations are described in greater detail in Section 33.6. Studies have shown that the cutting fluid gains access to the tool–chip interface by seeping from the *sides* of the chip (perpendicular to the page in Figs. 21.11 and 21.12), through the *capillary action* of the interlocking network of surface asperities in the interface.

Because of the small size of this capillary network, the cutting fluid should have a *small molecular size* and possess proper *wetting* (*surface tension*) characteristics. Therefore, for example, grease cannot be an effective lubricant in machining, whereas low-molecular-weight oils, suspended in water (*emulsions*), are very effective. Note that in discontinuous machining operations, cutting fluids have more access to tool–chip–workpiece interfaces, but then the tools are more susceptible to thermal shock.

EXAMPLE 22.1   Effects of Cutting Fluids on Machining

Given: A machining operation is being carried out with a cutting fluid that is an effective lubricant.

Find: Describe the changes in the cutting operation mechanics if the fluid supply is interrupted.

Solution: Since the cutting fluid is a good lubricant, the following chain of events will take place after the fluid is shut off:

1. Friction at the tool–chip interface will increase.
2. The shear angle will decrease, in accordance with Eq. (21.3).
3. The shear strain will increase, as seen from Eq. (21.2).
4. The chip will become thicker.
5. A built-up edge is likely to form.

As a result of these changes, the following events will occur:

1. The shear energy in the primary zone will increase.
2. The frictional energy in the secondary zone will increase.
3. The total energy will increase.
4. The temperature in the cutting zone will rise, causing greater tool wear.
5. Surface finish of the workpiece will begin to deteriorate, and dimensional accuracy may be difficult to maintain, because of the increased temperature and thermal expansion of the workpiece during machining.

Types of Cutting Fluids.   The characteristics and applications of metalworking fluids and their trends are described in Section 33.7. Briefly, four general types of cutting fluids are commonly used in machining operations:

1. **Oils,** also called *straight oils*, include mineral, animal, vegetable, compounded, and synthetic oils; typically are used for low-speed operations where temperature rise is not significant.
2. **Emulsions,** also called *soluble oils*, are a mixture of oil and water and additives; they generally are used for high-speed operations where the temperature rise is significant. The presence of water makes emulsions highly effective coolants, and the presence of oil reduces or eliminates the tendency of water to cause oxidation of workpiece surfaces.
3. **Semisynthetics** are chemical emulsions containing little mineral oil, diluted in water, and with additives that reduce the size of oil particles, thus making them more effective.
4. **Synthetics** are chemicals with additives, diluted in water, and containing no oil.

Because of the complex interactions among the cutting fluid, the workpiece materials, temperature, time, and cutting-process variables, the selection and application of fluids cannot be generalized. In Chapters 23 and 24, recommendations for cutting fluids for specific machining operations are given.

Methods of Cutting-fluid Application.   There are four basic methods of cutting-fluid applications in machining:

1. **Flooding.** This is the most common method, as shown in Fig. 22.12 and indicating good and poor flooding practices. Flow rates typically range from 10 L/min (3 gal/min) for single-point tools to 225 L/min (60 gal/min) per cutter for multiple-tooth cutters, as in milling. In some operations, such as drilling and milling, fluid pressures in the range from 700 to 14,000 kPa (100–2000 psi) also are used to flush away the chips produced, to prevent interfering with the operation.
2. **Mist.** This type of cooling supplies fluid to inaccessible areas, in a manner similar to using an aerosol can, and provides better visibility of the workpiece being machined, as compared with flood cooling. This method is particularly effective with water-based fluids and at air pressures ranging from 70 to 600 kPa (10–80 psi). However, it has limited cooling capacity, and requires venting to prevent the inhalation of airborne fluid particles by the machine operator and other personnel nearby.
3. **High-pressure systems.** With the increasing speed and power of computer-controlled machine tools, heat generation in machining has become a significant factor. Particularly effective is the use of high-pressure *refrigerated coolant systems* to increase the rate of heat removal. High pressures also are used in delivering the cutting fluid via specially designed nozzles that aim a powerful jet

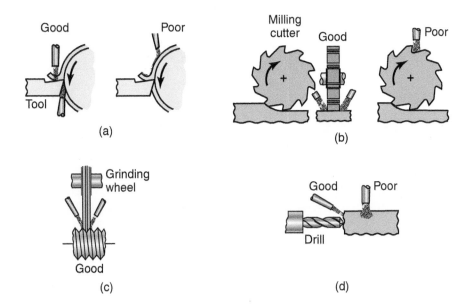

FIGURE 22.12   Schematic illustration of the methods of applying cutting fluids (by flooding) in various machining operations: (a) turning, (b) milling, (c) thread grinding, and (d) drilling.

of fluid to the cutting zone, particularly into the *clearance* or *relief face* of the tool (see Fig. 21.3). The pressures are usually in the range from 5.5 to 35 MPa (800–5000 psi), and also act as a chip breaker in situations where the chips produced would otherwise be long and continuous, interfering with the cutting operation. Proper cycling and continuous filtering of the fluid is essential to maintain workpiece surface quality.

A more recent design is shown in Fig. 22.13, which achieves good performance with more modest pressure requirements. This method has been found to be especially effective in machining titanium and other difficult-to-machine materials, where tool life can be increased by over 300%. Instead of applying coolant to the workpiece surface or chip at a distance remote from the cutting zone, the coolant is applied on the *side* of the insert. For controlled depths of cut, the temperature rise in the tool and chip can be reduced significantly, as seen in Fig. 22.13b.

4. **Through the cutting-tool system.** For a more effective application, narrow passages can be produced in cutting tools, as well as in toolholders, through which cutting fluids can be supplied under high pressure. Two applications of this method are (a) gun drilling, shown in Fig. 23.22; note the long, small hole through the body of the drill itself, and (b) boring bars, shown in Fig. 23.17a, where there is a long hole through the shank (toolholder), to which an insert is clamped. Similar designs have been developed for cutting tools and inserts, and for delivering cutting fluids through the spindle of the machine tool.

Effects of Cutting Fluids.   The selection of a cutting fluid should also include considerations such as its effects on

- Workpiece material
- Machine tool components
- Biological considerations
- The environment

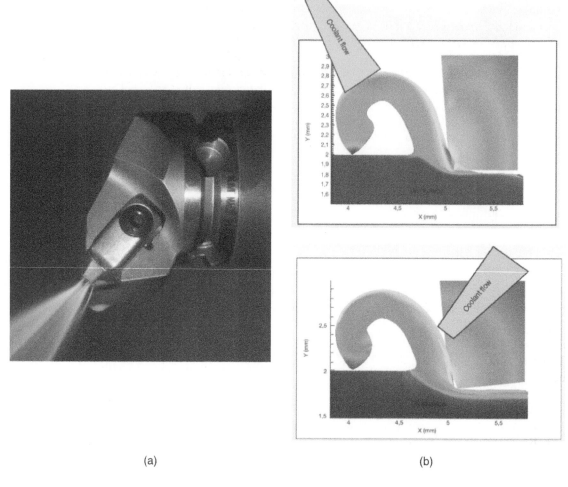

(a)   (b)

FIGURE 22.13 (a) A turning insert with coolant applied through the tool; (b) Comparison of temperature distributions for conventional and through-the-tool application. The workpiece material is Inconel 718; cutting speed = 180 m/min, feed = 0.020 in./rev, and tool coating = TiC. *Source:* Courtesy of Kennametal Inc.

In selecting an appropriate cutting fluid, one should consider whether the machined component may be subjected to stresses and adverse effects during its service life, possibly leading to stress-corrosion cracking. For example, (a) cutting fluids containing sulfur should not be used with nickel-based alloys and (b) fluids containing chlorine should not be used with titanium, because of increased corrosion. Moreover, machined parts should be cleaned and washed, in order to remove any cutting-fluid residue, as described in Section 34.16. Because the cleaning operation can be significant in time and cost, the trend is to use water-based, low-viscosity fluids for ease of cleaning and filtering the fluids. Because cutting fluids also may adversely affect the machine tool components, their compatibility with various metallic and nonmetallic materials in the machine also must be considered.

The *health effects* on machine operators, in contact with fluids, also should be of primary concern. Mist, fumes, smoke, and odors from cutting fluids can cause severe skin reactions and respiratory problems, especially in using fluids with chemical constituents such as sulfur, chlorine, phosphorus, hydrocarbons, biocides, and various additives. Much progress has been made in ensuring the safe use of cutting fluids in manufacturing facilities, including reducing or eliminating their use by considering the more recent trends in dry or near-dry machining techniques, as well as in the design of machine tools with enclosed working areas (see Figs. 25.2 and 25.12).

Cutting fluids, as well as other metalworking fluids used in manufacturing operations, may undergo chemical changes as they are used repeatedly over time. These changes may be due to environmental effects or to contamination from various sources, including metal chips, fine particles produced during machining, and *tramp oil* (oils from leaks in hydraulic systems, on sliding members of machines, and from lubricating systems for the machine tools). The changes can also involve the growth of microbes (bacteria, molds, and yeast), particularly in the presence of water, becoming an environmental hazard, and also adversely affecting the characteristics and effectiveness of the cutting fluids.

Several techniques, such as settling, skimming, centrifuging, and filtering, are available for clarifying used cutting fluids. Recycling involves treatment of the fluids with various additives, agents, biocides, and deodorizers, as well as water treatment (for water-based fluids). Disposal practices for these fluids must comply with federal, state, and local laws and regulations.

22.12.1   Near-dry and Dry Machining

For economic and environmental reasons, there has been a continuing worldwide trend, since the mid-1990s, to minimize or eliminate the use of metalworking fluids. This trend has led to the practice of *near-dry machining* (NDM), with significant benefits such as:

- Alleviating the environmental impact of using cutting fluids, improving air quality in manufacturing plants, and reducing health hazards
- Reducing the cost of machining operations, including the cost of maintenance, recycling, and disposal of cutting fluids

The significance of this approach becomes apparent when one notes that, in the United States alone, millions of gallons of metalworking fluids are consumed each year. Furthermore, it has been estimated that metalworking fluids constitute about 7–17% of the total machining cost.

The principle behind near-dry cutting is the application of a fine mist of an air–fluid mixture, containing a very small amount of cutting fluid, which may be reformulated to contain vegetable oil. The mixture is delivered to the cutting zone through the *spindle* of the machine tool, typically through a 1-mm-diameter nozzle and under a pressure of 600 kPa (85 psi). It is used at rates on the order of 1–100 cc/h, which is estimated to be, at most, one ten-thousandth of that used in flood cooling. Consequently, the process is also known as *minimum-quantity lubrication* (MQL).

Dry machining also is a viable alternative. With major advances in cutting tools, dry machining has been shown to be effective in various machining operations, especially turning, milling, and gear cutting, on steels, steel alloys, and cast irons, although generally not for aluminum alloys.

One of the functions of a metal-cutting fluid is to *flush* chips from the cutting zone. Although this function appears to be a challenge with dry machining, tool designs have been developed that allow the application of *pressurized air*, often through the tool shank. Although the compressed air does not serve as a lubricant, and provides only limited cooling capacity, it is very effective at clearing chips from the cutting zone.

Cryogenic Machining.   More recent developments in machining include the use of cryogenic gases, such as *nitrogen* or *carbon dioxide*, as a coolant. With small-diameter nozzles and at a temperature of $-200°C$ ($-320°F$), liquid nitrogen is injected into the cutting zone. Because of the reduced temperature, tool hardness is maintained and hence tool life is improved, thus allowing for higher cutting speeds. The chips also are less ductile, thus machinability is increased. There is no adverse environmental impact, and the nitrogen simply evaporates.

SUMMARY

- Cutting tool materials have a broad range of mechanical and physical properties, such as hot hardness, toughness, chemical stability and inertness, and resistance to chipping and wear. A wide variety of cutting-tool materials are now available, the most commonly used ones being high-speed steels, carbides, ceramics, cubic boron nitride, and diamond.

- Several tool coatings have been developed, resulting in major improvements in tool life, surface finish, and the economics of machining operations. Common coating materials are titanium nitride, titanium carbide, titanium carbonitride, and aluminum oxide. The trend is toward multiphase coatings for even better performance.

- The selection of appropriate tool materials depends not only on the material to be machined, but also on process parameters and the characteristics of the machine tool.

- Cutting fluids are important in machining operations, as they reduce friction, wear, cutting forces, and power requirements. Generally, slower cutting operations and those with high tool pressures require a fluid with good lubricating characteristics. In high-speed operations, where the temperature rise can be significant, fluids with good cooling capacity and some lubricity are required. The selection of cutting fluids must take into account their possible adverse effects on the machined parts, on machine tools and their components, on personnel, and on the environment.

KEY TERMS

Alumina-based ceramics	Cutting fluids	Multiphase coatings	Stellite
Carbides	Diamond coatings	Nanocrystalline	Titanium carbide
Cast-cobalt alloys	Diamond tools	Near-dry machining	Titanium nitride
Ceramics	Dry machining	Polycrystalline cubic	Tool costs
Cermets	Finishing cuts	boron nitride	Tool reconditioning
Chemical stability	Flooding	Polycrystalline diamond	Toughness
Chip breaker	High-speed steels	Reconditioning of tools	Tungsten carbide
Coated tools	Inserts	Roughing cuts	Uncoated carbides
Coolants	Lubricants	Sialon	Wear resistance
Cryogenic machining	Micrograin carbides	Silicon-nitride-based	Whisker-reinforced tools
Cubic boron nitride	Mist	ceramics	

BIBLIOGRAPHY

ASM Handbook, Vol. 16: **Machining**, ASM International, 1989.

ASM Specialty Handbook: **Tool Materials**, ASM International, 1995.

Astakhov, V.P., **Tribology of Metal Cutting**, Elsevier, 2007.

Astakhov, V.P., and Joksch, S., **Metalworking Fluids for Cutting and Grinding: Fundamentals and Recent Advances**, Woodhead, 2012.

Byers, J.P. (ed.), **Metalworking Fluids**, 2nd ed., CRC Press, 2006.

Jackson, M.J., and Morrell, J., **Machining with Nanomaterials**, Springer, 2009.

Komanduri, R., *Tool Materials*, in **Kirk–Othmer Encyclopedia of Chemical Technology**, 4th ed., Vol. 24, 1997.

Nachtman, E.S., and Kalpakjian, S., **Lubricants and Lubrication in Metalworking Operations**, Marcel Dekker, 1985.

Roberts, G.A., Krauss, G., and Kennedy, R., **Tool Steels**, 5th ed., ASM International, 1997.

Shaw, M.C., **Metal Cutting Principles**, 2nd ed., Oxford, 2005.

Smith, G.T., **Cutting Tool Technology: Industrial Handbook**, Springer, 2008.

Trent, E.M., and Wright, P.K., **Metal Cutting**, 4th ed., Butterworth-Heinemann, 2000.

REVIEW QUESTIONS

22.1 What are the major properties required of cutting-tool materials? Why?

22.2 What is the composition of a typical carbide tool?

22.3 Why were cutting-tool inserts developed?

22.4 Why are some tools coated? What are the common coating materials?

22.5 Explain the applications and limitations of ceramic tools.

22.6 List the major functions of cutting fluids.

22.7 Why is toughness important for cutting-tool materials?

22.8 Is the elastic modulus important for cutting-tool materials? Explain.

22.9 Explain how cutting fluids penetrate the tool–chip interface.

22.10 List the methods by which cutting fluids are typically applied in machining operations.

22.11 Describe the advantages and limitations of (a) single-crystal and (b) polycrystalline diamond tools.

22.12 What is a cermet? What are its advantages?

22.13 Explain the difference between M-series and T-series high-speed steels.

22.14 Why is cBN generally preferred over diamond for machining steels?

22.15 What are the advantages to dry machining?

QUALITATIVE PROBLEMS

22.16 Explain why so many different types of cutting-tool materials have been developed over the years. Why are they still being developed further?

22.17 Which tool-material properties are suitable for interrupted cutting operations? Why?

22.18 Describe the reasons for and advantages of coating cutting tools with multiple layers of different materials.

22.19 Make a list of the alloying elements used in high-speed steels. Explain what their functions are and why they are so effective in cutting tools.

22.20 As stated in Section 22.1, tool materials can have conflicting properties when used for machining operations. Describe your observations regarding this matter.

22.21 Explain the economic impact of the trend shown in Fig. 22.6.

22.22 Why does temperature have such an important effect on tool life?

22.23 Ceramic and cermet cutting tools have certain advantages over carbide tools. Why, then, are they not completely replacing carbide tools?

22.24 What precautions would you take in machining with brittle tool materials, especially ceramics? Explain.

22.25 Can cutting fluids have any adverse effects in machining? If so, what are they?

22.26 Describe the trends you observe in Table 22.2.

22.27 Why are chemical stability and inertness important in cutting tools?

22.28 Titanium-nitride coatings on tools reduce the coefficient of friction at the tool–chip interface. What is the significance of this property?

22.29 Describe the necessary conditions for optimal utilization of the capabilities of diamond and cubic-boron-nitride cutting tools.

22.30 Negative rake angles generally are preferred for ceramic, diamond, and cubic-boron-nitride tools. Why?

22.31 Do you think that there is a relationship between the cost of a cutting tool and its hot hardness? Explain.

22.32 Make a survey of the technical literature, and give some typical values of cutting speeds for high-speed steel tools and for a variety of workpiece materials.

22.33 In Table 22.1, the last two properties listed can be important to the life of a cutting tool. Why?

22.34 It has been stated that titanium-nitride coatings allow cutting speeds and feeds to be higher than those for uncoated tools. Survey the technical literature and prepare a table showing the percentage increase of speeds and feeds that would be made possible by coating the tools.

22.35 Note in Fig. 22.1 that all tool materials, especially carbides, have a wide range of hardnesses for a particular temperature. Describe each of the factors that are responsible for this wide range.

22.36 Referring to Table 22.1, state which tool materials would be suitable for interrupted cutting operations. Explain.

22.37 Which of the properties listed in Table 22.1 is, in your opinion, the least important in cutting tools? Explain.

22.38 If a drill bit is intended only for woodworking applications, what material is it most likely to be made from?

(Hint: Temperatures rarely rise to 400°C in woodworking.) Explain.

22.39 What are the consequences of a coating on a tool having a different coefficient of thermal expansion than the substrate material?

22.40 Discuss the relative advantages and limitations of near-dry machining. Consider all relevant technical and economic aspects.

22.41 Emulsion cutting fluids typically consist of 95% water and 5% soluble oil and chemical additives. Why is the ratio so unbalanced? Is the oil needed at all?

22.42 List and explain the considerations involved in determining whether a cutting tool should be reconditioned, recycled, or discarded after use.

QUANTITATIVE PROBLEMS

22.43 Review the contents of Table 22.1. Plot several curves to show relationships, if any, among parameters such as hardness, transverse rupture strength, and impact strength. Comment on your observations.

22.44 Obtain data on the thermal properties of various commonly used cutting fluids. Identify those which are basically effective coolants (such as water-based fluids) and those which are basically effective lubricants (such as oils).

22.45 The first column in Table 22.2 shows 10 properties that are important to cutting tools. For each of the tool materials listed in the table, add numerical data for each of these properties. Describe your observations, including any data that overlap.

SYNTHESIS, DESIGN, AND PROJECTS

22.46 Describe in detail your thoughts regarding the technical and economic factors involved in tool-material selection.

22.47 One of the principal concerns with coolants is degradation due to biological attack by bacteria. To prolong the life of a coolant, chemical biocides often are added, but these biocides greatly complicate the disposal of the coolant. Conduct a literature search concerning the latest developments in the use of environmentally benign biocides in cutting fluids.

22.48 How would you go about measuring the effectiveness of cutting fluids? Describe your method and explain any difficulties that you might encounter.

22.49 Contact several different suppliers of cutting tools, or search their websites. Make a list of the costs of typical cutting tools as a function of various sizes, shapes, and features.

22.50 There are several types of cutting-tool materials available today for machining operations, yet much research

and development is being carried out on all these materials. Discuss why you think such studies are being conducted.

22.51 Assume that you are in charge of a laboratory for developing new or improved cutting fluids. On the basis of the topics presented in this chapter and in Chapter 21, suggest a list of topics for your staff to investigate. Explain why you have chosen those topics.

22.52 Tool life could be greatly increased if an effective means of cooling and lubrication were developed. Design methods of delivering a cutting fluid to the cutting zone, and discuss the advantages and shortcomings of your design.

22.53 List the concerns you would have if you needed to economically machine carbon-fiber reinforced polymers or metal–matrix composites with graphite fibers in an aluminum matrix.

CHAPTER 23

Machining Processes: Turning and Hole Making

- With the preceding two chapters as background, this chapter describes machining processes that are capable of generating round external or internal shapes.
- The most common machine tool used for such operations is the lathe, which is available in several types and automated systems.
- The wide variety of operations that can be performed on lathes are then described in detail, including turning, drilling, profiling, facing, grooving, thread cutting, and knurling.
- The chapter also describes operations such as boring, drilling, reaming, and tapping, and the characteristics of the machine tools associated with these processes.

Typical parts made: Machine components; engine blocks and heads; parts with complex shapes, close tolerances, and good surface finish; and externally and internally threaded parts.

Alternative processes: Precision casting, powder metallurgy, powder-injection molding, abrasive machining, thread rolling, and rotary swaging.

23.1 Introduction

This chapter describes machining processes with the capability of producing parts that basically are round in shape. Typical products made are as small as miniature screws for the hinges of eyeglass frames, and as large as turbine shafts for hydroelectric power plants and rolls for rolling mills.

One of the most basic machining processes is **turning**, meaning that the part is rotated while it is being machined. The starting material is generally a workpiece that has been made by other processes, such as casting, forging, extrusion, drawing, or powder metallurgy, as described in Parts II and III. Turning processes, which typically are carried out on a **lathe** or by similar *machine tools*, are outlined in Fig. 23.1 and Table 23.1. These machines are highly versatile and capable of performing several machining operations that produce a wide variety of shapes, such as:

- **Turning:** to produce straight, conical, curved, or grooved workpieces (Figs. 23.1a through d), such as shafts, spindles, and pins
- **Facing:** to produce a flat surface at the end of the part and perpendicular to its axis (Fig. 23.1e); parts that are assembled with other components; face grooving for such applications as O-ring seats (Fig. 23.1f)

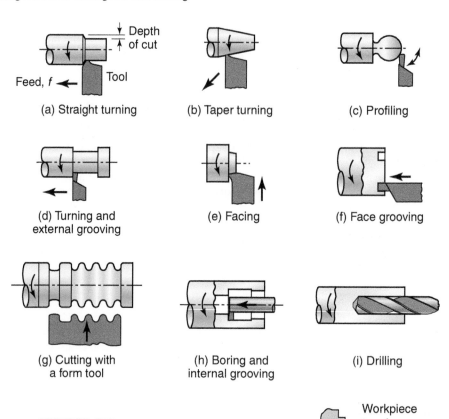

FIGURE 23.1 Miscellaneous operations that can be performed on a lathe; note that all parts are circular. The tools used, their shape, and the processing parameters are described throughout this chapter.

- **Cutting with form tools:** (Fig. 23.1g) to produce various axisymmetric shapes for functional or for aesthetic purposes
- **Boring:** to enlarge a hole or cylindrical cavity made by a previous process or to produce circular internal grooves (Fig. 23.1h)
- **Drilling:** to produce a hole (Fig. 23.1i), which then may be followed by boring it to improve its dimensional accuracy and surface finish
- **Parting:** also called **cutting off,** to remove a piece from the end of a part, as is done in the production of slugs or blanks for additional processing into discrete products (Fig. 23.1j)
- **Threading:** to produce external or internal threads (Fig. 23.1k)
- **Knurling:** to produce a regularly shaped roughness on cylindrical surfaces, as in making knobs and handles (Fig. 23.1l)

TABLE 23.1

General Characteristics of Machining Processes and Typical Dimensional Tolerances

Process	Characteristics	Typical dimensional tolerances, ±mm (in.)
Turning	Turning and facing operations on all types of materials, uses single-point or form tools; engine lathes require skilled labor; low production rate (but medium-to-high rate with turret lathes and automatic machines) requiring less-skilled labor	Fine: 0.025–0.13 (0.001–0.005) Rough: 0.13 (0.005)
Boring	Internal surfaces or profiles with characteristics similar to turning; stiffness of boring bar important to avoid chatter	0.025 (0.001)
Drilling	Round holes of various sizes and depths; high production rate; labor skill required depends on hole location and accuracy specified; requires boring and reaming for improved accuracy	0.075 (0.003)
Milling	Wide variety of shapes involving contours, flat surfaces, and slots; versatile; low-to-medium production rate; requires skilled labor	0.13–0.25 (0.005–0.01)
Planing	Large flat surfaces and straight contour profiles on long workpieces, low-quantity production, labor skill required depends on part shape	0.08–0.13 (0.003–0.005)
Shaping	Flat surfaces and straight contour profiles on relatively small workpieces; low-quantity production; labor skill required depends on part shape	0.05–0.13 (0.002–0.003)
Broaching	External and internal surfaces, slots, and contours; good surface finish; costly tooling; high production rate; labor skill required depends on part shape	0.025–0.15
Sawing	Straight and contour cuts on flat or structural shapes; not suitable for hard materials unless saw has carbide teeth or is coated with diamond; low production rate; generally low labor skill	0.8

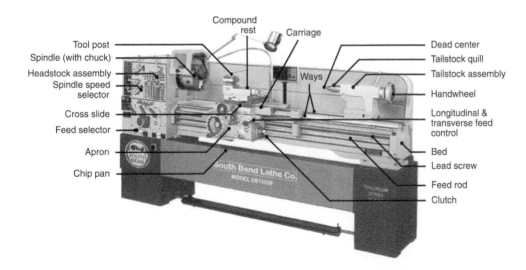

FIGURE 23.2   General view of a typical lathe, showing various components. *Source:* Courtesy of South Bend Lathe Co.

The cutting operations summarized above typically are performed on a *lathe* (Fig. 23.2), which is available in a wide variety of designs, sizes, capacities, and computer-controlled features, as described in Section 23.3 and Chapter 25. As shown in Figs. 21.2 and 23.3, turning is performed at various (a) rotational speeds, N, of the workpiece clamped in a spindle, (b) depths of cut, d, and (c) feeds, f, depending

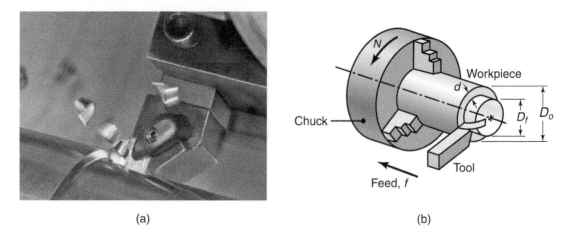

(a)

(b)

FIGURE 23.3 (a) A turning operation, showing insert and chip removal; the machine tool is traveling from right to left in this photograph. (b) Schematic illustration of the basic turning operation, showing depth of cut, d; feed, f; and spindle rotational speed, N, in rev/min. The cutting speed is the surface speed of the workpiece at the tool tip. *Source:* (a) Courtesy of Kennametal Inc.

QR Code 23.1 Turning with a wiper insert. (*Source: Courtesy of Sandvik Coromant*)

on the workpiece and cutting-tool materials, surface finish and dimensional accuracy required, and characteristics of the machine tool.

This chapter describes turning process parameters, cutting tools, process capabilities, and characteristics of the machine tools that are used to produce a variety of parts with round shapes. Design considerations to improve productivity for each group of processes also are outlined.

23.2 The Turning Process

The majority of turning operations involve the use of simple single-point cutting tools, with the geometry of a typical right-hand cutting tool shown in Figs. 21.10 and 23.4. As can be seen, such tools are described by a standardized nomenclature. Each group of workpiece materials has an optimum set of tool angles, which have been developed largely through experience over many years (Table 23.2).

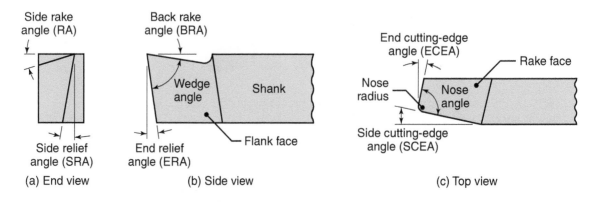

FIGURE 23.4 Designations for a right-hand cutting tool, meaning that the tool travels from right to left, as shown in Fig. 23.3b.

TABLE 23.2

General Recommendations for Tool Angles in Turning										
	High-speed steel					Carbide inserts				
Material	Back rake	Side rake	End relief	Side relief	Side and end cutting edge	Back rake	Side rake	End relief	Side relief	Side and end cutting edge
Aluminum and magnesium alloys	20	15	12	10	5	0	5	5	5	15
Copper alloys	5	10	8	8	5	0	5	5	5	15
Steels	10	12	5	5	15	−5	−5	5	5	15
Stainless steels	5	8–10	5	5	15	−5–0	−5–5	5	5	15
High-temperature alloys	0	10	5	5	15	5	0	5	5	45
Refractory alloys	0	20	5	5	5	0	0	5	5	15
Titanium alloys	0	5	5	5	15	−5	−5	5	5	5
Cast irons	5	10	5	5	15	−5	−5	5	5	15
Thermoplastics	0	0	20–30	15–20	10	0	0	20–30	15–20	10
Thermosets	0	0	20–30	15–20	10	0	15	5	5	15

The important process parameters that have a direct influence on machining processes, and the importance of controlling these parameters for optimized productivity, were described in Chapter 21. This section outlines important turning-process parameters, such as tool geometry and material-removal rate, and gives data for recommended cutting practices, including cutting-tool materials, depth of cut, feed, cutting speed, and use of cutting fluids.

Tool Geometry.   The various angles in a single-point cutting tool have important functions in machining operations. These angles are measured in a coordinate system consisting of the three major axes of the tool shank, as shown in Fig. 23.4.

- **Rake angle** is important in controlling both the direction of chip flow and the strength of the tool tip. Positive rake angles improve the cutting operation by reducing forces and temperatures, but they also result in a small included angle of the tool tip (see Figs. 21.3 and 23.4), possibly leading to premature tool chipping and failure, depending on the toughness of the tool material.
- **Side rake angle** is more important than the **back rake angle**, which usually controls the direction of chip flow; these angles typically are in the range from −5° to 5°.
- **Cutting-edge angle** affects chip formation, tool strength, and cutting forces to various degrees; typically, the cutting-edge angle is around 15°.
- **Relief angle** controls interference and rubbing at the tool–workpiece interface. If it is too large, the tool tip may chip off; if it is too small, flank wear may be excessive. Relief angles typically are 5°.
- **Nose radius** affects surface finish and tool-tip strength. The smaller the nose radius (meaning a sharp tool), the rougher the surface finish of the workpiece and the lower the strength of the tool; a large nose radii can, however, lead to tool *chatter*, as described in Section 25.4.

Material-removal Rate.   The *material-removal rate* (MRR) in turning is the volume of material removed per unit time, with the units of mm^3/min or in^3/min. Referring to Figs. 21.2 and 23.3, note that, for each revolution of the workpiece, a ring-shaped layer of material is removed, which has a cross-sectional area that equals the product of the distance the tool travels in one revolution (the feed, f) and the depth of cut, d. The volume of this ring is the product of the cross-sectional area, that is, $(f)(d)$, and the average circumference of the ring, πD_{avg}, where

$$D_{avg} = \frac{D_o + D_f}{2}.$$

For light cuts on large-diameter workpieces, the average diameter may be replaced by D_o.

The rotational speed of the workpiece is N, and the material removal rate per revolution is $(\pi)(D_{avg})(d)(f)$. Since there are N revolutions per minute, the removal rate is

$$MRR = \pi D_{avg} df N. \tag{23.1}$$

Note that Eq. (23.1) also can be written as

$$MRR = df V, \tag{23.2}$$

where V is the cutting speed and MRR has the same unit of mm³/min.

The cutting time, t, for a workpiece of length l can be calculated by noting that the tool travels at a feed rate of fN with units of (mm/rev)(rev/min) = mm/min. Since the distance traveled is l mm, the cutting time is

$$t = \frac{l}{fN}. \tag{23.3}$$

The foregoing equations and the terminology used are summarized in Table 23.3. The cutting time in Eq. (23.3) does not include the time required for *tool approach* and *retraction*. Because the time spent in noncutting cycles of a machining operation is unproductive, and thus adversely affects the overall economics, the time involved in approaching and retracting tools to and from the workpiece is an important consideration. Machine tools are designed and built to minimize this time. One method of accomplishing this aim is to *rapidly* traverse the tools during noncutting cycles, followed by a *slower* movement as the tool engages the workpiece.

Forces in Turning. The three principal forces acting on a cutting tool in turning are shown in Fig. 23.5. These forces are important in the design of machine tools, as well as in the deflection of tools and workpieces, particularly in precision-machining operations (see Section 25.7). It is essential that the machine tool and its components be able to withstand these forces, without undergoing significant deflections, vibrations, and chatter in the overall machining operation.

The **cutting force**, F_c, acts downward on the tool tip, and thus tends to deflect the tool downward and the workpiece upward. The cutting force supplies the energy required for the cutting operation, and it can be calculated with the data given in Table 21.2, from the energy per unit volume, described in Section 21.3. The product of the cutting force and its distance from the workpiece center determines the *torque* on the spindle. The product of the torque and the spindle speed then determines the *power* required in the turning operation.

The **thrust force**, F_t, acts in the longitudinal direction; it also is called the **feed force**, because it is in the feed direction of the tool. This force tends to push the tool toward the right and away from the chuck in Fig. 23.5. The **radial force**, F_r, acts in the radial direction and tends to push the tool away from the workpiece. Because of the several factors involved in the cutting process, forces F_t and F_r are difficult to calculate directly, and are usually determined experimentally.

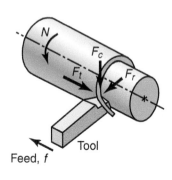

FIGURE 23.5 Forces acting on a cutting tool in turning. F_c is the cutting force, F_t is the thrust or feed force (in the direction of feed), and F_r is the radial force that tends to push the tool away from the workpiece being machined.

TABLE 23.3

Summary of Turning Parameters and Formulas

N = Rotational speed of the workpiece, rpm

f = Feed, mm/rev or in./rev

v = Feed rate, or linear speed of the tool along workpiece length, mm/min or in./min

  $= fN$

V = Surface speed of workpiece, m/min or ft/min

  $= \pi D_o N$ (for maximum speed)

  $= \pi D_{avg} N$ (for average speed)

l = Length of cut, mm or in.

D_o = Original diameter of workpiece, mm or in.

D_f = Final diameter of workpiece, mm or in.

D_{avg} = Average diameter of workpiece, mm or in.

  $= (D_o + D_f)/2$

d = Depth of cut, mm or in.

  $= (D_o - D_f)/2$

t = Cutting time, s or min

  $= l/fN$

MRR = mm^3/min or in^3/min

  $= \pi D_{avg} d f N$

Torque = N-m or lb-ft

  $= F_c D_{avg}/2$

Power = kW or hp

  $= (\text{Torque})(\omega)$, where $\omega = 2\pi N$ rad/min

Note: The units given are those that are commonly used; appropriate units must be used and checked in the formulas.

Video Solution 23.1 Turning Operations

QR Code 23.2 Roughing and finishing cuts in turning. (*Source:* Courtesy of Sandvik Coromant)

Roughing and Finishing Cuts. In machining, the usual procedure is to first perform one or more *roughing cuts*, typically at high feed rates and large depths of cut; thus, the material-removal rates are high, and there is little consideration for dimensional tolerance and surface roughness of the workpiece. These cuts are then followed by a *finishing cut*, typically performed at a lower feed and smaller depth of cut, in order to produce a good surface finish.

Tool Materials, Feeds, and Cutting Speeds. The general characteristics of cutting-tool materials have been described in Chapter 22. A broad range of applicable cutting speeds and feeds for these tool materials is given in Fig. 23.6, as a general guideline in turning operations. Specific recommendations regarding turning-process parameters for various workpiece materials and cutting tools are given in Table 23.4. These recommendations are based on experimental data, using standard turning arrangements. It is not uncommon to exceed these values in practice, especially with effective cutting fluids and a well-controlled process.

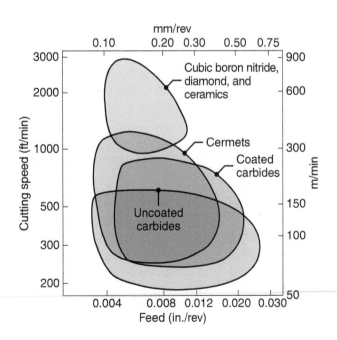

FIGURE 23.6 The range of applicable cutting speeds and feeds for a variety of tool materials.

TABLE 23.4

General Recommendations for Turning Operations (These Recommendations Are for Guidance Only, and Are Often Exceeded in Practice)

Workpiece material	Cutting tool	General-purpose starting conditions			Range for roughing and finishing		
		Depth of cut, mm (in.)	Feed, mm/rev (in./rev)	Cutting speed, m/min (ft/min)	Depth of cut, mm (in.)	Feed, mm/rev (in./rev)	Cutting speed, m/min (ft/min)
Low-C and free machining steels	Uncoated carbide	1.5–6.3 (0.06–0.25)	0.35 (0.014)	90 (300)	0.5–7.6 (0.02–0.30)	0.15–1.1 (0.006–0.045)	60–135 (200–450)
	Ceramic-coated carbide	"	"	245–275 (800–900)	"	"	180–495 (590–1600)
	Triple-coated carbide	"	"	185–200 (600–650)	"	"	90–245 (300–800)
	TiN-coated carbide	"	"	105–150 (350–500)	"	"	60–230 (200–750)
	Al$_2$O$_3$ ceramic	"	0.25 (0.010)	395–440 (1300–1450)	"	"	365–550 (1200–1800)
	Cermet	"	0.30 (0.012)	215–290 (700–950)	"	"	180–455 (590–1500)
Medium and high-C steels	Uncoated carbide	1.2–4.0 (0.05–0.20)	0.30 (0.012)	75 (250)	2.5–7.6 (0.10–0.30)	0.15–0.75 (0.006–0.03)	135–225 (440–725)
	Ceramic-coated carbide	"	"	185–230 (600–750)	"	"	120–410 (400–1350)
	Triple-coated carbide	"	"	120–150 (400–500)	"	"	75–215 (250–700)
	TiN-coated carbide	"	"	90–200 (300–650)	"	"	45–215 (150–700)
	Al$_2$O$_3$ ceramic	"	0.25 (0.010)	335 (1100)	"	"	245–455 (800–1500)
	Cermet	"	0.25 (0.010)	170–245 (550–800)	"	"	105–305 (350–1000)
Cast iron, gray	Uncoated carbide	1.25–6.3 (0.05–0.25)	0.32 (0.013)	90 (300)	0.4–12.7 (0.015–0.5)	0.1–0.75 (0.004–0.03)	75–185 (250–600)
	Ceramic-coated carbide	"	"	200 (650)	"	"	120–365 (400–1200)
	TiN-coated carbide	"	"	90–135 (300–450)	"	"	60–215 (200–700)
	Al$_2$O$_3$ ceramic	"	0.25 (0.010)	455–490 (1500–1600)	"	"	365–855 (1200–2800)
	SiN ceramic	"	0.32 (0.013)	730 (2400)	"	"	200–990 (650–3250)
	Polycrystalline cBN	"	"	1000 (3290)	"	"	200–1160 (650–3800)

Workpiece material	Cutting tool	Depth of cut mm (in.)	Feed mm/rev (in./rev)	Cutting speed m/min (ft/min)	Depth of cut mm (in.)	Feed mm/rev (in./rev)	Cutting speed m/min (ft/min)
Stainless steel, austenitic	Triple-coated carbide	1.5–4.4 (0.06–0.175)	0.35 (0.014)	150 (500)	0.5–12.7 (0.02–0.5)	0.08–0.75 (0.003–0.03)	75–230 (250–750)
	TiN-coated carbide	"	"	85–160 (275–525)	"	"	55–200 (175–650)
	Cermet	"	0.30 (0.012)	185–215 (600–700)	"	"	135–315 (450–1000)
High-temperature alloys, nickel based	Uncoated carbide	2.5 (0.10)	0.15 (0.006)	25–45 (75–150)	0.25–6.3 (0.01–0.25)	0.1–0.3 (0.004–0.012)	15–30 (50–100)
	Ceramic-coated carbide	"	"	45 (150)	"	"	20–60 (65–200)
	TiN-coated carbide	"	"	30–55 (95–175)	"	"	20–85 (60–275)
	Al_2O_3 ceramic	"	"	260 (850)	"	"	185–395 (600–1300)
	SiN ceramic	"	"	215 (700)	"	"	90–215 (300–700)
	Polycrystalline cBN	"	"	150 (500)	"	"	120–185 (400–600)
Titanium alloys	Uncoated carbide	1.0–3.8 (0.04–0.15)	0.15 (0.006)	35–60 (120–200)	0.25–6.3 (0.01–0.25)	0.1–0.4 (0.004–0.015)	10–75 (30–250)
	TiN-coated carbide	"	"	30–60 (100–200)	"	"	15–170 (50–550)
Aluminum alloys, Free machining	Uncoated carbide	1.5–5.0 (0.06–0.20)	0.45 (0.018)	490 (1600)	0.25–8.8 (0.01–0.35)	0.08–0.62 (0.003–0.025)	200–670 (650–2000)
	TiN-coated carbide	"	"	550 (1800)	"	"	60–915 (200–3000)
	Cermet	"	"	490 (1600)	"	"	215–795 (700–2600)
	Polycrystalline diamond	0.1–4.0	0.1–0.4	760 (2500)	"	"	1000–5000 (3200–16,250)
High silicon	Polycrystalline diamond	"	"	530 (1700)	"	"	365–915 (1200–3000)

(continued)

TABLE 23.4 (*continued*)

General Recommendations for Turning Operations (These Recommendations Are for Guidance Only, and Are Often Exceeded in Practice)

Workpiece material	Cutting tool	General-purpose starting conditions			Range for roughing and finishing		
		Depth of cut, mm (in.)	Feed, mm/rev (in./rev)	Cutting speed, m/min (ft/min)	Depth of cut, mm (in.)	Feed, mm/rev (in./rev)	Cutting speed, m/min (ft/min)
Copper alloys	Uncoated carbide	1.5–5.0 (0.06–0.20)	0.25 (0.010)	260 (850)	0.4–7.51 (0.015–0.3)	0.15–0.75 (0.006–0.03)	105–535 (350–1750)
	Ceramic-coated carbide	"	"	365 (1200)	"	"	215–670 (700–2200)
	Triple-coated carbide	"	"	215 (700)	"	"	90–305 (300–1000)
	TiN-coated carbide	"	"	90–275 (300–900)	"	"	45–455 (150–1500)
	Cermet	"	"	245–425 (800–1400)	"	"	200–610 (650–2000)
	Polycrystalline diamond	"	"	520 (1700)	0.05–2.0	0.03–0.3	400–1300 (1300–4200)
Tungsten alloys	Uncoated carbide	2.5 (0.10)	0.2 (0.008)	75 (250)	0.25–5.0 (0.01–0.2)	0.12–0.45 (0.005–0.018)	55–120 (175–400)
	TiN-coatedcarbide	"	"	85 (275)	"	"	60–150 (200–500)
Thermoplastics and thermosets	TiN-coated carbide	1.2 (0.05)	0.12 (0.005)	170 (550)	0.12–5.0 (0.005–0.20)	0.08–0.35 (0.003–0.015)	90–230 (300–750)
	Polycrystalline diamond	"	"	395 (1300)	"	"	250–730 (800–2400)
Composites, graphite reinforced	TiN-coated carbide	1.9 (0.075)	0.2 (0.008)	200 (650)	0.12–6.3 (0.005–0.25)	0.12–1.5 (0.005–0.06)	105–290 (350–950)
	Polycrystalline diamond	"	"	760 (2500)	"	"	550–1310 (1800–4300)

Source: Based on data from Kennametal Inc.

Note: Cutting speeds for high-speed steel tools are about one-half those for uncoated carbides.

CASE STUDY 23.1   Brake Disk Machining

An automotive brake manufacturer produces brake disks (See Fig. 23.7) by facing them on a lathe, using the process parameters in Table 23.5. The brake disks are made from a cast blank, machined on a lathe, and then mounting holes on the axle and cooling holes in the disk are produced on a CNC drill press. The material used is a gray cast iron (ASTM Class 25, see Table 12.4), using a silicon nitride insert. Unfortunately, this material can have very poor machinability because of insufficient aging or variations in composition. In addition, it is desired to modify the cutting conditions, in order to increase production rate.

Aluminum oxide (Al_2O_3) and polycrystalline cubic boron nitride (cBN) were investigated as alternate cutting-tool materials. As can be seen in Table 23.4, cBN is the only material that would allow for an increased cutting speed, compared to SiN for gray cast iron as the workpiece. Based on the recommendations in Table 23.4, the cutting parameters shown in Table 23.5 were selected.

With the cBN insert, it was found that the tool life could be dramatically increased to 4200 brake disks per tool edge, compared to only 40 with the silicon nitride, so that the higher cost of cBN could be economically justified as well. In addition, because of the longer life, the tool change time was greatly reduced, and the machine utilization increased from 82 to 94%. Thus, a change to polycrystalline cBN led to a simultaneous improvement in economy and production rate. Such dramatic improvements are not generally achieved, but gray cast iron is a target material for this cBN application.

Source: Courtesy of Kennametal Inc.

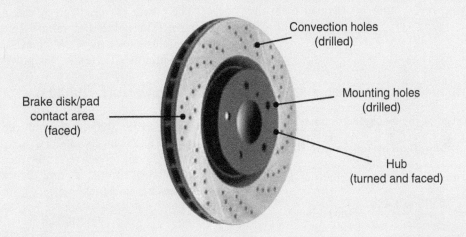

FIGURE 23.7   Brake disk, for Case Study 23.1.

TABLE 23.5

Process Parameter Comparison for SiN and cBN Tools in Facing of a Brake Disk		
	Cutting-tool material	
Parameter	SiN	cBN
Depth of cut, mm	1.5	2.0
Feed, mm/rev	0.5	0.4
Cutting speed, m/min	700	1000
Tool life, parts	40	4200
Machine utility, %	82	94

Cutting Fluids. Many metallic and nonmetallic materials can be machined without a cutting fluid, but in most cases the application of a cutting fluid can significantly improve the operation. General recommendations for cutting fluids, appropriate to various workpiece materials, are given in Table 23.6. However, recall the major current trend toward and the benefits of near-dry and dry machining, as described in Section 22.12.1.

EXAMPLE 23.1   Material-removal Rate and Cutting Force in Turning

Given: A 6-in.-long, 0.5-in.-diameter 304 stainless-steel rod is being reduced in diameter to 0.480 in. by turning it on a lathe. The spindle rotates at $N = 400$ rpm and the tool is traveling at an axial speed of 8 in./min.

Find: Calculate the cutting speed, material-removal rate, cutting time, power dissipated, and cutting force.

Solution: The cutting speed is the tangential speed of the workpiece. The maximum cutting speed is at the outer diameter, D_o, and is obtained from the equation

$$V = \pi D_o N.$$

Thus,

$$V = (\pi)(0.500)(400) = 628 \text{ in./min} = 52 \text{ ft/min.}$$

The cutting speed at the machined diameter is

$$V = (\pi)(0.480)(400) = 603 \text{ in./min} = 50 \text{ ft/min.}$$

From the information given, note that the depth of cut is

$$d = \frac{0.500 - 0.480}{2} = 0.010 \text{ in.}$$

and the feed is

$$f = \frac{8}{400} = 0.02 \text{ in./rev.}$$

According to Eq. (23.1), the material-removal rate is then

$$MRR = (\pi)(0.490)(0.010)(0.02)(400)$$
$$= 0.12 \text{ in}^3/\text{min.}$$

Equation (23.2) also can be used, in which case we find that $MRR = (0.01)(0.02)(52)(12) = 0.12$ in^3/min. The actual time to cut, according to Eq. (23.3), is

$$t = \frac{6}{(0.02)(400)} = 0.75 \text{ min.}$$

The power required can be calculated by referring to Table 21.2 and taking an average value for stainless steel as 4 W-s/mm^3 = 4/2.73 = 1.47 hp-min/in^3. Therefore, the power dissipated is

$$\text{Power} = (1.47)(0.123) = 0.181 \text{ hp.}$$

Since 1 hp = 396,000 in.-lb/min, the power dissipated is 71,700 in.-lb/min. The cutting force, F_c, is the tangential force exerted by the tool. Power is the product of torque, T, and the rotational speed in radians per unit time; hence,

$$T = \frac{71,700}{(2\pi)(400)} = 29 \text{ lb-in.}$$

The torque is $T = F_c D_{avg}/2$; hence

$$F_c = \frac{29}{0.490/2} = 118 \text{ lb.}$$

23.3   Lathes and Lathe Operations

Lathes generally are considered to be the oldest machine tools. Although woodworking lathes originally were developed during the period from 1000 to 1 B.C., metalworking lathes, with lead screws, were not built until the late 1700s. The most common lathe originally was called an *engine lathe*, because it was powered with overhead pulleys and belts from a nearby engine on the factory floor. Lathes became equipped with individual electric motors starting in the late 19th century.

The maximum spindle speed of lathes typically is around 4000 rpm, but may be only about 200 rpm for large lathes. For special applications, speeds may range from 10,000 to 40,000 rpm, or higher for very high-speed machining (see Section 25.5). The cost of lathes ranges from about $2000 for bench types to over $100,000 for larger units.

23.3.1 Lathe Components

Lathes are equipped with a variety of components and accessories, as shown in Fig. 23.2. Their basic features and functions are:

Bed. The bed supports all major components of the lathe; it has a large mass and is built rigidly, usually from gray or nodular cast iron. (See also Section 25.3 on new materials for machine-tool structures.) The top portion of the bed has two **ways**, with various cross-sections that are hardened and machined for wear resistance and good dimensional accuracy during turning. In a *gap-bed lathe*, a section of the bed in front of the headstock can be removed to accommodate larger diameter workpieces.

Carriage. The carriage, or *carriage assembly*, slides along the ways and consists of the *cross-slide*, *tool post*, and *apron*. The cutting tool is mounted on the *tool post*, usually with a *compound rest* that swivels for tool positioning and adjustment. The *cross-slide* moves radially in and out, controlling the radial position of the cutting tool in such operations as facing (see Fig. 23.1e). The *apron* is equipped with mechanisms for both manual and mechanized movement of the carriage and the cross-slide by means of the *lead screw*.

Headstock. The headstock is fixed to the bed and is equipped with motors, pulleys, and V-belts that supply power to a *spindle* and at various rotational speeds, which can be set through manually controlled selectors or by electrical controls. Most headstocks are equipped with a set of gears, and some have various drives to provide a *continuously variable* range of speed to the spindle. Headstocks have a *hollow spindle* on which work-holding devices (such as *chucks* and *collets*; see Section 23.3.2) are mounted, and long bars or tubing can be fed through them for various turning operations. The accuracy of the spindle is important for precision in turning, particularly in high-speed machining. Preloaded tapered or ball bearings typically are used to rigidly support the spindle.

Tailstock. The tailstock, which can slide along the ways and be clamped at any position, supports the other end of the workpiece. It is equipped with a *center*, which may be fixed (called *dead center*) or it may be free to rotate with the workpiece itself (*live center*). Drills and reamers (Sections 23.5 and 23.6) can be mounted on the tailstock *quill* (a hollow cylindrical piece with a tapered hole) to drill axial holes in the workpiece.

Feed Rod and Lead Screw. The feed rod is powered by a set of gears through the headstock. It rotates during the lathe operation, and provides movement to the carriage and the cross-slide by means of gears, a friction clutch, and a keyway along the length of the rod. Closing a *split nut* around the lead screw engages the rod with the carriage. The split nut is also used for cutting threads accurately.

TABLE 23.6

General Recommendations for Cutting Fluids for Machining (see also Section 33.7)

Material	Type of cutting fluid
Aluminum	D, MO, E, CSN
Beryllium	MO, E, CSN
Copper	D, E, CSN
Magnesium	D, MO
Nickel	MO, E, CSN
Refractory metals	MO, E, EP
Steels	
Carbon and low-alloy	D, MO, E, CSN, EP
Stainless	D, MO, E, CSN
Titanium	CSN, EP, MO
Zinc	C, MO, E, CSN
Zirconium	D, E, CSN

Note: CSN = chemical and synthetics; D = dry; E = emulsion; EP = extreme pressure; FO = fatty oil; and MO = mineral oil.

TABLE 23.7

Typical Capacities and Maximum Workpiece Dimensions for Machine Tools			
Machine tool	Maximum dimension (m)	Power (kW)	Maximum speed (rpm)
Lathes (swing/length)			
Bench	0.3/1	<1	3000
Engine	3/5	70	12000
Turret	0.5/1.5	60	6000
Automatic screw machines	0.1/0.3	20	10,000
Boring machines (work diameter/length)			
Vertical spindle	4/3	200	300
Horizontal spindle	1.5/2	70	2000
Drilling machines			
Bench and column (drill diameter)	0.1	10	12,000
Radial (column to spindle distance)	3	—	—
Numerical control (table travel)	4	—	—

Note: Larger capacities are available for special applications.

Lathe Specifications.   A lathe generally is specified by the following parameters:

- Its *swing*, the maximum diameter of the workpiece that can be accommodated (Table 23.7); this may be as much as 2 m (78 in.)
- The maximum distance between the headstock and tailstock centers
- The length of the bed

23.3.2 Work-holding Devices and Accessories

Work-holding devices are important, as they must hold the workpiece securely in place. As shown in Fig. 23.3, one end of the workpiece is clamped to the spindle of the lathe by a chuck, collet, face plate (see Fig. 23.8d), or mandrel.

A **chuck** usually is equipped with three or four *jaws*. *Three-jaw* chucks generally have a geared-scroll design that makes the jaws self-centering. They are used for round workpieces, such as bar stock, pipes, and tubing, and typically can be centered to within 0.025 mm (0.001 in.). *Four-jaw* chucks have jaws that can be moved and adjusted independently of each other; thus, for example, they can be used for square, rectangular, or odd-shaped workpieces. The jaws in some types of chucks can be reversed to permit clamping of hollow workpieces, such as pipes and tubing, either on their outside or inside surfaces. Also available are jaws made of low-carbon steel (called *soft jaws*) that can be machined into desired shapes. Because of their low strength and hardness, soft jaws conform to small irregularities on workpieces, thus resulting in better clamping. Chucks can be *power* or *manually actuated* using a chuck wrench.

Power chucks, actuated either pneumatically or hydraulically, are used in automated equipment for high production rates, including the loading of parts using industrial robots (Section 37.6). Also available are several types of power chucks with lever- or wedge-type mechanisms used to actuate the jaws. Chucks are available in various designs and sizes. Their selection depends on the type and speed of operation, workpiece size, production and dimensional accuracy requirements, and the jaw clamping forces required. By controlling the magnitude of the jaw forces, an operator can ensure that the part does not slip in the chuck during machining. High spindle speeds can significantly reduce jaw forces due to the effect of *centrifugal forces*.

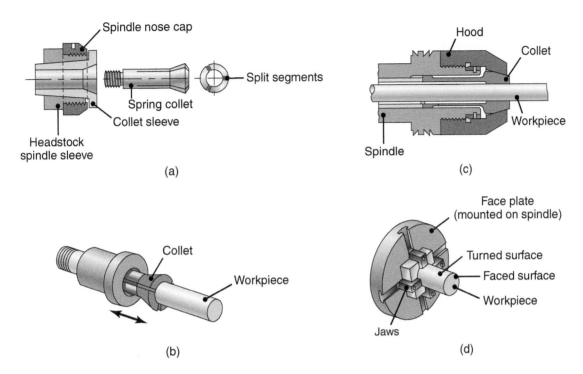

(a)

(c)

(b)

(d)

FIGURE 23.8 (a) and (b) Schematic illustrations of a draw-in type of collet. The workpiece is placed in the collet hole, and the conical surfaces of the collet are forced inward by pulling it with a draw bar into the sleeve. (c) A push-out type of collet. (d) Work holding of a workpiece on a face plate.

A **collet** is basically a longitudinally split, tapered bushing. The workpiece, generally with a maximum diameter of 1 in., is placed inside the collet, and the collet is pulled (*draw-in collet*; Figs. 23.8a and b) or pushed (*push-out collet*; Fig. 23.8c) mechanically into the spindle. The tapered surfaces shrink the segments of the collet radially, tightening them onto the workpiece. Collets are used for round workpieces as well as for other shapes. An advantage to using a collet, rather than a three- or four-jaw chuck, is that the collet grips nearly the entire circumference of the part, making the device particularly well suited for parts with small cross-sections.

Face plates are used for clamping irregularly shaped workpieces; they are round and have several slots and holes through which the workpiece is bolted or clamped (Fig. 23.8d). **Mandrels** (Fig. 23.9) are placed inside hollow or tubular workpieces,

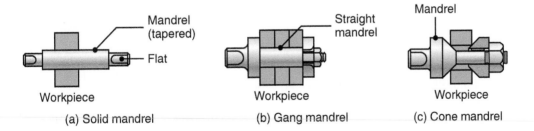

(a) Solid mandrel

(b) Gang mandrel

(c) Cone mandrel

FIGURE 23.9 Various types of mandrels to hold workpieces for turning; these mandrels usually are mounted between centers on a lathe. Note that in (a) both the cylindrical and the end faces of the workpiece can be machined, whereas in (b) and (c) only the cylindrical surfaces can be machined.

and are used to hold workpieces that require machining on both ends or on their cylindrical surfaces.

Accessories. Several devices are available as accessories and attachments for lathes. Among these devices are:

- Carriage and cross-slide stops, with various designs, to stop the carriage at a predetermined distance along the bed
- Devices for turning parts having various tapers
- Various attachments for milling, boring, drilling, thread cutting, gear-cutting, sawing, and grinding operations

23.3.3   Lathe Operations

In a typical turning operation, the workpiece is clamped by any one of the work-holding devices described previously. Long and slender parts must be supported by a *steady* rest placed on the bed, or by a *follow* rest, to keep the part from deflecting under the cutting forces. These rests usually are equipped with three adjustable fingers or rollers that support the workpiece while allowing it to rotate freely. Steady rests are clamped directly on the ways of the lathe (as in Fig. 23.2), whereas follow rests are clamped on the carriage and travel with it.

The cutting tool is attached to the tool post, which is driven by the lead screw. The cutting tool removes material by traveling along the bed. A *right-hand* tool travels toward the headstock, and a *left-hand* tool travels toward the tailstock. Facing operations are done by moving the tool radially inward with the cross-slide.

Form tools are used to machine various shapes on solid, round workpieces (Fig. 23.1g), by moving the tool radially inward while the part is rotating. Form cutting is not suitable for deep and narrow grooves or sharp corners, because of vibration and chatter, causing poor surface finish. As a rule, (a) the formed length of the part should not be greater than about 2.5 times the minimum diameter of the part, (b) the cutting speed should be set properly, and (c) cutting fluids should be used. The stiffness of the machine tools and work-holding devices also are important considerations.

Boring on a lathe is similar to turning, and is performed inside hollow workpieces or in a hole made previously by drilling or some other means. Out-of-shape round holes also can be straightened by boring. Boring large workpieces is described in Section 23.4.

Drilling (Section 23.5) can be performed on a lathe, by mounting the drill bit in a chuck or in the tailstock quill. The workpiece is clamped in a work holder on the headstock, and the drill bit is advanced by rotating the handwheel of the tailstock. The concentricity of the hole can be improved by subsequently boring the drilled hole. Drilled holes may be **reamed** (Section 23.6) on lathes in a manner similar to drilling, thus improving dimensional accuracy and surface finish.

The cutting tools for *parting, grooving, and thread cutting* are specially shaped for their particular purpose or are available as inserts. *Knurling* is performed on a lathe with hardened rolls (see Fig. 23.1l), in which the surface of the rolls is a replica of the profile to be generated. The rolls are pressed radially against the rotating workpiece while the tool moves axially along the part. Note that this is a forming operation and not machining, although it is done on a lathe.

23.3.4   Types of Lathes

Bench Lathes. As the name suggests, these lathes are placed on a workbench or a table. They have low power, are usually operated by hand feed, and are used to machine small workpieces. These are the simplest of lathes, and are commonly made for hobbyists or prototyping of small parts. *Toolroom bench lathes* have higher precision, enabling the machining of parts to close dimensional accuracy.

Special-purpose Lathes.   These lathes are used for such applications as railroad wheels, gun barrels, and rolling-mill rolls, with workpiece sizes as large as 1.7 m in diameter by 8 m in length (66 in. by 25 ft), and machine capacities as high as 450 kW (600 hp).

Tracer Lathes.   These lathes have special attachments that are capable of turning parts with various contours. Also called a *duplicating lathe* or *contouring lathe*, the cutting tool follows a path that duplicates the contour of a template, similar to a pencil following the shape of a stencil. These machine tools have largely been replaced by *numerical-control lathes* and *turning centers*, described in Section 25.2, although tracer attachments are available for many engine lathes.

Automatic Lathes.   Various mechanisms have been developed that enable machining operations on a lathe to follow a certain prescribed sequence. In a *fully automatic lathe*, parts are fed and removed automatically; in *semiautomatic* machines, these functions are performed by the operator, although the machining remains automatic. Automatic lathes, either with a horizontal or a vertical spindle, are suitable for medium- to high-volume production.

Lathes without tailstocks are called *chucking machines* or *chuckers*. They are used for machining individual pieces of regular or irregular shapes, and are either single- or multiple-spindle types. In another type of automatic lathe, the bar stock is fed periodically into the lathe, and after a part is machined, it is cut off from the end of the bar stock.

Automatic Bar Machines.   Also called **automatic screw machines**, these machine tools are designed for high-production-rate machining of screws and similar threaded parts. All operations are performed automatically, with tools attached to a special turret. After each part or screw is machined to finished dimensions, the bar stock is fed forward automatically through the hole in the spindle and then cut off. Automatic bar machines may be equipped with single or multiple spindles; capacities range from 3- to 150-mm (1/8- to 6-in.) diameter bar stock.

Single-spindle automatic bar machines are similar to turret lathes, and are equipped with various cam-operated mechanisms. In *Swiss-type automatics*, the cylindrical surface of the solid-bar stock is machined using a series of tools that move, radially and in the same plane, toward the workpiece. The bar stock is clamped close to the headstock spindle, thus minimizing deflections due to cutting forces. These machine tools are capable of high-precision machining of small-diameter parts.

Automatic bar machines are now equipped with computer numerical control, eliminating the use of cams, and the operation is programmed for a specific product (see Section 37.3).

Multiple-spindle automatic bar machines typically have four to eight spindles, arranged in a circle on a large drum, with each spindle carrying an individual workpiece. The cutting tools are arranged in various positions in the machine, and move in both axial and radial directions. Each part is machined in stages as it moves from one station to the next. Because all operations are carried out simultaneously, the cycle time per part is reduced.

Turret Lathes.   These machine tools are capable of performing multiple cutting operations, such as turning, boring, drilling, thread cutting, and facing (Fig. 23.9).

Several cutting tools, usually as many as six, are mounted on the hexagonal *main turret*, which is rotated after each specific cutting operation is completed. The lathe usually has a *square turret* on the cross-slide, equipped with as many as four cutting tools. The workpiece, generally a long, round bar stock, is advanced a preset distance through the chuck. After the part is machined, it is cut off by a tool mounted on the square turret, which moves radially into the workpiece. The rod then is advanced the same preset distance, and the next part is machined.

Turret lathes, either the bar type or the chucking type, are versatile, and the operations may be carried out either by hand, using the turnstile (*capstan wheel*), or automatically. Once set up properly, these machines do not require highly skilled operators. *Vertical turret lathes* also are available; they are more suitable for short, heavy workpieces with diameters as large as 1.2 m (48 in.).

The turret lathe shown in Fig. 23.10 is known as a **ram-type** turret lathe, in which the ram slides along a separate base on the saddle. The short stroke of the turret slide limits this machine to relatively short workpieces and light cuts in both small- and medium-quantity production. In another design, called the **saddle type**, the main turret is installed directly on the saddle, which slides along the bed. The length of the stroke is limited only by the length of the bed; this type of lathe is constructed more heavily, and is used to machine large workpieces. Because of the heavy weight of the components, saddle-type lathe operations are slower than ram-type operations.

Computer-controlled Lathes.   In the most advanced lathes, movement and control of the machine tool and its components are achieved by *computer numerical control* (CNC). The features of such a lathe are shown in Fig. 23.11a. These lathes generally are equipped with one or more turrets, and each turret is equipped with a variety of tools and performs several operations on different surfaces of the workpiece (Fig. 23.11b). Workpiece diameters may be as much as 1 m (36 in.).

To take advantage of new cutting-tool materials, computer-controlled lathes are designed to operate faster, and have higher power available compared with other lathes; they are equipped with *automatic tool changers* (ATCs). Their operations are reliably repeatable, maintain the desired dimensional accuracy, and require less skilled labor (once the machine is set up); they are suitable for low- to medium-volume production.

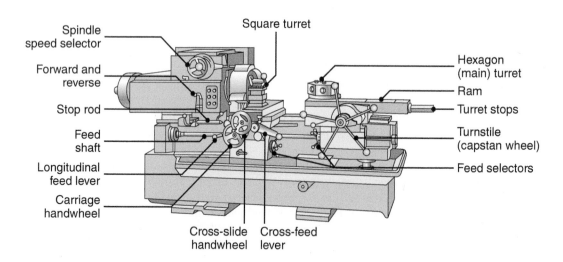

FIGURE 23.10   Schematic illustration of the components of a turret lathe; note the two turrets: square and hexagonal (main turret).

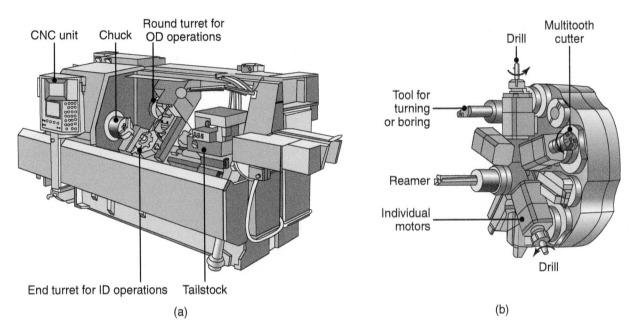

CNC unit   Chuck   Round turret for OD operations

End turret for ID operations   Tailstock

(a)

Drill   Multitooth cutter

Tool for turning or boring

Reamer

Individual motors

Drill

(b)

FIGURE 23.11   (a) A computer-numerical-control lathe; note the two turrets. These machines have higher power and spindle speed than other lathes in order to take advantage of new cutting tools with enhanced properties. (b) A typical turret equipped with 10 tools, some of which are powered.

EXAMPLE 23.2   Typical Parts Made on CNC Turning Machine Tools

The capabilities of CNC turning machine tools are illustrated in the parts shown in Fig. 23.12, indicating the workpiece material, the number of cutting tools used, and the machining times. These parts also can be made on manual or turret lathes, although not as effectively or consistently.

Source: Courtesy of Monarch Machine Tool Company.

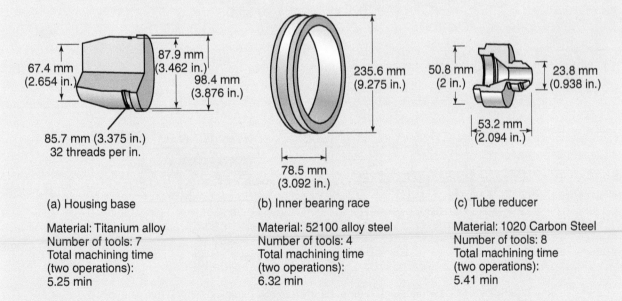

67.4 mm (2.654 in.)   87.9 mm (3.462 in.)   98.4 mm (3.876 in.)

85.7 mm (3.375 in.) 32 threads per in.

235.6 mm (9.275 in.)

78.5 mm (3.092 in.)

50.8 mm (2 in.)   23.8 mm (0.938 in.)

53.2 mm (2.094 in.)

(a) Housing base

Material: Titanium alloy
Number of tools: 7
Total machining time (two operations): 5.25 min

(b) Inner bearing race

Material: 52100 alloy steel
Number of tools: 4
Total machining time (two operations): 6.32 min

(c) Tube reducer

Material: 1020 Carbon Steel
Number of tools: 8
Total machining time (two operations): 5.41 min

FIGURE 23.12   Typical parts made on CNC lathes.

EXAMPLE 23.3   Machining of Complex Shapes

Note, in Example 23.2, that the parts are axisymmetric. The capabilities of CNC turning are further illustrated in Fig. 23.13, which shows three additional, more complex parts: a pump shaft, a crankshaft, and a tubular part with an internal rope thread. Descriptions of these parts are given below; as in most operations, machining of such parts consists of both roughing and finishing cuts:

1. *Pump shaft* (Fig. 23.13a). This part, as well as a wide variety of similar parts with external and internal features, including camshafts, was produced on a CNC lathe with two turrets. The lathe is similar in construction to the machine tool shown in Fig. 23.11a. Each turret can hold as many as eight tools. To produce this particular shape, the upper turret is programmed in such a manner that its radial movement is synchronized with the shaft's rotation (Fig. 23.13b).

   The spindle turning angle is monitored directly, a processor performs a high-speed calculation, and the CNC then issues a command to the cam turret in terms of that angle. The machine has absolute-position feedback, using a high-accuracy scale system. The CNC compares the actual value with the commanded one, then performs an automatic compensation, using a built-in learning function. The turret has a lightweight design for smooth operation, which also reduces inertial forces.

   The shaft may be made of aluminum or stainless steel. The machining parameters for aluminum are given in Table 23.8 (see Part a in the first column of the table). These parameters may be compared with the data given in Table 23.4, which has only a broad and approximate range as a guideline. The inserts were a K10 (C3) uncoated carbide with a compacted polycrystalline diamond (see Fig. 22.10). The OD machining in the table shown refers to the two straight cylindrical ends of the part. The total machining time for an aluminum shaft was 24 min; for stainless steel, it was 55 min, because the cutting speed for stainless steel is considerably lower than that for aluminum.

2. *Crankshaft* (Fig. 23.13c). This part is made of ductile (nodular) cast iron, and the machining parameters are shown in Part b of Table 23.7. The insert was K10 carbide. The machining time was 25 min, which is of the same order

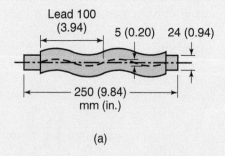

(a)

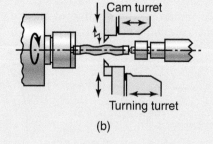

(b)

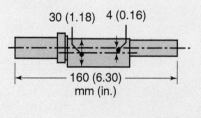

(c)

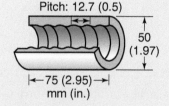

(d)

FIGURE 23.13 Examples of four parts with more complex shapes that can be produced on a CNC lathe.

TABLE 23.8

Machining Summary for Example 23.3

Operation	Speed (rpm)	Cutting speed	Depth of cut	Feed	Cutting tool
Parts a and b:					
Outer diameter (OD)					
Roughing	1150	160 m/min (525 ft/min)	3 mm (0.12 in.)	0.3 mm/rev (0.012 in./rev)	K10 (C3)
Finishing	1750	250 (820)	0.2 (0.008)	0.15 (0.006)	K10 (C3)
Lead					
Roughing	300	45 (148)	3 (0.12)	0.15 (0.006)	K10 (C3)
Finishing	300	45 (148)	0.1 (0.004)	0.15 (0.006)	Diamond compact
Part c: Eccentric shaft					
Roughing	200	5–11 (16–136)	1.5 (0.059)	0.2 (0.008)	K10 (C3)
Finishing	200	5–11 (16–136)	0.1 (0.004)	0.05 (0.0020)	K10 (C3)
Part d: Internal thread					
Roughing	800	70 (230)	1.6 (0.063)	0.15 (0.006)	Coated carbide
Finishing	800	70 (230)	0.1 (0.004)	0.15 (0.006)	Cermet

of magnitude as that for the pump shaft described above.

3. *Tubular part with internal rope threads* (Fig. 23.13d). This part, made of 304 stainless steel, was machined under the conditions given for Part d in Table 23.8. The starting blank was a straight tubular piece, similar to a bushing. The cutting tools were coated carbide and cermet. The boring bar was made of tungsten carbide, for increased stiffness, and, hence, improved dimensional accuracy and surface finish. For the threaded portion, the dimensional accuracy was ±0.05 mm (0.002 in.), with a surface finish of $R_a = 2.5$ μm (100 μin.).

The machining time for this part was 1.5 min, which is much shorter than those for the previous two parts. The reason is that (a) this part is shorter, (b) less material is removed, (c) it does not have the eccentricity features of the first two parts, so the radial movement of the cutting tool is not a function of the angular position of the part, and (d) the cutting speed is higher.

Source: Based on technical literature supplied by Okuma Corp.

23.3.5   Turning-process Capabilities

Relative *production rates* in turning, as well as in other machining operations described in the rest of this chapter and in Chapter 24, are shown in Table 23.9. These rates have an important bearing on productivity in machining operations. Note that there are major differences in the production rate among the processes listed in the table. These differences are due not only to the inherent characteristics of the processes and the machine tools, but also to various other factors, such as the setup times and the types and sizes of the workpieces to be machined.

The ratings given in Table 23.9 are relative, and there can be significant variations in special applications. For example, heat-treated, high-carbon cast-steel rolls for rolling mills can be machined on special lathes, using cermet tools and at material-removal rates as high as 6000 cm³/min (370 in³/min). Also called **high-removal-rate machining**, this process has at least two important requirements: (a) very high

TABLE 23.9

Typical Production Rates for Various Machining Operations

Operation	Rate
Turning	
Engine lathe	Very low to low
Tracer lathe	Low to medium
Turret lathe	Low to medium
Computer-controlled lathe	Low to medium
Single-spindle chuckers	Medium to high
Multiple-spindle chuckers	High to very high
Boring	Very low
Drilling	Low to medium
Milling	Low to medium
Planing	Very low
Gear cutting	Low to medium
Broaching	Medium to high
Sawing	Very low to low

Note: Production rates indicated are relative: *Very low* is about one or more parts per hour, *medium* is approximately 100 parts per hour, and *very high* is 1000 or more parts per hour.

machine-tool rigidity, to avoid chatter and associated tool breakage, and (b) high power, of up to 450 kW (600 hp).

The surface finish (Fig. 23.14) and dimensional accuracy (Fig. 23.15) obtained in turning and related operations depend on several factors, such as the characteristics and condition of the machine tool, stiffness, vibration and chatter, process parameters, tool geometry and wear, the use of cutting fluids, the machinability of the workpiece material, and operator skill. As a result, a wide range of surface finishes can be obtained, as shown in Fig. 27.4.

23.3.6 Design Considerations and Guidelines for Turning Operations

Several considerations are important in designing parts to be machined economically by turning operations. Because machining, in general, (a) takes considerable time, thus increasing the production cost, (b) wastes material, and (c) is not as economical as forming or shaping operations, and must be avoided as much as possible. When turning operations are necessary, the following general design guidelines should be considered:

1. Parts should be designed so that they can be fixtured and clamped easily into work-holding devices. Thin, slender workpieces are difficult to support properly and must be able to withstand clamping and cutting forces (see also *flexible fixturing*, Section 37.8).
2. The dimensional accuracy and surface finish specified should be as wide as permissible, but the part must still function properly.
3. Sharp corners, tapers, steps, and major dimensional variations in the part should be avoided.
4. Blanks to be machined should be as close to final dimensions as possible, such as by near-net-shape forming, so as to reduce production cycle time.
5. Parts should be designed so that cutting tools can travel across the workpiece without any obstruction.
6. Design features should be such that commercially available standard cutting tools, inserts, and toolholders can be used.
7. Workpiece materials should preferably be selected for their machinability (Section 21.7).

Guidelines for Turning Operations.   The following list outlines generally accepted guidelines for turning operations; see also Table 23.10 for probable causes for turning problems.

1. Minimize tool overhang
2. Support the workpiece rigidly
3. Use machine tools with high stiffness and high damping capacity
4. When tools begin to vibrate and chatter (Section 25.4), modify one or more of the process parameters, such as tool geometry, cutting speed, feed rate, depth of cut, and cutting fluid (see also adaptive control, Section 37.4)

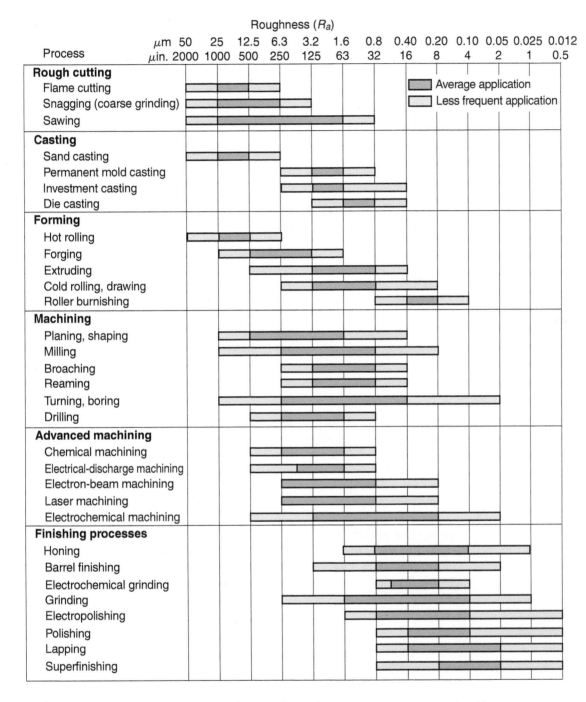

FIGURE 23.14   The range of surface roughnesses obtained in various processes; note the wide range within each group, especially in turning and boring.

23.3.7   Chip Collection Systems

The chips produced during machining must be collected and disposed of properly. The volume of chips produced can be very high, particularly in ultra-high-speed machining and high-removal-rate machining operations. For example, in a drilling

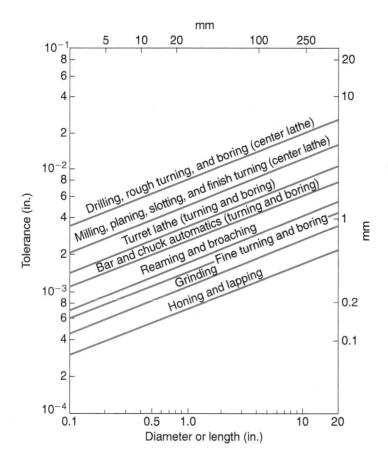

FIGURE 23.15 Range of dimensional tolerances obtained in various machining processes as a function of workpiece size; note that there is an order-of-magnitude difference between small and large workpieces.

operation on steel, during which only 1 in³ of metal is removed, the loose bulk volume of the chips can, depending on chip type (see Section 21.2.1), be in the range of 40–800 in³. Likewise, the milling of 1 in³ of steel produces 30–45 in³ of chips, while cast iron produces 7–15 in³ of chips.

Also called **chip management**, the system involves collecting chips from their source in the machine tool in an efficient manner and removing them from the work area. Long and stringy chips are more difficult to collect than short chips, which are produced by using tools with chipbreaker features (see Figs. 21.7 and 22.2). Thus, the type of chip produced must be an integral aspect of the chip-collecting system.

Chips can be collected by any of the following methods:

- Allowing gravity to drop them onto a steel conveyor belt
- Dragging the chips from a settling tank
- Using augers with feed screws, similar to those in meat grinders
- Using magnetic conveyors, for ferrous chips only
- Employing vacuum methods of chip removal

Modern machine tools are designed with automated chip-handling features. There may be a considerable amount of cutting fluid mixed with and coating the chips produced, thus proper filtration or draining is important. The cutting fluid and sludge can be separated, using chip wringers (centrifuges). Chip-processing systems usually require considerable floor space in the plant, and can cost from $60,000 for small shops to over $1 million for large plants.

TABLE 23.10

General Troubleshooting Guide for Turning Operations

Problem	Probable causes
Tool breakage	Tool material lacks toughness, improper tool angles, machine tool lacks stiffness, worn bearings and machine components, machining parameters too high
Excessive tool wear	Machining parameters too high, improper tool material, ineffective cutting fluid, improper tool angles
Rough surface finish	Built-up edge on tool; feed too high; tool too sharp, chipped, or worn; vibration and chatter
Dimensional variability	Lack of stiffness of machine tool and work-holding devices, excessive temperature rise, tool wear
Tool chatter	Lack of stiffness of machine tool and work-holding devices, excessive tool overhang, machining parameters not set properly

The collected chips may be recycled if it is determined to be economical to do so. Prior to their removal from a manufacturing plant, the large volume of chips can be reduced to as little as one-fifth of the loose volume by *compaction* (crushing) into briquettes or by *shredding*. Dry chips are more valuable for recycling, because of reduced environmental contamination. The method chosen for chip disposal depends on economics, as well as on compliance with local, state, and federal regulations. The trend now is to recycle all chips, as well as the used cutting fluids and the sludge.

23.3.8   Cutting Screw Threads

A *screw thread* may be defined as a ridge of uniform cross-section that follows a helical or spiral path on the outside or inside of a cylindrical (*straight thread*) or tapered surface (*tapered thread*). Machine screws, bolts, and nuts have straight threads, as do threaded rods for such applications as the lead screw in lathes and various machinery components (Fig. 23.2). Tapered threads commonly are used for water or gas pipes and plumbing supplies, which require a watertight or airtight connection. Threads may be *right handed* or *left handed*.

Threads traditionally have been machined, but they are increasingly being formed by **thread rolling** (described in Section 13.5). Rolled threads now constitute the largest quantity of external threaded parts produced. It also may be possible to cast threaded parts, but there are limitations to dimensional accuracy, surface finish, and minimum dimensions.

Threads can be machined, either externally or internally, with a cutting tool in a process called *thread cutting* or *threading*. External threads also may be cut with a *die* or by milling. Internal threads are typically produced by tapping, using a special threaded tool, called a *tap* (Section 23.7). Threads subsequently may be ground for high dimensional accuracy and surface finish for such applications as screw drives in machines.

Screw-thread Cutting on a Lathe.   A typical thread-cutting operation on a lathe is shown in Fig. 23.16a. The cutting tool, the shape of which depends on the type of thread to be cut, is mounted on a holder and moved along the length of the workpiece by the lead screw on the lathe. This movement is achieved by the engagement of a *split nut*, also called a *half nut*, inside the apron of the lathe (see Fig. 23.2).

The axial movement of the tool, in relation to the rotation of the workpiece, determines the lead of the screw thread (i.e., the axial distance moved in one complete revolution of the screw). For a fixed spindle speed, the slower the tool movement, the finer the thread will be. In thread cutting, the cutting tool may be fed radially into the workpiece, thus cutting both sides of the thread at the same time, as in form cutting described earlier; however, this method usually produces a poor surface finish.

A number of passes, in the sequence shown in Fig. 23.16b, generally are required to produce threads with good dimensional accuracy and surface finish. Figure 23.16c shows a carbide insert for screw-thread cutting (*threading insert*) for machining threads on a round shaft. Figure 23.16d shows an internal screw-thread cutting process. Except for small production runs, the thread-cutting process largely has been replaced by other methods, such as thread rolling, automatic screw machining, and the use of CNC lathes.

The production rate in cutting screw threads can be increased with tools called *die-head chasers* (Figs. 23.17a and b), which typically have four cutters with multiple teeth and can be adjusted radially. After the threads are cut, the cutters open automatically (thus the alternative name *self-opening die heads*) by rotating around

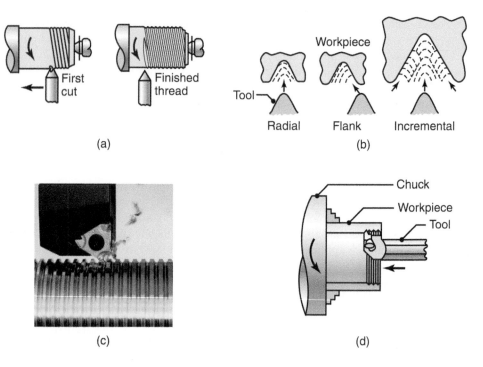

FIGURE 23.16 (a) Cutting screw threads on a lathe with a single-point cutting tool. (b) Cutting screw threads with a single-point tool in several passes, normally utilized for large threads. The small arrows in the figures show the direction of feed, and the broken lines show the position of the cutting tool as time progresses. In radial cutting, the tool is fed directly into the workpiece. In flank cutting, the tool is fed into the piece along the right face of the thread. In incremental cutting, the tool is fed first directly into the piece at the center of the thread, then at its sides, and finally into the root. (c) A typical coated-carbide insert in the process of cutting screw threads on a round shaft. (d) Cutting internal screw threads with a carbide insert. *Source:* (c) Courtesy of Iscar Metals, Inc.

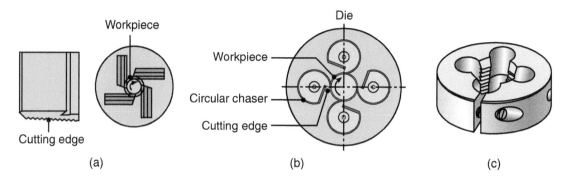

FIGURE 23.17 (a) Straight chaser for cutting threads on a lathe. (b) Circular chaser. (c) A solid threading die.

their axes to allow the part to be removed. *Solid-threading dies* (Fig. 23.17c) also are available for cutting straight or tapered screw threads.

Design Considerations for Screw Thread Machining. The design considerations that must be taken into account in order to produce high-quality and economical screw threads are:

- Designs should allow for the termination of threads before they reach a shoulder. Internal threads in blind holes should have an unthreaded length at the bottom. (The term *blind hole* refers to a hole that does not go through the thickness of the workpiece; see, for example, Fig. 23.1i.)
- Attempts should be made to eliminate shallow, blind, tapped holes.
- Chamfers should be specified at the ends of threaded sections, to minimize finlike threads with burrs.
- Threaded sections should not be interrupted with slots, holes, or other discontinuities.
- Standard threading tooling and inserts should be used as much as possible.
- Thin-walled parts should have sufficient thickness and strength to resist clamping and cutting forces. A common rule of thumb is that the minimum engagement length of a fastener should be 1.5 times the diameter.
- Parts should be designed so that all cutting operations can be completed in one setup.

23.4   Boring and Boring Machines

Boring enlarges a hole made previously by some other process or produces circular internal profiles in hollow workpieces (Fig. 23.1h). The cutting tools are similar to those used in turning, and are mounted on a *boring bar* (Fig. 23.18a), in order to reach the full length of the bore. It is essential that the boring bar be sufficiently stiff to minimize tool deflection and vibration, and thus maintain dimensional accuracy and surface finish. For this reason, a material with a high elastic modulus, such as tungsten carbide, is desirable. Boring bars also have been designed and built with capabilities for damping vibration (Fig. 23.18b).

Boring operations on relatively small workpieces can be carried out on lathes, whereas large workpieces are machined on **boring mills**. These machine tools are either horizontal or vertical, and are capable of performing various operations, such as turning, facing, grooving, and chamfering. In **horizontal boring mills**, the workpiece is mounted on a table that can move horizontally, in both the axial and radial directions. The cutting tool is mounted on a spindle that rotates in the headstock, and is capable of both vertical and longitudinal movements. Drills, reamers, taps, and milling cutters also can be mounted on the machine spindle. A **vertical boring mill** (Fig. 23.19) is similar to a lathe, has a vertical axis of workpiece rotation, and can accommodate workpieces with diameters as large as 2.5 m (98 in.).

The cutting tool is usually a single point, made of M2 or M3 high-speed steel, or P10 (C7) or P01 (C8) carbide. It is mounted on the tool head, which is capable of

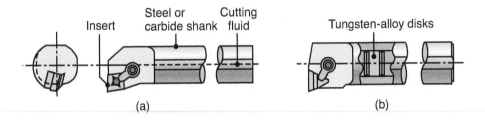

(a)   (b)

FIGURE 23.18   (a) Schematic illustration of a steel boring bar with a carbide insert; note the passageway in the bar for cutting fluid application. (b) Schematic illustration of a boring bar with tungsten-alloy "inertia disks" sealed in the bar to counteract vibration and chatter during boring; this system is effective for boring-bar length-to-diameter ratios of up to six.

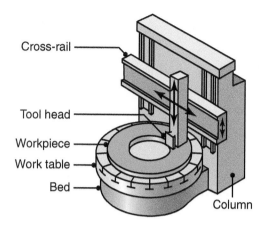

Cross-rail

Tool head

Workpiece

Work table

Bed

Column

FIGURE 23.19 Schematic illustration of a vertical boring mill; such a machine can accommodate workpiece sizes as large as 2.5 m (98 in.) in diameter.

vertical movement (for boring and turning) and radial movement (for facing), guided by the cross-rail; the head can be swiveled to produce conical (tapered) holes. Cutting speeds and feeds for boring are similar to those for turning (see Table 23.9).

Boring machines are available with a variety of features. Machine capacities range up to 150 kW (200 hp), and are available with computer numerical controls, allowing all movements of the machine to be programmed. Little operator involvement is required, and thus consistency and productivity are improved.

Design Considerations for Boring.   Guidelines for economical boring operations are similar to those for turning; additionally, the following factors should be considered:

- Whenever possible, through holes rather than blind holes should be specified. The term *blind hole* refers to a hole that does not extend through the thickness of the workpiece.
- The greater the length-to-bore-diameter ratio, the more difficult it is to hold dimensions, because of the deflections of the boring bar due to cutting forces, as well as the higher tendency for vibration and chatter.
- Interrupted internal surfaces, such as internal splines or radial holes that go through the thickness of the part, should be avoided.

23.5 Drilling, Drills, and Drilling Machines

When inspecting large or small products, it will be noted that the vast majority have several holes in them. **Hole making** is among the most important operations in manufacturing, and **drilling** is a major and common hole-making process; other processes for producing holes are punching (as described in Section 16.2) and various advanced machining processes (Chapter 27). The cost of hole making is among the highest machining costs in automotive engine production.

23.5.1 Drills

Drills typically have high length-to-diameter ratios (Fig. 23.20), hence they are capable of producing relatively deep holes. However, high ratios make drills somewhat flexible and prone to fracture or producing inaccurate holes; moreover, the chips produced within the hole present significant difficulties in their disposal and also ensuring cutting-fluid effectiveness.

Drills generally leave a *burr* on the bottom surface of the part upon breakthrough, often necessitating deburring operations (Section 26.8). Also, because of its rotary motion, drilling produces holes with walls that have *circumferential* marks; in contrast, punched holes have *longitudinal* marks (see Fig. 16.5a). This difference is significant in terms of the hole's fatigue properties, as described in Section 33.2.

The diameter of a hole produced by drilling is slightly larger than the drill diameter (*oversize*), as one can note by observing that a drill can easily be removed from the hole it has just produced. The amount of oversize depends on the quality of the drill, the equipment used, and on the machining practices employed. Furthermore, depending on their thermal properties, some metals and nonmetallic materials expand significantly due to the heat produced during drilling, thus the final hole diameter could be smaller than the drill diameter when the part cools down. For better surface

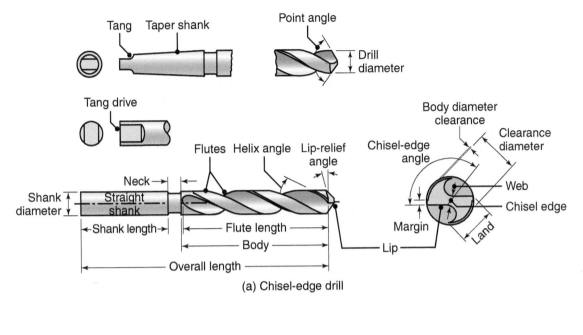

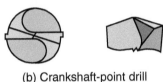

(b) Crankshaft-point drill

FIGURE 23.20   Two common types of drills: (a) Chisel-edge drill; the function of the pair of margins is to provide a bearing surface for the drill against walls of the hole as it penetrates into the workpiece. Drills with four margins (*double-margin*) are available for improved guidance and accuracy. Drills with chip-breaker features also are available. (b) Crankshaft-point drill; these drills have good centering ability, and because chips tend to break up easily, crankshaft drills are suitable for producing deep holes.

finish and dimensional accuracy, drilled holes may be subjected to subsequent operations, such as reaming and honing. The capabilities of drilling and boring operations are shown in Table 23.11.

Twist Drill.   The most common drill is the conventional *standard-point twist drill* (Fig. 23.20a). The geometry of the drill point is such that the normal rake angle and velocity of the cutting edge vary with the distance from the center of the drill. The main features of this drill are (with typical ranges of angles given in parentheses): (a) *point angle* (118°–135°), (b) *lip-relief angle* (7°–15°), (c) *chisel-edge angle* (125°–135°), and (d) *helix angle* (15°–30°).

Two spiral grooves, called *flutes*, run the length of the drill, and the chips produced are guided upward through these grooves. The grooves also serve as passageways to enable the cutting fluid to reach the cutting edges. Some drills have internal longitudinal holes (see, for example, the drill shown in Fig. 23.23a), through which cutting fluids are forced, thus improving

TABLE 23.11

General Capabilities of Drilling and Boring Operations			
		Hole depth/diameter	
Cutting tool	Diameter range (mm)	Typical	Maximum
Twist drill	0.5–150	8	50
Spade drill	25–150	30	100
Gun drill	2–50	100	300
Trepanning tool	40–250	10	100
Boring tool	3–1200	5	8

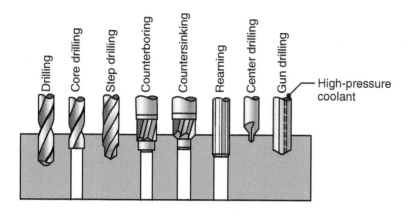

FIGURE 23.21   Various types of drilling and reaming operations.

lubrication and cooling, as well as washing away the chips. Drills are available with a **chip-breaker** feature, ground along the cutting edges. This feature is important in drilling with automated machinery, where continuous removal of long chips without operator assistance is essential.

The various angles on a drill have been developed through experience over many years, and are designed to produce accurate holes, minimize drilling forces and torque, and optimize drill life. Small changes in drill geometry can have a significant effect on a drill's performance, particularly in the chisel-edge region, which accounts for about 50% of the thrust force in drilling. For example, too small a lip relief angle (Fig. 23.20a) increases the thrust force, generates excessive heat, and increases drill wear. By contrast, too large an angle can cause chipping or breaking of the cutting edge.

In addition to conventional point drills, several other drill-point geometries have been developed to improve drill performance and increase the penetration rate. Special grinding techniques and equipment are used to produce these geometries.

Other Types of Drills.   Various types of drills are shown in Fig. 23.21. A *step drill* produces holes with two or more different diameters. A *core drill* is used to make an existing hole larger. *Counterboring* and *countersinking drills* produce depressions on the surface to accommodate the heads of screws and bolts below the workpiece surface. A *center drill* is short and is used to produce a hole at the end of a piece of stock, so that it may be mounted between centers of the headstock and the tailstock of a lathe (Fig. 23.2). A *spot drill* is used to spot (meaning to start) a hole at the desired location on a surface.

Spade drills (Fig. 23.22a) have removable tips or bits, and are used to produce large-diameter and deep holes. These drills have the advantages of higher stiffness

Video Solution 23.2 Drilling Operations

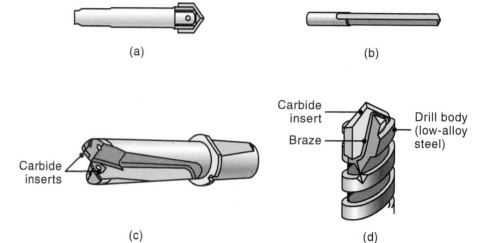

FIGURE 23.22   Various types of drills. (a) Spade drill; (b) straight-flute drill; (c) drill with indexable carbide inserts; and (d) drill with brazed-carbide tip.

(because of the absence of flutes in the body of the drill), ease of grinding the cutting edges, and lower cost. A similar drill is the *straight-flute drill* (Fig. 23.22b).

Solid carbide and carbide-tipped drills (Figs. 23.22c and d) are made for drilling hard materials, such as cast irons; high-temperature metals, such abrasive materials as concrete and brick (called *masonry drills*); and composite materials with abrasive fiber reinforcements, such as glass and graphite.

Gun Drilling.   Developed originally for drilling gun barrels, *gun drilling* is used for drilling deep holes; it requires a special drill, as shown in Fig. 23.23. The depth-to-diameter ratios of holes produced can be 300:1 or even higher. The thrust force (the radial force that tends to push the drill sideways) is balanced by bearing pads on the drill that slide along the inside surface of the hole. Consequently, a gun drill is *self-centering*, an important feature in drilling straight, deep holes. A variation of this process is **gun trepanning** (described below), which uses a cutting tool similar to a gun drill, except that the tool has a central hole.

Cutting speeds in gun drilling are usually high, and feeds are low. Tolerances typically are about 0.025 mm (0.001 in.). The cutting fluid is forced under high pressure through a longitudinal hole (passage) in the body of the drill (Fig. 23.22a). In addition to cooling and lubricating the workpiece, the fluid flushes out chips that otherwise would be trapped in the deep hole being drilled, and thus severely interfere with the drilling operation.

Trepanning.   In *trepanning* (from the Greek *trypanon*, meaning "boring a hole" or "auger,") the cutting tool (Fig. 23.24a) produces a hole by removing a disk-shaped piece (*core*), usually from flat plates. A hole is thus produced without reducing all of

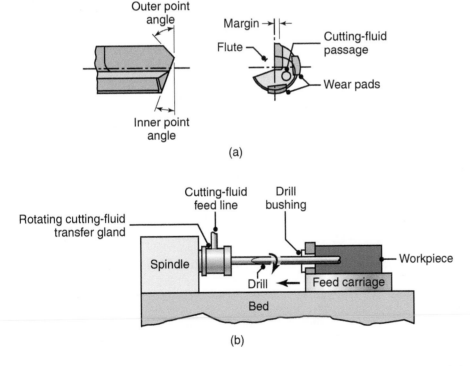

FIGURE 23.23   (a) A gun drill, showing various features. (b) Schematic illustration of the gun-drilling operation.

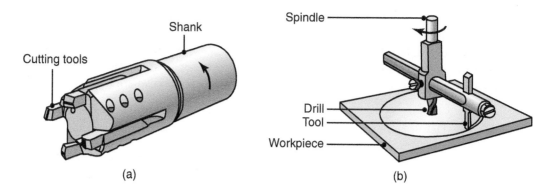

FIGURE 23.24   (a) Trepanning tool. (b) Trepanning with a drill-mounted single cutting tool.

the material that is removed to chips, as is the case in drilling. The trepanning process can be used to make disks up to 250 mm (10 in.) in diameter from flat sheets, plates, or structural members such as I-beams. It also can be used to make circular grooves in which O-rings are to be placed (similar to Fig. 23.1f). Trepanning can be carried out on lathes, drill presses, or other machine tools, using single-point or multipoint tools, as shown in Fig. 23.24b.

23.5.2   Material-removal Rate in Drilling

The *material-removal rate* (MRR) in drilling is the volume of material removed per unit time. For a drill with a diameter D, the cross-sectional area of the drilled hole is $\pi D^2/4$. The velocity of the drill perpendicular to the workpiece is the product of the feed, f (the distance the drill penetrates per unit revolution), and the rotational speed, N, where $N = V/\pi D$. Thus,

$$\text{MRR} = \left(\frac{\pi D^2}{4}\right) f N. \tag{23.4}$$

23.5.3   Thrust Force and Torque

The *thrust force* in drilling acts perpendicular to the hole axis; if this force is excessive, it can cause the drill to bend or break. An excessive thrust force also can distort the workpiece, particularly if it does not have sufficient stiffness (as is the case for thin sheet-metal structures) or it can cause the workpiece to slip into the work-holding fixture.

The thrust force depends on factors such as (a) the strength of the workpiece material, (b) feed, (c) rotational speed, (d) drill diameter, (e) drill geometry, and (f) cutting fluids. Thrust forces typically range from a few newtons for small drills to as high as 100 kN (23.5 klb) for drilling high-strength materials using large drills.

Torque.   A knowledge of the *torque* in drilling is essential for estimating the power requirement; however, because of the many factors involved, it is difficult to calculate. Torque can be estimated from the data given in Table 21.2, by noting that the power dissipated during drilling is the product of torque and rotational speed, and is also equal to the product of specific energy and material removal rate. Torque in drilling can be as high as 4000 N-m (3000 lb-ft).

EXAMPLE 23.4   Material-removal Rate and Torque in Drilling

Given: A hole is being drilled in a block of magnesium alloy with a 10-mm drill bit at a feed of 0.2 mm/rev and with the spindle running at $N = 800$ rpm.

Find: Calculate the material-removal rate and the torque on the drill.

Solution: The material-removal rate is calculated from Eq. (23.4):

$$\text{MRR} = \left[\frac{(\pi)(10)^2}{4} \right](0.2)(800)$$

$$= 12{,}570 \text{ mm}^3/\text{min} = 210 \text{ mm}^3/\text{s}.$$

Referring to Table 21.2, use an average unit power of 0.5 W-s/mm^3 for magnesium alloys. The power required is then

$$\text{Power} = (210)(0.5) = 105 \text{ W}$$

Power is the product of the torque on the drill and the rotational speed, which in this case is $(800)(2\pi)60 = 83.8$ radians per second. Noting that W = J/s and J = N-m,

$$T = \frac{105}{83.8} = 1.25 \text{ N-m}.$$

23.5.4 Drill Materials and Sizes

Drills usually are made of high-speed steels (M1, M7, and M10) and solid carbides or with carbide tips, typically made of K20 (C2) carbide, like those shown in Figs. 23.22c and d. Drills are now commonly coated with titanium nitride or titanium carbonitride, for increased wear resistance, as described in Section 22.5. Polycrystalline-diamond-coated drills are used for producing fastener holes in fiber-reinforced plastics. Because of their high wear resistance, several thousand holes can be drilled with little damage to the workpiece material.

Although there are continued developments, standard twist-drill sizes consist basically of the following series:

- **Numerical:** No. 97 (0.0059 in.) to No. 1 (0.228 in.)
- **Letter:** A (0.234 in.) to Z (0.413 in.)
- **Fractional:** Straight shank from $\frac{1}{64}$ to $1\frac{1}{4}$ (in $\frac{1}{64}$-in. increments) to $1\frac{1}{2}$ in. (in $\frac{1}{32}$-in. increments), and larger drills in larger increments. Taper shank from $\frac{1}{8}$ to $1\frac{3}{4}$ (in $\frac{1}{64}$ increments) to 3.5 in. (in $\frac{1}{16}$-in. increments)
- **Millimeter:** From 0.05 mm (0.002 in.) in increments of 0.01 mm

23.5.5 Drilling Practice

Drills and similar hole-making tools usually are held in *drill chucks*, which may be tightened with or without keys. Special chucks and collets, with various quick-change features that do not require stopping the spindle, are available for use on production machinery.

Because it does not have a centering action, a drill tends to "walk" on the workpiece surface at the beginning of the operation, a problem that is particularly severe with small-diameter long drills and can lead to failure. To start a hole properly, the drill should be guided, using fixtures, such as a bushing, to keep it from deflecting laterally. A small starting hole can be made with a *center drill*, usually with a point angle of 60°, or the drill point may be ground to an S shape (called *helical* or *spiral point*). This shape has a self-centering characteristic, thus eliminating the need for center drilling, and produces accurate holes and with improved drill life. These factors are particularly important in automated production with CNC machines, in which the usual practice is to use a *spot drill*. To keep the drill more centered, the point angles of the spot drill and of the drill are matched. Other alternatives for minimizing walking of the drill bit are to use a centering punch, to produce an initial impression where drilling starts, or else to directly incorporate dimples or other features into the cast or forged workpieces.

Drilling Recommendations.   Recommended ranges for drilling speeds and feeds are given in Table 23.12. The speed is the *surface speed* of the drill at its periphery; thus, a 0.5-in. (12.7-mm) drill rotating at 300 rpm has a surface speed of

$$V = \left(\frac{0.5}{2} \text{ in.}\right)(300 \text{ rev/min})(2\pi \text{ rad/rev})\left(\frac{1}{12} \text{ ft/in.}\right) = 39 \text{ ft/min} = 12 \text{ m/min.}$$

In drilling holes smaller than 1 mm (0.040 in.) in diameter, rotational speeds can range up to 30,000 rpm, depending on the workpiece material. The *feed* in drilling is the distance the drill travels into the workpiece per revolution. For example, Table 23.11 recommends that, for most workpiece materials, a drill 1.5 mm (0.060 in.) in diameter should have a feed of 0.025 mm/rev. If the speed column in the table indicates that the drill should rotate at, say, 2000 rpm, then the drill should travel into the workpiece at a linear speed of (0.025 mm/rev)(2000 rev/min) = 50 mm/min = 2 in./min.

Chip removal during drilling can be difficult, especially for deep holes in soft and ductile workpiece materials. The drill should be retracted periodically (called *pecking*) to remove chips that may have accumulated along the flutes. Otherwise, the drill may break because of excessive torque, or it may "walk" off location, and thus produce a misshaped hole. A general guide to the probable causes of problems in drilling operations is given in Table 23.13.

Drill Reconditioning.   Drills are *reconditioned* by grinding them, either manually or with special fixtures for better accuracy and productivity. Proper reconditioning of drills is important, particularly with automated manufacturing on CNC machines. Hand grinding is difficult, and requires considerable skill in order to

TABLE 23.12

General Recommendations for Speeds and Feeds in Drilling

			Drill diameter			
			Feed, mm/rev (in./rev)		Speed, rpm	
	Surface speed		1.5 mm (0.060 in.)	12.5 mm (0.5 in.)	1.5 mm (0.060 in.)	12.5 mm (0.5 in.)
Workpiece material	m/min	ft/min				
Aluminum alloys	30–120	100–400	0.025 (0.001)	0.30 (0.012)	6400–25,000	800–3000
Magnesium alloys	45–120	150–400	0.025 (0.001)	0.30 (0.012)	9600–25,000	1100–3000
Copper alloys	15–60	50–200	0.025 (0.001)	0.25 (0.010)	3200–12,000	400–1500
Steels	20–30	60–100	0.025 (0.001)	0.30 (0.012)	4300–6400	500–800
Stainless steels	10–20	40–60	0.025 (0.001)	0.18 (0.007)	2100–4300	250–500
Titanium alloys	6–20	20–60	0.010 (0.0004)	0.15 (0.006)	1300–4300	150–500
Cast irons	20–60	60–200	0.025 (0.001)	0.30 (0.012)	4300–12,000	500–1500
Thermoplastics	30–60	100–200	0.025 (0.001)	0.13 (0.005)	6400–12,000	800–1500
Thermosets	20–60	60–200	0.025 (0.001)	0.10 (0.004)	4300–12,000	500–1500

Note: As hole depth increases, speeds and feeds should be reduced; the selection of speeds and feeds also depends on the specific surface finish required.

TABLE 23.13

General Troubleshooting Guide for Drilling Operations	
Problem	Probable causes
Drill breakage	Dull drill, drill seizing in hole because of chips clogging flutes, feed too high, lip relief angle too small
Excessive drill wear	Cutting speed too high, ineffective cutting fluid, rake angle too high, drill burned and strength lost when drill was sharpened
Tapered hole	Drill misaligned or bent, lips not equal, web not central
Oversize hole	Same as previous entry, machine spindle loose, chisel edge not central, side force on workpiece
Poor hole surface finish	Dull drill, ineffective cutting fluid, welding of workpiece material on drill margin, improperly ground drill, improper alignment

produce symmetric cutting edges. Grinding on fixtures is accurate and is done on special computer-controlled grinders. Coated drills also can be recoated.

Measuring Drill Life. Drill life, as well as tap life (see Section 23.7), usually is measured by the number of holes drilled before the drill becomes dull and has to be reconditioned or replaced. *Drill life* can be determined experimentally, by first clamping a block of material on a suitable dynamometer or force transducer, and then drilling a number of holes while recording the torque or thrust force during each successive drilling operation. After a number of holes have been drilled, the torque and force begin to increase, because the drill is now becoming dull. Drill life is the number of holes drilled until this transition of the forces begins. Other techniques, such as monitoring vibration and acoustic emissions (Section 21.5.4), also can be used to determine drill life.

23.5.6   Drilling Machines

The most common drilling machine is the **drill press**, the major components of which are illustrated in Fig. 23.25a. The workpiece is placed on an adjustable table, either by clamping it directly into the slots and holes on the table or by using a vise which itself is clamped to the table. The drill is lowered manually by a handwheel or by power feed at preset rates. Manual feeding requires some skill in judging the appropriate feed rate.

Drill presses usually are designated by the largest workpiece diameter that can be accommodated on the table, and typically range from 150 to 1250 mm (6–50 in.). In order to maintain proper cutting speeds at the cutting edges of drills, the spindle speed on drilling machines has to be adjustable to accommodate different drill sizes. Adjustments are made by means of pulleys, gearboxes, or variable-speed motors.

The types of drilling machines range from simple bench-type drills, used to drill small-diameter holes, to large *radial drills* (Fig. 23.25b), which can accommodate large workpieces. The distance between the column and the spindle center can be as much as 3 m (10 ft). The drill head of *universal drilling machines* can be swiveled to drill holes at an angle. Drilling machines now include numerically controlled three-axis machines, in which the operations are performed automatically and in the desired sequence, with the use of a turret (Fig. 23.26), which can hold several different drilling tools.

Drilling machines with multiple spindles (**gang drilling**) are used for high-production-rate operations, and are capable of drilling, in one cycle, as many as 50 holes of varying sizes, depths, and locations. These machines also are used for

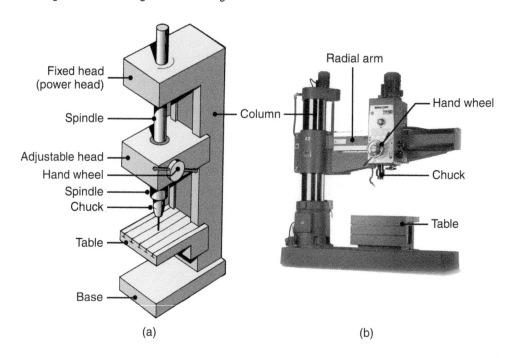

(a)   (b)

FIGURE 23.25   (a) Schematic illustration of a vertical drill press. (b) A radial drilling machine. *Source:* (b) Courtesy of Willis Machinery and Tools.

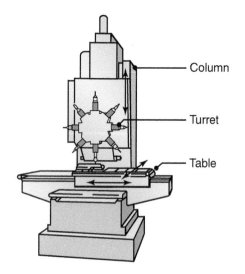

FIGURE 23.26 A three-axis computer numerical-control drilling machine; the turret holds as many as eight different tools, such as drills, taps, and reamers.

reaming and counterboring operations. Gang-drilling machines have largely been replaced with *numerical-control turret drilling machines.*

Work-holding devices for drilling are essential to ensure that the workpiece is located and clamped properly, to keep it from slipping or rotating during drilling. These devices are available in a variety of designs, with important features such as three-point locating for accuracy and three-dimensional work holding for secure fixturing. (See also Section 37.8.)

23.5.7 Design Considerations for Drilling

The basic design guidelines for drilling are:

- Designs should allow holes to be drilled on flat surfaces and perpendicular to the drill motion; otherwise, the drill tends to deflect and the hole will not be located accurately. Exit surfaces for the drill also should be flat.
- Design of hole bottoms should match standard drill-point angles, whenever possible; thus, flat bottoms or odd shapes should be avoided.
- Through holes are preferred over blind holes. If holes with large diameters are specified, the workpiece should have a preexisting hole, preferably made during fabrication of the part, such as by casting, powder metallurgy, or forming.
- Dimples should be provided when preexisting holes are not practical to produce, in order to reduce the tendency of the drill to walk.
- Parts should be designed so that all drilling can be performed with a minimum of fixturing and without the need to reposition the workpiece.

- Blind holes must be drilled deeper than subsequent reaming or tapping operations that may have to be performed.

23.6 Reaming and Reamers

Reaming is an operation used to (a) make an existing hole dimensionally more accurate than can be achieved by drilling alone and (b) improve its surface finish. The most accurate holes in workpieces generally are produced by the following sequence of operations:

1. Centering
2. Drilling
3. Boring
4. Reaming

For even better accuracy and surface finish, holes may be *burnished* or internally *ground* and *honed* (see Sections 26.4 and 26.7).

A *reamer* (Fig. 23.27a) is a multiple-cutting-edge tool, with straight or helically fluted edges that remove very little material. For soft metals, a reamer typically removes a minimum of 0.2 mm (0.008 in.) on the diameter of a drilled hole; for harder metals, about 0.13 mm (0.005 in.) is removed. Attempts to remove smaller layers can be detrimental, as the reamer may be damaged or the hole surface may become burnished (see also Fig. 21.22 as an analogy); in this case, honing would be preferred. In general, reamer speeds are one-half those of the same-size drill and three times the feed rate.

Hand reamers are straight or have a tapered end in the first third of their length. Various *machine reamers*, also called *chucking reamers* because they are mounted in a chuck and operated by a machine, are available in two types: (a) *Rose reamers* have cutting edges with wide margins and no relief (Fig. 23.27a); they remove considerable

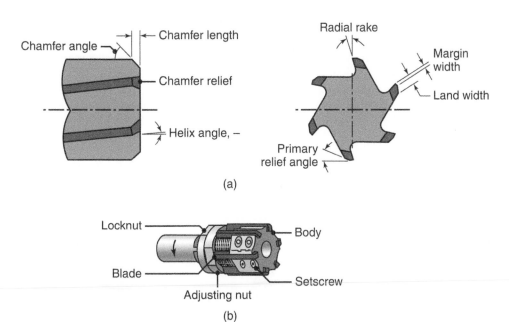

(a)

(b)

FIGURE 23.27   (a) Terminology for a helical reamer. (b) Inserted-blade adjustable reamer.

material and true up a hole for flute reaming. (b) *Fluted reamers* have small margins and relief, with a rake angle of about 5°; they usually are used for light cuts, of about 0.1 mm (0.004 in.) on the hole diameter.

Shell reamers are hollow and are mounted on an arbor and generally are used for holes larger than 20 mm (0.75 in.). *Expansion reamers* are adjustable, for small variations in hole size and also to compensate for wear of the reamer's cutting edges. *Adjustable reamers* (Fig. 23.27b) can be set for specific hole diameters, and are therefore versatile.

Reamers may be held rigidly, as in a chuck, or they may *float* in their holding fixtures to ensure alignment or to be *piloted* in guide bushings, placed above and below the workpiece. A further development in reaming consists of the *dreamer*, a tool that combines drilling and reaming. The tip of the tool first produces a hole by drilling, and the rest of the same tool performs a reaming operation. A similar development involves drilling and tapping in one stroke, using a single tool.

Reamers typically are made of high-speed steels (M1, M2, and M7) or solid carbides (K20, C2), or have carbide cutting edges. Reamer maintenance and reconditioning are important for hole accuracy and surface finish.

23.7 Tapping and Taps

Internal threads in workpieces can be produced by *tapping*, a *tap* being a chip-producing threading tool with multiple cutting teeth (Fig. 23.28a). Taps generally are available with two, three, or four flutes. The most common production tap is the two-flute spiral-point tap. The two-flute tap forces the chips into the hole, so that the tap needs to be retracted only at the end of the cut. Three-fluted taps are stronger, because more material is available in the flute. Tap sizes range up to 100 mm (4 in.), although larger threads can be machined in a mill or machining center (see Fig. 24.2f).

Tapered taps are designed to reduce the torque required for the tapping of through holes. *Bottoming taps* are for tapping blind holes to their full depth. *Collapsible taps* are used in large-diameter holes; after tapping has been completed, the tap is collapsed mechanically and is removed from the hole without having to rotate them in the hole, as do regular taps.

Chip removal can be a significant problem during tapping, because of the small clearances involved. If chips aren't removed properly, the torque increases significantly and can break the tap. The use of a cutting fluid and periodic reversal and

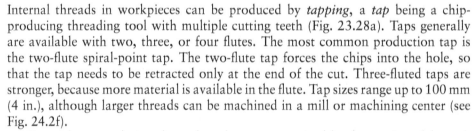

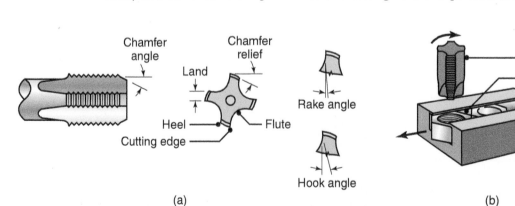

(a)   (b)

FIGURE 23.28   (a) Terminology for a tap. (b) Tapping of steel nuts in production.

removal of the tap from the hole are effective means of chip removal and of improving the quality of the tapped hole. For higher tapping productivity, drilling and tapping can be combined in a single operation (*drapping*) with a single tool. The tool has a drilling section at its tip, followed by a tapping section. Tapping may be done manually or with machines, such as (a) drilling machines, (b) lathes, (c) automatic screw machines, and (d) vertical CNC milling machines that combine the correct relative rotation and the longitudinal feed.

Special tapping machines also are available, with features for multiple tapping operations. Multiple-spindle tapping heads are used extensively, particularly in the automotive industry, where 30–40% of machining operations involve the tapping of holes. One simple system for the automatic tapping of nuts is shown in Fig. 23.28b.

Tap life can be determined with the same technique used to measure drill life. With proper lubrication, tap life may be as high as 10,000 holes. Taps usually are made of high-speed steels (M1, M2, M7, and M10). Productivity in tapping operations can be improved by *high-speed tapping*, with surface speeds as high as 100 m/min (350 ft/min). *Self-reversing* tapping systems are now in use with modern computer-controlled machine tools. Operating speeds can be as high as 5000 rpm, although actual cutting speeds in most applications are considerably lower. Cycle times typically are on the order of 1–2 s.

Some tapping systems now have capabilities for directing the cutting fluid to the cutting zone through the spindle and a hole in the tap, which also helps flush the chips out of the hole being tapped. **Chipless tapping** is a process of internal thread rolling using a forming tap, described in Section 13.5.

CASE STUDY 23.2   Bone Screw Retainer

A cervical spine implant is shown in Fig. 23.29a. In the event that a patient requires cervical bone fusion at one or more vertebral levels, this implant can act as an internal stabilizer by decreasing the amount of motion in the region, and thereby help promote a successful fusion. The plate is affixed to the anterior aspect of the spine, using bone screws that go through the plate and into the bone. The undersurface of the plate has a very rough surface, that helps hold the plate in place while the bone screws are being inserted.

One concern with this type of implant is the possibility of the bone screws loosening with time, due to normal and repetitive loading from the patient. In extreme cases, this can result in a screw backing out, with the head of the screw no longer flush with the plate, a condition that obviously is undesirable. The implant described here uses a retainer to prevent the bone screw from backing out away from the plate, as shown in the left half of Fig. 23.29b.

The retainer has a number of design features that are essential for it to function correctly, and without complicating the surgical procedure. To ease its use in surgery, the plate is provided with the retainers already in place, with the circular notches aligned with the bone screw holes. This arrangement allows the surgeon to insert the bone screws without interference from the retainer. Once the screws are inserted, the surgeon turns the retainer a few degrees so that each screw head is then captured. In order to ensure the retainer's proper orientation in the plate, the thread of its shank must start in the same axial location as point *S* in Fig. 23.29b.

The manufacturing steps followed to produce this part are shown in Fig. 23.29b. First, a 0.5-in. diameter Ti-6Al-4V rod is placed in a CNC lathe and faced. Then the threaded area is turned to the diameter necessary to machine the threads. The thread is turned on the shank, but over a longer length than is ultimately required, because of difficulties in obtaining high-quality threads at the start of machining. The cap then is turned to the required diameter, and a 0.10-in. radius is machined on the underside of the head. The part is removed, inspected, and placed in another CNC lathe, where it is faced to the specified length. The spherical radius in the cap is then machined, the center hole is drilled, and the hex head is broached. The cap is removed and inspected, and if the desired length has not been achieved, the cap is lapped (Section 26.7) to the final dimension.

(*continued*)

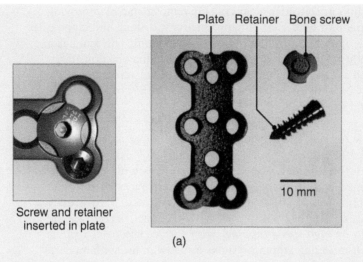

Plate   Retainer   Bone screw

Screw and retainer
inserted in plate

10 mm

(a)

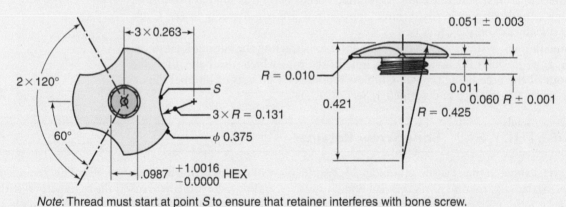

0.051 ± 0.003

←3 × 0.263→

2 × 120°

$R = 0.010$

S

3 × $R = 0.131$

ϕ 0.375

60°

0.421

0.011

0.060 R ± 0.001

$R = 0.425$

.0987 $^{+1.0016}_{-0.0000}$ HEX

Note: Thread must start at point *S* to ensure that retainer interferes with bone screw.

(b)

FIGURE 23.29   (a) A cervical spine implant, showing system components; (b) dimensions of the bone screw (all dimensions in inches).

At this point, the retainer is placed in a CNC milling machine, using a specially designed fixture that consists basically of a tapered and threaded hole. By carefully applying a predetermined torque on the retainer, when placing it into the fixture, the starting location of the threads can be controlled accurately. Once the cap is located in the fixture, the three circular notches are machined as per the drawing. The retainer is then deburred by tumbling to remove all sharp corners, and the bottom is grit blasted to match that of the underside of the plate. Finally, the parts are anodized (Section 34.10) and passivated to obtain the desired biocompatibility.

Source: Courtesy of J. Mankowski and B. Pyszka, Master Metal Engineering Inc., and C. Lyle and M. Handwerker, Wright Medical Technology, Inc.

SUMMARY

- Machining processes that typically produce external and internal circular profiles are turning, boring, drilling, and tapping. Because of the three-dimensional nature of these operations, chip movement and its control are important considerations.

Chip removal can be a significant problem, especially in drilling and tapping, and can lead to tool breakage.

- Optimization of each machining operation requires an understanding of the inter-relationships among design parameters (part shape, dimensional accuracy, and surface finish) and process parameters (cutting speed, feed, and depth of cut), tool material and shape, the use of cutting fluids, and the sequence of operations to be performed.
- The parts to be machined may have been produced by casting, forging, extrusion, or powder metallurgy. The closer the blank to be machined to the final shape desired (near-net shape), the fewer the number and extent of the subsequent machining processes required.

KEY TERMS

Automatic bar machine	Drill press	Knurling	Relief angle
Back rake angle	Dry machining	Lathes	Roughing cuts
Bed	Engine lathe	Lead screw	Screw threads
Boring	Face plate	Mandrel	Side rake angle
Boring mill	Facing	Material-removal rate	Tailstock
Carriage	Feed force	Nose radius	Tapping
Chip management	Feed rod	Parting	Threading
Chuck	Finishing cuts	Power chuck	Trepanning
Collet	Form tools	Rake angle	Turning
Cutting-edge angle	Gun drilling	Reamer	Turret lathe
Drilling	Headstock	Reaming	Twist drill
Drill life	Hole making	Reconditioning	

BIBLIOGRAPHY

ASM Handbook, Vol. 16: **Machining**, ASM International, 1989.

Boothroyd, G., and Knight, W.A., **Fundamentals of Machining and Machine Tools**, 3rd ed., Marcel Dekker, 2005.

Brown, J., **Advanced Machining Technology Handbook**, McGraw-Hill, 1998.

Byers, J.P. (ed.), **Metalworking Fluids**, 2nd ed., CRC Press, 2006.

Hoffman, E.G., **Jigs and Fixture Design**, 5th ed., Industrial Press, 2003.

Joshi, P.H., **Machine Tools Handbook**, McGraw-Hill, 2008.

Knight, W.A., and Boothroyd, G., **Fundamentals of Metal Machining and Machine Tools**, 3rd ed., Marcel Dekker, 2006.

Krar, S.F., and Check, A.F., **Technology of Machine Tools**, 6th ed., Glencoe Macmillan/McGraw-Hill, 2009.

Lopez, L.N., and Lamikiz, A. (eds.), **Machine Tools for High Performance Machining**, Springer, 2009.

Stout, K.J., Davis, E.J., and Sullivan, P.J., **Atlas of Machined Surfaces**, Chapman and Hall, 1990.

Walsh, R.A., **McGraw-Hill Machining and Metalworking Handbook**, 3rd ed., McGraw-Hill, 2006.

REVIEW QUESTIONS

23.1 Describe the types of machining operations that can be performed on a lathe.

23.2 What is turning? What kind of chips are produced by turning?

23.3 What is the thrust force in turning? What is the cutting force? Which is used to calculate the power required?

23.4 What are the components of a lathe?

23.5 (a) What is a tracer lathe? (b) What is an automatic bar machine?

23.6 Describe the operations that can be performed on a drill press.

23.7 Why were power chucks developed?

23.8 Explain why operations such as boring on a lathe and tapping are difficult.

23.9 Why are turret lathes typically equipped with more than one turret?

23.10 Describe the differences between boring a workpiece on a lathe and boring it on a horizontal boring mill.

23.11 How is drill life determined?

23.12 What is the difference between a conventional drill and a gun drill?

23.13 Why are reaming operations performed?

23.14 Explain the functions of the saddle on a lathe.

23.15 Describe the relative advantages of (a) self-opening and (b) solid-die heads for threading.

23.16 Explain how external threads are cut on a lathe.

23.17 What is the difference between a blind hole and a through hole? What is the significance of that difference?

QUALITATIVE PROBLEMS

23.18 Explain the reasoning behind the various design guidelines for turning.

23.19 Note that both the terms "tool strength" and "tool-material strength" have been used in the text. Do you think there is a difference between them? Explain.

23.20 (a) List and explain the factors that contribute to poor surface finish in the processes described in this chapter. (b) List the advantages and disadvantages of turning versus cold extruding a shaft.

23.21 Explain why the sequence of drilling, boring, and reaming produces a hole that is more accurate than drilling and reaming it only.

23.22 Why would machining operations be necessary even on net-shape or near-net-shape parts made by precision casting, forming, or powder metallurgy products, as described in preceding chapters? Explain.

23.23 A highly oxidized and uneven round bar is being turned on a lathe. Would you recommend a small or a large depth of cut? Explain.

23.24 Describe the difficulties that may be encountered in clamping a workpiece made of a soft metal in a three-jaw chuck.

23.25 (a) Does the force or torque in drilling change as the hole depth increases? Explain. (b) Drills usually have two flutes. Explain why.

23.26 Explain the similarities and differences in the design guidelines for turning and for boring.

23.27 Describe the advantages and applications of having a hollow spindle in the headstock of a lathe.

23.28 Assume that you are asked to perform a boring operation on a large-diameter hollow workpiece. Would you use a horizontal or a vertical boring mill? Explain.

23.29 Explain the reasons for the major trend that has been observed in producing threads by thread rolling as opposed to thread cutting. What would be the differences, if any, in the types of threads produced and in their performance characteristics?

23.30 Describe your observations concerning the contents of Tables 23.2 and 23.4, and explain why those particular recommendations are made.

23.31 The footnote to Table 23.12 states that as the hole diameter increases, speeds and feeds in drilling should be reduced. Explain why.

23.32 In modern manufacturing, which types of metal chips would be undesirable and why?

23.33 Sketch the tooling marks you would expect if a part was (a) turned; (b) reduced in diameter with a straight form tool; and (c) extruded.

23.34 What concerns would you have in turning a powder metal part, such as a shaft made by the Osprey process (see Fig. 17.21)?

23.35 The operational severity for reaming is much lower than that for tapping, even though they both are internal machining processes. Why?

23.36 Review Fig. 23.6, and comment on the factors involved in determining the height of the zones (cutting speed) for various tool materials.

23.37 Explain how gun drills remain centered during drilling. Why is there a hollow, longitudinal channel in a gun drill?

23.38 Comment on the magnitude of the wedge angle on the tool shown in Fig. 23.4.

23.39 If inserts are used in a drill bit (see Fig. 23.22), how important is the shank material? If so, what properties are important? Explain.

23.40 Refer to Fig. 23.11b, and in addition to the tools shown, describe other types of cutting tools that can be placed in toolholders to perform other machining operations.

QUANTITATIVE PROBLEMS

23.41 Calculate the same quantities as in Example 23.1 for high-strength titanium alloy and at $N = 700$ rpm.

23.42 Estimate the machining time required to rough turn a 0.50-m-long annealed copper-alloy round bar, from a 60-mm diameter to a 58-mm diameter, using a high-speed steel tool.

(See Table 23.4.) Estimate the time required for an uncoated carbide tool.

23.43 A high-strength cast-iron bar 8 in. in diameter is being turned on a lathe at a depth of cut $d = 0.050$ in. The lathe is equipped with a 15-hp electric motor and has a mechanical efficiency of 80%. The spindle speed is 500 rpm. Estimate the maximum feed that can be used before the lathe begins to stall.

23.44 A 0.30-in.-diameter drill is used on a drill press operating at 300 rpm. If the feed is 0.005 in./rev, what is the MRR? What is the MRR if the drill diameter is doubled?

23.45 In Example 23.4, assume that the workpiece material is high-strength aluminum alloy and the spindle is running at $N = 500$ rpm. Estimate the torque required for this operation.

23.46 For the data in Problem 23.45, calculate the power required.

23.47 A 6-in.-diameter aluminum cylinder 10 in. in length is to have its diameter reduced to 4.5 in. Using the typical machining conditions given in Table 23.4, estimate the machining time if a TiN-coated carbide tool is used.

23.48 A lathe is set up to machine a taper on a bar stock 120-mm in diameter; the taper is 1 mm per 10 mm. A cut is made with an initial depth of cut of 4 mm, at a feed rate of 0.250 mm/rev and at a spindle speed of 150 rpm. Calculate the average metal removal rate.

23.49 Assuming that the coefficient of friction is 0.25, calculate the maximum depth of cut for turning a hard aluminum alloy on a 20-hp lathe (with a mechanical efficiency of 80%) at a width of cut of 0.25 in., rake angle of 0°, and a cutting speed of 300 ft/min. What is your estimate of the material's shear strength?

23.50 A 3-in.-diameter, gray cast iron cylindrical part is to be turned on a lathe at 500 rpm. The depth of cut is 0.25 in. and the feed is 0.02 in./rev. What minimum horsepower is required for this operation?

23.51 Assume that you are an instructor covering the topics described in this chapter, and you are giving a quiz on the numerical aspects to test the understanding of the students. Prepare two quantitative problems and supply the answers.

SYNTHESIS, DESIGN, AND PROJECTS

23.52 Would you consider the machining processes described in this chapter as net-shape processes, thus requiring no further processing? Near-net-shape processing? Explain with appropriate examples.

23.53 Would it be difficult to use the machining processes described in this chapter on various soft nonmetallic or rubberlike materials? Explain your thoughts, commenting on the role of the physical and mechanical properties of such materials with respect to the machining operation, and any difficulties that may be encountered in producing the desired shapes and dimensional accuracies.

23.54 If a bolt breaks in a hole, it typically is removed by first drilling a hole in the bolt shank and then using a special tool to remove the bolt. Inspect such a tool and explain how it functions.

23.55 An important trend in machining operations is the increased use of flexible fixtures. Conduct a search on the Internet regarding these fixtures, and comment on their design and operation.

23.56 Review Fig. 23.8d, and explain if it would be possible to machine eccentric shafts, such as that shown in Fig. 23.13c, on the setup illustrated. What if the part is long compared with its cross-section? Explain.

23.57 Boring bars can be designed with internal damping capabilities, to reduce or eliminate vibration and chatter during machining (see Fig. 23.18). Referring to the technical literature, describe details of designs for such boring bars.

23.58 A large bolt is to be produced from extruded hexagonal bar stock by placing the hex stock into a chuck and machining the shank of the bolt by turning it on a lathe. List and explain the difficulties that may be involved in this operation.

23.59 Make a comprehensive table of the process capabilities of the machining operations described in this chapter. Using several columns, describe the machine tools involved, type of cutting tools and tool materials used, shapes of parts produced, typical maximum and minimum sizes, surface finish, dimensional tolerances, and production rates.

24

Machining Processes: Milling, Broaching, Sawing, Filing, and Gear Manufacturing

- This chapter begins with milling, one of the most versatile machining processes, in which a rotating cutter removes material while traveling along a desired path with respect to the workpiece.
- Several other machining processes are then described, such as planing, shaping, and broaching, in which either the cutting tool or the workpiece travels along a straight path, producing flat or profiled surfaces.
- Sawing processes are then covered, generally used for preparing blanks for subsequent operations, such as forming, welding, and machining. Also briefly discussed is filing, used to remove small amounts of material, usually from edges and corners.
- The chapter ends with descriptions of gear manufacturing by machining, the special cutters used, the equipment involved, and the quality and properties of the gears produced.

Typical parts made: Parts with complex external and internal features, splines, and gears.

Alternative processes: Die casting, precision casting, precision forging, powder metallurgy, powder-injection molding, creep-feed grinding, electrical discharge machining, rapid prototyping, and fabrication.

24.1 Introduction

In addition to producing parts with various external or internal round profiles, as described in Chapter 23, machining operations can produce many other complex shapes (Fig. 24.1). Although processes such as die casting, precision forging, and powder metallurgy also can produce parts with close tolerances and fine surface finish, it is often necessary to perform complex machining operations to satisfy design requirements and specifications.

The preceding chapter described machining processes that produce round shapes. While the processes and machinery covered in this chapter also can produce round, axisymmetric shapes, it is generally advisable to use the processes of Chapter 23 whenever possible, since the equipment is simpler, less expensive, and the processes are easier to set up and perform. This, however, cannot always be done, so the machining operations described in this chapter are often essential. In this chapter,

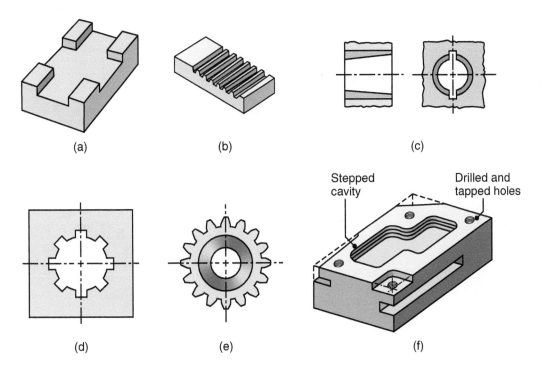

FIGURE 24.1 Typical parts and shapes that can be produced with the machining processes described in this chapter.

several cutting processes and machine tools that are capable of producing these shapes, using single-point, multitooth, and profiled cutting tools, are described (see also Table 23.1).

24.2   Milling and Milling Machines

Milling includes a number of highly versatile machining operations taking place in a variety of configurations (Fig. 24.2), with the use of a **milling cutter**, a multitooth cutter tool that produces a number of chips in one revolution (Fig. 24.3).

24.2.1   Peripheral Milling

In *peripheral milling*, also called *plain milling*, the axis of cutter rotation is parallel to the workpiece surface, as shown in Fig. 24.4. The cutter body, which generally is made of high-speed steel (Section 22.2), has a number of teeth along its circumference; each tooth acts like a single-point cutting tool. When the cutter is longer than the width of the cut, the operation is called **slab milling**.

Cutters for peripheral milling may have either *straight* or *helical teeth*, resulting in an orthogonal or oblique cutting action, respectively (see also Fig. 21.9). Helical teeth generally are preferred over straight teeth, because a tooth is always partially engaged with the workpiece as the cutter rotates. Consequently, the cutting force and the torque on the cutter are lower, resulting in a smoother milling operation and reduced chatter.

Conventional Milling and Climb Milling. Note in Fig. 24.5a that the cutter rotation can be either clockwise or counter-clockwise; this is significant in the milling

QR Code 24.1 Pocket Milling. (*Source:* Courtesy of Sandvik Coromant)

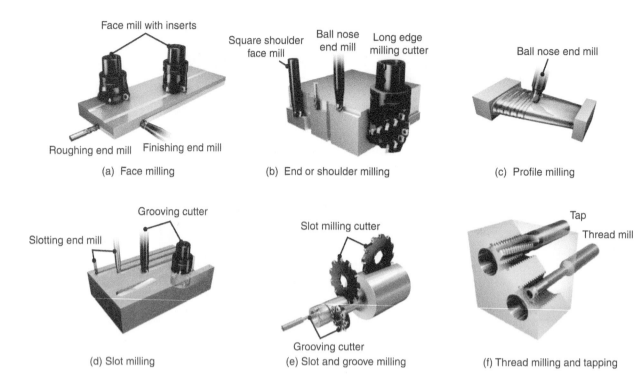

FIGURE 24.2   Some basic types of milling cutters and milling operations. *Source:* Courtesy of Sandvik Coromant.

FIGURE 24.3   The cutting action of a milling cutter using a number of inserts to remove metal. *Source:* Courtesy of Sandvik Coromant.

operation. In *conventional milling*, also called *up milling*, the maximum chip thickness is at the *end* of the cut as the tooth leaves the workpiece surface. Consequently, (a) tooth engagement is not a function of workpiece surface characteristics and (b) contamination or scale (oxide layer) on the surface does not adversely affect tool life. This is the more common method of milling, where the cutting operation is smooth. However, the cutter teeth must be sharp, as otherwise the tooth will rub against the surface being milled and smear it for some distance before it begins to engage and cut. There may also be a tendency for the cutter to chatter (Section 25.4), and the workpiece has a tendency to be pulled *upward* (because of the cutter rotation direction), thus necessitating proper clamping of the workpiece on the table of the machine.

In *climb milling*, also called *down milling*, cutting starts at the surface of the workpiece, where the chip is thickest. The advantage of this method is that the direction of rotation of the cutter will push the workpiece *downward*, thus holding the workpiece in place, a factor particularly important for slender parts. However, because of the resulting impact force when a tooth engages the workpiece, this operation must have a rigid work-holding setup, and gear backlash in the table feed mechanism must be eliminated. Climb milling is not suitable for machining of workpieces with surface scale, such as metals that have been hot worked, forged, or cast. The scale is hard and abrasive, and thus causes excessive wear and damage to the cutter teeth, and shortening tool life.

Milling Parameters.   The cutting speed, V, in peripheral milling is the surface speed of the cutter, or

$$V = \pi DN, \tag{24.1}$$

where D is the cutter diameter and N is the rotational speed of the cutter (Fig. 24.5).

Note from Fig. 24.3b that the thickness of the chip in slab milling will vary along its length, because of the relative longitudinal motion between the cutter and the workpiece. For a straight-tooth cutter, the approximate *undeformed chip thickness* (also called *chip depth of cut*), t_c, can be calculated from the equation

$$t_c = 2f\sqrt{\frac{d}{D}}, \tag{24.2}$$

where f is the feed per tooth of the cutter (the distance the workpiece travels per tooth of the cutter, in mm/tooth or in./tooth), and d is the depth of cut. As t_c becomes larger, the force on the cutter tooth will increase.

Feed per tooth is determined from the equation

$$f = \frac{v}{Nn}, \tag{24.3}$$

where v is the linear speed (also called *feed rate*) of the workpiece and n is the number of teeth on the cutter periphery.

The cutting time, t, is given by the equation

$$t = \frac{l + l_c}{v}, \tag{24.4}$$

where l is the length of the workpiece (Fig. 24.5c) and l_c is the horizontal extent of the cutter's first contact with the workpiece. Based on the assumption that $l_c \ll l$ (although this generally is not the case), the *material-removal rate* (MRR) is

$$\mathrm{MRR} = \frac{lwd}{t} = wdv, \tag{24.5}$$

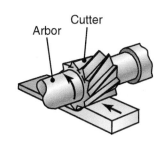

FIGURE 24.4   Schematic illustration of peripheral milling.

Video Solution 24.1 Time Required for Milling Operations

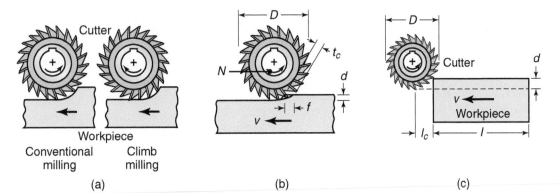

(a)   (b)   (c)

FIGURE 24.5   (a) Schematic illustration of conventional milling and climb milling. (b) Slab-milling operation showing depth of cut, d; feed per tooth, f; chip depth of cut, t_c, and workpiece speed, v. (c) Schematic illustration of cutter travel distance, l_c, to reach full depth of cut.

TABLE 24.1

Summary of Peripheral Milling Parameters and Formulas

N = Rotational speed of the milling cutter, rpm
F = Feed, mm/tooth or in./tooth
D = Cutter diameter, mm or in.
n = Number of teeth on cutter
v = Linear speed of the work piece or feed rate, mm/min or in./min
V = Surface speed of cutter, m/min or ft/min
   $= \pi DN$
f = Feed per tooth, mm/tooth or in./tooth
   $= v/Nn$
l = Length of cut, mm or in.
t = Cutting time, s or min
   $= (l + l_c)/v$, where l_c = extent of the cutter's first contact with the workpiece
MRR = mm³/min or in³/min
   $= wdv$, where w is the width of cut
Torque = N-m or lb-ft
   $= F_c D/2$
Power = kW or hp
   $= (\text{Torque})(\omega)$, where $\omega = 2\pi N$ radians/min

where w is the width of the cut, which, in slab milling, is equal to the width of the workpiece. The distance that the cutter travels in the *noncutting cycle* of the milling operation is an important economic consideration, and should be minimized by such means as faster travel of the machine tool components. The foregoing equations and the terminology used are summarized in Table 24.1.

The *power requirement* in peripheral milling can be measured and also calculated, but the *forces* acting on the cutter (tangential, radial, and axial; see also Fig. 23.5) are difficult to calculate because of the numerous variables involved, many of which pertain to the cutter geometry. These forces can be measured experimentally for a variety of milling conditions, and the *torque* on the cutter spindle can be calculated from the power (see Example 24.1). Although the torque is the product of the cutter radius and tangential force, the tangential force per tooth will depend on how many teeth are engaged at any moment during the cut.

EXAMPLE 24.1   Material-removal Rate, Power, Torque, and Cutting Time in Slab Milling

Given: A slab-milling operation is being carried out on a 12-in.-long, 4-in.-wide annealed mild-steel block at a feed $f = 0.01$ in./tooth and a depth of cut $d = \frac{1}{8}$ in. The cutter has a diameter of $D = 2$ in., 20 straight teeth, speed of $N = 100$ rpm, and, by definition, is wider than the block to be machined.

Find: Calculate the material-removal rate, estimate the power and torque required for this operation, and calculate the cutting time.

Solution: From the information given, the linear speed of the workpiece, v, can be calculated from Eq. (24.3):

$$v = fNn = (0.01)(100)(20) = 20 \text{ in./min.}$$

From Eq. (24.5), the material-removal rate is calculated to be

$$\text{MRR} = (4)\left(\frac{1}{8}\right)(20) = 10 \text{ in}^3/\text{min.}$$

Since the workpiece is annealed mild steel, the unit power is estimated from Table 21.2 as 1.1 hp-min/in³. Thus, the power required can be estimated as

$$\text{Power} = (1.1)(10) = 11 \text{ hp.}$$

The torque acting on the cutter spindle also can be calculated by noting that power is the product of torque and the spindle rotational speed (in radians per unit time). Therefore,

$$\text{Torque} = \frac{\text{Power}}{\text{Rotational speed}}$$

$$= \frac{(11 \text{ hp})(33,000 \text{ lb-ft/min-hp})}{(100 \text{ rpm})(2\pi)}$$

$$= 578 \text{ lb-ft.}$$

The cutting time is given by Eq. (24.4), in which the quantity l_c can be shown, from simple geometric relationships and for $D \gg d$, to be approximately equal to

$$l_c = \sqrt{Dd} = \sqrt{(2)\left(\frac{1}{8}\right)} = 0.5 \text{ in.}$$

Thus, the cutting time is

$$t = \frac{12 + 0.5}{20} = 0.625 \text{ min} = 37.5 \text{ s.}$$

24.2.2 Face Milling

In *face milling*, the cutter is mounted on a spindle having an axis of rotation perpendicular to the workpiece surface (Fig. 24.6b), and removes material in the manner shown in Fig. 24.6a. The cutter rotates at a rotational speed N, and the workpiece moves along a straight path and at a linear speed v. When the direction of cutter rotation is as shown in Fig. 24.6b, the operation is climb milling; when it is in the opposite direction (Fig. 24.6c), it is conventional milling. The cutting teeth, such as carbide inserts, are mounted on the cutter body as shown in Fig. 24.7 (see also Fig. 22.3c).

Because of the relative motion between the cutter teeth and the workpiece, face milling leaves *feed marks* on the machined surface (Fig. 24.8), similar to those left by turning operations as shown in Fig. 21.2. Note that the surface roughness of the workpiece depends on the corner geometry of the insert and the feed per tooth.

Video Solution 24.2 Mechanics of Face Milling Operations

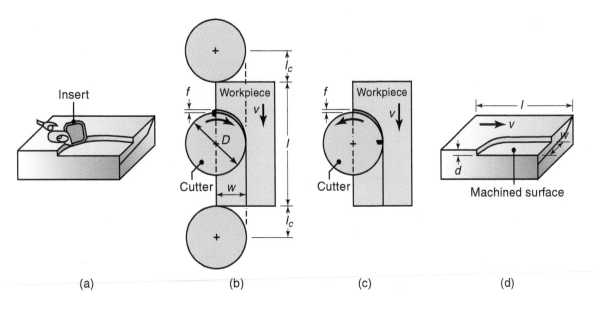

| (a) | (b) | (c) | (d) |

FIGURE 24.6 (a) Face-milling operation (with cutter removed) showing the action of a single insert; (b) climb milling; (c) conventional milling; and (d) dimensions in face milling. Note that the width of cut, w, is not necessarily the same as the cutter radius.

FIGURE 24.7 A face-milling cutter with indexable inserts. *Source:* Courtesy of Ingersoll Cutting Tool Company.

The terminology for a face-milling cutter, as well as the various angles, are shown in Fig. 24.9. As can be seen from the side view of the insert, in Fig. 24.10, the *lead angle* of the insert in face milling has a direct influence on the *undeformed chip thickness*, as it does in turning operations (see Fig. 23.3). As the lead angle (positive, as shown in Fig. 24.10b) increases, the undeformed chip thickness decreases (as does the chip thickness), and the length of contact, and hence chip width, increases. Note, however, that the cross-sectional area of the undeformed chip remains constant. The *lead angle* also influences the forces in milling. It can be seen that as the lead angle decreases, there is a smaller vertical-force component (i.e., the axial force on the cutter spindle). The lead angles for most face-milling cutters typically range from 0° to 45°.

A wide variety of milling cutters and inserts are available (see Figs. 24.2 and 24.11). The cutter diameter should be chosen such that it will not interfere with fixtures, work-holding devices, and other components in the setup. In a typical face-milling operation, the ratio of the cutter diameter, D, to the width of cut, w, should be no less than 3:2.

The relationship of cutter diameter to insert angles, and their position relative to the surface to be milled, is important, in that it will determine the angle at which an insert *enters* and *exits* the workpiece. Note in Fig. 24.6b for climb milling that, if the insert has zero axial and radial rake angles (see Fig. 24.9), the rake face of the insert engages the workpiece directly. As seen in Figs. 24.11a and b, however, the same insert may engage the workpiece at different angles, depending on the relative positions of the cutter and the workpiece width.

Note in Fig. 24.11a that the tip of the insert makes the first contact, thus there is a possibility for the cutting edge to chip off. In Fig. 24.11b, on the other hand, the first contacts (at entry, reentry, and the two exits) are at an angle and away from the tip of the insert; consequently, there is a lower tendency for the insert to fail, because the forces on the insert vary more slowly. Note from Fig. 24.9 that the radial and axial rake angles also will have an effect on this operation.

Figure 24.11c shows the exit angles for various cutter positions. Note that in the first two examples, the insert exits the workpiece at an angle, thus causing the force on the insert to reduce to zero at a slower rate (desirable for longer tool life) than in the third example, where the insert exits the workpiece suddenly (undesirable).

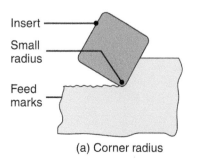

(a) Corner radius

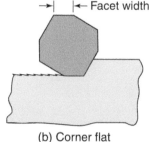

(b) Corner flat

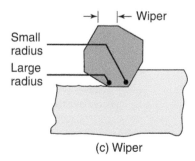

(c) Wiper

FIGURE 24.8 Schematic illustration of the effect of insert shape on feed marks on a face-milled surface: (a) small corner radius; (b) corner flat on insert; and (c) wiper, consisting of a small radius followed by a larger radius, resulting in smoother feed marks. (d) Feed marks due to various insert shapes.

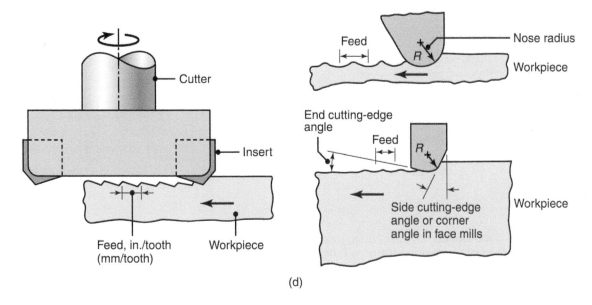

(d)

FIGURE 24.8 (*continued*)

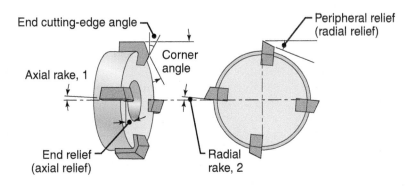

FIGURE 24.9 Terminology for a face-milling cutter.

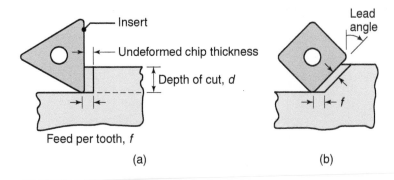

(a)   (b)

FIGURE 24.10 The effect of the lead angle on the undeformed chip thickness in face milling. Note that as the lead angle increases, the chip thickness decreases, but the length of contact (i.e., chip width) increases. The edges of the insert must be sufficiently long to accommodate the increase in contact length.

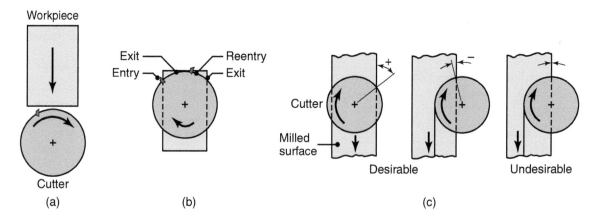

FIGURE 24.11   (a) Relative position of the cutter and insert as they first engage the workpiece in face milling. (b) Insert positions toward the end of cut. (c) Examples of exit angles of the insert, showing desirable (positive or negative angle) and undesirable (zero angle) positions. (In all figures, the cutter spindle is perpendicular to the page.)

EXAMPLE 24.2   Material-removal Rate, Power Required, and Cutting Time in Face Milling

Given: Refer to Fig. 24.6 and assume that $D = 150$ mm, $w = 60$ mm, $l = 500$ mm, $d = 3$ mm, $v = 0.6$ m/min, and $N = 100$ rpm. The cutter has 10 inserts, and the workpiece material is a high-strength aluminum alloy.

Find: Calculate the material-removal rate, cutting time, and feed per tooth, and estimate the power required.

Solution: First note that the cross-section of the cut is $wd = (60)(3) = 180$ mm^2. Then, noting that the workpiece speed, v, is 0.6 m/min = 600 mm/min, the material-removal rate (MRR) can be calculated as

$$\text{MRR} = (180)(600) = 108{,}000 \text{ mm}^3/\text{min.}$$

The cutting time is given by

$$t = \frac{l + 2l_c}{v}.$$

Note from Fig. 24.6 that, for this problem, $l_c = \frac{D}{2} = 75$ mm. The cutting time is therefore

$$t = \frac{500 + 150}{10} = 65 \text{ s} = 1.08 \text{ min.}$$

The feed per tooth can be obtained from Eq. (24.3), where $N = 100$ rpm = 1.67 rev/s, and hence

$$f = \frac{10}{(1.67)(10)} = 0.6 \text{ mm/tooth.}$$

For this material, let's estimate the unit power from Table 21.2 to be 1.1 Ws/mm^3; thus, the power is

$$\text{Power} = (1.1)(1800) = 1980 \text{ W} = 1.98 \text{ kW.}$$

24.2.3   End Milling

End milling is an important and common machining operation, because of its versatility and capability to produce various profiles and curved surfaces. The cutter, called an **end mill** (Fig. 24.12), has either a straight shank (for small cutter sizes) or a tapered shank (for larger sizes), and is mounted into the spindle of the milling machine. End mills may be made of high-speed steel, solid carbide, or with coated or uncoated carbide inserts, similar to those for face milling. The cutter usually rotates on an axis perpendicular to the workpiece surface, but it also can be tilted to conform to machine-tapered or curved surfaces.

End mills are available with hemispherical ends (*ball nose mills*; see Fig. 24.13) for the production of sculptured surfaces, such as on dies and molds; they can also be produced with a specific radius, a flat end, or with a chamfer. *Hollow end mills* have internal cutting teeth, and are used to machine the cylindrical surfaces of solid, round workpieces. End milling can produce a variety of surfaces at any depth, such as curved, stepped, and pocketed (Fig. 24.2f). The cutter can remove material on both its end and on its cylindrical cutting edges.

Vertical-spindle and horizontal-spindle milling machines (see Section 24.2.8), as well as machining centers (see Fig. 25.7), can all be used for end-milling workpieces of various sizes and shapes. The machines can be programmed such that the cutter can follow a complex set of paths in order to optimize the whole machining operation for higher productivity and at minimum cost.

High-speed End Milling.   *High-speed end milling* has become an important process, with numerous applications, such as the milling of large aluminum-alloy aerospace components and honeycomb structures. (See also *high-speed machining*, Section 25.5.) With spindle speeds in the range from 20,000 to 80,000 rpm, the machines must have high stiffness, usually requiring hydrostatic or air bearings, as well as high quality work-holding devices. The spindles have a rotational accuracy of 10 μm; thus the workpiece surfaces produced have very high accuracy. At such high rates of material removal, chip collection and disposal can be a significant problem, as discussed in Section 23.3.7.

Producing cavities in metalworking dies (called **die sinking**, such as in forging or in sheet-metal forming) also is done by high-speed end milling, often using TiAlN-coated ball-nose end mills (Fig. 24.13). The machines generally have *four-axis* or *five-axis* movement capabilities (see, for example, Fig. 24.21), but machining centers (Section 25.2) can add more axes for more complex geometries. Such machines are able to accommodate dies as large as 3 m × 6 m (9 ft × 18 ft) and weighing 60 tons, and costing over $2 million. The advantages of five-axis machines are that they (a) are capable of machining very complex shapes and in a single setup, (b) can use shorter cutting tools, thus reducing the tendency for vibration and chatter, and (c) enable drilling of holes at various compound angles.

FIGURE 24.12   A selection of end mills; the number of teeth and helix angle are selected based on whether a roughing or finishing cut will be made. *Source:* Courtesy of Kennametal Inc.

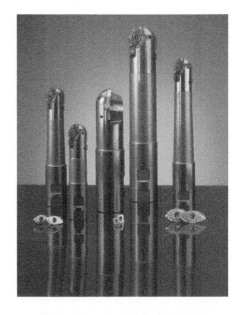

FIGURE 24.13   Ball nose end mills; these cutters can produce elaborate contours and are often used in the machining of dies and molds. (See also Fig. 24.2d.) *Source:* Courtesy of Dijet, Inc.

QR Code 24.2 Milling with ball nose end mills. (*Source:* Courtesy of Sandvik Coromant)

24.2.4   Other Milling Operations and Milling Cutters

Several other milling operations and cutters are used to machine workpieces. In **straddle milling**, two or more cutters are mounted on an arbor, and are used to simultaneously machine two parallel surfaces (Fig. 24.14a). **Form milling** produces

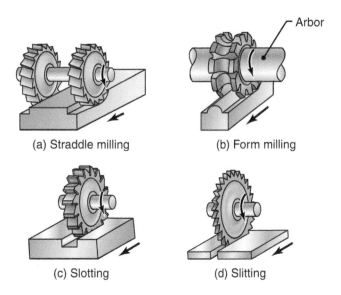

(a) Straddle milling   (b) Form milling

(c) Slotting   (d) Slitting

FIGURE 24.14 Cutters for (a) straddle milling; (b) form milling; (c) slotting; and (d) slitting with a milling cutter.

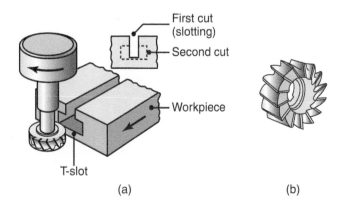

(a)   (b)

FIGURE 24.15 (a) T-slot cutting with a milling cutter. (b) A shell mill.

curved profiles, using cutters that have specially shaped teeth (Fig. 24.14b); such cutters are also used for cutting gear teeth, as described in Section 24.7.

Slotting and **slitting** operations are performed with *circular cutters*, as shown in Figs. 24.14c and d, respectively. The teeth may be staggered slightly, like those in a saw blade (Section 24.5), to provide clearance for the cutter width when machining deep slots. *Slitting saws* are relatively thin, usually less than 5 mm $\left(\frac{3}{16} \text{ in.}\right)$. *T-slot cutters* are used to mill T-slots, such as those found in machine-tool worktables for clamping workpieces. As shown in Fig. 24.15a, a slot is first milled with an end mill, and the cutter then machines the complete profile of the T-slot, in one pass.

Key seat cutters are used to make the semicylindrical or *Woodruff* key seats for shafts. *Angle milling cutters*, either single-angle or double-angle, are used to produce tapered surfaces with various angles. *Shell mills* (Fig. 24.15b) are hollow inside, and are mounted on a shank, thus allowing the same shank to be used for different-sized cutters. The use of shell mills is similar to that for end mills.

Milling with a single cutting tooth, mounted on a high-speed spindle, is known as *fly cutting*, and is generally used in simple face-milling and boring operations. This tool can be shaped as a single-point cutting tool, and can be placed in various radial positions on the spindle, in an arrangement similar to that shown in Fig. 23.24b.

24.2.5   Toolholders

The stiffness of toolholders and cutters is important for surface quality and in reducing vibration and chatter during milling operations. Milling cutters are classified as either arbor cutters or shank cutters. **Arbor cutters** are mounted on an *arbor* (see Figs. 24.14 and 24.18a), for operations such as peripheral, face, straddle, and form milling. In **shank-type cutters**, the cutter and the shank are made in one piece, the most common examples being end mills. Small end mills have straight shanks, but larger ones have tapered shanks, for better mounting in the machine spindle in order to resist the high forces and torque involved during cutting. Cutters with straight shanks are mounted in collet chucks or in special end-mill holders; those with tapered shanks are mounted in tapered toolholders. Hydraulic toolholders and arbors also are available.

24.2.6   Milling Process Capabilities

In addition to the various characteristics of the milling processes described thus far, milling process capabilities include such parameters as surface finish, dimensional tolerances, production rate, and cost considerations. Data on process capabilities are presented in Tables 23.1 and 23.8, Figs. 23.13 and 23.14, and Chapter 40.

The conventional ranges of cutting speeds and feeds for milling are given in Table 24.2, as guidelines. Depending on the workpiece material and its condition, cutting-tool material, and process parameters, cutting speeds vary widely in the range from 30 to 3000 m/min (90–10,000 ft/min). Feed per tooth typically ranges from about 0.1 mm (0.004 in.) to 0.5 mm (0.02 in.), and depths of cut are usually 1–8 mm (0.04–0.30 in.). For cutting-fluid recommendations, see Table 23.6.

A general **troubleshooting guide** for milling operations is given in Table 24.3; the last four items in this table are illustrated in Figs. 24.16 and 24.17. *Back striking* involves double feed marks, made by the trailing edge of the cutter. Note from Table 24.3 that some recommendations (such as changing milling parameters or cutting tools) are easier to accomplish than others (such as changing tool angles, cutter geometry, and the stiffness of spindles and work-holding devices).

TABLE 24.2

General Recommendations for Milling Operations (Note That These Values Are for a Particular Machining Geometry and Are Often Exceeded in Practice)

Material	Cutting tool	General-purpose starting conditions		Range of conditions	
		Feed, mm/tooth (in./tooth)	Speed, m/min (ft/min)	Feed, mm/tooth (in./tooth)	Speed, m/min (ft/min)
Low-carbon and free-machining steels	Uncoated carbide, coated carbide, cermets	0.13–0.20 (0.005–0.008)	100–472 (320–1550)	0.085–0.38 (0.003–0.015)	90–425 (300–1400)
Alloy steels					
Soft	Uncoated, coated, cermets	0.10–0.18 (0.004–0.007)	100–260 (360–860)	0.08–0.30 (0.003–0.012)	60–370 (200–1200)
Hard	Cermets, PcBN	0.10–0.15 (0.004–0.006)	90–220 (310–720)	0.08–0.25 (0.003–0.010)	75–460 (250–1500)
Cast iron, gray					
Soft	Uncoated, coated, cermets, SiN	0.10–0.20 (0.004–0.008)	160–440 (530–1440)	0.08–0.38 (0.003–0.015)	90–1370 (300–4500)
Hard	Cermets, SiN, PcBN	0.10–0.20 (0.004–0.008)	120–300 (400–960)	0.08–0.38 (0.003–0.015)	90–460 (300–1500)
Stainless steel, Austenitic	Uncoated, coated, cermets	0.13–0.18 (0.005–0.007)	120–370 (370–680)	0.08–0.38 (0.003–0.015)	90–500 (300–1800)
High-temperature alloys Nickel based	Uncoated, coated, cermets, SiN, PcBN	0.10–0.18 (0.004–0.007)	30–370 (100–1200)	0.08–0.38 (0.003–0.015)	30–550 (90–1800)
Titanium alloys	Uncoated, coated, cermets	0.13–0.15 (0.005–0.006)	50–60 (175–200)	0.08–0.38 (0.003–0.015)	40–140 (125–450)
Aluminum alloys					
Free machining	Uncoated, coated, PCD	0.13–0.23 (0.005–0.009)	1200–1460 (3920–4790)	0.08–0.46 (0.003–0.018)	300–3000 (1000–10,000)
High silicon	PCD	0.13 (0.005)	610 (2000)	0.08–0.38 (0.003–0.015)	370–910 (1200–3000)
Copper alloys	Uncoated, coated, PCD PCD	0.13–0.23 (0.005–0.009)	300–760 (1000–2500)	0.08–0.46 (0.003–0.018)	90–1070 (300–3500)
Plastics	Uncoated, coated, PCD PCD	0.13–0.23 (0.005–0.009)	270–460 (900–1500)	0.08–0.46 (0.003–0.018)	90–1370 (300–4500)

Source: Based on data from Kennametal, Inc.
Note: Depths of cut, d, usually are in the range of 1–8 mm (0.04–0.3 in.). PcBN: polycrystalline cubic-boron nitride; PCD: polycrystalline diamond. See also Table 23.4 for range of cutting speeds within tool material groups.

TABLE 24.3

General Troubleshooting Guide for Milling Operations	
Problem	Probable causes
Tool breakage	Tool material lacks toughness, improper tool angles, machining parameters too high
Excessive tool wear	Machining parameters too high, improper tool material, improper tool angles, improper cutting fluid
Rough surface finish	Feed per tooth too high, too few teeth on cutter, tool chipped or worn, built-up edge, vibration and chatter
Tolerances too broad	Lack of spindle and work-holding device stiffness, excessive temperature rise, dull tool, chips clogging cutter
Workpiece surface burnished	Dull tool, depth of cut too low, radial relief angle too small
Back striking	Dull cutting tools, tilt in cutter spindle, negative tool angles
Chatter marks	Insufficient stiffness of system; external vibrations; feed, depth of cut, and width of cut too large
Burr formation	Dull cutting edges or too much honing, incorrect angle of entry or exit, feed and depth of cut too high, incorrect insert shape
Breakout	Lead angle too low, incorrect cutting-edge geometry, incorrect angle of entry or exit, feed and depth of cut too high

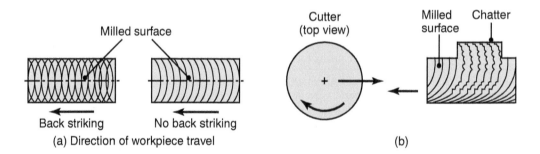

FIGURE 24.16   Workpiece surface features in face milling. (See also Fig. 24.8.)

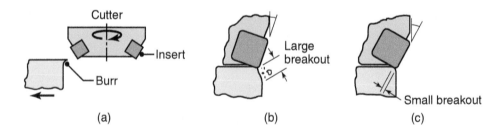

FIGURE 24.17   Edge defects in face milling: (a) burr formation along workpiece edge; (b) breakout along workpiece edge; and (c) how it can be avoided by increasing the lead angle (see also last row in Table 24.3).

24.2.7   Design and Operating Guidelines for Milling

The guidelines for turning and boring, given in Sections 23.3.6 and 23.4, are also generally applicable to milling operations. Additional factors relevant to milling operations include the following:

- Standard milling cutters should be used as much as possible, depending on part design features; costly special cutters should be avoided.
- Chamfers should preferably be specified, instead of radii, as it is difficult to smoothly match various intersecting surfaces if radii are specified.
- Internal cavities and pockets with sharp corners should be avoided, because of the difficulty of milling them, since cutting teeth or inserts have a finite edge radius. When possible, the corner radius should match the milling cutter geometry.
- Although small milling cutters can be used for milling any surface, they are less rugged and more susceptible to chatter and to tool breakage than large cutters.
- Workpieces should be sufficiently rigid to minimize deflections that may result from clamping and cutting forces.
- Workpieces should be designed so that they can be clamped or held in fixtures during machining. Furthermore, the fixturing should be designed to minimize the number of times that the part needs to be repositioned to complete the milling operation.

Guidelines for avoiding vibration and chatter in milling are similar to those for turning; in addition, the following practices should be considered:

- Cutters should be mounted as close to the spindle base as possible, in order to reduce tool deflections.
- Toolholders and fixturing devices should be as rigid as possible.
- In cases where vibration and chatter occur, tool shape and process conditions should be modified, including the use of cutters with fewer teeth or, whenever possible, with random tooth spacing (see Section 25.4).

24.2.8   Milling Machines

Because they are capable of performing a wide variety of cutting operations, milling machines are among the most versatile and useful of all machine tools. The first milling machine was built in 1820 by E. Whitney (1765–1825). A wide selection of milling machines with numerous features is now available, the most common of which are described below. These machines are now being rapidly replaced with *computer numerical-control* (CNC) *machines* and *machining centers*; although inexpensive, manually controlled machines are still widely used, especially for small production runs or for making prototypes. Modern machines are very versatile and capable of milling, drilling, boring, and tapping, with repeat and high accuracy (Fig. 24.20).

QR Code 24.3 Turbine Blade Milling. (*Source:* Courtesy of Sandvik Coromant)

Column-and-knee-type Machines. Used for general-purpose milling operations, *column-and-knee type machines* are the most common milling machines. The spindle on which the milling cutter is mounted may be *horizontal* (Fig. 24.18a), for peripheral milling, or *vertical*, for face and end milling, boring, and drilling operations (Fig. 24.18b). The basic components of these machines are:

- *Worktable:* on which the workpiece is clamped using T-slots; the table moves longitudinally relative to the saddle.
- *Saddle:* supports the table and can move in the transverse direction.
- *Knee:* supports the saddle and gives the table vertical movement, so that the depth of cut can be adjusted and workpieces with various heights can be accommodated.

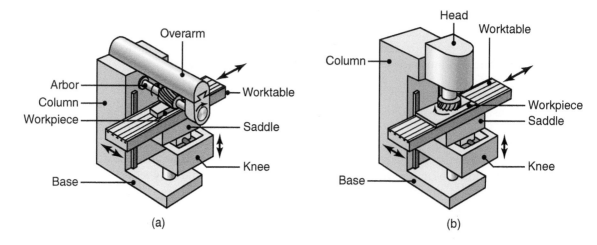

FIGURE 24.18 Schematic illustration of (a) a horizontal-spindle column-and-knee type milling machine and (b) vertical-spindle column-and-knee type milling machine. *Source:* After G. Boothroyd.

- *Overarm:* used on horizontal machines; it is adjustable to accommodate different arbor lengths.
- *Head:* contains the spindle and cutter holders; in vertical machines, the head may be fixed or can be adjusted vertically, and it can be swiveled in a vertical plane on the column for cutting tapered surfaces.

Plain milling machines have at least three axes of movement, with the motion usually imparted manually, either with a power screw actuator or by engaging powered actuators to the drive motor. In **universal column-and-knee milling machines,** the table can be swiveled on a horizontal plane; in this way, complex shapes, such as helical grooves at various angles, can be machined to produce such parts as gears, drills, taps, and cutters.

Bed-type Milling Machines. In *bed-type machines*, the worktable is mounted directly on the bed, which replaces the knee and moves only longitudinally (Fig. 24.19). Although not as versatile as other types, these machines have high stiffness and typically are used for high-production work. The spindles may be horizontal or vertical and of duplex or triplex types (with two or three spindles, respectively), for the simultaneous machining of two or three workpiece surfaces.

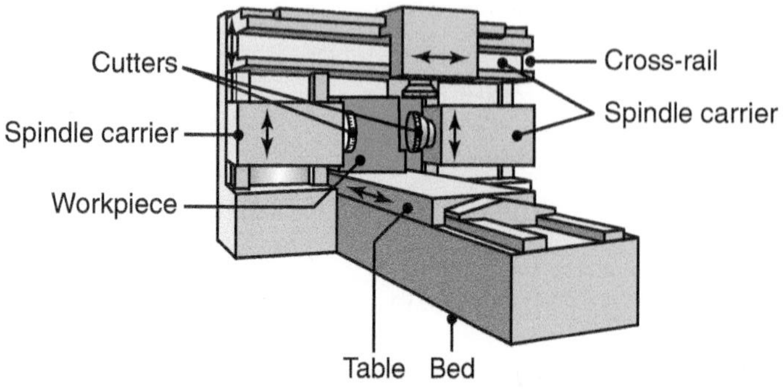

Other Types of Milling Machines. Several other types of milling machines are available (see also *machining centers,* Section 25.2). **Planer-type milling machines,** which are similar to bed-type machines, are equipped with several heads and cutters to mill different surfaces. They are typically used for heavy workpieces and are more efficient than simple planers (Section 24.3) when used for similar purposes. **Rotary-table machines** are similar to vertical milling machines, and are equipped with one or more heads for face-milling operations. Also available are **profile milling**

FIGURE 24.19 Schematic illustration of a bed-type milling machine.

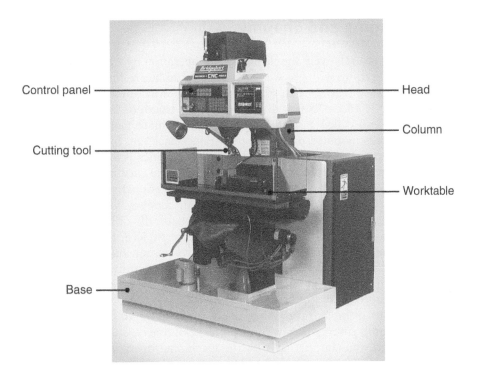

Control panel

Cutting tool

Base

Head

Column

Worktable

FIGURE 24.20 A computer numerical-control (CNC) vertical-spindle milling machine; this is one of the most versatile machine tools. The original vertical-spindle milling machine, used in job shops, is still referred to as a "Bridgeport," after its manufacturer in Bridgeport, Connecticut. *Source:* Courtesy of Bridgeport Machines Division, Textron, Inc.

machines, which have five axes of movement (Fig. 24.21); note the three linear and two angular movements of the machine components.

Work-holding Devices and Accessories.   The workpiece to be milled must be clamped securely to the worktable in order to resist cutting forces and prevent slipping during milling. Various fixtures and vises generally are used for this purpose. (See also Section 37.8 on *flexible fixturing*.) Mounted and clamped to the worktable, using the T-slots seen in Figs. 24.18a and b, vises are used for small production work on small parts, while *fixtures* are used for higher production work, and can be automated by various mechanical and hydraulic means.

   Accessories for milling machines include various fixtures and attachments for the machine head, as well as the worktable, that are designed to adapt them to different milling operations. The accessory that has been used most commonly, typically in job shops, is the *universal dividing (index) head*. Manually operated, this fixture rotates (*indexes*) the workpiece to specified angles between individual machining steps. Typically, it has been used to mill parts with polygonal surfaces and to machine gear teeth.

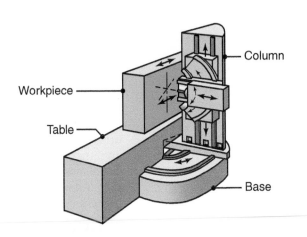

Column

Workpiece

Table

Base

FIGURE 24.21 Schematic illustration of a five-axis profile milling machine; note that there are three principal linear and two angular movements of machine components.

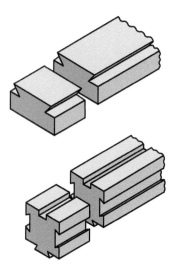

FIGURE 24.22 Typical parts that can be machined on a planer.

24.3 Planing and Shaping

Planing. This is a relatively simple machining operation by which flat surfaces, as well as cross-sections with grooves and notches, can be produced along the length of the workpiece (Fig. 24.22). Planing usually is done on large workpieces, as large as 25 m × 15 m (75 ft × 40 ft), although a length of 10 m is more typical. In a **planer**, also called a *scalper* when a layer is machined from a cast ingot, the workpiece is mounted on a table that travels back and forth along a straight path. A horizontal *cross-rail*, which can be moved vertically along the ways of the column, is equipped with one or more tool heads. The cutting tools are mounted on the heads, and the machining is done along a straight path. In order to prevent tool cutting edges from chipping, when they rub along a workpiece during the return stroke, tools are either tilted or lifted mechanically or hydraulically.

Because of the reciprocating motion of the workpiece, the noncutting time elapsed during the return stroke is significant. Consequently, these operations are neither efficient nor economical, except for low-quantity production, which is generally the case for large and long workpieces. The efficiency of the operation can be improved by equipping planers with toolholders and tools that cut in both directions of table travel. Also, because of the length of the workpiece, it is essential to equip cutting tools with chip breakers; otherwise, the chips produced can be very long, thus interfering with the machining operation and becoming a safety hazard.

Shaping. Machining by shaping is basically the same as by planing, except that it is the tool, and not the workpiece, that travels, and the workpieces are smaller, typically less than 1 m × 2 m (3 ft × 6 ft) of surface area. In a **horizontal shaper**, the cutting tool travels back and forth along a straight path. The tool is attached to the tool head, which is mounted on the ram; the ram has a reciprocating motion. In most machines, cutting is done during the forward movement of the ram (*push cut*); in others, it is done during the return stroke of the ram (*draw cut*). **Vertical shapers** (called **slotters**) are used to machine notches, keyways, and dies. Because of low production rates, only special-purpose shapers (such as gear shapers, Section 24.7.2) are in common use today.

24.4 Broaching and Broaching Machines

Broaching is similar to shaping with a long multiple-tooth cutter, and is used to machine internal and external surfaces, such as holes of circular, square, or irregular section; keyways; the teeth of internal gears; multiple spline holes; and flat surfaces (Fig. 24.23). In a typical **broach** (Fig. 24.24a), the total depth of material removed in one stroke is the sum of the depths of cut of each tooth of the broach. A large broach can remove material as deep as 38 mm (1.5 in.) in one stroke. Broaching is an important production process, and can produce parts with good surface finish and dimensional accuracy. It competes favorably with other machining processes, such as boring, milling, shaping, and reaming, to produce similar shapes. Although broaches can be expensive, the cost is justified with high-quantity production runs.

Broaches.   The terminology for a typical broach is given in Fig. 24.24b. The *rake* (hook) *angle* depends on the material cut (as it does in turning and other cutting operations), and usually ranges from 0° to 20°. The *clearance angle* is typically 1°–4°; finishing teeth have smaller angles; too small a clearance angle causes rubbing of the teeth against the broached surface. The *pitch* of the teeth depends on such

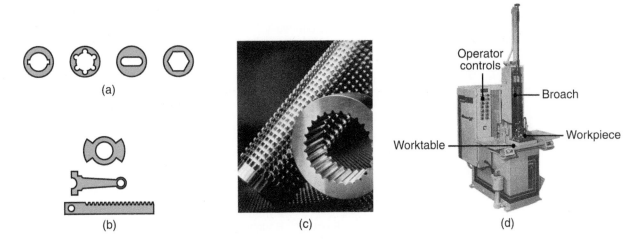

FIGURE 24.23 (a) Typical parts made by internal broaching. (b) Parts made by surface broaching. (c) A spline broach and internal spline used for a shaft coupling. (d) Vertical broaching machine. *Source:* (a) and (b) Courtesy of General Broach Company, (c) Courtesy of The Broachmasters, Inc., and (d) Courtesy of Ty Miles, Inc.

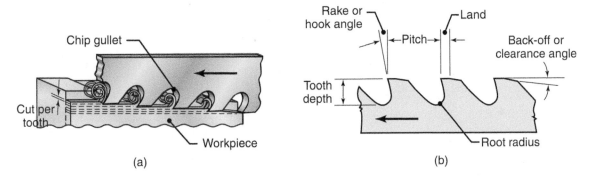

FIGURE 24.24 (a) Cutting action of a broach, showing various features. (b) Terminology for a broach.

factors as the length of the workpiece (length of cut), tooth strength, and size and shape of chips.

The tooth depth and pitch must be sufficiently large to accommodate the chips produced during broaching, particularly for long workpieces. At least two teeth should be in contact with the workpiece at all times. The following formula may be used to obtain the pitch for a broach to cut a surface of length l:

$$\text{Pitch} = k\sqrt{l}, \tag{24.6}$$

where k is a constant, equal to 1.76 when l is in mm and 0.35 when l is in inches. An average pitch for small broaches is in the range from 3.2 to 6.4 mm (0.125–0.25 in.), and for large ones it is in the range from 12.7 to 25 mm (0.5–1 in.). The depth of cut per tooth depends on the workpiece material and the surface finish required. It is usually in the range from 0.025 to 0.075 mm (0.001–0.003 in.) for medium-sized broaches, but can be larger than 0.25 mm (0.01 in.) for larger broaches.

Broaches are available with various tooth profiles, including some with *chip breakers* (Fig. 24.25). The variety of *surface* broaches include *slab* (for cutting flat

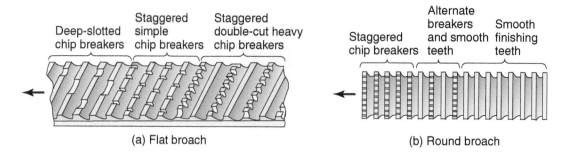

FIGURE 24.25 Chip breaker features on broaches.

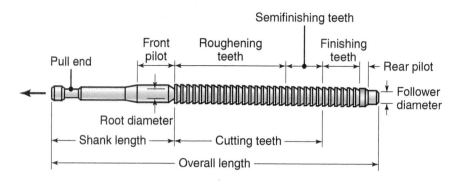

FIGURE 24.26 Terminology for a pull-type internal broach, used for enlarging long holes.

surfaces), *slot, contour, dovetail, pot* (for precision external shapes), and *straddle*. *Internal broach* types include *hole* (for close-tolerance holes, round shapes, and other shapes; Fig. 24.26), *keyway, internal gear,* and *rifling* (for gun barrels). Irregular internal shapes usually are broached by starting with a round hole drilled or bored in the workpiece.

Turn Broaching.   This process is typically used for broaching the bearing surfaces of crankshafts and similar parts. The crankshaft is rotated between centers, and the broach, which is equipped with multiple carbide inserts, passes tangentially across the bearing surfaces and removes material. Turn broaching is a combination of *shaving* and *skiving* (removing a thin layer of material with a specially shaped cutting tool). Straight as well as circular broaches can be used successfully in turn broaching. Machines that broach a number of crankshafts simultaneously also have been built.

Broaching Machines.   The machines for broaching are relatively simple in construction, have only linear motions, and usually are actuated hydraulically, although some are moved by crank, screw, or rack mechanisms. Several styles of machines are available, and sizes range from machines for making needlelike parts to those used for broaching gun barrels, including rifling (producing internal spiral grooves).

   Broaching machines either pull or push the broaches and are either horizontal or vertical. *Push broaches* usually are shorter, generally in the range from 150 to 350 mm (6–14 in.). *Pull broaches* tend to straighten the hole, whereas pushing permits the broach to follow any irregularity of the leader hole. Horizontal machines are capable of longer strokes. The *force* required to pull or push the broach depends on the (a) strength of the workpiece material, (b) total periphery of the cut, (c) cutting speed, (d) tooth profile, and (e) type of cutting fluid used. The pulling force capacities of broaching machines are as high as 0.9 MN (100 tons).

Process Parameters.   Cutting speeds for broaching may range from 1.5 m/min (5 ft/min) for high-strength alloys to as much as 30 m/min (100 ft/min) for aluminum and magnesium alloys. The most common broach materials are M2 and M7 high-speed steels, as well as carbide inserts. Smaller, high-speed steel blanks for making broaches can be produced by powder metallurgy techniques (Chapter 17) for better control of quality. Cutting fluids generally are recommended, especially for internal broaching.

Design Considerations.   Broaching, as with other machining processes, requires that certain guidelines be followed in order to obtain economical and high-quality production. The major requirements are:

- Blanks should be designed and prepared such that they can be securely clamped in broaching machines, and should have sufficient structural strength and stiffness to withstand the cutting forces during broaching.
- Keyways, splines, gear teeth, etc., should all have standard sizes and shapes, so as to allow the use of common broaches.
- Balanced cross-sections are preferable, to keep the broach from drifting laterally, thus maintaining close tolerances.
- Radii are difficult to broach and chamfers are preferred; inverted or dovetail splines should be avoided.
- Blind holes should be avoided whenever possible, but if necessary, there must be a relief at the end of the area to be broached.

CASE STUDY 24.1   Broaching Internal Splines

The part shown in Fig. 24.27 is made of nodular iron (65-45-15; Section 12.3.2), with internal splines, each 50 mm (2 in.) long. The splines have 19 involute teeth, with a pitch diameter of 63.52 mm (2.5009 in.). An M2 high-speed steel broach, with 63 teeth, a length of 1.448 m (57 in.), and a diameter the same as the pitch diameter, was used to produce the splines. The cut per tooth was 0.116 mm (0.00458 in.). The production rate was 63 pieces per hour. The number of parts per grind was 400, with a total broach life of about 6000 parts.

Source: Reprinted with permission of ASM International. All rights reserved. www.asminternational .org.

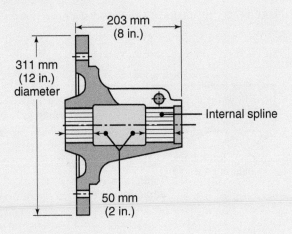

FIGURE 24.27   Example of a part with internal splines produced by broaching.

24.5 Sawing

Sawing is a common process dating back to around 1000 B.C. The cutting tool is a *blade* (the saw) having a series of small teeth, each tooth removing a small amount of material with each stroke or movement of the saw. This process can be used for all materials and is capable of producing various shapes (Fig. 24.28). Sawing is an efficient material-removal process and can produce near-net shapes from blanks. The width of the cut (called **kerf**) in sawing usually is small, so that the process wastes little material.

Typical saw-tooth and saw-blade configurations are shown in Fig. 24.29, where tooth spacing is generally in the range from 0.08 to 1.25 teeth per mm (2–32 per in.). A wide variety of sizes, tooth forms, tooth spacing, and blade thicknesses and widths

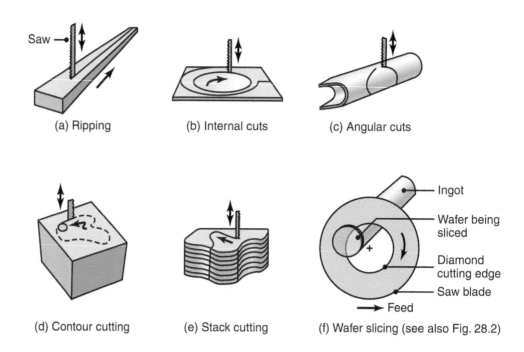

(a) Ripping   (b) Internal cuts   (c) Angular cuts

(d) Contour cutting   (e) Stack cutting   (f) Wafer slicing (see also Fig. 28.2)

FIGURE 24.28 Examples of various sawing operations.

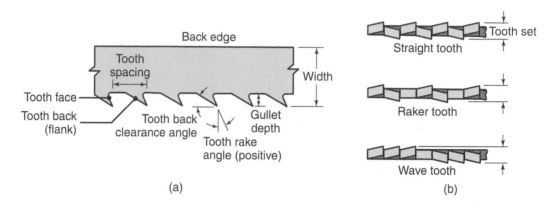

(a)   (b)

FIGURE 24.29 (a) Terminology for saw teeth. (b) Types of tooth sets on saw teeth, staggered to provide clearance for the saw blade to prevent binding during sawing.

are available. Saw blades generally are made from high-carbon and high-speed steels (M2 and M7); carbide or high-speed-steel tipped steel blades are used to saw harder materials and at higher speeds (Fig. 24.30).

The **tooth set** in a saw (Fig. 24.29b) is important in providing a sufficiently wide kerf for the blade to move freely in the workpiece, without binding or excessive frictional resistance, thus reducing the heat generated. Elevated temperatures can have adverse effects on the cut, especially in cutting thermoplastics, which soften rapidly when heated (see Fig. 7.11). The tooth set also allows the blade to track a path accurately, following the pattern to be cut without wandering. At least two or three teeth always should be engaged with the workpiece, in order to prevent *snagging* (catching of the saw tooth on the workpiece); this is the reason why thin materials, especially sheet metals, can be difficult to saw. The thinner the stock, the finer the saw teeth should be, and the greater the number of teeth per unit length of the saw. Cutting fluids generally are used to improve the quality of the cut and the life of the saw.

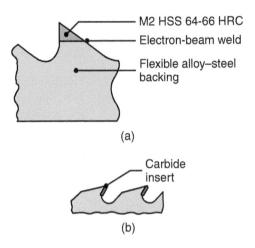

FIGURE 24.30 (a) High-speed steel teeth welded onto a steel blade. (b) Carbide inserts brazed on blade teeth.

Types of Saws.   *Hacksaws* have straight blades and reciprocating motions. Developed in the 1650s, they generally are used to cut off bars, rods, and structural shapes; they may be manual or power operated. Because cutting takes place during only one of the two reciprocating strokes, hacksaws are not as efficient as band saws. *Power hacksaw* blades are usually 1.2–2.5 mm (0.05–0.10 in.) thick, and up to 610 mm (24 in.) long. The rate of strokes ranges from 30 per minute for high-strength alloys to 180 per minute for carbon steels. The hacksaw frame in power hacksaws is weighted by various mechanisms, applying as much as 1.3 kN (300 lb) of force to the workpiece to improve the cutting rate. *Hand hacksaw* blades are thinner and shorter than power hacksaw blades, which have as many as 1.2 teeth per mm (32 per in.) for sawing sheet metal and thin tubing.

Circular saws, also called *cold saws* in cutting metal, generally are used for high-production-rate sawing, a process called *cutting off*. Cutting-off operations also can be carried out with thin, *abrasive* disks, as described in Section 26.4. Cold sawing is common in industry, particularly for cutting off large cross-sections. These saws are available with a variety of tooth profiles and sizes, and can be fed at any angle into the workpiece. In modern machines, cutting off with circular saws produces relatively smooth surfaces, with good thickness control and dimensional accuracy, because of the stiffness of the machines and of the saws. The inner-diameter cutting saw, shown in Fig. 24.28f, is used widely to cut single-crystal silicon wafers for microelectronic devices (Section 28.4).

Band saws have continuous, long, flexible blades. *Vertical band saws* are used for straight as well as *contour cutting* of flat sheets and other parts, supported on a horizontal table (Fig. 24.28d). Also available are computer-controlled band saws, with the capability of guiding the contour path automatically. *Power band saws* also are available; they have higher productivity than power hacksaws, because of their continuous cutting action. With high-speed steel blades, cutting speeds for sawing high-strength alloys are up to about 60 m/min (200 ft/min), and 120 m/min (400 ft/min) for carbon steels.

Blades and high-strength wire can be coated with diamond powder (**diamond-edged blades** and **diamond-wire saws**), so that the diamond particles act as cutting teeth (abrasive cutting); carbide particles also are used for this purpose. These blades and wires are suitable for sawing hard metallic, nonmetallic, and composite materials.

Wire diameters range from 13 mm (0.5 in.) for use in rock cutting to 0.08 mm (0.003 in.) for precision cutting. Hard materials also can be sawed with thin, abrasive disks and with advanced machining processes, described in Chapter 27.

Friction Sawing.   *Friction sawing* is a process in which a mild-steel blade or disk rubs against the workpiece, at speeds of up to 7600 m/min (25,000 ft/min). The frictional energy is converted into heat, which rapidly softens a narrow zone in the workpiece. The action of the blade, which can have teeth or notches for higher cutting efficiency, pulls and ejects the softened metal from the cutting zone. The heat generated in the workpiece produces a *heat-affected zone* (Section 30.9) on the cut surfaces; thus, the workpiece properties along the cut edges can be affected adversely by this process. Because only a small portion of the blade is engaged with the workpiece at any time, the blade itself cools rapidly as it passes through the air.

The friction-sawing process is suitable for hard ferrous metals and reinforced plastics, but not for nonferrous metals, because of their tendency to stick to the blade. Friction sawing also is commonly used to remove flash from castings. Friction-sawing disks as large as 1.8 m (6 ft) in diameter are used to cut off large steel sections.

CASE STUDY 24.2   Ping Golf Putters

In their efforts to develop high-end, top performing putters, engineers at Ping Golf, Inc., utilized advanced machining practices in their design and production processes for a new style of putter, called the Anser® series, shown in Fig. 24.31. In response to a unique set of design constraints, they had the task and goal of creating putters that would be practical for production quantities and also meet specific functional and aesthetic requirements.

One of the initial decisions concerned the selection of a proper material for the putter to meet its functional requirements. Four types of stainless steel (303, 304, 416, and 17-4 precipitation hardening; see Section 5.6) were considered for various property requirements, including machinability, durability, and the sound or feel of the particular putter material, another requirement that is unique to golf equipment. Among the materials evaluated, 303 stainless steel was chosen because it is a free-machining material (Section 21.7), indicating that in machining, it has smaller chips, lower power consumption, better surface finish, and improved tool life, thus allowing for increased machining speeds and hence higher productivity.

The next step of the project involved determining the optimum blank shape and the sequence of operations to be performed during its production. For this case, engineers chose to develop a slightly oversized forged blank (Chapter 14). A forging

was chosen because it provided a favorable internal grain structure, as opposed to a casting, which could result in porosity and an inconsistent surface finish after machining. The blank incorporated a machining allowance, whereby dimensions were specified approximately 0.050–0.075 in. (1.25–1.9 mm) larger in all directions than that of the final part.

The most challenging, and longest, task was developing the necessary programming and designing fixtures for each part of the putter. Beyond the common requirements of typical machined parts, including tight tolerances and repeatability, putters require an additional set of aesthetic specifications. In this case, both precise machining and the right overall appearance of the finished part were imperative. A machining technique, known as surfacing or contouring (commonly used in making injection molds), was used to machine most of the finished geometry. Although this operation required additional machining, it provided a superior finish on all surfaces and allowed machining of more complex geometries, thus adding value to the finished product.

As for all high-volume machined parts, repeatability was essential. Each forged blank was designed with a protrusion across the face of the putter, allowing for the initial locating surfaces, for ease of fixturing. A short machining operation

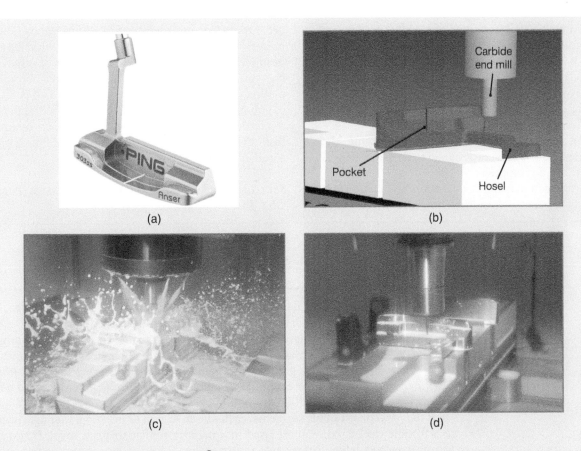

FIGURE 24.31 (a) The Ping Anser® golf putter; (b) CAD model of rough machining of the putter's outer surface; (c) rough machining on a vertical machining center; and (d) machining of the lettering in a vertical machining center; the operation was paused to take the photo, as normally the cutting zone is flooded with a coolant. *Source:* Printed with permission of Ping Golf, Inc., Phoenix, Arizona.

removed a small amount of material around the bar and produced three flat, square surfaces, as a reference location for the first primary machining operation.

Each putter required six different operations in order to machine all of its surfaces, and each operation was designed to provide locating surfaces for the next step in the manufacturing process. Several operations were set up, using a tombstone loading system (see Section 37.8) on a horizontal-spindle CNC milling machine. This method allowed machine operators to load and unload parts while other parts were being machined, thus significantly increasing the efficiency of the operation.

Modular fixturing and using tungsten-carbide cutting tools coated with TiAlN (Section 22.5.2) allowed for the quick changeover between right- and left-handed parts, as well as different putter models.

After the initial locating operation was completed, the parts were transferred to a three-axis vertical machining center (see, for example, Fig. 25.7) to cut the putter cavity. Since the forged blanks were near-net shape, the maximum radial depth of cut on most surfaces was 0.075 in., but the axial depth of cut of 1.5 in. inside the cavity of the putter was the most demanding milling operation (see Figs. 24.31b and c). The putter has small inside radii with a comparatively long depth ($7\times$ the diameter or greater).

A four-axis horizontal machining center (see, for example, Fig. 25.2) was used to reduce the number of setups in this operation. The rotary axis was used for creating the relatively complex geometry of the hosel (the socket for the shaft of the golf club). Since the hosel is relatively unsupported, chatter was the most complex challenge to

(*continued*)

overcome. Several iterations of spindle speeds were attempted in conjunction with upfront guidance from a simulation model. Modal analyses were conducted on the fixtured parts, in an attempt to identify and avoid the natural frequencies of the part or the fixture (see Section 25.4). The machines had spindle speeds ranging from 12,000 to 20,000 rpm, each having 30 hp. With the near-net-shape forging, the milling operations were designed to have low depths of cut, but high speed.

After each machining operation was completed, a small amount of hand finishing was necessary in order to produce a superior surface appearance. The putters were then lightly shot blasted (with glass bead media, Section 34.2) for the purpose of achieving surface consistency. A black, nickel–chrome plating (Section 34.9) was then applied to all parts, to enhance aesthetic appeal and protect the stainless steel from small dings and dents, and from corrosion from specific chemicals that might be encountered on a golf course.

Source: Based on D. Jones and D. Petersen, Ping Golf, Inc.

24.6 Filing

Filing involves the small-scale removal of material from a surface, corner, edge, or hole, including the removal of burrs (see Fig. 16.2). First developed around 1000 B.C., files usually are made of hardened steel, and are available in a variety of cross-sections, such as flat, round, half-round, square, and triangular. Files can have several tooth forms and coarseness grades. Although filing usually is done by hand, *filing machines*, with automatic features, are available for high production rates, and with files reciprocating at up to 500 strokes/min.

Band files consist of file segments, each about 75 mm (3 in.) long, that are riveted to a flexible steel band, and are used in a manner similar to band saws. *Disk-type files* also are available. *Rotary files* and *burs* (Fig. 24.32) are used for such applications as deburring, removing scale from surfaces, producing chamfers on parts, and removing small amounts of material in die making. These tools generally are conical, cylindrical, or spherical in shape, and have various tooth profiles. Their cutting action (similar to that of reamers, Section 23.6) removes small amounts of material at high rates. The rotational speed of burs ranges from 1500 rpm for cutting steels (using large burs) to as high as 45,000 rpm for magnesium (small burs).

24.7 Gear Manufacturing by Machining

Several processes for making gears or producing gear teeth on various components were described in Parts II and III, including casting, forging, extrusion, drawing, thread rolling, and powder metallurgy. *Blanking* of sheet metal also can be used for making thin gears, such as those used in mechanical watches, clocks, and similar mechanisms. Plastic gears can be made by such processes as casting (Chapter 11) and injection molding (Section 19.3).

Gears may be as small as those used in watches or as large as 9 m (30 ft) in diameter for rotating mobile crane superstructures and mining equipment. The dimensional accuracy and surface finish required for gear teeth depend on the intended use. Poor gear-tooth quality contributes to inefficient energy transmission, increased vibration and noise, and adversely affects the gear's frictional and wear characteristics. Submarine gears, for example,

(a) High-speed steel bur   (b) Carbide bur   (c) Rotary file

FIGURE 24.32 Types of burs used in burring operations.

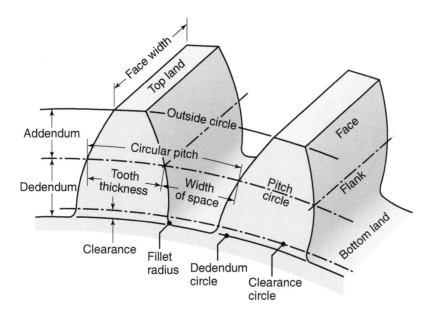

FIGURE 24.33 Nomenclature for an involute spur gear.

have to be of extremely high quality so as to reduce noise levels, thus helping to avoid detection.

The standard nomenclature for an involute spur gear is shown in Fig. 24.33. Starting with a wrought or cast gear blank, there are two basic methods of making gear teeth: form cutting and generating.

24.7.1 Form Cutting

In *form cutting*, the cutting tool is similar to a form-milling cutter made in the shape of the space between the gear teeth (Fig. 24.34a). The gear-tooth shape is reproduced by machining the gear blank around its periphery. The cutter travels axially along the length of the gear tooth, and at the appropriate depth, to produce the gear-tooth profile. After each tooth is cut, the cutter is withdrawn, the gear blank is rotated (*indexed*), and the cutter proceeds to cut another tooth; this process continues until all of the teeth are machined. Each cutter is designed to cut a range of numbers of teeth.

The precision of the form-cut tooth profile depends on the accuracy of the cutter and on the machine and its stiffness. Because the cutter has a fixed geometry, form cutting can be used only to produce gear teeth that have a constant width, that is, on spur or helical gears but not on bevel gears. Internal gears and gear teeth on straight surfaces, such as those in a rack and pinion, are form cut with a shaped cutter on a machine similar to a shaper.

Broaching also can be used to machine gear teeth and is particularly suitable for producing internal teeth. The broaching process is rapid and produces fine surface finish with high dimensional accuracy. However, because a different broach is required for each gear size (and broaches are expensive), this method is suitable almost exclusively for high-quantity production.

Although inefficient, form cutting also can be done on milling machines, with the cutter mounted on an arbor and the gear blank mounted in a dividing head. Gear teeth also may be cut on special machines with a single-point cutting tool, guided by

QR Code 24.4 Manufacturing cycle. (*Source:* Courtesy of the Forging Industry Association)

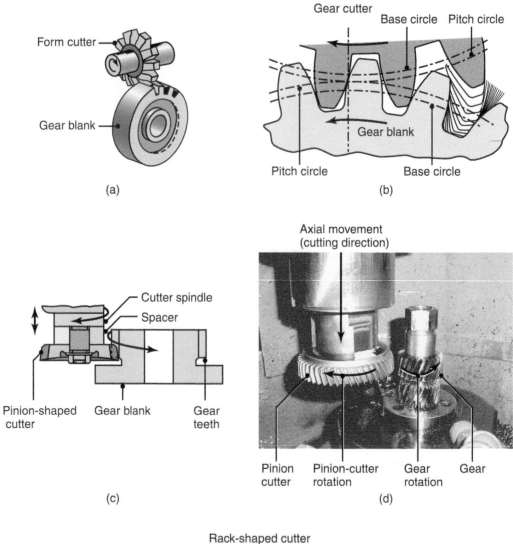

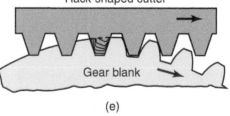

FIGURE 24.34 (a) Producing gear teeth on a blank by form cutting. (b) Schematic illustration of gear generating with a pinion-shaped gear cutter. (c) and (d) Gear generating in a gear shaper using a pinion-shaped cutter; note that the cutter reciprocates vertically. (e) Gear generating with rack-shaped cutter. *Source:* (d) Courtesy of Schafer Gear Works, Inc.

a *template* in the shape of the gear-tooth profile. Because the template can be made much larger than the gear tooth, dimensional accuracy is improved.

Form cutting is a relatively simple process and can be used for cutting gear teeth with various profiles. Nonetheless, it is a slow operation, and, furthermore, some types of machines require skilled labor. Machines with semiautomatic features can be used economically for form cutting on a limited-production basis. Generally, however, form cutting is suitable only for low-quantity production.

24.7.2 Gear Generating

The cutting tool used in *gear generating* may either be a pinion-shaped cutter, a rack-shaped straight cutter, or a hob.

1. A **pinion-shaped cutter** can be considered as one of the two gears in a conjugate pair, with the other being the gear blank (Fig. 24.34b). This type of cutter is used on vertical *gear shapers* (Figs. 24.34c and d). The cutter has an axis parallel to that of the gear blank, and rotates slowly with the blank at the same pitch–circle velocity and in an axial-reciprocating motion. A train of gears provides the required relative motion between the cutter shaft and the gear-blank shaft. Cutting may take place at either the downstroke or the upstroke of the machine. Because the clearance required for the cutter travel is small, gear shaping is suitable for gears that are located close to obstructing surfaces, such as a flange in the gear blank shown in Figs. 24.34c and d. This process can be used for low-quantity as well as high-quantity production.

2. On a **rack shaper,** the generating tool is a *segment of a rack* (Fig. 24.34e), which reciprocates parallel to the axis of the gear blank. Because it is not practical to have more than 6–12 teeth on a rack cutter, the cutter must be disengaged at suitable intervals, and returned to the starting point. The gear blank remains fixed during this operation.

3. A **hob** (Fig. 24.35) is basically a gear-cutting worm, or screw, made into a gear-generating tool by a series of longitudinal slots or gashes machined into it to form the cutting teeth. When hobbing a spur gear, the angle between the hob and gear-blank axes is 90° minus the lead angle at the hob threads. All motions in hobbing are rotary, and the hob and gear blank rotate continuously, much as two gears in mesh, until all of the teeth are cut.

Hobs are available with one, two, or three threads. For example, if the hob has a single thread and the gear is to have 40 teeth, the hob and the gear spindle must be geared together such that the hob makes 40 revolutions, while the gear blank makes 1 revolution. Similarly, if a double-threaded hob is used, the hob would make 20 revolutions to the gear blank's 1 revolution. In addition, the hob must be fed parallel to the gear axis for a distance greater than the face width of the gear tooth (Fig. 24.33), in order to produce straight teeth on spur gears. The same hobs and machines can be used to cut helical gears by tilting the axis of the hob spindle.

Because it produces a variety of gears at high rates and with good dimensional accuracy, gear hobbing is used extensively in industry. Although the process also is suitable for low-quantity production, it is most economical for medium- to high-quantity production.

Gear-generating machines also can produce spiral-bevel and hypoid gears. Like most other machine tools, modern gear-generating machines are computer controlled. *Multiaxis computer-controlled machines* are capable of generating many types and sizes of gears, using indexable milling cutters.

QR Code 24.5 Gear cutting. (*Source:* Courtesy of Sandvik Coromant)

24.7.3 Cutting Bevel Gears

Straight bevel gears generally are roughed out in one cut, with a form cutter, on machines that index automatically; the gear is then finished to the proper shape on a gear generator. The cutters reciprocate across the face of the bevel gear, as does the

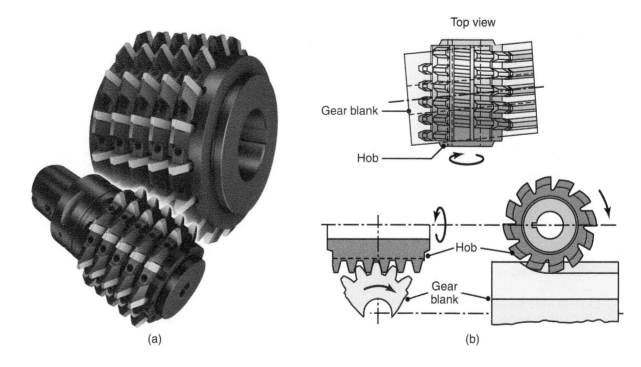

(a)

Top view

(b)

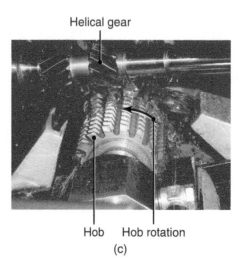

(c)

FIGURE 24.35   (a) Hobs, used to machine gear teeth; (b) schematic illustration of gear cutting with a hob; and (c) production of a worm gear by hobbing. *Source:* (a) Courtesy of Sandvik Coromant; and (c) Courtesy of Schafer Gear Works, Inc.

tool on a shaper (Fig. 24.36a). The machines for *spiral bevel gears* operate essentially on the same principle, in which the spiral cutter is basically a face-milling cutter, with a number of straight-sided cutting blades protruding from its periphery (Fig. 24.36b).

24.7.4   Gear-finishing Processes

As produced by any of the processes described, the surface finish and dimensional accuracy of gear teeth may still not be sufficient for certain specific applications. Moreover, the gears may be noisy, or their mechanical properties, especially fatigue

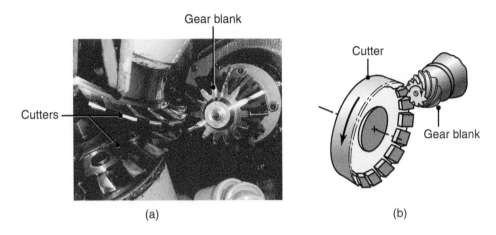

(a)

(b)

FIGURE 24.36   (a) Cutting a straight bevel-gear with two cutters. (b) Cutting a helical bevel gear. *Source:* Courtesy of Schafer Gear Works, South Bend, Indiana, USA.

life and wear resistance, may not be acceptable. Several *finishing processes* are available to improve the surface quality of the gears, the choice being dictated by the method of gear manufacture, the desired performance, and whether the gears have been hardened by heat treatment. As described in Chapter 4, heat treating can cause distortion of parts; consequently, for a precise gear-tooth profile, heat-treated gears typically are subjected to appropriate finishing operations.

Shaving.   The gear-shaving process involves a *cutter*, made in the exact shape of the finished tooth profile, which removes very small amounts of metal from the surface of the gear teeth. The cutter, which has a reciprocating motion, has teeth that are slotted or gashed at several points along its width, making the process similar to fine broaching. Shaving and burnishing (described next) can be performed only on gears with a hardness of 40 HRC or lower.

Although tooling is expensive and special machines are required, shaving is rapid and is the most commonly used process for gear finishing. It produces gear teeth with improved surface finish and good dimensional accuracy of the tooth profile. Shaved gears subsequently may be heat treated and then ground for improved hardness, wear resistance, and a more accurate tooth profile.

Burnishing.   The surface finish of gear teeth also can be improved by burnishing. Introduced in the 1960s, burnishing is basically a surface plastic-deformation process (see Section 34.2), using a special hardened, gear-shaped burnishing die, which subjects the tooth surfaces to a surface-rolling action (called **gear rolling**). The resulting cold working of the tooth surfaces not only improves the surface finish, but also induces compressive residual stresses on the surfaces of the gear teeth, thus improving their fatigue life. It has been shown, however, that burnishing does not significantly improve the dimensional accuracy of the gear tooth. With powder metallurgy gears, burnishing leads to surface densification and with a significant improvement in performance (see Section 17.5).

Grinding, Honing, and Lapping.   For the highest dimensional accuracy in tooth spacing and form, and surface finish, gear teeth subsequently may be ground, honed, and lapped, as described in Chapter 26. Specially dressed grinding wheels are used, for either forming or for generating gear-tooth surfaces. There are several types of grinders, with the single-index form grinder being the most commonly used. In **form**

QR Code 24.6 Cutting of a bevel gear. (*Source:* Courtesy of Sandvik Coromant)

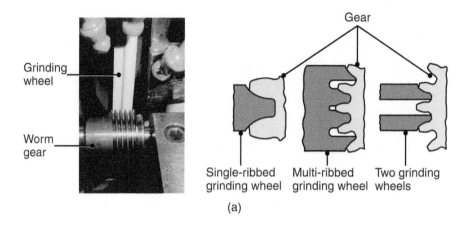

(a)

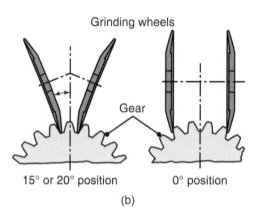

(b)

FIGURE 24.37 Finishing gears by grinding: (a) form grinding, with shaped grinding wheels and (b) grinding by generating, using two grinding wheels.

grinding, the shape of the grinding wheel is identical to that of the tooth spacing (Fig. 24.37a). In **generating,** the grinding wheel acts in a manner similar to the gear-generating cutter, described previously (Fig. 24.37b).

The **honing** tool is a plastic gear impregnated with fine abrasive particles. The honing process is faster than grinding, and is used to improve surface finish. To further improve the finish, ground gear teeth are **lapped,** using abrasive compounds with either (a) a gear-shaped lapping tool made of cast iron or bronze, or (b) a pair of mating gears that are run together. Although production rates are lower and costs are higher, these finishing operations are particularly suitable for producing hardened gears of very high quality, long life, and quiet operation.

24.7.5   Design Considerations and Economics of Gear Machining

Design considerations for gear-cutting operations may be summarized as follows:

- Gears should preferably be machined prior to their assembly on shafts; wide gears are more difficult to machine than narrow ones.
- Sufficient clearance should be provided between gear teeth and flanges, shoulders, and other features of the part, so that the cutting tool can machine without any interference.

- Blank design is important for proper fixturing and to ease cutting operations. Machining allowances must be provided in blanks, and if machining is to be followed by subsequent finishing operations, the part must still be oversized after machining; that is, it must have a finishing allowance after being machined.
- Spur gears are easier to machine than helical gears, which, in turn, are easier to machine than bevel gears or worm gears.
- Dimensional tolerances and standardized gear shapes are specified by industry standards. A gear quality number should be selected so that the gear has as wide a tolerance range as possible while still meeting performance requirements in service.

Economics.   As in all machining operations, the cost of gears increases rapidly with improved surface finish and gear quality. Figure 24.38 shows the relative manufacturing cost of gears as a function of quality, as specified by the AGMA (American Gear Manufacturers Association). The higher the number, the higher is the dimensional accuracy of the gear teeth. As noted in this figure, the manufacturing cost can vary by two orders of magnitude, depending on dimensional tolerances.

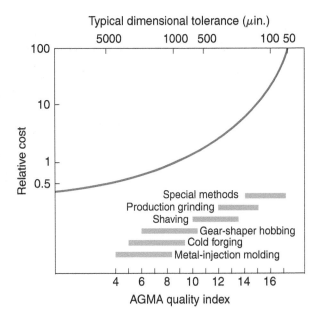

FIGURE 24.38   Gear manufacturing cost as a function of gear quality.

SUMMARY

- Various complex shapes can be machined by the processes described in this chapter. Milling is one of the most common machining processes, because it is capable of economically producing a variety of shapes and sizes from a large number of workpiece materials.
- Although these processes basically are similar to turning, drilling, and boring, and involve similar cutting mechanics, tool materials, and cutting fluids, most of the processes described in this chapter utilize multitooth cutters and tools at various axes with respect to the workpiece.
- Machine tools used to produce complex shapes are mostly computer controlled, having various dedicated features, and imparting much more flexibility in their application than traditional machine tools.
- Broaching is a method of accurately enlarging a round hole or other profiles in a workpiece. Sawing is the gradual removal of material by small teeth spaced on a saw, and is very versatile. Filing involves small-scale removal of material from a surface, especially the removal of burrs and sharp profiles.
- In addition to being produced by the various forming and shaping processes, gears also are produced by machining, either by form cutting or generating; the latter produces gears with better surface finish and higher dimensional accuracy. The quality of the gear-tooth profile is further improved by finishing operations, such as shaving, burnishing, grinding, honing, and lapping.

KEY TERMS

Arbor	Filing	Indexing	Sawing
Broaching	Fly cutting	Kerf	Shaping
Bur	Form cutting	Lapping	Shaving
Burnishing	Friction sawing	Milling	Slab milling
Climb milling	Gear generating	Planing	Tooth set
Die sinking	High-speed milling	Pull broach	Turn broaching
End milling	Hob	Push broach	Work holding
Face milling	Honing	Rack shaper	

BIBLIOGRAPHY

ASM Handbook, Vol. 16: **Machining**, ASM International, 1989.

Boothroyd, G., and Knight, W.A., **Fundamentals of Machining and Machine Tools**, 3rd ed., Marcel Dekker, 2005.

Brown, J., **Advanced Machining Technology Handbook**, McGraw-Hill, 1998.

Davim, J.P. (ed.), **Machining: Fundamentals and Recent Advances**, Springer, 2010.

Davis, J.R., (ed.), **Gear Materials, Properties and Manufacture**, ASM International, 2006.

Joshi, P.H., **Machine Tools Handbook**, McGraw-Hill, 2008.

Kibbe, R.R., Neely, J.E., White, W.T., and Meyer, R.O., **Machine Tool Practices**, 9th ed., Prentice Hall, 2009.

Krar, S.F., and Check, A.F., **Technology of Machine Tools**, 6th ed., Glencoe Macmillan/McGraw-Hill, 2009.

Lopez, L.N., and Lamikiz, A. (eds.), **Machine Tools for High Performance Machining**, Springer, 2009.

Radzevich, S.P., **Dudley's Handbook of Practical Gear Design and Manufacture**, 2nd ed., CRC Press, 2012.

Shaw, M.C., **Metal Cutting Principles**, 2nd ed., Oxford University Press, 2005.

Stephenson, D.A., and Agapiou, J.S., **Metal Cutting: Theory and Practice**, 2nd ed., Marcel Dekker, 2006.

Walsh, R.A., **McGraw-Hill Machining and Metalworking Handbook**, 3rd ed., McGraw-Hill, 2006.

REVIEW QUESTIONS

24.1 Explain why milling is such a versatile machining operation.

24.2 Describe a milling machine. How is it different from a drill press?

24.3 Describe the different types of cutters used in milling operations and give an application of each type.

24.4 Define the following: face milling, peripheral milling, shoulder milling, slot milling, thread milling.

24.5 Can threads be machined on a mill? Explain.

24.6 What is the difference between feed and feed per tooth? Can they ever be the same?

24.7 Explain the relative characteristics of climb milling and up milling.

24.8 Describe the geometric features of a broach and explain their functions.

24.9 What is a pull broach? A push broach?

24.10 Why is sawing a commonly used process? Why do some saw blades have staggered teeth? Explain.

24.11 What advantages do bed-type milling machines have over column-and-knee-type machines for production operations?

24.12 Explain why the axis of a hob is tilted with respect to the axis of the gear blank.

24.13 What is a shell mill? Why is it used?

24.14 Why is it difficult to saw thin sheet metals?

24.15 Of the processes depicted in Fig. 24.2, which is the most similar to hobbing?

24.16 Describe the tool motion during gear shaping.

24.17 When is filing necessary?

QUALITATIVE PROBLEMS

24.18 Would you consider the machining processes described in this chapter to be near-net or net-shape processing? Explain with appropriate examples.

24.19 Why is end milling such an important versatile process? Explain with examples.

24.20 List and explain factors that contribute to poor surface finish in the processes described in this chapter.

24.21 Are the feed marks left on the workpiece by a face-milling cutter true segments of a true circle? Explain, with appropriate sketches.

24.22 Explain why broaching crankshaft bearings is an attractive alternative to other machining processes.

24.23 Several guidelines are presented in this chapter for various cutting operations. Discuss the reasoning behind these guidelines.

24.24 What are the advantages of helical teeth over straight teeth on cutters for slab milling?

24.25 Explain why hacksaws are not as productive as band saws.

24.26 What similarities and differences are there in slitting with a milling cutter and with a saw?

24.27 Why do machined gears have to be subjected to finishing operations? Which of the finishing processes are not suitable for hardened gear teeth? Why?

24.28 How would you reduce the surface roughness shown in Fig. 24.8? Explain.

24.29 Why are machines such as the one shown in Fig. 24.20 so useful?

24.30 Comment on your observations concerning the designs illustrated in Fig. 24.23b and on the usefulness of broaching operations.

24.31 Explain how contour cutting could be started in a band saw, as shown in Fig. 24.28d.

24.32 In Fig. 24.30a, high-speed steel cutting teeth are welded to a steel blade. Would you recommend that the whole blade be made of high-speed steel? Explain your reasons.

24.33 Describe the parts and conditions under which broaching would be the preferred method of machining.

24.34 With appropriate sketches, explain the differences between and similarities among shaving, broaching, and turn-broaching operations.

24.35 Explain the reason that it is difficult to use friction sawing on nonferrous metals.

24.36 Would you recommend broaching a keyway on a gear blank before or after machining the gear teeth? Why?

QUANTITATIVE PROBLEMS

24.37 In milling operations, the total cutting time can be significantly influenced by (a) the magnitude of the noncutting distance, l_c, shown in Figs. 24.5 and 24.6, and (b) the ratio of width of cut, w, to the cutter diameter, D. Sketch several combinations of these parameters, give dimensions, select feeds and cutting speeds, etc., and determine the total cutting time. Comment on your observations.

24.38 A slab-milling operation is being performed at a specified cutting speed (surface speed of the cutter) and feed per tooth. Explain the procedure for determining the table speed required.

24.39 Show that the distance l_c in slab milling is approximately equal to $\sqrt{Dd}$ for situations where $D \gg d$. (See Fig. 24.5c.)

24.40 In Example 24.1, which of the quantities will be affected when the feed is increased to $f = 0.02$ in./tooth?

24.41 Calculate the chip depth of cut, t_c, and the torque in Example 24.1.

24.42 Estimate the time required to face mill a 10-in.-long, 1-in.-wide brass block with a 6-in.-diameter cutter with 10 high-speed steel inserts.

24.43 A 12-in.-long, 1-in.-thick plate is being cut on a band saw at 150 ft/min. The saw has 12 teeth per inch. If the feed per tooth is 0.003 inch, how long will it take to saw the plate along its length?

24.44 A single-thread hob is used to cut 40 teeth on a spur gear. The cutting speed is 120 ft/min and the hob is 3 in. in diameter. Calculate the rotational speed of the spur gear.

24.45 Assume that in the face-milling operation shown in Fig. 24.6 the workpiece dimensions are 4 in. by 10 in. The cutter is 6 in. in diameter, has eight teeth, and rotates at 300 rpm. The depth of cut is 0.125 in. and the feed is 0.005 in./tooth. Assume that the specific energy requirement for this material is 2 hp-min/in.^{3} and that only 75% of the cutter diameter is engaged during cutting. Calculate (a) the power required and (b) the material-removal rate.

24.46 A slab-milling operation will take place on a part 300 mm long and 40 mm wide. A helical cutter 75 mm in diameter with 10 teeth will be used. If the feed per tooth is 0.2 mm/tooth and the cutting speed is 0.75 m/s, find the machining time and metal-removal rate for removing 6 mm from the surface of the part.

24.47 Explain whether the feed marks left on the workpiece by a face-milling cutter (as shown in Fig. 24.16a) are segments of true circles. Describe the parameters you consider in answering this question.

24.48 In describing the broaching operations and the design of broaches, we have not given equations regarding feeds, speeds, and material-removal rates, as we have done in turning and milling operations. Review Fig. 24.24 and develop such equations.

SYNTHESIS, DESIGN, AND PROJECTS

24.49 The parts shown in Fig. 24.1 are to be machined from a rectangular blank. Suggest the machine tool(s) required, the fixtures needed, and the types and sequence of operations to be performed. Discuss your answer in terms of the workpiece material, such as aluminum versus stainless steel.

24.50 Would you prefer to machine the part in Fig. 24.1f from a preformed blank (near-net shape) rather than a rectangular blank? If so, how would you prepare such a blank? How would the number of parts required influence your answer?

24.51 If expanded honeycomb panels (see Section 16.12) were to be machined in a form-milling operation, what precautions would you take to keep the sheet metal from distorting due to tool forces? Think up as many solutions as you can.

24.52 Assume that you are an instructor covering the topics described in this chapter and you are giving a quiz on the numerical aspects to test the understanding of the students. Prepare two quantitative problems and supply the answers.

24.53 Suggest methods whereby milling cutters of various designs (including end mills) can incorporate carbide inserts.

24.54 Prepare a comprehensive table of the process capabilities of the machining processes described in this chapter. Using several columns, list the machines involved, types of tools and tool materials used, shapes of blanks and parts produced, typical maximum and minimum sizes, surface finish, dimensional tolerances, and production rates.

24.55 On the basis of the data developed in Problem 24.54, describe your thoughts regarding the procedure to be followed in determining what type of machine tool to select when machining a particular part.

24.56 Make a list of all the processes that can be used in manufacturing gears, including those described in Parts II and III of this text. For each process, describe the advantages, limitations, and quality of gears produced.

Machining Centers, Machine-Tool Structures, and Machining Economics

- This chapter presents the characteristics, types, and advantages of machining centers, and the concept of reconfigurable machine tools.
- Emphasis is placed on the importance of understanding the performance of machine tools, and their modules and components, particularly with regard to stiffness, vibration, chatter, and damping characteristics. These are important considerations not only for quality and dimensional accuracy, but also because of their influence on tool life, productivity, and the economics of machining operations.
- Presented next are high-speed machining, hard machining, and ultraprecision machining operations, topics that are strongly tied to the economics of machining.
- The chapter ends with a simple method of cost analysis for determining the conditions under which machining parameters can be selected, so that machining cost per piece or machining time per piece can be minimized.

25.1 Introduction

The preceding four chapters have described machining operations and machine tools, but have not emphasized the widespread integration of advanced computer technology and the flexibility it allows in manufacturing operations. Computers have dramatically improved the capabilities of machine tools, whereby they now have the capability of rapidly and repeatedly producing very complex part geometries. The program controlling a machine tool can incorporate changes in cutting conditions, compensate for tool wear, automatically change tools, and machine a workpiece without refixturing or having to transfer it to another machine tool, as had been the practice for many years.

In addition to advanced computer technologies, vibration and chatter and their avoidance, high-speed machining, hard machining, and advanced analysis of machining economics are now highly developed and have revolutionized machining operations.

25.2 Machining Centers

In describing the individual machining processes and machine tools in the preceding chapters, it was noted that each machine, regardless of how highly it is automated, is designed to perform basically the same type of operation, such as turning, boring,

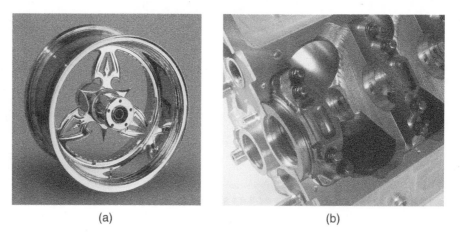

(a)   (b)

FIGURE 25.1   Examples of parts that can be machined on machining centers using various processes such as turning, facing, milling, drilling, boring, reaming, and threading; such parts ordinarily would require the use of a variety of machine tools to complete. (a) Forged motorcycle wheel, finish machined to tolerance and subsequently polished and coated. (b) Detailed view of an engine block, showing complex cavities, threaded holes, and planar surfaces. *Source:* (a) Courtesy of R.C. Components; (b) courtesy of Donovan Engineering, programming by N. Woodruff, and Photography by E. Dellis, Powersports Photography.

drilling, milling, broaching, planing, or shaping. It was also shown that most parts manufactured by the methods described throughout this book require further operations on their various surfaces before they are completed. Note, for example, that the parts shown in Fig. 25.1 have a variety of complex geometric features, and that all of the surfaces on these parts require a different type of machining operation to meet certain specific requirements concerning shapes, features, dimensional tolerances, and surface finish. Note also the following observations:

- Recall that some possibilities exist in *net-shape* or *near-net shape* production of these parts, depending on specific constraints on shapes, dimensional tolerances, detailed surface features, surface finish, and various mechanical and other properties to meet service requirements. Shaping processes that are candidates for such parts are precision casting, powder metallurgy, powder-injection molding, and precision forging. Even then, however, it is very likely that the parts will still require some additional finishing operations. For example, small-diameter deep holes, threaded holes, flat surfaces for sealing with gaskets, parts with very close dimensional tolerances, sharp corners and edges, and flat or curved surfaces, with different surface-finish requirements, will require further machining operations.
- If some machining is required, or if it is shown to be more economical to finish machining these parts to their final shapes, then it is obvious that none of the machine tools described in Chapters 23 and 24 could, *individually* and *completely*, produce the parts. Note also that traditionally, machining operations are performed by moving the workpiece from one machine tool to another, until all of the required machining operations are completed.

The Concept of Machining Centers.   The traditional method of machining parts by using different types of machine tools has been, and continues to be, a viable manufacturing method. This method can be highly automated in order to increase productivity, and in fact it is the principle behind **transfer lines**, also called *dedicated* *manufacturing lines* (DML), as described in Section 37.2.4. Commonly used in *high-volume* or *mass production*, transfer lines consist of several specific (*dedicated*) machine tools, arranged in a logical sequence. The workpiece, such as an automotive engine block, is moved from one station to another, with a specific machining operation performed at each station, after which it is transferred to the next machine for further specific machining operations.

There are situations, however, where transfer lines are not feasible or economical, particularly when the types of products to be processed change rapidly, due to factors such as product demand or changes in product shape or style. It is very costly and time-consuming to rearrange these machine tools to respond to the needs for the next and different production cycle. An important concept that addresses flexibility in manufacturing, developed in the late 1950s, is that of **machining centers.**

A machining center (Fig. 25.2) is an advanced computer-controlled machine tool that is capable of performing a variety of machining operations, on different surfaces and different orientations of a workpiece, without having to remove it from its work-holding device or fixture. The workpiece generally is stationary and the cutting tools rotate, as they do in such operations as milling, drilling, honing, and tapping. Whereas, in transfer lines or in traditional shops and factories, the workpiece is brought *to the machine*; in machining centers it is the machining operation that is brought *to the workpiece*.

The development of machining centers is related closely to advances in automation and computer control of machine tools, the details of which are described in Chapter 37. Recall that, as an example of the advances in modern lathes, Fig. 23.10 illustrates a numerically controlled lathe (**turning center**), with two turrets, each carrying several cutting tools.

Components of a Machining Center.   The workpiece in a machining center is placed on a **pallet**, or *module*, that can be moved and swiveled (oriented) in various directions (Fig. 25.3). After a particular machining operation has been completed, another operation begins, which may require reindexing of the workpiece on its pallet. After all of the machining operations have been completed, the pallet automatically moves away with the finished part, and another pallet, carrying another workpiece or

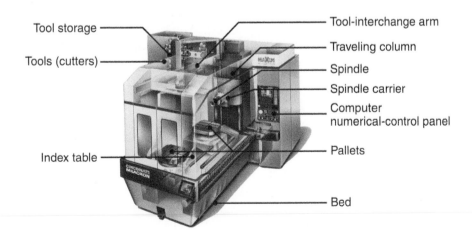

Tool storage — Tool-interchange arm — Traveling column — Spindle — Spindle carrier — Computer numerical-control panel — Pallets — Bed

Tools (cutters) —

Index table —

FIGURE 25.2 A horizontal-spindle machining center equipped with an automatic tool changer; tool magazines can store up to 200 cutting tools of various functions and sizes. *Source:* Courtesy of Cincinnati Milacron.

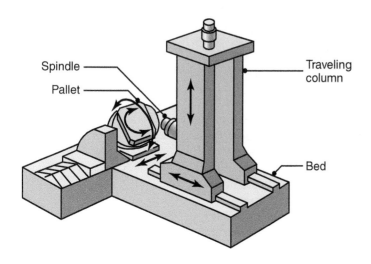

FIGURE 25.3 Schematic illustration of the principle of a five-axis machining center. Note that, in addition to possessing three linear movements (three axes), the pallet, which supports the workpiece, can be swiveled around two axes (hence a total of five axes), allowing the machining of complex shapes, such as those shown in Fig. 25.1. *Source:* Courtesy of Toyoda Machinery.

workpieces to be machined, is brought into position by an **automatic pallet changer** (Fig. 25.4). All movements are computer controlled, with pallet-changing cycle times on the order of only 10–30 s. Pallet stations are available with several pallets serving one machining center. The machines also can be equipped with various automatic features, such as part loading and unloading devices.

A machining center is equipped with a programmable **automatic tool changer** (ATC). Depending on the particular design, up to 100 cutting tools can be stored in a magazine, drum, or chain (*tool storage*). *Auxiliary* tool storage also is available on some special and large machining centers, raising the tool capacity to 200. The cutting tools are selected automatically for the shortest route to the machine spindle. The maximum dimensions that the cutting tools can reach around a workpiece in a machining center is known as the **work envelope**, a term that first was used in connection with industrial robots, as described in Section 37.6.

The **tool-exchange arm** shown in Fig. 25.5 is a common design; it swings around to pick up a particular tool and places it in the spindle; note that each tool has its own toolholder, which makes the transfer of cutting tools to the machine spindle highly efficient. Tools are identified by bar codes, QR codes, or coded tags attached directly to their toolholders. Tool-changing times are typically between 5 and 10 s, but may be up to 30 s for tools weighing up to 110 kg (250 lb), and less than one second for small tools.

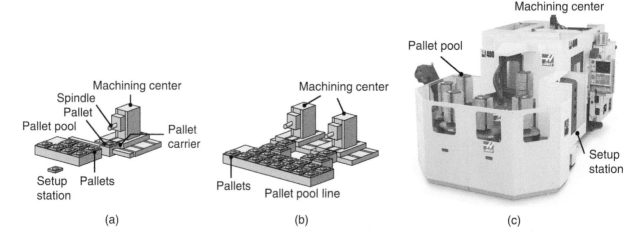

FIGURE 25.4 (a) Schematic illustration of the top view of a horizontal-spindle machining center showing the pallet pool, setup station for a pallet, pallet carrier, and an active pallet in operation (shown directly below the spindle of the machine). (b) Schematic illustration of two machining centers with a common pallet pool. (c) A pallet pool for a horizontal-spindle machining center; various other pallet arrangements are possible in such systems. *Source:* (a) and (b) Courtesy of Hitachi Seiki Co., Ltd.; (c) Courtesy of Haas Automation, Inc.

Machining centers may be equipped with a **tool-checking** and/or **part-checking station** that feeds information to the machine control system, so that it can compensate for any variations in tool settings or tool wear. **Touch probes** (Fig. 25.6) can be installed into a toolholder to determine workpiece-reference surfaces, for selection of tool settings and for online inspection of parts being machined.

Note in Fig. 25.6 that several surfaces can be contacted (see also *sensor technology*, Section 37.7), and that their relative positions are determined and stored in the database of the computer software. The data are then used to program tool paths (see, for example, Fig. 37.12) and to compensate for tool length, tool diameter, and for tool wear, in more advanced machine tools. Noncontact probes also can be used, and can measure dimensions, surface roughness, or temperature.

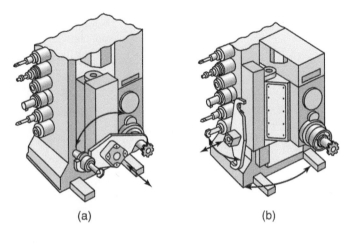

(a)   **(b)**

FIGURE 25.5   Swing-around tool changer on a horizontal-spindle machining center. (a) The tool-exchange arm is placing a toolholder with a cutting tool into the machine spindle; note the axial and rotational movement of the arm. (b) The arm is returning to its home position; note its rotation along a vertical axis after placing the tool and the two degrees of freedom in its home position.

25.2.1 Types of Machining Centers

There are various designs for machining centers. The two basic types are vertical spindle and horizontal spindle, although many machines are capable of operating along both axes.

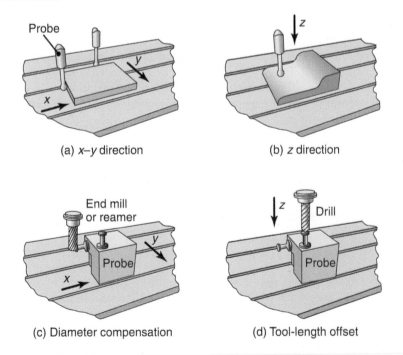

(a) *x–y* direction   (b) *z* direction

(c) Diameter compensation   (d) Tool-length offset

FIGURE 25.6   Touch probes used in machining centers for determining workpiece and tool positions and surfaces relative to the machine table or column. Touch probe determining (a) the $X-Y$ (horizontal) position of a workpiece; (b) the height of a horizontal surface; (c) the planar position of the surface of a cutter (e.g., for cutter–diameter compensation); and (d) the length of a tool for tool-length offset.

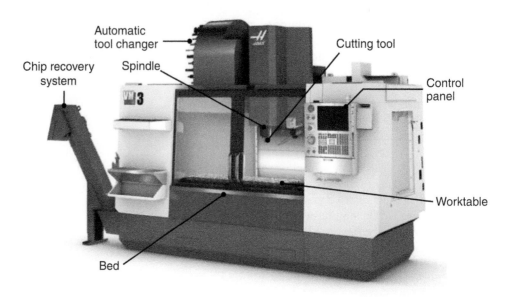

FIGURE 25.7 A vertical-spindle machining center; the tool changer is on the left of the machine, and has a 40 tool magazine. *Source:* Courtesy of Haas Automation, Inc.

Vertical-spindle Machining Centers. Also called *vertical machining centers* (VMC), these machines are capable of performing various machining operations on parts with deep cavities, as required in mold and die making (also called *die sinking*). A vertical-spindle machining center, which is similar to a vertical-spindle milling machine, is shown in Fig. 25.7. The tool magazine is on the left of the machine, and all operations and movements are directed and modified through the computer control panel, shown on the right. Because the thrust forces in vertical machining are directed downward, such machines have high stiffness and hence produce parts with good dimensional accuracy. VMCs generally are less expensive than horizontal-spindle machines of similar capacity.

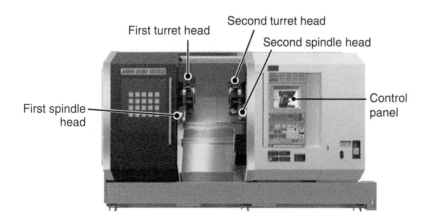

FIGURE 25.8 A computer numerical-controlled turning center. The two spindle heads and two turret heads make the machine very flexible in its machining capabilities; up to three turret heads are commercially available. *Source:* Courtesy of DMG/MORI SEIKI.

Horizontal-spindle Machining Centers. Also called *horizontal machining centers* (HMC), these machines are suitable for large as well as tall workpieces that require machining on a number of their surfaces. The pallet can be swiveled on different axes (e.g., see Fig. 25.3) to various angular positions.

Turning Centers. This is another category of horizontal-spindle machines, and basically are computer-controlled *lathes*, with several features. A multi-turret turning center is shown in Fig. 25.8. It is constructed with two horizontal spindles and two turrets, equipped with a variety of cutting tools used to perform several operations on a rotating workpiece. The turrets can be powered to allow for drilling or milling operations within the CNC turning center, and without the need to refixture the workpiece. For this reason, such machines are often referred to as **CNC Mill-turn Centers.**

Universal Machining Centers.   These machines are equipped with both vertical and horizontal spindles. They have a variety of features and are capable of simultaneously machining all of the surfaces of a workpiece, i.e., vertically, horizontally, and at a wide range of angles.

25.2.2   Characteristics and Capabilities of Machining Centers

The major characteristics of machining centers are summarized as:

- Machining centers are capable of handling a wide variety of part sizes and shapes efficiently, economically, repetitively, and with high dimensional accuracy and with tolerances on the order of ± 0.0025 mm (0.0001 in.).
- These machines are versatile and capable of quick changeover from one type of product to another.
- The time required for loading and unloading workpieces, changing tools, gaging of the part being machined, and troubleshooting is reduced. Because of the inherent flexibility of machining centers, the workpiece may not need to be refixtured during machining, referred to as the *one and done* approach. Productivity is improved, labor requirements (particularly skilled labor) are reduced, and production costs are minimized.
- These machines can be equipped with tool-condition monitoring devices for the detection of tool breakage and wear, as well as with probes for tool-wear compensation and tool positioning.
- In-process and postprocess gaging and inspection of machined workpieces are now features of machining centers.
- These machines are relatively compact and highly automated, and have advanced control systems, so one operator can attend to two or more machining centers at the same time, thus reducing labor costs.

Because of the high productivity of machining centers, large amounts of chips are produced and must be collected and disposed of properly (*chip management*, Section 23.3.7). Several system designs are available for *chip collection*, with one or more chain or spiral (screw) conveyors; they collect the chips along troughs in the machine and deliver them to a collecting point (see Fig. 25.7).

Machining centers are available in a wide variety of sizes and features. Typical capacities range up to 75 kW (100 hp). Maximum spindle speeds are usually in the range from 4000 to 8000 rpm, and some are as high as 75,000 rpm for special applications, using small-diameter cutters. Modern spindles can accelerate to a speed of 20,000 rpm in only 1.5 s. Some pallets are capable of supporting workpieces weighing as much as 7000 kg (15,000 lb), although even higher capacities are available for special applications. The cost of machining centers ranges from about $50,000 to $1 million and higher.

25.2.3   Selection of Machining Centers

Machining centers generally require significant capital expenditure; to be cost effective, they may have to be operated for more than one shift per day. Consequently, there must be sufficient and continued demand for parts to justify their purchase. Because of their inherent versatility, however, machining centers can be used to produce a wide range of products, particularly for *mass customization* or *just-in-time manufacturing*, as described in Section 39.5.

The selection of the type and size of machining centers depends on several factors, especially the following:

- Type of products, their size, and shape complexity
- Type of machining operations to be performed and the type and number of cutting tools required
- Dimensional accuracy specified
- Production rate required

CASE STUDY 25.1   Machining Outer Bearing Races on a Turning Center

Outer bearing races (Fig. 25.9) are machined on a turning center. The starting material is a hot-rolled 52100 steel tube, with 91 mm (3.592 in.) OD and 75.5 mm (2.976 in.) ID. The cutting speed is 95 m/min (313 ft/min) for all operations. All tools are carbide, including the cutoff tool (used in the last operation shown), which is 3.18 mm ($\frac{1}{8}$ in.), instead of 4.76 mm ($\frac{3}{16}$ in.) for the high-speed steel cutoff tool that formerly was used.

The amount of material saved by this change is significant, because the race width is small. The turning center was able to machine these races at high speeds and with repeatable tolerances of ±0.025 mm (0.001 in.). (See also Example 23.2.)

Source: Based on data from McGill Manufacturing Company.

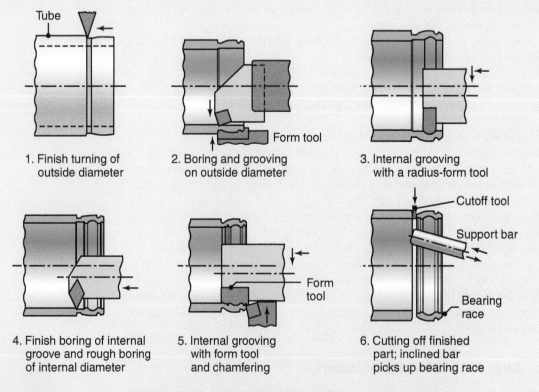

FIGURE 25.9   Steps in machining of outer bearing races.

25.2.4 Reconfigurable Machines and Systems

The need for the flexibility of manufacturing processes has led to the more recent concept of *reconfigurable machines*, consisting of various modules. The term reconfigurable stems from the fact that, by using advanced computer hardware and reconfigurable controllers, and utilizing advances in information management technologies, the machine components can be arranged and rearranged quickly into a number of configurations to meet specific production demands.

Figure 25.10 shows an example of how the basic machine-tool structure of a three-axis machining center can be reconfigured to become a *modular* machining center. With such flexibility, the machine can perform different machining operations while accommodating various workpiece sizes and part geometries. Another example is given in Fig. 25.11, where a five-axis (three linear and two rotational movements) machine can be reconfigured by assembling different modules.

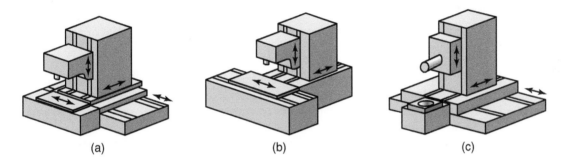

(a)   (b)   (c)

FIGURE 25.10 Schematic illustration of a reconfigurable modular machining center capable of accommodating workpieces of different shapes and sizes, and requiring different machining operations on their various surfaces. *Source:* After Y. Koren.

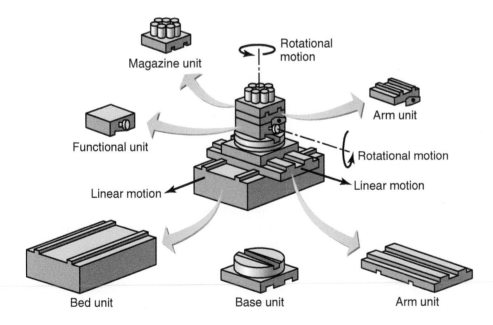

FIGURE 25.11 Schematic illustration of the assembly of different components of a reconfigurable machining center. *Source:* After Y. Koren.

Reconfigurable machines have the promise of (a) improving the productivity and efficiency of manufacturing operations, (b) reducing lead time for production, and (c) providing a cost-effective and rapid response to market demands (see also Chapter 39). These capabilities are significant, especially in view of the frequent introduction of new products into a highly competitive global marketplace, fluctuations in product demand and product mix, and unpredictable modifications in product design.

25.3 Machine-tool Structures

This section describes the materials and design aspects of machine-tool structures that are important in producing parts, with acceptable geometric features and dimensional and surface finish characteristics.

25.3.1 Materials

The following is a list of the materials that commonly have been used or are suitable for machine-tool structures.

- **Gray cast iron** was the first material used in machine-tool structures, and has the advantages of good damping capacity and low cost, but the disadvantage of being heavy. Most machine-tool structures are made of class 40 cast iron; some are made of class 50 (see Table 12.4).
- **Welded steel** structures (see Chapters 30 and 31) are lighter than cast-iron structures. Wrought steels, typically used in these structures, (a) are available in a wide range of section sizes and shapes, such as channels, angles, and tubes, (b) have good mechanical properties, (c) possess good manufacturing characteristics, such as formability, machinability, and weldability, and (d) have low cost. Structures made of steels can have high stiffness-to-weight ratios, by using cross-sections such as tubes and channels; in contrast, however, their damping capacity is very low.
- **Ceramic** components (Chapters 8 and 18) are used in advanced machine tools for their strength, stiffness, corrosion resistance, surface finish, and thermal stability. Ceramic components were first introduced in the 1980s. Spindles and bearings now can be made of silicon nitride, which has better friction and wear characteristics than traditional metallic materials. Furthermore, the low density of ceramics makes them suitable as the components of high-speed machinery that undergo rapid reciprocating or rotating movements, in which low inertial forces are desirable to maintain the system's stability, reduce inertial forces, and thus reduce the noncutting time in high-speed machining operations.
- **Composites** (Chapter 9) may consist of a polymer matrix, metal matrix, or ceramic matrix, with various reinforcing materials. The compositions can be tailored to provide appropriate mechanical properties in selected axes of the machine tool. Although they are presently expensive, composites are likely to become significant materials for high-accuracy, high-speed machining applications.
- **Granite–epoxy composites**, with a typical composition of 93% crushed granite and 7% epoxy binder, were first used in precision centerless and internal grinders in the early 1980s (see Section 26.4). These composite materials have several favorable properties: (a) good castability, thus allowing for design versatility in machine tools, (b) high stiffness-to-weight ratio, (c) thermal stability, (d) resistance to environmental degradation, and (e) good damping capacity.

- **Polymer concrete** is a mixture of crushed concrete and plastic (typically poly-methylmethacrylate), and easily can be cast into desired shapes for machine bases and various components. Although it has low stiffness (about one-third that of class 40 cast iron) and poor thermal conductivity, polymer concrete has good damping capacity and also can be used for sandwich construction with cast irons, thus combining the advantages of each type of material. Plain concrete can be poured into cast-iron machine-tool structures, to increase their mass and improve their damping capacity. Filling the cavities of machine bases with loose sand also has been demonstrated to be an effective means of improving damping capacity.

25.3.2 Machine-tool Design Considerations

Important considerations in machine tools generally involve the following factors:

- Design, materials, and construction
- Spindle materials and type of construction
- Thermal distortion of machine components
- Error compensation and the control of moving components along slideways

Stiffness. Stiffness, which is a major factor in the dimensional accuracy of a machine tool, is a function of (a) the elastic modulus of the materials used and (b) the geometry of the structural components, including the spindle, bearings, drive train, and slideways. Machine stiffness can be enhanced by design improvements, such as using diagonally arranged interior ribs.

Damping. Damping is a critical factor in reducing or eliminating vibration and chatter in machining operations. Principally, it involves (a) the types of materials used and (b) the type and number of joints (such as welded versus bolted) in the machine-tool structure. Cast irons and polymer-matrix composites have much better damping capacity than metals or ceramics; also, the greater the number of joints in a machine structure, the more damping there is.

Thermal Distortion. An important factor in machine tools is the thermal distortion of their components, which contributes significantly to their lack of precision.

There are two sources of heat in machine tools:

1. *Internal sources*, such as from bearings, ballscrews, machine ways, spindle motors, pumps, and servomotors, as well as from the cutting zone during machining (Section 21.4).
2. *External sources*, such as from cutting fluids, nearby furnaces, heaters, other nearby machines, sunlight, and fluctuations in ambient temperature from such sources as air-conditioning units, vents, or even someone opening or closing a door or a window.

These considerations are significant, particularly in **precision** and **ultraprecision machining** (Section 25.7), where dimensional tolerances and surface finish are now approaching the nanometer range. The machine tool used for these applications are equipped with the following features:

- Various thermal and geometric real-time error-compensating features, including (a) the modeling of heating and cooling and (b) electronic compensation for accurate ballscrew positions

- Gas or fluid hydrostatic spindle bearings, allowing tools to more easily achieve precise motions without encountering high friction or stick-slip phenomenon (Section 33.4)
- New designs for traction or friction drives, for smoother linear motion
- Extremely fine feed and position controls, using microactuators
- Fluid-circulation channels in the machine-tool base, for maintaining thermal stability

The structural components of the machine tool can be made of materials with high dimensional stability and low coefficient of thermal expansion, such as Super-Invar (Section 3.6), granite, ceramics, and composites. *Retrofitting* also is a viable option for enhancing the performance of older machines.

Assembly Techniques for Machine-tool Components. Traditionally, machine-tool components have been assembled using threaded fasteners and by welding (Part VI). Advanced assembly techniques now include integral casting and resin bonding. Steel guideways, with their higher stiffness, can be cast integrally over a cast-iron bed, using a hybrid casting technology. *Resin bonding* is being used to assemble machine tools, replacing mechanical fastening. Adhesives, described in Section 32.4, have favorable characteristics for machine-tool construction, as they do not require special preparation and are suitable for assembling both nonmetallic and metallic machine components.

Guideways. The preparation of guideways in machine tools traditionally has required significant effort. The plain cast-iron ways in machines, which is the most common material, require much care to achieve the required precision and service life. The movements of various components in a machine tool, along its various axes, usually have utilized high-precision *ballscrews*, *rotating-screw drives*, and *rotary motors*. This system of mechanical and electrical components has several unavoidable design characteristics, such as speed limitations, length restrictions, inertia effects, gear backlash and other errors, wear of the components, and low efficiency. Modern controls can compensate for these characteristics to achieve higher precision as discussed above.

Linear Motor Drives. A *linear motor* is like a typical rotary electric motor that has been rolled out (opened) flat. This is the same principle used in some high-speed ground transportation systems in which the cars are levitated by magnetic forces (Maglev). The sliding surfaces in these drives are separated by an air gap and, as a result, have very low friction and energy loss.

Linear motor drives in machine tools have important advantages:

- Simplicity and minimal maintenance, since there is one moving part and no mechanical linkages
- Smooth operation, better positioning accuracy, and repeatability, at as low as submicron ranges
- A wide range of linear speeds, from 1 μm/s to 5 m/s
- Acceleration rates of about 1–2 g (10–20 m/s^2), and as high as 4–10 g for smaller units
- The moving components do not undergo any wear, because there is no physical contact between the sliding surfaces of the machine

Machine Foundations. Foundation materials, their mass, and the manner in which they are installed in a plant are important considerations, as they help reduce vibration and do not adversely affect the performance of nearby machinery in the plant. For example, in the installation of a special grinder for high-precision grinding of 2.75-m (9-ft) diameter marine-propulsion gears, the concrete foundation was 6.7 m (22 ft) deep. Its large mass, combined with the machine base, reduced the amplitude of vibrations. Even better results can be obtained when a machine is installed on an *independent* concrete slab, that is isolated from the rest of the plant floor with shock-isolation devices.

25.3.3 Hexapod Machines

Developments in the design and materials used for machine-tool structures and their various components are taking place continually, with the purposes of (a) imparting machining flexibility to machine tools, (b) increasing their machining envelope (the space within which machining can be done), and (c) making them lighter. A truly innovative machine-tool structure is a self-contained octahedral (eight-sided) machine frame.

Referred to as **hexapods** (Fig. 25.12) or *parallel kinematic linked machines*, these machines have a design that is based on a mechanism called the *Stewart platform* (after D. Stewart), developed in 1966 and first used to position aircraft cockpit simulators. The main advantage of this system is that the links in the hexapod are loaded axially, thus the bending forces and lateral deflections are minimal, resulting in a very stiff structure.

The workpiece is mounted on a stationary table. Three pairs of *telescoping tubes* (called *struts* or *legs*), each with its own motor and equipped with ballscrews, are used to maneuver a rotating cutting-tool holder. While various features and curved surfaces are being machined, the controller automatically shortens some tubes and extends others, so that the cutter can follow a specified path around the workpiece. Six sets of coordinates are involved in these machines (hence the term *hexapod*, meaning "six legged"): three linear sets and three rotational sets. Every motion of the cutter,

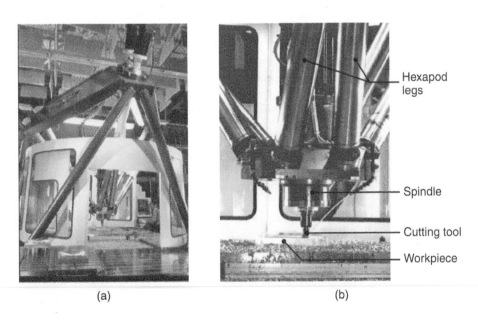

(a) (b)

FIGURE 25.12   (a) A hexapod machine tool, showing its major components. (b) A detailed view of the cutting tool in a hexapod machining center. *Source:* Courtesy of National Institute of Standards and Technology.

even a simple linear motion, is translated into six coordinated leg lengths, moving in real time. The motions of the legs are rapid; consequently, high accelerations and decelerations, with resulting high inertial forces, are involved.

These machines (a) have high stiffness, (b) are not as massive as machining centers, (c) have about one-third fewer parts than machining centers, (d) have a large machining envelope (hence greater access to the work zone), (e) are capable of maintaining the cutting tool perpendicular to the surface being machined (thus improving the machining operation), and (f) have high flexibility (with six degrees of freedom) in the production of parts with various geometries and sizes, without the need to refixture the work in progress. Unlike most machine tools, these are basically portable; in fact, *hexapod attachments* are now available so that a conventional machining center can easily be converted into a hexapod machine.

A limited number of hexapod machines have been built. In view of their potential as efficient machine tools, their performance is being evaluated continually regarding stiffness, thermal distortion, friction within the struts, dimensional accuracy, speed of operation, repeatability, and reliability.

25.4 Vibration and Chatter in Machining Operations

In describing machining processes and machine tools, it was noted on several occasions that *machine stiffness* is as important as any other parameter in machining. Low stiffness can cause *vibration* and *chatter* of the cutting tools and the machine components, and thus can have adverse effects on product quality. Vibration and chatter in machining are complex phenomena, and will be reviewed here briefly as a guide.

Uncontrolled vibration and chatter can result in:

- Poor surface finish, as shown in the right central region of Fig. 25.13
- Loss of dimensional accuracy of the workpiece
- Premature wear, chipping, and failure of the cutting tool, a critical consideration with brittle tool materials, such as ceramics, some carbides, and diamond
- Possible damage to the machine-tool components, from excessive vibration
- Objectionable noise, particularly if it is of high frequency, such as the squeal heard when turning brass on a lathe

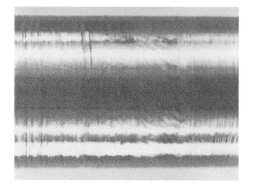

FIGURE 25.13 Chatter marks (right of center of photograph) on the surface of a turned part. *Source:* Courtesy of General Electric Company.

There are two basic types of vibration in machining: forced and self-excited.

Forced Vibration. Forced vibration generally is caused by some *periodic* applied force that develops in the machine tool, such as that from gear drives, imbalance of the machine-tool components, misalignment, and motors and pumps. In operations such as the milling or turning of a splined shaft, or a shaft with a keyway or radial hole, forced vibrations are caused by the periodic engagement of the cutting tool with the workpiece surface (see, for example, Figs. 24.9 and 24.14).

The basic solution to forced vibration is to *isolate* or remove the forcing element. For example, if the forcing frequency is at or near the natural frequency of a machine-tool system component, one of these two frequencies may be raised or lowered. The amplitude of vibration can be reduced by increasing the stiffness or by damping the system.

The cutting parameters generally do not appear to greatly influence the magnitude of forced vibrations; however, changing the cutting speed and the tool geometry can be helpful. It is also recognized that the source of vibrations can be minimized by changing the configuration of the machine-tool components, as may be done when the driving forces are close to, or act through, the *center of gravity* of a particular component. This approach will reduce the bending moment on the component, thus reducing deflections and improving dimensional accuracy.

Self-excited Vibration.   Generally called **chatter**, self-excited vibration is caused by the interaction of the chip-removal process with the structure of the machine tool. Self-excited vibrations usually have a very high amplitude, and are audible. Chatter typically begins with a disturbance in the cutting zone, such as by (a) the type of chips produced, (b) inhomogeneities in the workpiece material or its surface condition, and (c) variations in the frictional conditions at the tool–chip interface, as influenced by cutting fluids and their effectiveness.

The most important type of self-excited vibration is **regenerative chatter**, which is caused when a tool is cutting a surface that has a roughness or geometric disturbances developed from the previous cut (e.g., see Figs. 21.2 and 21.21). Thus, the depth of cut varies, and the resulting variations in the cutting force subject the tool to vibrations; the process continues repeatedly, hence the term *regenerative*. This type of vibration easily can be observed while driving a car over a rough road, the so-called *washboard effect*.

Self-excited vibrations generally can be controlled by:

- Increasing the *stiffness* and, especially, the *dynamic stiffness* of the system; the system includes not only the tool, tool holder, machine frame, etc., but also the *workpiece* and how it is supported
- *Damping* the system

Dynamic stiffness is defined as the ratio of the applied-force amplitude to the vibration amplitude. For example, recall that in the trepanning operation, shown in Fig. 23.23b, there are four machine components involved in the deflections that would cause vibrations: (a) spindle, (b) supporting arm for the cutting tool, (c) drill, and (d) cutting tool.

Experience and analysis of the system would indicate that, unless all of the machine components are sufficiently stiff, the trepanning operation likely will lead to chatter, beginning with torsional vibration around the spindle axis and twisting of the arm. Two similar examples are (a) long and slender drills that may undergo torsional vibrations and (b) cutting tools that are long or are not well supported, such as that shown schematically in Fig. 23.3.

Factors Influencing Chatter.   It has been observed that the tendency for chatter during machining is proportional to the cutting forces and the depth and width of the cut. Because cutting forces increase with strength (hence with hardness of the workpiece material), the tendency to chatter generally increases as hardness increases. Thus, aluminum and magnesium alloys, for example, have a lower tendency to chatter than do martensitic and precipitation-hardening stainless steels, nickel alloys, and high-temperature and refractory alloys.

Another important factor in chatter is the type of chip produced during cutting operations. Continuous chips involve basically steady cutting forces, and such chips generally do not cause chatter; discontinuous chips and serrated chips (Fig. 21.5), on the other hand, may do so. These types of chips are produced periodically, and the resulting force variations during cutting can thus cause chatter. Other factors that may contribute to chatter are the use of dull tools or cutters, lack of cutting fluids, and worn machine-tool ways and components.

Damping.   *Damping* is defined as the rate at which vibrations decay. This effect can be demonstrated on an automobile's shock absorbers, by pushing down on the car's front or rear end and observing how rapidly the vertical motion stops. Damping is a major factor in controlling machine-tool vibration and chatter; it consists of internal and external damping:

1. **Internal damping** results from the *energy loss* in materials during vibration; for example, composite materials have a higher damping capacity than gray cast iron, as shown in Fig. 25.14. The difference in the damping capacity of materials can easily be observed by striking them with a gavel and listening to the sound. For example, try striking a brass cymbal, a piece of concrete, and a piece of wood, and listen to the variations in their sound.

   Bolted joints in the structure of a machine tool also are a source of damping, their effectiveness depending on size, location, and the number of joints. Because friction dissipates energy, small relative movements along dry (unlubricated) joints increase damping. Because machine tools consist of a number of large and small components, assembled by various means, this type of damping is cumulative. Note in Fig. 25.15, for example, how overall damping increases as the number of components on a lathe and their contact areas increase. However, the overall stiffness of the machine tool will decrease as the number of joints increases. As described and illustrated in Fig. 23.17b, damping also can be accomplished by mechanical means, whereby energy is dissipated by the frictional resistance of the components within the structure of a boring bar.

2. **External damping** is accomplished with external dampers, similar to shock absorbers on automobiles or machinery. Special vibration absorbers have been developed and installed on machine tools for this purpose. Machinery can be installed on specially prepared floors and foundations to isolate forced vibrations, such as those from nearby machines on the same floor.

Guidelines for Reducing Vibration and Chatter.   It is evident from the foregoing discussion that a balance must be achieved between the increased stiffness of a machine tool and the desirability of increased damping, particularly in the

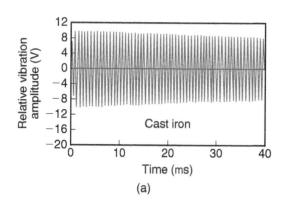
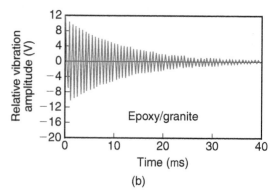

FIGURE 25.14 The relative damping capacity of (a) gray cast iron and (b) an epoxy–granite composite material. The vertical scale is the amplitude of vibration and the horizontal scale is time.

construction of high-precision machine tools. In various sections of Chapters 23 and 24, several guidelines were given for reducing vibration and chatter in machining operations. These basic guidelines may be summarized as:

- Minimize tool overhang
- Improve the stiffness of work-holding devices and support workpieces rigidly
- Modify tool and cutter geometry to minimize forces
- Change process parameters, such as cutting speed, feed, depth of cut, and cutting fluids
- Increase the stiffness of the machine tool and its components by improving their design, and by using larger cross-sections and materials with a higher elastic modulus
- Improve the damping capacity of the machine tool

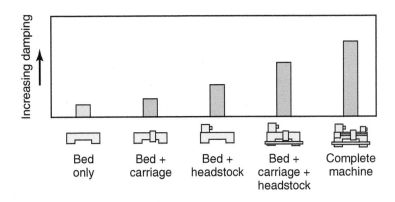

FIGURE 25.15   The damping of vibrations as a function of the number of components on a lathe. Joints dissipate energy; the greater the number of joints, the higher is the damping capacity of the machine. (See also Fig. 23.2.) *Source:* After J. Peters.

25.5   High-speed Machining

With continuing demands for higher productivity and lower production costs, the continuing trends are for increasing the cutting speed and the material-removal rate in machining operations, particularly in the aerospace and automotive industries.

The term "high" in *high-speed machining* (HSM) is somewhat relative; as a general guide, however, an approximate range of cutting speeds may be defined as:

1. **High speed:** 600–1800 m/min (2000–6000 ft/min)
2. **Very high speed:** 1800–18000 m/min (6000–60,000 ft/min)
3. **Ultrahigh speed:** Higher than 18,000 m/min

Spindle rotational speeds in machine tools now range up to 50,000 rpm, although the automotive industry generally has limited them to 15,000 rpm for better reliability and less downtime should a failure occur. The *spindle power* required in high-speed machining is generally on the order of 0.004 W/rpm (0.005 hp/rpm), much less than in traditional machining, which typically is in the range from 0.2 to 0.4 W/rpm (0.25–0.5 hp/rpm). Feed rates in high-speed machining are now up to 1 m/s (3 ft/s), and the acceleration rates of machine-tool components are very high.

Spindle designs for high speeds require *high stiffness* and *accuracy*, and generally involve an integral electric motor. The armature is built onto the shaft, and the stator is placed in the wall of the spindle housing. The bearings may be rolling elements or hydrostatic; the latter is more desirable because it requires less space than does the former. Because of *inertia* during the acceleration and deccelaration of machine-tool components, the use of lightweight materials, including ceramics and composite materials, is an important consideration.

The selection of appropriate cutting-tool materials is always a major consideration. On the basis of the discussions of tools and their selection in Chapter 22, and especially by reviewing Table 22.2, it is apparent that, depending on the workpiece material, multiphase coated carbides, ceramics, cubic-boron nitride, and diamond are all candidate tool materials for high-speed operations.

It also is important to note that high-speed machining should be considered primarily for operations in which **cutting time** is a significant portion of the total time in the overall machining operation. As described in Section 38.6 and Chapter 40, **noncutting time** and various other factors are important considerations in the overall assessment of the benefits of high-speed machining.

Studies have indicated that high-speed machining is economical for certain specific applications. As successful examples, it has been implemented in machining (a) aluminum structural components for aircraft; (b) submarine propellers 6 m (20 ft) in diameter, made of a nickel–aluminum–bronze alloy, and weighing 55,000 kg (50 tons); and (c) automotive engines, with 5–10 times the productivity of traditional machining. High-speed machining of complex three- and five-axis contours has been made possible by advances in CNC control technology, as described regarding *machining centers* in this chapter and in Chapter 37.

Another major factor in the adoption of high-speed machining has been the requirement to further improve dimensional tolerances. Note in Fig. 21.14 that as the cutting speed increases, a large percentage of the heat generated is removed by the chip, with the tool and the workpiece remaining closer to ambient temperature. This is beneficial, because there is no significant thermal expansion and thus warping of the workpiece during machining.

The important considerations in high-speed machining are summarized as:

1. Spindle design, for stiffness, accuracy, and balance at very high rotational speeds
2. Fast feed drives
3. Inertia of the components of the machine tool
4. Selection of appropriate cutting tools
5. Processing parameters and their computer control
6. Work-holding devices, which can withstand high centrifugal forces
7. Chip-removal systems, which are effective at very high rates of material removal

25.6   Hard Machining

It has been noted that as the hardness of the workpiece increases, its machinability decreases, and tool wear and fracture, surface finish, and surface integrity can become significant problems. However, it is still possible to machine hard metals and alloys by selecting an appropriate hard tool material and using machine tools with high stiffness, power, and precision.

An example is the finish machining of heat-treated steel (45–65 HRC) shafts, gears, pinions, and various automotive components, using polycrystalline cubic-boron nitride (PcBN), cermet, or ceramic cutting tools. Called *hard machining* or *hard turning*, this operation produces machined parts with good dimensional accuracy, surface finish (of as low as 25 μm or 10 μin.), and surface integrity. The important factors are the (a) available power, (b) static and dynamic stiffness of the machine tool and its spindle, and (c) work-holding devices and fixturing.

As described in Section 25.3, trends in the design and construction of modern machine tools, especially for hard machining, include the use of hydrostatic bearings for the spindles and slideways. The headstock and the slanted bed in the machines (see Fig. 23.11a) can be made of *epoxy–granite composite materials*, which have unique properties, such as high stiffness-to-weight ratio, thermal stability, and good damping capacity. Cutting-tool selection and edge preparation also are important to avoid premature failure in hard machining.

From technical, economic, and ecological considerations, hard turning has been found to compete successfully with the *grinding* process (Chapter 26). For instance, in some specific cases, hard turning has been shown to be three times faster than grinding, requiring fewer operations to finish the part, and utilizing five times less energy. A detailed comparative example of hard turning versus grinding is presented in Example 26.4.

25.7 Ultraprecision Machining

Beginning in the 1960s, increasing demands have been made concerning the precision manufacturing of components for computer, electronic, nuclear, and defense applications. Some specific examples include optical mirrors and lenses, fiber optic connection components, computer memory disks, metrology equipment of all kinds, and drums for photocopying machines. Surface-finish requirements are in the nanometer (10^{-9} m or 0.04 μin.) range, and dimensional tolerances and shape accuracies are in the micrometer (μm) and submicrometer range.

The trend toward ultraprecision manufacturing continues to grow. Modern **ultraprecision machine tools**, with advanced computer controls, can now position a cutting tool within an accuracy approaching 1 nm, as can be seen from Fig. 25.16. Also, note in this figure that higher precision is now being achieved by such processes as abrasive machining, ion-beam machining, and molecular manipulation.

The cutting tool for *ultraprecision machining* applications is almost exclusively a single-crystal diamond, where the process is called **diamond turning**. The diamond tool has a polished cutting edge, with a radius as small as a few nanometers. Wear of the diamond can be a significant problem, and more recent advances include **cryogenic diamond turning**, in which the tooling system is cooled by liquid nitrogen, to a temperature of about $-120°C$ ($-184°F$).

The workpiece materials for ultraprecision machining include copper alloys, aluminum alloys, silver, gold, electroless nickel, infrared materials, and plastics (acrylics). With depths of cut in the nm range, hard and brittle materials produce continuous chips, in a process known as **ductile-regime cutting** (see Section 26.3.4); deeper cuts in brittle materials produce discontinuous chips.

The machine tools for ultraprecision machining are built with very high precision and high stiffness of the machine, spindle, and work-holding devices. These machines have components that are made of structural materials with low thermal expansion and good dimensional stability (see Section 25.3). They are located in a dust-free environment (*clean rooms*; Section 28.2), where the temperature is controlled to within a fraction of one degree.

Vibrations from internal machine sources, as well as from external sources such as nearby machines on the same floor, must be avoided. Laser metrology (Section 35.5) is used for feed and position control, and

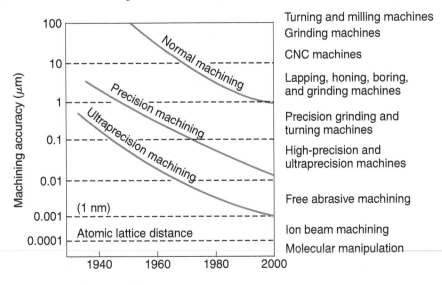

FIGURE 25.16 Improvements in machining accuracy over the years, using ultraprecision machining technologies. *Source:* After C.J. McKeown, N. Taniguchi, G. Byrne, D. Dornfeld, and B. Denkena.

the machines are equipped with highly advanced computer-control systems and with thermal and geometric error-compensating features.

General Considerations for Precision Machining.   There are several important factors in precision and ultraprecision machining and machine tools, somewhat similar to those in high-speed machining:

1. Machine-tool design, construction, and assembly, including the spindle, must provide stiffness, damping, and geometric accuracy
2. Motion control of various machine components, both linear and rotational
3. Thermal expansion of the machine tool, compensation for thermal expansion, and control of the machine-tool environment, especially ambient temperature
4. Real-time performance and control of the machine tool, and implementation of a tool-condition monitoring system

25.8 Machining Economics

Video Solution 25.1 Selection of Machining Operations

The material and process parameters that are relevant to efficient machining operations have been described in the preceding three chapters. In analyzing the *economics* of machining, several other factors also have to be considered. These factors include the costs involved in (a) the machine tools, work-holding devices and fixtures, and cutting tools; (b) labor and overhead associated with indirect costs; (c) the time required in setting up the machine for a particular operation; (d) material handling and movement, such as loading the blank and unloading the machined part; (e) gaging for dimensional accuracy and surface finish; and (f) cutting times and noncutting times.

Actual machining time is an important consideration, and recall also the discussion in Section 25.5 regarding the role of noncutting time in high-speed machining. Thus, unless noncutting time is a significant portion of the floor-to-floor time, high-speed machining should not be considered, unless it has other benefits.

Economic analysis is based on the ability to achieve a desired outcome, such as tolerance and surface finish, and, as such, requires that a machining process be robust and under control (see Section 36.5.1). For example, if a milling cutter is mounted such that the exposed spindle length varies randomly with every tool change, then this alone could result in high tolerances. The same analysis for different machine tools, whose dynamic stiffness and damping ability may differ (see Section 25.3), the use of cutters with different numbers of inserts, or loss of ambient temperature control, etc., all can result in variations that can significantly affect the ability to machine accurately.

Full-factorial design of experiments can be used to characterize the machine-tool/workpiece/operator system, but this approach is complex and has its own limitations. This section will assume that a process has been carefully designed to be robust, so that variations in these contributing factors can be ignored, and the effect of cutting speed on economics and productivity can be explored.

Minimizing Machining Cost per Piece.   As in all manufacturing processes and operations, the relevant parameters in machining can be selected and specified in such a manner that the *machining cost per piece*, as well as *machining time per piece*, is minimized. Various methods and approaches have been developed over the years to accomplish this goal, a task that has now become easier with the increasing use of computers and user-friendly software. In order for the results of the methods used to be reliable, however, it is essential that input data be accurate and up to date.

Described next is one of the simpler and more commonly used methods of analyzing machining costs and uses a *turning* operation to demonstrate the approach.

In machining a part by turning, the total machining cost per piece, C_p, is

$$C_p = C_m + C_s + C_l + C_t, \tag{25.1}$$

where

C_m = Machining cost
C_s = Cost of setting up for machining, including mounting the cutter, setting up fixtures, and preparing the machine tool for the operation
C_l = Cost of loading, unloading, and machine handling
C_t = Tooling cost, often only about 5% of the total machining operation; consequently, using the least expensive tool is not necessarily the proper way of reducing machining costs

The **machining cost** is given by

$$C_m = T_m (L_m + B_m), \tag{25.2}$$

where T_m is the machining time per piece, L_m is the labor cost of production personnel per hour, and B_m is the *burden rate*, or *overhead charge*, of the machine, including depreciation, maintenance, and indirect labor.

The **setup cost** is a fixed figure in dollars per piece. The **loading, unloading,** and **machine-handling cost** is

$$C_l = T_l (L_m + B_m), \tag{25.3}$$

where T_l is the time involved in loading and unloading the part, in changing speeds and feed rates, and making any other adjustments before machining. The **tooling cost** is

$$C_t = \frac{1}{N_i} [T_c (L_m + B_m) + D_i] + \frac{1}{N_f} [T_i (L_m + B_m)], \tag{25.4}$$

where N_i is the number of parts machined per cutting tool insert, N_f is the number of parts that can be produced per insert edge, T_c is the time required to change the insert, T_i is the time required to index the insert, and D_i is the depreciation of the insert, in dollars.

The time required to machine one part is

$$T_p = T_l + T_m + \frac{T_c}{N_i} + \frac{T_i}{N_f}, \tag{25.5}$$

where T_m has to be calculated for each particular operation on the part. For example, let's consider a turning operation, where the machining time (see Section 23.2) is given by

$$T_m = \frac{L}{fN} = \frac{\pi L D}{fV}, \tag{25.6}$$

where L is the length of cut, f is the feed, N is the angular speed (rpm) of the workpiece, D is the workpiece diameter, and V is the cutting speed. (Note that appropriate units must be used in all these equations.)

From Eq. (21.25) for tool life we have

$$T = \left(\frac{C}{V}\right)^{1/n}, \tag{25.7}$$

where T is the time, in minutes, required to reach a flank wear of certain dimension, after which the tool has to be reground or changed. Note that the tool may have to be replaced due to other reasons as well, such as crater wear, built-up edge, or nose wear. This analysis is restricted to *flank wear* as the important tool-failure criterion, but could be made more elaborate to include other variables. The number of pieces machined per insert edge follows from the Taylor equation, Eq. 21.25, as

$$N_f = \frac{T}{T_m},$$ (25.8)

and the number of pieces per insert is given by

$$N_i = mN_f = \frac{mT}{T_m}.$$ (25.9)

Sometimes not all of the edges are used before the insert is discarded; thus, it should be recognized that m corresponds to the number of edges that are actually used, not the number provided per insert. Combining Eqs. (25.6) through (25.9) yields

$$N_i = \frac{mfC^{1/n}}{\pi L D V^{(1/n)-1}}.$$ (25.10)

The cost per piece, C_p in Eq. (25.1), can now be defined in terms of several variables. To find the optimum cutting speed and the optimum tool life for **minimum cost**, C_p must be differentiated with respect to V and set to zero. Thus,

$$\frac{\partial C_p}{\partial V} = 0.$$ (25.11)

The *optimum cutting speed, V_o,* is

$$V_o = \frac{C(L_m + B_m)^n}{\left(\frac{1}{n} - 1\right)^n \left\{\frac{1}{m}[T_c(L_m + B_m) + D_i] + T_i(L_m + B_m)\right\}^n}$$ (25.12)

and the *optimum tool life, T_o,* is

$$T_o = \left(\frac{1}{n} - 1\right) \frac{\frac{1}{m}[T_c(L_m + B_m) + D_i] + T_i(L_m + B_m)}{L_m + B_m}.$$ (25.13)

To find the optimum cutting speed and the optimum tool life for **maximum production**, T_p must be differentiated with respect to V and set to zero. Thus,

$$\frac{\partial T_p}{\partial V} = 0.$$ (25.14)

The **optimum cutting speed** then is

$$V_o = \frac{C}{\left[\left(\frac{1}{n} - 1\right)\left(\frac{T_c}{m} + T_i\right)\right]^n},$$ (25.15)

and the **optimum tool life** is

$$T_o = \left(\frac{1}{n} - 1\right)\left(\frac{T_c}{m} + T_i\right). \qquad (25.16)$$

Qualitative plots of *minimum cost per piece* and *minimum time per piece* (hence the *maximum production rate*) are given in Figs. 25.17a and b. It should be noted that the cost of machining a part also depends on the surface finish required; the additional cost increases rapidly with finer surface finish, as shown in Fig. 26.35.

The preceding analysis indicates the importance of identifying all relevant parameters in a machining operation, determining various cost factors, obtaining relevant tool-life curves for the particular operation, and properly measuring the various time intervals involved in the overall operation. The importance of obtaining accurate data is shown in Fig. 25.17; note that small changes in cutting speed can have a significant effect on the minimum cost or minimum time per piece. The speeds

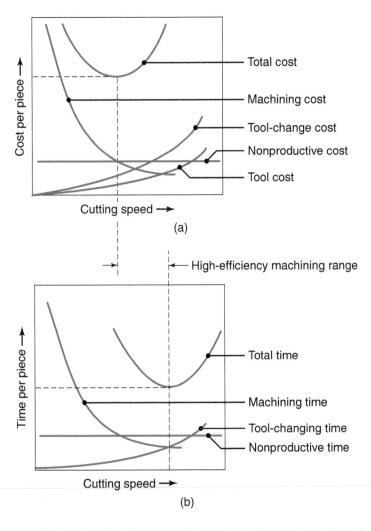

FIGURE 25.17   Graphs showing (a) cost per piece and (b) time per piece in machining; note the optimum speeds for both cost and time. The range between the two is known as the *high-efficiency machining range*.

and feeds recommended in Tables 23.4 and 24.2 generally lie in the *high-efficiency machining range*, which is between the speeds that yield the highest economy and highest production rate.

For many applications, such as finish machining of surfaces on soft metal castings, the machining cost per piece is fairly insensitive to cutting speed within this range; that is, the curve in Fig. 25.17 is fairly flat. With difficult-to-machine materials, however, as are routinely encountered in the medical products and aerospace industries, the cost per piece is very sensitive to cutting speed. Consequently, greater care is taken to ensure that machining takes place near the desired speed. Moreover, it should be recognized that the data given in Tables 23.4 and 24.2 are a summary for various tool and material grades; specific data is often available for machining particular alloys.

Such an economic analysis is typically done for all manufacturing operations, and it also can be a valuable tool for guiding process selection. For example, the cost per part in a sand-casting process to produce blanks, and in a machining operation to achieve final dimensional tolerances, can be calculated from an equation similar to Eq. (25.1), but including costs associated with sand casting, such as the cost of mold production, pattern depreciation, etc. A similar calculation can be made on a processing approach that uses powder metallurgy (thus increasing die and machinery costs), but requires less machining because of its ability to produce net-shape parts and with tighter tolerances, thereby reducing machining costs. A comparison of cost estimates can then help determine a processing strategy, as discussed in greater detail in Section 40.9.

SUMMARY

- Because they are versatile and capable of performing a variety of machining operations on small or large workpieces of various shapes, machining centers are now among the most important machine tools. Their selection depends on such factors as part complexity, the number and type of machining operations to be performed, the number of cutting tools required, and the dimensional accuracy and production rate specified.

- Vibration and chatter in machining are important considerations for workpiece dimensional accuracy, surface finish, and tool life. Stiffness and damping capacity of machine tools are major factors in controlling vibration and chatter.

- The economics of machining operations depends on factors such as nonproductive costs, machining costs, tool-change costs, and tool costs. Optimum cutting speeds can be determined for both minimum machining time per piece and minimum machining cost per piece.

KEY TERMS

Automatic pallet changer	Hard machining	Reconfigurable machines	Touch probes
Automatic tool changer	Hexapods	Regenerative chatter	Turning center
Chatter	High-efficiency machining range	Self-excited vibration	Ultraprecision machining
Chip collection	High-speed machining	Stiffness	Universal machining center
Damping	Machining center	Tool-exchange arm	Work envelope
Dynamic stiffness	Modular construction	Tool- and part-checking station	
Forced vibration	Pallet		

BIBLIOGRAPHY

Boothroyd, G., and Knight, W.A., **Fundamentals of Machining and Machine Tools**, 3rd ed., Marcel Dekker, 2005.

Chang, K., **Machining Dynamics: Fundamentals, Applications and Practices**, Springer, 2010.

Dashchenko, A.I. (ed.), **Reconfigurable Manufacturing Systems and Transformable Factories**, Springer, 2006.

Davim, J.P., **Machining of Hard Materials**, Springer, 2011.

Davim, J.P., **Machining of Complex Sculptured Surfaces**, Springer, 2012.

Erdel, B., **High-speed Machining**, *Society of Manufacturing Engineers*, 2003.

Gegg, B.C., Suh, C.S., and Luo, C.J., **Machine Tool Vibrations and Cutting Dynamics**, Springer, 2011.

Ito, Y., **Modular Design for Machine Tools**, McGraw-Hill, 2008.

Ito, Y., **Thermal Deformation in Machine Tools.**, McGraw-Hill, 2010.

Joshi, P.H., **Machine Tools Handbook**, McGraw-Hill, 2008.

Lopez de Lacalle, N., and Lamikiz, A. (eds.), **Machine Tools for High Performance Machining**, Springer, 2009.

Mickelson, D., **Hard Milling & High Speed Machining: Tools of Change**, Hanser Gardner, 2005.

Rivin, E.I., **Stiffness and Damping in Mechanical Design**, Marcel Dekker, 1999.

Schmitz, T.L., and Smith, K.S., **Machining Dynamics: Frequency Response to Improved Productivity**, Springer, 2008.

Zhang, D., **Parallel Robotic Machine Tools**, Springer, 2009.

REVIEW QUESTIONS

25.1 Describe the distinctive features of machining centers, and explain why these machines are so versatile.

25.2 Explain how the tooling system in a machining center functions. What are the typical tool-changing times?

25.3 Explain the trends in materials used for machine-tool structures.

25.4 Is there any difference between chatter and vibration? Explain.

25.5 What are the differences between forced and self-excited vibration?

25.6 Explain the importance of foundations in installing machine tools.

25.7 Explain why automated pallet changers and automatic tool changers are important parts of machining centers.

25.8 What types of materials are machine-tool bases typically made from? Why?

25.9 What is meant by the "modular" construction of machine tools?

25.10 What is a hexapod? What are its advantages?

25.11 What factors contribute to costs in machining operations?

25.12 List the reasons that temperature is important in machining operations.

QUALITATIVE PROBLEMS

25.13 Explain the technical and economic factors that led to the development of machining centers.

25.14 Spindle speeds in machining centers vary over a wide range. Explain why this is so, giving specific applications.

25.15 Explain the importance of stiffness and damping of machine tools. Describe how they are implemented.

25.16 Are there machining operations described in Chapters 23 and 24 that cannot be performed in machining and turning centers? Explain, with specific examples.

25.17 How important is the control of cutting-fluid temperature in operations performed in machining centers? Explain.

25.18 Review Fig. 25.10 on modular machining centers, and describe some workpieces and operations that would be suitable on such machines.

25.19 Review Fig. 25.15 and estimate the amount of damping you would expect in a hexapod. Is vibration a serious concern with hexapods? Explain.

25.20 Describe the adverse effects of vibration and chatter in machining operations.

25.21 Describe some specific situations in which thermal distortion of machine-tool components would be important.

25.22 Explain the differences in the functions of a turret and of a spindle in turning centers.

25.23 Explain how the pallet arrangements shown in Figs. 25.4a and b would be operated in using these machines on a shop floor.

25.24 Review the tool changer shown in Fig. 25.5. Are there any constraints on making their operations faster in order to reduce the tool changing time? Explain.

25.25 List the parameters that influence the temperature in metal cutting, and explain why and how they do so.

25.26 List and explain factors that contribute to poor surface finish in machining operations.

25.27 Can high-speed machining be performed without the use of cutting fluids? Explain.

25.28 In addition to the number of joints in a machine tool (see Fig. 25.15), what other factors influence the rate at which damping increases? Explain.

25.29 Describe types and sizes of workpieces that would not be suitable for machining on a machining center. Give specific examples.

25.30 Other than the fact that they each have a minimum, are the overall shapes and slopes of the total-cost and total-time curves in Fig. 25.17 important? Explain.

25.31 Explain the advantages and disadvantages of machine-tool frames made of gray-iron castings.

25.32 What are the advantages and disadvantages of (a) welded-steel frames, (b) bolted steel frames, and (c) adhesively bonded components of machine tools? Explain.

25.33 What would be the advantages and limitations of using concrete or polymer–concrete in machine tools?

25.34 Explain how you would go about reducing each of the cost factors in machining operations. What difficulties would you encounter in doing so?

25.35 Describe workpieces that would not be suitable for machining on a machining center. Give specific examples.

25.36 Give examples of forced vibration or self-excited vibration in general engineering practice.

QUANTITATIVE PROBLEMS

25.37 A machining-center spindle and tool extend 10 in. from their machine-tool frame. Calculate the temperature change that can be tolerated in order to maintain a tolerance of 0.001 in. in machining. Assume that the spindle is made of steel.

25.38 Using the data given in the example, estimate the time required to manufacture the parts in Example 25.1 with conventional machining and with high-speed machining.

25.39 A machining-center spindle and tool extend 12 in. from its machine-tool frame. What temperature change can be tolerated to maintain a tolerance of 0.0001 in. in machining? A tolerance of 0.001 in.? Assume that the spindle is made of steel.

25.40 In the production of a machined valve, the labor rate is $19.00 per hour, and the general overhead rate is $15.00 per hour. The tool is a ceramic insert with four faces and costs $25.00, takes five minutes to change and one minute to index. Estimate the optimum cutting speed from a cost perspective. Use $C = 100$ for V_O in m/min.

25.41 Estimate the optimum cutting speed in Problem 25.40 for maximum production.

25.42 Develop an equation for optimum cutting speed in face milling using a cutter with inserts.

25.43 Develop an equation for optimum cutting speed in turning, where the tool is a high-speed steel tool that can be reground periodically.

SYNTHESIS, DESIGN, AND PROJECTS

25.44 If you were the chief engineer in charge of the design of advanced machining and turning centers, what changes and improvements would you recommend on existing models? Explain.

25.45 Review the technical literature and outline the trends in the design of modern machine tools. Explain why there are those trends.

25.46 Make a list of components of machine tools that could be made of ceramics, and explain why ceramics would be suitable.

25.47 Survey the company literature from various machine-tool manufacturers, and prepare a comprehensive table indicating the capabilities, sizes, power, and costs of machining and turning centers. Comment on your observations.

25.48 The cost of machining and turning centers is considerably higher than for traditional machine tools. Since many operations performed by machining centers also can be done on conventional machines, how would you go about justifying the high cost of these centers? Explain, with appropriate examples.

25.49 In your experience using tools or other devices, you may have come across situations in which you experienced vibration and chatter. Describe your experience and explain how you would go about minimizing the vibration and chatter.

25.50 Describe your thoughts on whether or not it is feasible to include grinding operations (see Chapter 26) in machining centers. Explain the nature of any difficulties that may be encountered.

25.51 Is the accuracy and surface finish that can be achieved in a machining center a function of the number of inserts on a cutter? Explain.

25.52 The following experiment is designed to better demonstrate the effect of tool overhang on vibration and chatter: With a sharp tool, scrape the surface of a piece of soft metal by holding the tool with your arm fully outstretched. Repeat the experiment, this time holding the tool as close to the workpiece as possible. Describe your observations regarding the tendency for the tool to vibrate. Repeat the experiment with different types of metallic and nonmetallic materials.

25.53 Review the part in Fig. 25.1a and list the machining operations and machine tools you would recommend to produce this part.

Abrasive Machining and Finishing Operations

- Abrasive machining is important because of its capability to impart high dimensional accuracy and surface finish to parts.

- A wide variety of abrasive finishing processes is available, many of them based on the fundamental mechanism of abrasion.

- This chapter opens with a description of the grinding process, the mechanics of material removal, and the roles of types of abrasives and bonds in grinding wheels.

- Some abrasive machining operations, including polishing, buffing, honing, and sanding, require a bonded or coated abrasive; others, such as ultrasonic machining, lapping, abrasive flow machining, and electrochemical machining and grinding, use loose abrasives.

- These processes are described in detail, including their fundamental capabilities, applications, and design considerations.

- The chapter ends with a discussion of economic considerations for finishing operations.

Typical parts made: Any part requiring high dimensional accuracy and surface finish, such as ball and roller bearings, piston rings, valves, cams, gears, and tools and dies.

Alternative processes: Precision machining, electrical-discharge machining, electrochemical machining and grinding, and abrasive-jet machining.

26.1 Introduction

There are many situations in manufacturing where the processes described thus far cannot produce the required dimensional accuracy or surface finish for a part, or the workpiece material is too hard or too brittle to process. Consider, for example, the accuracy and fine surface finish required on ball bearings, pistons, valves, cylinders, cams, gears, molds and dies, and numerous precision components used in instrumentation. One of the most common methods for producing such demanding characteristics on parts is *abrasive machining*.

An **abrasive** is a small, hard particle having sharp edges and an irregular shape (Fig. 26.1), sand being the simplest example. It is capable of removing small amounts of material from a surface by a cutting process that produces tiny chips. Familiar applications of abrasives are *sandpaper* or *emery cloth*, used to smoothen

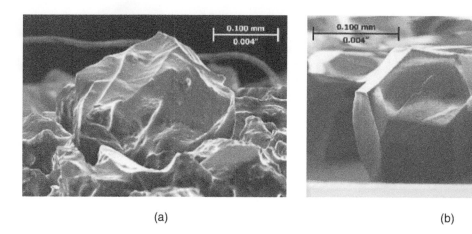

(a)   (b)

FIGURE 26.1 Typical abrasive grains; note the angular shape with sharp edges. (a) A single, 80-mesh Al_2O_3 grit in a freshly dressed grinding wheel; (b) An 80/100 mesh diamond grit; diamond and cubic boron nitride grains can be manufactured in various geometries, including the "blocky" shape shown. *Source:* Courtesy of J. Badger.

FIGURE 26.2 A variety of bonded abrasives used in abrasive-machining processes. *Source:* Courtesy of Norton/Saint Gobain Abrasives.

surfaces and remove sharp corners, and *grinding wheels*, as shown in Fig. 26.2, to sharpen knives, tools, or to impart good dimensional accuracy and surface finish on numerous components of products. Abrasives also are used to hone, lap, buff, and polish workpieces.

With the use of computer-controlled machines, abrasive processes and equipment are now capable of producing a wide variety of workpiece geometries, as can be seen in Fig. 26.3, and very fine dimensional accuracy and surface finishes, as shown in Figs. 23.14, 23.15, and Table 26.4. For example, dimensional tolerances on parts can now be less than 1 μm (40 μin.) and surface roughness can be as fine as 0.025 μm (1 μin.).

Because they are hard, abrasives also are used in *finishing processes* for heat-treated metals and alloys, and for very hard parts, in such applications as (a) finishing of ceramics and glasses, (b) cutting off lengths of bars, structural shapes, masonry, and concrete, (c) removing unwanted weld beads and spatter, (d) creating a very smooth and flat surface on silicon wafers in order to produce integrated circuits, (e) polishing bearings and races, and (f) cleaning surfaces with jets of air or water containing abrasive particles.

The chapter begins with a description of the characteristics of abrasives, along with their use in various abrasive material-removal processes. As with cutting operations, the mechanics of abrasive operations is described first. This knowledge is essential in assisting the establishment of the interrelationships between the (a) workpiece material and process variables and (b) dimensional accuracy, surface finish, and surface integrity of the parts produced.

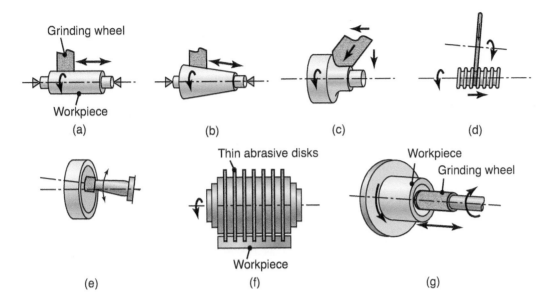

FIGURE 26.3   The types of workpieces and operations typical of grinding: (a) cylindrical surfaces; (b) conical surfaces; (c) fillets on a shaft; (d) helical profiles; (e) concave shape; (f) cutting off or slotting with thin wheels; and (g) internal grinding.

26.2   Abrasives and Bonded Abrasives

Abrasives that are used most commonly in abrasive-machining operations are:

Conventional abrasives

- *Aluminum oxide* (Al_2O_3)
- *Silicon carbide* (SiC)

Superabrasives

- *Cubic boron nitride* (cBN)
- *Diamond*

As described in Chapter 8, these abrasives are much harder than conventional cutting-tool materials, as may be seen by comparing Tables 22.1 and 26.1 (see also Fig. 2.15). Cubic boron nitride and diamond are listed as superabrasives because they are the two hardest materials known.

In addition to hardness, an important characteristic of abrasives is **friability**, defined as the ability of abrasive grains to fracture into smaller pieces. This property gives abrasives their *self-sharpening* characteristics, essential in maintaining their sharpness during use. High friability indicates low strength or low fracture resistance of the abrasive. Thus, a highly friable abrasive grain fragments more rapidly under grinding forces than one with low friability. For example, aluminum oxide has lower friability than silicon carbide and, correspondingly, a lower tendency to fragment.

The *shape* and *size* of the abrasive grain also affect its friability. For example, *blocky* grains, which are analogous to a negative rake angle in single-point cutting tools (as shown in Fig. 21.3), are less friable than less

TABLE 26.1

Ranges of Knoop Hardness for Various Materials and Abrasives	
Common glass	350–500
Flint, quartz	800–1100
Zirconium oxide	1000
Hardened steels	700–1300
Tungsten carbide	1800–2400
Aluminum oxide	2000–3000
Titanium nitride	2000
Titanium carbide	1800–3200
Silicon carbide	2100–3000
Boron carbide	2800
Cubic boron nitride	4000–5000
Diamond	7000–8000

blocky or platelike grains. Moreover, because the probability of defects decreases as grain size decreases, smaller grains are stronger and less friable than larger ones, a phenomenon known as *size effect*. (See also Section 26.3.)

Abrasive Types.   The abrasives commonly found in nature are *emery, corundum* (alumina), *quartz, garnet,* and *diamond.* Because in their natural state these abrasives generally contain impurities and possess nonuniform properties, their performance as an abrasive is inconsistent and unreliable; consequently, abrasives have been made *synthetically* for many years.

- **Aluminum oxide** was first made in 1893, and is produced by fusing bauxite, iron filings, and coke. Fused aluminum oxides are categorized as *dark* (less friable), *white* (very friable), or *single crystal.*
- **Seeded gel** was first introduced in 1987, and is the purest form of *unfused aluminum oxide.* Also known as *ceramic aluminum oxide,* it has a grain size on the order of 0.2 μm (coarse human hair is about 200 μm), which is much smaller than other types of commonly used abrasive grains. These grains are *sintered* (heating without melting; see Section 17.4) to form larger sizes. Because they are harder than fused alumina and have relatively high friability, seeded gels maintain their sharpness and are especially effective for difficult-to-grind materials.
- **Silicon carbide** was first discovered in 1891, and is made with silica sand and petroleum coke. Silicon carbides are classified as *black* (less friable) or *green* (more friable), and generally have higher friability than aluminum oxide; hence, they have a greater tendency to fracture and thus remain sharp.
- **Cubic boron nitride** was first developed in the 1970s; its properties and characteristics are described in Sections 8.2.3 and 22.7.
- **Diamond,** also known as *synthetic* or *industrial diamond,* was first used as an abrasive in 1955; its properties and characteristics are described in Sections 8.7 and 22.9.

Abrasive Grain Size.   As used in manufacturing operations, abrasives generally are very small when compared to the size of cutting tools and inserts, described in Chapters 21 and 22. They have sharp edges, thus allowing the removal of very small quantities of material from the workpiece surface, resulting in very fine surface finish and dimensional accuracy.

The size of an abrasive grain is identified by a **grit number**, which is a function of sieve size, whereby the smaller the grain size, the larger the grit number. For example, grit number 10 is typically regarded as very coarse, 100 as fine, and 500 as very fine. Sandpaper and emery cloth also are identified in this manner, as can readily be observed by noting the grit number printed on the back of an abrasive paper or cloth.

Compatibility of Abrasive versus Workpiece Material.   As in selecting cutting-tool materials for machining, the *affinity* of an abrasive grain to the workpiece material is an important consideration. The less the reactivity of the two materials, the less wear and dulling of the grains during grinding, thus making the operation more efficient and causing less damage to the workpiece surface (see Section 26.3.1 for details). As an example, because of its chemical affinity, diamond (which is a form of carbon, Section 8.7) cannot be used for grinding steels, since diamond dissolves in iron at the high temperatures that are encountered in grinding. Generally, the following recommendations are made with regard to selecting abrasives:

- **Aluminum oxide:** Carbon steels, ferrous alloys, and alloy steels
- **Silicon carbide:** Nonferrous metals, cast irons, carbides, ceramics, glass, and marble
- **Cubic boron nitride:** Steels and cast irons above 50 HRC hardness and high-temperature alloys
- **Diamond:** Ceramics, carbides, and some hardened steels where the hardness of diamond is more significant than its reactivity with the carbon in steel

26.2.1 Grinding Wheels

Each abrasive grain typically removes only a very small amount of material at a time; consequently, high material-removal rates can only be achieved if a very large number of these grains act together. This is done by using **bonded abrasives**, typically in the form of a grinding wheel, in which the abrasive grains are distributed and oriented randomly.

As shown schematically in Fig. 26.4, the abrasive grains in a grinding wheel are held together by a **bonding material** (Section 26.2.2), which acts as supporting posts or braces between the grains. In bonded abrasives, *porosity* is essential in order to provide clearance for the chips being produced, as otherwise there would be no space for the chips being produced, and thus would severely interfere with the grinding operation. Porosity can be observed by looking at the surface of a grinding wheel with a magnifying glass.

A very wide variety of types and sizes of abrasive wheels is available today. Some of the more commonly used types of grinding wheels for conventional abrasives are shown in Fig. 26.5; superabrasive wheels are shown in Fig. 26.6. Note that, due to their high cost, only a small volume of superabrasive material is used on the periphery of these wheels.

Bonded abrasives are marked with a standardized system of letters and numbers, indicating the type of abrasive, grain size, grade, structure, and bond type. Figure 26.7 shows the marking system for aluminum-oxide and silicon-carbide bonded abrasives; the marking system for diamond and cubic boron nitride bonded abrasives is shown in Fig. 26.8.

The cost of grinding wheels depends on the type and size of the wheel. Small wheels (up to about 25 mm, or 1 in., in diameter) cost approximately \$2 to \$15 for

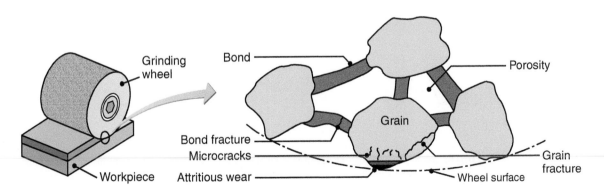

FIGURE 26.4 Schematic illustration of a physical model of a grinding wheel, showing its structure and its wear and fracture patterns.

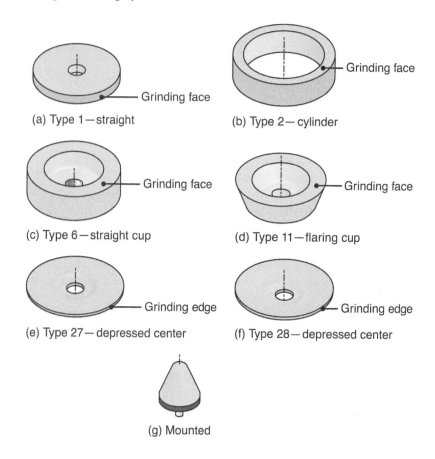

(a) Type 1—straight
Grinding face

(b) Type 2—cylinder
Grinding face

(c) Type 6—straight cup
Grinding face

(d) Type 11—flaring cup
Grinding face

(e) Type 27—depressed center
Grinding edge

(f) Type 28—depressed center
Grinding edge

(g) Mounted

FIGURE 26.5   Common types of grinding wheels made with conventional abrasives; note that each wheel has a specific grinding face; grinding on other surfaces is improper and unsafe.

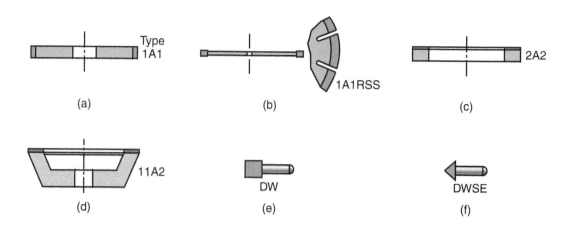

Type 1A1

1A1RSS

2A2

(a)

(b)

(c)

11A2

DW

DWSE

(d)

(e)

(f)

FIGURE 26.6   Examples of superabrasive wheel configurations; the annular regions (rims) are superabrasive grinding surfaces, and the wheel itself (core) generally is made of metal or composites. The bonding materials for the superabrasives are (a), (d), and (e) resinoid, metal, or vitrified; (b) metal; (c) vitrified; and (f) resinoid.

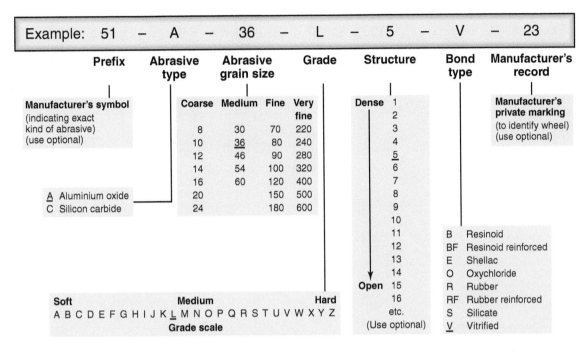

| Example: | 51 | – | A | – | 36 | – | L | – | 5 | – | V | – | 23 |

FIGURE 26.7 Standard marking system for aluminum-oxide and silicon-carbide bonded abrasives.

conventional abrasives, $30 to $100 for diamond, and $50 to $300 for cubic boron nitride wheels. For a large wheel of about 500 mm in diameter and 250 mm in width (20 in. × 10 in.), the wheel costs are $500 for conventional abrasives, $5,000 to $8,000 for diamond, and $20,000 for cubic boron nitride.

26.2.2   Bond Types

The common types of bonds used in bonded abrasives are:

Vitrified.   Also called *ceramic bonds, vitrified bonds* (from the Latin *vitrum* for glass; Section 8.4) are the most common and widely used material. The raw materials consist of feldspar (a crystalline mineral) and clays. They are mixed with the abrasives, moistened, and molded under pressure into the shape of grinding wheels. These "green" wheels, which are similar to powder metallurgy parts (Chapter 17), are then fired slowly up to a temperature of about 1250°C (2300°F), to fuse the glass and develop structural strength. The wheels are then cooled slowly (to avoid thermal cracking, due to temperature gradients), finished to size, inspected for quality and dimensional accuracy, and tested for any defects.

Wheels with vitrified bonds are strong, stiff, and resistant to oils, acids, and water; however, they are brittle and lack resistance to mechanical and thermal shock. To improve strength during their use, vitrified wheels also are made with steel backing plates or cups, for better structural support of the bonded abrasives. The color of the grinding wheel can be modified by adding various elements during its manufacture, so that wheels can be color coded for use with specific workpiece materials, such as ferrous, nonferrous, and ceramic.

Resinoid.   Resinoid bonding materials are *thermosetting resins*, and are available in a wide range of compositions and properties (Sections 7.4 and 7.7). Because the bond

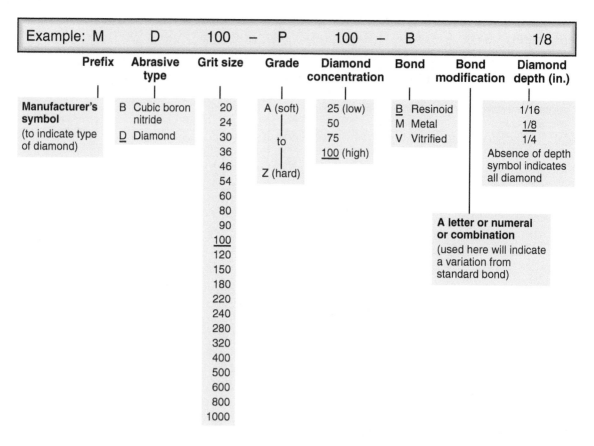

FIGURE 26.8 Standard marking system for cubic boron nitride and diamond bonded abrasives.

is an organic compound, wheels with *resinoid bonds* also are called **organic wheels**. The manufacturing technique for producing them consists basically of (a) mixing the abrasive with liquid or powdered phenolic resins and additives, (b) pressing the mixture into the shape of a grinding wheel, and (c) curing it at temperatures of about 175°C (350°F) to set the bond. In addition to pressing, *injection molding* also is used to manufacture grinding wheels (see Sections 17.3 and 19.3).

Because the elastic modulus of thermosetting resins is lower than that of glasses (see Table 2.2), resinoid wheels are more flexible than vitrified wheels. As a bonding material, *polyimide* (Section 7.7) also is used as a substitute for the phenolic resin; it is tougher and more resistant to higher temperatures.

Reinforced Wheels.   These wheels typically consist of one or more layers of *fiber-glass mats* (Section 8.4.2) of various mesh sizes. The fiberglass in this laminate structure provides reinforcement by way of retarding the disintegration of the wheel, rather than improving its strength, should the wheel fracture or break for some reason during use. Large-diameter resinoid wheels can be further supported using one or more internal rings, made of round steel bars inserted during molding of the wheel.

Thermoplastic.   In addition to thermosetting resins, thermoplastic bonds (Section 7.3) also are used in grinding wheels. Wheels are available with sol-gel abrasives bonded with thermoplastics.

Rubber.   The most flexible matrix used in abrasive wheels is rubber (Section 7.9). The manufacturing process consists of (a) mixing crude rubber, sulfur, and the abrasive grains together, (b) rolling the mixture into sheets, (c) cutting out disks of various diameters, and (d) heating the disks under pressure to vulcanize the rubber. Thin wheels can be made in this manner (called *cutoff blades*) and are used like saws for cutting-off operations.

Metal.   Using powder metallurgy techniques, the abrasive grains, usually diamond or cubic boron nitride, are bonded to the periphery of a metal wheel to depths of 6 mm (0.25 in.) or less, as illustrated in Fig. 26.5. Metal bonding is carried out under high pressure and temperature. The wheel itself (the core) may be made of aluminum, bronze, steel, ceramics, or composite materials, depending on such requirements as strength, stiffness, and dimensional stability. Superabrasive wheels may be *layered*, so that a single abrasive layer is plated or brazed to a metal wheel with a particular desired shape. Layered wheels are lower in cost, and are used for small production quantities.

26.2.3   Wheel Grade and Structure

The *grade* of a bonded abrasive is a measure of its bond strength, including both the type and the amount of bonding material in the wheel. Because strength and hardness are directly related (see Section 2.6.2), the grade is also referred to as the **hardness** of a bonded abrasive. Thus, for example, a hard wheel has a stronger bond and/or a larger amount of bonding material between the grains than a soft wheel.

   The *structure* of a bonded abrasive is a measure of its *porosity* (the spacing between the grains, as shown in Fig. 26.4). The structure ranges from *dense* to *open*, as shown in Fig. 26.7. Recall that some porosity is essential to provide clearance for the grinding chips, as otherwise they would interfere with the grinding operation.

26.3   The Grinding Process

Grinding is a chip-removal process that uses an individual abrasive grain as the cutting tool (Fig. 26.9a). The major differences between the action of an abrasive grain and that of a single-point cutting tool can be summarized as:

- The individual abrasive grains have *irregular shapes* (Fig. 26.1), and are spaced randomly along the periphery of the wheel (Fig. 26.10).
- The average rake angle of the grains is highly negative, typically $-60°$ or even less; consequently, grinding chips undergo much larger plastic deformation than they do in other machining processes. (See Section 21.2.)
- The radial positions of the grains (over the peripheral surface of a wheel) vary, thus not all grains are active during grinding.
- Surface speeds of grinding wheels (equivalent to cutting speeds) are very high, typically 20–30 m/s (4000–6000 ft/min), and can be as high as 150 m/s (30,000 ft/min) in high-speed grinding, using specially designed and manufactured wheels.

   The grinding process and its parameters can best be observed in the *surface-grinding* operation, shown schematically in Fig. 26.11. A straight grinding wheel

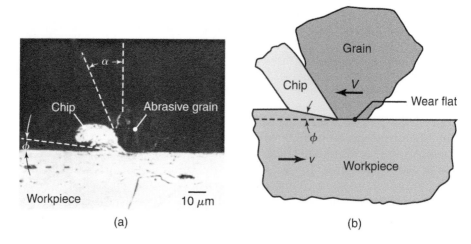

FIGURE 26.9 (a) Grinding chip being produced by a single abrasive grain; note the large negative rake angle of the grain. (b) Schematic illustration of chip formation by an abrasive grain with a wear flat; note the negative rake angle of the grain and the small shear angle. *Source:* (a) After M.E. Merchant.

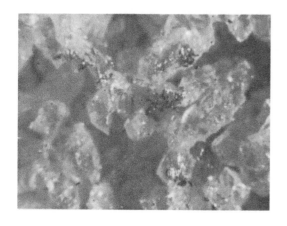

FIGURE 26.10 The surface of a grinding wheel (A46-J8V), showing abrasive grains, wheel porosity, wear flats on grains, and metal chips from the workpiece adhering to the grains; note the random distribution and shape of the abrasive grains. Magnification: 50×.

Video Solution 26.1 Material Removal Rate and Chip Thickness in Grinding

(Fig. 26.5a), with a diameter of D, removes a layer of metal at a depth d (called **wheel depth of cut**). An individual grain on the periphery of the wheel moves at a tangential velocity of V, while the workpiece moves at a velocity of v. Each abrasive grain produces a small chip, which has an *undeformed thickness* (**grain depth of cut**), t, and an *undeformed length*, l.

Typical chips from grinding operations are shown in Fig. 26.12. Note that the chips, just as in machining, are thin and long. From geometric relationships, it can be shown that the undeformed chip length in surface grinding (Fig. 26.11) is approximated by the equation

$$l = \sqrt{Dd} \qquad (26.1)$$

and the undeformed chip thickness, t, by

$$t = \sqrt{\left(\frac{4v}{VCr}\right)\sqrt{\left(\frac{d}{D}\right)}}, \qquad (26.2)$$

where C is the number of cutting points per unit area of the wheel periphery. Generally, C is estimated to be in the range from 0.1 to 10 per mm^2 (10^2–10^3 per in^2). The quantity r is the ratio of chip width to average undeformed chip thickness, and has an estimated value typically between 10 and 20.

As an example, l and t can be calculated for the following process parameters: Let $D = 200$ mm, $d = 0.05$ mm, $v = 30$ m/min, and $V = 1800$ m/min. Using the preceding formulas gives

$$l = \sqrt{(200)(0.05)} = 3.2 \text{ mm} = 0.13 \text{ in.}$$

Assuming that $C = 2$ per mm^2 and that $r = 15$ gives

$$t = \sqrt{\frac{(4)(30)}{(1800)(2)(15)}} \sqrt{\frac{0.05}{200}} = 0.006 \text{ mm}$$

$$= 0.00025 \text{ in.}$$

Because of plastic deformation during chip formation, the actual chip will be shorter and thicker than the values calculated. (See Figs. 26.9 and 26.12.) Note from this example that grinding chip's dimensions typically are much smaller than those generally obtained in metal-cutting operations, as described in Chapter 21.

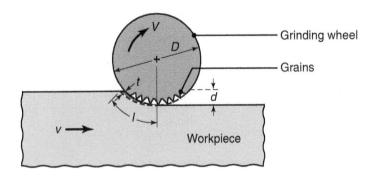

FIGURE 26.11 Schematic illustration of the surface-grinding process, showing various process variables; the figure depicts conventional (up) grinding.

Grinding Forces.   A knowledge of grinding forces is essential for

- Estimating power requirements
- Designing grinding machines and work-holding devices and fixtures
- Determining the deflections that the workpiece, as well as the grinding machine and its components, may undergo; deflections adversely affect dimensional accuracy, and are especially critical in precision and ultraprecision grinding

If it is assumed that the cutting force on the grain is proportional to the cross-sectional area of the undeformed chip, it can be shown that the **grain force** (acting tangential to the wheel) is a function of process variables:

$$\text{Grain force} \propto \left(\frac{v}{V}\sqrt{\frac{d}{D}}\right) (\text{UTS}). \tag{26.3}$$

It can be shown that, because of the small dimensions involved, forces in grinding are typically much smaller than those in the machining operations described in Chapters 23 and 24. The forces in grinding should be kept low, in order to avoid distortion and to maintain high dimensional accuracy of the workpiece.

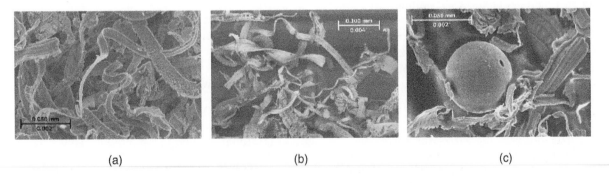

(a)   (b)   (c)

FIGURE 26.12 Typical chips, or *swarf*, from grinding operations. (a) Swarf from grinding a conventional HSS drill bit; (b) swarf from grinding nickel alloy; (c) swarf of cast iron, showing a melted globule among the chips. *Source: Courtesy of J. Badger.*

Video Solution 26.2 Mechanics of Grinding

TABLE 26.2

Approximate Specific-energy Requirements for Surface Grinding			
		Specific energy	
Workpiece material	Hardness	W-s/mm^3	hp-min/in^3
Aluminum	150 HB	7–27	2.5–10
Cast iron (class 40)	215 HB	12–60	4.5–22
Low-carbon steel (1020)	110 HB	14–68	5–25
Titanium alloy	300 HB	16–55	6–20
Tool steel (T15)	67 HRC	18–82	6.5–30

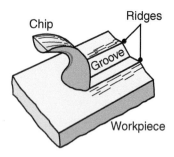

FIGURE 26.13 Chip formation and plowing of the workpiece surface by an abrasive grain.

Specific Energy. The *energy* dissipated in producing a grinding chip consists of the energy required for:

- *Plastic deformation* in chip formation
- *Plowing*, as shown by the ridges formed in Fig. 26.13
- *Friction*, caused by rubbing of the abrasive grain along the workpiece surface

Note in Fig. 26.9b that, after some use, the grains along the periphery of the wheel develop a **wear flat**, a phenomenon that is similar to *flank wear* in cutting tools, shown in Fig. 21.15. The wear flat continuously rubs along the ground surface, dissipates energy (because of friction), and thus makes the grinding operation less efficient.

Specific-energy in grinding is defined as the energy per unit volume of material ground from the workpiece surface, and is shown in Table 26.2. Note that the energy levels are much higher than those in machining operations (Table 21.2). This difference has been attributed to such factors as the presence of a wear flat, high negative rake angles of the abrasive grains (which require more energy; Section 21.3), and a possible contribution of the size effect (the smaller the chip, the higher the energy required to produce it; Section 1.5.1). Also, it has been observed that with effective lubrication, the specific energy in grinding can be reduced by a factor of four or more.

EXAMPLE 26.1   Forces in Surface Grinding

Given: Assume that a surface-grinding operation is being carried out on low-carbon steel, with a wheel of diameter $D = 10$ in. and rotating at $N = 4000$ rpm, and a width of cut of $w = 1$ in. The depth of cut is $d = 0.002$ in. and the feed rate of the workpiece, v, is 60 in./min.

Find: Calculate the *grinding force* (the force tangential to the wheel), F_c, and the *thrust force* (the force normal to the workpiece surface), F_n, using specific-energy data.

Solution: The material-removal rate (MRR) is determined as

$$MRR = dwv = (0.002)(1)(60) = 0.12 \text{ in}^3/\text{min}.$$

The power consumed is given by

$$\text{Power} = (u)(MRR),$$

where u is the specific energy, as obtained from Table 26.2. (See also Section 21.3.) For low-carbon steel, it is estimated to be 15 hp-min/in^3. Thus,

$$\text{Power} = (15)(0.12) = 1.8 \text{ hp}.$$

Because 1 hp = 33,000 ft-lb/min = 396,000 in.-lb/min,

$$\text{Power} = (1.8)(396,000) = 712,800 \text{ in.-lb/min}.$$

Since power is defined as

$$\text{Power} = T\omega,$$

where the torque $T = F_c D/2$ and ω is the rotational speed of the wheel, in radians per minute ($\omega = 2\pi N$). It then follows that

$$712,800 = (F_c)\left(\frac{10}{2}\right)(2\pi)(4000),$$

and hence, $F_c = 5.7$ lb. The thrust force, F_n, can be calculated directly; however, it also can be estimated by noting from experimental data in the technical literature that it is about 30% higher than the cutting force, F_c. Consequently,

$$F_n = (1.3)(5.7) = 7.4 \text{ lb}.$$

Temperature. The temperature rise in grinding is an important consideration because

- It can adversely affect the surface properties of the workpiece, including metallurgical changes
- The temperature rise can cause residual stresses on the workpiece
- Temperature gradients in the workpiece cause distortions due to thermal expansion and contraction of the workpiece surface, thus making it difficult to control dimensional accuracy

Video Solution 26.3 Temperatures in Grinding

The *surface-temperature rise* (ΔT) in grinding is related to process variables by the following expression:

$$\Delta T \propto D^{1/4} d^{3/4} \left(\frac{V}{v}\right)^{1/2}. \tag{26.4}$$

Thus, temperature increases with increasing depth of cut, d, wheel diameter, D, and wheel speed, V, and decreases with increasing workpiece speed, v. Note from this equation that the depth of cut has the largest exponent; hence, it has the greatest influence on temperature.

Although *peak temperatures* during grinding can reach $1600°C$ ($3000°F$), the time involved in producing a chip is on the order of microseconds, hence the chip produced may or may not melt. Because the chips carry away much of the heat generated, as do chips formed in high-speed machining processes (see Section 25.5), only a small fraction of the heat generated in grinding is conducted to the workpiece. If this was not the case, it would be very difficult to grind workpieces with sufficient dimensional accuracy and without causing any possible metallurgical changes in the workpiece.

Sparks. The sparks produced when grinding metals are actually chips that glow, due to the *exothermic* (heat producing) reaction of the hot chips with oxygen in the atmosphere. Sparks do not occur during grinding in an oxygen-free environment or when the workpiece material does not readily oxidize at elevated temperatures. The color, intensity, and shape of the sparks depend on the composition of the metal being ground. Charts are available that help identify the type of metal being ground, from the appearance of its sparks. If the heat generated due to exothermic reaction is sufficiently high, chips can melt, acquire a spherical shape (because of surface tension), and solidify as metal particles.

Tempering. An excessive temperature rise in grinding can cause *tempering* and *softening* of the workpiece surface. Process variables must therefore be selected carefully in order to avoid excessive temperature rise. The use of grinding fluids (Section 26.4) is an effective means of controlling temperature.

Burning.   Excessive temperature rise during grinding may burn the workpiece surface. A *burn* is characterized by a bluish color on ground steel surfaces, an indication that high temperature has caused oxidation of the workpiece. A burn, which can be detected by etching and metallurgical techniques, may not be objectionable in itself, unless surface layers have undergone *phase transformations* (Chapter 4). For example, if martensite forms in higher carbon steels from rapid cooling, it is called a *metallurgical burn*, which will adversely affect the surface properties of ground parts, and reduce surface ductility and toughness.

Heat Checking.   High temperatures in grinding may cause the workpiece surface to develop cracks, a condition known as *heat checking*. The cracks usually are perpendicular to the grinding direction, although under severe conditions, parallel cracks also may appear. As expected, such a surface lacks toughness and has low fatigue and corrosion resistance. Heat checking also occurs in dies during die casting, as described in Section 11.4.5.

Residual Stresses.   Temperature gradients within the workpiece during grinding are primarily responsible for the development of *residual stresses*. Grinding fluids and their method of application, as well as process parameters such as depth of cut and speeds, significantly influence the magnitude and type of residual stresses. Because of the adverse effect of tensile residual stresses on fatigue strength, process variables should be selected carefully. Residual stresses usually can be reduced by lowering wheel speed and increasing workpiece speed, a process called **low-stress grinding** or *gentle grinding*. Softer grade wheels, known as **free-cutting** grinding wheels, also may be used to reduce residual stresses.

26.3.1   Grinding-wheel Wear

Similar to the wear on cutting tools, grinding-wheel wear is an important consideration, because it adversely affects the shape and dimensional accuracy of ground surfaces. Wear of grinding wheels is caused by three different mechanisms: attritious grain wear, grain fracture, and bond fracture.

Attritious Grain Wear.   In *attritious wear*, which is similar to flank wear in cutting tools (see Fig. 21.15), the cutting edges of an originally sharp grain become dull and develop a *wear flat* (Fig. 26.9b). This type of wear involves both physical and chemical reactions, and is caused by the interaction of the grain with the workpiece material. These reactions are complex, and involve diffusion, chemical degradation or decomposition of the grain, fracture at a microscopic scale, plastic deformation, and melting.

Attritious wear is low when the two materials (i.e., the grain and the workpiece) are *chemically inert* with respect to each other, much like what has been observed with cutting tools (see Section 22.1). The more inert the materials, the lower is the tendency for reaction and adhesion to occur between the grain and the workpiece. Thus, for example, because aluminum oxide is relatively inert with respect to iron, its rate of attritious wear when it is used to grind steels is much lower than that of silicon carbide and diamond. By contrast, silicon carbide can dissolve in iron, hence it is not suitable for grinding steels. Cubic boron nitride has a higher inertness with respect to steels, hence is suitable for use as an abrasive.

Grain Fracture. Because abrasive grains are brittle, their fracture characteristics in grinding are significant. If the wear flat caused by attritious wear is excessive, the grain becomes dull and grinding becomes inefficient and produces undesirably high temperatures. Ideally, a dull grain should fracture or fragment at a moderate rate, so that new sharp cutting edges are produced continuously during grinding. This situation is equivalent to breaking a dull piece of chalk or a stone into two or more pieces in order to expose new sharp edges (see Section 26.2 on *friability*).

The selection of grain type and size for a particular application also depends on the attritious-wear rate. A grain–workpiece material combination that has a high attritious wear and low grain friability dulls grains and develops a large wear flat. Grinding then becomes inefficient, and surface damage, such as burning, is likely to occur.

Bond Fracture. The strength of the bond (*grade*) is a significant parameter in grinding. If the bond is too strong, for example, dull grains cannot be dislodged, which, in turn, prevents other sharp grains along the circumference of the grinding wheel from contacting the workpiece. Conversely, if the bond is too weak, the grains are dislodged easily, and the wear rate of the wheel increases. Maintaining dimensional accuracy then becomes difficult.

In general, softer bonds are recommended for harder materials, and for reducing residual stresses and thermal damage to the workpiece. Hard-grade wheels are used for softer materials, and for removing large amounts of material at high rates.

26.3.2 Grinding Ratio

Grinding-wheel wear is generally correlated with the amount of workpiece material ground by a parameter, called the *grinding ratio*, G, defined as

$$G = \frac{\text{Volume of material removed}}{\text{Volume of wheel wear}}. \tag{26.5}$$

In practice, grinding ratios vary widely, ranging from 2 to 200, and even higher, depending on the type of wheel, workpiece material, grinding fluid, and process parameters, such as depth of cut and speeds of the wheel and workpiece. It has been shown that effective grinding fluids can increase the grinding ratio by a factor of 10 or more, thus greatly improving wheel life.

During grinding, a particular wheel may **act soft** (thus exhibit high wear rate) or **act hard** (low wear rate), regardless of the wheel grade. Note, for example, that an ordinary pencil acts soft when writing on rough paper, but it acts hard when writing on soft paper, even though it is the same pencil. Acting hard or soft is a function of the force on the individual grain on the periphery of the wheel. The higher the force, the greater the tendency for the grains to fracture or be dislodged from the wheel surface, and the higher the wheel wear and thus the lower the grinding ratio.

Note from Eq. (26.3) that the grain force increases with the strength of the workpiece material, work speed, and depth of cut, and decreases with increasing wheel speed and wheel diameter; thus, a grinding wheel acts soft when v and d increase or when V and D decrease. Note also that attempting to obtain a high grinding ratio in practice (so as to extend wheel life) isn't always desirable, because high ratios may indicate grain dulling, thus possible surface damage to the workpiece. A lower ratio may be acceptable when an overall technical and economic analysis justifies it.

EXAMPLE 26.2   Action of a Grinding Wheel

Given: A surface-grinding operation is being carried out with the wheel running at a constant spindle speed. Assume that the depth of cut, d, remains constant and the wheel is dressed periodically (see Section 26.3.3).

Find: Will the wheel act soft or hard as the wheel wears down over time?

Solution: Referring to Eq. (26.3), note that the parameters that change over time in this operation are the wheel diameter, D, and its surface speed, V. As D becomes smaller, the relative grain force increases, thus the wheel acts softer. To accommodate the changes due to the reduction of wheel diameter over time or to make provisions for using wheels of different diameters, some grinding machines are equipped with variable-speed spindle motors.

26.3.3   Dressing, Truing, and Shaping of Grinding Wheels

Dressing is the process of (a) *conditioning*, producing *sharp new edges* on worn grains on the grinding surface of a wheel and (b) *truing*, producing a *true circle* on a wheel that has become out of round. Dressing is necessary when excessive attritious wear dulls the wheel, called **glazing** (because of the shiny appearance of the wheel surface), or when the wheel becomes *loaded* (see below). For softer wheels, truing and dressing are done separately, but for harder wheels, such as cBN, both are done in one operation.

Loading of a grinding wheel occurs when the porosities on the wheel surfaces (Fig. 26.10) become filled or clogged with chips from the workpiece. Loading can occur while grinding soft materials or from improper selection of wheels or process parameters. A loaded wheel grinds inefficiently and generates much frictional heat, resulting in surface damage and loss of dimensional accuracy of the workpiece.

The techniques used to dress grinding wheels are:

- A specially shaped *diamond-point tool* or *diamond cluster* is moved across the width of the grinding face of a rotating wheel, and removes a small layer from the wheel surface with each pass. This method can be performed either dry or wet, depending on whether the wheel is to be used dry or wet, respectively. In practice, however, the wear of the diamond with harder wheels can be significant, requiring the use of a diamond disk or cup wheel.
- A set of *star-shaped steel disks* is pressed manually against the wheel. Material is removed from the wheel surface by crushing the grains; as a result, this method produces a coarse surface on the wheel and is used only for rough grinding operations, on bench or pedestal grinders.
- *Abrasive sticks* may be used to dress grinding wheels, particularly softer wheels; however, this technique is not appropriate for precision grinding operations.
- Dressing techniques for metal-bonded diamond wheels involve the use of *electrical-discharge* and *electrochemical machining* techniques, as described in Chapter 27. These processes erode away very thin layers of the metal bond and thus expose new diamond cutting edges.
- Dressing for form grinding involves *crush dressing* or *crush forming*. The process consists of pressing a metal roll on the surface of the grinding wheel, which typically is a vitrified wheel. The roll (which usually is made of high-speed steel, tungsten carbide, or boron carbide) has a machined or ground profile on its periphery; thus, it reproduces a replica of this profile on the surface of the grinding wheel being dressed. (See Section 26.4.)

Dressing techniques and their frequency are important for quality, because they affect grinding forces and workpiece surface finish. Modern computer-controlled grinders are equipped with automatic dressing features, which dress the wheel continually as grinding progresses. The first contact of the dressing tool with the grinding wheel is very important, as it determines the nature of the new surface produced; this action usually is monitored precisely by using piezoelectric or acoustic-emission sensors (Section 37.7). Features such as vibration sensors, power monitors, and strain gages also are used in the dressing setup of high-precision grinding machines.

For a typical aluminum-oxide wheel, the depth removed during dressing is on the order of 5 to 15 μm (200–600 μin.), but for a cBN wheel, it is 2–10 μm (80–400 μin.). Consequently, modern dressing systems have a resolution as low as 0.25–1 μm (10–40 μin.).

Grinding wheels can be *shaped* to the form to be ground on the workpiece (Section 26.4). The grinding face on the Type 1 straight wheel shown in Fig. 26.5a is cylindrical, thus it produces a flat ground surface. The wheel surface also can be shaped into various forms by dressing it (Fig. 26.14a). Although templates have

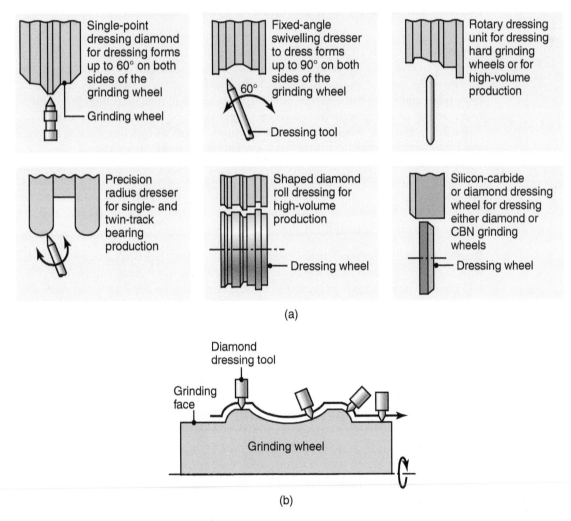

(a)

(b)

FIGURE 26.14 (a) Forms of grinding-wheel dressing. (b) Shaping the grinding face of a wheel by dressing it by computer control; note that the diamond dressing tool is normal to the surface at the point of contact with the wheel. *Source:* Courtesy of Okuma Corporation. Printed with permission.

TABLE 26.3

	Typical Ranges of Speeds and Feeds for Abrasive Processes			
Process variable	Grinding, conventional	Grinding, creep-feed	Polishing	Buffing
Wheel speed (m/min)	1500–3000	1500–3000	1500–2400	1800–3500
Work speed (m/min)	10–60	0.1–1	—	—
Feed (mm/pass)	0.01–0.05	1–6	—	—

been used for this purpose, modern grinders are equipped with computer-controlled shaping features. Unless it already has the desired form, the diamond dressing tool traverses the wheel face automatically along a certain prescribed path (Fig. 26.14b), and produces very accurate surfaces. Note in Fig. 26.14b that the axis of the diamond dressing tool remains normal to the grinding-wheel face at the point of contact.

26.3.4   Grindability of Materials and Wheel Selection

The term *grindability* of materials, like the terms *machinability* (Section 21.7) or *forgeability* (Section 14.5), is difficult to define precisely. Grindability is a general indicator of how easy it is to grind a material, and includes such considerations as the quality of the surface produced, surface finish, surface integrity, wheel wear, grinding cycle time, and overall economics of the operation. As in machinability, grindability of a material can be greatly enhanced by proper selection of process parameters (see Table 26.3), grinding wheels, and grinding fluids, as well as by using the appropriate machine characteristics, fixturing methods, and work-holding devices.

Grinding practices are now well established for a wide variety of metallic and non-metallic materials, including newly developed aerospace materials and composites. Specific recommendations for selecting wheels and appropriate process parameters for metals can be found in various handbooks, manufacturers' literature, and the references in the bibliography of this chapter.

Ductile-regime Grinding.   It has been shown that with light passes and machines with high stiffness and damping capacity, it is possible to produce *continuous* chips and good surface integrity in grinding brittle materials, such as ceramics (Fig. 26.13), a process known as *ductile-regime grinding*. This regime is useful because it produces fewer surface cracks and leads to better performance in fatigue and bearing applications. Ceramic chips are typically 1–10 μm (40–400 μin.) in size; they are more difficult to remove from grinding fluids than do metal chips, requiring the use of fine filters and special techniques.

26.4   Grinding Operations and Machines

The selection of a grinding process and machine for a particular application depends on the workpiece shape and features, size, ease of fixturing, and production rate required (Table 26.4). Modern grinding machines are computer controlled, and have such features as automatic workpiece loading and unloading, part clamping, and automatic dressing and wheel shaping. Grinders also can be equipped with probes and gages, for determining the relative position of the wheel and workpiece surfaces (see also Fig. 25.6), as well as with tactile-sensing features, whereby diamond dressing-tool breakage, for example, can be detected readily during the dressing cycle.

TABLE 26.4

General Characteristics of Abrasive Machining Processes and Machines

Process	Characteristics	Typical maximum dimensions, length and diameter (m)*
Surface grinding	Flat surfaces on most materials; production rate depends on table size and level of automation; labor skill depends on part complexity; production rate is high on vertical-spindle rotary-table machines	Reciprocating table L: 6 Rotary table D: 3
Cylindrical grinding	Round workpieces with stepped diameters; low production rate unless automated; low to medium labor skill	Workpiece D: 0.8, roll grinders D: 1.8, universal grinders D: 2.5
Centerless	Round and slender workpieces; high production rate; low to medium labor skill	Workpiece D: 0.8
Internal	Holes in workpiece; low production rate; low to medium labor skill	Hole D: 2
Honing	Holes in workpiece; low production rate; low labor skill	Spindle D: 1.2
Lapping	Flat, cylindrical, or curved workpieces; high production rate; low labor skill	Table D: 3.7
Chemical mechanical polishing	Flat surfaces, generally used for semiconductors for microelectronics or MEMS applications; moderate production rate; high labor skill.	D: 0.3
Abrasive flow machining	Used for deburring and finishing of complex geometries; low production rate; low labor skill	D: 0.3
Ultrasonic machining	Holes and cavities with various shapes; suitable for hard and brittle materials; medium labor skill	—

*Larger capacities are available for special applications.

Surface Grinding. *Surface grinding* (Fig. 26.15) generally involves the grinding of flat surfaces. Typically, the workpiece is held on a *magnetic chuck*, attached to the worktable of the grinder (Fig. 26.16); nonmagnetic materials are held by vises, vacuum chucks, or some other fixtures. In this operation, a straight wheel is mounted on the horizontal spindle of the surface grinder. Traverse grinding occurs as the table reciprocates longitudinally and is fed laterally (in the direction of the spindle axis) after each stroke.

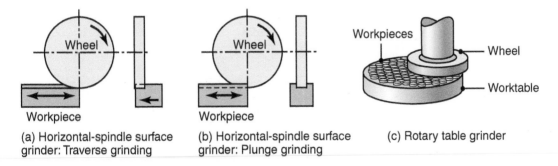

(a) Horizontal-spindle surface grinder: Traverse grinding

(b) Horizontal-spindle surface grinder: Plunge grinding

(c) Rotary table grinder

FIGURE 26.15 Schematic illustrations of various surface-grinding operations. (a) Traverse grinding with a horizontal-spindle surface grinder. (b) Plunge grinding with a horizontal-spindle surface grinder, producing a groove in the workpiece. (c) A vertical-spindle rotary-table grinder (also known as the *Blanchard* type).

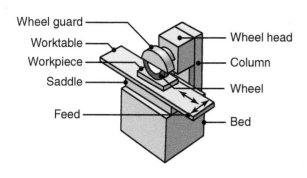

FIGURE 26.16   Schematic illustration of a horizontal-spindle surface grinder.

The movement of the grinding wheel may be along the surface of the workpiece (*traverse grinding*, *through-feed* grinding, or *cross feeding*), or the wheel may move radially into the workpiece (*plunge* grinding), as is the case when grinding a groove (Fig. 26.15b). Surface grinders make up the largest percentage of grinders used in industry, followed by bench grinders (usually with two wheels at each end of the spindle), cylindrical grinders, tool and cutter grinders, and internal grinders, as described below.

In addition to the surface grinder shown in Fig. 26.16, other types include *vertical spindles* and *rotary tables* (referred to as the *Blanchard* type, Fig. 26.15c). These configurations allow a number of pieces to be ground in one setup. Steel balls for ball bearings, for example, are ground in special setups and at high production rates (Fig. 26.17).

Cylindrical Grinding.   In *cylindrical grinding*, also called *center-type grinding* (Fig. 26.18; see also Fig. 26.3), the external cylindrical surfaces and shoulders of workpieces, such as crankshaft bearings, spindles, pins, and bearing rings, are ground. The rotating cylindrical workpiece reciprocates laterally along its axis, to cover the whole width to be ground. In *roll grinders*, used for large and long workpieces such as rolls for rolling mills (see Fig. 13.1), the grinding wheel reciprocates. These machines are capable of grinding rolls as large as 1.8 m (72 in.) in diameter.

The workpiece in cylindrical grinding is held between centers or in a chuck, or it is mounted on a faceplate in the headstock of the grinder. For straight cylindrical surfaces, the axes of rotation of the wheel and workpiece are parallel, and each is driven by separate motors and at different speeds. Long workpieces with two or more diameters also can be ground on cylindrical grinders. As *form grinding* and *plunge grinding*, the operation also can produce shapes in which the wheel is dressed to the workpiece form to be ground (Fig. 26.19).

Cylindrical grinders are identified by the maximum diameter and length of the workpiece that can be ground. In *universal grinders*, both the workpiece and the

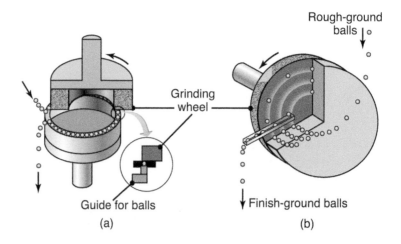

FIGURE 26.17   (a) Rough grinding of steel balls on a vertical-spindle grinder; the balls are guided by a special rotary fixture. (b) Finish grinding of balls in a multiple-groove fixture; the balls are ground to within 0.013 mm (0.0005 in.) of their final size.

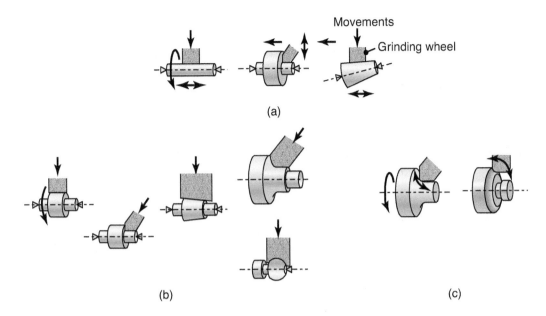

(a)

(b)   (c)

FIGURE 26.18 Examples of various cylindrical-grinding operations: (a) traverse grinding; (b) plunge grinding; and (c) profile grinding. *Source:* Courtesy of Okuma Corporation. Printed with permission.

wheel axes can be moved and swiveled around a horizontal plane, thus permitting the grinding of tapers and other shapes.

With computer control, *noncylindrical* parts, such as cams, also can be ground on rotating workpieces. As illustrated in Fig. 26.20, the workpiece spindle speed is synchronized, such that the radial distance, x, between the workpiece and the wheel axes is continuously varied to produce a particular shape, such as the one shown in the figure.

Thread grinding is done on cylindrical grinders, using specially dressed wheels matching the shape of the threads, as shown in Fig. 26.21. (See also *centerless grinding*.) Although expensive, threads

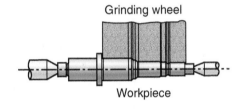

FIGURE 26.19 Plunge grinding of a workpiece on a cylindrical grinder with the wheel dressed to a stepped shape.

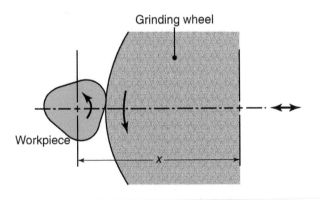

FIGURE 26.20 Schematic illustration of grinding a non-cylindrical part on a cylindrical grinder with computer controls to produce the shape. The part rotation and the distance x between centers are varied and synchronized to grind the particular workpiece shape.

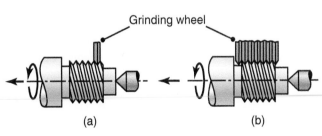

(a)   (b)

FIGURE 26.21 Thread grinding by (a) traverse and (b) plunge grinding.

produced by grinding are the most accurate of any manufacturing process, and have very fine surface finish. Typical applications requiring such threads include ballscrew mechanisms, used for precise movement of various machine components. The workpiece and wheel movements are synchronized to produce the pitch of the thread, usually in about six passes.

EXAMPLE 26.3   Cycle Patterns in Cylindrical Grinding

As in most grinding operations, the grinding wheel in cylindrical grinding typically makes several passes along a path, in order to produce the final geometry on the workpiece. Figure 26.22 illustrates the cycle patterns for producing various shapes on a multifunctional, computer-controlled precision grinder. The downward arrowheads with numbers in the figures indicate the beginning of the grinding cycle.

The determination of the optimum and most economical pattern for minimum cycle time depends on the volume of material to be removed, the shape of the part, and the process parameters. All the patterns shown are generated automatically by the software in the computer controls of the grinder.

Source: Based on Toyoda Machinery/JTEKT.

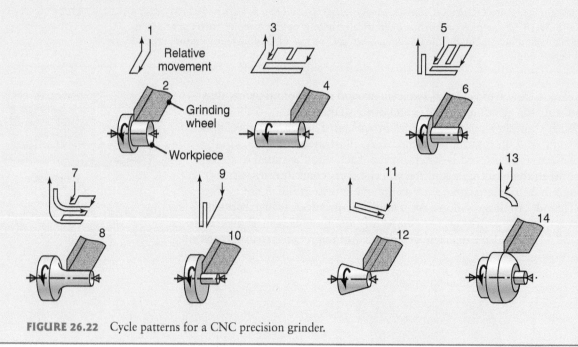

FIGURE 26.22   Cycle patterns for a CNC precision grinder.

Internal Grinding.   In *internal grinding* (Fig. 26.23), a small wheel is used to grind the inside diameter of the part, such as in bushings and bearing races. The workpiece is held in a rotating chuck, and the wheel rotates at 30,000 rpm or higher. Internal profiles also can be ground with *profile-dressed* wheels, that move radially into the workpiece. The headstock of internal grinders also can be swiveled on a horizontal plane to grind tapered holes.

Centerless Grinding.   *Centerless grinding* is a high-production process for grinding cylindrical surfaces; the workpiece is supported not by centers (hence the term "centerless") or chucks, but by a *blade*, as shown in Figs. 26.24a and b. Typical parts made by this process are roller bearings, piston pins, engine valves, camshafts, and similar components. Parts with diameters as small as 0.1 mm (0.004 in.) can be ground. Centerless grinders (Fig. 26.24d) are capable of wheel surface speeds on the order of 10,000 m/min (35,000 ft/min), typically using cubic boron nitride abrasive wheels.

In *through-feed grinding*, the workpiece is supported on a work-rest blade and is ground continuously (hence the term "through-feed") between two wheels (Fig. 26.24a). Grinding is done by the larger wheel, while the smaller wheel regulates the axial movement of the workpiece. The rubber-bonded *regulating wheel* is tilted and runs at a much slower speed of about one-twentieth of the grinding wheel speed.

Parts with variable diameters, such as bolts, valve tappets, and multiple-diameter shafts, can be ground by a process called *infeed* or *plunge grinding* (Fig. 26.24b), an operation similar to plunge or form grinding on cylindrical grinders. Tapered pieces are centerless ground by *end-feed grinding*. *Thread grinding* can be done at high production rate with centerless grinders, using specially dressed wheels. In *internal centerless grinding*, the workpiece is supported between three rolls and is ground internally; typical applications are sleeve-shaped parts and rings (Fig. 26.24c).

Creep-feed Grinding.   Although grinding traditionally has been associated with small rates of material removal (Table 26.3) and fine finishing operations, it can also be used for large-scale metal-removal operations. In *creep-feed grinding*, the wheel depth of cut, *d*, is as much as 6 mm (0.25 in.) and the workpiece speed is low (Fig. 26.25). The wheels are softer grade resin bonded and have an open structure, in order to keep workpiece temperatures low and improve surface finish.

The machines for creep-feed grinding have special features, such as power up to 225 kW (300 hp), high stiffness (because of the high forces due to the large depth of material removed), high damping capacity, variable spindle and worktable speeds, and ample capacity for grinding fluids. They can continuously dress the wheel, using a diamond-encrusted roll as the dressing tool.

Creep-feed grinding can be competitive with other machining processes, such as milling, broaching, and planning, and is economical for such specific applications as shaped punches, key seats, twist-drill flutes, the roots of turbine blades (Fig. 26.25c), and various complex parts made of superalloys. Because the wheel is dressed to the shape of the workpiece to be produced, the workpiece does not have to be previously shaped by milling, shaping, or broaching; near-net-shape castings and forgings are thus suitable for creep-feed grinding. Although a single grinding pass generally is sufficient, a second pass may be necessary for improved surface finish.

Heavy Stock Removal by Grinding.   Grinding also can be used for heavy stock removal by increasing process parameters, such as wheel depth of cut. This operation can be economical in specific applications and can compete favorably with machining processes, particularly milling, turning, and broaching. In this operation, surface finish is of secondary importance, the dimensional tolerances are on the same order as those obtained by most machining processes, and the grinding wheel (or belt) can be utilized to its fullest capabilities, while minimizing grinding cost per piece. Heavy stock removal by grinding is also performed on welds, castings, and forgings, to smoothen weld beads and remove flash.

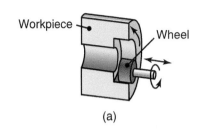

(a)

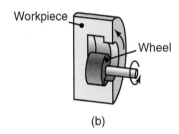

(b)

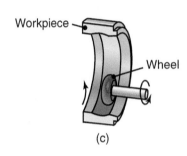

(c)

FIGURE 26.23   Schematic illustrations of internal grinding operations: (a) traverse grinding; (b) plunge grinding; and (c) profile grinding.

Through-feed grinding

Plunge grinding

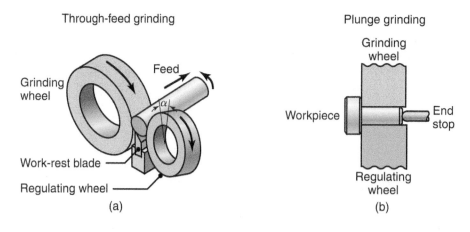

(a)

(b)

Internal grinding

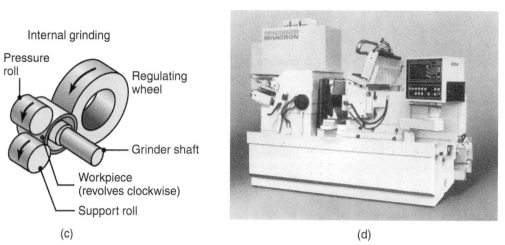

(c)

(d)

FIGURE 26.24   Schematic illustrations of centerless grinding operations: (a) through-feed grinding, (b) plunge grinding, (c) and internal grinding; and (d) a computer numerical-control cylindrical-grinding machine. *Source:* Courtesy of Cincinnati Milacron, Inc.

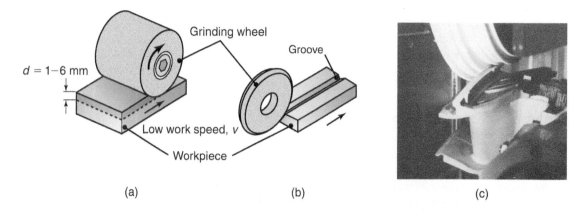

(a)

(b)

(c)

FIGURE 26.25   (a) Schematic illustration of the creep-feed grinding process; note the large wheel depth of cut, *d*. (b) A shaped groove produced on a flat surface by creep-feed grinding in one pass; groove depth is typically on the order of a few mm. (c) An example of creep-feed grinding with a shaped wheel; this operation also can be performed by some of the processes described in Chapter 27. *Source:* Courtesy of Blohm, Inc.

EXAMPLE 26.4   Grinding versus Hard Turning

In view of the discussions presented thus far, it is apparent that, in some specific applications, grinding and hard turning (described in Section 25.6) can be competitive in specific applications. Hard turning continues to be increasingly competitive with grinding, and dimensional tolerances and surface finish are approaching those obtained with grinding. Consider the case of *machining* of heat-treated steels, with hardness above 45 HRC, using a single-point polycrystalline cubic boron nitride tool, versus *grinding* these steels.

In comparing Tables 21.2 and 26.2, it will be noted that (a) turning requires much less energy than grinding, (b) thermal and other types of damage to the workpiece surface is less likely to occur in machining, (c) cutting fluids may not be necessary,

and (d) lathes are less expensive than grinders. Moreover, finishing operations, including finish grinding, can be performed on the turned part while it is still chucked in the lathe.

On the other hand, work-holding devices for large and especially slender workpieces during hard turning can present significant problems, because the cutting forces are higher than grinding forces. Furthermore, tool wear and its control can be a significant problem as compared with the automatic dressing of grinding wheels. It is evident that the competitive position of hard turning versus grinding must be evaluated individually for each application, in terms of product surface finish, integrity, quality and overall economics.

Several types of grinders are used for various operations:

- **Universal tool and cutter grinders** are used for grinding single-point or multipoint tools and cutters, including drills. They are equipped with special work-holding devices for accurate positioning of the tools to be ground. A variety of CNC tool grinders is available, making the operation simpler and faster and with consistent results. The cost of these grinders ranges from about $150,000 to $400,000.
- **Tool-post grinders** are self-contained units, and usually are attached to the tool post of a lathe (see Fig. 23.2). The tool is mounted on the headstock and is ground by moving the tool post. These grinders are versatile, but it is essential for the lathe components to be protected from abrasive debris.
- **Swing-frame grinders** are used in foundries for grinding large castings. Rough grinding of castings is called **snagging**, and is usually done on *floorstand grinders* using wheels as large as 0.9 m (36 in.) in diameter.
- **Portable grinders** are used for such operations as grinding off weld beads and for *cutting off*, using thin abrasive disks. They are driven either pneumatically, electrically, or with a flexible shaft connected to an electric motor or a gasoline engine.
- **Bench** and **pedestal grinders** are used for the routine offhand grinding of tools and small parts. They usually are equipped with two grinding wheels, mounted on the two ends of the shaft of an electric motor; generally, one wheel is coarse for rough grinding and the other is fine for finish grinding.

Grinding Fluids.   The functions of grinding fluids are similar to those of cutting fluids, described in Section 22.12. Although grinding and other abrasive-removal processes can be performed dry, the use of a fluid is important because it

- Reduces temperature rise in the workpiece
- Improves part surface finish and dimensional accuracy
- Improves the efficiency of the operation, by reducing wheel wear, reducing loading of the wheel, and lowering power consumption

TABLE 26.5

General Recommendations for Grinding Fluids	
Material	Grinding fluid
Aluminum	E, EP
Copper	CSN, E, MO + FO
Magnesium	D, MO
Nickel	CSN, EP
Refractory metals	EP
Steels	CSN, E
Titanium	CSN, E

Note: D = Dry; E = Emulsion; EP = Extreme pressure; CSN = Chemicals and synthetics; MO = Mineral oil; FO = Fatty oil. (See also Section 33.7.)

Grinding fluids typically are *water-based emulsions* for general grinding and *oils* for thread grinding (Table 26.5). They may be applied as a stream (flood) or as mist (a mixture of fluid and air). Because of the high surface speeds involved, an airstream (*air blanket*) around the periphery of the wheel may prevent the fluid from reaching the wheel–workpiece interface. Special nozzles, which conform to the shape of the cutting surface of the grinding wheel, have been designed whereby the grinding fluid is supplied under high pressure.

There can be a significant rise in the temperature of water-based grinding fluids as they remove heat from the grinding zone. This can cause the workpiece to expand, thus making it difficult to control dimensional accuracy. A common method, to maintain a more uniform workpiece temperature, is to use refrigerating systems (*chillers*), through which the grinding fluid is circulated continuously and maintained at a roughly constant temperature. As described in Section 22.12, the *biological* and *ecological* aspects of disposal, treatment, and recycling of metalworking fluids are important considerations in their selection and use. The practices employed must comply with federal, state, and local laws and regulations.

Grinding Chatter.   *Chatter* is particularly important in grinding because it adversely affects surface finish and wheel performance. Studying **chatter marks** on ground surfaces often can help identify their source, which may include (a) bearings and spindles of the grinding machine, (b) nonuniformities in the grinding wheel, as manufactured, (c) uneven wheel wear, (d) poor dressing techniques, (e) grinding wheels that are not balanced properly, and (f) external sources, such as nearby machinery. The grinding operation itself can cause *regenerative chatter*, as it does in machining, described in Section 25.4.

The important factors in controlling chatter are the stiffness of the grinding machine, the stiffness of work-holding devices, and damping of the system. General guidelines have been established to reduce the tendency for chatter in grinding, including especially (a) using soft-grade grinding wheels, (b) dressing the wheel frequently, (c) changing dressing techniques, (d) reducing the material-removal rate, and (e) supporting the workpiece rigidly.

Safety in Grinding Operations.   Because grinding wheels are brittle and rotate at high speeds, they can easily fracture. Certain procedures must be followed in their handling, storage, and use; failure to follow these procedures, and the instructions and warnings printed on individual wheel labels, may result in serious injury or fatality. Grinding wheels should be stored properly and protected from environmental extremes, such as high temperature or humidity. They should be inspected visually for cracks and damage prior to installing them on grinders. Vitrified wheels should be tested prior to their use by *ringing* them, that is, supporting them at the hole, tapping them gently, and listening to the sound; a damaged wheel will have a flat ring to it, similar to that of a cracked dinner plate.

Damage to a grinding wheel can severely reduce its **bursting speed**. Defined as the surface speed at which a freely rotating wheel bursts (*explodes*), the bursting speed (expressed in rpm) depends on the type of wheel, such as its bond, grade, and structure. In diamond and cBN wheels (Fig. 26.6), which are operated at high surface speeds, the type of core material used in the wheel affects the bursting speed. Metal cores, for example, have the highest bursting speed, typically on the order of about 250 m/s (800 ft/s).

26.5 Design Considerations for Grinding

Design considerations for grinding operations are basically similar to those for machining, as described in various sections in Chapters 23 and 24. In addition, specific attention should be given to the following:

- Parts to be ground should be designed so that they can be mounted securely, either in chucks, magnetic tables, or suitable fixtures and work-holding devices. Thin, straight, or tubular workpieces may distort during grinding, thus requiring special attention.
- If high dimensional accuracy is required, interrupted surfaces, such as holes and keyways, should be avoided, as they can cause vibrations and chatter.
- Parts for cylindrical grinding should be balanced, and long and slender designs should be avoided to minimize deflections. Fillets and corner radii should be as large as possible, or relief should be provided for them, by prior machining of these regions.
- In centerless grinding, short pieces may be difficult to grind accurately, because the blade may not support them. In through-feed grinding, only the largest diameter on the parts can be ground.
- The design of parts requiring accurate form grinding should be kept as simple as possible, to avoid frequent form dressing of the wheel.
- Deep and small holes, and blind holes, requiring internal grinding, should be avoided, or they should include a relief.

In general, part designs should require that a minimum amount of material be removed by grinding, except for creep-feed grinding. Moreover, in order to maintain good dimensional accuracy, designs preferably should allow for all grinding to be done without having to reposition the workpiece.

26.6 Ultrasonic Machining

In *ultrasonic machining* (UM), material is removed from a surface by *microchipping* and *erosion*, with loose, fine abrasive grains in a water slurry (Fig. 26.26a). The tip of the tool (called a *sonotrode*) vibrates at a frequency of 20 kHz and an amplitude of 0.0125–0.075 mm (0.0005–0.003 in.). The vibration imparts a high velocity to abrasive grains between the tool and the workpiece. The stress produced by the abrasive particles impacting the workpiece surface is high, because (a) the time of

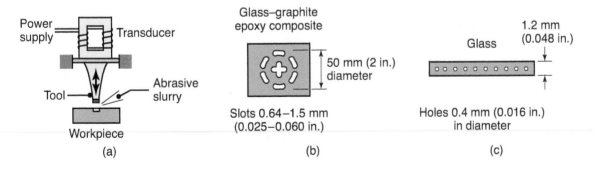

FIGURE 26.26 (a) Schematic illustration of the ultrasonic machining process. (b) and (c) Types of parts made by this process; note the small size of the holes produced.

contact between the particle and the surface is very short, on the order of 10 to 100 μs and (b) the area of contact is very small. In brittle materials, these impact stresses are sufficiently high to remove material from the workpiece surface.

The abrasive grains are typically boron carbide, although aluminum oxide or silicon carbide grains are also used, with sizes ranging from grit number 100 for roughing to grit number 1000 for finishing operations. The grains are carried in a water slurry, with concentrations of 20–60% by volume; the slurry also carries the debris away from the cutting zone. Ultrasonic machining is best suited for materials that are hard and brittle, such as ceramics, carbides, precious stones, and hardened steels; two examples are shown in Figs. 26.26b and c. A special tool is required for each shape to be produced, thus it is also called a *form tool*. The tip of the tool, which is attached to a transducer through the toolholder, usually is made of mild steel.

Rotary Ultrasonic Machining.   In this process, the abrasive slurry is replaced by a tool with metal-bonded diamond abrasives that are either impregnated or electroplated on the tool surface. The tool is vibrated ultrasonically and rotated at the same time, while being pressed against the workpiece surface at a constant pressure. The process is similar to a face-milling operation (see Fig. 24.5), but with the inserts being replaced with abrasives. The chips produced are washed away by a coolant, pumped through the core of the rotating tool. Rotary ultrasonic machining (RUM) is particularly effective in producing deep holes and at high material-removal rates in ceramic parts.

Design Considerations for Ultrasonic Machining.   The basic design guidelines for UM include the following:

- Avoid sharp profiles, corners, and radii, because they can be eroded by the abrasive slurry.
- Remember that holes produced will have some taper.
- Note that, because of the tendency of brittle materials to chip at the exit end of holes, the bottom of the parts should have a backup plate.

26.7   Finishing Operations

Several other processes utilize fine abrasive grains and are used as a final finishing operation. Because these operations can significantly affect production time and product cost, they should be specified only after due consideration of their additional costs versus benefits.

Coated Abrasives.   Common examples of *coated abrasives* are sandpaper and emery cloth; the majority of which are made of aluminum oxide, with silicon carbide and zirconia alumina making up the rest. Coated abrasives usually have a much more open structure than do grinding wheels, and their grains are more pointed and aggressive. The grains are deposited electrostatically on flexible backing materials, such as paper, cotton, rayon polyester, polynylon, and various blends.

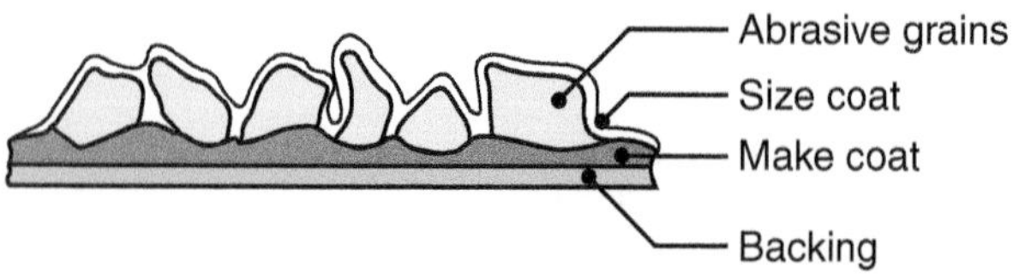

As shown in Fig. 26.27, the bonding material (matrix) typically is resin, which first is applied to the backing (called *make coat*); then the grains are bonded with a second layer (*size coat*). The grains have their long axes

- Abrasive grains
- Size coat
- Make coat
- Backing

FIGURE 26.27   Schematic illustration of the structure of a coated abrasive; sandpaper (developed in the 16th century) and emery cloth are common examples of coated abrasives.

aligned perpendicular to the plane of the backing, thus improving their cutting action. Coated abrasives are available as sheets, belts, and disks. They are used extensively to finish flat or curved surfaces of metallic and nonmetallic parts, metallographic specimens, and in woodworking.

Belt Grinding.   Coated abrasives also are used as *belts* for high-rate material removal and with good surface finish. *Belt grinding* is an important production process, and in some cases competes with and is preferred to conventional grinding operations. Belts with grit numbers ranging from 16 to 1500 (see Figs. 22.6 and 22.7) are available. Belt speeds are in the range of 700–1800 m/min (2500–6000 ft/min). Machines for abrasive-belt operations require proper belt support and must have rigid construction to minimize vibrations.

Conventional coated abrasives have abrasives randomly placed on their surface, and may consist of single or multiple layers of abrasives. An alternative surface is produced by **microreplication**, in which abrasives, in the shape of tiny aluminum-oxide pyramids, are placed in a predetermined orderly arrangement on the belt surface. When used on stainless steels and superalloys, they perform more consistently than conventional coated abrasives, and the temperatures involved are lower. Typical applications for finishing with coated abrasives include belt grinding of golf clubs, firearms, turbine blades, surgical implant, and medical and dental instruments.

EXAMPLE 26.5   Belt Grinding of Turbine Nozzle Vanes

The turbine nozzle vane shown in Fig. 26.28 was investment cast (Section 11.8) from a cobalt-based superalloy. To remove a thin diffusion layer from the root skirt and tip skirt sections of the vane, it was ground on a cloth-backed abrasive belt (60-grit aluminum oxide). The vanes were mounted on a fixture, and ground dry at a belt surface speed of 1800 m/min (6000 ft/min). The production rate was 93 s per piece. Each vane weighed 21.65 g before and 20.25 g after belt grinding, a reduction in weight of about 6.5%.

Source: Reprinted with permission of ASM International. All rights reserved. www.asminternational .org.

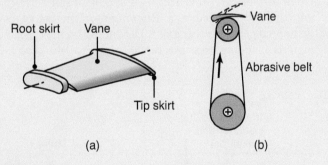

FIGURE 26.28   Turbine nozzle vane considered in Example 26.5.

Wire Brushing.   In this process, also called *power brushing*, the workpiece is held against a circular wire brush that rotates at speeds ranging from 1750 rpm for large wheels to 3500 rpm for small wheels. As they rub against it, the tips of the wire produce longitudinal scratches on the workpiece surface. Wire brushing is used to produce a fine or controlled surface texture. Performed under proper conditions, wire brushing also may be considered as a light material-removal process.

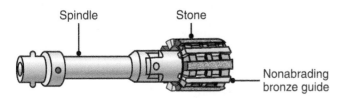

Spindle   Stone

Nonabrading
bronze guide

FIGURE 26.29 Schematic illustration of a honing tool used to improve the surface finish of bored or ground holes.

In addition to metal wires, polymeric wires (such as nylon, Section 7.6) embedded with abrasives can be used effectively. (See also *diamond wire saws*, Section 24.5.)

Honing. *Honing* is an operation that is used primarily to improve the surface finish of holes produced by processes such as boring, drilling, and internal grinding. The honing tool consists of a set of aluminum-oxide or silicon-carbide bonded abrasive sticks, usually called *stones* (Fig. 26.29). They are mounted on a mandrel that rotates in the hole, at surface speeds of 45–90 m/min (150–300 ft/min), applying a radial outward force on the hole surface. The stones can be adjusted radially for different hole sizes.

The tool has a reciprocating axial motion, which produces a crosshatched pattern on the surface of the hole. Oil- or water-based honing fluids generally are used to help flush away the debris and keep temperatures low. Honing is also done on external cylindrical or flat surfaces, and to manually remove sharp edges on cutting tools and inserts. (See Fig. 22.5.)

The quality of the surface finish produced by honing can be controlled by the type and size of abrasive used, the pressure applied, and speed. If not performed properly, honing can produce holes that are neither straight nor cylindrical, but rather in shapes that are bell mouthed, wavy, barrel shaped, or tapered.

Superfinishing. In this process, the pressure applied is very light and the motion of the honing stone has a short stroke. The motion of the stone is controlled so that the grains do not travel along the same path on the surface of the workpiece. Examples of external superfinishing of a round part are shown in Fig. 26.30.

Lapping. This is an operation for finishing flat, cylindrical, or curved surfaces. Generally, the *lap* (Fig. 26.31a) is relatively soft and porous, made of such materials as cast iron, copper, leather, or cloth. The abrasive particles either are embedded in the lap or may be carried in a slurry. Lapping of spherical objects and glass lenses is done with specially shaped laps. *Running-in* of mating gears can be done by lapping, as on hypoid gears for rear axles. Depending on the type and hardness of the workpiece material, lapping pressures range from 7 to 140 kPa (1–20 psi).

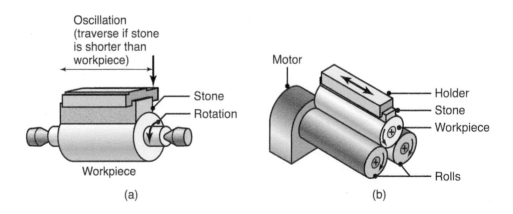

Oscillation
(traverse if stone
is shorter than
workpiece)

Motor

Stone
Rotation

Holder
Stone
Workpiece

Workpiece

Rolls

(a)   (b)

FIGURE 26.30 Schematic illustrations of the superfinishing process for a cylindrical part. (a) Cylindrical microhoning. (b) Centerless microhoning.

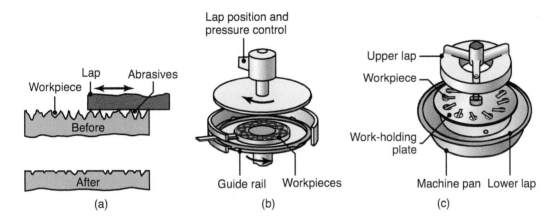

FIGURE 26.31 (a) Schematic illustration of the lapping process. (b) Production lapping on flat surfaces. (c) Production lapping on cylindrical surfaces.

Dimensional tolerances on the order of ±0.0004 mm (0.000015 in.) can be obtained in lapping by using fine abrasives (up to grit size 900), and the surface finish can be as smooth as 0.025–0.1 μm (1–4 μin.). Production lapping on flat or cylindrical pieces is done on machines similar to those shown in Figs. 26.31b and c.

Polishing. *Polishing* is a process that produces a smooth, lustrous surface finish. The basic mechanism involved in the polishing process is the softening and smearing of surface layers, by frictional heating developed during polishing, as well as by some very fine-scale abrasive removal from the workpiece surface. The shiny appearance commonly observed on polished surfaces results from a smearing action.

Polishing is done with disks or belts, made of fabric, leather, or felt, which typically are coated with fine powders of aluminum oxide or diamond. In *double-sided polishing*, pairs of pads are attached to the faces of platens that rotate horizontally and in opposite directions. Parts with irregular shapes, sharp corners, deep recesses, and sharp projections can be difficult to polish.

Chemical–mechanical Polishing. *Chemical–mechanical polishing* (CMP) is extremely important in the semiconductor industry (Chapter 28). This process, shown in Fig. 26.32, uses a suspension of abrasive particles in a water-based solution, with

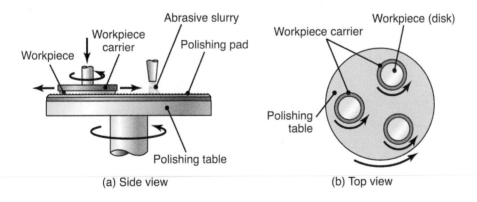

FIGURE 26.32 Schematic illustration of the chemical–mechanical polishing process. This process is used widely in the manufacture of silicon wafers and integrated circuits and also is known as *chemical–mechanical planarization*; for other materials and applications, more carriers and more disks per carrier are used.

a chemistry selected to cause controlled corrosion. Material removed from the workpiece surface is through combined actions of abrasion and corrosion. The surface produced has an exceptionally fine finish and is especially flat; for this reason, the process is often referred to as **chemical–mechanical planarization** (see Section 28.10).

A major application of this process is the polishing of silicon wafers (Section 28.4), in which case the primary function of CMP is to polish at the micrometer level. Therefore, to remove material evenly and across the whole wafer, the wafer is held face down on a rotating carrier, and is pressed against a polishing pad attached to a rotating disk, as shown in Fig. 26.32. The carrier and pad angular velocities are selected such that wear is uniform across the entire wafer surface. The angular velocities are adjusted such that there is a constant relative velocity between the carrier and the pad on the axis connecting their centers. The pad contains grooves intended to uniformly supply slurry to all wafers. Also, pad rotation ensures that a linear lay does not develop (see Section 33.3).

Specific abrasive and solution-chemistry combinations have been developed for the polishing of copper, silicon, silicon dioxide, aluminum, tungsten, and other metals. For silicon dioxide or silicon polishing, for example, an alkaline slurry of colloidal silica (SiO_2 particles in a KOH solution or in NH_4OH) is fed continuously to the pad-wafer interface.

Electropolishing.   Mirrorlike finishes can be obtained on metal surfaces by *electropolishing*, a process that is the reverse of electroplating (described in Section 34.9). Because there is no mechanical contact with the workpiece, this process is particularly suitable for polishing irregular shapes. The electrolyte attacks projections and peaks on the workpiece surface at a higher rate than the rest of the surface, producing a smooth surface. Electropolishing is also used for deburring operations (Section 26.8).

Polishing in Magnetic Fields.   In this technique, abrasive slurries are supported with magnetic fields. There are two basic methods:

1. In the **magnetic-float polishing** of ceramic balls, illustrated schematically in Fig. 26.33a, a magnetic fluid (containing abrasive grains and extremely fine ferromagnetic particles in a carrier fluid, such as water or kerosene) is filled in the chamber within a guide ring. The ceramic balls are located between a driveshaft and a float. The abrasive grains, the ceramic balls, and the float (made

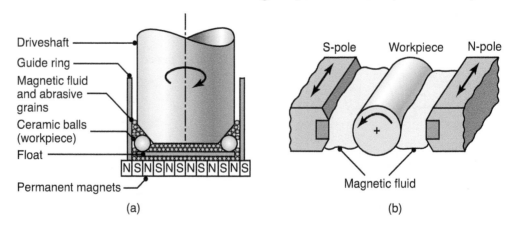

FIGURE 26.33 Schematic illustration of polishing of balls and rollers by magnetic fields. (a) Magnetic-float polishing of ceramic balls. (b) Magnetic-field-assisted polishing of rollers. *Source:* After R. Komanduri, M. Doc, and M. Fox.

of a nonmagnetic material) all are suspended by magnetic forces. The balls are pressed against the rotating driveshaft and are polished by the abrasive action. The forces applied by the abrasive particles on the balls are extremely small and controllable, hence the polishing action is very fine. Because polishing times are much lower than those involved in other polishing methods, this process is highly economical and the surfaces produced have few, if any, significant defects.

2. **Magnetic-field-assisted polishing** of ceramic rollers is illustrated in Fig. 26.33b. A ceramic or steel roller (as the workpiece) is clamped and rotated on a spindle. The magnetic poles are oscillated, introducing a vibratory motion to the magnetic–abrasive conglomerate, an action that polishes the cylindrical roller surface. Bearing steels of 63 HRC hardness have been mirror finished in 30 s with this process.

Buffing.   *Buffing* is similar to polishing, with the exception that an even finer surface finish is obtained using very fine abrasives, on soft disks that typically are made of cloth or hide. The abrasive is supplied externally from a stick of abrasive compound.

26.8   Deburring Operations

Burrs are thin ridges, usually triangular in shape, that develop along the edges of a workpiece from such operations as machining, shearing sheet metals (as in Figs. 16.2 and 16.3), and trimming of forgings and castings. Burrs can be detected by simple means, such as with a finger, toothpick, or cotton swab; visual inspection of burrs includes the use of magnifiers and microscopes. As yet there are no widely accepted standards for specifically defining a burr, partly because of the variety of burrs developed on parts.

Burrs have several disadvantages: (a) They may interfere with the mechanical assembly of parts, and can cause jamming and misalignment of parts, as well as cause short circuits in electrical components. (b) Because they are usually sharp, they can be a safety hazard to personnel handling the parts. (c) Burrs may reduce the fatigue life of components. (d) Sheet metal may have lower bendability if the burr is on the tensile side (see Section 16.2). On the other hand, burrs on thin drilled or tapped components, such as the tiny parts in mechanical watches and mechanisms, can provide additional thickness and improve the holding torque of screws.

Several **deburring processes** are available. Their cost-effectiveness depends on factors such as the extent of deburring required, part complexity, burr location, the number of parts to be deburred, floor space available, labor costs, and safety and environmental considerations. Deburring operations include:

1. Manual deburring, using files and scrapers. It is estimated, however, that manual deburring can contribute up to 10% of the cost of manufacturing the part
2. Mechanical deburring, by machining pieces such as cylindrical parts on a rotating spindle
3. Wire brushing or using rotary nylon brushes, consisting of filaments embedded with abrasive grits
4. Using abrasive belts
5. Ultrasonic machining
6. Electropolishing
7. Electrochemical machining
8. Magnetic–abrasive finishing
9. Vibratory finishing

10. Shot blasting or abrasive blasting
11. Abrasive-flow machining, such as extruding a semisolid abrasive slurry over the edges of the part
12. Thermal energy, using lasers or plasma

The last four processes are described next; other processes are covered elsewhere in this book.

Vibratory and Barrel Finishing.   These processes are used to remove burrs from large numbers of relatively small workpieces. This is a batch-type operation, in which specially shaped *abrasive pellets* of nonmetallic or metallic media (stones or balls) are placed in a container, along with the parts to be deburred. The container is then either *vibrated* or is *tumbled*, by various mechanical means. The impact of individual abrasives or metal particles removes the burrs and sharp edges from the parts. Depending on the application, this can be a *dry* or a *wet* process; liquid compounds may be added for such purposes as decreasing and adding corrosion resistance to the parts being deburred. When chemically active fluids and abrasives are used, this process is a form of chemical–mechanical polishing.

Shot Blasting.   Also called **grit blasting**, this process involves abrasive particles (usually sand) that are propelled by a high-velocity jet of air or by a rotating wheel, onto the surface of the workpiece. Shot blasting is particularly useful in deburring metallic and nonmetallic materials, and in stripping, cleaning, and removing surface oxides. The surfaces produced have a matte finish, but surface damage can result if the process parameters are not properly controlled. **Microabrasive blasting** consists of small-scale polishing and etching, using very fine abrasives, on bench-type units.

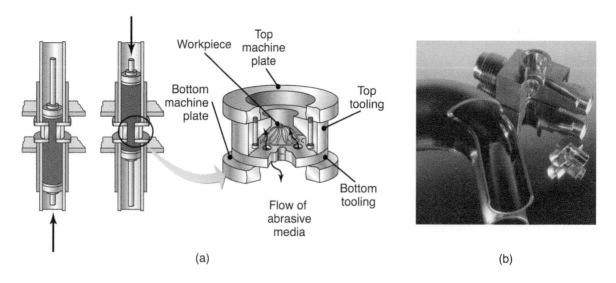

(a)   (b)

FIGURE 26.34   (a) Schematic illustration of abrasive-flow machining to deburr a turbine impeller; the arrows indicate movement of the abrasive media; note the special fixture, which is usually different for each part design. (b) Valve fittings treated by abrasive-flow machining to eliminate burrs and improve surface quality. *Source:* Courtesy of Kennametal Extrude Hone Corporation.

Abrasive-flow Machining. This process involves the use of abrasive grains, such as silicon carbide or diamond, that are mixed in a puttylike matrix and then forced back and forth through the openings and passageways in the workpiece. The movement of the abrasive matrix under pressure erodes away both burrs and sharp corners, and polishes the part. Abrasive-flow machining (AFM) is particularly suitable for workpieces with internal cavities, such as those produced by casting, that are inaccessible by other means. Pressures applied range from 0.7 to 22 MPa (100–3200 psi).

External surfaces also can be deburred with this method, by containing the workpiece within a fixture that directs the abrasive media to the edges and the areas to be deburred. The deburring of a turbine impeller by this process is illustrated in Fig. 26.34.

FIGURE 26.35 An example of thermal energy deburring. *Source:* Courtesy of Kennametal Extrude Hone Corporation.

In **microabrasive-flow machining,** the process mechanics is similar to those in ordinary abrasive-flow machining, but with much smaller abrasive media and less viscous carriers, which allows the media to flow through very small holes ranging from 50 μm (0.002 in.) to 750 μm (0.030 in.) diameter. Micro-AFM has been applied to the production of high-quality diesel fuel injectors and other fine nozzles, where a burr or rough surface finish could adversely affect flow quality.

Thermal Energy Deburring. This process consists of placing the part in a chamber that is then injected with a mixture of natural gas and oxygen. When the mixture is ignited, a burst of heat is produced at a temperature of about 3300°C (6000°F). The burrs are instantly heated and melt (see Fig. 26.35), while the temperature of the bulk part reaches only about 150°C (300°F). There are, however, drawbacks to this process: (a) Larger burrs tend to form beads after melting, (b) thin and slender parts may distort, and (c) the process does not polish or buff the workpiece surfaces, as do several other deburring processes.

FIGURE 26.36 A deburring operation on a robot-held die-cast part for an outboard motor housing; the operation uses a grinding wheel. Abrasive belts (Fig. 26.28) or flexible abrasive radial-wheel brushes also can be used for such operations. *Source:* Courtesy of Acme Manufacturing Company.

Robotic Deburring. Deburring and flash removal from finished products are being performed increasingly by *programmable robots* (Section 37.6), using a force-feedback system for controlling the path and rate of burr removal. This method eliminates tedious and expensive manual labor, and results in more consistent and repeatable deburring. An example is the robotic deburring (see also Section 37.6.3) of a die-cast outboard motor housing, shown in Fig. 26.36. In another application, the manual deburring of a double-helical gear for a helicopter gearbox was deburred in 150 min, whereas robotic deburring required 15 min.

26.9 Economics of Abrasive Machining and Finishing Operations

Abrasive machining and finishing operations often are necessary, because forming, shaping, and machining processes alone do not achieve high enough dimensional accuracy or smooth enough surfaces. Abrasive processes may be used both as a finishing and as well as a large-scale material-removal operation. For example, creep-feed grinding is an economical alternative to machining operations, such as milling or broaching, even though wheel wear is high.

Much progress has been made in automating the equipment involved in these operations, including the use of computer controls, sensors, process optimization, and robotic handling of parts. Consequently, labor costs and production times have been reduced, even though such machinery generally requires major capital investment.

Because they are additional operations, the processes described in this chapter can significantly affect product cost, especially since many of these processes are relatively slow. Moreover, as the surface-finish requirement increases, more operations may be necessary, further increasing production costs, as clearly seen in Fig. 26.37; note how rapidly the cost increases as surface finish is improved by such additional processes as grinding and honing.

The total cost of abrasive operations depends on various factors, such as part size, shape, surface finish, and dimensional accuracy required, as well as machinery, tooling, fixturing, and labor involved. Whereas machinery costs can be high for grinding, the costs for machinery for finishing processes are rather low. Grinding wheel costs are generally low compared with other aspects of the overall grinding operation. As stated in this chapter, however, the cost of grinding wheels can run into

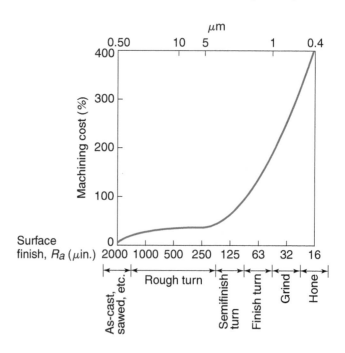

FIGURE 26.37 Increase in the cost of machining and finishing a part as a function of the surface finish required; this is the main reason that the surface finish specified on parts should not be any finer than is necessary for the part to function properly.

hundreds or even thousands of dollars, depending on their composition and size. The costs of finishing tools, such as those for honing and lapping, vary widely, and labor costs and operator skill depend greatly on how well the equipment is automated.

If finishing is likely to be an important factor in manufacturing a particular product, the conceptual and original design stages should involve an analysis of the level of surface finish and the dimensional accuracy required, and of whether they can be relaxed. Furthermore, all processes that precede finishing operations should be analyzed for their capability to produce more acceptable surface characteristics. This can be accomplished through the proper selection of tools, selection of process parameters, metalworking fluids, and the characteristics of the machine tools, their level of automation, computer controls, and of the work-holding devices involved.

SUMMARY

- Abrasive machining often is necessary and economical when workpiece hardness and strength is high, the materials are brittle, and surface finish and dimensional tolerance requirements are high.
- Conventional abrasives consist of aluminum oxide and silicon carbide; superabrasives consist of cubic boron nitride and diamond. The friability of abrasive grains is an important factor in their performance, as are the shape and size of the grains.
- Grinding wheels, also known as bonded abrasives in contrast to loose abrasives, consist of a combination of abrasive grains and bonding agents. Important characteristics of wheels are type of abrasive grain and bond, grade, and hardness. Wheels also may be reinforced to maintain their integrity, if and when a crack develops during their use.
- Grinding-wheel wear is an important consideration in the surface quality and integrity of the ground product. Dressing and truing of wheels are necessary operations, and are done by various techniques.
- A variety of abrasive-machining processes and machinery is available for surface, external, and internal grinding. Abrasive machining also is used for large-scale material-removal processes, such as creep-feed grinding, making it competitive with machining processes such as milling and turning.
- The selection of abrasives and process variables, including grinding fluids, is important in obtaining the desired surface finish and dimensional accuracy. Otherwise, damage to surfaces, such as burning, heat checking, detrimental residual stresses, and chatter, may develop.
- Several finishing operations are available for improving surface finish. Because they can significantly affect product cost, the appropriate selection and implementation of these operations is important.
- Deburring may be necessary for some finished components. Commonly used methods are vibratory finishing, barrel finishing, and shot blasting, although thermal energy and other methods also are available.

KEY TERMS

Abrasive-flow machining	Aluminum oxide	Belt grinding	Burning
Abrasives	Attritious wear	Bonded abrasives	Burr
	Barrel finishing	Buffing	Chatter marks

Chemical–mechanical
   polishing
Coated abrasives
Creep-feed grinding
Cubic boron nitride
Deburring
Diamond
Dressing
Ductile-regime
   grinding
Electropolishing
Finishing
Free-cutting wheels
Friability
Glazing

Grade
Grain depth of cut
Grain size
Grindability
Grinding
Grinding ratio
Grit number
Hardness of wheel
Heat checking
Honing
Lapping
Loading
Low-stress grinding
Magnetic-field-assisted
   polishing

Magnetic-float polishing
Metallurgical burn
Microabrasive-flow
   machining
Microreplication
Polishing
Reinforced wheels
Resinoid bond
Robotic deburring
Rotary ultrasonic
   machining
Seeded gel
Shot blasting
Silicon nitride
Snagging

Sonotrode
Sparks
Specific energy
Structure of wheel
Superabrasives
Superfinishing
Tempering
Truing
Ultrasonic machining
Vibratory finishing
Vitrified bond
Wear flat
Wheel depth of cut
Wire brushing

BIBLIOGRAPHY

Astashev, V.K., and Babitsky, V.I., **Ultrasonic Processes and Machines**, Springer, 2010.

Brown, J., **Advanced Machining Technology Handbook**, McGraw-Hill, 1998.

Gillespie, L.K., **Deburring and Edge Finishing Handbook**, Society of Manufacturing Engineers/American Society of Mechanical Engineers, 2000.

Hwa, L.S., **Chemical Mechanical Polishing in Silicon Processing**, Academic Press, 1999.

Jackson, M.J., and Davim, M.J., **Machining with Abrasives**, Springer, 2010.

Krar, S., and Ratterman, E., **Superabrasives: Grinding and Machining with CBN and Diamond**, McGraw-Hill, 1990.

Kuchle, A., **Manufacturing Processes 2: Grinding, Honing, Lapping**, Springer, 2009.

Malkin, S., and Guo, C., **Grinding Technology**, 2nd ed., Industrial Press, 2008.

Marinescu, I.D., Hitchiner, M., Uhlmann, E., and Rowe, W.B. (eds.), **Handbook of Machining with Grinding Wheels**, CRC Press, 2006.

Marinescu, I.D., Tonshoff, H.K., and Inasaki, I., **Handbook of Ceramic Grinding & Polishing**, William Andres, 2000.

Marinescu, I.D., Uhlmann, E., and Doi, T. (eds.), **Handbook of Lapping and Polishing**, CRC Press, 2006.

Oliver, M.R., **Chemical Mechanical Planarization of Semiconductor Materials**, Springer, 2004.

Rowe, W.B., **Principles of Modern Grinding Technology**, William Andrew, 2009.

Webster, J.A., Marinescu, I.D., and Trevor, T.D., **Abrasive Processes**, Marcel Dekker, 1999.

REVIEW QUESTIONS

26.1 What is an abrasive? What are superabrasives?

26.2 How is the size of an abrasive grain related to its number?

26.3 Why are most abrasives made synthetically?

26.4 Describe the structure of a grinding wheel and its features.

26.5 Explain the characteristics of each type of bond used in bonded abrasives.

26.6 What causes grinding sparks in grinding? Is it useful to observe them? Explain.

26.7 Define metallurgical burn.

26.8 Define (a) friability, (b) wear flat, (c) grinding ratio, (d) truing, and (e) dressing.

26.9 What is creep-feed grinding and what are its advantages?

26.10 How is centerless grinding different from cylindrical grinding?

26.11 What are the differences between coated and bonded abrasives?

26.12 What is the purpose of the slurry in chemical–mechanical polishing?

QUALITATIVE PROBLEMS

26.13 Explain why grinding operations may be necessary for components that have previously been machined.

26.14 Why is there such a wide variety of types, shapes, and sizes of grinding wheels?

26.15 Explain the reasons for the large difference between the specific energies involved in machining (Table 21.2) and in grinding (Table 26.2).

26.16 The grinding ratio, G, depends on the type of grinding wheel, workpiece hardness, wheel depth of cut, wheel and workpiece speeds, and the type of grinding fluid. Explain.

26.17 What are the consequences of allowing the temperature to rise during grinding? Explain.

26.18 Explain why speeds are much higher in grinding than in machining operations.

26.19 It was stated that ultrasonic machining is best suited for hard and brittle materials. Explain.

26.20 Explain why parts with irregular shapes, sharp corners, deep recesses, and sharp projections can be difficult to polish.

26.21 List the finishing operations commonly used in manufacturing operations. Why are they necessary? Explain why they should be minimized.

26.22 Referring to the preceding chapters on processing of materials, list the operations in which burrs can develop on workpieces.

26.23 Explain the reasons that so many deburring operations have been developed over the years.

26.24 What precautions should you take when grinding with high precision? Comment on the machine, process parameters, grinding wheel, and grinding fluids.

26.25 Describe the factors involved in a grinding wheel acting "soft" or acting "hard."

26.26 What factors could contribute to chatter in grinding? Explain.

26.27 Generally, it is recommended that, in grinding hardened steels, the grinding be wheel of a relatively soft grade. Explain.

26.28 In Fig. 26.5, the proper grinding faces are indicated for each type of wheel. Explain why the other surfaces of the wheels should not be used for grinding and what the consequences may be in doing so.

26.29 Describe the effects of a wear flat on the overall grinding operation.

26.30 What difficulties, if any, could you encounter in grinding thermoplastics? Thermosets? Ceramics?

26.31 Observe the cycle patterns shown in Fig. 26.22 and comment on why they follow those particular patterns.

26.32 Which of the processes described in this chapter are particularly suitable for workpieces made of (a) ceramics, (b) thermoplastics, and (c) thermosets? Why?

26.33 Grinding can produce a very fine surface finish on a workpiece. Is this finish necessarily an indication of the quality of a part? Explain.

26.34 Jewelry applications require the grinding of diamonds into desired shapes. How is this done, since diamond is the hardest material known?

26.35 List and explain factors that contribute to poor surface finish in the processes described in this chapter.

QUANTITATIVE PROBLEMS

26.36 Calculate the chip dimensions in surface grinding for the following process variables: $D = 10$ in., $d = 0.001$ in., $v = 100$ ft/min, $V = 5000$ ft/min, $C = 500$ per in^2, and $r = 20$.

26.37 If the strength of the workpiece material is doubled, what should be the percentage decrease in the wheel depth of cut, d, in order to maintain the same grain force, with all other variables being the same?

26.38 Assume that a surface-grinding operation is being carried out under the following conditions: $D = 200$ mm, $d = 0.1$ mm, $v = 40$ m/min, and $V = 2000$ m/min. These conditions are then changed to the following: $D = 150$ mm, $d = 0.1$ mm, $v = 30$ m/min, and $V = 2500$ m/min. How

different is the temperature rise from the rise that occurs with the initial conditions?

26.39 Estimate the percent increase in the cost of the grinding operation if the specification for the surface finish of a part is changed from 250 to 32 μin.

26.40 Assume that the energy cost for grinding an aluminum part with a specific energy requirement of 8 W-s/mm^3 is $1.50 per piece. What would be the energy cost of carrying out the same operation if the workpiece material were T15 tool steel?

26.41 In describing grinding processes, we have not given the type of equations regarding feeds, speeds, material-removal rates, total grinding time, etc., as we did in the turning and milling operations discussed in Chapters 23 and 24.

Study the quantitative relationships involved and develop such equations for grinding operations.

26.42 What would be the answers to Example 26.1 if the workpiece is high-strength titanium and the width of cut is $w = 0.75$ in.? Give your answers in newtons.

26.43 It is known that, in grinding, heat checking occurs when grinding is done with a spindle speed of 5000 rpm, a wheel diameter of 8 in., and a depth of cut of 0.0015 in. for a feed rate of 50 ft/min. For this reason, the standard operating procedure is to keep the spindle speed at 3500 rpm. If a new, 10-in.-diameter wheel is used, what spindle speed can be used before heat checking occurs? What spindle speed should be used to keep the same grinding

temperatures as those encountered with the existing operating conditions?

26.44 A grinding operation takes place with a 10-in. grinding wheel with a spindle speed of 4000 rpm. The workpiece feed rate is 50 ft/min and the depth of cut is 0.002 in. Contact thermometers record an approximate maximum temperature of 1800°F. If the workpiece is steel, what is the temperature if the speed is increased to 5000 rpm? What if the speed is 10,000 rpm?

26.45 Derive an expression for the angular velocity of the wafer shown in Fig. 26.30b as a function of the radius and angular velocity of the pad in chemical–mechanical polishing.

SYNTHESIS, DESIGN, AND PROJECTS

26.46 With appropriate sketches, describe the principles of various fixturing methods and devices that can be used for the processes described in this chapter.

26.47 Make a comprehensive table of the process capabilities of abrasive-machining operations. Using several columns, describe the features of the machines involved, the type of abrasive tools used, the shapes of blanks and parts produced, typical maximum and minimum sizes, surface finish, tolerances, and production rates.

26.48 Vitrified grinding wheels (also called ceramic wheels) use a glasslike bond to hold the abrasive grains together. Given your understanding of ceramic-part manufacture (as described in Chapter 18), list methods of producing vitrified wheels.

26.49 Assume that you are an instructor covering the topics described in this chapter and you are giving a quiz on the numerical aspects to test the understanding of the students. Prepare three quantitative problems and supply the answers.

26.50 Conduct a literature search, and explain how observing the color, brightness, and shape of sparks produced in grinding can be a useful guide to identifying the type of material being ground and its condition.

26.51 Visit a large hardware store and inspect the grinding wheels that are on display. Make a note of the markings on the wheels and, on the basis of the marking system shown in Fig. 26.6, comment on your observations, including the most common types of wheels available in the store.

26.52 Obtain a small grinding wheel or a piece of a large wheel. (a) Using a magnifier or a microscope, observe its surfaces and compare them with Fig. 26.9. (b) Rub the abrasive wheel by pressing it hard against a variety of flat metallic and nonmetallic materials. Describe your observations regarding the surfaces produced.

26.53 In reviewing the abrasive machining processes in this chapter, you will note that some use bonded abrasives while others involve loose abrasives. Make two separate lists for these processes and comment on your observations.

26.54 Obtain pieces of sandpaper and emery cloth of different coarseness. Using a magnifier or a microscope, observe their surface features and compare them with Fig. 26.25.

26.55 On the basis of the contents of this chapter, describe your thoughts on whether or not it would be possible to design and build a "grinding center." (See Chapter 25.) Comment on any difficulties that may be encountered in such machines and operations.

Advanced Machining Processes and Equipment

- It is often necessary to machine or finish products that are made of very hard or strong materials; in this case, the conventional machining and grinding strategies described thus far become impractical.
- This chapter describes advanced machining processes that are based on nonmechanical means of material removal. The chapter begins by examining chemical machining and photochemical blanking processes, in which material is removed through the corrosive action of a fluid.
- Electrochemical machining and grinding are then described, where material is removed by the action of an electrical power source and ion transfer inside an electrolytic fluid. Electrical-discharge machining removes material by melting small portions of the workpiece by sparks.
- Laser-beam and electron-beam machining processes, as well as water-jet and abrasive-jet machining operations, also are described, with examples of their unique applications.
- The chapter ends with a review of trends in hybrid machining operations and the economics of advanced machining processes.

Typical parts made: Skin panels for missiles and aircraft, turbine blades, nozzles, parts with complex cavities and small-diameter deep holes, dies, laser cutting of sheet metals, cutting of thick metallic and nonmetallic parts.

Alternative methods: Abrasive machining, ultrasonic machining, and precision machining.

27.1 Introduction

The machining processes described in the preceding chapters involved material removal by mechanical means of chip formation, abrasion, or microchipping. There are situations, however, in which mechanical methods are not satisfactory, economical, or even possible, for the following reasons:

- The **strength** and **hardness** of the workpiece material are very high, typically above 400 HB (see Fig. 2.15)
- The material is too **brittle** to be machined without damage to the part, typically the case with highly heat-treated alloys, glass, ceramics, and powder-metallurgy parts
- The workpiece is **too flexible** or **slender** to withstand forces involved in machining or grinding, or the parts are difficult to clamp in fixtures and workholding devices

FIGURE 27.1   Examples of parts made by advanced machining processes. (a) Samples of parts produced by water-jet cutting. (b) Turbine blade, produced by plunge electrical-discharge machining; the holes are produced by electrical-discharge machining. *Source:* (a) Courtesy of OMAX Corporation; (b) Courtesy of HI-TEK Mfg., Inc.

- The part **has a complex shape** (Fig. 27.1), including such features as internal and external profiles or holes with high length-to-diameter ratios in very hard materials
- Special **surface finish** and **dimensional tolerance** requirements that cannot be obtained by other manufacturing processes or are uneconomical
- The **temperature rise** during processing and **residual stresses** developed in the workpiece are not acceptable

Beginning with the 1950s, these difficulties led to the development of chemical, electrical, laser, and high-energy beams as energy sources for removing material from metallic or nonmetallic workpieces, as outlined in Table 27.1. Also called *nontraditional* or *unconventional* machining, these processes remove material not by producing chips, as in traditional machining and grinding, but by means such as chemical dissolution, etching, melting, evaporation, and hydrodynamic action, sometimes with the assistance of fine abrasive particles.

A major advantage of these processes is that their efficiency is independent of workpiece hardness. When selected and applied properly, advanced machining processes offer major technical and economic advantages over more traditional methods. This chapter describes these processes, including their characteristics, typical applications, limitations, product quality, dimensional accuracy, surface finish, and economics.

27.2 Chemical Machining

Chemical machining (CM) is based on the observation that chemicals attack and etch most materials, thereby removing small amounts of material from the surface. The CM process is carried out by chemical dissolution using **reagents** or **etchants,** such as acids and alkaline solutions. Chemical machining is the oldest of the advanced machining processes, and has been used in engraving metals and hard stones, in deburring, and in the production of printed-circuit boards and microelectronic devices (see Chapters 28 and 29).

TABLE 27.1

General Characteristics of Advanced Machining Processes

Process	Characteristics	Process parameters and typical material-removal rate or cutting speed
Chemical machining (CM)	Shallow removal on large flat or curved surfaces; blanking of thin sheets; low tooling and equipment cost; suitable for low-production runs	0.0025–0.1 mm/min. (0.0001–0.004 in./min)
Electrochemical machining (ECM)	Complex shapes with deep cavities; highest rate of material removal among other nontraditional processes; expensive tooling and equipment; high power consumption; medium-to-high production quantity	V: 5–25 D.C.; A: 1.5–8 A/mm^2; 2.5–12 mm/min (0.1–0.5 in./min), depending on current density
Electrochemical grinding (ECG)	Cutting off and sharpening hard materials, such as tungsten-carbide tools; also used as a honing process; higher removal rate than grinding	A: 1–3 A/mm^2; typically 25 mm^3/s (0.0016 in^3/s) per 1000 A
Electrical-discharge machining (EDM)	Shaping and cutting complex parts made of hard materials; some surface damage may result; also used as a grinding and cutting process; expensive tooling and equipment	V: 50–380; A: 0.1–500; typically 300 mm^3/min (0.02 in^3/min)
Wire electrical-discharge machining	Contour cutting of flat or curved surfaces; expensive equipment	Varies with material and thickness
Laser-beam machining (LBM)	Cutting and hole making on thin materials; heat-affected zone; does not require a vacuum; expensive equipment; consumes much energy	0.50–7.5 m/min (1.67–25 ft/min)
Laser Microjet	Water-jet guided laser uses a 25–100 μm diameter stream to mill or cut; large depth of field; little thermal damage from laser machining	Varies with material; up to 20 mm in silicon, 2 mm in stainless steel; up to 300 mm/s in 50 μm thick silicon.
Electron-beam machining (EBM)	Cutting and hole making on thin materials; very small holes and slots; heat-affected zone; requires a vacuum; expensive equipment	1–2 mm^3/min (0.004–0.008 in^3/h)
Water-jet machining (WJM)	Cutting all types of nonmetallic materials; suitable for contour cutting of flexible materials; no thermal damage; noisy	Varies considerably with material
Abrasive water-jet machining (AWJM)	Single-layer or multilayer cutting of metallic and nonmetallic materials	Up to 7.5 m/min (25 ft/min)
Abrasive-jet machining (AJM)	Cutting, slotting, deburring, etching, and cleaning of metallic and nonmetallic materials; tends to round off sharp edges; can be hazardous	Varies considerably with material

Chemical Milling.   In *chemical milling*, shallow cavities are produced on plates, sheets, forgings, and extrusions, generally for overall reduction of weight, as can be seen in Fig. 27.2. The process has been used on a wide variety of metals, with depths of removal up to 12 mm (0.5 in.). *Selective attack* by the chemical reagent on different areas of the workpiece surfaces is accomplished by removable layers of material, called *masking* (Fig. 27.3a), or by partial immersion of the part in the reagent.

The procedure for chemical milling consists of the following steps:

1. If the part to be machined has residual stresses from prior processing, the stresses first should be relieved in order to prevent warping after chemical milling.
2. The surfaces are degreased and cleaned thoroughly, to ensure both good adhesion of the masking material and uniform rate of material removal; scale from prior heat treatment also should be removed.

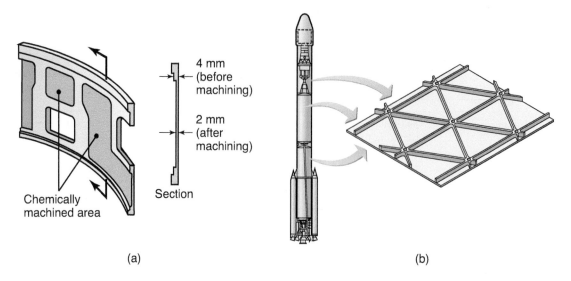

FIGURE 27.2 (a) Missile skin-panel section contoured by chemical milling to improve the stiffness-to-weight ratio of the part. (b) Weight reduction of space-launch vehicles by the chemical milling of aluminum-alloy plates. The plates are chemically milled after they have been formed into desired shapes by a process such as roll forming or stretch forming.

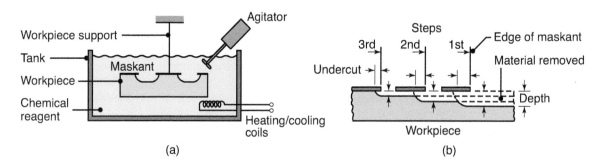

FIGURE 27.3 (a) Schematic illustration of the chemical-machining process; note that no forces or machine tools are involved in this process. (b) Stages in producing a profiled cavity by chemical machining; note the undercut.

3. The masking material, called **maskant**, is applied, typically using tapes or paints, although elastomers (rubber and neoprene) and plastics (polyvinyl chloride, polyethylene, and polystyrene) also are used. Note that the maskant should not react with the chemical reagent.

4. The areas in the maskant that will require etching is peeled off, using the scribe-and-peel technique.

5. The exposed surfaces are machined chemically, using etchants, such as sodium hydroxide (for aluminum), solutions of hydrochloric and nitric acids (for steels), and iron chloride (for stainless steels). Temperature control and agitation (stirring) of the etchant during chemical milling is important, in order to remove a uniform depth of material from the part.

6. After machining, the parts are washed thoroughly with water, to prevent further reactions with, or exposure to, any etchant residues.

7. The rest of the masking material is removed, and the part is cleaned and inspected. Note that although the maskant is unaffected by the reagent, it

can easily be dissolved by a different and appropriate type of solvent, such as acetone or Piranha (see Table 28.3).

8. Additional finishing operations may be performed on chemically milled parts, such as abrasive flow machining (see Section 26.8) or electroplating (Section 34.9).

9. This sequence of operations can be repeated to produce stepped cavities and various contours (Fig. 27.3b).

Chemical milling is used in the aerospace industry to remove shallow layers of material from large aircraft components, missile skin panels (Fig. 27.2), and extruded parts for airframes. Tank capacities for reagents are as large as 3.7 m × 15 m (12 ft × 50 ft). The process is also used to fabricate microelectronic devices, and often is referred to as **wet etching** (see Section 28.8.1). The ranges of surface finish and tolerance obtained by chemical machining and other machining processes are shown in Fig. 27.4.

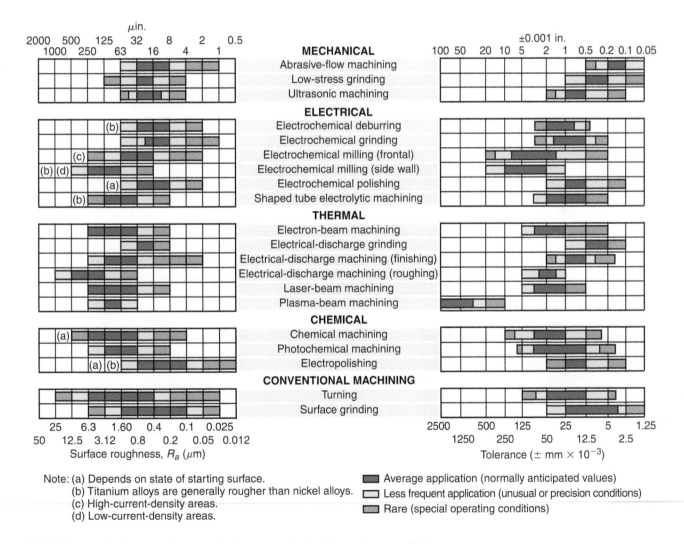

Note: (a) Depends on state of starting surface.
(b) Titanium alloys are generally rougher than nickel alloys.
(c) High-current-density areas.
(d) Low-current-density areas.

■ Average application (normally anticipated values)
□ Less frequent application (unusual or precision conditions)
■ Rare (special operating conditions)

FIGURE 27.4 Surface roughness and tolerances obtained in various machining processes; note the wide range within each process (see also Fig. 23.13). *Source:* Based on data from *Machining Data Handbook*, 3rd ed. Copyright 1980.

Some surface damage may result from chemical milling because of *preferential etching* and *intergranular attack*, which adversely affect surface properties. Chemical milling of welded and brazed structures also may result in uneven material removal, and castings may result in uneven surfaces, caused by porosity and property nonuniformities in the material.

Chemical Blanking.   *Chemical blanking* is similar to the blanking of sheet metals (Fig. 16.4). Typical applications for chemical blanking are the burr-free etching of printed circuit boards, decorative panels, and thin sheet-metal stampings, as well as the production of complex or very small shapes.

Photochemical Blanking.   *Photochemical blanking*, also called *photoetching*, is a modification of chemical milling. Material is removed, usually from flat thin sheet, by photographic techniques to produce a mask, followed by chemical machining. Complex, burr-free shapes can be blanked (Fig. 27.5) on metal foil as thin as 0.0025 mm (0.0001 in.). Sometimes called **photochemical machining**, this process is also used for etching, such as for electrical connectors or pattern plates for reflow (paste) soldering (Section 32.2.3).

The procedure in photochemical blanking consists of the following steps:

1. The design of the part to be blanked is prepared at a magnification of up to 100×; a photographic negative is then made and reduced to the size of the finished part (called **artwork**). Note that the original (enlarged) drawing allows inherent design errors to be reduced in size by the amount of reduction (such as 100×) for the final artwork image.
2. The sheet blank is coated with a photosensitive material (**photoresist**, and often called *emulsion*), by dipping, spraying, spin casting, or roller coating; it is then dried in an oven.
3. The negative is placed over the coated blank and exposed to ultraviolet light, which hardens the exposed areas.
4. The blank is developed, thus dissolving the unexposed areas.
5. The blank is then immersed into a bath of reagent (as in chemical milling) or is sprayed with the reagent, which etches away the exposed areas.
6. The masking material is removed, and the part is washed thoroughly with water, to remove all chemical residues.

FIGURE 27.5   Various parts made by chemical blanking; note the fine detail. *Source:* Courtesy of Buckbee-Mears, St. Paul, Minnesota.

The handling of chemical reagents requires precautions and special safety considerations, to protect the workers against exposure to both liquid chemicals and volatile chemicals. Furthermore, the disposal of chemical by-products from this process is a major drawback, although some by-products can be recycled.

Although skilled labor is required, tooling costs are low, the process can be automated, and is economical for medium- to high-production volume. Photochemical blanking is capable of making very small parts, in cases when traditional blanking dies (Section 16.2) are too difficult to produce. The process is also effective for blanking fragile workpieces and materials. Tolerances are on the order of 10% of the sheet thickness. Typical applications for photochemical blanking include fine metal screens, printed-circuit cards, electric-motor laminations, flat springs, and various components of miniaturized systems.

Design Considerations for Chemical Machining.   General design guidelines for chemical machining are:

- Designs involving sharp corners, deep and narrow cavities, severe tapers, folded seams, or porous workpiece materials should be avoided, because the etchant attacks all exposed surfaces continuously.
- Because the etchant attacks the material in both vertical and horizontal directions, undercuts may develop, as shown in Fig. 27.3, in the areas under the edges of the maskant.
- To improve production rate, the bulk of the workpiece preferably should be shaped by other and higher volume rate processes, such as machining, prior to chemical machining.
- Dimensional variations can occur, because of size changes in the deposited mask pattern due to humidity and temperature. These variations can be minimized by properly selecting artwork media and by controlling both the environment in which the artwork is generated and the production area in the plant.
- Product designs are now produced with computer-aided design systems (Chapter 38) and can be translated into a useful format for etching machinery.

27.3 Electrochemical Machining

Electrochemical machining (ECM) is basically the reverse of electroplating (Section 34.9). An **electrolyte** acts as the current carrier (Fig. 27.6), and the high flow rate of electrolyte in the tool–workpiece gap (typically 0.1–0.6 mm) washes metal ions away from the workpiece (*anode*) before they have a chance to plate onto the tool (*cathode*). Note that the cavity produced is the mating image of the tool shape.

The shaped tool, either in solid or tubular form, is generally made of brass, copper, bronze, or stainless steel. The electrolyte is a highly conductive inorganic fluid, such as an aqueous solution of sodium nitrate; it is pumped through the passages in the tool, at rates of 10–16 m/s (3–50 ft/s). A DC power supply, in the range from 10 to 25 V, maintains current densities, which, for most applications, are 20–200 A/cm^2 (130–1300 A/in^2) of active machined surface.

The *material-removal rate* (MRR) in electrochemical machining, for a current efficiency of 100%, may be estimated from the equation

$$MRR = CI, \qquad (27.1)$$

where MRR is in mm^3/min, I is the current in amperes, and C is a material constant with the unit of mm^3/A-min. For pure metals, C depends on the valence: the higher the valence, the lower is its value.

Machines having current capacities as high as 40,000 A, and as low as 5 A, are available. The penetration rate of the tool is proportional to the current density, and the material removal rate typically ranges between 1.5 and 4 mm^3 per A-min. Because the metal-removal rate is a function only of the ion exchange rate, it is not affected by the strength, hardness, or toughness of the workpiece, a characteristic that is common to the processes described in this chapter.

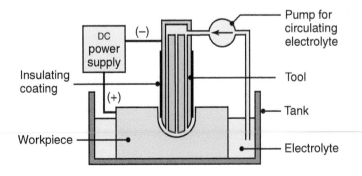

FIGURE 27.6 Schematic illustration of the electrochemical machining process.

Process Capabilities.   The basic concept of electrochemical machining developed rapidly beginning with the 1950s, whereupon it became an important manufacturing process. It generally is used to machine complex cavities and shapes in high-strength materials, particularly in the aerospace industry for the mass production of turbine blades, jet-engine parts, and nozzles (Fig. 27.7); other applications include the automotive (engines castings and gears) and medical industries.

Electrochemical machining also is used for machining and finishing forging-die cavities (called *die sinking*) and to produce small holes. Modifications of this process are used for turning, facing, milling, slotting, drilling, trepanning, and profiling operations, as well as in the production of continuous metal strips and webs. More recent applications of ECM include *micromachining* (Chapters 28 and 29) for the electronics industry.

An advance in ECM is *shaped-tube electrolytic machining* (STEM); it is used for producing small-diameter deep holes, as in turbine blades (Fig. 27.8). The electrolyte is acid-based, to make sure the worn metal is dissolved and carried away by the solution. The tool is a titanium tube, for corrosion resistance, coated with an electrically insulating resin to restrict the electrolytic action to the front surface of the electrode. Holes as small as 0.5 mm can be made, and at depth-to-diameter ratios as high as 300:1; larger holes can be produced by *electrolytic trepanning*, as shown in Fig. 27.8b.

The ECM process leaves a burr-free, bright surface, and it can also be used as a deburring operation. It does not cause any thermal damage to the part, and the absence of tool forces prevents distortion, especially in thin, flexible parts. Furthermore, there is no tool wear (since only hydrogen is generated at the cathode), and the process is capable of producing complex shapes. The mechanical properties of components made by ECM should, however, be compared carefully with those of

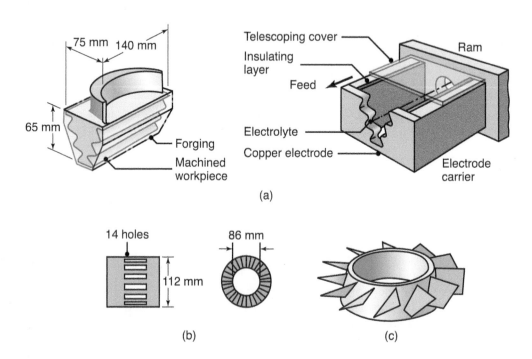

FIGURE 27.7   Typical parts made by electrochemical machining. (a) Turbine blade made of a nickel alloy of 360 HB; note the shape of the electrode on the right. (b) Thin slots on a 4340-steel roller-bearing cage. (c) Integral airfoils on a compressor disk.

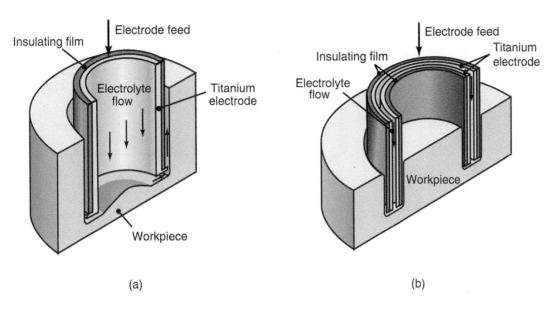

(a)   (b)

FIGURE 27.8 Shaped electrolytic machining operations: (a) shaped-tube electrolytic machining, used to make small holes with aspect ratios as large as 300:1; (b) electrolytic trepanning, used for larger diameter holes.

components made by other processes, to ensure that there has not been a significant compromise due to chemical reactions.

Electrochemical-machining systems now are available as *numerically controlled machining centers*, with the capability of high production rates, high flexibility of operation, and the maintenance of close dimensional tolerances. The ECM process also can be combined with electrical-discharge machining (EDM) on the same machine (called **hybrid machining**; see Section 27.10).

Design Considerations for Electrochemical Machining.   The following are general design guidelines for electrochemical machining:

- Electrochemical machining is not suited for producing sharp square corners or flat bottoms, because of the tendency for the electrolyte to erode away sharp profiles.
- Controlling the electrolyte flow may be difficult, hence irregular cavities may not be produced to the desired shape and with acceptable dimensional accuracy.
- Part designs should make provision for a small taper for holes and cavities to be machined.

CASE STUDY 27.1   Electrochemical Machining of a Biomedical Implant

A total knee-replacement system consists of a femoral and tibial implant, combined with an ultrahigh-molecular-weight polyethylene (UHMWPE) insert, as shown in Fig. 27.9a. The polyethylene has superior wear resistance and low friction against the cobalt–chrome alloy femoral implant.

The UHMWPE insert is compression molded (Section 19.7), and the metal implant is cast and ground on its external mating surfaces.

Designers of implants, manufacturing engineers, and clinicians have long been concerned particularly with the contact surface in the cavity

(*continued*)

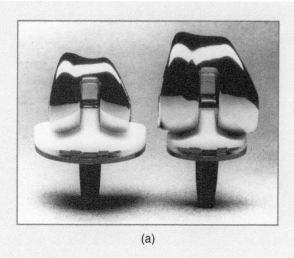

(a)

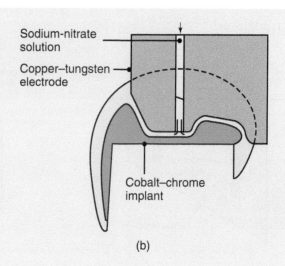

Sodium-nitrate solution

Copper–tungsten electrode

Cobalt–chrome implant

(b)

FIGURE 27.9   (a) Two total knee-replacement systems, showing metal implants (top pieces) with ultrahigh-molecular-weight polyethylene inserts (bottom pieces). (b) Cross-section of the ECM process as applied to the metal implant. *Source:* Courtesy of Biomet, Inc.

of the metal implant that mates with a protrusion on the polyethylene insert. As the knee articulates during its normal motion, the polyethylene slides against the metal part, becoming a potentially serious wear site (Section 33.5). This geometry is necessary to ensure lateral stability of the knee; that is, to prevent the knee from buckling sideways.

In order to produce a smooth surface, the grinding of the bearing surfaces of the metal implant, using both hand-held and cam-mounted grinders, was a procedure that had been followed for many years. However, grinding produced marginal repeatability and part quality. The interior surfaces of this part are extremely difficult to access for grinding, and the cobalt–chrome alloy is difficult to grind. Consequently, advanced machining processes, particularly electrochemical machining, were considered to be ideal candidates for this operation.

As shown in Fig. 27.9b, the current procedure consists of placing the metal implant in a fixture and bringing a tungsten electrode of the desired final contour in close proximity to the implant. The electrolyte is a sodium nitrate and water mixture, and is pumped through the tool, filling the gap between the tool and the implant. A power source (typically 10 V and 225 A) is applied, causing local electrochemical machining of the high spots on the implant surface, and producing a polished surface.

The electrolyte flow rate can be controlled in order to maximize surface quality. When the rate is too low, defects appear on the machined surface as localized dimples; if the flow rate is too high, machining times become longer. Typical machining times for this part are four to six minutes.

Source: Courtesy of T. Hershberger and R. Redman, Biomet, Inc.

27.3.1   Pulsed Electrochemical Machining

The *pulsed electrochemical machining* (PECM) process is a refinement of ECM; it uses very high current densities (on the order of 1 A/mm^2), but the current is *pulsed*, rather than direct current. The purpose of pulsing is to eliminate the need for high electrolyte flow rates, which limit the usefulness of ECM in die and mold making. Investigations have shown that PECM improves fatigue life compared to ECM, and the process does not have the characteristic recast layer on die and mold surfaces. The tolerances obtained typically are in the range from 20 to 100 μm.

Machines now can perform a combination of both EDM and PECM, thus the need to move the tool and workpiece between the two processes is eliminated. If these operations occur on separate machines, it is difficult to maintain precise alignment when moving the workpiece from the EDM to PECM. If misaligned significantly, all polishing will occur where the gap is smallest, and passivation (see Section 3.8) will occur where the gap is largest. Also, the process leaves metal residues suspended in the aqueous solution, which is harmful to the environment if disposed of without proper treatment.

The ECM process can be effective for *micromachining* (see section 29.2). The complete absence of tool wear implies that this process also can be used for precision electronic components, although the erosion problem due to stray current has to be overcome. ECM machines now have increased flexibility, by the implementation of numerical controls.

27.4   Electrochemical Grinding

Electrochemical grinding (ECG) combines electrochemical machining with conventional grinding. The equipment used is similar to a conventional grinder, except that the wheel is a rotating *cathode*, embedded with abrasive particles (Fig. 27.10a). The wheel is metal bonded with diamond or aluminum-oxide abrasives, and rotates at a surface speed from 1200 to 2000 m/min (4000–7000 ft/min).

The abrasives have two functions: (a) to serve as insulators between the wheel and the workpiece and (b) to mechanically remove electrolytic products from the working area. A flow of electrolyte solution, usually sodium nitrate, is provided for the electrochemical machining phase of the operation. Current densities range from 1 to 3 A/mm^2 (500–2000 A/in^2). The majority of metal removal in ECG is by electrolytic action, and, typically, less than 5% of the metal is removed by the abrasive action of the wheel; thus, wheel wear is very low and the workpiece remains cool. Finishing cuts usually are made by the grinding action, but only to produce a surface with good finish and dimensional accuracy.

The ECG process is suitable for applications similar to those for milling, grinding, and sawing (Fig. 27.10b), but it is not adaptable to cavity-sinking operations. EGC, which can be applied successfully to carbides and high-strength alloys, offers a distinct advantage over traditional diamond-wheel grinding of very hard materials, where wheel wear can be high. ECG machines are available with numerical controls, which improve dimensional accuracy and repeatability, and increase productivity.

Electrochemical honing combines the fine abrasive action of honing (Section 26.7) with electrochemical action. Although the equipment is costly, this process is as much as 5 times faster than conventional honing, and the tool lasts as much as 10 times longer. Electrochemical honing is used primarily for finishing internal cylindrical surfaces.

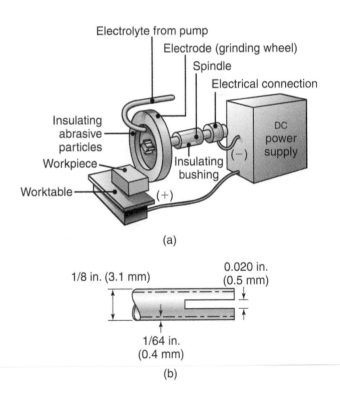

FIGURE 27.10 (a) Schematic illustration of the electrochemical-grinding process. (b) Thin slot produced on a round nickel-alloy (Inconel) tube by this process.

Design Considerations for Electrochemical Grinding. In addition to the design considerations already listed for electrochemical machining, ECG requires two additional ones:

- Designs should avoid sharp inside radii.
- If a surface is to be flat, it should be narrower than the width of the grinding wheel.

27.5 Electrical-discharge Machining

The principle of *electrical-discharge machining* (EDM), also called *electrodischarge* or *spark-erosion machining*, is based on the erosion of metals by spark discharges. Recall that when two current-carrying wires are allowed to touch each other, an arc is produced. When the point of contact between the two wires is examined closely, it will be noted that a small portion of the metal has been eroded away, leaving a small crater on the surface. Although this phenomenon has been known since the discovery of electricity, it was not until the 1940s that a machining process based on that principle was developed. The EDM process has become one of the most important and widely used production technologies in manufacturing.

Video Solution 27.1 Time
Required for EDM

Principle of Operation. The basic EDM system consists of a shaped tool (*electrode*) and the workpiece, connected to a DC power supply and placed in a **dielectric** (electrically nonconducting) fluid, as shown in Fig. 27.11a. When the potential difference between the tool and the workpiece is sufficiently high, the dielectric breaks down and a transient spark discharges through the fluid, removing a very small amount of metal from the workpiece surface. The capacitor discharge is repeated continuously, at rates between 200 and 500 kHz, with voltages usually ranging between 50 and 380 V and currents from 0.1 to 500 A. The volume of material removed per spark discharge is typically in the range from 10^{-6} to 10^{-4} mm^3 (10^{-10} to 10^{-8} in^3).

The EDM process can be used on any material that is an electrical conductor. Two important physical properties, which determine the volume of metal removed per discharge, are the melting point and the latent heat of melting of the workpiece material. As these quantities increase, the rate of material removal decreases. The material-removal rate can be estimated from the empirical formula

$$MRR = 4 \times 10^4 I T_w^{-1.23}, \tag{27.2}$$

where MRR is in mm^3/min, I is the current in amperes, and T_w is the melting point of the workpiece in °C.

The workpiece is fixtured within the tank containing the dielectric fluid, and its movements are controlled by numerically controlled systems. The gap between the tool and the workpiece (*overcut*) is critical; thus, the downward feed of the tool is controlled by a servomechanism, which automatically maintains a constant gap. The frequency of discharge or the energy per discharge, the voltage, and the current usually are varied to control the removal rate. The rate and surface roughness produced increase with increasing current density and with decreasing frequency of sparks.

Dielectric Fluids. The functions of the dielectric fluid are to:

1. Act as an insulator until the potential is sufficiently high
2. Provide a cooling medium
3. Act as a flushing medium and carry away the debris in the gap

The EDM machines are equipped with a pump and filtering system for the dielectric fluid. The most common dielectric fluids are mineral oils, although kerosene and

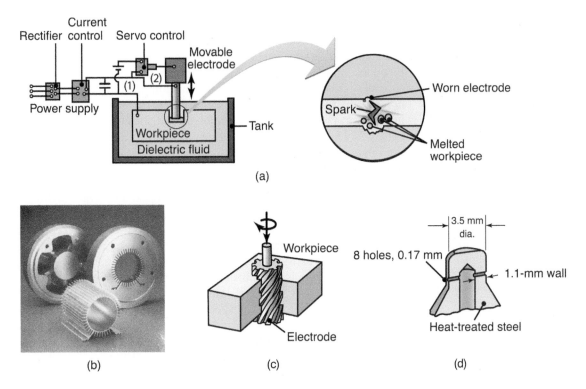

FIGURE 27.11   (a) Schematic illustration of the electrical-discharge machining process; this is one of the most widely used machining processes, particularly for die-sinking applications. (b) Examples of cavities produced by EDM, using shaped electrodes; the two round parts (rear) are the set of dies used in extruding the aluminum piece shown in front (see also Fig. 15.9b). (c) A spiral cavity produced by EDM using a slowly rotating electrode similar to a screw thread. (d) Holes in a fuel-injection nozzle made by EDM; the material is heat-treated steel. *Source:* (b) Courtesy of AGIE USA, Ltd.

distilled or deionized water also are used in specialized applications. Although more expensive, clear, low-viscosity fluids that make cleaning easier are also available.

Electrodes.   Electrodes for EDM usually are made of graphite, although brass, copper, or copper–tungsten alloys also are used. The tools can be shaped by forming, casting, powder metallurgy, or CNC machining techniques. Tungtsen-wire electrodes as small as 0.1 mm (0.005 in.) in diameter have been used to produce holes, with depth-to-diameter ratios of up to 400:1, a ratio that is much higher than those obtained by conventional methods (see Table 23.10).

The sparks in EDM also erode away the electrode, thus changing its geometry and adversely affecting the shape produced and its dimensional accuracy. *Wear ratio* is defined as the ratio of the volume of workpiece material removed to the volume of tool wear; it ranges from about 3:1 for metallic electrodes to as high as 100:1 for graphite electrodes. It has been shown that tool wear is related to the melting points of the materials involved: the higher the melting point of the electrode, the lower the wear rate. Consequently, graphite electrodes have the highest wear resistance; also, the higher the current, the higher is the wear. Tool wear can be minimized by reversing the polarity and using copper tools, a process called **no-wear EDM.** Care must be taken to control the process; it is possible for workpiece material to coat the electrode and change its shape.

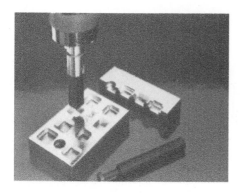

FIGURE 27.12 Stepped cavities produced with a square electrode by the EDM process. The workpiece moves in the two principal horizontal directions, and its motion is synchronized with the downward movement of the electrode to produce these cavities; also shown is a round electrode capable of producing round or elliptical cavities. *Source: Courtesy of AGIE USA, Ltd.*

Process Capabilities. Electrical-discharge machining has numerous applications, such as dies for forging, extrusion, die casting, injection molding, and large sheet-metal automotive-body components (produced in **die-sinking machining centers**, with computer numerical control). Other applications include machining small-diameter deep holes, using tungsten wire as the electrode; narrow slots in parts; cooling holes in superalloy turbine blades; and various intricate shapes (see Figs. 27.11b and c). Stepped cavities can be produced by controlling the relative movements of the workpiece in relation to the electrode (Fig. 27.12).

Blue Arc Machining. One variation of electrical discharge machining is the *Blue Arc* process, developed for roughing cuts of difficult-to-machine materials, especially nickel-based superalloys. The geometry of bladed disks (or *blisks*) used in aircraft engines is quite challenging to machine; the Blue Arc process removes most of the material to achieve a rough shape, which is then finish machined through conventional CNC milling. Blue arc uses an electrode and electrical discharge machining to remove material, but adds high-pressure fluid flushing to remove chips from the cutting zone. This approach has been shown to reduce energy consumption by over 30% compared to milling, and has reduced the cycle time to produce blisks from days to hours. Variations of Blue Arc machining exist for turning and grinding.

Because of the molten and resolidified (recast) surface structure developed, high rates of material removal produce a very rough surface finish, with poor surface integrity and low fatigue properties. Finishing cuts are therefore made at low removal rates, or the recast layer is subsequently removed by finishing operations. It also has been shown that surface finish can be improved by *oscillating* the electrode in a planetary motion, at amplitudes of 10–100 μm.

Design Considerations for EDM. The general design guidelines for electrical-discharge machining are:

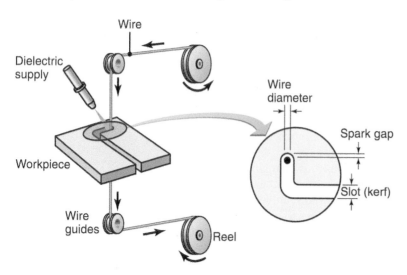

FIGURE 27.13 Schematic illustration of the wire EDM process; as many as 50 h of machining can be performed with one reel of wire, which is then recycled or discarded.

- Parts should be designed so that the required electrodes can be shaped economically.
- Deep slots and narrow openings should be avoided.
- For economic production, the surface finish specified should not be too fine.
- In order to achieve high production rate, the bulk of material removal should be done by conventional processes (called *roughing out*).

27.5.1   Wire EDM

A variation of EDM is *wire EDM*, or *electrical-discharge wire cutting* (Fig. 27.13). This process is similar to contour cutting with a band saw (illustrated in Fig. 24.28), in which a slowly moving wire travels along a prescribed path and cuts the work-piece. Figure 27.14a shows a thick plate being cut by this process, on a machine similar to that shown in Fig. 27.14b. Plates as thick as 300 mm (12 in.), and punches, tools, and dies made of hard metals, as well as intricate components for the electronics industry, can be cut by this process.

The wire travels at a constant velocity in the range from 0.15 to 9 m/min (6–360 in./min), and a constant gap (*kerf*) is maintained during the cut. The cutting speed generally is given in terms of the cross-sectional area cut per unit time. Typical examples are 32,000 mm²/h (50 in²/h) for 50-mm (2-in.) thick D2 tool steel, and 80,000 mm²/h (125 in²/h) for 150-mm (6-in.) thick aluminum. These removal rates indicate a linear cutting speed of 32,000/50 = 640 mm/h = 10.7 mm/min and 80,000/150 = 533 mm/h = 8.9 mm/min, respectively.

The wire is usually made of brass, copper, tungsten, or molybdenum; zinc- or brass-coated, multicoated, and steel-cored wires also are used. The wire diameter is typically about 0.30 mm (0.012 in.) for roughing cuts and 0.20 mm (0.008 in.) for finishing cuts. The wire should have high electrical conductivity and tensile strength, as the tension on it is typically 60% of its tensile strength. It usually is used only once, as it is relatively inexpensive compared with the type of operation it performs.

Multiaxis EDM wire-cutting machining centers are capable of producing three-dimensional shapes and are equipped with such features as:

- Computer controls, for controlling the cutting path of the wire (Fig. 27.14b) and its angle with respect to the workpiece plane
- Multiheads, for cutting two parts at the same time
- Features such as controls for preventing wire breakage
- Automatic self-threading features, in case of wire breakage
- Programmed machining strategies, to optimize the operation

Two-axis computer-controlled machines can produce cylindrical shapes, in a manner similar to a turning operation or cylindrical grinding. Modern wire EDM machines allow the control of the feed and take-up ends of the wire, in order to traverse independently in two principal directions, so that tapered parts also can be made.

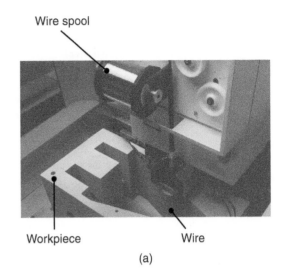

Wire spool

Workpiece   Wire

(a)

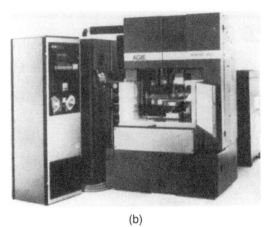

(b)

FIGURE 27.14 (a) Cutting a thick plate with wire EDM. (b) A computer-controlled wire EDM machine. *Source:* Courtesy of AGIE USA, Ltd.

27.5.2 Electrical-discharge Grinding

The grinding wheel in *electrical-discharge grinding* (EDG) is made of graphite or brass and contains no abrasives. Material is removed from the workpiece surface by spark discharges between the rotating wheel and the workpiece. This process is used primarily for grinding carbide tools and dies, but can also be used with fragile parts, such as surgical needles, thin-walled tubes, and honeycomb structures.

Material is removed by chemical action, with the electrical discharges from the graphite wheel breaking up the oxide film on the workpiece, and is washed away by the electrolyte flow. The material-removal rate can be estimated from the equation

$$MRR = KI, \qquad (27.3)$$

where MRR is in mm^3/min, I is the current in amperes, and K is a workpiece material factor in mm^3/A-min. For example, $K = 4$ for tungsten carbide and $K = 16$ for steel.

In **EDM sawing**, a setup similar to a band or circular saw (but without any teeth) is used, with the same electrical circuit for EDM. Narrow cuts can be made at high rates of metal removal. Because cutting forces are negligible, the process can be used on thin and slender components as well.

The EDG process can be combined with electrochemical grinding. Called **electrochemical-discharge grinding** (ECDG), the process uses a graphite wheel, and intermittent spark discharge, from alternating current or pulsed direct current, causes material removal. ECDC also commonly uses a highly conductive electrolyte instead of a dielectric fluid and uses lower voltages. This process is faster than EDG, but power consumption is higher.

27.6 Laser-beam Machining

In *laser-beam machining* (LBM), the source of energy is a **laser** (an acronym for *light amplification by stimulated emission of radiation*), which focuses optical energy on the surface of the workpiece (Fig. 27.15a). The highly focused, high-density energy source melts and evaporates portions of the workpiece in a controlled manner. This process, which does not require a vacuum, is used to machine a variety of metallic and nonmetallic materials.

There are several types of lasers used in manufacturing operations (Table 27.2):

1. **CO$_2$** (*pulsed* or *continuous wave*)
2. **Nd:YAG** (neodymium:yttrium–aluminum–garnet)
3. **Nd:glass, ruby**
4. **Diode**
5. **Excimer** (from the words *exc*ited and di*mer*, meaning two mers, or two molecules of the same chemical composition)

Important physical parameters in LBM are the *reflectivity* and *thermal conductivity* of the workpiece surface and its specific heat and latent heats of melting and

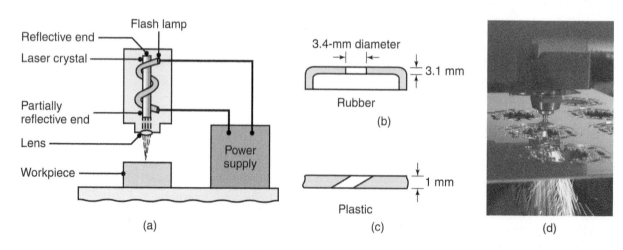

FIGURE 27.15 (a) Schematic illustration of the laser-beam machining process. (b) and (c) Examples of holes produced in nonmetallic parts by LBM. (d) Cutting sheet metal with a laser beam. *Source:* (d) Courtesy of ROFIN-SINAR Laser GmbH.

evaporation; the lower these quantities, the more efficient is the process. The cutting depth may be expressed as

$$t = \frac{CP}{vd} \qquad (27.4)$$

where t is the depth, C is a constant for the process, P is the power input, v is the cutting speed, and d is the laser-spot diameter. Peak energy densities of laser beams are in the range from 5 to 200 kW/mm^2.

The surface produced by LBM is usually rough and has a *heat-affected zone* (as described in Section 30.9), which, in critical applications, may have to be removed or heat-treated. Kerf width is an important consideration, as it is in other cutting processes, such as sawing, wire EDM, and electron-beam machining. In general, the smaller the kerf, the greater the accuracy and material utilization, and the lower the heat affected zone.

Laser beams may be used in combination with a gas (such as oxygen) stream, called **laser-beam torch**, to increase energy absorption for cutting sheet metals. *High-pressure, inert-gas* (nitrogen or argon) *assisted laser cutting* is used for stainless steel and aluminum; it leaves an oxide-free edge that can improve weldability of these metals. Gas streams also have the important function of blowing away molten and vaporized material from the workpiece surface.

TABLE 27.2

General Applications of Lasers in Manufacturing	
Application	Laser type
Cutting	
Metals	PCO$_2$, CWCO$_2$, Nd:YAG, ruby
Plastics	CWCO$_2$
Ceramics	PCO$_2$
Drilling	
Metals	PCO$_2$, Nd:YAG, Nd:glass, ruby
Plastics	Excimer
Marking	
Metals	PCO$_2$, Nd:YAG
Plastics	Excimer
Ceramics	Excimer
Surface treatment	CWCO$_2$
Welding	
Metals	PCO$_2$, CWCO$_2$, Nd:YAG, Nd:glass, ruby, diode
Plastics	Diode, Nd:YAG
Lithography	Excimer

Note: P = Pulsed, CW = Continuous wave, Nd:YAG = Neodymium:yttrium–aluminum–garnet.

Process Capabilities. Laser-beam machining is now widely used for hole making, trepanning, and cutting metals, nonmetallic materials, ceramics, and composite materials (Figs. 27.15b and c). The cleanliness of the operation has made laser-beam machining an attractive alternative to traditional machining methods. Holes as small as 0.005 mm (0.0002 in.), with depth-to-diameter ratios of 50:1 have been made in various materials, although a more practical minimum is 0.025 mm (0.001 in.). Steel plates as thick as 32 mm (1.25 in.) can be cut with laser beams.

Significant cost savings have been achieved by laser-beam machining, a process that is competing with electrical-discharge machining. It is being used increasingly in the electronics and automotive industries, and for composite materials. As two examples: the cooling holes in some vanes for the Boeing 747 jet engines, and bleeder holes for fuel-pump covers and lubrication holes in transmission hubs are produced through laser-beam machining. Lasers also are used for the following applications:

- **Welding** (Section 30.7)
- Small-scale and localized **heat treating** of metals and ceramics, to modify their surface mechanical and tribological properties
- **Marking** of parts, such as letters, numbers, and codes; note that marking also can be done by (a) punches, pins, styluses, and scroll rolls; (b) stamping; and (c) etching; although the equipment is more expensive than that used in other methods, laser marking and engraving has increasingly become common due to its accuracy, reproducibility, flexibility, ease of automation, and online application in manufacturing

The inherent *flexibility* of the laser-cutting process, including its *fiber-optic beam* delivery, simple fixturing, and low setup times, as well as the availability of multi-kW machines and two- and three-dimensional computer-controlled robotic laser-cutting systems, is a competitive and attractive feature. Laser cutting of sheets, for example, can successfully replace traditional punching processes, described in Chapter 16. Laser beams also can be combined with another process for improved overall efficiency, as described in Section 27.10 and shown in Example 27.1.

Design Considerations for LBM.   General design guidelines for laser-beam machining are:

- Sharp corners should be avoided, because they can be difficult to produce.
- Deep cuts will produce tapered walls.
- Dull and unpolished surfaces are preferable.
- There may be adverse effects on the properties of the machined materials, caused by high local temperatures and the heat-affected zone.

EXAMPLE 27.1   Combining Laser-beam Cutting and Punching of Sheet Metal

Laser cutting and punching processes have their respective advantages and limitations regarding both technical and economic aspects (see *hybrid machining*, Section 27.10). The advantages of *laser-beam cutting* generally are (a) flexibility of the operation, because hard tooling is not needed and there is no limitations to part size, (b) wide range of material thicknesses, (c) prototyping capability, and lot sizes can be as low as one, (d) materials and composites that otherwise might be cut with difficulty, and (e) complex geometries that can easily be programmed.

Drawbacks and advantages of *punching* include (a) large lot sizes in order to economically justify purchasing of tooling and equipment, (b) relatively simple shapes, (c) small range of part thicknesses, (d) fixed and limited punch geometries, even when using turrets, and (e) high production rate.

It is evident that the two processes cover different, but complementary, ranges. It is not difficult to visualize parts with some features that can be produced best by one process and other features that are best produced by the other process.

Machines have been designed and built in such a manner that the processes and fixturing can be utilized jointly to their full extent, yet without interfering with each other's operational boundaries. The purpose of combining them is to increase the overall efficiency and productivity of the manufacturing process, for parts that are within the capabilities of each of the two processes, similar to the concept of the machining centers described in Section 25.2. For example, turret-punch presses have been equipped with an integrated laser head; the machine can either punch or laser cut, but it cannot do both simultaneously.

Several factors must be taken into account in such a combination of two processes with respect to the characteristics of each operation: (a) the ranges of sizes, thicknesses, and shapes to be produced and how they are to be nested; (b) processing and setup times, including the loading, fixturing, and unloading of parts; (c) programming for cutting; and (d) the process capabilities of each method, including system dynamics, vibrations, and shock from punching that may disturb adjustments and alignments of the laser components.

Laser Microjet.   Laser Microjet®, illustrated in Fig. 27.16, uses a low-pressure laminar water stream to serve as a variable-length fiber optic cable to direct the laser and deliver laser power at the bottom of the kerf. This has an advantage in that the laser focus depth is very large, and high aspect ratio cuts can be produced. The water jet is produced by a sapphire or diamond nozzle with an opening of 25–100 μm, and exerts a force less than 0.1 N. In laser Microjet machining, material removal is due to the action of the laser, and the water provides cooling (reducing the heat affected zone, see Section 30.9), and prevents weld splatter from attaching to the workpiece.

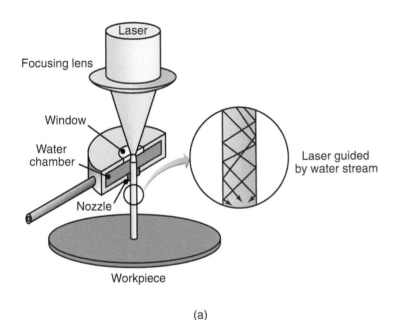

(a)

(b)

FIGURE 27.16   (a) Schematic illustration of the laser Microjet process. (b) A 25-mm-diameter wristwatch face produced from brass sheet by laser Microjet machining. *Source:* (b) Courtesy of Synova S.A.

The laser is typically an Nd:Yag laser, with micro- or nanosecond pulse duration and power between 10 and 200 W.

27.7 Electron-beam Machining

The energy source in *electron-beam machining* (EBM) is high-velocity electrons, which strike the workpiece surface and generate heat (Fig. 27.17). The machines utilize voltages in the range from 150 to 200 kV, to accelerate the electrons to 50–80% of the speed of light (300,000 km/s). Applications of this process are similar to those of laser-beam machining, except that, unlike lasers, EBM requires a *vacuum*; consequently, it is used much less frequently than laser-beam machining.

Electron-beam machining can be used for very accurate cutting of a wide variety of metals, and surface finish is better and the kerf is narrower than in other thermal cutting processes. (See also Section 30.6 on *electron-beam welding.*) However, the interaction of the electron beam with the workpiece surface produces hazardous X-rays, thus the equipment should be used only by highly trained personnel.

Design Considerations for EBM.   The guidelines for EBM generally are similar to those for LBM; additional considerations are:

- Because vacuum chambers have limited capacity, individual parts or batches should closely match the size of the vacuum chamber.
- If a part requires electron-beam machining on only a small portion of its volume, consideration should be given to manufacturing it as a number of smaller components, then assembling them after electron-beam machining.

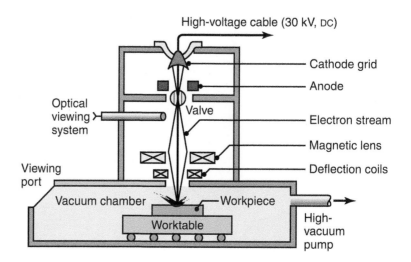

FIGURE 27.17 Schematic illustration of the electron-beam machining process; unlike LBM, this process requires a vacuum, so the workpiece size is limited to the size of the vacuum chamber.

Plasma-arc Cutting.   In plasma-arc cutting (PAC), *plasma beams* (ionized gas) are used to rapidly cut ferrous and nonferrous sheets and plates (see also Section 30.3). The temperatures generated are very high, on the order of 9400°C (17,000°F) in the torch when using oxygen as a plasma gas, thus material-removal rates are much higher than those associated with the EDM and LBM processes. The process is fast, the kerf width is small, parts can be machined with good reproducibility, and the surface finish is good; parts as thick as 150 mm (6 in.) can be cut. Plasma-arc cutting is now highly automated, using programmable controllers.

27.8   Water-jet Machining

The principle of *water-jet machining* (WJM), also called **hydrodynamic machining,** is based on the force that results from the momentum change of a stream of water. This force is sufficiently high to cut metallic and nonmetallic materials, as illustrated in Fig. 27.18. The water jet acts like a saw and cuts a narrow groove in the material. (See also *water-jet peening*, Section 34.2.)

A water-jet cutting machine is shown in Fig. 27.18b. A variety of materials can be cut, including plastics, fabrics, rubber, wood products, paper, leather, insulating materials, brick, and composite materials (Fig. 27.18c). A pressure level of about 400 MPa (60 ksi) is generally used for efficient operation, although pressures as high as 1400 MPa (200 ksi) can be generated. Jet-nozzle diameters typically range between 0.05 and 1 mm (0.002 and 0.040 in.).

Depending on the materials, thicknesses can range up to 25 mm (1 in.) and even higher. Vinyl and foam coverings for automobile dashboards, as well as some body panels, can be cut by multiple-axis, robot-guided water-jet machining equipment. Because it is an efficient and clean operation, as compared to other cutting processes, it is also used in the food-processing industry for cutting and slicing food products. The process also can be used for deburring.

The advantages of WJM are:

- Cuts can be started at any location without the need for predrilled holes
- No heat is produced

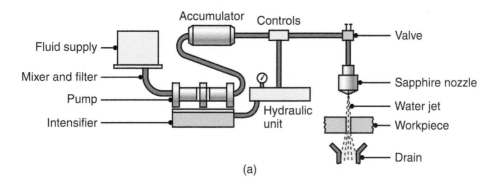

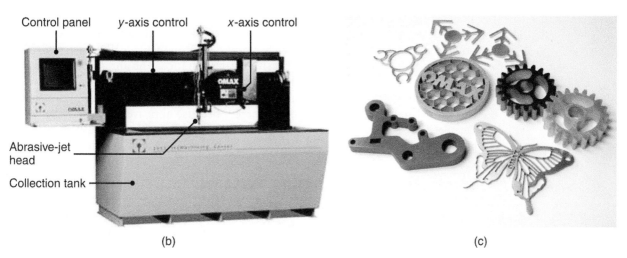

FIGURE 27.18   (a) Schematic illustration of the water-jet machining process. (b) A computer-controlled water-jet cutting machine. (c) Examples of various nonmetallic parts produced by the water-jet cutting process. *Source:* Courtesy of OMAX Corporation.

- No deflection of the rest of the workpiece takes place, thus the process is suitable for flexible materials
- Little wetting of the workpiece takes place
- The burr produced is minimal
- It is an environmentally safe manufacturing operation

Abrasive Water-jet Machining.   In *abrasive water-jet machining* (AWJM), the water jet contains abrasive particles, such as silicon carbide or aluminum oxide, which greatly increase the material-removal rate. Metallic, nonmetallic, and advanced composite materials of various thicknesses, in single layer or multilayers, can be cut. The optimum level of abrasives in the jet stream is controlled automatically in modern AWJM systems. Nozzles are typically made of rubies, sapphires, and diamond (Fig. 27.18a).

AWJM is particularly suitable for heat-sensitive materials that cannot be machined by processes in which heat is produced. Cutting speeds can be as high as 7.5 m/min (25 ft/min) for reinforced plastics, but are much lower for metals; consequently, the process may not be economical for applications requiring high production rates.

In making holes, the minimum size that can be produced satisfactorily to date is about 400 μm (0.015 in.), and maximum hole depth is on the order of 25 mm (1 in.).

With multiple-axis and robot-controlled machines, complex three-dimensional parts can be machined economically to finish dimensions.

27.9 Abrasive-jet Machining

In *abrasive-jet machining* (AJM), abrasive particles are propelled at the workpiece surface by a high-velocity jet of dry air, nitrogen, or carbon dioxide (Fig. 27.19). The impact of the particles develops a concentrated force (see also Section 26.6) sufficiently high to remove material, such as (a) cutting small holes, slots, or intricate patterns in very hard or brittle metallic and nonmetallic materials, (b) deburring or removing small flash from parts, (c) trimming and beveling of edges, (d) removing oxides and other surface films, and (e) cleaning of components with irregular surfaces.

The gas pressure is on the order of 850 kPa (125 psi), and the abrasive-jet velocity can be as high as 300 m/s (100 ft/s). The nozzles are usually made of tungsten carbide or sapphire, both of which have abrasive wear resistance. The abrasive size is in the range from 10 to 50 μm (400–2000 μin.). Because the flow of the free abrasives tends to round off corners, designs for abrasive-jet machining should avoid sharp corners. Also, holes made tend to be tapered, because abrasives preferentially wear the inlet side. There is some hazard involved in using this process, because of airborne particulates, a problem that does not exist with abrasive water-jet machining.

27.10 Hybrid Machining Systems

A more recent development in material removal processes is the concept of *hybrid machining systems*. Two or more individual machining processes are *combined* into one system, thus taking advantage of the capabilities of each process, increasing production speed, and thereby improving the efficiency of the operation. The system is able to handle a variety of materials, including metals, ceramics, polymers, and composites. Examples of hybrid machining systems include combinations and integration of the following processes:

1. Abrasive machining and electrochemical machining
2. Abrasive machining and electrical discharge machining
3. Abrasive machining and electrochemical finishing

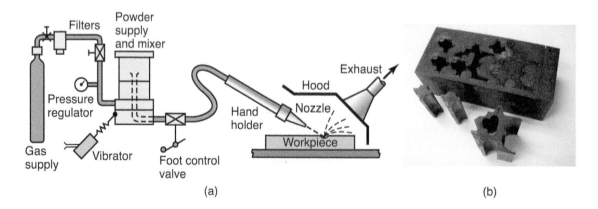

(a)   (b)

FIGURE 27.19   (a) Schematic illustration of the abrasive-jet machining process. (b) Examples of parts made by abrasive-jet machining, produced in 50-mm (2-in.) thick 304 stainless steel. *Source:* Courtesy of OMAX Corporation.

4. Water-jet cutting and wire EDM
5. High-speed milling, laser ablation, and abrasive blasting, as an example of *three* integrated processes
6. Machining and blasting
7. Electrochemical machining and electrical discharge machining (ECDM), also called electrochemical spark machining (ECSM)
8. Machining and forming processes, such as laser cutting and punching of sheet metal, described in Example 27.1
9. Combinations of various other forming, machining, and joining processes

The implementation of these concepts, and the development of appropriate machinery and control systems, present significant challenges. Important considerations include such factors as:

1. The workpiece material and its manufacturing characteristics; see, for example, Table I.3 in the General Introduction
2. Compatibility of processing parameters among the two or more processes to be integrated, such as speed, size, force, energy, and temperature
3. Cycle times of each individual operation involved and their synchronization
4. Safety considerations and possible adverse effects of the presence of various elements, such as abrasives, chemicals, wear particles, chips, and contaminants on the overall operation
5. Consequence of a failure in one of the stages in the system, since the operation involves sequential processes

27.11 Economics of Advanced Machining Processes

Advanced machining processes have unique applications, and are useful particularly for difficult-to-machine materials and for parts with complex internal and external profiles. The economic production run for a particular process depends on such factors as the (a) costs of tooling and equipment, (b) operating costs, (c) material-removal rate required, (d) level of operator skill required, and (e) secondary and finishing operations that subsequently may be necessary.

In chemical machining, an inherently a slow process, an important factor is the cost of reagents, maskants, and disposal, together with the cost of cleaning the parts. In electrical-discharge machining, the cost of electrodes and the need to periodically replace them can be significant. The rate of material removal and the production rate can vary significantly in processes described, as can be seen in Table 27.1. The cost of tooling and equipment also varies considerably, as does the operator skill required. The high capital investment for machines, such as those for electrical and high-energy-beam machining, especially when equipped with robotic control, has to be justified in terms of the production runs and the feasibility of manufacturing the same part by other means if at all possible.

CASE STUDY 27.2 Manufacture of Small Satellites

Satellites built in the early days of the Space Age (1960s) were very large, and those smaller than 1000 kg were very rare. Table 27.3 shows the classification of modern satellites by their mass. This case study describes the manufacture of propulsion systems for micro- and nanosatellites.

(continued)

There are several compelling reasons to reduce the size of satellites, none greater than the cost of putting the satellite into orbit. One of the main contributors to weight in a satellite is the propulsion system, which is essential for changing its orbit or correcting for any drift. Figure 27.20a shows the propulsion system for a microsatellite, incorporating several cold-gas microthrusters, a propellant storage tank, filters, and temperature and pressure sensors.

Selected components of the propulsion system are shown in Fig. 27.20b. Note that the production of these miniature parts would be difficult if made through conventional forming, casting, or machining technologies. Moreover, connecting the plumbing for all of the components would be very difficult, even with larger components, and almost impossible to perform inside a clean-room environment.

TABLE 27.3

Satellite Classification

Group name	Mass kg or g	Mass lb
Large satellite	>1000 kg	2200
Medium satellite	500–1000	1100–2200
Minisatellite	100–500	2200–1100
Small satellites		
Microsatellite	10–100	22–220
Nanosatellite	1–10	2.2–22
Picosatellite	0.1–1	0.22–2.2
Femtosatellite	<100 g	<0.22

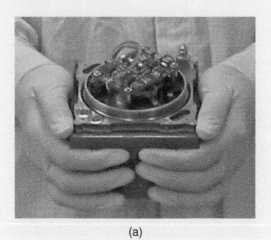

(a)

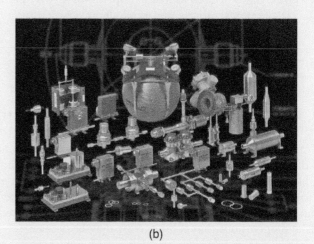

(b)

FIGURE 27.20 Propulsion system for a small satellite. (a) Miniaturized system suitable for a micro- or nanosatellite and (b) selected propulsion system components. *Source:* Courtesy of R. Hoppe, VACCO Industries, Inc.

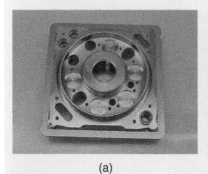

(a)

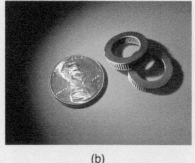

(b)

(c)

FIGURE 27.21 Photochemically etched and blanked components for micro- and nanosatellites. (a) Mounting board incorporating fluid flow channels in an integrated package; (b) microscale valve spring placed next to a U.S. penny; and (c) fuel filter. *Source:* Courtesy of R. Hoppe, VACCO Industries, Inc.

An attractive alternative is the production of an integrated system, with fluid connections made internally through a photochemically-etched and diffusion-bonded support, on which components are welded or fastened mechanically. Such a support is shown in Fig. 27.21, along with valve springs and filters that are produced through a combination of photochemical blanking diffusion-bonding processes.

Figure 27.22 depicts the manufacturing sequence involved. Titanium is commonly used for propulsion system components, because it has a high

FIGURE 27.22 Processing sequence for photochemical etching of microsatellite components: (a) Clean the raw material; (b) coat with photosensitive material; (c) expose with photographic tool; (d) develop a resist image; (e) etch; and (f) remove the resist. *Source:* Courtesy of R. Hoppe, VACCO Industries, Inc.

(*continued*)

strength-to-weight ratio, thus making possible lightweight designs. A mask is prepared, as described in Section 27.2, and the titanium is etched or blanked in a solution of hydrofluoric and nitric acid. Multiple layers of titanium are then diffusion bonded (see Section 31.7), to produce internal features such as flow channels.

Such fully integrated systems have resulted in the production of satellite propulsion systems that are less complex, more robust, and less massive than those in previous designs.

Source: Based on R. Hoppe, VACCO Industries, Inc.

SUMMARY

- Advanced machining processes have unique capabilities, utilizing chemical, electrochemical, electrical, and high-energy-beam sources of energy. The mechanical properties of the workpiece material are not significant, because these processes rely on mechanisms that do not involve the strength, hardness, ductility, or toughness of the material; rather, they involve physical, chemical, and electrical properties.

- Chemical and electrical methods of machining are suitable particularly for hard materials and complex part shapes. They do not produce forces (and therefore can be used for thin, slender, and flexible workpieces), significant temperatures, or residual stresses. However, the effects of these processes on surface integrity must be considered, as they can damage surfaces, thus reducing the fatigue life of the product.

- High-energy-beam machining processes basically utilize laser beams, electron beams, and plasma beams. They have important industrial applications, possess a high flexibility of operation with robotic controls, and are economically competitive with various other processes.

- Water-jet machining, abrasive water-jet machining, and abrasive-jet machining processes can be used for cutting as well as deburring operations. Because they do not utilize hard tooling, they have an inherent flexibility of operation.

- Hybrid machining processes offer possibilities for more efficient production of complex parts, by combining two or more processes and thus reducing production times.

KEY TERMS

Abrasive-jet machining	Electrical-discharge machining	Electron-beam machining	Photoresist
Abrasive water-jet machining	Electrochemical-discharge grinding	Etchant	Plasma-arc cutting
Blue arc machining	Electrochemical grinding	Hybrid machining	Plasma beams
Chemical blanking	Electrochemical honing	Hydrodynamic machining	Pulsed electrochemical machining
Chemical machining	Electrochemical machining	Laser	Reagent
Chemical milling	Electrode	Laser-beam machining	Shaped-tube electrolytic machining
Dielectric	Electrolyte	Laser Microjet	Undercut
Die sinking	Electrolytic trepanning	No-wear EDM	Water-jet machining
Electrical-discharge grinding		Photochemical blanking	Wire EDM
		Photochemical machining	
		Photoetching	

BIBLIOGRAPHY

Asibu, Jr., E.K., **Principles of Laser Materials Processing**, Wiley, 2009.

Brown, J., **Advanced Machining Technology Handbook**, McGraw-Hill, 1998.

Dahotre, N.B., and Samant, A., **Laser Machining of Advanced Materials**, CRC Press, 2011.

El-Hofy, H., **Advanced Machining Processes: Nontraditional and Hybrid Machining Processes**, McGraw-Hill, 2005.

Grzesik, W., **Advanced Machining Processes of Metallic Materials: Theory, Modelling and Applications**, Elsevier, 2008.

Guitrau, E.B., **The EDM Handbook**, Hanser, 2009.

Jameson, E.C., **Electrical Discharge Machining**, Society of Manufacturing Engineers, 2001.

Marinescou, I.D., Uhlmann, E., and Doi, T., **Handbook of Lapping and Polishing**, CRC Press, 2006.

Momber, A.W., and Kovacevic, R., **Principles of Abrasive Water Jet Machining**, Springer, 1998.

Steen, W.M., and Mazumder, J., **Laser Material Processing**, 4th ed., Springer, 2010.

REVIEW QUESTIONS

27.1 Describe the similarities and differences between chemical blanking and conventional blanking, using dies.

27.2 Name the processes involved in chemical machining. Describe briefly their principles.

27.3 Explain the difference between chemical machining and electrochemical machining.

27.4 What is the underlying principle of electrochemical grinding?

27.5 Explain how the EDM process is capable of producing complex shapes.

27.6 What are the important features of the Blue Arc machining process?

27.7 What are the capabilities of wire EDM? Could this process be used to make tapered parts? Explain.

27.8 Explain why laser Microjet has a large depth of field.

27.9 Describe the advantages of water-jet machining.

27.10 What is the difference between photochemical blanking and chemical blanking?

27.11 What type of workpiece is not suitable for laser-beam machining?

27.12 What is an undercut? Why must it be considered in chemical machining?

27.13 Explain the principle of hybrid machining.

QUALITATIVE PROBLEMS

27.14 Give technical and economic reasons that the processes described in this chapter might be preferred over those described in the preceding chapters.

27.15 Why is the preshaping or premachining of parts sometimes desirable in the processes described in this chapter?

27.16 Explain why the mechanical properties of workpiece materials are not significant in most of the processes described in this chapter.

27.17 List the processes that can produce shaped holes, that is, holes that are not circular.

27.18 List the advantages of laser Microjet over conventional laser machining.

27.19 Why has electrical-discharge machining become so widely used in industry?

27.20 Describe the types of parts that are suitable for wire EDM.

27.21 Which of the advanced machining processes would cause thermal damage? What is the consequence of such damage to workpieces?

27.22 Which of the processes described in this chapter require a vacuum? Explain why?

27.23 Describe your thoughts regarding the laser-beam machining of nonmetallic materials. Give several possible applications, including their advantages compared with other processes.

27.24 Are deburring operations still necessary for some parts made by advanced machining processes? Explain and give several specific examples.

27.25 List and explain factors that contribute to a poor surface finish in the processes described in this chapter.

27.26 What is the purpose of the abrasives in electrochemical grinding?

27.27 Which of the processes described in this chapter are suitable for producing very small and deep holes? Explain.

27.28 Is kerf width important in wire EDM? Explain.

27.29 Comment on your observations regarding Fig. 27.4.

27.30 Why may different advanced machining processes affect the fatigue strength of materials to different degrees?

QUANTITATIVE PROBLEMS

27.31 A 200-mm-deep hole that is 30 mm in diameter is being produced by electrochemical machining. A high production rate is more important than machined surface quality. Estimate the maximum current and the time required to perform this operation.

27.32 If the operation in Problem 27.31 were performed on an electrical-discharge machine, what would be the estimated machining time?

27.33 A cutting-off operation is being performed with a laser beam. The workpiece being cut is 0.5 in. thick and 15 in. long. If the kerf is $\frac{3}{32}$ in. wide, estimate the time required to perform this operation.

27.34 A 0.80-in.-thick copper plate is being machined by wire EDM. The wire moves at a speed of 4 ft/min and the kerf width is $\frac{1}{16}$ in. What is the required power? Note that it takes 1550 J (2100 ft-lb) to melt one gram of copper.

SYNTHESIS, DESIGN, AND PROJECTS

27.35 Explain why it is difficult to produce sharp profiles and corners with some of the processes described in this chapter.

27.36 Make a list of the processes described in this chapter in which the following properties are relevant: (a) mechanical, (b) chemical, (c) thermal, and (d) electrical. Are there processes in which two or more of these properties are important? Explain.

27.37 Would the processes described in this chapter be difficult to perform on various nonmetallic or rubberlike materials? Explain your thoughts, commenting on the influence of various physical and mechanical properties of workpiece materials, part geometries, etc.

27.38 Describe the types of parts that would be suitable for hybrid machining. Consider one such part and make a preliminary sketch for a hybrid machine to produce that part.

27.39 Describe your thoughts as to whether the processes described in (a) Chapters 13 through 16, and (b) Chapters 23 and 24 can be suitable for a hybrid system of making parts. Give a preliminary sketch of a machine for the two groups of processes listed.

27.40 Make a list of machining processes that may be suitable for each of the following materials: (a) ceramics, (b) cast iron, (c) thermoplastics, (d) thermosets, (e) diamond, and (f) annealed copper.

27.41 At what stage is the abrasive in abrasive water-jet machining introduced into the water jet? Survey the available literature, and then prepare a schematic illustration of the equipment involved.

27.42 How would you manufacture a large-diameter, conical, round metal disk with a thickness that decreases from the center outward? Make appropriate sketches.

27.43 Describe the similarities and differences among the various design guidelines for the processes described in this chapter.

27.44 Describe any workpiece size limitations in advanced machining processes. Give examples.

27.45 Suggest several design applications for the types of parts shown in Fig. 27.5. (See also Fig. 27.18c.)

27.46 Based on the topics covered in Parts III and IV, make a comprehensive table of hole-making processes. Describe the advantages and limitations of each method, and comment on the quality and surface integrity of the holes produced.

27.47 Review Example 27.1 and explain the relevant parameters involved; then design a system whereby both processes can be used in combination to produce parts from sheet metal.

27.48 Marking surfaces with numbers and letters for part-identification purposes can be done with a variety of mechanical and nonmechanical methods. Based on the processes described throughout this book thus far, make a list of these methods, explaining their advantages, limitations, and typical applications.

27.49 *Precision engineering* is a term that is used to describe manufacturing high-quality parts with close dimensional tolerances and good surface finish. Based on their process capabilities, make a list of advanced machining processes with decreasing order of the quality of parts produced. Comment on your observations.

27.50 With appropriate sketches, describe the principles of various work-holding methods and devices that can be used for the processes described in this chapter.

27.51 Make a table of the process capabilities of the advanced machining processes described in this chapter. Use several columns and describe the machines involved, the type of tools and tool materials used, the shapes of blanks and parts produced, the typical maximum and minimum sizes, surface finish, tolerances, and production rates.

27.52 One of the general concerns regarding advanced machining processes is that, in spite of their many advantages, they generally are slower than conventional machining operations. Conduct a survey of the speeds, machining times, and production rates involved, and prepare a table comparing their respective process capabilities.

27.53 It can be seen that several of the processes described in Part IV of this book can be employed, either singly or in combination, to make or finish dies for metalworking operations. Write a brief technical paper on these methods, describing their advantages, limitations, and typical applications.

Micromanufacturing and Fabrication of Microelectronic Devices

PART V

The importance of the topics covered in the following two chapters can best be appreciated by considering the manufacture of a simple metal spur gear. It is important to recall that some gears are designed to transmit *power*, such as those in gear boxes, yet others transmit *motion*, such as in rack and pinion mechanisms in automobile steering systems. If the gear is, say, 100 mm (4 in.) in diameter, it can be produced by traditional methods, such as starting with a cast or forged blank, and machining and grinding it to its final shape and dimensions. A gear that is only 2 mm (0.080 in.) in diameter, on the other hand, can be difficult to produce by these methods. If sufficiently thin, the gear could, for example, be made from sheet metal, by fine blanking, chemical etching, or electroforming.

If the gear is only a *few micrometers* in size, it can be produced by such techniques as optical lithography, wet and dry chemical etching, and related processes. A gear that is only a *nanometer* in diameter would, however, be extremely difficult to produce; indeed, such a gear would have only a few tens of atoms across its surface.

The challenges faced in producing gears of increasingly smaller sizes is highly informative, and can be put into proper perspective by referring to the illustration of length scales shown in Fig. V.1. Conventional manufacturing processes, described in Chapters 11 through 27, typically produce parts that are larger than a millimeter or so, and can be described as visible to the naked eye. The sizes of such parts generally are referred to as **macroscale**, the word "macro" being derived from the Greek *makros*, meaning "long," and the processing of such parts is known as **macromanufacturing**. Macroscale is the most developed and best understood size range from a design and manufacturing standpoint, with a very wide variety of processes available for producing components of that size. All of the examples and case studies given thus far have been examples of macromanufacturing.

Note that the gear shown in Fig. V.1 is a few tens of micrometers across, and thus fits into the category of **micromanufacturing**, developed mostly for electronic devices of all types, including computer processors and memory chips, sensors, and magnetic storage devices. For the most part, this type of manufacturing relies heavily on lithography, wet and dry etching, and coating techniques. Examples of products that rely on micromanufacturing techniques are a wide variety of sensors and probes (see Fig. V.2), ink-jet printing heads, microactuators, magnetic hard-drive heads, and such devices as computer processors and memory chips. Microscale mechanical devices are still a relatively new technology, but one that has been developing rapidly.

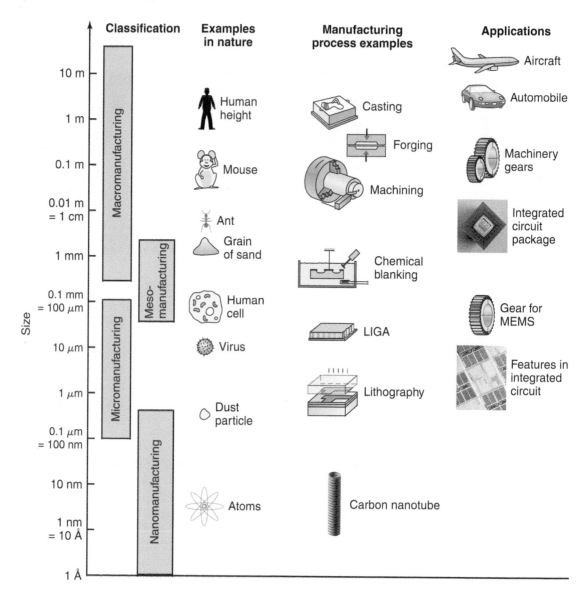

FIGURE V.1   Illustration of the regimes of macro-, meso-, micro-, and nanomanufacturing, the range of common sizes of parts, and the capabilities of manufacturing processes in producing these parts.

Mesomanufacturing overlaps macro- and micromanufacturing, as seen by the illustrations given in Fig. V.1. Examples of mesomanufacturing are extremely small motors, bearings, and components for miniature devices, such as hearing aids, stents, heart valves, and electronic toys, with components the same as the gear shown in Fig. V.1.

In **nanomanufacturing**, parts are produced at scales of one billionth of a meter, typically between 10^{-6} and 10^{-9} m in length; many of the features in integrated circuits are at this length scale. *Biomanufacturing*, covering such areas as molecularly engineered pharmaceutical products, genetic testing, gene therapy, and agricultural products, are at nanoscale level, and it is now recognized that many physical and

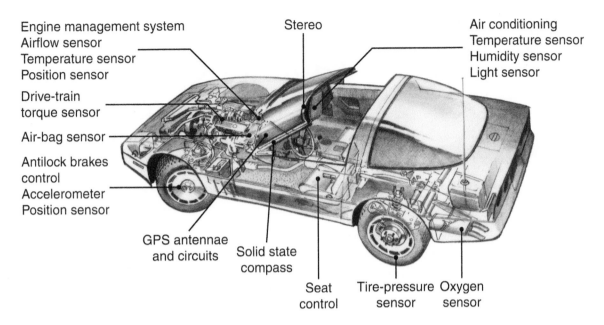

FIGURE V.2   Microelectronic and microelectromechanical devices and parts used in a typical automobile.

biological processes act at this scale and that nanomanufacturing holds much promise for future innovations.

In Chapter 28, the manufacture of silicon wafers and microelectronic devices is described, which include a wide variety of computer processors, memory devices, and integrated circuits. Controls, transportation, communications, engineering design and manufacturing, medicine, and entertainment all have been changed greatly by the wide availability of *metal-oxide-semiconductor* (MOS) *devices*, generally based on single-crystal silicon. Microelectronics is the best known and commercially important example of micromanufacturing, with some aspects of the applications exemplifying nanomanufacturing. The chapter also covers the techniques used in packaging and assembling integrated circuits onto printed circuit boards.

The production of microscale devices, that are mechanical and electrical in nature, is described in Chapter 29. Depending on their level of integration, these devices are called **micromechanical devices** or **microelectromechanical systems** (MEMS). While the historical origins of MEMS manufacture stem from the same processes used for microelectronic systems, and from identical processes and production sequences, still in use, several unique approaches also have been developed.

28 Fabrication of Microelectronic Devices

- This chapter presents the science and technologies involved in the production of integrated circuits, a product that has fundamentally changed our society.
- The chapter begins by describing silicon, the preferred material for most integrated circuits, and its unique properties that make it attractive. Beginning with a cast ingot, the operations required to produce a wafer are described.
- The production of patterns on wafers is next discussed, including the processes of lithography, wet and dry etching, and doping.
- Metallization and testing are then described, as are the approaches for obtaining electrical connections from integrated circuits to circuit boards.
- The chapter concludes with a description of the different packages used for integrated circuits.

Typical parts produced: Computer processors, memory chips, printed circuit boards, and integrated circuits of all types.

28.1 Introduction

Although semiconducting materials have been used in electronics for a long time (the word semiconductor first appeared in 1838), it was the invention of the *transistor* in 1947 that set the stage for what would become one of the greatest technological achievements in all of history. **Microelectronics** has played an increasing role ever since the *integrated circuit* (IC) technology became the foundation for calculators, wrist watches, controls for home appliances and automobiles, information systems, telecommunications, robotics, space travel, weaponry, and personal computers.

The major advantages of today's ICs are their very small size and low cost. As their fabrication technology has become more advanced, the size and cost of such devices as transistors, diodes, resistors, and capacitors continue to decrease, and the global market has become highly competitive. More and more components can now be put onto a **chip**, a very small piece of semiconducting material on which the circuit is fabricated.

Typical chips produced today have sizes that are as small as 0.5 mm × 0.5 mm and, in rare cases, can be more than 50 mm × 50 mm, if not an entire wafer. New technologies now allow densities in the range of 10 million devices per chip (Fig. 28.1), a magnitude that has been termed **very large scale integration** (VLSI). Some of the advanced ICs may contain more than 100 million devices, termed **ultralarge-scale**

QR Code 28.1 Tour of an integrated circuit fabrication facility. (*Source:* Courtesy of Intel)

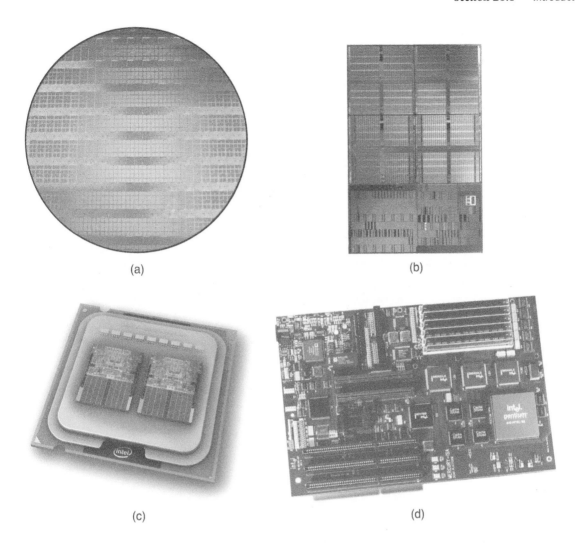

(a)

(b)

(c)

(d)

FIGURE 28.1   (a) A 300-mm (11.8-in.) wafer with a large number of dies fabricated onto its surface; (b) detail view of an Intel 45-nm chip, including a 153-Mbit SRAM (static random access memory) and logic test circuits; (c) image of the Intel® Itanium® 2 processor; and (d) Pentium® processor motherboard. *Source:* Courtesy of Intel Corporation.

integration (ULSI). The Intel® Itanium® processors, for example, have surpassed 2 billion transistors, and the Advanced Micro Devices Tahiti® graphic processing unit has surpassed 4.3 billion transistors.

Among more recent advances is **wafer-scale integration** (WSI), in which an entire silicon wafer is used to build a single device. This approach has been of greatest interest in the design of massively parallel supercomputers, including **three-dimensional integrated circuits** (3DICs), which use multiple layers of active circuits, maintaining connections both horizontally and vertically.

This chapter describes the processes that are currently in use in the fabrication of microelectronic devices and integrated circuits, following the basic sequence shown in Fig. 28.2. The major steps in fabricating a **metal-oxide-semiconductor field-effect transistor** (MOSFET), one of the dominant devices used in modern IC technology, are shown in Fig. 28.3.

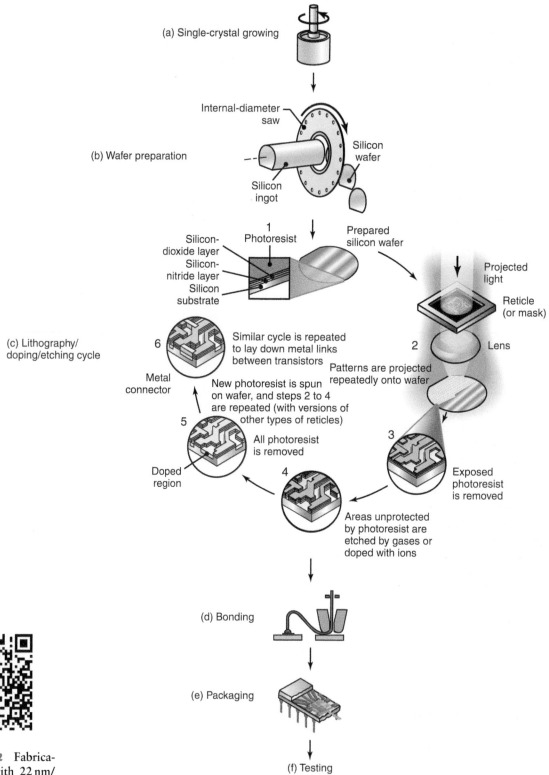

(a) Single-crystal growing

(b) Wafer preparation

Internal-diameter saw

Silicon ingot

Silicon wafer

(c) Lithography/ doping/etching cycle

1 Photoresist

Silicon-dioxide layer
Silicon-nitride layer
Silicon substrate

Prepared silicon wafer

Projected light

Reticle (or mask)

2   Lens

Patterns are projected repeatedly onto wafer

6

Similar cycle is repeated to lay down metal links between transistors

Metal connector

New photoresist is spun on wafer, and steps 2 to 4 are repeated (with versions of other types of reticles)

5

All photoresist is removed

Doped region

4

3

Exposed photoresist is removed

Areas unprotected by photoresist are etched by gases or doped with ions

(d) Bonding

(e) Packaging

(f) Testing

FIGURE 28.2   Outline of the general fabrication sequence for integrated circuits.

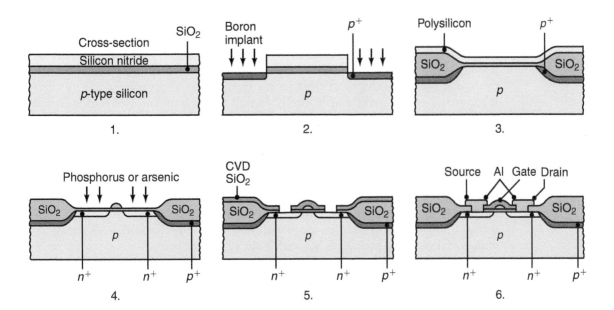

FIGURE 28.3   Cross-sectional views of the fabrication of a MOSFET transistor. *Source:* After R.C. Jaeger.

28.2   Clean Rooms

Clean rooms are essential for the production of integrated circuits, a fact that can be appreciated by noting the scale of manufacturing to be performed. Integrated circuits are typically a few millimeters in length, and the smallest features in a transistor on the circuit may be as small as a few tens of nanometers. This size range is smaller than particles that generally are not considered harmful, such as dust, smoke, and perfume; however, if these contaminants are present on the surface of a silicon wafer during its processing, they can seriously compromise the performance of the entire device.

There are various levels of clean rooms, defined by the **class** of the room. The size and the number of particles are significant in defining the class of a clean room, as shown in Fig. 28.4. The traditional classification system refers to the number of 0.5-μm or larger particles within a cubic foot of air. Thus, a Class-10 clean room has 10 or fewer such particles per cubic foot. This standard has been superseded by an ISO standard, but the traditional classification scheme is still widely used. Most clean rooms for microelectronics manufacturing range from Class 1 to Class 10; in comparison, the contamination level in modern hospitals is on the order of 10,000 particles per cubic foot.

To obtain controlled atmospheres that are free from particulate contamination, all ventilating air is passed through a *high-efficiency particulate air* (HEPA) filter. In addition, the air usually is

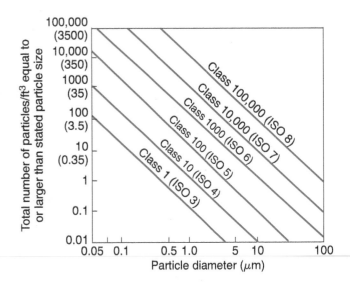

FIGURE 28.4   Allowable particle size counts for various clean-room classes; the numbers in parentheses on the ordinate are particle counts per cubic meter.

conditioned so that it is at 21°C (70°F) and 45% relative humidity. The largest source of contaminants in clean rooms are the workers themselves. Skin particles, hair, perfume, makeup, clothing, bacteria, and viruses are given off naturally by people, and in sufficiently large numbers to quickly compromise a Class-100 clean room. For these reasons, most clean rooms require special coverings, such as white laboratory coats, gloves, and hairnets, as well as the avoidance of perfumes and makeup. The most stringent clean rooms require full-body coverings, called *clean room smocks*. Other precautions include using (a) a ballpoint pen, instead of a pencil, to avoid objectionable graphite particles from pencils and (b) special clean-room paper, to prevent the accumulation of paper particles in the air.

Clean rooms are designed such that the cleanliness at critical *processing areas* is better than in the clean room in general. This is accomplished by always directing the filtered air *from top down* in the clean rooms, and from *floor to ceiling* in the service aisle, a goal that can be facilitated by laminar-flow hooded work areas. To minimize defects, no product is allowed in the service aisle.

28.3 Semiconductors and Silicon

As the name suggests, **semiconductor materials** have electrical properties that lie between those of conductors and insulators; they exhibit resistivities between 10^{-3} and 10^8 Ω-cm. Semiconductors have become the foundation for electronic devices, because their electrical properties can be altered when controlled amounts of selected impurity atoms are added to their crystal structures. These impurity atoms, also known as **dopants**, have either one more valence electron (*n*-type, or negative, dopant) or one less valence electron (*p*-type, or positive, dopant) than the atoms in the semiconductor lattice.

For silicon, which is a Group IV element in the Periodic Table, typical *n*-type and *p*-type dopants include, respectively, phosphorus (Group V) and boron (Group III). The electrical operation of semiconductor devices can thus be controlled through the creation of regions with different doping types and concentrations.

Although the earliest electronic devices were fabricated on *germanium*, **silicon** has become the industry standard. The abundance of alternative forms of silicon in the crust of the Earth is second only to that of oxygen, making it economically attractive. Silicon's main advantage over germanium is its large energy gap (1.1 eV), as compared with that of germanium (0.66 eV). This energy gap allows silicon-based devices to operate at temperatures of about 150°C (270°F), higher than devices fabricated on germanium, which operate at about 100°C (180°F).

Another important processing advantage of silicon is that its oxide (*silicon dioxide*, SiO_2) is an excellent electrical *insulator*, and can be used for both isolation and passivation (see Section 3.8) purposes. By contrast, germanium oxide is water soluble and thus unsuitable for electronic devices. Moreover, the oxidized form of silicon allows the production of **metal-oxide-semiconductor** (MOS) devices, which are the basis for MOS transistors. These materials are used in memory devices, processors, and other devices, and are by far the largest volume of semiconductor material produced worldwide.

Structure of Silicon.   The crystallographic structure of silicon is a diamond-type fcc structure, as shown in Fig. 28.5, along with the *Miller indices* of an fcc material. (Miller indices are a useful notation for identifying planes and directions within a unit cell; see also Section 1.3.) A crystallographic plane is defined by the reciprocal of its intercepts with the three axes. Because anisotropic etchants (see Section 28.8.1)

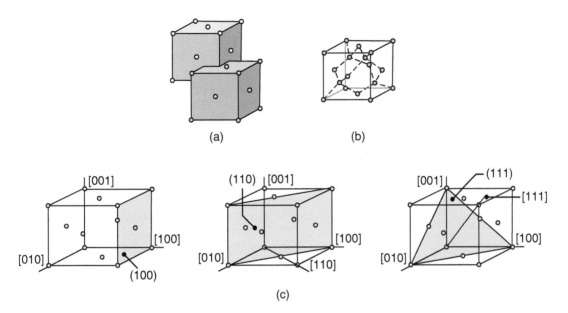

FIGURE 28.5 Crystallographic structure and Miller indices for silicon. (a) Construction of a diamond-type lattice from interpenetrating face-centered cubic cells; one of eight penetrating cells is shown. (b) Diamond-type lattice of silicon; the interiors have been shaded in color. (c) Miller indices for a cubic lattice.

preferentially remove material in certain crystallographic planes, the orientation of the silicon crystal in a wafer is an important consideration.

In spite of its advantages, however, silicon has a larger energy gap (1.1 eV) than germanium oxide and, therefore, has a higher maximum operating temperature (about 200°C; 400°F). This limitation has encouraged the development of *compound semiconductors*, specifically **gallium arsenide**. Its major advantage over silicon is its ability to emit light, thus allowing the fabrication of devices such as lasers and light-emitting diodes (LEDs).

Devices fabricated on gallium arsenide also have much higher operating speeds than those fabricated on silicon. Some of gallium arsenide's disadvantages, on the other hand, are its considerably higher cost, greater processing complications, and, most critically, the difficulty of growing high-quality oxide layers, the need for which is emphasized throughout the rest of this chapter.

28.4 Crystal Growing and Wafer Preparation

Silicon occurs naturally in the forms of silicon dioxide and various silicates. It must, however, undergo a series of purification steps in order to become the high-quality, defect-free, single-crystal material that is necessary for semiconductor device fabrication. The purification process begins by heating silica and carbon together in an electric furnace, which results in a 95–98% pure polycrystalline silicon. This material is converted to an alternative form, commonly trichlorosilane (a compound of silicon, hydrogen, and chlorine), which, in turn, is purified and decomposed in a high-temperature hydrogen atmosphere. The resulting product is extremely high quality *electronic-grade silicon* (EGS).

Single-crystal silicon usually is obtained through the **Czochralski, or CZ, process**, described in Section 11.5. The process utilizes a seed crystal that is dipped into

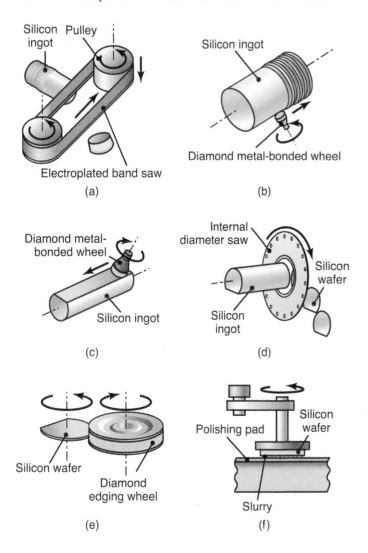

FIGURE 28.6 Finishing operations on a silicon ingot to produce wafers: (a) sawing the ends off the ingot; (b) grinding of the end and cylindrical surfaces of a silicon ingot; (c) machining of a notch or flat; (d) slicing of wafers; (e) end grinding of wafers; and (f) chemical–mechanical polishing of wafers.

a silicon melt, and is then pulled out slowly while being rotated. At this point, controlled amounts of impurities can be added to obtain a uniformly doped crystal. The result of the CZ process is a cylindrical single-crystal ingot, typically 100–300 mm (4–12 in.) in diameter and over 1 m (40 in.) in length. Because this technique does not allow for exact control of the ingot diameter, ingots are grown a few millimeters larger than the required size, and are ground to a desired diameter. Silicon **wafers** then are produced from silicon ingots, by a sequence of machining and finishing operations, illustrated in Fig. 28.6.

Next, the crystal is sliced into individual wafers, by using an inner-diameter diamond-encrusted blade (Fig. 24.28f), whereby a rotating, ring-shaped blade with its cutting edge on the inner diameter of the ring is utilized. While the substrate depth required for most electronic devices is no more than several microns, wafers typically are cut to a thickness of about 0.5 mm (0.02 in.). This thickness provides the physical support necessary for the absorption of temperature variations and the mechanical support needed during subsequent fabrication of the wafer.

The wafer is then ground along its edges using a diamond wheel. This operation gives the wafer a rounded profile, which is more resistant to chipping. Finally, the wafers must be polished and cleaned, to remove surface damage caused by the sawing process. This operation is commonly performed by *chemical–mechanical polishing*, also referred to as chemical–mechanical planarization, described in Section 26.7.

In order to properly control the manufacturing process, it is important to determine the *orientation* of the crystal in a wafer, which is done by notches or flats machined into them for identification, as shown in Fig. 28.7. Most commonly, the (100) or (111) plane of the crystal defines the wafer surface, although (110) surfaces also can be used for micromachining applications (Section 29.2). Wafers are also identified by a laser *scribe* mark, produced by the manufacturer. Laser scribing of information may take place on the front or on the back side of the wafer. The front side of some wafers has an exclusion edge area, 3–10 mm in size and reserved for the scribe information, such as lot numbers, orientation, and a wafer identification code unique to the particular manufacturer.

Wafers are typically processed in lots of 25 or 50, with 150–200 mm (6–8 in.) diameters each, or lots of 12–25 with 300-mm (12-in.) diameters each. In this way, they can be easily handled and transferred during subsequent processing steps. Because of the small device size and large wafer diameter, thousands of individual circuits can be placed on one wafer. Once processing is completed, the wafer is then sliced into individual **chips**, each containing one complete integrated circuit.

At this point, the single-crystal silicon wafer is ready for the fabrication of the integrated circuit or device. Fabrication takes place over the entire wafer surface, and thus many chips are produced at the same time, as shown in Fig. 28.1a. Because the number of chips that can be produced is dependent on the cross-sectional area of the wafer, advanced-circuit manufacturers have moved toward using larger single-crystal solid cylinders; 300-mm (12-in.) diameter wafers now are common, with 450-mm (18-in.) diameter wafers under development.

28.5   Film Deposition

Films are used extensively in microelectronic-device processing, particularly insulating and conducting types. Commonly deposited films include polysilicon, silicon nitride, silicon dioxide, and tungsten, titanium, and aluminum. In some cases, the wafers merely serve as a mechanical support, on which custom *epitaxial layers* are grown.

Epitaxy.   Defined as the growth of a vapor deposit, *epitaxy*, or *electrodeposit*, occurs when the crystal orientation of the deposit is related directly to the crystal orientation in the underlying crystalline substrate. The advantages of processing on these deposited films, instead of on the actual wafer surface, include

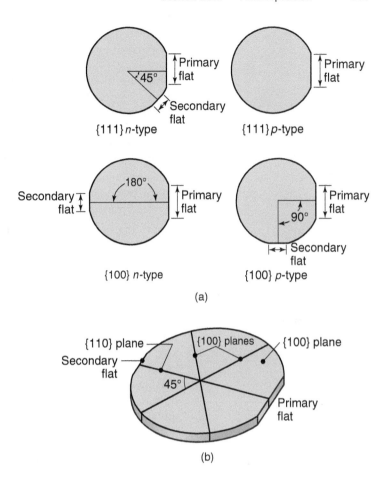

FIGURE 28.7   Identification of single-crystal wafers of silicon; this identification scheme is common for 150-mm (6-in.) diameter wafers, but notches are more common for larger wafers.

fewer impurities, especially carbon and oxygen, improved device performance, and the tailoring of material properties (which cannot be done on the wafers themselves).

Some of the major functions of deposited films are **masking** and protecting the semiconductor surface. In masking applications, the film must both inhibit the passage of dopants and concurrently display an ability to be etched into patterns of high resolution. Upon completion of device fabrication, films are applied to protect the underlying circuitry. Films used for masking and protecting include silicon dioxide, phosphosilicate glass (PSG), and silicon nitride. Each of these materials has distinct advantages, and they often are used in combination.

Conductive films are used primarily for device interconnection. These films must have a low electrical resistivity, be capable of carrying large currents, and be suitable for connection to terminal packaging leads with wire bonds. Generally, aluminum and copper are used for this purpose. Increasing circuit complexity has required up to six levels of conductive layers, all of which must be separated by insulating films.

Film Deposition.   Films can be *deposited* by several techniques, involving a variety of pressures, temperatures, and vacuum systems (see also Chapter 34):

- One of the oldest and simplest methods is **vacuum deposition**, used primarily for depositing metal films. The metal is first heated in a vacuum to its point of vaporization; upon evaporation, it forms a thin layer on the substrate surface. The heat of evaporation usually is generated by a heating filament or electron beam.

- **Sputtering** involves bombarding a target in a vacuum with high-energy ions, usually argon, aluminum, or copper. Sputtering systems generally include a DC power supply to produce the energized ions. As the ions impinge on the target, atoms are knocked off and are subsequently deposited on wafers mounted within the system. Although some argon may be trapped within the film, sputtering results in highly uniform coverage. Recent advances include using a radio-frequency power source (**RF sputtering**) and introducing magnetic fields (**magnetron sputtering**).
- In one of the most common techniques, **chemical-vapor deposition** (CVD), film is deposited by way of the reaction and/or decomposition of gaseous compounds. Using this technique, silicon dioxide is deposited routinely by the oxidation of silane or a chlorosilane. Figure 28.8a shows a continuous CVD reactor that operates at atmospheric pressure.

  A similar method that operates at lower pressures is referred to as **low-pressure chemical-vapor deposition** (LPCVD), as shown in Fig. 28.8b. Capable of coating hundreds of wafers at a time, this method results in a much higher production rate than that of atmospheric-pressure CVD, and provides superior film uniformity and with less consumption of carrier gases. The technique is commonly used for depositing polysilicon, silicon nitride, and silicon dioxide.
- **Plasma-enhanced chemical-vapor deposition** (PECVD) involves the processing of wafers in an RF plasma containing the source gases. This method has the advantage of maintaining a low wafer temperature during deposition.

Silicon **epitaxy** layers, in which the crystalline layer is formed, using the substrate as a seed crystal, can be grown by a variety of methods. If the silicon is deposited from the gaseous phase, the process is known as **vapor-phase epitaxy** (VPE). In another variation, the heated substrate is brought into contact with a liquid solution containing the material to be deposited, called **liquid-phase epitaxy** (LPE).

Another high-vacuum process, called **molecular-beam epitaxy** (MBE), utilizes evaporation to produce a thermal beam of molecules that are deposited on the heated substrate. This process results in a very high degree of purity. In addition, since the films are grown one atomic layer at a time, it is possible to have excellent control over doping profiles. This level of control is important especially in gallium-arsenide

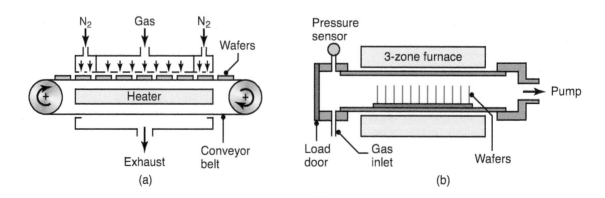

FIGURE 28.8 Schematic diagrams of a (a) continuous atmospheric-pressure CVD reactor and (b) low-pressure CVD. *Source: After S.M. Sze.*

technology; however, the MBE process has relatively low growth rates as compared to other conventional film-deposition techniques.

28.6 Oxidation

The term *oxidation* refers to the growth of an oxide layer as a result of the reaction of oxygen with the substrate material. Oxide films also can be formed by the previously described deposition techniques. Thermally grown oxides, described in this section, display a higher level of purity than deposited oxides, because they are grown directly from the high-quality substrate. However, methods of deposition must be used if the composition of the desired film is different from that of the substrate material.

Silicon dioxide is the most widely used oxide in IC technology today, and its excellent characteristics are one of the major reasons for the widespread use of silicon. Aside from its effectiveness in dopant masking and device isolation, silicon dioxide's most critical role is that of the *gate oxide* material. Silicon surfaces have an extremely high affinity for oxygen, and thus a freshly sawed slice of silicon will quickly grow a native oxide of 30–40 Å thickness.

- **Dry oxidation** is a relatively simple process, and is accomplished by elevating the substrate temperature, typically to about 750°–1100°C (1380°–2020°F), in an oxygen-rich environment. As a layer of oxide forms, the oxidizing agents must be able to pass through the oxide and reach the silicon surface, where the actual reaction takes place. Thus, an oxide layer does not continue to grow on top of itself, but rather, it grows from the silicon surface outward. Some of the silicon substrate is consumed in the oxidation process (Fig. 28.9).

  The ratio of oxide thickness to the amount of silicon consumed is found to be 1:0.44; thus, to obtain an oxide layer 1000 Å thick, approximately 440 Å of silicon will be consumed. This requirement does not present a problem, as substrates always are grown sufficiently thick. One important effect of the consumption of silicon is the rearrangement of dopants in the substrate near the interface. Because different impurities have different segregation coefficients or mobilities in silicon dioxide, some dopants become depleted away from the oxide interface while others pile up there. Consequently, processing parameters must be properly adjusted to compensate for this effect.

- **Wet oxidation** utilizes a water-vapor atmosphere as the agent. This method results in a considerably higher growth rate than that of dry oxidation, but it suffers from a lower oxide density and, therefore, a lower dielectric strength. The common practice in industry is to combine both dry and wet oxidation methods, by growing an oxide in a three-part layer: dry–wet–dry. This approach combines the advantages of wet oxidation's much higher growth rate and dry oxidation's high quality.

The two oxidation methods described are useful primarily for coating the entire silicon surface with oxide; however, it also may be necessary to oxidize only certain portions of the surface. The procedure of oxidizing only certain areas is called **selective oxidation,** and uses silicon nitride, which inhibits the passage of oxygen and water vapor. Thus, by covering certain areas with silicon nitride, the silicon under these areas remains unaffected while the uncovered areas are oxidized.

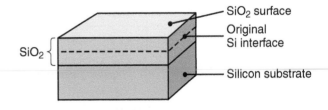

FIGURE 28.9 Growth of silicon dioxide, showing consumption of silicon. *Source:* After S.M. Sze.

TABLE 28.1

General Characteristics of Lithography Techniques

Method	Wavelength (nm)	Finest feature size (nm)
Ultraviolet (photolithography)	365	350
Deep UV	248	250
Extreme UV	10–20	30–100
X-ray	0.01–1	20–100
Electron beam	—	80
Immersion	193	11

Source: After P.K. Wright.

28.7 Lithography

Lithography is the process by which the geometric patterns that define devices are transferred to the substrate surface. A summary of lithographic techniques is given in Table 28.1, and a comparison of the basic lithography methods is shown in Fig. 28.10. The most common technique used today is **photolithography**, and most IC applications can be manufactured successfully with photolithography. *Electron-beam* and *X-ray lithography* are of great interest, because of their ability to transfer patterns with higher resolution, a necessary feature for the increased miniaturization of integrated circuits.

Photolithography. Photolithography uses a **reticle**, also called a **mask** or **photomask**, which is a glass or quartz plate with a pattern of the chip deposited onto it with a chromium film. The reticle image can be the same size as the desired structure on the chip, but it is often an enlarged image, usually 4× to 20× larger, although 10× magnification is the most common. The enlarged images

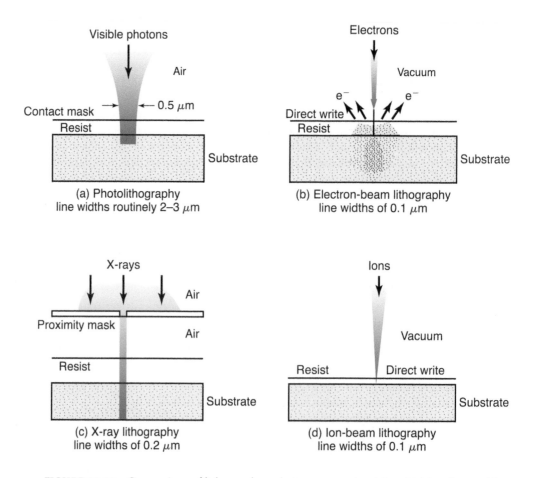

FIGURE 28.10 Comparison of lithography techniques; note that (a) and (c) involve masking to achieve pattern transfer, while (b) and (d) scribe the pattern without a mask, known as direct writing.

are then focused onto a wafer through a lens system, an operation that is referred to as *reduction lithography*.

In current practice, the lithographic process is applied to each microelectronic circuit as many as 25 times, each time using a different reticle to define the different areas of the working devices. Typically designed at several thousand times their final size, reticle patterns undergo a series of reductions before being applied permanently to a defect-free quartz plate. Computer-aided design (CAD; see Section 38.4) has had a major impact on reticle design and generation.

Cleanliness is especially important in lithography, and manufacturers now use robotics and specialized wafer-handling equipment in order to minimize contamination from dust and dirt. Once the film deposition process is completed and the desired reticle patterns have been generated, the wafer is cleaned and coated with a **photoresist** (PR), an organic polymer.

A photoresist consists of three principal components:

1. A polymer that changes its structure when exposed to radiation
2. A sensitizer that controls the reactions in the polymer
3. A solvent, to deliver the polymer in liquid form

Photoresist layers 0.5–2.5 μm (20–100 μin.) thick are produced by applying the photoresist to the substrate and then spinning it, at several thousand rpm, for 30 or 60 s to give uniform coverage (Fig. 28.11).

The next step in photolithography is **prebaking** the wafer, to remove the solvent from the photoresist and harden it. This step is carried out on a hot plate, heated to around 100°C. The pattern is transferred to the wafer through *stepper* or *step-and-scan* systems. With wafer steppers (Fig. 28.12a), the full image is exposed in one flash, and the reticle pattern is then refocused onto another adjacent section of the wafer. With step-and-scan systems (Fig. 28.12b), the exposing light source is focused into a line, and the reticle and wafer are translated simultaneously in opposite directions to transfer the pattern.

The wafer must be aligned carefully under the desired reticle. In this crucial step, called **registration**, the reticle must be aligned correctly with the previous layer on the wafer. Once the reticle is aligned, it is subjected to ultraviolet (UV) radiation. Upon development and removal of the exposed photoresist, a duplicate of the reticle

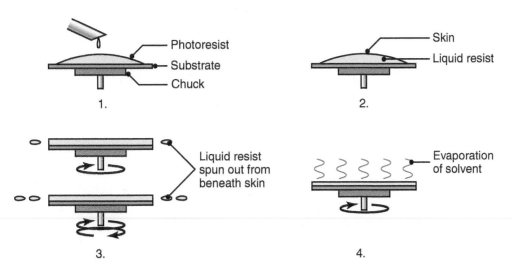

FIGURE 28.11   Spinning of an organic coating on a wafer.

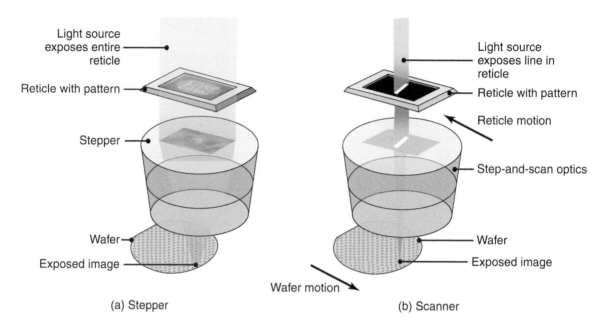

FIGURE 28.12 Schematic illustration of (a) wafer stepper technique for pattern transfer and (b) wafer step-and-scan technique.

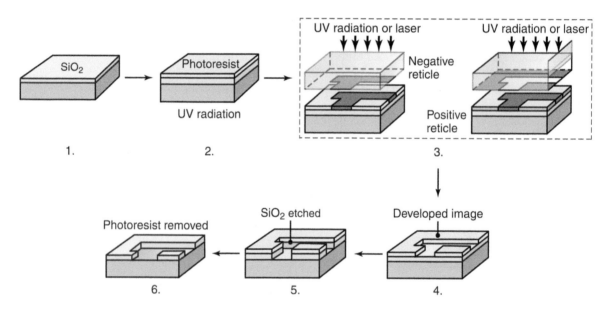

FIGURE 28.13 Pattern transfer by photolithography; note that the reticle in Step 3 can be a positive or negative image of the pattern.

pattern will appear in the photoresist layer. As seen in Fig. 28.13, the reticle can be a negative image or a positive image of the desired pattern. A positive reticle uses the UV radiation to break down the chains in the organic film, so that these films are removed preferentially by the developer. Positive masking is more common than negative masking because, with negative masking, the photoresist can swell and distort, thus making it unsuitable for small features. Newer negative photoresist materials do not have this problem.

Following the exposure and development sequence, **postbaking** the wafer drives off the solvent, and toughens and improves the adhesion of the remaining resist. In addition, a deep UV treatment (thereby baking the wafer to about 150°–200°C in ultraviolet light) can be used to further strengthen the resist against high-energy implants and dry etches. The underlying film not covered by the photoresist is then etched away (Section 28.8) or implanted (Section 28.9).

Following lithography, the developed photoresist must be removed, in a process called **stripping**. In *wet stripping*, the photoresist is dissolved by such solutions as acetone or strong acids. In this method, the solutions tend to lose potency in use. *Dry stripping* involves exposing the photoresist to an oxygen plasma, referred to as **ashing**. Dry stripping has become more common, because it (a) does not involve the disposal of consumed hazardous chemicals and (b) is easier to control and can result in exceptional surfaces.

One of the major issues in lithography is **line width**, which is the width of the smallest feature imprintable on the silicon surface, and is called **critical dimension (CD)**. As circuit densities have escalated over the years, device sizes and features have become smaller and smaller; today, commercially feasible minimum critical dimension is 32 nm, with a major trend to obtain 16 nm or even smaller.

Because pattern resolution and device miniaturization have been limited by the wavelength of the radiation source used, the need has arisen to move to wavelengths shorter than those in the UV range, such as deep UV wavelengths, extreme UV wavelengths, electron beams, and X-rays. In these technologies, the photoresist is replaced by a similar resist that is sensitive to a specific range of shorter wavelengths.

Pitch Splitting.   Multiexposure techniques have been developed to obtain higher resolution images than can be attained through conventional single-exposure lithography, and have been applied for sub-32 nm feature development. *Pitch splitting* is shown in Fig. 28.14, using conventional lithography in multiple stages. Recognizing that the spacing between features is the limiting dimension, pitch splitting involves breaking up the desired pattern into two complimentary portions and creating corresponding masks. By using two imaging steps, features can be developed in the substrate with twice the resolution of a single imaging step.

There are two forms of pitch splitting. In **double exposure** (DE), a mask exposes some of the desired trenches or regions in the photoresist, then a second mask is used to expose the remaining features (Fig. 28.14a). The photoresist is then exposed and the substrate is etched. **Double patterning** (DP) involves two sequential lithography and etch steps, so that it is sometimes referred to as the *LELE* (lithography-etch-lithography-etch) *process*.

Immersion Lithography.   The resolution of lithography systems can be increased by inserting a fluid with a high refractive index between the final lens and the wafer, an approach called *immersion lithography*. Water has been mainly used to date, and has been the main approach used to attain feature sizes below 45 nm. Fluids with a refractive index higher than that of water also are being investigated to increase the resolutions of immersion lithography.

Immersion lithography requires careful process controls, especially thermal controls, since any bubbles that develop in the water will result in defects due to distortion of the light source.

Extreme Ultraviolet Lithography.   The pattern resolution in photolithography and immersion lithography is ultimately limited by light diffraction. One of the means of reducing the effects of diffraction is to use ever shorter wavelengths. *Extreme ultraviolet lithography* (EUV) uses light at a wavelength of 13 nm, in order to obtain

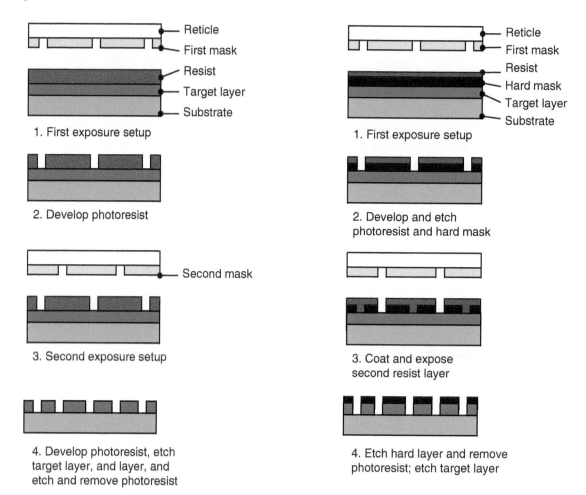

FIGURE 28.14 Pitch splitting lithography. (a) Double exposure (DE) process and (b) double patterning (DP) process, also known as the LELE (lithography-etch-lithography-etch) process.

features commonly in the range from 30 to 100 nm, but is expected to be useful for sub-25 nm features. Because glass lenses absorb some EUV light, the waves are focused through highly reflective molybdenum–silicon mirrors through the mask to the wafer surface.

X-Ray Lithography. Although photolithography is the most widely used lithography technique, it has fundamental resolution limitations associated with light diffraction. *X-ray lithography* is superior to photolithography, because of the shorter wavelength of the radiation and its very large depth of focus. These characteristics allow much finer patterns to be resolved, and make X-ray lithography far less susceptible to dust than photolithography. Moreover, the *aspect ratio* (defined as the ratio of depth to lateral dimension) can be higher than 100, whereas it is limited to around 10 with photolithography. However, to achieve this benefit, synchrotron radiation is required, which is expensive and available at only a few research laboratories.

Given the large capital investment required for a manufacturing facility, industry has preferred to refine and improve optical lithography, instead of investing new capital into X-ray-based production. Currently, X-ray lithography is not widespread, although the LIGA process, described in Section 29.3, fully exploits the benefits of this technique.

Electron-beam and Ion-beam Lithography.   Like X-ray lithography, *electron-beam* (e-beam) and *ion-beam* lithography are superior to photolithography, in terms of attainable resolutions. These two methods involve high current density in narrow electron or ion beams (known as *pencil sources*), which scan a pattern one pixel at a time onto a wafer. The masking is done by controlling the point-by-point transfer of the stored pattern, called *direct writing*, using software. These techniques have the advantages of accurate control of exposure over small areas of the wafer, large depth of focus, and low defect densities. Resolutions are limited to about 10 nm, because of electron scatter, although 2-nm resolutions have been reported for some materials.

It should be noted that the scan time significantly increases as the resolution improves, because more highly focused beams are required. The main drawback of these two techniques is that electron and ion beams have to be maintained in a vacuum, thus significantly increasing equipment complexity and production time. Moreover, the scan time for a wafer is much longer than that for other lithographic methods.

SCALPEL.   In the SCALPEL (*sc*attering with *a*ngular *l*imitation *p*rojection *e*lectron-beam *l*ithography) process (Fig. 28.15), a mask is first produced from about a 0.1-μm-thick membrane of silicon nitride, and then patterned with an approximately 50-nm-thick coating of tungsten. High-energy electrons pass through both the silicon nitride and the tungsten, but the tungsten scatters the electrons widely, whereas

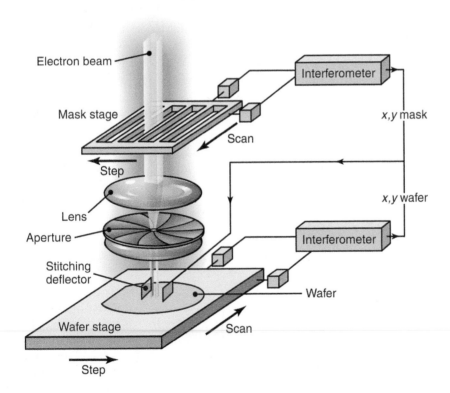

FIGURE 28.15   Schematic illustration of the SCALPEL process.

the silicon nitride results in very little scattering. An aperture blocks the scattered electrons, resulting in a high-quality image at the wafer.

The limitation to the SCALPEL process is the small-sized masks that are currently in use, but the process has high potential. Perhaps its most significant advantage is that energy does not need to be absorbed by the reticle; instead, it is blocked by the aperture, which is not as fragile or expensive as the reticle.

EXAMPLE 28.1   Moore's Law

G. Moore, an inventor of the integrated circuit and past chairman of Intel Corporation, observed in 1965 that the surface area of a single transistor is reduced by 50% every 12 months. In 1975, he revised this estimate to every two years; the resulting estimate is widely known as Moore's law, and it has been remarkably accurate. Figure 28.16 shows the historical progression of feature size in *dynamic random access memory* (DRAM) bits, as well as projected future developments. Looking ahead, however, one can note that there are some major impediments to the continued reliability of Moore's law. Among the more important ones are:

- To produce ever smaller features in a transistor requires that even more stringent manufacturing tolerances be achieved. For example, 180-nm line widths require ±14-nm dimensional tolerances, whereas 50-nm line widths require ±4 nm. Either requirement is especially problematic for the metal connection lines within the transistor.

- Smaller transistors can operate only if the dopant concentration is increased. Above a certain limit, however, the doping atoms cluster together. The result is that *p*-type and *n*-type silicon cannot be produced reliably at small length scales.
- The gate-switching energy of transistors has not been reduced at the same rate as their size; the result is increased power consumption in integrated circuits. This effect has a serious consequence, in that it is very difficult to dissipate the heat produced.
- At smaller length scales, microprocessors require lower voltages for proper operation. However, since the power consumption is still relatively high, very large currents are needed between the power-conversion devices and the central-processing units of modern microprocessors. These large currents result in resistive heating, thus compounding the heat-extraction problems.

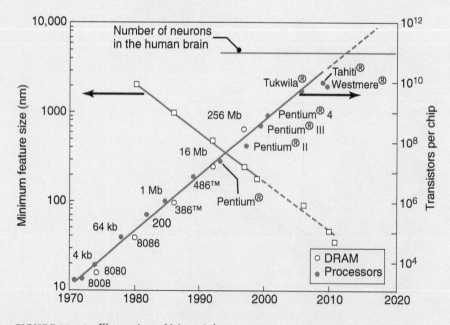

FIGURE 28.16   Illustration of Moore's law.

Much research continues to be directed toward overcoming these limitations. Moore's law was intended as a prediction of the short-term future of the semiconductor industry, and was put forward at a time when photolithography was the only option. During the four decades since the law was first stated, researchers often have identified seemingly insurmountable problems which, in turn, have been overcome.

Soft Lithography.   *Soft lithography* refers to several processes for pattern transfer, all of which require that a master mold be created by one of the standard lithography techniques described above. The master mold is then used to produce an elastomeric pattern, or stamp, as shown in Fig. 28.17. An elastomer that has commonly been used for the stamp is silicone rubber (polydimethylsiloxane, PDMS), because it is chemically inert, is not hygroscopic (it does not swell due to humidity), and has good thermal stability, strength, durability, and surface properties.

Several PDMS stamps can be produced using the same pattern, and each stamp can be used several times. Some of the common soft lithography processes are:

1. **Microcontact printing (μCP).** In *microcontact printing*, the PDMS stamp is coated with an "ink" and then pressed against a surface. The peaks of the pattern are in contact with the opposing surface, and a thin layer of the ink is transferred, often only one molecule thick (called a *self-assembled monolayer* or *boundary film*; see Section 33.6). This film can serve as a mask for selective wet etching, described below, or it can be used to impart a desired chemistry onto the surface.

2. **Microtransfer molding (μTM).** In this process, shown in Fig. 28.18a, the recesses in the PDMS mold are filled with a liquid polymer precursor, and then pressed against a surface. After the polymer has cured, the mold is peeled off, leaving behind a pattern suitable for further processing.

3. **Micromolding in capillaries.** Called MIMIC, in this technique (Fig. 28.18b) the PDMS stamp pattern consists of channels that use capillary action to wick a liquid into the stamp, either from the side of the stamp or from reservoirs within the stamp itself. The liquid can be a thermosetting polymer, a ceramic sol gel, or suspensions of solids within liquid solvents. Good pattern replication can be

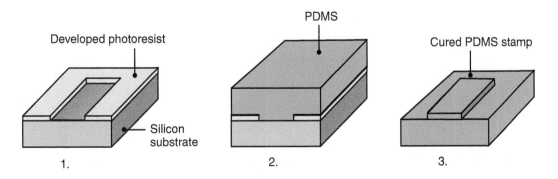

FIGURE 28.17 Production of a polydimethylsiloxane (PDMS) mold for soft lithography. 1. A developed photoresist is produced through standard lithography (see Fig. 28.13). 2. A PDMS stamp is cast over the photoresist. 3. The PDMS stamp is peeled off the substrate to produce a stamp. The stamp shown has been rotated to emphasize the replication of surface features; the master pattern can be used several times. *Source:* Based on Y. Xia and G.M. Whitesides.

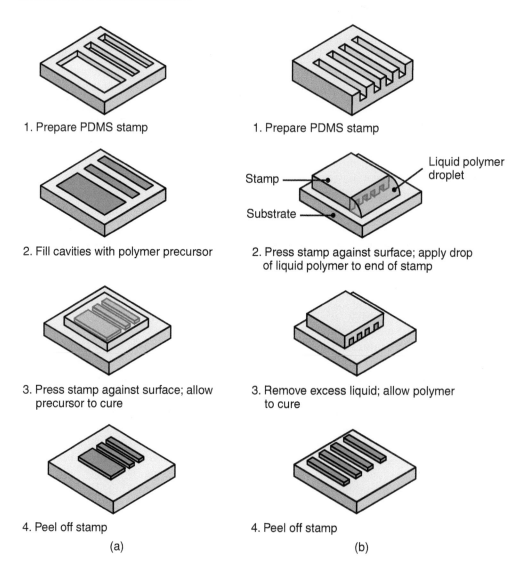

FIGURE 28.18 Soft lithography techniques. (a) Microtransfer molding (μTM) and (b) micromolding in capillaries (MIMIC). *Source:* After Y. Xia and G.M. Whitesides.

obtained, as long as the channel aspect ratio is moderate and the actual channel dimensions allow fluid flow. The dimensions required depend on the liquid used. The MIMIC process has been used to produce all-polymer field-effect transistors and diodes, and has various applications in sensors (Section 37.7).

28.8 Etching

Etching is the process by which entire films or particular sections of films are removed. One of the key criteria in this process is **selectivity**, that is, the ability to etch one material without etching another. In silicon technology, an etching process must etch the silicon-dioxide layer effectively, with minimal removal of either the underlying silicon or the resist material. In addition, polysilicon and metals must be etched into high-resolution lines with vertical wall profiles, and with minimal removal of either the underlying insulating film or the photoresist. Typical etch rates range from hundreds

TABLE 28.2

General Characteristics of Silicon Etching Operations

Wet etching	Temperature (°C)	Etch rate (μm/min)	{111}/{100} selectivity	Nitride etch rate (nm/min)	SiO_2 etch rate (nm/min)	p^{++} etch stop
Wet etching						
$HF:HNO_3:CH_3COOH$	25	1–20	—	Low	10–30	No
KOH	70–90	0.5–2	100:1	<1	10	Yes
Ethylene-diamine pyrocatechol (EDP)	115	0.75	35:1	0.1	0.2	Yes
$N(CH_3)_4OH$ (TMAH)	90	0.5–1.5	50:1	<0.1	<0.1	Yes
Dry (plasma) etching						
SF_6	0–100	0.1–0.5	—	200	10	No
SF_6/C_4F_8 (DRIE)	20–80	1–3	—	200	10	No

Source: Based on Kovacs, G.T.A., Maluf, N.I., and Peterson, K.E., "Bulk micromachining of silicon" in Integrated Sensors, Microactuators and Microsystems (MEMS), pp. 1536–1551, K.D. Wise (ed), Proceedings of the IEEE, v. 86, No. 8, August 1998.

to several thousands of angstroms per minute, and selectivities can range from 1:1 to 100:1. A summary of etching processes and etchants is given in Tables 28.2 and 28.3.

28.8.1   Wet Etching

Wet etching involves immersing the wafers in a liquid solution, usually acidic. A primary feature of most wet-etching operations is that they are *isotropic*; that is, they etch in all directions of the workpiece at the same rate. Isotropy results in *undercuts* beneath the mask material (see, for example, Figs. 27.3b and 28.19a), and thus limits the resolution of geometric features in the substrate.

Effective etching requires the following conditions:

1. Etchant transport to the surface
2. A chemical reaction
3. Transport of reaction products away from the surface
4. Ability to stop the etching process rapidly, known as *etch stop*, in order to obtain superior pattern transfer, usually by using an underlying layer with high selectivity

If the first or third condition listed above limits the speed of the process, agitation or stirring of the etchant can be employed to increase etching rates. If the second condition limits the speed of the process, the etching rate will strongly depend on temperature, etching material, and solution composition. Reliable etching thus requires both proper temperature control and repeatable stirring capability.

Isotropic Etchants.   These etchants are widely used for:

- Removing damaged surfaces
- Rounding sharply etched corners, to avoid stress concentrations
- Reducing roughness developed after anisotropic etching
- Creating structures in single-crystal slices
- Evaluating defects

Fabrication of microelectronic devices as well as microelectromechanical systems, described in Chapter 29, requires the precise machining of structures, done through masking. With isotropic etchants, however, masking can be a challenge, because the strong acids (a) etch aggressively at a rate of up to 50 μm/min, with an etchant of 66% HNO_3 and 34% HF, although etch rates of 0.1–1 μm/s are more typical,

Video Solution 28.1 Wet Etching

TABLE 28.3

Comparison of Etch Rates for Selected Etchants and Target Materials

Etchant	Target material	Etch rate (nm/min)[a]							
		Polysilicon n^+	Polysilicon, undoped	Silicon dioxide	Silicon nitride	Phospho-silicate glass, annealed	Aluminum	Titanium	Photoresist (OCG-820PR)
Wet etchants									
Concentrated HF (49%)	Silicon oxides	0	—	2300	14	3600	4.2	>1000	0
25:1 HF:H_2O	Silicon oxides	0	0	9.7	0.6	150	—	—	0
5:1 BHF[b]	Silicon oxides	9	2	100	0.9	440	140	>1000	0
Silicon etchant (126 HNO_3:60 H_2O:5 NH_4F)	Silicon	310	100	9	0.2	170	400	300	0
Aluminum etchant									
(16 H_3PO_4:1 HNO_3:1 HAc:2 H_2)	Aluminum	<1	<1	0	0	<1	660	0	0
Titanium etchant									
(20 H_2O:1 H_2O_2:1 HF)	Titanium	1.2	—	12	0.8	210	>10	880	0
Piranha (50 H_2SO_4:1 H_2O_2)	Metals and organics (cleaning off)	0	0	0	0	0	180	240	>10
Acetone (CH_3COOH)	Photoresist	0	0	0	0	0	0	0	>4000
Dry Etchants									
CF_4 + CHF_3 + He, 450 W	Silicon oxides	190	210	470	180	620	—	>1000	220
SF_6 + He, 100 W	Silicon nitrides	73	67	31	82	61	—	>1000	69
SF_6, 12.5 W	Thin silicon nitrides	170	280	110	280	140	—	>1000	310
O_2, 400 W	Ashing photoresist	0	0	0	0	0	0	0	340

Notes:

[a]Results are for fresh solutions at room temperature, unless noted. Actual etch rates will vary with temperature and prior use of solution, area of exposure of film, other materials present, film impurities, and microstructure.

[b]Buffered hydrofluoric acid (33% NH4F and 8.3% HF by weight).

Source: After K. Williams and R. Muller.

(b) produce rounded cavities, and (c) are etch rate that is highly sensitive to agitation, thus lateral and vertical features are difficult to control. Because the size of the features in an integrated circuit determines its performance, there is a strong need to produce extremely small and well-defined structures. Such small features cannot be attained through isotropic etching, because of the poor definition resulting from undercutting of masks.

Anisotropic Etching. This situation occurs when etching is strongly dependent on compositional or structural variations in the material. There are two basic types of anisotropic etching: *orientation-dependent etching* (ODE) and *vertical etching*. Orientation-dependent etching commonly occurs in a single crystal, when etching takes place at different rates in different directions, as shown in Fig. 28.19b. Most vertical etching is done with dry plasmas, as described later.

Anisotropic etchants produce geometric shapes, with walls that are defined by the crystallographic planes that resist the etchants. For example, Fig. 28.20 shows the vertical etch rate for silicon as a function of temperature. Note that the etching rate is more than one order of magnitude lower in the [111] crystal direction than in the other directions; therefore, well-defined walls can be obtained along the [111] crystal direction.

The **anisotropy ratio** for etching is defined by

$$AR = \frac{E_1}{E_2}, \qquad (28.1)$$

where E is the etch rate in the crystallographic direction of interest, and the two subscripts refer to two directions for the materials of interest. The anisotropy ratio is unity for isotropic etchants, but can be as high as 400/200/1 for $\langle 110 \rangle / \langle 100 \rangle / \langle 111 \rangle$ silicon. The $\langle 111 \rangle$ planes always etch at the lowest rate, but the $\langle 100 \rangle$ and $\langle 110 \rangle$ planes can be controlled through etchant chemistry.

Masking is also a concern in anisotropic etching, but for different reasons than those for isotropic etching. Anisotropic etching is slow, typically 3 μm/min, with anisotropic etching through a wafer taking as much as several hours. Silicon oxide may etch too rapidly to be used as a mask, hence a high-density silicon nitride mask may be needed.

Often, it is important to rapidly halt the etching process, especially when thin membranes are to be manufactured or features with very precise thickness control are required. Conceptually, rapid halting can be accomplished by removing the wafer from the etching solution. However, etching depends to a great extent on the ability to circulate fresh etchants to the desired locations. Since the circulation varies across a wafer's surface, this strategy for halting the etching process would lead to large variations in the etched depth.

The most common approach to obtain uniform feature sizes across a wafer is to use a boron etch stop (Fig. 28.21), whereby a boron layer is diffused or implanted into the silicon. Examples of common etch stops are the placement of a boron-doped layer beneath silicon or the placement of silicon dioxide (SiO_2) beneath silicon nitride (Si_3N_4). Because anisotropic etchants do not attack boron-doped silicon as aggressively as they do undoped silicon, surface features or membranes can be created by **back etching**.

Several etchant formulations have been developed, including hydrofluoric acid, phosphoric acid, mixtures of nitric acid and hydrofluoric acid, potassium hydrochloride, and mixtures of phosphoric acid, nitric acid, acetic acid, and water. Wafer *cleaning* is done with a solution consisting of sulfuric acid and peroxide, called *Piranha solution*, a trade name. Photoresist can be removed with these solutions, although acetone is commonly used for this purpose.

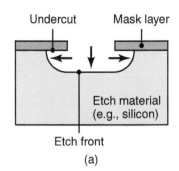

(a)

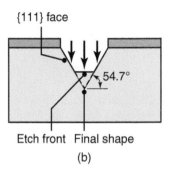

(b)

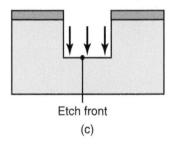

(c)

FIGURE 28.19 Etching directionality. (a) Isotropic etching; etch proceeds vertically and horizontally at approximately the same rate, with significant mask undercut. (b) Orientation-dependent etching (ODE); etch proceeds vertically, terminating on {111} crystal planes with little mask undercut. (c) Vertical etching; etch proceeds vertically with little mask undercut. *Source:* After K. Williams and R. Muller.

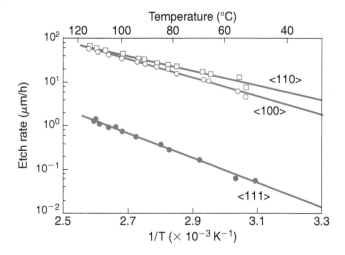

FIGURE 28.20 Etch rates of silicon in different crystallographic orientations, using ethylene-diamine/pyrocatechol-in-water as the solution. *Source:* After H. Seidel.

28.8.2 Dry Etching

Integrated circuits are now etched exclusively by *dry etching*, which involves the use of chemical reactants in a low-pressure system. In contrast to wet etching, dry etching can have a high degree of directionality, resulting in highly anisotropic etching profiles (Fig. 28.19c). Also, the dry-etching process requires only small amounts of the reactant gases, whereas the solutions used in wet etching have to be refreshed periodically.

Dry etching usually involves a plasma or discharge in areas of high electric and magnetic fields; any gases that are present are dissociated to form ions, photons, electrons, or highly reactive molecules. Table 28.2 lists some of the more common dry etchants, their target materials, and typical etch rates. There are several specialized dry-etching techniques.

Sputter Etching.   This process removes material by bombarding the surface with noble gas ions, usually Ar^+. The gas is ionized in the presence of a cathode and an anode (Fig. 28.22). If a silicon wafer is the target, the momentum transfer associated with the bombardment of atoms causes bond breakage and material to be ejected or sputtered. If the silicon chip is the substrate, then the material in the target is deposited onto the silicon, after it has been sputtered by the ionized gas.

The major concerns in sputter etching are:

- The ejected material can be redeposited onto the target, especially with large aspect ratios
- Sputtering can cause damage or excessive erosion of the material
- Sputter etching is not material selective, and because most materials sputter at about the same rate, masking is difficult
- The sputter etching process is slow, with etch rates limited to tens of nm/min
- The photoresist is difficult to remove

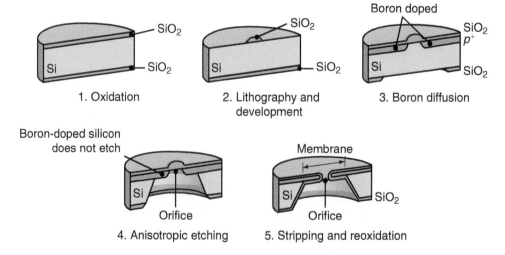

1. Oxidation

2. Lithography and development

3. Boron diffusion

4. Anisotropic etching

5. Stripping and reoxidation

FIGURE 28.21 Application of a boron etch stop and back etching to form a membrane and orifice. *Source:* Based on I. Brodie and J.J. Murray.

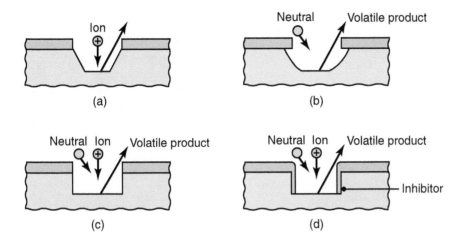

FIGURE 28.22 Machining profiles associated with different dry-etching techniques: (a) sputtering; (b) chemical; (c) ion-enhanced energetic; and (d) ion-enhanced inhibitor. *Source:* After M. Madou.

Reactive Plasma Etching. Also referred to as **dry chemical etching**, this process involves chlorine or fluorine ions (generated by RF excitation), and other molecular species, that diffuse into and chemically react with the substrate. As a result, a volatile compound is formed, which is then removed by a vacuum system. The mechanism of reactive plasma etching is illustrated in Fig. 28.23:

1. A reactive species, such as CF_4, is first produced, and it dissociates upon impact with energetic electrons, to produce fluorine atoms.
2. The reactive species then bombard and diffuse into the surface.
3. The reactive species chemically reacts to form a volatile compound.
4. The reactant desorbs from the surface.
5. It then diffuses into the bulk gas, where it is removed by a vacuum system.

Some reactants polymerize on the surface, thus requiring additional removal of material, either with oxygen in the plasma reactor or by an external *ashing* operation (see *dry stripping* in Section 28.7). The electrical charge of the reactive species is not high enough to cause damage through impact on the surface, hence no sputtering occurs; thus, the etching is isotropic and undercutting of the mask takes place (Fig. 28.19a).

Physical–chemical Etching. Processes such as *reactive ion-beam etching* (RIBE) and *chemically assisted ion-beam etching* (CAIBE) combine the advantages of physical and chemical etching. These processes use a chemically reactive species to drive material removal, but are assisted physically by the impact of ions onto the surface. In RIBE, also known as *deep reactive-ion etching* (DRIE), vertical trenches hundreds of nanometers deep can be produced by periodically interrupting the etching process and depositing a polymer layer, as shown in Fig. 28.23d.

In CAIBE, ion bombardment can assist dry chemical etching by:

- Making the surface more reactive
- Clearing the surface of reaction products and allowing the chemically reactive species access to the cleared areas

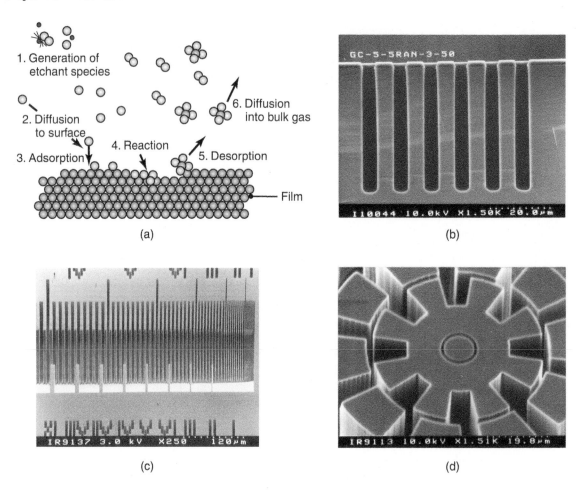

FIGURE 28.23   (a) Schematic illustration of reactive plasma etching. (b) Example of a deep reactive-ion etched trench; note the periodic undercuts, or scallops. (c) Near-vertical sidewalls produced through deep reactive-ion etching (DRIE), an anisotropic-etching process. (d) An example of cryogenic dry etching, showing a 145-μm deep structure etched into silicon with the use of a 2.0-μm-thick oxide masking layer; the substrate temperature was $-140°C$ during etching. *Source:* (a) Based on M. Madou. (b) through (d) After R. Kassing and I.W. Rangelow, University of Kassel, Germany.

- Providing the energy to drive surface chemical reactions; however, the neutral species do most of the etching

Physical–chemical etching is extremely useful because the ion bombardment is directional, so that etching is anisotropic. Also, the ion-bombardment energy is low and does not contribute much to mask removal, thus allowing the generation of near-vertical walls with very large aspect ratios. Since the ion bombardment does not remove material directly, masks can be used.

Cryogenic Dry Etching.   This method is used to obtain very deep features with vertical walls. The workpiece is first lowered to cryogenic temperatures, then the CAIBE process takes place. The very low temperatures involved ensure that insufficient energy is available for a surface chemical reaction to take place, unless ion bombardment is normal to the surface. Oblique impacts, such as those occurring on sidewalls in deep crevices, cannot drive the chemical reactions.

Because dry etching is not selective, etch stops cannot be applied directly and dry-etching reactions must be terminated when the target film is removed. This can be done by measuring the wavelength of light being emitted during a reaction; when the target film is removed, the wavelength of light emitted will change and can be detected with proper sensors.

EXAMPLE 28.2   Comparison of Wet and Dry Etching

Consider the case where a ⟨100⟩ wafer (see Fig. 28.5) has an oxide mask placed on it, in order to produce square or rectangular holes. The sides of the square are oriented precisely within the ⟨100⟩ direction of the wafer surface, as shown in Fig. 28.24.

Isotropic etching results in the cavity shown in Fig. 28.24a, and since etching occurs at constant rates in all directions, a rounded cavity that undercuts the mask is produced. An orientation-dependent etchant produces the cavity shown in Fig. 28.24b. Because etching is much faster in the ⟨100⟩ and ⟨110⟩ directions than in the ⟨111⟩ direction, sidewalls defined by the ⟨111⟩ plane are

generated. For silicon, these sidewalls are at an angle of 54.74° to the surface.

The effect of a larger mask or shorter etch time is shown in Fig. 28.24c. The resultant pit is defined by ⟨111⟩ sidewalls and by a bottom in the ⟨100⟩ direction parallel to the surface. A rectangular mask and the resulting pit are shown in Fig. 28.24d. Deep reactive-ion etching is depicted in Fig. 28.24e. Note that a polymer layer is deposited periodically onto the hole sidewalls, to allow for deep pockets, but scalloping (greatly exaggerated in the figure) is unavoidable. A hole resulting from CAIBE is shown in Fig. 28.24f.

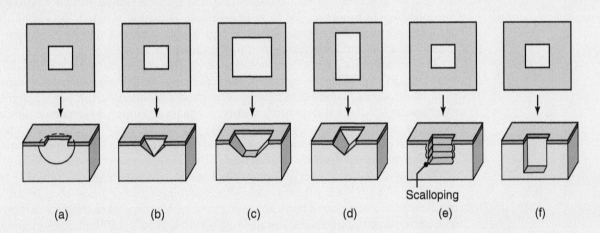

(a)   (b)   (c)   (d)   (e)   (f)

FIGURE 28.24 Holes generated from a square mask in (a) isotropic (wet) etching; (b) orientation-dependent etching (ODE); (c) ODE with a larger hole; (d) ODE of a rectangular hole; (e) deep reactive-ion etching; and (f) vertical etching. *Source:* After M. Madou.

28.9 Diffusion and Ion Implantation

Recall that the operation of microelectronic devices depends on regions that have different doping types and concentrations. The electrical characteristics of these regions are altered through the introduction of dopants into the substrate, by *diffusion* and *ion-implantation processes*. This step in the fabrication sequence is repeated several times, since many different regions of microelectronic devices must be defined.

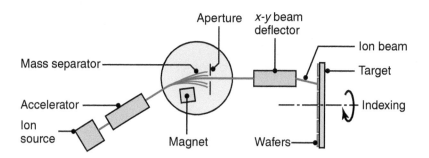

FIGURE 28.25   Schematic illustration of an apparatus for ion implantation.

In the diffusion process, the movement of atoms is a result of thermal excitation. Dopants can be introduced to the substrate surface in the form of a deposited film or the substrate can be placed in a vapor, containing the dopant source. The process takes place at temperatures usually 800°–1200°C (1500°–2200°F). Dopant movement within the substrate is strictly a function of temperature, time, and the diffusion coefficient (or *diffusivity*) of the dopant species, as well as the type and quality of the substrate material.

Because of the nature of diffusion, the dopant concentration is very high at the substrate surface, and drops off sharply away from the surface. To obtain a more uniform concentration within the substrate, the wafer is heated further to drive in the dopants, in a process called **drive-in diffusion**. Diffusion, whether desired or not, is highly isotropic and always occurs at high temperatures, a phenomenon that is taken into account during subsequent processing steps.

Ion implantation is a much more extensive process and requires specialized equipment (Fig. 28.25; see also Section 34.7). Implantation is accomplished by accelerating the ions through a high-voltage field of as much as 1 million electron volts, and then by choosing the desired dopant by means of a mass separator. In a manner similar to that of cathode-ray tubes, the beam is swept across the wafer by sets of deflection plates, thus ensuring uniformity of coverage of the substrate. The whole implantation operation must be performed in a vacuum.

The high-velocity impact of ions on the silicon surface damages the lattice structure and results in lower electron mobilities. Although this condition is undesirable, the damage can be repaired by an annealing step, which involves heating the substrate to relatively low temperatures, usually 400°–800°C (750°–1500°F), for a period of 15–30 min. Annealing provides the energy that the silicon lattice needs to rearrange and mend itself. Another important function of annealing is driving in the implanted dopants. Implantation alone imbeds the dopants less than half a micron below the silicon surface; annealing enables the dopants to diffuse to a more desirable depth of a few microns.

EXAMPLE 28.3   **Processing of a *p*-type Region in *n*-type Silicon**

Given: It is desired to create a *p*-type region within a sample of *n*-type silicon.

Find: Draw cross-sections of the sample at each processing step in order to accomplish this task.

Solution: Refer to Fig. 28.26. This simple device, known as a *pn-junction diode*, is the foundation for most semiconductor devices.

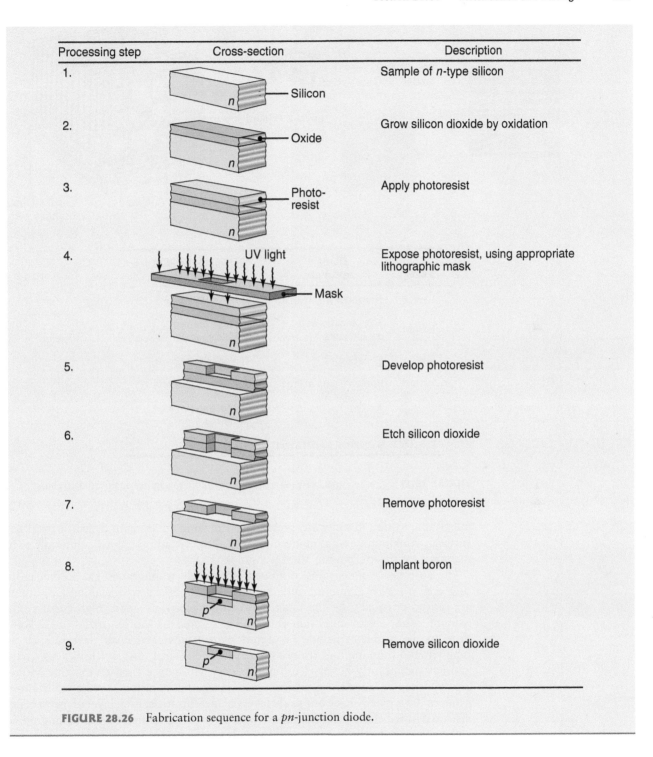

Processing step	Cross-section	Description
1.	Silicon	Sample of *n*-type silicon
2.	Oxide	Grow silicon dioxide by oxidation
3.	Photo-resist	Apply photoresist
4.	UV light / Mask	Expose photoresist, using appropriate lithographic mask
5.		Develop photoresist
6.		Etch silicon dioxide
7.		Remove photoresist
8.		Implant boron
9.		Remove silicon dioxide

FIGURE 28.26   Fabrication sequence for a *pn*-junction diode.

28.10   Metallization and Testing

The preceding sections focused on device fabrication. Generating a complete and functional integrated circuit requires that these devices be interconnected, and this must take place on a number of levels (Fig. 28.27). **Interconnections** are made using

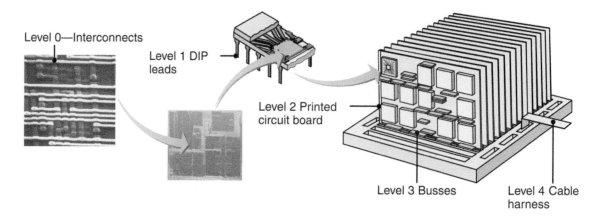

Level	Element example	Interconnection method
Level 0	Transistor within an IC	IC metallization
Level 1	ICs, other discrete components	Package leads or module interconnections
Level 2	IC packages	Printed circuit board
Level 3	Printed circuit boards	Connectors (busses)
Level 4	Chassis or box	Connectors/cable harnesses
Level 5	System, e.g., computer	

FIGURE 28.27 Connections between elements in the hierarchy for integrated circuits.

metals that exhibit low electrical resistance and good adhesion to dielectric insulator surfaces. Aluminum and aluminum–copper alloys remain the most commonly used materials for this purpose in VLSI technology today.

Because device dimensions continue to shrink, **electromigration** has become more of a concern with aluminum interconnects, a process by which aluminum atoms are moved physically by the impact of drifting electrons under high currents. In extreme cases, electromigration can lead to severed or shorted metal lines. Solutions to the problem include (a) addition of sandwiched metal layers, such as tungsten and titanium and (b) use of pure copper, which displays lower resistivity and has significantly better electromigration performance than aluminum. Metals are deposited by standard deposition techniques, an operation called **metallization**. Modern ICs typically have one to six layers of metallization, each layer of metal being insulated by a dielectric. Interconnection patterns are generated through lithographic and etching processes.

Planarization, that is, producing a planar surface of interlayer dielectrics, is critical to the reduction of metal shorts and the line width variation of the interconnect. A common method used to achieve a planar surface is a uniform oxide-etch process that smoothens out the peaks and valleys of the dielectric layer. Planarizing high-density interconnects has now become the process of **chemical–mechanical polishing** (CMP), described in Section 26.7. This process entails physically polishing the wafer surface, in a manner similar to that by which a disk or belt sander flattens the ridges

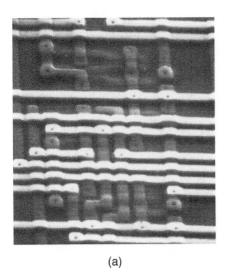

(a)

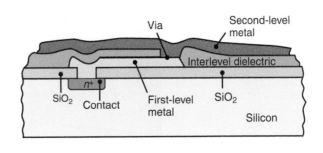

(b)

FIGURE 28.28   (a) Scanning-electron microscope (SEM) photograph of a two-level metal interconnect; note the varying surface topography. (b) Schematic illustration of a two-level metal interconnect structure. *Source:* (a) Courtesy of National Semiconductor Corporation. (b) After R.C. Jaeger.

in a piece of wood. A typical CMP process combines an abrasive medium with a polishing compound or slurry, and can polish a wafer to within 0.03 μm (1.2 μin.) of being perfectly flat, with an R_q roughness (see Section 4.4) on the order of 0.1 nm for a new silicon wafer.

Layers of metal are connected together by **vias,** and access to the devices on the substrate is achieved through **contacts** (Fig. 28.28). As devices continue to become smaller and faster, the size and speed of some chips have become limited by the metallization process itself. Wafer processing is completed upon application of a *passivation layer,* usually silicon nitride (Si_3N_4). The silicon nitride acts as a barrier to sodium ions and also provides excellent scratch resistance.

The next step is to test each of the individual circuits on the wafer (Fig. 28.29). Each chip, also known as a **die,** is tested by a computer-controlled probe platform, containing needlelike probes that access the bonding pads on the die. The probes are of two forms:

1. **Test patterns or structures:** The probe measures test structures, often outside of the active dice, placed in the scribe line (the empty space between dies); these probes consist of transistors and interconnect structures that measure various processing parameters, such as resistivity, contact resistance, and electromigration.
2. **Direct probe:** This approach involves 100% testing on the bond pads of each die.

The platform scans across the wafer and uses computer-generated timing waveforms, to test whether each circuit is functioning properly. If a chip is defective, it is marked with a drop of ink. Up to one-third of the cost of a microelectronic circuit can be incurred during this testing.

After the wafer-level testing is completed, back grinding may be done to remove a large amount of the original substrate. The final die thickness depends on the packaging requirement, but anywhere from 25 to 75% of the wafer thickness may be removed. After back grinding, each die is separated from the wafer. *Diamond*

FIGURE 28.29 A probe (top center) checking for defects in a wafer; an ink mark is placed on each defective die. *Source:* Courtesy of Intel Corp.

sawing is a commonly used separation technique and results in very straight edges, with minimal chipping and cracking damage. The chips are then sorted, the functional dice are sent on for packaging, and the inked dice are discarded.

28.11   Wire Bonding and Packaging

The working dice must be attached to a more rugged foundation to ensure reliability. One simple method is to fasten a die to its packaging material with epoxy cement; another method makes use of a eutectic bond, made by heating metal-alloy systems (see Section 4.3). One widely used mixture is 96.4% Au and 3.6% Si, which has a eutectic point at 370°C (700°F).

Once the chip has been attached to its substrate, it must be connected electrically to the package leads. This is accomplished by *wire bonding* very thin (25 μm diameter; 0.001 μin.) gold wires from the package leads to bonding pads, located around the perimeter or down the center of the die (Fig. 28.30). The bonding pads on the die are typically drawn at 75−100 μm (0.003−0.004 in.) per side, and the bond wires are attached by means of thermocompression, ultrasonic, or thermosonic techniques (Fig. 28.31).

The connected circuit is now ready for final packaging. The *packaging* process largely determines the overall cost of each completed IC since the circuits are mass produced on the wafer, but are then packaged individually. Packages are available in a wide variety of styles; the appropriate one must reflect the operating requirements. Consideration of a circuit's package includes the chip size, number of external leads, operating environment, heat dissipation, and power requirements; for example, ICs used for military and industrial applications require packages of particularly high strength, toughness, and resistance to temperature.

Packages are produced from polymers, metals, or ceramics. Metal containers are made from alloys such as Kovar (an iron−cobalt−nickel alloy with a low coefficient of expansion; see Section 3.6), which provide a hermetic seal and good thermal conductivity, but are limited in the number of leads that can be used. Ceramic packages usually are produced from aluminum oxide (Al_2O_3), are hermetic, and

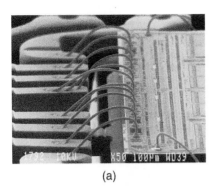
(a)
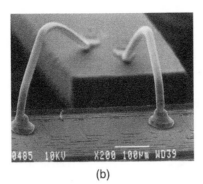
(b)
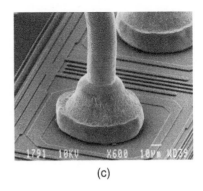
(c)

FIGURE 28.30 (a) SEM photograph of wire bonds connecting package leads (left-hand side) to die bonding pads. (b) and (c) Detailed views of (a). *Source:* Courtesy of Micron Technology, Inc.

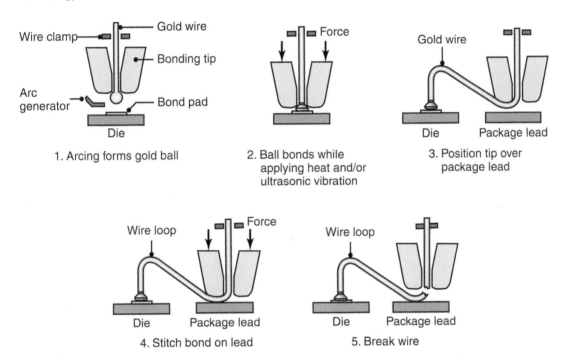

FIGURE 28.31 Schematic illustration of thermosonic welding of gold wires from package leads to bonding pads.

have good thermal conductivity, but have higher lead counts than metal packages; they are also more expensive. Plastic packages are inexpensive and have high lead counts, but they cannot withstand high temperatures and are not hermetic.

An older style of packaging is the **dual-in-line package** (DIP), shown schematically in Fig. 28.32a. Characterized by low cost and ease of handling, DIP packages are made of thermoplastics, epoxies, or ceramics, and can have from 2 to 500 external leads. Ceramic packages are designed for use over a broader temperature range and in high-reliability and military applications, but cost considerably more than plastic packages.

Figure 28.32b shows a *flat ceramic package* in which the package and all of the leads are in the same plane. This package style does not offer the ease of handling

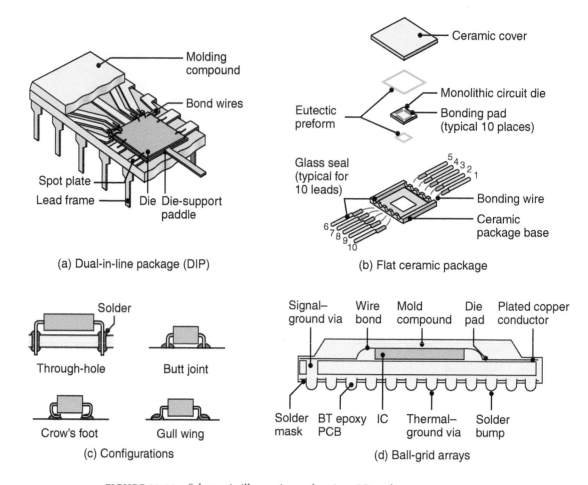

FIGURE 28.32   Schematic illustrations of various IC packages.

or the modular design of the DIP package. For that reason, it usually is affixed permanently to a multiple-level circuit board, in which the low profile of the flat package is necessary.

Surface-mount packages have become common for today's integrated circuits. Some examples are shown in Fig. 28.32c; note that the main difference among them is in the shape of the connectors. The DIP connection to the surface board is *via prongs*, which are inserted into corresponding holes, while a surface mount is soldered onto a specially fabricated pad or *land* (a raised solder platform for interconnections among components in a printed circuit board). Package size and layouts are selected from standard patterns, and usually require adhesive bonding of the package to the board, followed by **wave soldering** of the connections (see Section 32.3.3).

Faster and more versatile chips require increasingly tightly spaced connections. **Pin-grid arrays** (PGAs) use tightly packed pins that connect onto printed circuit boards by way of through-holes. PGAs and other in-line and surface-mount packages are, however, extremely susceptible to plastic deformation of the wires and legs, especially with small-diameter, closely spaced wires. One way of achieving tight packing of connections, and avoiding the difficulties of slender connections, is through **ball-grid arrays** (BGAs), as shown in Fig. 28.32d. This type of array has a solder-plated coating on a number of closely spaced metal balls on the underside of the package. The spacing between the balls can be as small as 50 μm (2000 μin.), but more commonly it is standardized as 1.0 mm (0.040 in.), 1.27 mm (0.050 in.), or 1.5 mm (0.060 in.).

Although BGAs can be designed with over 1000 connections, this is extremely rare and usually 200–300 connections are sufficient for demanding applications. By using the **reflow soldering** technique (see Section 32.3.2), the solder serves to center the BGA by surface tension, thus resulting in well-defined electrical connections for each ball. After the chip has been sealed in the package, it undergoes final testing. Because one of the main purposes of packaging is isolation from the environment, testing at this stage usually involves such factors as heat, humidity, mechanical shock, corrosion, and vibration. Destructive tests also are performed to determine the effectiveness of sealing.

Chip on Board.   *Chip on board* (COB) designs refer to the direct placement of chips onto an adhesive layer on a circuit board. Electrical connections are then made by wire bonding the chips directly to the pads on the circuit board. After wire bonding, final encapsulation with an epoxy is necessary, not only to attach the IC package more securely to the printed circuit board but also to transfer heat evenly during its operation.

Flip-chip on Board.   The flip-chip on board (FOB) technology, illustrated in Fig. 28.33, involves the direct placement of a chip with solder bumps onto an array of pads on the circuit board. The main advantage to flip chips, and ball-grid array packages, is that the space around the package, normally reserved for bond pads, is saved; thus, a higher level of miniaturization can be achieved.

System in Package.   A trend that allows for more compact devices involves incorporating more than one integrated circuit into a package. Figure 28.34 illustrates the major categories of *system-in-package* (SiP) designs. Although these packages can be integrated horizontally, vertical integration through stacked or embedded structures (Figs. 28.34b and c) has the advantage of achieving performance increases over conventional packages. These benefits have been described as "more than Moore" (see Example 28.1), although SiPs also have other advantages, such as (a) they present reduced size and less noise, (b) cross talk between chips can be better isolated, and (c) individual chips can be upgraded more easily. On the other hand, these packages are more complex, require higher power density and associated heat extraction, and are more expensive than conventional packages.

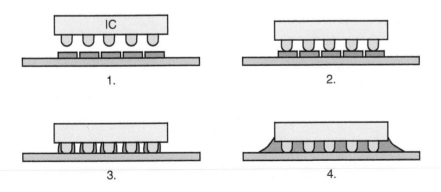

FIGURE 28.33   Illustration of flip-chip technology. Flip-chip package with 1. Solder-plated metal balls and pads on the printed circuit board. 2. Flux application and placement. 3. Reflow soldering. 4. Encapsulation.

SiP packages can, however, be made very simple by incorporating more than one chip inside a single package, as shown in Fig. 28.34a. To preserve area on a circuit board, chips and/or flip chips can be stacked and bonded to a circuit board, to produce three-dimensional integrated circuits, as illustrated in Fig. 28.34b. Here, an

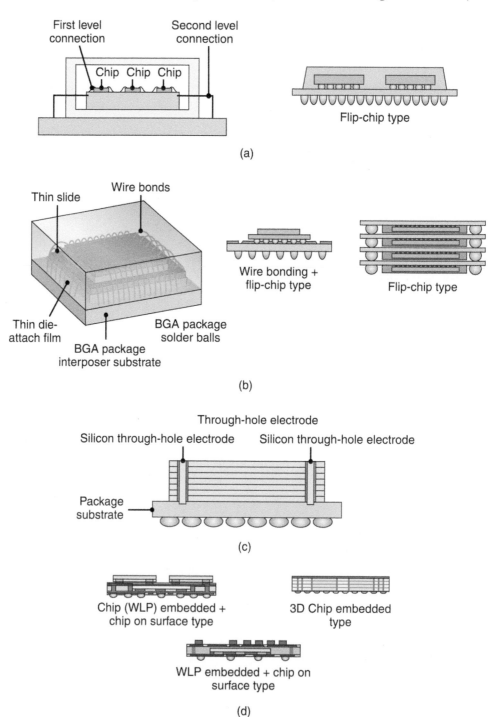

FIGURE 28.34   Major categories of system-in-package designs. (a) Horizontal placement, or multichip modules (MCMs); (b) interposer-type stacked structure; (c) interposerless stacked structure with through-silicon vias; and (d) embedded structure. WLP = wafer level package.

interposing layer, commonly an adhesive, separates the chips and electrically isolates adjacent layers. An alternative is to employ a so-called *interposerless* structure, using **through-silicon vias** (TSVs) instead of wire bonding, to provide electrical connections to all layers. TSVs are sometimes considered a packaging feature, but it has been noted that this is perhaps a case of 3D integration of a wafer, as shown in Fig. 28.34c.

28.12 Yield and Reliability

Yield is defined as the ratio of functional chips to the total number of chips produced. The overall yield of the total IC manufacturing process is the product of the wafer yield, bonding yield, packaging yield, and test yield. This quantity can range from only a few percent for new processes to more than 90% for mature manufacturing lines. Most yield loss occurs during wafer processing, due to its more complex nature. Wafers are commonly separated into regions of good and bad chips. Failures at this stage can arise from point defects (such as oxide pinholes), film contamination, metal particles, and area defects (such as uneven film deposition or nonuniformity of the etch).

A major concern about completed ICs is their **reliability** and **failure rate**. Since no device has an infinite lifetime, statistical methods are used to characterize the expected lifetimes and failure rates of microelectronic devices. The unit of failure rate is the FIT (failure in time), defined as *one failure per 1 billion device-hours*. However, complete systems may have millions of devices, so the overall failure rate in entire systems is correspondingly higher.

Equally important in failure analysis is determining the *failure mechanism*, that is, the actual process that causes the device to fail. Common failures due to processing involve:

Video Solution 28.2 Yield and Reliability of Integrated Circuits

- Diffusion regions: nonuniform current flow and junction breakdown
- Oxide layers: dielectric breakdown and accumulation of surface charge
- Lithography: uneven definition of features and mask misalignment
- Metal layers: poor contact and electromigration, resulting from high current densities
- Other failures, originating in improper chip mounting, poorly formed wire bonds, or loss of the package's hermetic seal

Because device lifetimes are very long, it is impractical to study device failure under normal operating conditions. One method of studying failures efficiently is **accelerated life testing**, which involves accelerating the conditions whose effects cause device breakdown. Cyclic variations in temperature, humidity, voltage, and current are used to stress the components. Chip mounting and packaging are strained by cyclical temperature variations. The statistical data taken from these tests are then used to predict device-failure modes and device life under normal operating conditions.

28.13 Printed Circuit Boards

Packaged ICs seldom are used alone; rather, they usually are combined with other ICs to serve as building blocks of a yet larger system. A *printed circuit board* (PCB) is the substrate for the final interconnections among all of the completed chips, and serves as the communication link between the outside world and the microelectronic circuitry within each packaged IC. In addition to possessing ICs, circuit boards usually contain discrete circuit components (such as resistors and capacitors), which take up too much "real estate" on the limited silicon surface, have special power-dissipation requirements, or cannot be implemented on a chip. Other common discrete components are inductors (that cannot be integrated onto the silicon surface), high-performance transistors, large capacitors, precision resistors, and crystals (for frequency control).

A PCB is basically a plastic (resin) material, containing several layers of copper foil (Fig. 28.35). *Single-sided PCBs* have copper tracks on only one side of an insulating substrate; *double-sided boards* have copper tracks on both sides. *Multilayered boards* also can be constructed from alternating layers of copper and insulator, but single-sided boards are the simplest form of circuit board.

Double-sided boards usually must have locations where electrical connectivity is established between the features on both sides of the board; this is accomplished with vias, as shown in Fig. 28.35. Multilayered boards can have partial, buried, or through-hole vias to allow for extremely flexible PCBs. Double-sided and multilayered boards are preferable, because IC packages can be bonded to both sides of the board, thus allowing for more compact designs.

The insulating material is usually an epoxy resin, 0.25–3 mm (0.01–0.12 in.) thick, reinforced with an epoxy-glass fiber, and is referred to as E-glass (see Section 9.2.1). The assembly is produced by impregnating sheets of glass fiber with epoxy, and pressing the layers together between hot plates or rolls. The heat and pressure cure the board, resulting in a stiff and strong basis for printed circuit boards.

Boards are first sheared to a desired size, and about 3-mm-diameter locating holes are then drilled or punched into the board's corners, to permit alignment and proper location of the board within the chip-insertion machines. Holes for vias and connections are punched or produced through CNC drilling (Section 37.3); stacks of boards can be drilled simultaneously to increase production rates.

The conductive patterns on circuit boards are defined by lithography, although originally they were produced through screen-printing technologies, hence the term *printed* circuit board or *printed wiring board* (PWB). In the *subtractive method*, a copper foil is bonded to the circuit board. The desired pattern on the board is defined

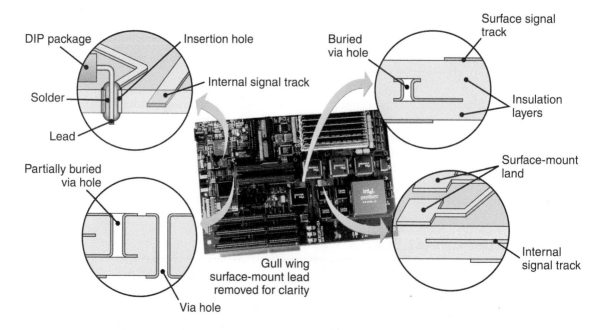

FIGURE 28.35 Printed circuit board structures and design features.

by a positive mask, developed through photolithography, and the remaining copper is removed through wet etching. In the *additive method*, a negative mask is placed directly onto an insulator substrate, to define the desired shape. Electroless plating and electroplating of copper serve to define the connections, tracks, and lands on the circuit board.

The ICs and other discrete components are then soldered to the board. This is the final step in making both the ICs and the microelectronic devices they contain into larger systems, through connections on PCBs. *Wave soldering* and *reflow paste soldering* (see Section 32.3.3 and Example 32.1) are the preferred methods of soldering ICs onto circuit boards.

Basic design considerations in laying out PCBs are:

1. Wave soldering should be used only on one side of the board; thus, all through-hole mounted components should be inserted from the same side of the board. Surface-mount devices placed on the insertion side of the board must be reflow soldered in place; surface-mount devices on the lead side can be wave soldered.
2. To allow good solder flow in wave soldering, IC packages should be laid out carefully on the PCB. Inserting the packages in the same direction is advantageous for automated placing, because random orientations can cause problems in the flow of solder across all of the connections.
3. The spacing of ICs is determined mainly by the need to remove heat during the operation. Sufficient clearance between packages and adjacent boards is thus required to allow forced airflow and heat convection.
4. There should be sufficient space also around each IC package to allow for reworking and repairing without disturbing adjacent devices.

SUMMARY

- The microelectronics industry continues to develop rapidly, and possibilities for new device concepts and circuit designs appear to be endless. The fabrication of microelectronic devices and integrated circuits involves several different types of processes, many of which have been adapted from those of other fields in manufacturing.
- A rough shape of single-crystal silicon is first obtained by the Czochralski process. This shape is ground to a solid cylinder of well-controlled dimensions, and a notch or flat is machined into the cylinder. The cylinder is then sliced into wafers, which are ground on their edges and subjected to chemical–mechanical polishing to complete the wafer.
- After bare wafers have been prepared, they undergo repeated oxidation or film deposition, and lithographic or etching steps to open windows in the oxide layer in order to access the silicon substrate.
- Wet etching is isotropic and relatively fast. Dry etching, using gas plasmas, is anisotropic and allows for more accurate lithography and large-scale integration of integrated circuits.
- After each of the processing cycles is completed, dopants are introduced into various regions of the silicon structure, through diffusion and ion implantation. The devices are then interconnected by multiple metal layers, and the completed circuit is packaged and made accessible through electrical connections.
- The packaged circuit and other discrete devices are then soldered to a printed circuit board for final installation.

KEY TERMS

Accelerated-life testing	Epitaxy	Microcontact printing	Selectivity
Bonding	Etching	Micromolding in	Semiconductor
Chemical–mechanical	Evaporation	capillaries	Silicon
polishing	Failure rate	Microtransfer molding	Soft lithography
Chemical-vapor	Film deposition	Oxidation	Sputtering
deposition	Flip-chip on board	Packaging	Surface-mount package
Chip	Gallium arsenide	Photoresist	System in package
Chip on board	Immersion lithography	Pitch splitting	Three-dimensional
Contacts	Integrated circuit	lithography	integrated circuits
Critical dimension	Ion implantation	Planarization	Very large scale
Czochralski process	LELE process	Postbaking	integration
Die	Line width	Prebaking	Vias
Diffusion	Lithography	Printed circuit board	Wafer
Dopants	Masking	Registration	Wafer-scale integration
Dry etching	Metal-oxide-	Reliability	Wet etching
Dry oxidation	semiconductor	Reticle	Wet oxidation
Dual-in-line package	field-effect transistor	SCALPEL	Wire bonding
Electromigration	Metallization	Selective oxidation	Yield

BIBLIOGRAPHY

Campbell, S.A., **The Science and Engineering of Micro-electronic Fabrication**, 2nd ed., Oxford University Press, 2001.

Doering, R., and Nishi, Y., **Handbook of Semiconductor Manufacturing Technology**, 2nd ed., CRC Press, 2008.

Geng, H., **Semiconductor Manufacturing Handbook**, McGraw-Hill, 2005.

Harper, C.A. (ed.), **Electronic Packaging and Interconnection Handbook**, 4th ed., McGraw-Hill, 2004.

Jackson, K.A., and Schroter, W. (eds.), **Handbook of Semiconductor Technology: Processing of Semiconductors**, 2 vols., Wiley, 2000.

Jaeger, R.C., **Introduction to Microelectronic Fabrication**, 2nd ed., Prentice Hall, 2001.

Khandpur, R.S., **Printed Circuit Boards: Design, Fabrication, and Assembly**, McGraw-Hill, 2005.

May, G.S., and Spanos, C.J., **Fundamentals of Semiconductor Manufacturing and Process Control**, Wiley, 2006.

Nishi, Y., and Doering, R., **Handbook of Semiconductor Manufacturing Technology**, 2nd ed., CRC Press, 2007.

Rizvi, S., **Handbook of Photomask Manufacturing Technology**, CRC Press, 2005.

Schroder, D.K., **Semiconductor Material and Device Characterization**, 3rd ed., Wiley-Interscience, 2006.

Ulrich, R.K., and Brown, W.D. (eds.), **Advanced Electronic Packaging**, Wiley, 2006.

Van Zant, P., **Microchip Fabrication**, 5th ed., McGraw-Hill, 2004.

Yoo, C.S., **Semiconductor Manufacturing Technology**, World Scientific Publishing, 2008.

REVIEW QUESTIONS

28.1 Define the terms wafer, chip, die, device, integrated circuit, line width, registration, surface mount, accelerated-life testing, and yield.

28.2 Why is silicon the semiconductor most used in IC technology?

28.3 What do the abbreviations BJT, MOSFET, VLSI, IC, CVD, CMP, LELE, and DIP stand for?

28.4 Explain the differences between wet and dry oxidation.

28.5 Explain the differences between wet and dry etching.

28.6 What are the purposes of prebaking and postbaking in lithography?

28.7 Define selectivity and isotropy, and their importance in relation to etching.

28.8 Compare the diffusion and ion-implantation processes.

28.9 Explain the difference between evaporation and sputtering.

28.10 What are the levels of interconnection?

28.11 Which is cleaner, a Class-10 or a Class-1 clean room?

28.12 Review Fig. 28.2 and describe the fabrication sequence for integrated circuits.

28.13 What is a via? Why is it important?

28.14 Describe how electrical connections are established between a die and a package.

28.15 What is a flip chip?

28.16 Describe the procedures of image splitting lithography and immersion lithography.

QUALITATIVE PROBLEMS

28.17 Comment on your observations regarding the contents of Fig. V.1.

28.18 Describe how *n*-type and *p*-type dopants differ.

28.19 How is silicon nitride used in oxidation?

28.20 How is epitaxy different from other techniques used for deposition? Explain.

28.21 Note that, in a horizontal epitaxial reactor (see Fig. P28.21), the wafers are placed on a stage (susceptor) that is tilted by a small amount, usually $1°-3°$. Explain why this is done.

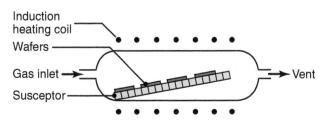

FIGURE P28.21

28.22 The table that follows describes three wafer-manufacturing changes: increasing the wafer diameter, reducing the chip size, and increasing process complexity. Complete the table by filling in "increase," "decrease," or "no change," and indicate the effect that each change would have on the wafer yield and on the overall number of functional chips.

Change	Wafer yield	Number of functional chips
Increase wafer diameter Reduce chip size Increase process complexity		

28.23 The speed of a transistor is directly proportional to the width of its polysilicon gate; thus, a narrower gate results in a faster transistor and a wider gate in a slower transistor. Knowing that the manufacturing process has a certain variation for the gate width (say, $\pm 0.1\ \mu\text{m}$), how would a designer modify the gate size of a critical circuit in order to minimize its variation in speed? Are there any negative effects of this change?

28.24 What is accelerated life testing? Why is it practiced?

28.25 Explain the difference between a die, a chip, and a wafer.

28.26 A common problem in ion implantation is channeling, in which the high-velocity ions travel deep into the material via channels along the crystallographic planes before finally being stopped. How could this effect be avoided? Explain.

28.27 Examine the hole profiles shown in Fig. P28.27 and explain how they might be produced.

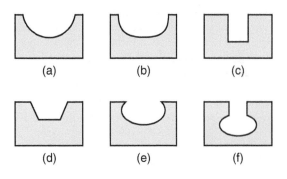

FIGURE P28.27

28.28 Referring to Fig. 28.24, sketch the shape of the holes generated from a circular mask.

QUANTITATIVE PROBLEMS

28.29 A certain wafer manufacturer produces two equal-sized wafers, one containing 500 chips and the other containing 200. After testing, it is observed that 50 chips on each wafer are defective. What are the yields of these two wafers? Can any relationship be drawn between chip size and yield?

28.30 A chlorine-based polysilicon etching process displays a polysilicon-to-resist selectivity of 5:1 and a polysilicon-to-oxide selectivity of 60:1. How much resist and exposed oxide will be consumed in etching 3500 Å of polysilicon? What would the polysilicon-to-oxide selectivity have to be in order to reduce the loss to only 40 Å of exposed oxide?

28.31 During a processing sequence, three silicon-dioxide layers are grown by oxidation to 2500 Å, 4000 Å, and 1500 Å, respectively. How much of the silicon substrate is consumed?

28.32 A certain design rule calls for metal lines to be no less than 2-μm wide. If a 1-μm thick metal layer is to be wet etched, what is the minimum photoresist width allowed (assuming that the wet etching is perfectly isotropic)? What would be the minimum photoresist width if a perfectly anisotropic dry-etching process is used?

28.33 Using Fig. 28.20, obtain mathematical expressions for the etch rate as a function of temperature.

28.34 If a square mask of side length 100 μm is placed on a {100} plane and oriented with a side in the ⟨100⟩ direction, how long will it take to etch a hole 5 μm deep at 80°C using ethylene–diamine/pyrocatechol? Sketch the resulting profile.

28.35 Obtain an expression for the width of the trench bottom as a function of time for the mask shown in Fig. 28.19.

SYNTHESIS, DESIGN, AND PROJECTS

28.36 Describe products that would not exist today without the knowledge and techniques described in this chapter. Explain.

28.37 Inspect various electronic and computer equipment, take them apart as much as you can, and identify components that may have been manufactured by the techniques described in this chapter.

28.38 Describe your understanding of the important features of clean rooms and how they are maintained.

28.39 Make a survey of the necessity for clean rooms in various industries, including the medical, pharmacological, and aerospace industries, and what their requirements are.

28.40 Review the technical literature, and give further details regarding the type and shape of the abrasive wheel used in the wafer-cutting process shown in Step 2 in Fig. 28.2. (See also Chapter 26.)

28.41 List and discuss the technologies that have enabled the manufacture of the products described in this chapter.

28.42 Estimate the time required to etch a spur gear blank from a 75-mm thick slug of silicon.

28.43 Microelectronic devices may be subjected to hostile environments, such as high temperature, humidity, and vibration, as well as physical abuse, such as being dropped onto a hard surface. Describe your thoughts on how you would go about testing these devices for their endurance under these conditions. Are there any industry standards regarding such tests? Explain.

28.44 Review the specific devices, shown in Fig. V.2. Choose any one of these devices, and investigate what they are, what their characteristics are, how they are manufactured, and what their costs are.

Fabrication of Microelectro-mechanical Devices and Systems; Nanoscale Manufacturing

- Many of the processes and materials used for manufacturing microelectronic devices are also used for manufacturing micromechanical devices and microelectromechanical systems; this chapter investigates topics in the production of very small mechanical and electromechanical products. The chapter begins with considerations of micromachining and surface machining of mechanical structures from silicon.

- The LIGA process and its variations are then described, along with micromolding, EFAB, and various other techniques for replicating small-scale mechanical devices.

- Solid free-form fabrication processes are sometimes suitable for the production of MEMS and MEMS devices.

- The chapter ends with a discussion of the emerging area of nanoscale manufacturing.

Typical parts made: Sensors, actuators, accelerometers, optical switches, ink-jet printing mechanisms, micromirrors, micromachines, and microdevices.

Alternative methods: Fine blanking, small scale machining, microforming.

29.1   Introduction

The preceding chapter dealt with the manufacture of integrated circuits and products that operate purely on electrical or electronic principles, called **microelectronic devices**. These semiconductor-based devices often have the common characteristic of extreme miniaturization. A large number of devices exist that are mechanical in nature and are of a similar size as microelectronic devices. A **micromechanical device** is a product that is purely mechanical in nature, and has dimensions between a few mm and atomic length scales, such as some very small gears and hinges.

A **microelectromechanical device** is a product that combines mechanical and electrical or electronic elements at these very small length scales. A **microelectromechanical system** (MEMS) is a microelectromechanical device that also incorporates an integrated electrical system into one product. Common examples of

FIGURE 29.1   SEM view of a micro mechanical system, the L3G4200DH accelerometer, used in popular smartphones. Sensors such as this accelerometer are perhaps the most common application of MEMS. *Source:* Courtesy of STMicroelectronics, Inc.

micromechanical devices are sensors of all types (Fig. 29.1). Microelectromechanical systems include accelerometers in mobile phones, gyroscopes and global positioning systems (GPS), air-bag sensors in automobiles, and digital micromirror devices. Parts made by **nanoscale manufacturing** generally have dimensions that are between 10^{-6} and 10^{-9} m, as described in Section 29.5.

Many of the materials and manufacturing methods and systems described in Chapter 28 also apply to the manufacture of microelectromechanical devices and systems. However, microelectronic devices are semiconductor-based, whereas micro-electromechanical devices and portions of MEMS do not have this restriction; thus, many more materials and processes are suitable for these applications. Regardless, silicon often is used because several highly advanced and reliable manufacturing processes using silicon have been developed for microelectronic applications. This chapter emphasizes the manufacturing processes that are applicable specifically to microelectromechanical devices and systems, but it should be realized that processes and concepts such as lithography, metallization, etching, coating, and packaging described in Chapter 28 still apply.

MEMS and MEMS devices are rapidly advancing, and new processes or variations on existing processes are continually being developed. A significant leap in the numbers of commercial MEMS devices occurred in the past few years as mobile phones and tablet computers, integrated accelerometers and gyroscopes into their products. However, it should be recognized that many of the processes described in this chapter have not yet become widespread, but are of interest to researchers and practitioners in MEMS, and hold great potential for future applications.

29.2 Micromachining of MEMS Devices

The topics described in the preceding chapter dealt with the manufacture of integrated circuits and products that operate purely on electrical or electronic principles. These processes also are suitable for manufacturing devices that incorporate mechanical elements or features. The following four types of devices can be made through the approach described in Fig. 28.2:

1. **Microelectronic devices** are semiconductor-based devices that often have the common characteristics associated with extreme miniaturization, and use electrical principles in their design.
2. **Micromechanical devices** are products that are purely mechanical in nature and have dimensions between atomic length scales and a few mm; very small gears and hinges are examples.
3. **Microelectromechanical devices** are products that combine mechanical and electrical or electronic elements at very small length scales; most sensors are examples of microelectromechanical devices.
4. **Microelectromechanical systems** are microelectromechanical devices that also incorporate an integrated electrical system in one product. Microelectromechanical systems are rare compared with microelectronic, micromechanical, or microelectromechanical devices, typical examples being air-bag sensors and digital micromirror devices.

The production of features from μm to mm in size is called *micromachining*. MEMS devices have been constructed from **polycrystalline silicon** (*polysilicon*) and **single-crystal silicon,** because the technologies for integrated-circuit manufacture, described in Chapter 28, are well developed and exploited for these devices; other new processes also have been developed that are compatible with the existing processing steps. The use of anisotropic etching techniques (Section 28.8.1) allows the fabrication of devices with well-defined walls and high aspect ratios; for this reason, some MEMS devices have been fabricated from single-crystal silicon.

One of the difficulties associated with the use of silicon for MEMS devices is the high adhesion encountered between components at small length scales and the associated rapid wear (Section 33.5). Most commercial devices are designed to avoid friction by, for example, using flexing springs instead of bearings. However, this approach complicates designs and makes some MEMS devices not feasible. Consequently, significant research is being conducted to identify materials and lubricants that provide reasonable life and performance, and that would allow sliding on the microscale without excessive wear.

Silicon carbide, diamond, and metals (such as aluminum, tungsten, and nickel) have been investigated as potential MEMS materials; various lubricants also have been investigated. It is known, for example, that surrounding the MEMS device in a silicone oil practically eliminates adhesive wear, but it also limits the performance of the device. Self-assembling layers of polymers also are being investigated, as well as novel and new materials with self-lubricating characteristics. However, the tribology of MEMS devices remains a main technological barrier to any further expansion of their already widespread use.

29.2.1   Bulk Micromachining

Until the early 1980s, *bulk micromachining* was the most common method of machining at micrometer scales. This process uses orientation-dependent etches on single-crystal silicon (see Fig. 28.15b), an approach that depends on wet etching (Section 28.8) into a surface and stopping on certain crystal faces, doped regions, and etchable films to form a desired structure. As an example of this process, consider the fabrication of the silicon cantilever shown in Fig. 29.2. Using the masking techniques described in Section 28.7, the process changes a rectangular patch of the *n*-type silicon substrate to *p*-type silicon, through boron doping. Etchants such as potassium hydroxide will not be able to remove heavily boron doped silicon; hence, this patch will not be etched.

A mask is then produced—for example, with silicon nitride on silicon. When etched with potassium hydroxide, the undoped silicon will be removed rapidly, while the mask and the doped patch will essentially be unaffected. Etching progresses until the (111) planes are exposed in the *n*-type silicon substrate; they undercut the patch, leaving a suspended cantilever (as shown in Fig. 29.2).

29.2.2   Surface Micromachining

Although bulk micromachining is useful for producing very simple shapes, it is restricted to single-crystal materials because polycrystalline materials will not wet etch at different rates in different directions. Many MEMS applications require the use of other materials or material combinations; hence, alternatives to bulk micromachining are needed. One such method is *surface micromachining*, the basic steps of which are illustrated for silicon devices in Fig. 29.3.

In surface micromachining, a spacer or sacrificial layer is deposited onto a silicon substrate coated with a thin dielectric layer, called an *isolation*, or *buffer*, *layer*. Phosphosilicate glass deposited by chemical-vapor deposition is the most common material for a spacer layer, because it etches very rapidly in hydrofluoric acid, a property that is useful in step 5. Step 2 in Fig. 29.3 shows the spacer layer after the application of masking and etching. At this stage, a structural thin film is deposited onto the spacer layer; the film can be polysilicon, metal, metal alloy, or a dielectric (step 3 in Fig. 29.3).

The structural film is then patterned, usually through dry etching, in order to maintain vertical walls and tight dimensional tolerances. Finally, wet etching of the sacrificial layer leaves a freestanding, three-dimensional structure, as shown in step 5 of Fig. 29.3. The wafer must be annealed to remove the residual stresses in the deposited metal before it is patterned, otherwise the structural film will severely warp once the spacer layer is removed.

Video Solution 29.1 Design and Manufacture of an Accelerometer

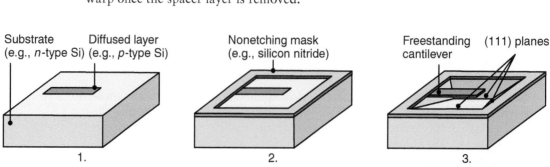

FIGURE 29.2 Schematic illustration of the steps in bulk micromachining. 1. Diffuse dopant in desired pattern. 2. Deposit and pattern-masking film. 3. Orientation-dependent etching (ODE) leaves behind a freestanding structure. *Source:* Courtesy of K.R. Williams.

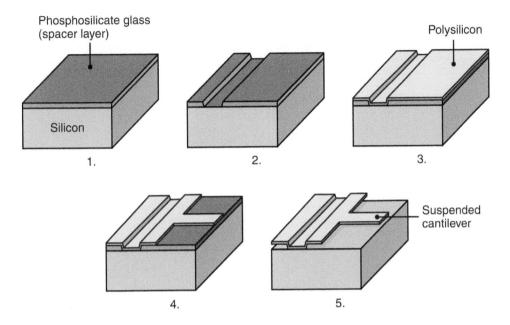

FIGURE 29.3 Schematic illustration of the steps in surface micromachining: 1. Deposition of a phosphosilicate glass (PSG) spacer layer. 2. Lithography and etching of spacer layer. 3. Deposition of polysilicon. 4. Lithography and etching of polysilicon. 5. Selective wet etching of PSG, leaving the silicon substrate and deposited polysilicon unaffected.

Figure 29.4 shows a microlamp that emits a white light when current passes through it; it has been produced through a combination of surface and bulk micromachining. The top patterned layer is a 2.2-μm layer of plasma-etched tungsten, forming a meandering filament and bond pad. The rectangular overhang is dry-etched silicon nitride. The steeply sloped layer is wet-hydrofluoric acid-etched phosphosilicate glass; the substrate is silicon, which is orientation-dependent etched.

The etchant used to remove the spacer layer must be selected carefully, as it must preferentially dissolve the spacer layer while leaving the dielectric, silicon, and structural film as intact as possible. With large features and narrow spacer layers, this task becomes very difficult, and etching can take many hours. To reduce the etching time, additional etched holes can be designed into the microstructures, to increase access of the etchant to the spacer layer.

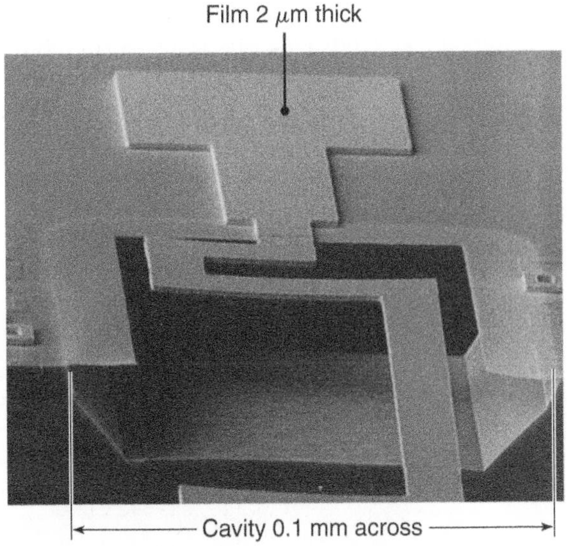

Another difficulty that must be overcome is **stiction** after wet etching, which can be described by considering the situation illustrated in Fig. 29.5. After the spacer layer has been removed, the liquid etchant is dried from the wafer surface. A meniscus forms between the layers, resulting in capillary forces that can deform the film and cause contraction of the substrate as the liquid evaporates. Since adhesive forces are more significant at small length scales, it is possible that the film may *stick* permanently to the surface; thus, the desired three-dimensional features will not be produced.

Film 2 μm thick

Cavity 0.1 mm across

FIGURE 29.4 A microlamp produced from a combination of bulk and surface micromachining processes. *Source:* Courtesy of K.R. Williams.

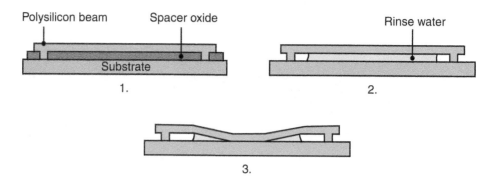

FIGURE 29.5   Stiction after wet etching: 1. Unreleased beam. 2. Released beam before drying. 3. Released beam pulled to the surface by capillary forces during drying. Once contact is made, interfacial adhesive forces prevent the beam from returning to its original shape. *Source:* After B. Bhushan.

EXAMPLE 29.1   Surface Micromachining of a Hinge

Surface micromachining is a widespread technology for the production of MEMS, with applications that include accelerometers, pressure sensors, micropumps, micromotors, actuators, and microscopic locking mechanisms. Often, these devices require very large vertical walls, which cannot be manufactured directly because the high vertical structure is difficult to deposit. This obstacle is overcome by machining large, flat horizontal structures, and then rotating or folding them into an upright position, as shown in Fig. 29.6.

Figure 29.6a shows a micromirror that has been inclined with respect to the surface on which it was manufactured; such systems can be used for reflecting light, that is oblique to a surface, onto detectors or toward other sensors. It is apparent that a device which has such depth, and has the aspect ratio of the deployed mirror, is very difficult to machine directly. Instead, it is easier to surface micromachine the mirror along with a linear actuator, and then to fold the mirror into a deployed position. In order to do so, special hinges (as shown in Fig. 29.6b) are integrated into the design.

Figure 29.7 shows the cross-section of a hinge during its manufacture. The following steps are involved in the production of the hinges:

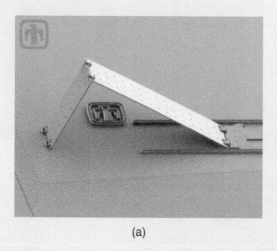

(a)

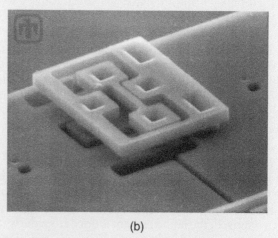

(b)

FIGURE 29.6   (a) SEM image of a deployed micromirror. (b) Detail of the micromirror hinge. *Source:* Courtesy of Sandia National Laboratories.

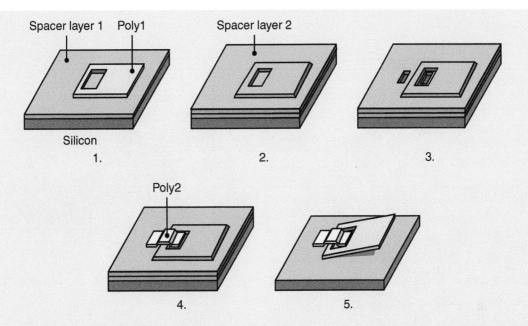

FIGURE 29.7 Schematic illustration of the steps required to manufacture a hinge. 1. Deposition of a phosphosilicate glass (PSG) spacer layer and polysilicon layer (see Fig. 29.3). 2. Deposition of a second spacer layer. 3. Selective etching of the PSG. 4. Deposition of polysilicon to form a staple for the hinge. 5. After selective wet etching of the PSG, the hinge can rotate.

1. A 2-μm-thick layer of phosphosilicate glass is first deposited onto the substrate material.
2. A 2-μm-thick layer of polysilicon (Poly1 in step 1 in Fig. 29.7) is deposited onto the PSG, patterned by photolithography, and dry etched to form the desired structural elements, including the hinge pins.
3. A second layer of sacrificial PSG, with a thickness of 0.5 μm, is deposited (step 2 in Fig. 29.7).
4. The connection locations are etched through both layers of PSG (step 3 in Fig. 29.7).

5. A second layer of polysilicon (Poly2 in step 4 in Fig. 29.7) is deposited, patterned, and etched.
6. The sacrificial layers of PSG are then removed by wet etching.

Hinges such as these have very high friction. Thus, if mirrors (as shown in Fig. 29.6) are manipulated manually and carefully with probe needles, they will remain in position. Often, such mirrors will be combined with linear actuators to precisely control their deployment.

CASE STUDY 29.1   Digital Micromirror Device

An example of a commercial MEMS-based product is the *digital pixel technology* (DPT) device, illustrated in Fig. 29.8. This device uses an array of *digital micromirror devices* (DMD) to project a digital image, as in movie theater projection systems or in nano projectors (Fig. 29.9). The aluminum mirrors can be tilted so that light is directed into or away from the optics that focus light onto a screen. That way, each mirror can represent a pixel of an image's resolution. The mirror allows light or dark pixels to be projected, but levels of gray also can be accommodated. Since the switching time is about 15 μs (which is much faster than the human eye can respond), the mirror will switch between the on and off states in order to reflect the proper dose of light to the optics.

The fabrication steps for producing the DMD device are shown in Fig. 29.10. This sequence is

(continued)

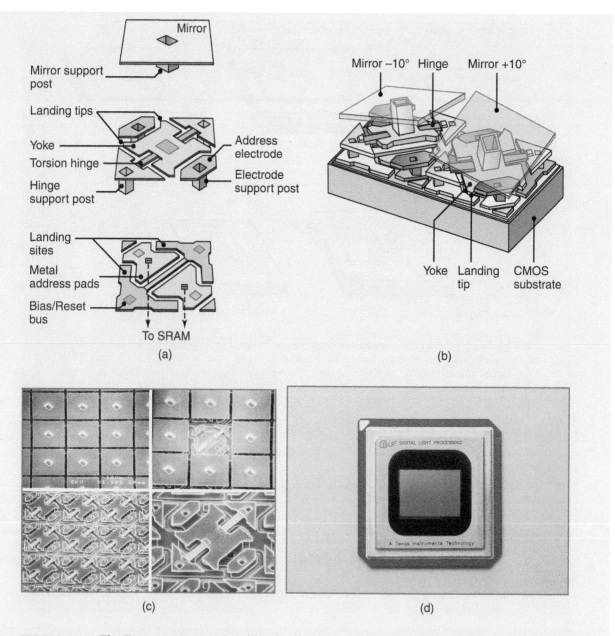

FIGURE 29.8 The Texas Instruments digital pixel technology (DPT) device. (a) Exploded view of a single digital micromirror device (DMD). (b) View of two adjacent DMD pixels. (c) Images of DMD arrays with some mirrors removed for clarity; each mirror measures approximately 17 μm (670 μin.) on a side. (d) A typical DPT device used for digital projection systems, high-definition televisions, and other image display systems. The device shown contains 1,310,720 micromirrors and measures less than 50 mm (2 in.) per side. *Source:* Courtesy of Texas Instruments.

similar to that of other surface micromachining operations, but has the following important differences:

- All micromachining steps take place at temperatures below 400°C, which is sufficiently low to

ensure that no damage occurs to the electronic circuit.

- A thick silicon-dioxide layer is deposited and is chemical–mechanical polished (Section 26.7) to

FIGURE 29.9 A prototype pico projector based on DPT.
Source: Courtesy of Texas Instruments Corp.

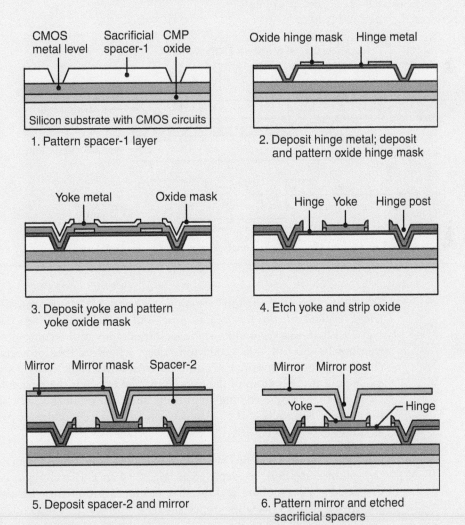

FIGURE 29.10 Manufacturing sequence for the Texas Instruments DMD device.

(*continued*)

provide an adequate foundation for the MEMS device.

- The landing pads and electrodes are produced from aluminum, which is deposited by sputtering.
- High reliability requires low stresses and high strength in the torsional hinge, which is produced from a proprietary aluminum alloy.
- The MEMS portion of the DMD is very delicate, and special care must be taken in separating the dies. When completed, a wafer saw (see Fig. 28.6c) cuts a trench along the edges of the DMD, which allows the individual dice to be broken apart at a later stage.
- A special step deposits a layer that prevents adhesion between the yoke and landing pads.

- The DMD is placed in a hermetically sealed ceramic package (Fig. 29.11) with an optical window.

An array of such mirrors represents a grayscale screen. Using three mirrors (one each for red, green, and blue light) for each pixel results in a color image with millions of discrete colors. Digital pixel technology is widely applied in digital projection systems, high-definition television, and other optical equipment. However, to produce the device shown in Fig. 29.8 requires much more than two-and-one-half-dimensional features, thus full three-dimensional, multipart assemblies have to be manufactured.

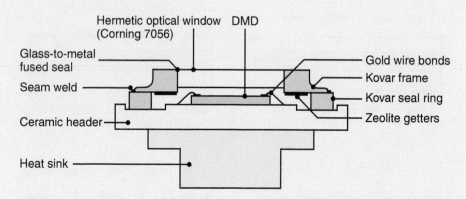

FIGURE 29.11   Ceramic flat-package construction used for the DMD device.

SCREAM.   Another method for making very deep MEMS structures is the SCREAM (*single-crystal silicon reactive etching and metallization*) process, depicted in Fig. 29.12. In this technique, standard lithography and etching processes produce trenches 10–50 μm (400–2000 μin.) deep, which are then protected by a layer of chemically vapor deposited silicon oxide. An anisotropic-etching step removes the oxide only at the bottom of the trench, and the trench is then extended through dry etching. An isotropic etching step (using sulfur hexafluoride, SF_6) laterally etches the exposed sidewalls at the bottom of the trench. This undercut (when it overlaps adjacent undercuts) releases the machined structures.

SIMPLE.   An alternative to SCREAM is SIMPLE (*silicon micromachining by single-step plasma etching*), as depicted in Fig. 29.13. This technique uses a chlorine-gas-based plasma-etching process that machines *p*-doped or lightly doped silicon anisotropically, but heavily *n*-doped silicon isotropically. A suspended MEMS device can thus be produced in one plasma-etching device, as shown in the figure.

Some of the concerns with the SIMPLE process are:

- The oxide mask is machined, although at a slower rate, by the chlorine-gas plasma; therefore, relatively thick oxide masks are required.

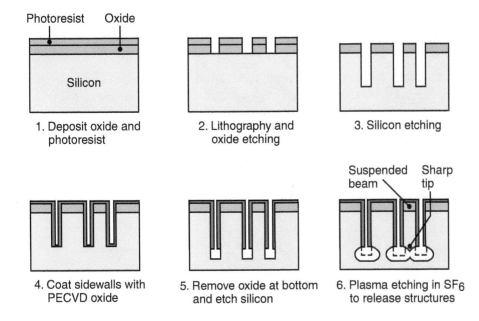

FIGURE 29.12   Steps in the SCREAM process. *Source:* After N. Maluf.

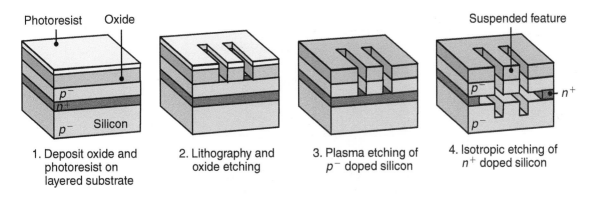

FIGURE 29.13   Schematic illustration of silicon micromachining by the single-step plasma etching (SIMPLE) process.

- The isotropic etch rate is low, typically 50 nm/min; consequently, this is a very slow process.
- The layer beneath the structures will have developed deep trenches, which may affect the motion of free-hanging structures.

Etching Combined with Diffusion Bonding. Tall structures can be produced in crystalline silicon through a combination of *silicon-diffusion bonding and deep reactive-ion etching* (SFB–DRIE), as illustrated in Fig. 29.14. First, a silicon wafer is prepared with an insulating oxide layer, with the deep trench areas defined by a standard lithography procedure. This step is followed by conventional wet or dry etching to form a large cavity. A second layer of silicon is then fusion bonded to the oxide layer; the second silicon layer can be ground and lapped to the desired

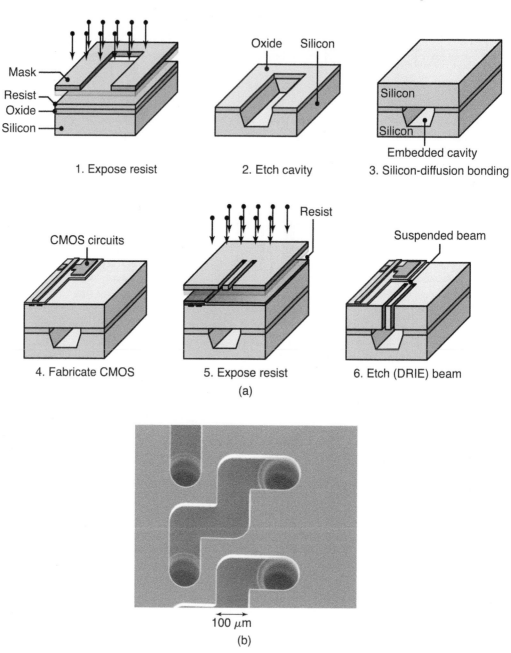

FIGURE 29.14 (a) Schematic illustration of silicon-diffusion bonding combined with deep reactive-ion etching to produce large, suspended cantilevers. (b) A microfluid-flow device manufactured by DRIE etching two separate wafers and then aligning and silicon-fusion bonding them together. Afterward, a Pyrex® layer (not shown) is anodically bonded over the top to provide a window to observe fluid flow. *Source:* (a) After N. Maluf. (b) Courtesy of K.R. Williams.

thickness, if necessary. At this stage, integrated circuitry is manufactured through the steps outlined in Fig. 28.2. A protective resist is applied and exposed, and the desired trenches are then etched by deep reactive-ion etching to the cavity in the first layer of silicon.

EXAMPLE 29.2   Operation and Fabrication Sequence for a Thermal Ink-jet Printer

Thermal ink-jet printers are among the most successful applications of MEMS to date. These printers operate by ejecting nano- or picoliters (10^{-12} l) of ink from a nozzle toward the paper. Ink-jet printers use a variety of designs, but silicon-machining technology is most applicable to high-resolution printers. Note that a resolution of 1200 dpi requires a nozzle spacing of approximately 20 μm.

The mode of operation of an ink-jet printer is shown in Fig. 29.15. When an ink droplet is to be generated and expelled, a tantalum resistor (placed below a nozzle) is heated, which makes a thin film of ink form a bubble within 5 μs, with internal pressures reaching 1.4 MPa (200 psi). The bubble then expands rapidly, and as a result, the fluid is forced rapidly out of the nozzle. Within 24 μs, the tail of the ink-jet droplet separates because of surface tension, the heat source is turned off, and the bubble collapses inside the nozzle. Within 50 μs, sufficient ink has been drawn into the nozzle from a reservoir to form the desired meniscus for the next droplet.

Traditional ink-jet printer heads have been made with electroformed nickel nozzles, produced separately from the integrated circuitry, thus requiring a bonding operation to attach these two components. With increasing printer resolution, it is more difficult to bond the components with a tolerance of less than a few micrometers. For this reason, single-component, or monolithic, fabrication is of interest.

The fabrication sequence for a monolithic ink-jet printer head is shown in Fig. 29.16. A silicon wafer is first prepared and coated with a phosphosilicate-glass (PSG) pattern and a low-stress silicon-nitride coating. The ink reservoir is obtained by isotropically etching the back side of the wafer, followed by PSG removal and enlargement of the reservoir. The required CMOS (complementary metal-oxide semiconductor) controlling circuitry is then produced, and a tantalum heater pad is deposited. The aluminum interconnection between the tantalum pad and the CMOS circuit is formed, and the nozzle is produced through laser ablation. An array of such nozzles can be placed inside an ink-jet printing head, and resolutions of 2400 dpi or higher can be achieved.

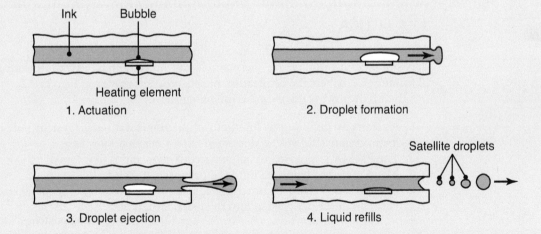

FIGURE 29.15 Sequence of operation of a thermal ink-jet printer. 1. Resistive heating element is turned on, rapidly vaporizing ink and forming a bubble. 2. Within 5 μs, the bubble has expanded and displaced liquid ink from the nozzle. 3. Surface tension breaks the ink stream into a bubble, which is discharged at high velocity; the heating element is turned off at this time, so that the bubble collapses as heat is transferred to the surrounding ink. 4. Within 24 μs an ink droplet (and some undesirable satellite droplets) are ejected, and surface tension of the ink draws more liquid from the reservoir. *Source: After F.-G. Tseng.*

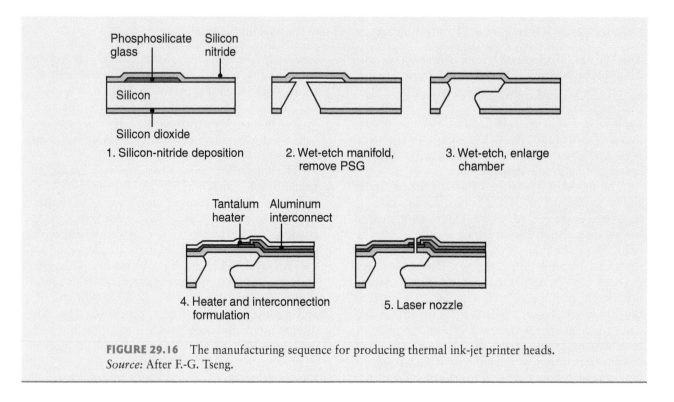

FIGURE 29.16 The manufacturing sequence for producing thermal ink-jet printer heads. *Source:* After F.-G. Tseng.

29.3 Electroforming-based Processes

29.3.1 LIGA

LIGA is a German acronym for the combined processes of X-ray lithography, electro-deposition, and molding (in German, X-ray *li*thographie, *g*alvanoformung, und *a*bformung). A schematic illustration of this process is given in Fig. 29.17.

The LIGA process involves the following steps:

1. A relatively thick (up to hundreds of micrometers) resist layer of polymethyl-methacrylate (PMMA) is deposited onto a primary substrate.
2. The PMMA is exposed to columnated X-rays and is developed.
3. Metal is electrodeposited onto the primary substrate.
4. The PMMA is removed or stripped, resulting in a freestanding metal structure.
5. Plastic is injection-molded into the metal structure.

Depending on the application, the final product from a LIGA process may consist of one of the following:

- A freestanding metal structure, resulting from the electrodeposition process
- A plastic injection-molded structure
- An investment-cast metal part, using the injection-molded structure as a blank
- A slip-cast ceramic part, produced using the injection-molded parts as the molds

The substrate used in LIGA is a conductor or a conductor-coated insulator. Examples of primary substrate materials include austenitic steel plate, silicon wafers with a titanium layer, and copper plated with gold, titanium, or nickel. Metal-plated

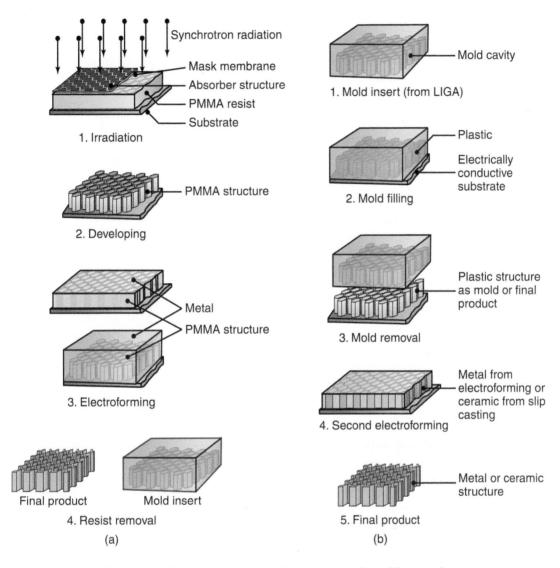

FIGURE 29.17 The LIGA (lithography, electrodeposition, and molding) technique. (a) Primary production of a metal final product or mold insert. (b) Use of the primary part for secondary operations or replication. *Source:* Based on data from IMM Institut für Mikrotechnik, Mainz, Germany.

ceramic and glass also have been used. The surface may be roughened by grit blasting to encourage good adhesion of the resist material.

Resist materials must have high X-ray sensitivity, dry- and wet-etching resistance when unexposed, and thermal stability. The most common resist material is polymethylmethacrylate, which has a very high molecular weight (more than 10^6 per mole; Section 7.2). The X-rays break the chemical bonds, leading to the production of free radicals and to a significantly reduced molecular weight in the exposed region. Organic solvents then preferentially dissolve the exposed PMMA in a wet-etching process. After development, the remaining three-dimensional structure is rinsed and dried, or it is spun and blasted with dry nitrogen.

Two newer forms of LIGA are **UV-LIGA** and **Silicon-LIGA**. In *UV-LIGA*, special photoresists are used, instead of PMMA, and they are exposed through ultraviolet lithography (Section 28.7). *Silicon-LIGA* uses deep reactive-ion-etched silicon (Section 28.8.2) as a preform for further operations. These processes, like the traditional X-ray-based LIGA, are used to replicate MEMS devices, but, unlike LIGA, they do not require the expensive columnated X-ray source for developing their patterns.

The electrodeposition of metal usually involves the electroplating of nickel (Section 34.9). The nickel is deposited onto exposed areas of the substrate; it fills the PMMA structure and can even coat the resist (Fig. 29.17a). Nickel is the preferred material because of the relative ease in electroplating with well-controlled deposition rates. Electroless plating of nickel also is possible, and the nickel can be deposited directly onto electrically insulating substrates. However, because nickel displays high wear rates in MEMS, significant research is being directed toward the use of other materials or coatings.

After the metal structure has been deposited, precision grinding removes either the substrate material or a layer of the deposited nickel. The process is referred to as *planarization* (Section 28.10). The need for planarization is obvious when it is recognized that three-dimensional MEMS devices require micrometer tolerances, on layers many hundreds of micrometers thick. Planarization is difficult to achieve, because conventional lapping leads to preferential removal of the soft PMMA and smearing of the metal. Planarization usually is accomplished with a diamond-lapping procedure (Section 26.7) referred to as *nanogrinding*. Here, a diamond-slurry-loaded, soft metal plate is used to remove material in order to maintain flatness within 1 μm (40 μin.) over a 75-mm (3-in.) diameter substrate.

If cross-linked, the PMMA resist is then exposed to synchrotron X-ray radiation, and removed by exposure to an oxygen plasma or through solvent extraction. The result is a metal structure, which may be processed further. Examples of freestanding metal structures produced through the electrodeposition of nickel are shown in Fig. 29.18.

The processing steps used to make freestanding metal structures are time consuming and expensive. The main advantage of LIGA is that these structures serve as molds for the rapid replication of submicron features through molding operations. The processes that can be used for producing micromolds are shown and compared in Table 29.1, where it can be seen that LIGA provides some clear advantages. Reaction injection molding, injection molding, and compression molding (described in Chapter 19) also have been used to make the micromolds.

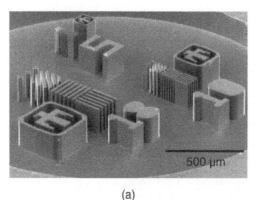

(a)

(b)

FIGURE 29.18   (a) Electroformed 200-μm-tall nickel structures and (b) detail of nickel lines and spaces. *Source*: Courtesy of T. Christenson, Sandia National Laboratories.

TABLE 29.1

Comparison of Micromold Manufacturing Techniques

Characteristic	Production technique		
	LIGA	Laser machining	EDM
Aspect ratio	10–50	10	up to 100
Surface roughness	<50 nm	100 nm	$0.3-1\mu m$
Accuracy	<1 μm	$1-3\mu m$	$1-5\mu m$
Mask required	Yes	No	No
Maximum height	$1-500\mu m$	$200-500\mu m$	μm to mm

Source: After L. Weber, W. Ehrfeld, H. Freimuth, M. Lacher, M. Lehr, P. Pech, and K.R. Williams.

EXAMPLE 29.3   Production of Rare-earth Magnets

A number of scaling issues in electromagnetic devices indicate that there is an advantage in using rare-earth magnets from the samarium cobalt (SmCo) and neodymium iron boron (NdFeB) families, which are available in powder form. These alloys are of interest because they can produce magnets that are one order of magnitude more powerful than conventional magnets (Table 29.2). Such materials can be used when effective miniature electromagnetic transducers are to be produced.

The processing steps involved in manufacturing these magnets are shown in Fig. 29.19. The PMMA mold is produced by exposure to X-ray radiation and solvent extraction. The rare-earth powders are mixed with a binder of epoxy and applied to

TABLE 29.2

Comparison of Properties of Permanent-magnet Materials

Material	Energy product (Gauss–Oersted $\times 10^{-6}$)
Carbon steel	0.20
36% Cobalt steel	0.65
Alnico I	1.4
Vicalloy I	1.0
Platinum–cobalt	6.5
$Nd_2Fe_{14}B$, fully dense	40
$Nd_2Fe_{14}B$, bonded	9

Source: Courtesy of T. Christenson, Sandia National Laboratories.

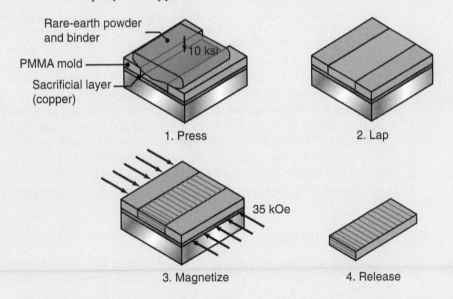

FIGURE 29.19 Fabrication process used to produce rare-earth magnets for microsensors. *Source:* Courtesy of T. Christenson, Sandia National Laboratories.

the mold through a combination of calendering (see Fig. 19.22) and pressing. After curing in a press at a pressure around 70 MPa (10 ksi), the substrate is planarized. The substrate is then subjected to a magnetizing field, of at least 35 kilo-oersteds (kOe),

in the desired orientation. Once the material has been magnetized, the PMMA substrate is dissolved, leaving behind the rare-earth magnets, as shown in Fig. 29.20.

(a)

(b)

FIGURE 29.20   SEM images of $Nd_2Fe_{14}B$ permanent magnets. The powder particle size ranges from 1 to 5 μm, and the binder is a methylene-chloride-resistant epoxy. Mild distortion is present in the image due to magnetic perturbation of the imaging electrons. Maximum energy products of 9 MGOe have been obtained with this process. *Source:* Courtesy of T. Christenson, Sandia National Laboratories.

29.3.2   Multilayer X-Ray Lithography

The LIGA technique is very powerful for producing MEMS devices with large aspect ratios and reproducible shapes. It is, however, often useful to obtain a multilayer stepped structure that cannot be made directly through LIGA. For nonoverhanging part geometries, direct plating can be applied. In this technique, a layer of electrodeposited metal with surrounding PMMA is produced, as previously described. A second layer of PMMA resist is then bonded to this structure and X-ray exposed, using an aligned X-ray mask.

Often, it is useful to have overhanging geometries within complex MEMS devices. A batch diffusion-bonding and release procedure has been developed for this purpose, as schematically illustrated in Fig. 29.21a. This process involves the preparation of two PMMA patterned and electroformed layers, with the PMMA subsequently removed. The wafers are then aligned, face to face, with guide pins that press-fit into complementary structures on the opposite surface. Finally, the substrates are joined in a hot press, and a sacrificial layer on one substrate is etched away, leaving behind one layer bonded to the other. An example of such a structure is shown in Fig. 29.21b.

29.3.3   HEXSIL

This process, illustrated in Fig. 29.21, combines *hex*agonal honeycomb structures, *sil*icon micromachining, and thin-film deposition to produce high-aspect-ratio, freestanding structures. HEXSIL can produce tall structures, with a shape definition that rivals that of structures produced by LIGA.

In HEXSIL, a deep trench is first produced in single-crystal silicon by dry etching, followed by shallow wet etching to make the trench walls smoother. The depth

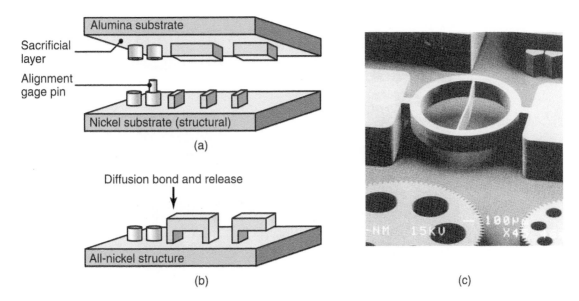

FIGURE 29.21 Multilevel MEMS fabrication through wafer-scale diffusion bonding. (a) Two wafers are aligned and assembled. (b) Resultant structure after diffusion bonding and removal of alumina substrate. (c) A suspended ring structure for measurement of tensile strain, formed by two-layer wafer-scale diffusion bonding. *Source:* (c) Courtesy of T. Christenson, Sandia National Laboratories.

of the trench matches the desired structure height, and is limited practically to around 100 μm. An oxide layer is then grown or deposited onto the silicon, followed by an undoped-polycrystalline silicon layer, which results in good mold filling and shape definition. A doped-silicon layer then follows, providing a resistive portion of the microdevice. Electroplated or electroless nickel plating is then deposited. Figure 29.22 shows various trench widths to demonstrate the different structures that can be produced in HEXSIL.

Microscale tweezers produced through the HEXSIL process are shown in Fig. 29.23. A thermally activated bar activates the tweezers, which have been used for microassembly and microsurgery applications.

29.3.4 MolTun

This process (short for *mol*ding of *tun*gsten) was developed in order to utilize the higher mass of tungsten in micromechanical devices and systems. In MolTun, a sacrificial oxide is patterned through lithography and then etched, but instead of electroforming, a layer of tungsten is deposited through chemical vapor deposition. Excess tungsten is then removed by chemical–mechanical polishing, which also ensures good control over layer thickness. Multiple layers of tungsten can be deposited to develop intricate geometries (Fig. 29.24).

MolTun has been used for micro mass-analysis systems and a large number of micro-scale latching relays, which take advantage of tungsten's higher strength compared to other typical MEMS materials. The depth of MolTun structures can be significantly larger than those produced through silicon micromachining; the mass analysis array in Fig. 29.24, for example, has a total thickness of around 25 μm.

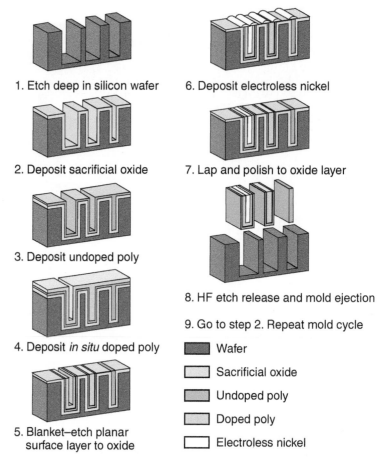

1. Etch deep in silicon wafer

2. Deposit sacrificial oxide

3. Deposit undoped poly

4. Deposit *in situ* doped poly

5. Blanket–etch planar
   surface layer to oxide

6. Deposit electroless nickel

7. Lap and polish to oxide layer

8. HF etch release and mold ejection

9. Go to step 2. Repeat mold cycle

■ Wafer

□ Sacrificial oxide

■ Undoped poly

□ Doped poly

□ Electroless nickel

FIGURE 29.22   Illustration of the hexagonal honeycomb structure, silicon micromachining, and thin-film deposition (the HEXSIL process).

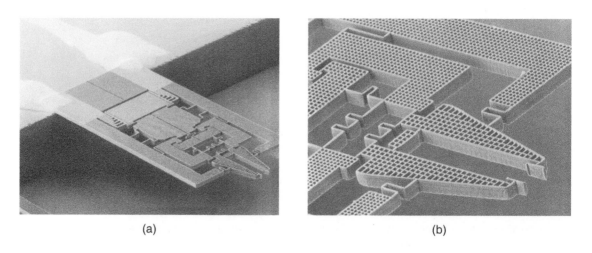

(a)   (b)

FIGURE 29.23   (a) SEM image of microscale tweezers used in microassembly and microsurgery applications. (b) Detailed view of gripper. *Source:* Courtesy of MEMS Precision Instruments.

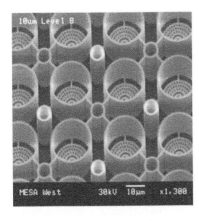
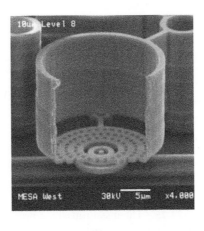

(a)   (b)

FIGURE 29.24   An array of micro-mass analysis systems consisting of cylindrical ion traps, constructed of 14 layers of molded tungsten, including 8 layers for the ring electrode. *Source:* Courtesy of Sandia National Laboratories.

29.4   Solid Free-form Fabrication of Devices

Solid free-form fabrication is another term for *rapid prototyping*, as described in Chapter 20. This method is unique in that complex three-dimensional structures are produced through additive manufacturing, as opposed to material removal. Many of the advances in rapid prototyping also are applicable to MEMS manufacture for processes with sufficiently high resolution. *Stereolithography* (Section 20.3.2) involves curing a liquid thermosetting polymer, using a photoinitiator and a highly focused light source. Conventional stereolithography uses layers between 75 and 500 μm in thickness, with a laser dot focused to a diameter of 0.05–0.25 mm.

Microstereolithography.   *Microstereolithography* uses the same basic approach as stereolithography; however, there are some important differences between the two processes, including the following:

- The laser is more highly focused, to a diameter as small as 1 μm, as compared with 10 to over 100 μm in stereolithography.
- Layer thicknesses are around 10 μm, which is an order of magnitude smaller than in stereolithography.
- The photopolymers used must have much lower viscosities, to ensure the formation of uniform layers.
- Support structures are not required in microstereolithography, since the smaller structures can be supported by the fluid.
- Parts with significant metal or ceramic content can be produced, by suspending nanoparticles in the liquid photopolymer.

The microstereolithography technique has several cost advantages, but the MEMS devices made by this method are difficult to integrate with the controlling circuitry.

Electrochemical Fabrication.   The solid free-form fabrication of MEMS devices using instant masking is known as *electrochemical fabrication* (EFAB). Instant masking is one EFAB technique for producing MEMS devices (Fig. 29.25). A mask of

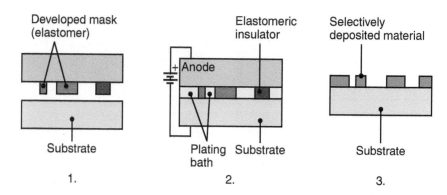

FIGURE 29.25 The instant-masking process: 1. Bare substrate. 2. During deposition, with the substrate and instant mask in contact. 3. The resulting pattern deposited. *Source:* Courtesy of Microfabrica.

elastomeric material is first produced through conventional photolithography techniques, described in Section 28.7. The mask is pressed against the substrate in an electrodeposition bath, so that the elastomer conforms to the substrate and excludes the plating solution in contact areas. Electrodeposition takes place in areas that are not masked, eventually producing a mirror image of the mask. By using a sacrificial filler, made of a second material, instant masking technology can produce complex three-dimensional shapes complete with overhangs, arches, and other features.

CASE STUDY 29.2 Accelerometer for Automotive Air Bags

Accelerometers based on lateral resonators represent the largest commercial application of surface micromachining today, and are used widely as sensors for automotive air-bag deployment systems. The sensor portion of such an accelerometer is shown in Fig. 29.26. A central mass is suspended over the substrate, but anchored through four slender beams, which act as springs to center the mass under static-equilibrium conditions. An acceleration of the car causes the mass to deflect, reducing or increasing the clearance between the fins on the mass and the stationary fingers on the substrate.

By measuring the electrical capacitance between the mass and fins, the deflection of the mass (and therefore the acceleration or deceleration of the system) can be directly measured. Figure 29.26 shows an arrangement for the measurement of acceleration in one direction, but commercial sensors employ several masses so that accelerations can be measured in multiple directions simultaneously.

Figure 29.27 shows a 50-g surface micromachined accelerometer (ADXL-50), with onboard signal conditioning and self-diagnostic electronics. The polysilicon sensing element (visible in the center of the die) occupies only 5% of the total die area, and the whole chip measures 500 μm × 625 μm (20 μin. × 25 μin.). The mass is approximately 0.3 μg, and the sensor has a measurement accuracy of 5% over the ±50-g range.

Fabrication of the accelerometer proved to be a challenge, since it required a *complementary metal-oxide-semiconductor* (CMOS) fabrication sequence to be integrated closely with a surface micromachining approach. Analog Devices, Inc., was able to modify a CMOS production technique to directly incorporate surface micromachining. In the sensor design, the *n*+ doped silicon underpasses connect the sensor area to the electronic circuitry, replacing the usual heat-sensitive aluminum connect lines. Most of the sensor processing is inserted into the

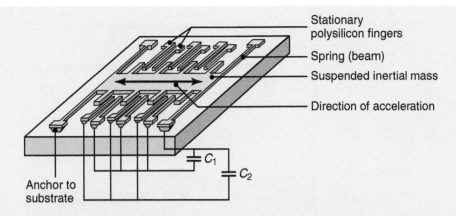

FIGURE 29.26   Schematic illustration of a microacceleration sensor. *Source:* After N. Maluf.

FIGURE 29.27   Photograph of Analog Devices' ADXL-50 accelerometer with a surface micromachined capacitive sensor (center), on-chip excitation, and self-test and signal-conditioning circuitry. The entire chip measures 0.500 mm × 0.625 mm. *Source:* After R.A. Core.

fabrication process right after a borosilicate-glass planarization process.

After the planarization, a designated sensor region, or *moat*, is cleared in the center of the die (step 1 in Fig. 29.28). A thin oxide is then deposited to passivate the $n+$ underpass connects, followed by a thin, low-pressure chemical-vapor deposited (LPCVD) nitride to act as an etch stop for the final

(continued)

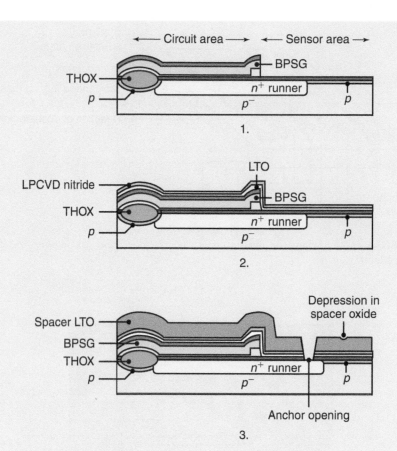

FIGURE 29.28 Preparation of IC chip for polysilicon. 1. Sensor area post-borophosphosilicate glass (BPSG) planarization and moat mask. 2. Blanket deposition of thin oxide and thin nitride layer. 3. Bumps and anchors made in low-temperature oxide (LTO) spacer layer. *Source:* From T.A. Core, et al., *Solid State Technol.*, v. 36, pp. 39–47, 1993. Printed by permission of PennWell Corporation.

polysilicon released etching (step 2 in Fig. 29.28). The spacer or sacrificial oxide used is a 1.6-μm (64-μin.) densified low-temperature oxide (LTO), deposited over the whole die (step 3 in Fig. 29.28).

In a first etching, small depressions (that will form bumps or dimples on the underside of the polysilicon sensor) are created in the LTO layer. These bumps will limit adhesive forces and sticking in the event that the sensor comes in contact with the substrate. A subsequent etching step cuts anchors into the spacer layer, to provide regions of electrical and mechanical contact (step 3 in Fig. 29.28). The 2-μm (80-μin.) thick sensor of polysilicon layer

is deposited, implanted, annealed, and patterned (step 1 in Fig. 29.29).

Metallization follows, starting with the removal of the sacrificial spacer oxide from the circuit area, along with the LPCVD nitride and LTO layer. A low-temperature oxide is deposited on the polysilicon-sensor part, and contact openings appear in the IC part of the die, where platinum is deposited to form platinum silicide (step 2 in Fig. 29.29). The trimmable thin-film material (TiW barrier metal) and Al–Cu interconnect metal are sputtered on and patterned in the IC area.

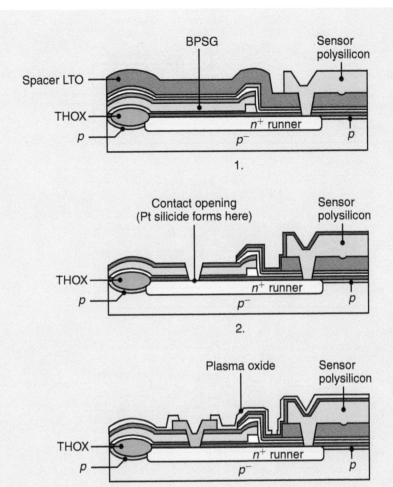

FIGURE 29.29 Polysilicon deposition and IC metallization. 1. Cross-sectional view after polysilicon deposition, implanting, annealing, and patterning. 2. Sensor area after removal of dielectrics from circuit area, contact mask, and platinum silicide. 3. Metallization scheme and plasma-oxide passivation and patterning. *Source:* From T.A. Core, et al., *Solid State Technol.*, v. 36, pp. 39–47, 1993. Printed by permission of PennWell Corporation.

The circuit area is then passivated in two separate deposition steps. First, plasma oxide is deposited and patterned (step 3 in Fig. 29.29), followed by a plasma nitride (step 1 in Fig. 29.30), to form a seal with the previously deposited LCVD nitride. The nitride acts as a hydrofluoric-acid barrier in the subsequent etch release in surface micromachining. The plasma oxide left on the sensor acts as an etch stop for the removal of the plasma nitride (step 1 in Fig. 29.30). The sensor area is then prepared for the final release etch. The dielectrics are removed from the sensor, and the final protective resist mask is applied. The photoresist protects the circuit area from the long-term buffered oxide etch (step 2 in Fig. 29.30). The final device cross-section is shown in step 3 in Fig. 29.30.

Source: Adapted from M. Madou, *Fundamentals of Microfabrication*, 2nd ed., CRC Press, 2002.

(*continued*)

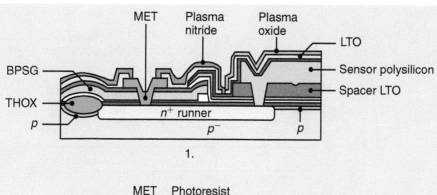

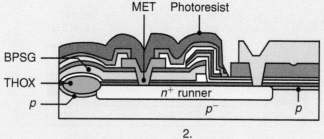

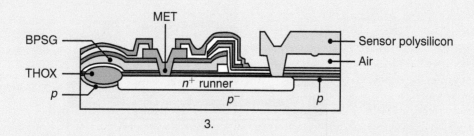

FIGURE 29.30 Prerelease preparation, and release. 1. Post-plasma nitride passivation and patterning. 2. Photoresist protection of the IC. 3. Freestanding, released polysilicon beam. *Source:* From T.A. Core, et al., *Solid State Technol.*, v. 36, pp. 39–47, 1993. Printed by permission of PennWell Corporation.

29.5 Nanoscale Manufacturing

In *nanomanufacturing*, parts are produced at nanometer length scales; the term usually refers to manufacturing below the micrometer scale, or between 10^{-9} and 10^{-6} m in length. Many of the features in integrated circuits are at this length scale, but very little else has significant relevance to manufacturing. Molecularly engineered medicines and other forms of biomanufacturing are the only commercial applications at present, except for some limited uses of carbon nanotubes (see Section 8.8). However, it has been recognized that many physical and biological processes act at this length scale; consequently, the approach holds much promise for future innovations.

Nanoscale manufacturing techniques are outlined in Table 29.3. Nanomanufacturing takes two basic approaches: top down and bottom up. **Top-down** approaches use *large building blocks* (such as a silicon wafer; see Fig. 28.2) and various manufacturing processes (such as lithography, and wet and plasma etching) to construct ever smaller features and products (microprocessors, sensors, and probes). At the other

TABLE 29.3

Comparison of Nanoscale Manufacturing Techniques

Characteristic	Top down			Both top down and bottom up	Bottom up	
Nanopatterning technique	Photo-lithography	Electron beam lithography	Nanoimprint lithography	Dip pen nanolithography	Microcontact printing	Scanning tunneling microscopy
Material flexibility	No	No	No	Yes	Yes	Limited
Resolution	~35 nm	~15 nm	~10 nm	14 nm	~100 nm	Atomic
Registration accuracy	High	High	High	Extremely high	Low	Extremely high
Speed	Very fast	Moderate	Fast	Slower, but scalable	Fast	Very slow
Cycle time	Weeks	Days	Days–week	Hours	Days–weeks	Days
Cost						
Purchase	>$10 M	>$1 M	>$500K	<$250 K	~$200 K	>$250 K
Operation	High	High	Moderate	Low	Moderate	Low

Source: Courtesy of NanoInk, Inc. Reprinted by permission.

extreme, **bottom-up** approaches use *small building blocks* (such as atoms, molecules, or clusters of atoms and molecules) to build up a structure. In theory, bottom-up approaches are similar to the additive manufacturing technologies described in Section 20.3. When placed in the context of nanomanufacturing, however, bottom-up approaches suggest the manipulation and construction of products are on an atomic or molecular scale.

Bottom-up approaches are widely used in nature (e.g., building cells is a fundamentally bottom-up approach), whereas conventional manufacturing has, for the most part, consisted of top-down approaches. In fact, there are presently no nanomanufactured products (excluding medicines and drugs "manufactured" by bacteria) that have demonstrated commercial viability.

Bottom-up approaches in various research applications can use atomic-force microscopy (AFM) for the manipulation of materials on the nanoscale. Figure 29.29 is an illustration of an atomic-force microscope. A probe (Fig. 29.31b) is mounted into the microscope, and a laser is reflected from a mirror on the back side of the probe so that it reflects onto a set of photosensors. Any vertical or torsional deflection of the cantilever is registered as a change in voltage on the photosensors. Atomic-force microscopes can have true atomic resolution of $<1 \times 10^{-10}$ m.

Atomic-force microscopes are widely used to measure the surface profile of very smooth surfaces (Section 33.3). Several approaches have been developed to allow nanoscale manufacturing processes to be performed on these microscopes. Some top-down approaches are:

- **Photolithography, electron-beam lithography, and nanoimprint (soft) litho-graphy.** These techniques are capable of top-down manufacture of structures, with resolution under 100 nm, as discussed in Section 28.7.
- **Nanolithography.** The probes used in atomic-force microscopy vary greatly in size, materials, and capabilities. The diamond-tipped stainless-steel cantilever shown in Fig. 29.31b has a tip radius of around 10 nm. By contacting and plowing across a surface, it can produce grooves up to a few μm thick. The spacing between lines depends on the groove depth needed.

- **Dip pen nanolithography.** This approach (Fig. 29.32) is used in an atomic-force microscope to transfer chemicals onto substrates. The process can produce lines as narrow as 10 nm. Dip pen nanolithography can be used with many parallel pens (Fig. 29.32b), typically made of silicon nitride and containing as many as 55,000 pens in a 1-cm^2 area. In a top-down approach, dip pen nanolithography is used to produce a mask suitable for lithography.

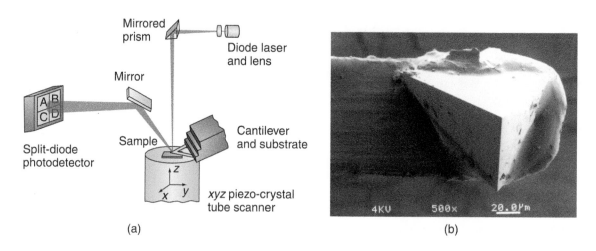

(a)   (b)

FIGURE 29.31   (a) Schematic illustration of an atomic-force microscope. A probe is mounted on a cylinder containing piezoelectric material; this arrangement allows translation of the probe in three dimensions. A laser, reflected from a mirror on the back of the probe onto a set of photosensors, allows measurement of the probe's location and monitoring of interactions with a sample surface. (b) Scanning-electron microscope image of a diamond-tipped stainless-steel cantilever suitable for nanolithography.

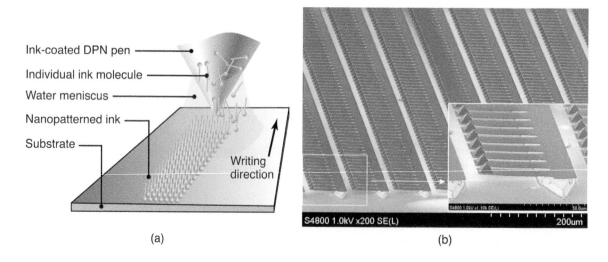

(a)   (b)

FIGURE 29.32   (a) Schematic illustration of dip pen nanolithography. (b) An array of pens used to produce identical patterns on surfaces. Commercial pen arrays can contain up to 55,000 pens; only a fraction of the available pens are shown. The inset highlights individual cantilevers, showing the 7.5-μm-high tips. *Source:* Courtesy of NanoInk, Inc.

Bottom-up approaches include:

- Dip pen nanolithography also can be a bottom-up approach, wherein the ink contains the material used to build the structure.
- *Microcontact printing* uses soft-lithography approaches, to deposit material on surfaces from which nanoscale structures can be produced.
- *Scanning tunneling microscopy* can be used to manipulate an atom on an atomically smooth surface (usually cleaved mica or quartz).

SUMMARY

- MEMS is relatively new and developing rapidly. Although most successful commercial MEMS applications are in the optics, printing, and sensor industries, the possibilities for new device concepts and circuit designs appear to be endless.
- MEMS devices are manufactured through techniques and with materials that, for the most part, have been pioneered in the microelectronics industry. Bulk and surface micromachining are processes that are well developed for single-crystal silicon.
- Specialized processes for MEMS include variations of machining, such as DRIE, SIMPLE, and SCREAM. These processes produce freestanding mechanical structures in silicon.
- Polymer MEMS can be manufactured through LIGA or microstereolithography. LIGA combines X-ray lithography and electroforming to produce three-dimensional structures. Related processes include multilayer X-ray lithography and HEXSIL.
- Nanoscale manufacturing is a relatively new area that has significant potential. The processes are typically bottom up, whereas conventional manufacturing is top down. Some lithography processes extend to the nanoscale, as does dip pen lithography. Materials such as carbon nanotubes have great potential for nanoscale devices.

KEY TERMS

Atomic-force microscope	HEXSIL	Multilayer X-ray	SIMPLE
Bulk micromachining	LIGA	lithography	Stiction
Diffusion bonding	MEMS	Planarization	Surface
Dip pen nanolithography	Micromachining	Sacrificial layer	micromachining
EFAB	Microstereolithography	SCREAM	UV-LIGA
Electrochemical fabrication	MolTun	Silicon-LIGA	

BIBLIOGRAPHY

Adams, T.M., and Layton, R.W., **Introductory MEMS: Fabrication and Applications**, Springer, 2009.

Allen, J.J., **Microelectromechanical System Design**, CRC Press, 2006.

Elwenspoek, M., and Jansen, H., **Silicon Micromachining**, 2nd ed., Cambridge University Press, 2004.

Elwenspoek, M., and Wiegerink, R., **Mechanical Microsensors**, Springer, 2001.

Fraden, J., **Handbook of Modern Sensors: Physics, Designs, and Applications**, 4th ed., Springer, 2010.

Gad-el-Hak, M. (ed.), **The MEMS Handbook**, 2nd ed. (3 vols.), CRC Press, 2006.

Ghodssi, R., and Lin, P. (eds.), **MEMS Materials and Processes Handbook**, Springer, 2011.

Harper, C.A. (ed.), **Electronic Packaging and Interconnection Handbook**, 4th ed., McGraw-Hill, 2004.

Hsu, T.-R., **MEMS & Microsystems: Design, Manufacture, and Nanoscale Engineering,** 2nd ed., Wiley, 2008.

Jha, A.R., **MEMS and Nanotechnology-based Sensors and Devices for Communications, Medical and Aerospace Applications,** CRC Press, 2008.

Kempe, V., **Inertial MEMS: Principles and Practice,** Cambridge, 2011.

Korvink, J., and Oliver, P. (eds.), **MEMS: A Practical Guide to Design, Analysis, and Applications,** William Andrew, 2005.

Leondes, C.T., **MEMS/NEMS Handbook** (5 vols.), Springer, 2007.

Liu, C., **Foundations of MEMS,** 2nd ed., Prentice Hall, 2011.

Madou, M.J., **Manufacturing Techniques for Microfabrication and Nanotechnology,** CRC Press, 2009.

Madou, M.J., **Applications of Microfabrication and Nanotechnology,** CRC Press, 2009.

Maluf, N., **An Introduction to Microelectromechanical Systems Engineering,** 2nd ed., Artech House, 2004.

Nishi, Y., and Doering, R. (eds.), **Handbook of Semiconductor Manufacturing Technologies,** 2nd ed., CRC Press, 2007.

Rockett, A., **The Materials Science of Semiconductors,** Springer, 2008.

Schaeffer, R., **Fundamentals of Laser Micromachining,** Taylor & Francis, 2012.

Sze, S.M., **Semiconductor Devices: Physics and Technology,** 3rd ed., Wiley, 2012.

Varadan, V.K., Jiang, X., and Varadan, V., **Microstereolithography and Other Fabrication Techniques for 3D MEMS,** Wiley, 2001.

REVIEW QUESTIONS

29.1 Define MEMS, SIMPLE, SCREAM, and HEXSIL.
29.2 Give three examples of common microelectromechanical systems.
29.3 Why is silicon often used with MEMS devices?
29.4 Describe bulk and surface micromachining.
29.5 What is the purpose of a spacer layer in surface micromachining?
29.6 What is the main limitation to successful application of MEMS?
29.7 What are common applications for MEMS and MEMS devices?
29.8 What is LIGA? What are its advantages?
29.9 What is a sacrificial layer?
29.10 Explain the differences between stereolithography and microstereolithography.
29.11 What is MolTun? What are its main advantages?
29.12 What is HEXSIL?
29.13 What do SIMPLE and SCREAM stand for?

QUALITATIVE PROBLEMS

29.14 Describe the difference between isotropic etching and anisotropic etching.
29.15 Lithography produces projected shapes, so true three-dimensional shapes are more difficult to produce. What lithography processes are best able to produce three-dimensional shapes, such as lenses? Explain.
29.16 Which process or processes in this chapter allow the fabrication of products from polymers?
29.17 What is the difference between chemically reactive ion etching and dry-plasma etching?
29.18 The MEMS devices discussed in this chapter are applicable to macroscale machine elements, such as spur gears, hinges, and beams. Which of the following machine elements can or cannot be applied to MEMS, and why? (a) Ball bearings, (b) bevel gears, (c) worm gears, (d) cams, (e) helical springs, (f) rivets, and (g) bolts.
29.19 Explain how you would produce a spur gear if its thickness was one-tenth of its diameter and its diameter was (a) 1 mm, (b) 10 mm, and (c) 100 mm.
29.20 List the advantages and disadvantages of surface micromachining compared with bulk micromachining.
29.21 What are the main limitations to the LIGA process? Explain.
29.22 Other than HEXSIL, what process can be used to make the microtweezers shown in Fig. 29.23? Explain.
29.23 Is there an advantage to using the MolTun process for other materials? Explain.

QUANTITATIVE PROBLEMS

29.24 The atomic-force microscope probe shown in Fig. 29.31 has a stainless steel cantilever that is 450 μm × 40 μm × 2 μm. Using equations from solid mechanics, estimate the stiffness of the cantilever, and the force required to deflect the end of the cantilever by 1 μm.

29.25 Estimate the natural frequency of the cantilever in Problem 29.24. *Hint:* See Problem 3.21.

29.26 Tapping-mode probes for the atomic-force microscope are produced from etched silicon and have typical dimensions of 125 μm in length, 30 μm in width, and 3 μm in thickness. Estimate the stiffness and natural frequency of such probes.

29.27 Using data from Chapter 28, derive the time needed to etch the hinge shown in Fig. 29.7 as a function of the hinge thickness.

29.28 It is desired to produce a 500 μm by 500 μm diaphragm, 25 μm thick, in a silicon wafer 250 μm thick.

Given that you will use a wet etching technique, with KOH in water with an etch rate of 1 μm/min, calculate the etching time and the dimensions of the mask opening that you would use on a (100) silicon wafer.

29.29 If the Reynolds number for water flow through a pipe is 2000, calculate the water velocity if the pipe diameter is (a) 10 mm; (b) 100 μm. Do you expect flow in MEMS devices to be turbulent or laminar? Explain.

SYNTHESIS, DESIGN, AND PROJECTS

29.30 List similarities and differences between IC technologies described in Chapter 28 and miniaturization technologies presented in this chapter.

29.31 Figure I.7b in the General Introduction shows a mirror that is suspended on a torsional beam, and can be inclined through electrostatic attraction by applying a voltage on either side of the micromirror at the bottom of the trench. Make a flowchart of the manufacturing operations required to produce this device.

29.32 Referring to Fig. 29.5, design an experiment to find the critical dimensions of an overhanging cantilever that will not stick to the substrate.

29.33 Design an accelerometer by using (a) the SCREAM process and (b) the HEXSIL process.

29.34 Design a micromachine or device that allows the direct measurement of the mechanical properties of a thin film.

29.35 Conduct a literature search and determine the smallest diameter hole that can be produced by (a) drilling, (b) punching, (c) water-jet cutting, (d) laser machining, (e) chemical etching, and (f) EDM.

29.36 Perform a literature search and write a one-page summary of applications in biomes.

Joining Processes and Equipment

Some products, such as paper clips, nails, steel balls for bearings, screws and bolts, are made of only one component. Almost all products, however, are assembled from components that have been manufactured as individual parts. Even relatively simple products consist of at least two different components, joined by various means. For example, (a) the eraser of an ordinary pencil is attached with a brass sleeve; (b) some kitchen knives have wooden or plastic handles that are attached to the metal blade with fasteners; and (c) cooking pots and pans have metal, plastic, or wooden handles and knobs that are attached to the pot by various methods.

On a much larger scale, observe power tools, washing machines, motorcycles, ships, and airplanes and how their numerous components are assembled and joined, so that they not only can function reliably but also are economical to produce. As shown in Table I.1 in the General Introduction, a rotary lawn mower has about 300 parts, a typical automobile consists of 15,000 components (Fig. VI.1), while a Boeing 747–400 aircraft has more than 6 million parts. In contrast, a Boeing 787 Dreamliner has fewer parts because its composite fuselage eliminates a large number of fasteners.

Joining is an all-inclusive term covering processes such as welding, brazing, soldering, adhesive bonding, and mechanical fastening. These processes are an essential

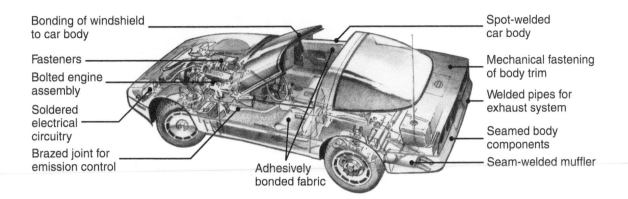

FIGURE VI.1 Various parts in a typical automobile that are assembled by the processes described in Part VI.

and important aspect of manufacturing and **assembly** operations, for one or more of the following reasons:

- Even a relatively simple product may be impossible to manufacture as a *single piece*. Consider, for example, the tubular construction shown in Fig. VI.2a. Assume that each of the arms of this product is 5 m (15 ft) long, the tubes are 100 mm (4 in.) in diameter, and their wall thickness is 1 mm (0.04 in.). After reviewing all of the manufacturing processes described in the preceding chapters, one would conclude that manufacturing this product in one piece would be impossible or uneconomical.
- A product such as a cooking pot with a handle is *easier* and *more economical* to manufacture as individual components, which are then assembled.
- Products such as appliances, hair dryers, and automobile engines must be designed so as to be able to be *taken apart* for maintenance or replacement of their worn or broken parts.
- *Different properties* may be desirable for functional purposes of the product. For example, surfaces subjected to friction, wear, corrosion, or environmental attack generally require characteristics that differ significantly from those of the components' bulk. Examples are (a) masonry drills with carbide cutting tips brazed to the shank of a drill (Fig. VI.2b), (b) automotive brake shoes, and (c) grinding wheels bonded to a metal backing (Section 26.2).
- *Transporting* the product in individual components and assembling them later may be easier and less costly than transporting the completed item. Note, for example, that metal or wood shelving, backyard grills, and large machinery are assembled after the components or subassemblies have been transported to the appropriate site.

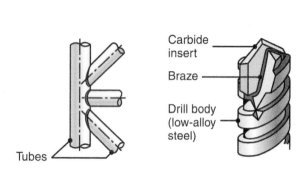

Carbide insert

Braze

Drill body (low-alloy steel)

Tubes

(a)   (b)

(c)

FIGURE VI.2 Examples of parts utilizing joining processes. (a) A tubular part fabricated by joining individual components; this product cannot be manufactured in one piece by any of the methods described in the previous chapters if it consists of thin-walled, large-diameter, tubular-shaped long arms. (b) A drill bit with a carbide cutting insert brazed to a steel shank—an example of a part in which two materials need to be joined for performance reasons. (c) Spot welding of automobile bodies.

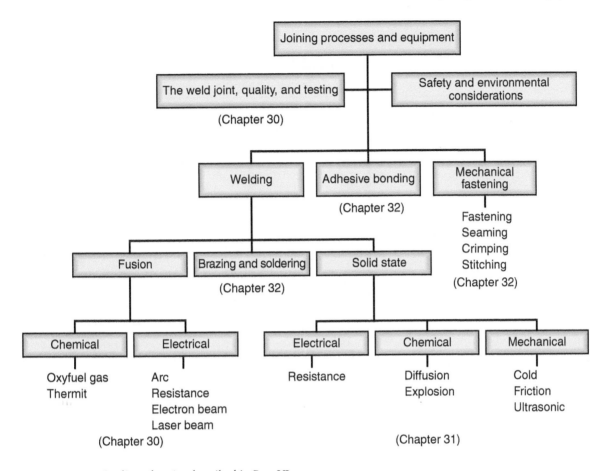

FIGURE VI.3   Outline of topics described in Part VI.

Although there are different ways of categorizing the wide variety of available joining processes, according to the American Welding Society (AWS) they fall into the following three major categories (Figs. VI.3 and I.7f):

- **Welding**
- **Adhesive bonding**
- **Mechanical fastening**

Table VI.1 lists the relative characteristics of various joining processes. Welding processes, in turn, are generally classified into three basic categories: Fusion Welding, Solid-state Welding, Brazing and Soldering.

As will be seen, some types of welding processes can be classified into both the fusion and the solid-state categories.

Fusion welding is defined as the melting together and coalescing of materials by means of heat, usually supplied by chemical or electrical means. Filler metals may or may not be used in fusion welding. This welding operation is composed of *consumable-* and *nonconsumable-electrode arc welding* and *high-energy-beam welding* processes. The welded joint undergoes important metallurgical and physical changes, which, in turn, have a major influence on the properties and performance of the welded component or structure. The terminology for some simple welded joints are illustrated in Fig. VI.4.

TABLE VI.1

Comparison of Various Joining Methods

Method	Strength	Design	Small parts	Large parts	Tolerances	Reliability	Ease of manufacture	Ease of inspection	Cost
Arc welding	1	2	3	1	3	1	2	2	2
Resistance welding	1	2	1	1	3	3	3	3	1
Brazing	1	1	1	1	3	1	3	2	3
Bolts and nuts	1	2	3	1	2	1	1	1	3
Rivets	1	2	3	1	1	1	3	1	2
Seaming and crimping	2	2	1	3	3	1	3	1	1
Adhesive bonding	3	1	1	2	3	2	3	3	2

Note: 1 = Very good; 2 = Good; 3 = Poor; For cost, 1 = Lowest.

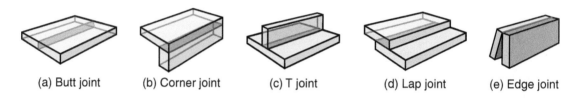

(a) Butt joint   (b) Corner joint   (c) T joint   (d) Lap joint   (e) Edge joint

FIGURE VI.4 Examples of joints that can be made through the various joining processes described in Chapters 30 through 32.

In **solid-state welding,** joining takes place without fusion; consequently, there is no liquid (molten) phase in the joint. The basic processes in this category are *diffusion bonding* and *cold, ultrasonic, friction, resistance, and explosion welding.* **Brazing** uses filler metals and involves lower temperatures than in welding. **Soldering** uses similar filler metals (*solders*) and involves even lower temperatures.

Adhesive bonding has unique applications requiring strength, sealing, thermal and electrical insulating, vibration damping, and resistance to corrosion between dissimilar metals. **Mechanical fastening** involves traditional methods of using various fasteners, such as bolts, nuts, and rivets. The **joining of plastics** can be accomplished by adhesive bonding, fusion by various external or internal heat sources, and mechanical fastening.

CHAPTER 30

Fusion-welding Processes

● ●

- This chapter describes fusion-welding processes, in which two pieces are joined together by the application of heat, which melts and fuses the interface; the operation is sometimes assisted by a filler metal.

- All fusion-welding processes are described in this chapter, beginning with oxyfuel–gas welding in which acetylene and oxygen provide the energy required for welding.

- Various arc-welding processes are then described, in which electrical energy and consumable or nonconsumable electrodes are used to produce the weld; specific processes examined include shielded metal-arc welding, flux-cored arc welding, gas tungsten-arc welding, submerged arc welding, and gas metal-arc welding.

- Welding with high-energy beams is then discussed, in which electron beams or lasers provide highly focused heat sources.

- The chapter concludes with a discussion of the nature of the weld joint, and includes weld quality, inspection, and testing procedures, along with a discussion of weld design practices and process selection.

30.1 Introduction

● ●

The welding processes described in this chapter involve partial melting and fusion between two members to be joined. Here, **fusion welding** is defined as *melting together and coalescing* materials by means of heat. *Filler metals*, which are metals added to the weld area during welding, also may be used. Welds made without the use of filler metals are known as *autogenous welds*.

This chapter covers the basic principles of each welding process; the equipment used; the relative advantages, limitations, and capabilities of the process; and the economic considerations affecting process selection (Table 30.1). The chapter continues with a description of the weld-zone features and the variety of discontinuities and defects that can exist in welded joints. The weldability of various ferrous and nonferrous metals and alloys are then reviewed. The chapter concludes with a discussion of design guidelines for welding, with several examples of good weld-design practices, and the economics of welding.

30.2 Oxyfuel–gas Welding

● ●

Oxyfuel–gas welding (OFW) is a general term to describe any welding process that uses a **fuel gas** combined with *oxygen* to produce a flame, which is the source of the heat required to melt the metals at the joint. The most common gas-welding

TABLE 30.1

General Characteristics of Fusion-welding Processes

Joining process	Operation	Advantage	Skill level required	Welding position	Current type	Distortion*	Typical cost of equipment ($)
Shielded metal arc	Manual	Portable and flexible	High	All	AC, DC	1–2	Low (1500+)
Submerged arc	Automatic	High deposition	Low to medium	Flat and horizontal	AC, DC	1–2	Medium (5000+)
Gas metal arc	Semiautomatic or automatic	Most metals	Low to high	All	DC	2–3	Medium (5000+)
Gas tungsten arc	Manual or automatic	Most metals	Low to high	All	AC, DC	2–3	Medium (2000+)
Flux-cored arc	Semiautomatic or automatic	High deposition	Low to high	All	DC	1–3	Medium (2000+)
Oxyfuel	Manual	Portable and flexible	High	All	—	2–4	Low (500+)
Electron beam, laser beam	Semiautomatic or automatic	Most metals	Medium to high	All	—	3–5	High (100,000– 1 million)
Thermit	Manual	Steels	Low	Flat and horizontal	—	2–4	Low (500+)

*1 = Highest; 5 = Lowest.

process uses *acetylene*; the process is known as *oxyacetylene–gas welding* (OAW) and is typically used for structural metal fabrication and repair work.

Developed in the early 1900s, OAW utilizes the heat generated by the combustion of acetylene gas (C_2H_2) in a mixture with oxygen. The heat is generated in accordance with a pair of chemical reactions. The primary combustion process, which occurs in the inner core of the flame (Fig. 30.1), involves the following reaction:

$$C_2H_2 + O_2 \rightarrow 2CO + H_2 + \text{Heat}. \qquad (30.1)$$

This reaction dissociates the acetylene into carbon monoxide and hydrogen and produces about one-third of the total heat generated in the flame. The secondary combustion process is

$$2CO + H_2 + 1.5O_2 \rightarrow 2CO_2 + H_2O + \text{Heat}. \qquad (30.2)$$

This reaction consists of the further burning of both the hydrogen and the carbon monoxide, and produces about two-thirds of the total heat. The temperatures developed in the flame can reach 3300°C (6000°F). Note from Eq. (30.2) that the reaction also produces water vapor.

Types of Flames. The proportion of acetylene and oxygen in the gas mixture is an important factor in oxyfuel–gas welding. At a ratio of 1:1 (i.e., when there is no excess oxygen), the flame is considered to be *neutral* (Fig. 30.1a). With a higher oxygen supply, the flame can be harmful (especially for steels), because it oxidizes the metal; for this reason, a flame with excess oxygen is known as an **oxidizing flame** (Fig. 30.1b). Only in the welding of copper and copper-based alloys is an oxidizing flame desirable, because in those situations, a thin protective layer of *slag* (compounds of oxides) forms over the molten metal. If the oxygen is insufficient for full combustion, the flame is known as a **reducing**, or **carburizing, flame** (Fig. 30.1c).

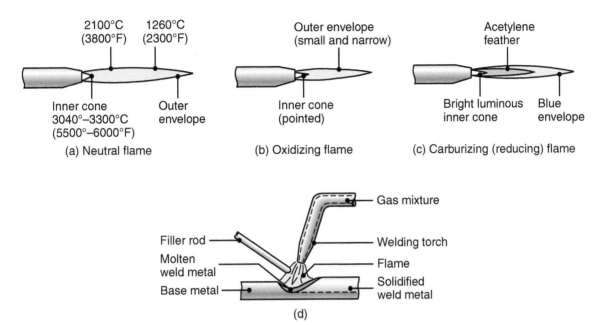

FIGURE 30.1 Three basic types of oxyacetylene flames used in oxyfuel–gas welding and cutting operations: (a) neutral flame; (b) oxidizing flame; and (c) carburizing, or reducing, flame. The gas mixture in (a) is basically equal volumes of oxygen and acetylene. (d) The principle of the oxyfuel–gas welding process.

The temperature of a reducing flame is lower, hence such a flame is suitable for applications requiring low heat, as in brazing and soldering (Chapter 32) and flame-hardening (Table 4.1) operations.

Other fuel gases, such as hydrogen and methylacetylene propadiene, also can be used in oxyfuel–gas welding. However, the temperatures developed by these gases are lower than those produced by acetylene, hence they are used for welding metals with low melting points, such as lead, and parts that are thin and small.

Filler Metals. *Filler metals* are used to supply additional metal to the weld zone during welding, and are available as **filler rods** or **wire** (Fig. 30.1d) and may be bare or coated with **flux**. The purpose of the flux is to retard oxidation of the surfaces of the parts being welded, by generating a gaseous shield around the weld zone. The flux also helps to dissolve and remove oxides and other substances from the weld zone, thus making the joint stronger. The *slag* developed (compounds of oxides, fluxes, and electrode-coating materials) protects the molten puddle of metal against oxidation as the weld cools.

Welding Practice and Equipment. Oxyfuel–gas welding can be used with most ferrous and nonferrous metals for almost any workpiece thickness, but the relatively low heat input limits the process to thicknesses of less than 6 mm (0.25 in.). Small joints made by this process may consist of a single-weld bead; deep-V groove joints are made in multiple passes. Cleaning the surface of each weld bead prior to depositing a second layer over it is important for joint strength and in avoiding defects (see Section 30.9). Wire brushes (hand or power) may be used for this purpose.

The equipment for oxyfuel–gas welding basically consists of a **welding torch**, connected by hoses to high-pressure gas cylinders and equipped with pressure gages and regulators (Fig. 30.2). The use of safety equipment, such as proper goggles

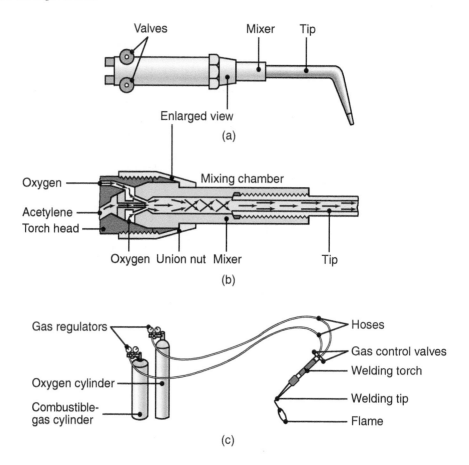

FIGURE 30.2 (a) General view of, and (b) cross-section of, a torch used in oxyacetylene welding. The acetylene valve is opened first; the gas is lit with a spark lighter or a pilot light. Then the oxygen valve is opened and the flame adjusted. (c) Basic equipment used in oxyfuel–gas welding. To ensure correct connections, all threads on acetylene fittings are left handed, whereas those for oxygen are right handed. Oxygen regulators usually are painted green and acetylene regulators red.

with shaded lenses, face shields, gloves, and protective clothing, is essential. Proper connection of the hoses to the cylinders also is an important safety issue; oxygen and acetylene cylinders have different threads, so that the hoses cannot be connected to the wrong cylinders. The low equipment cost is an attractive feature of this process. Although it can be mechanized, the operation is essentially manual, and therefore slow; however, the advantages of being portable, versatile, and economical for simple and low-quantity work.

Pressure-gas Welding.   In this method, the welding of two components starts with heating the interface by means of a torch, using typically an oxyacetylene–gas mixture (Fig. 30.3a). After the interface begins to melt, the torch is withdrawn. A force is then applied to press the two components together and is maintained until the interface solidifies. Note in Fig. 30.3b the formation of a flash, due to the upsetting of the joined ends of the two components.

Thermit Welding.   Also known as *thermite* or *exothermic welding*, and developed in 1895, *thermit welding* involves mixing a metal powder with a metal oxide, and

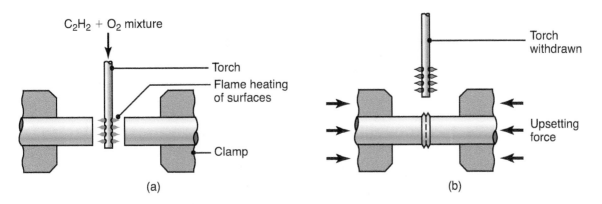

C₂H₂ + O₂ mixture

Torch

Flame heating
of surfaces

Clamp

(a)

Torch
withdrawn

Upsetting
force

(b)

FIGURE 30.3 Schematic illustration of the pressure-gas welding process: (a) before and (b) after; note the formation of a flash at the joint; later the flash can be trimmed off.

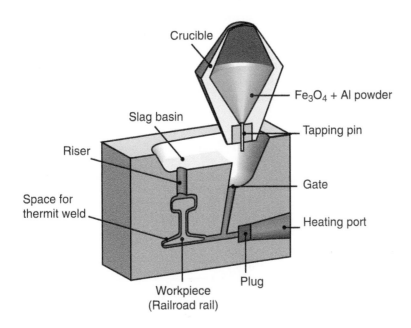

Crucible

Fe_3O_4 + Al powder

Slag basin

Tapping pin

Riser

Gate

Space for
thermit weld

Heating port

Plug

Workpiece
(Railroad rail)

FIGURE 30.4   Schematic illustration of thermit welding.

using a high-temperature ignition source to cause an oxidation–reduction reaction (Fig. 30.4). A common arrangement in this process is to use iron oxide (rust) powder in combination with aluminum powder; upon ignition by a magnesium fuse, the resulting chemical reaction forms aluminum oxide (Al_2O_3) and iron.

Temperatures can reach 2500°C (4500°F), melting the iron which subsequently flows into a pouring basin and then into a mold placed around the parts to be welded. The aluminum oxide floats to the slag basin because of its lower density. The features of a thermit welding mold are very similar to a casting mold (see Fig. 11.3). Note from Fig. 30.4 that a heating port is present, a feature that allows insertion of an oxyacetylene torch to preheat the workpieces and prevent weld cracks (see Section 30.9.1).

Several combinations of powder and oxide can be used in thermit welding, but aluminum powder combined with iron oxide is the most common, because of the widespread use of thermit welding for joining railroad rails. Some copper and magnesium oxides are often added to improve flammability. Other applications of thermit welding include the welding of large-diameter copper conductors using copper oxide and field repair of large equipment, such as locomotive axle frames.

30.3 Arc-welding Processes: Nonconsumable Electrode

In *arc welding*, developed in the mid-1800s, the heat required is obtained from electrical energy. The process involves either a *nonconsumable* or a *consumable electrode*. An AC or a DC power supply produces an arc between the tip of the electrode and the workpiece to be welded. The arc generates temperatures of about 30,000°C (54,000°F), much higher than those developed in oxyfuel–gas welding.

In *nonconsumable-electrode* welding processes, the electrode is typically a **tungsten electrode** (Fig. 30.5). Because of the high temperatures involved, an externally supplied shielding gas is necessary in order to prevent oxidation of the weld zone. Typically, *direct current* is used, and, as described below, its **polarity** (the direction of current flow) is important. The selection of current levels depends on such factors as the type of electrode, the metals to be welded, and the depth and width of the weld zone.

In **straight polarity**, also known as *direct-current electrode negative* (DCEN), the workpiece is positive (anode), and the electrode is negative (cathode). DCEN

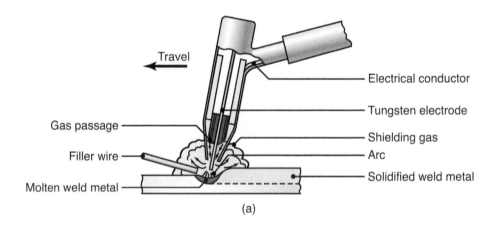

(a)

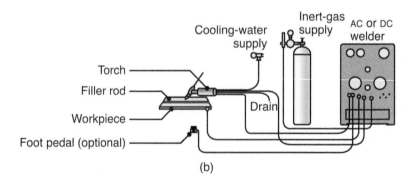

(b)

FIGURE 30.5   (a) The gas tungsten-arc welding process, formerly known as TIG (for tungsten inert gas) welding. (b) Equipment for gas tungsten-arc welding operations.

generally produces welds that are narrow and deep (Fig. 30.6a). In **reverse polarity**, also known as *direct-current electrode positive* (DCEP), the workpiece is negative and the electrode is positive. Weld penetration is less, and the weld zone is shallower and wider (Fig. 30.6b); consequently, DCEP is preferred for sheet metals and for joints with very wide gaps. In the **AC current** method, the arc pulsates rapidly. This method is suitable for welding thick sections and for using large-diameter electrodes at maximum currents (Fig. 30.6c).

Heat Transfer in Arc Welding. The heat input in arc welding is given by the equation

$$\frac{H}{l} = e\frac{VI}{v}, \tag{30.3}$$

where H is the heat input (J or BTU), l is the weld length, V is the voltage applied, I is the current (amperes), and v is the welding speed. The term e is the efficiency of the process, which varies from around 75% for shielded metal-arc welding to 90% for gas metal-arc and submerged-arc welding. The efficiency is an indication that not all of the available energy is beneficially used to melt the material: the heat is conducted through the workpiece, some is dissipated by radiation, and still more is lost by convection to the surrounding environment.

The heat input given by Eq. (30.3) melts a certain volume of material (usually the electrode or filler metal), and can also be expressed as

$$H = uV_m = uAl, \tag{30.4}$$

where u is the specific energy required for melting, V_m is the volume of metal melted, and A is the cross-section of the weld. Some typical values of u are given in Table 30.2. Equations (30.3) and (30.4) result in an expression of the welding speed as

$$v = e\frac{VI}{uA}. \tag{30.5}$$

Although these equations have been developed for arc welding, similar forms also can be obtained for other fusion-welding operations, while taking into account differences in weld geometry and process efficiency.

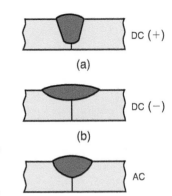

FIGURE 30.6 The effect of polarity and current type on weld beads: (a) DC current with straight polarity; (b) DC current with reverse polarity; and (c) AC current.

TABLE 30.2

Approximate Specific Energies Required to Melt a Unit Volume of Commonly Welded Metals

Material	Specific energy, u	
	J/mm^3	BTU/in^3
Aluminum and its alloys	2.9	41
Cast irons	7.8	112
Copper	6.1	87
Bronze (90Cu–10Sn)	4.2	59
Magnesium	2.9	42
Nickel	9.8	142
Steels	9.1–10.3	128–146
Stainless steels	9.3–9.6	133–137
Titanium	14.3	204

Note: 1 BTU = 1055; J = 778 ft-lb.

EXAMPLE 30.1   Welding Speed for Different Materials

Given: Consider a case in which a welding operation is being performed with $V = 20$ volts, $I = 200$ A, and the cross-sectional area of the weld bead is 30 mm^2.

Find: Estimate the welding speed if the workpiece and electrode are made of (a) aluminum, (b) carbon steel, and (c) titanium. Assume an efficiency of 75%.

Solution: For aluminum, from Table 30.2, the specific energy required is $u = 2.9$ J/mm^3. Therefore,

from Eq. (30.5),

$$v = e\frac{VI}{uA} = (0.75)\frac{(20)(200)}{(2.9)(30)} = 34.5 \text{ mm/s}.$$

Similarly, for carbon steel, u is estimated as 9.7 J/mm^3 (average of extreme values in the table), and thus $v = 10.3$ mm/s. For titanium, $u = 14.3$ J/mm^3, thus $v = 7.0$ mm/s.

Gas Tungsten-arc Welding.   In *gas tungsten-arc welding* (GTAW), formerly known as *TIG* (for "tungsten inert gas") *welding*, the filler metal is supplied from a **filler wire** (Fig. 30.5a). Because the tungsten electrode is not consumed in this operation, a constant and stable arc gap is maintained at a constant current level. The filler metals are similar to the metals to be welded, and flux is not used. The shielding gas is usually argon or helium, or a mixture of the two gases. Welding with GTAW may be done without filler metals, such as in welding close-fit joints.

Depending on the metals to be joined, the power supply is either DC at 200 A or AC at 500 A (Fig. 30.5b). In general, AC is preferred for aluminum and magnesium, because the cleaning action of AC removes oxides and improves weld quality. Thorium or zirconium may be used in the tungsten electrodes to improve their electron emission characteristics. The power supply ranges from 8 to 20 kW. Contamination of the tungsten electrode by the molten metal can be a significant problem, particularly in critical applications, because it can cause discontinuities in the weld; thus, contact of the electrode with the molten-metal pool should be avoided.

The GTAW process is used for a wide variety of applications and metals, particularly aluminum, magnesium, titanium, and the refractory metals; it is especially suitable for thin metals. The cost of the inert gas makes this process more expensive than SMAW, but it provides welds of very high quality and good surface finish. GTAW is used in a variety of critical applications, with a wide range of workpiece thicknesses and shapes, and the equipment is portable.

Plasma-arc Welding.   In *plasma-arc welding* (PAW), developed in the 1960s, a concentrated plasma arc is produced and directed toward the weld area. The arc is stable and reaches temperatures as high as 33,000°C (60,000°F). A **plasma** is an ionized hot gas composed of nearly equal numbers of electrons and ions. The plasma is initiated between the tungsten electrode and the orifice by a low-current pilot arc. What makes plasma-arc welding unlike other processes is that the plasma arc is concentrated, because it is forced through a relatively small orifice. Operating currents usually are below 100 A, but they can be higher for special applications. When a filler metal is used, it is fed into the arc, as is done in GTAW. Arc and weld-zone shielding is supplied by means of an outer-shielding ring and the use of gases, such as argon, helium, or mixtures.

There are two methods of plasma-arc welding:

- In the **transferred-arc** method (Fig. 30.7a), the workpiece being welded is part of the electrical circuit. The arc transfers from the electrode to the workpiece, hence the term *transferred*.
- In the **nontransferred** method (Fig. 30.7b), the arc occurs between the electrode and the nozzle, and the heat is carried to the workpiece by the plasma gas. This thermal-transfer mechanism is similar to that for an oxyfuel flame (see Section 30.2).

Compared with other arc-welding processes, plasma-arc welding has better arc stability, less thermal distortion, and higher energy concentration, thus permitting deeper and narrower welds. In addition, higher welding speeds, from 120 to 1000 mm/min (5–40 in./min), can be achieved. A variety of metals can be welded, with part thicknesses generally less than 6 mm (0.25 in.).

The high heat concentration can completely penetrate through the joint (known as the **keyhole technique**), with thicknesses as much as 20 mm (0.75 in.) for some titanium and aluminum alloys. In this technique, the force of the plasma arc displaces the molten metal and produces a hole at the leading edge of the weld pool. Plasma-arc

Video Solution 30.1 Speed in Welding

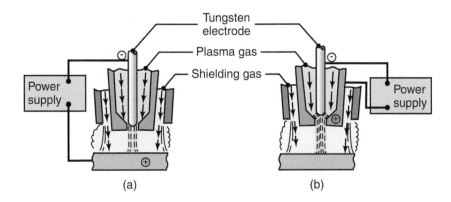

FIGURE 30.7 Two types of plasma-arc welding processes: (a) transferred and (b) non-transferred; deep and narrow welds can be made by these processes at high welding speeds.

welding (rather than the GTAW process) is often used for butt and lap joints, because of its higher energy concentration, better arc stability, and higher welding speeds. Proper training and skill are essential to use this equipment; safety considerations include protection against glare, spatter, and noise from the plasma arc.

Atomic-hydrogen Welding. In *atomic-hydrogen welding* (AHW), an arc is generated between two tungsten electrodes in a shielding atmosphere of hydrogen gas. The arc is maintained independently of the parts being welded. The hydrogen gas normally is diatomic (H_2) but where the temperatures are over 6000°C (11,000°F) near the arc, the hydrogen breaks down into its atomic form, thus simultaneously absorbing a large amount of heat from the arc. When the hydrogen strikes the relatively cold surface of the workpieces to be joined, it recombines into its diatomic form and rapidly releases the stored heat, reaching temperatures up to 4000°C. Thus, it is one of the few processes that can be used for welding tungsten. The energy in AHW can be easily varied by changing the distance between the arc stream and the workpiece surface.

30.4 Arc-welding Processes: Consumable Electrode

There are several consumable-electrode arc-welding processes, as described below.

30.4.1 Shielded Metal-arc Welding

Shielded metal-arc welding (SMAW) is one of the oldest, simplest, and most versatile joining processes; consequently, about 50% of all industrial and maintenance welding is performed by this process. The electric arc is generated by touching the tip of a *coated electrode* against the workpiece, and withdrawing it quickly to a distance sufficient to maintain the arc (Fig. 30.8a). The electrodes are in the shapes of thin, long round rods (hence the process also is referred to as *stick welding*) that are held manually.

The heat generated melts a portion of the electrode tip, its coating, and the base metal in the immediate arc area. The molten metal consists of a mixture of the base metal (the workpiece), the electrode metal, and substances from the coating on the electrode; this mixture forms the weld when it solidifies. The electrode coating

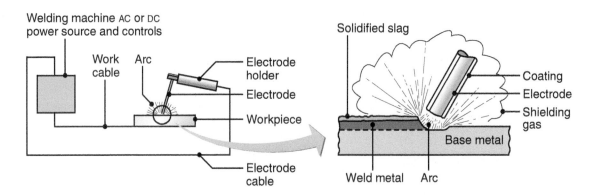

FIGURE 30.8 Schematic illustration of the shielded metal-arc welding process; about 50% of all large-scale industrial-welding operations use this process.

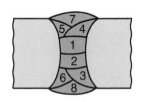

FIGURE 30.9 A deep weld, showing the buildup sequence of eight individual weld beads.

deoxidizes the weld area and provides a shielding gas, to protect it from oxygen in the environment.

A bare section at the end of the electrode is clamped to one terminal of the power source, while the other terminal is connected to the workpiece being welded (Fig. 30.8b). The current, which may be DC or AC, usually ranges from 50 to 300 A. For sheet-metal welding, DC is preferred because of the steady arc it produces. Power requirements generally are less than 10 kW.

The equipment consists of a power supply, cables, and an electrode holder. The SMAW process is commonly used in general construction, shipbuilding, pipelines, and maintenance work. It is especially useful for work in remote areas where a portable fuel-powered generator can be used as the power supply. This process is best suited for workpiece thicknesses of 3 to 19 mm (0.12–0.75 in.), although this range can easily be extended by skilled operators, using *multiple-pass* techniques (Fig. 30.9).

The multiple-pass approach requires that the slag be removed after each weld bead. Unless removed completely, the solidified slag can cause severe corrosion of the weld area (and thus lead to failure of the weld), and also prevents the fusion of weld layers and compromises the weld strength. The slag can be removed by, for example, wire brushing or chipping of the weld; consequently, both labor costs and material costs are high.

30.4.2 Submerged-arc Welding

In *submerged-arc welding* (SAW), the weld arc is shielded by a *granular flux*, consisting of lime, silica, manganese oxide, calcium fluoride, and other compounds. The flux is fed into the weld zone from a hopper by gravity flow through a nozzle (Fig. 30.10). The thick layer of flux completely covers the molten metal, and prevents spatter and sparks, and suppresses the intense ultraviolet radiation and fumes characteristic of the SMAW process. The flux also acts as a thermal insulator, by promoting deep penetration of heat into the workpiece.

The consumable electrode is a coil of bare round wire 1.5 to 10 mm ($\frac{1}{16}-\frac{3}{8}$ in.) in diameter, and is fed automatically through a tube (**welding gun**). Electric currents typically range from 300 to 2000 A, but multiple arc arrangements can be as high as 5000 A. The power supplies are usually connected to standard single- or three-phase power lines with a primary rating up to 440 V.

Because the flux is gravity fed, the SAW process is largely limited to welds in a flat or horizontal position having a backup piece. Circular welds can be made on pipes and cylinders, provided that they are rotated during welding. As Fig. 30.10

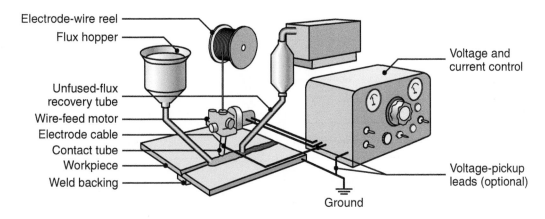

FIGURE 30.10 Schematic illustration of the submerged-arc welding process and equipment; the unfused flux is recovered and reused.

illustrates, the unfused flux can be recovered, treated, and reused; typically, 50 to 90% of the flux is recovered. SAW is automated and is used to weld a variety of carbon and alloy steel and stainless-steel sheets or plates at speeds as high as 5 m/min (16 ft/min), and occasionally is used for nickel-based alloys. The quality of the weld is very high, with good toughness, ductility, and uniformity of properties. The SAW process provides very high welding productivity, depositing 4 to 10 times the amount of weld metal per hour as the SMAW process. Typical applications include thick-plate welding for shipbuilding and for pressure vessels.

30.4.3 Gas Metal-arc Welding

In *gas metal-arc welding* (GMAW), developed in the 1950s and formerly called *metal inert-gas* (MIG) *welding*, the weld area is shielded by an effectively inert atmosphere of argon, helium, carbon dioxide, or various other gas mixtures (Fig. 30.11a). The consumable bare wire is automatically fed through a nozzle into the weld arc by a wire-feed drive motor (Fig. 30.11b). In addition to using inert shielding gases, deoxidizers usually are present in the electrode metal itself, in order to prevent oxidation of the molten-weld puddle. Multiple-weld layers also can be deposited at the joint.

Metal can be transferred by three methods:

1. **Spray transfer:** Small, molten metal droplets from the electrode are transferred to the weld area, at a rate of several hundred droplets per second; the transfer is spatter free and very stable. High DC currents and voltages and large-diameter electrodes are used, with argon or an argon-rich gas mixture as the shielding gas. The average current required in this process can be reduced by using a **pulsed arc**, superimposing high-amplitude pulses onto a low, steady current. The process can be used in all welding positions.

2. **Globular transfer:** Carbon-dioxide-rich gases are utilized, and globules are propelled by the forces of the electric-arc transfer of the metal, resulting in considerable spatter. High welding currents are used, making it possible for deeper weld penetration, and welding speeds are higher than in spray transfer. Heavier sections are commonly joined by this method.

3. **Short circuiting:** The metal is transferred in individual droplets (more than 50/s) as the electrode tip touches the molten weld metal and short circuits. Low currents and voltages are utilized, with carbon-dioxide-rich gases, and electrodes made of small-diameter wire; the power required is about 2 kW.

QR Code 30.1 Narrow groove tandem metal arc welding. (*Source:* Courtesy of EWI)

QR Code 30.2 GMAW of Titanium. (*Source:* Courtesy of EWI)

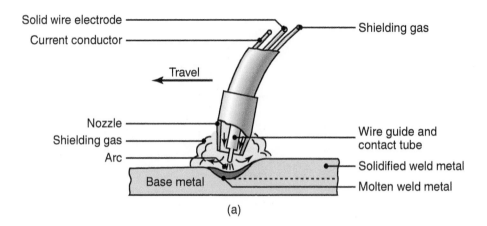

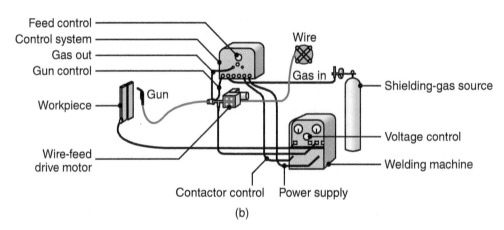

FIGURE 30.11 (a) Schematic illustration of the gas metal-arc welding process, formerly known as MIG (for metal inert-gas) welding. (b) Basic equipment used in gas metal-arc welding operations.

The temperatures generated in GMAW are relatively low; consequently, this method is suitable only for thin sheets and sections of less than 6 mm (0.25 in.); otherwise incomplete fusion may result. The operation, which is easy to perform, is commonly used for welding ferrous metals in thin sections. Pulsed-arc systems are used for thin ferrous and nonferrous metals.

The GMAW process is suitable for welding most ferrous and nonferrous metals, and is used extensively in the metal-fabrication industry. Because of the relatively simple nature of the process, the training of operators is easy. The process is versatile, rapid, and economical, and welding productivity is double that of the SMAW process. The GMAW process can easily be automated, and lends itself readily to robotics and to flexible manufacturing systems (Chapters 37 and 39).

30.4.4 Flux-cored Arc Welding

The *flux-cored arc welding* (FCAW) process, illustrated in Fig. 30.12, is similar to gas metal-arc welding, except that the electrode is tubular in shape and is filled with flux, hence the term *flux-cored*. Cored electrodes produce a more stable arc, improve the weld contour, and produce better mechanical properties of the joint. The flux in these electrodes is much more flexible than the brittle coating used on SMAW electrodes, thus the tubular electrode can be provided in long coiled lengths.

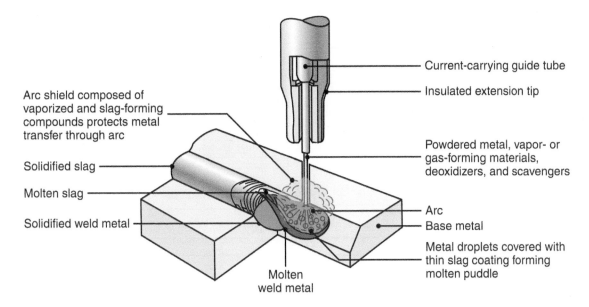

Current-carrying guide tube

Insulated extension tip

Arc shield composed of vaporized and slag-forming compounds protects metal transfer through arc

Powdered metal, vapor- or gas-forming materials, deoxidizers, and scavengers

Solidified slag

Molten slag

Solidified weld metal

Arc

Base metal

Metal droplets covered with thin slag coating forming molten puddle

Molten weld metal

FIGURE 30.12   Schematic illustration of the flux-cored arc-welding process; this operation is similar to gas metal-arc welding, shown in Fig. 30.11.

The electrodes are usually 0.5 to 4 mm (0.020–0.15 in.) in diameter, and the power required is about 20 kW. Self-shielded cored electrodes also are available; they do not require any external shielding gas, because they contain emissive fluxes that shield the weld area against the surrounding atmosphere. Small-diameter electrodes have made the welding of thinner materials not only possible but often preferable. Also, small-diameter electrodes make it relatively easy to weld parts at different positions, and the flux chemistry permits the welding of many metals.

The FCAW process combines the versatility of SMAW with the continuous and automatic electrode-feeding feature of GMAW. The process is economical and versatile, thus it is used for welding a variety of joints, mainly of steels, stainless steels, and nickel alloys. The higher weld-metal deposition rate of the FCAW process, as compared with that of GMAW, has led to its use in joining sections of all thicknesses. The use of *tubular electrodes* with very small diameters has extended the use of this process to workpieces of small section size.

A major advantage of FCAW is the ease with which specific weld-metal chemistries can be developed. By adding alloying elements to the flux core, virtually any alloy composition can be produced. The process is easy to automate and is readily adaptable to flexible manufacturing systems and robotics.

30.4.5   Electrogas Welding

Electrogas welding (EGW) is primarily used for welding the edges of sections, vertically and in one pass, with the pieces placed edge to edge (*butt joint*). It is classified as a *machine-welding* process, because it requires special equipment (Fig. 30.13). The weld metal is deposited into a weld cavity between the two pieces to be joined. The space is enclosed by two water-cooled copper *dams* (called *shoes*) to prevent the molten slag from running off; mechanical drives move the shoes upward. Circumferential welds, such as those on pipes, also are possible, with the workpiece rotating.

Single or multiple electrodes are fed through a conduit, and a continuous arc is maintained, using flux-cored electrodes at up to 750 A or solid electrodes at 400 A.

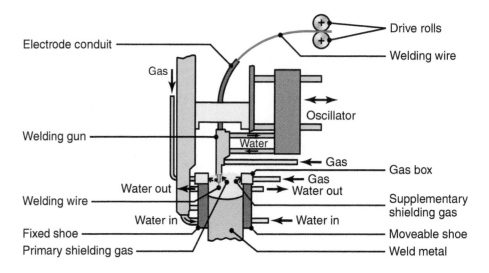

FIGURE 30.13 Schematic illustration of the electrogas welding process.

Power requirements are about 20 kW. Shielding is done by means of an inert gas, such as carbon dioxide, argon, or helium, depending on the type of material being welded. The gas may be provided either from an external source or from a flux-cored electrode, or from both.

Weld thickness ranges from 12 to 75 mm (0.5–3 in.) on steels, titanium, and aluminum alloys. Typical applications are in the construction of bridges, pressure vessels, thick-walled and large-diameter pipes, storage tanks, and ships. The equipment for this process is reliable and training for operators is relatively simple.

30.4.6 Electroslag Welding

Electroslag welding (ESW) and its applications are similar to electrogas welding (Fig. 30.14), the main difference is that the arc is started between the electrode tip and the bottom of the part to be welded. Flux is added, which then melts due to the heat of the arc. After the molten slag reaches the tip of the electrode, the arc is extinguished. Heat is produced continuously by the electrical resistance of the molten slag. Because the arc is extinguished, ESW is not strictly an arc-welding process. Single or multiple solid as well as flux-cored electrodes may be used. The guide may be nonconsumable (conventional method) or consumable.

Electroslag welding is capable of welding plates with thicknesses ranging from 50 mm to more than 900 mm (2–36 in.), and welding is done in one pass. The current required is about 600 A at 40 to 50 V, although higher currents are used for thick plates. The travel speed of the weld is in the range of 12–36 mm/min (0.5–1.5 in./min). This process is used for large structural-steel sections, such as heavy machinery, bridges, oil rigs, ships, and nuclear-reactor vessels. Weld quality is good.

30.5 Electrodes for Arc Welding

Electrodes for consumable arc-welding processes are classified according to the following properties:

- Strength of the deposited weld metal
- Current (AC or DC)
- Type of coating

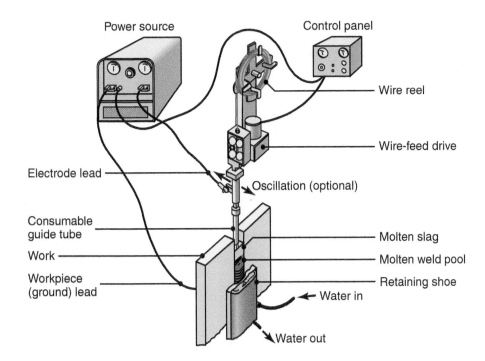

FIGURE 30.14   Equipment used for electroslag-welding operations.

Electrodes are identified by numbers and letters (Table 30.3), or by color code if the numbers and letters are too small to imprint. Typical coated-electrode dimensions are in the range of 150–460 mm (6–18 in.) in length and 1.5–8 mm ($\frac{1}{16} - \frac{5}{16}$ in.) in diameter.

Specifications for electrodes and filler metals, including dimensional tolerances, quality control procedures, and processes, are published by the American Welding Society (AWS) and the American National Standards Institute (ANSI). Some specifications are available in the Aerospace Materials Specifications (AMS) by the Society of Automotive Engineers (SAE). Electrodes are sold by weight and are available in a wide variety of sizes and specifications. Criteria for selection and recommendations for electrodes for a particular metal and its application can be found in supplier literature and in the various handbooks and references listed at the end of this chapter.

Electrode Coatings.   Electrodes are *coated* with claylike materials that include silicate binders and powdered materials, such as oxides, carbonates, fluorides, metal alloys, cotton cellulose, and wood flour. The coating, which is brittle and takes part in complex interactions during welding, has the following basic functions:

- Stabilize the arc
- Generate gases, to act as a shield against the surrounding atmosphere; the gases produced are carbon dioxide, water vapor, and small amounts of carbon monoxide and hydrogen
- Control the rate at which the electrode melts
- Act as a flux, to protect the weld against the formation of oxides, nitrides, and other inclusions and, with the resulting slag, to protect the molten-weld pool
- Add alloying elements, to the weld zone to enhance the properties of the joint— among these elements are deoxidizers to prevent the weld from becoming brittle

TABLE 30.3

Designations for Mild-steel Coated Electrodes

The prefix "E" designates arc welding electrode. The first two digits of four-digit numbers and the first three digits of five-digit numbers indicate minimum tensile strength:

E60XX	60,000 psi
E70XX	70,000 psi
E110XX	110,000 psi

The next-to-last digit indicates position:

EXX1X	All positions
EXX2X	Flat position and horizontal fillets

The last two digits together indicate the type of covering and the current to be used; the suffix (Example: EXXXX-A1) indicates the approximate alloy in the weld deposit:

–A1	0.5% Mo
–B1	0.5% Cr, 0.5% Mo
–B2	1.25% Cr, 0.5% Mo
–B3	2.25% Cr, 1% Mo
–B4	2% Cr, 0.5% Mo
–B5	0.5% Cr, 1% Mo
–C1	2.5% Ni
–C2	3.25% Ni
–C3	1% Ni, 0.35% Mo, 0.15% Cr
–D1 and D2	0.25–0.45% Mo, 1.75% Mn
–G	0.5% min. Ni, 0.3% min. Cr, 0.2% min. Mo, 0.1% min. V, 1% min. Mn (only one element required)

The deposited electrode coating or slag must be removed after each pass in order to ensure a good weld. Bare electrodes and wires, typically made of stainless steels and aluminum alloys, also are available, and are used as filler metals in various welding operations.

30.6 Electron-beam Welding

In *electron-beam welding* (EBW), developed in the 1960s, heat is generated by high-velocity, narrow-beam electrons; the kinetic energy of the electrons is converted into heat as they strike the workpiece to be welded. This process requires special equipment in order to focus the beam on the workpiece, typically in a vacuum. The higher the vacuum, the deeper the beam penetrates, and the greater is the depth-to-width ratio of the weld; thus, the methods are called EBW-HV (for "high vacuum") and EBW-MV (for "medium vacuum"); some materials may also be welded by EBW-NV (for "no vacuum").

Almost any metal can be welded and workpiece thicknesses range from foil to plate. Capacities of electron beam guns range up to 100 kW. The intense energy also is capable of producing holes in the workpiece (see **keyhole technique**, Section 30.3). Generally, no shielding gas, flux, or filler metal is required.

The EBW process has the capability of making high-quality welds that are almost parallel sided, are deep and narrow, and have small heat-affected zones (see Fig. 30.15 and Section 30.9). Depth-to-width ratios are in the range of 10–30. The size of welds made by EBW are much smaller than those of welds made by conventional processes. With the use of automation and servo controls, processing parameters can be controlled accurately, at welding speeds as high as 12 m/min (40 ft/min).

Almost any metal can be welded (butt or lap) with this process and at thicknesses up to 150 mm (6 in.). Distortion and shrinkage in the weld area are minimal, and weld quality is good. Typical applications include welding of aircraft, missile, nuclear, and electronic components, as well as gears and shafts for the automotive industry. Electron-beam welding equipment generates X-rays, hence proper monitoring and periodic maintenance are essential.

30.7 Laser-beam Welding

Laser-beam welding (LBW) utilizes a high-power laser beam as the source of heat to produce a fusion weld.

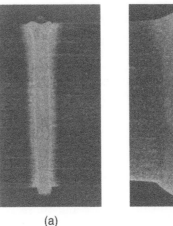 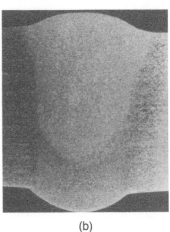

(a)   (b)

FIGURE 30.15 Comparison of (a) electron-beam weld and (b) gas tungsten arc weld. *Source:* Volume 3, *Welding Handbook, Welding Processes*, Part 2, Miami: American Welding Society, page 465. Used with permission.

Because it can be focused onto a very small area, the beam has high energy density and deep-penetrating capability. The laser beam can be directed, shaped, and focused precisely on the workpiece; laser spot diameters can be as small as 0.2 mm (0.008 in.). Consequently, this process is particularly suitable for welding deep and narrow joints (Fig. 30.15), with depth-to-width ratios typically ranging from 4 to 10.

Laser-beam welding has become extremely widespread and is now used in most industries. The laser beam may be **pulsed** (in milliseconds), with power levels up to 100 kW for such applications as the spot welding of thin materials. **Continuous** multi-kW laser systems are used for deep welds on thick sections.

Laser-beam welding produces welds of good quality with minimum shrinkage or distortion. Laser welds have good strength and are generally ductile and free of porosity. The process can be automated, to be used on a variety of materials, with thicknesses up to 25 mm (1 in.). As described in Section 16.2.2, *tailor-welded sheet-metal blanks* are joined principally by laser-beam welding, using robotics for precise control of the beam path.

Typical metals and alloys welded include aluminum, titanium, ferrous metals, copper, superalloys, and the refractory metals. Welding speeds range from 2.5 m/min (8 ft/min) to as high as 80 m/min (250 ft/min) for thin metals. Because of the nature of the process, welding can be done in otherwise inaccessible locations. As in other and similar automated welding systems, the operator skill required is minimal. Safety is particularly important in laser-beam welding, due to the extreme hazards to the eye as well as the skin; solid-state (YAG) lasers also are dangerous. (See Table 27.2 on types of lasers.)

The major advantages of LBW over EBW can be summarized as:

- A vacuum is not required, and the beam can be transmitted through air.
- Laser beams can be shaped, manipulated, and focused optically by means of fiber optics, thus the process can easily be automated.
- The beams do not generate X-rays.
- The quality of the weld is better than in EBW; there is less distortion and the weld has less tendency toward incomplete fusion, spatter, and porosity.

EXAMPLE 30.2   Laser Welding of Razor Blades

A close-up of the Gillette Sensor razor cartridge is shown in Fig. 30.16. Each of the two narrow, high-strength blades has 13 pinpoint welds, 11 of which can be seen (as darker spots, about 0.5 mm in diameter) on each blade in the photograph. The welds are made with an Nd:YAG laser equipped with fiber-optic delivery. This equipment provides very flexible beam manipulation and can target exact locations along the length of the blade. With a set of these machines, production is at a rate of 3 million welds per hour, with accurate and consistent weld quality.

Source: Based on Lumonics Corporation, Industrial Products Division.

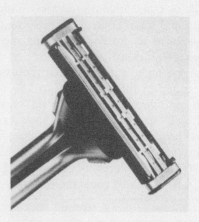

FIGURE 30.16 Detail of Gillette Sensor razor cartridge, showing laser spot welds. Courtesy of Procter & Gamble.

QR Code 30.3 Laser hybrid welding. (*Source:* Courtesy of EWI)

Laser GMAW.   *Laser GMAW* is an emerging hybrid welding technology, that combines the narrow heat-affected zone of laser welding with the high deposition rates of gas metal-arc welding. In this process, shown in Fig. 30.17, the laser is focused on the workpiece ahead of the GMAW arc, resulting in deep penetration and allowing high travel speeds. In addition, the process is able to bridge larger gaps than traditional laser welding, and the metallurgical quality of the weld is improved because of the presence of the shielding gas.

QR Code 30.4 Hybrid laser arc welding. (*Source:* Courtesy of EWI)

30.8   Cutting

In addition to being cut by mechanical means, as described in Part IV of this book, material can be cut into various contours by using a heat source, that melts and removes a narrow zone in the workpiece. The sources of heat can be torches, electric arcs, or lasers.

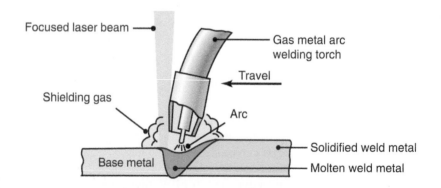

FIGURE 30.17   (a) Schematic illustration of the laser GMAW hybrid welding process. *Source:* Courtesy of Lincoln Electric.

Oxyfuel–gas Cutting.   *Oxyfuel–gas cutting* (OFC) is similar to oxyfuel–gas welding (Section 30.2), but the heat source is now used to *remove* a narrow zone from a metal plate or sheet (Fig. 30.18a). This process is particularly suitable for steels, where the basic reactions are

$$Fe + O \rightarrow FeO + Heat, \qquad (30.6)$$

$$3Fe + 2O_2 \rightarrow Fe_3O_4 + Heat, \qquad (30.7)$$

and

$$4Fe + 3O_2 \rightarrow 2Fe_2O_3 + Heat. \qquad (30.8)$$

The greatest heat is generated by the second reaction, with temperatures rising to about 870°C (1600°F). However, because this temperature is not sufficiently high, the workpiece is *preheated* with fuel gas, and oxygen is introduced later, as can be seen from the nozzle cross-section in Fig. 30.18a. The higher the carbon content of the steel, the higher is the preheating temperature required. Cutting takes place mainly by the oxidation (burning) of the steel; some melting also takes place. Cast irons and steel castings also can be cut by this method. The process generates a **kerf**, similar to that produced in sawing with a saw blade or by wire electrical-discharge machining (see Fig. 27.12). Kerf widths range from about 1.5 to 10 mm (0.06–0.4 in.), with good control of dimensional tolerances. However, distortion caused by uneven temperature distribution can be a problem in OFC.

The maximum thickness that can be cut by OFC depends mainly on the gases used. With oxyacetylene gas, for example, the maximum thickness is about 300 mm (12 in.); with oxyhydrogen, it is about 600 mm (24 in.). The flame leaves **drag lines** on the cut surface (Fig. 30.18b), resulting in a rougher surface than that produced by processes such as sawing, blanking, or other similar operations that use mechanical cutting tools. *Underwater cutting* is done with specially designed torches, that produce a blanket of compressed air between the flame and the surrounding water. Although long used for salvage and repair work, OFC can be used in manufacturing as well. Torches may be guided along specified paths manually, mechanically, or automatically by machines using programmable controllers and robots.

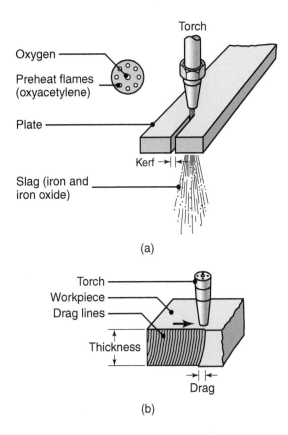

FIGURE 30.18   (a) Flame cutting of a steel plate with an oxyacetylene torch, and a cross-section of the torch nozzle. (b) Cross-section of a flame-cut plate, showing drag lines.

Arc Cutting.   *Arc-cutting* processes are based on the same principles as arc-welding processes. A variety of materials can be cut at high speeds by arc cutting, although as in welding, these processes also leave a heat-affected zone that needs to be taken into account, particularly in critical applications.

In **air carbon-arc cutting** (CAC-A), a carbon electrode is used and the molten metal is blown away by a high-velocity air jet. This process is used especially for gouging and *scarfing* (removal of metal from a surface). However, the process is noisy, and the molten metal can be blown substantial distances and cause safety hazards.

Plasma-arc cutting (PAC) produces the highest temperatures, and is used for the rapid cutting of nonferrous and stainless-steel plates. The cutting productivity of this process is higher than that of oxyfuel–gas methods. PAC produces a good surface finish and narrow kerfs, and is the most common cutting process utilizing programmable controllers employed in manufacturing today. *Electron beams* and *lasers* also are used for very accurately cutting a wide variety of metals, as was described in Sections 27.6 and 27.7. The surface finish is better than that of other thermal cutting processes, and the kerf is narrower.

30.9 The Weld Joint, Weld Quality, and Testing

Three distinct zones can be identified in a typical weld joint, as shown in Fig. 30.19:

1. Base metal
2. Heat-affected zone
3. Weld metal

The metallurgy and properties of the second and third zones depend strongly on the type of metals joined, the particular joining process, the filler metals used (if any), and welding process variables. Recall that a joint produced without using a filler metal is called *autogenous*, and its weld zone is composed of the *resolidified base metal*. A joint made with a filler metal has a central zone, called the *weld metal*, and is composed of a mixture of the base and the filler metals.

Solidification of the Weld Metal.   After the application of heat and the introduction of the filler metal (if any) into the weld zone, the weld joint is allowed to cool to ambient temperature. The solidification process is similar to that in casting (Section 10.2) and begins with the formation of *columnar (dendritic)* grains, as shown in Fig. 10.3. These grains are relatively long and they form parallel to the heat flow. Because metals are much better heat conductors than the surrounding air, the grains lie *parallel* to the plane of the two components being welded (Fig. 30.20a). In contrast, the grains in a *shallow* weld are as shown in Figs. 30.20b and c.

Grain structure and grain size depend on the specific metal alloy, the particular welding process employed, and the type of filler metal. Because it began in a molten state, the weld metal basically has a *cast structure*, and since it has cooled slowly, it has coarse grains. Consequently, this structure generally has low strength, toughness, and ductility. However, with proper selection of filler-metal composition or of heat treatments, following welding, the mechanical properties of the joint can be improved.

The resulting structure depends on the particular alloy, its composition, and the thermal cycling to which the joint is subjected. For example, cooling rates may be controlled and reduced by *preheating* the general weld area prior to welding. Preheating is important, particularly for metals having high thermal conductivity, such as aluminum and copper (Table 3.2). Without preheating, the heat produced during welding dissipates rapidly through the rest of the parts being joined.

Heat-affected Zone.   The *heat-affected zone* (HAZ) is within the base metal itself. It has a microstructure different from that of the base metal prior to its welding, because it has been temporarily subjected to elevated temperatures during welding. The portions of the base metal that are far away from the heat source

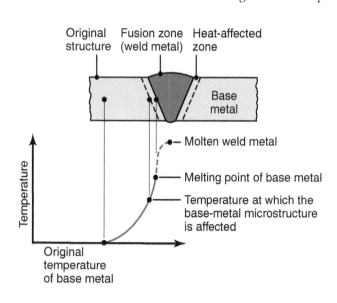

FIGURE 30.19   Characteristics of a typical fusion-weld zone in oxyfuel–gas and arc welding.

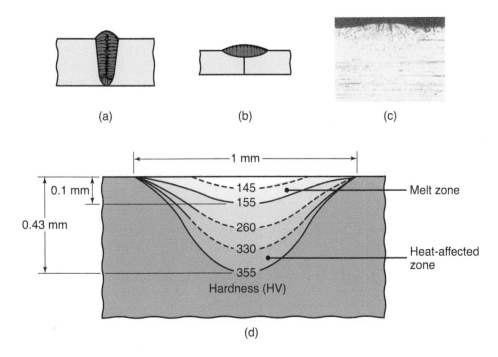

FIGURE 30.20   Grain structure in (a) a deep weld and (b) a shallow weld; note that the grains in the solidified weld metal are perpendicular to their interface with the base metal. (c) Weld bead on a cold-rolled nickel strip produced by a laser beam. (d) Microhardness (HV) profile across a weld bead.

do not undergo microstructural changes during welding, because of the far lower temperature to which they are subjected.

The properties and microstructure of the HAZ depend on (a) the rate of heat input and cooling and (b) the temperature to which this zone was raised. In addition to metallurgical factors (such as the original grain size, grain orientation, and degree of prior cold work), physical properties (such as the specific heat and thermal conductivity of the metals) also influence the size and characteristics of the HAZ.

The strength and hardness of the HAZ (Fig. 30.20d) depend partly on how the original strength and hardness of the base metal was developed prior to welding. As described in Chapters 2 and 4, they may have been developed by (a) cold working, (b) solid-solution strengthening, (c) precipitation hardening, or (d) various heat treatments. The effects of these strengthening methods are complex; the simplest to analyze are those in a base metal that has been cold worked, such as by cold rolling or cold forging.

The heat applied during welding *recrystallizes* the elongated grains of the cold-worked base metal. On the one hand, grains that are away from the weld metal will recrystallize into fine, equiaxed grains; on the other hand, grains close to the weld metal have been subjected to elevated temperatures for a longer time. Consequently, the grains will grow in size (*grain growth*, Section 1.7), and this region will be softer and have lower strength; such a joint will be weakest at its HAZ.

The effects of heat on the HAZ for joints made from dissimilar metals and for alloys strengthened by other methods are complex and are beyond the scope of this book. Details can be found in the more advanced references listed in the bibliography at the end of this chapter.

30.9.1 Weld Quality

As a result of a history of thermal cycling and its attendant microstructural changes, a welded joint may develop various **discontinuities**. Welding discontinuities also can be caused by an inadequate or careless application of welding technologies or by poor operator training. The major discontinuities that affect weld quality are described below.

Porosity. *Porosity* in welds may be caused by:

- Gases released during melting of the weld area but trapped during solidification
- Chemical reactions during welding
- Contaminants

Most welded joints contain some porosity, which is generally in the shape of spheres or of elongated pockets. (See also Section 10.6.1.) The distribution of porosity in the weld zone may be random or the porosity may be concentrated in a certain region in the zone. Porosity in welds can be reduced by the following practices:

- Proper selection of electrodes and filler metals
- Improved welding techniques, such as preheating the weld area or increasing the rate of heat input
- Proper cleaning and the prevention of contaminants from entering the weld zone
- Reduced welding speeds, to allow time for gas to escape

Slag Inclusions. *Slag inclusions* are compounds, such as oxides, fluxes, and electrode-coating materials, that are trapped in the weld zone. If shielding gases are not effective during welding, contamination from the environment also may contribute to such inclusions. Welding conditions also are important: with control of welding process parameters, the molten slag will float to the surface of the molten weld metal and thus will not become entrapped.

Slag inclusions can be prevented by implementing the following practices:

- Cleaning the weld-bead surface by means of a wire brush (hand or power) or a chipper before the next layer is deposited
- Providing sufficient shielding gas
- Redesigning the joint to permit sufficient space for proper manipulation of the puddle of molten weld metal

Incomplete Fusion and Penetration. *Incomplete fusion* produces poor weld beads, such as those shown in Fig. 30.21. A better weld can be obtained by implementing the following practices:

- Raising the temperature of the base metal
- Cleaning the weld area prior to welding
- Modifying the joint design and changing the type of electrode used
- Providing sufficient shielding gas

Incomplete penetration occurs when the depth of the welded joint is insufficient. Penetration can be improved by the following practices:

- Increasing the heat input
- Reducing the travel speed during welding
- Modifying the joint design
- Ensuring that the surfaces to be joined fit each other properly

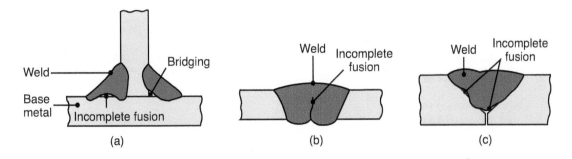

FIGURE 30.21   Examples of various discontinuities in fusion welds.

Weld Profile.   *Weld profile* is important not only because of its effects on the strength and appearance of the weld, but also because it can indicate incomplete fusion or the presence of slag inclusions in multiple-layer welds.

- **Underfilling** results when the joint is not filled with the proper amount of weld metal (Fig. 30.22a).
- **Undercutting** results from the melting away of the base metal and the subsequent generation of a groove in the shape of a sharp recess or notch (Fig. 30.22b). If it is deep or sharp, an undercut can act as a stress raiser and thus reduce the fatigue strength of the joint, and lead to premature failure.
- **Overlap** is a surface discontinuity (Fig. 30.22b), usually caused by poor welding practice or by selection of improper materials. Figure 30.22c shows a weld that would be considered to be good.

Cracks.   *Cracks* may occur in various locations and directions in the weld area. Typical types of cracks are *longitudinal, transverse, crater, underbead,* and *toe cracks* (Fig. 30.23). Cracks generally result from a combination of the following factors:

- Temperature gradients causing thermal stresses in the weld zone
- Variations in the composition of the weld zone causing different rates of contraction during cooling

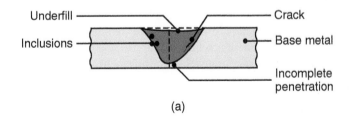

(a)

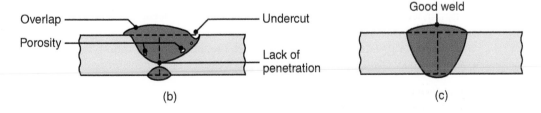

(b)   (c)

FIGURE 30.22   Examples of various defects in fusion welds and the cross-section of a good weld.

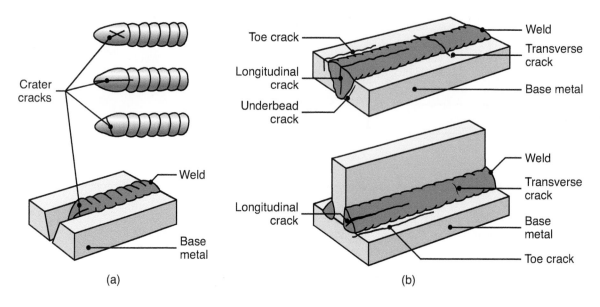

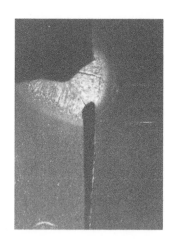

FIGURE 30.23 Types of cracks developed in welded joints; the cracks are caused by thermal stresses, similar to the development of hot tears in castings, as shown in Fig. 10.12.

FIGURE 30.24 Crack in a weld bead; the two welded components did not contract freely after the weld was completed.

- Embrittlement of grain boundaries (Section 1.5.2), caused by the segregation of such elements as sulfur to the grain boundaries, and occurring when the solid–liquid boundary moves as the weld metal begins to solidify
- Hydrogen embrittlement (Section 2.10.2)
- Inability of the weld metal to contract during cooling (Fig. 30.24), a situation similar to *hot tears* that develop in castings (Fig. 10.12), and is related to excessive restraint of the workpiece during the welding operation

Cracks also are classified as **hot cracks** (which occur while the joint is still at elevated temperatures) and **cold cracks** (which develop after the weld metal has solidified). The basic crack-prevention measures in welding are:

- Modify the joint design to minimize stresses developed from shrinkage during cooling
- Change the parameters, procedures, and sequence of the welding operation
- Preheat the components to be welded
- Avoid rapid cooling of the welded components

Lamellar Tears. In describing the anisotropy of plastically deformed metals in Section 1.5, it was stated that the workpiece is weaker when tested in its thickness direction, because of the alignment of nonmetallic impurities and inclusions (*stringers*). This condition is particularly observed in rolled plates and structural shapes. In welding such components, *lamellar tears* may develop, because of shrinkage of the restrained components of the structure during cooling. Such tears can be avoided by providing for shrinkage of the members or by modifying the joint design to make the weld bead penetrate the weaker component more deeply.

Surface Damage. Some of the hot metal may spatter during welding and be deposited as small droplets on adjacent surfaces. In arc-welding processes, the electrode inadvertently may touch the parts being welded at places other than the weld zone; such encounters are called **arc strikes**. The surface discontinuities thereby

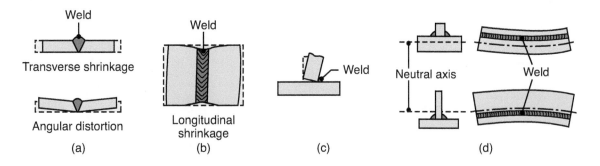

FIGURE 30.25 Distortion of parts after welding; distortion is caused by differential thermal expansion and contraction of different regions of the welded assembly.

produced may be objectionable for reasons of appearance or of subsequent use or assembly of the welded structure. If severe, these discontinuities may also adversely affect the properties of the welded structure, particularly for notch-sensitive metals. Using proper welding techniques and procedures is important in avoiding surface damage.

Residual Stresses.   Because of localized heating and cooling during welding, the expansion and contraction of the weld area causes *residual stresses* in the workpiece. (See also Section 2.11.) Residual stresses can lead to the following defects:

- Distortion, warping, and buckling of the welded parts (Fig. 30.25)
- Stress-corrosion cracking (Section 2.10.2)
- Further distortion if a portion of the welded structure is subsequently removed, such as by machining or sawing
- Reduced fatigue life of the welded structure

The type and distribution of residual stresses developed in welds is best described by referring to Fig. 30.26a. When two plates are being welded, a long narrow zone is subjected to elevated temperatures, while the plates, as a whole, are essentially at ambient temperature. After the weld is completed and as time elapses, the heat from the weld zone dissipates laterally into the plates, while the weld area begins to cool. Thus, the plates now begin to expand longitudinally, while the welded length begins to contract (Fig. 30.25).

If the plate is not constrained, it will warp, as shown in Fig. 30.25a. However, if the plate is not free to warp, it will develop residual stresses, which typically are distributed throughout the material, such as the stresses shown in Fig. 30.26b. Note that the magnitude of the compressive residual stresses in the plates diminishes to zero at the top and bottom surfaces of the welded plate. Because no external forces are acting on the welded plates, the tensile and compressive forces represented by these residual stresses must balance each other.

The sequence of events leading to the distortion of a simple tubular welded structure is shown in Fig. 30.27. Prior to welding, the structure is stress free, as shown in Fig. 30.27a, and it may be fairly rigid; also some fixturing may be present to support the structure as part of a larger assembly. During welding, the molten metal fills the gap between the surfaces to be joined and forms a weld bead. As the weld bead begins to solidify, both the weld bead and the surrounding material begin to cool down to

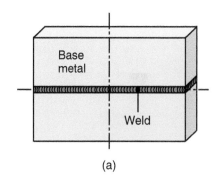

(a)

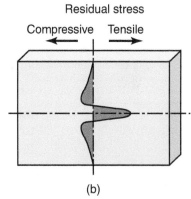

(b)

FIGURE 30.26 Residual stresses developed in a straight-butt joint; note that the residual stresses shown must be balanced internally since there are no external forces. (See also Fig. 2.29.)

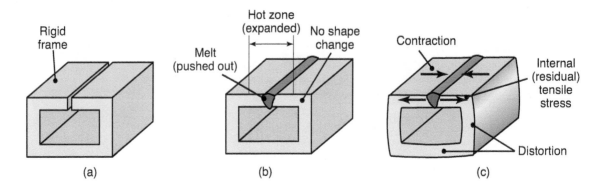

FIGURE 30.27   Distortion of a welded structure. *Source:* After J.A. Schey.

room temperature. As these materials cool, they tend to contract but are constrained by the rest of the weldment; as a result, the part distorts (Fig. 30.27c) and residual stresses develop.

The residual stresses produce the deformation shown in Fig. 30.27c, and put the weld and the heat-affected zone into a state of residual tension, which is not desirable for fatigue performance. In general, the HAZ is less fatigue resistant than the base metal. Because the residual stresses developed can be harmful, it is not unusual to stress relieve welds in highly stressed or fatigue-susceptible applications (see below). Recall that the weld itself may have porosity (see Fig. 30.22b), which also can act as a stress raiser and lead to fatigue crack growth.

In complex welded structures, residual-stress distributions are three dimensional and difficult to analyze. Note that the two plates shown in Fig. 30.26 were not restrained from movement; in other words, the plates were not an integral part of a larger structure. If, however, they were restrained, reaction stresses would develop, because the plates are not free to expand or contract. This is a situation that arises particularly in structures with high stiffness.

Stress Relieving of Welds.   The problems caused by residual stresses, such as distortion, buckling, and cracking, can be reduced by **preheating** the base metal or the parts to be welded. Preheating reduces distortion by reducing the cooling rate after welding and the level of thermal stresses developed (by lowering the elastic modulus). This technique also reduces shrinkage and possible cracking of the joint.

For optimum results, preheating temperatures and cooling rates must be controlled carefully in order to maintain acceptable strength and toughness in welded structures. The workpieces may be heated in several ways, including (a) in a furnace, (b) electrically (either resistively or inductively), or (c) by radiant lamps or a hot-air blast, especially for thin sections. The temperature and time required for stress relieving depend on the type of material and on the magnitude of the residual stresses developed.

Other methods of stress relieving include *peening, hammering,* or *surface rolling* (Section 34.2) of the weld-bead area. These techniques induce compressive residual stresses, which, in turn, lower or eliminate tensile residual stresses in the weld. For multilayer welds, the first and last layers should not be peened, in order to protect them against possible peening damage on the surface.

Residual stresses can also be relieved or reduced by *plastically* deforming the structure itself by a small amount. For example, this technique can be used in welded pressure vessels by pressurizing the vessels internally (*proof stressing*). In order to reduce the possibility of sudden fracture under high internal pressure, however, the weld must be made properly and must be free of notches and discontinuities, which can act as stress raisers.

In addition to being preheated for stress relieving, welds may be heat treated by various other techniques in order to modify other properties. These techniques include the annealing, normalizing, quenching, and tempering of steels and the solution treatment and aging of various alloys, as described in Chapter 4.

30.9.2 Weldability

The *weldability* of a metal is generally defined as its capability to be welded into a specific structure that has certain properties and characteristics, and will satisfactorily meet service requirements. Weldability involves a large number of variables, thus generalizations are difficult. Recall that material characteristics, such as alloying elements, impurities, inclusions, grain structure, and processing history, of both the base metal and the filler metal, are all important. For example, the weldability of steels decreases with increasing carbon content, because of martensite formation (which is hard and brittle; Section 4.7), and thus reduces the strength of the weld. Coated steel sheets (Chapter 34) also can present various challenges in welding, depending on the type and thickness of the coating.

Because of the effects of melting and solidification and of the associated micro-structural changes, a thorough consideration of the phase diagram and the response of the metal or alloy to sustained elevated temperatures is essential. Also influencing weldability are mechanical and physical properties: strength, toughness, ductility, notch sensitivity, elastic modulus, specific heat, melting point, thermal expansion, surface-tension characteristics of the molten metal, and corrosion resistance.

The preparation of *surfaces* for welding is important, as are the nature and pro-perties of surface-oxide films and of adsorbed gases (see also Section 33.2). The particular welding process employed significantly affects the temperatures developed and their distribution in the weld zone. Other factors that affect weldability are shielding gases, fluxes, moisture content of the coatings on electrodes, welding speed, welding position, cooling rate, and level of preheating, as well as such postwelding techniques as stress relieving and heat treating.

Weldability of Ferrous Materials

- *Plain-carbon steels:* Generally excellent for low-carbon steels, fair to good for medium-carbon steels, and poor for high-carbon steels
- *Low-alloy steels:* Similar to medium-carbon steels
- *High-alloy steels:* Generally good under well-controlled conditions
- *Stainless steels:* Generally weldable by various processes
- *Cast irons:* Generally weldable, although their weldability varies greatly

Weldability of Nonferrous Materials

- *Aluminum alloys:* Weldable at a high rate of heat input; an inert shielding gas and absence of moisture are important; aluminum alloys containing zinc or copper generally are considered unweldable
- *Copper alloys:* Depending on composition, generally weldable at a high rate of heat input; an inert shielding gas and absence of moisture are important
- *Magnesium alloys:* Weldable, using a protective shielding gas and fluxes
- *Nickel alloys:* Similar to stainless steels; absence of sulfur is undesirable
- *Titanium alloys:* Weldable, with proper use of shielding gases
- *Tantalum:* Similar to titanium

- *Tungsten:* Weldable, under well-controlled conditions
- *Molybdenum:* Similar to tungsten
- *Niobium (columbium):* Good weldability

30.9.3 Testing of Welds

As in all manufacturing processes, the quality of a welded joint is established by testing. Several standardized tests and test procedures have been established, and are available from organizations such as the American Society for Testing and Materials (ASTM), the American Welding Society (AWS), the American Society of Mechanical Engineers (ASME), the American Society of Civil Engineers (ASCE), and various federal agencies.

Welded joints may be tested either *destructively* or *nondestructively* (see also Sections 36.10 and 36.11). Each technique has certain capabilities and limitations, as well as process parameter sensitivity, reliability, and requirements for special equipment and operator skill.

Destructive Testing Techniques

- **Tension test.** *Longitudinal and transverse tension tests* are performed on specimens removed from actual welded joints and from the weld-metal area; stress–strain curves are then developed using the procedures described in Section 2.2. These curves indicate the yield strength, ultimate tensile strength, and ductility of the welded joint (elongation and reduction of area) in different locations and directions.
- **Tension-shear test.** The specimens in the *tension-shear test* (Fig. 30.28a) are prepared to simulate conditions to which actual welded joints are subjected. These specimens are subjected to tension so that the shear strength of the weld metal and the location of fracture can be determined.

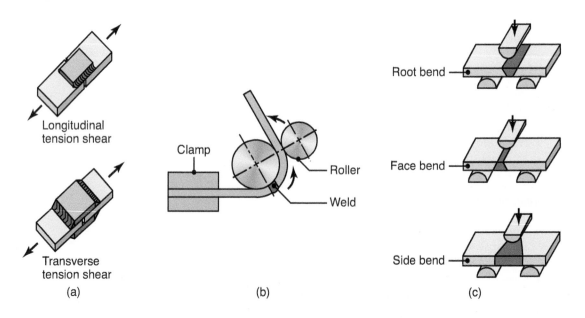

FIGURE 30.28 (a) Specimens for longitudinal tension-shear testing and for transfer tension-shear testing. (b) Wraparound bend-test method. (c) Three-point transverse bending of welded specimens.

- **Bend test.** Several bend tests have been developed to determine the ductility and strength of welded joints. In one common test, the welded specimen is bent around a fixture (*wraparound bend test*, Fig. 30.28b). In another test, the specimens are tested in *three-point transverse bending* (Fig. 30.28c; see also Fig. 2.11a). These tests help to determine the relative ductility and strength of welded joints.

- **Fracture toughness test.** Fracture toughness tests commonly utilize the impact testing techniques, described in Section 2.9. *Charpy V-notch* specimens are first prepared and tested for toughness. In the *drop-weight test*, the energy is supplied by a falling weight.

- **Creep and corrosion tests.** *Creep tests* (Section 2.8) are essential in determining the behavior of welded joints and structures subjected to elevated temperatures. Welded joints also may be tested for their resistance to *corrosion* (Section 3.8). Because of the difference in the composition and microstructure, *preferential corrosion* may take place in the weld zone.

Nondestructive Testing Techniques.   Welded structures often have to be tested *nondestructively* (Section 36.10), particularly for critical applications in which weld failure can be catastrophic, such as in pressure vessels, load-bearing structural members, and power plants. Nondestructive testing techniques for welded joints generally consist of the following methods:

- Visual
- Radiographic (X-rays)
- Magnetic-particle
- Liquid-penetrant
- Ultrasonic

As an example of another nondestructive method, testing for hardness distribution (see Section 2.6 and Figs. 16.3 and 30.20) in the weld zone also would be a useful indicator of weld strength and microstructural changes.

30.10   Joint Design and Process Selection

In describing individual welding processes, several examples were given regarding the types of welds and joints produced and their applications in various consumer and industrial products. Typical types of joints produced by welding, together with their terminology, are shown in Fig. 30.29. Standardized symbols commonly used in engineering to describe the types of welds are given in Fig. 30.30. These symbols identify the type of weld, the groove design, the weld size and length, the welding process, the sequence of operations, and other necessary information.

General design guidelines for welding are given in Fig. 30.31. Various other types of joint design are included in Chapters 31 and 32. Important design guidelines are summarized below.

- Product design should minimize the number of joints because, unless automated, welding can be time consuming and costly.
- Weld locations should be selected so as to avoid excessive local stresses or stress concentrations, as well as for better appearance.
- Weld location should be selected so as not to interfere with any subsequent processing of the joined components, or with their intended use.

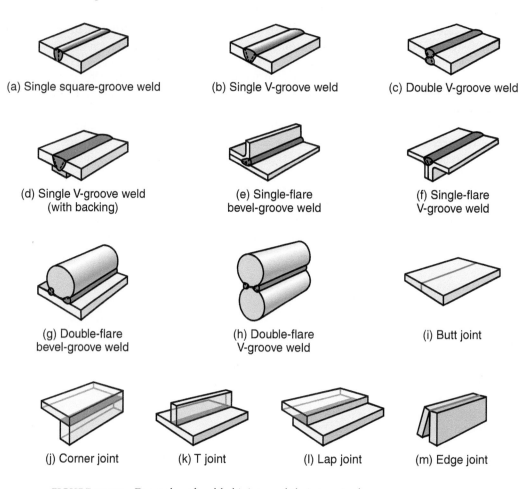

(a) Single square-groove weld

(b) Single V-groove weld

(c) Double V-groove weld

(d) Single V-groove weld
(with backing)

(e) Single-flare
bevel-groove weld

(f) Single-flare
V-groove weld

(g) Double-flare
bevel-groove weld

(h) Double-flare
V-groove weld

(i) Butt joint

(j) Corner joint

(k) T joint

(l) Lap joint

(m) Edge joint

FIGURE 30.29   Examples of welded joints and their terminology.

- The need for edge preparation should be minimized or avoided.
- Weld-bead size should be as small as possible while maintaining joint strength of the joint, in order to conserve weld metal and for better appearance.

Welding Process Selection.   In addition to considering welding process characteristics, capabilities, and material considerations, described thus far in this chapter, the selection of a weld joint and an appropriate process also involves the following considerations (see also Chapters 31 and 32):

- Configuration of the parts to be joined, joint design, thickness and size of the components, and number of joints required
- Methods used in manufacturing the components to be joined
- Types of materials involved
- Location, accessibility, and ease of joining
- Application and service requirements, such as type of loading, any stresses generated, and the environment
- Effects of distortion, warping, appearance, discoloration, and service
- Costs involved in edge preparation, joining, and postprocessing (including machining, grinding, and finishing operations)
- Costs of equipment, materials, labor and skills required, and of the overall operation

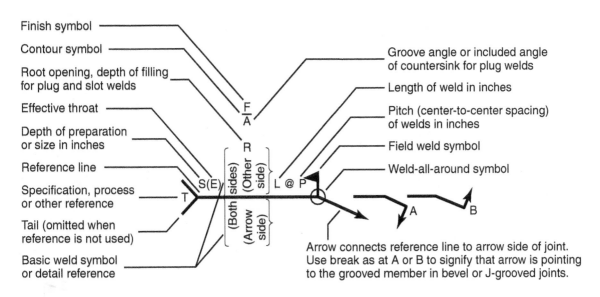

Basic arc- and gas-weld symbols							
Bead	Fillet	Plug or slot	Groove				
			Square	V	Bevel	U	J
⌒	△	▽	||	⌄	⌵	⌣	Ụ

Basic resistance-weld symbols			
Spot	Projection	Seam	Flash or upset
✳	✕	XXX	|

Finish symbol

Contour symbol

Root opening, depth of filling for plug and slot welds

Effective throat

Depth of preparation or size in inches

Reference line

Specification, process or other reference

Tail (omitted when reference is not used)

Basic weld symbol or detail reference

Groove angle or included angle of countersink for plug welds

Length of weld in inches

Pitch (center-to-center spacing) of welds in inches

Field weld symbol

Weld-all-around symbol

(Both sides) (Arrow side)   (Other side)

S(E)   R   L @ P   T

F / A

Arrow connects reference line to arrow side of joint. Use break as at A or B to signify that arrow is pointing to the grooved member in bevel or J-grooved joints.

FIGURE 30.30   Standard identification and symbols for welds.

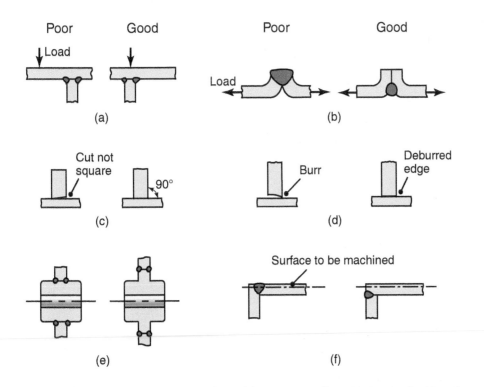

FIGURE 30.31   Some design guidelines for welds. *Source:* Bralla, J.G., *Design for Manufacturability Handbook*, 2nd ed., McGraw-Hill, 1999, ISBN No. 0-07-07139-X.

EXAMPLE 30.3   Weld Design Selection

Three different types of weld designs are shown in Fig. 30.32. The two vertical joints in Fig. 30.32a can be welded either externally or internally. Note that full-length external welding will take considerable time and will require more weld material than the alternative design, which consists of intermittent internal welds. Moreover, in the alternative method the appearance of the structure is improved and distortion is reduced.

In Fig. 30.32b, it can be shown that the design on the right can carry three times the moment M of the one on the left. Note also that both designs require the same amount of weld metal and welding time. In Fig. 30.32c, the weld on the left requires about twice the amount of weld material than does the design on the right. Moreover, because more material must be machined, the design on the left will require more time for edge preparation, and more base metal will be wasted.

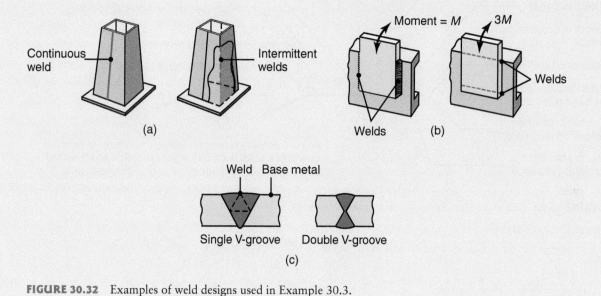

FIGURE 30.32   Examples of weld designs used in Example 30.3.

SUMMARY

- Oxyfuel–gas, arc, and high-energy-beam welding are among the most commonly used joining operations. Gas welding uses chemical energy; and to supply the necessary heat, arc and high-energy-beam welding use electrical energy.

- In all of the processes described, heat is used to bring the joint being welded to a liquid state. Shielding gases are used to protect the molten-weld pool and the weld area against oxidation. Filler metals may or may not be used in oxyfuel–gas and arc welding.

- The selection of a welding process for a particular operation depends on the workpiece material, its thickness and size, its shape complexity, the type of joint required, the strength required, and the change in product appearance caused by welding.

- A variety of welding equipment is available, much of which is now computer and robotics controlled, with programmable features.

- Cutting of metals also can be done by the processes based on oxyfuel–gas and arc welding. Cutting of steels occurs mainly through oxidation (burning) of the material. The highest temperatures for cutting are obtained by plasma-arc cutting.

- The welded joint consists of solidified metal and a heat-affected zone; each has a wide variation in microstructure and properties, depending on the metals joined and on the filler metals. The metallurgy of the welded joint is an important aspect of all welding processes, because it determines the strength, toughness, and quality of the joint.

- Discontinuities, such as porosity, inclusions, incomplete welds, tears, surface damage, and cracks, can develop in the weld zone. Residual stresses and relieving them are important considerations.

- The weldability of metals and alloys depends greatly on their composition, mechanical and physical properties, type of welding operation and process parameters employed, and the control of welding parameters.

- General guidelines are available for the selection of suitable and economical methods for a particular welding application.

KEY TERMS

Arc cutting	Electroslag welding	Laser-beam welding	Shielded metal-arc welding
Arc welding	Filler metal	Laser SMAW welding	Slag
Atomic-hydrogen welding	Flux	Neutral flame	Stick welding
Base metal	Flux-cored arc welding	Nonconsumable electrode	Submerged-arc welding
Carburizing flame	Fusion welding	Oxidizing flame	Tears
Coated electrode	Gas metal-arc welding	Oxyfuel–gas cutting	Thermit welding
Consumable electrode	Gas tungsten-arc welding	Oxyfuel–gas welding	Weld profile
Discontinuities	Heat-affected zone	Plasma-arc welding	Weld metal
Drag lines	Inclusions	Polarity	Weldability
Electrode	Joining	Porosity	Welding gun
Electrogas welding	Kerf	Reducing flame	
Electron-beam welding	Keyhole technique	Residual stresses	

BIBLIOGRAPHY

ASM Handbook, Vol. 6: **Welding, Brazing, and Soldering,** ASM International, 1993.

ASM Handbook, Vol. 6A: **Welding Fundamentals and Processes,** ASM International, 2011.

Cary, H.B., and Helzer, S., **Modern Welding Technology,** 6th ed., Prentice Hall, 2004.

Davies, A.C., **The Science and Practice of Welding,** 10th ed. (2 vols.), Cambridge University Press, 1993.

Duley, W.W., **Laser Welding,** Wiley, 1998.

Houldcroft, P.T., **Welding and Cutting: A Guide to Fusion Welding and Associated Cutting Processes,** Industrial Press, 2001.

Kou, S., **Welding Metallurgy,** 2nd ed., Wiley-Interscience, 2002.

Steen, W.M., and Mazumder, J., **Laser Material Processing,** 4th ed., Springer, 2010.

Welding Handbook, 9th ed. (3 vols.), American Welding Society, 2007.

Welding Inspection Handbook, American Welding Society, 2000.

Weman, K., **Welding Processes Handbook,** CRC Press, 2003.

REVIEW QUESTIONS

30.1 Describe fusion as it relates to welding operations.

30.2 Explain the features of neutralizing, reducing, and oxidizing flames. Why is a reducing flame so called?

30.3 What is stick welding?

30.4 Explain the basic principles of arc-welding processes.

30.5 Why is shielded metal-arc welding a commonly used process?

30.6 What keeps the weld bead on a steel surface from oxidizing (rusting) during welding?

30.7 Describe the functions and characteristics of electrodes. What functions do coatings have? How are electrodes classified?

30.8 What are the similarities and differences between consumable and nonconsumable electrodes?

30.9 What properties are useful for a shielding gas?

30.10 What are the advantages to thermit welding?

30.11 Explain where the energy is obtained in thermit welding.

30.12 Explain how cutting takes place when an oxyfuel–gas torch is used. How is underwater cutting done?

30.13 What is the purpose of flux? Why is it not needed in gas tungsten-arc welding?

30.14 What is meant by weld quality? Discuss the factors that influence it.

30.15 How is weldability defined?

30.16 Why are welding electrodes generally coated?

30.17 Describe the common types of discontinuities in welded joints.

30.18 What types of destructive tests are performed on welded joints?

30.19 Explain why hydrogen welding can be used to weld tungsten without melting the tungsten electrode.

30.20 What materials can be welded by laser SMAW hybrid welding?

QUALITATIVE PROBLEMS

30.21 Explain the reasons that so many different welding processes have been developed over the years.

30.22 It has been noted that heat transfer in gas-metal arc welding is higher than in shielded-metal arc welding. Explain why this would be the case. Which process would lead to more heat-affected zone cracking in hardened steels?

30.23 Explain why some joints may have to be preheated prior to welding.

30.24 Describe the role of filler metals in welding.

30.25 List the processes that can be performed with two electrodes. What are the advantages in using two electrodes?

30.26 What is the effect of the thermal conductivity of the workpiece on kerf width in oxyfuel–gas cutting? Explain.

30.27 Describe the differences between oxyfuel–gas cutting of ferrous and of nonferrous alloys. Which properties are significant?

30.28 Could you use oxyfuel–gas cutting for a stack of sheet metals? (*Note:* For stack cutting, see Fig. 24.25e.) Explain.

30.29 What are the advantages of electron-beam and laser-beam welding compared with arc welding?

30.30 Describe the methods by which discontinuities in welding can be avoided.

30.31 Explain the significance of the stiffness of the components being welded on both weld quality and part shape.

30.32 Comment on the factors that influence the size of the two weld beads shown in Fig. 30.15.

30.33 Which of the processes described in this chapter are not portable? Can they be made so? Explain.

30.34 Thermit welding is commonly used for welding railroad rails. List the reasons that make thermit welding attractive for this application. Review your list and create a list of products that would be suitable for thermit welding, and then identify any difficulties you would expect in applying thermit welding to that application.

30.35 Describe your observations concerning the contents of Table 30.1.

30.36 What determines whether a certain welding process can be used for workpieces in horizontal, vertical, or upside-down positions—or, for that matter, in any position? (See Table 30.1.) Explain and give examples of appropriate applications.

30.37 Comment on the factors involved in electrode selection in arc-welding processes.

30.38 In Table 30.1, the column on the distortion of welded components is ordered from lowest distortion to highest. Explain why the degree of distortion varies among different welding processes.

30.39 Explain the significance of residual stresses in welded structures.

30.40 Rank the processes described in this chapter in terms of (a) cost and (b) weld quality.

30.41 Must the filler metal be made of the same composition as the base metal that is to be welded? Explain.

30.42 What is weld spatter? What are its sources? How can spatter be controlled? Explain.

30.43 Describe your observations concerning Fig. 30.20.

30.44 If the materials to be welded are preheated, is the likelihood for porosity increased or decreased? Explain.

QUANTITATIVE PROBLEMS

30.45 Plot the hardness in Fig. 30.20d as a function of the distance from the top surface, and discuss your observations.

30.46 A welding operation will take place on carbon steel. The desired welding speed is around 0.8 in./s. If an arc-welding power supply is used with a voltage of 12 V, what current is needed if the weld width is to be 0.2 in.?

30.47 In Fig. 30.26b, assume that most of the top portion of the top piece is cut horizontally with a sharp saw. The residual stresses will now be disturbed and the part will change its shape, as was described in Section 2.11. For this case, how do you think the part will distort: curved downward or upward? Explain. (See also Fig. 2.30d.)

30.48 A welding operation takes place on an aluminum-alloy plate. A pipe 2.5 in. in diameter, with a 0.20 in. wall thickness and a 2 in. length, is butt welded onto an angle iron 6 in. × 6 in. × 0.25 in. thick. The angle iron is of an L cross-section and has a length of 1 ft. If the weld zone in a gas tungsten arc welding process is approximately 0.5 in. wide, what would be the temperature increase of the entire structure due to the heat input from welding only? What if the process were an electron-beam welding operation, with a bead width of 0.08 in.? Assume that the electrode requires 1500 J and the aluminum alloy requires 1200 J to melt one gram.

30.49 An arc welding operation is taking place on carbon steel. The desired welding speed is around 1 in./s. If the power supply is 10 V, what current is needed if the weld width is to be 0.25 in.?

30.50 In oxyacetylene, arc, and laser-beam cutting, the processes basically involve melting of the workpiece. If a 80.0 mm diameter hole is to be cut from a 250 mm diameter, 12 mm thick plate, plot the mean temperature rise in the blank as a function of kerf. Assume that one-half of the energy goes into the blank.

SYNTHESIS, DESIGN, AND PROJECTS

30.51 Comment on workpiece size and shape limitations for each of the processes described in this chapter.

30.52 Arc blow is a phenomenon where the magnetic field induced by the welding current, passing through the electrode and workpiece in shielded metal arc welding, interacts with the arc and causes severe weld splatter. Identify the variables that you think are important in arc blow. When arc blow is a problem, would you recommend minimizing it by using AC or DC power?

30.53 Review the types of welded joints shown in Fig. 30.29 and give an application for each.

30.54 Comment on the design guidelines given in various sections of this chapter.

30.55 You are asked to inspect a welded structure for a critical engineering application. Describe the procedure that you would follow in order to determine the safety of the structure.

30.56 Discuss the need for, and the role of, work-holding devices in the welding operations described in this chapter.

30.57 Make a list of welding processes that are suitable for producing (a) butt joints, where the weld is in the form of a line or line segment, (b) spot welds, and (c) both butt joints and spot welds. Comment on your observations.

30.58 Explain the factors that contribute to the differences in properties across a welded joint.

30.59 Explain why preheating the components to be welded is effective in reducing the likelihood of developing cracks.

30.60 Review the poor and good joint designs shown in Fig. 30.31, and explain why they are labeled so.

30.61 In building large ships, there is a need to weld thick and large sections of steel together to form a hull. Consider each of the welding operations discussed in this chapter, and list the benefits and drawbacks of that particular joining operation for this application.

30.62 Inspect various parts and components in (a) an automobile, (b) a major appliance, and (c) kitchen utensils, and explain which, if any, of the processes described in this chapter has been used in joining them.

30.63 Comment on whether there are common factors that affect the weldability, castability, formability, and machinability of metals, as described in various chapter of this book. Explain with appropriate examples.

30.64 If you find a flaw in a welded joint during inspection, how would you go about determining whether or not the flaw is significant?

30.65 Lattice booms for cranes are constructed from extruded cross-sections (see Fig. 15.2) that are welded together. Any warpage that causes such a boom to deviate from straightness will severely reduce its lifting capacity. Conduct a literature search on the approaches used to minimize distortion due to welding and how to correct it, specifically in the construction of lattice booms.

30.66 A common practice in repairing expensive broken or worn parts (such as may occur when a fragment is broken from a forging) is to fill the area with layers of weld beads and then to machine the part down to its original dimensions. Make a list of the precautions that you would suggest to someone who uses this approach.

30.67 Consider a butt joint that is to be welded. Sketch the weld shape you would expect for (a) SMAW, (b) laser welding, and (c) laser-SMAW hybrid welding. Indicate the size and shape of the heat-affected zone you would expect. Comment on your observations.

30.68 Make an outline of the general guidelines for safety in welding operations described in this chapter. For each of the operations, prepare a poster which effectively and concisely gives specific instructions for safe practices in welding (or cutting). Review the various publications of the National Safety Council and other similar organizations.

CHAPTER

31

Solid-State Welding Processes

- This chapter describes a family of joining processes in which the workpieces do not undergo a phase change, and if heat is used, it is generated internally.
- The chapter begins with a discussion of cold welding, followed by ultrasonic welding and the various forms of friction-welding processes.
- Resistance welding is then described, followed by explosion welding and diffusion bonding; these three processes have unique capabilities and applications suitable for a wide variety of materials, and can be automated for large-scale production.
- The chapter also examines special capabilities of diffusion bonding and joining processes that are combined with superplastic forming.
- Finally, economic considerations in welding are discussed.

31.1 Introduction

This chapter describes **solid-state welding** processes where joining takes place without fusion at the interface of the two parts to be welded. Unlike in the fusion-welding processes described in Chapter 30, in solid-state welding no liquid or molten phase is required for joining. The principle of solid-state welding is demonstrated best with the following example: If two clean surfaces are brought into close contact with each other under sufficient pressure, they form a bond and produce a joint. For a strong bond, it is essential that the interface be free of contaminants, such as oxide films, residues, metalworking fluids, and even adsorbed layers of gas.

Solid-state bonding involves one or more of the following parameters:

- **Heat:** Applying external heat increases *diffusion* (the transfer of atoms across an interface) and improves the strength of the weld between the two surfaces being joined, as occurs in *diffusion bonding*. Heat may be generated (a) internally by friction (as utilized in friction welding), (b) through electrical-resistance heating (as in *resistance-welding* processes, such as *spot welding*), and (c) externally by induction heating (as in *butt-welding* tubes).
- **Pressure:** The higher the pressure, the stronger is the resulting interface (as in *roll bonding* and *explosion welding*), where plastic deformation also occurs. Pressure and heat may be combined, as in *flash welding*, *stud welding*, and *resistance projection welding*.
- **Relative interfacial movements:** When relative motion (sliding) of the contacting surfaces (called *faying surfaces*) occurs (as in *ultrasonic welding*), even very

small amplitudes will disturb the interface, break up any oxide films present, and generate new, clean surfaces, thus improving the strength of the weld.

Most joining processes are now automated by robotics, vision systems, sensors, and adaptive and computer controls (described in Part VIII) for the purposes of cost reduction, consistency of operation, reliability of weld quality, and higher productivity. The costs involved in joining process are outlined in Section 31.8.

31.2 Cold Welding and Roll Bonding

In *cold welding* (CW), pressure is applied to the workpieces through dies or rolls. Because of the *plastic deformation* involved, it is necessary that at least one (but preferably both) of the mating parts be sufficiently ductile. Thus, cold welding is usually performed on nonferrous metals or on soft iron with little, if any, carbon content. Prior to welding, the interface is degreased, wire brushed, and wiped to remove oxide smudges. Applications include products made of wire and electrical connections.

During the joining of two dissimilar metals that are mutually soluble, brittle *intermetallic* compounds may form (Section 4.2.2), which will produce a weak and brittle joint; an example is the bonding of aluminum and steel. The best bond strength is obtained with two similar materials.

Roll Bonding. The pressure required for welding can be applied through a pair of rolls (Fig. 31.1), a process called *roll bonding* or *roll welding* (ROW). Developed in the 1960s, roll bonding is used for manufacturing some U.S. coins (see Example 31.1). Surface preparation is important for good interfacial strength. The operation also can be carried out at elevated temperatures (*hot roll bonding*).

Typical examples of this process are the *cladding* of (a) pure aluminum over precipitation-hardened aluminum-alloy sheet (Alclad, a trade name), for a corrosion-resistant surface with a strong inner core, typically used in the aerospace industry, (b) stainless steel over mild steel, for corrosion resistance, and (c) copper over steel, for coaxial cables. A common application of roll bonding is the production of *bimetallic strips* for thermostats and similar controls, using two layers of materials with different thermal-expansion coefficients (see Table 3.1). Bonding in only selected regions in the interface can be achieved by depositing a parting agent, such as graphite or ceramic, called *stop-off* (see Section 31.7).

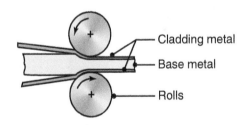

FIGURE 31.1 Schematic illustration of the roll bonding, or cladding, process.

EXAMPLE 31.1   Roll Bonding of the U.S. Quarter

The technique used for manufacturing composite U.S. quarters is the roll bonding of two outer layers of 75% Cu–25% Ni (cupronickel), where each layer is 1.2 mm (0.048 in.) thick, with an inner layer of pure copper 5.1 mm (0.20 in.) thick. To obtain good bond strength, the faying surfaces are cleaned chemically and wire brushed. First, the strips are rolled to a thickness of 2.29 mm (0.090 in.); a second rolling operation reduces the thickness to 1.36 mm (0.0535 in.). The strips thus undergo a total reduction in thickness of 82%.

Because of plastic deformation, there is a major increase in the surface area between the layers, which causes the generation of clean interfacial surfaces. This extension in surface area under the high pressure applied by the rolls, combined with the solid solubility of nickel in copper (see Section 4.2.1), produces a strong bond between the metal layers.

31.3   Ultrasonic Welding

In *ultrasonic welding* (USW), the faying surfaces of the two components are subjected to a normal force and oscillating shearing (tangential) stresses. The shearing stresses are applied by the tip of a **transducer** (Fig. 31.2a), which is similar to that used for ultrasonic machining (see Fig. 26.26a). The frequency of oscillation is generally in the range of 10–75 kHz, although a lower or higher frequency can be employed. Proper coupling between the transducer and the tip (called a **sonotrode**, from the words *sonic* and *electrode*; also called the *horn*) is important for efficient operation.

The shearing stresses cause plastic deformation at the interface of the two components, breaking up oxide films and contaminants, and thus allowing good contact and producing a strong solid-state bond. The temperature generated in the weld zone is usually one-third to one-half of the melting point (on the absolute scale) of the metals joined; consequently, neither melting nor fusion takes place. In certain situations, however, the temperature generated can be sufficiently high to cause metallurgical changes in the weld zone, thus affecting the strength of the bond.

The ultrasonic-welding process is versatile and reliable, and it can be used with a wide variety of metallic and nonmetallic materials, including dissimilar metals (as in bimetallic strips). It is used extensively for the joining of plastics (see Section 32.6), for packaging with foils, and in the automotive and consumer electronics industries for the lap welding of sheet, foil, and thin wire. The welding tip can be replaced with *rotating disks* (Fig. 31.2b) for the seam welding of structures in which one component is sheet, foil, or polymer-woven material (a process similar to *resistance seam welding*, Section 31.5.2).

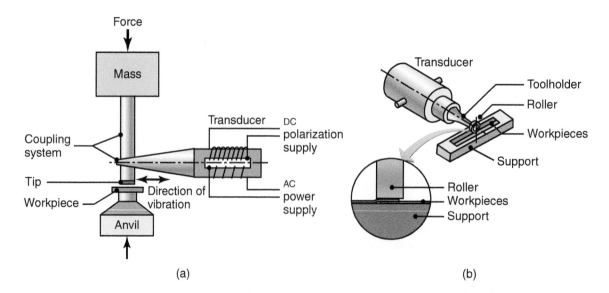

(a)   (b)

FIGURE 31.2   (a) Components of an ultrasonic-welding machine for making lap welds; the lateral vibrations of the tool tip cause plastic deformation and bonding at the interface of the workpieces. (b) Ultrasonic seam welding, using a roller as the *sonotrode*.

31.4 Friction Welding

In the joining processes described thus far, the energy required for welding is supplied from external sources, typically chemical, electrical, or ultrasonic energy. In *friction welding* (FRW), the heat required for welding is generated through friction at the interface of the two components being joined.

Developed in the 1940s, one of the workpiece components in friction welding remains stationary while the other is placed in a chuck or collet, and rotated at a constant peripheral speed as high as 15 m/s (3000 ft/min). The two members to be joined are then brought into contact under an axial force (Fig. 31.3). After sufficient contact is established, the rotating member is brought to a quick stop (so that the weld is not destroyed by shearing) while the axial force is increased. Oxides and other contaminants at the interface are expelled by the radially outward movement of the hot metal at the interface.

The pressure at the interface and the heat resulting from friction are sufficient for a strong joint to form. The weld zone is usually confined to a narrow region, and its size and shape depend on the (a) amount of heat generated, (b) thermal conductivity of the materials, (c) mechanical properties of the materials being joined at elevated temperatures, (d) rotational speed, and (e) axial pressure applied (Fig. 31.4). These factors must be controlled to obtain a uniform and strong joint.

Friction welding can be used to join a wide variety of materials, provided that one of the components has some rotational symmetry. Solid or tubular parts can be welded, with good joint strength. Solid steel bars up to 100 mm (4 in.) in diameter and pipes up to 250 mm (10 in.) in outside diameter have been friction welded successfully. Because of the combined heat and pressure, the interface in friction

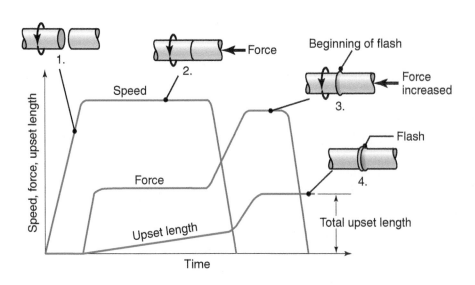

FIGURE 31.3 Sequence of operations in the friction-welding process: (1) The part on the left is rotated at high speed. (2) The part on the right is brought into contact with the part on the left under an axial force. (3) The axial force is increased, and the part on the left stops rotating; flash begins to form. (4) After a specified upset length or distance is achieved, the weld is completed. The *upset length* is the distance the two pieces move inward during welding after their initial contact; thus, the total length after welding is less than the sum of the lengths of the two pieces. The flash subsequently can be removed by machining or grinding.

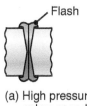

(a) High pressure or low speed

(b) Low pressure or high speed

(c) Optimum

FIGURE 31.4 Shape of the fusion zones in friction welding, as a function of the axial force applied and the rotational speed.

welding develops a *flash* by plastic deformation (upsetting) of the heated zone. This flash (if objectionable) can easily be removed by machining or grinding. Friction-welding machines are fully automated, and the operator skill required is minimal, once individual cycle times for the complete operation are set properly.

QR Code 31.1 Demonstration of inertia friction welding. (*Source:* Courtesy of Manufacturing Technology, Inc. [MTI])

QR Code 31.2 Demonstration of linear friction welding. (*Source:* Courtesy of Manufacturing Technology, Inc. [MTI])

QR Code 31.3 Demonstration of friction stir welding. (*Source:* Courtesy of Manufacturing Technology, Inc. [MTI])

Inertia Friction Welding. This process is a modification of friction welding, although the two terms have been used interchangeably. In *inertia friction welding*, the energy required for frictional heating is supplied by a flywheel. The flywheel is first accelerated to the proper speed, the two members are brought into contact, and an axial force is applied. As friction at the interface begins to slow the flywheel, the axial force is increased; the weld is completed when the flywheel comes to a stop. The timing of this sequence is important for good weld quality.

The rotating mass in inertia friction welding machines can be adjusted for applications requiring different levels of energy, where the levels depend on the workpiece size and properties. In one application of this process, 10-mm- (0.4-in.-) diameter shafts are welded to automotive turbocharger impellers, at a rate of one joint every 15 s.

Linear Friction Welding. In a further development of friction welding, the interface of the two components to be joined is subjected to a *linear reciprocating motion*, as opposed to a rotary motion. Thus, in *linear friction welding*, the components do not have to be circular or tubular in their cross-section. In this operation, one part is moved across the face of the other part by a balanced reciprocating mechanism. The process is capable of welding square or rectangular components, as well as round parts, made of metals or plastics.

In one application of this process, a rectangular titanium-alloy part was friction welded at a linear frequency of 25 Hz with an amplitude of ± 2 mm (0.08 in.) and under a pressure of 100 MPa (15,000 psi) acting on a 240 mm^2 (0.38 in^2) interface. Various other metal parts, with rectangular cross-sections as large as 50 mm × 20 mm (2 in. × 0.8 in.), have been welded successfully.

Friction Stir Welding. In the *friction stir welding* (FSW) process, developed in 1991, a third body (called a *probe*) is plunged into the joint and rubs against the two surfaces to be joined. The nonconsumable rotating probe is typically made of cubic boron nitride (Section 8.2.3), 5 to 6 mm in diameter and 5 mm high. (Fig. 31.5). The contact pressure causes frictional heating, raising the temperature to 230° to 260°C (450°–500°F). The tip of the rotating probe forces mixing or stirring of the material in the joint. No shielding gas or surface cleaning is required.

The thickness of the welded material can be as little as 1 mm and as much as 50 mm (2 in.), welded in a single pass. Aluminum, magnesium, nickel, copper, steel, stainless steel, and titanium have been welded successfully, and developments are taking place to extend FSW applications also to polymers and composite materials. The FSW process is now being applied to aerospace, automotive, shipbuilding, and military vehicles, using sheet or plates. With developments in rotating-tool design, other possible applications include inducing microstructural changes, refining grain in materials, and improving localized toughness in castings.

The welding equipment can be a conventional, vertical-spindle milling machine (see Fig. 24.15b), and the process is relatively easy to implement. For special applications, machinery dedicated to friction stir welding is also available (Fig. 31.5b). Welds produced by FSW have high quality, minimal pores, and a uniform material structure. Because the welds are produced with low heat input, there is low distortion and little microstructural changes.

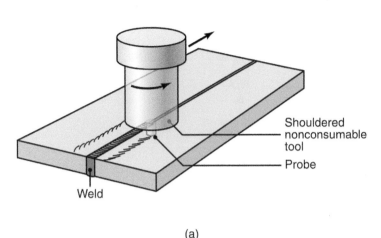

Shouldered
nonconsumable
tool

Probe

Weld

(a)

(b)

FIGURE 31.5   The friction-stir-welding process. (a) Schematic illustration of friction stir welding; aluminum-alloy plates up to 75 mm (3 in.) thick have been welded by this process. (b) Multi-axis friction stir welding machine for large workpieces, such as aircraft wing and fuselage structures, that can develop 67 kN (15,000 lb) axial forces, is powered by a 15 kW (20 hp) spindle motor, and can achieve welding speeds up to 1.8 m/s. *Source:* (b) Courtesy of Manufacturing Technology, Inc.

31.5   Resistance Welding

The category of *resistance welding* (RW) covers a number of processes in which the heat required for welding is produced by means of *electrical resistance* across the two components to be joined. These processes have major advantages, such as high-quality welds that do not require consumable electrodes, shielding gases, or flux, and can be produced quickly. Resistance welding lends itself very well to automation, and is often applied using welding robots (see Section 37.6).

The heat generated in resistance welding is given by the general expression

$$H = I^2Rt, \quad (31.1)$$

where

H = Heat generated in joules (watt-seconds)
I = Current (in amperes)
R = Resistance (in ohms)
t = Time of current flow (in seconds)

Equation (31.1) is often modified, so that it represents the actual heat energy available in the weld, by including a factor K, which compensates for the energy losses through conduction and radiation. The equation then becomes

$$H = I^2RtK, \quad (31.2)$$

where it can be noted that the value of K is less than unity.

Video Solution 31.1   Current in Resistance Welding

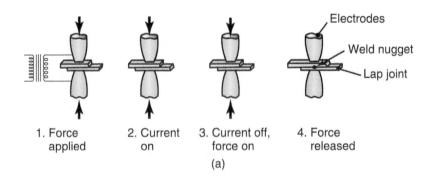

1. Force applied   2. Current on   3. Current off, force on   4. Force released

(a)

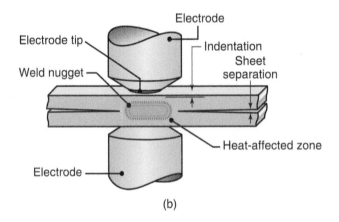

(b)

FIGURE 31.6 (a) Sequence of events in resistance spot welding of a lap joint. (b) Cross-section of a spot weld, showing the weld nugget and the indentation of the electrode on the sheet surfaces. This is one of the most commonly used processes in sheet-metal fabrication and in automotive-body assembly.

The *total resistance* is the sum of the following (see Fig. 31.6):

1. Resistances of the electrodes
2. Electrode–workpiece contact resistance
3. Resistances of the individual parts to be welded
4. Contact resistance between the two workpieces to be joined (faying surfaces)

The actual temperature rise in the joint depends on the specific heat and the thermal conductivity of the metals to be joined. For example, metals such as aluminum and copper have high thermal conductivity (see Table 3.1), thus they require high heat concentrations. Similar as well as dissimilar metals can be joined by resistance welding. The magnitude of the current in resistance-welding operations may be as high as 100,000 A, but the voltage is typically only 0.5 to 10 V. The strength of the bond depends on surface roughness and on the cleanliness of the mating surfaces. Oil films, paint, and thick oxide layers should therefore be removed prior to welding, although the presence of uniform, *thin* layers of oxide and other contaminants is not as critical.

Developed in the early 1900s, resistance-welding processes require specialized machinery, much of which is now operated by programmable computer control. Generally, the machinery is not portable, and the process is suitable primarily for use in manufacturing plants and machine shops. The operator skill required is minimal, particularly with modern machinery.

31.5.1   Resistance Spot Welding

In *resistance spot welding* (RSW), the tips of two opposing solid, cylindrical electrodes touch a lap joint of two sheet metals, and resistance heating produces a spot weld (Fig. 31.6a). In order to obtain a strong bond in the **weld nugget**, pressure is applied until the current is turned off and the weld has solidified. Accurate control and timing of the alternating current (AC) and of the pressure are essential in resistance welding. In the automotive industry, for example, the number of cycles ranges up to about 30 at a frequency of 60 Hz. (See also *high-frequency resistance welding* in Section 31.5.3)

The surfaces of a spot weld have a slightly discolored indentations. The weld nugget (Fig. 31.6b) may be up to 10 mm (0.375 in.) in diameter. Currents range from 3000 to 40,000 A, the current level depending on the materials being welded and their thicknesses; for example, the current is typically 10,000 A for steels and 13,000 A for aluminum. Electrodes generally are made of copper alloys and must have sufficient electrical conductivity and hot strength to maintain their shape.

The simplest and most commonly used resistance-welding process, spot welding, may be performed by means of single (most common) or multiple pairs of electrodes (as many as a hundred or more); the required pressure is supplied through mechanical or pneumatic means. **Rocker-arm-type** spot-welding machines normally are used for smaller parts; **press-type** machines are used for larger workpieces. The shape and surface condition of the electrode tip and the accessibility of the site are important factors in spot welding. A variety of electrode shapes are used for areas that are difficult to reach (Fig. 31.7).

Spot welding is widely used for fabricating sheet-metal parts; examples range from attaching handles to stainless-steel cookware (Fig. 31.8a) to spot-welding mufflers (Fig. 31.8b) and large sheet-metal structures. Modern spot-welding

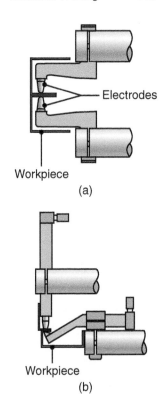

FIGURE 31.7   Two electrode designs for easy access to the components to be welded.

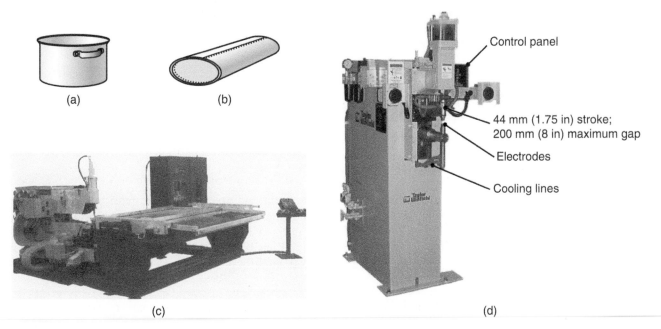

FIGURE 31.8   Spot-welded (a) cookware and (b) muffler. (c) A large automated spot-welding machine. The welding tip can move in three principal directions; sheets as large as 2.2 m × 0.55 m (88 in. × 22 in.) can be accommodated in this machine, with proper workpiece supports. (d) A spot welding machine. *Source:* (c) and (d) Courtesy of Taylor-Winfield Technologies, Inc.

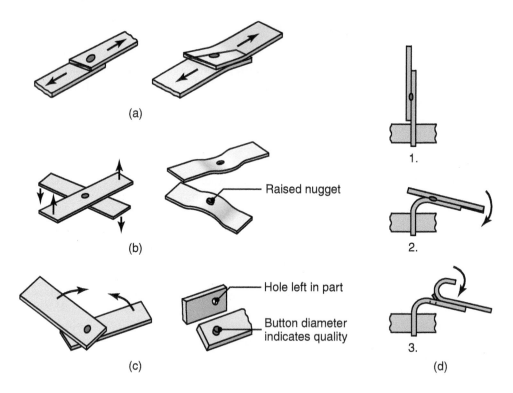

FIGURE 31.9 Test methods for spot welds: (a) tension-shear test, (b) cross-tension test, (c) twist test, and (d) peel test (see also Fig. 32.9).

equipment is computer controlled, for optimum timing of current and pressure, and the spot-welding guns are manipulated by programmable robots. Automobile bodies can have as many as 10,000 spot welds; they are welded at high rates using multiple electrodes (see Fig. I.9 in the General Introduction).

Testing Spot Welds.   Spot-welded joints may be tested for weld-nugget strength by means of the following techniques (Fig. 31.9):

- Tension-shear
- Cross-tension
- Twist
- Peel

Because they are easy to perform and are inexpensive, tension-shear tests are commonly used in fabricating facilities. The cross-tension and twist tests are capable of revealing flaws, cracks, and porosity in the weld area. The peel test is commonly used for thin sheets; after the joint has been bent and peeled, the shape and size of the torn-out weld nugget are evaluated.

EXAMPLE 31.2   Heat Generated in Spot Welding

Given: Assume that two 1-mm (0.04-in.) thick steel sheets are being spot-welded at a current of 5000 A and over a current flow time of 0.1 s by means of electrodes 5 mm (0.2 in.) in diameter.

Find: Estimate the heat generated and its distribution in the weld zone if the effective resistance in the operation is 200 $\mu\Omega$.

Solution: According to Eq. (31.1),

Heat $= (5000)^2(0.0002)(0.1) = 500$ J.

From the information given, the weld-nugget volume can be estimated to be 30 mm^3 (0.0018 in^3). Assume that the density for steel (Table 3.1) is 8000 kg/m^3; then the weld nugget has a mass of 0.24 g. The heat required to melt 1 g of steel is about 1400 J, so the heat required to melt the weld nugget is $(1400)(1400)(0.24) = 336$ J. The remaining heat (164 J) is dissipated into the metal surrounding the nugget.

31.5.2   Resistance Seam Welding

Resistance seam welding (RSEW) is a modification of spot welding wherein the electrodes are replaced by rotating wheels or rollers (Fig. 31.10a). Using a continuous AC power supply, the electrically conducting rollers produce a spot weld whenever the current reaches a sufficiently high level in the AC cycle. The typical welding speed is 1.5 m/min (60 in./min) for thin sheets.

With a sufficiently high frequency or slow traverse speed, these spot welds actually overlap into a continuous seam and produce a joint that is liquid- and gas-tight (Fig. 31.10b). The RSEW process is used to make the longitudinal seam of steel cans (for household products), mufflers, and gasoline tanks.

In **roll spot welding**, the current to the rolls is applied only intermittently, resulting in a series of spot welds that are at specified intervals along the length of the seam (Fig. 31.10c). In **mash seam welding** (Fig. 31.10d), the overlapping welds are about one to two times the sheet thickness, and the welded seam thickness is only about 90% of the original sheet thickness. This process is also used in producing *tailor-welded sheet-metal blanks*, which can be made by laser welding as well (see Section 16.2.2).

31.5.3   High-frequency Resistance Welding

High-frequency resistance welding (HFRW) is similar to seam welding, except that a high-frequency current of up to 450 kHz is employed. A typical application is the production of *butt-welded* tubing or pipe, where the current is conducted through two sliding contacts (Fig. 31.11a) to the edges of roll-formed tubes. The heated edges are then pressed together by passing the tube through a pair of squeeze rolls; the flash formed, if any, is then trimmed off.

Structural sections, such as I-beams, can be fabricated by HFRW, by welding the webs and flanges made from long, flat pieces. Spiral pipe and tubing, finned tubes

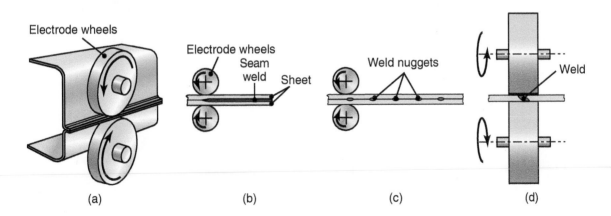

(a)   (b)   (c)   (d)

FIGURE 31.10   (a) Seam-welding process in which rotating rolls act as electrodes; (b) overlapping spots in a seam weld; (c) roll spot welds; and (d) mash seam welding.

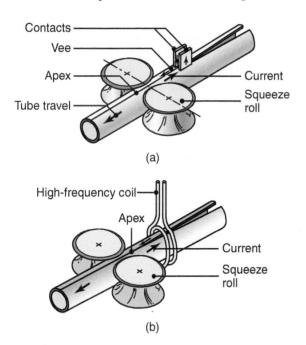

(a)

(b)

FIGURE 31.11 Two methods of high-frequency continuous butt welding of tubes.

for heat exchangers, and wheel rims also may be made by this technique. In another method, called **high-frequency induction welding** (HFIW), the roll-formed tube is subjected to high-frequency induction heating, as shown in Fig. 31.11b.

31.5.4 Resistance Projection Welding

In *resistance projection welding* (RPW), high electrical resistance at the joint is developed by embossing one or more projections (*dimples*; see Fig. 16.39) on one of the surfaces to be welded (Fig. 31.12). The projections may be round or oval for design or strength purposes. High localized temperatures are generated at the projections, which are in contact with the flat mating part. Typically made of copper-based alloys, the electrodes are large and flat, and are water cooled to keep their temperature low. Weld nuggets are similar to those in spot welding, and are formed as the electrodes exert pressure to soften and compress and flatten the projections.

Spot-welding equipment can be used for resistance projection welding by modifying the electrodes. Although the embossing of the workpieces adds to production cost, the operation produces several welds in one pass and extends electrode life; moreover, it is capable of welding metals of different thicknesses, such as a sheet welded over a plate. Nuts and bolts also can be welded to sheets and plates by this process (Figs. 31.12c and d), with projections that may be produced either by machining or forging. Joining a network of rods and wires (such as the ones making up metal baskets, grills (Fig. 31.12e), oven racks, and shopping carts) is considered resistance projection welding, because of the many small contact areas between crossing wires (grids).

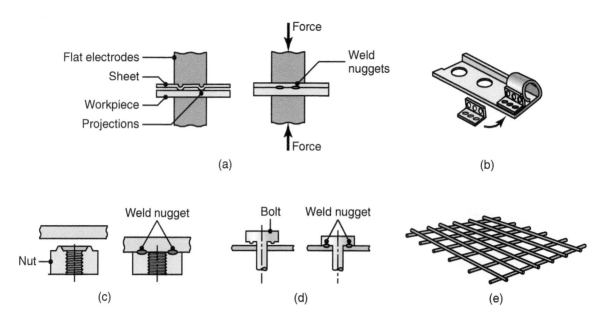

FIGURE 31.12 (a) Schematic illustration of resistance projection welding. (b) A welded bracket. (c) and (d) Projection welding of nuts or threaded bosses and studs. (e) Resistance-projection-welded grills.

31.5.5   Flash Welding

In *flash welding* (FW), also called **flash butt welding**, heat is generated very rapidly from the arc as the ends of the two members begin to make contact and develop an electrical arc at the joint (Fig. 31.13a). After the proper temperature is reached and the interface begins to soften, an axial force is applied at a controlled rate, producing a weld by plastic deformation of the joint. The mechanism involved is called *hot upsetting*, and the term *upset welding* (UW) also is used for this process. Some molten metal is expelled from the joint as a shower of sparks during the process, hence the name *flash welding*. Because of the presence of an arc, the process can also be classified as arc welding.

Impurities and contaminants are squeezed out during this operation, and a significant amount of material may be burned off during the welding process. The joint quality is good, and it may be machined later to further improve its appearance. The machines for flash welding usually are automated and large, and have a variety of power supplies ranging from 10 to 1500 kVA.

The FW process is suitable for end-to-end or edge-to-edge joining of strips and sheets of similar or dissimilar metals 0.2 to 25 mm (0.01–1 in.) thick, and for end-joining bars 1 to 75 mm (0.05–3 in.) in diameter. Thinner sections have a tendency to buckle under the axial force applied during welding. Rings made by forming (such as by the methods shown in Fig. 16.22) can be flash butt welded. In addition, the process is also used to repair broken band-saw blades (Section 24.5) using fixtures that are mounted on the band-saw frame.

The flash-welding process can be automated for reproducible welding operations. Typical applications are the joining of pipe and of tubular shapes for metal furniture, doors, and windows. The process also is used for welding the ends of sheets or coils of wire, in continuously operating rolling mills (Chapter 13) and in the feeding of

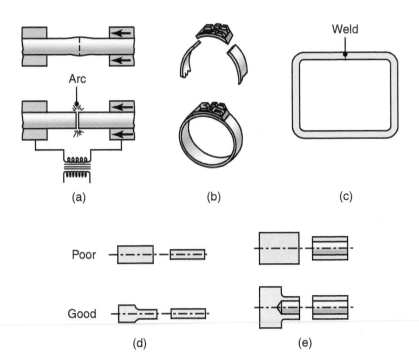

FIGURE 31.13   (a) Flash-welding process for end-to-end welding of solid rods or tubular parts. (b) and (c) Typical parts made by flash welding. (d) and (e) Some design guidelines for flash welding.

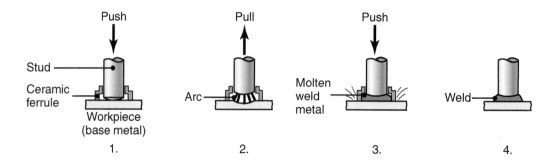

FIGURE 31.14 Sequence of operations in stud welding, commonly used for welding bars, threaded rods, and various fasteners onto metal plates.

wire-drawing equipment (Section 15.11). Some design guidelines for mating surfaces in flash welding are shown in Figs. 31.13d and e; note the importance of having uniform cross-sections at the joint.

31.5.6 Stud Welding

Stud welding (SW), also called *stud arc welding*, is similar to flash welding. The stud, which may be a threaded metal rod, hanger, or handle, serves as one of the electrodes while it is being joined to another component, usually a flat plate (Fig. 31.14). Polarity for aluminum is typically direct-current electrode positive (DCEP); for steel, it is direct-current electrode negative (DCEN).

In order to concentrate the heat generated, prevent oxidation, and retain the molten metal in the weld zone, a disposable ceramic ring (called *ferrule*) is placed around the joint. The equipment for stud welding can be automated, with various controls for arcing and for applying pressure. Portable stud-welding equipment also is available. Typical applications of stud welding include automobile bodies, electrical panels, and shipbuilding; the process is also used in building construction.

In **capacitor-discharge stud welding**, a DC arc is produced from a capacitor bank. No ferrule or flux is required, because the welding time is on the order of only 1 to 6 ms. The choice between this process and stud arc welding depends on such factors as the types of metals to be joined, the workpiece thickness and cross-section, the stud diameter, and the shape of the joint.

31.5.7 Percussion Welding

The resistance-welding processes already described usually employ an electrical transformer, to meet the power requirements; alternatively, the electrical energy for welding may be stored in a capacitor. *Percussion welding* (PEW) utilizes this technique, in which the power is discharged within 1 to 10 ms in order to develop localized high heat at the joint. The process is useful where heating of the components adjacent to the joint is to be avoided, as in electronic assemblies and electrical wires.

EXAMPLE 31.3   **Resistance Welding versus Laser-beam Welding in the Can-making Industry**

The cylindrical bodies of cans for food and for household products have been resistance seam welded (with a lap joint up the side of the can) for many years. Beginning in about 1987, laser-beam welding technology was introduced into the can-making industry. The joints are welded by lasers, with the same productivity as in resistance welding but with the following advantages:

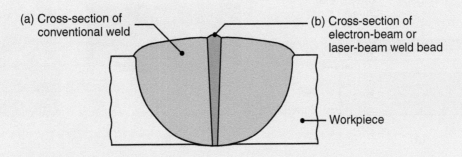

FIGURE 31.15 The relative sizes of the weld beads obtained by tungsten-arc and by electron-beam or laser-beam welding.

- As opposed to the lap joints suitable for resistance welding, laser welding utilizes butt joints; thus, some material is saved. Multiplied by the billions of cans made each year, this amount becomes a very significant savings.
- Because laser welds have a very narrow zone (Fig. 31.15; see also Fig. 30.15), the unprinted area on the can surface (called the printing margin) is greatly reduced. As a result, the can's appearance and its customer acceptance are improved.

- The resistance lap-welded joint can be subject to corrosion by the contents of the can (which can be acidic, such as tomato juice). This effect may change the taste and can cause a potential liability risk. A butt joint made by laser-beam welding eliminates this problem.

Source: Courtesy of G.F. Benedict.

31.6 Explosion Welding

In *explosion welding* (EXW), pressure is applied by detonating a layer of explosive that has been placed over one of the components being joined, called the *flyer plate* (Fig. 31.16). The contact pressures developed are extremely high (see also *explosive hardening*, Section 34.2), and the kinetic energy of the plate striking the mating component causes a wavy interface. This impact mechanically interlocks the two surfaces (Figs. 31.16b and c), so that pressure welding by plastic deformation also takes place. As a result, the bond strength from explosion welding is very high.

The explosive may be a flexible plastic sheet or cord or in granulated or liquid form, which is cast or pressed onto the flyer plate. The detonation speed is usually in the range from 2400 to 3600 m/s (8000–12,000 ft/s); it depends on the type of explosive, the thickness of the explosive layer, and the packing density of the layer. There is a minimum denotation speed necessary for welding to occur in this process. Detonation is carried out with a standard commercial blasting cap.

This process, developed in the 1960s, is particularly suitable for cladding a plate or a slab with a dissimilar metal. Plates as large as 6 m × 2 m (20 ft × 7 ft) have been clad explosively; they may then be rolled into thinner sections. Tubes and pipes can be joined to the holes in the header plates of boilers and heat exchangers by placing the explosive inside the tube; the explosion expands the tube. Explosion welding is inherently dangerous, thus it requires safe handling by well-trained and experienced personnel.

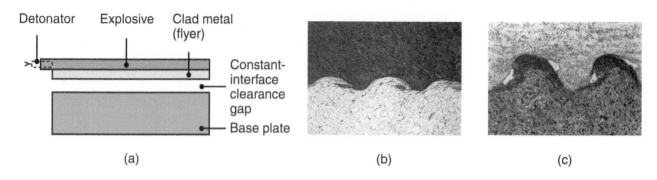

FIGURE 31.16   (a) Schematic illustration of the explosion-welding process. (b) Cross-section of explosion-welded joint: titanium (top) and low-carbon steel (bottom). (c) Iron–nickel alloy (top) and low-carbon steel (bottom).

31.7   Diffusion Bonding

Diffusion bonding, or **diffusion welding** (DFW), is a process in which the strength of the joint results primarily from diffusion (movement of atoms across the interface) and secondarily from plastic deformation of the faying surfaces. This process requires temperatures of about 0.5 T_m (where T_m is the melting point of the metal, on the absolute scale) in order to have a sufficiently high diffusion rate between the parts being joined (see also Sections 1.7 and 1.8).

The interface in diffusion welding has essentially the same physical and mechanical properties as the base metal; its strength depends on (a) pressure, (b) temperature, (c) time of contact, and (d) cleanliness of the faying surfaces. These requirements can be relaxed by using a filler metal at the interface. Depending on the materials joined, brittle intermetallic compounds may form at the interface, which may be avoided by first electroplating the surfaces with suitable metal alloys. In diffusion bonding, pressure may be applied by dead weights, a press, differential gas pressure, or the thermal expansion of the parts to be joined. The parts usually are heated in a furnace or by electrical resistance. High-pressure autoclaves also are used for bonding complex parts.

Although DFW was developed in the 1970s as a modern welding technology, the principle of diffusion bonding dates back centuries when goldsmiths bonded gold over copper, to create a product called **filled gold**. First, a thin layer of gold foil is placed over copper, and pressure is applied by a weight on top of the foil. The assembly is then placed in a furnace and left until a strong bond is developed, hence the process is also called *hot-pressure welding* (HPW).

Diffusion bonding generally is most suitable for joining dissimilar metals, and is also used for reactive metals (such as titanium, beryllium, zirconium, and refractory metal alloys) and for composite materials such as metal-matrix composites (Section 9.5). Diffusion bonding is an important mechanism of sintering in powder metallurgy (Section 17.4). Because diffusion involves migration of the atoms across the joint, DFW is slower than other welding processes.

Although diffusion welding is used for fabricating complex parts in low quantities for the aerospace, nuclear, and electronics industries, it has been automated to make it suitable and economical for moderate-volume production as well. Unless highly automated, considerable operator training and skill are required. Equipment cost is related approximately to the diffusion-bonded area, and is in the range of $3–$6/mm^2 ($2000–$4000/in^2).

EXAMPLE 31.4   Diffusion-bonding Applications

Diffusion bonding is especially suitable for such metals as titanium and the superalloys used in military aircraft. Design possibilities allow the conservation of expensive strategic materials and the reduction of

manufacturing costs. The military aircraft illustrated in Fig. 31.17 has more than 100 diffusion-bonded parts, some of which are shown.

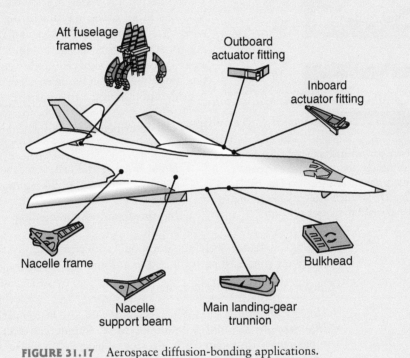

FIGURE 31.17   Aerospace diffusion-bonding applications.

Diffusion Bonding–Superplastic Forming.   Sheet-metal structures can be fabricated by combining *diffusion bonding* with *superplastic forming* (see also Section 16.10). Typical structures in which flat sheets are diffusion bonded and formed are shown in Fig. 31.18. After the diffusion bonding of selected locations on the sheets, the unbonded (*stop-off*) regions are expanded in a mold by air or fluid pressure. These structures are thin and have high stiffness-to-weight ratios, hence they are particularly useful in aircraft and aerospace applications.

Diffusion bonding–superplastic forming improves productivity by eliminating the number of parts in a structure, mechanical fasteners, labor, and cost. It produces parts with good dimensional accuracy and low residual stresses. First developed in the 1970s, this technology is now well advanced for titanium structures (typically using Ti-6Al-4V and 7475-T6) and various other alloys for aerospace applications.

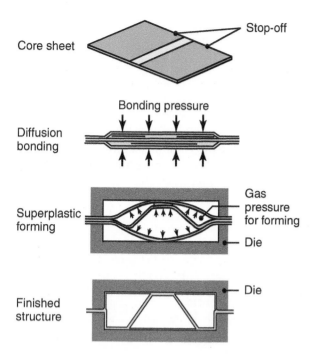

FIGURE 31.18 Sequence of operations in the fabrication of a structure by the diffusion bonding and superplastic forming of three originally flat sheets; see also Fig. 16.51. *Source:* (a) & (d) after D. Stephen and S.J. Swadling, (b) & (c) courtesy of Rockwell International Corp.

Other welding processes also can be used with post-welding superplastic forming of plates, notably friction welding and friction stir welding.

31.8 Economics of Welding Operations

The characteristics, advantages, and limitations of the welding processes described thus far have included brief discussions regarding welding costs. The relative costs of some selected processes are shown in Tables 30.1 and VI.1. As in all other manufacturing operations, costs in welding and joining processes can vary widely, depending on such factors as the equipment capacity, level of automation, labor skill required, weld quality, production rate, and preparation required, as well as on various other considerations specific to a particular joining operation.

Welding and joining costs for some common operations (all described throughout Chapters 30–32) may be summarized as:

- *High:* Brazing and fasteners (such as bolts and nuts), as they require hole-making operations and fastener costs
- *Intermediate:* Arc welding, riveting, adhesive bonding
- *Low:* Resistance welding, seaming, and crimping, as these operations are relatively simple to perform and to automate

Equipment costs for welding may be summarized as:

- *High* ($100,000–$200,000): Electron-beam and laser-beam welding
- *Intermediate* ($5,000–$50,000+): Spot, submerged arc, gas metal-arc, gas tungsten-arc, flux-cored arc, electrogas, electroslag, plasma arc, and ultrasonic welding
- *Low* ($500+): Shielded metal-arc and oxyfuel–gas welding

Labor costs in welding generally are higher than in other metalworking operations because of the operator skill, welding time, and the preparation required. Much also depends on the level of automation of the equipment employed, including the use of robotics and computer controls, programmed to follow a prescribed path (called *seam tracking*) during welding. It has been observed, for example, that in systems with robotic controls, the actual welding time reaches 80% of the total time, whereas in manual welding operations (see Table 30.1), the actual time spent by the operator on welding is only about 30% of the total time.

Labor costs may be summarized as:

- *High to intermediate*: Oxyfuel–gas welding and shielded metal-arc welding
- *High to low*: Electron-beam and laser-beam welding and flux-cored arc welding
- *Intermediate to low*: Submerged-arc welding

CASE STUDY 31.1   Friction Welding of Pistons

There has been a sustained effort among heavy-truck manufacturers to design and manufacture diesel engines with reduced emissions. A number of technologies have become more prevalent since the 1980s, reflecting the need for *green design* (see Sections I.4 and 40.4). Exhaust-gas recirculation (reintroduction of a portion of the spent exhaust gases into the intake stream of the engine) has become standard, and is known to reduce nitrous-oxide emissions. However, this strategy leads to less efficient combustion and lower component durability, because of the presence of abrasive-wear particles (see Section 33.5) and acids that are recirculated into the engine. To maintain and even improve efficiency, engine manufacturers have increased cylinder pressures and operating temperatures, which together lead to an even more demanding environment for engine components.

The traditional aluminum pistons in diesel engines (in the U.S. market) were found not to function reliably in modern engine designs. The problems identified with pistons were (a) a tendency to "mushroom" and fracture under the high firing pressures in the cylinder, (b) inadequate cooling of the piston, and (c) scuffing (wear) at the pin that joins the piston to the connecting rod. One solution, shown in Fig. 31.19, is a Monosteel piston, which has the following design attributes:

- The piston is made of steel (Chapter 5), which has higher strength and better high-temperature mechanical properties than the aluminum alloys (Section 6.2) previously used.
- A two-piece design allows the incorporation of an oil gallery, permitting circulation of cooling oil in the piston. One of the main advantages of the Monosteel design is the use of a very large gallery, resulting in effective heat removal from the piston. This design has been shown to reduce piston temperatures in the rim by around 40°C as compared with earlier piston designs.
- The piston's steel skirt is much more rigid than the aluminum skirt (because of steel's higher elastic modulus), resulting in smaller deformation and allowing for designs with tighter clearances. This feature results in a more stable piston (more concentric with the cylinder bore) with less oil consumption, thus further reducing harmful exhaust emissions.

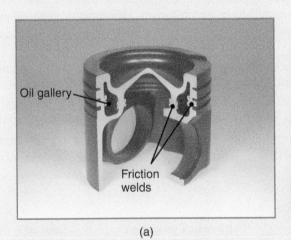

(a)   (b)

FIGURE 31.19   The Monosteel piston. (a) Cutaway view of the piston, showing the oil gallery and the friction-welded sections; (b) detail of the friction welds before the external flash is removed and cylindrical grooves are machined.

(*continued*)

Monosteel pistons are produced from two forged components, which are then machined prior to welding. The process used to join these components is inertia friction welding (Section 31.4), which has the following advantages in this particular application:

- The welding process leads to well-controlled, reliable, and repeatable high-quality welds.
- Friction welds are continuous and do not involve porosity, thereby producing a high-strength weld that seals the oil gallery.
- The welding process is fairly straightforward to optimize, with the main process variables being energy (or spindle speed for a given flywheel) and contact pressure.

Because it is entirely machine controlled, friction welding does not require operator intervention or expertise. Although the capital investment is significant, as compared with that of other appropriate welding technologies, weld quality and the ability to weld in this application is significantly more favorable.

The piston shown in Fig. 31.19 was produced on a 250-ton force capacity inertia friction welder, using a peripheral velocity of 7.5 m/s (1500 ft/min) and a contact pressure of 140 MPa (20 ksi); see Fig. 31.3. As can be seen, the weld zone has the optimum flash shape (Fig. 31.4); it is then removed from the exterior piston surface by a turning operation (Section 23.2), after which the piston skirt is ground (Section 26.3). Production times are typically at 40 to 60 s, but can be higher or lower depending on piston size.

Source: Courtesy of D. Adams, Manufacturing Technology, Inc., and K. Westbrooke, Federal Mogul, Inc.

SUMMARY

- In addition to the traditional joining processes of oxyfuel–gas and arc welding, several other joining processes that are based on producing a strong joint under pressure and/or heat also are available.
- Surface preparation and cleanliness are important in some of these processes. Pressure is applied mechanically or by explosives. Heat may be supplied externally (by electrical resistance or furnaces), or it may be generated internally (as in friction welding).
- Combining diffusion-bonding and superplastic-forming processes improves productivity and the capability to make complex parts economically.
- As in all manufacturing operations, certain hazards are inherent in welding operations. Some relate to the machinery and equipment used, others to the nature of the process itself (as in explosion welding). Proper safety precautions must always be taken in work areas.

KEY TERMS

Cold welding
Diffusion bonding (welding)
Explosion welding
Faying surfaces
Ferrule
Filled gold
Flash welding
Flyer plate
Friction stir welding
Friction welding
High-frequency resistance welding
Horn
Inertia friction welding
Linear friction welding
Percussion welding
Resistance projection welding
Resistance seam welding
Resistance spot welding
Resistance welding
Roll bonding
Roll spot welding
Roll welding
Seam welding
Solid-state welding
Sonotrode
Stud welding
Superplastic forming
Transducer
Ultrasonic welding
Weld nugget

BIBLIOGRAPHY

American Welding Society, **Welding Handbook**, 9th ed., 2011.

ASM Handbook, Vol. 6: **Welding, Brazing, and Soldering**, ASM International, 1993.

Bowditch, W.A., Bowditch, K.E., and Bowditch, M.A., **Welding Technology Fundamentals**, 4th ed., Goodheart–Willcox, 2009.

Cary, H.B., **Modern Welding Technology**, 6th ed., Prentice Hall, 2004.

Houldcroft, T., **Welding and Cutting: A Guide to Fusion Welding and Associated Cutting Processes**, Industrial Press, 2001.

Jeffus, L.F., **Welding: Principles and Applications**, 7th ed., Delmar Publishers, 2011.

Kou, S., **Welding Metallurgy**, 2nd ed., Wiley-Interscience, 2002.

Lagoda, T., **Life Estimation of Welded Joints**, Springer, 2008.

Mouser, J.D., **Welding Codes, Standards, and Specifications**, McGraw-Hill, 1997.

Nicholas, M.G., **Joining Processes: Introduction to Brazing and Diffusion Bonding**, Chapman & Hall, 1998.

Weman, K., **Welding Processes Handbook**, 2nd ed., Woodhead Publishing, 2012.

Zhang, H., and Senkara, J., **Resistance Welding: Fundamentals and Applications**, 2nd ed., CRC Press, 2011.

REVIEW QUESTIONS

31.1 Explain what is meant by solid-state welding.

31.2 What is cold welding? Why is it so called?

31.3 What is (a) a ferrule, (b) filled gold, and (c) a flyer plate?

31.4 What are faying surfaces in solid-state welding processes?

31.5 What is the basic principle of (a) ultrasonic welding and (b) diffusion bonding?

31.6 Explain how the heat is generated in the ultrasonic welding of (a) metals and (b) thermoplastics.

31.7 Describe the advantages and limitations of explosion welding.

31.8 Describe the principle of resistance-welding processes.

31.9 What materials would you recommend for resistance-welding electrodes?

31.10 What type of products are suitable for stud welding? Why?

31.11 What is the advantage of linear friction welding over inertia friction welding?

31.12 What are the main forms of friction welding?

31.13 Which processes in this chapter are applicable to polymers?

31.14 Describe how high-frequency butt welding operates.

31.15 What materials are typically used in diffusion bonding?

QUALITATIVE PROBLEMS

31.16 Make a list of processes in this chapter, ranking them according to (a) the pressure achieved, (b) the maximum temperature, and (c) suitability for bonding dissimilar materials.

31.17 Make a list of standard abbreviations for welding processes. For example, cold welding is CW and roll welding is ROW.

31.18 Explain the reasons why the processes described in this chapter were developed.

31.19 Explain the similarities and differences between the joining processes described in this chapter and those in Chapter 30.

31.20 Describe your observations concerning Figs. 31.16c and d.

31.21 Would you be concerned about the size of weld beads, such as those shown in Fig. 31.15? Explain.

31.22 What advantages does friction welding have over other methods described in this and in the preceding chapter?

31.23 List the process parameters that you think will affect the weld strength of a friction weld, and explain why you think those parameters are important.

31.24 Describe the significance of faying surfaces.

31.25 Discuss the factors that influence the strength of (a) a diffusion-bonded and (b) a cold-welded component.

31.26 What are the sources of heat for the processes described in this chapter?

31.27 Can the roll-bonding process be applied to a variety of part configurations? Explain.

31.28 Why is diffusion bonding, when combined with the superplastic forming of sheet metals, an attractive fabrication process? Does it have any limitations?

31.29 List and explain the factors involved in the strength of weld beads.

31.30 Give some of the reasons that spot welding is used commonly in automotive bodies and in large appliances.

31.31 Explain the significance of the magnitude of the pressure applied through the electrodes during a spot-welding operation.

31.32 Give some applications for (a) flash welding, (b) stud welding, and (c) percussion welding.

31.33 Discuss the need for, and role of, work-holding devices in the welding operations described in this chapter.

31.34 Inspect Fig. 31.4, and explain why those particular fusion-zone shapes are developed as a function of pressure and speed. Comment on the influence of material properties.

31.35 Could the process shown in Fig. 31.11 also be applicable to part shapes other than round? Explain, and give specific examples.

31.36 In spot-weld tests, what would be the reason for weld failure to occur at the locations shown in Fig. 31.9?

31.37 Can friction stir welding be used for powder metal parts? Explain.

31.38 Do any of the processes described in this chapter use a filler metal? Explain.

31.39 Which processes in this chapter are not affected by an oxide film? Explain.

31.40 Consider the situation where two round components are welded together. You believe that the components were friction welded, with the flash removed by machining. How could you confirm or disprove your suspicion?

31.41 Is there any advantage in preheating the workpieces in friction welding? Explain.

QUANTITATIVE PROBLEMS

31.42 The energy required in ultrasonic welding is found to be related to the product of workpiece thickness and hardness. Explain why this relationship exists.

31.43 Two flat copper sheets (each 1.0 mm thick) are being spot welded by the use of a current of 7000 A and a current flow time of 0.3 s. The electrodes are 4 mm in diameter. Estimate the heat generated in the weld zone. Assume that the resistance is 200 $\mu\Omega$.

31.44 Calculate the temperature rise in Problem 31.43, assuming that the heat generated is confined to the volume of material directly between the two round electrodes, and the temperature is distributed uniformly.

31.45 Calculate the range of allowable currents in Problem 31.43 if the temperature should be between 0.7 and 0.8 times the melting temperature of copper. Repeat this problem for carbon steel.

SYNTHESIS, DESIGN, AND PROJECTS

31.46 Comment on workpiece size and shape limitations (if any) for each of the processes described in this chapter.

31.47 Explain how you would fabricate the structures shown in Fig. 31.18 by methods other than diffusion bonding and superplastic forming.

31.48 Describe part shapes that cannot be joined by the processes described in this chapter. Give specific examples.

31.49 Comment on the feasibility of applying explosion welding in a factory environment.

31.50 Discuss your observations concerning the welding design guidelines illustrated in Figs. 31.13d and e.

31.51 Referring to Fig. 14.11b, could you use any of the processes described in Chapters 30 and 31 to make a large bolt by welding the head to the shank? Explain the advantages and limitations of this approach.

31.52 Explain how the projection-welded parts shown in Fig. 31.12 could be made by any of the processes described in this book.

31.53 Using a magnifier, inspect the periphery of coins such as the U.S. dime and nickel, and comment on your observations.

31.54 Describe the methods you would use for removing the flash from welds, such as those shown in Fig. 31.3. How would you automate these methods for a high-production facility?

31.55 In the roll-bonding process shown in Fig. 31.1, how would you go about ensuring that the interfaces are clean and free of contaminants, so that a good bond is developed? Explain.

31.56 Inspect several metal containers for household products and for food and beverages. Identify those which have utilized any of the processes described in this chapter. Describe your observations.

31.57 Inspect the sheet-metal body of an automobile, and comment on the size and frequency of the spot welds applied. How would you go about estimating the number of welds in an automobile body?

31.58 Alclad stock is made from 5182 aluminum alloy, and has both sides coated with a thin layer of pure aluminum. The 5182 provides high strength, while the outside layers of pure aluminum provide good corrosion resistance, because of their stable oxide film. Hence, Alclad is commonly used in aerospace structural applications. Investigate other common roll-bonded metals and their uses, and develop a summary table.

31.59 Design a test method for evaluating the bond strength in roll welding.

31.60 Review Fig. 31.3 and sketch the flash pattern you would expect if (a) two tubular parts were inertia friction welded; (b) two elliptical parts were inertia friction welded; and (c) a butt weld was created with linear friction welding.

31.61 Sketch the microstructure you would expect if a butt joint were created by (a) linear friction welding, (b) friction stir welding, (c) mash seam welding, and (d) flash welding.

32

Brazing, Soldering, Adhesive-bonding, and Mechanical Fastening Processes

- Brazing and soldering are different from welding, in that no diffusion takes place at the interface; bond strength depends on adhesive forces, which are often increased through the use of a proper filler metal to produce a strong joint.

- Brazing and soldering are differentiated by the temperature at which filler metals melt: Brazing takes place above 450°C (840°F) and produces stronger joints than soldering, while soldering involves lower temperatures. Soldering is widely applied in the electronics industry.

- Adhesive-bonding is versatile, and a wide variety of adhesives is available for numerous applications.

- Mechanical joining processes are then described, such as using bolts, nuts, rivets, snap fasteners, or shrink fits in assembly operations.

- The chapter ends with a discussion of economic considerations in joining operations.

32.1 Introduction

In most of the joining processes described in Chapters 30 and 31, the faying (mating) surfaces of the components are heated by various external or internal means, to cause fusion and bonding at the joint. However, what if the parts to be joined are fragile, intricate, made of two or more materials with very different characteristics and dimensions, or the components to be joined cannot withstand high temperatures, such as electronic components?

This chapter first describes two joining processes, *brazing* and *soldering*, that require lower temperatures than those used for fusion welding. Filler metals are placed in or supplied to the joint, and are melted by an external source of heat; upon solidification, a strong joint is obtained (Fig. 32.1). These two processes are distinguished arbitrarily by temperature. Temperatures for soldering are lower than those for brazing, and the strength of a soldered joint is much lower.

The chapter also describes the principles and types of *adhesive-bonding*. The ancient method of joining parts with animal-derived glues (typically employed in bookbinding, labeling, and packaging) has been developed into an important joining technology, for both metallic and nonmetallic materials. Modern adhesives are advanced polymers or composites, and are rarely animal-based. The joining process has wide applications in numerous consumer and industrial products, as well as in the aircraft and aerospace industries. Bonding materials such as thermoplastics,

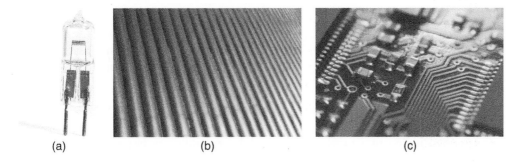

(a)   (b)   (c)

FIGURE 32.1 Examples of brazed and soldered joints. (a) Resistance-brazed light bulb filament; (b) brazed rocket tubing assembly; (c) soldered circuit board.

thermosets, ceramics, and glasses, either to each other or to other materials, present various challenges.

Although all of the joints described thus far are of a permanent nature, in many applications joined components have to be taken apart for purposes such as replacement of worn or broken components, general maintenance, or repair. Moreover, some joints are designed not to be permanent, but still must be strong. The obvious solution is then to use methods of *mechanical fastening*, such as using bolts, screws, nuts, and various special fasteners.

32.2 Brazing

Brazing is a joining process in which a **filler metal** is placed at the periphery or between the interfaces of the faying surfaces to be joined. The temperature is then raised sufficiently to melt the filler metal, but not the components (the base metal), as would be the case in fusion welding, described in Chapter 30. Brazing is derived from the word *brass* meaning "to harden;" the process was first used as far back as 3000 to 2000 B.C.

It will be noted that brazing is a *liquid–solid-state bonding* process. Upon cooling and solidification of the filler metal, a strong joint is obtained. Filler metals for brazing typically melt above 450°C (840°F), which is below the melting point (*solidus temperature*) of the metals to be joined (see, for example, Fig. 4.4).

Figure 32.2a shows a typical brazing operation, where a filler (*braze metal*) in the shape of wire is placed along the periphery of the components to be joined. Heat is then applied, by various external means, melting the braze metal and, by *capillary* action, filling the closely fitting space (called *joint clearance*) at the interfaces (Fig. 32.2b). In **braze welding**, filler metal (typically brass) is deposited at the

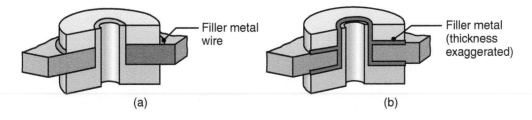

FIGURE 32.2 An example of furnace brazing (a) before and (b) after brazing. The filler metal is a shaped wire and the molten filler moves into the interfaces by capillary action, with the application of heat.

joint by a technique similar to oxyfuel–gas welding (see Fig. 30.1d). (For details, see Section 32.2.1.)

Examples of joints made by brazing and soldering are shown in Fig. 32.3. Intricate, lightweight shapes can be joined rapidly and with little distortion, good joint strength, and with dissimilar metals.

Filler Metals. Several filler metals are available, with a range of brazing temperatures as shown in Table 32.1. Note that, unlike those for the welding operations described in the two previous chapters, filler metals for brazing generally have compositions that are significantly different from those of the metals to be joined. They are available in a variety of shapes, such as wire, rod, ring, shim stock, and filings. The selection of the type of filler metal and its composition are important in order to avoid *embrittlement* of the joint by (a) grain-boundary penetration of liquid metal (Section 1.5.2); (b) the formation of *brittle intermetallic compounds* (Section 4.2.2); and (c) *galvanic corrosion* in the joint (Section 3.8).

Because of *diffusion* between the filler metal and the base metal, the mechanical and metallurgical properties of a brazed joint can change as a result of subsequent processing or during the service life of a brazed part. For example, when titanium is brazed with pure tin as the filler metal, it is possible for the tin to diffuse completely

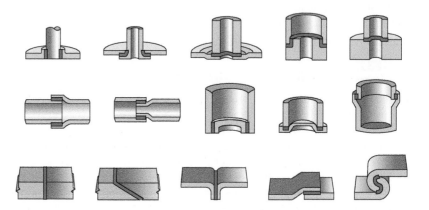

FIGURE 32.3 Joint designs commonly used in brazing operations. The clearance between the two parts being brazed is an important factor in joint strength; if the clearance is too small, the molten braze metal will not fully penetrate the interface, and if it is too large, there will be insufficient capillary action for the molten metal to fill the interface.

TABLE 32.1

Typical Filler Metals for Brazing Various Metals and Alloys

Base metal	Filler metal	Brazing temperature (°C)
Aluminum and its alloys	Aluminum–silicon	570–620
Magnesium alloys	Magnesium–aluminum	580–625
Copper and its alloys	Copper–phosphorus and gold–copper–phosphorus	700–925
Ferrous and nonferrous (except aluminum and magnesium)	Silver and copper alloys, copper–phosphorus, copper–zinc	620–1150
Iron-, nickel-, and cobalt-based alloys	Gold–copper and gold–paladium	900–1100
Stainless steels, nickel- and cobalt-based alloys	Nickel–silver	925–1200

into the titanium base metal when it is subjected to subsequent aging or to heat treatment; consequently, the joint no longer exists.

Fluxes. The use of a *flux* is essential in brazing, because it prevents oxidation and removes oxide films. Brazing fluxes generally are made of borax, boric acid, borates, fluorides, and chlorides. *Wetting agents* may be added to improve both the wetting characteristics of the molten filler metal and the capillary action.

It is essential that the surfaces to be brazed are clean and free from rust, oil, and other contaminants, in order (a) for effective wetting and distribution (spreading) of the molten filler metal in the joint interfaces and (b) to develop maximum bond strength. Grit blasting (Section 26.8) also may be used to improve the surface finish of the faying surfaces. Because they are corrosive, fluxes must be removed after brazing, usually done by washing with hot water.

Brazed Joint Strength. The strength of the brazed joint depends on (a) joint clearance, (b) joint area, and (c) the nature of the bond at the interfaces between the components and the filler metal. Joint clearances typically range from 0.025 to 0.2 mm (0.001–0.008 in.). As shown in Fig. 32.4, the smaller the gap, the higher is the *shear strength* of the joint. Note that there is an optimum gap for achieving maximum *tensile strength* of the joint. The shear strength can reach 800 MPa (120 ksi) by using brazing alloys containing silver (*silver solder*). Because clearances in brazing are very small, the roughness of the faying surfaces becomes important (see also Section 33.3).

32.2.1 Brazing Methods

As described below, the heating methods used in brazing identify the various processes.

Torch Brazing. The heat source in *torch brazing* (TB) is oxyfuel gas with a carburizing flame (see Fig. 30.1c). Brazing is performed by first heating the joint with the torch and then depositing the brazing rod or wire at the joint. Part thicknesses are typically in the range from 0.25 to 6 mm (0.01–0.25 in.). Torch brazing is difficult to control and requires skilled labor, but it can be automated as a production process by using multiple torches.

Furnace Brazing. The parts in *furnace brazing* (FB) are first cleaned and preloaded with brazing metal, in appropriate configurations; the assembly is then placed in a furnace, where it is heated uniformly. Furnaces may be either batch type, for complex shapes, or continuous type, for high production runs, especially for small parts with simple joint designs. Vacuum furnaces or neutral atmospheres are used for metals that react with the environment; hydrogen can be used to reduce oxides in metals that are not affected by hydrogen embrittlement (Section 2.10.2A). Skilled labor is not required, and complex shapes can be brazed because the whole assembly is heated uniformly in the furnace.

Induction Brazing. The source of heat in *induction brazing* (IB) is induction heating by high-frequency AC current. Parts are preloaded with filler metal and are placed near the induction coils for rapid heating (see Fig. 4.24). Unless a protective (neutral) atmosphere is utilized, fluxes generally are required. Part thicknesses typically are less than 3 mm (0.125 in.). Induction brazing is particularly suitable for continuous brazing of parts (Fig. 32.5).

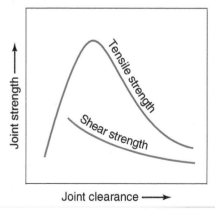

FIGURE 32.4 The effect of joint clearance on the tensile and shear strength of brazed joints; note that, unlike tensile strength, the shear strength continually decreases as the clearance increases.

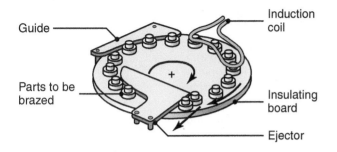

Guide

Parts to be
brazed

Induction
coil

Insulating
board

Ejector

FIGURE 32.5 Schematic illustration of a continuous induction brazing setup for increased productivity.

Resistance Brazing.   In *resistance brazing* (RB), the source of heat is the electrical resistance of the components to be brazed; electrodes are utilized in this method, as they are in resistance welding. Parts typically with thicknesses of 0.1 to 12 mm (0.004–0.5 in.) either are preloaded with filler metal or supplied externally during brazing. The operation is rapid, heating zones can be confined to very small areas, and the process can be automated to produce reliable and uniform joint quality.

Dip Brazing.   In *dip brazing* (DB), an assembly of two or more parts are joined by dipping in a bath of filler metal, or by immersing in a bath of molten salt; in the latter case, a filler metal needs to be part of the assembly. The molten salt acts as a flux, so that bonding occurs on oxide-free surfaces. The molten filler metal or the molten salt bath (Section 4.12) is at a temperature just above the melting point of the filler metal, so that all component surfaces are coated with the filler metal. Consequently, dip brazing in metal baths is typically used for small parts (such as sheet, wire, and fittings), usually less than 5 mm (0.2 in.) in thickness or diameter.

Depending on the size of the parts and the bath size, as many as 1000 joints can be made at one time. Dip brazing usually requires self-jigging (self-assembling) parts, but tack welding or pinning can be used; lap joints are preferred, although butt joints also can be produced.

Infrared Brazing.   The heat source in *infrared brazing* (IRB) is a high-intensity quartz lamp. The radiant energy is focused on the joint, and brazing can be carried out in a vacuum. *Microwave heating* also can be used. The process is particularly suitable for brazing very thin components, usually less than 1 mm (0.04 in.) thick, including metal honeycomb structures (Section 16.13).

Diffusion Brazing.   *Diffusion brazing* (DFB) is carried out in a furnace where, with proper control of temperature and time, the filler metal diffuses into the faying surfaces of the components to be joined. The brazing time required may range from 30 min to 24 h. This process is used for strong lap or butt joints and for difficult-to-join materials, but usually simple binary or three-metal alloys. More complex alloys may produce intermetallic compounds at the joint that can compromise joint strength. Because the rate of diffusion at the interface does not depend on the thickness of the components, part thicknesses may range from foil to as much as 50 mm (2 in.).

High-energy Beams.   For specialized and high-precision applications and with high-temperature metals and alloys, *electron-beam* or *laser-beam* heating may be used, as described in Sections 27.6 and 27.7.

Braze Welding.   The joint in *braze welding* is prepared as it is in fusion welding, described in Chapter 30. While an oxyacetylene torch with an oxidizing flame is used, filler metal is deposited at the joint (hence the term welding) rather than drawn in by capillary action. As a result, considerably more filler metal is used than in brazing. Temperatures in braze welding generally are lower than those in fusion welding, thus part distortion is minimal. The use of a flux is essential in this process. The principal use of braze welding is for maintenance and repair work, such as on ferrous castings and steel components, although the process can be automated for mass production.

32.2.2   Design for Brazing

As in all joining processes, *joint design* is important in brazing; some design guidelines are given in Fig. 32.6. Strong joints require a larger contact area for brazing than for welding. A variety of special fixtures and work-holding devices and fixtures (see also Section 37.8) may be required to hold the parts together during brazing; some will allow for thermal expansion and contraction during the brazing operation.

32.3   Soldering

In *soldering*, the filler metal (**solder**) melts at a relatively low temperature. As in brazing, the solder fills the joint, by capillary action, between closely fitting or closely placed components. Heat sources for soldering are typically soldering irons, torches, or ovens. The word "solder" is derived from the Latin *solidare*, meaning "to make solid." Soldering with copper–gold and tin–lead alloys was first practiced as far back as 4000 to 3000 B.C.

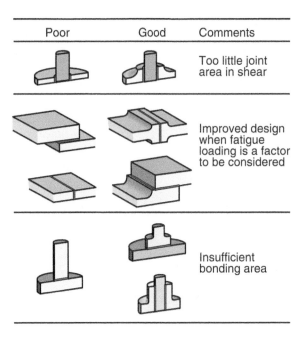

FIGURE 32.6   Examples of poor and good designs for brazing. *Source:* American Welding Society.

32.3.1   Types of Solders and Fluxes

Two important characteristics of solders are low surface tension and high wetting capability. Solders melt at the *eutectic point* of the solder alloy (see, for example, Fig. 4.8). Solders traditionally have been tin–lead alloys in various proportions. A solder of 61.9% Sn–38.1% Pb composition, for example, melts at 188°C (370°F), whereas tin melts at 232°C (450°F) and lead at 327°C (621°F). For special applications and higher joint strength, especially at elevated temperatures, other solder compositions are tin–zinc, lead–silver, cadmium–silver, and zinc–aluminum alloys (Table 32.2).

Because of the *toxicity* of lead and its adverse effects on the environment, **lead-free solders** are available. Since the European union prohibited the intentional addition of lead to consumer electronics in 2006, tin–silver–copper solders have come into wide use, with a typical composition of 96.5% tin, 3.0% silver, and 0.5% copper. A fourth element, such as zinc or manganese, is often added to provide desired mechanical or thermal characteristics. For nonelectrical applications, a large number of solders are available, and can also incorporate cadmium, gold, bismuth, and indium.

Fluxes are used in soldering, for the same purposes as they are in welding and brazing, as described in Section 32.2, and also serve to assist the wetting of surfaces by solder. Fluxes for soldering are generally of two types:

1. *Inorganic acids* or *salts*, such as zinc–ammonium-chloride solutions, which clean the surface rapidly. To avoid corrosion, the flux residues should be removed after soldering by washing the joint thoroughly with water.
2. Noncorrosive *resin-based fluxes*, used typically in electrical applications.

TABLE 32.2

A Selection of Common Solders and Their Typical Applications	
Tin–lead	General purpose
Tin–zinc	Aluminum
Lead–silver	Strength at higher than room temperature
Cadmium–silver	Strength at high temperatures
Zinc–aluminum	Aluminum, corrosion resistance
Tin–silver	Electronics
Tin–bismuth	Electronics

32.3.2 Soldering Techniques

The more common soldering techniques are:

1. **Torch soldering** (TS)
2. **Furnace soldering** (FS)
3. **Iron soldering** (INS)
4. **Induction soldering** (IS)
5. **Resistance soldering** (RS)
6. **Dip soldering** (DS)
7. **Infrared soldering** (IRS)

Other soldering techniques, for special applications, are:

8. **Ultrasonic soldering:** A transducer subjects the molten solder to ultrasonic cavitation; this action removes the oxide films from the surfaces to be joined and thus eliminates the need for a flux (hence also the term *fluxless soldering*).
9. **Reflow (paste) soldering** (RS)
10. **Wave soldering** (WS)

The last two techniques are widely used for bonding and packaging in *surface-mount technology*, as described in Section 28.11. Because they are significantly different from other soldering methods, they are described next in some detail.

Reflow Soldering.   *Solder pastes* are solder–metal particles held together by flux, binding, and wetting agents. The pastes are semisolid in consistency, have high viscosity, and thus are capable of maintaining their shape for relatively long periods. The paste is placed directly onto the joint or on flat objects for finer detail; it can be applied via a *screening* or *stenciling* technique, as shown in Fig. 32.7a. Stenciling is commonly used during the attachment of electrical components to printed circuit boards. An additional benefit of reflow soldering is that the high surface tension of the paste helps keep surface-mount packages aligned on their pads, a feature that improves the reliability of the solder joints. (See also Section 28.11.)

After the paste has been placed and the joint is assembled, it is heated in a furnace and soldering takes place. In reflow soldering, the product is heated in a controlled manner, so that the following sequence of events occurs:

1. Solvents present in the paste are evaporated
2. The flux in the paste is activated, and fluxing action occurs
3. The components are preheated
4. The solder particles are melted and wet the joint
5. The assembly is cooled at a low rate, to prevent thermal shock and possible fracture of the joint

Although this process appears to be straightforward, there are several variables for each stage, and good control over temperatures and durations must be maintained at each stage, in order to ensure proper joint strength.

Wave Soldering.   *Wave soldering* is a common technique for attaching circuit components to their boards (see Section 28.11). Although slowly being replaced by reflow soldering, this process is still widely used in industrial practice.

It is important to note that because the molten solder does not wet all surfaces, it will not stick to most polymer surfaces, and is easy to remove while molten. The solder wets metal surfaces and forms a good bond, but only when the metal is preheated to a certain temperature. Thus, wave soldering requires separate fluxing and preheating operations before it can be completed.

A typical wave-soldering operation is illustrated in Fig. 32.7b. A standing laminar wave of molten solder is first generated by a pump; preheated and prefluxed circuit

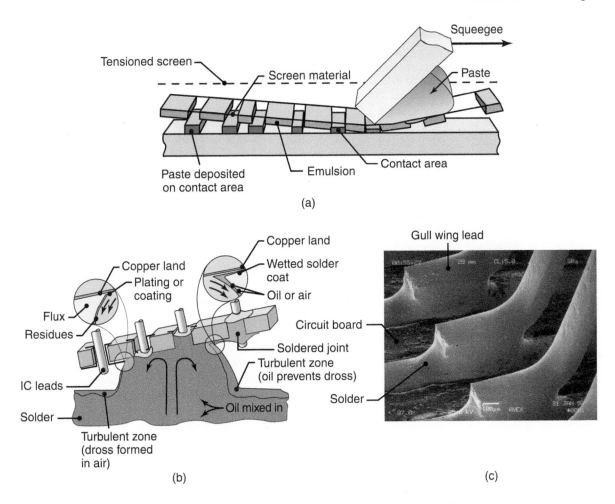

FIGURE 32.7 (a) Screening solder paste onto a printed circuit board in reflow soldering. (b) Schematic illustration of the wave-soldering process. (c) SEM image of a wave-soldered joint on a surface-mount device. *Source:* (a) After V. Solberg.

boards are then conveyed over the wave. The solder wets the exposed metal surfaces, but (a) it does not remain attached to the polymer package for the integrated circuits and (b) it does not adhere to the polymer-coated circuit boards. An *air knife* (a high-velocity jet of hot air; see also Section 34.11) blows excess solder away from the joint, to prevent bridging between adjacent leads.

When surface-mount packages are to be wave soldered, they must be bonded adhesively to the circuit board before soldering can commence. Bonding usually is accomplished by the following sequence: (1) screening or stenciling epoxy onto the boards, (2) placing the components in their proper locations, (3) curing the epoxy, (4) inverting the board, and (5) performing wave soldering. A scanning-electron microscope (SEM) photograph of a typical surface-mount joint is shown in Fig. 32.7c.

EXAMPLE 32.1   Soldering of Components onto a Printed Circuit Board

The computer and consumer electronics industries place extremely high demands on electronic components. Integrated circuits and other electronic devices are expected to function reliably for extended periods, during which they may be subjected to significant temperature variations and to vibration (see also Section 28.12). In recognition of this requirement, it is essential that solder joints be sufficiently strong and reliable, and also that the joints be applied extremely rapidly, using automated equipment.

A continuing trend in the computer and the consumer electronics industries is toward the reduction of chip sizes and increasing compactness of circuit boards. Further space savings are achieved by mounting integrated circuits into surface-mount packages, which allow tighter packing on a circuit board. More importantly, the technique allows components to be mounted on *both* sides of the board.

A challenging problem arises when a printed circuit board has both surface-mount and in-line circuits on the same board, and it is essential to solder all of the joints via a reliable, automated process. It is important to recognize that, for efficiency of assembly, all of the in-line circuits be inserted from one side of the board. Indeed, there is no performance requirement that would dictate otherwise, and this restriction greatly simplifies manufacturing.

The basic steps in soldering the connections on such a board are (see Figs. 32.7b and c):

1. Apply solder paste to one side of the board
2. Place the surface-mount packages onto the board, and insert in-line packages through the primary side of the board
3. Reflow the solder
4. Apply adhesive to the secondary side of the board
5. Using the adhesive, attach the surface-mount devices onto the secondary side
6. Cure the adhesive
7. Perform a wave-soldering operation on the secondary side, to produce an electrical attachment of the surface mounts and the in-line circuits to the board

Applying solder paste is done with chemically-etched stencils or screens, so that the paste is placed only onto the designated areas of a circuit board. (Stencils are used more widely for fine-pitched devices, as they produce a more uniform paste thickness.) Surface-mount circuit components are then placed on the board, and the board is heated in a furnace to around 200°C (400°F) to reflow the solder, and to form strong connections between the surface mount and the circuit board.

At this stage, the components with leads are inserted into the primary side of the board, their leads are crimped, and the board is flipped over. A dot of epoxy at the center of a surface mount component location is printed onto the board. The surface-mount packages are then placed onto the adhesive by high-speed automated, computer-controlled systems. The adhesive is then cured, the board is flipped, and wave soldering is done.

The wave-soldering operation simultaneously joins the surface-mount components to the secondary side, and it solders the leads of the in-line components from the board's primary side. The board is then cleaned and inspected prior to electronic quality checks.

32.3.3 Solderability

Solderability may be defined in a manner similar to weldability (Section 30.9.2). Special fluxes have been developed to improve the solderability of metals and alloys. As a general guide,

- *Copper, silver,* and *gold* are easy to solder
- *Iron* and *nickel* are more difficult to solder
- *Aluminum* and *stainless steels* are difficult to solder, because of their thin, strong oxide films
- *Steels, cast irons, titanium,* and *magnesium,* as well as *ceramics* and *graphite,* can be soldered by first plating them with suitable metallic elements to induce interfacial bonding; this method is similar to that used for joining carbides and ceramics (see Section 32.6.3). An example is *tinplate* (a common material used in making cans for food, which is steel sheet coated with tin, thus making it very easy to solder.

32.3.4 Soldering Applications and Design Guidelines

Soldering is used extensively in the electronics industry; note, however, that because soldering temperatures are relatively low, a soldered joint has very limited utility at

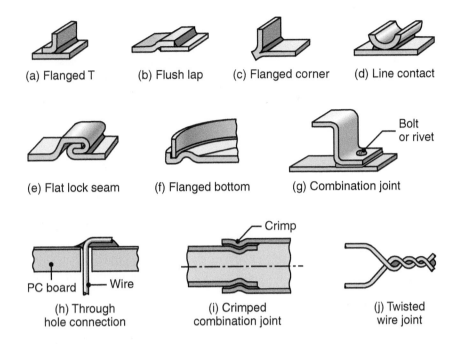

(a) Flanged T   (b) Flush lap   (c) Flanged corner   (d) Line contact

(e) Flat lock seam   (f) Flanged bottom   (g) Combination joint

(h) Through hole connection   (i) Crimped combination joint   (j) Twisted wire joint

FIGURE 32.8   Joint designs commonly used for soldering.

elevated temperatures. Moreover, since they generally do not have much strength, solders cannot be used for load-bearing (structural) members. Joint strength can be improved significantly by mechanical *interlocking* of the joint, as illustrated in Fig. 32.8.

Design guidelines for soldering are similar to those for brazing (see Section 32.2.2). Some frequently used joint designs are shown in Fig. 32.8. Note the importance of large contact surfaces (because of the low strength of solders) for developing sufficient joint strength in soldered products. Since the faying surfaces generally would be small, solders are rarely used to make butt joints.

Video Solution 32.1 Analysis of an Adhesive Joint

32.4   Adhesive-bonding

One of the most versatile joining process is the use of **adhesives** between two surfaces, generally using a rubber or polymer as the filler material. A common example of *adhesive-bonding* is plywood, where several layers of wood are bonded with wood glue. Modern plywood was developed in 1905, but the practice of adhesive-bonding of wood layers, using animal glue, dates back to 3500 B.C.

Adhesive-bonding has gained increased acceptance in manufacturing ever since its first use on a large scale: the assembly of load-bearing components in aircraft during World War II (1939–1945). Adhesives are available in various forms: liquid, paste, solution, emulsion, powder, tape, and film. When applied, adhesives typically are about 0.1 mm (0.004 in.) thick.

To meet the requirements of a particular application, an adhesive may require one or more of the following properties (Table 32.3):

QR Code 32.1 Adhesive Bonding Processing. (*Source:* Courtesy of EWI)

- Strength: shear and peel
- Toughness
- Resistance to various fluids and chemicals

TABLE 32.3

Typical Properties and Characteristics of Chemically Reactive Structural Adhesives

	Epoxy	Polyurethane	Modified acrylic	Cyanoacrylate	Anaerobic
Impact resistance	Poor	Excellent	Good	Poor	Fair
Tension-shear strength, MPa (10^3 psi)	15–22 (2.2–3.2)	12–20 (1.7–2.9)	20–30 (2.9–4.3)	18.9 (2.7)	17.5 (2.5)
Peel strength,* N/m (lb/in.)	<523 (3)	14,000 (80)	5250 (30)	<525 (3)	1750 (10)
Substrates bonded	Most	Most smooth, nonporous	Most smooth, nonporous	Most nonporous metals or plastics	Metals, glass, thermosets
Service temperature range, °C (°F)	−55 to 120 (−70 to 250)	−40 to 90 (−250 to 175)	−70 to 120 (−100 to 250)	−55 to 80 (−70 to 175)	−55 to 150 (−70 to 300)
Heat cure or mixing required	Yes	Yes	No	No	No
Solvent resistance	Excellent	Good	Good	Good	Excellent
Moisture resistance	Good–excellent	Fair	Good	Poor	Good
Gap limitation, mm (in.)	None	None	0.5 (0.02)	0.25 (0.01)	0.60 (0.025)
Odor	Mild	Mild	Strong	Moderate	Mild
Toxicity	Moderate	Moderate	Moderate	Low	Low
Flammability	Low	Low	High	Low	Low

*Peel strength varies widely, depending on surface preparation and quality.

- Resistance to environmental degradation, including heat and moisture
- Capability to wet the surfaces to be bonded

32.4.1   Types of Adhesives and Adhesive Systems

Several types of adhesives are available, and more continue to be developed that provide good joint strength, including fatigue strength (Table 32.4). The three basic types of adhesives are:

1. **Natural adhesives,** such as starch, soya flour, animal products, and dextrin (a gummy substance obtained from starch)
2. **Inorganic adhesives,** such as sodium silicate and magnesium oxychloride
3. **Synthetic organic adhesives,** which may be thermoplastics (used for nonstructural and some structural bonding) or thermosetting polymers (used primarily for structural bonding)

Because of their strength, synthetic organic adhesives are the most important adhesives in manufacturing operations, particularly for load-bearing applications. They are classified as:

- **Chemically reactive:** Polyurethanes, silicones, epoxies, cyanoacrylates, modified acrylics, phenolics, and polyimides; also included are anaerobics (which cure in the absence of oxygen), such as Loctite® for threaded fasteners; see also Case Study 32.1
- **Pressure sensitive:** Natural rubber, styrene–butadiene rubber, butyl rubber, nitrile rubber, and polyacrylates
- **Hot melt:** Thermoplastics (such as ethylene–vinyl acetate copolymers, polyolefins, polyamides, and polyester) and thermoplastic elastomers
- **Reactive hot melt:** A form of thermoplastic, with a thermoset portion (based on urethane's chemistry), with improved properties; also known as hot glue
- **Evaporative or diffusion:** Vinyls, acrylics, phenolics, polyurethanes, synthetic rubbers, and natural rubbers

TABLE 32.4

General Characteristics of Adhesives		
Type	Comments	Applications
Acrylic	Thermoplastic; quick setting; tough bond at room temperature; two components; good solvent; chemical and impact resistance; short work life; odorous; ventilation required	Fiberglass and steel sandwich bonds, tennis racquets, metal parts, and plastics
Anaerobic	Thermoset; easy to use; slow curing; bonds at room temperature; curing occurs in absence of air; will not cure where air contacts adherents; one component; not good on permeable surfaces	Close-fitting machine parts, such as shafts and pulleys, nuts and bolts, and bushings and pins
Epoxy	Thermoset; one or two components; tough bond; strongest of engineering adhesives; high tensile and low peel strengths; resists moisture and high temperature; difficult to use	Metal, ceramic, and rigid plastic parts
Cyanoacrylate	Thermoplastic; quick setting; tough bond at room temperature; easy to use; colorless	"Krazy Glue"; bonds most materials; especially useful for ceramics and plastics
Hot melt	Thermoplastic; quick setting; rigid or flexible bonds; easy to apply; brittle at low temperatures; based on ethylene vinyl acetate, polyolefins, polyamides, and polyesters	Bonds most materials; packaging, book binding, and metal can joints
Pressure sensitive	Thermoplastic; variable strength bonds; primer anchors adhesive to roll tape backing material—a release agent on the back of web permits unwinding; made of polyacrylate esters and various natural and synthetic rubbers	Tapes, labels, and stickers
Phenolic	Thermoset; oven cured; strong bond; high tensile and low impact strength; brittle; easy to use; cures by solvent evaporation	Acoustical padding, brake lining and clutch pads, abrasive grain bonding, and honeycomb structures
Silicone	Thermoset; slow curing; flexible; bonds at room temperature; high impact and peel strength; rubberlike	Gaskets and sealants
Formaldehyde (urea, melamine, phenol, resorcinol)	Thermoset; strong with wood bonds; urea is inexpensive, is available as powder or liquid, and requires a catalyst; melamine is more expensive, cures with heat, and the bond is waterproof; resorcinol forms a waterproof bond at room temperature. Types can be combined	Wood joints, plywood, and bonding
Urethane	Thermoset; bonds at room temperature or oven cure; good gap-filling qualities	Fiberglass body parts, rubber, and fabric
Water-based (animal, vegetable, rubbers)	Inexpensive, nontoxic, and nonflammable	Wood, paper, fabric, leather, and dry-seal envelopes

- **Film and tape:** Nylon, epoxies, elastomer epoxies, nitrile phenolics, vinyl phenolics, and polyimides
- **Delayed tack:** Styrene–butadiene copolymers, polyvinyl acetates, polystyrenes, and polyamides
- **Electrically** and **thermally conductive:** Epoxies, polyurethanes, silicones, and polyimides. Electrical conductivity is obtained by the addition of fillers, such as silver (used most commonly), copper, aluminum, and gold. Fillers that improve the electrical conductivity of adhesives generally also improve their thermal conductivity.

Adhesive Systems.   These systems may be classified on the basis of their specific chemistries:

- **Epoxy-based systems:** These systems have high strength and high-temperature properties, to as high as 200°C (400°F); typical applications include automotive brake linings and bonding agents for sand molds for casting.
- **Acrylics:** Because the adhesive acts as a solvent, these adhesives are more tolerant of contaminants on surfaces.
- **Anaerobic systems:** Curing of these adhesives is done under oxygen deprivation, and the bond is usually hard and brittle; curing times can be reduced by external heat or by ultraviolet (UV) radiation.
- **Cyanoacrylate:** The bond lines are thin and the bond sets within 5 to 40 s.
- **Urethanes:** These adhesives have high toughness and flexibility at room temperature; used widely as sealants.
- **Silicones:** Highly resistant to moisture and solvents, these adhesives have high impact and peel strength; however, curing times are typically in the range from 1 to 5 days.

Many of these adhesives can be combined to optimize their properties, such as the combinations of *epoxy–silicon, nitrile–phenolic,* and *epoxy–phenolic.* The least expensive adhesives are epoxies and phenolics, followed, in affordability, by polyurethanes, acrylics, silicones, and cyanoacrylates. Adhesives for high-temperature applications, in a range up to about 260°C (500°F), such as polyimides and polybenzimidazoles, are generally the most expensive. Most adhesives have an optimum temperature, ranging from about room temperature to about 200°C, for maximum shear strength.

32.4.2   Electrically Conducting Adhesives

Although the majority of adhesive-bonding applications require mechanical strength, a relatively recent advance is the development and application of electrically conducting adhesives to replace lead-based solder alloys, particularly in the electronics industry. These adhesives require curing or setting temperatures that are lower than those required for soldering. Applications of electrically conducting adhesives include calculators, remote controls, and control panels. In addition, there are high-density uses in electronic assemblies, liquid-crystal displays, pocket TVs, and electronic games.

In these adhesives, the polymer is the matrix and contains conducting metals (fillers) in such forms as flakes and particles (see also Section 7.3 on electrically conducting polymers). There is a minimum proportion of fillers necessary to make the adhesive electrically conducting; typically, in the range of 40 to 70% by volume.

The size, shape, and distribution of the metallic particles, the method of heat and pressure application, and the individual conducting particle contact geometry can be controlled to impart isotropic or anisotropic electrical conductivity to the adhesive. Metals used are typically silver, nickel, copper, and gold, as well as carbon. More recent developments include polymeric particles, such as polystyrene, coated with thin metallic films of silver or gold. Graphite also can be used as a filler, usually to produce an electrically conductive adhesive that is nonmagnetic, and can provide electromagnetic interference (EMI) shielding for electronic components. Matrix materials are generally epoxies, although thermoplastics also are used and are available as film or as paste.

32.4.3 Surface Preparation, Process Capabilities, and Applications

Surface preparation is very important in adhesive-bonding, as joint strength depends greatly on the absence of dirt, dust, oil, and various other contaminants. This dependence can be observed when one is attempting to apply an adhesive tape over a dusty or oily surface. Contaminants also affect the wetting ability of the adhesive and prevent an even spreading of the adhesive over the interface. Thick, weak, or loose oxide films on workpiece surfaces are detrimental to adhesive-bonding. On the other hand, a porous or a thin and strong oxide film may be desirable, particularly one with some surface roughness to improve adhesion or to introduce mechanical locking. However, the roughness must not be too high, because air may be trapped, in which case the joint strength is reduced. Various compounds and primers are available that modify surfaces to improve adhesive-bond strength. Liquid adhesives may be applied by brushes, sprayers, or rollers.

Process Capabilities.   Adhesives can be used for bonding a wide variety of similar and dissimilar metallic and nonmetallic materials and components with different shapes, sizes, and thicknesses. Adhesive-bonding also can be combined with mechanical joining methods (Section 32.5) to further improve bond strength. Joint design and bonding methods require care and skill. Special equipment is usually required, such as fixtures, presses, tooling, and autoclaves and ovens for curing.

Nondestructive inspection of the quality and strength of adhesively bonded components can be difficult. Some of the techniques described in Section 36.10, such as acoustic impact (tapping), holography, infrared detection, and ultrasonic testing, are effective testing methods for adhesive bonds.

Testing of Adhesives.   Recall that adhesives are most successful when they support shear stresses, and are less successful under other loading condition. Many adhesives are weak when loaded by tensile stresses. Recognizing that loadings can be complex, a large number of test configurations have been used to evaluate adhesives, depending on the particular application and the stresses encountered (Fig. 32.9). Tapered cantilever and wedge tests are particularly useful for high-strain-rate evaluations; wedge tests can result in combined shear and normal stresses when the two members have different thicknesses.

The most common test is the peel test, shown in Figs. 32.9b and 32.10, which also illustrates the strengths and limitations of adhesives. Note, for example, how easy it is to peel adhesive tape from a surface, yet it is very difficult to slide it along the surface. During peeling, the behavior of an adhesive may be brittle or it can be ductile and tough, thus requiring high forces to peel the adhesive from a surface.

Applications.   Major industries that use adhesive-bonding extensively are the aerospace, automotive, appliances, and building-products industries. Applications include automotive brake-lining assemblies, laminated windshield glass, appliances, helicopter blades, honeycomb structures, and aircraft bodies and control surfaces.

An important consideration in the use of adhesives in production is curing time, which can range

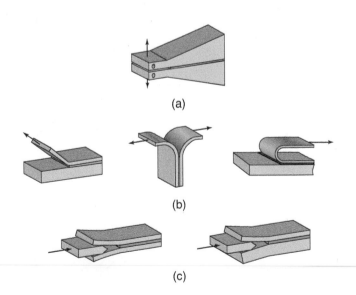

FIGURE 32.9   Common arrangements for evaluating adhesives: (a) tapered double cantilever beam, (b) peel test, and (c) wedge test.

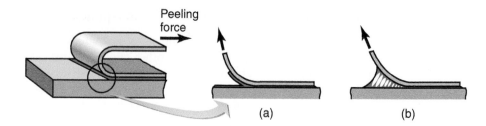

FIGURE 32.10 Characteristic behavior of (a) brittle and (b) tough adhesives in a peeling test; this test is similar to the peeling of adhesive tape from a solid surface.

from a few seconds (at high temperatures) to several hours (at room temperature), particularly for thermosetting adhesives. Thus, production rates can be low as compared with those of other joining processes. Moreover, adhesive bonds for structural applications rarely are suitable for service above 250°C (500°F).

The major advantages of adhesive bonding are:

- The interfacial bond has sufficient strength for structural applications, but is also used for nonstructural purposes, such as sealing, insulation, the prevention of electrochemical corrosion between dissimilar metals, and the reduction of vibration and of noise (by means of internal damping at the joints).
- Adhesive-bonding distributes the load at an interface, and thereby eliminates localized stresses that usually result from joining the components with mechanical fasteners, such as bolts and screws. Moreover, structural integrity of the sections is maintained (because no holes are required).
- The external appearance of the bonded components is unaffected.
- Very thin and fragile components can be bonded, without significant increase in their weight.
- Porous materials and materials of very different properties and sizes can be joined.

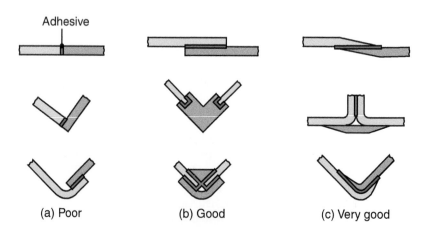

FIGURE 32.11 Various joint designs in adhesive-bonding; note that good designs require large contact areas between the members to be joined.

- Because adhesive-bonding is usually carried out at a temperature between room temperature and about 200°C (400°F), there is no significant distortion of the components or change in their original properties.

The major limitations of adhesive-bonding are:

- Limited range of service temperatures
- Bonding time can be long
- The need for great care in surface preparation
- Bonded joints are difficult to test nondestructively, particularly for large structures
- Limited reliability of adhesively-bonded structures during their service life, and significant concerns regarding hostile environmental conditions, such as degradation by temperature, oxidation, stress corrosion, radiation, or dissolution

The cost of adhesive-bonding depends on the particular operation. In many cases, however, the overall economics of the process make adhesive-bonding an attractive joining process, and sometimes it may be the only one that is feasible or practical. The cost of equipment varies greatly, depending on the size and type of operation.

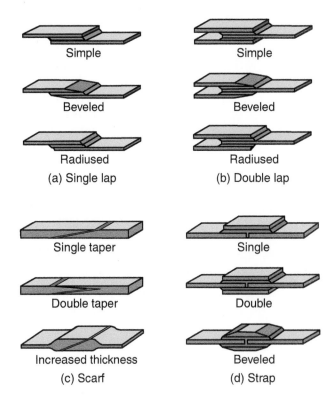

FIGURE 32.12 Desirable configurations for adhesively bonded joints.

32.4.4 Design for Adhesive-bonding

- Several joint designs for adhesive-bonding are shown in Figs. 32.11–32.13. They vary considerably in strength; hence, selection of the appropriate design is important and should include such considerations as the type of loading and the environment.
- Designs should ensure that joints are preferentially subjected only to compressive or shear forces, although limited tension can be supported. Peeling and cleavage should be avoided.
- Butt joints require large bonding surfaces; tapered (scarf) joints should be used when feasible. Simple lap joints tend to distort under tension, because of the force couple at the joint (see Fig. 31.9.). If this is a concern, double lap joints or straps can be used (Figs. 32.12b and d).
- The coefficients of thermal expansion (Table 3.1) of the components to be bonded should preferably be close to each other, in order to avoid internal stresses during adhesive-bonding. Situations in which thermal cycling can cause differential movement across the joint should be avoided.

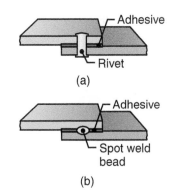

FIGURE 32.13 Two examples of combination joints, for purposes of improved strength, air of liquid tightness, and resistance to crevice corrosion.

32.5 Mechanical Fastening

Two or more components may have to be joined or fastened in such a way that they can be taken apart during the product's service life or life cycle. Numerous products, such as pens, shaft couplings, car wheels, appliances, engines, and bicycles, have components that are fastened mechanically. *Mechanical fastening* may be preferred over other methods for the following reasons:

- Ease of manufacturing
- Ease of assembly and transportation
- Ease of disassembly, maintenance, parts replacement, or repair
- Ease in creating designs that require movable joints such as hinges, sliding mechanisms, and adjustable components and fixtures
- Lower overall cost of manufacturing the product

The most common method of mechanical fastening is by the use of **fasteners**; these may be pins, rivets, or keys; *threaded fasteners*, such as bolts, nuts, screws, and studs; or other types, such as various integrated fasteners. Also known as **mechanical assembly**, mechanical fastening generally requires that the components have *holes* through which the fasteners are inserted. These joints may be subjected to both shear and tensile stresses, and thus should be designed to resist such forces.

Hole Preparation.   An important aspect of mechanical fastening is *hole preparation*. As described in Chapters 16, 23, and 27, a hole can be produced by several processes, such as punching, drilling, chemical and electrical means, and high-energy beams. Recall from Parts II and III that holes also may be *produced integrally* in the product during such processes as casting, forging, extrusion, and powder metallurgy. For improved accuracy and surface finish, many of these hole-making operations may be followed by finishing operations, such as shaving, deburring, reaming, and honing, as described in various sections of Part IV.

Because of the fundamental differences in their characteristics, each hole-making process produces holes with different surface finish, surface properties, and dimensional accuracy. The most significant influence of a hole in a solid body is its tendency to reduce the component's fatigue life, because of stress concentrations. Fatigue life can best be improved by inducing compressive residual stresses on hole surface in the hoop direction. These stresses usually are induced by pushing a round rod (*drift pin*) through the hole and expanding it by a very small amount. This operation plastically deforms the surface of the hole, in a manner similar to shot peening or in roller burnishing (Section 34.2).

Threaded Fasteners.   Bolts, screws, and studs are among the most commonly used threaded fasteners. Numerous standards and specifications, including thread dimensions, dimensional tolerances, pitch, strength, and the quality of the materials used to make these fasteners, are described in the references at the end of this chapter.

Bolts are used with through holes and depend on a nut to develop a preload; screws use a threaded hole, or they may be *self-tapping* (whereby the screw either cuts or forms the thread into the part to be fastened). The self-tapping method is particularly effective and economical in plastic products in which fastening does not require a tapped hole or a nut (see also Section 32.6). If the joint is to be subjected to vibration, such as in aircraft, engines, machinery, and appliances, several specially designed nuts and *lock washers* are available. They increase the frictional resistance in the torsional direction and thus inhibit any vibrational loosening of the fasteners.

Rivets.   The most common method of permanent or semipermanent mechanical joining is by *riveting* (Fig. 32.14). Hundreds of thousands of rivets may be used in the assembly of a large commercial aircraft. Rivets may be solid or tubular. Installing a solid rivet takes two steps: placing the rivet in the hole (usually punched or drilled) and deforming the end of its shank by upsetting it (*heading*; see Fig. 14.11). When a hole can only be accessed from one side, a *blind rivet* can be used, which uses a tubular rivet with an internal mandrel. When the rivet is inserted, the mandrel is pulled back toward the insertion end, resulting in a flared end that locks the rivet in place, as in Fig. 32.14c. Explosives can be placed within the rivet cavity and detonated to expand the end of the rivet.

The riveting operation can be performed manually or by mechanized means, including the use of programmable robots. Some design guidelines for riveting are illustrated in Fig. 32.15.

32.5.1 Various Fastening Methods

Numerous other techniques are used in joining and assembly applications.

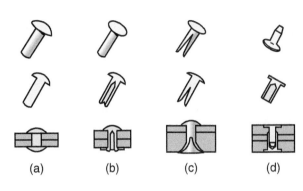

FIGURE 32.14 Examples of rivets: (a) solid, (b) tubular, (c) split or bifurcated, and (d) compression.

Metal Stitching and Stapling. The process of *metal stitching* and *stapling* (Fig. 32.16) is much like that of ordinary stapling of papers. The operation is fast, and it is particularly suitable for joining thin metallic and nonmetallic materials, including wood; a common example is the stapling of cardboard containers. In *clinching*, two or more materials are plastically deformed by a punch and die to produce an interlocking geometry. The fastener material must be sufficiently thin and ductile to withstand the large localized deformation in sharp 90° bends.

Seaming. *Seaming* (Fig. 32.17) is based on the simple principle of folding two thin pieces of material together, much like the joining of two pieces of paper by folding them at the corners. Common examples of seaming are found at the tops of beverage cans (see Fig. 16.31), containers for food and household products, and heating and air-conditioning ducts. In seaming, the materials should be capable of undergoing bending and folding at very small radii without cracking. The performance and reliability of seams may be improved by the addition of adhesives or polymeric coatings and sealing materials or by soldering; these approaches also make the seams impermeable.

Crimping. The *crimping* process is a method of joining without using fasteners. It can be done with beads or dimples (Fig. 32.18), which can be produced by shrinking or swaging operations (see Section 14.4). Crimping can be done on both tubular and flat components, provided that the materials are sufficiently thin and ductile in order

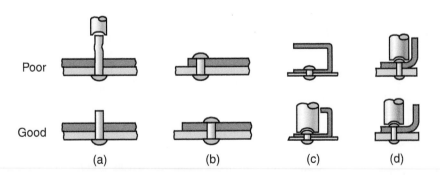

FIGURE 32.15 Design guidelines for riveting. (a) Exposed shank is too long; the result is buckling instead of upsetting. (b) Rivets should be placed sufficiently far from edges of the parts, to avoid stress concentrations. (c) Joined sections should allow ample clearance for the riveting tools. (d) Section curvature should not interfere with the riveting process. *Source:* After J.G. Bralla.

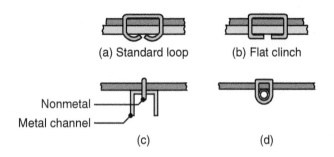

(a) Standard loop   (b) Flat clinch

Nonmetal ——
Metal channel ——

(c)   (d)

FIGURE 32.16   Typical examples of metal stitching.

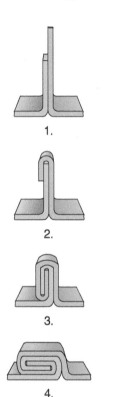

1.

2.

3.

4.

FIGURE 32.17   Stages in forming a double-lock seam.

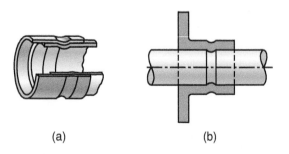

(a)   (b)

FIGURE 32.18   Two examples of mechanical joining by crimping.

to withstand the large localized deformations. Metal caps are fastened to glass bottles by crimping, just as some connectors are crimped to electrical wiring. To provide a stronger joint, crimping can also be performed using a sleeve around the parts to be joined.

Spring and Snap-in Fasteners. Several types of *spring and snap-in fasteners* are shown in Fig. 32.19. These fasteners are widely used in automotive bodies and household appliances; they are economical and permit easy and rapid component assembly. *Integrated snap fasteners* are increasingly common because they ease assembly, and can be molded at the same time as the part they are to fasten.

Shrink and Press Fits.   Components also may be assembled by shrink or press fitting. In *shrink fitting*, a component is heated so that it expands and can be mounted over a shaft or another component; upon cooling, it contracts and develops a high contact stress. Typical applications are assembling die components and mounting gears and cams onto shafts. In *press fitting*, one component is forced over another; when the components are designed properly, this process results in high joint strength.

Shape-memory Alloys.   The characteristics of these materials were described in Section 6.13. Because of their unique capability to recover their shape, *shape-memory alloys* can be used for fasteners. Several advanced applications include their use as couplings in the assembly of titanium-alloy tubing for aircraft.

32.5.2   Design for Mechanical Fastening

The design of mechanical joints requires a consideration of the type of loading to which the structure will be subjected and of the size and spacing of holes. General design guidelines for mechanical joining include the following (see also Section 37.10):

- It is generally less costly to use fewer, but larger, fasteners than to use a large number of small ones.
- Part assembly should be accomplished with a minimum number of fasteners.
- The fit between parts to be joined should be as loose as possible to facilitate the assembly process and reduce costs.
  - Fasteners of standard size should be used whenever possible.
  - Holes should not be too close to edges or corners, to avoid the possibility of tearing the material when it is subjected to external forces.

Compatibility of the fastener material with that of the components to be joined is important, as otherwise it may lead to *galvanic corrosion* (also known as *crevice corrosion*; see Section 3.8). For example, in a system in which a steel bolt or rivet is used to fasten copper sheets, the bolt is anodic and the copper plate is cathodic; this combination causes rapid corrosion and loss of joint strength. Aluminum or zinc fasteners on copper products react in a similar manner.

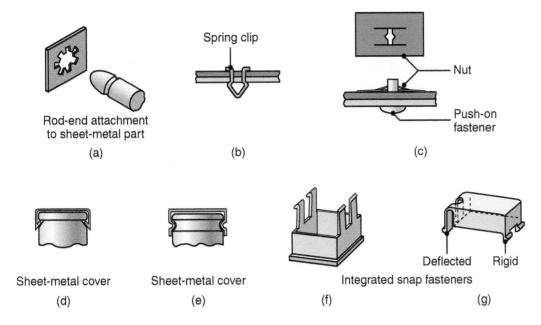

FIGURE 32.19   Examples of spring and snap-in fasteners, used to facilitate assembly.

32.6   Joining Plastics, Ceramics, and Glasses

Plastics can be joined by many of the methods already described for joining metals and nonmetallic materials, especially adhesive-bonding and mechanical fastening.

32.6.1   Joining Thermoplastics

Thermoplastics can be joined by thermal means, adhesive-bonding, solvent bonding, and mechanical fastening.

Thermal Methods.   Thermoplastics (Section 7.3) soften and melt as the temperature is increased. Consequently, they can be joined when heat is generated at the interface (from either an external or internal source), allowing **fusion** to take place. The heat softens the thermoplastic at the interface to a viscous or molten state, and ensures a good bond with the application of pressure.

About the low thermal conductivity of thermoplastics, however, the heat source may burn or char the surfaces of the components if applied at too high a rate. Burning or charring can then cause difficulties in developing sufficiently deep fusion for proper joint strength. *Oxidation* also can be a problem in joining some polymers, such as polyethylene, because it causes *degradation*; an inert shielding gas, such as nitrogen, can be used to prevent oxidation.

External heat sources may be chosen from among the following, depending on the compatibility of the polymers to be joined:

- **Hot air** or *inert gases*
- **Hot-tool welding** or *hot-plate welding*, where heated tools and dies are pressed against the surfaces to be joined, heating them by the interdiffusion of molecular chains; this process is commonly used in butt-welding of plastic pipes and tubing

- **Infrared radiation** (from high-intensity quartz heat lamps) is focused into a narrow beam onto the surfaces to be joined
- **Radio waves** are particularly useful for thin polymer films; frequencies are in the range of 100–500 Hz
- **Dielectric heating**, at frequencies of up to 100 MHz, are effective for through-heating of such polymers as nylon, polyvinyl chloride, polyurethane, and rubber
- **Electrical resistance** elements (such as wires or braids, or carbon-based tapes, sheets, and ropes) are placed at the interface to create heat by the passing of electrical current, a process known as *resistive-implant welding*. Alternatively, in induction welding, these elements at the interface may be subjected to radio-frequency exposure; in both cases, the elements at the interface must be compatible with the use of the joined product, because they are left in the weld zone
- **Lasers,** emitting defocused beams at low power to prevent degradation of the polymer

Internal heat sources are developed by the following means:

- **Ultrasonic welding** (Section 31.3) is the most commonly used process for thermoplastics, particularly amorphous polymers, such as acrylonitrile-butadiene-styrene (ABS) and high-impact polystyrene; frequencies are in the range of 20–40 kHz
- **Friction welding** (also called *spin welding*, for polymers) and *linear friction welding* (also called *vibration welding*) are particularly useful for joining polymers with a high degree of crystallinity, such as acetal, polyethylene, nylons, and polypropylene
- **Orbital welding** is similar to friction welding, except that the rotary motion of one component is in an orbital path

The fusion method is particularly effective with plastics that cannot be bonded easily using adhesives; plastics, such as PVC, polyethylene, polypropylene, acrylics, and ABS, can be joined in this manner. Specially designed portable fusion-sealing systems are used to allow in-field joining of plastic pipe, usually made of polyethylene and used for natural-gas delivery.

Coextruded multiple food wrappings consist of different types of films, bonded by heat during extrusion (Section 19.2.1). Each film has a different function; for example, one film may keep out moisture, another may keep out oxygen, and a third film may facilitate heat sealing during the packaging process. Some wrappings have as many as seven layers, all bonded together during production of the film.

Adhesive-bonding.   This method is best illustrated in joining of sections of PVC pipe (used extensively in plumbing systems) and ABS pipe (used in drain, waste, and vent systems). A primer that improves adhesion is first used to apply the adhesive to the connecting sleeve and pipe surfaces (a step much like that of using primers in painting), and then the pieces are pressed together.

Adhesive-bonding of polyethylene, polypropylene, and polytetrafluoroethylene (Teflon) can be difficult, because adhesives do not bond readily to them. The surfaces of parts made of these materials usually have to be treated chemically to improve bonding. The use of adhesive primers or double-sided adhesive tapes also is effective.

Mechanical Fastening.   This method is particularly effective for most thermoplastics (because of their inherent toughness and resilience) and for joining plastics to metals. Plastic or metal screws may be used; the use of self-tapping metal screws is a common practice. *Integrated snap fasteners* have gained wide acceptance for simplifying assembly operations; typical fastener geometries are shown in Figs. 32.19f and g. Because the fastener can be molded directly at the same time as the plastic is molded, it adds very little to the cost of assembly.

Solvent Bonding.   This method consists of the following sequence of steps:

1. Roughening the surfaces with an abrasive
2. Wiping and cleaning the surfaces with a solvent appropriate for the particular polymer
3. Pressing the surfaces and holding them together until sufficient joint strength is developed

Electromagnetic Bonding.   Thermoplastics also may be joined by *magnetic* means, by embedding tiny metal particles on the order of 1 μm (40 μin.) in the polymer. A high-frequency field then causes induction heating of the polymer and melts it at the interfaces to be joined.

32.6.2 Joining Thermosets

Thermosetting plastics, such as epoxy and phenolics, can be joined by the following techniques:

- **Threaded** or molded-in **inserts**
- **Mechanical fasteners**, particularly self-tapping screws and integrated snap fasteners
- **Solvent bonding**
- **Co-curing**, in which the two components to be joined are placed together and cured simultaneously
- **Adhesive-bonding**

CASE STUDY 32.1   Light Curing Acrylic Adhesives for Medical Products

Cobe Cardiovascular, Inc., is a manufacturer of blood collection and processing systems, as well as extracorporeal systems for cardiovascular surgery. The company, like many other device manufacturers, traditionally used solvents for bonding device components and subassemblies. However, several federal agencies began to encourage industries to avoid the use of solvents, and Cobe particularly wanted to eliminate its use of methylene chloride, for environmental and occupational safety reasons.

Toward this goal, the company began to redesign most of its assemblies to use light-curing (ultraviolet or visible) adhesives. Most of their devices were made of transparent plastics; consequently, its engineers needed clear adhesive bonds for aesthetic purposes and with no tendency for stress cracking or crazing.

As an example of a typical product, Cobe's blood salvage or collection reservoir is an oval polycarbonate device, approximately 300 mm (12 in.) tall, 200 mm (8 in.) in major diameter, and 100 mm (4 in.) deep (Fig. 32.20). The reservoir is a one-time use, disposable device; its purpose is to collect and hold the blood during open-heart or chest surgery or for arthroscopic and emergency room procedures.

Up to 3000 cc of blood may be stored in the reservoir while the blood awaits passage into a 250-cc centrifuge, which cleans the blood and returns it to the patient after the surgical procedure is completed. The collection reservoir consists of a clear, polycarbonate lid joined to a polycarbonate bucket. The joint is a tongue-and-groove configuration; the goal was to have a strong, elastic joint that could withstand repeated stresses with no chance of leakage.

(continued)

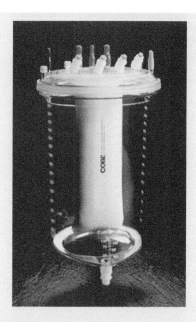

FIGURE 32.20 The Cobe Laboratories blood reservoir; the lid is bonded to the bowl with an airtight adhesive joint and tongue-in-groove joint. *Source:* Courtesy of Cobe Laboratories.

Light-cured acrylic adhesives offer a range of performance properties that make them well suited for this application because, first and foremost, they achieve high bond strength to the thermoplastics typically used to form medical-device housings. For example, Loctite® 3211 (see anaerobic adhesives, Section 32.4.1) achieves shear strengths of 11 MPa (1600 psi) on polycarbonate. As important as the initial shear strength may be, it is even more important that the adhesive be able to maintain the high bond strength after sterilization.

Another consideration that makes light-cured adhesives well suited for this application is their availability in formulations that allow them to withstand large strains prior to yielding; Loctite 3211, for example, yields at elongations in excess of 200%. Flexibility is critical, because the bonded joints are typically subjected to large amounts of bending and flexing when the devices are pressurized during qualification testing and their use. If an adhesive is too rigid, it will fail in this type of testing, even if it offers higher shear strength than a comparable and more flexible adhesive. Light-cured acrylics are widely available in formulations that meet international quality standard certification (ISO; see Section 36.6), which means that, when processed properly, they will not cause biocompatibility problems in the final assembly.

It also is important that the joint be designed properly to maximize performance. If the enclosure is bonded with a joint consisting of two flat faces in intimate contact, peel stresses (see Fig. 32.10) will be acting on the bond when the vessel is pressurized. These stresses are the most difficult type for an adhesive joint to withstand, because the entire load is concentrated on the leading edge of the joint. The tongue-and-groove design that the company adopted addressed this concern, with the groove acting as to hold and contain the adhesive during the dispensing operation.

When the parts are mated and the adhesive is cured, this design allows much of the load on the joint (when the device is pressurized) to be translated into shear and tensile forces (which the adhesive is much better suited to withstand). The gap between the tongue and the groove can vary widely, because most light-cured adhesives can quickly be cured to depths in excess of 5 mm (0.20 in.). This feature allows the manufacturer to have a robust joining process, meaning that wide dimensional tolerances can be accommodated.

With the new design and using this adhesive, the environmental concerns and the issues associated with solvent bonding were eliminated, with the accompanying benefits of a safer, faster, and more consistent bond. The light-curing adhesive provided the aesthetic-bond line the company wanted, one that was clear and barely perceptible. It also provided the structural strength required, and thus maintaining a competitive edge for the company in the marketplace.

Source: Courtesy of P.J. Courtney, Loctite Corporation.

32.6.3 Joining Ceramics and Glasses

A wide variety of ceramics and glasses are now available, with unique and important properties. Ceramics and glasses are used as products, as components of products, or as tools, molds, and dies. These materials often are assembled into components or subassemblies, and are joined either with the same type of material or with different metallic or nonmetallic materials. Generally, ceramics, glasses, and many similar materials can be joined by adhesive-bonding. A typical example is assembling broken ceramic pieces with a two-component epoxy, which is dispensed from two separate tubes and is mixed just prior to its application. Other joining methods include mechanical means, such as fasteners and spring or press fittings.

Ceramics.   As described in Chapter 8, ceramics have properties that are very different from metallic and nonmetallic materials, especially regarding stiffness, hardness, brittleness, resistance to high temperatures, and chemical inertness. Joining them to each other or to other metallic or nonmetallic materials requires special considerations; several highly specialized joining processes are now available.

A common technique, that is effective in joining difficult-to-bond combinations of materials, consists of first applying a coating of a material that bonds itself well to one or both components, thus acting as a bonding agent. For example, the surface of *alumina ceramics* can be *metallized*, as described in Section 34.5. In this technique, known as the *Mo–Mn* process, the ceramic part is first coated with a slurry of oxides of molybdenum and manganese. Next, the part is fired, forming a glassy layer on its surface. This layer is then plated with nickel; since the part now has a metallic surface, it can be brazed to a metal surface by means of an appropriate filler metal.

Tungsten carbide and titanium carbide can easily be brazed to other metals, because they both have a metallic matrix: WC has a matrix of cobalt and TiC has nickel–molybdenum alloy as a matrix (see Chapter 22). Common applications include brazing cubic boron nitride or diamond tips to carbide inserts (Figs. 22.10 and 22.11) and carbide tips to masonry drills (Fig. 23.22d). Depending on their particular structure, ceramics and metals also can be joined by *diffusion bonding*, although it may be necessary to place a metallic layer at the joint to make it stronger.

Ceramic components also can be joined or assembled together during their primary shaping process (Section 18.2); a common example is attaching handles to coffee mugs prior to firing them. Thus, shaping of the whole product is done *integrally* rather than as an additional operation after the part is already made.

Glasses.   As evidenced by the availability of numerous glass objects, glasses can easily be bonded to each other. This is commonly done by first heating and softening the surface to be joined, then pressing the two pieces together, and finally cooling them. Bonding glass to metals also is possible, because of diffusion of metal ions into the amorphous surface structure of glass; however, the differences in the coefficients of thermal expansion of the two materials must be taken into account.

32.7   Economics of Joining Operations

As in the economics of welding operations (described in Section 31.8), the joining processes discussed in this chapter depend greatly on several considerations. From Table VI.1, it can be seen that, in relative terms, the cost distribution for some of these processes:

- *Highest:* Brazing, bolts, nuts, and other fasteners
- *Intermediate:* Riveting and adhesive-bonding
- *Lowest:* Seaming and crimping

The variety of processes and the general costs involved are described below. For **brazing,**

- Manual brazing: The basic equipment costs about $300, but it can be over $50,000 for automated systems.

- Furnace brazing: Costs vary widely, ranging from about $2000 for simple batch furnaces to $300,000+ for continuous-vacuum furnaces.
- Induction brazing: For small units, the cost is about $10,000.
- Resistance brazing: Equipment costs range from $1000, for simple units, to more than $10,000 for larger, more complex units.
- Dip brazing: The cost of equipment varies widely, from $2000 to more than $200,000; the more expensive equipment includes various computer-control features.
- Infrared brazing: Equipment cost ranges from $500 to $30,000.
- Diffusion brazing: The cost of equipment ranges from $50,000 to $300,000.

Soldering.   The cost of soldering equipment depends on its complexity and on the level of automation. The cost ranges from less than $20 for manual soldering irons to more than $50,000 for automated equipment.

SUMMARY

- Joining processes that do not rely on fusion or pressure at the interfaces include brazing and soldering; instead, they utilize filler materials that require some temperature rise in the joint. They can be used to join dissimilar metals of intricate shapes and various thicknesses.
- Adhesive-bonding has gained increased acceptance in major industries, such as the aerospace and automotive industries. In addition to good bond strength, adhesives have other favorable characteristics, such as the ability to seal, insulate, prevent electrochemical corrosion between dissimilar metals, and reduce vibration and noise by means of internal damping in the bond. Surface preparation and joint design are important factors.
- Mechanical fastening is one of the most common joining methods. Bolts, screws, and nuts are typical fasteners for machine components and structures that are likely to be taken apart for maintenance and for ease of transportation.
- Rivets and various fasteners are semipermanent or permanent fasteners, used in buildings, bridges, and transportation equipment.
- Thermoplastics can be joined by fusion-welding techniques, adhesive-bonding, or mechanical fastening. Thermosets are usually joined by mechanical means, such as molded-in inserts and fasteners, or by solvent bonding. Ceramics can be joined by adhesive-bonding and metallizing techniques. Glasses are joined by heating the interfaces or by using adhesives.

KEY TERMS

Adhesive-bonding	Filler metal	Reflow soldering	Stapling
Braze welding	Flux	Rivet	Stitching
Brazing	Hole preparation	Seaming	Threaded fasteners
Crimping	Integrated snap fastener	Shrink fitting	Wave soldering
Electrically conducting adhesives	Lead-free solders	Snap-in fastener	
Fasteners	Mechanical fastening	Soldering	
	Press fitting	Solvent bonding	

BIBLIOGRAPHY

Adams, R.D. (ed.), **Adhesive-bonding: Science, Technology and Applications**, CRC Press, 2005.

Bath, J. (ed.), **Lead-Free Soldering**, Springer 2007.

Bickford, J.H., **Introduction to the Design and Behavior of Bolted Joints**, 4th ed., Marcel Dekker, 2007.

Brazing Handbook, 5th ed., American Welding Society, 2007.

Brockmann, W., Geiss, P.L., Klingen, J., and Schreoeder, K.B., **Adhesive-bonding: Materials, Applications and Technology**, Wiley, 2009.

Ebnesajjad, S., **Adhesives Technology Handbook**, 2nd ed., William Andrew, 2008.

Gourley, R., and Walker, C. (eds.), **Brazing and Soldering 2012**, American Society for Metals, 2012.

Hamrock, B.J., Jacobson, B., and Schmid, S.R., **Fundamentals of Machine Elements**, 2nd ed., McGraw-Hill, 2005.

Humpston, G., and Jacobson, D.M., **Principles of Soldering**, ASM International, 2004.

Jacobson, D.M., and Humpston, G., **Principles of Brazing**, ASM International, 2005.

Petrie, E.M., **Handbook of Adhesives and Sealants**, 2nd ed., McGraw-Hill, 2006.

Pizzi, A., and Mittal, K.L., **Handbook of Adhesive Technology**, 2nd ed., CRC Press, 2003.

Roberts, P., **Industrial Brazing Practice**, CRC Press, 2003.

Rotheiser, J., **Joining of Plastics: Handbook for Designers and Engineers**, 3rd ed., Hanser, 2009.

Schwartz, M.M., **Brazing**, 2nd ed., ASM International, 2003.

REVIEW QUESTIONS

32.1 What is the difference between brazing and braze welding?

32.2 Are fluxes necessary in brazing? If so, why?

32.3 Why is surface preparation important in adhesive-bonding?

32.4 What materials are typically used in solder?

32.5 Soldering is generally applied to thinner components. Explain why.

32.6 Explain the reasons why a variety of mechanical joining methods have been developed over the years.

32.7 List three brazing and three soldering techniques.

32.8 Describe the similarities and differences between the functions of a bolt and those of a rivet.

32.9 What precautions should be taken in the mechanical joining of dissimilar metals?

32.10 What difficulties are involved in joining plastics? Why?

32.11 What are the advantages of rivets?

32.12 What are the principles of (a) wave soldering and (b) reflow soldering?

32.13 What is a peel test? Why is it useful?

32.14 What is a combination joint?

32.15 What test methods are used to evaluate adhesives?

QUALITATIVE PROBLEMS

32.16 Describe some applications in manufacturing for single-sided and some for double-sided adhesive tapes.

32.17 Explain how adhesives can be made to be electrically conductive.

32.18 Comment on your observations concerning the joints shown in Figs. 32.3, 32.6, 32.10, and 32.11.

32.19 Give examples of combination joints other than those shown in Fig. 32.12.

32.20 Discuss the need for fixtures for holding workpieces in the joining processes described in this chapter.

32.21 Explain why adhesively bonded joints tend to be weak in peeling.

32.22 It is common practice to tin-plate electrical terminals to facilitate soldering. Why is it used?

32.23 Give three applications where adhesive-bonding is the best joining method.

32.24 How important is a close fit for two parts that are to be brazed? Explain.

32.25 If you are designing a joint that must be strong and also needs to be disassembled several times during the product's life, what kind of joint would you recommend? Explain.

32.26 Review Fig. 32.11 and explain why the examples under the 'Poor', 'Good' and 'Very good' have these classifications.

32.27 Rate lap, butt, and scarf joints in terms of joint strength. Explain your answers.

32.28 What are the advantages of integrated snap fasteners?

QUANTITATIVE PROBLEMS

32.29 Refer to the simple butt and lap joints shown at the top row of Fig. 32.10a. (a) Assuming that the area of the butt joint is 5 mm × 20 mm, and referring to the adhesive properties given in Table 32.3, estimate the minimum and maximum tensile force that this joint can withstand. (b) Estimate these forces for the lap joint, assuming that its area is 15 mm × 15 mm.

32.30 In Fig. 32.13a, assume that the cross-section of the lap joint is 20 mm × 20 mm, that the diameter of the solid rivet is 4 mm, and that the rivet is made of copper. Using the strongest adhesive shown in Table 32.3, estimate the maximum tensile force that this joint can withstand.

32.31 As shown in Fig. 32.15a, a rivet can buckle if it is too long. Referring to Chapter 14 on forging, determine the maximum length-to-diameter ratio of a rivet so that it would not buckle during riveting.

32.32 Figure 32.4 shows qualitatively the tensile and shear strength in brazing as a function of joint clearance. Search the technical literature, obtain data, and plot these curves quantitatively. Comment on your observations.

32.33 When manufacturing the fuselage of a commuter airplane, aluminum plates are adhesively bonded together with lap joints. Because the elastic deformation for a single plate differs from the deformation for two plates in a lap joint, the maximum shear stress in the adhesive is twice as high as the average shear stress. The shear strength of the adhesive is 20 MPa, the tensile strength of the aluminum plates is 95 MPa, and their thickness is 4.0 mm. Calculate the overlapping length needed to make the adhesive joint twice as strong as the aluminum plate.

SYNTHESIS, DESIGN, AND PROJECTS

32.34 Examine various household products and describe how their components were joined and assembled. Explain why those particular processes were used and not others.

32.35 Name several products that have been assembled by (a) seaming, (b) stitching, and (c) soldering.

32.36 Suggest methods of attaching a round bar (made of a thermosetting plastic) perpendicularly to a flat metal plate. Discuss their advantages and limitations.

32.37 Describe the tooling and equipment that would be necessary to perform the double-lock seaming operation shown in Fig. 32.17, starting with a thin, flat sheet.

32.38 Prepare a list of design guidelines for joining by the processes described in this chapter. Would these guidelines be common to most processes? Explain.

32.39 What joining methods would be suitable for assembling a thermoplastic cover over a metal frame? Assume that the cover is removed periodically, as is the lid of a coffee can.

32.40 Answer Problem 32.39, but for a cover made of (a) a thermoset, (b) a metal, and (c) a ceramic. Describe the factors involved in your selection of methods.

32.41 Comment on workpiece size and shape limitations, if any, for each of the processes described in this chapter.

32.42 Describe part shapes that cannot be joined by the processes covered in this chapter. Give specific examples.

32.43 Give examples of products in which rivets in a structure or in an assembly may have to be removed and later replaced by new rivets.

32.44 Visit a hardware store and investigate the geometry of the heads of screws that are permanent fasteners—that is, fasteners that can be screwed in, but not out.

32.45 Obtain a soldering iron and attempt to solder two wires together. First, try to apply the solder at the same time as you first put the soldering iron tip to the wires. Second, preheat the wires before applying the solder. Repeat the same procedure for a cool surface and a heated surface. Record your results and explain your findings.

32.46 Perform a literature search to determine the properties and types of adhesives used to affix artificial hips onto the human femur.

32.47 Review Fig. 32.9a and explain the shortcoming in using a constant thickness beam instead of a tapered double cantilever beam.

32.48 Review Fig. 32.9 and carefully sketch the stress distributions you expect in each geometry.

32.49 Design a joint to connect two 25 mm wide, 5 mm thick steel members. The overlap may be as much as 25 mm, and any one approach described in this chapter can be used.

32.50 For the same members in Problem 32.49, design a joint using threaded fasteners arranged in one row. Do you advise the use of one large fastener or many small fasteners? Explain.

32.51 For the same members in Problem 32.49, design a joint using a *combination* of joining techniques.

Surface Technology

Our first visual or tactile contact with the objects around us is through their *surfaces*. We can see and feel surface roughness, waviness, reflectivity, and various other features. The preceding chapters described the properties of materials and manufactured components mostly in terms of their *bulk* characteristics, such as strength, ductility, hardness, and toughness. Also included were some descriptions of the influences of surfaces on these properties, such as the effect of surface preparation on fatigue life and in joining processes, and the sensitivity of brittle materials to surface roughness, scratches, and various defects.

Machinery and their accessories typically have numerous members that slide against each other, such as bearings, slideways, pistons and cylinders, and tools and dies for cutting and forming operations. Close examination will reveal that some of these surfaces are

- Smooth while others are rough
- Slide against each other at high relative speeds while others move slowly
- Lubricated while others are dry
- Subjected to heavy loads while others support light loads
- Subjected to elevated temperatures while others are at room temperature

In addition to its geometric features, a **surface** constitutes a thin layer on the bulk material. The mechanical, physical, chemical, and metallurgical properties of a surface depend not only on the material and its processing history but also on the environment to which the surface has been exposed. Consequently, the surface of a manufactured part usually possesses properties and behavior that can be significantly different from those of its bulk.

Although the bulk material generally determines a component's overall mechanical properties, the component's surfaces directly influence the part's performance in the following areas (see Fig. VII.1):

- Appearance and geometric features of the part and their role in subsequent operations, such as welding, soldering, adhesive bonding, painting, and coating
- Resistance to corrosion
- Effectiveness of lubricants during the manufacturing process and throughout the part's service life

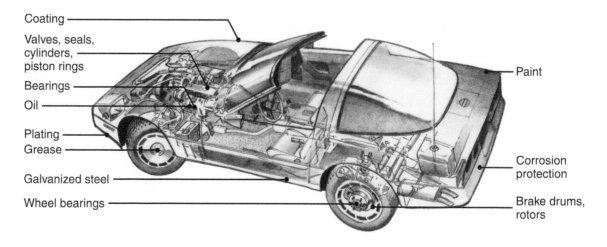

FIGURE VII.1   Components in a typical automobile that are related to the topics described in Part VII.

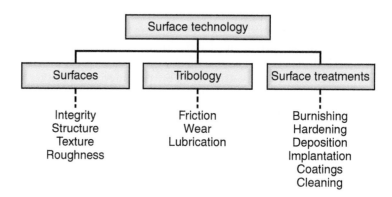

FIGURE VII.2   An outline of topics covered in Part VII.

- Crack initiation and propagation, as a result of surface such defects as roughness, scratches, seams, and heat-affected zones, which can lead to weakening and premature failure of the part, through fatigue
- Thermal and electrical conductivity of contacting bodies; rough surfaces, for example, have lower thermal and electrical conductivity than smooth surfaces
- Friction and wear of tools, molds, dies used in manufacturing, and of the products made

Following the outline shown in Fig. VII.2, Chapter 33 discusses surface characteristics in terms of their structure and topography. The material and process variables that influence the friction, wear, and lubrication of materials are then described. Chapter 34 describes the methods that can be used to modify surfaces for improved frictional behavior, effectiveness of lubricants, resistance to wear and corrosion, and surface finish and appearance.

Surface Roughness and Measurement; Friction, Wear, and Lubrication

CHAPTER **33**

- This chapter describes various features of surfaces that have a direct effect on both the selection of manufacturing processes and the service life of the parts produced.
- Surface features, such as roughness, texture, and lay, are discussed, as well as approaches used to quantitatively describe and measure surfaces.
- The chapter also examines the nature of friction, its role in manufacturing, and the factors involved in its magnitude.
- Wear and lubrication are then discussed, along with various approaches to minimizing wear.
- The chapter ends with a summary of commonly used lubricants and their selection for a particular manufacturing process and the materials involved.

33.1 Introduction

Surfaces are distinct entities, with properties that can be significantly different from those of the bulk; this is particularly the case for metals, because of surface-oxide layers, work-hardened layers, and various other features. Several defects can exist on a surface, depending on the manner in which the surface was generated. These defects, as well as various surface textures, can have a major influence on the surface integrity of workpieces, as well as of tools, molds, and dies.

The common methods of surface-roughness measurement in engineering practice are then described, including the instrumentation involved, followed by a brief description of surface-roughness requirements in engineering design and products. Because of their increasing importance in precision manufacturing and nanofabrication, three-dimensional surface measurements also are discussed.

The chapter then describes those aspects of *friction*, *wear*, and *lubrication*, collectively known as **tribology**, that are relevant to manufacturing processes and operations and to the service life of products. The nature of friction and wear for metallic and nonmetallic materials, and how they are influenced by various material and process variables, are then described. Wear has a major economic impact, as it has been estimated that in the United States alone the total cost of replacing worn parts is more than $100 billion per year.

Finally, the chapter describes the fundamentals of metalworking fluids, including the types, characteristics, and applications of commonly used liquid and solid lubricants and the lubrication practices employed, including the importance of biological and environmental considerations in the use, application, recycling, and ultimate disposal of metalworking fluids.

33.2   Surface Structure and Integrity

Upon close examination, it can be observed that the surface of a piece of metal generally consists of several layers, as illustrated in Fig. 33.1:

1. The *bulk* metal, also known as the metal **substrate**, has a structure that depends on the composition and processing history of the metal.

2. Above the bulk metal is a layer that usually has been deformed plastically and work hardened to a greater extent than the bulk during the manufacturing process. The depth and properties of the work-hardened layer, called **surface structure**, depend on such factors as the processing method employed and the effects of frictional sliding on the surface. If, for example, the surface has been produced by machining it with a dull and worn tool (see Fig. 21.22), or the surface has been subjected to sliding against tools and dies, the *work-hardened layer* will be relatively thick, and usually will also develop residual stresses (Section 2.11).

3. An *amorphous layer* (without crystalline structure; also called the *Beilby* layer, after G. Beilby) may also be present. Such an amorphous layer occurs when a surface is subjected to high heat or even melting (such as from high frictional forces), and then cooled rapidly, so that grains do not have time to form a crystalline structure (see Section 1.5). This can occur, for example, in flame cutting or high-stress grinding operations.

4. Unless the metal is processed and kept in an inert (oxygen-free) environment or is a noble metal (such as gold or platinum), an **oxide layer** forms over the work-hardened layer. The oxide layer is generally much harder than the base metal, hence it is more abrasive; as a result, it has important effects on friction, wear, and lubrication. Each metal forms its own unique oxide or oxides, and their behavior can be very complex. For example:

   - *Iron* has a surface oxide structure, with FeO adjacent to the bulk metal, followed by a layer of Fe_3O_4, and then a layer of Fe_2O_3 (which is exposed to the environment).
   - *Aluminum* has a dense, amorphous surface layer of Al_2O_3, with a thick, porous, and hydrated aluminum-oxide layer over it.
   - *Copper* has a bright, shiny surface when freshly scratched or machined; soon after, however, it develops a Cu_2O layer, which is covered with a layer of CuO; the latter layer gives copper its somewhat dull color.

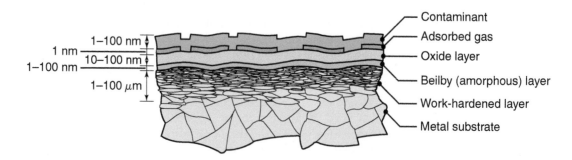

FIGURE 33.1   Schematic illustration of a cross-section of the surface structure of a metal; the thickness of the individual layers depends on both processing conditions and the processing environment. *Source:* Based on E. Rabinowicz and B. Bhushan.

- *Stainless steels* are "stainless" because they develop a protective layer of chromium oxide (by *passivation*, as described in Section 3.8).

5. Under normal environmental conditions, surface oxide layers generally are covered with *adsorbed* layers of gas and moisture.
6. Finally, the outermost surface of the metal may be covered with contaminants, such as dirt, dust, lubricant residues, cleaning-compound residues, and pollutants from the environment.

Surfaces thus have properties that generally are very different from those of the substrate material. The factors that pertain to the surface structures of the metals just described also are factors in the surface structure of plastics and ceramics (Chapters 7 and 8). The surface texture of these materials also depends, as with metals, on the method of production.

Surface Integrity.   *Surface integrity* describes not only the topological (geometric) features of surfaces and their physical and chemical properties, but also their mechanical and metallurgical properties and characteristics. Surface integrity is an important consideration in manufacturing operations, because it influences such properties as fatigue strength (see Fig. 2.29), resistance to corrosion, and service life.

Several *surface defects* caused by and produced during component manufacturing can be responsible for inadequate surface integrity. These defects usually are caused by a combination of factors, such as (a) defects in the raw or original material, (b) the method or methods by which the surface is produced, and (c) improper control of the process parameters, which can result in excessive stresses, temperatures, or surface deformation.

The following list gives general definitions of the major **surface defects** (in alphabetical order) found in practice:

- **Cracks** may be external or internal; those that require a magnification of $10\times$ or higher to be seen by the naked eye are called **microcracks**.
- **Craters** are shallow depressions.
- **Heat-affected zone** is that portion of a metal which is subjected to thermal cycling without melting, such as that shown in Fig. 30.19.
- **Inclusions** are small, nonmetallic elements or compounds in the material.
- **Intergranular attack** is the weakening of grain boundaries through liquid-metal embrittlement and corrosion (Section 1.5.2).
- **Laps, folds,** and **seams** are surface defects resulting from the overlapping of material during processing (see, for example, Fig. 14.17).
- **Metallurgical transformations** involve microstructural changes caused by temperature cycling of the material; these changes may consist of phase transformations, recrystallization, alloy depletion, decarburization, and molten and then recast, resolidified, or redeposited material.
- **Pits** are shallow surface depressions, usually the result of chemical or physical attack.
- **Residual stresses** on the surface are caused by nonuniform deformation and nonuniform temperature distribution (Section 2.11).
- **Splatter** is small resolidified molten metal particles deposited on a surface, as during welding.
- **Surface plastic deformation** is a severe surface deformation caused by high stresses due to such factors as friction, tool and die geometry, worn tools, and processing methods (see, for example, Fig. 21.21).

33.3 Surface Texture and Roughness

Regardless of the method of production, all surfaces have their own characteristics which, collectively, are referred to as *surface texture*. Although the description of surface texture as a geometrical property is complex, the following guidelines have been established for identifying surface texture in terms of well-defined and measurable quantities (Fig. 33.2):

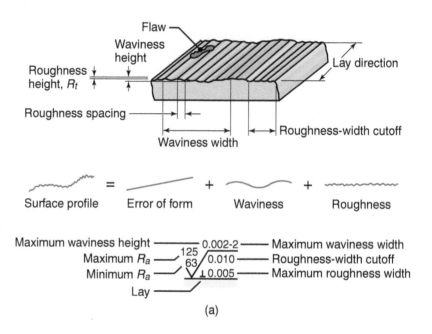

(a)

Lay symbol	Interpretation	Examples
=	Lay parallel to the line representing the surface to which the symbol is applied	
⊥	Lay perpendicular to the line representing the surface to which the symbol is applied	
X	Lay angular in both directions to line representing the surface to which symbol is applied	
P	Pitted, protuberant, porous, or particulate nondirectional lay	

(b)

FIGURE 33.2   (a) Standard terminology and symbols to describe surface finish; the quantities given are in microinches. (b) Common surface lay symbols.

- **Flaws** or **defects** are random irregularities, such as scratches, cracks, holes, depressions, seams, tears, or inclusions.
- **Lay (directionality)** is the direction of the predominant surface pattern, usually visible to the naked eye.
- **Roughness** is defined as closely spaced, irregular deviations on a small scale; it is expressed in terms of its height, width, and distance along the surface.
- **Waviness** is a recurrent deviation from a flat surface; it is measured and described in terms of the distance between adjacent crests of the waves (*waviness width*) and the height between the crests and valleys of the waves (*waviness height*).

Surface roughness is generally characterized by two methods. The **arithmetic mean value** (R_a) is based on the schematic illustration of a rough surface, as shown in Fig. 33.3, and is defined as

$$R_a = \frac{a + b + c + d + \cdots}{n},\qquad(33.1)$$

where all ordinates $a, b, c, \ldots$, are absolute values and n is the number of readings. The *units* generally used for surface roughness are μm (microns) or μin. Note that $1\ \mu m = 40\ \mu in.$ and $1\ \mu in. = 0.025\ \mu m$.

The **root-mean-square roughness** (R_q, formerly identified as RMS) is defined as

$$R_q = \sqrt{\frac{a^2 + b^2 + c^2 + d^2 + \cdots}{n}}.\qquad(33.2)$$

The datum line AB in Fig. 33.3 is located such that the sum of the areas above the line is equal to the sum of the areas below the line.

The **maximum roughness height** (R_t) is defined as the height from the deepest trough to the highest peak. It indicates how much material has to be removed in order to obtain a smooth surface, such as by polishing.

Because of its simplicity, the arithmetic mean value was adopted internationally in the mid-1950s and is used widely in engineering practice. Dividing Eq. (33.2) by Eq. (33.1) yields the ratio R_q/R_a, which, for typical surfaces produced by machining and finishing processes is 1.1 for cutting, 1.2 for grinding, and 1.4 for lapping and honing.

In general, a surface cannot be described by its R_a or R_q value alone, since these values are averages. Two surfaces may have the same roughness value, but have actual topographies that are very different. For example, a few deep troughs on an otherwise smooth surface will not affect the roughness values significantly. However, this type of surface profile can be significant in terms of friction, wear, and fatigue characteristics of a manufactured product. Consequently, it is important to analyze a surface in great detail, particularly for parts that are to be used in critical applications.

Symbols for Surface Roughness.   Limits for surface roughness are specified on technical drawings by symbols, typically shown around the check mark in the lower portion of Fig. 33.2a, and the values of these limits are placed to the left of the check mark. The symbols and their meanings concerning the lay are given in Fig. 33.2b; note that the symbol for the lay is placed at the lower right of the check mark. Symbols are used to describe a surface, specifying only its roughness, waviness, and lay; they do not include flaws. Therefore, whenever

Video Solution 33.1   Calculation of Surface Roughness

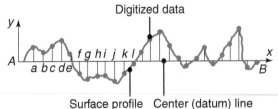

FIGURE 33.3 Coordinates used for surface-roughness measurement, defined by Eqs. (33.1) and (33.2).

necessary, a special note is included in technical specifications to describe the method or methods that should be used to inspect for surface flaws.

Measuring Surface Roughness.   Instruments called **surface profilometers** typically are used to measure and record surface roughness. A profilometer has a *diamond stylus* that travels along a straight line over the surface (Fig. 33.4a) and records periodic height measurements. The distance that the stylus travels is called the **cutoff,** which generally ranges from 0.08 to 25 mm (0.003–1 in.). A cutoff of 0.8 mm (0.03 in.) is typical for most engineering applications. The rule of thumb is that the cutoff must be large enough to include 10 to 15 roughness irregularities, as well as all surface waviness.

In order to highlight roughness, profilometer traces are recorded on an exaggerated vertical scale (a few orders of magnitude larger than the horizontal scale; see Fig. 33.4c–f); the magnitude of the scale on the recording instrument is called **gain.** Thus, the recorded profile is distorted significantly, and the surface will appear to be much rougher than it actually is. The recording instrument compensates for any surface waviness, and thus indicates only surface roughness.

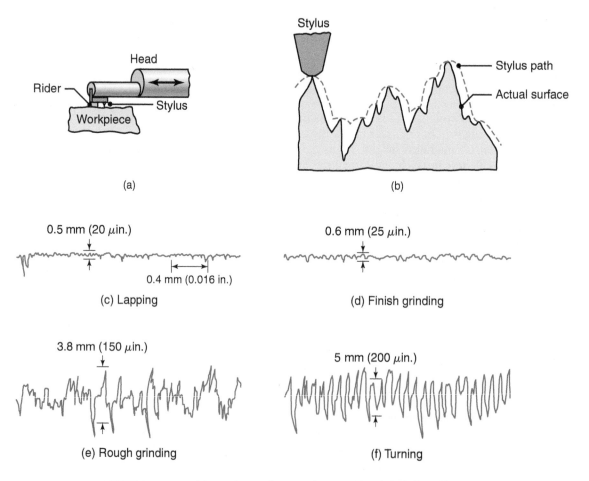

FIGURE 33.4   (a) Measuring surface roughness with a stylus; the rider supports the stylus and guards against damage. (b) Path of the stylus in surface-roughness measurements (broken line), compared with the actual roughness profile; note that the profile of the stylus path is smoother than that of the actual surface. (c)–(f) Typical surface profiles produced by various machining and surface-finishing processes; note the difference between the vertical and horizontal scales.

Because of the finite radius of the diamond stylus tip, the path of the stylus is different from the actual surface (note the path with the broken line in Fig. 33.4b), and the measured roughness is lower. The most commonly used stylus-tip diameter is 10 μm (400 μin.). The smaller the stylus diameter and the smoother the surface, the closer is the path of the stylus to the actual surface profile.

Three-dimensional Surface Measurement.   Because surface properties can vary significantly with the direction in which a profilometer trace is taken, there is often a need to measure *three-dimensional* surface profiles. In the simplest case, this can be done with a surface profilometer that has the capability of indexing a short distance between traces. A number of other alternatives have been developed, two of which are optical interferometers and atomic-force microscopes.

1. **Optical-interference microscopes** shine a light against a reflective surface and record the interference fringes that result from the incident and its reflected waves. This technique allows for a direct measurement of the surface slope over the area of interest. As the vertical distance between the sample and the interference objective is changed, the fringe patterns also change, thus allowing for surface height measurement.

2. **Atomic-force microscopes** (AFMs) are used to measure extremely smooth surfaces, and in some arrangements have the capability of distinguishing atoms on atomically smooth surfaces. In principle, an AFM is merely a very fine surface profilometer with a laser that is used to measure probe position. The surface profile can be measured with high accuracy and with vertical resolution on the atomic scale; scan areas can be on the order of 100 μm^2, although smaller areas are more common.

Surface Roughness in Engineering Practice.   Requirements for surface-roughness design in typical engineering applications vary by as much as two orders of magnitude. Some examples are:

- Bearing balls   0.025 μm (1 μin.)
- Crankshaft bearings   0.032 μm (13 μin.)
- Brake drums   1.6 μm (63 μin.)
- Clutch-disk faces   3.2 μm (125 μin.)
- Gage blocks and precision instruments   0.02 μm (0.8 μin.)

33.4 Friction

Friction plays an important role in manufacturing processes because of the relative motion and the friction forces that are always present between tool, die, and workpiece interfaces. Friction (a) *dissipates energy*, thus generating heat, which can have detrimental effects on an operation and (b) *impedes free movement* at interfaces, thus significantly affecting the flow and deformation of materials in metalworking processes. However, friction is not always undesirable; without friction, for example, it would be impossible to roll metals, control material flow in forming and shaping operations, clamp workpieces on machines, or hold drill bits in chucks.

There have been several explanations of friction. A commonly accepted theory of friction is the **adhesion theory**; it is based on the observation that two clean and dry surfaces, regardless of how smooth they are, contact each other at only a fraction of their apparent contact area (Fig. 33.5). The maximum slope of real

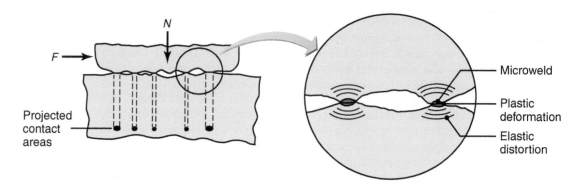

FIGURE 33.5   Schematic illustration of the interface of two bodies in contact, showing real areas of contact at the asperities; in engineering surfaces, the ratio of the apparent-to-real areas of contact can be as high as 4 to 5 orders of magnitude.

surfaces range typically from 5° to 15°, unless they are purposely manufactured to have high roughness. In such a situation, the normal (contact) load, N, is supported by minute **asperities**, which are small projections from the surface that are in contact with each other. The normal stresses at these asperities are therefore high, and thus can cause *plastic deformation* at the junctions, creating an *adhesive bond*. In other words, the asperities form microwelds, and it takes a certain force to shear the microweld. The cold pressure welding process (see Section 31.2), for example, is based on this principle.

Another theory of friction is the **abrasion theory**; it is based on the notion that an asperity from a hard surface (such as a tool or a die) *penetrates* and *plows* a softer surface (the workpiece), as discussed in Section 26.3. Plowing will cause displacement of the material and/or produce small chips or slivers, as in filing. In both situations described above, sliding between two bodies in contact is possible only if a *tangential force* is applied; this force is the **friction force**, F, required in order to shear the junctions or *plow* through the softer material.

The ratio F/N (Fig. 33.5) is the **coefficient of friction**, μ. Depending on the materials and processes involved, μ in manufacturing vary significantly. For example, in metal-forming processes, it typically ranges from about 0.03 in cold working to about 0.7 in hot working, and from about 0.5 to as much as 2 in machining operations.

Almost all of the energy dissipated in overcoming friction is converted into *heat*, which raises the surface temperature; a small fraction of the energy becomes *stored energy* (see Section 1.6) in the plastically deformed surfaces. The temperature increases (a) with increasing friction and sliding speed, (b) decreasing thermal conductivity, and (c) decreasing specific heat of the sliding materials (see also Section 21.4). The interface temperature may be so high as to soften and even melt the surfaces, and possibly cause microstructural changes in the materials involved. Note also that temperature affects the viscosity and other properties of lubricants, and high temperatures can cause breakdown of the lubricant.

Friction of Plastics.   Because their strength is low as compared with that of metals (Tables 2.2 and 7.1), *plastics* generally possess low frictional characteristics, especially polyimides, polyesters, and fluorocarbons (*Teflon*). This property makes plastics better than metals for bearings, gears, seals, prosthetic joints, and general friction-reducing applications, provided that the loads are not high. Because of this characteristic, polymers are sometimes described as *self lubricating*.

The factors involved in the friction of metals are generally applicable to polymers as well. In sliding, the plowing component of friction in thermoplastics and elastomers is a significant factor, because of their viscoelastic behavior (i.e., they exhibit both viscous and elastic behavior) and subsequent hysteresis loss (see Fig. 7.14). This condition can easily be simulated by dragging a dull nail across the surface of rubber, and observing how the rubber quickly recovers its shape.

An important factor in plastics applications is the effect of temperature rise at the sliding interfaces, caused by friction. As described in Section 7.3, thermoplastics rapidly lose their strength and become soft as temperature increases. Thus, if the temperature rise is not controlled, sliding surfaces can undergo permanent deformation and thermal degradation. The frictional behavior of various polymers on metals is similar to that of metals on metals. The well-known low friction of Teflon has been attributed to its molecular structure, which has no reactivity with metals; consequently, its adhesion is poor and its friction is low.

Friction of Ceramics.   The mechanics of friction for ceramics is similar to that of metals; thus, adhesion and plowing at interfaces contribute to the friction force in ceramics as well. Usually, however, adhesion is less important with ceramics because of their high hardness (see Fig. 2.15), whereby the real area of contact at sliding interfaces is small. Abrasion and plowing can be significant, especially when ceramics interface with softer materials.

Reducing Friction.   Friction can be reduced mainly through (a) the selection of materials that have low adhesion, such as carbides and ceramics, and (b) the use of surface films and coatings. *Lubricants* (such as oils) or solid films (such as graphite) interpose an adherent layer between the tool, die, and workpiece, which minimizes adhesion and interactions between two sliding bodies. Friction also can be reduced significantly by subjecting the tool- or die-workpiece interface to *ultrasonic vibrations*, typically at 20 kHz. The amplitude of the vibrations periodically separates the two surfaces and thus allows the lubricant to flow more freely into the interface.

Friction Measurement.   The coefficient of friction usually is determined experimentally, either during an actual manufacturing process or in simulated laboratory tests using small-scale specimens of various shapes. A test that has gained wide acceptance, particularly for bulk-deformation processes, is the **ring-compression test.** A flat ring is upset plastically between two flat platens (Fig. 33.6a). As its height is reduced, the ring expands radially outward. If friction at the interfaces is zero, both the inner and outer diameters of the ring expand as if it were a solid disk. With increasing friction, however, the internal diameter becomes smaller. For a particular reduction in height, there is a critical friction at which the internal diameter increases from its original diameter if μ is lower, and it decreases if μ is higher (Fig. 33.6b).

By measuring the change in the specimen's internal diameter and using the curves shown in Fig. 33.7 (obtained through theoretical analyses), the coefficient of friction can be determined. Note that each ring geometry and material has its own specific set of curves; the most common geometry of a specimen has an outer diameter/inner diameter/height proportion of 6:3:2. The actual size of the specimen is usually not relevant in these tests. Thus, once the percentage of reduction in internal diameter and height is known, the magnitude of μ can be determined from the appropriate chart.

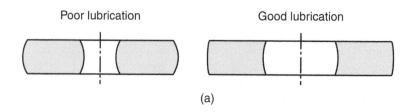

(a)

Video Solution 33.2 Ring Compression Tests

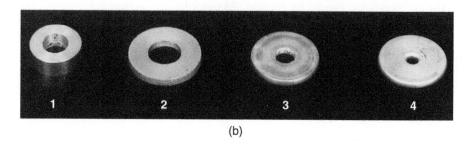

(b)

FIGURE 33.6   Ring-compression test between flat dies. (a) Effect of lubrication on type of ring-specimen barreling. (b) Test results: (1) original specimen and (2)–(4) increasing friction. *Source:* Based on A.T. Male and M.G. Cockcroft.

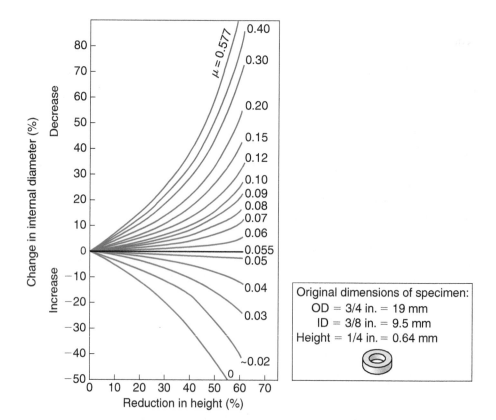

FIGURE 33.7   Chart to determine friction coefficient from a ring-compression test. Reduction in height and change in internal diameter of the ring are measured, then μ is read directly from this chart; for example, if the ring specimen is reduced in height by 40% and its internal diameter decreases by 10%, the coefficient of friction is 0.10.

EXAMPLE 33.1   Determination of Coefficient of Friction

Given: In a ring-compression test, a specimen 10 mm in height and with an outside diameter (OD) of 30 mm and an inner diameter (ID) of 15 mm is reduced in thickness by 50%.

Find: Determine the coefficient of friction, μ, if the OD is 38.9 mm after deformation.

Solution: First it is necessary to determine the new ID (which is obtained from volume constancy) as follows:

$$\text{Volume} = \frac{\pi}{4}\left(30^2 - 15^2\right)(10)$$

$$= \frac{\pi}{4}\left(38.9^2 - \text{ID}^2\right)(5).$$

From this equation, the new ID is calculated as 12.77 mm. Thus, the change in internal diameter is

$$\Delta\text{ID} = \frac{12.77 - 15}{15}$$

$$= -0.1487 \text{ or } 14.87\% \text{ (decrease).}$$

With a 50% reduction in height and a 14.87% reduction in internal diameter, the friction coefficient can then be obtained from Fig. 33.7 as $\mu = 0.09$.

33.5 Wear

The importance of *wear* is evident in the number of parts and components that continually have to be replaced or repaired in a wide variety of consumer and commercial products. *Wear plates*, placed in dies and sliding mechanisms where the loads are high, are an important component in some metalworking machinery. These plates, also known as wear parts, are expected to wear, but they can easily be replaced, and thus prevent more costly repairs.

Although wear generally alters a part's surface topography, and may result in severe surface damage, it can also have a beneficial effect. The *running-in* period for engines produces small particles of wear while removing the peaks from asperities, as can be seen in Fig. 33.8. Thus, under controlled conditions, wear can be regarded as a type of smoothing or polishing process. Note also that writing with an ordinary pencil or chalk is a wear process, and the words written actually consist of wear particles.

Described below are basic wear mechanisms relevant to manufacturing operations.

Adhesive Wear.   If a tangential force is applied to the model shown in Fig. 33.9, shearing can take place either (a) at the original interface of the two bodies or (b) along a path below or above the interface; in either case, sliding causes *adhesive wear*, also called *sliding wear*. Because of such factors as strain hardening at the asperity contacts, diffusion between the two bodies, and mutual solid solubility (Section 4.3) of the materials involved, the adhesive bonds formed at the asperity junctions often are stronger than the base metals themselves. Thus, during sliding, fracture usually follows a path in the weaker or softer component; that is how a wear fragment is generated. Although this fragment is typically attached to the harder component (the upper surface in Fig. 33.9c), it eventually becomes detached during further sliding at the interface and develops into a loose **wear particle**.

In more severe conditions, such as ones with high normal loads and strongly bonded asperities, adhesive wear is described as *scuffing, smearing, tearing, galling*, or *seizure*, and is called **severe wear**. Oxide layers on surfaces, however, have a great influence on adhesive wear, sometimes acting as a protective film, resulting in **mild wear**, which consists of small wear particles.

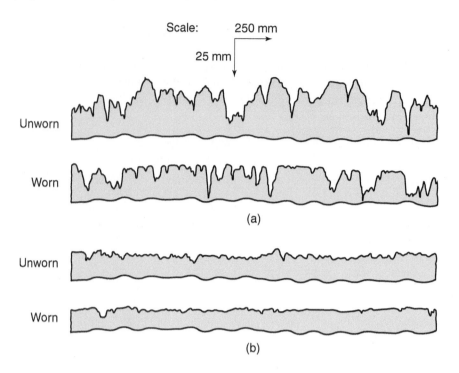

FIGURE 33.8 Changes in original (a) wire-brushed and (b) ground-surface profiles, after wear; note the difference in the vertical and horizontal scales. *Source:* Based on E. Wild and K.J. Mack.

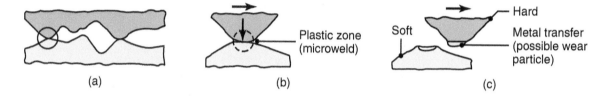

FIGURE 33.9 Schematic illustration of (a) two contacting asperities, (b) adhesion between two asperities, and (c) the formation of a wear particle.

Adhesive wear can be reduced by one or more of the following methods:

1. Selecting materials that do not develop strong adhesive bonds
2. Using a harder material as one member of the pair
3. Using materials that oxidize more easily
4. Applying hard coatings (see Chapter 34)
5. Coating one surface with a softer material, such as tin, silver, lead, or cadmium
6. Using an appropriate lubricant

Abrasive Wear.   This type of wear is caused by a hard, rough surface, or a surface containing hard protruding particles, sliding across another surface. As a result, *microchips* or *slivers* are produced as wear particles, thereby leaving grooves or scratches on the softer surface (Fig. 33.10). Processes such as filing, grinding, ultrasonic machining, and abrasive-jet and abrasive water-jet machining act in this manner. Unlike wear, which generally is not intended or desirable, the parameters in these processes are *controlled* to produce the desired surfaces and shape changes through the mechanism of wear.

There are two basic types of abrasive wear. In **two-body wear**, abrasive action takes place between two sliding surfaces or between abrasive (hard) particles and a solid body. This type of wear is the basis of **erosive wear**, such as occurs from the movement of slurries through pipes or sand particles acting on a ship's propeller. In **three-body wear**, an abrasive particle is present between two sliding solid bodies, such as a wear particle (contaminant) being carried by a lubricant. Such a situation indicates the importance of properly and periodically filtering lubricants in metalworking operations, as well as in machinery, automobile engines, and aircraft and helicopter engines.

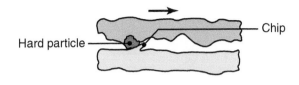

FIGURE 33.10 Schematic illustration of abrasive wear in sliding; longitudinal scratches on a surface usually indicate abrasive wear.

The **abrasive-wear resistance** of pure metals and ceramics has been found to be directly proportional to their hardness. Thus, abrasive wear can be reduced by increasing the hardness of materials (usually by heat treating) or by reducing the normal load. Elastomers and rubbers resist abrasive wear well, because they *deform elastically* and then recover when abrasive particles cross past over their surfaces. The best example is an automobile tire, which lasts thousands of miles, even though it is operated on paved or unpaved road surfaces, which generally are rough and abrasive; even hardened steels would not last long under such conditions.

Corrosive Wear.   Also known as *oxidation* or *chemical wear*, this type of wear is caused by chemical and electrochemical reactions between the surface and the environment. The fine corrosive products on the surface constitute the wear particles in corrosive wear. When the corrosive layer is destroyed or removed through sliding or abrasion, another layer begins to form, and the process of removal and corrosive-layer formation is repeated. Among corrosive media are water, seawater, oxygen, acids, chemicals, and atmospheric hydrogen sulfide and sulfur dioxide.

Corrosive wear can be reduced by:

- Selecting materials that will resist environmental attack
- Applying a coating
- Controlling the environment
- Reducing operating temperatures in order to lower the rate of chemical reaction.

Fatigue Wear.   Fatigue wear, also called *surface fatigue* or *surface-fracture wear*, is caused when surfaces are subjected to cyclic loading, such as rolling contact in bearings or in forging operations. The wear particles are usually formed through the mechanism of *spalling* or *pitting*. **Thermal fatigue** is another type of fatigue wear, whereby surface cracks are generated by thermal stresses from thermal cycling, as when a cool die is repeatedly brought in contact with hot workpieces. The individual cracks then join each other, and the surface begins to spall, a phenomenon similar to the development of potholes on roads. Thermal fatigue results in *heat checking* of molds and dies in die casting and hot working operations.

Fatigue wear can be reduced by:

- Lowering contact stresses
- Reducing thermal cycling
- Improving the quality of materials, by removing impurities, inclusions, and various other flaws that may act as local points for crack initiation and propagation

Several other types of wear also can be observed in manufacturing operations:

- **Erosion,** caused by loose particles abrading a surface
- **Fretting corrosion,** when interfaces are subjected to very small reciprocal movements

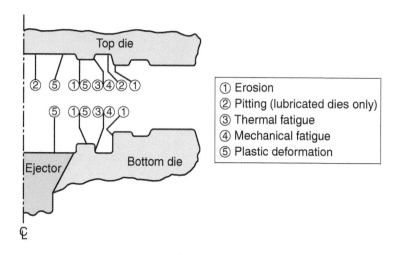

① Erosion
② Pitting (lubricated dies only)
③ Thermal fatigue
④ Mechanical fatigue
⑤ Plastic deformation

FIGURE 33.11   Types of wear observed in the cavity of a single pair of dies used for hot forging. *Source:* Based on T.A. Dean.

• **Impact wear,** which is removal of very small amounts of material from a surface through the impacting action of particles, similar to the mechanism of ultrasonic machining (Section 26.6)

In many situations in manufacturing, component wear is the result of a *combination* of different types of wear. Note, in Fig. 33.11, for example, that even in the same forging die, various types of wear take place in different locations of the die cavity. A similar situation also can exist in cutting tools, as shown in Fig. 21.18.

Wear of Thermoplastics.   The wear behavior of thermoplastics is similar to that of metals. Their abrasive-wear behavior depends partly on the ability of the polymer to deform and recover elastically, as in rubber and elastomers. Typical polymers with good wear resistance are polyimides, nylons, polycarbonate, polypropylene, acetals, and high-density polyethylene. These polymers are molded or machined, to make gears, pulleys, sprockets, and similar mechanical components. Because thermoplastics can be made with a wide variety of compositions, they also can be blended with internal lubricants (such as polytetrafluoroethylene, silicon, graphite, molybdenum disulfide, and rubber particles) that are interspersed within the polymer matrix.

Wear of Reinforced Plastics.   The wear resistance of reinforced plastics depends on the type, amount, and direction of reinforcement in the polymer matrix (see Chapter 9); carbon, glass, and aramid fibers all improve wear resistance. Wear usually takes place when the fibers are pulled away from the matrix, called *fiber pullout*. Wear is highest when the sliding direction is parallel to the fibers, because they can then be pulled out more easily. Long fibers increase the wear resistance of composites, because they (a) are more difficult to pull out and (b) prevent cracks in the matrix from propagating to the surface.

Wear of Ceramics.   When ceramics slide against metals, wear is caused by (a) small-scale deformation (plastic) and surface fracture (brittle), (b) plowing, (c) fatigue, and (d) surface chemical reactions. While sliding along each other, material from the surface of a metal body can be transferred to the oxide-type ceramic surface, forming metal oxides. Thus, sliding actually takes place between the metal and the metal-oxide surface.

33.6   Lubrication

Lubrication is used to reduce friction and wear, and dates back about four millenia. Egyptian chariot wheels, for example, were lubricated with beef tallow in 1400 B.C. A variety of oils were used for lubrication in metalworking operations, beginning in about 600 A.D. (see Table I.2).

As noted in various chapters, the surfaces of tools, molds, dies, and workpieces typically are subjected to (a) *force* and *contact pressure*, which ranges from very low to multiples of the yield stress of the workpiece material; (b) *relative speed*, from very low to very high; and (c) *temperature*, which generally ranges from ambient to melting. In addition to selecting appropriate materials and controlling process parameters to reduce friction and wear, **lubricants**, or, more generally, **metalworking fluids,** are applied widely.

Regimes of Lubrication.   There are four regimes of lubrication that are generally of interest in manufacturing operations (Fig. 33.12):

1. **Thick-film lubrication:** The surfaces are separated completely by a film of lubricant, and lubricant viscosity is an important factor. Such films can develop in some regions of the workpiece in high-speed operations, and also can develop from high-viscosity lubricants that become trapped at die–workpiece interfaces. A thick lubricant film results in a dull, grainy surface appearance on the workpiece after forming operations, with the degree of roughness varying with grain size. In operations such as coining and precision forging, trapped lubricants are undesirable; they prevent the blank from completely filling the die cavity.

2. **Thin-film lubrication:** As the load between the die and the workpiece increases or as the speed and viscosity of the metalworking fluid decrease, the lubricant film becomes thinner and the process is known as **thin-film lubrication.** This condition raises the friction at the sliding interfaces and results in slight wear.

3. **Mixed lubrication:** A significant portion of the load is carried by the physical contact between the asperities of the two contacting surfaces. The rest of the load is carried by the fluid film trapped in pockets, such as the valleys between asperities.

4. **Boundary lubrication:** The load is supported by contacting surfaces that are covered with a *boundary film* of lubricant (Fig. 33.12d)—a thin molecular lubricant layer. The film is attracted to the metal surfaces and prevents direct metal-to-metal contact of the two bodies, thus reducing wear. Boundary lubricants typically are natural oils, fats, fatty acids, esters, or soaps. However, boundary films can *break down* (a) as a result of desorption caused by high temperatures developed at the sliding interfaces or (b) by being rubbed off during sliding. Deprived of this protective film, the sliding metal surfaces then begin to wear and may also score severely.

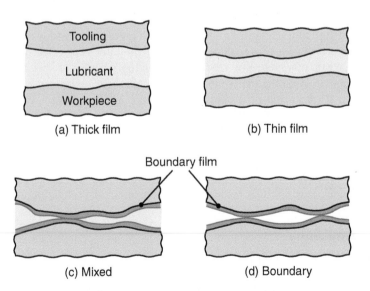

FIGURE 33.12 Regimes of lubrication generally occurring in metalworking operations. *Source:* Based on W.R.D. Wilson.

Other Considerations.   Note that the valleys in the surface of the contacting bodies (see Figs. 33.2a, 33.4, and 33.5) can serve as local reservoirs or pockets for lubricants, thereby supporting a substantial portion of the load. The workpiece, but not the die, should have the rougher surface; as the workpiece plastically deforms, its surface is flattened by the tooling and the lubricant is released or *percolated* from the surface. If the harder die surface is rough, there is no percolation effect, and the asperities, acting like a file, may damage the workpiece surface. The recommended surface roughness on most dies is about 0.4 μm (15 μin.).

The overall *geometry* of the interacting bodies is an important consideration in ensuring proper lubrication. The movement of the workpiece in the deformation zone, as occurs during wire drawing, extrusion, and rolling, should allow a supply of lubricant to be carried into the die–workpiece interface.

33.7 Metalworking Fluids and Their Selection

The functions of a *metalworking fluid* are to:

- *Reduce friction*, thus reducing force and energy requirements and any rise in temperature
- *Reduce wear*, thus reducing seizure and galling
- *Improve material flow* in tools, dies, and molds
- Act as a *thermal barrier* between the workpiece and its tool and die surfaces, thus preventing workpiece cooling in hot-working processes
- Act as a *release* or *parting agent*, a substance that helps in the removal or ejection of parts from dies and molds

Several types of metalworking fluids are now available, with diverse chemistries, properties, and characteristics that fulfill these requirements. (See also Section 22.12 on cutting fluids.)

33.7.1 Oils

Oils maintain high film strength on the surface of a metal, as it readily can be observed when trying to clean an oily surface. Although they are very effective in reducing friction and wear, oils have low thermal conductivity and low specific heat. Consequently, they do not effectively conduct away the heat generated by friction and plastic deformation during processing. Moreover, it is difficult and costly to remove oils from component surfaces that are to be painted or welded, and it is difficult to dispose of them. (See Section 34.16.)

The sources of oils may be (a) **mineral** (*petroleum* or *hydrocarbon*), (b) **animal**, or (c) **vegetable**. Oils may be *compounded* with any number of additives or with other oils. Compounding is used to change such properties as viscosity–temperature behavior, surface tension, heat resistance, and boundary-layer characteristics.

33.7.2 Emulsions

An *emulsion* is a mixture of two immiscible liquids (usually oil and water, in various proportions), along with additives. *Emulsifiers* are substances that prevent the dispersed droplets in a mixture from joining together, hence the term immiscible. Milky in appearance, emulsions also are known as **water-soluble oils** or **water-based** coolants, and are of two types. In *indirect emulsion*, water droplets are dispersed in the oil. In *direct emulsion*, mineral oil is dispersed in water in the form of very small droplets. Direct emulsions are important metalworking fluids, because the presence of water gives them high cooling capacity. They are particularly effective in high-speed machining (Section 25.5), where a severe temperature rise has detrimental effects on tool life, the surface integrity of workpieces, and the dimensional accuracy of parts.

33.7.3   Synthetic and Semisynthetic Solutions

Synthetic solutions are chemical fluids that contain inorganic and other chemicals, dissolved in water; they do not contain any mineral oils. Chemical agents are added to impart various properties. Semisynthetic solutions are basically synthetic solutions to which small amounts of emulsifiable oils have been added.

33.7.4   Soaps, Greases, and Waxes

Soaps typically are reaction products of sodium or potassium salts with fatty acids. Alkali soaps are soluble in water, but other metal soaps generally are insoluble. Soaps are effective boundary lubricants and can form thick film layers at die–workpiece interfaces, particularly when applied on conversion coatings for cold metalworking applications (Section 34.10).

Greases are solid or semisolid lubricants and generally consist of soaps, mineral oil, and various additives. They are highly viscous and adhere well to metal surfaces. Although used extensively in machinery, greases are of limited use in manufacturing processes. *Waxes* may be of animal or plant (*paraffin*) origin. Compared with greases, they are less "greasy" and are more brittle. Waxes are of limited use in metalworking operations, except as lubricants for copper and, in the form of a chlorinated paraffin, and for stainless steels and high-temperature alloys.

33.7.5   Additives

Metalworking fluids usually are *blended* with various additives, including oxidation inhibitors, rust-preventing agents, foam inhibitors, wetting agents, and antiseptics.

Sulfur, *chlorine*, and *phosphorus* are important additives to oils. Known as **extreme-pressure** (EP) **additives**, and used either singly or in combination, they react chemically with metal surfaces and form adherent surface films of metallic sulfides and chlorides. These films have low shear strength and good antiweld properties, thus can effectively reduce friction and wear. However, they may preferentially attack the cobalt binder in tungsten-carbide tools and dies (through *selective leaching*), causing changes in the surface roughness and integrity of those tools (see Section 22.4).

33.7.6   Solid Lubricants

Because of their unique properties and characteristics, several solid materials are used as lubricants in manufacturing operations.

Graphite.   The general properties of graphite were described in Section 8.6. Graphite (Section 8.6) is weak in shear along its *basal planes* (see Fig. 1.4), thus it has a low coefficient of friction in that direction. It can be an effective solid lubricant, particularly at elevated temperatures; however, friction is low only in the presence of air or moisture. Otherwise, friction is very high; in fact, graphite can be abrasive.

It can be applied either by rubbing it on surfaces or by making it part of a *colloidal* (dispersion of small particles) suspension in a liquid carrier, such as water, oil, or alcohol.

Molybdenum Disulfide.   A widely used lamellar solid lubricant, molybdenum disulfide (MoS_2) is somewhat similar in appearance to graphite. Unlike graphite, however, it has a high friction coefficient in an ambient environment. Molybdenum disulfide is used as a carrier for oils; it typically is applied by rubbing it on the workpiece surface.

Metallic and Polymeric Films.   Because of their low strength, thin layers of soft metals and polymer coatings also are used as solid lubricants. Suitable metals include lead, indium, cadmium, tin, and silver; polymers such as polytetrafluoroethylene, polyethylene, and methacrylates also are used. These coatings have limited applications because of their lack of strength under high contact stresses, especially at elevated temperatures.

Soft metals also are used to coat high-strength metals, such as steels, stainless steels, and high-temperature alloys. For example, copper or tin is chemically deposited on the surface of a metal before it is processed. If the oxide of a particular metal has low friction and is sufficiently thin, the oxide layer can serve as a solid lubricant, particularly at elevated temperatures (see also Section 15.3).

Glasses.   Although it is a solid material, glass becomes viscous at elevated temperatures, and thus it can serve as a liquid lubricant; viscosity is a function of temperature (but not of pressure) and depends on the type of glass (Section 8.4). Poor thermal conductivity also makes glass attractive, since it acts as a thermal barrier between hot workpieces and relatively cool dies. Glass lubrication is typically used in such applications as hot extrusion and hot forging.

Conversion Coatings.   Lubricants may not always adhere properly to workpiece surfaces, particularly under high normal and shearing stresses. Failure to adhere has detrimental effects in forging, extrusion, and the wire drawing of steels, stainless steels, and high-temperature alloys. For these applications, the workpiece surfaces are first transformed through a chemical reaction with acids, hence the term *conversion*. (See also Section 34.10.)

This reaction leaves a somewhat rough and spongy surface, which acts as a carrier for the lubricant. After treatment, any excess acid from the surface is removed, using borax or lime. A liquid lubricant, such as a soap, is then applied to the surface; it adheres to the surface and cannot be scraped off easily. *Zinc-phosphate* conversion coatings often are used on carbon and low-alloy steels. *Oxalate coatings* are used for stainless steels and high-temperature alloys.

33.7.7 Selection of Metalworking Fluids

Selecting a metalworking fluid for a particular application and workpiece material involves a consideration of several factors:

1. Specific manufacturing process
2. Workpiece material
3. Tool or die material
4. Processing parameters
5. Compatibility of the fluid with the tool and die materials and the workpiece
6. Surface preparation required

7. Method of applying the fluid
8. Removal of the fluid and cleaning of the workpiece after processing
9. Contamination of the fluid by other lubricants, such as those used to lubricate machinery
10. Storage and maintenance of fluids
11. Treatment of waste lubricant
12. Biological and environmental considerations
13. Costs involved in all of the factors listed above

The specific function of a metalworking fluid, whether it is primarily a *lubricant* or a *coolant*, also must be taken into account. Water-based fluids are very effective coolants but, as lubricants, they are not as effective as oils. It is estimated that water-based fluids are used in 80 to 90% of all machining operations.

Specific requirements for metalworking fluids are:

- They should not leave any harmful residues that could interfere with production operations.
- They should not stain or corrode the workpiece or the equipment.
- Periodic inspection is necessary to detect deterioration caused by bacterial growth, accumulation of oxides, chips, wear debris, and general degradation and breakdown due to temperature and time. The presence of wear particles is particularly important, because they cause damage to the system; proper inspection and filtering are thus essential.

After the completion of manufacturing operations, workpiece surfaces usually have lubricant residues; these should be removed prior to further processing, such as welding or painting. Oil-based lubricants are more difficult and expensive to remove than water-based fluids. Various cleaning solutions and techniques used for this purpose are described in Section 34.16.

Biological and Environmental Considerations.   These considerations are important factors in the selection of a metalworking fluid. Hazards may result if one contacts or inhales some of these fluids, such as inflammation of the skin (*dermatitis*) and long-term exposure to carcinogens. The improper disposal of metalworking fluids may cause adverse effects on the environment as well. To prevent or restrict the growth of microorganisms, such as bacteria, yeasts, molds, algae, and viruses, chemicals (*biocides*) are added to metalworking fluids.

Much progress has been made in developing environmentally safe (*green*) fluids and the technology and equipment for their proper treatment, recycling, and disposal. In the United States, laws and regulations concerning the manufacture, transportation, use, and disposal of metalworking fluids are promulgated by the U.S. Occupational Safety and Health Administration (OSHA), the National Institute for Occupational Safety and Health (NIOSH), and the Environmental Protection Agency (EPA).

SUMMARY

- Surfaces and their properties are as important as the bulk properties of materials. A surface not only has a particular shape, roughness, and appearance, but also has properties that can differ significantly from those of the bulk material.

- Surfaces are exposed to the environment and thus are subject to environmental attack. They also may come into contact with tools and dies (during processing) or with other components (during their service life).

- The geometric and material properties of surfaces can significantly affect their friction, wear, fatigue, corrosion, and electrical and thermal conductivity properties.

- The measurement and description of surface features, including their characteristics, are important aspects of manufacturing. The most common surface-roughness measurement is the arithmetic mean value. The instruments usually used to measure surface roughness include profilometers, optical interferometers, and atomic force microscopes.

- Friction and wear are among the most significant factors in processing materials. Much progress has been made in understanding these phenomena and identifying the factors that govern them.

- Other important factors are the affinity and solid solubility of the two materials in contact, the nature of surface films, the presence of contaminants, and process parameters, such as load, speed, and temperature.

- A wide variety of metalworking fluids, including oils, emulsions, synthetic solutions, and solid lubricants, is available for specific applications. Their selection and use requires a careful consideration of many factors regarding the workpiece and die materials and the particular manufacturing process.

- Metalworking fluids have various lubricating and cooling characteristics. Biological and environmental considerations also are important factors in their selection.

KEY TERMS

Abrasive wear	Flaw	Pit	Surface profilometer
Additives	Fretting corrosion	Plowing	Surface roughness
Adhesion	Friction force	Ring-compression test	Surface structure
Adhesive wear	Greases	Root-mean-square	Surface texture
Arithmetic mean value	Impact wear	average	Thick-film lubrication
Asperities	Lay	Running-in	Thin-film lubrication
Boundary lubrication	Lubricant	Selective leaching	Tribology
Coefficient of friction	Lubrication	Self lubricating	Ultrasonic vibrations
Compounded oils	Maximum roughness	Severe wear	Water-soluble oils
Conversion coatings	height	Soaps	Waviness
Coolant	Metalworking fluids	Solid lubricants	Waxes
Emulsion	Microwelds	Substrate	Wear
Extreme-pressure	Mixed lubrication	Surface defects	Wear parts
additives	Oils	Surface finish	
Fatigue wear	Oxide layer	Surface integrity	

BIBLIOGRAPHY

Astakhov, V.P., **Tribology of Metal Cutting**, Elsevier, 2007.

Basu, B., and Kalin, M., **Tribology of Ceramics and Composites**, Wiley-American Ceramic Society, 2011.

Bhushan, B., **Introduction to Tribology**, Wiley, 2002.

Bhushan, B., **Modern Tribology Handbook**, 2 vols., CRC Press, 2000.

Burakowski, T., and Wiershon, T., **Surface Engineering of Metals: Principles, Equipment, Technologies**, CRC Press, 1998.

Byers, J.P., **Metalworking Fluids**, 2nd ed., CRC Press, 2006.

Chattopadhyay, R., **Surface Wear: Analysis, Treatment, and Prevention**, ASM International, 2001.

Gohar, R., **Fundamentals of Tribology**, Imperial College Press, 2008.

Rabinowicz, E., **Friction and Wear of Materials**, 2nd ed., Wiley, 1995.

Stachowiak, G.W., **Wear: Materials, Mechanisms and Practice**, Wiley, 2006.

REVIEW QUESTIONS

33.1 What is tribology?

33.2 Explain what is meant by (a) surface texture and (b) surface integrity.

33.3 List and explain the types of defects typically found on surfaces.

33.4 Define the terms (a) roughness, and (b) waviness.

33.5 Explain why the results from a profilometer are not a true depiction of the actual surface.

33.6 Describe the features of the ring-compression test. Does it require the measurement of forces?

33.7 List the types of wear generally observed in engineering practice.

33.8 Define the terms wear, friction, and lubricant.

33.9 How can adhesive wear be reduced? Abrasive wear?

33.10 Explain the mechanisms through which a wear particle is formed from adhesive wear, and two- and three-body abrasive wear.

33.11 Explain the functions of a lubricant in manufacturing processes.

33.12 What is a grease? What is an emulsion?

33.13 What is the role of additives in metalworking fluids?

33.14 Describe the factors involved in lubricant selection.

33.15 Explain why graphite and molybdenum disulfide are effective solid lubricants.

QUALITATIVE PROBLEMS

33.16 Give several examples that show the importance of friction in manufacturing processes described in Parts III and IV.

33.17 Explain the significance of the fact that the hardness of metal oxides is generally much higher than that of the base metals themselves. Give some examples.

33.18 What factors would you consider in specifying the lay of a surface for a part? Explain.

33.19 Explain why identical surface-roughness values do not necessarily represent the same type of surface.

33.20 Why are the requirements for surface-roughness design in engineering applications so broad? Explain with specific examples.

33.21 What is the significance of surface-temperature rise resulting from friction? Give some examples based on topics covered in the preceding chapters.

33.22 Explain the causes of lay on surfaces.

33.23 Give several examples of how wear on molds, tools, and dies affects a manufacturing operation.

33.24 Comment on the surface roughness of various parts and components with which you are familiar. What types of parts exhibit the coarsest surface? What types exhibit the finest? Explain.

33.25 (a) Give two examples in which waviness on a surface would be desirable. (b) Give two examples in which it would be undesirable.

33.26 Do the same as for Problem 33.25, but for surface roughness.

33.27 Describe your observations regarding Fig. 33.7.

33.28 Give the reasons that an originally round specimen in a ring-compression test may become oval after it is upset.

33.29 Could the ring compression test be applied to rolling? Explain.

33.30 Explain the reason that the abrasive-wear resistance of a material is a function of its hardness.

33.31 On the basis of your own experience, make a list of parts and components that have to be replaced because of wear.

33.32 List the similarities and differences between adhesive and abrasive wear.

33.33 Explain why the types of wear shown in Fig. 33.11 occur in those particular locations in the forging die.

33.34 List the requirements of a lubricant.

33.35 List manufacturing operations in which high friction is desirable and those in which low friction is desirable.

33.36 List manufacturing operations in which high wear is desirable and those in which low wear is desirable.

33.37 It is observed that the coefficient of friction between the carriage and the ways on a lathe is 0.35. To reduce friction and wet the surfaces, kerosene (a very low viscosity fluid) is applied to the interface. Instead of reducing the friction, it is now measured to be 0.38. Provide an explanation for these results.

QUANTITATIVE PROBLEMS

33.38 Refer to the profile shown in Fig. 33.3, and offer some reasonable numerical values for the vertical distances from the centerline. Calculate the R_a and R_q values. Then give another set of values for the same general profile and calculate the same two quantities. Comment on your observations.

33.39 Obtain several different parts made of various materials, inspect their surfaces under an optical microscope at different magnifications, and make an educated guess as to what manufacturing process or finishing process was likely used to produce each of these parts. Explain your reasoning.

33.40 A surface with a triangular sawtooth roughness pattern has a peak-to-valley height of 4 μm. Find the R_a and R_q values.

33.41 Refer to Fig. 33.6b, and make measurements of the external and internal diameters (in the horizontal direction in the photograph) of the four specimens shown. Remembering that in plastic deformation the volume of the rings remains constant, estimate (a) the reduction in height and (b) the coefficient of friction for each of the three compressed specimens.

33.42 Using Fig. 33.7, make a plot of the coefficient of friction versus the change in internal diameter for a constant reduction in height of 35%.

33.43 Assume that in Example 33.1 the coefficient of friction is 0.16. If all other parameters remain the same, what is the new internal diameter of the specimen?

SYNTHESIS, DESIGN, AND PROJECTS

33.44 List the steps you would follow if you wished to reduce friction in a manufacturing process.

33.45 Discuss the tribological differences between ordinary machine elements (such as gears, cams, and bearings) and metalworking processes using tools, molds, and dies. Consider such factors as load, speed, and temperature.

33.46 Section 33.2 listed major surface defects. How would you go about determining whether or not each of these defects is a significant factor in a particular application?

33.47 Describe your own thoughts regarding biological and environmental considerations in the use of metalworking fluids.

33.48 Wear can have detrimental effects in manufacturing operations. Can you visualize situations in which wear could be beneficial? Explain, and give some examples.

33.49 Many parts in various appliances and automobiles have to be replaced because they were worn. Describe the methodology you would follow in determining the type(s) of wear these components have undergone.

33.50 In the second paragraph of the introduction to Part VII, five different sets of interfacial conditions were outlined, from (a) to (e). For each of these, give several examples from the manufacturing processes described in this book.

33.51 Describe your thoughts on the desirability of integrating surface-roughness measuring instruments into the machine tools described in Parts III and IV? How would you go about doing so, giving special consideration to the factory environment in which they are to be used? Make some preliminary sketches of such a system.

Surface Treatments, Coatings, and Cleaning

- Although material and process selection are critical aspects of manufacturing, often the surface properties of a component or a part also determine its performance.
- This chapter describes various surface-modification operations that can be performed on parts, for technical and aesthetic reasons.
- The chapter presents the surface treatment, cleaning, and coating processes that are commonly performed, and includes a discussion of mechanical surface treatments such as shot peening, laser peening, and roller burnishing, for imparting compressive residual stresses onto metal surfaces.
- Coating operations are then examined, including cladding, thermal-spray operations, physical and chemical vapor deposition, ion implantation, and electroplating; the benefits of diamond and diamond-like carbon coatings also are investigated.

34.1 Introduction

After a part is manufactured, some of its surfaces may have to be processed further in order to ensure that they have certain properties and characteristics. **Surface treatments** may be necessary in order to

- *Improve resistance to wear, erosion, and indentation,* such as for machine-tool slideways (Figs. 23.2 and 35.1), shafts, rolls, cams, and gears
- *Reduce friction,* such as on sliding surfaces of tools, dies, bearings, and machine ways
- *Reduce adhesion,* such as for electrical contacts
- *Improve resistance to corrosion and oxidation,* on sheet metals for automobile bodies, gas-turbine components, food packaging, and medical devices
- *Improve fatigue resistance,* of bearings and shafts with fillets
- *Rebuild surfaces,* on worn tools, dies, molds, and machine components
- *Modify surface texture,* appearance, dimensional accuracy, and frictional characteristics
- *Impart decorative features,* such as texture or color and texture

Numerous techniques are employed to impart these characteristics to various types of metallic, nonmetallic, and ceramic materials, and include mechanisms that involve (a) plastic deformation of the workpiece surface, (b) chemical reactions, (c) thermal means, (d) deposition, (e) implantation, and (f) organic coatings and paints. Some of these techniques also are used in the manufacture of semiconductor devices, as described in Chapter 28.

The chapter ends with a discussion of the methods used for cleaning manufactured surfaces, before the components are assembled into the completed product and made ready for service. Environmental considerations regarding the fluids used and the waste material from various surface-treatment processes also are included.

34.2 Mechanical Surface Treatments

Several approaches are used to mechanically improve the surface properties of manufactured components; the more common methods are:

Shot Peening.   In this process, the workpiece surface is impacted repeatedly with cast steel, glass, or ceramic balls (called *shot*), which make overlapping indentations on the surface. Using shot sizes that range from 0.125 to 5 mm (0.005–0.2 in.) in diameter, this action causes plastic surface deformation, at depths up to 1.25 mm (0.05 in.). Because the plastic deformation is not uniform throughout the part's thickness (see also Fig. 2.14c), shot peening causes compressive residual stresses on the surface, thus improving the fatigue life of the component. Unless the process parameters are controlled properly, the deformation can be so severe as to cause damage to the surface. The extent of deformation can be reduced by *gravity peening*, which involves larger shot sizes, but fewer impacts on the workpiece surface.

Shot peening is used extensively on shafts, gears, springs, oil-well drilling equipment, and turbine and compressor blades. Note, however, that if these parts are later subjected to high temperatures (such as gas turbine blades), the residual stress will begin to relax (*thermal relaxation*) and their beneficial effects will be diminished.

Laser Shot Peening.   In this process, also called *laser shock peening* (first developed in the mid-1960s but not commercialized until much later), the workpiece surface is subjected to *pulses* (planar laser shocks) from high-power lasers. This peening process produces compressive residual-stress layers that are typically 1 mm (0.04 in.) deep, with less than 1% of cold working taking place at the surface.

Laser shot peening has been applied successfully and reliably to jet-engine fan blades and to materials such as titanium, nickel alloys, and steels for improved fatigue resistance and some corrosion resistance. Laser intensities are on the order of 100–300 J/cm^2 and have a pulse duration of 10–50 ns. The basic limitation of laser shot peening for industrial, cost-effective applications is the high cost of the high-power lasers (up to 1 kW) that must operate at energy levels of 100 J/pulse.

Water-jet Peening.   In this process, a water jet, at pressures as high as 400 MPa (60,000 psi), impinges on the workpiece surface, inducing compressive residual stresses, and with surface and subsurface hardening at the same level as in shot peening. The water-jet peening process has been used successfully on steels and aluminum alloys. The control of process variables (such as jet pressure, jet velocity, nozzle design, and its distance from the surface) is important in order to avoid excessive surface roughness or damage.

Ultrasonic Peening.   This process uses a hand tool that is vibrated by a piezoelectric transducer, at a frequency of 22 kHz. A variety of heads can be used for different applications.

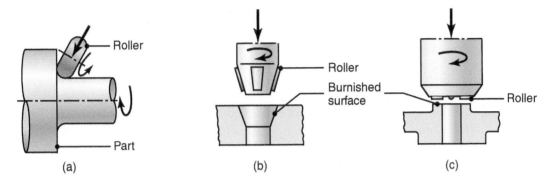

FIGURE 34.1 Burnishing tools and roller burnishing of (a) the fillet of a stepped shaft to induce compressive surface residual stresses for improved fatigue life; (b) a conical surface; and (c) a flat surface.

Roller Burnishing.   Also called *surface rolling*, the surface of the component is cold worked by the action of a hard and highly polished roller or set of rollers; the process is used on various flat, cylindrical, or conical surfaces (Fig. 34.1). Roller burnishing improves surface finish by removing scratches, tool marks, and pits, and induces beneficial compressive surface residual stresses. Consequently, corrosion resistance is improved, since corrosive products and residues cannot be entrapped. In a variation of this process, called *low-plasticity burnishing*, the roller travels only once over the surface, inducing minimal plastic deformation.

The internal cylindrical surfaces of holes also can be burnished by a process called **ballizing** or **ball burnishing**. In this operation, a smooth ball (slightly larger than the bore diameter) is pushed through the length of the hole.

Typically used on hydraulic-system components, seals, valves, spindles, and fillets on shafts, roller burnishing improves mechanical properties as well as surface finish. It can be used either by itself or in combination with other finishing processes (such as grinding, honing, and lapping) in which case, the finishing operation is performed after burnishing the part, in order to produce a smooth surface. The equipment can be mounted on various CNC machine tools, for improved productivity and consistency of performance. All types of soft or hard metals can be roller burnished.

Explosive Hardening.   In this process, the surfaces are subjected to high pressures through the placement and detonation of a layer of an explosive sheet, placed directly on the workpiece surface. The contact pressures developed can be as high as 35 GPa (5×10^6 psi), lasting about 2–3 μs. Significant increases in surface hardness can be achieved, with very little change (less than 5%) in the shape of the component. Railroad rail surfaces, for example, are explosively hardened.

34.3   Mechanical Plating and Cladding

Mechanical Plating.   In *mechanical plating* (also called *mechanical coating, impact plating*, or *peen plating*), fine metal particles are compacted over the workpiece surfaces by glass, ceramic, or porcelain beads, that are propelled by rotary means (such as tumbling). This process, which is basically one of cold-welding particles onto a surface, is typically used for hardened-steel parts, such as for automobiles, with plating thickness usually less than 25 μm (0.001 in.).

Cladding.   In this process, also called *clad bonding*, metals are bonded with a thin layer of corrosion-resistant metal through the application of pressure by rolls or other means (see Fig. 31.1). A typical example is the cladding of aluminum (*Alclad*), in which a pure or corrosion-resistant layer of aluminum alloy is clad over an aluminum-alloy body (core). The cladding layer is anodic to the core and usually has a thickness less than 10% of the total thickness.

Examples of cladding are 2024 aluminum clad with 1230 aluminum, and 3003, 6061, and 7178 aluminum clad with 7072 aluminum; other applications include steels clad with stainless-steel or nickel alloys. The cladding material may also be applied using dies (as in cladding steel wire with copper) or with explosives. Multiple-layer cladding is also utilized in special applications.

Laser cladding involves fusion of a wire or powder material over a substrate. It has been applied successfully to metals and ceramics, especially for enhanced friction and wear behavior of the components.

34.4   Case Hardening and Hard Facing

Surfaces also may be hardened by thermal means in order to improve their frictional and wear properties, as well as their resistance to indentation, erosion, abrasion, and corrosion. The most common methods are:

Case Hardening.   Traditional methods of case hardening (*carburizing, carbonitriding, cyaniding, nitriding, flame hardening,* and *induction hardening*) are described in Section 4.10 and summarized in Table 4.1. In addition to common heat sources (such as gas and electricity), an electron beam or laser beam also can be used as a heat source, for both metals and ceramics. Case hardening, as well as some of the other surface-treatment processes described in this chapter, induces residual stresses on surfaces, such as, for example, by the formation of martensite, which causes compressive residual stresses.

Hard Facing.   In this process, a relatively thick layer, edge, or point of wear-resistant hard metal is deposited on the workpiece surface by fusion-welding techniques (Chapter 30). Several layers, known as *weld overlay*, can be deposited to repair worn parts. Hard facing enhances the wear resistance of the materials, hence it is used in the manufacture of tools, dies, and various industrial components.

Spark Hardening.   Hard coatings of tungsten, chromium, or molybdenum carbides can be deposited by an electric arc, in a process called *spark hardening, electric spark hardening,* or *electrospark deposition*. The deposited layer is typically $250 \ \mu$m (0.01 in.) thick. Hard-facing alloys can be used as electrodes, rods, wires, or powder in spark hardening. Typical applications are for valve seats, oil-well drilling tools, and dies for hot metalworking.

34.5   Thermal Spraying

Thermal spraying is a series of processes in which coatings of various metals, alloys, carbides, ceramics, and polymers are applied to metal surfaces by a spray gun, with a stream heated by an oxyfuel flame, an electric arc, or a plasma arc. The earliest applications of thermal spraying, in the 1910s, involved metals, hence the term **metallizing**, which also has been used. The surfaces to be sprayed are first cleaned of oil and dirt, and then roughened by, for example, grit blasting, to improve their bond strength (see Section 26.8). The coating material can be in the form of wire, rod, or powder, and when the droplets or particles impact the workpiece, they solidify and bond to the surface.

Particle velocities typically range from a low of about 150 to 1000 m/s, but can be higher for special applications. Temperatures are in the range of 3000° to 8000°C (5500°–14,000°F). The sprayed coating is hard and wear resistant, with a layered structure of deposited material; however, the coating can have a porosity as high as 20%, due to entrapped air and oxide particles. Bond strength depends on the particular process and techniques used, and is mostly mechanical in nature (hence the importance of roughening the surface prior to spraying), but can also be metallurgical. Bond strength generally ranges from 7 to 80 MPa (1–12 ksi), depending on the particular process used.

Typical applications of thermal spraying include aircraft engine components (such as those used in rebuilding worn parts), structures, storage tanks, tank cars, rocket motor nozzles, and components that require resistance to wear and corrosion. In an automobile, thermal spraying is often applied to crankshafts, valves, fuel-injection nozzles, piston rings, and engine blocks. The process also is used in the gas and petrochemical industries, for the repair of worn parts, and to restore dimensional accuracy to parts that may have not been machined or formed properly.

The source of energy in thermal-spraying processes is of two types: chemical combustion and electrical.

1. **Combustion Spraying**

   - **Thermal wire spraying** (Fig. 34.2a): The oxyfuel flame melts the wire and deposits it on the surface. The bond is of medium strength, and the process is relatively inexpensive.
   - **Thermal metal powder spraying** (Fig. 34.2b): This process is similar to thermal wire spraying, but uses a metal powder instead of wire.
   - **Detonation gun**: Controlled and repeated explosions take place by means of an oxyfuel–gas mixture. The detonation gun has a performance similar to that of plasma.
   - **High-velocity oxyfuel–gas spraying** (HVOF): This process has characteristics that are similar to that of the detonation gun, but is less expensive.

2. **Electrical Spraying**

   - **Twin-wire arc**: An arc is formed between two consumable wire electrodes; the resulting bond has good strength, and the process is the least expensive.
   - **Plasma**: Conventional, high-energy, or vacuum (Fig. 34.2c) plasma produces temperatures on the order of 8300°C (15,000°F), and results in good bond strength with very low oxide content. **Low-pressure plasma spray** (LPPS) and **vacuum plasma spray** both produce coatings with high bond strength and with very low levels of porosity and surface oxides.

Cold Spraying.   The particles to be sprayed are at a lower temperature and are not melted; thus, oxidation is minimal. The spray jet in cold spraying is narrow and highly focused; it has very high impact velocities, thereby improving the bond strength of the particles on the surface.

34.6 Vapor Deposition

Vapor deposition is a process in which the substrate (workpiece surface) is subjected to chemical reactions by gases that contain chemical compounds of the material to be deposited. The coating thickness is usually a few microns, which is much less than

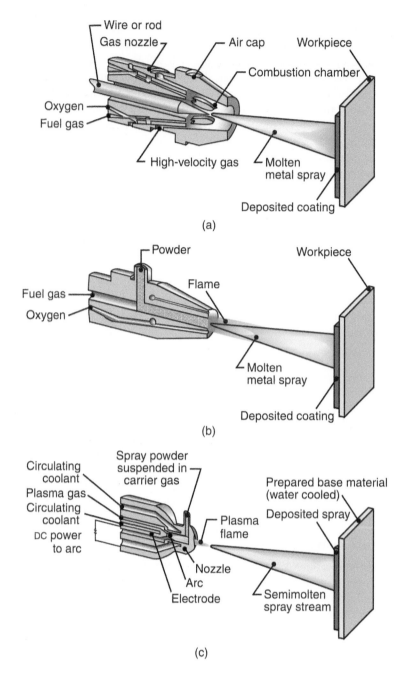

FIGURE 34.2   Schematic illustrations of thermal-spray operations: (a) thermal wire spray, (b) thermal metal-powder spray, and (c) plasma spray.

the thicknesses that result from the techniques described in Sections 34.2 and 34.3. The substrate may be metal, plastic, glass, or paper, and the deposited material may consist of metals, alloys, carbides, nitrides, borides, ceramics, or oxides. Control of the coating composition, thickness, and porosity is important. Typical applications for vapor deposition are coating of cutting tools, drills, reamers, milling cutters, punches, dies, and wear surfaces.

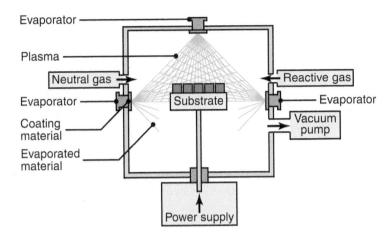

FIGURE 34.3 Schematic illustration of the physical-vapor-deposition process; note that there are three arc evaporators and the parts to be coated are placed on a tray inside the chamber.

There are two major vapor-deposition processes: physical vapor deposition and chemical vapor deposition.

34.6.1 Physical Vapor Deposition

The three basic types of *physical vapor deposition* (PVD) processes are (a) vacuum deposition or arc evaporation; (b) sputtering; and (c) ion plating. These processes are carried out in a high vacuum and at temperatures in the range from 200° to 500°C (400°–900°F). In PVD, the particles to be deposited are carried physically to the workpiece, rather than by chemical reactions, as in chemical vapor deposition.

Vacuum Deposition.   In vacuum deposition or evaporation, the metal is evaporated at a high temperature in a vacuum and is deposited on the substrate, which usually is at room temperature or slightly higher, for improved bonding. Coatings of uniform thickness can be deposited, even on complex shapes. In **arc deposition** (PV/ARC), the coating material (cathode) is evaporated by several arc evaporators (Fig. 34.3), using highly localized electric arcs. The arcs produce a highly reactive plasma, which consists of the ionized vapor of the coating material; the vapor condenses on the substrate (anode) and coats it. Applications of this process are both functional (oxidation-resistant coatings for high-temperature applications, electronics, and optics) and decorative (hardware, appliances, and jewelry). **Pulsed-laser** and **electron-beam deposition** are more recent, related processes in which the energy beams heat the target into a vapor.

Sputtering.   In *sputtering*, an electric field ionizes an inert gas (usually argon); the positive ions then bombard the coating material (cathode) and cause sputtering (ejection) of its atoms. The atoms condense on the workpiece, which is heated to improve bonding (Fig. 34.4). In **reactive sputtering**, the inert gas is replaced by a reactive gas (such as oxygen), in which case the atoms are oxidized and the oxides are deposited. Carbides and nitrides also are deposited by

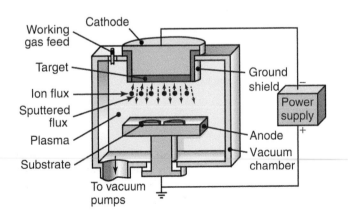

FIGURE 34.4 Schematic illustration of the sputtering process.

reactive sputtering. Alternatively, very thin polymer coatings can be deposited on metal and polymeric substrates, with a reactive gas, causing polymerization of the plasma. **Radio-frequency (RF) sputtering** is used for nonconductive materials, such as electrical insulators and semiconductor devices.

Ion Plating.   *Ion plating* is a generic term that describes various combined processes of sputtering and vacuum evaporation. An electric field causes a glow, generating a plasma (Fig. 34.5); the vaporized atoms are ionized only partially. **Ion-beam-enhanced (assisted) deposition** is capable of producing thin films, as coatings for semiconductor, tribological, and optical applications. Bulky parts can be coated in large chambers, using high-current power supplies of 15 kW and voltages of 100,000 DC. **Dual ion-beam deposition** is a hybrid coating technique that combines PVD with simultaneous ion-beam bombardment, resulting in good adhesion on metals, ceramics, and polymers. Ceramic bearings and dental instruments are examples of its applications.

34.6.2   Chemical Vapor Deposition

Chemical vapor deposition (CVD) is a *thermochemical* process (Fig. 34.6). In a typical application, such as coating cutting tools with titanium nitride (TiN; Section 22.5), the tools are placed on a graphite tray and heated to 950°–1050°C (1740°–1920°F), at atmospheric pressure and in an inert atmosphere. Titanium tetrachloride (a gas), hydrogen, and nitrogen are then introduced into the chamber. The chemical reactions form titanium nitride on the tool surfaces, with hydrogen chloride produced and exhausted from the reaction chamber. Because of its toxicity, however, this exhaust gas must be carefully cleaned, using exhaust scrubbers before being vented to the atmosphere. For a coating of titanium carbide, methane is substituted for the other gases.

Deposited CVD coatings usually are thicker than those obtained with PVD. A typical cycle is long, consisting of (a) three hours of heating, (b) four hours of coating, and (c) six to eight hours of cooling to room temperature. The thickness of the coating depends on the flow rates of the gases used, time, and temperature.

CVD is a very versatile process. Almost any material can be coated and any material can serve as a substrate, although bond strength will vary. This process

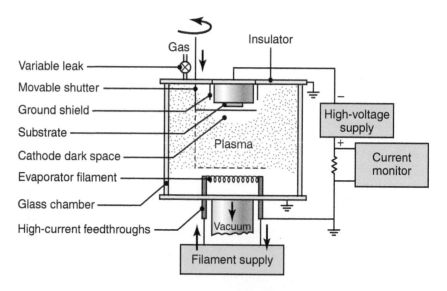

FIGURE 34.5   Schematic illustration of an ion-plating apparatus.

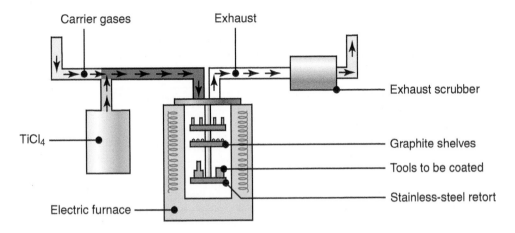

FIGURE 34.6   Schematic illustration of the chemical-vapor-deposition process; note that parts and tools to be coated are placed on trays inside the chamber.

is also used to produce diamond coatings without binders, unlike polycrystalline diamond films which use 1 to 10% binder materials. The **medium-temperature CVD** (MTCVD) technique results in a higher resistance to crack propagation than CVD.

34.7   Ion Implantation and Diffusion Coating

In *ion implantation*, ions (charged atoms) are introduced into the surface of the workpiece. The ions are accelerated in a vacuum, to such an extent that they penetrate the substrate to a depth of a few microns. Ion implantation (not to be confused with ion plating, Section 34.6.1) modifies surface properties, by increasing surface hardness and improving resistance to friction, wear, and corrosion. The process can be controlled accurately, and the surface can be masked to prevent ion implantation in unwanted locations.

Ion implantation is particularly effective on such materials as aluminum, titanium, stainless steels, tool and die steels, carbides, and chromium coatings. The process is typically used on cutting and forming tools, dies and molds, and metal prostheses, such as artificial hips and knees. When used in specific applications, such as semiconductors (Section 28.3), ion implantation is called **doping**, that is, alloying with small amounts of various elements.

Diffusion Coating.   This is a process in which an alloying element is diffused into the surface of the substrate (usually steel), altering its surface properties. The alloying elements can be supplied in solid, liquid, or gaseous states. The process has acquired different names, depending on the diffused element, as shown in Table 4.1, which lists various diffusion processes, such as *carburizing*, *nitriding*, and *boronizing*.

34.8   Laser Treatments

As described in various chapters of this book, lasers are having increasingly wider use in manufacturing processes (e.g., laser machining, forming, joining, rapid prototyping, and metrology) and surface engineering (laser peening, alloying, surface treatments, and texturing). Powerful, efficient, reliable, and less expensive lasers are now available for a variety of cost-effective surface treatments, as outlined in Fig. 34.7.

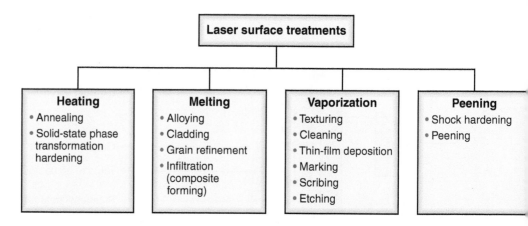

FIGURE 34.7 An outline of laser surface-engineering processes. *Source:* Based on N.B. Dahotre.

EXAMPLE 34.1   Applications of Laser Surface Engineering

Several applications of lasers in engineering practice are given in this example. The most commonly used lasers are Nd:YAG and CO_2; excimer lasers are generally used for surface texturing (see also Table 27.2).

1. *Localized surface hardening*

   - Cast irons: diesel-engine cylinder liners, automobile steering assemblies, and camshafts
   - Carbon steels: gears and electromechanical parts

2. *Surface alloying*

   - Alloy steels: bearing components
   - Stainless steels: diesel-engine valves and seat inserts

   - Tool and die steels: dies for forming and die casting

3. *Cladding*

   - Alloy steels: automotive valves and valve seats
   - Superalloys: turbine blades

4. *Ceramic coating*

   - Aluminum-silicon alloys: automotive-engine bore

5. *Surface texturing*

   - Metals, plastics, ceramics, and wood: all types of products

---

34.9 Electroplating, Electroless Plating, and Electroforming

Plating imparts resistance to wear, resistance to corrosion, high electrical conductivity, better appearance, and reflectivity.

Electroplating.   In *electroplating*, the workpiece (cathode) is plated with a different metal (anode), which is transferred through a water-based electrolytic solution (Fig. 34.8). Although the plating process involves a number of reactions, the process consists basically of the following sequence:

1. The metal ions from the anode are discharged by means of the potential energy from the external source of electricity, or are delivered in the form of metal salts.

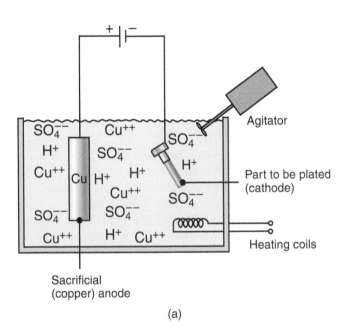

(a)   (b)

FIGURE 34.8 (a) Schematic illustration of the electroplating process. (b) Examples of electroplated parts. *Source:* Courtesy of BFG Manufacturing Service.

2. The metal ions are dissolved into the solution.
3. The metal ions are deposited on the cathode.

The volume of the plated metal can be calculated from the equation

$$\text{Volume} = cIt, \tag{34.1}$$

where I is the current, in amperes, t is time, and c is a constant that depends on the plate metal, the electrolyte, and the efficiency of the system; typically, it is in the range of 0.03–0.1 mm^3/amp-s. It can be noted that for the same volume of material deposited, the deposited thickness is inversely proportional to the surface area. The deposition rate is typically on the order of 75 μm/h, thus electroplating is a slow process. Thin plated layers are typically on the order of 1 μm (40 μin.); for thick layers, the plating can be as much as 500 μm.

Video Solution 34.1 Electroplating and Electroforming

The *plating solutions* are either strong acids or cyanide solutions. As the metal is plated from the solution, it has to be periodically replenished. This is accomplished through two principal methods: (a) salts of metals are occasionally added to the solution or (b) a *sacrificial anode*, of the metal to be plated, is used in the electroplating tank and dissolves at the same rate that the metal is deposited.

There are three basic methods of electroplating:

1. **Rack plating:** The parts to be plated are placed in a rack, which is then conveyed through a series of process tanks.
2. **Barrel plating:** Small parts are placed inside a permeable barrel, made mostly of wire mesh, which is then placed inside the processing tank(s). This operation is commonly performed with small parts, such as bolts, nuts, gears, and fittings. Electrolytic fluid can freely penetrate through the barrel and provide the metal for plating, and electrical contact is provided through the barrel and through contact with other parts.

3. **Brush processing:** the electrolytic fluid is pumped through a handheld brush with metal bristles. The workpiece can be very large, and the process is suitable for field repair or plating, and it can be used to apply coatings on large equipment without disassembling them.

Simple electroplating can be done in a single-process bath or tank, but more commonly, a sequence of operations is involved in a plating line. The following equipment and processes may be part of an electroplating operation:

- Chemical cleaning and degreasing tanks will be used to remove surface contaminants and enhance surface adhesion of the plated coating.
- The workpieces may be exposed to a strong acid bath (pickling solution) to eliminate or reduce the thickness of the oxide coating on the workpiece.
- A base coating may be applied. This may involve the same or a different metal than that of the final surface; for example, if the desired metal coating will not adhere well to the substrate, an intermediate coating can be applied. Also, if thick films are required, a plating tank can be used to quickly develop a film and a subsequent tank, with brightener additives in the electrolytic solution, is used to develop the final surface finish.
- A separate tank performs final electroplating.
- Rinse tanks will be used throughout the sequence.

Rinse tanks are essential for several reasons. Some plating is performed with cyanide salts, delivering the required metal ions. If any residue acid (such as that from a pickling tank) is conveyed to the cyanide-solution tank, poisonous hydrogen-cyanide gas is evolved. This is a significant safety concern, thus environmental controls are essential in plating facilities. Also, plating solution residue will contain some metal ions, and it is often desirable to recover those ions by capturing them in a rinse tank.

The rate of coating deposition depends on the local current density, which is not necessarily uniform on a part. Workpieces with complex shapes may require a modified geometry because of varying plating thicknesses, as can be seen in Fig. 34.9.

Common plating metals are chromium, nickel (for corrosion protection), cadmium, copper (corrosion resistance and electrical conductivity), and tin and zinc (corrosion protection, especially for sheet steel). **Chromium plating** is done by first plating the metal with copper, then with nickel, and finally with chromium. **Hard chromium plating** is done directly on the base metal, and results in a surface hardness of up to 70 HRC (see Fig. 2.14) and a thickness of about 0.05 mm (0.002 in.) or

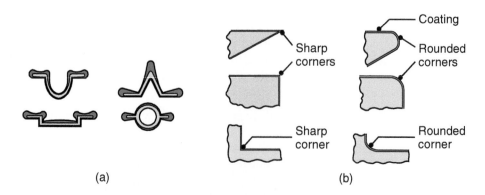

(a)   (b)

FIGURE 34.9 (a) Schematic illustration of nonuniform coatings (exaggerated) in electroplated parts. (b) Design guidelines for electroplating; note that sharp external and internal corners should be avoided for uniform plating thickness.

higher. This method is used to improve the resistance to wear and corrosion of tools, valve stems, hydraulic shafts, and diesel- and aircraft-engine cylinder liners.

Examples of electroplating include copper-plating aluminum wire and phenolic boards for printed circuits, chrome-plating hardware, tin-plating copper electrical terminals (for ease of soldering), galvanizing sheet metal (see also Section 34.11), and plating components such as metalworking dies that require resistance to wear and galling (cold welding of small pieces from the workpiece surface). Metals such as gold, silver, and platinum are important electroplating materials in the electronics and jewelry industries, for electrical contact and for decorative purposes, respectively.

Plastics, such as ABS, polypropylene, polysulfone, polycarbonate, polyester, and nylon, also can be electroplated. Because they are not electrically conductive, plastics must first be preplated, by a process such as electroless nickel plating. Parts to be coated may be simple or complex, and size is not a limitation.

Electroless Plating.   This process is carried out by a chemical reaction and without using an external source of electricity. The most common application utilizes nickel as the plating material, although copper also is used. In *electroless nickel plating*, nickel chloride (a metallic salt) is reduced (with sodium hypophosphite as the reducing agent) to nickel metal, which is then deposited on the workpiece. The hardness of nickel plating ranges between 425 and 575 HV; the plating can subsequently be heat treated to 1000 HV. The coating has excellent wear and corrosion resistance.

Cavities, recesses, and the inner surfaces of tubes can be plated successfully. Electroless plating also can be used with nonconductive materials, such as plastics and ceramics. The process is more expensive than electroplating, but unlike electroplating, the coating thickness of electroless plating is always uniform.

Electroforming.   A variation of electroplating, *electroforming* is a metal-fabricating process. Metal is electrodeposited on a *mandrel* (also called a *mold* or a *matrix*), which is then removed with an appropriate solvent; the coating itself thus becomes the product (Fig. 34.10). Both simple and complex shapes can be produced by

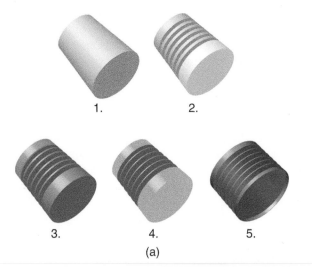

1.   2.

3.   4.   5.

(a)

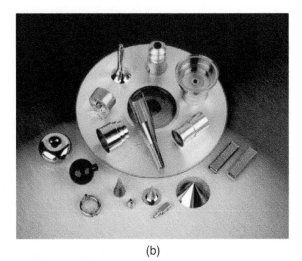

(b)

FIGURE 34.10   (a) Typical sequence in electroforming: (1) A mandrel is selected with the correct nominal size. (2) The desired geometry (in this case, that of a bellows) is machined into the mandrel. (3) The desired metal is electroplated onto the mandrel. (4) The plated material is trimmed, if necessary. (5) The mandrel is dissolved through chemical machining. (b) A collection of electroformed parts. *Source:* Courtesy of Servometer®—PMG LLC, Cedar Grove, NJ.

electroforming, with wall thicknesses as small as 0.025 mm (0.001 in.). Parts may weigh from a few grams to as much as 270 kg (600 lb).

Mandrels are made from a variety of materials, including (a) metallic, such as zinc or aluminum; (b) nonmetallic, which can be made electrically conductive with the appropriate coatings; and (c) low-melting alloys, wax, or plastics, all of which can be melted away or dissolved with suitable chemicals. Mandrels should be physically removable from the electroformed part without damaging it.

The electroforming process is particularly suitable for low production quantities or intricate parts (such as molds, dies, waveguides, nozzles, and bellows) made of nickel, copper, gold, and silver. The process is also suitable for aerospace, electronics, and electro-optics applications.

34.10   Conversion Coatings

Conversion coating, also called *chemical-reaction priming*, is the process of producing a coating that forms on metal surfaces as a result of chemical or electrochemical reactions. Oxides that naturally form on their surfaces (see Section 33.2) represent a form of conversion coating. Various metals (particularly steel, aluminum, and zinc) can be conversion coated.

Phosphates, chromates, and *oxalates* are used to produce conversion coatings, for such purposes as providing corrosion protection, prepainting, and decorative finishing. An important application is the conversion coating of workpieces to serve as lubricant carriers in cold-forming operations, particularly zinc-phosphate and oxalate coatings (see Section 33.7.6). The two common methods of coating are *immersion* and *spraying*.

Anodizing.   This is an oxidation process (*anodic oxidation*) in which the workpiece surfaces are converted to a hard and porous oxide layer, that provides corrosion resistance and a decorative finish. The workpiece is the anode in an electrolytic cell immersed in an acid bath, which results in chemical adsorption of oxygen from the bath. Organic dyes of various colors (usually black, red, bronze, gold, or gray) can be used to produce stable, durable surface films. Typical applications are aluminum furniture and utensils, architectural shapes, picture frames, keys, and sporting goods. Anodized surfaces also serve as a good base for painting, especially on aluminum, which otherwise is difficult to paint.

Coloring.   As the name implies, *coloring* involves processes that alter the color of metals, alloys, and ceramics. This change is caused by the conversion of surfaces (by chemical, electrochemical, or thermal processes) into such chemical compounds as oxides, chromates, and phosphates. An example is the *blackening* of iron and steels, a process that utilizes solutions of hot caustic soda, resulting in chemical reactions that produce a lustrous, black oxide film on surfaces.

34.11   Hot Dipping

In *hot dipping*, the workpiece (usually steel or iron) is dipped into a bath of molten metal, such as (a) zinc, for galvanized-steel sheet and plumbing supplies; (b) tin, for tinplate and tin cans for food containers; (c) aluminum, for aluminizing; and (d) *terne*, an alloy of lead with 10–20% tin. Hot-dipped coatings on discrete parts provide long-term corrosion resistance to galvanized pipes, plumbing supplies, and many other products.

A typical continuous *hot-dipped galvanizing line* for sheet steel is shown in Fig. 34.11. The rolled sheet is first cleaned electrolytically and scrubbed by brushing. The sheet is then annealed in a continuous furnace with controlled atmosphere and temperature, and dipped in molten zinc at about 450°C (840°F).

The thickness of the zinc coating is controlled by a wiping action from a stream of air or steam, called an *air knife* (also used in wave soldering; see Fig. 32.7b). Proper draining for the removal of excess coating materials is important for product quality. The service life depends on the thickness of the zinc coating and the environment to which the galvanized steel is exposed. Various **precoated sheet steels** are used extensively in automobile bodies.

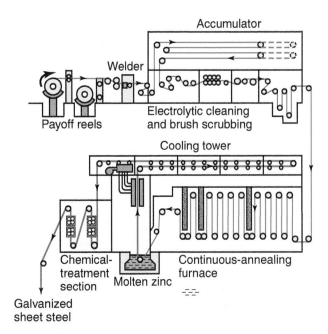

FIGURE 34.11   Flow line for the continuous hot-dipped galvanizing of sheet steel; the welder (upper left) is used to weld the ends of coils to maintain continuous material flow. *Source:* Courtesy of American Iron and Steel Institute.

34.12   Porcelain Enameling; Ceramic and Organic Coatings

Metals can be coated with a variety of glassy (*vitreous*) coatings to provide corrosion and electrical resistance, and for protection at elevated temperatures. These coatings usually are classified as **porcelain enamels**, and generally include enamels and ceramics. The root of the word "porcelain" is *porcellana*, in Italian meaning "marine shell." Note that the word *enamel* also is used as a term for glossy paints, indicating a smooth, hard coating.

Enamels.   Porcelain enamels are glassy inorganic coatings that consist of various metal oxides, and are available in various colors and transparencies. *Enameling* (which was a fully developed art by the Middle Ages) involves fusing the coating material to the substrate, at temperatures of 425° to 1000°C (800°–1800°F) to liquefy the oxides. The coating may be applied by dipping, spraying, or electrodeposition, and thicknesses are usually in the range of 0.05–0.6 mm (0.002–0.025 in.). The viscosity of the material can be controlled using binders, so that the coating adheres to vertical surfaces during application. Depending on their composition, enamels have varying resistances to alkali, acids, detergents, cleansers, and water.

Typical applications for porcelain enameling are household appliances, plumbing fixtures, chemical-processing equipment, signs, cookware, and jewelry; they are also used as protective coatings on jet-engine components. Metals coated are typically steels, cast iron, and aluminum. For chemical resistance, glasses are used as a lining, where the thickness of the glass is much greater than that of the enamel. **Glazing** is the application of glassy coatings onto ceramic wares to give them decorative finishes and to make them impervious to moisture.

Ceramic Coatings.   Ceramics, such as aluminum oxide and zirconium oxide, are applied to a substrate at room temperature by means of binders, then fired in a furnace to fuse the coating material. Usually applied using thermal-spraying techniques, these coatings act as thermal barriers for turbine blades, diesel-engine components, hot-extrusion dies, and nozzles for rocket motors, to extend the life of these components. They also are used for electrical-resistance applications to withstand repeated arcing.

Organic Coatings. Metal surfaces can be coated or precoated with a variety of organic coatings, films, and laminates, in order to improve appearance and corrosion resistance. Coatings are applied to the coil stock on continuous lines (see Fig. 13.10), with thicknesses generally in the range of 0.0025–0.2 mm (0.0001–0.008 in.). Organic coatings have a wide range of properties, such as flexibility, durability, hardness, resistance to abrasion and chemicals, color, texture, and gloss. Coated sheet metals are subsequently shaped into various products, such as cabinets, appliance housings, paneling, shelving, residential-building siding, gutters, and metal furniture.

Critical applications of organic coatings involve, for example, the protection of naval aircraft, which are subjected to high humidity, rain, seawater, pollutants, aviation fuel, deicing fluids, and battery acid, as well as being impacted by particles such as dust, gravel, stones, and deicing salts. For aluminum structures, organic coatings consist typically of an epoxy primer and a polyurethane topcoat.

EXAMPLE 34.2   Ceramic Coatings for High-temperature Applications

Table 34.1 shows various ceramic coatings and their typical applications at elevated temperatures. These coatings may be applied either singly or in layers, each layer having its own special properties, as is done in multiple-layer coated cutting tools (see Fig. 22.8).

TABLE 34.1

Ceramic Coatings Used for High-temperature Applications		
Property	Type of ceramic	Applications
Wear resistance	Chromium oxide, aluminum oxide, aluminum titania	Pumps, turbine shafts, seals, and compressor rods for the petroleum industry; plastics extruder barrels; extrusion dies
Thermal insulation	Zirconium oxide (yttria stabilized), zirconium oxide (calcia stabilized), magnesium zirconate	Fan blades, compressor blades, and seals for gas turbines; valves, pistons, and combustion heads for automotive engines
Electrical insulation	Magnesium aluminate, aluminum oxide	Induction coils, brazing fixtures, general electrical applications

---

34.13   Diamond Coating and Diamond-like Carbon

The properties of *diamond* that are relevant to manufacturing engineering were described in Section 8.7. Important advances have been made in *diamond coating* of metals, glass, ceramics, and plastics, using various techniques such as chemical vapor deposition, plasma-assisted vapor deposition, and ion-beam-enhanced deposition.

Examples of diamond-coated products are scratchproof windows, such as those used in aircraft and military vehicles for protection in sandstorms; cutting tools, such as inserts, drills, and end mills; wear faces of micrometers and calipers; surgical knives; razors; electronic and infrared heat seekers and sensors; light-emitting diodes; speakers for stereo systems; turbine blades; and fuel-injection nozzles.

Techniques also have been developed to produce **freestanding diamond films,** on the order of 1 mm (0.04 in.) thick and up to 125 mm (5 in.) in diameter. These films include smooth, optically clear diamond film, which is then laser cut to desired shapes, and brazed onto workpieces.

Studies also are continuing concerning growing of diamond films on crystalline-copper substrate, by implantation of carbon ions. An important application is in making computer chips (Chapter 28). Diamond can be doped to form p- and n-type ends on semiconductors to make transistors, and its high thermal conductivity allows the closer packing of chips than would be possible with silicon or gallium-arsenide chips, significantly increasing the speed of computers. Diamond is also an attractive material for MEMS devices (Chapter 29), because of its favorable friction and wear characteristics.

Diamond-like Carbon.   *Diamond-like carbon* (DLC) coatings, a few nanometers in thickness, are produced by a low-temperature, ion-beam-assisted deposition process. The structure of DLC is between that of diamond and graphite (Section 8.6). Less expensive than diamond films but with similar properties (such as low friction, high hardness, and chemical inertness, as well as having a smooth surface), DLC has applications in such areas as tools and dies, engine components, gears, bearings, MEMS devices, and microscale probes. As a coating on cutting tools, DLC has a hardness of about 5000 HV, as compared with about twice that for diamond.

34.14 Surface Texturing

Recall that each manufacturing process produces a certain surface texture, finish, and appearance. However, manufactured surfaces can be modified further by secondary operations for functional, optical, or aesthetic reasons.

Called *surface texturing*, the secondary operations generally consist of the following techniques:

- **Etching,** using chemicals or sputtering techniques
- **Electric arcs**
- **Lasers,** using excimer lasers with pulsed beams
- **Atomic oxygen,** reacting with surfaces to produce a fine, cone-like surface texture

The possible beneficial as well as adverse effects of these additional processes on material properties, and the performance of the textured parts are important considerations. Usually, the performance is related to appearance or tribological issues. A rough surface, for example, can help entrain a lubricant and reduce friction, but it can also increase friction in unlubricated contacts.

34.15 Painting

Because of its decorative and functional properties, such as environmental protection, low cost, relative ease of application, and the range of available colors, *paint* has been widely used for thousands of years, as a surface coating and decoration. Paints are generally classified as

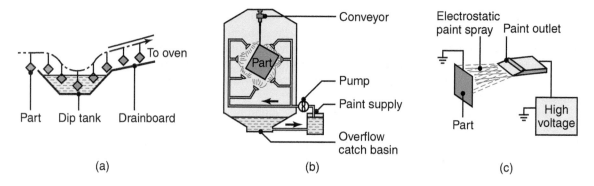

FIGURE 34.12 Methods of paint application: (a) dip coating, (b) flow coating, and (c) electrostatic spraying (used particularly for automotive bodies).

- **Enamels,** producing a smooth coat with a glossy or semiglossy appearance
- **Lacquers,** forming an adherent film by evaporation of a solvent
- **Water-based paints,** applied easily, but have a porous surface and absorb water, making them more difficult to clean

Paints are now available with good resistance to abrasion, high temperatures, and fading. Their selection depends on specific requirements, such as resistance to mechanical actions (abrasion, marring, impact, and flexing) and to chemical reactions (acids, solvents, detergents, alkali, fuels, staining, and general environmental attack).

Common methods of applying paint are dipping, brushing, rolling, and spraying (Fig. 34.12). In **electrocoating** or **electrostatic spraying**, paint particles are charged *electrostatically* and are attracted to surfaces to be painted, producing a uniformly adherent coating. Unlike paint losses in conventional spraying, which may be as much as 70% of the paint, the loss can be as little as 10% in electrostatic spraying. However, deep recesses and corners can be difficult to coat with this method. The use of *robotic controls* for guiding the spray nozzles is now in common practice (Section 37.6.3).

34.16 Cleaning of Surfaces

The importance of surfaces in manufacturing, and the effects of deposited or adsorbed layers of various contaminants on surface characteristics, have been stressed throughout many chapters. The word **clean** or *degree of cleanliness* of a surface is somewhat difficult to define; two common and simple tests are:

1. Observing whether water continuously and uniformly coats the surface, known as the *waterbreak* test; if water collects as individual droplets, the surface is not clean. (This phenomenon can easily be observed by wetting dinner plates that have been washed to different degrees of cleanliness.)
2. Wiping the surface with a clean white cloth and observing any residues on the cloth.

A clean surface can have both beneficial and detrimental effects. Although a surface that is not clean may reduce the tendency for adhesion in sliding (and thus reduce friction), cleanliness generally is essential for a more effective application of coatings, painting, adhesive bonding, brazing, and soldering, as well as for the reliable functioning of moving parts in machinery, assembly operations, and for food and beverage containers.

Cleaning involves the removal of solid, semisolid, or liquid contaminants from a surface, and is an important part of manufacturing operations and the economics of production. In manufacturing operations, the type of cleaning process required depends on the type of *metalworking-fluid residues* and *contaminants* to be removed. Water-based fluids, for example, are easier and less expensive to remove than oil-based fluids. Contaminants (also called *soils*) may consist of rust, scale, chips, various metallic and nonmetallic debris, metalworking fluids, solid lubricants, pigments, polishing and lapping compounds, and general environmental elements.

Basically, there are three types of cleaning methods:

Mechanical Cleaning. This operation consists of physically disturbing the contaminants, often with wire or fiber brushing, abrasive blasting, tumbling, or steam jets. Many of these processes are particularly effective in removing rust, scale, and other solid contaminants from surfaces. *Ultrasonic cleaning* is also in this category.

Electrolytic Cleaning. A charge is applied to the part to be cleaned, in an aqueous (often alkaline) cleaning solution; the charge results in bubbles of hydrogen or oxygen (depending on polarity) being released at the surface; the bubbles are abrasive and aid in the removal of contaminants.

Chemical Cleaning. Chemical cleaning usually involves the removal of oil and grease from surfaces; this operation consists of one or more of the following processes:

- **Solution:** The soil dissolves in the cleaning solution
- **Saponification:** A chemical reaction converts animal or vegetable oils into a soap that is soluble in water
- **Emulsification:** The cleaning solution reacts with the soil or lubricant deposits and forms an emulsion; the soil and the emulsifier then become suspended in the emulsion
- **Dispersion:** The concentration of soil on the surface is decreased by surface-active elements in the cleaning solution
- **Aggregation:** Lubricants are removed from the surface by various agents in the cleanser, and are collected as large dirt particles

Cleaning Fluids. Common cleaning fluids, used in conjunction with electrochemical processes for more effective cleaning, include:

- **Alkaline solutions:** A complex combination of water-soluble chemicals, alkaline solutions are the least expensive and most widely used cleaning fluids in manufacturing operations. Small parts may be cleaned in rotating drums or barrels. Most parts are cleaned on continuous conveyors by spraying them with the solution and rinsing them with water.
- **Emulsions:** Emulsions generally consist of kerosene and oil-in-water and various types of emulsifiers.
- **Solvents:** Petroleum solvents, chlorinated hydrocarbons, and mineral–spirits solvents generally are used for short runs. Fire and toxicity are major hazards.
- **Hot vapors:** Chlorinated solvents can be used to remove oil, grease, and wax. The solvent is first boiled in a container and then condensed and collected for reuse. This hot-vapor process, also known as **vapor degreasing**, is simple and

the cleaned parts are dry. The solvents are never diluted or made less effective by the dissolution of oil, since the oil does not evaporate.

- **Acids, salts, and mixtures of organic compounds:** These are effective in cleaning parts that are covered with heavy paste or oily deposits and rust.

Design Guidelines for Cleaning. Cleaning discrete parts with complex shapes can be difficult. Some basic design guidelines include (a) avoiding deep, blind holes; (b) making several smaller components instead of one large component, which may be difficult to clean; and (c) providing appropriate drain holes in the parts to be cleaned.

The *treatment* and *disposal* of cleaning fluids, as well as of various fluids and waste materials from the processes described in this chapter, are among the most important considerations for environmentally safe manufacturing operations. (See also Section I.4.)

SUMMARY

- Surface treatments are an important aspect of all manufacturing operations. They are used to impart specific mechanical, chemical, and physical properties, such as appearance, and corrosion, friction, wear, and fatigue resistance.

- The processes used include mechanical working and surface treatments, such as heat treatment, deposition, and plating. Surface coatings include enamels, nonmetallic materials, and paints.

- Clean surfaces can be important in the further processing, such as coating, painting, and welding, and in the use of the product. Cleaning can have a significant economic impact on manufacturing operations.

KEY TERMS

Anodizing	Diamond coating	Hard chromium plating	Porcelain enamel
Ballizing	Diamond-like carbon	Hard facing	Roller burnishing
Blackening	Diffusion coating	Hot dipping	Shot peening
Case hardening	Electroforming	Ion implantation	Spraying
Chemical cleaning	Electroless plating	Ion plating	Sputtering
Chemical vapor	Electroplating	Laser peening	Surface texturing
deposition	Enamel	Mechanical plating	Thermal spraying
Cladding	Explosive hardening	Metallizing	Vacuum evaporation
Cleaning fluids	Freestanding diamond	Painting	Vapor deposition
Coloring	film	Physical vapor	Waterbreak test
Conversion coating	Glazing	deposition	Water-jet peening

BIBLIOGRAPHY

ASM Handbook, Vol. 5, **Surface Engineering**, ASM International, 1994.

Bunshah, R.F. (ed.), **Handbook of Hard Coatings: Deposition Technologies, Properties and Applications**, William Andrew, 2002.

Davis, J.R. (ed.), **Surface Engineering for Corrosion and Wear Resistance**, ASM International and IOM Communications, 2001.

Davis, J.R. (ed.), **Surface Hardening of Steels**, ASM International, 2002.

Polak, T.A. (ed.), **Handbook of Surface Treatments and Coatings**, ASME Press, 2003.

Roberge, P.R., **Handbook of Corrosion Engineering**, 2nd ed, McGraw-Hill, 2012.

Schulze, V., **Modern Mechanical Surface Treatment**, Wiley-VCH, 2006.

Tracton, A.A. (ed.), **Coatings Technology Handbook**, 3rd ed., CRC Press, 2005.

REVIEW QUESTIONS

34.1 Explain why surface treatments may be necessary for various parts made by one or more processes.

34.2 What is shot peening? Why is it performed?

34.3 What are the advantages of roller burnishing?

34.4 Explain the difference between case hardening and hard facing.

34.5 Describe the principles of physical and chemical vapor deposition. What applications do these processes have?

34.6 Review Fig. 34.2 and describe the principles behind thermal spray.

34.7 What is electroplating? Why can it be hazardous?

34.8 What is the principle of electroforming? What are its advantages?

34.9 Explain the difference between electroplating and electroless plating.

34.10 How is hot dipping performed?

34.11 What is an air knife? How does it function?

34.12 Describe the common painting systems presently in use in industry.

34.13 What is a conversion coating? Why is it so called?

34.14 Describe the difference between thermal spraying and plasma spraying.

34.15 What is cladding, and why is it performed?

34.16 How are diamond coatings produced?

QUALITATIVE PROBLEMS

34.17 Describe how roller-burnishing processes induce compressive residual stresses on the surfaces of parts.

34.18 Explain why some parts may be coated with ceramics. Give some examples.

34.19 List and briefly describe five surface treatment techniques that use lasers.

34.20 Give examples of part designs that are suitable for hot-dip galvanizing.

34.21 Comment on your observations regarding Fig. 34.9.

34.22 It is well known that coatings may be removed or depleted during the service life of components, particularly at elevated temperatures. Describe the factors involved in the strength and durability of coatings.

34.23 Make a list of the coating processes described in this chapter, and classify them in relative terms as "thick" or "thin."

34.24 Sort the coating processes described in this chapter according to (a) maximum thickness generally achieved and (b) typical coating time.

34.25 Why is galvanizing important for automotive-body sheet metals?

34.26 Explain the principles involved in various techniques for applying paints.

QUANTITATIVE PROBLEMS

34.27 Taking a simple example, such as the parts shown in Fig. 34.1, estimate the force required for roller burnishing. (*Hint:* See Sections 2.6 and 14.4.)

34.28 Estimate the plating thickness in electroplating a 20-mm solid-metal ball using a current of 10 A and a plating time of 1.5 h. Assume that $c = 0.08$ in Eq. (34.1).

SYNTHESIS, DESIGN, AND PROJECTS

34.29 Which surface treatments are functional, and which are decorative? Are there any treatments that serve both functions? Explain.

34.30 An artificial implant has a porous surface area where it is expected that the bone will attach and grow into the implant. Without consulting the literature, make recommendations for producing a porous surface; then review the literature and describe the actual processes used.

34.31 If one is interested in obtaining a textured surface on a coated piece of metal, should one apply the coating first or apply the texture first? Explain.

34.32 It is known that a mirrorlike surface finish can be obtained by plating workpieces that are ground; that is, the surface finish improves after coating. Explain how this occurs.

34.33 It has been observed in practice that a thin layer of chrome plating, such as that on older model automobile bumpers, is better than a thick layer. Explain why, considering the effect of thickness on the tendency for cracking.

34.34 Outline the reasons that the topics described in this chapter are important in manufacturing processes and operations.

34.35 Shiny, metallic balloons have festive printed patterns that are produced by printing screens and are then plated onto the balloons. How can metallic coatings be plated onto a rubber sheet?

34.36 Because they evaporate, solvents and similar cleaning solutions have adverse environmental effects. Describe your thoughts on what modifications could be made to render cleaning solutions more environmentally friendly.

34.37 A roller-burnishing operation is performed on a shaft shoulder to increase fatigue life. It is noted that the resulting surface finish is poor, and a proposal is made to machine the surface layer to further improve fatigue life. Will this be advisable? Explain.

34.38 The shot-peening process can be demonstrated with a ball-peen hammer (in which one of the heads is round). Using such a hammer, make numerous indentations on the surface of a piece of aluminum sheet (a) 2 mm and (b) 10 mm thick, respectively, placed on a hard flat surface such as an anvil. Note that both pieces develop curvatures, but one becomes concave and the other convex. Describe your observations and explain why this happens. (*Hint:* See Fig. 2.14.)

34.39 Obtain several pieces of small metal parts (such as bolts, rods, and sheet metal) and perform the waterbreak test on them. Then clean the surfaces with various cleaning fluids and repeat the test. Describe your observations.

34.40 Inspect various products, such as small and large appliances, silverware, metal vases and boxes, kitchen utensils, and hand tools, and comment on the type of coatings they may have and the reasons why they are coated.

Engineering Metrology, Instrumentation, and Quality Assurance

The vast majority of manufactured parts are components or a subassembly of a product, and they must fit and be assembled properly so that the product performs its intended function during its service life. Dimensions and other surface features of a part are measured to ensure that it is manufactured consistently and within the specified range of dimensional tolerances. Note, for example, that (a) a piston must fit into a cylinder within specified tolerances, (b) a turbine blade must fit properly into its slot on a turbine disk, and (c) the slideways of a machine tool must be produced with a certain accuracy so that the parts produced on that machine are accurate within their specifications.

Measurement of relevant dimensions and features is an integral aspect of **interchangeable parts manufacturing,** which is the basic concept behind standardization and mass production. If, for example, a ball bearing in an appliance is worn and has to be replaced, all one has to do is purchase a similar one with the same specification or part number.

The first of the next two chapters describes the principles involved in and the various instruments and modern machines used for measuring dimensional features, such as length, angle, flatness, and roundness. Testing and inspecting parts are equally important aspects of manufacturing operations; thus, the methods used for the nondestructive and destructive testing of parts also are described.

Product quality is one of the most important aspects of manufacturing, as described in Chapter 36, which discusses the technological and economic importance of building quality *into* a product, rather than inspecting it *after* it is made, as had been done traditionally for centuries.

35

Engineering Metrology and Instrumentation

Video Solution 35.1 Thermal Expansion and Tolerances

- This chapter describes the importance of the measurement of manufactured parts, noting that measurement and their certification to a certain standard is essential to ensure their proper fit and operation.
- A wide variety of measurement techniques, gages, instruments, and machines is now available, as described and illustrated in this chapter.
- The chapter also describes features of automated measurement, ending with an introduction to the principles of dimensioning and tolerancing.

35.1 Introduction

Engineering metrology is defined as the measurement of dimensions, such as length, thickness, diameter, taper, angle, flatness, and profile. Note that these are *geometric* measures; mechanical and physical property measurements are not included in metrology. Consider, for example, the slideways for machine tools (Fig. 35.1; see also Figs. 23.2, 24.18, and 25.11); these components must have specific dimensions, angles, and flatness in order for the machine tool to function properly and within the specified dimensional accuracy.

Traditionally, measurements have been made after the part has been produced, an approach known as **postprocess inspection.** Here, the term *inspection* means checking the dimensions of what has been produced or is being produced, and determining whether those dimensions comply with the specified dimensional tolerances and other specifications. Today, measurements are being made *while* the part is produced on the machine, an approach known as **in-process, online,** or **real-time inspection.**

An important aspect of metrology is **dimensional tolerance,** defined as the permissible variation in the dimensions of a part. Tolerances are important because of their major role on part interchangeability, proper functioning of the product, and manufacturing costs; generally, however, the smaller the tolerance, the higher are the production costs.

35.2 Measurement Standards

Our earliest experience with measurement is usually with a simple *ruler*, also called a *rule*, to measure **length.** Rulers are used as a *standard* against which dimensions are measured. Traditionally, in the United States, the units *inch* and *foot* have been and continue to be used, which, originally, were based on parts of the human body (see Example 35.1); consequently, it was common to find significant variations in the length of 1 foot.

In most of the world, however, the *meter* is used as a length standard. Originally, 1 meter was defined as one ten-millionth of the distance between the North Pole and the equator. Later, meter was standardized as the distance between two scratches on a platinum-iridium bar, kept under controlled conditions in a building outside Paris. In 1960, the meter was officially defined as 1,650,763.73 wavelengths, in a vacuum, of the orange light given off by electrically excited krypton 86, a rare gas. The precision of this measurement was set as 1 part in 10^9. The meter is now an international standard of length in the Système International d'Unités (SI).

Numerous measuring instruments and devices are used in engineering metrology, each of which has its own resolution, precision, and other features. Two terms commonly used to describe the type and quality of an instrument are:

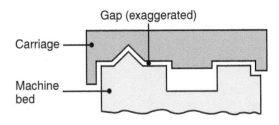

FIGURE 35.1 Cross-section of a machine-tool slideway; the width, depth, angles, and other dimensions all must be produced and measured accurately for the machine tool to function as designed.

1. **Resolution:** The smallest difference in dimensions that the measuring instrument can detect or distinguish; a wooden yardstick, for example, has far less resolution than a micrometer.
2. **Precision:** Sometimes incorrectly called accuracy, it is the degree to which the instrument gives repeated measurements of the same standard. For example, an aluminum ruler will expand or contract, depending on temperature variations in the environment in which it is used; thus, its precision can be affected even when being held by hand.

EXAMPLE 35.1   Length Measurements throughout History

Several standards for length measurement have been developed during the past 6000 years. A common standard in Egypt, around 4000 B.C., was the King's elbow, which was equivalent to 0.4633 m. One *elbow* was equal to 1.5 feet (or 2 hand spans, 6 hand widths, or 24 finger thicknesses). In 1101, King Henry I of England declared a new standard, called the *yard* (0.9144 m); it was the distance from his nose to the tip of his thumb.

During the Middle Ages, almost every kingdom and city established its own length standard, some of which had identical names. In 1528, the French physician Jean Fernel proposed the distance between Paris and Amiens, a city 120 km north of Paris, as a general length reference. During the 17th century, some scientists suggested that the length of a certain pendulum be used as a standard. In 1661, the British architect Sir Christopher Wren suggested that a pendulum with a period of one-half second be used. The Dutch mathematician Christian Huygens proposed a pendulum that had a length one-third of Wren's pendulum and a period of 1 second.

To put an end to the widespread confusion of length measurement, a definitive length standard began to be developed in 1790, in France, with the concept of a *mètre* (from the Greek word *metron*, meaning "measure"). A gage block 1 meter long was made of pure platinum with a rectangular cross-section, and was placed in the National Archives in Paris in 1799. Duplicates of this gage were made for other countries over the years.

During 1870–1872, international committees met and decided on an international meter standard. A new bar was made of 90% platinum and 10% iridium, with an X-shaped cross-section and overall dimensions of 20 mm × 20 mm. Three marks were engraved at each end of the bar; the standard meter was the distance between the central marks at each end, measured at 0°C. Today, an extremely accurate measurement is based on the *speed of light in a vacuum*, which is calculated by multiplying the wavelength of the standardized infrared beam of a laser by its frequency.

35.3 Geometric Features of Parts: Analog and Digital Measurements

The most commonly used quantities and geometric features in engineering practice are described in this section.

- **Length,** includes all linear dimensions of parts
- **Diameter,** includes outside and inside diameters
- **Roundness,** includes out-of-roundness, concentricity, and eccentricity
- **Depth,** such as that of drilled holes and cavities in dies and molds
- **Straightness,** such as that of shafts, bars, and tubing
- **Flatness,** such as that of machined, ground, and polished surfaces
- **Parallelism,** such as that of two shafts or slideways in a machine
- **Perpendicularity,** such as that of a threaded bar inserted into a flat plate
- **Angle,** includes internal and external angles
- **Profile,** such as curvatures of parts made by various processes

A wide variety of instruments and machines is now available to accurately and rapidly measure these quantities, either on stationary parts or on parts that are in continuous production. In engineering metrology, the words **instrument** and **gage** often are used interchangeably. Because of major global trends in automation and computer control of manufacturing operations (see Part IX), modern measuring equipment and instrumentation are now an *integral part of production machinery.* The implementation of **digital** instrumentation and developments in computer-integrated manufacturing have, together, led to the total integration of measurement technologies within manufacturing systems.

Temperature control is very important, particularly for making measurements with precision instruments. The standard measuring temperature is 20°C (68°F), and all gages are calibrated at this temperature. In the interest of accuracy, measurements should be taken in controlled environments maintaining the standard temperature, usually within ±0.3°C (0.5°F).

An **analog** instrument, such as a vernier caliper or micrometer (Fig. 35.2a), relies on the skill of the operator to properly interpolate and read the graduated scales. In contrast, in a digital micrometer, measurements are indicated directly (Fig. 35.2b). More importantly, digital equipment can easily be integrated into other equipment (Fig. 35.2c), including production machinery and systems for statistical process control (SPC), as described in Chapter 36.

35.4 Traditional Measuring Methods and Instruments

35.4.1 Line-graduated Instruments

Graduated means marked to indicate a certain quantity; these instruments are used for measuring lengths or angles.

Linear Measurement (Direct Reading)

- **Rules:** The simplest and most commonly used instrument for making linear measurements is a *steel rule (machinist's rule),* bar, or tape, with fractional or decimal graduations. Lengths are measured directly, to an accuracy that is limited to the nearest division, usually 1 mm (0.040 in.).

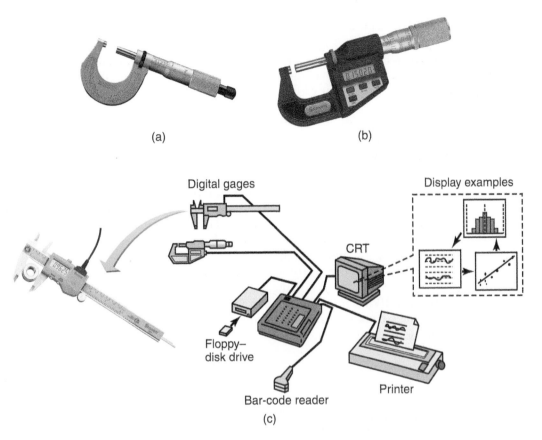

(a)   (b)

(c)

FIGURE 35.2 (a) A vernier (analog) micrometer. (b) A digital micrometer, with a range of 0 to 1 in. (0–25 mm) and a resolution of 1.25 μm (50 μin.); note that it is much easier and faster to read dimensions on this instrument than on analog micrometers. (c) Schematic illustration showing the integration of digital gages with microprocessors for real-time data acquisition for statistical process control (see Section 36.8). *Source:* Courtesy of L.S. Starrett Co.

- **Calipers:** Also called *caliper gages* and *vernier calipers* (named for P. Vernier, who lived in the 1600s), they have a graduated beam and a sliding *jaw*. *Digital calipers* are now in wide use.
- **Micrometers:** These instruments are commonly used for measuring the thickness and inside or outside dimensions of parts. *Digital micrometers* are equipped with readouts (Fig. 35.2b), either in metric or in English units. Micrometers also are available for measuring internal diameters (*inside micrometer*) and depths (*micrometer depth* gage, Fig. 35.3). The anvils on micrometers can be equipped with conical or ball contacts, to measure recesses, threaded-rod diameters, and wall thicknesses of tubes and curved sheets or plates.

FIGURE 35.3 A micrometer depth gage, with both digital and vernier gages. *Source:* Courtesy of L.S. Starrett Co.

Linear Measurement (Indirect Reading). These instruments typically are calipers and *dividers*, without any graduated scales, and are used to transfer the measured

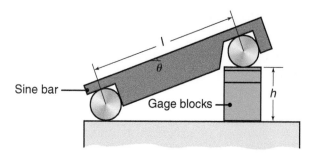

FIGURE 35.4   Schematic illustration of a sine bar.

size to a direct-reading instrument, such as a ruler. Because of the experience required to use them and their dependence on graduated scales, the accuracy of indirect-measurement tools is limited. *Telescoping gages* can be used for the indirect measurement of holes or cavities.

Angle Measurement

- **Bevel protractor:** This is a direct-reading instrument, similar to a common protractor, except that it has a movable element. The two blades of the protractor are placed in contact with the part being measured, and the angle is read directly on the vernier scale. Another common type of bevel protractor is the *combination square*, which is a steel rule equipped with devices for measuring 45° and 90° angles.
- **Sine bar:** A *sine bar* consists of a plate mounted on two solid cylinders (Fig. 35.4). One cylinder is a pivot for the plate; the other is located a fixed distance, commonly 10 in., from the first bar. By inserting gage blocks (Section 35.4.4) under the second bar, the angle of the plate can be varied. The relationship is given by

$$\sin\theta = \frac{h}{l} \tag{35.1}$$

- **Surface plates:** These plates are used to support both the part to be measured and the measuring instrument. They are produced to high flatness, with variations as little as 0.25 μm across the plate. The plates typically are made of cast iron or natural stones, such as granite. Granite plates have the desirable properties of being resistant to corrosion, are nonmagnetic, and have low thermal expansion, thereby minimizing thermal distortion.

Comparative Length Measurement.   Instruments used for measuring comparative lengths (also called *deviation-type* instruments) amplify and measure variations in the distance between two or more surfaces. These instruments, the most common example being a **dial indicator** (Fig. 35.5), compare dimensions, hence the term *comparative*. The indicator is set to zero, at a certain reference surface, and the instrument, or the surface to be measured (either external or internal), is brought into contact with the pointer. The movement of the indicator is read directly on the circular dial, to accuracies as high as 1 μm (40 μin.). Dial indicators with electrical and fluidic amplification mechanisms and with a digital readout also are available.

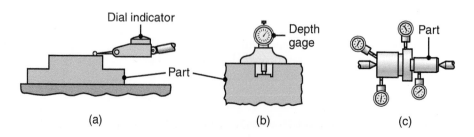

FIGURE 35.5   Three uses of dial indicators to measure (a) roundness and (b) depth, and (c) for multiple-dimension gaging of a part.

35.4.2   Measuring Geometric Features

Straightness.   *Straightness* can be checked with a straightedge or a dial indicator (Fig. 35.6). An *autocollimator*, which resembles a telescope with a light beam that bounces back from the object, is used to accurately measure small angular deviations on a flat surface. *Laser beams* are now widely used to align individual machine elements in the assembly of machine components.

Flatness.   *Flatness* can be measured by mechanical means, using a *surface plate* and a *dial indicator*. This method can be used to measure perpendicularity, which also can be measured by precision steel squares.

Another method for measuring flatness is **interferometry**, which uses an *optical flat*. This device is a glass or fused-quartz disk, with parallel flat surfaces, that is placed on the workpiece surface (Fig. 35.7a). When a *monochromatic* light beam (a beam with one wavelength) is aimed at the surface at an angle, the optical flat splits the light into two beams, appearing as light and dark fringes or bands (Fig. 35.7b).

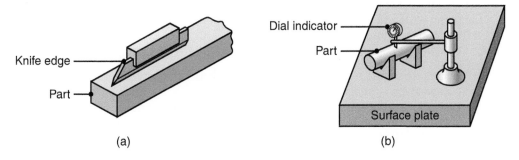

FIGURE 35.6   Measuring straightness manually with (a) a knife-edge rule and (b) a dial indicator. *Source:* Based on F.T. Farago.

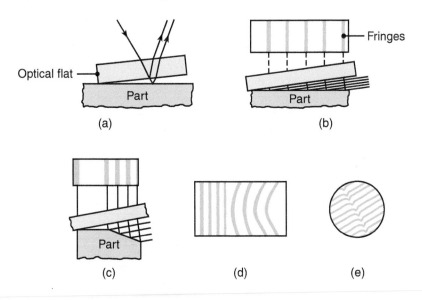

FIGURE 35.7   (a) Interferometry method for measuring flatness with an optical flat. (b) Fringes on a flat, inclined surface; an optical flat resting on a flat workpiece surface will not split the light beam, and no fringes will be present. (c) Fringes on a surface with two inclinations; note that the greater the incline, the closer together are the fringes. (d) Curved fringe patterns, indicating curvatures on the workpiece surface. (e) Fringe pattern indicating a scratch on the surface.

The number of fringes that appear is related to the distance between the surface of the part and the bottom surface of the optical flat (Fig. 35.7c). A truly flat surface (i.e., one in which the angle between the two surfaces is zero) will not split the light beam, and hence fringes will not appear. When surfaces are not flat, the fringes are curved (Fig. 35.7d). The interferometry method is also used for observing surface textures and scratches (Fig. 35.7e).

Diffraction gratings consist of two optical flat glasses of different lengths and closely spaced parallel lines scribed on their surfaces. The grating on the shorter glass is inclined slightly; as a result, *interference fringes* develop when viewed over the longer glass. The position of these fringes depends on the relative position of the two sets of glasses. With modern equipment and with the use of electronic counters and photoelectric sensors, a resolution of 2.5 μm (0.0001 in.) can be obtained with gratings having 40 lines/mm (1000 lines/in.).

Roundness.   This feature is described as a deviation from true roundness (mathematically, a *circle*). The term *out-of-roundness* (ovality) is actually more descriptive of the shape of the part (Fig. 35.8a) than the word *roundness*. True roundness is essential to the proper functioning of rotating shafts, bearing races, pistons, cylinders, and steel balls in bearings.

Methods of measuring roundness generally fall into two categories:

1. The round part is placed on a *V-block* or between centers (Figs. 35.8b and c, respectively), and is rotated while the point of a dial indicator is in continuous contact with the part's surface. After a full rotation of the workpiece, the difference between the maximum and minimum readings on the dial is noted, called the **total indicator reading** (TIR) or the **full indicator movement.** This method also can be used to measure the straightness (squareness) of end faces of shafts that are machined, such as in the facing operation shown in Fig. 23.1e.

2. In *circular tracing*, the part is placed on a platform and its roundness is measured by rotating the platform (Fig. 35.8d); alternatively, the probe can be rotated around a stationary part.

Profile.   *Profile* may be measured by such means as (a) comparing the surface with a template or profile gage (as in the measurement of radii and fillets) for conformity and (b) using several dial indicators or similar instruments. The best method, however, is using the advanced measuring machines, described in Section 35.5.

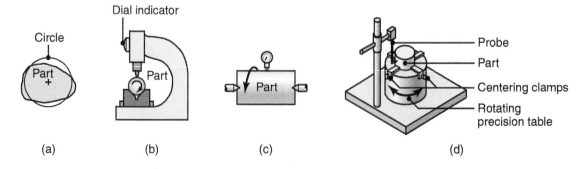

(a)   (b)   (c)   (d)

FIGURE 35.8   (a) Schematic illustration of out-of-roundness (exaggerated). Measuring roundness with (b) a V-block and dial indicator, (c) a round part supported on centers and rotated, and (d) circular tracing. *Source:* Based on F.T. Farago.

Screw Threads and Gear Teeth. Threads can be measured by means of *thread gages* of various designs, that compare the thread produced against a standard thread. Some of the commonly used gages are *threaded plug gages*, *screw-pitch gages*, micrometers with cone-shaped points, and *snap gages* (see Section 35.4.4) with anvils in the shape of threads. Gear teeth are measured with (a) instruments that are similar to dial indicators, (b) calipers (Fig. 35.9a), and (c) micrometers, using round pins or balls of various diameters (Fig. 35.9b). Advanced methods include the use of optical projectors and coordinate-measuring machines (Section 35.5).

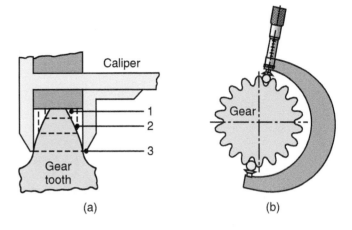

(a)   (b)

FIGURE 35.9 Measuring gear-tooth thickness and profile with (a) a gear-tooth caliper and (b) pins or balls and a micrometer. *Source:* Courtesy of American Gear Manufacturers Association.

35.4.3 Optical Contour Projectors

These instruments, also called **optical comparators**, were first developed in the 1940s to check the geometry of cutting tools for machining screw threads, but are now used for checking all profiles (Fig. 35.10). The part is mounted on a table or between centers, and the image is projected onto a screen, at magnifications of 100× or higher. Linear and angular measurements are made directly on the screen, which is marked with reference lines and circles. For angular measurements, the screen can be rotated.

35.4.4 Gages

Gage Blocks. *Gage blocks* are individual square, rectangular, or round blocks of various sizes. For general use, they are made from heat-treated and stress-relieved alloy steels. The better gage blocks are made of ceramics (often zirconia) and chromium carbide; unlike steels, these materials do not corrode, but they are brittle and must be handled carefully. *Angle blocks* are used for angular gaging. Gage blocks have a flatness within 1.25 μm (50 μin.). Environmental temperature control is essential when gages are used for high-precision measurements.

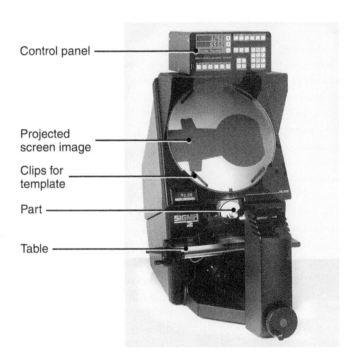

FIGURE 35.10 A horizontal-beam contour projector with a 16-in.-diameter screen, with 150-W tungsten halogen illumination. *Source:* Courtesy of L.S. Starrett Company, Precision Optical Division.

Fixed Gages. These gages are replicas of the shapes of the parts to be measured. Although *fixed* gages are easy to use and are inexpensive, they indicate only whether a part is too small or too large, as compared with an established standard. They have become less common, as they are incompatible with quality control methods (described in Section 36.7). Examples of fixed gages include:

- **Plug gages** are commonly used for holes (Figs. 35.11a and b). The *GO gage* is smaller than the *NOT GO* (or *NO GO*) gage, and slides into any hole that

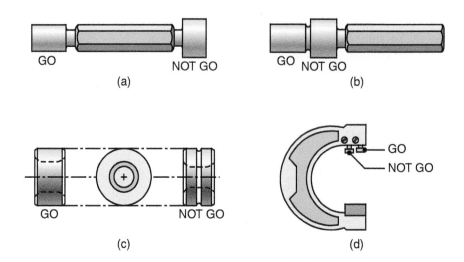

FIGURE 35.11   (a) Plug gage for holes, with *GO* and *NOT GO* on opposite ends of the gage. (b) Plug gage with *GO* and *NOT GO* on one end. (c) Plain ring gages for gaging round rods. (d) Snap gage with adjustable anvils.

has a dimension larger than the diameter of the gage. The *NOT GO* gage must not go into the same hole. Two gages are required for such measurements, although both may be on the same device, either at opposite ends or in two segments at one end (*step-type gage*). Plug gages also are available for measuring internal tapers, splines, and threads (in which the *GO* gage must screw into the threaded hole).

- **Ring gages** (Fig. 35.11c) are used to measure shafts and similar round parts. Ring *thread gages* are used to measure external threads.
- **Snap gages** (Fig. 35.11d) are commonly used to measure external dimensions. They are made with adjustable gaging surfaces, for use with parts that have different dimensions; one of the gaging surfaces can be set at a different gap from the other, thus making the device a one-unit *GO*-and-*NOT-GO* gage.

Air Gages.   The basic operation of an *air gage*, also called a **pneumatic gage**, is shown in Fig. 35.12a. The gage head (called air plug) has two or more holes, typically 1.25 mm (0.05 in.) in diameter, through which pressurized air (supplied by a constant-pressure line) escapes. The smaller the gap between the gage and the hole, the more difficult it is for the air to escape, and hence the higher is the back pressure. The back pressure, which is indicated by a pressure gage, is calibrated to measure the dimensional variations of holes.

Air gages can be rotated during use to indicate and measure any out-of-roundness of a hole. The outside diameters of parts (such as pins and shafts) also can be measured when the air plug is in the shape of a ring, slipped over the part. In cases where a ring is not suitable, a fork-shaped gage head (with the airholes at the tips) can be used (Fig. 35.12b). Various shapes of air heads, such as the conical head shown in Fig. 35.12c, can be made for use in specialized applications.

Air gages are easy to use, and the resolution can be as fine as 0.125 μm (5 μin.); if the surface roughness of the part is too high, however, the readings may be unreliable. Their noncontacting nature and the low pressure has the benefit of not distorting or damaging the measured part, as could be the case with mechanical gages.

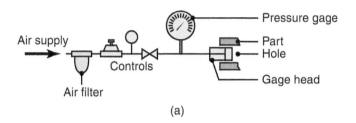

(a)

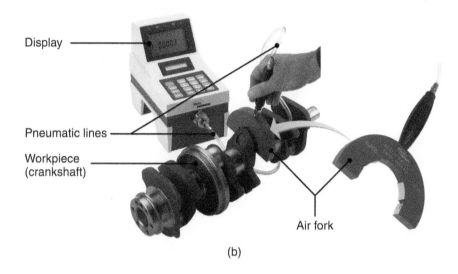

(b)

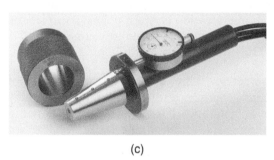

(c)

FIGURE 35.12   (a) Schematic illustration of the principle of an air gage. (b) Illustration of an air-gage system used in measuring the main-bearing dimension on a crankshaft. (c) A conical head for air gaging; note the three small airholes on the conical surface. *Source:* (b) Courtesy of Mahr Federal, Inc. (c) Courtesy of Stotz Gaging Co.

35.5   Modern Measuring Instruments and Machines

A wide variety of measuring instruments and gages has been developed, ranging from simple, hand-operated devices to computer-controlled machines with very large workspaces.

Electronic Gages.   Unlike mechanical systems, *electronic gages* sense the movement of the contacting pointer through changes in the electrical resistance of a strain gage, inductance, or capacitance. The electrical signals are then converted and digitally displayed as linear dimensions. A handheld electronic gage for measuring bore diameters is shown in Fig. 35.13. When its handle is squeezed slightly, the tool can be inserted

FIGURE 35.13 An electronic gage for measuring bore diameters; the measuring head is equipped with three carbide-tipped steel pins, for wear resistance. *Source:* Courtesy of TESA SA.

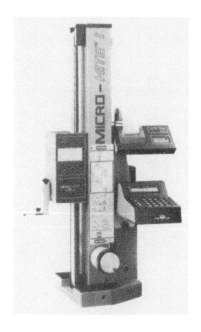

FIGURE 35.14 An electronic vertical length–measuring instrument, with a resolution of 1 μm (0.0001 in.). *Source:* Courtesy of TESA SA.

into the bore, and the bore diameter is read directly. A microprocessor-assisted electronic gage for measuring vertical length is shown in Fig. 35.14.

A commonly used electronic gage is the *linear-variable differential transformer* (LVDT), for measuring small displacements. The chemical-vapor-deposition (CVD; Section 34.6) coating on these gages has a wear resistance superior to that of steel or tungsten-carbide edges; it also resists corrosion. Although more expensive than other types, electronic gages have such advantages as ease of operation, rapid response, a digital readout, lower possibility of human error, versatility, flexibility, and the capability to be integrated into automated systems, through microprocessors and computers.

Laser Micrometers. In this instrument, a laser beam scans the workpiece (Fig. 35.15), typically at a rate of 350/s. Laser micrometers are capable of resolutions as high as 0.125 μm (5 μin.). They are suitable not only for stationary parts, but also for in-line measurement of stationary, rotating, or vibrating parts, as well as parts in continuous, high-speed production facilities. Moreover, because there is no physical contact, they can measure parts that are at elevated temperatures or are too flexible to be measured by other means. The laser beams can be of various types (such as scanning or rastoring for stationary parts), yielding **point cloud** descriptions of part surfaces. In a *point cloud*, a large number of surface points are measured and their coordinates stored; through interpolation between points, a surface is then defined. Laser micrometers are of the shadow type or are charge-coupled device (CCD) based for in-line measurement while a part is in production.

Laser Interferometry. This technique is used to check and calibrate machine tools for various geometric features during their assembly. The method has better accuracies than those of gages or indicators. Laser interferometers also are used to automatically compensate for positioning errors in coordinate-measuring machines and computer-numerical control machines.

Photoelectric Digital Length Measurement. This type of measurement is done by an instrument that can measure the overall dimensions, thickness, and depth of a variety of parts. Resolution settings can range from 5 to 0.01 μm.

35.5.1 Coordinate-measuring Machines

As schematically shown in Fig. 35.16a, a *coordinate-measuring machine* (CMM) basically consists of a platform on which the workpiece being measured is placed, and is then moved linearly or rotated. A *probe* (Fig. 35.16b; see also Fig. 25.6) is attached to a head that is capable of various movements and records all measurements. In addition to the tactile probe shown in the figure, other types of probes are scanning, laser (Fig. 35.16c), and vision probes, all of which are nontactile. A CMM for measurement of a typical part is shown in Fig. 35.16d.

CMMs are very versatile and capable of recording measurements of complex profiles, with high resolution (0.25 μm; 10 μin.) and high speed. They are built rigidly and ruggedly to resist environmental effects in manufacturing plants, such as temperature variations and vibration. They can be placed close to machine tools for more efficient measurement and rapid feedback, so that processing parameters

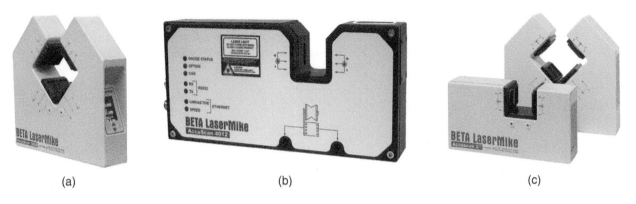

(a)   (b)   (c)

FIGURE 35.15 (a) and (b) Types of measurements made with a laser scan micrometer. (c) Two types of laser micrometers; note that the instrument in the front scans the part (placed in the opening) in one dimension; the other instrument scans the part in two dimensions. *Source:* Courtesy of BETA LaserMike.

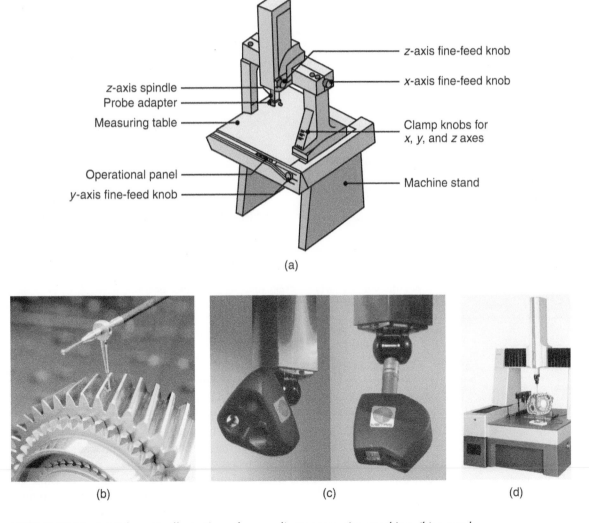

FIGURE 35.16 (a) Schematic illustration of a coordinate-measuring machine; (b) a touch signal probe; (c) examples of laser probes; (d) a coordinate-measuring machine, with a complex part being measured. *Source:* (b)–(d) Courtesy of Mitutoyo America Corporation.

are corrected before the next part is made. Although large machines can be expensive, most machines with a touch probe and computer-controlled three-dimensional movement are suitable for use in small shops, generally costing less than $20,000.

EXAMPLE 35.2   A Coordinate-measuring Machine for Car Bodies

A large horizontal CNC coordinate-measuring machine used to measure all dimensions of a car body is shown in Fig. 35.17. This machine has a measuring range of 6 m × 1.6 m × 2.4 m (236 in. × 63 in. × 94 in.) high, a resolution of 0.1 μm (4 μin.), and a measuring speed of 5 mm/s (0.2 in./s). The system has temperature compensation within a range from 16° to 26°C (60°–78°F), in order to maintain good measurement accuracy.

For efficient measurements, the machine has two heads, with touch-trigger probes that are controlled simultaneously and have full three-dimensional

movements. The probes are software controlled, and the machine is equipped with safety devices to prevent the probes from inadvertently hitting any part of the car body during their movements. The equipment, shown around the base of the machine, includes supporting hardware and software that controls all movements and records all measurements.

Source: Courtesy of Mitutoyo America Corporation.

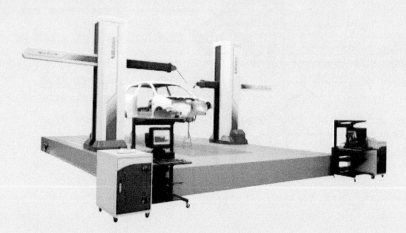

FIGURE 35.17 A large coordinate-measuring machine with two heads measuring various dimensions on a car body.

35.6 Automated Measurement

Automated measurement is based on various **online sensor systems** that continuously monitor the dimensions of parts while they are being made and, if necessary, use these measurements as input to make adjustments (see also Sections 36.12 and 37.7). Manufacturing cells and flexible manufacturing systems (see Chapter 39) all have led to the adoption of advanced measuring techniques and systems.

To appreciate the importance of online monitoring of dimensions, consider the following: If a machine tool has been producing a certain part with acceptable dimensions, what factors contribute to the subsequent deviation in the dimensions of the same part produced by the same machine? There are several technical, as well as human, factors that may be involved:

- *Variations* in the properties and dimensions of the incoming material
- *Distortion* of the machine, because of thermal effects caused by such factors as changes in the ambient temperature, deterioration of metalworking fluids, and changes in machine tool bearings and various components
- *Wear* of tools, dies, and molds
- *Human errors*

Due to one or more of these factors, the dimensions of parts made will vary, thus making continuous monitoring during production necessary.

35.7   General Characteristics and Selection of Measuring Instruments

The characteristics and quality of measuring instruments are generally described by various specific terms, defined as follows (in alphabetical order):

- **Accuracy:** The degree of agreement of the measured dimension with its true magnitude
- **Amplification**, also called **magnification:** The ratio of instrument output to the input dimension
- **Calibration:** The adjustment or setting of an instrument, to give readings that are accurate within a reference standard
- **Drift**, also called **stability:** An instrument's capability to maintain its calibration over time
- **Linearity:** The accuracy of the readings of an instrument over its full working range
- **Precision:** Degree to which an instrument gives repeated measurement of the same standard
- **Repeat accuracy:** The same as accuracy, but repeated several times
- **Resolution:** Smallest dimension that can be read on an instrument
- **Rule of 10** (*gage maker's rule*): An instrument or gage should be 10 times more accurate than the dimensional tolerances of the part being measured; a factor of 4 is known as the *mil standard rule*
- **Sensitivity:** Smallest difference in dimension that an instrument can distinguish or detect
- **Speed of response:** How rapidly an instrument indicates a measurement, particularly when a number of parts are measured in rapid succession

The selection of an appropriate measuring instrument for a particular application also depends on (a) the size and type of parts to be measured, (b) the environment, such as temperature, humidity, and dust, (c) operator skills required, and (d) the cost of equipment.

35.8   Geometric Dimensioning and Tolerancing

Individually manufactured parts eventually are assembled into products. It is often taken for granted that when, for example, a thousand lawn mowers are manufactured and assembled, each part of the mower will fit properly with its intended components. Likewise, when a broken bolt on an old machine is to be replaced, an identical bolt is purchased.

This is done with confidence, because the bolt is manufactured according to certain standards and specifications, and thus the dimensions of all similar bolts will vary by only a small, specified amount, called *tolerance*, that will not affect their function. In other words, all similar bolts are **interchangeable**.

Video Solution 35.2 Dimensions and Tolerances

Dimensional Tolerance.   *Dimensional tolerance*, from the Latin *tolerare*, meaning to "endure," is defined as the permissible or acceptable variation in the dimensions, such as height, width, depth, diameter, and angles, of a part. Tolerances are unavoidable, because it is virtually impossible and unnecessary to manufacture two parts that have precisely the same dimensions.

Moreover, because close dimensional tolerances can significantly increase the product cost (see Fig. 40.4), an unnecessarily narrow tolerance range is economically undesirable. For many parts, however, close tolerances *are* necessary for their proper functioning; examples are precision measuring instruments, hydraulic pistons, rolling-element bearings, and turbine blades for aircraft engines.

Measuring dimensions and features of parts rapidly and reliably can be a challenging task. For example, each of the 6 million parts of a Boeing 747-400 aircraft requires measuring about 25 features, representing a total of 150 million measurements. Surveys have shown that the dimensional tolerances on state-of-the-art manufactured parts are shrinking by a factor of 3 every 10 years, and that this trend will continue (see also Fig. 25.16). It is estimated that the accuracies of

- Conventional turning and milling machines (Chapters 23 and 24) will rise from the present 2 to 0.5 μm
- Diamond-wheel wafer-slicing machines for semiconductor fabrication (see Fig. 28.6d) to as low as 0.25 μm
- Precision diamond turning machines (Section 25.7) from 0.03–0.01 μm
- Ultraprecision ion-beam machines (Section 28.7) to less than 0.001 μm

Importance of Tolerance Control.   Surfaces that are free and not functional do not need close tolerance control; dimensional tolerances become important only when a part is to be assembled with another part. For example, the accuracy of the holes and the distance between the holes for a connecting rod (Fig. 14.8a) are far more critical than the rod's width and thickness at various locations along its length.

To recognize the importance of dimensional tolerances, consider the assembly of a simple round shaft (axle) and a wheel with a round hole. Assume that the shaft's diameter is 1 in. (Fig. 35.18). The wheel is a casting that has a 1-in. diameter hole

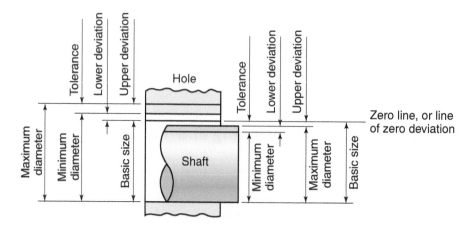

FIGURE 35.18   Basic size, deviation, and tolerance on a shaft, according to the ISO system.

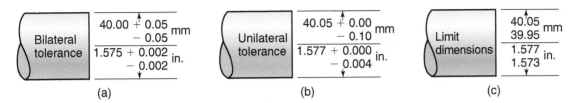

FIGURE 35.19 Various methods of assigning tolerances on a shaft: (a) bilateral tolerance, (b) unilateral tolerance, and (c) limit dimensions.

machined into it. Will the rod fit into the hole without having to force it or will it be loose in the hole? The 1-in. dimension is the **nominal size** of the shaft. If we produce a second shaft, at a different time or select one randomly from a large lot, each shaft will likely have a slightly different diameter (see also Chapter 36). Machines with the same setup may produce rods of slightly different diameters, depending on a number of factors, such as speed of operation, temperature, lubrication, and variations in the properties of the incoming material.

Certain terminology has been established over the years to clearly define these geometric quantities. One such system is the International Organization for Standardization (ISO) system, shown in Fig. 35.18. Note that both the shaft and the hole have minimum and maximum diameters, the difference being the tolerance for each member. A proper engineering drawing would specify these parameters with numerical values, as shown in Fig. 35.19.

The range of dimensional tolerances achievable in manufacturing processes is given in various figures and tables throughout this book. There is a general relationship between tolerances and part size (Fig. 35.20) and between tolerances and surface

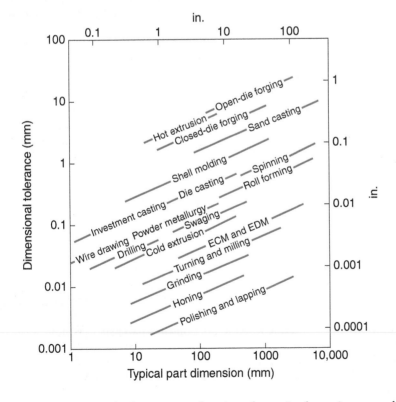

FIGURE 35.20 Dimensional tolerances as a function of part size for various manufacturing processes; note that because many factors are involved, there is a broad range for tolerances.

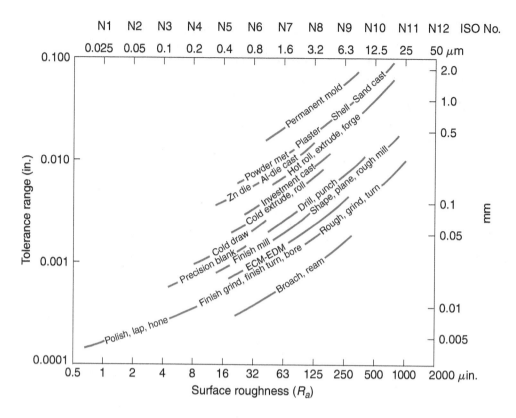

FIGURE 35.21   Dimensional tolerance range and surface roughness obtained in various manufacturing processes; these tolerances apply to a 25-mm (1-in.) workpiece dimension. *Source:* Based on J.A. Schey.

finish of parts manufactured by various processes (Fig. 35.21). Note the wide range of tolerances and surface finish obtained, and that the larger the part, the greater is its obtainable tolerance range.

Definitions.   Several terms are used to describe features of dimensional relationships between mating parts. Details of the definitions are available in the ANSI/ASME B4.2, ANSI/ASME Y14.5, and ISO/TC10/SC5 standards. The commonly used terms for geometric characteristics are defined briefly, as follows (in alphabetical order):

- **Allowance,** also called **functional dimension** or **sum dimension:** The specified difference in dimensions between mating parts
- **Basic size:** Dimension from which limits of size are derived, with the use of tolerances and allowances
- **Bilateral tolerance:** Deviation (plus or minus) from the basic size
- **Clearance:** The space between mating parts
- **Clearance fit:** Fit that allows for rotation or sliding between mating parts
- **Datum:** A theoretically exact axis, point, line, or plane
- **Feature:** A physically identifiable portion of a part, such as hole, slot, pin, or chamfer
- **Fit:** The range of looseness or tightness that can result from the application of a specific combination of allowance and tolerance in the design of mating-part features

- **Geometric tolerances:** Tolerances that involve shape features of the part
- **Hole-basis system,** also called **standard hole practice** or **basic hole system;** tolerances based on a zero line on the hole
- **Interference:** Negative clearance
- **Interference fit:** A fit having limits of size, so prescribed that an interference always results when mating parts are assembled
- **International tolerance** (IT) **grade:** A group of tolerances that vary with the basic size of the part, but provide the same relative level of accuracy within a grade
- **Limit dimension,** also called **limits:** The maximum and minimum dimensions of a part
- **Maximum material condition** (MMC): The condition whereby a feature of a certain size contains the maximum amount of material within the stated limits of that size
- **Nominal size:** An approximate dimension that is used for the purpose of general identification
- **Positional tolerancing:** A system of specifying the true position, size, and form of the features of a part, including allowable variations
- **Shaft basis system,** also called **standard shaft practice** or **basic shaft system;** tolerances based on a zero line on the shaft.
- **Transition fit:** A fit with small clearance or interference that allows for accurate location of mating parts

Type of feature	Type of tolerance	Characteristic	Symbol
Individual (no datum reference)	Form	Flatness	▱
		Straightness	—
		Circularity (roundness)	○
		Cylindricity	⌭
Individual or related	Profile	Profile of a line	⌒
		Profile of a surface	⌓
Related (datum reference required)	Orientation	Perpendicularity	⊥
		Angularity	∠
		Parallelism	//
	Location	Position	⊕
		Concentricity	◎
	Runout	Circular runout	∕
		Total runout	⌰

(a)

FIGURE 35.22   Geometric characteristic symbols to be indicated on engineering drawings of parts to be manufactured. *Source:* Courtesy of The American Society of Mechanical Engineers.

(*continued*)

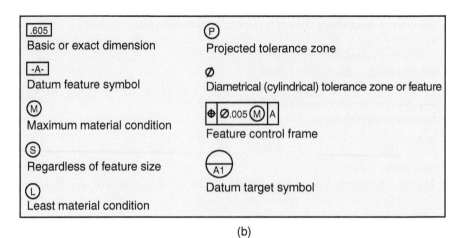

(b)

FIGURE 35.22 (*continued*)

TABLE 35.1

Classes of Fit			
Class	Description	Type	Applications
1	Loose	Clearance	Where accuracy is not essential, such as in building and mining equipment
2	Free	Clearance	In rotating journals with speeds of 600 rpm or greater, such as in engines and some automotive parts
3	Medium	Clearance	In rotating journals with speeds under 600 rpm, such as in precision machine tools and precise automotive parts
4	Snug	Clearance	Where small clearance is permissible and where mating parts are not intended to move freely under load
5	Wringing	Interference	Where light tapping with a hammer is necessary to assemble the parts
6	Tight	Interference	In semipermanent assemblies suitable for drive or shrink fits on light sections
7	Medium	Interference	Where considerable pressure is required for assembly and for shrink fits of medium sections; suitable for press fits on generator and motor armatures and for automotive wheels
8	Heavy force or shrink	Interference	Where considerable bonding between surfaces is required, such as locomotive wheels and heavy crankshaft disks of large engines

TABLE 35.2

Recommended Tolerances in mm for Classes of Fit				
Class	Allowance	Interference	Hub tolerance	Shaft tolerance
1	$0.0073d^{2/3}$	—	$0.0216d^{1/3}$	$0.0216d^{1/3}$
2	$0.0041d^{2/3}$	—	$0.0112d^{1/3}$	$0.0112d^{1/3}$
3	$0.0026d^{2/3}$	—	$0.0069d^{1/3}$	$0.0069d^{1/3}$
4	0.000	—	$0.0052d^{1/3}$	$0.0035d^{1/3}$
5	—	0.000	$0.0052d^{1/3}$	$0.0035d^{1/3}$
6	—	$0.00025d$	$0.0052d^{1/3}$	$0.0052d^{1/3}$
7	—	$0.0005d$	$0.0052d^{1/3}$	$0.0052d^{1/3}$
8	—	$0.0010d$	$0.0052d^{1/3}$	$0.0052d^{1/3}$

- **Unilateral tolerancing:** Deviation from the nominal dimension, in one direction only
- **Zero line:** Reference line along the basic size from which a range of tolerances and deviations are specified

Because the dimensions of holes are more difficult to control than those of shafts, the hole-basis system is commonly used for specifying tolerances in shaft and hole assemblies. The symbols used to indicate geometric characteristics are shown in Figs. 35.22a and b.

Limits and Fits. *Limits and fits* are essential in specifying dimensions for holes and shafts (see Tables 35.1 and 35.2). There are two standards on limits and fits, as described by the American National Standards Institute (see ANSI/ASME B4.1, B4.2, and B4.3). One standard is based on the traditional inch unit; the other is based on the metric unit and has been developed in greater detail.

SUMMARY

- Modern manufacturing technology requires measuring instrumentation with several features and characteristics.
- Various devices are available for measurements, from simple gage blocks to electronic gages with high resolution. Major advances have been made in automated measurement, linking measuring devices to microprocessors and computers for accurate in-process control of manufacturing operations.
- Reliable linking, monitoring, display, distribution, and manipulation of data are important factors, as are the significant costs involved in implementing them.
- Dimensional tolerances and their selection are important factors in manufacturing. The smaller the range of specified tolerances, the higher is the cost of production; thus, tolerances should be as broad as possible but should still maintain the functional requirements of the product.

KEY TERMS

Air gage
Analog instruments
Autocollimator
Bevel protractor
Comparative length-measuring instruments
Coordinate-measuring machine
Dial indicator
Diffraction gratings
Digital instruments
Dimensional tolerance
Electronic gages
Fits
Fixed gage
Gage block
Interferometry
Laser micrometer
Limits
Line-graduated instruments
Measurement standards
Micrometer
Optical contour projector
Optical flat
Plug gage
Pneumatic gage
Precision
Resolution
Ring gage
Sensitivity
Snap gage
Tolerance
Total indicator reading
Vernier caliper

BIBLIOGRAPHY

Bentley, J.P., **Principles of Measurement Systems**, 4th ed., Prentice Hall, 2005.
Bucher, J.L. (ed.), **The Metrology Handbook**, American Society for Quality, 2004.
Cogorno, G., **Geometric Dimensioning and Tolerancing for Mechanical Design**, 2nd ed., McGraw-Hill, 2011.
Curtis, M.A., **Handbook of Dimensional Measurement**, 4th ed., Industrial Press, 2007.
Hocken, R.J., and Pereira, P.H., **Coordinate Measuring Machines and Systems**, 2nd ed., CRC Press, 2011.
Krulikowski, A., **Fundamentals of Geometric Dimensioning and Tolerancing**, 3rd ed., Delmar, 2012.

Madsen, D.A., **Geometric Dimensioning and Tolerancing,** 7th ed., Goodheart-Wilcox, 2003.

Meadows, J.D., **Geometric Dimensioning and Tolerancing Handbook,** ASME, 2009.

Whitehouse, D.J., **Handbook of Surface and Nanometrology,** 2nd ed., CRC Press, 2010.

REVIEW QUESTIONS

35.1 What is metrology?

35.2 Explain how a meter is defined and measured.

35.3 Explain what is meant by standards for measurement.

35.4 What is the basic difference between direct-reading and indirect-reading linear measurements? Name the instruments used in each category.

35.5 What is meant by comparative length measurement?

35.6 Explain how flatness is measured. What is an optical flat?

35.7 Describe the principle of an optical comparator.

35.8 Why have coordinate measuring machines become important instruments?

35.9 What is the difference between a plug gage and a ring gage?

35.10 What are dimensional tolerances? Why is their control important?

35.11 Why is a sine bar known by that name?

35.12 Explain the difference between tolerance and allowance.

35.13 What is the difference between bilateral and unilateral tolerance?

35.14 How is straightness measured?

35.15 When is a clearance fit desirable? An interference fit?

35.16 What factors contribute to deviations in the dimensions of the same type of parts made by the same machine?

QUALITATIVE PROBLEMS

35.17 Why are the words "accuracy" and "precision" often incorrectly interchanged?

35.18 Review the following results from an archery competition. Indicate which of the targets display an archer with (a) precise placement of the arrows; (b) accurate placement of the arrows.

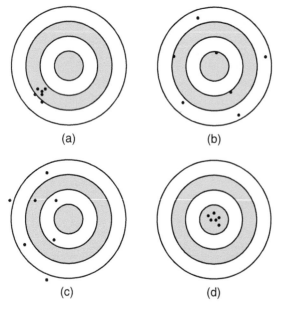

(a)   (b)

(c)   (d)

FIGURE P35.18

35.19 Why do manufacturing processes produce parts with a wide range of tolerances? Explain, giving several examples.

35.20 Explain the need for automated inspection.

35.21 Dimensional tolerances for nonmetallic parts usually are wider than for metallic parts. Explain why. Would this also be true for ceramics parts?

35.22 Comment on your observations regarding Fig. 35.21. Why does dimensional tolerance increase with increasing surface roughness?

35.23 Review Fig. 35.20, and comment on the range of tolerances and part dimensions produced by various manufacturing processes.

35.24 In the game of darts, is it better to be accurate or to be precise? Explain.

35.25 What are the advantages and limitations of *GO* and *NOT GO* gages?

35.26 Comment on your observations regarding Fig. 35.19.

35.27 What are gage blocks? Explain three methods that gage blocks can be used in metrology.

35.28 Why is it important to control temperature during the measurement of dimensions? Explain, with examples.

35.29 Describe the characteristics of electronic gages.

35.30 What method would you use to measure the thickness of a foam-rubber part? Explain.

35.31 Review Fig. 35.20 and explain why the dimensional tolerance is related to the part dimensions.

35.32 Review Fig. 35.21 and give reasons that there is a range of tolerance and surface roughness for each manufacturing process.

QUANTITATIVE PROBLEMS

35.33 Assume that a steel rule expands by 0.07% due to an increase in environmental temperature. What will be the indicated diameter of a shaft with a diameter of 1.200 in. at room temperature?

35.34 If the same steel rule as in Problem 35.33 is used to measure aluminum extrusions, what will be the indicated diameter at room temperature? What if the part were made of a thermoplastic?

35.35 A shaft must meet a design requirement of being at least 1.10 in. in diameter, but it can be 0.015 in. oversized. Express the shaft's tolerance as it would appear on an engineering drawing.

35.36 Review Table 35.2 and plot the tolerance of a hub and the shaft it mounts on as a function of diameter.

SYNTHESIS, DESIGN, AND PROJECTS

35.37 Describe your thoughts on the merits and limitations of digital measuring equipment over analog instruments. Give specific examples.

35.38 Take an ordinary vernier micrometer (see Fig. 35.2a) and a simple round rod. Ask five of your classmates to measure the diameter of the rod with this micrometer. Comment on your observations.

35.39 Obtain a digital micrometer and a steel ball of, say, $\frac{1}{4}$-in. diameter. Measure the diameter of the ball when it (a) has been placed in a freezer, (b) has been put into boiling water, and (c) when it has been held in your hand for different lengths of time. Note the variations, if any, of measured dimensions, and comment on them.

35.40 Repeat Problem 35.39, but with the following parts: (a) the plastic lid of a small jar, (b) a thermoset part such as the knob or handle from the lid of a saucepan, (c) a small juice glass, and (d) an ordinary rubber eraser.

35.41 What is the significance of the tests described in Problems 35.39 and 35.40?

35.42 Explain the relative advantages and limitations of a tactile probe versus a laser probe.

35.43 Make simple sketches of some forming and cutting machine tools (as described in Parts III and IV of the book)

and integrate them with the various types of measuring equipment described in this chapter. Comment on the possible difficulties involved in doing so.

35.44 Inspect various parts and components in consumer products, and comment on how tight dimensional tolerances have to be in order for these products to function properly.

35.45 As you know, very thin sheet-metal parts can distort differently when held from various locations and edges of the part, just as a thin paper plate or aluminum foil does. How, then, could you use a coordinate-measuring machine for "accurate" measurements? Explain.

35.46 Explain how you would justify the considerable cost of a coordinate-measuring machine such as that shown in Fig. 35.17.

35.47 Explain how you would measure the dimensions of an extruded hexagonal cross-section, and how you would indicate deviation from the hexagon's form.

35.48 Conduct an Internet search, and make a list of the method you would use for length measurement as a function of the length (*e.g.*, a few mm versus a km).

35.49 How are dimensions of MEMS devices measured?

36

Quality Assurance, Testing, and Inspection

- This chapter outlines the procedures used to ensure the manufacture of high-quality products. It describes the mathematical tools and inspection techniques that have been developed to produce high-quality products, including total quality management, Taguchi methods, and the Deming and Juran approaches.
- Statistical methods of quality control and control charts are then described, including acceptance sampling, in order to ensure that production meets quality standards.
- The chapter concludes with a discussion of the methods used in destructive and nondestructive testing of materials and products.

36.1 Introduction

Manufactured products develop certain external and internal characteristics that result, in part, from the type of production processes employed. External characteristics most commonly involve dimensions, size, and surface finish and integrity considerations. Internal characteristics include such defects as porosity, impurities, inclusions, phase transformations, residual stresses, embrittlement, cracks, and debonding of laminations in composite materials.

Some of these defects may exist in the original material (stock), while others are introduced or induced during the particular manufacturing operation. Before they are marketed, manufactured parts and products are inspected in order to

- Ensure dimensional accuracy, so that parts fit properly into other components during assembly; recall that a Boeing 747-400 aircraft has six million parts to be assembled.
- Identify products whose failure or malfunction may have serious implications, including bodily injury or even fatality. Typical examples are elevator cables, switches, brakes, grinding wheels, railroad wheels, welded joints, turbine blades, and pressure vessels.

Product quality always has been one of the most important aspects of manufacturing operations. In view of a global competitive market, *continuous improvement in quality* is a major priority; in Japan, the single term **kaizen** is used to signify *never-ending improvement*. *Quality must be built into a product* and not merely considered *after* the product already has been made. Thus, close cooperation and communication among design and manufacturing engineers, and the direct involvement and encouragement of company management are vital.

Major advances in quality engineering and productivity have been made over the years, largely because of the efforts of quality experts such as W.E. Deming, G. Taguchi, and J.M. Juran. The importance of the *quality, reliability*, and *safety* of products in a global economy is now internationally recognized, as evidenced by the establishment of various ISO and QSO standards and, nationally, by the Malcolm Baldrige National Quality Award in the United States.

36.2 Product Quality

What is quality? Unlike most technical terms, quality is difficult to define precisely; generally, it can be defined as a *product's fitness for use*. Thus, quality is a broad-based characteristic or property, and its factors consist not only of well-defined technical considerations, but also of subjective opinions. Several aspects of quality that generally are identified are: performance, durability, reliability, robustness, availability, cost, and serviceability, as well as aesthetics and perceived quality.

Consider, for example, the following: (a) The handle on a kitchen utensil is installed improperly or its handle discolors or cracks during its normal use, (b) a weighing scale functions erratically, (c) a vacuum cleaner requires frequent repairs, and (d) a machine tool cannot maintain the specified dimensional tolerances because of lack of stiffness or poor construction. These examples indicate that the product is of low quality. Thus, the general perception is that a high-quality product is one that performs its functions reliably over a long time, without breaking down or requiring repairs (see Table I.4 in the General Introduction). The level of quality that a manufacturer chooses for its products depends on the market for which the products are intended; low-quality, low-cost tools, for example, have their own market niche.

Contrary to general public perception, high-quality products do not necessarily cost more, especially considering the fact that poor-quality products

- Present difficulties in assembling and maintaining components
- Require in-field repairs (see Table I.5)
- Have the significant built-in cost of customer dissatisfaction (see Section 36.5)

As described in Section 40.9, the total product cost depends on several variables, including the level of automation in the manufacturing plant. There are many ways for engineers to review and modify overall product design and manufacturing processes, in order to minimize a product's cost without affecting its quality. Thus, quality standards are essentially a balance among several considerations; this balance is also called **return on quality** (ROQ), and usually includes some *limit* on the expected life of the product.

36.3 Quality Assurance

Quality assurance is the total effort made by a manufacturer to ensure that its products conform to a detailed set of specifications and standards. It can be defined as all actions necessary to ensure that quality requirements will be satisfied. **Quality control** is the set of operational techniques used to fulfill quality requirements.

The standards cover several types of parameters, such as dimensions, surface finish, tolerances, composition, and color, as well as mechanical, physical, and chemical properties and characteristics. In addition, standards usually are written to ensure proper *assembly*, using *interchangeable* defect-free components and resulting in a product that performs as intended by its designers.

A major aspect of quality assurance is the capability to *analyze* defects as they occur on the production line, and *promptly eliminate* or reduce them to acceptable levels. In an even broader sense, quality assurance involves *evaluating* the product and its customer satisfaction. The sum total of all these activities is referred to by terms such as as **total quality control** and **total quality management.**

In order to control quality, it is essential to be able to

- *Measure* the level of quality quantitatively
- *Identify* all of the material and process variables that can be controlled

The quality level built in during production can then be checked by *continuously* inspecting the product, in order to determine whether it meets the relevant specifications for dimensional tolerances, surface finish, defects, and various other characteristics.

36.4 Total Quality Management

Total quality management (TQM) is a *systems approach*, in that both management and employees make a concerted effort to consistently manufacture high-quality products; *defect prevention* rather than *defect detection* is the major goal.

Leadership and *teamwork* in the organization are essential to ensure that the goal of **continuous improvement** in manufacturing operations is paramount, because they *reduce product variability* and improve customer satisfaction. The TQM concept also requires *control of the processes*, and not the *control of parts produced*, so that process variability can be reduced and no defective parts are allowed to continue through the production line.

Quality Circle.   A *quality circle*, first established in Japan in 1962, consists of groups of employees (workers, supervisors, and managers) who volunteer to meet regularly to discuss how to improve and maintain product quality at *all* stages of the manufacturing operation. Worker involvement, responsibility, and creativity, as well as a team effort, are emphasized. Comprehensive *training* is provided so that the worker can become conscious of quality, and also be capable of analyzing statistical data, identifying the causes of poor quality, and taking immediate action to correct the situation. Experience has indicated that quality circles are more effective in *lean-manufacturing* environments, described in Section 39.6.

Quality Engineering as a Philosophy.   Experts in quality control have placed many of the quality-control concepts and methods into a larger perspective. Notable among these experts have been Deming, Juran, and Taguchi, whose philosophies of quality and product cost have had, and continue to have, a major impact on modern manufacturing.

36.4.1 Deming Methods

During World War II, the American statistician, W.E. Deming (1900–1993) and several others developed new methods of *statistical process control* for wartime-industry

TABLE 36.1

Deming's 14 Points

1. Create constancy of purpose toward improvement of product and service.
2. Adopt the new philosophy: refuse to accept defects.
3. Cease dependence on mass inspection to achieve quality.
4. End the practice of awarding business on the basis of price tag.
5. Improve the system of production and service constantly and forever, to improve quality and productivity and thus constantly decrease cost.
6. Institute training for the requirements of a particular task, and document the requirements for future training.
7. Institute leadership, as opposed to supervision.
8. Drive out fear so that everyone can work effectively.
9. Break down barriers between departments.
10. Eliminate slogans, exhortations, and targets for zero defects and new levels of productivity.
11. Eliminate quotas and management by numbers, or numerical goals. Substitute leadership.
12. Remove barriers that rob the hourly worker of pride of workmanship.
13. Institute a vigorous program of education and self-improvement.
14. Put everyone in the company to work to accomplish the transformation.

manufacturing plants. The methods arose from the recognition that there were *variations* (a) in the performance of machines and of people and (b) in the quality and dimensions of raw materials (stock). The efforts of these pioneers involved not only statistical methods of analysis, but also a new way of looking at manufacturing operations from the perspective of *improving quality while lowering costs*.

Deming recognized that manufacturing organizations are *systems* of management, workers, machines, and products. He placed great emphasis on communication, direct worker involvement, and education in statistics and modern manufacturing technology. His basic ideas are summarized in the well-known *14 points*, given in Table 36.1. These points are not to be seen as a checklist or menu of tasks; they are what Deming recognized as *characteristics of companies* that produce high-quality goods.

36.4.2 Juran Methods

A contemporary of Deming, J.M. Juran (1904–2008), an electrical engineer and management consultant, emphasized the importance of

- Recognizing quality at all levels of an organization, including upper management
- Fostering a responsive corporate culture
- Training all personnel in planning, control, and improvement of quality

The main concern of the top management in an organization is business and management, whereas those in quality control are basically concerned with technology. These different worlds have, in the past, often been at odds, and their conflicts have led to quality problems. Planners determine who the customers are and their needs. An organization's customers may be external (end users who purchase the product or service), or they may be internal (different parts of an organization that rely on other segments of the organization to supply them with products and services). The planners then develop product and process designs to respond to the customer's needs. The plans are turned over to those in charge of operations, who then become responsible for implementing both quality control and continued improvement in quality.

36.5   Taguchi Methods

G. Taguchi (1924–2012), an engineer and a statistician, proposed that high quality and low costs are achieved by combining engineering and statistical techniques to optimize product design and manufacturing processes. *Taguchi methods* is now a term that refers to the approaches he developed to manufacture high-quality products. One fundamental viewpoint put forward is the quality challenge facing manufacturers: Provide products that delight your customers, and to do so, manufacturers should offer products with the following product characteristics:

- High reliability
- Perform the desired functions well
- Good appearance
- Inexpensive
- Upgradeable
- Available in the quantities desired when needed
- Robust over their intended life (see Section 36.5.1)

These characteristics clearly are the goals of manufacturers striving to provide high-quality products. Although it is very challenging to actually provide all of these characteristics, excellence in manufacturing is undeniably a prerequisite.

Taguchi also contributed to the approaches that are used to document quality, recognizing that any deviation from the optimum state of a product represents a financial loss, because of such factors as reduced product life, performance, and economy. *Loss of quality* is defined as the *financial loss* to society after the product is shipped. Loss of quality results in the following problems:

- Poor quality leads to customer dissatisfaction
- Costs are incurred in servicing and repairing defective products, especially when such repairs have to be made in the field
- The manufacturer's credibility in the marketplace is diminished
- The manufacturer eventually loses its share of the market

The Taguchi methods of **quality engineering** emphasize the importance of

- **Enhancing cross-functional team interaction:** Design engineers and manufacturing engineers communicate with each other in a common language. They quantify the relationships between design requirements and manufacturing process selection.
- **Implementing experimental design:** The factors involved in a process or operation and their interactions are studied simultaneously.

In *experimental design*, the effects of controllable and uncontrollable variables on the product are identified. This approach minimizes variations in product dimensions and properties and, ultimately, brings the mean to the desired level. The methods used for experimental design are complex, and involve using *factorial design* and *orthogonal arrays*, both of which reduce the number of experiments required. These methods also are capable of identifying the effects of variables that cannot be controlled (called *noise*), such as changes in environmental conditions in a plant.

The use of factorial design and orthogonal arrays results in (a) the rapid *identification* of the controlling variables, referred to as *observing main effects*, and (b) the ability to determine the best method of process control. Control of these variables sometimes requires new equipment or major modifications to existing equipment. Thus, for example, variables affecting dimensional tolerances in machining a particular component can readily be identified and, whenever possible, the correct cutting speed, feed, cutting tool, and cutting fluids can be specified.

36.5.1   Robustness

Another aspect of quality, originally suggested by Taguchi, is *robustness*: A robust design, process, or system is one that continues to function, within acceptable parameters, despite variabilities (often unanticipated) in its environment. In other words, its outputs (such as its performance) have *minimal sensitivity* to its input variations (such as variations in environment, load, and power source).

In a robust design, for example, a part will function sufficiently well even if the loads applied, or their directions, exceed anticipated values. Likewise, a robust machine or a system will undergo minimal deterioration in performance even if it experiences variations in environmental conditions, such as temperature, humidity, air quality, and vibrations. A robust machine will also have no significant reduction in its performance over its life, whereas a less robust design will perform less efficiently as time passes.

As a simple illustration of a robust design, consider a sheet-metal mounting bracket to be attached to a wall with two bolts (Fig. 36.1a). The positioning of the two mounting holes on the bracket will include some error due to the manufacturing process involved; this error will prevent the top edge of the bracket from being perfectly horizontal.

A more robust design is shown in Fig. 36.1b, in which the mounting holes have been moved twice as far apart as in the original design. Even though the precision of hole location remains the same, and the manufacturing cost also is the same, the variability in the top edge of the bracket (from the horizontal) has now been reduced by one-half. If the bracket is subjected to vibration, however, the bolts may loosen over time. An even more robust design approach would then be to use an adhesive to hold the bolt threads in place, or to use a different type of fastener that would not loosen over time (see also Section 32.5).

36.5.2   Taguchi Loss Function

The *Taguchi loss function*, introduced in the early 1980s, is a tool for comparing quality on the basis of minimizing variations. It calculates the increasing loss to the company when the component deviates from the design objective. This function is defined as a parabola where one point is the cost of replacement (including shipping,

Video Solution 36.1 Taguchi Loss Function

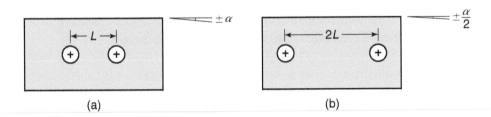

(a)   (b)

FIGURE 36.1 A simple example of robust design. (a) Location of two mounting holes on a sheet-metal bracket, where the deviation keeping the top surface of the bracket from being perfectly horizontal is $\pm \alpha$. (b) New locations of the two holes; the deviation keeping the top surface of the bracket from being perfectly horizontal is now reduced to $\pm \alpha/2$.

scrapping, and handling costs) at an extreme of the tolerances, while a second point corresponds to zero loss at the design objective.

Mathematically, the loss cost can be written as

$$\text{Loss cost} = k \left[(Y - T)^2 + \sigma^2 \right], \tag{36.1}$$

where Y is the mean value from manufacturing, T is the target value from design, σ is the standard deviation of parts from manufacturing (see Section 36.7), and k is a constant, defined as

$$k = \frac{\text{Replacement cost}}{(\text{LSL} - T)^2}, \tag{36.2}$$

where LSL is the lower specification limit. When the lower (LSL) and upper (USL) specification limits are the same distance from the mean (i.e., the tolerances are balanced), either of the limits can be used in this equation.

EXAMPLE 36.1   Production of Polymer Tubing

Given: High-quality polymer tubes are being produced for medical applications in which the target wall thickness is 2.6 mm, a USL is 3.2 mm, and an LSL is 2.0 mm (2.6 ± 0.6 mm). If the units are defective, they are replaced at a shipping-included cost of $10.00. The current process produces parts with a mean of 2.6 mm and a standard deviation of 0.2 mm. The current volume is 10,000 sections of tube per month. An improvement is being considered for the extruder heating system. This improvement will cut the variation in half, but it costs $50,000.

Find: Determine the Taguchi loss function and the payback period for the investment.

Solution: The quantities involved are: USL = 3.2 mm, LSL = 2.0 mm, $T = 2.6$ mm, $\sigma = 0.2$ mm, and $Y = 2.6$ mm.

The quantity k is given by Eq. (36.2) as

$$k = \frac{\$10.00}{(3.2 - 2.6)^2} = \$27.28.$$

The loss cost before the improvement is, from Eq. (36.1),

$$\text{Loss cost} = (27.78) \left[(2.6 - 2.6)^2 + 0.2^2 \right]$$
$$= \$1.11 \text{ per unit.}$$

After the improvement, the standard deviation is 0.1 mm; thus, the loss cost is

$$\text{Loss cost} = (27.78) \left[(2.6 - 2.6)^2 + 0.1^2 \right]$$
$$= \$0.28 \text{ per unit.}$$

The savings are then ($1.11−$0.28)(10,000)= $8300 per month. Thus, the payback period for the investment is $50,000/($8300/month) = 6.02 months.

CASE STUDY 36.1   Manufacture of Television Sets by Sony Corporation

Sony Corporation executives found a confusing situation in the mid-1980s. Television sets manufactured in Japanese production facilities sold faster than those produced in a San Diego, CA facility, even though they were produced from identical designs. There were no identifications to distinguish the sets made in Japan from those made in the United States, and there was no apparent reason for this discrepancy. However, investigations revealed that the sets produced in Japan were superior to the U.S. versions, as color sharpness was better and hues were more brilliant. Since they were on display in stores, consumers could easily detect and purchase the model that had the best picture.

Although the difference in picture quality was obvious, the reasons for the difference were not clear. A further point of confusion was the constant assurance that the San Diego facility had a total quality program in place, and that the plant was maintaining quality-control standards so that no defective parts were produced. Although the Japanese facility did not have a total quality program, there was an emphasis on reducing variations from part to part.

Further investigations found a typical pattern in an integrated circuit that was critical in affecting color density. The distribution of parts meeting the color-design objective is shown in Fig. 36.2a; the Taguchi loss function for these parts is shown in Fig. 36.2b. In the San Diego facility, where the number of defective parts was minimized (to zero in this case), a uniform distribution within the specification limits was achieved.

The Japanese facility actually produced parts outside of the design specification, but the standard deviation about the mean was lower. Using the Taguchi loss-function approach (see Example 36.1) made it clear that the San Diego facility lost about $1.33 per unit while the Japanese facility lost $0.44 per unit.

Traditional quality viewpoints would find a uniform distribution without defects to be superior to a distribution in which a few defects are produced but the majority of parts are closer to the design target values. Consumers, however, can readily detect which product is superior, and the marketplace proves that minimizing deviations is indeed a worthwhile quality goal.

Source: Based on D.M. Byrne and G. Taguchi.

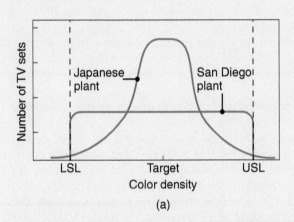

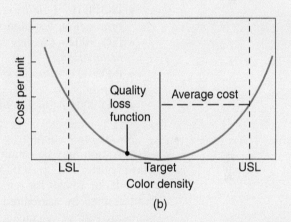

(a)   (b)

FIGURE 36.2 (a) Objective-function value distribution of color density for television sets. (b) Taguchi loss function, showing the average replacement cost per unit to correct quality problems. *Source:* Based on G. Taguchi.

EXAMPLE 36.2   Increasing Quality without Increasing the Cost of a Product

A manufacturer of clay tiles noticed that excessive scrap was being produced because of temperature variations in the kiln used to fire the tiles, thus adversely affecting the company's profits. The first solution the manufacturer considered was purchasing new kilns with better temperature controls; however, this solution would require a major capital investment. A study was then undertaken to determine whether modifications could be made in the composition of the clay so that it would

be less sensitive to temperature fluctuations during firing.

On the basis of factorial experiment design, in which the factors involved in a process and their interactions are studied simultaneously, it was found that increasing the lime content of the clay made the tiles less sensitive to temperature variations during firing. This modification (which was also the low-cost alternative) was implemented, reducing scrap substantially and improving tile quality.

36.6   The ISO and QS Standards

Customers worldwide are increasingly demanding high-quality products and services at low prices, and are looking for suppliers that can respond to this demand consistently and reliably. This trend in the global marketplace has, in turn, created the need for international conformity and consensus regarding the establishment of methods for quality control, reliability, and safety of products. In addition to these considerations, there are equally important concerns regarding the environment and quality of life that continue to be addressed.

36.6.1   The ISO 9000 Standard

First published in 1987 and revised in 1994, the ISO 9000 standard (**Quality Management and Quality Assurance Standards**) is a deliberately generic series of quality system-management standards. This standard has permanently influenced the manner in which manufacturing companies conduct business in world trade, and has become the world standard for quality.

The ISO 9000 series includes the following standards:

- **ISO 9001**—*Quality systems: Model for quality assurance in design/development, production, installation, and servicing*
- **ISO 9002**—*Quality systems: Model for quality assurance in production and installation*
- **ISO 9003**—*Quality systems: Model for quality assurance in final inspection and testing*
- **ISO 9004**—*Quality management and quality system elements: Guidelines*

Companies voluntarily register for these standards and are issued certificates. Registration may be sought generally for ISO 9001 or 9002, and some companies have registration up to ISO 9003. The 9004 standard is simply a guideline and not a model or a basis for registration. For certification, a company's plants are visited and audited by accredited and independent third-party teams, to certify that the standard's 20 key elements are in place and are functioning properly.

Depending on the extent to which a company fails to meet the requirements of the standard, registration may or may not be recommended at that time. The audit team does not advise or consult with the company on how to fix discrepancies, but merely describes the nature of the noncompliance. Periodic audits are required to maintain certification. The certification process can take from six months to a year or more and can cost tens of thousands of dollars, depending on the company's size, number of plants, and product line.

The ISO 9000 standard is not a product certification, but a **quality process certification**. Companies establish their own criteria and practices for quality. However, the documented quality system must be in compliance with the ISO 9000 standard; thus, a company cannot write into the system any criterion that opposes the intent of the standard. Registration symbolizes a company's commitment to conform to consistent practices, as specified by the company's own quality system (such as quality in design, development, production, installation, and servicing), including proper documentation of such practices. In this way, customers (including government agencies) are assured that the supplier of the product or service (which may or may not be within the same country) is following specified practices. In fact, manufacturing companies are themselves assured of such practices regarding their own suppliers that have ISO 9000 registration; thus, suppliers also must be registered.

36.6.2   The QS 9000 Standard

Jointly developed by Chrysler, Ford, and General Motors, the QS 9000 standard was first published in 1994. Prior to its development, each of these automotive companies had its own standard for quality system requirements. The ISO/TS 16949 standard superseded QS 9000 in 2002, and is intended to be applied across an entire supply chain. Tier I automotive suppliers have been required to obtain third-party registration to the standard.

36.6.3   The ISO 14000 Standard

ISO 14000 is a family of standards first published in 1996 and pertaining to international **environmental management systems** (EMS). It concerns the way an organization's activities affect the environment throughout the life of its products (see also Section I.6 in the General Introduction). These activities (a) may be internal or external to the organization, (b) range from production to ultimate disposal of the product after its useful life, and (c) include effects on the environment, such as pollution, waste generation and disposal, noise, depletion of natural resources, and energy use.

The ISO 14000 family of standards has several sections: Guidelines for Environmental Auditing, Environmental Assessment, Environmental Labels and Declarations, and Environmental Management. ISO 14001, *Environmental Management System Requirements,* consists of sections titled General Requirements, Environmental Policy, Planning, Implementation and Operation, Checking and Corrective Action, and Management Review.

36.7   Statistical Methods of Quality Control

Because of the numerous variables involved in manufacturing processes and operations, the implementation of *statistical methods of quality control* is essential. Some of the more commonly observed variables in manufacturing are:

- Cutting tools, dies, and molds undergo wear, thus part dimensions and surface characteristics vary over time.
- Machinery performs differently depending on its quality, age, condition, and level of maintenance; older machines tend to chatter and vibrate, can be difficult to adjust, and do not maintain tolerances.
- The effectiveness of metalworking fluids declines as they degrade; thus, tool and die life, surface finish and surface integrity of the workpiece, and forces and energy requirements are adversely affected.
- Environmental conditions, such as temperature, humidity, and air quality in the plant, may change from one hour to the next, affecting the performance of machines and workers.
- Different shipments, and at different times, of raw materials to a plant may have significantly different dimensions, properties, surface characteristics, and overall quality.
- Operator attention may vary during the day or from operator to operator.

Those events that occur *randomly*—that is, without any particular trend or pattern—are called **chance variations** or **special causes**; those that can be traced to *specific causes* are called **assignable variations** or **common causes.**

Although the existence of **variability** in production operations had been recognized for centuries, it was E. Whitney (1765–1825), an American inventor and arms manufacturer, who first understood its full significance, when he observed that *interchangeable parts* were indispensable to the mass production of firearms. Modern statistical concepts relevant to manufacturing engineering were first developed in the early 1900s, notably through the work of W.A. Shewhart (1891–1967), a physicist, engineer, and statistician.

36.7.1   Statistical Quality Control

Statistical quality control (SQC) involves the use of probability theory along with the testing of random subsets of parts produced to obtain an understanding of quality. To properly use statistical quality control, the following commonly used terms must first be defined:

- **Sample size:** The number of parts to be inspected in a sample. The properties of the parts in the sample are studied to gain information about the whole population.
- **Random sampling:** Taking a sample from a population or lot in which each item has an equal chance of being included in the sample.
- **Population:** The total number of individual parts of the same design from which samples are taken; also called the **universe.**
- **Lot size:** The size of a subset of the population. One or more lots can be considered subsets of the population and may also be considered as representative of the population.

The sample is inspected for several characteristics and features, such as tolerances, surface finish, and defects, using the instruments and techniques described in Chapter 35 and in Sections 36.10 and 36.11. These characteristics fall into two categories: (a) those which are measured quantitatively (*method of variables*) and (b) those which are measured qualitatively (*method of attributes*).

1. The **method of variables** is the *quantitative measurement* of the part's characteristics, such as dimensions, tolerances, surface finish, and physical or mechanical properties. The measurements are made for each of the units in the group under consideration, and the results are then compared against specifications.
2. The **method of attributes** involves observing the presence or absence of *qualitative characteristics* (such as external or internal defects in machined, formed, or welded parts, and dents in sheet-metal parts) in each of the units in the group under consideration. The sample size for attributes-type data generally is larger than that for variables-type data.

Consider measuring the diameters of machined shafts produced on a lathe. For a variety of reasons, described in this chapter, the diameters will vary. When the measured diameters of the turned shafts in a given population are listed, one or more parts will have the smallest diameter and one or more will have the largest diameter. The rest of the turned shafts will have diameters that lie between these two extremes.

All the diameter measurements can be grouped and plotted in a bar graph, called a *histogram*, representing the number of parts in each diameter group (Fig. 36.3a). The bars show a **distribution**, also called a **spread** or **dispersion** of the diameter

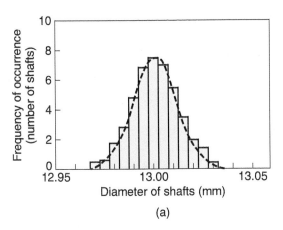

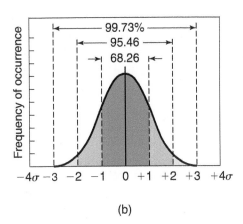

(a)   (b)

FIGURE 36.3   (a) A histogram of the number of shafts measured and their respective diameters; this type of curve is called frequency distribution. (b) A normal distribution curve indicating areas within each range of standard deviation; note that the greater the range, the higher is the percentage of parts that fall within it.

measurements. The *bell-shaped curve* in Fig. 36.3a is called **frequency distribution**, and shows the frequencies with which parts of each diameter are being produced.

Data from manufacturing processes often fit curves represented by a mathematically derived **normal-distribution curve** (Fig. 36.3b), called *Gaussian*, after K.F. Gauss (1777–1855), a German mathematician and physical scientist, who developed it on the basis of *probability*. The bell-shaped normal distribution curve fitted to the data shown in Fig. 36.3a has two features. First, it shows that most part diameters tend to cluster around an *average value* (**arithmetic mean**), and is designated as $\bar{x}$ and calculated from the expression

$$\bar{x} = \frac{x_1 + x_2 + x_3 + \cdots + x_n}{n}, \qquad (36.3)$$

where the numerator is the sum of all of the measured values (shaft diameters) and n is the number of measurements (number of shafts).

The second feature of this curve is its width, indicating the **dispersion** of the diameters measured; the wider the curve, the greater is the dispersion. The difference between the largest value and the smallest value is called the **range**, R:

$$R = x_{max} - x_{min}. \qquad (36.4)$$

The dispersion is estimated by the **standard deviation**, given by the expression

$$\sigma = \frac{\sqrt{(x_1 - \bar{x})^2 + (x_2 - \bar{x})^2 + \cdots + (x_n - \bar{x})^2}}{n - 1}, \qquad (36.5)$$

where x_i is the measured value for each part.

Note from the numerator in Eq. (36.5) that, as the curve widens, the standard deviation becomes greater, and that σ has the same units as x_i. Since the number of turned parts that fall within each group is known, the percentage of the total population represented by each group can be calculated. Thus, Fig. 36.3b shows that in the measurement of shaft diameters,

- 99.73% of the population falls within the range $\pm 3\sigma$,
- 95.46% within $\pm 2\sigma$, and
- 68.26% within $\pm 1\sigma$.

These quantities are only valid for distributions that are normal, as shown in Fig. 36.3, and are not skewed. It will be noted that only 0.27% fall outside the $\pm 3\sigma$ range, which means that there will be 2700 defective parts per one million parts produced. In modern manufacturing, that is not an acceptable rate, because at this level of defects no modern computer would function reliably.

36.7.2   Six Sigma

Six sigma is a set of statistical tools based on total quality management principles of *continually* measuring the quality of products and services. Although six sigma indicates 3.4 defective parts per million, it includes considerations such as understanding *process capabilities* (described in Section 36.8.2), delivering defect-free products, and thus ensuring customer satisfaction. This approach consists of a clear focus on (a) defining quality problems, (b) measuring relevant quantities, and (c) analyzing, controlling, and improving processes and operations.

As stated in Section 36.7.1, three sigma would result in 0.27% defective parts, an unacceptable rate in modern manufacturing. Also, in the service industries, at this rate 270 million incorrect credit-card transactions would be recorded each year in the United States alone. It has further been estimated that companies operating at three- to four-sigma levels would lose about 10–15% of their total revenue due to defects. Extensive efforts continue to be made to eliminate virtually all defects in products, processes, and services, resulting in savings estimated to be billions of dollars. Because of its major impact on business, six sigma is now widely recognized as a good management philosophy.

36.8   Statistical Process Control

If the number of parts that do not meet set standards begin to increase during a production run, it is essential to determine the cause (such as incoming materials, machine controls, degradation of metalworking fluids, operator boredom, or various other factors) and take appropriate action. Although this statement at first appears to be self-evident, it was only in the early 1950s that a systematic statistical approach was developed in order to guide operators in manufacturing plants.

The statistical approach advises the operator to take certain measures and actions, and tells the operator when to take them to avoid producing further defective parts. Known as *statistical process control* (SPC), this technique consists of the following elements:

- Control charts and control limits
- Capabilities of the particular manufacturing process
- Characteristics of the machinery involved

36.8.1   Shewhart Control Charts

The frequency distribution curve shown in Fig. 36.3b indicates a range of shaft diameters being produced that may fall beyond the design tolerance range. The same bell-shaped curve is shown in Fig. 36.4 but now includes the *specified tolerances* for the diameter of the turned shafts.

Video Solution 36.2 Shewhart Control Charts

Control charts graphically represent the variations of a process over time; they consist of data taken and plotted *during* production. Typically, there are two plots. The quantity $\bar{x}$ (Fig. 36.5a) is the average for each subset of samples taken and inspected—say, each subset consists of five parts. A sample size of between 2 and 10 parts is sufficiently accurate (although more parts are better), provided that the sample size is held constant throughout the inspection.

The frequency of sampling depends on the nature of the process; some processes may require continual sampling, whereas others may require only one sample per day. Quality-control analysts are best qualified to determine this frequency for a particular operation. Since the measurements in Fig. 36.5a are made consecutively, the abscissa of the control charts also represents time.

The solid horizontal line in this figure is the **average of averages (grand average)**, denoted as $\bar{\bar{x}}$, and represents the population mean. The upper and lower horizontal broken lines indicate the **control limits** for the process. The control limits are set on these charts according to statistical-control formulas designed to keep actual production within acceptable levels of variation. One common approach is to ensure that all parts are within three standard deviations of the mean ($\pm 3\sigma$).

The standard deviation also can be expressed as a function of range. Thus, for $\bar{x}$,

Upper control limit $(\text{UCL}_{\bar{x}})$

$$= \bar{x} + 3\sigma = \bar{\bar{x}} + A_2\bar{R} \quad (36.6)$$

and

Lower control limit $(\text{LCL}_{\bar{x}})$

$$= \bar{x} - 3\sigma = \bar{\bar{x}} - A_2\bar{R}, \quad (36.7)$$

where A_2 is obtained from Table 36.2 and $\bar{R}$ is the average of R values. The quantities $\bar{\bar{x}}$ and $\bar{R}$ are estimated from the measurements taken.

The control limits are calculated on the basis of the past production capability of the equipment itself, and are not associated with either design tolerance specifications or dimensions. They indicate the limits within which a certain percentage of measured values normally are expected to fall, because of the inherent variations of the process itself and upon which the limits are based. The major goal of statistical process control is to improve the manufacturing process with the aid of control charts so as to eliminate assignable causes. The control chart continually indicates progress in this area.

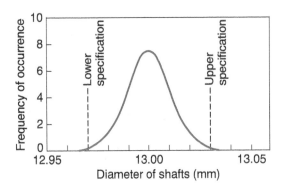

FIGURE 36.4 Frequency distribution curve, showing lower and upper specification limits.

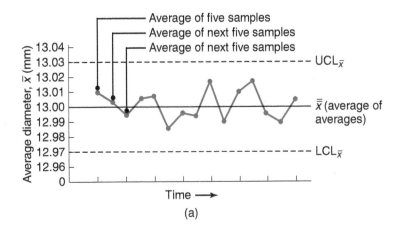

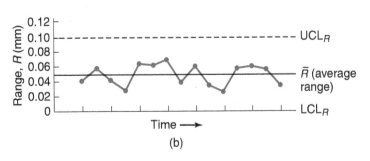

FIGURE 36.5 Control charts used in statistical quality control; the process is in good statistical control because all points fall within the lower and upper control limits. In this illustration, the sample size is 5 and the number of samples is 15.

TABLE 36.2

Constants for Control Charts

Sample size	A_2	D_4	D_3	d_2
2	1.880	3.267	0	1.128
3	1.023	2.575	0	1.693
4	0.729	2.282	0	2.059
5	0.577	2.115	0	2.326
6	0.483	2.004	0	2.534
7	0.419	1.924	0.078	2.704
8	0.373	1.864	0.136	2.847
9	0.337	1.816	0.184	2.970
10	0.308	1.777	0.223	3.078
12	0.266	1.716	0.284	3.258
15	0.223	1.652	0.348	3.472
20	0.180	1.586	0.414	3.735

The second control chart, shown in Fig. 36.5b, indicates the range, R, in each subset of samples. The solid horizontal line represents the average of R values in the lot, denoted as $\bar{R}$, and is a measure of the variability of the samples. The upper and lower control limits for R are obtained from the equations:

$$UCL_R = D_4\bar{R} \tag{36.8}$$

and

$$LCL_R = D_3\bar{R}, \tag{36.9}$$

where the constants D_4 and D_3 take on the values given in Table 36.2. The table also includes the constant d_2, which is used to estimate the standard deviation of the process distribution shown in Fig. 36.4 from the equation

$$\sigma = \frac{\bar{R}}{d_2}. \tag{36.10}$$

When the curve of a control chart is like the one shown in Fig. 36.5a, it is said that the process is in *good statistical control*, meaning that

- There is no discernible trend in the pattern of the curve
- The points (measured values) are random with time
- The points do not exceed the control limits

It can be seen that in curves such as those in Figs. 36.6a–c, there are certain *trends*. Note, for example, that in the middle of the curve in Fig. 36.6a, the diameter of the shafts is increasing with time, a reason for which may be a change in one of the process variables, such as wear of the cutting tool.

If the trend is consistently toward large diameters, as in the curve in Fig. 36.6b, with diameters hovering around the upper control limit, it could mean that the tool settings on the lathe may be incorrect and, as a result, the parts being turned are consistently too large. The curve in Fig. 36.6c shows two distinct trends, that may be due to such factors as a change in the properties of the incoming material or a change in the performance of the cutting fluid (e.g., its degradation). These situations place the process *out of control*; warning limits to this effect are sometimes set at $\pm 2\sigma$.

Analyzing patterns and trends in control charts requires considerable experience so that one may identify the specific cause(s) of an out-of-control situation. Among such causes may be one or more of those variables listed at the beginning of Section 36.7. *Overcontrol* of the manufacturing process (i.e., setting upper and lower control limits too close to each other, resulting in a smaller standard-deviation range) is another cause of out-of-control situations, and is the reason why control limits are calculated on the basis of *process variability* rather than on potentially inapplicable criteria.

It is evident that operator training is critical for the successful implementation of SPC on the shop floor. Once process control guidelines are established, operators also should have some responsibility for making adjustments in processes that are beginning to become out of control. The capabilities of individual operators also should be taken into account so that they are not overloaded with data input and thus fail to interpret the data properly.

This task is now greatly simplified through dedicated software. For example, digital readouts on electronic measuring devices are now directly integrated into a computer system for real-time SPC. Figure 35.2 shows such a multifunctional computer system in which the output from a digital caliper or micrometer is analyzed

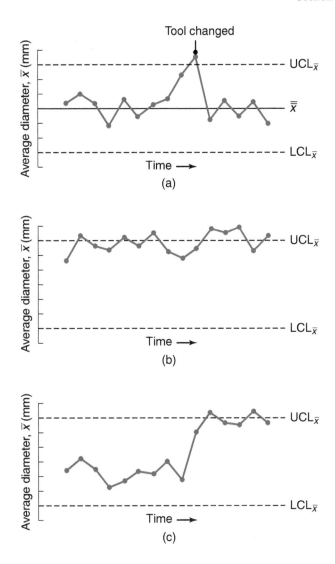

FIGURE 36.6   Control charts. (a) Process begins to become out of control because of such factors as tool wear (*drift*); the tool is changed and the process is now in statistical control. (b) Process parameters are not set properly; thus, all parts are around the upper control limit (*shift in mean*). (c) Process becomes out of control, because of factors such as a change in the properties of the incoming material (*shift in mean*).

by a microprocessor, in real time, and is displayed in several ways, such as frequency distribution curves and control charts.

36.8.2   Process Capability

Process capability is defined as the ability of a process to produce defect-free parts in controlled production. It indicates that a particular manufacturing process can produce parts consistently and repeatedly within specific limits of precision (Fig. 36.7). Various indices are used to determine process capability, describing the relationship between the variability of a process and the spread of lower and upper specification limits. Since a manufacturing process typically involves materials, machinery, and operators, each factor can be analyzed individually to identify a problem when process capabilities do not meet specification limits.

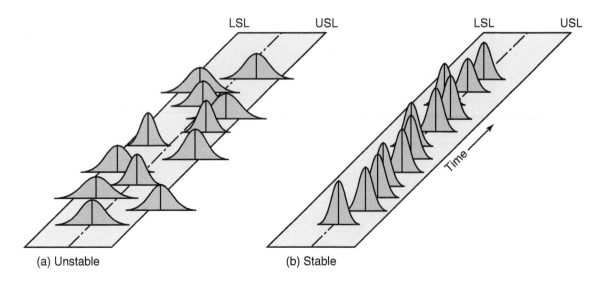

FIGURE 36.7 Illustration of processes that are (a) unstable or out of control and (b) stable or in control. Note in part (b) that all distributions have standard deviations that are lower than those of the distributions shown in part (a) and have means that are closer to the desired value. *Source:* Based on K. Crow.

EXAMPLE 36.3   Calculation of Control Limits and Standard Deviation

Given: The data given in Table 36.3 show length measurements (in in.) taken on a machined workpiece. The sample size is 5, and the number of samples is 10; thus, the total number of parts measured is 50. The quantity $\bar{x}$ is the average of five measurements in each sample.

Find: Determine the upper and lower control limits and the standard deviation for the population of machined parts.

Solution: The average of averages, $\bar{\bar{x}}$, is

$$\bar{\bar{x}} = \frac{44.296}{10} = 4.430 \text{ in.}$$

The average of the R values is

$$\bar{R} = \frac{1.03}{10} = 0.103 \text{ in.}$$

Since the sample size is five, the following constants can be determined from Table 36.2: $A_2 = 0.577$, $D_4 = 2.115$, and $D_3 = 0$. The control limits can now be calculated from Eqs. (36.4)–(36.7).

TABLE 36.3

Data for Example 36.3

Sample number	x_1	x_2	x_3	x_4	x_5	$\bar{x}$	R
1	4.46	4.40	4.44	4.46	4.43	4.438	0.06
2	4.45	4.43	4.47	4.39	4.40	4.428	0.08
3	4.38	4.48	4.42	4.42	4.35	4.410	0.13
4	4.42	4.44	4.53	4.49	4.35	4.446	0.18
5	4.42	4.45	4.43	4.44	4.41	4.430	0.04
6	4.44	4.45	4.44	4.39	4.40	4.424	0.06
7	4.39	4.41	4.42	4.46	4.47	4.430	0.08
8	4.45	4.41	4.43	4.41	4.50	4.440	0.09
9	4.44	4.46	4.30	4.38	4.49	4.414	0.19
10	4.42	4.43	4.37	4.47	4.49	4.436	0.12

Thus, for averages,

$$\text{UCL}_{\bar{x}} = 4.430 + (0.577)(0.103)$$

$$= 4.489 \text{ in.}$$

and

$$\text{LCL}_{\bar{x}} = 4.430 - (0.577)(0.103)$$

$$= 4.371 \text{ in.}$$

For ranges,

$$\text{UCL}_R = (2.115)(0.103) = 0.218 \text{ in.}$$

and

$$\text{LCL}_R = (0)(0.103) = 0 \text{ in.}$$

From Eq. (36.10), the standard deviation, σ, for the population, for a value of $d_2 = 2.326$, can be estimated as

$$\sigma = \frac{0.103}{2.326} = 0.044 \text{ in.}$$

36.8.3 Acceptance Sampling and Control

Acceptance sampling consists of taking a few random samples from a lot and inspecting them to judge whether the entire lot is acceptable or whether it should be rejected or reworked. Developed in the 1920s and used extensively during World War II for military hardware (MIL STD 105), this statistical technique is used widely. Acceptance sampling is particularly useful for inspecting high-production-rate parts when 100% inspection would be too costly. There are certain critical devices, such as pacemakers and prosthetic devices, for example, that must be subjected to 100% inspection.

A number of acceptance sampling plans have been prepared, for both military and national standards, on the basis of an acceptable, predetermined, and limiting percentage of nonconforming parts in the sample. If this percentage is exceeded, the entire lot is rejected or it is reworked, if economically feasible. Note that the actual number of samples, but not the percentages of the lot that are in the sample, can be significant in acceptance sampling.

The greater the number of samples taken from a lot, the greater is the chance that the sample will contain nonconforming parts, and the lower is the probability of the lot's acceptance. **Probability** is defined as the relative occurrence of an event. The probability of acceptance is obtained from various operating characteristics curves, one example of which is shown in Fig. 36.8.

The **acceptance quality level** (AQL) is commonly defined as the level at which there is a 95% acceptance probability for the lot. This percentage indicates to the manufacturer that 5% of the parts in the lot may be rejected by the consumer (**producer's risk**); likewise, the consumer knows that 95% of the parts are acceptable (**consumer's risk**).

The manufacturer can salvage those lots that do not meet the desired quality standards through a secondary rectifying inspection. In this method, a 100% inspection is made of the rejected lot, and the defective parts are removed. The process is time consuming and costly, and is an important incentive for the manufacturer to better control its production processes.

Acceptance sampling requires less time and fewer inspections than do other sampling methods. Consequently, inspection of the parts can be more detailed. Automated inspection techniques (Section 36.12) have been developed so that 100% inspection of all parts is indeed possible and inspection can also be economical.

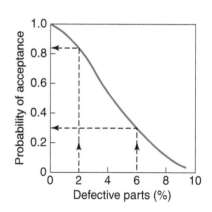

FIGURE 36.8 A typical operating characteristics curve used in acceptance sampling; the higher the percentage of defective parts, the lower is the probability of acceptance by the consumer.

36.9 Reliability of Products and Processes

All products eventually fail in some manner or other: Automobile tires become worn, electric motors burn out, water heaters begin to leak, dies and cutting tools wear out or break, and machinery stops functioning properly. **Product reliability** is generally defined as the probability that a product will perform its intended function, and without failure, in a given environment for a specified period of time while in normal use by the customer.

The more critical the application of a particular product, the higher its reliability must be. Thus, the reliability of an aircraft jet engine, a medical instrument, or an elevator cable, for example, must be much higher than that of a kitchen faucet or a mechanical pencil. From the topics described in this chapter, it can be noted that, as the quality of each component of a product increases, so, too, does the reliability of the whole product.

Predicting reliability involves complex mathematical relationships and calculations. The importance of predicting the reliability of the critical components of civilian or military aircraft is obvious. The reliability of an automated and computer-controlled high-speed production line, with all of its complex mechanical and electronic components, also is important, as its failure can result in major economic losses to the manufacturer.

Process reliability can be defined as the capability of a particular manufacturing process to operate predictably and smoothly over time. It is implicit that there must be no significant deterioration in performance, which otherwise would require downtime on machines, interrupt production, and result in major economic loss.

36.10 Nondestructive Testing

Nondestructive testing (NDT) is carried out in such a manner that product integrity and surface texture remain unchanged. The techniques employed generally require considerable operator skill and interpreting test results accurately, which may be a difficult task. The extensive use of computer graphics and other enhancement techniques, however, has significantly reduced the likelihood of human error. Current systems have various capabilities for data acquisition and for qualitative and quantitative inspection and analysis.

Liquid Penetrants.   In this technique, fluids are applied to the surfaces of the part and allowed to penetrate into cracks, seams, and pores (Fig. 36.9). By capillary action, the penetrant can seep into cracks as small as 0.1 μm (4 μin.) in width. Two common types of liquids used for this test are (a) *fluorescent penetrants*, with various sensitivities and which fluoresce under ultraviolet light and (b) *visible penetrants*, using dyes (usually red) that appear as bright outlines on the workpiece surface.

The liquid-penetrants method can be used to detect a variety of surface defects. The equipment is simple and easy to use, can be portable, and is less costly to operate than those of other methods. However, the method can detect only defects that are open to the surface or are external.

Magnetic-particle Inspection.   This technique consists of placing fine ferromagnetic particles on the surface of the part. The particles can be applied either dry or in a liquid carrier, such as water or oil. When the part is magnetized with a magnetic field, a discontinuity (defect) on the surface causes the particles to gather visibly

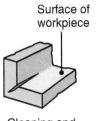
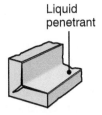

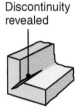

Surface of workpiece   Liquid penetrant   Developing agent   Discontinuity revealed

1. Cleaning and drying of surface

2. Application of liquid penetrant to surface

3. Water-wash removal of liquid penetrant from surface but not the defect

4. Applying developing agent

5. Inspection

FIGURE 36.9   Sequence of operations for liquid-penetrant inspection to detect the presence of cracks and other surface flaws in a workpiece. *Source:* Reprinted with permission of ASM International. All rights reserved. www.asminternational.org.

around the defect (Fig. 36.10). The ferromagnetic particles may be colored with pigments for better visibility on metal surfaces.

The defect then becomes a magnet, due to flux leakages where the magnetic-field lines are interrupted by the defect; this, in turn, creates a small-scale *N–S* pole at either side of the defect as field lines exit the surface. The particles generally take the shape and size of the defect. Subsurface defects also can be detected by this method, provided that they are not too deep. The magnetic fields can be generated with either DC or AC, and with yokes, bars, and coils. Subsurface defects can be detected best with DC. The magnetic-particle method can also be used on pure ferromagnetic materials, but the parts have to be demagnetized and cleaned after inspection. The equipment may be portable or stationary.

Ultrasonic Inspection.   In this technique, an ultrasonic beam travels through the part; an internal defect (such as a crack) interrupts the beam and reflects back a portion of the ultrasonic energy. The amplitude of the energy reflected and the time required for its return indicate the presence and location of any flaws in the workpiece.

The ultrasonic waves are generated by transducers, called *search units* or *probes*, available in various types and shapes. Transducers operate on the principle of *piezoelectricity* (Section 3.7), using materials such as quartz, lithium sulfate, or various ceramics. Most inspections are carried out at a frequency of 1–25 MHz. Couplants, such as water, oil, glycerin, and grease, are used to transmit the ultrasonic waves from the transducer to the test piece. The ultrasonic-inspection method has high penetrating power and sensitivity. It can also be used from various directions to inspect flaws in large parts, such as railroad wheels, pressure vessels, and die blocks. The method requires experienced personnel to properly conduct the inspection and to correctly interpret the results.

Acoustic Methods.   The **acoustic-emission technique** (see also Section 21.5.4) detects signals (high-frequency stress waves) generated by the workpiece itself during plastic deformation, crack initiation and propagation, phase transformation, and abrupt reorientation of grain boundaries. Bubble formation during the boiling of a liquid and friction and wear of sliding interfaces are other sources of acoustic signals.

Acoustic-emission inspection is usually performed by elastically stressing the part or structure, such as bending a beam, applying torque to a shaft, or internally pressurizing a vessel. Sensors, typically consisting of piezoelectric ceramic elements,

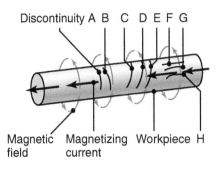

Discontinuity A B C D E F G

Magnetic field   Magnetizing current   Workpiece H

FIGURE 36.10   Schematic illustration of magnetic-particle inspection of a part with a defect in it. Cracks that are in a direction parallel to the magnetic field (such as discontinuity A) would not be detected, whereas the others shown would. Discontinuities F, G, and H are the easiest to detect. *Source:* Reprinted with permission of ASM International. All rights reserved. www.asminternational.org.

detect acoustic emissions. This method is particularly effective for continuous surveillance of load-bearing structures.

The **acoustic-impact technique** consists of tapping the surface of an object, listening to the signals produced, and analyzing them to detect discontinuities and flaws. The principle is basically the same as that employed when tapping walls, desktops, or countertops at various locations, with a finger or a hammer, and listening to the sound emitted. Vitrified grinding wheels (Section 26.2) are tested in a similar manner (called *ring test*) to detect cracks in the wheel, which may not be visible to the naked eye. The acoustic-impact technique is easy to perform and can be instrumented and automated.

Radiography.   *Radiography* involves X-ray inspection to detect such internal flaws as cracks and porosity. The technique detects differences in density within a part. Thus, for example, on an X-ray image, the metal surrounding a defect is typically denser, and thus shows up as lighter than the flaws. The source of radiation is generally an X-ray tube, and a visible, permanent image is made on a film or radiographic paper (Fig. 36.11a). **Fluoroscopes** also are used to produce X-ray images very quickly, and it is a real-time radiography technique that shows events as they are occurring. Radiography requires expensive equipment and proper interpretation of results, and can be a radiation hazard.

Three radiographic technics are:

- **Digital radiography.** The film is replaced by a linear array of detectors (Fig. 36.11b). The X-ray beam is collimated into a fan beam (compare Figs. 36.11a and b) and the workpiece is moved vertically. The detectors digitally sample the radiation, and the data are stored in computer memory; the monitor then displays the data as a two-dimensional image of the workpiece.

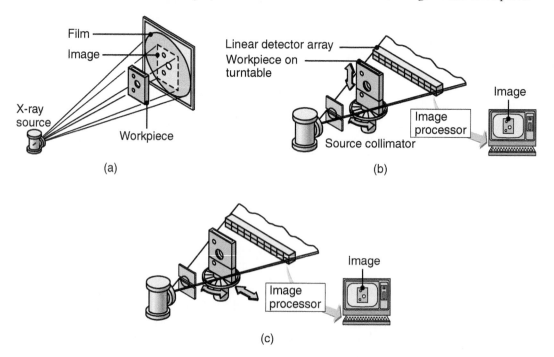

FIGURE 36.11   Three methods of radiographic inspection: (a) conventional radiography; (b) digital radiography; and (c) computed tomography. *Source:* Reprinted with permission of ASM International. All rights reserved. www.asminternational.org.

- **Computed tomography.** This technique is based on the same system as described for digital radiography, except that the workpiece is rotated along a vertical axis as it is being moved vertically (Fig. 36.11c), and the monitor produces X-ray images of thin cross-sections of the workpiece. The translation and rotation of the workpiece provide several angles from which to precisely view the object.
- **Computer-assisted tomography (CAT scan).** This technique is based on the same principle as above and is used widely in medical practice and diagnosis.

Eddy-current Inspection.   This method is based on the principle of *electromagnetic induction*. The part is placed in or adjacent to an electric coil through which alternating current (exciting current) flows, at frequencies of 60 Hz–6 MHz. The current causes eddy currents to flow in the part. Defects in the part impede and change the direction of the eddy currents (Fig. 36.12) and cause changes in the electromagnetic field. These changes then affect the exciting coil (inspection coil), the voltage of which is monitored to determine the presence of flaws. Inspection coils can be made in various sizes and shapes to suit the shape of the part being inspected. Parts must be conductive electrically, and flaw depths detected usually are limited to 13 mm (0.5 in.). The technique requires the use of a standard reference sample to set the sensitivity of the tester.

Thermal Inspection.   *Thermal inspection* involves using contact- or noncontact-type heat-sensing devices that detect temperature changes. Defects in the workpiece (such as cracks, and poor joints, debonded regions in laminated structures) cause a change in temperature distribution. In **thermographic inspection**, materials such as heat-sensitive paints and papers, liquid crystals, and other coatings are applied to the workpiece surface; any changes in their color or appearance indicate defects. The most common method of noncontact thermographic inspection uses infrared detectors (usually infrared scanning microscopes and cameras), which have a high response time and sensitivities of as low as 1°C (2°F). **Thermometric inspection** utilizes devices such as thermocouples, radiometers, and pyrometers, as well as materials such as waxlike crayons.

Holography.   The *holography technique* creates a three-dimensional image of the part by utilizing an optical system (Fig. 36.13). Generally used on simple shapes and highly polished surfaces, this technique records the image on a photographic film.

The use of holography has been extended to **holographic interferometry** for the inspection of parts with various shapes and surface features. Using double- and multiple-exposure techniques, while the part is being subjected to external forces or time-dependent variations, any changes in the images reveal defects in the part.

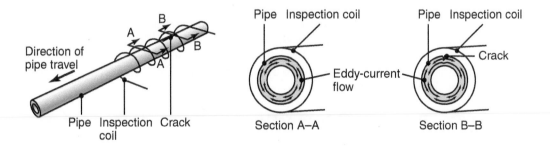

FIGURE 36.12   Changes in eddy-current flow caused by a defect in a part. *Source:* Reprinted with permission of ASM International. All rights reserved. www.asminternational.org.

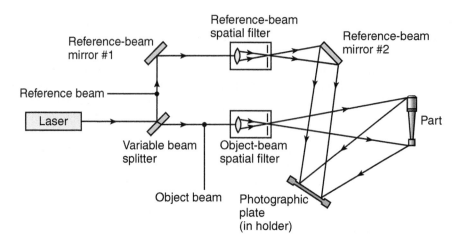

FIGURE 36.13 Schematic illustration of the basic optical system used in holography elements in radiography for detecting flaws. *Source:* Reprinted with permission of ASM International. All rights reserved. www.asminternational.org.

- **Acoustic holography.** Information on internal defects is obtained directly from the image of the interior of the part. In *liquid-surface acoustical holography*, the part and two ultrasonic transducers (one for the object beam and the other for the reference beam) are immersed in a water-filled tank. A holographic image is then obtained from the ripples in the tank.
- **Scanning acoustical holography.** Only one transducer is used and a hologram is produced by electronic-phase detection. In addition to being more sensitive, the equipment is usually portable and can accommodate very large workpieces by using a water column instead of a tank.

36.11 Destructive Testing

As the name suggests, the part tested by *destructive-testing* methods no longer retains its integrity, original shape, or surface characteristics. Mechanical test methods, described in Chapter 2, are all destructive, in that a sample or specimen has to be removed from the product in order to test it. Examples of other destructive tests include the speed testing of grinding wheels to determine their bursting speed (Section 26.4) and the high-pressure testing of pressure vessels to determine their bursting pressure.

Hardness tests that leave relatively large indentations (Figs. 2.13 and 2.14) also may be regarded as destructive testing. Microhardness tests, however, may be regarded as nondestructive because of the very small permanent indentations produced; this distinction is based on the assumption that the material is not notch sensitive (see Section 2.9), as typically is the case with brittle materials. Generally, most glasses, highly heat-treated metals, and ceramics are notch sensitive; consequently, a small indentation produced by the indenter can greatly reduce their strength and toughness significantly.

36.12 Automated Inspection

Traditionally, individual parts and subassemblies have been manufactured in batches, sent to inspection in quality-control rooms (**postprocess inspection**) and, if approved, placed into inventory. If the parts do not pass the quality inspection, they are either scrapped or are kept and used on the basis of having a certain acceptable deviation from the standard. Parts also may be inspected immediately after they are produced (**in-process inspection**).

In contrast, *automated inspection* uses a variety of sensor systems that monitor the relevant parameters during the manufacturing operation (**online inspection**). Using the measurements obtained, the process automatically corrects itself to produce acceptable parts. Thus, further inspection of the part at another location in the plant becomes unnecessary. The development of accurate sensors and advanced computer-control systems has greatly enabled automated inspection to be integrated into manufacturing operations (Chapters 37 and 38). Such integration ensures that no part is moved from one process to another (such as a turning operation followed by cylindrical grinding), unless the part is made correctly and meets the standards in the first operation.

Automated inspection is flexible and responsive to product design changes. Furthermore, because of automated equipment, less operator skill is required, productivity is increased, and parts have higher quality, dimensional accuracy, and reliability.

Sensors for Automated Inspection.   Continuing advances in *sensor technology*, described in Section 37.7, have made real-time monitoring of manufacturing processes feasible. Directly or indirectly, and with the use of various *probes*, sensors can detect dimensions, surface finish, temperature, force, power, vibration, tool wear, and the presence of defects.

Sensors operate on the principles of strain gages, inductance, capacitance, ultrasonics, acoustics, pneumatics, infrared radiation, optics, lasers, or various electronic gages. They may be *tactile* (touching) or *nontactile*. They are linked to microprocessors and computers for graphic data display (see also *programmable logic controllers*, Section 37.2.6). This capability allows rapid online adjustment of any processing parameter, thus resulting in the production of parts that consistently are within specified standards. Such systems already are standard equipment on machine tools (described in Part IV of this book).

SUMMARY

- Quality must be built into products. Quality assurance concerns various aspects of production, such as design, manufacturing, assembly, and especially inspection, at each step of production for conformance to specifications.
- The traditional approach of inspecting a part after it is made has largely been replaced by online and 100% inspection of all parts being manufactured.
- Statistical quality control and process control are indispensable in modern manufacturing; they are particularly important in the production of interchangeable parts and in the reduction of manufacturing costs.
- Although all quality-control approaches have their limits of applicability, the implementation of total quality management, the ISO and QSO 9000 standards, and the ISO 14000 standard are among the most significant developments in quality management in manufacturing.
- Several nondestructive and destructive testing techniques, each of which has its own applications, advantages, and limitations, are available for inspection of parts produced.

KEY TERMS

Acceptance quality level	Experimental design	Process reliability	Specification limits
Acceptance sampling	Factorial design	Product reliability	Standard deviation
Assignable variations	Frequency distribution	Producer's risk	Statistical process
Automated inspection	Grand average	Quality	control
Chance variations	ISO standards	Quality assurance	Statistical quality
Common cause	Juran methods	Quality circle	control
Consumer's risk	Kaizen	QS standards	Statistics
Continuous improvement	Lot size	Random sampling	Taguchi loss function
Control charts	Lower control limit	Range	Taguchi methods
Control limits	Method of attributes	Reliability	Total quality control
Defect prevention	Method of variables	Return on quality	Total quality
Deming methods	Nondestructive testing	Robustness	management
Destructive testing	Normal distribution	Sample size	Upper control limit
Dispersion	curve	Sensors	Variability
Distribution	Population	Shewhart control charts	
Environmental	Probability	Six sigma	
management systems	Process capability	Special cause	

BIBLIOGRAPHY

Allen, T.T., **Introduction to Engineering Statistics and Lean Sigma: Statistical Quality Control and Design of Experiments and Systems**, 2nd ed., Springer, 2010.

Bendell, A., **Taguchi Methods**, Springer, 2007.

Besterfield, D.H., **Quality Control**, 8th ed., Prentice Hall, 2008.

Breyfogl, F.W., **Implementing Six Sigma**, Wiley, 2003.

De Feo, J., and Juran, J.M., **Juran's Quality Handbook**, 6th ed., McGraw-Hill, 2010.

DeVor, R.E., Chang T., and Sutherland, J.W., **Statistical Quality Design and Control**, 2nd ed., Prentice Hall, 2006.

Fowlkes, W.Y., **Engineering Methods for Robust Product Design: Using Taguchi Methods in Technology and Product Development**, Pearson/Education, 2012.

Imai, M., **Gemba Kaizen: A Commonsense Approach to Continuous Improvement**, 2nd ed., McGraw-Hill, 2012.

Joglekar, A., **Statistical Methods for Six Sigma: In R&D and Manufacturing**, Wiley, 2004.

Montgomery, D.C., **Introduction to Statistical Quality Control**, 7th ed., Wiley, 2012.

Pyzdek, T., and Keller, P., **The Six Sigma Handbook**, 3rd ed., McGraw-Hill, 2009.

Ryan, T.P., **Modern Experimental Design**, Wiley, 2007.

Ryan, T.P., **Statistical Methods for Quality Engineering**, 3rd ed., Wiley, 2011.

Schilling, E.G., and Neubauer, D.V., **Acceptance Sampling in Quality Control**, 2nd ed., Chapman and Hall/CRC, 2009.

Smith, G.M., **Statistical Process Control and Quality Improvement**, 5th ed., Prentice Hall, 2003.

Taguchi, G., Chowdhury, S., and Wu, Y., **Taguchi's Quality Engineering Handbook**, Wiley-Interscience, 2004.

Wang, J.X., **Engineering Robust Designs with Six Sigma**, Prentice Hall, 2005.

Yang, G., **Life Cycle Reliability Engineering**, Wiley, 2007.

REVIEW QUESTIONS

36.1 Define the terms sample size, random sampling, population, and lot size.

36.2 What are chance variations?

36.3 Explain the difference between method of variables and method of attributes.

36.4 Define standard deviation. Why is it important in manufacturing?

36.5 Describe what is meant by statistical process control.

36.6 When is a process out of control? Explain.

36.7 Explain why control charts are developed. How are they used?

36.8 What is a loss function? How is it used?

36.9 What do control limits indicate?

36.10 Define process capability. How is it used?

36.11 What is acceptance sampling? Why was it developed?

36.12 Search the technical literature and explain the difference between series and parallel reliability.

36.13 What is meant by six-sigma quality?

36.14 Explain the difference between (a) probability and reliability and (b) robustness and reliability?

36.15 How are liquid penetrants used, and what can they detect?

36.16 Give three methods of nondestructive testing that measure material properties.

QUALITATIVE PROBLEMS

36.17 Explain why major efforts are continually being made to build quality into products.

36.18 Give examples of products for which 100% sampling is not possible or feasible.

36.19 What is the consequence of setting lower and upper specifications closer to the peak of the curve in Fig. 36.4?

36.20 Identify several factors that can cause a process to become out of control.

36.21 Describe situations in which the need for destructive testing techniques is unavoidable.

36.22 Which of the nondestructive inspection techniques are suitable for nonmetallic materials? Why?

36.23 What are the advantages of automated inspection? Why has it become an important part of manufacturing engineering?

36.24 Why is reliability important in manufacturing engineering? Give several examples.

36.25 Give examples of the acoustic-impact inspection technique other than those given in the chapter.

36.26 Explain why *GO* and *NOT GO* gages (see Section 35.4.4) are incompatible with the Taguchi philosophy.

36.27 Review Case Study 36.1 and list the important lessons that are illustrated.

36.28 List advantages and disadvantages of incorporating fluoroscopy as an in-line inspection tool.

36.29 Search the technical literature and give examples of robust design in addition to that shown in Fig. 36.1.

36.30 Will Eq. (36.10) yield exact results? Why or why not?

36.31 What is a Taguchi loss function? What is its significance?

QUANTITATIVE PROBLEMS

36.32 Beverage-can manufacturers try to achieve failure rates of less than one can in ten thousand. If this corresponds to n-sigma quality, find n.

36.33 Assume that in Example 36.3 the number of samples was 8 instead of 10. Using the top half of the data in Table 36.3, recalculate the control limits and the standard deviation. Compare your observations with the results obtained by using 10 samples.

36.34 Calculate the control limits for averages and ranges for (a) number of samples = 8, (b) $\bar{\bar{x}} = 65$, and (c) $R = 6$.

36.35 Calculate the control limits for (a) number of samples = 6, (b) $\bar{\bar{x}} = 36.5$, and (c) $\text{UCL}_R = 5.75$.

36.36 In an inspection with a sample size of 12 and a sample number of 40, it was found that the average range was 14 and the average of averages was 80. Calculate the control limits for averages and for ranges.

36.37 Determine the control limits for the data given in the following table:

x_1	x_2	x_3	x_4
0.57	0.61	0.50	0.55
0.59	0.55	0.60	0.58
0.55	0.50	0.55	0.51
0.54	0.57	0.50	0.50
0.58	0.58	0.60	0.56
0.60	0.61	0.55	0.61
0.58	0.55	0.61	0.53

36.38 The average of averages of a number of samples of size 9 was determined to be 124. The average range was 17.82 and the standard deviation was 4. The following measurements were taken in a sample: 121, 130, 125, 130, 119, 131, 135, 121, and 128. Is the process in control?

36.39 A manufacturer is ring rolling ball-bearing races (see Fig. 13.15a). The inner surface has a surface roughness specification of 0.10 ± 0.006 μm. Measurements taken from rolled rings indicate a mean roughness of 0.112 μm, with a standard deviation of 0.02 μm. Fifty thousand rings per month are manufactured, and the cost of rejecting a defective ring is $10.00. It is known that by changing lubricants to a special emulsion, the mean roughness could be made essentially equal to the design specification. What additional cost per month can be justified for the lubricant?

36.40 For the data in Problem 36.39, assume that the lubricant change can cause the manufacturing process to achieve a roughness of 0.10 ± 0.01 μm. What additional cost per month for the lubricant can be justified? If the lubricant did not add any new cost, would you recommend its use?

SYNTHESIS, DESIGN, AND PROJECTS

36.41 Which aspects of the quality-control concepts of Deming, Taguchi, and Juran would, in your opinion, be difficult to implement in a typical manufacturing facility? Why?

36.42 Describe your thought on whether products should be designed and built for a certain expected life. Would your answer depend on whether the products were consumer or industrial products? Explain.

36.43 Survey the available technical literature, contact various associations, and prepare a comprehensive table concerning the life expectancy of various consumer products.

36.44 Deming's 14 points have been formulated as the characteristics of a company that produces high-quality products, so that very few defects are produced. Could the same rules be used to minimize a company's carbon footprint? Explain.

36.45 Would it be desirable to incorporate nondestructive inspection techniques into metalworking machinery? Give a specific example, make a sketch of such a machine, and explain its features.

36.46 Name several material and process variables that can influence product quality in metal (a) casting, (b) forming, and (c) machining.

36.47 Identify the nondestructive techniques that are capable of detecting internal flaws and those which detect external flaws only.

36.48 Explain the difference between in-process and post-process inspection of manufactured parts. What trends are there in such inspections? Explain.

36.49 Review Table 36.1 and outline changes that would occur if your instructor incorporated all of Deming's 14 points.

36.50 Many components of products have a minimal effect on part robustness and quality. For example, the hinges in the glove compartment of an automobile do not have an impact on the owner's satisfaction, and the glove compartment is opened so infrequently that a robust design is easy to achieve. Would you advocate using Taguchi methods (such as loss functions) on this type of component? Explain.

36.51 You are instructed to design an automated inspection system for an orthogonal cutting operation (Section 21.2). What property or properties would you attempt to measure, and with what instruments/probes?

Manufacturing in a Competitive Environment

PART **IX**

In a highly competitive global marketplace for consumer and industrial goods, advances in manufacturing processes, machinery, tooling, and operations are being driven by goals that may be summarized as:

- Products must fully meet **design** and **service requirements, specifications,** and **standards.**
- Manufacturing activities must continually strive for higher levels of **quality** and **productivity;** quality must be built into the product at each stage of design and manufacture.
- Manufacturing processes and operations must have sufficient **flexibility** to respond rapidly to constantly changing market demands.
- The most **economical methods** of manufacturing must be explored and implemented.

Although numerical control of machine tools, beginning in the early 1950s, was a key factor in setting the stage for modern manufacturing, much of the progress in manufacturing activities stems from our ability to view these activities and operations as a *large system*, with often complex interactions among all of its components. In implementing a *systems approach* to manufacturing, various functions and activities, that for a long time had been separate and distinct entities, can now be *integrated* and *optimized*.

As the first of the four chapters in the final part of this book, Chapter 37 introduces the concept of *automation* and its implementation in terms of key developments in numerical control and, later, in computer numerical control. This introduction is followed by a description of the advances made in automation and controls, involving major topics such as as adaptive control, industrial robots, sensor technology, material handling and movement, and assembly systems, and how they are implemented in modern production.

Manufacturing systems and how their individual components and operations are integrated are described in Chapter 38, along with the critical role of computers, communications, and various *enabling technologies* as an aid to such activities as product design, engineering, manufacturing, and process planning. The enabling technologies include industrial robots, sensor technology, adaptive control, flexible fixturing, and assembly systems.

Computer-integrated manufacturing, with its various features, such as cellular manufacturing, flexible manufacturing systems, just-in-time production, lean manufacturing, and artificial intelligence, are then described in Chapter 39.

The aim of Chapter 40 is to highlight the importance of the numerous, and often complex, factors and their interactions that have a major impact on competitive manufacturing in a global marketplace. Among the factors involved are product design, quality, and product life cycle; selection of materials and processes and their substitution in production; process capabilities; and costs involved, including machinery, tooling, and labor.

Automation of Manufacturing Processes and Operations

- This chapter describes automation in all aspects of manufacturing processes and operations, by which parts are produced reliably, accurately, at high production rates, and economically. It begins with a description of the types of automation and their various applications.
- Flexibility in manufacturing through numerical control of machines is then discussed, with detailed descriptions of their important features.
- The chapter also investigates the different control strategies that can be used, including open-loop, closed-loop, and adaptive control.
- Industrial robots are then reviewed, including their capabilities and guidelines for applications. A discussion of sensor technology and its important applications follows.
- The chapter concludes with a comprehensive description of flexible fixturing and assembly systems and their design considerations.

37.1 Introduction

Until the early 1950s, most operations in a typical manufacturing plant were carried out on traditional machinery, such as lathes, milling machines, drill presses, and various equipment for forming, shaping, and joining materials. Such equipment generally lacked flexibility, and it required considerable skilled labor to produce parts with acceptable dimensional accuracy and surface characteristics. Moreover, each time a different product was to be manufactured, the machinery had to be retooled, fixtures had to be prepared or modified, and the movement of materials among various machines had to be rearranged. The development of new products and of parts with complex shapes required numerous trial-and-error attempts by the operator to set the proper processing parameters. Also, because of human involvement, making parts that were exactly alike was often difficult, time consuming, and costly.

These circumstances meant that processing methods generally were inefficient and that labor costs were a significant portion of the overall production cost. The necessity for reducing the labor share of product cost became increasingly apparent, as did the need to improve the efficiency and flexibility of manufacturing operations.

Productivity also became a major concern; generally defined as output per employee per hour, it basically measures operating efficiency. An efficient operation makes optimum use of all resources, such as materials, energy, capital, labor, machinery, and available technologies. With rapid advances in the science and technology of manufacturing, the efficiency of manufacturing operations began to improve and the percentage of total cost represented by labor began to decline.

In the past decade, *free trade zones* have proliferated across the entire globe. One effect of this development is that labor-intensive manufacturing operations or products, such as furniture, shoes, clothing, and assembly, often have been relocated to regions where labor costs are low (see Table I.7 in the General Introduction). Still, countries with higher labor costs continue to try to remain competitive, mainly by achieving significant increases in productivity.

The important elements in improving productivity have been *mechanization, automation,* and *control* of manufacturing equipment and systems, as well as widespread adoption of communications and software. **Mechanization** controls a machine or process with the use of various mechanical, hydraulic, pneumatic, or electrical devices; it reached its peak by the 1940s. In spite of the obvious benefits of mechanized operations, however, the worker would still be directly involved in a particular operation and would continually check a machine's performance. Consider, for example, the following situations:

- A cutting tool wears or fractures during a machining operation
- A part is overheated during its heat treatment
- The surface finish of a part begins to deteriorate during grinding
- Dimensional tolerances and springback become too large in sheet-metal forming

In each of these situations, the operator had to intervene and change one or more of the relevant process parameters and machine settings, a task that required considerable training and experience.

The next step in improving the efficiency of manufacturing operations was **automation**, a word coined in the mid-1940s by the U.S. automobile industry to indicate the **automatic handling and processing of parts** in and among production machines. Efficiency became further increased by rapid advances in automation and the development of several enabling technologies, largely through advances in **control systems,** with the help of increasingly powerful computers and software.

This chapter follows the outline shown in Fig. 37.1. It first reviews the history and principles of automation and how it has helped to *integrate* various key operations

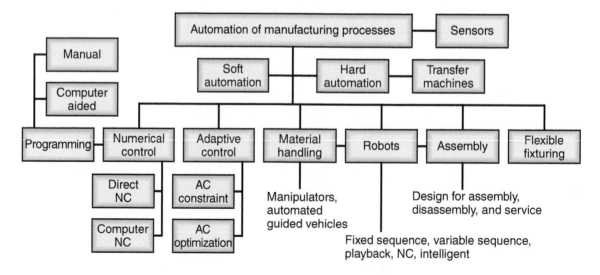

FIGURE 37.1   Outline of topics covered in this chapter.

and activities in a manufacturing plant. It then introduces the concept of the control of machines and systems through **numerical control** and **adaptive control** techniques. The chapter also describes how the important activity of material handling and movement has been developed into various systems, particularly those including the use of **industrial robots** to greatly improve handling efficiency.

The subject of **sensor technology** is then described; this is a topic that is an essential element in the control and optimization of machinery, processes, and systems. Significant developments in **flexible fixturing** and **assembly operations** also are covered; these methods are essential in advanced manufacturing technologies, particularly flexible manufacturing systems. Also included is a discussion of **design for assembly, disassembly, and service,** with specific recommendations to help improve the efficiency of each of these operations. The final topic of the chapter describes the **economics** of the equipment, processes, and manufacturing operations.

37.2 Automation

Automation, from the Greek word *automatos*, meaning "self acting," is generally defined as the process of enabling machines to follow a predetermined sequence of operations with little or no human intervention, and using specialized equipment and devices that perform and control processes and operations. Table 37.1 shows the development of automation throughout history. Full automation is achieved through

TABLE 37.1

History of the Automation of Manufacturing Processes	
Date	Development
1500–1600	Water power for metalworking; rolling mills for coinage strips
1600–1700	Hand lathe for wood; mechanical calculator
1700–1800	Boring, turning, and screw-cutting lathe; drill press
1800–1900	Copying lathe, turret lathe, universal milling machine; advanced mechanical calculators
1808	Sheet-metal cards with punched holes for automatic control of weaving patterns in looms
1863	Automatic piano player (Pianola)
1900–1920	Geared lathe; automatic screw machine; automatic bottle-making machine
1920	First use of the word robot
1920–1940	Transfer machines; mass production
1940	First electronic computing machine
1943	First digital electronic computer
1945	First use of the word automation
1947	Invention of the transistor
1952	First prototype numerical-control machine tool
1954	Development of the symbolic language APT (Automatically Programmed Tool); adaptive control
1957	Commercially available NC machine tools
1959	Integrated circuits; first use of the term group technology
1960s	Industrial robots
1965	Large-scale integrated circuits
1968	Programmable logic controllers
1970	First integrated manufacturing system; spot welding of automobile bodies with robots
1970s	Microprocessors; minicomputer-controlled robot; flexible manufacturing systems; group technology
1980s	Artificial intelligence; intelligent robots; smart sensors; untended manufacturing cells
1990s–2000s	Integrated manufacturing systems; intelligent and sensor-based machines; telecommunications and global manufacturing networks; fuzzy-logic devices; artificial neural networks; Internet tools; virtual environments; high-speed information systems
2010s	Cloud-based storage; MTConnect for information retrieval; three-dimensional geometry files; STEP-NC and autogenerated G-code

various devices, sensors, actuators, techniques, and equipment that are capable of (a) monitoring all aspects of, (b) making decisions concerning changes that should be made in, and (c) controlling all aspects of, the operation. Modern systems also have the capability to store data regarding these activities and communicate the data over a network to computers or storage devices for analysis.

Automation is an *evolutionary* rather than a revolutionary concept, and has been implemented especially in the following basic areas of manufacturing activities:

- **Production processes:** Machining, forging, cold extrusion, casting, powder metallurgy, and grinding operations.
- **Material handling and movement:** Materials and parts in various stages of completion (called *work in progress*) are moved throughout a plant by computer-controlled equipment, with little or no human guidance.
- **Inspection:** Parts are inspected automatically for dimensional accuracy, surface finish, quality, and various specific characteristics during their production (*in-process inspection*).
- **Assembly:** Individually manufactured parts and components are assembled automatically into subassemblies and then into assemblies to complete a product.
- **Packaging:** Products are packaged automatically for shipment.

37.2.1   Evolution of Automation

As shown in Table I.2, some metalworking processes were used as early as 4000 B.C. However, it was not until the beginning of the Industrial Revolution in the 1750s (also referred to as the *First Industrial Revolution*) that automation began to be introduced in the production of goods. The *Second Industrial Revolution* began in the mid 1950s, with advances in several areas.

Machine tools, such as turret lathes, automatic screw machines, and automatic glass bottle-making equipment, began to be developed in the late 1890s. Mass-production techniques and transfer machines were developed in the 1920s. These machines had *fixed* automatic mechanisms and were designed to produce *specific* parts, best represented by the automobile industry, which produced passenger cars at high rates and low cost; for example, by the end of the 1920s, Ford was producing more than one car per minute, and it could be easily afforded by the typical factory employee.

The major breakthrough in automation began with numerical control (NC) of machine tools in the early 1950s (see Section 37.3). Since this historic development, rapid progress has been made in automating almost all aspects of manufacturing, from the introduction of computers into automation, to computerized numerical control (CNC) and adaptive control (AC), to industrial robots, to computer-aided design, engineering, and manufacturing (CAD/CAE/CAM) and computer-integrated manufacturing (CIM) systems.

As described throughout various chapters, manufacturing involves different levels of automation, depending on the processes used, the products to be made, and production volumes required. Manufacturing systems include the following classifications, in order of increasing automation (see also Fig. 37.2):

- **Job shops:** These facilities use general-purpose machines and machining centers, with high levels of labor involvement.
- **Stand-alone NC production:** This method uses numerically controlled machines, but with significant operator–machine interaction.

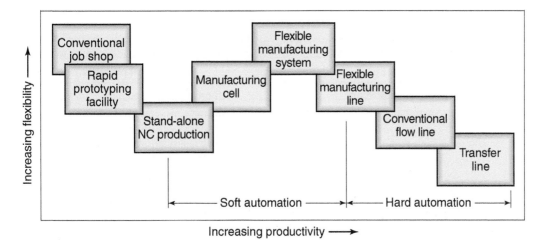

FIGURE 37.2   Flexibility and productivity of various manufacturing systems; note the overlap between the systems, due to the various levels of automation and computer control that are possible in each group. (See Chapter 39 for details.)

- **Manufacturing cells:** These cells use a cluster of machines with integrated computer control and flexible material handling, often with industrial robots.
- **Flexible manufacturing systems:** These systems use computer control of all aspects of manufacturing, the simultaneous incorporation of several manufacturing cells, and automated material-handling systems.
- **Flexible manufacturing lines:** These lines involve computer-controlled machinery, organized in production lines instead of cells. Part transfer is through hard automation and product flow is more limited than in flexible manufacturing systems, but the throughput is larger for higher production quantities.
- **Flow lines** and **transfer lines:** These lines consist of organized groupings of machinery, with automated material handling between machines. Because the goal is to produce only one type of part, the manufacturing line is designed with limited or no flexibility.

37.2.2   Implementation of Automation

Automation generally has the following primary goals:

- *Integrate* various aspects of manufacturing operations to improve product quality and uniformity, minimize cycle times and effort, and reduce labor costs.
- *Improve productivity* by reducing manufacturing costs through better control of production; parts are loaded, fed, and unloaded on machines more efficiently; machines are used more effectively; and production is organized more efficiently.
- *Improve quality* by using more repeatable processes.
- *Reduce human involvement*, boredom, and thus the possibility of human error.
- *Reduce workpiece damage* caused by the manual handling of parts.
- *Raise the level of safety* for personnel, especially under hazardous working conditions.
- *Economize on floor space* in the plant by arranging machines, material handling and movement, and auxiliary equipment more efficiently.

Automation and Production Quantity.   The production quantity is crucial in determining the type of machinery and the level of automation required to produce parts

TABLE 37.2

Approximate Annual Production Quantities		
Type of production	Number produced	Typical products
Experimental or prototype	1–10	All products
Piece or small-batch	10–5000	Aircraft, missiles, special machinery, dies, jewelry, and orthopedic implants
Batch or high-volume	5000–100,000	Trucks, agricultural machinery, jet engines, diesel engines, computer components, and sporting goods
Mass production	100,000 and over	Automobiles, appliances, fasteners, and food and beverage containers

economically. *Total production quantity* is defined as the total number of parts to be made, whereas *production rate* is defined as the number of parts produced per unit time. The production quantity is produced in batches of various *lot sizes*. The approximate and generally accepted ranges of production volume are shown in Table 37.2 for some typical applications. Note that, as expected, experimental or prototype products represent the lowest volume (see Chapter 20).

Job shops typically produce small quantities per year (Fig. 37.2), using various standard general-purpose machine tools (called *stand-alone machines*) or using *machining centers* (Chapter 25). The operations performed typically have high part variety, meaning that different parts can be produced in a short time without extensive changes in tooling or in operations. Machinery in job shops generally requires skilled labor to operate, and production quantities and rates are typically low; as a result, production cost per part is high (Fig. 37.3). When parts involve a large labor component, the production is called *labor intensive*.

Digital manufacturing or *rapid prototyping* has significantly transformed low-volume production. Along with the development of computer software with three-dimensional geometric modeling ability, piece parts can now be designed and manufactured with less effort and cost than previously possible. Operator/programmer skill is still quite high, but rapid prototyping facilities can achieve almost

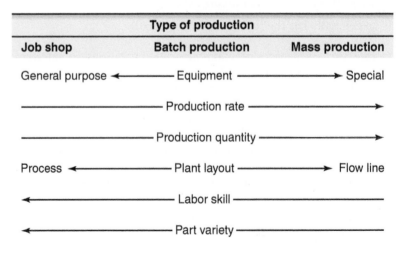

FIGURE 37.3 General characteristics of three types of production methods: job shop, batch, and mass production.

the same flexibility as job shops, being limited only in the number of materials that can be effectively processed.

- **Piece-part production** generally involves very small quantities, and is suitable for job shops; the majority of piece-part production is done in lot sizes of 50 or less.
- **Small-batch production** quantities typically are in the range of 10–100; the equipment consists of general-purpose machines and machining centers.
- **Batch production** usually involves lot sizes of 100–5000; the machinery is similar to that used for small-batch production but with specially designed fixtures for higher productivity.
- **Mass production** involves quantities typically over 100,000. It requires special-purpose machinery, called **dedicated machines,** and automated equipment for transferring materials and parts in progress. These production systems are organized for a specific type of product; consequently, they lack flexibility. Although the machinery, equipment, and specialized tooling are expensive, both the labor skills required and the labor costs are relatively low.

37.2.3   Applications of Automation

Automation can be applied to the manufacturing of all types of goods, from raw materials to finished products, and in all types of production, from job shops to large manufacturing facilities. The decision to automate a new or existing production facility includes the following considerations:

- Type of product manufactured
- Production quantity and rate of production required
- Particular phase of the manufacturing operation to be automated if not all phases are to be automated
- Level of skill in the available workforce
- Reliability and maintenance problems that may be associated with automated machinery and systems
- Economics of the whole operation

Because automation generally involves high initial equipment cost and requires a knowledge of operation and maintenance principles, a decision about the implementation of even low levels of automation must involve a careful study of the actual needs of an organization. In some situations, **selective automation,** rather than total automation, of a facility is desirable. As described in the rest of this final part of this book, there are several important and complex issues involved in making decisions about the appropriate level of automation. Generally, if a manufacturing facility is already automated, the operator skill level required is lower.

On the other hand, the application of computers and computer-controlled machinery in the manufacturing environment requires higher level of sophistication in the workforce than for hard automation. Also, automation requires greater skill in the maintenance and setup workers employed. Thus, it should be recognized that the skills and attributes of a plant's workforce are now different and are tied to the level of automation to be achieved.

37.2.4   Hard Automation

In *hard automation,* also called **fixed-position automation,** the machines are designed to produce a standard product, such as a gear, a shaft, or an engine block. Although product size and processing parameters, such as machining speed, feed, and depth of cut, can be modified, these machines are specialized and hence lack flexibility.

Because the machines are expensive to design and build, their economical use requires production of parts in very large quantities—for example, automotive engines. The machines, generally called *transfer machines* and consisting of *power-head production units* and *transfer mechanisms*, usually are built on the **modular (building-block) principle.** (See also Section 25.2.4.)

Power-head Production Units. Consisting of a frame or bed, electric drive motors, gearboxes, and tool spindles, these units are self contained; their components are available commercially, in various standard sizes and capacities. Because of this inherent modularity, and thus their adaptability and flexibility, they can easily be regrouped to produce a different part.

Transfer Machines. Typically consisting of two or more powerhead units, these machines can be arranged on the shop floor in linear, circular, or U-shaped patterns. Transfer machines also are used extensively in automated assembly, as described in Section 37.9. *Transfer mechanisms* are used to move the workpiece from one station to another in the machine or from one machine to another, in order to enable various operations to be performed on the part. Workpieces are transferred by such methods as (a) rails, along which the parts (usually placed on pallets) are pushed or pulled by various mechanisms (Fig. 37.4a), (b) rotary indexing tables (Fig. 37.4b), and (c) overhead conveyors.

Transfer Lines. Figure 37.5 shows a *transfer line*, or **flow line;** note the variety of operations performed. A common example of a large transfer line is in the production of automotive engine blocks, consisting of numerous machining operations at high rates. The weight and shape of the workpieces influence the arrangement of the individual machines, important for continuity of operation in the event of tool failure or machine breakdown in one or more of the units. *Buffer storage* features are incorporated in the machines to permit continued operation in such an event.

The transfer of parts from station to station is usually controlled by sensors and other devices. Tools on the machines can easily be changed using toolholders with *quick-change features.* The machines can be equipped with various automatic gaging and inspection systems, ensuring that the dimensions of a part produced in one station are within acceptable tolerances before that part is transferred to the next station along the line.

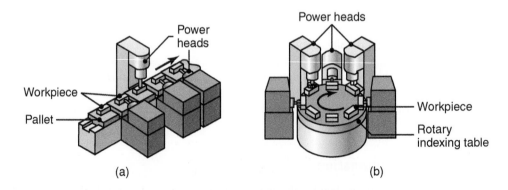

(a)   (b)

FIGURE 37.4 Two types of transfer mechanisms: (a) straight rails and (b) circular or rotary.

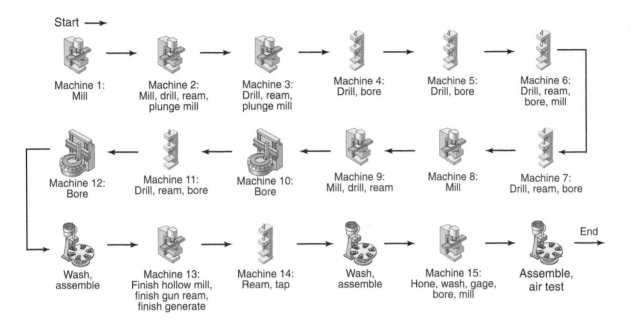

Start →

Machine 1:
Mill

Machine 2:
Mill, drill, ream,
plunge mill

Machine 3:
Drill, ream,
plunge mill

Machine 4:
Drill, bore

Machine 5:
Drill, bore

Machine 6:
Drill, ream,
bore, mill

Machine 12:
Bore

Machine 11:
Drill, ream, bore

Machine 10:
Bore

Machine 9:
Mill, drill, ream

Machine 8:
Mill

Machine 7:
Drill, ream, bore

Wash,
assemble

Machine 13:
Finish hollow mill,
finish gun ream,
finish generate

Machine 14:
Ream, tap

Wash,
assemble

Machine 15:
Hone, wash, gage,
bore, mill

Assemble,
air test

End →

FIGURE 37.5 Schematic illustration of a transfer line.

37.2.5 Soft Automation

Recall that hard automation generally involves mass-production machines that lack flexibility. Greater flexibility is achieved in *soft automation*, also called **flexible** or **programmable automation**, through the use of computer control of the machine and of its functions; thus, soft automation can produce parts with complex shapes. The machines can easily be reprogrammed to produce a part that has a shape or dimensions different from the one produced just prior to it. Advances in soft automation include the extensive use of modern computers, leading to the development of **flexible manufacturing systems,** with high levels of efficiency and productivity (Section 39.3).

37.2.6 Programmable Logic Controllers

The control of a production process in the proper sequence of operations, especially one involving groups of machines and material-handling equipment, has traditionally been performed by switches, relays, timers, counters, and similar hardwired devices that are based on mechanical, electromechanical, and pneumatic principles.

Beginning in 1968, *programmable logic controllers* (PLCs) were introduced to replace these hardwired devices. Because PLCs eliminate the need for relay control panels, and they can be reprogrammed and take less space, they have been adopted widely in manufacturing systems and operations. Their basic functions are (a) on–off, (b) motion, (c) sequential operations, and (d) feedback control. These controllers perform reliably in industrial environments and improve the overall efficiency of an operation. Although they have become less common in new installations (because of advances in numerical-control machines), PLCs still represent a very large installation base. Their sustained popularity is due to their low cost and the proliferation of powerful software to allow programming of PLCs from personal computers, which upload control programs via Ethernet or wireless communications.

Modern PLCs are often programmed in specialized versions of BASIC or C programming languages. *Microcomputers* are now used more often, because they are less expensive and are easier to program and to network. PLCs are also used in system control with high-speed digital-processing and communication capabilities.

37.2.7 Total Productive Maintenance and Total Productive Equipment Management

The management and maintenance of a wide variety of machines, equipment, and systems are among the important aspects affecting productivity in a manufacturing organization. The concepts of *total productive maintenance* (TPM) and *total productive equipment management* (TPEM) include the continued analysis of such factors as:

- Equipment breakdown and equipment problems
- Monitoring and improvement of equipment productivity
- Implementation of preventive and predictive maintenance
- Reduction in setup time, idle time, and cycle time
- Full utilization of machinery and equipment and the improvement of their effectiveness
- Reduction in product defects

As expected, *teamwork* is an important component of this activity and involves the full cooperation of machine operators, maintenance personnel, engineers, and the management of the organization. (See also **kaizen**, Section 36.1.)

37.3 Numerical Control

Numerical control (NC) is a method of controlling the movements of machine components by directly inserting coded instructions into the system. The system then automatically interprets these data and converts them to output signals that, in turn, control various machine components—for example, turning spindles on and off, changing tools, moving the workpiece or the tools along specific paths, and turning cutting fluids on and off.

The importance of numerical control in manufacturing can be illustrated by the following example: Assume that several holes are to be drilled on a part, in the positions shown in Fig. 37.6. In the traditional manual method of machining this part, the operator positions the drill bit with respect to the workpiece (see Fig. 23.25), using reference points given by any of the three methods shown in the figure; the operator then proceeds to drill the holes.

Assume now that 100 parts, all having exactly the same shape and dimensional accuracy, are to be drilled. Obviously, this operation is going to be tedious and time consuming, because the operator has to go through the same motions repeatedly. Moreover, the probability is high that, for a variety of reasons, some of the parts drilled will be different from others.

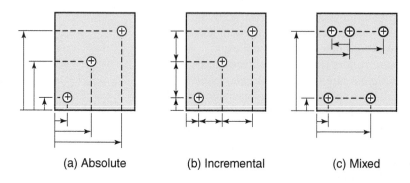

(a) Absolute   (b) Incremental   (c) Mixed

FIGURE 37.6 Positions of drilled holes in a workpiece. Three methods of measurements are shown: (a) Absolute dimensioning, referenced from one point at the lower left of the part; (b) incremental dimensioning, made sequentially from one hole to another; and (c) mixed dimensioning, a combination of both methods.

Assume further that during this production run, the order of processing these parts is changed and that 10 of the parts now require holes in different positions. The machinist now has to reposition the worktable, an operation that is time consuming and subject to error. Such operations can be performed easily by numerical-control machines (see Fig. 23.26), which are capable of producing parts repeatedly and accurately and of handling different parts simply by loading different part programs.

In numerical-control operations, data concerning all aspects of the machining operation, such as tool locations, speeds, feeds, and cutting fluids, are stored on hard disks. On the basis of input information, relays and other devices, known as *hardwired controls*, can be actuated to obtain a desired machine setup. Complex operations, such as turning a part having various contours or die sinking in a milling machine, are now carried out easily. NC machines are used extensively in small- and medium-quantity production (typically 500 or fewer parts) of a wide variety of parts, both in small shops and in large manufacturing facilities.

EXAMPLE 37.1   Historical Origin of Numerical Control

The basic concept behind numerical control appears to have been implemented in the early 1800s, when punched holes in sheet-metal cards were used to automatically control the movements of weaving machines. Needles were activated by sensing the presence or absence of a hole in the card. This invention was followed by automatic piano players (*Pianola*), in which the keys were activated by air flowing through holes punched in a perforated roll of paper.

The principle of numerically controlling the movements of machine tools was first conceived in the 1940s by J.T. Parsons (1913–2007), an American inventor, in his attempt to machine complex helicopter blades. The first prototype NC machine, built in 1952 at the Massachusetts Institute of Technology, was a vertical-spindle, two-axis copy-milling machine retrofitted with servomotors; the machining operations consisted of end milling and face milling (Chapter 24) on a thick aluminum plate.

The numerical data, to be punched into the paper tapes, were generated by a digital computer (another invention that was being developed at the same time at MIT). In the first experiments, parts were machined successfully, accurately, and repeatedly, without any operator intervention. On the basis of this success, the machine-tool industry began designing, building, and marketing NC machine tools. Later, these machines were equipped with computer-numerical controls (CNCs), with greater flexibility, accuracy, versatility, and ease of operation.

37.3.1   Computer Numerical Control

In the next step in the development of numerical control, the control hardware (mounted on the NC machine) was converted to local computer control by software.

- In **direct numerical control** (DNC), several machines were controlled directly (step by step) by a central mainframe computer. In this system, the operator had access to the central computer through a remote terminal. With DNC, the status of *all* machines in a manufacturing facility can be monitored and assessed from a central computer. However, DNC has a crucial disadvantage in that if the computer shuts down for some reason, all of the machines become inoperative.
- In **distributed numerical control**, a central computer serves as the control system over a number of individual CNC machines having onboard microcomputers. This approach was valuable when computers lacked the memory necessary to store all of the data for machining operations essential to produce a part.

DNC became less popular in the 1990s, although using a central computer is still common for collecting, storing, and analyzing data. A recent development is to use **cloud storage,** where files are stored on the servers of a third company and accessed, as needed, through Internet tools.

- **Computer numerical control** (CNC) is a system in which a control micro-computer is an integral part of a machine (*onboard computer*). The machine operator can program onboard computers, modify the programs directly, pre-pare programs for different parts, and store the programs. CNC systems are widely used today, because of the availability of (a) small computers with large memory, (b) low-cost programmable controllers and microprocessors, and (c) program-editing capabilities.

37.3.2 Principles of NC Machines

The basic elements and operation of a typical NC machine are shown in Fig. 37.7. The functional elements in numerical control and the components involved are:

- **Data input:** The numerical information is read and stored in computer memory.
- **Data processing:** The programs are read into the machine control unit for pro-cessing.
- **Data output:** This information is translated into commands (typically, pulsed commands) to the servomotor (Fig. 37.8). The servomotor then moves the work-table to specific positions, through linear or rotary movements, by means of stepping motors, leadscrews, or other similar devices.

Types of Control Circuits. An NC machine can be controlled through two types of circuits. In the **open-loop** system (Fig. 37.8a), the signals are sent to the servomotor by the controller, but the movements and final positions of the worktable are not checked for accuracy. In contrast, the **closed-loop** system (Fig. 37.8b) is equipped with various transducers, sensors, and counters that accurately measure the posi-tion of the worktable. Through *feedback control*, the position of the worktable is compared against the signal; the table movements terminate when the proper coordinates are reached.

Position measurement in NC machines is carried out through two methods (Fig. 37.9). In *indirect measuring systems, rotary encoders,* or *resolvers,* a rotary movement is converted to trans-lation. However, backlash (the play between two adjacent mating gear teeth), can significantly affect measurement accuracy. Position feedback mechanisms utilize various sensors that are based mainly on magnetic and photoelectric principles. In *direct measuring sys-tems,* a sensing device reads a graduated scale on the machine table, or slide, for linear movement (Fig. 37.9c). This system is more accu-rate because the scale is built into the machine, and thus backlash in the mechanisms is not significant.

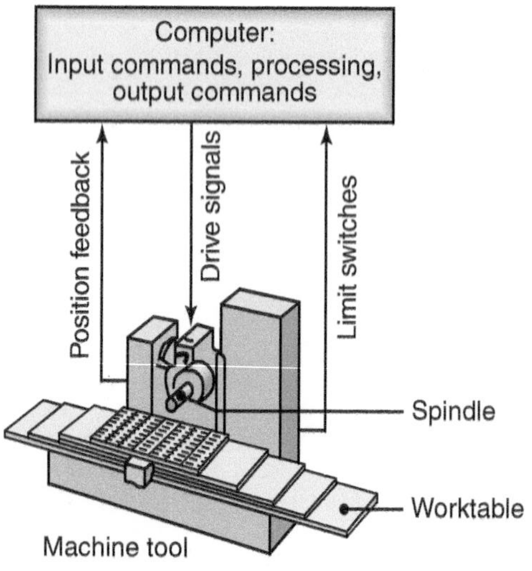

37.3.3 Types of Control Systems

There are two basic types of control systems in numerical control:

1. In a **point-to-point system,** also called a *positioning system,* each axis of the machine is driven separately by leadscrews and at different velocities, depending on the type of opera-tion. The tool moves initially at maximum velocity, in order to

FIGURE 37.7 Schematic illustration of the major components for position control on a numerical-control machine tool.

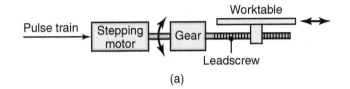

(a)

Video Solution 37.1 Open-loop and Closed-loop Control

FIGURE 37.8 Schematic illustration of the components of (a) an open-loop and (b) a closed-loop control system for a numerical-control machine. DAC = digital-to-analog converter.

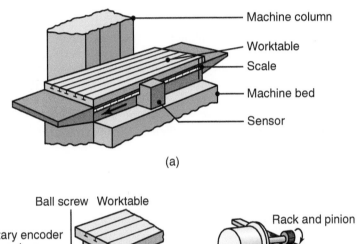

(a)

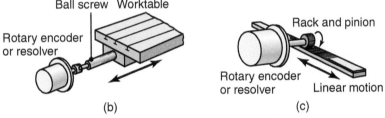

(b)   (c)

FIGURE 37.9 (a) Direct measurement of the linear displacement of a machine-tool worktable; (b) and (c) indirect measurement methods.

reduce nonproductive time, but then it decelerates as the tool approaches its numerically defined position. Thus, in an operation such as drilling a hole, the positioning and drilling take place *sequentially* (Fig. 37.10a).

After the hole is drilled, the tool retracts and moves rapidly to another specified position, and the operation is repeated. The tool path followed from one position to another is important in only one respect: it must be chosen to minimize the time of travel for better efficiency. Point-to-point systems are used mainly in drilling, punching, and straight milling operations.

2. In a **contouring system**, also known as a *continuous-path system*, the positioning and the operations are both performed along controlled paths, but at

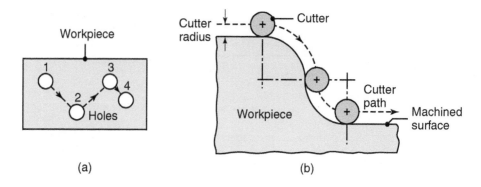

(a)   (b)

FIGURE 37.10 Movement of cutting tools in numerical-control machining. (a) Point-to-point, in which the drill bit drills a hole at position 1, is retracted and moved to position 2, and so on. (b) Continuous path by a milling cutter; note that the cutter path is compensated for by the cutter radius. The path also can be compensated for cutter wear.

different velocities. Because the tool cuts as it travels along a prescribed path (Fig. 37.10b), the control and synchronization of velocities and movements are important for accuracy. The contouring system is typically used on lathes, milling machines, grinders, welding machinery, and machining centers.

Interpolation. Movement of the tool along a path (*interpolation*) occurs incrementally, by one of several basic methods (Fig. 37.11). Examples of actual paths in drilling, boring, and milling operations are shown in Fig. 37.12. In all interpolations, the path controlled is that of the *center of rotation* of the tool. However, compensation for different types of tools, different diameters of tools, or for tool wear during machining also can be made in the NC program.

- **Linear interpolation.** The tool moves in a straight line from start to end (Fig. 37.11a) along two or three axes. Although, theoretically, all types of profiles can be produced by this method, by making the increments between the points small (Fig. 37.11b), a large amount of data has to be processed in order to do so.
- **Circular interpolation** (Fig. 37.11c). The inputs required for the path are the coordinates of the end points, the coordinates of the center of the circle and its radius, and the direction of the tool travel along the arc.
- **Parabolic interpolation** and **cubic interpolation.** The tool path is approximated by curves, based on higher-order mathematical equations. This method is effective in 5-axis machines and is useful in die-sinking operations, such as for

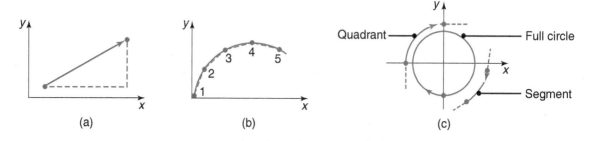

(a)   (b)   (c)

FIGURE 37.11 Types of interpolation in numerical control: (a) linear, (b) continuous path, approximated by incremental straight lines, and (c) circular.

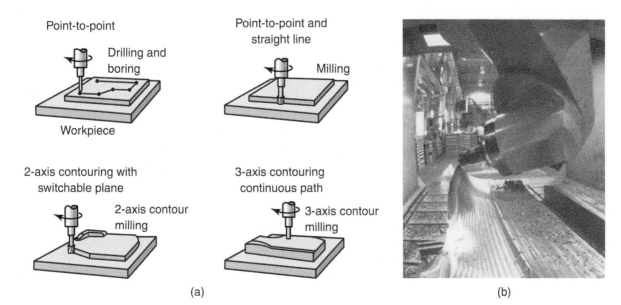

Point-to-point

Drilling and boring

Workpiece

Point-to-point and straight line

Milling

2-axis contouring with switchable plane

2-axis contour milling

3-axis contouring continuous path

3-axis contour milling

(a)

(b)

FIGURE 37.12 (a) Schematic illustration of drilling, boring, and milling along various paths. (b) Machining a sculptured surface on a 5-axis numerical-control machine tool. *Source:* Courtesy of The Ingersoll Milling Machine Co.

dies for sheet forming of automotive bodies. The interpolations also are used for the movements of industrial robots (Section 37.6).

37.3.4 Positioning Accuracy in Numerical Control

Positioning accuracy in numerical-control machines is defined by how accurately the machine can be positioned with respect to a certain coordinate system. *Repeat accuracy* is defined as the closeness of the agreement of the repeated position movements under the same operating conditions of the machine. *Resolution*, also called *sensitivity*, is the smallest increment of motion of the machine components (see also Section 35.7).

The *stiffness* of the machine tool and the *backlash* in gear drives and lead-screws are important factors in achieving dimensional accuracy. Backlash in modern machines is eliminated by using preloaded ball screws. Also, a rapid response to command signals requires that friction in machine slideways and inertia be minimized. The latter can be achieved by reducing the mass of moving components of the machine, such as, for example, by using lightweight materials, including ceramics (see also Section 25.3).

37.3.5 Advantages and Limitations of Numerical Control

Numerical control has the following advantages over conventional methods of machine control:

- Greater flexibility of operation, as well as the ability to produce complex shapes with good dimensional accuracy and repeatability; high production rates, productivity, and product quality; and lower scrap loss.
- Machine adjustments are easy to make.
- More operations can be performed with each setup, and the lead time required for setup and machining is less than the lead time required in conventional methods.

- Programs can be prepared rapidly and can be recalled at any time.
- Operator skill required is less than that for a qualified machinist and the operator has more time to attend to other tasks in the work area.

The major limitations of NC are (a) the relatively high initial cost of the equipment, (b) the need and cost for programming, and (c) the special maintenance required.

37.3.6 Programming for Numerical Control

A *program* for numerical control consists of a sequence of instructions that causes an NC machine to carry out a certain operation; machining operations of all forms are the most commonly applied. *Programming for NC* may be

- Performed by computer software, such as *G-Code generators*, that produce numerical code from geometry data files; such software is now very common, with numerous open-source codes available
- Done on the shop floor (rare but minor changes in G-code)
- Purchased from an outside source

The program contains the following instructions and commands:

1. *Geometric instructions* pertain to relative movements between the tool and the workpiece.
2. *Processing instructions* concern spindle speeds, feeds, cutting tools, cutting fluids, and so on.
3. *Travel instructions* pertain to the type of interpolation and to the speed of movement of the tool or the worktable.
4. *Switching instructions* concern the on–off position for coolant supplies, direction or lack of spindle rotation, tool changes, workpiece feeding, clamping, and so on.

Computer-aided part programming involves special *symbolic programming languages* that determine such coordinate points as corners, edges, and surfaces of the part. A **programming language** is a means of communicating with the computer and involves the use of symbolic characters. The programmer describes the component to be processed and the computer converts that description to commands for the NC machine. Several languages are available commercially and are manufacturer-specific. Programs for part programming are referred to as *macros*, and are similar to the BASIC programming language.

Complex parts can be machined with the use of graphics-based computer-aided machining programs. Standardized programming languages, such as STEP-NC and the older but still common G-Code, are used for communicating machining instructions to the CNC hardware.

The STEP-NC software is a standardized language that has been extended beyond machine tools, and incorporates models for milling, turning, and EDM. Plasma cutting and laser welding and cutting systems also have been developed for use with STEP-NC.

37.4 Adaptive Control

Adaptive control (AC) is basically a *dynamic-feedback* system in which the operating parameters automatically adapt themselves to conform to new circumstances; it is thus a logical extension of computer numerical control systems. Human reactions to occurrences in everyday life already contain dynamic-feedback control. For example, driving a car on a smooth road is relatively easy; on a rough road, however, it has to be steered in order to avoid potholes by *visually* and *continuously* observing the condition of the road directly ahead of the car. Moreover, the driver's body feels the car's rough movements and vibrations, and then reacts by changing the direction or the speed of the car to minimize these effects. An **adaptive controller** continuously checks road conditions, calculates an appropriate desired braking profile (e.g., an antilock brake system and traction control), and then uses feedback to implement it.

As described in Section 37.3, in manufacturing operations the part programmer sets the processing parameters, on the basis of the existing knowledge of the work-piece material and relevant data on the particular operation. In CNC machines, these parameters are held constant during a particular process cycle; in AC, however, the system is capable of automatic adjustments *during* the operation, through closed-loop feedback control (Fig. 37.13). Several adaptive-control systems are commercially available for a variety of applications.

Adaptive Control in Manufacturing.   The main purposes of adaptive control in manufacturing are to

- Optimize production rate
- Optimize product quality
- Minimize production cost

Consider a machining operation, such as turning on a lathe (Section 23.2). The adaptive control system senses, *in real time*, parameters such as cutting forces, spindle torque, temperature rise, tool wear, tool condition, and surface finish produced. The AC system converts this information into commands, modifies the process parameters to hold them within certain limits, and optimizes the machining operation.

Those systems which place a constraint on a process variable, such as force, torque, or temperature, are called **adaptive-control constraint** (ACC) systems.

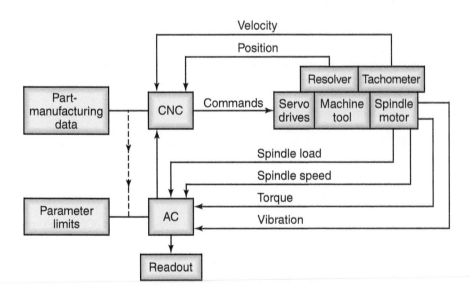

FIGURE 37.13   Schematic illustration of the application of adaptive control (AC) for a turning operation. The system monitors such parameters as cutting force, torque, and vibrations; if these parameters are excessive, it modifies process variables (such as feed and depth of cut) to bring the parameters back to acceptable levels.

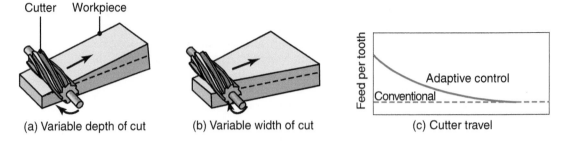

(a) Variable depth of cut   (b) Variable width of cut   (c) Cutter travel

FIGURE 37.14 An example of adaptive control in milling; as the depth of cut (a) or the width of cut (b) increases, the cutting forces and the torque increase. The system senses this increase and automatically reduces the feed (c) to avoid excessive forces or tool breakage in order to maintain cutting efficiency. *Source:* Based on Y. Koren.

Thus, for example, if the cutting force (and hence the torque) increase excessively (due, say, to the presence of a hard region in a casting), the system modifies the cutting speed or the feed, as necessary, in order to lower the cutting force to an acceptable level (Fig. 37.14).

Without AC or the direct intervention of the operator (as has been the case in traditional machining operations), high cutting forces may cause tool failure or the workpiece to deflect or distort excessively. As a result, workpiece dimensional accuracy and surface finish begin to deteriorate. Those systems that optimize an operation are called **adaptive-control optimization** (ACO) systems. Optimization may, for example, involve maximizing the material-removal rate between tool changes or improving the surface finish of the part.

Response time must be short for AC to be effective, particularly in high-speed machining operations (Section 25.5). Assume, for example, that a turning operation is being performed on a lathe at a spindle speed of 1000 rpm, and the tool suddenly breaks, adversely affecting the surface finish and dimensional accuracy of the part. Obviously, in order for the AC system to be effective, the sensing system must respond within a very short time; otherwise the damage to the workpiece may become extensive.

In adaptive control, *quantitative* relationships must therefore be established and coded in the computer software as mathematical models. If, for instance, the tool-wear rate in a machining operation is excessive (Section 21.5), the computer must be able to (a) calculate how much of a change in speed or in feed is necessary and (b) decide whether to increase or decrease the speed or the feed, in order to reduce the tool-wear rate to an acceptable level. The system also must be able to compensate for dimensional changes in the workpiece, due to such factors as tool wear and temperature rise (Fig. 37.15).

It is apparent that, coupled with CNC, adaptive control is a powerful tool in optimizing manufacturing operations. While it has significant demonstrable benefits with respect to quality and productivity, it has not been extensively applied to date. Most machine-tool control systems are geometry-based and do not incorporate adaptive control. The latest versions of STEP-NC programming languages, however, allow for deflection and wear management.

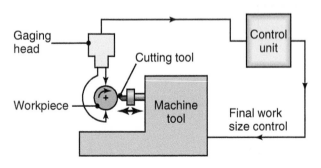

FIGURE 37.15 In-process inspection of a workpiece diameter in a turning operation; the system automatically adjusts the radial position of the cutting tool in order to produce the correct diameter.

37.5   Material Handling and Movement

During a typical manufacturing operation, raw materials and parts in progress are moved from storage to machines, from machine to machine, from assembly to inventory, and, finally, to shipment. For example, (a) workpieces are loaded on machines, as when a forging is mounted on a milling-machine bed for machining, (b) sheet metal is fed into a press for stamping, (c) parts are removed from one machine and loaded onto another, as when a machined forging is to be subsequently ground, for better surface finish and dimensional accuracy, and (d) finished parts are assembled into a final product.

Material handling is defined as the functions and systems associated with the transportation, storage, and control of materials and parts in the total manufacturing cycle of a product. The total time required for actual manufacturing operations depends on the part size and shape and on the set of operations to be performed. Note that idle time and the time required for transporting materials can constitute the majority of the time consumed, thus reducing productivity. Material handling must be an integral part of the planning, implementing, and control of manufacturing operations; moreover, it must be repeatable and predictable.

Plant layout is an important aspect of the orderly flow of materials and components throughout the manufacturing cycle. The time and distances required for moving raw materials and parts should be minimized (see *lean manufacturing*, Section 39.6). For parts requiring multiple operations, which typically is the case, equipment should be grouped around the operator or an industrial robot or robots (see also *cellular manufacturing*, Section 39.2).

Material-handling Methods.   Several factors must be considered in selecting a suitable material-handling method for a particular manufacturing operation:

1. Shape, weight, and characteristics of the parts
2. Distances involved and the position and orientation of the parts during their movement and at final destination
3. Conditions of the path along which the parts are to be transported
4. Level of automation, the controls needed, and any integration with other equipment and systems
5. Operator skill required
6. Economic considerations

For small-batch manufacturing operations, raw materials and parts can be handled and transported manually, but this method is generally costly; also, because it involves people, the practice can be unpredictable, unreliable, or unsafe. In contrast, in automated manufacturing plants, computer-controlled material and parts flow are implemented, resulting in improved productivity at lower labor costs.

Equipment.   The types of equipment that can be used for moving materials and parts in progress may consist of conveyors, rollers, carts, forklift trucks, self-powered monorails, and various mechanical, electrical, magnetic, pneumatic, and hydraulic devices and manipulators. **Manipulators** are designed to be controlled directly by the operator, or they can be automated for repeated operations, such as the loading and unloading of parts from machine tools, presses, and furnaces. Manipulators are capable of gripping and moving heavy parts and of orienting them, as necessary, between the manufacturing and assembly operations. Workpieces are transferred directly from machine to machine. Machinery combinations having the capability of

(a)

(b)

FIGURE 37.16 (a) An automated guided vehicle (Tugger type). This vehicle can be arranged in a variety of configurations to pull caster-mounted cars; it has a laser sensor to ensure that the vehicle operates safely around people and various obstructions. (b) An automated guided vehicle configured with forks for use in a warehouse. *Source:* Courtesy of Egemin, Inc.

conveying parts without the use of additional material-handling apparatus are called **integral transfer devices**.

Flexible material handling and movement, with real-time control, has become an integral part of modern manufacturing. Industrial robots, specially-designed pallets, and **automated guided vehicles** (AGV) are used extensively in *flexible manufacturing systems* to move parts and to orient them as required (Fig. 37.16). AGVs operate automatically along pathways in a plant, with in-floor wiring or tapes for optical scanning for guidance without operator intervention.

This transport system has high flexibility and is capable of random delivery to different workstations. It optimizes the movement of materials and parts in cases of congestion around workstations, machine breakdown (*downtime*), or the failure of one section of the production system.

The movements of AGVs are planned such that they interface with **automated storage/retrieval systems** (AS/RS), in order to utilize warehouse space efficiently and to reduce labor costs. However, these systems are now considered undesirable, because of the current focus on *minimal inventory* and on *just-in-time* production methods (Section 39.5).

Coding systems that have been developed to locate and identify parts in progress throughout the manufacturing system and to transfer them to their appropriate stations are outlined as:

- **Bar coding**, including **QR Codes** (as also used in this book), is the most widely used coding system and the least costly.
- **Magnetic strips** constitute the second most common system.
- **RF** (radio frequency) **tags** are sometimes used; although expensive, they do not require the clear line of sight essential in the previous two systems, and have a long range and are rewritable.
- **Acoustic waves**, **optical character recognition**, and **machine vision** are the principles of other identification systems.

37.6 Industrial Robots

QR Code 37.1 The most common industrial robot applications. (*Source:* © 2012, Courtesy of ABB)

The word **robot** was coined in 1920 by the Czech author K. Čapek, in his play *R.U.R.* (Rossum's Universal Robots); it is derived from the word *robota*, meaning "worker." An **industrial robot** has been described by the International Organization for Standardization (ISO) as a machine consisting of a mechanism including several degrees of freedom, often having the appearance of one or several arms ending in a wrist capable of holding a tool, a workpiece, or an inspection device. In particular, an industrial robot's control unit must use a memorizing method, and may also use sensing or adapting features to take into account environment and special circumstances.

Industrial robots were first introduced in the early 1960s. Computer-controlled robots were commercialized in the early 1970s, and the first robot controlled by a microcomputer appeared in 1974. Industrial robots were first employed in hazardous operations, such as the handling of toxic and radioactive materials and the loading and unloading of hot workpieces from furnaces in foundries.

Simple rule-of-thumb applications for robots are described as the three D's, for *dull, dirty,* and *dangerous* (a fourth D would be for *demeaning*) and the three H's, for *hot, heavy,* and *hazardous.* Industrial robots are now indispensable components in almost all manufacturing operations and have greatly improved productivity.

37.6.1   Robot Components

An industrial robot (Fig. 37.17) has a number of basic components.

Manipulator.   Also called an **arm and wrist,** the *manipulator* is a mechanical device that provides motions (trajectories) similar to those of a human arm and hand. The end of the wrist can reach a point in space having a specific set of coordinates and in a specific orientation. Most robots have six rotational joints; seven degrees of freedom (*redundant* robots) also are available for special applications.

End Effector.   The end of the wrist in a robot is equipped with an *end effector,* also called *end-of-arm tooling.* Depending on the type of operation, conventional end effectors may be equipped with any of the following devices (Fig. 37.18):

- Grippers, hooks, scoops, electromagnets, vacuum cups, and adhesive fingers for material handling (Fig. 37.19a)
- Spray guns for painting
- Attachments for spot and arc or laser welding (Fig. 37.19b), and for arc cutting
- Power tools, such as drills, nut drivers, and burrs (Figs. 37.18b and c)
- Measuring instruments (Fig. 37.18e)

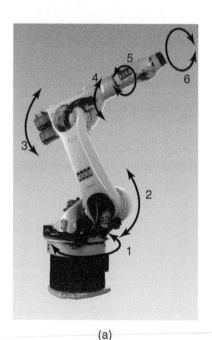
(a)

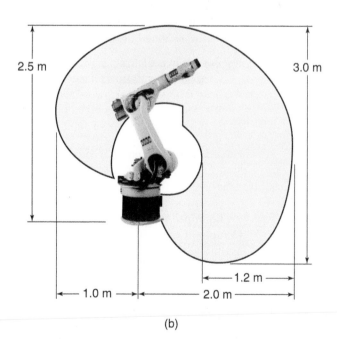
(b)

FIGURE 37.17   (a) Schematic illustration of a 6-axis KR-30 KUKA robot; the payload at the wrist is 30 kg and repeatability is ± 0.15 mm (0.006 in.). The robot has mechanical brakes on all of its axes, which are directly coupled to one another. (b) The work envelope of the robot, as viewed from the side. *Source:* Courtesy of KUKA Robotics Corp.

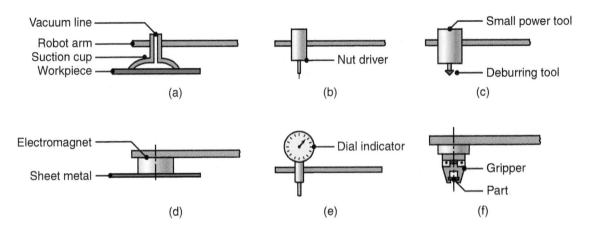

FIGURE 37.18   Types of tools and devices attached to end effectors to perform a variety of operations.

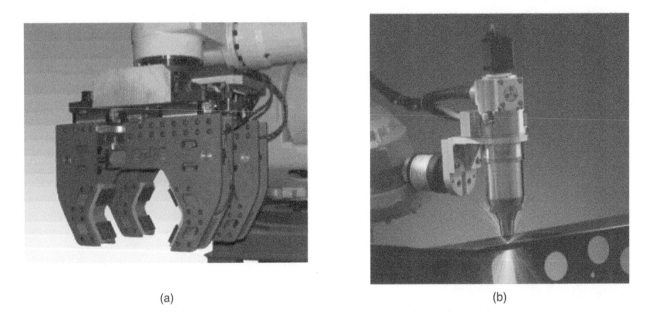

(a)   (b)

FIGURE 37.19   Examples of end effectors. (a) Gripper with four fingers and (b) a laser cutting head. *Source:* (a) Courtesy of RoboMatrix; (b) Courtesy of the Fabricators & Manufacturers Association.

Equipped with two or more fingers, mechanical grippers are used most commonly. **Compliant end effectors** are typically used to handle fragile materials or to facilitate assembly; they can use mechanisms that limit the force applied to the workpiece or they can be designed with a desired stiffness. The selection of an appropriate end effector for a specific application depends on such factors as the payload, environment, reliability, and cost; consequently, end effectors generally are custom made to meet specific handling requirements.

Power Supply. Each motion of the manipulator (linear or rotational) is controlled and regulated by independent actuators that use an electrical, pneumatic, or hydraulic power supply. Each source of energy and each type of motor has its own characteristics, applications, advantages, and limitations.

Controller.   Also known as the *control system*, the controller is the communications and information-processing system that gives commands for the movements of the robot; it is the brain of the robot and stores data to initiate and terminate movements of the manipulator. The controller is also the nervous system of the robot; it interfaces with computers and other equipment, such as manufacturing cells or assembly systems.

Feedback devices, such as transducers, are an important part of the control system. Robots with a fixed set of motions have *open-loop control*, in which commands are given and the robot arm goes through its motions. Unlike feedback in *closed-loop systems*, the accuracy of the movements in open-loop systems is not monitored; consequently, an open-loop system does not have a self-correcting capability.

Depending on the particular task, the *positioning repeatability* required may be as small as 0.050 mm (0.002 in.), as would be required in assembly operations for electronic printed circuit boards (Section 28.13). Accuracy and repeatability vary greatly with payload and with position within the work envelope.

37.6.2   Classification of Robots

Robots may be classified by basic type (Fig. 37.20):

1. **Cartesian**, or **rectilinear**
2. **Cylindrical**
3. **Spherical**, or **polar**
4. **Articulated**, *revolute, jointed,* or *anthropomorphic*

Robots may be attached permanently to the floor of a plant; they may also move along overhead rails (*gantry robots*) or be equipped with wheels to move along the factory floor (*mobile robots*).

Fixed-sequence and Variable-sequence Robots.   The *fixed-sequence robot*, also called a **pick-and-place robot,** is programmed for a specific sequence of operations; its movements are from point to point and the cycle is repeated continuously. These robots are simple and relatively inexpensive. The *variable-sequence robot* also is programmed for a specific sequence of operations, but it can be *reprogrammed* to perform a different sequence of operations.

Playback Robot.   An operator leads or walks the *playback robot* and its end effector through the desired path; in other words, the operator *teaches* the robot by showing it what to do. The robot records the path and sequence of the motions and can repeat them continually without further action or guidance by the operator. The

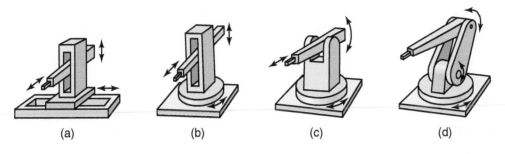

(a)   (b)   (c)   (d)

FIGURE 37.20   Four types of industrial robots: (a) Cartesian (rectilinear); (b) cylindrical; (c) spherical (polar); and (d) articulated (revolute, jointed, or anthropomorphic).

teach pendant robot utilizes handheld button boxes connected to the control panel; the pendants are used to control and guide the robot and its tooling through the work to be performed. These movements are then registered in the memory of the controller and are automatically reenacted by the robot whenever needed.

Numerically Controlled Robot.   This type of robot is programmed and operated much like a numerically controlled machine; it is servocontrolled by digital data and its sequence of movements can be modified with relative ease. As in NC machines, there are two basic types of control. (1) *Point-to-point* robots are easy to program and have a higher load-carrying capacity and a larger **work envelope,** the maximum extent or reach of the robot hand or working tool in all directions, as shown in Fig. 37.21. (2) *Continuous-path* robots have greater accuracy than point-to-point robots but they have lower load-carrying capacity.

Intelligent Robot.   The *intelligent robot,* also called a *sensory robot,* is capable of performing some of the functions and tasks carried out by humans. The robot is equipped with a variety of sensors with *visual (computer vision)* and *tactile* capabilities (Section 37.7). Much like humans, the robot observes and evaluates its immediate environment and its own proximity to other objects (especially machinery), by perception and *pattern recognition.* It then makes appropriate decisions for the next movement and proceeds accordingly.

Developments in intelligent robots include:

- Behaving more like humans and performing such tasks as moving among a variety of machines and equipment on the shop floor and avoiding collisions
- Recognizing, selecting, and properly gripping the correct raw material or workpiece for further processing
- Transporting a part from machine to machine
- Assembling components into subassemblies or into a final product

37.6.3   Applications and Selection of Robots

Major applications of industrial robots include:

- Material-handling operations can be performed reliably and repeatedly, thereby improving quality and reducing scrap losses. Some examples are (a) casting and

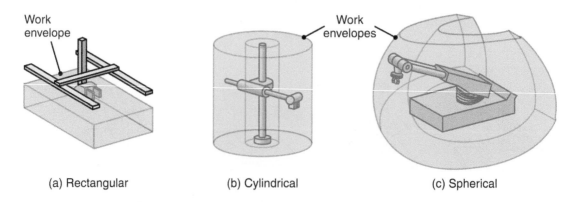

(a) Rectangular   (b) Cylindrical   (c) Spherical

FIGURE 37.21   Work envelopes for three types of robots; their selection depends on the particular application (see also Fig. 37.17b).

molding operations in which molten metal, raw materials, and parts in various stages of completion are handled without operator interference; (b) heat-treating operations in which parts are loaded and unloaded from furnaces and quench baths; and (c) operations in which parts are loaded and unloaded from presses and various other types of metalworking machinery.

- Spot welding of automobile and truck bodies, producing welds of good quality (Fig. 37.22a). Robots also perform other, similar operations, such as arc welding, arc cutting, and riveting.
- Operations such as deburring, grinding, and polishing can be done, by using appropriate tools attached to end effectors.
- Applying adhesives and sealants, as in the automobile frame shown in Fig. 37.22b.
- Spray painting (particularly of complex shapes) and cleaning operations are frequent applications, because the motions required for treating one piece are repeated accurately for the next piece.
- Automated assembly (Fig. 37.23).
- Inspection and gaging at speeds much higher than those that can be achieved by humans.

Robot Selection.   Factors that influence the selection of robots in manufacturing are:

- Load-carrying capacity
- Work envelope (see Figs. 37.17b and 37.21)
- Speed of movement
- Reliability
- Repeatability
- Arm configuration
- Degrees of freedom
- The control system

QR Code 37.2 Robots used to paint car bodies. (*Source:* © 2012, Courtesy of ABB)

(a)	(b)

FIGURE 37.22 Examples of industrial robot applications. (a) Spot welding automobile bodies. (b) Sealing joints of an automobile body. *Source:* (b) Courtesy of Cincinnati Milacron, Inc.

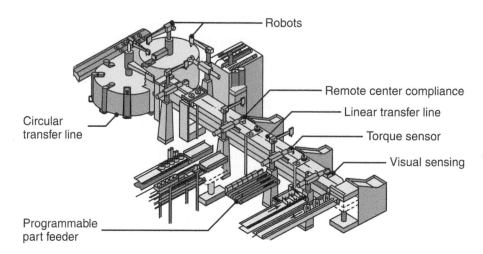

Robots

Remote center compliance

Linear transfer line

Torque sensor

Visual sensing

Circular transfer line

Programmable part feeder

FIGURE 37.23 Automated assembly operations using industrial robots and circular and linear transfer lines.

Economics. In addition to the technical factors, cost and benefit considerations are major aspects of robot selection and their use. Due to their increasing availability, reliability, and their reduced costs, intelligent robots are having a major impact on manufacturing operations.

Robot Safety. Depending on the size of the robot's work envelope, speed, and proximity to humans, safety considerations in a robot environment are important, particularly for programmers and maintenance personnel who are in direct physical interaction with robots. In addition, the movement of the robot with respect to other machinery requires a high level of reliability in order to avoid collisions and damage to nearby equipment. The robot's material-handling activities require the proper securing of raw materials and parts in its gripper at various stages in the production line.

CASE STUDY 37.1   Robotic Deburring of a Blow-molded Toboggan

Roboter Technologie, of Basel, Switzerland, produces high-quality toboggans and car seats made of plastic by injection molding (Section 19.3) or by blow molding (Section 19.4). After molding and while the part is cooling, holes have to be cut and deburred (Section 26.8). The deburring operation is ideal for a robot to perform but it is very difficult to automate. If a rotating burr tool is used, fumes and particles are generated that pose a health risk, and a nonrotating cutter requires that the robot allow deviations in its programmed path in order to accommodate shrinkage variations in the molded parts.

Roboter Technologie found a solution: a float-mounted tool (see *compliant end effectors*, Section 37.6.1) that can accommodate various cutter

blades. To remove burrs, the blade must maintain the correct cutting angle and a constant cutting force. This task is achieved using a KUKA KR-15 robot, which performs the cutting and deburring operations in a single step while also compensating for plastic shrinkage (Fig. 37.24).

As soon as the blow-molded part leaves the molding machine, an operator removes the flash and places the part on a rotary indexing table (Fig. 37.4b). When a part is rotated into the robot's workspace, the robot cuts and deburrs the holes; once the side parts have been cut and deburred, the fixture tilts the toboggan to a vertical position so that its top can be accessed. An automatic tool changer (Section 25.2) is used during the processing of each toboggan, to switch from a convex cutter (used for

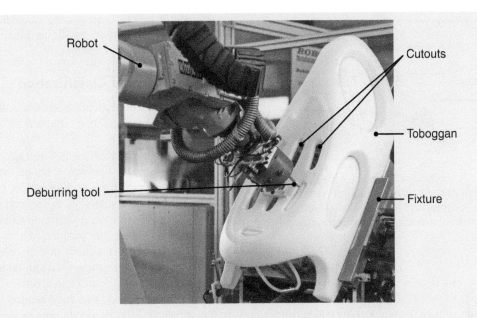

Robot

Cutouts

Toboggan

Deburring tool

Fixture

FIGURE 37.24   Robotic deburring of a blow-molded toboggan. *Source:* Courtesy of Kuka Robotics, Inc., and Roboter Technologie, GmbH. (See also Fig. 26.34.)

the cutouts) to a straight edge cutter in order to produce a smooth exterior contour.

The complex shape of a toboggan is a good example of the flexibility achievable by robots. The robot completes the machining operation in 40–50 s, as opposed to the cycle time of the blow molding process of 120 s; thus, the implementation of the robot is consistent with the goals of a pull system (see Section 39.5). The robot was successfully used in a hazardous and dirty environment, and it

eliminated tasks that were associated with work-related injuries, incurred by fume exposure and stresses on the wrist in manual deburring. Moreover, because of the higher quality of deburring and the elimination of rejected parts, the robot cell paid for itself within only three months.

Source: Courtesy of Kuka Robotics, Inc. and Roboter Technologie, GmbH.

37.7   Sensor Technology

A *sensor* is a device that produces a signal in response to its detecting or measuring a specific property, such as position, force, torque, pressure, chemistry, temperature, humidity, speed, acceleration, or vibration. Traditionally, actuators and switches have been used to set limits on the performance of machines, such as (a) stops on machine tools to restrict worktable movements, (b) pressure and temperature gages with automatic shutoff features, and (c) governors on engines to prevent excessive speed of operation.

Sensor technology is now an important aspect of manufacturing processes and systems, and essential for data acquisition, monitoring, communication, and computer control of machines and systems (Fig. 37.25). Sensors are necessary for the control of intelligent robots and are being developed with capabilities that resemble those of humans, called **smart sensors.**

Because they convert one quantity to another, sensors are often referred to as *transducers.* **Analog sensors** produce a signal, such as voltage, that is proportional to the measured quantity. **Digital sensors** have digital outputs that can be

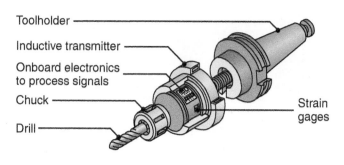

Toolholder

Inductive transmitter

Onboard electronics
to process signals

Chuck

Drill

Strain
gages

FIGURE 37.25 A toolholder equipped with thrust-force and torque sensors (*smart toolholder*), capable of continuously monitoring the cutting operation; such toolholders are essential for the adaptive control of manufacturing operations.

transferred directly to computers. *Analog-to-digital converters* (ADCs) are available for interfacing analog sensors with computers.

37.7.1 Sensor Classification

Sensors that are of greatest interest in manufacturing operations are generally classified as:

- **Mechanical** sensors measure such quantities as strain, mass, position, shape, velocity, force, torque, pressure, and vibration.
- **Chemical** sensors measure concentrations of target chemicals. A common example is the O_2 sensor in an automobile, which measures the amount of oxygen gas in the exhaust; this allows inference of the proper amounts of gas and fuel intake in order to minimize harmful emissions.
- **Electrical** sensors measure voltage, current, charge, and electrical conductivity.
- **Magnetic** sensors measure magnetic field, flux, and permeability.
- **Thermal** sensors measure temperature, flux, thermal conductivity, and specific heat.
- Other types of sensors are **acoustic, ultrasonic, optical, radiation, laser, and fiber optic.**

Depending on its application, a sensor may consist of metallic, nonmetallic, organic, or inorganic materials, as well as fluids, gases, plasmas, or semiconductors. Using the special characteristics of these materials, sensors convert the quantity or property measured to analog or digital output. The operation of an ordinary mercury thermometer, for example, is based on the difference between the thermal expansion of mercury and that of glass. Likewise, a machine part, a physical obstruction, or a barrier in a space can be detected by breaking the beam of light, sensed by a photoelectric cell. A *proximity sensor*, which senses and measures the distance between it and an object or a moving member of a machine, can be based on acoustics, magnetism, capacitance, or optics. Other types of sensors contact the object and then take appropriate action, usually by electromechanical means.

Tactile Sensing.   *Tactile* sensing involves the continuous sensing of variable contact forces, commonly by an array of sensors. Such a system is capable of operating within an arbitrary three-dimensional space. Fragile parts, such as eggs, thin glass objects, and electronic devices, can be handled by robots with *compliant (smart) end effectors*. These effectors can sense the force applied to the object being handled, using, for example, strain gages, piezoelectric devices, magnetic induction, ultrasonics, and optical systems of fiber optics and light-emitting diodes (LEDs).

The force that is sensed is monitored and controlled through closed-loop feedback devices. Compliant grippers, with force-feedback capabilities and sensory perception, are complex and require powerful computers. *Anthropomorphic end effectors* are designed to simulate the human hand and fingers, and to have the capability of sensing touch, force, and movement. The ideal tactile sensor must also sense *slip*, a normal capability of human fingers and hands; note, for example, how, even with closed eyes, one can sense if an object is beginning to slip from one's hand.

Visual Sensing.   In visual sensing, cameras scan an image, and the software processes the data. Its most important applications in manufacturing are *pattern*

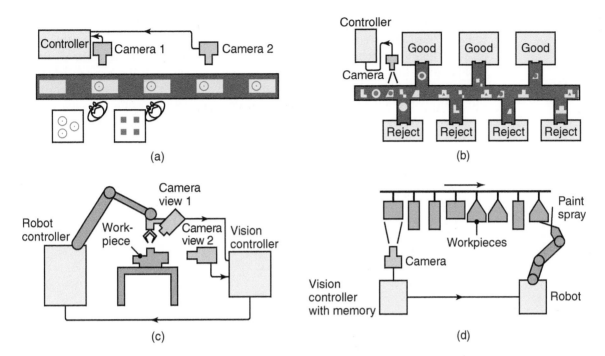

FIGURE 37.26 Examples of machine-vision applications. (a) In-line inspection of parts. (b) Identification of parts with various shapes, and inspection and rejection of defective parts. (c) Use of cameras to provide positional input to a robot relative to the workpiece. (d) Painting parts having different shapes and input from a camera; the system's memory allows the robot to identify the particular shape to be painted and to proceed with the correct movements of a paint spray attached to the end effector.

recognition (Fig. 37.26), edge detection, and transfer of information, such as with simple barcodes.

Machine vision commonly uses digital cameras that communicate with a computer through wireless, USB, or Ethernet connections. Scanning can take place in (a) one (line scan, such as with bar codes), (b) two (2D scanning, as with QR Codes), or (c) three dimensions (3D scanning, as with CT scanning or confocal cameras, as shown in Fig. 37.27). Three-dimensional scanning has become more common, with powerful software available to even allow simple digital cameras on smart phones to take three-dimensional data. Most manufacturing applications require 2D scanning, although 3D scanners are useful for geometry capture.

Machine vision is particularly useful (a) for parts that are otherwise inaccessible, (b) in hostile manufacturing environments, (c) for measuring a large number of small features, and (d) in situations where physical contact may cause damage to a part. Applications of machine vision include online, real-time inspection in sheet-metal stamping lines and sensors for machine tools that can sense tool offset and tool breakage, verify part placement and fixturing, and monitor surface finish.

FIGURE 37.27 The use of a 3D scanner to digitize the geometry of a casting (below) to generate a 3D data file describing the geometry (above). The data file can then be used for quality control, or it can be sent to a 3D printer to produce a part with the same geometry. *Source:* Courtesy of EMS-USA.

Applications. Several applications of machine vision in manufacturing are shown in Fig. 37.26. With visual sensing capabilities, end effectors are capable of picking up parts and grip them in the proper orientation and location. Machine vision is capable of in-line identification and inspection of parts and of rejecting defective ones. *Robust sensors* have been developed to withstand extremes of temperature, shock and vibration, humidity, corrosion, dust and various contaminants, fluids, electromagnetic radiation, and other interferences.

The *selection* of a sensor for a particular application depends on such factors as:

- The particular quantity to be measured or sensed
- The sensor's interaction with other components in the system
- The sensor's expected service life
- Level of sophistication
- Difficulties associated with its use
- The power source
- Cost

Smart Sensors. These sensors have the capability to perform a logic function, conduct two-way communication, make decisions, and take appropriate actions. The knowledge required to make a decision can be built into a *smart sensor*; in machining, for example, a computer chip with sensors can be programmed to turn a machine tool off when a cutting tool breaks. Likewise, a smart sensor can stop a mobile robot, or a robot arm, from accidentally coming into contact with an object or people by sensing quantities such as distance, heat, and noise.

37.7.2   Sensor Fusion

Sensor fusion basically involves the integration of several sensors in such a manner that individual data from each of the sensors (such as force, vibration, temperature, and dimension data) are combined to provide a higher level of information and reliability. A simple application of sensor fusion can be demonstrated when someone drinks from a cup of hot coffee. Although such a common event is often taken for granted, it can readily be seen that this activity involves data input from eyes, lips, tongue, and hands. Through the five basic senses of sight, hearing, smell, taste, and touch, there is now real-time monitoring of relative movements, positions, and temperatures. If the coffee is too hot, for example, the hand movement of the cup toward the lip is slowed and adjusted accordingly; note that the fingers and the hand also already sense the temperature, thus becoming an input into the control system.

The earliest engineering applications of sensor fusion were in robot movement control, missile flight tracking, and similar military applications, often when these activities involve movements that mimic human behavior. An important aspect in sensor fusion is *sensor validation*: the failure of any one sensor is detected, so that the control system maintains high reliability; thus, it is essential to receive redundant data from different sensors.

Sensor fusion has become practical and available at relatively low cost, largely because of the advances made in sensor size, quality, and technology, as well as continued developments in computer-control systems, artificial intelligence, expert systems, and artificial neural networks, all described in Chapter 39.

37.8   Flexible Fixturing

In describing *work-holding devices* for the manufacturing operations throughout this book, the words *clamp, jig,* and *fixture* often were used interchangeably and sometimes in pairs, such as in jigs and fixtures. Briefly,

- **Clamps** are simple multifunctional work-holding devices.
- **Jigs** have various reference surfaces and points for accurate alignment of parts or tools.
- **Fixtures** generally are designed for specific applications.

Other common work-holding devices are *chucks, collets,* and *mandrels.* Some work-holding devices, such as *power chucks,* are designed and operated at various levels of mechanization and automation, and are driven by mechanical, hydraulic, or electrical means. Work-holding devices generally have specific ranges of capacity. For example, (a) a particular collet can accommodate bars only within a certain range of diameters; (b) four-jaw chucks can accommodate square or prismatic workpieces having a certain range of dimensions; (c) *dedicated fixtures* are designed and made for specific workpiece shapes and dimensions and for specific tasks; and (d) if a workpiece has curved surfaces, the contacting surfaces of the jaws are shaped by machining them to conform to the workpiece surfaces, known as *machinable jaws.*

The emergence of flexible manufacturing systems has necessitated the design and use of work-holding devices and fixtures that have *built-in flexibility*. There are several methods of **flexible fixturing**, based on different principles, called **intelligent fixturing systems**. These devices are capable of quickly accommodating a range of part shapes and dimensions, without requiring extensive changes, adjustments, or operator intervention.

Modular Fixturing.   *Modular fixturing* is often used for small or moderate lot sizes, especially when the cost of dedicated fixtures and the time required to produce them would be difficult to justify. Complex workpieces can be located within machines using fixtures produced quickly from standard components, and can then be disassembled when a production run is completed. Modular fixtures usually are based on tooling plates or blocks, configured with grid holes or T-slots upon which a fixture is constructed.

To quickly produce a fixture, several standard components, such as locating pins, workpiece supports, V-blocks, clamps, springs, and adjustable stops can be mounted onto a *base plate* or *block*. Using computer-aided fixture-planning techniques, such fixtures can be designed, assembled, and modified using industrial robots. Compared with dedicated fixturing, modular fixturing has been shown to be low in cost, has a shorter lead time, has more easily repaired components, and possesses more intrinsic flexibility of application.

Tombstone Fixtures.   Also referred to as *pedestal-type fixtures, tombstone fixtures* have between two and six vertical faces (hence resembling tombstones) onto which workpieces can be mounted. These fixtures are typically used in automated or robot-assisted manufacturing. The machine tool performs the desired operations on one face of the workpiece, then flips or rotates the fixture to work on other surfaces. Tombstone fixtures are commonly used for higher volume production, typically in the automotive industry. (See also Case Study 24.2.)

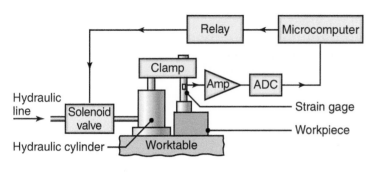

FIGURE 37.28   Schematic illustration of an adjustable-force clamping system; the clamping force is sensed by the strain gage, and the system automatically adjusts this force. *Source:* Based on P.K. Wright.

Bed-of-nails Device. This type of fixture consists of a series of *air-actuated pins* that conform to the shape of the external surfaces of the workpiece. Each pin moves as necessary to conform to the shape at its point of contact with the piece; the pins are then mechanically locked against the part. The device is compact, has high stiffness, and is reconfigurable.

Adjustable-force Clamping. A schematic illustration of this type of system is shown in Fig. 37.28. The strain gage mounted on the clamp senses the magnitude of the clamping force; the system then adjusts this force to keep the workpiece securely clamped for the particular application. It can also prevent excessive clamping forces that otherwise may damage the workpiece surface, particularly if it is soft or has a slender design.

Phase-change Materials. Other than by hard tooling, there are two methods capable of holding irregular-shaped or curved workpieces:

1. A *low-melting-point metal* is used as the clamping medium. Typically, an irregular-shaped workpiece is dipped into molten lead and allowed to set (similar to a wooden stick in a popsicle), a process that is similar to *insert molding* (Sections 11.3.5 and 19.3). After setting, the solidified lead block is clamped in a simple fixture. However, the possible adverse effect of such materials as lead on the workpiece to be clamped (due to *liquid–metal embrittlement*; see Section 1.5.2) must be considered.

2. The supporting medium is a *magnetorheological* (MR) or *electrorheological* (ER) fluid. In the MR method, the particles are ferromagnetic or paramagnetic nanoparticles, in a nonmagnetic fluid; surfactants are added to prevent the particles from settling. After the workpiece is immersed in the fluid, an external magnetic field is applied, whereby the particles are polarized and the behavior of the fluid changes from a liquid to that of a solid. The workpiece is later retrieved by removing the external magnetic field. In the ER method, the fluid is a suspension of fine dielectric particles in a liquid of low dielectric constant. Upon application of an electrical field, the liquid becomes a solid.

37.9 Assembly Systems

The individual parts and components produced by various manufacturing processes must be *assembled* into finished products. The total assembly operation is usually broken into individual assembly operations (*subassemblies*), with an operator assigned to carry out each step. Traditionally, assembly has involved much manual work and thus has contributed significantly to product cost.

Depending on the type of product, assembly costs can vary widely. For example, Apple iPhones cost $12.50–$30.00 in total labor, with a total cost of $200–$600. Assembly costs are 10–50% of the total cost of manufacturing, with the percentage of workers involved in assembly operations ranging from 20 to 60%. In developed countries, with high productivity and associated automation, the number of workers involved in assembly is on the low end of this range; in countries with inexpensive labor, the percentage is higher. As production costs and quantities of products to be assembled began to increase, the necessity for *automated assembly* became obvious. Beginning with the hand assembly of muskets with *interchangeable parts*, in the late 1700s and the early 1800s, assembly methods have vastly been improved upon over the years.

The first large-scale efficient application was the assembly of flywheel magnetos for the Model T Ford automobile. This experience eventually led to mass production of the automobile. The choice of an assembly method and system depends on the required production rate, the total quantity to be produced, the product's life cycle, the availability of labor, and cost.

Automated Assembly.   Recall that parts are manufactured within certain dimensional tolerance ranges. Taking ball bearings as an example, it is well known that, although they all have the same *nominal* dimensions, some balls in a lot will be smaller than others, although by a very small amount. Likewise, some bearing races will be smaller than others in the lot. There are two methods of assembly for such high-volume products.

In **random assembly**, the components are put together by selecting them randomly from the lots produced. In **selective assembly**, the balls and races are segregated (separated) by groups of sizes, from smallest to largest. The parts are then selected to mate properly; thus, the smallest diameter balls are mated with inner races having the largest outside diameter, and, likewise, with outer races having the smallest inside diameter.

Methods and Systems of Assembly.   There are three basic methods of assembly: manual, high-speed automatic, and robotic; they can be used individually or, as is the case in most applications, in combination. As shown in Fig. 37.29, an analysis of the product design must first be made to determine an appropriate and economical method of assembly:

1. **Manual assembly** uses relatively simple tools and generally is economical for small lots. Because of the dexterity of the human hand and fingers, and their capability for feedback through various senses, workers can manually assemble even complex parts without much difficulty. (In spite of the use of sophisticated mechanisms, robots, and computer controls, aligning and placing of a simple square peg into a square hole with small clearances can be a difficult task in automated assembly.)

2. **High-speed automated assembly** utilizes *transfer mechanisms* designed specially for assembly. Two examples are shown in Fig. 37.30, in which individual assembly is carried out on products that are *indexed* for proper positioning.

3. In **robotic assembly**, one or more general-purpose robots operate at a single workstation (Fig. 37.31) or they operate at a multistation assembly system.

There are three basic types of assembly systems:

1. **Synchronous systems.** In these *indexing* systems, individual parts are supplied and assembled at a constant rate at fixed individual stations. The rate of movement of the parts in this system is based on the station that takes the longest time to complete its portion of the assembly. The synchronous system is used primarily for high-volume, high-speed assembly of small products.

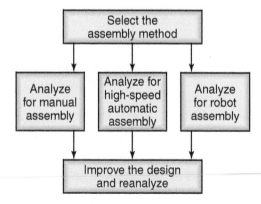

FIGURE 37.29 Stages in the design for assembly analysis. *Source:* Based on G. Boothroyd and P. Dewhurst.

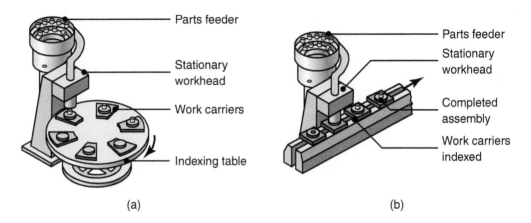

FIGURE 37.30 Transfer systems for automated assembly: (a) rotary indexing machine and (b) in-line indexing machine. *Source:* Based on G. Boothroyd.

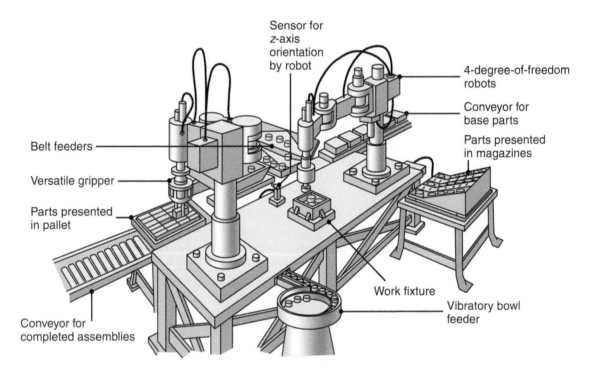

FIGURE 37.31 A two-arm robot assembly station. *Source:* Based on G. Boothroyd and P. Dewhurst.

Transfer systems move the partial assemblies from workstation to workstation by various mechanical means; two typical transfer systems (*rotary indexing* and *in-line indexing*) are shown in Fig. 37.30. These systems can operate in either a fully automatic or a semiautomatic mode; note, however, that a breakdown of one station will shut down the whole assembly operation.

Part feeders supply the individual parts to be assembled and place them on other components, which are mounted on work carriers or fixtures. The feeders move the individual parts by vibratory or other means through delivery chutes, and ensure their proper orientation by various ingenious means, some of which

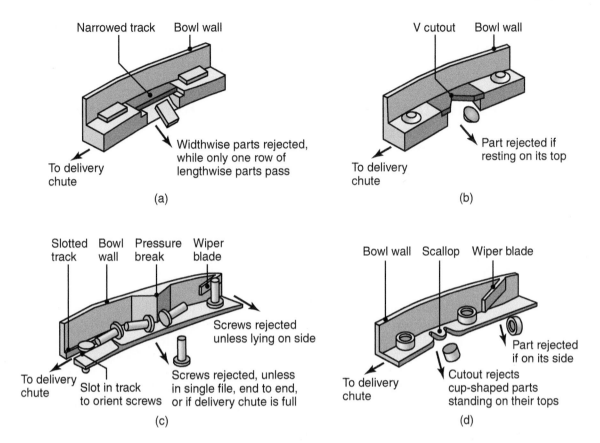

FIGURE 37.32 Examples of guides to ensure that parts are properly oriented for automated assembly. *Source:* Based on G. Boothroyd.

are shown in Fig. 37.32. Orienting parts properly and avoiding jamming are essential in all automated assembly operations.

2. **Nonsynchronous systems.** Each station operates independently, and any imbalance is accommodated in *buffer* (storage) between stations. The station continues operating until the next buffer is full or the previous buffer is empty. Also, if one station becomes inoperative, the assembly line continues to operate until all the parts in the buffer have been used up. Nonsynchronous systems are suitable for large assemblies with many parts to be assembled. Note that if the time required for the individual assembly operations vary significantly, the output will be constrained by the slowest station.

3. **Continuous systems.** The product is assembled while moving at a constant speed on pallets or similar workpiece carriers. The components to be assembled are brought to the product by various means, and their movements are synchronized with the continuous movement of the product. Typical applications of this system are in bottling and packaging plants, although the method also has been used on mass-production lines for automobiles and appliances.

Flexible Assembly Systems. Assembly systems generally are set up for a specific product line. They can, however, be modified for increased flexibility in order to assemble product lines that have a variety of product models. *Flexible assembly systems* (FAS) utilize computer controls, interchangeable and programmable workheads and feeding devices, coded pallets, and automated guiding devices. This system is capable of, for example, assembling up to a dozen different transmission and engine combinations and power steering and air-conditioning units.

37.10 Design Considerations for Fixturing, Assembly, Disassembly, and Servicing

As in many aspects of production, design of the devices and systems described above is an integral part of the total manufacturing operation.

37.10.1 Design for Fixturing

The proper design, construction, and operation of flexible work-holding devices and fixtures are essential to the efficient operation of advanced manufacturing systems. The major design issues involved are:

- Work-holding devices must position the workpiece automatically and accurately. They must maintain its location precisely and with sufficient clamping force to withstand the requirements for a particular manufacturing operation. Fixtures also should be able to accommodate parts repeatedly in the same position.
- Fixtures must have sufficient stiffness to resist, without excessive distortion, the normal and shear forces developed at the workpiece–fixture interfaces.
- The presence of loose machining or grinding chips and various other debris between the locating surfaces of the workpiece and the fixture can be a serious problem. Chips are most likely to be present where cutting fluids are used, as they tend to adhere to the wet surfaces due to surface-tension forces.
- A flexible fixture must be able to accommodate the parts to be made by different processes and ones with dimensions and surface features that vary from part to part. These considerations are even more important when the workpiece (a) is fragile or made of a brittle material, (b) is made of a relatively soft and flexible material, such as thermoplastics and elastomers, or (c) has a relatively soft coating on its contacting surfaces.
- Clamps and fixtures must have low profiles to avoid collision with cutting tools. Avoiding collisions is an important consideration in programming tool paths in machining operations.
- Flexible fixturing must meet special requirements in manufacturing cells and flexible manufacturing systems.
- Workpieces must be designed so as to allow locating and clamping within the fixture. Flanges, flats, or other locating surfaces should be incorporated into product design to simplify fixture design and to aid in part transfer into various machinery.

37.10.2 Design for Assembly, Disassembly, and Servicing

Design for Assembly.   While product design for manufacture is a matter of major interest, *design for assembly* (DFA) has attracted special attention (particularly design for automated assembly), because of the continued need to reduce assembly costs. In *manual assembly*, a major advantage is that humans can easily pick the correct parts from bulk, such as from a nearby bin, orient them properly, and insert them as necessary. In *high-speed automated assembly*, however, automatic handling generally requires that parts be separated from the bulk, conveyed by hoppers or vibratory feeders (Fig. 37.32), and assembled in their proper locations and orientations.

The principle of **poka-yoke** has often been applied in assembly. Poka-yoke is a Japanese term that refers to "mistake-proofing" or "fail-safing." This method also is applied to lean manufacturing systems (Section 39.6). With respect to assembly, this principle suggests that assembly operations must be designed so that operator-caused errors are unlikely or even impossible to occur. This approach requires a review of assembly operations and identification of potential problems, as well as corrective actions to minimize assembly errors.

Some of the general guidelines for design for assembly may be summarized as:

1. Reduce the number and variety of parts in a product. Simplify the product design and incorporate multiple functions into a single part, and design parts for easy insertion. Use common parts as much as possible. Consider subassemblies that would serve as modules.

2. Ensure that parts have a high degree of symmetry, such as round or square, or a high degree of asymmetry, such as oval or rectangular, so that they cannot be installed incorrectly and do not require locating, aligning, or adjusting.

3. Designs should allow parts to be assembled without any obstructions. There should be a direct line of sight. Assemblies should not have to be turned over for insertion of components.

4. Consider methods such as snap fits (see Fig. 32.19) to avoid the need for fasteners, such as bolts, nuts, and screws. If fasteners are used, their variety should be minimized and they should be spaced and located such that tools can be used without obstruction.

5. Part designs should consider such factors as size, shape, weight, flexibility, abrasiveness, and possible entanglement with other parts.

6. Assembly from two or more directions can be difficult. Parts should be inserted from a single direction, preferably vertically and from above in order to take advantage of gravity.

7. Products must be designed, or existing products redesigned, so that there are no physical obstructions to the free movement of parts during assembly. Sharp external and internal corners, for example, should be replaced with chamfers, tapers, or radii.

8. Color codes should be used on parts that may appear to be similar but are different. Letters or other symbols also can be provided to ensure proper part identification.

Robotic Assembly.   Design guidelines for robotic assembly include the following additional considerations:

- Parts should be designed so that they can be gripped and manipulated by the same gripper of the robot. Parts should be made available to the gripper in the proper orientation.
- Assembly that involves threaded fasteners (bolts, nuts, and screws) may be difficult to perform by robots; one exception is the use of self-threading screws for sheet metal, plastics, and wooden parts. Note that robots easily can handle snap fits, rivets, welds, and adhesives. The advances in compliant end effectors and dexterous manipulators have made robotic assembly even more attractive.

Evaluating Assembly Efficiency.   To evaluate *assembly efficiency*, each component of an assembly is evaluated with respect to its features that can affect both assembly itself and a baseline estimated time required to incorporate the part into the assembly. Note that assembly efficiency can also be measured for existing products. The assembly efficiency, η, is given by

$$\eta = \frac{Nt}{t_{\text{tot}}}, \tag{37.1}$$

where N is the number of parts, t_{tot} is the total assembly time, and t is the ideal assembly time for a small part that presents no difficulties in handling, orientation, or assembly; t is commonly taken to be 3 s. On the basis of Eq. (37.1), competing designs can thus be evaluated with respect to design for assembly. It has been observed that products that are in need of redesign to facilitate assembly usually have assembly efficiencies around 5–10%, while well-designed parts have efficiencies around 25%.

Design for Disassembly.   The manner and ease with which a product may be taken apart for maintenance or replacement of its parts is another important consideration in product design. Recall, for example, the difficulties one has in removing certain components from under the hood of some automobiles; similar difficulties exist in the disassembly of appliances and numerous other products.

The general approach to *design for disassembly* requires the consideration of factors that are similar to those for design for assembly. Analysis of computer or physical models of products and their components, with regard to disassembly, can generally indicate any potential problems, such as obstructions, size of passageways, lack of a line of sight, and the difficulty of firmly gripping and guiding components.

An important aspect of design for disassembly is how, after its life cycle (see Section 40.4), a product is to be taken apart for recycling, especially to salvage its more valuable components. Note, for example, that depending on their design and location, the type of tools used, and whether manual or power tools are used, (a) rivets will take longer to remove than screws or snap fits and (b) a bonded layer of valuable material on a component would be very difficult, if not impossible, to remove for recycling or reuse.

Obviously, the longer it takes to take components apart, the higher is the cost of doing so. It is then possible that this cost becomes prohibitive; consequently, the time required for disassembly also has to be studied and measured. Although the time depends on the manner in which disassembly is performed, some examples are: (a) cutting wire at 0.25 s, (b) disconnecting wire at 1.5 s, (c) effecting snap fits and clips at 1–3 s, and (d) loosening machine screws and bolts at 0.15–0.6 s per revolution. These quantities will of course greatly depend on the level of automation employed.

Design for Servicing.   *Design for servicing* is essentially based on the concept that the elements which are most likely to need servicing are at the *outer* layers of the product. In this way, individual parts are easier to reach and service, without the need to remove various other parts in order to do so. Thus, designing for assembly and disassembly should take into account the ease with which a product can be serviced and, if necessary, repaired.

37.11   Economic Considerations

As described in greater detail in Chapter 40, and as seen throughout many chapters in this book, there are numerous considerations involved in determining the overall *economics of production*. Because all production systems are essentially combinations of machines and people, important factors influencing the final decisions include:

- Type and cost of machinery, equipment, and tooling
- Cost of operation of the machinery
- Skill level and amount of labor required
- Production quantity desired

Recall also that lot size and production rate greatly influence the economics of production. Small quantities per year can be produced in job shops. However, the type of machinery in job shops generally requires skilled labor and the production quantity and rate are low; as a result, the cost per part can be high. Similarly, rapid prototyping facilities can be used for low production runs, and can be more economical if material requirements are compatible with the processing sequence.

At the other extreme is the production of very large quantities, using conventional flow lines and transfer lines and involving special-purpose machinery and equipment, specialized tooling, and computer-control systems. Although all of these components constitute major investments, both the level of skill required and the labor costs are relatively low, because of the high level of automation implemented. These production systems are, however, organized for a specific type of product and hence they lack flexibility.

Because most manufacturing operations are between the preceding two extremes, an appropriate decision must be made regarding the optimum level of automation to be implemented. In many situations, selective automation rather than total automation of a facility has been found to be cost effective.

SUMMARY

- Automation has been implemented in manufacturing processes, material handling, inspection, assembly, and packaging at increasing rates. There are several levels of automation, ranging from simple automation of machines to untended manufacturing cells.

- True automation began with the numerical control of machines, which offers flexibility of operation, lower cost, and ease of making different parts with less operator skill. Production quantity and rate are important factors in determining the economic levels of automation.

- Manufacturing operations are optimized further, both in quality and in cost, by adaptive control techniques, which continuously monitor an operation and quickly make necessary adjustments in the processing parameters.

- Major advances have been made in material handling, particularly with the implementation of industrial robots and automated guided vehicles.

- Sensors are essential in the implementation of modern technologies; a wide variety of sensors based on various principles have been developed and installed.

- Further advances include flexible fixturing and automated assembly techniques that reduce the need for worker intervention and lower manufacturing costs. Their effective and economic implementation requires that design for assembly, disassembly, and servicing be recognized as an important factor in the total design and manufacturing operations.

- Efficient and economic implementation of these techniques also requires that design for assembly, disassembly, and servicing be recognized as important elements in manufacturing.

KEY TERMS

Adaptive control	End effector	Open-loop control	Selective automation
Assembly	Feedback	Part programming	Sensor fusion
Automated guided vehicle	Flexible assembly systems	Poka-Yoke	Sensors
Automation	Flexible fixturing	Positioning	Smart sensors
Buffer	Hard automation	Power-head production units	Soft automation
Closed-loop control	Hardwired controls	Productivity	Stand-alone machines
Compliant end effectors	Industrial robot	Programmable controller	Tactile sensing
Computer numerical control	Intelligent robot	Programming language	Tombstone fixture
Computer vision	Interpolation	Random assembly	Total productive maintenance
Continuous path	Machine vision	Repeat accuracy	Transfer lines
Contouring	Manipulators	Resolution	Visual sensing
Control systems	Material handling	Robot	Work envelope
Dedicated machines	Mechanization	Selective assembly	
	Numerical control		

BIBLIOGRAPHY

Batchelor, B.G. (ed.), **Machine Vision Handbook**, Springer, 2012.

Boothroyd, G., Dewhurst, P., and Knight, W.A., **Product Design for Manufacture and Assembly**, 3rd ed., CRC Press, 2010.

Craig, J.J., **Introduction to Robotics: Mechanics and Control**, 3rd ed., Prentice Hall, 2004.

Davies, E.R., **Computer and Machine Vision**, 4th ed., Academic Press, 2012.

Fraden, J., **Handbook of Modern Sensors: Physics, Designs, and Applications**, 4th ed., Springer, 2010.

Hornberg, A., **Handbook of Machine Vision**, Wiley, 2007.

Ioannu, P.A., **Robust Adaptive Control**, Prentice Hall, 1995.

Kandray, D., **Programmable Automation Technologies: An Introduction to CNC, Robotics and PLCs**, Industrial Press, 2010.

Kurfess, T.R. (ed.), **Robotics and Automation Handbook**, CRC Press, 2004.

Mitchell, H.B., **Multi-Sensor Data Fusion: An Introduction,** Springer, 2007.

Quesada, R., **Computer Numerical Control Machining and Turning Centers**, Prentice Hall, 2004.

Rehg, J.A., **Introduction to Robotics in CIM Systems**, 5th ed., Prentice Hall, 2002.

Smid, P., **CNC Programming Handbook,** 2nd ed., Industrial Press, 2002.

Snyder, W.E., and Qi, H., **Machine Vision**, Cambridge University Press, 2010.

Soloman, S., **Sensors Handbook**, McGraw-Hill, 2009.

Stenerson, J., and Curran, K.S., **Computer Numerical Control: Operation and Programming**, 3rd ed., Prentice Hall, 2006.

Valentino, J.V., and Goldenberg, J., **Introduction to Computer Numerical Control**, 5th ed., Prentice Hall, 2012.

Zhang, J., **Practical Adaptive Control: Theory and Applications**, VDM Verlag, 2008.

REVIEW QUESTIONS

37.1 Describe the differences between mechanization and automation.

37.2 Explain the difference between hard and soft automation. Why are they so called?

37.3 What is productivity? Why is it important?

37.4 Explain the difference between a flexible manufacturing line and a transfer line.

37.5 Describe the principle of numerical control of machines.

37.6 Explain open-loop and closed-loop control circuits.

37.7 Describe the principle and purposes of adaptive control.

37.8 What factors have led to the development of automated guided vehicles?

37.9 What is a point-to-point control system? How is it different from a contouring system?

37.10 Describe the features of an industrial robot. Why are these features necessary?

37.11 List and describe the principles of various types of sensors.

37.12 Describe the concept of design for assembly. Why has it become an important factor in manufacturing?

37.13 Is it possible to have partial automation in assembly? Explain.

37.14 Explain the advantages of flexible fixturing.

37.15 How are robots programmed to follow a certain path?

37.16 What kind of end effectors are available for robots?

QUALITATIVE PROBLEMS

37.17 (a) Why is automation generally regarded as evolutionary rather than revolutionary? (b) Explain why it would be difficult to justify automation for small production runs.

37.18 Are there activities in manufacturing operations that cannot be automated? Explain.

37.19 What is a programmable logic controller? Why are they popular?

37.20 Explain the factors that have led to the development of numerical control.

37.21 Giving specific examples, discuss your observations concerning Fig. 37.2.

37.22 What are the relative advantages and limitations of the two arrangements for power heads shown in Fig. 37.4?

37.23 Discuss methods of online gaging of workpiece diameters in turning operations other than that shown in Fig. 37.15.

37.24 Are drilling and punching the only applications for the point-to-point system shown in Fig. 37.10a? Explain.

37.25 If three points were known on a straight line, is it better to use linear or circular interpolation? Explain.

37.26 What determines the number of robots in an automated assembly line such as that shown in Figs. 37.22a and 37.30?

37.27 Describe situations in which the shape and size of the work envelope of a robot (Fig. 37.21) can be critical.

37.28 Explain the difference between an automated guided vehicle and a self-guided vehicle.

37.29 Explain why sensors have become so essential in the development of automated manufacturing systems.

37.30 Table 37.2 shows a few examples of typical products for each category. Add several other examples to the table.

37.31 List applications for industrial robots.

37.32 What is meant by the term sensor fusion?

37.33 Describe applications of machine vision for specific parts that are similar to the examples shown in Fig. 37.26.

37.34 What is a tombstone fixture?

37.35 Sketch the workspace (envelope) of each of the robots shown in Fig. 37.20.

37.36 List the advantages and disadvantages of modular filtering.

QUANTITATIVE PROBLEMS

37.37 A spindle–bracket assembly uses the following parts: a steel spindle, two nylon bushings, a stamped steel bracket, and six screws and six nuts to attach the nylon bushings to the steel bracket, and thereby support the spindle. Compare this assembly with the spindle–bracket assembly shown in Problem 16.65, and estimate the assembly efficiency for each design.

37.38 Disassemble a simple ballpoint pen. Carefully measure the time it took for you to reassemble the pen, and calculate the assembly efficiency. Repeat the exercise for a mechanical pencil.

37.39 Examine Fig. 37.11b, and obtain an expression for the maximum error in approximating a circle with linear increments, as a function of the radius of the circle and the number of increments on the circumference of the circle.

37.40 Review Example 14.1, and develop open- and closed-loop control system equations for the force if the coefficient of friction is μ.

SYNTHESIS, DESIGN, AND PROJECTS

37.41 Refer to Part III of this book, and give an example of a metal forming operation that is suitable for adaptive control.

37.42 Describe possible applications for industrial robots not discussed in this chapter.

37.43 Design two different systems of mechanical grippers for two widely different applications.

37.44 Give some applications for the systems shown in Figs. 37.26a and c.

37.45 For a system similar to that shown in Fig. 37.28, design a flexible fixturing setup for a lathe chuck.

37.46 Give examples of products that are suitable for the three types of production shown in Fig. 37.3.

37.47 Describe situations in which tactile sensors would not be suitable. Explain why.

37.48 Are there situations in which machine vision cannot be applied properly and reliably? Explain.

37.49 Choose one machine each from Parts II through IV, and design a system in which sensor fusion can be used effectively.

37.50 Think of a product, and design a transfer line for it which is similar to that shown in Fig. 37.5. Specify the types and the number of machines required.

37.51 Describe your thoughts on the usefulness and applications of modular fixturing consisting of various individual clamps, pins, supports, and attachments, mounted on a base plate.

37.52 Inspect several household products and describe the manner in which they have been assembled. Comment on any product design changes you would make so that assembly, disassembly, and servicing are simpler and faster.

37.53 Inspect Table 37.1 on the history of automation, and describe your thoughts as to what new developments might be added to the bottom of the list in the near future.

37.54 Design a robot gripper that will pick up and place the following: (a) eggs, (b) an object made of foam rubber, (c) a metal ball with a very smooth and polished surface, (d) a newspaper, and (e) tableware, such as knives, spoons, and forks.

37.55 Design an end effector that can pick up and place (a) a marshmallow in the shape of an egg and (b) a steel casting in the shape of an egg.

37.56 Review the specifications of various numerical-control machines, and make a list of typical numbers for their (a) positioning accuracy, (b) repeat accuracy, and (c) resolution. Comment on your observations.

37.57 Describe the sensors that you use in a simple act, such as walking or throwing a ball.

37.58 Give an example of each of the design rules in Section 37.10.2.

37.59 Obtain an old toaster and disassemble it. Explain how you would go about reassembling it by automated assembly.

37.60 Conduct an Internet search and obtain product literature for four kinds of 3D scanners. Compare the price and capabilities of each.

37.61 Assume that you are asked to give a quiz to students on the contents of this chapter. Prepare five quantitative problems and five qualitative questions, and supply the answers.

Computer-aided Manufacturing

- Computers have fundamentally changed the product design and manufacturing enterprise; powerful computer software is now available to assist and integrate all engineering tasks.
- This chapter opens with a description of computer-aided design, in which the graphic description of parts is created and stored in software.
- The use of computers in the direct control of manufacturing processes and in computer-aided manufacturing is then discussed.
- The chapter then describes how software can allow the simulation of manufacturing processes and systems.
- Finally, a description of group technology is presented—an approach that is often built into CAD software, allowing the rapid recovery of previous design and manufacturing experience, and an essential tool for production flow analysis.

38.1 Introduction

The importance of product quality was emphasized in Chapter 36, along with the necessity for the commitment of a company to total quality management. Recall also the statements that *quality must be built into the product*, that high quality does not necessarily mean high costs, and that marketing poor-quality products can indeed be very costly to the manufacturer.

High quality is far more attainable and less expensive if design and manufacturing activities are properly integrated, rather than treated as separate activities. Integration can be performed, successfully and effectively, through *computer-aided design, engineering, manufacturing, process planning,* and *simulation of processes and systems*, as described throughout this chapter. The widespread availability of high-speed computers and powerful software has allowed computers to proliferate into all areas of manufacturing.

Computer technology is pervasive and exists at many levels. A part geometry can be programmed in CAD software, which is in itself a fairly complex computer program. The manufacture of a part can be achieved, for example, by programming it into G-code (Section 37.3.6), which uses another fairly complex computer program to translate geometric instructions into machine actions. Software is currently available and sufficiently powerful to integrate design with CNC programming activities and, thereby, streamline the design and manufacturing process. Indeed, every aspect of the modern manufacturing enterprise is currently associated with computers and software, and integration of the entire business through communication standards is now possible.

38.2 Manufacturing Systems

Manufacturing is a complex *system*, because it consists of many diverse physical and human elements. Some of these elements are difficult to predict and control, due to such factors as the supply and cost of raw materials, the impact of continually developing technologies, global market changes, and human behavior and performance. Ideally, a manufacturing system should be represented by mathematical and physical models that show the nature and extent of the interdependence of all relevant variables. In this way, the effects of a change or a disturbance that occurs anywhere in the system can be analyzed and necessary and timely adjustments can be made.

The supply of a particular raw material may, for example, decrease significantly due to global demands or for geopolitical reasons. Because, as a result, the raw material cost will rise (supply and demand), alternative materials have to be considered and selected. The selection must be made after a careful consideration of several factors, because such a change may have adverse effects on product quality, production rate, and manufacturing costs. The material selected, for example, may not be as easy to form, machine, or weld, and thus product integrity may suffer.

In a constantly changing global marketplace, the demand for a product also may fluctuate randomly and rapidly for a variety of reasons. As examples, note the downsizing of automobiles in response to rising fuel costs and the increasing popularity of gas–electric hybrids, fuel cells, and electrically powered vehicles. The manufacturing system must be able to produce the modified product in a relatively short lead time while minimizing large expenditures in new machinery and tooling that will be required.

Such a complex system can be difficult to analyze and model, largely because of a lack of comprehensive and reliable data on all of the variables involved. Moreover, it is difficult to correctly predict and control some of these variables, because (a) raw-material costs are difficult to predict accurately, (b) machine-tool characteristics, their performance, and their response to random external disturbances cannot be precisely modeled, and (c) human behavior and performance are even more difficult to model.

38.3 Computer-integrated Manufacturing

Computer-integrated manufacturing (CIM) involves the computerized integration of all aspects of product design, process planning, production, and distribution, as well as the management and operation of the whole manufacturing organization. CIM is a *methodology*, and its effectiveness critically depends on the use of a *large-scale integrated communications system*, consisting of computers, machines, and their controls, described in Section 39.7. Because CIM ideally should involve the total operation of an organization, it requires an extensive *database*, concerning the technical and business aspects of the operation. Consequently, if planned all at once, CIM can be prohibitively expensive, particularly for small and medium-size companies.

Comprehensive and long-range *strategic planning* covering all phases of the operation is thus essential in order to fully benefit from CIM. Such planning and the level of integration must take into account considerations such as (a) the mission, goals, and culture of the organization; (b) the availability of financial, technical, and human resources; and (c) the existing, as well as emerging, technologies in the areas of the products to be manufactured.

Subsystems of CIM.   Computer-integrated manufacturing systems comprise the following subsystems, which are integrated into a whole (Fig. 38.1):

1. Business planning and support
2. Product design
3. Manufacturing process planning
4. Process automation and control
5. Production-monitoring systems

The subsystems are designed, developed, and implemented in such a manner that the output of one subsystem serves as the input of another. Organizationally, the subsystems generally are divided into two functions:

- **Business-planning functions:** Forecasting, scheduling, material-requirements planning, invoicing, and accounting
- **Business-execution functions:** Production and process control, material handling, testing, and inspection of the system

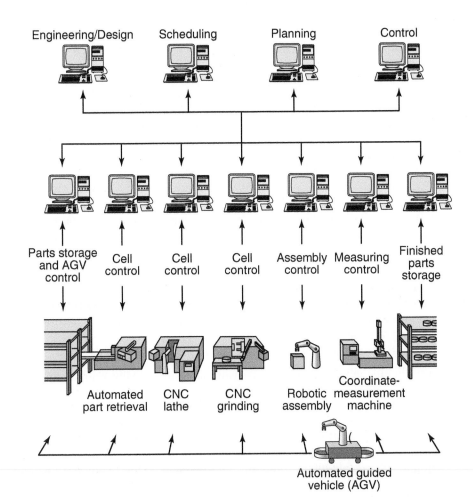

FIGURE 38.1   A schematic illustration of a computer-integrated manufacturing system; the manufacturing cells and their controls shown at the lower left are described in Section 39.2. *Source:* Based on U. Rembold.

If implemented properly, the major benefits of CIM are:

- Emphasis on *product quality* and uniformity, through better process control
- *Efficient* use of materials, machinery, and personnel and a major reduction in work-in-progress inventory, all of which improve productivity and lower product cost
- *Total control* of the production, schedules, and management of the entire manufacturing operation
- *Responsiveness* to shorter product life cycles, changing market demands, and global competition

38.3.1   Database

Databases consist of up-to-date, detailed, and accurate information relating to designs, products, processes, materials, machinery, production, finances, purchasing, sales, and marketing. This vast array of information is stored in computer memory and recalled or modified as necessary, either by individuals in the organization or by the CIM system itself. An efficient CIM system requires a *single database* to be shared by the entire manufacturing organization.

A database typically consists of the following items, some of which are classified as technical and others as nontechnical:

- **Product data:** Part shape, dimensions, and specifications
- **Data-management attributes:** Part number and revision level, including descriptions or keywords to assist in retrieving data
- **Production data:** Manufacturing processes employed
- **Operational data:** Scheduling, lot sizes, and assembly requirements
- **Resources data:** Capital, machines, equipment, tooling, personnel, and their capabilities

Databases are compiled by individuals in the organization, with input from various sensors in production machinery and equipment. Data are automatically collected by a **data-acquisition system** (DAS), which can track the number of parts being produced per unit of time and their dimensional accuracy, surface finish, weight, and other characteristics, at specified rates of sampling. The components of DAS include microprocessors, transducers, and analog-to-digital converters (ADC). Data-acquisition systems also are capable of analyzing data and transferring them to other computers for such purposes as statistical analysis, data presentation, and the forecasting of product demand.

Several factors are important in the use and implementation of databases:

1. They should be timely, accurate, easily accessible, easily shared, and user friendly.
2. Because they are used for a variety of purposes and by many people in an organization, databases must be flexible and responsive to the needs of different users.
3. CIM systems can be accessed by designers, manufacturing engineers, process planners, financial officers, and the management of the company, through appropriate access codes; companies must protect data against tampering or unauthorized use.
4. If problems arise with data accuracy or loss of data, the correct data should be recovered and restored.

38.4   Computer-aided Design and Engineering

Computer-aided design (CAD) involves the use of computers to create design drawings and product models (see also Fig. I.10 in the General Introduction). CAD is generally associated with **interactive computer graphics**, known as a **CAD system**. These systems are powerful tools and are used in the design and geometric modeling of components and products. The designer can easily conceptualize the part to be designed on a computer monitor, and can consider alternative designs or modify a particular design to quickly respond to specific design requirements.

There are several powerful commercially available programs to aid designers in geometry description and engineering analysis, such as SolidWorks, ProEngineer, CATIA, AutoCAD, Solid Edge, and VectorWorks. The software can help identify potential problems, such as excessive loads, deflections, or interference at mating surfaces when encountered during assembly. Information, such as a list of materials, specifications, and manufacturing instructions, also is stored in the CAD database. Using this information, the product designer can then analyze the manufacturing economics of alternative designs.

Computer-aided engineering (CAE) allows several applications to share the information in the database. These applications include, for example, (a) finite-element analysis of stresses, strains, deflections, and temperature distribution in structures and load-bearing members; (b) the generation, storage, and retrieval of NC data; and (c) the design of integrated circuits and various electronic devices.

38.4.1   Exchange Specifications

Because of the availability of a wide variety of CAD systems with different characteristics and supplied by different vendors, effective communication and exchange of data between these systems is essential. **Drawing exchange format** (DFX) was developed for use with *Autodesk®* and is still supported, but has been superseded by the DWG file format. Stereolithography (STL) formats are used to export three-dimensional geometries, initially only to *rapid-prototyping systems* (Chapter 20), but they now have become a format for data exchange between different CAD systems.

The necessity for a single, neutral format for better compatibility and for the transfer of more information than geometry alone is currently filled mainly by the **Initial Graphics Exchange Specification** (IGES). This format is used for translation in two directions (in and out of a system), and is also widely used for the translation of three-dimensional line and surface data. There are several variations of IGES in existence; the latest is version 5.3, published in 1996.

Another useful format is a solid-model-based standard, called the **Product Data Exchange Specification** (PDES), which is based on the Standard for the Exchange of Product model data (STEP) and developed by the International Standards Organization. PDES allows information on shape, design, manufacturing, quality assurance, testing, maintenance, etc., to be transferred between CAD systems. The increasing popularity of PDES and STEP have led to less frequent use of IGES.

38.4.2   Elements of CAD Systems

The design process using a CAD system consists of four stages: geometric modeling, design analysis and optimization, design review and evaluation, and database.

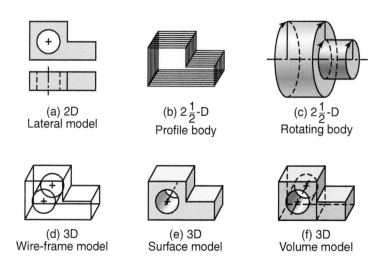

(a) 2D
Lateral model

(b) $2\frac{1}{2}$-D
Profile body

(c) $2\frac{1}{2}$-D
Rotating body

(d) 3D
Wire-frame model

(e) 3D
Surface model

(f) 3D
Volume model

FIGURE 38.2   Various types of modeling for CAD.

Geometric Modeling.   In *geometric modeling*, a physical object or any of its parts is described mathematically. The designer first constructs a geometric model by giving commands that create or modify lines, surfaces, solids, dimensions, and text. Together, these elements present an accurate and complete two- or three-dimensional representation of the object. The results are displayed and can be moved around on the screen, and any section can be magnified to view details.

The models in a CAD system can be presented in three ways (Fig. 38.2):

1. In **line representation**, also called **wire-frame** representation, all of the *edges* of the model are visible as solid lines. This image can, however, be ambiguous or difficult to visualize, particularly for complex shapes.

The three types of wire-frame representations are *two, two-and-one-half,* and *three dimensional*. A two-dimensional image shows the profile of the object, and a two-and-one-half-dimensional image can be obtained by a *translational sweep* (by moving the two-dimensional object along the *z*-axis). For round objects, a two-and-one-half-dimensional model can be generated simply by *rotating* a two-dimensional model around its axis.

2. In the **surface model**, all visible *surfaces* are shown; these models define surface features and edges of objects. CAD programs now use *Bezier* curves, B-splines, or nonuniform rational B-splines (NURBS) for surface modeling. Each of these approaches uses control points to define a polynomial curve or surface. A Bezier curve passes through the first and last vertex and uses the other control points to generate a blended curve. The drawback to Bezier curves is that any modification of one control point will affect the entire curve.

B-splines are blended piecewise polynomial curves where the modification of a control point affects only the curve in the area of the modification. Figure 38.3 shows examples of two-dimensional Bezier curves and B-splines. A *NURBS* is a special type of B-spline such that each control point has a weight associated with it.

3. In the **solid model**, all surfaces are shown, but the data describes the interior volume. Solid models can be constructed from (a) *swept volumes* (Fig. 38.2c and e) or by the techniques shown in Fig. 38.4; (b) *boundary representation* (B-rep), where surfaces are combined to develop a solid model (Fig. 38.4a); and (c) *constructive solid geometry* (CSG), where simple shapes, such as spheres, cubes, blocks, cylinders, and cones (called *primitives of solids*), are combined to develop a solid model (Fig. 38.4b).

Computer programs allow the user to select any combination of these primitives and their sizes and combine them into the desired solid model. Although solid models have such advantages as ease of design analysis and ease of preparation for manufacturing the part, they require more computer memory and processing time than the wire-frame and surface models shown in Fig. 38.2.

The standard for rapid prototyping machinery, the **STL file format** (an abbreviation for *stereolithography*, but also called *Standard Tessellation Language*), allows

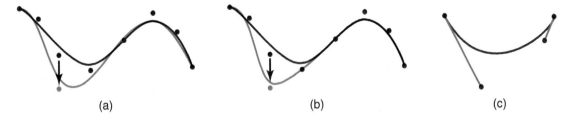

FIGURE 38.3 Types of splines. (a) A Bezier curve passes through the first and last control point, but it generates a curve from the other points; changing a control point modifies the entire curve. (b) A B-spline is constructed piecewise, so that changing a vertex affects the curve only in the vicinity of the changed control point. (c) A third-order (cubic) piecewise Bezier curve is constructed through two adjacent control points, with two other control points defining the slope of the curve at the endpoints. A third-order piecewise Bezier curve is continuous, but its slope may be discontinuous.

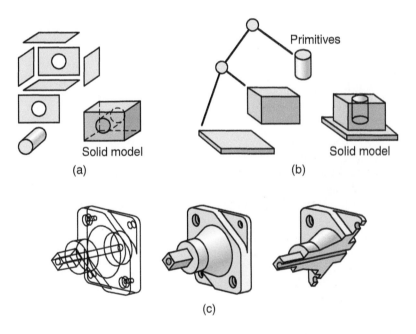

FIGURE 38.4 (a) Boundary representation of solids, showing the enclosing surfaces of the solid model and the generated solid model. (b) A solid model represented as compositions of solid primitives. (c) Three representations of the same part by CAD. *Source:* Based on P. Ranky.

for three-dimensional part descriptions. Basically, an STL file consists of a number of triangles that define the exterior surface (Fig. 38.5). With a sufficiently large number of triangles, the surface can be defined within a prescribed tolerance, although it requires a larger file size. With additive manufacturing, a part cross-section can be obtained at any height, and the resulting polygon is then used to plan the part (see Fig. 20.3). The use of STL in rapid prototyping, along with its easy implementation, has led to the use of this format in other applications as well, such as computer graphics and general CAD data transfer.

A special kind of solid model is a **parametric model**, where a part is stored not only in terms of a B-rep or CSG definition, but is derived from the dimensions and constraints that define the features (Fig. 38.6). Whenever a change is made, the part

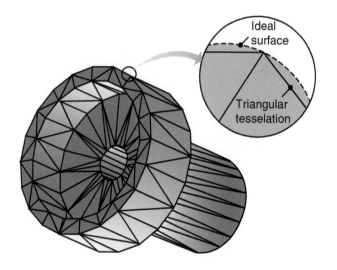

FIGURE 38.5 An example of an STL part description. Note that the surface is defined by a tessellation of triangles, and that there is an inherent error of form that occurs with curved surfaces; however, this can be brought to any desired tolerance by incorporating more triangles in the surface.

is recreated from these definitions, a feature that allows for simple and straightforward updates and changes to be made to the models.

The **octree representation** of a solid object is shown in Fig. 38.7; it is a three-dimensional analog to pixels on a television screen. Just as any area can be broken down into quadrants, any volume can be broken down into octants, which are then identified as solid, void, or partially filled. Partially filled *voxels* (from the words *vo*lume and pi*xels*) are broken into smaller octants and are reclassified. With increasing resolution, exceptional part detail can be achieved. Although this process may appear to be somewhat cumbersome, it allows for accurate description of complex surfaces.

Octree representation is used particularly in biomedical applications, such as modeling bone geometries, and also has been implemented in computer games where visual impact of objects at multiple scales is desired. For simpler applications, voxel representations at one resolution can be used.

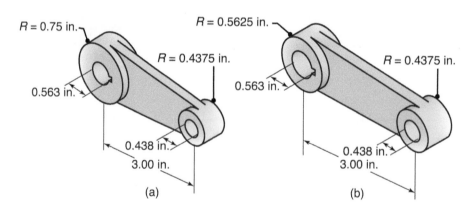

FIGURE 38.6 An example of parametric design; the dimensions of part features can easily be modified to quickly obtain an updated solid model.

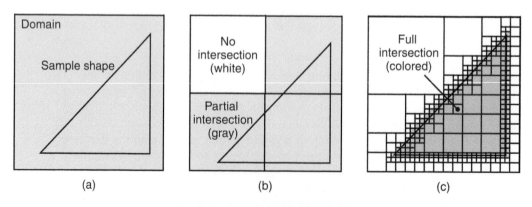

FIGURE 38.7 The octree representation of a solid object; any volume can be broken down into octants, which are then identified as solid, void, or partially filled. Shown is a two-dimensional version (quadtree) for the representation of shapes in a plane.

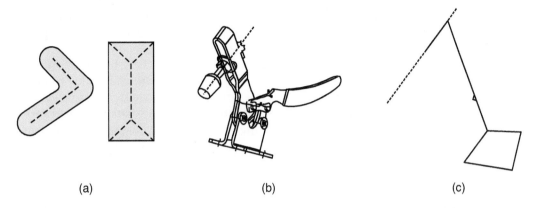

(a)   (b)   (c)

FIGURE 38.8   (a) Illustration of the skeleton data structure for solid objects; the skeleton is the dashed line in the object interior; (b) general view of a clamp; and (c) skeleton model. *Source*: S.D. Lockhart and C.M. Johnson, *Engineering Design Communication*, Prentice Hall, 2000.

A **skeleton** (Fig. 38.8) is the family of lines, planes, and curves that describe a part, but without the detail of surface models; it is commonly used for kinematic analysis of parts or assemblies. Conceptually, a skeleton can be constructed by fitting the largest circle (or sphere, for three-dimensional objects) within the geometry; the skeleton is the set of points that connect the centers of the circles or spheres. A continuing area of research involves using skeleton models instead of conventional surface or solid models, especially for complicated geometries such as in biomedical applications.

Design Analysis and Optimization.   After the geometric features of a particular design have been determined, the design is subjected to engineering analysis. This phase may, for example, consist of analyzing stresses, strains, deflections, vibrations, heat transfer, temperature distribution, or dimensional tolerances. Various software packages are now available, such as the finite element-based programs ABAQUS, ANSYS, NASTRAN, LS-DYNA, MARQ, and ALGOR, each having the capability to compute these quantities accurately and rapidly.

Because of the relative ease with which such analyses can be carried out, designers increasingly are willing to analyze a design more thoroughly before it is moved on to production. Experiments and measurements in the field nonetheless may be necessary to determine the actual effects of loads, temperature, and various other variables on the designed components.

Design Review and Evaluation.   An important design stage is design review and evaluation in order to check for any interference or excess gap between various components. The review is done in order to avoid difficulties either during assembly or in the use of the part and to determine whether moving members, such as linkages, are going to operate as intended. Software is available with animation capabilities to identify potential problems with moving members and other dynamic situations. During this stage, the part is dimensioned and toleranced precisely to the full degree required for manufacturing.

Database.   Many components, such as bolts and gears, either are standard components, mass produced according to a given design specification, or are identical to the parts used in previous designs. CAD systems thus have a built-in database management system that allows designers to locate, view, and adopt parts from a stock part library. These parts can be modeled parametrically to allow cost-effective updating of the part geometry. Some databases are available commercially, with extensive parts libraries; many vendors make their part libraries also available on the Internet.

38.5 Computer-aided Manufacturing

Computer-aided manufacturing (CAM) involves the use of computers to assist in all phases of manufacturing a product; it encompasses many of the technologies described in Chapter 37 and in this chapter. Because of their joint benefits, CAD and CAM are often combined into **CAD/CAM systems**. This combination allows the transfer of information from the design stage to the stage of planning for manufacture, without the necessity to reenter the data on part geometry manually.

The database developed during CAD is stored and further processed by CAM into the relevant data and instructions, for such purposes as operating and controlling production machinery, material-handling equipment, and automated testing and inspection for product quality. CAD/CAM systems also are capable of coding and classifying parts into groups that have similar design or manufacturing attributes, as described in Section 38.8.3.

Typical applications of CAD/CAM include:

- Programming for numerical control and industrial robots
- Design of dies and molds for casting in which, for example, shrinkage allowances are preprogrammed
- Dies for metalworking operations, such as complex dies for sheet forming and progressive dies for stamping
- Design of tooling and fixtures, and EDM electrodes
- Quality control and inspection, such as coordinate-measuring machines programmed on a CAD/CAM workstation
- Process planning and scheduling
- Plant layout

An important feature of CAD/CAM in machining operations, for example, is the capability to calculate and describe the *tool path* (see Figs. 20.3, 24.2, 26.14, and 26.22). The instructions (*programs*) are computer generated, and they can be modified by the programmer to optimize the tool path. The engineer or technician can then display and visually check the tool path for possible tool collisions with clamps, fixtures, or other interferences.

By standardizing product development and reducing design effort, tryout, and prototype work, CAD/CAM has made possible significantly reduced manufacturing costs and improved productivity. The two-engine Boeing 777 passenger airplane, for example, was designed completely by computer (known as **paperless design**), with 2000 workstations linked to eight computers. The plane was constructed *directly* from the CAD/CAM software that was developed (an enhanced CATIA system), and no prototypes or mock-ups were built, as were required for previous models. The cost for this development was on the order of $6 billion.

38.6 Computer-aided Process Planning

Process planning is basically concerned with selecting methods of production: tooling, fixtures, machinery, sequences of operations, and assembly; all of these diverse activities must be planned, which traditionally has been done by process planners.

ROUTING SHEET		
CUSTOMER'S NAME: Midwest Valve Co.	PART NAME: Valve body	
QUANTITY: 15	PART NO.: 302	
Operation No.	Description of operation	Equipment
10	Inspect forging, check hardness	Rockwell tester
20	Rough machine flanges	Lathe No. 5
30	Finish machine flanges	Lathe No. 5
40	Bore and counterbore hole	Boring mill No. 1
50	Turn internal grooves	Boring mill No. 1
60	Drill and tap holes	Drill press No. 2
70	Grind flange end faces	Grinder No. 2
80	Grind bore	Internal grinder No. 1
90	Clean	Vapor degreaser
100	Inspect	Ultrasonic tester

FIGURE 38.9   An example of a simple, traditional routing sheet. *Operation sheets* may include additional information on materials, tooling, the estimated time for each operation, processing parameters (such as cutting speeds and feeds), and other information; the routing sheet travels with the part from operation to operation. Routing sheets have been largely replaced with computer databases; this routing sheet illustrates the representative minimum content for a simple part.

The sequence of processes and operations to be performed, the machines to be used, the standard time for each operation, and similar information all are documented in a computer file. **Routing sheets** (Fig. 38.9) are the traditional means for storing manufacturing data, and are useful to demonstrate the type of data required.

   Computer-aided process planning (CAPP) accomplishes the complex task of process planning by viewing the total operation as an *integrated* system, so that the individual processing steps are coordinated and performed efficiently and reliably. CAPP is particularly effective in small-volume, high-variety parts production. Although extensive software and good coordination with CAD/CAM and other aspects of integrated manufacturing systems (described throughout the rest of this chapter) are required, CAPP is a powerful tool for efficiently planning and scheduling manufacturing operations.

38.6.1   Elements of CAPP Systems

There are two types of computer-aided process-planning systems.

Variant System.   Also called the **derivative system,** these computer files contain a standard process plan for the part to be produced. On the basis of its shape and its manufacturing characteristics, a search for a standard plan is then conducted in the database, using a specific code number for the part. The plan is retrieved, displayed for review, and printed as a routing sheet.

The *variant-process plan* includes such information as the types of tools and machines required, the sequence of operations to be performed, and the speeds, feeds, and time required for each sequence. Minor modifications of an existing process plan, which usually are necessary, also can be made. If the standard plan for a particular part is not in the computer files, a plan that is close to it and that has a similar code number and an existing routing sheet is retrieved. If a routing sheet does not exist for a new part, one is made and stored in computer memory.

Generative System.   In this system, a process plan is automatically generated on the basis of the same logical procedures that would be followed by a traditional process planner in making that particular part. However, the *generative system* is complex, because it must contain comprehensive and detailed information about (a) the part shape and dimensions, (b) process capabilities, (c) selection of manufacturing methods, machinery, and tools, and (d) the sequence of operations to be performed.

The generative system can create a new plan instead of having to use and modify an existing plan, as the variant system must do. Although generally used less commonly than the variant system, the generative system has such advantages as (a) flexibility and consistency in process planning for new parts and (b) higher overall planning quality, because of the capability of the decision logic in the system to optimize the planning and to utilize up-to-date manufacturing technology.

The process-planning capabilities of computers also can be integrated into the planning and control of production systems. These activities are a subsystem of computer-integrated manufacturing, as described in Section 38.3. Several functions can be performed, such as **capacity planning** for plants to meet production schedules, control of inventory, purchasing, and production scheduling.

38.6.2   Material-requirements Planning and Manufacturing Resource Planning

Computer-based systems for managing inventories and delivery schedules of raw materials and tools are called *material-requirements planning* (MRP) systems. Also regarded as a method of **inventory control**, MRP involves the keeping of complete records of inventories of materials, supplies, parts in various stages of production (called *work in progress*, WIP), orders, purchasing, and scheduling. Several files of data usually are involved in a master production schedule. These files pertain to the raw materials required (listed on a **bill of materials**), product structure levels (individual items that compose a product, such as components, subassemblies, and assemblies), and scheduling.

Manufacturing resource planning (MRP-II) controls all aspects of manufacturing planning through feedback. Although the system is complex, it is capable of final production scheduling, monitoring actual results in terms of performance and output, and comparing those results against the master production schedule.

38.6.3   Enterprise Resource Planning

Enterprise resource planning (ERP) is basically an extension of MRP-II. Although there are several variations, it is also a method for effective planning and control of all the resources needed in a business enterprise to take orders for products, produce them, ship them to the customer, and service them. ERP thus attempts to coordinate, optimize, and dynamically integrate all information sources and the widely diverse technical and financial activities in a manufacturing organization.

Effective implementation of ERP can be a challenging task because of

- The difficulties encountered in timely, effective, and reliable communication among all parties involved, especially in a global business enterprise
- The need for changing and evolving business practices, in an age where information systems and **e-commerce** (defined as buying and selling of products or services over electronic systems) have become highly relevant and important to the success of business organizations
- The need to meet extensive and specific hardware and software requirements for ERP. ERP-II is a more recent development that uses web-based tools to perform the tasks of ERP. These systems are intended to extend the ERP capabilities beyond the host organization, and allowing interaction and coordination across corporate entities

38.7 Computer Simulation of Manufacturing Processes and Systems

With increasing sophistication of computer hardware and software, *computer simulation of manufacturing processes and systems* has advanced rapidly. Simulation takes two basic forms:

1. It is a model of a specific operation, intended to determine the viability of a process or to optimize and improve its performance.
2. It models multiple processes and their interactions, to help process planners and plant designers in the layout of machinery and facilities.

Individual processes can be modeled using various mathematical schemes. Typical problems addressed are (a) **process viability**, such as the formability of sheet metal in a certain die and (b) **process optimization**, such as material flow in a forging in a given die to identify potential defects, or mold design in casting to eliminate hot spots, promote uniform cooling, and minimize defects. Finite-element analysis is increasingly being applied in software packages (called **process simulation**), that are available commercially and are inexpensive.

The simulation of an entire manufacturing system, involving multiple processes and equipment, helps plant engineers to organize machinery and to identify critical machinery elements. Such models also can assist manufacturing engineers with scheduling and routing by *discrete-event simulation*. Commercially available software packages often are used for these purposes, although dedicated software programs, written for a particular company, also can be developed.

EXAMPLE 38.1   Simulation of Plant-scale Manufacturing

Several examples and case studies presented in this book have focused upon the simulation of individual manufacturing processes and operations. The availability of low-cost, high-performance computer systems and the development of advanced software have allowed the simulation of *entire* manufacturing systems, and have led to the optimization of manufacturing and assembly operations.

As an example, *Digital Manufacturing Hub* software (Delmia Corporation) allows the simulation of manufacturing processes in three dimensions, including (a) the use of human manikins, to identify safety hazards, manufacturing problems, or bottlenecks; (b) improving machining accuracy; or (c) optimizing tooling organization (see Fig. 38.10a). Since simulation can be performed prior to building an assembly line, it can significantly reduce development times and cost.

Figure 38.10b illustrates the simulation of a robotic welding line in an automotive plant, where the motions of a robot can be simulated and collisions between neighboring robots or other machinery can be detected in a virtual environment. The program then can be modified to prevent such collisions before the welding line is actually put into operation. While this example is a powerful demonstration of the utility of system simulation, a more common application is to optimize the sequence of operations and organization of machinery, to reduce manufacturing costs.

The software also has the capability of conducting ergonomic analyses of various operations and machinery setups and, therefore, of identifying bottlenecks in the movement of parts, equipment, or personnel. The bottlenecks can then be relieved by the process planner, by adjusting the automated or manual procedures at these locations. Using such techniques, a Daimler-Chrysler facility in Rastatt, Germany, was able to balance its production lines so that each worker is productive an average of 85–95% of the time.

Another application of systems simulation is the planning of manufacturing operations to optimize production and to prepare for *just-in-time production* (Section 39.5). For example, if an automobile manufacturer needs to produce 1000 vehicles in a given time, production can be optimized by using certain strategies, such as distributing the number of vehicles with sunroofs throughout the day, or grouping vehicles by color so that the number of paint changes in paint booths is minimized.

With respect to just-in-time production, software, such as that produced by ILOG Corporation, can plan and schedule plant operations far enough in advance to allow ordering materials as they are needed, thus eliminating costly stockpiled inventory.

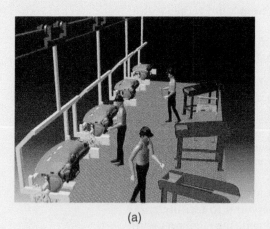

(a)

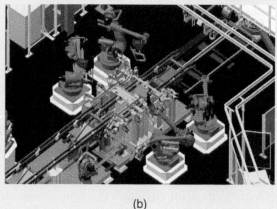

(b)

FIGURE 38.10 Simulation of plant-scale manufacturing operations. (a) The use of virtual manikins to evaluate the required motions and efficiency in manually assembling an automotive dashboard. (b) A robot welding line, where interactions of multiple robots and workpieces can be simulated in order to detect and avoid collisions and improve productivity. *Source:* Courtesy of Dassault Systemes. Printed with permission.

38.8 Group Technology

Group technology (GT) is a methodology that seeks to take advantage of the **design** and **processing similarities** among the parts to be produced. As illustrated in Fig. 38.11, these characteristics clearly suggest that major benefits can be obtained by **classifying** and **coding** the parts into *families*. One company found that by

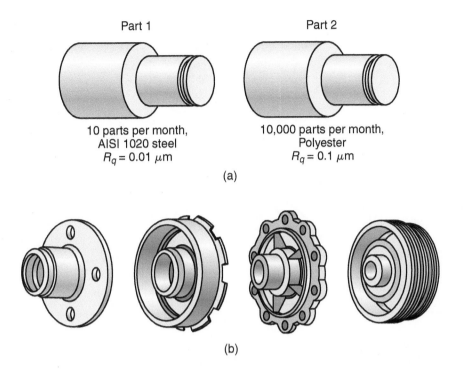

Part 1

Part 2

10 parts per month,
AISI 1020 steel
$R_q = 0.01\ \mu m$

10,000 parts per month,
Polyester
$R_q = 0.1\ \mu m$

(a)

(b)

FIGURE 38.11   Grouping parts according to their (a) geometric similarities and (b) manufacturing attributes.

disassembling each product into its individual components and then identifying the similar parts, 90% of the 3000 parts made fell into only five major families of parts.

A pump, for example, can be broken down into its basic components, such as the motor, housing, shaft, flanges, and seals. It will be noted that each of these components in a pump is basically the same in terms of its design and manufacturing characteristics; consequently, all shafts can be placed in one family of shafts or a family of seals, and so on. Group technology becomes especially attractive because of the ever-growing variety of products, which are often produced in batches. Since nearly 75% of manufacturing today is batch production, improving the efficiency of batch production becomes especially important.

Plant Layout.   A traditional product flow in batch manufacturing, called **functional layout**, is shown in Fig. 38.12a. Note that machines of the same type are arranged in groups, that is, groups of lathes, milling machines, drill presses, and grinders. It will be noted that in such a layout there is considerable random movement, as shown by the arrows indicating movement of materials and parts. Because it wastes time and effort, such an arrangement is obviously not efficient. A better product flow line that would take advantage of group technology is the **group layout**, shown in Fig. 38.12b (see also *cellular manufacturing*, Section 39.2). Note the greater simplicity and the decrease in the number of paths and movements among the machines.

38.8.1   Advantages of Group Technology

The major advantages of group technology are:

- It makes possible the standardization of part designs and the minimization of design duplication. New part designs can be developed from similar, yet previously used, designs, thus saving a significant amount of time and effort.

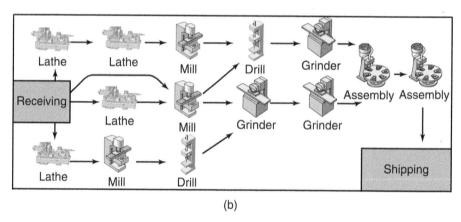

FIGURE 38.12 (a) Functional layout of machine tools in a traditional plant; arrows indicate the flow of materials and parts in various stages of completion. (b) Group-technology (cellular) layout. *Source:* Based on M.P. Groover.

Moreover, the product designer can quickly determine whether data on a similar part already exists in the computer files of the company.

- Data that reflect the experience of the designer and the manufacturing process planner are stored in the database; thus, a new and less experienced engineer can quickly benefit from that experience by retrieving any of the previous designs and process plans.

- Manufacturing costs can be more easily estimated and the relevant statistics on materials, processes, number of parts produced, and other factors can be more easily obtained.

- Process plans can be standardized and scheduled more efficiently, orders can be grouped for more efficient production, and thus machine utilization is improved. Setup times are reduced and parts are produced more efficiently, and with better and more consistent product quality. Similar tools, fixtures, and machinery are shared in the production of a family of parts.

- With the implementation of CAD/CAM, cellular manufacturing, and CIM, group technology is capable of greatly improving productivity and reducing costs, with batch production approaching the benefits of mass production. Depending on the level of implementation, savings in each of the various design and manufacturing phases can range from 5 to 75%.

38.8.2   Classification and Coding of Parts

In group technology, parts are identified and grouped into families by **classification and coding** (C/C) **systems**. This process is a critical and a complex first step, and is done according to the part's design attributes and manufacturing attributes (see Fig. 38.14).

Design Attributes.   These attributes pertain to similarities in geometric features and consist of

- External and internal shapes and dimensions
- Aspect ratios, such as length-to-width or length-to-diameter ratios
- Dimensional tolerances
- Surface finish
- Part functions

Manufacturing Attributes.   Group technology uses the similarities in the methods and sequence of the manufacturing operations performed on the part. Because the selection of a manufacturing process or processes depends on numerous factors, manufacturing and design attributes are interrelated. The manufacturing attributes of a part consist of

- Primary processes
- Secondary and finishing processes
- Dimensional tolerances and surface finish
- Sequence of operations performed
- Tools, dies, fixtures, and machinery
- Production quantity and production rate

Coding can be time consuming and considerable experience is required. It can be done simply by viewing the shapes of the parts in a generic way and then classifying the parts accordingly; for example, parts having rotational symmetry, parts having rectilinear shape, and parts having large surface-to-thickness ratios. A more thorough approach is to review all of the data and drawings concerning the design *and* manufacture of all of the parts.

Parts also may be classified by studying their production flow during the manufacturing cycle, an approach called **production flow analysis** (PFA). Recall from Section 38.6 that routing sheets clearly show process plans and the sequence of operations to be performed. One drawback to PFA, however, is that a particular routing sheet does not necessarily indicate that the total manufacturing operation is optimized.

38.8.3   Coding

The code for parts can be based on a company's own system of coding or it can be based on one of several classification and coding systems available in commercial software; often, coding is incorporated into CAD/CAM packages. Whether it was developed in-house or it was purchased, the coding system must be compatible with the company's other systems, such as NC machinery and CAPP systems.

The code structure for part families typically consists of numbers, letters, or a combination of the two. Each specific component of a product is assigned a code. This code may pertain to design attributes only (generally, fewer than 12 digits) or to manufacturing attributes only; most advanced systems include both, using as many as 30 digits. Coding may be done without input from the software user and displayed only if the information is requested. Commonly, design or manufacturing data retrieval can be based on keyword searches.

The three basic levels of coding vary in degree of complexity:

1. **Hierarchical coding.** Also called **monocode**, hierarchical coding interprets each succeeding digit on the basis of the value of the preceding digit. Each symbol amplifies the information contained in the preceding digit; therefore, a digit in the code cannot be interpreted alone. The advantage of this method is that a short code can contain a large amount of information; however, the method is difficult to apply in a computerized system.

2. **Polycodes.** In this method, also known as **chain-type** coding, each digit has its own interpretation, which does not depend on the preceding digit. This

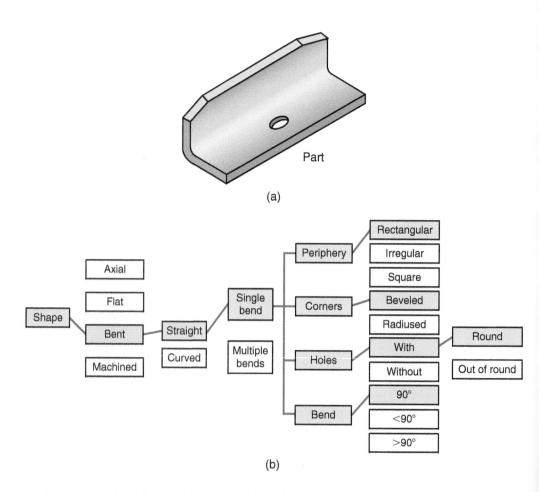

FIGURE 38.13 Decision-tree classification for a sheet-metal bracket. *Source:* Based on G.W. Millar.

structure tends to be relatively long, but it allows the identification of specific part attributes and is well suited to computer implementation.

3. **Decision-tree coding.** This type of coding, also called **hybrid coding**, is the most advanced and combines both design and manufacturing attributes (Fig. 38.13).

38.8.4 Coding Systems

Three major industrial coding systems are:

1. The **Opitz system** (after H. Opitz, 1905–1977) was the first comprehensive coding system developed. It consists basically of nine digits (12345 6789) representing design and manufacturing data (Fig. 38.14); four additional codes

		Form Code				Supplementary Code			
		1st digit Part class	**2nd digit** Main shape	**3rd digit** Rotational surface machining	**4th digit** Plane surface machining	**5th digit** Auxiliary holes, gear teeth, and forming	Digit 1	Digit 2	Digit 3	Digit 4
0	Rotational parts	$\frac{L}{D} < 0.5$	External shape, external shape elements	Internal shape, internal shape elements	Plane surface machining	Auxiliary holes	Dimension	Material	Original shape of raw material	Accuracy
1		$0.5 < \frac{L}{D} < 3$								
2		$\frac{L}{D} > 3$				Gear teeth				
3		$\frac{L}{D} < 2$ with deviation	Main shape	Rotational machining, internal and external shape elements		Auxiliary holes, gear teeth, and forming				
4		$\frac{L}{D} \leq 2$ with deviation								
5		Special								
6	Nonrotational parts	$\frac{A}{B} < 3, \frac{A}{C} > 4$ Flat parts	Main shape							
7		$\frac{A}{B} > 3$ Long parts	Main shape	Principal bores	Plane surface machining	Auxiliary holes, gear teeth, and forming				
8		$\frac{A}{B} < 3, \frac{A}{C} > 4$ Cubic parts	Main shape							
9		Special								

FIGURE 38.14 Classification and coding system according to Opitz, consisting of a form code of five digits and a supplementary code of four digits.

(1234) may be used to identify the type and sequence of production operations to be performed. The Opitz system has two drawbacks: (a) It is possible to have different codes for parts that have similar manufacturing attributes and (b) a number of parts with different shapes can have the same code.

2. The **multiClass system** was developed to help automate and standardize several design, production, and management functions; it involves up to 30 digits (Fig. 38.15). This system is used interactively with a computer that asks the user a number of questions; on the basis of the answers given, the computer automatically assigns a code number to the part.

3. The **KK-3 system** is a general-purpose system for parts that are to be machined or ground; this code uses a 21-digit decimal system. The code is much greater in length than the two previous codes described, but it classifies dimensions and dimensional ratios, such as the length-to-diameter ratio of the part. The structure of a KK-3 system for rotational components is shown in Fig. 38.16.

Production Flow Analysis.   One of the main benefits of GT is the design of manufacturing cells (Section 39.2). Consider the situation in Fig. 38.17, which is intended to show a highly simplified list of parts that are to be produced and the required machinery. As can be seen in this figure, the types of machines and variety of parts do not lend themselves to groupings of machines. By using GT, parts can be classified and codified, then sorted into logical groupings. Combined with production requirements, this approach allows the design of manufacturing cells to achieve the associated benefits of flexibility and utility.

For example, Fig. 38.17a is a spreadsheet showing a collection of parts produced by a pump manufacturer; the machines required to produce these parts are also

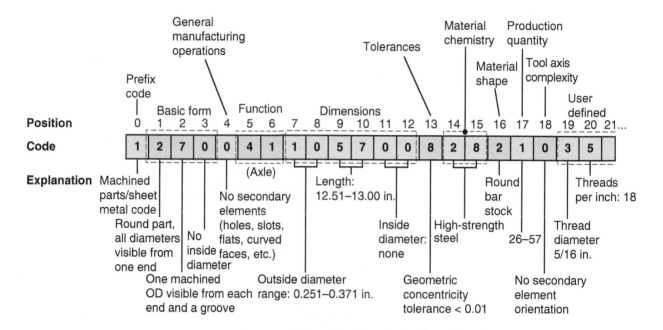

FIGURE 38.15   Typical multiClass code for a machined part.

Digit	Items	(Rotational component)
1	Part name	General classification
2		Detail classification
3	Materials	General classification
4		Detail classification
5	Major dimensions	Length
6		Diameter
7	Primary shapes and ratio of major dimensions	
8	External surface	External surface and outer primary shape
9		Concentric screw-threaded parts
10		Functional cutoff parts
11		Extraordinary shaped parts
12		Forming
13		Cylindrical surface
14	Internal surface	Internal primary shape
15		Internal curved surface
16		Internal flat and cylindrical surface
17	End surface	
18	Nonconcentric holes	Regularly located holes
19		Special holes
20	Noncutting process	
21	Accuracy	

(Digits 8–20 are grouped under the side label "Shape details and kinds of processes")

FIGURE 38.16 The structure of a KK-3 system for rotational components. *Source:* Courtesy of Japan Society for the Promotion of Machine Industry.

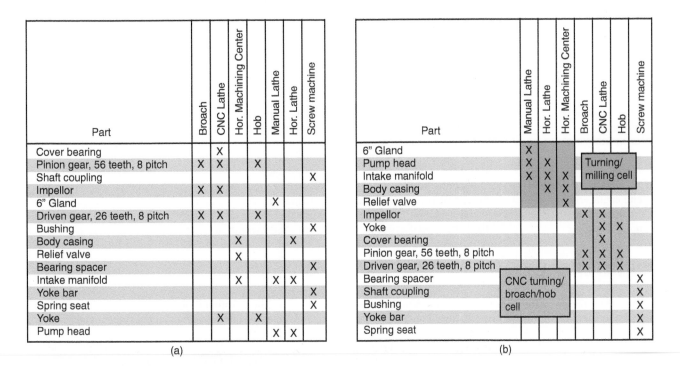

FIGURE 38.17 The use of group technology in organizing manufacturing cells. (a) Original spreadsheet of parts; no logical organization is apparent. (b) After grouping parts based on manufacturing attributes, logical machinery groupings are discernable.

indicated. From this spreadsheet, no intuitive machine groupings are apparent. With a reorganization of the data, the parts suggest logical groupings of machines, as shown in Fig. 38.17b. It should be recognized that these figures are highly simplified in order to demonstrate the significance of group technology in organizing cells; usually there are many more parts and processes (including metrology) that must be considered.

SUMMARY

- Integrated manufacturing systems can be implemented to various degrees to optimize operations, improve product quality, and reduce production costs.
- Computer-integrated manufacturing operations have become the most important means of improving productivity, responding to rapidly changing market demands, as well as improving the control of both manufacturing and the management functions of an organization.
- Advanced software and powerful computers now allow the description of part geometry in several different formats, including wire-frame, octree, surface models, solid models, skeletons, and boundary representations.
- Computers also are used to simulate manufacturing operations and systems, and to aid in the selection of manufacturing processes.
- Group technology is a powerful approach that allows the rapid recovery of previous design and manufacturing experiences, by encoding a part on the basis of its geometric features or manufacturing attributes. Several group-technology coding systems are available.

KEY TERMS

Business-execution function
Business-planning function
Classification and coding systems
Coding
Computer-aided design and engineering
Computer-aided manufacturing
Computer-aided process planning
Computer-integrated manufacturing
Computer simulation
Data-acquisition system
Database
Design attributes
Enterprise resource planning
Exchange specifications
Functional layout
Group layout
Group technology
Manufacturing attributes
Manufacturing resource planning
Material-requirements planning
Modeling
Octree representation
Paperless design
Process planning
Production flow analysis
Routing sheet
Solid model
Surface model
Wire frame

BIBLIOGRAPHY

Amirouche, F.M.L., **Principles of Computer-Aided Design and Manufacturing,** 2nd ed., Prentice Hall, 2004.
Chang, T.-C., Wysk, R.A., Wang, H.P., and Rembold, U., **Computer-Aided Manufacturing,** 3rd ed., Prentice Hall, 2007.
Chryssolouris, G., **Manufacturing Systems: Theory and Practice,** Springer, 2006.
Curry, G.L., and Feldman, R.M., **Manufacturing Systems Modeling and Analysis,** Springer, 2008.
Groover, M., **Automation, Production Systems and Computer-Integrated Manufacturing,** 3rd ed., Prentice Hall, 2007.
Parsaei, H., Leep, H., and Jeon, G., **The Principles of Group Technology and Cellular Manufacturing Systems,** Wiley, 2006.
Rehg, J.A., and Kraebber, H.W., **Computer-Integrated Manufacturing,** 3rd ed., Prentice Hall, 2004.
Rong, Y., and Huang, H., **Advanced Computer-Aided Fixture Design,** Academic Press, 2005.

REVIEW QUESTIONS

38.1 In what ways have computers had an impact on manufacturing?

38.2 Describe the benefits of computer-integrated manufacturing operations.

38.3 What is a database? Why is it necessary?

38.4 What is an STL file? What is it popularly used for?

38.5 What are the differences between the terms computer-aided and computer-integrated?

38.6 What are the advantages of CAD systems over traditional methods of design? Are there any limitations?

38.7 What do the following abbreviations mean: NURB, DAS, DFX, PDES?

38.8 What are the main advantages of CIM?

38.9 Describe the purposes of process planning. How are computers used in such planning?

38.10 Describe the features of a routing sheet. Why is it necessary?

38.11 What are the advantages of simulation of manufacturing lines?

38.12 What is group technology? Why was it developed? Explain its advantages.

38.13 Explain the three types of GT coding: hierarchical, polycode, and decision-tree.

38.14 Describe what is meant by the term manufacturing system. What are its benefits?

38.15 What does classification and coding mean in group technology?

38.16 What do the Opitz and KK-3 systems have in common?

38.17 Examine a cardboard box. Is this an example of a boundary representation, a solid model, or a profile body?

QUALITATIVE PROBLEMS

38.18 Describe your observations regarding Figs. 38.1 and 38.2.

38.19 Some software packages displace STL files as collections of surface quadrilaterals, not triangles. Explain why this can be done.

38.20 Give examples of primitives of solids other than those shown in Figs. 38.4a and b.

38.21 Describe your understanding of the octree representation in Fig. 38.7.

38.22 Explain the logic behind the arrangements shown in Fig. 38.12b.

38.23 What are the advantages of hierarchical coding?

38.24 What is the difference between a variant system and a generative system?

38.25 Referring to Fig. 38.3, explain the advantages of a third-order piecewise Bezier curve over a B-spline or a conventional Bezier curve.

38.26 Sketch a sphere using (a) wire fame; (b) 3D surface models; and (c) STL formats.

38.27 Describe situations that would require a design change at its larger end of the part in Fig. 38.6.

38.28 Describe your thoughts on the differences between e-commerce and traditional business practices.

SYNTHESIS, DESIGN, AND PROJECTS

38.29 How would you describe the principle of computer-aided manufacturing to an older worker in a manufacturing facility who is not familiar with computers?

38.30 Review various manufactured parts illustrated in this book, and group them in a manner similar to those shown in Fig. 38.11. Explain.

38.31 Think of a simple part and make a decision-tree chart similar to that shown in Fig. 38.13.

38.32 Review the machine arrangements in Fig. 38.12 and suggest changes that may improve the flow of materials and parts.

38.33 Think of a simple product and make a routing sheet, similar to that shown in Fig. 38.9. If the same part is given to another person, what is the likelihood that the routing sheet developed will be the same? Explain.

38.34 Review Fig. 38.9, and prepare a routing sheet for one of the following: (a) a spur gear, (b) a turbine blade, (c) a glass bottle, (c) an automotive connecting rod, and (d) a forging die.

38.35 Consider a large-scale kitchen, such as those used to prepare meals for a school, hospital, or large office building. Make a list of equipment necessary, arrange the equipment in a functional layout, and then arrange the equipment in a cellular layout. Describe your observations for flow lines when different foods are prepared.

38.36 Using drawing tools, carefully create a sketch of the end view of the part in Fig. 38.6a. Make the same sketch on a CAD program. Compare the time required to produce each sketch.

39

Computer-integrated Manufacturing Systems

- This chapter describes the influence of computer systems and communications networks on product development and manufacturing, and the integration of all related activities.
- The chapter begins by describing the principles of manned and untended manufacturing cells and their features, and how cells can be integrated into flexible manufacturing systems. The concept of holonic manufacturing and its applications also are reviewed.
- Just-in-time production, lean manufacturing, and communication systems are then described.
- The chapter concludes with a discussion of artificial intelligence and expert systems as applied to manufacturing.

39.1 Introduction

This chapter focuses on the **computer integration of manufacturing operations**, where *integration* means that manufacturing processes, operations, and their management are all treated as a *system*. A major advantage of such an approach is that machines, tooling, and manufacturing operations now acquire a built-in flexibility, called **flexible manufacturing**. As a result, the system is capable of (a) *rapidly responding* to changes in product types and fluctuating demands and (b) ensuring *on-time delivery* of products to the customer. Failure of on-time delivery in a highly competitive global environment can have major adverse effects on a company's operations and its success.

This chapter first describes the key elements that enable the execution of the functions necessary to achieve flexible manufacturing, beginning with *cellular manufacturing*, which is the basic unit of flexibility in the production of goods. Manufacturing cells can be broadened into *flexible manufacturing systems*, with major implications for production capabilities. Described next is *holonic manufacturing*, which is a strategy of organizing manufacturing units such as to achieve higher efficiency in operations.

The important concept and the benefits of *just-in-time production* is then examined, in which parts are produced "just in time" to be made into subassemblies, assemblies, and final products. This is a method that eliminates the need for inventories (a major financial burden on a company), as well as significantly saving on space and storage facilities.

Because of the necessity for and extensive use of computer controls, hardware, and software in all the activities just outlined, the planning and effective implementation of *communication networks* constitute critical components of the overall manufacturing operations. The chapter concludes with a review of *artificial intelligence*, which consists of expert systems, natural-language processing, machine vision, artificial neural networks, and fuzzy logic.

39.2 Cellular Manufacturing

A **manufacturing cell** is a small unit consisting of one or more workstations. A workstation usually contains either one machine (called a *single-machine cell*) or several machines (called a *group-machine cell*), with each machine performing a different operation. The machines can be modified, retooled, and regrouped for different product lines within the same family of parts.

QR Code 39.1 Fully automated piston rod welding cell. (*Source:* Courtesy of Manufacturing Technology, Inc. [MTI])

Cellular manufacturing has thus far been utilized primarily in machining, finishing, and sheet-metal-forming operations. The machine tools commonly used in the cells are lathes, milling machines, drills, grinders, and electrical-discharge machines; for sheet forming, the equipment typically consists of shearing, punching, bending, and other forming machines. The equipment may include special-purpose machines and CNC machines; automated inspection and testing equipment also are generally a part of a manufacturing cell.

The capabilities of cellular manufacturing typically involve the following operations:

- Loading and unloading raw materials and workpieces at workstations
- Changing tools at workstations
- Transferring workpieces and tooling between workstations
- Scheduling and controlling the total operation in the cell

In *manned* or attended machining cells, raw materials and parts can be moved and transferred manually by the operator (unless the parts are too heavy or the movements are too hazardous) or by an industrial robot (Section 37.6) located centrally in the cell.

Flexible Manufacturing Cells.   Manufacturing cells can be made flexible by using CNC machines and machining centers (Section 25.2), and by means of industrial robots or other mechanized systems for handling materials and parts in various stages of completion (Section 37.6).

An example of an attended *flexible manufacturing cell* (FMC) that involves several machining operations is illustrated in Fig. 39.1. Note that machining centers, equipped with automatic tool changers and tool magazines, have the capability of performing a wide variety of operations (see Section 25.2). A computer-controlled inspection station, with a coordinate-measuring machine or other flexible metrology equipment, can similarly inspect dimensions on a wide variety of parts. Thus, the organization of these machines into a cell can allow the manufacture of very different parts. With computer integration, a manufacturing cell can produce parts in batch sizes as small as one part, with negligible delay between batches. The actual delay is the time required to download new machining instructions.

Flexible manufacturing cells are usually *unmanned* (unattended) and their design and operation are more exacting than those for other cells. The selection of the machines and industrial robots, including the types and capabilities of their end effectors and control systems, is critical to the effective functioning of the FMC.

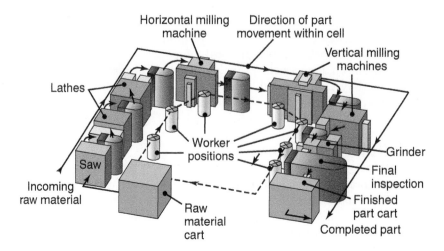

FIGURE 39.1 Schematic illustration of a manned flexible manufacturing cell, showing various machine tools and an inspection station. *Source:* Based on JT. Black.

As with other flexible manufacturing systems (Section 39.3), the cost of FMCs is very high (see Table 40.6); however, this disadvantage is outweighed by increased productivity, flexibility, and controllability.

Cell Design.   Because of the unique features of manufacturing cells, their design and placement requires efficient layout and organization of the plant, and the consideration of product flow lines (see *production flow analysis*, Section 38.8). The machines may be arranged along a line or in a U-shape, an L-shape, or a loop. Selecting the best machine and arrangement for material-handling equipment also involves taking into account such factors as the type of product, its shape, size, and weight, and production rate. In designing these cells, the likelihood of a significant change in demand for a particular part families must be considered, in order to ensure that the equipment involved has the required flexibility and capacity.

39.3   Flexible Manufacturing Systems

A *flexible manufacturing system* (FMS) integrates all of the major elements of production into a highly automated system (Fig. 39.2); a general view of an FMS installation in a plant is shown in Fig. 39.3. First utilized in the late 1960s, an FMS consists of (a) a number of manufacturing cells, each containing an industrial robot serving several CNC machines and (b) an automated material-handling system. An automated guided vehicle (AGV; Section 37.5) moves parts between machines and inspection stations, which could easily involve an industrial robot or integral transfer device. All of these activities are computer controlled, and different computer instructions can be downloaded for each successive part passing through a particular workstation. The system can handle a *variety of part configurations* and produce them *in any order*.

An FMS is capable of optimizing each step of the total operation. These steps may involve (a) one or more processes and operations, such as machining, grinding, cutting, forming, powder metallurgy, heat treating, and finishing, (b) the handling of raw materials and parts, (c) measurement and inspection, and (d) assembly. The most common applications of FMS to date have been in machining and assembly operations.

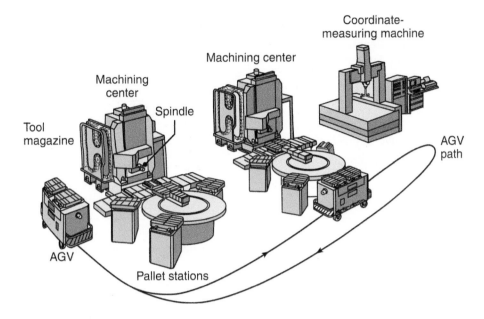

FIGURE 39.2   A schematic illustration of a flexible manufacturing system, showing two machining centers, a measuring and inspection station, and two automated guided vehicles. *Source:* Based on JT. Black.

FIGURE 39.3   A general view of a flexible manufacturing system in a plant, showing several machining centers and automated guided vehicles moving along the white line in the aisle. *Source:* Courtesy of Cincinnati Milacron, Inc.

FMS can be regarded as a system that combines the benefits of two systems: (1) the highly productive, but inflexible, transfer lines and (2) job-shop production, which can produce large product variety on stand-alone machines, but is inefficient (see also Fig. 37.2). The relative characteristics of transfer lines and FMS are shown in Table 39.1. Note that with an FMS, the time required for a changeover to a different

TABLE 39.1

Comparison of General Characteristics of Transfer Lines and Flexible Manufacturing Systems		
Characteristic	Transfer line	FMS
Part variety	Few	Infinite
Lot size	>100	1–50
Part-changing time	Long	Very short
Tool change	Manual	Automatic
Adaptive control	Difficult	Available
Inventory	High	Low
Production during breakdown	None	Partial
Justification for capital expenditure	Simple	Difficult

part is very short; the quick response to product and market-demand variations is a major attribute of FMS.

Compared with conventional manufacturing systems, FMS have the following major benefits:

- Parts can be produced in any order, in batch sizes as small as one, and at lower unit cost.
- Direct labor and inventories are reduced or eliminated.
- The lead times required for product changes are shorter.
- Because the system is self-correcting, production is more reliable and product quality is uniform.

However, FMS has the following drawbacks:

- Equipment costs are higher, and computer integration requires more highly trained operating personnel.
- FMS is generally not suitable for very low or very high production rates.

Elements of FMS.   The basic elements of a flexible manufacturing system are:

- Workstations and cells
- Automated handling and transport of materials and parts
- Control systems

The workstations are arranged to yield the highest efficiency in production, with an orderly flow of materials and parts in progress through the system. The types of machines in workstations depend on the type of production. For machining operations, for example, they usually consist of a variety of 3- to 5-axis machining centers, CNC lathes, milling machines, drill presses, and grinders (described in Part IV). Also included are various pieces of equipment, such as that for automated inspection (including coordinate-measuring machines), assembly, and cleaning. In addition to machining, other types of manufacturing operations suitable for FMS are sheet-metal forming, punching, shearing, and forging. FMS also may incorporate various equipment, trimming presses, heat-treating facilities, and cleaning equipment.

The flexibility of FMS requires that material-handling systems be computer-controlled and performed by automated guided vehicles, conveyors, and various transfer mechanisms. Thus, FMS are capable of transporting raw materials, blanks, and parts in various stages of completion to any machine, in random order, and at any time. *Prismatic parts* are usually moved on specially designed **pallets**, whereas those having *rotational symmetry* are usually moved by robots and various mechanical devices.

Scheduling.   In FMS, efficient machine utilization is essential; machines must not stand idle, thus proper *scheduling* and process planning are crucial. Unlike that in job shops, where a relatively rigid schedule is followed to perform a set of operations, scheduling for FMS is *dynamic*, meaning that it is capable of responding to quick changes in product type, thus it is responsive to real-time decisions. The scheduling system in FMS (a) specifies the types of operations to be performed on each part and (b) identifies the machines or manufacturing cells on which these operations are to take place.

No setup time is wasted in switching between manufacturing operations. The characteristics, performance, and reliability of each unit in the system must, however, be monitored to ensure that parts are of acceptable quality and dimensional accuracy before they move on to the next workstation.

Economic Justification of FMS.   FMS installations are very capital intensive, costing millions of dollars. Consequently, a thorough cost–benefit analysis must be conducted, which must include such factors as (a) the costs of capital, energy, materials, and labor; (b) the expected markets for the products to be manufactured; (c) any anticipated fluctuations in market demand and product type; and (d) the time and effort required for installing and debugging the whole system.

As can be in Fig. 37.2, the most effective FMS applications are in medium-quantity batch production. When a variety of parts is to be produced, FMS is suitable for production quantities typically up to 15,000 to 35,000 aggregate parts per year. For individual parts with the same configuration, production may, however, reach 100,000 parts per year. In contrast, high-volume, low-variety parts production is best obtained using transfer machines (*dedicated equipment*). Finally, low-volume, high-variety parts production can best be done on conventional standard machinery (with or without numerical control) or by using machining centers.

EXAMPLE 39.1   Flexible Manufacturing Systems in Large and Small Companies

Many manufacturers have long considered implementing a large-scale system in their facilities. After detailed review, and on the basis of the experience of other similar companies, however, most decide on smaller, simpler, modular, and less expensive systems that are more cost effective. These systems include flexible manufacturing cells (the cost of which would be on the order of a few hundred thousand dollars), stand-alone machining centers, and various CNC machine tools that are significantly easier to control than an FMS.

39.4   Holonic Manufacturing

Holonic manufacturing describes a unique organization of manufacturing units, whereby each component in a holonic manufacturing system is, at the same time, an independent or *whole* entity and a subservient *part* of a hierarchical organization. The word *holonic* is from the Greek *holos* (meaning "whole") and the suffix *on* (meaning "a part of").

Holonic organizational systems have been studied since the 1960s, and there are a number of examples in biological systems. Three fundamental features about these systems are:

1. Complex systems will evolve from simple systems much more rapidly if there are stable intermediate forms than if there are none; also, stable and complex systems require a hierarchical system for their evolution.

2. Holons are simultaneously self-contained wholes of their subordinated parts and dependent parts of other systems. Holons are autonomous and self-reliant units that have a degree of independence, and can handle contingencies without asking higher levels in the hierarchical system for instructions. At the same time, holons are subject to control from multiple sources of higher system levels.

3. A *holarchy* consists of (a) autonomous wholes in charge of their parts and (b) dependent parts controlled by higher levels of a hierarchy. Holarchies are coordinated according to their local environment.

In biological systems, hierarchies have the characteristics of (a) stability in the face of disturbances, (b) optimum use of available resources, and (c) a high level of flexibility when their environment changes. A *manufacturing holon* is an autonomous and cooperative building block of a manufacturing system for the production, storage, and transfer of objects or information. It consists of a control part and an optional physical-processing part; for example, a holon can be a combination of a CNC milling machine and an operator interacting via a suitable interface. A holon can also consist of other holons that provide the necessary processing, information, and human interfaces to the outside world, such as a group of manufacturing cells. Holarchies can be created and dissolved dynamically, depending on the current needs of the particular manufacturing process.

A holonic systems view of the manufacturing operation is one of creating a working manufacturing environment from the bottom up. Maximum flexibility can be achieved by providing intelligence within holons to (a) support all production and control functions required to complete production tasks and (b) manage the underlying equipment and systems. The manufacturing system can dynamically reconfigure itself into operational hierarchies, to optimally produce the desired products, with holons or elements being added or removed as needed.

Holarchical manufacturing systems rely on fast and effective communication among holons, as opposed to traditional hierarchical control where individual processing power is essential. Although a large number of specific arrangements and software algorithms have been proposed for holarchical systems, the general sequence of events can be outlined as follows:

1. A factory consists of a number of *resource holons*, available as separate entities in a resource pool. Available holons may, for example, consist of (a) a CNC milling machine and operator, (b) a CNC grinder and operator, or (c) a CNC lathe and operator.

2. Upon receipt of an order or a directive from higher levels in the factory hierarchical structure, an *order holon* is formed and begins communicating and negotiating with the available resource holons.

3. The negotiations lead to a self-organized grouping of resource holons, which are assigned on the basis of product requirements, resource holon availability, and customer requirements. For example, a given product may require a CNC lathe, a CNC grinder, and an automated inspection station, to organize it into a *production holon*.

4. In case of a breakdown, the unavailability of a particular machine or due to changing customer requirements, other holons from the resource pool can be added or subtracted as needed, allowing a reorganization of the production holon. Production bottlenecks can be identified and eliminated through communication and negotiation among the holons in the resource pool. Step 4 has been referred to as *plug and play*, a term borrowed from the computer industry, where hardware components are seamlessly integrated into a system.

39.5 Just-in-time Production

The *just-in-time* (JIT) *production* concept originated in the United States, with novel contributions by Ford Motor Co. and John Deere Corporation. It was first implemented on a large scale in 1953 at the Toyota Motor Company in Japan to eliminate waste of materials, machines, capital, manpower, and inventory throughout the manufacturing system. The JIT concept has the following goals:

- Receive supplies just in time to be used
- Produce parts just in time to be made into subassemblies
- Produce subassemblies just in time to be assembled into finished products
- Produce and deliver finished products just in time to be sold

In traditional manufacturing, the parts are made in batches, placed in inventory, and then used whenever necessary. This approach is known as a **push system**, meaning that parts are made *according to a schedule* and are placed in inventory, to be used whenever they are needed. In contrast, JIT is a **pull system**, meaning that parts are *produced to order*, and thus the production is matched with demand for the final assembly of products.

There are no stockpiles in JIT and the ideal production quantity is one, hence it is also called *zero inventory*, *stockless production*, and *demand scheduling*. Moreover, parts are inspected as they are manufactured and are used within a short period of time. In this way, a worker maintains continuous production control, immediately identifying defective parts, and reducing process variation to produce quality products.

Implementation of the JIT concept requires that all aspects of manufacturing operations be monitored and reviewed, so that all operations and resources which do not add value are eliminated. This approach emphasizes (a) pride and dedication in producing high-quality products, (b) elimination of idle resources, and (c) teamwork among workers, engineers, and management to promptly solve any problems that arise during production or assembly.

The ability to detect production problems as parts are being produced has been likened to what happens to the level of water (representing *inventory levels*) in a lake, covering a bed of boulders (representing *production problems*). When the water level is high (analogous to the high levels of inventory, associated with push production), the boulders are not exposed. By contrast, when the level is low (analogous to the low inventories, associated with pull production), the boulders are exposed and thus can promptly be identified and removed. This analogy indicates that high inventory levels can mask quality and production problems with parts that are already made and stockpiled.

The just-in-time concept requires the timely and reliable delivery of all supplies and parts from outside sources and/or from other divisions of a company, thus significantly reducing or eliminating in-plant inventory. Suppliers are expected to deliver, often on a daily basis, preinspected goods as they are needed for production and assembly. The reliability of supply can become a major concern, particularly in generally unpredictable major natural events, such as earthquakes, flooding, and nuclear disasters.

The JIT approach requires (a) reliable suppliers, (b) close cooperation and trust among the company and its vendors, and (c) a reliable system of transportation; also important for smoother operation is a reduction in the number of its suppliers.

Advantages of JIT.   The major advantages of just-in-time production are:

- Low inventory-carrying costs
- Rapid detection of defects in the production or the delivery of supplies and, hence, low scrap loss
- Reduced inspection and reworking of parts
- High-quality products made at low cost

Although there can be significant variations in performance, JIT production has resulted in reductions of 20–40% in product cost, 60–80% in inventory, up to 90% in rejection rates, 90% in lead times, and 50% in scrap, rework, and warranty costs. Increases of 30–50% in direct-labor productivity and of 60% in indirect-labor productivity also have been attained.

Kanban.   The implementation of JIT in Japan involved *kanban*, meaning "visible record." These records originally consisted of two types of cards (called *kanbans*, now replaced by bar-coded tags and other devices):

- The *production card*, which authorizes the production of one container or cart of identical, specified parts at a workstation
- The *conveyance card* or *move card*, which authorizes the transfer of one container or cart of parts from a particular workstation to the workstation where the parts will be used

The cards or tags contain information on the (a) type of part, (b) location where the card was issued, (c) part number, and (d) number of items in the container. The number of containers in circulation at any time is completely controlled and can be scheduled, as desired, for maximum production efficiency.

39.6 Lean Manufacturing

In a modern manufacturing environment, companies must be responsive to the needs of the customers and their specific requirements, and to fluctuating global market demands. At the same time, to ensure competitiveness, the manufacturing enterprise must be conducted with a minimum amount of wasted resources. This realization has lead to *lean production* or *lean manufacturing* strategies.

Lean manufacturing involves the following steps:

1. **Identify value.** The critical starting point for lean thinking is a recognition of *value*, which can be done only by a customer and considering a customer's product (see also Section 40.10.1). The goal of any organization is to produce a product that a customer wants, and at a desired price, location, time, and volume. Providing the wrong good or service produces waste, even if it is delivered efficiently. It is important to identify all of a manufacturer's activities from

the *viewpoint of the customer* and *optimize processes* to maximize added value. This viewpoint is critically important, because it helps identify whether or not a particular activity:

   a. Clearly adds value
   b. Adds no value, but cannot be avoided
   c. Adds no value, and can be avoided

2. **Identify value streams.** *Value stream* is the set of all actions required to produce a product, including

   a. Product design and development tasks, involving all actions from concept, to detailed design, and to production launch
   b. Information management tasks, involving order taking, detailed scheduling, and delivery
   c. Physical production tasks, by means of which raw materials progress to a finished product delivered to the customer

3. **Make the value stream flow.** *Flow* is achieved when parts encounter a minimum of idle time between any successive operation. It has been noted that flow is easiest to achieve in mass production, but it is more difficult for small-lot production. However, production in batches inherently involves part idle time; this should be avoided, thus just-in-time approaches are essential. In such cases the solution is to use manufacturing cells, where minimum time and effort are required to switch from one product to another; a product being manufactured encounters continuous flow regardless of the production rate.

   In addition to JIT approaches, establishing product flow through factories requires

   - Eliminating waiting time, which may be caused by unbalanced workloads, unplanned maintenance, or quality problems; thus, the efficiency of workers must be maximized at all times.
   - Leveling or balancing production. Production may vary at different times of the day or the day of the week; uneven production occurs when machines are underutilized, invariably leading to waiting time and waste. If production can be leveled, inventory is reduced, productivity increases, and process flow is easier to achieve. It should also be noted that balancing can occur on cell, line, and plant scales.
   - Eliminating unnecessary or additional operations and steps, because they represent costs.
   - Minimizing or eliminating movements of parts or products in plants, because it represents an activity that adds no value. This waste can, for example, be either eliminated or minimized by using machining cells or better plant layouts.
   - Performing motion and time studies to avoid unnecessary part or product movements, or to identify inefficient workers.
   - Eliminating part defects.
   - Avoiding a single-source supplier, especially in the unforeseen events of natural disasters and regional conflicts.

4. **Establish pull.** It has been observed that once value streams are flowing, significant savings are gained in terms of inventory reduction, as well as efficiency gains in product development, order processing, and production. In some cases, up to 90% savings in actual production time have been obtained. Under these circumstances, it is possible to establish *pull manufacturing* (Section 39.5),

where products are produced upon order by a customer or upstream machine, and not in batches that ultimately are unwanted and do not create value.

5. **Achieve perfection.** As described in Section 36.1, *kaizen* is used to signify continuous improvement, and clearly, there is a need for continuous improvement in all organizations. With lean manufacturing approaches, it has been found that continuous improvement can be accelerated, so that production without waste is possible. Moreover, upon the adoption of lean manufacturing principles, companies encounter an initial benefit, referred to as *kaikaku* or "radical improvement."

39.7 Communications Networks in Manufacturing

In order to maintain a high level of coordination and efficiency of operation in integrated manufacturing, an extensive, high-speed, and interactive *communications network* is essential. The **local area network** (LAN) is a hardware-and-software system, in which logically related groups of machines and equipment communicate with each other. A LAN links these groups to each other, bringing different phases of manufacturing into a unified operation.

A LAN can be very large and complex, linking hundreds or even thousands of machines and devices in several buildings of an organization. Various network layouts of fiber-optic or copper cables are typically used, over distances ranging from a few meters to as much as 32 km (20 mi); for longer distances, **wide area networks** (WANs) are used. Different types of networks can be linked or integrated through "gateways" and "bridges," often with the use of secure *file transfer protocols* (FTPs) over Internet connections. Several advanced network protocols, including *ipV6* (internet protocol version 6) and *Internet2*, have been implemented and applied to manufacturing networks. A *carrier-sense multiple access with collision detection* (CSMA/CD) system was developed in the 1970s and implemented in *Ethernet*, which is now the industry standard.

Conventional LANs require the routing of wires, often through masonry walls or other permanent structures, and require computers and machinery to remain stationary. **Wireless local area networks** (WLANs) allow equipment, such as mobile test stands or data-collection devices (e.g., bar-code readers), to easily maintain a network connection. A communication standard (IEEE 802.11) currently defines frequencies and specifications of signals, and two radio-frequency methods and one infrared method for WLANs. Although wireless networks are slower than those that are hardwired, their flexibility makes them desirable, especially in situations where slow tasks, such as machine monitoring, are the main application.

Personal area networks (PANs) are used for electronic devices, such as cellular telephones and personal data assistants, but are not as widespread for manufacturing applications. PANs are based on communications standards (such as Bluetooth, IrDA, and HomeRF), and are designed to allow data and voice communication over short distances. For example, a short-range Bluetooth device will allow communication over a 10-m (32-ft) distance. PANs are undergoing major changes, and communications standards are continually being refined.

Communications Standards.   Typically, one manufacturing cell is built, with machines and equipment purchased from one vendor, another cell, with machines purchased from another vendor, and a third, purchased from yet another vendor. As a result, a variety of programmable devices are involved and are driven by several

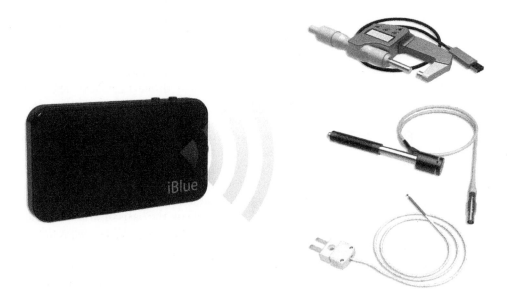

FIGURE 39.4 The iBlue communications device, with a micrometer (top), shore hardness probe (middle), and thermal probe (bottom). The iBlue is the same size as a mobile phone, and uses the MTConnect communications standard to communicate with instruments and machinery. *Source:* Courtesy of J. Neidig, ITAMCI, Inc.

computers and microprocessors purchased at various times from different vendors and having various capacities and levels of sophistication.

In 2008, **MTConnect** was demonstrated for the first time. *MTConnect* is a machine tool *communications protocol,* developed by the Association for Manufacturing Technology, which has quickly become an industry standard (aided by its royalty-free availability). MTConnect uses *hypertext transfer protocols* (http) for communication of data suitable for all machine-tool manufacturers. Software is widely available for retrieving data from machine tools, called *agents,* and is even available for tablet computers, smart phones, or specially-designed Bluetooth-based devices, such as iBlue (see Fig. 39.4). Collection of manufacturing data, tracking of machine utilization and production rates, and other forms of data are greatly enabled by this protocol, allowing for real-time plant management.

39.8 Artificial Intelligence

Artificial intelligence (AI) is that part of computer science concerned with systems that exhibit some characteristics that are usually associated with intelligence in humans: learning, reasoning, problem solving, recognizing patterns, and understanding language. The goal of AI is to simulate such behaviors on the computer. The art of bringing relevant principles and tools of AI to bear on difficult application problems is known as **knowledge engineering.**

Artificial intelligence has had a major effect on the design, automation, and overall economics of the manufacturing operation, largely because of advances made in computer-memory expansion (see VLSI chip design; Chapter 28) and decreasing costs. Artificial intelligence packages costing as much as a few thousand dollars have been developed, many of which can be run on inexpensive personal computers.

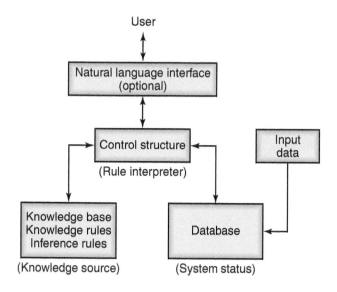

FIGURE 39.5 Basic structure of an expert system. The knowledge base consists of knowledge rules (general information about the problem) and inference rules (the way conclusions are reached); the results may be communicated to the user through the natural language interface.

Expert Systems. An *expert system* (ES), also called a **knowledge-based system**, generally is defined as an intelligent computer program that has the capability to solve difficult real-life problems by using **knowledge-based** and **inferential** procedures (Fig. 39.5). The goal of an expert system is to conduct an intellectually demanding task in the way that a human expert would.

The field of knowledge required to perform the task in question is called the **domain** of the expert system. Expert systems utilize a knowledge base containing facts, data, definitions, and assumptions. They also have the capability of adopting a **heuristic** approach: making good judgments on the basis of discovery and revelation and making high-probability guesses, just as a human expert would.

The knowledge base is expressed in computer code, usually in the form of **if-then rules**, and can generate a series of questions. The mechanism for using these rules to solve problems is called an **inference engine**. Expert systems can communicate with other computer software packages.

To construct expert systems for solving the complex design and manufacturing problems one encounters in real life, one needs a great deal of knowledge and a mechanism for manipulating that knowledge to create solutions. Because of the difficulty involved in (a) accurately modeling the many years of experience of an expert, or a team of experts, and (b) the complex inductive reasoning and decision-making capabilities of humans (including the capability to learn from mistakes), developing knowledge-based systems requires considerable time and effort.

Expert systems operate on a real-time basis, and their short reaction times provide rapid responses to problems. The programming languages most commonly used for these applications are C++, Java, and Python; other languages also are available. An important aspect of expert-system software is **shells** or **environments**, also called **framework systems**. These software packages are essentially expert-system outlines that allow a person to write specific applications to suit special needs; writing these programs requires considerable experience and time.

Several expert systems have been developed that utilize computers with various capabilities and for such specialized applications as:

- Problem diagnosis in various types of machines and equipment, and the determination of corrective actions
- Modeling and simulation of production facilities
- Computer-aided design, process planning, and production scheduling
- Management of a company's manufacturing strategy

Natural-language Processing. Traditionally, retrieving information from a database in computer memory has required the utilization of computer programmers to translate questions in natural language into "queries" in some machine language. *Natural language* interfaces with database systems, which are in continuous development, allow a user to retrieve information by entering English or other language commands in the form of simple, typed questions.

Software shells are available and are used in such applications as scheduling material flow in manufacturing plants and analyzing information in databases. Significant progress continually is being made on computer software that will have speech synthesis and recognition (**voice recognition**) capabilities, in order to eliminate the need to type commands on keyboards.

Machine Vision.   In *machine vision* (Section 37.7.1), computers and software, implementing artificial intelligence, are combined with cameras and other optical sensors (Fig. 39.6). These machines then perform such operations as inspecting, identifying, and sorting parts, as well as guiding robots (see *intelligent robots*; Section 37.6.2), operations that otherwise would require human intervention.

Artificial Neural Networks.   Although computers are much faster than the human brain at performing sequential tasks, humans are much better at pattern-based tasks that can be performed with parallel processing, such as recognizing features (on faces and in voices, even under noisy conditions), assessing situations quickly, and adjusting to new and dynamic conditions. These advantages also are due partly to the ability of humans to use all five senses (sight, hearing, smell, taste, and touch) in real time and simultaneously, called *data fusion*. The branch of AI called *artificial neural networks* (ANN) attempts to gain some of these capabilities through computer imitation of the way that data are processed by the human brain.

The human brain has about 100 billion linked **neurons** (cells that are the fundamental functional units of nerve tissue) and more than a thousand times that many connections. Each neuron performs exactly one simple task: it receives input signals from a fixed set of neurons; when those input signals are related in a certain way (specific to that particular neuron), it generates an electrochemical output signal that goes to a fixed set of neurons. It now is believed that human learning is accomplished by changes in the strengths of these signal connections among neurons.

Artificial neural networks are used in such applications as noise reduction (in telephones), speech recognition, and process control. For example, they can be used for predicting the surface finish of a workpiece obtained by end milling on the basis of input parameters, such as cutting force, torque, acoustic emission, and spindle acceleration.

Fuzzy Logic.   An element of AI having important applications in control systems and pattern recognition is *fuzzy logic*, also called *fuzzy models* or *fuzzy reasoning*. Introduced in 1965 and based on the observation that people can make good decisions on the basis of imprecise and nonnumeric information, fuzzy models basically are

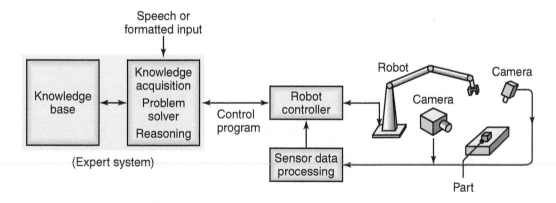

FIGURE 39.6   Illustration of an expert system as applied to an industrial robot guided by machine vision.

mathematical means of representing vagueness and imprecise information, hence the term "fuzzy."

These models have the ability to recognize, represent, manipulate, interpret, and utilize data and information that are vague or lack precision. Fuzzy models deal with reasoning and decision making at a level higher than those for neural networks. Typical linguistic examples are the words: *few, very, more or less, small, medium, extremely,* and *almost all.*

Fuzzy technologies and devices have been developed and successfully applied in such areas as robotics and motion control, image processing and machine vision, machine learning, and the design of intelligent systems. Some applications include (a) the automatic transmission in automobiles; (b) a washing machine that automatically adjusts the washing cycle for load size, type of fabric, and amount of dirt; and (c) a helicopter that obeys vocal commands to go forward, up, left, and right, and to hover and land.

39.9 Economic Considerations

The economic considerations in implementing the various computer-integrated activities described in this chapter are critical in view of their many complexities and the high costs involved. Flexible manufacturing system installations are very capital intensive, thus requiring a thorough cost–benefit analysis, including

- Costs of capital, energy, materials, and labor
- Expected markets for the products to be made
- Anticipated fluctuations in market demand and in the type of product
- Time and effort required for installing and debugging the system

Typically, an FMS system can take two to five years to install and a few months to debug. Although it requires few, if any, machine operators, the personnel in charge of the total operation must be trained and highly skilled; these personnel also include manufacturing engineers, computer programmers, and maintenance engineers. The most effective FMS applications have been in medium-volume batch production. When a variety of parts is to be produced, FMS is suitable for production volumes of 15,000–35,000 parts per year; for individual parts that are of the same configuration, production may reach 100,000 units per year. In contrast, high-volume, low-variety parts production is best obtained from transfer machines (see *dedicated equipment,* Section 37.2.4). Low-volume, high-variety parts production can best be done on conventional standard machinery, with or without NC, or by using machining centers (Section 25.2).

SUMMARY

- Integrated manufacturing systems are implemented to various degrees to optimize operations, improve product quality, and reduce costs.
- Computer-integrated manufacturing systems have become the most important means of improving productivity, responding to changing market demands, and enhancing the control of manufacturing and management functions.
- Advances in holonic manufacturing, just-in-time production, and communications networks are all essential elements in improving productivity.
- Lean manufacturing is intended to identify and eliminate waste, leading to improvements in product quality, customer satisfaction, and decreasing product cost.

- Artificial intelligence continues to create new opportunities in all aspects of manufacturing engineering and technology.
- Economic considerations in the design and implementation of computer-integrated manufacturing systems, especially flexible manufacturing systems, are particularly crucial because of the major capital expenditures required.

KEY TERMS

Artificial intelligence	Environments	Internet tools	Natural-language
Artificial neural networks	Ethernet	Just-in-time production	processing
Cellular manufacturing	Expert systems	Kanban	Pallet
Communications	Flexible manufacturing	Knowledge engineering	Pull system
network	Flow	Knowledge-based	Push system
Communications	Framework systems	system	Value
standard	Fuzzy logic	Lean manufacturing	Waste
Computer-integrated	Holonic manufacturing	Local area network	Wireless networks
manufacturing	If-then rules	Machine vision	Zero inventory
systems	Inference engine	Manufacturing cell	

BIBLIOGRAPHY

Black, J.R., **Lean Production**, Industrial Press, 2008.

Black, J.T., and Hunter, S.L., **Lean Manufacturing Systems and Cell Design**, Society of Manufacturing Engineers, 2003.

Bodek, N., **Kaikaku: The Power and Magic of Lean**, PCS Press, 2004.

Conner, G., **Lean Manufacturing for the Small Shop**, 2nd ed., Society of Manufacturing Engineers, 2008.

Deen, S.M. (ed.), **Agent-Based Manufacturing: Advances in the Holonic Approach**, Springer, 2003.

Giarratano, J.C., and Riley, G.D., **Expert Systems: Principles and Programming**, 4th ed., Course Technology, 2004.

Irani, S.A. (ed.), **Handbook of Cellular Manufacturing Systems**, Wiley, 1999.

Jackson, P., **Introduction of Expert Systems**, 3rd ed., Addison Wesley, 1998.

Koren, Y., **The Global Manufacturing Revolution: Product-Process-Business Integration and Reconfigurable Systems**, Wiley, 2010.

Liker, J., and Convys, G.L., **The Toyota Way to Lean Leadership**, McGraw-Hill, 2011.

Meyers, F.E., and Stewart, J.R., **Motion and Time Study for Lean Manufacturing**, 3rd ed., Prentice Hall, 2001.

Myerson, P., **Lean Supply Chain and Logistics Management**, McGraw-Hill, 2012.

Negnevitsky, M., **Artificial Intelligence: A Guide to Intelligent Systems**, 2nd ed., Addison Wesley, 2004.

Ortiz, C.A., **Kaizen Assembly: Designing, Constructing, and Managing a Lean Assembly Line**, CRC Press, 2006.

Padhy, N.P., **Artificial Intelligence and Intelligent Systems**, Oxford University Press, 2003.

Rehg, J.A., **Introduction to Robotics in CIM Systems**, 5th ed., Prentice Hall, 2002.

Rehg, J.A., and Kraebber, H.W., **Computer-Integrated Manufacturing**, 3rd ed., Prentice Hall, 2004.

Russell, S., and Norvig, P., **Artificial Intelligence: A Modern Approach**, 3rd ed., Prentice Hall, 2009.

Siler, W., and Buckley, J.J., **Fuzzy Expert Systems and Fuzzy Reasoning**, Wiley-Interscience, 2004.

Wilson, L., **How to Implement Lean Manufacturing**, McGraw-Hill, 2009.

REVIEW QUESTIONS

39.1 What is a manufacturing cell? Why was it developed?

39.2 Describe the basic principle of flexible manufacturing systems.

39.3 Why is a flexible manufacturing system capable of producing a wide range of lot sizes?

39.4 What are the benefits of just-in-time production? Why is it called a pull system?

39.5 Explain the function of a local area network.

39.6 What is an expert system?

39.7 What are the advantages of a communications standard?

39.8 What is MTConnect?

39.9 What is a WLAN? a PAN?

39.10 Describe your understanding of holonic manufacturing.

39.11 What is Kanban? Explain.

39.12 What is lean manufacturing?

39.13 What is a push system?

39.14 In the lean manufacturing concept, what is the meaning of a value stream?

39.15 Describe the elements of artificial intelligence. Is machine vision a part of it? Explain.

QUALITATIVE PROBLEMS

39.16 In what ways have computers had an impact on manufacturing? Explain.

39.17 What advantages are there in viewing manufacturing as a system? What are the components of a manufacturing system?

39.18 One restaurant makes sandwiches as they are ordered by customers. Another competing restaurant makes sandwiches in advance and sells them to customers from their inventory. Which is a pull, and which is a push system? Explain which restaurant makes the better sandwiches.

39.19 Discuss the benefits of computer-integrated manufacturing operations.

39.20 (a) Why is just-in-time production required in lean manufacturing? (b) What drawbacks are there to just-in-time?

39.21 Would machining centers be suitable in just-in-time production? Explain.

39.22 Give an example of a push system and of a pull system. Indicate the fundamental difference between the two methods.

39.23 What is fuzzy logic? Give three examples where you personally have made decisions based on fuzzy data.

39.24 What are the advantages to having level production across lines and with respect to time?

39.25 Is there a minimum to the number of machines in a manufacturing cell? Explain.

39.26 Are robots always a component of an FMC? Explain.

39.27 Are there any disadvantages to zero inventory? Explain.

39.28 Review Table 36.1 and identify the points that are consistent with lean manufacturing.

39.29 Give examples in manufacturing processes and operations in which artificial intelligence could be effective.

SYNTHESIS, DESIGN, AND PROJECTS

39.30 Think of a product line for a commonly used household item and design a manufacturing cell for making it. Describe the features of the machines and equipment involved.

39.31 What types of (a) products and (b) production machines would not be suitable for FMC? What design or manufacturing features make them unsuitable? Explain with examples.

39.32 Describe your opinions concerning the voice-recognition capabilities of future machines and controls.

39.33 Can a factory ever be completely untended? Explain.

39.34 Assume that you own a manufacturing company and you are aware that you have not taken full advantage of the technological advances in manufacturing. However, now you would like to do so, and you have the necessary capital. Describe how you would go about analyzing your company's needs and how you would plan to implement these technologies. Consider technical as well as human aspects.

39.35 How would you describe the benefits of FMS to an older worker in a manufacturing facility whose experience has been running only simple machine tools?

39.36 Artificial neural networks are particularly useful where problems are ill defined and the data are vague. Give examples in manufacturing where artificial neural networks can be useful.

39.37 It has been suggested by some that artificial intelligence systems ultimately will be able to replace the human brain. Do you agree? Explain.

39.38 Evaluate a process from a lean-production perspective. For example, observe the following closely, and identify, eliminate (when possible), or optimize the steps that produce waste in (a) preparing breakfast for a group of eight, (b) washing clothes or cars, (c) using Internet browsing software, and (d) studying for an exam, writing a brief report, or writing a term paper.

39.39 Pull can be achieved by working with one supplier and developing a balanced flow of products. However, it was stated that single-source suppliers should be avoided in the unforeseen events of natural disasters. Write a one-page paper explaining this paradox.

39.40 (a) Explain how you can make your study habits more lean. (b) Conduct an Internet search and list five software packages that incorporate MTConnect.

Product Design and Manufacturing in a Competitive Environment

40

- Manufacturing high-quality products at the lowest possible cost is critical in a global economy.
- This chapter describes the interrelated factors in product design, development, and manufacturing; it begins with a discussion of product design and life-cycle assessment in design and manufacturing.
- The importance of energy considerations in manufacturing and their role in production costs are then discussed.
- Material and process selection, together with their effects on design and manufacturing, are then described, followed by a discussion of the economic factors involved in making a product.
- Finally, the principle of value analysis is described, along with a discussion of how it can help optimize manufacturing operations and minimize product cost.

40.1 Introduction

In an increasingly competitive global marketplace, manufacturing high-quality products at the lowest possible cost requires an understanding of the often complex interrelationships among numerous factors. It was indicated throughout this text that

1. Product design and selection of materials and manufacturing processes are interrelated.
2. Designs are sometimes modified to (a) improve product performance, (b) strive for zero-based rejection and waste, (c) make them easier and faster to manufacture, and (d) consider new materials and processes that are continually being developed.

Because of the increasing variety of materials and manufacturing processes now available, the task of producing a high-quality product by selecting the best materials and the best processes, while, at the same time, minimizing costs, continues to be a major challenge, as well as an opportunity. The term **world class** is widely used to indicate high levels of product quality, signifying the fact that products must meet international standards and be acceptable and marketable worldwide.

World-class status, like product quality, is not a fixed target for a company to reach but rather a *moving target*, also known as **continued improvement**, rising to higher and higher levels. The *selection of materials* for products traditionally has required much experience; however, several databases and expert systems are now widely available that facilitate the selection process. Moreover, in reviewing the materials used in existing products, it readily becomes apparent that there are numerous opportunities for the *substitution of materials* for improved performance and, especially, cost savings.

In the production phase of a product, the *capabilities of manufacturing processes* must be properly assessed in the selection of an appropriate process or sequence of processes. As described throughout various chapters, there usually is more than one method of manufacturing the individual components of a product.

Increasingly important are also *life-cycle* assessment and *life-cycle engineering* of products, services, and systems, particularly regarding their potentially adverse global impact on the environment. A major emphasis now is on *sustainable manufacturing*, with the purpose of reducing or eliminating any and all adverse effects of manufacturing on the environment, while still allowing companies to continue to be profitable.

Although the *economics* of individual manufacturing processes has been described throughout the book, this chapter takes a broader view and summarizes the overall manufacturing cost factors. It also introduces cost-reduction methods, including *value analysis*, a powerful tool to evaluate the cost of each manufacturing step relative to its contribution to a product's value.

40.2   Product Design

Design for manufacture and assembly (DFMA) and *competitive aspects* of manufacturing have been highlighted throughout this text. Several guidelines for the selection of materials and manufacturing processes are given in the references listed in Table 40.1. Major advances are continually being made in design for manufacture and assembly, for which several software packages are now available. Although their use requires considerable training, these advances greatly help designers develop high-quality products with fewer components, thus reducing production time and assembly, as well as product cost.

40.2.1   Product Design Considerations

In addition to the design guidelines regarding individual manufacturing processes, there are general product design considerations (see also *robust design*, Section 36.5.1). Designers often must check and verify whether they have addressed considerations such as:

1. Have all alternative designs been thoroughly investigated?
2. Can the design be simplified and the number of components minimized, without adversely affecting the intended functions and performance?
3. Can the design be made smaller and lighter?
4. Are there unnecessary features in the product and, if so, can they be eliminated or combined with other features?
5. Have modular design and building-block concepts (see, e.g., Section 25.2.4) been considered for a family of similar products and for servicing, repair, and upgrading?
6. Are the specified dimensional tolerances and surface finish unnecessarily tight, thereby significantly increasing product cost, and, if so, can they be relaxed without any significant adverse effects?
7. Will the product be difficult or excessively time consuming to assemble or to disassemble, for maintenance, servicing, or recycling of some or all of its components?

TABLE 40.1

References to Various Topics in This Book (Page numbers are in parentheses)

Material Properties
Tables 2.1 (56), 2.2 (58), and 2.3 (61), and Figs. 2.5, 2.6, 2.7, 2.15, 2.16, and 2.17
Tables 3.1 (89) and 3.2 (90), and Figs. 3.1, 3.2, and 3.3
Tables 5.3 (138), 5.4 (139), and 5.6 (143)
Tables 6.3 through 6.10 (152–160)
Tables 7.1 (170) and 7.2 (178)
Tables 8.1 (197), 8.2 (201), and 8.3 (205), and Fig. 8.3
Tables 9.1 (217), 9.2 (218), and 9.4 (228), and Figs. 9.3, 9.5, and 9.7
Table 10.1 (249)

Table 11.3 (280)
Tables 12.3 (305), 12.4 (306), and 12.5 (306), and Fig. 12.5
Tables 16.2 (398), 16.3 (403), and 16.4 (415), and Fig. 16.14
Tables 17.3 (464), 17.4 (465), and 17.5 (465), and Fig. 17.11
Table 20.2 (539)
Tables 22.1 (602), 22.2 (603), 22.3 (603), and 22.5 (608), and Figs. 22.1 and 22.9
Table 26.1 (731)
Table 32.3 (944)

Manufacturing Characteristics of Materials
Table I.3 (14)
Table 4.1 (120)
Table 5.8 (145)
Table 6.2 (151)
Tables 12.1 (297) and 12.6 (308)

Table 14.3 (349)
Table 16.3 (403) and Fig. 16.34
Tables 17.1 (454) and 17.2 (462)
Tables 21.1 (568) and 21.2 (579)
Fig. 22.1 and Table 40.2 (1149)

Dimensional Tolerances and Surface Finish
Table 11.2 (259)
Table 23.1 (627) and Figs. 23.14 and 23.15
Fig. 25.16

Fig. 27.4
Figs. 35.20 and 35.21
Fig. 40.4

Capabilities of Manufacturing Processes
Tables 11.1 (257) and 11.2 (259)
Table III.1 (314)
Tables 14.1 (339) and 14.4 (356)
Table 16.1 (388)
Section 17.7 and Fig. 17.15
Table 18.1 (476)
Tables 19.1 (495) and 19.2 (530)
Table 20.1 (538)
Tables 23.1 (627), 23.7 (638), 23.9 (646), and 23.11 (653)
Tables 26.3 (746) and 26.4 (747)

Table 27.1 (771)
Tables 28.1 (810), 28.2 (819), and 28.3 (820), and Fig. 28.19
Tables 29.1 (857) and 29.3 (867)
Table VI.1 (876)
Table 30.1 (878)
Table 32.4 (945)
Table 37.2 (1064) and Fig. 37.3
Table 39.1 (1128)
Tables 40.4 (1151) and 40.5 (1155), and Figs. 40.2 and 40.3

Design Considerations in Processing
Abrasive processes: Section 26.5
Advanced machining: Various sections in Chapter 27
Casting: Section 12.2
Ceramics shaping: Section 18.5
Extrusion: Section 15.7
Forging: Section 14.6

Heat treating: Section 4.13
Joining processes: Various sections in Chapters 30–32
Machining: Sections in Chapters 23–24
Polymers processing: Section 19.15
Powder metallurgy: Section 17.6
Sheet-metal forming: Section 16.15

Costs and Economics
Tables I.5 (29), I.6 (31), I.7 (32), and Section I.10
Table 6.1 (151)
Section 12.4
Section 14.9
Section 16.16
Table 17.6 (469) and Section 17.7
Table 19.2 (530) and Section 19.16
Section 25.8

Section 26.9 and Fig. 26.37
Section 27.11
Section 31.8
Section 32.7
Section 37.11
Section 39.9
Table 40.6 (1159) and Section 40.10

8. Is the use of fasteners minimized, including their quantity and type variety?
9. Have environmental considerations been taken into account and incorporated into product design and material and process selection?
10. Have green design and life-cycle engineering principles been applied, including recycling and cradle-to-cradle considerations?
11. Can any of the design or manufacturing activities be outsourced, and can any currently outsourced activities be reshored?

40.2.2 Product Design and Quantity of Materials

Significant reductions in the *quantity* of materials required can be achieved by approaches such as (a) reducing the component's size or volume or (b) using materials with higher strength-to-weight or stiffness-to-weight ratios (see Fig. 3.2). The latter can be attained by improving and optimizing the product design and by selecting different cross-sections, such as those having a high moment of inertia (e.g., I-beams and channels) or by using tubular or hollow components instead of solid sections.

Implementing design changes may, however, present significant challenges in manufacturing; for example, consider the following:

1. Casting or molding thin cross-sections can present difficulties in die and mold filling and in meeting specified dimensional accuracy and surface finish (Section 12.2).
2. Forging of thin sections requires higher forces and can present difficulties (Section 14.3).
3. Impact extrusion of thin-walled parts can be difficult, especially when high dimensional accuracy and part symmetry are required (Section 15.4.1).
4. The formability of sheet metals may be reduced as their thickness decreases, also possibly leading to wrinkling of the part (Section 16.3).
5. Machining and grinding of thin workpieces may lead to part distortion, poor dimensional accuracy, and vibration and chatter (Section 26.5); consequently, advanced machining processes may have to be considered (Chapter 27).
6. Welding thin sheets or slender structures can cause significant distortion of parts and structures (Section 30.10).

Conversely, making parts with thick cross-sections can have their own adverse effects:

1. The production rate in die casting (Section 11.4.5) and injection molding (Section 19.3) can become slower, because of the increased cycle time required to allow sufficient time for the thicker regions to cool before removing the part from the die or mold.
2. Porosity can develop in thicker regions of castings (Fig. 10.14).
3. In die-cast parts, thinner sections will have lower strength per unit thickness as compared to thicker sections (Section 11.4.5).
4. Processing plastic parts requires increased cycle times as their thickness or volume increases, because of the longer time required for the parts to cool sufficiently to be removed from their molds (Chapter 19).
5. The bendability of sheet metals decreases as their thickness increases (Section 16.5).
6. In parts made by powder metallurgy, there can be significant variations in density, hence properties, throughout regions with varying thickness (Section 17.6).
7. Welding thick sections can present difficulties in the quality of the welded joint (Section 30.9).

40.3   Product Quality

Recall that the word *quality* (Section 36.2) is difficult to define precisely, largely because it includes not only well-defined technical characteristics but also human, and hence subjective, opinions. Generally, however, a high-quality product is considered to have at least the following characteristics: it (a) satisfies the needs and expectations of the customer, (b) has high reliability and thus functions well over its intended life, (c) has a pleasing appearance, and (d) is easy to install and maintain, and future improvements are easy to make and at low cost.

A major priority in product quality is the concept of **continuous improvement,** as exemplified by the Japanese term **kaizen,** meaning *never-ending improvement* (Section 36.1). Recall also that the *level* of quality a manufacturer chooses to impart to a particular product depends on the market for which the product is intended. Low-quality, low-cost products, such as hand tools, have their own market niche, just as there is a market for high-quality and high-end products, such as a Rolls-Royce automobile, a diamond-studded gold watch, high-precision machine tools and measuring instruments, and high-performance sports equipment.

Return on Quality.   In implementing quality into products, it is important to appreciate the concept of *return on quality* (ROQ), because of the following considerations:

- Quality must be viewed as an investment because of its long-term major influence on customer satisfaction.
- Any improvement in quality must be investigated with respect to the additional costs involved.
- There must be a certain limit to how much a manufacturer should spend on incremental quality improvements, especially when quality can be rather subjective.

Although customer satisfaction can be difficult to assess and include in cost calculations, when their satisfaction increases, customers are more likely to be retained and become repeat customers; this is unlikely if there are significant defects in the products they purchase. It is also important to note the fact that the relative costs involved in identifying and repairing defects in products grow by orders of magnitude, in accordance with the *rule of ten*, as shown in Table I.5.

40.4   Life-cycle Assessment and Sustainable Manufacturing

Life-cycle assessment (LCA) is defined, according to the ISO 14000 standard, as "a systematic set of procedures for compiling and examining the inputs and outputs of materials and energy, and the associated environmental impacts or burdens directly attributable to the functioning of a product, process, or service system throughout its entire life cycle." **Life cycle** involves consecutive and interlinked stages of a product or a service, from the very beginning of design and manufacture to its recycling or disposal; it includes

1. Extraction of natural resources
2. Processing of raw materials
3. Manufacturing of products
4. Transportation and distribution of the product to the customer

QR Code 40.1 The Sustainability Continuum. (*Source:* Courtesy of the Metal Powder Industries Federation)

5. Use, maintenance, and reuse of the product
6. Recovery, recycling, and reuse of components of the product or their disposal, which also include metalworking fluids, cleaning solvents, and various liquids used in heat-treating and plating processes

A product typically has several components made of a variety of metallic and nonmetallic materials, processed into individual parts, and then assembled. Thus, each component has its own life cycle, such as the tires in an automobile, rubber washers in faucets, bulbs in light fixtures, and belts in vacuum cleaners. Moreover, (a) some products, particularly those made of paper, cardboard, glass, or inexpensive plastics, are intentionally made to be *disposable* but nonetheless are now recyclable and (b) numerous other products are completely reusable.

Life-cycle Engineering. The major aim of *life-cycle engineering* (LCE), also called **green design** or **green engineering**, is to consider *reusing* and *recycling* the components of a product, beginning with the earliest stage of product design (see Fig. I.2). Although life-cycle analysis and engineering are comprehensive and powerful tools, their implementation can be challenging, time consuming, and costly, largely because of uncertainties (a) regarding materials, processes, and long-term effects, (b) the input data, and (c) the time required to collect reliable data, to properly assess the often-complex interrelationships among the numerous components of the whole system. Software is being developed to expedite these analyses, especially for the chemical and manufacturing industries, because of the higher potential for environmental and ecological impact; examples of such software include FeaturePlan and Teamcenter, which runs in a ProEngineer environment.

Cradle-to-cradle. The terms and concepts of *cradle-to-cradle* design (coined by W.R. Stahel in the 1970s and also called *CRC* or *regenerative design*) and *cradle-to-grave* (which ends with the disposal phase of products) are described in Section I.4. These are basically a holistic model for human activity with due respect for life and the well-being of future generations. (See also McDonough and Braungart in the Bibliography at the end of this chapter.)

Sustainable Manufacturing. As universally acknowledged, the natural resources on the planet Earth are limited, thus necessitating conservation of both materials and energy. The concept of *sustainable manufacturing* emphasizes the need for conserving resources, particularly through proper maintenance and reuse. Sustainable manufacturing is meant to meet the main purposes of (a) increasing the life cycle of products, (b) eliminating harm to the environment and the ecosystem, and (c) ensuring our collective well-being.

EXAMPLE 40.1   Sustainable Manufacturing of Nike Athletic Shoes

Among numerous examples, the production of Nike shoes illustrates the benefits of sustainable manufacturing. Athletic shoes are assembled using adhesives (Section 32.4). Up to around 1990, the adhesives used contained petroleum-based solvents, which pose health hazards to humans and contribute to petrochemical smog. The company cooperated with chemical suppliers to successfully develop a water-based adhesive technology, now used in the majority of shoe-assembly operations. As a result, solvent use in all manufacturing processes in Nike's subcontracted facilities in Asia has been greatly reduced.

The rubber outsoles of the athletic shoes are made by a process that results in significant amounts of extra rubber around the periphery of the sole, called *flashing* (similar to the flash shown in Figs. 14.6d and 19.17). With about 40 factories using thousands of molds and producing over a million outsoles a day, the flashing constitutes the largest chunk of waste in manufacturing of these shoes.

In order to reduce this significant waste, the company developed a technology that grinds the flashing into 500-μm rubber powder, which is then added back into the rubber mixture needed to make the outsole; with this approach, waste was reduced by 40%. Moreover, it was found that the mixed rubber had better abrasion resistance and durability, and its overall performance was higher than the best premium rubber.

40.5 Energy Consumption in Manufacturing

The manufacturing sector consumes approximately one-fourth of the annual global energy production; this number has fallen since it peaked at around 50% in the 1970s, because of major efforts to reduce waste and improve the efficiency of machinery and manufacturing operations. By far, the most common source of energy in manufacturing is electrical, produced from oil, natural gas, biofuels, coal, nuclear, wind, solar, and wave. Given such a large percentage and varied sources of energy, all of the concerns regarding energy availability and conservation still must be considered and addressed in manufacturing. Indeed, it is not likely that viable national energy policies can be developed and implemented without a central consideration of the manufacturing sector.

Process Energy Demand. The energy required to produce a particular part or component is determined, to a large extent, by its design, material, and the manufacturing operation, as well as the quality, condition, and age of the machinery and equipment. The energy requirement is relatively easy to calculate for a manufacturing process (see, for instance, Table 21.2), but becomes difficult when ancillary equipment is included in the final calculation. For example, pumps, fans, blowers, furnaces, and lights are all involved in manufacturing operations. Moreover, the efficiency of operation typically varies depending on a particular plant's practices and procedures, or from company to company. For example:

- Some manufacturing operations are more demanding from an energy standpoint than others, as shown in Fig. 40.1.
- Each manufacturing process has a range of performance; processing rates can vary greatly depending on, for example, the specified tolerances and surface finish specified, with tighter tolerances and smoother surfaces being the most time consuming and energy intensive (Figs. 40.2 and 40.3).
- Energy requirements for some processes are strongly related to the sequence of operations

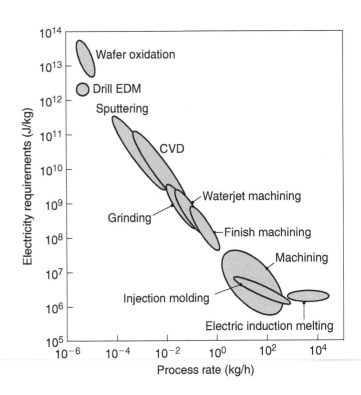

FIGURE 40.1 Specific energy requirements for various manufacturing processes, including ancillary equipment. *Source:* Based on S. Gutowski.

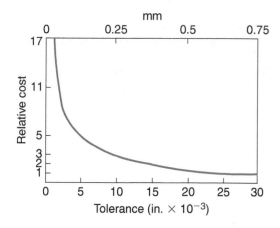

FIGURE 40.2 Dependence of manufacturing cost on dimensional tolerance.

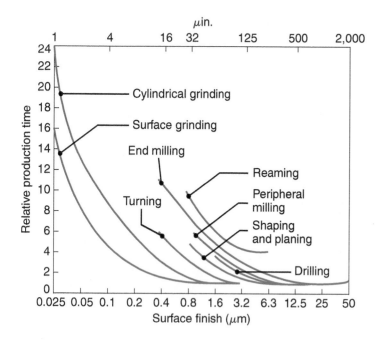

FIGURE 40.3 Relative production time as a function of surface finish produced by various manufacturing processes. (See also Fig. 26.35.)

performed, as when, for example, a machine is ready but not yet actively processing the material; thus, (a) in injection molding (Section 19.3), the mold has been installed in the machine but it has not yet been filled with the polymer or (b) when a workpiece is being repositioned in a fixture on a machining center (Section 25.2).

• Some machinery have continuously operating hydraulic pumps, while others will shut the pumps off during periods of inactivity.

• Workpieces in hot-working operations can be cooled by blowing cool air over them, and, in practice, the hot air is often vented to the atmosphere. The resulting hot air can, however, be used to preheat stock material for processing, heat the facility, or provide hot water.

Effect of Materials. Whether a material is mined, refined, and cast; synthesized from chemicals; or recycled from discarded products, energy is required to produce stock materials into forms that can be further processed. In many applications, it is noted that certain materials have a performance advantage over others. For example, titanium and aluminum alloys are obviously preferred to steel for aircraft, primarily because of their lower strength-to-weight and stiffness-to-weight ratios (see Fig. 3.2), allowing for lighter designs and associated fuel savings.

It is important to note, however, that different materials have very different energy requirements, and recycled materials have significantly less energy needs. Table 40.2 summarizes the energy required to produce various materials, and presents the data by mass and by volume. It has been noted that if energy or *carbon footprint* is divided by mass, there is a natural benefit to using heavier materials, that may not be justified in weight-constrained problems. On the other hand, if volume is a constraint, then the energy per unit weight may be a fair measure.

Note in Table 40.2 that metals require significant energy to be produced; they are generally reduced or extracted from their oxides, a process that is energy intensive. It has been estimated that 5% of the total energy consumption in the United States is used to produce aluminum (Section 6.2). It is not uncommon for the energy required to produce a material to be three or four orders of magnitude greater than the energy required to form it into its final shape. It is therefore understandable that the use of aluminum in automobiles is indeed difficult to justify without the sustained implementation of recycling.

40.6 Material Selection for Products

In *selecting materials* for a product, it is essential to have a clear understanding of the *functional requirements* for each of its individual components. The general criteria for selecting materials were described in Section I.5 of the General Introduction; this section discusses them in more specific detail.

General Properties of Materials. *Mechanical properties* (Chapter 2) include strength, toughness, ductility, stiffness, hardness, and resistance to fatigue, creep, and impact. *Physical properties* (Chapter 3) include density, melting point, specific heat, thermal and electrical conductivity, thermal expansion, and magnetic properties. *Chemical properties* of primary concern in manufacturing are their susceptibility to oxidation and corrosion, and their response to various surface-treatment processes (Chapter 34). The cost of corrosion, although largely hidden, is estimated to be on the order of 3% of the U.S. gross domestic product.

Material selection has become easier and faster because of the increasing availability of extensive computer databases that provide greater accessibility and accuracy. To facilitate the selection of materials, **expert-system software** (called **smart databases,** Section 39.8) have been developed. These systems are capable of identifying appropriate materials for a specific application, just as an expert or a team of experts would.

Shapes of Commercially Available Materials. After selecting appropriate materials, the next step is to determine the shapes and the sizes in which these materials are available *commercially* (Table 40.3). Depending on the type, materials generally are available as castings, extrusions, forgings, drawn rod and wire, rolled bars, plates, sheets, foil, and powder metals.

Purchasing materials in shapes that require the least amount of additional processing obviously is a major economic consideration; also relevant are such characteristics as surface finish, dimensional tolerances, straightness, and flatness (see, e.g., Figs. 23.14, 23.15, 27.4, and Table 11.2). Obviously, the better and the more consistent these characteristics are, the less additional processing will be required. If, for example, simple round shafts with good dimensional accuracy, roundness, straightness, and surface finish are desired, then round bars that are first turned (or drawn) and centerless-ground (Fig. 26.23) to the dimensions specified could be purchased.

Unless the facilities in a plant have the capability of economically producing round bars, it generally is less expensive to purchase them. If a stepped shaft (having different diameters along its length, as shown in Fig. IV.3) is required, a round bar with a diameter at least equal to the largest diameter of the stepped shaft could be purchased, then turned on a lathe or processed and formed by some other means in order to reduce its diameter.

TABLE 40.2

Energy Content of Selected Materials

Material	Energy content	
	MJ/kg	GJ/m³
Metals		
Aluminum		
From bauxite	300	810
Recycled	42.5	115
Cast iron	30–60	230–460
Copper		
From ore	105	942
Recycled	55.4	497
Lead	30	330
Magnesium	410	736
Steel		
From ore	55	429
Recycled	9.8	76.4
Zinc	70	380
Polymers		
Nylon 6,6	175	200
Polyethylene		
High density	105–120	100–115
Low density	80–100	75–95
Polystyrene	95–140	95–150
Polyvinyl chloride	67–90	90–150
Ceramics	1–50	4–100
Glasses	10–25	30–60
Wood	1.8–4.0	1.2–3.6

TABLE 40.3

Shapes of Commercially Available Materials

Material	Available as
Aluminum	B, F, I, P, S, T, W
Ceramics	B, p, s, T
Copper and brass	B, f, I, P, s, T, W
Elastomers	b, P, T
Glass	B, P, s, T, W
Graphite	B, P, s, T, W
Magnesium	B, I, P, S, T, w
Plastics	B, f, P, T, w
Precious metals	B, F, I, P, t, W
Steels and stainless steels	B, I, P, S, T, W
Zinc	F, I, P, W

Note: B = bar and rod; F = foil; I = ingots; P = plate and sheet; S = structural shapes; T = tubing; and W = wire; all metals are generally available as powders. Lowercase letters indicate limited availability.

Each manufacturing operation produces a part that has a specific shape, surface finish, and dimensional accuracy; consider the following examples:

- Castings generally have lower dimensional accuracy and a poorer surface finish than parts made by cold forging, cold extrusion, or powder metallurgy.
- Hot-rolled or hot-drawn products generally have a rougher surface finish and larger dimensional tolerances than cold-rolled or cold-drawn products.
- Extrusions have smaller tolerances than parts made by roll forming of sheet metal.
- Round bars machined on a lathe have a rougher surface finish than similar bars that are ground.
- The wall thickness of welded tubing is generally more uniform than that of seamless tubing.

Manufacturing Characteristics of Materials.   *Manufacturing characteristics* generally include castability, workability, formability, machinability, weldability, and hardenability by heat treatment. Recall also that the quality of the raw material (stock) can greatly influence its manufacturing properties. The following are typical examples:

- A bar with a longitudinal seam or lap will develop cracks during even simple upsetting or heading operations.
- Rods with internal defects, such as hard inclusions, will crack during further processing.
- Porous castings will develop a poor surface finish when subsequently machined for better dimensional accuracy.
- Parts that are heat treated nonuniformly, or cold-drawn bars that are not properly stress relieved, will distort during subsequent processing.
- Incoming stock that has significant variations in its composition and microstructure cannot be heat treated or machined consistently and uniformly.
- Sheet metal having variations in its cold-worked conditions will exhibit different degrees of springback during subsequent bending and other forming operations.
- If prelubricated sheet-metal blanks are supplied with nonuniform lubricant thickness and distribution, their formability, surface finish, and overall quality in subsequent stamping operations will be adversely affected.

Reliability of Material Supplies.   Several factors influence the *reliability of material supplies*: shortages of materials, strikes, geopolitics, and the reluctance of suppliers to produce materials in a particular shape or quality. Moreover, even though raw materials may generally be available throughout a country as a whole, they may not readily be available at a particular plant's location.

Recycling Considerations.   *Recycling* may be relatively simple for products such as scrap metal, plastic bottles, and aluminum cans; often, however, the individual components of a product must be taken apart and separated into groups. Obviously, if much effort and time has to be expended in doing so, recycling can indeed become prohibitively expensive. Some general guidelines to facilitate recycling are:

- Reduce the number of parts and types of materials in products
- Reduce the variety of product models
- Use a modular design to facilitate disassembly of a product

QR Code 40.2 Case studies of Green Engineering. (*Source:* Courtesy of Intel Corp.)

- For products made of plastic, use single types of polymers, as much as possible
- Mark plastic parts for ease of identification
- Avoid using coatings, paints, and plating; instead, use molded-in colors in plastic parts
- Avoid using adhesives, rivets, and other permanent joining and assembly methods; instead, use fasteners, especially snap-in fasteners

As one example of such an approach to recycling, one manufacturer of laser-jet printers reduced the number of parts in a cartridge by 32% and the variety of plastic materials used by 55%.

TABLE 40.4

Approximate Scrap Produced in Various Manufacturing Operations

Process	Scrap (%)
Machining	10–80
Hot forging	20–25
Sheet-metal forming	10–25
Hot extrusion	15
Permanent-mold casting	10
Powder metallurgy	<5
Rolling	<1

Cost of Materials and Processing.   Because of its processing history, the *unit cost* of a raw material (typically per unit weight) depends not only on the material itself, but also on its shape, size, and condition. Also, for example, (a) because more operations are involved in the production of thin wire than for a round rod, its unit cost is much higher; (b) powder metals generally are more expensive than bulk metals; and (c) the cost of materials typically decreases as the quantity purchased increases, especially for automotive companies that purchase materials in very large quantities.

The cost of a particular material is subject to fluctuations, caused by factors as simple as supply and demand or as complex as geopolitics. If a material used in a product is no longer cost competitive, alternative materials have to be selected. For example, (a) the copper shortage in the 1940s led the U.S. government to mint pennies from zinc-plated steel (see Table I.2) and (b) when the price of copper increased substantially during the 1960s, electrical wiring in homes was switched to aluminum; however, this substitution led to the redesign of terminals of switches and outlets in order to avoid excessive localized heating, because aluminum has a higher contact resistance at the junctions than copper.

Scrap.   The value of *scrap* (Table 40.4) is deducted from a material's cost in order to obtain the net material cost. Its value depends on the type of material and on demand for it, and typically it is 10–40% of the original cost of the material. Note that in machining, scrap can be very high, whereas such processes as rolling, ring rolling, and powder metallurgy produce the least scrap, hence they are called net- or near-net-shape processes.

40.7   Material Substitution

Substitution of materials plays a major role in the economics of manufacturing products. Automobile and aircraft manufacturing are typical examples of major industries in which the substitution of materials is an ongoing activity; a similar trend is evident in sporting goods and numerous other products. There are several reasons for substituting materials in products:

1. Reduce the costs of materials and processing them
2. Improve the efficiency of manufacturing and assembly operations
3. Improve the performance of the product, such as by reducing its weight and by improving its resistance to wear, fatigue, and corrosion
4. Increase stiffness-to-weight and strength-to-weight ratios
5. Reduce the need for periodic maintenance and repair
6. Reduce vulnerability to the unreliability of the supply of materials

7. Improve compliance with legislation and regulations prohibiting the use of certain materials, especially for health reasons
8. Improve robustness to reduce variations in performance or environmental sensitivity of the product
9. Improve ease of recycling

Substitution of Materials in the Automotive Industry

- Certain components of the metal body replaced with plastic or reinforced-plastic parts
- Metal bumpers, gears, pumps, fuel tanks, housings, covers, clamps, and various other components replaced with plastics or composites
- Carbon-steel chassis pillars replaced with TRIP or TWIP steels
- Structural steel components replaced with aluminum alloys
- Metallic engine components replaced with ceramic and composites parts
- All-metal drive shafts replaced with composite-material drive shafts
- Cast-iron engine blocks changed to cast-aluminum, forged crankshafts to cast crankshafts, and forged connecting rods to cast, powder metallurgy, or composite-material connecting rods

Substitution of Materials in the Aircraft and Aerospace Industries

- Conventional aluminum alloys (particularly 2000 and 7000 series) replaced with aluminum–lithium alloys, titanium alloys, polymer-reinforced composites, and glass-reinforced aluminum, because of the higher strength-to-weight ratios of these materials
- Forged parts replaced with powder metallurgy parts that are manufactured with better control of impurities and microstructure; the powder metallurgy parts also require less machining and produce less scrap of expensive materials
- Advanced composite materials and honeycomb structures replacing traditional aluminum airframe components, and metal-matrix composites replacing some of the aluminum and titanium in structural components

---

EXAMPLE 40.2 Material Substitution in Products

In the following list, the commonly available products can be made of either set of materials mentioned:

1. Baseball bat: Metal vs. wood
2. Hammer: Metal vs. reinforced-plastic or wood handle
3. Engine intake manifold: Metal vs plastic
4. Lawn chair: Cast-iron vs. aluminum
5. Light-switch plate: Plastic vs. sheet metal
6. Aircraft fuselage: Aluminum vs. composite materials
7. Pneumatic fittings: Zinc vs. copper

More recent examples of possible material substitutions are the following, some of which have been built and some are under various stages of consideration:

8. Automobile windows: Glass vs. polycarbonate
9. Pedestrian bridge: Steel vs. titanium
10. Bicycle: Metal vs. hardwood
11. Automobile bodies: Steel vs. aluminum
12. Truck bodies: Steel vs. stainless steel
13. Guitar: Wood vs. plastic or aluminum
14. Armor plate: Aluminum vs. steel or other high-strength alloys
15. Connections in IC packages: Copper vs. gold

40.8 Manufacturing Process Capabilities

Process capability is the ability of a particular manufacturing process to produce, under controlled production conditions, defect-free parts within certain limits of precision (see also Section 36.8.2). The capabilities of several manufacturing processes regarding their dimensional limits are shown in Fig. 40.4. Note, for example, that sand casting cannot produce thin parts, whereas cold rolling is capable of producing very thin materials, as also evidenced by such a common product as aluminum foil.

Equally important are the capabilities of various processes to meet stringent dimensional tolerance and surface-finish requirements, as shown in Figs. 23.14 and 40.4. Note, for example, how sand casting typically produces much rougher surfaces than does polishing. The importance of process control can be appreciated when one views the ranges attainable in the figure. Note, for instance, the large envelope for machining and finishing operations in Fig. 23.14, with roughness boundaries that, for a variety of reasons, span three orders of magnitude. Thus, if a turning operation is carried out on a poorly maintained old lathe or using inappropriate tools and processing parameters, then the tolerances and surface finish will, of course, be poor.

Dimensional Tolerances and Surface Finish. The *dimensional tolerances* and *surface finish* produced are particularly important in (a) subsequent assembly operations, because of possible difficulties in fitting the parts together during assembly and (b) the proper operation of machines and instruments, because their performance will affect tolerances and finish.

Closer tolerances and better surface finish can be achieved by subsequent and additional finishing operations (see Section 26.7) but at higher cost. Moreover the finer the surface finish specified, the longer is the manufacturing time. For example, it has been observed that in machining aircraft structural members, made of titanium

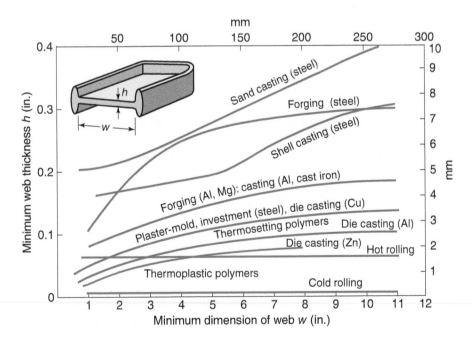

FIGURE 40.4 Manufacturing process capabilities for minimum part dimensions. *Source:* Based on J.A. Schey.

alloys, as much as 60% of the cost of machining may be expended in the final machining pass. Thus, unless otherwise required, parts should be specified with as rough a surface finish and as wide a dimensional tolerance as functionally and aesthetically will be acceptable.

Production Quantity.   Depending on the type of product, the production quantity, also known as *lot size*, varies widely. Bearings, bolts and nuts, metal or plastic containers, tires, automobiles, and lawn mowers, for example, are produced in very large quantities, whereas jet engines, diesel engines, locomotives, and medical equipment are produced in much more limited quantities. Production quantity also plays a significant role in process and equipment selection. In fact, an entire manufacturing discipline, called **economic order quantity**, is devoted to mathematically determining the optimum production quantity.

Production Rate.   An important factor in manufacturing process selection is the *production rate*, defined as the number of pieces to be produced per unit of time (hour, day, or year). The rate obviously can be increased by using multiple equipment and highly automated machinery. Die casting, powder metallurgy, deep drawing, wire drawing, and roll forming are high-production-rate operations; by contrast, sand casting, electrochemical machining, superplastic forming, adhesive and diffusion bonding, and the processing of reinforced plastics typically are relatively slow operations.

Lead Time.   *Lead time* is generally defined as the length of time between the receipt of an order for a product and its delivery to the customer. Depending on part size, material, and shape complexity and precision of the dies required, the lead time for such processes as forging, extrusion, die casting, roll forming, and sheet-metal forming can range from weeks to months. By contrast, processes such as machining, grinding, and advanced material-removal processes have significant built-in flexibility, due to the fact that they utilize machinery and tooling that can readily be adapted to most production requirements. Recall that machining centers, flexible manufacturing cells, and flexible manufacturing systems are all capable of responding rapidly and effectively to product changes and to production quantities. (See also *rapid prototyping* in Chapter 20.)

Robustness of Manufacturing Processes and Machinery.   *Robustness* was described in Section 36.5.1 as characterizing a design, a process, or a system that continues to function within acceptable parameters despite variabilities in its environment. To appreciate its importance in manufacturing processes, consider a situation in which a simple plastic gear is being produced by injection molding: it has been noticed that there are significant and unpredictable variations in quality as the gears are being produced. There are several well-understood variables and parameters in injection molding, including the effects of raw-material quality, speed of operation, and temperatures within the system. All these are independent variables, hence they can be controlled.

There are, however, certain other variables, called **noise**, that are largely beyond the control of the operator, such as (a) variations in ambient-temperature and humidity in the plant throughout the day, (b) dust in the air entering the plant from an open door, thus possibly contaminating the pellets being fed into the hoppers of the injection-molding machine, (c) variation in the mold temperature, depending on the delay between successive molding shots, and (d) variability in the performance of individual operators during different shifts. In order to maintain good product quality, it is essential to understand the effects, if any, of each element of noise in the operation, such as, for example: (a) Why and how does the ambient temperature affect the quality and surface characteristics of the molded gears? (b) Why and how does the dust coating on a pellet in the machine's hopper affect its behavior in the molding machine? As a result, certain control features may be instituted in the system.

40.9 Process Selection

Process selection is intimately related to the characteristics of the materials to be processed, as shown in Table 40.5.

Characteristics and Properties of Workpiece Materials. Recall that some materials can be processed at room temperature, whereas others require elevated temperatures, thus the need for certain furnaces, appropriate tooling, and various controls. Some materials are easy to process because they are soft and ductile; others are hard, brittle, and abrasive, thus requiring appropriate processing techniques, tooling, and equipment.

TABLE 40.5

General Characteristics of Manufacturing Processes for Various Metals and Alloys

	Carbon steels	Alloy steels	Stainless steels	Tool and die steels	Aluminum alloys	Magnesium alloys	Copper alloys	Nickel alloys	Titanium alloys	Refractory alloys
Casting										
Sand	1	1	1	2	1	1	1	1	2	1
Plaster	3	3	3	3	1	1	1	3	3	—
Ceramic	1	1	1	1	2	2	1	1	2	1
Investment	1	1	1	3	1	2	1	1	1	1
Permanent	2	2	3	3	1	1	1	3	3	—
Die	3	3	3	3	1	1	1	3	3	—
Forging										
Hot	1	1	1	1	1	1	1	1	1	1
Cold	1	1	1	3	1	2	1	3	3	—
Extrusion										
Hot	1	1	1	2	1	1	1	1	1	1
Cold	1	2	1	3	1	3	1	2	3	3
Impact	3	3	3	3	1	1	1	3	3	3
Rolling	1	1	1	3	1	1	1	1	1	2
Powder metallurgy	1	1	1	1	1	1	1	1	1	1
Sheet-metal forming	1	1	1	3	1	1	1	1	1	1
Machining	1	1	1	3	1	1	1	2	1	2
Chemical	1	2	1	2	1	1	1	2	2	2
ECM	3	1	2	1	3	3	2	1	1	1
EDM	3	2	2	1	3	3	2	2	2	1
Grinding	1	1	1	1	1	1	1	1	1	1
Welding	1	1	1	3	1	1	1	1	1	1

Note: 1 = generally processed by this method; 2 = can be processed by this method; 3 = usually not processed by this method.

As can be noted in Table 40.5, few materials have favorable manufacturing characteristics in all categories. A material that is castable or forgeable, for example, may later present difficulties in subsequent processing, such as machining, grinding, and polishing, that may be required for an acceptable surface finish and dimensional accuracy.

Moreover, materials have different responses to the rate of deformation (see *strain-rate sensitivity*, Sections 2.2.7 and 7.3) to which they are subjected. Consequently, the speed at which a particular machine is operated can affect product quality, including the development of external and internal defects. For example, impact extrusion or drop forging may not be appropriate for a certain material with high strain-rate sensitivity, whereas the same material will perform well in a hydraulic press or in direct extrusion.

Geometric Features.   Requirements for part shape, size, and thickness; dimensional tolerances; and surface finish may greatly influence the selection of a process or processes, as described throughout this and various other chapters in the book.

Production Rate and Quantity.   These requirements dictate process selection by way of the productivity of a process, machine, or system (see Section 40.7).

Process Selection Considerations.   The major factors involved in process selection may be summarized as:

1. Are some or all of the parts or components that are needed in a product commercially available as standard items?
2. Which components of the product have to be manufactured in the plant?
3. Is the tooling that is required available in the plant? If not, can it be purchased as a standard item?
4. Can group technology be implemented for parts that have similar geometric and manufacturing attributes?
5. Have all alternative manufacturing processes been investigated?
6. Are the methods selected economical for the type of material, the part shape to be produced, and the required production rate?
7. Can the requirements for dimensional tolerances, surface finish, and product quality be met consistently or can they be relaxed?
8. Can the part be produced to its final dimensions and surface characteristics without requiring additional processing or finishing operations?
9. Are all processing parameters optimized?
10. Is scrap produced, and if so, is it minimized? What is the value of the scrap?
11. Have all the automation and computer-control possibilities been explored for all phases of the total manufacturing cycle?
12. Are all in-line automated inspection techniques and quality control being implemented?

EXAMPLE 40.3   Process Substitution in Products

The following list gives some typical choices that can be made in process selection for the products listed. Recall that each pair of processes has its own advantages and limitations regarding such considerations as weight, life expectancy, durability, and cost.

1. Crankshaft: Forged vs. cast
2. Connecting rod: Forged vs. powder metallurgy
3. Hubcap: Sheet metal vs. cast
4. Automobile wheel: Forged vs. cast
5. Large gear: Machined vs. precision formed

6. Spur gear: Forged vs. powder metallurgy
7. Threaded fastener: Thread rolling vs. machining
8. Frying pan: Casting vs. stamping

9. Outdoor furniture: Formed aluminum tubing vs. cast iron
10. Machine-tool structures: Welding vs. mechanical fastening

---

EXAMPLE 40.4   Process Selection in Making a Simple Axisymmetric Part

Given: The simple part shown in Fig. 40.5a is to be produced in large quantities. It is 125 mm (5 in.) long and its large and small diameters are, respectively, 38 mm and 25 mm (1.5 in. and 1.0 in.). Assume that this part must be made of metal because of functional requirements such as strength, stiffness, hardness, and resistance to wear at elevated temperatures.

Find: Select a manufacturing process, and describe how you would organize the production facilities to manufacture a cost-competitive, high-quality product.

Solution: Recall that, as much as possible, parts should be produced at or near their final shape (net- or near-net-shape manufacturing), under an approach that largely eliminates much secondary processing and thus reduce the total manufacturing time and cost. Because its shape is relatively simple, this part can be made by (a) *casting*, (b) *powder metallurgy*, (c) *upsetting* or forging, (c) *extrusion*, (d) *machining*, or (e) *joining* two separate pieces together.

For net-shape production, the two suitable processes are casting and powder metallurgy; each of these two processes has its own characteristics, need for specific tooling, labor skill, and costs. The part can also be made by cold, warm, or hot forming. One method is upsetting (see heading, Fig. 14.12), in which a 25-mm (1-in.) round blank is placed in

a round die cavity to form the larger end. Another possibility is cold extrusion (a variation of closed-die forging, see Section 15.4), or partial direct extrusion of a 38-mm (1.5-in.) diameter blank to reduce its diameter to 25 mm. Note that each of these processes produces little or no material waste, an important factor in green manufacturing.

This part also can be made by machining a 38-mm-diameter bar stock to reduce the lower section to 25 mm. Machining, however, may require much more time than forming it, and a considerable amount of material inevitably will be wasted as metal chips (see Table 40.4). However, unlike net-shape processes, which generally require special dies, machining involves no special tooling, and the operation can be carried out easily on a CNC lathe and at high rates. Note also that the part can be made in two separate pieces and then joined by welding, brazing, or adhesive bonding.

After these initial considerations, it appears that if only a few parts are needed, machining this part is the most economical method. However, for a high production quantity and rate, as stated, producing this part by a heading operation or by cold extrusion would be an appropriate choice. Finally, note that if for some technical reason, the top and bottom portions of the part must be made of different materials, then the part can be made in two pieces, and joining them would be the most appropriate choice.

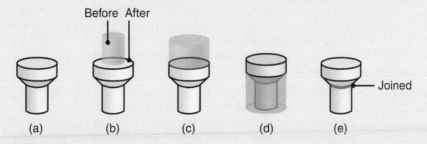

FIGURE 40.5   Various methods of making a simple part: (a) casting or powder metallurgy; (b) forging or upsetting; (c) extrusion; (d) machining; and (e) joining two pieces.

40.10 Manufacturing Costs and Cost Reduction

The *total cost* of a product typically consists of material, tooling, fixed, variable, direct-labor, and indirect-labor costs. As a general guide to the costs involved, see the sections on the economics of each chapter concerning individual groups of manufacturing processes and operations. Depending on the particular company and the type of products made, different methods of cost accounting may be used, with methodologies of accounting procedures that can be complex, and even controversial. Moreover, because of the numerous technical and operational factors involved, calculating individual cost factors can be challenging, time consuming, and not always reliable.

Costing systems, also called *cost justification*, typically include the following considerations: (a) intangible benefits of quality improvements and inventory reduction, (b) life-cycle costs, (c) machine usage, (d) cost of purchasing machinery compared with that of leasing it, (e) financial risks involved in implementing highly automated systems, and (f) implementation of new technologies and their impact on products.

Additionally, the costs to a manufacturer that are attributed directly to *product liability* continue to be a matter of major concern, and every product has a built-in cost to cover possible product liability claims. It has been estimated that liability suits against car manufacturers in the United States add about $500 to the indirect cost of an automobile, and because of the risks involved in its use, 20% of the price for a ladder is attributed to potential product liability costs.

Material Costs. Some cost data on materials are given in various tables throughout this book. Because of the different operations involved in producing raw materials (stock), their costs depend on the (a) type of material; (b) its processing history; and (c) its size, shape, and surface characteristics. For example, per unit weight:

- Drawn round bars are less expensive than bars that are ground to close tolerances and fine surface finish
- Bars with square cross-sections are more expensive than round bars
- Cold-rolled plate is more expensive than hot-rolled plate of the same thickness
- Thin wire is more expensive than thick wire
- Solid stock is much less expensive than metal powders of the same type

Tooling Costs. *Tooling costs* can be very high but they can be justified in high-volume production, such as automotive applications, where die costs can be $2 million or more. The expected life of tools and die, and their obsolescence because of product changes, also are major considerations.

Tooling costs are greatly influenced by the operation they perform; for example:

- Tooling cost for die casting is higher than that for sand casting
- Tooling costs for machining or grinding are much lower than those for powder metallurgy, forging, or extrusion
- Carbide tools are more expensive than high-speed steel tools, but tool life is longer
- If a part is to be made by spinning, the tooling cost for conventional spinning is much lower than that for shear spinning
- Tooling for rubber-forming processes is less expensive than the die sets (male and female) used for the deep drawing and forming of sheet metals

Fixed Costs. These costs include electric power, fuel, taxes on real estate, rent, insurance, and capital (including depreciation and interest). A company has to meet fixed costs regardless of whether or not it has made a product; thus, fixed costs are not sensitive to production volume.

Capital Costs.   These costs represent machinery, tooling, equipment, and investment in buildings and land. As can be seen in Table 40.5 the cost of machines and systems can vary widely, depending on numerous factors. In view of the generally high equipment costs (particularly those involving transfer lines and flexible-manufacturing cells and systems), high production quantities and rates are essential to justify such large expenditures, as well as to keep product costs at or below the all-important competitive level. Lower unit costs (cost per piece) can be achieved by continuous production, involving around-the-clock operation, as long as demand warrants it. Equipment maintenance also is essential to ensure high productivity, as any breakdown of machinery, leading to *downtime*, can be very costly, by as much as thousands of dollars per hour.

Direct-labor Costs.   These costs involve labor that is *directly* involved in manufacturing products, also known as *productive labor*; they include the costs of all labor, from the time raw materials are first handled to the time when the product is manufactured, a period generally referred to as *floor-to-floor time*. Direct-labor costs are calculated by multiplying the labor rate (the hourly wage, including benefits) by the amount of time that the worker spends producing the particular part.

The time required for producing a part depends not only on its specified size, shape, dimensional accuracy, and surface finish, but also on the workpiece material itself. For example, the cutting speeds for machining high-temperature alloys are lower than those for machining aluminum or plain-carbon steels. Labor costs in manufacturing and assembly vary greatly from country to country (see Table I.7 in the General Introduction).

It is thus understandable that most products one purchases today are either made or assembled in countries where labor costs are low. Note, for example, that the global market for bicycles in 2011 was $61 billion and 66% of the bicycles were made in China. Firms located in countries with high labor rates tend, on the other hand, to emphasize *high value added* manufacturing tasks or high automation levels and the associated increases in productivity, such that the labor component of the cost is significantly reduced.

For labor-intensive industries, such as clothing and textiles, steelmaking, petrochemicals, and chemical processing, manufacturers generally consider moving production to countries with a lower labor rate, a practice known as **outsourcing**. While this approach can be financially attractive, the cost savings anticipated may not always be realized, because of the following hidden costs associated with outsourcing:

- International shipping is far more involved and time consuming than domestic shipping. For example, it takes roughly four to six weeks for a container ship to bring a product from China to the United States or Europe, an interval that has continued to increase because of homeland security issues. Moreover, shipping costs can fluctuate significantly and in an unpredictable manner; it is therefore not possible to reliably predict or budget shipping costs.
- Lengthy shipping schedules indicate that the benefits of the just-in-time manufacturing approach and its associated

TABLE 40.6

Relative Costs for Machinery and Equipment	
Automatic screw machine	M–H
Boring mill, horizontal	M–H
Broaching	M–H
Deep drawing	M–H
Die casting	M–H
Drilling	L–M
Electrical-discharge machining	L–M
Electron-beam welding	M–H
Extruder, polymer	L–M
Extrusion press	M–H
Flexible manufacturing cell and system	H–VH
Forging	M–H
Fused–deposition modeling	L
Gas tungsten-arc welding	L
Gear shaping	L–H
Grinding	L–H
Headers	L–M
Honing, lapping	L–M
Injection molding	M–H
Laser-beam welding	M–H
Lathes	L–M
Machining center	L–M
Mechanical press	L–M
Milling	L–M
Powder-injection molding	M–H
Powder metallurgy	L–M
Powder metallurgy, HIP	M–H
Resistance spot welding	L–M
Ring rolling	M–H
Robots	L–M
Roll forming	L–M
Rubber forming	L–M
Sand casting	L–M
Spinning	L–M
Stereolithography	L–M
Stamping	L–M
Stretch forming	M–H
Transfer lines	H–VH
Ultrasonic welding	L–M

Note: L = low; M = medium; H = high; VH = very high. Costs vary greatly, depending on size, capacity, options, and level of automation and computer controls. See also the sections on economics in various chapters.

cost savings may not be realized. Moreover, because of the long shipping times and rigid schedules, design modifications cannot be made easily, and companies cannot readily address changes in the market or in demand. Thus, companies that outsource can lose agility, and may also have difficulties in following lean-manufacturing approaches.

- Legal systems are not as well established in countries with lower labor rates as they are in other countries. Procedures that are common in the United States and the European Union, such as accounting audits, protection of patented designs and intellectual property, and conflict resolution, are more difficult to enforce or obtain in other countries.
- Because payments typically are expected on the basis of units completed and sold, the consequences of defective products and their rates can become significant.
- There are various other hidden costs in outsourcing, such as increased paperwork and documentation, lower productivity from existing employees because of lower morale, and difficulties in communication.

For all these reasons, many manufacturers that have outsourced manufacturing activities have not been able to realize the hoped-for benefits; consequently, a trend toward *reshoring* manufacturing efforts has recently begun.

Indirect-labor Costs. These costs are generated in the servicing of the total manufacturing operation, consisting of such activities as supervision, maintenance, quality control, repair, engineering, research, and sales, as well as the cost of office staff. Because they do not contribute directly to the production of specific products, these costs are referred to as the *overhead* or *burden rate*, and are charged proportionally to all products. The personnel involved in these activities are categorized as **nonproductive labor**.

Manufacturing Costs and Production Quantity. One of the most significant factors in manufacturing costs is the *production quantity* (see Table 37.2). A large production quantity obviously requires high production rates which, in turn, require the use of **mass production** techniques that involve special machinery (*dedicated machinery*), and employ proportionally less direct labor. At the other extreme, a smaller production quantity usually means a larger direct-labor involvement.

Small-batch production usually involves general-purpose machines, such as lathes, milling machines, and hydraulic presses. The equipment is versatile, and parts with different shapes and sizes can be produced by appropriate changes in the tooling; however, direct-labor costs are high because these machines usually are operated by skilled labor.

In **medium-batch production**, the quantities are larger and general-purpose machines are equipped with various jigs and fixtures, or they can be computer controlled. To further reduce labor costs, machining centers and flexible-manufacturing systems are important alternatives. Generally, for quantities of 100,000 or more, the machines are designed for specific purposes (*dedicated*), and they perform a variety of specific operations with very little direct labor involved.

Cost Reduction. *Cost reduction* requires a study of how the costs described previously are interrelated, using *relative costs* as an important parameter. The unit cost of a product can vary widely. For example, some parts may be made from expensive materials but require very little processing, as in the case of minted gold coins; consequently, the cost of materials relative to that of direct labor is high.

By contrast, some products may require several complex and expensive production steps to process relatively inexpensive materials, such as carbon steels. An electric motor, for example, is made of relatively inexpensive materials, yet several different manufacturing operations are involved in making the housing, rotor, bearings, brushes, and various other components. Unless highly automated, assembly operations for such products can become a significant portion of the overall cost (see Section 37.9).

In the 1960s, labor accounted for as much as 40% of the production cost; today, it can be as low as 5%, depending on the type of product and level of automation (see Table I.6). Note also that the contribution of the design phase is only 5%, yet it is the *design* phase that generally has the largest influence on the quality and success of a product in the marketplace.

There are various opportunities for cost reduction, as have been discussed in various chapters throughout this book. Introducing more automated systems and the adoption of up-to-date technologies are an obvious means of reducing some costs. However, this approach must be undertaken after a thorough cost–benefit analysis, which requires reliable input data and consideration of all the technical as well as the human factors involved. Advanced technologies, some of which can be very costly to implement, should be considered only after a complete analysis of the more obvious cost factors, known as *return on investment* (ROI).

40.10.1   Value Analysis

Manufacturing adds *value* to materials as they become discrete parts, components, and products, and are then marketed. Because the value is added in individual stages during the creation of the product, the utilization of value analysis, also called *value engineering*, *value control*, and *value management*, is important. *Value analysis* is thus a system that evaluates each step in design, material and process selection, and operations, in order to manufacture a product that performs all of its intended functions and does so at the lowest possible cost.

In this analysis, developed at the General Electric Co. during World War II, a monetary value is established for each of two product attributes: (a) **use value**, reflecting the functions of the product, and (b) **esteem** or **prestige value**, reflecting the attractiveness of the product that makes its ownership desirable. The *value of a product* is then defined as

$$\text{Value} = \frac{\text{Product function and performance}}{\text{Product cost}} \qquad (40.1)$$

Although there are different versions, value analysis basically consists of the following six phases:

1. *Information phase*: Gathering data and determining costs
2. *Analysis phase*: Defining functions and identifying problems as well as opportunities
3. *Creativity phase*: Seeking ideas in order to respond to problems and opportunities without judging the value of each idea
4. *Evaluation phase*: Selecting the ideas to be developed and identifying the costs involved
5. *Implementation phase*: Presenting facts, costs, and values to the company management; developing a plan and to motivate positive action, all in order to obtain a commitment of the resources necessary to accomplish the task

6. *Review phase*: Reexamining the overall value-analysis process in order to make necessary adjustments

Value analysis is usually coordinated by a value engineer and conducted jointly by designers, manufacturing engineers, and quality-control, purchasing, and marketing personnel, and it must have the full support of a company's top management. The implementation of value analysis in manufacturing can result in such benefits as (a) significant cost reduction, (b) reduced lead times, (c) better product quality and performance, (d) a reduced time for manufacturing the product, and (e) reduced product weight and size.

SUMMARY

- Regardless of how well a product meets design specifications and quality standards, it also must meet economic criteria in order to be competitive in the domestic and global marketplace.

- Important considerations in product design and manufacturing include manufacturing characteristics of materials, product life expectancy, life-cycle engineering, and an awareness of minimizing any potential harm to the environment and the Earth's ecosystem.

- Substitution of materials, modification, and simplifying product design, and relaxing of dimensional tolerance and surface finish requirements are among important methods of cost reduction.

- The total cost of a product includes several elements, such as the costs of materials, tooling, capital, labor, and overhead. Material costs can be reduced through careful selection, without compromising design and service requirements, functions, specifications, or standards for good product quality.

- Labor costs generally are becoming an increasingly smaller percentage of production costs in highly industrialized countries. To counteract lower wages in some developing countries, labor costs can be further reduced through highly automated and computer-controlled equipment and operations.

KEY TERMS

Burden rate	Direct labor	Nonproductive labor	Reshoring
Capital costs	Downtime	Outsourcing	Return on investment
Cost–benefit analysis	Economic order quantity	Overhead	Scrap
Cost justification	Fixed costs	Process capabilities	Smart databases
Cost reduction	Floor-to-floor time	Production quantity	Sustainable manufacturing
Cradle-to-cradle	Indirect labor	Production rate	Value
Cradle-to-grave	Lead time	Recycling	Value analysis
Dedicated machines	Life-cycle assessment	Relative costs	World class

BIBLIOGRAPHY

Anderson, D.M., **Design for Manufacturability & Concurrent Engineering**, CIM Press, 2010.

Andrae, A.S.G., **Global Life Cycle Assessments of Material Shifts**, Springer, 2009.

Ashby, M.F., **Materials Selection in Mechanical Design**, 4th ed., Butterworth-Heinemann, 2010.

ASM Handbook, Vol. 20: **Materials Selection and Design**, ASM International, 1997.

Boothroyd, G., Dewhurst, P., and Knight, W., **Product Design for Manufacture and Assembly**, 3rd ed., CRC Press, 2010.

Bralla, J.G., **Design for Manufacturability Handbook**, 2nd ed., McGraw-Hill, 1999.

Cook, H.F., and Wissmann, L.A., **Value Driven Product Planning and Systems Engineering**, Springer, 2010.

Fiksel, J., **Design for the Environment**, 2nd ed., McGraw-Hill, 2011.

Giudice, F., La Rosa, G., and Risitano, A., **Product Design for the Environment**, CRC Press, 2006.

Harper, C.A. (ed.), **Handbook of Materials for Product Design**, McGraw-Hill, 2001.

Hundai, M. (ed.), **Mechanical Life Cycle Handbook**, CRC Press, 2001.

Kutz, M., **Environmentally Conscious Manufacturing**, Wiley, 2007.

Madu, C. (ed.), **Handbook of Environmentally Conscious Manufacturing**, Springer, 2011.

Magrab, E.B., Gupta, S.K., McCluskey, F.P., and Sandborn, P., **Integrated Product and Process Design and Development: The Product Realization Process**, 2nd ed., CRC Press, 2009.

Mangonon, P.C., **The Principles of Materials Selection for Engineering Design**, Prentice Hall, 1998.

McDonough, W., and Braungart, M., **Cradle to Cradle: Remaking the Way We Make Things**, North Point Press, 2002.

Poli, C., **Design for Manufacturing: A Structured Approach**, Butterworth-Heinemann, 2001.

Ribbens, J., **Simultaneous Engineering for New Product Development: Manufacturing Applications**, Wiley, 2000.

Seliger, G., (ed.), **Sustainability in Manufacturing**, Springer, 2010.

Swift, K.G., and Booker, J.D., **Process Selection: From Design to Manufacture**, 2nd ed., Butterworth-Heinemann, 2003.

Ulrich, K., and Eppinger, S., **Product Design and Development**, 5th ed., McGraw-Hill, 2011

Wenzel, H., Hauschild, M., and Alting, L., **Environmental Assessment of Products**, Springer, 2003.

REVIEW QUESTIONS

40.1 Explain what is meant by manufacturing properties of materials.

40.2 Why is material substitution an important aspect of manufacturing engineering?

40.3 What factors are involved in the selection of manufacturing processes? Explain why they are important.

40.4 How is production quantity significant in process selection? Explain.

40.5 List and describe the major costs involved in manufacturing.

40.6 Why does material selection influence energy requirements for products?

40.7 Describe life-cycle assessment and life-cycle engineering.

40.8 Define what is meant by economic order quantity.

40.9 Explain the difference between direct-labor cost and indirect-labor cost.

40.10 Describe your understanding of the following terms: (a) life expectancy, (b) life-cycle engineering, (c) sustainable manufacturing, and (d) green manufacturing.

40.11 What is the difference between production quantity and production rate?

40.12 Is there a significant difference between cradle-to-grave and cradle-to-cradle production? Explain.

40.13 How would you define value? Explain.

40.14 Define sustainable manufacturing.

40.15 What is the meaning and significance of the term return on investment? Explain.

QUALITATIVE PROBLEMS

40.16 Describe the major considerations involved in selecting materials for products.

40.17 What is meant by manufacturing process capabilities? Select four different manufacturing processes and describe their capabilities.

40.18 Comment on the magnitude and range of scrap shown in Table 40.4 and the reasons for the variations.

40.19 Explain why the value of the scrap produced in a manufacturing process depends on the type of material and processes involved.

40.20 Describe your observations concerning the information given in Table 6.1.

40.21 Other than the size of the machine, what factors are involved in the range of prices in each machine category shown in Table 40.6? Explain.

40.22 Explain why it takes different amounts of energy to produce different materials. Consider both the material and the processing history.

40.23 Refer to Table 40.2 and explain why it is essential to recycle aluminum and magnesium. From a life-cycle standpoint, explain why aluminum and magnesium should or should not be used in automobiles.

40.24 Explain how the high cost of some of the machinery listed in Table 40.6 can be justified.

40.25 On the basis of the topics covered in this book, explain the reasons for the relative positions of the curves shown in Fig. 40.2.

40.26 What factors are involved in the shape of the curve shown in Fig. 40.4? Explain.

40.27 Describe the problems that may have to be faced in reducing the quantity of materials in products. Give some examples.

40.28 Explain the reasons that there is a strong desire in industry to practice near-net-shape manufacturing.

40.29 State and explain your thoughts concerning cradle-to-cradle manufacturing.

40.30 Why is the amount of scrap produced in a manufacturing process important?

40.31 Discuss the advantages to long lead times, if any, in production.

40.32 Review Table 40.2 and estimate the carbon footprint of materials (mass of carbon produced per mass or volume of material) if the energy used to produce the material is obtained from (a) hydroelectric power, wind, or nuclear energy; (b) coal.

40.33 Explain why the larger the quantity per package of food products, the lower is the cost per unit weight.

40.34 List and explain the advantages and disadvantages of outsourcing manufacturing activities to countries with low labor costs.

SYNTHESIS, DESIGN, AND PROJECTS

40.35 As you can see, Table 40.5 lists only metals and their alloys. On the basis of the information given in various chapters in this book and in other sources, prepare a similar table for nonmetallic materials, including ceramics, plastics, reinforced plastics, and both metal-matrix and ceramic-matrix composite materials.

40.36 Is it always desirable to purchase stock that is close to the final dimensions of a part to be manufactured? Explain why or why not and give some examples.

40.37 What course of action would you take if the supply of a raw material selected for a product line becomes unreliable? Explain.

40.38 Estimate the position of the curves for the following processes in Fig. 40.4: (a) centerless grinding, (b) electrochemical machining, (c) chemical milling, and (d) extrusion.

40.39 Review Fig. I.2 in the General Introduction and present your own thoughts concerning the two flowcharts. Would you want to make any modifications, and if so, what would they be?

40.40 Over the years, numerous consumer products (such as rotary-dial telephones, analog radio tuners, turntables, and vacuum tubes) have become obsolete or nearly so, while many new products have entered the market. Make two lists: a comprehensive list of obsolete products that you can think of and a list of new products. Comment on the reasons for the changes you observe.

40.41 List and discuss the different manufacturing methods and systems that have enabled the manufacture of new products. (Recall that these products and systems are known as enabling technologies.)

40.42 Select three different products, and make a survey of the changes in their prices over the past 10 years. Discuss the possible reasons for the changes.

40.43 Describe your own thoughts concerning the replacement of aluminum beverage cans with those made of steel.

40.44 Select three different products commonly found in homes. State your opinions on (a) what materials were used in each product, (b) why those particular materials were chosen, (c) how the products were manufactured, and (d) why those particular processes were used.

40.45 Comment on the differences, if any, among the designs, materials, and processing and assembly methods used for making such products as hand tools and ladders for professional use and those for consumer use.

40.46 The cross-section of a jet engine is shown in Fig. 6.1. On the basis of the topics covered in this book, select any three individual components of such an engine and describe the materials and processes that you would use in making them in quantities of, say, 1,000.

40.47 Inspect some products around your home, and describe how you would go about taking them completely apart quickly and recycling their components. Comment on their design regarding the ease with which they can be disassembled.

40.48 What products do you know of that would be very difficult to disassemble for recycling purposes? Explain.

40.49 Conduct a literature search and perform a lifecycle assessment of a typical automobile. Estimate the amount of energy needed to produce the car from its raw materials, and compare this to the energy that is consumed by the car during its intended life of 160,000 km (100,000 miles). What recommendations would you make regarding the use of aluminum and magnesium instead of steel in cars?

40.50 Discuss the trade-offs involved in selecting between the two materials for each of the applications listed:

a. Sheet metal vs. reinforced-plastic chairs
b. Forged vs. cast crankshafts
c. Forged vs. powder metallurgy connecting rods
d. Plastic vs. sheet metal light-switch plates
e. Glass vs. metal water pitchers
f. Sheet-metal vs. cast hubcaps
g. Steel vs. copper nails
h. Wood vs. metal handles for hammers

Discuss also the typical conditions to which these products are subjected in their normal use.

40.51 Discuss the factors that influence the choice between the following pairs of processes to make the products indicated:

a. Sand casting vs. die casting of a fractional electric-motor housing
b. Machining vs. forming of a large-diameter bevel gear
c. Forging vs. powder metallurgy production of a cam
d. Casting vs. stamping a sheet-metal frying pan
e. Making outdoor summer furniture from aluminum tubing vs. cast iron
f. Welding vs. casting of machine-tool structures
g. Thread rolling vs. machining of a bolt for high-strength application
h. Thermoforming a plastic vs. molding a thermoset to make the blade for an inexpensive household fan

40.52 A dish-shaped part is to be produced; two methods are under consideration as illustrated in Fig. P40.52. The part can be formed by placing a flat piece of sheet metal between two dies and closing the dies to get the desired shape; the part can also be made by explosive forming as shown. (a) List the advantages and disadvantages to these options if the part is 2 m (80 in.) in diameter, and only 50 parts are required. (b) What other manufacturing processes would be suitable for producing this part?

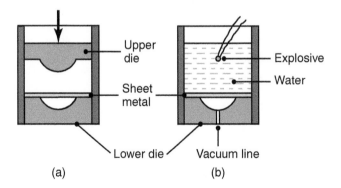

(a)   (b)

FIGURE P40.52

40.53 Figure P40.53 shows a sheet-metal part made of steel:

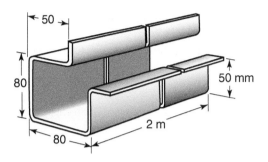

FIGURE P40.53

Discuss how this part could be made and how your selection of a manufacturing process may change (a) as the number of parts required increases from 10 to thousands and (b) as the length of the part increases from 2 m to 20 m.

40.54 The part shown in Fig. P40.54 is a carbon-steel segment (partial) gear:

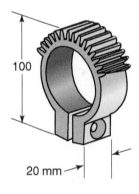

FIGURE P40.54

The small hole at the bottom is for clamping the part onto a round shaft, using a screw and a nut. Suggest a sequence of manufacturing processes to make this part. Consider such factors as the influence of the number of parts required, dimensional tolerances, and surface finish. Discuss such processes as machining from a bar stock, extrusion, forging, and powder metallurgy.

40.55–40.59 Review the products illustrated below and, for each component, describe your thoughts on the (a) material or materials that would be appropriate for use and (b) specific manufacturing process or processes that could be employed in making each component. Assume now that after presenting your suggestions, you are told that the product would be too expensive to produce as recommended. Suggest the changes that you could make to reduce the overall cost, explaining your reasons.

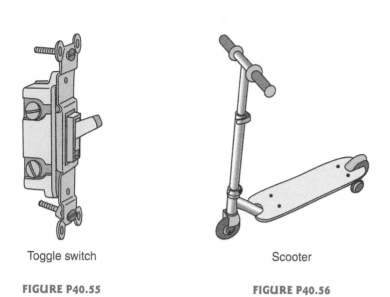

Toggle switch

FIGURE P40.55

Scooter

FIGURE P40.56

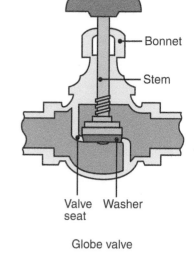

Globe valve

FIGURE P40.57

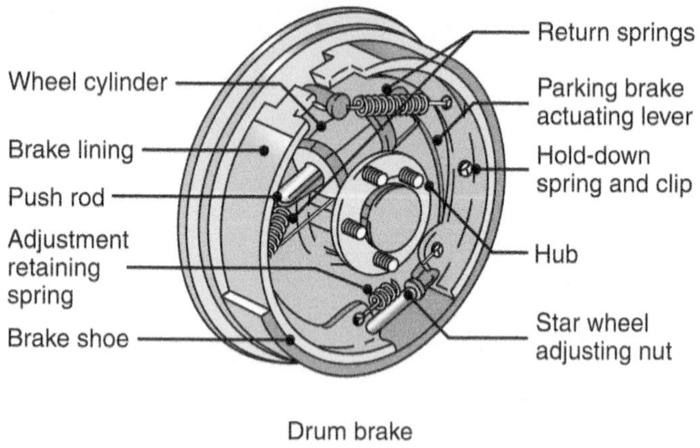

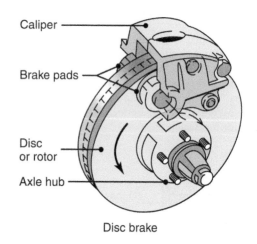

Drum brake

FIGURE P40.58

Disc brake

FIGURE P40.59

Index

LIST OF EXAMPLES

Chattel, Servant or Citizen.

Women's Status in Church, State and Society

Edited by Mary O'Dowd and Sabine Wichert.

HISTORICAL STUDIES XIX

Papers read before the xxist Irish Conference of Historians, held at
Queen's University of Belfast, 27–30 May 1993

The Institute of Irish Studies
The Queen's University of Belfast

Published 1995
The Institute of Irish Studies
The Queen's University of Belfast

British Library Cataloguing-in-Publication Data. A catalogue
record for this book is available from the British Library.

ISBN: 0 85389 567 8 HB
0 85389 576 7 PB

The editors wish to thank the following for their assistance in
the publication of this volume: the Department of Education
for Northern Ireland; Kate Newmann; Roseanne Donnelly;
the Publications Fund and the School of Modern History,
Queen's University, Belfast.

Miriam and Ira D. Wallach Division of Art, Prints and Photographs; The New York Public Library; Astor, Lenox and Tilden
Foundations are acknowledged for permission to use the prints
in chapter 11.

Printed by W & G Baird Ltd, Antrim
Cover design by Rodney Miller Associates

Previous Volumes in the Series

T.D. Williams (ed.), *Historical Studies I* (London: Bowes and Bowes 1958)

M. Roberts (ed.), *Historical Studies II* (London: Bowes and Bowes 1959)

J. Hogan (ed.), *Historical Studies III* (London: Bowes and Bowes 1961)

G. A. Hayes-McCoy (ed.), *Historical Studies IV* (London: Bowes and Bowes 1963)

J. L. McCracken (ed.), *Historical Studies V* (London: Bowes and Bowes 1965)

T. W. Moody (ed.), *Historical Studies VI* (London: Routledge & Kegan Paul 1968)

J. C. Beckett (ed.), *Historical Studies* VII (London: Routledge & Kegan Paul 1969)

T. D. Williams (ed.), *Historical Studies VIII* (Dublin: Gill & MacMillan 1971)

J. G. Barry (ed.), *Historical Studies IX* (Belfast: Blackstaff Press 1974)

G. A. Hayes-McCoy (ed.), *Historical Studies X* (Dublin: ICHS 1976)

T. W. Moody (ed.), *Nationality and the Pursuit of National Independence: Historical Studies XI* (Belfast: Appletree Press 1978)

A. C. Hepburn (ed.), *Minorities in History: Historical Studies XII* (London: Edward Arnold 1978)

D. W. Harkness and M. O'Dowd (eds), *The Town in Ireland: Historical Studies XIII* (Belfast: Appletree Press 1981)

J. I. McGuire and A. Cosgrove (eds), *Parliament and Community: Historical Studies XIV* (Belfast: Appletree Press 1983)

P. J. Corish (ed.), *Radicals, Rebels and Establishments: Historical Studies XV* (Belfast: Appletree Press 1985)

Tom Dunne (ed.), *The Writer as Witness: Literature as Historical Evidence: Historical Studies XVI* (Cork: Cork University Press 1987)

Ciarán Brady (ed.), *Ideology and the Historians: Historical Studies XVII* (Dublin: The Lilliput Press 1991)

T. G. Fraser and Keith Jeffery (eds), *Men, Women and War: Historical Studies XVIII* (Dublin: The Lilliput Press 1993)

Contributors

Maxine Berg: Lecturer in History, University of Warwick
Caitriona Clear: Lecturer in History, University College, Galway
Mary Daly: Professor of History, University College, Dublin
Barbara Gates: Professor of English Literature, University of Delaware
Maria Grever: Professor of History, University of Nijimegen
Joan Hoff: Professor of History, University of Indiana
Franca Iacovetta: Professor of History, University of Toronto
Peter Jupp: Professor of History, Queen's University, Belfast
Margaret MacCurtain: Formerly Lecturer in History, University College, Dublin
Elizabeth Meehan: Professor of Politics, Queen's University, Belfast
Christine Meek: Professor of History, Trinity College, Dublin
Grace Neville: Lecturer in French, University College, Cork
Muireann Ní Bhrolcháin: Lecturer in Irish, St Patrick's College, Maynooth
Máirín Ní Dhonnchadha: Assistant Professor in Dublin Institute of Advanced
  Studies
Donnchadh Ó Corráin: Professor of History, University College, Cork
Mary O'Dowd: Lecturer in History, Queen's University of Belfast
Kevin O'Neill: Professor of History, Boston College
Catherine Shannon: Professor of History, Westfield State College
Laura Strumingher: Provost, Hunter College, New York
Pauric Travers: Lecturer in History, St Patrick's College, Dublin
Joan Thirsk: Formerly Reader in Economic History, University of Oxford
Maryann Gialanella Valiulis: Director of Women's Studies, Trinity College,
  Dublin
Sabine Wichert: Lecturer in History, Queen's University, Belfast
Marian Yeates: Professor of History, Montana State University

Contents

Introduction

The theme of the twenty-first Irish Conference of Historians was the history of women. Held in Queen's University, Belfast in the spring of 1993, the programme of the conference was one of the most ambitious and most international sponsored by the Irish Committee of Historical Sciences. The conference organisers were motivated by two major considerations when choosing the theme. First, while the previous decade had witnessed the expansion and development of research into Irish women's history, the subject remained on the periphery of the Irish academic world. The conference was thus seen as an opportunity for historians in Ireland to become more aware of the academic validity and significance of women's history. Secondly, it was hoped that the conference would communicate to the Irish historical community the high quality and intellectual challenge of international research in the field of women's history. An unusually large number of international scholars were therefore invited to present papers. They included many leading authorities in the field. Not surprisingly, scholars from the United States of America predominated as it is there that women's history has been most developed, but historians from Canada, England, Germany, Italy and the Netherlands also delivered papers[1]. The conference attracted a large attendance and helped to generate a considerable number of worthwhile discussions on the role and status of women in the past. A striking feature of the debates was the concern of contributors to incorporate the experiences of Irish women into a wider, comparative context. A long term consequence of the event was thus the establishment of Irish women's history within the intellectual framework developed by scholars in other countries. In this and in other ways, the conference was generally considered to have marked a milestone in the development of women's history in Ireland, north and south.

The four day programme aimed to be as representative of current trends in research in women's history as possible. Invited speakers were therefore given a free choice in the selection of their topics. The result could have been a haphazard, disjointed collection of papers on many different subjects, but, as this volume demonstrates, there was a surprising coherence and inter-twining of themes between the conference papers, despite the different national histories in which they were located. This may have been coincidental, or an expression of common intellectual concerns at international level but it may also reflect the shared experiences of women, particularly in relation to the two institutions on which many papers concentrated, the church and the state.

The first four papers in the volume focus on the theory and historiography of women's history. Although varying in topic from the writing of female religious history in Ireland to the teaching of women's history in the Netherlands, the

---

1 In addition to the papers published in this volume, papers were presented by Mary Cullen, Natalie Zemon Davis, Olwen Hufton, Greta Jones, Belinda Loftus, Maria Luddy, Dympna McLoughlin, Mary Nash, Mary Beth Norton, Uta Ranke-Heinemann, Ursula Vogel and Marina Warner.

four studies share common concerns and themes. All explicitly or implicitly acknowledge the advances which have been made in women's history in the past twenty five years but all also accept that the subject still encounters problems. The most obvious problem is the struggle to win full academic recognition. Maria Grever's paper reports on the negative reception given to women's history by leading (and predominantly male) Dutch historians. Women's history has encountered similar resistance in other countries. Joan Thirsk documents the writings of a large number of women historians whose work has been neglected, a theme echoed by Maxine Berg who notes that most of Eileen Power's research on women was ignored by historians until very recently; and for Ireland, Margaret MacCurtain records the low status accorded to female religious archives by Irish historians.

Reservations concerning the nature and content of women's history are understandable in the context of the close connection between the political campaign for women's rights of the 1970s and the re-emergence of writing about women's role in the past. Some of the published work on women's history which dates from that time was imbued with political rhetoric and lacked scholarly detachment and objectivity. But in the past twenty years the study of women's historical role has advanced and become more intellectually sophisticated. Most research in women's history is now of an impeccable high academic standard. The refusal by some male scholars, such as the Dutch historians described by Maria Grever, to acknowledge developments in women's history and the stimulating contribution which it can make to historical studies is often a consequence of their failure to update their knowledge of current trends in the field; and so they maintain a stereo-typical view of the subject and its researchers.

Reluctant academic recognition is, however, only one of the problems which women's history has encountered and, as the subject wins more acceptance through integration into university and school curricula, it may no longer be the most serious. Far more complex and more difficult to resolve is the relationship between women's history and 'mainstream' interpretations of 'general history'. Integrating research on women's activities in the past into 'mainstream' interpretations has proved more problematic than the early advocates of women's history may have imagined. Even in countries where a considerable amount of historical research on women has been accumulated such as in the United States, integration has only been achieved in a haphazard and marginal fashion. Women's status or activities still tend to be dealt with in, at best, a separate chapter or, at worst, an occasional paragraph. Some women historians have suggested, as Grever points out, that integration should not be a priority for women's history. They advocate a 'sectarian' view of the subject arguing that it has its own intellectual integrity and does not require incorporation into mainstream interpretations of history to be academically valid. Other historians, following Joan Scott's influential thesis, have argued in favour of gender history as a means of integrating female and male historical experiences on an equal basis. The two arguments are not mutually exclusive. Many of the articles in this volume can be interpreted from either perspective. They advance our knowledge of women's role in the past in several areas while at the same time they provide the basis for a gendered debate on a number of different issues such as the early Christian church and the state in the twentieth century. Thus the collection provides an empirical indication of future developments in women's history. In twenty years' time gender history and women's history may not be presented in such opposing terms as they have been in recent years.

Another controversial and scarcely acknowledged problem encountered in the study of women's history which is raised by Joan Thirsk and Maria Grever is the gendered nature of intellectual study. Both writers imply that women and men have different intellectual interests. Thirsk compares the approaches of her male and female graduate students while Grever demonstrates that girls did better than boys in the new history curriculum in the Netherlands. The girls preferred the new approach which included the history of women while the boys did better in the older more traditional syllabus. The issue is a complex one on which psychologists are not yet agreed. The effects of nurturing and parental and, later, teacher influence on the respective intellectual choices of men and women need to be explored more thoroughly before firm conclusions can be reached. The issue has, however, profound implications for women's history and the Thirsk and Grever articles may initiate a wider debate on the subject.

The four writers who focus on theoretical issues share an optimism about the future of women's history, despite the problems which it is still encountering. As Margaret MacCurtain points out, research into women's history has led to new sources being discovered and new methodologies being developed. The essays in this volume amply demonstrate the wide range of materials being exploited by historians of women. They include many previously neglected or under-valued sources such as convent archives, the nineteenth century French prints of women analysed by Laura Struminger, the social welfare files used by Franca Iacovetta and the diaries of Mary Shackleton which Kevin O'Neill describes as possibly 'the most complete record of the private life of any individual who has lived in Ireland'. Astonishingly, as it now seems, this source, easily available in the National Library of Ireland, was neglected by historians until very recently. Many of the contributors also emphasise the need to re-evaluate familiar documents from a gender point of view. Thus Caitriona Clear points to the problem of using Irish census statistics for assessing the numbers of women at work and Grace Neville demonstrates the richness of the Irish Folklore Commission archives for studying the lives of rural women in nineteenth and twentieth-century Ireland.

One of the principal themes in the volume is the common and seemingly paradoxical attitudes of ecclesiastical and state institutions to women. The early Christian church devised laws to protect women against abuses in marriage, as documented by Christine Meek for medieval Lucca, or against participation in war which was the purpose behind Adomnán's law of the seventh century, as Máirín Ní Dhonnchadha notes. In formulating these laws the church found itself in opposition to local customs which were not so concerned to protect women. Donnchadh Ó Corráin's article analyses in detail the influence of the church on what have often been considered indigenous Irish laws. For the sake of political expediency Irish chiefs often ignored the church's teaching on marriage. Christine Meek also notes the strength of local marriage customs in Lucca which ran counter to church law.

But if the early church devised laws to protect women it also created an ideal of womanhood which in the long term bequeathed a legacy of restriction and confinement which was to have a powerful and lasting impact on European and Anglo-American society. Through the construction of the role model of Mary, the church's image of the ideal woman was defined by Mary's role as

wife and mother. The submissive and domesticated view of women was transferred into the political sphere as individual states in the early modern and
modern world incorporated a similar image of women into their legislative programmes. In the new Irish state of the early twentieth century, for example, the
ideal Christian woman became synonymous with the ideal Irish woman and
the new state's legislation severely curtailed the public role of women, as the
contributions of Maryann Valiulis, Pauric Travers and Caitriona Clear document. Nor was such convergence between catholic and national identity confined to Ireland. As Valiulis argues there are many similarities between Italy
and Ireland in this regard. Joan Hoff's analysis of abortion suggests that a similar situation exists in Poland. Mary Daly's study of the International Labour
Organisation indicates that the I.L.O. was also imbued with a similar traditional view of womanhood which it shared with left-wing groups in the middle
decades of the twentieth century. Operating in a different ideological context,
the Canadian government also tried to impose a restricting model of the ideal
Canadian woman on female emigrants from Eastern Europe, as Franca
Iacovetta's analysis demonstrates.

Elizabeth Meehan and Joan Hoff continue the story into the second half of
the twentieth century and analyse the contemporary debate about the role of
women in the state and definitions of citizenship. Meehan illustrates the way in
which this debate has been conducted in the European Union. The Union,
Meehan argues, offers women better prospects for equality in citizenship than
individual nation states. Her optimistic conclusion suggests that the European
Union may play a leading role in finally eradicating the submissive role model
for women in the legislation of European states.

There was and is no common response on the part of women to the restrictive legislation of churches and states. Some have opted to work within the
system, others openly to resist it or, at least, to campaign for changes in the
status of women. The varied responses of women are often influenced by their
economic and social background. Women of the same class have often more in
common with women from similar social backgrounds from different cultures
or different periods of time than with their contemporaries from a different
class. Thus the aristocratic women in early Christian Ireland, as described by
Muireann Ní Bhrolcháin, had a similar political role to aristocratic women in
pre-democratic Britain as documented by Peter Jupp. Both belonged to a small
elite group of families among whom marriage was often a political alliance and
in which women had a recognised political influence even if they held no political office. As a consequence of their birth and privileged position women in
such families could challenge the traditional view of women without arousing
the opposition of society.

In middle class society women's actions were more circumscribed. Male advocates of democracy or socialism were often slow to extend their idealism to
include women. Despite their political radicalism they often held very traditional views on women's role in society. Laura Strumingher points to one group
of women who were keen to extend the equality of the French Revolution to
women but she also illustrates the rather ambiguous view of the male illustra-
tor, Edouard de Beaumont, to their activities. In twentieth century Northern
Ireland there have been very few women political leaders, as Catherine Shannon's article points out. Partly out of frustration with the political scene women
have become involved in community groups and social action.

Elsewhere also, middle-class women often diverted their energy into similar areas such as charitable work or campaigning on selected issues. Barbara Gates and Marian Yeates demonstrate how this tendency was shared by women in England and in north America. Yeates also analyses the attempts by a number of nineteenth-century thinkers to develop an ideology which embraced both men and women. The debate has echoes in the twentieth-century discourse on citizenship as described by Joan Hoff and Elizabeth Meehan.

A prime factor in the slow change in attitudes to women's public role in modern times was the advent of industrialisation and the democratisation which it brought. By the early twentieth century the number of women who wanted or needed to work outside the home presented a problem for the ideologies of church and state which had not arisen when women were involved in the rural economy in a less democratic age. Divisions among women themselves on the issue also existed and were again related to their position within the social structure of their respective societies. The debate on whether or not women should work at night or in the mines, for instance, appears to have divided middle-class women who argued that women's principal duty was with family and home, from working-class women who demanded the right to work when and where they could, out of economic necessity. In catholic countries in particular, such as Ireland, Poland and Italy the state also held on relentlessly to the ideal of woman's place in the home. But increasingly in the twentieth century more and more women rejected the role allotted to them by church and state. Rather than live in a confined and restricted fashion in Ireland many single women opted for emigration. Ironically, their positive action in choosing emigration could in turn lead to a deeply misogynist critique of their behaviour in their home locality, as Grace Neville's research in the Irish Folklore Commission archives indicates.

The women's movement of the 1970s had a profound impact on attitudes to women as institutionalised in western states and churches. As Meehan's article demonstrates, the European Union encouraged member states to enact equal opportunity legislation. The movement also infiltrated the academy and women in the historical profession became aware that their own history had been subsumed into general or men's history and began the process of searching for their part in the history of humankind. The contributions in this volume show that considerable inroads have been made in this quest but also how very much more needs to be done, before any serious attempt at possible reintegration of the history of women into general history can be made.

1   The History Women

Joan Thirsk

T HE title of this paper echoes but alters the title of a book by John Kenyon, published in 1983, entitled *The history men*. That work was subtitled *The historical profession in England since the renaissance* and so on the face of it, it told a long and complete story about the writers of history. Kenyon noticed in the preface that 'many people do not realize what a comparatively recent development the writing by "professional historians" is'.[1] True enough, but in surveying the writers, he started from the seventeenth century, and while devoting most space to the nineteenth century when the professionalisation took place, he had barely a word anywhere for the women.

One woman who did receive Kenyon's attention in the second half of the eighteenth century, was Catharine Macaulay. In fact, he acknowledged her to have been a serious rival to David Hume and a reading of Bridget Hill's discerning recent biography of her shows, rather more clearly than did Kenyon, her strengths in comparison with Hume. Kenyon did, however, recognise two of her distinctive achievements: she explored the seventeenth century more deeply than did Hume; and she worked on original documents at the British Library and State Paper Office.[2] Moving into the nineteenth century and the era of professionalisation, Kenyon ignored all the women except Alice Stopford Green, who, as the wife of J. R. Green, received notice for bringing out one of her husband's volumes after his death, and then establishing an independent reputation as a historian with her *Town life in the fifteenth century* published in London in 1894. But she is robbed of any originality by Kenyon's summary statement that her book on town life was 'an extension of her husband's early work in urban history'.[3]

Over the years I have become aware of the large number of women writing history in the nineteenth century, before the subject was professionalised. They then continued in the early days of professionalisation as well. When it was a new field of investigation, lying wide open to anyone who was interested in exploring it, the women may well have been as numerous as the men. That observation, in itself, aroused curiosity for it offered yet another illustration of what I now boldly call *Thirsk's Law*, which is this: that whenever new openings have appeared on the English scene, whether in crafts, or in trade, and, in the modern world, in new academic endeavours, or in the setting up of new organisations in the cultural field, women have usually been prominent alongside the men, sometimes even outnumbering them. Certainly they have been involved

---

1   John Kenyon, *The history men. The historical profession in England since the renaissance* (London, 1983), p. x.
2   Bridget Hill, *The republican virago. The life and times of Catharine Macaulay, historian* (Oxford, 1992). Bridget and Christopher Hill have since then written an account of her impressive library ('Catharine Macaulay's *History* and her catalogue of tracts' in *The Seventeenth Century*, viii (2), 1993, pp 269–285).
3   Kenyon, *op. cit.*, pp 54–5, 163–4.

1

on a basis of equality with men. But that situation has lasted only until the venture has been satisfactorily and firmly established, when it has become institutionalised, formalised, and organised. Then, when the formal structure hardens, the direction, and the style as well, always fall under the control of men. I suspect that I have recently alighted on another example to add to my list, namely, the University Extension Service in the late nineteenth century. The many women, including Alice Stopford Green and Louise Creighton, giving history courses in the early days of the adult education movement, had already attracted my attention; and their preponderance is certainly implied in a derogatory reference in 1892 to 'the many petticoated lecturers who so obligingly lighten our darkness.'[4]

Maxine Berg has convincingly illustrated *Thirsk's Law* at work in the emergence of economic and social history as an academic subject of study in the universities. But I posit it as a general rule of life, and finding so many women writing history in the nineteenth century suggested that I was observing the same law in operation there. In that spirit I explored it further, and it is proving a very large and rewarding subject. Before completing a full investigation, I offer here some of the first insights.

A starting point for such a survey is an essay entitled ' Female historians' by E. Beresford Chancellor, published in a volume of *Literary diversions* in 1925. It was one of a number of essays by Chancellor which had originally appeared in journals and reviews and from internal evidence it must have been written soon after the 1914–18 war. The author was a Fellow of the Royal Historical Society and a Fellow of the Society of Architects, and his life spanned the years 1868–1937. He wrote mostly on London, its streets, its buildings and antiquities, though he also edited and annotated (with Sir Mayson Beeton) a volume on Defoe's *Tour of London* in 1928.[5] The essay on women historians gives a fresh view by an open-minded scholar, judging the scene in the 1920s. Chancellor started with the remark that women began to feature as historians in the early years of Queen Victoria's reign, and many, he said, were still household words. In fact, when naming them, he went back further still, to around 1820, so that we are entitled to call the roll of the 'History women' from the beginning of the nineteenth century, along with the 'History men' like Henry Hallam, T. B. Macaulay and John Lingard who launched the nineteenth century for Professor Kenyon. In fact, it is fair to regard the women as continuing the line of Catharine Macaulay, who died in 1791.

Who are these forgotten women? The first of the new century to be named by Chancellor was Lucy Aikin who wrote *Memoirs of the court of Queen Elizabeth* (2 vols, London, 1818), *Memoirs of the court of King James the first* (2 vols, London, 1822), and *Memoirs of the court of King Charles the first* (2 vols, London, 1833); and Elizabeth Ogilvie Benger whose work is described as being on a lower plane than that of Lucy Aikin, but who produced *Memoirs of the life of Anne Boleyn* (London, 1821), *Memoirs of Mary Queen of Scots* (2 vols, Lon-

---

4 Mark Reid, 'Our young historians' in *Macmillans Magazine*, 67 (1892–3), p. 9 (See also on p.98 a reference to 'the petticoated legions' of undergraduates in Oxford and Cambridge); Louise Creighton, *The economics of the household* (London, 1907). The prominence of women in adult education was doubtless strengthened by the fact that the university extension service was started from Cambridge university, partly on the initiative of the North of England Council for Promoting the Higher Education of Women. See Albert Mansbridge, *An adventure in working-class education, being the story of the Workers' Educational Association, 1903–15* (London, 1920), pp 6–7, 36–45.

5 E. Beresford Chancellor, *Literary diversions* (London, 1925), pp 81–6; *Who was who, 1929–40, sub nomine.*

don, 1823) and *Memoirs of Elizabeth Stuart, Queen of Bohemia* (London, 1825). Two women singled out for their writing on French history were Miss Pardoe on *Louis XIV* and Martha Walker Freer on the reign of Henry IV and the life of Margaret d'Angoulême and other works (seven in all).[6] Freer was writing in the 1850s and 1860s, and in her work on *Elizabeth de Valois and . . . the court of Philip II* (London, 1857) special notice is given in the title of the book to her use of 'numerous unpublished sources in the archives of France, Italy, and Spain', while the work on Marguerite d'Angoulême is described as being 'from numerous unpublished sources'. The works of these two were praised by Chancellor for their graphic descriptions of the periods treated, a quality which we shall see being stressed many times more in the writing of women.

Chancellor also dwelt on the documentation lying behind the work of the early history women, and among the early examples in the 1820s, he ranked Mrs Anthony Todd Thomson (Katherine Byerley), who 'exhibited far more research than was usually to be found in such works at this period'. Her *Memoirs of the court of Henry VIII* (2 vols, London, 1826) was described by the *Edinburgh Review*, as a work of much good sense and impartiality. It was followed by a *Life of Sir Walter Raleigh* (London, 1830); *Memoirs of Sarah, Duchess of Marlborough . . .* (2 vols, London, 1839); and, spanning the decades up to 1860, *Memoirs of the Jacobites of 1715 and 1745* (3 vols, London, 1845–6); *Memoirs of Viscountess Sundon . . .* (London, 1847); and *The life and times of George Villiers, Duke of Buckingham* (London, 1860). An author commended for her good work in earlier decades was Caroline Halstead, writing a *Life of Margaret Beaufort, Countess of Richmond* (London, 1839) and *Richard III* (London, 1844). Other authors described as 'lesser exponents of historical investigation', but having all 'done something more or less important in forwarding' historical knowledge were Miss Cooper on *The life and letters of Lady Arabella Stuart* (London, 1866), Emma Roberts on *Memoirs of the rival houses of York and Lancaster* (London, 1827), Mrs Marsh on the *Protestant reformation in France* (London, 1847), Mrs Matthew (i.e. Martha) Hall on *The queens before the conquest* (London, 1854), Emma Wilsher, on *Memoirs of the queen of Prussia* (London, 1858), and Miss Hookham on *The life and times of Margaret of Anjou* (2 vols, London, 1872).[7] Finally, Chancellor came to Agnes Strickland, and there for the first time appeared a better-remembered name. On account of her *Lives of the queens of England* (1840–8), Chancellor knew that she would be known to the majority of people. It had long been a standard work, he said, 'although later research has rather tended to remove it from the shelves of the more serious student' but it was 'a remarkable effort of industry and investigation, and, being written in an easy and pleasant style, did much to lay open the arcana of the annals of this country to the curious.' Chancellor then enumerated Strickland's many other books on the *Lives of the queens of Scotland and English princesses . . .* (with her sister, Elizabeth) (London, 1850–59); *Lives of the bachelor kings of England* (London, 1861); *Lives of the last four princesses of*

---

6 *History of the reign of Henry IV* (2 vols, London, 1860–63); *The life of Marguerite d'Angoulême, Queen of Navarre . . . from numerous unpublished sources* (2 vols, London, 1854). Julia S.H. Pardoe also wrote on the court and reign of Francis I, on Marie de Medici, books of travel, on Turkey and Portugal, oriental tales, and novels (See *British Library general catalogue of printed books to 1975* (London, 1984)).

7 Elizabeth Cooper also wrote *The life of Thomas Wentworth, Earl of Strafford and lord-lieutenant of Ireland* (2 vols, London, 1874) and *A popular history of America . . .* (London, 1865). Anne Marsh (later Marsh-Caldwell) wrote no other historical works, but published many novels and children's stories (*British Library general catalogue of printed books to 1975*). Miss Hookham was Mary Anne Hookham.

the royal house of Stuart (London, 1872); and *Letters of Mary Queen of Scots* (London, 1842). He did not underline her work in the archives, though he could have done (she went to great lengths to obtain access to them), but plainly that was in his mind, for he then went on to 'one who did great work among the hitherto unpublished archives of the country', namely, Mrs Mary Anne Everett Green (1818–1895). Mrs Everett Green had started with the same aspirations as Agnes Strickland. She wrote six volumes on the *Lives of the princesses of England from the Norman conquest* (London, 1849–55), and edited royal letters, including those of Princess Henrietta Maria (1857). But then she was asked to calendar documents in the Public Record Office, and ended up completing forty one volumes of Calendars of State Papers Domestic from the Tudors to the Commonwealth, including five magnificent volumes of the *Calendar of the Proceedings of the Committee for Compounding, 1643–1660* (7 parts, London, 1889–93).[8] Her grander plans for her own writing fell by the wayside, and the promised work on the lives of the queens of England of the house of Brunswick never appeared, though she lived until 1895.

Returning to historians, as opposed to editors of documents, Chancellor continued his list with Mary Berry, writing on *A comparative view of the social life of England and France from the restoration of Charles II . . .* (London, 1828); Cornelia Knight, author of an *Autobiography of Miss C. K., lady companion to the princess Charlotte of Wales* (2 vols, London, 1861); and Mrs Elizabeth Stone, writing *The chronicles of fashion from the time of Elizabeth* in two volumes which was published in London in 1845. His words used to describe this last work should be carefully weighed by those who believe that economic and social history started with Thorold Rogers, Arnold Toynbee, William Ashley, and William Cunningham.[9] Mrs Stone's volumes 'should not be forgotten,' he said,

for in that interesting work the writer has collected and presented in a very attractive form those features of the history of our country which tell not so much of the lives of kings or the actions of ministers, as of the life, habits and customs of the society over which they ruled; and which was, after all, the conception of what history should be, held by J. R. Green, one of the great masters of the art of historical disquisition.

And lest that remark be taken to mean that Mrs Stone followed the lead given her by J. R. Green, the dates of their works of history need to be underlined: Mrs Stone was writing in 1845, J. R. Green's *Short history of the English people* appeared in 1874.

Chancellor's essay is instructive in its attention to women, and its generous acknowledgment of their contribution to the writing of history. When investigated in the library catalogue, his authors prove to have been highly versatile women, writing, in addition to works of history, poetry, stories, school textbooks, and memoirs. Frequently, they were women of means, they had been well educated, and they had access to good libraries.[10] They were daughters and/or wives of clergymen, gentry, or middle-class families with bookish inter-

---

8 For a list of Green's publications see *British Library general catalogue of printed books to 1975.*
9 As in N. B. Harte, *The study of economic history* (London, 1971), pp xi–xxiv, and D. C. Coleman, *History and the economic past* (Oxford, 1987), pp 37–8.
10 Though not in the case of Elizabeth Ogilvie Benger, of whom it was said by Lucy Aikin in a memoir of her that she was so hungry for books, when living in Wells in her youth, that she used to gaze through a bookseller's window, reading title pages, and returning day after day in the hope that a page had been turned over, and she could read more (E. O. Benger, *Memoirs of the life of Anne Boleyn, with memoir of the author by Miss Aikin* (3rd edition, London, 1827), p.iv).

ests. Their fathers and husbands often took a positive view of the importance of educating women, or else their pursuits brought the whole family into contact with current political and historical debates. They were thus well placed to move into a field that was then without boundaries, and wide open to all.

Chancellor's choice of writers alighted noticeably on authors of biographies or memoirs, and editors of original documents. Thus his essay seemed silently to assemble some special features that distinguished the work of women historians of that time. This struck a chord with me, because I also had come to see qualities in women's work which distinguishes it from that of men. I first became aware of the differences when supervising men and women graduate students at Oxford. They made their own choices, free from much interference by me in the selection of subjects. Admittedly, a certain influence is exerted nowadays by postgraduate courses of training, instituted in recent years. But those pressures were not strongly exerted before the late 1970s. Choosing their themes in relative freedom, men preferred spacious surveys, often starting from large conceptual propositions, and they favoured statistical aggregation as a means of arriving at their conclusions. Women preferred to observe people at closer quarters, and had a keener eye for detail. They chose smaller groups for intimate study, though this in due course opened a window on a larger world. They did not shy away from far-reaching conclusions, but they arrived at them from a different direction. In other words, the biographical and documentary preferences of the history women, which are seen in the nineteenth century, still float to the surface. The historical conventions and themes have changed but the instinctive differences between women's and men's views of the world persist.

This case should not be overstated, or distorted by exaggerated generalisations. Plainly, when surveying a wide spectrum of humanity, the contrasts between men and women will be blurred in individual cases. Moreover, men and women influence each other. But, in general, differences can be identified between the historical interests and style of the history men and those of the history women, the men clustering at one end, and the women at the other end of a spectrum.

In his essay on 'Female historians' of the nineteenth century, Chancellor seemed to be saying, first, that they favoured the study of history through the lives of individuals, and, secondly, that they attached great importance to the study of original documents. I would go further and say that the two concerns went together, making the search for original documents the essential requirement for writing a graphic, well-authenticated, fully-rounded, and convincing history of people.

Women's interest in the study of the original documents was evident already in the work of Catharine Macaulay and it is most emphatic in the case of Mrs Everett Green. It is again evident, to an almost fanatical degree, in the work of Lady Theresa Lewis, writing her *Lives of the friends and contemporaries of Lord Chancellor Clarendon* (3 vols, London, 1852). In the preface she explained that

nothing ... has been stated without full reference to the authorities from which it is drawn, and in no case has information knowingly been accepted at second hand when the original was accessible, still less have the conjectures of the author been allowed to supply the place of authentic testimony.

Mrs Everett Green, in her turn, dedicated her *Letters of Queen Henrietta Maria* to Lady Lewis (in 1857), commending the excellence of her procedures.[11]

---

11 Lady Theresa Lewis, *Lives of the friends and contemporaries of Lord Chancellor Clarendon: illustrative of portraits in his gallery* (3 vols, London, 1852), i, p. v; Mary A. Everett Green, *Letters of Queen Henrietta Maria, including her private correspondence with Charles the First* ... (London, 1857), dedication.

The importance which the women attached to original documents is further underlined in the many volumes of correspondence which they edited, and in sidelights shed on their own lives, showing how much effort was put into gaining access to the archives. Agnes Strickland went to great trouble to get access to manuscripts in private hands in England and Scotland. In France Guizot gave her permission to consult state papers, and in Paris she kept two transcribers at work. She also went to the provinces in search of more manuscripts. Mrs Everett Green said in her *Lives of the princesses of England*, that she had investigated documents 'in almost every civilized country in Europe'.[12] Julia Cartwright (Mrs Henry Ady), writing *A life of Henrietta, daughter of Charles I and duchess of Orleans* (London, 1894), managed, a generation after Agnes Strickland, to get access to original archives in France through diplomatic intervention. A manuscript of a different kind, Celia Fiennes's *Travels through England on a side saddle in the time of William and Mary* (London, 1888) saw the first light of day through the editorship of the Hon. Mrs Emily Griffiths. Her opportunity came because her father, the Baron Saye and Sele, acquired the manuscript about 1885 and gave it to her. Other original documents edited by women in the nineteenth century make a long list, which cannot be enumerated here.

The role of documents for the history women was, before all else, to give them insights into the lives of people (not simply their public, but their private lives as well), and to conjure up the authentic atmosphere of their period. The literary force of this endeavour was highlighted for me when I examined in some detail the very different style of Edward Hasted (1732–1812), the historian of Kent. He was unusual in his time for his indefatigable search for original documents. No one could accuse him of neglecting the accessible archives in the British Library and the Public Record Office, and, in the opportunities available to him, he can be directly compared with Catharine Macaulay, for they were contemporaries, she being born in 1731 and he in 1732, though he lived more than twenty years longer than she. His use of the documents, however, was markedly different. For him the documents were the basis of a history of Kent, and they established events. His history was markedly deficient in curiosity about the personalities of people. When he said anything about individuals, the descriptions were wooden and totally failed to bring people to life.[13] By contrast, Catharine Macaulay was markedly perceptive and assured when seeking to portray character, personality, and idiosyncrasy.

The history women were clearly prominent in the writing of biographies in the nineteenth century. The history men favoured political, constitutional, and ecclesiastical surveys, in which high politics dominated; this was the male preference, or style. When women wrote in the same vein, the explanation often turns out to be that they were the wives or family members of male historians. In other words, they had absorbed other influences, and they undertook some tasks which had been handed their way by the menfolk. Thus Mrs S. R. Gardiner (Bertha Meriton Cordery) wrote on the *The French revolution, 1789–95* (London, 1883) and *The struggle against absolute monarchy, 1603–88* (London, 1877)

---

12 Agnes Strickland, *Lives of the queens of England from the Norman conquest* (6 vols, London, 1864 edn., (an edition that sold 11,000 copies)), p. xii; *ibid.* (Philadelphia edition, 1907, with biographical introduction by John Foster Kirk), pp xi–xii, xviii–xx; Una Pope-Hennessy, *Agnes Strickland. Biographer of the queens of England, 1796–1874* (London, 1940), p. 8; M. A. E. Green, *Lives of the princesses of England from the Norman conquest* (6 vols, London, 1849–55), i, p. vi.

13 Joan Thirsk, 'Hasted as historian' in *Archaeologia Cantiana*, cxi (1993), pp l–15.

and Mrs Mandell Creighton (Louise von Glehn) wrote on *England: a continental power from the conquest to Magna Charta, 1066–1216* (London, 1876).

The coupling of biography with the study of documents was no accident, for together they kept living people at the centre of the stage, and gave women's writing a lively quality. Even when stern editors lacked sympathy for a woman author, as did William Sonnenschein when commenting on a work by A. Mary F. Robinson (Madame Darmesteter), he admitted, after saying that the work had no great scientific value, that the text supplied 'sympathetic and vivid medieval pictures.'[14] Women's success lay in dramatising history, and in capturing and retaining its human scale. They used anecdotes to conjure up a living picture. They also had a keen sense of place: Strickland reckoned to visit the places mentioned in her writing, and made several journeys to Scotland in search of atmosphere when writing about Mary Queen of Scots. This explains how women also came to write so many history books for children. A list of such authors would bring hundreds more women writers to light.[15]

When we turn to the beginnings of economic and social history, the women's concern to focus on people in all aspects of their lives, rather than restricting attention to the public dimension, again brings women to the forefront, as pioneers in exploring the subject from their preferred direction. It was perhaps inevitable, given the angle from which they approached history, that they should be early investigators on this scene. Their lives were concerned with the domestic management of families and households, and so, if economic and social history was to become a subject for respectable academic study, one of its branches would, in fact, be transferring women's age-old preoccupations onto a wider stage. This transference was attempted when Mrs Elizabeth Stone wrote her *Chronicles of fashion* in 1845. She took for her subject what we would nowadays consider to be an aspect of social history. Her chapters dealt, in a somewhat high flown style, with banquets, food, and table manners, habitations, carriages, amusements, theatrical performances, costume, assemblies, horse racing, and watering places. But she clearly realised that she was battling against the tide of historical fashion for she defiantly wrote under the main title of her book, 'Lore which wig-crowned history scorns.'

Three years earlier, Mrs Stone had made a move into economic history, tackling it in more experimental fashion. The influences behind her venture can be seen in her own life story. She was the daughter of the proprietor of the *Manchester Chronicle*. This plainly kept her in contact with current economic and social developments. She was, in fact, a campaigner for women, as well as a historian, and had attempted to get the working hours of millinery apprentices shortened. She wrote a book on *The art of needlework* (London) in 1840, and then in 1843 *The young milliner* (London), a fictional account of real life which brought her face to face with the abject working conditions of millinery apprentices. Her indignation boiled over at the end of *Chronicles of fashion*, when

---

14 William Swan Sonnenschein, *The best books. A reader's guide* (London, 1891), p. 398. Mary Robinson became Madame Darmesteter and later Madame Duclaux. Her works include the *Letters from the Marchioness de Sévigné* (in French first (Paris, 1914), and then in English (London, 1927)), poems, a novel, a book of songs, a life of Emily Brontë, and a work on Froissart (see *British Library general catalogue of printed books to 1975*).

15 Among the most successful writers of children's books was Mrs Markham (the assumed name of Mrs Elizabeth Penrose of Bracebridge, Lincoln, second daughter of Edmund Cartwright, inventor of the power loom) whose histories of England, France, Germany, Greece, Rome, etc. passed through innumerable editions. Their writing had been prompted by her own children's desire to read history, and young Richard's difficulty in understanding David Hume (Mrs Markham, *History of England from the first invasion by the Romans . . .* (London, 1848 edn.), p. vii.

she issued a clarion call to the women of England. 'With you this power rests: with you it remains to decide whether still, without care and without remorse, the young and friendless of your own sex and your own country shall continue to be sacrificed to the demon FASHION!'[16]

The full dimension of Mrs Stones's economic insights, however, is seen in another work of hers in 1842. She was a proud resident of Manchester, hotly defending Lancashire against current opinion which labelled it a place of smoky chimneys, filthy streets, drunken men, immoral girls, and squalid children. She wrote two volumes, entitled *William Langshawe, the cotton lord* (London, 1842) as a novel, but it clearly rested on historical fact. Indeed, a footnote in volume one actually says 'this improbable-looking incident is Fact'. In short, her book gave, in 1842, the first glimmerings of a future for the study of economic and social history, through the working lives of individuals. This was long before the history men such as Thorold Rogers and William Cunningham, took up their pens. And when they did, they took economic history in a different direction, making it at first a study of national economic developments and government policy, not the history of men and women who lived under the regime.[17]

Maxine Berg has told the story of the professionalisation of economic history from the 1880s onwards through to the 1920s and 1930s, showing that the women were prominent and numerous at the beginning, though slowly their numbers fell away.[18] Outside the professional circle that was assembling in the 1880s and 1890s some women carried on the non-professional tradition, and their work continued to demonstrate a characteristically female starting point and view point. One of the first milestones in the writing of women's history was set in place by Miss Georgiana Hill, who had begun her author's career in the 1860s producing numerous, highly successful cookery books. They were at first, in 1860, directed at cottage kitchens, but they came to incorporate recipes from many different European countries. They sold cheaply and easily; her *Everybody's pudding book* (London, 1887), claimed to have sold ten thousand copies. She then went on to produce a *History of English dress from the Saxon period to the present day* (2 vols, London, 1893), and finally *Women in English life from medieval to modern times* (2 vols, London, 1896). Georgiana Hill was making the first exploration of women's history, running far ahead of current historical fashions. But being by then an experienced author, and having worked diligently on her primary sources, she was supremely confident, and put forward some spacious historical judgments that did not fail to arouse the indignation of the history men. Some angry reviews appeared. One reviewer felt 'throughout that lack of grip of the subject which is the necessary consequence of superficial study'. Another did not like her failure to stress the role of women as mothers and nurses, and feared that if women followed other pursuits, they would be accused in the end of 'the degradation of the race'.[19]

At the end of this survey of nineteenth-century women's writing on history,

---

16 E. Stone, *Chronicles of fashion from the time of Elizabeth to the early part of the nineteenth century in manners, amusements, banquets, costume, etc.* (2 vols, London, 1845), ii, p.455.

17 E. Stone, *William Langshawe, the cotton lord* (2 vols, London, 1842), i, p. 121. It is significant that R. H. Tawney in a robust and memorably-phrased memoir of George Unwin attributed to him the broadening of economic history beyond state policy to the study of people, but for Unwin the chief interest lay in people not as individuals, but operating in communities and associations (George Unwin, *Studies in economic history: the collected papers of George Unwin* (London, 1927), pp lxiiff. See also Unwin's own words, on pp 28ff).

18 Maxine Berg, 'The first women economic historians' in *Economic History Review*, xlv, 2 (1992), pp 308–29.

19 Anon., '*Women in English Life*, by Georgina (sic) Hill' in *The Athenaeum*, no. 3591, 22 Aug. 1896, pp 252–3; Anon., 'Women in English life' in *The Spectator*, vol. 77, 8 Aug. 1896, pp 177–9.

the strong conclusion emerges that the viewpoint, style, and subject matter of the history women were, and, I would say, still are different from those of the history men. In consequence, they did not, and do not now, always like each other's work. But they influence each other profoundly, and sooner or later they absorb each others' viewpoints with their own. The study of household, family, children, fashion, food, and consumerism were explored by the women in the nineteenth century, but not by men; they have now been accepted as worthy subjects on the agenda of economic and social history. At the same time, it has to be said that when the history men take up the history women's subjects, they carry them in a somewhat different direction. The work of Wrigley and Schofield, for example, shows how the subject of the family has been carried along a path that is more congenial to men than to women.[20] The subject of rural industries similarly has been diverted along different tracks, from the subject envisaged by this author when writing in 1958–9. Then rural industries were seen to be embedded in a dense thicket of economic and social circumstances which were peculiar to their own locality, and which had to be explored within a local, cultural framework, even while they were seen also in the broader economic landscape of national industries and commerce. The subject was transformed by the history men into 'proto-industrialisation', which was a disturbing simplification and dehumanisation of a complex story.[21] Even now we see the study of the consumer society being carried off by the history men, after being revived in the 1970s by the history women in England and America.

This appraisal of the differences between the history men and the history women is not intended to start a war between them, but has the constructive purpose of promoting a clearer understanding and appreciation of their differing contributions. For if the differences are generally acknowledged, then women's insights must be deemed essential if history is to be constantly refreshed from as many sources as possible. The propositions offered here, furthermore, give a rational, and hence consoling, explanation for some of the bleak facts of life as they unfold in our present-day historical world. They explain, for example, why economic history has moved away from its much broader span of interests in the early twentieth century, and accepts its dehumanised straitjacket now; the history men have imposed their preferences. It explains the present-day obsession of historians with statistical proofs, even when the available statistics are so meagre that they narrow the rich panorama of the scene, and oblige scholars to peer at it through the tiniest of windows. In short, the content of history has been tilted towards quantitative rather than qualitative analyses, and is dominated by masculine preferences. A better balance between the two viewpoints needs to be restored.

In the 1880s and 1890s an illuminating debate was conducted in the journals on the content of history. Critics realised that history was taking a new direction, examining causes and consequences, setting problems, whereas before it had described events dramatically. The new style was called 'scientific history', and was much disliked by many essayists. It was dull and arid, they said, and

---

20 See E. A. Wrigley and R. S. Schofield, *The population history of England, 1541–1871. A reconstruction* (Cambridge, 1981), for which the research required the reconstitution of families.

21 Compare Joan Thirsk, 'Industries in the countryside' in F. J. Fisher (ed.), *Essays in the economic and social history of Tudor and Stuart England*, (Cambridge, 1961), pp 70–88, with F. F. Mendels, 'Proto-industrialization: the first phase of the industrialization process' in *Journal of Economic History*, xxxii (1972), pp 241–61, and P. Kriedte, H. Medick, and J. Schlumbohm, *Industrialization before industrialization* (English transl., Cambridge, 1981).

they bemoaned the loss of history as an art. 'The field of history', said Augustine Birrell, 'should not merely be well tilled, but well peopled ... tell me their names that I may repeat them to my children'. Not for him was history to be filled with maxims, morals, theoretical studies, political philosophy, political economy, 'setting us all problems'. Lecky also saw the debate as a tussle between epic and scientific history, though he did not express a preference about the outcome. Froude, in contrast, defended the epic form: history, for him, was a stage in which the drama of humanity was played out.[22]

In this debate, the historians who were singled out as exemplars of the two styles were always men, and never women. Freeman was named as a scientific historian, Macaulay and Carlyle were cited as the epic writers : 'we find him (i.e. Carlyle)', says Birrell 'always treating even comparatively insignificant facts with a measure of reverence and handling them lovingly.' In fact, he was commending a style which the women also preferred, though they were passed over in silence. Ignoring the women had become the accepted convention, as was conspicuously evident in the work of A. W. Ward. In a magisterial chapter in the nineteenth-century volume of the *Cambridge history of English literature* (Cambridge, 1916), Sir Adolphus W. Ward devoted 69 pages to historians and biographers. But he spared only 58 lines, less than 3% of his text for the women. It is quite breathtaking when one then reads the memoirs of Alice Drayton Greenwood, and discovers that this lady, with a first class degree in history from Somerville College in 1888, did substantial work for his *Cambridge history of English literature* (she wrote three chapters in volume ii) and carried out innumerable grinding chores, for the last volume of the *Cambridge modern history* (Cambridge, 1910) which Ward also edited. She was responsible for the genealogical tables, lists of popes, governors of British colonies, chancellors of Germany, tables of parliaments and congresses, plus the indexes. As well as all that, Alice Greenwood wrote books of her own, and clearly might have written more. But as Edith Wilson, writing her memoir, remarked mildly: 'She was one of those students who get too little credit for their own work because of the time spent in helping others.'[23]

In the middle of this debate of the 1890s on the content of history, one writer devoted a whole essay to the 'Women historians'. Augustus Jessop explained how he had read them avidly in his youth, and plainly owed them much for stimulating his interest in history. Nevertheless, in his riper years he took the view that it must all have been 'poor stuff'. Surveying the work of the more recent history women, he had high praise for Agnes Strickland, Mary Everett Green, Kate Norgate, Lina Eckenstein, and Mary Bateson, but in a conclusion that was astonishingly at variance with what went before, he arrived at what he considered 'the most striking fact', namely, 'the immeasurable superiority of the men over the women in mere style,' and concluded that 'the present outlook seems to point to this that in the future the great builders-up, the great discoverers, the great thinkers, the great historians will not be women'. Such a

---

22 Augustine Birrell, 'The muse of history' in his *Obiter Dicta* (London, 1927 edn.), pp 260–78; Mark Reid, *op. cit.*, pp 91–8. See also Anon., [*recte* Arthur Ready], 'The real historian' in *Macmillan's Magazine*, 66 (1892), pp 221–7.

23 Edith C. Wilson, *In memoriam Alice Drayton Greenwood, 1862–1935* (privately printed, 1935). Alice Greenwood's *Lives of the Hanoverian queens of England* (2 vols, London, 1909–11) contains an entirely characteristic history woman's claim about the scrupulous use of documentary sources, and warns that if the actions of eminent men do not seem to tally with the accepted portraits, it is because she has used documents concerning their private lives (*ibid.*, ii, preface, pp ix–x).

conclusion is only explicable if one accepts the general proposition argued here that the styles and subject matter of the men and the women are, indeed, different. We might all then agree with Jessop that women do not generally choose to write about 'greatness'.[24]

It seemed that a debate on the content of history might be revived when Lawrence Stone wrote recently on the revival of narrative history. This is the twentieth-century term for what the nineteenth-century writers called 'epic history', in contrast with scientific history. But Stone did not trace the story so far back as to link it with the debate of the 1890s, for he examined it only from the 1930s onwards. And his is yet another survey of the history men, with only one reference to a woman, Natalie Zemon Davis.[25] It did not initiate fresh discussion. A second theme of the 1890s also returned when an article in *The Times*, on 14 November 1992, asked 'Why are so many of our modern academic historians so dull?' The article carried the headline, 'The dead hand of the history men', and it was, indeed, entirely concerned with the writings of men. This spirited challenge likewise failed to kindle a flame.[26] Perhaps it is high time that the women joined the debate, taking as their starting point a shrewd essay by Mrs Alice Stopford Green on 'Women's place in the world of letters', published in *The Nineteenth Century* in 1897 and reissued in a slim booklet in 1913. She was obviously concerned at the women's tendency to follow the style of men. The woman, she observed 'seeks safety in what is known in Nature as protective mimicry. Woman sails under any colour but her own as though in perilous days a racing yacht hoisted the black flag of the pirate to be in fashion with the wide world'; and again, 'the busy contrivances of women for adaptation and assimilation do tend to obliterate distinctions, and to rob their work of both the eccentricity which they fear and the originality they distrust. The tortoise's head is kept well under cover'. Mrs Green's essay is a thoughtful, sensitive analysis, which deserves careful reading. But, in summary, she saw women having 'a sense of values permanently different from that of the man', and she finished up her essay with a vision of the future in which women would 'open new horizons where men's vision has stopped short'.[27] This review of the history women, still only in its preliminary stages, endeavours to show how and where the writers of the nineteenth century opened up 'new horizons where men's vision stoppped short'. In many of their ventures they received at first condescending words of cool comment, if not adverse criticism, but as the decades have passed, the themes they broached have been drawn into the mainstream, have been accepted, and sometimes have been taken over by the history men. In order to arrive at a more balanced account of the historical profession in England since the Renaissance, it is necessary to call the roll of the history women, and recognise their innovations alongside those of the men. They have played a positive role in opening up new horizons, and its full dimensions need to be identified. Then the women's task must be to declare and proclaim that fact, for the record of the past teaches another lesson very plainly indeed, that if the history women do not do it, then the history men will not do it for them.

---

24 Augustus Jessop, 'Women as historians' in *Literature*, iv (1899), pp 41–2, 67–8.
25 Lawrence Stone, 'The revival of narrative: reflections on a new old history' in *Past and Present*, 85 (Nov. 1979), pp 3–24.
26 Daniel Johnson. 'The dead hand of the history men' in *The Times*, 14 Nov. 1992, p.18.
27 A. S. Green, *Woman's place in the world of letters* (London, 1913), pp 7–8, 10.

2 A Woman in History: Eileen Power and the Early Years of Social History and Women's History

Maxine Berg

WOMEN'S writing has been a subject of enduring interest, but it is only recently that women's history writing has attracted attention. This is despite the place which history was accorded in the arts during the eighteenth and early nineteenth centuries, when it was considered the highest form of literature after poetry. Clio was represented as a woman, but the great historians Thucydides, Gibbon, Hume, Macaulay, Ranke, Maitland, and others who are now remembered were men, and the historical profession, progressing through various schools of history – the German historical school, the *Annales* School, the students of Croce seemed to be masculine. Biographies of Toynbee, Namier, Bloch and Trevelyan have appeared in recent years. The history men, as Joan Thirsk has argued, dominate our view of the profession.[1]

When Trevelyan published his *English social history* in 1944 he dedicated it to the memory of Eileen Power, economic and social historian. Her story has not been told, yet she brought together high academic honours, an international reputation, and a popular literary following in ways the great male historians did not achieve. For Power, as a part of the academic elite, eschewed writing history only for such an elite. During the inter-war years she brought medieval history into the general culture and made social history a prominent part of the historical discipline. Eileen Power together with R.H. Tawney helped to create economic and social history through her teaching at the London School of Economics and her key role in founding the Economic History Society. She was the author of one of the best-known of medieval histories, *Medieval people*, and one of the first writers and teachers of women's history. Her work spanned a great range of medieval history, and her publications were major contributions not just to social and cultural history, but to comparative economic history.[2]

---

1 W.H. McNeill, *Arnold J. Toynbee. A life* (Oxford, 1989); Linda Colley, *Namier* (Cambridge, 1989); David Cannadine, *G.M. Trevelyan. A life in history* (London, 1992); J.P. Kenyon, *The history men* (London, 1983). See also pp 1–11 above.

2 Her books included *The Paycockes of Coggeshall* (London, 1920); *Medieval English nunneries* (Cambridge, 1922); *Medieval people* (London, 1924); *The Goodman of Paris* (London, 1928); *The wool trade in English medieval history* (posthumously Oxford, 1941). She was an editor and major contributor to R.H. Tawney and Power (eds), *Tudor economic documents* (London, 1924); Power and M.M. Postan (eds), *Studies in English trade in the fifteenth century* (London, 1933); and Power and Clapham, *The Cambridge economic history of Europe* (3 vols, Cambridge, 1941), vol. i. She contributed a highly regarded chapter for J. R. Tanner, C. W. Previté-Orton, Z. N. Brooke (eds), *The Cambridge medieval history* (8 vols, Cambridge, 1911–36), vii entitled 'Peasant life and rural conditions c. 1100 to 1500' in 1932. She also wrote a wide range of major articles; some of these were collected together by Postan in 1975 for the well-known, *Medieval women* (Cambridge, 1975).

12

Power was a professor by her early forties, and collected several honours, including two honorary doctorates of literature at the University of Manchester and Mount Holyoke College, and a corresponding membership of the Medieval Academy of America.

Eileen Power appears, at first sight, to be a unique and extraordinary case of mainstream academic achievement in her field. She was not rare among women in choosing to study history, but far more studied history and published on the subject than subsequently acquired any kind of position, and have since been forgotton. There was Lucie Varga, known less for her publications in the *Annales*,[3] than for her affair with Lucien Febvre. There was Alice Stopford Green, known less for the parts of the *History of England* which she wrote, and the other parts for which she did the research, than for being the wife of J.R. Green.[4] There was Frances Collier who was rewarded for her minor classic, *The family economy of the working classes in the cotton industry,* by being made the departmental secretary of the country's first economic history department at Manchester.[5]

Two recent articles in *History Workshop* – Peter Schottler's study of Lucie Varga, and Natalie Zemon Davis's on the women of the *Annales* School – highlight how many fine women historians turned into research assistants for major historical figures. Simonne Bloch and Paule Braudel were well known to be close collaborators in their husbands' projects. Many other women wrote major theses in colleges which women could attend during the inter-war years, but few ended up other than as school teachers, copy editors, research assistants and administrators on the *Annales* and the *Encyclopedie Francaise*.[6]

Their example was matched in England by the great research projects at the London School of Economics. W. H. Beveridge's monumental *Prices and wages* was based on the research work of five women assistants. The Webbs' historical projects absorbed another large group of female researchers – the best known was Amy Harrison, author of works on the history of the Factory Acts. Another research project was the monumental Victoria County History.[7]

This paper will not attempt to provide a full overview of Power's contribution to history, but will concentrate instead on her formation as a woman and scholar, and on her pioneering work in women's history.

Eileen Power (1889–1940) was born the eldest of three sisters just outside Manchester, in the exclusive suburb of Antrincham. Her family had Irish roots. Her paternal grandfather, the Reverend Philip Bennett Power (1822–1899) was from Waterford, and took a degree in Trinity College, Dublin and was ordained. He left Ireland shortly after, and lived in England the rest of his life, working as a curate. But he gave this up in his forties, and became well-known after that as a prolific evangelical tract writer.[8] There is no evidence that Eileen Power ever visited Ireland, but she perceived herself as influenced by this heritage, and

---

3 Natalie Zemon Davis, 'Women and the world of the *Annales*' in *History Workshop*, 33 (1992), pp 121–138; Peter Schottler, 'Lucie Varga: a central European refugee in the circle of the French "Annales", 1934–1941' in *History Workshop*, 33 (1992), pp 100–121.

4 See pp 1, 11 above.

5 Frances Collier, *The family economy of the working classes in the cotton industry, 1784–1833* (1921, new edn., Manchester, 1965).

6 Natalie Zemon Davis, *op. cit.*; Peter Schottler, *op. cit.*

7 See Maxine Berg, 'The first women economic historians' in *Economic History Review*, xlv, 2 (1992), pp 308–329.

8 I owe the information to Sir John Habakkuk. See also Rev. Philip Bennett Power, *Series of tracts* (Stirling, 1892–1910); and C.H. Unwin, 'Introduction' to P. B. Power, *Breviates or short texts and their teachings* (London, 1916). These collections and others published over this period numbered several hundred.

took an enduring interest in Irish history and literature.[9] Her father, Philip Ernest le Poer Power (b.1861) took an Oxford degree, and became a stock broker in Manchester and was dealing on a large scale by the time Eileen Power was born. He entertained some ideas of grandeur, for he added 'le Poer' to the family name. Her mother's family was middle class, and there were close connections with two spinster aunts and her maternal grandfather. This extended family was to be important, because Power's father was convicted of a massive fraud on the Manchester Stock Exchange in 1891, was declared a bankrupt in 1893, and served a five year prison term, and several others in the years afterwards for subsequent financial crime.[10] The family, tainted in their view, by a Victorian style scandal, moved to Bournemouth and lived for a time under an assumed name. All contact with the girls' father was cut off. The Power sisters' mother died of tuberculosis when Eileen Power was fourteen, and the sisters were then brought up by their spinster aunts and maternal grandfather. Their mother had wanted her daughters to attend the Oxford High School for Girls, so the family moved to Oxford, and leased a modest house for the time that the sisters were at the school.

The scandal had a deep emotional impact on Eileen Power; neither she nor her younger sister, Beryl ever saw their father afterwards and she perceived his crime as a dark shadow on her life. Her mother's death, followed by her grandfather's death shortly before she finished at university left her with responsibility for her younger sisters.

Power went to Girton College, Cambridge in 1907, and took a first in both parts of the Historical Tripos in 1910. She was taught history there by Winifred Mercier, the suffragist and educationalist, and by Ellen McArthur, a suffragist and economic historian. Through her tutors in Girton she formed the connection with economic history and the London School of Economics (L.S.E.) which was to set her future career. Power's time as a student in Cambridge brought close friendships with several students from much more cosmopolitan and intellectual backgrounds than her own. It also brought close contact with the suffrage politics then important in Cambridge, and especially in Girton.

Power's education at Girton in the Historical Tripos was already, by the time she came, a route to historical research and outward careers of some kind in a new expanding field, economic history. Women's education both in the college and in the wider education was fostered by William Cunningham, one of the first major economic historians. Ellen McArthur (1862–1927) was one of his students and collaborators on his *The growth of English industry and commerce in modern times* (6th edn., Cambridge, 1907). Another of his students and collaborators, Lilian Knowles, also taught at Girton, but went on by the time Power arrived, to hold the first lectureship in economic history at the London School of Economics. Both women helped to create a direct route, for Power and several other women of her generation, through Girton College and the Cambridge Historical Tripos to research at the L.S.E.

Power also drew great support from the suffrage movement. She had been brought up in a family of self-reliant women, determined on education for the three Power sisters. Her college and her tutors were prominent in the suffrage

---

9 See Power's reviews, 'The Dark Rosaleen'; 'Early Ireland'; 'Light and flashlight on the Irish question' in *The Nation and the Athenuem*, 20 Dec. 1924, 16 May 1925; Eileen Power, *Poems from the Irish selected by E. Power* (London, 1927); Eileen Power's correspondence also contains references to Irish poetry and fairytales.
10 *The Times*, Criminal Courts, 29 Apr. 1905.

campaigns. Girton established the first suffrage club in Cambridge in 1907.[11] Power's tutors, Winifred Mercier and Ellen McArthur were among the leading educationalists and suffragists of their generation.[12] But it was her friends and their families who were to provide a more important influence. Power's closest friends included Margery Garrette, later Spring-Rice, the niece of Elizabeth Garrette Anderson and Millicent Garrette Fawcett, and Karin Costelloe, sister of Ray Strachey, daughter of Mary Berenson and niece of Alys Russell. Eileen Power's friendships with women, especially Margery Garrette and M.G. Jones were passionate, life-long connections.

During her final year at Girton in 1910 and in the years immediately following Power spent large parts of the academic holidays at the Russell household where Alys Russell taught her German and took her to suffrage meetings. She described her initiation into speaking in the suffrage cause in 1910:

Last week Mrs. Russell took me off to a suffrage meeting in a village near here. I went to Court Place for the night and we bicycled 8 miles to a lovely Elizabethan house owned by a rich American artist – a charming woman. We had dinner (prepared by a French chef. Lor!), and then Mrs. Russell addressed a meeting from the Terrace. She speaks excellently. I had to speak too, which rather disconcerted me, as you know what an atrocious speaker I am at the best of times, and it was too dark for me even to use notes. Moreover a crowd of shy bucolic yokels fill me with more awe than the academic audiences of Girton. However desperation probably carried me through . . . we are probably going to do a little campaign among the villages near here in September.[13]

Power also spent time in the Garrette house in Suffolk, she was involved in organizing for the NUWSS in 1914. Her sister Beryl was organizing in the Eastern region and was speaking at meetings with Mrs Fawcett.[14] By 1918 her net was cast wider and she was lecturing for the League of Nations Union and the Workers Education Association.[15] Power wrote many times during this period to her friends and colleagues on marriage, women's education, experiences of sexism in the universities and on her own and other women's careers.[16] In answer to a request for information about herself from someone writing about former Girton students for the *Girls Realm*, she included instructions to put in 'that I was a strong feminist!'[17] Her reaction to the marriage of her friend Karin Costelloe to Adrian Stephen, Virginia Woolf's brother was 'heaven be praised he won't interfere with the philosophy.'[18]

Power's feminist interests and views were fostered first by her own upbringing and in the relatively isolated single sex and feminist environment of Girton College. Rosamund Lehmann's *Dusty Answer* conveyed a picture of 'black garments, grey, close-brushed intellectual heads, serious thin faces looking down the room'. This was quite at odds with the memories of Power's own students

11 Martin Pugh, *Women and the women's movement in Britain 1914–1959* (London, 1992), p. 60; Philippa Levine, 'Love, friendship and feminism in later nineteenth century England' in *Women's Studies International Forum*, 13 (1990), pp 630–78, especially p. 64.
12 Lynda Grier, *The life of Winifred Mercier* (Oxford, 1937); Martha Vicinus, *Independent women: work and community for single women 1850–1920* (London, 1985); M.B. Curran, 'Ellen Annette McArthur, 1862–1927' in *Girton Review*, 75 (1927), pp 2–4.
13 Power to Margery Garrette, 18 Aug. 1910 (Power papers in possession of Lady Cynthia Postan).
14 Power to Margery Garrette, 24 Sept. 1914; 16 Mar. 1915 (Power papers in possession of Lady Cynthia Postan).
15 Power to Garrette, 25 Sept. 1918 (Power papers in possession of Lady Cynthia Postan)
16 Letters to Margery Garrette (Power papers in possession of Lady Cynthia Postan); Letters to G.G. Coulton (Power papers, Girton College, Cambridge, Archives); Letters to Lilian Knowles (Eileen Power Personal File (London School of Economics)).
17 Power to Garrette, Feb. 1911 (Power papers in possession of Lady Cynthia Postan).
18 Power to Garrette, 25 Sept. 1914 (Power papers in possession of Lady Cynthia Postan).

who remembered poetry and musical evenings presided over by a charismatic and beautiful woman 'dressed like a princess out of a medieval story book,' and tutorials conducted by a young woman who 'sat at the edge of her chair, offering chocolates.'[19]

After Girton, Power was given a fellowship to spend a year at the Ecole des Chartes in Paris, and she studied there between 1910 and 1911 under the supervision of C.V. Langlois. The time allowed her to range widely over medieval and Renaissance art and literature. She wrote in the autumn of 1910;

I don't believe one can ever do good work on the history of a period without getting soaked in its literature, art and general atmosphere, and that is what I am doing now . . . I love being able to spread myself over what at college had to be irrelevancies – the art and literature and to feel that duty and pleasure coincide.[20]

She started a thesis there on Isabella of France, wife of Edward II (popularly known as the she-wolf), and described by Power as 'the most disreputable woman of her day – her young life was a perfect hotch potch of lovers and murders and plots.'[21] It was Langlois, the eminent literary and social historian who had also taught Marc Bloch who directed her toward the biographical topic. Hopes of pursuing the thesis at the Sorbonne were soon dashed by lack of money, and she returned home in the summer to apply for teaching jobs.

Shortly after, Power was told about the Shaw Fellowship at the London School of Economics; she applied and was awarded it for two years. This paid her £105 a year, and she also did some research work for Hubert Hall. On this she kept a set of rooms big enough to accommodate her sisters, Rhoda and Beryl, both now at university, during their term breaks. Money was always short in the family; the house in Oxford had been given up, and one of her aunts was doing secretarial work in London.

Under the terms of her fellowship, Power's historical research now took on new dimensions. From a biographical portrait of a queen she moved to broader work on medieval women. The subject set for her under the fellowship was 'The social and economic position of women in England during the xiii or perhaps xiv century'. She eventually narrowed the subject down to research on medieval English nunneries.

The historians at the L.S.E. including Lilian Knowles had clearly targeted women's history as an important area of research to foster. The Shaw fellowship, endowed by Charlotte Shaw in 1904 went to women four out of six times between 1904 and 1915, and these included Marion Phillips, a founder of the Fabian Women's Group and a Labour Party activist, and Alice Clark who used the time to research her *Working life of women in the seventeenth century* (London, 1919). Women research students featured prominently at the L.S.E. at the time, and were awarded over a third of the research studentships between 1896 and 1932. There is no evidence as to why women's history was a particular interest to the selection committee in 1911, but the Fabian Women's Group which included Charlotte Shaw, Marion Phillips, Mabel Atkinson and B.L. Hutchins was just then discussing women's economic situation and its historical back-

---

19 Rosamund Lehmann, *Dusty answer* (London, 1927; 1936 edition), p. 107; Dorothy Marshall, 'Unpublished memoirs'; author's interview with Dorothy Marshall, Jan. 1992; Mary Llewelyn Daviews and Theodora Calvert, 'Memoirs' (Girton College, Cambridge, Archives).
20 Power to Margery Garrette, 6 Nov. 1910 (Power papers held by Lady Cynthia Postan).
21 Power to Garrette, 26 Mar. 1911 (Power papers held by Lady Cynthia Postan).

ground. Knowles was clearly aware of these debates, took them up in her lectures and later in her textbook.

Power's literary and biographical tastes and background had not quite prepared her for this, and she expressed reservations: 'I am extremely perturbed because the whole thing has turned out very much more economic than I expected.'[22] But she compromised by narrowing the subject down to the social history of the nunneries. Power's research then took another turn into women's religious life, under iconoclastic supervision from Cambridge by the renowned anti-clerical historian of monastic life, G.G. Coulton.[23]

The first four chapters were submitted for a London M.A., and Power eventually published her first major book on this work, *Medieval English nunneries*.[24] Power's debt to Coulton was clear in this, her major work of women's history. The limitations of this history and her later women's history arose out of this early connection with Coulton, but it was also Langlois and Coulton who fostered her deep knowledge of archival and literary sources. Coulton was a rabid anti-catholic historian of monastic life who chose to focus not on spirituality, but on the social history of the monasteries. He had a mission in life, and this was to dispel the myth of medieval golden ages, and he applied this not just to the lives of the monks, but also to the medieval peasantry.[25] Power absorbed Coulton's approach to social history, but not his combative historical style.

Power's book broke free of former historical traditions in manorial history and medieval religious thought to present a highly readable account of the wealth distribution of convents, their economic activities and division of labour, and the social backgrounds, daily lives and careers of the nuns. These subjects are only now beginning to appear on the agenda of social and feminist historians. The book was no romantic story of powerful women's communities. Most of the convents of the later middle ages were small and poor. Her history was a long tale of financial embarrassment and managerial incompetence. Small communities of upper class women maintained high proportions of servants, but contributed little in the way of education, learning or spirituality. The nuns of whom she wrote produced no saintly women or great mystics; there were no successors to the learned Anglo-Saxon abbesses. These women did not copy and illuminate manuscripts, and no nunnery produced a chronicle. Their role as educators of the young was as poor as was their own education. There was little sustained adherence to their vows, and their moral lapses were the commonplaces of medieval literature. This was not material out of which to make great history, and Power was a sympathetic, but critical observer of the convents. She did not, however, make any serious effort to examine the constraints on these women's lives set by the church itself, nor the reasons for the failure of the nuns to overcome or even to challenge these. Nor did she examine the impact of contemporary economic change which had exacerbated the problems of small landlords, a role which many convents fulfilled in the rural economy.[26] Eileen Power's negative assessment of women's religious commu-

---

22 Power to Garrette, 26 Mar 1911 (Power papers held by Lady Cynthia Postan.)
23 Coulton is best known for his *Five Centuries of Religion* (4 vols, Cambridge, 1923–50).
24 Eileen Power, *Medieval English nunneries, c. 1275–1535* (Cambridge, 1922).
25 On Coulton see Gerald Christianson, 'G.G. Coulton: the medieval historian as controversialist' in *Catholic Historical Review*, vol. 57 (1971), pp 421–441.
26 'A melancholy chronicle' in *New Statesman*, vol. 20, 27 Jan. 1923, p. 412; 'Medieval people' in *Times Literary Supplement*, 11 Sept. 1924, p. 551; Bertha H. Putnam, 'Medieval English nunneries' in *American Historical Review*, vol. xxix (1924), pp 538–9.

nities continued to influence the perceptions of generations of historians to come; for a long time there was very little else written on the nunneries. Recent research has, however, challenged her perspectives and her uncritical and incomplete use of the sources available.[27]

The book was to a large extent still in Coulton's mould. It avoided issues of women's spirituality, and dealt only obliquely with the power structures within which these women's communities operated. Power's debt to Coulton was clear to some reviewers.[28] Coulton had not only supervised her, but published the book in the series he edited. Her book was even announced as Coulton's in *The New Statesman*, eliciting Power's remark, 'a view of the comparative importance of mother and obstetrician which looks like another injustice to women'. One reviewer even erased her existence from the book, assuming it was another of Coulton's anti-catholic outpourings.[29]

As soon as she finished her *Nunneries* Coulton suggested she work on the *Menagier of Paris*,[30] and he remained her closest contact for bibliographic material, criticism of her work and testimonials for the next several years. With this, however, his influence upon her work ended. She adopted Coulton's choice of social history as well as his synthetic literary style for a wide reading public, but for quite different reasons than his. While Coulton's intellectual concerns remained historiographical debate and anti-clericalism, Power's turned increasingly to feminism, pacifism and education.

Power returned to Girton in 1913 as Director of Studies in History, and stayed there for the next eight years. She wrote her *Nunneries* book during this period, and wrote more broadly on women's medieval history, including her widely read articles, 'Medieval ideas about women', and 'The position of women' in G. C. Crump and E.F. Jacob (eds), *The legacy of the middle ages* (Oxford 1926). She was awarded the Kahn Travelling Fellowship which funded her for a year's travel around the world during 1920–21. While she was in China she received a letter from Beveridge offering her a job at the L.S.E.[31] The L.S.E. offered Power the scope she had been longing for. Girton, by the time of the war had become claustrophobic to her, and she yearned for broader pastures in London. She wrote in 1917, 'I don't believe I was ever cut out for this sort of life: I die within me month by month . . . Of course the war makes it worse, for Cambridge is an awful place to be in just now – no one under forty! . . . It seems such a short life for all the best years of it to be spent here.'[32] Shortly after she accepted the lectureship at the L.S.E. she wrote to her old advisor,

I do find the L.S.E. a more stimulating place to work in and London a more congenial place to live in than Cambridge. My idea of life is to have enormous quantities of friends or to live alone. And I do not know whether Girton or the study of medieval nunneries did more to convince me that I was not born to live in a community.[33]

Eileen Power subsequently built her career at the L.S.E., became a professor there in 1931, and in 1947 married her former student and research assistant,

27 See the research of Marilyn Oliva to be published in *Medieval Prosopography* (Spring, 1995).
28 See some of the reviews of *Medieval English nunneries*, for example, 'A melancholy chronicle', *op. cit.*; 'Medieval English Nunneries', review by Bertha H. Putnam in *American Historial Review*, vol. xxix (1924), pp 538–9.
29 Power to G.G. Coulton, 9 Oct. 1922 (Power papers, Girton College, Cambridge).
30 Power to G.G. Coulton, 23 Dec. 1920 (Power papers, Girton College, Cambridge).
31 W.H. Beveridge Papers (British Library of Political Science).
32 Power to Garrette, 10 June 1917 (Power papers in possession of Lady Cynthia Postan).
33 Power to G.G. Coulton, 30 Jan. 1922 (Power Papers, Girton College, Cambridge).

M.M. Postan. When Power first came to the L.S.E. she described herself as half way through a book on medieval women.[34] The book never appeared, but it was planned, and several chapters were written and given as lectures or published separately as articles over the years. She became well-known for her lectures on working women, bourgeois women, medicine and midwifery. Her studies of the menagier's wife and the medieval nun appeared later in *Medieval people*. Her translation and edition of *The Goodman of Paris*, and her work on the family of clothiers, the Paycockes, included extended discussions of courtship, marriage and middle class women's domestic lives. Her work on women over this period included research on servants, their wages, duties and supervision, on medicine and midwives, and on prostitution in medieval France and England. She planned for a major section of her book to be focused on marriage, including not only marriage vows, betrothal and property holding, but also wife-beating and murder. She also planned a large-scale research project on women and the family from the middle ages to the eighteenth century.[35]

Power took a very pessimistic view of women's position in the upper classes and in the religious life of the middle ages. She was much more optimistic on the participation of women in the labour market, and in their significance in managing the household and family affairs.[36] Power has since been criticised for her failure to set out adequately the extent of women's subordination in spite of their contributions.[37] This criticism has, however, focused on one paper, 'The position of women' in Crump and Jacob, *The legacy of the middle ages*, which Power herself dismissed as 'one of those gossips about social life which ought to be bought by the yard at a department store.' Her preferred article on medieval ideas about women had been rejected by the editors because it 'wasn't sufficiently respectful of women, the church and the proprieties.'[38]

Why Power never published a book on medieval women is not known. Her lectures on the subject became well known not just in academic circles, but in the wider literary culture throughout the 1920s. Ray Strachey recalled her excitement at the prospect of a lecture by Eileen Power on 'Many evil women', and her great disappointment when she arrived to discover it was on 'Medieval women'. Long after Power's death, some of her best-known lectures were collected and published in a popular form in *Medieval women* edited by M.M. Postan.

At the L.S.E. the predominantly literary, cultural and social framework of Power's history changed under the influence of the social sciences and the historian with whom she worked closely to create the economic and social history course at the L.S.E., R.H. Tawney. The major political influence on her work from this time was her great commitment to the peace movement, and from the mid-thirties, to the anti-appeasement movement. Women's history was no longer the focus of her attention except as a part of the social and economic history which she now extended to international and comparative history. Her initiative in developing these was her special contribution to the peace movement.

---

34 Power to William Beveridge, 26 June 1921 (Eileen Power Personal File (London School of Economics)).
35 Power Papers, Cambridge University Library.
36 Power, 'The Menagier's wife, a Paris housewife in the fourteenth century' in Power, *Medieval people* (1924), London, 1986, pp. 96–120; Power, 'The working woman', in Power, *Medieval women*, edited by M.M. Postan (Cambridge, 1975), pp 62–70.
37 J.M. Bennett, '"History that stands still": Women's work in the European past' in *Feminist Studies*, 14 (1988), p. 270; Olwen Hufton, 'Women in history: early modern Europe' in *Past and Present*, 101 (1983), pp 38–41.
38 Power to G. G. Coulton, 5 Sept. 1920 (Power papers, Girton College, Cambridge, Archives).

She argued that European and world history, especially social history would help to create a community with common historical ideas and with a sense of the likeness between nations. This was the framework for her most famous work, *Medieval people*, and for her article, 'A plea of the middle ages'.

Medieval people was Eileen Power's most famous work. It went into ten editions, and since that last tenth edition, has been reprinted three times, the most recent in 1983 with an introduction by Richard Smith. *Medieval people* was her application of a methodology which conveyed the ideal types of medieval social structure as the personal histories of selected individuals. It was also the literary synthetic history she espoused throughout her reviewing in the *Nation and the Athenaeum* and the *Times Literary Supplement*. Bodo represented peasant life on a typical medieval estate; Marco Polo, Venetian trade with the East; Madam Eglentyne depicted monastic life and the convent; the menagier's wife, domestic life and medieval ideas about women. Thomas Betson illustrated the wool trade and the merchants of the staple; and Thomas Paycocke, the cloth industry in East Anglia.

Power's pacifist politics in the years during and following the First World War were practised in the creation of a new type of history – one which was international and comparative, and above all, social history. In a lecture she gave on the League of Nations, and later published in *Economica* in 1922 under the title 'A plea for the middle ages' she argued the parallels between the corporations and communalism of early medieval society and new federal principles for the League. She denounced the rise of the modern state which was both too large 'to fulfil man's need for an intense civil and political life, and too small to represent the actual internationalism of his economic and cultural life.' She argued that the League needed not only a new theory of the state, but the education of public opinion.

History as at present taught in the majority of schools is ... mainly a record of the conflicts between states, and the story of the growing interdependence of states is never told ... A greater insistence on social history ... would impress almost imperceptibly upon their minds the points of likeness between the men of all nations.[39]

In *Medieval people* Power started with the simple Carolingean peasant Bodo, and with the newest of all the branches of history, economic history. And with these she made her famous proclamation for a history of ordinary people, and a turning away from the glories of war.

We still praise famous men, for he would be a poor historian who could spare one of the great figures who have shed glory or romance upon the page of history; but we praise them with due recognition of the fact that not only great individuals, but people as a whole, unnamed and undistinguished masses of people, now sleeping in unknown graves, have also been concerned in the story. Our fathers that begat us have come to their own at last. As Acton put it, 'The great historian now takes his meals in the kitchen.' This book is chiefly concerned with the kitchens of History.[40]

From this work Power turned from the mid 1920s increasingly to medieval trade and industry. The great *Cambridge economic history of Europe* which she initiated and the first volume of which she edited with J.H. Clapham was perhaps the greatest testimony to the analytical, comparative and international history

39 'A plea for the middle ages'.
40 *Medieval people*, p. 19.

which she fostered at the L.S.E. But by this time another war was in view, and from her internationalist stance she moved to warn her fellow historians against appeasement of Hitler. Her lecture in 1938, 'The eve of the dark ages' set the parallels between Hitler's expansionism and the road to the fall of the Roman Empire. 'If historians were less obsessed with an outworn fetish of progress, less convinced that it is always for the best, they might be of greater help to their generation. The men of the Dark Ages had no such illusion; they knew what they had lost and the memory of Rome haunted their wistful minds like the dream of a golden age.'[41]

Power died suddenly in 1940, aged only 51. She held a central place in the creation of the new discipline of economic and social history, and in making this form of history a major part of the national culture during the inter-war years. Women's history was a significant part of this. Eileen Power's rise to a prominent place in the academic elite was made in a field and in a university still in their formative phases. They had not yet become professionalised or narrowed in scope as they were to be in the years following the Second World War. There was a greater openness for new subjects and for a broadly based academic constituency including many women. Power's own success was made on the building blocks created by other women researchers and academics in her field. Substantial numbers of these women either gave up research or worked in menial academic positions. But others did achieve academic appointments and promotions in much higher proportions than those gained by women in any time since the Second World War and up to the present.[42] Power benefited from the other women in her field, and made great efforts to continue the process through fostering her research students and her seminar which included many women. These included several, men as well as women, whose early academic careers were scarred by poor exam results.[43] Yet her legacy, and memories of her achievements faded in the years following the Second World War. Her subject was narrowed. It lost its political framework and the women historians largely disappeared.

Eileen Power was remembered finally as a woman, for her face, her clothes, her charm and her wit, and her history was forgotten. R.H. Tawney is credited with the origins of social history, and as such features as one of the History Men. But the woman in history with whom he worked has now been all but forgotten.

---

41 Power, 'The eve of the dark ages: a tract for the times', unpublished lecture, 1938.
42 See Berg, 'The first women economic historians', pp 308–9.
43 These included her research assistant and future husband M.M. Postan, Eleonora Carus-Wilson and Elizabeth Crittall. For evidence of her efforts on their behalf see Eileen Power, Personal File (London School of Economics); M. Chibnall, 'Eleanora Mary Carus-Wilson, 1899–1977' in *Proceedings of the British Academy*, xviii (1982), 503–20; interviews with Marjorie Chibnall and Elizabeth Crittall, 1990, 1992.

3   'Scolding Old Bags and Whining Hags'. Women's History and the Myth of Compatible Paradigms in History

Maria Grever

E XPERTS in women's history are often confronted with the question of how long it will take until their field of study will become a truly integrated part of the historical sciences. The question implies that in time the separate approach will become redundant. 'When do you expect this "fashion" to be over', journalists and scholars regularly ask, 'in about five or ten years?' During a recent academic forum a woman historian in all sincerity declared that in five years' time women's history would be superfluous. When asked if she then would agree to abolish all the existing academic infrastructure of this field, she was at a loss for words.[1]

These questions, which essentially stem from ignorance and disdain, will most likely never be asked of other experts. What's more, established historians who argue for the eventual abolition of women's history as a separate field would probably be insulted if they were asked when their field will become superfluous and, therefore, extinct. Dealing with this line of argument requires presence of mind and full understanding of the underlying bias which women's history has to endure within the academic discipline of history. At first sight it seems reasonable to acknowledge that special attention for women should be a temporary necessity only, since history is traditionally conceptualised as a unified story about a universal subject. However, as Joan Scott has pointed out, this supposed universality is achieved through a process of differentiation and marginalisation, resulting in the exclusion of women, the poor and ethnic and other minorities.[2]

The history of women's history is a never-ending story of assessing this field of scholarly inquiry by two standards. Consequently, the differences in perception and validation between women's history and other historical fields in effect confirm and enforce the existing asymmetry between women's history and so called mainstream history.[3] The asymmetry is continued by the constant creation of new differences. The double standard with respect to women's his-

---

1 The academic forum was about 'use and uselessness of women's history', at the University of Louvain, 26 Feb. 1992. This question also came up during an interview I had with the Dutch VPRO radio on 19 Nov. 1992. I responded by asking the interviewer whether he would ask a professor in social history when this field of study would be abolished.

2 Joan W. Scott, *Gender and the politics of history* (New York, 1988), pp 178–198.

3 See for theoretical reflections on the double standard of morals in sexual relations A. Komter, *De macht van de dubbele moraal. Verschil en gelijkheid in de verhouding tussen de seksen (The power of the double standard. Difference and equality in the relations of the sexes)* (Amsterdam, 1990).

tory is manifest in the way historians formulate inconsistent critiques of women's history and pick and choose any real or perceived difference which they can come up with to 'settle' whatever issue they happen to be 'debating' at the time; experts in women's history would certainly not get away with this.

Most of these arguments are hardly rational but emanate from the bias of hostile scholars. My paper focuses on the reasons why blunt questions, irrational arguments and stereo-typed images about women's history exist. I hope to explain the double standard in the historical profession with respect to the evaluation of women's history. The effects of measuring this field by a different standard can be demonstrated in particular by efforts in the Netherlands to institutionalise women's history at academic institutions and within secondary education. I will link these experiences with the position of the first professionally trained women historians in the Netherlands. The Dutch case might offer a recognisable way to reflect on the epistemological implications of women's history and the sociological practices of professionals. The last section of this paper suggests that the connection between the double standard and the universality claim of the historical sciences is a plausible explanation for the exclusion of women in history.

Women's History in Dutch Secondary Schools

In the Netherlands women's history is flourishing. Feminist historians have been well organised in the Women's History Association since 1976. Many books on women's history have been published since then; forums, congresses and exhibitions have been organised.[4] In the past ten years various dissertations about new aspects of women's history have appeared, and attention has been devoted to basic premises that challenge the dominant historiography. At almost every Dutch university at least one woman historian has been appointed especially for women's history.[5]

Outside the academic world in Dutch secondary education, women's history was introduced as a required part of the final examinations in history, in itself an elective course.[6] This meant that in 1990 and 1991 over 80,000 pupils in lower vocational schools up to the gymnasia were taught by some 2500 teachers about twentieth century women's history in the Netherlands and the United States. The central question of the subject-matter was: 'In what way did the position of women change in relation to the family, their participation in the labour market as well as national politics during the economic crisis of the thirties, the Second World War and the building of the welfare-state?' Particular attention was paid to processes of continuity and change for the period circa 1929 to

---

4 See also Francisca de Haan, 'Women's history behind the dykes: reflections on the situation in the Netherlands', in Karen Offen, Ruth Roach Pierson and Jane Rendall (eds), *Writing women's history: international perspectives* (London, 1991), pp 259–77; José Eijt, 'Women's history: "the take-off" of an important discipline. Developments in the Netherlands and Belgium since 1985' in *Historical research in the Low Countries 1985–1990* (The Hague, 1992), pp 76–88.

5 In 1993 a women's history course became a required part of the main curriculum for all students at the history department of the University of Nijmegen. It is also possible to proceed to an M.A. in women's history in Nijmegen.

6 Carla Wijers and I, sustained by a network of women historians, wrote the subject matter of the women's history examinations. See Maria Grever, '"Pivoting the centre": Women's history as a compulsory examination subject in all Dutch secondary schools in 1990 and 1991', in *Gender and History*, 3, no 1 (Spring 1991), pp 65–80.

1969, in both the Netherlands and the United States. Pupils learned about feminism in the 1920s, the attacks on women's labour in the 1930s, and the attempts
by Eleanor Roosevelt and Mary Macleod Bethune to feminise the New Deal
and eliminate racism. They also studied how Dutch women resisted or collaborated with the Germans during World War II, how American women worked in
factories during the war, only to be thrown out again after the war was over.
Furthermore, the pupils were taught about the Kinsey Report on sexuality and
the rise of the second feminist movement.[7]

All this was not achieved overnight. A conservative member of parliament
and three teachers' unions objected. There was also stiff resistance from some
history teachers in permanent education courses. Some of them complained
that women's history had not been part of their college training and rejected
this 'feminist indoctrination'. But since the final history examinations are compulsory, they had no choice but to teach women's history. Some of these teachers rushed through their lessons, never letting their pupils take advantage of
the wonderful books and audio-visual materials produced for this examination.

By now, much of the disagreement has disappeared as the course has proved
its value. Teachers experienced vivid discussions in the classroom and girls in
particular showed great interest, although they were sometimes teased by boys.
Pupils were amazed at the discrimination which women endured only a few
decades ago. Many of them, as well as the teachers, were fascinated by the
history of women. Throughout the process the board of the Dutch Association
of History Teachers supported us to overcome prejudices among its members
and enabled us to establish a special committee for women's history within the
association. This committee can guarantee the integration of gender into future
history examinations and promote women's history within the overall curriculum of secondary schools.

After the examination had taken place, the Association of History Teachers
organised a survey among 400 teachers in thirty regions, the first of its kind.[8]
Because the subject was so controversial (everybody had been afraid that wom-
en's history would not be 'neutral' or 'scientific')[9] the association was curious
about the experiences of the teachers. The results of the survey showed that
teachers had become interested in women's history. Several high schools organised a Women's History Day and invited elderly women to talk about their
lives. A few teachers even delivered lectures on gender for their colleagues.

Most surprising were the results of the examination itself. The subjects of the
examinations before 1990 were focused on the political history of statesmen and
parliament. Although 60% of the 40,000 high school pupils who select history in

---

7 See also the course book for history teachers on the women's history topic: Maria Grever and Carla Wijers,
   *Vrouwen in de twintigste eeuw. De positie van de vrouw in Nederland en de Verenigde Staten van Amerika,
   1929–1969* (Women of the 20th century. The position of women in the Netherlands and the United States,
   1929–1969) (Ijsselstein, 1988).
8 'Verslagen van reakties op het geschiedeniseindexamen VWO (14 verslagen) en Havo (16 verslagen) (Report of reactions on the history examination of the two highest levels of secondary schools gathered from 30
   groups of teachers) published in *Kleio*, 31 (July/August 1990), pp 34–37; *Intern verslag resultaten van de
   eindexamens 1990 van het Cito (Report of the Central Testing Institute about the final examinations)* (Arnhem,
   1990), pp 23–24; see also the journal of the Dutch Ministry of Education *Uitleg* (October 1990), pp 16–17.
9 Various groups investigated the women's history examinations: scholars in history and women's studies, feminists and educationalists. They all had their opinions, even though they were either not familiar with the
   examinations procedures or did not know anything about women's history. For instance many journals and
   newspapers published in advance interviews and gave their opinions. (See educational magazines *Nieuw
   Zicht* and *School*, and feminist magazines *Opzij* and *Dinamiek*).

their school curriculum are female, and girls are more interested in history than boys, their history examination scores have always been lower than those of the male pupils on all four high school levels.[10] However, this changed when the two subjects were 'The position of women in the Netherlands and the United States, 1929–1969' and 'The Netherlands during the Second World War'. For the first time female pupils of the highest secondary education level (VWO) had (statistically significant) better grades for the whole history examination. Male pupils had better overall grades at the three other levels (Havo, Mavo-C en Mavo-D). Considering the two subjects, there was a remarkable difference in achievement: girls had better grades for women's history in three of four high school levels, whereas boys at all levels had better grades for the Second World War. Interestingly, women's history was the more difficult topic for all pupils, although this was more so in the case of the boys than the girls.[11] (see tables)

Table 1. Pupils who selected history for their examination curriculum (*Statistiek van het VWO, Havo en Mavo-eindexamen*. The Hague, CBS, 1985–1991)

| | VWO (highest) | | | Havo | | | Mavo (lowest) | | |
%	tot	male	fem	tot	male	fem	tot	male	fem
1984	44	37	52	34	29	38	32	26	37
1985	44	37	51	32	27	36	31	25	36
1986	45	38	52	31	26	35	31	26	35
1987	43	37	50	30	25	34	30	26	34
1988	42	36	48	31	26	34	29	25	32
1989	43	37	48	29	24	34	27	24	31
1990	43	38	48	28	23	32	26	23	29
1991	43	37	49	27	22	30	27	23	30

Table 2. Percentage of pupils with low grades in history (*Statistiek van het VWO, Havo en Mavo-examen*. The Hague, CBS, 1985–1991)

| | VWO | | | HAVO | | | MAVO | | |
%	tot	male	fem	tot	male	fem	tot	male	fem
1984	4	3	5	9	6	12	5	2	7
1985	5	3	6	11	8	13	6	4	8
1986	4	3	5	9	6	11	7	5	9
1987	5	4	5	9	6	10	6	4	8
1988	6	5	6	8	5	10	7	3	10
1989	7	6	8	11	7	14	13	9	16
CSE with women's history									
1990	2	2	2	5	4	5	8	5	10
1991	4	4	4	11	12	11	7	4	9

10 *Statistiek van het VWO, Havo en Mavo-eindexamen* (Den Haag, CBS, 1985–1991). See also Jan Krol, 'De beleving van het geschiedenisonderwijs door meisjes en jongens (How girls and boys experience history education at secondary schools)' in *Kleio, 29 (Kleio-katern*, 1988), pp 8–14.
11 The Dutch Central Testing Institute (CITO) took a random sample of 8,084 pupils' examination scores. (*CITO report*, 1990, 1991).

Table 3. Differences in achievement between boys' and girls' final written examinations in history 1990 and 1991 (*CITO report,* 1990, 1991, and unpublished data of the psychometric service of the CITO)

	Difference in average p'value between boys and girls		
	two subjects	**Women's History**	**WWII**
1990			
VWO	girls 1,2% + (s)	girls 2,9% + (s)	boys 0,6% + (ns)
HAVO	boys 1,7% + (s)	girls 0,9% + (ds)	boys 4,2% + (s)
MAVO-D	boys 2,0% + (s)	girls 1,8% + (s)	boys 5,9% + (s)
MAVO-C	boys 3,6% + (s)	boys 2,0% + (s)	boys 5,3% + (s)
1991			
VWO	girls 1,3% + (s)	girls 2,2% + (s)	girls 0,5% + (ns)
HAVO	boys 0,2% + (ns)	girls 2,2% + (s)	boys 2,6% + (s)
MAVO-D	boys 1,9% + (s)	girls 0,8% + (ns)	boys 4,8% + (s)
MAVO-C	boys 2,0% + (s)	boys 0,6% + (ns)	boys 3,3% + (s)

Explanation symbols

VWO	=	highest level in Dutch secondary education
MAVO-C	=	lowest level in Dutch secondary education
girls x% +	=	grades girls better than boys
boys x% +	=	grades boys better than girls
(s)	=	significant (valid sample taken at random: generalizations are valid for total population)
(ns)	=	not significant
(ds)	=	dilute significant

At this stage we cannot conclude that there is a causal relationship between the different subjects of the history examinations and the records of boys and girls, although the results indicate a positive correlation. The National Testing Institute is continuing to research the differences between the results of boys and girls. What seems clear, however, is that several history teachers have come to realise that gender bias in history education, both in content and presentation, does exist. All this seems to justify the continued presence of women's history in secondary education. More attention to this subject may match the historical interest of girls and improve their grades as well. That does not mean that the political history of the state is not important for them or that women's history is only relevant for girls. Teachers have to be aware of the possible differences in historical interest between girls and boys, and confront them at times with subjects which are more or less geared to them.

The women's history examination revealed unmistakably the male perspective of the history curriculum at secondary schools. The fact that no one had ever protested about the 'one-sided' male history nor worried about the difficulties girls had identifying with the material, shows the double standard with which historical research and history education is approached. The fear of a 'feminist bias' or resistance among teachers and their unions once women's history became part of the examination, shows how deeply this double standard is rooted – the only concern seemed to be the historical interest of male pupils.

Ambivalence in the Academic Field

Though women's history is an exciting field in the Netherlands, there are drawbacks and persistent obstructions as well. Despite all the research, the number of academic publications and efforts to integrate those results into popularised books, the curricula of universities, colleges and secondary education, there is no chair in women's history in the Netherlands. Recently new chairs have been established in the history of mentalities, the history of journalism and even in the history of castles.[12] There are ten chairs in women's studies, none of which has an assignment in a history department. Only the women's studies chair of the Belle van Zuylen Institute in Amsterdam includes women's history.

Moreover, at the moment several historians feel that the integration of women's history into the historical profession has gone far enough. In the discussion about the new attainment targets for history and politics in basic education (the age group between 12 and 15) of Dutch high schools some history professors have argued in prestigious newspapers that it is time to stop for fear of corrupting the field's high standards.[13] The objection to women's history appears to stem from a fear of losing control over the historical territory. In that sense, the double standard functions as a strategy to counteract the emergence of both women historians and women's history. This impression was confirmed when the Dutch Historical Association organised its first national conference on women's history.

This conference, on 16 April 1992, was attended by well over one hundred historians.[14] There were hardly any male members present. Unfortunately, the debate after the long morning programme lost itself in general statements about women's history, ignoring the interesting lectures. An already disparate debate became chaotic when the president of the Historical Association, who chaired the conference, argued that there were many black holes in history, and that, therefore, women historians should not 'whine'. 'More work has to be done everywhere', he said, ignoring fundamental complaints about male bias in historical research. The public became noisy and some women left the room. Despite the ambivalent feelings of many women historians about this conference, good contacts were established with the board of the Dutch Historical Association. It was agreed that more women should become members of the association's board and committees. A little later, a woman historian was invited to join the editorial team of the journal of the Dutch Historical Association. For the second time in the journal's long history a woman scholar became one of its editors.[15]

---

12 For instance, regular chairs in the history of culture and mentalities have been installed since the 1980s, as well as special chairs in the same field in Maastricht in 1990, and Nijmegen and Groningen in 1992. A regular chair in the history of journalism has been installed in Amsterdam in 1986 and a special chair in the same field in Rotterdam in 1991. A special chair in 'kastelenkunde' (the historical knowledge of castles and country-houses) was installed in Utrecht in 1992, sponsored by the Dutch Foundation of Castles.

13 Prof.dr. H. Righart (Un. of Utrecht) in *HP/De Tijd* (1992) and dr. A. van Hooff (Un. of Nijmegen) in *NRC-Handelsblad* (1993) and *Historisch Nieuwsblad* (1993) have supported this view.

14 The conference had the not very exciting title: 'Vrouwengeschiedenis en de "gevestigde" geschiedwetenschap: een ontmoeting' ('Women's history and the 'established' historical sciences: an encounter'; my translation).

15 The present journal of the Dutch Historical Association, *Bijdragen en Mededelingen betreffende de Geschiedenis der Nederlanden (BMGN)* was established in 1969 as a merger of journals founded in 1837, 1846, 1877 and 1946. See for its history J.C.H. Blom, 'Een eeuwfeest? Een verkenning ter gelegenheid van de laatste aflevering van deel honderd van de *BMGN*' ('The centenary of the journal of the Dutch Historical Association'), in *BMGN*, 100, no. 4 (1985), pp 576–87. The first female member in the editorial team of the BMGN was Johanna Kossmann-Putto from 1976–1981; since 1993 the second female member is Elsbeth Locher-Scholten. I am grateful to Gees van der Plaat for her information on this subject.

However, although the president supported female participation he still stuck to his negative view of the women historians at the conference. In his annual address to the Historical Association six months later he considered the women's history conference to a certain extent a failure. He thought the lectures were very interesting and regretted the small male attendance, but he insisted that the debate was a failure because of those 'scolding old bags from women's history'.[16] The audience – the majority of whom were male – burst out laughing. Nobody protested. The president's complaint about the absence of men at the women's history conference and his sexist sneerings illustrate again the power of the double standard. The presence of a hundred women historians – a unique event in the history of the Dutch Historical Association – proved to be insufficient in his view and that of many others. Success depends on 'established', that is male, recognition.

Expressions like 'scolding old bags' fit into a long tradition of ridiculing women as historians while upholding gender-neutral standards for the historical profession. That tradition is certainly not an exclusively Dutch phenomenon.

The Emergence of the Historical Profession and Women Historians

Bonnie Smith, Natalie Zemon Davis, Jacqueline Goggin and others have pointed out that the process through which the modern genre of history emerged in western society was shaped by the perceived differences between the sexes.[17] The growing number of publications on this subject is gratifying. Until recently the existing body of research into the history of historiography paid little attention to the gender bias in historical consciousness, historical writing, and consequently against women historians.

Even in 1991 the editors of the *Historikerlexikon. Von der Antike bis zum 20. Jahrhundert*, an international lexicon on 530 historians, stated: 'That only two women historians could be inserted, reflects the male preponderance in the historical profession'.[18] The women included are Hedwig Hintze-Guggenheimer (1884–1942) and Lily Ross Taylor (1886–1969). There are no entries for Catharine Macaulay, Marguerite Thibert, Eileen Power, Mary Ritter Beard, Alice Clark or Annie Romein-Verschoor. Although we might appreciate the editors' admission of male dominance, the almost total absence of women historians gives cause for concern since it perpetuates the invisibility of women in history both as subjects and objects.

In my own research on the historian Johanna Naber and Dutch historiography (1860–1948) I found that gender boundaries within the historical field emerged with its professionalisation.[19] In fact the discipline was constructed without

---

16 P.W. Klein, 'Jaarrede van de voorzitter van het Nederlands Historisch Genootschap (. . .), 23 oktober 1992' ('Annual adress of the president') in *BMGN*, 108, no. 1 (1993), pp 186–9. See for his critique on the women's history conference (on request of the board here without the statement about the 'scolding old bags'), p. 186.

17 Scott, *Gender and the politics of history*, pp 178–98; Peter Schöttler, *Lucie Varga. Les autorités invisible* (Paris, 1991); Natalie Z. Davis, 'Women and the world of the Annales', in *History Workshop*, 33 (Spring 1992), pp 121–37; Jacqueline Goggin, 'Challenging sexual discrimination in the historical profession: Women historians and the American Historical Association, 1890–1940', in *American Historical Review* 97 (June 1992), pp 769–802; B.G. Smith, 'Historiography, objectivity, and the case of the abusive widow', in *History and Theory* (December 1992), pp 15–32.

18 R. vom Bruch und R.A. Müller (eds), *Historikerlexikon. Von der Antike bis zum 20. Jahrhundert* (München 1991), p.ix. My translation.

19 Maria Grever, 'The historical representation of spinsters by Johanna Naber', in Margret Brügmann a.o. (eds), *Who's afraid of femininity? Questions of identity* (Amsterdam, 1993), pp 75–86; and my forthcoming book, *Struggle against silence. Johanna Naber (1859–1941) and women's voice in history* (Hilversum, 1994).

women and sometimes in direct opposition to them. First of all, the professionalisation process meant an increased use of records and archives, access to which was often difficult for women. Secondly, the new emphasis on the use of empirical evidence in historical writing dramatically lowered the stature of the historical novel which was an important genre through which women expressed their historical interest. Thirdly, both academic associations and universities explicitly excluded women, as the famous writer and historian Trui Bosboom-Toussaint discovered in the 1860s and 1870s when she was not granted membership of two academic associations. Apart from this, women's colleges were unknown in the Netherlands unlike in many other countries. Until the last decade of the century, Dutch women historians did not receive any professional training outside the home. Meanwhile, historical writing was looking for academic recognition. This was eventually achieved in 1921 when history became a separate field of study at Dutch universities.

Simultanously Dutch women were emancipated and finally obtained their constitutional civil rights in 1922. In the first half of the twentieth century 81 Dutch women completed doctoral dissertations on historical subjects and sixty per cent of these became members of the Historical Association. Though women gained access to the academic community, the historical profession remained largely dominated by men and there is little evidence that they supported their female colleagues. The Historical Association, founded in 1845, showed hardly any interest in these women historians. There are no records that demonstrate a formal policy of excluding them (unlike the Society of Dutch Literature and the Provincial Society of Utrecht which did not allow women to become members until the 1890s and only then after heated debates). In 1901 three non-academic women historians (one of them was Naber) appeared on the Historical Association's membership rolls for the first time. Apparently, they were admitted without any discussion. Most of the female members however left after two or three years. To date only one woman historian, dr. Johanna Oudendijk, has been a board member of the Association (1957–1959).

Despite the access of women to the historical profession, inequities persisted. All the women historians – married or single, with or without children – occupied the lowest ranks within the profession. Only the Dutch Economic History Archive (NEHA), an association founded in 1914, allowed two professional women historians on its board. This new historical field offered more opportunities for women as Maxine Berg's research on women economic historians has also suggested.[20] As for the universities, dr. Elizabeth C. Visser (1908–1986) was in 1947 the first woman historian to be appointed at a Dutch university. She became a full professor in ancient history.

As the social practices of the historical profession and the validation of historical knowledge are intimately related, the organisational invisibility of women affected their marginality as historical subjects and resulted in the exclusion of women from history itself. History consolidated its authority by narrating the story of state power to which women had frequently no access. Political history is obviously still a dominant field of study. Moreover, the construction of 'significant history' by leading male historians implied the projection and confirmation of their existence as citizens, a privilege from which all women had been formally excluded until they won the vote.

---

20 M. Berg, 'The first women economic historians', in *Economic History Review*, xlv, no. 2 (1992), pp 308–29.

Dutch historians of the nineteenth century in particular were involved in the history of politics, which was actually interpreted as the history of the nation. They were eager to overcome the lack of national identity in the Netherlands, trying to present a 'reality' that would supersede the existing conflicting political interests of liberals, Roman catholics, protestants and socialists.[21] They argued that historical writing should be impartial and religious bias avoided. The process of professionalisation in the Netherlands was, therefore, stimulated by the political objective of mainly the liberals to achieve a peaceful co-existence between all parties.

Until the nineteenth century the historian was usually an amateur, an erudite person or a writer. In the course of professionalisation history started to be practised by members of a well-defined profession, that followed generally accepted rules and methods. They turned away from the rhetorical and literary historical writing of their predecessors and concentrated on research. The professional claims of history were guarded by men who 'served as the gatekeepers for the profession'.[22] The 'true' historian was male, and by definiton impartial. He faithfully presented the past and was history's spokesman, not of a particular party, sex or class, but of society as a whole. Women were not capable of 'truthful historical writing', as the Dutch historian, P.J. Blok argued in 1907:

It is my experience that the female brain misses impartiality and open-mindedness. It is by nature too partial, too superficial for historical writing, and bent upon one single fact instead of the cohesion of facts. (. . .) As an assistant archivist or an assistant librarian a woman can be very helpful to collect and classify data. However, I do not think she is suited for independent historical study of a high ranking scientific nature.[23]

This overtly sexist attitude was sustained in a metaphorical way. The historian was not only a white male professional, but he was also a conqueror or liberator of the records in archives. He was even a Prince Charming, according to the historian G.W. Kernkamp. In 1895 he compared the hidden history of trade with the 'sleeping beauty in the forest who is waiting for the wake-up kiss of the researcher'.[24] The male gatekeepers interpreted research and historical writing as a male enterprise, and referred to it in terms of fame, honour and eroticism. Curiously enough, although Clio is a female muse who had to be conquered, liberated or awakened it was mainly the history of men and male institutions that was revealed. And this representation claimed to reveal the truth of 'the universal history'.

In this male dominated intellectual atmosphere of the interbellum, professional women historians tried to express their historical fascination by studying archival sources, publishing history books, and organising historical expositions. Some of them wrote histories of women and feminism. As a result the International Archive for the Women's Movement was founded in Amsterdam in 1935. This Women's Archive functioned as a resort for research in women's history as women historians continued to be marginalised and the first stirrings of a feminist historiography were muted. In contrast with social and economic history, the breakthrough of a woman oriented history in the Netherlands failed at

---

21 Jo Tollebeek, *De toga van Fruin. Denken over geschiedenis in Nederland sinds 1860 (The robe of Fruin. Reflections on history in the Netherlands since 1860)* (Amsterdam, 1990).
22 Goggin, 'Challenging sexual discrimination in the historical profession', p. 780.
23 P.J. Blok, 'Vrouwelijke studenten' ('Female students'), in *Onze Eeuw (Our Century)*, 7 (1907), pp 447–62. All translations are mine.
24 Tollebeek, p. 138. Cf. Smith about Ranke in 'Historiography', p. 18.

the time. The concept of man as the representative human subject in the historical field was hardly susceptible to the kind of pluralisation that would include women.[25]

The 1970s brought a new generation of women historians and the emergence of women's history as a new historical field. In comparison with the pre-war period and the 1950s and 1960s, the position of women historians improved enormously. A far greater number of Dutch women historians now have appointments at universities in various fields, many of them in women's history. Apart from offering women more employment opportunities, this field also functioned as a general breeding ground for women historians. Several women went from their assignments in women's history to other subjects and now have tenure.

The Dream of a 'Complete' History

However, in the Netherlands many historians seem to have been afflicted by the dream of a 'complete' history, a story of universal reality which must tell us everything that is important to know. In my view this assumption is one of the main reasons why the women's history examination project met such persistent resistance and why women's history as a field is viewed with contempt by professional (female and male) historians. To them women's history signifies a peculiar subject that lies outside the realm of the established discipline. Or they deal with it superficially in terms of 'restoring women to history', like adding castles to medieval history. The underlying assumption is that women's history in itself can never be valid because it fails to narrate the 'whole story'. What is at stake is the refusal to accept the particularity of all human subjects. We notice here the power of the double standard within science, i.e. the common belief in the universal nature of the male perspective and the exclusion of the female as specific.

The effect of this double standard is an antipathy towards women's history or, as is more often the case, an incredible indifference towards it. This is illustrated by the ignorance of many historians concerning the field of women's history. They have read few of the recent journals, yearbooks or dissertations on women's history and cultivate images from fifteen years ago. A persistent 'misunderstanding' is that women's history is identical to the history of emancipation. Many 'outsiders' argue as if experts in women's history are constantly searching for proto-feminists in the past, as if we are not trained to detect tricky anachronisms. A professor in history in Utrecht became bored with women's history because it was a never ending story about 'liberating the oppressed'. He and others accused women's history of applying contemporary standards to the past (hodiecentrism). According to them there is no room for a romantic sense of the past in this politically committed field of women's history.[26]

The historian Bastiaan Bommeljé published an extensive review article in a prestigious newspaper about six books on women's studies, two of which focus on women's history. He concluded that the flourishing of women's studies is

---

25 Scott, *Gender and the politics of history*, p. 192.
26 Comment of Righart in *Dinamiek* 8 (1991), pp 12–13; J. Tollebeek, 'De drang naar hoger leven. Geschiedbeoefening en fin de siècle in Nederland' ('The urge for a higher life. Historical writing and fin de siècle in the Netherlands'), in *Tijdschrift voor sociale geschiedenis* 17, no. 3 (Aug. 1991), pp 249–70.

based on quicksand. Having a weak theoretical basis it does not constitute a discipline and there are, therefore, no starting-points to legitimate a separate approach. According to him the concept of gender is vague and there is no agreement concerning a definition. The theoretical debates are too abstract and consequently make women's studies useless and an enormous waste of government money. He appreciated that there were some excellent scholars in women's studies but thought it was a pity that they did not work in regular departments which would allow them to escape the climate of 'mutual admiration' and 'scientific in-breeding', as well as the 'sectarian character of women's studies'.[27]

Bommeljé's article is an interesting example of double standards. There is hardly consensus about the concept of class in social history, neither do social historians agree about the boundaries of their field. Professors in the history of journalism also recently questioned the principal content of their field.[28] So why is women's history singled out for its alleged 'vagueness'? Why is a separate approach of gender not legitimate? Why are journals on women's history often considered as 'sectarian periodicals' unlike those of other historical specialisms?[29] A separate approach guarantees attention for gender and stimulates intellectual creativity. Furthermore, the debates in women's history indicate the liveliness and the autonomy of the subject, and demonstrate that gender is a dynamic concept and does not simply have a singular meaning. The challenge is to employ gender as a historical category: historical with respect to the view of the historian and the changing relationship between the sexes in the past.

Finally, there are both external and intrinsic criteria for assessing women's history as a separate field. There are academic historians who consider themselves experts in women's history, who publish books on women's history for an academic community and a large national and international public. In addition to the expanding infrastructure, the field of women's history has developed a specific historical critique of science, specific theoretical notions and its own shifts of perspectives. For instance, whereas in the 1970s the notion of oppression was dominant with 'woman' and 'man' as supposedly coherent categories, the construction of these categories as cultural products in various contexts of meaning have been examined in the 1980s and 1990s.

The number of historians who refuse to acknowledge gender as an important historical category is at times discouraging. On a deeper level they may be reluctant to recognize that the different voices of women, blacks and the poor cannot be reduced to a single narrative. From this point of view women's history represents a different paradigm because it critically analyses categories often taken for granted, such as science, profession, history, progress, women, men, equality or difference.[30] The indifference towards gender calls to mind the

---

27 B. Bommeljé, 'Het misverstand vrouwenstudies' ('Women's studies, a misconception'), in *NRC*, 28 Mar. 1992. In his article, women's studies as a discipline and women's history as a subfield are taken together. See also his response to feminist critiques on his article in *NRC*, 11 Apr. 1992.
28 G. Borst and M. Prenger, 'In de loopgraven van persgeschiedenis' ('In the trenches of the history of journalism'), in *Historisch Nieuwsblad*, 1 (1992), pp 6–11. This article was an interview with four specialists in the history of journalism, two of whom are professors: Jan Blokker, Joan Hemels, Gerard Mulder and Frank van Vree.
29 See for instance the reviews on 'gender history' in *Historisch Nieuwsblad*, 1 (1992), p. 29.
30 See Scott, *Gender and the politics of history*, p. 196.

incommensurability thesis of Thomas Kuhn.[31] In his view adepts of different paradigms are generally not capable of convincing each other because the arguments advanced to support a paradigm always contain non-rational elements that go beyond logic or empiricism. A paradigm deeply influences the point of view of a scholar analysing scientific problems as some facts are crucial within one constellation while others are irrelevant or do not even exist. Because paradigms are ultimately based in metaphysics, they will by definition be incompatible, and debates between incompatible paradigms can only go so far. The world and its past can and should, therefore, be seen in different ways.

In his second edition of *The structure of scientific revolutions* (Chicago, 1970) Kuhn admitted that paradigms sometimes overlap, and that some of these may be incompatible but certainly not all of them. In addition he stated that the 'conversion' moment of scientists, when they switch to a new paradigm, is exaggerated. Yet, he maintained the role of normative judgements, especially when a choice between competitive scientific approaches is at stake. It seems to me that the transition from one paradigm to another always requires readiness to embrace the new paradigm, otherwise the research generated within the new framework will never be fully understood.

Because of a continuing inertia within the historical sciences toward gender, experts in women's history will have to demonstrate a great deal of perseverance to get acknowledgment for their field. Trying to convince historians who reject the gender paradigm in advance is a waste of energy. Therefore, we should concentrate on publishing high quality books on women's history while building a strong infrastructure for the field, and influencing research programmes, itself a tremendous task. We will then continue to experience the sensation of reconstructing women's histories from the archives and subsequently subverting male-made dichotomies such as 'scolding old bags' versus 'true historians'. I am sure that women historians of the past, such as Eileen Power and Johanna Naber, can inspire us in the way they dealt with the gendered double standard in the historical profession.[32]

---

31 Thomas Kuhn, *The Structure of scientific revolutions* (Chicago, 1970, 2nd ed.); K. van Berkel, 'De historisering van de wetenschapsfilosofie' ('The historizing of the philosophy of science'), in *Tijdschrift voor Geschiedenis (Journal of History)* 101 (1988), pp 525–45, p. 528.

32 I would like to thank Wim Roefs for his critical reading of this paper, and Angélique Janssens and Joke van der Leeuw-Roord for their steadfast support. I also benefited from the thoughtful comments of Berteke Waaldijk which helped me sharpen my argument. I am grateful to my students of the women's history seminar in Njimegen in 1993.

4 Late in the Field: Catholic Sisters in Twentieth-Century Ireland and the New Religious History

Margaret MacCurtain

THE most significant development of the 1980s has been the emergence of women's history. Its acceptance by the professional history establishment is evident from the theme of this, the 21st Irish Conference of Historians: 'The history of women'. Its entry into, and its reluctant acceptance by the world of academia in Ireland have given rise to new ways of looking at primary sources, innovative methods, and an interest by a reading and listening public that is unprecedented, certainly in my long span of recollection.[1] No summer school, no third level or university department of history can afford to ignore the treatment of women as an important field of studies. Yearly the volume of new scholarship angled on women accelerates as graduate schools take on the task of guiding theses in women's history, often without the training and specialised skills needed for this rapidly expanding field. High on the list of new tools is the issue of gender as a category of analysis and to it may be added in the 1990s those of class and ethnicity in the redefinition of 'difference'. A feminist analysis of power yields a different configuration of the past and in general, according to Linda Gordon in a recent survey of women's history in the United States, 'most women's historians consider themselves social historians, focusing more on private than on public experience, more on informal than on official sources of power'.[2]

Turning to catholic sisters, the chronology of writing twentieth-century nuns into Irish history is full of surprises. A cluster of studies in the first years of the twentieth century focused on subjects such as Foxford and its woollen mills founded by Agnes Morrogh-Bernard, Mother Arsenius,[3] and Irish convent industries including lace and linen weaving.[4] Then, somewhat later, Helena Concannon, wrote several studies of the Poor Clare nuns of Ireland establishing securely the spiritual traditions of contemplative nuns in Ireland.[5] Her work

---

1 Maria Luddy, Margaret MacCurtain and Mary O'Dowd, 'An agenda for women's history in Ireland, 1500–1900' in *Irish Historical Studies*, 28, no. 109 (May 1992), pp 1–37.
2 Linda Gordon, *U.S. women's history* (American Historical Association, 1991), p. 5.
3 Denis Gildea, *Mother Arsenius of Foxford* (Dublin, 1936); Bernie Joyce, *Agnes Morrogh-Bernard, 1842–1932, foundress of Foxford Woollen Mills* (Ballina, 1991).
4 Rosa Mulholland, 'Skibbereen Convent of Mercy linen weaving' in *Irish Monthly* 18 (March 1890) pp 145–8; Jane Houston-Almquist, *Mountmellick work* (Mountrath, 1985); Nellie Ó Cléirigh, *Carrickmacross lace* (Mountrath, 1985); see also, Alan S. Cole, *A renascence of the Irish art of lacemaking* (London, 1888); William T. McCarthy (ed.) *Irish rural life and industry* (Dublin, 1907).
5 Helena Concannon, *The Poor Clares in Ireland* (Dublin, 1929); 'Historic Galway convents, I – Poor Clare' in *Studies*, 38 (December 1949), pp 439–446.

contributes in a scholarly method to a history of Irish female spirituality. Two major biographies of foundresses, one by Roland Burke, S.J., *A Valiant Dublin woman: the history of George's Hill* (1940) and one by Thomas. J. Walsh, *Nano Nagle and the Presentation Sisters* (1959), exemplified the tenets of catholic historiography of the mid-century: namely, that historical biography, especially of those who were leaders, was 'the key to unlocking the past, and the object of this exercise was to understand more fully the history of the institution'.[6]

In 1965 Angela Bolster published her Ph.D. thesis 'The Sisters of Mercy in the Crimea', a revisionist study that upset admirers of Florence Nightingale. Bolster, with her emphasis on documentary sources, went on to edit *The letters of Catherine McAuley* and she has been one of the main influences behind the steady stream of postgraduate studies of Mercy Sisters since the seventies.[7] At least fifty post-graduate theses dealing with the role or activities of Irish religious women in the twentieth century have been researched in various university departments over the past fifteen years. Of these, five dealt with the Mercy Sisters in different regions of Ireland: four with the Presentation Order of nuns, three with the Loreto Institute, two with the Sisters of the Holy Faith, one each with the Dominicans, Ursulines, Sisters of St Louis, Poor Clare teaching sisters. Several more theses, including the Ph.D. dissertation of John Coolahan, professor of education at Maynooth College, treated the catholic sisters in the context of training colleges, and several more examined catholic sisters in their involvement with workhouses.[8] Biographies have appeared in the 1980s of Mother Kevin Kearney, founder of the Franciscan Missionary Sisters, Mother Mary Martin, foundress of the Medical Missionaries of Mary in 1937, and the nun of Calabar, Mary Charles Walker.[9] At undergraduate level sixty or more special studies have examined aspects of sisters' ministry, such as the work of the Salesian Sisters in Limerick; the St Louis teaching ministry in rural domestic economy schools; the Hospice Movement; and the Little Company of Mary.[10]

Then come critical published studies of which Joseph Robbins *From rejection to integration: A century of service of the Daughters of Charity to persons with a mental handicap* was a narrative study based on primary sources, whereas Mavis Arnold's *Children of the Poor Clares, the story of an orphanage,* looked at the darker side of orphanage life and care in the mid-century in a case-study of the fire disaster that consumed the lives of young orphans in an Irish country town.[11]

Given the solidity of the research which this author has sketched in a rudimentary fashion, a new perspective of how to record what has happened to catholic sisters in the twentieth century is coming into focus: the present and

---

6 Roland Burke, *A valiant Dublin woman: the story of George's Hill, 1766–1940* (Dublin, 1940); Thomas J. Walsh, *Nano Nagle and the Presentation Sisters* (Dublin, 1959); Jay P. Dolan, 'New directions in American catholic history' in Jay P. Dolan and James P. Wind, *New dimensions in American religious history* (Grand Rapids, 1992), p. 153.

7 Angela Bolster, *The Sisters of Mercy in the Crimean War* (Cork, 1965); *Catherine McAuley in her own words* (Dublin, 1978); *The correspondence of Catherine McAuley, 1827–41* (Cork, 1989).

8 Sheila Lunney, 'List of twentieth-century theses on women religious orders in Ireland at post-graduate and under-graduate levels' (forthcoming for the Irish Association for Research in Women's History bulletin, 1995).

9 Sheila O'Hara, *Dare to live. Mother M. Kevin Kearney* (Dublin, 1979); Mary Purcell, *To Africa with love: life of Mother Mary Martin* (Dublin, 1987); Colman Cooker, *Mary Charles Walker: the nun of Calabar* (Dublin, 1980).

10 Lunney, *op. cit.*

11 Joseph Robbins, *From rejection to integration: a century of service of the Daughters of Charity to persons with a mental handicap* (Dublin, 1992); Mavis Arnold, *Children of the Poor Clares, the story of an orphanage* (Belfast, 1985).

next generation of scholars are asking different questions of the past and are looking for wider sources and new methodologies in an attempt to answer questions, free now from the burden of writing apologetic and promotional church history (which was what the old approach required). What has changed in the last decade is the perspective of the trained historians of women using new as well as old tools of research.

The origins of the new religious history are linked with the developments of the 1970s. Great changes within the history profession reflected the growing acknowledgement of new methods and approaches to doing history. This was mainly brought about by the refinement of analytical, theoretical methods and with the introduction of computer technology and more efficient means of assessing quantitative material. Latterly with the techniques involved in information systems research, the use of multiple, qualitative methods for a particular research problem is at hand for collecting data. The work of Dr Eileen Trauth, business historian, in her exposition of the methods used for assessing the role of societal factors in Ireland's progress from an agrarian to an information society combines data collection methods, documentary analysis, and skills for designing a questionnaire and a set of questions for oral interviews based on triangulation and reflexivity. These tools of research, which can now be performed with the aid of computer analysis, may sound depressing to the older historian like myself but as Dr Eileen Trauth observes: 'the underlying assumption . . . is that the research problem should drive the choice of methods used'.[12]

The creation of the Social Science History Association in the 1970s in the United States does not have a counterpart in Ireland which may explain why influential studies analysing religious belief and behaviour such as Dr Máire Nic Ghiolla Phádraig's analysis of church-going practices;[13] Dr Tom Inglis's study *Moral monopoly: the catholic church in modern Irish society*;[14] Dr Tony Fahy's Ph.D. dissertation, 'Female asceticism in the catholic church: a case study of nuns in Ireland in the nineteenth century', are coming from the discipline of social science.[15] At this point of time, it is generally acknowledged that the rise of feminist history and the parallel study of gender history in the 1980s has influenced historians to seek connections between religious and non-religious variables. A feminist approach to a historical problem studies the group, the life-cycle, the recurrent. For example, Caitriona Clear in her *Nuns in nineteenth-century Ireland* took a sample of nine convents and examined nuns as women in their social and economic contexts.[16] Dr Jacinta Prunty, in her study of Holy Faith Sisters and their ministry of the poor of Dublin moves the scrutiny away from an individual as dramatic as Margaret Aylward, foundress of the Holy Faith Congregation, and examines the forces and interest-groups of women

---

12 Eileen M. Trauth and Barbara O'Connor, 'A study of the inter-action between information technology and society: an illustration of combined qualitative research methods' in Hans-Erik Nissen, Heinz K. Klein and Rudy Hirschein, *Information systems research, contemporary approaches and emergent traditions* (Amsterdam, 1991), p. 142.

13 Máire Nic Ghiolla Phádraig, 'Religion in Ireland: preliminary analysis' in *Social Studies*, 5, no. 2 (1976), pp 113-80; 'Religious practice and secularisation' in Pat Clancy, Sheila Drudy, Kathleen Lynch and Liam O'Dowd (eds), *Ireland: a sociological profile* (Dublin, 1986), pp 137–54.

14 Tom Inglis, *Moral monopoly: the catholic church in modern Irish society* (Dublin, 1987).

15 Tony Fahy, 'Nuns in the catholic church in Ireland in the nineteenth century' in Mary Cullen (ed.), *Girls don't do honours: Irish women in education in the nineteenth and twentieth centuries* (Dublin, 1987), pp 7–30; 'Female asceticism in the catholic church: a case-study of nuns in Ireland in the nineteenth century' (unpublished Ph.D. thesis, University of Illinois at Urbana-Champaign, 1982).

16 Caitriona Clear, *Nuns in nineteenth-century Ireland* (Dublin, 1987).

involved in the struggle for children's souls and welfare with some pertinent analysis of population and housing trends.[17]

Professor Margaret Susan Thompson, historian of American catholic nuns, author of many monographs and of the eagerly-awaited book, *The yoke of grace: American nuns and social change, 1808–1917*, distinguishes the new feminist religious history from traditional historiography by the following characteristics:'its critique of patriarchy, its analysis of the connection between ordination and power, its recognition of the pervasiveness and importance of unordained ministry and the roles of the laity, and fourthly its identification of transdenominational patterns.'[18] This is a challenging list for Irish historians schooled in the empirical pastures of the formal document. How does it work in practice? Keeping within the framework of 'new' religious history the problem of sources presents itself. Linda Gordon in her essay 'What's new in women's history' suggests that 'the question of what counts as evidence is far more substantive than methodology.' She continues in tones reminiscent of Marc Bloch: 'I consider as evidence material once thought of as outside history – gossip, menstruation, latrines.' She is, of course, in that essay arguing that rather than indulging in a tolerant acceptance of difference, historians 'integrate that experience as part of our whole approach to the study of women.'[19]

Of its nature, the new religious feminist history requires inter-disciplinarity. It cannot rely only on quantitative methodology or statistical data. Not only are these less available for women than for men but ecclesiastical archives in general preserve information on men and tend not to record women unless they have transgressed canon law. The reports of Margaret Anna Cusack, the nun of Kenmare that repose among the Kirby papers in the Irish College Archives in Rome are a prime example: the depositions concerning her eccentric behaviour and actions as filed in Propaganda Fide Archives in Rome which have been examined by Sr Avril Reynolds have proved to be incomplete, if not one-sided.[20]

When finishing doctoral research on aspects of the Irish Counter-Reformation thirty years ago, this author had to make a case for the validity of convent archives, setting them apart from the authoritative state archives and ecclesiastical archives used: the Public Records Office in London, the archives of the various ministries of foreign affairs in Madrid, Paris and Lisbon, Simancas in Castille and the Archivio Segreto Vaticane in Rome. The convent archives appear small and unimportant in the description of sources used. Yet at the oral examination of the completed thesis, the archives of Bom Sucesso Convent in Lisbon, founded in 1639 for Irish-born nuns exiled from Ireland was the area that aroused most interest.[21] The archives of religious congregations possess records of unique value for the historian. All convents are obliged by ecclesiastical law to appoint an annalist who may also be the archivist of the community.

---

17 Jacinta Prunty, 'The geography of poverty, Dublin 1850–1900; The social mission of the church with particular reference to Margaret Aylward and co-workers' (unpublished Ph.D. thesis, University College Dublin, 1992).

18 Margaret S. Thompson, 'Women, feminism, and the new religious history: catholic sisters as a case-study' in Philip R. Vandermeer and Robert R. Swierenga (eds), *Belief and behaviour: essays in the new religious history* (New Jersey, 1991), p. 138.

19 Linda Gordon, 'What's new in women's history' in Ineja Gunen (ed.), *A reader in feminist knowledge* (London, 1991), pp 78–9.

20 Avril Reynolds, 'From loneliness to loneliness: Margaret Anna Cusack, the nun of Kenmare' (paper presented to the Society for the History of Women Conference in University College, Dublin, February 1990).

21 Margaret MacCurtain, 'Women, education and learning in early modern Ireland' in Margaret MacCurtain and Mary O'Dowd (eds), *Women in early modern Ireland* (Edinburgh, 1991), pp 169–70.

Many communities have an unofficial or even a published history of their convent. Convent annals are mainly a chronology of the major events that occur in
the life of that community: they also chronicle personal details of the sisters
and their relatives, benefactors, distinguished or unusual visitors to the convent. Where a convent is responsible for the management and staffing of schools,
asylums, custodial institutes, hospitals, hostels and halls of residence, occasionally letters and less formal observations find their way into convent archives. A
second category of material in convent archives is correspondence. Mary
Peckham, in the course of her research on the re-emergence and early development of women's religious orders in Ireland from 1700 to 1870,[22] described for
the summer-school of the Association of Religious Archivists in Ireland the
kind of source material she came across: 'I have found contemporary accounts
of such major historical events as the near landing of the French in Bantry Bay
in December 1796, of the exile of women religious from nineteenth-century
France and their search for new homes in Irish convents, of Irish emigration, of
the founding of missions throughout the world and moving accounts of the
great Famine . . .'[23] The 1989 *Draft directory of the Irish religious archives* compiled for the members of the Association, lists the material available in the
archives of religious houses.[24] In the present century, Irish congregations of religious women worked with the Irish and British governments and with governments and civil service departments throughout the world. For much of this
century the involvement of female religious orders with state projects, education, hospitals, social welfare has been of staggering proportions: the closing in
1986 of Cariesfort Training College in Dublin managed by the Irish Sisters of
Mercy was the closing of a chapter of intense collaboration between the state
and religious women.[25] Elites and power players have not been examined to
any great extent in twentieth-century Ireland. Leaders of women's religious
orders were both, but that is not the stuff of feminist history. As Caitriona Clear
remarks:

The archives of religious congregations have much to offer the history of women, the
history of work and work practices. They throw light on the history of the family in
Ireland: it is notable that several siblings very often entered the religious life, often the
same congregation.[26]

From the historian's angle, one of the most significant developments has been
the changed attitude on the part of many religious communities towards the
preservation and administration of religious archives. Over the past decade the
summer courses organised by Miss Ailsa Holland and Mr Seamus Helferty of
the Archives Department, University College Dublin, have effected a dramatic
transformation in the preservation of religious archival material. According to
David Sheehy, archivist of the Dublin archdiocesan archives, religious archives

---

22 Mary L. Peckham, 'Catholic female congregations and religious change in Ireland, 1770–1870' (unpublished
   Ph.D. thesis, University of Wisconsin-Madison, 1993).
23 Mary L. Peckham, 'Religious archives: a rich source of social history' (paper presented to the Association of
   Religious Archivists of Ireland, Dublin in July 1989).
24 The Association of Religious Archivists of Ireland compiled a draft directory of Irish religious archives in
   1985, under the direction of Henry O'Shea, OSB. In 1992 the association changed its name to 'Association of
   Church Archivists of Ireland'.
25 Angela Bolster, *Catherine McAuley, her educational thought and its influence on the origin and development
   of an Irish training college: two centenary lectures* (Dublin, 1980).
26 Caitriona Clear, 'The archives of religious congregations: to what purpose?' (paper presented to the Association of Religious Archivists of Ireland, Dublin in July 1989).

have been the leading growth area in the Irish archival profession over the last decade.[27] In practical terms it means the historian no longer has to justify convent archives in terms of authenticity.

By the beginning of the twentieth century there were just over eight thousand nuns in the country with a total of thirty five religious orders. There were 368 convents, in architectural style and design, supplying a distinctive structure quite different from the Georgian or Victorian mansion or larger houses. In a comprehensive analysis of the numbers of religious (men and women) in Ireland in 1989 (using the computer methodology outlined in an earlier section of this paper), the Conference on Major Superiors has presented the researcher with figures drawn from census reports, monastic and convent archives, catholic directories and diocesan archives to present a multi-layered profile of the members of religious engaged in ministry in the catholic church in Ireland in 1989.[28] The total numbers of female religious in Ireland in 1989 was 11,415, of whom 405 were cloistered contemplatives. There were 128 religious congregations of women and twenty four monasteries of women contemplatives. The most visible change in the 1980s was the closing of thirty convents and the gradual transition of large communities to smaller groups: 141 community houses opened during the same period. The other dramatic change is the decline in vocation and the age-structure of nuns: 57 per cent were over sixty years of age at the time of the *Profile's* reckoning, 1989–90.[29]

Dr Suellen Hoy, in a forthcoming study, 'The journey out: the recruitment and emigration of Irish religious women to the United States, 1812–1913', notes that 10 per cent of the number of nuns in the U.S. in 1915 were Irish women who had emigrated as professed nuns, novices, postulants or aspirants (there were 40,000 nuns working in the U.S. in 1900, 175,000 in 1915).[30] Dr Hoy has placed her investigations in the context of women's experience of emigration and high on the research agenda of twentieth-century sisters in Ireland is a continuation of her work. According to J. J. Lee, the economic historian, of every 100 girls in Connaught aged between fifteen and nineteen years in 1946, 42 had left by 1951. The most the government would say in a memo from the Department of External Affairs was: 'the present high volume of emigration is due at least to causes other than economic necessity.' According to this estimate Ireland would boast the highest rate of female emigration of any European country between 1945 and 1960. 'If the comely maidens would laugh, it would be the bitter sweet laugh of liberation through emigration from a sterile society where the Bridies left behind would be glad to settle, their girlhood dreams dashed, for the Bowser Egans.'[31]

Lee does not pick up on his own statistics which he quoted in an earlier essay on women and the church since the famine, that by the 1941 census one out of every four hundred women was entering a convent. Moreover, admissions to novitiates climbed steadily, peaked around 1960, remained stable until 1972 and then declined decisively to the present situation where 2 per cent (227) of

27 David Sheehy in conversation with author, 20 May 1993.
28 Emmet Larkin, 'The devotional revolution in Ireland, 1850–75' in *American Historical Review*, 77, no. 3 (June 1972), pp 626, 664, 651 for nineteenth-century estimates. For late twentieth-century estimates see: Conference of major superiors of Ireland (eds), *Profile of religious in Ireland 1989/1990* (Dublin, 1990).
29 *Ibid.*, pp 1–97.
30 Suellen Hoy, 'The journey out: the recruitment and emigration of Irish religious women to the United States, 1812–1914' in *Journal of Women's History* (forthcoming).
31 Joseph J. Lee, *Ireland, 1912–1985, politics and society* (Cambridge, 1980), p. 377. See pp 187–214 below.

the 11,415 nuns in Ireland are under twenty-nine years of age.[32] Clearly these statistics contain investigations of much suggestibility. Why were Irish women leaving Ireland? Is there a connection between the restrictions on entry to the United States regulated in 1917 and the swing to other parts of the world? How does the historian account for the rise in female vocations in the century which offered women undreamt of opportunities in the growing professionalisation of work? How many of those emigrant women were missionary sisters?

Though the ministerial cabinet of the Irish Free State kept referring as late as the 1930s and 1940s to the freedom of movement of Irish women to emigrate and find work as domestic servants, the reality had been that since the 1890s, there was a growing acceptance of women into university courses and of their certification in most fields of business. They also made advances into the fields of professional medicine and nursing. Dr Edmund Hogan, in his study on the Irish missionary movement,[33] has a section on 'The development of medical missions'. There he deals with the winning of the initiatives by women missionaries and the Missionary Sisters Congregations, notably the Franciscan Missionaries of Mary, the Mercy Sisters in India, the Medical Missionaries of Mary and the Holy Rosary Sisters of Killeshandra, to qualify themselves for the medical care of women and children (as surgeons, obstetricians and gynaecologists). Spelt out, this involved a struggle first with the great medical schools, and then, a prolonged lobbying of Rome, in particular of the papacy, to allow catholic sisters to study and practise medicine and to take state qualifications in midwifery. Hogan's investigations are sensitive to what Professor Margaret Thompson discerns as one of the characteristics of the 'new feminist religious history': he recognises the actuality and need for unordained ministry, and demonstrates how laywomen and religious women worked closely together to achieve an objective. What also comes through in his study, though not overtly, is the patriarchal nature of the struggle where popes withheld dispensation and permission: not just from catholic women but also from enlightened missionary bishops. As an appendix to this study, Hogan prints in full Canon 489, which was issued in February 1936, 'Maternity training for missionary sisters'.[34] That, of course, was not the end of the story though it represents a victory for Teresa Keaveney, Mother Kevin, who in 1921 had opened a Midwifery Training School for her women in Uganda with Evelyn Connolly, a Dublin doctor, qualified in obstetrics. Keaveney herself had not succeeded in obtaining a dispensation to train in midwifery because of a Roman embargo.[35] The quest for the acquisition of professional qualifications such as those involved in surgery and obstetrics, the so-called 'forbidden skills', belongs to the area of public power, and ultimately to the area of church-state politics. The paradox of the missionary situation of the early twentieth century was that the tensions came from within the catholic structure of authority. The players were those at the top, leaders in their own domain. Three successive popes, Pius X, Benedict XV, and Pius XI, withheld dispensations (which is the normal procedure to get around a canon law). Missionary bishops of the calibre of Bishop Shanahan of Nigeria, Monsignor Wagner of Rawalpindi (who needed a hospital for women in purdah, run

32 Joseph J. Lee, 'Women and the church since the famine' in Margaret MacCurtain and Donncha Ó Corráin (eds), *Women in Irish society, the historical dimension* (Dublin, 1978), p. 40.
33 Edmund Hogan, *The Irish missionary movement, a historical survey, 1830–1980* (Dublin, 1992).
34 Canon 489: Maternity training for missionary sisters (Instruction, 5. Congreg. Prep. Fide, 11 February 1936), *Acta apostolical aedis*, 28, p. 208, quoted in full in Hogan, *op. cit.*, pp 195–6 (Appendix B).
35 Hogan, *op. cit.*, pp 113–20.

by women), and in the home diocese, Cardinal Bourne of Westminster, Archbishop Dougherty of Philadelphia, Cardinal Manning in Australia, all lobbied the Roman Congregation of Bishops and Regulars unsuccessfully. Fr John Blowick, founder of the Maynooth mission to China and a seasoned campaigner, organised a formidable constellation of women determined, like Teresa Keaveney, to end the embargo. Three laywomen, medical doctors all, Agnes McLaren, Margaret Lamont and Anna Dengel, came together with Teresa Keaveney, Mother Xavier Murphy, a Presentation Sister in South India, and Frances Moloney, later founder of the Columban Sisters. They were joined by two medical students, Marie Martin and Agnes Ryan. They sent Dr Agnes McLaren to speak on their behalf with the pope, Benedict XV, and other Roman officials. To no avail. Moloney went ahead and founded the Columban Sisters in 1922 'vowed to the medical care of the sick in non-Christian countries and whose members would be properly qualified in medicine, surgery and midwifery.'[36] In 1924, Mary Ryan began to set up the Sisters of the Holy Rosary Killeshandra[37] and then, in 1936, Marie Martin founded the Medical Missionaries of Mary after long and careful deliberation, an institute devoted exclusively to medical and related problems.[38] From the mid-1940s a familiar sight in student medical classes and midwifery courses was a soberly-clad young woman without adornment who took her place with her fellow students and as often as not took high grades in class results. Within fifty years of its foundation the Institute of the Medical Missionaries of Mary had some 450 members working internationally and an international hospital in Drogheda, County Louth, which has a maternity unit, and specialises in tropical diseases. The Sisters of St. Columban in their first thirty years listed nine doctors and forty two nurses among its members. In 1948 Sr Aquinas Monaghan went to Hong Kong and became a world-specialist in tuberculosis.[39]

This sketch of the politics of the women's campaign for medical training, where their resources were pitted against the highest authorities within the catholic church, is an important dimension of women's political activity in the twentieth century. It bears out a point made by Joan Wallach Scott in her article in *Past and present*:

> To ignore politics in the recovery of the female subject is to accept the reality of public/private distinctions and the separate or distinctive qualities of women's character and experience. It misses the chance not only to challenge the accuracy of binary distinctions between men and women in the past and present, but to expose the very political nature of a history written in those terms.[40]

Far less known is the history of the general nurse and her struggle for state recognition, and after that, the efforts to rescue midwifery from the irregularities of the untrained and place it on a professional footing, in line with the 1875 British parliamentary act for the training of midwives and the 1902 midwifery act of Britain which recognised the Coombe and Rotunda Hospitals as training

---

36 Margaret Fairburn, 'Missionary sisters of St. Columban' in *Studies*, 36 (December 1947) pp 451–60.
37 Mary Bridig, 'The congregation of the Missionary Sisters of the Holy Rosary. Origin and growth' in *Capuchin Annual* (Dublin, 1955), pp 254–376.
38 Mary Purcell, *To Africa with love: the biography of Mother Mary Martin* (Dublin, 1987).
39 Hogan, *op. cit.*, p. 120.
40 Joan W. Scott, 'Survey articles: women in history: the modern period' in *Past and present*, no. 101 (Nov. 1983), p. 156.

schools.[41] Legislation passed in 1918 made it incumbent on nurses to register as midwives after training. The history of nursing dates back to the nineteenth century and Pauline Scanlon's book: *The Irish nurse: a study of nurses in Ireland: history and education, 1718–1981*,[42] and Dr Ruth Barrington's *Health, medicine and politics in Ireland 1900–1970* are valuable for a general background.[43] The medical profession, however, is not a homogeneous one: it is based on hierarchy and, to some extent, is also class-based. The emergence of the nursing profession, and the endeavours to train the general nurse and to set up state qualifications in midwifery, belong to the mainstream of feminist history. Here we are focusing on the nun's story which had to confront the double patriarchy of hospital administration and church control in the person of the local bishop who chaired the board of trustees. Mercy Sisters and Irish Sisters of Charity nursed side by side in the cholera epidemics of the 1830s and 1840s. When the Sisters of Mercy agreed to take over the running of the Limerick Workhouse and thereafter the other state workhouses, they received assistance and training from the 'Florence Nightingale trained nurses'. There was a perception that had to be combated by catholic women, that the quality of recruitment to the general hospital was quite low and that in the workhouses the night-attendant, who was quite often the local midwife, was not always a savoury character.[44]

The gradual assumption of administration and of senior nursing posts by trained religious sisters, gave a credibility to catholic hospitals which the Irish Free State and the government in Northern Ireland was eager to recognise. Once the Cork Mercy Hospital, founded in 1854, began to provide state training for nuns in September 1911,[45] and then much later, by arrangement with Holles Street Maternity Hospital, midwifery training, the voluntary catholic hospitals became, to some extent, victims of their own success, as church and state jockeyed for control of policy and appointments.[46]

The need for sustained analysis on the nature of twentieth-century Irish feminism has been demonstrated by Mary Cullen in her essay 'How radical was Irish feminism'.[47] At this point in the development of women's history in Ireland, it is difficult to chart how the female nursing and medical professions divided on lines of class and gender divisions of labour.

A study of Irish women and the professions, 1900–1936 by Clare Eager brings the debate a step further:

Service was the watchword for Irish women who chose to work in such diverse professions as teaching, nursing and in the pursuit of the public good. With entrée into those professions assured after 1901, women's choice of career option determined the public perception of their role by a society becoming more and more conservative. One can

41 Jo Murphy-Lawless, 'The silencing of women in chidbirth, or let's hear it for Bartholomew and the boys' in *Women's Studies International Forum*, 11, no. 4 (1988), pp 293–8; Ian Campbell Ross (ed.), *Public virtue, public love: the early years of the Rotunda Lying-in hospital* (Dublin, 1986).
42 Pauline Scanlon, *The Irish nurse: a study of nurses in Ireland: history and education, 1718–1981* (Leitrim, 1991).
43 Ruth Barrington, *Health, medicine and politics in Ireland, 1900–1970* (Dublin, 1987).
44 Patricia Kelly, 'From workhouse to hospital: the role of the workhouse in medical relief to 1921' (unpublished M.A. thesis, University College Galway, 1972).
45 Emmanuel Browne, *Mansion of mercy: Cork Mercy Hospital* (Cork, 1988).
46 Emmanuel Browne, *A tale of two hospitals* (Cork, no date).
47 Mary Cullen, 'How radical was Irish feminism between 1860 and 1920?' in Patrick J. Corish (ed.), *Radicals, rebels and establishments* (Belfast, 1985), pp 185–201.

suggest that the 'feminisation' of the professions in Ireland began in 1901 and had, by 1936, become an integral part of the state.[48]

Conclusions

There is an impression that religious life for women in twentieth-century Ireland was sterile and narcissistic and that women crowded into convents only because of economic and social conditions. As has been noted, the flow of vocations requires a different kind of analysis and the research, on which to base a viable interpretation, has only begun. We need to hear the voices of women religious, the self which is no longer annalist but the subject of the testimony. The journals of nuns exist and the voices of religious women released into familiar speech with their families and friends need to be heard. There are letters in family collections that nuns have written which are records of private feelings and thoughts, others that provide glimpses of spirituality, nuances of alternative realities to console the sorrows of bereaved relatives and friends. Dr Angela Bourke's pioneering studies of Irish women's lament poetry supplies an original methodology for a study of the spirituality of religious women. In her essay 'More in anger than in sorrow: Irish women's lament poetry', Bourke states: 'the lament poetry considered in this essay contains messages that would have been intelligible to an inner circle of women but not necessarily to the rest of the audience.'[49] The resources of female Christian spirituality in Ireland (which flourished in the teeth of Irish patriarchy) are being explored by Mary Condren, first in her book, *The serpent and the goddess: women, religion and power in Celtic Ireland*, and in a subsequent series of essays which open up new fields of enquiry.[50] The work of Seamus Enright in examining the spirituality of Irish nuns is welcome, as are the new studies of Bridget.[51] The forthcoming volume four of *The field day anthology* promises to capture the voices of religious women as well as the religious voices of women, in hymn, in sermon, and in diary.

On a less exalted note and as a vigorous *pis-aller* the history-writing of religious women is a shared work, possibly a collaboration, certainly one which the lay-historian is invited to venture into and research. There are situations in the lives of religious women which require analysis of an objectivity not always available to those caught up in that way of life. Caitriona Clear and Tony Fahy have both drawn attention to the anomalous position of laysisters within the nineteenth-century Irish convent. It may come as a surprise to learn that the two-tiered division within convents was only abolished in the 1960s. In the same decade apostolic religious orders of women, what used to be known as the ac-

---

48 Clare Eager, 'Alice through the looking glass: Irish women in the professions, 1901–36' (paper presented to Boston College-Harvard University Irish Colloquium, 18 March 1993).

49 Angela Bourke, 'More in anger than in sorrow: Irish women's lament poetry' in Joan Newlon Radner (ed.), *Feminist messages: coding in women's folk culture* (Chicago, 1993), p. 161.

50 Mary Condren, *The serpent and the goddess: women, religion and power in Celtic Ireland* (San Francisco, 1989).

51 Seamus Enright, ongoing research for Ph.D. dissertation in Maynooth College, Ireland. See also Kim McCone, 'Brigit in the seventh century: a saint with three lives?' in *Peritia* 1 (1982), pp 107–45; Sean Connolly and Jean Michel Picard, 'Cogitosus: life of St Brigit' in *Journal of the Royal Society of Antiquaries*, no. 117 (1987), pp 11–27.

tive orders, discovered they were members of the laity.[52] By combining insights and methods drawn from a variety of disciplines, by making connections with the lives of women in faith groupings other than Roman catholicism, catholic sisters will become an integral part of Irish history and an indispensable part of the new religious history of Ireland.

---

52 Austin Flannery (ed.), *Vatican council II, the conciliar and post conciliar documents* (New York, 1975), p. 388.

5 Women and the Law in Early Ireland

Donnchadh Ó Corráin

I N the case of early Ireland we have no marriage charters, no records of law-suits concerning property, and thus virtually no prosopographical data about marital property in the broadest sense or its assignment. What we do have is very detailed treatment of christian marriage in the Latin law tracts[1] and, in the vernacular law, detailed treatment of divorce and the division of marital property in the case of divorce. We also have a good deal of information on a woman's legal position, the relationship with uxorilateral kindred, women's relationships and responsibilities in regard to children, women and personal injuries (including rape and marital violence).[2] This material occurs in extensive vernacular legal tracts written for the most part between 650 and 850, and these are equipped with an elaborate apparatus of gloss and commentary that refers sometimes to this period, and often to much later times. (Some of the commentaries are best regarded as law tracts in their own right that refer to a period later than that of the classical tracts.) These materials can be supplemented by reference to the contemporary genealogies (of which there is an abundance, though material on women is relatively thin),[3] the *Banshenchas* 'History of women' a twelfth-century tract, in prose and verse recensions, listing famous women and their marriages),[4] and to a very extensive vernacular literature, prose as well as poetry, including a fairly extensive wisdom literature.[5] I will refer briefly to one literary text: the introduction to the recension of *Táin Bó Cúalṅge* in the Book of Leinster, namely, the famous pillow-talk between king Ailill and queen Medb in Cruachain.

A word about the legal materials. These are not the unaltered records of a pagan past, the oral teaching of pagan lawyers, and thus an artifact of the Celtic culture, if not of a remote Indo-European antiquity. This peculiar understanding of Irish law held the field until recently amongst scholars who saw the texts

---

1 Collectio canonum Hibernensis, books xlv ('De quaestionibus mulierum') and xlvi ('De ratione matrimonii'), F. W. Wasserschleben (ed.), *Die irische Kanonensammlung* (2nd. ed., Leipzig), pp 180–95; Donnchadh Ó Corráin, 'Marriage in early Ireland' in Art Cosgrove (ed.), *Marriage in Ireland* (Dublin, 1985), pp 5-24 has references to the more recent literature.

2 Most of this material is conveniently collected and discussed in R. Thurneysen, D. A. Binchy et al (eds), *Studies in early Irish law* (Dublin, 1936). The central text, 'Cáin lánamna', is edited and translated into German by Thurneysen (*ibid.*, pp 1–80). All these texts have been re-edited in D. A. Binchy's monumental *Corpus iuris Hibernici* (Dublin, 1978) (hereafter CIH). These materials are ably discussed in Fergus Kelly, *A guide to early Irish law* (Dublin, 1988), pp 68–79, 134–37.

3 M. A. O'Brien, *Corpus genealogiarum Hiberniae* (Dublin, 1967) is an edition of some of the earliest materials.

4 M. E. Dobbs (ed.), 'The Ban-shenchas' in *Revue Celtique*, xlvii (1930), pp 283–339; xlviii (1931), pp 163–234; xlix (1932), pp 437–89; see below, pp 70–81.

5 For some more conjectural reflections on women in early Irish literature, see Charles Bowen, 'Great-bladdered Medb: mythology and invention in *Táin Bó Cuailnge*' in *Éire-Ireland*, x (1975), pp 14–34; Lisa Bitel, '"Conceived in sins, born in delights": stories of procreation in early Ireland' in *Journal of the History of Sexuality*, iii (1992), pp 181–202.

as pagan, with a light christian varnish.[6] In fact, the laws are the product of the self-confident and vigorous clerical culture of early Ireland that consciously created a christian law for a christian people.[7] The encounter between inherited legal ideas, the law of the Pentateuch in the hands of skilful exegetes, and the general christian inheritance in law led to rapid legal developments and a remarkable and independent jurisprudence. For this reason, I will be pointing to explicitly biblical, Roman, patristic and papal sources for some aspects of the law governing women. Here I propose to deal only with three points: the contractual nature of marriage and the lawyers' concern with property, the law concerning rape, and the rules about female inheritance.

The Contractual Nature of Marrriage

The lawyers writing in the vernacular discuss three principal kinds of marriage: *lánamnas comthinchuir* 'marriage of common contribution'; *lánamnas for ferthinchur* 'marriage on man contribution'; *lánamnas for bantinchur* 'marriage on woman contribution'. The concern of the lawyers is with property and status. This becomes evident even in the introduction to the main tract on marriage where they dwell on the general characteristics of relationships sharing a common life and a common economy (abbot and *manaig*, teacher and pupil, man and wife):

Comdíles do cách díb cia tarta di araile, cia imarbara cách díb ar araile cen elguin, cen taíde. As-renar aithgin cach díchmairc cairichther co troscud acht i n-eclais. Aithgin olchena cach díchmairc cairigther co imchim troiscthe nó élud dligid. Cach taíde, cach elguin, cach díchmairc cairigther follaighther, is cona díre do-bongar.[8]

[Equally exempt for each is whatever one of them may have given the other, whatever one of them may have used as against the other, without violent crime, without stealth. Everything taken without permission that is complained about is repaid by simple replacement of the object until the matter reaches the legal remedy of fasting,[9] except in the case of the church. Repayment by simple replacement of what is taken without permission and complained about is all that is required until evasion of the legal obligations arising from fasting or legal default. Anything taken by stealth, by violent crime, anything taken without permission that is complained about and ignored, is levied with its penalty fine.]

6 K. McCone, *Pagan past and christian present in early Irish literature*, Maynooth Monographs iii (Maynooth, 1990), pp 84–106.
7 Donnchadh Ó Corráin, 'Irish law and canon law', in Próinsías Ní Chatháin and Michael Richter (eds), *Irland und Europa: die Kirche im Frühmittelalter* (Stuttgart, 1984), pp 157–66; Donnchadh Ó Corráin, 'Irish vernacular law and the Old Testament' in Próinsías Ní Chatháin and Michael Richter (eds), *Irland und die Christenheit* (Stuttgart, 1987), pp 284–307; Donnchadh Ó Corráin, Liam Breatnach and Aidan Breen, 'The law of the Irish' in *Peritia*, iii (1984), pp 382–438; Liam Breatnach, 'Canon law and secular law in early Ireland: the significance of *Bretha nemed*' in *Peritia*, iii (1984), pp 439–59; Liam Breatnach, 'The ecclesiastical element in the Old-Irish legal tract *Cáin Fhuithirbe*' in *Peritia*, v (1986), pp 36–62; Liam Breatnach, 'The first third of *Bretha nemed*' in *Ériu*, xl (1989), pp 1–40; the role of the monastic men of learning as lawyers is discussed in Donnchadh Ó Corráin, 'Nationality and kingship in pre-Norman Ireland', in T. W. Moody (ed.), *Nationality and the pursuit of national independence: Historical Studies, xi* (Belfast, 1978), pp 1–35.
8 CIH 504; *Studies in early Irish law*, p. 3, §3. Thurneysen (*ibid.*, p. 4) makes the odd suggestion that the Irish lawyers list all such legal relationships 'because definition is foreign to the Irish'.
9 A formal hunger-strike by which a plaintiff forces a defendant (usually of higher rank) to initiate a legal settlement of his claim; for details see R. Thurneysen, 'Das Fasten beim Pfändungsverfahren' in *Zeitschrift für celtische Philologie*, xv (1925), pp 260–76; D. A. Binchy, 'A pre-christian survival in mediaeval Irish hagiography', in D. Whitelock et al (eds), *Ireland in early mediaeval Europe* (Cambridge, 1982), pp 165–78.

This sets out the basis of the marriage partnership: the couple are partners in a joint enterprise, in which they invest, in different proportions, and in which their property-nexus is regulated as in similar relationships.[10]

I cite the relevant texts in respect of the first two types of marriage (the third, that of the marriage of heiresses, is a special case, to be dealt with later):

Lánamnas comthinchuir: mad co tír 7 cethra 7 intreb 7 mad comsaír comthéchta a cuma lánamnusa – 7 is don bein sin as-berar bé cuitchernsa – nibi cor cor nechtar dá lína sech araile inge curu lesaigter a cumthus. It é-side in so: comul comair fri coibne téchta in tan nád bí occaib fadesin comobair trebtha; fochraic tíre; tionól cua; comull sollumun; síl cethra do luaig; lánad treib intreib; comul comsa; creic neich do-da-esaib do toischidib. Cach cunnrad cen díchell, sochur sochubus iarna coïr coitechta, co n-imaititiu i neoch crenar amal mbes selb neich renar and.[11]

[Union of common contribution: if it is a union with land and stock and household equipment, and if their condition of marital relationship is one of equal status and equal propriety – and a woman in such a situation is called a woman of joint dominion – no contract of either is a valid one without the consent of the other, except for contracts that benefit their establishment. These are: an agreement for common ploughing with proper kinsmen when they do not themselves have a full ploughing team; the leasing of land; getting together food for a coshering; getting food for feast-days; the buying of young stock; outfitting the household; making an agreement for joint husbandry; the purchase of any essentials that they lack. Every contract shall be without neglect, an advantageous contract, conscientious, in accordance with right and propriety, with acknowledgement on both sides that the ownership of what is acquired belongs to the person whose property was alienated to acquire it.]

Lánamnus mná for ferthinchur: is cor a chor in fir sech in ṁbein acht reic étaig 7 bíd; 7 rec bó 7 cẹrech mad bé n-urnadma nab cétmunnter. Mad bé cétmunterasa téchta, comaith 7 comcheniúil – sech is comcheniúil cach comaith – fo-fuasna-ide a churu uile mad baíth – ar ní-said dílse for diubirt ná fogurrud – conda-tathbongat a maic.[12]

[Union of a woman on a man's contribution: the man's contract is a valid contract without the wife's consent, except for the sale of clothing and food; and the sale of cattle and sheep if she is a duly contracted wife who is not a *cétmuinter* (principal wife). If she is a proper *cétmuinter* equally good and equally well bred – for everyone of equal goodness is of equal birth – she impugns all his contracts if they are foolish – for validity does not sit on cheating and on what is forcefully protested against – and her sureties annul them.]

The most formal type of marriage is a contract brought about by a procedure called *airnaidm* 'binding, tying' (the term is the verbal noun of the verb *ar-naisc* 'to bind'). The force of the pre-verb *ar* is a little uncertain: it may mean to 'bind forward' but it more probably means 'to bind publicly'. This is a formal contract between two families marked by the exchange of property. The term used for the property paid by the groom, in the first instance to the father or legal representative of the bride (what the anthropologists call bridewealth), is *coibche*. This word is an innovation that displaced the older term *tinscra*, associated with the older type of marital contract, *lánamnas for ferthinchur*. *Coibche* originally meant 'contract', and shifted semantically to mean 'marriage contract', and then the 'consideration' or principal external 'consideration' of the con-

---

10 Cf. the implicit comparison between divorce and the parting of a superior from his church (Ó Corráin, 'Irish law and canon law', pp 161–64).
11 CIH 505–06; *Studies in early Irish law*, pp 18–19, §5.
12 CIH 512; *Studies in early Irish law*, pp 45–46, §§21–22.

tract, the bridewealth, or, to use the Roman law term, *donatio propter nuptias*. This semantic shift underlines the contractual nature of the proceeding. *Airnaidm* completes the contract: it was not necessary for less formal kinds of marital relationships. Like other contracts, its execution is guaranteed by others.[13] The guarantors were usually more exalted personages than the parties to the contract. Others members of society, then, had an interest in seeing that the conditions of the contract were observed and had a legal role in suing out a woman's rights in marriage.

Two things are evident from the texts cited: the lawyers are preoccupied with the need of the couple to be free and equal and, in the most formal type of marriage, formal payments are made by both parties. These conditions derive from late Roman law – the constitution of Majorian of 458 (abrogated by Leo and Severus in 463), as interpreted by Pope Leo the Great in his letter to Rusticus, Bishop of Narbonne in 459. This letter is cited at length in the Collectio canonum Hibernensis, appears in the Hispana (first half of the seventh century), and in the fifth-century collection of Dionysius Exiguus.

Nuptiarum autem foedera inter ingenuos sunt legitima et inter aequales ... nisi forte illa mulier et ingenua facta et dotata legitime et publicis nuptiis honestata uideatur.[14]

It is evident that, in the most formal type of marriage, the Irish lawyers prescribe the two payments made in late Roman law and custom – *donatio propter nuptias* and *dos*. Hence the technical term *lánamnas comthinchuir* 'marriage of common contribution', implying not equality of total assets, but equality in marital payments in the way the *donatio propter nuptias* and *dos* were equal in late Roman usage. The third condition of Leo's Letter – *publicis nuptiis honestata* i.e. that the contract be public, and not clandestine – is met by the term *airnaidm* itself. By definition, this is a contract with witnesses and guarantors for both sides and the term may well mean 'public contract'.[15] It is clear, then, that *lánamnas comthinchuir* is essentially a christian Roman marriage and the more formal kind of *lánamnas for ferthinchur* has been reshaped by clerical thinking. Thurneysen and Binchy have both argued that *lánamnas comthinchuir*, the type of marriage treated of in greatest detail in *Cáin lánamna*, was the normal form of marriage in early Ireland.[16] If this is correct (and it seems to be), clerical reformation of the institution of marriage was both radical and successful.

13 The expression *conda-tathbongat a maic* 'and her sureties annul them' (cited above) indicates that the guarantors had an on-going role in the observance of the terms of the *airnaidm* rather than the simpler duty of seeing to it that the terms of its dissolution were equitable. cf. *Is mese cach ben a uccu im la mac beith a cáin fa la fine fa la fer a sliasta, acht cétmuinter. Ar is la cach cétmuinter téchta a cáin-side, manis-coirbet a anfolaid lánamnais; a n-am inda-coirbet, is ann is meise side imscartha fris* (CIH 443=R. Thurneysen, 'Irisches Recht', *Abhandlungen der preussischen Akademie der Wissenschaften*, Phil-Hist. Kl. 2., Jhrg 1934 (Berlin, 1931), p. 34, §34) 'Every woman is competent to decide whether the suing out of her rights should belong to her guarantor, her family, or the man she is sleeping with, except for a *cétmuinter* (principal wife) for her suing out of rights belongs to every proper husband (*cétmuinter*) unless his misdeeds pollute her. When they do, she is entitled to divorce him'.
14 *Patrologia Latina* 54, 1205–05; Collectio canonum Hibernensis, xlvi 19 (Wasserschelben, *Kanonensammlung*, p. 190).
15 A nice parallel for Irish law is found in *Lex Romana Burgundiorum* (dating from the beginning of the sixth century): *Nuptiae legitimae contrahuntur, si conventu parentum aut ingenuorum virorum, intercurrente nuptiali donatione, legitime celebrentur. Quodsi pares fuerint honestate personae, consensus perficit nuptias, sic tamen ut nuptialis donatio solenniter celebretur: aliter, filii exinde nati legitimorum locum obtinere non poterint* (Monumenta Germaniae Historica, Leges in Quart. Sect. I 2 37§§1–2). This is a mixture of late Roman law, the Novel of Majorian, and the Letter of Pope Leo I. Note that the semantic instability of *donatio/dos* also occurs in Irish material.
16 R. Thurneysen, *Studies in early Irish law*, p. 20 ('*Lánamnas comthinchuir* ist für den Verfasser offenbar gewissermassen die Normalehe; darum stellt er sie voran und behandelt sie ausführlicher als alle anderen'); D. A. Binchy, *ibid.*, p 210.

One of the more remarkable echoes of Roman law occurs in *Cáin lánamna* in regard to the improper repudiation of a wife and, again, it is concerned with property rather than morality.

Mad coibche fir bein do-rata cid dia sétaib fadesin, is dílis don chétmuintir in choibche sin má ógaid a mámu téchta i lánamnas {gloss: ni ime tucad bean tara cend im inddliged to denam}. Is fíachach cach adaltrach do-thét for cend cétmuintire: as-ren lóg n-enech na cétmuintire.[17]

[If he gives bridewealth to another woman, even from his own private property, that bridewealth is forfeit to his *cétmuinter* (principal wife) if she carries out her marital obligations {*gloss*: It was not because she did wrong that another woman was married 'over her head'.} Every secondary wife [literally: adulteress] who comes 'over the head' of a cé*tmuinter* is liable to penalty: she pays the honour-price of the *cétmuinter*.]

These provisions do not derive from any traditional Irish law but from rules of Roman law that occur in the *Codex Theodosianus* and subsequently in the various recensions of the *Lex Romana Visigothorum* (whence they may have reached Irish lawyers). That will be evident from the following citation from Codex Theodosianus III 16.1 (Mommsen, i 156):

In masculis etiam, si repudium mittant, haec tria crimina inquiri conveniet, si moecham vel medicamentariam vel conciliatricem repudiare voluerint. Nam si ab his criminibus liberam eiecerit, omnem dotem restituere debet et aliam non ducere. Quod si fecerit, priori coniugi facultas dabitur domum eius invadere et omnem dotem posterioris uxoris ad semet ipsam transferre pro iniuria sibi inlata.

[In the case of a man also if he should send a notice of divorce, inquiry shall be made as to the following three criminal charges, namely, if he wishes to divorce her as an adulteress, a sorceress, or a procuress. For if he should cast off a wife who is innocent of these crimes, he must restore her entire dowry, and he shall not marry another woman. But if he should do this, his former wife shall be given the right to enter his house by force and to transfer to herself the entire dowry of his later wife in recompense for the outrage inflicted upon her. [AD 331]]

These parallels with Roman law point to the syncretistic activities of the early Irish clerical lawyers in devising an innovative law of marriage for their christian society. It goes without saying that other inherited usages in regard to marriage were tolerated in Ireland – as they were elsewhere in early medieval Europe – and the second type of marriage, *lánamnas for ferthinchur*, seems to preserve more of the characteristics of native marriage and of marriage as it was generally understood in northern Europe.

The marriage of an heiress, *lánamnas for bantinchur*, is a special case but one in which the general rule that property determines marital status holds good. I cite some of the relevant texts:

Lánamnas fir for bantinchur: is i suidiu téit fer i n-uidiu mná 7 ben i n-uidiu fir. Mad fer fognama is nómad a h-arbim don fhir; 7 don saill mad cend comairle cuindrig muintire fri comairle comnirt . . . Acht is fer do-renar a h-inchaib na mná mad lé in tothchus uile, inge mad sofoltachu in fer oldaas in ben, nó mad cáidiu nó mad saíre no mad airmidnechu.[18]

[Union of a man on a woman's contribution: in that case, the husband goes in the track of the wife and wife in the track of the husband. If he is a man of service he receives [on the occasion of a divorce] a ninth of the corn; and of the salt meat if he is a "head of counsel" who controls the people of the household with advice of equal stand-

---

17 CIH 513; *Studies in early Irish law*, p. 49, §23.
18 CIH 515–16; *Studies in early Irish law*, p. 57, §29; p. 62, §31.

ing [to that of his wife] . . . But he is a husband who is paid honour-price in accordance with his wife's status if she holds all the property, unless he is more godly, more high-born or more estimable than she.]

Ar cach recht la Féiniu acht óentríar, is lethlóg a enech dia mnaí: fer cen seilb cen tothchus lasmbí bancomarba, a inchaib a mná dí-renar side; 7 fer in-etet tóin a mná tar crích, dí-renar a inchuib a mná; 7 cú glas, dí-renar side a inchuib a mná 7 is sí íccas a cinta, mad iarna urnadmaim nó aititen dia finib. It tualaing na téora mná so imoicheda cor a céle, connatát meisse recce ná crecce secha mná acht ní for-congrat.[19]

[In the case of all kinds of men in Irish law, except for three alone, their wives have half their honour-price: a man without land without property who is married to an heir-ess – she is paid honour-price according to the honour-price of his wife; a man who follows his wife's arse over a border[20] – he is paid honour-price according to the honour price of his wife; and a foreigner – he is paid honour-price according to the honour price of his wife; she pays for his offences if she is contracted in marriage or recognised by her family. These three women are capable of impugning the contracts of their husbands, so that these latter are not competent to sell or buy without their wives, but they can do only what their wives authorise.]

This is the whole point of the spirited dialogue in the *Táin Bó Cuailgñe*: king Ailill is a lackland and an outlander, and his spouse queen Medb is heiress-queen of Connacht and daughter of the high-king of Ireland – and thus his superior by birth. Even the term *coibche* is cleverly inverted in Medb's speech:[21]

"Tucusa cor 7 coibchi duit amal as dech téit do mnaí .i. timthach dá fher déc d'étuch, carpat trí secht cumal, comlethet t'aigthi do dergór, comthrom do riged clí do fhindruinni. Cipé imress méla 7 mertain 7 meraigecht fort, ní fhuil díri nó eneclann duit-siu ind acht na fil dam-sa", ar Medb, "dáig fer ar tincur mná atachomnaic."

["I gave you a contract and a bride-price as befits a woman, namely, the rainment of twelve men, a chariot worth thrice seven *cumala*, the breadth of your face in red gold, the weight of your left arm in white bronze. Whoever, brings shame and annoyance and confusion on you, you have no claim for compensation or for honour-price for it except what claim I have", said Medb, "for you are a man dependent on a woman's marriage-portion."]

Rape and Illicit Intercourse

Some texts deal with serious crimes against women – rape and illicit intercourse. Here it is possible only to treat the matter briefly and to point out the hand of the churchmen and the effect of the Old-Testament legal inheritance on Irish law.

Tá .uii. mná la Féniu ada dílsi ina frithigib nacon dlegar díre ná eneclainn ina sleith; ní tuillet fiachu ná éric ina forcur cibé do-d-róna: echlach oides a corp do chách co rogaib genus, ben ara-túaisi a sleith, ben con-ceil a forcor, ben for-curthar i cathuir ná fóccuir co ndichet do ráith, ben ara-foím immarmus do chind a céile, ben ara-dála fer cuice i muine no lige, ben ad-guid aitire Dé no duine i fomatu a cuirp, ben do-fairget ar decmuic. It é .uii. mná inso ada tualuing taburta a corp i fomatu lánamnuis acht ná methat a ngínmu. Ní berat comperta for fine nadbi tualuing somaine lánamnuis.[22]

19 CIH 427; Thurneysen, 'Irisches Recht', p. 64, §4; D. A. Binchy, 'The legal capacity of women in regard to contracts', in *Studies in early Irish law*, p. 215.
20 i.e. a man who marries a woman from another kingdom and goes to live with her.
21 Ernst Windisch (ed.), *Die altirische Heldensage Táin Bó Cúalnge* (Leipzig, 1905), pp 1–17; R. I. Best & M. A. O'Brien (eds), *Book of Leinster*, ii (Dublin, 1956), pp 261–63 [diplomatic edition]; Cecile O'Rahilly (ed. & trans.), *Táin Bó Cúalnge from the Book of Leinster* (Dublin, 1967), pp 1–3 (text), pp 137–40 (translation); summary, Rudolf Thurneysen, *Die irische Helden– und Königsage* (Halle, 1921), pp 242–43.
22 Heptad 47=CIH 42 (cf. 1845)=ALI v 272–77.

[There are seven women in Irish law who are liable in their encounters and who are not entitled to penalty or honour-price for their *sleith*;[23] they are not entitled to fine or body-fine for rape whosoever may have done so: a whore who offers her body to all, until she becomes chaste; a woman who observes that she is the victim of *sleith* [and does nothing about it]; a woman who conceals her rape; a woman who is raped in a town and who does not cry out until the rapist has got away; a woman who agrees to have illicit intercourse in despite of her husband; a woman who trysts with a man in the bushes or in bed; a woman who invokes a body-surety, cleric or lay, by the offer of sexual favours; a woman who offers herself for something trivial. These are seven women who are capable of giving their bodies in sexual intercourse, provided they do not fail in their duties.[24] Their children do not belong to the family and they are not entitled to the profits arising from cohabitation.]

The same provision in respect of rape is adverted to in *Gúbretha Caratniad*:

Rucus éricc do mnaí nadége oca forcor . . . ba deithbir ar ba i ndithruib forcorad.[25]

[I granted *éraic* (body fine) to a woman who did not cry out at her rape . . . it was proper because she was raped in a deserted place.]

This rule about rape (and *sleith* is legally only a variety of rape) derives not from any tradition of Irish law but from Hebrew law as set out in Deuteronomy 22:23–27:

si puellam virginem desponderit vir et invenerit eam aliquis in civitate et concubuerit cum illa educes utrumque ad portam civitatis illius et lapidibus obruentur puella quia non clamavit cum esset in civitate vir quia humiliavit uxorem proximi sui et auferes malum de medio tui sin autem in agro reppererit vir puellam quae desponsata est et adprehendens concubuerit cum illa ipse morietur solus puella nihil patietur nec est rea mortis quoniam sicut latro consurgit contra fratrem suam et occidit animam eius ita et puella perpessa est sola erat in agro clamavit et nullus adfuit qui liberaret eam.

Irish law took over the principle of Hebrew law but not its harsh sanctions. Something of the punishment for rape – that is, the legal penalty, as distinct from the possibly violent summary justice of kinsmen – can be found elsewhere.

Lánamnas éicne nó sleithe: ní téchtat ba acht comperta. 7 as-renar lánéraic i n-ingin macdacht 7 i mmacaillig ná diúlta cailli 7 i cétmuintir, lethéraic mad adaltracha – cen frithuide in so uile – co lánlóig einech bes sruithem fordo-bé do neoch diambi saindíles[26]

[Sexual union by rape or by stealth (*sleith*): they [the partners] possess nothing but offspring. Full wergild is paid for a virgin, for a young nun who does not reject her veil and for a *cétmuinter* (principal wife); half wergild for secondary wives – all this without the co-operation of the woman – together with the full honour-price of the man of highest rank who has authority over her of those to whom she specially belongs.]

Éraic, often translated 'wergild, body-fine' is the fixed penalty for homicide, namely, seven *cumals*. A *cumal*, literally 'a bond-maid', is a unit of account gen-

---

23 This term (the verbal noun of *selid 'creeps') is defined (DIL s. v.) as 'the act of surprising a sleeping woman, having intercourse with her'. It is better understood as non-consensual sexual intercourse with a woman who is sleeping, in a drunken stupor, or comatose for whatever reason.

24 The glossator (CIH 43) offers the tart comment: *amail robattar mna Tulcha Leis uair is ed fa gnathugad doib: cid mor do indlighid donedis, acht gu toirsidis do bleogain a mbo im eatra, a slainti doib* [such were the women of Tulach Léis, for their practice was that however much wrong-doing they did, they were reprieved provided they came home to milk the cows at milking time].

25 CIH 2197; R. Thurneysen, 'Aus dem irischen Recht III: die falschen Urteilsspürche Caratnia's' in *Zeitschrift für celtische Philologie*, xiv (1925), pp 302–76: p. 350 §39.

26 CIH 519; *Studies in early Irish law*, p. 71 §5

erally taken to be equal to three milch cows. The first penalty for the rape of a virgin, young nun or principal wife was 21 milch cows. The second is the honour-price of the man of highest rank who has authority over the woman – father or grandfather, guardian or, in the case of a young nun, her superior in religion. Honour-price varied according to status – from two-and-a-half milch cows in the case of a substantial farmer (*bó-aire*) to seven *cumals* in the case of a petty king, bishop or senior monastic superior. Therefore, the penalty for rape varied with the social connections and status of the victim.

A coarser indication of social life occurs elsewhere:

Rucus slán slithi mná óentama i tig midchuarda . . . Ba deithbir ar ba écóir ben i teglug cena céle 'co imchomét.[27]

[I dismissed the *sleith* of an unaccompanied woman in a feast-house. . . . It was correct for it was wrong for a woman to be in a house-party without her partner to watch over her.]

The commentator's observation is enlightening: *sleth cétamus, atá tucht ad-claid bein .i. cen éricc i saide dia fir* [*sleith*, then, there is a case that inculpates a woman i.e. no *éraic* is payable to her man on that account]. Another passage expresses the same attitude to a woman's contributory negligence:

Is and is díles sleith na mná gan éric .i. mása colladh do-rigni in ben i n-aenach nó a cuirmthech gan teist aga testugud, ní fuil éric ina sleith.[28]

[This is when *sleith* perpetrated on a woman is not actionable and without wergild i.e. if the woman fell asleep in an assembly or in an ale-house without a witness to testify there is no wergild for *sleith*.]

Women and Inheritance

The general rule is that daughters inherited real estate only in default of male siblings, and then they inherited only a life-interest in the estate which reverted to the nearest agnates. One of the most important sources on female inheritance is the difficult legal poem 'In-longat bantaid banchuru'. Myles Dillon edited and translated it in 1936[29] but much still remained dark. In 1993 Dr Thomas Charles-Edwards offered a fresh text and new translation that lightens much of its obscurity.[30] Earlier, Charles-Edwards dated this poem to the late sixth or the first half of the seventh century.[31] I should now much prefer the later date and perhaps one later still.

The poem envisages a case in which the smallest kin-group, the *gelfhine* (the common descendants of a grandfather), issues in an heiress (§1). *Ar-naisc finsruith finteda/ manip sesed imbera* [The head of the kindred binds forward family lands unless a sixth man acts (§2)]. What this means is that the surviving senior male of the *gelfhine* or, if he is dead or incapable, one of the ultimate

27 CIH 2198; Thurneysen, 'Gúbretha', p. 351, §40. I emend MS *teglaig* to *teglug*.

28 CIH 827.

29 Myles Dillon, 'The relationship of mother and son, of father and daughter, and the law of inheritance with regard to women' in *Studies in early Irish law*, pp 129–79; his edition and translation of the poem is on pp 135–59.

30 T. M. Charles-Edwards, *Early Irish and Welsh kinship* (Oxford, 1993), pp 516–19. I wish to express here my deep indebtedness to his work on this difficult tract and I depend on the text he has established. Reference will be to the sections of the poem in his edition: I call them stanzas for convenience.

31 T. M. Charles-Edwards, 'Kinship, status and the origin of the hide' in *Past & Present*, lvi (1972), pp 3–33.

heirs, a member of the *derbfhine* (the *sesed* of the poem) causes the inheriting female to enter into bonds that she will not alienate family land, namely, attempt to transmit it to her children by a non-kinsman, for she has a life-interest only and, on her death, the land should revert to her father's or grandfather's nearest male relatives. By executing such bonds, women may inherit land lawfully (and within these restrictions) (§3). Stanzas §§4–11 discuss the rules of inheritance generally. Stanza §12 states:

Ní mac bratas finteda/fine fri fót frithmesso/manip nessa fírchoibnius/máthair athair inorbae
[he is no son who takes kin-lands, lands that revert to the *fine*, unless a father capable of inheritance is nearer in true kinship than a mother]

This exceptionally succinct piece of legal writing can be interpreted as follows. When the *gelfhine* has none to inherit but an heiress, the land reverts on her death to the patrilateral kin and the heiress's offspring is excluded. Her son(s), however, can succeed if their father is nearer in relationship to the ultimate heirs than his mother i.e. he must be one of the ultimate heirs. Stanza §13 adds: *Orbae máthar munchoirche/a maic o laithib a ardimnai* [the land of a mother thus contracted in regard to property belongs to her son from the days of her testamentary disposition] – which may mean that a woman who has been under such property bonds and who concludes a marriage with the appropriate cousin may make her son an heir by a testamentary disposition.
But her offspring does not receive it all:

Do-aisic a leth immurgu/Dochum fine fírgriain/a leth n-aill a fírbrethaib/Síl a féola fodlaither (§13)
[However, he restores half of it/to the true kindred of the land/the other half, according to just judgements/is divided amongst the seed of her flesh.]

This provision limits the amount of the estate heritable by the heiress's offspring to half.
Another much later prose passage throws some more light on female inheritance:

Má tá comarba ferrdha ann nocho berann in ingen ní do díbad a athar do scuichthibh ná do annscuichthib acht lanna, ranna 7 bregda. Nó dano is na scuichthe do chomraind doib, 7 is as gabar esén: 'rannait ingena fri macu dlighthecha séta saindílsi athar ilchoraigh, genmothá orba n-athar urrannat maicne ciniudha caín'. Muna fhuil comarba ferrdha ann na scuichthi do breith di uile 7 na h-annscuichthi go fuba 7 ruba, nó a leth gen fuba 7 gen ruba.[32]
[If there is a male heir a daughter receives nothing of her father's inheritance of mobilia or immobilia except *lanna, rann,* and *bregda* [either utensils for female work or jewellery]. Or, otherwise, the mobilia are divided amongst them equally and that is taken from the maxim: "Daughters share with lawful sons possessions that are the personal property of a capable father, except for inherited paternal land that sons of fair kindred share." If there is no male heir, all the mobilia are given to her, and the immobilia with the obligation of military service or half without the obligation of military service.]

This contains an interesting distinction: kin-land is for male heirs only. Not so personal property – mobilia and immobilia – acquired by a successful father in his lifetime: his daughter may inherit that. Here Thurneysen sees the influence

---

32 CIH 736, 2039; Dillon, *op. cit.*, p. 133; Thurneysen, 'Irisches Recht', pp 30–31.

of canon law and cites the rulings: '*De eo quod dare debet pater hereditatem filiae inter fratres suos*' and '*De eo quod feminae dividunt hereditatem, non tamen principalem*',[33] but it it is, perhaps, better to see the Latin canon law and the vernacular law as parts of a single system produced by a single class of lawyers.

Now the rules of female inheritance set out in the ancient poem will, at first sight, give rise to problems for the churchmen because of the church's laws of consanguinity. Briefly, the church took over the prohibitions of Roman law and added to them. Taking its cue from Lev. 18:6 (*Omnis homo ad proximam sanguinis suae non accedat ut revelet turpitudinem eius*) the church forbade marriage with a widening circle of kindred: the sixth degree and some categories from the seventh. In the sixth century, Gaulish and Spanish synods forbade marriage within the sixth degree (thus excluding first and second cousins). Gregory the Great in his letter to Augustine of Canterbury (AD 601) absolutely ruled out marriage of first cousins and in a concession to the weakness in the faith of the Anglo-Saxons, allowed them for the time being to marry their second and third cousins.[34] When one applies these rules to the Irish situation, they pose a problem, especially in the matter of female inheritance. Irish canon law deals in detail with it:

*De his quod addunt auctores ecclesiae in feminis heredibus. Sinodus Hibernensis. Auctores ecclesiae hic multa addunt, ut feminae heredes dent ratas et stipulationes, ne transferatur hereditas ad alienos; Dominus enim dicit: Transibit hereditas earum fratribus patris sui, inde propinquis. Sciendum est, utrum dabunt partem Domino; si tacuerint propinqui earum, Domini erit, quod dabunt, sin autem, irritum erit. Sciendum est quid dabunt in testamentum, hoc est, vaccas, vestes et vasa. Sciendum est quid dabunt ministris, hoc est, partem de ovibus et lanam; si vero de propinquis fuerint ministri, dabunt eis aliquid de hereditate, et si ecclesiae habuerint partem (*vl. ecclesiam habuerint paternam*), dabunt ei de sua hereditate, et si genuerint filios viris suae cognationis dabunt hereditatem*[35]

[Concerning what the authorities of the church add in respect of female heirs. Irish Synod: The authorities of the church add much here, that female heirs should give guarantees and bonds lest they alienate the inheritance. The Lord says [paraphrase of Numbers 27:10, 11] their inheritance will go to their father's brethren and thus to their relatives. May they give a bequest to the Lord? If their kindred do not protest, what they will give will belong to the Lord. If not, the bequest is invalid. What may they give by testamentary disposition? Cows, clothes and vessels. What may they give to the servants? Portion of the sheep and the linen. If the servants be kin, they will grant them some of the inheritance. If they be part-owners of a church (*vl.* if they have a hereditary church) they will make a grant to it from the inheritance. If they have sons by men of their own kindred, they will transmit the inheritance to them.]

The provisions of vernacular law and of canon law evidently derive from the same source – the one mentioned in the canon law tract itself, namely, the church authorities. But if women married their patrilateral cousins in order to preserve a right of inheritance for their children and if parallel cousin marriage was used to consolidate family land in this way, was it not in breach of the universal teaching of the church in the sixth and seventh centuries and later? The Irish lawyers found the answer in the Old Testament and spelled it out in the canon law:

---

33 'Irisches Recht', p. 31; Collectio canonum Hibernensis, xxxii 17, 19 (Wasserschleben, *Kanonensammlung*, pp 115–16).

34 G. H. Joyce, *Christian marriage*, Heythrop Series 1 (London ,1933), pp 505–20.

35 Collectio canonum Hibernensis, xxxii 20 (Wasserschleben, *Kanonensammlung*, p. 116).

De eo quod feminae dividunt hereditatem, non tamen principalem. . . . Lex dicit: Filiae Selphat de tribu Manassen accesserunt ad Moysen in campestribus Moab dicentes: pater noster mortuus est, non habens filios, nec fuit in seditione Chore et Dathan, sed in suo peccato mortuus est, cur privamur hereditate ejus? Moyses retulit hanc questionem ad judicium Domini, qui dixit: Rem justam postulant filiae Selphat; date eis hereditatem in medio fratrum suorum. Sed Dominius praecipit, ut viris tribus suae nuberent, ne transferatur hereditas de tribu in tribum. In quo intelligendum est, quod Dominus ideo dixit: Nemo copuletur uxori nisi de tribu sua, ne hereditas transferatur de tribu in tribum[36]

[That women share the inheritance but not as the ultimate heirs . . . Scripture says [paraphrase of Numbers 27:1–11 and Josh. 17:3–6]: The daughters of Salphaad came to Moses in the plains of Moab saying: our father died in the desert nor did he take part in the sedition of Core and Dathan but he died in his own sin. And he had no sons. Why are we deprived of his inheritance? And Moses referred their cause to the judgement of the Lord, who said: The daughters of Salphaad demand a just thing. Give them an inheritance amongst their father's kindred. And the Lord commanded [Numbers 36:8–13] that they should marry men of their own tribe, so that the inheritance should not be transferred from tribe to tribe. From which is to be understood: let no man be joined to a wife not of his own tribe, lest the inheritance be transferred from tribe to tribe.]

When we compare Numbers 26:28–43 with Numbers 36:10–13 we find that the daughters of Salphaad married their father's brother's sons. This text, then, was used to legitimise the marriages of inheriting females with patrilateral kinsmen. This had two effects: women could inherit a life interest in their father's estates in kin-land and transmit an interest in the property to their offspring, and kin-lands were thus protected against alienation. But the lawyers go further and prescribe parallel cousin marriage as the preferred form of marriage, as it is in Semitic society.

In another recension of the Collectio canonum Hibernensis there is an additional book 'De tribu' that contains the following:

De eo quod omnia patris non habentis filium debentur viro filiae suae post mortem suam in una tribu.[37]

[That all the property of a man without sons should be given to the husband of his daughter who is of the same family.]

Evidently, the canonist translates Irish *fine* as Latin *tribus*, here as elsewhere – a satisfactory and accurate reading of the genealogies in Numbers. In support of this ruling, he cites Tobias 6:11–13:

. . . est hic Raguhel nomine propinquus vir de tribu tua et hic habet filiam nomine Sarram sed neque masculum neque feminam ullum habet praeter eam tibi detur omnis substantia eius et oportet te eam accipere coniugem.

[. . . Raguel, a near kinsman of thy tribe. And he hath a daughter named Sara: but he hath no son or other daughter beside her. All his substance is due to thee, and thou must take her to wife.]

---

36 Wasserschleben, *Kanonensammlung*, pp 115–16 (Collectio canonum Hibernensis, xxxii 19). Much the same rule occurs in Collectio canonum Hibernensis, xxxii 9 (Wasserschleben, *Kanonensammlung*, p. 112): *De hereditate non habentis filios servanda filiae ceterisque post eam heredibus* [That the inheritance of a man without sons is to be held by a daughter, and after her by the other heirs]. This is supported by a citation from Numbers 27:8–11: '*Homo cum mortuus fuerit absque filio ad filiam eius transibit hereditas, si filiam non habuerit habebit successores fratres suos, quod si fratres non fuerint dabitis hereditatem fratribus patris eius; sin autem nec patruos habuerit dabitur hereditas his qui ei proximi sunt*'.

37 Wasserschleben, *Kanonensammlung*, p. 171, note cc. The MSS that contain this book are: British Library, London, Cotton MS Otho E XIII (s. X in.); Bodleian Library, Oxford, Hatton 42 (s. IX); Rome, S. Maria Valicella XVIII (s. XI).

She is, of course, his father's brother's daughter. The leitmotiv of the book is the obligation to marry a woman of one's patrikin (1:9, 4:12–13, 6:16–19, 7:9–14). When Tobias married her, he received half of his father-in-law's property, and the balance later.

The provisions of the vernacular law and Latin canon law are, to a degree, different in detail, but they spring from the same principle. This lack of agreement may be due to diversity of regional custom or to changes that took place over time or to differences of opinion amongst the lawyers. However, it is evident that Hebrew law and Irish law (canon and vernacular) are at one in essentials in regard to female inheritance, and the latter is a borrowing of the former. How did this situation come about? Did the clerical lawyers succeed in replacing inherited native custom with Hebrew law? We know nothing at all about this hypothetical native custom that was evidently displaced by the work of the canonists, and we cannot even guess at its nature. When we first encounter it, the Irish law governing female inheritance (and parallel cousin marriage) is evidently a borrowing from the Old Testament and the Irish lawyers continued to adhere to this legal resolution of the problem of inheritance long after marriages of this kind were forbidden in the West. What, then, will appear to reformers – and to some modern historians – as a stubborn adherence to pagan practices is, in fact, an old-fashioned rule firmly based on Scripture. It is difficult to tell how old this rule is but I should not be surprised if it were in place by the end of the sixth century.

In the case of marriage law, the Latin canon law and the vernacular law are complimentary. *Cáin lánamna* is, for the most part, a practical guide to equity in regard to marital property and to its apportionment in the case of divorce rather than an ethical treatise as such. But it is evident from both *Cáin lánamna* and the Latin tracts that whilst the church prudently tolerated many different kinds of marital arrangements its aim was christian monogamy.[38] The *cétmuinter* ('principal wife') is so privileged that one may feel that for the lawyers she is the real wife (and women would have stood out for these privileges),[39] and the secondary wife, in *Cáin lánamna* and increasingly in later tracts, is termed *adaltrach* 'adulteress'.[40] Beside pragmatic provisions for life as it was actually lived in society, one can find a spirited defence of christian monogamy in the vernacular laws:

---

38 Collectio canonum Hibernensis, xlvi 14 ('*De eo, quod non accipienda uxor, vivente priore, licet adultera*'), 15 ('*De adulterio femiae non celando et penitentia ejus recipienda et alia uxore non ducenda*'), 18 ('*De concubinis non habendis cum legitima uxore*'), 31 ('*De omni adultero excommunicando*') (Wasserschleben, *Kanonensammlung*, pp 188, 190, 193).

39 To the privileges in the texts cited above one can add the canon law: *Non est dignus fideiussor fieri servus nec peregrinus nec brutus, nec monachus nisi imperante abbate, nec filius nisi imperante patre, nec femina nisi domina, virgo sancta* (Wasserschleben, *Kanonesammlung*, 122; cf. R. Thurneysen, 'Zu der Etymologie von ir. *ráth* 'Bürgschaft' und zu der irischen Kanonensammlung und den Triaden' in *Zeitschrift für celtische Philologie*, xviii (1930), pp 364–75: p. 368). The *domina* and the holy virgin are exceptionally the two women capable of acting as sureties, and *domina* is almost certain to be identified with the *bé cuitchernsa* ('woman of joint dominion') of *lánamnas comthinchuir* (Binchy, *Studies in early Irish law*, p. 233. A tract on status of c. AD 700, *Críth gablach*, represents the *bó-aire* (the characteritic substantial farmer) as living in legitimate marrige with a *cétmuinter* of his own class ('*a ben, ingen a chomgráid inna coir chétmuinterasa*'), the precise prescription of *Cáin lánamna*: see D. A. Binchy (ed.), *Crith gablach*, Mediaeval and Modern Series xi (Dublin, 1941), p. 8, §15.

40 *Studies in early Irish law*, p. 49, §23 (the bridewealth paid her, falls forfeit to the innocent *cétmuinter*); p. 71, §35 (the wergild for her rape or *sleith* is half that of a *cétmuinter*); N. Power, 'Classes of women described in the *Senchas Már*' in *Studies in early Irish law*, pp 84–89.

Ar nach rath érenar ni athcuirither corub comísel comarba uasal fir h-ísel 7 corub comuasal comarba ísel fri h-uasal, ar is é triar nad scara commaid co bás: céile fria thigerna iar ndígbáil tséd do dernuind, manach fria airchindech, cétmuinter dligthech fria céile iar n-urnaidm etar dá daingen, ar it díetarscarta iar comrac 7 comlebaid co ro scara lám fria taíb 7 cend fri coland 7 tenga fri comlabra. Amail nad scarat-saide co bás, ní arscara manach fria airchindech co saigid n-éca, ar ní arscara céile fria tigerna ná cétmuinter dligthech fria céile co n-adnacal díb línaib.[41]

[For every fief that is granted is not returned until a noble heir is as base as a base one and a base heir is as noble as a noble one, for these are the three that do not break their partnership until death: a client and his lord after taking chattels from his palm, a monastic tenant and his superior, and a legal first spouse and his/her partner after a marriage contract witnessed by two firm sureties, for they are indissoluble after sexual intercourse and sleeping together until the hand part from the side, the head from the body and the tongue from speech. As they do not part until death, so a monastic tenant does not part from his superior until death, for a client does not part from his lord or a legal first spouse from his/her partner until they are both in the grave.]

Other rules that govern aspects of women's lives may equally derive from clerical law-making. One of the more interesting of these concerns women's right to contract (or rather the lack of it):

Messom cundrada cuir ban. Air ní tualaing ben ro ria ní sech óen a cenn: ada-gair a athair i mbe ingen; ada-gair a cétmuinter i mbi bé cétmuintere; ada-gairet a mmeicc i mbi bé clainne; ada-gair fine i mbi bé fine, ada-gair eclais i mbi bé ecailse. Ní tualaing reicce ná creicce ná cuir ná cundruda sech óen a cenn, acht tabairt bes téchta d'óen a cenn cocur cen díchill.[42]

[The worst of transactions are women's contracts. For a woman is not capable of selling anything without the consent of one of those who has authority over her: her father looks after her when she is a girl; her *cétmuinter* looks after her when she is a *cétmuinter*; her sons look after her when she is a widow with children; her family looks after her when she is a woman of the family [a widow without living father, spouse or children]; the church looks after her when she is a woman of the church. She is not capable of selling or buying or contract or transaction without the consent of one of those who has authority over her, apart from a proper gift to one of those authorities with agreement and without neglect.]

Binchy saw these as inherited provisions and compared them with early Roman law and Indian law[43] but it is possible that one need not go so far afield or argue for separate development. Very likely, the Irish lawyers found these principles – as they found others – in Roman law and adapted them for their own purposes. It is surely interesting that in three of the texts that deal with women's lack of capacity to contract,[44] there are specific rules governing cases that involve the church or its dependent tenantry. Further research may show that the early Irish law in regard to women was more innovative – one might even say original – than scholars have thought.

---

41 CIH 2230–31.
42 CIH 443; R. Thurneysen, *Irisches Recht*, p. 35. §38; Binchy, *Studies in early Irish law*, pp 213–14.
43 *Studies in early Irish law*, p. 223.
44 For the others, see CIH 522 (=*Studies in early Irish law*, p. 212, §7) and CIH 351 (=R. Thurneysen, 'Aus dem irischen Recht, iv' in *Zeitschrift für celtische Philologie*, xvi (1926), p. 177, §12; p. 181, §12).

6 The *Lex Innocentium*: Adomnán's Law for Women, Clerics and Youths, 697 A.D

Máirín Ní Dhonnchadha

THE Law of Adomnán, titled *Cáin Adomnán* in Irish and *Lex Innocentium* in Latin, is unique among the laws of early medieval Ireland in that not only is the identity of the person at whose behest the law was proclaimed known, but also the exact year of its proclamation. The Law is also a great milestone in the written history of women in Ireland as the earliest surviving law concerned primarily with their welfare and very probably the first law with this focus to have been enacted in the country. Adomnán, abbot of Iona and head of the Columban *paruchia* from 679 until his death in 704, was an illustrious scholar and exegete as well as an extremely important figure in ecclesiastical and secular politics. The body of medieval writings by him and about him affords us a rare opportunity to conjecture the mind of the individual behind the Law and the thinking of the society to which he responded.[1] The Law survives today in a composite tract of which there are copies in two late manuscripts.[2] The Law proper is preceded in this tract by a Middle Irish introduction, dating from the late tenth or early eleventh century. The perspective on the Law found in this introduction is a valuable bridge between the time of its enactment roughly three hundred years earlier and a modern assessment one millennium later.

Before dealing with the formulation and concerns of *Cáin Adomnán*, it may be noted that this Law belongs to a body of legislation essentially different from *aurradas* or customary law. First of all, the Law was formally promulgated at a great synod in Birr in the year 697, and took its force from the fact of its enactment, not from time-honoured custom. Secondly, it introduced a jurisdiction specific to its own tenets and entirely without precedent in Irish law: it was to be effective over the whole island of Ireland and also over Britain. However, while the Law actually states that it was enacted *for feraib Érenn ocus Alpan* and *in Hibernia Britanniaque*, it is reasonable to suppose that only those areas of Britain over which the Columban *familia* had influence were targeted.[3] A further departure was that the embodiment of the Law in written form was essential to its instrumentality; a list contemporary with its promulgation of

---

1 For references to primary and secondary materials dealing with Adomnán and the Columban familia see the comprehensive study by Máire Herbert, *Iona, Kells and Derry: the history and hagiography of the monastic familia of Columba* (Oxford, 1988), Particular references or more recent work will be cited below.
2 The fifteenth/sixteenth-century manuscript (Bodleian Library, Oxford, Rawlinson B. 512, fos 48a–51v) and the seventeenth-century O'Clery manuscript (Bibliotheque Royale, Brussels, MS 2324–40, fos 76a–82b).
3 All translations below are from a new critical edition which I am preparing for publication (henceforth Law) but my numbering of the paragraphs corresponds to that in the earlier edition by K. Meyer, *Cáin Adamnáin* (Oxford, 1905); the above citations are from Law §§28, 34. For lack of space, all citations are kept to a minimum and given only in translation.

ninety-one ecclesiastical and secular guarantors was incorporated into the text of the Law. *Cáin*-laws such as the Law of Adomnán do not belong to a tradition which accommodates variant interpretations, or options in the precedents one may follow. While they are in force, they take precedence over any previous legislation that deals with the same matters. They are enacted for a specific period of time; the tenets of *Cáin Adomnán* are stated to be binding in perpetuity.[4] In *cáin*-law, the fines for offences for which there is a previous prescription in *aurradas* are greatly increased. A general principle in many cases is 'doubling of fines for every breach of *cáin*-law'. The Law of Adomnán legislates for a number of different offences against women, clerics and youths, ranging from murder to insulting imputation. Particular mention is made of violence perpetrated by people in large groups, extending to numbers of one thousand and more. Violation of church property is among the Law's secondary concerns. However, the draconian doubling of fines applies only in the case of offences against one of the three categories of persons discussed, that of women, but as we shall see, this category was its main concern.

The text of the Law defines it as 'a perpetual law for clerics, and women, and innocent youths until they are capable of killing a person and of taking their place in the *túath'*.[5] Underpinning the grouping of these three categories is the medieval concept of the just war and its corollary, the exemption of certain categories from the duty to participate even in justified military action. That this is the key to understanding the formulation of *Cáin Adomnán* finds support in the annalistic description of the Law as *Lex Innocentium*.[6] Although the Irish term *ennac* 'innocent' is used in the Law itself only of youths (*maic*), the annalistic references clearly indicate that all three categories are being subsumed under the one term. What *innocentia* connotes is that these categories have no obligation to take up arms, unlike *laici*, 'adult laymen', all of whom may be viewed as potential soldiers. In the case of male children, the exemption is usually temporary: the definition cited above clearly conveys that once adulthood has been reached, military service for just cause is expected of laymen. But if the 'innocent youths' are in fact clerical students, their exemption will continue in adult life. Provision for this is indicated in paragraph 35 of the text which speaks of the 'clerical student or the "innocent youth"'. Thus, while the term *macc* is used in other sources of boys and girls alike, the formulation in Adomnán's Law excludes the latter category.

The medieval Irish church absorbed the theories of what constituted just war and just cause, in particular as developed by Augustine. Augustine built on Cicero's idea that the precondition of a *justum bellum* was a *causa belli*: there had to be clear evidence that the offender had infringed pre-existing rights, public denunciation of the injustice and a formal declaration of war. Just war consisted not in a wilful exercise of violence but in an exercise to recover lost goods (*rebus repetitis*). A breach of contract in private law justified a civil suit to recover *damna* and *iniuriae*; similarly, the state or nation could proceed against an enemy who had injured its interests or undermined its territorial claims. Augustine took this definition further. For him, *iusta bella ulciscuntur iniurias*, 'just wars avenge injuries'. Therefore, one could avenge not only attacks on existing legal rights but also on the moral order. However, a just war had to be

---

4 *Law* §§28, 29, 34, 39.
5 *Law* §§34.
6 *A.U.* (*Chron. Scot., Ann. Rosc.*), s.a. 697.

waged on legitimate authority: lay christians should only take up arms when warfare was properly sanctioned and private individuals only in self-defence (and even this latter was disallowed by Augustine).[7]

It has been pointed out by Edward Ryan that in the first two centuries after Christ, the numbers of christians in Rome's imperial armies were few. It was implicit that baptised christians should avoid conscription if possible, but at no stage in the pre-Constantine era did the church issue a ruling against military service. Nevertheless, the question of how to reconcile civil obedience which included military service under the emperor with the tenets of the faith had to be addressed, as Christianity made substantial numbers of converts and Rome found it impossible to do without the services of its christian soldiers. The church's difficulty with christians enlisting did not arise from any rejection of the legitimacy of warfare but from fears that the soldier's duties and lifestyle would undermine his christian morals and conflict with the practice of his religion. The oath which bound him to promote the emperor's interests first and foremost might well lead to a conflict with christian imperatives, and even to killing fellow-christians for the avowal of their faith. Participation in emperor worship, sacrifice to the gods, and veneration of the *signa* of the legions were obviously unacceptable. Since soldiers could not marry, they might well be prey to sexual licentiousness. A small minority of early christians espoused pacifism from the beginning. With the clearly defined pacifist stance of Origen and Tertullian came the call for the removal of christians to the sidelines of imperial wars, their contribution to consist only of prayer for victory for the just party. However, once Constantine gave official recognition to Christianity, and the threat to soldiers practising their religion was removed, the church was free to support and define military service by christians and their active participation in just wars, which it duly did thereafter.[8]

Since Ireland was outside the ambit of imperial Rome, one does not expect to find arguments in support of christians avoiding military service in pagan legions. On the other hand, the duties of defence and prosecuting just wars will fall on the population at large, without differentiation by denomination, since christians cannot appeal to the existence of an organised military machine of non-christian base to serve in place of themselves. Irish sources proceed to deal with the question of just war, legitimate authority and immunity for certain categories on the assumption that the legitimate military will be recruited from the general – that is, christian – populace.

However, allowing for the different circumstances, there is a clear echo of the problems identified for early christians serving in pagan armies in the Irish formulations of proscribed forms of warring. It has been demonstrated that Latin *laicus* 'layman' was borrowed into Irish as *láech* which developed the sense 'warrior' when all laymen were perceived as potential arms-bearers.[9] *Laicus* assumed pagan connotations because of its inclusive use in reference to the entire non-clerical population, and since 'paganism in the seventh century was most prominently represented by groups more or less outlawed

---

7 This all too brief summary derives from F.H. Russell, *The just war in the middle ages* (Cambridge, 1975), pp 1–39.
8 See E.A. Ryan, 'The rejection of military service by the early christians' in *Theological Studies*, xiii (1952), pp 1–32.
9 P. Mac Cana, 'Two notes' in *Celtica 11*, (1976), pp 125–8; R. Sharpe, 'Hiberno-Latin *laicus*, Irish *láech* and the devil's men' in *Ériu*, 30 (1979), pp 75–92.

and practising brigandage, *laicus* comes to mean "brigand"'.[10] In Irish texts, the most commonly used synonyms of *laicus* in this sense are *díberg, féindid* (member of a *fían*-band, participator in *fíannas*), *mac báis* 'son of death', *mac mallachtan* 'son of malediction'. Various sources describe these categories as taking a *uotum mali* 'oath of evil-doing' which is binding until its malevolent injunctions have been fulfilled, entering into demonic clientship, wearing *signa diabolica* on their heads, harassing and preying upon peaceful christians, and so forth.[11]

The *fían* has been identified as 'a typical enough instance of the widely attested phenomenon of the "Männerbund" or association of wild young warrior-hunters that can be linked in various ways to systems of age-grading', continuing 'an Indo-European prototype with remarkably little alteration'.[12] A more contemporary resonance should perhaps be located for some of the references absorbed into this general formulation in an implicit opposition between christian and non-christian warfare, with the leaders of the latter cast in the mould of tyrannical kings and would-be usurpers who destroyed the *pax catholica*, a mould easily identifiable in Insular sources.[13] The aims of christian warfare are ultimately benevolent, to restore peace to the victimised and to bring the aggressors to charitable justice. Divine sanction is mediated by a christian ruler and victory assured by God if its cause is just. Christian soldiers fight under the sign of the cross, not under *signa diabolica*.

From the time of Constantine, various ordinances were passed to exempt churchmen from military service because it would interfere with their pastoral responsibilities. In Ireland, one can expect similar exemptions for clerics, but when dealing with lay-people, the focus is on the regulation of warfare. *Críth gablach* which dates from a period shortly after the Law of Adomnán describes the *aithech baitside* 'vassal of baptism' as one who is 'guiltless, innocent of theft, of plundering, of homicide except in the day of battle or in self-defence'.[14] Another law-tract, *Míadshlechta*, describes the type of strong farmer known as the *bó-aire* as 'one who does not kill a person except on the day of battle'.[15]

One Irish law concerned with the limits of licit and illicit warfare by kings is *Cáin Fhuithirbe*. This is the earliest of the *cánai* known to have been enacted, a Munster *cáin* drafted by a cleric or clerics and proclaimed sometime around the year 680 during the reign of the Munster king Finguine mac Con-cen-máthair. It survives in fragmentary form. One fragment of text contrasts illicit action when compensation has already been made with just cause when the culprits evade the king's call for redress; it is the king, presumably, who then orders a reprisal: 'Every spoiling by a host which is pursued after *cumals* have been given up is liable. Every burning which is carried out in the circuits of a king when he has bewailed evasion is immune'. This fragment also makes terse condemnation of 'intentional wounding on a battlefield . . . violent taking of hostages . . .

10 Sharpe, *op. cit.*, p. 88.
11 Sharpe, *op. cit.*, *passim*.
12 K. McCone, 'Werewolves, cyclopes, *díberga* and *fíanna*: juvenile delinquency in early Ireland' in *Cambridge Medieval Celtic Studies*, 12 (Winter, 1986), pp 1–22 (p. 22).
13 On Bede's views, for example, see T.M. Charles-Edwards, 'Bede, the Irish and the Britons' in *Celtica*, 13 (1983), pp 42–52.
14 D.A. Binchy, *Críth gablach* (Dublin, 1970), p. 6.
15 D.A. Binchy, *Corpus iuris hibernici* (Dublin, 1978), ii, p. 584, lines 27–8.

evading a hosting, a deserted hosting, wounding (treacherously) on (his) land
... a battle without the other side having just cause'.[16]

In the mid-seventh-century Hiberno-Latin text *De duodecim abusiuis saeculi*,
the ninth abuse details the reign of the *rex iniquus*.[17] This section was copied in
a somewhat garbled fashion into Book 25 of the *Collectio Canonum
Hibernensis*.[18] The *Collectio* survives in two recensions; in the earlier, A, the last
cited author is Theodore of Canterbury (died 690), in the later, B, Adomnán
himself. In one early manuscript, the work is attributed to Ruiben (died 725) of
Dairinis, an island in the Blackwater estuary near Youghal, and Cú Chuimne
(died 747), a cleric of Iona. Even if we did not have the *Collectio*'s citation from
Adomnán himself, or the early reference associating the work with Iona, the
more substantial matter of the extensive library of source-materials which must
be predicated for it and the work's broad ambition to provide a clear line of
guidance on issues on which there was diverse opinion, among seculars *and*
clerics, would point to Iona as the place of compilation and to Adomnán's influence in the enterprise, whether direct or indirect. Early in this century, a case
for Adomnán's authorship was mooted but not substantiated.[19] However, since
then, an extremely strong case has been made for the influence of Adomnán's
views on the compilers of the *Collectio* on the question of kingship. Much scholarly attention had already been paid to Adomnán's retrospective elevation of
certain of his earlier Uí Néill royal kinsmen in the *Uita Columbae* to the status
of monarchs of all Ireland, predestined or ordained by God. This has been seen
as a strategy to propagandise christian concepts of kingship in his own day, and
to underline for future generations the benefit he believed must continue to
accrue to the whole island from the hegemony of the Uí Néill kings.[20] Recent
work by Michael Enright has shown just how finely tuned an invocation of the
biblical precedents of Samuel, Saul and David in *I Kings* is Adomnán's paradigm for legitimate kingship with appropriate sacral ordination in the *Uita
Columbae*, for both Irish and Dál Riata kings.[21] Furthermore, Enright has demonstrated that Adomnán's thinking on christian kingship, his advocacy of the
church's central role in anointing and choosing kings, and his recourse to scrip-
tural precedent is reproduced by the compilers of the *Collectio*. 'Consequently',
he concludes, 'the references to a new consecration rite in *Vita* and *Collectio*
can not be treated as isolated, unconnected passages, but must be interpreted
as evidence of a common interest and at least a partially common goal'.[22] It is
reasonable to expect that other aspects of kingship and just rule as described in
the *Collectio* were in accord with Adomnán's own views.

As mentioned above, the text concerning the *rex iniquus* from *De duodecim
abusiuis saeculi* is cited in the *Collectio*. Chapter 3 of Book 25 (*De Regno*) states:
'The iniquity of an unjust king disrupts the peace of the peoples, raises impediments to government, diminishes the produce of the lands, leads to the deaths

16 For text, translation and discussion see L. Breatnach, 'The ecclesiastical element in the Old-Irish legal tract
     Cáin fhuithirbe' in *Peritia*, 5 (1986), pp 36–52; original of citations, *ibid.*, p. 39. The basic meaning of *cumal* is
     'female slave'; as a unit of value, it might be rendered in different currencies – slaves, livestock, grain, silver,
     etc.
17 S. Hellmann (ed.), *Psuedo-Cyprianus De XII abusivis saeculi* (Harnack and Schmidt, *Texte und Untersuchungen
     zur Geschichte der altchristlichen Literatur, xxxiv*) (Leipzig, 1909), pp 1–61.
18 H. Wasserschleben (ed.), *Die irische Kanonensammlung* (Leipzig , 1885).
19 See E.W.B. Nicholson, 'The origin of the "Hibernian" collection of Canons' in *Z.C.P.*, 3 (1901), pp 99–103.
20 For the most recent discussion see Herbert, *Iona, Kells and Derry*, pp 52–6.
21 M.J. Enright, *Iona, Tara and Soissons* (Berlin, New York, 1985), pp 1–106.
22 *Ibid.*, p. 48.

of loved ones, allows invasion of enemies into the border-territories . . .[23] This is reminiscent of the description of the effects of *fíannas* in the early eighth-century *Apgitir Chrábaid*: 'it contracts territories; it increases enmity; it annihilates life, it prolongs torments'.[24] Chapter 4 describes the reign of a just king: 'The justice of a king is the peace (*pax*) of the peoples, the protection of the *patria*, the immunity of the populace, the defence of the country, the care of the weak . . .'.[25] Book 55, *De Vera Innocentia*, deals succinctly with all christian subjects' obligation not to break the king's *pax*, contrasting the passing innocence of children who are not of an age to wound, with the adult's true innocence which consists in harming none, though he or she has the strength to do so.[26] Such strength, however, is depicted in Book 46 as unnatural in women: 'The name of 'man' (*vir*) comes from 'strength' (*virtus*), that is, for making war, working, defending, ruling and public speaking; the name 'woman' (*mulier*) on the other hand is derived from 'softness' (*mollitia*), that is, from fragility, weakness, lowliness and subjection'.[27]

Special ordinances were sometimes required to deal with the kind of crisis which would overwhelm the smooth working of customary law. In Irish sources the terms most frequently employed for these extraordinary decrees are *cáin*, *rechtgae* and *recht* and to a fair extent these terms are interchangeable; a treaty between territories is known as *cairde*. The law-tract *Córus Béscnai* says: 'Every overlord [pledges] on behalf of his people. Violence is abated by his good procedures of *cáin* and *rechtgae* and good custom and inter-territorial treaty'.[28] It is not surprising to find royal decrees to initiate reprisals contextualised with decrees to restore a ravaged populace to good order and an enhanced peace. *Críth gablach* describes the three types of *rechtgae* that it is proper for a king to enjoin on his people: 'a *rechtgae* when they have been defeated in battle that he may consolidate his peoples thereafter lest they destroy one another; and a *rechtgae* after a mortality; and a king's *rechtgae* such as the *rechtgae* of the king of Cashel in Munster. For there are three *rechtgaes* which it is proper for a king to bind on his peoples: a *rechtgae* for the expulsion of a foreign race, for example, against the Saxons; and a *rechtgae* for the cultivation of produce; and a *recht* to kindle the faith, such as the Law of Adomnán'.[29]

There is a neat homology between the three types of situations described in each of these two sentences. The first situation describes an aggressive war which calls for a reprisal; the second a mortality resulting from plague or famine, very possibly as a result of the first, and this requires organised action to produce food for the weakened community; the third ensures an improvement on the *status quo ante bellum*. While the '*rechtgae* of the king of Cashel in Munster' has not been identified (though I suggest *Cáin Fhuithirbe* as a possible candidate), the phraseology points to a *cáin*-law similar in type to the Law of Adomnán, proclaimed by a king or kings in conjunction with powerful ecclesiastical leaders; similar phraseology is found in the Irish annals for the series of *cáin*-laws enacted between the years 697 and 887.

It is possible to identify a link between some of the historical events underly-

---

23 Wasserschleben, *Kanonensammlung*, p. 77.
24 V. Hull, '*Apgitir chrábaid*: The alphabet of piety' in *Celtica*, 8 (1968), pp 60–1.
25 Wasserschleben, *Kanonensammlung*, p. 77.
26 *Ibid.*, p. 219.
27 *Ibid.*, pp 191–2.
28 Binchy, *Corpus iuris hibernici*, p. 526, lines 21–3.
29 Binchy, *Críth gablach*, pp 20–21.

ing this formulation in *Críth gablach*. D.A. Binchy was the first to point out that
the expulsion of the Saxons referred to an extraordinary raid on the territory of
Brega dispatched by Ecgfrith of Northumbria in the year 685.[30] The devastation it caused, and the year and location of the raid, are detailed in a number of
sources including the contemporary Irish annals, Bede's *Historia Ecclesiastica*
which he completed in 731, and the tenth-century Irish Life of Adomnán.[31]
One year later, a Pictish-Irish alliance defeated and killed Ecgfrith at
Nechtansmere. It has been convincingly demonstrated by Hermann Moisl that
the Uí Néill, Adomnán's kinsmen, took part in the alliance that revolted against
Ecgfrith in order to support the claim of his half-brother Aldfrith to the North-
umbrian kingship; Aldfrith was related through his mother to the Uí Néill. They
succeeded. Aldfrith was installed as king of Northumbria and, shortly afterwards, Adomnán came to him as ambassador *a sua gente* to request the return
of the captives taken on the raid on Brega.[32] In 687 he conducted the captives
home with great pomp and ceremony.

It can be argued that the enactment of *Cáin Adomnán* was the culmination
of Adomnán's efforts to shield certain sectors of the population from the kind
of violent warfare which ensued from raids such as that on Brega. By granting
permanent exemption from military service to clerics, youths and women, he
re-defined peace as a contract by which the entire populace – and foreigners –
were bound never to take action that would drive these categories to violence
even in self-defence. His Law elides the difference between women as agents
of violent action and women as its victims in an almost imperceptible shift.
Frederick Russell has pointed out that 'the legal foundation of the Roman just
war was the analysis of contractual obligation. The etymology of *pax* stems
from *pangere*, to make a pact or contract wherein the rights and duties of both
parties are specified'.[33] The primary meaning of *cáin* is 'regulation', the secondary one 'punishment (by fine or otherwise) for breach of regulation'.[34] The idea
of *cáin* as a contract binding different parties to peace is well brought out in the
frequent collocation of *síth* 'peace' and *cáin*, and in the compounds *síthcháin*
'peace-compact, peace' and *cáinchomracc* 'peace-convergence, amity'.

Various factors clearly indicate that Adomnán intended the tenets of his Law
to be effective. First, there is the collective strength of the guarantors whom he
co-opted to support it. The list of forty ecclesiastical guarantors is headed by
Flann Febla, the bishop of Armagh, the most powerful church figure in the
country after Adomnán. The list of fifty-one secular guarantors is headed by
Loingsech, Adomnán's Cenél Conaill kinsman who became the Uí Néill
overking in 696. One representative or more figures in the list for every major
kingship, over-kingship and monastic centre in the country. The list also includes Coeddi, bishop of Iona, Conamail, another cleric at Iona, Curetán, bishop
of Rosemarkie in Scotland, and Uuictberct, an English cleric living in the English colony at Clonmelsh near Carlow.[35] The guarantor-list confirms that at the

---

30 *Ibid.*, p. xiv.
31 *A.U., Chron. Scot., Ann. Clon.* s.a. 685; Bede, *Hist. ecc.* (eds) B. Colgrave and R.A.B. Mynors (Oxford, 1969)),
    iv, p. 26; M. Herbert and P. Ó Riain, *Betha Adamnáin. The Irish life of Adamnán* (London, 1988), pp 54–6.
32 H. Moisl, 'The Bernician royal dynasty and the Irish in the seventh century' in *Peritia*, 2 (1983), pp 120–4.
33 Russell, *The just war in the middle ages*, p. 4.
34 M.A. O'Brien, 'Etymologies and notes' in *Celtica*, 3 (1956), p. 172.
35 See further M. Ní Dhonnchadha, 'The guarantor list of *Cáin Adomnán*, 697' in *Peritia*, 1 (1982), pp 178–215.
    On the identity of Uuictberct see D. Ó Cróinín, 'Rath Melsigi, Willibrord and the earliest Echternach manu-
    scripts' in *Peritia*, 3 (1984), pp 21–6.

end of the seventh century, Adomnán was uniquely qualified to transcend the sectional interests of lesser secular and ecclesiastical federations and unite them in common support of this Law.[36] The choice of Birr as the site for its promulgation, a monastery in an area close to the borders of Munster, Connacht, southern Uí Néill and Leinster, drives home the point that he, with Loingsech's backing, could impose what was essentially an Uí Néill law over the entire country.

The opening paragraphs of the Law are framed in the form of disposition-clauses which incorporate the guarantor-list (§§28–9), and sanction-clauses of benediction and malediction (§§30–1), a form reminiscent of the Celtic charter and insinuating the legal force of such documents.[37] Certain modifications of customary law are spelt out, to preclude any recourse to the more lenient judgments of that system. Thus, one general principle is that 'everyone pays for the crimes of his own hand' (§43). If he cannot afford to do so – say, for example, in the case of murder – he forfeits his life and his kindred (*fine*) then makes payment for him. Whatever compromises were tolerated in practice, *cáin*-law states that no criminal is allowed to live while devolving the fines onto his kindred. Furthermore, in another apparent innovation of *cáin*-law, bystanders who do not take action to prevent a crime are unequivocally penalised as severely as the criminal, even to the death penalty (§§35, 47).[38] If a known murderer is not apprehended, liability for full fines devolves on his kindred, liability extending to fines and the death-penalty devolves on those who shelter him, and quarter-liability for fines devolves on the community for failing to bring him to justice. If no suspect can be named, liability for fines devolves on the community living in the area in which the body is found (§46). If grounds for suspicion exist, lot-casting is sufficient to establish guilt (§46). All these measures demanded that the entire population be vigilant in observing and prosecuting the Law, not just those directly affected as kin of the culprit or the victim. Every kin-group (*derbfhine*) and church throughout Ireland was to nominate at least one hostage-surety (*aitire*) who would be taken into custody if the criminal defaulted (§§39,53). To ensure rigour in prosecution, only clerics nominated by Adomnán's *familia* were to preside as judges (§37). All fines were to be met within five days of the bringing of a suit, as opposed to ten days in *aurradas* (§38).

In customary law, two main types of fine are paid to the victim's kin in the case of murder. The first is the fixed penalty for homicide – *éraicc* or *cró* – which is normally seven *cumals*, irrespective of rank. The second is a payment based on the honour-price of the various members of the victim's kin, paternal and maternal, who are entitled to compensation; where parents, brothers, sisters, aunts, uncles and cousins survive the deceased, the total sum may amount to a crippling debt. In the case of any offence other than murder, the main fine will be based on the honour-price of the victim, rather than the victim's kin.[39]

Penalties for almost every offence of which the Law of Adomnán treats are already specified in *aurradas*. In the case of *all* offences against clerics and youths, and offences against women other than murder, the procedure adopted in the Law is to add to the specified fine a superlevy of one-eighth, which then goes to

---

36 For background see Herbert, *Iona, Kells and Derry*, pp 47–56, 142–50.
37 On charter-form see W. Davies, 'The Latin charter-tradition in western Britain, Brittany and Ireland in the early medieval period', in D. Whitelock, R. McKitterick and D. Dumville, (eds), *Ireland in early medieval Europe* (Cambridge, 1982), pp 258–80.
38 On this see F. Kelly, *A guide to early Irish law* (Dublin, 1988), p 156.
39 For a detailed discussion see Kelly, *Guide*, pp 125–41.

Adomnán and his *familia*. This ensures that the Law does not encroach on the fines due under customary law to the kin, lord, or church of the victim. In the case of murder of women, however, there is a radical departure. The customary *éraicc* of seven *cumals* is doubled and the entire sum goes to Adomnán's *familia* (at their discretion, part of it may be returned to the enforcers): 'All have sworn, therefore, both layperson and cleric, to fulfil all of the Law of Adomnán until doom. They have offered the entire *éraicc*-fine of their woman-deaths to Adomnán, and to every heir who will be in his seat until doom. And Adomnán does not steal fines from lord or church or kin to whom there is due' (§29). While it is difficult to be certain, this last sentence presumably relates only to the payments based on honour-price which the victim's kin and guardians would normally expect to receive.

The appropriation of the entire *éraicc* for murder of women by the church in the person of Adomnán is one of the most remarkable aspects of this Law, and its import is difficult to interpret. Does it indicate that murder of women was acknowledged to reach such crisis level, on occasion at least, that political leaders accepted the necessity of church intervention and consequent loss of fines? Or rather, that murder of women was so rare that secular society was willing to forgo the *éraicc*? Two factors suggest that the former scenario is closer to the truth. The first is that the Law twice refers to situations where numbers up to three hundred are involved in slaying (§§33, 35), and once to situations where more than one thousand are involved (§35). The second is that the Law also specifies a fine for 'making use of women in a massacre or a muster or a raid' (§ 52). This does not suggest that women willingly engaged in warfare and, in fact, the evidence for women opting for the warrior life in medieval Ireland is insubstantial.[40] But, as noted above, the formulation of the Law as a *lex innocentium* elides the difference between women as agents and women as victims of violence. There is no shortage of evidence that women, as much as men, suffered in the endemic attacks on settlements, monastic and secular, throughout the period.

In the Middle-Irish introduction to the Law, we are told that Adomnán emancipated women from the double duty of fighting and performing servile work. In a neat piece of creative etymologising, the text says:

Cumalach ('Slavey') was the name for women until Adomnán came to emancipate them and this was the *cumalach* ('the *comla*-worker'): the woman for whom a pit was dug at the head of the sluice-gate (*comla*) so that it hid her nakedness. One end of the crossbar was supported by her until the grinding of the load was done. After coming out of that earth-hole, she dipped a candle four man-fists in thickness into a mound of butter or tallow and that candle remained on her palm until dividing and distributing and bed-making in the houses of kings and superiors had ended. That woman took no share in purse or basket, nor [did she live] in the one house with the house-master but in a cold hut outside the enclosure, lest any malevolence from land or sea befall her lord (§2).

The work which the finest of women used do was to proceed to battle and battlefield, division and encampment, killing and slaughter. On one side she carried her provision-bag, on the other side her infant. Her wooden pole upon her back, thirty feet in length. On one end of it [there was] an iron hook, which she would thrust into the opposite battalion at the hair of the other woman. Her husband behind her, a fence-stake in his hand, flogging her on to battle. For at that time it was a woman's head or two breasts which were taken as trophies (§3).

---

40 Ibid., pp 69–9.

That Adomnán was concerned with improving the general living conditions of women is confirmed in the the text of the Law. For in addition to legislating against the murder of women in battle, it also specifies the fines for their deaths in sites and situations in which they would have performed their daily work.

The sanction of the Law enjoins that payment in full fines is to be made for every woman that has been killed, whether a human had a part in it, or animals or dogs or fire or a ditch or a building. For in *cáin*-law every construction is to be paid for, including ditch and pit and bridge and hearth and step and pool and kiln and every hardship besides, if a woman should die because of it (§41; cf §42).

Fines are also specified for a range of physical injuries to women: a 'white blow' (which leaves no mark), a blow which leaves a livid mark or swelling, spilling of blood, a wound requiring a staunch, serious injury necessitating a leech (§44). Violent rape is the most serious sexual crime dealt with, but there is provision also for sexual harassment by touching a woman outside her clothing, or by putting a hand under her clothing (§50). Denying paternity or unjustly accusing a woman of fornication is severely penalised (§51) as is fathering a child on a woman without a prior and proper contract of marriage – irrespective of the woman's willingness (§52).

The Law exempts men from liability for women's deaths in one noteworthy case, that is, if a woman dies as a result of 'an act of God or proper lawful union' (§42; cf. §33), in other words, in childbirth within lawful christian marriage. In fact, one can discern a clear subtext in the Law whereby Adomnán calls on all laywomen (and men) to accommodate themselves to christian marriage and abandon unorthodox unions. Remarkably, the Law makes no reference what-soever to women in religious life. Early Irish ecclesiastical sources endorsed the idealised three-class categorisation of the faithful: all Christians were to accommodate themselves to one of the three categories of *virgines*, *penitentes* or those *in legitimo matrimonio*. The *virgines* were those who had not lost their virginity and who vowed themselves to perpetual celibacy. Those *in legitimo matrimonio* were to be monogamous, and celibate in periods strictly ordained by the church. The *penitentes*, 'penitent spouses', were those who took a vow of celibacy *after* a period as sexually active adults.[41] The force of Adomnán's Law is not said to extend to the entire lay-population, but only to the church's 'law-abiding laymen with their legitimate spouses who abide by the will of Adomnán and a proper, wise and holy confessor' (§34). By excluding all reference to vir-gin nuns and formerly married women who entered the church, Adomnán gives a new emphasis to women's *militia Christi*. Male 'soldiers of Christ' were ex-empted from military service under a temporal lord to wage spiritual war on the flesh and the devil. Adomnán's Law calls on women to serve not in ascetic denial but in marriage and motherhood.

The Law implies an opposition between men as takers of life (with just cause) and women as givers of life. The Middle-Irish introduction makes this opposi-tion explicit. For example, the perceived incompatibility of the two roles issues in an image beheld by Adomnán and his mother on a battlefield, which suppos-edly drove him eventually to enact his Law: 'Though they beheld the massacre, they did not see anything which they thought more pitiful than a woman's head

---

41 See further M. Ní Dhonnchadha, '*Caillech* and other terms for veiled women in medieval Irish texts' in *Éigse*, 28 (forthcoming).

on one bank and her body on the opposite bank with a child asleep at the breast, a stream of milk on one cheek and a stream of blood on the other' (§5). In the law-text proper, the role of women as givers of life is valorized through assimilation to the role of Mary, mother of Christ and consequently mother of the whole human race. There we are told that Adomnán received a directive from an angel of God who said:'You shall establish a law in Ireland and Britain for the sake of the mother of each one, because a mother has borne each one, and for the sake of Mary, mother of Jesus Christ, through whom the whole [human race] is . . . For great is the sin when anyone kills one who is mother, and sister to Christ's mother, and mother of Christ' (§33). The Middle-Irish introduction also collapses the distinction between 'woman' and 'mother'. The 'makings of a good woman' is equated with a mother, and 'a mother is a venerable treasure, a mother is a goodly treasure, mother of saints and bishops and righteous folk, an increase of the kingdom of heaven, earth's propagation' (§4). The exaltation of women as givers of life probably explains why the Law exempts women who are themselves guilty of murder from capital punishment, substituting the penalty of being alienated and cast adrift 'in a boat of one paddle', their fate thereafter left to God (§45).

Adomnán's and the Columban *familia*'s promotion of the Marian cult is well known. Adomnán devotes four chapters of his *De locis sanctis* to a discussion of a church of the Virgin in Jerusalem, another containing her tomb, a mantle which Mary is said to have woven, and an image of her which, to Adomnán's great indignation, was grossly violated by a Jew.[42] Cú Chuimne's famous hymn to Mary, '*Cantemus in omni die*', celebrates Mary's reversal of Eve's role by bringing salvation to humankind.[43] The earliest non-historicised images of the Virgin Mother and Child appear on the crosses of Iona and in the Book of Kells. What has received less attention is the Columban *familia's* engagement with the definition of christian marriage in the same period. Adomnán's Law offers no definition of what constitutes christian marriage. But Book 46 of the *Collectio canonum hibernensis* ('*De ratione matrimonii*') provides a blueprint which cuts a clear line through practices which the Iona theologians regarded as reprehensible although tolerated in other Irish legal sources arguably compiled by clerical writers. Both partners should be virgins until marriage and continent in marriage. One spouse should put aside the other for fornicating until such time as he or she has done penance, but the offended party may not take another partner or remarry; they should be reconciled after penance. The church is to excommunicate recalcitrant adulterers. The keeping of concubines is unacceptable at all times. A second marriage when one's spouse is dead is tolerated but not advocated. Widowers, if remarrying, should not marry widows. Exceptionally, if either spouse enters religion, the other may remarry, but a wife is exhorted to take a vow of celibacy. Another exceptional precept discriminating in men's favour allows that a man may take a second partner if his wife is infertile.[44] But on the whole, it can be said that christian marriage as defined in the *Collectio* disallows polygyny, concubinage, repudiation of wives, divorce and multiple, though serial, unions – all of which were common practice in Irish society, among the upper classes at least. Reference has been made

---

42 D. Meehan, *Adomnan's De Locis Sanctis* (Dublin, 1958), pp 48, 56–8, 118.
43 J.H. Bernard and R. Atkinson (eds), *The Irish Liber Hymnorum* (London, 1898), i, pp 32–4; for discussion of hymn see P. O'Dwyer, *Mary: a history of devotion in Ireland* (Dublin, 1988), pp 54–7.
44 Wasserschleben, *Kanonensammlung*, pp 185–95.

already to a citation from Adomnán which appears in B-recension of the *Collectio*, and the same citation also appears in the *Canones Adomnani*. It, too, disallows remarriage but significantly, casts the miscreant as a woman: 'Of a wife who is a harlot, thus the same man explained, that she will be a harlot who has cast off the yoke of her own husband, and is joined to a second husband or a third. Her husband shall not take another (wife) while she lives'.[45]

While Adomnán's *Uita Columbae* is ostensibly a biography of Colum Cille, it undoubtedly was also a vehicle for Adomnán's own views and, in it, we find his view of christian marriage presented with the endorsement of his predecessor. The longest narrative concerning a woman tells how Colum Cille miraculously restores a wife's affection for a husband whom she had found physically repulsive. The saint had asked:

> "Why, woman, do you attempt to put from you your own flesh? The Lord says, 'Two shall be in one flesh'. Therefore the flesh of your husband is your flesh". She replied: "I am ready to perform all things whatsoever what you may enjoin on me, however burdensome: save one thing, that you do not constrain me to sleep in one bed with Lugne. I do not refuse to carry on the whole management of the house; or, if you command it, even to cross the seas, and remain in some monastery of nuns". Then the saint said: "What you suggest cannot rightly be done. Since your husband is still alive, you are bound by the law of the husband; for it is forbidden that that should be separated, which God has lawfully joined".

The saint asked both husband and wife to fast and pray with him. The following day, she said: 'I know now that your prayer concerning me has been heard by God. For him whom I loathed yesterday I love today. In this past night, (how, I do not know) my heart has been changed in me from hate to love'.[46] The narrative immediately preceding this tells how Colum Cille miraculously relieved 'A certain young woman who was suffering, as a daughter of Eve, great and very hard pangs of childbirth'.[47] Yet another tells how Colum Cille induced a druid who kept a slave-girl to part from her.[48] Two further narratives tell how evil men met their deaths while lying on a couch with the harlot of the Apocalypse.[49] In the entire *Uita*, one finds only one narrative concerning a religious woman.[50]

The enactment of Adomnán's Law surely appears to have been motivated by benevolence towards women, true indignation at the hardships they endured, and some compassion even when they were guilty of murder. In return, Adomnán's expectation of them was that they should conform to the role and model of marriage which he advocated, presumably, to his thinking, as being in their best interests. Adomnán must also have held in respect women who opted for a single life, or life in religion, but there is no evidence that he was willing to see real power in their hands.

---

45 The B-recension is not published but the citation is quoted in Enright, *Iona, Tara, and Soissons*, p. 24, note 83; for the citation in the *Canones Adomnani* see L. Bieler, *The Irish penitentials* (Dublin, 1975), pp 178–9.
46 A.O. and M.O. Anderson (eds), *Adomnan's life of Columba* (Edinburgh, 1961) pp 437–41.
47 *Ibid.*, pp 435–7.
48 *Ibid.*, pp 399–405.
49 *Ibid.*, pp 291–3.
50 *Ibid.*, pp 337–9.

7   The *Banshenchas* Revisited

Muireann Ní Bhrolcháin

DESPITE the prominence of women in early Irish sagas and the consequent attention accorded them by modern commentators it is widely recognised that the historical documents of the medieval period give them scant space. A notable exception is the twelfth century *Banshenchas* or *The lore of women*, extant in both metrical and prose form which contains a detailed account of the marriages contracted by Irish aristocratic women up to and including the twelfth century, concluding with the very early Norman period.[1] It first appears in metrical form composed by Gilla Mo-Dutu Ó Caiside who is described as follows in the *Lebor Gabála Érenn* (*Book of Invasions*): 'Gilla Mo-Dutu ... he was blind and flat-faced and he never chanted falsehood or a crooked history.'[2]

In a colophon the poet informs us that he came from Ardbracken, County Meath and that the poem was written in 1147 on Devinish Island situated on Lower Lough Erne in County Fermanagh.

Exactly one thousand, one hundred and forty seven years of pure deeds have passed since the birth of Christ, a fitting memory, until I related the companies of women.

Gilla Mo-Dutu composed this poem with noble harmonies. He came from Ardbracken, he is not worried or distressed by minor matters in Devinish of the gracious speeches and of the clear-coloured holy arts.[3]

There are four extant copies of the poem. The first, in the contemporaneous Book of Leinster, seems abbreviated, particularly towards the end and the remainder is somewhat inaccurate compared with the other manuscripts: The Book of Uí Mhaine (fourteenth/fifteenth century), The Book of Lecan (fifteenth century) and finally the fourteenth-century manuscript known as G3 in the National Library of Ireland.[4]

Gilla Mo-Dutu purports to catalogue famous women, primarily Irish, from the time of Eve up to and including the date of the composition of the text. Apart from brief references to some obviously non-Irish biblical figures such as Adam, Eve, Noah and their families in the opening stanzas, the vast majority of characters are Irish with the inclusion of some Viking characters in the appropriate historical period.

The structure of the poem indicates that the author confined himself to cataloguing the wives, daughters and families of the reputed high-kings of Ireland.

---

1 Margaret Dobbs (ed.), 'The Ban-shenchus' in *Revue Celtique*, xlvii (1930), pp 283–339; Muireann Ní Bhrolcháin (ed.), '*An Banshenchas Filíochta*' (unpublished M.A. thesis, University College, Galway, 1977). Cf. Muireann Ní Bhrolcháin, 'The manuscript tradition of the Banshenchas' in *Ériu*, xxxiii (1982), pp 109–35.
2 R.A.S. Macalister (ed.), *Lebor Gabála Érenn* (Irish Texts Society, vol. 44, Dublin, 1942), pp 412–3.
3 Margaret Dobbs, *op cit.*, pp 315, 339.
4 *The Book of Leinster* ((eds) R.I. Best and M.A. O'Brien, Dublin, 1956); *The Book of Uí Mhaine* (introduction by R. A. Macalister, Dublin, 1942); *The Book of Lecan* (introduction by K. Mulchrone, Dublin, 1937).

He divides the poem into two separate sections, pre-Christian and Christian with the latter based directly on the extant lists of the high-kings. He thus pays little or no attention to the minor royal families who are treated at length in the prose version. Gilla Mo-Dutu comments that he does not intend to insult anyone by his composition: 'I omit the strict enumeration of the harlots and base offspring and evil women and people of low birth of the generations, so that there would be no true dishonour against the king by listing them before their people.'[5]

Gilla Mo-Dutu is also the reputed author of another long composition entitled *Éri og inis na Naem* (*Ireland the pure island of saints*) which is part of the *Lebor Gabála*, dated 1143 and enumerates the kings of the Christian era. It is possible that the *Banshenchas* was intended as its companion piece.[6] The *Banshenchas* contains a short eulogy to Tigernán Ua Ruairc, his overlord in Devinish, and may have been composed as a tribute to him or at his request. Tigernán's wife, Derborgaill, was the daughter of Murchad Ua Maeil Sechlainn, king of Meath where Ardbracken is situated. Moreover, the poem ends with a tribute to Tigernán, Derborgaill and her parents. Perhaps Gilla Mo-Dutu intended the poem to honour Derborgaill. It is, therefore, possible that she was the inspiration for this unusual metrical composition.[7]

The prose version is anonymous and appears to be an expansion of the original poem compiled between 1147 and 1169, just before Strongbow came to Ireland.[8] There are eight extant copies of the prose version and two are of particular importance in this context: the fifteenth-century Book of Lecan and the Book of Uí Mhaine written towards the end of the fourteenth and the beginning of the fifteenth centuries. Both contain a metrical version also.[9] These contribute far more information on the historical period (the eleventh and twelfth centuries particularly) and illustrate the marriages contracted by the contemporary Irish dynasties and contain genealogical material which demonstrates distinct regional bias.

The Book of Lecan also incorporates a long genealogical tract compiled by a scribe who had a particular interest in the Uí Maeil Sechlainn family of Meath. The anonymous scribe/author may have used an independent, local tract as his source, and some entries have been duplicated in an attempt to tabulate the members of this family. Despite the author's best efforts this was not fully achieved and no similar genealogical tract has come to light elsewhere. The Uí Mhaine prose copy has a strong Leinster flavour and a distinct interest in the family and marital connections of Diarmait Mac Murchada. It is the only manuscript which mentions the historic marriage of Aífe, daughter of Diarmait to Strongbow or Richard fitz Gilbert de Clare. This copy also blames Tiernán Ua Ruairc of Bréifne for the flight of Diarmait to England:

---

5 Omitted from Dobbs' edition.
6 *Lebor Gabála Érenn*, pp 540–65.
7 It has been suggested recently by Máire Ní Mhaonaigh that he may have based the poem on a shorter exemplar and added a substantial amount. Could he have added to his own original poem at a later date? Cf. Máire Ní Mhaonaigh, 'Bréifne Bias in *Cogad Gáedel re Gallaib*' in *Ériu*, 43 (1992), pp 154–56.
8 Margaret Dobbs, 'The Ban-shenchus' in *Revue Celtique*, xlviii (1931), pp 136–233; Muireann Ní Bhrolcháin (ed.), 'The prose *Banshenchas*' (unpublished Ph.D. thesis, University College, Galway, 1980). Indices to the texts were compiled by Margaret Dobbs in 'The Ban-shenchus' in *Revue Celtique*, xlix (1932), pp 437–89.
9 There are also copies in *The Book of Uí Mhaine*; *The Book of Ballymote*; in Trinity College Dublin, MS H 3 17; in the National Library in Brussels (written by Míchéal Ó Cléirigh in Louvain, January 1642) and in the National Library of Scotland in Edinburgh.

Mór . . . mother of Derborgaill wife of Tigernán Ua Ruairc who was the instigator of all, nearly all, the miseries of Ireland, that is by the expulsion of Diarmait, it happened that he journeyed over sea after a time and brought the foreigners with him to Ireland.[10]

Another entry mentions Diarmait's mother: 'Inchdelb daughter of Cernachan Ua Gairbita, mother of Diarmait son of Donnchad son of Murchad high king of Leinster and of the Foreigners and of most of all Ireland'[11] and also included are 'Aife wife of the Earl'[12] and 'Sadb wife of the watchman of Diarmait king of Leinster'.[13]

These tracts contain the most comprehensive view hitherto available of aristocratic marriages. Some women who are referred to in the annals are omitted by the *Banshenchas*, perhaps because they had no offspring and were not worthy of mention. The content of the prose tract is, therefore, frequently uncorroborated but when the information is verifiable it is unfailingly accurate. It seems reasonable to assume therefore, that the authority of the uncorroborated evidence may generally be accepted also.

There is very little or no comment on the women themselves in the prose texts, with the occasional exception. Gilla Mo-Dutu was a little more forthright in his views and comments perhaps due to the strictures and constraints of his complicated metre rather than to his own personal opinions. A typical entry is a simple bald statement of fact as follows: 'Caillech Dé daughter of Ua Eidin, mother of Ruaidrí Ua Conchobair and of Muirchertach and Tairrdelbach Ua Briain king of Munster.'[14]

The text is sparse, lacking any indication of the chronological order of marriages, divorces and subsequent remarriage which would make it easier to connect them to ongoing historical events. No data is given on fosterage, property, inheritance or status but the majority of the women included are from the upper echelon of Irish medieval society with some notable exceptions. The principal aim of these sections is to document the royal marriages and offspring of the women and they are very infrequently described as the mothers of other than royalty. Poets, bishops, coarbs or others of high rank seldom feature and the mothers usually had royal offspring as well. There are some exceptions to the rule such as: 'Caintigern daughter of Guaire Ua Lachtnáin of Teathba the learned man of Clonmacnoise mother of the two sons of Flann Ua Maeil Sechlainn .i. Murchad and Domnall.'[15]

The marriages seem to coincide with contemporary historical events and, as might be expected, women were probably used as pawns in the political manoeuvres of male relatives. Marriages would have been initiated and terminated by a father or brother to secure an existing alliance or to appease and pacify enemies. This was similar to the general European situation where royal marriage was also a political expedient. As the direction of the family's political ambition shifted so might the woman's husband. It may not have been advantageous to fall in love with a husband – he might not last long!

In these circumstances it is hardly surprising that church laws on consanguinity were ignored and that divorce laws were so liberal, facilitating the rapid

---

10 Dobbs, 'The Ban-shenchus' in *Revue Celtique*, xlviii (1931), p. 232. Part of the original entry is in Latin. Thanks to Dáibhí Ó Cróinín for translation assistance.
11 *Ibid.*, pp 198, 231.
12 *Ibid.*, p. 232.
13 *Ibid.*, pp 232, 233.
14 *Ibid.*, pp 191, 234.
15 *Ibid.*, pp 190, 197, 229.

turnover of both male and female spouses. These laws are frequently cited today as evidence of the greater autonomy of women under Brehon law but it is probable that only the wealthy could afford frequent divorce and it seems unlikely that a woman had much prospect of availing of it without the active support of her immediate family to whom she would normally return afterwards. Such information is unavailable from the *Banshenchas*.

Another fact worth noting is that, although the major dynastic families frequently intermarried, there are a large number of unions with relatively minor dynasties and families, mainly within the ambit of the larger dynasty itself. These probably strengthened relationships between major dynasties and satellite families under their influence but may also reflect fostering traditions. Or it may indicate a marriage of the woman's own choosing as Donnchadh Ó Corráin has suggested.[16] As consecutive marriages continued women may have had greater freedom to choose later husbands due to their increased power and wealth following various divorce settlements. In the *Banshenchas* the successive marriages of a woman are graded with the most prestigious marriages noted first.

Changing patterns also emerge within families according to status as instanced by the rise of the Uí Bhriain family of Munster where Cenn Éitig, father of Brian Ború, married Bé Binn daughter of the relatively obscure Urchad son of Murchad of West Connacht, but Brian himself married Gormlaith the daughter of the more prestigious Murchad son of Finn of Leinster. She is one of the most famous medieval women who also married both Mael Sechlainn of Meath and the Viking king of Dublin.[17]

Two main points will be considered below: (i) the relationship of marriages to current political events with particular attention paid to the Uí Bhriain and Uí Cheinnselaig after the death of Mael Sechlainn of Meath in 1022 and (ii) the multiple marriages contracted by women within this era. Reference shall be made to other families if necessary, however it seems prudent to confine the examination to better known individuals and not to involve too many disparate groups which might lead to confusion. Almost all the personages noted appear in the genealogies and the discussion is chronological.

Marriages and Contemporary Events

After the near massacre of the Uí Briain family of Munster in the battle of Clontarf in 1014, the strongest king to emerge was Diarmait mac Maeil na mBó of the Uí Cheinnselaig of Leinster. While establishing himself as king of Leinster he allied himself with Uí Eochada of Ulster on four different occasions. Diarmait had three wives according to the *Banshenchas* including Derbáil daughter of Uí Eochada.[18] Furthermore, Murchad son of Diarmait married Ben Ulad daughter of Niall Ua Eochada (apparently Derbáil's sister).[19]

Diarmait also married Derborgaill daughter of Donnchad son of Brian Ború; Donnchad was Diarmait's strongest political opponent. The marriage may have been intended to bring about some reconciliation, or perhaps it coincided with the submission by 'the son of Brian' (probably Donnchad), referred to in 1058

---

16 Donnchadh Ó Corráin, suggested this verbally to the author.
17 Dobbs, 'The Ban-shenchus' in *Revue Celtique*, xlviii (1931), pp 188, 227, 196.
18 *Annals of the kingdom of Ireland* (translated by John O'Donovan, 7 vols, Dublin, 1848–1851) (hereafter *AFM*), 1046, 1047, 1048, 1049; Dobbs, *op. cit.*, pp 48, 194.
19 *Ibid.*, pp 196, 197, 229.

in the annals. Diarmait's third wife was Coblaid daughter of Uí Néill of Mag-da-Chonn, a minor Leinster chieftain.[20]

Diarmait supported Tairrdelbach Ua Briain, Donnchad's nephew, during his disagreements with Donnchad and in Tairrdelbach's claim to become king of Munster. This may have been because, as *AFM* indicates, Tairrdelbach was Diarmait's fosterson and Bé Binn, Tairrdelbach's sister, married Donnchad the grandson of Diarmait mac Maeil na mBó.[21] Tairrdelbach's son, Diarmait, may in fact have been named after Diarmait of Leinster.

Diarmait Mac Murchada of the Uí Cheinnselaig, while trying to assert his rights as king of Leinster, supported Conchobar Ua Briain against Mac Carthaig of Munster. However Diarmait subsequently made peace with Mac Carthaig in 1139 and this event may be reflected in the marriage of Diarmait to Órlaith daughter of Cormac Mac Carthaig.[22] Diarmait's sister Dub Coblaid married Muirchertach Ua Maeil Sechlainn which may be connected to Diarmait's alliance in the years 1138 and 1142 with Murchad Ua Maeil Sechlainn Muirchertach's grandfather.[23] However, this alliance did not prevent Diarmait from participating in and profiting by some of the divisions of Meath which occurred sporadically during the twelfth century.[24]

Diarmait was no more loyal to any of the other families to whom he was connected by marriage. For example, he blinded, or had blinded, Domnall lord of the Uí Faeláin in 1141 despite his marriage to Sadb daughter of an Ua Faeláin. In the same year he blinded Murchad Ua Tuathail although he himself was married to Mór daughter of Muirchertach Ua Tuathail.[25]

Among the Uí Bhriain a similar pattern emerges. Tairrdelbach Ua Briain, son of Tadg, son of Brian Ború had three wives. Gormlaith is mentioned in the annals as the daughter of Ua Fógartaig and the annalists remark that Ua Fógartaig was a lord to Tairrdelbach. This is one of the localised marriages noted already as a feature of the marriage-patterns. Furthermore Tairrdelbach's daughter Derbáil who died in 1116 is also said to have married an Ua Fógartaig.[26]

A second wife of Tairrdelbach's who is listed in the *Banshenchas*, was Derborgaill daughter of Tadg, son of Gilla Pádraig of Ossory.[27] This marriage may have been arranged by Tairrdelbach's fosterfather, Diarmait mac Maeil na mBó, as the Osraige were allies of Diarmait. Tairrdelbach's own mother was Aífe daughter of Gilla Pádraig of Ossory, and her daughter was named after her. This name followed on through the generations, as, for example, when Diarmait Mac Murchada named his famous daughter Aífe in the mid-twelfth century.[28]

Tairrdelbach Ua Briain's third wife was Dub Coblaid daughter of Ua Lorcáin

20 'Annála Tigernach' (ed.) Whitley Stokes in *Revue Celtique,* 16 (1895), pp 374–479; 17 (1896), pp 50–77, 79–223, 229–312; 18 (1897), pp 313–449, 450–466. (hereafter *An. Tig.*); *Chronicum Scotorum. A chronicle of Irish affairs from the earliest times to A.D. 1135* (London, 1866) (hereafter *Chron. Scot*), 1058; Dobbs, 'The Banshenchus' in *Revue Celtique,* xlviii (1931), pp 190, 229.

21 *Ibid.,* pp 196, 197, 229. It has been suggested by Seán Duffy that Donnchad may have connived with Domnall son of Tadg, Bé Binn's brother. Cf. Seán Duffy, 'Irishmen and Islemen in the kingdoms of Dublin and Man' in *Ériu,* 43 (1992), pp 93–134.

22 Dobbs, 'The Ban-shenchus' in *Revue Celtique,* xlviii (1931), p. 234.

23 *Ibid.,* p. 233; *AFM,* 1138, 1142.

24 *AFM,* 1144, 1152, 1156.

25 *Ibid.,* pp 232, 233. His most notorious alliance was that arranged between Aífe and Strongbow.

26 *Ibid.,* pp 194, 195; *Annals of Loch Cé . . . from A.D. 1014 to A.D. 1590* (ed. W.M. Hennesy, 2 vols, London, 1871) (hereafter *ALC*); *Annals of Clonmacnoise from the earliest perioid to A.D. 1408* (ed. D. Murphy, Dublin, 1896) (hereafter *An.Clonm.*); *An. Tig.*, 1076; *AFM,* 1077.

27 *Ibid.,* p. 190.

28 *Ibid.,* pp 189, 228.

of the Uí Mhuiredaig in Leinster, and this marriage to a relatively minor Leinster chieftain may again indicate friendship or alliance with Diarmait mac Maeil na nBó.[29] Tairrdelbach's daughter Mór married both Mael Sechlainn Ua Maeil Sechlainn of Meath and Ruaidrí Ua Conchobair of Connacht, mirroring the policies pursued by her father during his reign. When Diarmait mac Maeil na mBó died in 1072, Mael Sechlainn of Meath submitted to Tairrdelbach who retained a strong hold on Meath afterwards. Tairrdelbach also attempted to play the principal combatants in Connacht, the Uí Conchobair and Uí Ruairc, against each other. Here again christian names are seen to move about within families. It was by means of Mor's marriage to Ruaidrí Ua Conchobair that the name of Tairrdelbach passed to the Uí Conchobair family.[30]

Mór's marriage to Ruaidrí Ua Conchobair was only one part of a double alliance between these two families. Her brother Diarmait son of Tairrdelbach married another Mór the daughter of Ruaidrí Ua Conchobair. His second wife was the daughter of Domnall Ua Gilla Pádraig of Ossory continuing the existing relationship between the two families referred to above.[31] This alliance would be further strengthened by the fact that Derbáil, Tairrdelbach's daughter became the mother of Gilla Pádraig Ruad of Ossory.[32]

Muirchertach son of Tairrdelbach married Derborgaill daughter of Lethlobar Ua Laidcnéin termed the 'high-king of the Airgialla', a relatively unimportant family.[33] This is the only marriage which I have found for him in contrast with many of his contemporaries. He succeeded Tairrdelbach on his death in 1086 and he was a dominant political figure until his illness in 1114. One of Muirchertach's major concerns was Meath, which he battled to keep under control. This may be reflected in the marriage of his daughter Mór to Murchad Ua Maeil Sechlainn.[34]

Muirchertach's son Domnall married Cacht daughter of Úgaire Ua Tuathail, a relatively minor political force, and also Ragnailt daughter of Gilla Pádraig following the pattern established by his aunt, uncle and grandfather before him. Ragnailt was also married to Donnchad of Uí Cheinnselaig.[35] Tairrdelbach son of Diarmait Ua Briain (Muirchertach's brother) married Sadb daughter of Donnchad son of Gilla Pádraig of Ossory.[36] The other two were Caillech Dé daughter of the Uí Eidin of Connacht and an unnamed daughter of Cormac Mac Carthaig of Munster.[37]

Tairrdelbach Ua Conchobair was married six times according to the combined evidence of the annals and the *Banshenchas* and his wives included two daughters of Domnall Mac Lochlainn of Ailech. Mór is mentioned in the annals and Derborgaill appears in the *Banshenchas* as the daughter of Domnall.[38]

---

29 *Ibid.*, p. 193.
30 *Ibid.*, pp 191, 234.
31 *Ibid.*, p. 196.
32 *Ibid.*, p. 194.
33 *Ibid.*
34 *Ibid.*, pp 191, 200. Mór was a popular name of the Uí Bhriain, unlikely to be revived today with the negative connotations of the adjective 'large'.
35 *Ibid.*, pp 198, 230, 232 and 189, 190. *The Annals of Inisfallen* mention two marriages arranged by Muirchertach between two unnamed daughters and French and Norse chieftains not found in the *Banshenchas*.
36 Dobbs, 'The Ban-shenchus' in *Revue Celtique*, xlviii (1931), pp 198, 231.
37 *Ibid.*, pp 191, 192, 195, 234.
38 *Ibid.*, p 191; *ALC*; *The Annals of Inisfallen* (ed. S. Mac Airt, Dublin, 1951) (hereafter *AI*); *Annála Uladh, Annals of Ulster* . . . (ed. W. M. Hennesy and B. MacCarthy, 4 vols, Dublin, 1887–1901) (hereafter *AU*), 1122; *AFM*, 1151. It appears that he married two sisters contrary to the laws of consanguinity but Derborgaill's *obit* of 1151 (*AFM*) seems much later than that of her brothers Niall (*AFM*, 1129) and Conchobar (*AFM*, 1136).

A third was Caillech Dé daughter of Ua Eidin of Connacht from a not particularly prestigious family. She also married Tairrdelbach Ua Briain as noted earlier.[39] Two of the two most powerful and famous men in Ireland were, therefore, Caillech Dé's husbands and she was the mother of Ruaidrí Ua Conchobair. The annals also mention Tairrdelbach Ua Conchobair's marriages to two daughters of the Uí Maeil Sechlainn family of Meath, Órlaith and Taillte. These may again be sisters and the daughters of Murchad Ua Maeil Sechlainn although this is not certain, but both marriages reflect the many attacks levelled by Tairrdelbach at Meath – twelve raids in a period of forty years.[40] (The unusual name Taillte appears again in the Uí Maeil Sechlainn family, Taillte the wife of Imar Ua Cathasaig of Brega and Taillte wife of Domnall son of Murchad Ua Fergail of Forthuatha.[41]) The annals also name Tairrdelbach's sixth wife as Dub Coblaid daughter of Mael Sechlainn son of Tadg Ua Maeil Ruanaid possibly of Mag Luirg in Roscommon.[42]

The *Banshenchas* cites one wife for Ruaidrí Ua Conchobair (Tairrdelbach's famous son), Dub Coblaid daughter of Ua Ruairc, presumably Tigernán Ua Ruairc. According to the *Annals of Clonmacnoise* she died in 1181. This may reflect a tentative alliance between Tigernán Ua Ruairc and Tairrdelbach Ua Conchobair. But it is more likely to coincide with the strong relationship which developed between Ruaidrí Ua Conchobair himself and Tigernán and the annals note nine instances of alliances within a period of thirteen years.[43] Despite the apparent divergence from the polygamy practised by his father Tairrdelbach, there is an intriguing comment in the *Annals of Connacht* in 1233 indicating that Ruaidrí had much in common with his father: 'the Pope offered him the title to (the kingship) of Ireland for himself and his seed forever, and likewise six wives if he would renounce the sin of adultery henceforth.'

Within the Uí Chonchobair family there are some more marriages which seem rather complex if the evidence may be believed. It appears that Aed Ua Conchobair who died in 1067 married two sisters, namely Caillech Caemgen daughter of Ócán Ua Fallamain of Fotharta in Leinster and Sadb daughter of Ua Fallamain.[44] His third wife was Etan daughter of Uí Egra of Sligo who died in 1104 according to the *Fragmentary Annals* but the *Banshenchas* states that she also married his son Ruaidrí.[45] It may be that she was Aed's youngest wife and therefore Ruaidrí's contemporary. Certainly she was not Ruaidrí's mother who has been mentioned above as Caillech Dé daughter of Uí Eidin of Connacht.

Multiple Marriages Contracted by Women

It is a well known fact that men married more than once in early Irish society and it is obvious from the material described above that they married up to six

---

39 *Ibid.*, pp 191, 234.
40 *An. Tig.*, 1114; *An. Tig.*, *AFM*, 1115 *An.Tig.*, 1117; *An.Tig.*, *AFM*, 1120; *An.Tig.*, 1124; *An.Tig.*, *AFM*, 1125; *ALC, AU*, 1127; *AFM*, 1128; *An.Clonm.*, 1129; *An.Tig.*, 1132; *An.Tig.*, *AFM*, 1140; *An.Tig.*, *AFM*, 1143; *An.Tig. AFM*, 1144; *An.Tig.*, 1145; *An.Tig*, 1148.
41 Not part of the *Banshenchas*.
42 *AFM*, 1168.
43 *AU, AFM*, 1159; *AFM*, 1160, 1161; *AU, AFM*, 1166; *AFM*, 1167, 1168, 1169; *AU, AFM*, 1170; *AFM*, 1171.
44 Dobbs, 'The Ban-shenchus' in *Revue Celtique*, xlviii (1931), pp 190, 193.
45 *Ibid.*, p. 191; 'Fragmentary annals . . .' (ed.), E.J. Gwynn in *R.I.A.Proc.*, xxxvii (1925–7), section C, pp 149–57.

times, as in the case of Tairrdelbach Ua Conchobair. It was also legally permitted at a certain period for a man to marry bigamously, which appears not to have been the case for women. In this context, one of the most interesting facts which emerges from the prose text of the *Banshenchas* is the number of marriages contracted by some of the women. These marriages are not recorded in the annals or genealogies.[46] One example illustrates this with particular clarity:

Derborgaill daughter of Tadg son of Gilla Pádraig king of Ossory, mother of Aed Ua Conchobair, and mother of the two sons of Tairrdelbach Ua Briain that is, Tadg and Muirchertach and mother of the two sons of Dúnlaing Ua Caellaide king of Ind Ochla, that is, Finn and Donn Sléibe the two sons of Dúnlaing Ua Caellaide king of Éile and Fotharta. And mother of Mael Sechlainn Ua Fógartaig and of Domnall Ua Lorcáin and of Sadb and Dub Leasa and Cacht the three daughters of Ua Lorcáin and of Órlaith daughter of Ua Domnaill.[47]

The christian names of all the husbands are not given, perhaps the identity of the offspring was more important, but it is clear that Derborgaill married six different men and had eleven children. Two husbands were from the leading dynasties of the time, Tadg Ua Conchobair and Tairrdelbach Ua Briain but the remainder were not so well-known: Dúnlaing Ua Caellaide and the families of Uí Fógartaig, Uí Lorcáin and Uí Domnaill. Although her death is recorded by both *AU* and *AFM* in 1098, *AFM* simply notes that she was the mother of Tadg and Muirchertach Ua Briain, her most prestigious offspring and thus the annalistic obit conceals her many marriages.

Derborgaill is unique in the *Banshenchas*. No other woman is documented as having six husbands but women contracted three, four and indeed five marriages. Dub Coblaid daughter of the Uí Lorcáin of Leinster had five husbands although her five offspring are again considered more important than the full names of the husbands. She was the mother of Diarmait son of Tairrdelbach Ua Briain of Munster, Úgaine son of Gilla Pádraig of Ossory, Comgaill son of Dúnlaing of Leinster, Órlaith daughter of Ua Gilla Comgaill and finally of Ben Féta daughter of Ua Domnaill. The same distinct mixture of distinguished and lesser-known families noted above emerges here also.[48]

Another woman, Dub Coblaid, half-sister to Diarmait Mac Murchada, had four husbands with offspring by each. She was the mother of Dúnlaing Ua Nualláin the king of the Fotharta of Leinster, Caillech Mór daughter of Muirchertach Ua Maeil Sechlainn of Meath, two sons of Dalbach Ua Domnaill king of Uí Fheilmeda of Leinster and of Crimthainn Mac Dalbaig.[49]

Some women were married twice or thrice but those twice-married are too numerous for enumeration. Gormlaith daughter of Mael Sechlainn Ua Maeil Sechlainn of Meath was the mother of Cú Ulad Ua Caindelbáin of Tuatha Laegaire and of Congalach Ua Conchobair of Uí Failgi and of Ruaidrí son of Conchobar Ua Flaithbertaig of West Connacht.[50] Aífe, the mother of Diarmait mac Maeil na mBó of Leinster, was also the mother of Aimirgin Ua Mórda of Laois and of Echdonn son of Dúnlaing of the Uí Dhúnlainge of Leinster.[51]

---

46 For the sake of clarity the names of women who married more than once are italicised in the genealogies and the names of their spouses are written alongside.
47 Dobbs, 'The Ban-shenchus' in *Revue Celtique*, xlviii (1931), p. 190. According to *AFM* and *AU*, she died in 1098.
48 *Ibid.*, p. 193.
49 *Ibid.*, pp 197, 229, 234.
50 *Ibid.*, p. 194.
51 *Ibid.*, pp 189, 228.

What emerges then from this cursory examination of the eleventh and twelfth-century section of the text is that the marriages echo, by and large, the contemporary political activity of powerful dynasties. It is probable that the liberal divorce laws of early Ireland were partly aimed at facilitating the political marriages and were availed of more for political expediency than for personal need. The multiple marriages of men, and particularly of women, are clearly visible in the *Banshenchas* in contrast with the annals and genealogies.

There appears to be an elite aristocratic circle of eleven families who intermarried and warred with each other in fairly equal proportions: Uí Bhriain of Munster, Uí Cheinnselaig of Leinster, Uí Chonchobair of Connacht, Uí Maeil Sechlainn of Meath and Mac Lochlainn of Ailech were the most prominent, followed by such names as Mac Gilla Pádraig of Ossory, Uí Ruairc of Bréifne, Uí Mórda of Laois, Uí Chonchobair of Uí Failgi, Mac Carthaig of Munster and Uí Tuathail of Leinster.[52]

---

52 There is obviously a fashion in personal names within the period as there is today and the most popular are, in order – Sadb, Cacht, Mór, Gormlaith, Aífe, Órlaith, Derbáil, Derborgaill, Ailbe, Dub Coblaid, Ragnailt and Bé Binn.

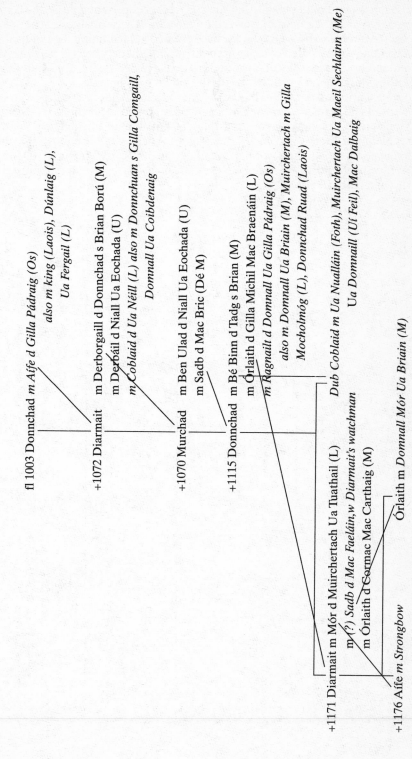

UÍ CHEINNSELAIG

fl 1003 Donnchad *m Aífe d Gilla Pádraig (Os)*
also m king (Laois), Dúnlaig (L),
Ua Fergail (L)

+1072 Diarmait   *m Derborgaill d Donnchad s Brian Ború (M)*
m Derbáil d Niall Ua Eochada (U)
m Coblaid d Ua Néill (L) also m Donnchuan s Gilla Comgaill,
Domnall Ua Coibdenaig

+1070 Murchad   *m Ben Ulad d Niall Ua Eochada (U)*
m Sadb d Mac Bric (Dé M)

+1115 Donnchad   *m Bé Binn d Tadg s Brian (M)*
m Órlaith d Gilla Míchíl Mac Braenáin (L)
m Ragnailt d Domnall Ua Gilla Pádraig (Os)
also m Domnall Ua Briain (M), Muirchertach m Gilla
Mocholmóg (L), Donnchad Ruad (Laois)

Dub Coblaid m Ua Nualláin (Foth), Muirchertach Ua Maeil Sechlainn (Me)
Ua Domnaill (Uí Feil), Mac Dalbaig

+1171 Diarmait m Mór d Muirchertach Ua Tuathail (L)
m (?) Sadb d Mac Faeláin,w Diarmait's watchman
m Órlaith d Cormac Mac Carthaig (M)

Órlaith m Domnall Mór Ua Briain (M)

+1176 Aífe m Strongbow

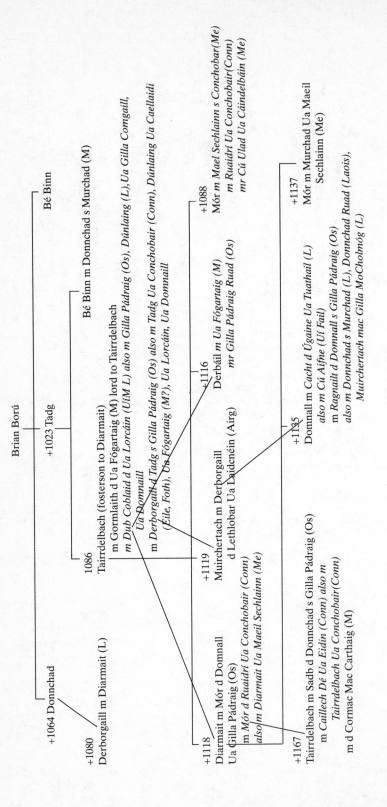

Uí BHRIAIN

Brian Ború

+1023 Tadg

Bé Binn

+1064 Donnchad

+1080
Derborgaill m Diarmait (L)

Bé Binn m Donnchad s Murchad (M)

1086
Tairrdelbach (fosterson to Diarmait)
m Gormlaith d Ua Fógartaig (M) lord to Tairrdelbach
m Dub Coblaid d Ua Lorcáin (UíM L) also m Gilla Pádraig (Os), Dúnlaing (L), Ua Gilla Comgaill,
Ua Domnaill
m Derborgaill d Tadg s Gilla Pádraig (Os) also m Tadg Ua Conchobair (Conn), Dúnlaing Ua Caelláidi
(Éile, Foth), Ua Fógartaig (M?), Ua Lorcáin, Ua Domnaill

+1119
Muirchertach m Derborgaill
d Lethlobar Ua Laidcnéin (Airg)

+1116
Derbáil m Ua Fógartaig (M)
mr Gilla Pádraig Ruad (Os)

+1088
Mór m Mael Sechlainn s Conchobar(Me)
m Ruaidrí Ua Conchobair(Conn)
mr Cú Ulad Ua Cáindelbáin (Me)

+1118
Diarmait m Mór d Domnall
Ua Gilla Pádraig (Os)
m Mór d Ruaidrí Ua Conchobair (Conn)
also m Diarmait Ua Maeil Sechlainn (Me)

+1135
Domnall m Cacht d Úgaire Ua Tuathail (L)
also m Cú Aifne (Uí Fail)
m Ragnailt d Domnall s Gilla Pádraig (Os)
also m Donnchad s Murchad (L), Donnchad Ruad (Laois),
Muirchertach mac Gilla MoCholmóg (L)

+1137
Mór m Murchad Ua Maeil
Sechlainn (Me)

+1167
Tairrdelbach m Sadb d Donnchad s Gilla Pádraig (Os)
m Caillech Dé Ua Eidin (Conn) also m
Tairrdelbach Ua Conchobair(Conn)
m d Cormac Mac Carthaig (M)

UÍ CHONCHOBAIR

+1030
Tadg m Derborgaill d Gilla Pádraig (Os)

+1067
Aed m Caillech Caemgen d Ócán Ua Fhallamain (Foth L)
      m Sadb d Ua Fallamain (Foth L)
      m *Etan d Ua Egra (Sligo) also m Murchad s Donnchad (M)*

+1118
Ruaidrí m *Mór d Tairrdelbach Ua Briain (M) also m Mael Sechlainn s Conchobar (Sligo)*
        m *Etan d Ua Egra (Sligo)*
        m d Ua Conaing (Uí Bhriúin Rátha)
        m d Echnechán Ua Cuinn (Muinter Gillgáin)

+1156
Tairrdelbach m *Caillech Dé d Ua Eidin (Conn) also m Tairrdelbach Ua Briain (M)*
            m Derborgaill d Domnall Mac Lochlainn (CE)
            m Mór d Domnall Mac Lochlainn (CE)
            m Órlaith d Murchad Ua Maeil Sechlainn (Me)
            m Taillte d Murchad Ua Maeil Sechlainn (Me)
            m Dub Coblaid d Mael Sechlainn son of Tadg Ua Maeil Ruanaid (Roscommon)

+1198
Ruaidrí m Dub Coblaid d Ua Ruairc (Bré)

Italicised names indicate those women who married more than once

Abbreviations

Airg	Airgialla	Me	Meath
Bré	Bréifne	M	Munster
CE	Cenél nEogain	Os	Osraige
Conn	Connacht	s	son
d	daughter	U	Ulster
Dé	Déise	Uí Fáil	Uí Fáilgi
Foth	Fotharta	Uí Feil	Uí Féilmeda
L	Leinster	Uí M	Uí Muiredaig
m	married		

8 Women, the Church and the Law: Matrimonial Litigation in Lucca under Bishop Nicolao Guinigi (1394–1435)

Christine Meek[1]

MARRIAGE in Lucca, as elsewhere in northern Italy, was governed by a combination of canon law, statutory provisions and local social customs. By the late fourteenth and early fifteenth centuries it had long been established in canon law that the essential element in a marriage was the free consent of the two parties, provided that there was no canonical impediment to their union.[2] In social custom, as revealed in statutes, diaries and family memoirs and in details provided by some of the litigants in matrimonial cases, marriage was very much a family affair. A father normally married off his daughter, or looking at it another way, was expected to provide a husband for her. He was certainly expected to provide a dowry. Consent could become rather lost in all this, although it was crucial in the church's eyes.

Marriage in normal social practice was very much a two-part affair. The marriage ceremony took place, with the man and woman declaring that they took each other as man and wife *per verba de presenti*, that is in the present as opposed to the future tense, with the placing of a ring on the bride's finger. Some weeks or months or even years later came the *transductio*, the formal transfer of the bride to her husband's house. This must have seemed to the couple and their friends and relatives to be the real marriage, but it was in fact the promise *per verba de presenti* that was binding, a fact that not all couples seem to have realised.

This paper illustrates these and other problems primarily from the records of the episcopal court in Lucca in the time of Nicolao Guinigi, who was bishop for over forty years, 1394–1435. The records of the bishop's court are not sufficiently complete to lend themselves to a statistical treatment, but they do allow a discussion of the character of matrimonial cases coming before the court and make possible some reflections on the nature of marriage and the effects which this had on both the men and women involved. The matrimonial cases coming before the bishop's court can be divided into two main categories: those concerned with whether or not a particular union constituted a valid marriage and those in which the validity of the marriage was not in dispute, but where problems had arisen between husband and wife.

---

1 I would like to express my thanks to the Arts and Social Sciences Benefaction Fund of Trinity College Dublin for a grant towards the cost of microfilm, which greatly facilitated the preparation of this paper.
2 M.M. Sheehan, 'Choice of marriage partner in the middle ages: development and mode of application of a theory of marriage' in *Studies in Medieval and Renaissance History*, n.s., i (1978), pp 3–33.

Marriage could be entered into very easily. Many marriages described in law-suits before the episcopal court were remarkably informal and impromptu. Despite canon law provisions, in no case was a priest said to be present or the ceremony to have taken place in or near a church. Sometimes a notary was present, but more frequently publicity was achieved by having the ceremony take place before lay witnesses and often in informal surroundings. One morn-ing in May 1403 monna Bionda, daughter of ser Niccolò da Buggiano went to ask the advice of Dino Guinigi, an elderly cousin of the ruler of Lucca, about a marriage proposed between herself and the merchant, Francesco Vinciguerra. Dino advised in favour and the couple were married the same day in the pres-ence only of Dino and his wife, Dino conducting what ceremonies there were.[3]

At just about the same date Fettoro Coluccini, a barber surgeon, and Tommaso Boccacci, an armourer, agreed on Fettoro's marriage to Tommaso's daughter Lisa, a young widow. They then proceeded to Tommaso's house, where Lisa agreed to the marriage, which took place immediately with bride and groom exchanging promises *per verba de presenti*. Probably neither Lisa nor her fa-ther realised that these promises were binding, since two days later she married another man. Fettoro was able to produce a number of witnesses to prove that the marriage to him had taken place first, and obtain a decision in his favour.[4]

The normal expectation after a marriage ceremony of this kind would have been that the bride remained in her previous dwelling until the proper arrange-ments could be made for the dowry, the trousseau and the *transductio*. When disputes arose the bridegroom tried to prove that he had been treated as the woman's husband by her and her family, that he had been invited to her house and had given her gifts, that he had held her hand or kissed her. In no case is it claimed that the marriage had been consummated, although that would obvi-ously have greatly strengthened the man's claim that they were man and wife.

Most marriages, no doubt, proceeded to the *transductio*, but if one or both parties changed their minds, problems arose, because it was the marriage cer-emony that was binding, even if the marriage had not been consummated and the couple had never lived together. One party could sue the other in the bishop's court either to have the marriage declared invalid and the plaintiff declared free to marry someone else, or alternatively to compel the defendant on pain of excommunication to consummate the marriage and cohabit with the plaintiff.

One of the most effective grounds for pleading that a marriage was invalid was that the bride was under the canonical age of consent at the time. Since the age of consent was twelve for girls and fourteen for boys this was remarkable. If the bride had been under the age of twelve at the time the marriage cer-emony took place, the marriage could be repudiated, whatever words she had been induced to pronounce. But the brides in these cases usually claim to have been acting under duress, as well as being under age. Thus Niera, daughter of the cathedral organist, messer Matteo Martini, declared that she had been in terror at her father's threats and still not quite eleven, when in March 1396 she had married Baldassino Jacobi of Pistoia, a mercer.[5] Maddalena, daughter of

3 Liber Causarum 35, ff. 7r–13r, 38v–41v, 31 May–23 June 1403. All documents cited are from the Archivio Arcivescovile in Lucca unless otherwise indicated.
4 Ibid., ff. 29r–30v, 35r–37v, 25 May–9 June 1403. Similar case in Actorum Liber 39, ff. 64r–65r, 21 Nov. 1414, where the dowry had been paid and the bride *transducta* by the second 'husband', but the decision neverthe-less went in favour of the first.
5 Liber Causarum 1396, ff. 68r–70r, 16 Oct. 1396.

Giannetto di Giovanni, claimed that her father had passed from fair words and persuasion to beatings to induce her to consent to marriage with Nanne di Nuto of Decimo, hitting her about the face, shoulders and back and dragging her by the hair and threatening her with still worse treatment.[6] Benvenuta, daughter of the late Domenico Colucci of Lucca, claimed that it was her mother who had forced her into marriage with Lorenzo di Francesco of Città di Castello, when she was a little girl *(juventula)* less than twelve years old; indeed it was later stated that she was only ten.[7]

These picturesque details should not be taken entirely at their face value. The girls are clearly stressing aspects of their stories that would render the marriages invalid in canon law. The cruel parents whom they allege beat them into submitting, accompany them to court, act for them and give evidence on their behalf. They do not challenge the system of arranged marriages as such, and Maddalena specifically required her father and mother to find her an alternative bridegroom.[8] Even the claim that they were under twelve years of age at the time of the marriage cannot necessarily be assumed to have been true. If Lorenzo admitted that his bride Benvenuta looked to be about ten at the time of their marriage, both Baldassino and Nanne denied that their brides were under twelve, Nanne with considerable force and corroborative detail.

It seems likely that in at least some of the cases the family had had second thoughts about the marriage. One possible indication of this is given in Niera's case, where in addition to claiming that she was below the canonical age and under duress, she said that she had been given to understand that Baldassino's social status and financial position were better than she subsequently found them to be. After the marriage had taken place, the bridegroom and the bride's father had appointed an arbitrator to fix the amount appropriate for the dowry and trousseau, and it seems likely that in the course of these negotiations the bride's family discovered that the marriage was less advantageous than it had previously supposed and used the claim that Niera was under age to go back on the match.[9]

The episcopal court was well aware of the possibility of deception and collusion. The vicar-general questioned Niera carefully about whether or not she was acting under compulsion in the claims she made in court. Monna Anna, who admitted to terrorising her daughter Benvenuta in order to induce her to say yes to marriage with Lorenzo di Francesco, was specifically asked if she was giving this testimony in order to cancel the marriage and she replied that she was telling the truth. She had wanted Benvenuta to marry Lorenzo because she wanted her to marry well, but the effort to compel her had turned out badly, and she now wanted her to be free of her equivocal position.[10]

But if the allegations in these cases cannot be taken entirely at face value, they do suggest certain general reflections. They demonstrate that girls were indeed married very young. Whether or not they were under the canonical age of twelve, they cannot have been all that much older. They refer to their physical appearance and obvious youth at the time of their appearance in court. The claim that girls could easily be threatened and bullied into consenting was not

---

6 Liber Causarum Civilium 37, ff. 51r–60v, 29 May–20 Sept. 1409.
7 Liber Causarum 1426–7, ff. 77v–78v, 2–31 July 1427, esp. f. 78v.
8 Liber Causarum Civilium 37, f. 54v, 3 July 1409.
9 Liber Causarum 1396, ff. 70r–70v, 23 Oct. 1396; Liber Causarum 1397, f. 22v, 31 March 1397.
10 Liber Causarum 1426–7, f. 83r, 31 July [1427 ?].

challenged as such, and was probably recognised as corresponding to social realities. All these cases involve girls; in the one case in which a boy repudiated a marriage on reaching canonical age no attempt was made to urge duress.[11]

Some of the same considerations apply to quite a large number of cases, where the contract of marriage was made not by the bride herself, but by kinsmen on her behalf. All these contracts were in the form of solemn promises by the bride's male and very occasionally female relatives, confirmed by oaths, to ensure that the woman married the man in question.[12] When these cases came to the bishop's court, each of these girls said much the same thing, that is that the promise had been made by others without her consent, and in some cases without her knowledge. It is probably significant that while three of the cases were begun by bridegrooms demanding fulfilment of the promises,[13] three were begun by the girls who sought to clarify their position and be free of any claim,[14] so that they could marry someone else. Two of the bridegrooms also sought to be freed from the promises they had made, since the bride would not consent and they wished to seek another marriage.[15] Again it is probably significant that virtually all these cases involve unwillingness on the part of the bride, but there is one case where a man was unwilling to fulfill a promise of marriage which his father had made on his behalf.[16]

Where a marriage had been contracted *per verba de presenti* by the two parties in person and where it could not be alleged that either party was under the canonical age, the marriage was binding. Only in extraordinary circumstances could a marriage *per verba de presenti* be declared invalid. In two cases women who had contracted marriage *per verba de presenti* claimed that this had been conditional and was not binding since the conditions had not been fulfilled. Monna Lemma, claimed that she had married Simone Betti only if and insofar as this was acceptable to her kinsmen and relatives and not otherwise. Simone denied this and the decision seems to have been going in his favour anyway, but monna Lemma greatly weakened her position by going through a form of marriage with a certain Giusto da Scilivano and beginning to cohabit with him, while the case was still pending, with the result that the original case became submerged in proceedings against her, and eventually also Giusto, for adultery, defiance of canon law and contempt of the bishop's court.[17] Maria alias Constantia, daughter of Johannes de Pomes de Spania, or Guasconia, alleged that her marriage to Piero de Valentia had been conditional on his obtaining her release from the Lucchese prison, where she was detained for a debt to a third party, and which he had presumably failed to do. The case was eventually decided in her favour, at least partly because Piero gave up the attempt to prove that the marriage was unconditional and renounced his claims.[18]

Bartolomea, daughter of Filippo Coluccini, obtained an annulment, when she and her father were able to prove that the fact that her bridegroom was

---

11 Liber Causarum 1397, ff. 26r–29r, 1 June–10 Sept. 1397.
12 Liber Reclamorum 27, f.21v, 26 June 1393 ; Liber Causarum 42, f. 76r, 8 Jan. 1420, f. 20v, 18 July 1420; Liber Causarum 43, unfoliated, 5 Feb. 1422; Liber Causarum 45, f.11r, 25 June 1426, f.32r, 24 May 1426; Liber Causarum 48, f. 51r, 21 Feb. 1435. A girl's mother involved, Liber Reclamorum 27, f. 62v, 2 May 1394 ; Liber Causarum 46, f. 24v, 1 October [1429].
13 Liber Reclamorum 27, f. 62v, 2 May 1394 ; Liber Causarum 42, f. 20v; Liber Causarum 45, f. 11r, 25 June 1426.
14 Liber Causarum 43, unfoliated, 5 Feb. 1422; Liber Causarum Spiritualium 48, f. 51r.
15 Liber Causarum 43, ff. 76r–76v, 2 May 1426; Liber Causarum 45, f. 32r, 24 May 1426.
16 Liber Causarum 29, f. 90v, 7 April 1399.
17 Liber Causarum 40, ff. 85r–86v, 29 July–20 Sept. 1419.
18 Liber Causarum 45, ff.48v–49r and *cedula* between folios 44 and 45, 21 Oct. 1427– 20 Jan. 1428.

afflicted with incurable leprosy had been deliberately concealed from them by the go-betweens.[19] Another Bartolomea obtained a separation, when she was able to prove that her husband, Simone Jacobi Davini of Pisa, had abandoned the Christian faith and become a Moslem while being held as a slave in Tunis, where he had also taken a Moslem wife.[20] Monna Bionda was able to escape from the marriage she had contracted with Francesco Vinciguerra by alleging that she had taken the veil. Despite serious doubts about the truth of this claim, the vicar-general ruled that, although the marriage between them had been valid, it was automatically dissolved by the monna Bionda's entry into religious life.[21]

There are two cases of marriages annulled for inability to consummate, one of them fairly straightforward, the other distinctly curious. Caterina, wife of Antonio Bartolomei, a weaver of velvets, obtained an annulment on the grounds of her husband's impotence, he consenting to the judgement.[22] The other case involved a couple, Giovanni Dominici and his wife Giuliana, who had earlier obtained an annulment on the grounds of the wife's inability to have sexual relations. Giovanni was given permission to remarry and Giuliana joined an order of Dominican tertiaries. However, she subsequently gave birth to an illegitimate daughter. She therefore petitioned to have her marriage to Giovanni declared valid after all and for him to be compelled to take her back as his wife, since the grounds for the dissolution of the marriage were clearly untrue. Although she had a good case in canon law, Giovanni not unnaturally took a different view, asking for the marriage to be annulled on the grounds of her fornication. The vicar-general ruled in favour of Giovanni.[23] Five years later in 1435 she was before the court again. Giovanni had died and she had appealed to the pope to be freed from her religous vow, especially as the organisation she had joined had never received official recognition. She was in fact freed and given leave to marry.[24]

There are surprisingly few cases involving alleged consanguinity, bigamy or precontract. There is in fact only one consanguinity case, although there are a number of dispensations for marriage within the prohibited degrees. The couple were treated remarkably sympathetically, their case being settled in a single day. They were made to take an oath that they were indeed third cousins, as they claimed, and that they were seeking an annulment in order to avoid mortal sin and not out of fraud, but they were not required to produce any witnesses, although their claim to be related within the prohibited degrees involved seven named persons besides themselves.[25]

The cases involving bigamy or pre-contract are not numerous, but those that do exist suggest a rather casual attitude to marriage on the part of those concerned. Giovanni Cecchi petitioned in 1394 for a declaration that the marriage he had contracted three years previously to Francesca, daughter of Vanne Gratia of Siena, be declared null, because he had subsequently discovered that she was already married at the time to a man called Coscino Giuntori. She agreed

19 Libri Antichi 82, f. 30r, 27 Aug. 1410. Evidence of witnesses including two doctors, ff. 26r–26v.
20 Liber Causarum 1424, ff. 117r–119v, 3–21 Nov. 1424.
21 Liber Causarum 35, ff. 10r–13r, 14–25 May, ff. 39r–41r, 16–23 June 1403.
22 Liber Causarum 48, ff. 102r–103v, 120r–120v, 25 June–12 Aug. 1436.
23 Liber Causarum 46, unfoliated, 4 Oct. 1430; J.A.Brundage, *Law, sex, and marriage in medieval Europe* (Chicago and London, 1987), pp 376–8.
24 Liber Causarum Spiritualium 48,ff. 116r–116v, 3 Aug. 1435.
25 Liber Causarum 1396, ff. 43r–44r, 15 May 1396.

that she had married Coscino in 1384 and had lived with him for four years, but had believed him to be dead when she married Giovanni.[26] Giovanni Nucchori of Saltocchio had a similar experience. He had married Divitia, daughter of the late Giovanni Pardelli of Corsena in August 1397, but by February 1398 was in court asking for an annulment, because he had by then discovered that she was already married to a certain Arrighetto Mannini of Coreglia. She admitted that she had married Arrighetto eighteen years previously and had lived with him for a while, but said she did not know he was still alive. In fact he was not only still alive, but still living in the same commune, and her 'marriage' to Giovanni Nucchori was therefore declared invalid.[27]

If men and women were treated equally, at any rate from a legal point of view, as far as matters like consent and pre-contract were concerned, this was not the case where problems arose within the marriage. Neither the secular nor the ecclesiastical authorities were entirely indifferent to misconduct by married men. Adultery was an offence under secular law for both men and women, and although prosecutions are rare, they do exist.[28] But all complaints in the bishop's court about adultery or desertion were made by husbands against their wives. No woman ever appears in court to complain that her husband has committed adultery or deserted her or even that he has ill-treated her or failed to provide for her, except in defence. A woman who left her husband was at a disadvantage, because he could obtain an injunction from the bishop's court ordering her to return within a short space of time on pain of excommunication. The wife had the right to appear in court to show cause why she was not obliged to return, but she was on the defensive from the very beginning. She was in a more unfavourable position than if she had simply been summonsed and a number of women asked that the injunction be converted to an ordinary summons.

In some cases the women had left their husbands for another man. Monna Pasqua had left her husband Giovanni Guglielmi of Pietrabuona and was living with a notary, ser Stefano Martini of Pescia.[29] Monna Riccha of Uzzano had left her husband and children for a priest, Antonio, who was a chaplain in Pescia,[30] and monna Tita, wife of a tanner of Pescia, was also said to be living with a priest.[31] Monna Onofria, wife of Lorenzo di Jacobo Ceci, was living with Giovanni di ser Niccolò, a prominent merchant in Lucca,[32] while Puccinello Guiduccini named several men as lovers of his wife, monna Contessa.[33] It is seldom clear what happened in these cases. The husbands of monna Pasqua, monna Riccha and monna Tita demanded their return and were therefore presumably willing to take them back. These women were threatened with excommunication or even actually excommunicated, and the case then disappears from the record.

Rather more is known of the cases of monna Contessa and monna Onofria. Monna Contessa agreed that she had committed adultery with several men in the four years of her marriage, but said that she had been impelled to this by the

26 Liber Reclamorum 27, ff. 15r–16r, 6 July 1394.
27 Liber Reclamorum 29, ff. 43r–44r, 26 Feb. 1398.
28 e.g. Archivio di Stato in Lucca [A.S.L.], Sentenze e Bandi 72, ff. 64r–65r.
29 Liber Causarum et Reclamorum 34, ff. 129r, 140r, 10 April 1402.
30 Ibid. ff. 149r–150r, 22 Jan. 1403.
31 Liber Causarum Spiritualium 48, f. 19v.
32 Liber Causarum 34, ff. 93r–95r, 7 Jan. 1401.
33 Liber Causarum 1396, ff. 15r–16r, 17 April 1396.

ill-treatment she had suffered at the hands of her husband, who beat and injured her and denied her the necessities of life, so that she had suffered hunger and thirst.[34] Monna Onofria also accused her husband of cruelly beating her and denying her the necessities of life, so that she said she had legitimate cause to commit adultery. Cruelty could not, of course, be accepted as a justification for adultery, and the vicar-general may have suspected collusion, since he insisted on witnesses being produced, despite the fact that monna Onofria had admitted adultery.[35] In both of these cases the husband was granted a separation.

There are well over a dozen other cases where a wife who had left her husband was ordered to return.[36] Most of them do not provide much detail. Two of the women are said to be living in the houses of Lucchese citizens, but as their names make it clear that the women are from the peasantry, it seems more likely that they had found a job in Lucca than that they were living in adultery. One of the citizens, ser Manuello Guccini, appeared in court as the woman's proctor.[37] Monna Oliva, wife of Jacobo Cini of Lunata, had returned to the house of her father, Jacobo Salvi, but the parents of a run-away wife were not permitted to harbour her, and he too was summoned to court and accused of keeping her from her husband.[38]

Where there is any indication of the women's motives for leaving their husbands the issues seem to be cruelty and to a lesser extent the husband's ability to provide for his wife. The father of monna Simona, as her proctor, contested the demand of her husband, Ghiberto Francisci, a shoemaker, that she return to him, saying that her husband was unable to earn enough to feed them both and that she was willing to return whenever he was able to support her and redeem her clothing, which he had pawned.[39] Several of the husbands promise to treat their wives well or are ordered to do so, which suggests that these were cases where wives had fled from cruel husbands.[40] This certainly seems to have been the case with monna Elizabetta, wife of Bartolomeo Pieri, whose marriage led to a ding-dong battle in the courts. Monna Elizabetta said that she could not return to Bartolomeo without risking her life. Bartolomeo denied this and said that all their problems were her fault, since 'sweet speech and humble actions by a wife make and keep a husband good-humoured.'[41] It is probably significant that the case was initiated by Bartolomeo's demand that she return and not by Elizabetta's complaints about his cruelty.

This leads to a consideration of cases one does *not* find in the episcopal records in Lucca, which studies of other areas, especially England, have led one to expect.[42] There are, for example, no cases of clandestine marriages, that is mar-

34 Liber Causarum 1396, f. 15v, 17 April 1396.
35 Liber Causarum 34, f. 93v, 7 Jan 1401.
36 In addition to the cases discussed below see Liber Reclamorum 27, f. 27v, 27 Aug. 1393, f. 31v, 10 Sept. 1393 ; Liber Reclamorum 29, f. 83v, 24 Jan. 1399 ; Liber Reclamorum 35, f. 77v, 21 July 1413, f. 78r, 18 Aug. 1412 ; Liber Causarum et Etiam 40, f. 8v, 12 Feb. 1418 ; Liber Causarum 43, f. 22v, 5 Aug. 1422 ; Liber Causarum 1425, f. 62v, 20 June 1425.
37 Liber Reclamorum 29, f. 47r, 15 March 1398. Other case see Liber Reclamorum 27, f. 13v, 28 Feb. 1393.
38 Liber . . . 1425, *cedula* ff. 64–65, 18 June 1425.
39 Liber Causarum 40, f. 8v, 12 Feb. 1418.
40 Liber Reclamorum 29, f. 83v, 24 Jan. 1399; Liber Causarum 38, ff. 12r–12v, 27 Feb. 1410.
41 '*dulce verbum et humilis actus uxoris reddit atque conservat maritum mansuetum*', Liber Causarum 46, f.66v. Their matrimonial problems first came to court in 1427, Liber Causarum 45, ff. 64v–65v, 86r–87v, 27 June–23 July 1427, with further disputes in 1432, Liber Causarum 46, ff. 63r–67r, 16 Feb.–12 May 1432.
42 R.H.Helmholz, *Marriage litigation in medieval England* (Cambridge, 1974); M.M. Sheehan 'The formation and stability of marriage in fourteenth-century England: evidence of an Ely register' in *Medieval Studies* 32 (1971), pp 228–263; A.Finch, '*Repulsa uxore sua*: marital difficulties and separation in the later middle ages' in *Continuity and Change* 8 (1993), pp 11–38.

riages with no witnesses present. There are very few cases of bigamy and none of multiple bigamy. There are no cases of brides insisting that reluctant bride-grooms be compelled to consummate the marriage and cohabit with them. It was perfectly possible to deal with marital problems without recourse to any court. There are a number of notarial documents in which a couple agree to live apart and make the necessary financial and maintenance agreements. There are inci-dental references in other documents which reveal couples who were not living together.[43]

There are many indications too of matrimonial irregularities that never reached the courts. Several children born to women by men other than their husbands ended up in the foundling hospital and there are wills in which the testator speaks of children born to him by a woman who was married to someone else at the time.[44] Cases of defamation, insults, quarrels and brawls often involve accusa-tions of immorality, especially against women.[45] There seem to have been numer-ous irregular couples. A husband had the right to take legal action against a man who seduced his wife or enticed her away from him,[46] or against anyone who attacked or insulted his wife, but this right applied only to husbands. The defend-ant was often able to deflect the charge, not by claiming that it was untrue, but by denying that the plaintiff was the woman's husband, so that he had no stand-ing in the case, which therefore collapsed.[47] Some of the cases which do appear in the bishop's court suggest a lack of scruple about marriage vows. A number of favourable decisions seem to have been obtained by false pretences, without the sucessful party apparently suffering any qualms of conscience.

Although the regulations about consanguinity, pre-contract and above all consent applied equally to men and women, there were distinct gender differ-ences in the type of case actually brought before the ecclesiastical court. Were there no women who were abandoned by their husbands or who found that the man they had married already had a wife? Of course, there were, but they did not apparently have recourse to the episcopal court. They tended to concen-trate on practical matters, obtaining support for themselves and their children and above all recovering their dowries, which meant pleading in the secular courts. No woman sued in the bishop's court for her husband to return to her. Virtually all reluctant spouses are female, whether the case takes the form of the bride trying to evade the marriage or the husband trying to enforce it.[48] This may reflect a view that it was hardly decent to pursue a reluctant bridegroom

---

43 Monna Santa was living apart from her husband maestro Giovanni Fasani, and with her son Giuntino, pre-sumably by a previous marriage, since his father was given as the late Giovanni Giuntini, when she made her will (25 Nov. 1415, A. S. L. Testamenti 9, f. 125 bis r).

44 A. S. L. Spedale di S. Luca 907, f.202r, 20 May 1416 (where the mother had died), f. 242r, 15 March 1426 (where the mother was still alive). Bonturellus Bonifatii in his will speaks of monna Buona as his '*legiptima concubina*'. They had a daughter and four sons, the eldest of whom was born while she still had a husband, but the others after she was free (Testamenti 7, ff. 8v–9r, 28 July 1418). The will of monna Caterina, wife of a silk weaver Michele Papi of Florence, speaks of her natural son, Ambrogio (Testamenti 9, f. 125bis v, 25 March 1415). The wills of male testators not infrequently mention illegitimate children.

45 e.g. A. S. L. Capitano del Popolo 21, unfoliated, 4 Oct. 1413. For two cases involving fixing animal horns to a citizen's door to dishonour him and his wife, see Capitano del Popolo 18, unfoliated, 16 March 1405, Capitano del Popolo 19, unfoliated, 7 Jan. 1412.

46 A man from Prato was fined £1800 for seducing a man's wife while living in their house and taking her with him when he left (Capitano del Popolo 29, ff. 172r–173r, 2 March 1425). For another case see Capitano del Popolo 21, unfoliated, 28 May 1412.

47 e.g. A. S. L. Sentenze e Bandi 18, unfoliated, 18 June 1356.

48 Cf. C. Donahue Jr., 'Female plaintiffs in marriage cases in the court of York in the later middle ages: what can we learn from the numbers?', in Sue Sheridan Walker (ed.), *Wife and widow in medieval England* (Ann Arbor, 1993), pp 183–213, especially pp 195–8.

through the courts, but it is more likely to be a consequence of the greater possibility of manipulating women in the family interest. One of the great weaknesses of laws to protect women or to give equal value to the need for their assent was always the possibility of using persuasion, pressure, and in the last resort force, to get them to agree to what their menfolk, whether their husbands or their own family, wanted.

9 'Almost a Gentlewoman': Gender and Adolescence in the Diary of Mary Shackleton

Kevin O'Neill

I N the *History of youth*, Michael Mitterauer points out that there are very important areas of darkness in our images of nineteenth-century European adolescence. Although Europe was an overwhelmingly rural world, we know little about the experience of rural youth. This is especially true of rural women: our notions of their adolescent experience are almost non-existent. Mitterauer notes:

> It seems to me to be of the greatest importance to pursue gender-specific factors in the lives of young people in earlier times. However, the necessary prerequisites for such an analysis are for the most part missing.[1]

One reason for this is that diaries, which are often the most useful source of information about adolescence, rarely survive from rural areas. This essay attempts to offer some evidence about the lives of rural women by making use of the diary of Mary Shackleton of Ballitore, Kildare.

This source is unusual in several regards. Begun in 1769 when Mary was eleven years old, the diary runs continuously until shortly before her death in 1826. Together with several hundred surviving letters, five manuscript volumes of poetry, a cure book, household accounts, records of the Religious Society of Friends, and several literary manuscripts, it may be the most complete record of the private life of any individual who has lived in Ireland. Her diary also represents one of the very few surviving European diaries that details the adolescent years of a young woman living in a rural community. As such, it offers unusual opportunities to explore the social construction and dynamics of gender, ethnicity, class, and adolescence.

The historical definition of adolescence is a complex subject. Despite the conflictive nature of the field, the works of Aries, Stone and Davis demonstrate that the very notion of adolescence as a transitional stage between the dependency of childhood and the independence of adulthood was historically determined.[2] This paper suggests that, at least by the late eighteenth century, many Irish adolescents experienced a prolonged period of transition between childhood and adult status that was characterised by social, economic and psycho-

---

1 Michael Mitterauer, *A history of youth* (translated by Graeme Dunphy, Cambridge, 1993).
2 Philippe Aries, *Centuries of childhood* (New York, 1962); Lawrence Stone, *The family, sex and marriage in England, 1500–1800* (London, 1977); Natalie Zemon Davis, *Society and culture in early modern France* (Stanford, 1975).

logical forms of apprenticeship; and that, while there were significant differences between male and female adolescent experiences, those differences were less dramatic than the previous literature on this subject would suggest.

Theodore Scharmann defines adolescence as:

... the step from the under-age, dependent child to the mature, responsible person, adjusted to reality, whose income and ability to support others is so far developed that he or she can go on to take the second great life decision, the choice of a partner, and can consider or achieve the founding of a family.[3]

This definition is especially useful because it provides both sensitivity to psychological factors and flexibility in considering economic factors. Scharman's emphasis on income and employment does tend to accentuate aspects of adolescence more powerful in male experience, but his notion of an 'adjustment to reality' has a very significant meaning for the construction of gender. Insofar as 'male' and 'female' are social and historical constructions, a very crucial part of the transmission of those constructions, as well as their coercive power, is contained within this adjustment. It is during this period that any individual variations from the perceived norms of maleness and femaleness will be either negotiated, confronted or surrendered.

This definition stresses social and economic components of adolescence, and hence positions us to explore adolescence as a historically rather than biologically determined process. It is important not to limit our notion of adolescence to the physical and psychological aspects of developing sexual maturity; for while it is painfully obvious that sexuality is at the centre of adolescence, too limited a focus on sexuality poses a risk that the simple and compelling relationship between gender and biology will lead us into a determinism that marginalises the economic and social aspects of this period of youth. In most societies, this stage of life was critical in the development of adolescent work and social skills. Only some of these skills were related to sexuality and marriage; and, while preparation for marriage was the critical task of youth, economic – not sexual – aspects of that education were the essential prerequisites for the achievement of adult status. For most Ballitore young people, the meaning of marriage centered on promised emotional support and the social and economic identity that it conveyed. Of course, sexual liberty was an aspect of emotional fulfillment, and procreation was closely related to adult identity, but the understanding which Mary Shackleton and her peers had of marriage was decidedly practical.

This preparation for marriage usually entailed a gradual process of growing independence rather than a quick, sharp break with either household or parents. Education, physical separation from the nuclear family, courtship, and growing awareness of alternative social, moral and geographic space all proceeded incrementally towards a state of social autonomy and maturity. However, the speed and continuity of these processes varied considerably across class and gender lines. One major cause of such variation may have been the relationship between work and gender. Mitterauer argues that as women rarely had apprenticeships and were less likely to take wage labour than males, they had more difficulty separating from hearth and home.

---

3 Theodore Scharmann, *Jugend in Arbeit und Beruf* (Munich, 1965), p.18.

In the biography of girls who never sought paid employment outside the home, the only working threshold was when as children they began to assist their mothers. Boundaries relating to the world of work seem as a rule to be ill-suited for the delineation of female youth in history, as they take their orientation from the patterns of male careers.[4]

Mary Shackleton's diary suggests that this may be true only as far as our ignorance masks the organisation of work within the rural household. She began her diary at age eleven, and her earliest entries indicate that she took work responsibilities from both her father and mother. Over the next several years she gradually became a central part of a complicated household economy.

From the age of eleven, she records helping with household chores, and especially with needle work. She was soon cutting shirts and assisting in butter making, jam making, cooking of various sorts, brewing, and the major household activity of washing clothes. Despite this activity, these subjects received little comment in the diary. She considered this sort of work so normal and ever-present that most references to it are of the sort 'worked as usual today', or the only slightly more informative entries such as 'washing today', 'made jam', 'mangling'. While this familiarity may be disappointing to the researcher, it underscores the normalcy of young women taking on work responsibilities from an early age, and could be seen as an apprenticeship for rural women who would one day manage a complex rural household.

Mary's preparation for adult life was not limited to this domestic sphere. Certainly one of her greatest advantages was the education that she received from her schoolmaster father. When she was at an early age her parents decided that, as she had an obvious academic aptitude, she alone of her sisters should attend the family school – as its only female student. This was difficult for Mary, who was often teased by older boys, because of both her gender and a speech defect that caused her great embarrassment. Still by the age of fifteen she was able to read French literature and to practise geometry with the most advanced scholars.

She received less formal types of education as well. Mary acted as an informal apprentice to her aunt, who was the local herbal healer. In helping to prepare and deliver medicines, she received an important generational transmission of knowledge and moral obligation. As her delivery routes took her though the village and beyond, she also entered into the households and lives of the poorest families of the surrounding countryside. These visitations not only involved her in the lives of a wide variety of places and people, but also gave her confidence, and a familiarity with suffering that prepared her for the difficult tasks she would face as a mother and as a female leader of a disordered community in the 1790s.

There is one peculiarity involving this apprenticeship as an herbalist that illustrates some of the interpretive problems involved in the use of the diary. While Mary lived with and helped her aunt, there were frequent references to this healing activity; but, with her aunt's death in 1778, all such references ceased. There is no indication that anyone – least of all Mary – had taken on these responsibilities. Then, without explanation, in 1787 she began to record episodes in which she had clearly stepped into her aunt's role as village doctress. The first reference was not a happy one:

---

4 Mitterauer, *op. cit.*, p. 37.

A poor woman today said to us "You gave me a cure for my daughter & she died on me, she was 15 years of age, & I have no other daughter, & it has turned me grey." However she did not charge us with her death.[5]

It is difficult to interpret the years of silence about healing: Was she assisting someone else? Was she perhaps carrying on herself without recording it in her diary – perhaps uncertain of the propriety of a youth serving in this capacity? Such an anxiety would be consistent with notions of healing as the proper work of a mature woman. In any case, the store of pharmacological knowledge with which she carried on this tradition gave her great pride and satisfaction; and her emergence as a 'doctress' (for such she was nominated by her rural neighbours in the 1780s) was an important milestone in her achievement of adult status.

There is other evidence of the breadth of Mary's preparation for the complex work world of rural women which helps to identify the normative gender boundaries of work in Ballitore and suggests that they were not immutable. From an early age, Mary helped her father and grandfather with farm work, particularly in hay-saving. Mary's enjoyment and talent in this field led her grandfather to make her a small hay fork so that she could manage this work more easily. The culmination of this informal apprenticeship came in the year before her marriage when – aided by her friend, Molly Bewley – she took charge of her father's hay-saving:

... It being proposed in our council yesterday that I should attempt the character of Steward. I begin this afternoon, & equipping myself for the purpose with my Straw hat ... [I] take my station in the Hay-field. My M.B. [Molly Bewley] by my side ... The men wonder & titter at me I believe.
12th day About 1/2 after 4 my MB & I awoke escorted by M[ary] Casey walked to Molly's ... Then set the men to work, & found myself as much entrusted in the work as if I was making a shirt ...[6]

These activities demonstrate the achievement before marriage of at least an intermediate status between dependence and adulthood. In reviewing her work-related training for life in this community, it is difficult to agree with Mitterauer that once a girl began to help her mother there were few 'markers' of development. Of course these experiences were hardly typical; but it would not be too heroic an assumption to suggest that more modest forms of practical training gave most rural young women a growing sense of their capacity as home and farm managers. Farming daughters could also find moments in which they were entrusted with adult work, such as taking charge of the butter making or the sale of eggs or poultry. For those of even the most humble economic backgrounds, parallel experience came in the form of domestic or farm service in the household of a neighbour. Mitterauer argues that such paid labour outside the familial household was often a critical step in adolescent development. Besides providing practical training and familiarity with a world more mercenary than the family, it also significantly weakened the bonds of dependency upon the family.[7]

Mary's experience, which did not include paid labour outside the family, suggests that we need to extend Mitteraruer's argument to include other forms of

---

5 National Library of Ireland (hereafter NLI), MS 9312, 16 Feb. 1787.
6 NLI, MS 9314, 11 August 1789.
7 Mitterauer, *A history of youth*, p.22.

living outside the family home. For Mary, there were social and at least quasi-economic equivalents of service that also served to loosen the bonds of the parental household. Like other young women of her milieu, she spent significant periods of time living in the households of extended family members as a companion or household assistant to an adult female relative. She spent several periods of a month or longer with her older married sister in Clonmel and also with various cousins in Cork, Waterford, Dublin and the north. A few of these visits were primarily social, or related to yearly or quarterly meetings of the Society of Friends, but others were to help a household in short supply of female labour, especially at the time of childbirth. These sojourns provided young women with valuable social and emotional experience. An entry from 1778 provides a glimpse of her first direct experience with childbirth which occurred while accompanying her mother to her sister's home in Clonmel.

At four o'clock in the morning Suzy Jessop came into the room in which my Mother & I stay "what is the matter" says my Mother, "The Mistress is ill" says she "& The Doctor is with her" My mother got up & desired me to do so, but I trembled so, I could scarce put on my clothes – as our room was behind my Sister's & a door opened into hers, I could hear her groans very plain – I hurried down stairs & stood by the Kitchen fire in all the agony of suspense as I never had been at such a place before, my fancy painted in the darkest shades . . .[8]

While Mary may not have helped her sister at this moment, over the next few weeks she was a major asset in the household. She was also learning important life lessons. This type of extended family assistance provided experience outside of her household and village that familiarised her with different modes of behaviour. This was essential to her development of a sense of community and of self-identity. On a more practical level, it added to her knowledge and experience of household and farm management. Perhaps most importantly, it gave her essential information regarding childbirth and infant care, a subject with which, as the second youngest child in her own family, she had no direct experience. These were often difficult periods for her, both because of homesickness, and the anxiety which her future role as a mother provoked.

This was no little anxiety for Mary in her adolescent years; but despite the several difficult births that she witnessed, her greatest anxieties focused not on childbirth itself, but upon her ability to nurse an infant. This anxiety represents one of the most recurrent themes in her writings of this time. It dates to her first thoughts about motherhood, and was given special meaning by her elder sister's difficulty in nursing her children. Her diaries contain numerous snatches of poetry on this theme, and it is one of the few subjects about which she copied published work into her diary. One example, a quotation taken from a journal poem describing an infant on the breast of a dying mother, provides some notion of the depth of her anxiety:

> From "the desolation of America, lately published"
> In vain for food, the dying infant cries;
> With ghostly visage & beseeching eyes:
> The pious mother doom'd to certain death
> For his dear sake, retains the fleeting breath;
> But while with fatal tenderness she drains

---

8 NLI, MS 9304, 26 May 1778.

> The milky treasure & exhausts her veins;
> She sees her breast deny its balmy flood
> And fills his guiltless mouth with streaming blood
> Then with a bursting heart from pain
> Look up to Heaven & on her babe expires.[9]

This concern for infant care remained important throughout her life. She had difficulty nursing her own children, and she would devote considerable time to helping other women in such circumstances. She also carried on a critique of the wet nursing system in her published work.

Independent of, but parallel to this growing domestic identity, Mary was also developing a public identity through her activity in the Society of Friends. The Society provided women much greater autonomy than other churches in Ireland; each meeting had its own separate and semi-autonomous women's meeting.[10] At the monthly women's meeting, women dealt with both spiritual and temporal issues of their community. For Mary, like many of her friends, these meetings provided an important corporate sense of the female community as well as role models of female leadership. They frequently offered young women their first opportunity for public speaking and, eventually, for public responsibility. In December 1782, Mary acted as the clerk of the Carlow monthly women's meeting. This was an important and challenging role. The clerk was responsible not only for recording the minutes of the meeting, but also for implementing decisions reached by the meeting. This entrance into the leadership of the Society was an important step towards gaining full status in her community. As with house and field work, this transition into the adult world occurred well before her marriage in 1790.

Another aspect of female autonomy that is central to our understanding of adolescence is the degree to which young, unmarried women could travel and associate with other young people, especially young men. Mitterauer identifies this as one of the main areas of differentiation between male and female experience. The evidence from Mary's diary suggests that while there was a difference in the degrees of freedom accorded males and females in eighteenth-century Ireland, the situation was not nearly as stark as Mitterauer suggests. His claim that, 'as a rule they were only permitted to go out in the evening if accompanied by a chaperon'[11] was manifestly untrue of the young women of Ballitore. The youth of Ballitore, male and female, did require permission for evening excursions; but for both males and females, permission was routinely granted. The youth of Ballitore went on numerous 'night walks' unescorted by adults, and contrary to Mitteraruer's suggestion that perceived sexual dangers were the cause of social restriction, on the one recorded occasion when permission for such a revel was denied, the reasons which adults gave had to do with concerns that the young men in the group might come into conflict with other youth.

There was, of course, sexual danger in this world, and these fears for male youth may indeed represent another part of the dynamic of sexual aggression. Sexual assaults upon women were not unusual occurrences in rural Kildare, and Mary records in a rather offhand way an attempted wife-stealing carried out by one of her childhood playmates and classmates:

---

9 NLI, MS 9304, 6 March 1778.
10 See Phyllis Mack, *Visionary women: ecstatic prophecy in seventeenth-century England* (Berkeley, California, 1992); Isabel Grubb, *Quakers In Ireland 1654–1900* (London, 1927) and Maurice J. Wigham, *The Irish Quakers* (Dublin, 1992).
11 Mitterauer, *A history of youth*, p. 44.

... Tom Johnson ... & another fellow going out at night met a man & his wife whom they were for taking from him – on his resistance Tom snapped a pistol & the other fellow knocked him down & almost killed him.[12]

Mary does not record the fate of this woman, nor any surprise over the occurrence. Instead, she expressed shock over the fate of Tom Johnson, and the failure of a Friends education to guide him towards correct behaviour. Whatever the realities of sexual violence in its many forms, it did not lead to the type of social restrictions upon young women that Mitterauer describes.

This is surprising, as by definition this was the period during which individuals encountered both their own developing sexuality and the social and cultural constraints that their community placed upon that sexuality. While we should not exaggerate the centrality of biology to the adolescent experience, this process of sexual acculturation is a central part of the social and psychological meaning of adolescence. This process is also the most powerful tool that the community possessed for transmitting and validating its sexual mores. This was especially true in a society where a prolonged period of courtship led to the development of a courting subculture that was important for the construction of gender, and especially for its transmission across generations. This youth culture acted as a reservoir of gender norms and customs that was particularly well insulated from adult society. The effect on male adolescent culture was often striking because its manifestations tended to be outwardly focused; but it was likely just as important in the development of female identity.

Mary recorded an emerging interest in young men shortly after beginning her diary at age 11. In 1773, at the age of fifteen, she noted her first relationship with a boy, a student in her father's school. Mary wrote several poems in efforts to express her feelings about the young man. Most of her juvenile poetry is suitably immature, and least interesting when it was most successful in copying the pastoral models that she followed. Yet occasionally, when the pastoral mode did not really fit her subject, she could produce poems that carry the force and meaning with which she invested them. The title of the very first of these poems about relationships offers more than a hint of the difficulties inherent in first love. To avoid detection and embarrassment, she gave the young man the pseudonym of 'Lysander.'

> Distress.
>
> ... Devoid of charm of dress or face,
>   Alas I plainly see
> He could not his affection place
>   On such a one as me.
>
> Or had I sparkling beams of wit
>   Which might ensnare alone,
> My tongue's debarr'd from shewing it.
>   But truly I have none ... [13]

The distress of this poem, written in August 1773, had by November turned to despair. She entitled her effort of that month 'Falsehood'. It ends:

---

12 NLI, MS 9302, 2 June 1776.
13 NLI, MS 23574.

> But now I'll tear thee from my bleeding heart
> Where thou too long hast claim'd the better part.
> Come Fortitude assist the bold design,
> And tear him from this feeble breast of mine . . .

Mary's distress was only temporary. She soon had regained her composure sufficiently to gain some poetic justice. In 'A Soliloquy', she alters her lover's pseudonym from 'Lysander' to 'Philander' as part of her sly retaliation, and in a subsequent poem she welcomes the return to schoolroom normalcy that the end of this relationship brought:

> The Welcome to Liberty.
>
> More welcome than returning light,
> To those who painful pass the night
>   Is Freedom unto me.
>   . . .
> 'Twas wond'rous kind to set me free,
> To school I'll go, nor think of thee,
>   Though thou before me stands,
> my Sums, my books, my globes I'll ply,
> Nor let my pen (to view thy eye)
>   Lie useless in my hands.
>
> Of Mars & Jupiter I'll tell,
> Venus & Saturn knowing well
>   More radiant than thy eyes,
> Which once I said, (but 'twas as a Lie)
> Outshine these lamps that hang on high
>   To light the sable skies.[14]

Unfortunately, if Mary really prized her freedom, she was about to enter into another relationship, one that was both more enduring and painful. The object of her attention in this case was another student at her father's school, William Hall. She was now sixteen years old and enjoyed considerable freedom to associate with the young men of Ballitore. An entry in her diary from December 1774 gives us an idea of the rather open and athletic nature of social interaction between these young men and women:

Tom Pim was sitting by me Sally made me change my seat very often. I was facing the fire which she was afraid would hurt my eyes so desired me to sit in the arm chair, I went to it for that purpose but Dick Jacob sat him on in it & pulled me in his lap, I did not like it at all & Cousin Sam & Jonathan Gotchell were facing us, so I desired him to let me up for I was not an infant therefore I did not chance to be nursed, all was in vain, till in his struggling to hold me he overturned the screen & then I got lose, after a bit they took leave & went, to my joy – . . .[15]

Her joy at the young men's departure should not mislead us into thinking that Mary did not like the rough-housing associated with this early courtship. The problem here was twofold. Her elder cousin Sam was an unwelcome observer; and the rough-houser was the wrong young man. Later in the same month she entered an even more confrontational situation, but in this case she approved of the young man's dynamic attention:

---

14 *Ibid.*
15 NLI, MS 9298, 18 Dec. 1774.

Billy Hall was pinching [Sally] & I undertook to be her champion, but instead of saving her, I drew him on us both, then he & I fell to it & slapped each other stoutly we had fine sport till Betty called the boys to eat . . .[16]

And on 31 December:

. . . Billy Hall did not slap as much but squelched my hand – he did not hurt me though – no indeed . . .[17]

And again on 5 January:

Billy Hall punched me, I punched him, am I sorry I hurted him? Why ? ask my heart.[18]

This relationship, however combative, was a happy one, but it was soon cast in shadows by William's impending departure for the army. Although William was a successful scholar, his father was unable to continue to finance his education, and William was signed on to a regiment. In a diary entry, Mary records both the pain this caused and her intense desire to hide her feelings from more mature folks – a constant feature of this youth courtship culture:

Betty says to me "Billy Hall is to go to Dublin in a few days." I immediately felt my face & neck all in a glow. "O" says I & went on reading a letter I had in my hand, but I knew not what I read . . . I affected indifference but I am a bad dissembler I faltered in my speech . . . What shall I do! I am in such anguish of mind alas should I never see him more

[At school] . . .when I had laid my cloak & bonnet on the table & cast my eyes over to Billy Hall who sat at Pat's desk they filled with tears, & did as often as I looked at him. I endeavoured to conceal my grief & did myself so much violence that I but seldom look'd at him . . . at so tender an age to expose so lovely a boy to such danger is cruel.

Later that day Mary and her sister had another encounter with Billy:

. . . I took Billy Halls measure & made him put his name on it

"It will be white-washed over" says he

"I will keep it in for thy sake" says I "Will thou forget us . . . when . . . among the Officers & soldiers?"

he said indeed he would not & I would fain believe him . . .

Sally & I powdered our heads Billy Hall . . . put on my cap: he look'd very pretty in it . . . My dear Creature pinch'd my arm, O that he would pinch it black & blue that I might keep the marks in ![19]

But William had to go. In poems written in March of that year, Mary tried to deal with his departure. The 'Maids of Sorrow' expressed the depth of her personal loss:

> . . . The turtle alone on a bough
>   Whose love has submitted to fate.
> Soft warbles her ne'er dying vow,
>   And lamenteth the loss of her mate.
> . . .
> To hide the full grief of my heart,
>   Alas I did vainly pretend,

---

16 NLI, MS 9298, 29 Dec. 1774.
17 NLI, MS 9298, 31 Dec. 1774.
18 NLI, MS 9298, 5 Jan. 1775.
19 NLI, MS 9298, 9 Jan. 1775.

> For how was I grieved to part,
>> With my loved school-fellow & friend . . .[20]

And in a poem entitled 'The Recollection' she expressed her fear for his safety:

> . . .
> ye Cannons that tumultuous roar;
>> And bathe the fields in blood,
> Ah do not riot in his gore,
>> Procure some other food . . .[21]

Unfortunately William Hall would indeed face the cannons of war. With several other Ballitore boys, he was sent to the Americas to deal with the rebellious colonists. They served and suffered from Long Island to Yorktown, returning to Ballitore with broken spirits and bodies.

Mary would form several similar relationships over the following decade before entering into the lifelong relationship with another former Ballitore schoolboy, William Leadbeater. There is much more material in the diary regarding these relationships, but the compelling nature of another type of relationship requires that we look at a different type of emotional engagement that was central to Mary's adolescent life. Like many of the adolescent women of Ballitore, she formed special or, as her contemporaries called them, 'partial' friendships with other young women. Mary established several such special friendships. One of them was especially intense, and though it cannot be amply treated in this essay, any discussion of her adolescent life would be incomplete if it is not at least introduced.[22]

Like most of the young women of Ballitore, Mary rarely, if ever, slept alone. During her childhood she shared a bed with her sisters and her aunt. Upon reaching adolescence she had a much wider range of bedfellows. In 1774 a cousin, Eliza Shannon, who had recently been orphaned, joined her household. As Mary reported it: '. . . My Cousin Betsy Shannon will now be *our* companion I am afraid, she will be so much bigger . . .'[23] The two quickly developed a special relationship that was to last for a decade. For three years the young women shared the same bed, and nearly all of their experiences. In 1777, the friendship was tested when Betsy was sent into service as a domestic servant to a Cork family. The parting was very difficult, and Mary has left a vivid poetic record of the relationship's intensity:

> To Eliza Shannon the day before she left Ballitore.
>
> . . . sad-to-morrow
>> Takes thee & all my joy away,
> In your stead comes grief & sorrow,
>> Alas unhappy woeful day –
>
> All our looks our grief discover
>> Low each hangs the droopy head
> Now we wish the parting over,
>> Now the cruel hour we dread

20 NLI, MS 23575.
21 *Ibid.*
22 I am currently writing a separate article about this type of relationship.
23 NLI, MS 9298, 22 Dec. 1775.

> Thy dear handkerchief, lov'd Maiden,
>   I for a precious relick bear.
> Nothing from my bosom bleeding,
>   Shall this precious relick tear.
>
> This when Memory sad shall soften
>   With full grief my flowing Eye,
> With my streaming tears I'll moisten
>   And with burning kisses dry . . .[24]

We need to exercise caution in interpreting this poem. For while it clearly describes an intense relationship, the precise romantic form it takes may represent a victory of genre over gender. Mary had no appropriate poetic models for same sex relationships. The only models she would likely have been familiar with were derivative from Greek warrior friendship poems – and these were clearly not very useful in describing the relationship of two Quaker women. She may have deployed a more familiar pastoral form that came closer to her need; perhaps not. In any case it would be an ahistorical imposition of our norms to assign a name of our liking to this relationship. She chose to call it a 'partial friendship', and we can do no better.

Whatever the exact nature of the relationship, it was central to her happiness. After eight months of separation, Mary learned in a letter that Betsy would be accompanying her new master to Dublin for the Friends' yearly meeting.

I meet these words, "Sam Neale his wife & c[ousin] Shannon set off for Dublin next 2nd day . . ." No Tongue or pen (I believe) can describe the sensations which passed through to my mind at that time, my grief & joy seem to struggle together, I thought if my grief was gone I would be the happiest creature on the world, however I made strange mistakes, forgot peoples names & took Leadbeater for Johney Hutchinson who is gone 2 years.[25]

This meeting in Dublin was a highly emotional moment:

In Dublin, she in Dublin, & have not seen Betsy this was my first reflection on waking up. Just as we were entering J. P['s] door says Cousin Sam "There are some young Women but I don't think Betty Shannon amongst them." I went from him & . . . I found myself grasping her dear hand & trembling all over, "don't kiss in the street" says Nancy Shannon. We took her advice & all marched into Barrington's shop where we clasped each other & laid our heads on each other's shoulder, while I burst into tears, she wept not, but I am certain felt an equal joy – & was ashamed at weeping . . .[26]

After dining separately they were reunited:

It was to be sure a delightful Evening as I sat by Betsy I could not help saying "If this be a Dream; it is the pleasantest I ever had" but it was sweet reality . . . No one could be happier than I with my Betsy. I went to bed – . . . & my mind incessantly wandered on the preceding day – . . . Oh delightful sincerity ! says I how thou shines in my Betsy Shannon in her I find true friendship – . . .

There would be other meetings and a constant exchange of letters in which support and affection were the major themes; though there were also frequent

---

24 NLI, MS 23576.
25 NLI, MS 9304, 21 April 1778.
26 *Ibid.*

chidings and apologies for not writing often enough. A letter from Betsy to Mary written shortly after her return to Cork underscores the intensity of the relationship and the difficulties of describing it in acceptable written form.

My dear dear Molly
   ... maybe thou would come to Cork, O if thou would what happiness would be in store for me: then I could say a thousand things to my beloved Molly that I cannot commit to paper. but it is time for me to say I got thy to [sic] dear letters and welcome the[y] were to me indeed: ....[27]

This most partial friendship came to an end with Betsy's marriage in 1783. These special relationships between female friends were of considerable importance in giving shape and voice to notions of female identity and autonomy by providing networks of support, information and intimacy outside of family control or male influence. As with the other aspects of female adolescence in Ballitore, these relationships helped young women develop the secure sense of themselves as individuals and as women that was a prerequisite for adult identity in their community.

Our understanding of female adolescence is still sketchy, but the following tentative conclusions are warranted for at least one Quaker female adolescence in eighteenth-century Ireland. Mary Shackleton and many of her peers:

1. received a practical education in household and farm management skills;
2. were free to travel to, and were responsible for helping female relations;
3. were free to associate with young men with only limited adult supervision;
4. had opportunities to achieve responsibility and leadership in the women's meetings of the Society of Friends; and
5. formed special personal relations with male and female peers.

Each of these areas of personal development provided Mary with valuable knowledge and space for personal development. Before her marriage in 1790, Mary Shackleton was well prepared for the demanding and difficult life that lay ahead. Her adolescence had provided a sense of autonomy, functionality and community. It prepared her for the perceived challenges of rural life, rather than for some perceived normative gentlewoman's life. She did not acquire the accoutrements of a lady, but all who knew her recognised that she was a substantial woman.

---

27 Friends Historical Library, Dublin, Ms Box 36, Leadbeater Shackleton Collection.

10 The Roles of Royal and Aristocratic Women in British Politics, *c.* 1782–1832

P.J. Jupp

THE purpose of this paper[1] is to explore the roles of royal and aristocratic women in British ministerial and parliamentary politics during a period sandwiched between two political upheavals: the constitutional crisis of 1782–84 when George III successfully asserted the influence of the crown against the whig party; and 1830–32 when the whigs successfully asserted the influence of the house of commons and public opinion in the interest of parliamentary reform. In keeping with at least one type of conference paper, the exploration is speculative and has no pretensions to being definitive. The British aristocracy, although small by European standards, consisted of between 7–10,000 noble and gentry families, a number sufficient in the present state of research to preclude definitive statements for many years to come. Furthermore, this paper is based principally, although not exclusively, on the published memoirs and correspondence of women as opposed to either the much larger number of published literary and philosophical works relevant to their role or the huge quantity of manuscript sources in the archives. That said, the published memoirs and correspondence for the period are exceptionally numerous and rich: they not only provide some ballast for speculation but also suggest that it was a time when the roles of women of this class were changing significantly.

There is little doubt that this was a phase in the history of high politics when conditions were particularly suited to women playing a role. The monarch, for example, still had an important part in the making and unmaking of governments and in shaping their policies. In the period from 1782 to the accession of Victoria in 1837, to cite just one example, George III, George IV and William IV contrived to oblige five governments to resign.[2] It was for this reason that parties in power courted the monarch and his or her entourage; and that parties in opposition courted the causes of disgruntled royals, especially if there was a chance that they might soon succeed to the throne. The prince of Wales was therefore the intermittent figurehead of the main opposition party from 1780 until he abandoned it on becoming regent in 1812, at which point the party intensified their support for the causes of his wife Caroline and their daughter Charlotte, both of whom he had treated abominably. Furthermore the monarchy as an institution was exceptionally large. This was principally due to George III and Queen Charlotte producing seven surviving sons and six daughters but

---

1 I am grateful to both the editors for their invitation to venture into what was for me relatively uncharted territory and particularly so to Sabine Wichert who first suggested this specific topic.
2 In 1783, 1801, 1807, 1832 and 1834.

was also the result of the number of mistresses established by his sons (about a dozen in this period) and the fact that each royal had his or her own household or court. In 1828, for example, the number of household officials was nearly three times greater than the Whitehall staff of all the ministerial departments put together.[3] For these reasons queens, princesses, mistresses and female household officials were not only exceptionally numerous but also had relatively easy access to those with effective power.

In the cases of governments and parliament, aristocrats commanded the vast majority of offices and perks, and kinship often determined where, exactly, they were located in the political spectrum. Moreover the spectrum was exceptionally diverse. Politics was driven by competition between two principal groups – the self-styled whigs and those who eventually styled themselves tories – but neither was sufficiently numerous to command a parliamentary majority on their own. As a consequence they had to seek the support of a plethora of other largely aristocratic groups. Parliamentary politics was therefore principally an aristocratic and family affair in which permutations of allegiance were frequent and the opportunities for aristocratic women to play a role, various.

However a fundamental shift in the sources of power and the substance of debate is detectable. At the beginning of the period a largely aristocratic debate was being conducted about the power and influence of the monarchy. Towards the end, professional politicians with weak aristocratic ties were guiding the debate towards how best to arbitrate between the aristocracy and the new social and economic interests seeking their place in the sun. 'Palace politics', as one historian has suggested, was giving way to 'real politics'.[4]

With regard to female attitudes to politics, it is useful to draw a distinction between the royals (I use this term in its widest sense) and aristocrats. In the former case, only one, Princess Charlotte, the heir presumptive from her birth in 1796 to her death in 1817, has left a record by which her views on politics can be judged. This reveals her to have been highly intelligent, well read, an accomplished pianist and with a passion for politics of a whiggish to liberal kind which she was determined to implement when she eventually succeeded to the throne.[5] These views were undoubtedly stimulated by her virtual imprisonment at Windsor Castle by her father and the encouragement of the leading whigs, but they seem to have flowed from personal conviction. Believing in the cause of liberty, particularly for Ireland, she was unique amongst the female royals for actually stating that she longed for the power to further it.[6]

By contrast, but as one might expect from those excluded from effective power but close to its source, aristocratic women had an ambivalent attitude to politics. On the one hand, a significant number stated that women were not suited to the subject. Some thought this was due to differences between male and female intellects. Mary Berry, the author and society host, expressed the view that women had livelier imaginations than men but lesser powers of what she

---

3 The offices of the members of Wellington's cabinet of January 1828 employed c.490 individuals; the members of the royal households of that date numbered c.1200.

4 Boyd Hilton, 'The political arts of Lord Liverpool' in *Transactions of the Royal Historical Society*, 5th series, vol. 38 (1988), pp 147–170, partic., pp 168–170.

5 A. Aspinall (ed.), *Letters of the Princess Charlotte 1811–1817* (London, 1949), *passim* but particularly the introduction, pp ix–xxii and 28, 95.

6 *Ibid.* p. 19 (to Miss Mercer Elphinstone, 30 Dec.[1811]); Lady Theresa Lewis (ed.), *Extracts of the journals and correspondence of Miss Berry 1783–1852* (3 vols, London, 1865, hereafter *Berry Corrs.*,), ii, pp 452–53.

called 'steady attention'.[7] Lady Charlotte Bury, novelist and one time lady in waiting to Princess Charlotte, wrote of women not having the 'strength and terseness ascribed to male intellect alone'.[8] Lady Louisa Stuart, a daughter of Lord Bute, prime minister in the early part of George III's reign, commented in 1830 that women should not play a part in public life because they had feelings that 'melt . . . resolution, impede . . . exertion, weaken . . . reason, combat . . . interest, overpower . . . prudence'.[9]

Others believed that politics were too violent or too pedantic to be suitable pastimes for women of quality. Virtually all the women whose memoirs I have consulted expressed their distaste for the violent terms in which men conducted their politics and were shocked by those women who became similarly afflicted. The Countess Granville, for example, in the course of explaining that political commitment was unfeminine, noted that the facial beauty of Lady Jersey, a devout tory, was adversely affected by 'a stern and political countenance' that she adopted when engaged in political discourse. And even worse, that when excited, she was given to unfeminine actions such as on one occasion grabbing Lord William Russell by the lapels and shouting at him: 'why should we have Germans to reign over us ?'.[10] To Lady Charlotte Bury such scenes would have proved that the political world was inimical to all that was 'intellectual or noble in our nature'[11] and that she was better occupied with what she called the 'illusory world' of her '*vie interieur*'.[12] It is therefore not suprising that when commenting on women who were active in politics, other women ascribed supposedly masculine characteristics to them. For example, Lady Shelley wrote that Mrs Arbuthnot, a confidante of tory ministers in the 1820s, had 'a man-like sense' and was 'devoid of womanly passions'[13] while Mrs Arbuthnot herself described the Duchess of Rutland's command over her husband's politics as being due to 'a masculine strength of mind'.[14]

On the other hand, despite the widespread view that women were not suited to politics, the evidence suggests a considerable interest in the subject. The most obvious reason is that politics was an aristocratic and family affair. In virtually all the aristocratic correspondence of this period the norm was for men to inform their female relations and friends of political news, for women to reciprocate, and when corresponding with each other, to comment on political events, at least *en passant*. But it was also due in many cases to education, reading and experience. Many aristocratic women were brought up in houses with substantial libraries, were educated by governesses, tutors and masters and travelled extensively in Europe, particularly in France and Italy. Virtually all the women who have been studied for the purposes of this paper were fluent in French and some in Italian too. Furthermore, many were widely read, particularly in the

---

7   *Berry Corrs.*, ii, p. 313.
8   *Berry Corrs.*, ii, p. 470 note.
9   Hon. James A. Home (ed.), *Lady Louisa Stuart* (Edinburgh, 1899), p. 166.
10 Hon. F. Leveson Gower (ed.), *Letters of Harriet Countess Granville 1810–1845* (2 vols, London, 1894), ii, pp 196–97; in 1828 she dreaded Lady Canning attacking her late husband's friend, Huskisson, for joining the Wellington administration as 'Personal abuse and attack would destroy my sympathy – unfeminine, unfeeling, unchristian.' (*ibid.*, ii, pp 9–10).
11 Lady Charlotte Bury, *The diary of a lady-in-waiting* (2 vols, London, 1908, hereafter *Lady Bury*), ii, p. 19. Lady Bury felt the plight of women in a male dominated society very strongly and even more so than others I have consulted. She deserves a biographer.
12 *Ibid.*, i, pp 52 –53.
13 Francis Bamford and the Duke of Wellington (eds), *The journal of Mrs Arbuthnot 1820–1832* (2 vols, London, 1950, hereafter *Arbuthnot Jnl.,*), i, p. xiv.
14 *Ibid.*, i, p. 430.

works of the French Enlightenment (Voltaire, Rousseau, Diderot, and Condorcet were favoured), in history and to some extent in English and Scottish moral philosophy and political economy.[15] In fact my impression is that many women of this class were more widely read, especially in Enlightenment literature, than those men who had entered the busyish world of politics immediately following repeated doses of the classics at school and university. Moreover there is some evidence that experience of politics in other, more accessible, settings stimulated a greater interest in parliament. A number of women who took a particularly active part in British politics had been struck by the easy access for women to the proceedings of the French parlements before the Revolution,[16] to the debates of the National Assembly during it, and to the seven year trial of Warren Hastings at Westminster Hall between 1788–95, one of the great political events of the time which was dominated by the stars of parliamentary debate and attended by large numbers of women.[17] Fanny Burney went to the trial at least seven times in 1790 and took pride in her place in the lord chamberlain's box.[18]

Whatever the causes, there is no doubting an increasing interest. Important developments in this respect were the very substantial increase in the reporting of the debates in the press during this period as a result of the waiving of reporting restrictions; and the comparable increase in the number of national and provincial newspapers. These developments provided this class of women with the means to learn very much more of what was said in parliament than previously, an opportunity which, to judge by the evidence, they accepted eagerly. In addition, increasing numbers appear to have braved the notorious ventilator room above the ceiling of the Commons through which they could see and hear the debates; and to have taken advantage of the slightly more comfortable spaces in the Lords which were behind the red curtains on each side of the throne and in the lord chancellor's room. The increased attendance of women in Commons and Lords is particularly noticeable after the conclusion of the great war in 1815 and most of all during the debates on catholic relief and parliamentary reform from 1828 to 1832. In 1829 Lady Holland reported that women were so numerous in the Lords that they were continually breaking through the curtains and 'are very troublesome upon the Throne, much to the annoyance of H[ouse]. of C[ommons]. [members] & Peers' eldest sons, who have a right to the Throne.'[19] At the height of the Reform Bill crisis, in October 1831, Harriet Ellis went to the House of Lords for the first time and stayed from 3.30 to 11.00 pm, 'a long time, but nothing to some of the ladies who go every night

15 See, for example, The Earl of Bessborough (ed.), *Lady Bessborough and her family circle* (London, 1940), p. 141; the Earl of Ilchester (ed.), *The journal of Elizabeth Lady Holland (1791–1811)* (2 vols, London, 1908), i, p. 192 where she records that between Christmas 1797 and June 1798 she read 'the D. of Marlbro's *Apology*, Burnet's *History*, ye XIII. *Satire of Juvenal*, Hearne's *Travels into N. America*, Smith on ye figure and complexion of ye human species, Bancroft on dying, some desultory chemistry, *Roderick Random*, *Lazarillo de Tormes*, Leti's *Life of Sixtus V*, various German and French plays, novels and trash, Cook's *Third Voyage*, Wolfe's *Ceylon*, part of Ulloa's *Voyage* [to South America], and some papers and ye memoirs of ye Exeter Society. Frequent dippings into Bayle, Montaigne, La Fontaine, Ariosto. Read ye three first books of Tasso; Ld.Orford's works.'
16 The Earl of Bessborough (ed.), *Lady Bessborough and her family circle* (London, 1940, hereafter *Lady Bessborough*), pp 24–25.
17 Emily J. Climenson (ed.), *Passages from the diaries of Mrs. Philip Lybbe Powys* (London, 1899), pp 232–33.
18 Joyce Hemlow et al (eds), *The journals and letters of Fanny Burney* (vols, 1- , Oxford 1972- , hereafter *Burney Jnl.*), i, (ed. Joyce Hemlow, 1972), pp 115 and n.3, 116.
19 The Earl of Ilchester (ed.), *Elizabeth, Lady Holland to her son 1821–1845* (London, 1946, hereafter Ilchester, *Lady Holland to her son*), p. 101.

and stay to the end.' She also noted with some surprise that Lady Jersey 'was seated among the reporters'.[20]

Another indication of interest is the number of diaries written by women that deal extensively with high politics and constitute some of the best records we have of the events they describe. Five of these are particularly important – those of Lady Holland, Lady Elizabeth Foster, Georgiana, Duchess of Devonshire, Mrs Harriet Arbuthnot and Princess Lieven – especially as they take their most detailed form at crucial moments in politics.[21] They certainly convey the impression that their authors sense that they could make their mark by providing a record for posterity – the only way, in the views of Mary Berry and Lady Charlotte Bury, that women could make their mark.[22]

What roles then did women play in politics ? As had traditionally been the case in political systems dominated by monarchies and landed aristocracies, some played a significant role as confidantes and advisers to, and as activists on behalf of, the male members of the predominant parties and factions. How many and whether they were more numerous and more influential in this, as opposed to other, periods are difficult questions to answer but my impression is that their number and influence were unusually high.

Of the royals and the royal mistresses, for example, several play an intermittent part in this respect. Amongst the royals, Queen Charlotte's influence was probably heightened somewhat following George III's mental collapse in 1788–89 and was sustained by its recurrence in 1801 and 1804.[23] Later, in 1831–32, Queen Adelaide, William IV's wife, attempted to muster opposition to the reform bills, measures in which she felt her husband too easily acquiesced.[24] Princess Charlotte looked forward to influence which never materialised and Queen Caroline, although supported by the whigs and a host of women in society at large, was interested solely in her own cause against her husband. In the case of mistresses, only those of George IV as opposed to those of his uncles and brother aspired to political influence. A number of these, such as the duchess of Devonshire, Mrs Fitzherbert and Lady Jersey were alleged to have occasionally exerted political influence – Mrs Fitzherbert's dislike of Fox, the whig leader, is supposed to have deterred the prince from championing the whigs's cause on the change of ministers in 1801[25] – but the two most influential were probably Lady Hertford and Lady Conyngham, his successive mistresses from about 1812 to his death in 1830. It was Lady Hertford, who amongst other political doings, composed some of the articles that appeared in his newspaper, *The Morning Herald*, from 1812. In these she sought to justify his politics, most notably his abandonment of the whigs on becoming regent.[26] Lady Conyngham's role is less easy to establish.

20 Maud, Lady Leconfield and John Gore (eds), *Three Howard sisters* (London, 1955), p. 212.
21 The sections of Georgiana, Duchess of Devonshire's journal that are referred to cover the periods 1788–89 and 1800–02 and are in the Devonshire MSS, Chatsworth, Derbyshire. The section dealing with 20 Nov. 1788–12. Jan. 1789 can be found in W. Sichel, *Sheridan* (2 vols, London, 1909), ii, pp 399–426; the relevant section of the journal of Lady Elizabeth Foster, Georgiana Devonshire's successor as Duchess, runs from Nov. 1788–Sept. 1799 – a typescript is in my possession and the original is in the possession of Hon. Jane Dormer; the Earl of Ilchester (ed.), *The journal of Elizabeth Lady Holland (1791–1811)* (2 vols, London, 1908); the journals of Mrs Arbuthnot and Princess Lieven are first referred to in footnotes 13 and 28 respectively.
22 *Berry Corrs.*, ii, p. 22 ; *Lady Bury*, ii, pp 143–44.
23 Hon. James A. Hope (ed.), *Lady Louisa Stuart* (Edinburgh, 1899), p. 147.
24 Ilchester, *Lady Holland to her son*, pp 120–21, 136.
25 The Duchess of Devonshire's Journal, Tuesday [20 Oct. 1801] (Devonshire MSS, Chatsworth, Derbyshire).
26 *Arbuthnot Jnl.*, ii, pp 299–300; Lady Holland thought she 'ruled with the instruments of inspiring him with hatred and suspicion of all about him, not a good mechanism for a King's mistress' (Ilchester, *Lady Holland to her son*, p. 8).

Some observers thought she was principally interested only in 'Loves and Graces' as one woman put it,[27] but there is evidence that she was a dab hand at political intrigue. For example, she was certainly an active member of the 'cottage coterie' that plotted against Canning's foreign policy after 1822.[28]

In the case of the whigs there were three leading female activists: Georgiana, Duchess of Devonshire, who having married the 5th duke in 1774 became a mistress of the prince of Wales and bore a child by Charles Grey, a rising star of the whigs in the 1790s; Lady Elizabeth Foster, who as a widow lived from the 1780s with the duke and duchess of Devonshire in a *ménage-à-trois* and became the duke's second wife in 1809; and Elizabeth, Lady Holland, who married the third baron in 1797 following her bearing his child and divorcing her first husband, Sir Godfrey Webster. All three became members of families at the very centre of whig politics. All three were highly partisan in the cause but as their diaries and extensive correspondence indicate, held distinctive views of their own on all major matters of policy and leadership. All three were close friends and confidantes of the dozen or so – and it was no more than that – who led the whigs at any given point in this period. They conversed freely with them on politics either *tete-à-tete* or in small groups and gave equally freely of their views. They corresponded with them, received them singly to give them advice or hear the gossip; and when the whigs were in office, which was but briefly in this period, were besieged by those seeking a post.[29] In short, they played a role at this level of activity which was indistinguishable from that of the leading men. Furthermore, as the whigs were out of office for all but a few years in this period, this kind of activity was particularly important.

As for the 'alarmists' and tories, four stand out: Mrs Crewe, the wife of John, subsequently Lord Crewe, who in the early 1790s gathered around her those who shared Burke's views about the threat of jacobinism; Lady Jersey, a mistress of George IV who became known as a 'virago' of the tories in the 1820s;[30] Lady Lyndhurst, who championed the most vehement opponents of parliamentary reform – 'the malignants';[31] and Mrs Arbuthnot, the wife of a minister in successive tory governments from 1809 to 1830. By far the most notable was Mrs Arbuthnot. A staunch partisan in the tory cause, she was given to focusing her attention on individuals, most notably Lord Castlereagh, Foreign Secretary and Leader of the House of Commons from 1812–22; and the duke of Wellington, cabinet minister in the 1820s and prime minister 1828–30. Both Castlereagh and Wellington reciprocated. Castlereagh visited her virtually every other day at her breakfast hour from 1820 until his suicide in 1822 – in her words 'to take his orders'.[32] Wellington became an even more frequent visitor and when apart they corresponded daily. In both cases politics was the dominant subject of discussion and both men gave freely of their views and plans – to a degree in Wellington's case which is astonishing. From 1822 he gave her sight of confidential letters from the king and from foreign courts and when prime minister forewarned her of every major initiative, sometimes well before he made his

---

27 Ilchester, *Lady Holland to her son*, p. 8.
28 H. Temperley (ed.), *The diary of Princess Lieven* (London, 1925, hereafter Temperley, *Lieven*), pp 61–62; *Arbuthnot Jnl.*, i, p. 78; ii, p. 258.
29 For example, the Duchess of Devonshire to Lord Hartington, 4 Feb. 1806 (Devonshire MSS, 5th Duke's Group, no. 1852, Chatsworth, Derbyshire).
30 Ilchester, *Lady Holland to her son*, p. 64, n.1.
31 *Ibid.* p. 118.
32 *Arbuthnot Jnl.*, ii, pp 176–77.

views known to his colleagues. This was the case, for example, with his plans to settle the Irish crisis in 1828.[33] She, for her part, spared them nothing of her views[34] and in Wellington's case was soon regarded by both colleagues and foreign diplomats as the best person through whom they could bring their influence to bear on him.[35] It is also interesting that she gradually came to resent the fact that as a woman she could not play a direct part. On one occasion in her journal she exclaims how she would love to be a man so that she could deliver her firm views of foreign policy in parliament;[36] and on another, in 1828, she regrets that she could not be Irish secretary for if she was, she would commit herself to business and become, unlike the actual incumbent, (and these are her own words), 'master of the whole subject'.[37]

However, perhaps the most influential woman in this role was Princess Lieven, the wife of the Russian ambassador in London from 1812.[38] Recognising early on that her husband was an effective administrator but an ineffective diplomat, she devoted herself to advancing Russian policy in London. By 1820 she had established herself as a member of the royal and the very highest whig and tory circles. From 1823 to 1825 she plotted with various royals to undermine Canning's foreign policy.[39] From 1825, at the specific request of the tsar, she acted in the opposite cause, to soothe relations between Canning and the king.[40] In 1827 she was influential in having the unstable Lord Dudley succeed Canning as foreign secretary.[41] During the 1828–30 Wellington ministry she simultaneously kept up close relations with the inexperienced foreign secretary, Lord Aberdeen, while cultivating pro-Russian views amongst the whigs. Throughout she kept up a correspondence with the whig leader, Lord Grey, and was influential in having him appoint Palmerston rather than Lord Lansdowne as his foreign secretary in 1830.[42]

Another prominent role played by women was that of host during the parallel social and parliamentary seasons which lasted normally from January to July. In this regard five points stand out. The first, outlined in a superb memoir by Lady Susan O'Brien in 1818 but confirmed by other evidence, is that the daily timetable of the season changed in the course of this period. In 1760, she remembered, parliament was usually up by 4.00 pm at which point society sat down to dinner. By 1818, however, parliament often did not begin its main business until 4.00 pm and debates continued late into the night. The consequences were that dinner time shifted to 7.00 or 8.00 pm and there was greater need for less formal events such as 'at homes', *soirées*, salons, assemblies and suppers which men could attend during a lull in parliamentary proceedings or afterwards.[43]

The second point is that the number of social events seems to have increased

---

33 *Ibid.* i, pp 187–88, 260, 272–73, ii; pp 198–99.
34 *Ibid.*, ii, p. 158
35 *Ibid.*, i, pp 390–391 where Herries clearly hopes that his views will be passed on; see also, ii, pp 14, 163–65, 170.
36 *Ibid.*, i, p. 381.
37 *Ibid.*, ii, p. 283.
38 This passage is based on Temperley, *Lieven* and Lionel G.Robinson (ed.), *Letters of Dorothea, Princess Lieven 1812–1834* (London, 1902, hereafter Robinson, *Lieven*), *passim*. Specific references are provided below.
39 Temperley, *Lieven*, pp 61–62.
40 *Ibid.*, p. 107.
41 *Ibid.*, pp 126–27 and Robinson, *Lieven*, p. 275.
42 Temperley, *Lieven*, p. 167 and Robinson, *Lieven*, pp 128, 156–58, 162, 196, 204, 244–45, 281.
43 The Countess of Ilchester and Lord Stavordale (eds), *The life and letters of Lady Sarah Lennox 1745–1826* (2 vols, London, 1901), ii, p. 291.

considerably. As Lady Susan put it: in 1760 assemblies and balls were few and 'one place in an evening was an engagement, and sufficient amusement' whereas in 1818 such events were so numerous that '2 or 3 of a night it is common to go to.'[44] A further indication of an increase is the establishing of fixed times for events hosted by particular individuals, presumably to avoid clashes. To quote two of a great many examples: in 1827 it was made known that Ladies Belfast and Gwydyr and Mrs Hope were 'at home' every week on Mondays, Wednesdays and Fridays respectively;[45] and in 1834 Lady Holland complained that she had been unable to go 'even *once* to Princess Lieven's Wednesdays, Lady Grey's Thursdays, Duchess of Dino's Fridays, or Mrs Baring's Saturdays' and that there had been a fracas between Lady Keith and Mrs Dawson over Sundays because Mrs Dawson had suddenly opened her doors and drained 'le ton' from Lady Keith's.[46] Finally, on this point, the evidence does suggest a large number and a considerable variety of such events. On any given day of the season, especially after Easter, there were numerous 'at homes', dinners, salons, assemblies, suppers, balls, as well as plays and operas at Covent Garden or Drury Lane where, incidentally, the boxes were often leased by women to women.[47]

The third point is that women appear to have organised the vast majority of these events, with the exception of the dinner party in which male hosts were more prominent and which many women said they did not like because of the convention that they retired from the table after pudding.[48] The most interesting events were the salons. In the 1770s there were a number of 'bluestocking' salons where cards and politics were banished and women and such men as were invited were restricted, in theory at least, to conversation about literary and moral subjects.[49] The famous ones were those of Mrs Montagu, Mrs Vesey, Mrs Boscawen and Mrs Thrale. However, it is my impression that although bluestocking salons continued in this period – Mrs Sotheby hosted one in the early 1800s[50] – they were superseded in number and significance by salons in which discussion of politics was encouraged, albeit in settings in which politicians of both sexes mixed with those from the literary and artistic worlds. In addition to the salons of Mary and Agnes Berry which continued throughout this period, mention is frequently made of the salons of Lady Melbourne and Mrs Crewe in the 1780s and '90s and later of Lady Hertford's, Lady Jersey's, Lady Holland's, Princess Lieven's and Mme. de Stael's, the latter being regarded by many women as almost a role model in view of her brilliant intellect and the fact that as Necker's daughter, she had played a significant part in French politics at the beginning of the Revolution and continued to do so as an émigré and writer afterwards.[51]

---

44 *Ibid.*, p. 291.
45 *Arbuthnot Jnl.*, ii, p. 80.
46 Ilchester, *Lady Holland to her son*, p. 145.
47 It is likely that there was a similar increase in social events outside London during the recesses.
48 One wonders whether the convention of women leaving the table to the men after dinner was a peculiarly British one as the future third Lord Lyttelton informed his wife in 1817 of a dinner in Paris where 'we got up all together, men and women, and went out, *donnant les bras*, after the foreign fashion, in less than half an hour after the dessert was served.' (The Hon. Mrs. Hugh Wyndham (ed.), *Correspondence of Sarah Spencer Lady Lyttelton 1787–1870* (London, 1912, hereafter *Lyttelton Corrs.*), p. 201).
49 Reginald Blunt, *Mrs Montagu, queen of the Blues* (2 vols, London, 1921), ii, pp 1–3; Hon. James A. Hope, *Lady Louisa Stuart* (Edinburgh, 1899), p. 156.
50 *Berry Corrs.*, ii, p. 454.
51 Mabel, Countess of Airlie, *In whig society* (London, 1921), p. 7; *Burney Jnl.*, ii (ed. Joyce Hemlow and Anthea Douglas), (Oxford, 1972), pp xi–xix, 16–19; Robinson, *Lieven*, p. xvii; *Berry Corrs.*, ii, p. 545; *Arbuthnot Jnl.*, i, p. 135; see also, *Lady Bury*, i, pp 143, 190.

Unfortunately we have only a modest amount of evidence of what actually happened in these salons, a problem compounded by the fact that it is sometimes difficult to distinguish salons from *soirées*, supper parties and small assemblies. The chief point, however, is that they were occasions when men and women mixed in relatively small numbers and where politics were a topic of open conversation. Perhaps the following description of a small-scale gathering at Mrs Crewe's in June 1792 might convey the flavour of such events. The company was Dr and Fanny Burney, Edmund, Mrs Burke and their son, Richard, Mr and Mrs Crewe and their son, John, Lord Loughborough (soon to be Pitt's lord chancellor), Miss French, Miss Townshend, William Elliot, a Burkean, and Thomas and Mrs Erskine (Thomas flirting at this time with radicalism). At one point in the evening the party was described by Fanny Burney as being disposed as follows: 'My Father, Mr Richard Burke, and Mr Eliot (sic) entered into some general discourse; Mr Burke took up a volume of Boileau, & read aloud, though to himself, & with a pleasure that soon made him seem to forget all intruders. Lord Loughborough joined Mrs Burke; Miss French & Miss Townshend & Miss Crewe sat quietly together & Mr Erskine, seating himself next to Mrs Crewe engrossed her entirely, yet talked loud enough for all to hear, who were not engaged themselves.' There was then some general conversation about the radical society 'The Friends of the People' and at Mrs Crewe's instigation, an examination of how one should define 'the People'. The evening closed with a reading of Samuel Rogers' recently published poem, 'The Pleasures of Memory' following which, in Fanny Burney's words, we 'retired in very serene good humour, I believe, with one another'.[52]

The fourth point is that there was a tendency for all these types of events to became increasingly partisan. There were therefore whig and tory assemblies or balls and a myriad of factional salons and *soirées* set up as need arose.[53] Some women disliked the partisan nature of their role but others such as Lady Spencer, who in 1811 set up a regular 'at home' for her relations and the Grenville party, relished the responsibility.[54]

Finally under this head, there is little doubt that these events played a considerable part in the lives of male politicians. The great expansion in the volume of government and parliamentary business in this period does not seem to have been accompanied by a concerted effort to masculinise the social context of public life. For the moment these events held sway. Of course the point is difficult to substantiate but here are some examples of the frequency with which three men whose diaries record such events partook of female hospitality:

George Canning: during his first sixteen months as an M.P. in 1793–95 he went to Mrs Crewe's suppers and salons 47 times; to Lady Payne's *soirées* 10 times; dined at Lady Charlotte Greville's 9 times, at Lady Malmesbury's 34 times and with the countess of Sutherland 36 times. In addition he went to innumerable plays and operas, usually taking a place if not a seat in a woman's box.[55]

William Windham: Secretary at War in 1797 when the war against France was going badly and he was very busy. In the 46 days between 1 February and 17

---

52 *Burney Jnl.*, i, pp 193–203.
53 For example, Miss Berry noted this in 1809–1811, see *Berry Corrs.*, ii, pp 378, 406–07, 467.
54 *Lyttelton Corrs.*, p. 121.
55 Peter Jupp (ed.), *The Letter-Journal of George Canning 1793–1795* (Camden Fourth Series, vol. 41, London, 1991). See the index under the names of the persons mentioned in the text.

April for which he left a record, he 'called' on women twice and went to one breakfast, three dinners, one party and one ball at their various requests.[56]

Lord Howick: son and heir of Lord Grey, the whig leader and just embarking on his political career. In June 1829 he attended five parties, four 'at homes' and one dinner given by women; and enjoyed a private play at Lady Salisbury's and went to one opera. In February 1830, to take another example at random, he 'called' eight times and attended two evening parties, two dinners, two balls and three plays at Covent garden – all at the invitation of women.[57]

Finally and briefly, what of electoral and pressure group politics? In the case of parliamentary elections, the possession of property was the basis of significant electoral interests with the result that elections were dominated by the aristocracy. It was this that enabled numerous aristocratic women to exercise an influence over their families' electoral interests in this period, particularly when their husbands and their heirs were unable to do so – indeed many brought electoral interest with them to their marriages. However, there was a considerable difference between managing an electoral interest from the big house, as many women of this period did, and canvassing the voters during a contested election. Most aristocratic women whose memoirs have been consulted for the purposes of this paper disliked contested elections because of the violence that often accompanied them.[58] Sarah Lyttelton regarded them as occasions of 'chaos, riots, drunkeness, bribes, and bustles'.[59] Mrs Arbuthnot said she was opposed to giving seats to populous towns like Manchester or Birmingham because elections would encourage violence amongst the electors.[60] It was probably for this reason that the number of cases of women canvassing urban constituencies is relatively small and that the famous case of Westminster in 1784 is exceptional. In that year about a dozen women of high rank canvassed the largest and most prestigious of English urban constituencies for the whigs and the tories and went to such lengths as conveying artisan voters in their own carriages. However, their activities scandalised aristocratic society to such an extent that the most famous participant, the duchess of Devonshire, declined to repeat the effort at the next election for Westminster in 1788.[61]

Pressure group politics, not surprisingly, found little or no support from women of this class. In fact the evidence that has been consulted suggests that such women were nervous about even very modest degrees of support for causes that could be construed as putting pressure on parliament. It is interesting, for example, that following the formation of a Ladies' Society in the early 1790s to support the cause of the French émigrés, Hannah More thought it appropriate to woo supporters by penning a piece justifying women taking such a part in public affairs on the grounds that benevolence (as opposed to politics) was the motive.[62] Moreover, in the case of the most popular and frequent way in

---

56 Mrs. Henry Baring (ed.), *The diary of the Right Hon. William Windham 1784–1810* (London, 1866), pp 351–357.

57 Lord Howick's journal for June 1829 and Feb. 1830 (Durham University Department of Palaeography and Diplomatic (Grey of Howick Collection), 3rd Earl Grey MSs., Box F 105).

58 See, for example, The Earl of Bessborough (ed.), *Lady Bessborough*, pp 2–3.

59 *Lyttelton Corrs.*, p. 210

60 *Arbuthnot Jnl.*, ii, p. 173.

61 Reginald Blunt, *Mrs Montagu, queen of the Blues* (2 vols, London, 1923), ii, pp 170–71; the Duchess of Devonshire to Lady Spencer, n.d. [Apr. 1784] (Devonshire MS 610.3, Chatsworth House, Derbyshire); the Earl of Bessborough (ed.), *Georgiana Duchess of Devonshire* (London, 1955) p. 132.

62 *Burney Jnl.*, iii (ed. Joyce Hemlow, Oxford, 1973), pp 14 and notes 4, 5 and 6–15; this scheme was extended to Ireland and raised £200 (*ibid.*, pp 18, 37).

which benevolence was expressed by such women collectively – the charity ball – considerable efforts were made to ensure that they were not regarded as political.[63]

To conclude. There appears to have been a widespread view on the part of royal and aristocratic women that they were not suited to politics in the form that they were familiar with. This was due to theories about the differences between men and women, to distaste for the violence or the pedantry of politics, and to what might be called a caste consciousness – that there were certain things which neither women nor men of quality should do if they were anxious to preserve their caste. On the other hand, many such women took a great deal of interest in day to day politics and possibly an increased interest by comparison with earlier times as a result of such developments as the publication of fuller proceedings of parliament in the press.

The roles that they played were roughly the same as those played by similar women both earlier and later. However, the considerable influence of the monarchy, the highly aristocratic and factional structures of high politics and the contemporaneous increase in parliamentary business combined to invest them with much more significance than hitherto. Furthermore it is tempting to regard political developments after 1832 as having reduced their significance. In the case of the monarchy, the accession of Victoria led, ironically, to women having fewer opportunities to influence events. After 1841 the political influence of the crown declined as did the number of female royals of all kinds. As for parliamentary politics, the growth of party organisation, the increase in male clubs, the tighter grip of central party organisations over the constituencies and in general, the developing notion of public life as a profession rather than a vocation – also appear to have reduced the significance of these roles. In this respect a thesis can be advanced about royal and aristocratic women which bears a resemblance to the thesis that Linda Colley and others have advanced on the part of women in general in this period: that their political role increased; but subsequently, was checked.[64]

---

63 See, for example, *Arbuthnot Jnl.,* ii, pp 22–23.
64 Linda Colley, *Britons* (London, 1992), chapters 6 and 8.

11   *Jolies Femmes*/Feisty Women: The Images of Edouard de Beaumont

Laura Strumingher Schor[1]

S EVERAL years ago while reading the *Voix des femmes* in search of a better understanding of women's activities in Paris during the revolutionary months of 1848, I spotted an article which described a legion of young women, nicely dressed and well-mannered, who met at the base of the column on the Place Vendôme. There they floated a banner in the national colours with the word *Vésuviennes* stitched in golden letters. At noon, the legion set out, observing great order, and marched to the Hôtel de Ville where they asked for the aid of the provisional government. The author of the article explained that the *Vésuviennes* were poor young women who lived in community in Belleville. Their regulations were strict; they guaranteed food and lodging and ten francs per month to each. The author went on to comment: 'We find the work of the *Vésuviennes* worthy of all our sympathy; but why this name *Vésuvienne*? Their youth, their devotion to the Republican cause certainly permits its use, but is it in the general interest?'[2]

This simple article piqued my curiosity so that I began to hunt for every reference to the *Vésuviennes* in the literature of the period. I found many articles, all humorous spoofs about young women who made up a quasi-military group. I also found several illustrations of *Vésuviennes* in *Charivari* (see Plates 10 and 11), the premier satirical journal of the period. Except for one drawing by Cham, the rest of the *Vésuviennes* in *Charivari* were by Edouard de Beaumont, a little-known artist whose work when mentioned was frequently compared to the giants of the last century – Daumier and Gavarni, and found to be derivative though not without charm.[3]

I was very happy to have stumbled across Beaumont because I hoped he might tell me something about the women of the last century. Given the paucity of material available to historians who wish to understand the thinking of women who left few records, every new source is precious. Though the lithographs were intended for a popular audience during the middle decades of the last century, their meanings were far from clear to me several generations after they were printed. I also wondered how the images were experienced by women who read and contributed to the *Voix des femmes*. I began discussing the images with friends and colleagues who were specialists in art history. Beatrice

---

1 This article is adapted from the catalogue, *Les Jolies femmes d'Edouard de Beaumont*, printed in conjunction with an exhibition of the same title at the Leubsdorf Gallery of Hunter College, April–May 1994.
2 See Laura S. Strumingher, 'The Vésuviennes: images of women warriors in 1848 and their significance for French history' in *History of European Ideas*, 8, no. 4/5 (1987), pp 451–488.
3 Gustave Kahn, *La femme dans la caricature francaise* (Paris, 1907), pp 247–254. Kahn calls Beaumount's woman, *la petite femme*.

Farwell, Elizabeth Childs, John House, Jane Roos and Patricia Mainardi have each reviewed the images and offered valuable suggestions. Others working in related fields, including Karen Offen, Jeanine Plottel, Jeffrey Rosen, and Ann Ilan-Alter were generous with their comments.

My greatest debt is, however, to Edouard de Beaumont himself. Though he certainly did not intend for his work to become a source for the study of women's history, his images have taught me about the women who took leading roles in the struggle for women's rights during Beaumont's lifetime. These activists who wrote articles, attended meetings, and signed petitions, were also viewers of images of women in the popular satiric journals of the day. Their response to the images, both humorous and self-conscious, formed part of the sub-text to their meetings and their efforts at reform. Some planned their meetings knowing that satiric images and humorous texts about them would appear in *Charivari* and the other popular newspapers. These images helped the leaders develop into public personalities. It brought them and their cause out of their private and sheltered worlds and into the political arena with its open hostility to women's demands for change.

In April of 1848 one of the young leaders of the women's movement, Marie Noémi Constant, wrote a letter to the editor of *Charivari* which was printed in the feminist daily *Voix des femmes*:

Citizen *Charivari*,

It is with regret that we read in yesterday's issue your malicious citation of a petition signed by several members of our society; in France, the witty have always been the advocates of the fair sex, so why do you attack us now? Oh, citizen *Charivari*, do you, by any chance, take us for bluestockings?

Do you, by any chance, take us for some old, recalcitrant shrews, wearing both a petticoat and a moustache?! That would be proof that you have no idea who we are!

Because, if such were your views of us, think again, citizen *Charivari*: far from being nameless individuals, prickly, wrathful, all puffed up, dishevelled, with inky fingers and quills behind our ears, we are simply women ... who read *Charivari* and like it ... in spite of it all!

Be assured, citizen *Charivari* of our sincere fraternity.

One of the principal writers,
Marie Nóemi.[4]

This spirited letter is direct confirmation that women read and liked *Charivari*. They were not, however, passive observers of the misogynistic content of the journal. Marie Nóemi Constant protested the confusion of the editors who mistook women who struggled for change with shrewish women who wore moustaches, had inky fingers and quills behind their ears. Constant did not want to be identified with the blue stocking images of Daumier, but these images did not horrify her and cause her to stop reading *Charivari*.

While none of Marie Noémi Constant's contemporaries left such clear comments on their reading habits, it is likely that many women in the leadership of the struggle for greater equality in 1848 read *Charivari*. Eugénie Niboyet, the editor of the *Voix des femmes* and target of much satiric commentary in *Charivari*, was certainly familiar with the newspaper. Delphine Gay Girardin, a journalist, and the wife of Emile de Girardin, a leading journalist and publisher of newspapers, was another likely reader. Adèle Esquiros, a political activist and

---

4 'Citizen Charivari' in *La Voix des femmes*, 24 April 1848.

author, and the wife of Alphonse Esquiros, also an activist, author and a lithographic printer, had many reasons for following *Charivari*.

Adèle Esquiros and Marie Noémi Constant both signed the constitution of the *Vésuviennes*, a radical document codifying rights and duties for male and female citizens including mandatory military service for both, mandatory household *labour* for both, and mandatory marriage. The constitution was printed in September 1848 following six months of *Vésuviennes* stories and images in *Charivari*. There is no question that the constitution was a direct response to the satire. In the preamble the authors explain the term *Vésuvienne*. Far from associating themselves with the incendiary eruptions of Mount Vesuvius which was the intention of the satirists who developed the term, these women turned the metaphor on its head. They assured the public that they took the name because of the regenerative power of the lava which flowed from a volcano. It was their aim to restore balance and harmony in a peaceful manner under the banner of *Vésuviennes*.[5]

While this interpretation would certainly have surprised the authors of the *Vésuviennes* articles and imagery including Beaumont, it illustrates again that the women who read *Charivari* were not a passive audience; on the contrary they interpreted the complex images and texts and ferreted out multiple meanings depending on their sense of the way the world functions.[6] Particularly with expressive art like caricature, there are ample ambiguities left for the reader to interpret.[7] Perhaps it was that very openness that permitted some women readers to find positive meanings in Beaumont's images.

Beaumont's images were published regularly in *Charivari* over a twenty-three year period. They appealed to a readership that was entertained by references to seasonal festivities – summer vacations in the countryside and masked balls during carnival, and to recurring contests of will between men and women – primarily older bourgeois men and young, pretty, poor women. The selection of the lithographs for this article was designed to include images from each of Beaumont's major series. To create some coherence, I have divided the images into five categories: The Countryside, The Trickster, Women Demand Change, The Thirteenth Arrondissement, and The Masked Ball. Despite the variety of settings and themes, Beaumont's women remain the same. Each was a *jolie femme* – round-shouldered, full-bosomed, tiny-waisted, and winsome. Perhaps Beaumont was unable to draw women in any other way. Whatever the reason, Beaumont's use of the pretty figure engaged in rebellious acts was a sharp break with contemporary images of women demanding changes in public and private life.

For twenty-three years, from 1843–1866, Beaumont contributed his views about women to *Charivari*, the popular satirical daily, which also exhibited the broader ranging works of Daumier, Gavarni, and Cham. Beaumont averaged fifty-three publications per year during this period;[8] all were about women. His best known series were *L'opéra au xixe siècle, Les jolies femmes de Paris, Quartier*

5 Struminger, '*The Vésuviennes*', *op.cit.*, p. 455.
6 See Lawrence Levine, 'The folklore of industrial society: popular culture and its audiences' in *American Historical Review*, vol. 97, no. 5 (Dec. 1992), pp 1373, 1381.
7 For an interesting analysis of how popular culture can be interpreted in different ways depending on the predisposition of the audience see Janice Radway, *Reading the romance: women, patriarchy and popular literature* (Chapel Hill, North Carolina, 1991).
8 For a complete listing of Beaumont's lithographs, see Laran, Jean, 'Edouard de Beaumont', *Inventaire du fonds francais après 1800*, vol. i (Paris, 1930).

de la Boule Rouge, Au bal masqué, Les Vésuviennes, La guerre des femmes, Les jolies Parisiennes, and *Ces petites dames.* These works are influenced by the work of Beaumont's contemporaries as well as by the actual events the artist witnessed. Because he concentrated on women in their relationships with other women, in their relationships with men, and in their quiet reflections, study of Beaumont's *oeuvre* has frequently been dismissed as derivative and of secondary importance by art historians.

Beaumont produced a new type of woman which he dubbed the *jolie femme de Paris* in an early series for *Charivari.* This *jolie femme* sometimes called *jolie Parisienne, petite dame,* or even *Vésuvienne,* is graceful, almost doll-like, but simultaneously exudes self-confidence. She is aware of her power, particularly of her power over men who want something from her. Beaumont's women were spunky and struggling and simultaneously irresistible. They provided his viewers with an alternative to the largely negative images of independent women contributed to *Charivari* by Gavarni, Daumier, Cham and others.

The Countryside

Beginning with *Aux eaux de Baden* in 1845, and continuing for twenty years in *A la campagne, Croquis d'été, Canotiers et canotières,* and *Plaisirs d'été,* Beaumont contributed more than one hundred images of summertime in the country to *Charivari.* This group of lithographs reflects the keen interest of *Charivari* readers, mostly urbanites, in life in rural settings. Whether they are reminiscing about their own past in the village or envisioning summer holidays away from Paris in fashionable Baden or artistic Dieppe, Beaumont's viewers were doubtless charmed by his light-hearted depictions of women confiding in each other, of men seeking to control women, and of women boasting of their conquests. In the countryside images, Beaumont opened questions about the relationships between men and women, and about women's culture; but as in all of his work, he refrained from providing absolute answers. The viewer is given an opportunity to fill in the gaps.

One such opportunity can be seen in 'A view by the seashore' (Plate 1), published by *Charivari* on 10 August, 1848. The scene is probably of a promontory at Dieppe where Beaumont and Dumas fils and many of their artistic friends spent several weeks in summer.[9] Parisian readers would note the sharp contrast between the revolution-scarred streets with overturned vehicles and dug-up paving stones in their neighbourhoods[10] and the scene pictured by Beaumont. The caption testifies to the reassuring permanence of the cliff and the sea and summer breezes, despite the upheavals of revolution in Paris. The women on the cliff look like the party-goers Beaumont drew in his *Opéra au xixe siècle* series one year earlier. The blowing curls covering the woman's face resemble a mask. Beaumont introduces an ironic note in this otherwise idyllic scene, while the two women concentrate on a view of the sea, the young man hopes for a view of petticoats and ankles. This type of joke is found frequently in *Charivari;* it certainly helped sell subscriptions to men. But, some of Beaumont's women

---

9 *The Dieppe connection: the town and its artists from Turner to Braque* (London, 1992), pp 9–12.
10 Beaumont contributed to a series of 17 illustrations of the devastation wreaked by the Cavaignac's army on working-class neighbourhoods in Paris. Some of the illustrations showed astonished middle-class men and women staring at the rubble. They appeared in *Charivari* from 1 July to 8 September 1848.

readers might have viewed this scene differently. They might have smiled at the joke but also wondered what were these young women seeking? Did they imagine that there was greater freedom beyond the seas?

Two years later Beaumont places a reminder of the revolution in a rural setting in *A la campagne* #11 (Plate 2). Here, Beaumont shows two easily identifiable characters, the hobo and the country girl, suggestively lighting each other's cigars, a pleasure definitely forbidden to proper women. Beaumont's readers would remember a similar drawing of a militant *Vésuvienne* and a guardsman published in Spring 1848.[11] The two smokers, standing in front of an obscured rural backdrop appear to be on their way to the city. The hobo carries his walking stick and the young woman has her traveling bonnet and umbrella. Beaumont depicted many emigrants to the city who yearned for a better life. But the real subject of this lithograph is caught by the ironic caption: 'The Smokers' Republic – the only one in which the motto – Liberty, Equality, Fraternity – is really practised.' Beaumont participated in the ironic representation of the calls of feminists for an expansion of the ideals of the revolution to include women. But while his contemporaries ridiculed women who sought change,[12] Beaumont's treatment of women's struggle for rights could be read as humorous; he was never hostile.

But Beaumont knew that young people, especially women, also faced danger from more experienced men who lured them to Paris. In *Fariboles* #64 (Plate 3) he uses a bucolic backdrop to illustrate an encounter between a bourgeois man, identified by his top hat, and a village woman. The man comments on the ease with which he could transform an innocent woman into a woman of easy virtue. 'Well, well but the little one is very nice . . . All she needs to become a fashionable lady are camellias . . . and shoes.' The association of camellias and a life of debauchery would have been clear in 1852, the year that Dumas fils' *La dame aux camélias* opened to great acclaim. But Beaumont's lithograph introduces something new. Here the bourgeois man inspects the young woman; she in turn, inspects him. The reciprocal glance creates a balance not found in Dumas' play which evokes sympathy for its dying heroine. In this drawing Beaumont implies that women as well as men hold power.

On 18 August 1860, *Charivari* printed a lithograph (plate 4) in which the countryside is present only in a dream and in a painting on the wall. Here Beaumont presents a dialogue between two women who have traded their virtue for a comfortable apartment with bourgeois furnishings. Both are enjoying the fruit of their labour, one by playing the piano, the other by indulging in opium. Beaumont's caption allows him to introduce the belief widely held by his contemporaries that opium freed the imagination.[13] The smoker dreams that she still lives in her village and that she is recognised as an innocent maiden, crowned with roses. Her companion is amazed that she can even imagine such things! Though the dialogue is naive, even childish, Beaumont's women are appealing; their conversation is fresh and amusing. While some women readers may have been concerned about the morals of these kept women, others were

---

11 *Entre deux camarades, Vésuviennes* #12, appeared in *Charivari* on 14 June 1848. See Strumingher, 'Vésuviennes', *op. cit.*, p. 476.
12 Daumier's *Les femmes socialistes* were printed in *Charivari* in 1849. For an explanation of Daumier's misogyny see Daumier, *Intellectuelles et femmes socialistes*, preface by Francois Partuier (Paris, 1974).
13 Elizabeth Childs, 'Honoré Daumier and the exotic vision: studies in French culture and caricature, 1830–1870' (Ph. D. dissertation, Columbia Univ. Press, 1989; Ann Arbor: U.M.I., 1992), p. 244.

FARIBOLES

Chez Aubert, Pl. de la Bourse   Imp. Aubert & Cie

Plate 1

FARIBOLES I, #10

Une vue au bord de la mer.
A view by the seashore.

CHARIVARI 10 August 1848

Hunter College Collection

A LA CAMPAGNE

Chez Aubert, Pl. de la Bourse   Imp. Aubert & Cie

Plate 2

A LA CAMPAGNE I, #11

République des fumeurs! – La seule dans laquelle on pratique véritablement la devise:
Liberté, égalité, fraternité.

Smokers' Republic! The only one in which the motto – Liberté, Égalité, Fraternité is
really practised.

CHARIVARI 7 June 1850

New York Public Library

FARIBOLES

Maison Martinet, r. Vivienne 41 et 11 r. du Coq Sᵗ Honoré. Paris.   Imp. Ch. Trinocq, Cour des Miracles, 9. Paris.

Plate 3

FARIBOLES I, #64

Tiens tiens, mais elle est très gentille cette petite . . . il ne lui manque, pour devenir une femme à la mode, que des camélias . . . et des souliers!

Well, well, but the little one is very nice . . . all she needs to become a fashionable lady are camellias . . . and shoes!

CHARIVARI 27 April 1852

Hunter College Collection

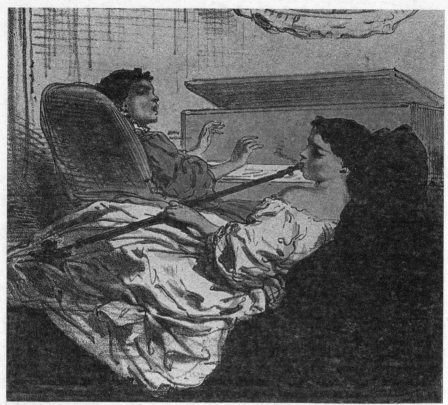

Mᵒⁿ Martinet, 172, r. Rívoli et 41, r. Vivienne.   Lith. Destouches, 28, r. Paradis Pʳᵉ Paris.

Plate 4

CES PETITES DAMES I, #3

– *Quel plaisir trouves-tu à fumer de l'opium?*
– *Je fais des songes charmants . . ., je rêve que j'habite
  encore mon village, que je suis une honnête jeune
  fille et qu'on me couronne rosière!*
– *Peut-on rêver de choses aussi invraisemblables!*

– What pleasure do you find in smoking opium?
– I have enchanting dreams . . . I dream that I still live
  in my village, that I am an honest girl and that I'm
  made queen of maidens!
– How can you dream such incredible things!

CHARIVARI 18 August 1860

Hunter College Collection

certainly charmed by the wit. They may have seen humour in some of the statements and appreciated the survival skills demonstrated in others. This glimpse at the confidential discussions is brief, but it presents a world in which dreams and sorrows are shared by women.

The Trickster

The Trickster images express Beaumont's admiration for women's cunning, especially when it relates to women who trick men out of money. This group of cartoons appear to have been influenced by an earlier form of popular culture, the oral *fabliaux*. In these tales of French peasant life, clever women frequently overcome difficult circumstances by resorting to their natural wit and cunning.[14] Triumphing over the hardships of life by eating a splendid meal was the central focus of many peasant oral tales.[15] This old tradition entered the popular culture of the nineteenth century through the new medium best adopted to mass communication in the city, the political cartoon printed in the daily press.[16] The trickster images did not stop with the desire for immediate gratification in the form of food. Beaumont modified the form to include women whose trickery extended to other areas of life, women who knew that clear thinking and quick wit were necessary to extract what was needed from men.

Whether they are at a masked ball or immersed in a lake, Beaumont's women appear to always need a square meal. While peasant stories involved preparing succulent capons, the nineteenth-century version of this obsession with food usually included a restaurant and finding a man to pay the bill. In the next four images (Plate 5, 6, 7, and 8) men become targets of opportunity in a world in which women can not earn enough money as seamstresses or flowermakers. In planning how to pay for a meal, the tricksters agree that a full stomach is more important than who pays the bill. Beaumont's poor women are equally prepared to dupe a waiter, a restaurant owner, or a Russian prince. One woman plans to faint with hunger to achieve her goal (Plate 7). Another (Plate 8) asks a waiter to add her old restaurant debts to the bill of her new bourgeois suitor. Because the object of all these plots is sustenance, the women are probably perceived as clever and witty, even humorous, rather than as immoral.

In the final drawing of this type, *Ces petites dames* #8, (Plate 9) Beaumont presented a beautiful woman receiving instructions on how to shoot pigeons. Her response, 'Hunting pigeons, I know that better than you!', is the ultimate retort of the self-confident trickster. The rifle-holding woman would have reminded Beaumont's viewers of the militant *Vésuviennes* whom he popularised in 1848. But it is the caption that reveals Beaumont's keen empathy with women in his audience who would have enjoyed the chance to make such an honest and witty response.

---

14 Mary Jane Schenck, *The Fabliaux: tales of wit and deception* (Philadelphia, 1987), p. 12.
15 Robert Darnton, *The great cat massacre* (New York, 1984), p. 34.
16 See also Natalie Zemon Davis, 'Toward mixtures and margins' in *American Historical Review*, vol. 97, no.5 (Dec. 1992), p. 1410. I am indebted to Jeanine Plottel for reminding me of the rich literary tradition of tricksters pervasive from the time of Rabelais, through Molière, to Prèvost, and continuing with Beaumont's contemporary Nèrimèe.

AU BAL MASQUÉ

M^{on} Martinet172, r. Rívoli et 41, r. Vivienne   Lith Destouches 28, r. Paradis P^{re} Paris.

Plate 5

AU BAL MASQUÉ IV, #32

– *Dis-donc, si ces messieurs qui nous ont donné*
  rendez-vous ici n'allaient pas venir . . ., que ferions-nous?
– *C'est bien simple, nous souperions sans eux!*
– *Mais qui paiera l'addition?*
– *Nous inviterons le garçon à souper . . . et comme il n'y*
  aura que lui en fait d'homme . . . il sera bien obligé de payer!

– Say, if these men who told us to meet them here didn't
  show up, what would we do?
– Very simple, we'd have supper without them!
– But who will pay the bill?
– We will invite the waiter . . . and since he will be the only
  man at the table . . . he'll have no choice but to pay!

CHARIVARI 26 March 1860

Hunter College Collection

CES PETITES DAMES

M^{on} Martinet, Paris   Lith. Destouches, Paris.

Plate 6

CES PETITES DAMES IV, #8

– *Nous avons quartre francs cinquante; en ne prenant qu'un potage, une matelotte, un homard un filet Chateaubriand et un peu de Champagne, nous n'arriverons jamais à payer notre dîner.*
– *Si fait, supprimons le pourboire au garçon et empruntons 20 francs au restaurateur.*

– We have four francs fifty; if we only order soup, fishstew, lobster, Chateaubriand fillet and a little bit of champagne, we'll never be able to pay for our dinner.
– Oh yes, let's not give a tip to the waiter and borrow 20 francs from the owner.

CHARIVARI 12 September 1865

Armand Hammer Collection

AU BAL MASQUÉ

maison Martinet, 172, r. Rivoli et 41 r. Vivienne.   Lith Destouches, 28, r. Paradis P^re Paris.

Plate 7

AU BAL MASQUÉ IV, #38

– *Fanny, serre-moi davantage . . .*
– *Mais madame se trouvera mal!*
– *Tant mieux . . . je ferai croire que c'est d'inanition . . .,*
  et un Prince russe compatissant m'emmènera souper!

– Fanny, make it tighter . . .
– But madam will faint! . . .
– Good . . . I'll make believe that it's with hunger . . .,
  and a compassionate Russian Prince will buy me supper!

CHARIVARI 21 January 1861

Hunter College Collection

AU BAL MASQUÉ

A de Vresse F.d^r r. Rivoli 55

Lith Destouches 28, r. Paradis P^{re}

Plate 8

AU BAL MASQUÉ I, #6

*Garçon, ne manquez pas de porter sur l'addition le 11
déjeûners et les 14 dîners que je vous dois: j'ai amené un
bon jeune homme.*

Waiter, do not fail to add onto the check the eleven
lunches and the fourteen dinners that I owe you. I
brought a young gentleman.

CHARIVARI 5 March 1866

Armand Hammer Collection

CES PETITES DAMES

M^{on} Martinet, 172, r. Rivoli et 41, r. Vivienne   Lith Destouches 28, r. Paradis P^{re}

Plate 9

CES PETITES DAMES III, #8

- *Écoutez bien . . . pour les Pigeons . . .*
- *La chasse aux Pigeons, je connais ça mieux que vous!*

- Now listen . . . as for pigeons . . .
- Hunting pigeons, I know that better than you!

CHARIVARI 4 October 1864

Armand Hammer Collection

Women Demand Change

1848 marked a new beginning for the women's rights movement in France. Leaders like Eugénie Niboyet, Jeanne Deroin, Desirée Gay, Pauline Roland, and Suzanne Voilquin energetically pursued a broad agenda to expand women's political, legal, educational, and economic rights.[17] Beaumont devoted more attention to women demanding change than any of his contemporaries. He created *Les Vésuviennes*, images of militant women who enlisted in the Guards; *La Guerre des femmes*, images of male-female power struggles; and *Les Femmes en révolution*. His treatment of the subject was also more sympathetic than that of Daumier, Cham, or Gustave Doré or of the satirical articles in *Charivari* which ridiculed women who joined clubs. Beaumont's work implicitly refuted the article in *La Liberté* which stated: 'Woman is consecrated to obedience and silence; to confiding love and modest humble devotion. This is a much more beautiful position than that of woman voter, woman guardsman, woman legislator.'[18]

Looking at Beaumont's depiction of the *Vésuviennes*, one is struck by the fact that here a woman is both a guardsman and charming (Plate 10). Madame Coquardeau[19] teasingly thumbs her nose at her husband who has forbidden her to fulfill her military duty. 'Mâme Coquardeau, I forbid you to go to arms . . . It makes no sense to leave me here with three kids to watch and no bottle!' Her posture is suggestive; her uniform is reminiscent of the stevedore costume, a preferred outfit for wild devotees of the masked ball.[20] The husband, on the other hand, looks like a bewildered hobo. His inability to handle the babies shows him as ineffectual; the situation is humorous, not alarming. A life of silence and obedience does not seem very attractive to Beaumont's women. The *Vésuviennes* are serious about their mission, but they also seem to be having a good time.

Beaumont took up another family topic, the corporal chastisement of wives by husbands, in *Les Vésuviennes* #8 (Plate 11) but in this drawing the practice is boldly reversed.[21] The decisive kick placed by the *Vésuvienne* on the *derrière* of a man who did not take her seriously is reminiscent of *fabliaux* lore in which women get even with men. In view of laws which permitted husbands to discipline their wives, and required wives to obey their husbands, this kick was a cartoon corrective to a world in which men were legally always on top.

In the final drawing depicting women's demand for change, Beaumont shows a moment of triumph for women in the setting of the masked ball (Plate 12). The caption reads: 'Hurrah! . . . at last woman ceases to be oppressed . . . long live the droll and jolly republic!' As in the previous image of the two smokers (Plate 2) where liberty, equality and fraternity holds true only in the imaginary smokers' republic, here women are enjoying their superior role only during the fantasy of the ball. Both of these images allow male readers to be amused by the pretty women, while providing women readers with the opportunity to imagine achieving their goals.

---

17 See Claire G. Moses, *French feminism in the nineteenth century* (New York, 1984), pp 89–149.
18 This statement in *La Liberté* provoked an angry response from *La Voix des femmes* on 15 June 1848.
19 For Gavarni, Coquardeau represents the cuckold, victimised by his wife or mistress, or both. Bourgeois and middle-aged, he is at once suspicious and naive, and hopefully out of mood with the prevailing high spirits of carnival (Nancy Olson, *The carnival lithographs* (New Haven, Yale, 1979), p.36).
20 Gavarni created the *débardeur* (stevedore) costume based on the uniform of the men who loaded the barges on the Seine. This costume was admired for its nonconformity to modern dress. (See Olson *op. cit.*, p. 18).
21 See Roderick Phillips, 'Women, neighborhood and family in the late eighteenth century' in *French Historical Studies*, vol. 18, no. 1 (Spring 1993), pp 1–12.

LES VÉSUVIENNES

Chez Aubert Pl de la Bourse.   Imp. Aubert & Comp^ie

Plate 10

LES VÉSUVIENNES, #1

*Mâme Coquardeau, j'te défends d'aller au rappel . . . n'y
a pas d'bon sens de me laisser comme ça avec trois enfants
sur les bras . . . et pas de biberon!*

Ma'am Coquardeau, I forbid you to go to arms . . . It makes
no sense to leave me here with three kids to watch and no
bottle!

CHARIVARI 1 May 1848

New York Public Library

LES VÉSUVIENNES

Chez Aubert Pl. de la Bourse.   Imp. Aubert & Cie

Plate 11

LES VÉSUVIENNES, #8

Danger d'insulter une femme armée.
It's dangerous to insult an armed woman!

CHARIVARI 1 June 1848

Hunter College Collection

AU BAL MASQUÉ

Chez Aubert Pl. de la Bourse 29   Imp. Aubert & Cie

Plate 12

AU BAL MASQUÉ I, #10

A la bonne heure . . . v'la enfin la femme qui cesse d'être opprimée . . .
vive la république drôlatique et joviale! . . .

Hurrah! . . . at last, woman ceases to be oppressed . . .
long live the droll and jolly republic! . . .

CHARIVARI 21 February 1849

New York Public Library

Thirteenth Arrondissement

Before the reorganisation of Paris by Haussman, the city had only twelve *arrondissements*. The Thirteenth was a euphemism created by contemporaries to denote a meeting place for seductive and willing women. Its centre was the poor neighbourhoods of Paris where seamstresses, laundresses, and flowermakers lived in garrets under the eaves, alternately, freezing in winter and boiling in summer. Bourgeois men came to the Thirteenth to seek out poor women for an evening's entertainment. These women were forced by hunger or the need to help aged parents to supplement their meagre earnings by working nights as well. These young women, frequently country girls who were new-comers to Paris, became the subject of literary and artistic works which covered up the real conditions of their lives by romanticising them. The truth about their conditions was made known to the public by A.J.B. Parent-Duchâtelet in a path-breaking study, but the images of the artists and journalists were better received than the dry statistics of the medical observer.[22]

Henry Monnier introduced the image of the adorable *grisette*, frail and loving, to a large audience in 1827.[23] The charm of the loyal and unassuming *grisette* captivated Parisian audiences until she was replaced by the more worldly *lorette* in the mid-1840s. Gavarni made the *lorette*, a denizen of the neighbourhood surrounding the church of Notre Dame de Lorette, a greedy and manipulative woman who was more interested in money than in love.[24] She required expensive gifts like cashmere shawls and elegant furniture and paintings from her lovers.

Beaumont also illustrated the personal relations between the *jolie femme* and her clients. He demonstrates that when they are in their prime, these women are a cunning and successful group. In one image (Plate 13) Beaumont shows a bourgeois man on his knees begging forgiveness from his young mistress who eyes him with a dubious look. Should she forgive him? She calls him a despot. What was his misdeed? Could he have suspected her of faithlessness? Many of the images in the *Quartier de la Boule Rouge* attest to the infidelity of both men and women. Two months after the previous image appeared in *Charivari*, Beaumont contributed an image which illustrated that the man's doubts were justified (Plate 14). Here we see the young woman quickly dispatching a boy-friend or another client while the bourgeois busies himself looking at a painting. Here, Beaumont, the artist, pokes fun at the bourgeois who is occupied studying a painting while his mistress arranges to meet another man. These two drawings of the same couple appearing within two months of each other in the same series indicate the deceit and treachery in the relationships between young women and their casual lovers who are older and bourgeois and very likely married.

Beaumont brings his public back to the reality of life for women in the Thirteenth Arrondissement on 3 December 1850 (Plate 15). Against a somber backdrop we meet two young women admiring an infant whom they are about to give to a wet-nurse. Looking at the baby's eyes and nose they seek a clue to the parentage. Who fathered the baby, Ernest or Adolphe? In reality it matters

---

22 A.J.B. Parent-Duchâtelet, *De la prostitution dans la ville de Paris* (Paris, 1836 and 1857), p. xix.
23 Edith Melcher, *The Life and Times of Henry Monnier* (Cambridge, Harvard 1950), pp 42–49.
24 Thérèse Stamm, *Gavarni and the critics* (Ann Arbor, Michigan, 1981), p. 145. Champfleury observed: 'Le rôle de la femme, suivant Gavarni, consiste à endiabler l'homme et à le faire sauter comme les écus.'

QUARTIER DE LA BOULE ROUGE

Chez Aubert, Pl. de la Bourse.   Imp. Aubert & Cⁱᵉ

Plate 13

QUARTIER DE LA BOULE ROUGE, #3

Je ne sais pas trop si je dois vous pardonner . . . vous êtes un despote . . . un tyran!

I'm not sure if I should forgive you . . . you're a despot . . . a tyrant!

CHARIVARI 18 September 1847

Hunter College Collection

QUARTIER DE LA BOULE ROUGE

Chez Aubert Pl. de la Bourse.   Imp. Aubert & Cie

Plate 14

QUARTIER DE LA BOULE ROUGE, #15

Un visite intempestive.

An untimely visit.

CHARIVARI 13 November 1847

Hunter College Collection

FARIBOLES

Chez Aubert & Cⁱᵉ Pl. de la Bourse, 29. Paris.   Imp. de Mᵉ Vᵉ Aubert, 5. r. de l'Abbaye, Paris.

Plate 15

FARIBOLES I, #38

– *N'est ce pas qu'il a tout à fait les yeux d'Ernest!*
– *Oui, mais il me semble qu'il a le nez d'Adolphe!*

– Doesn't he just have Ernest's eyes!
– Yes, but I think he has Adolphe's nose!

CHARIVARI 3 December 1850

Hunter College Collection

very little since neither will be able or inclined to support the child. The cost of the nurse will be born by the continued *labour* of the mother who may receive some help from her women friends. The expressions on the faces of the three women indicate routine acceptance of their situation, illustrating the banality of the out-of-wedlock birth and the uncertainty of paternity. The real situation of the *jolies femmes* could not be more vivid.

Three months later Beaumont shows what happens to *jolies femmes* who are down on their luck; they emigrate to California. This clever and beautifully drawn lithograph (Plate 16) joins the excitement of the day about gold discovery in California to the stereotype of the period that women were interested in men only for their gold. Beaumont also extends the *lorette* metaphor to include the religious idea of a pilgrimage of the faithful. These pilgrims are faithful to themselves and to their struggle to survive.

In a final comment on infidelity, Beaumont takes us away from the ball and from the homes of the *jolies femmes* to the comfortable sitting room of the bourgeois (Plate 17). In the disjointed dialogue between husband and wife, Beaumont points to the infidelity of married bourgeois women. The viewer may sympathise with the bourgeois wife whose husband surely frequents the *Quartier de la Boule Rouge* for female companionship while burying himself behind a newspaper at home. The neglected wife resorts to buying herself presents, exhibiting tastes not unlike those of the *jolies femmes*, and to daydreaming about a rendez-vous with Monsieur Edmond.

Beaumont shows that while infidelity meets the economic needs of poor women, it also answers the need for intimacy of bourgeois women. The problems caused by husbands who were totally absorbed with their businesses leaving their wives to get into trouble by daydreaming was the subject of literature destined for middle-class women.[25] But while the stories were laden with strong moral overtones urging the wives to be content as managers of their homes and mothers to their children, Beaumont presents the ideas of this deceitful wife without moral overtone. While Madame is dressed as a proper lady and feigns interest in her husband's reading, she plans her own activities like any *jolie femme*. By failing to express disapproval of Madame, Beaumont implicitly recognises her dilemma and allows her to dream her own dreams. By using the same type of woman in all of his work, Beaumont suggests a similarity in the conditions and in the struggle of all women of his day.

The Masked Ball

Beaumont began to draw costumed revelers in 1848 and continued through 1866, drawing a total of 206 lithographs depicting men and women at the ball. This large and varied production permitted Beaumont to comment on gender roles, class issues, urban and country mores, all enacted in relief against the backdrop of the most colourful and popular amusement of his day. Originally balls were segregated by class and often by geographic origin, but by mid-century the opera ball which cut across class lines was the most popular. There, on the Rue Le Peletier at the old opera house, restored decades earlier by Beaumont's father, bourgeois men escorted their mistresses, while masked bour-

---

25 Laura S. Strumingher, 'Mythes et réalités de la condition féminine à travers la presse féministe lyonnaise des années 1830' in *Cahiers d'histoire*, iv, (1976), pp 408–424.

FARIBOLES

Chez Aubert & Cᵗᵉ Pl. de la Bourse 29. Paris.   Imp. de Mᵉ Vᵉ Aubert, 5. r. de l'Abbaye Paris

Plate 16

FARIBOLES I, #46

Jeunes fidèles de la Paroisse Notre Dame de Lorette
partant en pèlerinage en Californie.

Young believers from the parish of Our Lady of Lorette
going on a pilgrimage to California.

CHARIVARI 17 March 1851

Hunter College Collection

CROQUIS PARISIENS

Mᵒⁿ Martinet, 146, r. Rivoli et 41, r. Vivienne.   Lith. Destouches, 38, r. Paradis Pʳᵉ Paris.

Plate 17

CROQUIS PARISIENS IV, #20

Une Après-Dinée Conjugale

Monsieur: (lisant tout haut le journal) – On comprend que ces motifs rendent bien difficile,
. . . quant à present, . . . une reprise sérieuse sur les actions de chemins de fer. Aussi la
lourdeur de ces valeurs et leur abandon par la spéculation s'expliquent-ils parfaitement.
Nous avons publié vendredi les recettes brutes de la semaine dernière sur toutes les lignes.
Les recettes kilométriques ont présenté les résultats suivants . . . (Monsieur s'arrête, éternue,
se mouche et continue sa lecture)
Madame: (se parlant tout bas) – Demain à une heure je vais me commander un chapeau
en velours, puis me choisir une robe en brocatelle, . . . de là, j'irai au bois pour retrouver
M. Edmond, qui doit m'attendre auprès de la cascade . . . (Madame s'arrête, s'endort et
rêve le reste)

Monsieur and Madame After Dinner

Monsieur (reading the newspaper aloud) – It is understood that these motives render
quite difficult . . . as for now, . . . a healthy recovery of railway shares. The securities'
slackness and their abandon by speculators is then perfectly understandable. On Friday,
we published last week's gross revenues for all (railway) lines: a breakdown of revenues
per kilometer shows the following figures . . . (Monsieur sneezes blows his nose and
continues to read)
Madame (to herself) – Tomorrow at one o'clock, I shall order myself a velvet hat and
choose a 'brocatelle' dress, . . . then I'll go on to the park and meet Edmond, who will be
waiting for me by the waterfall . . . (Madame stops, falls asleep and dreams the rest)

CHARIVARI 9 December 1856

Armand Hammer Collection

geois women flirted incognito. Using the setting of the ball, Beaumont exhibited women's continuing struggle to make ends meet and to gain respect.

One of the earliest and most striking ball images appeared in *Charivari* in November 1849 (Plate 18). It showed a woman dressed as a stevedore with additional plumage on her képi-style hat. Her costume was a development of the *Vésuvienne* costume Beaumont had popularised the previous year. Facing her was a man dressed as a 'savage' with a feathered headdress and a tomahawk behind his back neatly balancing the elaborate bustle attached to the stevedore's belt. The costume of the savage, based on the American Indian prototypes, was a standard of the day.[26] Beaumont's use of the icon enabled him to contrast the delicate beauty of the woman with the strength yet brutishness of the man. He gives the caption to the stevedore: 'You a savage? You are not more savage than I.' In this line, like in the verbal challenge of the pigeon hunters discussed above (Plate 9), Beaumont shows the bravado of the woman who struggles for survival. She is willing to engage in a contest with the savage because he provides her with the best chance of survival. She baits him to get him interested. His glance appears menacing, but she has captured his attention.

In an image drawn a decade later, Beaumont shifts focus from a male-female exchange to one between two women. In this image (Plate 19) the rivalry between two women is expressed through their comments on costumes chosen for the ball. The lady criticises the stevedore for not being respectable; the stevedore costume carried with it a reputation for wanton behaviour. The stevedore responds with verve that the woman dressed as a lady is the more camouflaged. The exchange reflects the growing ability of poor women to pass themselves off as ladies because of the availability of ready-made ladies' clothing. Unlike most of the Trickster images described above, the stevedore and the lady are rivals; they do not collaborate to achieve their goals. In a similar drawing, Beaumont presents another stevedore and a masked lady competing for the attention of Adolphe (Plate 20). The stevedore thumbs her nose to indicate a complete disregard for convention. There is no gender solidarity in this image, only a ruthless desire to succeed.

Around the same time Beaumont published another image in the *Au bal masqué* series which showed an alert 'lady' making sure that her escort will deliver the supper he has promised (Plate 21). The clown urges Turlurette to lift her leg higher in dance; she refuses explaining that she does not want to be arrested and as a consequence miss supper. Readers of *Charivari* would have known of the extensive legislation governing dance halls, including restrictions on various dances thought to provoke riots.[27] An earlier Beaumont dealt with the same theme of leg lifting, but here the stevedore warns the *pierrot* to behave himself (Plate 22).

Over a decade later (in 1861), Beaumont reassured his readers with a tender portrait of a hobo lighting a cigar for a young woman (Plate 23). This couple bears strong resemblance to the countryside lithograph celebrating the republic of smokers (Plate 2) published in June 1850. Like the image, the characters are also eleven years older. The innocent young country girl of the smokers' republic no longer has her country pipe; she claims that it broke during the

---

26 Childs, *op. cit.*, pp 183–184. Childs points out that the outfit usually included a club, topknot of hair, a feather headdress, hoop earrings or nose rings, a grass skirt or loin cloths and elabourate body paint to imitate tattooing.
27 François Gasnault, *Guinguettes et lorettes: Bals publics et danse sociale à Paris entre 1830 et 1870* (Paris, 1976), p.23.

AU BAL MASQUÉ

Chez Aubert Pl. de la Bourse, 29.   Imp. Aubert & Cie

Plate 18

AU BAL MASQUÉ I, #11

Toi sauvage? . . . t'es pas plus sauvage que moi!

You a savage? . . . you're not more of a savage than I!

CHARIVARI 13 November 1849

New York Public Library

AU BAL MASQUÉ

maison Martine, 172, r. Vivienne   Lith. Destouches, 28. r. Paradis P^re Paris.

Plate 19

AU BAL MASQUÉ IV, #15

– *Comment, une femme qui se respecte peut-elle venir
ici déguisée en débardeur!*
– *T'es encore bien plus déguisée que moi, puisque tu
t'es mise en femme du monde!*

– How can a woman with self-respect come here
disguised as a stevedore.
– You're even more disguised than I am, since you
come here as a lady!

CHARIVARI 11 January 1860

Hunter College Collection

AU BAL MASQUÉ

maison Martine, 172, r. Rivoli et r. Vivienne   Lith. Destouches, 28. r. Paradis Pre Paris.

Plate 20

AU BAL MASQUÉ V, #9

– *La première fois que je te verrai parler à Adolphe, tu . . .*
– *As-tu fini tes manières! . . . tu m'en demanderas raison . . .*
peut-être? . . . je ne te crains pas moi . . . Ah! mais c'est que
j'ai reçu une bonne éducation, moi . . . j'ai eu un professeur
qui m'a appris à mettre les poings sur les nez . . .

– The first time that I see you speak with Adolphe, you . . .
– Will you stop it! What, you'll demand satisfaction for it,
perhaps? I'm not scared of you one bit . . . Hey, I got myself
a good education . . . I had a teacher who taught me to
thumb my nose . . .

CHARIVARI 28 December 1860

Hunter College Collection

AU BAL MASQUÉ

M^{on} Martinet r. Rivoli et 41, r. Vivienne   Lith. Destouches. Paris.

Plate 21

AU BAL MASQUÉ V, #6

– *Voyons, Turlurette . . ., lève donc mieux la jambe que ça!*
– *Farceur . . ., on voit bein qu'il est trois heures du matin . . .,*
  tu veux me faire mettre au violon pour ne pas avoir à
  m'emmener souper!

– Come on, Turlurette . . ., lift your leg better than that!
– You clown, I can tell it's three o'clock in the morning . . .,
  you want to get me locked up so you won't have to buy me
  supper! . . .

CHARIVARI 25 December 1860

Hunter College Collection

AU BAL MASQUÉ

Chez Aubert Pl. de la Bourse.   Imp. Aubert & Cᶦᵉ

Plate 22

AU BAL MASQUÉ I, #25

Un Pierrot Blanc

Fais-donc attention! . . . pour un Pierrot réactionnaire, tu vas trop en avant!

A White Pierrot

Watch where you're going! For a reactionary Pierrot, you're too forward!

CHARIVARI 14 February 1850

New York Public Library

AU BAL MASQUÉ

M^{on} Martine, 172, r. Rivoli et 41, r. Vivienne   Lith. Destouches, 28, r. Paradis P^{re} Paris.

Plate 23

AU BAL MASQUÉ V, #16

– *Comment! vous, ma p'tite dame . . ., vous fumez des cigares?*
– *Oh! ne m'en parlez pas . . ., j'en suis bien contrariée moi-même . . . mais j'y suis bien forcée! . . . car en dansant le grand galop de la fin du bal, j'ai cassé ma pipe qui était dans ma poche!*

– What! the little lady smokes cigars?
– Tell me about it! I don't like it either . . . but I'm forced to do it, since when I was dancing the great gallopade at the end of the ball, I broke my pipe that was in my pocket!

CHARIVARI 14 February 1861

Hunter College Collection

gallopade, a wild dance which announced the end of the ball. It's possible that she gave up smoking a pipe and replaced it with a cigar because of the more modern and daring connotations attached to cigar smoking for women. One wonders what else has broken in the passage of the innocent girl to *jolie femme*. The visual representation is of a woman who has had a rough time and harbours no illusions. But her acceptance of the light from the hobo is a poignant illustration of a moment of friendship amid the wild atmosphere preceding the deprivations of Lent.

The masked ball images provided Beaumont with a colourful and exciting backdrop on which to juxtapose the icons of the period – the stevedore, the *pierrot*, and the *loge*, with the realities of the day – legislation which regulated the dance halls, salesmen who sought to take advantage of the would-be stevedore, and the persistent struggle for survival which faced the poor women of Paris. His ability to show amusing and serious aspects of the Masked Ball and of each of the themes sketched above – the Countryside, the Trickster, Women Demand Change, and Thirteenth Arrondissement, make Beaumont's images fresh and vital more than a century after his death.

The twenty-three years during which Beaumont contributed lithographs to *Charivari* was a period in which the traditional relationships between French men and women, codified early in the century by Napoleon, were severely challenged. The discontent with the status quo came from women of diverse backgrounds. Poor women were primarily interested in the right to a job and wages which would support them without the need to resort to prostitution to supplement their income. Lower middle-class women were particularly interested in education as an avenue to escape poverty through teaching jobs. Middle-class women were interested in civil and political rights as well as in extending the right to higher education to women. All of these women struggled to find solutions to common problems like divorce and balancing the needs of families with the need or desire to work in larger arenas. These issues were raised in women's newspapers beginning with the *Conseiller des femmes* edited by Eugénie Niboyet in 1834.[28] They are also found in the writings of each of the leading women authors of the period – Daniel Stern, Louise Colet, George Sand, Anaïs Segalas, and Delphine Gay Girardin.

The revolution of February 1848 was the catalyst that united the distinct groups of women and propelled them into action. Their newspaper, the *Voix des femmes*, was the focal point for the creation of a women's reading room, for collections of emergency relief for unemployed women, for efforts at creating employment bureaus and employment opportunities for women, for efforts to secure the right to vote and to extend opportunities for higher education for women, for demands to change the divorce laws, and finally, for a petition to award midwives civil servant status and hence decent salaries. The women marched, signed petitions, and staged meetings that grew in size to accommodate hundreds of women and men.[29]

The reaction criticising women's efforts to expand their horizons and to protect their human rights was already evident in the work of Daumier prior to 1848. His *Bas bleus* series of shrewish images of women writers was printed four years before the revolution took place. This series was an early response to demands which had been simmering for at least a decade. Daumier's *Divorceuses*

---

28 Strumingher, 'Mythes . . .', *op. cit.*, pp 414–418.
29 Moses, *op. cit.*, pp 127–149.

and *Femmes socialistes* appeared shortly after the revolution and completed his attack on women who questioned the traditional family. The appealing *grisette* made famous by Monnier was another response to women's demands for change. The *grisette* created a sense of immutability about women and poverty and vulnerability that was a clear negative response to women's demand for change. Gavarni's introduction of the *lorette* was, if anything, less sympathetic; the *lorette* was too much inspired by Guizot's famous dictum – *enrichissez-vous* – to be appreciated by her viewers. Indeed, Gavarni's later series of the *Lorettes vieillies* show them to share some of the same miserable qualities as Daumier's hags.

While Beaumont was certainly not a proponent of women's rights, his contribution to the discussion of women's struggle for change is a departure from that of his contemporaries in several ways. First, he devotes almost all of his 1,273 lithographs to women. Second, he presents his women as struggling, clever and attractive at the same time. Third, he appears to encourage their struggle and to enjoy their successes, however ephemeral. Fourth, he shows women engaged in conversations with women friends about women's concerns – whether they are sharing dreams or planning tactics, or trying to ascertain the paternity of a child. Beaumont's images allowed women who struggled to change the status quo to see themselves in a different and more positive light.

In seeking to understand how Beaumont's audience responded to his images and to the captions which were an integral part of the presentation, it is important to recall the multi-valued context in which these works were published. The audience responded both to the real conditions of women and to the satiric images and articles which appeared in the popular press. Beaumont's images of *jolies femmes* and of stevedores and masked ladies were familiar icons to the regular readers of *Charivari* and the other satirical journals of the day. But Beaumont used these images to express a positive view about the struggle of witty and beautiful women to gain a measure of control over their lives. Those women most engaged in struggle themselves would not have missed the visual support offered them by Beaumont. Others, including art historians, who persisted for years in only seeing Beaumont's work as derivative of Gavarni, missed the unique contribution he made to the art of his day and to the readers of *Charivari*.

12 Writing for the Birds.
Women Founders of the Society
for the Protection of Birds

Barbara T. Gates

T HIS essay is not a tribute to the increasingly famous female Victorian sages, like George Eliot or Florence Nightingale, nor social reformers, like Josephine Butler or Frances Power Cobbe, but to a little-heralded group of women who have affected your own life every time you have driven by or stepped into a marsh—the English women who founded the Society for the Protection of Birds. To find these women in their own habitat, imagine yourself at afternoon tea in 1889. You might be at the home of Mrs Robert Williamson, in Didsbury, Manchester, discussing the slaughter of American egrets to supply the voracious appetite for feathered millinery. You would then be with acquaintances who had just formed a fledgling society to protect such birds. Or you might be in Croydon, at Mrs Edward Phillips', sipping your tea in the company of W. H. Hudson along with other members of the newly established Fur, Fin and Feather Club. In either case, you would probably be headed for membership in a soon-to-be combined group that would take the name of the Society for the Protection of Birds (SPB).

That group would establish its headquarters in London and would be run by women whose female leadership was unprecedented in the establishment and continuation of protective or scientific societies. Power Cobbe's Victoria Street Society, which protested against vivisection, was founded in 1876 by five women and eight men, but throughout most of its Victorian history it was dominated by male presidents and vice-presidents, while the Royal Society for the Prevention of Cruelty to Animals prevented women from sitting on its general council until 1896. The new SPB was different. In its first year, all of its officers were women. Eliza Phillips, with whom you have just had tea, became the organisation's vice-president and publications editor, a position she held through many tempestuous years of propagandising until her death in 1916. She did the annual reports as well as selecting the pamphlets to be reproduced and editing the house organ, *Bird notes and news*. Winifred, Duchess of Portland, became the Society's first and only president for sixty five years, and Margaretta Lemon, its first honorary secretary. Because of legal intricacies, Lemon's husband, a lawyer, took over her title when the society became incorporated by royal charter in 1904 and became the RSPB, but Lemon continued virtually to run the organisation for forty years. Among her duties were recorder, secretary and editor of publications. Thus until Phillips' death in 1916, Lemon and Phillips controlled the pen at the RSPB, and it is their particular skill as polemicists and writers and editors of protest literature that I shall detail here.

When these two women began their campaign to save wild birds, protest and concern had been underway for a quarter of a century. In the 1860s, repeated use of sea birds as hat decorations had led to the passage of the Sea Birds Protection Act of 1869. Unfortunately, this measure seemed only to encourage the use of other kinds of birds in millinery, particularly imported birds. In just one year in the 1880s, over 400,000 West Indian and Brazilian birds and 350,000 East Indian birds were sold on the London market. Not just feathers, but pieces of birds and even entire animals were used to decorate hats. By the 1880s and 1890s, for example, it was fashionable to set the heads of owls, with false glass eyes staring out into space, into the crowns of ladies' chapeaux. A cartoon from *The Westminster Gazette* in 1901, with an owled hat, named a 'killing hat', spoofs the ends to which women were believed to go in their quest for originality. Coupled with the owl is a pair of stork legs, and coupled with the illustration, a poem:

> I have found out a gift for my fair–
> A pair of stork legs–think of that!
> If they do look absurd
> That's the fault of the bird,
> Not to grow legs more fit for a hat.[1]

Ten years before this cartoon was published, Lemon and Phillips had set to work on the same cause, not in fun, but in deadly earnest. In her history of the RSPB, Lemon tells us that new members of their society took a pledge and received a white card of membership that enjoined them in gold letters to: 'discourage the wanton destruction of Birds, and interest themselves generally in their protection' and to 'refrain from wearing the feathers of any bird not killed for purposes of food, the ostrich excepted.'[2] Ostrich feathers were taken during molts and considered fair game. Lemon felt so strongly about women who were wearing plumed hats that in the early years of the society she took note of all the ladies in her church who were wearing them. Early on Monday mornings she would then send each of them a letter indicating the cruelties of feather-hunting. To appeal to their sentiment, she would include discussion of the starving young birds whose breeding-plumaged parents had been killed for the women's gratification. Not all of the church-goers might have been convinced, but by the end of its first year, the SPB had over 5,000 members.

Nevertheless, Lemon and Phillips had a huge public relations task ahead of them. In the first two years of the Society's life, they had hoped to enrol enough women to halt the trade in feathers. By 1891, they knew they were in for a long fight. For every hundred women who joined their group, hundreds of thousands continued to wear plumes. Just how Lemon and Phillips engineered their task of written protest and persuasion can be detected in the pages of the early pamphlets they issued, before their society was incorporated and before *Bird notes and news* came into being in 1903.

I would like to turn to aspects of the rhetoric of several of those pamphlets now. I have confined myself to pamphlets written by Lemon or Phillips and those that reveal their editorial skill in selecting other work to complement their own. So much as is possible, since not all of the pamphlets were dated and

---

1 For the information in this paragraph I am indebted to Robin W. Doughty, *Feather fashions and bird preservation: a study in nature protection* (Berkeley, California, 1975).
2 Mrs Frank E. Lemon, 'The story of the R.S.P.B.' in *Bird notes and news*, 20, no. 5 (Spring 1943), p. 68.

many were reissued and reprinted, I have also assumed a rough chronology that will show the drift of the Society's aims from its early years until the turn of the century.

Phillips, always the more trenchant of the two writers, set out her concerns in the 1891 leaflet, *Destruction of ornamental-plumaged birds*.[3] The piece was sent to the queen that same year and was accepted with the somewhat curt notation that 'Her Majesty's attention had been already directed to this subject.' All the same, in an important way, the queen was a proper recipient, for Phillips was aiming to prove that bird protection was a woman's question and hoping to gain the support of the most revered woman in the realm. In the pamphlet, Phillips assumes the voice of a woman speaking to women. 'It is our vanity,' she says, 'that stimulates the greed of commerce, and our money that tempts bird-slaughterers to continue their cruel work at home and abroad.' She particularly castigates those women who are mothers with nurseries at home—a reference intended to recall the nestling birds which she has just discussed—who wear bird feathers 'even when engaged in public worship of the Creator of the beautiful and useful life of which they are inciting the continued destruction.' And she violently puts down all women who speak in public 'with their heads bedecked with stuffed birds,' implying that their messages must be as dead or their heads as stuffed as their headgear.

If Phillips' primary appeal was to women, her primary tactic was to shame them. But further appeals were made via carefully selected authorities. Buttressing her own powerful language with that of well-respected Englishmen, Phillips cites the president of the British Ornithologists' Union, Lord Lilford, himself a very outspoken man, on the 'disfigurement of women's heads' effected by the wearers of 'keeper's-gibbets,' Phillips' own loaded word for the current hat style made up of trapped, then executed birds. Phillips' strategy here shows one of the ways she chose to reinforce her arguments: she takes her own phrase and like a ventriloquist projects it from another more prestigious person. In this way she not only authenticates her own prose but validates her female outrage without seeming unfemininely harsh. After quoting Lilford, she goes on to cite the prominent writer W. H. Hudson on the horrors of bird warehouses, 'where,' according to Hudson, 'it [was] possible for a person to walk ankle deep—literally to wade—in bright-plumaged bird skins, and see them piled shoulder high on each side of him.' Next she appeals to aesthetics and common sense: birds are beautiful and birds are useful, especially as insect eaters. Orchards and gardens need them, as agriculturalists in England and in America will attest. Her final stirring appeal is directed to the clergy of England whom she enjoins to speak out from their pulpits and condemn those who neglect the 'duty of righteous and merciful dealing with every living creature, as inseparable from the dominion given by God to man.'

I have reviewed and quoted from this early pamphlet at some length because it is typical of the Society's first efforts at persuasion. The women of the SPB were hard on other women because they felt responsible for reforming their sisters. They did not see bird preservation as a feminist issue. And because they were women speaking to women, they spent less time writing about the feathered military—another favourite target of bird protectors—than they did about feathered women. For them, if women were the victims of a culture that de-

---

3 Leaflet no. 1 (London: Society for the Protection of Birds, 1891).

manded inutility and conspicuous consumption, they were nonetheless also people who might be exhorted and persuaded to alter their ways through intelligent and passionate address. Because men too aimed to protect birds, they used the names and words of their most famous members—like Hudson—to bolster their own arguments. And they appealed to the religious sense of the Victorian public in their quest for justice for birds.

In support of her viewpoints, Phillips arranged for pamphlets complementary to *Destruction to ornamental-plumaged birds* also to be printed and disseminated. Among those early SPB leaflets were Hudson's paper on 'Osprey, or egrets and aigrettes' – osprey and aigrettes being inaccurate names for egret feathers when turned into hat materials[4] – and Hudson's 1893 letter to *The Times* (17 October) suggesting that cannibals had more respect for God-given creatures than did 'Ruskin-reading' civilised persons of the nineteenth century. Also among the early leaflets was one by a Reverend H. Greene, used to strengthen Phillips' exhortation for the clergy to severely criticise all feather-wearers. If Phillips cited male authorities, Greene, in his essay entitled 'As in a mirror', cited women, recalling that E.V.B. once noticed that bird hats were great levellers: there were no shoddy birds killed for the trade involving the lower classes because there was 'no such thing to be found as an ill-made seagull'.[5] I bring in Hudson and Greene for the same reasons that Phillips did—to indicate that despite the appeals to women and the importance of women founders to the SPB, the movement to save birds was never exclusively a women's movement. Animal rights over-rode women's rights here, as in other early environmental protection movements.

Lemon's hand was nearly as visible in the first pamphlets as was Phillips'. Lemon had begun her interest in bird welfare early. After reading Eliza Brightwen's *Wild nature won by kindness* (London, 1890) she had become determined to do something about the plight of birds slaughtered during mating season, a cause she would always espouse. Brightwen was a popular writer on natural history subjects and an expert on exotic pets who was respected as an authority on animals in her day. Like Brightwen, Lemon began her crusading hopefully, almost sentimentally. Not surprisingly, an essay by Brightwen also became one of the early choices for an SPB leaflet.

In her autobiography, *Eliza Brightwen: the life and thoughts of a naturalist*, Brightwen recounts that she was ready to support efforts at bird preservation because she found the suffering of birds victimised for their feathers all but unbearable. She describes an idyllic Florida nesting sight, then redescribes it 'invaded by a gang of men, bang go the guns, the little tender, loving mothers cannot bear to leave their young and hover close around them. On goes the slaughter, and with regardless cruelty the skin of the back and the wings are torn off the poor birds whilst they are still alive.'[6] In the essay reprinted by the RSPB, Brightwen reveals equivalent sympathy and sentiment for mother birds but with comparable vividness delineates the cruelties of electrifying tired song-birds on trap-like telegraph wires deliberately placed along the French coast to entice them in their first landfall after a longish flight over water.[7]

Lemon's own early work as pamphleteer was on behalf of the bird of para-

4 *Feathered Women*, leaflet no.10 (London: Society for the Protection of Birds, no date).
5 Leaflet no. 2 (London: Society for the Protection of Birds, no date), p. 9.
6 Eliza Brightwen, *The life and thoughts of a naturalist* (London, 1909), p. 135.
7 *A talk about birds*, leaflet no. 9 (London: Society for the Protection of Birds, no date).

dise. Aware in 1895 that exceptionally large numbers of these rare birds were killed for the plume trade and aware, too, of the enormity of the task of halting that trade, she spoke out with utter forthrightness, eschewing the Brightwen-like sentimentality that she had once so admired. At this stage of Lemon's career, the facts themselves were glaring enough: living birds were no longer capable of reproducing the feathers that the trade required because most specimens were not 'allowed to live long enough to reach maturity, the full plumage of the male bird requiring several years for its development!'[8] In an 1898 footnote to the revised 1895 pamphlet, she pointed out that on six auction days during 1898, 34,860 single birds of paradise, and 45 packages were disposed of at the London commercial sale rooms. Despite a decade's efforts at bird preservation, Lemon often seems to have felt that the SPB's work was losing ground.

But back to 1895. By then, Phillips too had changed her tone – from trenchant to angry. In that year, her pamphlet entitled *Mixed Plumes*[9] disdainfully described the bouquet-like sprays that then adorned women's hats. Phillips would no longer mince words. 'Nuptial plumage' is 'torn' from parent birds, and the people wearing such plumes should be not only 'feathered', but 'tarred'. The hats women wore might no longer be 'stuffed carcasses', but a look at the newly fashionable bunched plumes indicated to Phillips that 'the tar brush, suggested [by her] for the wearers, seems to have been employed upon the feathers', and that women seen in the boxes at the opera were adorned with a hair ornament 'exactly like the sort of brush servants use to clean lamp-chimneys with'. In *Mixed Plumes* we can also see how Phillips and Lemon collaborated. Phillips refers to Lemon on the birds of paradise to insinuate her point that the newly beloved feather sprays are drawn from the most precious of species.

Phillips concluded her essay with a direct appeal to the princess of Wales to serve as a model for other women and desist in wearing mixed plumes. This time she was read and heeded. *The Evening Standard* (21 November 1895) reiterated her plea to the princess and cited Coleridge's moral from the 'Ancient mariner' in her defence:'He prayeth well who loveth well, both man and bird and beast.' In fact, the royals did begin to listen to the Society and by the time of the royal charter (1904), Princess Alexandra would quit wearing feathers. Earlier, in 1899, Queen Victoria had ordered her regiments to cease wearing 'ospreys'.

Still, by the end of the century, there was much left to be done. Lemon's history recalls the pride with which the young SPB welcomed Linley Sambourne's 1892 *Punch* cartoon of a female harpy, dressed in full Victorian feather regalia, about to snare a tiny songbird in her talons – a piece done in honor of the fledgling Society. But Sambourne was still at it in 1899, when he felt moved to offer another *Punch* cartoon in support of the discouraging and long-continued attempts at egret preservation. Here he represented familiar images from the essays I have discussed: baby egrets starving in the nest, birds plummeting from the sky, and a haughty, unheeding woman crowned in feathers and holding a new chapeau of dead birds.[10]

All of this was still occurring when Margaretta Lemon addressed the International Congress of Women in Westminster in 1899.[11] By century's end, she

8 *The Bird of Paradise*, leaflet no. 20 (London: Society for the Protection of Birds, 1895, 1898).
9 Leaflet no. 22 (London: Society for the Protection of Birds, 1895).
10 Lemon, 'The story of the R..S.P.B.' in *Bird notes and news*, 20, no. 5 (Spring 1943).
11 Printed as *Dress in relation to animal life*, leaflet no. 33 (London: Society for the Protection of Birds, 1899).

used yet another tone of voice – a tired certainty along with a ring of authority. Now a woman to be reckoned with, she presented herself to her audience as an expert with a belief that her cause was not 'only a sentimental one, but a serious economic one.' Vermin, she noted, were increasing because of owl slaughter, and women were not guiltless. Current laws protecting birds were virtually ignored. Seabirds and Florida egrets were dying by hordes. Birds of paradise were diminishing. Mother birds were still ripped away from their young. And then, once again practising a kind of gendered ventriloquism when it came to real nastiness, an SPB leader would crown an argument with the words of an indignant Lord Lilford: 'the fittest place for any wilful destroyer of an owl was an asylum for idiots.' With all of these reminders before them, Lemon asked the women of the congress to become 'citizens of the world', an epithet she put before them along with the word 'extinction'. Women – 'good' women – had become exterminators rather than nurturers.

Throughout, Lemon's rhetoric was hyperbolic and dramatic. Toward the end of her talk, she called out the plea, 'when it is too late, man (and woman) will discover what a poor, worthless, uninhabitable place this world is without the birds.' And she concluded by presenting a painting which was a personal gift to the Society, one valorising woman's compassion toward birds, George Watts' 'Shuddering angel'. As she looked at it, Lemon recalled Watts' dedication 'to all who love the beautiful and mourn over the senseless and cruel destruction of bird life and beauty.' She then read a poem, which she adapted from *Punch*:

> Feathers deck the hat and bonnet;
> Though the plumage seemeth fair,
> Angels as they look upon it
> See but slaughter in the air.
> Many a fashion gives employment,
> Unto thousands needing bread;
> This to add to your enjoyment,
> Means the dying and the dead.

> Wear the hat without the feather,
> All ye women, kind and true,
> Birds enjoy the summer weather
> And the sea as much as you.
> There's the riband, silk, or jewel;
> Fashion's whims are oft absurd,
> This is execrably cruel,
> Leave his feathers to the bird.

Dramatic though it might be, I cannot quite leave the story here, at the International Congress of Women, or, tempting as it might be, take it back full-circle and bring you back to tea. In 1899, the day for tea parties with respect to birds was over, as I hope Lemon and Phillips have shown. The two worked on together – with the help of a new addition in 1900. She was Linda Gardiner, a brilliant stylist and contributor to the growing store of pamphlets, but there is not space to tell her story here. For the story of women writing for the RSPB did go on. Politicians rather than the hoped-for clergy became the women's primary allies in a long battle for bird preservation. Despite such events as the famous sandwich-board protests illustrating the 'life of the egret' that took place in the streets of London in 1911, it would be 1921, five years after Phillips' death, before the Plumage Act would become operative. And it would be another forty-three years before the indefatigable Lemon would, for her, by then, old companion, *Bird news and notes*, write the history of the society she had so dedicated herself to for over half a century.

13　Jane Addams and the Great Railway Strike of 1894: Toward a [Reconstructed] Universalist Ethic

Marian Yeates

IN studying the social thought of American reformer Jane Addams, I was struck by similarities in Addams' notion of social ethics and that of contemporary theorist, Seyla Benhabib. In particular, I noted features of Hull House that reflect principles of what Benhabib calls an interactive, communicative ethic – a social ethic which holds within it the possibility of arbitrating divergent claims among culturally diverse individuals or communities. Indeed, Hull House could serve as a working model of what Benhabib calls a 'participationist' ethical community. To highlight similarities, I will contrast Hull House with George Pullman's industrial community. Further, I argue that the railroad strike of 1894 brought two distinct models into collision, and the resulting impact was instrumental in shaping Addams' emerging social thought.

In her recent book, *Situating the self*, Benhabib sifts through the rubble of the once-proud, universalist ethical tradition discarding those principles ravaged by post-modern, communitarian, and feminist criticism, searching the wreckage to reclaim those principles which may survive, perhaps thrive, in a pluralistic, intellectual environment. Aided by post-modern methodology, Benhabib makes two claims. First, she asserts all truth is discursive which means all truth resides in texts; and, subjects construct texts from the negotiable symbols available in language and culture. Second, subjects are finite and embodied. Consequently no truth can exist outside the subject's finitude and physical embodiment. Given this agenda, Benhabib declares the need to alter radically the 'moral point of view' by discarding absolutist assertions and replacing them with practical, 'interactive' reason set in discursive, interactionist terms. In calling for 'interactive reason' as the starting point for constructing a 'moral point of view,' Benhabib departs from Jürgen Habermas who seeks consensus through interactive reason. Rather, Benhabib argues that the possibility of achieving consensus among the multiple claims of self-interested individuals in a pluralistic society is a 'counterfactual illusion.' Rather, she prefers Hannah Arendt's notion of 'representative thinking' to achieve, not consensus, but 'understanding.'[1] Basically, Benhabib's strategy, as suggested by the title, *Situating the self*, is to situate the moral self in its peculiar context of gender and community, and from there, arm the individual, moral self with universalist principles with which to shape her discursive environment from which both ethics – i.e. the rules of

---

1 Seyla Benhabib, *Situating the self: gender, community and postmodernism in contemporary ethics* (New York, 1992), pp 5–7.

conduct governing institutions and individuals, and truth, i.e. the recognisable facts that constitute reality – emerge.

Benhabib's strategy creates a subtle but important shift in perspective. When bound together in interactive dialogue, individuals understand that they share the need to maintain relationships among parties; and further, all parties understand they must reach agreement from which further dialogue can proceed. Benhabib rejects the universalist hope of discovering a 'general interest' upon which to form a common ethic. What she retains is the notion of a 'universalizability procedure' understood as a 'reversing of perspectives and the willingness to reason from the other's point of view.' This procedure demonstrates a 'readiness to seek understanding in an open-ended moral conversation.'[2] The resulting shift places relationship rather than principle at the heart of any ethical project.

Holding this theoretical framework in mind, let us turn to Jane Addams. In the winter of 1889, Jane Addams and Ellen Starr launched their scheme to open a 'social settlement' in Chicago's 19th Ward. Using Starr's social contacts, for six months the two young women visited groups within Chicago's philanthropic community soliciting support for a scheme which at that point was largely undefined. Their general idea was to establish a social settlement drawn from London's Toynbee Hall, but beyond that, they had no clear idea of what they hoped to achieve, and further, resisted imposing any ideological agenda upon themselves. The open-ended character of the project was reflected in Addams' letters of the period. In February 1889, she wrote to her sister, Mary Linn, the following: 'I have been writing a good deal on the "scheme" this week, not to publish but to clarify my own mind . . .'[3] Biographer, James Linn, supposed the purpose of Hull House was 'not to provide a higher civic and social life for anybody.' Rather the purpose was to develop '*a center for the development* of such a life' (emphasis by author).[4] By refusing to impose a preconceived ideal, Addams worked to foster a 'social spirit' that facilitated 'social intercourse' between the employer and employee classes. Further, she insisted that the social spirit was to be the 'undercurrent of the life of Hull House,' and allowed that spirit to go in 'whatever direction that stream might take.'[5]

The open-ended nature of the 'scheme' was the core of Addams' social ethic. This was a highly dialogic model predicated upon the need to construct concrete channels of expression between individuals, groups, and classes to act as outlets for human expression. Addams borrowed the channel metaphor from Edward Caird who described Abraham Lincoln as a president 'content merely to dig the channels through which the moral life of his country men might flow.'[6] In a sense, Hull House was a 'channel of expression' between representatives of the philanthropic community and residents of the 19th Ward. Labour unions were potential 'channels of expression' between labour and capital. The Chicago Woman's Club, the Civic Federation of Chicago, and municipal government were other institutional structures that potentially could channel resources to bridge the chasm between isolated groups. The opening of meaningful 'chan-

---

2 *Ibid.*, p. 9.
3 Jane Addams to Mary Linn, 26 February 1889 (Mary Lynn McCree (ed.), *The Jane Addams papers, 1860–1960* Microform (Ann Arbor, Michigan, 1984)).
4 James Weber Linn, *Jane Addams: a biography* (New York, 1935), p. 111.
5 C. R. Henderson, *Social settlements* (New York, 1899), p. 50.
6 Jane Addams, *Twenty years at Hull-House* (Urbana, 1990), p. 25.

nels of expression' through institutional links, both public and private was the operating principle behind Addams' notion of 'social ethics'. With each presentation, ideas were exchanged, new points of view assimilated. As she gathered information, her own ideas became more definite while her confidence grew. In a letter dated 19 February 1889, Jane told of a reaction to the 'scheme'. She wrote, 'I was almost immediately requested to begin and talked for about fifteen minutes as well as I could. Some of the older ladies cried'. In spite of their initial success, Addams quickly assured her older sister that 'Our heads are not turned by the first flush of success.'[7]

In developing a 'narrative strategy' for the scheme, Addams targeted her audience – the women of the Chicago Woman's Club. This powerful woman's association dominated the social, cultural and philanthropic world of Chicago in the 1880s. The club at that time was made up of an alliance of over five-hundred women drawn from the social elite, literary circles, business and professional women, as well as 'plain wives and daughters.'[8] From its inception in 1886, the club declared its intention to tackle the 'knotty' problems of the day on behalf of women and children. By committing energy and resources to the social ills of the day, the Club became 'a potent factor for good, and a force that should not be despised.'[9] While visiting the Chicago Exposition in 1893, Julian Ralph described the Chicago Woman's Club for *Harper's Monthly*. According to Ralph, the club represented an alliance of the 'rich and stylish' with 'forceful women' of the community whose 'astonishing activity has worked wonders in that city.' Their commitment to reform was unquestioned; and they were also able to raise funds. In one venture, the Woman's Club raised $168,000 in just six weeks to build a woman's dormitory at the University of Chicago. Through the Reform Committee, the Club organised campaigns on behalf of humanity, 'especially feminine humanity.' In comparing the Chicago group with others, Ralph observed that even more than Washington, Chicago women 'respect talent, and vie with one another to honor those who have any standing in the World of Intellect.'[10] This characteristic would hold Jane Addams in good stead.

To cultivate channels of expression among the residents of Hull House and the neighbourhood, Addams and her associates showed a willingness to remain open to new ideas. Through daily experience, Addams learned that the 'poor' shared many values in common with the middle class, many were well-educated, many possessed native skills and traditions that enriched the social life of the neighbourhood. One of the first activities at Hull House was a reading of George Eliot's *Romolo* attended by young women who followed the story with 'unflagging interest.'[11] In this first encounter, Addams discovered that art and culture provided an invaluable means for bringing people together in order to stimulate discussion of common needs. Likewise, Addams grew to appreciate the degree of learning that many of the poor possessed. The 'first resident' at Hull House was a 'charming' old lady who presented five consecutive readings from Hawthorne, 'interspersing the magic tales most delightfully with recollections of the elusive and fascinating author'. In her younger days, the woman

---

7 Jane Addams to Mary Linn, 12 Feb. 1889 (McCree, *op. cit.*).

8 Julian Ralph, 'Chicago's gentle side' in *Harper's new monthly magazine,* June–Nov. 1893, pp 286–297.

9 Henriette Greenbaum Frank and Amalie Hofer Jerome (compilers), *Annals of the Chicago woman's club for the first forty years of its organization, 1876–1916* (Chicago, 1916), p. 14.

10 *Ibid.*, pp 286–298.

11 Addams, *Twenty years at Hull-House*, p. 61.

was a resident of Brook Farm and had come to Hull House because she wished to live in an atmosphere where 'idealism ran high'.[12] During the winter of 1889, Addams spent a good deal of time meeting new people, and speaking to various groups trying to interest them in settlement work. In making speeches about the needs of the neighbourhood, biographer James Linn reported that Addams always took a neighbour with her to 'check any tendency to hasty generalisations' by forcing on her the consciousness that she had at least one auditor who knew the conditions more intimately than she.[13]

In addition to learning from the residents, Addams recognised the importance of translating cultural signs. In an essay entitled 'The devil baby at Hull House', Addams recounted a strange incident. One day, quite unexpectedly, three Italian women came to Hull House demanding to see the devil baby which they believed was kept there. These women knew exactly what he looked like: cloven hoofs, pointed ears, a diminutive tail. Moreover, they knew that as soon as he was born, he began to speak and was 'most shockingly profane'.[14] No amount of persuasion could convince these women that no such creature lived there. The women left undissuaded.

For the next six weeks, 'a veritable multitude' came to Hull House demanding to see the devil baby. Apparently there were several versions of the devil baby story. In the Italian version of the story, a pious Italian girl married an atheist who tore a holy picture from the bedroom wall swearing he would rather have a devil in the house, whereupon, the devil incarnated himself in her child. A Jewish version told of a father of six daughters, who swore that if the seventh were another girl, he would sooner have the devil; whereupon the devil baby promptly appeared. People of all ages, cultures, and educational backgrounds came to Hull House as if on a pilgrimage to see the devil baby.

The spectacle at first disturbed, perhaps disgusted, Addams. She wrote: 'I will confess that, as the empty show went on day after day, I quite revolted against such a vapid manifestation of an admirable human trait.' However, rather than condemning, she struggled to make sense of the event noting that 'whenever I heard the high eager voices of old women, I was irresistibly interested, and left anything I might be doing in order to listen to them'.[15] By listening to the stories of the old women, Addams realised that the birth of a devil baby was a tangible mark of divine displeasure against men. The legend persisted through the centuries because of its 'taming effects upon recalcitrant husbands and fathers'.[16] Women of the neighbourhood helped their men to see the sign, each woman hoping to 'tame her mate and to make him a better father to her children'. From this experience, Addams realised that women of the neighbourhood were 'establishing rules of conduct as best they may, to counteract the base temptation of a man's world'. Further, she concluded that in order to protect women from domestic abuse, the women of the 19th Ward needed protective measures stronger than the devil baby.

In an early description of Hull House, Addams described what she believed a social settlement was meant to be. 'It aims to gather to itself . . . whatever social life its neighbourhood may afford, to focus, and give form to that

---

12 *Ibid.*, p. 101.
13 James Weber Linn, *Jane Addams: A biography* (New York, 1935), p. 112.
14 Jane Addams, 'The devil baby at Hull-House' in *Atlantic Classics* (second series) (Boston, 1918), pp 52–77.
15 *Ibid.*, p. 55.
16 *Ibid.*, p. 64.

life . . .'[17] In daily operations, programmes were instituted based upon the needs of neighbourhood residents. As needs became known, programmes proliferated to include political clubs for men, social evenings for women, a boarding house, bakery, gymnasium and coffee house. In a sense, Hull House served as a neighbourhood, community-action centre and clearing house for municipal services. In initiating new programmes, Addams wrote that experience taught them 'not to hold to preconceived ideas of what the neighbourhood ought to have, but to keep ourselves in readiness to modify and adapt our undertakings as we discovered those things which the neighborhood was ready to accept'.[18]

Hull House was opened, Addams declared, 'on the theory that the dependence of classes on each other is reciprocal'; and that as 'the social relation is essentially a reciprocal relation, it gave a form of expression that has peculiar value'.[19] The 'peculiar value' of social expression arising from daily intercourse among residents and patrons of Hull House formed the basis of an interactive, communicative ethic. This ethic was contextual, situated, and bounded by the neighbourhood. Its value was site-specific, meaning value was negotiable only within the confines of its particular environment. Value could not be abstracted from its context to circulate in economies outside its own.

As described by Benhabib, Hull House presented a 'participationist' vision of community. This model allows for the individual, in dialogue with others, to exercise the individual rights of *'political agency and efficacy'* (emphasis by author).[20] It asserts that loss of political agency is caused, not by the separation of the personal from the political, but rather, from the 'contraction between various spheres which diminishes one's agency in one sphere on the basis of one's position in another sphere'.[21] To regain political agency, the contradictions between spheres must be eliminated. Thus the participationist model advocates the elimination of contradictions by means of non-exclusive principles of membership, or, by developing avenues of expression through which agency is exercised both within and among spheres.[22] This strategy is particularly important for marginally enfranchised workers and women denied political agency by virtue of their assigned position in 'non-propertied' spheres.

The participationist model claims agency is not an abstract right. Therefore, agency cannot be granted from an external, absolute source; rather, agency is a faculty to be exercised – to be cultivated – by embodied individuals in daily intercourse with one another. Likewise, values, both communal and individual, are cultivated through exercise of moral faculties rather than pronounced by external authority. Following this model, Hull House was a site for the exercise of agency carried on through a myriad of schemes instituted to draw disenfranchised individuals in from the street, and, in Addams' words, give form to their communal life. The political clubs, educational programmes, and neighbourhood projects served as channels of expression for the exercise of individual, political agency. From the exercise of agency, communal values emerged. In this model, values are cultivated through process rather than dictated by demand.

If Hull House represented a participationist model of a interactive, ethical

---

17 Jane Addams, 'Outgrowth of Toynbee Hall', Speech delivered to the Chicago Woman's Club Board, Dec. 1891 (McCree, *op. cit.,* p.8).
18 Addams, *Twenty years at Hull-House*, p. 79.
19 Henderson, *Social settlements*, pp 1–2.
20 Benhabib, *Situating the self*, p. 77.
21 *Ibid.*
22 *Ibid.*, p. 78.

community as described by Benhabib, a few miles down the road George Pullman's industrial community represented a model of a different sort. George Pullman was by all accounts an adventuresome businessman who embodied the ideals of an American entrepreneur. In 1855, George Pullman arrived in Chicago and quickly established himself as a contractor who specialised in shoring up sinking buildings and roads along Lake Michigan. On a trip to the east coast in 1859, he spent the night in a sleeping car and began thinking of how the design could be improved so as to be more comfortable. Returning to Chicago, he rented a machine shop and began working out the design of his sixteen-wheel car. The new car was superior to others in terms of comfort, but was also more expensive. Pullman's Pioneer cost $18,000 whereas a standard coach car cost $4,000. Convincing a railroad president to try the new cars was a hard sell, given the price, but after much effort, George Pullman persuaded James F. Joy of the Michigan Central to try his new sleeper cars and Pullman's Pioneer became the industry standard. In spite of higher costs, Pullman cars consistently out-sold conventional models. Sensing the commercial potential of mass-produced comfort, Pullman declared that 'people are always willing to pay for the best, provided they get the worth of their money'.[23] He was fond of insisting that the roughest man when placed in a room elegantly furnished becomes more 'refined' as a result of his physical environment.[24] From this experience, George Pullman learned that people were very greatly influenced by their physical surroundings.

While his sixteen-wheeled sleeper set the standard of comfort in the industry, Pullman's vision extended beyond railroad cars. Following the strikes of 1877, Pullman became convinced that industry must take the initiative in alleviating the dangerous gap between capital and labour. He also saw that much labour discontent arose from urban tenements where workers endured poor conditions, and were prone to contamination by radical agitators. In 1878, as Pullman began plans for his model city, his vision expanded to combine a state-of-the-art manufacturing complex with a model, industrial city in which to house company workers. From the beginning, the model town was designed to take care of all workers' needs, provide a 'refined' environment, and still make a 6% return. Pullman explained that his scheme was 'simplicity itself' – 'we are the landlord and employers'.[25] The heart of the 'Pullman system' lay in its ability to combine the aesthetic with the functional and still turn a profit.

In selecting a site to build his new factory/model town complex, Pullman looked to the undeveloped suburbs of Chicago where workers would be distanced from the corruption of the city, and where land prices were cheap. The 3,500 acre site south of Chicago along the Calumet River seemed ideal. All aspects of the construction were co-ordinated to maximise productivity – even the sewage from the town sewage plant was pumped to the village farm to be used as fertiliser. Sanitary facilities were of a high standard; care was taken to make the town aesthetically pleasing; a shopping arcade, library, theatre and church on the site kept residents close to home. All services normally supplied by municipal government were performed by the company leaving manage-

---

23 C. W. Tyler, 'The rise of the Pullman Company' in Mrs Duane Doty (ed.), *The town of Pullman* (Pullman, Illinois, 1893), pp 21–22.
24 *Ibid.*, p. 23.
25 Stanley Buder, *Pullman: An experiment in industrial order and community planning, 1880–1930* (New York, 1967), p. vii.

ment free to run the town without interference from local politicians. The town was designed to attract the best class of mechanics and 'exclude the baneful influences'. It was believed that such a policy would result in the greatest measure of economic success. So great was George Pullman's faith in the redemptive power of environment that he believed that by providing his workers with the needs of body as well as soul, 'the disturbing conditions of strikes and other troubles that periodically convulse the world of labour would not be found here'.[26] The Pullman System was a unilateral and costless gesture by capital to help labour aid both itself and the company in the process. According to Stanley Buder, this was the 'genius' of the Pullman System.[27] The Pullman experiment represented an imaginative, perhaps even radical attempt, to apply a thoroughly rationalised model to an industrial setting.

In contrast to the interactive, 'participationist' mode of Hull House, Pullman can be seen as an 'integrationist' ethical community based upon consensus, assumptions of 'common interest', and abstract notions of a common good. Using Benhabib's distinction between an 'integrationist' and 'participationist' model, the distinction between Hull House and Pullman town becomes more pronounced. According to the integrationist model, community is established by recovery, or revitalisation, of a coherent value scheme based upon religion, civic interests, or codes of civility that promote solidarity, if not friendship. Communal cohesion depends upon uniform acceptance and enforcement of values designed to shape moral character to complement communal values. Missing from this model are spaces for exchanging views, forums for public discussion and channels for worker input. Following this model, when workers signed on with Pullman, they entered a world where all the decisions had been made, and would continue to be made, so as to maximise the efficiency of the whole. They became an interchangeable part in a great industrial machine that was carefully co-ordinated by the top echelons of the company. The choice was to stay or go; but if they chose to stay, they had to follow the programme, not just on the job, but in their private lives as well.

Denied a voice in community affairs, resentment grew in the town. While the streets were cleaned, the garbage cleared, and flowers bloomed perpetually in company gardens, the heavy-hand of the corporation invaded every aspect of workers' lives. Pastor William H. Carwardine described the atmosphere:

The corporation is everything and everywhere. The corporation trims your lawn and attends to your trees; the corporation sweeps your street, and sends a man around to pick up every cigar stump, every bit of paper, every straw and leaf; the corporation puts two barrels in your back yard, one for ashes and one for refuse of the kitchen; the corporation has the ashes and refuse hauled away; the corporation provides you with new barrels when the others are worn out; the corporation does practically everything but sweep your room and make your bed, and the corporation expects you to enjoy it and hold your tongue.[28]

In addition to having little voice in local affairs, the dominance of the company in every aspect of the workers' lives led to an 'all-pervading feeling of insecurity'.[29]

---

26 *Ibid.*, p. viii.
27 *Ibid.*, p. 45.
28 William H. Carwardine, 'The Pullman strike' in Leon Stein (ed.), *The Pullman strike* (New York, 1969), p. 24.
29 Richard T. Ely, 'Pullman: a social study' in *Harpers' New Monthly Magazine*, Feb. 1885, pp 452–466.

Others expressed the same sentiment. Writing a feature story for *Harper's Monthly* in 1885, Richard Ely observed that workers were free to exercise political franchise in national elections, but had no say in local political issues.[30] Pullman repeatedly resisted efforts to annex the town into neighbouring Hyde Park fearing higher taxes and loss of control.[31] Fear of company retaliation, especially in hard times, curtailed criticism so that residents dared not speak out openly. While Richard Ely found workers had many grievances, he found no one in Pullman who would 'give expression to them in print over his own name'.[32] Residents were made aware that they were expected to appear in public properly attired. Likewise, residents were expected to maintain their homes so as not to encourage 'sloth'. If it were discovered that families 'accustomed to filth and squalor' could not conform, company inspectors would visit to threaten fines. This lead to a sense that workers were on display for the sake of the company image.[33]

Long-standing fear and resentment festering among the workers was aggravated by financial distress brought on by the depression of 1893. As the depression deepened, Pullman scrambled to cut wages and work time, but offered no compensatory cuts in rents or utility rates, insisting on his usual 6% return. By April of 1894, 15% of the town's units stood empty; workers owed the company $70,000 in back rent.[34] Reverend Carwardine observed that when the long winter of 1894 passed, the workers had been 'so ground between the upper mill stone of "low wages" and the nether mill stone of "high rents" . . . and the system of surveillance that seems to be indigenous to the very atmosphere of the place that they were in no condition to be trifled with by the Company'.[35] On 9 May 1894 a delegation of forty three workers met with the company to discuss worker grievances centring on wage disputes, discriminatory work assignments, and the critical issue of rents. Initially, Thomas Wickes, the second vice-president met with the workers and promised to investigate all charges of unfair practices, but gave no satisfaction on the issue of wages or rent. Later, George Pullman joined the group and read from a prepared statement. He elaborated what Wickes had said, stressing that the company was producing railroad cars at a loss. On the issue of rents, he remained adamant, arguing that the company was just like any other landlord and could not consider tenants' salaries in setting rents. He cautioned the workers not to confuse the company's roles as employer and landlord, and claimed that the two had nothing to do with each other. He maintained the town's well-being was very important to him and concluded by remarking that he thought of workers as 'his children'.[36] The next day, the workers of Pullman's model city walked off the job igniting one of the most bloody, violent strikes in American history.

In 1893, Jane Addams was appointed to the Industrial Committee of the Chicago Civic Federation. As a member of that committee, Addams was one of the first civic representatives to meet with workers and management to begin arbitration. Other members were unavailable, so Addams went alone to serve as arbiter, walking the two miles from Hull House to Pullman. During the initial

30 Richard T. Ely, *Ground under our feet* (New York, 1938), p. 169.
31 Buder, *Pullman: an experiment in industrial order and community planning*, p. 109.
32 Ely, 'Pullman: a social study', p. 464.
33 Buder, *Pullman: an experiment in industrial order and community planning*, p. 95.
34 *Ibid.*, p. 148.
35 Carwardine, 'The Pullman strike', p. 32.
36 Buder, *op. cit.*, p.156.

meeting, Addams confronted the highly-charged issue of rents hoping to diffuse tensions and establish a basis for discussion of more complex issues. When both sides presented their positions, Addams proposed that a real estate board be appointed to make an estimate of comparable suburban rents to determine the fairness of rental rates in Pullman. To this seemingly innocuous proposal, George Pullman issued his now-famous dictum: 'There is nothing to arbitrate'.[37] In refusing to consider any possibility of adjustment, Addams saw the futility of further talks. Infuriated by George Pullman's response, she left in disgust, retracing her steps back along the dusty, unpaved road to her home at Hull House.

Although she had no further direct involvement in the strike, Addams followed events throughout the summer with pained interest. In pondering her experience with George Pullman, later that year, Addams wrote an essay, entitled, 'A modern tragedy: an analysis of the Pullman strike' in which she began to set down formally her developing ideas on social ethics. Much of the essay is reflective of the contrast between the 'participationist' model being developed at Hull House, and the 'integrationist' model of Pullman. Addams opened by stating that 'perhaps one of the most subtle tragedies of these latter days was that enacted during the summer of 1894'.[38] While ancient tragedy implied maladjustment between individuals, this modern tragedy resulted from the 'maladjustment between two large bodies of men, an employing company and a mass of employees'.[39] According to Addams, the source of this maladjustment was rooted in the failure to allow workers a voice in shaping their industrial world. By denying a voice, Pullman destroyed the channel for dialogue thereby frustrating the social relationship between parties. To resolve this maladjustment, Addams suggested that society work from the 'family relationship which we all have shared, and concerning which our code of ethics is somewhat settled, towards the industrial relationship, in which we also bear a part, but concerning which our ethics are so uncertain'.[40]

In refusing to extend the higher, social standard of care to industrial relations, Pullman denied workers the sign of respect and esteem accorded to members of the public sphere by failing to take their demands seriously, thereby inferring their status as unemancipated adults. Acting as a parent, Pullman assumed that he knew their needs better than they, and rather than wishing them to express themselves, 'he denied to them the simple right of trade organization ... which would have been ... the merest preliminary to an attempt at associated expression'.[41] The tyrannical pre-emption of 'absolute authority' over the town lay in the imposition of 'one will' that directed 'the energies of many others, without regard to their desires, and having in view only the commercial results'.[42] Wounded by a perceived lack of gratitude, Pullman, like Lear, lost the power to recognise the good in what the workers were attempting to do. By failing thus, Pullman 'failed to catch the great moral lessons' of the times.[43] Like Cordelia, the worker had awakened to a 'wider existence' of associational soli-

---

37 Addams' testimony on her role in the Pullman Strike was recorded in the United States Strike Commission Report on the Chicago Strike of June–July, 1894 S. Ex. Doc., No. 7 (Serial set 3276), 53rd Congress, 3rd Sess., 1895, pp 654–658.
38 Jane Addams, 'A modern tragedy: an analysis of the Pullman strike' (McCree, *op. cit.*, p. 1).
39 *Ibid.*
40 *Ibid.*, p. 2.
41 *Ibid.*, p. 5.
42 *Ibid.*
43 *Ibid.*, p. 4.

darity which recognised an ethic of brotherhood, sacrifice, and subordination of individual and trade interest to the good of the working class.[44] In failing to sense the change in his daughter, Lear failed the 'test' which comes sooner or later to all parents – the test of acknowledging that the relation between parent and child had become a relation between adults.[45] Like the domestic tragedy of Lear, the industrial tragedy at Pullman resulted from the failure to acknowledge the first awkward attempts of a maturing working class to express their 'new code of ethics'.[46]

Throughout the winter of 1894-95, Addams submitted the essay to various journals for publication, but was uniformly rejected. Charles Lamb suggested 'A modern tragedy' represented a certain 'excess of participation'. George Pullman apparently offered a 'more emphatic' appraisal that could not be reprinted.[47] Addams noted that because she was perceived as supporting the strikers, Hull House 'lost many friends'.[48] Even Eugene Debs did not like the essay. He called it 'just another attempt to put out a fire with rosewater'.[49] One of the few enthusiasts was John Dewey who wrote, 'It is quite impossible to say anything in the way of criticism . . . upon the Pullman paper . . . it is one of the greatest things I have read both as to its form and its ethical philosophy'.[50] The essay was not published until 1912, when it appeared in revised form under the title, *A modern Lear.*

In a more recent appraisal, revisionist historian, Rivka Lissak, has taken a critical view of Addams contending that the agenda at Hull House was assimilationist. Lissak argued that in their zeal to Americanise properly alien workers, Hull House patrons stripped immigrants of their ethnic traditions, disarming class consciousness in order to diffuse hostility towards the privileged classes.[51] While the merits of Addams' social ethics will continue to be debated, Lissak's charge misses a central theme in Addams' social thought. Addams stressed repeatedly the belief that human relationships were reciprocal; that exchanges between individuals, communities, even nations inevitably produced modification in both; that through interaction, the fundamental nature of each would be altered by the encounter. Addams' critique of George Pullman reflected Shakespeare's critique of King Lear – i.e. each stubbornly clung to the fantasy that somehow he could achieve immortality by reproducing the sacred image of the same in his children, whether natural or industrial. Perhaps if Lear had sired sons rather than daughters, or if George Pullman had built a model community of Harvard business graduates rather than urban workers, their dreams of immortality could more easily have been realised. But this was not the case. The industrial world was growing more diverse. There were new players on the scene demanding new rules for old games; new actors seeking a new stage; new voices creating their own channels of expression. As an educated, but disenfranchised, single woman, this was a reality Jane Addams not only accepted, but welcomed.

---

44 *Ibid.*, p. 8.
45 *Ibid.*, p. 9.
46 *Ibid.*
47 Addams, *Twenty years*, p. 128.
48 *Ibid.*, p. 134.
49 James Weber Linn, *Jane Addams: a biography*, p. 167.
50 John Dewey to Jane Addams, 19 Jan. 1896 (McCree, *op. cit.*).
51 Rivka Shpak Lissak, *Pluralism and progressives: Hull House and the new immigrants, 1890–1919* (Chicago, 1989), p. 22.

In 1902, Jane Addams published her first full-length work entitled, *Democracy and social ethics*. Addams opened the treatise declaring that each generation has its own test by which its moral achievements would be judged. The 'test' must necessarily include that which has been attained by past generations, but if it contained no more, 'we shall fail to go forward, thinking complacently that we had "arrived" when in reality we have not yet started'.[52] In Addams' view, the test by which her generation would be judged was the 'social test'. The stern questions that would be asked were not related to personal or family relations. That, she asserted, had been the test of a former generation. The social test of her generation asked that the privileged extend to the larger social arena the same individual standard of care granted to family and friends.

If the test were to extend standards of personal care to the larger society, the question was: How could this be done? Addams and others of her generation shared the belief that the social problems of an industrialised society had grown too complex, too diverse, too massive to find resolution in traditional approaches. Corporate America produced miracles of production, technology and wealth, yet had failed to offer a corresponding programme for social development. The disjunction between the 'progress' of industry and the 'poverty' of the social order was evident everywhere. To heal the disjunction, Addams insisted that 'identification with the common lot' marked the starting point for constructing a modern 'social ethic'.[53]

Between George Pullman's now-famous declaration, 'There is nothing to arbitrate', and Addams' idea that there is 'nothing that we cannot modify or adapt' lies a century of ethical thinking. Jane Addams opened Hull House in 1889 hoping to make a difference to the residents of Chicago's 19th ward – and she did. Hull House was a site of change – but the point is, all parties to the transaction emerged enriched, or depleted, whatever the interpretation, but nonetheless, changed. Addams envisioned Hull House as the arbiter of difference, a facilitator of change for both sides. It was this vulnerability to change, this willingness to accept new elements, even at the risk of rendering an old, familiar world irrevocably different that distinguishes Addams' work.

In opening the twentieth century, Addams was asked to consider the question: What was the greatest menace of this new century's progress? In considering her response, Addams observed the following:

America does not yet believe that each soul has within itself a tremendous power, which because we distrust it, has not been awakened. Our democracy has not succeeded because it has not been thoroughly tried. Our philanthropy allows itself to belittle the human individual in order to help him, and in doing so, lowers human nature. We have a way of believing that if any great thing is to be done, it must be done by commercial means; and, moral energy is good as long as it is applied to individuals and families, but not as a great national force. We fear and distrust alien peoples because we distrust human energy and the power of human thought.

Thus, Addams concluded, the greatest menace of the twentieth century was 'lack of faith in the people'.[54]

Whether enlightened or misguided, Addams maintained a stubborn faith that by cultivating proper channels for the expression of human feeling, men and

---

52 Jane Addams, *Democracy and social ethics* (ed. Anne Firor Scott, Cambridge, Massachusetts, 1964), p. 2.
53 *Ibid.*, p. 1.
54 Jane Addams, 'What is the greatest menace of the twentieth century?' First printed in the *Minutes of the Sunset Club*, 14 Feb. 1901 (McCree, *op. cit.*).

women of all races, classes and religions can discover that they are more alike than different, that they have more to respect than to resent, and possibly, they can find more to love than to hate. Nearly one hundred years later, as we look over the devastation of this century, of this city, perhaps even here, we can still summon the courage to believe that Addams' vision could actually be realised.

14   Neither Feminist nor Flapper: the Ecclesiastical Construction of the Ideal Irish Woman

Maryann Valiulis

IN the early years of the Free State, after the turmoil and turbulence of the Anglo-Irish war of independence and the civil war, there was an ongoing debate between political and ecclesiastical authorities on the one hand, and middle-class feminists on the other, over women's relationship to, and role in the new state.

Irish feminists maintained that they were full citizens of the state, that they had a right to inclusion within the body politic on the same terms as men. This, they argued, was guaranteed in the constitution, and was, moreover, a right which women had earned during the revolutionary struggle. They also contended that, as women, they had a special contribution to make to the political life of the state.

Political and ecclesiastical leaders argued that women needed to be returned to their rightful position within the home – a position some had vacated during the revolutionary struggle. Returning women to the home, these authorities declared, was essential to the stability of the family, the state, and a catholic society. Public duties simply drew women away from their proper domestic sphere and gave them access to an arena in which they neither belonged nor were needed.

This conflict over women's roles was a theme in Irish society throughout the years of the Free State. Respective governments brought in gendered legislation which restricted women's access to the public sphere and increasingly curtailed their freedom.[1] Ecclesiastical leaders sanctioned and legitimated the political restrictions.

In advocating and supporting – indeed applauding – restrictive gender legislation, catholic leaders constructed a particular identity for women. This article examines the virtues and attributes of the ideal Irish catholic woman as defined by the prescriptive ecclesiastical literature of the period. Secondly, it assesses the significance of this ideal for Irish society. What becomes apparent in focusing on women and the construction of a feminine identity is that the years of the Irish Free State were not simply a time of turning inward, of creating if not

---

1 In 1924 and 1927, the Cosgrave government brought in legislation to restrict women's right to serve on juries; in 1925, women's right to sit for all examinations in the civil service was curtailed; in 1932, compulsory retirement was introduced for married women teachers and eventually applied to the entire civil service; in 1935, the government assumed the right to limit the employment of women in any given industry; in 1937, the constitution defined women's role in the state exclusively in terms of the hearth and home.

celebrating a parochial, puritanical Irish catholic state. Rather the ideas of ecclesiastical leaders are part of a broader, more general movement of European conservatism, as a brief comparison to fascist Italy will demonstrate. In fact, the construction of the ideal Irish catholic woman, which church leaders claimed was in keeping with the tradition of the ancient Gaelic state, in reality took its lead more from papal encyclicals than from the early Irish annals.

Any description of the ideal Irish catholic woman of the 1920s and 1930s must begin in that bastion of domesticity, the home. Overwhelmingly, political and ecclesiastical leaders in the Irish Free State constructed an identity for Irish women solely in domestic terms – women were mothers, women were wives. More than that, women were subordinate to their husbands, inferior to men. That was the 'natural' hierarchical order of the world.

Certainly, the dominant political belief was that the proper function of women was motherhood, that their place was in the home, tending to the needs of their husband and children – from Kevin O'Higgins's assertion that the natural and normal role for women was that of bearers of children and keepers of the home and only abnormal women thought otherwise[2] to de Valera's 1937 constitution which clearly situated women in the home. One example from the political discourse of the period – de Valera's eulogy on the death of Margaret Pearse – can serve not only as an example of the dominant political beliefs regarding women, but also as an excellent statement of the views of both political and ecclesiastical leaders. There was little noticeable difference.

Margaret Pearse, mother of 1916 leaders Patrick and Willie, died in April 1932. Her death received a great deal of attention and she was valorised as the ideal mother, the ideal woman. De Valera began his eulogy by noting that:

But for the fame of her sons the noble woman at whose grave we are gathered would, perhaps never have been heard of outside the narrow circle of her personal friends. Her modesty would have kept her out of the public eye. Yet it was from her that ... [her sons] learnt that ardent love for Ireland and for Gaelic culture and tradition that became the passion of their lives. It was from her that they inherited the strength of soul that made them resolute and unshrinking in the career they foresaw would end in death.[3]

De Valera then went on to note that Margaret Pearse must have known long in advance of her son's desire for martyrdom and must have

suffered in advance the sorrow of his death. When the time of parting came she too was prepared. This loving and tender woman resisted the promptings of her mother's heart; she did not seek to hold her sons back. She bade them go ... she bore bravely the sorrow of their death. As she once said, she knew that her boys had done right and that she too had done right in giving them for their country.[4]

After the death of her sons, de Valera noted approvingly that Margaret Pearse saw her role as being 'to hold what they upheld'. She fulfilled this role, he claimed, without bitterness and complaint, but rather with courage, charity and cheerfulness.[5]

Here was the ideal Irish woman. She was first and foremost a mother, who inculcated in her children, her sons in particular, a love of country, of Gaelic

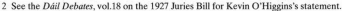

2 See the *Dáil Debates*, vol.18 on the 1927 Juries Bill for Kevin O'Higgins's statement.
3 *Irish Press*, 27 Apr. 1932.
4 *Ibid.*
5 *Ibid.*

culture and traditions, of freedom for Ireland. This was the Irish version of republican motherhood which surfaced after both the American Revolution and the French Revolution. Women's role was to produce sons and educate them in the nationalist tradition to be good and virtuous citizens of the new state.

De Valera describes at length the particular attributes that make up the Irish republican mother. Like Mary who understood that her son Jesus must die, so too did Margaret Pearse understand that her sons must die, putting the good of the nation above her own motherly desires. Self-sacrifice is thus an integral part of her character. Indeed de Valera paints the picture of the Irish *pieta*, the noble, suffering mother holding the dead son – in this case two dead sons.

The ideal woman is also passive. She has no work of her own to do, but rather fulfils the wishes of her sons. She performs her role in public not with an agenda of her own, but rather as a living vessel through which the dead may speak. All of this manifests an air of self-effacement, of meekness, of indirectness. What it lacks is passion, vitality, independence and assertiveness.

De Valera's picture is one which could have been drawn by any catholic bishop or priest. Ecclesiastical discourse of the period of the 1920s and 1930s – that is, the lenten pastorals, the popular catholic press, more scholarly catholic writing, the propaganda of various catholic lay organisations – all supported, legitimised and gave moral sanction to this image of women. Drawing heavily from the papal encyclicals of the period, especially *Casti connubi* published late in 1930, and from the long tradition of the subordination of women in catholic teachings, catholic leaders denied women a public identity, casting them solely in terms of domesticity. As one popular catholic publication proclaimed:

Her [woman's] natural qualities fit her more for the activities and life of the home; . . . woman's gifts point to her as the manager of the household, the educator of the children, and the principal source of brightness and sympathy and love which all seek in the domestic circle. It is the woman's special function, too, to maintain a high ideal of purity and goodness among the members of the family; and to impart to the home that element of aestheticism and beauty which does so much to brighten and elevate human life.[6]

This adds another dimension to de Valera's picture. The ideal Irish catholic woman was also pure and good with a particular appreciation for the beautiful, the pleasing. Implicit in this statement was a reiteration of the belief in woman as the angel in the house who creates a haven to which men can retreat after their sordid dealings in the world of political and economic power.

This public arena wherein political and economic power resided was no place for women, ecclesiastical discourse maintained. Any attempt by women to leave their domestic confines would wreak havoc not only on the home but on the nation as well. As one catholic publication noted, woman has but one vocation:

the one for which nature had admirably suited her . . . that of wife and mother. The woman's duties in this regard especially that of bringing up the children, are of such far-reaching importance for the nation and the race, that the need of safeguarding them must outweigh almost every other consideration.[7]

---

6 Edward Cahill, 'Notes on Christian Sociology' in *Irish Monthly*, Oct. 1924.
7 *Irish Monthly*, 1925.

Such beliefs certainly made the feminist demand for full citizenship seem particularly threatening. The feminist view, said some in the catholic church, lays 'claim to equality which is foreign to her [woman's] nature',[8] and which is based on a 'new (unChristian) conception of society'.[9] Ecclesiastical leaders saw women's citizenship as fraught with danger. Some within the catholic church believed that women should not have even been granted the vote because it was

inconsistent with the Christian ideal of the intimate union between husband and wife that they should exercise the political franchise as distinct units, and be thus enabled by law even to take opposite sides on public issues.[10]

Nor should they participate in other aspects of public life. During the debate over the issue of jury service in the 1920s, church leaders argued that to suggest

that married women should be called upon for the duties of jurors is manifestly inconsistent with their home duties; and that any women be eligible to act as jurors in certain types of criminal cases is contrary to the Christian ideal of female modesty. Hence it is desirable that women be exempted from that duty.[11]

According to church teachings, those who desired a public identity for women were offering nothing but false liberty, an exaggerated and distorted notion of equality of rights.[12]

It was not only political duties which posed a threat to women's continued domesticity. There were economic snares as well. Ignoring the fact that most Irish women worked because they were obliged to out of economic necessity, church leaders proclaimed it

the duty of a Christian State to remedy, by prudent legislation, the abuses which have driven an excessive number of women into industrial employment outside the home.... In a Christian State women should be excluded even by law from occupations unbecoming or dangerous to female modesty. The employment of wives or mothers in factories or outside their own household should be strictly limited by legislation. Girls should not be employed away from their homes or in work other than domestic until they have reached a sufficiently mature age, so that they be not exposed too soon to external dangers to their modesty; and that they have sufficient time before leaving home to become acquainted with household work.[13]

Political and economic power were thus to be left in the hands of men.

But the catholic church was concerned with more than just a glorification of motherhood, a sanctification of the cult of domesticity. Throughout the 1920s, there were other clearly articulated themes prevalent in ecclesiastical discourse: the pursuit of pleasure, the evils of modern dress, modern dance, the cinema, and indecent literature. Quite often, the bishops, for example, would thunder about the 'lure of exotic dances, extravagance and immodesty in dress, and the craze for hectic pleasures of every kind'[14] – all of which they believed were destroying traditional Irish catholic life.

Women were central to this discourse – as symbols of the nation, as innocent

---

8 W.P.Mac Donagh, 'The position of women in modern life' in *Irish Monthly,* June 1939.
9 Edward Cahill, S.J., 'Notes on Christian sociology,' in *Irish Monthly,* Dec. 1924.
10 Edward Cahill, S.J., 'Social status of women' in *Irish Monthly,* Jan. 1925.
11 *Ibid.*
12 *Cork Examiner,* 26 Jan. 1931.
13 Edward Cahill, S.J., 'Notes on Christian sociology'.
14 Editorial, *The Irish Catholic,* 5 March 1927.

victims of modern trends, and as purveyors of immorality. This level of argument was clearly anti-emancipationist, seeing in the stereo-typical flapper of the period the incarnation of immorality. The flapper was juxtaposed to the young girl who was innocent and vulnerable. Ecclesiastical discourse thus defined women in the traditional Madonna/Eve split – a dualism which is an integral part of catholic teaching.

Women were associated with both national identity and the moral health of the nation. Ecclesiastical discourse explicitly tied together nationalism and catholicism, arguing that a return to catholic standards would bring about the return of a traditional Gaelic nation.

According to the bishops, what was at stake in this discussion was the very self-definition of the Irish people. To many in the church, one fundamental basic characteristic of Irish society, of Irish national being was purity.[15] Purity was primarily cast as a woman's responsibility, a woman's crowning glory. Women were thus critical to Irish self-definition and any rejection of traditional standards of purity endangered Ireland's definition of self. As the 1925 statement of the Irish bishops meeting in Maynooth asserted:

There is a danger of losing the name which the chivalrous honour of Irish boys and the Christian reserve of Irish maidens has won for Ireland. If our people part with the character that gave rise to the name, we lose with it much of our national strength . . . Purity is strength and purity and faith go together. Both virtues are in danger these times, but purity is more directly assailed than faith.[16]

If there were any doubt about who had to bear responsibility for this state of affairs, societies like the Catholic Truth Society, made it abundantly clear who was to blame:

The women of Ireland, heretofore, renowned for their virtue and honour, go about furnished with the paint-pot, the lip-stick, . . . and many of them have acquired the habit of intemperance, perhaps one of the sequels to their lately adopted vogue of smoking. A so-called dress performance or dance today showed some of our Irish girls in such scanty drapery as could only be exceeded in the slave markets of pagan countries.[17]

Or as one bishop more moderately noted: 'They could not have a clean and noble race till woman was restored to her former dignity'.[18]

Purity was, therefore, a primary characteristic of the ideal Irish women. Purity meant sexual purity. It also meant eschewing make-up, not smoking or drinking, and modesty in dress and demeanour. Modesty was an important adjunct of purity. It also was a virtue which was being threatened by revealing fashions, suggestive dances, and the like: 'The cult of sex is everywhere. Sex is blazoned on our fashion plates, palpitates in our novels, revels in our ball-rooms . . .'[19]

Purity and modesty, however, were not simply about sexual behaviour. There were implications for cultural nationalism as well, especially as regards fashion and dance. The bishops regularly exhorted mothers to dress their daughters in Irish fabrics – heavy, solid tweeds which covered rather than draped the body and in 'an Irish standard of dress instead of imitating those foreign importations which offend Christian refinement'.[20] Similarly, the bishops exhorted their

---

15 *Irish Monthly*, Nov. 1925.
16 *Cork Examiner*, 30 Nov. 1925.
17 *Irish Independent*, 13 Oct. 1926.
18 *Irish Independent*, 28 Feb. 1927.
19 *Irish Monthly*, March 1926.
20 *Cork Examiner*, 28 Feb. 1927.

flocks to engage in traditional Irish dances which were not sexually provocative and which had the added virtue that they could not be danced for long
hours at a time.[21] Why, the bishops asked, 'should this ancient catholic nation
copy and ape what is worst in the foreigner?'[22]

All would be well – or at least significantly better – if women were pure and
modest, if mothers raised their children by Irish catholic standards, that is dressed
their children in Irish fashions and insisted that they dance only Irish dances.
The moral climate of society would be redeemed and the national culture revi-
talised. To women then fell the onerous responsibility for the moral and cultural life of the country – leaving men free, not surprisingly, to pursue political
and economic power.

But there was a problem. Women seemed to be failing in this role. What was
particularly appalling to ecclesiastical leaders was that in catholic Ireland which
they believed had a history of noble and virtuous mothers, there were now to
be found mothers

who shirked or neglected their duty to their children. . . . There were mothers who preferred the fashionable and crowded thoroughfare to their own quiet home; there were
mothers who preferred talking on a platform or in a council chamber to chatting with
their children in the nursery . . .[23]

To these mothers, the bishops said:

Do not forget that you are Irish mothers; do not forget your glorious traditions . . . Appear seldom on the promenade, and sit oftener by the cradles; come down from the
platform and attend to the cot; talk less with your gossipers, pray more with you child.[24]

Throughout the 1920s, the bishops warned mothers time and time again that
they were failing in their responsibilities, that they were allowing their daughters to go to dances, immodestly dressed, unchaperoned and unprotected. To
these mothers the message was clear. They were shirking their duty and imper-
illing both the spiritual lives of their children and the spiritual life of the nation.

But it would be misleading to think that the bishops were only concerned
about sexual morality and Gaelic resuscitation. Much more is going on in this
discussion about dress, dance, cinemas, make-up and the like. What the prel-
ates, the priests, the propagandists were also concerned about was the possibility for emancipation inherent in these modern amusements and the levelling
effect which these modern amusements had on women in particular and the
population in general. Modern dances, for example, were ones in which anyone
who could pay the price of admission was free to attend. Modern dress defied
hierarchy and did not signify one's place in the class structure – as did nineteenth-century fashion. As one historian has noted:

. . . the battle over fashion was also about social rank. Fashion, as Georg Simmel observed, signals the cohesiveness of those belonging to the same social circles, at the
same time as it closes off these circles to those of inferior social rank.[25]

However, in the 1920s and 1930s, 'the dress styles, body movements, and beauty
canons associated with them made women similar enough in outward appear-

21 *Cork Examiner*, 30 Nov. 1925.
22 *Irish Independent*, 2 Dec. 1924.
23 *Irish Independent*, 25 Oct. 1924.
24 *Irish Independent*, 25 Oct. 1924.
25 Victoria de Grazia, *How fascism ruled Italy* (Berkeley, California, 1992), p. 222.

ance so that, as snobby mothers opined, "you couldn't tell what sort she was until she opened her mouth."[26]

Thus modern dress and modern dance threatened the traditional hierarchies of Irish society.

Objections to the cinema, the pictures, were equally complex. Certainly, some films offended catholic sexual morality. But ecclesiastical objection also centred on the fact that the cinema was noted throughout Europe as a very female pastime and the movie theatre as a female public space. Not only did the church object to public spaces for women, they particularly objected to those which were uncensored, unchaperoned, unprotected. Equally significant, these films vividly portrayed alluring lifestyles which according to religious leaders left young women 'sick with discontent at the grim contrast presented by the realities of their own drab lives'.[27] Thus young women were venturing out in a 'public space' only to be morally corrupted and culturally dissatisfied.

This complexity of concerns was demonstrated in the 1934 controversy surrounding the decision of the National Athletic and Cycling Association to allow women to participate in the same athletic meetings as men. This did not mean that men and women would compete in the same event. Rather it meant that at men's athletic meetings women would have their own contests, their own events. It was a highly controversial decision.

One of the first into the dispute was Dr McQuaid, then president of Blackrock College. McQuaid sent a letter of protest to the Athletic Association, characterising their decision as 'uncatholic and un-Irish' and stating unequivocally '... that no boy from my college will take part in any athletic meeting controlled by your Organisation, at which women will compete, *no matter what attire they may adopt.*'[28] [emphasis added]. It was not, therefore, a question only of modesty in dress, but of venturing out into public arenas which had been the traditional preserve of men. To compete in these events, therefore, would be a violation of the prescribed gender hierarchy.

Later that month, McQuaid again wrote to the Association explaining his position in more detail:

It is un-Irish, for, that mixed athletics is a social abuse outraging our rightful national tradition is a statement that requires not proof, but only some reflection. It is un-Catholic . . . mixed athletics are a moral abuse, formally reprobated by the Sovereign Pontiff, Pius XI.[29]

Specifically, McQuaid cited the pope's position that 'Christian modesty of girls must be, in a special way, safeguarded, for it is supremely unbecoming that they should flaunt themselves and display themselves before the eyes of all.'[30]

McQuaid's sentiments were echoed by others in the community. Some on the athletic board supported his objections, but for quite different reasons. For example, one member said that 'men members of the association had been disgruntled because women members who had won medals had set themselves up

---

26 *Ibid.*
27 *Irish Monthly*, Feb. 1925.
28 Letter from Dr McQuaid to Honorable Secretary of the National Athletic and Cycling Association, 6 Feb. 1934 (Dublin Diocesan Archives, McQuaid Papers, General Correspondence, 1933–1937).
29 Letter from Dr McQuaid to the National Athletic and Cycling Association, 23 Feb. 1934 (Dublin Diocesan Archives, McQuaid Papers, General Correspondence, 1933–1937).
30 *Ibid.*

to be as good as the men'.[31] Under attack from McQuaid and others, the council eventually rescinded its original position and concluded that if women wanted an athletic association, they should form it themselves.

From this conflict, another dimension in the construction of Irish womanhood emerges. True Irish women are not only pure, virtuous and modest, they are demure and deferential, not wanting to flaunt themselves, to call attention to themselves – regardless of their talents. They do not venture into public space. They do not want to compete with men, to be on equal terms with men, but rather accept their separate and subordinate role. In essence, they know their place. They are neither feminists nor flappers.

Interestingly, McQuaid offers no reference for his statement that mixed athletics is antithetical to the national tradition. He simply asserts its validity. Catholic leaders simply made the equation of catholic and Irish and that which violated catholic doctrine also had to be at odds with the national tradition.

Overall, the construction of the ideal Irish woman revealed much about catholic ecclesiastical leaders in particular and Irish society in general. First, the picture which ecclesiastical discourse constructs – pure, modest, deferential, respectful of hierarchy, unassuming, content with one's station in life – represents the ideal of a pre-modern society. In societies which have modernised, the emphasis is on equality of opportunity with merit superseding birth, by a decline of deference based on hereditary status, by the creation of political consciousness, and by the growth of functional specialisation. The ideal which the bishops were advocating reinforced those attributes which were antithetical to a modern or modernising society.

Historically, the catholic church saw the forces of modernisation as a threat to their power and influence – as evidenced in a number of papal encyclicals.[32] In this instance, the Irish catholic bishops chose women as the group with which to make their stand not simply against the modern world as is often said, but against the forces of modernisation. Because they were primarily defined in domestic, private terms, because the church believed women were more malleable with their 'more emotional temperament and . . . weaker personality', their 'natural gifts of sympathy and love', their 'keener sensitiveness', their 'special aptitude to promote the happiness of domestic life', church leaders identified women with a pre-modern way of life. The ideal Irish woman thus represented a bulwark against modernisation. Given their views about women's inferiority and subordinate status, it is indeed ironic that the church turned to women to ward off the attack, to beat back the forces of modernisation, to save the traditional way of life.[33]

This stance against modernisation also gave added zeal to the catholic church's anti-feminist and anti-emancipationist position. As I have argued elsewhere, by embracing and advocating the concept of woman as citizen, feminists aligned themselves with the forces of modernity, that is with the ideas of a post-French Revolution modern political order which was based on a constitution, was open and democratic, espoused the idea of equality and merit, and embodied the

---

31 *Irish Press*, 12 Mar. 1934.
32 See for example Pope Pius IX' s *Syllabus of errors* issued in 1864.
33 The Irish situation was not unique, however. Bonnie Smith makes this same point about the catholic church and upper middle-class women in the second half of the nineteenth century in France (Bonnie Smith, *Ladies of the leisure class* (Princeton, 1981)).

notion of progress.[34] This alternative vision of womanhood clearly threatened to undermine the ability of the church to use women as a barrier against modernisation.

This was equally true of modern amusements and pastimes. Modern dance, the cinema, mixed athletics were associated with the emancipation of women, the desire of some women to broaden their cultural horizons, explore the physical freedom of athletics, enjoy the physicality of their bodies, break through the gender and class hierarchies which had imprisoned women for so long. These attitudes would clearly weaken women's identification with the pre-modern world.

The anti-modernisation stance of the catholic church, therefore, dictated a very narrow, very restrictive gender ideology. The dominant ecclesiastical discourse refused to countenance a more complex identity for women, to acknowledge that other options existed for women outside of the ideal Irish woman.

But other options did exist for women. One reason, in fact, that the prescriptive literature was so uncompromising was because so many women were refusing to be bound by this restrictive image. Church leaders were hostile to expanding definitions of women's role because they felt under siege, because they knew that the reality of women's experiences, of women's lives was much more complex, much more challenging than their definition would admit.

Women's lives clearly transcended the single domestic dimension of the ideal constructed by ecclesiastical discourse. The reality was that increasing numbers of women worked outside the home. A significant number of women never married. Women continued to emigrate in increasing numbers. Women were exploring their sexuality, were having children outside of marriage. Women were going to dances, wearing imported fashions and going to films – often enough for the complaint to be heard that they were never at home. Women were agitating for political rights, demanding a public identity. In essence, women were modern actors in a modernising society. Thus, the church's construction of womanhood was, on one level, a response to women's changing life style – a statement of disapproval, an acknowledgment that what they defined as traditional Irish catholic culture and traditional Irish catholic virtues were indeed under attack.

Was ecclesiastical discourse then simply a denial of reality, a discursive attempt to create an ideal which had no bearing on reality? To dismiss it thus would be to overlook an important consequence. Ecclesiastical discourse on the ideal women gave an important moral justification to political restrictions against women. For example, because women were not supposed to work outside the home, the government need not have any qualms about restricting women's right to the highest levels of the civil service, or of giving the Minister for Industry and Commerce the power to limit the number of women in any given industry. Because women had no power to claim political rights and a public identity, women were effectively denied the right to serve on juries. Because bearing children was the primary and defining function of women, women were denied any access to birth control information. Thus Irish ecclesiastical discourse about the ideal woman had very direct and very real consequences for women in the Irish Free State.

---

34 Maryann Gialanella Valiulis, 'Defining their role in the new state: Irishwomen's protest against the Juries Act of 1927' in *Canadian Journal of Irish Studies*, 18, no. 1 (July 1992), pp 43–60.

Second, the ecclesiastical construction of the ideal Irish woman was important because of the perspective it provided. When viewed through the lens of gender ideology, Irish ecclesiastical discourse emerges not as a unique and isolated phenomenon, but, more accurately, as a part of a general conservative movement throughout Europe. For example, the parallels with Italy in the inter-war period are striking. This is certainly not to suggest that Ireland was fascist or to gloss over the extremely significant differences between the two countries. The point is rather that both fascist Italy and the Irish Free State took their gender ideology from the same source – the papacy – and hence created a socially conservative image of women. Irish conservatism, while fitting the specific needs of Ireland at the time, also emerged as part of a broader European phenomenon. It was not simply the parochial reaction of a country which turned inward after independence.[35]

Similarities abound. In both countries, for example, women were defined primarily as mothers – in keeping with catholic doctrine:

... the way women as a group were singled out in the policies of the period depended on a vision which located them primarily in the family as mothers. ...The most concentratedly offensive initiatives were those that attempted to keep Italian women confined to a destiny of reproduction and nurturing in a family seen as the fundamental nucleus of the Nation.[36]

There are other similarities. The question of modern dress, for example, was as hotly debated in Italy as in Ireland. Involved were the same questions of modesty, of purity and the connection with nationalism. Inter-war Italy, much like inter-war Ireland, was

... the natural home of papal "crusades for purity" and of Sunday hell and brimstone sermons about slackening public morality. ... the National Committee for Cleaning Up Fashion ... led a two-year "universal uprising" against the "horrid vice", the ... "shameful wound" of indecent and scandalous dress. ... several thousand small town girl parishioners vowed to renounce cosmetics, wear sleeves to cover their elbows, and hem their skirts no more than ten centimeters from the ankle down. ... By the late 1920's, major Italian churches had posted signs banning "immodest dress".[37]

There was a similar crusade in Ireland. The Mary Immaculate Modest Dress and Deportment Crusade was aimed at 'stamping out what is mannish and immodest', and girls promised not to wear short or suggestive dresses, not to bear their arms or wear attire which was sexually provocative.[38]

Nationalist concerns also mingled with moral prohibitions in Italy as well as in Ireland. The 'horrible reptile of foreign fashion' obviously tied into nationalism. So too did the fact 'that several billion lira annually were expended on

---

35 This impression has been primarily conveyed through the study of censorship. Noted Irish writers, disgusted by censorship and Irish conservatism, painted the picture of Irish society as inward-looking, parochial and self-absorbed. However, once the focus is broadened, different interpretations emerge. The re-construction of gender ideology demonstrates, as I have indicated, that Irish conservatism is part of a wider European trend. Moreover, an evaluation of the political activities of successive Free State governments reveal their desire to have Ireland play its role on an international stage – be it in the British Commonwealth or the League of Nations.

36 Lesley Caldwell, 'Reproducers of the nation: women and the family in fascist policy' in David Forgacs (ed.), *Rethinking Italian fascism* (London, 1986), pp 110–111.

37 Victoria de Grazia, *How fascism ruled women*, pp 205–206.

38 *Mary Immaculate Training College Annual* (Limerick, 1927), p. 36. The Crusade specifically said girls were not to wear dresses less than four inches below the knee. It also covered behaviour and outlawed smoking, loud talking and the like, suggestive or immodest dances, alcoholic drinks, improper cinema-shows and 'anything opposed to modesty in what relates to sea-bathing ...' (*ibid.*, p. 36.).

luxury imports [which] spurred campaigns to "buy Italian". Propaganda berated the upper classes for their "foreignophilia".'[39]

Similarities are also apparent on the question of athletics. Initially, the fascists had supported women's participation in athletics. However, when the emancipationist implications became apparent and when the papacy condemned it in 1928, the fascists changed their position. Mussolini even demanded that 'Italian women give up even the mildest athletic pursuits as being too virile'.[40]

Moreover, as in Ireland, ecclesiastical discourse was used as moral justification for restrictions against women. Defining women primarily as mothers and wives '. . . provided the justification for the introduction of severely restrictive policies for the vast masses of women.'[41] Thus, under fascism, there was an attempt to limit women's participation in the workforce. In 1927, women's salaries were cut fifty per cent. In the 1930s, the number of women working in government offices was limited to ten per cent and women were limited to the lower sections of the civil service – typists, stenographers and the like[42] – the very jobs which Ernest Blythe singled out for women in the Irish civil service in 1925. Women could be limited to the lower ranks because this was not their primary vocation in life.

The ideal Irish woman then – the self-sacrificing mother whose world was bound by the confines of her home, a woman who was pure, modest, who valued traditional culture, especially that of dress and dance, a woman who inculcated these virtues in her daughters and nationalist ideology in her sons, a woman who knew and accepted her place in society – served the purposes of the ruling Irish male elite. Political and ecclesiastical leaders sought to re-establish gender boundaries and hierarchy after the promise of equality and the experience of freedom of the revolutionary period.

However, the ideal Irish woman could also be found in cities and villages throughout Italy, indeed, one suspects, throughout Europe – wherever Roman catholicism reigned supreme. Neither feminist nor flapper, the ideal woman as constructed by catholic ecclesiastical leaders stood as a potent discursive symbol of European womanhood in the inter-war years.

39 Victoria de Grazia, *How fascism ruled women*, p.222.
40 Bonnie Smith, *Changing Lives* (Lexington, Massachussetts, 1989), p.461.
41 Lesley Caldwell, 'Reproducers of the nation: women and the family in fascist policy,' in *Rethinking Italian fascism*, pp 110–111.
42 Bonnie Smith, *Changing Lives*, p. 461.

15 'The Women Can Not be Blamed':The Commission on Vocational Organisation, Feminism and 'Home-makers' in Independent Ireland in the 1930s and '40s

Caitriona Clear

THE war in Europe was the result of masculine statesmanship, Lucy Kingston of the National Council of Women in Ireland told the Commission on Vocational Organisation in 1940; the women, she said, could not be blamed. This remark was made as a result of intense provocation by the chairman of the session in which her organisation, the Joint Committee of Women's Societies and Social Workers, represented by Mrs P. S. O'Hegarty, and the Catholic Federation of Women Secondary School Teachers, represented by Mrs Vera Dempsey, gave evidence together. The three organisations were trying to explain what they meant by the term 'home-maker' and what they saw as these women's interests, and, as will be shown later, their explanations were somewhat vague.[1] This paper explores some of the contradictions and exclusions in public perceptions of women doing household work in their own homes in Ireland in the 1930s and '40s. By 'public' I mean any utterance, definition and opinion in the public domain, made by politicians, civil servants, journalists and women's organisations. The 'woman of the house', as she was commonly known, was somebody who had or who shared primary responsibility in a household (house, room, caravan, tent) for the everyday re-ordering of the environment, the gathering and preparation of food – in short, for the ongoing life-maintenance work without which health and well-being are impossible. In addition many of these workers were also responsible for childbearing and childrearing.

The Commission on Vocational Organisation was set up in 1939 and it heard evidence from a large number of organisations – professional bodies, trade unions, voluntary organisations and interest groups of various kinds. The purpose of the Commission was to investigate vocational organisation in Ireland, following on the principles and ideas put forward in the papal encyclical of 1931, *Quadrogesimo anno*.[2] The evidence which was given to the Commission gives valuable insight into the ideals, aspirations, and strategies of all these groups.

---

1 Minutes of evidence of the Commission on Vocational Organisation, 12 Dec. 1940 (National Library of Ireland (hereafter NLI), MS 930, vol. 9, pp 3065–3080).
2 For a full discussion of vocationalism and its application in Ireland see J. H.Whyte, *Church and state in modern Ireland* (Dublin, 1971, 1980); and J.Lee, 'Aspects of corporatist thought in Ireland: the Commission on Vocational Organization 1939–43' in A.Cosgrove and D. McCartney (eds), *Studies in Irish History* (Dublin, 1979).

Irish women's organisations had a hard job in the 1920s and '30s to defend the political and civil equality which had been promised them in 1916 and guaranteed to them in the 1922 constitution.[3] These decades saw ongoing debate about women's work and women's 'role'.There was a strong perception that new opportunities were appearing every day for women in the field of paid work, and this gave rise to both approval and alarm. 'Infinite romance and adventure' was an *Irish Press* reporter's description, in 1931, of the female occupational tables in the census, and the picture she believed they drew of women's gradual but thorough permeation of all areas of economic life. Woman had 'found her place in the world,' said a Dr Quin in 1945 – all the more reason, he argued, why there should be a comprehensive maternity service.[4] There was a rise of just over 2% in the 1936 census in the proportion of the adult female population who were engaged in paid employment (An adult female was 12 and over in 1926 and 14 and over in 1936). Was it alarm at this development which prompted Taoiseach Eamon de Valera to insert Article 41.2 in his 1937 constitution, an article which aroused the opposition of women's organisations?

In particular the State recognises that by her life within the home woman gives to the State a support without which the common good cannot be achieved.
The State shall therefore endeavour to ensure that mothers shall not be obliged by economic necessity to engage in labour to the neglect of their duties outside the home.[5]

According to T.P. O'Neill, de Valera put this in because he had been reading Ivy Pinchbeck's pioneering *Women workers and the industrial revolution*, and feared that the appalling maternal ill health and infant mortality of early industrialisation might happen as a result of the industrialising drive of the 1930s.[6] De Valera would, at any rate, have had to explain himself to nationalist feminists, most of whom had supported him since 1922. One way of preventing high maternal and infant mortality among mothers in the paid workforce would have been to follow the French example and introduce statutory maternity benefits, breast-feeding bonuses and so on, or government could have brought in protective legislation for mothers and expectant mothers in dangerous and unhealthy work.[7] As it turned out the only statutory marriage bars affecting women were in national teaching and the public service, and maternal and infant mortality rates remained unacceptably high, according to many doctors, and not only among women who worked outside the home.[8]

It is hard to know where all this worry about women deserting the home was coming from. What spurred Fr Hunter Guthrie, for example, to write in 1940

3 M.Clancy,'Aspects of women's contribution to Oireachtas debate in the Irish Free State 1922–37' in M. Luddy and C. Murphy (eds), *Women surviving: studies in Irish women's history in the nineteenth and twentieth centuries* (Dublin, 1990) pp 206–32.
4 'Irish women in industry' in *Irish Press*, 5 Sept. 1931 (first issue); J. Quin, 'A suggested maternity service for Éire' in *Irish Journal of Medical Science* (6th series, 1945), p.19.
5 *Bunreacht na hÉireann* (Dublin, 1937).
6 Census occupational tables 1926, 1936; conversation with T.P.O'Neill, December 1987 concerning de Valera and Ivy Pinchbeck, *Women workers and the industrial revolution* (London, 1932).
7 Karen Offen, 'Body politics: women, work and the politics of motherhood in France 1920–1950' in G.Bock and P.Thane (eds), *Maternity and gender policies: women and the rise of European welfare states 1880s–1950s* (London, 1991, pp 138–59); Susan Pedersen, *Family, dependence, and the origins of the welfare state. Britain and France, 1914–1945* (Cambridge, 1993); S. Koven and S. Michel (eds), *Mothers of a new world: maternalist politics and the origins of welfare states* (London, 1993).
8 On the marriage bar correspondence between the Department of Finance and the Department of Education, see National Archives, Dublin, S 7985 A, B. See also evidence of Irish National Teachers Organization to Commission on Vocational Organization (NLI, MS 923, vol. 2). On health, see Ruth Barrington, *Health, medicine and politics in Ireland 1900–1970* (Dublin, 1987).

that women's insistence on a public role was the root cause of many social ills?[9] This anxiety was common all over Europe in the inter-war period, and it was based on alarm at the presence of married women in the paid workforce, but there was less basis for this anxiety in Ireland than anywhere else. The vast majority of married Irish women did not work outside the home for wages, and from 1926 to 1961 unmistakeably the largest single block of adult females in the census was women described by the census as 'engaged in home duties'. These women made up 49% of all adult females in 1926, 51.5% in 1936, 54.5% in 1946, rising to 60% in 1961.[10] Furthermore, only 3% of married women were formally in the paid workforce, according to the 1946 census.[11] These census figures embody several perceptions of women of the house. Firstly people, almost all of them women, who are counted as 'engaged in home duties' are in the 'Not gainfully occupied' section of the census, along with invalids, students and retired people. This might seem to be sensible, in that women of the house are not necessarily paid a wage in respect of the work they perform in the house, but then, neither were many farmers' assisting relatives paid formally for the work they performed around the farm. The Irish Countrywomen's Association in 1940, and Hugh Brody writing about a small western community in the 1960s, noted that farmers' sons and daughters received pocket-money, their keep, and for some sons the prospect of inheriting, for other sons and daughters perhaps an education or a dowry.[12] Farmers' wives – as opposed to other assisting relatives were counted by the census as not gainfully occupied, despite the fact that they seem to have had more control than their 'occupied' relatives over certain key remunerative areas of farm work – the poultry and dairy produce. Many farmers' wives, given the age difference between spouses, also had the prospect of inheriting. Who was 'engaged in home duties' when the farmer's wife became the farmer?

Secondly, by noting only principal occupations, the census gives the impression that people in the paid workforce did not do any extra work – this hides the domestic workload of married and single women who were gainfully occupied. James Deeny's survey of married mill-workers in Lurgan in 1940 noted the daily struggle of women who alternated between home and work, never really resting – the same must have been true of working-class married women workers in the Free State.[13]

Thirdly, the census enumerators obviously had some fixed, if not very well-defined idea of what constituted 'home duties', for they noted in the 1926 census and thereafter, that in households of five members and less, only one person could validly be counted as 'engaged in home duties'. No stipulations of age and health were made about either the family members or those looking after them. Somebody like Mary Healy, a former domestic servant who married a carpenter and settled in a Tipperary village in 1942, could have done with

9 Hunter Guthrie S.J., 'Woman's role in the modern world' in *Irish Monthly*, May 1941, pp 246–52; Mary Hayden, 'Woman's role in the modern world' in *Irish Monthly*, Aug. 1941, pp 392–402.

10 Occupational tables, censuses 1926, 1936, 1946, 1961. See under 'Persons not gainfully occupied' for women engaged in home duties.

11 This statistic is cited in the *Commission on emigration and other population problems, 1948–54* (Dublin, 1954), p. 81.

12 Irish Countrywomen's Association evidence to the Commission on Vocational Organisation (NLI MS 930, vol 9, pp 2928–48); Hugh Brody, *Innishkillane* (New York, 1973).

13 James Deeny, 'Poverty as a cause of ill-health' in *Statistical and Social Inquiry Journal of Ireland*, vol. xvi (1939–40), pp 75–89; see also Deeny's memoirs *To cure and to care: memoirs of a chief medical officer* (Dublin, 1989).

somebody to help her when she had two small children still in nappies, in a house without running water and a husband who had to work all hours of the day with the county council to support his family.[14] Had Mary Healy's mother come to live with her, the census would not have considered that there was enough work for the two women. Yet, had Mary Healy employed one, or several domestic servants, whatever the size of the household, she would still have been regarded by the census as 'engaged in home duties'. Finally, women of the house of the lower-middle and working classes – those who could not afford servants – often relied heavily upon children – usually girls – to mind younger children, for long periods of time. This was valuable caretaking and the work consisted of young workers preventing toddlers and small children from drowning, poisoning themselves, falling under vehicles and so on. Nan Joyce, who grew up on the roads of Ireland in the 1940s, remembers never being able to play freely because she always had to mind a younger one or two.[15] Mothers who had nobody to help them at home could not, according to Sheila Greene, a social worker writing in *The Irish Housewife* in 1946, be expected to supervise their children every hour of the day; she recommended supervised playgroups in disadvantaged areas.[16] Photographs of the period often show small children dragging or carrying children half their size, or even bigger.[17] These child workers do not feature in the census at all.

Despite statistical evidence to the contrary, fears about mothers 'deserting' their duties seem to have been part of a more general social anxiety about the rising public profile of women in paid work and in public life, however low this profile might seem to us today. There was a noticeable increase in the proportion of women working in white-collar clerical jobs in the 1920s and 1930s, and some increase in women's industrial employment also.[18] Fears about the breakup of the 'home' merged with resentment at so-called 'pin money girls', or 'flappers', at a time of high male unemployment. Still there was some approval for women working outside the home. Mary Hayden, in an equally long and equally prominent reply to Hunter Guthrie, also in 1940, argued that women's involvement in public life enhanced public morality, and insisted that, while the home was woman's sphere, woman's sphere was not the home, i.e. was not bounded by the home. Hayden, like many other prominent women of her generation – Vera Brittain and Winifred Holtby in Britain, for example – believed that women were uniquely suited to household and domestic responsibilities, but that public rights and duties, including paid work, should be tacked onto this.[19] By the end of the 1930s, however, many women in public, in Ireland and elsewhere, were demanding that the state take some responsibility for childcare, or recognise the work of the woman of the house in some way.[20] The Commission on

---

14 Mary Healy, *For the poor and for the gentry: Mary Healy remembers her life* (Dublin, 1989).
15 Nan Joyce, with Anna Farmar, *Traveller* (Dublin, 1985), pp 15–16.
16 Sheila Greene, 'Youth and crime' in *Irish Housewife*, no. 1 (1946), pp 36–39.
17 See for example, photograph captioned 'Happy in spite of all: children at play in a Dublin tenement street' with article entitled D. Macardle, 'Some Irish mothers and their children' in *Irish Press*, 14 Sept. 1931.
18 Census occupational tables. Between 1926 and 1936, the proportion of the female workforce employed in industrial work rose from 9.4% to 10.3%, women clerks and typists rose from 4.8% to 6.6%, and women in the professions – mainly teaching, nursing and the religious life – rose slightly also, from 8.5% to 9.2%.
19 Mary Hayden, 'Woman's role'. On British feminists between the wars, see Martin Pugh, *Women and the women's movement in Britain 1914–1959* (London, 1992), and C. Dyhouse, *Feminism and the family in England 1880–1939* (Oxford, 1989).
20 See Bock and Thane, *op. cit.* and Koven and Michel, *op. cit.*

Vocational Organisation gave some of these organisations a chance to state their views.

The Irish Countrywomen's Association was (and is) the largest women's organisation in the country, with branches in 17 counties and about 2500 members in 1940.[21] Countrywomen, according to Muriel Gahan and others giving evidence, needed to be represented vocationally in a separate category both from farmers – thought of as male – and from other women working in the home, because they were both producers and consumers, and it was impossible to distinguish their remunerative work from their home and life maintenance work. A countrywoman was defined as anybody who lived in the country, in a farm of whatever size, in a labourer's family, or in a town of under 3000 inhabitants. Every woman resident in these areas could and should, they argued, benefit from the Association's activities – lectures and classes in everything from fowl-rearing to beekeeping, from business organisation to home management skills. Although the delegates to the Commission insisted on the fact of social equality within the ICA, they admitted that organisation was slow in the poorer agricultural areas of the country. Then as now the kind of people who joined this or any group were likely to be those who had a margin of leisure; many members, it was reported, cycled over ten miles to meetings. Although the ICA co-operated with other rural development groups like Muintir na Tíre on community initiatives like group water schemes and rural electrification, they held fast, in the face of the Commission's fears of the overlapping of groups with similar functions, to the need for a separate women's organisation. In meetings, they said, women received valuable training for public life, and besides, in mixed organisations – they were tactful but insistent on this point – men did all the talking. They also told the frankly incredulous commissioners that women of all ages were equal within the ICA. 'Surely a grandmother would not allow a girl of 16 to talk up to her?' one of them asked – this incredulity suggesting that the hierarchy of age was quite strong at this time. A shocked 'Good gracious no!' was the delegation's reply to the suggestion that their organisation's activity might be 'a form of rural slumming', and when asked if any of the women they worked with objected to the designation 'countrywomen', they claimed: 'They are charmed to be called countrywomen. They are living in the country and they love the country. All they want is a little bit more pocket money'.[22]

The charge of slumming was also advanced, this time more forcefully, against the three women's organisations noted earlier: the National Council of Women, the Joint Committee of Women's Societies and Social Workers, and the Catholic Federation of Women Secondary School Teachers, who gave evidence in a joint session to the Commission some weeks later. These organisations recommended firstly that those women whom they called 'home-makers' have a voice in government and be represented as a vocational group, and secondly that a wide-ranging network of support organisations for women with home responsibilities be set up – baby advice clubs, day nurseries for children whose mothers worked outside the home, social organisations for general support and ad-

---

21 A full history of the ICA has yet to be written, but some of the early writings and information on the organisation and its precursor, the United Irishwomen, are collected in Pat Bolger (ed.), *And see her beauty shining there: the story of the Irish Countrywomen's Association* (Dublin, 1986). Information on membership is taken from the ICA's submission to the Commission on Vocational Organisation (NLI, MS 923, vol. 196).

22 *Ibid.*

vice. They also suggested that the voluntary social work which was carried out by many women who did not work for wages outside the home, be recognised in some way by government.

Bishop Michael Browne, the chairman of the Commission for Vocational Organisation, chaired this session, and he set a tone of confrontation from the beginning, constantly interrupting the women in mid-sentence, and flatly contradicting them. It would be a mistake, though, to let Browne's rudeness deflect our attention from the fact that the women's definitions were sometimes problematic. Browne charged the women, in two separate exchanges, with being 'philanthropic ladies of leisure' performing 'slum work'. They retorted to the first charge, that most members of their organisations worked for a living 'in various professions and trades'. They were not, therefore, referring to themselves primarily as 'home-makers', and throughout the session they referred to 'home-makers' in the third person. They also stated that voluntary work in the community with women who needed support and encouragement and advice was part of a 'home-maker's' job. In this way they were identifying two kinds of 'home-maker' – one who ran baby clubs and supervised, through paid or unpaid work, other services, and the other, who availed of these services. Would the former have been in a position of some authority over the latter? Certainly she would have had superior 'expertise', and access to resources. The women's organisations did not challenge the 'slum work' point, apart from retorting, 'It is not slum work'. Pushed to a definition of 'home-maker', the delegates said that a 'home-maker' was anybody who had charge of a house, married or single, male or female, yet they used the words 'home-maker' and 'woman' interchangeably, and furthermore, defined 'home-maker' interests largely as the interests of mothers of young children. The urgent need to reduce infant mortality and childhood disease no doubt prompted this prioritisation, but there were 'home-makers' with older children, women in sibling households, women caring for aging parents or for younger siblings, for nieces and nephews and so on.[23]

Browne's hostility to the women's definitions and demands can only be understood as stemming from the contemporary fear of women in public life, because the demands were socially conservative and in keeping with what some historians have called the 'maternalist politics' of the early twentieth century – demands made by educated women of the upper-middle classes on behalf of working-class women, in organisations and initiatives which, while they often actually saved lives and improved the quality of life, gave the former authority over the latter and a high political profile also.[24] When Mrs. P.S.O'Hegarty began to explain why there was a need to educate mothers in baby clubs (advice centres for mothers with babies), Browne interrupted her: 'But it may be that the only home-makers in Ireland at present are doing their work remarkably well?', suggesting that these women were a dying breed. Even though delegates explained patiently and at length that day nurseries were for the children of mothers who worked outside the home from economic need, he described the nurseries as 'plans to abolish the mother's home as the ideal place for looking after babies.' Anyway, he said, surely health education and advice about childcare

---

23 Commission on Vocational Organisation (NLI, MS 930, vol. 9, pp 3065–80).
24 See Bock and Thane, *op. cit.* and Koven and Michel, *op. cit.* and for a critical appraisal of the concept of maternalism see Jane Lewis, 'Gender, the family and women's agency in the building of welfare states: the British case' in *Social History*, vol. 19, no. 1 (Jan., 1994), pp 37–55.

should be provided by the Department of Education.This was an odd suggestion from an advocate of vocationalism: that a centralised government department take over such a delicate responsibility, and it suggests that the thought of
women – any women – telling other women what to do, caused Browne some
panic. He put it to the delegates that their organisations were 'in fact, feminist-
suffragist' and when they replied, with, one imagines, weary courage, that yes,
they were, he then claimed triumphantly that they had no authority whatso-
ever to ask for rights and services for 'home-makers', since all they were interested in was getting women out of the home to work.

The kind of obvious unfriendliness shown by Browne to the women's organisations was also experienced from some quarters by the Irish Housewives' Association, a campaigning organisation set up by Hilda Tweedy and others in
1942. The interests of 'housewives' were broadly defined, encompassing everything from price and quality control of foodstuffs to clean milk and clean streets,
living accommodation and public health. Made up largely of urban, middle-
class protestants, it contained many first-wave feminists among its founding
members and friends.[25] The first issue of the IHA's annual publication, *The Irish
Housewife* appeared in 1946, with articles on subjects as diverse as rational
house design, conditions in national schools, and women in local government.
Susan Manning wrote an editorial on the problems facing 'the housewife of today'; these included travelling on crowded buses, waiting in queues for basic
foodstuffs (Emergency rationing had not ended), having to carry home all the
groceries oneself, the absence of good, reasonably priced restaurants, and, a
contemporary detail which might make us look more kindly on the wrapped
sliced pan, the loaf of bread left on the windowsill by the breadman, dirty from
being handled so much. Housewives, she concludes, 'must organise not only for
themselves, but for their poorer neighbours'.[26] This is what the IHA did, and its
concerns were in no way limited by its class composition, but one might ask,
were there no 'housewives' among the 'poorer neighbours'? Housewives are, in
Manning's definition, middle to lower-middle-class urban women. To point this
out is in no way to minimise the problems faced by such women working in
their own homes – they usually carried the burden of family budgeting, and
more and they, like rural women, had to do without the helping hands of younger
sisters or older daughters, who were by this time spending longer at school,
working for wages outside the home, or emigrating. The ongoing decline in the
proportion of single women 'engaged in home duties' began to accelerate from
the mid-1940s and cannot be entirely accounted for by the dying out of sibling
and 'carer' households. It was paralleled by an even sharper decline in the
number of female assisting relatives on farms.[27] Were women in charge of households becoming more and more short-handed and isolated from other women
in the 1940s?

In the end, the Commission on Vocational Organisation when it reported in
1943, recommended that women working in their own households be organ-

---

25 Hilda Tweedy, *A link in the chain: the story of the Irish Housewives' Association 1942–1992* (Dublin, 1992).
26 *The Irish Housewife*, vol.1, no.1 (1946), foreword by Susan Manning.
27 Census occupational tables, agriculture; the number of female assisting relatives on farms fell by 33% be-
    tween 1926 and 1946. Single (not married nor widowed) women 'engaged in home duties' fell from 24% of all
    women thus engaged in 1926, to 21% in 1946, but by 1951 their proportion to the total women thus described
    by the census had fallen to 11.1%, which suggests particularly rapid change in the late 1940s. Much more
    work needs to be done on these figures and on their relationship to one another. These are crude statistics,
    but they give some indication of trends.

ised in some way, so as to exert pressure on government on matters directly
relevant to them – public health, education, prices, household design, resources
of water and power, and public morality. There was, however, no permanent
place for 'home-makers' on the proposed National Vocational Assembly, although there was a provision for their co-option if this was thought to be necessary. This caused at least two of the commissioners, Maire McGeehin and G.L.C.
Crampton, to point out in an addendum to the Report, that 'home-makers'
made up 25% of the adult population in the 1936 census, and that a minimum
of 5 seats would be necessary to give them representation proportionate to
their size on the proposed Assembly.[28] As it happened, the Commission's recommendations were ignored by government and the Assembly never came into
being, but the Commission's perspective on women's household work shows
their evaluation of its importance, and crucially, the very faint consultative voice
they were prepared to allow its representatives.

Hanna Sheehy Skeffington wrote an article for the first issue of *The Irish
Housewife* in 1946, an article published after her death. In it she applauded the
Irish Housewives' Association, but lamented the use of the word housewife –
why, she asked, was there no word or phrase in English to compare with the
French *menagere*, or the Irish *bean a' tí*. Still, she said, it was no harm to use a
term like housewife, because such 'man-made language' was a salutary reminder
of 'how little free we really are'.[29] My preferred term for women performing
work in their own homes is 'women of the house', not only because it is the
most inclusive term imaginable, taking in women working outside the home
who had domestic responsibilities, married, single and widowed, urban and rural women, but because it was, and is (in the singular), the term most often used
in ordinary conversation. If 'women of the house' (i.e. the plural) sounds strange
this is no harm, because it reminds us that these workers were seldom, if ever,
referred to in the plural, and that they had no collective identity to speak of.
Given this, it is hardly any wonder that all public attempts to formulate collective definitions of women of the house ran into difficulties in Ireland in the
1930s and '40s, whether the designation was 'home-maker', 'countrywoman' or
'housewife'. These terms were inadequate and exclusive, but they represented
an attempt to qualify the term 'woman'. Fr Coyne, one of the more sympathetic
commissioners at the joint session of the three organisations, suggested to the
delegates, when they were trying to define a 'home-maker', that they 'should
not put forward being a woman as a specialised vocation.'[30] Everybody present,
however, and everyone who made public statements either supporting or opposing women's political and civil equality in this period, slipped into generalisations about women's nature and social function. Lucy Kingston's tirade against
masculine statesmanship was sparked off by Browne's boorishness, but recourse
to moral-superiority arguments had its own pitfalls. Browne retorted that the
women who had reared these statesmen must bear some of the responsibility.
Feminists in Ireland in this period were torn between arguing for rights and
services for women on the basis of strict political and civil equality with men, or
on that of difference from them. This was a dilemma faced by all nineteenth-
and twentieth-century feminists, and one which feminists still face.

---

28 *Commission on Vocational Organisation Report* (Dublin, 1943), pp 279–81, 414–5, 471–4, 481–6.
29 'Random reflections on housewives, their ways and works' in *Irish Housewife*, vol. 1, no. 1 (1946).
30 Evidence to the Commission on Vocational Organisation (NLI, MS 930, vol. 9, p. 307–8).

16 Emigration and Gender: The Case of Ireland, 1922–60

Pauric Travers

E MIGRATION has generally been seen in male terms: the destitute or adventurous young man who leaves his homeland, driven out by poverty or attracted by the prospect of economic improvement in a new land.[1] Insofar as women are considered it is as dependents who accompany or later join their menfolk. Where the existence of large female emigration flows has not been ignored, they have been seen as secondary movements, dependent on the original primary migration of young men. According to the traditional model, women did not *choose* to emigrate: they chose to accompany their menfolk or to rejoin them even though this meant 'abandoning environments they love.'[2] As one influential study concluded, 'women generally emigrate to create or reunite a family.'[3]

While a large part of female emigration can certainly be seen in these terms, the overall picture is much more complex. For example, female immigration in the United States has exceeded male since the 1930s. In the case of Irish emigration, the preponderance of female emigrants is even longer established. Between 1871 and 1971, net female emigration (i.e. gross emigration minus immigration) exceeded that of male. Despite this, historians have paid remarkably little attention to Irish female emigration, particularly in the period after 1922.

Although there have been a number of pioneering works on Irish female emigration in recent times, much more work is necessary on the origins of Irish female emigrants, why they emigrated, where they emigrated to and how they adjusted to their new environments.[4] It might then be possible to move from a woman centred analysis towards a gendered understanding of the emigration process. Recent studies elsewhere have concluded that ethnicity, class and gender interact in the process of migration. Irish migration studies have tended to concentrate on the first two and virtually ignore the third.[5] It is not simply a question of female migration having been ignored or under-rated. The exclu-

---

1 A more detailed examination of many of the issues discussed in this paper can be found in P. Travers, 'Irish female emigration 1922–71' in P. Flanagan (ed.), *Irish women and Irish migration* (forthcoming, Leicester 1995).

2 See, for example, E.S. Lee's influential article 'A theory of emigration' in *Demography*, iii (1966), pp 47–57.

3 M. F. Houston, R. G. Kramer and J. M. Barrett,'Female preponderance of immigration to the United States since 1930' in *International Migration Review*, xviii (1983), p. 919.

4 See for example, David Fitzpatrick, 'A share of the honeycomb': education, emigration and Irishwomen' in *Continuity and Change*, i, no. 2 (1986), pp 217–34; H.R. Diner, *Erin's daughters in America : Irish immigrant women in the nineteenth century* (Baltimore, 1983); P. Jackson, 'Women in nineteenth century Irish emigration' in *International Migration Review*, xix (1984), pp 1004–20. For the period after 1922, see Mary Lennon et al, *Across the water: Irish women's lives in Britain* (London, 1988); Joy Rudd, 'The emigration of Irish women' in *Social Studies*, 9 (Spring 1987), pp 1–11.

5 For a splendid discussion of these issues see Silvia Pedraza, 'Women and migration: the social consequences of gender' in *Annual Review of Sociology*, xvii (1991), pp 302–25.

sion of gender has produced a flawed understanding of the wider process of migration, both male and female.[6]

This paper aims to focus particularly on the factors which gave rise to large scale female emigration from the south of Ireland in the forty years after independence in 1922. Female emigration held up a mirror to Irish society. The main concern here is with what that mirror reveals about Irish society and the position of women within it.

The continuance of large scale emigration in this period posed a significant challenge to the new Irish state. The traditional nationalist intrepretation of emigration which saw it as a side effect of British colonialism and Irish emigrants as exiles became increasingly untenable. The process of coming to a more realistic appraisal of the causes of emigration was a difficult and traumatic one, not least because it involved confronting deep-rooted forces in Irish society.

In the century from 1871 to 1971, net female emigration outnumbered male overall and in six decades, between 1871 and 1901, the balance was fairly even. There was a female predominance between 1901–11, 1926–36, 1946–51 and 1961–71. During the period of the two world wars, the outflow of female emigrants reduced considerably while that of males increased because of recruitment and the outflow of men to Britain to fill the vacuum left by British recruitment. During the 1950s, a sharp decline in male employment in Ireland (male employment fell by 135,000 whereas female declined only by 35,000) contributed to a substantial excess of male over female emigration.[7]

It is worth noting that if one excludes inheriting sons from the calculations, there is little significant variation between the numbers of male and female emigrants. For the century as a whole, the annual average net emigration for males was 15,707 whereas for females it was 15,983. This excess of females is small and might disappear altogether if one allowed for the likely inflation of female totals caused by the larger male presence in reverse emigration. The Commission on Emigration (1948–54) found it impossible to produce reliable figures for gross emigration (i.e. the total of persons leaving the country to take up permanent residence abroad) because of the difficulty of differentiating between emigrants and other travellers and of acquiring and maintaining continuous and accurate records of movements by sea and air. It therefore relied mainly on net emigration figures. These were based on population changes between censuses, adjusted to take into account births and deaths. The net emigration figures failed to take into account reverse emigration i.e. returning emigrants. In the Irish case this was small but it was more popular with men than women.[8] Thus the representation of men in gross outflows is underestimated. Moreover, the under-registration of deaths, particularly of women, up to the 1890s, also served to inflate net female emigration figures for the earlier period.[9] However, whatever adjustments are made, the rate of female

6 On women's history versus gender history, see Cliona Murphy, 'Women's history: feminist history or gender history?' in *Irish Review*, xii (1992), pp 21–7.

7 National Economic and Social Council, *The economic and social implications of emigration* (Dublin, 1991) – hereafter *Economic Implications* – pp 67–9. The economic upturn in the 1960s benefitted men more so than women thus helping to reverse the balance. In the 1970s, there was a net inward migration with male emigrants returning to Ireland in greater numbers than females. With resumption of large scale emigration in the 1980s, the initial outflow was strongly male rather than female.

8 The rate of return emigration varied but averaged at c.10%. See Mark Wyman, *Round trip to America: the immigrants return to Europe 1880–1930* (London, 1993), pp 10–12.

9 *Commission on Emigration and Other Population Problems, 1984–54. Reports* (Dublin, 1954) – hereafter CEOP – pp 115–17. I am indebted to Dr David Fitzpatrick for bringing the discrepancy in the registration of deaths to my attention.

emigration remains significant, not least in the context of the assumption that emigration is a predominantly male phenomenon. The fact that male emigrants returned to Ireland in greater numbers than female is itself significant.

A significant characteristic of Irish emigration since the famine is that is has been predominantly youthful, the majority of emigrants falling within the 15–24 age group. During the period 1924–39, 50 percent of emigrants fell into this category. This continued during the Second World War, when emigration was largely to Britain to work in war industries but there was a noticeable increase in the emigration of older males. This was not true of female emigration which remained decisively youthful. Family emigration dried up but it reappeared in the 1950s whereas emigration among older males declined.[10]

During the period covered by this paper, female emigrants were on average younger than male while fewer older women than men emigrated.[11] A surprisingly large proportion of women emigrated between the age of 15 and 19. Irish female emigrants were on average younger than those from other comparable European countries such as Spain and Italy. The most likely reason for this is the large numbers of Irish women who went to work in domestic service. Average age of employees in domestic service was lower than in the unskilled employment which most Irish male emigrants gravitated towards. While domestic service declined as an occupation after the Second World War, it remained throughout this period the major employment for Irish female emigrants. The closer monitoring and regulation of emigration to Britain during and immediately after the war has allowed for more detailed analysis of these years.

While the relative youth of female emigrants can be explained, it should not be underestimated. It highlights the inadequacy of traditional models of emigration: women going to work in domestic service were not part of family emigration, they were not rejoining 'their menfolk', and they cannot reasonably be considered secondary emigrants. In human terms, their relative youth must have made the emotional turmoil of the emigration process for Irish female emigrants correspondingly greater.

Table I. Classification of Intended Empoyment of Females Granted Travel Permits to Britain, Jun. 1948 –Dec. 1951[12]

Employment	Numbers	%
Domestic Service	21,903	53.7
Nursing	5,578	13.7
Agriculture	689	1.7
Clerical	1,237	3.0
Factory Work etc	11,412	28.0
Total	40,819	100.0

10 *Economic Implications*, pp 70–1; D. Hannan, 'Emigration 1946–71', unpublished Thomas Davis lecture, October 1972.
11 See *Economic Implications*, Table 3.5, p. 73.
12 *CEOP* p. 323. A large proportion of the women who emigrated to jobs in domestic service had already been in domestic service in Ireland. For position regarding travel permits, see Emigration for Employment, memorandum by Department of Social Welfare for Commission on Emigration (Marsh papers, Trinity College, Dublin, MS 8302, pp 70–1).

In terms of geographical origins the pattern for female emigration corresponds roughly with that of men. Connacht and the north west were the worst hit. Male emigration from urban centres, especially Dublin, was slightly higher than female while from rural areas it was slightly lower. Internal migration was much more popular with women, especially better educated women, than it was with men. It would seem that women were more eager to leave rural areas than urban whether because of availability of employment or more conducive social conditions. Of the 122,954 women who sought travel permits to go to work in Britain between 1943 and 1951, 12.5 percent lived in Dublin county or borough, 19.5 percent came from other urban districts while 68 percent came from rural districts.[13]

The United States was the preferred destination of Irish emigrants until 1929 when Britain became the most popular destination, followed by the USA, Canada, Australia and other 'overseas' destinations. Female emigrants were slightly more likely to go to Britain than male. Between 1926 and 1951, approximately 180,000 female emigrants went from Ireland to Britain whereas just over 52,000 went to the United States. Of the 101,338 female migrants between 1924 and 1952 who went to 'overseas' destinations (i.e. outside Europe), 86 percent went to the United States, 5.3 percent to Canada, 4 percent to Australia and 4.8 percent to other countries. A smaller proportion of women than men went to Canada and Australia.[14]

The relative youth of the female emigrants, their rural background and their undoubted frustration with rural life must certainly have shaped their experience. So too did the fact that so many emigrated as individuals rather than as part of a family, which makes them different from female emigrants from other parts of Europe. Little or no effort was made to prepare emigrants for their new life. There was a reluctance to argue in favour of pre-emigration education lest it appear that emigration was being condoned or encouraged. A proposal by the Commission on Emigration to establish emigrant advice centres and to prepare emigrants for the conditions they were likely to meet was largely ignored. Emigrants continued to make do as best they could.[15]

There was a contradiction between the reality of large scale emigration and the official state ethos. The latter saw emigration as evil, dangerous, even anti-social. Female emigrants, as one commentator put it, 'risked plunging into the sinks of iniquity and exchanging the spotless purity of their Irish homes for the pagan turpitude of a modern Babylon'.[16] The catholic church, in particular, highlighted the moral danger posed to young women who preferred the 'kitchens, factories and dancehalls of other lands' and fled from 'the green fields of Ireland to the grey streets of an alien underworld'.[17] How many of the female emigrants really considered London or New York an alien underworld is a matter worth investigating. What is certain is that the vast majority implicitly rejected Bishop Lucey of Cork's position that urban life was by definition corrupt and that their abandonment of the rural life was 'socially and morally indefensible'.[18]

---

13 *Economic Implications*, pp 75–6; *CEOP*, pp 321 and 7; Mary Daly, 'Internal migration', unpublished Thomas Davis lecture, December 1972.
14 *CEOP*, pp 25 and 317.
15 See address by Peadar O'Donnell to Symposium on Emigration, Dublin, May 1954 in *Irish Press*, 10 May 1954. For some rather dubious advice to Irish female emigrants see Gertrude Gaffney, *Emigration to England: what you should know; advice to Irish girls* (Dublin, 1937).
16 *Irish Times*, 25 Mar. 1948.
17 *Irish Press*, 23 Apr. 1948; Pastoral letter by Dr Lyons of Kilmore, Feb. 1948, *Irish Press*, 9 Feb. 1948.
18 *CEOP*, Minority Report, pp 335–42.

Undoubtedly, male and female emigration were both largely a product of the same general factors but some of these impinged more forcefully on women than men. There was also a range of additional factors relating to women's status which all played a part in the decision to emigrate. Having examined this question for six years, the Commission on Emigration concluded in 1954 that the fundamental cause of emigration was economic. However, it added that the decision to emigrate arose from an interplay of factors, social, political, economic and psychological.[19] Young people were more demanding; they sought a better standard of life and an escape from 'the drabness of the average Irish village' with its 'frustrations, inhibitions and its sterile outlook'.[20]

These conclusions were based partly on surveys of emigrants conducted for the Commission. Intending emigrants were interviewed and surveys were undertaken of social conditions in different parts of the country. The causes of emigration offered by the people themselves are tabulated. The reasons cited ranged from general poverty to restlessness and absence of freedom, from the impossibility of getting farmers to marry to the poor social conditions, the monotony of rural life and the attractions of life in urban centres abroad. The remedies offered by the people included planning, decentralisation, running water and lavatories. Many emigrants cited hardship as a reason for emigration but they also cited the opportunities offered by emigration.[21]

A survey of intending emigrants in County Cavan during the 1960s found that a majority of girls with secondary school education would emigrate even if a job was available.[22] In 1948 when Eamon de Valera was faced with explaining why two thirds of emigrants were women, he conceded that there were reasons other than employment which persuaded women to emigrate and confessed with resignation that '. . . we cannot corral the people and say "You must not go out". That is an interference with human liberty and would not be justified except in times of grave emergency.'[23]

In the same month, Dr McNamee, Bishop of Ardagh and Clonmacnoise in a lenten pastoral also admitted that emigration was a product of disillusionment as much as unemployment and warned parents that they should 'weigh carefully and prayerfully before permitting much less encouraging their children to join in the present wild exodus from their native land.'[24]

A study of Irish emigrants living in London in the early 1970s confirmed that a higher proportion of women cited reasons other than employment as a reason for emigration. Dissatisfaction with their life and status in Ireland was identified as a more important factor for women than for men. Arising from this, a larger proportion of women than men indicated that they did not intend to return to Ireland. No doubt, partly because of this, women were found to have integrated better than men into their new society.[25] A more recent survey of post-war female emigrants confirms all these findings with respondents citing

---

19 *CEOP*, p. 153.
20 *Irish Times*, 25 Mar. 1948.
21 Rural Survey: County Clare (Marsh Papers, TCD, MS 8301, SSI, p.17).
22 See D. Hannan, *Rural exodus: a study of the forces influencing the large scale migration of Irish youth* (Dublin, 1971); C. O'Gráda, 'On two aspects of post-war Irish emigration', Centre for Economic Research, UCD, Working paper no. 31; Mary Daly, 'Internal migration'.
23 De Valera was speaking at Bray during the 1948 election campaign (*Irish Press*, 3 Feb. 1948).
24 *Irish Press*, 16 Feb. 1948.
25 M. O'Briain, 'Irish immigrants in London' (unpublished M.Phil thesis, the City University, London, 1981), cited in J. Rudd, 'The emigration of Irish women', pp 1–11. For a good recent discussion of Irish immigrant integration in Britain in the last two generations, see *Economic Implications*, pp 161–214.

employment as a major determinant of their decision to emigrate but also other factors including restlessness, dissatisfaction with their lot, poor social conditions, marriage and the influence of an emigrant network. A minority admitted to unhappiness within their extended family or frustration with social or religious aspects of Irish society.[26]

In its evidence to the Commission on Emigration, the Irish Housewives' Association asserted that female emigration was a product of poor conditions on small farms, the fact that women were dismissed from public appointments on marriage and the inferior status of women in Irish society.[27] The official encapsulation of that status was the 1937 constitution which enshrined an idealisation of the family and the aspiration that 'there may be established on the land in economic security as many families as in the circumstances shall be practicable' (Article 45.2). No clauses in the constitution attracted more attention at the time than those which related to women. The constitution reflected a particularly Victorian view of the role of women as the home-maker. In the section on fundamental rights, all citizens were proclaimed equal before the law. De Valera initially omitted the phrase 'without distinction of sex' which appeared in the Free State constitution of 1922 but added a new qualification which alarmed many women: 'This shall not be held to mean that the state shall not in its enactments have due regard to differences of capacity, physical and moral, and of social function'.

The articles which followed stated *inter alia* that the state recognised that

by her life within the home, woman gives to the State a support without which the common good cannot be achieved. The State shall, therefore, endeavour to ensure that mothers shall not be obliged by economic necessity to engage in labour to the neglect of their duties in the home.

In the Dáil debate on the constitution, one critic pointed to the reality that with so many women emigrating, these sections were little more than 'pious resolutions'.[28]

The strength of the female reaction against the constitution is remarkable.[29] The campaign was led by the National University Women Graduates' Association, the Irish Women Worker's Union, the Standing Committee on Legislation Affecting Women and the Joint Committee of Women's Societies and Social Workers representing more than fourteen women's groups.[30] Mary Hayden claimed that women's rights as citizens, their civil status and their right to work were to be left to the whim of government ministers. Mary S. Kettle of the Joint Committee alleged that the government was striking at the rights of women under the cloak of the constitution. So vociferous was the attack that the *Irish*

---

26 See Travers, 'Female emigration'.
27 Rural Surveys: Co.Clare (Marsh Papers, TCD, Ms 8301, SS4(b), p.2). These views met with a mixed response from the Commission. They were vigorously rejected by Bishop Lucey who flatly stated that they did not stand up to cross-examination.
28 *Bunreacht na hÉireann* (Dublin, 1937); M. Moynihan (ed.), *Speeches and statements by Eamon De Valera 1917–1973* (Dublin, 1980), pp 326–7; P. Travers, 'Irish responses to emigration' in O. MacDonagh and W.F. Mandle (eds), *Irish Australian studies* (Canberra, 1989), p.323.
29 On women and the constitution, see Yvonne Scannell, 'The constitution and the role of women', in B. Farrell (ed.), *De Valera's constitution and ours* (Dublin, 1988), pp 123–35; Mary Clancy, 'Aspects of women's contribution to the Oireachtas debate in the Irish Free State, 1922–37', in Maria Luddy and Cliona Murphy (eds), *Women surviving: studies in Irish women's history in the nineteenth and twentieth Centuries* (Dublin, 1989), pp 206–32; Mary McGinty, 'A study of the campaign for and against the enactment of the 1937 constitution' (unpublished M.A. thesis, University College Galway, 1987); Marie O'Neill, *Jenny Wyse Power* (Dublin, 1993).
30 *Irish Press*, 11 May 1937.

Press was forced to come to de Valera's defence in a series of editorials which accused some of his critics of 'gratuitous libel'.[31] Even after de Valera met delegations from the Joint Committee and the Women Graduates' Association, and promised to consider their reservations, the storm continued. A joint committee representing republican women's groups including Cumann na mBan, Mná na Poblachta, Cumann na Poblachta and Sinn Féin added their voice to the protest. A campaign was organised to lobby TDs and to get women's groups in Ireland and abroad to take up the issue.[32]

Gertrude Gaffney accused the constitution of sounding the death knell of the working women. De Valera, she said,

had always been a reactionary where women were concerned. He dislikes and distrusts us as a sex, and his aim ever since he came into office has been to put us in what he considers is our place, and to keep us there . . . If he had more contact with the average working man and woman he would know that ninety percent of the women who work for their living in this country do so because they must.[33]

The women's campaign against the provisions of the constitution achieved a modest success in persuading the normally unyielding de Valera to drop the reference to the 'inadequate strength of women' and to reintroduce the Free State reference to 'without distinction of sex'. It also provides evidence of a considerable level of organisation and politicisation particularly on the part of middle-class, professional women in Ireland. Indeed, some conservative critics who urged de Valera not to give in to the 'clamour of these suffragettes' saw the campaign as evidence of the wisdom of the constitutional provisions and the need to make the articles on the family and women in the home mandatory by guaranteeing a family wage. That, argued the influential Jesuit, Fr Cahill, was the only way to deal with female emigration.[34]

The campaign against the constitution was not an isolated one. Many of the groups involved had been active on issues of concern to women in earlier years and they continued their involvement afterwards. The Joint Committee of Women's Societies under the formidable leadership of women such as W.R. O'Hegarty is a good example. The Committee campaigned on issues as diverse as pervasive state censorship, equal pay for women, the marriage bar in the civil service which forced women to retire on marriage, the exclusion of women from key positions in the civil service, Garda Siochána and prison service, the inadequacy of welfare provisions, and general health issues. The response from the state was invariably obstructive, as for example on the equal pay issue. Even though the Department of Finance admitted that lower payment of women civil servants might discourage many able women, it strenuously opposed all demands to end discrimination in rates paid for equal work.[35]

An even better example of the unfriendly nature of the state apparatus as far as women were concerned is the campaign to have women included in the Garda Siochána. The women's police issue was first raised with the Minister for Jus-

---

31 *Ibid.*, 11, 12, 13 May 1937; *Irish Independent*, 12 May 1937.
32 W. G. Hassard to de Valera, 22 May 1937; Louie Bennett to de Valera, 18 May 1937; D. Macardle to de Valera, 21 May 1937 (National Archives, Dublin (hereafter NA), D/T). See also various letters in the NA, D/T file, Women: position under constitution, 1937, S9880.
33 'A woman's view of the constitution' in *Irish Independent*, 7 May 1937.
34 Cahill to de Valera, 23 May 1937; J.J. Walsh to de Valera, 15 May 1937 (NA, D/T S9880); B.B. Waters, letter to *Irish Times*, 22 May 1937.
35 Memorandum on Equal Pay, circulated 23 Nov. 1934 (N.A. D/T S6834/A).

tice in 1926. In 1930, the Cork Council of Women raised it with the president. The following year, the Carrigan Committee on the Criminal Law recommended inclusion of women in the police. In 1933, the League of Nations raised the issue. In 1936, the Joint Committee met the deputy commissioner of the Garda and pressed the case but the Minister for Justice refused requests to meet a delegation. Undaunted, the Joint Committee drew up a scheme and another delegation met the assistant commissioner in 1938. In 1939 the Joint Committee won the support of forty public bodies including Dublin Corporation, but the Taoiseach and Minister for Justice refused even to meet them to discuss the issue.[36]

In 1939, 1940, 1945, 1953, 1954 and 1955, the Joint Committee raised the matter again. In 1951, an official inquiry recommended the appointment of women police officers and in 1955 the issue reached the floor of the Dáil. What was envisaged initially was the recruitment of a small number of women officers who would handle mainly sensitive women's issues. Provison was finally made for women police officers in 1958 but on a lower rate of pay than their male counterparts and with compulsory retirement on marriage.[37] It is doubtful whether exclusion from the police caused many Irish women to emigrate but it is symptomatic of the unfriendly nature of Irish society and particularly the state apparatus as far as many women were concerned. At a time when the expectations of women were increasing, the state remained insensitive if not hostile.

While a majority of Irish women emigrants went into domestic service, it would be wrong to assume that they were all unskilled or uneducated. As the period progressed, a growing proportion enrolled as nurses in Britain. This reflects a trend towards their being better educated. Throughout the period, the educational qualifications of female emigrants were on average superior to those of their male counterparts. As the Land Commission inspectors testified, for many farmers' daughters, secondary school education was a significant catalyst in the decision to abandon the farming life. Such women might have aspired to employment in the civil service and related employment had conditions been more conducive.[38]

Lack of employment was a significant factor in female as in male emigration. In Northern Ireland where employment prospects were greater, the rate of female emigration was much lower.[39] For some women, however, it was as much the attitude to work which prompted emigration as its scarcity. As one female emigrant remembered: 'If a woman got married and then started to work in the 1930s, she'd be criticised all over the place. "God, she didn't make much of a match, he can't even keep her." It never dawned on anybody that a person might like to go out to work.'[40]

Another significant factor in the decision of many young women to emigrate was the late age and the low rate of marriage in Ireland.[41] Although the average

---

36 S.A. Roche to Mr Ó Muimhneacháin, 23 Nov. 1939; Memorandum by W.R. O'Hegarty, 1 Nov. 1939, Women in Garda Síochána, 1938–58 (NA, D/T S16210).
37 Minister's Brief for Dáil Question, 9 Nov. 1954; Department of Justice Memorandum, September 1957 (NA, D/T S16210).
38 *Economic Implications*, p.83; *CEOP*, p. 323.
39 *CEOP*, p.120.
40 Quoted in Mary Lennon, et al, *Across the water*, p.33.
41 For marriage in rural Ireland, see K.H. Connell, 'Peasant marriage in Ireland: its structure and development since the Famine' in *Economic History Review*, xiv (1962), pp 504–23.

age at marriage declined during this period, it remained high compared to other countries. In 1945–6, the average age of marriage for women was 28.0 for women and 33.1 for men. In England and Wales at the same time the average age for women was 26.6 and for men 29.7.[42]

The low rate of marriage which was one of the lowest in the world throughout the period was even more important. A side effect of the famine, late marriage persisted in Ireland until the 1960s. From an annual average rate of 5.10 per thousand population in the late 1860s, the marriage rate declined to 4.02 in the 1880s but increased again, reaching 5.11 in 1911–20. During the first decade of the Irish Free State, it declined to 4.76 but then increased gradually, reaching 5.59 in the decade spanning the Second World War before declining again during the 1950s. In 1936, one in four women remained finally unmarried as compared with one in eight in 1841. Irish marriage rates fell far short of comparable countries. In 1936, the number of married women under 45 in the Irish Free State was 25 percent lower than Northern Ireland and just half of the United States. By the 1950s, Ireland had both the lowest marriage rate and the highest celibacy rate in Europe with one woman in four and one man in three over the age of 55 unmarried. The Free State had the lowest marriage rate of twenty three countries investigated by the Commission on Emigration.

The low rate of marriage was a particular problem in rural areas where farmers and their sons showed a remarkable reluctance to marry. The typical marriage practices among farmers in rural Ireland were described by one observer in the 1940s as follows:

What invariably happens in the case of these people is the eldest son is retained at home say at 18 years of age to assist the father on the farm who is say now 50 years of age. The son's term of service, usually ending when the last member of the family has been done for, can be taken to be anything from 20 to 25 years. He is then anything from 40 to 45 years of age, a settled middle-aged man, careless about marriage and with ambitioin failing. If he does happen to marry he will probably choose a partner about his own age, oftentimes ending in a childless marriage.[43]

Table 2. Married Women Under 45 Years per 1,000 of Population in 1936[44]

Irish Free State	73	Italy	118
Finland	96	Canada	119
Northern Ireland	97	Australia	120
Norway	97	England & Wales	123
Scotland	105	South Africa	125
Sweden	108	Denmark	126
Portugal	110	Poland	133
Switzerland	110	France	139
Netherlands	115	Belgium	141
New Zealand	116	USA	145
Latvia	118		

42 R.C. Geary, 'Some reflections on Irish population questions' in *Studies* (Summer 1954), p.177. Geary was a member of the Commission on Emigration.

43 T.P. O'Cathán, Jan. 1944 (NA, S13413/1). In fact, the eldest son did not invariably inherit. And if he married, he was as likely to take a younger bride, thus skipping a generation.

44 Report of Inter-Departmental Committee on the Question of Making Available on Farms a Second Dwellinghouse, with a View to Removing Certain Difficulties in the Way of Early Marriage (hereafter Dower House Committee), Nov. 1944 (NA, D/T S13413/1, p.10).

The reluctance of farmers to marry was matched by a reluctance on the part of young women to marry into the farming life. As a series of agricultural inspectors testified in the 1940s, there was a growing impatience with backward social conditions which led many women to choose emigration as a form of escape. R. M. Duncan reported that there was a general distaste for the drudgery of life on small farms but even where the standard of comfort left nothing to be desired women were still unwilling to marry farmers. In the case of the better class of farmers, 'the girls have generally been sent to good secondary schools and will not contemplate becoming farmers' wives even where an adequate staff of servants is available'.[45] They were influenced by the lure of jobs in the towns where conditions were better and the added attractions included the cinema which provided a glimpse of a different kind of life. Again and again in these reports, the cinema is identified as a significant factor in luring women away from rural areas and increasing their expectations.

The low rate of marriage affected large farms even more than small. The larger the farm the greater the number of sons and daughters who remained unmarried. In 1936 for instance, 53 percent of all daughters in the age group 20–34 on farms 30 acres and under were single, whereas the figure was 57 percent for farms 30–50 acres or more, 60 percent for farms 50–200 acres and 62 percent for farms over 200 acres. T. P. O'Cathán, another agricultural inspector, concluded that there was 'something wrong with the natural sequence of marriage by the larger class of farmer'.[46]

The late age of marriage among larger farmers suggests that poverty and poor social conditions alone were not the only factors involved. Although it was declining, the dowry system was still widespread. It contributed to the problem by restricting choice and giving the parents undue influence.[47] Above all, as a government report on the subject concluded in the 1940s, 'the most important single factor causing the postponement of marriage among farmers is the unwillingness of the head of the household to retire from his occupation with advancing years.'[48] One rural clergyman described as 'debilitating and stagnatory' the effect of this state of affairs which led to frequent absurdities such as a twenty-five-year-old son having to ask for the price of a smoke or the price of a ticket to the cinema or a dance.[49] The inevitable frustration goes a long way towards explaining the high rate of emigration among daughters and non-inheriting sons.

Paradoxically, marriage was central to the maintenance of the Irish rural economy. The new wife provided labour and the heir to whom the farm would ultimately be transferred. As K. H. Connell remarked in 1962, 'marriage was likely to be contemplated not when a man needed a wife but when the land needed a woman'.[50] The deeply entrenched commitment to the maintenance of the family inheritance and the avoidance of subdivision proved stronger than the official state commitment to establish as many people as possible on the land. The latter policy was intended to end emigration whereas the former de-

---

45 R.M. Duncan to Land Commission, 1 Jan. 1944 (NA, D/T 13413/1). Duncan's impression was accurate. Daughters of large farmers, especially those with a dowry were more inclined to marry the salaried or professional classes (Dower House Report, p.12).

46 Dower House Report, p. 7; T.P. O'Cathán to the Land Commission, January 1944, *loc. cit.*

47 Rural Survey: Co. Clare (Marsh Papers, TCD, Ms 8301, SSI pp 16–17 and 30–1).

48 *Ibid.*, pp 11–12.

49 Rev. T.H. McFall, Piltown, County Waterford to Taoiseach, 4 Dec. 1943 (NA, D/T 13413/A).

50 K.H. Connell, 'Peasant marriage', p.503.

manded it. Emigrants, particularly female emigrants, were victims of both. Political rhetoric or official government policy never acquired sufficient force to counter-balance these central features of the rural economy.

In the absence of viable alternative employment locally, emigration was an inevitable result of large family size and inheritance practices. The size of the flow depended to a large extent on economic circumstances elsewhere. The economic depression of the 1930s and the popularity of protectionist policies kept emigration in check and helped de Valera pursue his economic experiment. But when, in the 1940s and 1950s, the floodgates opened, the crisis which developed threatened to submerge de Valera's vision *and* the rural economy.

Insofar as the extent of female emigration was the subject of public comment in these years, it was in the context of marriage and the moral welfare of the emigrants. His supporters warned de Valera that female emigration was 'a danger to the race' while his critics sought to make capital out of the issue. In January 1948, Daniel Morrissey TD told his constituents in a public letter that Ireland had 'failed to retain any percentage of the cream of young Ireland to build up the nation.'[51] In Morrissey's view, some emigration was inevitable and the loss of some emigrants was to be more lamented than others, not least if it threatened to deplete the 'brood stock'. Dr T. F. O'Higgins in Westmeath in August 1947 voiced a widespread fear: 'Could anybody', he asked, 'contemplate a more serious national situation...[with] a steady outward flow of young women so that many parishes have not a single young girl left.'[52]

By the 1950s it was widely accepted that drastic economic reform was necessary. It was also accepted that changes in marriage practices were necessary. However such changes were not easy to bring about. Successive governments between 1922 and 1958 considered piecemeal schemes aimed at improving social conditions and particularly at making life in rural Ireland more attractive for young women. One way of achieving that which was widely discussed in government circles for many years was the promotion of earlier marriages. In 1937, the Executive Council of the Irish Free State considered but rejected a scheme from a West Cork parish priest to introduce a marriage bounty as a means of encouraging young women to marry into small holdings.[53] It was argued that even if the scheme proved expensive to implement, the state would be getting good value as 'the young girls would be kept at home and would settle down with the young men, working the small holdings that were at present not cared for.' Similar proposals for what would in effect have been a state dowry surfaced periodically. In 1954, Robert K. Flynn of Chicago wrote to de Valera proposing a graduated scheme under which the younger the groom the larger the bounty, on the basis that 'the raging fires of young love were more productive than the flukey flames of advancing age.'[54]

By the 1950s, de Valera had grown increasingly disillusioned about the possibility of stemming emigration by changing marriage practices. His attitude had been soured by the failure of a pet project of his own, the provision of dower or second houses on farms to encourage the marriage of sons who normally postponed marriage until they inherited the farm on the death of their father. De

---

51 Joe O'Flanagan to de Valera, 12 June 1944, 7 July 1947 (NA, D/T 13413/A); *Irish Times*, 7, 12 July 1947; *Midland Tribune*, 10 Jan. 1948.
52 *Westmeath Independent*, 2 Aug. 1947.
53 'Small Farmers in West Cork: Economic Conditions', 23 Feb. 1937; 'Marriage Bounty', March 1937 (NA, D/T S9696, 9645).
54 Robert K Flynn to de Valera, 17 March 1954 (NA, D/T S9645).

Valera was convinced that there was a direct link between emigration and the late age of marriage. He also came to accept that the success of his party's policy of establishing as many families as possible on the land was dependent on persuading young women to marry into small farms. He raised the possibility of state subsidised dower houses in the late 1920s and again in the 1930s. In 1943, he was responsible for establishing an inter-departmental committee to prepare such a scheme but it did not get beyond the drawing board.[55] He reopened the issue again in 1947 and 1951.[56]

The failure of the dower house scheme despite the strong sponsorship of de Valera illustrates the way in which key elements of the rural economy outweighed government policy or the desire to end emigration. The main objection to the dower house proposal was that two houses on a holding might lead to subdivision of the holding. This was specifically forbidden in various land acts and ran counter to the policy pursued by the Land Commission almost since its foundation. When the 1943 committee sought the views of agricultural inspectors from different parts of the country as to the possible impact of such a scheme, this point was made repeatedly. F. W. Stock from Galway suggested that the provision of a dower house would be a 'direct and irresistible incentive to permanent subdivision of the holding'. For that reason it was 'too dangerous an experiment to be fostered by the State'.[57]

The hostile response to de Valera's attempts to promote a dower house scheme revealed that the maintenance intact of the family holding was of supreme value in rural Ireland, a priority outweighing all other considerations, including the fight against emigration and the preferences of the Taoiseach. J. F. Glynn, a land commissioner from Tipperary warned melodramatically that the scheme would 'strike at the foundations of family life as lived heretofore in Ireland.' Normal family life did not require providing a 'love nest' for the newly weds so that they 'could spend a protracted honeymoon billing and cooing until such times as they realised that they had also to earn a living.'[58]

Although the dower house proposal was unsuitable for small farms which could not support two families, it did help to highlight the frustration and friction which could be generated within extended families living in over-crowded and substandard accommodation in rural areas. M. J. Curley admitted that 'a normal married woman still able to work demands a home of her own.' A woman who chose to marry into a farm had to contend with enormous difficulties. There was liable to be 'continual friction due to the overlap of the spheres of two queen-bees confined to the same hive.' The balance in the relationship was more likely to be in favour of the older couple as they normally retained ownership of the holding rather than signing it over. Thus, as Curley admitted, 'the clash of wills can become almost intolerable particularly for those who were most enterprising.'[59] These were precisely the people who were likely to seek the escape offered by emigration.

The strong evidence presented in the reports of the Land Commission inspectors, of women in rural Ireland being frustrated and unhappy with their lot

---

55 *Irish Press*, 4 Dec. 1943; *Farmer's Gazette*, 11 Dec. 1943; D. Glavin to De Valera, 13 Dec. 1943 (NA, D/T S13413/A).
56 *Irish Press*, 5 July 1947; *Dáil Debates*, 2 July 1947, vol. 107, pp 608–9, 661–2, 735–9; *Sunday Press*, 4 March 1951; Report of Fianna Fáil Dower House Committee, 11 June 1953 (NA, D/T 13413/A).
57 F.W. Stock, 17 Jan. 1944 (NA, D/T 13413/A).
58 J.F. Glynn, 4 Jan. 1944, *loc.cit.*
59 M.J. Curley, Jan. 1944, *loc. cit.*

was supported a decade later by the Commission on Emigration and Other Population Problems established in 1948. This was the most ambitious and extensive attempt by the government of the Irish Free State to investigate the roots of emigration. Despite the fact that the majority of emigrants at the time were women, the Commission included only two women, a fact strongly condemned by the Irish Housewives' Association.[60] Although it did not produce any cure for emigration, the Commission did contribute to a more realistic understanding as to its causes and it did confirm that more women than men were emigrating.[61]

The experience of Irish female migrants in their new environments lies beyond the scope of this paper but it does provide some clue to their state of mind on leaving Ireland. Given their dissatisfaction with rural life, it is hardly surprising that female emigrants congregated in cities in even larger proportions than male emigrants. Only a small fraction settled in rural areas. It is ironic that such large numbers found themselves in domestic service in Britain and the United States, especially in the pre-war period, at a time when domestic service in Ireland was on the decline. For many, domestic service was the only employment available. However, given their reasons for emigrating in the first place, the common assumption that these women, or Irish female emigrants generally were docile, subservient or deferential and brought these attributes to their new societies is at least questionable.

The 1991 National Economic and Social Council investigation of the causes of contemporary Irish emigration concluded that gender alone is not a good indicator of a likelihood to emigrate but that it is a factor of some significance. That is certainly the case in Ireland between 1922 and 1960. There is little doubt that ethnicity, gender and class interact in the process of emigration. Irish women emigrated in relatively greater numbers than their European sisters. Poverty and employment were important factors. However, the high rate of emigration among the daughters of larger farmers points to the necessity for a broader approach. That approach must include gender as one of the key interacting causes.[62]

---

60 *Irish Press* 26 Apr. 1948. See NA, CAB S14/249A/1. For a more extensive account of the Commission, see P. Travers, 'Irish responses to emigration 1922–60'. For the Irish Housewives' Association, see Hilda Tweedy, *A link in the chain: the story of the Irish Housewives' Association 1942–1992* (Dublin, 1992).
61 *CEOP*, pp 150–1.
62 Pedraza, 'Women and migration', p.303; *Economic Implications*, p. 160.

17 Dark Lady of the Archives: Towards an Analysis of Women and Emigration to North America in Irish Folklore

Grace Neville

B ETWEEN 1885 and 1920, an estimated seven hundred thousand Irishwomen emigrated to the United States.[1] Who were they? Why did they leave? Where did they come from and where were they going? What impression did their new home make on them? How well did they settle there? With a few notable exceptions,[2] historians of Irish emigration have, to date, focused on male emigration and male experiences. Female emigrants, despite their arguably greater numbers, generally flicker like shadows at the edges of their perception; in Eavan Boland's telling phrase: 'they are, they have always been outside history. They keep their distance.'[3]

If, for a wide variety of reasons, the basic material needed by the historian of emigration – reliable statistics and census returns, first-hand accounts of the experiences to be studied – is less than plentiful in this area, then first-hand accounts by female Irish emigrants are rarer still. Thus, in this context, the 1.5 million pages that constitute the archives of the Irish Folklore Commission are an invaluable source of first-hand material.[4] These archives (housed in the Department of Folklore, University College, Dublin) consist largely of interviews collected in the 1930s and somewhat later by approximately two thousand collectors from forty thousand informants, overwhelmingly male and elderly, living throughout rural Ireland. The result is a veritable Aladdin's cave packed with an astonishing range of eclectic material including mythology, recent history, folk customs and beliefs, song and dance, accounts of folk habitations and furniture, sports and pastimes and memories of life long ago: a kind of Finnegan's Wake for the common man. As many of the informants were born shortly after the Great Famine, their memories and experiences date largely from the second half of the nineteenth and the early twentieth centuries. To date, much of the analysis carried out on this corpus has concentrated on the mythological content, although this is changing as more and more scholars realise its extraor-

---

1 Janet A. Nolan, *Ourselves alone: women's emigration from Ireland 1885–1920* (Lexington, 1989), p.100.
2 In addition to Janet Nolan's work, see Hasia Diner, *Erin's daughters in America: Irish immigrant women in the nineteenth century* (Baltimore, 1983); Pauline Jackson, 'Women in nineteenth century Irish emigration' in *International Migration Review*, xix (Winter, 1984), pp 1004–20; Ann Rossiter, 'Bringing the margins into the centre: a review of aspects of Irish women's emigration' in Seán Hutton and Paul Stewart (eds), *Ireland's histories: aspects of state, society and ideology* (London, 1991); Robert E. Kennedy, *The Irish: emigration, marriage and fertility* (Berkeley, California, 1973); Joseph J.Lee, *Ireland, 1912–1985* (Cambridge, 1989), pp 375–7.
3 Eavan Boland, *Outside history: selected poems, 1980–1990* (London, 1991), p. 50.
4 For a useful account of the function and contents of the archives of the Irish Folklore Commission, see, 'The Department of Irish Folklore and its archives' in *Celtic Cultures Newsletter*, v (December 1987), pp 28–32.

dinary potential as source material, for example in the area of women's studies.[5] The Irish Folklore Commission archive has not been used by historians of female Irish emigration, perhaps because most of it is in pre-standard Irish. This paper examines the experiences of Irish women emigrants as seen through the information in the archive. It emphasises gender specific aspects of emigration, leaving aside, as far as possible, aspects of the experience that were identical for both sexes.

Who left? Most female emigrants featured in the archives were young, single, unattached and often alone. The uniqueness of this pattern in the wider context of female emigration from Europe to North America has been stressed by historians and analysts elsewhere. Female emigrants from Ireland seem to have been younger on average than their male counterparts: around sixteen or seventeen ('as soon as she came fit') for girls and twenty for boys.[6]

Why did they leave? Because, according to the archives, there was little or no adequately paid work for them at home: 'what ever chance men had of staying at home at the Head and earning their living by fishing the women had none'[7] or because of some aspect of their marital situation: their husband or fiancé had emigrated some time before and had sent the passage money for them to follow on which sometimes meant that they left their children behind in Ireland.[8] Some were recorded as leaving in order to follow a man they fancied: 'Girls sometimes went after a fellow they had a wild strong notion of [in love with the boy]'.[9] Others left in order to avoid a man who fancied them – in some cases their husbands. The archives cite married women running away to the United States i.e. using America as a refuge, a step on the road to Irish-style divorce: 'Sam['s] wife was running away to America. Sam followed her and coaxed her back';[10] '. . . he and his wife had some differences and she ran from him to America. He followed her and searched the States upside down for her and advertised in the principal papers. No trace of her and no replies'.[11] Other women departed because they were pregnant or had had an illegitimate child and emigrated under pressure from close relations. The father of the illegitimate child might be willing to marry the mother in the United States though not in Ireland for fear of disgrace.[12] Other couples emigrated because they had married or were eloping despite their family's opposition. In Ireland, the girl would have no dowry and the man no accommodation. Life, they had decided, would be kinder in the United States than at home. Such couples often emigrated if they could afford the price of the crossing.[13] Some emigrated in order not to spoil the marriage chances of their elder sister who had not yet married and who, because she was the eldest, was expected to marry first.[14]

How did they leave? While the passage westwards was usually pre-paid by a relative or friend 'on the other side', we sometimes encounter women who put

---

5 See, for example, the work of folklorists Áine O'Connor, *Child murderess and dead child traditions* (Helsinki, 1992); Fionnuala Nic Suibhne, 'Cuntas ar ghnéithe de shaol an bhaineannaigh as insint bhéil fhaisnéiseoirí mná ó Chúige Uladh' (unpublished M.A. thesis University College, Cork, 1992).
6 Irish Folklore Commission Archive (hereafer IFC), 1411/391. See also IFC, 1411/20 for an example of a twelve-year-old female emigrant; and pp 187–99 above.
7 IFC, 1411/3, 206, 280, 391.
8 IFC, 1411/320, 374, 213.
9 IFC, 1411/396.
10 IFC, 1362/142.
11 IFC, 591/462.
12 IFC, 1411/396. See also 1411/377; 1403/101.
13 IFC, 160/228–9.
14 IFC, 1411/40–1.

together the price of the crossing themselves. Witness the frenetic determination of the woman featured in the following passage. The narrator recalls saving hay with her in North Cork:

As fast as I could pike it up she had it up and the súgáns on it [. . .] "Th'anam on diabhal," says I, "but you're a great girl." "Well," says she, "I have one a month here for nine months and as soon as that's up I'm going over the Western Ocean." Everyone that could scrape the money together at all them times used to go away to America.[15]

While we learn virtually nothing about the preparation given to male emigrants for their new life in the United States (because there was none?), we learn a little of the preparation afforded to girls. The accounts of women's lives in the United States to be gleaned from the archives are mostly negative and off-putting. They underline the Irishwoman's vulnerablility from the moment of arrival as well as the unhelpfulness and bias of the surrounding community. Could these scare tactics be construed as 'preparation', aimed at frightening women into staying at home, thus stemming the flood of female emigration or, at the very least, if they insisted on emigrating, at frightening them into 'minding themselves'? Religion, too, featured in the little preparation given to female emigrants. A priest saying Sunday mass before the long sea journey westwards: 'had a special word of encouragement for the young Irish girls and told them never to miss saying three Hail Marys every morning to Our Lady, and they would find that she would look after and protect them.'[16]

'American wakes' were held for most emigrants, male and female. Yet even at these rites of passage, male and female roles and experiences differed significantly, reflecting the separate spheres inhabited by men and women at the time as well as the different expectations made of them. Even the gifts that formed an integral part of these ceremonies were gender specific: men paid for and brought alcohol which they then proceeded to drink ('the men supplied the drink only'),[17] whereas women's role as nurturers was echoed in the somewhat more durable and altruistic gifts they provided: 'All the women who attended the Wake also brought something, a present for the emigrant, such as cakes, milk, or something like that [. . .] perhaps butter . . .'[18] '. . . In the times before that when the sailing ships were going the women of the parish would be busy baking oat bread for them to eat on the boat for at that time you had to take your barrell of stuff with you or go hungry.'[19]

Gender differences are reflected even in the beverages consumed at the wake: 'a barrell or half barrel of porter for the men and tea and wine for the women'.[20] The men entertained the onlookers with acrobatics or went to buy extra porter while the women were more restrained: they were the audience or they 'got tay ready'. They also acted as a brake on proceedings: late in the night, 'the girls were getting tired an' bate out'. This change of pace and mood heralded the moment of final leavetaking.[21] However, this active/passive split could be reversed at the dancing that took place during the wake:

15 IFC, 107/462.
16 IFC, 1430/226–7. Religion shades into superstition in the 'preparation' of the young women for their new lives as we read of love charms involving menstral blood and pubic hair (IFC, 1411/413).
17 IFC, 1430/216.
18 *Ibid*.
19 IFC, 1411/5. See also 1411/113.
20 IFC, 1430/228. See also 1430/217.
21 IFC, 514/63, 64.

very often the boys would stand in the middle of the floor and wouldn't ask the girls out to dance. Two men would go around and make the girls get out on the floor. Thus every girl was compelled to get out to dance. When she did go out on the floor, she could go to any of the boys she liked and dance with him. No girl was let sit down but had to dance. Apparently the boys were shy and wouldn't ask the girls to dance.[22]

Thus, the girls were made responsible for curing the boys' shyness! On a more serious note, one cannot help wondering what experience of socialisation with the opposite sex these young people had had and how they managed to cope with life in America if the boys could not even pluck up the courage to ask girls whom they had known all their lives out to dance.

The prominence given to women in the emigrant songs that marked these ceremonies is worth noting. The main characters in these maudlin songs ('Noreen Bawn', 'The maid of Coolmore ', 'My charming Kate O'Neill', 'Nora McShane', 'My lovely Irish rose', 'The maid of Galway town', 'Annie Laurie') tended to be women grieving because their lovers were emigrating without them or because they themselves, having emigrated and then returned, were now confronting sickness, heartache, tragedy and death. Cliché outdid cliché to produce 'the saddest songs they could think of', ballads which must have had a traumatic effect on people already in agony.[23] One tells of a broken-hearted girl begging a departing emigrant to carry shamrock to her only brother, now in America: the shamrock comes from the grave of their mother who died of a broken heart at the loss of her son and was watered lovingly by the girl's tears![24] Perhaps men's sorrow at leaving or being left could somehow be vicariously exorcised by such lamentations. Perhaps the paroxysms of anguish they must have produced could somehow numb the listeners and thus help anaesthetise the pain of the leavetaking. However, their depiction of women as eternal losers – victims of emigration whether they stayed or whether they left – must have provided cold comfort indeed for female listeners. As one perceptive female informant remarks: 'imagine if you were going away the next morning and hear a song like that wouldn't it put you out of your mind.'[25]

Although most of the features of the long journey westwards were experienced equally by both sexes, differences did exist. Men sometimes worked their way westwards by helping on the ship. I have encountered no instance of women doing likewise. Pregnancy, childbirth and even the death of a baby could mark the crossing. It was also believed that pregnant emigrants could be barred from entering the United States: 'the child would be a burden on the state'.[26] Unmarried motherhood, could, it was believed, be a fatal handicap for a would-be immigrant, even if the baby in question had already died and even without the incriminating evidence of a baby. The loss of her reputation would debar her from entry: 'girls suspected of having been pregnant were not allowed enter the United States'.[27] Fear of deportation must have added enormously to the stress of the situation especially since for many of these women, there literally was no going back. It was precisely to avoid the stigma of unmarried motherhood in Ireland that they had boarded the emigrant ship in the first place. Nor could

22 IFC, 1430/217.
23 IFC, 1411/7, 47, 74, 113, 137, 136, 220, 230; IFC, 514/61.
24 IFC, 1411/49.
25 IFC, 1411/51.
26 IFC, 1411/392. See also 1411/393–4, 414.
27 IFC, 1411/392.

bluff help in this instance. The New York immigration authorities could, it was believed, determine whether a woman was pregnant or not: 'I was told that girls were made to stand on ice and from that they [the immigration authorities] could tell someway whether the girl was in the family way or not.'[28]

On arrival in New York, another hurdle awaited immigrants, especially women. Here, immigrants were usually 'claimed' by a relative or friend.[29] The danger especially to women of having no one to claim them on arrival is repeatedly stressed. Alone in New York, they might disappear: 'especially girls were a great worry to their people for at that time people might lure them away after landing [in America] and that would be the last you would hear tell of them.'[30] The women, it was believed, could fall foul of slave traders, pimps, burglars, rapists or they might be drawn into prostitution.[31] This portrayal of the newly arrived women as passive and child-like, is strengthened, albeit unconsciously, in the following passage, by the use of passive verbs that define the woman as the inert object of male actions:

When my mother landed in New York her brother was there to claim her, but he had no letter from her and she had no papers, and he wouldn't be allowed to take her away: not till he went up the town and got two men to support him. Girls were being taken into the slave traffic then.[32]

One might be tempted to dispute these man-made impressions of utter female vulnerability. After all, these were young – often very young – women who had just successfully crossed the Atlantic for an unknown continent though many of them were unaccompanied and had never travelled anywhere before. In the case of a woman described as 'destitute when she got to the far side, but her husband's people with her gave her what kept her', we learn that her husband had died during the crossing and that: 'to fulfil the law his wife S[o]phia had to put her hand to his corpse before it was put overboard.'[33]

Whereas male immigrants are said to have gone into a wide range of jobs such as labourers, tradesmen, tram drivers, saloon managers, cowboys, peddlars, shipyard workers, miners, oil refinery workers, farm hands and even a turf-cutter in New York,[34] most women went into some form of domestic service: in hotels, restaurants or, overwhelmingly, as servants in private houses.[35] This was clearly an occupation in which most of them had served a long apprenticeship, either informally in their own homes or formally as domestics in the homes of others. So many Irish women worked in domestic service that 'Bridget' became the generic term for a female servant in the United States. For young, unattached, single Irishwomen, this occupation offered several advantages: familiarity as it mirrored aspects of the life they had known at home; security as it provided previously arranged board and lodgings and the opportunity to save most of their wages, as outgoings were negligible. Domestic service was not, however, always or even usually an enviable profession and anti-Irish bias was

---

28 IFC, 1411/392.
29 IFC, 1411/394.
30 IFC, 1411/43. See also 1411/253, 352, 356.
31 IFC, 1411/350–1, 353–6.
32 IFC, 1362/143. See also 1433/414.
33 IFC, 1365/83.
34 See G. Neville, 'Westward bound: emigration to North America in the Irish Folklore Commission Archives' in *Études Irlandaises*, xvii (June 1992), p. 204.
35 See IFC, 486/112–4 for a woman who worked in a paper mill and later a cotton mill.

rife there. One Irish housemaid was dismissed because: 'the people said they couldn't keep her because they didn't want the children to pick up her English.'[36] The harsh reality of domestic service for many Irishwomen in American homes was known to some in Ireland. Consequently, around 1900, the clergy began to rail against emigration, especially female emigration: 'They used to tell the women they'd have to work down in a kitchen underground, where they'd never see daylight.'[37]

The whole question of women and emigration touches, of course, not just female emigrants but the people left behind. In the archives which I have examined, daughters more than sons are lamented and missed by their parents. Of an American travelling in Ireland, we read: 'He was talking to an old Irishman one day. When the old fellow heard the tourist was from America, he asked him: "Would you know my daughter? She's out there this good many years"'[38] Elsewhere, we read of the decision of a widower of over eighty years to emigrate to America in order to be with his daughter. Such a decision is understandable given how invaluable a devoted daughter could be to ageing parents.[39]

Mothers more than fathers expressed grief at the loss of their children. At her child's departure: 'the mother would put her arms round their neck and sometimes hold on and on until somebody would take her away.'[40] One account graphically tells of a mother accompanying her daughter to the train station: 'She had never seen a train before. The train came into the station, blowing its horn, and the woman began cursing it: "Bad luck to you [. . .] you're taking me daughter away from me."'[41] Elsewhere we read: 'One old woman had three daughters, all of whom had gone to America, and the poor old woman was making herself ill with weeping and mourning for them.'[42] Another account tells of the desperation of yet another mother reduced to getting 'the old woman to "read the rake" [necromantic reading or predicting of future from hieroglyphics in fire-ashes] to see if he [her son] had ever landed at all.'[43]

Again, I have located more accounts of mothers than fathers accompanying offspring to the station or the port. Mothers could feel cut off not just from their offspring but from neighbours and peers who might have been expected to comfort them. Thus, we read of neighbours' indifference towards the mother of fourteen children faced with the departure of her eldest son for the United States. A neighbour, an old man, fears that frost might ruin his potatoes. The old woman's priorities are different: she reposts that this would be a lesser calamity than the departure of her eldest child for the United States.[44]

Other women affected by emigration include wives and fiancées left behind and women whose husbands (or fiancés) had emigrated first and who are left behind (sometimes indefinitely), awaiting the call to America: 'when a married man left he would not, as a rule, take his wife with him but when he got settled down and got a job on the other side he would send for her to come out.'[45] One

---

36 IFC, 1430/223.
37 IFC, 1430/220. See also pp 189–90 above.
38 IFC, 1430/240.
39 IFC, 1430/213.
40 IFC, 1411/224–5.
41 IFC, 1430/229.
42 IFC, 1430/213.
43 IFC, 1411/397.
44 IFC, 434/234–5; 1365/89; 1430/229.
45 IFC, 1411/3. See also 1411/21, 61, 149–50, 207, 319, 320, 374.

wonders how these emigration widows managed to cope with a whole variety of problems as they were left alone for upwards of a decade with their children in Ireland.[46] One woman, tired of waiting for over ten years, married someone else. Another woman was abandoned by her husband who fled to America since he felt tricked into marrying her. A pregnant woman was left by her husband who decided on impulse at a few hours notice to emigrate alone. Indeed, pregnant women were not infrequently abandoned as their men fled to America. Another woman was in the process of eloping only to be abandoned by her fiancé at the port of Larne. (She decided to emigrate accompanied instead by the man who had driven them to the port and who, with little or no notice, took her lover's ticket!)[47] One of the most poignant victims of female emigration recorded in the archives was, curiously, the second wife of a man whose first wife had abandoned him and gone to the United States. Though the man followed her and searched for her as best he could, he did not find her. He returned home and married again. Some time later, his first wife reappeared. The second family were then effectively ostracised by the local clergy in Limerick, and when the second wife was about to have one of her many babies, the priests forced her to go to a protestant nursing home. The narrator concludes ominously: 'the family is prospering but all say that "the curse is still on the place". We know not what that curse may be.'[48] At least in the case of female emigrants, one feels that they had some control over their destinies; the worst of all worlds was surely experienced by those mothers, wives and lovers left behind, drifting, helpless, their fate decided by others, usually men, who in many instances were not just absent but clearly, like Lot, had no intention of ever looking back.

It would appear fom the fragmented evidence to be gleaned from the archives that women more than men were interested in the fate of emigrants. As one woman said (about American wakes): 'they go back [. . .] longer than I remember or my mother or her mother.'[49] In so far as lifelines were thrown from Ireland to emigrants in the United States, this was usually ensured, it appears, by women who sent presents of food, knitted gifts and letters (sometimes dictated) to their emigrant children: 'it was nearly always the mother who wrote these letters and if she could not write some of the family or a neighbour wrote for her and put down exactly what she told them.'[50] An illiterate grandmother would oblige her grandson to answer her son's letters to her by return of post. In this context it is perhaps significant that 'American letters' seem to be addressed to mothers rather than to fathers. Indeed, in order to illustrate the formulaic nature of such writings, one informant says that they usually started with the salutation 'My Dear Mother'.[51] Interestingly, responsibility for communicating the contents of these letters far and wide throughout the community was taken by the emigrant's mother, to such an extent that: 'when night would come the house would be full to hear Johnny's letter [. . .] the woman had read the letter so often that she would know it off by heart [. . .] so well was she practised.'[52]

---

46 IFC, 1411/309.
47 IFC, 510/516, 1411/82, 376, 396; 1430/263; 1403/101. .
48 IFC, 591/462ff.
49 IFC, 1411/74.
50 IFC, 1411/240, 330; 1405/568.
51 IFC, 1411/139. See also 1411/241; 1430/231.
52 IFC, 1411/124–5. See also 1411/31, 53.

The archives abound with countless illustrations of the generosity and self-sacrifice of female emigrants towards their families back home:

I knew girls and they would have a cup on the side board in their boarding house and every week they would put a few cents into it until they would have so many dollars saved. When they would have five or six saved they would send it home. Then those at home thought that it was such a great country and that all in it had "lashings and lavings" of money.[53]

One woman paid the passage to the United States of 'five or six of her sister's children': as if she was not even counting. Another sent home the price of dowries for not one but two of her daughters as well as money to buy more land. Nor was such generosity reserved for relatives. One woman sent the fare to the United States to a female neighbour in order to rescue her from ill-treatment by her employer. Elsewhere, a woman described as a 'black [i.e. total] stranger' promised to send the passage money for another girl who was being badly-treated by her employers.[54] Of a daughter who emigrated, we read:

every month after that girl landed the pound note came to Margret Rid's [i.e. her mother]. Margret used to say: "We're getting a pound a month: we can put on the kettle now." Another girl emigrated: the monthly amount increased to two pounds. Others went. The family house was slated and general improvements made.[55]

Other emigrant daughters elsewhere 'raised [added another storey to] the parents' home'.[56] One cannot help thinking that for the families left behind, these emigrant women were regarded as endless sources of dollars and kudos to be milked for all they could get. Of a female emigrant we read: 'she never then after that forgot him'. This euphemism refers solely to the fact that she regularly sent home dollars to the person in question, as if the acts of remembering and forgetting were mere financial transactions, as if those who remained at home had an unspoken right to the hard-earned dollars of those who left.[57] Ironically, through their generosity, these selfless sisters and daughters continued to shore up the economic system that could not support them in the first place and which, without their lavish remittances, might have been forced into reforms that could have staunched the tide of subsequent female emigration.

Historians suggest that for most men and women, the decision to leave was prompted by a lack of work opportunities at home and the vision of the United States as a land where a bright future through plentiful, well-paid work was a real possibility. This was compounded, for women, by a realisation of the declining status of women in general and of single women in particular.[58] Demographic and economic changes in the post-Famine period as well as changing marriage and inheritance patterns, the rural exodus, changes in farm practices, rising expectations especially among women encouraged by their longer schooling than men and growing access to the outside world in the form of postal

---

53 IFC, 1411/142–3. See also 1411/33.
54 IFC, 1430/295, 213; 1411/160–1, 398–9; 1362/139.
55 IFC, 1411/399.
56 IFC, 1390/224.
57 IFC, 1411/345.
58 See *inter alia*, J.J. Lee, 'Women and the church since the Famine' in Margaret MacCurtain and Donncha Ó Corráin (eds), *Women in Irish society: the historical dimension* (Dublin, 1978), pp 37–45; David Fitzpatrick, 'The modernisation of the Irish female' in P. O'Flanagan et al (eds), *Rural Ireland 1600–1900: modernisation and change* (Cork, 1987), pp 162–80.

services, cheap magazines and even 'American letters' telling of fame and fortune beyond the waves, combined in different ways to convince women that emigration to the United States was the best and, in many cases, the only way they had to acceed to adulthood (viewed as financial independence, marriage and motherhood) rather than remain in prolonged childhood in Ireland, celibate, unemployed and dependent on a male relative.

In the archives, we are led to believe that men's motivation in emigrating was wide and varied, doubtless a reflection of what was seen as their multi-faceted personalities and social roles. They left in order to avoid the law, because they were political activists or criminals or for family and financial considerations. However, despite the wide and complex range of reasons for female emigration advanced by historians, those quoted in the archives are few. Unlike men, women were perceived to have emigrated because of some aspect of their marital status or potential as marriage partners. Above all many women were reported to have emigrated in order to put together a dowry – and returned as soon as this was done. This unanimity deserves further analysis. It is undoubtedly true that in the post-Famine period, dowries acquired unprecedented importance.[59] Since men usually dowried only one daughter or one sister, the others had to fend for themselves or remain single. Thus, in the archives, we are repeatedly told that once women had earned a dowry in the United States, their successful return home to a desirable lifestyle (and the status of married woman) was guaranteed:

Most of the girls went to America with the idea of making enough money to enable them to marry, because their chances of marrying without a fortune would be very slight. One must understand that in the rural parts of Ireland, even to this day, a woman has a very small chance of marriage, unless she has a fortune.[60]

Men intending to return to Ireland had to acquire the price of the purchase and maintenance of a farm. Even those who managed to do this were not always successful back in Ireland: 'some men also came home and bought land, and settled down. But very often those who bought land had not sufficient capital to keep it going and had to sell out and go back to America.'[61] Women, on the other hand, had to put together merely the price of a dowry (a smaller amount that could be earned in just a few years), in order to ensure their successful return. These financially independent women came home ostensibly on holidays or for some other reason but in reality to find a husband, or so the archives tell us. This mission was inevitably successful. The archives provide innumerable accounts of the marriages of these returned women. Of a woman who had been a maid in a doctor's house in New York for six years before returning home 'on holidays' we learn: 'She had made enough money to enable her to make a good match. She had literally made her fortune (in the sense of dowry) in the States and then came home to marry.'[62] Another woman returned home in bad health, on her doctor's recommendation. She afterwards married and settled down in the County Westmeath.[63] Some had even given a great deal of thought to their return: we read of girls who used to send home money to a

59 Nolan, *Ourselves alone*, p. 67.
60 IFC, 1430/205. See also 1430/222.
61 IFC, 1430/222.
62 IFC, 1430/201.
63 IFC, 1430/293.

local matchmaker who put it in trust for them for their return and eventual marriage. In other words, preparations for their marriage were already quite advanced: the trifling matter of their choice of husband could be dealt with later![64]

The effect which these returned emigrants and their as yet unattached dowries had on the home community are described in detail. Understandably, the men left behind were not indifferent to the 'fortunes' of these women who were now free to pick and choose a husband in a way that would have been quite out of the question had they not emigrated: 'The men at home were very eager to marry these returned Americans'; 'When [a woman] returned from America with a few hundred pounds or even a hundred, she had no difficulty whatever in getting a good match, and marrying into a good sized farm. She could pick and choose them.'[65] Significantly, in this passage, 'them' could refer to farms or husbands, for in reality these women married both. On the other hand, and with remarkable symmetry, returned male emigrants with American fortunes could now, exceptionally, aspire to the hands of the wealthiest of the local girls: 'The man who married a woman with a farm had either to have as big a farm himself, or its equivalent in cash. So many of the returned male emigrants brought back enough money to enable them to make a good match.'[66]

In other words, there was now a larger pool of potential husbands for local, high-status women. Some returned female emigrants are said to have enhanced their chances of making a good match by pretending to be richer than they were. Little wonder that these women, whatever size their fortunes, became known in some parts of the country as 'Redeemers'.[67] The marriage prospects of women who had not emigrated were weakened as a result of the presence of these 'Redeemers'. Even if their fathers or brothers could dowry them, they had to compete (for competition it was) with these returned emigrants, former friends and classmates perhaps, and their as yet floating dollars. This pitted woman against woman in a contest in which the prize was some local man. Local girls who had not emigrated and whose families could not dowry them now had the poorest marriage prospects of all, relegated to the back of the queue for husbands largely because of the emigration of others.

The subsequent fate of the dowries of returned female emigrants is worth noting. Passing out of the women's control, they were handed over to their future fathers-in-law and used to build up the wealth of the husband and his family.[68] They bought more land, renovated houses and, significantly, were end-lessly recycled as dowries for their sisters-in-law and other women: 'The eldest son of a farmer has to 'fortune' his sister with the fortune obtained with his wife.'[69] This money had originally liberated. It now ensnared. Earned often with difficulty and even heartache in a foreign country, it now went like the mailed remittances into propping up the patriarchy that had been unable to support the women in the first place and ensured that other women would consequently, like them, and for the same reasons, be forced onto the emigrant ship.

---

64 IFC, 1411/416.
65 IFC, 1430/220, 237. See also 1411/416.
66 IFC, 1430/206.
67 IFC, 1411/347–8; 1430/238.
68 For a discussion of peasant marriage involving remarkably similar dowry practices in late nineteenth century France see Pierre Bourdieu, 'Célibat et condition paysanne' in *Études Rurales*, v, vi (April, Sept. 1962), pp 32–135.
69 IFC, 1430/206.

Historians have shown that in the vast majority of cases, Irish emigrants, male and female, stayed in the United States.[70] One might also posit that most women stayed in America because life there held far more attractions for them than for their male counterparts. As one narrator states, remembering comments of returned emigrants: 'the reliable returned Yank said in effect that if men (not girls) worked as hard and attended to their business at home there would be no need to go to America. Girls they said could do much better in America than at home.'[71] Or as another put it more pithily: 'this [America] is a great country for women and dogs but no good for men and horses.'[72] The archives occasionally suggest that women were voting with their feet as chain emigration threatened to drain the country of its young women: 'my mother's name was Doherty and likely she would have gone away too only she got married.'[73] An old Donegal woman names her nine sisters who emigrated, like a litany for the dear departed. Another recalls her sisters' fury when she refused to join them in America. They clearly expected her to prefer life in America with them to life in Ireland with her fiancé.[74] Despite the favourable accounts of the emigrants' experiences, the overwhelming impression given by the archives was that female emigrants returned in great numbers: 'I never knew many men who came home to stay but I did know plenty of women.'[75] Returned emigrants were reported to be rushing to marry men they would have spurned prior to emigrating, bottlewashers, *scriosain*, ugly men, beggars, small farmers, even the owners of a 'humble house something next to a pig's *cró*', anyone at all, in fact.[76]

This contention that women returned willingly and in droves deserves further analysis. It is true that returned women emigrants occasionally remark, without explaining why, that they 'did not like America'.[77] The archives, however, were the work not of historians, anthropologists or statisticians but of mostly elderly people living throughout rural Ireland and reporting as witnesses on their personal experiences and memories. Thus, understandably, women who returned feature more prominently than those who did not. This may (partly) explain the mistaken impression that women returned in great numbers. There were, however, other reasons. These may relate to the informants' mental landscape, their *mentalité*, their self-image, and the vision they had of Irishmen, Irishwomen and of Ireland. Historians suggest that women left for a whole variety of reasons to do with the declining status of women in general and of single women in particular. Barring disasters, many of them clearly had no intention of ever returning to Ireland. In the archives, however, one sometimes suspects informants of recounting less what they had seen than what they wanted to see. Unlike men, women are depicted largely as one-dimensional characters with just one aim in life: to get married locally i.e. to someone not unlike the informant or the collector. The acquisition of a dowry thus 'explains' the crisscrossing of the Atlantic by these young women hungry to get married in order to avoid the leper-like status of spinster, of man-less woman. Furthermore, male collectors and informants may not have wished to acknowledge their part in

---

70 Nolan, *Ourselves alone*, p.71.
71 IFC, 1403/108.
72 I FC, 1411/52.
73 IFC, 1411/20.
74 IFC, 1411/39, 205–6.
75 IFC, 1411/162. See also 1411/37, 132, 183, 196, 237, 238.
76 IFC, 1411/346, 3, 37, 255, 237, 346, 162, 196.
77 IFC, 1411/96. See also 1411/93, 140.

the women's decision to leave. To say that these women were dowry-hunting is reassuring to men. It confirms their image of themselves as the centre of the universe; their view of single women as incomplete beings seeking through marriage and pocketfuls of dollars to buy their way in from the margins, their vision of Irishmen as the only mates these women could ever want and of Ireland as the only place in which these women would wish to live. The 'dowry theory' of emigration is also flattering to men. It places them effortlessly on a pedestal, the prize for the hardest (or, more accurately, richest) trier, preferable to the alternative suggestion that the women's decision had nothing to do with men in general or with them in particular or, worse still, that it had: that, consciously or not, it was precisely to find a different kind of mate from those available at home that these women had fled in the first place.

Irrespective of the marriage element, it is interesting to note that in the archives, apart from a handful of female emigrants who kept 'the old ways' going in the United States (e.g. a Rathlin Island woman who smoked – presumably pipes; and a woman who scolded her American-born child in Irish,[78]), most female emigrants are seen to be more anxious to become acculturated to the American way of life, more eager to be 'modern' than their male counterparts. Being modern and American meant, at one level, sounding modern i.e. speaking English. Thus, we learn of women who, as a first step, anglicised their name from McQuilkin to Wilkinson, and of a native Irish speaking woman who pretended to have no Irish after she emigrated to America.[79] It was not enough to sound American, however: one also had to look American, at least if one was a woman. Hence, the following advice from a returned female emigrant to the mother of a girl about to head westwards:

"Don't bother dressing her up, for her clothes will only be burned when she gets to the far side anyway." (Burned, I was told, because the style and fashion were entirely wrong for America).[80]

It is interesting to note that the offending clothes would not be simply discarded or recycled but burned/obliterated: there are hints here of some kind of purification ritual, the eradication/purging of the girl's past. Not only did these women sound and look American, they also acted American, at least according to one priest who berated Irishwomen in America for not attending mass.[81] It would appear from the same source that Irishmen were no more assiduous in their churchgoing than their female counterparts but they did not bear the full brunt of the cleric's displeasure.

Money was not all that they brought or sent back to Ireland. According to the archives: informants often criticise returned female emigrants for being 'uppity' ('great "swells" of girls'),[82] for no longer knowing their 'proper'/subservient place, having learnt different manners and risen 'above their station' in the United States.[83] While for some informants everything returned emigrants, male or female, did was wrong (everything about them grated: their perhaps unavoidably acquired foreign accents, clothes and manners), returned female emigrants were 'more wrong' still.

---

78 IFC, 1390/225.
79 IFC, 1390/224, 227; 53/188–9.
80 IFC, 1362/139.
81 IFC, 1105/176.
82 IFC, 1411/18.
83 IFC, 84/249–50.

They brought their version of modernity with them, too. Once back in Ireland, some of them pretended – to local disbelief – to have forgotten all their Irish (for them a symbol of backwardness) whereas their returned male emigrants had no such aversion to their mother tongue.[84] The love-hate relationship of the locals to these returned emigrants is crystallised in their attitudes to and interest in their American clothes. This was especially true in the case of local women looking at the clothes of returned female emigrants:

a woman of course would have more 'style' on coming home than a man and all the other women would manoeuvre into a good position at the chapel gate to insure [sic] that they would get a look at the Yankee. Then when they went home they would discuss her clothes from shoes to hat.[85]

Little wonder, therefore, that a departing son felt that the promise of the 'grand clothes [. . .] he would send her the minute he would land' in America might console his grieving mother.[86] Little by little, through American influence, local women began to favour capes rather than shawls[87], and indeed, emigrant women continued to make their 'new ways' felt from the other side of the Atlantic through the clothes parcels which they sent home.[88]

The archives show little sympathy for what are seen as the pretensions of these emigrants, especially female ones. One girl is lambasted by the narrator for ignoring him when their paths cross in America and for speaking English. He reminds her that he knew her when she used to pick potatoes off the rockface. In other words, she must never be allowed to forget her destitute youth, as if her former poverty were some kind of moral transgression to be eternally paraded before all, never forgiven and certainly never forgotten. Could it be that, in this as in other instances, the suspicion that the woman may have risen higher than the narrator and thus transgressed some 'natural' order is what rankles most? In any case, his criticism has the desired effect which is gleefully recalled: '*las sí le náire*' (she lit up with shame).[89] The frustration of informants who suspected that they had been left behind not just by their peers but by life in general may explain such expressions of begrudgery especially towards female emigrants.

The stereo-typed vision of women held by informants and collectors in the archives is, of course, as old as time and frequently self-contradictory. Women emigrants are vulnerable and in need of male protection despite plentiful evidence that they could fend for themselves in an often hostile and even dangerous environment, escaping through sheer ingenuity from burglars and potential murderers, not to mention wreaking vengeance on racist employers, something most of the (male) informants had never done. Women are wily, resourceful and deceitful especially in money matters. They live on their wits, even resorting occasionally to stealing from their employers. Their only aim in life is to get married. This is why they emigrated and why they later came back.[90]

Many of the jokes and humorous stories in the archives are told at the expense of women by men. (As previously noted, most of the collectors and in-

---

84 IFC, 1219/60.
85 IFC, 1411/190. See also 1411/161. 296.
86 IFC, 1411/28.
87 IFC, 1411/82.
88 IFC, 1403/107.
89 IFC, 53/188–9.
90 IFC, 1411/355, 251–2, 352, 408; IFC 179/146–50.

formants in the archives were men, men talking to other men.) Added to that is the stereo-type in the informants' minds of the Irish as dirty, lying and stupid as well as occasionally vulgar.[91] One of the many ironies here is the fact that the informants themselves are Irish. Hence the suggestion that many female emigrants were dirty, stupid and faintly ridiculous. To be Irish and female is thus to live at the intersection of the worst of all worlds, at the bottom of all social ladders. The prejudice of the informants is also infused with the traditional dual standard concerning men's and women's behaviour. Thus, shock which is expressed at the discovery of a female Irish prostitute in New York is unconsciously juxtaposed with the matter-of-fact announcement that her brother frequented brothels. Likewise, we are encouraged elsewhere to disapprove of a returned female emigrant who had pretended to be richer than she was in order to ensnare a husband, but the local man thus lured into marrying her in the hope of striking it rich earns the narrator's pity rather than his criticism.[92]

Despite all the tales of leaving, of lives uprooted and disjointed, of hopes realised and dashed, one is struck above all in this vast opus quite simply by the silence, the great silence surrounding these female emigrants, all hundreds of thousands of them. While both in the archives and in later historical analyses, they are said to have emigrated in greater numbers than men, the least one could say is that in this vast collection they do not hog the limelight. Most of them are eclipsed by their menfolk. If, as is generally acknowledged, naming is a defining act, it is significant that many of these women are nameless, being referred to as their brother's sister, their father's daughter or their husband's wife. The term 'married people', one discovers, means 'married men' , even when the informant is a woman. Of a woman who emigrated after a bizarre incident remembered in detail, the narrator adds: 'I never heard her name.'[93] The female composers of emigrant songs remain nameless though their songs are remembered verbatim.[94] Accounts of leave-takings emphasise drinking and even hooliganism: male rather than female activities. Men's jobs, men's lives in the United States are described in more detail than women's. Even when an Irishwoman succeeds in America, as in the case of an artist, her name still does not warrant a mention, unlike that of her arguably less successful father whose name and occupation are given.[95] If, as the archives proclaim, the women returned in droves why are they not better recorded in the archives? And as for the women in the lives of the many returned male emigrants whose testimony fills these pages, it is often a case of *cherchez la femme.* Their presence is sensed rather than seen. One narrator cannot even remember whether a particular man took his wife with him when he emigrated.[96] Of a man who sent for his wife a year or two after leaving, we read: 'he made a good bit of money in America but he came home again and stayed till he died.'[97] What happened to his wife? One informant simply mentions, in passing, without further comment, the five daughters and five sons he left behind on emigrating.[98] Over four hundred pages on emigration were elicited by a series of thirteen questions none of

---

91 IFC, 1411/408; 696/369–70; 1430/239–40.
92 IFC, 1411/347–8.
93 IFC, 1390/231.
94 IFC, 72/39–41; 85/210–2.
95 IFC, 1390/221.
96 IFC, 1411/62.
97 IFC, 1411/61.
98 IFC, 1105/178.

which, however, have any gender-specific slant.[99] It would be too easy to blame the collectors for asking the 'wrong' questions and for concentrating on the 'wrong' issues. Sensibilities change over the decades and the archives are, inevitably, a product of their time. Perhaps in the conventions existing when these archives were compiled, the correct questions were asked.

One regrets that more time was not devoted to these women and to women generally in the archives not least because many of them were such striking and accurate observers of the world around them. The culture shock that all emigrants must have experienced not just on Ellis Island but so much sooner is vividly expressed by an Irish-speaking Kerry woman. A whole lifetime after the event in question, she still remembered her astonishment as a young emigrant on seeing oranges for the first time not in the United States but in the port of Queenstown, in County Cork.[100] An eighty-three year old woman remembers vividly her thrill over sixty years earlier when, on the voyage westwards, she was able to dip into a barrel of delicious crackers any time she wished.[101] The understandable jealousy felt by so many of those who never got as far as Ellis Island is expressed by a Cavan woman who, on first sighting a local house built with American money, quipped: 'See what Brooklyn money has done [. . .] it has built the tower of Babylon.'[102]

Women's first-hand testimony, relatively rare though it may be, is shot through with such colour and sharpness that one longs to interview them again and to elicit different information from them this time. Thus, in this as in other areas, further analysis of the vast and unique stream of consciousness that constitutes the archives of the Irish Folklore Commission could help to move women back from the shadows, from the gaps between the lines, into history.[103]

99 IFC, 1411.
100 IFC, 486/105–6.
101 IFC, 1411/89.
102 IFC, 1430/233.
103 I would like to express my gratitude to Professor Bo Almqvist and the Irish Folklore Commission, University College, Dublin for permission to cite material from the archives. I also wish to thank Elizabeth Steiner-Scott for her assistance.

18 'Fanaticism and Excess' or 'the Defence of Just Causes': the International Labour Organisation and Women's Protective Legislation in the Inter-war Years

Mary E. Daly

IN 1923, under the heading, 'Women at Odds on Industrial Equality Plans' U.S. feminist Crystal Eastman wrote that protective legislation, controlling women's working conditions was 'the issue on which (I believe), feminists mainly will divide in the next decade or two'.[1] Eastman was reporting on the 9th World Congress of the International Women's Suffrage Assocation, in Rome where a debate on equal pay and the right to work had exposed serious divisions between delegates who favoured absolute equality in the workplace and those who believed that women were in need of legal protection specific to their gender. The result was, what Eastman described as a 'not very successful compromise' resolution.

That no special regulations for women's work, different from regulations for men, should be imposed contrary to the wishes of the women concerned; that laws relating to women, as mothers, should be so framed as not to handicap them in their economic position, and that all future labour regulations should tend towards equality of men and women.[2]

Eastman's claim that protective labour legislation would prove a divisive issue in coming years was broadly correct. In the U.S. it handicapped efforts to draft an Equal Rights Amendment and split the ranks of American feminists, with the National Women's Party opting for equality and opposing sex-based protective legislation whereas groups such as the Women's Trade Union League and the Women's Bureau – which had pioneered research into the working conditions of American women – favoured protection.[3] In Britain the equal rights lobby was led by the Open Door Council, frequently dismissed as merely a group of middle and upper-class women, out of touch with the realities of working-class life; women's organisations linked with the trade union movement such as the Permanent Mixed Committee of Women's Industrial Organisations and individuals such as Labour M.P. Margaret Bondfield, Britain's first woman cabinet minister upheld protective legislation.[4]

Following the divisive debate in Rome, the equal rights faction prepared a

---

1 Unidentified press cutting (Schlesinger Library, Harvard University (hereafter SL), WRC M-93 (2), p. 207).
2 *Ibid.*
3 Nancy F. Cott, *The grounding of modern feminism* (New Haven, 1987), pp 117–142.
4 Martin Pugh, *Women and the women's movement in Britain, 1914–1959* (London, 1992), pp 99, 29.

draft report for submission to the 1926 Congress of the International Women's Suffrage Alliance in Paris[5] where a motion favouring equal rights was carried. The issue of protective legislation also surfaced at meetings of the International Council of Women, the 1928 Pan-Pacific Women's Congress and other gatherings too numerous to mention.[6] Between 1923 and the early 1930s few international women's congresses, irrespective of their primary purpose, met without debating the question of gender-specific protective legislation.

This preoccupation reflects a refocusing of the feminist political agenda, with suffrage receding from its hitherto dominant place and forcing the women's movement to confront a changing role in post-war society. Protective legislation also came to the fore because of the prominence assigned it by the International Labour Organisation, whose inaugural conference in Washington in 1920 included a draft convention prohibiting women from working at night, or before and after childbirth or in 'unhealthy processes'.[7] While the prohibition relating to childbirth aroused no controversy, some delegates argued that a ban on working in 'unhealthy processes' should apply to both sexes. The proposal to ban women from night-work was to prove the most controversial of the organisation's conventions. Successive directors' reports devote considerable attention to the matter, with opponents of the Convention described as 'extremists in feminist organisations', motivated by 'fanaticism and excess',[8] language not generally associated with the annual reports of international organisations. The ILO headquarters in Geneva became the focus for considerable lobbying by women's groups, with the Open Door Council – which expanded to become an international organisation in 1929[9] – opening a Geneva office for this purpose.

The ILO offered an influential platform for debating the economic role of women. By 1929 it had fifty-five member states, and being open to non-members of the League of Nations, these included both the U.S.A. and defeated belligerents in World War One.[10] Although legislation banning night-work for women was already on the statute books in many countries, its inclusion as one of the first ILO Conventions gave the prohibition added significance.

The ILO's preoccupation with protecting or restricting women's working conditions reflects the organisation's links with the trade union movement; it was established in response to their insistence that international economic reform was essential to the maintenance of long-term peace in the aftermath of World War One. In November 1914, several years before the U.S. entered the war, Samuel Gompers, general secretary of the AFL (American Federation of Labor), proposed that an international labour meeting be held simultaneously with a post-war peace conference with the 'aim of protecting the interests of the toilers and thereby assisting in laying foundations of a more lasting peace'.[11] The positive response to this proposal after the Armistice, and to similar ideas emanating from French and British trade unionists reflects the enhanced status of the trade union movement as a result of their involvement in the manage-

---

5 International Woman Suffrage Alliance. Committee for like Conditions of Work for Men and Women (SL, A-1116, Box 12, 221).
6 See *International Labour Office Director's Report 1927*.
7 George N. Barnes, *History of the International Labour Office* (London, 1926), pp 58–9.
8 *ILO Director's Reports, 1927, 1930*.
9 Else Luders, 'The effects of German labour legislation on employment possibilities for women' in *International Labour Review*, Sept. 1929.
10 Lewis L. Lorwin, *The International Labour Movement. History, policies and outlook* (New York, 1953), p. 128.
11 Quoted in John N. Horne, *Labour at war. France and Britain, 1914–1918* (Oxford, 1991), pp 308–9.

ment of wartime economies. In both the U.K. and France it also reflected a belief that the introduction and enforcement of international norms for working conditions, together with some mechanism for labour market regulation, were essential, if hard-won advances in wages and working conditions were not to be undermined by competition from low cost workers in other countries.[12] In February 1919 a Commission for International Labour Legislation met in Paris, parallel with the peace conference, to draw up proposals for an international labour organisation and for guideline conventions.[13] Chaired by Samuel Gompers of the AFL with major inputs from trade unionists and officials from the ministries of Labour in Britain and France, the quasi-corporatist structure which was proposed for the new organisation, with national delegations representing government, employers and trade unions reflected the dramatic growth of such contacts in wartime Britain and France.[14] The Commission followed in the footsteps of previous international labour conferences, including the Congress of Berlin in 1890 and the Berne Conference of 1906[15] by devoting considerable attention to protective legislation affecting women and children. Areas listed as needing attention in the preamble to the Commission's report were 'the regulation of hours of work, prevention of unemployment, the provision of a living wage, the protection of the worker against sickness, disease and injury arising out of his employment, the protection of child and female labour, provision for old age and injury', protection of the interests of migrant workers and freedom of association for workers.[16]

II

The long-standing interest shown by trade unionists in measures designed to protect women's working conditions has been viewed as evidence of patriarchy: a desire to restore male authority in the household which had been undermined by the Industrial Revolution, though an alternative point of view sees it as a cohesive effort on the part of the working-class family to improve overall living conditions.[17] The widespread support shown by international labour groups can be read both as an indication that the separate spheres ideal was widely upheld throughout Europe and as evidence that protective legislation applying to women and children was one of the few topics on which workers from different countries could agree, because it entailed no concessions on the part of the male trade unionists involved. The 1906 Berne conference produced only two draft conventions: one prohibiting the use of white phosphorous in the production of matches – which provoked considerable controversy – and a widely-accepted ban on women working at night.[18] Thus conventions regulat-

---

12 Horne, *Labour at war*, p. 322.
13 James T. Shotwell, (ed.), *The origins of the International Labour Organization* (2 vols, New York, 1934), i, pp 84–198.
14 Horne, *op. cit.*, pp 267–84.
15 For details of these and many other international labour conferences see John W. Follows, *Antecedents of the International Labour Organization* (Oxford, 1951), pp 120–139 and 162–169.
16 Shotwell, *The origins of the International Labour Organisation*, i, p. 132.
17 Heidi Hartman, 'Capitalism, patriarchy and job segregation by sex' in M Blaxall and B Regan (eds), *Women and the workplace* (New York, 1976); Jane Humphries, 'Protective legislation, the capitalist state and working class men: the case of the 1842 Mines Regulation Act' in *Feminist Review*, 7 (Spring 1981), pp 1–33.
18 Ernest Mahaim, 'The historical and social importance of international labor legislation' in Shotwell (ed.), *op. cit.*, i, p. 10. and Lorwin, *International Labour Movement*, p. 164.

ing women's work served to conceal a lack of trade-union action on other international issues; women (and children) became code words for threats to high pay and working-class respectability, agreement on measures to control their working hours became a token of international solidarity which often concealed wider divisions. By the late nineteenth century working-class respectability was grounded in a commitment to separate spheres, with men earning adequate wages to support a family and women concentrating on domestic duties. There is no reason to assume that such attitudes had changed by 1920. On the contrary the disruption of the war years and the influx of women into traditional male occupations such as engineering plants and transport services may have increased the sense of insecurity among working men and reinforced their commitment to restore traditional gender roles. Restrictions on women working at night had been lifted during the war years in order to maximise labour resources and facilitate the redeployment of women into non-traditional occupations; their reimposition and strengthening, as in the 1920 British Women and Young Persons Act which banned night work by women and younger workers signalled a return to the past. Women's war work in non-traditional occupations had been monitored for evidence of ill-health or other adverse affects; the report of the *Health of Munitions Workers Committee*, which referred to the unfavourable effect of long hours on workers was cited in British trade union circles as justifying an eight-hour day for all workers and a return to a pre-war ban on night-work for women.[19] However the scientific basis of this and other reports concerning the impact of work on women's health is highly dubious: studies claiming fatigue among female munitions workers fail to examine a control group of housewives, charwomen or women in traditional employment. In 1919, evidence submitted to a government inquiry concluded that there was no apparent evidence of injury to the health of women or children who had worked in wartime employment.[20]

III

The founding fathers of the ILO however shared the wider trade union viewpoint that working women were in need of protection. Trade unionists dominated the Labour Commission which determined the functions of the ILO; the first director, Albert Thomas, was a former leader of the French socialist trade union, the SFIO and wartime Minister for Munitions; trade unionists had lobbied for his appointment. Thomas, who exercised virtually complete control over the hiring of ILO personnel in turn recruited many of the early staff from the ranks of his former associates.[21] His trade union background would invariably have conditioned Thomas' attitude towards women in the workforce; as wartime Minister for Munitions, he had been concerned to placate the concerns expressed by French workers' representatives. Many of these focused on the influx of women into traditional male employment. Although explicit statements of Thomas's attitude towards working women for these years are not to

19 Irene Osgood Andrews and Margarett A Hobbs, *Economic effects of the world war upon women and children in Great Britain* (Oxford, 1921), p. 8.
20 *Report of War Cabinet Committee on women in Industry* 1919 xxxi, cd. 135. Memo by Dr Janet Campbell.
21 Martin Fine, 'Albert Thomas: A reformer's vision of modernization, 1914–22' in *Journal of Contemporary History* 12 (1977), pp 545–564.

hand, they are unlikely to have diverged radically from those of the French socialist movement which was strongly committed to the ideal of separate spheres.[22]

There is no evidence of any specific female contribution to earlier international labour conferences. In a footnote to his account of the 1906 Berne Convention, published in 1935, Ernest Mahaim, Belgian academic, labour expert and a founding father of the ILO noted (with apparent relief) that 'the propaganda of extreme feminism had not at that time reached the pitch which it has today'.[23] In March 1919 however a delegation representing several international women's organisations, including the International Council of Women presented its demands to the International Labour Commission. The joint manifesto started from the principle of equality and demanded equal pay, equality of employment opportunities, the suppression of night work where possible without discriminating against women and the representation of women in all discussions concerning the organisation of work and specifically regarding matters pertaining to women's work. A subsequent submission from the International Council of Working Women reiterated the demand for equal treatment and suggested that gender-specific working restrictions should only be imposed following consultation with women workers on the advice of women professionals and only in cases where alternative remedies such as improved factory safety conditions were deemed futile;[24] this however was bitterly opposed by Mary Anderson, head of the U.S. Department of Labor's Women's Bureau,[25] a foretaste of divisions to come.

The response of the International Labour Commission to these submissions was one of compromise: the draft articles establishing the ILO specifically stated that women could serve as delegates on equal terms with men and an amendment to the original articles, which was adopted, provided that when questions specifically affecting women were being considered at ILO Conferences at least one of the International Labour Office's advisers should be a woman. They also promised to appoint an unspecified percentage of women to the ILO staff.[26] Proposals for a convention limiting the working hours of women were dropped in favour of one advocating an eight-hour day for all workers;[27] however although a commitment in principle to equal pay for men and women was included in article 427 of the treaty,[28] it was not among the draft conventions considered or adopted at the Washington conference in the autumn of 1919. These were limited to restrictions on female and child employment; the eight-hour day and the adoption of measures dealing with unemployment.

ILO procedures provided for draft conventions to be ratified by a delegate Conference; following such approval, they should then be adopted by individual member states. By 1925 although the Washington Convention prohibiting night work for women had been ratified by fifteen countries, it was facing increasing opposition. In response the International Labour Office adopted a two-pronged

22 James F. McMillan, *Housewife or harlot. The place of women in French society, 1870–1940* (New York, 1981), pp 13–16.
23 Ernest Mahaim, 'The historical and social importance of international labor legislation' in Shotwell (ed.), *op. cit.*, i, p. 10.
24 Ki-Tcheng, *La femme et la Societe des Nations* (Paris, 1928), pp 106–110.
25 Alice Kessler-Harris, *Out to work. A history of wage-earning women in the United States* (Oxford, 1982), p. 207.
26 Shotwell, *op. cit.*, i, p. 140; Ki-Tcheng, *La Femme*, pp 110–111.
27 Shotwell, *op. cit.*, i, p. 178.
28 *ILO Director's Report, 1929.*

approach: actively attacking its critics and lobbying for support from sympathetic women's organisations. In 1926 a woman member of the ILO staff was dispatched to attend the quinquennial assembly of the International Council of Women in Washington and a liaison committee was established in Geneva to maintain permanent links with the International Council of Women; formal links were also established with the International Women's TUC.[29] Groups who opposed the Convention were dismissed, in successive annual reports as extremists who represented 'the more exclusive section of feminists'[30] rather than the rank and file of working women; the 1929 report specifically welcomed the defeat of the ERA amendment in the U.S. while lamenting the apparently growing strength of 'extremists' in Britain in the guise of the Open Door Council. Passages advocating the Convention invariably spoke of women workers as fragile victims; the 1926 report spoke of 'sentiments of sympathy, of pity, of human solidarity which presided at the origin of the movement for legal protection'; the 1929 report stated bluntly: 'There are many women at present working in factories; they require protection'. Determined efforts were made to emphasise class divisions within the women's movement: middle-class advocates of equality were contrasted with protection-seeking working women; demands from women for entry to the professions on equal terms with men were juxtaposed with the 'hardship and sufferings of industrial women workers'. Assertions were made that women suffered health problems specific to their sex as a result of the combined effects of working conditions and maternity; however, despite cursory references to statistics none were cited to support this argument.

The ILO repeatedly cited trade union support for protective legislation, together with statements suggesting that the Washington Convention provided much less protection for women workers than women trade unionists demanded. The organisation appears to have made a conscious effort to avoid the controversy being viewed as a male/female confrontation. Several annual reports relied on the manifesto favouring protection for women, published in Britain by the Permanent Mixed Committee of Women's Industrial Organisations, a body claiming to represent the views of over 1 million working women, to justify the ILO stance; statements from female trade unionists supporting the Convention were cited. The British manifesto reiterated the long-standing views of the (male) trade union movement that the improved working conditions which male workers derived from trade union organisation could only be obtained by women workers via legislation; it further argued that women's employment was in need of official regulation because women were less fitted to carrying out certain dangerous tasks. The enforcement of protective legislation limited to women was seen as paving the way for its extension to men.[31] Most intriguing was the resort among advocates of protection to the traditional (and accurate argument) that working women carried a double burden of paid work and domestic duties; in this case however night work, rather than work *per se* constituted the double burden – as if the domestic duties of women who engaged in paid employment during the day vanished by magic.[32] In all statements supporting the

---

29 *ILO Director's Report, 1926.*
30 *ILO Director's Report, 1929.*
31 *ILO Director's Report, 1927, 1928.*
32 E.g. Proceedings of International Labour Conference, 17 June 1930: contribution of Miss Mohrke, German Workers' Advisor.

Convention, regardless of their origin, women workers were invariably viewed as housekeepers, wives and mothers, never as individuals: there were repeated claims that the lack of rest resulting from night work upset household routine and injured children. Although this argument would seem equally applicable to men, this was never stated; instead it was emphasised that unlike men, women were unable to resist industrial fatigue 'as they cannot be freed from household duties'. If women's work challenged the idealised family structure; women's night work posed a double challenge – even if we ignore the fact that it might facilitate female access to traditional male jobs and could be regarded as posing certain sexual dangers. Working a night-shift was obviously not in keeping with the role model outlined by one French trade unionist in 1917: a woman should stay at home though she could go out with her *compagnon* when he has finished his day's work.[33]

The debate at the ILO in the early 1930s on a proposed revision of the Washington Convention further underlined the importance attached to woman's domestic role. Belgian textile firms sought to alter the hours when women's night-work was banned, from the existing 10 p.m to 5 a.m to 11 p.m. to 6 a.m. Opposing this motion, Gertrud Hanna a German delegate, argued that

A great number of women, whether married or not, have to share in household duties particularly in the morning. They have to help their husbands or their younger brothers or sisters or their children to get ready for work; so that the change does not mean that they will have an hour extra for rest but would have to get up at the same time and would therefore lose an hour's rest.[34]

Opposition to protective legislation which singled out women was led by the Open Door Council, though the equal-rights viewpoint also received strong support from the International Alliance of Women for Suffrage and Equal Citizenship. The campaign of equal-rights feminists escalated following the ILO's adoption in 1928 of a convention on Wage-Fixing Machinery – to enforce mininum wages – which excluded provisions for equal pay. The Open Door Council contrasted ILO efforts to prohibit night work by women with its lack of commitment to equal pay. The International Alliance of Women for Suffrage and Equal Citizenship also protested at the ILO's separation of the issue of minimum wages from that of equal pay. Feminist beliefs that the director's attitudes owed more to ideology than to scientific evidence was a major cause of grievance. Miss Hesselgren, a Swedish government representative to the 1929 conference protested at the ILO's automatic assumption that women workers were in need of protective legislation and urged that the organisation commission 'an unbiassed investigation as to how far women still need special regulation'.[35] In June 1931 Mrs Eugene Boschet, a workers' representative from Austria voiced the need for precise information about the impact of both protective legislation and social insurance regulations on women's employment and unemployment, particularly given the deepening economic crisis.[36] Faced with ILO proposals to restrict the weights which women should be permitted to lift in the course of employment, a further extension of sex-based legislation, the Inter-

---

33 Cited in J. L. Robert, 'La CGT et la famille ouvriere, 1914–18, premiere approche' in *Le mouvement social* (July–Sept. 1981), p. 59.
34 International Labour Conference, 15 June 1931.
35 International Labour Conference, 12 June 1929.
36 International Labour Conference, 6 June 1931.

national Women's Medical Association, argued for controls based not on sex, but on physique.[37] The equal rights camp consistently dismissed the protectionist case as one owing more to tradition than to science: they called for research to be carried out on the actual impact of night-work on women's health and on the family as a substitute for generalised assertions. Others questioned whether the views expressed at annual ILO Conferences were representative of the voices of working women. Mrs Eugene Wasniewska, advisor to the Polish Workers' Delegation and an opponent of the Night-Work Convention protested at the paucity of women delegates, the failure to appoint a woman president (the office rotated annually) and the absence of female representation from many committees.[38]

Equal rights opposition, which rested on the dual issues of the need for research into women's working conditions and demands for adequate representation of their point of view came to a head in 1931 when the original Convention became due for amendment. Proposed amendments related to detail rather than substance: ending the exclusion from night work for women holding supervisory or management positions and an alteration in the hours covered by the Convention at the behest of the Belgian textile industry. While neither modification posed a fundamental challenge to the Convention, they were initially opposed by the International Labour Office, who feared, according to one Irish official, that any revision, however minor, would open the door to further alterations and 'the demands for absolute equality between men and women might be further advanced'.[39] The secretariat successfully thwarted the efforts of Mrs Wasniewska to use the occasion to press for the establishment of an advisory committee – with women members included – to examine the impact of protective legislation on women's paid work. The chairman ruled against the tabling of her motion, deeming the matter not to be urgent. A similar resolution tabled by a male workers' delegate from Chile M. Arancibia was also excluded by the director, Albert Thomas. When Arancibia subsequently arranged for Thomas to receive a delegation from women's groups opposed to the Convention (consisting of Mrs Pethick-Lawrence of Open Door International; Madaleine Doty of the Women's International League for Peace and Freedom together with Alice Paul and Marta Vergara of the Inter-American Commission of Women), Thomas terminated the meeting abruptly before any of the women had spoken.[40]

Despite these repulses, opposition to the Convention appeared to be having some effect by the autumn of 1931. Mrs Wasniewska resubmitted her resolution for consideration at the 1932 conference, coupling protective legislation with the problem of falling female employment, the impact of rationalisation and with allegations that the exclusion of women from certain jobs was driving them to prostitution. Despite the fact that she cited no evidence in support of the latter assertion, – a practice for which she and other women had criticised the director – this was an emotionally- effective issue to raise given the involvement of both the ILO and the League of Nations in controlling prostitution and trafficking in women[41] and the extent to which the protectionist camp had

---

37 *ILO Director's Report*, 1929.
38 International Labour Conference, 20 June 1930.
39 Partial Revision of Convention concerning the employment of Women at night, 18th session of ILO (National Archives Dublin, (henceforth NA) Department of Labour, TIL 18/4/A).
40 Marta Vergara 'Women fight at International Labour Conference for Equal Economic Rights' (SL, A 116, Box ii, Folder 219).
41 Ki-Tcheng, *op. cit.*, pp 49, 65–88.

rested their case on moral considerations. The director's office welcomed the resolution on the grounds that it would 'allow it, so to speak, to take its bearing as regards the activities concerning women's work and the possibility of developing those activities' and proposed that the ILO governing body should instruct the director to prepare a report on the Convention for the 1932 conference.[42] In fact the 1932 report is silent on the matter; Thomas died suddenly and with his departure protective legislation faded from prominence. The 1933 discussion of women's work focused on pressures to exclude married women from the labour market as a solution to mass unemployment; the research of ILO official Marguerite Thibert which demonstrated the futility of such measures in reducing unemployment[43] contrasts sharply with the lack of similar research concerning night-work and women.

Revision of the Night-Work Convention was given greater urgency in 1934, following a decision of the International Court of Justice in the Hague, that the Convention applied to women holding managerial and supervisory posts, contrary to the practices of many member states. Amendments excluding women in supervisory and managerial positions from the Convention's ambit and altering the hours covered by the Night-Work Convention were carried, in marked contrast to the 1931 session, though not without opposition.[44] Women's employment attracted little attention at the ILO for the remainder of the decade: the Night-Work Convention was not reconsidered until 1948 when some minor amendments were unsuccessfully proposed; the number of female delegates appears to have declined, as did the number of women speakers. This may reflect a tendency to reduce the size of national delegations as a result of straitened circumstances. In 1934 the Irish government responded to a request from the International Alliance of Women for Suffrage and Equal Citizenship that women be included in all ILO delegations, with a statement that the official Irish delegation was to be limited to four. In the interests of economy no advisers were being sent:[45] women frequently appeared in Geneva in the capacity of advisers rather than delegates.

Superficially the debate over women's night-work, with its representation of women as wives, mothers and housekeepers echoes earlier impositions of protective legislation such as the ban on women in coal-mining.[46] However unlike in the nineteenth century when women's views went largely unheard, on this occasion they are active on both sides of the debate. Critics of the equal right's lobby alleged that they did not represent the views of working women. The ILO Directorate consistently sought legitimacy by referring to the support among the trade union movement for protective legislation. Although the extent of that support was challenged by Miss Hesselgren a Swedish government advisor who emphasised the opposition of Danish and Swedish women trade union members,[47] it remains the case that the majority of trade unionists favoured a ban on night-work by women. Male support for a ban on night-work can perhaps be dismissed as designed to protect spheres of employment from female encroachment; the attitudes of women merit closer scrutiny.

---

42 Extract from Supplementary Report of the Director of the ILO, 12 Oct 1931 (SL A 116 Box 11, Folder 219).
43 Marguerite Thibert, 'The economic depression and the employment of women I, II' in *International Labour Review*, Apr. 1933; May 1933.
44 International Labour Conference, 15 June 1934.
45 NA, Depatrment of Labour, TIL 18/4/A;
46 Humphreys, *op. cit.*
47 International Labour Conference, 12 June 1928.

The most vocal support for the ILO Directive came from European christian trade unions who became a particular target of the director's attempts to secure the Convention's ratification. Christian trade unions were commited to ending all factory work by married women, – 'the most intolerable and most serious evil of the present organisation of society' – according to the German Christian Union of Textile Workers; any restriction on women's factory work was regarded as a step in that direction.[48] Otherwise opposition focused on the threat which night work posed to women's childcare and domestic duties, though some of this hostility reflected a more general fear of rationalisation and increasing work intensity. The introduction of an eight-hour day had made two-shift operations feasible in many industries – bringing the spectre of over-production and reduced control by workers over the production process. The ban on women's night work ruled out two-shift operations in all plants employing women. British workers' opposition to women's night-work originated among the textile unions who saw shift-working as exacerbating problems of over-capacity.[49] The request by Belgian textile producers for an alternation in the hours covered by the Convention to facilitate two-shift operations was strongly opposed by British trade unionists. In other industries male workers effectively hid behind women's skirts: the ban on night-work for women prevented them from being forced to work at night.

Whether the hostility voiced by trade unionists accurately reflected the views of working women remains in doubt. The question of female representation: who could claim to speak for women, and specifically for working-class women is an important sub-text in the discourse. Albert Thomas assumed that trade unions constituted the representative voice of working women – not a surprising stance given his own trade-union origins – but evidence suggests that women within the trade union movement lacked both the independence and resources to diverge from male trade union orthodoxy.[50] Alice Kessler-Harris sees U.S. Women's Bureau chief Mary Anderson as inextricably tied to the trade union position which favoured protective legislation because trade union support was essential to ensure the Bureau's viability.[51] In the address which she was prevented from delivering to Albert Thomas, Mrs Emmeline Pethick-Lawrence argued that women who were organised together with men in trade unions occupied an inferior position, 'over-shadowed and dominated by the traditions and old-fashioned ideas of women's sphere that are brought over by men from the past'. Whereas Thomas and upholders of protective legislation claimed that the issue was relevant only to working and working-class women, Pethick-Lawrence argued that by stereotyping the position of women as one of 'permanent inferiority', the Washington Convention was relevant to the interests of all women. On this basis she argued the right of the wider women's movement to be consulted.[52] However there may have been some truth in the allegations that women's groups which supported equal rights only voiced the concerns of more prosperous and better-qualified women. A letter from the International Alliance of Women for Suffrage and Equal Citizenship, addressed to all gov-

48 Cited in *ILO Director's Report*, 1927.
49 'Protection of women and children. British legislation affecting women and young persons' in *International Labour Review*, Jan. 1921.
50 Alice Cook, Val R. Lorwin and Arlene Kaplan Daniels (eds), *Women and trade unions in eleven industrialized countries* (Philadelphia, 1984), introduction by Alice Cook, p. 17.
51 Kessler-Harris, *op. cit.*, pp 209–10.
52 Vergara, 'Women fight at international labour conference for equal economic rights', p. 3.

ernments of member states of the ILO pressed that at least one woman be included in each national delegation 'most particularly women belonging to women's professional organisations'. Although the women who lobbied the Irish government in favour of the Convention's revision referred to the problems which it posed for female laundry workers, the delegation consisted exclusively of representatives of university women's groups.[53]

The extent to which either faction represented the views of working-class working women remains in some doubt because, as in the nineteenth century, we know little of the views of women directly affected by the measures. However the fragmentary surviving evidence suggests that given adequate pay and working conditions, night-work proved popular among working women with family commitments. Single women workers were less enthusiastic. A study of the effects of a two-shift system on textile workers in Britain showed that married women preferred shift-work i.e. either early or late, because it allowed more time for family duties; more extensive evidence confirms the popularity of night-work among American working women.[54] Night-work and shift work posed no immediate physical dangers to women: the injury rate on shift-work was identical to that among day workers.[55] Else Luders, a Labour Ministry official in Berlin claimed that during the war, when the prohibition was lifted on women working at night, women had actively sought night-work because they could combine it with childcare. However she regarded such popularity as justification for banning women from night-work[56] – determined to protect women from themselves. Far from destroying the family, night-work would appear to have offered women with families the possibility of meeting both the economic and emotional needs which working-class family life dictated. Women trade union leaders, or labour ministry officials such as Else Luders and Mary Anderson, who sought to prevent such women working at night may be condemned for viewing working-class women as dependents rather than as consenting adults, for failing to understand their need to combine paid employment and family duties, and for imposing models of famiy behaviour appropriate to middle-class or respectable working-class families on those who found them to be financially unattainable.

Supporters of protective legislation invariably argued that the prohibition of night-work had little effect on female employment or unemployment. A similar conclusion was reached by a British government study which was published in 1930. However the report is so unsound methodologically, and so riddled with contradictions that its findings should be dismissed; the document is primarily of value as an example of the extent to which the conclusions of an apparently-scientific study of women's work were distorted by ideology. The report claimed that legislative restrictions had little effect on women's employment, despite citing numerous case histories such as a jute mill which replaced all but six of its female workforce with men, because it was deemed unfair to have men on permanent night shift; laundries which hired men for night-work; hosiery firms where men operated the circular knitting machine at night, de-

---

53 Partial revisions of Convention regarding the employment of women at night, 18th session of ILO (NA, Department of Labour, TIL 18/4/A).
54 Kessler-Harris, *op. cit.*, pp 192–5.
55 H.M. Vernon, 'The development of the two-shift system in Great Britain' in *International Labour Review*, Feb. 1934.
56 Else Luders, 'The effects of German labour legislation on employment possibilities for women' in *International Labour Review*, Sept., 1929.

spite the fact that women had monopolised this activity 'since the beginning'.[57] An inquiry conducted by the Women's Bureau of the U.S. Department of Labor reached a similar conclusion, despite estimating that at least 60,000 women – who were frequently 'at the cutting edge of job opportunities' had lost their jobs as a result of the ban on night-work.[58] A subsequent report by the Women's Bureau took refuge in assertions that night-work was in decline for both men and women and that employers would not hire women for night-work even if the legislative ban was repealed.[59]

That said, there is no precise picture of the extent to which women's employment options were reduced by the ban on night-work. The most frequently-cited instances concern women print workers and the majority of infringements of the night-work prohibition in Ireland related to the printing industry. Printing was an industry which offered women the prospect of skilled and relatively well-paid employment; it was also an industry which was characterised in the nineteenth and early twentieth century by some of the most divisive disputes over gender demarcation. During the war there was considerable substitution of female for male workers.[60] The renewed ban on night-work helped to reassert male dominance. In countries where the legislation applied to transport workers, women conductors were seriously affected. In textiles, the women's industrial sector par excellence, the expanding production of man-made fibres brought a need for continuous production and a rising percentage of men among textile workers during the inter-war years. While the ban on night-work may not have been the sole factor accounting for that development, it was a contributory factor. The British Home Office report, while denying that the legislation had any impact on women's employment cited cases in jute, laundries, cotton doubling, tinplating, hosiery, lace making, artificial silk and bakeries where men worked by night at tasks performed by women during the day, or where a female workforce was replaced by men or boys to enable a night-shift to be worked. These examples suggest that the total number of women potentially affected was considerable.

In Britain the prohibition on women supervisors working at night excluded women from jobs which they had held in the expanding electricity industry, an industry which came to be dominated by men; in Ireland those affected included laundry supervisors and chemistry graduates who were excluded from government scholarships to the Glasgow School of Sugar Technology because they were unable to hold supervisory control posts in sugar factories where night-work was common.[61]

The final point worth noting is the divergence in national opinions on this issue. Scandinavian countries generally opposed the Night-Work Convention and restrictions on women's night-work were either non-existent, as in Norway and Denmark, or severely contested as in Sweden. These attitudes may reflect

57 *A study of the factors which have operated in the past and those which are operating now to determine the distribution of women in industry.* December 1929. 1930 Cmd 3508.

58 Kessler-Harris, *op. cit.*, p. 211.

59 Mary Elizabeth Pidgeon, *Women in the economy of the United States of America* (U.S. Department of Labor; Washington, 1937), pp 10–11.

60 Ellic Howe and John Child, *The Society of London Bookbinders, 1780–1951* (London, 1952), p. 258; Sian Reynolds, 'Women in the printing and paper trades in Edwardian Scotland' in Eleanor Gordon and Esther Breitanbach (eds), *The world is ill-divided, women's work in Scotland in the nineteenth century* (Edinburgh. 1990), pp 49–69.

61 Partial revisions of Convention concerning employment of women at night. 18th Session of ILO (N.A. Department of Labour TIL 18/4/A).

their later industrial revolution, the relative dearth of 'dark satanic' textile mills employing numerous women and children in the early years of industrialisation, or perhaps the philosophical basis of the dominant labour party.[62] In contrast strong support for protective legislation in catholic countries reflects the impact of catholic social teaching and the strongly-articulated message about the evils of women working outside the home, though such attitudes were not limited to catholic countries.

It would be naive to argue that the Washington Convention banning night-work by women played a key role in determining the nature of their work during the inter-war years. In many countries national legislation which predated the Convention imposed more stringent controls on women's work. Nor is the debate entirely new: the division in the women's ranks between equal rights advocates and those favouring protection marks a continuation of earlier divisions between the allegedly middle and upper-class suffragists and those seeking to improve conditions of working women. The confrontation is of primary interest for the insights it affords into some of the problems facing historians writing about women, notably the difficulty in disentangling facts from rhetoric: the extent to which, as Alice Kessler-Harris states, 'politics masqueraded behind a facade of statistical data'.[63] One of the most notable contributions of the International Labour Organisation has been its compilation of comparative data on numerous aspects of employment, unemployment, wages and working conditions. Yet the debate over night-work among women is notable for the almost complete absence of hard information, and, in the case of the British Home Office, for the serious distortion of what little exists.

Finally, although the degree of support among working women for a ban on night-work is open to question, it is equally evident that the Open Door discourse with its focus on equal rights failed to appeal to such women. For most women working outside the home, night-work was not a matter of equal rights but an instrument to be used in the battle for survival.

62 On Scandinavia see, Ida Blom, 'The struggle for women's suffrage in Norway, 1885–1913' in *Scandinavian Journal of History*, 5 (1980), pp 3–22; Gunnar Qvist, 'Policy towards women and the women's struggle in Sweden' in *Scandinavian Journal of History*, 5 (1980); Rita Liljestrom 'Sweden' in Sheila B. Kamer and Alfred J. Kah, (eds), *Family policy: government and families in fourteen countries* (New York, 1978), pp 28–31; Jennifer Schirmer, *The limits of reform, women, capital and welfare* (Cambridge, Massachussetts, 1982).
63 Kessler-Harris, *op. cit.*, p. 211.

19 Gender Ideologies, Social Experts and Immigrant Women in Early Post-World War Two Canada: Ontario, 1940s–'60s

Franca Iacovetta

I N spite of the recent attention which immigrant and minority women in Canada have received from scholars, government officials, and media experts, they remain largely faceless women, as othered. Often portrayed as isolated women trapped inside households, insular ethnic colonies or female job ghettos, immigrant women, despite remarkably diverse racial-ethnic, class and socio-religious backgrounds, generally are lumped together as women who fall outside the mainstream of Canadian society. We are accustomed to thinking of immigrant and refugee women as being acted upon by wider socio-economic forces, sexist and racist immigration policies, segmented labour markets, and domineering husbands and fathers. Rarely have Canadians asked if the politics, world view, activities, and demands of immigrant women have affected the society of their adoption. How, for instance, have they affected the households, neighbourhoods, workplaces, and wider communities of post-war Ontario, the Canadian province that has long been the most popular target of immigrants and refugees entering Canada? During the early post-world war era – between 1946 and 1965 – some 52 percent of Canada's two million newcomers chose Ontario. While almost forty percent were British nationals, more than one half of the immigrants came from continental Europe. (By contrast, only a tiny minority of immigrants of colour had entered Canada by 1965.) Among the adult immigrants who came in these years, women comprised well over forty percent of the total.[1]

The marginalisation of minority women in society is replicated in the academy. Despite the recent proliferation of sophisticated works both in immigrant and women's history in Canada, studies of immigrant women, especially non-English-speaking women, remain scanty and they occupy a position outside the main bodies of works on Canadian immigration and Canadian women's history.[2] As Ruth Roach Pierson has astutely observed, the histories of immi-

---

1 The total number and proportions of British and European immigrants to Ontario 1946–65 are as follows: British Isles: 475,934 (38%); western and northern Europe: 273,411 (22%); southern and eastern Europe: 310,235 (25%); Asian: 14,053 (1%). My calculations derived from data contained in Canada, Immigration Branch, *Annual Reports* (1946–65). Thanks to Andrew Boyd for collecting the data.
2 For a more detailed examination of the Canadian literature, several historiographical essays might be consulted, including: Roberto Perin, 'Clio as an ethnic: the third force in Canadian historiography' in *Canadian Historical Review* 64 (1983), pp 441–67; Bettina Bradbury, 'Women's history and working class history' in *Labour/Le Travail* 19 (1987), pp 22–43; introduction in Franca Iacovetta and Mariana Valverde (eds), *Gender conflicts: new essays in women's history* (Toronto, 1992); Franca Iacovetta, 'Manly militants, cosy communities and defiant domestics: writing about immigrants in Canadian history' in *Labour/Le Travail* forthcoming.

grant and minority women, like those of lesbians, 'have been marginalized through publication in separate anthologies and through insufficient integration into the mainstream of Canadian women's history.'[3] Critical debates that have preoccupied Canadian feminist historians, such as gender formation, sexuality, and life cycle approaches, generally have not been informed by the experiences of women who did not belong to the dominant English-Canadian or (in Quebec) French-Canadian majority. To date, historians of immigrant women have focused almost exclusively on the subject of paid and unpaid work, ignoring other crucial aspects of ethnic women's lives, including, for example, their sexual history. In an immigrant nation like Canada, whose multi-ethnic and multi-racial reality has long been a critical factor influencing economic, social, intellectual, and political developments, historians of women need to ask serious questions not only about the particular experiences of specific groups of minority women – such specialised studies are now under way – but also about how we can fully integrate ethnicity and race into all of our analyses.

In my new, comparative research on European immigrant women in early post-World War Two Ontario, I am trying to do that. I want to consider the diversity of Ontario's post-1945 female immigrants and the influences that the arrival of tens of thousands of European 'foreign' females – Holocaust survivors, Baltic and other east European Displaced Persons, and volunteer immigrants from southern Europe and elsewhere – had for post-war Ontario society.[4] I am a specialist in Italian working-class immigrant women, and so this work represents a departure from earlier work and a desire to move beyond the single-ethnic-group or ethnic-community approach that has dominated recent immigration history. While it has yielded important results and helped give face and voice to immigrants, the ethnic-community approach also has had a tendency to encourage an insular perspective, one focusing exclusively on the members of that ethnic group and ignoring the 'others' who inhabited the houses, shops, and streets of 'Little Italy', 'Polish Town' and so on. Such an approach tends to down play immigrant encounters with members of the host society, be they doctors, government welfare workers, employers, or Anglo-Celtic neighbours.

Dealing more specifically with immigrant women, I want also to challenge, or at least modify, the victim model that dominates the literature, feminist and otherwise, on minority women in Canada. By suggesting that immigrant women could make a difference to the lives of their family members, neighbours, and kin, and that their actions and demands could affect the character of Ontario's post-war workplaces and communities, I am not implying that we substitute a model of victimisation for one of unlimited agency. Economic and racial discrimination as well as conditions of scarcity were central to the lives of On-

---

3 Ruth Roach Pierson, 'Colonization and Canadian women's history,' in *Journal of Women's History* (1992), pp 134–56. On the United States, see, for example, Donna Gabaccia's insightful 'Immigrant women. Nowhere at home?' in *Journal of American Ethnic History* (Fall 1990).

4 This paper draws on material contained in two preliminary essays that explore aspects of the larger themes raised here: my 'Making new Canadians: social workers, women, and the reshaping of immigrant families' in Iacovetta and Valverde (eds), *Gender Conflicts*; 'Remaking their lives: women immigrant, survivors and refugees in early post war Ontario' in Joy Parr (ed.), *Women of Ontario 1940s–1970s* (working title) forthcoming. My work on Italians includes: 'From Contadina to woman worker: southern Italian working women in Toronto,' in Jean Burnet (ed.), *Looking into my sister's eyes: an exploration in women's history* (Toronto, 1986); *Such hardworking people: Italian immigrants in postwar Toronto* (Montreal and Kingston, 1992); Karen Dubinsky and Franca Iacovetta, 'Murder, womanly virtue and motherhood: the case of Angelina Napolitano, 1911–22' in *Canadian Historical Review* (Dec. 1991), pp 505–31.

tario's post-war female newcomers, many of whom were also trying to remake their lives in the aftermath of horrific wartime experiences. Nor were immigrant women, many of whom arrived with few resources and/or few marketable industrial skills, entirely free to choose how they might remake their lives here. On the contrary, their experiences involved multiple encounters with members of the host society, including public health officials, social workers, and zealous reformers active in immigrant reception work. Such encounters, moreover, involved relations between unequal partners. The 'hosts' almost always had the upper hand, with respect to language, and to cultural and material resources, and their assistance was offered not simply in the hope of ensuring the successful adjustment of newcomers. They were also bent on exerting a permanent 'Canadianising' (read: middle-class Anglo-Celtic) influence over 'foreigners', including women. Far from simply embracing or acquiescing in such pressures, however, immigrant women responded in various ways as they sought to adapt to, confront, and at times, even overcome countless challenges and constraints. The nature of the encounter between immigrant and host was complex; its dynamics deserve close study.

Given the propensity of post-war Canadian social experts to down play deliberately the national, ethno-racial, political, and class differences among the European newcomers – the label 'New Canadian' was a useful shorthand in this regard – it is important to note the diversity of the European women who entered Ontario by 1965. They differed with respect to class, social, cultural, political, linguistic, marital, and religious backgrounds, and in regard to their pre-migration and wartime experiences. No more was this true than of the protestant, catholic, and Jewish women who hailed from central, eastern, and southern Europe. While Ontario long had attracted a variety of immigrants, the early post-war years brought an unprecedented number of continental Europeans: they included Jews and catholics, leftists, liberals, and virulent anti-communists from Soviet-controlled homelands, resistance fighters and Nazi collaborators.

Note, for instance, the remarkably different pre-migration and post-migration trajectories of three Ontario women whose stories have been recorded. Sarah Ginsberg, a Jewish survivor from Poland, had spent the war years in a woman's slave labour camp in Czechoslovakia, working long hours in the factory and daily witnessing Nazi guards beat Jews. At the end of the war, Ginsberg, along with her mother and sister, spent weeks trying to locate family members who, it turned out, had not survived the Nazi death camps. They then made their way to a Displaced Person's Camp in Germany. By 1953, Sarah had arrived in Toronto, where for years she was employed in the city's garment trade.

While the Second World War dramatically affected the lives of southern Europeans like Italian-born Nina Laurenza, such women had not had to face a genocide. Laurenza spent the war years raising a child and running the family farm in her Calabrian village, where she resided next door to her parents. Like much of the Italian south, Laurenza's village had been periodically bombed throughout the war and she spent the war years worrying about the safety of her husband-soldier. But he returned home safely from the war and by 1954, the couple had settled in Toronto. The migration had occurred in a ragged fashion: he went first, as a farm labourer on a Canadian government, one-year contract, and she and her child followed two years later.

Protestant and catholic women living in the east and central European nations that had fallen to Hitler were also spared the Final Solution, but many were directly affected by the war. Eva Kogler, a Polish-born, Ukrainian woman

who eventually settled in the northwestern Ontario town of Port Arthur (later,
Thunder Bay), for instance, had been a young, single woman when war had
broken out in Europe. It disrupted her education and separated her from her
family. Transported by the invading Nazis from Poland to Vienna, Kogler was
posted to domestic duty with a well-placed German family. After the war, she
returned home to find her village reduced to 'ashes'. On the move at the end of
the war, she eventually arrived at a Displaced Persons camp in Czechoslovakia,
where she befriended and eventually married a Czech man. A few years later,
she resettled in Port Arthur in northwestern Ontario. Following a two-year separation, Kogler had joined her husband, who initially gained entry into Canada
by means of a labour contract with a northern Ontario logging company.[5]

Ontario's work places, markets, neighbourhoods, English classes, and other
social arenas, then, not only drew together women from different backgrounds
but, in some cases, brought together women who recently had been bitter
enemies. Sometimes, it caused tensions. In other cases, women had the opportunity to get to know women whom they had once been taught to treat as
enemies.

Patriotic and socially conscious Ontarians were propelled into action by the
anticipation and then arrival of hundreds of thousands of newcomers. The newcomers provoked into action a loose collection of middle-class professionals,
government officials, social workers, journalists, women's groups, service organisations, and ethnic Canadians of various political stripes. They debated
questions of citizenship, democracy, and appropriate family values, mobilised
resources for 'receiving' the newcomers and exerting a 'Canadianising' influence over them. Immigrant women figured prominently in the scenarios of enthusiastic Ontarians; as the 'gatekeepers' they were keen to encourage wives
and mothers of young children to adopt Canadian ways in everything from
cooking, to household furnishings, to child-rearing methods. Women's groups
conducted tours of immigrant women through middle-class homes with the intention of encouraging the newcomers to aspire to middle-class standards of
living and taste, while a whole range of government and volunteer agencies ran
movie nights and citizenship classes that extolled the virtues of Canada's parliamentary and democratic traditions. Nutritional experts even advised changes
in 'ethnic' diets on the grounds that too much of the family budget was going to
purchase fresh foods and lean meats. Thus, the Italians were encouraged to buy
'cheaper' cuts of meat, the Portuguese to replace fresh fish with the frozen variety, and various immigrant groups to rely on canned fruits and vegetables.[6]
The immigrants were entering Ontario at a time when constructs such as 'the
family', 'democracy', and 'citizenship' were being defined and redefined in light
of the catastrophic events of the great depression and the Second World War.
Acting as powerful indicators were the major ideological currents of the day,
namely the cold war and a family-centred gender paradigm that idealised
nuclear families in which the male breadwinner supported a housekeeping wife
and children. As the world entered the atomic age and cold war, there was
much anxiety about the crisis facing Canadian families and, moreover, consid-

---

5 In regard to personal interviews, I have used fictitious names and occasionally modified tiny details in order
  to ensure privacy. Interview with Sarah Ginsberg, care of Paula Draper, Director, Oral History Project of
  Canadian survivors; Nina Luarenza interview by author (Toronto); Eva Kogler interview by author (Thunder
  Bay).
6 For further details on these and other examples of reception work see especially my 'Remaking their lives.'

erable discussion regarding the need to ensure the political and social integration of the European immigrants. Social workers and other professional experts expressed fears that after two decades of depression and war, Canadians did not know what 'normal' family life was. In an address entitled, 'The ordinary family in extraordinary times', a social worker with the Family Welfare Association of Montreal provided a gendered analysis of her times:

[F]amily life has been taking a beating during the past 20 years . . . [T]he great depression . . . [meant the] postponement of marriage for many young people because of lack of jobs, which was extremely serious, both biologically and psychologically . . . Many husbands appeared to be failing in their role as husbands and fathers appeared to be failing in their role as breadwinners and consequently in the eyes of their children to be total failures . . . Then came the war . . . Couples were flung apart to have widely different experiences. When the war was over, they had to begin all over again, but the situation was not the same . . . For men there was the difficulty of settling down to a humdrum life. For women too, who during the war had to rally and do many things that they had never done before, there was a letdown . . . A very special hazard to family life has been the extreme housing shortage [so that] young people have been unable to find a place of their own . . . Overcrowding . . . in the homes of relatives does not offer a natural opportunity for family life to grow and has contributed greatly to the development of strained and broken family relationships and subsequently less useful citizens.[7]

Such fears, as scholars have shown for the Canadian and U.S. contexts, reflected an underlying concern to protect and bolster the ideal model of the middle-class, nuclear family, one in which gender relations were firmly rooted in the arrangement described above.[8] It was often expressed in the cry that Canadians return to the 'normal' family of the past, even though that was largely a mythical construct. Concerns were voiced about increasing rates of divorce and juvenile delinquency, an increase of married women in the paid workforce, the lack of affordable housing for newly married couples and subsequent doubling up, and anxieties and tensions related to the cold war. All were viewed as powerful indicators that the family was in need of bolstering.

The twin metaphors of family and democracy also underlined middle-class Ontarians' discourses regarding and responses to the new immigrants, especially women. Fears of 'dysfunctional' Canadians unfamiliar with so-called 'normal' family life were compounded by the anticipation of potentially more disfunctional newcomers – the casualties of war, refugee camps, and Soviet aggression – who would require extensive material support and counselling but might never turn into useful Canadian citizens. Referring to the 'many immigrants' from 'so many countries', one writer warned:

[A] problem in regard to the mental health of our immigrants does exist and must be faced. Otherwise there is a real possibility that the immigrant population, many of them survivors of terrible experiences and most of them having at least some difficulty in adjusting to a strange environment and to the new ideas and customs, may form a pool

---

7 Elinor G, Barnstead, 'The ordinary family in extraordinary times' in *Social Worker*, 16:2 (Dec. 1947), pp 1–9.
8 See, for example, Denise Riley, 'Some peculiarities in social policy concerning women in wartime and postwar Britain' in Margaret Randolph Higonnet et al (eds), *Behind the lines: gender and the two world wars* (New Haven, 1987); Elaine Tyler May, *Homeward bound: American families in the cold war era* (New York, 1988); Veronica Strong-Boag, 'Home dreams: women and the suburban experiment in Canada, 1945–60' in *Canadian Historical Review*, 72:4 (Dec. 1991), pp 471–504; Iacovetta, 'Making new Canadians'; Ruth Roach Pierson, 'Gender and the unemployment debates in Canada, 1934–1940' in *Labour/le Travail*, 25 (Spring 1990), pp 77–103.

of mental ill-health that will add greatly to our already impressive total problem in this respect.[9]

The tendency among newcomers to live in multiple or extended family arrangements was also viewed with disapproval. Such views emerged, for example, in the discourses surrounding the Displaced Persons. The plight of the DPs, especially the children, was held up by many as a reminder of the havoc that war wreaked on family life and the misery experienced by those without families. The Canadian Association of Social Workers, an organisation representing professional social workers in the country, strenuously lobbied the Canadian government on the refugees' behalf. But the same association also insisted that these people, now accustomed to living in chaotic 'mass' groupings of 'little Europes,' be settled in Canada in separate 'family' groups of nuclear households so as to facilitate their resumption of normal family life. Careful screening of communists was urged as well. The twin influences of family and democracy also shaped the parameters of immigrant reception work and social services. Active Ontarians, whether volunteers or professionals, were pleased not only to find immigrants contributing to the post-war economy by filling critical jobs, but they applauded successful 'New Canadians' – that is, women and men who became citizens, couples who set up shop or became homeowners, immigrant housewives who attended English, cooking, and citizenship classes. Such people were held up as the beneficiaries of the grand opportunities available in the democratic west.[10]

More to the point, many social experts who served an immigrant clientele, especially counsellors, saw the process of assisting their clients' adjustment as largely a process of re-shaping 'deviant' or otherwise 'foreign' families according to the prevailing, North American model that combined patriarchal ideals about family with the notion of a companionate marriage. That Canadian professionals assumed that European women suffered from more deeply entrenched patriarchal cultures is clearly evident in the words of Benjamin Schlesinger, a lecturer at the University of Toronto's social work school who explained that 'social workers are faced with interpreting community standards to the *New Canadian.*' Such community standards included a more egalitarian view of the family than Europeans were considered to possess – one that placed limitations on a husband's domineering attitude towards his wife, including his right to beat her, and a more indulgent form of child rearing. Thus, while social workers such as Schlesinger could claim the merits of a 'socio-cultural approach' that acknowledged the role of culture in explaining certain behavioural patterns, they were nevertheless determined that the newcomers, especially the southern and eastern European families, required reforming. They were concerned not so much to eliminate the patriarchal organisation of such families as to alter their character and give them a more 'modern' or North American bias. Immigrant men were expected not only to acquire sole responsibility for the economic support of the family, but to be companionable partners to their wives. Married women were to adapt to Canadian standards of womanhood and motherhood and thereby provide well-adjusted homes for their husband and chil-

---

9 Brian Cahill, 'Do immigrants bring a mental health problem to Canada' in *Saturday Night*, 22 June 1957.
10 For example, see: M. Ross, 'How the immigrant becomes Canadian' in *Saturday Night*, 23 Dec. 1957; C.W. Woodside, 'Problem in integration: the New Canadians' in *ibid.*, 8 Jan. 1955; J. Kage, 'Immigration and social service' in *Canadian Welfare*, 1 March 1949; immigration issue, *Food For Thought*, Jan. 1953; R. Williams, 'How to keep red hands off our new Canadians' in *Financial Post*, 12 March 1949.

dren. Social experts were convinced that as the gatekeepers of modern, socially progressive democratic society, they had the right to intervene in the lives of needy or unruly immigrant clients. As Schlesinger put it, the newcomers had come to a country that sanctioned 'the community's right to protect children, to regulate family disorganisation, and to interfere in difficult family relations.'[11]

The debates, activities, and campaigns of active Ontarians represent only one side of the story. Social experts could not determine absolutely the adjustments of newly arrived immigrant women from Europe. The demands, strategies, responses, and experiences of immigrant women must also be taken into account, and indeed the immigrant woman's role was critical in determining the outcome of the programmes and campaigns staged for her supposed benefit. For their part, immigrant women did not always act according to a script that others had written, and their resistance and indifference to, or co-operation with, the plans of Ontarians to remake them into Canadians, helped shape post-war relations between newcomer and host. The presence of these women certainly made a difference to the character of post-war Ontario society. Huge numbers of immigrant women put pressures on existing social services, and English classes, libraries, medical care, baby clinics, job assistance programmes largely depended upon the women's willingness to participate as well as the nature (and length) of their participation.

The efforts of the professional and volunteer caregivers could be thwarted by the immigrant women themselves, many of whom tapped such services while at the same time withstanding pressures to transform entirely their behaviour or beliefs. The kind of tensions that could result from such encounters between newcomer and host can be seen in the instance of a Toronto baby clinic serving Italian mothers. The public health nurses involved in running the city clinic, which was run out of the basement of a catholic church parish with a significant Italian immigrant congregation, were continually frustrated by the fact that the Italian mothers who attended the clinic made use of the practical forms of aid they offered, such as free vaccination shots and milk supplies, but did not stick around long enough to attend the prescriptive lectures delivered on Canadian child-rearing practices. It was also noted that the Italian mothers were suspicious of 'English' nurses, whom they avoided until their own baby was being weighed or examined. This situation sent the public health nurse supervising the clinic into a tailspin. In her exasperation, she attributed the problem to the narrow-minded and suspicious qualities of the Italian character, and insisted that the Italian mothers had to be convinced that the nurses had their best interests at heart. 'It is an essential part of their integration into the Canadian way of life,' she wrote, 'and the sooner it is explained to them, the better.' That the Italian women saw the clinic as serving an important, practical need, explains why, the ire of the nurses notwithstanding, the clinic remained a going concern and the city committed resources to keeping it staffed and supplied.[12]

In order to capture some sense of the complex nature of the encounters that occurred between social experts and immigrant women, I examined 320 confidential case files of the International Institute of Metropolitan Toronto, an im-

---

11 Benjamin Schlesinger, 'Socio-cultural elements in casework – the Canadian scene' in *Social Worker*, 30:1 (Jan. 1962). For the United States see especially Linda Gordon's *Heroes of their own lives: the politics and history of family violence* (New York, 1980).
12 A. Cecilia Pope to Doris Clark, 15 Apr. 1958, with enclosed report, 'Civic Action in a Well Baby Clinic' (City of Toronto Archives, Social Planning Collection).

migrant aid agency serving thousands of non-British immigrants in Toronto, the single most popular target among Canada's post-war immigrants.[13] A charitable organisation, the Institute belonged to an elaborate web of private and public agencies and services that included family and children's aid societies, volunteers', women's, religious, and ethnic groups, and government welfare, employment and immigration offices. The Institute performed several functions, including reception work. It mostly involved hosting cultural, education, and recreational activities designed to inform newcomers about Canada and to permit mingling between immigrants and Torontonians so as to encourage immigrant adjustment to Canadian ways. At the reception centre, for instance, staff organised a variety of cultural, educational, and recreational activities – including movies, afternoon teas, bridge games, and Amateur Nights showcasing talented immigrant musicians – that were designed to offer newcomers an opportunity not only to learn about Canada and to meet one another but also to get to know the Anglo-Celtic Torontonians whom the Institute staff sought to attract to these events. In this way it was hoped that, as one staff member put it, 'the newcomers may be encouraged to become adjusted to the new culture and add to it by meeting, observing, and understanding their Canadian friends.'[14]

A staff of professional and volunteer social workers, some of them actually immigrants, also offered counselling services. While the vast majority of the clients were Europeans from a peasant or rural artisanal family background who had joined the ranks of the urban working class, a small minority was made up of skilled trades people or former professionals who now found themselves out of work or unable to transfer their skills to Canada. The latter group included mostly east European refugees.

It is the case files that this service generated that most graphically illustrate some of the patterns I have been describing. The files also offer the historian a way of approaching immigrant women's lives that avoids the dichotomy of victims/heroines. If the case files that I examined are in any way typical, we can conclude that not only were social workers serving immigrant female clients profoundly influenced by the predominant gender ideologies of the day but also that they were usually not successful in significantly reshaping their clients' lives. The failure largely reflected the limited contact caseworkers had with their clients. Most clients appeared, then quickly disappeared. Moreover, many of the women clients adopted a highly selective approach towards the services offered by the agency. Far from exhibiting an inclination to adopt fully Canadian ways, most clients responded in pragmatic fashion, taking advantage of its services in order to meet their own needs as they defined them.

Some very brief illustrations will have to suffice. That the Institute counsellors operated within the gender paradigms of the era is clearly evident in the cases involving women who had been deserted or were the victims of domestic violence. One wife abuse case that first came before the Institute counsellors in the spring of 1958 involved a recent Hungarian refugee woman who initially came to the agency hoping to find work and locate a day-care centre for her two small children. But it soon became clear that an apparently simple job placement case masked a turbulent history of marital cruelty. The husband's continual beatings, his hiring of private detectives in an effort to discredit her as a

---

13 For further details on the agency and my analysis of the case files see my 'Making new Canadians.'
14 MU6474, File: Reception Centre 1960, Progress Report nd (c 1960)(Archives of Ontario (AO), International Institute of Metropolitan Toronto (IIMT)).

mother, and his threats of child custody suits – all led the woman to attempt suicide by winter 1958. However, by the time the lengthy case file comes to an end in spring 1959, the woman appears to have taken matters in her own hands. She has left her husband, found a job, and placed her children temporarily in a catholic orphanage. Throughout much of this crisis, however, the Institute counsellors, though genuinely concerned for their client, nevertheless adopted a family reconciliation approach, even contacting the husband to notify him about the benefits of family counselling. In seeking to preserve this family, they hoped to reform the husband and thus re-shape the dynamics of this deviant family along a more companionate model. Moreover, the sympathy they showered on the woman reflected their conviction that she was indeed a 'good mother', as one caseworker described her, and thus a worthy client.[15] Not all women clients were viewed as sympathetically. Single women living with male lovers or women who demanded better job postings than those they received with the help of the agency, were quickly dubbed 'ungrateful,' 'choosy,' or in need of considerable case work counselling.

While Institute counsellors hoped to play a more profound role in all of their clients' private lives, they often found themselves confined simply to providing their female clients with material aid – handing out food and clothing vouchers or referring them to the local Welfare or Immigration Office for emergency handouts. Apart from trying to weed out fraudulent requests – a major preoccupation indeed! – the staff had little difficulty processing these cases. Staff members did, however, express frustration at not being able to recruit more of the immigrant housewives into their English classes, civics lectures, and other programmes designed to have a greater influence over them.

Finally, the caseworkers were at times compelled by their women clients to agree to assist them in ways that they, as the experts, did not necessarily condone. Their toleration of the immigrant married woman worker – a figure that did not conform to the hegemonic ideal of womanhood – is a case in point. With husbands unable to earn a 'family wage' sufficient to support the entire family, the majority of the married immigrant woman who came to the Institute wanted a job and help finding it. Counsellors made it clear they considered it unfortunate that married women, especially mothers, went out to work, but, significantly, they showed a far less rigid adherence to predominant gender ideologies on this score than we might have expected. Generally case workers worked hard trying to secure jobs for such women, although it is interesting that almost without exception the agency placed women in low paying, low skilled jobs such as domestic service and factory work. Several factors might account for their response. The smooth economic adjustment of immigrant families took temporary precedence over concerns that men and women fill their prescribed roles. Caseworkers also appeared to assume that such arrangements would be temporary. On several occasions, for example, they encouraged a distraught, unemployed husband to rely on his wife's earnings while he attended trade school in order to ensure that eventually he could take on his duty as bread winner. It would also be worth noting that, notwithstanding the dominant rhetoric of the day regarding the ideal stay-at-home mom, a good proportion of the caseworkers were themselves not only working married women, but also refugees and immigrants – recruited because of their language skills.

---

15 This case is contained in Confidential Case Files, Series E-3 (AO, IIMT).

Despite differences in class between them and their clients, these European women counsellors might have felt some empathy for their compatriots who came to them for help in getting paid work.

These, then, are some of the social arenas that immigrant women in early post-war Ontario inhabited. Many other arenas – including department stores, family-based ethnic businesses, factories, neighbourhoods, and volunteer and professional organisations – need to be explored if we are to enlarge our understanding of the thousands of immigrant and refugee women who in the two decades after World War Two made Ontario, Canada, their home.

20   Women in Northern Ireland

Catherine B. Shannon

THE origins and nature of the conflict that has engulfed Northern Ireland since 1968 has been analysed extensively, making writing about the 'troubles' one of the few growth industries in the economically devastated province. Although a complete bibliography now would contain over 7,000 items, until quite recently the impact of the political violence on the role and status of Northern Irish women has been virtually ignored.[1] The reasons for this lacuna are not hard to appreciate. They derive from the same subtle, often unconscious, yet pervasive patriarchal influences that dominated most historical and social science researchers and methodology until the 1970s and early 1980s. This silence has been compounded by the extreme traditionalism of Ulster's churches and schools as well as by an economic structure that keeps women who work in sex-segregated, low-paying, and often part-time jobs. Approximately 54% of economically active women are employed, but 18% of these are in part-time jobs, and 18% of part-timers work less than sixteen hours, thus disqualifying them from pension or maternity rights.[2] Moreover, the political necessity to document the discrimination against northern catholics as well as to analyse the attitudes of catholics and protestants regarding the controversial issue of the constitutional link with Britain delayed much social science research on the gross inequalities of gender that continue to characterise Northern Irish society. Fortunately, considerable progress has been made to close this gap by the publication since 1989 of a dozen or more articles written by university academics and surveys commissioned by the Equal Opportunities Commission.[3]

From House Rights to Civil Rights

A glaring example of northern women being written out of the history of the 'troubles' is the failure of most historical accounts to mention that the very first protests against discriminatory housing policies by Unionist dominated local government councils were initiated by women.

Although women from the Springtown estate in Derry had disrupted meetings of the Londonderry City Council to protest inadequate housing for Derry's

---

1 John Whyte, *Interpreting Northern Ireland* (Oxford, 1991).
2 Pamela Montgomery and Celia Davis, 'A woman's place in Northern Ireland' in Peter Stringer and Gillian Robinson (eds), *Social attitudes in Northern Ireland* (1990–91 edition, Belfast, 1991), pp 74–78. See also Celia Davies and Eithne McLaughlin (eds), *Women, employment and social policy in Northern Ireland: a problem postponed* (Belfast, 1991).
3 See especially Monica McWilliams, 'Women in Northern Ireland: an overview' and Hazel Morrisey, 'Economic change and the position of women in Northern Ireland' in Eamonn Hughes (ed.), *Culture and politics in Northern Ireland, 1960–1990* (London, 1991); Carmel Roulston, 'Women on the margin: the women's movement in Northern Ireland, 1973–1988' in *Science and Society*, vol. 53, no. 2 (1989), pp 219–236.

catholics as early as November 1959,[4] it was the protest organised in spring and summer 1963 by forty young catholic Dungannon women that helped unleash a chain of events that eventually led to the founding of the Northern Irish Civil Rights Association. Similar to thousands of working-class catholic women in Northern Ireland, these Dungannon women were struggling to raise their families in woefully over-crowded, expensive and often unsanitary private houses and flats. The severity of conditions in Dungannon was evident in local newspaper reports of rat-infested quarters and of as many as eight catholic couples living in one house where they shared two cookers and two toilets for a rent of £27 per week.[5] Although the Dungannon Urban District Council had built at least 194 houses since the Second World War, allocations consistently went to newly married protestant couples and in 58% of the cases to protestants from outside the constituency, leaving married Dungannon catholic couples with young children on the council housing waiting list for as long as eight to ten years.[6]

Inspired partly by the contemporary civil rights movement in the United States, these young women began to question the systematic denial of decent public housing for Dungannon's working-class catholics. Under the initial leadership of Mrs Angela McCrystal, the group drafted a petition documenting their housing grievances and submitted it to the council at its regular meeting on 13 May 1963. Convinced by the negative reaction of the councillors that nothing would be done, Angela McCrystal and her associates called a public meeting in St. Patrick's Hall, formed the Homeless Citizens League and began a series of pickets of the council meetings.[7] These demonstrations featured mothers dressed in their Sunday best, pushing baby prams with many toddlers in tow. Looking back at the newspaper pictures and accounts, these events appear quaintly naive and innocent. Yet these women, led primarily by Angela McCrystal and Patricia McCluskey, the wife of a local catholic doctor who joined the H.C.L. shortly after its formation, began collecting irrefutable evidence proving the substance of the original allegations that the Dungannon Urban Council discriminated against catholics in housing allocation.

These actions by ordinary Dungannon women raised a potentially explosive political issue since qualification for the local government franchise in Northern Ireland depended upon being a registered householder. In those areas, mostly west of the River Bann, where protestants almost equalled or outnumbered the catholic population, as in Derry, Unionist councils used housing allocations to preserve their political monopoly. In eventually taking their housing grievance to Stormont, these women mounted a direct challenge to one of the main bulwarks of forty years of protestant Unionist domination and catholic Nationalist exclusion. Their assertiveness on this fundamental right to adequate housing was in stark contrast to the irredentist rhetoric and abstentionist profile that had characterised the local nationalist councillors as well as the six Nationalist members at Stormont for years.

At the end of August and inspired further by the wide media coverage during

---

4 Frank Curran, *Countdown to disaster* (Dublin,1986), p. 19; author's interview with Mary Nelis, July 1985, June 1986.
5 *Dungannon Observer*, 11 May, 16 Sept. 1963.
6 *Ibid.*, 21 Sept. 1963; see also Bob Purdie, *Politics in the streets: the origins of the civil rights movement in Northern Ireland* (Belfast, 1990), p. 87.
7 *Dungannon Observer*, 18 May, 15 June, 13 July 1963; author's interviews with Angela McCrystal, 13 June 1993 and Patricia McCluskey, 14 May 1992.

the summer of United States civil rights marches which culminated in the Martin Luther King-led Washington march of 28 August,[8] a few members of the Homeless Citizens League took more decisive action after the Dungannon Council refused a request from H.C.L. members to make available to homeless catholics pre-fabricated housing at Fairmount Park then being vacated by protestant families moving to a newly built council estate. A squat was organised which eventually involved thirty-seven women and their families. In some instances, the vacating protestant tenants handed over their keys to the incoming squatters instead of council officials.[9] Within a week the Minister of Health and Local Government, Mr William Morgan, agreed to receive a delegation of the Homeless Citizens League at Stormont. The combination of widespread coverage of the squat by the Belfast media and the detailed nature of the statistics presented by the delegation was sufficient to force Stormont to put pressure on the Dungannon Council who were obliged to abandon their plans to eject forcibly the squatters, and make provision for homeless catholics more quickly by rushing the completion of a new Ballygawley housing estate in the nationalist West ward.[10]

The coverage given to the actions of the Homeless Citizens League in the summer and fall of 1963 elicited a deluge of supporting letters from across Northern Ireland which indicated that the minority's acquiescence in the face of Unionist misrule was wearing thin. This led the McCluskeys and their associates to examine the related issues of discrimination against catholics in public employment and how widespread gerrymandering of local electoral districts and the local electoral franchise of business owners virtually guaranteed Unionist domination of local councils even in areas where the catholic population outnumbered the protestant minority. In order to address these issues on a province-wide basis, the Campaign for Social Justice was formed in January 1964 by a committee of twelve that included Olive Scott, Maura Mullally and Patricia McCluskey.[11] Over the next few years the Campaign for Social Justice, largely managed by Mrs McCluskey and her husband Conn, systematically collected and published data that proved widespread discrimination against catholics in housing and public employment in other communities such as Derry, Enniskillen and Lurgan. They disseminated their findings to interested groups and public libraries throughout Ireland, the United Kingdom and the United States, one of the first significant occasions being Patricia McCluskey's presentation of C.S.J. data to the National Council of Civil Liberties meeting in London on 13 March 1965. In addition, the C.S.J. members encouraged those of the minority population who qualified to get on the voting register. Mrs McCluskey and Mrs McCrystal were particularly active in efforts to have Dungannon's catholics on the register for the next local elections. In May 1964 a slate of seven C.S.J. candidates, which included Patricia McCluskey, Angela McCrystal and Brid McAleer, ran for the Dungannon council on the slogan of 'Vote for Justice, Vote for the Team.' With a 97% voter turnout, four of the seven, including Patricia McCluskey were elected.[12] Mrs Brid Rodgers, who later became a leading member of the Social Democratic and Labour Party, began her political

---

8 *Dungannon Observer*, 18 May, 21 Sept. 1963.
9 Purdie, *Politics in the streets*, p. 88.
10 *Ibid.*, pp 88–9; Conn McCluskey, *Up off their knees* (Dublin, 1989), pp 12–13.
11 *Ibid.*, pp 15–17. The CSJ remained basically a catholic organisation despite initial efforts to recruit liberal protestants who had expressed sympathy with its aims.
12 McCluskey, *op. cit.*, p. 13.

involvement when she collected data on Lurgan for the C.S.J. As a result her husband's dental practice fell by 40% when his protestant patients boycotted his practice.[13]

Appreciating that the Government of Ireland Act of 1920 vested ultimate and specific responsibility in Westminster to protect the northern catholic minority from discrimination and injustice by any northern Irish authorities, Mrs McCluskey presented the C.S.J. data to Conservative Prime Minister Alec Douglas Home and the Labour Party opposition leader, Harold Wilson, in a protracted correspondence in the summer of 1964. Neither party leader was willing to do anything despite the extensive documentation of discrimination and Stormont's indifference contained in four C.S.J. publications.[14]

Meanwhile, a group of sixty Labour back-benchers and party regulars long concerned about Northern Ireland and supplied with C.S.J. data established the Campaign for Democracy in Ulster in early 1965. Mrs McCluskey spoke on the housing question at their inaugural meeting in the House of Commons the following July. Under the leadership of Paul Rose, the Labour MP for Blackley and Paddy Byrne, the C.D.U. eventually had over 100 parliamentary supporters. The group supported the C.S.J. demands for full parliamentary inquiries into the Stormont administration, with particular reference to discrimination in housing, public employment and the use of police powers. In April 1967 the C.D.U. sent a fact-finding delegation to Northern Ireland and found the injustices there comparable or worse than existed in India and South Africa.[15] The expectations of Northern Irish catholics that redress might come from London proved temporary for it was only after the outbreak of violence in early 1969 that Westminster launched parliamentary investigations of the Stormont regime.

By raising the explosive housing issue and showing that redress could not be achieved either through the highest legislative or judicial authorities, the women of the Homeless Citizens League and in the Campaign for Social Justice played a significant role in heightening the catholic sense of grievance and in preparing large segments of the northern catholic community to move beyond personal and local grievances to embrace the broader political ideology of civil rights. Thus the founding of the Northern Irish Civil Rights Association is as much the result of their groundwork as of the various trade union, republican and civil liberties groups that joined in establishing the organisation in April 1967. Within 18 months large segments of the catholic community were fully mobilised behind the programme and marching tactics of N.I.C.R.A. The courageous example of the Homeless Citizens League and the women associated with the C.S.J. was reflected in the involvement of a number of women in the initial activities of the Northern Irish Civil Rights Association. Patricia McCluskey and Brid Rodgers of C.S.J. as well as trade unionist leader Betty Sinclair, and a cadre of young women students, including Bernadette Devlin, Eilis McDermott, Inez McCormack, Ann Hope and Madge Davison were to have prominent roles in various civil rights campaigns over the next four years.[16]

13 Author's interview with Brid Rodgers, Jan. 1987.

14 McCluskey, *Up off their knees*, pp 22–3. The four publications were *The plain truth; Londonderry: one man, no vote; Northern Ireland: why justice cannot be done* and *Northern Ireland: legal aid to oppose discrimination, not likely*. A second edition of *The plain truth* was published in 1969 and reprinted again in 1972. Over 100,000 copies were distributed up to 1975.

15 McCluskey, *Up off their knees*, p. 29.

16 William Van Voris, *Violence in Ulster; an oral documentary* (Amherst, 1975); author's interviews with Bernadette Devlin, 16 July 1992; with Ann Hope, 4 June 1993; with Inez McCormack, Jan. 1987; with Madge Davison, 20 June 1988; with Brid Rodgers, Jan. 1987; with Patricia McCluskey, Apr., 1992.

The Early Troubles, 1969–1975

The high hopes of the civil rights marches of 1968–69 were soon crushed in an escalating cycle of sectarian rioting, police brutality, rising republican and loyalist para-military terrorism and by a repressive policy of counter-terrorism first by the Stormont administration and subsequently by the British government. The human costs since 1968 have been staggering. In a population of one and a half million up to 1991 there have been approximately 3,100 killings, 30,000 serious injuries, 32,000 shootings, 13, 391 explosions, and countless instances of intimidation. Just over two hundred of the dead have been women who ranged in age from infancy to the very elderly. They include women from every walk of life: the daughter of a judge, the wife of another judge, seven police women and a Roman catholic nun, Sister Catherine Dunne, who was killed by an I.R.A. landmine which also killed three policemen on the Killylea Road near Armagh on 24 July 1990. Marie Wilson, a young protestant nurse is remembered widely as the innocent victim of the November 1987 Enniskillen bombing. In April 1992 a twenty-six year old catholic mother of two, Philomena Hanna, was assassinated apparently by the loyalist Ulster Freedom Fighters while she was working at a Belfast chemist shop where she served customers from both the catholic Falls Road area and the protestant Shankill.

Gender has not exempted politically active women from assassination as the murders of two prominent republican supporters, Maire Drumm and Miriam Daly, in the late 1970s demonstrated. In the early 'eighties an unsuccessful attempt to assassinate Bernadette Devlin left her severely wounded. In October 1992, a twenty-nine year old Sinn Féin local government councillor and law student, Sheena Campbell, was gunned down within the shadows of Queen's University. Women married to politicians, policemen, judges, and even lawyers lived with the daily reality that their husband's political convictions or occupation might have resulted in a sudden and premature widowhood.

A poignant expression of the communal and personal pain these grim facts have inflicted came from the late Kathy Harkin, a dedicated community worker from Derry when she observed:

> War is ultimately male. It is the women who as givers of life suffer in a war when the lives they have produced are destroyed. It is the women who are left to grieve, to mourn, to nurse the sick and the wounded, to keep families together, to keep society 'normal'. They act as a buffer in war, trying to protect their children from it. And when the war is over it is the women who are left to pick up and rebuild the communities – so it is the women who make the bigger sacrifice.[17]

While Harkin's conflation of war primarily with maleness is problematic historically, her observation is significant nonetheless because it goes beyond the usual portrayal of northern women as mere victims, and suggests what ordinary women have done individually, collectively and, usually outside the official and public patriarchal structures, to protect and to minimise the impact of political violence on their families and communities. It is this aspect of northern women's experience that the remainder of this article will emphasise, first by focusing upon the responses of catholic women to the deteriorating political situa-

---

17 E. Fairweather, M. McFadyean, R. McDonough, *Only the rivers run free: Northern Ireland, women's war* (London, 1984), p. 36.

tion from 1969 to 1974, then by examining the response of protestant women to escalating Provisional I. R. A. violence after 1970, and finally to some instances where northern women have crossed the political and religious divide to co-operate in pursuit of common goals.

The response of catholic women to the besiegement of their communities and the internment of their men was vigorous and often very inventive. For instance, during the August 1969 'battle of the Bogside', women devised home-made gas masks from nappies soaked in lemon-juice, vinegar and water to dis-tribute to those being bombarded by the police with C. S. gas. Eileen Doherty, wife of a prominent Bogside leader, recalls converting her William Street kitchen into a virtual cafeteria to feed the male defenders.[18] The youth who rained pet-rol bombs down on the police from the top of Rossville Flats got their supplies from teen-aged girls who lugged crates of milk bottles and stones up to the roof.[19]

The first curfew imposed by the British army to execute an arms search in the catholic lower Falls Road area in early July 1970 left women imprisoned in their homes and unable to shop for two and a half days. The curfew was broken when women from other parts of the city arrived with supplies of badly needed bread and milk for their friends and relatives. After going straight through and over the barricades and distributing these goods, to the dismay of the soldiers and their officers, the women then marched around the district for a couple of hours singing civil rights songs.[20] The resentment which these searches caused in conjunction with the fears spawned by various sectarian riots since January 1969 brought new recruits to the I. R. A. and the newly formed Provisional I. R. A., bringing the estimated republican activists to 800 by December 1970. The response of the Stormont government, now led by Brian Faulkner, was the introduction of internment without trial on 9 August 1971.

Internment proved a huge security and political blunder. Outdated police intelligence enabled most active republicans to escape the initial round-up. The partisan nature of the initial operation, which neglected to lift known loyalist para-militaries, further incensed the nationalist community. Subsequent swoops in catholic areas over the next few years were hampered greatly by women who warned of imminent security searches by banging dust-bin lids and blowing whistles as soon as armoured vehicles entered catholic estates. Meanwhile, women were the principal instigators and organisers of the rate and rent strike that was endorsed on 10 August by a broad spectrum of nationalists to protest against the internment policy. This exercise in civil disobedience, adopted by 30,000 families in public housing estates occupied by the nationalist popula-tion, put considerable financial pressure upon the government. For instance, in West Belfast's Turf Lodge estate, the 411 families that put rent strike notices in their windows, cost the government £1233 per week in lost revenue. The fol-lowing December the usual postal revenues fell considerably when women or-ganised a Christmas card boycott. In the long run, the rent and rate strike proved costly because the original advice of the organisers from the Civil Rights Asso-ciation to have women organise street committees to collect and deposit rents in banks, was set aside when S.D.L.P. politicians advocated a rent spend

---

18 Author's interview with E. Doherty, July 1985.
19 Jonathan Bardon, *The history of Ulster* (Belfast, 1992), pp 668–70. 43,000 milk bottles were lost from one dairy during the 'battle of the Bogside.'
20 Author's interviews with Madge Davison, Inez McCormack, *op. cit.*

approach, advice which many, but not all families followed. Thus the strike ultimately caused increased hardship when the short-lived power-sharing executive and later the British government deducted arrears from social security benefits and in some cases from pay cheques. However ill-advised in hindsight, the rent strike was the only option that working-class women, traditional controllers of the family purse strings, had available to protest the fundamental violation of civil rights that the internment policy represented.[21]

Many women participated in weekly marches protesting internment over the next six months, the most infamous being that on 30 January in Derry after which 14 unarmed demonstrators were shot by soldiers from the British Paratrooper Regiment. Lost from most historical accounts is the fact that, despite a government ban on marches and grave fears of a potential repetition of Derry's Bloody Sunday, the Newry march of 20,000 held one week later, was led by seven young women civil rights activists including Bernadette Devlin, Eilish McDermott, Madge Davison and Ann Hope who had been wounded in Derry the previous Sunday.[22]

The abolition of the Stormont parliament and the imposition of direct rule from London in late March 1972 brought little relief to women in nationalist areas. Over the next three years, in their efforts to defeat an increasingly active Provisional I. R. A., the security forces conducted brutally invasive early morning house searches which left homes completely ransacked and often were accompanied by the beating of men in front of their families. In 1973, there were 73,000 searches in a community of 400,000 households.[23]

In Belfast's working-class estates, women organised 'hen patrols' to act as lookouts and raise the traditional dustbin alarm upon the arrival of the army and police. Often working in groups of ten, they followed the steps of the patrolling soldiers, not simply to harass them but to monitor their conduct and to try to prevent confrontations between soldiers and teenagers that could easily become violent. In October 1976 after an innocent thirteen-year-old boy, Brian Stewart, was shot by a British soldier in the midst of a *mêlée* between patrolling soldiers and Turf Lodge youths, women in Turf Lodge kept the army out of their district for a month by organising road blocks.[24] Less public actions to protect the youth of their community were taken by women in Derry. In the early 1970s, a number of Derry women stood daily guard at army check points while their children were going to and returning from school to prevent harassment by soldiers. Since the early 1970s women in Derry's Peace and Reconciliation group have acted as observers to insure that the legal rights of young men detained for questioning are upheld. Their mediation has been credited with reducing considerably the tensions between Derry's youth and the security officials in recent years.

By 1978 with 3000 men in jail, Northern Ireland had the highest and youngest male imprisonment rate in Europe.[25] The cancellation of political status for

21 Madge Davison interview and written memoir in Linen Hall Library, Belfast, Women's Collection and Political Collection, fiche 362.
22 Author's interview with Madge Davison, 20 June 1988; *Newry Reporter*, 10 Feb. 1972; *Irish Times*,7 Feb. 1972.
23 1600 catholics and 200 protestant men were interned and imprisoned without trial between 1971 and 1975. For statistical information see especially J. Darby (ed.), *Northern Ireland, the background to the conflict* (Belfast, 1986); *New Ireland Forum Report on the cost of violence relating to the Northern Irish crisis* (Dublin, 1983) and Brendan O'Leary and John McGarry (eds), *The politics of antagonism* (London, 1993), pp 8–44.
24 Author's interview with Lily Fitzsimons, July 1987; M. Ward and M. T. McGivern, 'Images of women in Northern Ireland' in *Cranebag*, vol. 4, no. 1 (19), pp 66–72.
25 An equivalent figure in the United States would be 1 million.

paramilitary prisoners on l March 1976 provided yet another catalyst for political activism among catholic women.[26] A few Ballymurphy women formed the Relatives' Action Committee in February 1976 and within a year there were nine branches, members of which were the chief organisers and participants in protests over conditions in the Maze Prison in the late 1970s during the blanket protest and no wash campaign by republican prisoners. Many of the women also demonstrated every Saturday at Dublin's General Post Office, an obviously symbolic site, in an effort to engage southern sympathy for the prisoners' plight. Although the Relatives' Action Committee was eventually absorbed into Sinn Féin, its women were especially prominent in the public demonstrations occasioned by the 1981 hunger strike. It was largely women visiting their husbands, sons and brothers in the Maze who sustained communication between the prisoners and the Provisional's Army Council and Sinn Féin during this tense period from January to August 1981.[27] In the 1980s women were in the forefront of organising public demonstrations and launching appeals against the use of the uncorroborated testimony of supergrass informers to obtain convictions. Indeed, there was co-operation between women from republican and loyalist areas in the Justice for Lifers organisation to have the sentences of men so convicted reviewed and to pressure the government to end this practice to obtain convictions.[28]

Generalisations about motivation are always problematic, and especially when evidence was being gathered in the midst of continuing conflict. However, the evidence suggests that from 1968–1974, their traditional roles as guardians of their families, rather than ideological factors, initially motivated women in nationalist areas to undertake the actions described. These responses share similarities with patterns of behaviour that Karen Offen has described as 'relational feminism' in her analysis of women's roles in European revolutionary situations in the late nineteenth century.[29] While it is widely acknowledged that the cult of Mary generally provides a passive role model for catholic women, in its Irish manifestation the Marian cult often strongly emphasises a positive role for the Virgin Mother as intercessor to secure salvation. Thus many mothers may have been inspired to engage in activities that previously they would have deemed inappropriate or beyond their capabilities. Recitations of the rosary were part of the weekly picketing ritual at the Cookstown police station in the early l970s and were a frequent feature of the street corner meetings during the Hunger Strike of 1981. Mary Nelis, a Sinn Féin supporter in Derry, who had two sons in the Maze in the mid-1970s, including one on the dirty protest, risked wide community opprobrium when she and two other women stood barefoot and clad only in a blanket outside Bishop Edward Daly's residence to protest what they perceived as the church's indifference to the withdrawal of political status. She recalled feeling '. . . like Mary, the Mother of Jesus doing the stations of the Cross . . . on the hill of Calvery.'[30] The west Belfast gable painting that depicts a fallen Provisional I. R. A. soldier, draped with the tricolour and

---

26 David Beresford, *Ten men dead* (London, 1987); Fairweather, McDonough and McFadyean, *Only the rivers run free*.

27 Beresford, *op.cit.*, p. 30.

28 *Fortnight*, 5 May l986. Over 600 arrests had been made based on the statements of 25 supergrass informers by May l986.

29 Karen Offen, 'Liberty, equality and justice for women: the theory and practice of feminism in nineteenth-century Europe' in R. Bridenthal, Claudia Koonz and Susan Stuard (eds), *Becoming visible: women in European history* (2nd edn., Boston, 1987), pp 335–73.

30 Author's interview with Mary Nelis, July 1985, June 1986.

held by his mother in the fashion of Michelangelo's Pieta surely is more a call to activism than passivity. The careful maintenance of the mural suggests the extent to which republican leaders appreciate and use its potential for encouraging female activism. The historical resonance of this motif can be traced to similar representations in Dublin following the execution of Patrick Pearse, the leader of the 1916 rebellion, and examples of communal recitations of the rosary during the war of independence from 1919–21.

On the other hand, younger republican women serving in Armagh in the early 1980's found a more satisfying role model in the mythical Celtic goddess Scuthath who had tutored the famous Cúchulainn in his martial skills. More recently, there is evidence of great interest among republican and Sinn Féin women in Mary Condren's portrayal of pre-christian Gaelic goddesses as figures of independence, power and authority.[31]

Undoubtedly, some catholic women, especially those with family links to the old I. R.A. became activists as a result of more conscious political convictions regarding the basic legitimacy of the northern state and the administration of its police and judicial system. Others were politicised by the very nature of the trauma which their communities endured from 1970 to 1975 and by the grip that a prison culture held over urban catholic neighbourhoods once large numbers of their men had been interned or convicted. The newly revived republican movement lost no time in efforts to enlist this female support. As early as January 1971 *Republican News* was encouraging young nationalist women to follow the examples of Maude Gonne and Countess Markievicz by joining Cumann na mBan. Home-made leaflets seeking women recruits to the organisation were distributed frequently in West Belfast's nationalist estates in the early 1970s.[32] At present, evidence attesting the exact number of young women who joined Cumann na mBan or even moved on to be regular members of the I. R.A. is not available. By 1976 there were 236 women held in Armagh gaol as convicted prisoners or internees. Over the years the convictions for republican women have been chiefly for aiding and abetting terrorist activity by storing guns and ammunition, keeping safe houses and washing clothes and serving as lookouts. According to a high ranking security official the fire-bombing campaign in London's department stores in winter 1991–92 was executed largely by women republicans. The more lethal para-military activity that Mairéad Farrell was allegedly involved in prior to her being shot by the S.A.S. in Gibraltar in 1988 seems to be the exception rather than the rule for women in the I. R. A.[33]

The deteriorating political situation between 1969–1974 was not without serious impact on many working-class protestant areas, especially in Belfast. Research on loyalist women is still at the preliminary stage, but there is evidence that some provided support services for various loyalist para-military organisations. Loyalist women were active at various locations during the Ulster Worker's Council strike in 1974 that helped to topple the power-sharing executive. Indeed, the Ulster Defence Association had a women's division until 1974 when it was disbanded subsequent to the conviction of three of its members for the murder of Ann Ogilby.[34] In the late 1980s, a woman who had joined the organi-

---

31 Marguertia D'Arcy, *Tell them everything* (London, 1981); discussion at Mary Ann McCracken Historical Society Conference, March 1992.
32 Samples of such recruiting leaflets are in the Linenhall Library Political Collection.
33 The identity of this Northern Ireland Office official is confidential.
34 Fairweather, McDonough and McFadyean, *Only the rivers run free*, p.283.

sation after being intimidated out of her Suffolk home after internment, was prominent in organising support services for families of loyalist prisoners and in efforts to curb criminal racketeering by some of its members. She eventually left the organisation and presently there is little evidence of active involvement by women.[35]

Women in Party Politics:

Although Bernadette Devlin of People's Democracy had moved onto the main political stage when she won the mid-Ulster Westminster seat as a Unity candidate in April 1969, it was not until the mid-seventies that a small number of women from the nationalist areas became involved in formal party politics. Their party choices reflected the historic divisions within Irish nationalism, i. e. between constitutionalism on the one hand, and physical force as an acceptable means to win independence for the entire island of Ireland on the other. The Social Democratic and Labour Party founded in August 1970 represents the first tradition while Sinn Féin represents the latter.

Female activists within Sinn Féin have tended to be women whose male relatives or friends were interned from 1971 to 1975, or who witnessed as teenagers their fellow catholics being intimidated and burned out of their homes and shops in mixed neighbourhoods. The memories of those experiences are deeply imbedded in the psyche of Sinn Féin supporters.[36] Unemployment, poverty and well-founded fears for their own security keeps many active Sinn Féin women as well as men physically and socially isolated in their own neighbourhoods.

By the early 1980s, younger Sinn Féin women, especially former prisoners, empowered by their roles in sustaining community solidarity and partly by the international feminist ferment, began to seek positions of leadership within the movement and to demand that women's issues be addressed alongside the constitutional issue in the party programme. In 1983 a separate Women's Department was established and women were guaranteed eight out of thirty-two seats on the Sinn Féin national executive. The adoption of feminist planks has not been without stress and strain as evidenced by the party's reversal in 1986 of a previous resolution recognising a woman's right to choose an abortion. Knowledgeable observers contend that there is often a huge gap between stated party policy on women's equality and its acceptance by the male rank and file in daily life.[37]

Women who became members of the Social Democratic and Labour Party in the mid to late 1970s generally had more financial security and education than those that joined Sinn Féin. Often their experience of working outside of Northern Ireland or having grown up in mixed neighbourhoods prior to 1969 were factors in their decisions to support a constitutional approach. The majority of S.D.L.P. women currently holding council seats in 1992 joined the party at this time and did a great deal of service work in fund raising and canvassing prior to

---

35 *Ibid.*, pp 281, 302, 309; author's confidential interviews with U.D.A woman, 9 July 1987, 28 July 1989.
36 Fionnuala O'Connor, *In search of a state: catholics in Northern Ireland* (Belfast, 1993). Interviews therein attest strongly to these memories.
37 See Catherine B. Shannon, 'Catholic women and the Northern Irish troubles' in Yonah Alexander and Alan O'Day (eds), *Ireland's terrorist trauma* (London, 1989), pp 234–248; see also Marie Mulholland, 'Between a rock and a hard place' in *Unfinished revolution: essays on the Irish women's movement* (Belfast, 1989), p.35.

running for council seats. Brid Rodgers is the most prominent female S.D.L.P. politician. She served as party general secretary in the early 1980s, was a delegate to the New Ireland Forum in 1983 and twice has contested the Upper Bann Westminster constituency against Ulster Unionist incumbents. She served two terms on Craigavon Council for a constituency that includes the protestant stronghold of Portadown. She worked diligently to persuade her vulnerable catholic constituents in Portadown's housing estates not to look to the Provisional I. R. A. for protection against loyalist attacks during times of heightened sectarian tensions in the aftermath of the Anglo-Irish Agreement and during the season of Orange parades in 1986 and 1987.

In the last four years, a few younger women members with university degrees have been active in party management and policy and are working hard to have the S.D.L.P. platform address the interests and needs of women, admittedly a difficult task given the conservative middle-class catholic attitudes predominant among its male leaders and membership. One female councillor acknowledged the need to attract more working-class members to the party, but indicated that such efforts by two women in her constituency were abandoned when bomb threats on their homes were made. Similarly, a twenty-five-year-old woman living near the interface of the Falls Road and the Shankill Road, who as a teenager had worked for Sinn Féin and was interrogated by the police at sixteen, was severely harassed by female neighbours when she worked for the S.D.L.P. in the Westminster election of 1987. Her support was especially courageous given that unquestioning loyalty for Sinn Féin and the Provisional I. R. A. is seen by many of her neighbours as the only defence against loyalist para-militaries from the nearby Shankill Road. She attributed her political migration to her conviction that the republican campaign was doing nothing to alleviate her own unemployment or to provide a better future for her small child. Although she did not become a party member, she volunteered her services again in the hotly contested 1992 election in which the S.D.L.P. candidate, Dr Joe Hendron, defeated the Sinn Féin President Gerry Adams for the West Belfast seat. Electoral returns from the last two Westminster elections indicate that the S.D.L.P. holds the support of nationalist voters by a margin of 2 to 1 over Sinn Féin. Recent surveys examining political attitudes by gender suggesting no marked differentials between male and female attitudes on the constitutional question would indicate that the 2 to 1 margin probably holds for women in the nationalist communities.[38]

Meanwhile, the small non-sectarian Alliance Party, which supports the link with Britain, provides space for catholic women who reject both the S.D.L.P. and Sinn Féin. Many of these women are middle class and have had better than average education. Quite a few active Alliance women members were drawn to the party as the only logical political choice resulting from their being in mixed marriages. Protestant members, the religious majority in Alliance, reject the politics of the main Unionist parties as essentially sectarian. The Alliance party is distinguished by the largest number of women members and had the highest percentage of women candidates in the 1989 local government elections and again in 1992 when 33% of its candidates were women. One quarter of its 44 councillors are women. Alliance was the party of choice for 7% of women polled in a 1991 survey whereas Sinn Féin received only 1% support.[39]

---

38 Valerie Morgan, 'Bridging the divide' in *Social attitudes survey*, 1991–2 (Belfast, 1992), p. 137.
39 *Ibid.*

Formal party involvement by women in the protestant community has grown much more slowly. This is not surprising given the Calvinist theology that infuses Ulster presbyterianism, the majority sect of northern protestants. After 1922 involvement of aristocratic and upper-middle-class women in the Ulster Unionist Party that had characterised the era from 1910 to 1922 decreased steadily, so that by the 1960s the Unionist Party structures had a very male ethos.[40] In the 1970s and early 1980s the majority of protestant women, feeling besieged by the Provisional I. R. A. and threatened by absorption into the Irish Republic, were content to leave politics and the defence of their state to the traditional male leadership. The fragmentation of the protestant church structure may have contributed to isolation and a reluctance to pursue political and social grievances through the type of collective action that nationalist women have used. Survey data from the mid-1980s showed middle-class protestant women over thirty-five as especially content with traditional privatised roles and adverse to active political participation.[41] Women prominent in the two main unionist parties have tended to be the wives or daughters of male politicians, such as Hazel Bradford in the Official Unionists, or Rhonda Paisley and Iris Robinson in the Democratic Unionist Party. In the 1993 local government elections women constituted 14% of Official Unionist and 10% of Democratic Unionist candidates. (These percentages are comparable to the ratio of women candidates for the S.D.L.P which had 16% and Sinn Féin 11.5%. The latter figure was down 4% from 1989, undoubtedly a result of the frequency of assassination attempts on Sinn Féin councillors in recent years.) Yet a recent survey showed a larger number of protestant than catholic women willing to indicate firm support for a political party.[42] The significance of this finding in relation to earlier evidence of lack of political participation is not entirely clear.

Nonetheless, during the past four years there has been a rising political consciousness among protestant working-class women which is putting pressure on both the Official Unionist Party and the Democratic Unionist Party to address critical issues for women, especially in the areas of childcare, education and training and social security benefits. Increasingly they have become aware that main-line unionism has little to offer their class and gender. As one woman active in the protestant Shankill community put it in 1991,

> . . . The protestant community just wasn't aware of the discrimination, but they are now. We were told we were God's own people. Stormont would look after us and all the rest of it. We don't believe that any longer. We haven't believed that for thirty years and we are fighting back. I don't feel discrimination is any less on the protestant side than it is on the catholic side, I think all women suffer the same discrimination whether it be by the church or by the state or whatever.[43]

Similar to their catholic sisters, they face immense obstacles in the conservative values shared by the male catholic and protestant church leaders and politicians who constitute a deeply entrenched ecumenical patriarchy that will not be easily defeated. Despite their bitter differences on the constitutional question, there was an unanimous opposition among the male leaders of the S.D.L.P.,

---

40 Diane Urquhart, 'The Ulster Women's Unionist Council, 1911–1940' (unpublished M.A. dissertation, Queen's University, Belfast, 1991).
41 John D. Kramer and Carol Curry, *Attitudes toward women in Northern Ireland*, (Belfast, 1985).
42 Morgan, *op. cit.*, p. 137.
43 Author's interview with Mina Wardell, 20 Nov. 1991.

the Official Unionists and Rev. Ian Paisley's Democratic Unionist Party toward the siting of a Brook Clinic in Belfast during 1991–92. Brook Clinics provide family planning services and sex education to young girls. The need for this service is underscored by the sky-rocketing rate of teenage pregnancy with 75% of mothers under twenty being single.[44] Clinic opponents fear that the more lenient criteria for abortion that pertain in the rest of the United Kingdom will be extended to Northern Ireland. However, with considerable cross-community and cross-class support from northern women the clinic opened in September 1992. On the opening day, the Reverend Ian Paisley arrived to preach his message of damnation, but he beat a hasty retreat after a number of women outside the entrance started dancing to the hymn 'Rock of ages' when it came blaring over his sound system.[45]

Despite signs of rising political consciousness, the number of women actively involved in formal party politics remains very small. Westminster and European parliament seats remain a male monopoly, and women constitute only 11% of local government councillors. Women candidates in the last local elections dropped by 15 from 14% in 1989 to 12% in 1993. Only 34% of northern women are willing to indicate a particular party allegiance while the 47% who are unwilling to state a party preference undoubtedly reflects the irrelevance of both orange and green nationalism to women's pressing daily needs.[46] Many working-class women express frustration that the absence of a northern Irish labour party prohibits them from registering their opposition to the conservative economic and social policies of the British Tories. Meanwhile, many middle-class feminists are involved in campaigns for legal and social reforms to remove the disabilities which all northern women suffer in areas of wage, job and educational discrimination and for better protection from domestic violence. Such women have been especially active in efforts to ensure that the full benefits of the European Social Charter be extended to Northern Irish women.[47]

Even for women who are politically interested, the pressure of family duties, lack of money and sufficient encouragement from the male dominated party structures present serious impediments to active participation even at the council level. Yet lack of involvement in formal party politics does not necessarily translate into political indifference or ignorance. A number of recent conferences such as the Worker's Education Association's 'Women in the 1990s' and surveys indicate that northern women's political views have become more varied, complex, sophisticated, and indeed flexible than is often assumed. Factors of class, educational attainment, age, geographic location of residence and personal experience of the 'troubles' are far greater influences on their political priorities and party choices than passive acceptance of the nationalist or unionist loyalties of their menfolk or districts. The diversity of women's experiences and political views even within the small confines of Belfast, let alone the geographic breadth of the province, is great. For instance, a middle-class Roman catholic woman living in the Malone Road area would find life only four or five miles away in west Belfast's Ballymurphy district quite alien to her experience and values. Likewise, a catholic nationalist women living in Portadown would feel far more threatened by sectarianism than her counterpart in Derry. The

44 *Social trends: annual report of social and economic statistics*, HMSO (Belfast, 1992), Table 24.
45 Talk by Inez McCormack to Boston Public Library Symposium on Irish Women, 9 Mar. 1993.
46 Valerie Morgan, *op. cit.*, p. 137.
47 Bronagh Hinds (ed.), *Northern Ireland women's European platform* (Belfast, 1992).

experience of a protestant woman living in Tiger's Bay in north Belfast would have little in common with a protestant woman living in Holywood or Cultra where the political conflict rarely intrudes on the daily lives of its upper-middle-class residents.[48]

Reaching for Common Ground:

Over the past decade there have been many instances where northern women, while acknowledging their political and religious differences, have co-operated on issues of concern to all women. One such instance arose in the early 1980s when the largely female N.U.P.E., whose membership spans both communities, was able to work through the contentious political implications of the strip-searching of female republican prisoners and condemn the practice as a violation of the sexual and personal integrity of all women.[49]

Poverty is the common reality that approximately 80,000 catholic and protestant Belfast women share as widows, divorcees, single parents and wives of prisoners and unemployed men. Their frustration with the irrelevance of party politics is often expressed by the slogan 'Neither of the two flags can you fry up in the pan for breakfast.' It was in this spirit that the Women's Information Group was established in 1980 to enable protestant and catholic women from different sections of the city to meet and acquire information regarding medical benefits, social security, locations of shelters for victims of domestic violence and opportunities for adult education. Their monthly meetings at different locations in the city represented a pioneering effort to break down the ideological prisons and sectarian isolation that has characterised working-class areas since the early 1970s.

Since the mid-1980s women have been particularly active in conducting investigations and surveys to document deplorable housing conditions in working-class areas of Belfast. Evidence collected by women living in Divis Flats and Turf Lodge showed the impact of poorly built and badly maintained housing on their families' health and helped bring about the demolition of Divis and renovations in Turf Lodge. In the last five years, similar research projects have been compiled by women living in Andersonstown, the Upper Shankill, Ballybean and Ardoyne which highlight the impact of poverty, unemployment and inadequate education, housing and health care on their families and communities.[50] Such investigations are no longer confined to Belfast as rural women are now banding together to document their needs and lobby for increased financial support.[51]

Recently women have co-operated with professional researchers to document the pervasiveness of domestic violence across the province and in every social stratum. The incidence is especially high in the disadvantaged urban areas where the presence of armed men can put women in vulnerable positions if they call the police into their communities. The heightened publicity given to

---

48 See O'Connor, *In search of a state* for examples of this in the catholic community.
49 Author's interview with Inez McCormack, Jan. 1987.
50 Ardoyne Women's Research Project, *Unheard voices: women's needs in Ardoyne* (Belfast, 1992); Alison Rocks and Edel Teague, *Women in Andersonstown* (Belfast, 1988); *Women in the Upper Shankill* (Belfast, 1988).
51 See for example, Avila Kilmurray and Carmel Bradley, *Rural women in south Armagh: needs and aspirations* (Belfast, 1989, rept. 1991).

this issue and determined lobbying has led the police to institute training pro-grammes to prepare officers to deal more effectively with the problem.[52]

Lack of adequate childcare facilities for children under four is a huge prob-lem confronting women who want to work or acquire training and further edu-cation. Northern Ireland has only half the nursery places per capita as are pro-vided in the rest of the United Kingdom. This leaves it at the very bottom of the list in comparison to the European Community member states.

Who fills this childcare gap? As always it is women working either through churches or local women's centres who have established mother and toddler groups enabling young mothers in disadvantaged areas to escape the isolation of their homes. Often participation in these groups has been the catalyst for young mothers who have few educational qualifications to enrol in personal development, assertiveness training, literacy and computer training courses. Many of these courses are organised by the Women's Education Project. In-creasingly women completing their programmes enrol at further education col-leges and some eventually go onto university level education. Reflective of the low priority given to women's needs by government bureaucracy is that despite a decade of success, the Women's Education Project struggles to obtain ad-equate funding and resources to meet the growing demands for its services.

The amount of voluntary work that women in Northern Ireland do locally for the elderly, the handicapped and those affected by the political violence is impressive. It ranges from counselling the widows of policemen, teaching in prisons and helping the families of prisoners cope with emotionally draining and time-consuming prison visits to providing respite help and training for young mothers with handicapped and special needs children. Women have been at the forefront in organising programmes aimed at keeping teenage boys away from potential recruitment by para-militaries, and in the more widely known reconciliation work at Corrymeela and with organisations like the Peace Peo-ple or Families Against Intimidation. To a large extent this is work that knows no political and sectarian boundaries and should be emulated by a wider repre-sentation of Northern Irish men than is now the case.

The increasing willingness of Northern Irish women to work for common goals irrespective of their political differences was graphically illustrated in 1990 when the Unionist dominated Belfast City Council threatened to remove its grant from the Falls Road Women's Centre which is situated in one of the strong-est pockets of Sinn Féin's support. Protests against this obviously partisan deci-sion were forthcoming even from predominantly protestant women's groups at the Shankill's Hummingbird Centre and the Ballybean Women's Centre. The independent Unionist councillor Elizabeth Seawright, whose husband George had been assassinated by the Provisional I. R. A. in December 1987, was espe-cially vocal in protesting this decision and worked with Sinn Féin women to get the funds restored. It is significant that the contacts this episode established between the Falls Road and the Hummingbird centres ultimately resulted in their joint invitation to President Mary Robinson to come to Belfast. Her Feb-ruary 1992 visit during which she met women from all sections of the commu-nity was the first by an Irish president to Northern Ireland. The co-operation begun in arranging President Robinson's visit also led to the establishment of

---

52 Monica McWilliams and Joan McKiernan, *Bringing it out in the open. Domestic violence in Northern Ireland* (HMSO, Belfast, 1993).

the Women's Support Network, a clearing house of information for 192 women's groups and the services they offer across the province.[53]

No less than in the United States, feminism remains a problematic word to many Northern Irish women, particularly for those in the working class. Yet it is undeniable that a feminist consciousness is being nurtured there. While this does not necessarily guarantee progress in ending the bitter political and religious divisions, it has the potential to foster a wider atmosphere of trust and an ability to deal with difference. As one high ranking security official observed to me in 1992, 'Without the women, we might have a Yugoslavia situation here.'

Gerda Lerner in her recent book defines feminist consciousness as

... the awareness of women that they belong to a subordinate group; that they have suffered wrongs as a group; that their condition of subordination is not natural, but is societally determined; that they must join with other women to remedy these wrongs; and finally that they must and can provide an alternative vision of societal organisation in which women as well as men will enjoy autonomy and self-determination.[54]

Over the last decade, women in Northern Ireland have gone a long way down this road. They still have miles to go, but there are grounds for hope and confidence that their continuing efforts are making a significant contribution to improving life for the entire community in Northern Ireland.

---

53 Ruth Taillon (ed.), *1992 directory of women's voluntary organisations in Northern Ireland* (Belfast, 1992).
54 Gerda Lerner, *The creation of a feminist consciousnesss* (New York, 1993), p. 274.

21   Comparative Analysis of Abortion in Ireland, Poland and the United States

Joan Hoff

THIS essay argues that despite cultural and national differences, women are not full citizens in Ireland, Poland, and the United States (nor are they in most other western, democratic countries). There are three reasons for this second-class citizenship of most women world wide: 1) women cannot meet the standards set by male definitions of citizenship, *except on male terms*; 2) women do not have the same reproductive and sexual rights as men; and 3) women do not have adequate legal protection against physical abuse (one manifestation of physical abuse I maintain is forcing women to carry unwanted foetuses to term). It is my contention that anti-choice arguments usually reflect civic or religious attitudes that implicitly condone or encourage second-class citizenship status.

To demonstrate how abortion policy in all three countries contributes to the position of women as citizens, this essay focuses the relationship between church and state and the role played by language in the debates over abortion, including the much heralded contemporary use of 'family' or 'christian' values to mask anti-woman sentiments in religion, legislation, and education. It concludes with a review of Irish abortion policy through to 1993.

Church and State Relationships

In both Ireland and Poland there has been a much longer and closer association between a single church and the state than in colonial America or the United States. In both Ireland and Poland the church has been an integral part of intellectual life and the struggle for nationhood. Both countries also have more homogeneous populations than the United States. Denied status as independent countries since the end of the eighteenth century when other nation states began to emerge in the rest of Europe, colonisation prevailed in the case of Ireland and partition in the case of Poland. In each instance, vulnerable geographical locations played a large role in the denial of nationhood to the Irish and Polish peoples: Ireland's proximity to Britain and Poland's sandwich-like position between Germany and Russia. In all three countries, however, the striving for modern nationhood began in the last quarter of the eighteenth century – with only the United States succeeding in its struggle for independence at that time.[1]

---

1 Patrick Clancy, Mary Kelly, Jerzy Wiatr, and Ryszard Zoltaniecki (eds), *Ireland and Poland: comparative perspectives* (Dublin, 1992), pp xi, 3, 15.

In the long and often bloody battle for traditional state status the catholic church has, of course, played important roles in both Ireland and Poland for centuries. This convergence between catholic and national identity still characterises both countries – a convergence that gave (and gives) the church a role in intellectual and everyday life that is difficult to comprehend for those of us who reside in nations where the connection between religion and nationalism is much more tenuous, if it exists at all.[2] Coming from the United States it is particularly difficult to appreciate fully the ways in which religion permeates the very fabric of both societies – and for legitimate historical reasons.

What is significant about the relationship of church and state in Ireland and Poland with respect to the issue of abortion, is not *how they differ* from the United States where we pride ourselves on our separation of church and state, but *how different* the current role of the catholic church in state affairs is in both countries. Since the 1970s, the church in Ireland has played an increasingly reactive role, while in Poland, because of its recent alliance with Solidarity in the 1980s and now under liberation, the church is playing a more aggressive role in state policy – initiating change rather than simply responding to it.[3] (Although it is by no means a completely accurate comparison, there is a similarity between the role played by the church in Ireland in the 1920s and 1930s and the one currently being played by the church in Poland in the 1980s and 1990s.)

Another anomaly in the role played by the church in Ireland and Poland can be seen in its relationship to the political economies of both countries. The church remains opposed to the accruement of wealth in each nation – an attitude that has made its philosophical views about capitalism problematic, to say the least. Consequently, the church in Poland opposed the imposition of communism in 1948 not only because it represented secular totalitarianism, but also because of its economic policies – even though the state-run economy 'facilitated upward mobility and satisfied material needs' on the part of the masses of Polish people better than previous domination or brief periods of independence in the past had.[4] Since liberation the Polish catholic church has supported privatisation of the economy (as long as it does not produce ostentatious wealth or encroach on the property confiscated under communism to which the church is now laying claim) even though this form of 'primitive capitalism' will initially impoverish many and make living conditions worse.

Likewise, in Ireland the church has not opposed the market economy inherited from colonial times, even though capitalism in Ireland cannot be considered a success since it has never overcome massive unemployment (in 1993 its 20 percent unemployment was the second highest among EU nations) which, in turn, has produced mass migration of Irish citizens abroad and numerous social inequalities at home. In other words, Poland was viewed as one of the most important and 'successful' of the socialist countries before liberation, while Ireland's practice of capitalism has, after a boom period in the 1960s, unfavourably compared to that in other western European nations. Yet the Poles with the support of the church chafed and rebelled against their relative material 'success' of socialism, while the Irish and the church seemingly have accepted

---

2 *Ibid.*, pp 289–291, 303–305
3 *Ibid.*, p. xi, 293–295, 307–309, 311–312
4 *Ibid.*, p. xxi.

the inequities of their 'floundering' version of capitalism. In neither country, regardless of socio-economic system, has the church taken the initiative to foster gender equality.

My point is that the catholic church has placed less emphasis on the economic well-being of people in both countries than it has on saving their souls – usually at the expense of women's rights, especially during periods of newly won liberation and independence. Contrary to United States government propaganda, capitalism usually exacerbates the existence of (or return to) sexist public policies in countries where democracy is weak because it flourishes best when only a modicum of democratic practices exists. Consequently, the state, in conjunction with the church, in both Ireland and Poland is at the moment cautiously pro-capitalist from an economic point of view and, therefore, quite accepting of the gender and economic inequalities – also characteristics of capitalism – that ensure women remain second-class citizens as they do in the United States.

Despite America's long-standing constitutional proclamation separating church and state, since abortion was declared legal in 1973, courts and legislative bodies have responded to a variety of traditional and fundamentalist religious groups who oppose abortion – not unlike the way courts and legislative bodies responded to religious pressure at the end of the last century. At the turn of both the twentieth and twenty-first centuries, therefore, it is possible to argue that anti-woman religious campaigns have occurred in part in response to successful women's movements, in part because of impending 'millennium anxieties', but mainly because of socially constructed cultural and religious mores that continue to be reflected in the political and juridical systems of the United States. Only at the end of this century, however, has the American tradition of religious freedom so distorted American politics that it temporarily threatens to tear the country apart.[5] For example, conservatives may use the abortion issue to thwart passage of a national health insurance programme under the Clinton administration if this medical procedure is covered.

The controversy over abortion in the United States is more publicly passionate and violent than in either Ireland or Poland because free speech about sexuality is much more common and handguns are increasingly being used to settle all kinds of disputes ranging from drug deals to traffic altercations. In all three countries, however, the rhetorical aspects of the debate is bordering on the hopeless because it has been side-tracked by religion into the metaphysical issue of whether the foetus is a human person, with one side automatically equating abortion with murder. This position brooks no compromise even though some of the most vocal opponents of abortion are sometimes hoisted on their own petard by the 'doctrine of double effect'. According to the way the 'double effect principle' is defined in Ireland (and applies in other countries) pro-lifers almost always admit that abortion is permissible to save the mother's life. An excellent example of the 'doctrine of double effect' occurred in the United States when George Bush and Dan Quayle, his running mate in the 1992 presidential

---

5 Ronald Dworkin, *Life's dominion: an argument about abortion and euthanasia* (New York, 1993). In making this point Dworkin argues that abortion and euthanasia are too complex for public debate because politics promotes such emotive, polarising discussion that there can be no compromise. I disagree because he is accepting the current debate in the United States at face value as though it could not be cast in any other than moral terms by either side; and ignoring the ability of technology to defuse the entire issue very rapidly through abortion-inducing chemicals and hormones administered in the privacy of the home, such as the abortifacient RU–486.

campaign, both as pro-lifers, said that they would support their daughters or granddaughters if they decided to have an abortion – clearly indicating that they did not think the foetus was in any sense a human being or that they irrevocably equated abortion with murder.[6]

Under the 'double effect principle' the foetus is not accorded the same protection as a born individual and hence there is a contradiction (or potentially division) within the pro-life argument that has not effectively been exploited by pro-choice advocates. By this I mean that it should be possible to find ways to negotiate with anti-abortionists who are willing to consider abortion under certain cases of necessity, such as rape, incest, or when the life of the mother is at risk. Even the 'double effect principle' does not recognise the rights and responsibilities of full citizenship for women, but it does move the argument out of the metaphysical dead end of whether the foetus is a human being. At present the anti-abortion legislation enacted or proposed in all three countries says that pregnant women cannot be granted moral agency to choose how to live – not just for nine months of their lives – but for the rest of their lives. With the exception of Ireland and Poland all other European nations allow abortion virtually on request during the first three months of pregnancy. The same was true of the United States, until recent supreme court decisions began to allow states to place restrictions on first trimester abortions.[7]

The Language of Abortion

Language is important to the discussion of abortion in all three countries because it has the literal and figurative power to construct social reality. So we hear of such terms as pro-life, pro-choice, and voluntary motherhood versus involuntary motherhood being bandied about. Non-viable foetuses are arbitrarily endowed with the same constitutional rights as viable adult females, as though the real lives of living women were not at stake. Sometimes women are left with nightmare choices about aborting non-viable foetuses or practising infanticide by letting viable new born babies die, and yet the public language with which to conceptualise and contextualise abortion remains limited – even in countries like the United States, France, and England where abortion is legal. The democratic right of citizenship as an adult to choose to control one's own body from forced pregnancies (or forced abortions) has no reality because the 'foetus is promoted to the status of "baby", [or child] while women are "demoted" to the materially undervalued condition of "mother".'[8] Thus, language remains key. The terms chosen by countries to discuss abortion, contraception, and reproductive health matters in general usually reflect their laws and customs.

The abortion debate has often been cast in terms of 'mother v. foetus', as if they were two separate entities. This is a false dichotomy if there ever was one.

---

6 William Duncan, 'Abortion, divorce and the debate about liberty,' in Anthony Whelan (ed.), *Law & liberty in Ireland* (Dublin, 1993), p. 124; Barbara Hinkson Craig and David M. O'Brien, *Abortion and American politics* (Chatham, New Jersey, 1933), pp 35–68 and Dworkin, *Life's dominion*, pp 3–24. Dworkin seems unaware that the Irish lawyers have already identified this 'double effect principle' and in the proposed amendment to the Irish constitution in 1992 the wording did give equal protection to the foetus. See footnote 20 below for the exact wording.

7 *The Irish Times*, 29 May 1993; *International Herald Tribune*, 15 May 1990; 29–30 May 1993, pp 1, 4.

8 Ailbhe Smyth (ed.), *The abortion papers* (Dublin, 1993), pp 7–8.

There can be no foetus without a womb, and there can be no womb without a woman. There is only one viable entity present in pregnancy and that is the woman because her womb is an integral part of her. This spurious separation allows religious and government spokespersons to claim that women do not really own their wombs – that the state does or churches do. As a result, women are accorded responsibility to care for any children they bear, but not the freedom of informed choice as citizens to control their own reproductive capacity. To give women such control in the name of citizenship seems to imply to those who want to distinguish women from their wombs that *they might get out of control as women* unless this part of their bodies is regulated by male-dominated institutions. In all three countries under discussion the public debate over abortion implies that women are irresponsible; *that all pregnant women will choose abortion*, when contemporary statistics strongly suggest that it is an act of last resort for most.

Moreover, foetalcentric or childcentric language leads to oxymoronic references to non-viable foetuses as 'developing members of the family',[9] while women are reduced to mothers by both sides as though this was their only or primary task in life. As long as abortion continues to be discussed in terms of motherhood, women will continue to be viewed as mere mobile incubators ('walking wombs') or transportation vessels for non-viable foetuses which have been designated by anti-abortionists as having independent and equal rights before birth with their mothers.

The fact that no western legal system has ever granted property or inheritance rights to these non-viable, dependent entities before birth and that few religions recognising baptism accord this sacrament to other than full-term, live babies is lost in the recent attempt to resurrect non-viable life forms into viable human beings before they can live outside a woman's body. Such arguments about the 'humanness' of unborn foetuses were not made on religious or scientific grounds until very recently.[10] That they have never made legal sense goes without saying, as western forms of jurisprudence clearly attest. In the United States at least, arguments about 'quality' of life are becoming part of the legal discourse at both ends of life because of the current controversy over

---

9 *New York Times*, 24 May 1991 (quoting the United Catholic Conference, the official organisation of American Roman catholic bishops).
10 Both the religious and medical reasons for opposing abortion are of very recent origin; that is, they date back at best to the end of the last century and in some instances only appeared in this century. Official religious opposition to abortion in the first two trimesters of pregnancy (or first 24 weeks) did not emerge among any of the major denominations, including the catholic church, until the last quarter of the nineteenth century. And in the United States the catholic church did not join the anti-contraceptive/anti-abortion crusade until well into the twentieth century (1968 to be exact) because it feared even more discrimination against catholic immigrants if it did because abortion was widely used as a means of birth control by white, middle and upper-middle-class protestants until their ministers began to fear the higher birth rates among non-protestant immigrants. So opposition to abortion procedures before what was commonly known as 'quickening' or detection *by women* of movement in their wombs – usually placed at the end of the second trimester or six months into the average pregnancy – has not existed from time immemorial as a principal teaching of the catholic church, nor did it exist from the reformation among major protestant denominations. Under Pius IX abortion *at any stage during pregnancy* became part of canon law, *but not until 1917*. The church still does not baptise miscarriages or still births carried to term, indicating that it does not consider them 'living creatures with souls.' (See Maureen Muldoon, *The abortion debate in the United States and Canada: a source book* (New York, 1991), pp 159–160; and Marilyn French, *The war against women* (London, 1993; reprint of 1992 Summit Books edition), p.85). Anti-abortion campaigns began in the United States, at least, as a reaction to demands by women for greater rights in society and also as a convenient, expedient collaboration among protestant groups and the medical profession at the end of the century to control what we now know was the emerging recognition of female (and male) sexuality as other than simply a function of reproduction. Today opposition to abortion is

abortion and euthanasia.[11] At the moment, however, this is too subtle an argument to be tolerated in the intolerant climate created by religious dogma on both issues.

Unfortunately most of the terminology used in the debates over abortion have been coined by the anti-abortionists (or in reaction to them) in the United States and England and then exported to other countries where they cannot always be easily translated for linguistic and cultural reasons. The term 'pro-life' was their supreme achievement because it implied that their opponents were 'anti-life,' even though, in the United States at least, it is usually the pro-life groups which are 'anti' most viable forms of life because they usually are the first to support American military actions abroad, capital punishment, and to oppose welfare and child care legislation. So it is often said that the American pro-life advocates believe life begins with conception and ends at birth. In all three countries the proponents of access to abortion were forced to adopt 'pro-choice' as their slogan, even though as a group they are usually 'for' more viable forms of life than their opponents.

The result is, that at the end of the twentieth century, *women as citizens* have been misrepresented or silenced, and almost always devalued rather than empowered in the public discourse over abortion. That is why I prefer the more controversial but less confusing terms, pro- and anti-abortion(ists) or at the very least pro- and anti-choice and I will use these two sets of label interchangeably throughout this essay. I also prefer to recast the debate (where it belongs) in the language of citizenship in order to transform the United Nations and EU concept of human rights from a feminist perspective. Among other things this would mean taking into account viable women's lives and the particular types

---

no longer supported by the international medical community or by most members of the medical profession in the United States, but it is supported even more strongly by religious groups for a reason very similar to the one that their ancestors reacted to: namely, the demand on the part of women for greater rights and responsibilities and to what were then contemporary conditions, including something we also face today the phenomenon known as 'millennium anxiety'. These concerns have little to do with religion, *per se* but have a great deal to do with preventing women from becoming full citizens on the grounds that unlike men they do not have the right to control their reproductive and sexual functions; that men, but not women, can be trusted with this fundamental right of citizenship.

Other reasons that one hears so often in the abortion debate are medical or scientific reasons against it because of the once again very recent assertion that life begins at conception. Doctors supported state legislation in the nineteenth-century anti-abortion movement in order to assert their own authority over pregnant women's bodies in their battle with midwives, but not on the grounds that life began at conception. These laws ostensibly 'protected' women from 'unscrupulous practitioners and the use of poisons, and the foetus *post-quickening abortions*.' There is no major body of scientific literature that has ever in the past or today claimed that human life as we know it as human personhood which is linked to capacity for consciousness and decision and hence agency and viability exists at the moment of conception (See Muldoon, *Abortion debate*, p. 159 (quotation, emphasis added); and Ruth Riddick, *The right to choose: questions of feminist morality* (Dublin, 1990), p.18). All of these functions of 'live' citizenship can only take place outside of the womb. Individual claims to the humanness of the egg at the moment of impregnation can be found but I stress there is no major body of medical literature that has ever claimed this to be true for one simple reason: even modern science cannot sustain the viability of human life outside the womb if the foetus weighs less than 500 grams. When and if medical science progresses to the point where this is possible then arguments about the viability of the foetus will change. But this type of viability does not now exist. All this quasi-scientific argument has produced is 'quarter-of-a million dollar foetuses' (this is how they are referred to by the medical profession in the United States because that is what it costs to try to save them – that is, to make them viable until they can function as viable humans without life support systems) – and most of these 'quarter-of-a-million-dollar' foetuses cannot now be saved. It is this unscientific claim about life at conception that allows the anti-abortionist to argue that non-viable foetuses have the same rights as viable girls or women.

11   Ruth Colker, *Abortion & dialogue: pro-choice, pro-life, & American law* (Bloomington, 1992); Mary Ann Glendon, *Abortion and divorce in western law: American failures, European challenges* (Cambridge, Massachusetts, 1987); and Dworkin, *Life's dominion*.

of global violation of female life, including absolute or partial denial of, or arbitrary enforcement of abortion.[12]

As if these euphemisms and oxymorons were not enough, the terms, 'family' or 'christian' values have come into prominence in the 1990s. For example, christian values in Poland correspond very closely to the Vatican Code on Family Values. In September 1990 the Polish catholic church, on its own authority initiated mandatory religious classes in public schools, and on 2 March 1993, Polish radio and TV were told by the senate to respect mandatory christian values. However, as of the summer of 1993 the parliament had not funded the council set up to monitor both private and public stations. In contrast, the church has been more circumspect, but no less firm, in Ireland with respect to its control over religious education in denominational schools, often leaving it to groups representing parents such as the Congress of Catholic Secondary School Parents to speak for it. Nonetheless, in a critical response to a policy paper on education issued by the Irish minister of education, the position of the church remained that 'curriculum and policies of the school[s] must be consonant with catholic teaching and tradition.' In the United States 'family values' featured prominently in the 1992 presidential campaign even though politicians and the public divided over what constituted a family in modern society.[13]

What is usually forgotten about the use of such phrases as 'family' or 'christian' values is that the attempt to regiment private morality by churches and governments and political parties is a sure sign that there no longer is national ethical consensus about them. This difference between a consensual ethic necessary for establishing a civil society and the assumption that consensus exists or can be enforced where private morals are concerned is a painful dilemma that all new nation states face. It also reflects the old story of locking the barn door after the horse has gone. It is time that the sexist (and in the United States, racist) assumptions of these terms be recognised once and for all.

Recent Developments in Irish Abortion Policy

The Republic of Ireland stands alone among EU countries in not granting women the right to abortion, except those acceptable to catholic theology under the doctrine of 'indirect intent', usually meaning only in the case of ectopic

---

12 I realise that the reformulation of human rights from a feminist perspective is a difficult task, but the rapes of women as part of the 'ethnic cleansing' in Bosnia-Herzegovina have, unfortunately, created the opportunity to internationalise the issue of rape as a war crime and to highlight other gendered violence directed towards women. Up to now such atrocities have been either ignored as exaggerated wartime stories, or simply viewed as a 'normal' part of warfare and, hence, seldom given credence in post-war legal documents and trials. This was most clearly evidenced by the way the mass rapes of Bangladeshi women by Pakistani soldiers in 1971 were quickly forgotten by the international community. It would also help if the United States could at long last be forced to ratify the major UN human rights treaties and conventions (some of which apply very specifically to women) (See Joan Hoff and Christie Farnham, 'Rape, murder, and plain old neglect' in *Journal of Women's History*, 5 (Spring 1993), pp 6–8; Elizabeth F. Defeis, 'International Trends' in *Journal of Women's History*, 3 (Spring 1991), pp 90–107; *International Herald Tribune*, 8 Dec. 1992; *Newsweek* (international edition), 4 Jan. 1993, pp 32–37; 'List of rape/death camps in Bosnia-Herzegovina,' and 'General report' on the need to protect women and children as victims and refugees, both dated 28 Sept. 1992 and issued by the Croatian women's group 'Tresnjevika'; and Charlotte Bunch and Roxanna Carrillo, *Gender violence: a development and human rights issue* (Dublin, 1992), pp 12–23).

13 Wanda Nowicka, 'Two steps back: Poland's new abortion law,' paper presented at European Population Conference, Mar. 1993 in *Journal of Women's History*, 5 (Winter 1994), pp 151–55; *The Irish Times*, 7 Apr. 1993; 2 May 1993, p. 7; *Irish Independent*, 11 Apr. 1993; and Joan Hoff and Christie Farnham, 'Sexist and racist: the post-cold war world's emphasis on family values' in *Journal of Women's History*, 4 (Fall 1992), pp 6–9.

pregnancies and other emergency conditions (such as cancer of the uterus), although even these are referred to as something else in medical reports. 'Direct' versus 'indirect' abortions have no standing in Irish law, where the distinction is usually between 'unlawful' and 'not lawful'.[14] As a result, however, of unofficial church-imposed restrictions against 'direct' abortions, it is estimated that at least 5,000 women from the Irish Republic travel each year to England to obtain this medical procedure.[15] Women seeking abortion in Northern Ireland must do the same because the British Abortion Act of 1967 does not apply to that portion of the United Kingdom. It could logically be asked why so little international attention has been focused on this equally egregious situation for women living in the North. It is just another anomaly among many where (until the recent peace) mafia-style violence and financial arrangements among loyalist and republican militants overshadowed the injustice experienced by *all* women in Northern Ireland, regardless of their political or religious orientation.[16]

The refusal of the state and religious groups in all the counties of Ireland to recognise officially that Irish catholic and protestant women are forced to have abortions abroad and to assume responsibility for this phenomenon has created a surrealistic atmosphere when it comes to discussing abortion throughout the island of Ireland. If anything, discussion is becoming more open, if not any less euphemistic, in the Republic of Ireland than in Northern Ireland and England where silence prevails for all the obvious political reasons over the denial of equal rights to abortion for women in the north. For example, in the November 1992 general election in the south the male-generated euphemism 'substantive issue' dominated the discussion about the merits of adding three new amendments dealing with abortion to the Irish constitution. One of the few voices of clarity during this campaign was that of Ailbhe Smyth, director of the women's studies programme at University College Dublin, who repeatedly pointed out that Irish women do not go to England to have a 'substantive issue'; they go to have an abortion.[17] Irish women know this even if their politicians do not.

The history of abortion in the Irish Republic and related reproductive health services is riddled with ironies. Abortion has been illegal since the 1861 Offences Against the Persons Act and it was rigidly enforced in the Irish Free State with the Dáil passing reinforcing legislation in 1929 against publishing, selling or distributing literature about birth control, and in 1935 contraception was criminalised. The only recourse for Irish women was to practise infanticide

---

14 Gerard Hogan, 'Law, liberty and the abortion controversy' in Anthony Whelan (ed.), *Law and liberty in Ireland* (Dublin, 1993), pp 113–114.
15 Paula Snyder, *European women's almanac* (New York, 1992), p. 193. Annual figures for Irish women from the south travelling to England for abortion range from the government's 'official' estimate of 4,000 to 7,000 or 8,000. Figures in an unpublished March 1993 paper by Ruth Riddick, director of the Open Line Counselling in Dublin, based on data from the Office of Population Census and Surveys show an almost steady increase in abortions from 3,320 in 1980 to 4,154 in 1991; a total of 45,456 for the decade or an average of 3,788 annually. Since the 1983 anti-abortion amendment to the Irish constitution, it is estimated that a minimum of 36,000 women and more likely around 70,000 have obtained abortions in England. The difficulty in keeping accurate statistics stems from the fact that Irish women often stay with relatives and use local addresses when signing into abortion clinics, and neither Irish nor British census takers or other government officials gathering demographic data try to find out about such women. Those who have abortions abroad literally do not 'exist' in Ireland (See: Smyth, *Abortion papers*, p. 20–21, 40–46, 130–137).
16 Scott Anderson, 'Making a killing: the high cost of peace in Northern Ireland' in *Harper's*, Feb. 1994, pp 45–54.
17 Ailbhe Smyth, 'Abortion: a woman's right to life', address to UCD Pro-Choice Action Group, 22 Oct. 1992.

(most often of illegitimate children) in the last half of the nineteenth century and increasingly to resort to illegal backstreet abortions in the early decades of the twentieth. The Irish government also paid no attention to court decisions and parliamentary acts in England that liberalised abortion in 1937 and again in 1967 with the British Abortion Act which legalised abortions during the first 24 weeks or first two trimesters. In 1991 the British parliament permitted abortions through the third trimester if the foetus might be born seriously handicapped or if the mother's health was at risk. The trek of pregnant women from Ireland to England to have abortions really began in the late 1930s, not the late 1960s as is usually stated, although such travel was interrupted due to commuting restrictions imposed during World War II. As a result, backstreet abortions and illegitimate birth rates soared during the war years.[18]

In 1979 there was no legal contraception in the Irish Republic because the pill could only be prescribed by doctors as a 'cycle regulator'. In that year the Dáil passed a bill designed to limit the buying of prescription contraceptives to married couples, and the then Minister for Health Charles Haughey coined another euphemism by calling this 'an Irish solution to an Irish problem.' In 1985 eighteen year-olds regardless of marital status were allowed to buy nonmedical contraceptives although their distribution was restricted to pharmacies and clinics.[19] In the interim, however, in 1983 sixty-seven percent of those voting supported an amendment to the Irish constitution that granted the nonviable unborn equal rights with the viable born and more than ever before women began to use English abortions as a form of contraception. As a result of this 1983 amendment to the constitution, the Irish government and people proclaimed that in theory a non-viable foetus had an 'equal' citizenship 'right to life' as a viable female adult.[20]

The 1983 referendum had been initiated by anti-abortionists who wanted to seal the process of social change with a constitutional amendment. At that time they borrowed terminology and tactics from their counterparts in England and the United States. Although a small group of Dublin feminists had formed a Women's Right to Choose group and established the Irish Pregnancy Counselling Centre in 1980, there was no single-minded feminist demand for abortion in the early 1980s. Instead, Irish feminists supported improved social and reproductive services with respect to maternity leave, illegitimacy, divorce, contraceptive information and access, and married women's domicile rights.[21] On the face of it the anti-abortionists succeeded because the eighth amendment to the Irish constitution effectively prohibited the Dáil from enacting legislation to legalise abortion *without another referendum*.

In the process of achieving this constitutional success, the anti-abortionists unintentionally opened up a public debate on the very topic few wanted discussed: abortion, or the so-called 'substantive issue'. Ireland was founded in

---

18 Jenny Beale, *Women in Ireland: voices of change* (Dublin,1986), p. 105; and Pauline Conroy Jackson, 'Outside the jurisdiction: Irish women seeking abortion,' in Smyth, *Abortion papers*, pp 125–137; and *International Herald Tribune*, 14 Apr. 1993. It should be noted that many Irish women had already been going to England to have their babies, if they were going to give them up for adoption, and so the exodus for abortions in the 1930s did not set a new precedent for travel on the part of Irish women with crisis pregnancies.

19 Alpha Connelly (ed.), *Gender and the law in Ireland* (Dublin, 1993), pp 131–135; and *The Irish Times*, 3 May 1993.

20 The exact wording of Article 40.3.3 is: 'The State acknowledges the right to life of the unborn, and with due regard to the equal right of the mother, guarantees in its laws to respect and, as far as practicable, by its laws to defend and vindicate that right.'

21 Beale, *Women in Ireland*, p. 113.

1921, after all, on the assumption or notion that there was national ethical consensus and that what the Irish have in common as a people is more important than their moral differences. Even though in 1977 seventy-four percent of Irish catholics agreed that abortion was always wrong and another 21 percent said it was generally wrong, the 1983 debate revealed deep divisions that have increased since then – much to the consternation of those anti-abortionists who had not wanted more discussion, but rather more silence on the question of abortion.[22] Thus, the original terms of the debate in the 1980s were set by the anti-abortionists – not in the name of broad civil or human rights (around which an ethical consensus can be based), but in the most narrow moral terms that always break down at the level of private practice. They confused the public consensus on statehood for a private one on morality.

In 1987 the high court issued a preliminary order closing pregnancy counselling services in Ireland and in 1988 the supreme court permanently restrained dissemination of abortion information. In October 1991 the European court of justice in Luxembourg ruled in the Grogan student case that Irish women had an explicit right to travel within the EU to obtain abortions because Article 59 or the Treaty of Rome conferred the right of freedom of movement to obtain medical services. A year later the European court of human rights in Strasbourg found that in the Open Door case the Irish high court had breached Article 10 of the European Convention when it barred women from receiving information about abortion.[23] There the situation remained until two things occurred in the spring and summer of 1992 – one received international publicity while the other went almost unnoticed except by Irish feminists.

International protest greeted the supreme court decision on 6 February 1992 prohibiting a fourteen-year-old girl who had been raped to travel to England for an abortion. This in turn prompted the equally publicised and extraordinary decision in *Attorney general v X & Others* on 5 March 1992.[24] A majority of the five persons on the Irish supreme court declared that abortion was legal in Ireland when there existed 'a real and substantial risk to the life, as distinct from the health, of the mother', or when the woman was potentially suicidal, as in the case of the fourteen-year-old. In so ruling, however, the supreme court relied on domestic constitutional law rather than on any precedent established under the EU Convention on Human Rights. While EU laws are binding on the Irish government, they cannot constitutionally be used in Irish courts unless specific implementing legislation is passed. So Ireland's legal relationship with the EU remains in a tenuous state, to say the least, where women's rights are concerned.

Just to be on the safe side, however, on 18 June 1992 when the Irish voted to approve the Maastricht Treaty, they did so with the little known Protocol 17 (inserted quietly by the conservative Fianna Fáil Irish government) that nothing in European Union treaties could contravene or supersede Article 40.3.3 of

---

22 *Ibid.*, pp 113, 116–117; and Fintan O'Toole, address at University College Dublin, 21 Apr. 1993.
23 Riddick, 'Abortion and the law in the Republic of Ireland'. The two Irish decisions overturned by these international courts of law were: *Society for the Protection of the Unborn [SPUC] (Irl) Ltd v Stephen Grogan & Others*, 1989 IR 753, and *AG (at the relation of SPUC (Irl) Ltd) v Open Door Counselling Ltd & Dublin Well Woman Centre Ltd*, 1988 IR 593.
24 An important, but usually overlooked aspect of the X case is that it not only involved a fourteen-year-old girl, but also her family and several members of the gardai – all of whom supported her decision. The fact that the 'others' in the case were all adults showed how anti-family the lower court decision had been in a country that purports to honour family values.

the Irish constitution. This Protocol appeared to obviate any force that EU law and the United Nations Convention on Human Rights might have over abortion and related matters inside Ireland, even after the Maastricht Treaty went into effect. Ireland's still murky relationship with the EU thus refuelled the debate on abortion, once the existence of Protocol 17 was revealed.

The stage was thus set for the last of the major debates over abortion in Ireland in what unexpectedly turned out to be a general election instead of simply a vote on three referenda. One provided for women's right to travel for abortion abroad; another for their right to information about abortion. The most controversial one represented an attempt by the anti-choice forces to override the decision in the X Case. It stated: 'It shall be unlawful to terminate the life of an unborn unless such termination is necessary to save the life, as distinct from the health, of the mother which gives rise to a real and substantial risk to her life, not being a risk of self-destruction.' Translated, this proposed 56-word amendment to the Irish constitution represented an 'assault on the integrity' of Irish women by attempting to differentiate between the lives and health of female (but obviously not male) citizens.[25] Fortunately, by a percentage that reversed the 1983 vote, two thirds of those voting rejected this amendment, but it is impossible to tell how many of those who opposed it were pro-life and how many were pro-choice. However, thirty-three percent of the electorate still agreed that women who are arbitrarily assigned the responsibility of raising children *should not* be granted the responsibility of deciding whether to have them in the first place. One of the posters in a Dublin pro-choice march in November 1992 pointed out this obvious contradiction with the slogan: 'If you can't trust me with choice, how can you trust me with a child?'

Unlike 1983, church opinion was divided on the third referendum. In a statement issued by the Catholic Bishops' Conference just before the election and sent by the primate of all Ireland, Cardinal Cathal Daly, to be read at all masses just before the vote, abortion was condemned as the 'deliberate and direct destruction of unborn human life', but parishioners were told they could vote either way. A few individual bishops urged a 'no' vote, but Cardinal Daly supported the official 'no stand' position of the country's 35 bishops.[26]

Anti-choice groups voted against it because it was too lenient on abortion and seemed to sanction 'direct' abortion, while pro-choice groups said that it was too restrictive in that it appeared to reverse aspects of the X case. So the lack of national consensus in Ireland continues on abortion. Public opinion, however, had dramatically changed since the 1977 poll cited above. In October 1992, sixty-four percent favoured granting abortion under 'special circumstances'.[27] Until the Dáil passes legislation defining conditions under which abortions can and cannot be performed in Ireland, *it remains legal under the terms established in the X case*, although no doctor in his or her right mind would perform one, nor is anyone about to open an abortion clinic in Ireland. The role of the state in distributing information about abortion is still contested (as it also is in Poland) despite the passage of the right to information amendment to the constitution. Anti-abortion, anti-woman attitudes remain so ingrained in public and private discourse that the lack of consensus over the 'substantive

---

25 *The Irish Times*, 13 Oct. 1992 (quoting the Irish Council for Civil Liberties).
26 *International Herald Tribune*, 23 Nov. 1992; 13 Nov. 1992, and 23 Nov. 1992
27 Hogan, 'Law, liberty and the abortion controversy,' p. 118; and *The Irish Times*, 6 Oct. 1992. Of the 64 percent in this poll, 16 percent believed abortion should be available on demand, while 48 percent favoured it only in 'special circumstances'.

issue' since 1983 has yet to bring the Irish Republic into line with other EU nations.

Moreover, family planning in Ireland is the worst among developed nations. The Health (Family Planning) (Amendment) Act of 1992 redefined contraceptives to include non-medical contraceptives, thus allowing oral contraceptives and spermicides to be sold from controlled outlets. After the conservative Fianna Fáil government passed a watered-down reform of the restrictive legislation governing sale of condoms in the summer of 1992, they still could not be purchased through vending machines. In May 1993 the Irish Minister of Health Brendan Howlin announced the introduction of legislation to legalise dispensing condoms by vending machines in 'appropriate' public places and removing the age limits for purchase. Although Howlin said that he had not consulted with the catholic hierarchy, he noted that in order to achieve this liberalisation over the objections of the church, the bill 'excludes condoms from the legal definition of contraceptives,' and justifies their use in the name of an AIDS awareness campaign, *not on behalf of family planning.* Although Ireland has only a little over 300 AIDS cases, most of them male homosexual and drug users, it is worth noting that it was an anti-AIDS initiative which prompted this government support of condoms, not the need of hundreds of thousands of Irish women (and men) long-denied the right to adequate contraception. The new Family Planning Bill passed the Dáil unanimously on 4 June 1993.[28]

The Catholic Bishops' Conference through its principal spokesperson Dr Joseph Duffy, the bishop of Clogher, announced on the day before its passage that the new legislation ending the inadequate 'Irish solution to an Irish problem', could give the impression that society approved an abuse of God's gift of sexuality. In the face of the AIDS prevention argument and the lack of opposition in the Dáil to the Family Planning Bill making condoms available through vending machines, the church has lamely fallen back on its old shibboleth that 'people can die through the false belief that condoms guarantee "safe sex,"' and that only permanently monogamous relationships between two faithful individuals can prevent AIDS. In 1993 the Dáil also decriminalised suicide and homosexuality for those seventeen or older (while continuing to impose severe penalties on female prostitutes). In that year the Dáil appeared favourably disposed toward granting married women half ownership in the homes they occupy with their husbands and to extending laws against domestic violence to those co-habiting, but not married. Moreover, the church is facing the inevitability of a referendum on divorce – another procedure currently banned by the Irish constitution since 1937. What is so blatantly anti-woman in the current debate in the Republic of Ireland over divorce is that libertarian arguments are made in favour of it on the grounds that such 'legal restraints . . . should not be imposed upon mature and competent adults', because half of these 'adults' happen to be male. The same liberal legalism is not applied to mature and competent adult women when it comes to abortion. On all these issues it is clear that there is no longer a consensus in Ireland about living by christian principles.[29]

---

28 Connelly, *Gender and the law in Ireland*, pp 135–136; and *Irish Independent*, 19 Oct. 1992; *The Irish Times*, 11 Nov. 1992; 25 May 1993 and 4 June 1993. As of May, 1993, Ireland had recorded a total of 341 cases of AIDS, with 150 deaths (approximately the same figures apply to the single city of Edinburgh, Scotland), although 1,368 Irish people had tested positive for HIV.

29 Duncan, 'Abortion, divorce and the debate about liberty', pp 121–122; *The Irish Times*, 2 June 1993; 3 June 1993 (quotation), 5 June 1993; and Desmond Fennell, *Nice people & rednecks: Ireland in the 1980s* (Dublin, 1986), pp 109–110.

So the church is left in the reactive position of maintaining that while laws cannot make people moral, they should not make moral living more difficult by encouraging immorality, and this argument is falling increasingly on indifferent ears inside the Republic of Ireland. However, the Irish supreme court on 20 July 1993 upheld a ruling that prevented the lifting of the injunction still in existence against those groups who originally brought the law suits to the attention of international courts. Thus, Well Woman and Open Door Counselling cannot distribute abortion information despite the 1992 approval of a referendum making abortion information available.[30]

Impact of Irish Abortion Policy on Feminists and Female Legal Status

While Irish feminists were galvanised by the referenda on abortion in the autumn of 1992, they too now seem to be retrenching and rethinking their tactics and strategies. They seem to have entered a 'down' period after the November 1992 election even though they did not suffer defeat over the abortion referendum as did Polish women when the government refused to honour a successful petition campaign calling for a national election on abortion. (Under the American electoral system, national referenda on specific issues are not possible.) Moreover, the overall position of Irish women is stronger than that of Polish women because they do have several national councils and commissions representing them and strong pro-choice grass roots organisations still in existence. The Second Report of the Commission on the Status of Women, issued in February 1993, avoided the issue of abortion but made liberal recommendations on all other legal aspects of women's lives in Ireland, much to the consternation of most male commentators.[31]

Despite recent liberal developments in legislation on contraceptives, suicide, homosexuals, domicile rights, and with the promise of a referendum on divorce, feminists in Ireland remain sceptical that these are simply sops being thrown out so that the then conservative/labour coalition government could indefinitely delay taking up the scarcely mentioned 'substantive issue' of abortion. In fact, an appearance of the Irish attorney general Harry Whelehan before the United Nations Committee on Human Rights in July 1993 seemed to confirm these suspicions. Extolling Ireland's recent or pending liberal legislative developments he mentioned abortion *only* in relation to the information referendum passed in November 1992. The fact that the Irish Commission for Justice and Peace established by the catholic church and headed by Father Michael O'Flaherty is one of the two major Non-Government Organisations representing Ireland at these hearings does not augur well for making access to abortion part of the human rights improvements necessary to bring the country into line with the United Nations Covenant on Civil and Political Rights which it signed in 1989.[32] Moreover, if only the X case is made the basis for such legislation, what is passed by the Dáil will inevitably be restrictive. Radical feminists insist, therefore, that repeal of the eighth amendment to the constitution has to re-

30 *New York Times*, 31 July 1993.
31 *Irish Press*, 11 Feb. 1993.
32 *The Irish Times*, 13 July 1993.

main their goal because nothing less will bring Ireland into line with most other European nations on the abortion issue. This would necessitate still another national referendum.

In Ireland, the economic and legal position of women is technically much weaker than that of Polish women (and American women) because fewer of them work outside the home and neither divorce nor abortion was available as of 1993. However, in the unlikely event that abortion legislation does emerge under any coalition government, it probably will not revert completely to pre-X case restrictions. The Irish supreme court had one member on it in 1994 who participated in the X case decision. So any other pronouncement by the court on abortion may not be as 'liberal' as in the X case. Moreover, unlike in the United States where supreme court decisions can be voided by congressional legislation, this is not possible in Ireland. Only referenda to the Irish constitution can change a pronouncement by the Irish supreme court on matters involving constitutional law. So appealing to the supreme court on feminist issues is much more 'dangerous' in Ireland than in the United States, and in Poland as it is the body that decides constitutional questions. The Irish supreme court ruling in July 1993 preventing clinics from informing women about where they can obtain abortions abroad in defiance of the favourable popular vote approving such abortion information is an indication of the problems Irish feminists face from their country's court system at all levels.[33]

Despite the fact that the country remains out of step with most industrialised European countries on almost every aspect of female sexuality, the situation for women in Ireland is more hopeful at the moment than in Poland, although it does not have even the modicum of supreme court protection for abortion that exists in the United States. The catholic church appears to be reacting defensively to attempts to liberalise abortion and other issues relating to reproductive health in the Republic, while the opposite is true in Poland. Moreover, the church in Ireland in the 1990s has no gaping political vacuum to fill as it does in Poland because the multi-party system of proportional representation provides stability, and under the conservative/labour government, the Republic saw the greatest participation of women in the parliament and in ministries in its seventy year history. 'When I was in Ireland I felt freer than I do here,' one non-catholic Polish woman was quoted as saying. 'There you feel change is coming; here you get the opposite feeling.' Finally, unlike Poland, Ireland's female president Mary Robinson, represents a positive model for women of all walks of life and is generally considered to be pro-choice.[34]

---

33 *New York Times*, 21 July 1993, Nov. 1994. See also footnote 28 above.
34 Bogna Piotrowska quoted in *The Observer*, 2 May 1993. As a member of the Irish senate in 1971 Robinson tried to liberalise the country's anti-contraception law and her career as a barrister was characterised by 'campaigns to legalise contraception, divorce, homosexuality and freedom of information on abortion.' (See *The Irish Times*, 4 June 1993; *Ms. Magazine*, May/June 1992, pp 16–19 (quotation at 16); and *Town and Country*, Feb. 1993, pp 60ff).

22 Women and Citizenship in the European Union

Elizabeth Meehan[1]

I have to confess that to speak to an audience of historians is somewhat intimidating for a political scientist. Historians, particularly those of the Maurice Cowling school of high politics, can be scathing about what political scientists do – especially those who, as in my case, have connections with the political science of Nuffield College. Nothing much can be said, it seems, about modern institutions without the benefits of access to many private archives. And, it seems, the institutions of the European Community (now the European Union – EU), as distinct from processes of integration or disintegration, can be boring. I agree with the latter in the sense that the way in which the Community has been explained in the last thirty years by political scientists and economists often does not inspire excitement.

In my own work, from which this paper comes, I have tried to argue that the Community is interesting precisely because it has to be seen in a different light from received wisdom; that is, it is something which cannot adequately be explained by existing categories of domestic and inter-governmental politics. One example of this is that, as a result of its existence, a new conception of citizenship is emerging that is, as in the Roman Empire, dualistic in that rights associated with citizenship are formulated and guaranteed through more than one political or legal authority. At least I can say that, in order to do what I felt necessary, I had to turn, if not to history, then to the history of political ideas.

In this paper, I concentrate on what these innovations mean for women. To begin with, I outline defects in the way in which women's citizenship is formulated in liberal democratic theory and then explain the view that liberal 'fictions' of equality are not eliminated in the EU but are replicated on a grander scale. My argument, however, is that the critics are too pessimistic. Thus, in my third section, I suggest that, because of the nature of politics as a human activity and because of the nature of this new polity, whose like we have not known before, the EU can be used to promote a set of policies that are friendlier to women's interests, as defined by themselves, than those defined by men in the social contract or liberal equal-rights traditions.

Problems of Liberal-Democratic Theory

The roots of what is wrong for women with the idea of the social contract and with the blindness to this of liberal reformers, are ancient. Aristotle believed

---

1 The author would like to thank those who contributed ideas at the conference and Ronan Fanning for commenting on the subsequent redraft.

that the dispositions of women rendered them incapable of rational thought about the public interest; their place was in the family where they were to create havens for men during retreats from public life. An association of femaleness with nature, partly to be subdued, partly the source of emotional sustenance, continued throughout the middle ages. It contrasted with an association between maleness and both military and civic virtue, the civilization of nature and with the fostering of an antithesis between rationality and emotion.[2] As Ursula Vogel[3] and Joan Hoff have argued, maleness was constitutive of citizenship; women could not be citizens because they were not men. And the 'generality of the public realm', epitomised by 'manly virtue', could be sustained only by insisting on its separateness from the 'particularity' of women and domestic concerns.[4] The impact of ideas about women's nature on their political rights was compounded by their limited right to property, itself a general qualification for political participation.

At the beginning of the French Revolution, there was, briefly, to be a woman's revolution as well as a class one – despite the influence of Rousseau, whose views on women were hardly different from the medievalists. But the feminism of the revolutionaries became attenuated and, under Napoleon, pre-revolutionary patriarchal powers over women were re-established.[5] In the century after the French Revolution, John Stuart Mill was one of the few male advocates of women's rights. He and his collaborator, friend and wife, Harriet Taylor, argued that we could not know that women were by nature different from men; if they seemed different, it might be because of artificial, man-made constrictions on their development. In any case, if men might differ naturally from one another and still be equal citizens, it was illogical to withhold civil and political rights from women on the ground that they were naturally different.

The triumph of the social contract way of thinking in political liberalism is commonly seen as a total revolution against patriarchal rule as theorised by Robert Filmer; that is, against political power being associated with the headship of families superimposed by the 'highest' family, the royal family. Many people have believed that the shift from status to contract as a way of ordering society – with its apparently gender-neutral concepts, such as citizen, individual, person and so on – means that women could be and have become citizens in the fullest sense. But Carole Pateman argues that the revolution was only half a one and must be understood as one of brothers against fathers. Women were brought into society, not as individuals, but as legal dependants – as wife, mother, sister or daughter. The meanings of citizenship and the qualifying virtues remained male, despite the universalism of terms in which they are described.[6] The discrepancy between rhetoric and reality is exemplified for Pateman in the terms of the marriage contract and by how men relax the central liberal tenet of consent when it comes to judging questions of rape and sexual violence.[7]

---

2 Genevieve Lloyd, 'The man of reason. 'Male' and 'female'', in *Western philosophy* (London, 1987).

3 Ursula Vogel, 'Is citizenship gender-specific?' in Ursula Vogel and Michael Moran (eds), *The frontiers of citizenship* (Basingstoke, 1991).

4 Iris Young, 'Polity and group difference: a critique of the ideal of universal citizenship' in Ethics, 99, no. 2 (Summer, 1989) pp 250–74; Will Kymlicka, *Contemporary political philosophy. An introduction* (Oxford, 1991), pp 251–6.

5 Darline Gay Levy and Harriet Branson Applewhite, 'Women and political revolution in Paris' in Renate Bridenthal, Claudia Koonz and Susan Stuard (eds), *Becoming visible. Women in European history* (2nd edition, Boston, 1987), pp 279–306; Vogel, *op cit.*

6 Carole Pateman, *The sexual contract* (Oxford, 1988).

7 Though called a contract, it is not an agreement between free and equal partners and, uniquely for a 'contract', subjugates the body and property of one person to another.

Pateman's analysis is essentially political but a concept of patriarchy also appears in analyses that are primarily economic, particularly Marxist feminism. Though I have no time to explain socialist understandings of patriarchy, I do need to comment briefly on the material conditions that are thought to be a prerequisite for the exercise of the legal and political rights of citizenship. Once, as in the ideas of John Locke, but even to the opponents of the reforming idealists, lack of property or an inability 'to fend for oneself' was a disqualification from citizenship. In the twentieth century, it has come to be thought, instead, that the poor could be given the material means of subsistence through which they could become rational and disinterested enough to be allowed to participate in politics.[8] And, more than that, there is the now classic idea of T. H. Marshall that the right to belong to workers' organisations and the right to an assured minimum standard of living and welfare, form an indispensible part of any effective conception of citizenship. His conclusion was accepted for some decades by those with an expedient concern, not for rights, but for the maintenance of public order.

However, the foundational principles of welfare states also rest upon Aristotelian conceptions of the public and private realms and the natures of men and women – explicitly so in their first manifestations. This can be seen in, for example, the Beveridge and Bowlby reports on health and education in the United Kingdom, the Irish constitution and the first legal interpretations of the American Fair Labor Standards Act. Again, according to Pateman and others, this is disguised by subsequent modifications in the surface language of politics.[9] Modern advocates of social rights, she argues, wrongly take 'fraternity' to be a synonym for communality. More recent use of gender-neutral terms such as 'head of household' or 'breadwinner' and 'worker' disguise the reality that men and women are not in the same position to claim the same levels of pay and benefits which are the 'floor' for their equal autonomy in legal and political life. This gives rise to an insistence by women that what they need is not necessarily the same treatment as men but treatment that takes account of *both* their similarity to men and their different situations.[10] It seems to have become normal in the Scandinavian countries for women and men to expect and enjoy a mix of policies based on an appropriate combination of the same and specific treatment.[11] That this approach is thought to be less firmly entrenched in the European Community is one of the reasons for opposition among Norwegian and Danish women to potential or actual membership. It is to this grander 'fiction' of equality that I now turn.

Fictions of Equal Citizenship on a Grander Scale

Scandinavian conceptions of social rights are based, at least more than in other democracies, on need and the idea that needs arise from individual circum-

---

8 Desmond King and Jeremy Waldron, 'Citizenship, social citizenship and the defence of welfare provision' in *British Journal of Political Science*, 18 (1988), pp 415–43.

9 Carole Pateman, *The disorder of women* (Oxford, 1989).

10 This can be a fraught position because anti-feminists have always justified their position on grounds of difference and because it can appear that feminism is also slipping into a similar biological essentialism.

11 Birte Siim, 'Welfare state, gender politics and equality policies: women's citizenship in the Scandinavian welfare states' in Elizabeth Meehan and Selma Sevenhuijsen (eds), *Equality politics and gender* (London, 1991), pp 175–194.

stances not the well-being of conventional family units. But EU rights originate in economic policy objectives, tend to be based on insurance contributions – which, because of different earning powers, result in unequal benefits – and are interpreted in the equal treatment tradition which is usually blind to the asymmetry of family relations.

This is not to belittle the impact of equal treatment, either within states or the Union. Despite my earlier criticisms, the claim of equal political rights has had revolutionary effects in so-called nation-states[12] and the equal treatment principles of the European Court of Justice have been said to have had significant effects on sexual politics in many member states (especially Ireland and Greece). The court has moved decisively against the preferences of governments by redefining economic instruments into what it calls fundamental rights and asserting that it is an overriding duty of all Community institutions, national and common, to uphold them. The rights of women workers have been placed firmly in this category.

But the court's main impact has been inevitably on the situations of men and women as workers because rights arise from the market-based objective in the Treaty of Rome of the free movement of labour. And, it can be argued, seemingly gender-neutral terms, such as 'migrant', 'member of a worker's family', 'citizen of the union' are as misleading as their counterparts in political theory. When it comes to the distinctive interests of women in matters relating to motherhood and income maintenance, where lie – as Marxist and socialist feminists point out – the basic inequalities that make formal, legal and political equality a fiction, the court has been more ambivalent than it is in cases involving the comparable situations and rights of women and men at work.

In one sense, the court's interpretations of motherhood could be seen as acknowledging one strand of feminism; that is, the view that motherhood is virtually the only female role that society values and that, as such, it merits specific guarantees. In another sense, the court can be seen as wholly traditional, thus inhibiting more flexibility about who inhabits the public and private realms.

The former is illustrated in cases of discrimination against pregnant women workers. Here, the court has expressed its specific concern for women more forcefully than the English courts. In England, it has been held that, since a man cannot become pregnant, a pregnant woman has no point of comparison unless analogous treatment of a sick male can be brought in. The contrasting position of the European Court of Justice is that, since pregnancy is unique to women, discrimination on grounds of pregnancy is direct sex discrimination; there is no need to bring in other factors.[13] The English rulings would satisfy equal rights feminists but those who emphasise the need not to overlook relevant differences might find the European rulings more congenial. However, the latter rulings might also be construed as embodying a patriarchal outlook, if considered in the light of other decisions.

Examples of rulings based on traditional assumptions include the court's rejection of a Commission claim that Italy was contravening Community law by

---

12 The early French Revolution was noted before. Abigail Adams wrote to her husband when he was helping to frame the American constitution that he could expect a rebellion if he ignored the rights of women (Eve Cary and Kathleen Willert Peratis, *Women and the law* (Skokie, Illinois, 1977). See also Juliet Mitchell, 'Women and equality' in Anne Phillips (ed.), *Feminism and equality* (Oxford, 1987), pp 24–43.

13 177/88 Dekker v Stichting Vormingscentrum voor Jonge Volwassenen (VIV-Centrum) Plus, weekly proceedings 23/90, 8.11.90; C-179/88 Hertz v Aldi Marked A/S, weekly proceedings 21/89, 6.11.90.

not allowing parental leave to adoptive fathers as well as mothers.[14] And it upheld a German decision that an unmarried father who looked after his child
was not entitled to leave and an allowance granted to mothers.[15] In both cases,
the court referred to the special need to ensure the protection of women and to
the distinctive importance of bonding between mother and child. Though, on
the face of it, this standard of justice might please 'difference' feminists, it has
also been criticised because the ruling also stated that Community rules about
equal treatment in paid employment were 'not intended to deal with matters of
family organisation or to "alter the division of responsibility between parents"'.[16]

Though it has strongly asserted the rights of some women under Community
social security rules, the court has also been reluctant to upset traditional sex
roles in its general jurisprudence in this field. It has stated categorically that the
two Social Security Directives are not intended to address the general condition of sex equality but only the situations of men and women as workers.[17]

The fact that women and men are not in employment which leads to similar
lifetime earnings and, hence, similar levels of benefits, may be thought to be
disappearing. On the face of it, changes in the patterns of industry and employment, already in train and likely to accelerate in the Single European Market,
combined with demographic trends, seem favourable to women.[18] Traditional
industries associated with male employment are declining while service industries are growing (including fortune telling in Italy!). But, as Jackson points
out, new technologies in banking, where there is a high density of female employment, may reduce their opportunities unless there is appropriate training.[19]
Generally, work is being organised into 'a-typical' forms with more 'flexible'
contracts, more sophisticated types of home-working, more self-employment
and more small enterprises. Although a-typical work is still a small proportion
of total Community employment, it is a much more significant dimension of
women's employment, analysed separately, and is becoming more so.[20]

Over a quarter of the Community's female workforce works part-time [more
among married women] and many more do this involuntarily than had previously been estimated; in Spain and Greece, 10 percent and 34 percent, respectively, of working women are unpaid because they work in family businesses
but in most countries women are more likely than men to be home-workers or
'assisting relatives' as their paid employment. While pressures on women to
accept a-typical employment contracts are, according to Jackson, stronger in
southern Europe, women are more likely than men throughout the Community to be on temporary contracts and the rate at which they are unwillingly in

---

14   163/82 Commission v Italy [1983] ECR 3273.
15   184/83 Hofmann v Barmer Ersatzkasse [1984] ECR 3042.
16   Ruth Nielsen and Erika Szyszczak, *The social dimension of the European Community* (Copenhagen, 1991), p
     106.
17   Joined cases 48/88, 106/88, 107/88 Achterberg-te Riele, Bernsen-Gustin and Egbers-Reuvers v Sociale
     Verzekeringsbank, weekly proceedings 15/89, 27.6.89.
18   Catherine Hakim, 'On the margins of Europe? The social policy implications of women's marginal work' in
     Margaret O'Brien, Linda Hantrais and Steen Mangen (eds), *Women, equal opportunities and welfare* (Aston
     University: The Cross-National Research Group, Cross-National Research Papers. New Series: The Implica-
     tions of 1992 for Social Policy, 1990), pp 21–28; Hazel Morrissey, 'Women and the impact of 1992' in Paul
     Hainsworth (ed.), *Towards 1992. Europe at the crossroads* (Jordanstown: University of Ulster, 1989), pp 37–
     46; Pauline Jackson, *1992. The impact of the completion of the internal market on women in the European
     Community* (Working Paper prepared for the Equal Opportunities Unit of Directorate Generale V of the
     European Commission, V/506/90-EN. Brussels, 1990).
19   Jackson, *op. cit.*
20   *Ibid.*, pp 44–9.

a-typical contracts grew throughout the 1980s in half the member states. Jackson predicts considerably more part-time, fixed-term or agency employment in the service sector.[21]

Though such forms of employment appear to help women who have to be 'flexible' in managing two roles, home-working usually escapes protection and success in transforming it into self-employment or a co-operative small business will require familiarity with new technologies and management planning. Investment in the remains of skilled manufacturing may be concentrated in the 'golden triangle' of Germany, France and the Benelux countries, thus requiring mobility. Low or semi-skilled manufacturing might increasingly be carried out in countries with poor levels of welfare, education and training. Without appropriate measures, labour markets segmented by sex may be reinforced – doubly so, within and across countries.

Yet, according to some writers, work, not need, is an increasingly prominent criterion of entitlement to directly related benefits and to others that affect domestic welfare.[22] Moreover, they argue that the idea of work in this connection is based on a model of male, full-time employment. This compounds the problematic way in which some member states are equalising social security entitlements; that is, their methods of eliminating direct discrimination increase indirect discrimination, which is also supposed to be unlawful. Substituting benefits labelled as for husbands or fathers by neutral ones for 'heads of household' or 'breadwinners' does not tackle the problem. Since men and women are rarely similarly situated in the labour market from which access to and levels of benefit are derived, it seldom will be more rational for women than men to be designated as 'head of household'.[23] As Luckhaus puts it, equal 'participation in the social security system remains entirely contingent on [women] being able to establish themselves, like men, as full-time paid workers'.[24] Most writers suggest that women cannot enjoy equal social rights unless social security schemes are 'individualised' not in the labelling of rights but in how needs are calculated.

For example, Mangen argues for a more flexible definition of work that would enable socially important activity that was not subject to a formal labour contract to qualify for insurance credits. He points out that such a settlement might 'institutionalize sexism' since there are no serious moves afoot to encourage men to take up 'substantially more of the roles of housewife, child carer, care of the elderly and disabled'; but women would be guaranteed 'a certain level of protection that, in many countries, is still denied to them'.[25]

---

21 *Ibid.*
22 Janet Finch, 'Women, equal opportunities and welfare in the European Community: some questions and issues' in Margaret O'Brien, Linda Hantrais and Steen Mangen (eds), *Women, equal opportunities and welfare*, pp 1–6; Hakim, *op. cit*; Linda Luckhaus, 'The social security directive: its impact on part-time work' in Margaret O'Brien, Linda Hantrais and Steen Mangen (eds), *op. cit.*, pp 11–20.
23 Ina Sjerps, 'Indirect discrimination in social security in the Netherlands: demands of the Dutch women's movement' in Mary Buckley and Malcolm Anderson (eds), *Women, equality and the law* (Basingstoke, 1988), pp 95–106; Catherine Hoskyns, '"Give us equal pay and we'll open our own doors" – a study of the impact in the Federal Republic of Germany and the Republic of Ireland of the European Community's policy on women's rights' (*idem*, pp 33–55); Catherine Hoskyns and Linda Luckhaus, 'The European community directive on equal treatment in social security' in *Policy and Politics*, 17, no. 4 (1989), pp 321–336; Jane Miller, 'Social security, equality and women in the UK' , *idem*, pp 311–320; Linda Luckhaus, *op. cit.*
24 Linda Luckhaus, *op cit.*, p 14.
25 Steen Mangen, 'The implications of 1992 for social policy; social insurance' in Steen Mangen, Linda Hantrais and Margaret O'Brien (eds), *The implications of 1992 for social insurance* (Aston University: Cross national Research Group, Cross-National Research Papers. New Series: The Implications of 1992 for Social Policy, 1990), pp 6–7.

The pessimism of these analyses appears to provide support for the idea that the EU is a fiction on a grander scale. But this can be countered in two ways. One is the nature of politics as a human activity and the other is the more pluralist structure of the Community compared to that of the nation-state. I turn to these in my concluding section.

The Nature of Politics and the New Fora for Political Activity

A friend described my inaugural lecture on citizenship and the European Community as post-modernist. Post-modernism contains the idea that our categories are not fixed and have to be understood in their contexts. Though I eschew the label because of the relativism or futility that crude versions of post-modernism can lead to, I can see what was meant because part of the thesis of my lecture was that there is no single 'truth' about the meaning of citizenship. But, I also share the view of, for example, Anne Phillips and Avtar Brah, that human beings are agents who can construct meanings that are appropriate to their aspirations.[26] Thus, politics must be, among other things, a struggle over which meanings are to be used as a basis for public policy and social practice. I give two illustrations in which women have been effective in such a struggle.

The first is Birte Siim's account of women, citizenship and Scandinavian welfare states.[27] She argues that measures devised for other reasons – class cohesion – have provided the means through which women have become able to be political actors with the power to define their own needs, including the capacity to say where equal treatment is appropriate and when specific treatment would be better.

The other is Vivien Hart's account of reinterpretations of the American Fair Labor Standards Act. She points out that in 1936 there was similar criticism of the concept of worker in the United States. That it was assumed to imply a model of male employment in manufacturing meant that agricultural work and domestic service were not accepted as 'work' and were deemed to be governed, not by contracts of employment, but by the conventions of familial relationships. Thus farm workers – mainly black males – and domestic servants – black and white females – were not protected by this American 'charter' of employment rights. But she also points out that, inspired by a political culture of constitutionalism and rights, the unprotected were able to use the pluralistic institutions of the United States to get their situations redefined and, thereby, protected by the Act. It took a long time but their success encourages Hart to be more optimistic than the critics noted above about the possibility of women using the new channels provided by Community institutions.[28]

In other accounts of public policy and civil rights in the United States, it has been argued that benefits or entitlements may be symbolic. Governments may pass laws designed to protect 'victims' without any tangible distributive or redistributive consequences. In his seminal work, Edelman argued that this

---

26 Anne Phillips, 'Must feminists give up on liberal democracy?' in *Political Studies*, xl (special issue on 'Prospects for democracy', edited by David Held, 1992), pp 68–81; Avtar Brah, 'Difference, diversity and differentiation' in James Donald and Ali Rattani (eds), *Race, culture and identity* (London, 1992), pp 126–145.
27 Siim, *op cit.*
28 Vivien Hart, 'The right to a fair wage: American experience and the European Community Charter of the Fundamental Social Rights of Workers' in Vivien Hart and Shannon Stimpson (eds), *Writing a national identity: political, economic and cultural perspectives on the written constitution* (Manchester, 1993), pp 106–124.

happened because agencies were 'captured' by the very groups whose behaviour was supposed to be regulated in the interest of the intended beneficiaries; for example, railway magnates 'captured' the Interstate Commerce Commission which was supposed to have maintained fair prices for travellers and the sellers of agricultural products.[29] Sometimes, symbolic laws may satisfy the 'victims' and sometimes they may create expectations for further action. For expectations to be transformed into tangible consequences, it is necessary for the 'victims' to enter into a bargaining relationship with public administrators.[30] If such a policy-network is to succeed, administrators must play an interventionist role in respect of the 'victims'. What has been called 'bland neutrality'[31] on the part of administrators as between, say, large corporations and racial minorities cannot lead to improvements for racial minority employees. Administrators need to counter-balance the pre-existing inequality of power by helping the supposed beneficiaries of legislation with special efforts to inform them of their rights and how to use the law. Heclo argues that in the last days of the New Deal ascendancy, Democrat presidents tried to ensure that civil rights and anti-poverty agencies were headed and staffed by people who were sympathetic to the then prevailing version of the liberal public philosophy.[32]

There is some evidence that the European Commission has tried to fulfil a role that is similar to that of American public administrators in the 1960s and 1970s by establishing links between itself and national and cross-national trades unions and women's organisations. And there is evidence that the sympathetic 'insiders' have been able to inscribe concepts and meanings in Community rules that reflect the wishes of citizens more than those of their governments. For example, Catherine Hoskyns refers to communication between Commission officials and members of voluntary groups that is more politicised, in the sense of publicising rights and opposition to proposals, than in systems where the civil service is supposedly neutral. She also refers to the partly successful efforts of officials to include in Community policy references to indirect discrimination and positive action, based on knowledge of the thinking of people outside governments.[33] Another example of the capacity of the Commission to absorb more feminist ways of thinking is provided by Julia Edwards and Linda McKie. They argue that the Community's approach to childcare is based on a meaning of equality which is not the narrow one of individual equal treatment but one which implies the promotion of group justice.[34] Peter Moss argues that regret that the Commission had to give way to member states by issuing a non-binding recommendation on childcare, instead of a binding directive, should not be felt too deeply.[35] The European Childcare Network has already brought

29 Murray Edelman, *The symbolic uses of politics* (Urbana, Illinois, 1964).
30 *Ibid.*; Jo Freeman, *The politics of women's liberation* (New York, 1975).
31 Alfred Blumrosen, 'Toward effective administration of new regulatory statutes' in *Administrative Law Review* (Parts I and II, Winter and Spring 1977), pp 90–120, 209–37.
32 Hugh Heclo, *A government of strangers* (Washington, 1977); Hugh Heclo, 'Issue networks and the executive establishment' in Anthony King (ed.), *The new American system* (Washington, 1978), pp 87–124.
33 Catherine Hoskyns, 'Women's equality and the European Community' in *Feminist Review*, 20 (1985), pp 71–88; Catherine Hoskyns, 'Women, European law and transnational politics' in *International Journal of the Sociology of the Law*, 14 (1986), pp 299–315.
34 Julia Edwards and Linda McKie, 'Equal opportunities and public policy: an agenda for change' in *Public Policy and Administration*, 8 no. 2 (Summer, 1993), pp 54–67.
35 Peter Moss, 'Childcare and equality of opportunity' in Linda Hantrais, Steen Mangen and Margaret O'Brien (eds), *Caring and the welfare state in the 1990s* (Aston University: The Cross-National Research Group, Cross-National Research Papers. New Series: The Implications of 1992 for Social Policy, 1990), pp 23–31.

about much greater awareness in each country of provisions in others and will continue to push common policy into more concrete forms.

Such networks are sponsored and funded by the Commission. Financial support also exists for outside groups and lobbies such as the Centre for Research on European Women, the European Network of Women and the European Women's Lobby; much written information is disseminated (for example, in newsletters like *Women of Europe*) and information offices have been set up in the member states to deal specifically with rights (in addition to the general presence of the Commission in national capitals and major cities). A vast number of committees exists, composed of Commission staff and national nominees, to monitor policy implementation, a process which reveals defects and can lead to reforms. The existence of channels through which to influence Community policies means that Community nationals have some of the secondary political rights associated with metics in a hospitable country. Some primary political rights based, not on working status but humanity, will have to be agreed upon now, following the ratification of the Maastricht Treaty of Union.

In conclusion, let me refer to some comparable arguments that are taking place in feminist theory, and in the way that radical democrats are conceptualising the European Community. Anne Phillips provides a way of transcending what can be seen as an impasse in Carole Pateman's accounts of women and citizenship and between feminists who emphasise sexual difference and those who prefer to retain something of similarity. She reminds us that radical and Marxist feminisms are useful correctives to the liberal fallacy of slipping from saying that differences should not determine who has political rights to saying that differences do not matter at all. But she goes on to point out that, since politics is an activity distinct from other things that we do, we have to accept that good mothers are not necessarily good citizens, though they may be both. Their identities arising from their womanhood are not irrelevant to citizenship since participation by anyone tends to stem from organisation based on social roles. But it is only through participation and interaction that they can contribute to the public policies that impinge upon them and would do so inevitably even if they retreated from public life, albeit for radical rather than traditional reasons. The mark of a genuinely 'woman-friendly' polity would be an assured place on the public agenda for general questions about the domestic division of labour.[36]

The renewal of emphasis on public politics in feminist thinking has a counterpart in radical theories of democracy. Chantal Mouffe and her contributors argue that the political authority regulating a democratic 'public space' cannot remain neutral but must ensure a place for all legitimate interests in the interplay of dissent, discussion, negotiation and decision-making.[37] These are the kinds of ideas used by Etienne Tassin to examine the European Community. He argues that the Community cannot be the precondition of a public space in the communitarian sense because, like multi-ethnic states, it does not embody a cohesive, common original identity. But, he argues, it can be what he calls 'a politically constituted public space in which the plurality of political interests,

---

36 Anne Phillips, 'Citizenship and feminist politics' in Geof Andrews (ed.), *Citizenship* (London, 1991), pp 76–88.
37 Chantal Mouffe, 'Democratic citizenship and the political community' in Chantal Mouffe (ed.), *Dimensions of radical democracy. Pluralism, citizenship, community* (London, 1992), pp 1–14.

feelings, wills, initiatives, judgements, decisions and actions come face to face'.[38] In these radical theories, there is a paradox in that, while the political authority has the responsibility for determining the boundaries of the public space and discussion in it, it must be constantly open to challenge in these matters from citizens – if it is to retain its status as democratic. Though the Community is defective in many ways – for different reasons, I would argue, from those put forward by the opponents of Maastricht – and though it is complicated and cumbersome, it is more open to challenge by women than would be the case if they had to rely on their national polities alone.[39]

---

38 Etienne Tassin, 'Europe: a political community?' in *idem*, pp 169–192.
39 It is notable that all the opponents of the Maastricht Treaty seem to try to frighten its supporters, or the indifferent, by painting a picture of a deadening uniformity. Unlike Tassin, they also try to argue that it cannot work because the people's of Europe are diverse and, therefore, that efforts to make it work must entail coerced uniformity.

Index